Handbook of Plant And Crop Stress, Fourth Edition

Handbook of Plant and Crop Stress,
Fourth Edition

Edited by
Mohammad Pessarakli

CRC Press
Taylor & Francis Group
Boca Raton London New York

CRC Press is an imprint of the
Taylor & Francis Group, an **informa** business

CRC Press
Taylor & Francis Group
6000 Broken Sound Parkway NW, Suite 300
Boca Raton, FL 33487-2742

First issued in paperback 2021

Library of Congress Cataloging-in-Publication Data

Names: Pessarakli, Mohammad, 1948- editor.
Title: Handbook of plant and crop stress / editor: Mohammad Pessarakli.
Description: Fourth edition. | Boca Raton, FL : CRC Press, Taylor & Francis
Group, 2019. | Includes index.
Identifiers: LCCN 2019009679| ISBN 9780815390824 (hardback : alk. paper) |
ISBN 9781351104609 (ebook)
Subjects: LCSH: Crops--Effect of stress on. | Plants--Effect of stress on. |
Crops--Effect of salts on. | Plants--Effect of salts on.
Classification: LCC SB112.5 .H36 2019 | DDC 630--dc23
LC record available at https://lccn.loc.gov/2019009679

Visit the Taylor & Francis Web site at
http://www.taylorandfrancis.com

and the CRC Press Web site at
http://www.crcpress.com

In memory of my beloved parents, Fatemeh and Vahab, who, regretfully, did not live to see this work and my other works, which in no small part resulted from their gift of many years of unconditional love to me.

Contents

SECTION I   Soil Salinity and Sodicity Problems

SECTION II   Plants/Crops Tolerance Mechanisms and Stressful Conditions

SECTION III Plants and Crops Responses: Physiology, Cellular and Molecular Biology, Microbiological Aspects, and Whole Plant Responses under Salt, Drought, Heat, Cold Temperature, Light, Nutrients, and Other Stressful Conditions

SECTION VIII   Examples of Empirical Investigations of Specific Plants and Crops Grown under Salt, Drought, and Other Environmental Stress Conditions

SECTION IX   Future Promises: Improving Plant and Crop Adaptation/ Tolerance and Cultivation under Stressful Conditions

SECTION X Beneficial Aspects of Stress on Plants/Crops

Preface

The dynamic and ever-expanding knowledge of environmental stresses and their effects on plants and crops has resulted in the compilation of a large volume of information since the third edition of the *Handbook of Plant and Crop Stress* was prepared and presented to scientists and professionals. This fact necessitated that this unique comprehensive source of information be revised again and all the new findings in this field be included in the new edition of this book. Like the first, second, and third editions, the new edition of the *Handbook of Plant and Crop Stress* is also a unique, most comprehensive, and complete collection of the issues of stress imposed on plants/crops.

More than 80% of the materials of the new edition of the *Handbook of Plant and Crop Stress* are entirely new, and these are included in this volume under new titles. The other 20% of the old materials have been updated. Therefore, overall, over 90% of this book is new, and it appears that a totally new volume has emerged.

Of the total 50 chapters, only 10 chapters are from the third edition, but substantially revised and updated. The other 40 chapters are entirely new.

Since the early 1900s, soil/plant scientists have observed that plant growth and crop yields decreased under salinity, drought, and/or other environmental stress conditions. The reduction in plant growth was reported to be a result of modification of the physiological process and environmental conditions that control growth. Stresses imposed on plants due to pollution or the application of agrichemicals have recently attracted the attention of scientists, investigators, and environmentalists in agriculture or related areas. The mechanisms by which salinity, drought, high/low temperatures or heat, high/low pH, high/low light, nutrient deficiency, pollution, agrichemicals, or any other abiotic or biotic stresses affect plant metabolism, thereby reducing plant growth and development, are still not completely understood. Among plant physiological processes, the change in nutrient uptake and metabolism induced by salt, drought, and/or other stress factors is commonly accepted among scientists as one of the most important factors responsible for abnormal plant metabolism, reduced growth, and decreased crop yield. It is vital to minimize the effects of salt, drought, extreme temperatures, extreme pH, extreme light, pollution, agrichemicals, or any other environmental stresses on plant growth and crop yield. Thus, a greater awareness of these stress factors and their related problems is essential for scientists, growers, and all involved in the field of agriculture.

This handbook is a comprehensive, up-to-date reference book effectively addressing issues and concerns related to plant and crop stress. While many reference books are in circulation about soil salinity, sodicity, specific plant/crop salt and drought stress, pollution, and other environmental stresses, these all exist in relative isolation from each other, covering only one specific topic.

Efficiency and effectiveness in solving plant and crop stress problems are dependent on the accountability and coordination of all the factors and their interrelationships involved in plant/crop stress.

While previous authors have indeed competently covered the many areas separately, the areas are, nonetheless, interrelated and should be covered comprehensively in a single text. Thus, the purpose of this book is to fill this niche.

This new and updated edition of the *Handbook of Plant and Crop Stress*, 4th Edition, has been prepared by about a hundred contributors from among the most competent and knowledgeable scientists, specialists, and researchers in agriculture from 20 countries. It is intended to serve as a resource for both lecture and independent purposes. Scientists, agriculture researchers, agriculture practitioners, and students will benefit from this unique comprehensive guide. This technical book covers plant stress problems from the soil to the atmosphere.

As with other fields, accessibility of knowledge is among the most critical factors involved with crop stress problems. Without due consideration of all the elements contributing to a specific crop stress problem, it is unlikely that a permanent solution will be achieved. For this reason, as many of the factors as possible are included in this handbook. To further facilitate the accessibility of the desired information in the areas of plant/crop stress covered in this collection, the volume has been divided into 10 sections: "Soil Salinity and Sodicity Problems"; "Plant/Crop Tolerance Mechanisms and Stressful Conditions"; "Plant and Crop Responses: Physiology, Cellular, and Molecular Biology, Microbiological Aspects, and Whole Plant Responses under Salt, Drought, Heat, Temperature, Light, Nutrients, and Other Stressful Conditions"; "Plant and Crop Responses under Pollution and Heavy Metal Stresses"; "Plant and Crop Responses under Biotic Stress"; "Genetic Factors and Plant/Crop Genomics under Stress Conditions"; "Plant/Crop Breeding under Stress Conditions"; "Examples of Empirical Investigations of Specific Plants and Crops Grown under Salt, Drought, and Other Environmental Stress Conditions"; "Future Promises: Improving Plant and Crop Adaptation/Tolerance and Cultivation under Stressful Conditions"; and "Beneficial Aspects of Stress on Plants/Crops." Each of these sections consists of one or more chapters to discuss, independently, as many aspects of stress as possible.

The "Soil" section consists of one chapter entitled "Soil Salinity and Sodicity as Particular Plant/Crop Stress Factors." This chapter explains soil as a medium of plant/crop growth, soil salinity and sodicity problems, and the effects of soil salinity and sodicity on plant/crop growth.

The "Plant and Crop Tolerance" section consists of 11 chapters: "Roles and Mechanisms of Rhizobacteria in Regulating Plant Tolerance to Abiotic Stress"; "Physiological, Biochemical, and Molecular Mechanisms Regulating Post-drought Stress Recovery in Grass Species"; "Regulatory Mechanisms for Stress-induced Leaf Senescence"; "Mechanisms of Salt Tolerance in Submerged Aquatic Macrophytes"; "Oxidative Stress in Plants: Production, Metabolism, and Biological Roles of Reactive Oxygen Species (ROS)"; "Oxidative Stress and Antioxidative Defense System in Plants Growing under Abiotic Stresses"; "Plant Biochemical Mechanisms for the Maintenance of Oxidative Stress under Control Conditions"; "Role of Proline and Other Osmoregulatory Compounds in Plant Responses to Abiotic Stresses"; "Role of Dehydrins in Plant Stress Response"; "Strigolactone Plant Hormone Role in Plant Stress Responses"; and "Plant Abiotic Stress Proteomics: An Insight into Plant Stress Response at Proteome Level." These chapters address plants' and crops' mechanisms of stress tolerance.

The next section, "Plant and Crop Responses: Physiology, Cellular and Molecular Biology, Microbiological Aspects, and Whole Plant Responses under Salt, Drought, Heat, Cold Temperature, Light, Nutrients, and Other Stressful Conditions," includes the following 11 chapters: "Responses of Photosynthetic Apparatus to Salt Stress: Structure, Function, and Protection"; "Responses of Plants to Stresses of the Sonoran Desert"; "Stresses in Pasture Areas in South-Central Apennines, Italy, and Evolution at Landscape Level"; "Turfgrass Nutrient Management under Stresses: A Part of Integrated Stress Management (ISM)"; "Nutrient Management of Golf Course Putting Greens under Stress"; "Molecular Chaperones and Stress Tolerance in Plants"; "Phytohormone Homeostasis and Crosstalk Effects in Response to Drought Stress"; "Heliotropism: Plants Follow the Sun under High Light and Heat Stress Conditions"; "Carbon Metabolic Pathways and Relationships with Plant Stress"; "Protein Synthesis by Plants under Stressful Conditions"; and "Ultraviolet Effects on Plants: Harmful or Beneficial?" Each of these chapters presents in-depth information, separately, on each of these topics.

The section dealing with "Plant and Crop Responses under Pollution and Heavy Metal Stresses" consists of three chapters: "Plant Heavy Metal Interactions and Pollution Stress"; "Plant Responses to Stress Induced by Toxic Metals and their Nanoforms"; and "Turfgrass Hyper-accumulative Characteristics to Alleviate Heavy and Toxic Metals Stress." These chapters provide detailed information on plants/crops influenced by pollution generated from either the soil, water, or the atmosphere as well as heavy metal toxicity in soils.

The section entitled "Plant and Crop Responses under Biotic Stress and Agrichemicals" consists of one chapter: "How Crops Stress Weeds." This chapter discusses how the interactions between weeds and agrichemicals and plants/crops counter stress and the potential problems caused by the application of agrichemicals to plants/crops.

"Genetic Factors and Plant/Crop Genomics under Stress Conditions" is a section containing two chapters: "Candidate Gene Expression Involved in Plant Osmotic Tolerance" and "Drought-induced Gene Expression Reprogramming Associated with Plant Metabolic Alterations and Adaptation." These chapters present detailed and comprehensive information on the available materials on these subjects.

A new section entitled "Plant/Crop Breeding under Stress Conditions" contains three chapters: "Marker Assisted Breeding for Disease Resistance in Legume Vegetable Crops";

"Breeding for Improved Crop Resistance to Osmotic Stress"; and "Breeding for Improved Plant–Symbiont Thermotolerance and Symbiotic Performance by Regulating Heat Shock Proteins, RNA Binding Proteins, and Chaperones." These chapters present the recent detailed information on these subjects.

Several examples of "Empirical Investigations of Specific Plants and Crops Grown under Salt, Drought, and Other Environmental Stress Conditions" are presented in a section consisting of eight chapters, presenting various plants and crops with different degrees of stress tolerance. These chapters are "Abiotic Stress Impact and Tolerance of Natural Sweetener Plant Stevia"; "Responses of Green Beans (*Phaseolus vulgaris* L.) in Terms of Dry Matter Production, Nitrogen Uptake, and Water Absorption under Salt Stress Conditions"; "Growth Responses of Pepper Plant (*Capsicum annuum* L.) in Terms of Biomass Production and Water Uptake under Deficit Irrigation system, Mild Water Stress Condition"; "Effects of Salinity Stress on Tomato Plants and the Possibility of its Mitigation"; "Water Stress Effects on Growth and Physiology of Corn"; "Moisture Stress and Its Effects on Forage Production Systems"; "Responses of Medicinal Plants to Abiotic Stresses"; and "Citrus Plant Botanic Characteristics and its Abiotic and Biotic Stress."

"Future Promises: Improving Plant and Crop Adaptation/Tolerance and Cultivation under Stressful Conditions" is a section that presents evidence and guidance on plants and crops that can be used under stressful conditions. This section consists of the following nine chapters: "Improving Crop Resistance to Abiotic Stresses through Seed Invigoration"; "Drought Resistance of Tropical Forage Grasses: Opening a Fertile Ground for Innovative Research"; "Drought Resistance of Common Bean: Water Spending and Water Saving Plant Ideotypes"; "Moringa and Tamarind: Potential Drought Tolerant Perennial Crops"; "Relationship of Medicinal Plants and Environmental Stresses: Advantages and Disadvantages"; "The Role of Beneficial Elements in Mitigation of Plant Osmotic Stress"; "The Role of Grafting by Vegetable Crops for Reducing Biotic and Abiotic Stresses"; "Why Root Morphology Is Expected to Be a Key Factor for Crop Salt Tolerance?"; and "Improving Plant Yield and Quality under Normal and Stressful Conditions by Modifying the Interactive Signaling and Metabolic Pathways and Metabolic Interaction Networks."

The important subject of "Beneficial Aspects of Stress on Plants/Crops," which has received very little attention, is also included in this book. This section consists of a unique chapter entitled "Beneficial Effects of Various Environmental Stresses on Vegetables and Medicinal Plants for the Production of High Value-added Plants," which presents available information on this subject.

Numerous figures and tables are exhibited in this technical guide to facilitate comprehension of the presented materials. Thousands of index words are also included in this volume to further increase the accessibility of the desired information.

Mohammad Pessarakli

Acknowledgments

I would like to express my appreciation for the secretarial and administrative assistance that I received from the secretarial and administrative staff of the School of Plant Sciences, College of Agriculture and Life Sciences, the University of Arizona. The encouraging words of several of my colleagues, particularly Dr. Dennis T. Ray, University Distinguished Professor, which are always greatly appreciated, have certainly been a driving force for the successful completion of this project.

In addition, I sincerely acknowledge Ms. Randy Brehm (Senior Editor, CRC Press/Taylor and Francis Group), whose professionalism, patience, hard work, and proactive methods helped in the completion of this project as well as my previous book projects. This job would not have been completed as smoothly and rapidly without her valuable support and efforts.

I am indebted to Ms. Laura Piedrahita (Senior Project Coordinator, CRC Press/Taylor and Francis Group) for her professional and careful handling of this volume. I would also like to acknowledge the eye for detail, sincere efforts, and hard work put in by the copy editors and the project editor.

The collective efforts and invaluable contributions of numerous experts in the field of Plant/Crop Stress made it possible to produce this unique source, which presents comprehensive information on this subject. Each and every one of these contributors and their contributions are greatly appreciated.

Last, but not least, I thank my wife, Vinca, a high school science teacher, and my son, Dr. Mahdi Pessarakli, MD, who supported me during the course of this work.

Mohammad Pessarakli

Editor

Dr. Mohammad Pessarakli, the editor, is a professor in the School of Plant Sciences, College of Agriculture and Life Sciences at the University of Arizona, Tucson. His work at the University of Arizona includes research and extension services as well as teaching courses in Turfgrass Science, Management, and Stress Physiology; he is currently teaching Plants and Our World to large classes of over 200 students each semester. He is the editor of the *Handbook of Plant and Crop Stress*, the *Handbook of Plant and Crop Physiology* (both titles formerly published by Marcel Dekker, Inc. and currently by Taylor and Francis Group, CRC Press), *Handbook of Photosynthesis*, *Handbook of Turfgrass Management and Physiology*, and *Handbook of Cucurbits*. He has written 30 book chapters and is editor-in-chief of the *Advances in Plants & Agriculture Research* (*APAR*) Journal, an editorial board member of the *Journal of Plant Nutrition and Communications in Soil Science and Plant Analysis* as well as the *Journal of Agricultural Technology*, a member of the Book Review Committee of the Crop Science Society of America, and a reviewer of the *Crop Science, Agronomy, Soil Science Society of America*, and *HortScience* journals. He is the author or co-author of over 200 journal articles. Dr. Pessarakli is an active member of the Agronomy Society of America, the Crop Science Society of America, and the Soil Science Society of America, among others. He is an executive board member of the American Association of University Professors (AAUP), Arizona Chapter. Dr. Pessarakli is a well-known internationally recognized scientist and scholar and an esteemed member (invited) of Sterling Who's Who, Marques Who's Who, Strathmore Who's Who, Madison Who's Who, and Continental Who's Who as well as numerous honor societies (i.e., Phi Kappa Phi, Gamma Sigma Delta, Pi Lambda Theta, and Alpha Alpha Chapter). He is a Certified Professional Agronomist and Certified Professional Soil Scientist (CPAg/SS), designated by the American Registry of the Certified Professionals in Agronomy, Crop Science, and Soil Science. Dr. Pessarakli is a United Nations Consultant in Agriculture for underdeveloped countries. He received the BS degree (1977) in Environmental Resources in Agriculture, the MS degree (1978) in Soil Management and Crop Production from Arizona State University, Tempe, and the PhD degree (1981) in Soil and Water Science from the University of Arizona, Tucson. Dr. Pessarakli's environmental stress research work and expertise on plants and crops are internationally recognized.

For more information about Dr. Pessarakli, the editor, please visit

https://cals.arizona.edu/spls/content/mohammad
https://cals.arizona.edu/spls/people/faculty

Contributors

Albert T. Adjesiwor
Department of Plant Sciences
University of Wyoming
Laramie, Wyoming

Fatemeh Aflaki
Department of Biology
Faculty of Science and Literature
Bolu Abant Izzet Baysal University
Bolu, Turkey

Tahmina Akter
Department of Biochemistry and Molecular Biology
Graduate School of Science and Engineering
Saitama University
Saitama, Japan

E.L. Apostolova
Institute of Biophysics and Biomedical Engineering
Bulgarian Academy of Sciences
Sofia, Bulgaria

Diego G. Arias
Instituto de Agrobiotecnología del Litoral (IAL, UNL
    CONICET) & FBCB
Laboratorio de Enzimología Molecular
Santa Fe, Argentina

Omid Askari-Khorasgani
Young Researchers and Elite Club Isfahan (Khorasgan)
    Branch
Islamic Azad University
Isfahan, Iran

Christian Baldwin
The Scotts Miracle-Gro Company
Marysville, Ohio

S.M.A. Barsa
Department of Agronomy
University of Agriculture
Faisalabad, Pakistan

Rodríguez-Garay Benjamín
Centro de Investigación y Asistencia en Tecnología y Diseño
    del Estado de Jalisco
Jalisco, Mexico

Frank Bethea
Coca-Cola Company
Atlanta, Georgia

A.K. Biswal
Center for Bioenergy Innovation
Oak Ridge National Laboratory
Oak Ridge, Tennessee
and
Complex Carbohydrate Research Center
University of Georgia
Athens, Georgia
and
Department of Biochemistry and Molecular Biology
University of Georgia
Athens, Georgia

Juan Andrés Cardoso
International Center for Tropical Agriculture (CIAT)
Cali, Colombia

Christopher Catanzaro
Department of Agriculture
Virginia State University
Petersburg, Virginia

Rakesh Kumar Chahota
Department of Agricultural Biotechnology
CSK Himachal Pradesh Krishi Vishvavidyalaya
Palampur (H.P.), India

Cathryn Chapman
Department of Plant Biology
Rutgers University
New Brunswick, New Jersey

Luana Circelli
Department of Agriculture, Environmental, Food Science
University of Molise
Campobasso, Italy

Claudio Colombo
Department of Agriculture, Environmental, Food Science
University of Molise
Campobasso, Italy

Thomas W. Crawford, Jr.
Global Agronomy, LLC

Jack Dekker
New Weed Biology Laboratory
Portland, Oregon

Erika Di Iorio
Department of Agriculture, Environmental, Food Science
University of Molise
Campobasso, Italy

Rama Shanker Dubey
Department of Biochemistry
Institute of Science, Banaras Hindu University
Varanasi, India

Ali Akbar Ebadi
Rice Research Institute of Iran
Agricultural Research, Education and Extension
    Organization (AREEO)
Rasht, Iran

Masarovičová Elena
Department of Soil Science
Faculty of Natural Sciences
Comenius University in Bratislava
Bratislava, Slovakia

William Errickson
Rutgers University
New Brunswick, New Jersey

M. Farooq
Department of Crop Sciences
College of Agricultural and Marine Sciences
Sultan Qaboos University
Al-Khoud, Oman
and
Department of Agronomy
University of Agriculture
Faisalabad, Pakistan
and
The UWA Institute of Agriculture and School of Agriculture
    & Environment
The University of Western Australia
Perth, Australia

Antonella Fatica
Department of Agriculture, Environmental, Food Science
University of Molise
Campobasso, Italy

Carlos M. Figueroa
Instituto de Agrobiotecnología del Litoral
Universidad Nacional del Litoral, CONICET
Facultad de Bioquímica y Ciencias Biológicas
Santa Fe, Argentina

Anwar H. Gilani
Department of Biological and Biomedical Sciences
The Aga Khan University Medical College
Karachi, Pakistan

Sergio A. Guerrero
Instituto de Agrobiotecnología del Litoral (IAL, UNL
    CONICET) & FBCB
Laboratorio de Enzimología Molecular
Santa Fe, Argentina

Bingru Huang
Ralph Geiger Endowed Chair Professor in Turfgrass Science
Director, Graduate Program in Plant Biology
Department of Plant Biology and Pathology
Rutgers University
New Brunswick, New Jersey

Alberto A. Iglesias
Instituto de Agrobiotecnología del Litoral (IAL, UNL
    CONICET) & FBCB
Laboratorio de Enzimología Molecular
Santa Fe, Argentina

M. Anowarul Islam
Department of Plant Sciences
University of Wyoming
Laramie, Wyoming

Vykouková Ivana
Department of Soil Science
Faculty of Natural Sciences
Comenius University in Bratislava
Bratislava, Slovakia

Ambuj Bhushan Jha
Department of Plant Sciences
College of Agriculture and Bioresources
University of Saskatchewan
Saskatoon, Saskatchewan, Canada

Jampílek Josef
Department of Analytical Chemistry
Faculty of Natural Sciences
Comenius University in Bratislava
Bratislava, Slovakia
and
Regional Centre of Advanced Technologies and Materials
Faculty of Science
Palacky University
Olomouc, Czech Republic

Noémi Kappel
Department of Vegetable and Mushroom Growing
Faculty of Horticultural Science
Szent István University
Gödöllő, Hungary

Kráľová Katarína
Institute of Chemistry
Faculty of Natural Sciences
Comenius University in Bratislava
Bratislava, Slovakia

Mojtaba Kordrostami
Department of Plant Biotechnology
Faculty of Agricultural Sciences
University of Guilan
Rasht, Iran

Klára Kosová
Department of Genetics and Plant Breeding
Crop Research Institute
Prague, Czech Republic

Zuzana Kriššáková
Department of Soil Science
Faculty of Natural Sciences
Comenius University in Bratislava
Bratislava, Slovakia

Hong Li
Haikou Cigar Research Institute
China Tobacco, Hainan Provincial Bureau
Hainan, China
and
Chinese Academy of Tropical Agricultural Sciences
Environment and Plant Protection Institute
Hainan Province, China

Haibo Liu
Department of Plant and Environmental Sciences
Clemson University
Clemson, South Carolina

Rincón-Hernández Manuel
Centro de Investigación y Asistencia en Tecnología y Diseño
    del Estado de Jalisco
Jalisco, Mexico

Mohammad Mafakheri
Department of Horticultural Science
Faculty of Agricultural Sciences
University of Guilan
Rasht, Iran

Yehouda Marcus
School of Plant Sciences and Food Security
Tel Aviv University
Tel Aviv, Israel

Sara Mardani
Department of Agricultural and Biosystems Engineering
South Dakota State University
Brookings, South Dakota

Barranco-Guzmán Angel Martín
Centro de Investigación y Asistencia en Tecnología y Diseño
    del Estado de Jalisco
Jalisco, Mexico

Rachel McDaniel
Department of Agricultural and Biosystems Engineering
South Dakota State University
Brookings, South Dakota

Maryam Mozafariyan Meimandi
Department of Vegetable and Mushroom Growing
Faculty of Horticultural Science
Szent István University
Gödöllő, Hungary

Nick Menchyk
Department of Urban Horticulture and Design
Farmingdale State College
Farmingdale, New York

Romina I. Minen
Instituto de Agrobiotecnología del Litoral
Universidad Nacional del Litoral, CONICET
Facultad de Bioquímica y Ciencias Biológicas
Santa Fe, Argentina

A.N. Misra
Khallikote University
Odisha, India
Center for Life Sciences
Central University of Jharkhand
Jharkhand, India

M. Misra
Khallikote University
Odisha, India

Ali Akbar Mozafari
Department of Horticulture
Faculty of Agriculture
University of Kurdistan
Sanandaj, Iran

Hitoshi Nakamoto
Department of Biochemistry and Molecular Biology
Graduate School of Science and Engineering
Saitama University
Saitama, Japan

Satya S.S. Narina
Department of Agriculture
Virginia State University
Petersburg, Virginia

Abdelaziz Nilahyane
Department of Plant Sciences
University of Wyoming
Laramie, Wyoming

Caleb Patrick
SePRO Research & Technology Campus
Whitakers, North Carolina

Arman Pazuki
Department of Biology
Faculty of Science and Literature
Bolu Abant Izzet Baysal University
Bolu, Turkey

Claudia V. Piattoni
Instituto de Agrobiotecnología del Litoral (IAL, UNL
    CONICET) & FBCB
Laboratorio de Enzimología Molecular
Santa Fe, Argentina

Florencio E. Podestá
Centro de Estudios Fotosintéticos y Bioquímicos
Universidad Nacional de Rosario, CONICET
Facultad de Ciencias Bioquímicas y Farmacéuticas
Rosario, Argentina

Jose A. Polania
International Center for Tropical Agriculture (CIAT)
Cali, Colombia
and
Departamento Biologia Molecular de Plantas
Instituto de Biotecnologia
Universidad Nacional Autonoma de Mexico
Morelos, Mexico

Rout Nutan Prasad
Centro de Investigación y Asistencia en Tecnología y Diseño
    del Estado de Jalisco
Jalisco, Mexico

Shaw Birendra Prasad
Institute of Life Sciences
Bhubaneswar, India

Ilja Tom Prášil
Department of Genetics and Plant Breeding
Crop Research Institute
Prague, Czech Republic

Babak Rabiei
Department of Agronomy & Plant Breeding
Faculty of Agricultural Sciences
University of Guilan
Rasht, Iran

Idupulapati M. Rao
International Center for Tropical Agriculture (CIAT)
Cali, Colombia
and
Plant Polymer Research Unit
National Center for Agricultural Utilization Research
Agricultural Research Service
United States Department of Agriculture
Peoria, Illinois

Abdul Rehman
Department of Crop Sciences and Biotechnology
Dankook University
Cheonan-si, Korea

Stephanie Rossi
Rutgers University
New Brunswick, New Jersey

Armin Saed-Moucheshi
Department of Plant Production and Plant Breeding
College of Agriculture
Shiraz University
Shiraz, Iran

Amir Hossein Saeidnejad
Department of Agriculture
Payame Noor University (PNU)
Tehran, Iran

Elisabetta Salimei
Department of Agriculture, Environmental, Food Science
University of Molise
Campobasso, Italy

Uwe Schleiff
Independent Expert for Irrigation and Salinity – Fertilizer
    and Crops – Soils and Environment
Wolfenbuettel, Germany

Bhallan Singh Sekhon
Department of Vegetable Science & Floriculture
CSK Himachal Pradesh Krishi Vishvavidyalaya
Palampur (H.P.), India

Ehsan Shakeri
Department of Plant Production and Plant Breeding
College of Agriculture
Shiraz University
Shiraz, Iran

Akhilesh Sharma
Department of Vegetable Science & Floriculture
CSK Himachal Pradesh Krishi Vishvavidyalaya
Palampur, India

Pallavi Sharma
Department of Life Sciences
Central University of Jharkhand
Ranchi, India

Kadambot H.M. Siddique
The UWA Institute of Agriculture and School of Agriculture
 & Environment
The University of Western Australia
Perth, Australia

Fatemeh Sohrabi
Department of Plant Biotechnology
College of Agriculture
Shiraz University
Shiraz, Iran

M. Stefanov
Institute of Biophysics and Biomedical Engineering
Bulgarian Academy of Sciences
Sofia, Bulgaria

I. Szabolcs
Research Institute for Soil Science and Agricultural
 Chemistry of the Hungarian Academy of Science
Budapest, Hungary

Jacob Taylor
Department of Plant and Environmental Sciences
Clemson University
Clemson, South Carolina

Satoru Tsukagoshi
Center for Environment
Health and Field Sciences
Chiba University
Chiba, Japan

Milan Oldřich Urban
Laboratory of Plant Stress Biology and Biotechnology
Division of Crop Genetics and Breeding
Crop Research Institute
Prague, Czechia

Gómez-Entzin Veronica
Centro de Investigación y Asistencia en Tecnología y Diseño
 del Estado de Jalisco
Jalisco, Mexico

Pavel Vítámvás
Department of Genetics and Plant Breeding
Crop Research Institute
Prague, Czech Republic

A. Wahid
Department of Botany
University of Agriculture
Faisalabad, Pakistan

Wataru Yamori
Department of Biological Sciences
Graduate School of Science
The University of Tokyo
Tokyo, Japan

Section I

Soil Salinity and Sodicity Problems

1 Soil Salinity and Sodicity as Particular Plant/Crop Stress Factors

Mohammad Pessarakli and I. Szabolcs

CONTENTS

1.1 INTRODUCTION

Years of research have gone into the impact of soil salinity and sodicity as a plant stress factor. Soil salinity and sodicity can potentially affect several traits in plants, including nutrient absorption and use, water uptake and plant water relations, physiological processes such as seed germination and emergence, plant growth rates, photosynthesis, fresh and dry matter production, yield and yield components, and eventually, plant senescence. Soil salinity and sodicity problems usually occur in climates with little to no rain, generally when evapotranspiration is higher than precipitation, so that salts accumulate at the soil surface. In dry and semi-dry areas, factors such as low precipitation, high evaporation, and high temperature as well as poor management of water resources result in increased soil salinity. Saline soils affect the development of plants because of the osmotic stress caused by the high salt concentration in both the liquid phase and the solid phase. Also, chlorides as toxic elements are a compound of saline soil that hinders the development of crops. Sodic soils, also known as alkaline soils, show unfavorable physical conditions (low permeability, high dispersion, low aggregation and flocculation), which negatively affect the water and air movement within the soil. Soil salinity and sodicity are among the major agricultural problems limiting plant growth and development throughout the world [2–4, 7–9, 14, 15, 17–20, 25], [31, 33, 34, 36, 39, 43, 44, 46, 47, 49], [51–53, 56, 58, 60, 71, 72, 74, 75], [80–84, 87–89, 93, 94, 99, 100], [102, 104–111, 114, 115, 120–127], [133, 136, 139–141, 143–146, 148], [151, 153, 154, 156, 185, 187, 191–193, 198–202], [207, 208, 210, 211, 215, 216, 223–228], [231–233, 237, 239–257, 259, 264–268, 270]. Salinity and sodicity problems in agriculture have an ancient history and have now become a very difficult problem in agricultural and farming activities [270]. The salinity and sodicity problems are more severe, and the situation is catastrophic, especially in arid regions with high soil salinity, hot and dry conditions, and limited water resources [1, 10, 13, 21, 24, 28, 101, 242, 255]. The use of Landsat data (LD), satellite-based detection (SBD), remote sensing (RS), and global positioning system (GPS) technologies has made the identification and measurement of salt-affected land easy and relatively accurate [1, 13, 35, 66, 78, 85, 91, 92, 138, 149, 194, 211, 261]. These problems are of particular concern for countries whose economies rely to a great extent on agriculture [214, 222].

Salinity and sodicity problems are common in arid and semiarid regions, where rainfall is insufficient to leach salts and excess sodium ions out of the rhizosphere. In addition, these areas often have high evaporation rates, which can encourage an increase in salt concentration at the soil surface. The arid and semiarid regions include almost one-third of the world's land [139, 203, 215]. According to the Food and Agricultural Organization (FAO) [66] of the United Nations, the total salt-affected area of the world has been estimated at over 800 million hectares.

The presence of a caliche horizon and/or a cemented hardpan layer at varying depths, plus insufficient precipitation for leaching, often adds to the salt accumulation in these soils.

Newly established irrigation projects, with improper planning and management practices, may also add salts to soils [158].

Soil salinity and sodicity problems are present in nearly every irrigated area of the world and also occur on nonirrigated croplands and rangelands. Thus, virtually no land is immune to salinization. Therefore, to sustain life on earth, it is vital and urgent to control these problems and find new ways to use these extensive saline and sodic soils and water resources, at least for agricultural purposes. Reclamation, or at least minimization of the effect of salinity and/or sodicity, is important and necessary. In this respect, the proper use of water for both plant growth and soil salinity and sodicity control is probably of the greatest importance.

3

The main focus of this introductory chapter is to summarize general information on salt-affected (saline and sodic) soils and factors influencing their formation and reclamation, and to discuss salinity and sodicity as plant/crop stress factors.

1.2 THE SIGNIFICANCE OF SOILS WITH RESPECT TO CROP STRESS

Inasmuch as all crops are grown on soils, soil properties have substantial influence on the living conditions of plants and crops. In nature, usually, particular plant species grow on specific soils. Thus, specific relationships exist between a particular soil and the vegetation cover of that specific soil. For example, Kreeb et al. [116] investigated soil and vegetation relationships associated with alkaline–saline soil surfaces.

Plant development and successful crop production require proper soil conditions, including adequate water and nutrient supply. Unfavorable soil conditions, environmental stress [3, 45, 101, 169, 236, 242, 245], salinity and/or sodicity [2, 7–9, 15, 18–20, 23, 25, 27, 28], [31–34, 38, 39, 42, 43, 49, 52, 58], [63, 64, 72, 79, 81, 84, 87–89, 93, 94, 99], [102, 104, 108–110, 114, 116, 122–125, 127], [136, 145, 148, 153, 154, 171–178, 182, 185, 187, 191], [201, 205, 213, 215, 218, 226, 228, 231, 236], [237, 239, 241, 243, 245–247, 249–253], [255, 256, 263–266], and inadequate nutrient supply [189, 258] have an adverse effect on the life of plants, sometimes seriously hindering their effective production. Salinity stress in soil or water, especially in hot, arid regions, could seriously limit plant growth and reduce yield.

Based on these facts, we can speak of stress factors originating in the soil, i.e., unfavorable soil conditions that cause, or contribute to, the stress factors plants and crops are exposed to.

It is impossible to list all, or even most of, such factors in a short introductory chapter. Therefore, the authors will limit the range of this chapter to a general description of soil behavior and its function in nature and production as well as an outline of one of the most serious factors in the origin of salt-affected soils. For more in-depth information regarding salt-affected soils, readers are referred to the more comprehensive available sources [10, 12, 28, 30, 40, 41, 48, 53–55, 59, 61], [66, 76, 77, 85, 86, 88, 90, 97, 98], [105, 106, 112, 113, 128, 133, 134, 137, 138], [142, 147, 150, 152, 157, 188], [194–197, 200, 203, 206, 210, 212, 218, 221–223], [226, 229, 231–235, 255, 260, 264, 272].

1.3 THE PLACE AND ROLE OF THE SOIL IN NATURE

It is generally accepted that the soil is a substantial part of the environment, comprising different substances and forming a special kind of ecosystem, with various properties and attributes, inside a given ecosystem. It is also accepted that the soil of the continents is of high diversity, which is dealt with by several branches of soil science, e.g., taxonomy, classification, survey, mapping, etc.

The soil, or the pedosphere, which is an environmental synonym for the soils of a given territory, has a specific place in nature. It is a natural body, similar to rocks, waters, or biota in the sense that they, too, have their own materials, mass and energy fluxes, development, and regularities. This fact should be mentioned because, not only in newspapers but also in technical literature, soils are frequently treated either as living substances or as non-biological substances. Neither of these approaches is correct, because one of the characteristics of the soil is its complexity, the fact that it contains both living and non-living substances, formed due to both biotic and abiotic processes.

The soil as a natural body is inseparable from the rocks and the crust of weathering on the surface of the continents from which it has developed, on the one hand, and from biological processes, on the other hand. The main characteristic that distinguishes soil from rock is the result of biological processes: the production of organic matter by the activities of micro-organisms, plants, invertebrates and other animals, and finally, human beings, which transforms rock into soil, capable of supplying plants and crops with nutrients and water and serving as an anchor for their establishment and maintenance on the land.

The processes of soil formation started concurrently with the appearance of life on the continents and continued during the billions of years of interactions between living substances and rocks under the influence of climatic conditions, with particular regard to the action of water, vegetation cover, organisms (both macro and micro), geomorphological patterns, and the time factor. As a result of their interactions, specific mass and energy fluxes formed the different soil types in various environmental conditions.

With the appearance of the human race on the face of the earth, even changes in the environment became different. Due to human activities, natural processes affected by biotic and abiotic factors accelerated, and several others, which had been unknown or minimal before, developed.

The role of soils in nature is complex and multi-faceted, including biospheric, hydrospheric, and lithospheric functions. Their interaction is illustrated in Figure 1.1 [234]. Figure 1.1 clearly shows that the soil is a specific body related to the ecosystem. Even the word "soil" is very often used as a synonym of "ecosystem" when characterizing the ecological conditions in a certain place. If we want to be precise, we must agree that the ecosystem includes the pedon; in other words, the soil. However, the soil includes different phases (solid, liquid, gaseous), living and nonliving substances, plants, animals, and microbes, and has its own energy and material fluxes. Therefore, it can be considered to be an ecosystem in its own right. In this respect, when speaking of soils in regard to their plant cover, we can consider the soils of a given location as the basis, ladder, and foothold; for instance, in savannas, or in the tropical belt, a well-defined plant cover develops, and very often, the soil properties promote or limit the living conditions of certain plant species or associations.

Based on these considerations, it can be accepted that certain soil types, when discussed as the habitat for certain plant

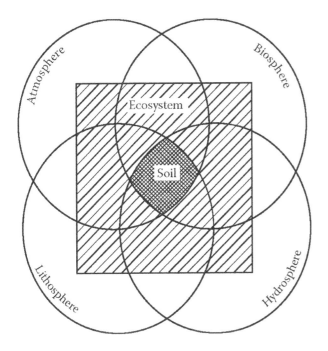

FIGURE 1.1 Schematic diagram of the interaction of lithosphere, atmosphere, biosphere, hydrosphere, ecosystems, and soils. (From Szabolcs, 1989.)

associations, often give their name to the ecosystem of the plant association concerned, as the pedon includes, apart from the plants, most of the components of the ecosystem.

Evidently, the soil, as a specific natural entity, is far from being identical with the vegetation, and, despite their close correlation, direct conversion between soil types and vegetation is almost impossible. Still, there are soil types that more or less determine the ecological function for certain types of vegetation, either by providing beneficial conditions for their development or by limiting the ecological conditions for other types of vegetation.

This is perhaps best demonstrated in the case of salt-affected soils, where high electrolyte contents and extreme pH conditions limit the development of the majority of plants and serve as a habitat only for species that can survive or tolerate the unfavorable conditions caused by the salinity and alkalinity of the soil. For example, the grass *Leptochloa fusca*, which grows vigorously on salt-affected soils, can tolerate extremely saline and sodic (alkaline) conditions [117]. This species is also well adapted to the waterlogging encountered on saline and sodic (alkaline) soils. Saltgrass (*Distichlis spicata*) is another example of a highly salt-tolerant plant species that grows vigorously on saline and sodic soils [132, 159–164, 168, 171, 173–179]. In fact, intensive investigations of this plant species by the senior author of this chapter and his co-workers found that this grass performed better than the control when some salt was added to it during its establishment period, and so far, it has been the most salt- and drought-tolerant species compared with the other highly salt-tolerant halophytes that have been tested by this investigator [159–164, 168, 171,

173–179]. Other investigators [57, 70, 76, 116, 188, 209, 213] have also reported on the soil and vegetation relationships by which specific plant types are adapted to and growing in specific habitats. In such respects, salt-affected soils can be considered as habitats or ecosystems for halophytes, and if we agree on this, correlations can be found between the different types of salt-affected soils and their flora and fauna as components of the ecosystem.

To cast light on both the theoretical and the practical aspects of such considerations, it is necessary to describe briefly the properties and grouping of salt-affected soils with regard to the possibilities of the occurrence and distribution of halophytes and xerophytes developing on them.

1.4 EXTENT AND GLOBAL DISTRIBUTION OF SALT-AFFECTED SOILS

Nearly 10% of the total land surface is covered with different types of salt-affected soils. Table 1.1 demonstrates the distribution of salt-affected soils in the world [113]. Table 1.1 shows that no continent on our globe is free from salt-affected soils. They are not only distributed in deserts and semi-desert regions [13, 28, 98, 101, 158, 162–168, 179–181, 197, 213, 272], but also frequently occur in fertile alluvial plains, river valleys and coastal areas, close to densely populated areas and irrigation systems [40, 55, 59, 65, 66, 133, 134, 157, 201–204, 209, 210, 234, 235].

Figure 1.2 shows the distribution of salt-affected soils throughout the world [235].

1.5 DEVELOPMENT AND GROUPING OF SALT-AFFECTED SOILS AND PARTICULAR PLANT/CROP GROWTH STRESS FACTORS

The United Nations Environment Programme estimates that approximately 20% of agricultural lands and 50% of croplands in the world are salt-affected. Soil salinization is

---

TABLE 1.1

Salt-Affected Soils on the Continents and Subcontinents (Million ha)

Continent	Area (Million ha)
North America	15.7
Mexico and Central America	2.0
South America	129.2
Africa	80.5
South Asia	87.6
North and Central Asia	211.7
South-east Asia	20.0
Australasia	357.3
Europe	50.8
Total	954.8

Source: Kovda, V.A. and Szabolcs, I., *Agrokemia es Talajtan*, Suppl., 1979. With permission.

FIGURE 1.2 Global distribution of the salt-affected soils. (From Szabolcs 1991.)

reducing the area that can be used for agriculture by 1–2% every year, hitting hardest in the arid and semi-arid regions [121]. Despite the fact that the properties and attributes of salt-affected soils have been well known for a long time, it is appropriate to give a brief definition of this group of soils right at the start, because salinity and sodicity (alkalinity) as well as acidity of soils are substantial stress factors, seriously affecting the productivity of the land and plant/crop responses [2, 3, 7–9, 14, 16–25, 28], [31, 33, 34, 36, 38, 39, 41–46, 49, 51–53], [56–58, 60, 62–64, 67–69, 71–75], [79–84, 87–89, 93–96, 99, 100], [102–111, 114, 116, 118, 120, 122–126], [128, 129, 133, 139–141, 143–148], [150, 151, 153, 154, 156, 172], [185, 187, 189–193, 198–201, 204, 205], [207, 208, 210, 215, 216, 218], [223–228, 231, 234–247, 249–259, 263–271].

Salt-affected (i.e., saline, saline–sodic, and sodic) soils usually have low biological activity both because of the osmotic and ionic effects of salts and due to limitation of carbonaceous substrates. Rao and Pathak [196] reported that microbial growth was depressed in sodic (alkaline) soils due to, at least in part, limitation in carbon substrate (carbon stress) and in saline soils due to salt stress.

For detailed information on the formation of salt-affected soils and their characterization and classification, readers are referred to Asmamaw et al. [28], Pessarakli [157], Szabolcs [234, 235], and Zia-ur-Rehman et al. [272].

Salt-affected soils can be characterized as soils formed under the dominant influence of different salts in their solid or liquid phases, which will then have a decisive influence on the development, the characteristics, the physical, chemical, and biological properties, and eventually, the fertility of the soil. Whenever and wherever this phenomenon occurs, it produces specific formations of soils where the high electrolyte concentration and its consequences overshadow the former soil-forming processes or former soil properties and environmental conditions, often radically changing them.

High electrolyte concentration is the only common feature of all salt-affected soils. Their chemistry, morphology, pH, and many other properties may be different depending on the character of salinization and/or alkalization.

Salt-affected soils, in the broader sense, can be divided into the following groups:

1. Saline soils that develop under the influence of electrolytes of sodium salts with nearly neutral reactions (predominantly sodium sulfate [Na_2SO_4], sodium chloride [$NaCl$], and occasionally sodium nitrate [$NaNO_3$]). These soils occur mainly in arid and semi-arid regions and form a major part of all the salt-affected soils of the world.

   High contents of soluble salts accumulated in these soils can significantly decrease their value and productivity. Salinization of soils is one of the most serious environmental problems in world agriculture. The problem of salinity is characterized by an excess of soluble salts and is common in arid and semi-arid lands, where it has been naturally formed under the prevailing climatic conditions and due to lower precipitation, higher rates of evapotranspiration, and lack of leaching water. Although more frequent in arid lands, saline soils are also present in areas where salinity is caused by poor quality of irrigation water. Soil salinity often causes constraints to plant/crop growth; hence, the productivity of a major category of plants, called *glycophytes*, is severely affected. Therefore, soil salinity is a serious concern in agriculture. Combination with other edaphic/environmental factors (precipitation, temperature, flooding, soil profile, and the water table) exacerbates the catastrophe synergistically. Improper irrigation systems and leaching fractions added by land clearing and deforestation have been noted as the major causes of soil salinity [88, 124, 231, 255, 261]. Saline soil induces physiological and metabolic disturbances in plants, affecting their development, growth, yield, and quality. Plants are adversely affected by salinity; as a result, seed germination, survival percentage, morphological characteristics, development and yield, and their components are substantially reduced under saline conditions [98]. In general, the photosynthesis and respiration rates of plants are decreased under salt stress conditions.

2. Sodic (alkaline) soils that develop under the influence of electrolytes capable of alkaline hydrolysis (mainly sodium carbonate [Na_2CO_3] and sodium bicarbonate [$NaHCO_3$] and occasionally sodium silicate [Na_2SiO_3] and sodium bisilicate [$NaHSiO_3$]). This is often recognized as an excess of sodium ions (Na^+) (sodicity or alkalinity) that imparts life-threatening consequences to plants due to mal-textured soil, hindering porosity and aeration, and leading to a physiological water deficit.

This group is widespread in practically all the climatic regions, from the humid tropics to beyond the polar circles, and the total salt content is usually lower than that of saline soils, sometimes even strongly sodic (alkaline). Virgin sodic (alkaline) soils have a high pH and high exchangeable sodium (Na) percentage (ESP) and are often barren.

Sodic soils exhibit poor physical conditions that adversely influence water and air movement in the soils. Sodicity causes soil erodibility and impairs plant growth [142, 157].

3. Salt-affected soils that mostly develop due to the presence of calcium sulfate ($CaSO4$) (gypsiferous soils) or, rarely, in the presence of calcium chloride ($CaCl2$). Gypsiferous soils can mainly be found in the arid and semi-arid regions of North America, North Africa, the Near, Middle and Far East, and Australia.

4. Salt-affected soils that develop under the influence of magnesium salts. This group occurs in arid, semi-arid, and even semi-humid regions and has particular significance, especially those soils that have a heavy texture.

5. Acid-sulfate soils whose salt content is composed mainly of Al2/SO4/3 and Fe2/SO4/3. This type of salt-affected soil broadly extends in the tidal marsh areas along the seashores of all the continents. These soils are particularly common in northern Europe, the western and eastern coastlines of Africa, along the coastline of south-east India, etc., and develop on sulfurous marine sediments.

Inland acid-sulfate soils can also be found in different areas of the world, such as the western territories of the United States, Asia Minor, and China. Such soils developed as a result of fluvial glacial processes and have had no connection with seashores in recent geological times.

Evidently, the different groups of salt-affected soils have diverse physico-chemical and biological properties besides the one they have in common, i.e., a comparatively high electrolyte content.

The grouping of the salt-affected soils and their properties causing plant and crop stress are presented in Table 1.2.

The five groups in Table 1.2 represent the formations of different salt-affected soils described earlier, indicating their chemical types, the environmental conditions where they dominate or occur, the mechanism of their main adverse effect on production, and the basic methods of their reclamation. For detailed information on the formation, reclamation, characterization, and classification of salt-affected soils, see Asmamaw et al. [28], Pessarakli [157], and Szabolcs [234, 235].

TABLE 1.2
Grouping of Salt-Affected Soils and Their Properties Causing Plant and Crop Stress

Types of Salt-Affected Soils	Electrolyte(s) Causing Salinity and/or Sodicity	Environment	Properties Causing Plant and Crop Stress	Methods for Reclamation
Saline	Sodium chloride an d sulfate (in extreme cases nitrate)	Arid and semi-arid	High osmotic pressure of soil solution, toxic effect of chlorides	Removal of excess salt (leaching)
Sodic	Sodium ions capable of alkaline hydrolysis	Semi-arid, semi-humid, and humid	High (alkaline) pH, poor water physical conditions	Lowering or neutralizing the high pH by chemical amendments
Magnesium	Magnesium ions	Semi-arid and semi-humid	Toxic effect, high osmotic pressure, Ca deficiency	Chemical amendments, leaching
Gypsiferous	Calcium ions (mainly $CaSO_4$)	Semi-arid and arid	Low (acidic) pH toxic effect	Alkaline amendments
Acid sulfate	Fe1Tic and aluminum ions (mainly sulfates)	Seashores and lagoons with heavy, sulfate-containing sediments, diluvial inland slopes and depressions	High acidity and the toxic effect of aluminum	Liming

Source: Szabolcs, I., *Salt-affected Soils*, CRC Press, Boca Raton, 1989. With permission.

In Table 1.2, the adverse properties of different salt-affected soils causing crop stress are also included. From these, it is clear that in various groups, different properties are responsible for hindering the development of plants and crops by causing stress.

In saline soils, it is the high salt concentration in the solid and liquid phases that results in high osmotic pressure, hindering the normal development of plants; the stress factor is the salinity, with all its disadvantageous consequences for plant life. Apart from this, some compounds of the salt content of these soils, e.g., chlorides as toxic elements, also act as one of the stress factors.

In sodic (alkaline) soils, as a rule, it is not the high salt concentration but the sodic (alkaline) pH value that is the stress factor, particularly in cases where there is a high concentration of sodium carbonate in the solid and liquid phases of the soil. The high pH hinders the biological functions of crops and limits their development.

In another group of sodic (alkaline) soils, which sometimes do not have a very alkaline pH value (solonetz type), the comparatively low concentration of sodium salts capable of sodic (alkaline) hydrolysis constitutes a stress factor through its action, resulting in poor physical properties of water in the soil. As a consequence of this phenomenon, the wilting point in the soil increases, and the plants suffer from water deficiency, even in wet soils, due to the swelling of clay saturated with sodium ions (Na^+).

In magnesium soils, which have not been adequately studied, the combination of a toxic effect, calcium deficiency, and poor soil physical properties are the stress factors.

In gypsiferous soils, the acidic pH, and sometimes the toxic effect of the high gypsum content, contributes to the appearance of stress factors for plant and crop life in regions with large areas of intensely gypsiferous soils.

In acid-sulfate soils, the very high acidity, with a pH sometimes below 2, causes stress due to all the adverse effects of extreme acidity. Furthermore, the high aluminum content of the soil solution has an intensely toxic effect.

Apart from this, the temporary or permanent waterlogging in such soils acts as a stress factor hindering the normal air and nutrient regime, necessary for plant life, in these soils.

Besides the salt-affected soils that develop due to natural soil-forming processes, the so-called *secondary salt-affected soils* have increasing importance in both scientific and practical terms. Secondary salt-affected soils are those that have been salinized due to human factors, mainly as a consequence of improper methods of irrigation. The extent of secondary salt-affected soils is sizeable, and this adverse process is as old as irrigated agriculture itself. Ancient civilizations in Mesopotamia, China, and pre-Columbian America fell as a consequence of the salinization of irrigated land. The process is also advancing vigorously at present, and more than half of all the irrigated lands in the world are under the influence of secondary salinization and/or alkalization.

When speaking of human-made factors of salinization, we also have to mention potential salt-affected soils, which are not salt-affected at present but in the case of increased irrigation, deforestation, overgrazing, and other human activities, can and will be salinized unless the necessary preventive procedures are undertaken in due time. No global records are available of the extent of potential salt-affected soils; however, the area that they cover is larger than that of existing salt-affected soils.

Secondary salt-affected soils can be divided into the following two categories:

1. Secondary formation of salt-affected soils caused by irrigation.

    Despite negative experiences, the salinization of irrigated and surrounding areas has not diminished. On the contrary, it is still on the increase.

    According to the estimates of the FAO and UNESCO (the United Nations Educational, Scientific and Cultural Organization), as many as half of all the existing irrigation systems of the world are, more or less, under the influence of secondary salinization, alkalization, and waterlogging. This phenomenon is very common not only in old irrigation systems but also in areas where irrigation has only recently begun.

    According to the estimates of these agencies, 10 million hectares of irrigated land are abandoned yearly because of the adverse effects of salinity due to irrigation, mainly secondary salinization and alkalinization.

    The mentioned losses and damages are not evenly distributed among the irrigating countries. In some of them, the damage may be relatively small, while in others, it actually constitutes the major problem in agriculture or even in the national economy of the country in question. In this respect, unfortunately, there are countless sad examples. In Pakistan, Ahmad [5] carried out statistical analyses in respect of secondary salinized land. According to his data, out of 35 million acres (approximately 16 million hectares) of total irrigated territory, salinized areas accounted for 5.3 million acres (approximately 2.4 million hectares) after a few years of irrigation. He indicated, among the causes of secondary salinization in Pakistan, the joint effects of irrigation and groundwater. According to Zavaleta [263], practically all irrigated alluvial soils in Peru show features of salinity and sodicity (alkalinity). It is known from FAO reports [65, 66] and the papers of Kovda [112] that more than 40% of irrigated soils in Iraq and Iran are affected by secondary salinization. In a country report on salinity in Syria, FAO [65] estimated the adverse effects of salinity as follows:

    (a) In more than 20,000 ha, salinity developed to a level where these soils had to be taken out of cultivation, and the loss is estimated at a total of 30,000 tons of cotton per year.

(b) In about 30,000 ha, the yield decreased by 50%, and the total loss is estimated at 20,000 tons of cotton per year.

(c) In about 60,000 ha, the yield decreased by 20%, and the total loss is estimated at about 18,000 tons of cotton per year.

At present, no continent is free from the occurrence of this very serious phenomenon. In Argentina, 50% of the 40,000 ha of land irrigated in the nineteenth century are now salinized. In Australia, secondary salinization and alkalization take place in the valley of the Murray River, and in Northern Victoria, 80,000 ha have been affected. The same phenomena can be observed in Alberta, Canada, and similar processes have been recorded in the northern states of the United States, where irrigation was introduced much later than in the dry west. It is noteworthy that these last examples, and many other irrigated regions, are far from being arid areas, and the majority of salt accumulations are associated with the sodium salts capable of sodic (alkaline) hydrolysis and not with the neutral sodium salts that we are familiar with in desert and semi-desert areas.

The more recent reports of the FAO [66] of the United Nations estimated the salt-affected areas due to irrigation in developing countries, including Iran, Iraq, Pakistan, and Syria, to be much higher than in the previous reports.

2. Secondary formation of salt-affected soils caused by human activities other than irrigation.

When speaking of secondary salinization, most people have irrigation and drainage in mind. However, there are also other anthropogenic factors causing this adverse phenomenon. It is true that the majority of secondary salt-affected soils develop as a result of improper methods of irrigation, but there are other human effects that more and more often trigger this process in many places, both in arid and humid areas.

Some of these anthropogenic processes are, including but not limited to, the following:

(a) Overgrazing

This process occurs mainly in arid and semi-arid regions, where the natural soil cover is poor and scarcely satisfies the fodder requirement of rather extensive animal husbandry. If, due to overgrazing, the natural vegetation is sparse or annihilated, progressive salinization develops, and step by step, the scarcity of the plant cover becomes increasingly pronounced. Sometimes, the process ends in desertification, because even the poor pasture diminishes, and no other fodder resources are available. According to Theunissen [238], the gradual decline in the ecological condition of natural pastures as a result of overgrazing and the application of

insufficient management decisions, coupled with the detrimental effects of long-term drought, has left extensive areas of high-potential grazing land in southern Africa in urgent need of restoration. However, due to the limited number of grasses currently available for rehabilitating and restoring the vast number of different habitats encountered, the selection of indigenous grasses suitable for the restoration of denuded areas in the arid and semiarid grasslands of southern Africa has been initiated.

(b) Deforestation in semi-humid and semi-arid areas

Particularly in the last few decades, it has become evident that deforestation results in many tropical and subtropical countries in the salinization and alkalization of soils due to the effects of soil migration in both the upper and the lower layers. In Southeast India, for example, vast territories of former forest land became intensely saline and sodic (alkaline) within a few years after the annihilation of the woods. Similar phenomena occurred in the forest steppe areas in Russia, Iran, Eastern and Central Europe, Latin America, etc.

(c) Salinization caused by contamination with chemicals

Despite the fact that the amount of chemicals applied in agriculture is practically negligible in comparison to the salt content of several soils, we have considered the fact that this kind of salinization more and more often occurs in modern intensive agricultural production, particularly in greenhouses and intensive farming systems. When production takes place in semi-closed systems (e.g., greenhouses), where the chemicals applied will not be removed regularly, the accumulation of salts or their components becomes possible in the upper layer of the soil, resulting in salinity and sodicity (alkalinity). In Japan, the Netherlands, and other countries with intensive agriculture, and particularly horticulture, this type of salinization is appearing more and more frequently, causing serious losses of crop yields.

(d) Accumulation of air-borne or water-borne salts

Due to the concentration of industrial plants, the emitted chemical compounds may accumulate in the soil, and if their concentration is high enough, they result in salt accumulation in the upper layer of the soil.

A similar phenomenon appears when, due to water system regulations, sludge water disposal, and other hydrotechnical measures, water with considerable salt concentration contaminates the upper soil layer, causing salinization and/or alkalization.

1.6 RECLAMATION OF SALT-AFFECTED SOILS AND RELIEF OR ELIMINATION OF PARTICULAR PLANT/CROP STRESS FACTORS

The major environmental stresses caused by soil salinity and sodicity existed long before agricultural practices began. Soil salinity and sodicity have a substantial effect in reducing agricultural production worldwide [3, 7–9, 14, 17–29, 31–34, 36, 39], [42–44, 46, 47, 49, 51–53, 56–58, 60], [63, 64, 67, 68, 71–75, 79–84, 87–89, 93], [94, 99, 100, 102–115, 122–130, 133, 136, 139], [141, 143–148, 150, 151, 153, 154, 156], [170–178, 182, 185, 187, 189–193, 198], [201, 202, 205, 207, 208, 210], [215, 216, 223–228, 231, 232, 238–259, 264–271]. This has a major impact on increasing food and feed insecurity globally, particularly in developing countries that are more prone and vulnerable to salinization and desertification due to lack of advanced technology, inadequate education, and other socioeconomic and technological problems. Population growth and increasing demand for food and agricultural products necessitate using salt-affected soils and marginal lands for food production. These soils are needed for the extension of agriculture, and hence, reclamation is required. Reclamation is needed on the millions of hectares of slowly permeable salt-affected (i.e., saline–sodic and sodic) soils throughout the world [3, 10, 11, 16, 30, 37, 38, 40, 41, 48, 53, 55], [61, 62, 66, 68, 76, 77, 86, 88, 90, 97, 98], [107, 108, 111, 117, 118, 124, 128], [133, 134, 137, 142, 147, 150, 153, 157, 158], [179, 186, 188, 190, 195–197, 204, 207, 210], [212, 217–221, 223, 229, 231, 232, 238, 260, 267, 272].

Different techniques of reclamation and preventive measures or management practices are used for the reclamation of salt-affected soils and to reduce the salt content of the growth medium or to find more stress-tolerant plant/crop species and cultivars via genetic engineering to combat salinity stress. These management practices aim to enable plants to grow in saline and sodic conditions to use salt-affected soils for agricultural practices and food production [3, 4, 6, 10–12, 14, 16, 17, 21–27, 29, 30, 32, 36–38], [40–47, 50, 53, 55–57, 60–64, 67–71], [73–77, 79, 80, 82–84, 86–88, 90, 95–98, 100], [105–107, 110, 117, 118, 121, 124, 127–130], [133–135, 137, 139, 140, 142–144], [147, 148, 150, 152, 155–158, 186, 188–191], [193, 195–200, 202, 203, 206–208, 210], [212, 216–225, 227–233, 236, 238, 240], [244, 251, 254, 257, 260, 265, 266, 268, 272]. Saline soils are usually reclaimed by leaching the salts out of the soil through irrigation and drainage systems, whereas the reclamation of sodic (alkaline) soils requires the application of chemical amendments followed by the leaching process.

The present recommendations for the reclamation the salt-affected soils are usually based only on relatively simple and often empirical relationships. Various amendments and management strategies have been used for the reclamation of salt-affected soils. To evaluate particular reclamation strategies, some specific considerations should be noted, as follows:

1. The quantity of water needed
2. The quality of water needed
3. The quantity of amendments to be used
4. The type(s) of amendment(s) to be used
5. The time required for reclamation to be completed

Chemical reactions such as cation exchange, precipitation, and dissolution of solid phases (reclamation amendments) and the soil hydraulic properties and corresponding changes in the water flow and solute transport rates must be considered [221].

Among the various reclamation practices, usually, a combination of added gypsum amendment and crop rotation has proved to be best.

The reclamation of salt-affected (saline–sodic and sodic) soils by chemical amendments has become cost intensive and requires high capital investment, and it is not always a practical solution to the problem of soil salinity and sodicity. Therefore, biotic approaches such as the cultivation of salinity- and sodicity-tolerant plants and crops on salt-affected soils (*saline agriculture*) may be another alternative. The use of plant growth–promoting bacteria as one of the tools and biological methods for salinity stress management, salinity alleviation, and the reclamation of saline soils has been reported by several investigators [27, 73, 121, 130, 186, 218]. Salinity decreases the yield of many crops. The use of plant growth–promoting rhizobacteria (PGPR), beneficial bacteria that live in the plant root zone named the *rhizosphere*, is one of the solutions to solve this issue, thus remediating salinity problems and reclaiming saline soils [121]. Soil salinity, especially in arid and semi-arid regions, is one of the main limiting factors for plant growth, development, and production. Alfalfa (*Medicago sativa* L.) is cultivated on a vast scale in these regions, where its growth is substantially reduced by soil salinity. To find salinity-tolerant cultivars suitable for the saline soils of the arid regions, Ashrafi et al. [27] conducted an experiment to evaluate the efficacy of the inoculation of arbuscular mycorrhizal fungus (AMF), *Glomus mosseae*, or alfalfa rhizobia *Sinorhizobium meliloti* (R) into seed in the development of salinity tolerance of different alfalfa cultivars at a variety of salinity levels. These investigators reported that inoculation of alfalfa seed with AMF or R, especially double inoculation, caused a considerable increase in alfalfa yield under both saline and non-saline conditions by increasing colonization, nodulation, and nutrient uptake.

Among different reclamation techniques, the use of various soil amendments (both organic and inorganic) has been practiced for the reclamation of salt-affected soils. Recently, biochar (solid carbonaceous residue, produced under oxygen-free or oxygen-limited conditions at temperatures ranging from 300 to 1000 °C) has attracted considerable attention as a soil amendment [106, 152]. Evidence has shown that biochar addition is effective in improving the physical, chemical, and biological properties of salt-affected soils. Biochar application enhances the fertility of soil, as it improves the soil cation exchange capacity and water holding capacity (WHC), and in turn, improves the nutrient holding capacity of the soil [152, 155]. Ali et al. [11] reported that the application of biochar increased plant growth, biomass, and yield under either drought and/or salt stress and also increased photosynthesis, nutrient uptake, and modified gas exchange characteristics in drought- and salt-stressed plants. These investigators also reported that under drought stress, biochar increased the WHC of soil and improved the physical and biological properties of soils, while under salt stress, biochar decreased sodium ion (Na^+) uptake and increased potassium ion (K^+) uptake by plants. The biochar-mediated increase in the salt tolerance of plants is primarily associated with an improvement in soil properties, thus increasing plant water status, reducing Na^+ uptake, increasing the uptake of minerals, and regulating stomatal conductance and phytohormones. Farhangi-Abriz and Torabian [67] studied changes in antioxidant enzyme and osmotic adjustment in bean seedlings treated with biochar under salt stress and reported that biochar alleviated the negative impacts of salt stress in bean plants. However, some studies have found an increase in soil salinity and sodicity at high rates of biochar application. Panel et al. [152] pointed out the high cost associated with the production of biochar and its high application rates as significant challenges to the widespread use of biochar in salt-affected areas. The relatively limited information on the long-term behavior of salt-affected soils subjected to biochar applications is another drawback to its use in the reclamation of saline and sodic soils.

Different salinity- and sodicity-tolerant plant types and species (i.e., grasses [37, 42, 62, 76, 86, 103, 107, 110, 116–120], [131, 132, 159–171, 173–185, 188–190], [209, 213, 238, 250, 252, 264], agronomic crops [6, 21, 22, 26, 29, 36, 60, 69–71], [104, 108, 110, 114, 126, 127, 140, 153, 156], [203, 215, 224–228, 253, 256, 264–271], and forest species and trees [14, 41, 89, 219, 220, 230, 241, 258] have been cultivated by several investigators for reclamation purposes. These plants can mobilize the native lime (calcium carbonate, $CaCO_3$) in these soils through root action, a substitute for the chemical approach. Qadir et al. [188], studying the combination of chemical amendments and biological (using plants) reclamation techniques, reported that soil treated with gypsum at a high rate (100% GR) removed the greatest amount of Na^+ from the soil columns and resulted in a marked decrease in soil salinity (EC) and sodicity, sodium absorption ratio (SAR), and exchangeable sodium percentage (ESP). The performance of grass treatment in enhancing the leaching of Na^+ was between the gypsum treatments.

According to Kumar [117] and Qadir [188], the grass *Leptochloa fusca* is very useful and effective in the reclamation of salt-affected soils. This plant can tolerate extremely saline and sodic (alkaline) conditions. Since its growth is not affected by gypsum application, planting with *Leptochloa* is an alternative biological rather than chemical method for the reclamation of sodic (alkaline) soils. This plant is also well adapted to the waterlogging encountered on saline and sodic (alkaline) soils. The plant improves the soil's physical, chemical, and biological properties, so that within 2 or 3 years, many commercial and forage crops can be grown on the soil [117]. *Leptochloa* excretes salts through specialized glands and is, therefore, reasonably palatable to farm animals. It must be noted that because of its vigorous growth on sodic (alkaline) soils, *Leptochloa* does not allow satisfactory growth of companion plant species, especially in the initial years of soil reclamation.

Subramaniam and Babu [230] also used a forest shrub species for the reclamation of sodic soils. According to these investigators [230], *Sophora mollis*, which is a shrub to medium-sized tree and is used for both fodder and firewood, can be used in the reclamation of sodic (alkaline) soils.

Kilic et al. [111] investigated the salt-removing capacity of purslane (*Portulaca oleracea* L.) by studying different stress criteria and by tracking its salt removal from germination to harvest. The results of their study showed that purslane could cumulatively remove considerable amounts of salt from the soil if it is practical to cultivate it as an intercrop all year round.

Saltgrass (*Distichlis spicata*), which has been found as the only vegetation cover on a highly sodic (alkaline) soil in Wilcox Playa, AZ, United States [178], can also be very effective in the reclamation of saline and sodic soils. As mentioned earlier, the senior author of this chapter and his co-workers found this grass to be a very highly salt-tolerant plant species that grows vigorously on saline and sodic soils [132, 159–164, 168, 171, 173–179]. Compared with the other highly salt-tolerant halophytes that have been tested by this investigator [165–167, 169, 170, 175, 180–184], so far, this grass has proved the most salt and drought tolerant of all the tested species.

Although it is slow, a definite improvement is achieved in the physico-chemical properties of salt-affected soils by encouraging the growth of vegetation on such lands. Tree species in general are effective in improving the soil properties, as reflected by the changes in physico-chemical characteristics of the soil, such as bulk density (BD), WHC, hydraulic conductivity (HC), pH, EC, OC, N, and exchangeable cations (Na^+ and Ca^{++}) [219].

Due to the low biological activity and depressed microbial growth of salt-affected (i.e., saline, saline–sodic, and sodic) soils, there is a need for applying organic amendments (i.e., plant residue or manure) during sodic (alkaline) soil reclamation. In the reclamation of saline soils, organic amendments must be applied following the leaching process.

Kumar et al. [118] conducted a study of a combination of biological and chemical reclamation on a highly sodic (alkaline) soil. These investigators [118] found that rice produced satisfactory yields in the first year of gypsum application, but sorghum and *Sesbania* yields were very poor. The yield of *Leptochloa* was not affected by gypsum application. In their crop rotation practice, Kumar et al. [118] reported that the green forage yield of sorghum was greatest when sorghum followed *Leptochloa*, grown for 2 years, and the harvested grass was left to decompose on the site.

In a biological reclamation study of saline soils, Helalia et al. [86] reported that amshot grass significantly reduced the soil salinity compared with either ponding or gypsum application, and this grass produced a higher fresh yield than clover cultivated in such soils.

These findings indicate that biological reclamation with salinity- or sodicity-tolerant plants (i.e., *Leptochloa*, grasses, shrubs, or trees) is a suitable substitute for chemical reclamation with gypsum, and the former has an economic advantage over the latter.

Yildirim et al. [257] evaluated the effects of selected biologicals on direct-seeded and transplanted squash plant growth and mineral content under salinity stress. These investigators reported that salinity negatively affected the growth of squash; however, biological treatments significantly increased the fresh weight compared with non-treated plants that were under salt stress. They also found that biological treatments increased the uptake of potassium compared with the non-treated control in both direct-seeded and transplanted squash. Based on their results, these investigators concluded that alteration of mineral uptake may be one mechanism for the alleviation of salt stress, and the use of biological treatments may provide a means of facilitating plant growth under salt stress conditions.

Compost, or any other organic material, is recommended for use during the reclamation process of salt-affected soils. The results of a field experiment conducted by Avnimelech et al. [30] verified that compost application improved both physical and chemical conditions of saline and sodic (alkaline) soils. Compost application to such soils is expected to release acids, which would ultimately lead to the replacement of exchangeable sodium by calcium. In addition, compost application would stabilize the soil structure and enhance plant growth. These investigators [30] found that the application of municipal solid waste compost was equivalent, or even superior, to the addition of gypsum, the most common amendment used to reclaim sodic (alkaline) soils. This was evident from the substantial increase in crop yields. The combined application of compost and gypsum raised yields to levels equal to those of the commercial fields.

In a field experiment, Batra et al. [37] compared the microbiological and chemical amelioration of a highly deteriorated sodic (alkaline) soil using two reclamation technologies:

1. Growing Karnal grass (*Leptochloa fusca*) as a first crop with no chemical amendment (biological reclamation)
2. Applying gypsum as a chemical amendment for different crop rotations

These investigators [37] reported that the microbiological properties changed more than the chemical properties of sodic (alkali) soil as the time period advanced.

In a biological reclamation study carried out on saline soils, Apte and Thomas [16] found that a brackish-water, nitrogen-fixing cyanobacterium, *Anabaena torulosa*, could successfully grow and fix nitrogen on moderately saline soils (EC of 5 to 8.5 dS m^{-1}). These investigators [16] reported that the cyanobacterium exhibited high rates of nitrogen fixation and substantially enriched the nitrogen status of saline soils. However, permanent removal of Na$^+$ from saline soils using cyanobacteria or any other microorganisms may not be possible, since Na$^+$ is released back into the soil subsequent to the death and decay of cyanobacteria or other microorganisms. The amelioration of soil salinity by simultaneous application of *Anabaena torulosa* during crop growth seems to be an attractive possibility for reclamation, especially since it can

also supplement the nitrogen requirement of the crops growing on these soils.

Blue-green algae that tolerate excess Na and grow extensively on the soil surface in wet seasons were found effective in the reclamation of sodic soils [195]. However, a permanent reclamation of such soils using only blue-green algae as a biological amendment to achieve sodic (alkaline) soil reclamation is neither possible nor comparable to an effective chemical amendment such as gypsum.

In the reclamation process of saline soils, de Villiers et al. [57] compared different annual and perennial species. Of the six species tested, the perennials seemed to be more effective and better suited for rehabilitation purposes under saline soil conditions.

The type of chemical compound being used also influences the reclamation process of salt-affected soils. Sharma and Upadhyay [217] reported that among the up-to-date known chemical compounds, cyclohexathiazenium chloride $(S_6N_4)^{2+}Cl^{2-}$ is the best and most suitable chemical to reclaim sodic (alkaline) soil at any pH.

When good-quality water is not available for leaching the salts out of the soil, low-quality water can be used for the initial stages of reclamation. In this regard, Singh and Bajwa [223] studied the effects of gypsum and sodic irrigation on the precipitation of Ca^{++} and removal of Na$^+$ from a sodic soil reclaimed with different levels of gypsum and growth of rice in a greenhouse experiment. Dubey and Mondal [61] also used low-quality, saline water in conjunction with organic and inorganic amendments for the initial stages of reclamation of sodic soils. Using low-quality water, Joshi and Dhir [98] evaluated the rehabilitation of degraded sodic soils using residual sodium carbonate water (low-quality water) combined with gypsum treatment and found that the combination treatment was effective in lowering the soil SAR and improved the water infiltration rate. In the first year of gypsum treatment, it was possible to establish the crop. In the second year, moderate productions of wheat (2610 kg ha^{-1}) and raga (*Brassica* sp.) (2000 kg ha^{-1}) were obtained [98].

Using the most common technique, an irrigation water and drainage system, for the reclamation of salt-affected soils, the results of an investigation carried out by Millette et al. [133] demonstrated the ability of fall irrigation to leach salts from the surface soil during a period of low consumptive use, which could lead to reclamation. Long-term monitoring would be required to determine whether a further and permanent decline in salinity could be achieved.

Concerning other reclamation materials and techniques, the results of Jones et al. [97] indicate that acid whey is effective in reclaiming sodic soil by lowering ESP, SAR, and pH and by improving infiltration rate. Rao and Leeds Harrison [197] used simulation models for the desalinization of a drained two-layered saline soil using surface irrigation for different water management practices to increase leaching efficiency. Based on image elements and their correlation with the ground features, Rao et al. [194] suggested categorizing sodic soils into moderately and strongly sodic groups. The delineation thus made would help the execution of a reclamation program for sodic soils at the study sites. Abdel-Hamid et al. [1] monitored soil salinity in the northern Nile delta of Egypt using data collected via Landsat and geographic information system (GIS). The collected data were used in making recommendations for the reclamation of the saline soils of the Nile delta area.

The vast area of salt-affected soils still remains a burden for societies, particularly undeveloped countries, in need of adequate resources to reclaim them, with the available technology involving initial heavy investments. The process of degradation due to reckless destruction of vegetation can be reversed by the re-establishment of vegetative cover, which results in a slow but definite improvement in such soils. This phenomenon has been frequently demonstrated by various parameters influencing the soil welfare in several investigations showing positive signs of improvement in terms of both physical and chemical properties of salt-affected soils. Such soils should, therefore, be brought under any type of vegetation cover (i.e., sods, shrubs, trees) if they are not found to be economical for regular farming and growing agronomic crops, and they should be taken care of by the community for posterity [219].

Even for the execution of reclamation processes, the status and behavior of nutrients in salt-affected soils (i.e., saline–sodic and sodic soils) during reclamation by crop rotation and chemical amendment requires a comprehensive assessment. This is because usually, during the leaching process of the soluble salts and the exchangeable sodium, some soil nutrients are also lost and leached out of the soil. In this regard, several investigators [41, 53, 189, 233, 258] have studied nutrient status and behavior during reclamation processes. Swarup et al. [233] reported the effect of gypsum on the behavior of soil phosphorus during the reclamation of a sodic soil. According to Bhojvaid et al. [41], soil nutrient status under a tree plantation was higher than that of the non-sodic farm soil. This finding confirms that successful tree plantation may restore the productivity and fertility of highly degraded sodic soils.

Regardless of the techniques used in the reclamation of salt-affected soils, post-reclamation management practices, i.e., the proper choice of crops, crop rotation, method of irrigation, the quality and quantity of water used for irrigation and reclamation, fertilization, and the economics of reclamation, must be taken into consideration and followed to achieve successful results.

1.7 CONCLUDING REMARKS

In this chapter, information has been given on the important functions of the soil in relation to soil-originated stress factors for plant and crop growth and development as well as a little more detailed information on particular problems related to salt-affected soils and their formation and reclamation.

The properties of the stress factors for plant and crop growth originating in soil are diverse and multi-faceted. We know comparatively little about the up-to-date orientation, particularly for finding methods to improve the situation and ensure better plant and crop growth and development. Therefore, target-oriented studies on the different kinds of soil-originated stress factors for plant and crop growth and development are necessary so that the complex correlations and actions in the soil–plant–water system can be revealed with the aim of a better characterization of stress factors, on the one hand, and improving environmental and production conditions, on the other hand. For this to be successful, it is necessary to understand how plants/crops respond to salinity and sodicity, the relative tolerances of different plants/crops and their sensitivity at different stages of growth, and how different soils and environmental conditions affect salt-stressed plants. By following these guidelines, agricultural practitioners and growers can expect to be more successful in any future growing practices and endeavors for more food and feed production to guarantee future food security for a growing and expanding population.

REFERENCES

1. Abdel-Hamid, M.A., D. Shrestha, and C. Valenzuela. 1992. Delineating Mapping and Monitoring of Soil Salinity in the Northern Nile Delta Egypt Using Landsat Data and a Geographic Information System. *Egyptian Journal of Soil Science*, 32(3):463481.
2. Abdul Qados, A.M.S. 2011. Effect of Salt Stress on Plant Growth and Metabolism of Bean Plant *Vicia faba* (L.). *Journal of the Saudi Society of Agricultural Sciences*, 10(1):7–15.
3. Adcock, D., A.M. McNeill, G.K. McDonald, and R.D. Armstrong. 2007. Subsoil Constraints to Crop Production on Neutral and Alkaline Soils in South-eastern Australia: A Review of Current Knowledge and Management Strategies. *Australian Journal of Experimental Agriculture*, 47(11):1245–1261.
4. Afzal, I., S.M.A. Basra, A. Hameed, and M. Farooq. 2006. Physiological Enhancements for Alleviation of Salt Stress in Wheat. *Pakistan Journal of Botany*, 38(5):1649–1659.
5. Ahmad, N. 1965. A Review of Salinity-Alkalinity Status of Irrigated Soils of West Pakistan. *Agrokemia es Talajtan*, 14(Suppl.):117–154.
6. Ahmad, S., J.D.H. Keatinge, A. Ali, and B.R. Khan. 1992. Selection of Barley Lines Suitable for Spring Sowing in the Arid Highlands of Baluchistan. *Sarhad Journal of Agriculture, Pakistan*, 8(1):49–56.
7. Ahmadi, A., Y. Emam, and M. Pessarakli. 2009. Response of Various Cultivars of Wheat and Maize to Salinity Stress. *Journal of Agriculture, Food, and Environment (JAFE)*, 7(1):123–128.
8. Akinci, S., K. Yilmaz, and I.E. Akinci. 2004. Response of Tomato (*Lycopersicon esculentum* Mill.) to Salinity in the Early Growth Stages for Agricultural Cultivation in Saline Environments. *Journal of Environmental Biology*, 25(3):351–357.
9. Al-Busaidi, A., T. Yamamoto, M. Inoue, M. Irshad, Y. Mori, and S. Tanaka. 2007. Effects of Seawater Salinity on Salt Accumulation and Barley (*Hordeum vulgare* L.) Growth under Different Meteorological Conditions. *Journal of Food Agriculture and Environment (JAFE)*, 5(2):270–279.
10. Ali, M. 2011. Management of Salt-Affected Soils. *Practice of Irrigation and On-farm Water Management*, 2:271–325.
11. Ali, S., M. Rizwan, M.F. Qayyum, Y.S. Ok, M. Ibrahim, M. Riaz, and A.N. Shahzad. 2017. Biochar Soil Amendment on Alleviation of Drought and Salt Stress in Plants: A Critical Review. *Environmental Science and Pollution Research International*, 24(14):12700–12712.

12. Alimova, R.Kh., R.Z. Kopp, A.A. Agazmkhodzhayev, and A. Dusmukhamedov. 1993. Stabilization of Saline Soils of the Aral Sea Coastal Region by Complex Additives. *Eurasian Soil Science*, 25(5):8388.

13. Allbed, A. and L. Kumar. 2013. Soil Salinity Mapping and Monitoring in Arid and Semi-Arid Regions Using Remote Sensing Technology: A Review. *Advances in Remote Sensing*, 2(4):373–385.

14. Altman, A. 2003. From Plant Tissue Culture to Biotechnology: Scientific Revolutions, Abiotic Stress Tolerance, and Forestry. *In Vitro Cellular and Developmental Biology-Plant*, 39(2):75–84.

15. Amirinejad, A.A., M. Sayyari, F. Ghanbari, and S. Kordi. 2017. Salicylic Acid Improves Salinity-Alkalinity Tolerance in Pepper (*Capsicum annuum* L.). *Advances in Horticultural Science*, 31(3):157–163.

16. Apte, S.K. and J. Thomas. 1997. Possible Amelioration of Coastal Soil Salinity Using Halotolerant Nitrogen Fixing Cyanobacteria. *Plant and Soil*, 189(2):205–211.

17. Arzani, A. 2008. Improving Salinity Tolerance in Crop Plants: A Biotechnological View. *In-Vitro Cellular and Developmental Biology-Plant*, 44(5):373–383.

18. Asch, F. and M.C.S. Wopereis. 2001. Responses of Field-grown Irrigated Rice Cultivars to Varying Levels of Floodwater Salinity in a Semi-arid Environment. *Field Crops Research*, 70(2):127–137.

19. Asch, F., M. Dingkuhn, and K. Dorffling. 2000. Salinity Increases CO_2 Assimilation but Reduces Growth in Field-grown, Irrigated Rice. *Plant and Soil*, 218(1–2):1–10.

20. Asfaw, K.G. 2011. The Response of Some Haricot Bean (*Phaseolus vulgaris*) Varieties for Salt Stress during Germination and Seedling Stage. *Current Research Journal of Biological Sciences*, 3(4):282–288.

21. Ashraf, M. 1994. Breeding for Salinity Tolerance in Plants. *Critical Review, Plant Sciences*, 13:17–42.

22. Ashraf, M. 2004. Some Important Physiological Selection Criteria for Salt Tolerance in Plants. *Flora*, 199:361–376.

24. Ashraf, M. and M.R. Foolad. 2005. Pre-sowing Seed Treatment – A Shotgun Approach to Improve Germination, Plant Growth, and Crop Yield under Saline and Non-saline Conditions. *Advances in Agronomy*, 88(Special Issue):223–271.

25. Ashraf, M. and A. Orooj. 2006. Salt Stress Effects on Growth, Ion Accumulation and Seed Oil Concentration in an Arid Zone Traditional Medicinal Plant Ajwain (*Trachyspermum ammi*) [L.] Sprague. *Journal of Arid Environments*, 64(2):209–220.

23. Ashraf, M., S.M. Akhtar, N. Imtiaz, and A. Ali. 2017. Salinization/Sodification of Soil and Physiological Dynamics of Sunflower Irrigated with Saline-sodic Water Amending by Potassium and Farm Yard Manure. *Journal of Water Reuse and Desalination*, 7(4):476–487.

26. Ashraf, M., H.R. Athar, P.J.C. Harris, and T.R. Kwon. 2008. Some Prospective Strategies for Improving Crop Salt Tolerance. *Advanced Agronomy*, 97:45–110.

27. Ashrafi, E., M. Zahedi, and J. Razmjoo. 2014. Co-inoculations of Arbuscular Mycorrhizal Fungi and Rhizobia under Salinity in Alfalfa. *Soil Science and Plant Nutrition*, 60(5):619–629.

28. Asmamaw, M., H. Ashenafi, and A. Gezai Abera. 2018. Characterization and Classification of Salt Affected Soils and Irrigation Water in Tendaho Sugarcane Production Farm, North-eastern Rift Valley of Ethiopia. *African Journal of Agricultural Research*, 13(9):403–411.

29. Athar, H.R., A. Khan, and M. Ashraf. 2008. Exogenously Applied Ascorbic Acid Alleviates Salt-Induced Oxidative Stress in Wheat. *Environmental and Experimental Botany*, 63:224–231.

30. Avnimelech, Y., D. Shkedy, M. Kochva, and Y. Yotal. 1994. The Use of Compost for the Reclamation of Saline and Alkaline Soils. *Compost Science and Utilization*, 2(3):611.

31. Awasthi, P., H. Karki, K. Bargali, and S.S. Bargali. 2016. Germination and Seedling Growth of Pulse Crop (*Vigna* spp.) as Affected by Soil Salt Stress. *Current Agriculture Research Journal*, 4(2):159–170.

32. Aydin, A., C. Kant, and M. Turan. 2012. Humic Acid Application Alleviate Salinity Stress of Bean (*Phaseolus vulgaris* L.) Plants Decreasing Membrane Leakage. *African Journal of Agricultural Research*, 7(7):1073–1086.

33. Azizpour, K., M.R. Shakiba, N.A. Khosh Kholg Sima, H. Alyari, M. Mogaddam, E. Esfandiari, and M. Pessarakli. 2010. Physiological Response of Spring Durum Wheat Genotypes to Salinity. *Journal of Plant Nutrition*, 33(6):859–873.

34. Bahaji, A., I. Mateu, A. Sanz, and M.J. Cornejo. 2002. Common and Distinctive Responses of Rice Seedlings to Saline- and Osmotically-Generated Stress. *Plant Growth Regulation*, 38(1):83–94.

35. Bai, L., C. Wang, S. Zang, Y. Zhang, Q. Hao, and Y. Wu. 2016. Remote Sensing of Soil Alkalinity and Salinity in the Wuyu'er-Shuangyang River Basin, Northeast China. *Remote Sensing*, 8(2):163.

36. Bao, A.K., S.M. Wang, G.Q. Wu, J.J. Xi, J.L. Zhang, and C.M. Wang. 2009. Over-expression of the Arabidopsis H^+-PPase Enhanced Resistance to Salt and Drought Stress in Transgenic Alfalfa (*Medicago sativa* L.). *Plant Science*, 176(2):232–240.

37. Batra, L., A. Kumar, M.C. Manna, and R. Chhabra. 1997. Microbiological and Chemical Amelioration of Alkaline Soil by Growing Karnal Grass and Gypsum Application. *Experimental Agriculture*, 33(4):389397.

38. Bauder, J.W. and T.A. Brock. 1992. Crop Species Amendment and Water Quality Effects on Selected Soil Physical Properties. *Soil Science Society of America Journal*, 56(4):12921298.

39. Ben-Gal, A. and U. Shani. 2002. Yield, Transpiration and Growth of Tomatoes under Combined Excess Boron and Salinity Stress. *Plant and Soil*, 247(2):211–221.

40. Bennett, D.R. 1990. Reclamation of Saline Soils Adjacent to Rehabilitated Irrigation Canals. *Canadian Agricultural Engineering*, 32(1):116.

41. Bhojvaid, P.P., V.R. Timmer, and G. Singh. 1996. Reclaiming Sodic Soils for Wheat Production by *Prosopis juliflora* (Swartz) DC Afforestation in India. *Agroforestry Systems*, 34(2):139150.

42. Bhuiyan, M., A. Raman, D. Hodgkins, D. Mitchell, and H. Nicol. 2015. Physiological Response and Ion Accumulation in Two Grasses, One Legume, and One Saltbush under Soil Water and Salinity Stress. *Ecohydrology*, 8:1547–1559.

43. Blanco, F.F., Folegatti, M.V., Gheyi, H.R., and P.D. Fernandes. 2008. Growth and Yield of Corn Irrigated with Saline Water. *Scientia Agricola*, 65(6):574–580.

44. Bochow, H., S.F. El-Sayed, H. Junge, A. Stavropoulou, and G. Schmiedeknecht. 2001. Use of *Bacillus subtilis* as Biocontrol Agent. IV. Salt-stress Tolerance Induction by *Bacillus subtilis* FZB24 Seed Treatment in Tropical Vegetable Field Crops, and Its Mode of Action. *Zeitschrift Fur Pflanzenkrankheiten Und Pflanzenschutz – Journal of Plant Diseases and Protection*, 108(1):21–30.

45. Bohnert, H.J., D.E. Nelson, and R.G. Jensen. 1995. Adaptations to Environmental Stresses. *Plant Cell*, 7(7):10991111.

46. Bonilla, I., A. El-Hamdaoui, and L. Bolanos. 2004. Boron and Calcium Increase *Pisum sativum* Seed Germination and Seedling Development under Salt Stress. *Plant and Soil*, 267(1–2):97–107.

47. Borsani, O., J. Cuartero, J.A. Fernandez, V. Valpuesta, and M.A. Botella. 2001. Identification of Two Loci in Tomato Reveals Distinct Mechanisms for Salt Tolerance. *Plant Cell*, 13(4):873–887.

48. Callaghan, M.V., E.E. Cey, and L.R. Bentley. 2014. Hydraulic Conductivity Dynamics during Salt Leaching of a Sodic, Structured Subsoil. *Soil Science Society of America Journal*, 78:1563–1574.

49. Campos, C.A.B., P.D. Fernandes, H.R. Gheyi, F.F. Blanco, and S.A.F. Campos. 2006. Yield and Fruit Quality of Industrial Tomato under Saline Irrigation. *Scientia Agricola*, 63(2):146–152.

50. Canellas, L.P., F.L. Olivares, N.O. Aguiar, D.L. Jones, A. Nebbioso, P. Mazzei, and A. Piccolo. 2015. Humic and Fulvic Acids as Biostimulants in Horticulture – A Review. *Scientia Horticulturae*, 196:15–27.

51. Cantrell, I.C. and R.G. Linderman. 2001. Preinoculation of Lettuce and Onion with VA Mycorrhizal Fungi Reduces Deleterious Effects of Soil Salinity. *Plant and Soil*, 233(2):269–281.

52. Chaparzadeh, N., Y. Aftabi, M. Dolati, F. Mehrnejad, and M. Pessarakli. 2014. Salinity Tolerance Ranking of Various Wheat Landraces from West of Uremia Lake in Iran by Using Physiological Parameters. *Journal of Plant Nutrition*, 37:1025–1039.

53. Chauhan, R.P.S. 1995. Effect of Amendments of Sodic Soil Reclamation and Yield and Nutrient Uptake of Rice (*Oryza sativa*) under Rice Fallow Rice System. *Indian Journal of Agricultural Science*, 65(6):438–441.

54. Chernousenko, G.I. and S.S. Kurbatskaya. 2017. Soil Salinization in Different Natural Zones of Intermontane Depressions in Tuva. *Eurasian Soil Science*, 50(11):1255–1270.

55. Dahiya, I.S. and R. Anlauf. 1990. Sodic Soils in India, Their Reclamation and Management. *Zeitschrift Fuer Kulturtechnik und Landentwicklung*, 31(1):26–34.

56. Dasgan, H.Y., H. Aktas, K. Abak, and I. Cakmak. 2002. Determination of Screening Techniques to Salinity Tolerance in Tomatoes and Investigation of Genotype Responses. *Plant Science*, 163(4):695–703.

57. de Villiers, A.J., M.W. van Rooyen, G.K. Theron, and A.S. Claassens. 1997. Tolerance of Six Namaqualand Pioneer Species to Saline Soil Conditions. *South African Journal of Plant and Soil*, 14(1):38–42.

58. di Caterina, R., M.M. Giuliani, T. Rotunno, A. de Caro, and Z.Flagella. 2007. Influence of Salt Stress on Seed Yield and Oil Quality of Two Sunflower Hybrids. *Analysis of Applied Biology*, 151(2):145–154.

59. Dinc, U., S. Senol, S. Kapur, M. Sari, M.R. Derici, and M. Sayin. 1991. Formation, Distribution, and Chemical Properties of Saline and Alkaline Soils of the Cukurova Region Southern Turkey. *Catena*, 18(2):173–183.

60. Djilianov, D., E. Prinsen, S. Oden, H. van Onckelen, and J. Muller. 2003. Nodulation under Salt Stress of Alfalfa Lines Obtained after In Vitro Selection for Osmotic Tolerance. *Plant Science*, 165(4):887–894.

61. Dubey, S.K. and R.C. Mondal. 1993. Sodic Reclamation with Saline Water in Conjunction with Organic and Inorganic Amendments. *Arid Soil Research and Rehabilitation*, 7(3):219–231.

62. Epstein, E., J.D. Norlyn, D.W. Rush, R.K. Kingsbury, D.B. Kelley, and A.F. Warna. 1980. Saline Culture of Crops: A Genetic Approach. *Science*, 210:339–404.

63. Erdal, S., M.Genisel, H.Turk, and Z. Gorcek. 2012. Effects of Progesterone Application on Antioxidant Enzyme Activities and K^+/Na^+ Ratio in Bean Seeds Exposed to Salt Stress. *Toxicology and Industrial Health*, 28(10):942–946.

64. Fahad, S., S. Hussain, and A. Matloob. 2015. Phytohormones and Plant Responses to Salinity Stress: A Review. *Plant Growth Regulation*, 75(2):391–404.

65. FAO. 1971. Irrigation and Drainage Paper 7, Rome.

66. FAO. 2005. *Global Network on Integrated Soil Management for Sustainable Use of Salt-Affected Soils*. Rome, Italy: FAO Land and Plant Nutrition Management Service. www.fao.org/ag/agl/agll/spush

67. Farhangi-Abriz, S. and S. Torabian. 2017. Antioxidant Enzyme and Osmotic Adjustment Changes in Bean Seedlings as Affected by Biochar under Salt Stress. *Journal of Ecotoxicology and Environmental Safety*, 137:64–70.

68. Fileccia, V., P. Ruisi, R. Ingraffia, D. Giambalvo, F.A. Salvatore, and F. Martinelli. 2017. Arbuscular Mycorrhizal Symbiosis Mitigates the Negative Effects of Salinity on Durum Wheat. *PLoS One*, 12(9): e0184158.

69. Flowers, T.J. 2004. Improving Crop Salt Tolerance. *Journal of Experimental Botany*, 55(96):307–319.

70. Flowers, T.J. and T.D. Colmer. 2015. Plant Salt Tolerance: Adaptations in Halophytes. *Annals of Botany*, 115(3):327–331.

71. Foolad, M.R. 2004. Recent Advances in Genetics of Salt Tolerance in Tomato. *Plant Cell Tissue and Organ Culture*, 76(2):101–119.

72. Forieri, H., U. Hildebrandt, and M. Rostás. 2016. Salinity Stress Effects on Direct and Indirect Defense Metabolites in Maize. *Environmental and Experimental Botany*, 122:68–77.

73. Forni, C., D. Duca, and B.R. Glick. 2017. Mechanisms of Plant Response to Salt and Drought Stress and Their Alteration by Rhizobacteria. *Plant Soil*, 410:335.

74. Gao, S.M., H.W. Zhang, Y. Tian, F. Li, Z.J. Zhang, X.Y. Lu, X.L. Chen, and R.F. Huang. 2008. Expression of TERF1 in Rice Regulates Expression of Stress-Responsive Genes and Enhances Tolerance to Drought and High-Salinity. *Plant Cell Reports*, 27(11):1787–1795.

75. Ge, H., N. Zhao, Y. Miao, M. Chen, and X. Wang. 2016. Research Progress on Salt Tolerance of Faba Bean. *Agricultural Science & Technology*, 17(3):569–572.

76. Gharaibeh, M., N. Eltaif, and A. Albalasmeh. 2011. Reclamation of Highly Calcareous Saline Sodic Soil Using *Atriplex halimus* and Byproduct Gypsum. *International Journal of Phytoremediation*, 13(9):873–883.

77. Ghosh, S., P. Lockwood, N. Hulugalle, H. Daniel, P. Kristiansen, and K. Dodd. 2010. Changes in Properties of Sodic Australian Vertisols with Application of Organic Waste Products. *Soil Science Society of America Journal*, 74:153–160.

78. Goldshleger, N., E. Ben-Dor, R. Lugassi, and G. Eshel. 2010. Soil Degradation Monitoring by Remote Sensing: Examples with Three Degradation Processes. *Soil Science Society of America Journal*, 74:1433–1445.

79. Golldack, D., I. Lüking, and O. Yang. 2011. Plant Tolerance to Drought and Salinity: Stress Regulating Transcription Factors and Their Functional Significance in the Cellular Transcriptional Network. *Plant Cell Reports*, 30(8):1383–1391.

80. Greenway, H. and R. Munns. 1980. Mechanisms of Salt Tolerance in Non-halophytes. *Annual Review of Plant Physiology*, 31:149–190.

81. Grieve, C.M., L.E. Francois, and J.A. Poss. 2001. Effect of Salt Stress during Early Seedling Growth on Phenology and Yield of Spring Wheat. *Cereal Research Communications*, 29(1–2):167–174.

82. Gulnaz, A., J. Iqbal, S. Farooq, and F. Azam. 1999. Seed Treatment with Growth Regulators and Crop Productivity. I. 2,4-D as an Inducer of Salinity-Tolerance in Wheat (*Triticum aestivum* L.). *Plant and Soil*, 210(2):209–217.

83. Hanafy, M.S., A.El-banna, H.M.Schumacher, H.J.Jacobsen, and F.S. Hassan. 2013. Enhanced Tolerance to Drought and Salt Stresses in Transgenic Faba Bean (*Vicia faba* L.) Plants by Heterologous Expression of the PR10a Gene from Potato. *Plant Cell Reports*, 32(5):663–674.

84. Hasanuzzaman, M., K. Nahar, and M. Fujita. 2013. Plant Response to Salt Stress and Role of Exogenous Protectants to Mitigate Salt-induced Damages. In: P. Ahmad, M. Azooz, and M.Prasad (eds.), *Ecophysiology and Responses of Plants under Salt Stress.*, New York, NY: Springer.

85. Heilig, J., J. Kempenich, J. Doolittle, E.C. Brevik, and M. Ulmer. 2011. Evaluation of Electromagnetic Induction to Characterize and Map Sodium-affected Soils in the Northern Great Plains. *Soil Horizons*, 52(3):77.

86. Helalia, A.M., S. El-Amir, S.T. Abou-Zeid, and K.F. Zaghloul. 1992. Bio-reclamation of Saline-sodic Soil by Amshot Grass in Northern Egypt. *Soil and Tillage Research*, 22(12):109–116.

87. Hernansez, L., O. Loyola-Gonzalez, B. Valle, J. Martinez, L. Diaz-Lopez, C. Aragon, and J.C. Lorenzo. 2015. Identification of Discriminant Factors after Exposure of Maize and Common Bean Plantlets to Abiotic Stresses. *Notulae Botanicae Horti Agrobotanici Cluj-Napoca*, 43(2):589.

88. Hijikata, N. 2019. On-site Use of Reclaimed Greywater. In: N.Funamizu (ed.), *Resource-Oriented Agro-sanitation Systems*, pp. 243–268, Tokyo: Springer.

89. Hokmabadi, H., K. Arzani, and P.F. Grierson. 2005. Growth, Chemical Composition, and Carbon Isotope Discrimination of Pistachio (*Pistacia vera* L.) Rootstock Seedlings in Response to Salinity. *Australian Journal of Agricultural Research*, 56(2):135–144.

90. Ilyas, M., R.H. Qureshi, and M.A. Qadir. 1997. Chemical Changes in a Saline-sodic Soil after Gypsum Application and Cropping. *Soil Technology*, 10(3):247–260.

91. Iqbal, F. 2011. Detection of Salt Affected Soil in Rice-Wheat Area Using Satellite Image. *African Journal of Agricultural Research*, 6:4973–4982.

92. Iqbal, S. and N. Mastorakis. 2015. Soil Salinity Detection Using Remote Sensing (RS) Data. In: *Advances in Environmental Science and Energy Planning*, pp. 277–281. ISBN 978-1-61804-280-4

93. Izadi, M.H., J. Rabbani, Y. Emam, M. Pessarakli, and A. Tahmasebi. 2014. Effects of Salinity Stress on Physiological Performance of Various Wheat and Barley Cultivars. *Journal of Plant Nutrition*, 37:520–531.

94. Janmohammadi, M., A. Abbasi, and N. Sabaghnia. 2012. Influence of NaCl Treatments on Growth and Biochemical Parameters of Castor Bean (*Ricinus communis* L.). *Acta Agriculturae Slovenica*, 99(1):31.

95. Javid, M.G., A. Sorooshzadeh, F. Moradi, M. Sanavy, S.A. Mohammad, and I. Allahdadi. 2011. The Role of Phytohormones in Alleviating Salt Stress in Crop Plants. *Australian Journal of Crop Science*, 5(6):726–734 [online].

96. Jithesh, M.N., S.R. Prashanth, K.R. Sivaprakash, and A.K. Parida. 2006. Antioxidative Response Mechanisms in Halophytes: Their Role in Stress Defense. *Journal of Genetics*, 85(3):237–254.

97. Jones, S.B., C.W. Robbins, and C.L. Hansen. 1993. Sodic Soil Reclamation Using Cottage Cheese Acid Whey. *Arid Soil Research and Rehabilitation*, 7(1):5161.

98. Joshi, D.C. and R.P. Dhir. 1991. Rehabilitation of Degraded Sodic Soils in an Arid Environment by Using Residual Sodium Carbonate Water for Irrigation. *Arid Soil Research and Rehabilitation*, 5(3):175–185.

99. Jouyban, Z. 2012. The Effects of Salt Stress on Plant Growth. *Technical Journal of Engineering and Applied Sciences*, 2: 7–10.

100. Kant, C., A. Aydin, and M. Turan. 2008. Ameliorative Effect of Hydro Gel Substrate on Growth, Inorganic Ions, Proline, and Nitrate Contents of Bean under Salinity Stress. *Journal of Plant Nutrition*, 31(8):1420–1439.

101. Kar, A. 2018. Desertification Causes and Effects, 48p. In: D. Bartlett and R. Singh (eds.), *Exploring Natural Hazards, A Case Study Approach.* New York: Chapman & Hall/CRC Press. eBook ISBN: 9781351681230.

102. Karami Chame, S., B. Khalil-Tahmasbi, P. ShahMahmoodi, A. Abdollahi, A. Fathi, S.J. Seyed Mousavi, M. Hossein Abadi, S. Ghoreishi, and S. Bahamin. 2016. Effects of Salinity Stress, Salicylic Acid and Pseudomonas on the Physiological Characteristics and Yield of Seed Beans (*Phaseolus vulgaris* L.). *Scientia Agriculturae*, 14(2):234–238.

103. Karimi, I.Y.M., S.S. Kurup, M.A.M.A. Salem, A.J. Cheruth, F.T. Purayil, S. Subramaniam, and M. Pessarakli. 2018. Evaluation of Bermuda and Paspalum Grass Types for Urban Landscapes under Saline Water Irrigation. *Journal of Plant Nutrition*, 41(7):888–902.

104. Katerji, N., J.W. van Hoorn, A. Hamdy, and M. Mastrorilli. 2003. Salinity Effect on Crop Development and Yield, Analysis of Salt Tolerance According to Several Classification Methods. *Agricultural Water Management*, 62(1):37–66.

105. Kaushik, A., N. Saini, S. Jain, P. Rana, R.K. Singh, and R.K. Jain. 2003. Genetic Analysis of a CSR10 (Indica) x Taraori Basmati F-3 Population Segregating for Salt Tolerance Using ISSR Markers. *Euphytica*, 134(2):231–238.

106. Khadem, A. and F. Raiesi. 2019. Response of Soil Alkaline Phosphatase to Biochar Amendments: Changes in Kinetic and Thermodynamic Characteristics. *Geoderma*, 337(1 March):44–54.

107. Khattak, R.A., K. Haroon, and D. Muhammad. 2013. Mechanism(s) of Humic Acid Induced Beneficial Effects in Salt-affected Soils. *Scientific Research and Essays*, 8(21):932–939.

108. Khosh Kholgh Sima N.A., H. Askari, H. Hadavand Mirzaei, and M. Pessarakli. 2009. Genotype-dependent Differential Responses of Three Forage Species to Ca Supplement in Saline Conditions. *Journal of Plant Nutrition*, 32(4):579–597.

109. Khosh Kholgh Sima, N.A., S. Tale Ahmad, R.A. Alitabar, A. Mottaghi, and M. Pessarakli. 2012. Interactive Effects of Salinity and Phosphorus Nutrition on Physiological Responses of Two Barley Species. *Journal of Plant Nutrition*, 35:1411–1428.

110. Khosh Kholgh Sima, N.A., S. Tale Ahmad, and M. Pessarakli. 2013. Comparative Study of Different Salts (Sodium Chloride, Sodium Sulfate, Potassium Chloride, and Potassium Sulfate) on Growth of Forage Species. *Journal of Plant Nutrition*, 36(2):214–230.

111. Kilic, C.C., Y.S. Kukul, and D. Anac. 2008. Performance of Purslane (*Portulaca oleracea* L.) as a Salt-removing Crop. *Agricultural Water Management*, 95(7):854–858.

112. Kovda, V.A. 1980. *Problems of Combating Salinizaton of Irrigated Soils*. UNEP.

113. Kovda, V.A. and I. Szabolcs. 1979. Modelling of Soil Salinization and Alkalization. *Agrokemia es Talajtan*.

114. Koyro, H.W. 2006. Effect of Salinity on Growth, Photosynthesis, Water Relations and Solute Composition of the Potential Cash Crop Halophyte *Plantago coronopus* (L.). *Environmental and Experimental Botany*, 56(2):136–146.

115. Koyro, H.W. and S.S. Eisa. 2008. Effect of Salinity on Composition, Viability and Germination of Seeds of *Chenopodium quinoa* Willd. *Plant and Soil*, 302(1–2):79–90.

116. Kreeb, K.H., R.D.B. Whalley, and J.L. Charley. 1995. Some Investigations into Soil and Vegetation Relationships Associated with Alkaline Saline Soil Surfaces in the Walcha Area, Northern Tablelands, New South Wales. *Australian Journal of Agricultural Research*, 46(1):209–224.

117. Kumar, A. 1996. Use of *Leptochloa fusca* for the Improvement of Salt Affected Soils. *Experimental Agriculture, India*, 32(2):143–149.

118. Kumar, A., L. Batra, and R. Chhabra. 1994. Forage Yield of Sorghum and Winter Clovers as Affected by Biological and Chemical Reclamation of a Highly Alkaline Soil. *Experimental Agriculture, India*, 30(3):343–348.

119. Kurup, S.S., A.W. Al Amouri, M.A.M.A. Salem, and M. Pessarakli. 2014. Identification of Saline Tolerant Turfgrass Species with Optimum Turf Quality for Urban Landscaping in the United Arab Emirates. *Acta Horticulturae*, 1051:117–125.

120. Kurup, S.S., M.A.M.A. Salem, A.J. Cheruth, S. Subramaniam, F. Thayale Purayil, A. Wahed Al Amouri, and M. Pessarakli. 2017. Changes in Antioxidant Enzyme Activity in Turfgrass Cultivars under Various Saline Water Irrigation Levels to Suit Landscapes under Arid Regions. *Communications in Soil Science and Plant Analysis Journal*, 48(17):1989–2001.

121. Lade, P.D. and H. Agron. 2014. Plant-growth-promoting Rhizobacteria to Improve Crop Growth in Saline Soils: A Review. *Sustainable Development*, 34(4):737–752.

122. Lee, M.K. and M.W. van Lersel. 2008. Sodium Chloride Effects on Growth, Morphology, and Physiology of Chrysanthemum (*Chrysanthemum xmorifolium*). *Hortscience*, 43(6):1888–1891.

123. Lei, Y., Q. Liu, C. Hettenhausen, G. Cao, Q. Tan, W. Zhao, and J. Wu. 2017. Salt-tolerant and -sensitive Alfalfa (*Medicago sativa* L.) Cultivars Have Large Variations in Defense Responses to the Lepidopteran Insect *Spodoptera litura* under Normal and Salt Stress Condition. *PLoS One*, 12(7).

124. Leuther, F., S. Schlüter, R. Wallach, and H.-J. Vogel. 2019. Structure and Hydraulic Properties in Soils under Long-term Irrigation with Treated Wastewater. *Geoderma*, 333(1 January):90–98.

125. Liu, H.R., G.W. Sun, L.J. Dong, L.Q. Yang, S.N. Yu, S.L. Zhang, and J.F. Liu. 2017. Physiological and Molecular Responses to Drought and Salinity in Soybean. *Biologia Plantarum*, 1(3):557–564.

127. Maggio, A., S. de Pascale, G. Angelino, C. Ruggiero, and G. Barbieri. 2004. Physiological Response of Tomato to Saline Irrigation in Long-term Salinized Soils. *European Journal of Agronomy*, 21(2):149–159.

126. Maggio, A., P.M. Hasegawa, R.A. Bressan, M.F. Consiglio, and R.J. Joly. 2001. Unravelling the Functional Relationship between Root Anatomy and Stress Tolerance. *Australian Journal of Plant Physiology*, 28(10):999–1004.

128. Mahdy, A.M. 2011. Comparative Effects of Different Soil Amendments on Amelioration of Sodic Soil. *Soil and Water Research*, 6(4):205–216.

129. Maliro, M.F.A., D. McNeil, B. Redden, J.F. Kollmorgen, and C. Pittock. 2008. Sampling Strategies and Screening of Chickpea (*Cicer arietinum* L.) Germplasm for Salt Tolerance. *Genetic Resources and Crop Evolution*, 55(1):53–63.

130. Maqshoof, A., A. Zahir, H. Naeem, and M. Asghar. 2011. Inducing Salt Tolerance in Mung Bean through Co-inoculation with Rhizobia and Plant-growth-promoting Rhizobacteria Containing 1-aminocyclopropane-1-carboxylate deaminase. *Canadian Journal of Microbiology*, 57(7):578–589.

131. Marcum, K.B. and M. Pessarakli. 2006. Salinity Tolerance and Salt Gland Excretion Activity of Bermudagrass Turf Cultivars. *Crop Science*, 46(6):2571–2574.

132. Marcum, K.B., M. Pessarakli, and D.M. Kopec. 2005. Relative Salinity Tolerance of 21 Turf-type Desert Saltgrasses Compared to Bermudagrass. *HortScience*, 40(3):827–829.

133. Millette, D., C. Madramootoo, and G.D. Buckland. 1993. Salt Removal in a Saline Soil Using Fall Irrigation under Subsurface Grid Drainage. *Canadian Agricultural Engineering Journal*, 35(1):1–9.

134. Minashina, N.G. 1996. Effectiveness of Drainage in Developing Saline Soils under Irrigated Agriculture during the Last Thirty Years. *Eurasian Soil Science*, 28(2):7790.

135. Miransari, M., H. Riahi, F. Eftekhar, A. Minaie, and D.L. Smith. 2013. Improving Soybean (*Glycine max* L.) N2 Fixation under Stress. *Journal of Plant Growth Regulation*, 32(4):909–921.

136. Mittal, S., N. Kumari, and V. Sharma. 2012. Differential Response of Salt Stress on *Brassica juncea*: Photosynthetic Performance, Pigment, Proline, D1 and Antioxidant Enzymes. *Plant Physiology and Biochemistry*, 54:17–26.

137. Miyamoto, S. and C. Enriquez. 1990. Comparative Effects of Chemical Amendments on Salt and Sodium Leaching. *Irrigation Science*, 11(2):8392.

138. Muller, S.J. and A.V. Niekerk. 2016. An Evaluation of Supervised Classifiers for Indirectly Detecting Salt-affected Areas at Irrigation Scheme Level. *International Journal of Applied Earth Observation and Geoinformation*, 49:138–150.

139. Munns, R. 2002. Utilizing Genetic Resources to Enhance Productivity of Salt-prone Land. *CAB Review: Perspectives in Agriculture, Veterinary Science, Nutrition, and Natural Resources*, 2(9).

140. Munns, R., R.A. James, and A. Lauchli. 2006. Approaches to Increasing the Salt Tolerance of Wheat and Other Cereals. *Journal of Experimental Botany*, 57(5):1025–1043.

141. Murtaza, B., G. Murtaza, M. Saqib, and A. Khaliq. 2014. Efficiency of Nitrogen Use in Rice Wheat Cropping System in Salt-affected Soils with Contrasting Texture. *Pakistan Journal of Agricultural Science*, 51:431–441.

142. Nadler, A., G.J. Levy, R. Keren, and H. Eisenberg. 1996. Sodic Calcareous Soil Reclamation as Affected by Water Chemical Composition and Flow Rate. *Soil Science Society of America Journal*, 60(1):252–257.

143. Nahar, K., M. Hasanuzzaman, M. Alam, and M. Fujita. 2015. Roles of Exogenous Glutathione in Antioxidant Defense System and Methylglyoxal Detoxification during Salt Stress in Mung Bean. *Biologia Plantarum*, 59(4):745–756.

144. Nahar, K., M. Hasanuzzaman, A. Rahman, M. Alam, J. Al-Mahmud, T. Suzuki, and M. Fujita. 2016. Polyamines Confer Salt Tolerance in Mung Bean (*Vigna radiata* L.) by Reducing Sodium Uptake, Improving Nutrient Homeostasis, Antioxidant Defense, and Methylglyoxal Detoxification Systems. *Frontiers in Plant Science*, 7:1104.

145. Negrão, S., S.M. Schmöckel, and M. Tester. 2017. Evaluating Physiological Responses of Plants to Salinity Stress. *Annals of Botany*, 119(1):1–11.

146. Netondo, G.W., J.C. Onyango, and E. Beck. 2004. Sorghum and Salinity: I. Response of Growth, Water Relations, and Ion Accumulation to NaCl Salinity. *Crop Science*, 44(3):797–805.

147. Nguyen, D.N. 2012. Plant Availability of Water in Soils Being Reclaimed from the Saline Sodic State. Unpublished Ph.D. Dissertation, Adelaide University, Adelaide Australia.

148. Nguyen, P.D., C.L. Ho, J.A. Harikrishna, M.C.V.L. Wong, and R.A. Rahim. 2007. Functional Screening for Salinity Tolerant Genes from *Acanthus ebracteatus* Vahl Using *Escherichia coli* as a Host. *Trees*, 21(5):515–520.

149. Nwer, B., H. Zurqani, and E. Rhoma. 2014. The Use of Remote Sensing and Geographic Information System for Soil Salinity Monitoring in Libya. *GSTF Journal of Geological Sciences*, 1(1).

150. Ouni, Y., T. Ghnayaa, F. Montemurro, C. Abdellya, and A. Lakhdara. 2014. The Role of Humic Substances in Mitigating the Harmful Effects of Soil Salinity and Improve Plant Productivity. *International Journal of Plant Production*, 8:353–374.

151. Pakar, N., H. Pirasteh-Anosheh, Y. Emam, and M. Pessarakli. 2016. Barley Growth, Yield, Antioxidant Enzymes and Ions Accumulation Affected by PGRs under Salinity Stress. *Journal of Plant Nutrition*, 39(10):1372–1379.

152. PanelS., S. Dahlawi, A. Naeem, Z. Rengel, and R. Naidu. 2018. Biochar Application for the Remediation of Salt-affected Soils: Challenges and Opportunities. *Science of the Total Environment*, 625:320–335.

153. Papiernik, S.K., C.M. Grieve, S.M. Lesch, and S.R. Yates. 2005. Effects of Salinity, Imazethapyr, and Chlorimuron Application on Soybean Growth and Yield. *Communications in Soil Science and Plant Analysis*, 36(7–8):951–967.

154. Parihar, P., S. Singh, R. Singh, V.P. Singh, and S.M. Prasad. 2014. Effect of Salinity Stress on Plants and Its Tolerance Strategies: A Review. *Environmental Science and Pollution Research*, 22(16):2056–4075.

155. Patel, A., P. Khare, and D.D. Patra. 2017. Biochar Mitigates Salinity Stress in Plants. In: V. Shukla, S. Kumar, and N. Kumar (eds.), *Plant Adaptation Strategies in Changing Environment*, pp. 153–182, , Singapore: Springer.

156. Paul, D. and S. Nair. 2008. Stress Adaptations in a Plant Growth Promoting Rhizobacterium (PGPR) with Increasing Salinity in the Coastal Agricultural Soils. *Journal of Basic Microbiology*, 48(5):378–384.

157. Pessarakli, M. 1991a. Formation of Saline and Sodic Soils and Their Reclamation. *Journal of Environmental Science and Health*, A26(7):1303–1320.

158. Pessarakli, M. 1991b. Water Utilization and Soil Salinity Control in Arid-zone Agriculture. *Communications in Soil Science and Plant Analysis*, 22(17–18):1787–1796.

159. Pessarakli, M. 2005a. Supergrass: Drought-tolerant Turf Might Be Adaptable for Golf Course Use. *Golfweek's SuperNews Magazine*, November 16, 2005, p. 21 and cover page. www.supernewsmag.com/news/golfweek/supernews/20051116/p21.asp?st=p21_s1.htm

160. Pessarakli, M. 2005b. Gardener's Delight: Low-maintenance Grass. *Tucson Citizen*, Arizona, Newspaper Article, September 15, 2005, Tucson, AZ, United States. Gardener's delight: Low-maintenance grass, www.tucsoncitizen.com/

161. Pessarakli, M. 2007. Saltgrass (*Distichlis spicata*), a Potential Future Turfgrass Species with Minimum Maintenance/Management Cultural Practices. In: M. Pessarakli (ed.), *Handbook of Turfgrass Management and Physiology*, pp. 603–615, Florida: CRC Press, Taylor & Francis.

162. Pessarakli, M. 2010. Saltgrass, a High Salt and Drought Tolerant Species for Sustainable Agriculture in Desert Regions. In *Proceedings of the 4th International Conference on Water Resources and Arid Environments (ICWRAE 4)*, Vol. 1, pp. 551–561, Riyadh, Saudi Arabia.

163. Pessarakli, M. 2011. Saltgrass, a High Salt and Drought Tolerant Species for Sustainable Agriculture in Desert Regions. *International Journal of Water Resources and Arid Environments*, 1(1):55–64.

164. Pessarakli, M. 2014a. Saltgrass, a True Halophytic Plant Species for Sustainable Agriculture in Desert Regions. *Research on Crop Ecophysiology*, 9/1(1):1–11.

165. Pessarakli, M. 2014b. Using Bermudagrass (*Cynodon dactylon* L.) in Urban Desert Landscaping and as a Forage Crop for Sustainable Agriculture in Arid Regions and Combating Desertification. In: *Proceedings of the 6th International Conference on Water Resources and Arid Environments (ICWRAE 6)*, Vol. 1, pp. 242–248, Riyadh, Saudi Arabia.

166. Pessarakli, M. 2015. Using Bermudagrass (*Cynodon dactylon* L.) in Urban Desert Landscaping and as a Forage Crop for Sustainable Agriculture in Arid Regions and Combating Desertification. *International Journal of Water Resources and Arid Environments*, 4(1):8–14.

167. Pessarakli, M. 2016a. Screening Various Cultivars of Seashore Paspalum (*Paspalum vaginatum* Swartz) for Salt Tolerance for Potential Use as a Cover Plant in Combatting Desertification. In: *Proceedings of the 7th International Conference on Water Resources and Arid Environments (ICWRAE 7)*, Vol. 1, pp. 509–516, Riyadh, Saudi Arabia.

168. Pessarakli, M. 2016b. Saltgrass, a Minimum Water and Nutrient Requirement Halophytic Plant Species for Sustainable Agriculture in Desert Regions. *Journal of Earth, Environment and Health Sciences*, 2(1):21–27.

169. Pessarakli, M. 2017. Growth Responses of Sacaton Grass (*Sporobolus airoides* Torr.) and Seashore Paspalum (*Paspalum vaginatum* Swartz) under Prolonged Drought Stress Condition. *Advances in Plants & Agriculture Research, MedCrave*, 7(4):00261.

170. Pessarakli, M., D.D. Breshears, J. Walworth, J.P. Field, and D.J. Law. 2017. Candidate Halophytic Grasses for Addressing Land Degradation: Shoot Responses of *Sporobolus airoides* and *Paspalum vaginatum* to Weekly Increasing NaCl Concentration. *Arid Land Research & Management Journal*, 31(2):169–181.

171. Pessarakli, M., N. Gessler, and D.M. Kopec. 2008. Growth Responses of Saltgrass (*Distichlis spicata*) under Sodium Chloride (NaCl) Salinity Stress. *USGA Turfgrass and Environmental Research Online*, October 15, 2008, 7(20):1–7. http://turf.lib.msu.edu/tero/v02/n14.pdf

172. Pessarakli, M., M.Haghighi, and A. Sheibanirad. 2015. Plant Responses under Environmental Stress Conditions. *Advances in Plants & Agriculture Research Journal*, 2(6):00073. http://medcraveonline.com/APAR/APAR-02-00073.pdf

173. Pessarakli, M., M.A. Harivandi, D.M. Kopec, and D.T. Ray. 2012. Growth Responses and Nitrogen Uptake by Saltgrass (*Distichlis spicata* L.), a Halophytic Plant Species, under Salt Stress, Using the 15N Technique. *International Journal of Agronomy*, 2012: Article ID 896971, 9 pages, doi

174. Pessarakli, M. and D.M. Kopec. 2005. Responses of Twelve Inland Saltgrass Accessions to Salt Stress. *USGA Turfgrass and Environmental Research Online*, 4(20):1–5.

175. Pessarakli, M. and D.M. Kopec. 2008. Establishment of Three Warm-season Grasses under Salinity Stress. *Acta HortScience*, ISHS, 783:29–37.

176. Pessarakli, M. and D.M. Kopec. 2010. Growth Responses and Nitrogen Uptake of Saltgrass (*Distichlis spicata* L.), a True Halophyte, under Salinity Stress Conditions using 15N

Technique. In: *Proceedings of the International Conference on Management of Soils and Ground Water Salinization in Arid Regions*, Vol. 2, pp. 1–11, Muscat, Sultanate of Oman.

177. Pessarakli, M., D.M. Kopec, and D.T. Ray. 2011. Growth Responses of Various Saltgrass (*Distichlis spicata*) Clones under Salt Stress Conditions. *Journal of Food, Agriculture, and Environment (JFAE)*, 9(3&4):660–664.

178. Pessarakli, M., K.B. Marcum, and D.M. Kopec. 2005. Growth Responses and Nitrogen-15 Absorption of Desert Saltgrass (*Distichlis spicata*) to Salinity Stress. *Journal of Plant Nutrition*, 28(8):1441–1452.

179. Pessarakli, M. and K.B. Marcum. 2013. *Distichlis spicata* – A Salt and Drought Tolerant Plant Species with Minimum Water Requirements for Sustainable Agriculture in Desert Regions and Biological Reclamation of Desert Saline Soils. In: F.K. Taha, S.A. Shahid, and M.A. Abdelfattah (eds.), *Developments in Soil Salinity Assessment and Reclamation: Innovative Thinking and Use of Marginal Soil and Water Resources in Irrigated Agriculture*, pp. 383–396. © Springer Science & Business Media Dordrecht 2013. International Center for Bio-saline Agriculture, Dubai, UAE.

180. Pessarakli, M. and D.E. McMillan. 2013. Seashore Paspalum, a High Salinity Stress Tolerant Halophytic Plant Species for Sustainable Agriculture in Desert Regions and Combating Desertification. In: *Proceedings of the 5th International Conference on Water Resources and Arid Environments (ICWRAE 5)*, Vol. 1, pp. 488–495, Riyadh, Saudi Arabia.

181. Pessarakli, M. and D.E. McMillan. 2014. Seashore Paspalum, a High Salinity Stress Tolerant Halophytic Plant Species for Sustainable Agriculture in Desert Regions and Combating Desertification. *International Journal of Water Resources and Arid Environments*, 3(1):35–42.

182. Pessarakli, M. and H. Touchane. 2006. Growth Responses of Bermudagrass and Seashore Paspalum under Various Levels of Sodium Chloride Stress. *Journal of Agriculture, Food, and Environment (JAFE)*, 4(3&4):240–243.

183. Pessarakli, M. and H. Touchane. 2010. Biological Technique in Combating Desertification Processes Using a True Halophytic Plant. In: *Proceedings of the 4th International Conference on Water Resources and Arid Environments (ICWRAE 4)*, Vol. 1, pp. 545–550, Riyadh, Saudi Arabia.

184. Pessarakli, M. and H. Touchane. 2011. Biological Technique in Combating Desertification Processes Using a True Halophytic Plant. *International Journal of Water Resources and Arid Environments*, 1(5):360–365.

185. Piwowarczyk, B., K. Tokarz, and I. Kamińska. 2016. Responses of Grass Pea Seedlings to Salinity Stress in In Vitro Culture Conditions. *Plant Cell Tissue Organ Culture*, 124:227.

186. Porcel, A. and J. Ruiz-Lozano. 2012. Salinity Stress Alleviation Using Arbuscular Mycorrhizal Fungi. A Review. *Agronomy for Sustainable Development*, 32(1):181–200.

187. Pushpavalli, R., J. Quealy, T.D. Colmer, N.C. Turner, K.H.M. Siddique, M.V. Rao, and V. Vade. 2016. Salt Stress Delayed Flowering and Reduced Reproductive Success of Chickpea (*Cicer arietinum* L.), a Response Associated with Na+ Accumulation in Leaves. *Journal of Agronomy and Crop Science*, 202(2):125–138.

188. Qadir, M.A., R.H. Qureshi, and N. Ahmad. 1996. Reclamation of a Saline-sodic Soil by Gypsum and *Leptochloa fusca*. *Geoderma*, 74(34):207–217.

189. Qadir, M.A., R.H. Qureshi, and N. Ahmad. 1997. Nutrient Availability in a Calcareous Saline-sodic Soil during Vegetative Bioremediation. *Arid Soil Research and Rehabilitation*, 11(4):343–352.

190. Qadir, M.A., R.H. Qureshi, N. Ahmad, and M. Ilyas. 1996. Salt-tolerant Forage Cultivation on a Saline-sodic Field for Biomass Production and Soil Reclamation. *Land Degradation and Development*, 7(2):11–18.

191. Quesada, V., S. Garcia-Martinez, P. Piqueras, M.R. Ponce, and J.L. Micol. 2002. Genetic Architecture of NaCl Tolerance in Arabidopsis. *Plant Physiology*, 130(2):951–963.

192. Radi, A.A., F.A. Farghaly, and A.M. Hamada. 2013. Physiological and Biochemical Responses of Salt-tolerant and Salt-sensitive Wheat and Bean Cultivars to Salinity. *Journal of Biology and Earth Sciences*, 3(1): Online.

193. Rady, M.M., C.Bhavya Varma, and S.M. Howladar. 2013. Common Bean (*Phaseolus vulgaris* L.) Seedlings Overcome NaCl Stress as a Result of Presoaking in *Moringa oleifera* Leaf Extract. *Scientia Horticulturae*, 162(23):63–70.

194. Rao, B.R.M., R.S. Dwivedi, L. Venkataratnam, T. Ravishankar, S.S. Thammappa, G.P. Bhargawa, and A.N. Singh. 1991. Mapping the Magnitude of Sodicity in Part of the IndoGangetic Plains of Uttar Pradesh Northern India Using Landsat Data. *International Journal of Remote Sensing*, 12(3):419–425.

195. Rao, D.L.N. and R.G. Burns. 1991. The Influence of Blue-green Algae on the Biological Amelioration of Alkali Soils. *Biology and Fertility of Soils*, 11(4):306–312.

196. Rao, D.L.N. and H. Pathak. 1996. Ameliorative Influence of Organic Matter on Biological Activity of Salt-affected Soils. *Arid Soil Research and Rehabilitation*, 10(4):311–319.

197. Rao, K.V.G.K and P.B. Leeds Harrison. 1991. Desalinization with Subsurface Drainage. *Agriculture Water Management*, 19(4):303–312.

198. Rathinasabapathi, B. 2000. Metabolic Engineering for Stress Tolerance: Installing Osmoprotectant Synthesis Pathways. *Analysis of Botany*, 86(4):709–716.

199. Rawat, L., Y. Singh, N. Shukla, and J. Kumar. 2011. Alleviation of the Adverse Effects of Salinity Stress in Wheat (*Triticum aestivum* L.) by Seed Biopriming with Salinity Tolerant Isolates of *Trichoderma harzianum*. *Plant and Soil*, 347(1/2):387–400.

200. Redly, M. 1996. Contribution of ISSS Subcommission on Salt Affected Soils to the Study and Utilization of Saline and Alkaline Soils (1964–1994). *Pochvovedenie*, 8(7):923–928.

201. Refahi, A. and A.R. Shahsavar. 2017. Effects of Salinity Stress on Certain Morphological Traits and Antioxidant Enzymes of Two *Carica papaya* Cultivars in Hydroponic Culture. *Advances in Horticultural Science*, 31(2):131–139.

202. Reynolds, M.P., A. Mujeeb-Kazi, and M. Sawkins. 2005. Prospects for Utilizing Plant-adaptive Mechanisms to Improve Wheat and other Crops in Drought- and Salinity-prone Environments. *Analysis of Applied Biology*, 146(2):239–259.

203. Rogers, M.E., A.D. Craig, R. Munns, T.D. Colmer, P.G.H. Nichols, C.V. Malcolm, E.G. Barrett-Lennard, et al. 2005. The Potential for Developing Fodder Plants for the Salt-affected Areas of Southern and Eastern Australia: An Overview. *Australian Journal of Experimental Agriculture*, 45:301–329.

204. Rogobete, G., D. Tarau, D. Dicu, and R. Bertici. 2013. Capillary and Solute Transport in Swelling and Shrinking Soils. *Soil Forming Factors and Processes from Temperate Zone*, 12(2):53–59.

205. Sanchez, P.L., M.-K. Chen, M. Pessarakli, H.J. Hill, M.A. Gore, and M.A. Jenks. 2014. Effects of Temperature and Salinity on Germination of Non-pelleted and Pelleted Guayule (*Parthenium argentatum* A. Gray) Seeds. *Journal of Industrial Crops and Products*, 55:90–96.

206. Sappor, D.K., B.A. Osei, and M.R. Ahmed. 2017. Reclaiming Sodium Affected Soil: The Potential of Organic Amendments. *International Journal of Plant & Soil Science*, 16(2):1–11.

207. Saqib, M., C. Zorb, and S. Schubert. 2008. Silicon-mediated Improvement in the Salt Resistance of Wheat (*Triticum aestivum*) Results from Increased Sodium Exclusion and Resistance to Oxidative Stress. *Functional Plant Biology*, 35(7):633–639.

208. Savvas, D., D. Giotis, E. Chatzieustratiou, M. Bakea, and G. Patakioutas. 2009. Silicon Supply in Soilless Cultivations of Zucchini Alleviates Stress Induced by Salinity and Powdery Mildew Infections. *Environmental and Experimental Botany*, 65(1):11–17.

209. Sayed, O.H. 1995. Edaphic Gradients and Species Attributes Influencing Plant Distribution in Littoral Salt Marshes of Qatar. *Qatar University Science Journal*, 14(2):257–262.

210. Schwabe, K.A., I. Kan, and K.C. Knapp. 2006. Drain Water Management for Salinity Mitigation in Irrigated Agriculture. *American Journal of Agricultural Economy*, 88(1):133–149.

211. Seifert, C., J.I. Ortiz-Monasterio, and D.B. Lobell. 2011. Satellite-based Detection of Salinity and Sodicity Impacts on Wheat Production in the Mexicali Valley. *Soil Science Society of America Journal*, 75(2):699.

212. Selassie, T.G., J.J. Jurinak, and L.M. Dudley. 1992. Saline and Saline-sodic Soil Reclamation First Order Kinetic Model. *Soil Science*, 154(1):17.

213. Sepehry, A., D. Akhzari, M. Pessarakli, and H. Barani. 2012. Studying the Effects of Salinity, Aridity and Grazing Stress on the Growth of Various Halophytic Plant Species (*Agropyron elongatum*, *Kochia prostrata* and *Puccinellia distans*). *World Applied Sciences Journal*, 17(10):1278–1286.

214. Shah, G.M. 2014. Socio-economic Impact of Climate Change in Aral Sea Basin. *The Journal of Central Asian Studies*, 21(1):48–57.

215. Shani, U. and L.M. Dudley. 2001. Field Studies of Crop Response to Water and Salt Stress. *Soil Science Society of America Journal*, 65(5):1522–1528.

216. Shannon, M.C. 1998. Adaptation of Plants to Salinity. *Advances in Agronomy*, 60:75–119.

217. Sharma, H.K. and V.K. Upadhyay. 1994. A Note on the pH Variations in Alkaline Soil of Aligarh District as Influenced by an Inorganic Heterocycle (S–6N–4)–2 + Cl–2. *Advances in Plant Science*, 7(1):6871.

218. Shrivastava, P. and R. Kumar. 2015. Soil Salinity: A Serious Environmental Issue and Plant Growth Promoting Bacteria as One of the Tools for Its Alleviation. *Saudi Journal of Biological Sciences*, 22(2):123–131.

219. Shukla, A.K. and P.N. Misra. 1993. Improvement of Sodic Soil under Tree Cover. *Indian Forester*, 119(1):4352.

220. Siddiqui, K.M. 1994. Tree Planting for Sustainable Use of Soil and Water with Special Reference to the Problem of Salinity. *Pakistan Journal of Forestry*, 44(3):97–102.

221. Simunek, J. and D.L. Suarez. 1997. Sodic Soil Reclamation Using Multicomponent Transport Modeling. *Journal of Irrigation and Drainage Engineering, ASCE*, 123(5):367–376.

222. Singh, A.P. and A.R. Singh. 2013. Seasonal Changes in the Physico-chemical Attributes of Salt Affected Habitat. *India Journal of Science Research*, 4(1):105–115.

223. Singh, H. and M.S. Bajwa. 1991. Effect of Sodic Irrigation and Gypsum on the Reclamation of Sodic Soil and Growth of Rice and Wheat Plants. *Agriculture Water Management*, 20(2):163–172.

224. Singla-Pareek, S.L., M.K. Reddy, and S.K. Sopory. 2003. Genetic Engineering of the Glyoxalase Pathway in Tobacco Leads to Enhanced Salinity Tolerance. *Proceedings of the National Academy of Sciences of the United States of America*, 100(25):14672–14677.

225. Singla-Pareek, S.L., S.K. Yadav, A. Pareek, M.K. Reddy, and S.K. Sopory. 2008. Enhancing Salt Tolerance in a Crop Plant by Overexpression of Glyoxalase II. *Transgenic Research*, 17(2):171–180.

226. Srivastava, P.K., M. Gupta, A. Panday, N. Singh, and S.K. Tewan. 2014. Effects of Sodicity Induced Changes on Soil Physical Properties on Paddy Root Growth. *Journal of Plant, Soil and Environment*, 60(4):165–169.

227. Srivastava, S., B. Fristensky, and N.N.V. Kav. 2004. Constitutive Expression of a PR10 Protein Enhances the Germination of *Brassica napus* under Saline Conditions. *Plant and Cell Physiology*, 45(9):1320–1324.

228. Steppuhn, H., M.T. van Genuchten, and C.M. Grieve. 2005. Root-zone Salinity: I. Selecting a Product-yield Index and Response Function for Crop Tolerance. *Crop Science*, 45(1):209–220.

229. Stewart, D.P.C. 1992. Reclamation of Saline-sodic Soils Adjacent to the Heathcote River, Christchurch. *New Zealand Natural Sciences*, 19:45–52.

230. Subramaniam, B. and C.R. Babu. 1994. New Nodulating Legumes of Potential Agricultural and Forestry Value from Subtropical Himalayan Ecosystems. *Biological Agriculture and Horticulture*, 10(4):297–302.

231. Sun, G., Y. Zhu, M. Ye, J. Yang, Z. Qu, W. Mao, and J. Wu. 2019. Development and Application of Long-term Root Zone Salt Balance Model for Predicting Soil Salinity in Arid Shallow Water Table Area. *Agricultural Water Management*, 213(1 March):486–498.

232. Sun, Z., Y. Wang, F. Mou, Y. Tian, L. Chen, S. Zhang, Q. Jiang, and X. Li. 2015. Genome-wide Small RNA Analysis of Soybean Reveals Auxin-responsive microRNAs That Are Differentially Expressed in Response to Salt Stress in Root Apex. *Frontiers in Plant Science*, 6:1273.

233. Swarup, A., S. Adhikari, and A.K. Biswas. 1994. Effect of Gypsum on the Behavior of Soil Phosphorus during Reclamation of a Sodic Soil. *Journal of Indian Society of Soil Science*, 42(4):543–547.

234. Szabolcs, I. 1989. *Salt-affected Soils.*, Boca Raton: CRC Press.

235. Szabolcs, I. 1991. *Soil Salinity and Biodiversity, The Biodiversity of Microorganisms and Invertebrates: Its Role in Sustainable Agriculture*. D.L. Hawksworth (ed.), CAB International.

236. Takeoka, Y., A. Al-Mamun, T. Wada, and P.B. Kaufman. 1992. *Developments in Crop Science Vol. 22. Reproductive Adaptation of Rice to Environmental Stress*. Tokyo, Japan: Japan Scientific Societies Press; Amsterdam, Netherlands, New York, NY: Elsevier Science Publishing.

237. Tavakkoli, E., J. Paull, P. Rengasamy, and G.K. McDonald. 2012. Comparing Genotypic Variation in Faba Bean (*Vicia faba* L.) in Response to Salinity in Hydroponic and Field Experiments. *Field Crops Research*, 127:99–108.

238. Theunissen, J.D. 1997. Selection of Suitable Ecotypes within *Digitaria eriantha* for Reclamation and Restoration of Disturbed Areas in Southern Africa. *Journal of Arid Environment*, 35(3):429–439.

239. Tunçtürk, M., R. Tunçtürk, B. Yildirim, and V. Çiftçi. 2011. Effect of Salinity Stress on Plant Fresh Weight and Nutrient Composition of Some Canola (*Brassica napus* L.) Cultivars. *African Journal of Biotechnology*, 10(10): 1827–1832.

240. Umar, S., I. Diva, N.A. Anjum, M. Iqbal, A. Iqbal, and E. Pereira. 2011. Potassium-induced Alleviation of Salinity Stress in *Brassica campestris* L. *The Central European Journal of Biology*, 6(6):1054–1063.

241. van Hoorn, J.W., N. Katerji, A. Hamdy, and M. Mastrorilli. 2001. Effect of Salinity on Yield and Nitrogen Uptake of Four Grain Legumes and on Biological Nitrogen Contribution from the Soil. *Agricultural Water Management*, 51(2):87–98.

242. van Oort, P.A.J. 2018. Mapping Abiotic Stresses for Rice in Africa: Drought, Cold, Iron Toxicity, Salinity and Sodicity. *Field Crops Research Journal*, 219:55–75.

243. Veatch, M.E., S.E. Smith, and G. Vandemark. 2004. Shoot Biomass Production among Accessions of *Medicago truncatula* Exposed to NaCl. *Crop Science*, 44(3):1008–1013.

244. Verma, D., S.L. Singla-Pareek, D. Rajagopal, M.K. Reddy, and S.K. Sopory. 2007. Functional Validation of a Novel Isoform of Na+/H+ Antiporter from *Pennisetum glaucum* for Enhancing Salinity Tolerance in Rice. *Journal of Biosciences*, 32(3):621–628.

245. Vierling, E. and J.A. Kimpel. 1992. Plant Responses to Environmental Stress. *Current Opinion in Biotechnology*, 3(2):164–170.

246. Villora, G., D.A. Moreno, G. Pulgar, and L.M. Romero. 1999. Zucchini Growth, Yield, and Fruit Quality in Response to Sodium Chloride Stress. *Journal of Plant Nutrition*, 22(6):855–861.

247. Waheed, A., I.A. Hafiz, G. Qadir, G. Murtaza, T. Mahmood, and M. Ashraf. 2006. Effect of Salinity on Germination, Growth, Yield, Ionic Balance and Solute Composition of Pigeon Pea (*Cajanus cajan* L., Millsp). *Pakistan Journal of Botany*, 38(4):1103–1117.

248. Wahid, A., M. Perveen, S. Gelani, and S.M.A. Basra. 2007. Pretreatment of Seed with H_2O_2 Improves Salt Tolerance of Wheat Seedlings by Alleviation of Oxidative Damage and Expression of Stress Proteins. *Journal of Plant Physiology*, 164(3):283–294.

249. Wang, D. and M.C. Shannon. 1999. Emergence and Seedling Growth of Soybean Cultivars and Maturity Groups under Salinity. *Plant and Soil*, 214(1–2):117–124.

250. Wang, D., J.A. Poss, T.J. Donovan, M.C. Shannon, and S.M. Lesch. 2002. Biophysical Properties and Biomass Production of Elephant Grass under Saline Conditions. *Journal of Arid Environments*, 52(4):447–456.

251. Weisany, W., Y. Sohrabi, G. Heidari, A. Siosemardeh, and K. Ghassemi-Golezani. 2012. Changes in Antioxidant Enzymes Activity and Plant Performance by Salinity Stress and Zinc Application in Soybean ('*Glycine max*' L.). *Plant Omics*, 5(2):60–67 [online].

252. Wilson, C. and J.J. Read. 2006. Effect of Mixed-salt Salinity on Growth and Ion Relations of a Barnyardgrass Species. *Journal of Plant Nutrition*, 29(10):1741–1753.

253. Wilson, C., X. Liu, S.M. Lesch, and D.L. Suarez. 2006. Growth Response of Major US Cowpea Cultivars. I. Biomass Accumulation and Salt Tolerance. *Hortscience*, 41(1):225–230.

254. Winicov, I. and D.R. Bastola. 1999. Transgenic Overexpression of the Transcription Factor Alfin1 Enhances Expression of the Endogenous MsPRP2 Gene in Alfalfa and Improves Salinity Tolerance of the Plants. *Plant Physiology*, 120(2):473–480.

255. Yadav, S., M. Irfan, A. Ahmad, and S. Hayat. 2011. Causes of Salinity and Plant Manifestations to Salt Stress: A Review. *Journal of Environmental Biology*, 32:667–685.

256. Yarami Ali, N. and R. Sepaskhah. 2015. Physiological Growth and Gas Exchange Response of Saffron (*Crocus sativus* L.) to Irrigation Water Salinity, Manure Application and Planting Method. *Agricultural Water Management*, 154:43–51.

257. Yildirim, E., A.G. Taylor, and T.D. Spittler. 2006. Ameliorative Effects of Biological Treatments on Growth of Squash Plants under Salt Stress. *Scientia Horticulturae*, 111(1):1–6.

258. Yobterik, A.C. and V.R. Timmer. 1994. Nitrogen Mineralization of Agroforestry Tree Mulches under Saline Soil Conditions. In: R.B. Bryan (ed.), *Advances in Geoecology, 27, Soil Erosion, Land Degradation and Social Transition*, pp. 181–194, Destedt, Germany: Catena Verlag.

259. You, M.P., T.D. Colmer, and M.J. Barbetti. 2011. Salinity Drives Host Reaction in *Phaseolus vulgaris* (Common Bean) to *Macrophomina phaseolina*. *Functional Plant Biology*, 38(12):984–992.

260. Zahow, M.F. and C. Amrhein. 1992. Reclamation of a Saline Sodic Soil Using Synthetic Polymers and Gypsum. *Soil Science Society of America Journal*, 56(4):1257–1260.

261. Zalacáin, D., S. Martínez-Pérez, R. Bienes, A. García-Díaz, and A. Sastre-Merlín. 2019. Salt Accumulation in Soils and Plants under Reclaimed Water Irrigation in Urban Parks of Madrid (Spain). *Agricultural Water Management*, 213(1 March):468–476.

262. Zare, E., J. Huang, F.M. Santos, and J. Triantafilis. 2015. Mapping Salinity in Three Dimensions Using a DUALEM-421 and Electromagnetic Inversion Software. *Soil Science Society of America Journal*, 79(6):1729.

263. Zavaleta, G.G. 1965. The Nature of Saline and Alkaline Soils of the Peruvian Coastal Zone. *Agrokemia es Talajtan*, 14(Suppl.):415–425.

264. Zehra, A. and M.A. Khan. 2007. Comparative Effect of NaCl and Sea Salt on Germination of Halophytic Grass *Phragmites karka* at Different Temperature Regimes. *Pakistan Journal of Botany*, 39(5):1681–1694.

265. Zhang, H., J. Cui, T. Cao, J. Zhang, Q. Liu, and H. Liu. 2011. Response to Salt Stress and Assessment of Salt Tolerability of Soybean Varieties in Emergence and Seedling Stages. *Shengtai Xuebao/Acta Ecologica Sinica*, 31(10):2805–2812.

266. Zhang, H.X., J.N. Hodson, J.P. Williams, and E. Blumwald. 2001. Engineering Salt-tolerant *Brassica* Plants: Characterization of Yield and Seed Oil Quality in Transgenic Plants with Increased Vacuolar Sodium Accumulation. *Proceedings of the National Academy of Sciences of the United States of America*, 98(22):12832–12836.

267. Zhang, Q.T., M. Inoue, K. Inosako, M. Irshad, K. Kondo, G.Y. Qiu, and S.P. Wang. 2008. Ameliorative Effect of Mulching on Water Use Efficiency of Swiss Chard and Salt Accumulation under Saline Irrigation. *Journal of Food Agriculture and Environment*, 6(3–4):480–485.

268. Zhao, G.Q., B.L. Ma, and C.Z. Ren. 2007. Growth, Gas Exchange, Chlorophyll Fluorescence, and Ion Content of Naked Oat in Response to Salinity. *Crop Science*, 47(1):123–131.

269. Zhu, H., G.H. Ding, K. Fang, F.G. Zhao, and P. Qin. 2006. New Perspective on the Mechanism of Alleviating Salt Stress by Spermidine in Barley Seedlings. *Plant Growth Regulation*, 49(2–3):147–156.

270. Zhu, J.K. 2001. Plant Salt Tolerance. *Trends in Plant Sciences*, 6:66–71.

271. Zhu, M., S. Shabala, L. Shabala, Y. Fan, and M.X. Zho. 2016. Evaluating Predictive Values of Various Physiological Indices for Salinity Stress Tolerance in Wheat. *Journal of Agronomy and Crop Science*, 202(2):115–124.

272. Zia-ur-Rehman, M., G. Murtaza, M. Farooq Qayyum, M. Saqib Saifullah, and J. Akhtar. 2017. Salt-affected Soils: Sources, Genesis and Management. In: M. Sabir, J. Akhtar, and K.R. Hakeem (eds.), *Soil Science Concepts and Applications*, pp. 191–216., Faisalabad, Pakistan: University of Agricultural Science Press.

Section II

Plants/Crops Tolerance Mechanisms and Stressful Conditions

2 Roles and Mechanisms of Rhizobacteria in Regulating Plant Tolerance to Abiotic Stress

William Errickson and Bingru Huang

CONTENTS

2.1 INTRODUCTION

Plants growing in natural environments or cultivated agricultural systems form interactions with microorganisms. The rhizosphere contains a high percentage of microbial diversity and abundance, with plants secreting root exudates that serve as substrates and attractants for bacterial and fungal organisms (Dakora and Phillips, 2002; Chaparro et al., 2014). Some bacteria develop symbiotic relationships with plants, which promote plant performance and survival in a wide range of environmental conditions, including root-colonizing plant growth-promoting rhizobacteria (PGPR) (Lugtenberg and Kamilova, 2009; Yang et al., 2009; Berg et al., 2014; Liu and Zhang, 2015; Poupin et al., 2016). Endophytic PGPR in the microbial communities form closer associations with the plant host than free-living bacteria in the rhizosphere, as they colonize the internal tissues and can directly regulate cellular processes of their host plant (Conn et al., 1997; Chanway et al., 2000).

Incorporating naturally occurring PGPR in agricultural and horticultural systems has arisen as a promising strategy to improve plant tolerance to abiotic and biotic stresses. PGPR have been shown to promote plant growth in various plant species under diverse environmental conditions, such as salt stress (Mayak et al., 2004a; Sergeeva et al., 2006; Cheng et al., 2007; Nadeem et al., 2007; Saravanakumar and Samiyappan, 2007), drought (Mayak et al., 2004b; Arshad et al., 2008; Zahir et al., 2008; Belimov et al., 2009; Shakir et al., 2012), flooding (Grichko and Glick, 2001), heat (Bensalim et al., 1998), and metal toxicity (Burd et al., 1998, 2000; Belimov et al., 2001; Farwell et al., 2007; Rodriguez et al., 2008). This is an environment-friendly and sustainable approach to improving plant growth and stress tolerance with low inputs and environmental impact.

This chapter overviews the current literature on biological effects of rhizobacteria on plant stress tolerance and the underlying mechanisms involved in rhizobacteria–plant symbiosis and growth regulation.

2.2 PLANT GROWTH-PROMOTING RHIZOBACTERIA

A large variety of bacteria, including *Achromobacter*, *Azospirillum*, *Variovorax*, *Bacillus*, *Burkholderia*, *Herbaspirillum*, *Enterobacter*, *Azotobacter*, and *Pseudomonas*, exhibit growth-promoting properties by colonizing various plant species (Table 2.1). Some PGPR may have a broad host range, and others may only colonize specific plant species and cultivars (Bashan and Holguin, 1998). Tomato plants inoculated with *Achromobacter piechaudii* demonstrated increased tolerance to salinity stress, quantified by increases in biomass and water use efficiency (WUE), as well as a reduction in endogenous ethylene levels compared with non-inoculated plants (Mayak et al., 2004a). Increased biomass and reduced ethylene levels were also observed in canola plants inoculated with *Pseudomonas putida* and exposed to salinity stress (Cheng et al., 2007), while wheat plants subjected to salinity stress exhibited increased shoot and root biomass when inoculated with *Bacillus* sp., *Burkholderia cepacia*, *Enterobacter* sp., or *Paenibacillus* sp. (Upadhyay et al., 2011). Pepper plants grown in high salt concentrations had increased shoot and root length as well as increased shoot and root biomass when inoculated with *Brevibacterium iodinum*, *Bacillus licheniformis*, or *Zhihengliuela alba* (Siddikee et al., 2011), while inoculation

TABLE 2.1

Recent Studies Evaluating Bacteria-Mediated Tolerance to Abiotic Stress

	Bacteria	Plant Species	Stress	Growth Promotion	Mechanism	Source
1	*Pseudomonas fluorescens, Enterobacter hormaechei, Pseudomonas migulae*	Foxtail millet (*Setaria italica*)	Drought	Increased germination and seedling growth	ACC deaminase, EPS	Niu et al. (2018)
2	*Achromobacter piechaudii*	Tomato (*Solanum lycopersicum*), pepper (*Capsicum annuum*)	Drought	Increased biomass, reduced ethylene	ACC deaminase	Mayak et al. (2004b)
3	*Pseudomonas putida, Pseudomonas fluorescens*	Pea (*Pisum sativum*)	Drought	Increased biomass and grain yield	ACC deaminase	Arshad et al. (2008)
4	*Pseudomonas fluorescens*	Pea (*Pisum sativum*)	Drought	Increased biomass, shoot and root length, number of leaves, WUE	ACC deaminase	Zahir et al. (2008)
5	*Variovorax paradoxus*	Pea (*Pisum sativum*)	Drought	Increased yield and nutrition	ACC deaminase	Belimov et al. (2009)
6	*Mitsuaria* sp., *Burkholderia* sp.	Maize (*Zea mays*), Arabidopsis (*Arabidopsis thaliana*)	Drought	Improved root architecture, reduced evapotranspiration, increased proline levels and antioxidant activity	ACC deaminase, EPS	Huang et al. (2017)
7	*Pseudomonas putida*	Sunflower (*Helianthus* sp.)	Drought	Increased seedling survival, biomass	Ammonia, IAA, oxalic acid (OA), P solubilization, Fe siderophores, EPS	Sandhya et al. (2009)
8	*Achromobacter xylosoxidans, Serratia ureilytica, Herbaspirillum seropedicae, Ochrobactrum rhizosphaerae*	Tulsi (*Ocimum sanctum*)	Flooding	Increased shoot growth, yield, chlorophyll content, reduced ethylene	ACC deaminase, IAA	Barnawal et al. (2012)
9	*Enterobacter cloacae, Pseudomonas putida*	Tomato (*Solanum lycopersicum*)	Flooding	Increased biomass, reduced ethylene	ACC deaminase	Grichko and Glick (2001)
10	*Pseudomonas* sp., *Alcaligenes xylosoxidans, Variovorax paradoxus*	Canola (*Brassica napus*), Indian mustard (*Brassica juncea*)	Metal toxicity (cadmium)	Increased biomass and root elongation, reduced ethylene	ACC deaminase, IAA, P solubilization	Belimov et al. (2001)
11	*Kluyvera ascorbata*	Canola (*Brassica napus*), tomato (*Solanum lycopersicum*)	Metal toxicity (nickel)	Increased biomass, reduced ethylene	ACC deaminase, Fe siderophores	Burd et al. (1998)
12	*Pseudomonas putida*	Canola (*Brassica napus*)	Metal toxicity (nickel)	Increased biomass and shoot length	ACC deaminase	Farwell et al. (2007)
13	*Pseudomonas putida*	Canola (*Brassica napus*)	Metal toxicity (nickel)	Increased biomass, nickel uptake	ACC deaminase, IAA, Fe siderophores	Rodriguez et al. (2008)
14	*Paenibacillus yonginensis*	Ginseng (*Panax ginseng*)	Salt	Osmolyte production, ion pump activation, increased total sugars, ROS scavenging	IAA, Fe siderophores, P solubilization	Sukweenadhi et al. (2018)
15	*Pseudomonas fluorescens*	Peanut (*Arachis hypogaea*)	Salt	Increased yield, root length, shoot length, germination percentage	ACC deaminase	Saravanakumar and Samiyappan (2006)
16	*Bacillus amyloliquefaciens*	Rice (*Oryza sativa*)	Salt	Increased shoot length, root length, dry weight, chlorophyll, proline	ACC deaminase, IAA, P solubilization	Nautiyal et al. (2013)
17	*Brevibacterium iodinum, Bacillus licheniformis, Zhihengliuela alba*	Pepper (*Capsicum annuum*)	Salt	Increased shoot and root length, shoot and root biomass	ACC deaminase, EPS	Siddikee et al. (2011)

(Continued)

TABLE 2.1 (CONTINUED)
Recent Studies Evaluating Bacteria-Mediated Tolerance to Abiotic Stress

	Bacteria	Plant Species	Stress	Growth Promotion	Mechanism	Source
18	*Bacillus* sp., *Burkholderia cepacia, Enterobacter* sp., *Paenibacillus* sp.	*Wheat (Triticum aestivum)*	Salt	Increased shoot and root biomass	EPS, IAA, P solubilization	Upadhyay et al. (2011)
19	*Achromobacter piechaudii*	*Tomato (Solanum lycopersicum)*	Salt	Increased biomass and water use efficiency (WUE), reduced ethylene	ACC deaminase	Mayak et al. (2004a)
20	*Pseudomonas putida*	*Canola (Brassica napus)*	Salt	Increased biomass, reduced ethylene	ACC deaminase	Cheng et al. (2007)

with *Bacillus amyloliquefaciens* resulted in increased shoot length, root length, dry weight, chlorophyll content, and proline content in rice (Nautiyal et al., 2013). Increased yield, root length, shoot length, and germination were also observed in peanut (*Arachis hypogaea*) plants inoculated with *Pseudomonas fluorescens* under salinity stress conditions (Saravanakumar and Samiyappan, 2006), and ginseng (*Panax ginseng*) plants inoculated with *Paenibacillus yonginensis* and grown in high levels of NaCl demonstrated increases in osmolyte production, ion pump activation, total sugar content, and scavenging of reactive oxygen species (ROS) (Sukweenadhi et al., 2018). Inoculation of pea plants (*Pisum sativum*) with *P. fluorescens* resulted in plants with increased biomass, shoot and root length, leaves per plant, and WUE in plants exposed to drought stress compared with non-inoculated controls (Zahir et al., 2008). Pea plants under drought stress conditions also demonstrated increased biomass and grain yield when inoculated with *P. fluorescens* and *P. putida* (Arshad et al., 2008), in addition to increased yield and nutrient content when inoculated with *Variovorax paradoxus* (Belimov et al., 2009). Tomato and pepper seedlings exposed to drought stress had increased biomass and reduced ethylene concentrations when inoculated with *Achromobacter piechaudii* compared with non-inoculated controls (Mayak et al., 2004b), while foxtail millet (*Setaria italica*) inoculated with *P. fluorescens*, *Enterobacter hormaechei*, or *P. migulae* displayed increased germination and seedling growth under drought stress conditions (Niu et al., 2018). *Achromobacter xylosoxidans, Serratia ureilytica, Herbaspirillum seropedicae,* and *Ochrobactrum rhizosphaerae* were all effective at improving the growth of tulsi (*Ocimum sanctum*) under flooding stress conditions by increasing shoot growth, yield, and chlorophyll content while reducing endogenous ethylene concentrations compared with non-inoculated plants (Barnawal et al., 2012). Increased biomass and reduced ethylene levels were also observed in tomato plants exposed to flooding stress when inoculated with *Enterobacter cloacae* or *P. putida* (Grichko and Glick, 2001). Several strains of PGPR have also proved effective for reducing heavy metal toxicity, including *P. putida*, which increased biomass, shoot length, and nickel uptake in inoculated canola plants exposed to high levels of nickel (Farwell et al., 2007;

Rodriguez et al., 2008). Inoculation of canola and tomato plants with *Kluyvera ascorbata* also resulted in increased biomass and reduced ethylene levels under nickel toxicity when compared with non-inoculated plants (Burd et al., 1998). Tolerance to cadmium toxicity in canola and Indian mustard was further increased by inoculation with *Pseudomonas* sp., *Alcaligenes xylosoxidans*, or *V. paradoxus*, with inoculated plants showing increases in biomass and root elongation, as well as decreased ethylene levels compared with plants growing without PGPR (Belimov et al., 2001).

PGPR regulation of plant growth and stress tolerance involves several mechanisms, including enhancing nutrient acquisition of plants from the environment, such as nitrogen fixation, excreting exopolysaccharides into the plant-growth medium, and stimulation of plant growth by altering plant metabolism through the microbial production of hormones or regulating endogenous hormone levels within the host plant (Glick, 1999, 2007a, 2007b, 2012; Lugtenberg and Kamilova, 2009).

2.3 BIOLOGICAL ROLES AND MECHANISMS OF PGPR REGULATION OF PLANT GROWTH AND STRESS TOLERANCE

The major mechanisms or metabolic pathways imparting the growth-promoting effects of PGPR are discussed in the following sections.

2.3.1 PGPR REGULATION OF PLANT NUTRITION

A direct mechanism by which PGPR can facilitate plant growth and tolerance to abiotic stress is through the increased acquisition of nutrients such as nitrogen, phosphorus, potassium, and iron. The best-studied symbiotic association for nutrient acquisition is the interaction between leguminous plants and *Rhizobium* for nitrogen fixation (Hirsch et al., 2001; Matamoros et al., 2003). However, other free-living and endophytic diazotrophs, including *Burkholderia, Azoarcus, Desulfuromonas, Geobacter, Azospirillum, Bacillus, Herbaspirillum, and Nitrobacter,* can also form associations with plants to encourage growth by making atmospheric N available for assimilation (Franche et al., 2009; Bashan and de-Bashan, 2010; Bahulikar et al., 2014). Some PGPR also have the ability to increase the solubility of phosphorus and potassium, increasing the availability of these nutrients to the plant (Han et al., 2006; Tarafdar and Gharu, 2006; Lugtenberg and Kamilova, 2009; Richardson and Simpson, 2011). The production of iron siderophores is employed by various PGPR to chelate iron, thus making it more absorbable by the plants' roots (Crowley et al., 1992; Radzki et al., 2013). Enhanced uptake of Fe, Zn, Mg, Ca, and K by crop plants inoculated with strains of *Pseudomonas* and *Acinetobacter* has also been reported (Khan, 2005). By improving plant nutrition, PGPR can increase physiological function and growth during periods of stress while reducing the need for synthetic fertilizers.

For plants to be able to assimilate atmospheric nitrogen (N_2), it must be converted to ammonia (NH_3) through bacterial nitrogen fixation. This is conducted via nitrogenase enzymes, which are composed of two components, each of which is oxygen sensitive. Component I is an association of two α-protein subunits, two β-protein subunits, 24 molecules of iron, two molecules of molybdenum, and an iron–molybdenum cofactor (FeMoCo). Component II contains two α-protein subunits that differ from those of component I in addition to several iron molecules. The reduction of N_2 to NH_3 is an energy-intensive process due to the stability of the triple bond in N_2 and thus requires the two components, a magnesium–ATP complex, and reducing agents for the reaction to occur (Glick, 2015). Component I catalyzes the reaction, while component II transfers electrons to component I. There are 15–20 accessory proteins that are involved in this process, many of which function to transfer electrons to component II and to synthesize the iron–molybdenum cofactor in component I. The complex nature of this process involves the coordinated effort of multiple genes, collectively referred to as *nif* genes (Glick, 2015). These genes are arranged in a single cluster in diazotrophs, occupying 24 kb, with seven operons, and encoding 20 different proteins. The *nifA* and *nifL* genes provide the regulatory control, with the NifA protein serving as a positive regulator by binding to the promoters of all other *nif* operons. The NifL protein is a negative regulator that binds to NifA and blocks it from binding to the *nif* promoters (Glick, 2015).

While considerable research has focused on understanding bacteria-mediated nitrogen fixation and assimilation, attention must also be directed toward bacteria-mediated phosphorus solubilization. In addition to being a major component of plant nutrition, P is also the most limiting factor in N_2 fixation by *Rhizobium* in leguminous plants (MacDermott, 2000; Gyaneshwar et al., 2002). Although many soils contain abundant amounts of total phosphorus, it often exists in forms that are not accessible for plant uptake, as soluble P is highly reactive with Ca, Fe, or Al, ultimately leading to P precipitation. Some of the most plant-available forms of P exist as orthophosphate anions (HPO_4^{2-} and $H_2PO_4^{1-}$), which are often only present in low concentrations in the soil solution. Transport proteins that facilitate the uptake of orthophosphate have high

expression in root hair cells (Mitsukawa et al., 1997), resulting in plant assimilation of P and a depletion of orthophosphate in the rhizosphere. Organic anions, such as citrate and oxalate, have been shown to be effective at mobilizing P bound in humic–metal complexes in the soil, thus making it more plant available (Gerke, 1993). Microbial-mediated P dynamics play a complex and essential role in facilitating adequate plant nutrition and contributing to abiotic stress tolerance. While multiple genera of bacteria possess the capacity to increase P availability to plants, there are various mechanisms through which this is achieved.

Microorganisms in the rhizosphere can both metabolize and release organic anions that facilitate P mobilization and plant uptake (Jones et al., 2003). Organic P can constitute as much as 50% of total P in the soil (Barber, 1995), with microbial turnover responsible for the bulk of dissolved organic P (largely present as nucleic acids and phospholipids) (Macklon et al., 1997). Phosphatase activity is higher in the rhizosphere, functioning in the hydrolysis of organic P, with microbially mediated mineralization of organic P contributing to the pool of plant-available P in the rhizosphere (Frossard et al., 2000). Microorganisms can also produce phytases, which can hydrolyze phytate, another major form of organic P in the soil (Alexander, 1977). An additional mechanism of P solubilization, especially in Gram negative bacteria, occurs via the secretion of gluconic acid (Kim et al., 1998). Bacteria use glucose dehydrogenase (GDH) and the cofactor pyrroloquinoline (PQQ) in the oxidative metabolism of glucose to produce gluconic acid, which functions to increase P availability to plants. Chabot et al. (1996) demonstrated increased growth of maize and lettuce when inoculated with P-solubilizing *Rhizobium leguminosarum*, with P solubilization resulting from microbial production of 2-ketogluconic acid influencing pH changes in the soil. The total capacity of bacteria-mediated P solubilization is dependent on the strains present as well as the specific P substrates available. *Bacillus megaterium* is one of the most studied P-solubilizing bacteria, and while it is effective at mineralizing P from organic phosphates, it cannot solubilize mineral P from existing complexes (Asea et al., 1988; Kucey, 1988). When *B. megaterium* was compared with other *Bacillus* species for their ability to solubilize P from a variety of substrates, *B. megaterium* was found to produce the highest concentration of soluble P from a fish bone substrate (Saied et al., 2018). Furthermore, even bacteria that do not directly solubilize P can play a role in increasing plant P availability, as P-assimilating bacteria can take up P with low solubility in the soil through high-affinity transporters. When these bacteria die, the P becomes available to the plants through mineralization.

2.3.2 Interactions of PGPR and Abiotic Stress Involving Exopolysaccharides

Bacteria produce diverse polysaccharides that coat their cell walls and are excreted into the environment. These substances are referred to as *exopolysaccharides* and play a valuable role in improving both bacterial and plant tolerance to various abiotic stresses, including drought and salinity, by effectively coating bacterial cells and plant tissues that are sensitive to desiccation in a polysaccharide film (Sandhya et al., 2009; Upadhyay, 2011). Exopolysaccharide-producing bacteria have been found in symbiotic relationships with many plants, including *Arabidopsis thaliana* (Huang et al., 2017), foxtail millet (Niu et al., 2018), and native cacti from a semi-arid region in Brazil (Kavamura et al., 2013).

Drought stress has been found to increase bacterial exopolysaccharide (EPS) production and to change the concentration and composition of EPS, functioning as a bacterial survival mechanism in conditions with reduced moisture and creating rhizosheath in conjunction with plant roots (Vardharajula and Ali, 2014; Naseem et al., 2018). As soil moisture decreases, increased EPS production composed of high–molecular weight carbohydrate complexes protects the bacteria from drought stress while increasing soil water holding capacity and aggregation (Vardharajula and Ali, 2014). Maize inoculated with EPS-producing strains of *Proteus*, *Psuedomonas*, and *Alcaligenes* demonstrated increased relative water content, protein, sugar, and proline content under drought stress conditions (Naseem and Bano, 2014). Inoculation with EPS-producing *Mitsuaria* sp. and *Burkholderia* sp. also enhanced drought tolerance in maize and *Arabidopsis thaliana* by altering root architecture, reducing evapotranspiration, and increasing proline levels and antioxidant activity (Huang et al., 2017). Scanning electron microscope images of sunflower roots inoculated with EPS-producing *P. putida* revealed the formation of biofilms on the root surfaces, which contributed to increased seedling survival rate and plant biomass in drought stress conditions (Sandhya et al., 2009). Wheat seedlings inoculated with EPS-producing PGPR strains and subjected to salinity stress had increased root and shoot biomass and reduced Na^+ uptake compared with non-inoculated plants (Upadhyay, 2011). *P. fluorescens* isolated from the roots of foxtail millet was found to have high 1-aminocyclopropane-1-carboxylate (ACC) deaminase and EPS-producing activity and was able to confer drought tolerance by stimulating seed germination and growth as well as increasing soil moisture and enhancing the root-adhering soil/root tissue ratio (Niu et al., 2018). EPS production and desiccation tolerance in the PGPR strain *P. fluorescens* are controlled by the algU-mucA-mucb gene cluster, in which algU is hypothesized to act as an on-off switch that drives the expression of defense-related genes, thus facilitating survival of the bacteria and growth of the plant in environments where moisture is a limitation (Schnider-Keel et al., 2001).

2.3.3 Promotive Effects on Plant Growth and Stress Tolerance through Hormonal Regulation

Phytohormones, such as auxin (IAA), cytokinin (CK), gibberellic acid (GA), abscisic acid (ABA), and ethylene, can have various impacts on plant growth and stress tolerance. PGPR can play a role in directly synthesizing these compounds or regulating their endogenous levels within the plant.

IAA is a major plant hormone that is involved in growth and development throughout the life of a plant, including shoot and root elongation and lateral root initiation, as well as cell division, elongation, and differentiation (Abel and Theologis, 1996). Under osmotic stress, including drought and salinity conditions, endogenous levels of IAA are reduced in leaves and roots (Du et al., 2013; Liu et al., 2015). Additionally, Du et al. (2013) observed decreased expression of YUC auxin biosynthesis genes in rice under osmotic stress. The role of auxin in osmotic stress tolerance was further clarified when the *Agrobacterium tumefaciens* tryptophan monooxygenase (iaaM) gene was overexpressed in *Arabidopsis*, resulting in overproduction of auxin and increased drought tolerance. Comparatively, a yuc1/yuc2/yuc6 triple mutant, with lower auxin biosynthesis, demonstrated reduced drought tolerance (Shi et al., 2014). Shi et al. (2014) also attributed the drought tolerance benefits of auxin to changes in root architecture, ABA gene expression, and ROS metabolism. Approximately 85% of all soil bacteria are able to synthesize IAA from tryptophan produced in the roots of plants, and many PGPR also possess this capability. IAA-producing *Burkholderia* sp. and *Pseudomonas* sp. were found to improve root growth in *Arabidopsis thaliana*, with increased branching of longer, thinner roots (Jiang et al., 2012), suggesting a morphological change in root architecture that is beneficial in drought stress conditions. Wheat seedlings inoculated with IAA-producing *Pseudomonas* sp. were found to have increased root growth and reduced dormancy under salinity stress conditions when compared with non-inoculated controls (Egamberdieva, 2009). Increased root and shoot growth were also observed in wheat seedlings inoculated with an IAA-producing strain of *Bacillus altitudinis* under iron stress (Sun et al., 2017). *Klebsiella variicola*, another IAA-producing PGPR, promoted growth in soybean plants subjected to flooding stress by initiating adventitious root formation, increasing chlorophyll content, and improving photosynthetic efficiency (Kim et al., 2017).

GA plays many important roles in plant growth and development, including stimulation of cell division and seed germination as well as stem and root elongation, although its role in stress tolerance is somewhat complex. Increased GA signaling and biosynthesis has been shown to increase plant growth under shade stress and submergence through an activated escape response, while reduced GA signaling contributes to diminished growth in plants exposed to various abiotic stresses, including cold, salinity, and osmotic stress (Franklin, 2008; Bailey-Serres and Voesenek, 2010; Colebrook et al., 2014). However, while reduced GA levels contribute to a reduced plant stature, they are also related to increased abiotic stress tolerance and survival (Achard et al., 2006; Colebrook et al., 2014). As such, appropriate regulation of GA for a specific abiotic stress is essential. Many species of bacteria from within the genera of *Acetobacter*, *Bacillus*, *Acinetobacter*, *Bradyrhizobium*, *Arthrobacter*, *Burkholderia*, *Pseudomonas*, and *Rhizobium* produce GA (Glick, 2015). Corn seedlings inoculated with GA-producing *Azospirillum lipoferum* demonstrated increased root growth and increased root hair growth under non-stress conditions (Fulchieri et. al., 1993). Shazad et al. (2016) found that inoculation of rice seedlings with GA-producing *Bacillus amyloliquefaciens* resulted in increased levels of endogenous salicylic acid and reduced levels of abscisic acid and jasmonic acid, suggesting the potential for improved stress tolerance through phytohormone regulation. However, a comprehensive understanding of the direct effects of PGPR-influenced GA modulation on stress tolerance remains incomplete.

CKs promote cell growth and division while inhibiting aging and senescence in the plant. The role of CKs in promoting abiotic stress tolerance is complex. Under heat stress, endogenous CK levels are decreased in the plant (Cheikh and Jones, 1994), with the current understanding suggesting that genes regulating CK biosynthesis are upregulated for a short period of time at the initial onset of stress and then decrease under moderate stress or increase further under severe stress (Nishiyama et al., 2011; Zwack and Rashotte, 2015). Studies that have evaluated the effects of exogenous CK application on abiotic stress have had mixed results, including reduced salt tolerance in beans with foliar application (Kirkham et al., 1974), increased salt tolerance in wheat grown in supplemented media and potato pretreated with CKs (Naqvi et al., 1982; Abdullah and Ahmad, 1990), and increased drought tolerance in bean, but decreased drought tolerance in sugar beet when irrigated with CKs (Pospisilova and Batkova, 2004). Exogenous applications of CKs have delayed leaf senescence in creeping bentgrass under heat stress while encouraging higher chlorophyll content and photochemical efficiency in addition to increased antioxidant function and reduced oxidative damage (Liu et al., 2002; Veerasamy et al., 2007; Xu and Huang, 2009). Transcriptome analyses of plant responses to CKs have shown upregulation of enzymes that counter oxidative stress, including peroxidase and glutathione transferase (Brenner and Schmülling, 2012; Bhargava et al., 2013). Production of CKs has been observed for several strains of PGPR, including *Azotobacter* spp., *Rhizobium* spp., *Pantoea agglomerans*, *Rhodospirillum rubrum*, *P. fluorescens*, *Bacillus subtilis*, and *Paenibacillus polymyxa* (Glick, 2015), while *Bacillus megaterium* has been found to induce expression of genes related to CK receptors in *Arabidopsis*, facilitating growth promotion (Ortíz-Castro et al., 2008). Alfalfa that was inoculated with strains of *Sinorhizobium* that had been engineered to overproduce CK demonstrated increased drought tolerance compared with non-inoculated plants (Xu et al., 2012). Understanding endogenous CK biosynthesis, signaling, and transcriptional regulation of stress-related genes may help to clarify the relationship between bacterially produced or regulated CK and stress tolerance. Crosstalk between CK and ABA also has a pronounced influence on the mechanisms involved in abiotic stress tolerance.

ABA plays an important role in conferring stress tolerance in plants largely through the regulation of WUE. Under drought stress conditions, ABA triggers stomatal closure, minimizing water loss through transpiration, and encourages root cell elongation, facilitating post-drought recovery (Daszkowska-Golec, 2016). ABA also induces antioxidant

defense genes, including superoxide dismutase (SOD), glutathione peroxidase (GPX), ascorbate peroxidase (APX), and catalase (CAT), which can prevent damage from osmotic or ionic stress by scavenging ROS (Bharti et. al., 2016). Suzuki et al. (2016) found that ABA was required for tolerance to a combination of heat and salt stress in *Arabidopsis thaliana* by observing that many transcripts associated with ABA were upregulated during the combined stress conditions and that mutants deficient in ABA metabolism were more susceptible to heat and salt stress than wild type plants. Several strains of PGPR have demonstrated an ability to produce ABA and/or regulate endogenous levels of plant ABA, including *Achromobacter xylosoxidans*, *Bacillus licheniformis*, *B. pumilus*, *B. subtilis*, *Brevibacterium halotolerans*, *Lysinibacillus fusiformis*, *Pseudomonas putida*, and *Paenibacillus yonginensis* (Sgroy et al., 2009; Glick, 2014; Sukweenadhi et al., 2018). Increased salinity tolerance was observed in *Panax ginseng* inoculated with *P. yonginensis* through the induction of ABA biosynthetic genes as well as an increase in ROS scavenging enzymes, proline content, ion transport, total sugars, and root hair formation (Sukweenadhi et al., 2018). Inoculation with *Bacillus amyloliquefaciens* also increased salinity tolerance in rice through ABA modulation, with inoculated plants demonstrating increased chlorophyll content, shoot length, root length, fresh weight, and dry weight compared with non-inoculated plants (Shahzad, 2016). Additionally, Bharti (2016) found that inoculation of wheat with *Dietzia natronolimnaea* improved salinity stress tolerance through the regulation of stress-responsive genes related to ABA.

Ethylene is produced in low concentrations in plants under optimal growth conditions, but its production can be stimulated to detrimental levels by various abiotic stresses, including heat, drought, and salinity stress (Stearns and Glick, 2003; Glick, 2005). Stress-induced accumulation of ethylene occurs in all tissues and organs of a plant and particularly in roots directly exposed to salt stress (Xu and Zou, 1993; Gomez-Cadenas et al., 1996). While an initial induction of ethylene synthesis at low concertation (typically below 0.1 μL L^{-1}) is thought to trigger plant protective responses (Stearns and Glick, 2003; Pierik et al., 2006), an excessive amount of ethylene accumulation during a prolonged period or severe stress can cause damage in plants, such as the induction and acceleration of leaf senescence, chlorosis, and abscission, thereby causing overall inhibitory effects on plant growth and yield (Morgan and Drew, 1997; Stearns and Glick, 2003; Pierik et al., 2006). The ethylene precursor ACC plays similar biological roles to those of ethylene, as it also induces leaf senescence and inhibits root cell elongation (Le et al., 2001; De Cnodder et al., 2005; Staal et al., 2011; Markakis et al., 2012). In studies with tomato and cucumber plants, increasing salt concentrations enhanced ethylene production rates, which caused premature leaf senescence (Feng and Barker, 1993; Helmy et al., 1994). Increased ethylene production also inhibits root elongation, while inhibitors of ethylene biosynthesis or action restore root elongation at low water potential (Spollen et al., 1997). Ethylene inhibition of cell elongation has been reported for lateral roots and adventitious roots as well as leaves (Visser et al., 1997; Pierik et al., 1999). High concentrations of ethylene and the ethylene precursor ACC have been found to inhibit root cell elongation by affecting cell-wall properties (Le et al., 2001; De Cnodder et al., 2005; Staal et al., 2011; Markakis et al., 2012). ACC induces apoplastic alkalinization in root cells, which leads to a decrease in the activity of cell-wall loosening agents (Staal et al., 2011). Markakis et al. (2012) reported that ACC downregulated the expression of several genes coding for cell-wall loosening proteins. At the cellular level, a high concentration of ethylene causes a reduction in membrane fluidity (Thompson et al., 1982; Coker et al., 1985) and increases phospholipid turnover in membranes (Drory et al., 1992) and solute leakage from plant cells (Eze et al., 1986; Drory et al., 1995). Ethylene and ACC can exacerbate the adverse effects of abiotic stress that trigger the ethylene response.

Considering the adverse effects of stress-induced production of ethylene and ACC on plant growth, it is anticipated that any approaches, either chemical or biological, that can suppress ethylene synthesis or action and lower ACC levels in plants should be effective for alleviating ethylene-related stress damage and improving stress tolerance. Various studies have found that the exogenous application of chemical inhibitors of ethylene synthesis to reduce ethylene accumulation and block the action or responses of ethylene can effectively promote shoot and root growth under abiotic stress conditions (Xu and Huang, 2007; Glick, 2014).

Some endophytic PGPR are particularly effective in suppressing ethylene production within the host plant by breaking down the ethylene precursor ACC using the bacterial ACC deaminase enzyme (ACCd). ACCd-rhizobacteria include a large variety of bacterial species, which are known to promote shoot and root growth as well as plant tolerance to abiotic stresses at morphological and physiological levels. The ACCd-rhizobacteria offer non-toxic and environment-friendly benefits to mitigate the adverse effects of stress-induced ethylene accumulation in plants. However, the mechanisms of ACCd-rhizobacteria interaction with plants and how ACCd-rhizobacteria may promote plant tolerance to abiotic stress are largely unknown. Further research in these areas will advance our understanding of how endophytic rhizobacteria interact with plant hosts, enabling the promotion of plant growth and the underlying factors controlling successful colonization of the bacteria in the host plant, which is important for improving PGPR efficacy.

The ACCd enzyme cleaves the ethylene precursor ACC to α-ketobutyrate and ammonia, which the bacteria use as a nitrogen source. As a result, this process can lower the endogenous level of ethylene within the plant tissue colonized by ACCd-rhizobacteria (Glick, 2005). One of the most pronounced effects of ACCd-rhizobacteria colonization is stimulation of root elongation (Belimov et al., 2009). The study of ACCd-producing *Burkholderia phytofirmans* PsJN mutants lacking the gene controlling ACCd has provided direct evidence for the growth-promoting effects of ACCd, which has shown that mutant bacteria lacking ACCd activity had no effect on root elongation in canola seedlings (Sessitsch et al., 2005; Sun et al., 2009). Endophytic rhizobacteria cause not only local effects on roots but also systemic effects on shoots via long-distance ACC signaling (Chen et al., 2013). The inoculation of plants with ACCd-rhizobacteria can also increase leaf relative water content (Nadeem et al., 2007, 2010; Ahmad et al., 2013), chlorophyll content (Mayak et al., 2004a, 2004b; Nadeem et al., 2007; Bal et al., 2013), and WUE (Mayak et al., 2004a, 2004b) under salinity stress. On the other hand, AcdS minus mutant strains without the functional acdS gene (encoding ACC deaminase) were not effective for alleviating ethylene inhibition of plant growth (Li et al., 2000; Sun et al., 2009). Ample evidence has demonstrated the positive roles of ACCd-bacteria in regulating plant growth and stress adaptation based on the phenotypic and physiological assessment of host plants (Verhagen et al., 2004; Wang et al., 2005; Lopez-Bucio et al., 2007; Cartieaux et al., 2008; Contesto et al., 2010; Schwachtje et al., 2012; Poupin et al., 2013; Zamioudis et al., 2013).

Burkholderia species are among the most abundant ACCd-rhizobacteria inhabiting the rhizosphere and endosphere. *Burkholderia* are present in diverse plant species and adapt to a range of environmental conditions. Different species in the *Burkholderia* genus have been isolated from roots, leaves, stems, and fruits in both dicot and monocot plant species, including *B. phytofirmans* from onion roots (Nowak and Shulaev, 2003), *B. cepacia* from branches of citrus (Araujo et al., 2002) and from roots of rice (Singh et al., 2006), *B. silvatlantica* from leaves of sugar cane (Perin et al., 2006), *B. pyrrocinia* from the stems of lodgepole pine (Bal et al., 2012), and *B. tropica* from the stems and fruits of pineapple (Magalhaes Cruz et al., 2001). The presence of *Burkholderia* throughout the plant tissues and in such diverse plant species demonstrates the wide host range of *Burkholderia* spp. and the endophytic nature of these bacteria in plant internal tissues with great potential for wide adaptation in agricultural ecosystems.

To date, *B. phytofirmans* PsJN is the most studied species/strain among the *Burkholderia* genus for growth-promoting effects. *B. phytofirmans* PsJN was found to promote both shoot (biomass) and root growth (increased biomass, number of lateral and adventitious roots) in a large number of plant species, including dicots and monocots, and herbaceous and woody plant species (Compant et al., 2005; Sessitsch et al., 2005; Ait Barka et al., 2006; Bordiec et al., 2011; Da et al., 2012; Fernandez et al., 2012; Kim et al., 2012; Naveed et al., 2014), such as tomato (Nowak et al., 2004), maize, and sorghum (Bevivino et al., 2000). *B. phytofirmans* also plays a role in plant adaptation to abiotic stresses. Ait Barka (2006) reported increased growth and physiological activities of *Vitis vinifera* by inoculation with *B. phytofirmans* PsJN at low temperature (4 °), suggesting the positive effects of the bacteria on cold tolerance. *B. phytofirmans* PsJN also enhanced salt stress tolerance (Pinedo et al., 2015; Cheng et al., 2016). However, the molecular and metabolic mechanisms of how *B. phytofirmans* PsJN may regulate plant growth and stress tolerance are yet to be determined.

Ethylene signaling and interplay with stress-related transcription factors also may play a crucial role in ACCd-bacteria-mediated plant stress responses. The ethylene-responsive element binding proteins (EREBP) in the ethylene response pathway show altered expression in response to stresses (Broekaert et al., 2006). Other important stress-signaling transcription factors include protein kinases; MAP-kinases MPK6 and MPK3 have been reported to phosphorylate ACC synthase, thereby affecting ethylene biosynthesis (Liu and Zhang, 2004). ACCd-rhizobacteria may alter stress-related transcriptional factors and downstream genes, proteins, and metabolites responsible for growth promotion and stress tolerance.

Transcriptomics is a comprehensive technology for sequencing all RNA transcripts of an organism, which enables the analysis of gene expression in its entirety and allows the identification of key molecular factors regulating plant growth and stress responses. Wang et al. (2005) analyzed the transcriptional changes induced by *P. fluorescens* strain FPT9601-T5 in *Arabidopsis* shoots and found many transcripts associated with stress defense were upregulated, such as genes involved in carbohydrate metabolism and oxidative burst. van de Mortel et al. (2012) analyzed the transcriptomic changes in *Arabidopsis* induced by *P. fluorescens* SS101 (Pf. SS101) and found that a large number of genes responsive to the inoculation in roots and leaves belong to functional categories in general stress response, root morphogenesis, metal ion transport, and secondary metabolism. Several genes were upregulated by ACCd-producing *Enterobacter cloacae* inoculation, including cdc48 and eIF3. cdc48 is highly expressed in proliferating cells of shoots and roots, regulating cell division and growth (Feiler et al., 1995), while eIF3 is necessary for protein production (Hontzeas et al., 2004). The upregulation of cdc48 and eIF3, which are involved in protein synthesis and cell division and growth, may contribute to the enhanced root elongation observed in *E. cloacae*–inoculated canola roots (Hontzeas et al., 2004). Inoculation with *B. phytofirmans* PsJN upregulated genes for ROS scavenging (Ascorbate Peroxidase 2) and detoxification (Glyoxalase I 7) (Pinedo et al., 2015). The transcriptional response to PsJN inoculation varies between plant organs, with more rapid induction of stress-responsive genes in roots than in shoots in *Arabidopsis* inoculated with *B. phytofirmans* PsJN (Pinedo et al., 2015). Hontzeas et al. (2004) examined changes in gene expression over time in canola roots treated with *E. cloacae* UW4 and an AcdS (ACCd structural gene) knockout mutant of *E. cloacae* UW4 and reported that ACCd-bacterium colonization repressed the stress-induced expression of stress response genes, including RAP1. RAP1 contains a glycine-rich domain common to many RNA-binding proteins involved in plant stress responses, which is typically induced by ethylene-producing stresses, such as cold, wounding, and flooding.

Transcriptional analysis in the vast majority of stress-related studies is focused on the first hours after inoculation or stress exposure, but little is known about the molecular changes after a long-term exposure (Geng et al., 2013; Kim et al., 2014). In addition, the transcriptional responses to PGPR differ between plant species (Stearns et al., 2012) and also largely depend on the bacterial partner (Loon, 2007). Straub et al. (2013) found that very few gene transcripts in

the *Herbaspirillum frisingense*–induced transcriptome profile of *Miscanthus* were common to other grass species, suggesting that endophyte strain–specific and plant species–specific responses exist during interaction with diverse bacterial species.

Proteomics is the large-scale identification and quantification of proteins and associated biological functions and metabolic pathways. A few proteomic studies have been conducted to determine endophyte-induced changes in plant protein expression and/or plant-induced changes in endophyte protein expression (Bestel-Corre et al., 2004). Banaei-Asl et al. (2016) found that *P. fluorescens* inoculation increased the abundance of proteins related to glycolysis, the tricarboxylic acid cycle, and amino acid metabolism, and attributed the plant proteomic changes to improved salt tolerance in canola. Another study with *P. fluorescens* inoculation in canola exposed to salt stress conditions also found an increased abundance of proteins related to amino acid metabolism and glycolysis (Banaei-Asla et al., 2015). Increases in the abundance of proteins involved in photosynthesis, anti-oxidative processes, transportation across membranes, and pathogenesis-related responses under salt stress were reported in canola inoculated with *P. putida* UW4 (Cheng et al., 2012). *P. putida* UW4 inoculation in cucumber roots under hypoxia stress led to changes in the abundance of 1735 protein spots in roots, which were identified to exhibit putative functions in transcription, protein synthesis, signal transduction, carbohydrate and nitrogen metabolism, stress defense, antioxidant binding, and others. Proteins involved in plant growth and development, such as xyloglucan endotransglycosylase, and defense, such as peroxidases, glutathione S-transferases, and kinases, exhibited increases in abundance in rice due to the colonization of *Bacillus cereus* (Wang et al., 2012).

Metabolomics is a powerful approach to the discovery of metabolites and metabolic pathways involved in the symbiotic interaction between the plant host and microbes (Feussner and Polle, 2015). Metabolic pathways of the host plant regulated by PGPR have not been well explored. A few studies have described the modifications in the content of primary and secondary metabolites following inoculation. van de Mortel et al. (2012) analyzed the metabolome of *Arabidopsis* plants inoculated with *P. fluorescens* SS101 and identified 46 and 13 metabolites differentially accumulated in roots and leaves, respectively, due to the inoculation, with increased content of some metabolites in both organs, including glucosinolates, phytoalexin scopoletin glucoside, d-gluconate, and indole-3-carboxylic acid β-d-glucopyranosyl ester. The inoculation of maize with *P. putida* KT2440 resulted in the accumulation of several phospholipids, particularly diacylglycerophosphocholine (Planchamp et al., 2015). Changes in the content of secondary metabolites, such as 2-hydroxy-4,7-dimethoxy-1,4-benzoxazin-3-one glucoside, were reported in maize inoculated with *P. fluorescens* F113 (Walker et al., 2011, 2012).

Suppression of the endogenous levels of ethylene in plant tissues due to the colonization of ACCd-rhizobacteria may rapidly activate the plant defense system, leading to a primed or faster response to abiotic stresses, and cause long-term reprogramming in the transcriptome, proteome, and metabolome of the host plant, ultimately protecting plants from ethylene-mediated stress damage. However, the stress defense mechanisms activated in the plant by ACCd-rhizobacteria inoculation are largely unknown. This can be addressed by an integrated approach combining transcriptomics, proteomics, and metabolomics, which has proved to be a powerful tool for unraveling molecular factors and metabolic pathways involved in stress adaptation to abiotic stresses and providing insightful guidance for developing strategies to improve plant productivity and stress tolerance.

2.3.4 HORMONAL CROSSTALK INVOLVED IN PGPR REGULATION OF PLANT GROWTH

The crosstalk between hormones is known to play key roles in plant stress tolerance. Among the hormones, the interaction of ethylene with auxin is well characterized. Auxin is known as a positive regulator and ethylene as a negative regulator of root elongation, and ethylene can inhibit auxin transport and signal transduction (Prayitno et al., 2006). The elevated ethylene synthesis in response to stresses can suppress IAA transport from leaves to roots, restricting root growth; on the other hand, the reduction in ethylene concentration in plant roots exposed to drought or salinity stress by the colonization of ACCd-rhizobacteria can relieve the repression of auxin transport and functions by ethylene, thereby promoting root growth under stress conditions (Dharmasiri and Estelle, 2004). The positive effects of *B. phytofirmans* PsJN on shoot growth have been associated with increases in endogenous IAA and GA levels, whereas increases in root growth were related to endogenous increases in IAA and trans-zeatin CK levels (Kurepin et al., 2015a). ACCd-producing *Bacillus subtilis* GB03 inoculation promoted *Arabidopsis* shoot growth and enhanced photosynthetic efficiency and chlorophyll content, while it inhibited ABA biosynthesis (Zhang et al., 2008). Kurepin et al. (2015b) reported that inoculation of potato plants with *B. phytofirmans* PsJN strain caused significant increases in the endogenous salicylic acid (SA) levels in shoots (1.5-fold) and roots (4.0-fold). Inoculation of potato plants with *B. phytofirmans* PsJN also increased IAA levels in roots (Kurepin et al., 2015a) as well as root jasmonic acid (JA) levels and ABA levels (Kurepin et al., 2015b). SA, JA, and ABA, along with ethylene, are considered as stress hormones, which are typically induced by stresses, while IAA, GA, and CKs are growth-promoting hormones, which typically are repressed by stress. These studies suggest that the regulation of plant growth and stress tolerance by ACCd-rhizobacteria may involve hormonal crosstalk. The systematic mechanisms by which ACCd-bacteria that inhibit ethylene synthesis may interact with stress hormones (SA, JA, and ABA) and growth-promoting hormones (IAA, GA, and CK) to enable the positive effects on the shoot and root growth under abiotic stress conditions or enhance stress tolerance deserve further investigation.

Transcript analyses of plants inoculated with different PGPR species with ACCd activities have shown changes (upregulation or downregulation) of different genes involved in the biosynthesis and signaling of ethylene and other hormones, such as IAA, GA, ABA, SA, and JA. Auxin-responsive genes, such as a putative auxin-regulated protein, a xyloglucan endo-1,4-β-d-glucanase precursor, and the heat-/auxin-/ethylene-/wounding-induced small protein, were induced in *Arabidopsis* by ACCd-producing *P. thivervalensis* MLG45 and *P. fluorescens* FPT9601-T5 (Cartieaux et al., 2003; Wang et al., 2005). *Arabidopsis* inoculated with *B. phytofirmans* PsJN demonstrated that ACCd-bacteria may regulate the expression of genes related to the auxin signaling pathway, facilitating plant tolerance to salt stress (Poupin et al., 2013). Auxin pathway induction was also found by inoculation with *Bacillus subtilis* (Lakshmanan et al., 2013). Auxin biosynthesis genes were upregulated during the inoculation of canola with *P. putida* UW4 (Stearns et al., 2012). The transcription of genes related to ABA signaling was downregulated by *B. phytofirmans* PsJN in *Arabidopsis* (Pinedo et al., 2015). The expression of GA signaling genes was also downregulated by ACCd-rhizobacteria inoculation (Glick et al., 2007). The response of JA signaling genes to ACCd-rhizobacteria inoculation varies with the time of inoculation. Straub et al. (2013) found that JA signaling genes were upregulated 3 h after inoculation and repressed at 3 weeks after inoculation of *Miscanthus sinensis* seedlings with ACCd-producing *Herbaspirillum frisingense* GSF30T. van de Mortel et al. (2012) analyzed the transcriptomic changes in *Arabidopsis* induced by *P. fluorescens* SS101 (Pf.SS101) and found upregulated SA signaling genes in roots and leaves. Verhagen et al. (2004) reported that root inoculation of *Arabidopsis* with *P. fluorescens* strain WCS417r resulted in the upregulation of genes involved in JA and ethylene signaling in leaves. However, other studies reported that *Arabidopsis* colonized by *P. fluorescens* WCS417r, *P. fluorescens* FPT9601-T5, and *P. chlororaphis* O6 demonstrated downregulation of genes involved in the ethylene signaling pathway (Wang et al., 2005; Cho et al., 2013). Straub et al. (2013) reported that ethylene receptors were upregulated, while ethylene response factors were repressed, in roots of *M. sinensis* inoculated with *H. frisingense* GSF30T, which is aligned with the effects of *H. frisingense* in reducing ethylene levels in plant roots by ACC deaminase competing with the plant ACC oxidase for ACC (Glick, 2005). As shown earlier, ACCd-rhizobacteria inoculation causes various transcriptional changes in hormone synthesis and signaling pathways, but the key molecular factors involved in the crosstalk of ethylene with other hormones by ACCd-bacteria inoculation that may confer the promotive effects of ACCd-rhizobacteria on stress tolerance warrant systematic investigation.

2.4 CONCLUSIONS AND FUTURE RESEARCH PERSPECTIVES

Physiological functions are well documented for a variety of PGPR species and strains, although the extent of the growth-promoting effects and impact on plant stress tolerance varies widely with bacteria species, host plant species, and stress types and severity levels. Mechanisms for hormonal regulation of phytostimulation and nitrogen fixation for plant nutrition have started to be unraveled using transcriptomic, proteomic, or metabolomic analysis, through the identification of critical metabolic pathways controlling the intimate interaction of the endophytic bacteria and their host. Future studies focusing on the simultaneous analysis of gene and protein expression patterns, as well as metabolic profiling in symbiotic microbes and host plants by integrating genomic information, will discover novel insights into the molecular mechanisms of PGPR regulation of plant stress tolerance and the symbiotic interactions.

BIBLIOGRAPHY

Abdullah, Z., & Ahmad, R. (1990). Effect of pre-and post-kinetin treatments on salt tolerance of different potato cultivars growing on saline soils. *Journal of Agronomy and Crop Science*, *165*(2–3), 94–102.

Abel, S., & Theologis, A. (1996). Early genes and auxin action. *Plant Physiology*, *111*(1), 9–17.

Achard, P., Cheng, H., De Grauwe, L., Decat, J., Schoutteten, H., Moritz, T., & Harberd, N. P. (2006). Integration of plant responses to environmentally activated phytohormonal signals. *Science*, *311*(5757), 91–94.

Ahmad, M., Zahir, Z. A., Khalid, M., Nazli, F., & Arshad, M. (2013). Efficacy of *Rhizobium* and *Pseudomonas* species to improve physiology, ionic balance and quality of mung bean under salt-affected conditions on farmer's fields. *Plant Physiology and Biochemistry*, *63*, 170–176.

Ait Barka, E., Nowak, J., & Clement, C. (2006). Enhancement of chilling resistance of inoculated grapevine plantlets with a plant growth-promoting rhizobacterium, *Burkholderia phytofirmans* strain PsJN. *Applied and Environmental Microbiology*, *72*(11), 7246.

Alberton, D., Müller-Santos, M., Brusamarello-Santos, L. C. C., Valdameri, G., Cordeiro, F. A., Yates, M. G., & de Souza, E. M. (2013). Comparative proteomics analysis of the rice roots colonized by *Herbaspirillum seropedicae* strain SmR1 reveals induction of the methionine recycling in the plant host. *Journal of Proteome Research*, *12*(11), 4757.

Alexander, M., Cook, A. M., & Daughton, C. G. (1977). Induction of microbial metabolism of organophosphorus compounds. US Dept of the Army; Cornell University, Ithaca, NY.

Amaral, F., Bueno, J., Hermes, V., & Arisi, A. (2014). Gene expression analysis of maize seedlings (DKB240 variety) inoculated with plant growth promoting bacterium *Herbaspirillum seropedicae*. *Symbiosis*, *62*(1), 41–50.

Araujo, W. L., Marcon, J., Maccheroni, W., Jr., van Elsas, J. D., van Vuurde, J. W. L., & Azevedo, J. L. (2002). Diversity of endophytic bacterial populations and their interaction with *Xylella fastidiosa* in citrus plants. *Applied and Environmental Microbiology*, *68*(10), 4906.

Arshad, M., Shaharoona, B., & Mahmood, T. (2008). Inoculation with *Pseudomonas* spp. containing ACC-deaminase partially eliminates the effects of drought stress on growth, yield, and ripening of pea (*Pisum sativum* L.). *Pedosphere*, *18*(5), 611–620.

Asea, P., Kucey, R., & Stewart, J. (1988). Inorganic phosphate solubilization by two *Penicillium* species in solution culture and soil. *Soil Biology and Biochemistry*, *20*(4), 459–464.

Bahulikar, R., Torres-Jerez, I., Worley, E., Craven, K., & Udvardi, M. (2014). Diversity of nitrogen-fixing bacteria associated with switchgrass in the native tallgrass prairie of northern Oklahoma. *Applied and Environmental Microbiology*, *80*(18), 5636–5643.

Bailey-Serres, J., & Voesenek, L. (2010). Life in the balance: a signaling network controlling survival of flooding. *Current Opinion in Plant Biology*, *13*(5), 489–494.

Bal, A., Anand, R., Berge, O., & Chanway, C. P. (2012). Isolation and identification of diazotrophic bacteria from internal tissues of *Pinus contorta* and *Thuja plicata*. *Canadian Journal of Forest Research*, *42*(4), 807–813.

Bal, H. B., Nayak, L., Das, S., & Adhya, T. K. (2013). Isolation of ACC deaminase producing PGPR from rice rhizosphere and evaluating their plant growth promoting activity under salt stress. *Plant and Soil*, *366*(1–2), 93–105.

Banaei-Asl, F., Bandehagh, A., Uliaei, E. D., Farajzadeh, D., Sakata, K., Mustafa, G., & Komatsu, S. (2015). Proteomic analysis of canola root inoculated with bacteria under salt stress. *Journal of Proteomics*, *124*(C), 88–111.

Banaei-Asl, F., Farajzadeh, D., Bandehagh, A., & Komatsu, S. (2016). Comprehensive proteomic analysis of canola leaf inoculated with a plant growth-promoting bacterium, *Pseudomonas fluorescens*, under salt stress. *BBA – Proteins and Proteomics*, *1864*(9), 1222–1236.

Barber, S. (1995). *Soil nutrient bioavailability: a mechanistic approach* (2nd ed.). New York: Wiley.

Barnawal, D., Bharti, N., Maji, D., Chanotiya, C., & Kalra, A. (2012). 1-Aminocyclopropane-1-carboxylic acid (ACC) deaminase-containing rhizobacteria protect *Ocimum sanctum* plants during waterlogging stress via reduced ethylene generation. *Plant Physiology and Biochemistry*, *58*(C), 227–235.

Bashan, Y., & Holguin, G. (1998). Proposal for the division of plant growth-promoting rhizobacteria into two classifications: biocontrol-PGPB (plant growth-promoting bacteria) and PGPB. *Soil Biology and Biochemistry*, *30*(8), 1225–1228.

Bashan, Y., & de-Bashan, L. E. (2010). How the plant growth-promoting bacterium *Azospirillum* promotes plant growth – a critical assessment. *Advances in Agronomy*, *108*, 77–122.

Belimov, A. A., Dodd, I. C., Hontzeas, N., Theobald, J. C., Safronova, V. I., & Davies, W. J. (2009). Rhizosphere bacteria containing 1-aminocyclopropane-1-carboxylate deaminase increase yield of plants grown in drying soil via both local and systemic hormone signaling. *New Phytologist*, *181*(2), 413–423.

Belimov, A. A., Safronova, V. I., Sergeyeva, T. A., Egorova, T. N., Matveyeva, V. A., Tsyganov, V. E., & Stepanok, V. V. (2001). Characterization of plant growth promoting rhizobacteria isolated from polluted soils and containing 1-aminocyclopropane-1-carboxylate deaminase. *Canadian Journal of Microbiology*, *47*(7), 642–652.

Bensalim, S., Nowak, J., & Asiedu, S. (1998). A plant growth promoting rhizobacterium and temperature effects on performance of 18 clones of potato. *American Journal of Potato Research*, *75*(3), 145–152.

Berg, G., Grube, M., Schloter, M., & Smalla, K. (2014). Unraveling the plant microbiome: looking back and future perspectives. *Frontiers in Microbiology*, *5*, 148.

Bestel-Corre, G., Dumas-Gaudot, E., & Gianinazzi, S. (2004). Proteomics as a tool to monitor plant-microbe endosymbioses in the rhizosphere. *Mycorrhiza*, *14*(1), 1–10.

Bevivino, A., Dalmastri, C., Tabacchioni, S., & Chiarini, L. (2000). Efficacy of *Burkholderia cepacia* MCI 7 in disease suppression and growth promotion of maize. *Biology and Fertility of Soils*, *31*(3), 225–231.

Bhargava, A., Clabaugh, I., To, J. P., Maxwell, B. B., Chiang, Y. H., Schaller, E. G., & Kieber, J. J. (2013). Identification of cytokinin responsive genes using microarray meta-analysis and RNA-seq in *Arabidopsis thaliana*. *Plant Physiology*, *162*(1), 272–294.

Bharti, N., Pandey, S. S., Barnawal, D., Patel, V. K., & Kalra, A. (2016). Plant growth promoting rhizobacteria *Dietzia natronolimnaea* modulates the expression of stress responsive genes providing protection of wheat from salinity stress. *Scientific Reports*, 6, 34768.

Bordiec, S., Paquis, S., Lacroix, H., Dhondt, S., Ait Barka, E., Kauffmann, S., & Dorey, S. (2011). Comparative analysis of defense responses induced by the endophytic plant growth-promoting rhizobacterium strain PsJN and the non-host bacterium pv. in grapevine cell suspensions. *Journal of Experimental Botany*, 62(2), 595–603.

Brenner, W., & Schmülling, T. (2012). Transcript profiling of cytokinin action in *Arabidopsis* roots and shoots discovers largely similar but also organ-specific responses. *BMC Plant Biology*, 12(1), 112.

Broekaert, W. F., Delaur, S. L., De Bolle, M. F. C., & Cammue, B. P. A. (2006). The role of ethylene in host-pathogen interactions. *Annual Review of Phytopathology*, 44(1), 393–416.

Brusamarello-Santos, L. C., Gilard, F., Brulé, L., Quilleré, I., Gourion, B., Ratet, P., & Hirel, B. (2017). Metabolic profiling of two maize (*Zea mays* L.) inbred lines inoculated with the nitrogen fixing plant-interacting bacteria *Herbaspirillum seropedicae* and *Azospirillum brasilense*. *PLoS ONE*, 12(3), e0174576.

Burd, G. I., Dixon, D. G., & Glick, B. R. (1998). A plant growth-promoting bacterium that decreases nickel toxicity in seedlings. *Applied and Environmental Microbiology*, 64(10), 3663–3668.

Burd, G., Dixon, D., & Glick, B. (2000). Plant growth-promoting bacteria that decrease heavy metal toxicity in plants. *Canadian Journal of Microbiology*, 46(3), 237–245.

Cartieaux, F., Contesto, C., Gallou, A., Desbrosses, G., Kopka, J., Taconnat, L., & Touraine, B. (2008). Simultaneous interaction of *Arabidopsis thaliana* with *Bradyrhizobium* sp strain ORS278 and *Pseudomonas syringae* pv. tomato DC3000 leads to complex transcriptome changes. *Molecular Plant-Microbe Interactions*, 21(2), 244–259.

Cartieaux, F., Thibaud, M. C., Zimmerli, L., Lessard, P., Sarrobert, C., David, P., & Nussaume, L. (2003). Transcriptome analysis of *Arabidopsis* colonized by a plant-growth promoting rhizobacterium reveals a general effect on disease resistance. *Plant Journal*, 36(2), 177–188.

Chabot, R., Antoun, H., Kloepper, J., & Beauchamp, C. (1996). Root colonization of maize and lettuce by bioluminescent *Rhizobium leguminosarum* biovar phaseoli. *Applied and Environmental Microbiology*, 62(8), 2767–2772.

Chanway, C. P., Shishido, M., Nairn, J., Jungwirth, S., Markham, J., Xiao, G., & Holl, F. B. (2000). Endophytic colonization and field responses of hybrid spruce seedlings after inoculation with plant growth-promoting rhizobacteria. *Forest Ecology and Management*, 133(1), 81–88.

Chaparro, J., Badri, D., & Vivanco, J. (2014). Rhizosphere microbiome assemblage is affected by plant development. *The ISME Journal*, 8(4), 790–803.

Cheikh, N., & Jones, R. J. (1994). Disruption of maize kernel growth and development by heat stress (role of cytokinin/abscisic acid balance). *Plant Physiology*, 106(1), 45–51.

Chen, L., Dodd, I. C., Theobald, J. C., Belimov, A. A., & Davies, W. J. (2013). The rhizobacterium *Variovorax paradoxus* 5C–2, containing ACC deaminase, promotes growth and development of *Arabidopsis thaliana* via an ethylene-dependent pathway. *Journal of Experimental Botany*, 64(6), 1565–1573.

Cheng, L., Zhang, N., & Huang, B. (2016). Effects of 1-aminocyclopropane-1-carboxylate-deaminase-producing bacteria on perennial ryegrass growth and physiological responses to salinity stress. *Journal of the American Society for Horticultural Science*, 141(3), 233–241.

Cheng, Z., Park, E., & Glick, B. R. (2007). 1-Aminocyclopropane-1-carboxylate deaminase from *Pseudomonas putida* UW4 facilitates the growth of canola in the presence of salt. *Canadian Journal of Microbiology*, 53(7), 912–918.

Cheng, Z., Woody, O. Z., McConkey, B. J., & Glick, B. R. (2012). Combined effects of the plant growth-promoting bacterium *Pseudomonas putida* UW4 and salinity stress on the *Brassica napus* proteome. *Applied Soil Ecology*, 61, 255–263.

Cho, S. M., Kang, B. R., Han, S. H., Anderson, A. J., Park, J.-Y., Lee, Y.-H., & Kim, Y. C. (2008). 2R,3R-butanediol, a bacterial volatile produced by *Pseudomonas chlororaphis* O6, is involved in induction of systemic tolerance to drought in *Arabidopsis thaliana*. *Molecular Plant-Microbe Interactions*, 21(8), 1067.

Coker, T., Mayak, S., & Thompson, J. E. (1985). Effect of water stress on ethylene production and on membrane microviscosity in carnation flowers. *Scientia Horticulturae*, 27(3), 317–324.

Colebrook, E. H., Thomas, S. G., Phillips, A. L., & Hedden, P. (2014). The role of gibberellin signaling in plant responses to abiotic stress. *Journal of Experimental Biology*, 217(1), 67–75.

Compant, S., Duffy, B., Nowak, J., Clément, C., & Barka, E. A. (2005). Use of plant growth-promoting bacteria for biocontrol of plant diseases: principles, mechanisms of action, and future prospects. *Applied and Environmental Microbiology*, 71(9), 4951–4959.

Conn, K. L., Lazarovits, G., & Nowak, J. (1997). A gnotobiotic bioassay for studying interactions between potatoes and plant growth-promoting rhizobacteria. *Canadian Journal of Microbiology*, 43(9), 801–808.

Contesto, C., Milesi, S., Mantelin, S., Zancarini, A., Desbrosses, G., Varoquaux, F., & Touraine, B. (2010). The auxin-signaling pathway is required for the lateral root response of *Arabidopsis* to the rhizobacterium *Phyllobacterium brassicacearum*. *Planta*, 232(6), 1455–1470.

Crowley, D. E., Römheld, V., & Marschner, H. (1992). Root-microbial effects on plant iron uptake from siderophores and phytosiderophores. *Plant Soil*, 142(1), 1–7.

Dakora, F., & Phillips, D. (2002). Root exudates as mediators of mineral acquisition in low-nutrient environments. *Plant and Soil*, 245(1), 35–47.

Daszkowska-Golec, A. (2016). The role of abscisic acid in drought stress: how ABA helps plants to cope with drought stress. In *Drought Stress Tolerance in Plants, Vol 2* (pp. 123–151). Cham: Springer.

De Cnodder, T., Vissenberg, K., Van Der Straeten, D., & Verbelen, J. P. (2005). Regulation of cell length in the *Arabidopsis thaliana* root by the ethylene precursor 1-aminocyclopropane-1-carboxylic acid: a matter of apoplastic reactions. *New Phytologist*, 168(3), 541–550.

Dharmasiri, N., & Estelle, M. (2004). Auxin signaling and regulated protein degradation. *Trends in Plant Science*, 9(6), 302.

Drory, A., Beja-Tal, S., Borochov, A., Gindin, E., & Mayak, S. (1995). Transient water stress in cut carnation flowers: effects of cycloheximide. *Scientia Horticulturae*, 64(3), 167–175.

Drory, A., Borochov, A., & Mayak, S. (1992). Transient water stress and phospholipid turnover in carnation flowers. *Journal of Plant Physiology*, 140(1), 116–120.

Du, H., Liu, H., & Xiong, L. (2013). Endogenous auxin and jasmonic acid levels are differentially modulated by abiotic stresses in rice. *Frontiers in Plant Science*, 4, 397.

Egamberdieva, D. (2009). Alleviation of salt stress by plant growth regulators and IAA producing bacteria in wheat. *Acta Physiologiae Plantarum*, 31(4), 861–864.

Eze, J. M. O., Mayak, S., Thompson, J. E., & Dumbroff, E. B. (1986). Senescence in cut carnation flowers: temporal and physiological relationships among water status, ethylene, abscisic acid and membrane permeability. *Physiologia Plantarum*, 68(2), 323–328.

Farwell, A. J., Vesely, S., Nero, V., Rodriguez, H., McCormack, K., Shah, S., & Glick, B. R. (2007). Tolerance of transgenic canola plants (*Brassica napus*) amended with plant growth-promoting bacteria to flooding stress at a metal-contaminated field site. *Environmental Pollution*, *147*(3), 540–545.

Feiler, H. S., Desprez, T., Santoni, V., Kronenberger, J., Caboche, M., & Traas, J. (1995). The higher plant *Arabidopsis thaliana* encodes a functional CDC48 homologue which is highly expressed in dividing and expanding cells. *The EMBO Journal*, *14*(22), 5626.

Feng, J., & Barker, A. V. (1993). Polyamine concentration and ethylene evolution in tomato plants under nutritional stress. *HortScience: A Publication of the American Society for Horticultural Science*, *2*, 109–110.

Fernandez, O., Theocharis, A., Bordiec, S., Feil, R., Jacquens, L., Clément, C., & Barka, E. A. (2012). *Burkholderia phytofirmans* PsJN acclimates grapevine to cold by modulating carbohydrate metabolism. *Molecular Plant-Microbe Interactions*, *25*(4), 496.

Ferrari, C., Amaral, F., Bueno, J., Scariot, M., Valentim-Neto, P., & Arisi, A. (2014). Expressed proteins of *Herbaspirillum seropedicae* in maize (DKB240) roots-bacteria interaction revealed using proteomics. *Applied Biochemistry and Biotechnology*, *174*(6), 2267–2277.

Feussner, I., & Polle, A. (2015). What the transcriptome does not tell – proteomics and metabolomics are closer to the plants' pathophenotype. *Current Opinion in Plant Biology*, *26*, 26–31.

Franche, C., Lindström, K., & Elmerich, C. (2009). Nitrogen-fixing bacteria associated with leguminous and non-leguminous plants. *Plant and Soil*, *321*(1), 35–59.

Franklin, K. A. (2008). Shade avoidance. *New Phytologist*, *179*(4), 930–944.

Frossard, E., Condron, L., Oberson, A., Sinaj, S., & Fardeau, J. (2000). Processes governing phosphorus availability in temperate soils. *Journal of Environmental Quality*, *29*(1), 15.

Fulchieri, M., Lucangeli, C., & Bottini, R. (1993). Inoculation with *Azospirillum lipoferum* affects growth and gibberellin status of corn seedling roots. *Plant and Cell Physiology*, *34*(8), 1305–1309.

Geng, Y., Wu, R., Wee, C. W., Xie, F., Wei, X., Chan, P. M. Y., & Dinneny, J. R. (2013). A spatio-temporal understanding of growth regulation during the salt stress response in *Arabidopsis*. *The Plant Cell*, *25*(6), 2132–2154.

Gerke, J. (1993). Phosphate adsorption by humic/Fe-oxide mixtures aged at pH 4 and 7 and by poorly ordered Fe-oxide. *Geoderma*, *59*(1–4), 279–288.

Glick, B. (2015). *Beneficial plant-bacterial interactions* (pp. 1–243). Cham: Springer International.

Glick, B. R. (1999). *Biochemical and genetic mechanisms used by plant growth promoting bacteria*. London: ICP.

Glick, B. R. (2005). Modulation of plant ethylene levels by the bacterial enzyme ACC deaminase. *FEMS Microbiology Letters*, *251*(1), 1–7.

Glick, B. R. (2014). Bacteria with ACC deaminase can promote plant growth and help to feed the world. *Microbiological Research*, *169*(1), 30–39.

Glick, B., Cheng, Z., Czarny, J., & Duan, J. (2007). Promotion of plant growth by ACC deaminase-producing soil bacteria. *European Journal of Plant Pathology*, *119*(3), 329–339.

Gomez-Cadenas, A., Tadeo, F. R., Talon, M., & Primo-Millo, E. (1996). Leaf abscission induced by ethylene in water-stressed intact seedlings of cleopatra mandarin requires previous abscisic acid accumulation in roots. *Plant Physiology*, *112*(1), 401.

Grichko, V. P., & Glick, B. R. (2001). Amelioration of flooding stress by ACC deaminase-containing plant growth-promoting bacteria. *Plant Physiology and Biochemistry*, *39*(1), 11–17.

Gyaneshwar, P., Naresh Kumar, G., Parekh, L., & Poole, P. (2002). Role of soil microorganisms in improving P nutrition of plants. *Plant and Soil*, *245*(1), 83–93.

Han, H., Supanjani, K., & Lee, K. (2006). Effect of co-inoculation with phosphate and potassium solubilizing bacteria on mineral uptake and growth of pepper and cucumber. *Plant, Soil and Environment*, *52*(3), 130–136.

Helmy, Y. H., El-Abd, S. O., Abou Hadid, A., El-Beltagy, M. S., & El-Beltagy, A. (1994). Ethylene production from tomato and cucumber plants under saline conditions. *Egyptian Journal of Horticulture*, *21*(2), 153–160.

Hirsch, A. M., Lum, M. R., & Downie, J. A. (2001). What makes the rhizobia-legume symbiosis so special? *Plant Physiology*, *127*(4), 1484.

Hontzeas, N., Saleh, S. S., & Glick, B. R. (2004). Changes in gene expression in canola roots induced by ACC-deaminase-containing plant-growth-promoting bacteria. *Molecular Plant-Microbe Interactions*, *17*(8), 865.

Huang, X., Zhou, D., Lapsansky, E., Reardon, K., Guo, J., Andales, M., & Manter, D. (2017). *Mitsuaria* sp. and *Burkholderia* sp. from *Arabidopsis* rhizosphere enhance drought tolerance in *Arabidopsis thaliana* and maize (*Zea mays* L.). *Plant and Soil*, *419*(1), 523–539.

Jiang, Y., Wu, Y., Xu, W., Cheng, Y., Chen, J., Xu, L., & Li, H. (2012). IAA-producing bacteria and bacterial-feeding nematodes promote *Arabidopsis thaliana* root growth in natural soil. *European Journal of Soil Biology*, *52*, 20–26.

Jones, D., Dennis, P., Owen, A., & van Hees, P. (2003). Organic acid behavior in soils – misconceptions and knowledge gaps. *Plant and Soil*, *248*(1), 31–41.

Kavamura, V., Santos, S., Silva, J., Parma, M., Avila, L., Visconti, A., & Melo, I. (2013). Screening of Brazilian cacti rhizobacteria for plant growth promotion under drought. *Microbiological Research*, *168*(4), 183–191.

Khan, A. G. (2005). Role of soil microbes in the rhizospheres of plants growing on trace metal contaminated soils in phytoremediation. *Journal of Trace Elements in Medicine and Biology*, *18*, 355–364.

Kim, A., Shahzad, R., Kang, S., Seo, C., Park, Y., Park, H., & Lee, I. (2017). IAA-producing *Klebsiella variicola* AY13 reprograms soybean growth during flooding stress. *Journal of Crop Science and Biotechnology*, *20*(4), 235–242.

Kim, C. K., Lim, H. M., Na, J. K., Choi, J. W., Sohn, S. H., Park, S. C., & Kim, D. Y. (2014). A multistep screening method to identify genes using evolutionary transcriptome of plants. *Evolutionary Bioinformatics*, *10*, 69–78.

Kim, K., Jordan, D., & Krishnan, H. (1998). Expression of genes from *Rahnella aquatilis* that are necessary for mineral phosphate solubilization in *Escherichia coli*. *FEMS Microbiology Letters*, *159*(1), 121–127.

Kim, S., Lowman, S., Hou, G., Nowak, J., Flinn, B., & Mei, C. (2012). Growth promotion and colonization of switchgrass (*Panicum virgatum*) cv. Alamo by bacterial endophyte *Burkholderia phytofirmans* strain PsJN. *Biotechnology for Biofuels*, *5*, 37.

Kirkham, M. B., Gardner, W. R., & Gerloff, G. C. (1974). Internal water status of kinetin-treated, salt-stressed plants. *Plant Physiology*, *53*(2), 241–243.

Kucey, R. (1988). Effect of *Penicillium bilaji* on the solubility and uptake of P and micronutrients from soil by wheat. *Canadian Journal of Soil Science*, *68*(2), 261–270.

Kurepin, L., Park, J., Lazarovits, G., & Bernards, M. (2015a). *Burkholderia phytofirmans*-induced shoot and root growth promotion is associated with endogenous changes in plant growth hormone levels. *Plant Growth Regulation*, *75*(1), 199–207.

Kurepin, L., Park, J., Lazarovits, G., & Hüner, N. (2015b). Involvement of plant stress hormones in *Burkholderia phytofirmans*-induced shoot and root growth promotion. *Plant Growth Regulation*, 77(2), 179–187.

Lakshmanan, V., Castaneda, R., Rudrappa, T., & Bais, H. (2013). Root transcriptome analysis of *Arabidopsis thaliana* exposed to beneficial *Bacillus subtilis* FB17 rhizobacteria revealed genes for bacterial recruitment and plant defense independent of malate efflux. *Planta*, 238(4), 657–668.

Le, J., Vandenbussche, F., Van Der Straeten, D., & Verbelen, J. P. (2001). In the early response of *Arabidopsis* roots to ethylene, cell elongation is up- and down-regulated and uncoupled from differentiation. *Plant Physiology*, 125(2), 519.

Li, J., Ovakim, D. H., Charles, T. C., & Glick, B. R. (2000). An ACC deaminase minus mutant of *Enterobacter cloacae* UW4 no longer promotes root elongation. *Current Microbiology*, 41(2), 101–105.

Liu, W., Li, R., Han, T., Cai, W., Fu, Z., & Lu, Y. (2015). Salt stress reduces root meristem size by nitric oxide-mediated modulation of auxin accumulation and signaling in *Arabidopsis*. *Plant Physiology*, 168(1), 343–356.

Liu, X. M., & Zhang, H. (2015). The effects of bacterial volatile emissions on plant abiotic stress tolerance. *Frontiers in Plant Science*, 6, 774.

Liu, X., Huang, B., & Banowetz, G. (2002). Cytokinin effects on creeping bentgrass responses to heat stress. *Crop Science*, 42(2), 457–465.

Liu, Y., & Zhang, S. (2004). Phosphorylation of 1-aminocyclopropane-1-carboxylic acid synthase by MPK6, a stress-responsive mitogen-activated protein kinase, induces ethylene biosynthesis in *Arabidopsis*. *The Plant Cell*, 16(12), 3386.

Loon, L. (2007). Plant responses to plant growth-promoting rhizobacteria. *European Journal of Plant Pathology*, 119(3), 243–254.

Lopez-Bucio, J., Campos-Cuevas, J., Hernandez-Calderon, E., Velasquez-Becerra, C., Farias-Rodriguez, R., Macias-Rodriguez, L. I., & Valencia-Cantero, E. (2007). *Bacillus megaterium* rhizobacteria promote growth and alter root-system architecture through an auxin- and ethylene-independent signaling mechanism in *Arabidopsis thaliana*. *Molecular Plant-Microbe Interactions*, 20(2), 207–217.

Lugtenberg, B., & Kamilova, F. (2009). Plant-growth-promoting rhizobacteria. *Annual Review of Microbiology*, 63(1), 541–556.

Macklon, A., Grayston, S., Shand, C., Sim, A., Sellars, S., & Ord, B. (1997). Uptake and transport of phosphorus by *Agrostis capillaris* seedlings from rapidly hydrolysed organic sources extracted from 32 P-labelled bacterial cultures. *Plant and Soil*, 190(1), 163–167.

Magalhaes Cruz, L., Maltempi de Souza, E., Weber, O. B., Baldani, J. I., Dobereiner, J., & de Oliveira Pedrosa, F. (2001). 16S Ribosomal DNA characterization of nitrogen-fixing bacteria isolated from banana (*Musa* spp.) and pineapple (*Ananas comosus* (L.) Merril). *Applied and Environmental Microbiology*, 67(5), 2375.

Markakis, M. N., De Cnodder, T., Lewandowski, M., Simon, D., Boron, A., Balcerowicz, D., & Vissenberg, K. (2012). Identification of genes involved in the ACC-mediated control of root cell elongation in *Arabidopsis thaliana*. *BMC Plant Biology*, 12(1), 208.

Matamoros, M. A., Dalton, D. A., Ramos, J., & Clemente, M. R. (2003). Biochemistry and molecular biology of antioxidants in the rhizobia-legume symbiosis1,2. *Plant Physiology*, 133(2), 499–509.

Mayak, S., Tirosh, T., & Glick, B. R. (2004a). Plant growth-promoting bacteria confer resistance in tomato plants to salt stress. *Plant Physiology and Biochemistry*, 42(6), 565–572.

Mayak, S., Tirosh, T., & Glick, B. R. (2004b). Plant growth-promoting bacteria that confer resistance to water stress in tomatoes and peppers. *Plant Science*, 166(2), 525–530.

McDermott, T. R. (2000). Phosphorus assimilation and regulation in *Rhizobia*. In *Prokaryotic nitrogen fixation: a model system for the analysis of a biological process* (pp. 529–548). Horizon Scientific Press.

Mitsukawa, N., Okumura, S., Shirano, Y., Sato, S., Kato, T., Harashima, S., & Shibata, D. (1997). Overexpression of an *Arabidopsis thaliana* high-affinity phosphate transporter gene in tobacco cultured cells enhances cell growth under phosphate-limited conditions. *Proceedings of the National Academy of Sciences of the United States of America*, 94(13), 7098–7102.

Morgan, P. W., & Drew, M. C. (1997). Ethylene and plant responses to stress. *Physiologia Plantarum*, 100(3), 620–630.

Nadeem, S. M., Zahir, Z. A., Naveed, M., & Arshad, M. (2007). Preliminary investigations on inducing salt tolerance in maize through inoculation with rhizobacteria containing ACC deaminase activity. *Canadian Journal of Microbiology*, 53(10), 1141–1149.

Nadeem, S. M., Zahir, Z. A., Naveed, M., Asghar, H. N., & Arshad, M. (2010). Rhizobacteria capable of producing ACC-deaminase may mitigate salt stress in wheat. *Soil Science Society of America Journal*, 74(2), 533–542.

Naqvi, S. S. M., Ansari, R., & Khanzada, A. N. (1982). Response of salt-stressed wheat seedlings to kinetin. *Plant Science Letters*, 26(2–3), 279–283.

Naseem, H., & Bano, A. (2014). Role of plant growth-promoting rhizobacteria and their exopolysaccharides in drought tolerance of maize. *Journal of Plant Interactions, Plant-Microorganism Interactions*, 9(1), 689–701.

Naseem, H., Ahsan, M., Shahid, M., & Khan, N. (2018). Exopolysaccharides producing rhizobacteria and their role in plant growth and drought tolerance. *Journal of Basic Microbiology*, 58(12), 1009–1022.

Nautiyal, C., Srivastava, S., Chauhan, P., Seem, K., Mishra, A., & Sopory, S. (2013). Plant growth-promoting bacteria *Bacillus amyloliquefaciens* NBRISN13 modulates gene expression profile of leaf and rhizosphere community in rice during salt stress. *Plant Physiology and Biochemistry*, 66, 1–9.

Naveed, M., Mitter, B., Yousaf, S., Pastar, M., Afzal, M., & Sessitsch, A. (2014). The endophyte *Enterobacter sp.* FD17: a maize growth enhancer selected based on rigorous testing of plant beneficial traits and colonization characteristics. *Biology and Fertility of Soils*, 50(2), 249–262.

Nishiyama, R., Watanabe, Y., Fujita, Y., Le, D. T., Kojima, M., Werner, T., & Sakakibara, H. (2011). Analysis of cytokinin mutants and regulation of cytokinin metabolic genes reveals important regulatory roles of cytokinins in drought, salt and abscisic acid responses, and abscisic acid biosynthesis. *The Plant Cell*, 23(6), 2169–2183.

Niu, X., Song, L., Xiao, Y., & Ge, W. (2017). Drought-tolerant plant growth-promoting rhizobacteria associated with foxtail millet in a semi-arid agroecosystem and their potential in alleviating drought stress. *Frontiers in Microbiology*, 8, 2580.

Nogueira, E., Vinagre, F., Hana Paula, M., Vargas, C., Padua, V., da Silva, F., & Adriana Silva, H. (2001). Expression of sugarcane genes induced by inoculation with *Gluconacetobacter diazotrophicus* and *Herbaspirillum rubrisubalbicans*. *Genetics and Molecular Biology*, 24, 199–206.

Nowak, J., & Shulaev, V. (2003). Priming for transplant stress resistance in in vitro propagation. *In Vitro Cellular & Developmental Biology – Plant*, 39(2), 107–124.

Nowak, J., Sharma, V. K., & A'Hearn, E. (2004). Endophyte enhancement of transplant performance in tomato, cucumber and sweet pepper. *Acta Horticulturae*, *631*, 253–263.

Ortíz-Castro, R., Valencia-Cantero, E., & López-Bucio, J. (2008). Plant growth promotion by *Bacillus megaterium* involves cytokinin signaling. *Plant Signaling & Behavior*, *3*(4), 263–265.

Perin, L., Martínez-Aguilar, L., Paredes-Valdez, G., Baldani, J. I., Estrada-de Los Santos, P., Reis, V. M., & Caballero-Mellado, J. (2006). *Burkholderia silvatlantica* sp. nov., a diazotrophic bacterium associated with sugar cane and maize. *International Journal of Systematic and Evolutionary Microbiology*, *56*(Pt 8), 1931.

Pierik, R., Tholen, D., Poorter, H., Visser, E. J. W., & Voesenek, L. A. C. J. (2006). The Janus face of ethylene: growth inhibition and stimulation. *Trends in Plant Science*, *11*(4), 176–183.

Pierik, R., Verkerke, W., Voesenek, R. A. C. J., Blom, K. W. P. M., & Visser, E. J. W. (1999). Thick root syndrome in cucumber (*Cucumis sativus* L.): a description of the phenomenon and an investigation of the role of ethylene. *Annals of Botany*, *84*(6), 755–762.

Pinedo, I., Ledger, T., Greve, M., & Poupin, M. (2015). *Burkholderia phytofirmans* PsJN induces long-term metabolic and transcriptional changes involved in *Arabidopsis thaliana* salt tolerance. *Frontiers in Plant Science*, *6*, 466.

Planchamp, C., Glauser, G., & Mauch-Mani, B. (2015). Root inoculation with *Pseudomonas putida* KT2440 induces transcriptional and metabolic changes and systemic resistance in maize plants. *Frontiers in Plant Science*, *5*, 719.

Pospíšilová, J., & Baťková, P. (2004). Effects of pre-treatments with abscisic acid and/or benzyladenine on gas exchange of French bean, sugar beet, and maize leaves during water stress and after rehydration. *Biologia Plantarum*, *48*(3), 395–399.

Poupin, M. J., Greve, M., Carmona, V., & Pinedo, I. (2016). A complex molecular interplay of auxin and ethylene signaling pathways is involved in *Arabidopsis* growth promotion by *Burkholderia phytofirmans* PsJN. *Frontiers in Plant Science*: *Plant Microbe Interactions*, *7*, 492.

Poupin, M., Timmermann, T., Vega, A., Zuniga, A., & Gonzalez, B. (2013). Effects of the plant growth-promoting bacterium *Burkholderia phytofirmans* PsJN throughout the life cycle of *Arabidopsis thaliana*. *PLoS ONE*, *8*(7), e69435.

Prayitno, J., Rolfe, B. G., & Mathesius, U. (2006). The ethylene-insensitive sickle mutant of *Medicago truncatula* shows altered auxin transport regulation during nodulation. *Plant Physiology*, *142*(1), 168–180.

Radzki, W., Gutierrez Mañero, F., Algar, E., Lucas García, J., García-Villaraco, A., & Ramos Solano, B. (2013). Bacterial siderophores efficiently provide iron to iron-starved tomato plants in hydroponics culture. *Antonie van Leeuwenhoek*, *104*(3), 321–330.

Richardson, A., & Simpson, R. (2011). Soil microorganisms mediating phosphorus availability update on microbial phosphorus. *Plant Physiology*, *156*(3), 989–996.

Rocha, F., Papini-Terzi, F., Nishiyama, M., Vencio, R., Vicentini, R., Duarte, R., & Souza, G. (2007). Signal transduction-related responses to phytohormones and environmental challenges in sugarcane. *BMC Genomics*, *8*, 71.

Rodriguez, H., Vessely, S., Shah, S., & Glick, B. (2008). Effect of a nickel-tolerant ACC deaminase-producing *Pseudomonas* strain on growth of non-transformed and transgenic canola plants. *Current Microbiology*, *57*(2), 170–174.

Saeid, A., Prochownik, E., & Dobrowolska-Iwanek, J. (2018). Phosphorus solubilization by *Bacillus* species. *Molecules*, *23*(11), 2897.

Sandhya, V., Skz, A., GroverM., Reddy, G., & Venkateswarlu, B. (2009). Alleviation of drought stress effects in sunflower seedlings by the exopolysaccharides producing *Pseudomonas putida* strain GAP-P45. *Biology and Fertility of Soils*, *46*(1), 17–26.

Saravanakumar, D., & Samiyappan, R. (2007). ACC deaminase from *Pseudomonas fluorescens* mediated saline resistance in groundnut (*Arachis hypogea*) plants. *Journal of Applied Microbiology*, *102*(5), 1283–1292.

Schnider-Keel, U., Lejbolle, K., Baehler, E., Haas, D., & Keel, C. (2001). The sigma factor AlgU controls exopolysaccharide production and tolerance towards desiccation and osmotic stress in the biocontrol agent *Pseudomonas fluorescens* CHA0. *Applied and Environmental Microbiology*, *67*(12), 5683–5693.

Schwachtje, J., Karojet, S., Kunz, S., Brouwer, S., & Van Dongen, J. T. (2012). Plant-growth promoting effect of newly isolated rhizobacteria varies between two *Arabidopsis* ecotypes. *Plant Signaling & Behavior*, *7*(6), 623–627.

Schwachtje, J., Karojet, S., Thormählen, I., Bernholz, C., Kunz, S., Brouwer, S., & van Dongen, J. T. (2011). A naturally associated rhizobacterium of *Arabidopsis thaliana* induces a starvation-like transcriptional response while promoting growth (*Arabidopsis* growth promotion by rhizobacteria). *PLoS ONE*, *6*(12), e29382.

Sergeeva, E., Shah, S., & Glick, B. R. (2006). Growth of transgenic canola (*Brassica napus* cv. Westar) expressing a bacterial 1-aminocyclopropane-1-carboxylate (ACC) deaminase gene on high concentrations of salt. *World Journal of Microbiology and Biotechnology*, *22*(3), 277–282.

Sessitsch, A., Coenye, T., Sturz, A. V., Vandamme, P., Barka, E. A., Salles, J. F., & Nowak, J. (2005). *Burkholderia phytofirmans* sp. nov., a novel plant-associated bacterium with plant-beneficial properties. *International Journal of Systematic and Evolutionary Microbiology*, *55*(3), 1187–1192.

Sgroy, V., Cassán, F., Masciarelli, O., Del Papa, M. F., Lagares, A., & Luna, V. (2009). Isolation and characterization of endophytic plant growth-promoting (PGPB) or stress homeostasis-regulating (PSHB) bacteria associated to the halophyte *Prosopis strombulifera*. *Applied Microbiology and Biotechnology*, *85*(2), 371–381.

Shahzad, R., Waqas, M., Khan, A., Asaf, S., Khan, M., Kang, S., & Lee, I. (2016). Seed-borne endophytic *Bacillus amyloliquefaciens* RWL-1 produces gibberellins and regulates endogenous phytohormones of *Oryza sativa*. *Plant Physiology and Biochemistry*, *106*(C), 236–243.

Shakir, M. A., Bano, A., & Arshad, M. (2012). Rhizosphere bacteria containing ACC-deaminase conferred drought tolerance in wheat grown under semi-arid climate. *Soil and Environment*, *31*(1), 108–112.

Shi, H., Chen, L., Ye, T., Liu, X., Ding, K., & Chan, Z. (2014). Modulation of auxin content in *Arabidopsis* confers improved drought stress resistance. *Plant Physiology and Biochemistry*, *82*, 209–217.

Siddikee, M., Glick, B., Chauhan, P., Yim, W., & Sa, T. (2011). Enhancement of growth and salt tolerance of red pepper seedlings (*Capsicum annuum* L.) by regulating stress ethylene synthesis with halotolerant bacteria containing 1-aminocyclopropane-1-carboxylic acid deaminase activity. *Plant Physiology and Biochemistry*, *49*(4), 427–434.

Singh, R., Mishra, R., Jaiswal, H., Kumar, V., Pandey, S., Rao, S., & Annapurna, K. (2006). Isolation and identification of natural endophytic rhizobia from rice (*Oryza sativa* L.) through rDNA PCR-RFLP and sequence analysis. *Current Microbiology*, *52*(5), 345–349.

Spollen, W., Lenoble, M. E., & Sharp, R. (1997). Regulation of root ethylene production by accumulation of endogenous ABA at low water potentials. *Plant Physiology, 114*(3), 446.

Staal, M., De Cnodder, T., Simon, D., Vandenbussche, F., Van Der Straeten, D., Verbelen, J.-P., & Vissenberg, K. (2011). Apoplastic alkalinization is instrumental for the inhibition of cell elongation in the *Arabidopsis* root by the ethylene precursor 1-aminocyclopropane-1-carboxylic acid. *Plant Physiology, 155*(4), 2049–2055.

Stearns, J. C., & Glick, B. R. (2003). Transgenic plants with altered ethylene biosynthesis or perception. *Biotechnology Advances, 21*(3), 193–210.

Stearns, J., Woody, O., McConkey, B., & Glick, B. (2012). Effects of bacterial ACC deaminase on *Brassica napus* gene expression. *Molecular Plant-Microbe Interactions, 25*(5), 668–676.

Straub, D., Yang, H., Liu, Y., Tsap, T., & Ludewig, U. (2013). Root ethylene signaling is involved in *Miscanthus sinensis* growth promotion by the bacterial endophyte *Herbaspirillum frisingense* GSF30(T). *Journal of Experimental Botany, 64*(14), 4603.

Sukweenadhi, J., Balusamy, S. R., Kim, Y. J., Lee, C. H., Kim, Y. J., Koh, S. C., & Yang, D. C. (2018). A growth-promoting bacteria, *Paenibacillus yonginensis* DCY84T enhanced salt stress tolerance by activating defense-related systems in *Panax ginseng*. *Frontiers in Plant Science*, 9, 813.

Sun, Y., Cheng, Z., & Glick, B. R. (2009). The presence of a 1-aminocyclopropane-1-carboxylate (ACC) deaminase deletion mutation alters the physiology of the endophytic plant growth-promoting bacterium *Burkholderia phytofirmans* PsJN. *FEMS Microbiology Letters, 296*(1), 131–136.

Sun, Z., Liu, K., Zhang, J., Zhang, Y., Xu, K., Yu, D., & Li, C. (2017). IAA producing *Bacillus altitudinis* alleviates iron stress in *Triticum aestivum* L. seedlings by both bioleaching of iron and up-regulation of genes encoding ferritins. *Plant and Soil, 419*(1), 1–11.

Suzuki, N., Bassil, E., Hamilton, J. S., Inupakutika, M. A., Zandalinas, S. I., Tripathy, D., & Nakano, R. (2016). ABA is required for plant acclimation to a combination of salt and heat stress. *PLoS One, 11*(1), e0147625.

Tarafdar, J., & Gharu, A. (2006). Mobilization of organic and poorly soluble phosphates by *Chaetomium globosum*. *Applied Soil Ecology, 32*(3), 273–283.

Thiebaut, F., Rojas, C. A., Grativol, C., Motta, M. R., Vieira, T., Regulski, M., & Ferreira, P. C. G. (2014). Genome-wide identification of microRNA and siRNA responsive to endophytic beneficial diazotrophic bacteria in maize. *BMC Genomics, 15*, 766.

Thompson, J. E., Shinitzky, M., & Halevy, A. H. (1982). Acceleration of membrane senescence in cut carnation flowers by treatment with ethylene. *Plant Physiology, 69*(4), 859–863.

Upadhyay, S., Singh, J., & Singh, D. (2011). Exopolysaccharide-producing plant growth-promoting rhizobacteria under salinity condition. *Pedosphere, 21*(2), 214–222.

van de Mortel, J. E., de Vos, R. C., Dekkers, E., Pineda, A., Guillod, L., Bouwmeester, K., & Raaijmakers, J. M. (2012). Metabolic and transcriptomic changes induced in *Arabidopsis* by the rhizobacterium *Pseudomonas fluorescens* SS101. *Plant Physiology, 160*(4), 2173–2188.

Vardharajula, S., & Ali, S. K. Z. (2014). Exopolysaccharide production by drought tolerant *Bacillus* spp. and effect on soil aggregation under drought stress. *The Journal of Microbiology, Biotechnology and Food Sciences, 4*(1), 51–57.

Veerasamy, M., He, Y., & Huang, B. (2007). Leaf senescence and protein metabolism in creeping bentgrass exposed to heat stress and treated with cytokinins. *Journal of the American Society for Horticultural Science, 132*(4), 467–472.

Verhagen, B. W., Glazebrook, J., Zhu, T., Chang, H. S., Van Loon, L. C., & Pieterse, C. M. (2004). The transcriptome of rhizobacteria-induced systemic resistance in *Arabidopsis*. *Molecular Plant-Microbe Interactions, 17*(8), 895–908.

Visser, E. J. W., Nabben, R. H. M., Blom, C. W. P. M., & Voesenek, L. A. C. J. (1997). Elongation by primary lateral roots and adventitious roots during conditions of hypoxia and high ethylene concentrations. *Plant, Cell & Environment, 20*(5), 647–653.

Walker, V., Bertrand, C., Bellvert, F., Moënne-Loccoz, Y., Bally, R., & Comte, G. (2011). Host plant secondary metabolite profiling shows a complex, strain-dependent response of maize to plant growth-promoting rhizobacteria of the genus *Azospirillum*. *The New Phytologist, 189*(2), 494.

Walker, V., Couillerot, O., Von Felten, A., Bellvert, F., Jansa, J., Maurhofer, M., & Comte, G. (2012). Variation of secondary metabolite levels in maize seedling roots induced by inoculation with *Azospirillum*, *Pseudomonas* and *Glomus* consortium under field conditions. *Plant and Soil, 356*, 151–163.

Wang, W., Chen, L. N., Wu, H., Zang, H., Gao, S., Yang, Y., & Gao, X. (2013). Comparative proteomic analysis of rice seedlings in response to inoculation with *Bacillus cereus*. *Letters in Applied Microbiology, 56*(3), 208–215.

Wang, Y., Ohara, Y., Nakayashiki, H., Tosa, Y., & Mayama, S. (2005). Microarray analysis of the gene expression profile induced by the endophytic plant growth-promoting rhizobacteria, *Pseudomonas fluorescens* FPT9601–T5 in *Arabidopsis*. *Molecular Plant-Microbe Interactions, 18*(5), 385–396.

Xu, C., & Zou, Q. (1993). Effect of drought on lipoxygenase activity, ethylene and ethane production in leaves of soybean plants. *Acta Botanica Sinica* (suppl), 31–37.

Xu, J., Li, X. L., & Luo, L. (2012). Effects of engineered *Sinorhizobium meliloti* on cytokinin synthesis and tolerance of alfalfa to extreme drought stress. *Applied and Environmental Microbiology, 78*(22), 8056–8061.

Xu, Y., & Huang, B. (2007). Heat-induced leaf senescence and hormonal changes for thermal bentgrass and turf-type bentgrass species differing in heat tolerance. *Journal of the American Society for Horticultural Science, 132*(2), 185–192.

Xu, Y., & Huang, B. (2009). Effects of foliar-applied ethylene inhibitor and synthetic cytokinin on creeping bentgrass to enhance heat tolerance. *Crop Science, 49*(5), 1876–1884.

Yang, J., Kloepper, J. W., & Ryu, C.-M. (2009). Rhizosphere bacteria help plants tolerate abiotic stress. *Trends in Plant Science, 14*(1), 1–4.

Zahir, Z. A., Munir, A., Asghar, H. N., Shaharoona, B., & Arshad, M. (2008). Effectiveness of rhizobacteria containing ACC deaminase for growth promotion of peas (*Pisum sativum*) under drought conditions. *Journal of Microbiology and Biotechnology, 18*(5), 958–963.

Zamioudis, C., Mastranesti, P., Dhonukshe, P., Blilou, I., & Pieterse, C. M. J. (2013). Unraveling root developmental programs initiated by beneficial *Pseudomonas* spp. bacteria. *Plant Physiology, 162*(1), 304.

Zhang, H., Xie, X., Kim, M. S., Kornyeyev, D. A., Holaday, S., & Paré, P. W. (2008). Soil bacteria augment *Arabidopsis* photosynthesis by decreasing glucose sensing and abscisic acid levels in planta. *Plant Journal, 56*(2), 264–273.

Zwack, P. J., & Rashotte, A. M. (2015). Interactions between cytokinin signaling and abiotic stress responses. *Journal of Experimental Botany, 66*(16), 4863–4871.

3 Physiological, Biochemical and Molecular Mechanisms Regulating Post-Drought Stress Recovery in Grass Species

Cathryn Chapman and Bingru Huang

CONTENTS

3.1 INTRODUCTION

Drought stress is an important environmental factor that can negatively impact the growth and development of many plant species. In recent decades, there has been an increase in the frequency and duration of drought events, which is of great agricultural concern worldwide (Karl et al., 2008, 2009; Mishra and Singh, 2010; IPCC, 2014). The demand for water continues to grow, and there is a constant struggle for optimum resources. Drought stress causes various damages, including decreases in leaf water relations, membrane stability and metabolic activities (Hsiao, 1973; Levitt, 1980; Nilsen and Orcutt, 1996). Because an adequate supply of water is essential to the normal functioning of plants, its limited availability due to lack of precipitation can have detrimental ecological and economic effects.

One of the major plant species adversely affected by drought stress is the grass family, Poaceae. The Poaceae family, which is one of the largest flowering plant families, is comprised of an estimated 11,000–12,000 different species (Kellogg, 2015; Soreng et al., 2017). Included are various annual cereal species, such as maize (*Zea* spp.), rice (*Oryza* spp.), wheat (*Triticum* spp.) and sorghum (*Sorghum* spp.) (USDA, 2018) which are cultivated as food staples. Various perennial grass species are widely cultivated or grown in natural areas, such as those used as forage, turfgrass and biofuels (USDA, 2018). Grass species in arid and semi-arid regions can be adversely affected by a moderate or severe drought stress level brought on by lack of water availability, which makes it critical for understanding the mechanisms that these plants have developed in order to survive during drought events. While there is a great interest within this particular area, there should be an equal interest in learning more about the mechanisms behind regrowth and re-establishment after the drought stress

is relieved, or rather when the level of precipitation resumes to that of a more sustainable level, after which any water use restrictions can be removed to allow for normal irrigation once again.

Strategies for plant drought resistance is an area of research that has been widely studied across a variety of grass species, including wheat (Morgan, 1983; Ritchie et al., 1990; Gunasekera and Berkowitz, 1992; Rampino et al., 2006; Han et al., 2015; Abid et al., 2018), switchgrass (*Panicum virgatum*) (Sun et al., 2012; Meyer et al., 2014; Liu et al., 2015) and Kentucky bluegrass (*Poa pratensis*) (Perdomo et al., 1996; Wang et al., 2003; Hu et al., 2010; Xu et al., 2011a, 2011b; Xu et al., 2013), to name a few. Indeed, it is important for grass species to be able to survive and tolerate long-term drought, but it is also critical that they can recover from drought stress through regrowth and re-establishment once the stress is removed (Levitt, 1980; Chaves and Oliveira, 2004; Lopes et al., 2011). In other words, grass plants need to be able to maintain normal functions under varying levels of water deficits, and then rapidly recover after the stress is relieved. The ability of grass species to withstand drought and recover is the key to ensuring a more sustainable grass crop for future generations.

This chapter will discuss physiological, biochemical and molecular factors associated with the recuperative ability of grass species from drought stress by addressing the questions on how drought-resistance traits may affect post-drought recovery, and identifying the specific mechanisms involved in post-drought stress recovery that enable plants to regenerate or re-establish and restore normal functions?

The following sections will discuss the details of important drought-resistance mechanisms and their relation to post-drought recovery mechanisms on a physiological, biochemical and molecular level.

3.2 CARBOHYDRATE ACCUMULATION AND REMOBILIZATION

A major area of post-drought recovery regrowth potential involves the use of carbohydrate reserves that are stored within tissues of certain plant organs during drought, which provides a recovering plant with the necessary resources for rapid regrowth. Under prolonged drought conditions where photosynthesis is limited, carbohydrates can accumulate due to a decrease in respiration and overall decrease in carbohydrate demand caused by growth inhibition (Levitt, 1980; Chaves, 1991; Nilsen and Orcutt, 1996; Thomas and James, 1999; Chaves and Oliveira, 2004; Fry and Huang, 2004; Huang et al., 2014). Managed turfgrasses that are used on home lawns or sports fields are a classic example of a type of grass that can store carbohydrate reserves within meristematic tissues during drought stress and then utilize them for rapid regrowth of tissues upon re-watering. In order to gain a better understanding of this type of post-drought stress recovery mechanism, however, it is relevant to understand the growth habits of turfgrass species. Unlike other plant species, whose growing points, or meristems, are at the tips of the plant, turfgrass species have meristematic tissue located near the soil level at the base of the plant, within an organ called the crown (Beard, 1973; Turgeon, 2012; Emmons and Rossi, 2016). This feature of turfgrass, as well as other perennial grasses, is what allows them to be cut down or mowed so long as the meristematic region remains intact and uninjured (Turgeon, 2012; Emmons and Rossi, 2016). Cells located in the meristematic region can divide and enlarge, thus providing overall plant growth.

In addition to the primary lateral shoots, or tillers, that grow from crowns, some turfgrass species generate new growth from stolons or rhizomes, which can form shoots and roots from their nodes to produce new above-ground daughter plants using axillary buds (Beard, 1973; Turgeon, 2012). The spreading growth habit of stolons or rhizomes can also strengthen the ability to recuperate from drought stress, allowing the turfgrass to spread laterally to fill in bare spots that may have been damaged during drought.

Since growth can be inhibited while a plant is experiencing drought, it would seem that such damages may subsequently inhibit the growth of new or established daughter plants once re-watering occurs. However, perennial grasses, such as turfgrass, have developed other mechanisms to rapidly recover using their important growing points that contain meristematic tissues. In this case, maintaining the viability of the meristematic tissues within stolons, rhizomes and crowns during drought stress is very important for continued plant survival (Chai et al., 2010; Sarath et al., 2014). The post-drought recovery mechanism that perennial grass species utilize is that they store carbohydrates during stress in various plant structures and are then able to remobilize the carbohydrates and send them to axillary buds in order to make new roots and shoots when the stress is relieved (Beard, 1973; Huang and Fu, 2000; Huang and Gao, 2000; DaCosta and Huang, 2006a, 2006b; Chai et al., 2010; Yang et al., 2013). Thus, research has focused on finding ways to ensure these organs are protected

so that they can allow the turfgrass to recuperate from the effects of drought. Additionally, by protecting the axillary buds of stolons and rhizomes during periods of drought, they can contribute to active growth and development of new daughter plants once water becomes available again.

It is worth noting that there are differences in drought recuperative ability of various turfgrass species. For instance, compared to bunch-type grasses that lack stolons or rhizomes, such as perennial ryegrass (*Lolium perenne*), stoloniferous and rhizomatous turfgrass species, such as creeping bentgrass (*Agrostis stolonifera*) and Kentucky bluegrass, respectively, may have a greater potential to survive during longer periods of drought and also have the ability to make new daughter plants (with roots and shoots) more rapidly after the stress is relieved, and they are re-watered (Beard, 1973).

In a previous study, which measured total nonstructural carbohydrate (TNC) content in the rhizomes of Kentucky bluegrass (KBG) and in the crowns of perennial ryegrass (PRG), it was found that the recuperative potential of PRG might be less than that of KBG, as exhibited by the TNC content decrease in the crowns of PRG during drought, when compared to the non-stressed control (Chai et al., 2010). According to this report (Chai et al., 2010), storage of carbohydrates in the rhizomes of KBG increased in comparison to the non-stressed control during drought stress, and then decreased upon re-watering, most likely because they were remobilizing to form new tissues, which was shown by an increase in the production of both new roots and new tillers by the end of the recovery period in comparison to PRG. It seems that due to the reduced carbohydrate storage in the crowns of PRG during drought stress, these plants, in comparison to KBG, had fewer reserves to remobilize upon re-watering and, therefore, could not form new roots or shoots as quickly or efficiently.

This trend of accumulating sugars during drought stress, including water-soluble carbohydrates, and subsequent improved growth potential upon re-watering has also been observed in various other grass species, including drought-tolerant populations of *Dactylis glomerata* (Volaire, 1994, 1995), rice (*Oryza sativa*) (Mostajeran and Rahimi-Eichi, 2009), barley (*Hordeum vulgare*) (Sicher et al., 2012) and wheat (*Triticum aestivum*) (Gupta et al., 2011). Studies have attributed the increase in water-soluble carbohydrates, such as sucrose, during drought stress, to the decreases in acid invertase which controls sucrose breakdown, and increased activities of two sucrose synthesizing enzymes, either sucrose synthase or sucrose-phosphate synthase (Kaur et al., 2007; Fu et al., 2010; Yang et al., 2013). Although certain water-soluble carbohydrates are most often associated with drought tolerance, fructans are the major contributor to regrowth of tissues during re-watering after drought (Volaire, 1994, 1995; Yang et al., 2013).

Of course, the concept of regrowth from stored reserves can also be applied to other grass species in addition to the aforementioned C_3 grass species. For example, many C_4 perennial grasses, such as switchgrass and miscanthus (*Miscanthus* x *giganteus*), must also utilize their storage organs to overcome adverse environmental conditions, including drought, by maintaining viable meristems (Sarath et al., 2014). This recovery mechanism, although not unique to drought stress, is made possible by the induction of a commonly known drought escape mechanism called dormancy, which prevents growth of the meristems during the adverse conditions, and ultimately stores reserves and resumes growth using those reserves once the stress is relieved (Sarath et al., 2014).

3.3 OSMOLYTE ACCUMULATION AND PROTECTIVE BENEFITS

Plants also accumulate other compounds in addition to the accumulation of some forms of carbohydrates during drought stress which may contribute to the post-drought recovery of plants. Under drought stress, plants accumulate various solutes in order to maintain cell turgor pressure and volume known as osmotic adjustment, which protect plant tissues from dehydration damages (Turner and Jones, 1980; Hale and Orcutt, 1987; Nilsen and Orcutt, 1996; Hopkins and Hüner, 2004). Common osmoregulants include inorganic ions, such as potassium, calcium, or sodium, organic compounds, such as sugars (sucrose, glucose and fructose), organic acids (malic or citric acid), sugar alcohols (mannitol or sorbitol) and amino acids (proline or glycine betaine). These compounds may accumulate within the vacuole and the cytoplasm, lowering the osmotic potential of plant tissues. When a plant accumulates a higher concentration of solutes, the more negative the osmotic potential becomes, which essentially results in higher osmotic adjustment ability and greater water retention potential when plants are exposed to drought stress.

Osmotic adjustment does not only maintain cell turgor, but also has the functions of protecting membranes, proteins, cellular organelles and enzymes from dehydration damages (Hoekstra et al., 2001; Oliver et al., 2010). Osmotic adjustment capacity can play a major role in a plant's ability for recovery once re-watering occurs. For instance, during shorter periods of drought, osmotic adjustment allows stomata to remain open through its maintenance of turgor pressure, allowing carbon gain and cell growth to occur despite the low water potential brought on by mild drought stress (Nilsen and Orcutt, 1996). Through their ability to tolerate drought stressed conditions and maintain carbon balance via turgor pressure, plants can have overall greater recovery potential upon re-watering by their ability to continue necessary metabolic functions despite the lack of water availability.

Osmotic adjustment has been shown to help improve post-drought recovery in various plant species, and many studies have correlated greater osmotic adjustment ability with greater post-drought recovery potential, as shown in studies with analysis of metabolite content in plants during drought and re-watering. Many metabolites, including glycine, isoleucine, asparagine, threonine, spermine, proline, adenosine, fructose and valine, found in leaves and roots of maize (*Zea mays*) that were responsive to drought stress were also responsive to re-watering (Sun et al., 2013). Additionally, the accumulation of amino acids, which are nitrogen enriched

compounds, has shown to be a key factor in promoting recovery upon re-watering. In studies on maize where many amino acids accumulated during drought stress, it was found that upon re-watering there was a decrease in isoleucine, leucine, valine and proline levels, suggesting that accumulated nitrogen from these amino acids might be used as a substrate for protein replacement during recovery (Sun et al., 2016). The overall accumulation and subsequent decrease in the levels of various osmolytes during drought stress and recovery has also had positive outcomes for improvements in synthesizing new biomass upon re-watering (Warren et al., 2012), and can be a major component in plant recovery through its assistance in resuming metabolic functions, such as photosynthesis and stomatal conductance (Souza et al., 2004; Foster et al., 2015). It is thought that drought-injured photochemical systems involve important repair processes to resume photosynthesis upon re-watering (Miyashita et al., 2005) and various abiotic stress-tolerance-related metabolites, such as the polyamine spermine, may be utilized for this purpose. For example, spermine can elicit repair processes that help prevent further damage to membranes or photosynthetic functions (Yamaguchi et al., 2007; Farooq et al., 2009; Zhou and Yu, 2010), and increases in photosynthesis are often seen during re-watering (Sun et al., 2013).

The benefits of osmotic adjustment and the accumulation of solutes for post-drought recovery have generally been positive in numerous studies; however, some have had conflicting results stating that this process occurs at the expense of yield. In the case of agricultural crops, where a large and profitable yield is the main goal, osmotic adjustment in this light could then be seen as a negative mechanism. There has been some controversy (Munns, 1988; Serraj and Sinclair, 2002) as to whether or not osmolyte accumulation benefits the recovery of growth or gas exchange directly, as seen by a poor correspondence between osmotic adjustment and yield at the end of the post-drought recovery period. This may lead some to think that this type of drought tolerance mechanism may only be useful in terms of post-drought recovery for turfgrass or other perennial forage grass species where a high yield is not necessarily as desirable. However, it is evident that there are many protective benefits that osmolytes can have, which may lead to rapid resumption of metabolic functions upon re-watering, such as increased photosynthetic activity (Chaves et al., 2009) or signaling of important compounds, which would most likely lead to overall improvements in crop biomass yield and should be considered a positive outcome for overall plant survival.

3.4 OXIDATIVE STRESS AND ANTIOXIDANT PROTECTION

One of the unfortunate effects of drought stress is oxidative stress, or the increased production of harmful reactive oxygen species (ROS) within plants, which include radicals such as superoxide or hydroxyl, and non-radicals such as hydrogen peroxide and singlet oxygen. Under mild stress conditions, ROS are formed, but can quickly be quenched by antioxidants, including either enzymatic antioxidants, such as superoxide dismutase (SOD), peroxidase (POD), catalase (CAT), or ascorbate peroxidase (APX), or non-enzymatic antioxidants, such as ascorbate, glucose, or tocopherols (Smirnoff, 1993; Chaves and Oliveira, 2004; Ahmad et al., 2008; Koyro et al., 2012). However, during prolonged and severe drought stress, the production of ROS is typically greater as a result of the overall limitation in photosynthesis, which causes decreased intercellular CO_2 and subsequent limited carbon reduction by the Calvin Cycle (Hsu and Kao, 2003; Chaves and Oliveira, 2004; Ahmad et al., 2008). More specifically, the decreased photosynthetic activities result in an increased transfer of electrons to oxygen (Smirnoff, 1993; Navari-Izzo and Rascio, 1999). Then, in a chain reaction, more ROS will be formed in multiple sites in addition to the chloroplast, including mitochondria or peroxisomes. Injury to these organelles can include the following: damages to lipids (through oxidation of unsaturated fatty acids, in a process called lipid peroxidation, as well as membrane leakage of solutes); proteins (through loss of function or modifications); or DNA (through mutations and other lethal genetic effects) (Møller, 2001; Foyer and Noctor, 2003; Apel and Hirt, 2004; Foyer and Noctor, 2005; Ahmad et al., 2008). The enzymatic and non-enzymatic antioxidants can work either alone or in combination to scavenge and detoxify ROS (Smirnoff, 1993; Apel and Hirt, 2004; Foyer and Noctor, 2005; Ahmad et al., 2008; Koyro et al., 2012).

Some amino acids also play positive roles in antioxidant functions when accumulated during drought stress (Smirnoff and Cumbes, 1989; Ashraf and Foolad, 2007; Molinari et al., 2007). Because of these benefits, certain amino acids can be applied exogenously to plants to aid in drought stress tolerance and, ultimately, in post-drought recovery. One example is proline, which is a well-known amino acid that accumulates in a variety of plant species during drought. Increases in endogenous proline content have been associated with increases in leaf hydration levels, better growth and yield and antioxidant defense (Szabados and Savouré, 2010; Anjum et al., 2012; An et al., 2013; Bandurska et al., 2017; Zegaoui et al., 2017). Proline levels accumulate during drought stress and subsequently decrease upon recovery (Krasensky and Jonak, 2012), a trend that is often seen with osmoregulants when the stress is relieved. For example, in cotton (*Gossypium hirsutum*) during drought stress, the accumulation of drought-induced proline has also been associated with increases in Δ^1-Pyrroline-5-carboxylate synthetase (P5CS) and Δ^1-Pyrroline-5-carboxylate reductase (P5CR), two major enzymes in the biosynthetic pathway for proline synthesis (Parida et al., 2008). Other studies have also seen increases in P5CS and proline synthesis in cowpea (*Vigna unguiculata*) (Zegaoui et al., 2017) and grass species, such as barley (Bandurska et al., 2017). It has also been noted that there are decreases in proline dehydrogenase (PDH), a proline degrading enzyme, during drought stress (Parida et al., 2008; Zegaoui et al., 2017). Upon recovery, the levels of P5CS and PDH decreased and increased, respectively, thus shedding more light on the regulation of these enzymes for controlling proline levels during drought stress and post-drought recovery.

GABA (γ – Aminobutyric acid) is another amino acid that has been found to accumulate in plant species under drought stress, which has also been associated with better growth and yield, as well as increases in chlorophyll content and decreases in oxidative damage (Shelp et al., 1999; De Diego et al., 2013; Bown and Shelp, 2016; Yong et al., 2017). GABA is a non-protein amino acid that plays a role in carbon/nitrogen balance, and can also regulate proline metabolism and assist proline in ROS scavenging (Bouché and Fromm, 2004; Signorelli et al., 2015). When exogenously applied, both proline and GABA have been known to have positive effects for alleviating damages due to drought in wheat (Farooq et al., 2017), creeping bentgrass (Li et al., 2017) and perennial ryegrass (Krishnan et al., 2013), and increases in endogenous concentrations of each, respectively, are often seen. Studies have found that in addition to the improvements in plant–water relations and lesser oxidative damage that the exogenous application of GABA can provide, it was also accompanied by up-regulation of genes encoding antioxidant enzymes, such as CAT, SOD and APX, as well as up-regulation of stress-protective genes, such as the following: genes that are involved in stress defense response signaling (mitogen-activated protein kinase [*MAPK1*]); ROS neutralization (Metallothionein [*MT1*]); and defense proteins (heat shock protein [*HSP70*]) (Li et al., 2018). Although the complete mechanisms behind proline- and GABA-assisted drought tolerance and post-drought recovery are not fully understood, exogenous application of these particular osmolytes deserves further attention in order to learn more about their potential to prevent irreversible damages to membranes or proteins that would have otherwise been caused by ROS from oxidative stress.

3.5 HORMONE METABOLISM AND PLANT GROWTH REGULATION

Drought stress can have a major impact on hormonal metabolism, and changes in hormone levels are often observed. For instance, during the initial effects of drought, when the plant begins to wilt, there can be decreases in growth promoters, such as cytokinins (CK), gibberellins (GA) and auxin, and increases in growth inhibitors, such as abscisic acid (ABA) and ethylene (Nilsen and Orcutt, 1996; Gupta, 2005). The effects of drought can be altered through exogenous application of hormones or plant growth regulators (PGRs), which can have benefits for promoting both drought stress tolerance and recovery upon re-watering. Therefore, the following section will discuss some key players in the hormone regulation of drought tolerance and post-drought recovery.

ABA is one of the most commonly known drought stress-induced hormones. It has been widely studied that under stressful environmental conditions, such as during initial drought responses, ABA production will increase and will reduce stomatal aperture in order to limit transpirational water loss so that the plant tissue can regain turgor (Hiron and Wright, 1973; Hsiao, 1973; Aspinall, 1980; Hale and Orcutt, 1987; Gupta, 2005; Tuteja, 2007). ABA production during drought stress is a vital drought avoidance mechanism, as expressed by its contribution to maintaining a favorable water status via either controlling water uptake through an extensive root system, or via reduced water loss from leaves (Levitt, 1980; Nilsen and Orcutt, 1996). In fact, studies on Kentucky bluegrass have shown that although ABA levels are much higher in drought-susceptible cultivars in comparison to drought-tolerant cultivars, it is thought that the drought tolerance is actually related to stomatal sensitivity of accumulated endogenous ABA content during mild drought stress, which was exhibited by less severe declines in stomatal conductance, photosynthetic rate and leaf water potential for the more drought-tolerant cultivars (Wang and Huang, 2003). Ultimately, this allows drought-tolerant plants to conserve more water and survive longer under drought stressed conditions. Additionally, due to the stomatal sensitivity, it can mean that even a low accumulation of ABA in drought-tolerant cultivars may lead to better maintenance of photosynthesis under short term drought and allow for more dry matter accumulation that can help survive longer drought periods and promote more rapid recovery (Wang and Huang, 2003).

Although the level of drought-induced ABA and its specific role may differ by species, the severity of the drought and/or plant developmental stage, it is generally understood that the increased ABA is related to chemical signaling, which can occur between the roots and the shoots in response to drought stress. This has been studied by others in depth (Levitt, 1980; Reid and Wample, 1985; Nilsen and Orcutt, 1996; Davies et al., 2002; Gupta, 2005). However, the question remains as to how regulation of ABA can increase the recuperative ability of plants after drought stress. In order to answer this question, it is important to think of how this drought avoidance mechanism can be utilized as more of a drought tolerance mechanism, which would promote survival and post-drought recovery. One way of doing so is to use ABA as a PGR to initiate earlier stomatal closure. This can be especially helpful for drought-susceptible cultivars of plants, which may have stomata that are not as sensitive to ABA levels compared to drought-tolerant cultivars. On an agricultural scale, this can be helpful for drought-susceptible grass crop species to promote earlier stomatal closure and, thus, more tolerance which can aid in post-drought recovery. In fact, when ABA is applied exogenously, both drought-tolerant and drought-susceptible cultivars respond more quickly to the application, closing stomata more rapidly and ultimately enhancing plant–water relations by inducing osmotic adjustment and causing less cell membrane damage, which has been seen in Kentucky bluegrass (Wang et al., 2003) and wheat (Kirkham, 1983). Additionally, low concentrations of exogenously applied ABA have been shown to induce antioxidative defense responses in maize seedlings (Jiang and Zhang, 2001) and Bermuda grass (*Cynodon dactylon*) (Lu et al., 2009).

During prolonged drought stress the stomata may remain closed, further limiting necessary metabolic functions. Since plants may have different rates at which stomata re-open upon re-watering, one way to promote the rapid resumption of photosynthesis and subsequent plant recovery could be with the use of another PGR, such as CK. For instance, post-drought recovery foliar treatment of synthetic cytokinin, 6-benzyl-aminopurine (6-BA), was able to recover Kentucky bluegrass plants from drought stress by enhancing stomatal re-opening and initiating photosynthesis (Hu et al., 2012a, 2012b). It has also been reported that the exogenous application of 6-BA promoted tiller growth and regeneration after re-watering, and induced endogenous accumulation of CK, which caused a decrease in the content of ABA (Hu et al., 2012a). In addition to the rapid post-drought recovery benefits that accompany the endogenous ratios or balance of CK to ABA upon re-watering (DaCosta and Huang, 2007), it is important to note that exogenous application of CK could have many benefits at the whole plant level during adverse environmental conditions, as observed in wheat (Monakhova and Chernyad'Ev, 2004; Kumari et al., 2018) and maize (Blackman and Davies, 1983; Wang et al., 2016). Overall, the rapid re-opening of stomata due to CK is critical for the resumption of gas exchange and photosynthetic activities that facilitate post-drought recovery, which makes this mechanism very applicable to the agricultural system for overall improved yield upon re-watering.

In addition to PGRs being utilized for the resumption of photosynthetic activities to promote growth upon re-watering, their use for promoting rapid regrowth in general is also significant. For instance, endogenous GA is known for promoting cell elongation and vertical shoot growth (Hopkins and Hüner, 2004; Gupta and Chakrabarty, 2013; Huang et al., 2014; Small and Degenhardt, 2018). In addition to the studies already discussed that have shown improved drought tolerance and post-drought recovery growth potential as a result of exogenous applications of CK, it has been found that exogenous applications of GA in combination with CK at the vegetative stage can also promote increases in plant biomass and grain yield in maize due to better maintenance of photosynthesis (Akter et al., 2014). This can be a very promising aspect for more sustainable post-drought recovery growth potential.

Application of PGRs is beneficial for modifying physiological responses to drought stress, which can eventually lead to adaptations that are sustainable for drought tolerance and subsequent recovery upon re-watering. Further, the growth effects of using PGR application can be enhanced through the use of a nitrogen source application upon re-watering. Application of nitrogen is a post-stress recovery practice that is often used by turfgrass managers to facilitate more rapid growth and re-greening of turfgrass (Beard, 1973). Nitrogen application in combination with CK has also shown to improve drought tolerance through enhanced antioxidant metabolism in creeping bentgrass (Chang et al., 2016). Therefore, perhaps using GA and CK for cell elongation and cell division for lateral root and shoot growth, in combination with a growth-promoting nitrogen source, could be the key for further enhancing drought tolerance and post-drought recovery. Such knowledge could be of great significance for developing effective post-drought management programs, taking advantage of the fact that these growth-promoting hormones and nutrients can enhance rapid regrowth or re-establishment of grass crop species during re-watering periods in areas with limited irrigation or rainfall.

3.6 CONCLUSIONS AND FUTURE RESEARCH PERSPECTIVES

Drought stress is a major factor limiting the growth and productivity of both annual grass crop species and perennial grasses, while rapid recovery from drought damages is critically important for the drought survival of perennial species. Rapid regeneration of new shoots and roots enable re-establishment of new plant stands and resumes growth and functions once water becomes available due to rainfall or irrigation. Drought-resistance traits and post-drought recovery mechanisms discussed in this chapter should provide a better understanding of how grass plants can survive, overcome and resume productivity in an ever-changing, unpredictable environment. Enhancing survival during drought stress through drought-resistance mechanisms is essential not only for maintaining functional metabolic processes, but also to ensure the plant will be able to recover upon re-watering. It is worth noting, however, that although there have been many studies done on the effects of drought and water deficit on different grass species and their coping strategies, the mechanisms behind their recovery from drought stress still deserves further investigation.

REFERENCES

Abid, M., Ali, S., Qi, L.K., Zahoor, R., Tian, Z., Jiang, D., Snider, J.L. and Dai, T. 2018. Physiological and biochemical changes during drought and recovery periods at tillering and jointing stages in wheat (*Triticum aestivum* L.). *Scientific Reports*, 8(1), 4615.

Ahmad, P., Sarwat, M. and Sharma, S. 2008. Reactive oxygen species, antioxidants and signaling in plants. *Journal of Plant Biology*, 51(3), 167–173.

Akter, N., Islam, M.R., Karim, M.A. and Hossain, T. 2014. Alleviation of drought stress in maize by exogenous application of gibberellic acid and cytokinin. *Journal of Crop Science and Biotechnology*, 17(1), 41–48.

An, Y., Zhang, M., Liu, G., Han, R. and Liang, Z. 2013. Proline accumulation in leaves of *Periploca sepium* via both biosynthesis up-regulation and transport during recovery from severe drought. *PloS one*, 8(7), 1–10.

Anjum, S.A., Farooq, M., Xie, X.Y., Liu, X.J. and Ijaz, M.F. 2012. Antioxidant defense system and proline accumulation enables hot pepper to perform better under drought. *Scientia Horticulturae*, 140, 66–73.

Apel, K. and Hirt, H. 2004. Reactive oxygen species: metabolism, oxidative stress, and signal transduction. *Annual Review of Plant Biology*, 55, 373–399.

Ashraf, M.F.M.R. and Foolad, M. 2007. Roles of glycine betaine and proline in improving plant abiotic stress resistance. *Environmental and Experimental Botany*, 59(2), 206–216.

Aspinall, D. 1980. Role of abscisic acid and other hormones in adaptation to water stress. In: *Adaptation of plants to water and high temperature stress*, ed. N.C. Turner and P.J. Kramer, pp. 155–172, New York: John Wiley & Sons, Inc.

Bandurska, H., Niedziela, J., Pietrowska-Borek, M., Nuc, K., Chadzinikolau, T., and Radzikowska, D. 2017. Regulation of proline biosynthesis and resistance to drought stress in two

barley (*Hordeum vulgare* L.) genotypes of different origin. *Plant Physiology and Biochemistry*, 118, 427–437.

Beard, J.B. 1973. *Turfgrass: science and culture*. Prentice Hall, Englewood Cliffs, NJ.

Blackman, P.G. and Davies, W.J. 1983. The effects of cytokinins and ABA on stomatal behaviour of maize and *Commelina*. *Journal of Experimental Botany*, 34(12), 1619–1626.

Bouché, N. and Fromm, H. 2004. GABA in plants: just a metabolite? *Trends in Plant Science*, 9(3), 110–115.

Bown, A.W. and Shelp, B.J. 2016. Plant GABA: not just a metabolite. *Trends in Plant Science*, 21(10), 811–813.

Chai, Q., Jin, F., Merewitz, E., and Huang, B. 2010. Growth and physiological traits associated with drought survival and post-drought recovery in perennial turfgrass species. *Journal of the American Society for Horticultural Science*, 135, 125–133.

Chang, Z., Liu, Y., Dong, H., Teng, K., Han, L. and Zhang, X. 2016. Effects of cytokinin and nitrogen on drought tolerance of creeping bentgrass. *PloS one*, 11(4), 1–19.

Chaves, M.M. 1991. Effects of water deficits on carbon assimilation. *Journal of Experimental Botany*, 42(1), 1–16.

Chaves, M.M., Flexas, J. and Pinheiro, C. 2009. Photosynthesis under drought and salt stress: regulation mechanisms from whole plant to cell. *Annals of Botany 103*, 551–560.

Chaves, M.M. and Oliveira, M.M. 2004. Mechanisms underlying plant resilience to water deficits: prospects for water-saving agriculture. *Journal of Experimental Botany*, 55(407), 2365–2384.

DaCosta, M. and Huang, B. 2006a. Changes in carbon partitioning and accumulation patterns during drought and recovery for colonial bentgrass, creeping bentgrass, and velvet bentgrass. *Journal of the American Society for Horticultural Science*, 131, 484–490.

DaCosta, M. and Huang, B. 2006b. Osmotic adjustment associated with variation in bentgrass tolerance to drought stress. *Journal of the American Society for Horticultural Science*, 131, 338–344.

DaCosta, M. and Huang, B. 2007. Drought survival and recuperative ability of bentgrass species associated with changes in abscisic acid and cytokinin production. *Journal of the American Society for Horticultural Science*, 132(1), 60–66.

Davies, W.J., Wilkinson, S. and Loveys, B. 2002. Stomatal control by chemical signalling and the exploitation of this mechanism to increase water use efficiency in agriculture. *New Phytologist*, 153(3), 449–460.

De Diego, N., Sampedro, M.C., Barrio, R.J., Saiz-Fernández, I., Moncaleán, P. and Lacuesta, M. 2013. Solute accumulation and elastic modulus changes in six radiata pine breeds exposed to drought. *Tree Physiology*, 33(1), 69–80.

Emmons, R. and Rossi, F. 2016. *Turfgrass science and management*. Stamford, CT: Cengage Learning.

Farooq, M., Nawaz, A., Chaudhry, M.A.M., Indrasti, R. and Rehman, A. 2017. Improving resistance against terminal drought in bread wheat by exogenous application of proline and gamma-aminobutyric acid. *Journal of Agronomy and Crop Science*, 203(6), 464–472.

Farooq, M., Wahid, A. and Lee, D.J. 2009. Exogenously applied polyamines increase drought tolerance of rice by improving leaf water status, photosynthesis and membrane properties. *Acta Physiologiae Plantarum*, 31(5), 937–945.

Foster, K., Lambers, H., Real, D., Ramankutty, P., Cawthray, G.R. and Ryan, M.H. 2015. Drought resistance and recovery in mature *Bituminaria bituminosa* var. *albomarginata*. *Annals of Applied Biology*, 166(1), 154–169.

Foyer, C.H. and Noctor, G. 2003. Redox sensing and signalling associated with reactive oxygen in chloroplasts, peroxisomes and mitochondria. *Physiologia Plantarum*, 119(3), 355–364.

Foyer, C.H. and Noctor, G. 2005. Redox homeostasis and antioxidant signaling: a metabolic interface between stress perception and physiological responses. *The Plant Cell*, 17(7), 1866–1875.

Fry, J. and Huang, B. 2004. *Applied turfgrass science and physiology*. Hoboken, NJ: John Wiley & Sons, Inc.

Fu, J., Huang, B. and Fry, J. 2010. Osmotic potential, sucrose level, and activity of sucrose metabolic enzymes in tall fescue in response to deficit irrigation. *Journal of the American Society for Horticultural Science*, 135(6), 506–510.

Gunasekera, D. and Berkowitz, G.A. 1992. Evaluation of contrasting cellular-level acclimation responses to leaf water deficits in three wheat genotypes. *Plant Science*, 86(1), 1–12.

Gupta, U.S. 2005. *Physiology of stressed crops: hormone relations*. Enfield: Science Publishers, Inc.

Gupta, R. and Chakrabarty, S.K. 2013. Gibberellic acid in plant: still a mystery unresolved. *Plant Signaling & Behavior*, 8(9), 1–5.

Gupta, A.K., Kaur, K. and Kaur, N. 2011. Stem reserve mobilization and sink activity in wheat under drought conditions. *American Journal of Plant Sciences*, 2(1), 70–77.

Hale, M.G. and Orcutt, D.M. 1987. *The physiology of plants under stress*. New York: John Wiley & Sons, Inc.

Han, H., Tian, Z., Fan, Y., Cui, Y., Cai, J., Jiang, D., Cao, W. and Dai, T. 2015. Water-deficit treatment followed by re-watering stimulates seminal root growth associated with hormone balance and photosynthesis in wheat (*Triticum aestivum* L.) seedlings. *Plant Growth Regulation*, 77(2), 201–210.

Hiron, R.W.P. and Wright, S.T.C. 1973. The role of endogenous abscisic acid in the response of plants to stress. *Journal of Experimental Botany*, 24(4), 769–780.

Hoekstra, F.A., Golovina, E.A. and Buitink, J. 2001. Mechanisms of plant desiccation tolerance. *Trends in Plant Science*, 6(9), 431–438.

Hopkins, W.G. and Hüner, N. 2004. *Introduction to plant physiology*, 3rd edition. Hoboken, NJ: John Wiley & Sons, Inc.

Hsiao, T.C. 1973. Plant responses to water stress. *Annual Review of Plant Physiology*, 24(1), 519–570.

Hsu, S.Y. and Kao, C.H. 2003. The protective effect of free radical scavengers and metal chelators on polyethylene glycol-treated rice leaves. *Biologia Plantarum*, 46(4), 617–619.

Hu, L., Wang, Z. and Huang, B. 2010. Diffusion limitations and metabolic factors associated with inhibition and recovery of photosynthesis from drought stress in a C3 perennial grass species. *Physiologia Plantarum*, 139(1), 93–106.

Hu, L., Wang, Z. and Huang, B. 2012a. Growth and physiological recovery of Kentucky bluegrass from drought stress as affected by a synthetic cytokinin 6-benzylaminopurine. *Crop Science*, 52(5), 2332–2340.

Hu, L., Wang, Z. and Huang, B. 2012b. Effects of cytokinin and potassium on stomatal and photosynthetic recovery of Kentucky bluegrass from drought stress. *Crop Science*, 53(1), 221–231.

Huang, B., DaCosta, M. and Jiang, Y. 2014. Research advances in mechanisms of turfgrass tolerance to abiotic stresses: from physiology to molecular biology. *Critical Reviews in Plant Sciences*, 33(2-3), 141–189.

Huang, B. and Fu, J. 2000. Photosynthesis, respiration, and carbon allocation of two cool-season perennial grasses in response to surface soil drying. *Plant and Soil*, 227(1-2), 17–26.

Huang, B. and Gao, H. 2000. Root physiological characteristics associated with drought resistance in tall fescue cultivars. *Crop Science*, 40(1), 196–203.

IPCC, 2014. *Climate Change 2014: Synthesis report. Contribution of working groups I, II and III to the fifth assessment report of the intergovernmental panel on climate Change*, eds. Core Writing Team, R.K. Pachauri and L.A. Meyer. IPCC, Geneva, Switzerland.

Jiang, M. and Zhang, J. 2001. Effect of abscisic acid on active oxygen species, antioxidative defence system and oxidative damage in leaves of maize seedlings. *Plant and Cell Physiology*, 42(11), 1265–1273.

Karl, T.R., Meehl, G.A., Miller, C.D., Hassol, S.J., Waple, A.M. and Murray, W.L. eds. 2008. *Weather and climate extremes in a changing climate*. US Climate Change Science Program.

Karl, T.R., Melillo, J.M., Peterson, T.C. and Hassol, S.J. eds. 2009. *Global climate change impacts in the United States*. Cambridge University Press.

Kaur, K., Gupta, A.K. and Kaur, N. 2007. Effect of water deficit on carbohydrate status and enzymes of carbohydrate metabolism in seedlings of wheat cultivars. *Indian Journal of Biochemistry and Biophysics*, 44, 223–230.

Kellogg, E.A. 2015, Flowering plants. Monocots: Poaceae. In: *The families and genera of vascular plants*, ed. K. Kubitski, pp. 1–416. Cham: Springer International.

Kirkham, M.B. 1983. Effect of ABA on the water relations of winter-wheat cultivars varying in drought resistance. *Physiologia Plantarum*, 59(1), 153–157.

Koyro, H.W., Ahmad, P. and Geissler, N. 2012. Abiotic stress responses in plants: an overview. In: *Environmental adaptations and stress tolerance of plants in the era of climate change*, pp. 1–28, New York: Springer.

Krasensky, J. and Jonak, C. 2012. Drought, salt, and temperature stress-induced metabolic rearrangements and regulatory networks. *Journal of Experimental Botany*, 63(4), 1593–1608.

Krishnan, S., Laskowski, K., Shukla, V. and Merewitz, E.B. 2013. Mitigation of drought stress damage by exogenous application of a non-protein amino acid γ–aminobutyric acid on perennial ryegrass. *Journal of the American Society for Horticultural Science*, 138(5), 358–366.

Kumari, S., Kumar, S. and Prakash, P. 2018. Exogenous application of cytokinin (6-BAP) ameliorates the adverse effect of combined drought and high temperature stress in wheat seedling. *Journal of Pharmacognosy and Phytochemistry*, 7(1), 1176–1180.

Levitt, J. 1980. *Responses of plants to environmental stresses: Water, radiation, salt, and other stresses*, Vol. 1, 2nd edition. New York: Academic Press.

Li, Z., Peng, Y. and Huang, B. 2018. Alteration of transcripts of stress-protective genes and transcriptional factors by γ-Aminobutyric Acid (GABA) associated with improved heat and drought tolerance in creeping bentgrass (*Agrostis stolonifera*). *International Journal of Molecular Sciences*, 19(6), 1–17.

Li, Z., Yu, J., Peng, Y. and Huang, B. 2017. Metabolic pathways regulated by abscisic acid, salicylic acid and γ-aminobutyric acid in association with improved drought tolerance in creeping bentgrass (*Agrostis stolonifera*). *Physiologia Plantarum*, 159(1), 42–58.

Liu, Y., Zhang, X., Tran, H., Shan, L., Kim, J., Childs, K., Ervin, E.H., Frazier, T. and Zhao, B. 2015. Assessment of drought tolerance of 49 switchgrass (*Panicum virgatum*) genotypes using physiological and morphological parameters. *Biotechnology for Biofuels*, 8(1), 152.

Lopes, M.S., Araus, J.L., Van Heerden, P.D. and Foyer, C.H. 2011. Enhancing drought tolerance in C4 crops. *Journal of Experimental Botany*, 62(9), 3135–3153.

Lu, S., Su, W., Li, H. and Guo, Z. 2009. Abscisic acid improves drought tolerance of triploid bermudagrass and involves H_2O_2-and NO-induced antioxidant enzyme activities. *Plant Physiology and Biochemistry*, 47(2), 132–138.

Meyer, E., Aspinwall, M.J., Lowry, D.B., Palacio-Mejía, J.D., Logan, T.L., Fay, P.A. and Juenger, T.E. 2014. Integrating transcriptional, metabolomic, and physiological responses to drought stress and recovery in switchgrass (*Panicum virgatum* L.). *BMC Genomics*, *15*(1), 527.

Mishra, A.K. and Singh, V.P. 2010. A review of drought concepts. *Journal of Hydrology*, *391*(1-2), 202–216.

Miyashita, K., Tanakamaru, S., Maitani, T. and Kimura, K. 2005. Recovery responses of photosynthesis, transpiration, and stomatal conductance in kidney bean following drought stress. *Environmental and Experimental Botany*, *53*(2), 205–214.

Molinari, H.B.C., Marur, C.J., Daros, E., De Campos, M.K.F., De Carvalho, J.F.R.P., Filho, J.C.B., Pereira, L.F.P. and Vieira, L.G.E. 2007. Evaluation of the stress-inducible production of proline in transgenic sugarcane (*Saccharum* spp.): osmotic adjustment, chlorophyll fluorescence and oxidative stress. *Physiologia Plantarum*, *130*(2), 218–229.

Møller, I.M. 2001. Plant mitochondria and oxidative stress: electron transport, NADPH turnover, and metabolism of reactive oxygen species. *Annual Review of Plant Biology*, *52*(1), 561–591.

Monakhova, O.F. and Chernyad'Ev, I.I. 2004. Effects of cytokinin preparations on the stability of the photosynthetic apparatus of two wheat cultivars experiencing water deficiency. *Applied Biochemistry and Microbiology*, *40*(6), 573–580.

Morgan, J.M. 1983. Osmoregulation as a selection criterion for drought tolerance in wheat. *Australian Journal of Agricultural Research*, *34*(6), 607–614.

Mostajeran, A. and Rahimi-Eichi, V. 2009. Effects of drought stress on growth and yield of rice (*Oryza sativa* L.) cultivars and accumulation of proline and soluble sugars in sheath and blades of their different ages leaves. *Agriculture and Environmental Science*, *5*(2), 264–272.

Munns, R. 1988. Why measure osmotic adjustment? *Functional Plant Biology*, *15*(6), 717–726.

Navari-Izzo, F. and Rascio, N. 1999. Plant response to water-deficit conditions. In: *Handbook of plant and crop stress*, 2nd edition, ed. M. Pessarakli, pp. 231–270, New York: Marcel Dekker, Inc.

Nilsen, E.T. and Orcutt, D.M. 1996. *Physiology of plants under stress: abiotic factors*, New York: John Wiley & Sons, Inc.

Oliver, M.J., Cushman, J.C. and Koster, K.L. 2010. Dehydration tolerance in plants. In: *Plant stress tolerance*, ed. R. Sunkar, pp. 3–24, New York: Humana Press.

Parida, A.K., Dagaonkar, V.S., Phalak, M.S. and Aurangabadkar, L.P. 2008. Differential responses of the enzymes involved in proline biosynthesis and degradation in drought tolerant and sensitive cotton genotypes during drought stress and recovery. *Acta Physiologiae Plantarum*, *30*(5), 619–627.

Perdomo, P., Murphy, J.A. and Berkowitz, G.A. 1996. Physiological changes associated with performance of Kentucky bluegrass cultivars during summer stress. *HortScience*, *31*(7), 1182–1186.

Rampino, P., Pataleo, S., Gerardi, C., Mita, G. and Perrotta, C. 2006. Drought stress response in wheat: physiological and molecular analysis of resistant and sensitive genotypes. *Plant, Cell & Environment*, *29*(12), 2143–2152.

Reid, D.M. and Wample, R.L. 1985. Water relations and plant hormones. In: *Hormonal regulation of development III: role of environmental factors*, pp. 513–578, Berlin: Springer-Verlag.

Ritchie, S.W., Nguyen, H.T. and Holaday, A.S. 1990. Leaf water content and gas-exchange parameters of two wheat genotypes differing in drought resistance. *Crop Science*, *30*(1), 105–111.

Sarath, G., Baird, L.M. and Mitchell, R.B. 2014. Senescence, dormancy and tillering in perennial C4 grasses. *Plant Science*, *217*, 140–151.

Serraj, R. and Sinclair, T.R. 2002. Osmolyte accumulation: can it really help increase crop yield under drought conditions? *Plant, Cell & Environment*, *25*(2), 333–341.

Shelp, B.J., Bown, A.W. and McLean, M.D, 1999. Metabolism and functions of gamma-aminobutyric acid. *Trends in Plant Science*, *4*(11), 446–452.

Sicher, R.C., Timlin, D. and Bailey, B. 2012. Responses of growth and primary metabolism of water-stressed barley roots to rehydration. *Journal of Plant Physiology*, *169*(7), 686–695.

Signorelli, S., Dans, P.D., Coitiño, E.L., Borsani, O. and Monza, J. 2015. Connecting proline and γ-aminobutyric acid in stressed plants through non-enzymatic reactions. *PLoS One*, *10*(3), 1–14.

Small, C.C. and Degenhardt, D. 2018. Plant growth regulators for enhancing revegetation success in reclamation: a review. *Ecological Engineering*, *118*, 43–51.

Smirnoff, N. 1993. The role of active oxygen in the response of plants to water deficit and desiccation. *New Phytologist*, *125*(1), 27–58.

Smirnoff, N. and Cumbes, Q.J. 1989. Hydroxyl radical scavenging activity of compatible solutes. *Phytochemistry*, *28*(4), 1057–1060.

Soreng, R.J., Peterson, P.M., Romaschenko, K., Davidse, G., Teisher, J.K., Clark, L.G., Barberá, P., Gillespie, L.J. and Zuloaga, F.O. 2017. A worldwide phylogenetic classification of the Poaceae (Gramineae) II: an update and a comparison of two 2015 classifications. *Journal of Systematics and Evolution*, *55*(4), 259–290.

Souza, R.P., Machado, E.C., Silva, J.A.B., Lagôa, A.M.M.A. and Silveira, J.A.G. 2004. Photosynthetic gas exchange, chlorophyll fluorescence and some associated metabolic changes in cowpea (*Vigna unguiculata*) during water stress and recovery. *Environmental and Experimental Botany*, *51*(1), 45–56.

Sun, C.B., Fan, X.W., Hu, H.Y., Liang, Y., Huang, Z.B., Pan, J.L., Wang, L. and Li, Y.Z. 2013. Pivotal metabolic pathways related to water deficit tolerance and growth recovery of whole maize plant. *Plant Omics*, *6*(6), 377–387.

Sun, C., Gao, X., Chen, X., Fu, J. and Zhang, Y. 2016. Metabolic and growth responses of maize to successive drought and rewatering cycles. *Agricultural Water Management*, *172*, 62–73.

Sun, G., Stewart Jr, C.N., Xiao, P. and Zhang, B. 2012. MicroRNA expression analysis in the cellulosic biofuel crop switchgrass (*Panicum virgatum*) under abiotic stress. *PLoS One*, *7*(3), 1–7.

Szabados, L. and Savouré, A. 2010. Proline: a multifunctional amino acid. *Trends in Plant Science*, *15*(2), 89–97.

Thomas, H. and James, A.R. 1999. Partitioning of sugars in *Lolium perenne* (perennial ryegrass) during drought and on rewatering. *The New Phytologist*, *142*(2), 295–305.

Turgeon, A.J. 2012. *Turfgrass management*, 9th edition. Boston, MA: Prentice-Hall Inc.

Turner, N.C. and Jones, M.M. 1980. Turgor maintenance by osmotic adjustment: a review and evaluation. In: *Adaptation of plants to water and high temperature stress*, ed. N.C. Turner and P.J. Kramer, pp. 87–103. New York: John Wiley & Sons, Inc.

Tuteja, N., 2007. Abscisic acid and abiotic stress signaling. *Plant Signaling & Behavior*, *2*(3), 135–138.

USDA. 2018. Classification for kingdom plantae down to family *Poaceae*. *Plants Database*.

Volaire, F. 1994. Effects of summer drought and spring defoliation on carbohydrate reserves, persistence and recovery of two populations of cocksfoot (*Dactylis glomerata*) in a Mediterranean environment. *The Journal of Agricultural Science*, *122*(2), 207–215.

Volaire, F. 1995. Growth, carbohydrate reserves and drought survival strategies of contrasting *Dactylis glomerata* populations in a Mediterranean environment. *Journal of Applied Ecology*, 56–66.

Wang, Z. and Huang, B. 2003. Genotypic variation in abscisic acid accumulation, water relations, and gas exchange for Kentucky bluegrass exposed to drought stress. *Journal of the American Society for Horticultural Science*, *128*(3), 349–355.

Wang, Z., Huang, B. and Xu, Q. 2003. Effects of abscisic acid on drought responses of Kentucky bluegrass. *Journal of the American Society for Horticultural Science*, *128*(1), 36–41.

Wang, X.L., Wang, J.J., Sun, R.H., Hou, X.G., Zhao, W., Shi, J., Zhang, Y.F., Qi, L., Li, X.L., Dong, P.H. and Zhang, L.X. 2016. Correlation of the corn compensatory growth mechanism after post-drought rewatering with cytokinin induced by root nitrate absorption. *Agricultural Water Management*, *166*, 77–85.

Warren, C.R., Aranda, I. and Cano, F.J. 2012. Metabolomics demonstrates divergent responses of two *Eucalyptus* species to water stress. *Metabolomics*, *8*(2), 186–200.

Xu, L., Han, L. and Huang, B. 2011a. Antioxidant enzyme activities and gene expression patterns in leaves of Kentucky bluegrass in response to drought and post-drought recovery. *Journal of the American Society for Horticultural Science*, *136*(4), 247–255.

Xu, L., Han, L. and Huang, B., 2011b. Membrane fatty acid composition and saturation levels associated with leaf dehydration tolerance and post-drought rehydration in Kentucky bluegrass. *Crop Science*, *51*(1), 273–281.

Xu, L., Yu, J., Han, L. and Huang, B. 2013. Photosynthetic enzyme activities and gene expression associated with drought tolerance and post-drought recovery in Kentucky bluegrass. *Environmental and Experimental Botany*, *89*, 28–35.

Yamaguchi, K., Takahashi, Y., Berberich, T., Imai, A., Takahashi, T., Michael, A.J. and Kusano, T. 2007. A protective role for the polyamine spermine against drought stress in *Arabidopsis*. *Biochemical and Biophysical Research Communications*, *352*(2), 486–490.

Yang, Z., Xu, L., Yu, J., DaCosta, M. and Huang, B. 2013. Changes in carbohydrate metabolism in two Kentucky bluegrass cultivars during drought stress and recovery. *Journal of the American Society for Horticultural Science*, *138*(1), 24–30.

Yong, B., Xie, H., Li, Z., Li, Y.P., Zhang, Y., Nie, G., Zhang, X.Q., Ma, X., Huang, L.K., Yan, Y.H. and Peng, Y. 2017. Exogenous application of GABA improves PEG-induced drought tolerance positively associated with GABA-shunt, polyamines, and proline metabolism in white clover. *Frontiers in Physiology*, *8*, 1–15.

Zegaoui, Z., Planchais, S., Cabassa, C., Djebbar, R., Belbachir, O.A. and Carol, P. 2017. Variation in relative water content, proline accumulation and stress gene expression in two cowpea landraces under drought. *Journal of Plant Physiology*, *218*, 26–34.

Zhou, Q. and Yu, B. 2010. Changes in content of free, conjugated and bound polyamines and osmotic adjustment in adaptation of vetiver grass to water deficit. *Plant Physiology and Biochemistry*, *48*(6), 417–425.

4 Regulatory Mechanisms for Stress-Induced Leaf Senescence

Stephanie Rossi and Bingru Huang

CONTENTS

4.1 INTRODUCTION

Leaf senescence is a naturally-occurring process during the life cycle of plants but can be induced or accelerated prematurely by environmental stresses. At the onset of leaf senescence, macromolecules, including chlorophyll, proteins, lipids, and DNA, are degraded, which can lead to leaf death and a decline in plant productivity (Buchanan-Wollaston, 1997). Cellular membranes, which are responsible for cell structure, cell-to-cell signaling, and compound transport, are also degraded (Taiz and Zeiger, 2002). Catabolites generated from the processes of leaf senescence may be transported or remobilized from aging leaves to other parts of the plant to support the growth of non-senescent organs during natural leaf senescence. Although leaf senescence is an important survival mechanism for the entire plant system, premature leaf senescence induced by environmental stress can lead to the reduction of plant growth and productivity.

The Earth is being faced with ongoing fluctuations in precipitation events, rises in average ambient temperatures, and the existence of other extreme weather activities, all of which are complicating crop management techniques. The agricultural and horticultural plant industries seek to control leaf senescence, as it reduces yield and biomass productivity of crops and contributes to postharvest declines in those that are transported and stored. As the worldwide human population continues to rise, it will be necessary to understand approaches for regulating abiotic stress-induced leaf senescence in order to maintain adequate food production to meet global demands. Studying the effects of abiotic stresses such as drought, salinity, elevated temperature, freezing, and high light conditions, on the processes of leaf senescence will be important for improving plant productivity under unfavorable environmental conditions.

This chapter overviews the current literature on metabolic processes and mechanisms regulating stress-induced leaf senescence. The mechanisms responsible for leaf chlorosis during stress-induced senescence are highlighted, with emphasis placed on chlorophyll metabolism. Other senescence-related processes, including proteolysis, amino acid metabolism, and oxidative metabolism related to membrane activity and stability are also discussed.

4.2 CHLOROPHYLL METABOLISM ASSOCIATED WITH STRESS-INDUCED LEAF SENESCENCE

Chlorophyll is the predominant pigment responsible for the green color of leaves and functions in light harvesting and conversion of light energy to carbohydrates. Each chlorophyll molecule is composed of four nitrogen atoms arranged in a porphyrin ring around one magnesium ion, bearing a phytol side chain (Fiedor et al., 2008). Since chlorophyll is the pigment most responsible for light absorption and harvesting, photosynthetic activity is reduced as chlorophyll content declines in senescent leaves, ultimately affecting carbon metabolism and energy production.

Chlorophyll content may decline in senescent leaves due to the inhibition of chlorophyll synthesis or acceleration of chlorophyll degradation. In efforts to maintain high quality and productivity in crops and other ornamental plants, many studies have examined the pathway of chlorophyll degradation, placing much focus on determining the factors associated with loss of leaf greenness. Many researchers have attempted to distinguish which chlorophyll enzymes are most likely associated with chlorophyll loss and how they correlate to it, but there are still some discrepancies regarding which processes are key to chlorosis.

Chlorophyllase (CHLASE) is the first enzyme performing in the catalytic degradation of chlorophyll proteins and is located in the thylakoid membrane of chloroplasts (Matile et al., 1997). Chlorophyll degradation initiates when CHLASE separates the phytol tail from the chlorophyll molecule to yield chlorophyllide. Following dephytylation of chlorophyll, Mg-dechelatase removes the Mg atom from chlorophyllide to generate pheophytin *a*. Pheophytin *a* is converted to pheophorbide *a* by the enzyme pheophytin pheophorbide hydrolase (pheophytinase; PPH) (Schelbert et al., 2009). Pheophorbide *a* oxygenase (PAO) severs the porphyrin ring of chlorophyllide by oxidizing pheophorbide *a*, which leads to the production of red chlorophyll catabolite (RCC) as well as other colorless metabolites (Hörtensteiner et al., 1998). The RCC formed in the reaction by which PAO oxidizes pheophorbide *a* is highly reactive and must be reduced by red chlorophyll catabolite reductase (RCCR) in order to prevent cell death. Once RCC is reduced, primary fluorescent chlorophyll catabolites (pFCCs) are formed and converted into non-fluorescent chlorophyll catabolites (NCCs) (Rodoni et al., 1997; Oberhuber et al., 2003). NCCs are not degraded any further for the purpose of remobilizing nitrogen (Hinder et al., 1996); therefore, it is likely that although plants degrade other cellular products to provide nitrogen for developing tissues, they undergo chlorophyll degradation to protect themselves from the highly reactive nature of chlorophyll byproducts.

Findings from many studies have shown that each of these enzymes may be important for the degradation of chlorophyll to proceed during leaf senescence. During prolonged heat stress, increased CHLASE enzyme activity and gene expression, along with loss of chlorophyll content, have been reported in different plant species, such as bentgrass (*Agrostis* spp.) (Jespersen et al., 2016; Rossi et al., 2017). In senescent barley (*Hordeum vulgare* L.) leaves, elevation of CHLASE activity corresponded with decreasing chlorophyll levels (Rodriguez et al., 1987). Conversely, CHLASE activity in snap bean (*Phaseolus vulgaris* L.) was highest just prior to leaf senescence but decreased as leaves yellowed, suggesting that CHLASE may not have been responsible for the loss of leaf greenness (Fang et al., 1998). As chlorophyll content declined in senescent canola (*Brassica napus* L.) under dark conditions, Mg-dechelatase activity decreased (Vicentini et al., 1995a). The efforts of Schelbert et al. (2009) showed that Arabidopsis (*Arabidopsis thaliana*) mutants unable to synthesize PPH remained green during leaf senescence and accumulated pheophytin, suggesting that leaf chlorosis may occur following dephytylation of pheophytin by PPH. Other studies on Yali pear (*Pyrus bretschneideri* Rehd. cv. "Yali"), broccoli (*Brassica oleracea*), and hybrid bentgrass (*A. capillaris* × *A. stolonifera*) during senescence have shown that the PPH gene is upregulated when chlorophyll content declines (Büchert et al., 2011; Cheng and Guan, 2014; Jespersen et al., 2016). Expression levels of the gene, PPH, and two others encoding the chlorophyll-degrading enzymes, PAO and RCCR, were elevated in senescent leaves of Chinese flowering cabbage (*Brassica rapa* var. parachinensis) (Zhang et al., 2011). PAO is only present in plant tissues that are chlorotic and can be measured at high levels once senescent leaves have yellowed (Ginsburg et al., 1994). Studies on various plants that exhibit the stay-green trait have suggested that PAO positively contributes to leaf senescence. In meadow fescue (*Festuca pratensis* Huds.) mutants with a trait for maintaining

chlorophyll, CHLASE and Mg-dechelatase remained active, while PAO activity was inhibited (Vicentini et al., 1995b). Similar results were seen for senescent peas (*Pisum sativum*) that maintained green color during leaf aging (Thomas et al., 1996). Additionally, a study showed that PAO gene expression is expressed at a higher level during salt-induced leaf senescence, implying that the catabolic enzyme may be responsible for salt-accelerated leaf senescence (Xiao et al., 2015). Disparities among studies regarding which chlorophyll degradation enzyme are essential for chlorosis to occur during leaf senescence may be attributed to differences among the species studied.

While multiple studies have reported that chlorophyll degradation is responsible for the loss of green color in leaf tissue, there is evidence that chlorosis may also be induced by suppression of chlorophyll synthesis. The first pyrrolic structure of chlorophyll biosynthesis, porphobilinogen, is produced when two molecules of δ-Aminolevulinic acid (δ-ALA) are condensed by ALA dehydrogenase. ALA dehyrogenase declined in sunflower (*Helianthus annuus*) leaves exposed to severe salinity stress while chlorophyllase activity was simultaneously inhibited, suggesting that both chlorophyll synthesis and degradation are negatively affected by abiotic stresses (Santos, 2004). Once porphobilinogen is generated, four molecules of the compound are joined in a linear fashion by porphobilinogen deaminase (PBGD) to produce hydroxymethylbilane. Previous research showed that in poinsettia (*Euphorbia pulcherrima*) and pepper (*Capsicum annuum*), PBGD activity was inhibited in senescent leaves but active in developing leaves, indicating that a decrease in PBGD activity occurs prior to leaf chlorosis (Frydman and Frydman, 1979). When hydroxymethylbilane is arranged into a ring structure during chlorophyll synthesis, a series of decarboxylation reactions occur, and the resultant uroporphyrinogen III is converted into protoporphyrinogen IX. Protoporphyrinogen oxidase facilitates dehydrogenation of protoporphyrinogen IX into protoporphyrin IX, a direct precursor of chlorophyll. Mg-chelatase inserts Mg^{2+} into protoporphyrin IX, and the product is methylated to produce protochlorophyllide, a molecule containing the five rings that are signature of all chlorophylls (Castelfranco and Beale, 1983; Steccanella et al., 2015). Enzyme activities of ALA dehydrogenase, porphobilinogen deaminase, and Mg-chelatase decreased in cucumber (*Cucumis sativus* L. cv poinsette) seedlings under heat and chilling stress, indicating the sensitivity of these enzymes to temperature extremes (Tewari and Tripathy, 1998). Protochlorophyllide oxidoreductase (POR) catalyzes the reduction of protochlorophyllide into chlorophyllide in the presence of light (Gabruk and Mysliwa-Kurdziel, 2015). When the enzyme chlorophyll synthetase catalyzes the esterification of a side chain in chlorophyllide, a phytol tail is added to the molecule, and chlorophyll is formed (Von Wettstein et al., 1995). Although many works have focused on elucidating the pathway of chlorophyll synthesis, there is no consensus regarding which step may be rate-limiting under abiotic stresses. An early study proposed that inhibition of proteins responsible for δ-ALA formation prevents chlorophyll from being synthesized (Nadler and Granick, 1970). Papenbrock et al. (2000) suggested that decreases in Mg-chelatase activity may negatively affect activities of other chlorophyll-synthesizing enzymes further downstream in the pathway, while other research exhibited that limiting Mg-chelatase itself is directly responsible for lowered rates of chlorophyll synthesis (Zhang et al., 2006). Further research regarding how abiotic stresses impact enzymes associated with chlorophyll biosynthesis will be necessary in the future in order to determine which are key in limiting leaf chlorophyll content.

4.3 REACTIVE OXYGEN SPECIES AND MEMBRANE PEROXIDATION RELATED TO STRESS-INDUCED LEAF SENESCENCE

Oxidative stress, generated by excessive production of reactive oxygen species (ROS), can accelerate leaf senescence by impacting the integrity of thylakoid membranes within chloroplasts. Under high light conditions, the quantity of photons harvested by leaves during photosynthesis may be above the optimal rate, allowing excess energy to be deposited on the electron transport chain and in turn, exciting chlorophyll a ($^1Chl^*$). As $^1Chl^*$ accumulates and is not reduced to its ground state, the highly reactive triplet chlorophyll ($^3Chl^*$) forms and transfers its energy to O_2 in photosystem II, generating singlet oxygen ($^1O_2$) (Pintó-Marijuan and Munné-Bosch, 2014). When excess electrons leak from photosystem II to O_2 at the same time as unreduced oxygen compounds are released, certain ROS, such as superoxide radical ($O_2^{.-}$) and hydrogen peroxide (H_2O_2) may form. If electrons are generated more rapidly than NADPH can be fixed to NADP during the process of carbon assimilation in photosynthesis, ROS are also generated in photosystem I. At this site, the reduction of O_2 by ferredoxin also leads to formation of $O_2^{.-}$, a product that can readily oxidize several molecules within chloroplasts (Asada, 2006). It is well-known that ROS are generated by plants in response to abiotic stresses for the purpose of initiating an internal stress defense system; however, when oxidative byproducts from the excitement of chlorophyll compounds accumulate, membranes and their integral proteins are damaged, causing a myriad of effects that ultimately lead to leaf senescence and eventual plant death.

In cell and organelle membranes, phospholipids are arranged in a bilayer, which provides a barrier that separates cellular materials from the surrounding environment and offers protection to components critical for metabolic function. Integral proteins embedded within this bilayer facilitate the transport of substances, accounting for the selective permeability of membranes. Additionally, these proteins play a role in photosynthesis and electron transport and function in ATP production. In healthy cells, the membrane typically exists in a fluid state, where the lipid bilayer is flexible and capable of transporting substances in and out of cells effectively; however, membranes of aged and stressed cells rigidify as unsaturated fatty acids are broken down, causing permeability to increase and ion leakage to occur (Barber and Thompson, 1980). As cellular membranes are disrupted, macromolecules, such as nucleic acids, proteins, and lipids, are degraded. Oxidative stress imposed by ROS disrupts membrane integrity. Lipid peroxidation results from exposure to the ROS that accumulate in senescent tissues when a stress is especially severe or prolonged. Malondialdehyde (MDA) is a highly reactive byproduct released as unsaturated fatty acids undergo peroxidation (Halliwel and Gutteridge, 2015). Several studies have shown that abiotic stresses increase lipid peroxidation in plants, inducing leaf senescence. In alfalfa (*Medicago sativa*) and wheat (*Triticum aestivum* L.) under drought stress, functionality of the antioxidant scavenging system decreased, leading to accumulation of MDA and a subsequent increase in lipid peroxidation (Irigoyen et al., 1992; Baisak et al., 1994). Salinity stress was shown to have a similar effect in rice and cotton (*Gossypium hirsutum*), where leaf senescence was accelerated following the destabilization of cell membranes following MDA accumulation (Lutts et al., 1996; Zhang et al., 2006). In addition to their effects on the structural integrity of thylakoid membranes, ROS impact photosynthesis. Since photosystems II and I are protein complexes embedded within the thylakoid membrane, they are rendered dysfunctional by oxidation of the membrane. As the proteins comprising photosystems II and I are denatured, and chlorophyll proteins are degraded, photosynthesis and CO_2 fixation are inhibited, leading to a reduction in plant growth and productivity. At temperatures greater than 35°C, wheat exhibited leaf senescence as a delay in photosynthetic activity resulting from inhibition of electron transport and degradation of chloroplasts (Harding et al., 1990). Rice (*Oryza sativa* L.) exposed to salinity stress was shown to have lower concentrations of chlorophyll and other proteins as well as disruption of membrane integrity, indicating that salt stress-induced leaf senescence occurred (Lutts et al., 1996). Oxidative stress-induced leaf senescence is associated with a reduction in the functionality of the photosynthetic apparatus and disability of plants to harvest light for energy.

When ROS induced by abiotic stresses are not countered by the effects of antioxidants, plants undergo leaf senescence more rapidly. Antioxidants are free radical scavengers that react with ROS, reducing them to less reactive compounds. Superoxide dismutase (SOD) initially acts on $O_2^{.-}$ to yield H_2O_2 and O_2. The H_2O_2 is scavenged by both SOD and catalase (CAT), preventing H_2O_2 from combining with $O_2^{.-}$ to form the dangerously reactive hydroxyl radical (.OH) (Scandalios, 1993). The remaining H_2O_2 is reduced to H_2O by hydrogen peroxidase (POD), catalase (CAT), and ascorbate peroxidase (APX), rendering it incapable of oxidizing cellular components (Foyer and Halliwell, 1976; Nakano and Asada, 1980; Nakano and Asada, 1981). ROS present in higher concentrations than antioxidants has been shown to greatly impact the performance of several important crops. Grain yield of foxtail millet (*Setaria italica* (L.) P. Beauv) under drought stress was reduced by an increase in ROS activity that impacted photosynthetic machinery, as indicated by the observed reduction in chlorophyll content (Dai et al., 2012). Sunflower (*Helianthus annuus* L.) maintained under high-temperature stress had elevated levels of H_2O_2 and decreased antioxidant activity, likely accounting for the measured reductions in leaf photosynthetic rate (De la Haba et al., 2014). The abilities of the antioxidants SOD and CAT to scavenge ROS were also inhibited in leaves of creeping bentgrass (*Agrostis stolonifera* L.) under heat stress, which had chlorophyll content and higher electrolyte leakage, indicating the occurrence of lipid peroxidation and senescence (Liu and Huang, 2000). Because of the consequences that ROS have on the acceleration of leaf senescence, it is necessary to determine the ways in which plants adequately control their damaging effects. Studying antioxidant potential to understand the mechanisms behind ROS-quenching will help in the development of plants that stay green under abiotic stress conditions.

4.4 AMINO ACID AND PROTEIN METABOLISM ASSOCIATED WITH STRESS-INDUCED LEAF SENESCENCE

During leaf senescence, proteins are typically broken down, with a majority of the degradation activity occurring in chloroplasts, adversely affecting photosynthesis (Hörtensteiner and Feller, 2002). Within the plant cell, proteolysis mainly occurs through the ubiquitin/26S proteasome pathway, senescence-associated vacuoles (SAVs), and proteases (Roberts et al., 2012; Díaz-Mendoza, 2014). In the ubiquitin/26S proteasome pathway, proteins that are destined for breakdown are tagged with ubiquitin and subsequently degraded (Vierstra, 2009). The transcription factors (AhUBC2, GmUBC2, and AtUBC32) related to managing the activities of ubiquitin enzymes were upregulated in response to osmotic stress (Zhou et al., 2010; Wan et al., 2011; Cui et al., 2012). Overexpression of ubiquitin was shown to confer drought and salinity tolerance to plants under osmotic stress (Guo et al., 2008). The decline in levels of proteins through the course of leaf senescence is accompanied by an increase in the activity of proteases as a result of elevated expression of their gene transcripts (Smart et al., 1995; Drake et al., 1996). Specifically, cysteine proteases are encoded by senescence-associated genes (SAGs), and it has been shown that *SAG12* is only expressed during senescence (Gan and Amasino, 1995). Abiotic stresses are known to induce and accelerate the activities of proteases, leading to the advancement of leaf senescence (Al-Khatib and Paulsen, 1984; Taiz and Zeiger, 2002). When the gene encoding the chloroplast protease CND41 was overexpressed in tobacco leaves, Rubisco was degraded during senescence, while silencing it prevented leaf senescence (Kato et al., 2005). In *Capsicum annuum* L., silencing the cysteine protease, CaCP, enhanced salt and drought tolerance (Xiao et al., 2014). In spinach (*Spinacia oleracea*) and *Ramonda serbica*, the activity of the serine protease, subtilase, was elevated under osmotic stress-induced senescence (Srivastava et al., 2009; Kidrič et al., 2014). SAVs are specifically located within chloroplasts and function in the degradation of photosynthetic proteins during leaf senescence (Otegui et al., 2005; Martínez et al., 2008). *SAG12* was shown to only be expressed within SAVs, vesicles that are produced in greater amounts as leaf senescence advances, suggesting that the gene may play an important role in the degradation of light-harvesting proteins (Otegui et al., 2005; Carrión et al., 2013). During proteolysis, partially degraded proteins must be transported to the vacuole from other cellular locations. There is evidence that some photosynthetic proteins are processed in the central vacuole in addition to SAVs (Avila-Ospina, 2014). Within the vacuole, proteins are finally degraded into amino acids and smaller peptides by proteases, and nutrients are remobilized to actively growing tissues (Hörtensteiner and Feller, 2002).

When proteins are degraded during leaf senescence, ammonium is generated as a byproduct. Since this form of nitrogen can become toxic to plants in excess, it must be reassimilated and mobilized to actively growing parts of the plant. Glutamine and asparagine, containing two nitrogen atoms in each molecule, are the two main amino acids from which nitrogen is transported. Prior to leaf senescence, glutamine synthetase (GS2) and ferredoxin-dependent glutamine oxoglutarate aminotransferase (Fd-GOGAT) complete this process within the chloroplast; however, once chloroplasts are degraded, the pathway occurs in the cytosol, with glutamate synthetase isoforms (GS1) compensating for the degraded GS2 (Sakurai et al., 1996). In the glutamate synthase cycle, the transfer of ammonium to glutamate is catalyzed by glutamine synthetase (GS1), yielding glutamine (Taiz and Zeiger, 2002). The nitrogen produced as a byproduct of this reaction is transferred to 2-oxoglutarate, generating two molecules of glutamate (Taiz and Zeiger, 2002). Alternatively, glutamate can be generated or deaminated by the enzyme glutamate dehydrogenase (GDH); however, the GS/GOGAT cycle is more effective in synthesizing glutamate (Taiz and Zeiger, 2002). Assimilation of nitrogen into glutamate and glutamine is critical for the biosynthesis of nitrogen-containing products, such as chlorophylls, nucleotides, and amino acids (Lea and Ireland, 1999).

There is evidence that the process of nitrogen assimilation and relocalization during leaf senescence may be negatively impacted by abiotic stress, in turn affecting the generation of other amino acids. Under salt-induced leaf senescence, expression of a gene associated with GDH was elevated in the presence of ROS, leading to GDH-mediated detoxification of the amine compounds generated by proteolysis (Skopelitis et al., 2006). Wheat under low salinity stress experienced an increase in GS activity; however, under high salinity stress, GS activity decreased while that of GDH was elevated (Wang et al., 2007). Proline accumulates in response to osmotic stresses, such as drought or salinity, in plants. In the biosynthesis of proline from glutamate, a reaction catalyzed by $\Delta^{'1}$-Pyrroline-5-carboxylate synthase (P5CS) produces pyrroline-5-carboxylate (P5C), which is in turn reduced by 1-pyrroline-5-carboxylate reductase (P5CR), yielding proline (Hu et al., 1992; Delauney and Verma, 1993). Wang et al. (2007) additionally determined that proline accumulated under a high level of salinity stress, while glutamate concentration was increased under both high and low levels. Under heat stress, GS, GOGAT, and P5CR enzyme activities declined, indicating that high temperatures may be damaging to nitrogen assimilation (Cui et al., 2006). These findings suggest that the efficient reuse of amino acids resulting from prote-olysis may help improve plant tolerance to abiotic stresses that accelerate protein degradation by manipulating protein metabolism to suppress leaf senescence.

4.5 HORMONE METABOLISM AND ABIOTIC STRESS-INDUCED LEAF SENESCENCE

Phytohormones play various roles in regulating plant growth and development, including leaf senescence. The effects of various hormones on the process of leaf senescence have been widely studied, with a primary focus on cytokinin, a senescence inhibitor, and ethylene, a promoter. Cytokinin and gibberellic acid (GA) are hormones that delay the onset of leaf senescence, while ethylene and abscisic acid (ABA) promote it. While auxin was widely believed to be inhibitory to leaf senescence in the past, its role in the regulation of this process is ambiguous, as several recent studies have suggested its function in the activation of senescence-promoting genes. The functions of hormones in regulating the development of leaf tissues are complex and vary with specific hormones, leaf ages, and environmental factors that affect leaf senescence.

Cytokinin is a hormone that mainly induces cell division in plants. Many studies have shown that endogenous cytokinin levels decrease in leaf tissue as senescence proceeds, and it has been suggested that once cytokinin content declines to a certain point, initiation of senescence is triggered (Noodén et al., 1990). Leaf senescence has been mitigated by exogenous application of cytokinin, and it has been proven that treatment with the hormone improves tolerance to abiotic stresses in various plant species. Under heat stress, creeping bentgrass treated with cytokinin, zeatin riboside (ZR) through foliar application exhibited delayed leaf senescence (Liu and Huang, 2002; Veerasamy et al., 2007). Application of 6-benzyladenine (6-BA) to perennial ryegrass (*Lolium perenne*) under salinity stress suppressed leaf senescence (Ma et al., 2016). The exact mechanisms by which exogenously applied cytokinins prevent the onset and progression of senescence are unknown, but research has suggested that the hormone may inhibit chlorophyll breakdown, allowing plants to remain photosynthetically active (Talla et al., 2016). Cytokinins are capable of inhibiting protein degradation, which normally occurs at a higher rate during leaf senescence, and protect membrane integrity by decreasing the activities of lipoxygenase, a membrane-degrading enzyme (Grossman and Ya'acovy, 1978; Legocka and Szweykowska, 1981). It is likely that cytokinins may delay senescence by preventing the structure of membranes from becoming interrupted, allowing for better retention of ions and solutes. Manipulating the expression of cytokinin synthesis genes may also serve as a mechanism by which leaf senescence can be controlled in plants exposed to stress. In transgenic plants expressing the *IPT* gene for cytokinin biosynthesis, multiple symptoms of senescence are suppressed, such as loss of chlorophyll and photosynthetic activity and breakdown of proteins, suggesting that genetic modification may prevent senescence at an early phase of the process (Wingler et al., 1998). Its applications in alleviating

the effects of abiotic stress-induced leaf senescence have been documented. *IPT*-tagged tobacco (*Nicotiana tabacum*) and lettuce (*Lactuca sativa* L.) under drought stress increased endogenous cytokinin levels and photosynthetic performance, which delayed leaf senescence and enhanced drought tolerance (McCabe et al., 2001; Rivero et al., 2009). Tall fescue (*Festuca arundinacea*) and creeping bentgrass tagged with *IPT* produced more tillers and presented higher freezing and heat tolerance, respectively (Hu et al., 2005; Xu et al., 2009). Research involving *IPT* has demonstrated the gene's clear function in the suppression of stress-induced leaf senescence. Genes involved in cytokinin perception also affect the process of leaf senescence. In Arabidopsis, cytokinin is sensed by histidine kinase receptors AHK2, AHK3, and AHK4, and the signal is transmitted to Arabidopsis response regulator (ARR) proteins to activate transcription of genes induced or suppressed by cytokinin (Kim et al., 2006). Kim et al. (2006) showed that overexpression of *ARR2* in Arabidopsis delayed leaf senescence. Furthermore, senescence was accelerated in loss-of-function *AHK3* Arabidopsis mutants, while gain-of-function mutants exhibited suppressed leaf senescence (Kim et al., 2006). These results indicate that the AHK3 receptor may specifically function in cytokinin-mediated regulation of leaf senescence. Furthermore, it is likely that cytokinins regulate the induction or suppression of genes associated with the signaling of other hormones involved in the leaf senescence process (Brenner et al., 2005). GA is developed in young, actively growing leaves, buds, roots, and shoots and is responsible for cell elongation. An initial study examining the effects of GA on leaf senescence showed that the hormone delayed the process by inhibiting RNA synthesis and allowing for the maintenance of chlorophyll levels (Fletcher and Osborne, 1965). Additionally, GA applied exogenously to nasturtium (*Tropaeolum majus*) and *Rumex* species inhibited the degradation of chlorophyll and other proteins during leaf senescence (Beevers, 1966; Goldthwaite and Laetsch, 1968). Exogenous application of GA to leaves of *Paris polyphylla* also decreased the activities of chlorophyll-degrading enzymes and hindered senescence by allowing GA to accumulate endogenously (Li et al., 2010). While a majority of studies reported the senescence-inhibiting properties of GA, some recent studies found that leaf senescence was promoted by this hormone. In Arabidopsis mutants lacking DELLA, a protein that represses the signaling of GA, leaf senescence was hastened but was delayed by overexpressing the gene (Chen et al., 2017). Furthermore, GA-receptor mutants also showed suppression of leaf senescence (Chen et al., 2017). The reasons for the inconsistency of GA effects on leaf senescence are unknown. Furthermore, the specific effects of GA on leaf senescence in plants under abiotic stresses have not been well-documented and deserve further examination.

ABA accumulates within plants during leaf senescence and is produced in response to environmental stresses, such as salinity, drought, and chilling (Gepstein and Thimann, 1980; Xiong and Zhu, 2003). The ABA biosynthesis gene, *NCED*, was shown to be upregulated in plants exposed to drought conditions (Tan et al., 1997; Qin and Zeevaart, 1999; Chernys and Zeevaart,

2000; Iuchi et al., 2001). Yang et al. (2011) determined that the transcription factor, *VNI2*, was also upregulated following exposure to salinity stress and ABA. Furthermore, overexpression of *VNI2* inhibited leaf senescence (Yang et al., 2011). It has been shown that exogenous application of ABA enhances the gene expression of *SAG*s, accelerating leaf chlorosis (Gao et al., 2016). Expression of *SAG*s is promoted when ABA activates SnRK2s, which in turn phosphorylate RAV1 and ABFs (Furihata et al., 2006). Specifically, the transcription factors ABF2, ABF3, and ABF4, all induce the activity of NYE1, a gene responsible for chlorophyll degradation (Gao et al., 2016). During senescence, *SAG113* is upregulated, and when the gene is knocked out, the process of leaf senescence is delayed (Zhang and Gan, 2012). Further research has shown that function of the ABA-responsive receptor kinase RPK1 is critical for plant perception of ABA signals and that mutants lacking RPK1 did not undergo leaf senescence when supplied with the hormone (Lee et al., 2011). ABA signaling is inhibited by the cytokinin receptor kinases AHK2 and AHK3 in Arabidopsis exposed to drought, salinity, and chilling stress, which accelerates leaf senescence (Tran et al., 2007; Jeon et al., 2010). ABA may act alone or interact with other hormones affecting leaf senescence.

Ethylene is produced in response to various abiotic stresses, such as heat, drought, salinity, chilling, and flooding (Johnson and Ecker, 1998). Ethylene is best-known for its ability to accelerate leaf senescence. In *ETR1-1* and *EIN2* Arabidopsis mutants insensitive to ethylene, leaf senescence was inhibited (Grbić and Bleecker, 1995; Oh et al., 1997). EIN3, a factor involved in ethylene signaling, accelerates leaf senescence by promoting the expression of *NYC1*, *NYE1*, and *PAO*, which are chlorophyll degradation genes (Qiu et al., 2015). During leaf senescence, genes responsible for ethylene synthesis are also upregulated (Hunter et al., 1999). When EIN3 interacts with ORE1, a senescence promoter, expression of the ethylene synthesis gene, *ACS2*, is enhanced (Qiu et al., 2015). Furthermore, repression of aminocyclopropane-1-carboxylic acid (ACC) oxidase, an enzyme functioning in ethylene biosynthesis, in tomato (*Lycopersicon esculentum* Mill.) inhibited leaf senescence (John et al., 1995). Inhibition of ACC synthase gene promoted salinity and heat tolerance by lowering ethylene response and production (Dong et al., 2011). Additionally, loss of expression in the ACC synthase gene of maize (*Zea mays*) led to a reduction in leaf senescence during drought stress (Young et al., 2004). Certain genes involved in ethylene signaling may also have functions during abiotic stress-induced leaf senescence. Profiling of transcripts present in senescent Arabidopsis leaves indicated the presence of mitogen-activated protein kinase 6 (MPK6), a transcription factor involved in ethylene signaling (Ouaked et al., 2003; Guo et al., 2004). MPK6 activity has been shown to increase during drought and chilling stress, suggesting that its accumulation may be responsible for the advancement of leaf senescence (Ichimura et al., 2000; Droillard et al., 2002). It is possible that by manipulating the genes associated with ethylene metabolism, new lines of plants that do not senesce as readily under abiotic stress may be generated. The effects of ethylene on plants under abiotic stresses have been implicated in multiple studies involving ethylene inhibitors. Under heat stress, aminoethoxyvinylglycine (AVG) and 1-Methylcyclopropene (1-MCP), two ethylene inhibitors, mitigated leaf senescence in creeping bentgrass and soybean, respectively (Yu and Huang, 2009; Djanaguiraman and Prasad, 2010; Jespersen et al., 2015; Jespersen and Huang, 2015). Treatment of rice under flooding stress with 1-MCP decreased the activity of CHLASE, allowing plants to maintain higher chlorophyll content, and suppressed leaf senescence (Ella et al., 2003). Lipoxygenase is known to have a role in ethylene biosynthesis, and application of lipoxygenase inhibitors was shown to suppress the production of ethylene, delaying senescence (Ferrie et al., 1994; Sheng et al., 2000). The results of these studies all suggest that ethylene production is enhanced under abiotic stress and that the effects of leaf senescence may be alleviated by genetic manipulation or utilization of ethylene inhibitors.

Auxins promote the elongation of cells and are synthesized in actively developing plant tissues, such as shoots and roots. Although the effects of auxin on leaf senescence have not been as widely examined as those of other hormones, there is research suggesting that it delays the onset of the process when applied exogenously to plants (Sacher, 1957). It is likely that auxin may regulate leaf senescence by controlling certain genes associated with aging. Under drought stress, foliar application of the auxin, indole-3-acetic acid (IAA), alleviated leaf senescence in white clover (*Trifolium repens*) by lowering transcriptional activity of the senescence-activated genes *SAG101* and *SAG102* (Li et al., 2018). Additionally, transcription of *SAG12* was suppressed by exogenous application of auxin (Noh and Amasino, 1999). Mutations in *ARF1* and *ARF2*, two auxin response factor genes, induced auxin signaling and delayed leaf senescence (Ellis et al., 2005; Lim et al., 2010). Overexpression of the key auxin synthesis gene, *YUCCA6*, inhibited leaf senescence by increasing endogenous levels of auxin (Kim et al., 2011). It has been shown that a majority of genes functioning in the transport of auxin are inhibited during senescence (van der Graaff et al., 2006; Lin et al., 2015). Conversely, recent studies have indicated that auxin may induce genes that promote leaf senescence. Overexpressing the auxin-activated genes, *SAUR36* and *GmSARK*, in an Arabidopsis mutant and soybean (*Glycine max*), respectively, accelerated leaf senescence, while inhibiting their expression delayed the process (Li et al., 2006; Xu et al., 2011; Hou et al., 2013). The reasons for contradictory results regarding how auxin affects leaf senescence are still unknown. Future research may explore the mechanisms by which auxin regulates stress-induced leaf senescence. Elucidating the mechanisms of hormone signaling in leaves during different stages of senescence will be critical for understanding how to regulate the aging response of plants under abiotic stress conditions.

4.6 APPROACHES TO SUPPRESS OR ALLEVIATE STRESS-INDUCED LEAF SENESCENCE

As discussed above, ROS accumulation is a major factor accelerating leaf senescence and threatens the functionality of many metabolic activities within plants. Engineering plants to have an altered expression of genes encoding antioxidant enzymes may be a strategy for developing lines of plants tolerant abiotic stress-induced leaf senescence. The use of non-enzymatic antioxidants can also be effective for mitigating oxidation-induced leaf senescence. Some antioxidants function in photochemical quenching, a mechanism by which excess energy on photo-excited compounds, such as light-harvesting pigments, is dissipated. Anthocyanins are pigment compounds that quench excess light energy under photoinhibition and have also been shown to confer tolerance to multiple stresses, such as chilling and drought, when present in leaves (Chalker-Scott, 1999; Sillie and Heterington, 1999; Pietrini et al., 2002; Farrant et al., 2003). Research has shown that anthocyanins are effective ROS scavengers and may interrupt the deleterious effects of free radicals on nutrient remobilization, allowing plants to sequester more nitrogen during senescence (Yamasaki et al., 1996; Hoch et al., 2001). Zeaxanthin, antheraxanthin, and violaxanthin are three carotenoids that protect excited chlorophyll molecules through the process of quenching. Through the xanthophyll cycle, oxygen is removed from violaxanthin, yielding antheraxanthin. When the de-epoxidation of antheraxanthin occurs, zeaxanthin is generated (Taiz and Zeiger, 2002). Zeaxanthin has been shown to alleviate heat stress in leaves by stabilizing thylakoid membranes that become more fluid under high temperatures (Havaux et al., 1996; Tardy and Havaux, 1997). Conversely, B-carotenes grant membranes greater fluidity and have been shown to be present in elevated quantities in membranes of plants exposed to chilling stress (Huner et al., 1984). Antioxidants have been shown to effectively mitigate the effects of leaf senescence in plants under various abiotic stresses. As environmental conditions become more severe, it may be necessary to use them to inhibit leaf chlorosis, as they function as membrane stabilizers and in protecting the photosynthetic apparatus from ROS damage.

Polyamines have been shown to have a beneficial role in the protection of membrane structures in plants and are involved in plant responses to leaf senescence as well as tolerance to various stresses, such as chilling, drought, salinity, and heat (Murkowski, 2001; Hummel et al., 2004; Liu et al., 2007; Yamaguchi et al., 2007; Duan et al., 2008; He et al., 2008; Gill and Tuteja, 2010). Kumar et al. (2008) found that treating plants with inhibitors of polyamine biosynthesis resulted in decreased stress tolerance, but that supplying polyamines exogenously reversed this effect, indicating their positive roles as suppressors of leaf senescence. Additionally, exogenous application of polyamines was found to lower the concentrations of H_2O_2 and O^- radicals, indicating their roles as antioxidant scavengers (Yiu et al., 2009). Wheat exposed to low temperatures accumulated putrescine and spermidine, two polyamines, in leaves, and alfalfa under chilling stress was also shown to have increased putrescine content (Nadeau et al., 1987; Kovács et al., 2006). It is possible that maintaining higher endogenous polyamine content may alleviate stress-induced leaf senescence in plants by alleviating the effects of membrane damage (Borrell et al., 1997). While many studies have reported the effects of the exogenous application of polyamines on mitigating various abiotic stresses, some have indicated that the transformation of certain genes involved with polyamine biosynthesis may confer tolerance to stresses. The gene AtADC2 has been shown to be critical for putrescine biosynthesis under various stresses, such as salinity, drought, chilling, and heat, as well as under optimal conditions, indicating that its manipulation in plants may lead to the production of genetic lines that better tolerate saline conditions (Urano et al., 2004; Prabhavathi and Rajam, 2007). Because polyamines have the ability to scavenge free radicals, their abilities to mitigate the effects of leaf chlorosis should be further examined in senescent plants under abiotic stress.

Glutamate is produced through the GS/GOGAT cycle and serves as the compound that initiates the synthesis of other amino acids that are important in regulating abiotic stress tolerance, such as y-aminobutyric acid (GABA). In addition to GABA being derived from glutamate, there is evidence that as the polyamine putrescine accumulates under abiotic stress, it is converted into GABA (Shelp et al., 2012). GABA is an important metabolite in plant cells, as it regulates carbon to nitrogen balance, maintains the pH of the cytosol, and has a role in oxidative stress signaling (Fait et al., 2008). Endogenous quantities of GABA are elevated in plants under abiotic stress, indicating the importance of GABA in conferring tolerance to adverse environmental conditions (Lee et al., 2011). When applied exogenously to plants under heat and drought stress, GABA increases proline levels and suppresses leaf senescence by supporting cell membrane structure, elevating leaf water content, and enhancing antioxidant potential (Nayyar et al., 2014; Li et al., 2016; Vijayakumari and Puthur, 2016). Proline is also important in the regulation of tolerance to abiotic stresses, especially to those that affect water potential, because of its properties as an osmoregulant. When proline accumulates under stress, it acts as an antioxidant in ROS scavenging (Shevyakova et al., 2009). During periods of stress, the P5CS gene, responsible for proline biosynthesis, is activated, leading to accumulation of proline (Hu et al., 1992). It is possible that in the future, modifying the P5CS gene in plants would promote osmotic regulation and mitigate the effects of leaf senescence, allowing for the development of transgenic lines tolerant to drought and salinity stress. Exogenous application of amino acids, other than GABA, has been shown to cause proline accumulation while reducing the permeability of membranes and improving ion uptake (Rai and Kumari, 1983; Handa et al., 1986; Khanna and Rai, 1998). In addition to proline, there are other amino acids that may confer tolerance to abiotic stress in plants. It was shown that proline, alanine, threonine, glycine, phenylalanine, lysine, tryptophan, leucine, and arginine prevented stomatal opening, while aspartate, glutamate, methionine, histidine, glutamine, and asparagine induced it (Rai and Sharma, 1991). It may be possible that the amino acids that reduce stomatal aperture may function in the regulation of drought stress by preventing water loss through transpiration. On the other hand, those that open stomata might offer protection against heat stress by allowing for evaporative cooling through transpiration. Therefore, foliar application of amino acids may be a useful way of controlling abiotic stress-induced leaf senescence in the future.

4.7　CONCLUSIONS

Abiotic stress-induced leaf senescence can lead to loss of crop yield. In order to ensure that plant production is maintained as environmental conditions become stressful, it will be necessary to develop strategies for controlling leaf senescence. Stress-induced leaf senescence may be mitigated by exogenous application of hormones, growth regulators, and antioxidant compounds. Additionally, the modification of critical genes, such as those involved in pigment or protein synthesis, ROS scavenging, hormone metabolism, and stress responses is another effective approach for developing new plant lines that are able to stay green under adverse conditions. The many research efforts made in the past have enabled us to better understand the mechanisms behind mitigating abiotic stress-induced leaf senescence and have laid the groundwork for the current works aiming to develop stay-green plants. By continuing to research plant abiotic stress responses and the mechanisms by which plants tolerate unfavorable environmental conditions, it will be possible to determine how to alleviate the effects of leaf senescence and improve the quality and productivity of plants.

REFERENCES

Al-Khatib, K., and G. M. Paulsen. 1984. "Mode of high temperature injury to wheat during grain development." *Physiologia Plantarum* 61, no. 3: 363–368.

Asada, K. 2006. "Production and scavenging of reactive oxygen species in chloroplasts and their functions." *Plant Physiology* 141, no. 2: 391–396.

Avila-Ospina, L., M. Moison, K. Yoshimoto, and C. Masclaux-Daubresse. 2014. "Autophagy, plant senescence, and nutrient recycling." *Journal of Experimental Botany* 65, no. 14: 3799–3811.

Baisak, R., D. Rana, P. B. B. Acharya, and M. Kar. 1994. "Alterations in the activities of active oxygen scavenging enzymes of wheat leaves subjected to water stress." *Plant and Cell Physiology* 35, no. 3: 489–495.

Barber, R. F., and J. E. Thompson. 1980. "Senescence-dependent increase in the permeability of liposomes prepared from bean cotyledon membranes." *Journal of Experimental Botany* 31, no. 5: 1305–1313.

Beevers, L. 1966. "Effect of gibberellic acid on the senescence of leaf discs of nasturtium (*Tropaeolum majus*)." *Plant Physiology* 41, no. 6: 1074.

Borrell, A., L. Carbonell, R. Farras, P. Puig-Parellada, and A. F. Tiburcio. 1997. "Polyamines inhibit lipid peroxidation in senescing oat leaves." *Physiologia Plantarum* 99, no. 3: 385–390.

Brenner, W. G., G. A. Romanov, I. Köllmer, L. Bürkle, and T. Schmülling. 2005. "Immediate-early and delayed cytokinin response genes of Arabidopsis thaliana identified by genome-wide expression profiling reveal novel cytokinin-sensitive processes and suggest cytokinin action through transcriptional cascades." *The Plant Journal* 44, no. 2: 314–333.

Buchanan-Wollaston, V. 1997. "The molecular biology of leaf senescence." *Journal of Experimental Botany* 48, no. 2: 181–199.

Büchert, A. M., P. M. Civello, and G. A. Martínez. 2011. "Chlorophyllase versus pheophytinase as candidates for chlorophyll dephytilation during senescence of broccoli." *Journal of Plant Physiology* 168, no. 4: 337–343.

Carrión, C. A., M. L. Costa, D. E. Martínez, C. Mohr, K. Humbeck, and J. J. Guiamet. 2013. "In vivo inhibition of cysteine proteases provides evidence for the involvement of 'senescence-associated vacuoles' in chloroplast protein degradation during dark-induced senescence of tobacco leaves." *Journal of Experimental Botany* 64, no. 16: 4967–4980.

Castelfranco, P. A., and S. I. Beale. 1983. "Chlorophyll biosynthesis: recent advances and areas of current interest." *Annual Review of Plant Physiology* 34, no. 1: 241–276.

Chalker-Scott, L. 1999. "Environmental significance of anthocyanins in plant stress responses." *Photochemistry and Photobiology* 70, no. 1: 1–9.

Chen, L., S. Xiang, Y. Chen, D. Li, and D. Yu. 2017. "Arabidopsis WRKY45 interacts with the DELLA protein RGL1 to positively regulate age-triggered leaf senescence." *Molecular Plant* 10, no. 9: 1174–1189.

Cheng, Y., and J. Guan. 2014. "Involvement of pheophytinase in ethylene-mediated chlorophyll degradation in the peel of harvested 'Yali' pear." *Journal of Plant Growth Regulation* 33, no. 2: 364–372.

Chernys, J. T., and J. A. D. Zeevaart. 2000. "Characterization of the 9-cis-epoxycarotenoid dioxygenase gene family and the regulation of abscisic acid biosynthesis in avocado." *Plant Physiology* 124, no. 1: 343–354.

Cui, F., L. Liu, Q. Zhao, Z. Zhang, Q. Li, B. Lin, Y. Wu, S. Tang, and Q. Xie. 2012. "Arabidopsis ubiquitin conjugase UBC32 is an ERAD component that functions in brassinosteroid-mediated salt stress tolerance." *The Plant Cell*

Cui, L., R. Cao, J. Li, L. Zhang, and J. Wang. 2006. "High temperature effects on ammonium assimilation in leaves of two *Festuca arundinacea* cultivars with different heat susceptibility." *Plant Growth Regulation* 49, no. 2–3: 127–136.

Dai, H.-P., C.-J. Shan, A.-Z. Wei, T. Yang, W.-Q. Sa, and B.-L. Feng. 2012. "Leaf senescence and photosynthesis in foxtail millet ['*Setaria italica*' (L.) P. Beauv] varieties exposed to drought conditions." *Australian Journal of Crop Science* 6, no. 2: 232.

Davletova, S., L. Rizhsky, H. Liang, Z. Shengqiang, D. J. Oliver, J. Coutu, V. Shulaev, Karen Schlauch, and R. Mittler. 2005. "Cytosolic ascorbate peroxidase 1 is a central component of the reactive oxygen gene network of Arabidopsis." *The Plant Cell* 17, no. 1: 268–281.

De la Haba, P., L. De la Mata, E. Molina, and E. Agüera. 2014. "High temperature promotes early senescence in primary leaves of sunflower (*Helianthus annuus* L.) plants." *Canadian Journal of Plant Science* 94, no. 4: 659–669.

Delauney, A. J., and D. P. S. Verma. 1993. "Proline biosynthesis and osmoregulation in plants." *The Plant Journal* 4, no. 2: 215–223.

Díaz-Mendoza, M., B. Velasco-Arroyo, P. González-Melendi, M. Martínez, and I. Díaz. 2014. "C1A cysteine protease–cystatin interactions in leaf senescence." *Journal of Experimental Botany* 65, no. 14: 3825–3833.

Djanaguiraman, M., and P. V. V. Prasad. 2010. "Ethylene production under high temperature stress causes premature leaf senescence in soybean." *Functional Plant Biology* 37, no. 11: 1071–1084.

Dong, H., Z. Zhen, J. Peng, L. Chang, Q. Gong, and N. N. Wang. 2011. "Loss of ACS7 confers abiotic stress tolerance by modulating ABA sensitivity and accumulation in Arabidopsis." *Journal of Experimental Botany* 62, no. 14: 4875–4887.

Drake, R., I. John, A. Farrell, W. Cooper, W. Schuch, and D. Grierson. 1996. "Isolation and analysis of cDNAs encoding tomato cysteine proteases expressed during leaf senescence." *Plant Molecular Biology* 30, no. 4: 755–767.

Droillard, M.-J., M. Boudsocq, H. Barbier-Brygoo, and C. Laurière. 2002. "Different protein kinase families are activated by osmotic stresses in *Arabidopsis thaliana* cell suspensions." *FEBS Letters* 527, no. 1–3: 43–50.

Duan, J., J. Li, S. Guo, and Y. Kang. 2008. "Exogenous spermidine affects polyamine metabolism in salinity-stressed *Cucumis sativus* roots and enhances short-term salinity tolerance." *Journal of Plant Physiology* 165, no. 15: 1620–1635.

Ella, E. S., N. Kawano, Y. Yamauchi, K. Tanaka, and A. M. Ismail. 2003. "Blocking ethylene perception enhances flooding tolerance in rice seedlings." *Functional Plant Biology* 30, no. 7: 813–819.

Ellis, C. M., P. Nagpal, J. C. Young, G. Hagen, T. J. Guilfoyle, and J. W. Reed. 2005. "AUXIN RESPONSE FACTOR1 and AUXIN RESPONSE FACTOR2 regulate senescence and floral organ abscission in *Arabidopsis thaliana*." *Development* 132, no. 20: 4563–4574.

Fait, A., H. Fromm, D. Walter, G. Galili, and A. R. Fernie. 2008. "Highway or byway: the metabolic role of the GABA shunt in plants." *Trends in Plant Science* 13, no. 1: 14–19.

Fang, Z., J. C. Bouwkamp, and T. Solomos. 1998. "Chlorophyllase activities and chlorophyll degradation during leaf senescence in non-yellowing mutant and wild type of *Phaseolus vulgaris* L." *Journal of Experimental Botany* 49, no. 320: 503–510.

Farrant, J. M., C. V. Willigen, D. A. Loffell, S. Bartsch, and A. Whittaker. 2003. "An investigation into the role of light during desiccation of three angiosperm resurrection plants." *Plant, Cell & Environment* 26, no. 8: 1275–1286.

Ferrie, B. J., N. Beaudoin, W. Burkhart, C. G. Bowsher, and S. J. Rothstein. 1994. "The cloning of two tomato lipoxygenase genes and their differential expression during fruit ripening." *Plant Physiology* 106, no. 1: 109–118.

Fiedor, L., A. Kania, B. Myśliwa-Kurdziel, Ł. Orzeł, and G. Stochel. 2008. "Understanding chlorophylls: central magnesium ion and phytyl as structural determinants." *Biochimica et Biophysica Acta (BBA)-Bioenergetics* 1777, no. 12: 1491–1500.

Fletcher, R. A., and D. J. Osborne. 1965. "Regulation of protein and nucleic acid synthesis by gibberellin during leaf senescence." *Nature* 207, no. 5002: 1176.

Foyer, C. H., and B. Halliwell. 1976. "The presence of glutathione and glutathione reductase in chloroplasts: a proposed role in ascorbic acid metabolism." *Planta* 133, no. 1: 21–25.

Frydman, R. B., and B. Frydman. 1979. "Disappearance of porphobilinogen deaminase activity in leaves before the onset of senescence." *Plant Physiology* 63, no. 6: 1154–1157.

Furihata, T., K. Maruyama, Y. Fujita, T. Umezawa, R. Yoshida, K. Shinozaki, and K. Yamaguchi-Shinozaki. 2006. "Abscisic acid-dependent multisite phosphorylation regulates the activity of a transcription activator AREB1." *Proceedings of the National Academy of Sciences* 103, no. 6: 1988–1993.

Gabruk, M., and B. Mysliwa-Kurdziel. 2015. "Light-dependent protochlorophyllide oxidoreductase: phylogeny, regulation, and catalytic properties." *Biochemistry* 54, no. 34: 5255–5262.

Gan, S., and R. M. Amasino. 1995. "Inhibition of leaf senescence by autoregulated production of cytokinin." *Science* 270, no. 5244: 1986–1988.

Gao, S., J. Gao, X. Zhu, Y. Song, Z. Li, G. Ren, X. Zhou, and B. Kuai. 2016. "ABF2, ABF3, and ABF4 promote ABA-mediated chlorophyll degradation and leaf senescence by transcriptional activation of chlorophyll catabolic genes and senescence-associated genes in Arabidopsis." *Molecular Plant* 9, no. 9: 1272–1285.

Gepstein, S., and K. V. Thimann. 1980. "Changes in the abscisic acid content of oat leaves during senescence." *Proceedings of the National Academy of Sciences* 77, no. 4: 2050–2053.

Gill, S. S., and N. Tuteja. 2010. "Polyamines and abiotic stress tolerance in plants." *Plant Signaling & Behavior* 5, no. 1: 26–33.

Ginsburg, S., M. Schellenberg, and P. Matile. 1994. "Cleavage of chlorophyll-porphyrin (requirement for reduced ferredoxin and oxygen)." *Plant Physiology* 105, no. 2: 545–554.

Goldthwaite, J. J., and W. M. Laetsch. 1968. "Control of senescence in Rumex leaf discs by gibberellic acid." *Plant Physiology* 43, no. 11: 1855–1858.

Grbić, V., and A. B. Bleecker. 1995. "Ethylene regulates the timing of leaf senescence in Arabidopsis." *The Plant Journal* 8, no. 4: 595–602.

Grossman, S., and L. Ya'acovy. 1978. "Lowering of endogenous lipoxygenase activity in *Pisum sativum* foliage by cytokinin as related to senescence." *Physiologia Plantarum* 43, no. 4: 359–362.

Guo, Q., J. Zhang, Q. Gao, S. Xing, F. Li, and W. Wang. 2008. "Drought tolerance through overexpression of monoubiquitin in transgenic tobacco." *Journal of Plant Physiology* 165, no. 16: 1745–1755.

Guo, Y., Z. Cai, and S. Gan. 2004. "Transcriptome of Arabidopsis leaf senescence." *Plant, Cell & Environment* 27, no. 5: 521–549.

Gupta, A. S., J. L. Heinen, A. S. Holaday, J. J. Burke, and R. D. Allen. 1993. "Increased resistance to oxidative stress in transgenic plants that overexpress chloroplastic Cu/Zn superoxide dismutase." *Proceedings of the National Academy of Sciences* 90, no. 4: 1629–1633.

Halliwell, B., and J. M. C. Gutteridge. 2015. *Free Radicals in Biology and Medicine.* Oxford University Press, Oxford, UK.

Handa, S., A. K. Handa, P. M. Hasegawa, and R. A. Bressan. 1986. "Proline accumulation and the adaptation of cultured plant cells to water stress." *Plant Physiology* 80, no. 4: 938–945.

Harding, S. A., J. A. Guikema, and G. M. Paulsen. 1990. "Photosynthetic decline from high temperature stress during maturation of wheat: I. Interaction with senescence processes." *Plant Physiology* 92, no. 3: 648–653.

Havaux, M., F. Tardy, J. Ravenel, D. Chanu, and P. Parot. 1996. "Thylakoid membrane stability to heat stress studied by flash spectroscopic measurements of the electrochromic shift in intact potato leaves: influence of the xanthophyll content." *Plant, Cell & Environment* 19, no. 12: 1359–1368.

He, L., Y. Ban, H. Inoue, N. Matsuda, J. Liu, and T. Moriguchi. 2008. "Enhancement of spermidine content and antioxidant capacity in transgenic pear shoots overexpressing apple spermidine synthase in response to salinity and hyperosmosis." *Phytochemistry* 69, no. 11: 2133–2141.

Hinder, B., M. Schellenberg, S. Rodoni, S. Ginsburg, E. Vogt, E. Martinoia, P. Matile, and S. Hörtensteiner. 1996. "How plants dispose of chlorophyll catabolites directly energized uptake of tetrapyrrolic breakdown products into isolated vacuoles." *Journal of Biological Chemistry* 271, no. 44: 27233–27236.

Hoch, W. A., E. L. Zeldin, and B. H. McCown. 2001. "Physiological significance of anthocyanins during autumnal leaf senescence." *Tree Physiology* 21, no. 1: 1–8.

Hou, K., W. Wu, and S.-S. Gan. 2013. "SAUR36, a small auxin up RNA gene, is involved in the promotion of leaf senescence in Arabidopsis." *Plant Physiology* 161, no. 2: 1002–1009.

Hörtensteiner, S., K. L. Wüthrich, P. Matile, K.-H. Ongania, and B. Kräutler. 1998. "The key step in chlorophyll breakdown in higher plants cleavage of pheophorbide a macrocycle by a monooxygenase." *Journal of Biological Chemistry* 273, no. 25: 15335–15339.

Hörtensteiner, S., and U. Feller. 2002. "Nitrogen metabolism and remobilization during senescence." *Journal of Experimental Botany* 53, no. 370: 927–937.

Hu, C. A., A. J. Delauney, and D. P. Verma. 1992. "A bifunctional enzyme (delta 1-pyrroline-5-carboxylate synthetase) catalyzes the first two steps in proline biosynthesis in plants." *Proceedings of the National Academy of Sciences* 89, no. 19: 9354–9358.

Hu, Y., W. Jia, J. Wang, Y. Zhang, L. Yang, and Z. Lin. 2005. "Transgenic tall fescue containing the Agrobacterium tumefaciens ipt gene shows enhanced cold tolerance." *Plant Cell Reports* 23, no. 10–11: 705–709.

Hummel, I., A. E. Amrani, G. Gouesbet, F. Hennion, and I. Couée. 2004. "Involvement of polyamines in the interacting effects of low temperature and mineral supply on *Pringlea antiscorbutica* (Kerguelen cabbage) seedlings." *Journal of Experimental Botany* 55, no. 399: 1125–1134.

Huner, N. P. A., B. Elfman, M. Krol, and A. McIntosh. 1984. "Growth and development at cold-hardening temperatures. Chloroplast ultrastructure, pigment content, and composition." *Canadian Journal of Botany* 62, no. 1: 53–60.

Hunter, D. A., S. D. Yoo, S. M. Butcher, and M. T. McManus. 1999. "Expression of 1-aminocyclopropane-1-carboxylate oxidase during leaf ontogeny in white clover." *Plant Physiology* 120, no. 1: 131–142.

Ichimura, K., T. Mizoguchi, R. Yoshida, T. Yuasa, and K. Shinozaki. 2000. "Various abiotic stresses rapidly activate Arabidopsis MAP kinases ATMPK4 and ATMPK6." *The Plant Journal* 24, no. 5: 655–665.

Irigoyen, J. J., D. W. Emerich, and M. Sánchez-Diaz. 1992. "Alfalfa leaf senescence induced by drought stress: photosynthesis, hydrogen peroxide metabolism, lipid peroxidation and ethylene evolution." *Physiologia Plantarum* 84, no. 1: 67–72.

Iuchi, S., M. Kobayashi, T. Taji, M. Naramoto, M. Seki, T. Kato, S. Tabata, Y. Kakubari, K. Yamaguchi-Shinozaki, and K. Shinozaki. 2001. "Regulation of drought tolerance by gene manipulation of 9-cis-epoxycarotenoid dioxygenase, a key enzyme in abscisic acid biosynthesis in Arabidopsis." *The Plant Journal* 27, no. 4: 325–333.

Jeon, J., N. Y. Kim, S. Kim, N. Y. Kang, O. Novák, S.-J. Ku, C. Cho et al. 2010. "A subset of cytokinin two-component signaling system plays a role in cold temperature stress response in Arabidopsis." *Journal of Biological Chemistry* 285, no. 30: 23371–23386.

Jespersen, D., and B. Huang. 2015. "Proteins associated with heat-induced leaf senescence in creeping bentgrass as affected by foliar application of nitrogen, cytokinins, and an ethylene inhibitor." *Proteomics* 15, no. 4: 798–812.

Jespersen, D., J. Zhang, and B. Huang. 2016. "Chlorophyll loss associated with heat-induced senescence in bentgrass." *Plant Science* 249: 1–12.

Jespersen, D., J. Yu, and B. Huang. 2015. "Metabolite responses to exogenous application of nitrogen, cytokinin, and ethylene inhibitors in relation to heat-induced senescence in creeping bentgrass." *PLoS One* 10, no. 3: e0123744.

John, I., R. Drake, A. Farrell, W. Cooper, P. Lee, P. Horton, and D. Grierson. 1995. "Delayed leaf senescence in ethylene-deficient ACC-oxidase antisense tomato plants: molecular and physiological analysis." *The Plant Journal* 7, no. 3: 483–490.

Johnson, P. R., and J. R. Ecker. 1998. "The ethylene gas signal transduction pathway: a molecular perspective." *Annual Review of Genetics* 32, no. 1: 227–254.

Kato, Y., Y. Yamamoto, S. Murakami, and F. Sato. 2005. "Post-translational regulation of CND41 protease activity in senescent tobacco leaves." *Planta* 222, no. 4: 643–651.

Khanna, S., and V. K. Rai. 1998. "Changes in proline level in response to osmotic stress and exogenous amino acids in *Raphanus sativus* L. seedlings." *Acta Physiologiae Plantarum* 20, no. 4: 393–397.

Kidrič, M., J. Sabotič, and B. Stevanović. 2014. "Desiccation tolerance of the resurrection plant *Ramonda serbica* is associated with dehydration-dependent changes in levels of proteolytic activities." *Journal of Plant Physiology* 171, no. 12: 998–1002.

Kim, H. J., H. Ryu, S. H. Hong, H. R. Woo, P. O. Lim, I. C. Lee, J. Sheen, H. G. Nam, and I. Hwang. 2006. "Cytokinin-mediated control of leaf longevity by AHK3 through phosphorylation of ARR2 in Arabidopsis." *Proceedings of the National Academy of Sciences* 103, no. 3: 814–819.

Kim, J. I., A. S. Murphy, D. Baek, S.-W. Lee, D.-J. Yun, R. A. Bressan, and M. L. Narasimhan. 2011. "YUCCA6 over-expression demonstrates auxin function in delaying leaf senescence in *Arabidopsis thaliana*." *Journal of Experimental Botany* 62, no. 11: 3981–3992.

Kovács, Z., L. Simon-Sarkadi, A. Szűcs, and G. Kocsy. 2010. "Differential effects of cold, osmotic stress and abscisic acid on polyamine accumulation in wheat." *Amino Acids* 38, no. 2: 623–631.

Kumar, V., P. Giridhar, A. Chandrashekar, and G. A. Ravishankar. 2008. "Polyamines influence morphogenesis and caffeine biosynthesis in in vitro cultures of *Coffea* canephora P. ex Fr." *Acta Physiologiae Plantarum* 30, no. 2: 217–223.

Lea, P. J., and R. J. Ireland. 1999. "Nitrogen metabolism in higher plants." In *Plant Amino Acids: Biochemistry and Biotechnology*. Marcel Dekker, New York, NY: 1–47.

Lee, I. C., S. W. Hong, S. S. Whang, P. O. Lim, H. G. Nam, and J. C. Koo. 2011. "Age-dependent action of an ABA-inducible receptor kinase, RPK1, as a positive regulator of senescence in Arabidopsis leaves." *Plant and Cell Physiology* 52, no. 4: 651–662.

Lee, O.-R., G. Sathiyaraj, Y.-J. Kim, J.-G. In, W.-S. Kwon, J.-H. Kim, and D.-C. Yang. 2011. "Defense genes induced by pathogens and abiotic stresses in Panax ginseng CA Meyer." *Journal of Ginseng Research* 35, no. 1: 1–11.

Legocka, J., and A. Szweykowska. 1981. "The role of cytokinins in the development and metabolism of barley leaves III. The effect on the RNA metabolism in various cell compartments during senescence." *Zeitschrift für Pflanzenphysiologie* 102, no. 4: 363–374.

Li, X.-P., R. Gan, P.-L. Li, Y.-Y. Ma, L.-W. Zhang, R. Zhang, Y. Wang, and N. N. Wang. 2006. "Identification and functional characterization of a leucine-rich repeat receptor-like kinase gene that is involved in regulation of soybean leaf senescence." *Plant Molecular Biology* 61, no. 6: 829–844.

Li, J. R., K. Yu, J. R. Wei, Q. Ma, B. Q. Wang, and D. Yu. 2010. "Gibberellin retards chlorophyll degradation during senescence of *Paris polyphylla*." *Biologia Plantarum* 54, no. 2: 395–399.

Li, Z., L. Zhang, Y. Yu, R. Quan, Z. Zhang, H. Zhang, and R. Huang. 2011. "The ethylene response factor AtERF11 that is transcriptionally modulated by the bZIP transcription factor HY5 is a crucial repressor for ethylene biosynthesis in Arabidopsis." *The Plant Journal* 68, no. 1: 88–99.

Li, Z., Y. Li, Y. Zhang, B. Cheng, Y. Peng, X. Zhang, X. Ma, L. Huang, and Y. Yan. 2018. "Indole-3-acetic acid modulates phytohormones and polyamines metabolism associated with the tolerance to water stress in white clover." *Plant Physiology and Biochemistry* 129: 251–263.

Li, Z., Y. Peng, and B. Huang. 2016. "Physiological effects of γ-aminobutyric acid application on improving heat and drought tolerance in creeping bentgrass." *Journal of the American Society for Horticultural Science* 141, no. 1: 76–84.

Lim, P. O., I. C. Lee, J. Kim, H. J. Kim, J. S. Ryu, H. R. Woo, and H. G. Nam. 2010. "Auxin response factor 2 (ARF2) plays a major role in regulating auxin-mediated leaf longevity." *Journal of Experimental Botany* 61, no. 5: 1419–1430.

Lin, M., C. Pang, S. Fan, M. Song, H. Wei, and S. Yu. 2015. "Global analysis of the *Gossypium hirsutum* L. Transcriptome during leaf senescence by RNA-Seq." *BMC Plant Biology* 15, no. 1: 43.

Liu, J.-H., H. Kitashiba, J. Wang, Y. Ban, and T. Moriguchi. 2007. "Polyamines and their ability to provide environmental stress tolerance to plants." *Plant Biotechnology* 24, no. 1: 117–126.

Liu, X., and B. Huang. 2002. "Cytokinin effects on creeping bentgrass response to heat stress." *Crop Science* 42, no. 2: 466–472.

Liu, X., and B. Huang. 2000. "Heat stress injury in relation to membrane lipid peroxidation in creeping bentgrass." *Crop Science* 40, no. 2: 503–510.

Lutts, S., J. M. Kinet, and J. Bouharmont. 1996. "NaCl-induced senescence in leaves of rice (Oryza sativa L.) cultivars differing in salinity resistance." *Annals of Botany* 78, no. 3: 389–398.

Ma, X., J. Zhang, and B. Huang. 2016. "Cytokinin-mitigation of salt-induced leaf senescence in perennial ryegrass involving the activation of antioxidant systems and ionic balance." *Environmental and Experimental Botany* 125: 1–11.

Martínez, D. E., M. L. Costa, F. M. Gomez, M. S. Otegui, and J. J. Guiamet. 2008. "'Senescence-associated vacuoles' are involved in the degradation of chloroplast proteins in tobacco leaves." *The Plant Journal* 56, no. 2: 196–206.

Matile, P., M. Schellenberg, and F. Vicentini. 1997. "Localization of chlorophyllase in the chloroplast envelope." *Planta* 201, no. 1: 96–99.

McCabe, M. S., L. C. Garratt, F. Schepers, W. J. R. M. Jordi, G. M. Stoopen, E. Davelaar, J. Hans, A. van Rhijn, J. B. Power, and M. R. Davey. 2001. "Effects of PSAG12-IPT gene expression on development and senescence in transgenic lettuce." *Plant Physiology* 127, no. 2: 505–516.

Murkowski, A. 2001. "Heat stress and spermidine: effect on chlorophyll fluorescence in tomato plants." *Biologia Plantarum* 44, no. 1: 53–57.

Nadler, K., and S. Granick. 1970. "Controls on chlorophyll synthesis in barley." *Plant Physiology* 46, no. 2: 240–246.

Nakano, Y., and K. Asada. 1980. "Spinach chloroplasts scavenge hydrogen peroxide on illumination." *Plant and Cell Physiology* 21, no. 8: 1295–1307.

Nakano, Y., and K. Asada. 1981. "Hydrogen peroxide is scavenged by ascorbate-specific peroxidase in spinach chloroplasts." *Plant and Cell Physiology* 22, no. 5: 867–880.

Nayyar, H., R. Kaur, S. Kaur, and R. Singh. 2014. "γ-Aminobutyric acid (GABA) imparts partial protection from heat stress injury to rice seedlings by improving leaf turgor and upregulating osmoprotectants and antioxidants." *Journal of Plant Growth Regulation* 33, no. 2: 408–419.

Noh, Y., and R. M. Amasino. 1999. "Identification of a promoter region responsible for the senescence-specific expression of SAG12." *Plant Molecular Biology* 41, no. 2: 181–194.

Noodén, L. D., S. Singh, and D. S. Letham. 1990. "Correlation of xylem sap cytokinin levels with monocarpic senescence in soybean." *Plant Physiology* 93, no. 1: 33–39.

Oberhuber, M., J. Berghold, K. Breuker, S. Hörtensteiner, and B. Kraütler. 2003. "Breakdown of chlorophyll: a nonenzymatic reaction accounts for the formation of the colorless 'nonfluorescent' chlorophyll catabolites." *Proceedings of the National Academy of Sciences* 100, no. 12: 6910–6915.

Oh, S. A., J.-H. Park, G. I. Lee, K. H. Paek, S. K. Park, and H. G. Nam. 1997. "Identification of three genetic loci controlling leaf senescence in Arabidopsis thaliana." *The Plant Journal* 12, no. 3: 527–535.

Otegui, M. S., Y.-S. Noh, D. E. Martínez, M. G. V. Petroff, L. A. Staehelin, R. M. Amasino, and J. J. Guiamet. 2005. "Senescence-associated vacuoles with intense proteolytic activity develop in leaves of Arabidopsis and soybean." *The Plant Journal* 41, no. 6: 831–844.

Ouaked, F., W. Rozhon, D. Lecourieux, and H. Hirt. 2003. "A MAPK pathway mediates ethylene signaling in plants." *The EMBO Journal* 22, no. 6: 1282–1288.

Papenbrock, J., E. Pfündel, H.-P. Mock, and B. Grimm. 2000. "Decreased and increased expression of the subunit CHL I diminishes Mg chelatase activity and reduces chlorophyll synthesis in transgenic tobacco plants." *The Plant Journal* 22, no. 2: 155–164.

Pietrini, F., M. A. Iannelli, and A. Massacci. 2002. "Anthocyanin accumulation in the illuminated surface of maize leaves enhances protection from photo-inhibitory risks at low temperature, without further limitation to photosynthesis." *Plant, Cell & Environment* 25, no. 10: 1251–1259.

Pintó-Marijuan, M., and S. Munné-Bosch. 2014. "Photo-oxidative stress markers as a measure of abiotic stress-induced leaf senescence: advantages and limitations." *Journal of Experimental Botany* 65, no. 14: 3845–3857.

Prabhavathi, V. R., and M. V. Rajam. 2007. "Polyamine accumulation in transgenic eggplant enhances tolerance to multiple abiotic stresses and fungal resistance." *Plant Biotechnology* 24, no. 3: 273–282.

Qin, X., and J. A. D. Zeevaart. 1999. "The 9-cis-epoxycarotenoid cleavage reaction is the key regulatory step of abscisic acid biosynthesis in water-stressed bean." *Proceedings of the National Academy of Sciences* 96, no. 26: 15354–15361.

Qiu, K., Z. Li, Z. Yang, J. Chen, S. Wu, X. Zhu, S. Gao et al. 2015. "EIN3 and ORE1 accelerate degreening during ethylene-mediated leaf senescence by directly activating chlorophyll catabolic genes in Arabidopsis." *PLoS Genetics* 11, no. 7: e1005399.

Rai, V. K., and A. Kumari. 1983. "Modulation of membrane permeability by amino acids in Vinca petals." *Experientia* 39, no. 3: 301–303.

Rai, V. K., and U. D. Sharma. 1991. "Amino acids can modulate ABA induced stomatal closure, stomatal resistance and K⁺ fluxes in *Vicia faba* leaves." *Beiträge zur Biologie der Pflanzen* 66: 393–405.

Rivero, R. M., V. Shulaev, and E. Blumwald. 2009. "Cytokinin-dependent photorespiration and the protection of photosynthesis during water deficit." *Plant Physiology* 150, no. 3: 1530–1540.

Roberts, I. N., C. Caputo, M. V. Criado, and C. Funk. 2012. "Senescence-associated proteases in plants." *Physiologia Plantarum* 145, no. 1: 130–139.

Rodoni, S., W. Muhlecker, M. Anderl, B. Krautler, D. Moser, H. Thomas, P. Matile, and S. Hörtensteiner. 1997. "Chlorophyll breakdown in senescent chloroplasts (cleavage of pheophorbide a in two enzymic steps)." *Plant Physiology* 115, no. 2: 669–676.

Rodriguez, M. T., M. P. González, and J. M. Linares. 1987. "Degradation of chlorophyll and chlorophyllase activity in senescing barley leaves." *Journal of Plant Physiology* 129, no. 3–4: 369–374.

Rossi, S., P. Burgess, D. Jespersen, and B. Huang. 2017. "Heat-induced leaf senescence associated with Chlorophyll metabolism in Bentgrass lines differing in heat tolerance." *Crop Science* 57, no. 1: S-169.

Sacher, J. A. 1957. "Relationship between auxin and membrane-integrity in tissue senescence and abscission." *Science* 125, no. 3259: 1199–1200.

Sakurai, N., T. Hayakawa, T. Nakamura, and T. Yamaya. 1996. "Changes in the cellular localization of cytosolic glutamine synthetase protein in vascular bundles of rice leaves at various stages of development." *Planta* 200, no. 3: 306–311.

Santos, C. V. 2004. "Regulation of chlorophyll biosynthesis and degradation by salt stress in sunflower leaves." *Scientia Horticulturae* 103, no. 1: 93–99.

Scandalios, J. G. 1993. "Oxygen stress and superoxide dismutases." *Plant Physiology* 101, no. 1: 7.

Schelbert, S., S. Aubry, B. Burla, B. Agne, F. K. K. Krupinska, and S. Hörtensteiner. 2009. "Pheophytin pheophorbide hydrolase (pheophytinase) is involved in chlorophyll breakdown during leaf senescence in Arabidopsis." *The Plant Cell* 21, no. 3: 767–785.

Shelp, B. J., G. G. Bozzo, C. P. Trobacher, A. Zarei, K. L. Deyman, and C. J. Brikis. 2012. "Hypothesis/review: contribution of putrescine to 4-aminobutyrate (GABA) production in response to abiotic stress." *Plant Science* 193: 130–135.

Sheng, J., Y. Luo, and H. Wainwright. 2000. "Studies on lipoxygenase and the formation of ethylene in tomato." *The Journal of Horticultural Science and Biotechnology* 75, no. 1: 69–71.

Shevyakova, N. I., E. A. Bakulina, and V. V. Kuznetsov. 2009. "Proline antioxidant role in the common ice plant subjected to salinity and paraquat treatment inducing oxidative stress." *Russian Journal of Plant Physiology* 56, no. 5: 663–669.

Skopelitis, D. S., N. V. Paranychianakis, K. A. Paschalidis, E. D. Pliakonis, I. D. Delis, D. I. Yakoumakis, A. Kouvarakis, A. K. Papadakis, E. G. Stephanou, and K. A. Roubelakis-Angelakis. 2006. "Abiotic stress generates ROS that signal expression of anionic glutamate dehydrogenases to form glutamate for proline synthesis in tobacco and grapevine." *The Plant Cell* 18, no. 10: 2767–2781.

Smart, C. M., S. E. Hosken, H. Thomas, J. A. Greaves, B. G. Blair, and W. Schuch. 1995. "The timing of maize leaf senescence and characterisation of senescence-related cDNAs." *Physiologia Plantarum* 93, no. 4: 673–682.

Srivastava, A. K., J. S. Nair, D. Bendigeri, A. Vijaykumar, N. K. Ramaswamy, and S. F. D'Souza. 2009. "Purification and characterization of a salinity induced alkaline protease from isolated spinach chloroplasts." *Acta Physiologiae Plantarum* 31, no. 1: 187.

Steccanella, V., M. Hansson, and P. E. Jensen. 2015. "Linking chlorophyll biosynthesis to a dynamic plastoquinone pool." *Plant Physiology and Biochemistry* 97: 207–216.

Taiz, L., and E. Zeiger. 2002. *Plant Physiology.* Sinauer Associates, Sunderland.

Talla, S. K., M. Panigrahy, S. Kappara, P. Nirosha, S. Neelamraju, and R. Ramanan. 2016. "Cytokinin delays dark-induced senescence in rice by maintaining the chlorophyll cycle and photosynthetic complexes." *Journal of Experimental Botany* 67, no. 6: 1839–1851.

Tan, B. C., S. H. Schwartz, J. A. D. Zeevaart, and D. R. McCarty. 1997. "Genetic control of abscisic acid biosynthesis in maize." *Proceedings of the National Academy of Sciences* 94, no. 22: 12235–12240.

Tanaka, Y., T. Sano, M. Tamaoki, N. Nakajima, N. Kondo, and S. Hasezawa. 2005. "Ethylene inhibits abscisic acid-induced stomatal closure in Arabidopsis." *Plant Physiology* 138, no. 4: 2337–2343.

Tardy, F., and M. Havaux. 1997. "Thylakoid membrane fluidity and thermostability during the operation of the xanthophyll cycle in higher-plant chloroplasts." *Biochimica et Biophysica Acta (BBA)-Biomembranes* 1330, no. 2: 179–193.

Tran, L.-S. P. Urao, T. Qin, F. Maruyama, K. Kakimoto, T., K. Shinozakiand K. Yamaguchi-Shinozaki. 2007. "Functional analysis of AHK1/ATHK1 and cytokinin receptor histidine

kinases in response to abscisic acid, drought, and salt stress in Arabidopsis." *Proceedings of the National Academy of Sciences* 104, no. 51: 20623–20628.

Tewari, A. K., and B. C. Tripathy. 1998. "Temperature-stress-induced impairment of chlorophyll biosynthetic reactions in cucumber and wheat." *Plant Physiology* 117, no. 3: 851–858.

Thomas, H., M. Schellenberg, F. Vicentini, and P. Matile. 1996. "Gregor Mendel's green and yellow pea seeds." *Botanica Acta* 109, no. 1: 3–4.

Urano, K., Y. Yoshiba, T. Nanjo, T. I. K. Yamaguchi-Shinozaki, and K. Shinozaki. 2004. "Arabidopsis stress-inducible gene for arginine decarboxylase AtADC2 is required for accumulation of putrescine in salt tolerance." *Biochemical and Biophysical Research Communications* 313, no. 2: 369–375.

van der Graaff, E., R. Schwacke, A. Schneider, M. Desimone, U.-I. Flügge, and R. Kunze. 2006. "Transcription analysis of Arabidopsis membrane transporters and hormone pathways during developmental and induced leaf senescence." *Plant Physiology* 141, no. 2: 776–792.

Veerasamy, M., Y. He, and B. Huang. 2007. "Leaf senescence and protein metabolism in creeping bentgrass exposed to heat stress and treated with cytokinins." *Journal of the American Society for Horticultural Science* 132, no. 4: 467–472.

Vicentini, F., F. Iten, and P. Matile. 1995. "Development of an assay for Mg-dechelatase of oilseed rape cotyledons, using chlorophyllin as the substrate." *Physiologia Plantarum* 94, no. 1: 57–63.

Vicentini, F., S. Hörtensteiner, M. Schellenberg, H. Thomas, and P. Matile. 1995. "Chlorophyll breakdown in senescent leaves identification of the biochemical lesion in a stay-green genotype of *Festuca pratensis* Huds." *New Phytologist* 129, no. 2: 247–252.

Vierstra, R. D. 2009. "The ubiquitin–26S proteasome system at the nexus of plant biology." *Nature Reviews Molecular Cell Biology* 10, no. 6: 385.

Vijayakumari, K., and J. T. Puthur. 2016. "γ-Aminobutyric acid (GABA) priming enhances the osmotic stress tolerance in *Piper nigrum* Linn. Plants subjected to PEG-induced stress." *Plant Growth Regulation* 78, no. 1: 57–67.

Von Wettstein, D., S. Gough, and C. G. Kannangara. 1995. "Chlorophyll biosynthesis." *The Plant Cell* 7, no. 7: 1039.

Wan, X., A. Mo, S. Liu, L. Yang, and L. Li. 2011. "Constitutive expression of a peanut ubiquitin-conjugating enzyme gene in Arabidopsis confers improved water-stress tolerance through regulation of stress-responsive gene expression." *Journal of Bioscience and Bioengineering* 111, no. 4: 478–484.

Wang, Z.-Q., Y.-Z. Yuan, J.-Q. Ou, Q.-H. Lin, and C.-F. Zhang. 2007. "Glutamine synthetase and glutamate dehydrogenase contribute differentially to proline accumulation in leaves of wheat (*Triticum aestivum*) seedlings exposed to different salinity." *Journal of Plant Physiology* 164, no. 6: 695–701.

Wingler, A., A. von Schaewen, R. C. Leegood, P. J. Lea, and W. P. Quick. 1998. "Regulation of leaf senescence by cytokinin, sugars, and light: effects on NADH-dependent hydroxypyruvate reductase." *Plant Physiology* 116, no. 1: 329–335.

Xiao, H.-J., K.-K. Liu, D.-W. Li, M. H. Arisha, W.-G. Chai, and Z.-H. Gong. 2015. "Cloning and characterization of the pepper CaPAO gene for defense responses to salt-induced leaf senescence." *BMC Biotechnology* 15, no. 1: 100.

Xiao, H.-J., Y.-X. Yin, W.-G. Chai, and Z.-H. Gong. 2014. "Silencing of the CaCP gene delays salt-and osmotic-induced leaf senescence in *Capsicum annuum* L." *International Journal of Molecular Sciences* 15, no. 5: 8316–8334.

Xiong, L, and J.-K. Zhu. 2003. "Regulation of abscisic acid biosynthesis." *Plant Physiology* 133, no. 1: 29–36.

Xu, F., T. Meng, P. Li, Y. Yu, Y. Cui, Y. Wang, Q. Gong, and N. N. Wang. 2011. "A soybean dual-specificity kinase GmSARK and its Arabidopsis homologue regulate leaf senescence through synergistic actions of auxin and ethylene." *Plant Physiology* 157 no. 4: 2131–2153.

Xu, Y., J. Tian, T. Gianfagna, and B. Huang. 2009. "Effects of SAG12-ipt expression on cytokinin production, growth and senescence of creeping bentgrass (*Agrostis stolonifera* L.) under heat stress." *Plant Growth Regulation* 57, no. 3: 281.

Yamaguchi, K., Y. Takahashi, T. Berberich, A. Imai, T. Takahashi, A. J. Michael, and T. Kusano. 2007. "A protective role for the polyamine spermine against drought stress in Arabidopsis." *Biochemical and Biophysical Research Communications* 352, no. 2: 486–490.

Yamasaki, H., H. Uefuji, and Y. Sakihama. 1996. "Bleaching of the red anthocyanin induced by superoxide radical." *Archives of Biochemistry and Biophysics* 332, no. 1: 183–186.

Yang, S.-D., P. J. Seo, H.-K. Yoon, and C.-M. Park. 2011. "The Arabidopsis NAC transcription factor VNI2 integrates abscisic acid signals into leaf senescence via the COR/RD genes." *The Plant Cell* 23 no. 6: 2155–2168.

Yiu, J.-C., L.-D. Juang, D. Y.-T. Fang, C.-W. Liu, and S.-J. Wu. 2009. "Exogenous putrescine reduces flooding-induced oxidative damage by increasing the antioxidant properties of Welsh onion." *Scientia Horticulturae* 120, no. 3: 306–314.

Young, T. E., R. B. Meeley, and D. R. Gallie. 2004. "ACC synthase expression regulates leaf performance and drought tolerance in maize." *The Plant Journal* 40, no. 5: 813–825.

Zhang, H., J. Li, J.-H. Yoo, S.-C. Yoo, S.-H. Cho, H.-J. Koh, H. S. Seo, and N.-C. Paek. 2006. "Rice Chlorina-1 and Chlorina-9 encode ChlD and ChlI subunits of Mg-chelatase, a key enzyme for chlorophyll synthesis and chloroplast development." *Plant Molecular Biology* 62, no. 3: 325–337.

Zhang, H. J., H. Z. Dong, W. J. Li, and D. M. Zhang. 2012. "Effects of soil salinity and plant density on yield and leaf senescence of field-grown cotton." *Journal of Agronomy and Crop Science* 198, no. 1: 27–37.

Zhang, K., and S.-S. Gan. 2012. "An abscisic acid-AtNAP transcription factor-SAG113 protein phosphatase 2C regulatory chain for controlling dehydration in senescing Arabidopsis leaves." *Plant Physiology* 158, no. 2: 961–969.

Zhang, X., Z. Zhang, J. Li, L. Wu, J. Guo, L. Ouyang, Y. Xia, X. Huang, and X. Pang. 2011. "Correlation of leaf senescence and gene expression/activities of chlorophyll degradation enzymes in harvested Chinese flowering cabbage (*Brassica rapa* var. parachinensis)." *Journal of Plant Physiology* 168, no. 17: 2081–2087.

Zhou, G.-A., R.-Z. Chang, and L.-J. Qiu. 2010. "Overexpression of soybean ubiquitin-conjugating enzyme gene GmUBC2 confers enhanced drought and salt tolerance through modulating abiotic stress-responsive gene expression in Arabidopsis." *Plant Molecular Biology* 72, no. 4–5: 357–367.

5 Mechanisms of Salt Tolerance in Submerged Aquatic Macrophytes

Rout Nutan Prasad, Shaw Birendra Prasad, and Rodríguez-Garay Benjamín

CONTENTS

5.1 INTRODUCTION

Salinity is well known among the environmental or abiotic stresses that affect the normal physiological functioning of an organism. The physiological and biochemical disturbances imposed by salinity in plant systems result in a huge loss in agricultural production worldwide (Munns and Tester, 2008). In terms of the magnitude of the salinity problem in agriculture, it has been estimated that more than 80 million ha of agricultural lands, representing 40% of total irrigated land worldwide, are affected because of salinization (Xiong and Zhu, 2001), and more than 50% of all arable lands are likely to suffer from soil salinization due to irrigation and climate change by 2050 (Wang et al., 2003). Among the crops, the cultivation of rice, which is grown as a paddy crop worldwide, is the most severely affected by salinity. The reason for the severe damaging effect of salinity on rice cultivation could be

because of its semi-aquatic nature, requiring a large amount of water for vegetative growth compared with other crops. The adverse effect of salinity on rice cultivation is against the backdrop of the requirement for its increase in production to 800 million tons from the current figure of ~500 million tons (Virk et al., 2004) to meet the demand of an estimated 9.6 billion people by 2050 (FAO, 2009). In fact, the demand for an increase in production of all crops is increasing day by day, and it will be necessary to increase agricultural production by 87% by the year 2050 (Kromdijk and Long, 2016); therefore, solutions are required to the agricultural loss caused by salinity.

Salinity is a condition of excess salt in the soil or in the water. Among the abiotic stresses, salinity is the most complex in action, as it involves both ionic and osmotic components that increase its potential toxicity compared with other environmental factors. In terms of chemistry, *salt* refers to the crystals of an electrolyte that when dissolved in water, break down into a cation, represented by a metal ion, and an anion, in the form of simple or compound ions of non-metals. Salts are widely present on the earth. The best-known example is the salt present in the sea. The landmass is also contaminated with salt to various degrees (Pasternak, 1987; Szabolcs, 1994). In seawater, the cationic species are mostly represented by Na^+, constituting nearly two-thirds of the total positive mass. This is followed by Mg^{2+}, Ca^{2+}, and K^+ in decreasing order. The anion species are constituted by Cl^- and to a negligible extent by SO_4^{2-}. As far as the landmass is concerned, both cationic and anionic constituents vary greatly. In landmass flooded by seawater, salinity is mainly constituted by Na^+ and Cl^-. Sodic soil, on the other hand, has comparatively greater representation of Ca^{2+}, Mg^{2+}, and SO_4^{2-}. Their relative levels, however, vary greatly depending on the soil. The level of Na^+ nevertheless remains the highest among the cations. In contrast, among the anions, the major partner could be either of the two, Cl^- or SO_4^{2-} (Curtion et al., 1993; Egan and Unger, 1998).

Apart from primary saline soil, secondary salinity is the major threat to the agricultural process, and increases with each and every irrigation of the agricultural land. It has been estimated that out of the total area of human-induced degraded soil, which was 1964 mha, the degradation due to salinity was 77 mha (Table 5.1) (Ghassemi et al., 1995; Munns and Richards, 1996). Munns and Tester (2008) also reported that as many as 1.5 million ha of agricultural land are taken out of production each year as a result of secondary salinization. Plants have developed a variety of powers of adaptation to

abiotic stresses, including salinity stress, involving cascades of signaling networks, which ultimately lead to changes in the expression of many genes determining the fate of cellular, biochemical, and physiological processes important from the viewpoint of tolerance to the stress (Vinocur and Altman, 2005). However, the mechanistic details of salt tolerance are exceedingly complex and defy facile definition, although many genes encoding effector and regulatory proteins have been identified, including those involved in the maintenance of ion homeostasis, osmotic protection through compatible solute accumulation, protection against oxidative stress, polyamine accumulation, and transcriptional regulation (Gupta and Huang, 2014). Thus, what constitutes tolerance or resistance to salinity appears to have many facets: water imbalance, ion homeostasis, nutritional disorders, osmotic adjustment, generation of oxidative radicals, membrane disorder, and genotoxicity, among many others (Hasegawa et al., 2000; Munns, 2002; Zhu, 2007; Carillo et al., 2011). All of them are equally important in enabling plants to tolerate a saline environment, but certainly, the resistance that a plant shows to salinity is not a result of additive effects of these physiological/biochemical processes; it is probably their coordinated functioning that imparts the tolerance. The involvement of all of them is probably a must for making a plant tolerant to salinity. The level of complexity making a plant tolerant to salinity may, however, be dependent on the habitat it is colonizing. In terrestrial plants, the complexity could be greater than in aquatic plants, as in the former, only the roots are exposed to salinity, while in the latter, the whole plant is exposed to the stress.

Aquatic macrophytes are defined as aquatic plants that can be observed by the naked eye and may be floating, submerged, or emergent (Hasan and Rina, 2009). Aquatic plants contribute to the maintenance of key functions of ecosystems, including biodiversity, and cater to many needs of human society (Bornette and Puijalon, 2009). They have a wide range of habitats, including estuaries, rivers, lakes, ponds, natural depressions, and ditches, and are used for animal and human food, filtration of water, extraction of value-added products, reduction of the nutrient load of water, and hindrance to the movement of fish (Carpenter and Lodge, 1986; Brix, 1994; Burks et al., 2006; Jeppesen et al., 2012; Brogan and Relyea, 2015, 2017). Apart from this, they are excellent models for the study of salinity stress, as the whole plant is submerged or fully exposed to stress conditions and experiences the variable saline environment constantly, compared with terrestrial plants, in which only the roots are exposed to the saline environment (Figure 5.1). The aim of this chapter is to present the information available on the mechanistic details of salt tolerance in aquatic macrophytes, especially in submerged aquatic macrophytes.

TABLE 5.1
Global Estimate of Secondary Salinization in the World's Irrigated Soils

Country	Total Land Area Cropped (Mha)	Area Irrigated		Salt-affected Irrigated Land	
		Mha	%	Mha	%
China	97	45	46	6.7	15
India	169	42	25	7.0	17
Soviet Union	233	21	9	3.7	18
United States	190	18	10	4.2	23
Pakistan	21	16	78	4.2	26
Iran	15	6	39	1.7	30
Thailand	20	4	20	0.4	10
Egypt	3	3	100	0.9	33
Australia	47	2	4	0.2	9
Argentina	36	2	5	0.6	34
South Africa	13	1	9	0.1	9
Subtotal	843	159	19	29.6	20
World	1474	227	15	45.4	20

Source: from Ghassemi, F., Jakeman, A.J., and Nix, H.A., *Salinisation of Land and Water Resources: Human Causes, Extent, Management and Case Studies*, UNSW Press, Sydney, Australia, and CAB International, Wallingford, UK, 1995. With permission.

5.2 STRESS CONCEPT AND SALINITY STRESS

5.2.1 CONCEPT OF STRESS

The concept of stress, as described by Selye (1950), is the "state of the whole organism or its partial mobilization affected by a strong external action." In general, for plants, anything that

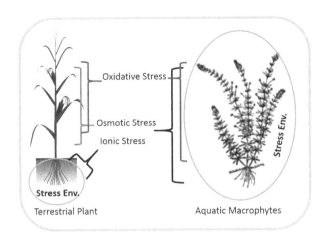

FIGURE 5.1 Saline exposure and salinity impact on terrestrial and aquatic plants.

affects or restricts their metabolism, growth, and development is regarded as stress. If the stress is living in nature, such as pathogens, insects, etc., then it is termed *biotic stress*, and when a non-living cause provokes the stress, the stress is called *abiotic stress*. Abiotic stresses are prevalent in nature, and some of the most important among these are wind, drought, salinity, heat, light, etc. All the environmental stresses significantly affect agricultural production in general. Plants, by virtue of their stationary status, cannot migrate to avoid unfavorable fluctuations in their environment as animals can, and hence, they must change their metabolic activity suitably to cope with the changing environment or perish. The resulting changes in their metabolism, called *stress responses*, may enable the plant to survive under the condition of stress, either for a short time only, known as *acclimation*, or the changes induced may be good enough to support continuous growth of the plant, known as *adaptation*. Lichtenthaler (1998) differentiated stress responses among plants into four phases: (1) the response phase, referring to the deviation from normal function, decline in activity, and imbalance of metabolism; (2) the restitution phase, referring to the initiation of adaptation and repair processes and reactivation of normal activity; (3) the stage of exhaustion (long-term stress), during which the plant becomes exhausted because of the intensity of the stress and unable to adapt, resulting in chronic damage, and (4) the regeneration phase, referring to full or partial regeneration and restoration of the physiological function when the stress is removed.

5.2.2 SALINITY STRESS

Plants require at least some salts for growth, and based on their requirement of salt for healthy growth, they have been classified into glycophytes and halophytes. Glycophytes do not survive under high-salinity conditions, whereas halophytes require the presence of salt in their environment for healthy growth. The actual concept of glycophytes and halophytes is, however, a little different: glycophytes are plants of non-saline or "sweet" environments and tend to accumulate

more K+ (glycophites), showing a high K+/Na+ ratio, whereas halophytes are plants of saline environments and accumulate more Na+ (halophiles) and thus, show a low K+/Na+ ratio. The requirement of glycophytes for salt is 10 mM or less, whereas for halophytes, the requirement varies from 20 to 500 mM (Flowers et al., 1977).

One of the oldest attempts to explain the mechanism of salt toxicity in the plant was that postulated at the end of the nineteenth century by Schimpes, who said that an excess of salts in the growth medium affects plant growth in the same way as drought stress, and this was termed *physiological drought stress* or *physiological dryness*. This, however, was questioned in the 1980s when the work related to salt toxicity really progressed. Nevertheless, this physiological dryness or salt-induced water deficit is still considered to be among the critical factors resulting in toxicity (Greenway and Munns, 1983; Jones, 1985; Kalaji and Pietkiewitcz, 1993).

There are many postulations about the mechanism of salt stress effects in plants (Jones, 1985; Kalaji and Pietkiewitcz, 1993). These, however, may be categorized into three types: 1) primary, which pertains to ion imbalance; 2) secondary, which is the result of a deficiency of essential ions and induction of water and osmotic stress, and 3) tertiary, which reflects toxicity, leading to death (Figure 5.2).

Salinity may affect a plant at various stages of its life, at both morphological and cellular levels, reducing plant growth and yield (Francois, 1994; Wilson et al., 2000; Munns and Tester, 2008; Liang et al., 2014, 2018), germination of seeds (Catalan et al., 1994; Lutts et al., 1995; Meloni et al., 2008), and growth of roots and shoots (Kafkafi and Bernstein, 1996; Acosta-Motos et al., 2015a). Among the physiological processes, both photosynthesis and respiration, the basic life processes in plants, are significantly affected by salt treatment (Brown et al., 1987; Belkhodja et al., 1999; Mäkelä et al., 1999; Munns et al., 2006; Hodges et al., 2016). Osmoregulation, which is the regulation of osmotic potential within a cell required for proper functioning of the biochemical process (Morant-Avice et al., 1998; Reddy and Iyenger, 1999; Munns, 2005; Tang et al., 2015a), is also severely affected by salt stress. Plants maintain the optimal cellular osmotic potential through osmotic adjustment brought about by the accumulation of organic molecules (Horie et al., 2011), which may be osmolytes such as organic acids, nitrogen compounds, and carbohydrates (Flower et al., 1977). During salinity stress, there occurs a decrease in total organic acids (Parez-Alfocea et al., 1994; Acosta-Motos et al., 2017) and an increase of proline and glycine betaine contents and the activities of proline and glycine betaine biosynthesizing enzymes (LaRosa et al., 1991; Williamson and Slocum, 1992; Rhodes and Hanson, 1993; Zhang et al., 1995; Santa-Cruz et al., 1997; Yoshiba et al., 1997; Ashraf and Fooland, 2007; Szabados and Savoure, 2010; Tang et al., 2015b).

The involvement of carbohydrates in the salt stress response is relatively less well studied (Karsten et al., 1991; Perez-Alfoceae and Larther, 1995; Hodges et al., 2016). One of the well-studied responses to salinity stress is that of the antioxidative machinery. The existing reports suggest that H_2O_2 and O^{2-} play an essential role in NaCl injury (Singha and

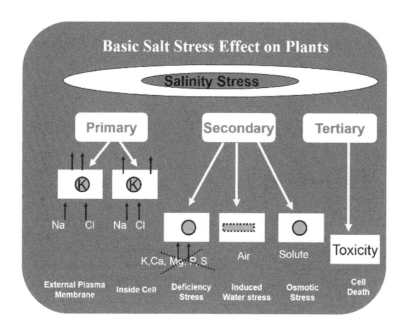

FIGURE 5.2 Basic salt stress effect on plants.

Choudhuri, 1990; Asada, 1992; Wang et al., 2016). The antioxidative enzymes such as superoxide dismutase (SOD), catalase, monodehydroascorbate reductase, ascorbate peroxidase (APX), and glutathione reductase (GR) constitute the antioxidative machinery, which is responsible for the removal of oxidant molecules such as H_2O_2 and O^{2-} (Tanaka et al., 1999; Kangasjärvi et al., 2012; Acosta-Motos et al., 2015b; Wang, et al., 2016).

5.3 AQUATIC MACROPHYTES

Aquatic macrophytes are photosynthetic plants in aquatic systems with a well-developed root and shoot system. These may be submerged, emergent, or free-floating. In all the categories, many aquatic macrophytes have been reported. Most of them are rooted, with the vegetative parts predominantly submerged, in both marine and freshwater ecosystems (Hasan and Rina, 2009), and their identification and taxonomical characterization are based on the manual by Fassett (2006).

5.3.1 Distribution and Importance

The aquatic macrophytes are distributed all over the world, except in some very deep and cold-water lakes in polar environments. Their distribution predominates in coastal lagoons and shallow water bodies along with freshwater lakes and rivers (Barnes, 1980; Kjerfve, 1994). Most of them belong to the families Ceratophyllaceae, Haloragaceae, Hydrocharitaceae, Nymphaeaceae, and Potamogetonaceae (Fassett, 2006; Hasan and Rina, 2009) (Table 5.2).

The distribution of aquatic macrophytes depends on two major factors: water quality and quantity. The depth of the water has the same importance as salinity, temperature, water current, sediment, nutrients, human activity, etc. Their distribution is also temporal because of changes in the flow of

current, sediment, nutrients, and salinity. It has been demonstrated that salinity is the primary variable for the distribution and abundance of macrophytes in water bodies (Lirman et al., 2008; Obrador and Pretus, 2008; Schubert et al., 2011). Changes in turbidity, sedimentation, and total nitrogen content are due to wind and water current or water discharge, and these comprise the second level of important factors that determine the availability and distribution of aquatic macrophytes in water bodies, as they reduce light availability for plants (Charpentier et al., 2005; Lawson et al., 2007; Millet et al., 2010). Since the availability of light at the bottom of water bodies is inversely related to their depth, the latter also greatly influences the availability and distribution of aquatic macrophytes, particularly submerged macrophytes (Vincent et al., 2006; Christia and Papastergiadou, 2007). Eutrophication is another factor that plays an important role in the availability and distribution of macrophytes, as nutrients support their growth (Leruste et al., 2016). As macrophytes experience varied salinity conditions in different habitats, the mechanism by which salt tolerance is achieved is very different among them. Besides, a macrophyte may face varied salinity conditions with seasonal variations in rainfall and temperature, and hence, the mechanism of salt tolerance is very different in such a plant than in one facing a uniform salinity regime during the whole year (Figure 5.3).

Aquatic plants contribute greatly to the maintenance of essential functions and related biodiversity in aquatic ecosystems and take care of the needs of human societies (Bornette and Puijalon, 2009). They are very important in supporting fisheries and are also used for aquaculture exploitation, recreation, and attracting tourism (Perez-Ruzafa et al., 2011). They also serve as feed for livestock and as sources of valuable carbohydrates, pigments, and polyunsaturated fatty acids (CEVA, 2005; Evans and Critchley, 2014), feed additives (Allen et al., 2001), and shrimp feeds (CEVA, 2005).

TABLE 5.2
Distribution of Some Submerged Aquatic Macrophytes

Scientific Name	Family	Common Name	Distribution (Predominate)	Reference
Blyxa lancifolia	Hydrocharitaceae	Blyxa	Japan, Australia Taiwan, China, Indonesia.	Govaerts (2018)
Cabomba caroliniana	Nymphaeaceae	Fanwort	United States, Australia, Japan, Netherlands.	Hassler (2018)
Ceratophyllum demersum/C. submersum	Ceratophyllaceae	Hornwort/Coontail	Netherlands, Germany, France, Sweden.	Hassler (2018)
Chara sp.	Characeae	Chara	Netherlands, Sweden, Germany, France, United States.	Beaune et al. (2018)
Elodea canadensis	Hydrocharitaceae	Canadian pondweed	United Kingdom, Netherlands, Germany, Sweden, France.	Govaerts (2018)
E. densa	Hydrocharitaceae	Brazilian pondweed	Japan, Brazil, United States, Germany, Russia.	Govaerts (2018)
E. michx	Hydrocharitaceae	Pondweed	Netherlands, United Kingdom, Germany, France.	Govaerts (2018)
Hydrilla verticillata	Hydrocharitaceae	Oxygen weed	United States, Korea, Australia, Japan, United Kingdom.	Govaerts (2018)
Myriophyllum aquaticum	Haloragaceae	Water milfoil	United States, Belgium, United Kingdom, France.	Hassler (2018)
M. exalbescens	Haloragaceae	Water milfoil	Canada, United States, Greenland, Sweden, Mexico.	Hassler (2018)
M. spicatum	Haloragaceae	Eurasian water milfoil	Netherlands, Germany, France, Sweden, United Kingdom.	Hassler (2018)
Najas guadalupensis	Hydrocharitaceae	Water velvet/ Najas	Japan, Indonesia, Taiwan, Philippines, India.	Govaerts (2018)
N. indica	Hydrocharitaceae	Water velvet/ Najas	Japan, Indonesia, Taiwan, Philippines, India	Govaerts (2018)
N. guadalupensis	Hydrocharitaceae	Water velvet/ Najas	United States, Mexico, Canada, Costa Rica, Brazil.	Govaerts (2018)
N. marina	Hydrocharitaceae	Water velvet/ Najas	France, Germany, Sweden, Netherlands, Finland.	Govaerts (2018)
Ottelia alismoides	Hydrocharitaceae	Ottelia	Japan, Australia, China, Korea.	Govaerts (2018)
Potamogeton crispus	Potamogetonaceae	Curlyleaf pondweed	Germany, Netherlands, France, United Kingdom, United States.	Govaerts (2018)
P. gramineus	Potamogetonaceae	Pondweed	Sweden, United Kingdom, Finland, Norway.	Govaerts (2018)
P. nodosus	Potamogetonaceae	Longleaf pondweed	France, United States, Spain, Germany, Netherlands	Govaerts (2018)
P. pectinatus	Potamogetonaceae	Sago pondweed	Netherlands, Germany, France, Belgium, Spain.	Govaerts (2018)
Ruppia maritima	Potamogetonaceae	Ruppia	Sweden, United Kingdom, Norway, United States, Spain.	Govaerts (2018)
Utricularia bulgaria	Nymphaeaceae	Bladderwort	Germany, Sweden, Netherlands, Finland, France.	Hassler (2018)
Vallisneria americana	Hydrocharitaceae	Eelgrass	United States, Canada, Australia, Mexico.	Govaerts (2018)
V. spiralis	Hydrocharitaceae	Eelgrass	France, Australia, Germany, New Zealand.	Govaerts (2018)

Source: Information based on Hasan, M.R. and Rina, C, Use of algae and aquatic macrophytes as feed in small-scale aquaculture: a review (No. 531), Food and Agriculture Organization of the United Nations (FAO), 2009 (www.gbif.org/species/search).

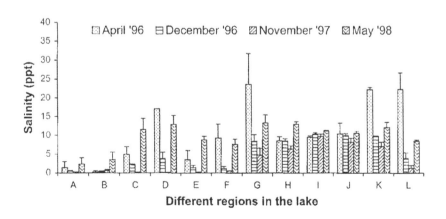

FIGURE 5.3 Spatial and temporal distribution of aquatic macrophytes in Chilika Lake, India. A to L are different spatial locations in the lake.

As the salinity problem has always been seen vis-à-vis agricultural losses, and as agricultural plant species are invariably terrestrial in nature, experiments related to salt tolerance have mostly been carried out in these plants. Work on aquatic macrophytes, in which salinity tolerance is more prevalent than in terrestrial plants, has been scant. The responses of terrestrial plants to salt treatment, however, could be complex in nature, since the organ of exposure is only the root, but the effect is also carried to the shoot and the rest of the plant; hence, the result could be misleading. Therefore, submerged aquatic macrophytes are the best models for studying the effects of salinity and the tolerance mechanisms in plants.

5.3.2 Salinity Stress in Aquatic Macrophytes

Macrophytes are aquatic angiosperms, but those that are submerged are also represented by macroalgae. These are very important for maintaining the biological quality of water bodies and for determining the ecological status of transitional waters, including coastal lagoons (Le Fur et al., 2018). Their spatial and temporal distribution depends greatly on the salinity condition of the water bodies (Bertness et al., 1992; Shaw et al., 2000; Sim et al., 2006; Watt et al., 2007; Goodman et al., 2010). Salinity greatly influences the growth and physiological processes of macrophytes (Haller et al., 1974; Warwick and Bailey, 1997; Rout and Shaw, 2001a; Macek and Rejmankova, 2007; Abraham, 2010; Izzati, 2016), including seed production (Kim, 2013), net photosynthesis (Jampeetong and Brix, 2009), and respiration (Lauer et al, 2011). Salinity also induces oxidative stress and lipid peroxidation in macrophytes (Jampeetong and Brix, 2009; Abraham and Dhar, 2010).

5.3.2.1 Effect on Growth

Reduction in the growth rate of both freshwater and saline water macrophytes has been reported on exposure to increasing salinity (Rozema, 1976; Van der Brick and Van der Velde, 1993; Adams and Bale, 1994). Like terrestrial glycophytes, freshwater macrophytes also show a wide range of salt sensitivity (James and Hart, 1993; Van der Brick and Van der Velde, 1993), but salinity above 1000 mg/L^{-1}, in general, is detrimental to them (Hart et al., 1990, 1991; Hart and James, 1993). Only 52% of plants (*Eleocharis acuta*, *Potamogeton tricarinatus*, and *Triglochin procera*) survived after 72 days of exposure to salinity of 7000 mg L^{-1} (Hart and James, 1993). It has also been documented that growth and biomass production are reduced in *Vallisneria americana* (Boustany et al., 2010) and *Potamogeton* species with decreased flowering (Van den Brink and Van der Velde, 1993). The growth rate is also reduced in *Spirodela polyrrhiza* (Leblebici et al., 2011), *Hydrilla verticillata* (Panda and Khan, 2004), and *Chara aspera* (Blindow et al., 2003) on exposure to salinity. Salinity also affects the length and number of roots and shoots (Haller et al., 1974; Upadhyay and Panda, 2005; Boustany et al., 2015; Borgnis and Boyer, 2016; Regan, 2017). Similarly to terrestrial plants, reduction in the growth of aquatic plants in response to salinity treatment is accompanied by an increase in root:shoot ratio (Ven den Brink and Van der Velde, 1993).

5.3.2.2 Effect on Photosynthesis and Respiration

Photosynthesis, the biochemical process of CO_2 assimilation, is known to be sensitive to many environmental stresses (Dubey, 1977). The process is essential and unique to plants, as it provides the substrates to drive all other metabolic activities, the cumulative effect of which leads to the growth of the plant. In fact, growth and photosynthesis are intimately related. The effect of environmental stress on the growth and final yield is generally expected to be first reflected by some alterations in the composition and/or levels of photosynthetic pigments. Pigments are primary components for photosynthesis. The levels of chlorophylls have been reported to both increase (Morale et al., 1992; Gonzalez and Moreno, 1997) and decrease (Salama et al., 1994; Del Zoppo et al., 1999) in aquatic macrophytes in response to exposure to salinity. Reports on the changes in relative levels of the pigments (Chl *a* to Chl *b*) in response to salinity are also variable, with some work showing an increase (Salama et al., 1994) and some showing no change (Morals et al., 1992; Del Zoppo et al., 1999) irrespective of the natural tolerance of the plants to salinity. A decrease in the level of chlorophylls has been observed in response to salinity in aquatic macrophytes such as *Hydrilla verticillata* (Twilley and Barko, 1990; Rout et al., 1997), *Myriophyllum spicatum*, *Potamogeton perfoliatus*, and *V. americana* (Twilley and Barko, 1990; Boustany et al., 2010), *Najas indica* and *N. graminea* (Rout et al., 1997), *Salvinia natans* (Jampeetong and Brix, 2009), *Pistia stratiotes* and *Salvinia molesta* (Upadhyay and Panda, 2005), *Lythrum salicaria* (Zhao et al., 2014), *Elodea canadensis*, *Myriophyllum aquaticum* and *Ludwigia grandiflora* (Thouvenot et al., 2015), *Ceratophyllum demersum* (Dogan and Demirors, 2018), and *Myriophyllum spicatum* (Gomes et al., 2017). On the other hand, *Salvinia minima* showed no significant changes in Chl *a* and Chl *b* concentrations on exposure to salinity (Al-Hamdani, 2008). However, the effect of salinity on various photosynthetic pigments is not uniform. This is reflected in reports on the effect of salinity on the ratios of various pigments. Exposure to salinity has been found to lead to an increase in the Chl *a*/Chl *b* ratio but a decrease in the chlorophyll/carotenoid ratio in *Chara* (Rout et al., 1997; Blindow et al., 2003). The difference in the effect of salinity on the levels of various photosynthetic pigments could be because of the difference in the roles they play in photosynthesis and their association with the light-harvesting complex. It is well known that carotenoids have a protective function against photooxidation (Krinsky, 1984; Siefermann-Harms, 1987), and although their concentration level decreases during salt stress (Upadhyay and Panda, 2005; Thouvenot et al., 2015; Gomes et al., 2017; Dogan and Demirors, 2018), the level increases relative to that of chlorophylls. In fact, carotenoids play the role of a quencher of excitation energy under conditions of inhibited carbon reduction (Barber and Anderson, 1992).

In plants, including aquatic macrophytes, reduction of photosynthesis in response to high salinity is a considerable stress factor (Haller et al., 1974; Rout and Shaw, 2001a; Blindow et al., 2003; Jampeetong and Brix, 2009). The inhibition of photosynthesis could be because of salt accumulation in leaves and due to stomatal closure induced by the reduction of water potential (Munns, 2002; Jampeetong and Brix, 2009). Aquatic macrophytes have a less significant change in photosynthetic activity compared with terrestrial plants and hence, are able to tolerate changes of salinity regime without significant photosynthetic stress symptoms (Rout and Shaw, 2001a; Thouvenot et al., 2012).

Respiration is an essential physiological process involving the breakdown of carbohydrate with simultaneous consumption of O_2 and production and release of CO_2. The response of the respiratory process to salinity in plants, particularly aquatic macrophytes, is, however, not very clear. This is because so far, 37% of studies have reported an increase, 34% have reported decreases, and 29% have reported no influence on the respiratory rate in response to salinity (Jacoby et al., 2011). In the aquatic macrophyte *Vallisneria americana*, the respiratory demand increases to nearly double the normal rate (Lauer et al., 2011).

5.3.2.3 Induction of Osmotic Stress and Ion Imbalance

A plant experiencing salt stress suffers in two ways: (1) osmotic stress (reduced water availability) and (2) ion imbalance (accumulation of inorganic ions, mostly Na^+). Excess Na^+ induces dehydration, which can damage proteins and membranes. An excess of ions also results in ion displacement, in which the accumulating ions displace inorganic ions in the proteins and membranes, resulting in changes of their properties, which renders them non-functional, reduces their functional potential, or alters their function *per se* (Garcia et al., 1997). Na^+ and Cl^- are essential for plants, but excessive Na^+ and Cl^- concentrations compete with other essential ions and directly affect their absorption (Stoeva and Kaymakanova, 2008; Iqbal et al., 2015; Abdallah et al., 2016), reducing stomatal opening and thereby, leading to decreases in intracellular CO_2 concentration (Munns and Tester, 2008).

Aquatic macrophyte species exhibit very different patterns of ion accumulation. Some species are unaffected by salinity. Na^+ and K^+ ion contents determined for leaves of different ages showed species-specific variations in Na^+/K^+ ratios, which in general, increased more in halophytes than in glycophytes. The halophytes may be capable of absorbing Na^+ into leaf vacuoles, the osmotic potential of which could be balanced by an increase in the concentration of organic compatible solutes in the cytoplasm (Warwick and Bailey, 1997; Rout and Shaw, 2001a; Masood et al., 2006; van Kempen et al., 2013). The reports available show the level of Ca, K, and Mg to decrease with increasing NaCl concentration in *Salvinia auriculata* (Gomes et al., 2017) and *S. natans* (Jampeetong and Brix, 2009). A supplement of essential ions, such as Ca^{2+}, however, reduces the increase in uptake of Na^+ and the ratio of Na^+/K^+ (Rout and Shaw, 2001a). It is an established fact that failure to maintain a favorable K^+ to Na^+ ratio can inhibit enzyme functions in plants (Greenway and Munns, 1983), including aquatic macrophytes (Abraham and Dhar, 2010). High salinity stress results in the development of toxicity symptoms,

such as reduced absorption of the macronutrients N, P, and K and the micronutrients Zn, Cu, and Fe and enhanced Mn^{2+} content in the macrophyte *Ceratophyllum demersum* (Dogan and Demirors, 2018).

5.3.2.4 Induction of Oxidative Stress

Salinity-induced oxidative damage to the plants occurs through reactive oxygen species (ROS) (Rout and Shaw, 2001b; He and Zhu, 2008; Aghaleh et al., 2009). H_2O_2 accumulation has been considered as a sign of oxidative stress, as its reaction with $O_2\bullet-$ leads to the formation of the highly reactive hydroxyl radical ($\bullet OH$), causing peroxidative damage to biomolecules (Shaw et al., 2004; Kim et al., 2005). The NaCl-induced increase in the cellular H_2O_2 content in many aquatic macrophytes is circumstantial evidence of the generation of ROS under salt stress (Shaw et al., 2004; Upadhyay and Panda, 2005; Mallik et al., 2011). The generation of ROS must be solely due to the disturbances in the chloroplast metabolism caused by NaCl (Rout and Shaw, 2001b). Malondialdehyde (MDA) is the result of oxidative damage to the membrane through lipid peroxidation, and lipid peroxidation levels increased significantly in many aquatic macrophytes during exposure to salinity (Upadhyay and Panda, 2005; Lauer et al, 2011; Zhao et al., 2014; Dogan and Demirors, 2018).

5.4 SALINITY TOLERANCE MECHANISM IN AQUATIC MACROPHYTES

Plants have very distinct abilities to tolerate salinity stress; they are classified based on their tolerance as either glycophytes or halophytes (Acosta-Motos et al., 2017). The mechanism of salt tolerance in aquatic macrophytes includes ion compartmentation and/or exclusion, osmotic adjustment, and antioxidant defenses (Munns, 2005; Abraham and Dhar, 2010; van Kempen et al., 2013; Tang et al., 2015a).

5.4.1 Ionic Regulation and Interaction

One of the most important aspects of salt tolerance is the regulation of inorganic ion transport across the membrane (Flowers et al., 1977; Munns et al., 2006; Hamed et al., 2007). The regulation of ion transport across the plasma membrane in plants is to limit the salt uptake and reduce the salt concentration in the cytoplasm and cell wall. Plants having a higher capability for ion compartmentation and/or exclusion are called halophytes; these are more tolerant to salinity stress than those not having this feature (Munns et al., 2006). In this context, it has been observed that the intracellular Cl^- concentration in plants parallels the fluctuation of salinity, but that of K^+ and Na^+ varies depending on the plant species (Kirst, 1990), suggesting that tolerance to salinity probably lies in the regulation of the transport of Na^+ and K^+. The levels of K^+ in the cell must be maintained, and a high ratio of K^+ to Na^+ is required for the survival of plants under saline conditions (Epstein et al., 1980; Greenway and Munns, 1983). Most importantly, the maintenance of the intracellular levels of K^+, Ca^{2+}, and Mg^{2+} is necessary for salt tolerance in plants (Rout and Shaw, 2001b).

The ionic interaction in aquatic macrophytes is completely different compared with that in land plants; the whole plant (shoot and root) is exposed to the saline environment, and the entry of Na⁺ probably also occurs through routes other than the outward rectifying K⁺ channels (Murata et al., 1994). A simple positive correlation between sodium exclusion and salt tolerance was observed only with lower concentrations of salt treatment in many aquatic plants (Rout and Shaw, 2001a; Masood et al., 2006; Kim, 2013), whereas higher salt treatment had just the opposite effect, and this was species-specific (Rout and Shaw, 2001a). The leaf age may not be a factor in absorbing Na⁺ from the water column directly, and some species may be capable of absorbing Na⁺ into leaf vacuoles, the osmotic potential of which could be balanced by a high concentration of compatible osmolytes in the leaf cell cytoplasm (Warwick and Bailey, 1997).

K⁺ influx is highly inhibited in the presence of NaCl, and a tolerant plant is less affected than a sensitive one (Warwick and Bailey, 1997; Rout and Shaw, 2001a; Masood et al., 2006; Jampeetong and Brix, 2009). A high Na⁺/K⁺ ratio at high salinity indicates the lack of a mechanism to maintain ionic homeostasis in the cells (Jampeetong and Brix, 2009; Gomes et al., 2017), although the K⁺/Na⁺ ratio in some cases may not support any relationship between salt tolerance and the K⁺/Na⁺ ratio (Rout and Shaw, 2001a), which could be due to leakage of the ion through the outward rectifying potassium channels (Murata et al., 1994). In principle, a salt-tolerant plant accumulates a lower level of Na⁺ ions in the cytoplasm and shows lower electrolyte leakage than a sensitive plant (Masood et al., 2006). Efficient modulation of antioxidant enzymes, coupled with regulation of ion transport, plays an essential role in the induction of salt tolerance (Abraham and Dhar, 2010). A lower decrease in Ca²⁺ and Mg²⁺ content in macrophytes under salt stress implies that they have a higher salt tolerance capability (Rout and Shaw, 2001a; Jampeetong and Brix, 2009). It has been observed that an external supply of calcium can restore the intracellular K⁺ level and improve the salt tolerance capability of a plant, particularly in the halophytes (Kim, 2013).

5.4.2 Osmotic Adjustment or Regulation

Plants exposed to salinity try to balance their water potential equivalent to the surroundings. This regulation of osmotic potential within a cell by accumulation or removal of solutes from the cytoplasm until the cytoplasmic osmotic potential approximately equals the osmotic potential of the medium surrounding the cell is called *osmoregulation* (Heuer and Feigin, 1993; Hasegawa et al., 2000). The presence of salt decreases the osmotic potential of the environment and turgidity of the cells, and the tolerance mechanism is triggered, with accumulation of ions in the cell vacuole and compatible solutes in the cytoplasm for adaptation to the saline environment (Hasegawa et al., 2000; Jaleel et al., 2007). The compounds reported to be involved in osmotic adjustment are carbohydrates (sucrose, sorbitol, mannitol, glycerol, arabinitol, and pinitol) (Bohnert and Jensen, 1996; Hare et al., 1999); nitrogenous compounds (proteins, betaine, glutamate, aspartate, glycine, proline, and 4-gamma aminobutyric acid); and organic acids (malate and oxalate) (Lamosa et al., 1998; Kinnersley and Turano, 2000).

5.4.2.1 Free Amino Acids

The free amino acid content increased significantly in two salt-tolerant aquatic macrophytes, *N. indica* and *N. graminea*, and decreased in the salt-sensitive *H. verticillata*, which advocates a positive role of free amino acids in salinity tolerance (Rout, 2001). It is also reported that salinity leads to a significant increase in free amino acids and a significant decrease in asparagine, glutamine, and gamma-aminobutyric acid in *Azolla filiculoides* and *Anabaena azollae*. The high glutamine concentrations and a linear relationship between proline concentrations and glutamine/glutamate ratios in the plants grown on high NaCl probably contribute to osmotic adjustment in *Azolla filiculoides* and *Anabaena azollae* (van Kempen et al., 2013).

5.4.2.2 Proline and Its Biosynthesis

The biosynthesis and accumulation of proline is a nearly universal response of plants to osmotic stress induced by salinity, and its function as an osmoprotectant under drought and salinity stress has been widely advocated (Perez-Alfoceae and Larher, 1995; Martinez et al., 1996; Patnaik and Debata, 1997; Petrusa and Winicov, 1997; Tang et al., 2015b). Proline can be synthesized from either glutamate (in cytosol) or ornithine (in mitochondria) (Figure 5.4). The first step of proline biosynthesis from glutamate is catalyzed by Δ1-pyrroline-5-carboxylate synthetase (P5CS), which is a bifunctional enzyme with apparent activities of γ-glutamyl kinase (γ-GK) and glutamic acid-5-semialdehyde (GSA) dehydrogenase.

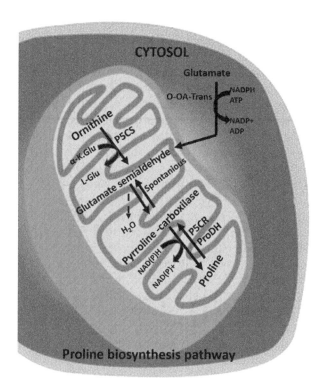

FIGURE 5.4 Proline biosynthesis in the plant cell.

GSA spontaneously cyclizes to Δ1-pyrroline-5-carboxylate (P5C), which forms a common intermediate and is reduced to proline by Δ1-pyrroline-5-carboxylate reductase (P5CR), the final step in proline biosynthesis (Planchet et al., 2014). The expression of both P5CS and P5CR has been reported to be enhanced in response to salinity (Williamson and Slocum, 1992; Liu and Zhu, 1997).

Proline in *Azolla microphylla* pre-exposed to salinity increases significantly as compared with plants exposed directly to salinity, suggesting that the compound may have an adaptive effect to salinity in plants and probably accounts for osmoregulation and salt tolerance (Abraham and Dhar, 2010). An increase in proline content has also been reported in some macrophytes in response to salinity (Zhao et al., 2014; Gomes et al., 2017). In *Azolla filiculoides*, however, proline accumulation seemed to be triggered by the decrease in moisture content as well as by an increase in the salt concentration of the medium (van Kempen et al., 2013). Dogan and Demirors (2018) in their study with *C. demersum* found that NaCl stress induced an increase in proline and Na content in a correlated manner in the plant tissues, supporting an indirect protective function of proline against the stress. The protective function of proline against NaCl stress is also supported by the observed low accumulation of the compound in *Salvinia natans* accompanied by its inability to sequester Na$^+$ and Cl$^-$ in the vacuole (Jampeetong and Brix, 2009). In a comparative study involving aquatic macrophytes varying in salt tolerance, the accumulation of proline was found to be maximal in the salt-sensitive plant *H. verticillata* compared with the moderately tolerant *Najas indica* and the highly tolerant *Najas graminea* under low-salinity treatments (Rout and Shaw, 1998, 2001a). However, *N. indica* and *N. graminea* accumulated proline to a much higher level compared with *H. verticillata* on exposure to high NaCl and seawater salinity (SWS) treatments, suggesting that maintenance of accumulation of the compound in high-salinity treatment conditions is essential for making a plant tolerant to salinity (Rout and Shaw, 1998, 2001a; Tripathi et al., 2007). In addition, it may be that proline might have a limited effect on salt tolerance in this group of macrophytes. It is often postulated that the increase in the cellular content of proline in plants in general on exposure to salinity could be a result of degradation of proteins. This may be true for salt-sensitive macrophytes such as *H. verticillata* that show an increase in proline content greater than that in salt-tolerant plants under low-level salt treatment. However, the fact that proline is synthesized *de novo* and accumulates in tissue on exposure of the macrophytes to salinity is reflected by an increase in the activity of the proline-biosynthesizing enzyme P5CR in response to salt treatment. Salt-tolerant *N. graminea* showed a higher activity level of the enzyme compared with the salt-sensitive *H. verticillata* and *N. indica*. The salt-induced *de novo* synthesis of proline is further reflected by the fact that the activity level of P5CR decreased in response to salinity in the salt-sensitive macrophyte compared with the tolerant ones and did not show any correlation with the level of proline accumulation (Rout, 2001).

Unlike P5CR, the control level activity of the proline-synthesizing enzyme P5CS differed more significantly in the macrophytes in response to the salt treatment; the salt-tolerant *N. graminea* had a much higher activity level of P5CS compared with the salt-sensitive *H. verticillata* and the moderately tolerant *N. indica*. A high expression of P5CS in response to salt treatment was also observed in *N. graminea*, but the transcript abundance did not correlate with the level of proline accumulation (Tripathi et al., 2007). This may be due to the high basal activity level of the enzyme being sufficient to promote high proline synthesis.

5.4.2.3 Glycine Betaine

Among the organic solutes synthesized during salt stress, the quaternary ammonium compound glycine betaine (also known as betaine) is considered to be the most effective osmolyte (Le Rudulier et al., 1984), and it has been widely studied (Storey et al., 1977; Weretilnyk and Hanson, 1989). Higher plants from several families (e.g., Chenopodiaceae, Poaceae, Asteraceae, etc.) accumulate betaine in response to salt stress or water deficit (Csonka and Hanson, 1991; Rhodes and Hanson, 1993). There are many findings indicating that in plants and in other organisms, betaine acts as a nontoxic protective cytoplasmic osmolyte, allowing normal metabolic function to continue in cells by maintaining the water balance between the plant cell and its environment (Robinson and Jones, 1986) and by stabilizing macromolecules during cellular dehydration and at high salt concentrations (Incharoensakdi et al., 1986; Murata et al., 1992; Santoro et al., 1992; Mohanty et al., 1993; Papageorgiou and Murata, 1995).

Betaine is synthesized by a two-step oxidation of choline in both eukaryotes and prokaryotes involving either one or two enzymes (Ikuta et al., 1977; Hanson et al., 1985; Andresen et al., 1988). In higher plants, the biosynthetic pathway has been described for a number of species (Hanson et al., 1985; Rhodes and Hanson, 1993). In the first step, choline is oxidized to betaine aldehyde by a ferredoxin-dependent choline monooxygenase (CMO) (Brouquisse et al., 1989). In the second step, betaine aldehyde is converted into betaine by an NAD$^+$-dependent betaine aldehyde dehydrogenase (BADH) (Weigel et al., 1986). The activity of these two enzymes is localized in the chloroplast stroma, but some BADH activity is also found in the cytoplasm (Weigel et al., 1986).

Betaine accumulation has been reported in members of several families of higher plants, such as Chenopodiaceae, Asteraceae, Poaceae, etc. (Storey et al., 1977; Weretilnyk and Hanson, 1989). The relatively salt-tolerant cereals sorghum, barley, and wheat can accumulate up to about 75–100 μmol betaine g^{-1} dry weight (Ladyman et al., 1983; Grieve and Maas, 1984). Concomitantly with an increase in accumulation of betaine, the level of BADH also increases in these plants on salt stress (Weigel et al., 1986; Weretilnyk and Hanson, 1989). Also, the level of BADH transcript has been found to increase in leaves of spinach (Weretilnyk and Hanson, 1989) and sugar beet (McCue and Hanson, 1992) under salt stress. Maize, which is more susceptible to salt and drought stress when compared with other crop plants, shows much less glycine betaine accumulation on exposure to salinity.

There are considerable species-specific variations with regard to betaine accumulation in plants under salt stress.

The accumulation could be a result of an increase in transcription of BADH and/or CHO genes, or translation of their respective proteins, or both. The possible cause of such variation in accumulation is, however, not known with certainty, and the rate-limiting step in betaine accumulation is yet to be demonstrated and established. The reason is possibly lack of sufficient work on the expression of the genes involved in its synthesis and regulation of their expression, particularly involving a diverse group of plants. Reports on the accumulation of quaternary ammonium compounds and expression of the BADH gene in aquatic plants under salt stress is, however, scant. A study with the aquatic macrophytes *N. indica* and *N. graminea* revealed no change in the cellular level of betaine in the plants on exposure to salinity (Tripathi et al., 2007). The enzyme BADH was also poorly expressed (Tripathi et al., 2007). The halophytes, nevertheless, show considerable expression of quaternary ammonium compounds and expression of BADH under salt stress (Gharat and Shaw, 2015).

5.4.2.4 Polyamines

Polyamines (PAs) are nitrogen compounds that are present in virtually all living beings, both prokaryotes and eukaryotes. *Polyamines* is a generic term for diamines (1,3 diaminopropane, putrescine, and cadaverine), triamines (norspermidine, spermidine, aminopropylcadaverine, and homospermidine), tetraamines (norspermine, spermine, thermospermine, and canavalmine), pentaamines (caldopentamine and homocaldopentamine), and hexamines (caldohexamine and homocaldohexamine) (Kuehn et al., 1990a). Among them, the most abundant are putrescine, spermine, and spermidine (common polyamines). In plants, PAs are a fundamental part in many physiological processes, including flower formation, embryogenesis, DNA and RNA protection, membrane protection, fruit development, and organogenesis, among many others (Kuehn et al., 1990b; Tiburcio et al., 2014).

The biosynthesis of common PAs is initiated with the formation of putrescine from ornithine or arginine, catalyzed by ornithine decarboxylase and by arginine decarboxylase, respectively. The formation of non-common PAs, so considered because of their limited distribution in nature, is initiated from 1,3 diaminopropane, which is derived from the action of polyamine oxidase (PAO) and serves as a substrate for repeated reactions that aggregate an aminopropyl group donated by decarboxylated S-adenosylmethionine (Kuehn et al., 1990a).

The uncommon polyamines, comprised in their entirety of the repetition of aminopropyl groups, such as norspermidine, norspermine, and caldopentamine, were reported for the first time in higher plants in studies carried on drought-tolerant alfalfa, *Medicago sativa* L., and in heat-tolerant cotton, *Gossypium hirsutum* L. and *G. barbadense* L. Norspermidine, norspermine, and caldopentamine were found in alfalfa shoot meristems under drought stress, while caldopentamine was found in cell suspensions subjected to drought stress. On the other hand, norspermine, norspermidine, caldopentamine, and homocaldopentamine were found in cell suspensions of *G. barbadense* under heat stress (Rodriguez-Garay et al.,

1989; Kuehn et al., 1990b). In Figure 5.5, the biosynthetic routes for common and uncommon PAs are depicted in a very simple way. For a more detailed biosynthetic pathway of PAs, see the review by Alcazar et al. (2010).

Accumulation of PAs has been observed in many plants in response to various types of abiotic stress, including high concentrations of salt and heavy metals, heat, cold, drought, and lack of oxygen in the roots due to water flooding (hypoxia), among others (Groppa and Benavides, 2008). The application of exogenous polyamines has been found to confer tolerance to various types of stress in plants (Kuznetsov and Shevyakova, 2007). Endogenous and/or exogenous PAs have been shown to modulate stress-activated ROS homeostasis and oxidative damage by enhancing the antioxidant enzyme activities and antioxidants in marine plants (Kumar et al., 2017).

Both marine and freshwater plants seem to share PA metabolic pathways similar to those found in land plants. However, more studies are required for a greater understanding of PA metabolism in marine macrophytes, as this process is not well studied compared with that in land plants (see Figure 12.1 in Kumar et al., (2017). Nevertheless, a study on the aquatic macrophyte *Potamogeton crispus* suggested a definite role of putrescine in abiotic stress tolerance, as the level of proline and putrescine increased when the plant was exposed to high concentrations of Pb (Xu et al., 2011). In the same way, Polo et al. (2015) found an important role of putrescine, spermidine, and spermine in conferring tolerance on the brown macroalga *Sargassum cymosum* against high concentrations of salinity and exposure to UV radiation under laboratory conditions.

5.4.3 Antioxidant Defenses and Enzymes

The growth of plants under high exogenous salt concentrations leads to the production of ROS,, which is highly damaging to cells and cellular components, such as pigments, lipids, proteins, and nucleic acids, leading to oxidative damage (Ghorbanli et al., 2004; Kim et al., 2005). Also, salinity induces ROS production in aquatic macrophytes, which leads to oxidative damage to them (Rout and Shaw, 2001b; He and Zhu, 2008; Aghaleh et al., 2009). The plant has well-armed enzymatic and non-enzymatic antioxidant systems to scavenge the free oxygen radicals generated during stress conditions (Corpas et al., 2015). These are composed of antioxidative enzymes such as SOD, APX, and catalase (CAT), which play a key role in the removal of ROS produced in aquatic macrophytes either as by-products of normal cell metabolism or as a result of a disturbance in the cell metabolic processes under abiotic stresses (Shaw et al., 2004; Abraham and Dhar, 2010; Mallik et al., 2011). In addition to the enzymes, there are also non-enzymatic antioxidants, such as ascorbic acid (vitamin C [Asc]), glutathione (GSH), tocopherols (vitamin E), and carotenes (provitamin A), that are involved in scavenging ROS and promoting plant recovery from oxidative damage (Gupta el al., 2018).

The primary production of ROS is in the chloroplasts, mitochondria, and peroxisomes, and these are superoxide radicals ($O_2 \bullet -$), hydrogen peroxide (H_2O_2), hydroxyl radicals

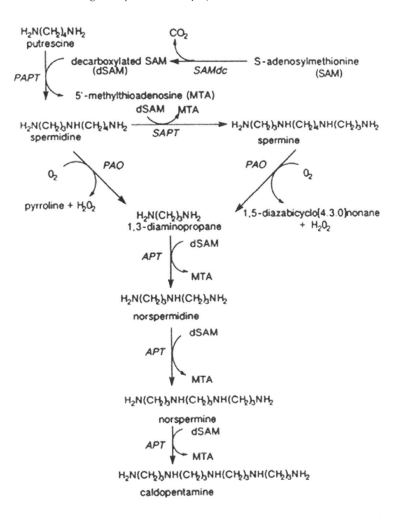

FIGURE 5.5 Biosynthetic pathways in higher plants known for the common polyamines and proposed for the uncommon polyamines initiating with 1,3-diaminopropane. PAPT, putrescine aminopropyltransferase; SAPT, spermidine aminopropyltransferase. (Republished with permission of the American Society of Plant Biologists, from: [Novel Occurrence of Uncommon Polyamines in Higher Plants, Kuehn et al., Plant Physiology 94(3): 855-857 (1990)]; permission conveyed through Copyright Clearance Center, Inc.)

(^{1}OH), and singlet oxygen ($^1O_2$). The flow of electrons from the reduced ferredoxin to oxygen in chloroplasts and from ubiquinone to oxygen in mitochondria results in the formation of $O_2\bullet-$. SOD acts on $O_2\bullet-$ to produce H_2O_2, and for the healthy growth of plants, it is essential that both these ROS are scavenged, which is done by the enzymes catalase and peroxidases (Asada, 2006; Corpas et al., 2015; Gupta et al., 2018). The oxidative radical formation and antioxidant protection system in the cell are summarized in Table 5.3.

5.4.3.1 Superoxide Dismutase (SOD)

The enzyme superoxide dismutase (SOD; EC 1.15.1.1) was discovered in 1969. It catalyzes the dismutation of superoxide radicals into molecular oxygen and H_2O_2 (McCord and Fridovich, 1969). In plants, there are three isoforms of SOD, depending on the prosthetic metals in their active sites: copper and zinc (Cu/Zn-SODs); manganese (Mn-SODs); and iron (Fe-SODs) (Wang et al., 2016; Gupta et al., 2018). The involvement of SOD in salinity stress tolerance in aquatic macrophytes has been well documented in many plants. Higher production of H_2O_2, together with an increase in the

activity of SOD, has been demonstrated in many studies, and the increase in their level has been found to be almost reciprocal to the salinity treatments (Upadhyay and Panda, 2005; Abraham, 2010), indicating their likely involvement in salt tolerance. Furthermore, Rout and Shaw (2001a) reported that the acceleration in the activity of SOD in response to NaCl was highest in salt-tolerant *N. indica* (moderately) and *N. graminea* and lowest in the sensitive plant *H. verticillata*, indicating the possible involvement of the enzyme in the salt tolerance process. In a similar study, Masood et al. (2006) observed the diverse response of SOD to salinity treatment in macrophytes, and the activity of the enzyme was higher in the tolerant plant compared with the sensitive ones. The importance of SOD in salt tolerance is also reflected from the fact that salinity treatment also results in the synthesis of new isoforms of SOD, which could be necessary to cope with the stress condition (Rout, 2001; Mallick et al., 2011).

5.4.3.2 Catalase

Catalase (EC 1.11.1.6) is located in the peroxisomes of plants. It catalyzes the breakdown of H_2O_2 into water and O_2 (Asada,

TABLE 5.3

Subcellular Localization of the Antioxidant Mechanism of the Cell

Subcellular Location	Type of Active Oxygen Species	Source of Active Oxygen Species	Enzymatic Scavenging Systems	Products	Non-enzymatic Scavenging Systems
Chloroplast	Superoxide, H_2O_2	Photosystem II, Enzymatic	SOD, Peroxidases	H_2O_2, Glutathione, $NADP^+$, dehydroascorbate	Ferredoxin, carotenoids, xanthophylls, α-tocopherol, glutathione, cysteine, hydroxyquinones, ascorbate (vitamin C)
Mitochondria	Superoxide	Electron transport	SOD	H_2O_2	
	H_2O_2	Enzymatic	Peroxidase	H_2O_2 + oxidized donor	
			Catalase	$H_2O_2 + O_2$	
Cytosol	Superoxide	Enzymatic	SOD	H_2O_2	
	H_2O_2	Enzymatic	Catalase	$H_2O + O_2$	
			Peroxidase	H_2O + oxidized donor	
Glyoxysomes	H_2O_2	β-oxidation	Catalase	$H_2O + O_2$	
	H_2O_2	Photorespiration	Catalase	$H_2O + O_2$	
Extracellular	Superoxide	Enzymatic	None known	None known	
	H_2O_2	Enzymatic	Peroxidase	Lignin, suberin, hydroxyproline	

Source: Modified from Asada, K., *Plant Physiology*, 141(2), 391–396, 2006.

2006; Kumar et al., 2017; Zhao et al., 2014). Enhanced production of H_2O_2 in response to salt treatment has been found to be accompanied by an increase in the activity of catalase in plants, indicating a positive role of the enzyme in the salt tolerance of the plant (Upadhyay and Panda, 2005; Abraham, 2010; Abraham and Dhar, 2010). However, the activity of catalase was found to decrease in response to NaCl and SWS in both the salt-tolerant macrophytes *N. indica* (moderately) and N. *graminea* and the salt-sensitive macrophyte H. *verticillata*, which is against the postulated role of the enzyme in the salt tolerance process, at least in aquatic macrophytes. The importance of the enzyme in the salt tolerance process, nevertheless, cannot be undermined, as the studies by Mallick et al. (2011) and Rout (2001), showed the appearance of new isoforms of the enzyme in response to the salt treatment.

5.4.3.3 Ascorbate Peroxidase

Ascorbate peroxidase (EC 1.11.1.11) is also a scavenger of H_2O_2 but differs in action from catalase in requiring an electron donor for reducing the H_2O_2. This enzyme has a high degree of specificity for ascorbate (Asc) as the electron donor (Asada, 1992; Asada et al., 1993). This has been reported to exist in three isomeric forms: thylakoid bound (tApx), stromal (sApx), and cytosolic (cApx) (Chen and Asada, 1992; Miyake and Asada, 1992). Stimulation of APX activity has been shown to be associated with salt tolerance in many plants (Sudhakar et al., 2001; Yazici et al., 2007). In the aquatic macrophyte *A. microphylla*, the activities of APX are considerably enhanced in response to salinity stress, and pre-exposure of the plant to ascorbate results in a better capacity of the plant to decompose H_2O_2 than in plants directly exposed to salinity (Abraham, 2010; Abraham and Dhar, 2010). The ability of plants pre-exposed to ascorbate to better tolerate salinity than those that do not tolerate indicates an important role of APX

in salt tolerance in plants, at least in aquatic macrophytes. It has been observed that salt-tolerant macrophytes such as *N. graminea* and *N. indica* have a greater basal activity of APX than the salt-sensitive macrophyte *H. verticillata*, further indicating the importance of the enzyme in salt tolerance in aquatic macrophytes (Rout, 2001; Rout and Shaw, 2001b).

5.4.3.4 Peroxidase

Peroxidase (POX), like APX, is a heme-containing enzyme, but unlike APX, POX is capable of accepting an electron from a very wide range of substances for catalyzing the breakdown of H_2O_2. Artificial substrates include guaiacol, benzidine, o-dionisidine, etc. Hence, the enzyme is also called *non-specific peroxidase*, having several electron donors (Van Huystee, 1987). Like the other antioxidative enzymes, POX also seems to play an important role in salt tolerance in aquatic macrophytes. This view stems from the fact that the salt-tolerant macrophyte *Pistia* shows a higher activity level when compared with the salt-sensitive macrophyte *Salvinia* (Upadhyay and Panda, 2005). Similarly, the salt-tolerant macrophytes *N. graminea* and *N. indica* have a higher basal activity of POD compared with the salt-sensitive macrophyte *H. verticillata* (Rout and Shaw, 2001b). Besides, salt treatment induces the synthesis of new isoforms of POD (Rout, 2001), probably to enhance its capacity to cope with the salt-induced generation of H_2O_2 (Rout, 2001).

5.4.4 Non-Enzymatic Antioxidants

Non-enzymatic antioxidants, such as ascorbic acid, glutathione, and phenolic compounds, have also been shown to play an important role in overcoming oxidative stress in plants (Foyer, 1994), in parallel with the antioxidative enzymes. It has been reported that ascorbate and glutathione contents increase in aquatic macrophytes in response to exposure to salinity (Dogan and Demirors, 2018), and the increase is higher in the salt-tolerant *Pistia* than in the sensitive *Salvinia*. An increase in the cellular contents of non-enzymatic antioxidants has also been reported in these aquatic macrophytes in response to exposure to salinity (Upadhyay and Panda, 2005).

5.5 CONCLUSIONS AND PERSPECTIVES

The ever-increasing human population pressure needs the development of new cultivars of crop plants with enhanced salt tolerance to bring unused or poorly used agricultural land into proper cultivation. Conventional approaches for breeding crops for salt tolerance have so far been unsuccessful (Flowers and Yeo, 1995). This is due to the quantitative nature of the salt tolerance trait (Fooland and Jones, 1993). Therefore, a complete understanding of the underlying mechanisms conferring salt tolerance is required, which may allow the development of salt-tolerant varieties of crops through genetic engineering and conventional genetic protocols by using proper selection procedures. With regard to understanding the mechanism of salt tolerance, a common approach is to compare and identify the distinguishable responses in salt-adapted and non-adapted glycophytes. Another approach is to identify the responses of halophytes to salt stress. Both glycophytes and halophytes are available as terrestrial plants. However, research using these as model organisms has not been able to provide universally accepted mechanistic details of salt tolerance. For example, the accumulation of glycine betaine is considered to be an important component imparting salt tolerance in terrestrial plants, whereas such accumulation is lacking in the salt-tolerant submerged aquatic macrophyte *Najas graminea* (Rout and Shaw, 2001b). Hence, the accumulation of glycine betaine in terrestrial plants could be considered as a response that is not related to salt tolerance. The difference in response of terrestrial plants and aquatic macrophytes could be because of the organizational complexities of the former. In this context, the aquatic macrophytes, particularly the submerged ones, offer good models to work with, as their response to salinity is expected to be similar in both root and shoot. The lack of interest in using aquatic macrophytes as model plants for salt response studies is probably due to the requirement for huge space in the laboratory for their culture and more importantly, the unavailability of protocols for their genetic manipulation, which is highly necessary for establishing a fact or proving a hypothesis in biological research. Nevertheless, aquatic macrophytes offer a great opportunity for conducting research on issues related to salt tolerance in plants, as they have an anatomical organization similar to that of terrestrial plants, and all their organs remain exposed to the same environment. In addition, there are salt-tolerant macrophytes, such as *Najas graminea*, which can survive and grow even in the absence of salt. Moreover, salt tolerance is more prevalent in aquatic macrophytes compared with terrestrial plants, and hence, their use as model plants in salt response studies for understanding the mechanistic details of salt tolerance deserves serious consideration.

ACKNOWLEDGMENTS

This research was carried out with the support of Project PlanTECC CONACyT-Mexico No. 2018-293362.

REFERENCES

Abdallah, S. B., Aung, B., Amyot, L., Lalin, I., Lachâal, M., Karray-Bouraoui, N., & Hannoufa, A. (2016). Salt stress (NaCl) affects plant growth and branch pathways of carotenoid and flavonoid biosyntheses in Solanum nigrum. *Acta Physiologiae Plantarum*, 38(3): 72.

Abraham, G. (2010). Antioxidant enzyme status in *Azolla microphylla* in relation to salinity and possibilities of environmental monitoring. *Thin Solid Films*, 519(3): 1240–1243.

Abraham, G., & Dhar, D. W. (2010). Induction of salt tolerance in *Azolla microphylla* Kaulf through modulation of antioxidant enzymes and ion transport. *Protoplasma*, 245(1–4): 105–111.

Acosta-Motos, J. R., Diaz-Vivancos, P., Álvarez, S., Fernández-García, N., Sánchez-Blanco, M. J., & Hernández, J. A. (2015a). NaCl-induced physiological and biochemical adaptative mechanisms in the ornamental *Myrtus communis* L. plants. *Journal of Plant Physiology*, 183: 41–51.

Acosta-Motos, J. R., Diaz-Vivancos, P., Álvarez, S., Fernández-García, N., Sanchez-Blanco, M. J., & Hernández, J. A. (2015b). Physiological and biochemical mechanisms of the ornamental *Eugenia myrtifolia* L. plants for coping with NaCl stress and recovery. *Planta*, 242(4): 829–846.

Acosta-Motos, J. R., Ortuño, M. F., Bernal-Vicente, A., Diaz-Vivancos, P., Sanchez-Blanco, M. J., & Hernandez, J. A. (2017). Plant responses to salt stress: adaptive mechanisms. *Agronomy*, 7(1): 18.

Adams, J. B., & Bate, G. C. (1994). The effect of salinity and inundation on the estuarine macrophyte Sarcocornia perennis (Mill.) AJ Scott. *Aquatic Botany*, 47(3–4): 341–348.

Aghaleh, M., Niknam, V., Ebrahimzadeh, H., & Razavi, K. (2009). Salt stress effects on growth, pigments, proteins and lipid peroxidation in *Salicornia persica* and S. *europaea*. *Biologia Plantarum*, 53(2): 243–248.

Alcázar, R., Altabella, T., Marco, F., Bortolotti, C., Reymond, M., Csaba Koncz, C., Pedro Carrasco, P., & Tiburcio, A. F. (2010). Polyamines: molecules with regulatory functions in plant abiotic stress tolerance. *Planta*, 231: 1237–1249.

Al-Hamdani, S. (2008). Influence of different sodium chloride concentrations on selected physiological responses of Salvinia. *Journal of Aquatic Plant Management*, 46: 172–175.

Allen, V. G., Pond, K. R., Saker, K. E., Fontenot, J. P., Bagley, C. P., Ivy, R. L., & Dettle, T. M. (2001). Tasco-Forage: III. Influence of a seaweed extract on performance, monocyte immune cell response, and carcass characteristics in feedlot-finished steers. *Journal of Animal Science*, 79(4): 1032–1040.

Andresen, P. A., Kaasen, I., Styrvold, O. B., & Boulnois, G. (1988). Molecular cloning, physical mapping and expression of the bet genes governing the osmoregulatory choline-glycine betaine pathway of *Escherichia coli*. *Microbiology*, 134(6): 1737–1746.

Asada, K. (1992). Production and scavenging of active oxygen in chloroplast. In: Scandlios, J. G. (Ed.). *Molecular Biology of Free Radical Scavenging Systems*, New York, Cold Spring Harbor Lab Press, p. 173.

Asada, K., Heber, U., & Schreiber, U. (1993). Electron flow to intersystem chain from stromal components and cyclic electron flow in maize chloroplast as detected in intact leaves by monitoring redox changes of p700 and chlorophyll fluorescence. *Plant Cell Physiology*, 34: 39–50.

Asada, K. (1999). The water-water cycle in chloroplasts: scavenging of active oxygens and dissipation of excess photons. *Annual Review of Plant Biology*, 50(1): 601–639.

Asada, K. (2006). Production and scavenging of reactive oxygen species in chloroplasts and their functions. *Plant Physiology*, 141(2): 391–396.

Ashraf, M. F. M. R., & Foolad, M. (2007). Roles of glycine betaine and proline in improving plant abiotic stress resistance. *Environmental and Experimental Botany*, 59(2): 206–216.

Barber, J., & Andersson, B. (1992). Too much of a good thing: light can be bad for photosynthesis. *Trends in Biochemical Sciences*, 17(2): 61–66.

Barnes, R. S. K. (1980). *Coastal Lagoons: The Natural History of a Neglected Habitat*, Cambridge University Press, Cambridge.

Beaune, D., Sellier, Y., Lambert, É., & Grandjean, F. (2018). The use of *Chara* spp. (Charales: Characeae) as a bioindicator of physico-chemical habitat suitability for an endangered crayfish *Austropotamobius pallipes* in lentic waters. *Aquatic Conservation: Marine and Freshwater Ecosystems*, 28(2): 506–511.

Belkhodja, R., Morales, F., Abadia, A., Medrano, H., & Abadia, J. (1999). Effect of salinity on chlorophyll fluorescence and photosynthesis of barley (*Hordeum vulgare* L) grown under a triple-line-source sprinkler system in the field. *Photosynthetica*, 36: 375–387.

Bertness, M. D., Gough, L., & Shumway, S. W. (1992). Salt tolerances and the distribution of fugitive salt marsh plants. *Ecology*, 73(5): 1842–1851.

Blindow, I., Dietrich, J., Möllmann, N., & Schubert, H. (2003). Growth, photosynthesis and fertility of *Chara aspera* under different light and salinity conditions. *Aquatic Botany*, 76(3): 213–234.

Blum, A. (1988). *Plant Breeding for Stress Environment*, CRC Press, Inc, Boca Raton, FL.

Bohnert, H. J., & Jensen, R. G. (1996). Metabolic engineering for increased salt tolerance–the next step. *Functional Plant Biology*, 23(5): 661–667.

Borgnis, E., & Boyer, K. E. (2016). Salinity tolerance and competition drive distributions of native and invasive submerged aquatic vegetation in the Upper San Francisco Estuary. *Estuaries and Coasts*, 39(3): 707–717.

Bornette, G., & Puijalon, S. (2009): Macrophytes: ecology of aquatic plants. In: *Encyclopedia of Life Sciences*, Wiley J & Sons, Chichester, pp. 1–9.

Boustany, R. G., Michot, T. C., & Moss, R. F. (2010). Effects of salinity and light on biomass and growth of *Vallisneria americana* from Lower St. Johns River, FL, USA. *Wetlands Ecology and Management*, 18(2): 203–217.

Boustany, R. G., Michot, T. C., & Moss, R. F. (2015). Effect of nutrients and salinity pulses on biomass and growth of *Vallisneria americana* in lower St Johns River, FL, USA. *Royal Society Open Science*, 2(2): 140053.

Brix, H. (1994). Functions of macrophytes in constructed wetlands. *Water Science and Technology*, 29(4): 71–78.

Brogan, W. R. III, & Relyea, R. A. (2015). Submerged macrophytes mitigate direct and indirect insecticide effects in freshwater communities. *PLoS One*, 10(5): e0126677.

Brogan, W. R. III, & Relyea, R. A. (2017). Multiple mitigation mechanisms: effects of submerged plants on the toxicity of nine insecticides to aquatic animals. *Environmental Pollution*, 220: 688–695.

Brouquisse, R., Weigel, P., Rhodes, D., Yocum, C. F., & Hanson, A. D. (1989). Evidence for a ferredoxin-dependent choline monooxygenase from spinach chloroplast stroma. *Plant Physiology*, 90(1): 322–329.

Brown, S., Day, D. A., & Critchley, C. (1987). Salt tolerance-does leaf respiration have a contribution to make? In: Moore, A. L., & Beechey, R. B. (Eds.). *Plant Mitochondria: Structure, Functional and Physiological Aspects*, Plenum Press, New York, p. 393.

Burks, R. L., Mulderij, G., Gross, E., Jones, I., Jacobsen, L., Jeppesen, E., & Van Donk, E. (2006). Center stage: the crucial role of macrophytes in regulating trophic interactions in shallow lake wetlands. In *Wetlands: Functioning, Biodiversity Conservation, and Restoration*.

Carillo, P., Annunziata, M. G., Pontecorvo, G., Fuggi, A., & Woodrow, P. (2011). Salinity stress and salt tolerance. In: *Abiotic Stress in Plants-Mechanisms and Adaptations*. InTech, Rijeka, Croatia.

Carpenter, S. R., & Lodge, D. M. (1986). Effects of submersed macrophytes on ecosystem processes. *Aquatic Botany*, 26: 341–370.

Catalan, L., Balzarini, M., Taleisnik, E., Sereno, R., & Karlin, U. (1994). Effects of salinity on germination and seedling growth of *Prosopis flexuosa* (D. C.). *Forest Ecology and Management*, 63: 347.

CEVA. (2005). Algues et alimentation animale. Algorythmes, No. 72. https://www.feedipedia.org/node/17854 (accessed on November 21, 2018).

Charpentier, A., Grillas, P., Lescuyer, F., Coulet, E., & Auby, I. (2005). Spatio-temporal dynamics of a *Zostera noltii* dominated community over a period of fluctuating salinity in a shallow lagoon, Southern France. *Estuarine, Coastal and Shelf Science*, 64(2–3): 307–315.

Chen, G.-X., & Asada, K. (1992). Inactivation of ascorbate peroxidase by thiols requires hydrogen peroxide. *Plant Cell Physiology*, 33: 117.

Christia, C., & Papastergiadou, E. S. (2007). Spatial and temporal variations of aquatic macrophytes and water quality in six coastal lagoons of western Greece. *Belgian Journal of Botany*, 39–50.

Csonka, L. N., & Hanson, A. D. (1991). Prokaryotic osmoregulation: genetics and physiology. *Annual Reviews in Microbiology*, 45(1): 569–606.

Curtion, D., Steppuhn, H., & Selles, F. (1993). Plant response to sulphate and chloride salinity: growth and ionic relations. *Soil Science Society of America Journal*, 57(5): 1304–1310.

Del Zoppo, M., Galleschi, L., Onnis, A., Pardossi, A., & Saviozzi, F. (1999). Effect of salinity on water relations, sodium accumulation, chlorophyll content and proteolytic enzymes in a wild wheat. *Biologia Plantarum*, 42(1): 97.

Dittami, S. M., Gravot, A., Renault, D., Sophie Goulitquer, S., Eggert, A., Bouchereau, A., Boyen, C., & Tonon, T. (2011). Integrative analysis of metabolite and transcript abundance during the short-term response to saline and oxidative stress in the brown alga *Ectocarpus siliculosuspce*. *Plant, Cell and Environment*, 34: 629–642.

Dogan, M., & Demirors Saygideger, S. (2018). Physiological effects of NaCl on *Ceratophyllum demersum* L., a submerged rootless aquatic macrophyte. *Iranian Journal of Fisheries Sciences*, 17(2): 346–356.

Dubey, R. S. (1997). Photosynthesis in plants under stressful conditions. In: Pessarakli, M. (Ed.). *Hand Book of Photosynthesis*, Marcel Dekker, New York, p. 859.

Egan, T. P., & Ungar, I. A. (1998). Effect of different salts of sodium and potassium on the growth of *Atriplex prostrata* (Chenopodiaceae). *Journal of Plant Nutrition*, 21(10): 2193–2205.

Epstein, E., Norlyn, J. D., Rush, D. W., Kingsbury, R. W., Kelley, D. B., Cunningham, G. A., & Wrona, A. F. (1980). Saline culture of crops: a genetic approach. *Science*, 210(4468): 399–404.

Evans, F. D., & Critchley, A. T. (2014). Seaweeds for animal production use. *Journal of Applied Phycology*, 26(2): 891–899.

Fassett, N. C. (2006). *A Manual of Aquatic Plants*. University of Wisconsin Press.

Flowers, T. J., & Yeo, A. R. (1995). Breeding for salinity resistance in crop plants: where next? *Functional Plant Biology*, 22(6): 875–884.

Flowers, T. J., Troke, P. F., & Yeo, A. R. (1977). The mechanism of salt tolerance in halophytes. *Annual Review of Plant Physiology*, 28(1): 89–121.

Food and Agriculture Organization (FAO) of the United Nations (UN) (2012). Food Balance Sheets (Datafile). Food and Agriculture Organization, Rome. Available from URL: via faostat.fao.org (accessed on June 27, 2012).

Foolad, M. R., & Jones, R. A. (1993). Mapping salt-tolerance genes in tomato (*Lycopersicon esculentum*) using trait-based marker analysis. *Theoretical and Applied Genetics*, 87(1–2): 184–192.

Foyer, C. H., Lelandais, M., & Kunert, K. J. (1994). Photooxidative stress in plants. *Physiologia Plantarum*, 92(4): 696–717.

Francois, L. E. (1994). Growth, seed yield, and oil content of canola grown under saline conditions. *Agronomy Journal*, 86(2): 233–237.

Garcia, A. B., Engler, J. D. A., Iyer, S., Gerats, T., Van Montagu, M., & Caplan, A. B. (1997). Effects of osmoprotectants upon NaCl stress in rice. *Plant Physiology*, 115(1): 159–169.

Gharat, S. A., & Shaw, B. P. (2015). NaCl induced changes in the ionic and osmotic components in rice cultivars vis-a-vis that in a natural halophyte. *ORYZA-An International Journal on Rice*, 52(1): 46–53.

Ghassemi, F., Jakeman, A. J., & Nix, H. A. (1995). *Salinisation of Land and Water Resources: Human Causes, Extent, Management and Case Studies*. UNSW Press, Sydney, Australia, and CAB International, Wallingford, UK.

Ghorbanli, M., Ebrahimzadeh, H., & Sharifi, M. (2004). Effects of NaCl and mycorrhizal fungi on antioxidative enzymes in soybean. *Biologia Plantarum*, 48(4): 575–581.

Gomes, M. A. D. C., Pestana, I. A., Santa-Catarina, C., Hauser-Davis, R. A., & Suzuki, M. S. (2017). Salinity effects on photosynthetic pigments, proline, biomass and nitric oxide in *Salvinia auriculata* Aubl. *Acta Limnologica Brasiliensia*, 29.

González-Moreno, S., Gómez-Barrera, J., Perales, H., & Moreno-Sánchez, R. (1997). Multiple effects of salinity on photosynthesis of the protist *Euglena gracilis*. *Physiologia Plantarum*, 101(4): 777–786.

Goodman, A. M., Ganf, G. G., Dandy, G. C., Maier, H. R., & Gibbs, M. S. (2010). The response of freshwater plants to salinity pulses. *Aquatic Botany*, 93(2): 59–67.

Govaerts, R. (Ed.). (2018). For a full list of reviewers see: http://apps.kew.org/wcsp/compilersReviewers.do (2018). WCSP: World Checklist of Selected Plant Families (version Sep 2014). In: Roskov, Y., Abucay, L., Orrell, T., Nicolson, D., Kunze, T., Flann, C., Bailly, N., Kirk, P., Bourgoin, T., DeWalt, R. E., Decock, W., & De Wever, A., (Eds.). Species 2000 & ITIS Catalogue *of Life, 2015 Annual Checklist*. Species 2000: Naturalis, Leiden, the Netherlands. ISSN 2405-8858. Digital resource at www. catalogueoflife.org/col.

Greenway, H., & Munns, R. (1983). Interactions between growth, uptake of Cl– and Na+, and water relations of plants in saline environments. II. Highly vacuolated cells. *Plant, Cell & Environment*, 6(7): 575–589.

Grieve, C. M., & Maas, E. V. (1984). Betaine accumulation in salt-stressed sorghum. *Physiologia Plantarum*, 61(2): 167–171.

Grillas, P., van Wijck, C., & Bonis, A. (1993). The effect of salinity on the dominance-diversity relations of experimental coastal macrophyte communities. *Journal of Vegetation Science*, 4(4): 453–460.

Groppa, M. D., & Benavides, M. P. (2008). Polyamines and abiotic stress: recent advances. *Amino Acids*, 34(1): 35–45.

Gupta, B., & Huang, B. (2014). Mechanism of salinity tolerance in plants: physiological, biochemical, and molecular characterization. *International Journal of Genomics*, 2014.

Gupta, D. K., Palma, J. M., & Corpas, F. J. (Eds.). (2018). *Antioxidants and Antioxidant Enzymes in Higher Plants*. Springer, USA.

Haller, W. T., Sutton, D. L., & Barlowe, W. C. (1974). Effects of salinity on growth of several aquatic macrophytes. *Ecology*, 55(4): 891–894.

Hamed, K. B., Castagna, A., Salem, E., Ranieri, A., & Abdelly, C. (2007). Sea fennel (Crithmum maritimum L.) under salinity conditions: a comparison of leaf and root antioxidant responses. *Plant Growth Regulation*, 53(3): 185–194.

Hanson, A. D., May, A. M., Grumet, R., Bode, J., Jamieson, G. C., & Rhodes, D. (1985). Betaine synthesis in chenopods: localization in chloroplasts. *Proceedings of the National Academy of Sciences*, 82(11): 3678–3682.

Hare, P. D., Cress, W. A., & Van Staden, J. (1999). Proline synthesis and degradation: a model system for elucidating stress-related signal transduction. *Journal of Experimental Botany*, 50(333): 413–434.

Hart, B. T., Bailey, P., Edwards, R., Hortle, K., James, K., McMahon, A., Meredith, C., & Swading, K. (1990). Effects of salinity on river, stream and wetland ecosystems in Victoria, Australia. *Water Research*, 24(9): 1103–1117.

Hart, B. T., Bailey, P., Edwards, R., Hortle, K., James, K., McMahon, A., Meredith, C., & Swading K. (1991). A review of the salt sensitivity of the Australian freshwater biota. *Hydrobiologia*, 210(1–2): 105–144.

Hasan, M. R., & Rina, C. (2009). Use of algae and aquatic macrophytes as feed in small-scale aquaculture: a review (No. 531). Food and Agriculture Organization of the United Nations (FAO).

Hasegawa, P. M., Bressan, R. A., Zhu, J. K., & Bohnert, H. J. (2000). Plant cellular and molecular responses to high salinity. *Annual Review of Plant Biology*, 51(1), 463–499.

Hassler, M. (2018). World Plants: Synonymic Checklists of the Vascular Plants of the World (version Jan 2015). In: Roskov, Y., Abucay, L., Orrell, T., Nicolson, D., Kunze, T., Flann, C., Bailly, N., Kirk, P., Bourgoin, T., DeWalt, R.E., Decock, W., & De Wever, A., (Eds.) *Species 2000 & ITIS Catalogue of Life, 2015 Annual Checklist*. Species 2000: Naturalis, Leiden, the Netherlands. ISSN 2405-8858. Digital resource at www. catalogueoflife.org/col.

Corpas, F. J., Gupta, D. K., & Palma, J. M. (2015). Production sites of reactive oxygen species (ROS) in organelles from plant cells. In: *Reactive Oxygen Species and Oxidative Damage in Plants under Stress*, Springer, Cham, pp. 1–22.

He, Y., & Zhu, Z. J. (2008). Exogenous salicylic acid alleviates NaCl toxicity and increases antioxidative enzyme activity in *Lycopersicon esculentum*. *Biologia Plantarum*, 52(4): 792.

Heuer, B., & Feigin, A. (1993). Interactive effects of chloride and nitrate on photosynthesis and related growth parameters in tomatoes. *Photosynthetica*, 28: 549.

Hodges, M., Dellero, Y., Keech, O., Betti, M., Raghavendra, A. S., Sage, R., & Weber, A. P. (2016). Perspectives for a better understanding of the metabolic integration of photorespiration within a complex plant primary metabolism network. *Journal of Experimental Botany*, 67(10): 3015–3026.

Horie, T., Kaneko, T., Sugimoto, G., Sasano, S., Panda, S. K., Shibasaka, M., & Katsuhara, M. (2011). Mechanisms of water transport mediated by PIP aquaporins and their regulation via phosphorylation events under salinity stress in barley roots. *Plant and Cell Physiology*, 52(4): 663–675.

Huystee, R. V. (1987). Some molecular aspects of plant peroxidase biosynthetic studies. *Annual Review of Plant Physiology*, 38(1): 205–219.

Ikuta, S., Imamura, S., Msaki, H., & Horiuti, Y. (1977). Purification and characterization of choline oxidase from *Arthrobacter globiformis*. *The Journal of Biochemistry*, 82(6): 1741–1749.

Incharoensakdi, A., Takabe, T., & Akazawa, T. (1986). Effect of beta-ine on enzyme activity and subunit interaction of ribulose-1, 5-bisphosphate carboxylase/oxygenase from *Aphanothece halophytica*. *Plant Physiology*, 81(4): 1044–1049.

Iqbal, N., Umar, S., & Khan, N. A. (2015). Nitrogen availability regulates proline and ethylene production and alleviates salinity stress in mustard (*Brassica juncea*). *Journal of Plant Physiology*, 178: 84–91.

Izzati, M. (2016). Salt Tolerance of Several Aquatic Plants. In: *Proceedings International Conference on Global Resource Conservation* (Vol. 6, No. 1).

Jacoby, R. P., Taylor, N. L., & Millar, A. H. (2011). The role of mito-chondrial respiration in salinity tolerance. *Trends in Plant Science*, 16(11): 614–623.

Jaleel, C. A., Gopi, R., Manivannan, P., & Panneerselvam, R. (2007). Responses of antioxidant defense system of *Catharanthus roseus* (L.) G. Don. to paclobutrazol treatment under salinity. *Acta Physiologiae Plantarum*, 29(3): 205–209.

James, K. R., & Hart, B. T. (1993). Effect of salinity on four fresh-water macrophytes. *Marine and Freshwater Research*, 44(5): 769–777.

James, K. R., Hart, B. T., Bailey, P. C., & Blinn, D. W. (2009). Impact of secondary salinisation on freshwater ecosystems: effect of experimentally increased salinity on an intermittent floodplain wetland. *Marine and Freshwater Research*, 60(3): 246–258.

Jampeetong, A., & Brix, H. (2009). Effects of NaCl salinity on growth, morphology, photosynthesis and proline accumula-tion of *Salvinia natans*. *Aquatic Botany*, 91(3): 181–186.

Jeppesen, E., Sondergaard, E., Sondergaard, M., & Christofferson, K. (2012). *The Structuring Role of Submerged Macrophytes in Lakes*. (Vol. 131 of Ecological Studies Series). Springer Science & Business Media, New York.

Jones, R. W. (1985). Salt tolerance in plants. *Chemistry in Britain*, 21(5): 454.

Kafkafi, U., & Bernstein, N. (1996). Root growth under salinity stress. In: Waisel, Y., Eshel, A., & Kafkafi, U. (Eds.). *Plant Roots, The Hidden Half*, 2nd edition, Marcle Dekker Inc, New York, p. 435.

Kalaji, M. H., & Pietkiewicz, S. (1993). Salinity effects on plant growth and other physiological processes. *Acta Physiologiae Plantarum*, 15(2): 89–124.

Kangasjärvi, S., Neukermans, J., Li, S., Aro, E. M., & Noctor, G. (2012). Photosynthesis, photorespiration, and light signalling in defence responses. *Journal of Experimental Botany*, 63(4): 1619–1636.

Karsten, U., Wiencke, C., & Kirst, G. O. (1991). The effect of salinity changes upon the physiology of eulittoral green macroalgae from Antarctica and southern Chile: II intracellular inor-ganic ions and organic compounds. *Journal of Experimental Botany*, 42(12): 1533–1539.

Kim, D. H. (2013). The influence of salinity on various life stages of *Ruppia tuberos* a and implications for its distribution in the Coorong (Doctor Thesis, South Australia). Available at https ://digital.library.adelaide.edu.au/dspace/bitstream/2440/825 35/9/01front.pdf

Kim, R. H., Smith, P. D., Aleyasin, H., Hayley, S., Mount, M. P., Pownall, S., & Westaway, D. (2005). Hypersensitivity of DJ-1-deficient mice to 1-methyl-4-phenyl-1, 2, 3, 6-tetrahy-dropyrindine (MPTP) and oxidative stress. *Proceedings of the National Academy of Sciences*, 102(14): 5215–5220.

Kinnersley, A. M., & Turano, F. J. (2000). Gamma aminobutyric acid (GABA) and plant responses to stress. *Critical Reviews in Plant Sciences*, 19(6): 479–509.

Kirst, G. O. (1990). Salinity tolerance of eukaryotic marine algae. *Annual Review of Plant Biology*, 41(1): 21–53.

Kjerfve, B. (1994). Coastal lagoons. In: Kjerfve, B. (Ed.). Coastal Lagoon Processes, Elsevier, Amsterdam, pp. 1–8.

Krinsky, N. I. (1984). Biology and photobiology of singlet oxygen. In: Bors, W., Saran, M., & Tait, D. (Eds.). *Oxygen Radicals in Chemistry and Biology*, Walter de Gruyter and Co, New York, p. 453.

Kromdijk, J., & Long, S. P. (2016). One crop breeding cycle from starvation? How engineering crop photosynthesis for ris-ing CO2 and temperature could be one important route to alleviation. *Proceedings of the Royal Society B*, 283(1826): 20152578.

Kuehn, G. D., Bagga, S., Rodriguez-Garay, B., & Phillips, G. C. (1990a). Biosynthesis of uncommon polyamines in higher plants and their relationship to abiotic stress responses. In: H.E. Flores, R.N. Arteca, and J.C. Shanon (Eds.). Polyamines and Ethylene: Biochemistry, physiology and interactions. *American Society of Plant Physiologists*, Rockville, MD, pp. 190–202.

Kuehn, G. D., Rodríguez-Garay, B., & Phillips, G. C. (1990b). Novel occurrence of uncommon polyamines in higher plants. *Plant Physiology*, 94: 855–857.

Kumar, M., Kumari, P., Reddy, C. R. K., & Jha, B. (2014). Salinity and desiccation induced oxidative stress acclimation in sea-weeds. In: Nathalie, B. (Ed.). *Advances in Botanical Research*, Academic Press, pp. 91–123.

Kumar, M., Kuzhiumparambil, U., Pernice, M., Jiang, Z., & Ralph, P. J. (2016). Metabolomics: an emerging frontier of systems biology in marine macrophytes. *Algal Research*, 16: 76–92.

Kumar, M., Kuzhiumparambil, U., Ralph, P. J., & Contreras-Porcia, L. (2017). Polyamines: stress metabolite in marine macro-phytes. In: Rastogi, R. P., Madamwar, D., & Pandey, A. (Eds.). *Algal Green Chemistry: Recent Progress in Biotechnology*, Elsevier, Amsterdam, Netherlands, pp. 243–255.

Kuznetsov, V. V. & Shevyakova, N. I. (2007). Polyamines and stress tolerance of plants. *Plant Stress*, 1(1): 50–71.

Ladyman, J. A. R., Ditz, K. M., Grumet, R., & Hanson, A. D. (1983). Genotypic variation for glycinebetaine accumulation by cul-tivated and wild barley in relation to water stress 1. *Crop Science*, 23(3): 465–468.

Lamosa, P., Martins, L. O., Da Costa, M. S., & Santos, H. (1998). Effects of temperature, salinity, and medium composi-tion on compatible solute accumulation by *Thermococcus spp*. *Applied and Environmental Microbiology*, 64(10): 3591–3598.

LaRosa, P. C., Rhodes, D., Rhodes, J. C., Bressan, R. A., & Csonka, L. N. (1991). Elevated accumulation of proline in NaCl-adapted tobacco cells is not due to altered Δ1-pyrroline-5-carboxylate reductase. *Plant Physiology*, 96(1): 245–250.

Lauer, N., Yeager, M., Kahn, A. E., Dobberfuhl, D. R., & Ross, C. (2011). The effects of short term salinity exposure on the sublethal stress response of *Vallisneria americana* Michx. (Hydrocharitaceae). *Aquatic Botany*, 95(3): 207–213.

Lawson, S. E., Wiberg, P. L., McGlathery, K. J., & Fugate, D. C. (2007). Wind-driven sediment suspension controls light avail-ability in a shallow coastal lagoon. *Estuaries and Coasts*, 30(1): 102–112.

Le Fur, I., De Wit, R., Plus, M., Oheix, J., Simier, M., & Ouisse, V. (2018). Submerged benthic macrophytes in Mediterranean lagoons: distribution patterns in relation to water chemistry and depth. *Hydrobiologia*, 808(1): 175–200.

Le Rudulier, D., Strom, A. R., Dandekar, A. M., Smith, L. T., & Valentine, R. C. (1984). Molecular biology of osmoregulation. *Science*, 224(4653): 1064–1068.

Leblebici, Z., Aksoy, A., & Duman, F. (2011). Influence of salinity on the growth and heavy metal accumulation capacity of *Spirodela polyrrhiza* (Lemnaceae). *Turkish Journal of Biology*, 35(2): 215–220.

Leruste, A., Malet, N., Munaron, D., Derolez, V., Hatey, E., Collos, Y., & Bec, B. (2016). First steps of ecological restoration in Mediterranean lagoons: shifts in phytoplankton communities. *Estuarine, Coastal and Shelf Science*, 180: 190–203.

Liang, W., Cui, W., Ma, X., Wang, G., & Huang, Z. (2014). Function of wheat Ta-UnP gene in enhancing salt tolerance in transgenic Arabidopsis and rice. *Biochemical and Biophysical Research Communications*, 450(1): 794–801.

Liang, W., Ma, X., Wan, P., & Liu, L. (2018). Plant salt-tolerance mechanism: a review. *Biochemical and Biophysical Research Communications*, 495(1): 286–291.

Lichtenthaler, H. K. (1998). Stress concept in plants. In: *Stress of Life: From Molecules to Man an Introduction, Annals of New York Academy Science* (Vol. 851), The New York Academy of Sciences, New York, p. 187.

Lirman, D., Deangelo, G., Serafy, J., Hazra, A., Hazra, D. S., Herlan, J., Luo, J., Bellmund, S., Wang, J., & Clausing, R. (2008). Seasonal changes in the abundance and distribution of submerged aquatic vegetation in a highly managed coastal lagoon. *Hydrobiologia*, 596(1): 105–120.

Littles, C. J. (2005). Effects of rapid salinity change on submersed aquatic plants (Doctoral dissertation, University of Florida).

Liu, J., & Zhu, J. K. (1997). Prolineaccumulation and salt-stress-induced gene expression in a salt-hypersensitive mutant of Arabidopsis. *Plant Physiology*, 114: 591–596.

Lutts, S., Kinet, J. M., & Bouharmont, J. (1995). Changes in plant response to NaCl during development of rice (*Oryza sativa* L.) varieties differing in salinity resistance. *Journal of Experimental Botany*, 46(12): 1843–1852.

Macek, P., & Rejmánková, E. (2007). Response of emergent macrophytes to experimental nutrient and salinity additions. *Functional Ecology*, 21(3): 478–488.

Mäkelä, P., Kontturi, M., Pehu, E., & Somersalo, S. (1999). Photosynthetic response of drought and salt stressed tomato and turnip rape plants to foliar applied glycinebetaine. *Physiologia Plantarum*, 105(1): 45–50.

Mallik, S., Nayak, M., Sahu, B. B., Panigrahi, A. K., & Shaw, B. P. (2011). Response of antioxidant enzymes to high NaCl concentration in different salt-tolerant plants. *Biologia Plantarum*, 55(1): 191–195.

Martinez, C. A., Maestri, M., & Lani, E. G. (1996). *In vivo* salt tolerance and proline accumulation in Andean potato (*Solanum* sp). differing in frost resistance. *Plant Science*, 116: 177.

Masood, A., Shah, N. A., Zeeshan, M., & Abraham, G. (2006). Differential response of antioxidant enzymes to salinity stress in two varieties of Azolla (*Azolla pinnata* and *Azolla filiculoides*). *Environmental and Experimental Botany*, 58(1–3): 216–222.

McCord, J. M., & Fridovich, I. (1969). Superoxide dismutase an enzymic function for erythrocuprein (hemocuprein). *Journal of Biological Chemistry*, 244(22): 6049–6055.

McCue, K. F., & Hanson, A. D. (1992). Salt-inducible betaine aldehyde dehydrogenase from sugar beet: cDNA cloning and expression. *Plant Molecular Biology*, 18(1): 1–11.

Meloni, D. A., Gulotta, M. R., & Martínez, C. A. (2008). Salinity tolerance in Schinopsis quebracho colorado: seed germination, growth, ion relations and metabolic responses. *Journal of Arid Environments*, 72(10): 1785–1792.

Millet, B., Robert, C., Grillas, P., Coughlan, C., & Banas, D. (2010). Numerical modelling of vertical suspended solids concentrations and irradiance in a turbid shallow system (Vaccares, Se France). *Hydrobiologia*, 638(1): 161–179.

Miyake, C., & Asada, K. (1992). Thylakoid-bound ascorbate peroxidase in spinach chloroplasts and photoreduction of its primary oxidation product monodehydroascorbate radicals in thylakoids. *Plant and Cell Physiology*, 33(5): 541–553.

Mohanty, P., Hayashi, H., Papageorgiou, G. C., & Murata, N. (1993). Stabilization of the Mn-cluster of the oxygen-evolving complex by glycinebetaine. *Biochimica et Biophysica Acta (BBA)-Bioenergetics*, 1144(1): 92–96.

Morales, F., Abadía, A., Gómez-Aparisi, J., & Abadía, J. (1992). Effects of combined NaCl and CaCl2 salinity on photosynthetic parameters of barley grown in nutrient solution. *Physiologia Plantarum*, 86(3): 419–426.

Morant-Avice, A., Pradier, E., & Houchi, R. (1998). Osmotic adjustment in triticales grown in presence of NaCl. *Biologia Plantarum*, 41(2): 227–234.

Munns, R., & Richards, R. A. (1996). Improving crop productivity in saline soils. In: *Proceedings of the 2nd International Crop Science Congress*, 453.

Munns, R. (2002). Salinity, growth and phytohormones. In: *Salinity: Environment-Plants-Molecules*, Springer, Dordrecht, pp. 271–290.

Munns, R. (2005). Genes and salt tolerance: bringing them together. *New Phytologist*, 167(3): 645–663.

Munns, R., James, R. A., & Läuchli, A. (2006). Approaches to increasing the salt tolerance of wheat and other cereals. *Journal of Experimental Botany*, 57(5): 1025–1043.

Murata, N., Mohanty, P. S., Hayashi, H., & Papageorgiou, G. C. (1992). Glycinebetaine stabilizes the association of extrinsic proteins with the photosynthetic oxygen evolving complex. *FEBS Letters*, 296(2): 187–189.

Munns, R., & Tester, M. (2008). Mechanisms of salinity tolerance. *Annual Review of Plant Biology*, 59: 651–681.

Obrador, B., & Pretus, J. L. (2008). Light regime and components of turbidity in a Mediterranean coastal lagoon. *Estuarine, Coastal and Shelf Science*, 77(1): 123–133.

Panda, S. K., & Khan, M. H. (2004). Changes in growth and superoxide dismutase activity in *Hydrilla verticillata* L. under abiotic stress. *Brazilian Journal of Plant Physiology*, 16(2): 115–118.

Papageorgiou, G. C., & Murata, N. (1995). The unusually strong stabilizing effects of glycine betaine on the structure and function of the oxygen-evolving photosystem II complex. *Photosynthesis Research*, 44(3): 243–252.

Parida, A., Das, A. B., & Das, P. (2002). NaCl stress causes changes in photosynthetic pigments, proteins, and other metabolic components in the leaves of a true mangrove, *Bruguiera parviflora*, in hydroponic cultures. *Journal of Plant Biology*, 45(1): 28–36.

Pasternak, D. (1987). Salt tolerance and crop production-a comprehensive approach. *Annual Review of Phytopathology*, 25(1): 271–291.

Patnaik, J., & Debata, B. K. (1997). In vitro selection of NaCl tolerant callus lines of *Cymbopogon martinii* (Roxb.) Wats. *Plant Science*, 124(2): 203–210.

Pérez-Alfocea, F., & Larher, F. (1995). Sucrose and proline accumulation and sugar efflux in tomato leaf discs affected by NaCl and polyethylene glycol 6000 iso-osmotic stresses. *Plant Science*, 107(1): 9–15.

Perez-Alfocea, F., Santa-Cruz, A., Guerrier, G., & Bolarin, M. C. (1994). NaCl stress-induced organic solute changes on leaves and calli of *Lycopersicon esculentum, L. pennellii* and their interspecific hybrid. *Journal of Plant Physiology*, 143(1): 106–111.

Pérez-Ruzafa, A., Marcos, C., Pérez-Ruzafa, I. M., & Pérez-Marcos, M. (2011). Coastal lagoons: "transitional ecosystems" between transitional and coastal waters. *Journal of Coastal Conservation*, 15(3): 369–392.

Petrusa, L. M., & Winicov, I. (1997). Proline status in salt-tolerant and salt-sensitive alfalfa cell lines and plants in response to NaCl. *Plant Physiology and Biochemistry*, 35: 303–310. (France).

Planchet, E., Verdu, I., Delahaie, J., Cukier, C., Girard, C., Morère-Le Paven, M. C., & Limami, A. M. (2014). Abscisic acid-induced nitric oxide and proline accumulation in independent pathways under water-deficit stress during seedling establishment in *Medicago truncatula*. *Journal of Experimental Botany*, 65(8): 2161–2170.

Polo, L. K., Felix, M. R. L., Kreusch, M., Pereira, D. T., Costa, G. B., Simioni, C., de Paula, Martins, R., Latini, A., Floh, E. S. I., Chow, F., Ramlov, F., Maraschin, M., Bouzon, Z. L., & Schmidt, É. C. (2015). Metabolic profile of the brown macroalga *Sargassum cymosum* (Phaeophyceae, Fucales) under laboratory UV radiation and salinity conditions. *Journal of Applied Phycology*, 27(2): 887–899.

Reddy, M. P. & Iyenger, E. R. R. (1999). Crop responses to salt stress: seawater application and prospects. In: Pessarakli, M. (Ed.). *Handbook of Plant and Crop Stress*, Marcel Dekker Inc, New York-Basel, p. 1041.

Regan, S. M. (2017). Factors affecting monoecious hydrilla (*Hydrilla verticillata*) in dynamic systems. (MS Thesis, North Carolina State University, Raleigh, North Carolina).

Rhodes, D., & Hanson, A. D. (1993). Quaternary ammonium and tertiary sulfonium compounds in higher plants. *Annual Review of Plant Biology*, 44(1): 357–384.

Robinson, S. P., & Jones, G. P. (1986). Accumulation of glycinebetaine in chloroplasts provides osmotic adjustment during salt stress. *Functional Plant Biology*, 13(5): 659–668.

Rodriguez-Garay, B., Phillips, G. C., & Kuehn, G. D. (1989). Detection of norspermidine and norspermine in *Medicago sativa* L. (alfalfa). *Plant Physiology*, 89(2): 525–529.

Rout, N. P. (2001). Effects of salinity on the metabolic activities of aquatic macrophytes. Thesis submitted to the Utkal University (India) for the degree of PhD in Science (Botany).

Rout, N. P., & Shaw, B. P. (1998). Salinity tolerance in aquatic macrophytes: probable role of proline, the enzymes involved in its synthesis and C4 type of metabolism. *Plant Science*, 136(2): 121–130.

Rout, N. P., & Shaw, B. P. (2001a). Salt tolerance in aquatic macrophytes: ionic relation and interaction. *Biologia Plantarum*, 44(1): 95–99.

Rout, N. P., & Shaw, B. P. (2001b): Salt tolerance in aquatic macrophytes: possible involvement of the antioxidative enzymes. *Plant Science*, 160(3): 415–423.

Rout, N. P., Tripathi, S. B., & Shaw, B. P. (1997). Effect of salinity on chlorophyll and proline contents in three aquatic macrophytes. *Biologia Plantarum*, 40(3): 453–458.

Rozema, J. (1976). An ecophysiological study on the response to salt of four halophytic and glycophytic Juncus species. *Flora*, 165(2): 197–209.

Salama, S., Trivedi, S., Busheva, M., Arafa, A. A., Garab, G., & Erdei, L. (1994). Effects of NaCl salinity on growth, cation accumulation, chloroplast structure and function in wheat cultivars differing in salt tolerance. *Journal of Plant Physiology*, 144(2): 241–247.

Santa-Cruz, A., Estan, M. T., Rus, A., Bolarin, M. C., & Acosta, M. (1997). Effects of NaCl and mannitol iso-osmotic stresses on the free polyamine levels in leaf discs of tomato species differing in salt tolerance. *Journal of Plant Physiology*, 151(6): 754–758.

Santoro, M. M., Liu, Y., Khan, S. M., Hou, L. X., & Bolen, D. W. (1992). Increased thermal stability of proteins in the presence of naturally occurring osmolytes. *Biochemistry*, 31(23): 5278–5283.

Schubert, H., Feuerpfeil, P., Marquardt, R., Telesh, I., & Skarlato, S. (2011). Macroalgal diversity along the Baltic Sea salinity gradient challenges Remane's species-minimum concept. *Marine Pollution Bulletin*, 62(9): 1948–1956.

Selye, H. (1936). A syndrome produced by various nucuous agents. *Nature*, 138: 32.

Shaw, B. P., Rout, N. P., Barman, B. C., Choudhury, S. B., & Rao, K. H. (2000). Distribution of macrophytic vegetation in relation to salinity in the Chilka lake, a lagoon along east coast of India. *Indian Journal of Marine Sciences*, 29: 144–148.

Shaw, B. P., Sahu, S. K., & Mishra, R. K. (2004). Heavy metal induced oxidative damage in terrestrial plants. In: M. N. V. Prasad (Ed.). *Heavy Metal Stress in Plants: From Biomolecules to Ecosystems*, 2nd edition, , Springer-Verlag, New York, pp. 84–126.

Shields, E. C., & Moore, K. A. (2016). Effects of sediment and salinity on the growth and competitive abilities of three submersed macrophytes. *Aquatic Botany*, 132: 24–29.

Siefermann-Harms, D. (1987). The light-harvesting and protective functions of carotenoids in photosynthetic membranes, *Physiologia Plantarum*, 69: 561.

Sim, L. L., Chambers, J. M., & Davis, J. A. (2006). Ecological regime shifts in salinised wetland systems. I. Salinity thresholds for the loss of submerged macrophytes. *Hydrobiologia*, 573(1): 89–107.

Singha, S. & Choudhury, M. A. (1990). Effect of salinity (NaCl) stress on H2O2 metabolism in *Vigna* and *Oryza* seedlings. *Biochemie und Physiologie der Pflanzen*, 186(1): 69–74.

Stoeva, N., & Kaymakanova, M. (2008). Effect of salt stress on the growth and photosynthesis rate of bean plants (*Phaseolus vulgaris* L.). *Journal of Central European Agriculture*, 9(3): 385–391.

Stoler, A., Sudol, K., Mruzek, J., & Relyea, R. (2018). Interactive effects of road salt and sediment disturbance on the productivity of seven common aquatic macrophytes. *Freshwater Biology*, 63(7): 709–720.

Storey, R., Ahmad, N., & Jones, R. W. (1977). Taxonomic and ecological aspects of the distribution of glycinebetaine and related compounds in plants. *Oecologia*, 27(4): 319–332.

Sudhakar, C., Lakshmi, A., & Giridarakumar, S. (2001). Changes in the antioxidant enzyme efficacy in two high yielding genotypes of mulberry (*Morus alba* L.) under NaCl salinity. *Plant Science*, 161(3): 613–619.

Szabados, L., & Savoure, A. (2010). Proline: a multifunctional amino acid. *Trends in Plant Science*, 15(2): 89–97.

Szabolcs, I (1994). Soils and Salinisation. In: M. Pessrakali (Ed.). *Handbook of Plant and Crop Stress*, Marcel Dekker, New York, p. 3.

Tanaka, Y., Hibino, T., Hayashi, Y., Tanaka, A., Kishitani, S., Takabe, T., & Yokota, S. (1999). Salt tolerance of transgenic rice overexpressing yeast mitochondrial Mn-SOD in chloroplasts. *Plant Science*, 148(2): 131–138.

Tang, X., Mu, X., Shao, H., Wang, H., & Brestic, M. (2015a). Global plant-responding mechanisms to salt stress: physiological and molecular levels and implications in biotechnology. *Critical Reviews in Biotechnology*, 35(4): 425–437.

Tang, X., Mu, X., Shao, H., Wang, H., & Brestic, M. (2015b). Global plant-responding mechanisms to salt stress: Physiological and molecular Szabados, L.; Savouré, A. Proline: A multifunctional amino acid. *Trends in Plant Science*, 15: 89–97.

Thouvenot, L., Deleu, C., Berardocco, S., Haury, J., & Thiébaut, G. (2015). Characterization of the salt stress vulnerability of three invasive freshwater plant species using a metabolic profiling approach. *Journal of Plant Physiology*, 175: 113–121.

Thouvenot, L., Haury, J., & Thiébaut, G. (2012). Responses of two invasive macrophyte species to salt. *Hydrobiologia*, 686(1): 213–223.

Tiburcio, A. F., Altabella, T., Bitrián, M., & Alcázar, R. (2014). The roles of polyamines during the lifespan of plants: from development to stress. *Planta*, 240: 1–18.

Tripathi, S. B., Gurumurthi, K., Panigrahi, A. K., & Shaw, B. P. (2007). Salinity induced changes in proline and betaine contents and synthesis in two aquatic macrophytes differing in salt tolerance. *Biologia Plantarum*, 51(1): 110.

Twilley, R. R., & Barko, J. W. (1990). The growth of submersed macrophytes under experimental salinity and light conditions. *Estuaries*, 13(3): 311–321.

Upadhyay, R. K., & Panda, S. K. (2005). Salt tolerance of two aquatic macrophytes, *Pistia stratiotes* and *Salvinia molesta*. *Biologia Plantarum*, 49(1): 157–159.

Van den Brink, F. W. B., & Van der Velde, G. (1993). Growth and morphology of four freshwater macrophytes under the impact of the raised salinity level of the lower Rhine. *Aquatic Botany*, 45(4): 285–297.

van Kempen, M. M., Smolders, A. J., Bögemann, G. M., Lamers, L. L., Visser, E. J., & Roelofs, J. G. (2013). Responses of the *Azolla filiculoides* Stras.–Stras.–*Anabaena azollae* Lam. association to elevated sodium chloride concentrations: amino acids as indicators for salt stress and tipping point. *Aquatic Botany*, 106: 20–28.

Vincent, C., Mouillot, D., Lauret, M., Do Chi, T., Troussellier, M., & Aliaume, C. (2006). Contribution of exotic species, environmental factors and spatial components to the macrophyte assemblages in a Mediterranean lagoon (Thau lagoon, Southern France). *Ecological Modelling*, 193(1–2): 119–131.

Vinocur, B., & Altman, A. (2005). Recent advances in engineering plant tolerance to abiotic stress: achievements and limitations. *Current Opinion in Biotechnology*, 16(2): 123–132.

Virk, P. S., Khush, G. S., & Peng, S. (2004). Breeding to enhance yield potential of rice at IRRI: the ideotype approach. *International Rice Research Notes*, 29(1): 5–9.

Wang, F., Liu, J., Zhou, L., Pan, G., Li, Z., & Cheng, F. (2016). Senescence-specific change in ROS scavenging enzyme activities and regulation of various SOD isozymes to ROS levels in psf mutant rice leaves. *Plant Physiology and Biochemistry*, 109: 248–261.

Wang, W., Vinocur, B., & Altman, A. (2003). Plant responses to drought, salinity and extreme temperatures: towards genetic engineering for stress tolerance. *Planta*, 218(1): 1–14.

Warwick, N. W., & Bailey, P. C. (1997). The effect of increasing salinity on the growth and ion content of three non-halophytic wetland macrophytes. *Aquatic Botany*, 58(1): 73–88.

Watt, S. C. L., García-Berthou, E., & Vilar, L. (2007). The influence of water level and salinity on plant assemblages of a seasonally flooded Mediterranean wetland. *Plant Ecology*, 189(1): 71.

Weigel, P., Weretilnyk, E. A., & Hanson, A. D. (1986). Betaine aldehyde oxidation by spinach chloroplasts. *Plant Physiology*, 82(3): 753–759.

Weretilnyk, E. A., & Hanson, A. D. (1989). Betaine aldehyde dehydrogenase from spinach leaves: purification, in vitro translation of the mRNA, and regulation by salinity. *Archives of Biochemistry and Biophysics*, 271(1): 56–63.

Williamson, C. L., & Slocum, R. D. (1992). Molecular cloning and evidence for osmoregulation of the Δ1-pyrroline-5-carboxylate reductase (*proC*) gene in pea (*Pisum sativum* L.). *Plant Physiology*, 100(3): 1464–1470.

Wilson, C., Lesch, S. M., & Grieve, C. M. (2000). Growth stage modulates salinity tolerance of New Zealand spinach (*Tetragonia tetragonioides*, Pall.) and red orach (*Atriplex hortensis* L.). *Annals of Botany*, 85(4): 501–509.

Wu, L., & Guo, X. (2002). Selenium accumulation in submerged aquatic macrophytes *Potamogeton pectinatus* L. and *Ruppia maritima* L. from water with elevated chloride and sulfate salinity. *Ecotoxicology and Environmental Safety*, 51(1): 22–27.

Xiong, L., & Zhu, J. K. (2001). Abiotic stress signal transduction in plants: molecular and genetic perspectives. *Physiologia Plantarum*, 112(2): 152–166.

Xu, Y., Shi, G. X., Ding, C. X., & Xu, X. Y. (2011). Polyamine metabolism and physiological responses of *Potamogeton crispus* leaves under lead stress. *Russian Journal of Plant Physiology*, 58(3): 460–466.

Yazici, I., Türkan, I., Sekmen, A. H., & Demiral, T. (2007). Salinity tolerance of purslane (*Portulaca oleracea* L.) is achieved by enhanced antioxidative system, lower level of lipid peroxidation and proline accumulation. *Environmental and Experimental Botany*, 61(1): 49–57.

Yoshiba, Y., Kiyosue, T., Nakashima, K., Yamaguchi-Shinozaki, K., & Shinozaki, K. (1997). Regulation of levels of proline as an osmolyte in plants under water stress. *Plant and Cell Physiology*, 38(10): 1095–1102.

Zhang, C. S., Lu, Q., & Verma, D. P. S. (1995). Removal of feedback inhibition of Δ1-pyrroline-5-carboxylate synthetase, a bifunctional enzyme catalyzing the first two steps of proline biosynthesis in plants. *Journal of Biological Chemistry*, 270(35): 20491–20496.

Zhao, H., Wang, F., Ji, M., & Yang, J. (2014). Effects of salinity on removal of nitrogen and phosphorus from eutrophic saline water in planted *Lythrum salicaria* L. microcosm systems. *Desalination and Water Treatment*, 52(34–36): 6655–6663.

Zhu, J. K. (2007). Plant salt stress. In: *Encyclopedia of Life Sciences*, John Wiley and Sons, pp. 1–3.

Zingelwa, N. S., & Wooldridge, J. (2016). Tolerance of macrophytes and grasses to sodium and chemical oxygen demand in winery wastewater. *South African Journal of Enology and Viticulture*, 30(2): 117–123.

6 Oxidative Stress in Plants
Production, Metabolism, and Biological Roles of Reactive Oxygen Species

Mojtaba Kordrostami, Babak Rabiei, and Ali Akbar Ebadi

CONTENTS

6.1 INTRODUCTION

Any changes in the amount and intensity of abiotic factors (temperature, relative humidity, light intensity, and nutrients), which leads to a change in the normal form of plant physiology, is known as stress (Pandey et al., 2017). Biologically, stress is the result of an inconsistency that tends to prevent the normal systems from operating (Negrão, Schmöckel, and Tester, 2017). Due to any kind of stress, reactive oxygen species (ROS) are produced (Schieber and Chandel, 2014). These species have high energy or an extra electron charge due to electron excitation. More precisely, under stress conditions, homeostasis of intracellular oxidation-reduction is impaired, which results in the formation of oxidative stress and reactive oxygen species (Nita and Grzybowski, 2016). Oxidative stress is produced by increasing the number of reactive oxygen species or free radicals, which results in early aging, increased permeability, ions leakage from the cell membranes, and reduced photosynthesis in plants (Sharma et al., 2012). Free radicals result in cellular damage through lipid peroxidation (mainly cell membranes) and the blocking of natural antioxidants. By measuring the malondialdehyde (MDA), which is the result of lipid peroxidation, we can find the amount of

stress in the plant cells (Grotto et al., 2009). Under oxidative stress conditions, by producing oxygen intermediates, which are relatively reduced or energy-intensive forms of atmospheric oxygen (O_2), the plant finds itself under stress conditions and activates a variety of defense systems such as antioxidants to protect against stress (Carvalho, 2008). The basis for measuring the amount of stress created in the plant, in many cases, is the measurement of the activity of the peroxidase enzymes, such as hydrogen peroxide (H_2O_2). In the case of the excessive accumulation of these activated oxygens, a variety of cell damage such as DNA damage, membrane lipid peroxidation, RNA damage, protein oxidation, and enzymatic inhibition, occurs in the cell (Saini et al., 2018). Free oxygen radicals or lipid peroxidation reactions in the plant membrane will selectively break up unsaturated fatty acids and accumulate hydrocarbons, aldehydes, and the like (Frankel, 1984). Therefore, in some studies, in order to determine the effect of environmental stresses on the membrane of plant cells, the amount of lipid peroxidation products, such as malondialdehyde (MDA) or hydrogen peroxide (H_2O_2), is measured and their results suggest the involvement of free oxygen radicals in response to stress (Birben et al., 2012).

6.2 THE PRODUCTION OF REACTIVE OXYGEN SPECIES IN THE PLANT

Research has shown that environmental stresses, by disrupting favorable conditions, cause metabolic disorders in plant cells. One of the main causes of these disorders is an increase in the production of ROS (Zandalinas and Mittler, 2018). Under normal conditions, four electrons are needed to reduce oxygen completely and convert it into water (Alberts et al., 2002). When the main path of electron transport is blocked, another sub-path for activating electron transfer is activated. In this pathway, unlike the main pathway, where the electrons were transferred to oxygen at once, electrons are transported individually to the oxygen. As a result of this, atmospheric oxygen is incompletely reduced, and by taking one, two, and three electrons, superoxide, hydrogen peroxide, and hydroxyl radicals are produced, respectively (Kordrostami, Rabiei, and Kumleh, 2017). The other form of ROS is singlet oxygen. In this form, electrons go into higher orbits and the molecule is excited (Khan and Wilson, 1995). A wide range of molecular impairments that lead to the physiological damage to plants under environmental stresses can be attributed to the production of reactive oxygen species. From the excitation of the molecular oxygen, singlet oxygen or excited oxygen are generated, and by reducing one, two, and three electrons, they will be converted to O_2^-, H_2O_2, and OH, respectively (Sharma et al., 2012). A Hydroxyl radical is able to react with any substrate, because it is very active, and the cell does not have an enzymatic mechanism to scavenge it. Because of this, the cell only lends itself to mechanisms that prevent its formation (Lobo et al., 2010). A superoxide anion is a non-radical ROS molecule which is often produced at the Photosystem II reaction center, and despite its short half-life, it can act as a molecular messenger through diffusion to the outer membrane of thylakoids or through secondary messengers such as carotenoid oxidation. Despite its toxic effects, this molecule triggers signaling cascades and causes programmed cell death (PCD) (Ramel et al., 2012). If this energy is not inhibited by carotenoids, O_2^- can flow energy among other molecules and can cause damage to them, including the high-peroxidation of unsaturated fatty acids (Das and Roychoudhury, 2014). The conversion of one ROS to another ROS is one of the important chemical characteristics of ROS (Suzuki et al., 2011).

Exposure to environmental stresses may increase the production of ROS and the oxidative stress, and eventually accelerate the damage to membranes, proteins, and DNA, changes in growth and physiological structure of the cells, and destroying enzymes. In addition, $O_2^{\bullet-}$ in chloroplasts is produced by Mehler reaction by transferring an electron from iron-sulfur to O_2. In this way, two $O_2^{\bullet-}$ and two H^+ in the reaction catalyzed by SOD, produce H_2O_2 and H_2O molecules. H_2O_2 conversion into $HO^{\bullet}$ is catalyzed by various bivalent cations, especially Fe^{2+} (Fenton reaction) (Sabater and Martín, 2013). The radicals direct the reactions that cause DNA damage, lipid peroxidation, and damage the membrane proteins and macro proteins in the cell, including chlorophylls and enzymes (Roy and Basu, 2009).

Under stress conditions, the rate of photosynthesis decreases, which indicates that the photosynthesis apparatus has been affected by stress (Gururani, Venkatesh, and Tran, 2015). Under osmotic stress conditions (salinity and drought), the stomata are closed, and subsequently the CO_2 concentration in the mesophilic tissue is reduced, and following this state the photosynthetic dark reactions (Calvin cycle) are interrupted and the products of light reactions, which include ATP and NADPH, are not consumed (Kreslavski et al., 2013). In such a situation, due to the lack of NADPH oxidation, the use of $NADP^+$ decreases for electron reception. Therefore, the oxygen molecule acts as the electron substitute receptor in the electron transport chain and leads to the formation of ROS (Pospíšil, 2009). These radicals cause molecular damage, cellular side effects on plants, and ultimately plant death (Sharma et al., 2012). In the cells, NADPH is produced by the light reaction, electron transport chain, oxidative pentose phosphate cycle, oxidoreductase reaction, Krebs cycle, and oxidoreductase enzymes activity (Berg, Tymoczko, and Stryer, 2012). Therefore, when the NADPH:$NADP^+$ ratio increases, the plant activates the pathway to modify the above ratio by purging the hydrogen potential (Zandalinas and Mittler, 2018).

6.3 CAN REACTIVE OXYGEN SPECIES (ROS) BE CONSIDERED AS SECONDARY MESSENGERS?

The reactive oxygen species play a dual role in the plants. At low concentrations, they act as messengers. Different studies have suggested that the reactive oxygen species act as messengers in biotic and abiotic stresses (Das and Roychoudhury, 2014). Lower concentrations of ROS are involved in cellular signal transduction, adaptability, and cross-resistance, while high concentrations of ROS are highly harmful to plants (Sewelam, Kazan, and Schenk, 2016). It is believed that, among all the ROS molecules, H_2O_2 is released freely through biological membranes (Birben et al., 2012). H_2O_2 in chloroplasts can directly affect the function of extracellular messenger components that play a potential role in the systemic reactions of the plants against excessive light (Mullineaux and Karpinski, 2002). H_2O_2 is a relatively stable ROS, which remains unchanged at a physiological pH (Sewelam et al., 2014). This molecule plays a major role in initiating the expression of antioxidant enzymes' genes (GR, CAT, SOD, APX) in photoprotection (Racchi, 2013). Under excessive light conditions, H_2O_2 is known to cause a secondary defense line and neutralize photo-inhibition due to its action as an intracellular message circulating from the chloroplast to the cytosol (Petrov and Van Breusegem, 2012). Hence, H_2O_2 is known as an intracellular and intercellular systemic message that is involved in excessive light adaptable plants (Mullineaux and Baker, 2010). The signaling mechanisms interacting with the ROS as described above (to pass ROS molecules outside of the chloroplast to multiply the message) depend on the expression of the nuclear gene (Surpin, Larkin, and Chory, 2002). Thus, it is believed that the electron transfer chain begins with NADPH in the stroma, traverses the chloroplast membrane and ends as a final receptor (in the outer portion of the chloroplastic surface) after reacting with oxygen (Mittler, 2002). The involvement of ROS has been shown to interfere with the regulation of stomatal closure and auxin-associated cellular reactions (Sierla et al., 2016), which are the most frequent cycles that may be related to the interference of ROS as the initiator of systemic symptoms in plants under oxidative stress conditions (Mittler, 2002). The use of transgenic plants with a complete analysis of the ROS producing and scavenging systems by the use of genomics and proteomics solves the role of ROS in the signal transduction in the plant (Saini et al., 2018).

6.4 CELLULAR DEFENSE MECHANISMS AGAINST ROS

As mentioned above, the reactive oxygen species play a dual role in the plants. At low concentrations, they act as messengers. Under the oxidative stress conditions, by producing oxygen intermediates, which are relatively reduced or energy-intensive forms of atmospheric oxygen (O_2), the plant finds itself under stress conditions and activates a variety of defense systems such as antioxidants to protect against stress (Carvalho, 2008). Reactive oxygen species are the inevitable products of cellular metabolism and are even produced under favorable environmental conditions (Mittler, 2002). Environmental stresses affect growth, plant physiological structure, protein synthesis, enzymatic and non-enzymatic activities, cellular respiration and metabolism (Shabala, 2017). Exposure to the stress may increase the production of various ROS, damage to membranes, proteins, and DNA, inducing changes in the growth and physiological structure of the cell and destroying enzymes (Das and Roychoudhury, 2014). Fortunately, plants have special defense mechanisms that enable them to scavenge ROS or prevent their formation. Antioxidants are molecules that prevent the function of the free radicals and prevent cell damage. They can give electrons to the free radicals, transform them into a sustainable form, and prevent their destructive effects (Ahmad et al., 2010). Antioxidant defense acts to protect the cells against the dangerous effects of active oxygen species. During oxidative stress, various events occur in plants which include increasing ROS production, increasing the expression of antioxidant genes, and increasing the capacity of scavengers, which can lead to increased plant stress tolerance (Ighodaro and Akinloye, 2017). The degree of antioxidant activity and the level of antioxidants in the plants depend on plant species, developmental stage, metabolic conditions, duration and intensity of stress (Caverzan, Casassola, and Brammer, 2016). The ROS scavenging mechanisms exist in all plants, and plants produce an effective antioxidant system against stressors (Das and Roychoudhury, 2014). Defense mechanisms often involve the use of antioxidant enzymes (superoxide dismutase, catalase, ascorbate peroxidase, glutathione reductase, glutathione peroxidase, dehydroascorbate reductase, etc.), and antioxidant molecules (vitamin E, ascorbates, glutathione, etc.) (Ighodaro and Akinloye, 2017).

6.5 NON-ENZYMATIC ANTIOXIDANTS

The harmful effects of ROS on cellular macromolecules can be reduced by the activity of antioxidant compounds such as ascorbic acid, glutathione, and carotenoids.

6.5.1 Ascorbic Acid

L-Ascorbic acid is a metabolic agent found abundantly in plant cells, which in some cases reaches up to 10% of the carbohydrate content of the plant (Hancock et al., 2003). Ascorbic acid plays an important role in the physiology of plant stress and plant growth and development (Hossain et al., 2017). Recently, there is increasing evidence that ascorbic acid plays an important role in the scavenging of reactive oxygen species in plants (Ye et al., 2011). Ascorbic acid has the capability of directly scavenging several ROS (including, singlet oxygen and hydroxyl radical) (Das and Roychoudhury, 2014). These researchers also recognized ascorbic acid as an antioxidant (α-tocopherol) protecting agent (bound to the membrane in a reduced state). In addition, ascorbic acid (as a cofactor in the xanthophyllic cycle) plays a major role in photoprotection (Jahns and Holzwarth, 2012). It has been observed that ascorbic acid is required as a cofactor for the violaxanthin-de-epoxidase enzyme (Urzica and Adler, 2012). The results of a study carried out by (Conklin, 2001) showed that the mutant plants lacking ascorbic acid, due to the decrease in the rate of this de-epoxidation reaction, had fewer levels of non-photochemical quenching.

6.5.2 Glutathione

Glutathione is a water-soluble non-proteinaceous compound and a low molecular weight sulfur tripeptide (De Kok et al., 1986). The sequential oxidation-reduction of NADPH, ascorbate, and glutathione, can act as a potential H_2O_2 scavenger (produced by the photo-oxidative stress in chloroplasts) (Foyer and Noctor, 2011). These reactions are generally referred to as an ascorbate-glutathione cycle. This cycle is performed in most of the cellular organs such as mitochondria, chloroplasts, and cytoplasm, which have been co-formed with glutathione reductase, dehydroascorbate reductase, ascorbate peroxidase enzymes, and ascorbate and glutathione (Jiménez et al., 1997). The mentioned cycle is one of the most important ways to decompose hydrogen peroxide into water. Hydrogen is required to restore hydrogen peroxide by H^+, NADPH; thus, the NADPH:$NADP^+$ ratio is moderated in most of the cellular organs by performing this cycle (Carvalho and Amâncio, 2002). Glutathione reductase (GR, EC 1.6.4.2) is a flavoenzyme found in all living organisms. This enzyme reduces glutathione disulfide (GSSG) to glutathione (GSH) along with NADPH oxidation, which is one of the major reactions to the system of scavenging reactive oxygen species in plants (Tsai et al., 2005). Reduced glutathione (GSH) is required for many cellular functions. Unlike the oxidized forms of many primary and secondary metabolites that can react with ROS, GSSG is rapidly reduced by GR in the important organs

(Noctor et al., 2012). Intracellular GSSG concentrations increase as a result of severe stresses (Emri, Pócsi and Szentirmai, 1997). Reconstruction of GSH by GR is a highly efficient process that scatters the energy and helps in setting the ATP:NADPH ratio when the CO_2 fixation is limited (Yousuf et al., 2012).

Currently, glutathione levels in plant cells are used as indicators of oxidation stress in higher plants (Tausz, Šircelj, and Grill, 2004). Also, various studies have shown that glutathione is an antioxidant in the mitochondria, cytosol, peroxisomes, and nucleus (Noctor et al., 2004).

6.5.3 Carotenoids

Carotenoids are one of the most important tetra-terpene group members (Arvayo-Enríquez et al., 2013). They are highly lipophilic compounds and are always associated with biological membranes. Carotenoids in chloroplasts are considered as an important pigment for photosynthesis (Ashraf and Harris, 2013). They also protect the plant against UV radiation. Most carotenoids are derived from a 40-carbon chain of which can be considered as a molecular skeleton. The configuration of carotenoids in nature is trans-type (Goodwin, 1980). Carotenoids play an important role in protecting plant cells against excessive light and, thus, cause the excess energy to be absorbed (Edreva, 2005). Hence, carotenoids are powerful receptors for ROS and protecting pigments and fats from oxidative damage.

6.6 ENZYMATIC ANTIOXIDANTS

The ROS scavenging enzymes are located in different parts of the plant cell (Tripathy and Oelmüller, 2012). The ROS scavenging enzymes in plant cells include catalase, superoxide dismutase, monodehydroascorbate, ascorbate peroxidase, dehydro-ascorbate reductase, glutathione reductase and glutathione peroxidase.

6.6.1 Superoxide Dismutase (SOD: EC 1.15.1.1)

Superoxide dismutase (SOD) is on the front line of antioxidant defense. Superoxide is known to be one of the major ROS in the cell, which causes changes in the nature of enzymes, lipid peroxidation, and DNA denaturation (Fukai and Ushio-Fukai, 2011). Superoxide dismutase is an enzyme capable of scavenging O_2^- and is the first enzymatic barrier for aerobic organisms. It is important to adjust its amount to modify the effects of oxidative stress (Racchi, 2013). When the ROS increases in the cell, the chain reaction begins, in which SOD converts two molecules of superoxide into oxygen and hydrogen peroxide (Bowler, Montagu and Inzé, 1992). Superoxide dismutase scavenges the first molecule produced from the reduction of the mono-valent oxygen (superoxide radical). For this reason, SOD is considered as the front line of antioxidant defense. The enzyme is a metalloenzyme (metalloprotein) encoded by a small gene family and divided into three groups based on the metal found in the active site (Shiriga et al., 2014). In plants, Cu/ZnSOD is generally present in the cytosol, FeSOD

and Cu/ZnSOD are present in chloroplasts, and MnSOD is present in mitochondria (Foyer et al., 1997). It seems that the role of the superoxide dismutase enzyme in the scavenging of superoxide is very important and is one of the most important antioxidants in the defense against reactive oxygen species. Because the phospholipid membrane is impermeable to superoxide radical, the superoxide dismutase enzyme is known to be the primary defense against free oxygen radicals; in fact, it acts as an antioxidant and protects cellular components (Fukai and Ushio-Fukai, 2011). Various studies have shown that the expression of chloroplastic SODs is a successful way to improve plant resistance to different sources of oxidative stress (Grene, 2002; Gill and Tuteja, 2010a).

6.6.2 Catalase (CAT, EC 1.11.1.6)

Catalase is one of the most effective enzymes produced in all living organisms under stress conditions (Racchi, 2013). This enzyme has a direct effect on hydrogen peroxide, which reduces its toxic effects. In fact, catalase uses hydrogen peroxide as a precursor and, by rapidly scavenging this molecule, inhibits its destructive effects (Gill and Tuteja, 2010b). The H_2O_2 molecule is scavenged by catalase without the need for energy, but only at high concentrations of H_2O_2, and its low concentrations are scavenged by other peroxidases (in collaboration with potent reducing agents such as glutathione and ascorbate) (Scandalios, 1997). Catalase activity reduces two H_2O_2 molecules to H_2O and O_2 (Anjum et al., 2016). Catalase is mainly located in the peroxisome, so its activity range is limited and when the amount of ROS increases under the stress conditions, it participates in scavenging (Ighodaro and Akinloye, 2017). The most important characteristic of catalase compared to other enzymes is that it does not require a reducer to catalyze the reaction (Mhamdi et al., 2010). Catalase is one of the first antioxidant enzymes that was discovered and characterized (Mhamdi, Noctor and Baker, 2012). Catalase is expressed with high intensity, especially in all types of plant cells, and as a result is an integral part of the plant's antioxidant system (Sofo et al., 2015).

6.6.3 Peroxidase (POD, EC 1.11.1.x)

Peroxidase is a large group of antioxidant enzymes that are produced in plants in response to biotic and abiotic stresses (You and Chan, 2015). Peroxidase typically catalyzes the oxidation-reduction reaction between H_2O_2 as an electron receptor and a large variety of phenolic substrates, ascorbic acid, aromatic amines, and cytochrome C and acts as ROS scavenging enzymes (Sharma et al., 2012). The peroxidase enzyme breaks down H_2O_2 in the cell and, thus, prevents ROS production, and with increasing levels of activity of this enzyme, the plant is less damaged by ROS. In other words, the peroxidase, by converting H_2O_2 into water and O_2, can neutralize the harmful effects of H_2O_2 (Das and Roychoudhury, 2014). Plant peroxidase can catalyze the reduction of H_2O_2 in their usual activity cycles. This group of enzymes is present in the cytosol, vacuole, chloroplasts, and apoplast (Sofo et al.,

2015). Peroxidase activity can be easily identified throughout the life cycle of different plants from the early stages of germination to the senescence stage by controlling the cell elongation, defense mechanisms, and several other functions (Niu and Liao, 2016). The enzyme has been found extensively in animals, plants, and microbes, and is involved in many cellular processes such as auxin metabolism, wood formation, response to environmental stresses, etc. Guaiacol peroxidase (EC 1.11.1.7) reacts with a large number of substrates, including guaiacol, pyrogallol, and phenol, which are used to measure the activity of the enzyme, but its specific substrate is hydrogen peroxide (Mika and Lüthje, 2003). In these enzymes, the nature of electron-donating depends to a large extent on the enzyme's structure and phenolic compounds such as guaiacol act as an electron donor to H_2O Guaiacol peroxidase is widely accepted as a stress enzyme and can act as an effective reducing agent for O_2 intermediary forms and peroxy radical under stress conditions (Blokhina, Virolainen and Fagerstedt, 2003).

6.6.4 GLUTATHIONE REDUCTASE AND GLUTATHIONE-S-TRANSFERASE

Glutathione reductase completes the Asada–Halliwell pathway through the re-production of a glutathione reservoir using NADPH as an electron donor (Perl-Treves and Perl, 2002). Although most of the glutathione reductase activity has been studied in chloroplasts, the mitochondrial forms of this enzyme are also described in cytosolic isoenzymes in plant cells. It has also been reported that glutathione reductase glyoxalase ΙΙ reduces the oxidized glutathione into reduced glutathione and helps to maintain the reduced glutathione: oxidized glutathione ratio in the plant cell (Foyer et al., 1997).

Glutathione-s-transferase is another important enzyme that protects plant cells through the scavenging of the harmful compounds (Gill and Tuteja, 2010b). It has been found that glutathione-s-transferase catalyzes the binding of glutathione to various toxic compounds in the cell. It has also been shown that glutathione-s-transferases play an important role in protecting plants from various biotic and abiotic stresses, including xenobiotic toxins, ultraviolet radiation, and photo-oxidation stress (Pellegrini et al., 2015). Also, products of oxidative damage are bound to the reduced glutathione by the glutathione-s-transferase enzyme and are scavenged. Therefore, glutathione-S-transferase is considered a key enzyme in maintaining glutathione self-preservation (Zeng, Lu and Wang, 2005).

6.6.5 PEROXIREDOXINS (PRX) AND ALTERNATIVE OXIDASE (AOX)

Peroxiredoxin has the ability to react directly with H_2O_2 and convert it to water by supplying the necessary hydrogen (Dietz et al., 2006). In this way, PRX becomes its oxidized form and establishes a thiol bond between the two molecules. In order to continue the activity of this cycle, the oxidized form of PRX should be converted again to its reduced form (Miller

et al., 2010). This action is done by thioredoxin (TRX) and glutaredoxin (GRX). Peroxiredoxin acts in important organs such as chloroplasts, mitochondria, and cytosols. Evidence suggests that there is a complex interaction between the TRX and GRX systems in plants (Foyer and Noctor, 2009). The oxidized disulfide form of TRX is reduced directly by NADPH and thioredoxin reductase, while GRX is reduced by glutathione (GSH) using NADPH electron. The cytosolic and mitochondrial TRX requires NADPH-dependent thioredoxin reductase: NTR for the reduction (Foyer and Shigeoka, 2011). While the reduction of chloroplastic TRX is performed by ferredoxin/TRX reductase: FTR, TRX plays an essential role in oxidative stress tolerance in plants. They deal with oxidative damage, with electron transfer, to reduce the reductase required to scavenge lipid hydroperoxides and, thus, repair the oxidized protein. They are involved in the ROS scavenging by regulating antioxidant enzymes (Schürmann and Buchanan, 2008).

Alternative oxidase is another pathway to scavenge the ROS. As already known, mitochondria are the engines for producing cellular energy in plants (Xu, Yuan, and Lin, 2011). They use the electron transfer chain to oxidize NADPH and produce a proton stimulus within the mitochondrial membrane, of which ADP is phosphorylated to ATP (Schertl and Braun, 2014). Another consequence of the activity of the electron transport chain is the transfer of an electron to oxygen and the production of superoxide. These reactive oxygen species may be converted to other ROS, including H_2O_2, by the activity of superoxide dismutase (SOD) enzyme (Zandalinas and Mittler, 2018). In this case, the amount of electron leakage increases, which increases the membrane potential and further reduction of the electron transport chain components. The plant mitochondria have a mechanism that can oxidize additional NADPH without the production of ATP to prevent over-reduction of the respiratory chain (Finnegan, Soole, and Umbach, 2004). The alternative oxidase pathway is one of the important ways to reduce the production of various active oxygen species in chloroplast and mitochondrial organs (Carvalho, 2008). In this path, oxygen is still the final receptor of the electron. But, instead of electrons being transmitted one by one, all of the four electrons needed to completely recover oxygen are absorbed by it and turn it into water. This action is done by the alternative oxidase enzyme. Activating this alternative pathway, like other inhibitor paths, prevents the formation of ROS and reduces the pressure of the available electrons on the electron transport chain (Maxwell, Wang, and McIntosh, 1999). By activating the alternative oxidase, the cell can reduce the production of hydrogen peroxide (Purvis, 1997). The AOX enzyme is a cyanide-resistant oxidase that catalyzes oxygen reduction in the mitochondrial electron transfer chain by receiving electrons from the ubiquinone (Vanlerberghe and McIntosh, 1996). Similar to the AOX function (reducing ATP production), uncoupling proteins (UCP) are also a membrane protein that blocks the proton slope. Uncoupling proteins were first discovered in animals; later they were detected in plants (Blokhina and Fagerstedt, 2010). In plants and animals, uncoupling proteins are coded in the nucleus by polygenic families. Studies have shown that UCP activity is stimulated by ROS. Although UCP activity may lead to heat generation, it is believed that its main role is not heat production in the plant. It is also reported that the UCP is able to operate when ROS levels increase (Borecký and Vercesi, 2005).

6.6.6 Mehler Cycle (Water–Water Cycle)

As its name suggests, it begins and ends with water. This cycle is performed only in chloroplasts. In this cycle, in photosystem I, instead of electron transfer from ferredoxin to $NADP^+$, this electron is transferred to oxygen and converts it to superoxide (Exposito-Rodriguez et al., 2017). Superoxide radical is immediately converted to hydrogen peroxide by the superoxide dismutase enzyme. The resulting hydrogen peroxide is completely degraded by the ascorbate peroxidase enzymes and converted to water. These reactions are carried out continuously, thus, preventing the distribution of active oxygen species. This will reduce the damage caused by the ROS (Slesak et al., 2007).

6.7 CONCLUSIONS

ROS are unstable, small, and highly reactive molecules that can oxidize DNA, proteins, and lipids. ROS is formed by an incomplete regeneration of an oxygen electron. ROS contains oxygen anions, free radicals, including superoxide and hydroxyl radicals and high oxides such as hydrogen peroxide (H_2O_2). ROS play a dual role in the plants. At low concentrations, they act as messengers. High concentrations of ROS are highly harmful to plants, while, in lower concentrations, reactive oxygen species are involved in cellular signal transduction, adaptability, and cross-resistance. Plants use different mechanisms to tolerate ROS oxidative stress. The result of the research indicated that the activity of a set of antioxidant enzymes could lead to detoxification of reactive oxygen species. Therefore, for a better understanding the ROS detoxifier factors, a comprehensive study should be done on the antioxidant system.

REFERENCES

Ahmad, P, CA Jaleel, MA Salem, G Nabi, and S Sharma. 2010. "Roles of enzymatic and nonenzymatic antioxidants in plants during abiotic stress." *Critical Reviews in Biotechnology* 30 (3):161–175.

Alberts, B, A Johnson, J Lewis, M Raff, K Roberts, and P Walter. 2002. "Electron-transport chains and their proton pumps." *Molecular Biology of the Cell* 4.

Anjum, NA, P Sharma, SS Gill, M Hasanuzzaman, EA Khan, K Kachhap, AA Mohamed, P Thangavel, GD Devi, and P Vasudhevan. 2016. "Catalase and ascorbate peroxidase—representative H_2O_2-detoxifying heme enzymes in plants." *Environmental Science and Pollution Research* 23 (19):19002–19029.

Arvayo-Enríquez, H, I Mondaca-Fernández, P Gortárez-Moroyoqui, J López-Cervantes, and R Rodríguez-Ramírez. 2013. "Carotenoids extraction and quantification: a review." *Analytical Methods* 5 (12):2916–2924.

Ashraf, MH, and PJC Harris. 2013. "Photosynthesis under stressful environments: an overview." *Photosynthetica* 51 (2):163–190.

Berg, JM, JL Tymoczko, and L Stryer. 2012. "The Calvin cycle and the pentose phosphate pathway." *Biochemistry* 6:565–591.

Birben, E, UM Sahiner, C Sackesen, S Erzurum, and O Kalayci. 2012. "Oxidative stress and antioxidant defense." *World Allergy Organization Journal* 5 (1):9.

Blokhina, O, and KV Fagerstedt. 2010. "Reactive oxygen species and nitric oxide in plant mitochondria: origin and redundant regulatory systems." *Physiologia Plantarum* 138 (4):447–462.

Blokhina, O, E Virolainen, and KV Fagerstedt. 2003. "Antioxidants, oxidative damage and oxygen deprivation stress: a review." *Annals of Botany* 91 (2):179–194.

Borecký, J, and AE Vercesi. 2005. "Plant uncoupling mitochondrial protein and alternative oxidase: energy metabolism and stress." *Bioscience Reports* 25 (3–4):271–286.

Bowler, C, M Montagu, and D Inzé. 1992. "Superoxide dismutase and stress tolerance." *Annual Review of Plant Biology* 43 (1):83–116.

Carvalho, LC, and S Amâncio. 2002. "Antioxidant defence system in plantlets transferred from in vitro to ex vitro: effects of increasing light intensity and CO_2 concentration." *Plant Science* 162 (1):33–40.

de Carvalho, MHC. 2008. "Drought stress and reactive oxygen species." *Plant Signaling & Behavior* 3:156–165.

Caverzan, A, A Casassola, and SP Brammer. 2016. "Antioxidant responses of wheat plants under stress." *Genetics and Molecular Biology* 39 (1):1–6.

Conklin, PL. 2001. "Recent advances in the role and biosynthesis of ascorbic acid in plants." *Plant, Cell & Environment* 24 (4):383–394.

Das, K, and A Roychoudhury. 2014. "Reactive oxygen species (ROS) and response of antioxidants as ROS-scavengers during environmental stress in plants." *Frontiers in Environmental Science* 2:53.

De Kok, LJ, FM Maas, J Godeke, AB Haaksma, and PJC Kuiper. 1986. "Glutathione, a tripeptide which may function as a temporary storage compound of excessive reduced sulphur in H_2S fumigated spinach plants." In *Fundamental, Ecological and Agricultural Aspects of Nitrogen Metabolism in Higher Plants*, 203–206. Berlin: Springer.

Dietz, K-J, S Jacob, M-L Oelze, M Laxa, V Tognetti, SMN de Miranda, M Baier, and I Finkemeier. 2006. "The function of peroxiredoxins in plant organelle redox metabolism." *Journal of Experimental Botany* 57 (8):1697–1709.

Edreva, A. 2005. "Generation and scavenging of reactive oxygen species in chloroplasts: a submolecular approach." *Agriculture, Ecosystems & Environment* 106 (2–3):119–133.

Emri, T, I Pócsi, and A Szentirmai. 1997. "Glutathione metabolism and protection against oxidative stress caused by peroxides in *Penicillium chrysogenum*." *Free Radical Biology and Medicine* 23 (5):809–814.

Exposito-Rodriguez, M, PP Laissue, G Yvon-Durocher, N Smirnoff, and PM Mullineaux. 2017. "Photosynthesis-dependent H_2O_2 transfer from chloroplasts to nuclei provides a high-light signalling mechanism." *Nature Communications* 8 (1):49.

Finnegan, PM, KL Soole, and AL Umbach. 2004. "Alternative mitochondrial electron transport proteins in higher plants." In *Plant Mitochondria: From Genome to Function*, 163–230. Springer.

Foyer, CH, H Lopez-Delgado, JF Dat, and IM Scott. 1997. "Hydrogen peroxide-and glutathione-associated mechanisms of acclimatory stress tolerance and signalling." *Physiologia Plantarum* 100 (2):241–254.

Foyer, CH, and G Noctor. 2009. "Redox regulation in photosynthetic organisms: signaling, acclimation, and practical implications." *Antioxidants & Redox Signaling* 11 (4):861–905.

Foyer, CH, and G Noctor. 2011. "Ascorbate and glutathione: the heart of the redox hub." *Plant Physiology* 155 (1):2–18.

Foyer, CH, and S Shigeoka. 2011. "Understanding oxidative stress and antioxidant functions to enhance photosynthesis." *Plant Physiology* 155 (1):93–100.

Frankel, EN. 1984. "Chemistry of free radical and singlet oxidation of lipids." *Progress in Lipid Research* 23 (4):197–221.

Fukai, T, and M Ushio-Fukai. 2011. "Superoxide dismutases: role in redox signaling, vascular function, and diseases." *Antioxidants & Redox Signaling* 15 (6):1583–1606.

Gill, SS, and N Tuteja. 2010a. "Reactive oxygen species and antioxidant machinery in abiotic stress tolerance in crop plants." *Plant Physiology and Biochemistry* 48 (12):909–930.

Gill, SS, and N Tuteja. 2010b. "Reactive oxygen species and antioxidant machinery in abiotic stress tolerance in crop plants." *Plant Physiology and Biochemistry* 48 (12):909–930.

Goodwin, TW. 1980. "Biogeochemistry of Carotenoids." In *The Biochemistry of the Carotenoids*, 346–349. Berlin: Springer.

Grene, R. 2002. "Oxidative stress and acclimation mechanisms in plants." *The Arabidopsis Book*/American Society of Plant Biologists 1.

Grotto, D, LS Maria, J Valentini, C Paniz, G Schmitt, SC Garcia, VJ Pomblum, JBT Rocha, and M Farina. 2009. "Importance of the lipid peroxidation biomarkers and methodological aspects for malondialdehyde quantification." *Quimica Nova* 32 (1):169–174.

Gururani, MA, J Venkatesh, and LSP Tran. 2015. "Regulation of photosynthesis during abiotic stress-induced photoinhibition." *Molecular Plant* 8 (9):1304–1320.

Hancock, RD, D McRae, S Haupt, and R Viola. 2003. "Synthesis of L-ascorbic acid in the phloem." *BMC Plant Biology* 3 (1):7.

Hossain, MA, S Munné-Bosch, DJ Burritt, P Diaz-Vivancos, M Fujita, and A Lorence. 2017. *Ascorbic Acid in Plant Growth, Development and Stress Tolerance*: Springer.

Ighodaro, OM, and OA Akinloye. 2017. "First line defence antioxidants-superoxide dismutase (SOD), catalase (CAT) and glutathione peroxidase (GPX): Their fundamental role in the entire antioxidant defence grid." *Alexandria Journal of Medicine*.

Jahns, P, and AR Holzwarth. 2012. "The role of the xanthophyll cycle and of lutein in photoprotection of photosystem II." *Biochimica et Biophysica Acta (BBA)-Bioenergetics* 1817 (1):182–193.

Jiménez, A, JA Hernandez, LA Del Rio, and F Sevilla. 1997. "Ascorbate-Glutathione cycle in mitochondria and peroxisomes of pea leaves: changes induced by leaf senescence." *Phyton, Annales Rei Botanicae, Horn* 37:101–108.

Khan, AU, and T Wilson. 1995. "Reactive oxygen species as cellular messengers." *Chemistry & Biology* 2 (7):437–445.

Kordrostami, M, B Rabiei, and HH Kumleh. 2017. "Biochemical, physiological and molecular evaluation of rice cultivars differing in salt tolerance at the seedling stage." *Physiology and Molecular Biology of Plants* 23 (3):529–544.

Kreslavski, VD, AA Zorina, DA Los, IR Fomina, and SI Allakhverdiev. 2013. "Molecular mechanisms of stress resistance of photosynthetic machinery." In *Molecular Stress Physiology of Plants*, 21–51. Berlin: Springer.

Lobo, V, A Patil, A Phatak, and N Chandra. 2010. "Free radicals, antioxidants and functional foods: Impact on human health." *Pharmacognosy Reviews* 4 (8):118.

Maxwell, DP, Y Wang, and L McIntosh. 1999. "The alternative oxidase lowers mitochondrial reactive oxygen production in plant cells." *Proceedings of the National Academy of Sciences* 96 (14):8271–8276.

Mhamdi, A, G Noctor, and A Baker. 2012. "Plant catalases: peroxisomal redox guardians." *Archives of Biochemistry and Biophysics* 525 (2):181–194.

Mhamdi, A, G Queval, S Chaouch, S Vanderauwera, F Van Breusegem, and G Noctor. 2010. "Catalase function in plants: a focus on Arabidopsis mutants as stress-mimic models." *Journal of Experimental Botany* 61 (15):4197–4220.

Mika, A, and S Lüthje. 2003. "Properties of guaiacol peroxidase activities isolated from corn root plasma membranes." *Plant Physiology* 132 (3):1489–1498.

Miller, G, N Suzuki, S Ciftci-Yilmaz, and R Mittler. 2010. "Reactive oxygen species homeostasis and signalling during drought and salinity stresses." *Plant, Cell & Environment* 33 (4):453–467.

Mittler, R. 2002. "Oxidative stress, antioxidants and stress tolerance." *Trends in Plant Science* 7 (9):405–410.

Mullineaux, P, and S Karpinski. 2002. "Signal transduction in response to excess light: getting out of the chloroplast." *Current Opinion in Plant Biology* 5 (1):43–48.

Mullineaux, PM, and NR Baker. 2010. "Oxidative stress: antagonistic signaling for acclimation or cell death?" *Plant Physiology* 154 (2):521–525.

Negrão, S, SM Schmöckel, and M Tester. 2017. "Evaluating physiological responses of plants to salinity stress." *Annals of Botany* 119 (1):1–11.

Nita, M, and A Grzybowski. 2016. "The role of the reactive oxygen species and oxidative stress in the pathomechanism of the age-related ocular diseases and other pathologies of the anterior and posterior eye segments in adults." *Oxidative Medicine and Cellular Longevity* 2016.

Niu, L, and W Liao. 2016. "Hydrogen peroxide signaling in plant development and abiotic responses: crosstalk with nitric oxide and calcium." *Frontiers in Plant Science* 7:230.

Noctor, G, C Dutilleul, R De Paepe, and CH Foyer. 2004. "Use of mitochondrial electron transport mutants to evaluate the effects of redox state on photosynthesis, stress tolerance and the integration of carbon/nitrogen metabolism." *Journal of Experimental Botany* 55 (394):49–57.

Noctor, G, A Mhamdi, S Chaouch, Y Han, J Neukermans, B Marquez-Garcia, G Queval, and CH Foyer. 2012. "Glutathione in plants: an integrated overview." *Plant, Cell & Environment* 35 (2):454–484.

Pandey, P, V Irulappan, MV Bagavathiannan, and M Senthil-Kumar. 2017. "Impact of combined abiotic and biotic stresses on plant growth and avenues for crop improvement by exploiting physio-morphological traits." *Frontiers in Plant Science* 8:537.

Pellegrini, E, A Campanella, M Tonelli, C Nali, and G Lorenzini. 2015. "Reactive oxygen species and antioxidant machinery in *Liriodendrum tulipifera* plants exposed to ozone." 28th Task Force Meeting ICP-Vegetation.

Perl-Treves, R, and A Perl. 2002. *Oxidative Stress: An Introduction, Oxidative Stress in Plants*. Baton Rouge: CRC Press.

Petrov, VD, and F Van Breusegem. 2012. "Hydrogen peroxide—a central hub for information flow in plant cells." *AoB Plants* 2012.

Pospíšil, P. 2009. "Production of reactive oxygen species by photosystem II." *Biochimica et Biophysica Acta (BBA)-Bioenergetics* 1787 (10):1151–1160.

Purvis, AC. 1997. "Role of the alternative oxidase in limiting superoxide production by plant mitochondria." *Physiologia Plantarum* 100 (1):165–170.

Racchi, ML. 2013. "Antioxidant defenses in plants with attention to Prunus and Citrus spp." *Antioxidants* 2 (4):340–369.

Ramel, F, S Birtic, C Ginies, L Soubigou-Taconnat, C Triantaphylidès, and M Havaux. 2012. "Carotenoid oxidation products are stress signals that mediate gene responses to singlet oxygen in plants." *Proceedings of the National Academy of Sciences* 109 (14):5535–5540.

Roy, B, and AK Basu. 2009. *Abiotic Stress Tolerance in Crop Plants: Breeding and Biotechnology*. New India Publishing.

Sabater, B, and M Martín. 2013. "Hypothesis: increase of the ratio singlet oxygen plus superoxide radical to hydrogen peroxide changes stress defense response to programmed leaf death." *Frontiers in Plant Science* 4:479.

Saini, P, M Gani, JJ Kaur, LC Godara, C Singh, SS Chauhan, RM Francies, A Bhardwaj, NB Kumar, and MK Ghosh. 2018. "Reactive oxygen species (ROS): A way to stress survival in plants." In *Abiotic Stress-Mediated Sensing and Signaling in Plants: An Omics Perspective*, 127–153. Berlin: Springer.

Scandalios, JG. 1997. "Molecular genetics of superoxide dismutases in plants." *Oxidative Stress and the Molecular Biology of Antioxidative Defenses*:527–568.

Schertl, P, and H-P Braun. 2014. "Respiratory electron transfer pathways in plant mitochondria." *Frontiers in Plant Science* 5:163.

Schieber, M, and NS Chandel. 2014. "ROS function in redox signaling and oxidative stress." *Current Biology* 24 (10):R453–R462.

Schürmann, P, and BB Buchanan. 2008. "The ferredoxin/thioredoxin system of oxygenic photosynthesis." *Antioxidants & Redox Signaling* 10 (7):1235–1274.

Sewelam, N, N Jaspert, K Van Der Kelen, VB Tognetti, J Schmitz, H Frerigmann, E Stahl, J Zeier, F Van Breusegem, and VG Maurino. 2014. "Spatial H_2O_2 signaling specificity: H_2O_2 from chloroplasts and peroxisomes modulates the plant transcriptome differentially." *Molecular Plant* 7 (7):1191–1210.

Sewelam, N, K Kazan, and PM Schenk. 2016. "Global plant stress signaling: reactive oxygen species at the cross-road." *Frontiers in Plant Science* 7:187.

Shabala, S. 2017. *Plant Stress Physiology*: CABI, Wallingford, UK.

Sharma, P, AB Jha, RS Dubey, and M Pessarakli. 2012. "Reactive oxygen species, oxidative damage, and antioxidative defense mechanism in plants under stressful conditions." *Journal of Botany* 2012.

Shiriga, K, R Sharma, K Kumar, SK Yadav, F Hossain, and N Thirunavukkarasu. 2014. "Expression pattern of superoxide dismutase under drought stress in maize." *International Journal of Innovative Research in Science, Engineering and Technology* 3 (4):11333–11337.

Sierla, M, C Waszczak, T Vahisalu, and J Kangasjärvi. 2016. "Reactive oxygen species in the regulation of stomatal movements." *Plant Physiology* 171 (3):1569–1580.

Slesak, I, M Libik, B Karpinska, S Karpinski, and Z Miszalski. 2007. "The role of hydrogen peroxide in regulation of plant metabolism and cellular signalling in response to environmental stresses." *Acta Biochimica Polonica* 54 (1):39.

Sofo, A, A Scopa, M Nuzzaci, and A Vitti. 2015. "Ascorbate peroxidase and catalase activities and their genetic regulation in plants subjected to drought and salinity stresses." *International Journal of Molecular Sciences* 16 (6):13561–13578.

Surpin, M, RM Larkin, and J Chory. 2002. "Signal transduction between the chloroplast and the nucleus." *The Plant Cell* 14 (suppl 1):S327–S338.

Suzuki, N, G Miller, J Morales, V Shulaev, MA Torres, and R Mittler. 2011. "Respiratory burst oxidases: the engines of ROS signaling." *Current Opinion in Plant Biology* 14 (6):691–699.

Tausz, M, H Šircelj, and D Grill. 2004. "The glutathione system as a stress marker in plant ecophysiology: is a stress-response concept valid?" *Journal of Experimental Botany* 55 (404):1955–1962.

Tripathy, BC, and R Oelmüller. 2012. "Reactive oxygen species generation and signaling in plants." *Plant Signaling & Behavior* 7 (12):1621–1633.

Tsai, Y-C, C-Y Hong, L-F Liu, and CH Kao. 2005. "Expression of ascorbate peroxidase and glutathione reductase in roots of rice seedlings in response to NaCl and H_2O_2." *Journal of Plant Physiology* 162 (3):291–299.

Urzica, EI, LN Adler, MD Page, CL Linster, MA Arbing, D Casero, M Pellegrini, SS Merchant, and SG Clarke. 2012. "Impact of oxidative stress on ascorbate biosynthesis in *Chlamydomonas* via regulation of the *VTC2* gene encoding a GDP-L-galactose phosphorylase." *Journal of Biological Chemistry* 287 (17):14234–14245.

Vanlerberghe, GC, and L McIntosh. 1996. "Signals regulating the expression of the nuclear gene encoding alternative oxidase of plant mitochondria." *Plant Physiology* 111 (2):589–595.

Xu, F, S Yuan, and H-H Lin. 2011. "Response of mitochondrial alternative oxidase (AOX) to light signals." *Plant Signaling & Behavior* 6 (1):55–58.

Ye, N, G Zhu, Y Liu, A Zhang, Yi Li, R Liu, L Shi, L Jia, and J Zhang. 2011. "Ascorbic acid and reactive oxygen species are involved in the inhibition of seed germination by abscisic acid in rice seeds." *Journal of Experimental Botany* 63 (5):1809–1822.

You, J, and Z Chan. 2015. "ROS regulation during abiotic stress responses in crop plants." *Frontiers in Plant Science* 6:1092.

Yousuf, PY, KUR Hakeem, R Chandna, and P Ahmad. 2012. "Role of glutathione reductase in plant abiotic stress." In *Abiotic Stress Responses in Plants*, 149–158. Springer.

Zandalinas, SI, and R Mittler. 2018. "ROS-induced ROS release in plant and animal cells." *Free Radical Biology and Medicine* 122:21–27.

Zeng, Q-Y, H Lu, and X-R Wang. 2005. "Molecular characterization of a glutathione transferase from Pinus tabulaeformis (Pinaceae)." *Biochimie* 87 (5):445–455.

7 Oxidative Stress and Antioxidative Defense System in Plants Growing under Abiotic Stresses

Pallavi Sharma, Ambuj Bhushan Jha, and Rama Shanker Dubey

CONTENTS

7.1 INTRODUCTION

Due to their sessile lifestyle, plants are exposed to drastic environmental conditions. The production of reactive oxygen species (ROS), an unavoidable product of aerobic metabolism, increases in plant cells facing stressful environmental conditions. ROS include highly reactive free radicals and non-radical molecules, e.g., superoxide anion ($O_2^{\bullet-}$), hydroxyl radical ($^\bullet OH$), hydrogen peroxide (H_2O_2), and singlet oxygen ($^1O_2$). The reduction of molecular oxygen (O_2) in stepwise fashion or exposure to high energy causes the production of ROS in several organelles of plant cells (Figure 7.1). During abiotic stresses, ROS are produced due to the disruption of various metabolic pathways (metabolic ROS), such as respiration and photosynthesis, and as part of the abiotic stress-response signal transduction network in plants (Suzuki et al., 2012; Choudhury et al., 2016). Metabolic ROS and signaling ROS can directly modify the redox state and hence, the function of regulatory proteins, and alter transcription and translation (Miller et al., 2010; Foyer and Noctor, 2016). Signaling ROS have been implicated in reducing the level of metabolic ROS by activating an acclimation response that reduces the effect of stresses on metabolic pathways. At low concentrations, ROS serve as important regulators of growth and defense pathways (Kärkönen, 2015). ROS also play an important role in lignification and other cross-linking processes in the cell wall (Bradley et al., 1992; Ogawa et al., 1997; Desikin et al.,

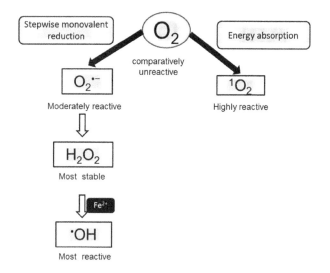

FIGURE 7.1 Formation of reactive oxygen species in plants. Activation of O_2 occurs by two different mechanisms. Stepwise monovalent reduction of O_2 leads to the formation of $O_2{}^{\bullet-}$, H_2O_2, and $^{\bullet}OH$, whereas energy transfer to O_2 leads to the formation of $^1O_2$.

2001; Mittler, 2002). The production and signaling of ROS in association with the action of brassinosteroids, auxin, gibberellins, ethylene, abscisic acid, strigolactones, jasmonic acid, and salicylic acid (SA) have been shown to play an important role in the coordinated regulation of growth and stress tolerance. Multiple points of reciprocal control and integration nodes involving Ca^{2+}-dependent processes and mitogen-activated protein kinase phosphorylation cascades have been identified in both local and systemic cross-talk of ROS and hormone signaling pathways (Xia et al., 2015).

ROS are regarded as indicators of cellular stress, as they act as secondary messengers in the stress-responsive signal transduction pathway (Choudhury et al., 2013; Das and Roychoudhury, 2014; Foyer et al., 2017). However, excessive production of ROS under abiotic and biotic stresses such as drought, salt stress, chilling, heat shock, excess levels of heavy metals, anaerobiosis, UV-B radiation, and gaseous pollutants causes disruption of cellular homeostasis (Yan et al., 1996; Pellinen et al., 1999; Shah et al., 2001; Mittler, 2002; Pasqualini et al., 2002; Mittler et al., 2004; Sharma and Dubey, 2005a,b; Suzuki and Mittler, 2006; Esfandiari et al., 2008; Hu et al., 2008; Han et al., 2009; Kumutha et al., 2009; Tanou et al., 2009; Dixon and Stockwell, 2014; Anjum et al., 2015; Devi and Giridhar, 2015; Dikilitas et al., 2015; Gill et al., 2015; Taibi et al., 2016). The increased generation of ROS during the period of abiotic stresses can pose a risk to cells by causing increased lipid peroxidation, protein oxidation, nucleic acid damage, enzyme inhibition, programmed cell death (PCD) pathway activation, and eventually, cell death (Mittler, 2002; Pasqualini et al., 2003; Verma and Dubey, 2003; Meriga et al., 2004; Sharma and Dubey, 2005a,b, 2007; Petrov et al., 2015) (Table 7.1).

It is proposed that whether ROS act as signal transducing or damaging molecules is mainly dependent on their quality (primary or secondary ROS) and second, on their quantity

TABLE 7.1

Abiotic Stresses That Overproduce ROS and Cause Oxidative Damage to Different Plant Species

Abiotic Stress	Plants	References
Drought	*Oryza sativa*	Sharma and Dubey (2005)
	Triticum aestivum	Esfandiari et al. (2008)
	Glycine max	Devi and Giridhar (2015)
Salinity	*Sorghum bicolor*	Hefny and Abdel-kader (2009)
	Fragaria ananassa	Tanou et al. (2009)
	Phaseolus vulgaris	Taibi et al. (2016)
Heat	*Hordeum vulgare*	El-Shintinawy et al. (2004)
	Lilium longiflorum	Yin et al. (2008)
	Triticum aestivum	Hameed et al. (2012)
Chilling	*Zea mays*	Prasad et al. (1994)
	Cucumis sativus	Hu et al. (2008)
	Sorghum bicolor	Guo et al. (2016)
Heavy metals	*Oryza sativa, Nicotiana tabacum, Hibiscus cannabinus*	Shah et al. (2001)
	Oryza sativa	Verma and Dubey (2003)
	Oryza sativa	Sharma and Dubey (2007)
	Oryza sativa	Maheshwari and Dubey (2009)
	Oryza sativa	Mishra et al. (2011)
	Oryza sativa	Li et al. (2013)
	Oryza sativa	Zrobek-Sokolnik et al. (2009)
Anaerobiosis	*Hordeum vulgare*	Kumutha et al. (2009)
	Zea mays	Jamei et al. (2009)
	Cajanus cajan	Bansal and Srivastava (2015)
Gaseous pollutants	*Betula pendula*	Pellinen et al. (1999)
	Medicago truncatula	Puckette et al. (2008)
	Oryza sativa	Ueda et al. (2013)
UV-B radiation	*Momordica charantia*	Agarwal and Shaheen (2007)
	Picea asperata	Han et al. (2009)
	Glycine max	Li et al. (2012)

(Weidinger and Kozlov, 2015). As ROS play multifunctional roles, it is essential for cells to control the level of ROS strictly to avoid any injury caused by oxidative stress without removing them completely. Plant cells scavenge or detoxify excess ROS through an efficient antioxidative system consisting of non-enzymic as well as enzymic antioxidants (Noctor and Foyer, 1998). Within the cell, ascorbate (AsA), glutathione (GSH), tocopherols, carotenoids, and phenolics are regarded as potent non-enzymic antioxidants, whereas guaiacol peroxidase (GPX), catalase (CAT), superoxide dismutase (SOD), and enzymes participating in the ascorbate-glutathione (AsA-GSH) cycle, e.g., ascorbate peroxidase (APX), glutathione reductase (GR), monodehydroascorbate reductase (MDHAR), and dehydroascorbate reductase (DHAR), are important enzymic antioxidants (Noctor and Foyer, 1998). However, the balance between production and removal of ROS is disrupted

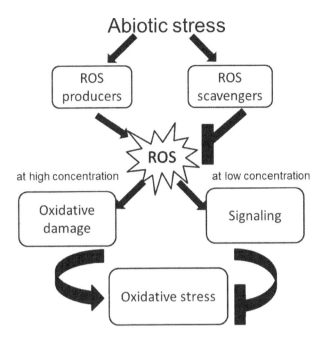

FIGURE 7.2 Action of reactive oxygen species in plants under abiotic stresses. Abiotic stresses cause increased production of ROS in plants. Plant cells try to scavenge excess ROS through an efficient antioxidative system. At low concentrations, ROS serve as important regulators of growth and defense pathways. At excessive levels, ROS cause oxidative stress.

under stressful conditions (Figure 7.2). The variation of antioxidant levels and activity establishes a key response for enduring hostile environmental conditions (Foyer et al., 1994; Noctor et al., 2002; Pitzschke and Hirt, 2006). Many groups have observed enhanced antioxidant enzyme activities in plants to combat oxidative stress induced by various stresses such as drought (Sharma and Dubey, 2005a; Devi and Giridhar, 2015; Huseynova, et al., 2015), soil salinity (Tsai et al., 2005; Tanou et al., 2009; Khaliq et al., 2015), high temperature (Almeselmani et al., 2006; Yin et al., 2008; Sgobba et al., 2015), chilling (Bafeel and Ibrahim, 2008; Radyuk et al., 2009; Guo et al., 2016), metal toxicity (Shah et al., 2001; Verma and Dubey, 2003; Sharma and Dubey, 2007; Gupta et al., 2009; Maheshwari and Dubey, 2009; Mishra et al., 2011; Malar et al., 2016), anaerobiosis (Lin et al., 2008; Kumutha et al., 2009; Zong et al., 2015), gaseous pollutants (Scebba et al., 2003; Aguiar-Silva et al., 2016), and UV-B radiation (Agarwal and Shaheen, 2007; Czégény et al., 2013). The inability of cells' antioxidant response to combat excess ROS is referred to as *oxidative stress*. It is a component of many abiotic stresses, which ultimately leads to yield reduction and death in plants. In plant cells, the detoxification of ROS is facilitated by various metabolic adaptations that decrease ROS production and maintain the level of free transient metals, such as Fe^{2+}, to block the production of the highly toxic $^{\bullet}OH$ via the Fenton reaction (Halliwell and Gutteridge, 2015). High antioxidant capacity has been linked to enhanced tolerance of plants to these abiotic stresses (Zaefyzadeh et al., 2009). It has been proposed that various abiotic stresses lead to different ROS signatures in different cellular compartments that ensure the

specificity of the adaptation process in response to the actual stress encountered by the plants (Choudhury et al., 2016).

Differential gene expression analysis in tolerant and sensitive genotypes can reveal the genetic basis of abiotic stress tolerance and identification of antioxidant genes involved in tolerance to a particular stress. The overexpression of antioxidant genes offers an opportunity to develop plants with improved tolerance to abiotic stresses. With the advancement of molecular tools, significant progress has been achieved in generating improved oxidative stress–tolerant crops (Allen et al., 1997; Lee et al., 2007a; Ashraf et al., 2008; Prashanth et al., 2008; Wang et al., 2017; Yin et al., 2017; Suekawa et al., 2018). However, as the process of ROS detoxification is very complex, altering one component may or may not affect the ability of the whole pathway (Tseng et al., 2007; Lee et al., 2009). Further, synergistic effects on stress tolerance due to overexpression of combinations of antioxidant enzymes have been seen (Lee et al., 2007a,b; Shafi et al., 2015; Yan et al., 2016; Shafi et al., 2017). Therefore, increased emphasis is being given to producing transgenic plants overexpressing genes associated with more than one antioxidant to achieve tolerance to multiple environmental stresses.

The present review focuses on abiotic stress–induced ROS production, oxidative damage caused by ROS, and the role of the antioxidative defense system in plants growing under different abiotic stresses. The progress made in the last few decades in improving abiotic stress tolerance in crop plants using genes responsible for the synthesis of antioxidants involving biotechnological approaches has also been discussed.

7.2 REACTIVE OXYGEN SPECIES AND INDUCTION OF OXIDATIVE STRESS

The oxygen molecule is used as the terminal electron acceptor in all aerobic organisms, including plants. To fully reduce O_2 to H_2O in one step, a high activation energy is required. Therefore, at ambient temperature, this reaction proceeds very slowly. When O_2 is involved in electron transfer chemical reactions or exposed to high energy, it is transformed to different chemical species such as $O_2^{\bullet-}$, $^{\bullet}OH$, H_2O_2, $^1O_2$, etc. collectively known as ROS. Some of the ROS are known to act as important signaling molecules that modify the expression of genes and alter the activity of specific defense proteins. However, all ROS at high concentrations are very damaging to organisms. Figure 7.1 shows various ROS, which are produced either due to stepwise reduction of O_2 or as a result of energy absorption. In plants, ROS are constantly formed by the unavoidable outflow of electrons onto O_2 from the electron transport chain operating in mitochondria, chloroplasts, and the plasma membrane (Pinto et al., 2003) or as a byproduct of respiration, photosynthesis, and other metabolic pathways (Foyer and Harbinson, 1994; Apel and Hirt, 2004; Tsukamoto et al., 2005). ROS disturb the homeostasis of the organism by oxidatively damaging biomolecules such as membrane lipids, proteins, chlorophylls, and nucleic acids (Kanazawa et al., 2000; Cassells and Curry, 2001; Jiang and Zhang, 2001;

Sharma and Dubey, 2005a, 2007; Konieczny et al., 2008; Maheshwari and Dubey, 2009).

Molecular O_2 is paramagnetic and comparatively unreactive in the ground state because of the presence of two electrons that are unpaired with parallel spin (Apel and Hirt, 2004). Two different mechanisms are known to be involved in activation of O_2: (i) monovalent reduction and (ii) reversal of spin of one of the unpaired electrons due to the absorption of adequate energy. Monovalent reduction of O_2 in a stepwise manner forms $O_2^{\bullet-}$, H_2O_2, and $^{\bullet}OH$, whereas absorption of energy by O_2 forms singlet oxygen, $^1O_2$ (Figure 7.1). Singlet oxygen, which is a very harmful ROS, reacts with most biomolecules at diffusion-controlled rates but is much more reactive toward organic molecules (Foyer and Harbinson, 1994). It cannot last for more than 100 μs in a non-polar environment and 4 μs in water (Foyer and Harbinson, 1994). An excessive amount of $^1O_2$ leads to oxidative stress and causes toxic effects because of the oxidation of lipids, DNA, and proteins (Agnez-Lima and Menck, 2012). Compared with other ROS, $^1O_2$ has received less attention; however, it has been observed that this species may be involved in programmed cell death, the acclimation process, and responses of plants to light (Triantaphylides and Havaux, 2009; Agnez-Lima and Menck, 2012).

$O_2^{\bullet-}$ is a nucleophilic reactant and is produced if one electron is added to ground state oxygen. Plasma membrane–bound nicotinamide adenine dinucleotide phosphate (NADPH)-oxidases use cytosolic NADPH to produce $O_2^{\bullet-}$ under abiotic stress conditions. $O_2^{\bullet-}$ is a short-lived, moderately reactive ROS having a half-life of approx. 2–4 μs. $O_2^{\bullet-}$ possesses both oxidizing and reducing properties (Halliwell, 1977). With regard to molecules of biological importance, $O_2^{\bullet-}$ can oxidize sulfur compounds, O-diphenols, ascorbic acid (Mishra, 1974; Miller and MacDowell, 1975), cytochrome C (McCord et al., 1977), and metal ions and metal complexes (McCord and Day, 1978). H_2O_2 is produced in the SOD-catalyzed disproportionation of $O_2^{\bullet-}$ or from the reduction of $O_2^{\bullet-}$ by AsA, manganese ions, or ferredoxin (Hideg, 1997). H_2O_2 is one of the most stable and important ROS, which regulates basic acclimatization and defense processes in plants (Slesak et al., 2007). It has been shown to act as a signal transduction molecule in several developmental processes. Nevertheless, at high concentrations, it causes oxidative stress, marked by increased lipid peroxidation and alteration of membrane permeability (Imlay, 2003). The main source of $^{\bullet}OH$ in biological systems is the decomposition of H_2O_2 in the Haber–Weiss reaction. This reaction is increased when a transition metal, such as Fe^{2+} (Fenton reaction), is present (Hideg, 1997). $^{\bullet}OH$ is the most reactive of all ROS. It has the capability to react with all biological molecules and can initiate self-perpetuating lipid peroxidation, protein damage, and membrane destruction (Foyer et al., 1997). As there is no enzymatic mechanism to remove this highly reactive ROS in cells, its excess production would ultimately cause cell death (Pinto et al., 2003). H_2O_2 itself is not very reactive, but in the presence of metal reductants, it forms highly reactive $^{\bullet}OH$, which is a strong oxidizing agent (Boo and Jung, 1999).

Under normal growth conditions, the formation of ROS is low in cells (Polle, 2001). However, various stressful

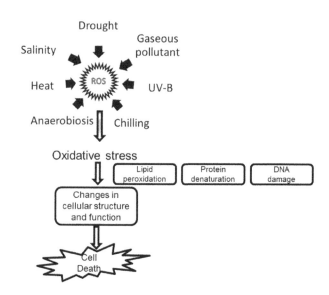

FIGURE 7.3 Abiotic stresses, production of ROS, and cellular damage in plants. Different abiotic stresses such as drought, salinity, extreme temperatures, heavy metals, anaerobiosis, gaseous pollutants, and UV-B radiation cause enhanced production of ROS in the tissues. These ROS cause oxidative damage to lipids, proteins and nucleic acids, which in turn, cause changes in cellular structure and function and ultimately cell death.

environmental conditions that disrupt cellular homeostasis enhance the production of ROS (up to 720 μM s^{-1} $O_2^{\bullet-}$ and a steady-state level of 5–15 μM H_2O_2) (Polle, 2001). When the ROS level exceeds the scavenging capacity of defense mechanisms, and more prooxidative reactions occur than antioxidative reactions, it is said that cells are under *oxidative stress* (Grzegorz, 1997). Figure 7.3 shows enhanced production of ROS in plants growing under various abiotic stresses, including drought, salinity, extremes of temperature, excess levels of metals, anaerobiosis, gaseous pollutants, and UV-B radiation. Enhanced levels of these ROS cause oxidative damage to biomolecules such as membrane lipids, proteins, enzymes, nucleic acids, chloroplast pigments, etc. (Dat et al., 2000; Mittler, 2002; Pasqualini et al., 2003; Meriga et al., 2004; Posmyk et al., 2005; Sharma and Dubey, 2005a, 2007). As primary ROS are scavenged easily by antioxidant enzymes such as SOD and CAT, and their reactions with biomolecules are reversible, they are ideal for intracellular signaling (Weidinger and Kozlov, 2015). Toxic reactive secondary species ($^{\bullet}OH$), which is produced by the reaction of primary species with transition metals and are not well controlled, cause irreversible damage to biomolecules. Along with primary ROS, they are necessary for the damaging effects of ROS. Iron-catalyzed ROS are now recognized as the main initiators and mediators causing cell death in various organisms (Dixon and Stockwell, 2014).

7.2.1 Lipid Peroxidation

Lipid peroxidation is regarded as the primary molecular mechanism associated with cellular oxidative damage (Repetto et al., 2012). Both free radicals and enzymes can cause initiation of lipid peroxidation in cellular and organellar membranes. $^1O_2$ reacts with conjugated double bonds, e.g., double bonds present in polyunsaturated fatty acids (PUFAs) (Rinalducci et al., 2008; reviewed by Braconi et al., 2011). Enhanced peroxidation of lipids has been observed in plants growing under stresses such as drought (Sharma and Dubey, 2005a; Esfandiari et al., 2008; Devi and Giridhar, 2015), salinity (Tanou et al., 2009; Ahmad et al., 2014; Taibi et al., 2016), chilling (O'Kane et al., 1996; Guo et al., 2016), heat (El-Shintinawy et al., 2004; Hameed et al., 2012), metal toxicity (Shah et al., 2001; Verma and Dubey, 2003; Sharma and Dubey, 2007; Maheshwari and Dubey, 2009; Gajewaska et al., 2012; Li et al., 2013; Sytar et al., 2013), anaerobiosis (Blokhina et al., 1999; Kumutha et al., 2009; Steffens et al., 2013), UV-B exposure (Yannarelli et al., 2006; Li et al., 2012), and gaseous pollutants (Puckette et al., 2007; Ueda et al., 2013). Increased lipid peroxidation has been observed with increased production of ROS under these stresses. Lipid peroxidation not only directly affects normal cellular functioning but also aggravates the oxidative stress by the generation of lipid-derived radicals (Montillet et al., 2005). The level of lipid peroxides has been commonly used as an indicator of free radical–mediated damage to cell membranes under stressful conditions. Malondialdehyde (MDA) which is one of the final products of the peroxidation of unsaturated fatty acids in phospholipids, is associated with cell membrane damage (Halliwell and Gutteridge, 1989). Yamauchi and coworkers (2008) reported protein modifications in heat-stressed plants due to MDA produced from peroxidized linolenic acid in chloroplasts. Linolenic acid–deficient mutants (*fad3fad7fad8* triple mutant) revealed that linolenic acid could be a primary source of protein modification by MDA in heat-stressed plants (Yamauchi et al., 2008).

Phospholipids are necessary components of the biomembrane surrounding cells and cellular structures, e.g., mitochondria and nucleus, and thus, damage to phospholipids can affect cell viability (Woessmann et al., 1999). Phospholipids are quite sensitive to attack by oxidants, including •OH. Various types of aldehydes, some of which are highly reactive, are formed due to peroxidation of PUFA present in biomembranes. PUFA peroxidation by ROS can cause chain breakage and thus, can increase the fluidity and permeability of the membrane. Two common sites at which oxygen free radical attack on phospholipid molecules occur are (i) unsaturated bonds present between two carbon atoms and (ii) the ester linkage between fatty acid and glycerol. Decrease in membrane stability or enhancement in membrane permeability has been associated with the extent of ROS-led lipid peroxidation (Sairam et al., 2002).

7.2.2 PROTEIN MODIFICATION

ROS attack on proteins causes site-specific modification of amino acids, peptide chain fragmentation, aggregation of cross-linked reaction products, modified electric charge, and enhanced susceptibility of proteins to proteolysis. ROS may cause direct as well as indirect modification of proteins. Direct modification through nitrosylation, carbonylation, di-sulfide

bond formation, and glutathionylation causes modulation of a protein's activity. Proteins can be modified indirectly by conjugation with breakdown products of fatty acid peroxidation (Yamauchi et al., 2008). $O_2^{•-}$ primarily reacts with Fe–S centers of proteins (Rinalducci et al., 2008; reviewed by Braconi et al., 2011). Due to excess ROS formation, tissues injured due to oxidative stress usually contain enhanced concentrations of carbonylated proteins (Dean et al., 1993). Proteins are regarded as major targets of oxidative modifications in abiotically stressed plants (Cramer et al., 2011; Anjum et al., 2015), and enhanced modification of proteins has been reported in plants under various stresses (Rao et al., 1995; Sharma and Dubey, 2005a, 2007; Maheshwari and Dubey, 2009; Tanou et al., 2009; Bhoomika et al., 2014). Increased protein carbonylation was reported in the leaves of mutant (*tasg1*) and wild-type cultivars of *Triticum aestivum* grown under drought stress (Tian et al., 2013). Nitration, carbonylation, and S-nitrosylation have been observed in acclimation to salinity stress in *Citrus aurantium* (Tanou et al., 2012). Increased protein carbonylation causes loss of protein function due to O_3 exposure in *Glycine max* leaves (Qiu et al., 2008).

The amino acids in a peptide differ in their susceptibility to attack by ROS, and various forms of ROS differ in their potential reactivity. •OH and alkoxyl radicals are mainly involved in the oxidation of proteins. Protein oxidation in plant mitochondria acts as a stress indicator (Moller and Kristensen, 2004). Sulfur-containing amino acids and thiol groups specifically are very susceptible sites for attack by ROS. Several heavy metals, such as Cd, Pb, and Hg, have been shown to cause depletion of protein-bound thiol groups (Stohs and Bagchi, 1995). During oxidative stress, unwanted disulfide bonds are produced in the normal cytosol-resident proteins (Cabiscol et al., 2000). ROS can form a thiyl radical by abstracting H atom from a cysteine residue, and this can cross-link to another thiyl radical to form a disulfide bridge. Alternatively, oxygen can add to a methionine to form a methionine sulfoxide derivative. A protein methionine sulfoxide reductase has been detected in pea chloroplasts (Ferguson and Burke, 1992). This enzyme reduces the methionyl sulfoxide back to methionyl residues in the presence of thioredoxin (Brot and Weissbach, 1982). Some forms of free radical attack on proteins are not reversible; for example, oxidation of iron–sulfur centers by $O_2^{•-}$ destroys enzymatic function (Gardner and Fridovich, 1991). In these cases, the metal (Fe) binds to a divalent cation-binding site on the protein. The metal (Fe) then reacts in a Fenton reaction to form •OH, which rapidly oxidizes an amino acid residue at or near the cation-binding site of the protein (Stadtman, 1986). Among amino acids, tyrosine and tryptophan are most prone to irreversible oxidation by ROS (Berlett and Stadtman, 1997; Davies, 2003; Stadtman and Levine, 2003; Todorovski et al., 2011). Tyrosine is readily cross-linked to form bityrosine products (Davies, 1987; Malencik and Anderson, 2003). In tryptophan, ROS generally targets the indole ring, whereby tryptophan oxidation products, including 5-hydroxy-tryptophan (5-HTP), oxindolylalanine (Oia), dioxindolylalanine, N-formylkynurenine (NFK), and kynurenine (Kyn) are formed (Garrison, 1987;

Simat and Steinhart, 1998). Histidine, lysine, proline, arginine, threonine, and serine form carbonyl groups on oxidation (Stadtman, 1986). With the finding that these amino acids are oxidized to carbonyl derivatives, several methods have been developed to detect the carbonyl content of proteins and used to measure protein damage (Cabiscol et al., 2000). Enzymes in which these amino acids are located at positions that are critical to enzyme activity will become inactivated by the interaction of ROS.

Oxidized proteins serve as better substrates for proteolytic digestion, and proteolytic pathways could provide a valuable line of "secondary antioxidant defense" (Cabiscol et al., 2000). A strong correlation has been demonstrated between increased hydrophobicity on the protein surface and the recognition and proteolytic degradation of oxidatively modified proteins. Many intracellular proteins are degraded by multicatalytic proteinase complexes, also called *proteasomes*, in a non-lysosomal pathway (Cabiscol et al., 2000; Jung et al., 2014). It has been suggested that protein oxidation could predispose the protein to ubiquitination, which in turn, would be a target for proteasomal degradation. It seems that after a certain degree of oxidative damage, further damage causes a decrease in proteolytic susceptibility. Several studies have revealed that heavily oxidized proteins, extensively cross-linked and aggregated, are not only poor substrates for degradation but also can inhibit proteases from degrading other oxidized proteins (Grune et al., 1997). The removal of damaged proteins is necessary to prevent their accumulation, which could compromise the correct metabolism of any cell exposed to oxidative stress.

In plants, cellular compartments such as cytosol, chloroplasts, peroxisomes, nucleus, and mitochondria may exhibit the presence of carbonylated proteins (Bartoli et al., 2004; Job et al., 2005; Rajjou et al., 2008). The highest concentration of oxidatively modified proteins has been reported in mitochondria of *T. aestivum* leaves (Bartoli et al., 2004) and legume nodules (Matamoros et al., 2013). Specific proteins appear to be particularly vulnerable to oxidative carbonylation in the matrix of plant mitochondria; these include several enzymes of the Krebs cycle, glycine decarboxylase, SOD, and heat shock proteins (HSPs). Among the techniques developed for the identification of oxidative stress–induced modifications of proteins, the so-called *redox proteome*, proteomics appears to be the best-suited approach. Oxidative stress leaves different footprints in the cell in the form of different oxidatively modified components, and by using the redox proteome, it will be possible to decipher the potential roles played by ROS-induced modifications in stressed cells (Rinalducci et al., 2008).

7.2.3 DNA Damage

Modification of DNA due to exposure to ROS is regarded as an early event involved in the damage caused due to oxidative stress agents. As DNA is the genetic material of the cell, any damage to the DNA can lead to alteration (i.e., mutation) in the encoded proteins, which may cause malfunction or total inactivation of these proteins. Thus, it is essential for the viability of the cell that the DNA remains intact. ROS are considered as a major source of DNA damage (Imlay and Linn, 1986), which can cause breakage of strands, nucleotide removal, and various modifications in the organic bases of the nucleotides (Cadet and Wagner, 2013). Alteration in the nucleotides of one strand can cause mismatches with the other strand nucleotides, leading to subsequent mutations. Although cells have developed repair mechanisms to correct naturally occurring changes in the DNA, additional or excessive changes caused by ROS or other agents can lead to permanent damage to the DNA with potentially detrimental effects for the cell (Chen et al., 2002). Mitochondrial and chloroplast DNA are more susceptible to oxidative damage than nuclear DNA due to the lack of protective protein histones, and proximity to the ROS-producing systems in the former (Richter, 1992). Enhanced DNA degradation has been observed in plants exposed to various environmental stresses, such as salinity (Liu et al., 2000; Dikilitas et al., 2015), metal toxicity (Meriga et al., 2004), gaseous pollutants (Pasqualini et al., 2003), and UV-B radiation (Gill et al., 2015).

Both the sugar and base moieties are susceptible to oxidation by ROS. Ravant et al. (2000) demonstrated that singlet oxygen is directly involved in the oxidation of DNA by forming 8-oxo-7,8-dihydro-2′-deoxyguanosine (8-oxodG). $^1O_2$ interacts with DNA molecules and may lead to the transversion of guanine to thymine due to 8-oxodG formation (Agnez-Lima and Menck, 2012). The degradation of bases produces numerous products, including 8-hydroxyguanine, hydroxymethyl urea, urea, thymine glycol, thymine, and adenine ring-opened and saturated products (Tsuboi et al., 1998). Around 100 oxidatively produced base lesions and 2-deoxyribose modifications, including thymidine hydroperoxides and diastereomeric nucleosides, which are formed initially, have been isolated and identified (Cadet et al., 2010, 2012). 8-Hydroxyguanine is the most commonly observed product. The primary cause of single-strand breaks is oxidation of the sugar moiety by the •OH. Under physiological conditions, neither H_2O_2 alone nor $O_2^{•-}$ can cause *in vitro* strand breakage. Therefore, it was concluded that the toxicity associated with these ROS *in vivo* is most likely as a result of Fenton reaction. Imlay and Linn (1986) concluded from their studies with *Escherichia coli* mutants that a Fenton-active metal is probably chelated to the phosphodiester linkage in DNA. When this metal is reduced by NADP(H) or $O_2^{•-}$, it reacts with H_2O_2 to form •OH (Imlay and Linn, 1986). •OH then oxidizes an adjacent sugar or base, causing breakage of the DNA chain. When •OH attacks either DNA or proteins associated with it, DNA–protein cross-links are formed (Oleinick et al., 1986). DNA–protein cross-links cannot be readily repaired and may be lethal if replication or transcription precedes repair.

•OH reacts with free carbohydrates, such as sugars, and polyols (Smirnoff and Cumbes, 1989). The oxidation of sugars with •OH often releases formic acid as the main breakdown product (Isbell et al., 1973). Plant cell wall polysaccharides have been shown to be susceptible to oxidative scission mediated by •OH *in vitro* under physiologically relevant conditions (Fry, 1998).

7.3 ANTIOXIDATIVE DEFENSE SYSTEM IN PLANTS

Adverse environmental conditions perturb the balance between the production and the scavenging of ROS, leading to rapid accumulation of ROS in the cells (Foyer et al., 1994; Noctor et al., 2002; Pitzschke and Hirt, 2006). At low concentrations, ROS play an important role in the process of acclimation of plants to abiotic stresses. Enhanced ROS production during abiotic stresses causes oxidative damage to lipids, proteins, and nucleic acids. Plants possess a complex antioxidative defense system consisting of non-enzymatic and enzymatic components to avoid the oxidative damage caused due to ROS. A detailed account of these components has been presented in Figure 7.4. Various cellular compartments of plants, such as chloroplasts, mitochondria, and peroxisomes, contain their own set of ROS-generating as well as scavenging pathways. These pathways from different organelles are coordinated (Pang and Wang, 2008; Choudhury et al., 2016).

7.3.1 NON-ENZYMATIC DEFENSE SYSTEM

Non-enzymic antioxidants include AsA, GSH, tocopherol, carotenoids, and phenolic compounds, which play key roles in the defense of plants against oxidative damage. Ascorbate and GSH are major cellular redox buffers. Both the concentration and the redox state of GSH and AsA influence the sensitivity of plants to stresses (Jozefczak et al., 2015). Mutants with decreased AsA, GSH, and tocopherol content appear to be hypersensitive to stresses (Howden et al., 1995; Conklin et al., 1996; Huang et al., 2005; Semchuk et al., 2009). Alterations in GSH/glutathione disulfide (GSSG) and AsA/dehydroascorbate (DHA) ratios have been associated with redox signaling (Foyer and Noctor, 2005; Jozefczak et al., 2012; Zagorchev et al., 2013). It has been proposed that several components of antioxidative defense (ascorbate, glutathione, carotenoids, tocopherols and other isoprenoids, flavonoids, and enzymatic

antioxidants), rather than a single antioxidant, help plants to withstand environmental stress (Munné-Bosch, 2005).

7.3.1.1 Ascorbate

Ascorbate is an abundant, water-soluble, low–molecular weight antioxidant present in plants. It is synthesized by four pathways involving D-mannose/L-galactose, L-gulose, D-galacturonate, and myo-inositol as key precursors. Ascorbate has a key role in defense against oxidative stress and is particularly abundant in photosynthetic tissues (Smirnoff et al., 2004; Höller et al., 2015; Kao, 2015; Yactayo-Chang et al., 2018). Along with other members of the antioxidant network, it functions in controlling the concentration of ROS (Colville and Smirnoff, 2008). It has been shown to play an important role in several physiological processes, including growth, differentiation, and metabolism in plants. Ascorbate acts as an enzyme cofactor and a modulator of cell expansion and division, flowering time, and gene regulation. It is also involved in the signal transduction pathway (Ortiz-Espín et al., 2017). To perform various physiological functions, the cellular content of AsA needs to be tightly regulated in response to various environmental conditions. Regulation of AsA biosynthesis occurs at different levels (Bulley and Laing, 2016; Ishikawa et al., 2018).

In plants, ascorbate exists in the reduced state (AsA) as well as in oxidized forms (such as monodehydroascorbate [MDHA] and DHA) (Ortiz-Espín et al., 2017). More than 90% of AsA is localized in the cytoplasm, but unlike other soluble antioxidants, a substantial portion is exported to the apoplast, where it is present in millimolar concentrations. Apoplastic AsA is believed to represent the first line of defense against potentially damaging external oxidants (Barnes et al., 2002). It protects critical macromolecules from oxidative damage. Ascorbate has a key role in the removal of H_2O_2 during abiotic stress via the AsA–GSH cycle (Pinto et al., 2003; Bartoli et al., 2017; Ortiz-Espín et al., 2017; Suekawa et al., 2017). It also reacts directly with $O_2^{•-}$, H_2O_2, or the tocopheroxyl radical

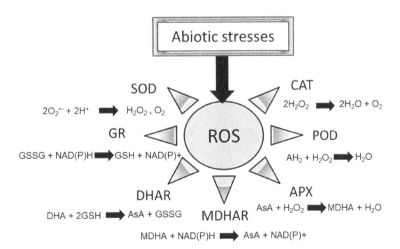

FIGURE 7.4 Reactions catalyzed by enzymes involved in the antioxidative defense system in plants. Efficient scavenging of ROS produced during abiotic stresses requires concerted action of the antioxidant enzymes superoxide dismutase (SOD), catalase (CAT), peroxidase (POD), ascorbate peroxidase (APX), monodehydroascorbate reductase (MDHAR), dehydroascorbate reductase (DHAR), and glutathione reductase (GR). The reactions catalyzed by individual enzymes are shown.

to form MDHA and DHA. The oxidation of AsA occurs in two sequential steps, first producing MDHA and subsequently DHA. The MDHA radical is also a primary product of APX reaction in the chloroplasts (Asada, 1997). MDHA can either spontaneously dismutate or be reduced to AsA by the NADP(H)-dependent enzyme monodehydroascorbate reductase (MDHAR) (Miyake and Asada, 1994). If MDHA is not rapidly re-reduced to AsA, the MDHA radical spontaneously disproportionates to AsA and DHA. DHA is also highly unstable at pH values greater than 6.0 and is decomposed to tartrate and oxalate (Noctor and Foyer, 1998). To prevent this, DHA is rapidly reduced to AsA by the enzyme DHAR using reducing equivalents from GSH (Asada, 1996; Suekawa et al., 2017). The level of reduced ascorbate has been reported to alter in response to drought (Sharma and Dubey, 2005a; Esfandiari et al., 2008; Mekki et al., 2016), salinity (Hernandez et al., 2001; Taibi et al., 2016), chilling (Radyuk et al., 2009), metal toxicity (Sharma and Dubey, 2007; Maheshwari and Dubey, 2009), anaerobiosis (Lin et al., 2004), gaseous pollutants (Ranieri et al., 1997; Scebba et al., 2003; Severino et al., 2007), UV-B radiation (Agarwal and Saheen, 2007), etc. The level of AsA under environmental stresses depends on the balance between the rates and capacity of AsA biosynthesis and turnover related to antioxidant demand (Chaves et al., 2002). The enzyme ascorbate oxidase (AO) converts AsA to MDHA with the simultaneous reduction of oxygen to water. It is proposed to play a key role in signaling between the external environment and the cell, and AO gene expression has been shown to respond to wounding, plant hormones, and stresses (Stevens et al., 2017).

An increased concentration of AsA in cells confers tolerance to multiple abiotic stresses, such as water deficit, salt, cold, and heat. Mutant studies have shown the importance of AsA in abiotic stress tolerance in plants. A low intrinsic AsA level in the *vtc-1* mutant under salt stress induced a dramatic decrease in the reduced form of AsA, which resulted in enhanced ROS content in the *vtc-1* mutants (Huang et al., 2005). Similarly, the AsA-deficient mutant *vtc-1* appeared to be more sensitive to supplementary UV-B treatment than wild-type plants (Gao and Zhang, 2008). GDP-L-galactose phosphorylase knockout reduced the level of AsA by 80% in rice, and the concentration of AsA showed a negative correlation with lipid peroxidation in foliar tissues under Zn deficiency and ozone stress (Höller et al., 2015).

7.3.1.2 Glutathione

The reduced form of the tripeptide glutathione (γ-glutamyl-cysteinyl-glycine [GSH]) is a low–molecular weight, non-protein thiol that is present abundantly in most plant species. GSH is synthesized in the cytosol and chloroplasts of plant cells by compartment-specific isoforms of γ-glutamyl-cysteinyl synthetase (γ-ECS) and glutathione synthetase (GS). Besides serving as a key antioxidant within the cell, GSH acts as a redox buffer and is involved in the transport and storage of the reduced form of sulfur. It is involved in the detoxification of xenobiotics, and it serves as a precursor for the synthesis of phytochelatins, which are crucial in controlling cellular heavy metal concentrations by chelating and sequestering them in

vacuoles (Foyer et al., 2001). It also acts as a regulator of epigenetic and genetic functions (Diaz-Vivancos et al., 2015).

Due to its reducing power, GSH plays an important role in diverse biological processes to maintain cellular homeostasis, such as cell growth/division, metabolic reducing reactions, transmembrane transport, receptor actions, and synthesis of proteins and nucleic acids (Foyer et al., 1997). GSH functions as an antioxidant in many ways. It can react chemically with $O_2^{\bullet-}$, $^{\bullet}OH$, and H_2O_2 and therefore, can function directly as a free radical scavenger. GSH can protect macromolecules (i.e., proteins, lipids, and DNA) either by the formation of adducts directly with reactive electrophiles (glutathiolation) or by acting as a proton donor in the presence of ROS or organic free radicals, yielding GSSG (Asada, 1994). Under unstressed conditions, GSSG is reduced efficiently back to GSH by the action of the enzyme glutathione reductase (GR), present in cytosol, mitochondria, and chloroplasts. In extremely stressful conditions, the rate of GSH oxidation exceeds that of GSH reduction, and the ratio of GSH/GSSG decreases (Foyer et al., 2001). GSH recycles AsA from its oxidized to its reduced form, catalyzed by the enzyme DHAR (Loewus, 1988). GSH can also reduce DHA by a non-enzymic mechanism at pH > 7 and at GSH concentrations greater than 1 mM. This may be a significant pathway in chloroplasts, where in the presence of light, pH remains around 8, and GSH concentration may be as high as 5 mM (Foyer and Halliwell, 1976). The enzyme GR uses NADPH as a cofactor to reduce GSSG back to two molecules of GSH. The reductant GSH, together with its oxidized sulfur–sulfur linked form (GSSG), forms a very potent redox couple and is considered to be the major redox buffer of the cell. Oxidative stress results in the formation of GSSG at the expense of GSH. This shift in the ratio of GSH/GSSG would change the redox status to a less negative potential. If this potential rose too much, it would either stimulate or impede redox-sensitive important cellular processes such as signal transduction pathways, calcium release, enzyme activation, etc. (Watanabe et al., 1972; Ernest and Kim, 1973). Therefore, the generation and maintenance of a reduced GSH pool, either by *de novo* synthesis or via recycling by GR, using NADPH as a cofactor and electron donor, is very important for the cell. The role of GSH in the antioxidative defense system provides a rationale for its use as a stress marker. A time-course analysis to monitor the level of GSH during stress response indicated that an initial stress response was related to changes in the GSH redox state, whereas acclimation was marked by increased GSH concentrations, increased GSH synthesizing enzyme activities, and/or a more reduced redox state of GSH (Tausz et al., 2004). The later was interpreted as overcompensation leading to enhanced regeneration of GSH. Glutathione accumulation has been observed in various plants subjected to abiotic stresses such as drought, salinity, cold, heat, herbicides, UV light, and air pollutants (Waśkiewicz et al., 2014). Enhanced glutathione has been shown to play important role in stress tolerance and global translational changes in *Arabidopsis* (Cheng et al., 2015).

Deteriorative effects on strong stress impacts were related to progressive oxidation of the GSH pool. When apple trees were subjected to progressive drought, the initial response was a slight oxidation of the GSH pool followed by increased GSH concentration. When the stress increased, GSH concentration dropped, and the redox state moved toward the oxidized form (Tausz et al., 2004). During light stress, total and free glutathione content increased quickly in *Arabidopsis* (Choudhury et al., 2018). Similarly to drought stress, an altered ratio of GSH/GSSG has been observed in plants under various stresses, such as salinity (Hefny and Abdel-Kader, 2009), chilling (Radyuk et al., 2009), metal toxicity (Sharma and Dubey, 2007; Maheshwari and Dubey, 2009), anaerobiosis (Lin et al., 2004), and gaseous pollutants (Scebba et al., 2003). Measurement of glutathione redox potential *in vivo* revealed that inside the cell, the environment is more reducing in comparison to the GSH/GSSG ratio in tissue extracts. A redox potential of about −300 mV was observed in the nuclei and cytosol of non-dividing cells. This was maintained even in cells subjected to oxidative stress by several mechanisms. Diaz-Vivancos et al. (2015) proposed that controlled ROS production associated with glutathione-mediated signaling events is characteristic of viable cells under stressful conditions. However, the importance of the GSH system relative to other components of the antioxidative defense system, as well as relative to stress avoidance strategies, remains to be established (Tausz et al., 2004).

7.3.1.3 Tocopherols and Carotenoids

Tocopherols and carotenoids are also important non-enzymic antioxidants in the plant cell. Tocopherols (α, β, γ, and δ) represent a group of lipophilic antioxidants that are synthesized only by photosynthetic organisms. They scavenge oxygen free radicals, lipid peroxy radicals, and $^1O_2$ (Diplock et al., 1989). The fully substituted benzoquinone ring and fully reduced phytyl chain of tocopherol act as antioxidants in redox interactions with $^1O_2$ (Fryer, 1992; Halliwell and Gutteridge, 1999). Tocopherol also acts as a membrane stabilizer. It is widely believed that the protection of pigments and proteins of photosynthetic system and PUFA from oxidative damage caused by ROS is the main function of tocopherols. Tocopherol coordinates with other antioxidants, such as ascorbate, and interacts with many phytohormones, e.g., abscisic acid (ABA), ethylene, SA, and jasmonic acid (Hasanuzzaman et al., 2014). The alteration in α-tocopherol levels resulting from altered expression of genes related to the α-tocopherol pathway, degradation and recycling, depends on the level of stress and the sensitivity of plant species to stress. An increase in α-tocopherol concentration has been associated with plant stress tolerance, whereas reduced levels favor oxidative damage (Munné-Bosch, 2005). Mutants of *Arabidopsis thaliana* with T-DNA insertions in tocopherol biosynthesis genes, tocopherol cyclase (*vte1*), and gamma-tocopherol methyltransferase (*vte4*) showed higher concentrations of protein carbonyl groups and GSSG compared with the wild type, indicating the state of oxidative stress in the tissues (Semchuk et al., 2009). The accumulation of tocopherol differs greatly between different plant species and different plant parts as well (Hasanuzzaman et al., 2014). The

accumulation of α-tocopherol has been shown to induce tolerance to chilling, water deficit, and salinity in different plant species (Yamaguchi-Shinozaki and Shinozaki, 1994; Munne-Bosch et al., 1999; Guo et al., 2006; Bafeel and Ibrahim, 2008). The most important role of tocopherol in mitigating abiotic stress-induced damage is its ability to detoxify ROS by scavenging or quenching lipid peroxides, oxygen radicals, or singlet oxygen. Although various studies have been done to examine the role of tocopherol in abiotic stress tolerance, the mechanisms through which tocopherol confers abiotic stress tolerance are still elusive (Hasanuzzaman et al., 2014).

Carotenoids serve as accessory pigments in light harvesting during photosynthesis, but perhaps a more important role of carotenoids is their ability to detoxify various forms of activated oxygen species (Young, 1991). Carotenoids also serve as precursors to signaling molecules that influence plant development and biotic/abiotic stress responses (Li et al., 2008). The ability of carotenoids to scavenge or to prevent or minimize the production of triplet chlorophyll may be accounted for by their chemical specificity. Carotenoids contain a chain of isoprenic residues bearing numerous conjugated double bonds, which allows easy energy uptake from excited molecules and dissipation of excess energy as heat (Edge et al., 1997; Mittler, 2002). Water stress significantly affects the concentration of some carotenoids in leaves of selected accessions of African eggplant (Mibei et al., 2017). Yildiz-Aktas and coworkers (2009) observed a higher content of carotenoids and antiradical capacity and lower MDA level in drought-tolerant genotypes of cotton compared with the sensitive genotypes in conditions of normal water supply. Hence, carotenoids are correlated with drought tolerance in cotton plants (Yildiz-Aktas et al., 2009). The first committed step in the plastid-localized biosynthetic pathway of carotenoids is mediated by the nuclear-encoded phytoene synthase (PSY). PSY1 was found to have a role in heat stress tolerance (Li et al., 2008). Orange genes, which play a role in carotenoid accumulation, have been isolated from several plant species, and their functions have been intensively investigated. The orange gene (*IbOr*) of sweet potato helps sustain carotenoid homeostasis to improve plant tolerance to environmental stress. *IbOr* interacts directly with PSY, a key enzyme involved in carotenoid biosynthesis, under stress conditions, resulting in increased carotenoid accumulation and abiotic stress tolerance. Transgenic sweet potato plants overexpressing *IbOr* showed enhanced tolerance to high temperatures (47 °C). These findings indicate that *IbOr* protects plants from environmental stresses not only by controlling carotenoid biosynthesis but also by directly stabilizing photosystem II (Kim et al., 2018).

7.3.1.4 Phenolic Compounds

The phenolic compounds flavonoids, phenylpropanoids, and phenolic acids serve as antioxidants (Urquiaga and Leighton, 2000). Phenolics contain an aromatic ring with –OH or OCH_3 substituents, which together contribute to their biological activity. There are two major phenolic classes: hydroxycinnamic acids and flavonoids. The concentration of phenolic compounds increases under abiotic stresses, and this increase shows a correlation with increased activity of enzymes participating in phenolics metabolism, indicating *de novo* synthesis of phenolics under abiotic stresses (Diaz et al., 2001). In contrast, some reports indicate that the increase in the level of phenolic compounds is primarily because of conjugate hydrolysis and not due to *de novo* synthesis (Parry et al., 1994). Flavonoids act as antioxidants and UV-B protectants. They can directly scavenge O_2^-, H_2O_2, •OH, singlet oxygen, or peroxy radicals *in vitro*. Most of the flavonoids outperform the well-known antioxidants, AsA and α-tocopherol, in *in vitro* antioxidant assays because of their strong capacity to donate electrons or hydrogen atoms. However, the antioxidant function of flavonoids in plants is still a matter of debate (Hernandez et al., 2009). During heavy metal stress, phenolic compounds can act as metal chelators, and on the other hand, phenolics can directly scavenge molecular species of active oxygen. Phenolics, especially flavonoids and phenylpropanoids, are oxidized by peroxidase and act in the H_2O_2-scavenging phenolic/AsA/peroxidase (POD) system. Their antioxidant action resides mainly in their chemical structure. There is some evidence of induction of phenolic metabolism in plants as a response to multiple stresses (including heavy metal stress) (Michalak, 2006). When the influence of a high copper sulfate concentration on lipid peroxidation and the contents of total phenolic compounds in roots of lentil (*Lens culinaris* Medic.) *cvs* Krak and Tina were investigated, it was observed that ROS could serve as a common signal for acclimation to Cu^{2+} stress and could cause accumulation of total phenolic compounds in dark-grown roots (Janas et al., 2009). Based on experiments conducted in a mutant with a single gene defect, which led to a block in the synthesis of a group of flavonoids, and wild-type *Arabidopsis*, Lois and Buchanan (1994) suggested an important role of flavonoids in protection from the damaging effects of UV-B light. Overaccumulation of flavonoids has been proposed to enhance tolerance to oxidative and drought stress. Nakabayashi et al. (2014) observed that overaccumulation of anthocyanin reduced ROS accumulation *in vivo* under these stresses, confirming the role of flavonoids in increasing stress tolerance in crops.

7.3.2 Enzymatic Defense System

The enzymatic components of the antioxidative defense system comprise several antioxidant enzymes, including SOD, CAT, guaiacol peroxidase (GPX), and the enzymes of the AsA–GSH cycle: APX, MDHAR, DHAR, and GR (Noctor and Foyer, 1998). These enzymes operate in different subcellular compartments and respond in concert when cells are exposed to oxidative stress. In plants, the decrease of one H_2O_2-scavenging enzyme leads to an increase in the activity of another H_2O_2-decomposing enzyme (Apel and Hirt, 2004). A detailed account of the reactions catalyzed by various antioxidant enzymes is presented in Figure 7.5.

7.3.2.1 Superoxide Dismutase

Superoxide dismutase is the most important antioxidant enzyme in plants. SOD, along with other necessary

downstream components, is needed for the full detoxification of ROS (Giannopolitis and Ries, 1977; Gill et al., 2015). Within the cell, SOD serves as the first line of defense against ROS (Alscher et al., 2002), belongs to the group of metalloenzymes, and catalyzes the disproportionation of $O_2^{•-}$ to O_2 and H_2O_2. This enzyme neutralizes the very reactive $O_2^{•-}$ produced in different compartments of plant cells into O_2 and H_2O_2 at a rate 10^4 times faster than the spontaneous dismutation reaction (Fridovich, 1975). Since SOD is present in all aerobic organisms and in most of the subcellular compartments that generate activated oxygen, it has been assumed that SOD has a central role in defense against oxidative stress (Scandalios, 1993). While all compartments of the cell are possible sites for $O_2^{•-}$ formation, chloroplasts, mitochondria, and peroxisomes are thought to be the most important generators of ROS (Alscher et al., 2002). In plant systems, three isozymes of SOD, namely, Cu/ZnSOD, MnSOD, and FeSOD, have been reported (Fridovich, 1989; Racchi et al., 2001). All forms of SOD are nuclear encoded and targeted to their respective subcellular compartments by an amino terminal targeting sequence (Bowler et al., 1992). Cu/ZnSOD is predominantly found in cytosol, chloroplasts, and mitochondria, and MnSOD in mitochondria and peroxisomes, whereas FeSOD is predominantly present in chloroplasts (del Rio et al., 1998; Arora et al., 2002). Cu/ZnSOD is cyanide sensitive, whereas the other two (MnSOD and FeSOD) are cyanide insensitive (del Rio et al., 1998). Eukaryotic Cu/ZnSOD isoforms are dimers, whereas MnSOD isoforms of mitochondria are tetramers (Scandalios, 1993).

The activity of SOD has been shown to increase under diverse stress situations such as drought, salinity, heat, chilling, metal toxicity, anaerobiosis, exposure to ozone, UV-B, etc. (Bowler et al., 1992; Rao et al., 1996; Lee and Lee, 2000; Verma and Dubey, 2003; Sharma and Dubey, 2005a, 2007; Almeselmani et al., 2006; Yin et al., 2008; Han et al., 2009; Hefny and Abdel-Kader, 2009; Kumutha et al., 2009; Maheshwari and Dubey, 2009; de Deus et al., 2015), and the increased activity is often correlated with tolerance of the plant against these stresses (Liu et al., 2015; Rohman et al., 2016; Esfandiari and Gohari, 2017; Kumar et al., 2017; Li et al., 2017; Zhao et al., 2018). A decreased level of SOD was observed in drought-sensitive durum wheat landraces, whereas in resistant and moderately resistant durum wheat landraces, SOD remained unchanged or increased. Therefore, it was suggested that SOD can be used as an indirect selection criterion for screening drought-resistant plants (Zaefyzadeh et al., 2009). Overproduction of SOD has been reported to result in enhanced oxidative stress tolerance in many plant species (McKersie et al., 1993; Perl et al., 1993; Foyer et al., 1994; Slooten et al., 1995a; Kaouthar et al., 2016; Zhang et al., 2017). The *SiCSD* gene from *Saussurea involucrata* Kar. expressing Cu/ZnSOD has been proposed to act as a positive regulator of drought and cold stress by mitigating oxidative damage (Zhang et al., 2017). In tobacco, overexpression of this gene led to enhanced drought, freezing, and oxidative stress tolerance; increased survival rate, relative water content, and photosynthesis efficiency; higher APX, CAT,

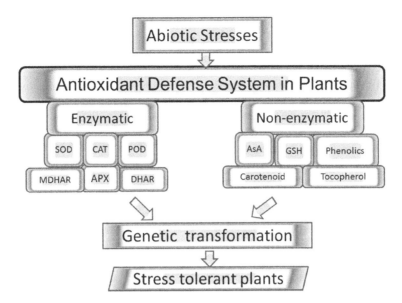

FIGURE 7.5 Antioxidant defense system comprising enzymatic and non-enzymatic components. Under abiotic stresses, increased levels of non-enzymic antioxidants and antioxidant enzymes are needed for efficient quenching of ROS. These antioxidants, singly or in combination, can be successfully used as attractive targets to produce abiotic stress–tolerant plants using biotechnological approaches.

and SOD activities; and reduced ion leakage and malondialdehyde content in comparison to wild type (Zhang et al., 2017). Similarly, a novel MnSOD (*TdMnSOD*) from durum wheat, overexpressed in yeast cells, enhanced tolerance to salt-, cold-, osmotic-, and H_2O_2-induced oxidative stresses by transforming yeast cells. In comparison to non-transformed plants, enhanced activities of total SOD, CAT, and POD were observed in the three transgenic lines subjected to abiotic stress (Kaouthar et al., 2016). The transgenic *A. thaliana* line with high FeSOD activities showed enhanced tolerance to oxidative stress and increased growth rates compared with wild-type plants (Van Breusegem et al., 1999). The expression of a cyanobacterial (*Nostoc flagelliforme*) FeSOD (*NfFeSOD*) into *Sedum alfredii*, which is a heavy metal hyperaccumulator, showed reduced metal accumulation, reduced plant growth, abnormal root architecture, and reduced photosynthetic efficiency. However, increased biomass growth was observed due to long-term high temperature and osmotic stresses. It was suggested that ectopic overexpression of *NfFeSOD* led to a significant increase in APX activity, which in turn, caused a sharp reduction in H_2O_2 concentration (Gao et al., 2016).

7.3.2.2 Catalase

Catalase, an antioxidant enzyme involved in H_2O_2 scavenging, plays an important role during abiotic and biotic stresses (Anjum et al., 2016). Among antioxidant enzymes, CAT was the first enzyme to be discovered and characterized. It directly converts H_2O_2 into H_2O and $^1/_2$ O_2 and does not need a reducing equivalent (Mhamdi et al., 2010; Demidchik, 2015). Therefore, when cells are stressed for energy and are rapidly generating H_2O_2 through catabolic processes, H_2O_2 is degraded by CAT in an energy-efficient manner (Mallick and Mohn, 2000). CAT is a tetrameric heme-containing enzyme found in all aerobic organisms. H_2O_2 has been implicated in many stress conditions. In plants, CAT scavenges

H_2O_2 generated during mitochondrial electron transport, β-oxidation of fatty acids, and most importantly, during photorespiratory oxidation (Scandalios et al., 1997). Different classes of CAT have been identified based on their expression profile. A nomenclature has been adopted for this classification (Willekens et al., 1994). Class I CATs are expressed in photosynthetic tissues and are regulated by light. Class II CATs are expressed at high levels in vascular tissues, whereas Class III CATs are highly abundant in seeds and young seedlings (Willekens et al., 1994). CATs are very sensitive to light and have a very rapid turnover rate similar to that of the D1 protein of PS II (Hertwig et al., 1992). This may be as a result of light absorption by the heme group or perhaps, inactivation due to H_2O_2. Abiotic stresses cause either enhancement or depletion of CAT activity (Feierabend et al., 1992; Fu and Huang, 2001; El-Shintinawy et al., 2004; Sharma and Dubey, 2005a; Moussa and Abdel-Aziz, 2008; Han et al., 2009; Kumutha et al., 2009; Noreen and Ashraf, 2009). The properties of CATs suggest that the enzyme is inefficient in removing low concentrations of H_2O_2 (Arora et al., 2002).

CATs function as a cellular sink for the removal of H_2O_2, as evidenced by the complementation of CAT deficiency by exogenous application of CAT, and the comparison of CAT-deficient and control leaf discs in removing external H_2O_2. Sofo et al. (2015) reviewed the role and genetic regulation of CAT in plants subjected to salinity and water stress. Stress analysis revealed increased susceptibility of CAT-deficient plants to paraquat, salt, and ozone, but not to chilling (Willekens et al., 1997). Transgenic tobacco plants having 10% wild-type CAT activity showed no visible disorders in low light, but in elevated light, they rapidly developed white necrotic lesions on the leaves. Leaf necrosis in such plants was correlated with accumulation of GSSG and a fourfold decrease in AsA, indicating that CAT is critical for maintaining the redox balance during oxidative stress (Willekens et al., 1997). Palatnik and coworkers (2002), while investigating the antioxidant status in a CAT-deficient mutant of barley RPr79/4 and in its mother line cv. Maris Mink, observed that seedlings of the CAT-deficient mutant grown in a growth chamber under a 14 h photoperiod (200 µmol quanta m^{-2} s^{-1}) showed higher concentrations of GSH and APX as compared with wild-type plants. Tobacco plants deficient in CAT did not show greater H_2O_2 accumulation in comparison to the wild line in response to Cd treatment due to CAT deficiency being counterbalanced by enhanced APX and GPX activities and decreased basal NADPH oxidase-like enzyme activity (Iannone et al., 2015). Anjum et al. (2016) reviewed cross-talk between CAT and APX, another H_2O_2-scavenging enzyme. An additional mitochondrial MnSOD isoenzyme was also detected in RPr79/4. When seedlings of the CAT-deficient mutant were grown at higher light intensities (370 µmol quanta m^{-2} s^{-1}), a Cu/ZnSOD isoform and the cytosolic glutamine synthetase isoenzyme were concomitantly induced. Taken together, these results suggest that several defense mechanisms operating in different subcellular compartments respond in concert to compensate for CAT deficiency in barley seedlings exposed to oxidative stress (Palatnik et al., 2002). Physical interaction of glycolate oxidase (GLO) occurs with CAT, and this association–dissociation of CAT and GLO seems to serve as a specific mechanism to alter H_2O_2 levels in rice in response to environmental stress (Zhang et al., 2016). SA-induced H_2O_2 accumulation likely results from SA-induced GLO–CAT dissociation.

7.3.2.3 Guaiacol Peroxidase

PODs belong to a large family of enzymes that occur ubiquitously in fungi, plants, and vertebrates and convert H_2O_2 into H_2O and O_2 (Mika and Luthje, 2003). Classical secretory plant peroxidases that commonly use guaiacol as a reducing substrate are also referred to as *guaiacol peroxidases* (GPX). These proteins usually contain a ferriprotoporphyrin IX prosthetic group that oxidizes several substrates in the presence of H_2O_2 (Vianello et al., 1997). Most secretory plant peroxidases are glycosylated (Johansson et al., 1992). These enzymes have four conserved disulfide bridges and contain two structural Ca^{2+} ions (Schuller et al., 1996). Many isoenzymes of GPX exist in plant tissues localized in vacuoles, the cell wall, and the cytosol (Asada, 1992). GPX oxidizes a wide range of organic substrates, and the oxidation processes are associated with many important biosynthetic processes. GPXs have been proposed to participate in lignification of the cell wall, degradation of indole-3-acetic acid, biosynthesis of ethylene, wound healing, and defense against pathogens (Kobayashi et al., 1996). Both intracellular and extracellular PODs have an important role in the antioxidative response of plant cells (Siegel and Siegel, 1986). PODs can function as effective quenchers of reactive intermediary forms of O_2 and peroxy radicals under stressed conditions (Vangronsveld and Clijsters, 1994). Various stressful environmental conditions have been shown to induce the activity of PODs (Rao et al., 1996; Radotic et al., 2000; Shah et al., 2001; El-baky et al., 2003; Harinasut et al., 2003; Verma and Dubey, 2003; Sharma

and Dubey, 2005a, 2007; Moussa and Abdel-Aziz, 2008; Han et al., 2009; Maheshwari and Dubey, 2009). PODs have been used as biochemical markers for various types of abiotic and biotic stresses (Castillo et al., 1992). Radotic and coworkers (2000) correlated increased activity of PODs with oxidative reactions under metal toxicity conditions and suggested their potential as a biomarker for sublethal metal toxicity in plants.

7.3.2.4 Enzymes of Ascorbate–Glutathione Cycle

Efficient scavenging/destruction of ROS generated during abiotic stresses requires the action of several antioxidant enzymes. The AsA–GSH cycle, also referred to as the Halliwell–Asada pathway, present in at least four different subcellular locations (cytosol, chloroplast, mitochondria, and peroxisomes), scavenges H_2O_2. The AsA–GSH cycle involves successive oxidation and reduction of AsA, GSH, and NADPH, catalyzed by the enzymes APX, MDHAR, DHAR, and GR (Figure 7.6). APX uses two molecules of AsA to reduce H_2O_2 to water with the concomitant generation of two molecules of MDHA. APX is a member of the Class I superfamily of heme peroxidases (Welinder, 1992). APX has a higher affinity for H_2O_2. In higher plants, five chemically and enzymatically distinct isoenzymes of APX have been found, differing in their subcellular localization and amino acid sequences. These are the cytosolic, stromal, thylakoidal, mitochondrial, and peroxisomal isoforms (Jimenez et al., 1997). The classification is based on sequence comparisons rather than on function. APX isoenzymes also differ in their molecular weight, substrate specificity, pH optima, and stability (Wang et al., 1999). Cytosolic APX has been characterized in photosynthetic as well as non-photosynthetic tissues

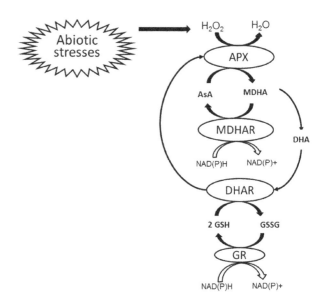

FIGURE 7.6 Components of the ascorbate-glutathione cycle play an important role in decomposition of H_2O_2 in plants under abiotic stresses. APX reduces H_2O_2 to H_2O using AsA, which generates MDHA. MDHA can be reduced to AsA by MDHAR. If not reduced rapidly, MDHA is disproportionated to AsA and DHA. DHA is reduced to AsA by DHAR using GSH as the reducing agent. GSH is converted to GSSG, which in turn, is reduced by GR using NADPH.

(Madhusudhan et al., 2003). The chloroplastic and cytosolic APX isoforms are specific for AsA as electron donor, and the cytosolic isoenzymes are less sensitive to depletion of AsA than the chloroplastic isoenzymes, including stromal and thylakoid-bound enzymes (Ishikawa et al., 1998). Many workers have reported enhanced expression of APX in response to abiotic stresses such as drought, salinity, heat, chilling, metal toxicity, anaerobiosis, UV irradiation, gaseous pollutants, etc. (Rao et al., 1996; Boo and Jung, 1999; Sharma and Dubey, 2005a, 2007; Han et al., 2009; Hefny and Abdel-Kader, 2009; Kumutha et al., 2009; Locato et al., 2009; Maheshwari and Dubey, 2009; Radyuk et al., 2009). The APX1 enzyme has been shown to play a key role when drought and heat stress are applied in *Arabidopsis* (Koussevitzky et al., 2008). The importance of APX in stress defense was demonstrated in APX-antisense transgenic tobacco, which was highly susceptible to ozone injury compared with the wild-type plants (Orvar and Ellis, 1997). Overexpression of a cytosolic APX gene derived from pea (*Pisum sativum* L.) in transgenic tomato plants (*Lycopersicon esculentum* L.) resulted in lower electrolyte leakage (20% to 23%) than in wild type (44%) after exposure to 4°C. Visual assessment of transgenic and wild-type lines exposed to salinity stress (200 or 250 mM) confirmed that overexpression of APX minimized leaf damage. Moreover, APX activity was nearly 25- and 10-fold higher in the leaves of transgenic plants in response to chilling and salt stresses, respectively. The results substantiate that increased levels of APX activity brought about by overexpression of a cytosolic APX gene may play an important role in ameliorating oxidative injury induced by chilling and salt stress (Wang et al., 2005b).

In plant cells, MDHAR and DHAR primarily recycle AsA via the AsA–GSH cycle for reducing ROS generated in leaves during photosynthesis (Suekawa et al., 2017). MDHAR isoforms, which represent multigenic families, are present in several subcellular locations (Leterrier and Cagnac, 2018). Different isoforms of MDHAR act in specific conditions, both during stress, causing excessive production of ROS, and in physiological processes, e.g., photosynthesis and the development of seedlings (Leterrier and Cagnac, 2018). MDHA radical, produced in APX-catalyzed reactions, has a short lifetime, and if not rapidly reduced, it disproportionates to AsA and DHA (Ushimaru et al., 1997). Within the cell, such as at the plasmalemma or at the thylakoid membrane, MDHA can be reduced directly to AsA. The electron donor for MDHA reduction may be b-type cytochrome, reduced ferredoxin, or NAD(P)H. The reaction is catalyzed by the enzyme MDHAR, which is found in several cellular compartments (Miyake and Asada, 1994). The isoenzymes of MDHAR are present both inside and outside the plastids (De Leonardis et al., 1995) and in shoots, roots, and dry seeds as well as fruits (Ushimaru et al., 1997; Leterrier et al., 2005). Various research groups have shown increased activity of MDHAR in plants subjected to abiotic stresses (Boo and Jung, 1999; Sharma and Dubey, 2005a, 2007; Locato et al., 2009; Maheshwari and Dubey, 2009). The presence of different regulatory motifs in the promoter region of the *MDAR1* gene has been suggested to

be responsible for the distinct responses of plants to various stress conditions. The functional analysis of MDHAR isoforms present in the different cell compartments in pea plants grown under eight stress conditions (continuous light, high light intensity, continuous dark, mechanical wounding, low and high temperatures, excess Cd, and application of the herbicide 2,4-dichlorophenoxyacetic acid) revealed a significant induction of MDHAR activity by high light intensity and Cd. On the other hand, expression studies demonstrated differential expression patterns of peroxisomal *MDAR 1* transcript in pea plants grown under these stressful conditions. These findings suggest that the peroxisomal *MDAR 1* has a differential regulation, which could be indicative of its specific function in peroxisomes (Leterrier et al., 2005). To evaluate the role of MDHAR isoform 3 (OsMDHAR3) in rice, transgenic rice plants overexpressing *mdhar3* with silenced NAD(P)H domain were used. These plants showed decreased AsA pool, redox homeostasis imbalance, reduced environmental adaptation ability, enhanced ion leakage and levels of hydroperoxide, reduced AsA/DHA ratio and chlorophyll content, and reduced biomass and yield (Kim et al., 2017).

Despite the possibility of enzymic and non-enzymic regeneration of AsA directly from MDHA, some DHA is always produced when AsA is oxidized in leaves and other tissues. DHA is reduced to AsA by the action of DHAR using GSH as the reducing substrate (Ushimaru et al., 1997). DHAR is a key component of the AsA recycling system. DHAR is a monomeric thiol enzyme abundantly found in dry seeds, roots, and etiolated as well as green shoots. Three functional DHAR genes have been reported to be encoded in the *Arabidopsis* genome. Abiotic stresses such as drought, metal toxicity, chilling, ozone exposure, etc. increase the activity of the DHAR in plants (Boo and Jung, 1999; Hernandez et al., 2001; Sharma and Dubey, 2005a, 2007; Yoshida et al., 2006; Locato et al., 2009; Maheshwari and Dubey, 2009). An *Arabidopsis* mutant completely lacking cytosolic DHAR activity was found to be highly ozone sensitive. The amounts of total AsA and GSH were similar in both lines, but the amount of apoplastic AsA in the mutant was 61.5% lower. These results indicate that the apoplastic AsA, which is generated through the reduction of DHA by cytosolic DHAR, is important for ozone tolerance (Yoshida et al., 2006). Consistent upregulation of the gene encoding cytosolic DHAR was found in *Lotus japonicus*, which was found to be more tolerant to salt stress than other legumes. This upregulation of DHAR was correlated with its role in AsA recycling in the apoplast (Rubio et al., 2009).

A reaction catalyzed by DHAR generates GSSG, which in turn, is re-reduced to GSH using NADPH in a reaction catalyzed by the enzyme GR. GR is an NAD(P)H-dependent enzyme ubiquitously present in mesophyll cells. Although it is located in chloroplasts, cytosol, and mitochondria, around 80% of GR activity in photosynthetic tissues is accounted for by chloroplastic isoforms (Edwards et al., 1990). The catalytic mechanism involves two steps. First, the flavin moiety is reduced by NADPH, the flavin is oxidized, and a redox-active disulfide bridge is reduced to produce a thiolate anion and a cysteine. The second step involves the reduction of

GSSG via thiol–disulfide interchange reactions (Ghisla and Massey, 1989). If the reduced enzyme is not re-oxidized by GSSG, it can suffer a reversible inactivation. Several authors have reported increased activity of this enzyme under abiotic stresses (Rao et al., 1996; Hernandez et al., 2001; Sharma and Dubey, 2005a, 2007; Yoshida et al., 2006; Kumutha et al., 2009; Locato et al., 2009; Maheshwari and Dubey, 2009). Pastori and Trippi (1992) observed correlation between oxidative stress resistance and the activity of GR and suggested that oxidative stress caused by paraquat or H_2O_2 could stimulate GR *de novo* synthesis, probably at the level of translation by preexisting mRNA.

In plants, glutathione S-transferases (GSTs), which are multifunctional proteins, constitute the major part of the antioxidant enzyme network and cellular detoxification system (Liu et al., 2013). The GPX family consists of many isoenzymes with specific localizations and different expression patterns in different tissues and under environmental stresses (Bela et al., 2015). GPXs may represent a link between the glutathione- and the thioredoxin-based system. Due to Cys in their active center, GR, GPX, and some GST isoenzymes are regulated by redox status (Csiszár et al., 2016).

7.4 LEVEL OF ROS, EXTENT OF OXIDATIVE STRESS, AND STATUS OF ANTIOXIDATIVE DEFENSE SYSTEM UNDER VARIOUS ABIOTIC STRESSES

Oxidative stress is one of the major limiting factors in crop productivity. Various stressful environmental conditions, such as drought, salinity, extreme temperatures, excessive levels of metals in the soil, anaerobiosis, gaseous pollutants, UV-B radiation, etc., affect the physiological processes of plants and lead to the production of ROS, which in turn, cause oxidative damage at the cellular level (Radotic et al., 2000; Abarca et al., 2001). Active oxygen species must be rapidly processed if oxidative damage is to be averted. An efficient antioxidative defense system can scavenge overproduced ROS in plants. The lifetime of ROS within the cellular environment is determined by the efficiency of the antioxidative system (both non-enzymatic and enzymatic antioxidants), which provides crucial protection against oxidative damage (Noctor and Foyer, 1998). This section describes the production of ROS, the induction of oxidative stress, and the roles of individual components of the antioxidative defense system under various abiotic stresses.

7.4.1 DROUGHT

Among stressful environmental conditions, drought serves as the major constraint limiting crop production in arid and semi-arid areas of the world. The growth and primary production of plants are severely reduced by water deficit (Tambussi et al., 2000; Fatima et al., 2005). Inhibition of carbon dioxide (CO_2) assimilation, coupled with the changes in photosystem activities and photosynthetic transport capacity under drought stress, results in accelerated production of ROS via the chloroplast Mehler reaction (Asada, 1999). ROS enhancement under drought stress functions as an alarm signal that triggers acclimatory/defense responses by specific signal transduction pathways that involve H_2O_2 as a secondary messenger. ROS signaling is linked to ABA, Ca^{2+} fluxes, and sugar sensing, and is likely to be involved both upstream and downstream of the ABA-dependent signaling pathways under drought stress. Nevertheless, if drought stress is prolonged beyond a certain extent, ROS production will overwhelm the scavenging action of the antioxidant system, resulting in extensive cellular damage (de Carvalho, 2008). The root is the first plant organ to sense a reduced water supply. Signals are generally sent from the roots to the leaves through the xylem sap, and the phytohormone ABA is considered to be one of the major root-to-shoot stress signals. After reaching the leaves, stress signals trigger stomatal closure, and the plant shifts to a water-saving strategy. By adjusting stomatal opening, plants control water loss by reducing the transpiration flux, but the entrance of CO_2 is also reduced concomitantly. The limited supply of CO_2 to the chloroplasts causes a lack of electron acceptors. As a result, the electron transport chain is over-reduced, and the production of ROS increases. The limitation on CO_2 fixation reduces $NADP^+$ regeneration, leading to an over-reduction of the photosynthetic electron transport chain. Thus, during water deficit, there is a considerable potential for increased accumulation of $O_2^{\bullet-}$ and H_2O_2 resulting from the increased rate of O_2 photoreduction in chloroplasts (Robinson and Bunce, 2000). A higher leakage of electrons to O_2 by the Mehler reaction has been observed in plants subjected to drought stress. Biehler and Fock (1996) reported 50% higher leakage of photosynthetic electrons to the Mehler reaction in drought-stressed wheat plants compared with unstressed plants. Increased formation of ROS, lipid peroxidation, and protein modification have been observed in several plant species subjected to drought stress (Boo and Jung, 1999; Sharma and Dubey, 2005a; Bai et al., 2006; Esfandiari et al., 2008; Pandey et al., 2016). The sequence of events in plant tissues subjected to drought includes (1) increased production of ROS, (2) increased expression of genes for antioxidant functions, and (3) increased levels of non-enzymic and enzymic antioxidants, etc., resulting in (4) tolerance against drought stress (Mano et al., 2002). Drought stress increases the specific activity of some of the antioxidant enzymes and also induces the synthesis of new isoenzymes to overcome the increased oxidative stress (Srivalli et al., 2003; Sharma and Dubey, 2005a). Enhanced activity of enzymes of the antioxidative defense system has been reported under drought stress in *Zea mays* (Jagtap and Bhargava, 1995), *Oryza sativa* (Boo and Jung, 1999; Sharma and Dubey, 2005a; Pandey et al., 2014), *Triticum durum* (Sgherri et al., 2000), and *Camellia sinensis* (Rajamickam et al., 2005) plants. The degree to which the activities of antioxidant enzymes and the amount of antioxidants increase in drought-stressed plants appears to be extremely variable among different plant species and even between different cultivars of the same species. Reddy and coworkers (2004) observed that among five mulberry (*Morus alba* L.) cultivars (BC2-59, K-2, MR-2, S-13, and

TR-10) subjected to drought, *cvs* S-13 and BC2-59 had efficient antioxidative characteristics, which could provide better protection against oxidative stress in leaves under water-limiting conditions. Compared with a drought-sensitive maize genotype (Trihybrid 321), a drought-tolerant maize genotype (Giza 2) exhibited lower accumulation of MDA and H_2O_2 and showed increased activities of the antioxidant enzymes SOD, CAT, and POD under water stress (Moussa and Abdel-Aziz, 2008). Differential responses of the antioxidative enzymes SOD and POD to progressive drought stress were also found between two contrasting populations (wet climate and dry climate population) of *Populus cathayana* Rehder. One of the reasons for the difference in drought tolerance was suggested to be the increased capacity of the antioxidative system to scavenge ROS and the consequent suppressed level of lipid peroxidation under drought conditions (Xiao et al., 2008a). The response of a plant species to drought depends not only on the species' inherent "strategy" but also on the duration and severity of the drought period. The activities of SOD and CAT increased in response to severe water deficit in mature leaves of *Populus deltoides × nigra* clones "Luisa Avanzo" and "Dorskamp." For both clones, three different SOD isoforms, MnSOD, FeSOD, and Cu/ZnSOD, were detected in varying amounts depending on drought intensity (Marron et al., 2006). An increase in the amount of AsA and GSH was observed as the level of drought was intensified in two widely cultivated varieties of wheat in Iran, Sab. and N. Sar. (Esfandiari et al., 2008). In rice plants, an increase in the capacity of the AsA regeneration system by *de novo* synthesis of MDHAR, DHAR, and GR has been shown to be one of the primary responses to water deficit so as to mitigate oxidative stress (Boo and Jung, 1999; Sharma and Dubey, 2005a). APX serves as an important component of the antioxidative defense system under drought conditions (Sharma and Dubey, 2005a).

Al-Ghamdi (2009) showed the ability of wheat plants to acclimatize and induce the antioxidant defense system under drought stress. The drought-acclimated leaves exhibited a systematic increase in the activity of H_2O_2-scavenging enzymes, particularly APX and CAT, and maintained an adequate AsA redox pool by efficient functioning of the APX enzyme. As a result, less membrane injury and a lower MDA content were observed in drought-acclimated plants (Al-Ghamdi, 2009). Under severe water stress, the MDA content increased, while the relative water content (RWC) and chlorophyll index decreased significantly in all the wheat genotypes studied. Antioxidant enzyme activities (SOD and POD) increased with the decrease in osmotic potential in drought-tolerant genotypes C306 and AKAW3717. Moreover, the transcript profile of MnSOD was upregulated significantly and was consistent with the trend of the variation in SOD activity, which suggests that MnSOD might play an important role in drought tolerance (Sheoran et al., 2015). Drought stress decreased mung bean leaf area and dry weight and enhanced ROS, lipid peroxidation, and activity of the lipoxygenase enzyme. It decreased RWC, leaf chlorophyll, ascorbate, the GSH/GSSG ratio, and the activities of CAT, MDHAR, and DHAR, and caused increased contents of GSH, GSSG, and proline and

activities of APX and GST (Nahar et al., 2015). The tolerance of *Oudneya* under drought conditions was reported to be mainly due to the efficiency of the AsA–GSH cycle. Talbi et al. (2015) suggested that *Oudneya* has the ability to activate an antioxidative defense system involved in the regulation of drought tolerance, probably involving H_2O_2-dependent signals.

The oil palm hybrid BRS Manicoré, which is more tolerant to drought than BRS C 2501, showed higher total glutathione, ascorbate, and carotenoid concentrations regardless of watering regimes, suggesting that it tolerates severe drought stress by activating multiple antioxidants (Silva et al., 2016). Under these conditions, physiological status indicators, ROS production, and antioxidant system activity were evaluated. Under drought stress, pea plants in the flowering phase showed accumulation of ROS, decreased stomatal conductance and photosynthetic quantum yield, increased activities of SOD, GR, APX, and increased total ascorbate content (Furlan et al., 2016).

7.4.2 SALINITY

Soil salinity is a major constraint to crop productivity, especially in the arid and semi-arid regions of the world. High concentrations of soluble salts in the soil environment cause nutrient imbalance, water deficit, and toxicity of salt ions in growing plants. Therefore, plants growing in saline soils encounter two types of stress: osmotic stress and ion toxicity (Lin and Kao, 2001; Munns, 2002). High salt concentrations normally impair cellular electron transport and lead to overproduction of the ROS $O_2^{\bullet-}$, $^{\bullet}OH$, H_2O_2, and $^1O_2$. Several researchers have demonstrated that salinity stress results in excessive generation of ROS (Tanou et al., 2009; Fan et al., 2015). Some workers suggest that stomatal closure on salt stress may limit the entry of CO_2, which in turn, may cause over-reduction of the photosynthetic electron transport system. Elevated CO_2 mitigates the oxidative stress caused by salinity, involving lower ROS generation and better maintenance of redox homeostasis as a consequence of higher assimilation rates and lower photorespiration (Perez-Lopez et al., 2009). NaCl-induced production of H_2O_2 was possibly linked to NAD(P)H-oxidase and amine oxidase regulation, signaled by nitric oxide (NO), SA, protein kinase, and Ca^{2+} channel activity (Tanou et al., 2009). In several plant species, including halophytes, NADPH oxidase–led generation of ROS has been shown to play an important role in the regulation of adaptation to salinity stress. Respiratory burst oxidase homolog (RBOH) proteins are synergistically activated by the binding of Ca^{2+} to EF-hand motifs as well as Ca^{2+}-dependent phosphorylation (Kurusu et al., 2015). Gémes et al. (2016) suggested that a feed-forward ROS amplification loop was formed by NADPH oxidase and the apoplastic polyamine oxidase (PAO), which induce oxidative stress and execute programmed cell death. Enhanced production of ROS under salinity stress induces phytotoxic reactions such as lipid peroxidation, protein degradation, and DNA mutation (Hefny and Abdel-Kader, 2009; Tanou et al., 2009). NaCl induced oxidative stress in strawberry leaves, as evidenced by $H_2O_2/O_2^{\bullet-}$ accumulation and an increase in lipid peroxidation and carbonyl-group content. NaCl caused oxidation of AsA and GSH redox pairs and inhibition of the activities of the ROS-metabolizing enzymes CAT, APX, and GR (Tanou et al., 2009). In olive plants (*Olea europaea*), salinity induced NADPH-producing dehydrogenases to recycle NADPH, necessary for protection against oxidative damage. The NADP-dehydrogenases appear to be key antioxidative enzymes in olive plants under salt stress (Valderrama et al., 2006).

The effect of salt stress depends on the organ of the plant, the developmental stage of the plant, and the genotype of the plant species, as well as the intensity and duration of the stress. Pronounced organ specificity of antioxidant defense system functioning was observed in *Plantago major* L. plants subjected to NaCl stress. The roots were characterized by high constitutive activities of SOD and three forms of peroxidase, and lower CAT activity. Unlike the leaves, the roots of *P. major* under saline conditions possessed higher activity of the antioxidant system, protecting the plants from the injurious action of oxidative stress and thereby providing survival to the plant species under stressful conditions (Radyukina et al., 2009). Salinity levels of 50 and 100 mM NaCl induced a significant increase in SOD activity, GSH levels, and carotenoid concentrations in all tolerant genotypes and the local genotype of forage sorghum seedlings compared with the sensitive group. It was suggested that these antioxidants could be considered as selection criteria for salt tolerance in sorghum species (Hefny and Abdel-Kader, 2009). Antioxidant systems depend on cultivars and cropping seasons. The antioxidant system in salt-stressed summer crop cultivars of "House Momotaro" was attributed to the enzymatic reactions of APX and GR, while in the salt-stressed "Mini Carol" crop, it was attributed to the non-enzymatic reactions of AsA and GSH. In the winter crop, the antioxidant systems were not influenced by salt stress in either cultivar. The seasonal and cultivar-specific differences in salt-induced changes in the antioxidant systems may result from differences in antioxidant capacities and the interaction between salt stress and growth conditions such as temperature and solar radiation (Zushi and Matsuzoe, 2009). The changes in the leaf apoplastic antioxidant defenses in response to NaCl stress in two pea (*Pisum sativum*) cvs, Lincoln and Puget, revealed that the cultivars had differences in antioxidative capacity, and the NaCl-induced response in the apoplast, and in the symplast from pea cv. Puget in comparison with pea cv. Lincoln, contributed to better protection of pea plants against salt stress (Hernandez et al., 2001).

Using DNA microarray, Kawasaki and coworkers (2001) showed salt-induced upregulation of several antioxidant genes in rice roots. Studies suggest that CAT, POD, and SOD isoenzymes can serve as useful markers in the analysis of gene functions and metabolic regulations, including salt-tolerance characteristics (Mittal and Dubey, 1991; Piqueras et al., 1996; El-baky et al., 2003). The specific activities and patterns of CAT, POD, and SOD isoenzymes are altered significantly in plants subjected to salinity stress. The induction of a new SOD isozyme was found in salt-tolerant embryonic callus

cultures of lemon (*Citrus limon* L. Burn) (Piqueras et al., 1996). Rice seedlings differing in salt tolerance possess a constitutively different number of peroxidase isoforms in non-salinized seedlings. When these seedlings were raised under NaCl salinity, certain new isoforms of peroxidases appeared, and the intensities of some of the preexisting isoenzymes increased. In 15-day-old seedlings of a salt-tolerant rice cv. CSR-1, three isoenzymes were observed in roots and five in shoots, whereas in a salt-sensitive cv. Ratna, six isoenzymes were observed in the roots as well as in shoots (Mittal and Dubey, 1991). Similarly, electrophoretic bands of POD and CAT isoenzymes of onion cultivars were found to vary in numbers and relative concentrations due to salt stress. New protein bands and the characteristic CAT and POD isoenzyme banding patterns have been suggested as a biochemical marker for the selection of salt-tolerant plants (El-baky et al., 2003).

Salt stress markedly enhances the activities of SOD and POD and the levels of total phenolics and γ- and δ-tocopherols, and causes a decrease in the level of total soluble proteins and CAT activity, while the internal levels of H_2O_2 remained unaffected in nine genetically diverse pea cultivars (Noreen and Ashraf, 2009). The tolerance of *Physcomitrella patens* to high-salinity environments makes it an ideal candidate for studying the molecular mechanisms by which plants respond to salinity stresses. Differential genomic and proteomic screenings carried out in these plants showed that plants responded to salinity stress by upregulating a large number of genes involved in the antioxidant defense system (Wang et al., 2008), suggesting that the antioxidative system plays a crucial role in protecting cells from oxidative damage following exposure to salinity stress in *P. patens* plants. Taibi et al. (2016) suggested that salt stress induced oxidative stress in beans and that this was primarily due to a reduced enzymatic defense system. Bose et al. (2014) suggested that salt-tolerant species with efficient Na^+ exclusion mechanisms do not allow excessive ROS production and hence, may not require a high level of antioxidant activity. They also suggested that H_2O_2 "signatures" may operate in plant signaling networks. It was proposed that in halophytes, higher SOD levels cause quick induction of the H_2O_2 "signature" and initiate adaptive responses. The basal level of H_2O_2 decreases due to the activity of enzymatic antioxidants once the signaling has been initiated. Non-enzymatic antioxidants have been proposed as the only effective way to stop the harmful effects of $\bullet OH$ on cellular structures.

7.4.3 Heat

High temperature is becoming one of the important abiotic stresses limiting plant growth and productivity, especially as the global temperature is likely to increase by 1.5–4.5 °C by 2050 (Houghton et al., 2001). Heat stress causes an array of physiological, biochemical, and molecular alterations in plants, influencing plant growth and development. The damage to plants exposed to heat stress has been ascribed to inhibition of photosynthesis, damage to cell membranes, senescence, and cell death (Xu et al., 2006). ROS could play a key role in mediating important signal transduction events. The rates and cellular sites of ROS production during high temperature or heat stress could play a central role in stress perception and protection. ROS levels, as well as ROS signals, are thought to be controlled by the ROS gene network of plants. It is likely that in plants, this network is interlinked with the different networks that control heat stress acclimation and tolerance. Suzuki and Mittler (2006) proposed a model for the involvement of ROS in heat stress sensing and defense. Heat stress leads to overproduction of ROS in cells (Bukhov et al., 1999; Yin et al., 2008). Ribulose bisphosphate carboxylase/oxygenase (Rubisco) can lead to the production of H_2O_2 as a result of its oxygenase reactions, and such H_2O_2 production substantially increases with higher temperature (Kim and Portis, 2004). Considerable work has revealed high temperature–induced oxidative damage in plants (El-Shintinawy et al., 2004; Yin et al., 2008). Lipid peroxidation level was directly correlated with temperature and exposure time in the seedlings of two Egyptian cultivars of barley (Giza 124 and 125). Heat shock caused an increase in the electrical conductivity of cell membranes and increased MDA content coupled with the disappearance of the polyunsaturated linolenic acid ($C_{18:3}$) in the seedlings, reflecting the peroxidation of membrane lipids, which led to the loss of membrane permeability (El-Shintinawy et al., 2004). Plant survival under heat stress requires the activation of proper defense mechanisms to avoid oxidative stress. The heat-tolerant wheat genotypes HD 2815 and HDR-77 showed relatively higher SOD, APX, GR, CAT, and POD activities compared with the sensitive genotypes PBW 343, PBW 175, and HD 2865 under high-temperature stress. Genotypes showing high activity of various antioxidant enzymes also showed minimal reduction in chlorophyll content and a lower membrane injury index, indicating the amelioration of high-temperature stress–induced oxidative stress by antioxidant enzymes (Almeselmani et al., 2006).

The antioxidant response depends on the severity and duration of the stress. When the effects of high temperature on antioxidant enzymes were investigated in three mulberry (*Morus alba* L.) cvs, K-2, MR-2, and BC2–59, maintained at 40 °C for 120, 240, and 360 min, the activities of SOD, CAT, POD, APX, and GR increased with duration of heat stress in all three cultivars. Cultivar BC2-59 showed the most efficient antioxidant system among the three cultivars, which was able to prevent oxidative damage to the leaves caused by high-temperature stress (Chaitanya et al., 2002). El-Shintinawy and coworkers (2004) showed that heat induced distinct and significant changes in the activities of antioxidant enzymes in barley. SOD and POD activities were progressively enhanced by moderate and elevated heat doses, but the most elevated temperature (45 °C for 8 h) caused a decrease in the activities of both enzymes. In contrast, CAT activity was reduced by all heat shocks (El-Shintinawy et al., 2004). In lily (*Lilium longiflorum* L.) plants, exposure to 37 °C and 42 °C for 10 h caused stimulation of the activities of the antioxidant enzymes SOD, POD, CAT, APX, and GR and elevated levels of AsA and GSH, which resulted in low levels of $O_2^{\bullet-}$ and H_2O_2 concentrations.

However, after 10 h exposure at 47 °C, the activities of SOD, APX, and GR as well as the concentration of GSH were similar to those in controls, while the activities of POD and CAT and the AsA concentration declined significantly as compared with the control, with a concomitant increase in $O_2^{\bullet-}$ and H_2O_2 concentrations. In addition, such heat-induced effects on antioxidant enzymes were also observed with SOD and POD isoforms, as Cu/ZnSOD maintained high stability under heat stress, whereas the activity of POD isoforms decreased with the duration of heat stress, especially at 47 °C. Oxidative damage induced by heat stress was related to the changes in antioxidant enzyme activities and levels of antioxidants (Yin et al., 2008).

In plants, the AsA-GSH cycle plays a pivotal role in controlling ROS levels and cellular redox homeostasis. When AsA-GSH cycle enzymes were analyzed in the cytosol, mitochondria, and plastids of tobacco Bright Yellow-2 (TBY-2) cultured cells subjected to two different heat shocks, it was observed that moderate heat shock (35 °C) did not affect cell viability, whereas exposure of cells to 55 °C led to heat shock–induced PCD. In relation to the AsA-GSH cycle, the three analyzed compartments appeared to have specific enzymatic profiles, which were diversely altered by the heat shock. The cytosol contained the highest activity of all AsA-GSH cycle enzymes; in particular, the cytosolic APX was found to be the most versatile enzyme, the activity of which was enhanced after moderate heat shock and declined during PCD induction, whereas other APX isoenzymes were affected only in the cells undergoing PCD (Locato et al., 2009).

TBY-2 cells subjected to short-term heat stress (SHS) suddenly and transiently enhanced antioxidant systems and HSPs, maintaining redox homeostasis and avoiding oxidative damage. Under prolonged heat stress, the antioxidant enzymes are inactivated, resulting in increased H_2O_2 production, lipid peroxidation, and protein oxidation. When TBY-2 cells were pretreated with galactone-γ-lactone (an ascorbate and glutathione precursor) before long-term heat stress treatment, cell growth and cell viability were better than in untreated cells, showing the key role of antioxidants in basal thermotolerance (Sgobba et al., 2015).

Among 37 moth bean (*Vigna aconitifolia*) genotypes (32 mutants and 5 varieties), six were categorized as tolerant, 13 as moderately tolerant, and 18 as susceptible for accession on the basis of wilting symptoms in response to heat stress. Short-term heat stress led to a significant overaccumulation of proline and total sugar and increased activity of GPX, CAT, and SOD. However, only GPX activity was found to be associated with the level of thermotolerance (Harsh et al., 2016).

7.4.4 CHILLING

Chilling stress is a key factor that affects the survival of plants and their geographical distribution. Susceptibility to chilling injury prevents the cultivation of many crops in regions where temperatures can drop far below the optimal growth temperatures (Ercoli et al., 2004). Chilling stress causes both pre-harvest and post-harvest damage to plants, which results in enormous yield losses every year. Evidence suggests that chilling stress in plants leads to increased formation of ROS in both roots and leaves. Chilling conditions cause an imbalance between light absorption and light use by inhibiting Calvin–Benson cycle activity (Logan et al., 2006). Enhanced photosynthetic electron flux to O_2 and over-reduction of the respiratory electron transport chain, resulting in ROS accumulation, has been shown in cucumber leaves subjected to chilling (Hu et al., 2008). Oxidative stress has been suggested to be a significant factor in relation to chilling injury in plants. Chilling causes enhanced production of ROS, slows down metabolism, and causes peroxidation of membrane lipids, resulting in a significant increase in MDA content. Chilling of *Arabidopsis thaliana* (L.) Heynh callus tissues to 4 °C led to conditions of oxidative stress, as indicated by increased levels of the products of peroxidative damage to cell membranes (O'Kane et al., 1996). Evidence for chilling-induced oxidative stress has also been observed in maize seedlings (Prasad et al., 1994).

Responses to chilling-induced oxidative stress include alteration in the activities of enzymes of the antioxidant defense system. In leaves of strawberry plantlets under low-temperature treatment, rates of $O_2^{\bullet-}$ generation and contents of H_2O_2 and MDA increased, while the activities of the antioxidant enzymes SOD, CAT, POD, and APX gradually increased to a certain degree and decreased thereafter. Therefore, it is suggested that low-temperature stress triggers increased production of ROS in plants, and early accumulation of ROS might lead to increased activity of the antioxidant defense system. If the duration of chilling stress is too long, the defense system may not remove overproduced ROS effectively, which may result in severe damage or even cell death (Yong et al., 2008). Chilling stress preferentially enhances the activities of the antioxidative enzymes SOD, APX, GR, and GPX, whereas the activity of CAT decreases in the leaves of cucumber seedlings (Lee and Lee, 2000). Non-enzymic antioxidants (AsA, GSH, and carotenoids) also play an important role in cold response. Under cold stress conditions, the content of low–molecular weight antioxidants, particularly of reduced AsA, increases in barley seedlings. After termination of the stress, the contents of total AsA, GSH, and carotenoids are reduced to a level close to initial. Although the level of reduced AsA declines, it remains at a higher level than initial. Therefore, reduced AsA has been suggested to be an important component in plant cell defense during low-temperature treatment (Radyuk et al., 2009).

ROS-producing and scavenging systems are found in various organelles. Radyuk and coworkers (2009) observed cytoplasmic SOD to be more active than its chloroplast isoforms in green barley (*Hordeum vulgare* L.) seedlings, suggesting that the initiation of the oxidative process under low-temperature stress occurred more actively in the cytosol than in the chloroplasts. The timing of response of different antioxidant enzyme isoforms to chilling appeared to be different in leaves of cucumber seedlings. Out of five APX isoforms present in leaves, the intensities of APX-4 and APX-5 were enhanced due to chilling stress, whereas APX-3 intensity increased in

the post-stress periods after chilling stress. The expression of MnSOD-2 and MnSOD-4 was enhanced only after 48 h of the post-stress period. All six GR isoforms showed increased intensity in stressed plants compared with control and post-stressed plants (Lee and Lee, 2000).

In lucerne leaves, after dark chilling treatment, a marked increase in the level of H_2O_2 and MDA was observed (Bafeel and Ibrahim, 2008). After a recovery period, MDA content decreased significantly due to the increase of phenolic compounds, which suppressed lipid peroxidation. Also, the redox properties of alpha-tocopherol play an important role in adsorbing and neutralizing free radicals and provide some forms of antioxidant protection. The activity of SOD increased sharply with the imposition of chilling stress, whereas CAT, APX, and GR activities slightly increased with the imposition of chilling. During the recovery period, the activities of CAT, APX, and GR increased significantly, which could possibly restrict the recycling of active oxygen species associated with chilling stress (Bafeel and Ibrahim, 2008). Radyuk and coworkers (2009) concluded that APX and CAT play an important role in plant cell defense against low temperatures in barley seedlings, whereas reduced GR and SOD activities are especially important during the poststress period.

Acclimation to prolonged chilling stress can be achieved by briefly pre-exposing the plants to low non-freezing temperatures, a process called *cold acclimation* (CA). For a chilling-sensitive crop such as maize, low-temperature treatment during the early stages of development can be detrimental to subsequent crop establishment and productivity (Stewart et al., 1990; Greaves, 1996). In maize seedlings, cold acclimation induces CAT activity, which seems to play a major role, along with other antioxidant enzymes, in inducing chilling tolerance (Prasad, 1997).

Two subspecies of cultivated rice (*Oryza sativa*) varieties from Asia, i.e., japonica cv. Nipponbare and Indica cv. 93-11, showed significant biochemical alterations related to antioxidative metabolism when subjected to chilling treatment. At the mid-treatment stage, a higher content of ROS and antioxidation-related compounds was observed in Nipponbare seedlings compared with 93-11. Also, the antioxidation-related compounds accumulated in Nipponbare earlier than in 93-11. RNA-seq revealed a significant contribution of ROS-mediated gene regulation, rather than the C-repeat binding factor/dehydration-responsive-element binding factor (CBF/DREB) regulon, to the more vigorous transcriptional stress response in Nipponbare (Zhang et al., 2016).

7.4.5 METAL TOXICITY

High concentrations of essential as well as non-essential metals in the soil environment, arising from mining and industrial activities, disposal of sewage sludge or soil acidification, and the use of pesticides and fertilizers, continue to be a serious risk for plant health (Barcelo and Poschenrieder, 1990). Soil contamination with a variety of metals has become a global problem leading to losses in agricultural yield (Salt et al., 1995). These metals are taken up from the soil by growing plants, and one of the consequences of the presence of the toxic metals within the plant tissues is the formation of free radical species, which can be initiated either directly or indirectly by the metals, consequently leading to oxidative damage to different cell constituents (Gallego et al., 2002). Metal toxicity for living organisms involves oxidative and/or genotoxic mechanisms (Briat and Lebrun, 1999). Enhanced generation of ROS can overwhelm cells' intrinsic antioxidant defenses and may result in a condition known as *oxidative stress*. Recent studies indicate that transition metals act as catalysts in the oxidative reactions of biological macromolecules; therefore, the toxicities associated with these metals might be due to oxidative tissue damage. Both redox-active as well as redox-inactive metals cause an increase in the production of ROS such as $O_2^{\bullet-}$, $^{\bullet}OH$, and H_2O_2. Redox-active metals, such as iron, copper, and chromium, undergo redox cycling, whereas redox-inactive metals, such as lead, cadmium, mercury, and others, deplete cells' major antioxidants, particularly thiol-containing antioxidants and enzymes (Gallego et al., 1996; Weckx and Clijsters, 1996; Yamamoto et al., 1997; Shah et al., 2001; Verma and Dubey, 2003; Sharma and Dubey, 2007; Maheshwari and Dubey, 2009). If metal-induced production of active oxygen species is not adequately counterbalanced by cellular antioxidants, oxidative damage to lipids, proteins, and nucleic acids ensues (Sharma and Dubey, 2007; Sandalio et al., 2009; Sharma and Dietz, 2009; Kumar et al., 2018). A significant decline in protein thiol content is observed when rice seedlings are subjected to Al or Ni toxicity (Sharma and Dubey, 2007; Maheshwari and Dubey, 2009). In addition to the use of plant seedlings or adult plants, another interesting approach has been the use of *in vitro* cell cultures to study plant responses to abiotic stresses, more importantly, metal toxicities (Gomes-Júnior et al., 2006). The induction of ROS production by heavy metals (cadmium and zinc) in *Nicotiana tabacum* L. cv. TBY-2 cells in suspension cultures showed properties comparable to the elicitor-induced oxidative burst in other plant cells (Zrobek-Sokolnik et al., 2009). These heavy metals generated $O_2^{\bullet-}$ and H_2O_2. The effects of CAT, N,N-diethyldithiocarbamate (DDC), and SOD on the heavy metal–induced ROS production indicated that it occurs outside the cells and that at least part of the H_2O_2 is produced by dismutation of the $O_2^{\bullet-}$. The results suggested that the enzyme responsible for cadmium- and zinc-induced ROS generation in tobacco cells contains a flavocytochrome (Zrobek-Sokolnik et al., 2009).

The increased activity of antioxidative enzymes in metal-stressed plants appears to serve as an important component of the antioxidant defense mechanism of plants to combat metal-induced oxidative injury (Shah et al., 2001). An antioxidant system consisting of several non-enzymic and enzymic components is activated in the cells as a response to metal stress (Gallego et al., 1996; Shah et al., 2001; Verma and Dubey, 2003). The activation of antioxidant enzymes such as POD, SOD, APX, and enzymes of the AsA-GSH cycle due to the toxicity of metals such as Cd, Pb, and Al has been reported in plants by various groups of workers (Cakmak and Horst, 1991; Shah et al., 2001; Verma and Dubey, 2003; Sharma and

Dubey, 2007). However, the results suggest that the activation of antioxidant enzymes in response to oxidative stress induced by metals is not enough to confer tolerance to metal accumulation. The responses of plants to heavy metal exposure vary depending on the plant species, tissue, and stage of development, and the type of metal and its concentration. One of the key responses includes the triggering of a series of defense mechanisms that involve enzymatic and non-enzymatic components (Gratão et al., 2005). Among various metals, such as Cd, Pb, Al, and the metalloid As, different effects have been observed on plant metabolism; although roots are regarded as the main sites of metal accumulation, the oxidative damage was found to be more detrimental in the leaves (Erdei et al., 2002; Jha and Dubey, 2004; Sharma and Dubey, 2007; Pandey et al., 2016).

Cu^{2+} ions are redox active and catalyze Fenton-type reactions producing $^{\bullet}OH$ (Elstner et al., 1988). Lipid peroxides also originate from the induction of the enzyme lipoxygenase in the presence of Cu^{2+} (Somashekaraiah et al., 1992). This enzyme is known to initiate lipid peroxidation. Iron has a pivotal and dual role in free-radical chemistry in all organisms. Free Fe can participate in Fenton reactions and can catalyze the generation of $^{\bullet}OH$ and other toxic oxygen species. On the other hand, Fe is a constituent of the antioxidant enzymes CAT, APX, GPX, and FeSOD (Becana et al., 1998). When plants are exposed to a variety of adverse environmental conditions, including chilling, high light, drought, etc., oxidative stress occurs primarily due to the increase in free-radical production mediated by catalytic Fe (Arora et al., 2002) and also due to a decrease in the antioxidative defense system (Arora et al., 2002).

Arsenic toxicity has been known for centuries and has recently received increased attention because of its chronic and epidemic effects on human health (Abernathy et al., 1999). Plants normally take up arsenic predominantly in trivalent (As(III)) and pentavalent (As(V)) forms, which are known to interfere with various metabolic pathways in cells, such as interaction with sulfhydryl groups, replacement of phosphate in ATP, and excessive production of ROS. Singh and coworkers (2007) showed that As induces oxidative stress resulting from enhanced lipid peroxidation in mung bean (*Phaseolus aureus* Roxb.). The accumulation of MDA in seedlings increased significantly with increasing arsenic concentration (both As(III) and As(V)). However, oxidative stress was more pronounced with As(III) treatment. Bandyopadhyay (2016) revealed the major role of ROS in As genotoxicity. As treatments significantly increased the activities of antioxidative enzymes (SOD and GR) and the contents of antioxidant metabolites (GSH and AsA) in Indian mustard, the increase being dependent on exposure time. The increase in the activity of CAT was not significant. It was concluded that Indian mustard was able to detoxify the low As level by induction of the antioxidant defense mechanism (Khan et al., 2009). The upregulation of APX and POD has also been reported in As(III)/As(V) treatments. Both GSH and Cys imparted enhanced tolerance to seedlings against arsenic stress. Seedling growth improved, while the level of MDA declined significantly when As(III)/

As(V) treatments were supplemented with GSH and Cys, suggesting GSH- and Cys-mediated protection against oxidative stress (Shri et al., 2009). While As(V) predominantly stimulates antioxidant enzyme activity, As(III) primarily causes enhanced levels of thiols (Srivastava et al., 2007). Mishra et al. (2011) reported increased formation of ROS and enhanced lipid peroxidation in arsenite-treated rice seedlings. Further, arsenite treatment increased non-enzymatic as well as enzymatic antioxidants and the synthesis of thiols and phytochelatins, which can be important components in alleviating As-induced oxidative damage. As led to modification of the root histological structure and damage to macromolecules. GST, which was markedly increased on exposure to As, is proposed as a suitable biomarker of arsenic contamination in peanut (Bianucci et al., 2017).

Cadmium is a potent heavy metal pollutant of the environment. It is a heavy metal of widespread occurrence. Cadmium induces oxidative stress in several plant species and increases the activity of enzymes of the antioxidant defense system (Shah et al., 2001; Metwally et al., 2003). The Cd-induced ROS burst is partially mediated by NADPH oxidase (Chmielowska-Bąk et al., 2018). The increase in NADPH oxidase activity was correlated over time with a significant accumulation of ROS. The NADPH oxidase inhibitor diphenyleneiodonium chloride (DPI) affected the expression of two Cd-inducible genes encoding DOF1 and MYBZ2 transcription factors (Chmielowska-Bąk et al., 2018). Using an *Arabidopsis* mutant, it has been shown that RBOHs differentially regulate ROS metabolism, redox homeostasis, and nutrient balance and could be of potential interest in biotechnology for the phytoremediation of polluted soils (Gupta et al., 2017). Lipids that contain phosphate groups are essential components of membranes that surround the cell as well as other cellular structures, such as the chloroplasts, mitochondria, and nucleus. Cd-induced oxidative damage involves peroxidation of PUFA of membrane lipids due to ROS generated by Cd (Lin et al., 2007). Lipid peroxidation eventually increases membrane fluidity and membrane permeability. Elevated levels of lipid peroxides were observed in Cd-stressed rice seedlings (Shah et al., 2001). Under 500 µM Cd treatment, about a 1.4- to 1.6-fold increase in MDA content was observed, indicating enhanced peroxidation of lipids due to Cd exposure (Shah et al., 2001). Increased production of ROS under Cd toxicity serves as a major source of DNA damage, leading to stand breakage, removal of nucleotides, and a variety of modifications in organic bases of nucleotides (Sarkar, 1995). A distinct pattern of DNA fragmentation, typical for PCD, was observed in Cd-exposed tobacco cells (Fojtova and Kovarik, 2000). DNA damage was suggested to be an important mechanism of Cd phytotoxicity in *Vicia faba* plants (Lin et al., 2007). Cadmium treatment decreases chlorophyll and heme levels in mungbean seedlings by the induction of lipoxygenase with the simultaneous inhibition of the antioxidative enzymes SOD and CAT (Somashekaraiah et al., 1992). Such inhibition results from binding of the metal to the important sulfhydryl group of enzymes, which increases the phytotoxic action of metals (Van Assche and Clijsters, 1990). In pea plants, Cd

toxicity induced carbonylation in proteins, and the extent of carbonylation was greater in peroxisomes than in the whole plant (Romero-Puertas et al., 2002). Increased protein oxidation was observed in cucumber seedlings (*Cucumis sativus* L.) grown in increasing Cd levels. Importance of the enzymatic and non-enzymatic antioxidant system in response to Cd toxicity has been shown by several authors. Enhancement in the activities of SOD, CAT, POD and GR is observed in Cd-stressed rice seedlings (Shah et al., 2001). The activity of another antioxidative enzyme CAT increased in rice seedlings grown at moderately toxic Cd (100 µM) level, whereas with highly toxic Cd (500 µM) level a marked inhibition in CAT activity was noted (Shah et al., 2001). Decline in CAT activity in plants growing under higher levels of Cd was suggested due to inhibition of enzyme synthesis or a change in assembly of enzyme subunits (Shah et al., 2001). Barley seedlings exposed to 25 µM Cd showed increased activity of H_2O_2-metabolising enzymes CAT and APX (Metwally et al., 2003). Apparently, it is the oxidative stress induced by Cd that enhances the activities of stress-related enzymes by increasing the levels of free radicals and peroxides. Activity of APX was inhibited while the activities of CAT and SOD were increased in cucumber seedlings (*Cucumis sativus* L.) grown in increasing levels of Cd. Simultaneously, AsA and non-protein thiol groups were also found to increase (Gonçalves et al., 2007). A comparison of closely related plant species with different degrees of sensitivity to toxic metals has established a link between the degree of plant tolerance to metals and the level of antioxidants (Sharma and Dietz, 2009). Nouairi and coworkers (2009) showed that *Brassica juncea* plants possessed greater potential for Cd accumulation and tolerance than *Brassica napus* plants. *B. napus* exhibited an increased level of lipid peroxidation on Cd exposure, whereas in treated *B. juncea* plants, the MDA content remained unchanged. In *B. napus*, with the exception of GPX, the activity levels of some antioxidant enzymes involved in the detoxification of ROS, including SOD, CAT, GR, and APX, decreased drastically at high Cd concentrations. By contrast, in leaves of Cd-exposed *B. juncea* plants, there was little or no change in the activities of the antioxidative enzymes. Analysis of the profile of the anionic isoenzymes of GPX revealed qualitative changes occurring during Cd exposure for both species. Mung bean genotypes (Pusa 9531, Pusa 9072, Pusa Vishal, and PS-16) treated with Cd (0, 25, 50, and 100 mg/kg soil) showed a differential response to Cd concentrations; Pusa 9531 was identified as Cd tolerant, whereas PS-16 was Cd susceptible. The results revealed the presence of a strong antioxidant defense system (elevated activities of SOD, CAT, APX, and GR and increased amounts of AsA and GSH) in the Cd-tolerant genotype (Pusa 9531) for providing adequate protection against oxidative stress induced by Cd (Anjum et al., 2008). In maize, Cd or As led to enhanced ROS content, osmolyte accumulation, and activities of antioxidant enzymes compared with controls. The combined application of Cd + As caused greater reduction in the plant biomass of both maize cultivars. The maize cultivar Dong Dan 80 showed less heavy metal–induced oxidative damage, which was associated with lower ROS production, higher antioxidant activities, and greater accumulation of osmolytes (Anjum et al., 2016).

Aluminum is a major constraint reducing crop productivity in acid soils throughout the world (Kochian, 1995; Pereira et al., 2006). Although Al itself is not a transition metal and cannot catalyze redox reactions, the involvement of oxidative stress in Al toxicity has been suggested in many plant species. In black gram pulse crops, Al^{3+} induces excessive ROS production, which leads to cellular damage, root injury, stunted root growth, and other metabolic shifts. Prolonged Al^{3+} stress can activate an aluminum detoxification defense mechanism. It was proposed that ROS played a key role as signaling molecules for aluminum stress in black gram (Chowra et al., 2017). Even without an external supply of Fe, enhanced peroxidation of lipids is observed due to Al in pea (*Pisum sativum*) roots (Yamamoto et al., 2001). Al ions have a strong affinity for biomembranes and cause the rigidification of membranes (Deleers et al., 1986), which seems to facilitate the radical chain reaction enhancing the peroxidation of lipids in phospholipid liposomes (Oteiza, 1994), soybean (*Glycine max*) roots (Cakmak and Horst, 1991), and cultured tobacco (*Nicotiana tabacum*) cells (Ono et al., 1995; Yamamoto et al., 1997). Al-induced peroxidation of lipids leads to loss of plasma membrane integrity and eventually cell death in cultured tobacco cells (Yamaguchi et al., 1999). Boscolo and coworkers (2003) observed a higher degree of protein oxidation in maize roots, compared with lipid peroxidation, in Al-treated plants and suggested that in maize roots, proteins are the primary target of damage due to ROS under Al toxicity. Increased oxidation of proteins and increased proteolysis have been reported in rice seedlings subjected to Al toxicity (Sharma and Dubey, 2007; Bhoomika et al., 2014). Al treatment led to DNA fragmentation in rice seedlings (Meriga et al., 2004). Signaling pathways involving H_2O_2 cause PCD in heavy metal–stressed plants. Cells treated with 100 μM $AlCl_3$ showed typical features of PCD, such as nuclear and cytoplasmic condensation, in tomato (*Lycopersicon esculentum* Mill.) suspension cells. Cell death was suppressed by the application of antioxidants and by inhibitors of phospholipase C (PLC), phospholipase D (PLD), and ethylene signaling pathways. The results suggest that low concentrations of heavy metal ions stimulate both PLC and PLD signaling pathways, leading to the production of ROS and subsequent cell death executed by caspase-like proteases (Yakimova et al., 2007).

As Al induces the expression of diverse genes in plant species such as wheat, maize, sugarcane, tobacco, and *Arabidopsis*, and many of these genes encode antioxidant enzymes such as GST, peroxidase, and SOD (Ezaki et al., 2000; Simonovicova et al., 2004), a strong correlation has been suggested between Al toxicity and oxidative stress in plants (Richards et al., 1998; Boscolo et al., 2003; Watt, 2004). Ezaki and coworkers (2000) confirmed this hypothesis when they showed that overexpression of some Al-induced genes in transgenic *Arabidopsis* plants conferred oxidative stress resistance. Meriga and coworkers (2004) found a close inverse relationship between decreased root growth and increased Al accumulation, lipid peroxidation, SOD

and POD activities, and DNA damage in rice plants. Higher activities of SOD and POD in Al-stressed rice seedlings suggested that these enzymes could serve as efficient free radical scavengers to minimize the adverse effects of lipid peroxidation and could contribute to the maintenance of membrane structure and integrity of rice plants under Al toxicity. Al toxicity is associated with the induction of oxidative stress in rice plants, and among antioxidative enzymes, SOD, guaiacol POD, and cytosolic APX appear to serve as important components of the antioxidative defense mechanism under Al toxicity (Sharma and Dubey, 2007; Pandey et al., 2016). Polyacrylamide gel electrophoresis (PAGE) confirmed the increased activity as well as the appearance of new isoenzymes of APX in Al-stressed seedlings. Immunoblot analysis revealed that changes in the activities of APX are due to changes in the amounts of enzyme protein (Sharma and Dubey, 2007). Transgenic *Arabidopsis thaliana* plants overexpressing enzyme GR synthesized higher amounts of GSH and were tolerant to Al (Yin et al., 2017).

Lead is a major heavy metal of antiquity and has gained considerable importance as a potent environmental pollutant (Sharma and Dubey, 2005; Srivastava et al., 2014). One of the phytotoxic effects of Pb appears to be the induction of oxidative stress in growing plant parts due to enhanced production of ROS, resulting in unbalanced cellular redox status. Pb is not an oxido-reducing metal like iron; therefore, the oxidative stress induced by Pb in plants appears to be an indirect effect of Pb toxicity, leading to the production of ROS, enhancing the pro-oxidant status of cell by reducing the pool of reduced GSH, activating calcium-dependent systems, and affecting iron-mediated processes (Pinto et al., 2003). Although the ROS-generating processes are slow under normal conditions, Pb accelerates them (Verma and Dubey, 2003). Such production depends on the intensity of the stress, repeated stress periods, and the species and age of plants (Asada, 1994; Verma and Dubey, 2003). Pea plants treated with Pb showed increased levels of the ROS $O_2^{•-}$ and H_2O_2 in the roots, and the increase was proportional to metal concentration (Malecka et al., 2009). Lipid peroxidation, which is regarded as an indicator of oxidative damage, involves oxidative degradation of PUFA residues of membranes (Girotti, 1990). Pb ions induce lipid peroxidation, decrease the level of saturated fatty acids, and increase the content of unsaturated fatty acids of membranes in several plant species (Halliwell and Gutteridge, 1999; Verma and Dubey, 2003; Reddy et al., 2005; Srivastava et al., 2014). The MDA level was reported to increase concomitantly with the increasing level of H_2O_2 produced in pea plants treated with Pb (Malecka et al., 2009). When rice (*Oryza sativa*) seedlings were raised in sand cultures under 500 and 1000 μM $Pb(NO_3)_2$ in the medium, during a growth period of 5–20 days, about a 21–177% increase in the level of lipid peroxides was observed, indicating that Pb induces oxidative stress in rice plants (Verma and Dubey, 2003).

The importance of the antioxidant defense mechanism for tolerating oxidative stress caused by Pb has been shown by various research groups. The activities of the antioxidative enzymes SODs, CAT, and APX and the levels of low-molecular antioxidants, particularly GSH, homoglutathione (h-GSH), and cysteine, were reported to increase concomitantly with increasing Pb concentration and duration of treatment (Malecka et al., 2009; Srivastava et al., 2014). The redox state (GSH/GSSG) of root cells dropped proportionally to stress intensity (Malecka et al., 2009). Pb-treated rice seedlings showed an increase in the activities of the enzymes SOD, GPX, APX, and GR compared with untreated controls (Verma and Dubey, 2003). Treatment of rice seedlings with 1000 μM Pb caused a decrease in the intensity of two preexisting CAT isoforms in shoots of rice seedlings. It is shown that Pb induces oxidative stress in growing rice plants and that SOD, peroxidases, and GR could serve as important components of the antioxidative defense mechanism against Pb-induced oxidative injury in rice plants (Verma and Dubey, 2003). Chloroplasts isolated from spinach seedlings treated with $PbCl_2$ showed an increase in both ROS and MDA content and a reduction in photosynthesis and in the activities of the antioxidant defense system (SOD, CAT, APX, GPX, and GSH content), indicating that the spinach chloroplasts underwent a stress condition due to oxidative attack. The results imply that spinach chloroplasts are not able to tolerate the oxidative stress induced by Pb^{2+}, possibly due to lack of an effective antioxidant defense mechanism (Xiao et al., 2008b).

Nickel is an essential nutrient for plants. However, the amount of Ni required for the normal growth of plants is very low. Nickel toxicity has been associated with oxidative stress in plants. Ni treatment enhanced electron flow through complex IV in *Vigna radiata* seed cotyledons. The altered antioxidant activity could be correlated with ROS generation and subsequent alteration in seed metabolism (Yadav et al., 2018). Nickel-treated rice seedlings showed increased rates of $O_2^{\bullet-}$ production; elevated levels of H_2O_2 and thiobarbituric acid reactive substances (TBARS), demonstrating enhanced lipid peroxidation; and a decline in protein thiol levels, indicative of increased protein oxidation, compared with controls (Maheshwari and Dubey, 2009). Gajewska and Skłodowska (2007) observed that despite prolonged increases in $O_2^{\bullet-}$ and H_2O_2 levels, oxidative damage measured in terms of lipid peroxidation did not occur in the leaves of Ni-treated wheat plants.

With progressively higher Ni concentrations, non-protein thiol and AsA levels increased, whereas the level of low–molecular weight thiols (such as GSH and hydroxyl-methyl GSH), the ratio of these thiols to their corresponding disulfides, and the ratio of AsA to DHA declined in the rice seedlings (Maheshwari and Dubey, 2009). The activities of all isoforms of SOD (Cu/ZnSOD, MnSOD, and FeSOd) GPX and APX were reported to increase in Ni-treated rice seedlings, while no clear alteration in CAT activity could be observed (Maheshwari and Dubey, 2009). An excess supply of Ni was shown to inhibit CAT activity and induce POD, APX, and SOD activities in maize. The localization of isoforms of these enzymes on native gels also revealed increases in the intensities of preexisting bands. The enhanced activities of peroxidase, APX, and SOD, however, did not appear to be sufficient to ameliorate the effects of excessively generated ROS due

to excess supply of Ni (Kumar et al., 2007). The activity of the AsA-GSH cycle enzymes MDHAR, DHAR, and GR significantly increased in Ni-treated seedlings (Maheshwari and Dubey, 2009). Exposure of *Nicotiana tabacum* cv. BY-2 cell suspensions to Ni led to alterations in isoenzymes of SOD, CAT, and GR. Activity staining analysis revealed that CAT activity plays a major role in the early response to Ni-induced oxidative stress, particularly when the Ni concentration used is low, while a specific GR isoenzyme appears to respond to the Ni-induced oxidative stress when a much higher Ni concentration is used to induce the stress for the same period of treatment (Pompeu et al., 2008). These results illustrate the importance and advantages of determining individual isoenzyme activities. Differential responses of SOD, CAT, and GR isoenzymes are associated with specific physiological phenomena of the cells due to the specific organellar localization of the isoenzymes (Pompeu et al., 2008). It has been observed that Ni-induced oxidative stress in rice seedlings can be reduced by modulating the antioxidative defense system with the use of aqueous extracts of the bark of *Terminalia arjuna* trees (Rajpoot et al., 2016).

In roots and shoots of wheat seedlings treated with Ni, a considerable increase in H_2O_2 concentration was observed (Gajewska and Sklodowska, 2008). The application of Ni led to many-fold enhanced activities of POD, APX, and GST in the shoots, whereas the activities were not significantly altered in the roots (Gajewska and Sklodowska, 2008). In comparison with *N. tabacum*, the oxidative damage was lower in *Aquilegia bertolonii* roots due to high endogenous activities of CAT and to a lesser extent, SOD (Boominathan and Doran, 2002). Exposure of rice (*Oryza sativa* L., cv. yangliangyou 6) plants to excess Ni concentrations resulted in a decline in root and shoot lengths as well as fresh weight (FW) and dry weight (DW), and depletion in the contents of photosynthetic pigments. Ni caused increased accumulation of hydrogen peroxide and MDA content. Ni stress decreased SOD activity but increased the activities of CAT and POD and the contents of AsA and GSH in both roots and shoots. Ni stress also triggered proline accumulation and decreased the contents of soluble protein and soluble sugars. The accumulation of osmolyte under Ni stress appeared to provide additional defense against oxidative stress (Rizwan et al., 2017).

7.4.6 ANAEROBIOSIS

Anaerobic stress is thought to be a major stress factor in the growth of crop plants in waterlogged soils (Armstrong, 1979; Sasidharan et al., 2018). Oxygen deficiency causes stomatal closure (Crawford, 1978), which reduces CO_2 availability in the leaves and inhibits carbon fixation. Thus, excessive excitation energy in chloroplasts could increase the generation of ROS and induce oxidative stress (Gossett et al., 1994). An excessive accumulation of $O_2^{\bullet-}$ and H_2O_2 is observed in plants subjected to anaerobic stress (Yan et al., 1996; Kumutha et al., 2009; Sairam et al., 2009). Under anoxic conditions, the production of ROS increases in mitochondria and chloroplasts (Sasidharan et al., 2018). Roots suffer from periodic or prolonged deprivation of oxygen, which leads to diminished respiration at the level of electron transport. The lack of a suitable electron acceptor leads to saturated redox chains, accumulation of NAD(P)H, and a decline in the generation of ATP. Flooding stress also contributes to the accumulation of acetaldehyde and other compounds derived from anaerobic metabolism, which could be susceptible to degradation, yielding H_2O_2 as an end product (Blokhina et al., 2003). The presence of H_2O_2 in the apoplast and in association with the plasma membrane has been visualized by transmission electron microscopy under hypoxic conditions in yellow flag iris (*Iris pseudacorus*), rice, and anoxia-intolerant wheat and garden iris (*Iris germanica*) (Blokhina et al., 2001). In the anoxia-tolerant species, the response was delayed in time, and in highly tolerant *I. pseudacorus*, plasma membrane–associated H_2O_2 was detected only after 45 d of oxygen deprivation.

Under flooding, plants show oxidative damage conditions due to the excess ROS production (Sasidharan et al., 2018). ROS interfere with normal metabolism through the oxidation of proteins, nucleic acids, and DNA and alter membrane integrity. Under flooding, plants intensify the antioxidant defense system, both enzymatic and non-enzymatic antioxidants, which scavenge excess ROS (Hasanuzzaman et al., 2017).

Waterlogging-induced production of H_2O_2 is also involved in the signaling and induction of various defense-related genes leading to the synthesis of proteins/enzymes imparting hypoxia tolerance (Sairam et al., 2009). Excessive production of ROS can oxidize biological molecules, such as DNA, proteins, and lipids, causing the malfunction of these molecules (Richter and Schweizer, 1997). Oxidative stress is an integral part of oxygen deprivation stress. Enhanced lipid peroxidation products have been observed in plants growing under low-oxygen conditions (Yan et al., 1996; Chirkova et al., 1998; Blokhina et al., 1999). An increased level of TBARS was observed at 4 and 6 d of waterlogging in pigeon pea plants over control plants, probably due to activation of diphenyleneiodonium-sensitive NADPH oxidase (Kumutha et al., 2009). The injury to biological lipids by ROS, as indicated by MDA content, was clearly detected in the waterlogged barley plants, but with a distinct difference between the Xiumai 3 (tolerant) and Gerdner (sensitive) genotypes, implying a possible difference in the capacity of their defense system against ROS (Zhang et al., 2007). Thus, the sensitivity of plants to environmental stress, including waterlogging, has been considered to be associated with antioxidative ability (Crawford, 1978; Foyer et al., 1994). Jamei and coworkers (2009) showed that excessive accumulation of H_2O_2 and $O_2^{\bullet-}$ induced membrane damage and lipid peroxidation in leaves of *Zea mays* subjected to hypoxia, which was as a result of reduced activity of SOD.

When the activities of antioxidant enzymes were examined for waterlogging tolerance of pigeon pea (*Cajanus cajan* L. Halls) genotypes ICP 301 (tolerant) and Pusa 207 (sensitive), it was observed that the activities of enzymes SOD, APX, GR, and CAT increased under waterlogging in both the cultivars and that comparatively greater antioxidant enzyme activities, resulting in lower oxidative stress, were observed

in the tolerant genotype ICP 301 compared with the sensitive genotype Pusa 207, suggesting that greater activities of antioxidative enzymes represent one of the factors determining higher tolerance of pigeon pea plants to flooding (Kumutha et al., 2009). Lin and coworkers (2008) observed that increased APX activity and increased levels of total GSH and oxidized and total AsA at different days of flooding confer flooding tolerance on sweet potato plants (*Ipomoea batatas* L. Lam. "Tainung 57"). When the eggplant genotypes EG117 and EG203 and the tomato genotypes L4422 and TNVEG6 were subjected to seven flooding treatments, the activity of APX in roots significantly increased during the period of continuous waterlogging (Lin et al., 2004). Slight increases in total AsA, reduced AsA, GSH, and total GSH contents in the roots were also observed throughout the entire waterlogging period. However, the activities of CAT, SOD, and GR, and the contents of AsA, GSSG, and α-tocopherol in the roots, were unaffected by waterlogging (Lin et al., 2004). Genotypes responded differently to oxidative injury according to the activity of their different antioxidative components. Following waterlogging treatments, APX activity in eggplants was generally higher than in tomatoes. Limited or less efficient APX in tomatoes led to accumulation of H_2O_2. It was concluded that increased APX activity could provide increased tolerance to waterlogging in plants (Lin et al., 2004).

Rice seedlings raised under water for 6 days showed reduced activities of SOD, APX, MDHAR, DHAR, GR, and CAT compared with seedlings raised in air for the same time period (Ushimaru et al., 1992). AsA and GSH were present in submerged seedlings at nearly the same levels as those found in aerobically grown controls. When submerged seedlings were exposed to air, the activities of the six antioxidative enzymes exceeded the levels in aerobically grown controls during 24 h of adaptation to air. The level of AsA increased slightly, but the level of GSH showed a rapid increase, reaching seven times that in aerobically grown controls within 12 h of adaptation to air. Therefore, in this case, the development of the antioxidative defense system appeared to consist of two steps: a rapid increase in the level of GSH and a subsequent slow increase in the activities of antioxidant enzymes (Ushimaru et al., 1992). The antioxidant systems of root and leaf cells behave differentially with regard to changes in oxygen concentration. In leaves, low-mass antioxidants play the major role in ROS detoxification, while in roots, increased alternative oxidase capacity supports the antioxidant systems, possibly by preventing ROS formation (Skutnik and Rychter, 2009).

When the responses of antioxidant enzymes to different depths of waterlogging in creeping bentgrass (*Agrostis stolonifera* L.) roots were examined, it was observed that SOD and APX were mainly involved in waterlogging-induced antioxidant responses and that partial waterlogging could also significantly affect root antioxidant activities (Lin et al., 2004). The ability to maintain a balance between the formation and the detoxification of ROS appears likely to contribute to the increased survival potential and the tolerance of the roots against oxidative stress (Lin et al., 2004).

Waterlogging has detrimental effects on crop productivity. In pigeon pea (*Cajanus cajan* (L.) Millsp.) plants, the activities of the enzymes SOD, APX, CAT, POD, and polyphenol oxidase (PPO) were higher in the waterlogging-resistant cultivar ICPL-84023 compared with the waterlogging-sensitive cultivar MAL-18. POD and PPO activity increased immediately (3 h) after the imposition of waterlogging stress, whereas changes in SOD activity were variable at different stages of observation. CAT activity increased significantly at 9 h and then decreased. APX activity did not change significantly under waterlogging (Bansal and Srivastava, 2015).

7.4.7 GASEOUS POLLUTANTS

An increase in the concentration of gaseous pollutants such as O_3, SO_2, CO, and NO_2 is detrimental to plant and human health (Lee et al., 2017). Ozone, the most abundant air pollutant, is hazardous to plants because of its strong oxidizing potential. Ozone concentration and exposure time determine the extent of toxicity and consequently, the severity of injury in plants. Ozone induces the formation of ROS in plants. Tropospheric ozone damages crop plants and forests by entering leaf mesophyll tissues through the stomata and rapidly generating the ROS $O_2^{\bullet-}$, $^{\bullet}OH$, and H_2O_2 at the cell perimeter (Pellinen et al., 1999; Pasqualini et al., 2002). Ozone enters plants through the stomata and is rapidly broken down to various ROS at the cell-wall interface. Ozone induces active ROS production. Ozone-induced accumulation of H_2O_2 is observed initially on the plasma membrane and cell wall in birch (*Betula pendula*) (Pellinen et al., 1999). Experiments with inhibitors of possible sources for H_2O_2 production in the cell wall suggested that both NADPH-dependent superoxide synthase and the cell-wall POD are involved in such H_2O_2 production. The H_2O_2 production continued in the cytoplasm, mitochondria, and peroxisomes when the ozone exposure was over, but not in chloroplasts. The timing of mitochondrial H_2O_2 accumulation coincided with the first symptoms of visible damage, and at the same time, the mitochondria showed disintegration of the matrix. These responses may not be directly connected with defense against oxidative stress but may, rather, indicate changes in oxidative balance within the cells that affect mitochondrial metabolism and the homeostasis of the whole cell, possibly leading to the induction of PCD (Pellinen et al., 1999). In sensitive plant species, even a few hours of exposure to this potent oxidant leads to severe oxidative stress that manifests the symptoms of visible cell death. Two mechanisms have been suggested for the injury brought about by ozone or ozone-induced ROS production: (i) modification of proteins (Rao et al., 1995), lipids (Calatayud and Barreno, 2001), and nucleic acids (Rousseaux et al., 1999) and (ii) activation of the PCD pathway, similar to the pathogen-induced hypersensitive response (Overmyer et al., 2005). It has been postulated that ozone or ozone-induced ROS may locally cause necrotic cell death, which in turn, can trigger the PCD pathway in surrounding tissues. An acute exposure to 150 ppb ozone decreased stomatal conductance to 60–70% of its initial value within 9–12 min in *Arabidopsis* plants

(Kollist et al., 2007). The transient decrease was absent in the ABA-insensitive mutant *abi2*, defective in a class 2C protein phosphatase. This provides *in vivo* confirmation that the early transient decrease in stomatal conductance is not a result of physical damage by the ROS formed from ozone breakdown but reflects the biological action of ROS, transduced through a signaling cascade (Kollist et al., 2007). Highly significant interactions between ozone damage and levels of ROS, AsA, GSH, and lipid peroxidation were observed in *Medicago truncatula* accessions from various geographical regions. There were significant differences among the accessions for these traits before and after the end of ozone fumigation, suggesting that multiple physiological and biochemical mechanisms may govern ozone tolerance or sensitivity (Puckette et al., 2007). Acute ozone treatment (300 nL L^{-1} for 6 h) led to an ROS burst in the sensitive genotype Jemalong 6 h post-fumigation, whereas in the resistant genotype JE154, the increase in ROS levels was much lower. Unique and shared transcriptional responses in an ozone-resistant and sensitive accession exemplify the complexity of oxidative signaling in plants (Puckette et al., 2007). It has been speculated that plants sensitive to acute ozone have an impairment in the perception of the initial signals generated by the action of this oxidant. This, in turn, leads to a delayed transcriptional response in the ozone-sensitive plants. In resistant plants, the rapid and sustained activation of several signaling pathways enables the deployment of multiple mechanisms for minimizing the toxic effect of this reactive molecule (Puckette et al., 2008).

In living organisms, ROS, directly or indirectly derived from ozone exposure, are scavenged by enzymatic and non-enzymatic antioxidant defensive mechanisms whose overall responsibility is to preserve cell structures and macromolecules from oxidative damage. These defenses are essentially those also involved in detoxifying the ROS inevitably produced by the metabolism of organisms living in an oxygen atmosphere (Iriti and Faoro, 2008). Studies on enzymatic and non-enzymatic antioxidants in two species belonging to a natural ecosystem, *Trifolium repens* L. (cv. Sonja) and *T. pratense* L. (cv. Milvus) in response to an acute ozone dose, revealed that *T. repens* had higher constitutive levels of the defense molecules compared with the other species. Moreover, it proved to be more prone to activating the antioxidant systems in response to ozone, showing a higher degree of sensitivity to the pollutant (Scebba et al., 2003). Rao and coworkers (1996) observed that *A. thaliana* genotype Landsberg erecta was capable of metabolizing ozone-induced activated oxygen by invoking the enzymes of the SOD/AsA-GSH cycle. Ozone exposure not only enhanced SOD, POD, GR, and APX activity but also modified the substrate affinity of both GR and APX (Rao et al., 1996). Ozone impacts on the yield and quality of crops are influenced by changing climatic conditions, increasing atmospheric CO_2, and altered emission patterns (Fuhrer, 2009). To study the relation between ozone sensitivity and leaf concentrations of antioxidants (AsA, total phenolics, and total antioxidant capacity), Severino and coworkers (2007) exposed ozone-sensitive (NC-S clone) and resistant plants (NC-R clone) of *T. repens* and *Centaurea jacea* to moderate ozone concentrations in ambient air. The NC-R clone showed the highest concentrations of antioxidants, with 50–70% more AsA than NC-S. NC-R had about five times more AsA in the young leaves and nine times more in the older leaves than *Centaurea*. In a fumigation experiment with acute ozone stress (100 nL L^{-1}), the antioxidant levels changed profoundly. The ozone-injured leaves of NC-S had six to eight times more total phenolics than uninjured leaves. Generally, older leaves had lower antioxidant concentrations and were more prone to ozone injury than younger leaves. Low AsA concentrations were more closely related to the appearance of visible ozone injury than the other antioxidant parameters.

Sulfur dioxide (SO_2) is a gaseous pollutant that adversely affects the growth and metabolism of plants (Lee et al., 2017). SO_2 penetrates into the leaves through the stomata. It dissolves in the aqueous medium surrounding the plant cells, generating the toxic molecular species sulfite and bisulfite. The detoxification reaction of sulfite to sulfate, which takes place by reactions initiated by light, is mediated by the photosynthetic electron transport chain (Asada, 1980) and leads to the formation of $O_2^{\bullet-}$, $^{\bullet}OH$, and H_2O_2 (Asada, 1980). These highly oxidant molecular species, together with the toxic sulfite, can damage the lipids and proteins of cell membranes. The damage to leaves by SO_2 in the presence of light has been attributed to the oxidative chain reaction of sulfite initiated by $O_2^{\bullet-}$ produced as a result of the Mehler reaction in chloroplasts (Asada and Takahashi, 1987). Exposure to SO_2 of plants of two wheat (*Triticum aestivum* L.) cvs, "Mec" and "Chiarano," each with a different sensitivity to SO_2, revealed that the different sensitivities to SO_2 of the two cultivars were due to differential ability to maintain elevated levels of AsA rather than increasing detoxifying enzyme activities (Ranieri et al., 1997). A significant increase in AsA, GSH, and their redox state was observed in plants exposed to high CO_2 and SO_2 compared with plants exposed to SO_2 alone. The absence of the negative effects of SO_2 in the presence of high CO_2 has been correlated with a high redox state of AsA and GSH (Rao and Dekok, 1994). At low concentration, SO_2 is beneficial to plants, but high concentrations of SO_2 can cause serious damage to plants within a few hours (Lee et al., 2017).

7.4.8 UV-B Radiation

The most damaging part of the UV spectrum reaching the earth's atmosphere is UV-B (290–320 nm). The increase in solar UV-B radiation reaching the earth's surface due to depletion of the stratospheric ozone layer has become a major global concern (Blumthaler and Ambach, 1990; Xiong and Day, 2001). UV-B impairs several biochemical and physiological processes, including gene expression (Booij-James et al., 2000). UV-B radiation causes enhanced generation of ROS, reduced photosynthesis, decreased protein synthesis, and impaired chloroplast functions (Strid et al., 1994; Booij-James et al., 2000; Han et al., 2009; Kohler et al., 2017). UV-B causes oxidative stress through excessive production of the ROS $O_2^{\bullet-}$, $^{\bullet}OH$, and H_2O_2 (Strid et al., 1994; Han et al., 2009), which in turn, cause enhanced lipid and protein

oxidation (Mittler, 2002; Babu et al., 2003; Frohnmeyer and Staiger, 2003; Yannarelli et al., 2006; Lu et al., 2009). The mechanisms of ROS generation by UV are not well understood (Green and Fluhr, 1995; Hideg et al., 2002). Rao and coworkers (1996) demonstrated that UV-B radiation enhances the activated oxygen species by increasing membrane-localized NADPH oxidase activity and decreasing CAT activity (Rao et al., 1996). Plants have to adapt to the deleterious effects of UV-B radiation, because they are dependent on sunlight for photosynthesis and therefore, cannot avoid exposure to UV-B radiation. Besides antioxidative enzymatic scavengers for ROS, such as SOD, POD, CAT, and APX, and non-enzymatic antioxidants such as AsA and GSH, plants possess substances such as UV-B-absorbing compounds and carotenoids to keep the balance between the production and removal of ROS (Costa et al., 2002; Han et al., 2009). In *Picea asperata* seedlings, enhanced UV-B (30%) induced the overproduction of ROS, leading to oxidative stress. UV-B increased the efficiency of an antioxidant defense system consisting of UV-B-absorbing compounds, carotenoids, and the antioxidant enzymes SOD, APX, CAT, and POD (Han et al., 2009). UV-B exposure preferentially enhanced GPX, APX, and POD specific to coniferyl alcohol and modified the substrate affinity of APX in *A. thaliana*. New isoforms of peroxidases and APX were synthesized in the flavonoid-deficient mutant transparent testa (*tt5*) of *A. thaliana* (Rao et al., 1996). Leaves from bitter gourd (*Momordica charantia* L) subjected to UV-B at three different stages of plant growth (pre-flowering, flowering, and post-flowering) revealed that the activities of SOD, CAT, POD, PPO, and GR and the concentrations of AsA, H_2O_2, and TBARS were elevated under UV-B exposure at all growth stages with the exception of H_2O_2 concentration at the post-flowering stage. It is suggested that *M. charantia* exhibits a protection mechanism against oxidative damage by maintaining a highly induced antioxidant system under UV-B stresses (Agarwal and Shaheen, 2007). Numerous studies have shown that large differences in UV-B sensitivity exist among plant species and even between cvs of the same species (Teramura and Murali, 1986; Barnes et al., 1993). Santos and coworkers (1999) observed the influence of UV-B radiation on the activity of SOD and the number and amounts of isoforms in leaves of the C_3 plants potato and wheat and the C_4 plants maize and sorghum. The total specific activity of SOD increased significantly in wheat, maize, and potato, whereas a decline was induced in sorghum. Native gels revealed that UV-B caused preferential changes of the SOD isoforms in all plants. The rise in SOD activity in maize, potato, and wheat was correlated with the UV-B tolerance of these crops, and the sensitivity of sorghum to UV-B was associated with the decrease in SOD activity.

Gao and Zhang (2008) studied response of an AsA-deficient *Arabidopsis thaliana* mutant, *vtc1*, to short-term increased UV-B exposure. After UV-B supplementation, *vtc1* mutants showed an increase in H_2O_2 content and production of TBARS and a decrease in chlorophyll content and chlorophyll fluorescence parameters. A reduced ratio of GSH/total GSH and an increased ratio of DHA/total AsA were observed in the *vtc1* mutants compared with the wild-type plants. In addition, the enzymes responsible for ROS scavenging, such as SOD, CAT, and APX, and enzymes responsible for the regeneration of AsA and GSH (including MDHAR, DHAR, and GR) had insufficient activity in the *vtc1* mutants compared with the wild-type plants. These results suggest that the AsA-deficient mutant *vtc1* is more sensitive than wild-type plants to supplementary UV-B treatment, and AsA can be considered as an important antioxidant for UV-B radiation (Gao and Zhang, 2008). Kohler and coworkers (2017) showed that UV-B exposure of the Antarctic vascular plant *Deschampsia antarctica* Desv. led to a sudden increase in the level of ROS and an increase in the activities of antioxidant enzymes. The high level of antioxidant enzymes in *D. antarctica* plants is associated with their UV-B tolerance (Kohler et al., 2017).

Many studies are focused on the elucidation of response mechanisms of plants to stresses applied individually; however, some abiotic stresses often occur simultaneously under field conditions, and the response to a combination of two different stresses is specific and cannot be determined from the stresses applied individually. Rivero et al. (2014) observed a specific response of plants to a combination of stresses, which included the accumulation of glycine, betaine, and trehalose. The accumulation of these compounds under the combined stresses led to a lower Na$^+$/K$^+$ ratio and a better performance of photosynthesis and cell water status in comparison with salinity alone. Heat stress, salinity stress, or a combination of both stresses led to an increase in oxidative stress. A metabolomic study revealed that under conditions where more flavonols were accumulated compared with hydroxycinnamic acids, the oxidative damage was lower, revealing the importance of flavonols as powerful antioxidants (Martinez et al., 2016).

Subjecting the drought-sensitive cotton cultivar 84-S and the drought-tolerant cultivar M-503 to the combined effects of drought and heat led to a greater decrease of relative growth rate and increase in lipid peroxidation in 84-S, suggesting that it is more sensitive than M-503 to combined stress. The sensitivity of 84-S to combined stress was associated with decreased activities of CAT and POD, resulting in higher H$_2$O$_2$ accumulation and oxidative stress–induced lipid peroxidation. The combined stress tolerance of M-503 was associated with proline accumulation, induced CAT and POD, and its ability to maintain constitutive activities of SOD and APX. The proline content of drought-resistant M-503 was greatly enhanced under drought and the combination of drought and heat treatments as compared with 84-S (Sekmen et al., 2014).

7.5 PRODUCTION OF ABIOTIC STRESS–TOLERANT TRANSGENIC CROP PLANTS USING COMPONENTS OF ANTIOXIDATIVE DEFENSE SYSTEM

Oxidative stress is a key component involved in the causation of damage to crop plants exposed to abiotic stresses. Maintaining redox balance is one of the crucial requirements for cells to endure abiotic stresses. Stress tolerance in plants can be enhanced by manipulating the level of ROS (You and Chan, 2015; Chakradhar et al., 2017). Tolerance to several abiotic stresses has been well correlated with an enhanced level of components of the antioxidant defense system. Transgenic plants offer novel possibilities to achieve complete understanding of the roles of different enzymes and biomolecules involved in protection against various types of abiotic stress. Genetic engineering approaches have gained considerable ground in terms of improving the traits of many crops within the shortest possible time period. Transgenic plants overexpressing enzymatic or non-enzymatic components of the antioxidant defense system, singly or in combination, offer attractive possibilities for producing abiotic stress–tolerant plants using biotechnological approaches (Chakradhar et al., 2017).

Enhancement of the capacity of the antioxidative defense system using gene transfer technology has been shown to enhance the tolerance of crop plants to abiotic stresses, with improved performance and productivity under these stresses. Table 7.2 provides an overview of abiotic stress–tolerant plants produced by overexpressing different enzymatic components of the antioxidative defense mechanism. SOD serves as the first enzyme in the chain of enzymatic components of the antioxidative defense system (Luis et al., 2018). Increased tolerance to some environmental stresses has been achieved by overexpressing SOD in target plants (Matters and Scandalios, 1986; Slooten et al., 1995a). Tolerance to drought (Badawi et al., 2004; Wang et al., 2005a), salinity (Liu et al., 2003; Prashanth et al., 2008; Zhang et al., 2008), low-temperature stress (Gupta et al., 1993; McKersie et al., 1993; Lee et al., 2009), and ozone (Van Camp et al., 1994) has been achieved by overexpressing SOD. Overexpression of wheat *TaSOD1.1* and *TaSOD1.2* genes increased SOD activities and decreased MDA content, resulting in less over-oxidation of the cellular membrane system and enhancement of physiological functions, with improved low-temperature stress tolerance and NaCl tolerance in transgenic tobacco plants (Zhang et al. 2008, 2009). Upregulation of target genes in transgenic plants at the transcriptional level enhanced the salt tolerance capacity of the plants (Zhang et al., 2008). Halophytic plants such as mangroves have been reported to have a high level of SOD activity, which plays a major role in defending the mangrove species against severe abiotic stresses. Sod1, a cDNA encoding a cytosolic Cu/ZnSOD from the mangrove plant *Avicennia marina*, was expressed in rice. The transgenic plants withstood salinity stress of 150 mM of NaCl for a period of 8 days, while the untransformed control plants wilted at the end of the stress treatment in hydroponics. The transgenic plants also revealed better tolerance to drought stress in comparison to untransformed control plants (Prashanth et al., 2008). A chimeric gene consisting of the coding sequence for cytosolic Cu/ZnSODs from *Oryza sativa* fused to the chloroplast transit sequence from *Arabidopsis thaliana* GR was used for generating transgenic tobacco plants. The first generation of the transgenic lines showed enhanced tolerance to salt and water stresses over the wild type, suggesting that the overexpressed Cu/ZnSOD enhances the chloroplast antioxidant system (Badawi et al., 2004).

TABLE 7.2

Overexpression of Antioxidant Enzymes to Produce Abiotic Stress Tolerant Plants

Genes	Gene Source	Transgenic Plant	Tolerance	References
SOD	Triticum aestivum	Nicotiana tabacum	Salinity	Zhang et al. (2008)
Cu/ZnSOD	Arachis hypogaea	Nicotiana tabacum	Dehydration and salinity	Negi et al. (2015)
	Saussurea involucrata	Nicotiana tabacum	Drought, cold, and oxidative stress	Zhang et al. (2017)
Fe-SOD	Arabidopsis thaliana	Medicago sativa	Freezing	McKersie et al. (2000)
	Nostoc flagelliforme	Sedum alfredii	High temperature and osmotic stress	Gao et al. (2016)
Mn-SOD	Arabidopsis thaliana	Arabidopsis thaliana	Salinity	Wang et al. (2004)
	Triticum durum	Arabidopsis thaliana	Salt, cold, osmotic, and H_2O_2-induced oxidative stress	Kaouthar et al. (2016)
CAT	Escherichia coli	Lycopersicon esculentum	Drought and chilling	Mohamed et al. (2003)
	Escherichia coli	Oryza sativa	Salinity	Moriwaki et al. (2008)
	Brassica juncea	Nicotiana tabacum	Cadmium	Guan et al. (2009)
APX	Pisum sativum	Lycopersicon esculentum	Salinity and chilling	Wang et al. (2005b)
	Puccinellia tenuiflora	Arabidopsis thaliana	Salinity	Guan et al. (2015)
	Brassica campestris	Arabidopsis thaliana	Heat	Chiang et al. (2015)
Cytosolic APX	Oryza sativa	Arabidopsis thaliana	Salinity	Lu et al. (2007)
	Prunus domestica	Pisum sativum	Water stress	Diaz-Vivancos et al. (2016)
	Camellia azalea	Nicotiana tabacum	Heat and cold	Wang et al. (2017)
Thylakoid-bound APX	Lycopersicon esculentum	Nicotiana tabacum	Salinity	Sun et al. (2009)
Peroxisomal APX	Hordeum vulgare	Arabidopsis thaliana	Salinity	Xu et al. (2008a)
	Hordeum vulgare	Arabidopsis thaliana	Zinc and cadmium	Xu et al. (2008b)
MDHAR	Arabidopsis thaliana	Nicotiana tabacum	Drought, salinity, ozone	Eltayeb et al. (2007)
	Brassica rapa	Arabidopsis thaliana	Freezing	Shin et al. (2013)
DHAR	Homo sapiens	Nicotiana tabacum	Salinity and chilling	Kwon et al. (2003)
	Oryza sativa	Arabidopsis thaliana	Salinity	Ushimaru et al. (2006)
GST/Glutathione POD	Suaeda salsa	Nicotiana tabacum	Thermal and salt stress	Roxas et al. (2000)
Glutathione POD	Chlamydomonas W80	Nicotiana tabacum	Salinity and chilling	Yoshimura et al. (2004)
GST	Nicotiana tabacum	Arabidopsis thaliana	Salinity	Qi et al. (2004)
	Juglans regia	Nicotiana tabacum	Chilling	Yang et al. (2016)
GR	Arabidopsis thaliana	Arabidopsis thaliana	Aluminum	Yin et al. (2017)

Overexpression of *Arachis hypogaea* (*AhCuZnSOD*) in tobacco plants led to enhanced SOD activity and tolerance to dehydration and salinity stress, as revealed by better germination of seeds and higher chlorophyll content under these stresses. Under stress conditions, transgenic plants showed less electrolyte leakage, less MDA, H_2O_2, and $O_2^{\bullet-}$ accumulation, and higher relative water content and antioxidant enzyme activity compared with wild-type plants, indicating that CuSOD plays a major role in mitigating oxidative stress induced by various environmental stresses (Negi et al., 2015).

Transgenic *Arabidopsis* plants overexpressing MnSOD showed enhanced tolerance to salt stress (Wang et al., 2004). When MnSOD from *Nicotiana plumbaginifolia* was targeted to tobacco mitochondria, only a minor effect on ozone tolerance was observed. However, overproduction of SOD in the chloroplasts resulted in a three- to fourfold reduction of visible ozone injury in transgenic tobacco plants (Van Camp et al., 1994). Overexpression of FeSOD increases $O_2^{\bullet-}$-scavenging capacity and thereby improves the oxidative stress tolerance of plants. Transgenic alfalfa plants transformed with *Arabidopsis* FeSOD with a chloroplast transit peptide showed

a novel FeSOD in native polyacrylamide gel electrophoresis (PAGE). As no detectable difference in the pattern of primary freezing injury, as shown by vital staining, or additional accumulation of carbohydrates in field-acclimated roots of the transgenic alfalfa plants could be observed, it was suggested that FeSOD overexpression reduced secondary injury symptoms and thereby, enhanced recovery from stresses experienced during winter (McKersie et al., 2000).

Attempts have been made to improve the stress tolerance of many crop plants by manipulating CAT genes (Mohamed et al., 2003; Nagamiya et al., 2007; Moriwaki et al., 2008; Guan et al., 2009). Salinity stress is a major constraint to the productivity of cereals. In an attempt to improve the salt tolerance of rice, a CAT gene of *Escherichia coli*, *katE*, was introduced into japonica rice cv. Nipponbare and indica rice cv. BR5. The resultant transgenic rice plants constitutively expressing *katE* were more tolerant to NaCl than wild-type plants (Nagamiya et al., 2007; Moriwaki et al., 2008). When the bacterial CAT gene was overexpressed in tomato chloroplasts, the transgenic plants had increased tolerance to photooxidative stress imposed by drought stress or chilling stress

(Mohamed et al., 2003). CAT has also been used to protect plants from Cd-induced oxidative stress caused by ROS. When a CAT gene from *Brassica juncea* was introduced into tobacco, wild-type plants became chlorotic and almost dead while transgenic tobacco plants remained green and phenotypically normal under 100 mM Cd treatment (Guan et al., 2009). Tolerance to various abiotic stresses was obtained by expressing a novel catalase, *TdCAT1*, from durum wheat, in yeast cells and *Arabidopsis*. Higher CAT and POD activities and lower H_2O_2 concentration were observed in transgenic lines (Feki et al., 2015).

Ascorbate peroxidase plays an important role in scavenging ROS by degrading H_2O_2 in plants. The use of APX genes for plant transformation has led to the development of transgenic plants with enhanced tolerance to oxidative stress (Slooten et al., 1995b; Wang et al., 2005b; Duan et al., 2006; Xu et al., 2008a,b). Overexpressing a cytosolic APX from *Arabidopsis thaliana* in the chloroplasts of *Nicotiana tabacum* cv. SR1 resulted in significantly enhanced H_2O_2-scavenging capacity (Slooten et al., 1995b). Different isoforms of APX in cytosol have different functional roles in rice. Transgenic lines overexpressing *OsAPXb* showed higher salt tolerance than *OsAPXa* transgenic *Arabidopsis* lines. An enhanced active oxygen scavenging system was suggested to protect plants from salt stress by equilibrating H_2O_2 metabolism in these lines. Overproduction of OsAPXb enhanced and maintained a higher level of APX activity than OsAPXa in transgenic *Arabidopsis* during treatment with different concentrations of NaCl. The findings suggest that the rice cytosolic *OsAPXb* gene has a more functional role than *OsAPXa* in the improvement of salt tolerance in transgenic plants (Lu et al., 2007). Thylakoid-bound APX and peroxisomal type APX (pAPX) also play an important role in protection against various abiotic stress–induced oxidative stresses (Xu et al., 2008a,b; Sun et al., 2009). When the thylakoid-bound APX gene (*LetAPX*) from tomato was overexpressed in tobacco, improved salt tolerance was observed in transgenic plants (Sun et al., 2009). Transgenic *A. thaliana* plants carrying a pAPX gene (*HvAPX1*) from barley were found to be more tolerant to salt stress than the wild type. The salt tolerance in transgenic plants was due to the maintenance and reestablishment of cellular ion homeostasis but to reduction of oxidative stress injury (Xu et al., 2008a). Transgenic rice plants overexpressing *HvAPX1* under excess Cd were significantly more tolerant to Cd stress and accumulated more Cd compared with the wild-type plants (Duan et al., 2006). *HvAPX1* also plays an important role in protection against excess Zn-induced oxidative stress and appears to be a novel candidate gene for developing high-biomass Zn-tolerant plants for phytoremediation of Zn-polluted environments (Xu et al., 2008b). The mechanism of Zn tolerance in transgenic plants was suggested to be due to reduced oxidative stress damage. OsAPX1 overexpressed in transgenic rice plants showed enhanced spikelet fertility under cold stress (Sato et al., 2011). OsAPX2-overexpressing plants showed increased APX activity, decreased H_2O_2 and MDA levels, and enhanced drought stress tolerance at the booting stage, as indicated by increased spikelet fertility (Zhang et al.,

2013). Overexpression of *PutAPX* from *Puccinellia tenuiflora*, which can grow in extreme saline–alkali soil, in *A. thaliana* enhanced salinity tolerance and reduced lipid peroxidation (Guan et al., 2015). Overexpression of BcAPX (*Brassica campestris*, Bc) in *Arabidopsis* enhanced heat tolerance by eliminating H_2O_2 and led to enhanced high temperature–stress tolerance (Chiang et al., 2015). Four copies of the cytosolic ascorbate peroxidase (*cytapx*), when expressed in plum plants, showed a marked effect at the physiological, biochemical, genetic, and proteomic levels, enhancing water stress tolerance (Diaz-Vivancos et al., 2016). Chiang et al. (2017) reported that overexpression of *LcAPX* from *Luffa cylindrica* and *SmAPX* from *Solanum melongena* enhanced flood tolerance, probably by eliminating H_2O_2, in *Arabidopsis* plants. The *PcAPX* gene from *Populus tomentosa*, when expressed in transgenic tobacco plants, led to drought, salt, and oxidative stress tolerance, evident by increased chlorophyll and lower levels of MDA (Cao et al., 2017). Overexpression of the *Camellia azalea* APX (*CaAPX*) gene showed elevated ROS-scavenging activities, which led to marked cold and heat tolerance and revealed a role of CaAPX in ROS signaling during temperature stresses (Wang et al., 2017).

The enzyme MDHAR is crucial for the regeneration of AsA and essential for maintaining a reduced pool of AsA. Overexpression of *MDHAR* has been shown to minimize the deleterious effects of environmental stresses (Eltayeb et al., 2007; Suekawa et al., 2018). Transgenic tobacco plants overexpressing *Arabidopsis thaliana MDHAR* gene (*AtMDAR1*) in the cytosol exhibited up to 2.1-fold higher MDHAR activity and 2.2-fold higher level of reduced AsA compared with non-transformed control plants. The transgenic plants showed enhanced stress tolerance in terms of significantly higher net photosynthesis rates under ozone, salt, and PEG stresses and greater photosystem II effective quantum yield under ozone and salt stresses. Furthermore, these transgenic plants exhibited significantly lower H_2O_2 level when tested under salt stress (Eltayeb et al., 2007). These results demonstrate that an overexpressed level of *MDHAR* confers enhanced tolerance to ozone, salt, and PEG stress (Eltayeb et al., 2007).

The enzyme DHAR is assumed to be critical for AsA recycling. Manipulation of DHAR expression is important for genetic engineering of stress-tolerant plants (Kwon et al., 2003; Ushimaru et al., 2006; Amako and Ushimaru, 2009). Expression of rice DHAR in transgenic *Arabidopsis thaliana* enhanced resistance to salt stress (Ushimaru et al., 2006). Similarly, enhanced tolerance to NaCl and low temperature was observed when a human DHAR was overexpressed in tobacco plants. Transgenic plants showed higher DHAR and GR activities than non-transgenic plants, and the ratio of AsA/DHA increased from 0.21 to 0.48, even though total AsA content was not significantly changed. When tobacco leaf discs were subjected to oxidative stress, reduction in membrane damage relative to non-transgenic plants was observed. Furthermore, transgenic plants showed enhanced tolerance to low temperature and NaCl compared with non-transgenic plants (Kwon et al., 2003). Overexpression of the gene *AtGR1* encoding GR in *Arabidopsis* led to higher GSH levels and GSH/GSSG ratio and improved Al tolerance (Yin et al., 2017). Glutathione peroxidase reduces H_2O_2 to H_2O by oxidizing GSH. Transgenic tobacco seedlings overexpressing GST with glutathione peroxidase activity resulted in enhanced growth under a variety of stressful conditions (Roxas et al., 2000). Similarly, overexpression of the GST gene of *Suaeda salsa* enhanced the salt tolerance of transgenic *Arabidopsis* plants (Qi et al., 2004). Voulgari et al. (2016) showed that the *PvGSTU3-3* gene from *Phaseolus vulgaris*, when expressed in tobacco plants, mediated physiological pathways that enhanced salt tolerance. The GST gene from *Juglans regia* (*JrGSTTau1*), when expressed in tobacco, improved the chilling tolerance of tobacco plants, indicating that *JrGSTTau1* is a candidate gene for potential application in molecular breeding to enhance plant abiotic stress tolerance (Yang et al., 2016). Increased GSH-dependent peroxide scavenging and alterations in GSH and AsA metabolism were suggested to reduce abiotic stress–induced oxidative damage in GST-overexpressing transgenic plants (Roxas et al., 2000; Qi et al., 2004). Glutathione peroxidase–like protein, over-expressed either in the cytosol or in chloroplasts of tobacco plants, increased tolerance to oxidative stress caused by various stressful conditions by removing unsaturated fatty acid hydroperoxides generated in cellular membranes (Yoshimura et al., 2004). Rice plants overexpressing *OsGS* encoding glutathione synthetase showed improved stress tolerance and productivity under paddy field conditions (Park et al., 2017).

Flavonoids are a class of plant secondary metabolites. Increased flavonoid content has been used to generate potato tubers with modified antioxidant capacities. The single-gene overexpression or simultaneous expression of genes encoding chalcone synthase (CHS), chalcone isomerase (CHI), and dihydroflavonol reductase (DFR) resulted in a significant increase in phenolic acids and anthocyanins accompanied by a decrease in starch and glucose levels in transgenic plants. The flavonoid-enriched plants showed improved antioxidant capacity; however, the participation of other compounds (which are not yet recognized) was also suggested to render antioxidant potential to the plants (Lukaszewicz et al., 2004). Overexpression of *Arabidopsis AtROS1*, a repressor of silencing, in tobacco plants showed enhanced salinity stress tolerance and increased demethylation in promoters and coding regions of genes encoding CHS, flavonol synthase, CHI, flavanone 3-hydroxylase, anthocyanidin synthase and dihydroflavonol 4-reductase of the flavonoid biosynthetic pathway, and GST, APX, GPX, and GR (Bharti et al., 2015). Higher expression levels of the genes encoding enzymes of the flavonoid biosynthetic and antioxidant pathways were suggested to be responsible for increased salt tolerance. In plants, the xanthophyll cycle (the reversible interconversion of two carotenoids, violaxanthin and zeaxanthin) is a promising target for genetic engineering to enhance stress tolerance. In *Arabidopsis thaliana*, overexpression of the *chy*B gene, which encodes β-carotene hydroxylase, an enzyme in the zeaxanthin biosynthetic pathway, causes a specific twofold increase in the size of the xanthophyll cycle pool. The plants are more tolerant to conditions of high light and high temperature, as shown by reduced leaf necrosis, reduced production of the stress indicator anthocyanin, and reduced lipid peroxidation. Stress protection was suggested to be due to the function of zeaxanthin in preventing oxidative damage to membranes (Davison et al., 2002). In view of the antioxidant properties of α-tocopherol, it is suggested that overexpression of α-tocopherol can increase the tolerance of plants to oxidative stress caused by abiotic stresses (Liu et al., 2008). Tocopherol cyclase (VTE1, encoded by the *VTE1* gene) catalyzes the penultimate step of tocopherol synthesis. Transgenic tobacco plants overexpressing VTE1 from *Arabidopsis* showed decreased lipid peroxidation, electrolyte leakage, and H_2O_2 content compared with the wild type when exposed to drought conditions (Liu et al., 2008).

Although it is possible to confer a certain degree of tolerance to a particular stress by overexpressing a single component of antioxidant defense system, only limited improvement in stress tolerance has been observed in these plants (Lee et al., 2009). As the ROS detoxification system is very complex, overexpressing one enzyme may or may not change the capacity of the whole pathway. Therefore, increases in one component may not result in an overall increase in protection (Lee et al., 2009). For example, Rubio and coworkers (2002) found no improvement in oxidative or environmental stress tolerance in transgenic alfalfa overexpressing SOD. Similarly, rice plants overexpressing *SodCc1*, encoding Cu/ZnSOD, did not show tolerance to oxidative stress. Further, a wilting assay also demonstrated no improvement in tolerance to either cold or drought (Lee et al., 2009). Overexpression of the MDHAR gene from non-heading Chinese cabbage (*Brassica campestris* ssp. chinensis Makino) showed reduced growth of tobacco plants, which altered the expression of the genes in the D-mannose/L-galactose pathway and the ascorbate glutathione pathway (Ren et al., 2014). The expression of combinations of antioxidant enzymes in transgenic plants has been shown to have synergistic effects on stress tolerance (Tseng et al., 2008). Chinese cabbage overexpressing both SOD and CAT in cytosol showed considerable tolerance to SO_2 damage (Tseng et al., 2008). Transgenic plants with an enhanced tolerance to multiple environmental stresses have been developed by overexpressing the genes of both SOD and APX in the chloroplasts (Kwon et al., 2002; Lee et al., 2007a; Lim et al., 2007; Kwak et al., 2009). Lee and coworkers (2007b) showed that simultaneous expression of multiple antioxidant enzymes, such as Cu/ZnSOD, APX, and DHAR, in chloroplasts is more effective than single or double expression for developing transgenic plants with enhanced tolerance to multiple environmental stresses. Expression of *CuZnSOD* and *APX* in salt-sensitive sweet potato led to enhanced salt tolerance (Yan et al., 2016). Overexpression of *RaAPX* (*Rheum australe*) and *PaSOD* (*Potentilla atrosanguinea*) in *Arabidopsis* promoted growth, biomass production, and yield under salt stress. It was suggested that salt stress–induced ROS are converted into H_2O_2 by SOD, and APX maintains its optimum concentration (Shafi et al., 2015). Similarly, in transgenic potato, overexpression of APX and SOD genes conferred salinity tolerance (Shafi et al., 2017).

7.6 CONCLUSIONS AND FUTURE PROSPECTS

The increased production of ROS in various cellular compartments is one of the inevitable consequences of aerobic metabolism. ROS play two divergent roles in plants. At low concentration, ROS serve as important regulators of growth and as signaling molecules for the activation of defense responses under stresses, whereas at high concentrations, they cause damage to cellular components such as lipids, proteins, and nucleic acids, disturb homeostasis, and in turn, cause severe damage to cell viability. Oxidative stress as a result of enhanced ROS production is a common manifestation of environmental stresses in plants. Various abiotic stressful environmental conditions severely limit agricultural productivity across the world. As ROS play multifunctional roles, it is essential for cells to control the level of ROS strictly to avoid any injury caused by oxidative stress without removing them completely. Plants possess an efficient antioxidative system consisting of non-enzymic antioxidants such as GSH, AsA, tocopherols, and carotenoids as well as enzymic antioxidants such as CAT, POD, and SOD and enzymes of the AsA-GSH cycle such as APX, MDHAR, DHAR, and GR. The antioxidant responses of plants depend not only on the species' inherent strategy but also on the tissues concerned and the duration and severity of the stress period. To improve abiotic stress tolerance in plants using genetic engineering techniques, molecular and cellular knowledge of antioxidative metabolism is very important. Naturally occurring abiotic stress–tolerant plants serve as attractive objects for such studies. Higher levels of the non-enzymic antioxidants AsA, GSH, and tocopherol and higher activities of the antioxidative enzymes SOD, CAT, peroxidases, MDHAR, DHAR, and GR have often been associated with tolerance to many abiotic stresses in various plants. Encouraging results have been obtained in producing abiotic stress–tolerant crop plants using biotechnological approaches, by increasing the synthesis of non-enzymic antioxidants or by overexpressing enzymes of the antioxidant defense system. However, due to the complex nature of stress tolerance, and due to the fact that in field conditions, plants are often confronted with more than one abiotic stress at any given time, overexpression of only one component of the antioxidant defense system may not always be helpful in conferring tolerance to a particular stress or multiple stresses. Further, it may disturb the balanced interaction among various components and metabolic pathways. In certain crops, results indicate that overexpression of a particular antioxidant component may provide tolerance to more than one abiotic stress. Therefore, future work employing high expression levels of various antioxidants in important crop plants is necessary to produce crop plants with enhanced levels of tolerance to multiple abiotic stresses.

BIBLIOGRAPHY

Abarca, D., M. Roldan, M. Martin, and B. Sabater. 2001. *Arabidopsis thaliana* ecotype cultivar shows an increased tolerance to photo-oxidative stress and contains a new chloroplastic copper/zinc SOD isoenzyme. *J. Exp. Bot.* 52:1417–1425.

Abernathy, C. O., Y. Liu, D. Longfellow et al. 1999. Arsenic: health effects, mechanisms of actions, and research issues. *Environ. Health Perspect.* 107:593–597.

Agarwal, S. and R. Shaheen. 2007. Stimulation of antioxidant system and lipid peroxidation by abiotic stresses in leaves of *Momordica charantia*. *Braz. J. Plant Physiol.* 19:149–161.

Agnez-Lima, L. F., J. T. Melo, A. E. Silva et al. 2012. DNA damage by singlet oxygen and cellular protective mechanisms. *Mutat. Res. Rev. Mutat. Res.* 751:15–28.

Aguiar-Silva, C., S. E. Brandão, M. Domingos, and P. Bulbovas. 2016. Antioxidant responses of Atlantic Forest native tree species as indicators of increasing tolerance to oxidative stress when they are exposed to air pollutants and seasonal tropical climate. *Ecol. Indic.* 63:154–164.

Ahmad, P., M. Ozturk, S. Sharma, and S. Gucel. 2014. Effect of sodium carbonate-induced salinity–alkalinity on some key osmoprotectants, protein profile, antioxidant enzymes, and lipid peroxidation in two mulberry (*Morus alba* L.) cultivars. *J. Plant Interact.* 9:460–467.

Al-Ghamdi, A. A. 2009. Evaluation of oxidative stress tolerance in two wheat (*Triticum aestivum*) cultivars in response to drought. *Int. J. Agric. Biol.* 11:7–12.

Allen, R. D., R. P. Webb, and S. A. Schake. 1997. Use of transgenic plants to study antioxidant defenses. *Free Rad. Biol. Med.* 23:472–479.

Almeselmani, M., P. S. Deshmukh, R. K. Sairam, S. R. Kushwaha, and T. P. Singh. 2006. Protective role of antioxidant enzymes under high temperature stress. *Plant Sci.* 171:382–388.

Alscher, R. G., N. Erturk, and L. S. Heath. 2002. Role of superoxide dismutases (SODs) in controlling oxidative stress in plants. *J. Exp. Bot.* 53:1331–1341.

Amako, K. and T. Ushimaru. 2009. Dehydroascorbate reductase and salt stress. *CAB Rev. Perspec. Agric. Vet. Sci. Nutr. Natur. Resour.* 4:1–7.

Anjum, N. A., P. Sharma, S. S. Gill et al. 2016. Catalase and ascorbate peroxidase-representative H_2O_2-detoxifying heme enzymes in plants. *Environ. Sci. Pollut. Res. Int.* 23:19002–19029.

Anjum, N. A., A. Sofo, A. Scopa et al. 2015. Lipids and proteins—major targets of oxidative modifications in abiotic stressed plants. *Environ. Sci. Pollut. Res. Int.* 22:4099–4121.

Anjum, N. A., S. Umar, A. Ahmad, and M. Iqbal. 2008. Responses of components of antioxidant system in moongbean genotypes to cadmium stress. *Commun. Soil Sci. Plant Anal.* 39:2469–2483.

Apel, K. and H. Hirt. 2004. Reactive oxygen species: metabolism, oxidative stress, and signal transduction. *Annu. Rev. Plant Physiol. Plant Mol. Biol.* 55:373–399.

Armstrong, W. 1979. Aeration in higher plants. *Adv. Bot. Res.* 7:225–232.

Arora, A., R. K. Sairam, and G. C. Srivastava. 2002. Oxidative stress and antioxidative system in plants. *Curr. Sci.* 10:1227–1238.

Asada, K. 1980. Formation and scavenging of superoxides in chloroplasts, with relation to injury by sulfur dioxide. Res. Rep. *Natl. Inst. Environ. Stud.* 11:165–179.

Asada, K. 1992. Ascorbate peroxidase—a hydrogen peroxide-scavenging enzyme in plants. *Physiol. Plant.* 85:235–241.

Asada, K. 1994. Production and action of active oxygen species in photosynthetic tissues. In *Causes of Photooxidative Stress and Amelioration of Defense Systems in Plants*, eds. C. H. Foyer and P. M. Mullineaux, 77–104. Boca Raton: CRC Press.

Asada, K. 1996. Radical production and scavenging in the chloroplasts. In *Photosynthesis and the Environment*, ed. N. R. Baker, 123–150. Dordrecht: Kluwer Academic Press.

Asada, K. 1997. The role of ascorbate peroxidase and monodehydro-ascorbate reductase in H_2O_2 scavenging in plants. In *Oxidative Stress and the Molecular Biology of Antioxidant Defenses*, ed. J. G. Scandalios, 715–735. New York: Cold Spring Harbor Laboratory Press.

Asada, K. 1999. The water-water cycle in chloroplasts: scavenging of active oxygens and dissipation of excess photons. *Annu. Rev. Plant Physiol. Plant. Mol. Biol.* 50:601–639.

Asada, K. and M. Takahashi. 1987. Production and scavenging of active oxygen in photosynthesis. In *Photoinhibition*, eds. D. J. Kyle, C. B. Osmond, and C. J. Arntzen, 227–287. Amsterdam: Elsevier.

Ashraf, M., H. R. Athar, P. J. C. Harris, and T. R. Kwon. 2008. Some prospective strategies for improving crop salt tolerance. *Adv. Agron.* 97:45–110.

Babu, T. S., T. A. Akhtar, M. A. Lampi, S. Tripuranthakam, D. G. Dixon, and B. M. Greenberg. 2003. Similar stress responses are elicited by copper and ultraviolet radiation in the aquatic plant *Lemna gibba*: implication of reactive oxygen species as common signals. *Plant Cell Physiol.* 44:1320–1329.

Badawi, G. H., Y. Yamauchi, E. Shimada et al. 2004. Enhanced tolerance to salt stress and water deficit by overexpressing SOD in tobacco (*Nicotiana tabacum*) chloroplasts. *Plant Sci.* 166:919–928.

Bafeel, S. O. and M. M. Ibrahim. 2008. Antioxidants and accumulation of alpha-tocopherol induced chilling tolerance in *Medicago sativa. Int. Agric. Biol.* 10:593–598.

Bai, L. P., F. G. Sui, T. D. Ge, Z. H. Sun, Y. Y. Lu, and G. S. Zhou. 2006. Effect of soil drought stress on leaf water status, membrane permeability and enzymatic antioxidant system of maize. *Pedosphere* 16:326–332.

Bandyopadhyay, A. 2016. Role of oxidative stress in arsenic (III) induced genotoxicity in cells of meristematic tissue of *Allium cepa*: an *in vivo* study. *Mater. Today* 3:3194–3199.

Bansal, R. and J. P. Srivastava. 2015. Antioxidative responses to short term waterlogging stress in pigeon pea. *Indian J. Plant Physiol.* 20:182–185.

Barcelo, J. and Ch. Poschenrieder. 1990. Plant water relations as affected by heavy metal stress: a review. *J. Plant Nutr.* 13:1–37.

Barnes, J. D., Y. Zheng, and T. M. Lyons. 2002. Plant resistance to ozone: the role of ascorbate. In *Air Pollution and Plant Biotechnology*, eds. K. Omasa, H. Saji, S. Youssefian, and N. Kondo, 235–254. Tokyo: Springer-Verlag.

Barnes, P. W., S. Maggard, S. R. Holman, and B. S. Vergara. 1993. Intraspecific variation in sensitivity to UV-B radiation in rice. *Crop Sci.* 33:1041–1046.

Bartoli, C. G., A. Buet, G. G. Grozeff, A. Galatro, and M. Simontacchi. 2017. Ascorbate-glutathione cycle and abiotic stress tolerance in plants. In *Ascorbic Acid in Plant Growth, Development and Stress Tolerance*, 177–200. Springer, Cham.

Bartoli, C. G., F. Gomez, D. E. Martínez, and J. J. Guiamet. 2004. Mitochondria are the main target for oxidative damage in leaves of wheat (*Triticum aestivum* L.). *J. Exp. Bot.* 55:1663–1669.

Becana, M., J. F. Moran, and I. Iturbe-Ormaetxe. 1998. Iron-dependent oxygen free radical generation in plants subjected to environmental stress: toxicity and antioxidant protection. *Plant Soil* 201:137–147.

Bela, K., E. Horváth, Á. Gallé, L. Szabados, I. Tari, and J. Csiszár. 2015. Plant glutathione peroxidases: emerging role of the antioxidant enzymes in plant development and stress responses. *J. Plant Physiol.* 176:192–201.

Berlett, B. S. and E. R. Stadtman. 1997. Protein oxidation in aging, disease, and oxidative stress. *J. Biol. Chem.* 272:20313–20316.

Bharti, P., M. Mahajan, A. K. Vishwakarma, J. Bhardwaj, and S. K. Yadav. 2015. AtROS1 overexpression provides evidence for epigenetic regulation of genes encoding enzymes of flavonoid biosynthesis and antioxidant pathways during salt stress in transgenic tobacco. *J. Exp. Bot.* 66:5959–5969.

Bhoomika, K., S. Pyngrope, and R. S. Dubey. 2014. Effect of aluminum on protein oxidation, non-protein thiols and protease activity in seedlings of rice cultivars differing in aluminum tolerance. *J. Plant Physiol.* 171:497–508.

Bianucci, E., A. Furlan, M. del Carmen Tordable, L. E. Hernández, R. O. Carpena-Ruiz, and S. Castro. 2017. Antioxidant responses of peanut roots exposed to realistic groundwater doses of arsenate: identification of glutathione S-transferase as a suitable biomarker for metalloid toxicity. *Chemosphere* 181:551–561.

Biehler, K. and H. Fock. 1996. Evidence for the contribution of the Mehler-peroxidase reaction in dissipating excess electrons in drought-stressed wheat. *Plant Physiol.* 112:265–272.

Blokhina, O. B., T. V. Chirkova, and K. V. Fagerstedt. 2001. Anoxic stress leads to hydrogen peroxide formation in plant cells. *J. Exp. Bot.* 52:1–12.

Blokhina, O. B., K. V. Fagerstedt, and T. V. Chirkova. 1999. Relationships between lipid peroxidation and anoxia tolerance in a range of species during post-anoxic reaeration. *Physiol. Plant.* 105:625–632.

Blokhina, O. B., E. Virolainenand, and K. V. Fagerstedt. 2003. Antioxidants, oxidative damage and oxygen deprivation stress: a review. *Ann. Bot.* 91:179–194.

Blumthaler, M. and W. Ambach. 1990. Indication of increasing solar ultraviolet-B radiation flux in alpine regions. *Science* 248:206–208.

Boo, Y. C. and J. Jung. 1999. Water deficit-induced oxidative stress and antioxidant defenses in rice plants. *J. Plant Physiol.* 155:255–261.

Booij-James, I. S., S. K. Dube, M. A. K. Jansen, M. Edelman, and A. K. Mattoo. 2000. Ultraviolet-B radiation impacts light-mediated turnover of the photosystem II reaction center heterodimer in *Arabidopsis* mutants altered in phenolic metabolism. *Plant Physiol.* 124:1275–1283.

Boominathan, R. and P. M. Doran. 2002. Ni-induced oxidative stress in roots of the Ni hyper accumulator, *Alyssum bertolonii. New Phytol.* 156:205–215.

Boscolo, P. R. S., M. Menossi, and R. A. Jorge. 2003. Aluminium-induced oxidative stress in maize. *Phytochemistry* 62:181–189.

Bose, J., A. Rodrigo-Moreno, and S. Shabala. 2014. ROS homeostasis in halophytes in the context of salinity stress tolerance. *J. Exp. Bot.* 65:1241–1257.

Bowler, C., M. Van Montagu, and D. Inze. 1992. Superoxide dismutase and stress tolerance. *Annu. Rev. Plant Physiol. Plant Mol. Biol.* 43:83–116.

Braconi, D., G. Bernardini, and A. Santucci. 2011. Linking protein oxidation to environmental pollutants: redox proteomic approaches. *J. Proteomics* 74: 2324–2337.

Bradley, D. J., P. Kjellbom, and C. J. Lamb. 1992. Elicitor- and wound-induced oxidative cross-linking of a proline-rich plant cell wall protein: a novel, rapid defense response. *Cell* 70:21–30.

Briat, J. F. and M. Lebrun. 1999. Plant responses to metal toxicity. *C. R. Acad. Sci. III–Life Sci.* 322:4354.

Brot, N. and H. Weissbach. 1982. The biochemistry of methionine sulfoxide residues in proteins. *Trends Biochem. Sci.* 7:137–139.

Bukhov, N. G., C. Wiese, S. Neimanis, and U. Heber. 1999. Heat sensitivity of chloroplasts and leaves: leakage of protons from thylakoids and reversible activation of cyclic electron transport. *Photosynth. Res.* 59:81–93.

Bulley, S. and W. Laing. 2016. The regulation of ascorbate biosynthesis. *Curr. Opin. Plant Biol.* 33:15–22.

Cabiscol, E., E. Oiulats, P. Echave, E. Herrero, and J. Ros. 2000. Oxidative stress promotes specific protein damage in *Saccharomyces cerevisiae. J. Biol. Chem.* 275:27393–27398.

Cadet, J. and J. R. Wagner. 2013. DNA base damage by reactive oxygen species, oxidizing agents, and UV radiation. *Cold Spring Harb. Perspect. Biol.* 5:a012559.

Cadet, J., T. Douki, and J. L. Ravanat. 2010. Oxidatively generated base damage to cellular DNA. *Free Radic. Biol. Med.* 49:9–21.

Cadet, J., J. L. Ravanat, M. TavernaPorro, H. Menoni, and D. Angelov. 2012. Oxidatively generated complex DNA damage: tandem and clustered lesions. *Cancer Lett.* 327:5–15.

Cakmak, I. and W. J. Horst. 1991. Effect of aluminium on lipid peroxidation, superoxide dismutase, catalase, and peroxidase activities in root tips of soybean (*Glycine max*). *Plant Physiol.* 83:463–468.

Calatayud, A. and E. Barreno. 2001. Chlorophyll a fluorescence, antioxidant enzymes and lipid peroxidation in tomato in response to ozone and benomyl. *Environ. Pollut.* 115:283–289.

Cao, S., X. H. Du, L. H. Li et al. 2017. Overexpression of *Populus tomentosa* cytosolic ascorbate peroxidase enhances abiotic stress tolerance in tobacco plants. *Russ. J. Plant Physiol.* 64:224–234.

Cassells, A. C. and R. F. Curry. 2001. Oxidative stress and physiological, epigenetic and genetic variability in plant tissue culture: implications for micropropagators and genetic engineers. *Plant Cell Tissue Organ Cult.* 64:145–157.

Castillo, T., D. R. Koop, S. Kamimura, G. Triadafilopoulos, and H. Tsukamoto. 1992. Role of cytochrome P-450 2E1 in ethanol-, carbon tetrachloride- and iron-dependent microsomal lipid peroxidation. *Hepatology* 16:992–996.

Chaitanya, K. V., D. Sundar, S. Masilamani, and A. R. Reddy. 2002. Variation in heat stress-induced antioxidant enzyme activities among three mulberry cultivars. *Plant Growth Regul.* 36:175–180.

Chakradhar, T., S. Mahanty, R. A. Reddy, K. Divya, P. S. Reddy, and M. K. Reddy. 2017. Biotechnological perspective of reactive oxygen species (ROS)-mediated stress tolerance in plants. In *Reactive Oxygen Species and Antioxidant Systems in Plants: Role and Regulation under Abiotic Stress*, 53–87. Singapore: Springer.

Chaves, M. M., J. S. Pereira, J. Maroco et al. 2002. How plants cope with water stress in the field. Photosynthesis and growth. *Annals Bot.* 89:907–916.

Chen, D., G. Cao, and T. Hastings. 2002. Age-dependent decline of DNA repair activity for oxidative lesions in rat brain mitochondria. *J. Neurochem.* 81:1273–1284.

Cheng, M. C., K. Ko, W. L. Chang, W. C. Kuo, G. H. Chen, and T. P. Lin. 2015. Increased glutathione contributes to stress tolerance and global translational changes in *Arabidopsis. Plant J.* 83:926–939.

Chiang, C. M., C. C. Chen, S. P. Chen et al. 2017. Overexpression of the ascorbate peroxidase gene from eggplant and sponge gourd enhances flood tolerance in transgenic *Arabidopsis. J. Plant Res.* 130:373–386.

Chiang, C. M., H. L. Chien, L. F. O. Chen et al. 2015. Overexpression of the genes coding ascorbate peroxidase from *Brassica campestris* enhances heat tolerance in transgenic *Arabidopsis thaliana. Biol. Plant.* 59:305–315.

Chirkova, T. V., L. O. Novitskaya, and O. B. Blokhina. 1998. Lipid peroxidation and antioxidant systems under anoxia in plants differing in their tolerance to oxygen deficiency. *Russ. J. Plant Physiol.* 45:55–62.

Chmielowska-Bąk, J., K. Izbiańska, A. Ekner-Grzyb, M. Bayar, and J. Deckert. 2018. Cadmium stress leads to rapid increase in RNA oxidative modifications in soybean seedlings. *Front Plant Sci.* 8:2219.

Choudhury, F. K., A. R. Devireddy, R. K. Azad, V. Shulaev, and R. Mittler. 2018. Rapid accumulation of glutathione during light stress in *Arabidopsis. Plant Cell Physiol.* doi. 10.1093/pcp/pcy101.

Choudhury, F. K., R. M. Rivero, E. Blumwald, and R. Mittler. 2017. Reactive oxygen species, abiotic stress and stress combination. *Plant J.* 90:856–867.

Choudhury, S., P. Panda, L. Sahoo, and S. K. Panda. 2013. Reactive oxygen species signaling in plants under abiotic stress. *Plant Signal Behav.* 8:e23681.

Chowra, U., E. Yanase, H. Koyama, and S. K. Panda. 2017. Aluminium-induced excessive ROS causes cellular damage and metabolic shifts in black gram *Vigna mungo* (L.) Hepper. *Protoplasma* 254:293–302.

Colville, L. and N. Smirnoff. 2008. Antioxidant status, peroxidase activity, and PR protein transcript levels in ascorbate-deficient *Arabidopsis thaliana vtc* mutants. *J. Exp. Bot.* 59:3857–3868.

Conklin, P. L., E. H. Williams, and R. L. Last. 1996. Environmental stress sensitivity of an ascorbic acid-deficient *Arabidopsis* mutant. *Proc. Natl. Acad. Sci. USA* 93:9970–9974.

Costa, H., S. M. Gallego, and M. L. Tomaro. 2002. Effect of UV-B radiation on antioxidant defense system in sunflower cotyledons. *Plant Sci.* 162:939–945.

Cramer, G. R., K. Urano, S. Delrot, M. Pezzotti, and K. Shinozaki. 2011. Effects of abiotic stress on plants: a systems biology perspective. *BMC Plant Biol.* 11:163.

Crawford, R. M. M. 1978. Metabolic adaptations to anoxia. In *Plant Life in Anaerobic Environments*, eds. D. D. Hook, and R. M. M. Crawford, 119–136. Michigan: Ann Arbor Science Publishers.

Csiszár, J., E. Horváth, K. Bela, and Á. Gallé. 2016. glutathione-related enzyme system: glutathione reductase (GR), glutathione transferases (GSTs) and glutathione peroxidases (GPXs). In *Redox State as a Central Regulator of Plant-Cell Stress Responses*, 137–158. Cham: Springer.

Das, K. and A. Roychoudhury. 2014. Reactive oxygen species (ROS) and response of antioxidants as ROS-scavengers during environmental stress in plants. *Front. Environ. Sci.* 2:53.

Dat, J., S. Vandenbeele, E. Vranova, M. Van Montagu, D. Inze, and F. Van Breusegm. 2000. Dual action of the active oxygen species during plant stress responses. *Cell. Mol. Life Sci.* 57:779–795.

Davies, K. J. A. 1987. Protein damage and degradation by oxygen radicals. I General aspects. *J. Biol. Chem.* 162:9895–9901.

Davies, M. J. 2003. Singlet oxygen-mediated damage to proteins and its consequences. *Biochem. Biophys. Res. Commun.* 305:761–770.

Davison, P. A., C. N. Hunter, and P. Horton. 2002. Overexpression of beta-carotene hydroxylase enhances stress tolerance in *Arabidopsis. Nature* 418:203–206.

de Carvalho, M. H. C. 2008. Drought stress and reactive oxygen species production, scavenging and signalling. *Plant Signal Behav.* 3:156–165.

de Deus, K. E., A. C. Lanna, F. R. M. Abreu et al. 2015. Molecular and biochemical characterization of superoxide dismutase (SOD) in upland rice under drought. *Aust. J. Crop Sci.* 9:744–753.

De Leonardis, S., G. De Lorenzo, G. Borraccino, and S. Dipierro. 1995. A specific ascorbate free radical reductase isoenzyme participates in the regeneration of ascorbate for scavenging toxic oxygen species in potato tuber mitochondria. *Plant Physiol.* 109:847–851.

Dean, R. T., S. Gieseg, and M. Davies. 1993. Reactive species and their accumulation on radical-damaged proteins. *Trends Biol. Sci.* 18:437–441.

del Rio, L. A., G. M. Pastori, J. M. Sandalio, and J. A. Hernandez. 1998. The activated oxygen role of peroxisome in senescence. *Plant Physiol.* 116:1195–1200.

Deleers, M., J. P. Servais, and E. Wülfert. 1986. Neurotoxic cations induce membrane rigidification and membrane fusion at micromolar concentrations. *Biochim. Biophys. Acta* 855: 271–276.

Demidchik, V. 2015. Mechanisms of oxidative stress in plants: from classical chemistry to cell biology. *Environ. Exp. Bot.* 109:212–228.

Desikin, R., S. A. H. Mackerness, J. T. Hancock, and S. J. Neill. 2001. Regulation of the *Arabidopsis* transcriptome by oxidative stress. *Plant Physiol.* 127:159–172.

Devi, M. A. and P. Giridhar. 2015. Variations in physiological response, lipid peroxidation, antioxidant enzyme activities, proline and isoflavones content in soybean varieties subjected to drought stress. *Proc. Natl. Acad. Sci. India Sect. B Biol. Sci.* 85:35–44.

Díaz, J., A. Bernal, F. Pomar, and F. Merino. 2001. Induction of shikimate dehydrogenase and peroxidase in pepper (*Capsicum annuum* L.) seedlings in response to copper stress and its relation to lignification. *Plant Sci.* 161:179–188.

Diaz-Vivancos, P., A. de Simone, G.Kiddle, and C. H. Foyer. 2015. Glutathione-linking cell proliferation to oxidative stress. *Free Radical Biol. Med.* 89:1154–1164.

Diaz-Vivancos, P., L.Faize, E.Nicolás, M. J.Clemente-Moreno, R.Bru-Martinez, L. Burgos, and J. A. Hernández. 2016. Transformation of plum plants with a cytosolic ascorbate peroxidase transgene leads to enhanced water stress tolerance. *Annals Bot.* 117:1121–1131.

Dikilitas, M., S. Karakas, and P. Ahmad. 2015. Effect of lead on plant and human DNA damages and its impact on the environment. In *Plant Metal Interaction*, 41–67. Elsevier.

Diplock, A. T., L. J. Machlin, L. Packer, and W. A. Pryor. 1989. Vitamin E: biochemistry and health implications. *Ann. N. Y. Acad. Sci.* 570:372–378.

Dixon, S. J. and B. R. Stockwell. 2014. The role of iron and reactive oxygen species in cell death. *Nat. Chem. Biol.* 10:9.

Duan, S. R., W. M. Shi, and J. R. Wang. 2006. Overexpressing peroxisomal APX gene in rice enhanced tolerance to cadmium stress. *Acta Pedologica Sinica* 43:111–116.

Edge, R., D. J. McGarvey, and T. G. Truscott. 1997. The carotenoids as anti-oxidants—a review. *J. Photochem. Photobiol. B-Biol.* 41:189–200.

Edwards, E. A., S. Rawsthorne, and P. M. Mullineaux. 1990. Subcellular distribution of multiple forms of glutathione reductase in leaves of pea (*Pisum sativum* L.). *Planta* 180:278–284.

El-baky, A., H. Hanaa, A. Amal, and M. M. Hussein. 2003. Influence of salinity on lipid peroxidation, antioxidant enzymes and electrophoretic patterns of protein and isoenzymes in leaves of some onion cultivars. *Asian J. Plant Sci.* 2:633–638.

El-Shintinawy F., M. K. H. Ebrahim, N. Sewelam, and M. N. El-Shourbagy. 2004. Activity of photosystem 2, lipid peroxidation, and the enzymatic antioxidant protective system in heat shocked barley seedlings. *Photosynthetica* 42:15–21.

Elstner, E. F., G. A. Wagner, and W. Schütz. 1988. Activated oxygen in green plants in relation to stress situation. *Curr. Top. Plant Biochem. Physiol.* 7:159–187.

Eltayeb, A. E., N. Kawano, G. H. Badawi et al. 2007. Overexpression of monodehydroascorbate reductase in transgenic tobacco confers enhanced tolerance to ozone, salt and polyethylene glycol stresses. *Planta* 225:1255–1264.

Ercoli, L., M. Mariotti, A. Masoni, and I. Arduini. 2004. Growth responses of sorghum plants to chilling temperature and duration of exposure. *Eur. J. Agron.* 21:93–103.

Erdei, S., A. Hegedus, G. Hauptmann, J. Szalai, and G. Horvath. 2002. Heavy metal induced physiological changes in the antioxidative response system. *Acta Biol. Szegediensis* 46:89–90.

Ernest, M. J. and K. H. Kim. 1973. Regulation of rat liver glycogen synthetase. Reversible inactivation of glycogen synthetase D by sulfhydryl-disulfide exchange. *J. Biol. Chem.* 248: 1550–1555.

Esfandiari, E. and G. Gohari. 2017. Response of ROS-scavenging systems to salinity stress in two different wheat (*Triticum aestivum* L.) cultivars. *Not. Bot. Horti Agrobot. Cluj Napoca* 45:287–291.

Esfandiari, E., M. R. Shakiba, S. A. Mahboob, H. Alyari, and S. Shahabivand. 2008. The effect of water stress on the antioxidant content, protective enzyme activities, proline content and lipid peroxidation in wheat seedling. *Pak J. Biol. Sci.* 11:1916–1922.

Ezaki, B., R. C. Gardner, Y. Ezaki, and H. Matsumoto. 2000. Expression of aluminium-induced genes in transgenic *Arabidopsis* plants can ameliorate aluminium stress and/or oxidative stress. *Plant Physiol.* 122:657–665.

Fan, Y., J. Bose, M. Zhou, and S. Shabala. 2015. ROS production, scavenging, and signaling under salinity stress. In *Managing Salt Tolerance in Plants: Molecular and Genomic Perspectives*, eds. S. H. Wani and M. A. Hossain, 187–199. Boca Raton: CRC Press.

Fatima, S., A. H. A. Farooqi, and R. S. Sangwan. 2005. Water stress mediated modulation in essential oil, proline and polypeptide profile in *Palmarosa* and *Citronella java*. *Physiol. Mol. Biol. Plants* 11:153–156.

Feierabend, J., C. Schaan, and B. Hertwig. 1992. Photoinactivation of catalase occurs under both high- and low-temperature stress conditions and accompanies photoinhibition of photosystem II. *Plant Physiol.* 100:1554–1561.

Feki, K., Y. Kamoun, R. B. Mahmoud, A. Farhat-Khemakhem, A. Gargouri, and F. Brini. 2015. Multiple abiotic stress tolerance of the transformants yeast cells and the transgenic *Arabidopsis* plants expressing a novel durum wheat catalase. *Plant Physiol. Biochem.* 97:420–431.

Ferguson, D. L. and J. J. Burke. 1992. A new method of measuring protein-methionine-S-oxide reductase activity. *Plant Physiol.* 100:529–532.

Fojtova, M. and A. Kovarik. 2000. Genotoxic effect of cadmium is associated with apoptotic changes in tobacco cells. *Plant Cell Environ.* 23:531–537.

Foyer, C. H. and B. Halliwell. 1976. The presence of glutathione and glutathione reductase in chloroplasts: a proposed role in ascorbate metabolism. *Planta* 133:21–25.

Foyer, C. H. and J. Harbinson. 1994. Oxygen metabolism and the regulation of photosynthetic electron transport. In *Causes of Photooxidative Stresses and Amelioration of Defense Systems in Plants*, eds. C. H. Foyer and P. Mullineaux, 1–42. Boca Raton: CRC Press.

Foyer, C. H. and G. Noctor. 2005. Redox homeostasis and antioxidant signaling: a metabolic interface between stress perception and physiological responses. *Plant Cell* 17:1866–1875.

Foyer, C. H., P. Descourvieres, and K. J. Kunert. 1994. Protection against oxygen radical: an important defense mechanism studied in transgenic plants. *Plant Cell Environ.* 17:507–523.

Foyer, C. H., H. Lopez-Delgado, J. F. Dat, and I. M. Scott. 1997. Hydrogen peroxide and glutathione-associated mechanisms of acclimatory stress tolerance and signaling. *Physiol Plant.* 100:241–254.

Foyer, C. H., A. V. Ruban, and G. Noctor. 2017. Viewing oxidative stress through the lens of oxidative signalling rather than damage. *Biochem. J.* 474:877–883.

Foyer, C. H., F. L. Theodoulou, and S. Delrot. 2001. The functions of intercellular and intracellular glutathione transport systems. *Trends Plant Sci.* 6:486–492.

Fridovich, I. 1975. Superoxide dismutases. *Annu. Rev. Biochem.* 44:147–159.

Fridovich, I. 1989. Superoxide dismutase: an adaptation to a paramagnetic gas. *J. Biol. Chem.* 264:7761–7764.

Frohnmeyer, H. and D. Staiger. 2003. Ultraviolet-B radiation-mediated responses in plants, balancing damage and protection. *Plant Physiol.* 133:1420–1428.

Fry, S. C. 1998. Oxidative scission of plant cell wall polysaccharides by ascorbate-induced hydroxyl radicals. *Biochem. J.* 332:507–515.

Fryer, M. J. 1992. The antioxidant effect of thylakoid vitamin-E (α-tocopherol). *Plant Cell Environ.* 15:381–392.

Fu, J. and B. Huang. 2001. Involvement of antioxidants and lipid peroxidation in the adaptation of two cool season grasses to localized drought stress. *Environ. Exp. Bot.* 45:105–114.

Fuhrer, J. 2009. Ozone risk for crops and pastures in present and future climates. *Naturwissenschaften* 96:173–194.

Furlan, A., E. Bianucci, M. del Carmen Tordable, A. Kleinert, A. Valentine, and S. Castro. 2016. Dynamic responses of photosynthesis and the antioxidant system during a drought and rehydration cycle in peanut plants. *Funct. Plant Biol.* 43:337–345.

Gajewska, E. and M. Skłodowska. 2007. Effect of nickel on ROS content and antioxidative enzyme activities in wheat leaves. *Biometals* 20:27–36.

Gajewska, E., P. Bernat, J. Długoński, and M. Skłodowska. 2012. Effect of nickel on membrane integrity, lipid peroxidation and fatty acid composition in wheat seedlings. *J. Agron. Crop Sci.* 198:286–294.

Gajewska, E. and M. Skłodowska. 2008. Differential biochemical responses of wheat shoots and roots to nickel stress: antioxidative reactions and proline accumulation. *Plant Growth Regul.* 54:179–188.

Gallego, S. M., M. P. Benavides, and M. L. Tomaro. 1996. Effect of heavy metal ion excess on sunflower leaves: evidence for involvement of oxidative stress. *Plant Sci.* 121:151–159.

Gallego, S. M., M. P. Benavides, and M. L. Tomaro. 2002. Involvement of an antioxidant defence system in the adaptive response to heavy metal ions in *Helianthus annuus* L. cells. *Plant Growth Regul.* 36:267–273.

Gao, Q. and L. Zhang. 2008. Ultraviolet-B-induced oxidative stress and antioxidant defense system responses in ascorbate-deficient *vtc1* mutants of *Arabidopsis thaliana*. *J. Plant Physiol.* 165:138–148.

Gao, X., W. L. Ai, H. Gong et al. 2016. Transgenic NfFeSOD *Sedum alfredii* plants exhibited profound growth impairments and better relative tolerance to long-term abiotic stresses. *Plant Biotechnol. Rep.* 10:117–128.

Gardner, P. R. and I. Fridovich. 1991. Superoxide sensitivity of the *Escherichia coli* 6-phosphogluconate dehydratase. *J. Biol. Chem.* 266:1478–1483.

Garrison, W. M. 1987. Reaction mechanisms in the radiolysis of peptides, polypeptides, and proteins. *Chem. Rev.* 87:381–398.

Gémes, K., Y. J. Kim, K. Y. Park et al. 2016. An NADPH-oxidase/polyamine oxidase feedback loop controls oxidative burst under salinity. *Plant Physiol.* 172:1418–1431.

Giannopolitis, C. N. and S. K. Ries 1977. Superoxide dismutases: I. Occurrence in higher plants. *Plant Physiol.* 59:309–314.

Gill, S. S., N. A. Anjum, R. Gill et al. 2015. Superoxide dismutase—mentor of abiotic stress tolerance in crop plants. *Environ. Sci. Pollut. Res.* 22:10375–10394.

Girotti, A. W. 1990. Photodynamic lipid peroxidation in biological systems. *Photochem. Photobiol.* 51:497–509.

Gomes-Júnior, R. A., C. A. Moldes, F. S. Delite et al. 2006. Antioxidant metabolism of coffee cell suspension cultures in response to cadmium. *Chemosphere* 65:1330–1337.

Gonçalves, J. F., A. G. Becker, D. Cargnelutti et al., 2007. Cadmium toxicity causes oxidative stress and induces response of the antioxidant system in cucumber seedlings. *Braz. J. Plant Physiol.* 19:223–232.

Gossett, D. R., E. P. Millhollon, and M. C. Lucas. 1994. Antioxidant response to NaCl stress in salt-tolerant and salt-sensitive cultivars of cotton. *Crop Sci.* 34:706–714.

Gratão, P. L., A. Polle, P. J. Lea, and R. A. Azevedo. 2005. Making the life of heavy metal-stressed plants a little easier. *Funct. Plant Biol.* 32:481–494.

Greaves, A. 1996. Improving suboptimal temperature tolerance in maize—the search for variation. *J. Exp. Bot.* 47:307–323.

Green, R. and R. Fluhr. 1995. UV-B induced PR-1 accumulation is mediated by active oxygen species. *Plant Cell* 7:203–212.

Grune, T., T. Reinheckel, and K. J. A. Davies. 1997. Degradation of oxidized proteins in mammalian cells. *FASEB J.* 11:526–534.

Grzegorz, B. 1997. Oxidative stress in plants. *Acta Physiol. Plant.* 19:47–64.

Guan, Q., Wang, Z., Wang, X., Takano, T. and S. Liu. 2015. A peroxisomal APX from *Puccinellia tenuiflora* improves the abiotic stress tolerance of transgenic *Arabidopsis thaliana* through decreasing of H_2O_2 accumulation. *J. Plant Physiol.* 175:183–191.

Guan, Z. Q., T. Y. Chai, Y. X. Zhang, J. Xu, and W. Wei. 2009. Enhancement of Cd tolerance in transgenic tobacco plants overexpressing a Cd-induced CAT cDNA. *Chemosphere* 76:623–630.

Guo, J., X. Liu, X. Li, S. Chen, Z. Jin, and G. Liu. 2006. Overexpression of VTE1 from *Arabidopsis* resulting in high vitamin E accumulation and salt stress tolerance increase in tobacco plant. *J. Appl. Environ. Biol.* 12:468–471.

Guo, Y., S. Liu, Z. Yang, S. Tian, and N. Sui. 2016. Responses of unsaturated fatty acid in membrane lipid and antioxidant enzymes to chilling stress in sweet sorghum (*Sorghum bicolor* (L.) Moench) seedling. *J. Agri. Sci.* 8:71.

Gupta, A. S., J. L. Heinen, A. S. Holaday, J. J. Burke, and R. D. Allen. 1993. Increased resistance to oxidative stress in transgenic plants that overexpress chloroplastic Cu/Zn superoxidedismutase. *Proc. Natl. Acad. Sci. U S A* 90:1629–1633.

Gupta, D. K., L. B. Pena, M. C. Romero-Puertas, A. Hernández, M. Inouhe, and L. M. Sandalio. 2017. NADPH oxidases differentially regulate ROS metabolism and nutrient uptake under cadmium toxicity. *Plant Cell Environ.* 40:509–526.

Gupta, M., P. Sharma, N. B. Sarin, and A. K. Sinha. 2009. Differential response of arsenic stress in two varieties of *Brassica juncea* L. *Chemosphere* 74:1201–1208.

Halliwell, B. 1977. Generation of hydrogen peroxide, superoxide and hydroxyl radicals during the oxidation of dihydroxyfumaric acid by peroxidase. *Biochem. J.* 163:441–448.

Halliwell, B. and J. M. C. Gutteridge. 1989. *Free Radicals in Biology and Medicine*, 2nd ed. Oxford: Oxford University. Press.

Halliwell, B. and J. M. C. Gutteridge. 1999. *Free Radicals in Biology and Medicine*, 3rd ed. Oxford: Oxford University. Press.

Halliwell, B. and J. M. Gutteridge. 2015. *Free Radicals in Biology and Medicine*. USA: Oxford University Press.

Hameed, A., M. Goher, and N. Iqbal. 2012. Heat stress-induced cell death, changes in antioxidants, lipid peroxidation, and protease activity in wheat leaves. *J. Plant Growth Regul.* 31:283–291.

Han, C., Q. Liu, and Y. Yang. 2009. Short-term effects of experimental warming and enhanced ultraviolet-B radiation on photosynthesis and antioxidant defense of *Picea asperata* seedlings. *Plant Growth Regul.* 58:153–162.

Harinasut, P., D. Poonsopa, K. Roengmongkol, and R. Charoensataporn. 2003. Salinity effects on antioxidant enzymes in mulberry cultivar. *Sci. Asia* 29:109–113.

Harsh, A., Y. K. Sharma, U. Joshi et al. 2016. Effect of short-term heat stress on total sugars, proline and some antioxidant enzymes in moth bean (*Vigna aconitifolia*). *Ann. Agric. Sci.* 61:57–64.

Hasanuzzaman, M., J. Al Mahmud, K. Nahar et al. 2017. Responses, adaptation, and ROS metabolism in plants exposed to water-logging stress. In *Reactive Oxygen Species and Antioxidant Systems in Plants: Role and Regulation under Abiotic Stress*, eds. M. Khan and N. Khan, 257–281. Singapore: Springer.

Hasanuzzaman, M., K. Nahar, and M. Fujita. 2014. Role of tocopherol (vitamin E) in plants: abiotic stress tolerance and beyond. In *Emerging Technologies and Management of Crop Stress Tolerance, Volume 2: A Sustainable Approach*, eds. P. Ahmad and S. Rasool, 267–289. Academic Press.

Hefny, M. and D. Z. Abdel-Kader. 2009. Antioxidant-enzyme system as selection criteria for salt tolerance in forage sorghum genotypes (*Sorghum bicolor* L. Moench). In *Salinity and Water Stress*, eds. M. Ashraf, M. Ozturk, and H. R. Athar, 25–36. Netherlands: Springer.

Hernandez, I., L. Alegre, F. Van Breusegem, and S. Munne-Bosch. 2009. How relevant are flavonoids as antioxidants in plants? *Trends Plant Sci.* 14:125–132.

Hernandez, J. A., M. A. Ferrer, A. Jimenez, A. R. Barcelo, and F. Sevilla. 2001. Antioxidant systems and $O_2\bullet-/H_2O_2$ production in the apoplast of pea leaves. Its relation with salt-induced necrotic lesions in minor veins. *Plant Physiol.* 127:817–831.

Hertwig, B., P. Streb, and J. Feierabend. 1992. Light dependence of catalase synthesis and degradation in leaves and the influence of interfering stress conditions. *Plant Physiol.* 100:1547–1553.

Hideg, E. 1997. Free radical production in photosynthesis under stress conditions. In *Handbook of Photosynthesis*, ed. M. Pessarakli, 911–930. New York: Marcel Dekker.

Hideg, E., C. Barta, T. Kalai, M. Vass, K. Hideg, and K. Asada. 2002. Detection of singlet oxygen and superoxide with fluorescence sensors in leaves under stress by photoinhibition or UV radiation. *Plant Cell Physiol.* 43:1154–1164.

Höhn, T. J. and T. Grune. 2014. The proteasome and the degradation of oxidized proteins: Part III—Redox regulation of the proteasomal system. *Redox Biol.* 2:388–394.

Höller, S., Y. Ueda, L. Wu et al. 2015. Ascorbate biosynthesis and its involvement in stress tolerance and plant development in rice (*Oryza sativa* L.). *Plant Mol. Biol.* 88:545–560.

Houghton, J. T., Y. Ding, D. J. Griggs et al. 2001. *Climate Change 2001: The Scientific Basis; Contribution of Working Group I to the Third Assessment Report of the Intergovernmental Panel on Climate Change*. Cambridge and New York: Cambridge University. Press.

Howden, R., C. R. Andersen, P. B. Goldsbrough, and C. S. Cobbett. 1995. A cadmium-sensitive, glutathione-deficient mutant of *Arabidopsis thaliana*. *Plant Physiol.* 107:1067–1073.

Hu, W. H., X. S. Song, K. Shi, X. J. Xia, Y. H. Zhou, and J. Q. Yu. 2008. Changes in electron transport, superoxide dismutase and ascorbate peroxidase isoenzymes in chloroplasts and mitochondria of cucumber leaves as influenced by chilling. *Photosynthetica* 46:581–588.

Huang, C., W. He, J. Guo, X. Chang, P. Su, and L. Zhang. 2005. Increased sensitivity to salt stress in an ascorbate-deficient *Arabidopsis* mutant. *J. Exp. Bot.* 56:3041–3049.

Huseynova, I. M., D. R. Aliyeva, A. C. Mammadov, and J. A. Aliyev. 2015. Hydrogen peroxide generation and antioxidant enzyme activities in the leaves and roots of wheat cultivars subjected to long-term soil drought stress. *Photosynth. Res.* 125:279–289.

Iannone, M. F., M. D. Groppa, and M. P. Benavides. 2015. Cadmium induces different biochemical responses in wild type and catalase-deficient tobacco plants. *Environ. Exp. Bot.* 109:201–211.

Imlay, J. A. 2003. Pathways of oxidative damage. *Ann. Rev. Microbiol.* 57:395–418.

Imlay, J. A. and S. Linn. 1986. DNA damage and oxygen radical toxicity. *Science* 240:1302–1309.

Iriti, M. and F. Faoro. 2008. Oxidative stress, the paradigm of ozone toxicity in plants and animals. *Water Air Soil Pollut.* 187:285–301.

Isbell, H. S., H. L. Frush, and E. T. Martin. 1973. Reaction of carbohydrates with hydroperoxides, Part I. Oxidation of aldoses with sodium peroxide. *Carbohydr. Res.* 26:287–295.

Ishikawa, T., T. Maruta, K. Yoshimura, and N. Smirnoff. 2018. Biosynthesis and regulation of ascorbic acid in plants. In *Antioxidants and Antioxidant Enzymes in Higher Plants*, eds. D. Gupta, J. Palma, and F. Corpas, 163–179. Cham: Springer.

Ishikawa, T., K. Yoshimura, K. Sakai, M. Tamoi, T. Takeda, and S. Shigeoka. 1998. Molecular characterization and physiological role of a glyoxysome-bound ascorbate peroxidase from spinach. *Plant Cell Physiol.* 39:23–34.

Jagtap, V. and S. Bhargava. 1995. Variation in antioxidant metabolism of drought-tolerant and drought-susceptible varieties of *Sorghum bicolor* (L.) Moench exposed to high light, low water and high temperature stress. *J. Plant Physiol.* 145:195–197.

Jamei, R., R. Heidari, J. Khara, and S. Zare. 2009. Hypoxia induced changes in the lipid peroxidation, membrane permeability, reactive oxygen species generation, and antioxidative response systems in *Zea mays* leaves. *Turk. J. Biol.* 33:45–52.

Janas, K. M., R. Amarowicz, J. Z. Tomaszewska, A. Kosińska, and M. M. Posmyk. 2009. Induction of phenolic compounds in two dark-grown lentil cultivars with different tolerance to copper ions. *Acta Physiol. Plant.* 31:587–595.

Jha, A. B. and R. S. Dubey. 2004. Arsenic exposure alters the activities of key nitrogen assimilatory enzymes in growing rice seedlings. *Plant Growth Regul.* 43:259–268.

Jiang, M. and J. Zhang. 2001. Effect of abscisic acid on active oxygen species, antioxidative defense system and oxidative damage in leaves of maize seedlings. *Plant Cell Physiol.* 42:1265–1273.

Jimenez, A., J. A. Hernandez, L. A. del Rio, and F. Sevilla. 1997. Evidence for the presence of the ascorbate-glutathione cycle in mitochondria and peroxisomes of pea leaves. *Plant Physiol.* 114:275–284.

Job, C., L. Rajjou, Y. Lovigny, M. Belghazi, and D. Job. 2005. Patterns of protein oxidation in *Arabidopsis* seeds and during germination. *Plant Physiol.* 138:790–802.

Johansson, A., S. K. Rasmussen, J. E. Harthill, and K. G. Welinder. 1992. cDNA, amino acid and carbohydrate sequence of barley seed-specific peroxidase BP1. *Plant Mol. Biol.* 18:1151–1161.

Jozefczak, M., E. Keunen, H. Schat et al. 2014. Differential response of *Arabidopsis* leaves and roots to cadmium: glutathione-related chelating capacity vs antioxidant capacity. *Plant Physiol. Biochem.* 83:1–9.

Jozefczak, M., T. Remans, J. Vangronsveld, and A. Cuypers. 2012. Glutathione is a key player in metal-induced oxidative stress defenses. *Int. J. Mol. Sci.* 13:3145–3175.

Jung, T., A. Höhn, and T. Grune. 2014. The proteasome and the degradation of oxidized proteins: Part III—Redox regulation of the proteasomal system. *Redox Biol.* 2:388–394.

Kanazawa, S., S. Sano, T. Koshiba, and T. Ushimaru. 2000. Changes in antioxidative enzymes in cucumber cotyledons during natural senescence: comparison with those during dark-induced senescence. *Physiol. Plant.* 109:211–216.

Kao, C. H. 2015. Role of L-ascorbic acid in rice plants. *Crop Environ. Bioinform.* 12:1–7.

Kaouthar, F., F. K. Ameny, K. Yosra, S. Walid, G. Ali, and B. Faical. 2016. Responses of transgenic *Arabidopsis* plants and recombinant yeast cells expressing a novel durum wheat manganese superoxide dismutase TdMnSOD to various abiotic stresses. *J. Plant Physiol.* 198:56–68.

Kärkönen, A. and K. Kuchitsu. 2015. Reactive oxygen species in cell wall metabolism and development in plants. *Phytochemistry* 112:22–32.

Kawasaki, S., C. Borchert, M. Deyholos et al. 2001. Gene expression profiles during the initial phase of salt stress in rice. *Plant Cell* 13:889–906.

Khaliq, A., M. Zia-ul-Haq, F. Ali et al. 2015. Salinity tolerance in wheat cultivars is related to enhanced activities of enzymatic antioxidants and reduced lipid peroxidation. *CLEAN-Soil Air Water* 43:1248–1258.

Khan, E. A., M. Misra, P. Sharma, and A. N. Misra. 2017. Effects of exogenous nitric oxide on protein, proline, MDA contents and antioxidative system in pea seedling (*Pisum sativum* L.) *Research J. Pharm. Tech.*, 10: 3137–3142.

Khan, I., A. Ahmad, and M. Iqbal. 2009. Modulation of antioxidant defence system for arsenic detoxification in Indian mustard. *Ecotoxicol. Environ. Safety* 72:626–634.

Kim, H. S., C. Y. Ji, C. J. Lee, S. E. Kim, S. C. Park, and S. S. Kwak. 2018. Orange: a target gene for regulating carotenoid homeostasis and increasing plant tolerance to environmental stress in marginal lands. *J. Exp. Bot.* 69:3393–3400.

Kim, J. J., Y. S. Kim, S. I. Park et al. 2017. Cytosolic monodehydroascorbate reductase gene affects stress adaptation and grain yield under paddy field conditions in *Oryza sativa* L. japonica. *Mol. Breed.* 37:118.

Kim, K. and A. Portis. 2004. Oxygen-dependent H₂O₂ production by Rubisco. *FEBS Lett.* 571:124–128.

Kobayashi, K., Y. Kumazawa, K. Miwa, and S. Yamanaka. 1996. ε-(γ-Glutamyl) lysine cross-links of spore coat proteins and transglutaminase activity in *Bacillus subtilis*. *FEMS Microbiol. Lett.* 144:157–160.

Kochian, L. V. 1995. Cellular mechanisms of aluminium resistance in plants. *Annu. Rev. Plant Physiol.* 46:237–270.

Köhler, H., R. A. Contreras, M. Pizzaro, R. Cortés-Antíquera, and G. E. Zúñiga. 2017. Antioxidant responses induced by UVB radiation in *Deschampsia antarctica* Desv. *Front Plant Sci.* 8: 921.

Kollist, T., H. Moldau, B. Rasulov et al. 2007. A novel device detects a rapid ozone-induced transient stomatal closure in intact *Arabidopsis* and its absence in abi2 mutant. *Physiol. Plant.* 129:796–803.

Konieczny, R., M. Libik, M. Tuleja, E. Niewiadomska, and Z. Miszalski. 2008. Oxidative events during *in vitro* regeneration of sunflower. *Acta Physiol. Plant.* 30:71–79.

Koussevitzky, S., N. Suzuki, S. Huntington et al. 2008. Ascorbate peroxidase 1 plays a key role in the response of *Arabidopsis thaliana* to stress combination. *J. Biol. Chem.* 283:34197–34203.

Kumar, P., R. K. Tewari, and P. N. Sharma. 2007. Excess nickel-induced changes in antioxidative processes in maize leaves. *J. Plant Nutr. Soil Sci.* 170:796–802.

Kumar, S., P. Sharma, M. Misra, and A. N. Misra. (2018) Lead induced root and shoot growth reduction in wheat (*Triticum aestivum* L.) is due to increase in membrane lipid peroxidation. *J. Pharmacogn. Phytochem.* 2018; SP1: 2080–2083 UGC S. No. 45051.

Kumutha, D., K. Ezhilmathi, R. K. Sairam, G. C. Srivastava, P. S. Deshmukh, and R. C. Meena. 2009. Waterlogging induced oxidative stress and antioxidant activity in *pigeonpea* genotypes. *Biol. Plant.* 53:75–84.

Kurusu, T., K. Kuchitsu, and Y. Tada. 2015. Plant signaling networks involving Ca2⁺ and Rboh/Nox-mediated ROS production under salinity stress. *Front. Plant Sci.* 6:427.

Kwak, S. S., S. Lim, L. Tang, S. Y. Kwon, and H. S. Lee. 2009. Enhanced tolerance of transgenic crops expressing both SOD and APX in chloroplasts to multiple environmental stress. In *Salinity and Water Stress*, eds. M. Ashraf, M. Ozturk, and H. R. Athar, 197–203. Netherlands: Springer.

Kwon, S. Y., S. M. Choi, Y. O. Ahn et al. 2003. Enhanced stress-tolerance of transgenic tobacco plants expressing a human dehydroascorbate reductase gene. *J. Plant Physiol.* 160:347–353.

Kwon, S. Y., Y. J. Jeong, H. S. Lee et al. 2002. Enhanced tolerances of transgenic tobacco plants expressing both superoxide dismutase and ascorbate peroxidase in chloroplasts against methyl viologen-mediated oxidative stress. *Plant Cell Environ.* 25:873–882.

Lee, D. H. and C. B. Lee. 2000. Chilling stress-induced changes of antioxidant enzymes in the leaves of cucumber: in gel enzyme activity assays. *Plant Sci.* 159:75–85.

Lee, H. K., I. Khaine, M. J. Kwak, J. H. Jang, T. Y. Lee and J. K. Lee. 2017. The relationship between SO₂ exposure and plant physiology: a mini review. *Hortic. Environ. Biotechnol.* 58:523–529.

Lee, S. C., S. Y. Kwon, and S. R. Kim. 2009. Ectopic expression of a chilling-responsive CuZn superoxide dismutase gene, *SodCc1*, in transgenic rice (*Oryza sativa* L.). *J. Plant Biol.* 52:154–160.

Lee, S. H., N. Ahsan, K. W. Lee et al. 2007a. Simultaneous overexpression of both CuZn superoxide dismutase and ascorbate peroxidase in transgenic tall fescue plants confers increased tolerance to a wide range of abiotic stresses. *J. Plant Physiol.* 164:1626–1638.

Lee, Y. P., S. H. Kim, J. W. Bang, H. S. Lee, S. S. Kwak, and S. Y. Kwon. 2007b. Enhanced tolerance to oxidative stress in transgenic tobacco plants expressing three antioxidant enzymes in chloroplasts. *Plant Cell Rep.* 26:591–598.

Leterrier, M. and O. Cagnac. 2018. Function of the various MDAR isoforms in higher plants. In *Antioxidants and Antioxidant Enzymes in Higher Plants*, eds. D. K. Gupta, J. M. Palma, and F. J. Corpas, 83–94. Cham: Springer.

Leterrier, M., F. J. Corpas, J. B. Barroso, L. M. Sandalio, and L. A. Del Rio. 2005. Peroxisomal monodehydroascorbate reductase. Genomic clone characterization and functional analysis under environmental stress conditions. *Plant Physiol.* 138:2111–2123.

Li, F. T., J. M. Qi, G. Y. Zhang et al. 2013. Effect of cadmium stress on the growth, antioxidative enzymes and lipid peroxidation in two kenaf (*Hibiscus cannabinus* L.) plant seedlings. *J. Integr. Agric.* 12:610–620.

Li, F., R. Vallabhaneni, J. Yu, T. Rocheford, and E. T. Wurtzel. 2008. The maize phytoene synthase gene family: overlapping roles for carotenogenesis in endosperm, photomorphogenesis, and thermal stress-tolerance. *Plant Physiol.* 147 1334–1346.

Li, X., L. Zhang, Y. Li, L. Ma, N. Bu, and C. Ma. 2012. Changes in photosynthesis, antioxidant enzymes and lipid peroxidation in soybean seedlings exposed to UV-B radiation and/or Cd. *Plant Soil* 352:377–387.

Lim, S., Y. H. Kim, S. H. Kim et al. 2007. Enhanced tolerance of transgenic sweetpotato plants that express both CuZnSOD and APX in chloroplasts to methyl viologen-mediated oxidative stress and chilling. *Mol. Breed.* 19:227–239.

Lin, A., X. Zhang, M. Chen, and Q. Cao. 2007. Oxidative stress and DNA damages induced by cadmium accumulation. *J. Environ. Sci.* 19:596–602.

Lin, C. C. and C. H. Kao. 2001. Relative importance of Na⁺, Cl⁻, and abscisic acid in NaCl induced inhibition of root growth of rice seedlings. *Plant Soil* 237:165–171.

Lin, K. H., C. C. Tsou, S. Y. Hwang, L. F. O. Chen, and H. F. Lo. 2008. Paclobutrazol leads to enhanced antioxidative protection of sweet potato under flooding stress. *Bot. Stud.* 49:9–18.

Lin, K. H., C. C. Weng, H. F. Lo, and J. T. Chen. 2004. Study of the root antioxidative system of tomatoes and eggplant under waterlogged conditions. *Plant Sci.* 167:355–365.

Liu, D., Liu, Y., Rao, J., Wang, G., Li, H., Ge, F. and C. Chen. 2013. Overexpression of the glutathione S-transferase gene from *Pyrus pyrifolia* fruit improves tolerance to abiotic stress in transgenic tobacco plants. *Mol. Biol.* 47:515–523.

Liu, M., J. Chen, Z. Guo, and S. Lu. 2017. Differential responses of polyamines and antioxidants to drought in a centipedegrass mutant in comparison to its wild type plants. *Front. Plant Sci.* 8:792.

Liu, T., J. van Staden, and W. A. Cress. 2000. Salinity induced nuclear and DNA degradation in meristematic cells of soybean (*Glycine max* L.) roots. *Plant Growth Regul.* 30:49–54.

Liu, X., X. Hua, J. Guo et al. 2008. Enhanced tolerance to drought stress in transgenic tobacco plants overexpressing *VTE1* for increased tocopherol production from *Arabidopsis thaliana*. *Biotechnol. Lett.* 30:1275–1280.

Liu, X. P., Y. C. Zhao, D. E. Huang, W. D. Li, and Q. S. Yuan. 2003. Metabolism of active oxygen and change of cell defence enzyme in potato transferred with Cu, Zn-SOD gene under NaCl stress. *Plant Protection* 29:21–24.

Locato, V., M. C. de Pinto, and L. De Gara. 2009. Different involvement of the mitochondrial, plastidial and cytosolic ascorbate-glutathione redox enzymes in heat shock responses. *Physiol. Plant.* 135:296–306.

Loewus, F. A. 1988. Ascorbic acid and its metabolic products. In *The Biochemistry of Plants*, ed. J. Preiss, 85–107. New York: Academic Press.

Logan, B. A., D. Kornyeyev, J. Hardison, and A. S. Holaday. 2006. The role of antioxidant enzymes in photoprotection. *Photosynth. Res.* 88:119–132.

Lois, R. and B. B. Buchanan. 1994. Severe sensitivity to ultraviolet radiation in an *Arabidopsis* mutant deficient in flavonoid accumulation. *Planta* 194:504–509.

Lu, Y., B. Duan, X. Zhang, H. Korpelainen, and C. Li. 2009. Differences in growth and physiological traits of *Populus cathayana* populations as affected by enhanced UV-B radiation and exogenous ABA. *Environ. Exp. Bot.* 66:100–109.

Lu, Z. Q., D. L. Liu, and S. K. Liu. 2007. Two rice cytosolic APXs differentially improve salt tolerance in transgenic *Arabidopsis*. *Plant Cell Rep.* 26:1909–1917.

Lukaszewicz, M., I. Matysiak-Kata, J. Skala, I. Fecka, W. Cisowski, and J. Szopa. 2004. Antioxidant capacity manipulation in transgenic potato tuber by changes in phenolic compounds content. *J. Agric. Food Chem.* 52:1526–1533.

Madhusudhan, R., T. Ishikawa, Y. Sawa, S. Shigeoka, and H. Shibata. 2003. Characterization of an ascorbate peroxidase in plastids of tobacco. *Physiol. Plant.* 117:550–557.

Maheshwari, R. and R. S. Dubey. 2009. Nickel-induced oxidative stress and the role of antioxidant defence in rice seedlings. *Plant Growth Regul.* 59:37–49.

Malar, S., S. S. Vikram, P. J. Favas, and V. Perumal. 2016. Lead heavy metal toxicity induced changes on growth and antioxidative enzymes level in water hyacinths [*Eichhornia crassipes* (Mart.)]. *Bot. Stud.* 55:54.

Małecka, A., M. Derba-Maceluch, K. Kaczorowska, A. Piechalak, and B. Tomaszewska. 2009. Reactive oxygen species production and antioxidative defense system in pea root tissues treated with lead ions: mitochondrial and peroxisomal level. *Acta Physiol. Plant.* 31:1053–1063.

Malencik, D. A. and S. R. Anderson. 2003. Dityrosine as a product of oxidative stress and fluorescent probe. *Amino Acids* 25:233–247.

Mallick, N. and F. H. Mohn. 2000. Reactive oxygen species: response of algal cells. *J. Plant Physiol.* 157:183–193.

Mano, J., Y. Torii, S. Hayashi et al. 2002. The NADPH:quinone oxidoreductase P1-ζ-crystallin in *Arabidopsis* catalyzes the α, β-hydrogenation of 2-alkenals: detoxication of the lipid peroxide-derived reactive aldehydes. *Plant Cell Physiol.* 43:1445–1455.

Marron, N., S. Maury, C. Rinaldi, and F. Brignolas. 2006. Impact of drought and leaf development stage on enzymatic antioxidant system of two *Populus deltoides* × *nigra* clones. *Ann. For. Sci.* 63:323–327.

Martinez, V., T. C. Mestre, F. Rubio et al. 2016. Accumulation of flavonols over hydroxycinnamic acids favors oxidative damage protection under abiotic stress. *Front. Plant Sci.* 7:838.

Matamoros, M. A., N. Fernández-García, S. Wienkoop, J. Loscos, A. Saiz, and M. Becana. 2013. Mitochondria are an early target of oxidative modifications in senescing legume nodules. *New Phytol.* 197:873–885.

Matters, G. L. and J. Scandalios. 1986. Effect of the free radical-generating herbicide paraquat on the expression of the superoxide dismutase (SOD) genes in maize. *Biochim. Biophys. Acta.* 882:29–38.

McCord, J. M. and E. D. Day. 1978. Superoxide-dependent production of hydroxyl radical catalyzed by iron-EDTA complex. *FEBS Lett.* 86:139–142.

McCord, J. M., J. D. Crapo, and I. Fridovich. 1977. Superoxide dismutase assay. A review of methodology. In *Superoxide and Superoxide Dismutase*, eds. A. M. Michelson, J. M. McCord, and I. Fridovich, 11–17. London: Academic Press.

McKersie, B. D., J. Murnaghan, K. S. Jones, and S. R. Bowley. 2000. Iron-superoxide dismutase expression in transgenic alfalfa increases winter survival without a detectable increase in photosynthetic oxidative stress tolerance. *Plant Physiol.* 122:1427–1437.

McKersie, B. D., Y. R. Chen, M. Debeus et al. 1993. Superoxide-dismutase enhances tolerance of freezing stress in transgenic alfalfa (*Medicago sativa* L.). *Plant Physiol.* 103:1155–1163.

Mekki, B. E. D., H. A. Hussien, and H. Salem. 2016. Role of glutathione, ascorbic acid and α-tocopherol in alleviation of drought stress in cotton plants. *Int. J. Chemtech Res.* 8:1573–1581.

Meriga, B., B. K. Reddy, K. R. Rao, L. A. Reddy, and P. B. K. Kishor. 2004. Aluminium-induced production of oxygen radicals, lipid peroxidation and DNA damage in seedlings of rice (*Oryza sativa*). *J. Plant Physiol.* 161:63–68.

Metwally, A., I. Finkemeier, M. Georgi, and K. J. Dietz. 2003. Salicylic acid alleviates the cadmium toxicity in barley seedlings. *Plant Physiol.* 132:272–281.

Mhamdi, A. and G. Noctor. 2016. High CO_2 primes plant biotic stress defences through redox-linked pathways. *Plant Physiol.* 172:929–942.

Mibei, E. K., J. Ambuko, J. J. Giovannoni, A. N. Onyango, and W. O. Owino. 2017. Carotenoid profiling of the leaves of selected African eggplant accessions subjected to drought stress. *Food Sci. Nutr.* 5:113–122.

Michalak, A. 2006. Phenolic compounds and their antioxidant activity in plants growing under heavy metal stress. *Polish J. Environ. Stud.* 15:523–530.

Mika, A. and S. Luthje. 2003. Properties of guaiacol peroxidase activities isolated from corn root plasma membranes. *Plant Physiol.* 132:1489–1498.

Miller, G., N. Suzuki, S. Ciftci-Yilmaz, and R. Mittler. 2010. Reactive oxygen species homeostasis and signalling during drought and salinity stresses. *Plant Cell Environ.* 33:453–467.

Miller, R. W. and F. D. H. MacDowell. 1975. The tiron free radical as a sensitive indicator of chloroplastic photoautooxidation. *Biochim. Biophys. Acta* 387:176–187.

Mishra, H. P. 1974. Generation of superoxide free radical during the autooxidation of thiols. *J. Biol. Chem.* 249:2151–2155.

Mishra, S., A. B. Jha, and R. S. Dubey. 2011. Arsenite treatment induces oxidative stress, upregulates antioxidant system and causes phytochelatin synthesis in rice seedlings. *Protoplasma* 248:565–577.

Mittal, R. and R. S. Dubey. 1991. Behaviour of peroxidases in rice: changes in enzymatic activity and isoforms in relation to salt tolerance. *Plant Physiol. Biochem.* 29:31–40.

Mittler, R. 2002. Oxidative stress, antioxidants and stress tolerance. *Trends Plant Sci.* 7:405–410.

Mittler, R., S. Vanderauwera, M. Gollery, and F. Van Breusegem. 2004. Reactive oxygen gene network of plants. *Trends Plant Sci.* 9: 490–498.

Miyake, C. and K. Asada. 1994. Ferredoxin-dependent photoreduction of the monodehydroascorbate radical in spinach thylakoids. *Plant Cell Physiol.* 35:539–549.

Mohamed, E. A., T. Iwaki, I. Munir, M. Tamoi, S. Shigeoka, and A. Wadano. 2003. Overexpression of bacterial catalase in tomato leaf chloroplasts enhances photo-oxidative stress tolerance. *Plant Cell Environ.* 26:2037–2046.

Moller, I. M. and B. K. Kristensen. 2004. Protein oxidation in plant mitochondria as a stress indicator. *Photochem. Photobiol. Sci.* 3:730–735.

Montillet, J. L., S. Chamnongpol, C. Rustérucci et al. 2005. Fatty acid hydroperoxides and H_2O_2 in the execution of hypersensitive cell death in tobacco leaves. *Plant Physiol.* 138:1516–1526.

Moriwaki, T., Y. Yamamoto, T. Aida et al. 2008. Overexpression of the *Escherichia coli* CAT gene, katE, enhances tolerance to salinity stress in the transgenic indica rice cultivar, BR5. *Plant Biotechnol. Rep.* 2:41–46.

Moussa, H. R. and S. M. Abdel-Aziz. 2008. Comparative response of drought tolerant and drought sensitive maize genotypes to water stress. *Aust. J. Crop Sci.* 1:31–36.

Munné-Bosch, S. 2005. The role of α-tocopherol in plant stress tolerance. *J. Plant Physiol.* 162:743–748.

Munné-Bosch, S., K. Schwarz, and L. Alegre. 1999. Enhanced formation of alpha-tocopherol and highly oxidized abietane diterpenes in water-stressed rosemary plants. *Plant Physiol.* 121:1047–1052.

Munns, R. 2002. Comparative physiology of salt and water stress. *Plant Cell Environ.* 25:239–250.

Nagamiya, K., T. Motohashi, K. Nakao et al. 2007. Enhancement of salt tolerance in transgenic rice expressing an *Escherichia coli* catalase gene, kat E. *Plant Biotechnol. Rep.* 1:49–55.

Nahar, K., M. Hasanuzzaman, M. Alam, and M. Fujita. 2015. Glutathione-induced drought stress tolerance in mung bean: coordinated roles of the antioxidant defence and methylglyoxal detoxification systems. *AoB Plants* 1:7.

Nakabayashi, R., K. Yonekura-Sakakibara, K. Urano et al. 2014. Enhancement of oxidative and drought tolerance in *Arabidopsis* by overaccumulation of antioxidant flavonoids. *Plant J.* 77: 367–379.

Negi, N. P., D. C. Shrivastava, V. Sharma, and N. B. Sarin. 2015. Overexpression of CuZnSOD from *Arachis hypogaea* alleviates salinity and drought stress in tobacco. *Plant Cell Rep.* 34:1109–1126.

Noctor, G. and C. Foyer. 1998. Ascorbate and glutathione: keeping active oxygen under control. *Annu. Rev. Plant Physiol. Plant Mol. Biol.* 49:249–279.

Noctor, G. and C. H. Foyer. 2016. Intracellular redox compartmentation and ROS-related communication in regulation and signaling. *Plant Physiol.* 171:1581–1592.

Noctor, G., S. Veljovic-Jovanovic, S. Driscoll, L. Novitskaya, and C. H. Foyer. 2002. Drought and oxidative load in the leaves of C3 plants: a predominant role for photorespiration? *Ann. Bot.* 89: 841–850.

Noreen, Z. and M. Ashraf. 2009. Assessment of variation in antioxidative defense system in salt-treated pea (*Pisum sativum*) cultivars and its putative use as salinity tolerance markers. *J. Plant Physiol.* 166(16):1764–1774.

Nouairi, I., W. Ben Ammar, N. Ben Youssef, D. D. Ben Miled, M. Ghorbal, and M. Zarrouk. 2009. Antioxidant defense system in leaves of Indian mustard (*Brassica juncea*) and rape (*Brassica napus*) under cadmium stress. *Acta Physiol. Plant.* 31:237–247.

Ogawa, K., S. Kanematsu, and K. Asada. 1997. Generation of superoxide anion and localization of CuZn-SOD in the vascular tissue of spinach hypocotyls: their association with lignifications. *Plant Cell Physiol.* 38:1118–1126.

O'Kane, D., V. Gill, P. Boyd, and R. Burdon. 1996. Chilling, oxidative stress and antioxidant responses in *Arabidopsis thaliana* callus. *Planta* 198:371–377.

Oleinick, N. L., S. Chiu, N. Ramakrishman, and L. Xue. 1986. The formation, identification, and significance of DNA-protein cross-links in mammalian cells. *Brit. J. Cancer* 55:135–140.

Ono, K., Y. Yamamoto, A. Hachiya, and H. Matsumoto. 1995. Synergistic inhibition of growth by aluminium and iron of tobacco (*Nicotiana tabacum* L.) cells in suspension culture. *Plant Cell Physiol.* 36:115–125.

Ortiz-Espín, A., A. Sánchez-Guerrero, F. Sevilla, and A. Jiménez. 2017. The role of ascorbate in plant growth and development. In *Ascorbic Acid in Plant Growth, Development and Stress Tolerance*, 25–45. Cham: Springer.

Orvar, B. L. and B. E. Ellis. 1997. Transgenic tobacco plants expressing antisense RNA for cytosolic ascorbate peroxidase show increased susceptibility to ozone injury. *Plant J.* 11:1297–1305.

Oteiza, P. I. 1994. A mechanism for the stimulatory effect of aluminium on iron-induced lipid peroxidation. *Arch. Biochem. Biophys.* 308:374–379.

Overmyer, K., M. Brosché, R. Pellinen et al. 2005. Ozone-induced programmed cell death in the *Arabidopsis* radical-induced cell death1 mutant. *Plant Physiol.* 137:1092–1104.

Palatnik, J. F., E. M. Valle, M. L. Federico et al. 2002. Status of antioxidant metabolites and enzymes in a catalase-deficient mutant of barley (*Hordeum vulgare* L.). *Plant Sci.* 162:363–371.

Pandey, P., R. K. Srivastava, and R. S. Dubey, 2014. Water deficit and aluminum tolerance are associated with a high antioxidative enzyme capacity in Indica rice seedlings. *Protoplasma* 251:147–160.

Pandey, P., R. K. Srivastava, R. Rajpoot et al. 2016. Water deficit and aluminum interactive effects on generation of reactive oxygen species and responses of antioxidative enzymes in the seedlings of two rice cultivars differing in stress tolerance. *Environ. Sci. Pollut. Res. Int.* 23(2):1516–1528.

Pang, C. H. and B. S. Wang. 2008. Oxidative stress and salt tolerance in plants. In *Progress in Botany*, eds. U. Lüttge, W. Beyschlag, and J. Murata, 231–245. Heidelberg: Springer.

Park, S. I., Y. S. Kim, J. J. Kim et al. 2017. Improved stress tolerance and productivity in transgenic rice plants constitutively expressing the *Oryza sativa* glutathione synthetase OsGS under paddy field conditions. *J. Plant Physiol.* 215:39–47.

Parry, A. D., S. A. Tiller, and R. Edwards. 1994. The effects of heavy metals and root immersion on isoflavonoid metabolism in alfalfa (*Medicago sativa* L.). *Plant Physiol.* 106:195–202.

Pasqualini, S., G. Della Torre, F. Ferranti et al. 2002. Salicylic acid modulates ozone-induced hypersensitive cell death in tobacco plants. *Physiol. Plant.* 115:204–212.

Pasqualini, S., C. Piccioni, L. Reale, L. Ederli, G. Della Torre, and F. Ferranti. 2003. Ozone-induced cell death in Bel W3 plants: the role of programmed cell death in lesion formation. *Plant Physiol.* 133:1122–1234.

Pastori, G. M. and V. S. Trippi. 1992. Oxidative stress induces high rate of glutathione-reductase synthesis in a drought-resistant maize strain. *Plant Cell Physiol.* 33:957–961.

Pellinen, R., T. Palva, and J. Kangasjärvi. 1999. Subcellular localization of ozone-induced hydrogen peroxide production in birch (*Betula pendula*) leaf cells. *Plant J.* 20:349–356.

Pereira, L. B., L. A. Tabaldi, J. F. Gonçalves et al. 2006. Effect of aluminum on δ-aminolevulinic acid dehydratase (ALA-D) and the development of cucumber (*Cucumis sativus*). *Environ. Exp. Bot.* 57:106–115.

Perez-Lopez, U., A. Robredo, M. Lacuesta et al. 2009. The oxidative stress caused by salinity in two barley cultivars is mitigated by elevated CO₂. *Physiol. Plant.* 135:29–42.

Perl, A., R. Perl-Treves, S. Galili et al. 1993. Enhanced oxidative-stress defense in transgenic potato expressing tomato Cu, Zn superoxide dismutases. *Theor. Appl. Genet.* 85:568–576.

Petrov, V., J. Hille, B. Mueller-Roeber, and T. S. Gechev. 2015. ROS-mediated abiotic stress-induced programmed cell death in plants. *Front. Plant Sci.* 6:69.

Pinto, E., T. C. S. Sigaud-Kutner, M. A. S. Leitao, O. K. Okamoto, D. Morse, and P. Colepicolo. 2003. Heavy metal-induced oxidative stress in algae. *J. Phycol.* 39:1008–1013.

Piqueras, A., J. L. Hernandez, E. Olmos, F. Sevilla, and E. Hellin. 1996. Changes in antioxidant enzymes and organic solutes associated with adaptation of citrus cells to salt stress. *Plant Cell Tissue Organ Cult.* 45:53–60.

Pitzschke, A. and H. Hirt. 2006. Mitogen-activated protein kinases and reactive oxygen species signalling in plants. *Plant Physiol.* 141:351–356.

Polle, A. 2001. Dissecting the superoxide dismutase-ascorbate-glutathione-pathway in chloroplasts by metabolic modeling. Computer simulations as a step towards flux analysis. *Plant Physiol.* 126:445–462.

Pompeu, G. B., P. L. Gratão, V. A. Vitorello, and R. A. Azevedo. 2008. Antioxidant isoenzyme responses to nickel-induced stress in tobacco cell suspension culture. *Sci. Agric.* 65:548–452.

Posmyk, M. M., C. Bailly, K. Szafranska, K. M. Janas, and F. Corbineau. 2005. Antioxidant enzymes and flavonoids in chilled soybean (*Glycine max* (L.) Merr.) seedlings. *J. Plant Physiol.* 162:403–412.

Prasad, T. K. 1997. Role of catalase in inducing chilling tolerance in pre-emergent maize seedlings. *Plant Physiol.* 114:1369–1376.

Prasad, T. K., M. D. Anderson, B. A. Martin, and C. R. Stewart. 1994. Evidence for chilling-induced oxidative stress in maize seedlings and a regulatory role for hydrogen peroxide. *Plant Cell* 6:65–74.

Prashanth, S. R., V. Sadhasivam, and A. Parida. 2008. Overexpression of cytosolic copper/zinc SOD from a mangrove plant *Avicennia mayina* in indica rice var Pusa Basmati-1 confers abiotic stress tolerance. *Transgenic Res.* 17:281–291.

Puckette, M. C., Y. Tang, and R. Mahalingam. 2008. Transcriptomic changes induced by acute ozone in resistant and sensitive *Medicago truncatula* accessions. *BMC Plant Biol.* 8:46.

Puckette, M. C., H. Weng, and R. Mahalingam. 2007. Physiological and biochemical responses to acute ozone-induced oxidative stress in *Medicago truncatula*. *Plant Physiol. Biochem.* 45:70–79.

Qi, Y. C., S. M. Zhang, L. P. Wang, M. D. Wang, and H. Zhang. 2004. Overexpression of *GST* accelerates the growth of transgenic *Arabidopsis* under the salt stress. *J. Plant Physiol. Mol. Biol.* 30:517–522.

Qiu, Q. S., J. L. Huber, F. L. Booker et al. 2008. Increased protein carbonylation in leaves of *Arabidopsis* and soybean in response to elevated [CO₂]. *Photosynth. Res.* 97:155–166.

Racchi, M. L., F. Bagnoli, I. Balla, and S. Danti. 2001. Differential activity of catalase and superoxide dismutase in seedlings and *in vitro* micropropagated oak (*Quercus robus* L.). *Plant Cell Rep.* 20:169–174.

Radotic, K., T. Ducic, and D. Mutavedzic. 2000. Changes in peroxidase activity and isoenzymes in spruce needles after exposure to different concentrations of cadmium. *Environ. Exp. Bot.* 44:105–113.

Radyuk, M. S., I. N. Domanskaya, R. A. Shcherbakov, and N. V. Shalygo. 2009. Effect of low above-zero temperature on the content of low-molecular antioxidants and activities of antioxidant enzymes in green barley leaves. *Russ. J. Plant Physiol.* 56:175–180.

Radyukina, N. L., S. Mapelli, Y. V. Ivanov, A. V. Kartashov, I. Brambilla, and V. V. Kuznetsov. 2009. Homeostasis of polyamines and antioxidant systems in roots and leaves of *Plantago major* under salt stress. *Russ. J. Plant Physiol.* 56:323–331.

Rajamickam, P., S. N. Meenakshi, R. S. Kumar, S. D. Joshi, and B. Ramasubramanian. 2005. Water deficit induced oxidative damage in tea (*Camellia sinesis*) plants. *J. Plant Physiol.* 162:413–419.

Rajjou, L., K. Gallardo, C. Job, and D. Job. 2008. Proteome analysis for the study of developmental processes in plants. *Plant Proteomics Ann. Plant Rev.* 28:151–184.

Ranieri, A., A. Castagna, G. Lorenzini, and G. F. Soldatini. 1997. Changes in thylakoid protein patterns and antioxidant levels in two wheat cultivars with different sensitivity to sulfur dioxide. *Environ. Exp. Bot.* 37:125–135.

Rao, M. V. and L. J. Dekok. 1994. Interactive effects of high CO₂ and SO₂ on growth and antioxidant levels in wheat. *Phyton Annal. Rei Bot.* 34:279–290.

Rao, M. V., B. A. Hale, and D. P. Ormrod. 1995. Amelioration of ozone-induced oxidative damage in wheat plants grown under high carbon dioxide (role of antioxidant enzymes). *Plant Physiol.* 109:421–432.

Rao, M. V., G. Paliyath, and D. P. Ormrod. 1996. Ultraviolet-B- and ozone-induced biochemical changes in antioxidant enzymes of *Arabidopsis thaliana*. *Plant Physiol.* 110:125–136.

Ravanat, J. L., P. Di Mascio, G. R. Martinez, M. H. Medeiros, and J. Cadet. 2000. Singlet oxygen induces oxidation of cellular DNA. *J. Biol. Chem.* 275:40601–40604.

Reddy, A. R., K. V. Chaitanya, P. P. Jutur, and K. Sumithra. 2004. Differential antioxidative responses to water stress among five mulberry (*Morus alba* L.) cultivars. *Environ. Exp. Bot.* 52:33–42.

Reddy, A. M., S. G. Kumar, G. Jyothsnakumari, S. Thimmanaik, and C. Sudhakar. 2005. Lead induced changes in antioxidant metabolism of horsegram (*Macrotyloma uniflorum* (Lam.) Verdc.) and bengalgram (*Cicer arietinum* L.). *Chemosphere* 60:97–104.

Ren, J., W. Duan, Z. Chen et al. 2015. Overexpression of the monodehydroascorbate reductase gene from non-heading Chinese cabbage reduces ascorbate level and growth in transgenic tobacco. *Plant Mol. Biol. Rep.* 33:881–892.

Repetto, M., J. Semprine, and A. Boveris. 2012. Lipid peroxidation: chemical mechanism, biological implications and analytical determination. In *Lipid Peroxidation*. InTech Open Access Publisher.

Richards, K. D., E. J. Schott, Y. K. Sharma, K. R. Davis, and R. C. Gardner. 1998. Aluminium induces oxidative stress genes in *Arabidopsis thaliana*. *Plant Physiol*. 116:409–418.

Richter, C. 1992. Reactive oxygen and DNA damage in mitochondria. *Mutat. Res*. 275:249–255.

Richter, C. and M. Schweizer. 1997. Oxidative stress in mitochondria. In *Oxidative Stress and the Molecular Biology of Antioxidant Defenses*, ed. J. G. Scandalios, 169–200. New York: Cold Spring Harbor Laboratory Press.

Rinalducci, S., L. Murgiano, and L. Zolla. 2008. Redox proteomics: basic principles and future perspectives for the detection of protein oxidation in plants. *J. Exp. Bot*. 59:3781–3801.

Rivero, R. M., T. C. Mestre, R. Mittler, F. Rubio, F. Garcia-Sanchez, and V. Martinez. 2014. The combined effect of salinity and heat reveals a specific physiological, biochemical and molecular response in tomato plants. *Plant Cell Environ*. 37:1059–1073.

Rizwan, M., M. Imtiaz, Z. Dai et al. 2017. Nickel stressed responses of rice in Ni subcellular distribution, antioxidant production, and osmolyte accumulation. *Environ. Sci. Pollut. Res*. 24:20587–20598.

Robinson, J. M. and J. A. Bunce. 2000. Influence of drought-induced water stress on soybean and spinach leaf ascorbate-dehydroascorbate level and redox status. *Int. J. Plant Sci*. 161:271–279.

Romero-Puertas, M. C., J. M. Palma, M. Gómez, L. A. del Río, and L. M. Sandalio. 2002. Cadmium causes the oxidative modification of proteins in pea plants. *Plant Cell Environ*. 25:677–686.

Rousseaux, M. C., C. L. Ballaré, C. V. Giordano et al. 1999. Ozone depletion and UV-B radiation: impact on plant DNA damage in southern South America. *Proc. Natl. Acad. Sci. U S A* 96:15310–15315.

Roxas, V. P., S. A. Lodhi, D. K. Garrett, J. R. Mahan, and R. D. Allen. 2000. Stress tolerance in transgenic tobacco seedlings that overexpress glutathione S-transferase/glutathione peroxidase. *Plant Cell Physiol*. 41:1229–1234.

Rubio, M. C., P. Bustos-Sanmamed, M. R. Clemente, and M. Becana. 2009. Effects of salt stress on the expression of antioxidant genes and proteins in the model legume *Lotus japonicus*. *New Phytol*. 181:851–859.

Rubio, M. C., E. M. Gonzalez, F. R. Minchin et al. 2002. Effects of water stress on antioxidant enzymes of leaves and nodules of transgenic alfalfa overexpressing superoxide dismutases. *Physiol. Plant*. 115:531–540.

Sairam, R. K., D. Kumutha, and K. Ezhilmathi. 2009. Waterlogging tolerance: nonsymbiotic haemoglobin-nitric oxide homeostasis and antioxidants. *Curr. Sci*. 96:674–682.

Sairam, R. K., K. V. Rao, and G. C. Srivastava. 2002. Differential response of wheat genotypes to long term salinity stress in relation to oxidative stress, antioxidant activity and osmolyte concentration. *Plant Sci*. 163:1037–1046.

Salt, D. E., M. Blaylock, N. P. B. A. Kumar, D. Viatcheslav, B. D. Ensley, I. Chet and I. Raskin. 1995. Phytoremediation: a novel strategy for the removal of toxic metals from the environment using plants. *Biotechnology* 13:468–473.

Sandalio, L. M., M. Rodríguez-Serrano, L. A. del Río, and M. C. Romero-Puertas. 2009. Reactive oxygen species and signaling in cadmium toxicity. In *Reactive Oxygen Species in Plant Signaling*, eds. L. A. Rio and A. Puppo, 175–189. Heidelberg: Springer.

Santos, I., J. Almeida, and R. Salema. 1999. The influence of UV-B radiation on the superoxide dismutase of maize, potato, sorghum, and wheat leaves. *Can. J. Bot*. 77:70–76.

Sarkar, B. 1995. Metal replacement in DNA-binding zinc finger proteins and its relevance to mutagenicity and carcinogenicity through free radical generation. *Nutrition* 11:646–649.

Sasidharan, R., S. Hartman, Z. Liu, S. Martopawiro, N. Sajeev, H. Veen, E. Yeung, and L. A. C. J. Voesenek. 2018. Signal dynamics and interactions during flooding stress. *Plant Physiol*. 176(2):1106–1117.

Sato, Y., Y. Masuta, K. Saito, S. Murayama, and K. Ozawa. 2011. Enhanced chilling tolerance at the booting stage in rice by transgenic overexpression of the ascorbate peroxidase gene, OsAPXa. *Plant Cell Rep*. 30:399–406.

Scandalios, J. G. 1993. Oxygen stress and superoxide dismutase. *Plant Physiol*. 101:7–12.

Scandalios, J. G., L. Guan, and A. N. Polidoros. 1997. Catalases in plants: gene structure, properties, regulation and expression. In *Oxidative Stress and the Molecular Biology of Antioxidant Defenses*, ed. J. G. Scandalios, 343–406. New York: Cold Spring Harbor Laboratory Press.

Scebba, F., G. Soldatini, and A. Ranieri. 2003. Ozone differentially affects physiological and biochemical responses of two clover species; *Trifolium repens* and *Trifolium pratense*. *Environ. Pollut*. 123:209–216.

Schuller, D. J., N. Ban, R. B. Huystee, A. McPherson, and T. L. Poulos. 1996. The crystal structure of peanut peroxidase. *Structure* 4:311–321.

Sekmen, A. H., R. Ozgur, B. Uzilday, and I. Turkan. 2014. Reactive oxygen species scavenging capacities of cotton (*Gossypium hirsutum*) cultivars under combined drought and heat induced oxidative stress. *Environ. Exp. Bot*. 99:141–149.

Semchuk, N. M., O. V. Lushchak, J. Falk et al. 2009. Inactivation of genes, encoding tocopherol biosynthetic pathway enzymes, results in oxidative stress in outdoor grown *Arabidopsis thaliana*. *Plant Physiol. Biochem*. 47:384–390.

Severino, J. F., K. Stich, and G. Soja. 2007. Ozone stress and antioxidant substances in *Trifolium repens* and *Centaurea jacea* leaves. *Environ. Pollut*. 146:707–714.

Sgherri, C., B. Stevanovic, and F. Navari-Izzo. 2000. Role of phenolic acids during dehydration and rehydration of *Ramonda serbica*. *Plant Physiol. Biochem*. 38:S196.

Sgobba, A., A. Paradiso, S. Dipierro, L. De Gara, and M. C. de Pinto. 2015. Changes in antioxidants are critical in determining cell responses to short-and long-term heat stress. *Physiol. Plant*. 153:68–78.

Shafi, A., R. Chauhan, T. Gill et al. 2015. Expression of SOD and APX genes positively regulates secondary cell wall biosynthesis and promotes plant growth and yield in *Arabidopsis* under salt stress. *Plant Mol. Biol*. 87:615–631.

Shafi, A., A. K. Pal, V. Sharma et al. 2017. Transgenic potato plants overexpressing SOD and APX exhibit enhanced lignification and starch biosynthesis with improved salt stress tolerance. *Plant Mol. Biol. Rep*. 35:504–518.

Shah, K., R. G. Kumar, S. Verma, and R. S. Dubey. 2001. Effect of cadmium on lipid peroxidation, superoxide anion generation and activities of antioxidant enzymes in growing rice seedlings. *Plant Sci*. 161:1135–1144.

Sharma, P. and R. S. Dubey. 2005a. Drought induces oxidative stress and enhances the activities of antioxidant enzymes in growing rice seedlings. *Plant Growth Regul*. 46:209–221.

Sharma, P. and R. S. Dubey. 2005b. Lead toxicity in plants. *Braz. J. Plant Physiol*. 17:35–52.

Sharma, P. and R. S. Dubey. 2007. Involvement of oxidative stress and role of antioxidative defense system in growing rice seedlings exposed to toxic levels of aluminium. *Plant Cell Rep.* 26:2027–2038.

Sharma, P., V. K.Srivastava, A. K. Gautam, A.Rupashree, R. Singh S. Abhijita, S. Kumar, A. B. Jha, and A. N. Misra. 2017. Mechanisms of metalloid uptake, transport, toxicity and tolerance in plants. In *Emerging Trends in Plant Physiology for Sustainable Crop Production*, eds. A. Zafar, A. K. Tiwari, and P. Kumar. New Jersey: Apple Academic Press.

Sharma, S. S. and K. J. Dietz. 2009. The relationship between metal toxicity and cellular redox imbalance. *Trends Plant Sci.* 14:43–50.

Sheoran, S., V. Thakur, S. Narwal et al. 2015. Differential activity and expression profile of antioxidant enzymes and physiological changes in wheat (*Triticum aestivum* L.) under drought. *Appl. Biochem. Biotechnol.* 177:1282–1298.

Shri, M., S. Kumar, D. Chakrabarty et al. 2009. Effect of arsenic on growth, oxidative stress, and antioxidant system in rice seedlings. *Ecotoxicol. Environ. Safety* 72:1102–1110.

Siegel, B. Z. and S. M. Siegel. 1986. Differential substrate specificity among peroxidases: a functional view of phyletic relations. In *Molecular and Physiological Aspects of Plant Peroxidases*, eds. H.

Silva, P. A., I. V. Oliveira, K. C. Rodrigues et al. 2016. Leaf gas exchange and multiple enzymatic and non-enzymatic antioxidant strategies related to drought tolerance in two oil palm hybrids. *Trees* 30:203–214.

Simat, T. J. and H. Steinhart. 1998. Oxidation of free tryptophan and tryptophan residues in peptides and proteins. *J. Agric. Food Chem.* 46:490–498.

Simonovicova, M., L. Tamas, J. Huttova, and I. Mistrík. 2004. Effect of aluminium on oxidative stress related enzymes activities in barley roots. *Biol. Plant.* 48:261–266.

Singh, H. P., D. R. Batish, R. K. Kohlo, and K. Arora. 2007. Arsenic-induced root growth inhibition in mung bean (*Phaseolus aureus* Roxb.) is due to oxidative stress resulting from enhanced lipid peroxidation. *Plant Growth Regul.* 53:65–73.

Skutnik, M. and A. M. Rychter. 2009. Differential response of antioxidant systems in leaves and roots of barley subjected to anoxia and post-anoxia. *J. Plant Physiol.* 166:926–937.

Slesak, I., M. Libik, B. Karpinska, S. Karpinski, and Z. Miszalski. 2007. The role of hydrogen peroxide in regulation of plant metabolism and cellular signalling in response to environmental stresses. *Acta Biochim. Pol.* 54:39–50.

Slooten, L., K. Capiau, S. Kushnir, M. van Montagu, and D. Inze. 1995b. Enhancement of oxidative stress tolerance in transgenic tobacco plants overexpressing APX in the chloroplasts. Xth International Photosynthesis Congress, Montpellier, France, 315–318.

Slooten, L., K. Capiau, W. Van Camp, M. Van Montagu, C. Sybesma, and D. Inze. 1995a. Factors affecting the enhancement of oxidative stress tolerance in transgenic tobacco overexpressing manganese superoxide dismutase in the chloroplasts. *Plant Physiol.* 107:737–750.

Smirnoff, N. and Q. J. Cumbes. 1989. Hydroxyl radical scavenging activity of compatible solutes. *Phytochemistry* 28:1057–1060.

Smirnoff, N., J. A. Running, and S. Gatzek. 2004. Ascorbate biosynthesis: a diversity of pathways. In *Vitamin C: Its Functions and Biochemistry in Animals and Plants*, eds. H. Asard, J. M. May, and N. Smirnoff, 7–29. New York: BIOS Scientific Publishers.

Sofo, A., A. Scopa, M. Nuzzaci, and A. Vitti. 2015. Ascorbate peroxidase and catalase activities and their genetic regulation in plants subjected to drought and salinity stresses. *Int. J. Mol. Sci.* 16:13561–13578.

Somashekaraiah, B. V., K. Padmaja, and A. R. K. Prasad. 1992. Phytotoxicity of cadmium ions on germinating seedlings of mung bean (*Phaseolus vulgaris*): involvement of lipid peroxides in chlorophyll degradation. *Physiol. Plant.* 85:85–89.

Srivalli, B., G. Sharma, and R. Khanna-Chopra. 2003. Antioxidative defense system in an upland rice cultivar subjected to increasing intensity of water stress followed by recovery. *Physiol. Plant.* 119:503–512.

Srivastava, R. K., P. Pandey, R. Rajpoot, A. Rani, and R. S. Dubey. 2014. Cadmium and lead interactive effects on oxidative stress and antioxidative responses in rice seedlings. *Protoplasma* 251:1047–1065.

Srivastava, S., S. Mishra, R. D. Tripathi, S. Dwivedi, P. K. Trivedi, and P. K. Tandon. 2007. Phytochelatins and antioxidant systems respond differentially during arsenite and arsenate stress in *Hydrilla verticillata* (L.f.) Royle. *Environ. Sci. Technol.* 41:2930–2936.

Stadtman, E. R. 1986. Oxidation of proteins by mixed-function oxidation systems: implication in protein turnover, aging and neutrophil function. *Trends Biochem. Sci.* 11:11–12.

Stadtman, E. R. and R. L. Levine. 2003. Free radical-mediated oxidation of free amino acids and amino acid residues in proteins. *Amino Acids* 25:207–218.

Steffens, B., Steffen-Heins, A. and M. Sauter. 2013. Reactive oxygen species mediate growth and death in submerged plants. *Front. Plant Sci.* 4:179.

Stevens, R., V. Truffault, P. Baldet, and H. Gautier. 2017. Ascorbate oxidase in plant growth, development, and stress tolerance. In *Ascorbic Acid in Plant Growth, Development and Stress Tolerance*, eds. M. Hossain, S. Munné-Bosch, D. Burritt, P. Diaz-Vivancos, M. Fujita, and A. Lorence, 273–295. Cham: Springer.

Stewart, C. R., B. A. Martin, L. Reding, and S. Cerwick. 1990. Respiration and alternative oxidase in corn seedling tissues during germination at different temperatures. *Plant Physiol.* 92: 755–760.

Stohs, S. J. and D. Bagchi. 1995. Oxidative mechanisms in the toxicity of metal ions. *Free Radical Biol. Med.* 18:321–336.

Strid, A., W. S. Chow, and J. M. Anderson. 1994. UV-B damage and protection at the molecular level in plants. *Photosynth. Res.* 39:475–489.

Suekawa, M., T. Kondo, Y. Fujikawa, and M. Esaka. 2017. Regulation of ascorbic acid biosynthesis in plants. In *Ascorbic Acid in Plant Growth, Development and Stress Tolerance*, 157–176. Cham: Springer.

Sun, W. H., F. Li, D. F. Shu, X. C. Dong, X. M. Yang, and Q. W. Meng. 2009. Tobacco plants transformed with tomato sense *LetAPX* enhanced salt tolerance. *Sci. Agric. Sinica* 42:1165–1171.

Suzuki, N. and R. Mittler. 2006. Reactive oxygen species and temperature stresses: a delicate balance between signaling and destruction. *Physiol. Plant.* 126:45–51.

Suzuki, N., S. Koussevitzky, R. Mittler, and G. Miller. 2012. ROS and redox signalling in the response of plants to abiotic stress. *Plant Cell Environ.* 35:259–270.

Sytar, O., A. Kumar, D. Latowski, P. Kuczynska, K. Strzałka, and M. N. V. Prasad. 2013. Heavy metal-induced oxidative damage, defense reactions, and detoxification mechanisms in plants. *Acta Physiol. Plant.* 35:985–999.

Taibi, K., F. Taïbi, L. A. Abderrahim, A. Ennajah, M. Belkhodja, and J. M. Mulet. 2016. Effect of salt stress on growth, chlorophyll content, lipid peroxidation and antioxidant defence systems in *Phaseolus vulgaris* L. *S. Afr. J. Bot.* 105:306–312.

Talbi, S., M. C. Romero-Puertas, A. Hernández, L. Terrón, A. Ferchichi, and L. M. Sandalio. 2015. Drought tolerance in a Saharian plant *Oudneya africana*: role of antioxidant defences. *Environ. Exp. Bot.* 111:114–126.

Tambussi, E. A., C. G. Bartoli, J. Beltrano, J. J. Guiamet, and J. L. Araus. 2000. Oxidative damage to thylakoid proteins in water-stressed leaves of wheat (*Triticum aestivum*). *Physiol. Plant.* 108:398–404.

Tanou, G., P. Filippou, M. Belghazi et al. 2012. Oxidative and nitrosative-based signaling and associated post-translational modifications orchestrate the acclimation of citrus plants to salinity stress. *Plant J.* 72: 585–599.

Tanou, G., A. Molassiotis, and G. Diamantidis. 2009. Induction of reactive oxygen species and necrotic death-like destruction in strawberry leaves by salinity. *Environ. Exp. Bot.* 65:270–281.

Tausz, M., H. Sircelj, and D. Grill. 2004. The glutathione system as a stress marker in plant ecophysiology: is a stress-response concept valid? *J. Exp. Bot.* 55:1955–1962.

Teramura, A. H. and N. S. Murali. 1986. Intraspecific differences in growth and yield of soybean exposed to ultraviolet-B radiation under greenhouse and field conditions. *Environ. Exp. Bot.* 26:89–95.

Tian, F., J. Gong, J. Zhang et al. 2013. Enhanced stability of thylakoid membrane proteins and antioxidant competence contribute to drought stress resistance in the tasg1 wheat stay-green mutant. *J. Exp. Bot.* 64:1509–1520.

Todorovski, T., M. Fedorova, and R. Hoffmann. 2011. Mass spectrometric characterization of peptides containing different oxidized tryptophan residues. *J. Mass. Spectrom.* 46:1030–1038.

Triantaphylides, C. and M. Havaux. 2009. Singlet oxygen in plants: production, detoxification and signaling. *Trends Plant Sci.* 14:219–228.

Tsai, Y. C., C. Y. Hong, L. F. Liu, and C. H. Kao. 2005. Expression of ascorbate peroxidase and glutathione reductase in roots of rice seedlings in response to NaCl and H_2O_2. *J. Plant Physiol.* 162:291–299.

Tseng, M. J., C. W. Liu, and J. C. Yiu. 2007. Enhanced tolerance to sulfur dioxide and salt stress of transgenic Chinese cabbage plants expressing both superoxide dismutase and catalase in chloroplasts. *Plant Physiol. Biochem.* 45:822–833.

Tseng, M. J., C. W. Liu, and J. C. Yiu. 2008. Tolerance to sulfur dioxide in transgenic Chinese cabbage transformed with both the superoxide dismutase containing manganese and catalase genes of *Escherichia coli*. *Sci. Hort.* 115:101–110.

Tsuboi, H., K. Kouda, H. Takeuchi et al. 1998. 8-Hydroxydeoxyguanosine in urine as an index of oxidative damage to DNA in the evaluation of atopic dermatitis. *Br. J. Dermatol.* 138:1033–1035.

Tsukamoto, S., S. Morita, E. Hirano, H. Yokoi, T. Masumura, and K. Tanaka. 2005. A novel *cis*-element that is responsive to oxidative stress regulates three antioxidant defense genes in rice. *Plant Physiol.* 137:317–327.

Ueda, Y., N. Uehara, H. Sasaki, K. Kobayashi, and T. Yamakawa. 2013. Impacts of acute ozone stress on superoxide dismutase (SOD) expression and reactive oxygen species (ROS) formation in rice leaves. *Plant Physiol. Biochem.* 70:396–402.

Urquiaga, I. and F. Leighton. 2000. Plant polyphenol antioxidants and oxidative stress. *Biol. Res.* 33:55–64.

Ushimaru, T., Y. Maki, S. Sano, K. Koshiba, K. Asada, and H. Tsuji. 1997. Induction of enzymes involved in the ascorbate-dependent antioxidative system, namely ascorbate peroxidase, monodehydroascorbate reductase and dehydroascorbate reductase, after exposure to air of rice (*Oryza sativa*) seedlings germinated under water. *Plant Cell Physiol.* 38:541–549.

Ushimaru, T., T. Nakagawa, Y. Fujioka et al. 2006. Transgenic *Arabidopsis* plants expressing the rice dehydroascorbate reductase gene are resistant to salt stress. *J. Plant Physiol.* 163:1179–1184.

Ushimaru, T., M. Shibasaka, and H. Tsuji. 1992. Development of the $O_2^{\bullet-}$-detoxification system during adaptation to air of submerged rice seedlings. *Plant Cell Physiol.* 33:1065–1071.

Valderrama, R., F. J. Corpas, A. Carreras et al. 2006. The dehydrogenase-mediated recycling of NADPH is a key antioxidant system against salt-induced oxidative stress in olive plants. *Plant Cell Environ.* 29:1449–1459.

Van Assche, F. and H. Clijsters. 1990. Effects of metals on enzyme activity in plants. *Plant Cell Environ.* 13:195–206.

Van Breuseguem, F., L. Slooten, J. M. Stassart et al. 1999. Overproduction of *Arabidopsis thaliana* FeSOD confers oxidative stress tolerance to transgenic maize. *Plant Cell Physiol.* 40: 515–523.

Van Camp, W., H. Willekens, C. Bowler et al. 1994. Elevated levels of superoxide-dismutase protect transgenic plants against ozone damage. *Bio/Technology* 12:165–168.

Vangronsveld, J. and H. Clijsters. 1994. Toxic effects of metals. In *Plants and the Chemical Elements. Biochemistry, Uptake, Tolerance and Toxicity*, ed. M. E. Farago, 150–177. Weinheim, Germany: VCH.

Verma, S. and R. S. Dubey. 2003. Lead toxicity induces lipid peroxidation and alters the activities of antioxidant enzymes in growing rice plants. *Plant Sci.* 164:645–655.

Vianello, A., M. Zancani, G. Nagy, and F. Macrì. 1997. Guaiacol peroxidase associated to soybean root plasma membranes oxidizes ascorbate. *J. Plant Physiol.* 150:573–577.

Wang, F. Z., Q. B. Wang, S. Y. Kwon, S. S. Kwak, and W. A. Su. 2005a. Enhanced drought tolerance of transgenic rice plants expressing a pea manganese SOD. *J. Plant Physiol.* 162:465–472.

Wang, J., B. Wu, H. Yin et al. 2017. Overexpression of CaAPX induces orchestrated reactive oxygen scavenging and enhances cold and heat tolerances in tobacco. *BioMed Res. Int.* 2017, Article ID 4049534, 15 pages, doi 10.1155/2017/4049534.

Wang, J., H. Zhang, and R. D. Allen. 1999. Overexpression of an *Arabidopsis* peroxisomal ascorbate peroxidase gene in tobacco increases protection against oxidative stress. *Plant Cell Physiol.* 40:725–732.

Wang, X., P. Yang, Q. Gao et al. 2008. Proteomic analysis of the response to high-salinity stress in *Physcomitrella patens*. *Planta* 228:167–177.

Wang, Y., M. Wisniewski, R. Meilan, M. Cui, R. Webb, and L. Fuchigami. 2005b. Overexpression of cytosolic ascorbate peroxidase in tomato confers tolerance to chilling and salt stress. *J. Am. Soc. Hort. Sci.* 130:167–173.

Wang, Y., Y. Ying, J. Chen, and X. Wang. 2004. Transgenic *Arabidopsis* overexpressing Mn-SOD enhanced salt-tolerance. *Plant Sci.* 167:671–677.

Waśkiewicz, A., G. Olimpia, S. Kinga, and G. Piotr. 2014. Role of glutathione in abiotic stress tolerance. In *Oxidative Damage to Plants: Antioxidant Networks and Signaling*, 149–181. Academic Press.

Watanabe, A., K. Tabeta, and K. Kosaka. 1972. Glutathione-dependent interconversion of microheterogeneous forms of glucose-6-phosphate dehydrogenase in rat liver. *J. Biochem.* 72:695–701.

Watt, D. A. 2004. Aluminium-responsive genes in sugarcane: identification and analysis of expression under oxidative stress. *J. Exp. Bot.* 385:1163–1174.

Weckx, J. E. J. and H. M. M. Clijsters. 1996. Oxidative damage and defense mechanisms in primary leaves of *Phaseolus vulgaris* as a result of root assimilation of toxic amounts of copper. *Physiol. Plant.* 96:506–512.

Weidinger, A. and A.V. Kozlov. 2015. Biological activities of reactive oxygen and nitrogen species: oxidative stress versus signal transduction. *Biomolecules* 5:472–484.

Welinder, K. G. 1992. Superfamily of plant, fungal and bacterial peroxidases. *Curr. Opin. Struc. Biol.* 2:388–393.

Willekens, H., S. Chamnongpol, M. Davey et al. 1997. Catalase is a sink for H_2O_2 and is indispensable for stress defense in C-3 plants. *EMBO J.* 16:4806–4816.

Willekens, H., C. Langebartels, C. Tire, M. Van Montagu, and D. Inze. 1994. Differential expression of catalase genes in *Nicotiana plumbaginifolia* (L.). *Proc. Natl. Acad. Sci. U S A* 91:10450–10454.

Woessmann, W., Y. H. Meng, and N. F. Mivechi. 1999. An essential role for mitogen-activated protein kinases, ERKs, in preventing heat-induced cell death. *J. Cell Biochem.* 74:648–662.

Xia, X. J., Y. H. Zhou, K. Shi, J. Zhou, C. H. Foyer, and J. Q. Yu 2015. Interplay between reactive oxygen species and hormones in the control of plant development and stress tolerance. *J. Exp. Bot.* 66:2839–2856.

Xiao, W., H. Hao, L. Xiaoqing et al. 2008b. Oxidative stress induced by lead in chloroplast of spinach. *Biol. Trace Elem. Res.* 126:257–268.

Xiao, X., X. Xu, and F. Yang. 2008a. Adaptive responses to progressive drought stress in two *Populus cathayana* populations. *Silva Fenn.* 42:705–719.

Xiong, F. S. and T. A. Day. 2001. Effect of solar ultraviolet-B radiation during springtime ozone depletion on photosynthesis and biomass production of Antarctic vascular plants. *Plant Physiol.* 125:738–751.

Xu, W. F., W. M. Shi, F. Liu, A. Ueda, and T. Takabe. 2008b. Enhanced zinc and cadmium tolerance and accumulation in transgenic *Arabidopsis* plants constitutively overexpressing a barley gene (*HvAPX1*) that encodes a peroxisomal APX. *Can. J. Bot.* 86:567–575.

Xu, W. F., W. M. Shi, A. Ueda, and T. Takabe. 2008a. Mechanisms of salt tolerance in transgenic *Arabidopsis thaliana* carrying a peroxisomal APX gene from barley. *Pedosphere* 18:486–495.

Xu, W. Y., S. N. Zhang, J. Zhang, Z. C. Zhang, and X. L. Hou. 2006. Effects of heat stress on growth and membrane damage of diploid and tetraploid *Raphanus sativus* L. *J. Nanjing Agric. Univ.* 29:43–47.

Yactayo-Chang J. P., L. M. Acosta-Gamboa, N. Nepal, and A. Lorence. 2017. The role of plant high-throughput phenotyping in the characterization of the response of high ascorbate plants to abiotic stresses. In *Ascorbic Acid in Plant Growth, Development and Stress Tolerance*, eds. M. Hossain, S. Munné-Bosch, D. Burritt, P. Diaz-Vivancos, M. Fujita, and A. Lorence, 321–354. Cham: Springer.

Yadav, N., K. Vati, S. K. Agarwal, and S. Sharma. 2018. Nickel toxicity, altering ROS scavenging mechanism and impairing sugar based signaling in *Vigna radiata* (L.) Wilczek during germination and seedling establishment. *Trop. Plant Biol.* 11:65–77.

Yakimova, E. T., V. M. Kapchina-Toteva, and E. J. Woltering. 2007. Signal transduction events in aluminum-induced cell death in tomato suspension cells. *J. Plant Physiol.* 164:702–708.

Yamaguchi, Y., Y. Yamamoto, and H. Matsumoto. 1999. Cell death process initiated by a combination of aluminium and iron in suspension-cultured tobacco cells (*Nicotiana tabacum*): apoptosis-like cell death mediated by calcium and proteinase. *Soil Sci. Plant Nutr.* 45:647–657.

Yamaguchi-Shinozaki, K. and K. Shinozaki. 1994. A novel *cis*-acting element in an *Arabidopsis* gene is involved in responsiveness to drought, low temperature, or high-salt stress. *Plant Cell* 6:251–264.

Yamamoto, Y., A. Hachiya, and H. Matsumoto. 1997. Oxidative damage to membranes by a combination of aluminium and iron in suspension-cultured tobacco cells. *Plant Cell Physiol.* 35:573–585.

Yamamoto, Y., Y. Kobayashi, and H. Matsumoto. 2001. Lipid peroxidation is an early symptom triggered by aluminium, but not the primary cause of elongation inhibition in pea roots. *Plant Physiol.* 125:199–208.

Yamauchi, Y., A. Furutera, K. Seki, Y. Toyoda, K. Tanaka, and Y. Sugimoto. 2008. Malondialdehyde generated from peroxidized linolenic acid causes protein modification in heat-stressed plants. *Plant Physiol. Biochem.* 46:786–793.

Yan, B., Q. Dai, X. Liu, S. Huang, and Z. Wang. 1996. Flooding-induced membrane damage, lipid oxidation and activated oxygen generation in corn leaves. *Plant Soil* 179:261–268.

Yan, H., Q. Li, S. C. Park, X. Wang, Y. J. Liu, Y. G. Zhang, W. Tang, M. Kou, and D. F. Ma. 2016. Overexpression of CuZnSOD and APX enhance salt stress tolerance in sweet potato. *Plant Physiol. Biochem.* 109: 20–27.

Yang, G., Z. Xu, S. Peng, Y. Sun, C. Jia, and M. Zhai. 2016. In planta characterization of a tau class glutathione S-transferase gene from *Juglans regia* (JrGSTTau1) involved in chilling tolerance. *Plant Cell Rep.* 35:681–692.

Yannarelli, G. G., G. O. Noriega, A. Batlle, and M. L. Tomaro. 2006. Heme oxygenase up-regulation in ultraviolet-B irradiated soybean plants involves reactive oxygen species. *Planta* 224:1154–1162.

Yildiz-Aktas, L., S. Dagnon, A. Gurel, E. Gesheva, and A. Edreva. 2009. Drought tolerance in cotton: involvement of non-enzymatic ROS-scavenging compounds. *J. Agron. Crop Sci.* 195: 247–253.

Yin, H., Q. M. Chen, and M. F. Yi. 2008. Effects of short-term heat stress on oxidative damage and responses of antioxidant system in *Lilium longiflorum. Plant Growth Regul.* 54:45–54.

Yin, L., J. I. Mano, K. Tanaka et al. 2017. High level of reduced glutathione contributes to detoxification of lipid peroxide-derived reactive carbonyl species in transgenic *Arabidopsis* overexpressing glutathione reductase under aluminum stress. *Physiol. Plant.* 161:211–223.

Yong, Z., T. Haoru, and L. Ya. 2008. Variation in antioxidant enzyme activities of two strawberry cultivars with short-term low temperature stress. *World J. Agric. Sci.* 4:458–462.

Yoshida, S., M. Tamaoki, T. Shikano et al. 2006. Cytosolic dehydroascorbate reductase is important for ozone tolerance in *Arabidopsis thaliana. Plant Cell Physiol.* 47:304–308.

Yoshimura, K., K. Miyao, A. Gaber et al. 2004. Enhancement of stress tolerance in transgenic tobacco plants overexpressing *Chlamydomonas* glutathione peroxidase in chloroplasts or cytosol. *Plant J.* 37:21–33.

You, J. and Z. Chan. 2015. ROS regulation during abiotic stress responses in crop plants. *Front. Plant Sci.* 6:1092.

Young, A. J. 1991. The photoprotective role of carotenoids in higher plants. *Physiol. Plant.* 83:702–708.

Zaefyzadeh, M., R. A. Quliyev, S. M. Babayeva, and M. A. Abbasov. 2009. The effect of the interaction between genotypes and drought stress on the superoxide dismutase and chlorophyll content in durum wheat landraces. *Turk. J. Biol.* 33:1–7.

Zagorchev, L., C. E. Seal, I. Kranner, and M. Odjakova. 2013. A central role for thiols in plant tolerance to abiotic stress. *Int. J. Mol. Sci.* 14:7405–7432.

Zhang, G. P., K. Tanakamaru, J. Abe, and S. Morita. 2007. Influence of waterlogging on some anti-oxidative enzymatic activities of two barley genotypes differing in anoxia tolerance. *Acta Physiol. Plant.* 29:171–176.

Zhang, H. N., J. T. Gu, W. J. Lu, C. D. Li, and K. Xiao. 2009. Improvement of low-temperature stress tolerant capacities in transgenic tobacco plants from overexpression of wheat TaSOD1.1 and TaSOD1.2 genes. *Sci. Agric. Sinica* 42:10–16.

Zhang, H. N., X. J. Li, C. D. Li, and K. Xiao. 2008. Effects of overexpression of wheat superoxide dismutase (SOD) genes on salt tolerant capability in tobacco. *Acta Agron. Sinica* 34:1403–1408.

Zhang, L., L. Sun, L. Zhang et al. 2017. A Cu/Zn superoxide dismutase gene from *Saussurea involucrata* Kar. and Kir., SiCSD, enhances drought, cold, and oxidative stress in transgenic tobacco. *Can. J. Plant Sci.* 97:816–826.

Zhang, Z., X. Li, L. Cui, S. Meng, N. Ye, and X. Peng. 2017. Catalytic and functional aspects of different isozymes of glycolate oxidase in rice. *BMC Plant Biol.* 17:135.

Zhang, Z., Q. Zhang, J. Wu et al. 2013. Gene knockout study reveals that cytosolic ascorbate peroxidase 2 (OsAPX2) plays a critical role in growth and reproduction in rice under drought, salt and cold stresses. *PLoS One* 8:57472.

Zhao, Q., L. Zhou, J. Liu et al. 2018. Relationship of ROS accumulation and superoxide dismutase isozymes in developing anther with floret fertility of rice under heat stress. *Plant Physiol. Biochem.* 122:90–101.

Zong, J., Y. Gao, J. Chen et al. 2015. Growth and enzymatic activity of four warm-season turfgrass species exposed to waterlogging. *J. Amer. Soc. Hortic. Sci.* 140:151–162.

Zrobek-Sokolnik, A., H. Asard, K. Gorska-Koplinska, and R. J. Gorecki. 2009. Cadmium and zinc-mediated oxidative burst in tobacco BY-2 cell suspension cultures. *Acta Physiol. Plant.* 31: 43–49.

Zushi, K. and N. Matsuzoe. 2009. Seasonal and cultivar differences in salt-induced changes in antioxidant system in tomato. *Sci. Hort.* 120:181–187.

8 Plant Biochemical Mechanisms for the Maintenance of Oxidative Stress under Control Conditions

Diego G. Arias, Claudia V. Piattoni, Sergio A. Guerrero, and Alberto A. Iglesias

CONTENTS

8.1   OXIDATIVE REACTIVE SPECIES PARADIGM: REDOX SIGNAL TRANSDUCTION VS. ANTIOXIDANT STRESS RESPONSE

The presence of O_2 in the earth's atmosphere originated over 2.2 billion years ago as a consequence of the evolution of oxygenic photosynthetic activity by cyanobacteria. After this, several organisms began to evolve a complex redox metabolism to cope with oxidative stress, giving them the capacity not only to tolerate O_2 but also to use it for metabolic transformation and biosynthesis (Halliwell, 2006; Slesak et al., 2007). It is remarkable that for aerobic organisms, O_2 is both beneficial and damaging. It is beneficial in its role as an essential electron exchanger for respiration and photosynthesis, besides which O_2 is widely used for signal transduction. Conversely, O_2 can cause dysfunction of cell components by irreversible modifications to DNA, proteins, sugars, and lipids. Hence, there is a balance/imbalance between the two sides, which is critical for cell functionality and survival (Slesak et al., 2007; Cerny et al., 2018; Moldogazieva et al., 2018).

In an aerobic atmosphere, a delicate homeostasis between oxidative and reductive reactions must be maintained inside cells. Plants and all other aerobic life forms evolved different antioxidative systems to protect cell components and ensure the use of O_2 in metabolic pathways. The presence of intracellular oxidative and reducing species determines a transitory redox environment, defined as the summation of the reduction potential and the reducing capacity of the linked intracellular

redox couples (Schafer and Buettner, 2001). Cells are exposed to a redox stress when the redox environment undergoes imbalance with enhanced levels of reduced or oxidized species. Both oxidative and reductive stress can trigger redox cascades that affect the oxidized/reduced status of biomolecules. Imbalances in the cellular redox environment can alter signal transduction, the synthesis of macromolecules (DNA, RNA, and proteins), enzyme activation, and even the regulation of the cell cycle. Thereby, the redox environment critically determines whether a cell will proliferate, differentiate, or perish (Moller, Jensen, and Hansson, 2007; Neill et al., 2002; Moldogazieva et al., 2018; Knuesting and Scheibe, 2018; Yang et al., 2018; Siqueira et al., 2018). For a clear scenario, we schematize in Figure 8.1 different agents that can cause stress conditions and even molecular damage to develop, resulting in a situation that challenges the cell to trigger defense mechanisms. Furthermore, simplified maps of different metabolic routes related to the flux of redox equivalents operating under physiological conditions as well as situations of coping with oxidative stress are detailed in Figures 8.2 and 8.3. A recent review article (Jones and Sies, 2015) about the redox code is recommended for a broad view of the redox networks operative in different cells.

Reductive (NADPH) and energetic (ATP) equivalents are mainly generated during respiration and photosynthesis, both processes involving electron transport chains and O_2 and generating different reactive oxygen species (ROS) as byproducts (Apel and Hirt, 2004; Buchanan and Balmer, 2005) (Figure 8.1).

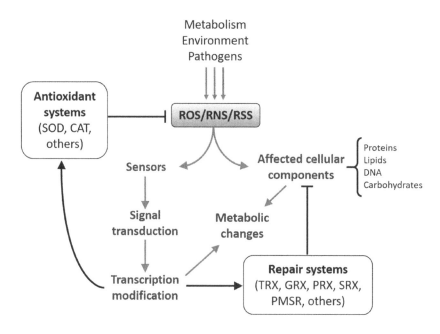

FIGURE 8.1   Redox homeostasis, redox signaling, and oxidative stress. The scheme shows a probable progression of the cellular response elicited by synthesis of oxidant species. Different modifications triggered by oxidants (marked in gray) are interrelated to increase the activity of antioxidant systems (marked in black), which might struggle to return the system to homeostasis.

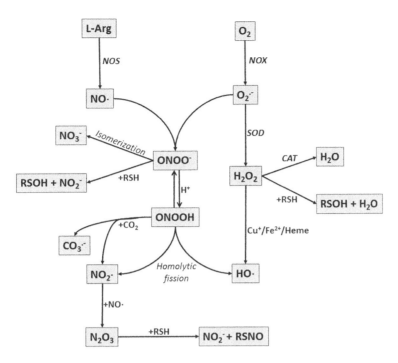

FIGURE 8.2 Molecular pathways of RNS and ROS generation. Nitric oxide (NO·) is formed from L-arginine and molecular oxygen (O_2) by the activity of various isoforms of NO synthase (NOS). The superoxide radical (O_2^-) is formed during cellular metabolism or as a product of the enzymatic activities of NADPH oxidase (NOX). O_2^- is dismutated by superoxide dismutase (SOD) enzymes to hydrogen peroxide (H_2O_2), which can either be detoxified to water by catalase or be converted to the hydroxyl radical (OH·) in the presence of metal (heme, copper, or iron-mediated Fenton reaction). NO· and O_2^- spontaneously and rapidly react to form the strong oxidant peroxynitrite ($ONOO^-$). $ONOO^-$ can be detoxified by isomerization to nitrate (NO_3^-) or may form secondary radicals through homolytic fission (rupture of a covalent bond) or through reaction with carbon dioxide (CO_2) of its conjugated acid peroxynitrous acid (ONOOH), yielding the carbonate radical (CO_3^-), the nitrogen dioxide radical (NO_2^-), or the OH· radical. Since NO_2 is a good one-electron oxidant, N_2O_3 has some ionic character ($NO^+–NO_2^-$). In the presence of a nucleophile, such as a thiol (RSH), N_2O_3 can thus transfer its nitrosonium group (NO^+) and generate a nitrosothiol (RSNO). Both $ONOO^-$ and H_2O_2 can react with RSH and generate sulfenic acid (RSOH).

The term ROS describes different potentially damaging oxygen species, such as free radicals containing one or more unpaired electrons, including superoxide anion radical (O_2^-), hydroxyl radical (·OH), per-hydroxyl radical ($HO_2^·$), and singlet oxygen ($^1O_2$). Some non-radical derivatives, such as hydrogen peroxide (H_2O_2), are also classified as ROS (Foyer et al., 2009). Plants may also be exposed to reactive nitrogen species (RNS) and reactive sulfur species (RSS). The major RNS compounds in the cell are nitric oxide (·NO), peroxynitrite ($ONOO^-$), nitrosoglutathione (GSNO), and metal–NO adducts (Moller, Jensen, and Hansson, 2007; Knuesting and Scheibe, 2018; Mittler, 2017). Other components that are major air pollutants and can cause oxidative modifications are ozone (O_3), nitric dioxide (NO_2), and sulfur dioxide (SO_2) (Moller, Jensen, and Hansson, 2007; Knuesting and Scheibe, 2018; Mittler, 2017).

To cope with the many oxidative species, aerobic organisms have developed antioxidant networks composed of different antioxidant systems (Figure 8.1). Any compound capable of quenching ROS without the generation (by self-conversion) of destructive radicals is named an *antioxidant* (Noctor and Foyer, 1998). Antioxidant networks are composed of numerous proteins, enzymes, and metabolites able to function in synchrony to ameliorate oxidative stress situations. For plant antioxidative networks, (i) catalase (CAT),

peroxiredoxin (PRX), thioredoxin (TRX), glutaredoxin (GRX), and sulfiredoxin (SRX) are identified as the main protein/enzyme components; and (ii) ascorbate, reduced glutathione (GSH), tocopherol, and NAD(P)H are key metabolic intermediaries. These antioxidant components act in association to form different groups of antioxidant systems (Sevilla et al., 2015; Stroher and Millar, 2012). Redox-active metabolites and signaling in plants are also relevant to the response to metal nutrient deficiency (Rajniak et al., 2018).

Because they are aerobic organisms performing oxygenic photosynthesis, plant cells mainly operate in oxidative environments. In addition, the sessile lifestyle characteristic of plants results in their exposure to a range of harmful biotic and abiotic stresses. These oxidative situations can trigger an oxidative burst as well as generating signal transducers (Figure 8.1). The state of oxidative stress depends on many factors, such as (i) the type of ROS that are produced, (ii) the concentration and the place where ROS are generated, (iii) the interaction of these molecules with other components in the organism, and (iv) the developmental period and the previous circumstances experienced by cells (Moller, Jensen, and Hansson, 2007; Wang et al., 2013). For example, an increase in levels of H_2O_2 is a signal of change in the milieu perceived by plants. Being a diffusible signal-transducer molecule, H_2O_2 warns the metabolism of the presence of threats (both biotic

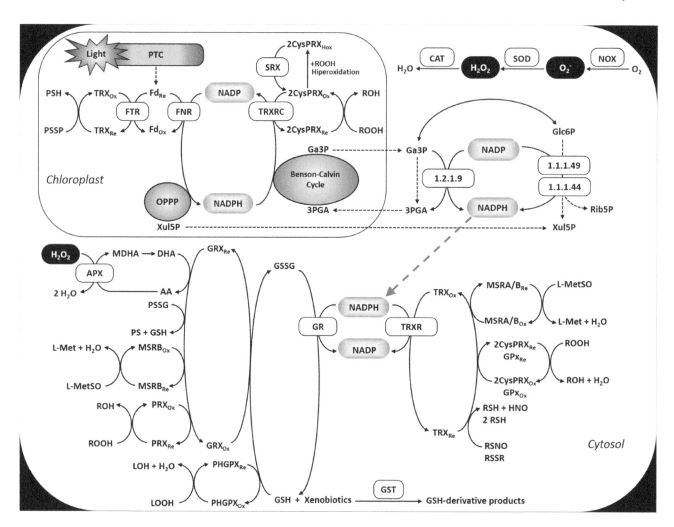

FIGURE 8.3 Cross-talk between antioxidant systems. The scheme shows fluxes between main redox components in plants. O_2: molecular oxygen, O_2^-: superoxide anion, H_2O_2: hydrogen peroxide, H_2O: water, NOX: NADPH oxidase, SOD: superoxide dismutase, CAT: catalase, PTC: photosynthetic transport chain, Fd: ferredoxin, FNR: ferredoxin-NADP oxido-reductase, FTR: ferredoxin-thioredoxin oxido-reductase, TRX: thioredoxin, PSH: thiol-protein, PSSP: disulfide-protein, TRXRC: thioredoxin reductase C, 2-CysPRX: typical two-cysteine peroxiredoxin, ROOH: peroxide, ROH: alcohol, SRX: sulfiredoxin, Ga3P: glyceraldehyde-3-phosphate, 3PGA: 3-phosphoglycerate, (1.2.1.9): non-phosphorylating glyceraldehyde-3-phosphate dehydrogenase, Glc6P: glucose-6-phosphate, (1.1.1.49): Glc6P dehydrogenase, (1.1.1.44): 6-phosphogluconate dehydrogenase, Rib5P: ribose-5-phosphate, Xul5P: xylulose-5-phosphate, OPPP: oxidative pentose phosphate pathway, GR: glutathione reductase, GSSG: oxidized glutathione, GSH: reduced glutathione, GRX: glutaredoxin, DHA: dehydroascorbate, AA: ascorbic acid, APX: ascorbate peroxidase, MDHA: monodehydroascobate, PSSG: glutathionyl-protein, MSRB: B-type methionine sulfoxide reductase, L-MetSO: L-methionine sulfoxide, L-Met: L-methionine, PRX: peroxiredoxin, PHGPX: phospholipid hydroperoxide glutathione peroxidase, LOOH: lipid hydroperoxide, LOH: lipid-alcohol, TRXR: thioredoxin reductase, MSRA: A-type methionine sulfoxide reductase, GPX: glutathione peroxidase, RSNO: *S*-nitrosothiols, RSSR: low–molecular weight disulfide, RSH: low–molecular weight thiol, HNO: nitroxyl, GST: glutathione-S-transferase.

and abiotic) (Noctor and Foyer, 1998; Cerny et al., 2018). When the plant response is deficient and unable to cope with the threat, oxidative stress results.

8.2 VERSATILITY OF OXYGEN, NITROGEN, AND SULFUR DERIVATIVE REACTIVE SPECIES

The incomplete reduction of oxygen, nitrogen, or sulfur generates oxidant species (Giles, Tasker, and Jacob, 2001; Nordberg and Arner, 2001) (Figure 8.2). The generation of these harmful molecules is a process linked to metabolism and to the interaction of the cell with its environment (Apel and Hirt,

2004) (Figure 8.1). Oxidant species exhibit different degrees of reactivity. Singlet oxygen, hydroxyl radical, peroxynitrite, and sulfenic acid are highly reactive molecules, while superoxide anion and hydrogen peroxide are less reactive (Giles and Jacob, 2002; Nordberg and Arner, 2001; Patel et al., 1999; Cerny et al., 2018). ROS, RNS, and RSS have the potential to cause damage to cells by oxidizing many biomolecules (Figure 8.1), which can modify their structures and/or cause the loss of their respective biological functions (Apel and Hirt, 2004; Friguet, 2006). The fact that oxidant species can have roles as signaling molecules is gaining consensus. Such signals enable the control of plant development in connection with external environments and in the defense against pathogens (Forman, Fukuto, and Torres, 2004; Foyer and Noctor, 2005; Pauly et al., 2006; Vranova, Inze, and Van Breusegem, 2002; Mittler, 2017).

8.2.1 REACTIVE OXYGEN SPECIES

ROS include a number of chemically reactive molecules (both inorganic and organic) derived from oxygen: oxygen ions, free radicals, and peroxides (Forman, Fukuto, and Torres, 2004). All these species are highly reactive due to the presence of unpaired electrons in a valence shell. ROS are generated as natural products of the metabolism of oxygen and have important roles in cell signaling, oxidative stress, and the cell's response to such damaging conditions (Hancock, 1997; Neill et al., 2002; Kotchoni and Gachomo, 2006; Turkan, 2015) (A recent review on the subject of ROS and plant signaling is recommended to readers; Waszczak, Carmody, and Kangasjarvi, 2018). Next, we detail the most common intracellular forms of ROS.

8.2.1.1 Singlet Oxygen

Singlet oxygen ($^1O_2$) is a diamagnetic form of molecular oxygen that is differentiated (for example, by its chemical properties) from triplet oxygen ($^3O_2$, the most stable form of molecular oxygen) (Halliwell, 2006). An input of energy able to rearrange electrons in valence shells generates $^1O_2$, a compound that even although it is not a free radical, has oxidizing properties, because its spin restriction is affected. $^1O_2$ is able to directly oxidize proteins, DNA, and lipids (Triantaphylides and Havaux, 2009). Insufficient energy dissipation during photosynthesis can lead to the formation of the chlorophyll triplet state, which can transfer its excitation energy to ground-state O_2, generating $^1O_2$ (Triantaphylides and Havaux, 2009; Halliwell, 2006). This originates oxidizing conditions for many components of the chloroplast and can trigger cell death. Plants have molecules able to quench $^1O_2$, including carotenoids, tocopherols, plastoquinones, ascorbate, vitamin B6, and flavonoids (Triantaphylides and Havaux, 2009). Alternatively, $^1O_2$ can also act as a signaling molecule, thus playing a critical role in plant responses to light stress (Halliwell, 2006; Triantaphylides and Havaux, 2009).

8.2.1.2 Superoxide Anion

Superoxide anion ($O_2{}^-$) is a free radical produced by the one-electron reduction of O_2, which occurs widely in nature (Vranova, Inze, and Van Breusegem, 2002). This anion is paramagnetic (as is the case for O_2) (Apel and Hirt, 2004; Forman, Fukuto, and Torres, 2004) and relatively poor as a reactive species, being unable to penetrate lipid membranes, so that it remains trapped within the compartment where it is generated (Nordberg and Arner, 2001). The generation of $O_2{}^-$ is a spontaneous process taking place in electron-rich aerobic environments (such as the vicinity of the mitochondrial inner membrane) (Nordberg and Arner, 2001). It can be produced endogenously by flavoenzymes, such as NADPH oxidase, lipoxygenase, and cyclooxygenase (Kotchoni and Gachomo,

2006). Biologically, $O_2{}^-$ plays relevant roles in the redox signaling response to different stimuli and in defense against pathogens (Kotchoni and Gachomo, 2006). Superoxide dismutase (SOD) is an enzyme found in almost all living organisms, which reduces the toxicity of $O_2{}^-$ with high efficiency by catalyzing the dismutation of two molecules of the anion to hydrogen peroxide and O_2 (Apel and Hirt, 2004; Knuesting and Scheibe, 2018).

8.2.1.3 Hydrogen Peroxide

Hydrogen peroxide (H_2O_2) is not a free radical. However, the peroxide is damaging to cells because of its remarkable diffusion capacity (Neill et al., 2002) and ability to penetrate biological membranes (Bienert, Schjoerring, and Jahn, 2006; Apel and Hirt, 2004; Neill et al., 2002). H_2O_2 is moderately reactive and relatively stable (with a half-life of 10^{-3} s), being an intermediate in the formation of other ROS (such as HOCl [hypochlorous acid] and OH [formed via the oxidation of transition metals; see later]) (Nordberg and Arner, 2001). By itself, H_2O_2 represents an intracellular signal that induces a wide range of molecular, biochemical, and physiological responses within cells (Neill et al., 2002; Hancock, 1997; Noctor, Reichheld, and Foyer, 2018). It can also inactivate enzymes through oxidation of their thiol groups. Potential sources of H_2O_2 in plants include $O_2{}^-$ dismutation by SOD, NADPH oxidase, cell wall peroxidase, amino oxidase, oxalate oxidase, and flavin-containing oxidase (Neill et al., 2002). The peroxide is also generated by induction following exposure to many abiotic and biotic stimuli: for example, UV irradiation, excess of excitation energy, exposure to ozone, phytohormones (i.e., abscisic acid [ABA]), dehydration, wounding, and pathogen attack (Neill et al., 2002; Apel and Hirt, 2004; Kotchoni and Gachomo, 2006). Catalases, glutathione peroxidases, ascorbate peroxidases, and peroxiredoxins are the main antioxidant enzymes involved in the removal of H_2O_2 in cells (Noctor and Foyer, 1998; Forman, Fukuto, and Torres, 2004; Maruta et al., 2016).

8.2.1.4 Hydroxyl Radical

Hydroxyl radical ($\cdot$OH) is a highly hazardous compound to organisms because of its strong reactivity toward biomolecules (Nordberg and Arner, 2001; Halliwell, 2006). Its very short half-life (approx. 10^{-9} s), which is faster than its diffusion rate, restricts the reactivity of the radical species to oxidizable compounds in its vicinity. Unlike $O_2{}^-$, which can be detoxified by SOD, $\cdot$OH cannot be eliminated enzymatically (Halliwell, 2006). It can damage almost all types of macromolecules: carbohydrates, nucleic acids (mutations), lipids (peroxidation), and amino acids (e.g., conversion of phenylalanine to m- and o-tyrosine) (Halliwell, 2006; Vranova, Inze, and Van Breusegem, 2002; Triantaphylides and Havaux, 2009). Glutathione and ascorbate protect important cell structures from $\cdot$OH (Noctor and Foyer, 1998). This radical is formed from H_2O_2 via the Fenton reaction (Reaction 8.1), which is catalyzed by metal ions (Fe^{2+} or Cu^+) that are complexed with different proteins or other molecules (Nordberg and Arner, 2001). As detailed by Reaction 8.2,

O_2^- is functionally connected with the Fenton reaction by recycling metal ions (Nordberg and Arner, 2001; Vranova, Inze, and Van Breusegem, 2002). The sum of Reactions 8.1 and 8.2 is the Haber–Weiss reaction (Reaction 8.3), which points to the importance of transition metals in the formation of ·OH (Vranova, Inze, and Van Breusegem, 2002; Vigani and Murgia, 2018).

$$Fe^{2+} / Cu^+ + H_2O_2 \rightarrow Fe^{3+} / Cu^{2+} + \cdot OH + OH^- \quad (8.1)$$

$$Fe^{3+} / Cu^{2+} + O_2 \rightarrow Fe^{2+} / Cu^+ + O_2 \quad (8.2)$$

$$H_2O_2 + O_2^- \rightarrow OH + OH^- + O_2 \quad (8.3)$$

8.2.2 Reactive Nitrogen Species

RNS are a group of reactive molecules derived from nitric oxide (·NO, which is produced enzymatically by inducible ·NO synthase) (Besson-Bard, Pugin, and Wendehenne, 2008). Analogously to ROS, RNS cause so-called *nitrosative stress*. In plants, RNS are also continuously produced in plants as byproducts of aerobic metabolism or in response to stress (Besson-Bard, Pugin, and Wendehenne, 2008; Shapiro, 2005; Fancy, Bahlmann, and Loake, 2017). Oxidative and nitrosative stress cause the production of ·NO-derived species such as peroxynitrite (ONOO⁻) and nitrogen dioxide (·NO₂) (Droge, 2002; Squadrito and Pryor, 1998). These compounds participate in the nitration of proteins (mainly tyrosine and cysteine residues) and lipids (unsaturated fatty acids), which constitute biologically relevant redox signals and injury events (Neill et al., 2002; Droge, 2002; Squadrito and Pryor, 1998; Moldogazieva et al., 2018; Mata-Perez et al., 2018).

8.2.2.1 Nitric Oxide

Despite having an unpaired electron, ·NO does not directly react with most biomolecules (Patel et al., 1999; Halliwell, 2006). It functions as a free radical scavenger, since it easily reacts with free radicals (i.e., peroxyl and alkyl radicals). In a clear example, ·NO has been shown to inhibit lipid peroxidation in cell membranes (Patel et al., 1999). However, the production of large amounts of O_2^- in combination with ·NO becomes highly cytotoxic, as the two species react to give peroxynitrite (OONO⁻) (Nordberg and Arner, 2001; Fancy, Bahlmann, and Loake, 2017). Four routes are found in plants to produce ·NO: (i) via L-arginine-dependent NOS (although this pathway has been reported in plants (Figure 8.2), no gene, cDNA, or protein related to NOS has been isolated to date) (Corpas et al., 2004; Corpas et al., 2006; Valderrama et al., 2007); (ii) mediated by plasma membrane–bound nitrate reductase (Corpas, 2004); (iii) through the electron transport chain in mitochondria (Besson-Bard, Pugin, and Wendehenne, 2008); and (iv) by non-enzymatic reactions (Besson-Bard, Pugin, and Wendehenne, 2008; Turkan, 2015).

The ability to be an intracellular messenger, stimulating guanylate cyclase and protein kinases, gives ·NO its main physiological functionality (Shapiro, 2005; Besson-Bard, Pugin, and Wendehenne, 2008). It acts as a signaling molecule to cope with oxidative stress and also plays a role in plant–pathogen interactions (Besson-Bard, Pugin, and Wendehenne, 2008; Shapiro, 2005). Since ·NO is able to cross cell membranes, it transmits signals between cells (Patel et al., 1999). S-nitrosylation is a biologically important reaction of ·NO that converts thiol groups (including cysteine residues in proteins) into S-nitrosothiols (RSNO; see later for further details about these compounds) (Squadrito and Pryor, 1998; Forman, Fukuto, and Torres, 2004; Zaffagnini et al., 2016), which is a mechanism for dynamic post-translational regulation of all major classes of proteins (Giustarini et al., 2004). Excessive production of ·NO is counteracted by its conjugation with glutathione. The resulting product (S-nitrosoglutathione adduct, abbreviated GSNO) can be cleaved directly by thioredoxin systems, with the consumption of NADPH and release of GSH and ·NO (Giustarini et al., 2004; Nordberg and Arner, 2001; Zaffagnini et al., 2016). In summary, the involvement of ·NO in cells' redox balance is multifaceted, and it behaves more as an antioxidant than as an oxidant.

8.2.2.2 Peroxynitrite

Peroxynitrite (ONOO⁻) is a non-radical oxidant and nitrating agent that reacts with many biomolecules in one- or two-electron reactions. It is a damaging agent that affects a wide number of cell components, mainly DNA, lipids, and proteins (Nordberg and Arner, 2001; Landino, 2008). Its action on proteins involves modification of tyrosine and cysteine residues (Landino, 2008; Besson-Bard, Pugin, and Wendehenne, 2008). The formation of ONOO⁻ *in vivo* has been described to occur by reaction between O_2^- and ·NO (Nordberg and Arner, 2001; Halliwell, 2006; Besson-Bard, Pugin, and Wendehenne, 2008). Also, ONOO⁻ readily reacts with CO_2 to form nitroso peroxocarboxylate (ONOOCO₂⁻), a highly reactive compound that can generate the potent oxidizing agents ·NO₂ and ·CO₃⁻ (carbonate radical) (Valderrama et al., 2007; Nordberg and Arner, 2001; Halliwell, 2006). At physiological pH, ONOO⁻ is rapidly protonated to peroxynitrous acid (ONOOH) (Figure 8.2) (Halliwell, 2006). It can directly damage macromolecules (Landino, 2008; Valderrama et al., 2007), but it is hazardous because undergoes homolytic fission to give noxious products, ·OH and ·NO₂, or rearranges to nitrate (NO₃⁻) (Squadrito and Pryor, 1998).

8.2.2.3 S-nitrosothiols

S-nitrosothiols (RSNO) may cause many of the biological effects of ·NO, mainly producing S-nitrosylation of proteins (Forman, Fukuto, and Torres, 2004). These compounds are donors of nitrosonium ion (NO⁺) and ·NO, which constitute signaling molecules in plant cells (Zhang and Hogg, 2005). The addition of a nitroso group to the sulfur atom of cysteines in proteins is known as S-nitrosation or S-nitrosylation (Zhang and Hogg, 2005; Forman, Fukuto, and Torres, 2004; Zaffagnini et al., 2016), and it functions as a mechanism of post-translational modification (analogous to phosphorylation) (Giustarini et al., 2004; Zhang and Hogg, 2005). Proteins modified by the formation of S-nitrosothiol adducts exhibit changes in activity, interaction with other molecules, or subcellular location (Giustarini et al., 2004; Zhang and Hogg, 2005; Lamotte et al., 2015).

8.2.3 Reactive Sulfur Species

ROS and RNS are primary oxidizing agents generated after oxidative stress (Apel and Hirt, 2004; Forman, Fukuto and Torres, 2004), although a second generation of reactive species functions by having a different spectrum of redox activity and biological targets (Giles, Tasker and Jacob, 2001). In this context, sulfur is usually considered as part of the cellular antioxidant systems, and many experimental data support the formation of RSS (with suppressor properties resembling ROS and RNS) under conditions of oxidative stress (Giles and Jacob, 2002; Jacob, Lancaster and Giles, 2004). Thiols, as well as disulfides, are promptly oxidized to species with sulfur in higher oxidation states (Jacob, Lancaster, and Giles, 2004), giving rise to sulfenyl radicals (RS·), disulfides (RSSR), sulfenic acids (RSO$^-$), disulfide-S oxides (RS(O)SR), and disulfide-S dioxides (RS(O$_2$)SR) (Giles and Jacob, 2002; Gruhlke and Slusarenko, 2012). These species modulate the redox status of biological sulfur molecules by reaction with low–molecular weight thiols and thiol moieties present in proteins. Examples include reports demonstrating that RS(O)SR are able to inactivate glyceraldehyde-3-phosphate dehydrogenase and alcohol dehydrogenase, and also that RSS induce zinc release from zinc-finger domains (Giles and Jacob, 2002; Jacob, Lancaster, and Giles, 2004).

8.3 MODIFICATION OF CELLULAR COMPONENTS. OXIDATIVE DAMAGE AND SIGNAL PERCEPTION

Figure 8.1 shows that exposure to physiological concentrations of ROS induces changes in metabolism, triggering cell responses for repair, adaptation, or transformation (Forman, Fukuto, and Torres, 2004). ROS represent potentially toxic and mutagenic agents (Smirnoff, 2005), targeting all major groups of biomolecules for damage (Halliwell, 2006; Zagorchev et al., 2013). The oxidative reactions are mainly irreversible and in some cases, induce specific repair mechanisms (Moller, Jensen, and Hansson, 2007). Despite their deleterious effects, reactive species also exert roles as indicators of oxidative stress and as signaling molecules. Redox signaling involves responses following a reversible mechanism specific to redox reactions (Forman, Fukuto, and Torres, 2004; Rouhier et al., 2015), involving reactive species as second messengers involved in signal transduction.

8.3.1 Effects on DNA, Lipids and Proteins

8.3.1.1 DNA

Reactive species act as mutagenic agents by chemically modifying nucleotide bases in DNA in different ways (Banerjee, 2008). The main alterations include cleavage of DNA, DNA–protein cross-linking, oxidation of guanine to 8-hydroxyguanine, and conjugation with polyunsaturated fatty acids (PUFA) (Moller, Jensen, and Hansson, 2007; Mueller and Berger, 2009). Besides mutations, oxidative DNA modifications can lead to changes in the methylation of cytosines, which is important for regulating gene expression (Moller, Jensen, and Hansson, 2007). These modifications are due to reactions with ROS, especially ·OH (exhibiting the highest reactivity) and $^1O_2$ (primarily attacking guanine) (Britt, 1996; Nordberg and Arner, 2001). (In contrast, H_2O_2 and $O_2{}^-$ are not reactive at all with DNA.) Under a deficit of repair systems (to restore intact DNA), a mutation will be established after erroneous base pairing during replication (Nordberg and Arner, 2001). A number of mechanisms are available for repairing mutated DNA both in the nucleus and in the mitochondria (Kimura et al., 2004), including direct reversion of the damage, replacement of the base, and replacement of the whole nucleotide (Britt, 1996; Banerjee, 2008; Roldan-Arjona and Ariza, 2009).

8.3.1.2 Lipids

Linoleic acid (18:2) and linolenic acid (18:3) are the major (PUFA) in plant membranes (Smirnoff, 2005). The presence of double bonds makes PUFA suitable targets for peroxidation by free radical (particularly $^1O_2$ and ·OH) action, with the emergence of a complex mixture of lipid hydroperoxides (LOOH) (Moller, Jensen, and Hansson, 2007; Mueller and Berger, 2009). Lipid peroxidation is probably the most explored area of research concerning ROS. Extensive PUFA peroxidation affects membrane functionality, leading to a decrease in fluidity, an increase in leakiness, and associated damage to membrane proteins (Moller, Jensen, and Hansson, 2007). The products of PUFA peroxidation are able to interact with DNA and proteins (Nordberg and Arner, 2001; Banerjee, 2008). In plant cells, some PUFA oxidation products behave as secondary messengers either directly or after further modification exerted by enzymes (Moller, Jensen, and Hansson, 2007; Mata-Perez et al., 2018). Phospholipid-hydroperoxide glutathione peroxidase is a member of the glutathione peroxidase family mainly involved in LOOH detoxification (Banerjee, 2008). This enzyme can directly transfer reductive equivalents from glutathione (GSH) to LOOH (Smirnoff, 2005).

8.3.1.3 Proteins

Protein oxidation is defined here as the covalent modification of proteins induced or mediated by ROS, RNS, or byproducts of oxidative stress (Banerjee, 2008). The process is widespread in stressed cells, and its determination is of utility as a diagnostic marker of the degree of oxidative damage (Moller and Kristensen, 2004; 2006). Protein oxidation is essentially irreversible except in a few cases (involving sulfur-containing amino acids) that are reversible (Forman, Fukuto, and Torres, 2004; Halliwell, 2006; Ruiz-May et al., 2018). The reaction of a polypeptide with oxidant species generates four main types of modification: (i) side chain oxidation, mainly cysteine (Kiley and Storz, 2004), methionine (Tarrago, Laugier, and Rey, 2009), and tryptophan oxidation (Moller and Kristensen,

2006); (ii) carbonylation (Banerjee, 2008); (iii) nitrosylation (Landino, 2008); and (iv) interaction with products of PUFA oxidation (Moller, Jensen, and Hansson, 2007; Mata-Perez et al., 2018). These modifications can occur in different degrees, from reducing functionality to producing denatured (non-functional) proteins.

The thiol group in cysteine can be oxidized to achieve different species, depending on the concentration and type of the reactive oxidant agent as well as the number of cysteinyl moieties concerned. The formation of a disulfide (R_1-S-S-R_2, also named cystine) between two cysteine residues is a relevant mechanism for metabolic redox regulation (Balmer et al., 2004; Buchanan and Balmer, 2005; Ruiz-May et al., 2018). Intra- or intermolecular disulfide bonds can be formed between thiol side chains. In many cases, thioredoxin (Balmer et al., 2004; Montrichard et al., 2009) or glutaredoxin (Lillig, Berndt, and Holmgren, 2008; Rouhier, Lemaire, and Jacquot, 2008) systems are able to reverse the process and reduce cystine to cysteine. A large number of potential thioredoxin-regulated proteins have been reported in cytosol, chloroplasts, and mitochondria (Montrichard et al., 2009). The thiol group of cysteine can be also oxidized to higher levels via cysteine sulfenic acid (R-SOH) and further, to cysteine sulfinic acid (R-SO$_2$H) (Trost et al., 2017). These oxidation states can be enzymatically reversed via the combined action of sulfiredoxin followed by thioredoxin or GSH and are probably involved in signaling pathways (Liu et al., 2006; Rey et al., 2007). The oxidation of cysteine to the highest level of cysteic acid (R-SO$_3$H) is irreversible and highly damaging (Kiley and Storz, 2004). Cysteine can also form mixed disulfides, primarily with GSH, which functions to protect the cysteinyl moiety from undergoing further oxidation (Rouhier, Lemaire, and Jacquot, 2008). Glutaredoxins and GSH can reduce mixed disulfides (Rouhier, Lemaire, and Jacquot, 2008).

The oxidation of methionine to methionine sulfoxide (MetSO) is another reversible modification in proteins (Banerjee, 2008). This sulfoxidation process is reversed in a reaction catalyzed by MetSO reductase (MSR) using thioredoxin or glutaredoxin as the reductant (Banerjee, 2008; Rey and Tarrago, 2018) (see Section 6.2 for further details). Reversible methionine sulfoxidation could constitute a key regulatory mechanism in cells (Rouhier et al., 2008). It has been proposed that some peripheral methionine residues could act as endogenous antioxidants by protecting sensitive functional domains in the protein while helping to remove ROS (Petropoulos and Friguet, 2006; Rouhier, Lemaire, and Jacquot, 2008). The more extensive oxidation of methionine to methionine sulfone (MetSO$_2$) appears to be irreversible and destructive of protein structure and function (Banerjee, 2008).

Nitrosylation and carbonylation represent the other two redox post-translational modifications of proteins operating in cells. The former is the covalent attachment of ·NO to the thiol moiety in cysteine, which potentially regulates the function of proteins (Lamattina et al., 2003; Besson-Bard, Pugin, and Wendehenne, 2008). Thiols present in a protein (and in GSH) can react with ·NO derivatives to produce a range of products: disulfides and sulfenic, sulfinic, and sulfonic acids as well as S-nitrosothiols (Shapiro, 2005; Dalle-Donne et al., 2007; Besson-Bard, Pugin, and Wendehenne, 2008; Zaffagnini et al., 2016). Carbonylation is an oxidative modification of proteins that proceeds irreversibly to give rise to derivative products commonly found in cells (Moller, Jensen, and Hansson, 2007). The oxidation of a number of amino acids (mainly arginine, histidine, lysine, proline, threonine, and tryptophan) produces free carbonyl groups (Banerjee, 2008; Kolbert et al., 2017). These carbonyl derivatives can modify the biological properties of macromolecules by reacting with (i) other side chain groups in amino acids, generating polypeptide cross-linking, or (ii) other biomolecules, such as DNA and PUFA (Banerjee, 2008; Cattaruzza and Hecker, 2008).

8.3.2 PROTEIN RECOVERY, REPLACEMENT, OR REMOVAL

An oxidized protein can be repaired or degraded depending on the degree and type (reversible or irreversible) of chemical modification exerted (Friguet, 2006; Moller, Jensen, and Hansson, 2007). Some of the oxidations of cysteine and methionine can be reversed (Banerjee, 2008). In this way, the thioredoxin system is effective for the reduction of both disulfide bridges and cysteine sulfenic acid (Schurmann and Jacquot, 2000; Smirnoff, 2005), whereas glutathione-dependent glutaredoxins reduce both disulfide bridges and low–molecular weight mixed disulfides, such as glutathione adducts (Smirnoff, 2005). Some examples show that the oxidation of methionine residues could contribute to the impairment of protein function, including the loss of activity in enzymes (Moller, Jensen, and Hansson, 2007). In some cases, the reversal of such oxidation by MSRA or MSRB allows recovery of functionality (Boschi-Muller et al., 2005; Banerjee, 2008; Tarrago, Laugier, and Rey, 2009; Rey and Tarrago, 2018; Ruiz-May et al., 2018). MSR enzymes have been described as a very important repair system for oxidized proteins, thus contributing to the regulation of cellular redox homeostasis (Moskovitz, 2005; Tarrago, Laugier, and Rey, 2009).

On the other hand, irreversible oxidation products are derived from amino acids that are frequently hydroxylated and carbonylated (Banerjee, 2008). Oxidized proteins are generally less active and less thermostable, with surface exposure of hydrophobic amino acids (Moller, Jensen, and Hansson, 2007). Protein damage can also occur due to the formation of adducts with peroxidized lipid derivatives (Moller, Jensen, and Hansson, 2007) or via oxidation by glycation (also known as *non-enzymatic glycosylation*, which is the result of covalent binding of a sugar [such as fructose or glucose] to a protein or lipid molecule without the controlling action of an enzyme). The latter processes lead to the formation of glycosylic adducts such as pentosidine or carboxy-methyl-lysine (McPherson, Shilton, and Walton, 1988; Ruiz-May et al., 2018).

Oxidized proteins are degraded relatively quickly, probably because a change in conformation exposes more hydrophobic residues, which are better recognized by proteases (Starke-Reed and Oliver, 1989). However, massive protein damage can generate toxic aggregates, which are not only resistant to proteolysis but able to inhibit proteases that degrade other oxidized proteins (Costa, Quintanilha, and Moradas-Ferreira, 2007). The accumulation of oxidized protein aggregates has been proposed to be a consequence of increased oxidative damage and/or decreased degradation and repair, being a hallmark of cellular aging (Friguet, 2006). The proteasome has been described as the main proteolytic pathway implicated in the degradation of oxidized polypeptides and the general turnover of proteins operating in the cytosol and the nucleus (Moller, Jensen, and Hansson, 2007). Also, the Lon protease (a family of ATP-dependent proteases) has been shown to selectively degrade oxidized proteins within the mitochondrial matrix (Moller, Jensen, and Hansson, 2007; Friguet, Bulteau, and Petropoulos, 2008; Ruiz-May et al., 2018).

8.3.3 Proteins Involved in Perception of Reactive Species

Proteins participating in the perception of reactive species and signal transduction have been characterized as polypeptides exhibiting high sensitivity to reversible oxidation processes, such as thiol/disulfide exchange (Kiley and Storz, 2004; Bindoli, Fukuto, and Forman, 2008). The reversible characteristics exhibited by most of the oxidized forms have led to the suggestion that thiol modification could be functional for signal transduction (by analogy to protein phosphorylation/dephosphorylation; Bindoli, Fukuto, and Forman, 2008). There are several examples of proteins whose activities are modulated by thiol oxidation and reduction, but sensors must have some specific characteristic enabling them to propagate this signal. A classical example is the OxyR transcription factor (Wang et al., 2008), which up-regulates peroxide defenses in *Escherichia coli* and other bacteria. Two critical cysteines in OxyR are oxidized to form an intramolecular disulfide bridge in cells exposed to peroxide stress (Kiley and Storz, 2004). Disulfide bond formation is associated with a conformational change that affects binding to DNA, allowing OxyR to activate the transcription of genes encoding for catalase, alkylhydroperoxide reductase, and other enzymes that decompose H_2O_2. Once levels of H_2O_2 are normalized, OxyR is reduced to reset the system (Banerjee, 2008).

Many biotic and abiotic factors can induce the generation of different ROS. Therefore, it might be expected that perception of the oxidative signal or stress would involve diverse cellular components. The knowledge (at the molecular level) of oxidative signal transduction pathways (including perception) in plants is far from complete. Still, some clues have been established following studies about redox-regulated proteins and proteomic analysis. It is known that exogenous H_2O_2 activates oxidative signal-inducible I kinase (OXI1) and mitogen-activated protein kinases (MAPK), but little is known about proteins that directly react with H_2O_2 (Hancock, 1997; Van Ruyskensvelde, Van Breusegem, and Van Der Kelen, 2018). A proteomic approach with iodoacetamide-based fluorescence tagging of proteins in conjunction with mass spectrometric analysis was used to identify potential targets of H_2O_2 in the cytosol of *Arabidopsis thaliana*. The cytosolic

phosphorylating glyceraldehyde-3-phosphate dehydrogenase (Ga3PDHase, EC 1.2.1.12) was identified, by this procedure, as the most prominent modified protein.

Studies *in vitro* on redox regulation of *Arabidopsis* Ga3PDHase have shown that it is inactivated after oxidation with H_2O_2, oxidized glutathione (GSSG), and GSNO; the oxidation is reversed by dithiothreitol (DTT) and GSH (Hancock et al., 2005; Zaffagnini et al., 2013). We have kinetically characterized the oxidation of Ga3PDHase from wheat (*Triticum aestivum*) with different reactive oxidant species. Interestingly, oxidation of the enzyme with H_2O_2 was significantly higher than with sodium nitroprusside or GSSG (Piattoni, Guerrero, and Iglesias, 2013). On the other hand, important intracellular reducing systems, such as those involving thioredoxin or glutaredoxin, were effective for reversing damage caused by oxidants to Ga3PDHase. The possibility of reversal was limited by the degree of enzyme oxidation, because extensive oxidative damage caused the irreversible formation of protein aggregates. The enzyme is also a target of post-translational modification by phosphorylation, after which it is inactivated and degraded (Piattoni et al., 2017).

After proteomic studies, it was proposed that Ga3PDHase could be involved in H_2O_2 perception (Hancock et al., 2005; Zaffagnini et al., 2013), which is in agreement with experiments developed in our laboratory suggesting a physiological redox regulation (thus, by sensing of H_2O_2) of Ga3PDHase. Recently (Schneider et al., 2018), it has been demonstrated that in *A. thaliana*, Ga3PDHase displays redox-dependent changes in subcellular localizations (cytosol, mitochondria, and nucleus) and activity, exhibiting moonlighting properties. Current information reported in plants accords with that from animals, in which it was demonstrated that Ga3PDHase is a multi-functional protein, with many roles besides its enzymatic activity in glycolysis (Schneider et al., 2018). Effectively, the enzyme participates in nuclear events, including gene transcription, RNA transport, DNA replication, and initiation of apoptotic cell death on nitrosylation (Zaffagnini et al., 2013).

Even when irreversible protein oxidation damages structure and function, recent reports have demonstrated that the induced structural changes confer, on certain polypeptides, the capacity to act as oxidative stress sensors (Dietz et al., 2006; Liebthal, Maynard, and Dietz, 2018). In this way, oxidized proteins become involved in many signal transduction routes (for example, apoptosis activation), or they display "non-conventional" enzymatic activities (now acting as chaperones or transcriptional factors). An example of the latter (found in plants, mammals, and some bacteria) is the over-oxidation of 2-Cys typical peroxiredoxin (2-CysPRX) (Smirnoff, 2005) (2-CysPRX are peroxidases lacking prosthetic groups that mediate in the defense against oxidative stress by reducing H_2O_2 and alkyl hydroperoxides.) This enzymatic activity relies on the high reactivity of two conserved cysteines, whose modification entails the usual thiol/disulfide exchange and further irreversible oxidation to sulfinic or sulfonic acid (Banerjee, 2008; Hall, Karplus, and Poole, 2009; Liebthal, Maynard and Dietz, 2018). These changes may induce other post-translational modifications, such as phosphorylation and acetylation, which contribute to the formation of high–molecular mass quaternary structures with chaperone-like activity (Aran et al., 2009; Barranco-Medina, Lazaro, and Dietz, 2009; Liebthal, Maynard, and Dietz, 2018). The latter structures further protect from thermal denaturation and act as protein activity modulators (Barranco-Medina, Lazaro, and Dietz, 2009; Hall, Karplus, and Poole, 2009).

8.4 SYSTEMS FOR REDUCING POWER GENERATION IN PLANTS

Inside cells, the reducing power required for all detoxifying and regenerative systems comes, directly or indirectly, from either NADPH or ferredoxin. Plants exhibit key characteristics: i) cells are highly compartmentalized, with the presence of plastids; ii) two different kinds of tissues exist, performing either photosynthesis or heterotrophic metabolism. The routes operating to produce NAPDH in a plant cell are shown schematically in Figure 8.3.

8.4.1 GENERATION OF NADPH IN PLASTIDS

In plants, we find different types of plastids: (i) chloroplasts, found in photosynthetic tissues; (ii) amyloplasts, leucoplasts, and chromoplasts, present in non-photosynthetic tissues (Piattoni et al., 2011). The diversity among plastids includes specialized functions and metabolic capacities. Furthermore, chloroplast metabolism differs during light and dark periods. The production of NADPH in plastids is associated with either photoreduction of $NADP^+$ during the day, by the photosynthetic electron flow, or generation by the oxidative pentose phosphate pathway (OPPP) during the night or in non-photosynthetic tissues.

In the light, chloroplasts actively perform photosynthesis, maintaining high NADPH as one key metabolite. Using solar energy, the photosynthetic electron transport chain transfers electrons from water to $NADP^+$ ($E^{\circ\prime}_{O2/H2O} = 815$ mV and $E^{\circ\prime}_{NADP+/NADPH} = -340$ mV, $E^{\circ\prime}$ being the midpoint oxidation-reduction potential relative to the standard hydrogen electrode at pH 7.0). The process is driven by two successive photochemical reactions involving two photosystems: photosystem II (PSII) and photosystem I (PSI). Both photosystems are connected via an electron transport chain composed of the cytochrome b_6f complex, an integral membrane component. Finally, reduced ferredoxin (Fd) reduces $NADP^+$ to NADPH, catalyzed by the enzyme ferredoxin-$NADP^+$ oxidoreductase. Electron transfer from water to $NADP^+$ is coupled to proton translocation across the thylakoid membrane, which generates the electrochemical potential gradient thermodynamically allowing the synthesis of ATP (Ziegler, 2000).

At night (as well in non-photosynthetic tissues), NADPH inside plastids is generated mainly by the OPPP, with oxidation, interconversion, and rearrangement of sugars–phosphate pools (Golowatzki, 1962). Two phases are distinguished: (i) one oxidative (phase 1), where in two consecutive reactions, glucose-6-phosphate (Glc6P) is converted to ribulose-5-phosphate (Rub5P) with the net production of NADPH, and (ii)

one non-oxidative (phase 2), involving the interconversion of pentose-P, hexose-P, and triose-P intermediaries to regenerate Glc6P in the cycle and connect it with glycolysis. In contrast to animal and yeast cells, plants have two operative OPPPs: one complete pathway inside the chloroplast and another incomplete one in the cytosol (Anderson and Advani, 1970; Bailey-Serres and Nguyen, 1992; Eicks et al., 2002; Kruger and von Schaewen, 2003). Despite the physical separation, both OPPPs are interconnected by specific solute transporters found in the chloroplast envelope (Eicks et al., 2002; Weber, Schwacke, and Flugge, 2005). The oxidation of Glc6P by one specific dehydrogenase (Glc6PDHase, EC 1.1.1.49) to form 6-phosphoglucono-δ-lactone and NADPH is the first reaction of phase 1. The lactone is hydrolyzed by a specific lactonase to yield 6-phosphogluconate (6PG), which then is oxidized and decarboxylated by 6PGDHase (EC 1.1.1.44), producing Rub5P and NADPH (Schnarrenberger, Tetour, and Herbert, 1975). Phase 2 includes a reversible set of interconversions between phosphorylated 3-, 4-, 5-, 6-, and 7-carbon sugars, mediated by ribose-5-P isomerase (Rib5P isomerase), Rub5P 3-epimerase, transaldolase (TA), and transketolase (TK) (Schnarrenberger, Tetour, and Herbert, 1975; Debnam et al., 1997).

Reversible enzyme reactions (except that involving TA) of the OPPP are amphibolic and shared with the Benson–Calvin cycle (Sharkey, 2018; Scholl and Nickelsen, 2015). Glc6PDHase is a strategic control point to regulate the Glc6P partitioning between glycolysis and the OPPP. Detailed studies evidenced that the activity of chloroplastidic Glc6PDHase is under coarse and fine regulatory control. The enzyme is sensitive to the NADPH:$NADP^+$ ratio, pH, Mg^{2+}, and levels of Glc6P . Also, a thioredoxin-dependent redox mechanism activates Glc6PDHase in the dark (when the NADPH:$NADP^+$ is low) (Buchanan and Balmer, 2005; Jablonsky, Bauwe, and Wolkenhauer, 2011). Several chloroplast enzymes have been reported to be regulated by reversible thioldisulfide interchange (Michelet et al., 2013). During the light period, covalent redox modification mediated by the ferredoxin–thioredoxin system leads to reductive activation of several stromal target enzymes: fructose-1,6-bisphosphatase, NADP-malate dehydrogenase, phosphoribulokinase, and NADP-dependent Ga3PDHase, among others. Chloroplastidic Glc6PDHase is regulated in an opposite way by the same system, the enzyme being inactive in its reduced state (light period) and inactive when oxidized (in the dark) (Buchanan and Balmer, 2005; Michelet et al., 2013). This opposite regulation prevents futile cycles, such as simultaneous carbohydrate synthesis in the Benson–Calvin cycle and catabolism by the OPPP. Thus, in accordance with its physiological role in chloroplasts, Glc6PDHase is active only at night, when photosynthetic generation of NADPH turns off (Buchanan and Balmer, 2005; Michelet et al., 2013).

8.4.2 GENERATION OF NADPH IN THE CYTOSOL OF PLANT CELLS

In the cytosol of plant cells, the production of NADPH exhibits a higher degree of complexity than in animals and yeasts; the extent to which the OPPP operates in the former is questionable. Both enzymes of phase 1, Glc6PDHase (Michelet et al., 2013) and 6PGDHase (Bailey-Serres and Nguyen, 1992), were identified in the cytosol of many plant cells; but the organization of the reversible, non-oxidative phase 2 of the pathway is far less clear. Some studies support the absence of Rub5P 3-epimerase, Rib5P isomerase, TK, and TA in the cytosol of spinach, pea, and maize leaf cells (Michelet et al., 2013; Sharkey, 2018). Similar results were reported after analysis of the complete genome from *A. thaliana* in search of the predicted subcellular localization of multiple copies of genes encoding for enzymes of the OPPP (Kruger and von Schaewen, 2003). Such an analysis identified putative cytosolic and plastidic forms of Glc6PDHase, 6PGDHase, Rib5P isomerase, and Rub5P 3-epimerase; whereas only plastidic localization was assigned for TK and TA. In other words, *Arabidopsis* has the genetic capacity to convert cytosolic Rub5P to Rib5P and Xyl5P, but any further rearrangement of the carbon backbone to regenerate fructose-6-P and triose-P only occurs within plastids. Similar results were reported by the group of M. J. Emes when they analyzed the subcellular distribution of OPPP in *Brassica napus* embryos by measuring activity of the different enzymes (Hutchings, Rawsthorne, and Emes, 2005). In the light of this information, the extent to which the OPPP is operative in the cytosol of plant cells is doubtful (Siqueira et al., 2018). To further use cytosolic pentose-P by phase 2 of the OPPP, transport of the metabolite across the plastid envelope would be required. Related to this, a phosphate translocator that preferentially mediates the counter-exchange of Xul5P, triose-P, and Pi has been identified in the inner membrane of the chloroplast envelope (Eicks et al., 2002).

As in plastids, Glc6PDHase seems to catalyze the key regulatory step in the cytosolic OPPP, although the characterization of the regulatory properties for this cytosolic enzyme is poor. An early report indicated that both chloroplastidic and cytosolic Glc6PDHases are inactivated in response to light and DTT (Anderson, Ng, and Park, 1974). More recent works detailed that the two cytosolic isoforms of G6PDHase exhibit differences in modulation, with one being slightly activated by reductants and the other one exhibiting insensitivity to redox signaling (Michelet et al., 2013; Considine and Foyer, 2014; Knuesting and Scheibe, 2018). Studies with *Arabidopsis* single and double null mutants of cytosolic Glc6PDHases revealed the occurrence of alternative, compensatory mechanisms to supply NADPH in the cytosol (Wakao, Andre, and Benning, 2008). Since the double mutant contained increased seed oil and mass, it was speculated that metabolic rearrangements may make possible an increase in the flux of carbon through glycolysis.

In addition to the incomplete OPPP, other enzymes are involved in the generation of NADPH in the cytosol of plant cells. A particular case is non-phosphorylating glyceraldehyde-3-phosphate dehydrogenase (np-Ga3PDHase; EC 1.2.1.9), which catalyzes the irreversible oxidation of Ga3P to 3-phosphoglycerate (3-PGA), specifically using NADP$^+$ and generating NADPH according to the reaction (Iglesias and Losada, 1988; Piattoni et al., 2011).

The np-Ga3PDHase has been identified as a member of the aldehyde dehydrogenase superfamily (Habenicht, 1997), and its presence is restricted to some specialized eubacteria (Habenicht, 1997), archaebacteria, and the cytosol of green algae and higher plants (Iglesias and Losada, 1988; Bustos and Iglesias, 2002; Iglesias and Podestá, 2005). The enzyme is coded by a single gene (*gapN*) and arranged in a tetrameric structure of about 200 kDa (Gomez Casati, Sesma, and Iglesias, 2000; Bustos and Iglesias, 2002). In photosynthetic cells, np-Ga3PDHase is involved in a shuttle system for the indirect export of photosynthetically generated NADPH from the chloroplast to the cytosol (Gomez Casati, Sesma, and Iglesias, 2000; Piattoni, Guerrero, and Iglesias, 2013; Piattoni et al., 2017), as detailed in Figure 8.3. This transport involves the triose-P/Pi translocator of the chloroplast envelope. On the other hand, the enzyme plays a key role in plants accumulating acyclic polyols (such as in celery) by supplying the NADPH necessary for the synthesis of the reduced sugars (Piattoni, Guerrero, and Iglesias, 2013; Piattoni et al., 2017). In non-green cells, the enzyme would couple NADPH production necessary for anabolism with glycolysis (Habenicht, 1997).

The presence of np-Ga3PDHase in the cytosol of plant cells constitutes an alternative to a glycolytic step, whereby Ga3P can be metabolized to 3-PGA either by a couple of phosphorylating, NAD-dependent Ga3PDHase (EC 1.2.1.12) and 3-PGA kinase (EC 2.7.2.3) or by np-Ga3PDHase (Plaxton, 1996; Zaffagnini et al., 2013). The first alternative generates NADH and ATP, whereas the second route produces NADPH but not ATP (Iglesias and Losada, 1988). This branch point in glycolysis critically determines the production of energetic or reductive power within the cell, which indicates that it should be under regulation (Plaxton, 1996). In fact, it has been reported that np-Ga3PDHase is a target for post-translational regulation by phosphorylation in non-photosynthetic plant cells (Bustos and Iglesias, 2002). The phosphorylated enzyme exhibits distinctive kinetic properties after interaction with 14–3-3 regulatory proteins (Bustos and Iglesias, 2006; 2005). Phosphorylation is mediated by an SnF1-related protein kinase, which modifies serine-404 of np-Ga3PDHase in wheat endosperm (Piattoni et al., 2011).

It has been evidenced that np-Ga3PDHase plays a pivotal role in carbon and energy metabolism in plants under both physiological and oxidative stress conditions (Bustos, Bustamante, and Iglesias, 2008; Piattoni, Guerrero, and Iglesias, 2013). An *Arabidopsis* mutant lacking np-Ga3PDHase was characterized as having altered morphology of the siliques, inhibited glycolytic flux, reduced CO_2 fixation capacity, and enhanced oxidative stress (Rius et al., 2006). Besides, the absence of np-Ga3PDHase elicited induction of Glc6PDHase activity, which produced an increase in levels of NADPH. When these results are analyzed together with those obtained with *Arabidopsis* mutants lacking cytosolic Glc6PDHases (described earlier), it potentially confirms the hypothesis that plants compensate for the disruption of one pathway generating NADPH in the cytosol by boosting enzymes involved in alternative routes operating within this cell compartment.

The non-photosynthetic isoform of NADP-malic enzyme (NADP-ME; L-malate:NADP oxidoreductase [oxaloacetate decarboxylating], EC 1.1.1.40) is another enzyme able to produce NADPH in the cytosol of plant cells. NADP-ME is an enzyme widely distributed and involved in different metabolic pathways. It catalyzes the oxidative decarboxylation of L-malate to yield pyruvate, CO_2, and NADPH in the presence of a divalent cation (Drincovich, Casati, and Andreo, 2001). Cytosolic isoforms of NADP-ME have been linked to plant defense responses and to lignin biosynthesis by providing NADPH, as well as to the control of pH by balancing the synthesis and degradation of malate (Drincovich, Casati, and Andreo, 2001). NADP-ME present in the cytosol of non-photosynthetic plant cells has been characterized with respect to kinetics, regulation, and function (Drincovich, Casati, and Andreo, 2001). Other studies report that genetically transformed *Arabidopsis* plants overexpressing rice cytosolic NADP-ME exhibit increased levels in the cytosolic NADPH:NADP+ ratio associated with salt tolerance (Weber, Schwacke, and Flugge, 2005).

With the background detailed here, it could be concluded that NADPH production in the cytosol of plant cells is not strictly dependent on a particular (single) system (Figure 8.3). Rather, it can be pictured as an integrated metabolic network, where different primary metabolic routes interact to guarantee (and maintain under control) the supply of reducing power demands of the cell and/or to respond to oxidative stress conditions.

8.5 SYSTEMS SCAVENGING OXIDATIVE SPECIES IN PLANT CELLS

8.5.1 ANTIOXIDANT MOLECULES AND REDOX COFACTORS

8.5.1.1 Glutathione

The tripeptide γ-glutamyl-cysteinyl-glycine, named *reduced glutathione* (GSH), interchanges with its disulfide oxidized form, GSSG, which critically affects normal cellular function. The standard reduction potential ($E^{\circ\prime}$ of −240 mV) of the GSSG/GSH couple determines that GSH is a moderate reducing agent (Banerjee, 2008). Levels of GSH are maintained by the action of glutathione reductase (GR), a pyridine nucleotide disulfide reductase that catalyzes reduction equivalents from NADPH to GSSG (Smirnoff, 2005). GSH is synthesized *de novo* in two ATP-dependent steps mediated by glutamate cysteine ligase and γ-glutamyltranspeptidase. Certain plants contain tripeptide homologues of GSH, which differ by having another amino acid instead of glycine at the carboxy-terminus (Noctor and Foyer, 1998). GSH is primarily found in eukaryotes and Gram-negative bacteria, although its use has also been characterized in a select number of Gram-positive prokaryotes (Banerjee, 2008). Approximately 90% of the intracellular GSH pool resides in the cytoplasm, with the remainder localized in organelles (mitochondria, endoplasmic reticulum, and nucleus) (Noctor and Foyer, 1998). In plant cells, GSH is synthesized in the cytosol and chloroplasts (Smirnoff, 2005; Diaz-Vivancos et al., 2015).

GSH is the major non-protein thiol in plants, and its physiological relevance encompasses sulfur metabolism and antioxidant defense (Scheibe, 1991; Smirnoff, 2005). As a product of primary sulfur metabolism, GSH constitutes a form of transport and storage of reduced sulfur and plays a key role as antioxidant and redox buffer (Buchanan and Balmer, 2005; Banerjee, 2008; Hernandez et al., 2015). Besides, GSH is relevant for phytochelatin synthesis; which is critical for the detoxification of heavy metals (mainly cadmium and nickel) (Noctor and Foyer, 1998; Abhilash, Jamil, and Singh, 2009; Hernandez et al., 2015). GSH is a substrate of many enzymes, most notably glutathione-S-transferases (GST) (Smirnoff, 2005; Kulinskii and Kolesnichenko, 2009), glutathione peroxidase (GPX) (Dietz et al., 2006), and phospholipids hydroperoxide glutathione peroxidase (PHGPX) (Smirnoff, 2005), involved in the detoxification of xenobiotics and the reduction of H_2O_2 and LOOH, respectively. Also, GSH acts as an electron donor for glutaredoxins (GRX) in the mixed-disulfides reduction of proteins (Lillig, Berndt, and Holmgren, 2008).

8.5.1.2 Ascorbate

Ascorbate (L-*threo*-hex-2-enono-1,4-lactone, or vitamin C) is well known for its radical-scavenging capacity and in terms of small, water-soluble molecules, is the major antioxidant in plants (Smirnoff, 2005). The moderately positive standard redox potential of $E^{\circ\prime} = 280$ mV makes of the ascorbate/monodehydroascorbate couple a highly operative one-electron donor to many enzymes, in particular oxygenases and hydroxylases (Noctor and Foyer, 1998). At physiological pH, ascorbic acid (AA; $pK_1 = 4.2$ and $pK_2 = 11.8$) is predominantly in the ascorbate anion state (Banerjee, 2008). The latter anion readily loses an electron from its ene-diol group to produce the monodehydroascorbate (MDHA) radical and further oxidizes to dehydroascorbate (DHA) (Banerjee, 2008). The conjugated structure of the free-radical, one-electron MDHA (with a five-atom lactone ring containing an ene-diol group) stabilizes after delocalization of the unpaired electron (Noctor and Foyer, 1998). The capacity of ascorbate as a one-electron donor plus the relatively low reactivity of MDHA radical sustains the biological utility of this system as an antioxidant and free radical scavenger. Oddly enough, ascorbate can also function as a pro-oxidant by reducing metal ions (Cu^{2+} and Fe^{3+}). The Fe^{2+} ion thus formed can then catalyze generation of OH from H_2O_2 by the Fenton reaction (see Reaction 2.1) (Halliwell, 2006; Banerjee, 2008).

With exception of guinea pigs and primates, all animals and plants can synthesize AA (Noctor and Foyer, 1998). In plants, ascorbate can reach millimolar concentrations in both photosynthetic and non-photosynthetic tissues. Synthesis occurs in two steps: oxidation of the sugar L-galactose via L-galactono-1,4-lactone in the Smirnoff–Wheeler–Running pathway, which involves the action of NAD-dependent L-galactose dehydrogenase and L-galactono-1,4-lactone dehydrogenase (Smirnoff and Wheeler, 2000; Linster and Clarke, 2008). L-galactose is available as GDP-L-galactose after epimerization from GDP-mannose (Noctor and Foyer,

1998; Smirnoff and Wheeler, 2000). Ascorbate catabolism converts L-galactose into oxalate, tartrate, and threonine, albeit the enzymes involved in cleavage of the ascorbate C-skeleton to generate such products have not been identified (Noctor and Foyer, 1998).

In plant cells, ascorbate is recycled by direct reduction of DHA by two molecules of GSH (Figure 8.3), a thermodynamically feasible reaction that has been identified in cell-free systems (Noctor and Foyer, 1998; Banerjee, 2008; Meyer, 2008). Plants also contain proteins with GSH-dependent DHA reductase activity (Smirnoff and Pallanca, 1996; Meyer, 2008) and two enzymes that catalyze ascorbate oxidation: ascorbate peroxidase (APX) (Smirnoff, 2005) and ascorbate oxidase (AO) (Dawson, Strothkamp, and Krul, 1975). APX catalyzes the ascorbate-dependent reduction of H_2O_2 to H_2O (Shigeoka et al., 2002), playing a central role associated with the antioxidant function of ascorbate in plants (Dawson, Strothkamp, and Krul, 1975). Concerning AO, the enzyme is a member of the blue copper oxidase family and a glycoprotein localized at the apoplastic level (Pignocchi and Foyer, 2003); its activity is highest in the growing tissues of cucurbit fruits (Arrigoni, 1994).

8.5.1.3 α-Tocopherol

α-Tocopherol is a potent lipid-soluble antioxidant synthesized exclusively by plants, constituting the major biologically active form (out of eight) of vitamin E (Banerjee, 2008). Seeds are rich in this antioxidant, which is linked with triacylglycerols in the oil bodies and probably in glyoxysomes, where they are oxidized during germination. In addition, tocopherol is highly abundant in thylakoid membranes containing PUFA, where it is in close proximity with ROS produced during photosynthesis (Zingg, 2007). The hydrophobic phytyl side chain makes α-tocopherol lipid soluble, allowing the antioxidant to be anchored to the membrane (Smirnoff and Wheeler, 2000; Miret and Munne-Bosch, 2015).

The antioxidant capacity of α-tocopherol comes from its capacity to quench free radicals by forming a stable radical itself (Zingg, 2007). The chromanol ring can donate a single electron with the generation of a resonance-stabilized tocopheroxyl (also named chromanoxyl) radical, which supports its capacity to avoid lipid peroxidation chain reactions. The reaction between lipid peroxyl radicals and tocopherol produces hydroperoxides, which may be reduced by PHGPX (Shao et al., 2008). Ubiquinol, AA, and (indirectly) dihydrolipoic acid are antioxidants able to reduce the tocopheroxyl radical (Banerjee, 2008). Furthermore, a putative non-antioxidant function has been attributed to α-tocopherol, whereby it stabilizes the cellular membrane structure after interaction with PUFA (Smirnoff, 2005; Miret and Munne-Bosch, 2015).

8.5.1.4 Carotenoids

As the most abundant pigmented plant-derived compounds, carotenoids impart red and yellow colors to fruits and vegetables (Lu and Li, 2008). These compounds are isoprenoids, consisting primarily of a structure of eight isoprene units, with 3 to 15 conjugated double bonds in a rigid backbone, which confer antioxidant properties. They are able to undergo some chemical modifications after cyclization of the carbon skeleton at one or both ends (Britton, 1989; Banerjee, 2008). In plants, carotenoids are essential, being directly involved in photosynthesis and reactions scavenging the vast amounts of ROS produced in chloroplasts (Lu and Li, 2008). Examples are as follows: (i) oxygen-containing carotenoids (xanthophylls) are integral components of PSII and the light-harvesting complex (Banerjee, 2008); (ii) zeaxanthin is implicated in non-photochemical quenching of excitation energy in PSII, in which excess energy in PSII is transferred to the carotenoid and re-radiated as heat (Ivanov et al., 2008); and (iii) several carotenoids participate in defense against photoinduced damage (Smirnoff, 2005). Both α-tocopherol and ascorbic acid can reduce carotenoid radicals, which supports the synergistic interactions reported to occur between these three important antioxidants (Smirnoff, 2005; Banerjee, 2008).

8.5.1.5 Flavonoids

Flavonoids are a group of polyphenolic compounds produced by secondary metabolism in plants (Banerjee, 2008). The basic structure of plant-derived flavonoids shows two aromatic benzene rings linked through three carbons that can form an oxygenated heterocycle; quercetin is the most abundant flavonoid of this class (Smirnoff, 2005). Other examples are (i) isoflavones (phytoestrogens), of high abundance in legumes (Veitch, 2009); (ii) hesperidin (a glycoside of the flavanone hesperetin), quercitrin, rutin (two glycosides of the flavonol quercetin), and the flavone tangeritin found in citrus (Gattuso et al., 2007); (iii) kaempferol and catechins (catechin, epicatechin, epicatechin gallate, and epigallocatechin gallate) as main components in green tea (Butt and Sultan, 2009); and (iv) resveratrol, present in grape skin, mainly the red variety (Dohadwala and Vita, 2009). The chelation of redox-active metals is the principal antioxidant activity of flavonoids in preventing peroxyl radical and lipid peroxidation, scavenging hydroxyl and peroxyl radicals, and quenching superoxide radicals and singlet oxygen (Banerjee, 2008).

8.5.1.6 NAD(P)⁺

In all living organisms, nicotinamide adenine dinucleotide (NAD^+) and its phosphorylated derivative ($NADP^+$) are the universal energy carriers performing reversible two-electron (specifically one hydride) transfer in a variety of essential metabolic reactions (Pollak, Dolle, and Ziegler, 2007). Worthy of mention is a recent forum editorial in a prestigious journal in this area (Fessel and Oldham, 2018a), which refers to these dinucleotides as the redox currency of the cell, which has gained explosive research interest in recent years (see also Fessel and Oldham (2018b). The coenzyme NAD^+ (the oxidized form) participates mainly in oxidative reactions (for example, in glycolysis) by accepting electrons from energy-rich substrates. Alternatively, NADPH (the reduced form) is the main electron donor in reductive biosynthetic reactions and has an important role in oxidant/antioxidant systems in plant cells, being a cofactor of NADPH-dependent enzymes such as NADPH oxidase and GR (Banerjee, 2008). The $E^{o'}$ for the two-electron reduction of the $NAD^+/NADH$ and $NADP^+/NADPH$ couples are similar (about -320 mV) (Schafer and Buettner, 2001). It is worth mentioning that NAD^+ also participates in reactions that are not related to redox metabolism. This is the case when NAD^+ is involved in (i) protein modification processes, such as poly-ADP-ribosylation of transcription factors in the nucleus (Ziegler, 2000); (ii) signaling pathways, as a precursor of the second messenger cyclic-ADP-ribose (Guse, 2004); and (iii) reactions catalyzed by sirtuins, a group of enzymes that use the dinucleotide as a substrate to remove acetyl groups from histones (in the regulation of transcription, apoptosis, and stress resistance) (Blander and Guarente, 2004; Trapp and Jung, 2006).

Two metabolic pathways in plants synthesize NAD^+. It is produced either *de novo*, from tryptophan or aspartic acid, or in salvage pathways, by recycling preformed components such as nicotinamide (Katoh et al., 2006; Banerjee, 2008). $NADP^+$ is the product of phosphorylation of the hydroxyl moiety in the $2'$ position of the ribose ring in NAD^+ (Banerjee, 2008). In chloroplasts, $NADP^+$ is reduced by ferredoxin-$NADP^+$ reductase in the last step of the photosynthetic electron chain. The photogenerated NADPH is then used as reducing power for carbohydrate biosynthesis in the Benson–Calvin cycle (Smirnoff, 2005). The balance between oxidized and reduced forms is the $NAD(P)^+/NAD(P)H$ ratio, which is a key component of what is named the *redox state* of a cell, a measurement reflecting the metabolic activities and health of cells (Schafer and Buettner, 2001).

8.5.1.7 Flavins

Flavin mononucleotide (FMN) and flavin adenine dinucleotide (FAD) partake as non-covalently or covalently bound redox cofactors in a series of enzyme-catalyzed reactions. A characteristic of flavins is the ability to couple one- and two-electron transfer between substrates and different electron carriers. Hence, flavins are able to equilibrate the reaction between quinone (oxidized), semiquinone (one-electron reduced), and hydroquinone (two-electron reduced) species with reversible electron transfer occurring across the isoalloxazine ring. The $E^{\circ\prime}$ of free FAD is -219 mV in solution (Banerjee, 2008), whereas it ranges between $+100$ and -400 mV for enzyme-bound flavin (Schafer and Buettner, 2001). The latter contributes to the involvement of many oxidoreductases (called *flavoenzymes* or *flavoproteins*) in a large diversity of reactions, namely dehydrogenation, electron transfer, dehalogenation, hydroxylation, luminescence, DNA repair, and disulfide reduction (Rouhier, Gelhaye, and Jacquot, 2002; Smirnoff, 2005; Banerjee, 2008).

The environment of the active site and the substrate/product complexation affect the reactivity of enzyme-bound reduced flavin with O_2^- (Miura, 2001). Such a reaction proceeds initially through a one-electron reduction of oxygen to generate a flavin semiquinone and O_2^- pair, which is followed by events leading to the formation of H_2O_2 (Miura, 2001). The capacity of reduced flavins to carry out the one-electron reduction of O_2 to O_2^- involves them in intracellular oxidative stress and pathogen defense (Yoshioka et al., 2009). Also,

some flavoenzymes participate in regulatory and signaling pathways; as in the example of the apoptotic inducing factor, which is a mitochondrial flavoenzyme with NADH oxidase and DNA-binding activities (Fleury, Mignotte, and Vayssiere, 2002; Le Bras et al., 2005). Another class of flavoenzymes includes the structurally related cryptochromes and DNA photolyases that are activated by blue light (Muller and Carell, 2009). Cryptochromes are implicated in photosensitive signaling pathways that set the circadian clock and regulate plant growth and development. Phototropins are another group of light-sensitive flavoproteins found in plants, which are concerned with adaptive responses to blue light, thus altering a wide variety of physiological processes (plant development, seed germination, and phototropism, among others) (Takagi, 2003; Demarsy and Fankhauser, 2009).

8.5.2 Antioxidant Enzymes

8.5.2.1 Thioredoxin System

Thioredoxins (TRXs) are small redox proteins (usually around 12 kDa) having two cysteine residues in the active site within the conserved WCG/PPC motif. TRX can exist either in reduced (dithiol) or in oxidized (disulfide) form (Banerjee, 2008). TRX is functional as a thiol reductant due to the fast reaction between its reduced form and disulfide substrates. This latter is linked to the reductive power of NADPH and ferredoxin systems to reduce disulfides in target proteins (Smirnoff, 2005; Smiri and Missaoui, 2014; Sevilla et al., 2015). The redox potential of TRXs is critical for activity, with the value found for plant forms ranging between -285 and -350 mV (Buchanan and Balmer, 2005). Plants have an oddly high number of genes coding for TRXs implicated in photosynthetic regulation. For example, the *A. thaliana* genome encodes at least 22 TRX isoforms, which are classified into six groups with different localization: f, m, x and y in chloroplasts; o in mitochondria; and h in several cell compartments (cytosol, mitochondria, and endoplasmic reticulum) (Gelhaye et al., 2005; Montrichard et al., 2009; Sevilla et al., 2015; Hagglund et al., 2016). TRX can control the activity of enzymes, receptors, and transcription factors due to its protein disulfide reductase activity. Two examples are: (i) photosynthetic enzymes are controlled by specific TRXs via light; and (ii) malate dehydrogenase and fructose bisphosphatase are specific for targets of TRX_m and TRX_f, respectively (Smirnoff, 2005; Hagglund et al., 2016).

TRXs were identified as the first thiol–disulfide exchange proteins in plants (Gelhaye et al., 2005). They have a common tertiary protein structure (known as the *thioredoxin fold*), which has the same domain structure found in many other thiol proteins (GRX, GST, GPX, PRX, and protein disulfide isomerase) (Martin, 1995). Reduced TRX is able to directly reduce protein disulfides by fast thiol–disulfide interchange reactions (Bindoli, Fukuto, and Forman, 2008). First, reduced TRX non-covalently docks to a target protein through a hydrophobic interaction surface and hydrogen bonds. Then, nucleophilic attack on the thiolate of the N-terminal cysteine (contained in the CXXC active domain) on the target disulfide performs a thiol–disulfide exchange reaction to generate a transient protein–protein complex disulfide intermediate. Lastly, intramolecular attack on the C-terminal cysteine cleaves the disulfide, releasing oxidized TRX and the reduced target protein (Buchanan and Balmer, 2005; Banerjee, 2008; Sevilla et al., 2015).

As illustrated by Figure 8.3, TRXs are components of two redox systems found in different plant cell compartments (Smirnoff, 2005). The ferredoxin system is located in chloroplasts and composed of ferredoxin (Fd, an iron-sulfur protein), ferredoxin–thioredoxin reductase (FTR), and TRX f, m, x, and y (Schurmann and Jacquot, 2000; Gelhaye et al., 2005). In this system, electrons flow via thiol–disulfide exchange intermediates from Fd (reduced in the light via PSI) to target proteins according to the sequence Light → Fd → FTR → TRX → target protein (Scheibe, 1991; Buchanan and Balmer, 2005). The system is converted to its oxidized form in the dark by different agents: O_2, oxidized TRX, GSSG, or ROS (Montrichard et al., 2009). Accordingly, the chloroplast differs from the cytoplasm in undergoing changes from a reductive state in the light to a more oxidative one in the dark (Buchanan and Balmer, 2005). The FTR is the central enzyme of the ferredoxin system, able to transfer a redox signal received from Fd to TRX by a unique mechanism involving a 4Fe-4S cluster and a disulfide bridge. In this way, FTR transforms an "electron signal" to a "thiol signal" (Buchanan et al., 2002; Schurmann, 2003; Rouhier et al., 2015; Sevilla et al., 2015).

The NADPH-dependent TRX system is composed of an FAD-containing enzyme and an NADPH-dependent thioredoxin reductase (NTR, a low–molecular weight thioredoxin reductase) (Gelhaye et al., 2005). In *A. thaliana*, FTR exists as two forms (A and B, coded by two distinct genes) that have been localized in either the cytoplasm or the mitochondria (Meyer, Reichheld, and Vignols, 2005; Meyer et al., 2008). The enzymes transfer electrons from NADPH to TRX_h (in the cytoplasm) or to TRX_h and TRX_o in mitochondria (following a thiol–disulfide exchange mechanism) according to the sequence NADPH → NTR → TRX → target protein (Buchanan and Balmer, 2005; Gelhaye et al., 2005; Smirnoff, 2005). In addition, it has been reported that chloroplasts contain a modified type of NTR, named NTRC, which presents a TRX domain in the C-terminal extension (Kirchsteiger et al., 2009). NTRC functions as a complete NADPH-dependent TRX system, having the capacity to transfer reducing equivalents from NADPH to BAS1 (a plastidic 2CysPRX) (Moon et al., 2006; Alkhalfioui, Renard, and Montrichard, 2007; Hashida et al., 2018).

The number of potential functions of the TRX system remains to be confirmed in several cases. Apart from its demonstrated involvement in dark/light regulation of photosynthesis, this system participates in responses to environmental stresses (Gelhaye et al., 2005). Exemplifying the latter, it is known that the TRX system is involved in ROS detoxification through several PRX isoforms, and GPX and MetSO reduction by means of MSRA/B proteins (Montrichard et al., 2009; Rey and Tarrago, 2018) (see Figure 8.3).

8.5.2.2 Glutathione-Dependent System

GSH is a low–molecular weight component playing a key role in the protection of plant cells against oxidative stress (Smirnoff, 2005). As shown in Figure 8.3, in this process, GSH is converted to its oxidized product, GSSG (Noctor and Foyer, 1998). A specific flavoenzyme of the disulfide oxidoreductase family (GR) catalyzes the reduction of GSSG to GSH using NADPH as electron donor (Rouhier, Lemaire, and Jacquot, 2008). Most GRs are homodimeric proteins with the active site made up by both subunits (Rybus-Kalinowska et al., 2009). Three domains are identified in the enzyme subunit: the FAD, the $NADP^+$, and the interface domain. The GSSG binding site is a bridge between the interface domain of one subunit and the FAD domain of the adjacent subunit (Banerjee, 2008; Rybus-Kalinowska et al., 2009; Gill et al., 2013).

Alternatively, GSSG can produce glutathionylation or protein thiolation by reaction with protein thiols to give rise to protein–glutathione mixed disulfides (PSSG) (Dalle-Donne et al., 2007; Rouhier, Lemaire, and Jacquot, 2008). The reverse process, protein dethiolation, takes place in cells after the involvement of GSH-dependent enzymes named *glutaredoxins* (GRXs or thioltransferases) (Dalle-Donne et al., 2007; Stroher and Millar, 2012). Glutathionylation can either inactivate or activate several enzymes (Buchanan and Balmer, 2005), due to which this regulated and reversible process is a tool to modulate certain metabolic pathways and to exert cell signaling (Buchanan and Balmer, 2005). The glutathione system (formed by NADPH, GR, GSH, and GRX) often functions in parallel with the TRX system in regulating redox homeostasis in the cell (Gill et al., 2013).

GRX is a member of the thiol-disulfide oxidoreductase enzyme family and an important component of the GSH system (Lillig, Berndt, and Holmgren, 2008). It is a small (10 to 24 kDa) protein that catalyzes the reduction of proteins that have been thiolated by GSH (PSSG) (Meyer et al., 2008; Stroher and Millar, 2012). The reduction of PSSG is performed by GSH, which is oxidized to GSSG and converted back to GSH via the recycling system of NADPH and GR (this is an electron transfer path: NADPH → GR → GSH → GRX) (Smirnoff, 2005). GRX can reduce protein disulfides (PSSP) using a dithiol mechanism, and it is also able to reduce PSSG or low–molecular weight dithiols (GSSR) via either a monothiol or a dithiol mechanism (Lillig, Berndt, and Holmgren, 2008). Many (about 30) GRX isoforms have been found in *A. thaliana* (Meyer et al., 2008). According to their redox-active center, GRXs are divided into three main classes: CPYC, CGFS, and CCX[C/S] types (Rouhier, Gelhaye, and Jacquot, 2004; Stroher and Millar, 2012). CC-type GRXs are only found in higher plants (Meyer et al., 2008). The redox potential of GRXs ranges between −190 and −230 mV, and they act as electron carriers in the glutathione-dependent synthesis of deoxyribonucleotides by ribonucleotide reductase (Lillig, Berndt, and Holmgren, 2008). Additionally, GRXs participate in antioxidant defense, performing reduction of DHA (Holmgren and Aslund, 1995) and serving as electron donors for several PRXs (Meyer et al., 2008) as well for MSRB (Tarrago, Laugier, and Rey, 2009). Besides their function in antioxidant defense, GRXs are able to bind iron-sulfur clusters and to deliver the clusters to enzymes on demand (Rouhier, Lemaire, and Jacquot, 2008). In *Arabidopsis*, GRXs were found to be involved in flower development and salicylic acid signaling (Meyer et al., 2008; Rouhier, Lemaire, and Jacquot, 2008; Stroher and Millar, 2012).

8.5.2.3 Peroxiredoxin

Peroxiredoxins (PRXs) are a family of thiol-based antioxidant enzymes, ubiquitously found in nature, that catalyze the redox reaction

$$2RSH + ROOH \rightarrow RSSR + ROH + H_2O$$

Compounds (ROOH) that are thus reduced include H_2O_2, a wide variety of organic hydroperoxides (from *tert*-butyl hydroperoxide, cumene hydroperoxide, and fatty acid peroxides to complex phosphatidyl choline peroxides), and $ONOO^-$ at the expense of thiol substrates according to the equation (Banerjee, 2008; Sevilla et al., 2015). PRXs belong to the group of proteins that have the thioredoxin fold, having the active site Cys in a TPXC motif (Dietz, 2003; Sevilla et al., 2015). PRXs are involved not only in the detoxification of peroxides but also in plant-specific functions, as revealed by their involvement in photosynthesis, phloem metabolism, environmental stress, pathogen resistance, proliferation, differentiation, and apoptotic pathways through both known and unknown mechanisms (Dietz, 2003; Smirnoff, 2005; Liebthal, Maynard, and Dietz, 2018).

Mechanistically, three classes of PRXs are identified: typical 2-CysPRX, atypical 2-CysPRX, and 1-CysPRX (Rouhier and Jacquot, 2002; Poole, 2007). These proteins present the same basic catalytic mechanism, whereby a redox-active cysteine (the peroxidatic cysteine or $R-S_PH$) in the active site is oxidized to a sulfenic acid ($R-S_pOH$) by the oxidant substrate. Regeneration of the oxidized cysteine differentiates three mechanistic classes. Typical and atypical 2-CysPRXs present a second reactive cysteine, named *resolving cysteine* ($R-S_RH$), which generates an intersubunit disulfide bond with the sulfenic acid form of the peroxidatic cysteine (to generate $R-S_p-S_R-R$) prior to reduction. This mechanism is followed by the typical 2-CysPRX with the $R-S_RH$ on a partner subunit (Dietz et al., 2006). Also, the atypical 2-CysPRX performs the same chemical mechanism (disulfide bond formation) for catalysis, except that both, the $R-S_RH$ and the $R-S_pH$ pertain to the same monomer, thus forming an intrasubunit disulfide bond prior to reduction (Smirnoff, 2005). The third type of PRXs, the 1-CysPRX, bypasses the disulfide bond formation prior to reduction (Banerjee, 2008). The mechanism for recycling 2-CysPRX involves TRX, GRX, or another CXXC-containing redox module protein (Dietz, 2003) such as CDSP32 (Rey et al., 2005), NTRC (Kirchsteiger et al., 2009), or cyclophilin (Dietz, 2007). The pathways for 1-Cys peroxiredoxin reduction are less known, although several studies have demonstrated that both ascorbate and GSH participate as reducing substrates (Dietz, 2003). The redox potential of plastidic PRXs ranges between −307 and −325 mV (Dietz et al., 2006; Liebthal, Maynard, and Dietz, 2018).

The *A. thaliana* genome contains 10 *prx* genes, with the codified proteins belonging to four different groups: 1-CysPRX, typical 2-CysPRX, PRXQ (atypical 2-CysPRX), and type II PRX (atypical 2-Cys-PRX) (Smirnoff, 2005). Some of the PRX transcripts have been detected in all analyzed plant tissues, although the expression of each individual member of the *prx* gene family may differ significantly. The accumulation of transcripts encoding chloroplastic PRX shows a correlation with chlorophyll both in tissue distribution and during leaf development (Dietz, 2003).

Even when numerous studies give PRXs an exclusive role as antioxidants by reducing peroxides, other works report alternative functions exerted by these proteins (Rouhier and Jacquot, 2002). Some evidence supports the idea that H_2O_2 is used as a signaling molecule to regulate a variety of important cellular functions (Neill et al., 2002; Noctor 2006; Vranova, Inze, and Van Breusegem, 2002). As abundant and ubiquitous peroxidases, the activity of PRXs is central in the regulation of these functions. It has been suggested that regulation of the peroxidase and chaperone activities of these multifunctional proteins is of relevance for the H_2O_2-mediated signal transduction in plants (Hall, Karplus, and Poole 2009; Liebthal, Maynard, and Dietz, 2018). Moreover, the functionality of some PRXs (particularly the typical 2-CysPRX) is regulated by over-oxidation (Banerjee, 2008), nitrosylation (Romero-Puertas et al., 2007), and/or phosphorylation (Aran et al., 2009). Therefore, PRXs are probably playing a key role in modulating and determining the intensity of the response of ROS–RNS regulatory networks.

PRXs modulate plant cell signaling by several mechanisms (Dietz et al., 2006). For example, it is known that PRXs are involved in (i) modulating H_2O_2 concentration (H_2O_2 is a major ROS signal during adaptation and development by modifying the nuclear gene expression; Foyer and Noctor, 2005); (ii) altering lipid hydroperoxide levels (PRX can change the content of lipid peroxide–derived oxylipins, which in turn, also modulate plant responses at the level of gene expression (Mueller and Berger, 2009)]; (iii) detoxifying $ONOO^-$ (which mediates nitration of proteins; Aran et al., 2009); (iv) "spending" electron donors (thus affecting the redox state of other targets); and (v) regulating the redox state of interacting thiol proteins (PRXs indirectly influence the redox state and associated activity of other targets that are regulated by redox interactions with these donors). PRXs may be considered as peroxide sensors, since they transmit the information of high peroxide concentrations to other thiol proteins (Dietz, 2003). They can also undergo conformational changes giving them the ability to bind to non-thiol proteins (2-CysPRX may act as molecular chaperones under oxidative conditions, interacting with non-thiol proteins and other cell structures; Hall, Karplus, and Poole, 2009; Aran et al., 2009; Liebthal, Maynard, and Dietz, 2018).

8.5.2.4 Superoxide Dismutase

Superoxide dismutases (SODs) are metalloproteins that constitute a relevant antioxidant defense in photosynthetic cells, since they catalyze the dismutation of O_2^- to produce O_2 and H_2O_2 (Smirnoff, 2005). *In vivo*, SOD represents the first line of defense against ROS, giving protection to cells and tissues from oxidative destruction (Hancock, 1997). Plant cells have several forms of SOD, which mainly vary in the metal cofactor they contain: copper and zinc (CuZnSOD), manganese (MnSOD), or iron (FeSOD). CuZnSODs are generally homodimeric enzymes (~32 kDa) containing one Cu^{2+} and one Zn^{2+} per subunit (Tainer et al., 1983). CuZnSODs localize in the cytosol as well as in many organelles (Hart et al., 1999). Cu^{2+} constitutes the catalytic center, while Zn^{2+} provides a structural supporting role (Hart et al., 1999). MnSODs localize in the mitochondrial matrix in animals and plants (Jackson et al., 1978) and are homotetrameric proteins with subunit molecular mass of 23 kDa (Jackson et al., 1978). FeSODs are constitutive enzymes found in plastids (Bridges and Salin, 1981).

In biological systems, $O_2^{.-}$ can undergo dismutation (velocity ~10^5 $M^{-1}s^{-1}$ at pH 7.0) (Banerjee, 2008), and it reacts even faster with other biological radicals, such as NO, generating a most toxic species, $ONOO^-$ (Valderrama et al., 2007). The latter gives a picture of the essential biological role played by SOD and the importance of the fact that it has the fastest turnover number of any known enzyme (~10^9 $M^{-1}s^{-1}$), which indicates that this reaction is only limited by the collision frequency between SOD and the substrate $O_2^{.-}$ (Banerjee, 2008).

8.5.2.5 Glutathione S-transferase

Glutathione S-transferases (GSTs) catalyze the nucleophilic attack of GSH on an electrophilic substrate (Smirnoff, 2005), with many of them also exhibiting GPX activity. GSTs are mainly involved in xenobiotic detoxification (Figure 8.3), although they also participate in different reactions of biosynthetic and catabolic pathways, thus playing a relevant function in redox metabolism by reducing ROOH and DHA (Banerjee, 2008). It is doubtful whether they are involved in other reactions related to stress signaling and protein binding (Noctor and Foyer, 1998). All GSTs from plants were primarily characterized as homodimers of 50 kDa, being classified into four major groups: phi-, tau-, zeta-, and theta-GST (Banerjee, 2008). Two additional groups were identified in *A. thaliana* (Dixon, Davis, and Edwards, 2002). A fifth group, omega-GST, is formed by four putative small DHARs with an active site represented by the CPFC/S motif (Smirnoff, 2005; Dixon, Davis, and Edwards, 2002). DHAR regenerates ascorbate from DHA using GSH as electron donor (Figure 8.3). An additional sixth subgroup, lambda-GST, comprises two members that exhibit no DHAR activity but are active as GSH-dependent thiol transferases, suggesting that they might function in the dethiolation of *S*-glutathionylated proteins accumulated under oxidative stress situations (Dixon, Davis, and Edwards, 2002; Rouhier, Lemaire, and Jacquot, 2008; Rezaei et al., 2013).

8.5.2.6 Ascorbate Peroxidase

Ascorbate peroxidase (APX) is a hemoprotein that catalyzes the detoxification of H_2O_2 using ascorbate as reducing substrate according to the reaction

$$ascorbate + H_2O_2 \rightarrow dehydroascorbate + 2H_2O$$

In contrast to the wide distribution of catalase and GPX, APX appears to be restricted to plants, algae, and some protozoa (Noctor and Foyer, 1998; Wilkinson et al., 2002). Initially, APX activity was detected in chloroplasts (Kelly and Latzko, 1979; Shigeoka, Nakano, and Kitaoka, 1980), but later, its presence was evidenced in almost every compartment of the plant cell, with an active participation in detoxification of H_2O_2 as part of the ascorbate–glutathione or Asada–Halliwell–Foyer pathway (Hiner et al., 2002). Three forms of chloroplastidic APX were reported: thylakoid, stromal, and soluble APX (localized in the lumen) (Smirnoff, 2000; Considine and Foyer, 2014). In *Arabidopsis*, nine different genes code for APX enzymes, with four of the proteins localizing in chloroplasts (Kubo et al., 1992). Two stromal APXs are dually targeted to the stroma and the mitochondrial intermembrane space (Smirnoff 2000; Maruta et al., 2016), whereas two APXs are targeted to peroxisomes and glyoxysomes (Lisenbee, Heinze, and Trelease, 2003), and the other two localize in cytoplasm (Santos et al., 1996). There are two microsomal APXs that are able to bind to the external surface of glyoxysomes or to be transported into peroxisomes (Lisenbee, Heinze, and Trelease, 2003).

APX functionality is dependent on the availability of ascorbate or alternatively, of GSH (Smirnoff, 2005). Cellular pools of these antioxidants are maintained in the reduced form by different enzymatic systems: mainly GR, DHAR, GRX, and GST (Buchanan and Balmer, 2005; Gill et al., 2013; Stroher and Millar, 2012). APXs exhibit a higher affinity for H_2O_2 than for organic peroxides, so that they actively participate in the removal of a key redox signaling molecule (Reddy et al., 2009; Apel and Hirt, 2004; Pauly et al., 2006). Studies with transgenic plants have afforded strong support to establish that APX is an important defense enzyme implicated in the elimination of H_2O_2 (Smirnoff, 2005; Maruta et al., 2016). In a more integrated scenario, APX acts in concert with other H_2O_2-regulation enzymes to (i) balance different plant cellular systems that generate the peroxide, as is the case for NADPH oxidase in pathogen response, and (ii) control the levels of H_2O_2 used for signaling during stress, either biotic or abiotic (Kotchoni and Gachomo, 2006; Apel and Hirt, 2004).

8.5.3 Cross-Talk between Antioxidant Systems

As described earlier, plants have developed efficient enzymatic systems to prevent/alleviate damage generated by ROS and RNS. The redox cellular status critically determines the signaling and regulation of several metabolic and cellular processes; including the regulation of enzymatic activities, gene expression, growth differentiation, pathogen resistance, and apoptosis. The TRX- and GSH-dependent systems constitute major routes involved in the maintenance of intracellular redox homeostasis. These systems do not act to prevent, repair and regulate oxidative damage in an individual and isolated manner but through interaction with other components (enzymatic and non-enzymatic). This interactive action is depicted in Figure 8.3, illustrating the flux of reductive/oxidative equivalents between the main plant redox components. In such a scenario, different antioxidant and repair systems interact to exert specific functions related to redox homeostasis inside the cell. Worthy of being highlighted in this picture is the function of NADPH, acting as a central donor supplier of reducing equivalents for all the different reducing systems.

8.6 OXIDATIVE DAMAGE REPAIR SYSTEMS OPERATING IN PLANTS

All reactive species can modify different macromolecules and generate conditions for cellular damage (Moskovitz, 2005). For this reason, cells have many protective systems helping to eliminate (or at least diminish) injurious molecular species or to repair the oxidative injury (Banerjee, 2008). Considering proteins, certain amino acids are highly susceptible to oxidation: cysteine and methionine are the two most sensitive; followed by histidine, tryptophan, and tyrosine (Moller, Jensen, and Hansson, 2007). The oxidation of proteins can generate conformational changes and (in some cases) loss of function (Friguet, 2006). Plants have two main systems for protein rescue: sulfiredoxins and methionine sulfoxide reductases (Smirnoff, 2005).

8.6.1 SULFIREDOXINS

As detailed in Section 8.5.2, typical 2-CysPRXs from eukaryotes are susceptible to hyperoxidation by an excess of oxidizing substrate (Kiley and Storz, 2004; Sevilla et al., 2015). As a result of this, the $R-S_pOH$ (formed during the catalytic cycle) generates the $R-S_pO_2H$ (a more oxidized sulfur state by two electrons) (Rouhier and Jacquot, 2002; Liebthal, Maynard, and Dietz, 2018). The capacity of this hyperoxidation pathway to regulate PRX activity could provide a tool for cells to control levels of H_2O_2 generated locally. Accordingly, the system operates by maintaining concentrations of the peroxide in ranges compatible with its role of redox signal (Rouhier, Gelhaye, and Jacquot, 2004). Another view speculates that PRX hyperoxidation would produce a multimeric (high–molecular mass) form of the protein; according to this view, chaperone activity aids in cell recovery from oxidative stress (Hall, Karplus, and Poole, 2009; Aran et al., 2009). In plants, 2-CysPRXs constitute the most abundant PRXs and are principally located in chloroplasts (Dietz et al., 2006). Lower oxidation states of cysteine (for example, disulfides) are readily reversible; but higher oxidation states (such as $R-S_pO_2H$) were once considered irreversible in a biological scenario (Moller, Jensen, and Hansson, 2007). This picture was revisited with the discovery of sulfiredoxin (SRX), a protein able to reduce $R-S_pO_2H$ back to $R-S_pOH$ by a mechanism requiring ATP (Biteau, Labarre, and Toledano, 2003; Chang et al., 2004; Liebthal, Maynard, and Dietz, 2018).

Plant SRXs present high identity with the orthologous proteins from yeasts and humans, all sharing the conserved signature sequence and residues essential for catalysis (Liu et al., 2006). In addition, SRXs from plants possess a distinctive transit peptide directing them to the chloroplast; this is reinforced by reports showing that the SRX-GFP fusion protein is targeted to the chloroplast in *Arabidopsis* mesophyll protoplast

(Rey et al., 2007; Sevilla et al., 2015). Several experiments have pointed to the expression of SRX genes in both vegetative and reproductive organs. Concerning SRX transcript, it was also characterized that (i) the highest amount is detected in leaves (Banerjee, 2008) and (ii) the level is considerably augmented under oxidative stress, in parallel with enhanced transcription of 2-CysPRX (which is essential in maintaining the chloroplast redox balance) (Dietz et al., 2006; Rey et al., 2007; Liebthal, Maynard, and Dietz, 2018).

SRX reduces hyperoxidized PRX in the presence of Mg^{2+} by transferring the phosphoryl group from ATP to the peroxiredoxin sulfinic acid, thus generating a sulfinic phosphoryl ester, which is the first step required for activation of this species (Jonsson, Johnson, and Lowther, 2009; Sevilla et al., 2015). The catalytic mechanism includes nucleophilic attack by the thiolate of the essential cysteine in SRX on the sulfinic phosphoryl ester, giving rise to linkage between the two proteins through the thiosulfinate (Jonsson et al., 2008). It follows the breakdown of the complex by reaction with another attacking thiolate to generate a disulfide bridge in SRX (Roussel et al., 2008). TRX, and possibly GSH, could serve as reductants in the repairing process (Roussel et al., 2009; Park et al., 2009). On the other hand, numerous studies support a differential reactivity of SRX toward various hyperoxidized PRX targets (Jonsson and Lowther, 2007; Sevilla et al., 2015).

8.6.2 METHIONINE SULFOXIDE REDUCTASES

Methionine (Met) oxidation leads to the formation of two S- and R-diastereoisomers of methionine sulfoxide (MetSO) that are reduced back to Met by methionine sulfoxide reductases (MSRs) A and B, respectively (Figure 8.4). We presented earlier (see Section 8.3.2) a general view on the oxidation of sulfur-containing amino acids in proteins and the role of repairing enzymes such as MSRs. Plants have multiple forms of both MSRA and MSRB, which are targeted to different cellular compartments (Tarrago, Laugier, and Rey, 2009). The multiple locations of MSRs indicate that the reduction of oxidized methionine residues in proteins takes place independently within the different subcellular locations (Rouhier et al., 2007; Rouhier et al., 2006; Rey and Tarrago, 2018). In *Arabidopsis*, the MSR family consists of 14 members, predicted to be localized in plastids, cytosol, and endoplasmic reticulum. The subcellular localization of some of them was proved experimentally, as reported for (i) the three plastidial MSRA4, MSRB1, and MSRB2 isoforms; (ii) MSRB7 and MSRB8 in cytosol; and (iii) MSRB3 in endoplasmic reticulum. In addition, the overexpression of MSRB2 in *A. thaliana* transgenic plants generates enhanced tolerance to cellular oxidative damage during long nights (Bechtold, Murphy, and Mullineaux, 2004; Romero et al., 2004). A direct antioxidant function was first attributed to these enzymes in the elimination of ROS via cyclic oxidation of Met in proteins and reduction by MSRs. Published data showed that in plants, the control of Met redox status is a key step in signaling pathways. Interestingly, phytohormones (such as ABA, jasmonic acid, salicylic acid, and ethylene) play more complex roles in

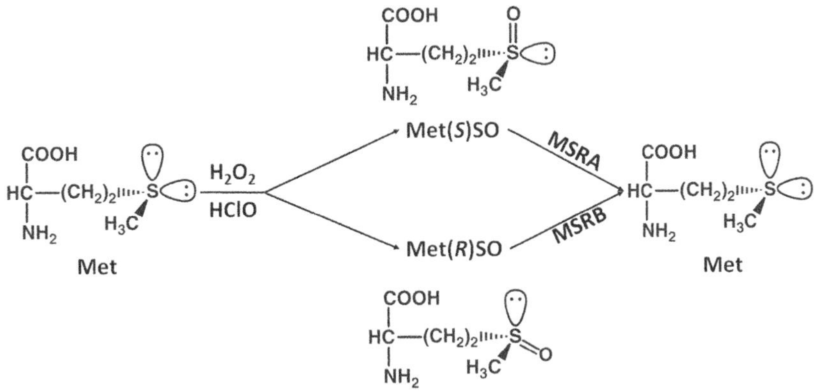

FIGURE 8.4 The sulfur atom in methionine (Met) is susceptible to oxidation by ROS (such as H_2O_2 and HClO). The oxidized form of methionine occurs in the form of two diastereomers, methionine-S-sulfoxide (Met(S)SO) and methionine-R-sulfoxide (Met(R)SO), which are reduced by MSRA and MSRB, respectively.

regulating MSR gene expression, probably in relation to both environmental conditions and the development stage.

Several studies indicate that MSRA and MSRB share a similar reaction mechanism (Kauffmann, Aubry, and Favier, 2005; Boschi-Muller, Gand, and Branlant, 2008), which proceeds in a series of steps. First, the catalytic cysteine thiolate (situated at the N-terminus in MSRA and the C-terminus in MSRB) attacks the sulfoxide, releasing methionine and producing a sulfenic acid intermediate on the cysteine. Second, the sulfenic acid intermediate is attacked by the resolving cysteine, generating an intramolecular disulfide bond. Finally, TRX (or other electron donors, such as GRX or GSH) reduces the intramolecular disulfide, regenerating MSR (Tarrago et al., 2009). Most MSRBs also contain a single zinc atom that supports a structural role in these proteins (Kauffmann, Aubry, and Favier, 2005). In addition, some MSRBs lack the resolving cysteine, suggesting that the sulfenic acid intermediate can be directly reduced by TRX or GSH (linking to GRX) (Vieira Dos Santos et al., 2007) or CDSP32 (Rey et al., 2005).

Modified lines of plant models and crop species indicate that MSRs play protective roles in abiotic and biotic environmental constraints, as well as in the control of the aging process, as shown in seeds subjected to adverse conditions. In their protective role against oxidative damage in plant proteins, MSRs act to preserve the activity of stress-responsive effectors such as GSTs and chaperones (Rey and Tarrago, 2018).

In response to biotic and abiotic stresses, plants induce a complex array of pathways. H_2O_2 has been implicated as the signaling molecule, assembling the downstream components of signaling pathways. Evidence has been provided for the existence of a mechanism by which plants perceive H_2O_2 associated with methionine oxidation (Emes, 2009). Previous data indicate that MSRs fulfill key signaling roles via interplay with Ca^{2+}- and phosphorylation-dependent cascades, thus transmitting ROS-dependent transduction pathways. The data support the idea that the chemical oxidation of methionine residues by H_2O_2 at key hydrophobic positions within canonical phosphorylation motifs in numerous

enzymes inhibits protein kinase binding. This inhibition of kinases (including the calcium-dependent protein kinase and AMP-activated protein kinase families) can be reversed by MSR *in vivo* (Hardin et al., 2009). It has been demonstrated that this mechanism is directly linked to oxidative signals (via a reversible process catalyzed by MSR) by means of changes in protein phosphorylation, contributing to the knowledge of the perception of redox signaling in plants (Rey and Tarrago, 2018).

8.7 PROTEIN S-NITROSYLATION IN PLANTS: AN OVERVIEW

Due to their unique physicochemical properties, cysteines participate in catalytic reactions, serve as metal ligands, and are also susceptible to various post-translational modifications (Ruiz-May et al., 2018). Whereas free cysteines have a thiol pKa value of about 8.3, in proteins, some cysteines (defined as reactive) possess a lower pKa (from 3.0 to 7.0) (Akter et al., 2015). In TRXs and GRXs, the lowering of the pKa results from the protein microenvironment, as these thiolates are stabilized by proximal positively charged amino acids, by specific hydrogen bonding, and/or by a dipole effect (Knuesting and Scheibe, 2018). At physiological pH (~7.0), these residues will be predominantly found as thiolates, which are much stronger nucleophiles than thiol groups. Consequently, proteins containing these reactive cysteines can undergo many different oxidation states in response to different redox signals. Many post-translational modifications are generated on protein thiol groups via oxidation with ROS, RNS, and RSS, particularly H_2O_2, ·NO, and H_2S (Fancy, Bahlmann, and Loake, 2017; Noctor, Reichheld, and Foyer, 2018; Trost et al., 2017; Lu et al., 2013) (Figure 8.5).

Plant cells are exposed to ROS and RNS produced during different metabolic processes, since there are many target macromolecules whose structures and/or functionalities can be modified by the action of other species. Protein oxidation can bring about conformational changes and in some cases, loss of function. Nitric oxide (NO) is now recognized as a key regulator of plant physiological processes (Romero-Puertas, Rodriguez-Serrano, and Sandalio, 2013; Noctor, Reichheld,

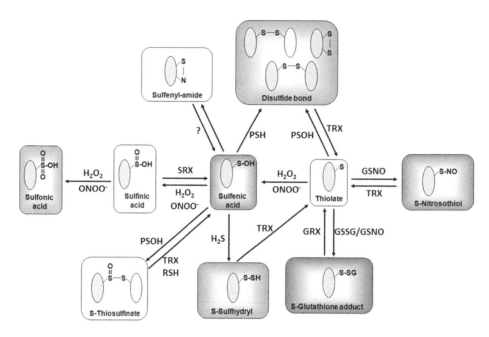

FIGURE 8.5 Principal oxidative modifications of cysteine residues and their reduction pathways. Reactive cysteine residues mostly exist in thiolate forms at physiological pH and can form a sulfenic acid (RSOH) by reacting with H_2O_2. This sulfenic acid is an intermediate for most other redox post-translational modifications, forming intra- or intermolecular disulfide bonds, glutathione adducts in the presence of GSH (or GSSG), sulfenyl-amides, persulfides on reaction with hydrogen sulfide, thiosulfinates by reacting with another sulfenic acid, and sulfinic (RSO_2H) and sulfonic (RSO_3H) acids by further reacting with H_2O_2. Other possible glutathionylation pathways could perform between a thiolate and nitrosoglutathione (GSNO) reaction. Most of these redox post-translational modifications are reversible, and the enzymatic reduction of these different oxidation forms is essentially achieved by GRX (mainly glutathionylated proteins but also some disulfides) and TRX (disulfides and possibly persulfides, nitrosothiols, and some glutathionylated proteins). In addition to specifically reducing sulfinic acids formed on peroxiredoxins, SRX may also catalyze PRX deglutathionylation. The reduction pathway of thiosulfinates to sulfenyl-amides is unclear, but glutathione and subsequently GRXs might be involved.

and Foyer, 2018). Several components of the signaling pathways relaying NO effects in plants, including second messengers, protein kinases, phytohormones, and target genes, have been characterized (Romero-Puertas, Rodriguez-Serrano, and Sandalio, 2013).

There is evidence that NO partly operates through post-translational modification of proteins, notably via S-nitrosylation and tyrosine nitration. The addition of a nitroso group (NO) to the sulfur atom of a protein cysteine residue is known as S-nitrosylation (a redox reversible post-translational modification). The formation of S-nitrosothiols (RSNOs) occurs through its reaction with a thiyl radical (RS·) formed from one-electron oxidation of thiolates via NO_2. Additionally, NO_2 can react with NO to form N_2O_3, which can combine with cysteine thiolate to form the RSNO (Romero-Puertas, Rodriguez-Serrano, and Sandalio, 2013; Knuesting and Scheibe, 2018). S-nitrosylated proteins can play an important role in signal, translational, and metabolic regulation processes. These modified proteins can transmit NO bioactivity and regulate the function of other proteins by means of mechanisms analogous to phosphorylation, which are known as *trans-nitrosylation* (Zaffagnini et al., 2016; Ruiz-May et al., 2018; Sevilla et al., 2015). These redox post-translational modifications lead to changes in biological activity of modified proteins, interactions between proteins, or subcellular localization of target proteins. For example,

Arabidopsis cytoplasmic glyceraldehyde-3-phosphate dehydrogenase activity is reversibly inhibited by nitrosylation of catalytic Cys[149] mediated either chemically with a strong NO donor or by trans-nitrosylation with GSNO. GSNO is found to trigger both GAPDH nitrosylation and glutathionylation, although nitrosylation is greatly prominent. The extent of GAPDH nitrosylation is dependent on the [GSH]/[GSNO] ratio (Zaffagnini et al., 2013).

The reverse of S-nitrosylation is denitrosylation, principally an enzymically controlled process. Multiple enzymes have been described to date, which fall into two main classes mediating the denitrosylation of protein and low–molecular mass RSNOs, respectively (Ruiz-May et al., 2018; Zaffagnini et al., 2016). Denitrosylation has been described as fundamental to the reversibility of protein S-nitrosylation. The TRX system is involved in this process in several organisms, including plants, with direct activity towards S-nitrosylated proteins. This denitrosylating activity is carried out via the catalytic Cys residues of TRX. Thus, SNO-protein denitrosylation/trans-nitrosylation implies an additional function for the TRX system as a key regulating point in metabolic and cellular process.

There is abundant evidence that TRX can be modified by S-nitrosylation in extra cysteine residues that do not directly take part in disulfide reductase activity to generate a nitrosyl form (TRX-SNO). In addition, it has been proved that TRX can control the S-nitrosylation of other target proteins

(Turkan, 2015; Moldogazieva et al., 2018). This control is carried out by either specifically eliminating the NO group in a process of specific denitrosylation or acting as a donor of NO by means of a specific trans-nitrosylation reaction (Turkan, 2015; Moldogazieva et al., 2018). On the other hand, GRX is a type of redox protein that uses GSH as a redox cofactor. In plant cells, GRXs have been proved not only to catalyze the reduction of mixed disulfides with GSH but also to modulate protein S-nitrosylation (via GSNO reduction) (Stroher and Millar, 2012). The role of NO as a major regulator of plant physiological functions has become increasingly evident. S-nitrosylation/denitrosylation is currently accepted as a critical redox-mediated regulation process in plant cells.

S-Nitrosoglutathione (GSNO) is an endogenous RSNO that plays a critical role in NO signaling and is a source of bio-available NO. In oxygenated media, formation of RSNOs is stimulated by the oxidation of NO to N_2O_3, with some evidence suggesting that both exogenous NO and endogenously derived NO from nitric oxide synthases can react with GSH to form GSNO (Knuesting and Scheibe, 2018; Fancy, Bahlmann, and Loake, 2017). The generation of GSNO can serve as a stable and mobile NO pool, which can effectively transduce NO signaling. The GSNO contributes to protein S-nitrosylation and S-glutathionylation, which is thought to be part of the signaling transduction. Unlike other low–molecular mass messengers that bind to and activate target cellular receptors, NO signaling is mediated by a coordinating complex between NO and transition metals or target cellular proteins, often via S-nitrosylation of cysteine residues (Turkan, 2015).

Nitrosoglutathione reductase (GSNOR, an NADH-dependent S-nitrosoglutathione reductase belonging to the family of class III alcohol dehydrogenases) accelerates the decomposition of S-nitrosoglutathione (GSNO) and SNO-proteins in equilibrium with GSNO. The enzyme is highly conserved from bacteria to humans and catalyzes the NADH-dependent reduction of GSNO, generating GSSG and NH_4^+ (Rizza and Filomeni, 2017; Lindermayr, 2018). Cellular homeostasis of GSNO, a major cache of NO bioactivity in plants, is under its control. GSNOR is a key regulator of S-nitrosothiol metabolism and is involved in plant responses to abiotic and biotic stresses. In higher plants, the first report on GSNO reductase activity involved Arabidopsis. Its activity was confirmed by functional complementation of the hypersensitivity to GSNO of a yeast mutant with impaired GSNO metabolism. Then, this protein was purified and characterized in several plant species, including pea seeds, maize, and rice. Thereafter, this activity began to attract the attention of many plant researchers, because its capacity to metabolize GSNO opens a new perspective on the metabolism of NO in plants under physiological and environmental stress (Lindermayr, 2018).

Thanks to the development of dedicated proteomic approaches (in particular, the use of the biotin switch technique [BST] combined with mass spectrometry) (Figure 8.6), hundreds of plant protein candidates for S-nitrosylation have been identified (Qin, Dey, and Daaka, 2013). Functional studies focused on specific proteins provided comprehensive views of how this post-translation modification impacts the structure and function of proteins and more generally, the way by which NO can regulate biological plant processes. The BST has become a mainstay assay for detecting S-nitrosylated proteins in complex biological systems. In a three-step procedure, nitrosylated cysteines are converted to biotinylated cysteines.

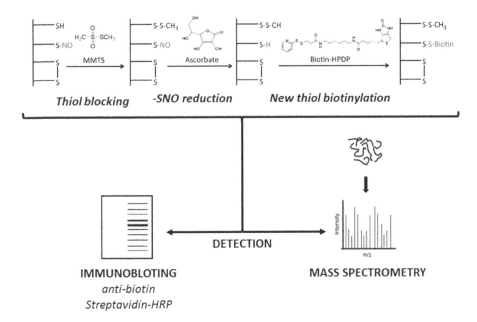

FIGURE 8.6 Representative schematic diagram of the biotin switch assay. Proteins that have been previously S-nitrosylated can be labeled with biotin-HPDP. Step 1 involves blocking off all free thiols with methyl methanethiosulfonate (MMTS). Step 2 involves the reduction of all S-nitrosylated thiols with ascorbate. Reduced thiols, which were originally S-nitrosylated, are then labeled with biotin-HPDP. Biotinylated proteins can then be detected by immunoblotting or can be purified by avidin-affinity chromatography. The isolated proteins can be identified by mass spectrometry.

Biotinylated proteins can then be detected by immunoblotting or can be purified by avidin-affinity chromatography (Lamotte et al., 2015). Initially, free (unmodified) thiol groups in the proteins are chemically blocked (using alkylating reagents such as iodoacetamide or oxidation with *S*-methyl methanethiosulfonate [MMTS]). Subsequently, the *S*-nitrosothiol group in cysteine residues is selectively reduced with ascorbate, generating new thiol groups. Then, these new thiol groups are biotinylated chemically (using the biotin-maleimide reagent) for later detection (with streptavidin-HRP) or isolation of the proteins that present them (with streptavidin-agarose resin). The isolated proteins can be identified by mass spectrometry (Lamotte et al., 2015; Astier et al., 2011).

8.8   CONCLUDING REMARKS

As aerobic organisms, plants take advantage of the redox potential of oxygen, using it in metabolic pathways. Antioxidant compounds and ROS/RNS/RSS comprise important interacting systems with different functions in higher plants; which make high redox flexibility feasible for the organism. Antioxidants are not passive bystanders in such a cross-talk; instead, they function as key signaling components that make up a dynamic metabolic interface between cell stress perception and physiological responses (Scandalios, 2005). Normal metabolic conditions produce ROS and RNS as byproducts that are hazardous, causing undesirable effects. Plants have developed defense mechanisms against the latter, including antioxidant enzymes and free radical scavengers. Under certain situations, such as biotic or abiotic stress as well as in the course of development, ROS and/or RNS synthesis is induced, generating cell signals that have complex downstream effects on both primary and secondary metabolism. The latter could result in different situations: (i) the cell responds to the ROS/RNS signal, making appropriate adjustments and returning to physiological conditions; or (ii) ROS/RNS production exceeds the capacity of antioxidant systems, cells are not able to return to homeostasis, oxidative damage prevails, and genetically programmed cell suicide events are triggered.

It is now clear that most cells can adapt to oxidative situations by altering global gene-expression patterns, including the transcription and translation of genes encoding antioxidant proteins (some of them with enzymatic activity), and/or by post-translational changes operating on proteins that could modify key regulators of redox responses (Waszczak, Carmody, and Kangasjarvi, 2018; Noctor, Reichheld, and Foyer, 2018). In plants, H_2O_2 has been identified as a second messenger for signals triggered by ROS, and a better picture has been gained of the understanding of metabolic changes operating in consequence. Moreover, the role of ·NO and the problematic of RNS have been clearly shown. Although major advances have been reached, areas related to the regulation of gene expression, the modulation of enzyme activity, and the coordination/redirectioning of metabolic fluxes require intensive research to complete the comprehension of how higher plants maintain levels of oxidative species under control. Post-genomics tools, including proteomics and metabolomics,

are emerging molecular tools of high utility in this field and are critical for advancements in integrative approaches. The challenge is complex, but the finding of responses to different open questions is predicted to be of high relevance. A complete understanding of the functioning of redox systems developed by plants during evolution is a key prerequisite for the design of manageable procedures to improve plant productivity under different environmental scenarios.

ACKNOWLEDGMENTS

Work in our laboratory received grants from ANPCyT, CONICET, and UNL. DGA, SAG, and AAI are Research Career Members from CONICET.

REFERENCES

Abhilash, P. C., S. Jamil, and N. Singh. 2009. Transgenic plants for enhanced biodegradation and phytoremediation of organic xenobiotics. *Biotechnol Adv* 27 (4):474–488.

Akter, S., J. Huang, C. Waszczak, S. Jacques, K. Gevaert, F. Van Breusegem, and J. Messens. 2015. Cysteines under ROS attack in plants: a proteomics view. *J Exp Bot* 66 (10):2935–2944.

Alkhalfioui, F., M. Renard, and F. Montrichard. 2007. Unique properties of NADP-thioredoxin reductase C in legumes. *J Exp Bot* 58 (5):969–978.

Anderson, L. E. and V. R. Advani. 1970. Chloroplast and cytoplasmic enzymes: three distinct isoenzymes associated with the reductive pentose phosphate cycle. *Plant Physiol* 45 (5):583–585.

Anderson, L. E., T. C. Ng, and K. E. Park. 1974. Inactivation of pea leaf chloroplastic and cytoplasmic glucose 6-phosphate dehydrogenases by light and dithiothreitol. *Plant Physiol* 53 (6):835–839.

Apel, K. and H. Hirt. 2004. Reactive oxygen species: metabolism, oxidative stress, and signal transduction. *Annu Rev Plant Biol* 55:373–399.

Aran, M., D. S. Ferrero, E. Pagano, and R. A. Wolosiuk. 2009. Typical 2-Cys peroxiredoxins—modulation by covalent transformations and noncovalent interactions. *FEBS J* 276 (9):2478–2493.

Arrigoni, O. 1994. Ascorbate system in plant development. *J Bioenerg Biomembr* 26 (4):407–419.

Astier, J., S. Rasul, E. Koen, H. Manzoor, A. Besson-Bard, O. Lamotte, S. Jeandroz, J. Durner, C. Lindermayr, and D. Wendehenne. 2011. S-nitrosylation: an emerging post-translational protein modification in plants. *Plant Sci* 181 (5):527–533.

Bailey-Serres, J. and M. T. Nguyen. 1992. Purification and characterization of cytosolic 6-phosphogluconate dehydrogenase isozymes from maize. *Plant Physiol* 100 (3):1580–1583.

Balmer, Y., W. H. Vensel, C. K. Tanaka, W. J. Hurkman, E. Gelhaye, N. Rouhier, J. P. Jacquot, et al. 2004. Thioredoxin links redox to the regulation of fundamental processes of plant mitochondria. *Proc Natl Acad Sci U S A* 101 (8):2642–2647.

Banerjee, R. 2008. *Redox Biochemistry*. USA: Wiley-Interscience.

Barranco-Medina, S., J. J. Lazaro, and K. J. Dietz. 2009. The oligomeric conformation of peroxiredoxins links redox state to function. *FEBS Lett* 583 (12):1809–1816.

Bechtold, U., D. J. Murphy, and P. M. Mullineaux. 2004. Arabidopsis peptide methionine sulfoxide reductase2 prevents cellular oxidative damage in long nights. *Plant Cell* 16 (4):908–919.

Besson-Bard, A., A. Pugin, and D. Wendehenne. 2008. New insights into nitric oxide signaling in plants. *Annu Rev Plant Biol* 59:21–39.

Bienert, G. P., J. K. Schjoerring, and T. P. Jahn. 2006. Membrane transport of hydrogen peroxide. *Biochim Biophys Acta* 1758 (8):994–1003.

Bindoli, A., J. M. Fukuto, and H. J. Forman. 2008. Thiol chemistry in peroxidase catalysis and redox signaling. *Antioxid Redox Signal* 10 (9):1549–1564.

Biteau, B., J. Labarre, and M. B. Toledano. 2003. ATP-dependent reduction of cysteine-sulphinic acid by *S. cerevisiae* sulphiredoxin. *Nature* 425 (6961):980–984.

Blander, G. and L. Guarente. 2004. The Sir2 family of protein deacetylases. *Annu Rev Biochem* 73:417–435.

Boschi-Muller, S., A. Gand, and G. Branlant. 2008. The methionine sulfoxide reductases: catalysis and substrate specificities. *Arch Biochem Biophys* 474 (2):266–273.

Boschi-Muller, S., A. Olry, M. Antoine, and G. Branlant. 2005. The enzymology and biochemistry of methionine sulfoxide reductases. *Biochim Biophys Acta* 1703 (2):231–238.

Bridges, S. M. and M. L. Salin. 1981. Distribution of iron-containing superoxide dismutase in vascular plants. *Plant Physiol* 68 (2):275–278.

Britt, A. B. 1996. DNA damage and repair in plants. *Annu Rev Plant Physiol Plant Mol Biol* 47:75–100.

Britton, G. 1989. Carotenoids and polyterpenoids. *Nat Prod Rep* 6 (4):359–392.

Buchanan, B. B. and Y. Balmer. 2005. Redox regulation: a broadening horizon. *Annu Rev Plant Biol* 56:187–220.

Buchanan, B. B., P. Schurmann, R. A. Wolosiuk, and J. P. Jacquot. 2002. The ferredoxin/thioredoxin system: from discovery to molecular structures and beyond. *Photosynth Res* 73 (1–3):215–222.

Bustos, D. M., C. A. Bustamante, and A. A. Iglesias. 2008. Involvement of non-phosphorylating glyceraldehyde-3-phosphate dehydrogenase in response to oxidative stress. *J Plant Physiol* 165 (4):456–461.

Bustos, D. M. and A. A. Iglesias. 2002. Non-phosphorylating glyceraldehyde-3-phosphate dehydrogenase is post-translationally phosphorylated in heterotrophic cells of wheat (*Triticum aestivum*). *FEBS Lett* 530 (1-3):169–173.

Bustos, D. M. and A. A. Iglesias. 2005. A model for the interaction between plant GAPN and 14-3-3zeta using protein-protein docking calculations, electrostatic potentials and kinetics. *J Mol Graph Model* 23 (6):490–502.

Bustos, D. M. and A. A. Iglesias. 2006. Intrinsic disorder is a key characteristic in partners that bind 14-3-3 proteins. *Proteins* 63 (1):35–42.

Butt, M. S. and M. T. Sultan. 2009. Green tea: nature's defense against malignancies. *Crit Rev Food Sci Nutr* 49 (5):463–473.

Cattaruzza, M. and M. Hecker. 2008. Protein carbonylation and decarboylation: a new twist to the complex response of vascular cells to oxidative stress. *Circ Res* 102 (3):273–274.

Cerny, M., H. Habanova, M. Berka, M. Luklova, and B. Brzobohaty. 2018. Hydrogen peroxide: its role in plant biology and crosstalk with signalling networks. *Int J Mol Sci* 19 (9): pii: E2812. doi: 10.3390/ijms19092812.

Considine, M. J. and C. H. Foyer. 2014. Redox regulation of plant development. *Antioxid Redox Signal* 21 (9):1305–1326.

Corpas, F. J. 2004. Enzymatic sources of nitric oxide in plant cells—beyond one protein—one function. *New Phytologist* 162:246–247.

Corpas, F. J., J. B. Barroso, A. Carreras, M. Quiros, A. M. Leon, M. C. Romero-Puertas, F. J. Esteban, et al. 2004. Cellular and subcellular localization of endogenous nitric oxide in young and senescent pea plants. *Plant Physiol* 136 (1):2722–2733.

Corpas, F. J., J. B. Barroso, A. Carreras, R. Valderrama, J. M. Palma, A. M. Leon, L. M. Sandalio, and L. A. del Rio. 2006. Constitutive arginine-dependent nitric oxide synthase activity in different organs of pea seedlings during plant development. *Planta* 224 (2):246–254.

Costa, V., A. Quintanilha, and P. Moradas-Ferreira. 2007. Protein oxidation, repair mechanisms and proteolysis in *Saccharomyces cerevisiae*. *IUBMB Life* 59 (4–5):293–298.

Chang, T. S., W. Jeong, H. A. Woo, S. M. Lee, S. Park, and S. G. Rhee. 2004. Characterization of mammalian sulfiredoxin and its reactivation of hyperoxidized peroxiredoxin through reduction of cysteine sulfinic acid in the active site to cysteine. *J Biol Chem* 279 (49):50994–51001.

Dalle-Donne, I., R. Rossi, D. Giustarini, R. Colombo, and A. Milzani. 2007. S-glutathionylation in protein redox regulation. *Free Radic Biol Med* 43 (6):883–898.

Dawson, C. R., K. G. Strothkamp, and K. G. Krul. 1975. Ascorbate oxidase and related copper proteins. *Ann N Y Acad Sci* 258:209–220.

Debnam, P. M., G. Shearer, L. Blackwood, and D. H. Kohl. 1997. Evidence for channeling of intermediates in the oxidative pentose phosphate pathway by soybean and pea nodule extracts, yeast extracts, and purified yeast enzymes. *Eur J Biochem* 246 (2):283–290.

Demarsy, E. and C. Fankhauser. 2009. Higher plants use LOV to perceive blue light. *Curr Opin Plant Biol* 12 (1):69–74.

Diaz-Vivancos, P., A. de Simone, G. Kiddle, and C. H. Foyer. 2015. Glutathione—linking cell proliferation to oxidative stress. *Free Radic Biol Med* 89:1154–1164.

Dietz, K. J. 2003. Plant peroxiredoxins. *Annu Rev Plant Biol* 54:93–107.

Dietz, K. J. 2007. The dual function of plant peroxiredoxins in antioxidant defence and redox signaling. *Subcell Biochem* 44:267–294.

Dietz, K. J., S. Jacob, M. L. Oelze, M. Laxa, V. Tognetti, S. M. de Miranda, M. Baier, and I. Finkemeier. 2006. The function of peroxiredoxins in plant organelle redox metabolism. *J Exp Bot* 57 (8):1697–1709.

Dixon, D. P., B. G. Davis, and R. Edwards. 2002. Functional divergence in the glutathione transferase superfamily in plants. Identification of two classes with putative functions in redox homeostasis in *Arabidopsis thaliana*. *J Biol Chem* 277 (34):30859–30869.

Dohadwala, M. M. and J. A. Vita. 2009. Grapes and cardiovascular disease. *J Nutr* 139 (9):1788S–1793S.

Drincovich, M. F., P. Casati, and C. S. Andreo. 2001. NADP-malic enzyme from plants: a ubiquitous enzyme involved in different metabolic pathways. *FEBS Lett* 490 (1–2):1–6.

Droge, W. 2002. Free radicals in the physiological control of cell function. *Physiol Rev* 82 (1):47–95.

Eicks, M., V. Maurino, S. Knappe, U. I. Flugge, and K. Fischer. 2002. The plastidic pentose phosphate translocator represents a link between the cytosolic and the plastidic pentose phosphate pathways in plants. *Plant Physiol* 128 (2):512–522.

Emes, M. J. 2009. Oxidation of methionine residues: the missing link between stress and signalling responses in plants. *Biochem J* 422 (2):e1–e2.

Fancy, N. N., A. K. Bahlmann, and G. J. Loake. 2017. Nitric oxide function in plant abiotic stress. *Plant Cell Environ* 40 (4):462–472.

Fessel, J. P. and W. M. Oldham. 2018a. Nicotine adenine dinucleotides: the redox currency of the cell. *Antioxid Redox Signal* 28 (3):165–166.

Fessel, J. P. and W. M. Oldham. 2018b. Pyridine dinucleotides from molecules to man. *Antioxid Redox Signal* 28 (3):180–212.

Fleury, C., B. Mignotte, and J. L. Vayssiere. 2002. Mitochondrial reactive oxygen species in cell death signaling. *Biochimie* 84 (2–3):131–141.

Forman, H. J., J. M. Fukuto, and M. Torres. 2004. Redox signaling: thiol chemistry defines which reactive oxygen and nitrogen species can act as second messengers. *Am J Physiol Cell Physiol* 287 (2):C246–C256.

Foyer, C. H. and G. Noctor. 2005. Redox homeostasis and antioxidant signaling: a metabolic interface between stress perception and physiological responses. *Plant Cell* 17 (7):1866–1875.

Foyer, C. H., G. Noctor, B. Buchanan, K. J. Dietz, and T. Pfannschmidt. 2009. Redox regulation in photosynthetic organisms: signaling, acclimation, and practical implications. *Antioxid Redox Signal* 11 (4):861–905.

Friguet, B. 2006. Oxidized protein degradation and repair in ageing and oxidative stress. *FEBS Lett* 580 (12):2910–2916.

Friguet, B., A. L. Bulteau, and I. Petropoulos. 2008. Mitochondrial protein quality control: implications in ageing. *Biotechnol J* 3 (6):757–764.

Gattuso, G., D. Barreca, C. Gargiulli, U. Leuzzi, and C. Caristi. 2007. Flavonoid composition of citrus juices. *Molecules* 12 (8):1641–1673.

Gelhaye, E., N. Rouhier, N. Navrot, and J. P. Jacquot. 2005. The plant thioredoxin system. *Cell Mol Life Sci* 62 (1):24–35.

Giles, G. I. and C. Jacob. 2002. Reactive sulfur species: an emerging concept in oxidative stress. *Biol Chem* 383 (3–4):375–388.

Giles, G. I., K. M. Tasker, and C. Jacob. 2001. Hypothesis: the role of reactive sulfur species in oxidative stress. *Free Radic Biol Med* 31 (10):1279–1283.

Gill, S. S., N. A. Anjum, M. Hasanuzzaman, R. Gill, D. K. Trivedi, I. Ahmad, E. Pereira, and N. Tuteja. 2013. Glutathione and glutathione reductase: a boon in disguise for plant abiotic stress defense operations. *Plant Physiol Biochem* 70:204–212.

Giustarini, D., R. Rossi, A. Milzani, R. Colombo, and I. Dalle-Donne. 2004. S-glutathionylation: from redox regulation of protein functions to human diseases. *J Cell Mol Med* 8 (2):201–212.

Golowatzki, I. D. 1962. [The role of glycolysis and the pentose cycle in the metabolism of phosphorylated monosaccharides in animal erythrocytes]. *Acta Biol Med Ger* 9:323–331.

Gomez Casati, D. F., J. I. Sesma, and A. A. Iglesias. 2000. Structural and kinetic characterization of NADP-dependent, non-phosphorylating glyceraldehyde-3-phosphate dehydrogenase from celery leaves. *Plant Sci* 154 (2):107–115.

Gruhlke, M. C. and A. J. Slusarenko. 2012. The biology of reactive sulfur species (RSS). *Plant Physiol Biochem* 59:98–107.

Guse, A. H. 2004. Regulation of calcium signaling by the second messenger cyclic adenosine diphosphoribose (cADPR). *Curr Mol Med* 4 (3):239–248.

Habenicht, A. 1997. The non-phosphorylating glyceraldehyde-3-phosphate dehydrogenase: biochemistry, structure, occurrence and evolution. *Biol Chem* 378 (12):1413–1419.

Hagglund, P., C. Finnie, H. Yano, A. Shahpiri, B. B. Buchanan, A. Henriksen, and B. Svensson. 2016. Seed thioredoxin h. *Biochim Biophys Acta* 1864 (8):974–982.

Hall, A., P. A. Karplus, and L. B. Poole. 2009. Typical 2-Cys peroxiredoxins—structures, mechanisms and functions. *FEBS J* 276 (9):2469–2477.

Halliwell, B. 2006. Reactive species and antioxidants. Redox biology is a fundamental theme of aerobic life. *Plant Physiol* 141 (2):312–322.

Hancock, J. T. 1997. Superoxide, hydrogen peroxide and nitric oxide as signalling molecules: their production and role in disease. *Br J Biomed Sci* 54 (1):38–46.

Hancock, J. T., D. Henson, M. Nyirenda, R. Desikan, J. Harrison, M. Lewis, J. Hughes, and S. J. Neill. 2005. Proteomic identification of glyceraldehyde 3-phosphate dehydrogenase as an inhibitory target of hydrogen peroxide in *Arabidopsis*. *Plant Physiol Biochem* 43 (9):828–835.

Hardin, S. C., C. T. Larue, M. H. Oh, V. Jain, and S. C. Huber. 2009. Coupling oxidative signals to protein phosphorylation via methionine oxidation in *Arabidopsis*. *Biochem J* 422 (2):305–312.

Hart, P. J., M. M. Balbirnie, N. L. Ogihara, A. M. Nersissian, M. S. Weiss, J. S. Valentine, and D. Eisenberg. 1999. A structure-based mechanism for copper-zinc superoxide dismutase. *Biochemistry* 38 (7):2167–2178.

Hashida, S. N., A. Miyagi, M. Nishiyama, K. Yoshida, T. Hisabori, and M. Kawai-Yamada. 2018. Ferredoxin/thioredoxin system plays an important role in the chloroplastic NADP status of *Arabidopsis*. *Plant J* 95 (6):947–960.

Hernandez, L. E., J. Sobrino-Plata, M. B. Montero-Palmero, S. Carrasco-Gil, M. L. Flores-Caceres, C. Ortega-Villasante, and C. Escobar. 2015. Contribution of glutathione to the control of cellular redox homeostasis under toxic metal and metalloid stress. *J Exp Bot* 66 (10):2901–2911.

Hiner, A. N., E. L. Raven, R. N. Thorneley, F. Garcia-Canovas, and J. N. Rodriguez-Lopez. 2002. Mechanisms of compound I formation in heme peroxidases. *J Inorg Biochem* 91 (1):27–34.

Holmgren, A. and F. Aslund. 1995. Glutaredoxin. *Methods Enzymol* 252:283–292.

Hutchings, D., S. Rawsthorne, and M. J. Emes. 2005. Fatty acid synthesis and the oxidative pentose phosphate pathway in developing embryos of oilseed rape (*Brassica napus* L.). *J Exp Bot* 56 (412):577–585.

Iglesias, A. A. and M. Losada. 1988. Purification and kinetic and structural properties of spinach leaf NADP-dependent non-phosphorylating glyceraldehyde-3-phosphate dehydrogenase. *Arch Biochem Biophys* 260 (2):830–840.

Iglesias, A. A. and F. E. Podestá. 2005. Photosynthate Formation and Partitioning in Crop Plants. In *Handbook of Photosynthesis*, 2nd Edition, edited by M. Pessarakli, pp. 525–545. CRC Press, Taylor & Francis Group, Boca Raton.

Ivanov, A. G., P. V. Sane, V. Hurry, G. Oquist, and N. P. Huner. 2008. Photosystem II reaction centre quenching: mechanisms and physiological role. *Photosynth Res* 98 (1–3):565–574.

Jablonsky, J., H. Bauwe, and O. Wolkenhauer. 2011. Modeling the Calvin-Benson cycle. *BMC Syst Biol* 5:185.

Jackson, C., J. Dench, A. L. Moore, B. Halliwell, C. H. Foyer, and D. O. Hall. 1978. Subcellular localisation and identification of superoxide dismutase in the leaves of higher plants. *Eur J Biochem* 91 (2):339–344.

Jacob, C., J. R. Lancaster, and G. I. Giles. 2004. Reactive sulphur species in oxidative signal transduction. *Biochem Soc Trans* 32 (Pt 6):1015–1017.

Jones, D. P. and H. Sies. 2015. The redox code. *Antioxid Redox Signal* 23 (9):734–746.

Jonsson, T. J., L. C. Johnson, and W. T. Lowther. 2009. Protein engineering of the quaternary sulfiredoxin-peroxiredoxin enzyme-substrate complex reveals the molecular basis for cysteine sulfinic acid phosphorylation. *J Biol Chem* 284 (48):33305–33310.

Jonsson, T. J. and W. T. Lowther. 2007. The peroxiredoxin repair proteins. *Subcell Biochem* 44:115–141.

Jonsson, T. J., M. S. Murray, L. C. Johnson, and W. T. Lowther. 2008. Reduction of cysteine sulfinic acid in peroxiredoxin by sulfiredoxin proceeds directly through a sulfinic phosphoryl ester intermediate. *J Biol Chem* 283 (35):23846–23851.

Katoh, A., K. Uenohara, M. Akita, and T. Hashimoto. 2006. Early steps in the biosynthesis of NAD in *Arabidopsis* start with aspartate and occur in the plastid. *Plant Physiol* 141 (3):851–857.

Kauffmann, B., A. Aubry, and F. Favier. 2005. The three-dimensional structures of peptide methionine sulfoxide reductases: current knowledge and open questions. *Biochim Biophys Acta* 1703 (2):249–260.

Kelly, G. J. and E. Latzko. 1979. Soluble ascorbate peroxidase: detection in plants and use in vitamin C estimation. *Naturwissenschaften* 66 (12):617–619.

Kiley, P. J. and G. Storz. 2004. Exploiting thiol modifications. *PLoS Biol* 2 (11):e400.

Kimura, S., Y. Tahira, T. Ishibashi, Y. Mori, T. Mori, J. Hashimoto, and K. Sakaguchi. 2004. DNA repair in higher plants; photoreactivation is the major DNA repair pathway in non-proliferating cells while excision repair (nucleotide excision repair and base excision repair) is active in proliferating cells. *Nucleic Acids Res* 32 (9):2760–2767.

Kirchsteiger, K., P. Pulido, M. Gonzalez, and F. J. Cejudo. 2009. NADPH thioredoxin reductase C controls the redox status of chloroplast 2-Cys peroxiredoxins in *Arabidopsis thaliana*. *Mol Plant* 2 (2):298–307.

Knuesting, J. and R. Scheibe. 2018. Small molecules govern thiol redox switches. *Trends Plant Sci* 23 (9):769–782.

Kolbert, Z., G. Feigl, A. Borde, A. Molnar, and L. Erdei. 2017. Protein tyrosine nitration in plants: present knowledge, computational prediction and future perspectives. *Plant Physiol Biochem* 113:56–63.

Kotchoni, S. O. and E. W. Gachomo. 2006. The reactive oxygen species network pathways: an essential prerequisite for perception of pathogen attack and the acquired disease resistance in plants. *J Biosci* 31 (3):389–404.

Kruger, N. J. and A. von Schaewen. 2003. The oxidative pentose phosphate pathway: structure and organisation. *Curr Opin Plant Biol* 6 (3):236–246.

Kubo, A., H. Saji, K. Tanaka, K. Tanaka, and N. Kondo. 1992. Cloning and sequencing of a cDNA encoding ascorbate peroxidase from *Arabidopsis thaliana*. *Plant Mol Biol* 18 (4):691–701.

Kulinskii, V. I. and L. S. Kolesnichenko. 2009. Glutathione system. I. Synthesis, transport, glutathione transferases, glutathione peroxidases. *Biomed Khim* 55 (3):255–277.

Lamattina, L., C. Garcia-Mata, M. Graziano, and G. Pagnussat. 2003. Nitric oxide: the versatility of an extensive signal molecule. *Annu Rev Plant Biol* 54:109–136.

Lamotte, O., J. B. Bertoldo, A. Besson-Bard, C. Rosnoblet, S. Aime, S. Hichami, H. Terenzi, and D. Wendehenne. 2015. Protein S-nitrosylation: specificity and identification strategies in plants. *Front Chem* 2:114.

Landino, L. M. 2008. Protein thiol modification by peroxynitrite anion and nitric oxide donors. *Methods Enzymol* 440:95–109.

Le Bras, M., M. V. Clement, S. Pervaiz, and C. Brenner. 2005. Reactive oxygen species and the mitochondrial signaling pathway of cell death. *Histol Histopathol* 20 (1):205–219.

Liebthal, M., D. Maynard, and K. J. Dietz. 2018. Peroxiredoxins and redox signaling in plants. *Antioxid Redox Signal* 28 (7):609–624.

Lillig, C. H., C. Berndt, and A. Holmgren. 2008. Glutaredoxin systems. *Biochim Biophys Acta* 1780 (11):1304–1317.

Lindermayr, C. 2018. Crosstalk between reactive oxygen species and nitric oxide in plants: key role of S-nitrosoglutathione reductase. *Free Radic Biol Med* 122:110–115.

Linster, C. L. and S. G. Clarke. 2008. L-Ascorbate biosynthesis in higher plants: the role of VTC2. *Trends Plant Sci* 13 (11):567–573.

Lisenbee, C. S., M. Heinze, and R. N. Trelease. 2003. Peroxisomal ascorbate peroxidase resides within a subdomain of rough endoplasmic reticulum in wild-type *Arabidopsis* cells. *Plant Physiol* 132 (2):870–882.

Liu, X. P., X. Y. Liu, J. Zhang, Z. L. Xia, X. Liu, H. J. Qin, and D. W. Wang. 2006. Molecular and functional characterization of sulfiredoxin homologs from higher plants. *Cell Res* 16 (3):287–296.

Lu, C., A. Kavalier, E. Lukyanov, and S. S. Gross. 2013. S-sulfhydration/desulfhydration and S-nitrosylation/denitrosylation: a common paradigm for gasotransmitter signaling by H_2S and NO. *Methods* 62 (2):177–181.

Lu, S. and L. Li. 2008. Carotenoid metabolism: biosynthesis, regulation, and beyond. *J Integr Plant Biol* 50 (7):778–785.

Martin, J. L. 1995. Thioredoxin—a fold for all reasons. *Structure* 3 (3):245–250.

Maruta, T., Y. Sawa, S. Shigeoka, and T. Ishikawa. 2016. Diversity and evolution of ascorbate peroxidase functions in chloroplasts: more than just a classical antioxidant enzyme? *Plant Cell Physiol* 57 (7):1377–1386.

Mata-Perez, C., M. N. Padilla, B. Sanchez-Calvo, J. C. Begara-Morales, R. Valderrama, M. Chaki, and J. B. Barroso. 2018. Biological properties of nitro-fatty acids in plants. *Nitric Oxide* pii: S1089-8603(17)30286-0. doi: 10.1016/j.niox.2018.03.011.

McPherson, J. D., B. H. Shilton, and D. J. Walton. 1988. Role of fructose in glycation and cross-linking of proteins. *Biochemistry* 27 (6):1901–1907.

Meyer, A. J. 2008. The integration of glutathione homeostasis and redox signaling. *J Plant Physiol* 165 (13):1390–1403.

Meyer, Y., J. P. Reichheld, and F. Vignols. 2005. Thioredoxins in *Arabidopsis* and other plants. *Photosynth Res* 86 (3):419–433.

Meyer, Y., W. Siala, T. Bashandy, C. Riondet, F. Vignols, and J. P. Reichheld. 2008. Glutaredoxins and thioredoxins in plants. *Biochim Biophys Acta* 1783 (4):589–600.

Michelet, L., M. Zaffagnini, S. Morisse, F. Sparla, M. E. Perez-Perez, F. Francia, A. Danon, et al. 2013. Redox regulation of the Calvin-Benson cycle: something old, something new. *Front Plant Sci* 4:470.

Miret, J. A. and S. Munne-Bosch. 2015. Redox signaling and stress tolerance in plants: a focus on vitamin E. *Ann N Y Acad Sci* 1340:29–38.

Mittler, R. 2017. ROS are good. *Trends Plant Sci* 22 (1):11–19.

Miura, R. 2001. Versatility and specificity in flavoenzymes: control mechanisms of flavin reactivity. *Chem Rec* 1 (3):183–194.

Moldogazieva, N. T., I. M. Mokhosoev, N. B. Feldman, and S. V. Lutsenko. 2018. ROS and RNS signalling: adaptive redox switches through oxidative/nitrosative protein modifications. *Free Radic Res* 52 (5):507–543.

Moller, I. M. and B. K. Kristensen. 2004. Protein oxidation in plant mitochondria as a stress indicator. *Photochem Photobiol Sci* 3 (8):730–735.

Moller, I. M., P. E. Jensen, and A. Hansson. 2007. Oxidative modifications to cellular components in plants. *Annu Rev Plant Biol* 58:459–481.

Moller, I. M. and B. K. Kristensen. 2006. Protein oxidation in plant mitochondria detected as oxidized tryptophan. *Free Radic Biol Med* 40 (3):430–435.

Montrichard, F., F. Alkhalfioui, H. Yano, W. H. Vensel, W. J. Hurkman, and B. B. Buchanan. 2009. Thioredoxin targets in plants: the first 30 years. *J Proteomics* 72 (3):452–474.

Moon, J. C., H. H. Jang, H. B. Chae, J. R. Lee, S. Y. Lee, Y. J. Jung, M. R. Shin, et al.. 2006. The C-type *Arabidopsis* thioredoxin reductase ANTR-C acts as an electron donor to 2-Cys peroxiredoxins in chloroplasts. *Biochem Biophys Res Commun* 348 (2):478–484.

Moskovitz, J. 2005. Methionine sulfoxide reductases: ubiquitous enzymes involved in antioxidant defense, protein regulation, and prevention of aging-associated diseases. *Biochim Biophys Acta* 1703 (2):213–219.

Mueller, M. J. and S. Berger. 2009. Reactive electrophilic oxylipins: pattern recognition and signalling. *Phytochemistry* 70 (13–14):1511–1521.

Muller, M. and T. Carell. 2009. Structural biology of DNA photolyases and cryptochromes. *Curr Opin Struct Biol* 19 (3):277–285.

Neill, S. J., R. Desikan, A. Clarke, R. D. Hurst, and J. T. Hancock. 2002. Hydrogen peroxide and nitric oxide as signalling molecules in plants. *J Exp Bot* 53 (372):1237–1247.

Noctor, G. 2006. Metabolic signalling in defence and stress: the central roles of soluble redox couples. *Plant Cell Environ* 29 (3):409–425.

Noctor, G. and C. H. Foyer. 1998. Ascorbate and glutathione: keeping active oxygen under control. *Annu Rev Plant Physiol Plant Mol Biol* 49:249–279.

Noctor, G., J. P. Reichheld, and C. H. Foyer. 2018. ROS-related redox regulation and signaling in plants. *Semin Cell Dev Biol* 80:3–12.

Nordberg, J. and E. S. Arner. 2001. Reactive oxygen species, antioxidants, and the mammalian thioredoxin system. *Free Radic Biol Med* 31 (11):1287–1312.

Park, J. W., J. J. Mieyal, S. G. Rhee, and P. B. Chock. 2009. Deglutathionylation of 2-Cys peroxiredoxin is specifically catalyzed by sulfiredoxin. *J Biol Chem* 284 (35):23364–23374.

Patel, R. P., J. McAndrew, H. Sellak, C. R. White, H. Jo, B. A. Freeman, and V. M. Darley-Usmar. 1999. Biological aspects of reactive nitrogen species. *Biochim Biophys Acta* 1411 (2–3):385–400.

Pauly, N., C. Pucciariello, K. Mandon, G. Innocenti, A. Jamet, E. Baudouin, D. Herouart, P. Frendo, and A. Puppo. 2006. Reactive oxygen and nitrogen species and glutathione: key players in the legume-Rhizobium symbiosis. *J Exp Bot* 57 (8):1769–1776.

Petropoulos, I. and B. Friguet. 2006. Maintenance of proteins and aging: the role of oxidized protein repair. *Free Radic Res* 40 (12):1269–1276.

Piattoni, C. V., D. M. Bustos, S. A. Guerrero, and A. A. Iglesias. 2011. Nonphosphorylating glyceraldehyde-3-phosphate dehydrogenase is phosphorylated in wheat endosperm at serine-404 by an SNF1-related protein kinase allosterically inhibited by ribose-5-phosphate. *Plant Physiol* 156 (3):1337–1350.

Piattoni, C. V., D. M. L. Ferrero, I. Dellaferrera, A. Vegetti, and A. A. Iglesias. 2017. Cytosolic glyceraldehyde-3-phosphate dehydrogenase is phosphorylated during seed development. *Front Plant Sci* 8:522.

Piattoni, C. V., S. A. Guerrero, and A. A. Iglesias. 2013. A differential redox regulation of the pathways metabolizing glyceraldehyde-3-phosphate tunes the production of reducing power in the cytosol of plant cells. *Int J Mol Sci* 14 (4):8073–8092.

Pignocchi, C. and C. H. Foyer. 2003. Apoplastic ascorbate metabolism and its role in the regulation of cell signalling. *Curr Opin Plant Biol* 6 (4):379–389.

Plaxton, W. C. 1996. The organization and regulation of plant glycolysis. *Annu Rev Plant Physiol Plant Mol Biol* 47:185–214.

Pollak, N., C. Dolle, and M. Ziegler. 2007. The power to reduce: pyridine nucleotides—small molecules with a multitude of functions. *Biochem J* 402 (2):205–218.

Poole, L. B. 2007. The catalytic mechanism of peroxiredoxins. *Subcell Biochem* 44:61–81.

Qin, Y., A. Dey, and Y. Daaka. 2013. Protein s-nitrosylation measurement. *Methods Enzymol* 522:409–425.

Rajniak, J., R. F. H. Giehl, E. Chang, I. Murgia, N. von Wiren, and E. S. Sattely. 2018. Biosynthesis of redox-active metabolites in response to iron deficiency in plants. *Nat Chem Biol* 14 (5):442–450.

Reddy, R. A., B. Kumar, P. S. Reddy, R. N. Mishra, S. Mahanty, T. Kaul, S. Nair, S. K. Sopory, and M. K. Reddy. 2009. Molecular cloning and characterization of genes encoding *Pennisetum*

glaucum ascorbate peroxidase and heat-shock factor: interlinking oxidative and heat-stress responses. *J Plant Physiol* 166 (15):1646–1659.

Rey, P., N. Becuwe, M. B. Barrault, D. Rumeau, M. Havaux, B. Biteau, and M. B. Toledano. 2007. The *Arabidopsis thaliana* sulfiredoxin is a plastidic cysteine-sulfinic acid reductase involved in the photooxidative stress response. *Plant J* 49 (3):505–514.

Rey, P., S. Cuine, F. Eymery, J. Garin, M. Court, J. P. Jacquot, N. Rouhier, and M. Broin. 2005. Analysis of the proteins targeted by CDSP32, a plastidic thioredoxin participating in oxidative stress responses. *Plant J* 41 (1):31–42.

Rey, P. and L. Tarrago. 2018. Physiological roles of plant methionine sulfoxide reductases in redox homeostasis and signaling. *Antioxidants (Basel)* 7 (9) pii: E114. doi: 10.3390/antiox7090114.

Rezaei, M. K., Z. S. Shobbar, M. Shahbazi, R. Abedini, and S. Zare. 2013. Glutathione S-transferase (GST) family in barley: identification of members, enzyme activity, and gene expression pattern. *J Plant Physiol* 170 (14):1277–1284.

Rius, S. P., P. Casati, A. A. Iglesias, and D. F. Gomez-Casati. 2006. Characterization of an *Arabidopsis thaliana* mutant lacking a cytosolic non-phosphorylating glyceraldehyde-3-phosphate dehydrogenase. *Plant Mol Biol* 61 (6):945–957.

Rizza, S. and G. Filomeni. 2017. Chronicles of a reductase: biochemistry, genetics and physio-pathological role of GSNOR. *Free Radic Biol Med* 110:19–30.

Romero, H. M., B. S. Berlett, P. J. Jensen, E. J. Pell, and M. Tien. 2004. Investigations into the role of the plastidial peptide methionine sulfoxide reductase in response to oxidative stress in *Arabidopsis*. *Plant Physiol* 136 (3):3784–3794.

Roldan-Arjona, T. and R. R. Ariza. 2009. Repair and tolerance of oxidative DNA damage in plants. *Mutat Res* 681 (2–3):169–179.

Romero-Puertas, M. C., M. Laxa, A. Matte, F. Zaninotto, I. Finkemeier, A. M. Jones, M. Perazzolli, E. Vandelle, K. J. Dietz, and M. Delledonne. 2007. S-nitrosylation of peroxiredoxin II E promotes peroxynitrite-mediated tyrosine nitration. *Plant Cell* 19 (12):4120–4130.

Romero-Puertas, M. C., M. Rodriguez-Serrano, and L. M. Sandalio. 2013. Protein S-nitrosylation in plants under abiotic stress: an overview. *Front Plant Sci* 4:373.

Rouhier, N. and J. P. Jacquot. 2002. Plant peroxiredoxins: alternative hydroperoxide scavenging enzymes. *Photosynth Res* 74 (3):259–268.

Rouhier, N., D. Cerveau, J. Couturier, J. P. Reichheld, and P. Rey. 2015. Involvement of thiol-based mechanisms in plant development. *Biochim Biophys Acta* 1850 (8):1479–1496.

Rouhier, N., E. Gelhaye, and J. P. Jacquot. 2002. Redox control by dithiol-disulfide exchange in plants: II. The cytosolic and mitochondrial systems. *Ann N Y Acad Sci* 973:520–528.

Rouhier, N., E. Gelhaye, and J. P. Jacquot. 2004. Plant glutaredoxins: still mysterious reducing systems. *Cell Mol Life Sci* 61 (11):1266–1277.

Rouhier, N., B. Kauffmann, F. Tete-Favier, P. Palladino, P. Gans, G. Branlant, J. P. Jacquot, and S. Boschi-Muller. 2007. Functional and structural aspects of poplar cytosolic and plastidial type a methionine sulfoxide reductases. *J Biol Chem* 282 (5):3367–3378.

Rouhier, N., C. S. Koh, E. Gelhaye, C. Corbier, F. Favier, C. Didierjean, and J. P. Jacquot. 2008. Redox based anti-oxidant systems in plants: biochemical and structural analyses. *Biochim Biophys Acta* 1780 (11):1249–1260.

Rouhier, N., S. D. Lemaire, and J. P. Jacquot. 2008. The role of glutathione in photosynthetic organisms: emerging functions for glutaredoxins and glutathionylation. *Annu Rev Plant Biol* 59:143–166.

Rouhier, N., C. Vieira Dos Santos, L. Tarrago, and P. Rey. 2006. Plant methionine sulfoxide reductase A and B multigenic families. *Photosynth Res* 89 (2–3):247–262.

Roussel, X., G. Bechade, A. Kriznik, A. Van Dorsselaer, S. Sanglier-Cianferani, G. Branlant, and S. Rahuel-Clermont. 2008. Evidence for the formation of a covalent thiosulfinate intermediate with peroxiredoxin in the catalytic mechanism of sulfiredoxin. *J Biol Chem* 283 (33):22371–22382.

Roussel, X., A. Kriznik, C. Richard, S. Rahuel-Clermont, and G. Branlant. 2009. The catalytic mechanism of sulfiredoxin from *Saccharomyces cerevisiae* passes through an oxidized disulfide sulfiredoxin intermediate that is reduced by thioredoxin. *J Biol Chem* 284 (48):33048–33055.

Ruiz-May, E., A. Segura-Cabrera, J. M. Elizalde-Contreras, L. M. Shannon, and V. M. Loyola-Vargas. 2018. A recent advance in the intracellular and extracellular redox post-translational modification of proteins in plants. *J Mol Recognit* 32 (1):e2754. doi: 10.1002/jmr.2754.

Rybus-Kalinowska, B., K. Zwirska-Korczala, M. Kalinowski, M. Kukla, E. Birkner, and J. Jochem. 2009. [Activity of antioxidative enzymes and concentration of malondialdehyde as oxidative status markers in women with non-autoimmunological subclinical hyperthyroidism]. *Endokrynol Pol* 60 (3):199–202.

Santos, M., H. Gousseau, C. Lister, C. Foyer, G. Creissen, and P. Mullineaux. 1996. Cytosolic ascorbate peroxidase from *Arabidopsis thaliana* L. is encoded by a small multigene family. *Planta* 198 (1):64–69.

Scandalios, J. G. 2005. Oxidative stress: molecular perception and transduction of signals triggering antioxidant gene defenses. *Braz J Med Biol Res* 38 (7):995–1014.

Schafer, F. Q. and G. R. Buettner. 2001. Redox environment of the cell as viewed through the redox state of the glutathione disulfide/glutathione couple. *Free Radic Biol Med* 30 (11):1191–1212.

Scheibe, R. 1991. Redox-modulation of chloroplast enzymes: a common principle for individual control. *Plant Physiol* 96 (1):1–3.

Schnarrenberger, C., M. Tetour, and M. Herbert. 1975. Development and intracellular distribution of enzymes of the oxidative pentose phosphate cycle in radish cotyledons. *Plant Physiol* 56 (6):836–840.

Schneider, M., J. Knuesting, O. Birkholz, J. J. Heinisch, and R. Scheibe. 2018. Cytosolic GAPDH as a redox-dependent regulator of energy metabolism. *BMC Plant Biol* 18 (1):184.

Scholl, R. and K. Nickelsen. 2015. Discovery of causal mechanisms: oxidative phosphorylation and the Calvin-Benson cycle. *Hist Philos Life Sci* 37 (2):180–209.

Schurmann, P. 2003. Redox signaling in the chloroplast: the ferredoxin/thioredoxin system. *Antioxid Redox Signal* 5 (1):69–78.

Schurmann, P. and J. P. Jacquot. 2000. Plant thioredoxin systems revisited. *Annu Rev Plant Physiol Plant Mol Biol* 51:371–400.

Sevilla, F., D. Camejo, A. Ortiz-Espin, A. Calderon, J. J. Lazaro, and A. Jimenez. 2015. The thioredoxin/peroxiredoxin/sulfiredoxin system: current overview on its redox function in plants and regulation by reactive oxygen and nitrogen species. *J Exp Bot* 66 (10):2945–2955.

Shao, H. B., L. Y. Chu, M. A. Shao, C. A. Jaleel, and H. M. Mi. 2008. Higher plant antioxidants and redox signaling under environmental stresses. *C R Biol* 331 (6):433–441.

Shapiro, A. D. 2005. Nitric oxide signaling in plants. *Vitam Horm* 72:339–398.

Sharkey, T. D. 2018. Discovery of the canonical Calvin-Benson cycle. *Photosynth Res.* doi 10.1007/s11120-018-0600-2

Shigeoka, S., T. Ishikawa, M. Tamoi, Y. Miyagawa, T. Takeda, Y. Yabuta, and K. Yoshimura. 2002. Regulation and function of ascorbate peroxidase isoenzymes. *J Exp Bot* 53 (372):1305–1319.

Shigeoka, S., Y. Nakano, and S. Kitaoka. 1980. Metabolism of hydrogen peroxide in *Euglena gracilis* Z by L-ascorbic acid peroxidase. *Biochem J* 186 (1):377–380.

Siqueira, J. A., P. Hardoim, P. C. G. Ferreira, A. Nunes-Nesi, and A. S. Hemerly. 2018. Unraveling interfaces between energy metabolism and cell cycle in plants. *Trends Plant Sci* 23 (8):731–747.

Slesak, I., M. Libik, B. Karpinska, S. Karpinski, and Z. Miszalski. 2007. The role of hydrogen peroxide in regulation of plant metabolism and cellular signalling in response to environmental stresses. *Acta Biochim Pol* 54 (1):39–50.

Smiri, M. and T. Missaoui. 2014. The role of ferredoxin:thioredoxin reductase/thioredoxin m in seed germination and the connection between this system and copper ion toxicity. *J Plant Physiol* 171 (17):1664–1670.

Smirnoff, N. 2000. Ascorbate biosynthesis and function in photoprotection. *Philos Trans R Soc Lond B Biol Sci* 355 (1402):1455–1464.

Smirnoff, N. 2005. *Antioxidants and Reactive Oxygen Species in Plants.* UK: Blackwell.

Smirnoff, N. and J. E. Pallanca. 1996. Ascorbate metabolism in relation to oxidative stress. *Biochem Soc Trans* 24 (2):472–478.

Smirnoff, N. and G. L. Wheeler. 2000. Ascorbic acid in plants: biosynthesis and function. *Crit Rev Biochem Mol Biol* 35 (4):291–314.

Squadrito, G. L. and W. A. Pryor. 1998. Oxidative chemistry of nitric oxide: the roles of superoxide, peroxynitrite, and carbon dioxide. *Free Radic Biol Med* 25 (4–5):392–403.

Starke-Reed, P. E. and C. N. Oliver. 1989. Protein oxidation and proteolysis during aging and oxidative stress. *Arch Biochem Biophys* 275 (2):559–567.

Stroher, E. and A. H. Millar. 2012. The biological roles of glutaredoxins. *Biochem J* 446 (3):333–348.

Tainer, J. A., E. D. Getzoff, J. S. Richardson, and D. C. Richardson. 1983. Structure and mechanism of copper, zinc superoxide dismutase. *Nature* 306 (5940):284–287.

Takagi, S. 2003. Actin-based photo-orientation movement of chloroplasts in plant cells. *J Exp Biol* 206 (Pt 12):1963–1969.

Tarrago, L., E. Laugier, and P. Rey. 2009. Protein-repairing methionine sulfoxide reductases in photosynthetic organisms: gene organization, reduction mechanisms, and physiological roles. *Mol Plant* 2 (2):202–217.

Tarrago, L., E. Laugier, M. Zaffagnini, C. Marchand, P. Le Marechal, N. Rouhier, S. D. Lemaire, and P. Rey. 2009. Regeneration mechanisms of *Arabidopsis thaliana* methionine sulfoxide reductases B by glutaredoxins and thioredoxins. *J Biol Chem* 284 (28):18963–18971.

Trapp, J. and M. Jung. 2006. The role of NAD+ dependent histone deacetylases (sirtuins) in ageing. *Curr Drug Targets* 7 (11):1553–1560.

Triantaphylides, C. and M. Havaux. 2009. Singlet oxygen in plants: production, detoxification and signaling. *Trends Plant Sci* 14 (4):219–228.

Trost, P., S. Fermani, M. Calvaresi, and M. Zaffagnini. 2017. Biochemical basis of sulphenomics: how protein sulphenic acids may be stabilized by the protein microenvironment. *Plant Cell Environ* 40 (4):483–490.

Turkan, I. 2015. ROS and RNS: key signalling molecules in plants. *J Exp Bot* 69 (14):3313–3315.

Valderrama, R., F. J. Corpas, A. Carreras, A. Fernandez-Ocana, M. Chaki, F. Luque, M. V. Gomez-Rodriguez, P. Colmenero-Varea, L. A. Del Rio, and J. B. Barroso. 2007. Nitrosative stress in plants. *FEBS Lett* 581 (3):453–461.

Van Ruyskensvelde, V., F. Van Breusegem, and K. Van Der Kelen. 2018. Post-transcriptional regulation of the oxidative stress response in plants. *Free Radic Biol Med* 122:181–192.

Veitch, N. C. 2009. Isoflavonoids of the leguminosae. *Nat Prod Rep* 26 (6):776–802.

Vieira Dos Santos, C., E. Laugier, L. Tarrago, V. Massot, E. Issakidis-Bourguet, N. Rouhier, and P. Rey. 2007. Specificity of thioredoxins and glutaredoxins as electron donors to two distinct classes of *Arabidopsis* plastidial methionine sulfoxide reductases B. *FEBS Lett* 581 (23):4371–4376.

Vigani, G. and I. Murgia. 2018. Iron-requiring enzymes in the spotlight of oxygen. *Trends Plant Sci* 23 (10):874–882.

Vranova, E., D. Inze, and F. Van Breusegem. 2002. Signal transduction during oxidative stress. *J Exp Bot* 53 (372):1227–1236.

Wakao, S., C. Andre, and C. Benning. 2008. Functional analyses of cytosolic glucose-6-phosphate dehydrogenases and their contribution to seed oil accumulation in *Arabidopsis*. *Plant Physiol* 146 (1):277–288.

Wang, B., Q. Shi, Y. Ouyang, and Y. Chen. 2008. [Progress in oxyR regulon—the bacterial antioxidant defense system—a review]. *Wei Sheng Wu Xue Bao* 48 (11):1556–1561.

Wang, Y., A. Lin, G. J. Loake, and C. Chu. 2013. H_2O_2-induced leaf cell death and the crosstalk of reactive nitric/oxygen species. *J Integr Plant Biol* 55 (3):202–208.

Waszczak, C., M. Carmody, and J. Kangasjarvi. 2018. Reactive oxygen species in plant signaling. *Ann Rev Plant Biol* 69:209–236.

Weber, A. P., R. Schwacke, and U. I. Flugge. 2005. Solute transporters of the plastid envelope membrane. *Annu Rev Plant Biol* 56:133–164.

Wilkinson, S. R., S. O. Obado, I. L. Mauricio, and J. M. Kelly. 2002. *Trypanosoma cruzi* expresses a plant-like ascorbate-dependent hemoperoxidase localized to the endoplasmic reticulum. *Proc Natl Acad Sci U S A* 99 (21):13453–13458.

Yang, S., Q. Yu, Y. Zhang, Y. Jia, S. Wan, X. Kong, and Z. Ding. 2018. ROS: the fine-tuner of plant stem cell fate. *Trends Plant Sci* 23 (10):850–853.

Yoshioka, H., S. Asai, M. Yoshioka, and M. Kobayashi. 2009. Molecular mechanisms of generation for nitric oxide and reactive oxygen species, and role of the radical burst in plant immunity. *Mol Cells* 28 (4):321–329.

Zaffagnini, M., M. De Mia, S. Morisse, N. Di Giacinto, C. H. Marchand, A. Maes, S. D. Lemaire, and P. Trost. 2016. Protein S-nitrosylation in photosynthetic organisms: a comprehensive overview with future perspectives. *Biochim Biophys Acta* 1864 (8):952–966.

Zaffagnini, M., S. Fermani, A. Costa, S. D. Lemaire, and P. Trost. 2013. Plant cytoplasmic GAPDH: redox post-translational modifications and moonlighting properties. *Front Plant Sci* 4:450.

Zaffagnini, M., S. Morisse, M. Bedhomme, C. H. Marchand, M. Festa, N. Rouhier, S. D. Lemaire, and P. Trost. 2013. Mechanisms of nitrosylation and denitrosylation of cytoplasmic glyceraldehyde-3-phosphate dehydrogenase from *Arabidopsis thaliana*. *J Biol Chem* 288 (31):22777–22789.

Zagorchev, L., C. E. Seal, I. Kranner, and M. Odjakova. 2013. A central role for thiols in plant tolerance to abiotic stress. *Int J Mol Sci* 14 (4):7405–7432 .

Zhang, Y. and N. Hogg. 2005. S-Nitrosothiols: cellular formation and transport. *Free Radic Biol Med* 38 (7):831–838.

Ziegler, M. 2000. New functions of a long-known molecule. Emerging roles of NAD in cellular signaling. *Eur J Biochem* 267 (6):1550–1564.

Zingg, J. M. 2007. Vitamin E: an overview of major research directions. *Mol Aspects Med* 28 (5–6):400–422.

9 Role of Proline and Other Osmoregulatory Compounds in Plant Responses to Abiotic Stresses

Ehsan Shakeri, Ali Akbar Mozafari, Fatemeh Sohrabi, and Armin Saed-Moucheshi

CONTENTS

9.1 INTRODUCTION

A variety of protective mechanisms have evolved in plants to allow them to acclimatize to unfavorable environmental conditions for survival and growth. It has been reported that amino acids, such as proline, alanine, valine, isoleucine, glutamic acid, arginine, serine, glycine, aspartic acid, amides such as glutamine and asparagine, and non-protein amino acids such as γ-aminobutyric acid, pipecolic acid, citrulline, and ornithine are accumulated in higher plants under different abiotic stress conditions [1], and thus may play a role in stress tolerance. The cellular source of these amino acids is the primary metabolic pathways [2]. Their main role is to increase the ability of cells to retain water without affecting the normal metabolism.

9.2 PROLINE

Proline (Pro) is a primary protectant against osmotic stress which plays an important role in coordinating the osmotic potential between cytosol and vacuolar (balancing capacity as an osmolyte) [3]. Proline accumulation is believed to play an adaptive role in plant stress tolerance. In plants, the precursor for proline biosynthesis is L-glutamic acid. It has been proposed that proline acts as a compatible osmolyte and as a way to store carbon and nitrogen [4]. Proline has also been proposed as functioning as molecular chaperone stabilizing the structure of proteins, and proline accumulation can provide a way to buffer cytosolic pH and to balance cell redox

status. Proline accumulation may be part of the stress signal influencing adaptive responses [5]. Proline may also act as a component of signal transduction pathways that regulate stress-responsive genes in addition to its previously described osmoprotective roles, thereby improving tolerance to salt stress [6].

Proline also increases cellular osmolarity (turgor pressure) that provides the turgor necessary for cell expansion under stress conditions [7]. The level of proline accumulation in plants under stress conditions varies from species to species and can be 100 times greater than that in control plants [8]. Accumulation of proline under stress in many plant species has been correlated with stress tolerance, and its concentration has been shown to be generally higher in stress-tolerant than in stress-sensitive plants [9]. Besides acting as an excellent osmolyte, proline plays a major role as a metal chelator, antioxidative defense molecule, and signaling molecule during stress. It imparts stress tolerance by modulating mitochondrial functions, which influences cell proliferation, triggers specific gene expressions, stabilizes membranes and thereby prevents electrolyte leakage and brings the concentrations of reactive oxygen species (ROS) within normal ranges leading to stress recovery. Foliar application of proline under heavy metal stress has shown to be an effective method for reducing the toxicity of metals by activating protective mechanisms in plants [10]. In general, proline plays multifarious roles, including adaptation, recovery, and signaling when it comes to combating stress in plants. It is well documented that the accumulation of proline is a response of plants to increased noxious elements [11]. Among these, the sodium ion is known as the most prominent one. It is reported that the noxious effects of higher Na+ concentrations could be inhibited by higher rates of proline. This might be a mechanism for salt tolerance in some plant cultivars under stress prevailing conditions [2]. Although proline accumulation in the leaves protects against salt toxicity, some authors suggested that proline does not play an important role as osmoregulator in water deficit stress [12, 13].

9.3 PROLINE AND ION HOMEOSTASIS

The role of proline in ion homeostasis at high salinity has been reported by several studies [14, 15]. Also, it is widely reported that proline could significantly reduce plant Na+ and Cl-, and increase K+ as well as Ca2+ under salt stress, thereby enhancing tolerance to salinity. For instance, Lone, Kueh, Wyn Jones, and Bright [16] indicated that exogenous proline application increasing barley shoot elongation under NaCl stress was associated with a decrease in shoot Cl- and Na+ accumulation. Moreover, over-expressing the P5CS gene in transgenic sugarcane, as a result of proline accumulation, reduced leaf Na+; therefore, the transgenic plants were able to cope with saline conditions [17]. Also, proline application under salt stress reduced the Na+/K+ ratio in rice [18]. The role of proline in ion homeostasis might be through its uninterrupted protection of transporter systems and membranes [19]. In this regard, it is crystal clear that proline acts in protective functions, ion homeostasis, and osmotic adjustment. Accordingly, proline can be a selection criterion for the stress tolerance of several plant species [20].

9.4 PROLINE BIOSYNTHESIS

The biosynthesis of proline in plants is adjusted by either the glutamate pathway or the ornithine pathway. The glutamate pathway accounts for major proline accumulation during stress. The synthesized proline is from glutamic acid via pyrroline-5-carboxylate (P5C) by two successive reductions catalyzed by P5C synthase (P5CS; EC 2.7.2.11) and P5C reductase (P5CR; EC 1.5.1.2) (Figure 9.1). P5CS catalyzes the activation of glutamate by phosphorylation and also the reduction of the labile intermediate γ-glutamyl phosphate into glutamate semialdehyde (GSA) which spontaneously cyclizes to d-pyrroline-5-carboxylate (P5C), and then finally reduced to proline by d1-pyrroline-5-carboxylate reductase (P5CR) [21]. The two plant isoforms, P5CS1 and P5CS2, function in the cytosol, which can probably be shifted from P5CS1 to the chloroplasts under abiotic stresses [22]. In accordance with this, Lehmann, Funck, Szabados and Rentsch [23] demonstrated that proline is synthesized in both the chloroplast and the cytoplasm under stressful conditions and is subsequently transported to, and degraded, in the root mitochondrion (Figure 9.1). In the ornithine pathway, it is transaminated to P5C by ornithine-d-aminotransferase (OAT; EC 2.6.1.13) via an intermediate pyrroline-2-carboxylate, but it is notable that this pathway is

still questionable [21]. It looks as if the ornithine pathway is predominantly in mitochondria under a high nitrogen supply; whereas, the glutamate pathway is activated during osmotic stress and nitrogen scarcity [23]. It appears that possibly in relation to N-nutritional status, regulation of proline synthesis could be different in plant species under stress conditions. Based on this hypothesis, the glutamate pathway is mainly activated in the grasses, and in the legumes, the ornithine pathway is predominant [22]. It is also reported that proline biosynthesis in not only regulated by enzymes, but also by other factors [11].

In fact, proline transportation along the plant parts and among cellular organelles have also been suggested as the most important factors. In the halophyte *Limonium latifolium*, proline was sequestered to vacuoles in non-stressed plants; whereas, proline accumulation was found in the cytosol of salt-stressed plants [24], indicating transportation of proline from the vacuole to the cytosol. A unique salt-inducible proline transporter was isolated from barley and found to be expressed strongly in root cap cells under salt stress [25], pointing to its role in the root tip region when plants are exposed to salt stress. Regarding the low capacity of roots in proline biosynthesis, proline accumulation is perhaps regulated by the rates of its transport and utilization. In support of this, Lehmann, Funck, Szabados, and Rentsch [23] reported that proline transporters, plasma membrane-localized transporters, were shown to facilitate the uptake of proline in plants, indicating a role in the import of proline into the roots. Additionally, AtProT1 expression was also found in the phloem cells, indicative of a role in the long-distance transport of proline as well as other solutes [24]. Based on the above findings, it is expected that proline content under stress conditions is related to transport processes.

9.5 PROLINE CATABOLISM

Proline catabolism occurs predominately in mitochondria by proline dehydrogenase (PDH) or proline oxidase (POX) producing P5C, which converted to glutamate by P5C dehydrogenase (P5CDH) (Figure 9.1) [26]. The PDH is encoded by two genes, whereas a single P5CDH gene has been identified in Arabidopsis and tobacco [24]. The PDH and P5CDH use FAD and NAD as electron acceptors and generate FADH and NADH, respectively, and deliver electrons for mitochondrial respiration [4]. Thus, the biosynthetic enzymes (P5CS1, P5CS2, and P5CR) were predicted to be localized in the cytosol and chloroplasts, whereas mitochondrial localization was predicted for the enzymes, such as PDH1, PDH2, P5CDH, and OAT, involved in proline catabolism [27].

9.6 REACTIVE OXYGEN SPECIES SCAVENGING

It has been reported that different stresses cause osmotic stress, and plants accumulate osmotically active compounds to lower the osmotic potential [9]. Induction of osmotic stress is responsible for the oxidative stress caused by reactive

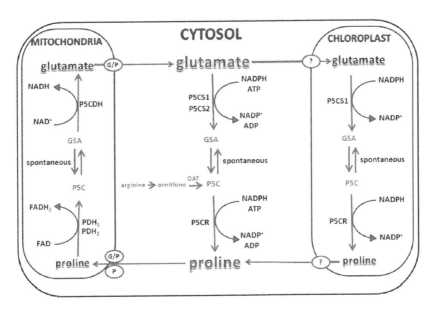

FIGURE 9.1   Schematic overview of proline biosynthesis in the cytosol or the chloroplast, and proline catabolism in the mitochondria. G/P, mitochondrial glutamate/proline antiporter; P, mitochondrial proline transporter.

oxygen species such as superoxide radical ($O_2{}^{\bullet-}$), hydrogen peroxide (H_2O_2) and hydroxyl radicals ($^{\bullet}OH$) [17]. Indeed, ROS are the byproduct of normal cellular metabolism and have important roles in cell signaling and transduction [28]. However, the excessive accumulation of ROS caused by different stresses could lead to cytotoxic oxidative damages to membrane lipids, proteins, and nucleic acids, and ultimately to cellular structure [23]. Also, ROS are highly reactive and can damage nucleic acids, and photosynthetic components [28]. The harmful effect of ROS is the oxidative attack on proteins, resulting in site-specific amino acid modifications, fragmentation of the peptide chain, aggregation of cross-linked reaction products and increased susceptibility to proteolysis [22]. The degree of damage by ROS depends on the balance between the product of ROS and its removal by the antioxidant scavenging systems [26, 29]. The main sources of ROS production in plants are chloroplasts, mitochondria, and peroxisomes (Figure 9.2). Plants have developed various protective mechanisms to eliminate or reduce the deleterious effects of ROS through antioxidant enzymes such as superoxide dismutase (SOD), catalase (CAT), ascorbate peroxidase (APX), glutathione reductase (GR), and non-enzymatic antioxidants (ascorbate, glutathione, and tocopherol), etc. [23]. It was proposed that proline is an effective hydroxyl radical scavenger [1]. Proline could also deactivate $1O_2$ by physical quenching [20]. Mechanistic evidence was collected using density functional theory coupled with the polarizable continuum model (DFT/PCM) to sustain the role for proline as a protective $OH^{\bullet}$ scavenger under stress conditions in plants. It was proposed proline–proline cycle operation (Figure 9.2) in which proline captures the first $OH^{\bullet}$ by H-abstraction followed by the second H-abstraction which also captures another $OH^{\bullet}$ yielding P5C which is, then, recycled back to proline by the action of the P5CR/NADPH enzymatic system [29].

9.7   EXOGENOUS PROLINE APPLICATION

It is widely reported that application of proline could enhance the tolerance of various plant species to different environmental stresses [10, 13, 15, 16]. These reports indicated that application of proline increased 1000-grains weight, grain, and essential oil yield, and water use efficiency in genotypes of fennel under drought-stress conditions. Also, proline application in date palm (*Phoenix dactylifera* L.) alleviated the oxidative damage induced by cadmium (Cd) accumulation and established better levels of plant growth, water status, and photosynthetic activity. Moreover, proline-treated plants showed high antioxidant enzyme activity (superoxide dismutase, catalase, and glutathione peroxidase) in roots and leaves as compared to Cd-treated plants [23]. Yu, Lin, Fan and Lu [30] demonstrated that exogenous proline was able to effectively reduce hexavalent chromium (Cr (VI))-induced lipid peroxidation in rice seedlings. Moreover, Rady, Taha and Mahdi [31] have found enhancing growth, productivity, and anatomy of lupine varieties under salt stress by exogenous application of proline. In fact, the mitigating effect of proline in this investigation was related to increased carotenoids, chlorophylls, total soluble sugars, and endogenous proline. In agreement with these results, exogenous proline supplementation in salt-treated rice plants, in addition to reducing the Na^+/K^+ ratio, increased proline production and P5CS activity as well as elevated P5CR transcript levels but decreased the activity of the antioxidant enzymes [1].

9.8   EFFECTIVE CONCENTRATIONS OF EXOGENOUS PROLINE

Reports in the literature strongly suggest that exogenous proline application with low concentration would be more efficient in inducing tolerance to abiotic stresses in different plant

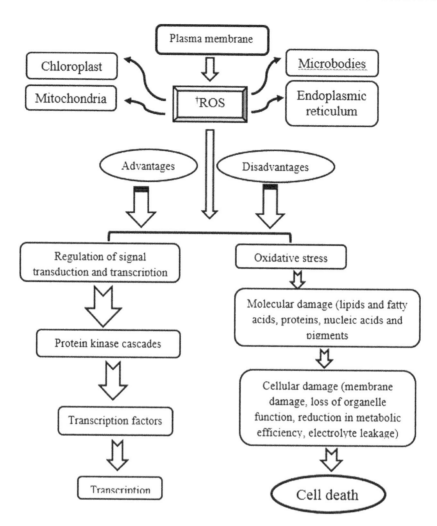

FIGURE 9.2 Sites of reactive oxygen species and the biological consequences leading to a variety of physiological dysfunctions that can lead to cell death. [†]: Reactive oxygen species.

species [26]. Therefore, it is essential to determine the optimal concentration of proline [3]. For example, Cuin and Shabala [32] indicated that in contrast to high proline concentration, low (0.5–5 mM) concentrations of exogenously supplied proline significantly reduced NaCl-induced K^+ efflux from barley roots. In fact, the authors concluded that high intracellular concentrations of compatible solutes are not required for an amelioratory role. In agreement with this, Rady, Taha and Mahdi [31] demonstrated that although proline at a low concentration (20, 30 mM) ameliorated the detrimental effects of salinity, higher concentrations (40, 50 mM) of proline had toxic effects on growth of rice seedlings. In accordance, application of proline in low concentrations (50 mM) was more effective than 100 mM, in alleviating the deleterious impact of salinity on the growth of groundnut [7]. Further, Lone, Kueh, Wyn Jones, and Bright [16] demonstrated that low concentrations of proline could mitigate the adverse effects of salinity in soaking rice seeds, while high proline concentration was injurious. Moreover, Heuer [33] found that an external supplementation of a low concentration (1 mM) of proline resulted in the higher K concentration in salt-stressed tomato leaves. In addition, an exogenous application of 10 mM proline stimulated growth,

increased activities of antioxidant enzymes and H^+-ATPase, increased proline and K levels, whereas reduced Na and lipid peroxidation in *Eurya emarginata* under salt stress [20].

It appears that a key impact of proline is increasing the antioxidant activities which contribute to improved tolerance. In view of this, several studies have attributed exogenous proline application to enhancing the activity of antioxidative enzymes in chickpeas [34, 35], lentil [36], sugarcane [37], and olives [38], under various stress conditions (Table 9.1). Researchers have suggested that proline could contribute to detoxification of O^{2-} radicals by increasing the activity of SOD in *Solanum nigrum* under Cd stress. The activities of APX, monodehydroascorbate reductase, and dehydroascorbate reductase, which are components of the ascorbate–glutathione cycle, are significantly enhanced by exogenous proline in tobacco cultures exposed to salinity stress [1, 26]. Some studies showed an increase in the content of ascorbate and glutathione, reduced glutathione/disulfide ratio, and higher activity of APX, GR, and CAT together with a decrease in H_2O_2 and malondialdehyde content in salt-treated mung beans that had been pre-treated with proline (Table 9.1). Reports also revealed that plants may over-accumulate proline and mitigate

TABLE 9.1

Exogenous Application of Proline and Its Protective Role in Plants Under Abiotic Stresses

Plant	Effects	Reference
Chickpea (*Cicer arietinum*)	Reduced oxidative injury by elevating enzymatic and non-enzymatic antioxidants	[35]
Mung bean (*Phaseolus vulgaris*)	Stimulated components of the ascorbate–glutathione cycle, increased activities of antioxidative enzymes, decreased lipid peroxidation and H_2O_2 content	[39]
Melon (*Cucumis melo*)	Increased fresh and dry masses, increased PN, Fv/Fm, and Chl content, reduced O_2 and the H_2O_2 content, enhanced activities of antioxidative enzymes	[40]
Melon (*Cucumis melo*)	Induced salt tolerance through plasma membrane protection, increased Ca^{2+}, K^+, N and decreased Na^+	[41]
Sea daffodil (*Pancratium maritimum*)	Protected protein turnover machinery against stress-damage and up-regulated stress-protective proteins	[42]
Lentil (*Lens culinaris*)	Induced upregulation of glutathione transferase and glyoxalase	[36]
Wheat (*Triticum aestivum*)	Improved shoot and root fresh and dry masses, shoot length, and grain yield	[43]
Tobacco (*Nicotiana tabacum*)	Increased fresh mass, enhanced activities of antioxidative enzymes	[44]
Rice (*Oryza sativa*)	Increased salt tolerance via reduced Na^+/K^+ ratio and increased transcript levels of proline biosynthesizing genes	[18]
Sugarcane (*Saccharum officinarum*)	Alleviated salt-induced oxidative stress by improving guaicol peroxidase activity	[37]
Olive (*Olea europaea*)	Modulated antioxidative defense system, increased photosynthetic activity, and plant growth maintaining a suitable plant water status	[45]
Rice (*Oryza sativa*)	Increased plant growth and N_2 assimilation enzyme activities	[46]

PN – Net Photosynthesis Rate; FV/Fm – Maximum Quantum Yield of Photosystem II Photochemistry; Chl – Chlorophyll

the deleterious effects of drought-induced oxidative stress by lowering lipid peroxidation [33]. In transgenic citrus plants carrying the heterologous gene P5CS112A, a high endogenous proline caused an increase in transcription of cytosolic APX and chloroplast GR, and Cu/Zn SOD isoforms [28]. In addition to enhancing the activity of antioxidant enzymes, proline is also known to increase some beneficial compounds such as enzyme protection and, also, could increase the activity of other enzymes [11].

9.9 SEED GERMINATION AND PROLINE

Although numerous scientists have studied exogenous proline improving the salt tolerance of plants, most of these studies concern using proline in medium or foliar-spraying. Heuer [33] indicated that increasing the level of free proline enhanced radicle emergence in Arabidopsis seeds and feedback inhibition of proline synthesis by exogenous proline decreased germination rate. The author [Heuer [33] is only one author.] studied dehydrogenase enzymes participating in the oxidative pentose phosphate pathway (OPPP) during the seed germination of Arabidopsis and observed the activation of this pathway as well as a fourfold increase in proline content before the radicle emergence. When the AtP5CS1 gene was inserted in an antisense orientation into Arabidopsis, the radicle emergence was delayed. It was inferred that proline

synthesis might replenish the NADP pool and, thus, activate OPPP. The results indicate a functional relationship between elevated proline biosynthesis and increased OPPP activity and hence coupling both pathways might be important in stimulating tolerance to salinity. These results are in agreement with the findings of Kavi Kishor, and Sreenivasulu [47] who suggested that OPPP is crucial for enhancing seed germination, and there is a link between proline and OPPP. Also, the results of Khan, Fatima, Ghani, Nadeem, Aziz, Hussain, and Ikram [13] showed that application of proline (30 mM) improved rice germination and seedling growth under salt stress, whereas higher concentrations (40 or 50 mM) resulted in reduced seedling growth and lowered K^+/Na^+ ratio. In addition, tobacco plants transformed with a P5CS dramatically increased the proline level, which resulted in improved germination and growth of tobacco seedlings under salt stress [33]. Improving the germination status of rice under salt stress was observed by using exogenous proline [36]. Moreover, the germination energy and relative germination energy of salt-stressed rice improved with increasing proline concentrations in the range of 5–45 mM. Also, soaking with 15 mM/L and 30 mM/L proline significantly improved amylase activity (e.g., alpha-amylase, beta-amylase, and total amylase) in rice under salt stress [48]. Subsequently, researchers [8, 17] found that osmopriming improved *Brassica napus* seed germination and salinity tolerance during post-priming germination and seedling establishment. Also, they concluded that this germination performance is highly related to proline accumulation as a result of the strong upregulation of the P5CSA gene, down-regulation of the PDH gene, and accumulation of hydrogen peroxide [8, 17].

In light of the available information, it seems that proline synthesis plays an important role in promoting germination. In fact, it appears that soaking seeds with proline can relieve the inhibition effect of stresses on seed germination. One idea is that soaking seeds with proline could improve the internal osmotic potential of seeds and enable them to maintain an adequate water supply under stress conditions.

9.10 GENETIC MANIPULATION OF PROLINE METABOLISM

In plants, numerous key enzymes in metabolic pathways are largely encoded by redundant genes, which are probably produced by gene duplication events during the evolution of the organism. Gene duplication generates two functionally identical copies of the gene that perform a redundant action immediately following the duplication event, and are often followed by categorized changes causing alterations in transcriptional regulation and contributing to the evolution of functional divergence. Figure 9.3 summarizes the genetic manipulations of proline metabolic pathways in plants that have been performed in scientific literature [49].

In Arabidopsis, knockout of the P5CS1 gene which encodes P5CS1 impairs proline synthesis leading to hypersensitivity to salt stress [47], indicating the importance of this enzyme for proline accumulation and enhancing tolerance to salinity. In this study, Arabidopsis P5CS1 insertion mutants had only 10% proline relative to the wild type and, also, more ROS and lipid peroxidation, suggestive of a proline role in scavenging ROS and, hence, lowering lipid peroxidation. Similarly, rice seedlings resulting from a knockout mutation in OsP5CS2 were more sensitive to salt stress than the wild types [33]. Moreover, over-expression of Vigna P5CS in sorghum [31] and tobacco [35] caused a higher yield and root biomass under salt stress. Also, the activities and expression of P5CS were found to be significantly increased in a cactus pear that lead to an enhanced proline accumulation under salt stress [16]. Accordingly, when rice plants were transformed with P5CS

cDNA and subjected to salt stress, overproduction of the P5CS enzyme resulted in proline accumulation and enhanced salinity tolerance [35]. Additional support comes from the finding that salt stress steadily increases the expression of the P5CS gene in line with proline content in a salt-tolerant rice line more than a sensitive one [48]. Furthermore, P5CS over-expression elevated free proline content in transgenic tobacco [21]. Also, tobacco plants transformed with a P5CS dramatically increased proline level, which resulted in the improved germination and growth of tobacco seedlings under salt stress [1]. Consistently, over-expression of the P5CS gene in tobacco resulted in 10- to 18-fold more proline in transgenes than in the control plants [21], resulting in improved salt tolerance. Overproduction of proline in this work enhanced root biomass and flower development in transgenic plants in response to osmotic stress. Additionally, transcript analysis demonstrated that the P5CS gene had a higher constitutive expression level in *Thellungiella salsuginea* than in Arabidopsis, and was induced more rapidly under salinity [6].

It is reported that the effects of genetic manipulation of proline synthesis can be plant species specific. Over-expressing the soybean P5CR gene in transgenic tobacco did not increase osmotolerance [49]. The over-expression of the P5CS encoding gene in transgenic tobacco plants resulted in increased proline production and conferred tolerance of these plants to osmotic stress, confirming that P5CS is of key importance for the biosynthesis of proline in plants [35]. Field trials in South Africa with P5CR transgenic soybean lines supported improved drought performance and higher heat tolerance compared to the wild type cultivars [14]. However, mutants displaying higher proline accumulation can also be salt hypersensitive [16]. It has been shown that transcriptional control of the P5CS gene is important for the regulation of the accumulation of proline during osmotic stress in plants. Studies of the transcriptional regulation of genes involved in proline synthesis confirmed developmental regulation. In young Arabidopsis plants, beta-glucuronidase (GUS) analysis of an AtP5R (Arabidopsis P5CR gene) promoter revealed high expression in apical meristem and young leaf, in the root meristem, secondary root primordia, and root vascular cylinder. In young leaf, high P5R expression could be detected all over the leaf blade, while in old leaves expression was restricted to the veins, hydathodes, guard cells, and base of trichomes

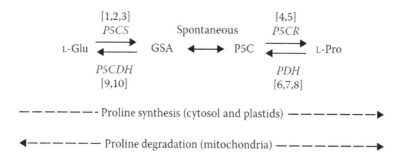

FIGURE 9.3 Genetic manipulations of proline metabolic pathways in plants. The main precursor of proline synthesis is L-glutamic acid (L-Glu). L-Glu is first reduced to glutamate semialdehyde, which spontaneously cyclizes to pyrroline-5-carboxylate (P5C), by P5C synthase (P5CS). The second reduction, of P5C to Pro, is catalyzed by P5C reductase (P5CR).

[1]. In flowering plants, high AtP5R expression could be detected in rapidly dividing cells, such as root meristem, and cells or tissues undergoing changes in water potential, such as hydathode, guard cell, ovule, developing seed, and pollen grain. AtHAL3a, a HAL3 homolog gene, in *Arabidopsis thaliana* over-expressing transgenic plants exhibit improved salt and sorbitol tolerance [16]. Over-expression of NtHAL3a, improved salt, osmotic, and lithium tolerance in cultured tobacco cells. Proline degradation can also be manipulated and depends on prior proline transport to mitochondria. Over-expression of PDH in *A. thaliana* did not result in morphological abnormalities, probably because proline homeostasis relies on regulated transport between cell compartments [18].

9.11 PROLINE AND TOLERANCE TO ENVIRONMENTAL STRESSES

Although proline accumulation is a common response to a wide range of stresses [13, 22, 28] which contribute substantially to the cytoplasmic osmotic adjustment [4, 5, 12], the relationship between proline accumulation and stress tolerance is not clear [10, 16, 37]. For example, though several researchers have already found that salinity tolerance is normally associated with proline accumulation[11, 14], some studies have demonstrated that salt-sensitive cultivars exhibited a greater accumulation of proline [22, 25, 31, 34]. For example, while in salt-tolerant alfalfa plants, proline concentration in the root rapidly doubled under salt stress, in salt-sensitive plants the response was slow [11]. Similarly, salt-tolerant genotypes of sesame cultivars accumulated more proline in response to salinity than the salt-sensitive genotypes. Moreover, proline was not involved in the osmotic adjustment of different plant species [22, 30, 31, 33] in response to salt treatment. Also, other authors reported that proline accumulation and distribution during shoot and leaf development in two sorghum genotypes contrasting in salt tolerance was a reaction to salt stress and not a plant response associated with tolerance. In fact, they suggested that proline accumulation did not appear to be related to salt tolerance in sorghum cultivars, and might be the result of plant responses to damage caused by stress [5, 21, 47]. Bavei, Shiran, and Arzani [50] observed positive relationships between proline content and fresh weight of sorghum plants under salinity stress. Therefore, they concluded that proline accumulation is not merely a symptom of salt-stress injury in sorghum, but rather it can be used as a criterion for salt tolerance. However, it seems that since proline is also accumulated under other stresses, such as high temperature, drought, and starvation, it cannot be used as an exclusive criterion for salt tolerance [8]. Hussain Wani, Brajendra Singh, Haribhushan, and Iqbal Mir [51] found that proline accumulation was not an essential part of the protection mechanism against salinity and drought in one variety of safflower, whereas proline accumulation was increased in the other varieties. While Hayat, Hayat, Alyemeni, Wani, Pichtel, and Ahmad [1] reported that species and cultivars which are classified as salinity-tolerant accumulate higher levels of proline; contrarily, a negative correlation between salinity tolerance and accumulation of proline have been found [20]. Also, in rice plants grown under salt stress, the accumulation of proline in the leaf was deemed to be a symptom of salt injury rather than an indication of salt tolerance [34]. In fact, the amount of proline accumulation in sensitive cultivars was greater than that in salt-tolerant cultivars. Signorelli, Corpas, Rodríguez-Ruiz, Valderrama, Barroso, Borsani, and Monza [29] indicated that because proline concentration increased as NaCl increased, it would not be appropriate to use this parameter to establish differences in sensitivity to salt stress among the three genotypes.

Also, it appears that proline accumulation is only a result of salt injury rather than an adaptation and acclimation to salinity conditions. On the other hand, the accumulation of one mole of proline usually needs 41 mole of ATP consumption [47], and this process usually occurs at the expense of plant growth. In general, it is controversial to say that the hyperaccumulation of proline is essential for improving salinity tolerance, because maybe it is just a symptom of salt stress.

In contrast to the above findings, there are strong correlations between stress tolerance and proline accumulation in plants. For example, Kaushal, Gupta, Bhandhari, Kumar, Thakur, and Nayyar [35] indicated that salt-tolerant rice lines accumulated more proline than the salt-sensitive ones under salt stress. Also, high salinity remarkably increased proline accumulation in the leaves of a salt-tolerant rice cultivar greater than that in the sensitive one [10]. Similarly, salt-tolerant brassica [13] and tomato [33] cultivars accumulated higher proline than the salt-sensitive cultivars in response to salinity treatments. In accordance, Zahedi, Hooman, and Saed-Moucheshi [20] concluded that since proline content decreased in the roots of the salt-sensitive rice cultivar, but increased in the tolerant one under salt stress, it was considered as an indicator of stress tolerance, not a symptom of injury. In agreement, Saed-Moucheshi, Heidari, Zarei, Emam, and Pessarakli [9] found a greater proline accumulation in drought-tolerant cultivars of wheat than in the drought-sensitive ones. Inconsistent to the above findings, proline shrank.

9.12 CONCLUDING REMARKS

Most plant species naturally accumulate proline and other compounds (for example, glycine betaine) as major organic osmolytes subject to altered abiotic stresses. These compounds are thought to play a role in the adaptation to stress conditions, as well as mediating osmotic adjustment for protecting cellular and subcellular constructions in stressed plants. Nevertheless, different plant species normally accumulate an altered amount of proline and other osmoregulators under different conditions. Consequently, different methodologies have been suggested and tested to increase the concentrations of these compounds under abiotic stress conditions to increase their adaptability and stress tolerance. In this chapter, some of the basic subjects related to these osmoregulators were discussed. Also, the mechanism and pathways for proline production and their roles in abiotic stresses were assessed. Additionally, this chapter considered some contradictions related to the influence of proline under stress conditions. Although numerous studies have been conducted related to proline as an osmoregulator, there is yet some gaps; for example, complete pathways and compounds that regulate the expression of this compound require more studies. Moreover, some plants respond better to a higher concentration of proline and other osmoregulators, which should be studied further, and different methods of genetic manipulation for altered pathways related to the production of these regulators should be tested and, considering their influence on stress tolerance in important plants, are highly recommended.

REFERENCES

1. Hayat, S., Hayat, Q., Alyemeni, M.N., Wani, A.S., Pichtel, J., and Ahmad, A. 2012. Role of Proline under Changing Environments: A Review. *Plant Signaling and Behavior* 7, (11):1456–466.
2. Slama, I., Abdelly, C., Bouchereau, A., Flowers, T., and Savoure, A. 2015. Diversity, Distribution and Roles of Osmoprotective Compounds Accumulated in Halophytes under Abiotic Stress. *Annals of Botany* 115, (3):433–447.
3. Ashraf, M. and Foolad, M. 2007. Roles of Glycine Betaine and Proline in Improving Plant Abiotic Stress Resistance. *Environmental Experimental Botany* 59, (2):206–216.
4. Ashraf, M., Akram, N., Al-Qurainy, F., and Foolad, M. 2011. Drought Tolerance: Roles of Organic Osmolytes, Growth Regulators, and Mineral Nutrients. In *Advances in Agronomy*, Elsevier, 249–296.
5. Saed-Moucheshi, A., Pakniyat, H., Pirasteh-Anosheh, H., and Azooz, M. 2014. *Role of Ros as Signaling Molecules in Plants*, Springer, New York, NY, p: 49.
6. Tadayyon, A., Nikneshan, P., and Pessarakli, M. 2018. Effects of Drought Stress on Concentration of Macro-and Micro-Nutrients in Castor (Ricinus Communis L.) *Plant. Journal of Plant Nutrition* 41, (3):304–310.
7. Júnior, D.F., Gaion, L., Júnior, G.S., Santos, D., and Carvalho, R. 2018. Drought-Induced Proline Synthesis Depends on Root-to-Shoot Communication Mediated by Light Perception. *Acta Physiologiae Plantarum* 40, (1):15.
8. Hunt, L., Amsbury, S., Baillie, A.L., Movahedi, M., Mitchell, A., Afsharinafar, M., Swarup, K., Denyer, T., Hobbs, J., and Swarup, R. 2017. Formation of the Stomatal Outer Cuticular Ledge Requires a Guard Cell Wall Proline-Rich Protein. *Plant Physiology* 42:715–729.
9. Saed-Moucheshi, A., Heidari, B., Zarei, M., Emam, Y., and Pessarakli, M. 2013. Changes in Antioxidant Enzymes Activity and Physiological Traits of Wheat Cultivars in Response to Arbuscular Mycorrhizal Symbiosis in Different Water Regimes. *Iran Agricultural Research* 31, (2):35–50.
10. Demiralay, M., Aluntas, C., Sezgin, A., and Terzi, R. 2017. Application of Proline to Root Medium Is More Effective for Amelioration of Photosynthetic Damages as Compared to Foliar Spraying or Seed Soaking in Maize Seedlings under Short-Term Drought. *Turkish Journal of Biology* 41, (4):649–660.
11. Khan, F., Niaz, K., Hassan, F.I., and Abdollahi, M. 2017. An Evidence-Based Review of the Genotoxic and Reproductive Effects of Sulfur Mustard. *Archives of Toxicology* 91, (3):1143–1156.
12. Abdallah, H.K., Abbas, M.K., and Hassan, A.E. 2017. Effect of Proline and Glycine Betaine in Improving Vegetative Growth of Washington Navel Orange (Citrus Sinensis L.) under Salinity Conditions. *Kufa Journal for Agricultural Science* 9, (1):1–30.
13. Khan, A., Fatima, H., Ghani, A., Nadeem, M., Aziz, A., Hussain, M., and Ikram, M. 2018. Improving Salinity Tolerance in Brassica (Brassica Napus Var. Bsa and Brassica Campestris Var. Toria) by Exogenous Application of Proline and Glycine Betaine. *Pakistan Journal of Scientific Industrial Research* 61, (1):1–9.
14. de Freitas, P.A.F., de Souza Miranda, R., Marques, E.C., Prisco, J.T., and Gomes-Filho, E. 2018. Salt Tolerance Induced by Exogenous Proline in Maize Is Related to Low Oxidative Damage and Favorable Ionic Homeostasis. *Journal of Plant Growth Regulation*:1–14.
15. Csiszár, J., Brunner, S., Horváth, E., Bela, K., Ködmön, P., Riyazuddin, R., Gallé, Á., Hurton, Á., Papdi, C., and Szabados, L. 2018. Exogenously Applied Salicylic Acid Maintains Redox Homeostasis in Salt-Stressed Arabidopsis Grl Mutants Expressing Cytosolic Rogfp1. *Plant Growth Regulation* 24:1–14.
16. Lone, M., Kueh, J., Wyn Jones, R., and Bright, S. 1987. Influence of Proline and Glycinebetaine on Salt Tolerance of Cultured Barley Embryos. *Journal of Experimental Botany* 38, (3):479–490.
17. Guerzoni, J.T.S., Belintani, N.G., Moreira, R.M.P., Hoshino, A.A., Domingues, D.S., Bespalhok Filho, J.C., and Vieira, L.G.E. 2014. Stress-Induced Δ1-Pyrroline-5-Carboxylate Synthetase (P5cs) Gene Confers Tolerance to Salt Stress in Transgenic Sugarcane. *Acta Physiologiae Plantarum* 36, (9):2309–2319.
18. Nounjan, N., Nghia, P.T., and Theerakulpisut, P. 2012. Exogenous Proline and Trehalose Promote Recovery of Rice Seedlings from Salt-Stress and Differentially Modulate Antioxidant Enzymes and Expression of Related Genes. *Journal of Plant Physiology* 169, (6):596–604.
19. de Freitas, P.A.F., de Souza Miranda, R., Marques, E.C., Prisco, J.T., and Gomes-Filho, E. 2018. Salt Tolerance Induced by Exogenous Proline in Maize Is Related to Low Oxidative Damage and Favorable Ionic Homeostasis. *Journal of Plant Growth Regulation* 27:1–14.
20. Zahedi, M.B., Hooman, R., and Saed-Moucheshi, A. 2016 Evaluation of Antioxidant Enzymes, Lipid Peroxidation and Proline Content as Selection Criteria for Grain Yield under Water Deficit Stress in Barley. *Journal of Applied Biological Sciences* (1):71–78.
21. Kaur, G. and Asthir, B. 2015. Proline: A Key Player in Plant Abiotic Stress Tolerance. *Biologia Plantarum* 59, (4):609–619.
22. Liang, X., Zhang, L., Natarajan, S.K., and Becker, D.F. 2013. Proline Mechanisms of Stress Survival. *Antioxidants Redox Signaling* 19, (9):998–1011.
23. Lehmann, S., Funck, D., Szabados, L., and Rentsch, D. 2010. Proline Metabolism and Transport in Plant Development. *Amino Acids* 39, (4):949–962.
24. Gagneul, D., Aïnouche, A., Duhazé, C., Lugan, R., Larher, F.R., and Bouchereau, A. 2007. A Reassessment of the Function of the So-Called Compatible Solutes in the Halophytic Plumbaginaceae Limonium Latifolium. *Plant Physiology* 144, (3):1598–1611.
25. Saed-Moocheshi, A., Shekoofa, A., Sadeghi, H., and Pessarakli, M. 2014. Drought and Salt Stress Mitigation by Seed Priming with Kno3 and Urea in Various Maize Hybrids: An Experimental Approach Based on Enhancing Antioxidant Responses. *Journal of Plant Nutrition* 37, (5):674–689.
26. Mansour, M.M.F. and Ali, E.F. 2017. Evaluation of Proline Functions in Saline Conditions. *Phytochemistry* 140:52–68.
27. Ren, Y., Miao, M., Meng, Y., Cao, J., Fan, T., Yue, J., Xiao, F., Liu, Y., and Cao, S. 2018. Dfr1-Mediated Inhibition of Proline Degradation Pathway Regulates Drought and Freezing Tolerance in Arabidopsis. *Cell Reports* 23, (13):3960–3974.

28. Saed-Moucheshi, A., Shekoofa, A., and Pessarakli, M. 2014. Reactive Oxygen Species (Ros) Generation and Detoxifying in Plants. *Journal of Plant Nutrition* 37, (10):1573–1585.

29. Signorelli, S., Corpas, F.J., Rodríguez-Ruiz, M., Valderrama, R., Barroso, J.B., Borsani, O., and Monza, J. 2018. Drought Stress Triggers the Accumulation of No and Snos in Cortical Cells of Lotus Japonicus L. Roots and the Nitration of Proteins with Relevant Metabolic Function. *Environmental Experimental Botany* 98:1201–1224.

30. Yu, X.-Z., Lin, Y.-J., Fan, W.-J., and Lu, M.-R. 2017. The Role of Exogenous Proline in Amelioration of Lipid Peroxidation in Rice Seedlings Exposed to Cr (Vi). *International Biodeterioration Biodegradation* 123:106–112.

31. Rady, M.M., Taha, R.S., and Mahdi, A.H. 2016. Proline Enhances Growth, Productivity and Anatomy of Two Varieties of Lupinus Termis L. Grown under Salt Stress. *South African Journal of Botany* 102:221–227.

32. Cuin, T.A. and Shabala, S. 2005. Exogenously Supplied Compatible Solutes Rapidly Ameliorate Nacl-Induced Potassium Efflux from Barley Roots. *Plant Cell Physiology* 46, (12):1924–1933.

33. Heuer, B. 2003. Influence of Exogenous Application of Proline and Glycinebetaine on Growth of Salt-Stressed Tomato Plants. *Plant Science* 165, (4):693–699.

34. Moucheshi, A., Heidari, B., and Dadkhodaie, A. 2011. Genetic Variation and Agronomic Evaluation of Chickpea Cultivars for Grain Yield and Its Components under Irrigated and Rainfed Growing Conditions. *Iran Agricultural Research* 29, (2):39–50.

35. Kaushal, N., Gupta, K., Bhandhari, K., Kumar, S., Thakur, P., and Nayyar, H. 2011. Proline Induces Heat Tolerance in Chickpea (Cicer Arietinum L.) Plants by Protecting Vital Enzymes of Carbon and Antioxidative Metabolism. *Physiology Molecular Biology of Plants* 17, (3):203.

36. Molla, M.R., Ali, M.R., Hasanuzzaman, M., Al-Mamun, M.H., Ahmed, A., Nazim-ud-Dowla, M., and Rohman, M.M. 2014. Exogenous Proline and Betaine-Induced Upregulation of Glutathione Transferase and Glyoxalase I in Lentil (Lens Culinaris) under Drought Stress. *Notulae Botanicae Horti Agrobotanici Cluj-Napoca* 42, (1):73–80.

37. Patade, V.Y., Lokhande, V.H., and Suprasanna, P. 2014. Exogenous Application of Proline Alleviates Salt Induced Oxidative Stress More Efficiently Than Glycine Betaine in Sugarcane Cultured Cells. *Sugar Tech* 16, (1):22–29.

38. Ben Ahmed, C., Ben Rouina, B., Sensoy, S., Boukhriss, M., and Ben Abdullah, F. 2010. Exogenous Proline Effects on Photosynthetic Performance and Antioxidant Defense System of Young Olive Tree. *Journal of Agricultural Food Chemistry* 58, (7):4216–4222.

39. Aggarwal, M., Sharma, S., Kaur, N., Pathania, D., Bhandhari, K., Kaushal, N., Kaur, R., Singh, K., Srivastava, A., and Nayyar, H. 2011. Exogenous Proline Application Reduces Phytotoxic Effects of Selenium by Minimising Oxidative Stress and Improves Growth in Bean (Phaseolus Vulgaris L.) Seedlings. *Biological Trace Element Research* 140, (3):354–367.

40. Yan, Z., Guo, S., Shu, S., Sun, J., and Tezuka, T. 2011. Effects of Proline on Photosynthesis, Root Reactive Oxygen Species (Ros) Metabolism in Two Melon Cultivars (Cucumis Melo L.) under Nacl Stress. *African Journal of Biotechnology* 10, (80):18381–18390.

41. Kaya, C., Tuna, A.L., Ashraf, M., and Altunlu, H. 2007. Improved Salt Tolerance of Melon (Cucumis Melo L.) by the Addition of Proline and Potassium Nitrate. *Environmental Experimental Botany* 60, (3):397–403.

42. Khedr, A.H.A., Abbas, M.A., Wahid, A.A.A., Quick, W.P., and Abogadallah, G.M. 2003. Proline Induces the Expression of Salt-Stress-Responsive Proteins and May Improve the Adaptation of Pancratium Maritimum L. To Salt-Stress. *Journal of Experimental Botany* 54, (392):2553–2562.

43. Kamran, M., Shahbaz, M., Ashraf, M., and Akram, N.A. 2009. Alleviation of Drought-Induced Adverse Effects in Spring Wheat (Triticum Aestivum L.) Using Proline as a Pre-Sowing Seed Treatment. *Pakistan Journal of Botany* 41, (2):621–632.

44. Hoque, M.A., Okuma, E., Banu, M.N.A., Nakamura, Y., Shimoishi, Y., and Murata, Y. 2007. Exogenous Proline Mitigates the Detrimental Effects of Salt Stress More Than Exogenous Betaine by Increasing Antioxidant Enzyme Activities. *Journal of Plant Physiology* 164, (5):553–561.

45. Ahmad, P., Jaleel, C., and Sharma, S. 2010. Antioxidant Defense System, Lipid Peroxidation, Proline-Metabolizing Enzymes, and Biochemical Activities in Two Morus Alba Genotypes Subjected to Nacl Stress. *Russian Journal of Plant Physiology* 57, (4):509–517.

46. Teh, C.-Y., Shaharuddin, N.A., Ho, C.-L., and Mahmood, M. 2016. Exogenous Proline Significantly Affects the Plant Growth and Nitrogen Assimilation Enzymes Activities in Rice (Oryza Sativa) under Salt Stress. *Acta Physiologiae Plantarum* 38, (6):151.

47. Kishor, K., Polvarap, B., and Sreenivasulu, N. 2014. Is Proline Accumulation Per Se Correlated with Stress Tolerance or Is Proline Homeostasis a More Critical Issue? *Plant, Cell Environmental Experimental Botany* 37, (2):300–311.

48. Singh, M., Singh, A., Nehal, N., and Sharma, N. 2018. Effect of Proline on Germination and Seedling Growth of Rice (Oryza Sativa L.) under Salt Stress. *Journal of Pharmacognosy Phytochemistry* 7, (1):2449–2452.

49. Verbruggen, N. and Hermans, C. 2008. Proline Accumulation in Plants: A Review. *Amino Acids* 35, (4):753–759.

50. Bavei, V., Shiran, B., and Arzani, A. 2011. Evaluation of Salinity Tolerance in Sorghum (Sorghum Bicolor L.) Using Ion Accumulation, Proline and Peroxidase Criteria. *Plant Growth Regulation* 64, (3):275–285.

51. Hussain Wani, S., Brajendra Singh, N., Haribhushan, A., and Iqbal Mir, J. 2013. Compatible Solute Engineering in Plants for Abiotic Stress Tolerance-Role of Glycine Betaine. *Current Genomics* 14, (3):157–165.

10 Role of Dehydrins in Plant Stress Response

Klára Kosová, Ilja Tom Prášil, and Pavel Vítámvás

CONTENTS

10.1 INTRODUCTION: DEFINITION OF DEHYDRINS, THEIR STRUCTURE AND FUNCTIONS

10.1.1 DEFINITION OF DEHYDRINS

Dehydrins are highly hydrophilic, well-soluble proteins which belong to a large family of Late embryogenesis-abundant (LEA) proteins whose name comes from Galau et al. (1986) who studied them for the first time in cotton (*Gossypium hirsutum* L.) embryos (Dure et al., 1981; Galau and Dure, 1981). They are classified as group 2 LEA (or LEA II) proteins (Bray, 1993; Ingram and Bartels, 1996) or LEA-D11 proteins according to one dehydrin member in a cotton embryo (Dure et al., 1989). In the Pfam database of protein domains (http://www.sanger.ac.uk/Software/Pfam; Bateman et al., 2004), they are simply named dehydrins and have a Pfam number – PF00257. They have a relatively high glycine content (greater than 6%) and a hydrophilicity index (Kyte and Doolittle, 1982) greater than 1, thus they can be classified as hydrophilins (Garay-Arroyo et al., 2000; Battaglia et al., 2008). The first reported dehydrin proteins were RAB21, a protein induced by salt and osmotic stress in rice (*Oryza sativa* L.) (Mundy and Chua, 1988), and D-11, a protein accumulating in maturating embryo of cotton (*Gossypium hirsutum* L.) (Baker et al., 1988). At the end of the 1980s, dehydrins were defined as 'dehydration-induced proteins' according to their mode of expression (Close et al., 1989). Later, as their sequence characteristics became available, dehydrins were re-defined on the basis of their sequential motifs. They were newly defined as proteins possessing at least one copy of a conserved lysine-rich amino acid sequence, a K-segment, in their molecules (Close, 1996,1997). Due to this definition, based on the presence of a unique amino acid motif, dehydrins can be easily detected by a specific primary antibody raised against the K-segment (Close et al.,1993).

10.1.2 Classification of Dehydrins

The K-segment is a lysine-rich amino acid (aa) sequence present in 1 to 11 copies near to the C-terminus of the dehydrin molecules. The consensus amino acid sequence of the K-segment in angiosperms is EKKGIME/DKIKEKLPG. Apart from the K-segment, dehydrins can possess other conserved sequential motifs: the tyrosine-rich Y-segment (consensus (V/T)D(E/Q)YGNP) near to the N-terminus and the serine-rich S-segment. S-segment is formed by a stretch of 4 to 10 serine residues which are a part of a conserved sequence $LHRSGS_{4-10}(E/D)_3$. This sequence motif can be distinguished by casein kinase II (CK_2)-type kinases and plays an important role in the phosphorylation of serine residues in the S-segment (Svensson et al., 2002; Battaglia et al., 2008). According to the presence of the K-, S- and Y-segments, dehydrins can be divided into five structural sub-groups: K_n, SK_n, K_nS, Y_xK_n, and Y_xSK_n where 'x' and 'y' can represent different numbers (Close, 1996, 1997; Campbell and Close, 1997). In addition, dehydrin molecules can contain less conserved sequential motifs rich in glycine and hydrophilic amino acids called Φ-segments. Some dehydrins also contain histidine-rich sequence motifs underlying their ion-binding and radical-scavenging activities (Hara et al., 2013).

An alternative classification of dehydrins (group 2 LEA proteins) has recently been proposed by Wise (2003) who divided them into three sub-groups: 2a, 2b, and 2c, according to protein hydrophilicity, which predicted the secondary structure and content of aromatic and charged amino acid residues. It is interesting that this classification corresponds to different protein-stress induction patterns: sub-groups 2b and 2c (especially 2c) include predominantly cold-inducible dehydrins while sub-group 2a includes dehydrins not inducible by cold stress.

10.1.3 Structure of Dehydrins

10.1.3.1 Primary Structure

In general, dehydrin aa sequences contain relatively large proportions of glycine (G) and hydrophilic aa, especially threonine (T). In contrast, they nearly lack cysteine (C) and tryptophan (W). These unique sequential chracteristics determine the high hydrophilicity of dehydrin molecules. Dehydrins are well-soluble in various kinds of aquaeous buffers and remain soluble in these buffers even after boiling. This fact can be used for their enrichment in plant tissue extracts (Campbell and Close, 1997; Close, 1997). Dehydrins also exhibit enhanced affinity to detergents such as sodium dodecyl sulfate (SDS). They bind unusually high proportions of SDS when compared with most protein molecules

and therefore they appear larger on SDS-PAGE gels (Their electrophoretic mobility is shifted to values typical for larger proteins on SDS-PAGE gels, i.e., their molecular weights determined on the basis of electrophoretic mobility correspond to the proteins with ca. 24% higher molecular weight when compared with the values calculated from a dehydrin aa sequence) (Close, 1997; Ismail et al., 1999a).

10.1.3.2 Secondary and Tertiary Structure

In aquaeous solutions, dehydrin molecules are present in the conformation of a random coil; i.e., their molecules form the maximum number of hydrogen bonds with neighboring water molecules (intermolecular hydrogen bonds) and a minimum of number of hydrogen bonds between different amino acid residues within protein molecules (intramolecular hydrogen bonds). Due to the low proportion of intramolecular hydrogen bonds, dehydrin molecules are intrinsically unstructured. Dehydrins share many features with other types of intrinsically disordered/unstructured proteins (IDPs/IUPs) (Tompa, 2002; Tompa et al., 2005): they contain high proportions of hydrophilic aa and change their conformation according to the changes in their ambient micro-environment. Based on several experimental studies (Lisse et al., 1996; Danyluk et al., 1998; Ismail et al., 1999a; Hara et al., 2001; Kovacs et al., 2008; Mouillon et al., 2008), it was confirmed that the decrease in dehydrin hydration status (loss of water molecules in their ambient micro-environment) or addition of high amounts of compatible solutes such as glycerol, detergents such as SDS or salts such as NaCl into dehydrin aquaeous solution leads to conformational changes which can be monitored by the technique of far-UV circular dichroism (CD). It was found out that under the conditions of reduced hydration, the regions of the K-segments form left-handed class A2 amphipathic α-helices; i.e., they form intramolecular hydrogen bonds between different aa residues within the same protein molecules instead of intermolecular hydrogen bonds between aa residues and surrounding water molecules which predominate in random coil conformation. Random coil conformation is relatively symmetrical, thus it does not deviate the plane of linearly polarized light beams significantly; whereas, α-helices are highly asymmetrical, and thus deviate the plane of linearly polarized light. The amphipathicity of the resulting α-helices is determined by the aa sequence of the K-segment. When α-helix is formed, negatively-charged aa (aa with acidic pI, e.g., D and E) lie on one side of the helix, hydrophobic aa (nonpolar aa, e.g., I or L) lie on the opposite side of the helix, and positively-charged aa (aa with basic pI, e.g., K) lie on the polar–nonpolar interface (Svensson et al., 2002; Saavedra et al., 2006). Recently, several studies indicated that metal ions such as copper or zinc or interaction with membranes induce reversible transitions of dehydrin molecules from disordered to ordered state (Rahman et al., 2010, 2011; Mu et al., 2011).

Some researchers (e.g., Soulages et al., 2003) suggest that some dehydrins (e.g., recombinant GmDHN1 from soybean *Glycine max*) do not form classical α-helical structures after addition of detergents; instead, they rather reveal a left-handed extended poly (L-proline)-type II (PII) helical conformation.

Soulages et al. (2003) emphasized that extended PII helical structures contain relatively higher portions of aa residues capable of forming intermolecular hydrogen bonds in comparison with 'classical α-helices' as protein secondary conformation motifs, and thus the extended PII conformations enable dehydrin interaction with membrane surfaces. It should be noted that extended PII helical conformations are typically formed by those protein molecules which have a high content of G (contain G repeats – every third aa is G) and P such as animal collagen. Hughes and Graether (2011) proposed crucial roles for both φ- and K-segments for dehydrin molecular functions: φ-segments are crucial for maintenance of intrinsically unstructured character and act as 'molecular shields' to prevent partially denatured proteins from interacting with one another, whereas the amphipathic K-segments may help to localize dehydrin molecules near both hydrophilic and hydrophobic surfaces such as active enzyme sites or biomembrane surfaces.

10.1.3.3 Relationships between Dehydrin Structure and Functions

The changes in protein conformation result also in changes in protein function. This phenomenon, which is characteristic for IDPs/IUPs, is called 'moonlighting' (Tompa, 2002; Tompa et al., 2005). In the case of IDPs/IUPs, the changes in protein ambient micro-environment, such as availability of water molecules, result in protein conformational and functional changes. The flexibility of protein conformation results in versatility of their functions which mirror the changes in the protein ambient micro-environment. In the following text, the major dehydrin protein functions including their hydrophilic, membrane-binding, chaperone, cryoprotective, ion-binding, radical-scavenging, and nucleic acid binding functions are briefly described.

Hydrophilic functions: In a well-hydrated state, individual macromolecules like proteins or phospholipids in biomembranes are surrounded by a thin layer of highly ordered water molecules bound to the protein or membrane surface via intermolecular hydrogen bonds ('water envelope'). Therefore, the protein or phospholipid macromolecules do not interact with each other directly. During dehydration, the 'water envelope' disrupts and the macromolecules can then come into mutual interaction (Israelachvili and Wennerström, 1996). The amphipathic α-helices can interact with partly dehydrated surfaces of various other proteins and also with surfaces of biomembranes. It has been proposed by Ingram and Bartels (1996) that several K-segments in one dehydrin molecule can form bundles when present in α-helical conformation thus enhancing their amphipathic character in protein–protein or protein–biomembrane interactions. The binding of dehydrin molecules to the partly dehydrated surface of other protein molecule enhances the formation of amphipathic α-helices in a dehydrin molecule and protects the other protein molecule from further loss of water envelope which can lead to irreversible changes in the protein conformation, i.e., protein denaturation. It has been suggested that these interactions between partly dehydrated surfaces of dehydrin molecules and partly

dehydrated surfaces of other protein molecules and/or biomembranes (observed e.g., by Koag et al., 2003 in case of maize DHN1), which are enabled by the amphipathic nature of the α-helices, present the basis of dehydrin protective functions such as chaperone function, 'molecular shield' function and cryoprotective function.

Membrane-binding functions: Recently, Clarke et al. (2015) found out that a small K_2-type dehydrin from *Vitis riparia* interacts with phosphatidic acid present in liposomes via its K-segments thus preventing membrane aggregation under freeze-thaw cycles. Amphipathic α-2 helices formed by K-segments are presumably involved in dehydrin interaction with membranes as reported for *A. thaliana Lti30* since negatively-charged lipid heads attract positively-charged lysine-rich sequences in K-segments leading to α-2 helix formation (Eriksson et al., 2016).

Chaperone functions: Kovacs et al. (2008) investigated chaperone activities of two dehydrin proteins isolated from *Arabidopsis thaliana*, ERD10 and ERD14, and described their protective effects against thermal aggregation of citrate synthase, firefly luciferase, inactivation of lysozyme and thermal inactivation of alcohol dehydrogenase. However, it turns out that the interactions between dehydrin molecules and other protein molecules are rather non-specific protein–protein interactions when compared with the interactions of classical chaperones (e.g., small heat-shock proteins). Therefore, some authors describe these dehydrin protective functions based on non-specific protein–protein interactions as 'molecular shield' (Tunnacliffe and Wise, 2007). Under environmental conditions when cells lose water, the relative spatial (three-dimensional) proportions among individual intracellular complexes do also change. These spatial changes are also harmful since they can lead to undesirable interactions and aggregation and denaturation of several proteins and membranaceous complexes. It has been proposed by several scientists (Tunnacliffe and Wise, 2007; Battaglia et al., 2008) that dehydrins can accumulate to relatively large amounts in various compartments inside the cells under the conditions associated with cellular dehydration. Thus, they can simply act as 'space-fillers'; i.e., they can participate in keeping the original, non-harmful distances among individual subcellular complexes. (During dehydration, cell volume diminishes, and this process can lead to several undesirable protein interactions resulting in protein denaturation.) Due to their largely unfolded state, capability to bind water and high level of accumulation, dehydrins can mimic the adverse effects of cellular dehydration and help in keeping the original cell volume, thus preventing structural collapse in the cell. This phenomenon, i.e., accumulation of various kinds of intrinsically unstructured hydrophilic proteins together with low-molecular compatible solutes, is sometimes termed 'macromolecular crowding' (Ellis, 2001a, 2001b; Mouillon et al., 2008).

Cryoprotective functions: Cryoprotective function, i.e., the ability to protect enzymes from the loss of their activity under freeze-thaw conditions, was first reported by Lin and Thomashow (1992) for the COR15a protein, a LEA-3 protein from *A. thaliana* using an LDH assay (LDH; lactate dehydrogenase EC 1.1.1.27.) developed by Carpenter and Crowe (1988). Since then, a cryoprotective activity has been reported for several dehydrin proteins, e.g., COR85 from spinach (Kazuoka and Oeda, 1994), WCS120 from common wheat (Houde et al., 1995), PCA60 from peach (Wisniewski et al., 1999), CuCOR19 from *Citrus unshiu* (Hara et al., 2001), DHN5 from barley (Bravo et al., 2003), and others. Recently, recombinant proteins from the WCS120 family were reported to be efficient in the cryopreservation of mammalian liver and pancreatic cells (Chow-Shi-Yee et al., 2016). As shown by Reyes et al. (2008), the presence of K-segments is essential for dehydrin cryoprotective activity. Similarly, Drira et al. (2013) and Yang et al. (2015) have demonstrated the crucial function of K-segments for a protective function of durum wheat DHN5 and common wheat WZY2 (both YSK_2), respectively, on retaining LDH and β-glucosidase activities *in vitro* due to a comparison of a wild-type protein with truncated forms lacking one or both copies a of K-segment. Moreover, Hughes and Graether (2011) revealed that not only K-segments but also φ regions are required to ensure dehydrin cryoprotective activity in an LDH assay.

Ion-binding and radical-scavenging functions: Some dehydrins, which contain relatively large amounts of histidine (H), arginine (R), and other reactive aa residues on the surface of their folded molecules, such as CuCOR15 and CuCOR19 from Satsuma mandarin (*Citrus unshiu*), also exhibit reactive oxygen species (ROS) scavenging and metal ion-binding properties. Both functions are mediated by direct interactions between the amino acid residue and the ROS species (superoxide anion radical O_2^{-}; singlet oxygen $^1O_2$; hydroxyl radical $HO^{\cdot}$; hydrogen peroxide H_2O_2) or the metal ion (Co^{2+}; Cu^{2+}; Fe^{2+}; Fe^{3+}; Ni^{2+}; Zn^{2+}). The interactions of the aa residue with ROS lead to oxidation of the residue; the interactions with metal ions lead to the formation of covalent bonds. Binding of free metal ions prevents the intracellular compounds from excessive ROS formation since free metal ions act as catalyzers of synthesis of various ROS. For example, a copper binding function as reported for *Arabidopsis* KS type dehydrin is caused by an interaction of histidine residues with Cu^{2+} and leads to a reduction of Cu-promoted ROS formation (Hara et al., 2013) and restoration of Cu-inhibited LDH activity (Hara et al., 2016). Dehydrins can thus also function as antioxidants (e.g., CuCOR15 and CuCOR19 in a Satsuma mandarin (*Citrus unshiu*) as reported by Hara et al. (2001, 2005)), chaperones (Hara et al., 2016), ion sequestrants (e.g., VCaB45 in celery (*Apium graveolens*)) vacuoles which binds Ca^{2+} (Heyen et al., 2002) or metal ion transporters in plant phloem sap (e.g., ITP protein from castor bean (*Ricinus communis*)) which binds Fe^{2+} and Fe^{3+} (Krüger et al., 2002).

Nucleic acid binding functions: Some dehydrins such as CuCOR15 (K_2S) from *Citrus unshiu* or VrDHN1 (Y_2K) from *Vigna radiata* reveal nucleic acid binding activities (DNA, tRNA) which are stimulated upon Zn^{2+} or Ni^{2+} addition (Hara et al., 2009; Lin et al., 2012). These results indicate that some dehydrin proteins can protect nucleic acids during stress treatments or seed maturation.

Besides dehydrin interactions with other molecules such as lipids, nucleic acids, and other proteins, dehydrin–dehydrin

interactions leading to the formation of dehydrin homo- and heterodimers were recently reported for acidic dehydrins in *Opuntia streptacantha* and *Arabidopsis thaliana*, respectively (Hernández-Sánchez et al., 2014, 2017).

Further details on dehydrin molecular structure and their functions determined by their structure can be found in recent reviews by Hara (2010), Hanin et al. (2011), Graether and Boddington (2014), and Liu et al. (2017).

10.1.4 Dehydrins and Dehydrin-Like Proteins within the Plant Kingdom and in Other Organisms

From the end of the 1980s until now, hundreds of dehydrin proteins have been described not only in dicotyledonous and monocotyledonous angiosperms such as cotton and rice but also in other plant taxons including gymnosperms, ferns, lycopods, and mosses. Proteins induced by osmotic stress which contain conserved sequential motifs resembling the dehydrin K-segment have also been found in various groups of cyanobacteria (Close and Lammers, 1993), algae (Li et al., 1998), in mycelia of nectrotrophic pathogen *Alternaria brassicicola* (Pochon et al., 2013) or white truffle *Tuber borchii* from Ascomycota (Abba et al., 2006), and in invertebrates, which can survive cellular desiccation in an 'anhydrobiosis state,' e.g., some nematodes (*Aphelenchus avenae*, *Caenorhabditis elegans*) and bdelloid rotifers (Browne et al., 2002; Tunnacliffe and Wise, 2007). The proteins, which resemble dehydrins distantly due to the presence of sequential motifs which can form amphipathic α-helices analogous to the dehydrin K-segment, also contain α-synucleins, proteins found in animal neurons where they bind to vesicles with neurotransmitters (Ismail et al., 1999a; Souza et al., 2000).

Dehydrin-like proteins in cyanobacteria and algae usually accumulate inside the cells under conditions associated with osmotic stress. It was proposed that dehydrins protect the cells against excessive water loss which is important namely for those algae that inhabit intertidal zones where they have to face periodical desiccation. Li et al. (1998) detected several dehydrin-like proteins in five species of fucoid algae (*Phaeophyceae*) living in the intertidal zone; unlike dehydrins in higher plants, these proteins were denatured by boiling.

In seedless plants, i.e., mosses, lycopods, and ferns, dehydrins have been characterized in several species. In contrast to gymnosperms and angiosperms, only one copy of the dehydrin gene per genome was found in all these plant taxons (groups). Dehydrins are thus present as single copy genes in genomes of seedless plants. In mosses, dehydrin genes were sequenced in a desiccation-tolerant moss, *Tortula ruralis* (Velten and Oliver, 2001), and in *Physcomitrella patens* (Saavedra et al., 2006). Velten and Oliver found out that during desiccation of *Tortula ruralis*, the level of dehydrin mRNA increases. The researchers have hypothesized that during rehydration, the dehydrin protein Tr288 becomes translated and helps to restore intracellular complexes. Tr288 is therefore also called 'rehydrin.' Near to the C-terminus, a conserved sequence which can form an amphipathic α-helical structure similar to the α-helices formed by angiosperm dehydrins occurs.

The dehydrin protein characterized in *P. patens*, PpDHNA, has probably a protective function in osmotic stresses since a *P. patens* knockout mutant *DHNA* reveals impaired ability to tolerate these factors (impaired growth on medium containing NaCl or mannitol). In contrast to *Tr288*, *PpDHNA* is also up-regulated by ABA. (In *Tortula ruralis*, endogenous ABA was not detected; Oliver et al., 1998.) Dehydrins were also detected in a poikilohydric fern, *Polypodium virginianum*, (Reynolds and Bewley, 1993) and in a resurrection lycopod, *Selaginella lepidophylla* (Iturriaga et al., 2006).

In all gymnosperm and angiosperm species studied so far, more than one copy of the dehydrin gene per genome was found. In other words, dehydrins are present as multi-copy genes in genomes of both gymnosperm and angiosperm plants. The different dehydrin genes identified in one plant species usually belong to different structural sub-groups and reveal a different mode of expression. Dehydrins in gymnosperms differ from their counterparts in angiosperms in the aa sequence of the K-segment. The consensus aa sequence of the K-segment for gymnosperms is (Q/E)K(P/A)G(M/L)LDKIK(A/Q)(K/M)(I/L)PG while for angiosperms is EKKGIMDKIKEKLPG. The first report about detection of dehydrins in gymnosperms comes from Close et al. (1993) who found a positive reaction on a polyclonal primary antibody raised against the angiosperm consensus K-segment in *Ginkgo biloba* and *Pinus taeda*. Later, Jarvis et al. (1996) detected dehydrins in Douglas fir (*Pseudotsuga menziesii*). Until now, dehydrins have also been identified in white spruce (*Picea glauca*) (Richard et al., 2000), Scots pine (*Pinus sylvestris*) (Kontunen-Soppela et al., 2000; Wachowiak et al., 2009) in Norway spruce (*Picea abies*) (Yakovlev et al., 2008), and in cypress (*Cupressus sempervirens*) (Pedron et al., 2009). Recently, dehydrin proteins lacking a complete K-segment sequence were described in Pinaceae; instead, two gymnosperm-specific sequence motifs were described in Pinaceae dehydrins (Perdiguero et al., 2012, 2014).

Until now, hundreds of dehydrin genes have been sequenced in different angiosperms including both dicotyledonous and monocotyledonous species. Among dicotyledons, dehydrins have already been characterized in small herbaceous plants such as the thale cress (*Arabidopsis thaliana*) (or mouse-ear cress) as well as in large, long-living woody species such as silver birch (*Betula pendula*) (Puhakainen et al., 2004a), downy birch (*Betula pubescens*) (Welling et al., 2004), beech (*Fagus sylvatica*) (Jiménez et al., 2008), poplar (*Populus tremula*) (Renaut et al., 2005), blueberry (*Vaccinium corymbosum*) (Levi et al., 1999), Satsuma mandarin (*Citrus unshiu*) (Hara et al., 1999), peach (*Prunus persica*) (Arora and Wisniewski, 1994), apple tree (*Malus domestica*) (Wisniewski et al., 2008), *Rhododendron catawbiense* (Wei et al., 2005; Peng et al., 2008), red-osier dogwood (*Cornus sericea*) (Karlson et al., 2003a, b), pistachio (*Pistacia vera*) (Yakubov et al., 2005), and others.

Several reviews on the relationships between dehydrin protein structure and their biological functions were published recently (e.g., Graether and Boddington, 2014; Liu et al., 2017) where readers can obtain additional information on this versatile protein family.

10.2 DEHYDRINS AND PLANT STRESS RESPONSE

10.2.1 EXPRESSION OF DEHYDRINS UNDER ENVIRONMENTAL STRESS FACTORS

Dehydrin proteins can be found in small amounts in young plant organs grown under optimum growth conditions and exhibiting rapid cell division or cell elongation, e.g., root tips, elongating stems, petioles, etc. (Nylander et al., 2001; Rorat et al., 2004). Analogously to other LEA proteins, dehydrins also accumulate in plant embryos in later stages of their development (embryo maturation, and desiccation). Apart from these natural physiological processes, dehydrins become expressed when plants are exposed to various stress factors which are accompanied by cellular dehydration. Thus, dehydrins can be induced by drought, osmotic stress, enhanced salinity, stresses associated with abscisic acid (ABA) signaling, low-temperature stresses (cold and frost), and also biotic stresses associated with water loss and jasmonic acid signaling such as wounding. All these stresses have a dehydration component. In addition, metal ion-binding properties have been reported for some dehydrins (CuCOR15 – Hara et al., 2005), which can thus function as chelators and seem to confer plant tolerance to heavy-metal stress (Xu et al., 2008). Dehydrins encoded by one plant genome are usually expressed in different plant tissues (organs) and in response to different developmental and environmental cues; see e.g., expression studies in *A. thaliana* by Nylander et al. (2001), Bies-Ethève et al. (2008), Hundertmark and Hincha (2008) revealed 51 *Lea* genes, 10 of them dehydrins; in rice, by Wang et al. (2007) 34 *Lea* genes, 8 of them dehydrins, were identified in japonica rice Nipponbare; in barley (*Hordeum vulgare*), by Choi et al. (1999) or a more recent work by Tommasini et al. (2008), 13 dehydrin genes were identified.

Under stress conditions, dehydrins accumulate in various compartments inside the cells – nucleus (PCA60 from peach – Wisniewski et al., 1999; WCS120 from common wheat – Houde et al., 1995; phosphorylated forms of maize RAB17 and DHN1 – Goday et al., 1994; Koag et al., 2003; phosphorylated TAS14 in salt-stressed tomato – Godoy et al., 1994; 24-kDa dehydrin in red-osier dogwood (*Cornus sericea*) – Karlson et al., 2003b; AmDHN1a from gray mangrove (*Avicennia marina*) – Mehta et al., 2009), nucleolus and euchromatin (TAS14 in salt-stressed tomato (*Lycopersicon esculentum*) – Godoy et al., 1994). Recently, Hara et al. (2009) reported that CuCOR15, a histidine-rich dehydrin protein, has Zn^{2+} dependent DNA and an RNA binding affinity. Unlike Cys_2His_2 zinc-finger transcription factors, the binding affinity of CuCOR15 is not sequence-specific. Thus, Hara et al. (2009) have proposed a general protective function for CuCOR15. Recently, at least three sequence motifs occurring in dehydrins were proposed to function as nuclear localization signals (NLS): a phosphorylated serine tract in YxSKn and SKn-type dehydrins (Brini et al., 2007a), an RRKK motif preceding the K-segment in Kn-type dehydrins (Malik et al., 2017), and a histidine-rich motif present in dehydrins with ion-binding properties (Hernández-Sánchez et al., 2015).

Furthermore, dehydrins have been localized to semiautonomous organelles such as the outer membrane of mitochondria (Borovskii et al., 2000, 2002; Hara et al., 2003; Carjuzaa et al., 2008; Rurek, 2010), chloroplasts (PCA60 from peach (*Prunus persica*) – Wisniewski et al., 1999; *HaDHN1* transcripts in sunflower (*Helianthus annuus*) – Natali et al., 2007), as well as immature proplastids (Carjuzaa et al., 2008), storage protein bodies and starch-rich amyloplasts (24-kDa dehydrin from downy birch (*Betula pubescens*) – Rinne et al., 1999). Dehydrins have also been localized to various membranaceous structures such as endoplasmic reticulum (CAP85 from spinach – Neven et al., 1993), rough endoplasmic reticulum cisternae (Carjuzaa et al., 2008), vacuole (VCaB45 from celery – Heyen et al., 2002), membranes of lipid vesicles containing acidic phospholipids (maize DHN1 – Koag et al., 2003), membranes of protein and lipid bodies (Egerton-Warburton et al., 1997), inner side of plasmalemma (WCOR410 from common wheat – Danyluk et al., 1998), or plasmodesmata of vascular cambium cells (24-kDa dehydrin in *Cornus sericea* – Karlson et al., 2003b). Abu-Abied et al. (2006) have detected *A. thaliana* ERD10 in association with actin stress fibers in transformed rat fibroblasts. They have also reported a protective function of GFP-ERD10 on the actin cytoskeleton in latrunculin-treated leaves of *Nicotiana benthamiana*. Taken together, it can be stated that dehydrins are intracellular proteins induced by stresses resulting in cellular dehydration. Until now, dehydrins have not been detected in cell walls.

At the tissue and organ level, dehydrins accumulate mainly in those tissues and organs which are exposed to rapid changes in water content, i.e., outer leaves of inflorescence buds, epidermis in herbs or bark in woody plants, stomatal guard cells (e.g., Rab18 from *A. thaliana* – Nylander et al., 2001; *HaDHN1* transcripts from sunflower (*Helianthus annuus*) – Natali et al., 2007), adventitious root primordia, root apices (e.g., Lti29/ERD10 and ERD14 in *A. thaliana* – Nylander et al., 2001) and vascular tissues. In vascular tissues, dehydrins usually accumulate in living cells in the vicinity of vascular elements, especially in xylem ray parenchyma cells and phloem ray parenchyma cells (Godoy et al., 1994; Bravo et al., 1999; Wisniewski et al., 1999, 2008; Yakubov et al., 2005; Natali et al., 2007; Yakovlev et al., 2008).

For some dehydrin genes, an alternative splicing was reported – e.g., *VvDHN-1a* and *VvDHN-1b* transcripts of the *VvDHN1* gene in grapevines (*Vitis vinifera*) (Xiao and Nassuth, 2006), *AmDHN1* and *AmDHN1a* transcripts of a dehydrin gene in gray mangrove (*Avicennia marina*) (Mehta et al., 2009), or *OsLea7* and *OsLea8* genes in japonica rice (*Oryza sativa* cv. japonica) (Wang et al., 2007). For some dehydrin proteins, two types of post-translational modifications – phosphorylation and O-glycosylation – under stress conditions have been reported. While phosphorylation has been quite widely reported, especially for dehydrins with the S-segment, and several functions for phosphorylated dehydrin proteins have been proposed. O-glycosylation has been reported only for some dehydrins in blueberry and pistachio (Levi et al., 1999; Yakubov et al., 2005) and no specific functions for glycosylated forms of these proteins have been suggested. Phosphorylation of the S-segment by casein kinase II (CK2)-type kinases has been shown to be associated with dehydrin

translocation from cytoplasm into nucleus (e.g., RAB17 in maize (*Zea mays*) – Goday et al., (1994)); Jensen et al., (1998); Riera et al., (2004); TAS14 in tomato (*Lycopersicon esculentum*) – Godoy et al., (1994); DHN5 in durum wheat (*Triticum turgidum* ssp. *durum*) – Brini et al., (2007a); however, nuclear localization has also been reported for some dehydrins that lack the S-segment (e.g., wheat WCS120 or peach PCA60; Houde et al., 1995; Wisniewski et al., 1999). For *A. thaliana* acidic dehydrins COR47, ERD10, and ERD14, and for celery (*Apium graveolens*) vacuole-located dehydrin VCaB45, it has been reported by Alsheikh et al. (2003, 2005) and by Heyen et al. (2002), respectively, that phosphorylation is necessary for Ca^{2+} binding properties of these proteins. Recently, *in vitro* phosphorylation has also been reported for K_n-type dehydrin Lti30 from *A. thaliana* by Eriksson and Harryson (2009). The phosphorylation of K_n-type dehydrins is provided by a different kinase than phosphorylation of dehydrins with an S-segment.

When dehydrin expression at transcript (mRNA) and protein level is compared, several differences can be found. Zhu et al. (2000) investigated the expression of the barley *DHN5* gene at the transcript and protein level in plants grown under field conditions and found out that either only transcript or only protein molecules can be present in the sample. Different dynamics of dehydrin transcript and protein accumulation during a time-course of a stress treatment has also been described. For example, Ganeshan et al. (2008) compared wheat dehydrin (*Wcs120*, *Wcor410*) expression at transcript and protein levels during a 98-day-cold acclimation (CA) and found out that the peak in transcript accumulation precedes the peak in protein accumulation during the time-course of CA treatment (2 days of CA versus 56 days of CA at transcript and protein levels, respectively).

Recently, dehydrin expression in various plant species and their functions under different stress conditions have been reviewed by Svensson et al. (2002), Allagulova (2003), Rorat (2006), and specifically for cold stress by Kosová et al. (2007). Basic characteristics of dehydrins induced by major abiotic stresses of low temperature, drought, and salinity in wheat and barley plants were recently summarized in a mini-review by Kosová et al. (2014). In the NCBI database (http://www.ncbi.nlm.nih.gov/), 5124 dehydrin protein sequences from plant species were available in April 2018. In the following paragraphs, expression of dehydrins and their specific functions in different plants under different stress conditions will be characterized according to the following scheme: First, brief characteristics of each stress factor will be given. Second, signaling pathways which cooperate in the induction of dehydrin expression under a given stress factor will be characterized. Third, examples of dehydrins expressed in various plant species under a given stress factor will be listed, and their specific functions (if known) will be discussed.

10.2.2 Drought Stress, Evaporation

10.2.2.1 Physiological Aspects of Drought Stress

Drought is one of the major abiotic stress factors which significantly limits plant growth and agronomic production in many areas worldwide. Drought, i.e., a shortage of rainfall, leads to a decrease in soil water potential (gravimetric soil water content), which is also often accompanied by a decrease in air relative humidity resulting in an enhanced rate of leaf transpiration. The decreased soil water potential draws water from an ambient environment with higher water potential including plant cells. Plant acclimation to drought stress which tries to eliminate excessive water loss lies in osmotic adjustment, i.e., a decrease of cell water potential in order to diminish the difference in water potential between the plant cell and the ambient soil. Osmotic adjustment is associated with the accumulation of low-molecular well-soluble metabolites collectively called compatible solutes (low-molecular saccharides – monosaccharides glucose, fructose, disaccharides – sucrose, oligosaccharides raffinose, stachyose and verbascose; sugar alcohols mannitol, pinitol, sorbitol; quaternary ammonium compounds called betaines – alanine betaine, glycine betaine; imino acid proline; polyamines spermine, spermidine, putrescine) or relatively high-molecular hydrophilic proteins inside the cells. Among hydrophilic proteins (hydrophilins), several LEA proteins including dehydrins accumulate to relatively high extents (amounts) in various plant cells during the process of osmotic adjustment. Osmotic adjustment lies in fact in the decrease of one major component of water potential – osmotic potential, which is determined by concentration (activity) of dissolved compounds in a solute (in fact, osmotic potential is defined as a negative value of osmotic pressure). Other components of cell water potential, i.e., pressure (or turgor) potential, which is determined by a hydrostatic pressure of cytosol against cell wall, gravitational potential, which is determined by position of water in a gravitational field, and matrix potential, which is determined by adhesive forces of water to cell surfaces, are not significantly affected by accumulation of compatible solutes.

In soil water potential (relative air humidity), as well as in leaf water potential, diurnal changes have been observed, with maximum values before dawn and minimum values at midday (noon). It is becoming evident that these changes are mirrored in the expression of several drought-inducible hydrophilic proteins including dehydrins (e.g., Cellier et al., 2000).

10.2.2.2 Signaling Pathways Involved in Drought-Inducible Dehydrin Gene Expression

Drought-inducible dehydrins contain ABA-responsive elements (ABRE), C-repeat/drought-responsive/low-temperature-responsive elements (CRT/DRE/LTRE), Myeloblastosis (MYB) and Myelocytomatosis (MYC) regulatory elements in their promoter regions. Their expression is thus regulated by both ABA-dependent and ABA-independent signaling pathways. ABA-dependent signaling pathways include either bZIP transcription activators named ABFs or AREBs (ABRE-binding factors), which bind to ABRE elements, homologues of A. thaliana CBF4/DREB1D transcription activator, which bind to CRT/DRE/LTRE elements, and MYBFs and MYCFs, which bind to MYB and MYC promoter elements. ABA-independent signaling pathways include homologues of A. thaliana DREB2A and DREB2B transcription activators, which bind to CRT/DRE/LTRE elements (for review

on stress signaling pathways and regulatory elements in the promoters of Cor/Lea genes, see, e.g., Bray, 1997; Shinozaki and Yamaguchi-Shinozaki, 1997, 2000; Zhu, 2002; Shinozaki et al., 2003; Chinnusamy et al., 2004; Yamaguchi-Shinozaki and Shinozaki, 2005, 2006).

10.2.2.3 Dehydrin Expression under Drought

10.2.2.3.1 Dehydrin Expression in Mature Seeds Under Drought

Dehydrins as LEA proteins accumulate in plant seeds which undergo desiccation during a physiological process of maturation drying. For example, Jiménez et al. (2008) and Kalemba et al. (2009) reported accumulation of dehydrins during the desiccation of beech (Fagus sylvatica) embryos and seeds. It is well known that mature plant embryos in seeds which undergo desiccation during their maturation are tolerant to drought. The seeds that undergo maturation drying are called orthodox seeds (e.g., barley, beech, maize, or rice). In contrast, seeds of some species, e.g., tropical wetland plants such as gray mangrove (Avicennia marina), but also some temperate woody plants such as horse chestnut (Aesculus hippocastanum) or oak (Quercus sp.), do not undergo maturation drying. They remain metabolically active throughout embryo maturation and can germinate before shedding from the parent plant. These seeds are called recalcitrant seeds. (To get more information about seed recalcitrance, see e.g., Berjak and Pammenter, 2008). Some researchers have shown that these two physiologically different types of seeds – orthodox and recalcitrant – differ also in accumulation of LEA proteins including dehydrins. Farrant et al. (1992) have reported an absence of dehydrins in late stages of seed development in gray mangrove (Avicennia marina) recalcitrant seeds. Similarly, Han et al. (1997) have shown that the susceptibility to desiccation in recalcitrant seeds of Trichilia dregeana may be due to their inability to accumulate sufficient amounts of dehydrins, especially after the beginning of germination. In contrast to these studies, no differences in dehydrin levels were found between desiccation-tolerant orthodox seeds of rice (Oryza sativa) and recalcitrant seeds of Zizania palustris during germination (Bradford and Chandler, 1992). In young seedlings, an ABA-induced dehydrin expression was observed in tolerant Cynanchum komarovii plantlets upon severe dehydration stress (Yang et al., 2007). It is well known that dehydrin accumulation in plant embryos and seedlings is regulated by ABA. However, dehydrin Mat1 in soybean (Glycine max) seedlings is induced by dehydration, but not by ABA – that is quite unusual for dehydrin expressed under drought conditions (Whitsitt et al., 1997).

10.2.2.3.2 Dehydrin Expression in Plants Under Drought

Dehydrins play an important role not only in maturating plant embryos but also in the later stages of plant development during the plant's response to stress factors associated with cellular desiccation. According to tolerance to cellular desiccation, plants can be divided into two major groups – poikilohydric plants and homoiohydric plants.

Poikilohydric plants are generally much more tolerant to cellular dehydration than homoiohydric plants due to small vacuoles and relatively low vacuole to cytoplasm ratio when compared with homoiohydric plants (for more on poikilohydry and homoiohydry in plants, see the review by Proctor and Tuba, 2002).

Poikilohydric plants include predominantly various desiccation-tolerant seedless plants such as desiccation-tolerant mosses, lycopods from the genus *Selaginella*, ferns from the genus *Polypodium*, etc. However, they also include a few desiccation-tolerant angiosperms which have cells with small vacuoles and can recover after up to 95% (practically total) water loss. These plants are sometimes called 'resurrection plants' – e.g., *Craterostigma plantagineum*, *Ramonda serbica*, etc. (Scott, 2000).

In desiccation-tolerant moss, *Tortula ruralis*, dehydrin Tr288 with several repeats of K-segments in its C–terminal domain was detected by Velten and Oliver (2001). These researchers found out that during dehydration, particles of Tr288 mRNP accumulate in desiccating cell cytoplasm. The researchers have hypothesized that Tr288 protein could become expressed (synthesized) during subsequent cellular rehydration rather than during dehydration. Therefore, Tr288 can be better characterized as a 'rehydrin.' Dehydrins have also been described in *Polypodium virginianum* (Reynolds and Bewley, 1993) and in *Selaginella lepidophylla* (Iturriaga et al., 2006). In *Craterostigma plantagineum*, an *in vitro* unfolded dehydrin Dsp16 has been studied (Piatkowski et al., 1990; Lisse et al., 1996).

Dehydrins also accumulate in homoiohydric plants when they are exposed (subjected) to drought stress. Several studies have shown that dehydrin accumulation during plant vegetative growth correlates with their drought tolerance. Labhilili et al. (1995) and Brini et al. (2007a) have shown that dehydrin accumulation confers tolerance to drought and salt stress in durum wheat (*Triticum turgidum* ssp. *durum*). The latter authors have also shown nuclear localization of a stress-inducible dehydrin protein DHN5, and they have also detected differences in the phosphorylation pattern of DHN5 between two Tunisian cultivars of durum wheat. A sensitive cultivar revealed only a low level of dehydrin phosphorylation while a highly tolerant cultivar exhibited a high level of dehydrin phosphorylation. DHN5 from durum wheat is Y_xSK_n-type dehydrin sequentially similar to bread wheat Rab15, barley DHN3, DHN4 and DHN9, or maize Rab17. In wild barley (*Hordeum vulgare* ssp. *spontaneum*), Suprunova et al. (2004) have observed expression of several Y_xSK_n-type dehydrins (especially *DHN1* and *DHN6*, but also *DHN3* and *DHN9*) associated with enhanced drought tolerance.

Similarly, Pelah et al. (1997) have observed a positive relationship between dehydrin accumulation and drought tolerance in aspen (*Populus tremula*). Analogously, Tabaei-Aghdaei et al. (2000) compared dehydrin protein accumulation in crown tissues of two wheatgrass species revealing different levels of drought tolerance – a drought-tolerant *Agropyron desertorum* and a less tolerant *Lophopyrum elongatum* – under drought stress (6 days without watering) and

the researchers detected much higher levels of dehydrin polypeptides in the drought-tolerant *A. desertorum* than in *L. elongatum*. Volaire (2002) found association of 22–24-kDa dehydrin accumulation with enhanced tolerance to drought and acquisition of summer dormancy in leaf bases of two contrasting varieties of cocksfoot (*Dactylis glomerata*), one of Mediterranean Moroccan origin (a drought-tolerant variety), and the other of oceanic French origin (a drought-susceptible variety). In contrast, no differences were observed at the transcript level (Volaire et al., 1998). These results suggest the existence of different regulatory mechanisms between *Dactylis glomerata* varieties with different drought tolerance at the post-transcriptional level.

Cellier et al. (1998) observed higher transcript levels of dehydrin genes *HaDHN1* and *HaDHN2* in a drought-tolerant cultivar of sunflower (*Helianthus annuus*) when compared with a drought-susceptible one. The accumulation of *HaDHN1* and *HaDHN2* transcripts correlated with plant leaf water potential when the cultivars were compared under the same gravimetric soil water content. Cellier et al. (2000) also observed diurnal changes in *HaDHN1* gene expression in sunflower plants subjected to drought with maximum *HaDHN1* expression at midday (noon) which mirrored the changes in leaf water potential. In contrast, *HaDHN2* gene showed no significant diurnal changes in its expression pattern.

10.2.3 SALINITY AND OSMOTIC STRESS

10.2.3.1 Brief Characteristics of Salinity Stress

High salinity means enhanced salt ion concentrations in soil water which is taken up by plants. Plants, as well as other organisms, actively regulate cytoplasmic (intracellular) ion concentrations, especially concentrations of Na^+, K^+, Ca^{2+}, and Cl^- and phosphate ions via ATP-dependent ion channels (pumps) in plasmalemma, tonoplast, and membranes of the endoplasmic reticulum. According to their tolerance to enhanced cytoplasmic ion concentrations, plants can generally be divided into glycophytes, which do not tolerate enhanced (increased) cytoplasmic ion concentrations, and halophytes which tolerate (or even require) enhanced cytoplasmic ion concentrations. Glycophytes include the most economically important plant crop species such as barley, wheat, maize, or rice, but also a model plant *Arabidopsis thaliana*. Halophytes include mostly plants living on sea coasts where they are in permanent contact with salt water – e.g., *Salsola kali*, *Salicornia*, or *Arabidopsis* salt-tolerant relative *Thellungiella halophila* which is used as a model plant for studies of plant salt tolerance mechanisms (Amtmann, 2009).

In the case of glycophytes, high ion concentrations in soil water result in the decreased osmotic potential of soil solution relative to cytoplasmic osmotic potential and this difference triggers water from plant cell cytoplasm. Thus, high salinity has a similar effect on glycophytes as drought stress. Therefore, compatible solutes such as proline, sugar alcohols, polyamines, low-molecular saccharides (products of starch degradation such as maltose, glucose), as well as hydrophilic proteins (dehydrins) accumulate in the cell cytoplasm upon salt stress.

10.2.3.2 Salt-Inducible Dehydrins

Dehydrin expression under salt stress has predominantly been studied in several economically important glycophytes whose production is strongly reduced by enhanced soil salinity. Only a few studies have been carried out on relatively salt-tolerant plant species such as *Avicennia marina* or *Lophopyrum elongatum* (Tabaei-Aghdaei et al., 2000; Mehta et al., 2009).

Godoy et al. (1994) have studied tissue- and subcellular localization of dehydrin TAS14 in salt-stressed tomatoes (*Lycopersicon esculentum*) and found its major accumulation in nucleus and nucleolus in the cells of xylem ray parenchyma neighboring the vascular tissue. The protein accumulated especially in stems and leaves and only a transient accumulation was detected in roots. Godoy et al. (1994) also found out that TAS14 is a phosphoprotein *in vivo* and that it could be phosphorylated by both casein kinase II and cAMP-dependent protein kinase *in vitro*.

Moons et al. (1995) detected several dehydrins and LEA 3 proteins of various molecular mass (dehydrins of 24, 35, and 50 kDa) in roots of salt-tolerant indica rice (*Oryza sativa* cv. indica) varieties Pokkali and Nona Bokra in response to salt stress (50 mM NaCl) and exogenous ABA treatment (20 or 100 μM ABA; pH 5.6). Similarly, Gulick and Dvořák (1992) have detected the expression of one dehydrin in roots of salt-tolerant wheatgrass species *Lophopyrum elongatum* in response to salt stress (250 mM NaCl). Masmoudi et al. (2001) found differences in the expression of the *DHN-5* gene (*Tdsi-5* cDNA clone) in roots and leaves of two cultivars of durum wheat (*Triticum turgidum* ssp. *durum*) with marked differences in their salt tolerance when the seedling plants were exposed to salt stress (compost soil supplemented with 200 mM NaCl). Later, Brini et al. (2007a) also reported differences in the DHN5 phosphorylation pattern between the same cultivars exposed to the same type of salt stress with large portions of phosphorylated forms of DHN5 in a tolerant cultivar and much fewer amounts of phosphorylated DHN5 in a salt-susceptible one. In contrast, Tabaei-Aghdaei et al. (2000) detected no significant differences in the accumulation of dehydrin polypeptides in crown tissues of two wheatgrass species – a salt-tolerant *Lophopyrum elongatum* and a less tolerant *Agropyron desertorum* – when the plants were subjected to a high-salt stress (daily exposure to 75 or 150 mM NaCl for 6 days).

The accumulation of dehydrins in response to salt and osmotic stress has also been reported in woody plants. In poplar (*Populus euramericana*), enhanced levels of dehydrin transcript of *PeuDHN1* (an SK_n-type dehydrin) were observed by Caruso et al. (2002) in response to salt stress (1.36 gL^{-1} NaCl) or osmotic stress (hydroponic cultivation with 50 gL^{-1} PEG 6000). The level of expression of this gene seemed to correlate with leaf osmotic potential. Mehta et al. (2009) found a rapid up-regulation of *AmDHN1* gene (a YSK_n type dehydrin) by 500 mM NaCl in gray mangrove (*Avicennia marina*), a pan-tropical mangrove species inhabiting coastal locations with excess salinity.

10.2.4 Dehydrins and Low-Temperature Stress (Cold and Frost)

10.2.4.1 Brief Characteristics of Low-Temperature Stresses (Cold and Frost)

Biennial and perennial plants inhabiting locations in higher latitudes and/or altitudes have to cope with relatively long periods of low temperatures (cold and frost) during winter. Annuals which survive winter as seeds often have to resist low temperatures as seedlings in early spring. In higher latitudes, periods of low temperatures are associated with short-day photoperiods which are sensed by plants and function as a signal which precedes low-temperature stress (e.g., in the initiation of winter dormancy in woody perennials). Plant winter hardiness has several components, e.g., desiccation tolerance, tolerance to the effects of snow cover, tolerance to flooding or repeated freeze-thaw cycles. Frost tolerance, i.e., plant ability to withstand adverse effects of low-temperature stress (especially effects of sub-zero temperatures, i.e., frost), presents a major component of plant winter hardiness. According to tolerance to low temperatures, plants can generally be divided into three groups: 1) chilling-susceptible plants which are susceptible to low above-zero temperatures (chilling is usually defined as a range between 0 and +12°C (+15°C)) – these plants include many crops of tropical and sub-tropical origin, such as cucumber (*Cucumis sativus*), tomato (*Lycopersicon esculentum*), maize (*Zea mays*), rice (*Oryza sativa*), cotton (*Gossypium hirsutum*), tropical legume crops such as cowpea (*Vigna unguiculata*) and chickpea (*Cicer arietinum*); 2) chilling-tolerant, but frost-sensitive plants, which are tolerant to chilling, but severely damaged by frost; and 3) frost-tolerant plants which can survive freezing temperatures in a vegetative stage – these include especially winter varieties of *Triticeae* cereals, i.e., winter barley (*Hordeum vulgare*), winter wheat (*Triticum aestivum*) or winter rye (*Secale cereale*) (Levitt, 1980; Sakai and Larcher, 1987).

However, in most frost-tolerant plants, frost tolerance is usually not a constitutive trait, but it can be induced by low, but above-zero temperatures (cold). This adaptive process is called cold acclimation. Different aspects of CA are reviewed in Levitt (1980), Sakai and Larcher (1987), Guy (1990), Pearce (1999), Thomashow (1999), Xin and Browse (2000), and others. The CA process leads to enhancement of plant winter hardiness and frost tolerance. Frost, i.e., the freezing of water – formation and growth of ice crystal nuclei – is accompanied by cellular dehydration. Therefore, plant cold acclimation leads to adaptive responses in order to minimize the adverse effects of cellular dehydration. These responses are associated with the accumulation of compatible solutes and hydrophilic proteins including dehydrins in various intracellular compartments.

10.2.4.2 Signaling Pathways Involved in Dehydrin Expression under Cold and Frost

Similarly to drought stress, dehydrins expressed under cold are regulated by both ABA-dependent and ABA-independent regulatory pathways. ABA-dependent pathways include

ABRE regulatory elements in dehydrin promoters which function as binding sites for bZIP ABRE-binding transcription factors (AREBs or ABFs). The main ABA-independent cold-inducible signaling pathway is the CBF/DREB1 pathway. The CBF/DREB1 transcription activators possess the AP2 DNA-binding domain flanked by two unique sequence motifs (called signature motifs) which can bind to CRT/DRE/LTRE elements in the dehydrin (and other *Cor* genes) promoter region. The up-regulation of *Cor* gene expression as a consequence of CBF gene overexpression was first reported by Jaglo-Ottosen et al. (1998) in *A. thaliana*. In the genome of *A. thaliana*, three predominantly cold-inducible *CBF/DREB1* genes – *CBF1/DREB1B*, *CBF2/DREB1C*, and *CBF3/DREB1A* – were characterized which are tandemly arranged on chromosome four. (The fourth member of the *CBF/DREB1* gene family – *CBF4/DREB1D* – is predominantly drought- and ABA-inducible gene – Haake et al., 2002). Currently, homologues of *A. thaliana CBF/DREB1* genes have been sequenced and further characterized in several other species, especially in cereals from the tribe *Triticeae* (e.g., Skinner et al., 2005,2006; Miller et al., 2006; Badawi et al., 2007; Francia et al., 2007) where the majority of *CBF* genes was mapped to the *Frost-resistance 2* (*Fr-2*) locus, previously characterized as a major QTL determining frost tolerance level and *Cor* gene expression in *Triticeae* (e.g., Vágújfalvi et al., 2000, 2003). The confirmation of the AP2 domain appears to be essential for the binding of the CBFs to CRT/DRE/LTRE promoter elements and the activation of *Cor* gene expression. Recently, Knox et al. (2008) described an allelic variation in *TmCBF12* gene located to *Fr-A^m2* locus in einkorn wheat (*Triticum monococcum*). A freezing-susceptible line of *T. monococcum* possesses *TmCBF12* allele with a deletion of a few aa residues in the AP2 domain of the protein. As a consequence, TmCBF12 cannot bind to CRT/DRE/LTRE elements in *Cor* gene promoters.

Recently, Bassett et al. (2009) reported conformational transitions in the promoter regions of cold-inducible peach dehydrin genes *PpDHN1* and *PpDHN3*. They hypothesized that these cold-inducible conformational transitions may strongly influence the cold-inducible expression of these dehydrins.

Vernalization-induced alterations in cold-induced dehydrin gene expression: It has been long known that fulfillment of vernalization requirement in winter cereals, but also in other plants such as *Arabidopsis thaliana*, results in a decrease of plant ability to induce frost tolerance when subjected to cold. Recently, molecular backgrounds of the relationships between the expression of vernalization-induced genes such as VRN1 in winter cereals and SOC1 in *A. thaliana*, and downregulation of cold-inducible pathways including CBF/DREB1 TFs and their downstream genes including COR/LEA proteins belonging to dehydrins. Negative correlations between the expression of SOC1 and CBFs plus COR/LEA genes were reported in *A. thaliana* following vernalization fulfillment (Seo et al., 2009) where a direct physical interaction between SOC1 and CBF promoter sequences was reported. Similarly, negative correlations were found between VRN1 gene expression and CBFs plus WCS120 gene expression in vernalized

winter wheat (Danyluk et al., 2003; Dhillon et al., 2010), and a possibility of VRN1 protein interaction with CBF promoter regions was predicted in barley based on chromatin immunoprecipitation approach (Deng et al., 2015). At the protein level, decreased WCS120 accumulation was found in vernalized winter wheat Mironovskaya 808 subjected to cold (2°C) following a reacclimation period (17°C) leading to dehydrin protein degradation when compared to non-vernalized winter wheat plants subjected to a reacclimation treatment (Vítámvás and Prášil, 2008).

10.2.4.3 Low-Temperature-Inducible Dehydrins and Their Features

Low-temperature induced dehydrins exhibit various protective functions: a cryoprotective function, antifreeze activity, cold-inducible phosphorylation, metal ion-binding, and ROS scavenging activities.

A cryoprotective activity has been reported for several cold-inducible dehydrins, e.g., COR85 from spinach (Kazuoka and Oeda, 1994), WCS120 from common wheat (Houde et al., 1995), PCA60 from peach (Wisniewski et al., 1999), CuCOR19 from *Citrus unshiu* (Hara et al., 2001), DHN5 from barley (Bravo et al., 2003). When a LDH assay is used, these proteins exhibit significant cryoprotective activity, which is sometimes stronger than the protective activity of several widely-used sugar cryoprotectants such as sucrose or cryoprotective proteins like bovine serum albumin (BSA). A protective effect on the activity of α-amylase (EC 3.2.1.1.) under cold was reported by Rinne et al. (1999) for a 24-kDa dehydrin from downy birch (*Betula pubescens*).

An antifreeze activity, i.e., modification of ice crystal growth and thermal hysteresis during freezing of aqueous protein solution, was first reported for PCA60, a Y_2K_9 dehydrin isolated from the bark tissue of peach (*Prunus persica*; Arora and Wisniewski, 1994; Wisniewski et al., 1999) and encoded by the *PpDHN1* gene. At high concentrations (micromolar to millimolar) of antifreeze proteins, ice crystal growth is inhibited along the *a*-axes while the *c*-axis becomes the preferred direction of growth. As a consequence, instead of flat round crystals formed by ice crystal nuclei in pure water, ice crystals of a shape of hexagonal bipyramids are formed in aqueous solutions containing proteins with an antifreeze activity. Moreover, the proteins with an antifreeze activity also decrease the freezing point of aqueous solutions in which they are dissolved; this phenomenon is called thermal hysteresis. In plants, antifreeze activity was first reported in 1992 for a specific group of proteins named antifreeze proteins (AFPs) (Griffith et al., 1992). PCA60 thus shares this function with AFPs which also accumulate in the bark of woody perennials during winter dormancy (for review on AFPs, see, e.g., Griffith and Yaish, 2004). Unlike dehydrins, AFPs are apoplastic proteins that accumulate in intercellular spaces where they inhibit growth and recrystallization of ice. Recently, a boiling-stable antifreeze protein revealing a sequence homology to dehydrins (containing several lysine-rich sequences and a Y-segment) was isolated from cold-acclimated bark tissue of *Forsythia suspensa* (Simpson et al., 2005).

Cold-induced phosphorylation of the S-segment has been reported for acidic SK_n-type dehydrins COR47, ERD10, and ERD14 in *A. thaliana* (Alsheikh et al., 2003, 2005). When phosphorylated, these proteins can bind Ca^{2+} ions, and Alsheikh et al. have speculated that they may act as calcium buffers or sugar chaperones under these conditions. It is also well known that Ca^{2+} ions act as second messengers in cold-inducible signaling pathways.

Radical-scavenging and metal ion-binding properties of dehydrin proteins have already been discussed above (Section 10.1.3.3).

10.2.4.3.1 Chilling-Induced Dehydrin Expression in Mature Seeds

An increase in the dehydrin transcript or protein level has also been observed in seeds exposed to chilling; e.g., the *FsDHN1* gene in European beech (*Fagus sylvatica*; Jiménez et al., 2008). These researchers have also found out that the expression of the *FsDHN1* gene in beech seeds is strongly up-regulated by ABA treatment. Ismail et al. (1999b) have reported for an extremely chilling-susceptible tropical legume crop *Vigna unguiculata* that seeds of a relatively chilling-tolerant line 1393-2-11 contained DHN1 (Y_2K-type), which positively correlated with seedling emergence under mild cold conditions (14°C), while the DHN1 was absent in genetically-related, but chilling-susceptible line 1393-2-1. Yao et al. (2005) have shown that expression of dehydrin genes *BnDHN1* and *BjDHN1* (both homologous Y_3SK_2 dehydrins) in cold-treated seeds of *Brassica napus* and *Brassica juncea*, respectively, enhances plant cold tolerance during seedling emergence. These findings are analogous to the situation in *Arabidopsis thaliana* where the RAB18 protein (also Y_2SK_2 dehydrin) is known to accumulate in mature seeds and to confer seedling tolerance to cold stress (Lång and Palva, 1992; Nylander et al., 2001; Hundertmark and Hincha, 2008).

10.2.4.3.2 Dehydrin Expression in Plants Upon Cold

In many plants, accumulation of dehydrin transcripts and/or proteins has been observed during cold acclimation process which often precedes frost in nature. During CA, dehydrin accumulation was reported for various intracellular compartments, e.g., plasmalemma in the case of WCOR410 in common wheat *Triticum aestivum* (Danyluk et al., 1998), endoplasmic reticulum in the case of CAP85 in spinach (Neven et al., 1993), nucleus in the case of WCS120 in common wheat (Houde et al., 1995) or PCA60 in peach (*Prunus persica*), chloroplasts in the case of peach PCA60 (Wisniewski et al., 1999), amyloplasts in the case of 24-kDa dehydrin from downy birch *Betula pubescens* (Rinne et al., 1999), mitochondria in the case of dehydrins in wheat (Borovskii et al., 2000, 2002). At the tissue level, dehydrin accumulation under cold stress was observed predominantly in epidermal cells and in xylem ray parenchyma cells, i.e., in tissues in the neighborhood of large extracellular spaces where ice crystals can be formed during freezing. This tissue distribution was observed for cold-inducible dehydrins in both dicotyledons and monocotyledons; it was reported for *Arabidopsis thaliana* (Nylander et al., 2001) as well as for

barley (Bravo et al., 1999). In citrus trees, some dehydrins have been reported to accumulate in fruit flavedo tissue after a rapid temperature change (chilling after a brief heat shock). This pattern of dehydrin induction was reported for CsDHN in oranges (*Citrus sinensis*) and for COR15 and CpDHN in grapefruits (*Citrus paradisi*) (Porat et al., 2002, 2004).

Cold-inducible dehydrins have been studied in many herbaceous and woody plants. Currently, the most studied plant is *Arabidopsis thaliana*, a total of 51 *Lea* genes have been identified (Hundertmark and Hincha, 2008), 10 of them represent dehydrins (Nylander et al., 2001). For five dehydrins, *Cor47*, *Erd10/Lti29*, *Erd14*, *Lti30/Xero2*, and *Rab18*, a cold-inducible expression has been reported (Nylander et al., 2001). Three cold-inducible dehydrin genes, *Cor47*, *Erd10/Lti29*, and *Erd14*, encode acidic SK_n-type proteins while *Lti30* encodes a K_n-type protein and *Rab18* encodes predominantly drought- and ABA-inducible (but also cold-inducible) basic Y_nSK_n-type protein. While the first four dehydrin proteins accumulate only in various non-seed tissues (predominantly vascular tissues, but also root tips in the case of ERD10, or anthers in case of Lti30), RAB18 also accumulates in mature embryos.

A positive effect of dehydrin transcript and/or protein accumulation upon CA conditions on plant tolerance to cold and/or frost has been reported for many cold-inducible dehydrins in various plant species, e.g., for CpDHN1 in *Cicer pinnatifidum*, which is a cold-tolerant wild relative of chickpea (*Cicer arietinum*) (Bhattarai and Fettig, 2005), CAS15 in alfalfa (*Medicago sativa*) (Monroy et al., 1993), CAS15 and CAS18 dehydrins in cell suspension culture of alfalfa (*Medicago falcata*) (Wolfraim et al., 1993), WCS120 in common wheat (*Triticum aestivum*) (Houde et al., 1992a, b; Vítámvás et al., 2007), DHN5 in barley (Close et al., 1995; Kosová et al., 2008), etc.

Effects of a broad range of suboptimum temperatures on dehydrin expression and protein accumulation: Studies aimed at an investigation of threshold induction temperatures leading to an accumulation of dehydrin transcripts or proteins in cereals (wheat, barley) revealed that highly frost-tolerant winter wheat and barley cultivars (Mironovskaya 808, Šárka, Dicktoo, Odesskij 31) started inducing acquired frost tolerance determined as lethal temperature for 50% of the sample (LT50) and accumulating cold-inducible dehydrin proteins (WCS120, DHN5) at higher temperatures (17–15°C) than the less tolerant winter ones (Fowler, 2008; Vítámvás et al., 2010; Kosová et al., 2013) which may correspond to earlier induction of dehydrins in the highly frost-tolerant winter cereals during autumn. Dehydrin transcript or protein accumulation at mild suboptimum temperatures (15–17°C) thus might be used as a preselection tool for the screening of highly frost-tolerant winter wheat or barley cultivars with respect to less tolerant ones without the necessity to subject the plants to long-term cold acclimation treatments.

10.2.4.3.3 Dehydrin Expression in Woody Plants during Winter Dormancy

In woody perennials, dehydrins accumulate predominantly during winter dormancy. Winter dormancy is a complex physiological process that is aimed at the enhancement of winter hardiness (overwintering) in woody perennials, and that is induced by low temperatures (both cold and frost) and short-day photoperiods during autumn. Processes of cold acclimation and winter dormancy in woody plants were reviewed recently by Welling and Palva (2006). The process of winter dormancy acquisition is associated with several changes in plant metabolism including the accumulation of various hydrophilic proteins (dehydrins).

Wisniewski et al. (2008) studied expression of dehydrin EST homologous to the *Arabidopsis Xero2* gene in young apple trees (*Malus × domestica* cv. 'Royal Gala') under cold (24 h at 5°C) and found a high increase in dehydrin expression in bark and xylem tissues and only a mild enhancement in leaves. Therefore, it can be concluded that dehydrin expression in woody plants subjected to cold is tissue-specific. In bark tissues, several hydrophilic proteins (e.g., bark storage proteins, 70 kDa heat-shock proteins, dehydrins) accumulate under cold acclimation (e.g., Wetzel et al., 1989; Wisniewski et al., 1996; Pagter et al., 2008).

For several dehydrins accumulating during winter dormancy, induction by short-day photoperiods has been reported; e.g., the *BpLti36* gene in silver birch (*Betula pendula*) (Puhakainen et al., 2004a), *BpuDHN1* (a dominant effect of SD induction) and *BpuDHN2* (only a small effect of SD induction) genes in pubescent (downy) birch (*Betula pubescens*) (Welling et al., 2004), *BbDHN7* gene in blueberry (*Vaccinium corymbosum*) (Dhanaraj et al., 2005), spliced forms of *DHN-1* genes in *Vitis riparia* and *Vitis vinifera* (Xiao and Nassuth, 2006), 24-kDa dehydrin in red-osier dogwood (*Cornus sericea*) (Karlson et al., 2003a, 2003b), or *PaDHN1* and *PaDHN6* genes in Norway spruce (*Picea abies*) (Yakovlev et al., 2008). Signaling pathways involved in short-day regulated dehydrin gene expression remain to be characterized although Welling and Palva (2008) have recently identified four homologues of *CBF* genes in silver birch (*Betula pendula*), named *BpCBF1* to *BpCBF4*, involved in the regulation of *BpLti36* gene expression.

Dehydrins which accumulate in bark tissues during winter dormancy exhibit often protective functions. For example, the 24-kDa dehydrin from *Betula pubescens*, which accumulates in the vicinity of storage protein bodies and starch-rich amyloplasts under cold, exhibits a protective effect on α-amylase (EC 3.2.1.1.) enzymatic activity *in vitro*. Similarly, dehydrin RcDHN5 from *Rhododendron catawbiense* can protect LDH activity *in vitro* (Peng et al., 2008).

Dehydrins also accumulate in buds during winter dormancy. Yakubov et al. (2005) detected dehydrin protein, PV-DHN, in the outer leaves of inflorescence buds in pistachio (*Pistacia vera*), grown in Negev Desert highlands, during December and January. Yakovlev et al. (2008) studied changes in dehydrin gene expression in buds of Norway spruce (*Picea abies*) during bud winter dormancy and subsequent flushing. During bud flushing, the expression of some *PaDHN* genes (*PaDHN1, PaDHN4.6, PaDHN5, PaDHN6, PaDHN2,* and *PaDHN3*) declines while the expression of other *PaDHN* genes (*PaCAP1, PaDHN4.2,* and *PaDHN7*) does not change significantly. These results can suggest that dehydrins from the first group may be related to bud winter hardiness while dehydrins from the second group may have general protective functions.

10.2.5 Dehydrins and Heavy-Metal Stress

Recently, metal ion-binding properties have been found in some dehydrins, e.g., in CuCOR15 from *Citrus unshiu* (Hara et al., 2005). Dehydrins may thus function as chelators analogously to metallothioneins and phytochelatins, although different amino acid residues are involved in metal ion binding. Instead of cysteine thiol groups responsible for metal ion-binding in metallothioneins and phytochelatins, histidine was reported to bind several metal ions: Co^{2+}; Cu^{2+}; Fe^{2+}; Fe^{3+}; Ni^{2+}; Zn^{2+} in CuCOR15. Recently, some papers showing a positive effect of dehydrin expression on plant heavy-metal tolerance have been published. Xu et al. (2008) found out that expression of BjDHN2 and BjDHN3 proteins from *Brassica juncea* in transgenic tobacco resulted in plants which accumulated higher concentrations of Cd^{2+} and Zn^{2+} in their root system with respect to wild-type plants when they were exposed to stress conditions (100 μM $CdCl_2$ or 200 μM $ZnCl_2$ for 10 days) and thus the transformants exhibited higher tolerance to heavy-metal stress. In contrast, *Brassica juncea* plants with inhibited expression of *BjDHN3* (via expression of antisense *BjDHN3* mRNA) revealed increased electrolyte leakage and reduced accumulation of Cd^{2+} and Zn^{2+} which indicates that they were not able to cope with heavy-metal stress. Zhang et al. (2006) have observed increased expression of an SK_n-type dehydrin *PvSR3* transcripts in roots of bean (*Phaseolus vulgaris*) in response to various heavy metals including cadmium and mercury (in the form of 0.2% (w/v) $CdCl_2$ or $HgCl_2$). Tamás et al. (2006) have observed enhanced expression of *DHN4* in barley roots in response to aluminum (10 mM $AlCl_3$ pH 4.0).

Recently, Hara et al. (2009) reported for CuCOR15 that increased concentrations of Zn^{2+} lead to the interaction of this histidine-rich dehydrin with DNA and RNA. They have shown that the CuCOR15 nucleic acid binding affinity is not sequence-specific; thus, they have hypothesized that binding of CuCOR15 to nucleic acids may have general protective functions under stress conditions.

10.2.6 Dehydrins and Biotic Stresses (Wounding)

Wounding, i.e., mechanical damage of plant tissues by herbivores or (sucking apparatus of) insects, is a common biotic stress that most plants have to face every day. Wounding is associated with cellular damage that leads to water loss. Thus, wounding can also be regarded as dehydration stress. In wounding stress signaling, salicylic acid (a derivative of 3-hydroxy-3-phenylpropanoic acid), jasmonic acid (a derivative of 12-oxo-*cis*, *cis*-10,1-phytodienoic acid) and its methyl ester methyl jasmonate play important roles.

For several dehydrins, e.g., CpDHN1 (Y_2K dehydrin) from *Cicer pinnatifidum* (Bhattarai and Fettig, 2005) or for PgDHN1 (S_8K_4 dehydrin) from white spruce *Picea glauca*

(Richard et al., 2000), induction of gene expression by jasmonic acid and methyl jasmonate was reported. Rouse et al. (1996) carried out a promoter analysis of *A. thaliana* cold-inducible K-type dehydrin gene *Lti30* (*Xero 2*) using a GUS reporter gene and they concluded that the *Lti30* promoter also revealed induction by wounding (damage of plant roots by forceps) among other treatments. In *Boea crassifolia*, expression of dehydrin *BcDHN2* is induced upon wounding via salicylic acid and methyl jasmonate signals (Shen et al., 2004). Recently, Sun et al. (2009) have observed a positive effect of low concentrations of exogenous salicylic acid (up to 0.25 mM) on the expression of drought-inducible dehydrins in barley seedlings subjected to drought stress. In contrast, higher concentrations of salicylic acid (0.25–0.50 mM) have led to a decrease of dehydrin expression under the same growth conditions (water stress). Dehydrins may also play an important role in plant defense mechanisms. Turco et al. (2004) have reported expression of several dehydrin proteins in drought-tolerant oak species *Quercus ilex* in response to infection with *Phytophthora cinnamomi*.

10.3 POSSIBILITIES OF THE USE OF DEHYDRINS FOR IMPROVEMENT OF PLANT TOLERANCE TO STRESS

Along with increasing knowledge of dehydrin protective functions in plant acclimation to various stress factors, recently, quite often, attempts have been made to exploit dehydrins for improvement of the stress tolerance of economically important plants, especially crops. These studies include both transgenic techniques (expression of a given dehydrin gene in other organisms or simply a modification of a dehydrin promoter sequence in order to enhance dehydrin gene expression) as well as the selection of crop varieties with an enhanced level of dehydrin expression which often (but not always) correlates with an enhanced level of stress tolerance.

10.3.1 Transgenic Studies

Several transgenic studies have proven to have a positive effect on dehydrin gene expression and/or dehydrin protein accumulation for plant stress tolerance. Studies carried out by Saavedra et al. (2006) on the moss *Physcomitrella patens*, which serves as a model organism due to the unique possibility to 'knockout' (make non-functional, disrupt) its genes via homologous recombination, have shown that a *P. patens* knockout mutant, which has only the dehydrin gene, *PpDHNA*, disrupted, reveals an impaired ability to recover after salt and osmotic stress (treatment either with 0.5 M NaCl or 0.9 M mannitol for 14 days).

Low-temperature stress: In contrast, overexpression of several dehydrin genes under a CaMV 35 S promoter (constructs pT9 containing *Cor47* and *Rab18* genes and pT10 containing *Lti29* (=*Erd10*) and *Lti30* genes) in *A. thaliana* led to enhanced plant tolerance to cold stress (Puhakainen et al., 2004b). Analogously, Peng et al. (2008) isolated the *RcDHN5* gene (an SK$_2$ acidic dehydrin) from frost-tolerant

Rhododendron catawbiense, and they have shown that its expression in *A. thaliana* led to the enhancement of frost tolerance. Similarly, Yin et al. (2006) concluded that expression of DHN24 protein from wild potato (*Solanum sogarandinum*) in cucumber (*Cucumis sativus*) led to the enhancement of frost tolerance under cold (4°C). Similarly, studies that used a dehydrin transgene expressed in a stress-susceptible plant have reported enhanced tolerance to stress in the transformed plant. For example, Hara et al. (2003) reported that expression of CuCOR19 from *Citrus unshiu* in tobacco mitochondria led to reduced lipid peroxidation. Houde et al. (2004) found out that expression of WCOR410 from common wheat in a strawberry led to the enhancement of the strawberry's frost tolerance.

Drought, salinity, and osmotic stress: Brini et al. (2007b) observed that expression of DHN5 protein from durum wheat (*Triticum turgidum* ssp. *durum*) in *A. thaliana* led to the increase in *A. thaliana* salt and osmotic stress tolerance. Saibi et al. (2015, 2016) found out that durum wheat DHN5 improves *A. thaliana* salinity tolerance due to regulation of proline metabolism leading to enhanced expression of pyrroline-5-carboxylate synthase (P5CS) and ROS scavenging enzymes catalase (CAT), peroxidase (POD), and superoxide dismutase (SOD) as well as via the modulating activities of some proteases, namely cysteinyl protease and aspartyl protase, in transgenic plants. RoyChoudhury et al. (2007) observed enhanced tolerance to drought and salt stress in tobacco plants transformed with the *Rab16A* (=*Rab21*) gene from salt-tolerant indica rice variety Pokkali. The transformed plants exhibited reduced levels of H$_2$O$_2$ and lipid peroxidation as well as lower accumulation of Na$^+$ and greater accumulation of K$^+$ when subjected to severe salt stress (200 mM NaCl). Similarly, Cheng et al. (2002) have shown that overexpression of wheat dehydrin PMA80 and wheat LEA I protein PMA1959 enhances rice tolerance to drought and salt stress. Figueras et al. (2004) reported a positive effect of the overexpression of maize Rab17 in *A. thaliana* on osmotic stress tolerance of the transformants. Xie et al. (2012) demonstrated that ectopic expression of alfalfa dehydrin MtCAS31 enhances drought tolerance in transgenic *Arabidopsis* due to a reduction in stomatal density leading to decreased stomatal transpiration. Reduced stomatal density in transgenic *Arabidopsis* overexpressing MtCAS31 is most probably underlined by an interaction of MtCAS31 with AtICE1 transcriptional regulator leading to alterations in stomatal development. Transgenic tomato overexpressing ShDHN, an SK$_3$ dehydrin from *Solanum habrochaites*, reveals enhanced levels of proline and antioxidant enzymes CAT and SOD indicating a pleiotropic effect of ShDHN overexpression on plant response to drought, salinity, and osmotic stress (Liu et al., 2015). Overexpression of four dehydrins *PmLEA10*, *PmLEA19*, *PmLEA20*, and *PmLEA29* from *Prunus mume* enhanced cold and drought tolerance in transgenic tobacco (Bao et al., 2017). Expression of ShDHN from *Solanum habrochaites* in tomato plants led to enhanced relative water content (RWC) and decreased ROS and malondialdehyde (MDA) levels and indicated less membrane damage in transgenic tomato (Liu et al., 2015).

Silencing of CaDHN1 in pepper (*Capsicum annuum*) leads to decreased chlorophyll content and reduced POD and Mn-SOD levels (Chen et al., 2015). In contrast, Iturriaga et al. (1992) did not recorded any improvement of drought tolerance determined by ion leakage tests of transgenic tobacco plants expressing several LEA proteins including dehydrins from desiccation-tolerant *Craterostigma plantagineum* when the transformants' leaf disks were subjected to 30% PEG 6000 solution.

Heavy-metal stress: Xu et al. (2008) found out that expression of BjDHN2 and BjDHN3 proteins from *Brassica juncea* in transgenic tobacco resulted in plants which could accumulate higher concentrations of Cd^{2+} and Zn^{2+} in their root system with respect to wild-type plants when they are exposed to stress conditions (100 μM $CdCl_2$ or 200 μM $ZnCl_2$ for 10 days), and thus the transformants exhibit higher tolerance to heavy-metal stress. In contrast, *Brassica juncea* plants with an inhibited expression of *BjDHN3* (via expression of anti-sense *BjDHN3* mRNA) revealed increased electrolyte leakage and reduced accumulation of Cd^{2+} and Zn^{2+} which indicates that they were not able to cope with heavy-metal stress.

10.3.2 Dehydrins as Markers of Plant Stress Tolerance

A positive effect of dehydrin gene expression or dehydrin protein accumulation on plant stress tolerance was reported not only by studies using transgenosis, but also by several other studies which compared varieties or cultivars of economically important plants which differ in their level of stress tolerance. Based on these studies, it is becoming evident that dehydrins can be used as indirect indicators of plant stress tolerance (for more on the direct and indirect methods of determination of plant stress tolerance, see, e.g., Prášil et al., 2007).

Plant tolerance to the effects of a certain stress factor can be associated with allelic variations in dehydrin sequences which can result in accumulation or absence of a dehydrin protein in certain plant tissues or organs. Ismail et al. (1999b) studied chilling tolerance during the process of seedling emergence in an extremely chilling-susceptible tropical legume crop *Vigna unguiculata*, and they described a presence of DHN1 protein (Y_2K-type dehydrin of 35 kDa) in seeds of a chilling-tolerant line 1393-2-11 which correlated positively with seedling emergence at 14°C. In contrast, the DHN1 protein was absent in the seeds of a genetically-related but chilling-susceptible line 1393-2-1.

However, the differences in plant stress tolerance are much more often determined by quantitative differences in dehydrin gene expression or dehydrin protein accumulation. It has repeatedly been proven that the rate of dehydrin transcript or protein accumulation positively correlates with the level of plant stress tolerance. In studies dealing with drought stress, Pelah et al. (1997) found a correlation between drought tolerance and the accumulation of dehydrin proteins in *Populus popularis*. Similarly, Labhilili et al. (1995) found a correlation between the level of dehydrin transcript accumulation and drought tolerance in two differently tolerant cultivars of

durum wheat (*T. turgidum* ssp. *durum*). Cellier et al. (1998) compared two lines of sunflower (*Helianthus annuus*) differing in their drought tolerance and found differences in their leaf water potential under the same gravimetric soil water content. When both lines were compared, the tolerant line had a higher leaf water potential under the same gravimetric soil water content and also exhibited a higher level of expression of two dehydrin genes, *HaDHN1* and *HaDHN2*, not only under the same gravimetric soil water content, but also when plants with the same leaf water potential were compared. The tolerant line also exhibited a higher level of *HaDHN2* gene expression after application of exogenous ABA. Analogously, Tabaei-Aghdaei et al. (2000) compared dehydrin protein accumulation in drought-tolerant *Agropyron desertorum* and in the less tolerant *Lophopyrum elongatum* under drought stress (6 days without watering) and detected much higher levels of dehydrin polypeptides in the drought-tolerant *A. desertorum* than in *L. elongatum*.

Brini et al. (2007a) studied DHN5 accumulation in two cultivars of Tunisian durum wheat (*Triticum turgidum* ssp. *durum*) with different levels of drought tolerance. They found differences not only in the accumulation of DHN5 but also in the pattern of DHN5 phosphorylation. The tolerant cultivar showed a high level of DHN5 phosphorylation while a sensitive cultivar nearly lacked phosphorylated forms of DHN5. So it can be postulated that different patterns of dehydrin post-translational modifications – namely phosphorylation – may underlie the differences in cultivar tolerance to stress.

Quantitative differences in dehydrin gene expression and dehydrin protein accumulation with respect to the low-temperature stress (cold and frost) have been studied especially in economically important cereals from the tribe *Triticeae* which are grown in temperate climates. In bread wheat (*T. aestivum*), Houde et al. (1992b) described a correlation between the accumulation of dehydrin proteins from the WCS120 family and the level of plant acquired frost tolerance. At the beginning of these studies, only the differences in dehydrin expression between cultivars (lines) with contrasting levels of acquired frost tolerance (i.e., spring cultivars versus winter ones) were studied (e.g., Zhu et al., 2000). However, Vítámvás et al. (2007) have shown that two winter wheat cultivars with a different level of acquired frost tolerance (Mironovskaya 808 and Bezostaya 1) can be distinguished according to the level of accumulation of WCS120 proteins after a 3-week cold treatment (2°C). Kosová et al. (2008) studied accumulation of the cold-inducible barley orthologue of the WCS120 protein, DHN5, and found a correlation between DHN5 accumulation and the level of acquired frost tolerance in a set of 21 barley cultivars of different geographical origin and growth habit (facultative, spring, winter) after a 3-week cold treatment. However, Kosová et al. (unpublished) demonstrated that the correlation between DHN5 accumulation and acquired frost tolerance or winter survival rate can be obtained only in those barley plants that are still in the vegetative phase of their development.

Stress-tolerant and stress-susceptible related plant species or cultivars belonging to one plant species can be distinguished according to the level of dehydrin expression not only under stress conditions but also under non-stress (or mild stress) growth conditions or in the early phases of stress treatment. This observation was reported by Xiao and Nassuth (2006) on dehydrin transcript of *DHN-1* in leaves and buds of two grapevine species *Vitis vinifera* and *Vitis riparia*, which differ in their frost tolerance. The tolerant *Vitis riparia* revealed a higher level of *DHN1* under control conditions and also an earlier increase in *DHN1* level under stress conditions than the sensitive *Vitis vinifera*. In *Triticeae*, it has also been shown by several authors that cultivars of wheat or barley with different levels of frost tolerance can be distinguished according to dehydrin (or other *Cor* gene) expression not only at cold temperatures typical of a cold acclimation process (5–2°C) but also at mildly cold temperatures (17–9°C). In the first studies published by Crosatti et al. (1995, 1996), spring versus winter cultivars of barley were distinguished according to the level of expression of the chloroplast-located *Cor14b* gene (a *Lea-3* gene). Later, Vágújfalvi et al. (2000) distinguished highly frost-tolerant winter wheats Albidum and Uljanovka from the less tolerant winter wheat Cheyenne according to *Cor14b* gene expression using the plants cultivated at a 18/13°C (day/night) regimen. Vítámvás et al. (2010) and (Kosová et al. 2013) used dehydrins from the *Wcs120* gene family and its barley ortho-logue DHN5 to screen frost tolerance of winter wheat and barley cultivars and they were able to distinguish differently frost-tolerant winter wheats and barleys grown at a broad temperature range from 20 to 5°C at a WCS120 or DHN5 accumulation level. Moreover, the level of accumulation of WCS120 proteins corresponded well with the level of plant winter survival. Thus it seems possible to use the level of WCS120 accumulation in wheat plants grown under mildly cold temperatures (17–9°C) as a means for estimation of plant winter survival. These results could enhance the pre-screening procedures in the breeding programs aimed at the improvement of wheat frost tolerance (or winter survival) immensely since the plants can be grown under relatively high temperatures; thus relatively large amounts of plant material can be obtained much faster when compared with plant growth under cold.

10.4 CONCLUDING REMARKS

With increasing data from diverse research fields, dehydrins appear to be an amazingly versatile group of LEA proteins presumably due to their intrinsically unstructured character. They exhibit myriads of functions in plant reactions on various stress factors, including drought, high-salinity stress, low-temperature stress, heavy-metal stress, and also some biotic stresses such as wounding. Several studies carried out on both normal (non-transgenic) and transgenic plants have reported a positive effect of dehydrin transcript or protein (expression) accumulation on plant tolerance to various stress factors. Future research studies will surely bring new insights into the functions of these fascinating proteins, and we can hope that these studies will significantly contribute to our better understanding of the roles which these stunning proteins play in the conundrum of plant stress response mechanisms.

ACKNOWLEDGMENTS

This work was supported by the Czech Ministry of Agriculture (MZe CR) institutional project MZe-RO0418 and project QK1710302, respectively.

REFERENCES

Abba, S., S. Ghignone, and P. Bonfante. 2006. A dehydration-inducible gene in the truffle *Tuber borchii* identifies a novel group of dehydrins. *BMC Genomics.* 7:39–53.

Abu-Abied, M., L. Golomb, E. Belausov, S. Huang, B. Geiger, Z. Kam, C.J. Staiger, and E. Sadot. 2006. Identification of plant cytoskeleton-interacting proteins by screening for actin stress fiber association in mammalian fibroblasts. *Plant J.* 48:367–379.

Allagulova, Ch.R., F.R. Gimalov, F.M. Shakirova, and V.A. Vakhitov. 2003. The plant dehydrins: structure and putative functions. *Biochemistry.* 68:945–951.

Alsheikh, M.K., B.J. Heyen, and S.K. Randall. 2003. Ion binding properties of the dehydrin ERD14 are dependent upon phosphorylation. *J. Biol. Chem.* 278:40882–408829.

Alsheikh, M.K., J.T. Svensson, and S.K. Randall. 2005. Phosphorylation regulated ion-binding is a property shared by the acidic subclass dehydrins. *Plant Cell Environ.* 28:1114–1122.

Amtmann, A. 2009. Learning from evolution: *Thellungiella* generates new knowledge on essential and critical components of abiotic stress tolerance in plants. *Mol. Plant.* 2:3–12.

Arora, R., and M.E. Wisniewski. 1994. Cold acclimation in genetically related (sibling) deciduous and evergreen peach (*Prunus persica* [L.] Batsch). II. A 60-kilodalton bark protein in cold-acclimated tissues of peach is heat stable and related to the dehydrin family of proteins. *Plant Physiol.* 105:95–101.

Badawi, M., J. Danyluk, B. Boucho, M. Houde, and F. Sarhan. 2007. The *CBF* gene family in hexaploid wheat and its relationship to the phylogenetic complexity of cereal *CBFs. Mol. Genet. Genomics.* 277:533–554.

Baker, J., C. Steele, and L. Dure III.. 1988. Sequence and characterization of 6 LEA proteins and their genes from cotton. *Seed Sci. Res.* 5:185–193.

Bao, F., D. Du, Y. An, W. Yang, J. Wang, T. Cheng, and Q. Zhang. 2017. Overexpression of *Prunus mume* dehydrin genes in tobacco enhances tolerance to cold and drought. *Front. Plant Sci.* 8:151.

Bassett, C.L., M.E. Wisniewski, T.S. Artlip, G. Richart, J.L. Norelli, and R.E.Jr. Farrell. 2009. Comparative expression and transcript initiation of three peach dehydrin genes. *Planta.* 230:107–118.

Bateman, A., L. Coin, R. Durbin, R.D. Finn, V. Hollich, S. Griffiths-Jones, A. Khanna, M. Marshall, S. Moxon, E.L.L. Sonnhammer, D.J. Studholme, C. Yeats, and S.R. Eddy. 2004. The Pfam protein families database. *Nucleic Acids Res.* 32:D138–D141.

Battaglia, M., Y. Olvera-Carrillo, A. Garciarrubio, F. Campos, and A.A. Covarrubias. 2008. The enigmatic LEA proteins and other hydrophilins. *Plant Physiol.* 148:6–24.

Berjak, P., and N.W. Pammenter. 2008. From *Avicennia* to *Zizania*: seed recalcitrance in perspective. *Ann. Bot.* 101:213–228.

Bhattarai, T., and S. Fettig. 2005. Isolation and characterization of a dehydrin gene from *Cicer pinnatifidum*, a drought-resistant wild relative of chickpea. *Physiol. Plant.* 123:452–458.

Bies-Ethève, N., P. Gaubier-Comella, A. Debures, E. Lasserre, E. Jobet, M. Raynal, R. Cooke, and M. Delseny. 2008. Inventory, evolution and expression profiling diversity of the LEA (late embryogenesis abundant) protein gene family in *Arabidopsis thaliana. Plant Mol. Biol.* 67:107–124.

Borovskii, G.B., I.V. Stupnikova, A.I. Antipina, C.A. Downs, and V.K. Voinikov. 2000. Accumulation of dehydrin-like proteins in the mitochondria of cold-treated plants. *J. Plant Physiol.* 156:797–800.

Borovskii, G.B., I.V. Stupnikova, A.I. Antipina, S.V. Vladimirova, and V.K. Voinikov. 2002. Accumulation of dehydrin-like proteins in the mitochondria of cereals in response to cold, freezing, drought and ABA treatment. *BMC Plant Biol.* 2:5.

Bradford, K.J., and P.M. Chandler. 1992. Expression of 'dehydrin-like' proteins in embryos and seedlings of *Zizania palustris* and *Oryza sativa* during dehydration. *Plant Physiol.* 99:488–494.

Bravo, L.A., T.J. Close, L.J. Corcuera, and C.L. Guy. 1999. Characterization of an 80-kDa dehydrin-like protein in barley responsive to cold acclimation. *Physiol. Plant.* 106:177–183.

Bravo, L.A., J. Gallardo, A. Navarrete, N. Olave, J. Martínez, M. Alberdi, T.J. Close, and L.J. Corcuera. 2003. Cryoprotective activity of a cold-induced dehydrin purified from barley. *Physiol. Plant.* 118:262–269.

Bray, E.A. 1993. Molecular responses to water deficit. *Plant Physiol.* 103:1035–1040.

Bray, E.A. 1997. Plant responses to water deficit. *Trends Plant Sci.* 2:48–54.

Brini, F., M. Hanin, V. Lumbreras, S. Irar, M. Pagès, and K. Masmoudi. 2007a. Functional characterization of DHN-5, a dehydrin showing a differential phosphorylation pattern in two Tunisian durum wheat (*Triticum durum* Desf.) varieties with marked differences in salt and drought tolerance. *Plant Sci.* 172:20–28.

Brini, F., M. Hanin, V. Lumbreras, I. Amara, H. Khoudi, A. Hassairi, M. Pagès, and K. Masmoudi. 2007b. Overexpression of wheat dehydrin DHN-5 enhances tolerance to salt and osmotic stress in *Arabidopsis thaliana*. *Plant Cell Rep.* 26:2017–2026.

Browne, J.A., A. Tunnacliffe, and A.M. Burnell. 2002. Anhydrobiosis – Plant desiccation gene found in a nematode. *Nature.* 416:38.

Campbell, S.A., and T.J. Close. 1997. Dehydrins: genes, proteins, and associations with phenotypic traits. *New Phytol.* 137:61–74.

Carjuzaa, P., M. Castellión, A.J. Distéfano, M. del Vas, and S. Maldonado. 2008. Detection and subcellular localization of dehydrin-like proteins in quinoa (*Chenopodium quinoa* Willd.) embryos. *Protoplasma.* 233:149–156.

Carpenter, J.F., and J.H. Crowe. 1988. The mechanism of cryoprotection of proteins by solutes. *Cryobiology.* 25:244–255.

Caruso, A., D. Morabito, F. Delmotte, G. Kahlem, and S. Carpin. 2002. Dehydrin induction during drought and osmotic stress in *Populus*. *Plant Physiol. Biochem.* 40:1033–1042.

Cellier, F., G. Conéjéro, J.-C. Breitler, and F. Casse. 1998. Molecular and physiological resonses to water deficit in drought-tolerant and drought-sensitive lines of sunflower. *Plant Physiol.* 116:319–328.

Cellier, F., G. Conéjéro, and F. Casse. 2000. Dehydrin transcript fluctuations during a day/night cycle in drought-stressed sunflower. *J. Exp. Bot.* 51:299–304.

Chen, R.G., H. Jing, W.L. Guo, S.B. Wang, F. Ma, B.G. Pan, and Z.H. Gong. 2015. Silencing of dehydrin CaDHN1 diminishes tolerance to multiple abiotic stresses in *Capsicum annuum*. Plant Cell Rep. 34:2189–2200.

Cheng, Z., J. Targolli, X. Huang, and R. Wu. 2002. Wheat LEA genes, PMA80 and PMA1959, enhance dehydration tolerance of transgenic rice (*Oryza sativa* L.). *Mol. Breed.* 10:71–82.

Chinnusamy, V., K. Schumaker, and J.K. Zhu. 2004. Molecular genetic perspectives on cross-talk and specificity in abiotic stress signalling in plants. *J. Exp. Bot.* 55:225–236.

Choi, D.-W., B. Zhu, and T.J. Close. 1999. The barley (*Hordeum vulgare* L.) dehydrin multigene family: sequences, allele types, chromosome assignments, and expression characteristics of 11 *Dhn* genes of cv. Dicktoo. *Theor. Appl. Genet.* 98:1234–1247.

Chow-Shi-Yee, M., J.G. Briard, M. Grondin, D.A. Averill-Bates, R.N. Ben, and F. Ouellet. 2016. Inhibition of ice recrystallization and cryoprotective activity of wheat proteins in liver and pancreatic cells. *Protein Sci.* 25:974–986.

Clarke, M.W., K.F. Boddington, J.M. Warnica, J.Atkinson, S. Mc Kenna, J. Medge, C.H. Barker, and S.P. Graether. 2015. Structural and functional insights into the cryoprotection of membranes by the intrinsically disordered dehydrins. *J. Biol. Chem.* 290:26900–26913.

Close, T.J. 1996. Dehydrins: Emergence of a biochemical role of a family of plant dehydration proteins. *Physiol. Plant.* 97:795–803.

Close, T.J. 1997. Dehydrins: A commonalty in the response of plants to dehydration and low temperature. *Physiol. Plant.* 100:291–296.

Close, T.J., R.D. Fenton, and F. Moonan. 1993. A view of plant dehydrins using antibodies specific to the carboxy terminal peptide. *Plant Mol. Biol.* 23:279–286.

Close, T.J., A.A. Kortt, and P.M. Chandler. 1989. A cDNA-based comparison of dehydration-induced proteins (dehydrins) in barley and corn. *Plant Mol. Biol.* 13:95–108.

Close, T.J., and P.J. Lammers. 1993. An osmotic stress protein of cyanobacteria is immunologically related to plant dehydrins. *Plant Physiol.* 101:773–779.

Close, T.J., N.C. Meyer, and J. Radik. 1995. Nucleotide sequence of a gene encoding a 58.5-kilodaltoon barley dehydrin that lacks a serine tract. Plant Gene Register. *Plant Physiol.* 107:289–290.

Crosatti, C., E. Nevo, A.M. Stanca, and L. Cattivelli. 1996. Genetic analysis of the accumulation of COR14 proteins in wild (*Hordeum spontaneum*) and cultivated (*Hordeum vulgare*) barley. *Theor. Appl. Genet.* 93:975–981.

Crosatti, C., C. Soncini, A.M. Stanca, and L. Cattivelli. 1995. The accumulation of a cold-regulated chloroplastic protein is light-dependent. *Planta.* 196:458–463.

Danyluk, J., A. Perron, M. Houde, A. Limin, B. Fowler, N. Benhamou, and F. Sarhan. 1998. Accumulation of an acidic dehydrin in the vicinity of the plasma membrane during cold acclimation of wheat. *Plant Cell* 10:623–638.

Danyluk, J., N.A.Kane, G. Breton, A.E.Limin, D.B. Fowler, and F.Sarhan 2003. TaVRT–1, a putative transcription factor associated with vegetative to reproductive transition in cereals. *Plant Physiology.* 132:1849–1860.

Deng, W.W., M.C. Casao, P. Wang, K. Sato, P.M. Hayes, E.J. Finnegan, and B. Trevaskis. 2015. Direct links between the vernalization response and other key traits of cereal crops. *Nat Commu.* 6:5882.

Dhanaraj, A.L., J.P. Slovin, and L.J. Rowland. 2005. Isolation of a cDNA clone and characterization of expression of the highly abundant, cold acclimation-associated 14 kDa dehydrin of blueberry. *Plant Sci.* 168:949–957.

Dhillon, T., S.P. Pearce, E.J. Stockinger, A. Distelfeld, C. Li, A.K. Knox, I. Vashegyi, *et al.* 2010. Regulation of freezing tolerance and fl owering in cereals: the *VRN–1* connection. *Plant Physiol.* 153:1846–1858.

Drira, M., W. Saibi, F. Brini, A. Gargouri, K. Masmoudi, and M. Hanin. 2013. The K-segments of the wheat dehydrin DHN-5 are essential for the protection of lactate dehydrogenase and β-glucosidase activities *in vitro*. *Mol. Biotechnol.* 54:643–650.

Dure, L. III, S.C. Greenway, and G.A. Galau. 1981. Developmental biochemistry of cottonseed embryogenesis and germination: changing messenger ribonucleic acid populations as shown by *in vitro* and *in vivo* protein synthesis. *Biochemistry.* 20:4162–4168.

Dure, L. III, M. Crouch, J.J. Harada, T. Ho, J. Mundy, R.S. Quatrano, T.L. Thomas, and Z.R. Sung. 1989. Common amino acid sequence domains among the LEA proteins of higher plants. *Plant Mol. Biol.* 12:475–486.

Egerton-Warburton, L.M., R.A. Balsamo, and T.J. Close. 1997. Temporal accumulation and ultrastructural localization of dehydrins in *Zea mays*. *Physiol. Plant.* 101:545–555.

Ellis, R.J. 2001a. Macromolecular crowding: an important but neglected aspect of the intracellular environment. *Curr. Opin. Struct. Biol.* 11:114–119.

Ellis, R.J. 2001b. Macromolecular crowding: obvious but underappreciated. *Trends Biochem. Sci.* 26:597–604.

Eriksson, S.K., and P. Harryson. 2009. Phosphorylation of the dehydrins: consequences for structure, lipid binding and kinase specificity. In: *International Conference on Plant Abiotic Stress Tolerance*. Poster. 8–11th February 2009, Vienna, Austria.

Eriksson, S., N. Eremina, A. Barth, J. Danielsson, and P. Harryson. 2016. Membrane-induced folding of the plant stress dehydrin *Lti30*. *Plant Physiol.* 171:932–943.

Farrant, J.M., P. Berjak, and N.W. Pammenter. 1992. Proteins in development and germination of a desiccation sensitive (recalcitrant) seed species. *Plant Growth Regulation.* 11:257–265.

Figueras, M., J. Pujal, A. Saleh, R. Save, M. Pagès, and A. Goday. 2004. Maize Rab17 overexpression in *Arabidopsis* plants promotes osmotic stress tolerance. *Ann. Appl. Biol.* 144:251–257.

Fowler, D.B. 2008. Cold acclimation threshold induction temperatures in cereals. *Crop Sci.* 48(3):1147–1154.

Francia, E., D. Barabaschi, A. Tondelli, G. Laidò, F. Rizza, A.M. Stanca, M. Busconi, C. Fogher, E.J. Stockinger, and N. Pecchioni. 2007. Fine mapping of a Hv*CBF* gene cluster at the frost resistance locus *Fr–H₂* in barley. *Theor. Appl. Genet.* 115:1083–1091.

Galau, G.A., and L. III. Dure. 1981. Developmental biochemistry of cottonseed embryogenesis and germination: changing messenger ribonucleic acid populations as shown by reciprocal heterologous complementary deoxyribonucleic acid-messenger ribonucleic acid hybridization. *Biochemistry.* 20:4169–4178.

Galau, G.A., D.W. Hughes, and L. Dure. III. 1986. Abscisic acid induction of cloned cotton late embryogenesis-abundant (Lea) mRNAs. *Plant Mol. Biol.* 7:155–170.

Ganeshan, S., P. Vítámvás, D.B. Fowler, and R.N. Chibbar. 2008. Quantitative expression analysis of selected *COR* genes reveals their differential expression in leaf and crown tissues of wheat (*Triticum aestivum* L.) during an extended low temperature acclimation regimen. *J. Exp. Bot.* 59:2393–2402.

Garay-Arroyo, A., J.M. Colmenero-Flores, A. Garciarrubio, and A.A. Covarrubias. 2000. Highly hydrophilic proteins in prokaryotes and eukaryotes are common during conditions of water deficit. *J. Biol. Chem.* 275:5668–5674.

Goday, A., A.B. Jensen, F.A. Culiáñez-Macià, M.M. Albà, M. Figueras, J. Serratosa, M. Torrent, and M. Pagès. 1994. The maize abscisic acid-responsive protein Rab17 is located in the nucleus and interacts with nuclear-localization signals. *Plant Cell.* 6:351–360.

Godoy, J.A., R. Lunar, S. Torres-Schumann, J. Moreno, R.M. Rodrigo, and J.A. Pintor-Toro. 1994. Expression, tissue distribution and subcellular localization of dehydrin TAS14 in salt-stressed tomato plants. *Plant Mol. Biol.* 26:1921–1934.

Graether, S.P., and K.F. Boddington, 2014. Disorder and function: a review of the dehydrin protein family. *Front. Plant Sci.* 5:576.

Griffith, M., P. Ala, D.S.C. Yang, W.-C. Hon, and B.A. Moffatt. 1992. Antifreeze protein produced endogenously in winter rye leaves. *Plant Physiol.* 100:593–596.

Griffith, M., and M.W.F. Yaish. 2004. Antifreeze proteins in overwintering plants: a tale of two activities. *Trends Plant Sci.* 9:399–405.

Gulick, P.J., and J. Dvořák. 1992. Coordinate gene response to salt stress in *Lophopyrum elongatum*. *Plant Physiol.* 100:1384–1388.

Guy, C.L. 1990. Cold acclimation and freezing stress tolerance: role of protein metabolism. *Annu. Rev. Plant Physiol. Plant Mol. Biol.* 41:187–223.

Haake, V., D. Cook, J.L. Riechmann, O. Pineda, M.F. Thomashow, and J.Z. Zhang. 2002. Transcription factor CBF4 is a regulator of drought adaptation in *Arabidopsis. Plant Physiol.* 130:639–648.

Han, B., P. Berjak, N. Pammenter, J. Farrant, and A.R. Kermode. 1997. The recalcitrant plant species, *Castanospermum australe* and *Trichilia dregeana*, differ in their ability to produce dehydrin-related polypeptides during seed maturation and in response to ABA or water-deficit-related stresses. *J. Exp. Bot.* 48:1717–1726.

Hanin, M., F. Brini, C. Ebel, Y.Toda , S. Takeda, and K. Masmoudi. 2011. Plant dehydrins and stress tolerance: Versatile proteins for complex mechanisms. *Plant Signalling Behaviour.* 6:10.

Hara, M. 2010. The multifunctionality of dehydrins: An overview. *Plant Signalling Behaviour.* 5:503–508.

Hara, M., M. Fujinaga, and T. Kuboi. 2005. Metal binding by citrus dehydrin with histidine-rich domains. *J. Exp. Bot.* 56:2695–2703.

Hara, M., M. Kondo, and T. Kato. 2013. A KS-type dehydrin and its related domains reduce Cu-promoted radical generation and the histidine residues contribute to the radical-reducing activities. *J. Exp. Bot.* 64:1615–1624.

Hara, M., S. Monna, T. Murata, T. Nakano, S. Amano, M. Nachbar, and H. Wätzig. 2016. The *Arabidopsis* KS-type dehydrin recovers lactate dehydrogenase activity inhibited by copper with the contribution of His residues. *Plant Sci.* 245:135–142.

Hara, M., Y. Shinoda, Y. Tanaka, and T. Kuboi. 2009. DNA binding of citrus dehydrin promoted by zinc ion. *Plant Cell Environ.* 32:532–541.

Hara, M., S. Terashima, T. Fukaya, and T. Kuboi. 2003. Enhancement of cold tolerance and inhibition of lipid peroxidation by citrus dehydrin in transgenic tobacco. *Planta.* 217:290–298.

Hara, M., S. Terashima, and T. Kuboi. 2001. Characterization and cryoprotective activity of cold-responsive dehydrin from *Citrus unshiu. J. Plant Physiol.* 158:1333–1339.

Hara, M., Y. Wakasugi, Y. Ikoma, M. Yano, K. Ogawa, and T. Kuboi. 1999. cDNA sequence and expression of a cold-responsive gene in *Citrus unshiu. Biosci. Biotechnol. Biochem.* 63:433–437.

Hernández-Sánchez, I.E., D.M. Martynowicz, A.A. Rodríguez-Hernández, M.B. Pérez-Morales, S.P. Graether, and J.F. Jiménez-Bremont. 2014. A dehydrin-dehydrin interaction: the case of SK3 from Opuntia streptacantha. *Front. Plant Sci.* 5:520.

Hernández-Sánchez, I.E., I. Maruri-López, A. Ferrando, J. Carbonell, S.P. Graether, and J.F. Jiménez-Bremont. 2015. Nuclear localization of the dehydrin OpsDHN1 is determined by histidine-rich motif. *Front. Plant Sci.* 6:702.

Hernández-Sánchez, I.E., I. Maruri-López, S.P. Graether, and J.F. Jiménez-Bremont. 2017. In vivo evidence for homo- and heterodimeric interactions of *Arabidopsis thaliana* dehydrins AtCOR47, AtERD10, and AtRAB18. *Sci. Rep.* 7:17036.

Heyen, B.J., M.K. Alsheikh, E.A. Smith, C.F. Torvik, D.F. Seals, and S.K. Randall. 2002. The calcium-binding activity of a vacuole-associated, dehydrin-like protein is regulated by phosphorylation. *Plant Physiol.* 130:675–687.

Houde, M., S. Dallaire, D. N'Dong, and F. Sarhan. 2004. Overexpression of the acidic dehydrin WCOR410 improves freezing tolerance in transgenic strawberry leaves. *Plant Biotech. J.* 2:381–387.

Houde, M., C. Daniel, M. Lachapelle, F. Allard, S. Laliberté, and F. Sarhan. 1995. Immunolocalization of freezing-tolerance-associated proteins in the cytoplasm and nucleoplasm of wheat crown tissues. *Plant J.* 8:583–593.

Houde, M., J. Danyluk, J.-F. Laliberte, E. Rassart, R.S. Dhindsa, and F. Sarhan. 1992a. Cloning, characterization, and expression of a cDNA encoding a 50-kilodalton protein specifically induced by cold acclimaiton in wheat. *Plant Physiol.* 99:1381–1387.

Houde, M., R.S. Dhindsa, and F. Sarhan. 1992b. A molecular marker to select for freezing tolerance in *Gramineae*. *Mol. Gen. Genet.* 234:43–48.

Hughes, S., and S.P. Graether. 2011. Cryoprotective mechanism of a small intrinsically disordered dehydrin protein. *Protein Sci.* 20:42–50.

Hundertmark, M., and D.K. Hincha. 2008. LEA (Late Embryogenesis Abundant) proteins and their encoding genes in *Arabidopsis thaliana*. *BMC Genomics.* 9:118–139.

Ingram, J., and D. Bartels. 1996. The molecular basis of dehydration tolerance in plants. *Annu. Rev. Plant Physiol. Plant Mol. Biol.* 47:377–403.

Ismail, A.M., A.E. Hall, and T.J. Close. 1999a. Purification and partial characterization of a dehydrin involved in chilling tolerance during seedling emergence of cowpea. *Plant Physiol.* 120:237–244.

Ismail, A.M., A.E. Hall, and T.J. Close. 1999b. Allelic variation of a dehydrin gene cosegregates with chilling tolerance during seedling emergence. *Proc. Nat. Acad. Sci. USA.* 96:13566–13570.

Israelachvili, J., and H. Wennerström. 1996. Role of hydration and water structure in biological and colloidal interactions. *Nature.* 379:219–225.

Iturriaga, G., M.A.F. Cushman, and J.C. Cushman. 2006. An EST catalogue from the resurrection plant *Selaginella lepidophylla* reveals abiotic stress-adaptive genes. *Plant Sci.* 170:1173–1184.

Iturriaga, G., K. Schneider, F. Salamini, and D. Bartels. 1992. Expression of desiccationrelated proteins from the resurrection plant *Craterostigma plantagineum* in transgenic tobacco. *Plant Mol. Biol.* 20:555–258.

Jaglo-Ottosen, K.R., S.J. Gilmour, D.G. Zarka, O. Schabenberger, and M.F. Thomashow. 1998. *Arabidopsis CBF1* overexpression induces *COR* genes and enhances freezing tolerance. *Science.* 280:104–106.

Jarvis, S.B., M.A. Taylor, M.R. MacLeod, and H.V. Davies. 1996. Cloning and characterisation of the cDNA clones of three genes that are differentially expressed during dormancy-breakage in the seeds of Douglas fir (*Pseudotsuga menziesii*). *J. Plant Physiol.* 147:559–566.

Jensen, A.B., A. Goday, M. Figueras, A.C. Jessop, and M. Pagès. 1998. Phosphorylation mediates the nuclear targeting of the maize Rab17 protein. *Plant J.* 13:691–697.

Jiménez, J.Á., A. Alonso-Ramírez, and C. Nicolás. 2008. Two cDNA clones (*FsDhn1* and *FsClo1*) up-regulated by ABA are involved in drought responses in *Fagus sylvatica* L. seeds. *J. Plant Physiol.* 165:1798–1807.

Kalemba E.M., F. Janowiak, and S. Pukacka. 2009. Desiccation tolerance acquisition in developing beech (*Fagus sylvatica* L.) seeds: the contribution of dehydrin-like protein. *Trees.* 23:305–315.

Karlson, D.T., Y. Zeng, V.E. Stirm, R.J. Joly, and E.N. Ashworth. 2003a. Photoperiodic regulation of a 24-kDa dehydrin-like protein in red-osier dogwood (*Cornus sericea* L.) xylem with relation to freeze-tolerance. *Plant Cell Physiol.* 44:25–34.

Karlson, D.T., T. Fujino, S. Kimura, K. Baba, T. Itoh, and E.N. Ashworth. 2003b. Novel plasmodesmata association of dehydrin-like proteins in cold-acclimated red-osier dogwood (*Cornus sericea*). *Tree Physiol.* 23:759–767.

Kazuoka, T., and K. Oeda. 1994. Purification and characterization of COR85–oligomeric complex from cold-acclimated spinach. *Plant Cell Physiol.* 35:601–611.

Knox, A.K., C. Li, A. Vágújfalvi, G. Galiba, E.J. Stockinger, and J. Dubcovsky. 2008. Identification of candidate *CBF* genes for the frost tolerance locus *Fr–Am2* in *Triticum monococcum*. *Plant Mol. Biol.* 67:257–270.

Koag, M.-C., R.D. Fenton, S. Wilkens, and T.J. Close. 2003. The binding of maize DHN1 to lipid vesicles. Gain of structure and lipid specificity. *Plant Physiol.* 131:309–316.

Kontunen-Soppela, S., K. Taulavuori, E. Taulavuori, P. Lähdesmäki, and K. Laine. 2000. Soluble proteins and dehydrins in nitrogen-fertilized Scots pine seedlings during deacclimation and the onset of growth. *Physiol. Plant.* 109:404–409.

Kosová, K., L. Holková, I.T. Prášil, P. Prášilová, M. Bradáčová, P. Vítámvás, and V. Čapková. 2008. Expression of dehydrin 5 during the development of frost tolerance in barley (*Hordeum vulgare*). *J. Plant Physiol.* 165:1142–1151.

Kosová, K., P. Vítámvás, and I.T. Prášil. 2007. The role of dehydrins in plant response to cold. *Biol. Plant.* 51:601–617.

Kosová, K., P. Vítámvás, P. Prášilová, and I.T. Prášil. 2013. Accumulation of WCS120 and DHN5 proteins in differently frost-tolerant wheat and barley cultivars grown under a broad temperature scale. *Biol. Plant.* 57:105–112.

Kosová, K., P. Vítámvás, and I.T. Prášil. 2014. Wheat and barley dehydrins under cold, drought, and salinity – what can LEA-II proteins tell us about plant stress response? *Front. Plant Sci.* 5:343.

Kovacs, D., E. Kalmar, Z. Torok, and P. Tompa. 2008. Chaperone activity of ERD10 and ERD14, two disordered stress-related plant proteins. *Plant Physiol.* 147:381–390.

Krüger, C., O. Berkowitz, U.W. Stephan, and R. Hell. 2002. A metal-binding member of the late embryogenesis abundant protein family transports iron in the phloem of *Ricinus communis* L. *J. Biol. Chem.* 277:25062–25069.

Kyte, J., and R.F. Doolittle. 1982. A simple method for displaying the hydropathic character of a protein. *J. Mol. Biol.* 157:105–132.

Labhilili, M., P. Joudrier, and M.-F. Gautier. 1995. Characterization of cDNAs encoding *Triticum durum* dehydrins and their expression patterns in cultivars that differ in drought tolerance. *Plant Sci.* 112:219–230.

Lång, V., and E.T. Palva. 1992. The expression of a *rab*-related gene, *rab18*, is induced by abscisic acid during the cold-acclimation process of *Arabidopsis thaliana* (L.) Heynh. *Plant Mol. Biol.* 20:951–962.

Levi, A., G.R. Panta, C.M. Parmentier, M.M. Muthalif, R. Arora, S. Shanker, and L.J. Rowland. 1999. Complementary DNA cloning, sequencing and expression of an unusual dehydrin from blueberry floral buds. *Physiol. Plant.* 107:98–109.

Levitt, J. 1980. *Responses of Pants to Environmental Stress. Chilling, Freezing and High Temperature Stresses.* Vol. 1. New York, NY: Academic Press.

Li, R., S.H. Brawley, and T.J. Close. 1998. Proteins immunologically related to dehydrins in fucoid algae. *J. Phycol.* 34:642–650.

Lin, C., and M.F. Thomashow. 1992. A cold-regulated *Arabidopsis* gene encodes a polypeptide having potent cryoprotective activity. *Biochem. Biophys. Res. Commun.* 183:1103–1108.

Lin, C.H., P.H. Peng, C.Y. Ko, A.H. Markhart, and T.Y. Lin. 2012. Characterization of a novel Y2K-type dehydrin *VrDhn1* from *Vigna radiata*. *Plant Cell Physiol.* 53:930–942.

Lisse, T., D. Bartels, H.R. Kalbitzer, and R. Jaenicke. 1996. The recombinant dehydrin-like desiccation stress protein from the resurrection plant *Craterostigma plantagineum* displays no defined three-dimensional structure in its native state. *Biol. Chem.* 377:555–561.

Liu, H., C.Y. Yu, H.X. Li, T.T. Wang, J.H. Zhang, X. Wang, and Z.B. Ye. 2015. Overexpression of *ShDHN*, a dehydrin gene from *Solanum habrochaites* enhances tolerance to multiple abiotic stresses in tomato. *Plant Sci.* 231:198–211.

Liu, Y., Q. Song, D. Li, X. Yang, and D. Li. 2017. Multifunctional roles of plant dehydrins in response to environmental stresses. *Front. Plant Sci.* 8:1018.

Malik, A.A., M. Veltri, K.F. Boddington, K.K. Singh, and S.P. Graether. 2017. Genome analysis of conserved dehydrin motifs in vascular plants. *Front. Plant Sci.* 8:709.

Masmoudi, K., F. Brini, A. Hassairi, and R. Ellouz. 2001. Isolation and characterization of a differentially expressed sequence tag from *Triticum durum* salt-stressed roots. *Plant Physiol. Biochem.* 39:971–979.

Mehta, P.A., K.C. Rebala, G. Venkataraman, and A. Parida. 2009. A diurnally regulated dehydrin from *Avicennia marina* that shows nucleo-cytoplasmic localization and is phosphorylated by Casein kinase II in vitro. *Plant Physiol. Biochem.* 47:701–709.

Miller, A.K., G. Galiba, and J. Dubcovsky. 2006. A cluster of 11 *CBF* transcription factors is located at the frost tolerance locus *Fr-Am2* in *Triticum monococcum*. *Mol. Genet. Genomics.* 275:193–203.

Monroy, A.F., Y. Castonguay, S. Laberge, F. Sarhan, L.P. Vezina, and R.S. Dhindsa. 1993. A new cold-induced alfalfa gene is associated with enhanced hardening at subzero temperature. *Plant Physiol.* 102:873–879.

Moons, A., G. Bauw, E. Prinsen, M. Van Montagu, and D. Van Der Straeten. 1995. Molecular and physiological responses to abscisic acid and salts in roots of salt-sensitive and salt-tolerant Indica rice varieties. *Plant Physiol.* 107:177–186.

Mouillon, J.-M., S.K. Eriksson, and P. Harryson. 2008. Mimicking the plant cell interior under water stress by macromolecular crowding: disordered dehydrin proteins are highly resistant to structural collapse. *Plant Physiol.* 148:1925–1937.

Mu, P., D. Feng, J. Su, Y. Zhang, J Dai, H Jin, B. Liu, Y. He, K. Qi, H. Wang, and J. Wang. 2011. Cu2+ triggers reversible aggregation of a disordered His-rich dehydrin MpDhn12 from *Musa paradisiaca*. *J. Biochem.* 150:491–499.

Mundy, J., and N.-H. Chua. 1988. Abscisic acid and water-stress induce the expression of a novel rice gene. *EMBO J.* 7:2279–2286.

Natali, L., T. Giordani, B. Lercari, P. Maestrini, R. Cozza, T. Pangaro, P. Vernieri, F. Martinelli, and A. Cavallini. 2007. Light induces expression of a dehydrin-encoding gene during seedling de-etiolation in sunflower (*Helianthus annuus* L.). *J. Plant Physiol.* 164:263–273.

Neven, L.G., D.W. Haskell, A. Hofig, Q.-B. Li, and C.L. Guy. 1993. Characterization of a spinach gene responsive to low temperature and water stress. *Plant Mol. Biol.* 21:291–305.

Nylander, M., J. Svensson, E.T. Palva, and B.V. Welin. 2001. Stress-induced accumulation and tissue-specific localization of dehydrins in *Arabidopsis thaliana*. *Plant Mol. Biol.* 45:263–279.

Oliver, M.J., A.J. Wood, and P. O'Mahony. 1998. 'To dryness and beyond' – preparation for the dried state and rehydration in vegetative desiccation-tolerant plants. *Plant Growth Regul.* 24:193–201.

Pagter, M., C.R. Jensen, K.K. Petersen, F. Liu, and R. Arora. 2008. Changes in carbohydrates, ABA and bark proteins during seasonal cold acclimation and deacclimation in *Hydrangea* species differing in cold hardiness. *Physiol. Plant.* 134:473–485.

Pearce, R.S. 1999. Molecular analysis of acclimation to cold. *Plant Growth Regul.* 29:47–76.

Pedron, L., P. Baldi, A.M. Hietala, and N. La Porta. 2009. Genotype-specific regulation of cold-responsive genes in cypress (*Cupressus sempervirens* L.). *Gene.* 437:45–53.

Pelah, D., W. Wang, A. Altman, O. Shoseyov, and D. Bartels. 1997. Differential accumulation of water stress-related proteins, sucrose synthase and soluble sugars in *Populus* species that differ in their water stress response. *Physiol. Plant.* 99:153–159.

Peng, Y., J.L. Reyes, H. Wei, Y. Yang, D. Karlson, A.A. Covarrubias, S.L. Krebs, A. Fessehaie, and R. Arora. 2008. *RcDhn5*, a cold acclimation-responsive dehydrin from *Rhododendron catawbiense* rescues enzyme activity from dehydration effects *in vitro* and enhances freezing tolerance in *RcDhn5*-overexpressing *Arabidopsis* plants. *Physiol. Plant.* 134:583–597.

Perdiguero, P., M.C. Barbero, M.T. Cervera, A. Soto, and C. Collada. 2012. Novel conserved segments are associated with difefrential expression patterns for Pinaceae dehydrins. *Planta* 236:1863–1874.

Perdiguero, P., C. Collada, and A. Soto. 2014. Novel dehydrins lacking complete K-segments in Pinaceae. The exception rather than the rule. *Front. Plant Sci.* 5:682.

Piatkowski, D., K. Schneider, F. Salamini, and D. Bartels. 1990. Characterization of five abscisic acid-responsive complementary DNA clones isolated from the desiccation-tolerant plant *Craterostigma plantagineum* and their relationship to other water-stress genes. *Plant Physiol.* 94:1682–1688.

Pochon, S., P. Simoneau, S. Pigné, S. Balidas, N. Bataillé-Simoneau, C. Campion, E. Jaspard, B. Calmes, B. Hamon, R. Berruyer, M. Juchaux, and T. Guillemette. 2013. Dehydrin-like proteins in the necrotrophic fungus *Alternaria brassicicola* have a role in plant pathogenesis and stress response. *PLoS One* 8:e75143.

Porat, R. Pasentsis, K. Rozentzvieg, D. Gerasopoulos, D. Falara, V. Samach, A. Lurie, S. , and A.K. Kanellis. 2004. Isolation of a dehydrin cDNA from orange and grapefruit citrus fruit that is specifically induced by the combination of heat followed by chilling temperatures. *Physiol. Plant.* 120:256–264.

Porat, R., D. Pavoncello, S. Lurie, and T.G. McCollum. 2002. Identification of a grapefruit cDNA belonging to a unique class of citrus dehydrins and characterization of its expression patterns under temperature stress conditions. *Physiol. Plant.* 115:598–603.

Prášil, I.T., P. Prášilová, and P. Mařík. 2007. Comparative study of direct and indirect evaluations of frost tolerance in barley. *Field Crops Res.* 102:1–8.

Proctor, M.C.F., and Z. Tuba. 2002. Poikilohydry and homoiohydry: an antithesis or a spectrum of possibilities? Tansley review no. 141. *New Phytol.* 156:327–349.

Puhakainen, T., Ch. Li, M. Boije-Malm, J. Kangasjärvi, P. Heino, and E.T. Palva. 2004a. Short-day potentiation of low temperature-induced gene expression of a C-repeat-binding factor-controlled gene during cold acclimation in silver birch. *Plant Physiol.* 136:4299–4307.

Puhakainen, T., M.V. Hess, P. Mäkelä, J. Svensson, P., Heino, and E.T. Palva. 2004b. Overexpression of multiple dehydrin genes enhances tolerance to freezing stress in *Arabidopsis*. *Plant Mol. Biol.* 54:743–753.

Rahman, L.N., L. Chen, S. Nazim, V.V. Bamm, M.W. Yaish, B.A. Moffatt, J.R. Dutcher, and G. Harauz. 2010. Interactions of intrinsically disordered *Thellungiella salsuginea* dehydrins TsDHN-1 and TsDHN-2 with membranes – synergistic effects of lipid composition and temperature on secondary structure. *Biochem. Cell Biol.* 88:791–807.

Rahman, L.N., V.V. Bamm, J.A. Voyer, G.S. Smith, L. Chen, M.W. Yaish, B.A. Moffatt, J.R. Dutcher, and G. Harauz. 2011. Zinc induces disorder-to-order transitions in free and membrane-associated *Thellungiella salsuginea* dehydrins TsDHN-1 and TsDHN-2: a solution CD and solid-state ATR-FTIR study. *Amino Acids.* 40:1485–1502.

Renaut, J., L. Hoffmann, and J.-F. Hausman. 2005. Biochemical and physiological mechanisms related to cold acclimation and enhanced freezing tolerance in poplar plantlets. *Physiol. Plant.* 125:82–94.

Reyes, J.L., F. Campos, H. Wei, R. Arora, Y. Yang, D.T. Karlson, and A.A. Covarrubias. 2008. Functional dissection of hydrophilins during *in vitro* freeze protection. *Plant Cell Environ.* 31:1781–1790.

Reynolds, T.L., and J.D. Bewley. 1993. Characterization of protein synthetic changes in a desiccation-tolerant fern, *Polypodium virginianum*. Comparison of the effects of drying, rehydration and abscisic acid. *J. Exp. Bot.* 44:921–928.

Richard, S., M.-J. Morency, C. Drevet, L. Jouanin, and A. Séguin. 2000. Isolation and characterization of a dehydrin gene from white spruce induced upon wounding, drought and cold stresses. *Plant Mol. Biol.* 43:1–10.

Riera, M., M. Figueras, C. Lopez, A. Goday, and M. Pagès. 2004. Protein kinase CK2 modulates developmental functions of the abscisic acid responsive protein Rab17 from maize. *Proc. Natl. Acad. Sci. USA.* 101:9879–9884.

Rinne, P.L.H., P.L.M. Kaikuranta, L.H.W. van der Plas, and C. van der Schoot. 1999. Dehydrins in cold-acclimated apices of birch (*Betula pubescens* Ehrh.): production, localization and potential role in rescuing enzyme function during dehydration. *Planta.* 209:377–388.

Rorat, T. 2006. Plant dehydrins – tissue location, structure and function. *Cell. Mol. Biol. Lett.* 11:536–556.

Rorat, T., W.J. Grygorowicz, W. Irzykowski, and P. Rey. 2004. Expression of KS-type dehydrins is primarily regulated by factors related to organ type and leaf developmental stage during vegetative growth. *Planta.* 218:878–885.

Rouse, D.T., R. Marotta, and R.W. Parish. 1996. Promoter and expression studies on an *Arabidopsis thaliana* dehydrin gene. *FEBS Lett.* 381:252–256.

RoyChoudhury, A., C. Roy, and D.N. Sengupta. 2007. Transgenic tobacco plants overexpressing the heterologous *lea* gene *Rab16A* from rice during high salt and water deficit display enhanced tolerance to salinity stress. *Plant Cell Rep.* 26:1839–1859.

Rurek, M. 2010. Diverse accumulation of several dehydrin-like proteins in cauliflower (*Brassica oleracea* var. *botrytis*), *Arabidopsis thaliana* and yellow lupin (*Lupinus luteus*) mitochondria under cold and heat stress. *BMC Plant Biol.* 10:181.

Saavedra, L., J. Svensson, V. Carballo, D. Izmendi, B. Welin, and S. Vidal. 2006. A dehydrin gene in *Physcomitrella patens* is required for salt and osmotic stress tolerance. *Plant J.* 45:237–249.

Saibi, W., K. Feki, R. Ben Mahmoud, and F. Brini. 2015. Durum wheat dehydrin (DHN-5) confers salinity tolerance to transgenic *Arabidopsis* plants through the regulation of proline metabolism and ROS scavenging system. *Planta.* 242:1187–1194.

Saibi, W., N. Zouari, K. Masmoudi, and F. Brini. 2016. Role of the durum wheat dehydrin in the function of proteases conferring salinity tolerance in *Arabidopsis thaliana* transgenic lines. *Int. J. Biol. Macromol.* 85:311–316.

Sakai, A., and W. Larcher. 1987. *Frost Survival of Plants. Responses and Adaptation to Freezing Stress.* 2nd Edition. Springer Verlag Berlin – Heidelberg – New York, NY – London – Paris – Tokyo.

Scott, P. 2000. Resurrection plants and the secrets of eternal leaf. *Ann. Bot.* 85:159–166.

Seo, E., H. Lee, J. Jeon, H. Park, J. Kim, Y.-S. Noh, and I. Lee, 2009. Crosstalk between cold response and flowering in *Arabidopsis* is mediated through the flowering-time gene *SOC1* and its upstream negative regulator *FLC*. *The Plant Cell.* 21:3185–3197.

Shen, Y., M.-J. Tang, Y.-L. Hu, and Z.-P. Lin. 2004. Isolation and characterization of a dehydrin-like gene from drought-tolerant *Boea crassifolia*. *Plant Sci.* 166:1167–1175.

Shinozaki, K., and K. Yamaguchi-Shinozaki. 1997. Gene expression and signal transduction in water-stress response. *Plant Physiol.* 115:327–334.

Shinozaki, K., and K. Yamaguchi-Shinozaki. 2000. Molecular responses to dehydration and low temperature: differences and cross-talk between two stress signaling pathways. *Curr. Opin. Plant Biol.* 3:217–223.

Shinozaki, K., K. Yamaguchi-Shinozaki, and M. Seki. 2003. Regulatory network of gene expression in the drought and cold stress responses. *Curr. Opin. Plant Biol.* 6:410–417.

Simpson, D.J., M. Smallwood, S. Twigg, C.J. Doucet, J. Ross, and D.J. Bowles. 2005. Purification and characterisation of an antifreeze protein from *Forsythia suspensa* (L.). *Cryobiology.* 51:230–234.

Skinner, J.S., P. Szücs, J. von Zitzewitz, L. Marquez-Cedillo, T. Filichkin, E.J. Stockinger, M.F. Thomashow, T.H.H. Chen, and P.M. Hayes. 2006. Mapping of barley homologs to genes that regulate low temperature tolerance in *Arabidopsis*. *Theor. Appl. Genet.* 112:832–842.

Skinner, J.S., J. von Zitzewitz, P. Szücs, L. Marquez-Cedillo, T. Filichkin, K. Amundsen, E.J. Stockinger, M.F. Thomashow, T.H.H. Chen, and P.M. Hayes. 2005. Structural, functional, and phylogenetic characterization of a large *CBF* gene family in barley. *Plant Mol. Biol.* 59:533–551.

Soulages, J.L., K. Kim, E.L. Arrese, C. Walters, and J.C. Cushman. 2003. Conformation of a group 2 late embryogenesis abundant protein from soybean. Evidence of poly (L-proline)-type II structure. *Plant Physiol.* 131:963–975.

Souza, J.M., B.I. Giasson, V.M.-Y. Lee, and H. Ischiropoulos. 2000. Chaperone-like activity of synucleins. *FEBS Lett.* 474:116–119.

Sun, X., D.H. Xi, H. Feng, J.B. Du, T. Lei, H.G. Liang, and H.H. Lin. 2009. The dual effects of salicylic acid on dehydrin accumulation in water-stressed barley seedlings. *Russ. J. Plant Physiol.* 56:348–354.

Suprunova, T., T. Krugman, T. Fahima, G. Chen, I. Shams, A. Korol, and E. Nevo. 2004. Differential expression of dehydrin genes in wild barley, *Hordeum spontaneum*, associated with resistance to water deficit. *Plant Cell Environ.* 27:1297–1308.

Svensson, J., A.M. Ismail, E.T. Palva, and T.J. Close. 2002. Dehydrins. In *Sensing, Signalling and Cell Adaptation*, ed. K.B. Storey, and J.M. Storey, 155–171. Elsevier Science.

Tabaei-Aghdaei, S.R., P. Harrison, and R.S. Pearce. 2000. Expression of dehydration-stress-related genes in the crowns of wheatgrass species [*Lophopyrum elongatum* (Host) A. Love and *Agropyron desertorum* (Fisch. ex Link.) Schult.] having contrasting acclimation to salt, cold and drought. *Plant Cell Environ.* 23:561–571.

Tamás, L., J. Huttová, I. Mistrík, M. Šimonovičová, and B. Široká. 2006. Aluminium-induced drought and oxidative stress in barley roots. *J. Plant Physiol.* 163:781–784.

Thomashow, M.F. 1999. Plant cold acclimation: Freezing tolerance genes and regulatory mechanisms. *Annu. Rev. Plant Physiol. Plant Mol. Biol.* 50:571–599.

Tommasini, L., J.T. Svensson, E.M. Rodriguez, A. Wahid, M. Malatrasi, K. Kato, S. Wanamaker, J. Resnik, and T.J. Close. 2008. Dehydrin gene expression provides an indicator of low temperature and drought stress: transcriptome-based analysis of barley (*Hordeum vulgare* L.). *Funct. Integr. Genom.* 8:387–405.

Tompa, P. 2002. Intrinsically unstructured proteins. *Trends Biochem. Sci.* 27:527–533.

Tompa, P., C. Szász, and L. Buday. 2005. Structural disorder throws new light on moonlighting. *Trends Biochem. Sci.* 30:484–489.

Tunnacliffe, A., and M.J. Wise. 2007. The continuing conundrum of the LEA proteins. *Naturwissenschaften.* 94:791–812.

Turco, E., T.J. Close, R.D. Fenton, and A. Ragazzi. 2004. Synthesis of dehydrin-like proteins in *Quercus ilex* L. and *Quercus cerris* L. seedlings subjected to water stress and infection with *Phytophthora cinnamomi*. *Physiol. Mol. Plant Pathol.* 65:137–144.

Vágújfalvi, A., C. Crosatti, G. Galiba, J. Dubcovsky, and L. Cattivelli. 2000. Two loci on wheat chromosome 5A regulate the differential cold-dependent expression of the *cor14b* gene in frost-tolerant and frost-sensitive genotypes. *Mol. Gen. Genet.* 263:194–200.

Vágújfalvi, A., G. Galiba, L. Cattivelli, and J. Dubcovsky. 2003. The cold-regulated transcriptional activator *Cbf3* is linked to the frost-tolerance locus *Fr-A2* on wheat chromosome 5A. *Mol. Genet. Genomics.* 269:60–67.

Velten, J., and M.J. Oliver. 2001. Tr.288, a rehydrin with a dehydrin twist. *Plant Mol. Biol.* 45:713–722.

Vítámvás, P., G. Saalbach, I.T. Prášil, V. Čapková, J. Opatrná, and A. Jahoor. 2007. WCS120 protein family and proteins soluble upon boiling in cold-acclimated winter wheat. *J. Plant Physiol.* 164:1197–1207.

Vítámvás, P. and I.T. Prášil,. 2008. WCS120 protein family and frost tolerance during cold acclimation, deacclimation and reacclimation of winter wheat. *Plant Physiol Biochem.* 46:970–976.

Vítámvás P., K. Kosová, P. Prášilová, and I.T. Prášil. 2010. Accumulation of WCS120 protein in wheat cultivars grown at 9°C or 17°C in relation to their winter survival. *Plant Breeding.* 129:611–616.

Volaire, F. 2002. Drought survival, summer dormancy and dehydrin accumulation in contrasting cultivars of *Dactylis glomerata*. *Physiol. Plant.* 116:42–51.

Volaire, F., H. Thomas, N., Bertagne, E. Bourgeois, M.-F. Gautier, and F. Lelièvre. 1998. Survival and recovery of perennial forage grasses under prolonged Mediterranean drought. II. Water status, solute accumulation, abscisic acid concentration and accumulation of dehydrin transcripts in bases of immature leaves. *New Phytol.* 140:451–460.

Wachowiak, W., P.A. Balk, and O. Savolainen. 2009. Search for nucleotide diversity patterns of local adaptation in dehydrins and other cold-related candidate genes in Scots pine (*Pinus sylvestris* L.). *Tree Genet. Genomes.* 5:117–132.

Wang, X.-S., H.-B. Zhu, G.-L. Jin, H.-L. Liu, W.-R. Wu, and J. Zhu. 2007. Genome-scale identification and analysis of *LEA* genes in rice (*Oryza sativa* L.). *Plant Sci.* 172:414–420.

Wei H., A.L. Dhanaraj, L.J. Rowland, Y. Fu, S.L. Krebs, and R. Arora. 2005. Comparative analysis of expressed sequence tags from cold-acclimated and non-acclimated leaves of *Rhododendron catawbiense* Michx. *Planta.* 221:406–416.

Welling, A., and E.T. Palva. 2006. Molecular control of cold acclimation in trees. *Physiol. Plant.* 127:167–181.

Welling, A., and E.T. Palva. 2008. Involvement of CBF transcription factors in winter hardiness in birch. *Plant Physiol.* 147:1199–1211.

Welling, A., P. Rinne, A. Viherä-Aarnio, S. Kontunen-Soppela, P. Heino, and E.T. Palva. 2004. Photoperiod and temperature differentially regulate the expression of two dehydrin genes during overwintering of birch (*Betula pubescens* Ehrh.). *J. Exp. Bot.* 55:507–516.

Wetzel, S., C. Demmers, and J.S. Greenwood. 1989. Seasonally fluctuating bark proteins are a potential form of nitrogen storage in three temperate hardwoods. *Planta* 178:275–281.

Whitsitt, M.S., R.G. Collins, and J.E. Mullet. 1997. Modulation of dehydration tolerance in soybean seedlings. *Plant Physiol.* 114:917–925.

Wise, M.J. 2003. LEAping to conclusions: a computational reanalysis of late embryogenesis abundant proteins and their possible roles. *BMC Bioinformatics.* 4:52.

Wisniewski, M., T.J. Close, T. Artlip, and R. Arora. 1996. Seasonal patterns of dehydrins and 70-kDa heat-shock proteins in bark tissues of eight species of woody plants. *Physiol. Plant.* 96:496–505.

Wisniewski, M., C. Bassett, J. Norelli, D. Macarisin, T. Artlip, K. Gasic, and S. Korban. 2008. Expressed sequence tag analysis of the response of apple (*Malus×domestica* 'Royal Gala') to low temperature and water deficit. *Physiol. Plant.* 133:298–317.

Wisniewski, M., R. Webb, R. Balsamo, T.J. Close, X.-M. Yu, and M. Griffith. 1999. Purification, immunolocalization, cryoprotective, and antifreeze activity of PCA60: a dehydrin from peach (*Prunus persica*). *Physiol. Plant.* 105:600–608.

Wolfraim, L.A., R. Langis, H. Tyson, and R.S. Dhindsa. 1993. cDNA sequence, expression, and transcript stability of a cold acclimation-specific gene, *cas18*, of alfalfa (*Medicago falcata*) cells. *Plant Physiol.* 101:1275–1282.

Xiao, H., and A. Nassuth. 2006. Stress- and development-induced expression of spliced and unspliced transcripts from two highly similar dehydrin 1 genes in *V. riparia* and *V. vinifera*. *Plant Cell Rep.* 25:968–977.

Xie, C., R.X. Zhang, Y.T. Qu, Z.Y. Miao, Y.Q. Zhang, X.Y. Shen, T. Wang, and J. Dong. 2012. Overexpression of MtCAS31 enhances drought tolerance in transgenic *Arabidopsis* by reducing stomatal density. *New Phytol.* 195:124–135.

Xin, Z., and J. Browse. 2000. Cold comfort farm: the acclimation of plants to freezing temperatures. *Plant Cell Environ.* 23:893–902.

Xu, J., Y.X. Zhang, W. Wei, L. Han, Z.Q. Guan, Z. Wang, and T.Y. Chai. 2008. *BjDHNs* confer heavy-metal tolerance in plants. *Mol. Biotechnol.* 38:91–98.

Yakovlev, I.A., D.K.A. Asante, C.G. Fossdal, J. Partanen, O. Junttila, and Ø. Johnsen. 2008. Dehydrins expression related to timing of bud burst in Norway spruce. *Planta.* 228:459–472.

Yakubov, B., O. Barazani, A. Shachack, L.J. Rowland, O. Shoseyov, and A. Golan-Goldhirsh. 2005. Cloning and expression of a dehydrin-like protein from *Pistacia vera* L. *Trees.* 19:224–230.

Yamaguchi-Shinozaki, K., and K. Shinozaki. 2005. Organization of *cis*-acting regulatory elements in osmotic- and cold-stress-responsive promoters. *Trends Plant Sci.* 10:88–94.

Yamaguchi-Shinozaki, K., and K. Shinozaki. 2006. Transcriptional regulatory networks in cellular responses and tolerance to dehydration and cold stresses. *Annu. Rev. Plant Biol.* 57:781–803.

Yang, L., C.L. Yu, F. Shi, Y.Q. Wei, C.C. Wang, H.T. Hu, and C.G. Cheng. 2007. Effects of abscisic acid on growth and dehydration tolerance of *Cynanchum komarovii* seedlings. *Plant Growth Regul.* 51:177–184.

Yang, W., L. Zhang, H. Lv, H. Li, Y. Zhang, Y. Xu, and J. Yu. 2015. The K-segments of wheat dehydrin WZY2 are essential for its protective functions under temperature stress. *Front. Plant Sci.* 6:406.

Yao, K., K.M. Lockhart, and J.J. Kalanack. 2005. Cloning of dehydrin coding sequences from *Brassica juncea* and *Brassica napus* and their low temperature-inducible expression in germinating seeds. *Plant Physiol. Biochem.* 43:83–89.

Yin, Z., T. Rorat, B.M. Szabala, A. Ziółkowska, and S. Malepszy. 2006. Expression of a *Solanum sogarandinum* SK3-type dehydrin enhances cold tolerance in transgenic cucumber seedlings. *Plant Sci.* 170:1164–1172.

Zhang, Y., J. Li, F. Yu, L. Cong, L. Wang, G. Burkard, and T. Chai. 2006. Cloning and expression analysis of SKn-type dehydrin gene from bean in response to heavy metals. *Mol. Biotechnol.* 32:205–217.

Zhu, B., D.-W. Choi, R. Fenton, and T.J. Close. 2000. Expression of the barley dehydrin multigene family and the development of freezing tolerance. *Mol. Gen. Genet.* 264:145–153.

Zhu, J.K. 2002. Salt and drought stress signal transduction in plants. *Annu. Rev. Plant Biol.* 53:247–273.

11 Strigolactone Plant Hormone's Role in Plant Stress Responses

Fatemeh Aflaki, Arman Pazuki, and Mohammad Pessarakli

CONTENTS

11.1 INTRODUCTION

As sessile organisms, environmental conditions have far-reaching and critical effects on plants. Any conditions adversely influencing a plant's growth and development is generally considered to be stress. Plant stresses are placed in three basic categories based on their duration (short- and long-term), intensity (high- and low-intensity), and type (biotic and abiotic stresses). Biotic and abiotic stresses are two common groups which have been widely studied. Drought (water deficit), salinity/sodicity, nutrient-deficiency, light, water (logging/flooding), temperature (cold, frost, and heat) stresses can be representative examples of abiotic stresses. Apart from abiotic stress-induced damages, plants may encounter tensions caused by biotic stresses, which can be instigated by living organisms, including pathogens (bacteria, viruses, fungi, and nematodes), plants (weeds), and pests. Facing environmental stresses can trigger some reactions in plants, i.e., signaling mechanisms, in which various components like plant hormones play leading roles (Pandey et al. 2016). Along with the major classes of hormones, i.e., auxins, ethylene (ET), cytokinins (CKs), gibberellins (GAs), abscisic acid (ABA), and brassinosteroids, some signaling molecules should also be pointed out, e.g., jasmonic acid, salicylic acid, and karrikins (Ren and Dai 2012; Munné-Bosch and Müller 2013). As a result of a decade's research, nowadays a new signaling molecule is universally recognized which is called strigolactone (SL). SL is a carotenoid derivative serving a dual function, as both an endogenous and exogenous signaling molecule (Foo et al. 2013; Pandey et al. 2016). Due to possessing the essential properties of phytohormones, i.e., acting in low concentration, having importance in plant growth and development, local or distal site of action, and crosstalk, SL is classified as a phytohormone (Gray 2004; Pandey et al. 2016). In this chapter, the hormone, its function, biosynthesis, and transduction, are briefly explained. Then, SL roles in plant responses to stresses are indicated to provide an integrated view of the newly discovered multifaceted phytohormone. Thus far, a limited number of research studying in-depth the roles of SL in stressful conditions is available which implies the absolute necessity of further research on this aspect of strigolactones.

11.2 STRIGOLACTONE

11.2.1 DISCOVERY

In 1966, SL was initially identified and isolated in pure crystalline compounds from the root exudates of cotton as a germination stimulator for a parasitic weed, *Striga lutea* (witchweed). Later, in 1972, strigol was structurally elucidated. They were given the name "Strigolactones" by Butler in 1995 (Zwanenburg and Pospíšil 2013).

11.2.2 STRIGOLACTONE BIOSYNTHESIS PATHWAY

Carotenoid is a prerequisite compound for SL biosynthesis. SL is a sesquiterpene lactone containing three isoprene units and a lactone ring. For ascertaining the fact that SL is a carotenoid

All-trans-β-carotene

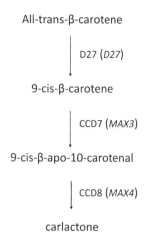

| D27 (*D27*) |

9-cis-β-carotene

| CCD7 (*MAX3*) |

9-cis-β-apo-10-carotenal

| CCD8 (*MAX4*) |

carlactone

FIGURE 11.1 The strigolactone-biosynthesis pathway in plastids.

derivative, two different sets of experiments were conducted. First, wild-type plants were treated with fluridone, a systemic herbicide (a carotenoid biosynthesis inhibitor). Second, maize carotenoid biosynthesis mutants were examined. In both pieces of research, SL level was remarkably low in the roots' exudates. It can be interpreted that the methyl-D-erythritol 4-phosphate (MEP) pathway is involved in SL biosynthesis, since interruption to the MEP pathway (by using fluridone) almost blocked carotenoid biosynthesis and subsequently SL biosynthesis. Therefore, the carotenoid biosynthesis pathway is required for SL. Carotenoid, as one of the MEP pathway products, enters the SL-biosynthesis pathway in plastids. In the SL-biosynthesis pathway, all-trans-β-carotene is converted to 9-cis-β-carotene, next to 9-cis-β-apo-10-carotenal, and carlactone by three enzymes, D27 (*D27*), CCD7 (*MAX3*), and CCD8 (*MAX4*), respectively. Carlactone (CL) is a mobile product, which moves into cytoplasm. CL contains A and D rings plus an enol-ether bridge; however, it lacks B and C rings (Figure 11.1). Subsequently, CL is converted to carlactonic acid (CLA) by *MAX1*-encoded cytochrome P450 (CYP450). Therefore, *MAX1* acts downstream of *MAX3* and *MAX4*. Afterward, CLA can be a precursor for methyl carlactonoate (MeCLA) or ent-2'-epi-5-deoxystrigol (4-deoxyorobanchol, 4DO). Thus, SL as the end product of the mentioned pathway is produced in the cytoplasm (Matusova et al. 2005; Cordoba et al. 2009; Goulet and Klee 2010; Alder et al. 2012; Pulido et al. 2012; Brewer et al. 2013; Umehara et al. 2015).

11.2.3 STRIGOLACTONE STRUCTURE

Primarily, SL structure consists of four rings. These four rings are grouped into two main moieties. One is butenolide and the other is tricyclic lactone. Butenolide is called D ring which bonds by an enol-ether bridge to the tricyclic lactone part, including three rings (A, B, and C). Substitution on A–B rings can differ from one SL to the other. It means that they can be modified by hydroxyl, methyl, or acetyloxyl groups on A and B rings (Marzec and Muszynska 2015; Umehara et al. 2015). On the other hand, there is another SL classification based on stereochemistry of the B–C ring junction, strigol- and orobanchol-like. Strigol-like SL includes a β-oriented C ring, whereas, orobanchol-like one has an α-oriented C ring (Xie et al. 2013; Scaffidi et al. 2014). Based on a structure–activity relationship (SAR), indicating dependency between a molecule chemical structure and its bio-activity, C and D rings as well as stereochemistry of D ring are crucial for SL bio-activity (Figure 11.2) (Akiyama et al. 2010; Zwanenburg and Pospíšil 2013).

11.2.4 STRIGOLACTONE FUNCTIONS

The crucial function of SL was gradually clarified. On the whole, SL function can be considered in two regions: the rhizosphere and the internal structure of plants. The recognized pathway was discovered through studying the harmful role of SL for crops. SLs were first discovered as secondary metabolites which result in the germination of root parasitic seeds. Afterward, SL-induced germinating weeds produce a haustorium; the tissue connecting a parasite to a host (Kuijt 1965), to attach to the root of the host plants and draw the water and the nutrients from their sap. Consequently, the crops may suffer a serious yield loss (Estabrook and Yoder 1998; Matusova et al. 2005). SL can exude from the roots of dicots, monocots, and even primitive plants, i.e., mosses, stoneworts, and liverworts (Delaux et al. 2012; Brewer et al. 2013).

The other role of SL was found in the rhizosphere as a hyphal branching regulator of arbuscular mycorrhizal fungus (AMF) (Cameron et al. 2013). Hence, SL may be influential in the symbiosis between plants and glomeromycota fungi.

After decades of searching, Gomez-Roldan et al. (2008) unmasked the internal role of SL as a new class of phytohormone. First, the inhibitory effect of SL on shoot branching and bud outgrowth was detected (Gomez-Roldan et al. 2008). The results of many further in-depth studies into mutant plants

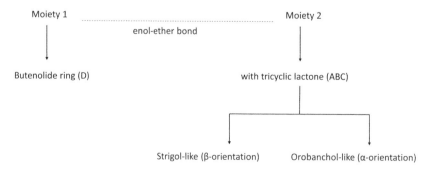

FIGURE 11.2 Strigolactone classification based on molecular structure.

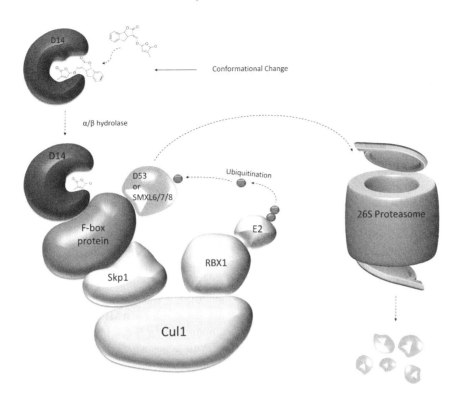

Conformational Change

α/β hydrolase

Ubiquitination

FIGURE 11.3 (See color insert.) SL's signal transduction pathway. The picture shows D14, a non-canonical hormone receptor for SL, and F-box protein as a part of SCF-ubiquitin ligase protein complex (Skp1-Cullin-F-box), which contributes to ubiquitination and subsequent degradation of SMXL6–8 by 26S proteasome.

supported its role in shoot branching inhibition, i.e., more axillary growth 1 to 4 (*MAX1–4*) in *Arabidopsis*, dwarf and high tillering dwarf (*d/htd*) in rice, decreased apical dominance 1 (*dad1*) in *Petunia hybrida* and ramosus 1 to 5 (*rms1–5*) in *Pisum sativum* (Umehara et al. 2008; Ruyter-Spira et al. 2013). Further comprehensive pieces of research showed more of SL roles. Besides the preliminary observation about the hormone effect on shoot branching and bud outgrowth, SL play an influential role in root growth, development, leaf shape, internode elongation, senescence, secondary growth, and stress responses through SL-signal transduction (Kapulnik et al. 2011; Koltai 2011; de Saint Germain et al. 2013; Liu et al. 2013; Bu et al. 2014; Ha et al. 2014; Zheng et al. 2015; Pandey et al. 2016).

11.2.5 Strigolactone's Signal Transduction Pathway

The aim of SL signal transduction pathway is suggested to be the degradation of SL repressors, e.g., D53 in rice, and SUPPRESSOR OF *MAX2*-LIKE6–8 (SMXL6–8) as D3 orthologous in *A. thaliana* through the ubiquitin-proteasome system (UPS) (Jiang et al. 2013). SMXL6–8 have a highly conserved ethylene-responsive element binding factor-associated amphiphilic repression (EAR) motif which interacts with TOPLESS and TOPLESS-RELATED PROTEINS (TPR2) as transcriptional co-repressors to suppress SL (Causier et al. 2012).

F-box protein MAX2 and DWARF14 (AtD14) are critical for SL signal transduction pathway. AtD14, a member of α/β hydrolase superfamily, is a non-canonical hormone receptor for SL in *A. thaliana*. Direct equivalents of AtD14 in rice and

petunia are D14 and DAD2, respectively (Zhou et al. 2013). D14, as an SL receptor, has a hydrophobic active site, which is composed of a conserved serine 96, histidine 247, and aspartate 218 (Ser-His-Asp) catalytic triad. The cooperation between Ser-His-Asp catalytic triad (in D14) and α/β hydrolase is required for a nucleophilic attack in which D ring separated from the other three rings of SL and attaches to D14 (Scaffidi et al. 2012). First, D ring transiently attaches to Ser 96. Later, it covalently bonds to His 247, leading to the formation of a CLIM (covalently linked intermediate molecule) which is non-canonical and stable. Next, a conformational change in AtD14 is induced by complexing with a *MAX2*-encoded protein (F-box protein) (Zhao et al. 2013). F-box protein is a part of the SCF-ubiquitin ligase protein complex (Skp1-Cullin-F-box) which contributes to the ubiquitination and subsequent degradation of SMXL6–8. As a result, SLs can carry out their functions as phytohormones and signaling molecules in stressful conditions (Figure 11.3) (Jiang et al. 2013; Zhao et al. 2013; De Cuyper et al. 2017).

11.3 SL MEDIATES ADAPTIVE RESPONSES TO ABIOTIC STRESSES

11.3.1 Strigolactone and Nutrient Stress

Nutrient stress is due to the availability and accessibility of micro-, and macronutrients in the rhizosphere. Typically, two sets of problems can be attributed to nutrient stress, i.e., the deficiency or excess of micro-, and macronutrients.

The macronutrient's importance lies in the demanded quantities for plant growth and development which are, in comparison with micronutrients, in large supply. Nitrogen (N) and phosphorus (P) are major macro elements for plants with a diverse range of functions. Even though about 78% of the atmosphere volume is occupied by N_2, due to the strong triple covalent bond, the molecule is not absorbable for plants. Therefore, only by N-fixing symbiosis it is accessible for plants. Deficiency of N, as one of the critical macronutrients for plant growth and development, may produce destructive effects through a reduction in chlorophyll content, photosynthetic rate, stomatal conductance, phosphoenolpyruvate carboxylate (PEPcase) activity, leaf area and an increase in senescence (Zhao et al. 2005). The only chemical form of P utilizable for plants is orthophosphate, which is generally called phosphate. Nowadays, P availability is restricted due to nonrenewable resources and becoming out-of-reach by bacteria converting phosphate to organic forms (López-Arredondo and Herrera-Estrella 2012). Since both N and P deficiencies have profound impacts on plant growth and development, they can be viewed as nutrient stress.

Plant hormones are of critical importance in order to mitigate stress damages (Javid et al. 2011). Thus far, it has been shown that SL may alleviate suffering from nutrient stresses, e.g., N and P shortage (Marzec et al. 2013). It has been demonstrated that SL biosynthesis, exudation, and perception increase under nutrient stresses to regulate appropriate responses, which include architecture modification and cooperation with other organisms (Yoneyama et al. 2007). Architecture modification comprises two major parts, i.e., alternation in above-ground and below-ground architectures. Above-ground modification, i.e., shoot branching inhibition, increases in low P condition. The modification was evidenced by SL-signaling and SL-biosynthesis mutants in which shoot branching was not inhibited under low P (Xi et al. 2015). The underlying reason for shoot inhibition in a P deficiency condition can be related to preserving a limited amount of P for already existing shoots. Additionally, as a consequence of the immobility of P, root modification seems necessary to reach an area containing P. It has been revealed that SL may positively affect root modification, including an increase in lateral roots (LRs), and root hairs (RHs) besides repressing primary root development (Ruyter-Spira et al. 2011). As a piece of evidence, applying an AbamineSG (an ABA biosynthesis inhibitor, targeting the enzyme catalyzing oxidative cleavage of 9-cis-epoxycarotenoids) resulted in the inhibition of SL biosynthesis (López-Ráez et al. 2010; Marzec et al. 2013) and decreased RHs under P deficiency. Thus, SL contributes to alleviating nutrient stress by communicating with other organisms, i.e., plants, bacteria, and fungi (Marzec et al. 2013).

Arbuscular mycorrhizal fungi are a type of fungi which can establish a symbiotic relationship with plants. AMF are clearly recognizable through two particular structures, vesicle, and arbuscule. AMF penetrate into root tissue (epidermis) by intracellular hyphae. Then, haustorium goes into the cortex part of the root and produces arbuscule. In this way, a mycorrhizae symbiosis can be established between plants (80% of land plants) and fungi (Van Der Heijden et al. 2008). In the mentioned symbiosis, plant roots provide the required carbohydrates for fungi. On the other hand, macronutrients can be provided by fungi for plants (Yoneyama et al. 2012; Van Der Heijden et al. 2017). Therefore, more mycorrhizal symbiosis means more chance of macronutrient availability.

In this regard, if SL has any influences on it, it may be a favorable opportunity to minimize the destruction of nutrient stresses. In research done on a *Lotus japonicus* root exudate, it has been shown that 5-deoxystrigol isolated from the root extract can stimulate the hyphal branching of AMF as a consequence of both cell proliferation and increasing the metabolism of the mitochondrion. In addition, it has been indicated that apart from an increase in hyphal branching, an increase in AMF colonization may be the other effect of SL (Marzec et al. 2013). It was proved that SL synthesis mutant plants were not colonized by AMF. In brief, it can be stated that a low amount of P or N can induce transcription of SL-biosynthesis enzyme genes which produce SL and as a result boost mycorrhizal symbiosis (López-Ráez et al. 2008; López-Ráez et al. 2009; Brewer et al. 2013).

The other communication that may be established under nutrient stressful conditions is between stressed plants and N-fixing bacteria. Through N-fixing bacteria, inorganic N, which is not absorbable by plants, is fixed and can be turned into ammonium and/or nitrate in usable forms (Xin-Hua et al. 2003). SL has shown the ability to increase the number of root nodules associated with the N-fixing bacteria. It has been indicated that even though SLs can promote nodulation in plant roots, they are not necessary for their formation (Soto et al. 2010; Foo and Davies 2011). Indeed, SL, by extending the number of nodules, may allow N-fixing bacteria to provide more N in stressful conditions. Therefore, SL can considerably modify plant architecture (shoot and root development) under nutrient-deficient conditions.

11.3.2 Strigolactone and Salinity Stress

One of the major environmental stresses is salinity, particularly in arid and semi-arid areas, which imposes a limitation on plant growth, and subsequently causes serious plant productivity and yield losses. Around 20% of the total cultivated lands and 50% of irrigated agricultural lands are severely affected by salinity (Sairam and Tyagi 2004). Salinity is the excessive salts accumulated in the soil surrounding plants. Soil salinization causes, for example, are irrigation with poor drainage, poor quality of water, and drought.

One of the reasons behind the adverse effects of salinity is oxidative damage (Saed-Moucheshi et al. 2014) caused by stress-produced reactive oxygen species (ROS) (Munns and Tester 2008; Heidari et al. 2012). For instance, ROS produced under salinity can damage polyunsaturated fatty acids in the membrane of plants by oxidizing them. The end result of the process is the formation of malondialdehyde (MDA) (Gunes et al. 2007). One of the problem-solving approaches that enable plants to lessen the destructive effects of ROS is antioxidants such as ROS scavengers, e.g., superoxide dismutase (SOD) and peroxidases (POD) (Miller et al. 2010; Wang et al. 2014). A decrease in photosynthesis rate and reduced CO_2 diffusion are the other major problems that plants encounter under salinity conditions (Flexas et al. 2004; Li et al. 2010).

Plant hormones are pivotal substances that modulate adaptive responses in stressful conditions, e.g., salinity stress (Peleg and Blumwald 2011). According to the previous intensive studies concerning the role of phytohormones in abiotic conditions, it has been demonstrated that ABA, ethylene, JA, AS, and IAA are major hormones involved in the responses to salinity (Shibli et al. 2007; Wasternack 2007; Cutler et al. 2010; Popko et al. 2010; Peleg and Blumwald 2011). Among these hormones, ABA is a stress hormone that induces 1-aminocyclopropane-1-carboxylic acid (ACC), as an ethylene precursor, in the plants under salinity stress (Zheng et al. 2015). Recent studies have focused on understanding the capacity of SL for inducing salinity-related responses.

Based on research, rac-GR24, a synthetic SL, application increased shoot and root growth of the plants in salinity conditions as compared with control (the salinity-exposed plants, no-hormonal application). Generally, physiological processes such as photosynthetic rate, stomatal conductance, transpiration rate, chlorophyll fluorescence, and chlorophyll content decrease in the plants exposed to salt stress. However, exogenous application of rac-GR24 in salinity conditions ameliorates the extent of damages to the mentioned processes (Visentin et al. 2016; Ma et al. 2017). On the other hand, increased amounts of lipid peroxidation in salinity-exposed/treated plants can be alleviated by rac-GR24. The synthetic SL can significantly reduce MDA and lipid peroxidation, which can be explained by the effect of SL on antioxidant enzyme activity. Rac-GR24 increases POD and SOD activity as important scavengers in stressed plants (Sedaghat et al. 2017). Therefore, SL can have a marked influence over scavenging salinity-generated ROS, e.g., superoxide radicals, hydrogen peroxide, and perhydroxyl radical (Banerjee and Roychoudhury 2018).

Root and shoot biomass can dramatically decline under saline stress due to salinity-posed ionic and osmotic stresses (Hu et al. 2016). However, rac-GR24 by positively influencing the responses (morphological, physiological, and biochemical) that are likely to affect shoot and root interaction in salinity can prevent them from declining (Ma et al. 2017). The other damage that can be alleviated through applied SL is the damage to photosynthesis which normally reduces its rate in salinity condition. The reduction can be attributed to impaired stomatal closure and lower content of leaf chlorophyll (Flexas et al. 2004; Ma et al. 2017).

It has been shown that SL can perform an ameliorating role within plants in salinity stress through interaction with the other phytohormones. For instance, stomatal and root/shoot development/function can be improved by SL–ABA and SL–IAA crosstalks, respectively.

11.3.3 Strigolactone and Drought Stress

Drought stress can be defined as a period without sufficient available water (either rainfall or irrigation) for plants which can lead to dehydration through the impairment of water intake and continuous water loss. Drought stress, as one of the most common abiotic stresses, by causing serious harm to plant growth and development decreases their productivity level (Jaleel et al. 2009). Generally, the status of leaf water and both stomatal density and closure are relevant and critical factors contributing to drought tolerance in plants (Bu et al. 2014). Plants in response to a damaging condition, such as drought stress, go through an elaborate system integrating multiple hormonal pathways (Ha et al. 2014). Researchers by following a predictable pattern of SL in regulating drought responses demonstrated exogenously applied SL could rescue drought-sensitive phenotypes in SL-deficient plants (MAX3 and MAX4) and can enhance drought tolerance in wild-type plants. In other words, MAX mutants showed high sensitivity to drought stress (Mishra et al. 2017). Expression of drought-induced genes revealed that drought condition could act as a trigger mechanism for MAX3 and MAX4 genes. Based on the microarray and RT-qPCR studies, MAX3 and MAX4 expression significantly increased in the leaves of the plants exposed to water stress and dehydration. Therefore, through activating SL-biosynthesis and -signaling, a successful adaptation may develop in drought conditions (Bu et al. 2014; Ha et al. 2014).

From a broad point of view, SL can be a positive regulator for drought stress-induced responses for plants through ABA-dependent and -independent pathways. Severe dehydration in MAX mutants can be due to the ABA-dependent pathway, which acts by losing water both rapidly and in large amounts. The underlying reasons for more and rapid water loss are higher stomatal density, slower ABA-mediated stomatal closure, and lower expression of ABA-inducible genes. In normal conditions, turgor pressure in guard cells is regulated by ABA, which induces stomatal closure when the plant is facing water stress and controls transpiration rate (Boursiac et al. 2013). This means that stomatal development and function are modulated by crosstalk between SL and ABA (Bu et al. 2014). It has been shown that in MAX mutants in addition to having denser stomata, and reduced stomatal sensitivity to ABA, the restricted capacity for transporting ABA to guard cells has an influential role in the faster wilting of plants in drought conditions (Saeed et al. 2017). It might be explained by research done on MAX2 mutants in which the genes encoding ABA importer proteins were down-regulated (Liu et al. 2015). Thus, stomatal cells can show dissimilar responses to ABA based on SL situation (Bu et al. 2014). An increased SL level in shoots might help to control leaf transpiration under drought stress via crosstalk with ABA (Liu et al. 2015).

In addition to the role of SL and ABA to coordinate appropriate plant responses to drought stress, SL may involve in the ABA-independent mechanism. It has been demonstrated that in MAX mutant plants, photosynthesis rate increases in drought conditions due to the up-regulation of the genes involved in photosynthesis which leads to a higher sensitivity to drought stress (Ha, et al. 2014). Interestingly, photosynthesis-related genes in wild-type plants containing SL are down-regulated which makes the plants moderately adaptable to drought stress (Pinheiro and Chaves 2010).

It seems that plant responses to drought stress are complex functions which are regulated by crosstalk among various hormones (Munné-Bosch and Müller 2013), such as ABA, CK, and SL, which modulate stomatal closure and leaf senescence. ABA as a primary phytohormone triggers short-term and long-term responses to drought (Zhang et al. 2006), and SL plays a role in promoting the senescence of leaves and stomatal closure. In contrast, CK can act as a negative regulator of both stomatal closure (Farber et al. 2016) and senescence (Zwack and Rashotte 2013; Talla et al. 2016).

Based on scientific evidence, it is suggested that gene expression regulation in drought stress conditions modulated by SL can be critical for plants to adapt accordingly. Generally, reduction of SL in stressed plants may lead to a drought sensitive phenotype that to a large extent is caused by lowering guard cell sensitivity to ABA and a more wilting phenotype (Liu et al. 2015).

11.3.4 STRIGOLACTONE AND OSMOTIC STRESS

A change in solute concentration around plant cells which alters osmotic pressure results in a swift water movement across the cell membrane. Osmotic stress can result from abiotic stresses, i.e., salinity, drought, and freezing stresses (Xiong and Zhu 2002). Comparable to the other mentioned stresses, osmotic stress can lead to a loss in plant yield through tightly constraining plant growth and development (García-Morales et al. 2018).

A research group explored the interaction of SL and osmotic stress, indicated that both SL accumulation and exudation were considerably lessened in osmotically stressed plants (Liu et al. 2015). As was previously discussed, SL level generally increases in the shoots of the plants under stressful conditions. However, in a study on osmotic stress, SL-depletion was observed in the roots. The lower amount of SL in the roots of plants after exposure to osmotic stress might be explained by the probable organ-specific functioning of SL (Visentin et al. 2016). For instance, an increase in SL level in the shoots of the plants under stressful conditions, e.g., drought, can be a representative case in the organ-targeted feature of SL. It has been postulated that the major organ of SL production can be switched from root to another organ, shoot, according to environmental conditions and the main plant organ suffering from and responding to stress (Liu et al. 2015). Acropetal movement of SL and/or a shifted SL-production site may be partly or mainly associated with ABA. The dramatic rise in SL levels in the shoot can be a signal for plants to close the stomata in a collaborative process with ABA. SL functions and interactions with other processes are intricate. In spite of the complexity, it seems that in osmotic stress-triggered responses the action part of SL is in the shoot, not the root, and in crosstalk with ABA

(Marzec and Muszynska 2015), which is the main stress hormone and endogenous messenger (Christmann et al. 2006). Principally, it can be considered that based on various regulations of SL biosynthesis, they are tissue- and organ-specific as well as external condition-specific responses (Marzec and Muszynska 2015; Visentin et al. 2016).

11.4 STRIGOLACTONE-MEDIATED RESPONSES TO BIOTIC STRESSES

Biotic stress can lead to severe losses in plants by the damages caused by herbivore attack and/or pathogens. In general, hormones effectively contribute to the responses of plants (resistance or susceptibility) to pathogens (Ma and Ma 2016). The main phytohormones that play roles in inducing effective defense responses to pathogens are salicylic acid (SA), jasmonates (JAs), and ABA (Okada et al. 2014; Zhang et al. 2017). The response of pathogen-attacked plants can also be modulated by brassinosteroids, gibberellins, auxins, and ET (Robert-Seilaniantz et al. 2011). Therefore, as an AMF hyphal branching stimulator, SL may directly or indirectly involve in plant defense signaling and affect fungal pathogenesis.

An experiment done on an SL-deficient tomato mutant, Slccd8, and wild-type tomato infected by pathogenic fungi, i.e., *Botrytis cinerea* and *Alternaria alternata*, indicated that a couple of days after infecting the plants with the fungi, the Slccd8 mutant showed more susceptibility (40% for both pathogens) and severe symptoms as than the wild-type tomatoes (up to 15%). It was observed that SL-deficient tomatoes (Slccd8) were increasingly susceptible to the pathogenic fungi regardless of the inoculum source, i.e., either germinating spores or developed hyphae. The results imply that SL can be associated with plant responses to pathogens. Later, to make clear whether SL effects are direct or not, the research group studied the severity of infections on leaf disks. One disk was soaked in water as a control, and the other three disks were completely soaked in a solution with 10 μl of three different concentrations of rac-GR24, including 10^{-6}, 10^{-7}, and 10^{-8} M. Afterward, the disks were placed in a dish containing *B. cinerea* in the middle (25°C, high humidity, for 5 days). The result showed no change in the pattern of *B. cinerea* growth. It can be interpreted that SL has no direct effects on both spore germination and pathogen development (Torres-Vera et al. 2014). To determine that there is a causal connection between susceptibility of SL-mutant tomato (Slccd8) and some hormones, such as JA, SA, and ABA, changes in hormone level were monitored. Based on the results, the content of JA, SA, and ABA in the Slccd8 mutant plant declined by 75, 50, and 35%, respectively. In addition to fungus pathosystem, it has been shown that strigolactones can have effects on bacterial pathosystem as well (Stes et al. 2015). To sum up, it seems that SL can have a key role in the defense system of plants against pathogens. However, it is mediated indirectly regarding the crosstalk between SL and the main hormones involved in plant defense responses (Torres-Vera et al. 2014). Therefore, SL via interacting with the other hormones, especially

JA, auxin, and cytokinin (Stes et al. 2015) can strengthen defense systems of plants.

11.5 STRIGOLACTONE EFFECTS ON MICRORNAS AND TRANSCRIPTION FACTORS

TFs play critical roles in response to a vast range of environmental stresses (biotic and abiotic stresses) via regulating the pattern of gene expression (Tran and Mochida 2010; Afrin et al. 2015). Understanding the role of TFs in activating genes can help to clarify the genes' functions (Khan et al. 2018).

SL can have diverse effects on the responses to stresses that can be explained by their differential regulation. Dissimilarity in SL levels is due to the variable regulations of SL-biosynthesis genes that can be changed as a result of TFs of almost all phytohormones (Marzec 2016). By studying the promoters of the genes involved in the SL-biosynthesis pathway encoding D27, CCD7, CCD8, and CYT P450, it has been made clear that there are 55 motifs and binding sites, for the TFs. Among the identified motifs, 26 were related to hormonal regulation, 23 to abiotic stresses, 19 to growth and development, 12 responded to light, 7 to biotic stresses and 6 to metabolism. Thus far, on the whole, 30 of the identified binding sites in SL promoters are induced by stresses, like WBOXATNPR1 and ASF1MOTIFCAMV motifs that may be induced under biotic stress (Marzec and Muszynska 2015).

MicroRNAs (miRNAs) are non-coding RNAs with a length of approximately 22 nucleotides which can play a major role in responding to stresses, commonly in coordination with TFs (Andolina et al. 2017). miRNAs as a post-transcriptional modification can function through degradation of mRNA, repression of translation, repression of TFs and gene expression (Hajdarpašić and Ruggenthaler 2012; Sunkar et al. 2012). About 69 target sequences have been recognized for different miRNAs in the genes encoding SL proteins in rice (Leung and Sharp 2010).

Responses to stresses can be divided into two levels, transcriptional and post-transcriptional levels mostly by TFs and miRNAs, respectively (Guerra et al. 2015; Kucherenko and Shcherbata 2018). Transcriptional modifications are appropriate for persistent stresses in order to slow long-lasting adaptation and post-transcriptional modifications for transient ones (Kucherenko and Shcherbata 2018). Regardless of the modification type, SL can interact with both.

11.6 CONCLUSIONS

To sum up, even though SLs have diverse functions, e.g., in parasitic seed germination, mycorrhizal fungi symbiosis, plant growth and development, plant-pathogen interaction, their potential in abiotic and, particularly, biotic stresses has not been well recognized. In spite of a few numbers of research studies, the effects of SL on environmental stresses are largely unknown. Thus, more intensive studies are needed in the future to provide a comprehensive explanation for SL

role in the adaptation to stress. Up until now, however, it has been demonstrated that SL can modulate stress tolerance through interaction with miRNAs and TFs. The presence of possible crosstalk between SL and other phytohormones is a key concept that can explain SL-induced responses to abiotic and biotic stresses. Another outstanding contribution of SL to alleviating stress can be made by SL-induced symbiosis with AMF and N-fixing rhizobial bacteria. To thoroughly understand about SL functions and interactions in alleviating stresses, more in-depth research should be carried out.

REFERENCES

Afrin, S., Zhu, J., Cao, H., Huang, J., Xiu, H., Luo, T., and Luo, Z. 2015. Molecular cloning and expression profile of an abiotic stress and hormone responsive MYB transcription factor gene from *Panax ginseng. Acta Biochimica et Biophysica Sinica*, 47(4): 267–277

Akiyama, K., Ogasawara, S., Ito, S., and Hayashi, H. 2010. Structural requirements of strigolactones for hyphal branching in AM fungi. *Plant and Cell Physiology*, 51(7): 1104–1117

Alder, A., Jamil, M., Marzorati, M., Bruno, M., Vermathen, M., Bigler, P., Ghisla, S., Bouwmeester, H., Beyer, P., and Al-Babili, S. 2012. The path from β-carotene to carlactone, a strigolactone-like plant hormone. *Science*, 335(6074): 1348–1351

Andolina, D., and Di Segni Matteo, R.V. 2017. MiRNA-34 and stress response. *Oncotarget*, 8(4): 5658–5659

Banerjee, A., and Roychoudhury, A. 2018. Strigolactones: multilevel regulation of biosynthesis and diverse responses in plant abiotic stresses. *Acta Physiologiae Plantarum*, 40(5): 86

Boursiac, Y., Léran, S., Corratgé-Faillie, C., Gojon, A., Krouk, G., and Lacombe, B. 2013. ABA transport and transporters. *Trends in Plant Science*, 18(6): 325–333

Brewer, P.B., Koltai, H., and Beveridge, C.A. 2013. Diverse roles of strigolactones in plant development. *Molecular Plant*, 6(1): 18–28

Bu, Q., Lv, T., Shen, H., Luong, P., Wang, J., Wang, Z., Huang, Z., Xiao, L., Engineer, C., Kim, T.H., and Schroeder, J.I. 2014. Regulation of drought tolerance by the F-box protein MAX2 in Arabidopsis. *Plant Physiology*, 164(1): 424–439

Cameron, D.D., Neal, A.L., van Wees, S.C., and Ton, J. 2013. Mycorrhiza-induced resistance: more than the sum of its parts? *Trends in Plant Science*, 18(10): 539–545

Causier, B., Ashworth, M., Guo, W., and Davies, B. 2012. The TOPLESS interactome: a framework for gene repression in Arabidopsis. *Plant Physiology*, 158(1): 423–438

Christmann, A., Moes, D., Himmelbach, A., Yang, Y., Tang, Y., and Grill, E. 2006. Integration of abscisic acid signalling into plant responses. *Plant Biology*, 8(3): 314–325

Cordoba, E., Salmi, M., and León, P. 2009. Unravelling the regulatory mechanisms that modulate the MEP pathway in higher plants. *Journal of Experimental Botany*, 60(10): 2933–2943

Cutler, S.R., Rodriguez, P.L., Finkelstein, R.R., and Abrams, S.R. 2010. Abscisic acid: emergence of a core signaling network. *Annual Review of Plant Biology*, 61: 651–679

De Cuyper, C., Struk, S., Braem, L., Gevaert, K., De Jaeger, G., and Goormachtig, S. 2017. Strigolactones, karrikins and beyond. *Plant, Cell & Environment*, 40(9): 1691–1703

de Saint Germain, A., Ligerot, Y., Dun, E.A., Pillot, J.P., Ross, J.J., Beveridge, C.A., and Rameau, C. 2013. Strigolactones stimulate internode elongation independently of gibberellins. *Plant Physiology*, 163: 1012–1025

Delaux, P.M., Xie, X., Timme, R.E., Puech-Pages, V., Dunand, C., Lecompte, E., Delwiche, C.F., Yoneyama, K., Bécard, G., and Séjalon-Delmas, N. 2012. Origin of strigolactones in the green lineage. *New Phytologist*, 195(4): 857–871

Estabrook, E.M., and Yoder, J.I. 1998. Plant-plant communications: rhizosphere signaling between parasitic angiosperms and their hosts. *Plant Physiology*, 116(1): 1–7

Farber, M., Attia, Z., and Weiss, D. 2016. Cytokinin activity increases stomatal density and transpiration rate in tomato. *Journal of Experimental Botany*, 67(22): 6351–6362

Flexas, J., Bota, J., Loreto, F., Cornic, G., and Sharkey, T.D. 2004. Diffusive and metabolic limitations to photosynthesis under drought and salinity in C3 plants. *Plant Biology*, 6(3): 269–279

Foo, E., and Davies, N.W. 2011. Strigolactones promote nodulation in pea. *Planta*, 234(5): 1073–1081

Foo, E., Yoneyama, K., Hugill, C., Quittenden, L.J., and Reid, J.B. 2013. Strigolactones: internal and external signals in plant symbioses? *Plant Signaling & Behavior*, 8: e23168

García-Morales, S., Gómez-Merino, F.C., Trejo-Téllez, L.I., Tavitas-Fuentes, L., and Hernández-Aragón, L. 2018. Osmotic stress affects growth, content of chlorophyll, abscisic acid, Na+, and K+, and expression of novel NAC genes in contrasting rice cultivars. *Biologia Plantarum*, 62(2): 307–317

Gomez-Roldan, V., Fermas, S., Brewer, P.B., Puech-Pagès, V., Dun, E.A., Pillot, J.P., Letisse, F., Matusova, R., Danoun, S., Portais, J.C., and Bouwmeester, H. 2008. Strigolactone inhibition of shoot branching. *Nature*, 455(7210): 189–194

Goulet, C., and Klee, H.J. 2010. Climbing the branches of the strigolactones pathway one discovery at a time. *Plant physiology*, 154(2): 493–496

Gray, W.M. 2004. Hormonal regulation of plant growth and development. *PLoS Biology*, 2(9): e311

Guerra, D., Crosatti, C., Khoshro, H.H., Mastrangelo, A.M., Mica, E., and Mazzucotelli, E. 2015. Post-transcriptional and post-translational regulations of drought and heat response in plants: a spider's web of mechanisms. *Frontiers in Plant Science*, 6: 57

Gunes, A., Inal, A., Alpaslan, M., Eraslan, F., Bagci, E.G., and Cicek, N. 2007. Salicylic acid induced changes on some physiological parameters symptomatic for oxidative stress and mineral nutrition in maize (*Zea mays* L.) grown under salinity. *Journal of Plant Physiology*, 164(6): 728–736

Ha, C.V., Leyva-González, M.A., Osakabe, Y., Tran, U.T., Nishiyama, R., Watanabe, Y., Tanaka, M., Seki, M., Yamaguchi, S., Van Dong, N., and Yamaguchi-Shinozaki, K. 2014. Positive regulatory role of strigolactone in plant responses to drought and salt stress. *Proceedings of the National Academy of Sciences*, 111(2): 851–856

Hajdarpašić, A., and Ruggenthaler, P. 2012. Analysis of miRNA expression under stress in *Arabidopsis thaliana*. *Bosnian Journal of Basic Medical Sciences*, 12(3): 169–176

Heidari, B., Pessarakli, M., Dadkhodaie, A., and Daneshnia, N. 2012. Reactive oxygen species-mediated functions in plants under environmental stresses. *Journal of Agricultural Science and Technology B*, 2(2): 159–168

Hu, Y., Xia, S., Su, Y., Wang, H., Luo, W., Su, S., and Xiao, L. 2016. Brassinolide increases potato root growth in vitro in a dose-dependent way and alleviates salinity stress. *BioMed Research International*, 2016: ID 8231873

Jaleel, C.A., Manivannan, P., Wahid, A., Farooq, M., Al-Juburi, H.J., Somasundaram, R., and Panneerselvam, R. 2009. Drought stress in plants: a review on morphological characteristics and pigments composition. *International Journal of Agriculture & Biology*, 11(1): 100–105

Javid, M.G., Sorooshzadeh, A., Moradi, F., Modarres Sanavy, S.A.M., and Allahdadi, I., 2011. The role of phytohormones in alleviating salt stress in crop plants. *Australian Journal of Crop Science*, 5(6): 726–734

Jiang, L., Liu, X., Xiong, G., Liu, H., Chen, F., Wang, L., Meng, X., Liu, G., Yu, H., Yuan, Y., and Yi, W. 2013. DWARF 53 acts as a repressor of strigolactone signalling in rice. *Nature*, 504: 401–405

Kapulnik, Y., Delaux, P.M., Resnick, N., Mayzlish-Gati, E., Wininger, S., Bhattacharya, C., Séjalon-Delmas, N., Combier, J.P., Bécard, G., Belausov, E., Beeckman, T., Dor, E., Hershenhorn, J., and Koltai, H. 2011. Strigolactones affect lateral root formation and root-hair elongation in Arabidopsis. *Planta*, 233(1): 209–216

Khan, S.A., Li, M.Z., Wang, S.M., and Yin, H.J. 2018. Revisiting the role of plant transcription factors in the battle against abiotic stress. *International Journal of Molecular Sciences*, 19(6): 1634

Koltai, H. 2011. Strigolactones are regulators of root development. *New Phytologist*, 190(3): 545–549

Kucherenko, M.M., and Shcherbata, H.R. 2018. miRNA targeting and alternative splicing in the stress response – events hosted by membrane-less compartments. *Journal of Cell Science*, 131(4): jcs202002

Kuijt, J. 1965. On the nature and action of the *Santalalean haustorium*, as exemplified by *Phthirusa* and *Antidaphne* (Loranthaceae). *Acta Botanica Neerlandica*, 14(3): 278–307

Leung, A.K., and Sharp, P.A. 2010. MicroRNA functions in stress responses. *Molecular Cell*, 40(2): 205–215

Li, G., Wan, S., Zhou, J., Yang, Z., and Qin, P. 2010. Leaf chlorophyll fluorescence, hyperspectral reflectance, pigments content, malondialdehyde and proline accumulation responses of castor bean (*Ricinus communis* L.) seedlings to salt stress levels. *Industrial Crops and Products*, 31(1): 13–19

Liu, J., He, H., Vitali, M., Visentin, I., Charnikhova, T., Haider, I., Schubert, A., Ruyter-Spira, C., Bouwmeester, H.J., Lovisolo, C., and Cardinale, F. 2015. Osmotic stress represses strigolactone biosynthesis in *Lotus japonicus* roots: exploring the interaction between strigolactones and ABA under abiotic stress. *Planta*, 241(6): 1435–1451

Liu, J., Novero, M., Charnikhova, T., Ferrandino, A., Schubert, A., Ruyter-Spira, C., Bonfante, P., Lovisolo, C., Bouwmeester, H.J., and Cardinale, F. 2013. CAROTENOID CLEAVAGE DIOXYGENASE 7 modulates plant growth, reproduction, senescence, and determinate nodulation in the model legume *Lotus japonicus*. *Journal of Experimental Botany*, 64(7): 1967–1981

López-Arredondo, D.L., and Herrera-Estrella, L. 2012. Engineering phosphorus metabolism in plants to produce a dual fertilization and weed control system. *Nature Biotechnology*, 30(9): 889–893

López-Ráez, J.A., Charnikhova, T., Gómez-Roldán, V., Matusova, R., Kohlen, W., De Vos, R., Verstappen, F., Puech-Pages, V., Bécard, G., Mulder, P., and Bouwmeester, H. 2008. Tomato strigolactones are derived from carotenoids and their biosynthesis is promoted by phosphate starvation. *New Phytologist*, 178(4): 863–874

López-Ráez, J.A., Kohlen, W., Charnikhova, T., Mulder, P., Undas, A.K., Sergeant, M.J., Verstappen, F., Bugg, T.D., Thompson, A.J., Ruyter-Spira, C., and Bouwmeester, H. 2010. Does abscisic acid affect strigolactone biosynthesis? *New Phytologist*, 187(2): 343–354

López-Ráez, J.A., Matusova, R., Cardoso, C., Jamil, M., Charnikhova, T., Kohlen, W., Ruyter-Spira, C., Verstappen, F., and Bouwmeester, H. 2009. Strigolactones: ecological significance and use as a target for parasitic plant control. *Pest Management Science*, 65(5): 471–477

Ma, K.W., and Ma, W. 2016. Phytohormone pathways as targets of pathogens to facilitate infection. *Plant Molecular Biology*, 91(6): 713–725

Ma, N., Hu, C., Wan, L., Hu, Q., Xiong, J., and Zhang, C. 2017. Strigolactones improve plant growth, photosynthesis, and alleviate oxidative stress under salinity in rapeseed (*Brassica napus* L.) by regulating gene expression. *Frontiers in Plant Science*, 8: 1671

Marzec, M. 2016. Perception and signaling of strigolactones. *Frontiers in Plant Science*, 7: 1260

Marzec, M., and Muszynska, A. 2015. In silico analysis of the genes encoding proteins that are involved in the biosynthesis of the RMS/MAX/D pathway revealed new roles of strigolactones in plants. *International Journal of Molecular Sciences*, 16(4): 6757–6782

Marzec, M., Muszynska, A., and Gruszka, D. 2013. The role of strigolactones in nutrient-stress responses in plants. *International Journal of Molecular Sciences*, 14(5): 9286–9304

Matusova, R., Rani, K., Verstappen, F.W., Franssen, M.C., Beale, M.H., and Bouwmeester, H.J. 2005. The strigolactone germination stimulants of the plant-parasitic *Striga* and *Orobanche* spp. are derived from the carotenoid pathway. *Plant Physiology*, 139(2): 920–934

Miller, G., Suzuki, N., Ciftci-Yilmaz, S., and Mittler, R. 2010. Reactive oxygen species homeostasis and signalling during drought and salinity stresses. *Plant, Cell & Environment*, 33(4): 453–467

Mishra, S., Upadhyay, S., and Shukla, R.K. 2017. The role of strigolactones and their potential cross-talk under hostile ecological conditions in plants. *Frontiers in Physiology*, 7: 691

Munné-Bosch, S., and Müller, M. 2013. Hormonal cross-talk in plant development and stress responses. *Frontiers in Plant Science*, 4: 529

Munns, R., and Tester, M. 2008. Mechanisms of salinity tolerance. *Annual Review of Plant Biology*, 59: 651–681

Okada, K., Abe, H., and Arimura, G.I. 2014. Jasmonates induce both defense responses and communication in monocotyledonous and dicotyledonous plants. *Plant and Cell Physiology*, 56(1): 16–27

Pandey, A., Sharma, M., and Pandey, G.K. 2016. Emerging roles of strigolactones in plant responses to stress and development. *Frontiers in Plant Science*, 7: 434

Peleg, Z., and Blumwald, E. 2011. Hormone balance and abiotic stress tolerance in crop plants. *Current Opinion in Plant Biology*, 14(3): 290–295

Pinheiro, C., and Chaves, M.M. 2010. Photosynthesis and drought: can we make metabolic connections from available data? *Journal of Experimental Botany*, 62(3): 869–882

Popko, J., Hänsch, R., Mendel, R.R., Polle, A., and Teichmann, T. 2010. The role of abscisic acid and auxin in the response of poplar to abiotic stress. *Plant Biology*, 12(2): 242–258

Pulido, P., Perello, C., and Rodriguez-Concepcion, M. 2012. New insights into plant isoprenoid metabolism. *Molecular Plant*, 5(5): 964–967

Ren, C.G., and Dai, C.C. 2012. Jasmonic acid is involved in the signaling pathway for fungal endophyte-induced volatile oil accumulation of *Atractylodes lancea* plantlets. *BMC Plant Biology*, 12: 128

Robert-Seilaniantz, A., Grant, M., and Jones, J.D. 2011. Hormone crosstalk in plant disease and defense: more than just jasmonate-salicylate antagonism. *Annual Review of Phytopathology*, 49: 317–343

Ruyter-Spira, C., Al-Babili, S., Van Der Krol, S., and Bouwmeester, H. 2013. The biology of strigolactones. *Trends in Plant Science*, 18(2): 72–83

Ruyter-Spira, C., Kohlen, W., Charnikhova, T., van Zeijl, A., van Bezouwen, L., de Ruijter, N., Cardoso, C., Lopez-Raez, J.A., Matusova, R., Bours, R., Verstappen, F., and Bouwmeester, H.

2011. Physiological effects of the synthetic strigolactone analog GR24 on root system architecture in Arabidopsis: another belowground role for strigolactones? *Plant Physiology*, 155(2): 721–734

Saed-Moucheshi, A., Shekoofa, A., and Pessarakli, M. 2014. Reactive oxygen species (ROS) generation and detoxifying in plants. *Journal of Plant Nutrition*, 37(10): 1573–1585

Saeed, W., Naseem, S., and Ali, Z. 2017. Strigolactones biosynthesis and their role in abiotic stress resilience in plants: a critical review. *Frontiers in Plant Science*, 8: 1487

Sairam, R.K., and Tyagi, A. 2004. Physiology and molecular biology of salinity stress tolerance in plants. *Current Science*, 86: 407–421

Scaffidi, A., Waters, M., Sun, Y.K., Skelton, B.W., Dixon, K.W., Ghisalberti, E.L., Flematti, G., and Smith, S. 2014. Strigolactone hormones and their stereoisomers signal through two related receptor proteins to induce different physiological responses in Arabidopsis. *Plant Physiology*, 165: 1221–1232

Scaffidi, A., Waters, M.T., Bond, C.S., Dixon, K.W., Smith, S.M., Ghisalberti, E.L., and Flematti, G.R. 2012. Exploring the molecular mechanism of karrikins and strigolactones. *Bioorganic & Medicinal Chemistry Letters*, 22(11): 3743–3746

Sedaghat, M., Tahmasebi-Sarvestani, Z., Emam, Y., and Mokhtassi-Bidgoli, A. 2017. Physiological and antioxidant responses of winter wheat cultivars to strigolactone and salicylic acid in drought. *Plant Physiology and Biochemistry*, 119: 59–69

Shibli, R.A., Kushad, M., Yousef, G.G., and Lila, M.A. 2007. Physiological and biochemical responses of tomato microshoots to induced salinity stress with associated ethylene accumulation. *Plant Growth Regulation*, 51(2): 159–169

Soto, M.J., Fernández-Aparicio, M., Castellanos-Morales, V., García-Garrido, J.M., Ocampo, J.A., Delgado, M.J., and Vierheilig, H. 2010. First indications for the involvement of strigolactones on nodule formation in alfalfa (*Medicago sativa*). *Soil Biology and Biochemistry*, 42(2): 383–385

Stes, E., Depuydt, S., De Keyser, A., Matthys, C., Audenaert, K., Yoneyama, K., Werbrouck, S., Goormachtig, S., and Vereecke, D. 2015. Strigolactones as an auxiliary hormonal defence mechanism against leafy gall syndrome in *Arabidopsis thaliana*. *Journal of Experimental Botany*, 66(16): 5123–5134

Sunkar, R., Li, Y.F., and Jagadeeswaran, G. 2012. Functions of microRNAs in plant stress responses. *Trends in Plant Science*, 17(4): 196–203

Talla, S.K., Panigrahy, M., Kappara, S., Nirosha, P., Neelamraju, S., and Ramanan, R. 2016. Cytokinin delays dark-induced senescence in rice by maintaining the chlorophyll cycle and photosynthetic complexes. *Journal of Experimental Botany*, 67(6): 1839–1851

Torres-Vera, R., García, J.M., Pozo, M.J., and López-Ráez, J.A. 2014. Do strigolactones contribute to plant defence? *Molecular Plant Pathology*, 15(2): 211–216

Tran, L.S.P., and Mochida, K. 2010. Identification and prediction of abiotic stress responsive transcription factors involved in abiotic stress signaling in soybean. *Plant Signaling & Behavior*, 5(3): 255–257

Umehara, M., Cao, M., Akiyama, K., Akatsu, T., Seto, Y., Hanada, A., Li, W., Takeda-Kamiya, N., Morimoto, Y., and Yamaguchi, S. 2015. Structural requirements of strigolactones for shoot branching inhibition in rice and Arabidopsis. *Plant and Cell Physiology*, 56(6): 1059–1072

Umehara, M., Hanada, A., Yoshida, S., Akiyama, K., Arite, T., Takeda-Kamiya, N., Magome, H., Kamiya, Y., Shirasu, K., Yoneyama, K., Kyozuka, J., and Yamaguchi, S. 2008. Inhibition of shoot branching by new terpenoid plant hormones. *Nature*, 455(7210): 195–200

Van Der Heijden, M.G.A., Bardgett, R.D., and Van Straalen, N.M. 2008. The unseen majority: soil microbes as drivers of plant diversity and productivity in terrestrial ecosystems. *Ecology Letters*, 11(3): 296–310

Van Der Heijden, M.G., Dombrowski, N., and Schlaeppi, K. 2017. Continuum of root–fungal symbioses for plant nutrition. *Proceedings of the National Academy of Sciences*, 114(44): 11574–11576

Visentin, I., Vitali, M., Ferrero, M., Zhang, Y., Ruyter-Spira, C., Novák, O., Strnad, M., Lovisolo, C., Schubert, A., and Cardinale, F. 2016. Low levels of strigolactones in roots as a component of the systemic signal of drought stress in tomato. *New Phytologist*, 212(4): 954–963

Wang, H.M., Xiao, X.R., Yang, M.Y., Gao, Z.L., Zang, J., Fu, X.M., and Chen, Y.H. 2014. Effects of salt stress on antioxidant defense system in the root of *Kandelia candel*. *Botanical Studies*, 55(1): 57

Wasternack, C. 2007. Jasmonates: an update on biosynthesis, signal transduction and action in plant stress response, growth and development. *Annals of Botany*, 100(4): 681–697

Xi, L., Wen, C., Fang, S., Chen, X., Nie, J., Chu, J., Yuan, C., Yan, C., Ma, N., and Zhao, L. 2015. Impacts of strigolactone on shoot branching under phosphate starvation in chrysanthemum (*Dendranthema grandiflorum* cv. Jinba). *Frontiers in Plant Science*, 6: 694

Xie, X., Yoneyama, K., Kisugi, T., Uchida, K., Ito, S., Akiyama, K., Hayashi, H., Yokota, T., Nomura, T., and Yoneyama, K. 2013. Confirming stereochemical structures of strigolactones produced by rice and tobacco. *Molecular Plant*, 6(1): 153–163

Xin-Hua, H., Critchley, C., and Bledsoe, C. 2003. Nitrogen transfer within and between plants through common mycorrhizal networks (CMNs). *Critical Reviews in Plant Sciences*, 22(6): 531–567

Xiong, L., and Zhu, J.K. 2002. Molecular and genetic aspects of plant responses to osmotic stress. *Plant, Cell & Environment*, 25(2): 131–139

Yoneyama, K., Xie, X., Kim, H.I., Kisugi, T., Nomura, T., Sekimoto, H., Yokota, T., and Yoneyama, K. 2012. How do nitrogen and phosphorus deficiencies affect strigolactone production and exudation? *Planta*, 235(6): 1197–1207

Yoneyama, K., Xie, X., Kusumoto, D., Sekimoto, H., Sugimoto, Y., Takeuchi, Y., and Yoneyama, K. 2007. Nitrogen deficiency as well as phosphorus deficiency in sorghum promotes the production and exudation of 5-deoxystrigol, the host recognition signal for arbuscular mycorrhizal fungi and root parasites. *Planta*, 227(1): 125–132

Zhang, J., Jia, W., Yang, J., and Ismail, A.M. 2006. Role of ABA in integrating plant responses to drought and salt stresses. *Field Crops Research*, 97(1): 111–119

Zhang, L., Zhang, F., Melotto, M., Yao, J., and He, S.Y. 2017. Jasmonate signaling and manipulation by pathogens and insects. *Journal of Experimental Botany*, 68(6): 1371–1385

Zhao, D., Reddy, K.R., Kakani, V.G., and Reddy, V.R. 2005. Nitrogen deficiency effects on plant growth, leaf photosynthesis, and hyperspectral reflectance properties of sorghum. *European Journal of Agronomy*, 22(4): 391–403

Zhao, L.H., Zhou, X.E., Wu, Z.S., Yi, W., Xu, Y., Li, S., Xu, T.H., Liu, Y., Chen, R.Z., Kovach, A., and Kang, Y. 2013. Crystal structures of two phytohormone signal-transducing α/β hydrolases: karrikin-signaling KAI2 and strigolactone-signaling DWARF14. *Cell Research*, 23(3): 436–439

Zheng, L., Meng, Y., Ma, J., Zhao, X., Cheng, T., Ji, J., Chang, E., Meng, C., Deng, N., Chen, L., and Shi, S. 2015. Transcriptomic analysis reveals importance of ROS and phytohormones in response to short-term salinity stress in *Populus tomentosa*. *Frontiers in Plant Science*, 6: 678

Zhou, F., Lin, Q., Zhu, L., Ren, Y., Zhou, K., Shabek, N., Wu, F., Mao, H., Dong, W., Gan, L., and Ma, W. 2013. D14–SCF D3-dependent degradation of D53 regulates strigolactone signalling. *Nature*, 504(7480): 406–410

Zwack, P.J., and Rashotte, A.M. 2013. Cytokinin inhibition of leaf senescence. *Plant Signaling & Behavior*, 8(7): p.e24737

Zwanenburg, B., and Pospíšil, T. 2013. Structure and activity of strigolactones: new plant hormones with a rich future. *Molecular Plant*, 6(1): 38–62

12 Plant Abiotic Stress Proteomics
An Insight into Plant Stress Response at Proteome Level

Klára Kosová, Milan Oldřich Urban, Pavel Vítámvás, and Ilja Tom Prášil

CONTENTS

12.1 INTRODUCTION

Stress can be defined as any environmental factor that adversely affects plant growth, development, and the final yield in crops. Plants can respond to stress stimulus in several ways, including an escape response, i.e., survival of stress treatments (stress periods) in a metabolically inactive stage such as seeds, or an active stress response, underlying either stress avoidance or stress tolerance. Stress avoidance means activation of plant biological mechanisms leading to maintenance of unstressed (control) conditions at plant cell level, while stress tolerance means an active plant acclimation to altered conditions at the cellular level (e.g., decreased cell water content, increased salt ion levels) inducing an enhanced stress resistance to an altered environment (Levitt, 1980).

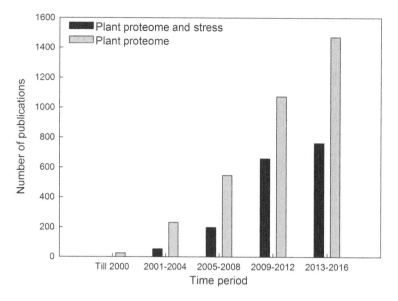

FIGURE 12.1 Number of publications found in the Web of Science database as a reply to a query on "plant proteome and stress" as compared with just "plant proteome."

Proteins play a crucial role in plant stress response, since they actively regulate plant response at epigenome, transcriptome, proteome, and metabolome levels as well as directly forming plant bodies (structural proteins). Protein structures and biological functions reflect the diversity and versatility of life. Unlike genes and transcripts, proteins are directly involved in plant stress response and actively shape plant phenotype. In the new millennium, the boost of novel high-throughput separation techniques based either on two-dimensional (2-D) electrophoresis (2DE or 2D-difference gel electrophoresis [DIGE]; a so-called *gel-based approach*) or liquid chromatography (nanoLC; a so-called *gel-free approach*) coupled with protein identification by mass spectrometry (MS) techniques (matrix assisted laser desorption ionization tandem time-of-flight [MALDI-TOF/TOF], electrospray ionization tandem mass spectrometry [ESI-MS/MS], or others), together with the publication of whole genome sequences of model plants and major crops (reviewed in Kosová et al., 2015), enabled researchers to separate and identify the individual components of complex mixtures such as protein extracts from biological samples. The boost of high-throughput proteomic techniques thus enabled study of the complex mixtures of proteins isolated from biological samples, the proteome. The term *proteome* was introduced in 1994 by Australian postdoc Mark Wilkins as "a protein complement of the genome," and it means the sum of all proteins in the given organism at a given time point (Wasinger et al., 1995). Unlike the genome, which is a static structure inherited from parents to offspring, the proteome is a dynamic structure due to alterations in gene expression, and its actual composition reflects plant developmental stage and health status as well as impacts of environmental conditions, including environmental stress. Moreover, the diversity and versatility of the proteome with respect to the genome are increased due to the important modulating role of posttranscriptional and posttranslational

modifications, i.e., modifications of pre-mRNA transcripts determining protein primary sequence and modifications of amino acid side chains in protein molecules determining protein biological functions. An example is the role of alternative pre-mRNA splicing of HAB1 protein phosphatase 2C (HAB1 PP2C) involved in abscisic acid (ABA)-mediated responses to environmental stress (Laloum et al., 2018).

Plant proteomics research aimed at the study of the impacts of environmental stresses on plants has experienced a boost in the last two decades. The number of research papers published on the topic of "plant proteome and stress" has risen geometrically from the year 2000 onwards, according to Web of Science (Figure 12.1); currently, more than 150 original papers and reviews are published each year on this topic. The topic of plant abiotic stress proteomics has been summarized in several review papers, including general reviews such as those by Jorrin-Novo et al. (2009) and Kosová et al. (2011) as well as more specialized reviews such as those on subcellular proteomics (Hossain et al., 2012) or aimed at individual stresses, such as low-temperature stress (Janmohammadi et al., 2015; Johnová et al., 2016), dehydration stresses (Johnová et al., 2016), heavy metal stress (Ahsan et al., 2009; Hossain and Komatsu, 2013), or individual crop plants such as wheat (Komatsu et al., 2014), wheat and barley (Kosová et al., 2014b, 2016), or soybean under flooding (Yin and Komatsu, 2017), among others. Increasing data on identified proteins led to the development of protein databases providing information on protein sequences as well as additional tools for the determination of basic protein characteristics based on amino acid composition, such as pI and molecular weight (MW) determination, etc. The major protein sequence databases include the National Center for Biotechnology Information (NCBI) protein database (www.ncbi.nlm.nih.gov), Uniprot (www.uniprot.org), and SwissProt (www.web.expasy.org) as well as specialized databases focused on important crops with sequenced and annotated genomes, such as rice (Rice Annotation Project Database; http://rapdb.dna.aff.rc.go.jp), soybean (Phytozome database), and others. Furthermore, protein databases providing information on protein biological functions include the KEGG pathway database (www.genome.jp/kegg/pathway), providing information on metabolic pathways the protein is involved in; Brenda Enzymes (www.brenda-enzymes.org), providing information on biological reactions catalyzed by a given enzyme; and functional databases such as MapMan Ontology (https://mapmangabipd.org) or the Gene Ontology database (www.geneontology.org), providing information on protein biological functions using three criteria: cellular localization, molecular function, and biological process.

The aim of this chapter is to provide a summary of the basic characteristics of the plant stress response at the proteome level, including general and specific characteristics of the major abiotic stress factors (drought, flooding, salinity, heavy metals, and extreme temperatures—cold, frost, heat). The dynamics of the plant stress response, including the alarm, acclimation, resistance, exhaustion, and recovery phases, are discussed, and specific features for each phase are characterized. Stress-induced alterations at the level of the individual proteins' functional groups (signaling and signal transduction, gene expression, protein metabolism, energy metabolism, carbohydrate metabolism, the metabolism of other compounds such as S-adenosylmethionine, phytohormones, etc., transport proteins, and structural and regulatory proteins) are briefly summarized with a focus on the differential response between stress-tolerant and stress-susceptible crop genotypes. Differential roles of protein isoforms and posttranslational modifications (PTMs) are discussed. Finally, studies aimed at the determination of protein biological functions under stress conditions, including transgenic studies involving heterologous transformation as well as protein–protein interaction studies, are briefly summarized. The possible use of the results of proteomic studies in crop breeding for improved abiotic stress tolerance is discussed. In conclusion, future perspectives of plant abiotic stress proteomics are outlined in a broader context of plant biology.

12.2 DYNAMICS OF PLANT STRESS RESPONSE

Plant stress response is a dynamic process, in which specific phases with differential stress resistance and a unique proteome composition can be distinguished: an alarm phase, an acclimation phase, a resistance phase, an exhaustion phase, and a recovery phase.

The alarm phase represents an initial phase of the plant stress response associated with stress signal perception and stress-induced disturbances in cellular homeostasis. Enhanced levels of transient metabolites that act as signaling molecules as well as reversible phosphorylation events can be found. The phase lasts from a few hours to a maximum of a few days.

The acclimation phase can be characterized by active adjustment of cellular metabolism to altered environmental conditions, which is aimed at gaining maximum stress resistance. It can be characterized by alterations in energy metabolism aimed at minimizing the risk of oxidative damage while adjusting to enhanced energy requirements, as well as the biosynthesis of several novel compounds, including novel proteins and metabolites such as osmotically active compounds, reactive oxygen species (ROS)-scavenging and detoxification enzymes, etc. The acclimation phase usually lasts several days, since it includes profound reorganization of the whole cellular proteome (metabolome) to adjust to the altered environment and to induce maximum stress resistance.

The resistance phase can be characterized by a maximum adjustment to the altered environment, leading to maximum stress resistance. However, the maintenance of the maximum stress resistance places increased demands on energy metabolism; thus, it can lead to exhaustion, which is characterized by stress-induced damage of cellular structures and disturbances in cellular metabolism, leading to a decrease in stress resistance. The exhaustion phase can occur when the stress lasts too long or is so severe that the plant cannot compensate for the adverse effects of stress.

The recovery phase follows stress treatment after the restoration of optimum growth conditions, and it can be characterized by the establishment of a novel homeostasis in the altered environment. At proteome level, it can be characterized by enhanced levels of several enzymes of energy metabolism, which are necessary to meet the increased requirements for energy. For example, an enhanced abundance of eight out of 12 glycolytic enzymes was found in drought-tolerant Australian wheat Excalibur 24 h after rewatering, indicating enhanced energy demand during the recovery process (Ford et al., 2011). Moreover, proteomic studies can represent a more subtle tool, complementing the results of physiological measurements, which can provide deeper insight into the molecular processes underlying stress acclimation as well as stress recovery. For example, a study on tomato plants subjected to 19 days of water withholding (drought stress) followed by 6 days of recovery indicated restoration of control levels of important characteristics and metabolites (ABA, chlorophylls, proline) at the whole plant level; however, the results of chloroplast proteome analysis, such as decreased levels of photosystem components (plastocyanin and chlorophyll a-b binding proteins), Calvin cycle enzymes (phosphoglycerokinase [PGK] and phosphoribulokinase [PRK]) and ATP synthase subunits (ATP synthase CF1α,γ) revealed that the process of establishment of a novel homeostasis was still ongoing (Tamburino et al., 2017).

12.3 STRESS DESCRIPTION: ENVIRONMENT-RELATED VERSUS PLANT-RELATED CHARACTERISTICS

Plants show considerable inter- and intra-species variability regarding plant response to various environmental stimuli. The same ambient conditions can have different impacts on a plant's cellular microenvironment depending on the plant's genetic mechanisms to cope with stress, its growth stage, and its health status. Therefore, it is important to characterize a given stress experiment not only by using environment-related characteristics (e.g., soil water capacity, or salt ion concentration in nutrient solution) but also by using plant-related characteristics (e.g., relative water content or water saturation deficit, or salt ion concentration in plant tissue). Relatively tolerant plants as compared with relatively susceptible ones often show a so-called delayed stress onset; i.e., adverse alterations in their physiological characteristics start later than in susceptible plants (Lawlor, 2013).

It has to be kept in mind that "stress tolerance" and "stress susceptibility" are relative terms, which depend on the given set of genotypes and the plant's growth stage and health status as well as on the given environmental conditions. Proteomic studies aimed at a comparison of "stress-tolerant" versus "stress-susceptible" plants revealed significant differences in cellular proteome composition not only under stress treatment but even under control conditions. Tolerant genotypes usually show constitutively enhanced levels of several stress-related proteins, such as protective proteins, hydrophilic proteins, ROS-scavenging and detoxification proteins, etc., thus showing fewer changes in their proteome composition under stress. For example, a comparative study by Pang et al. (2010) on *Arabidopsis thaliana*, a glycophyte, and *Thellungiella halophila*, a halophyte, subjected to 150 mM NaCl showed 88 differentially abundant proteins in response to salinity in *A. thaliana* but only 37 differentially abundant proteins in response to salinity in *T. halophila*. A study on threshold induction temperatures regarding cold-inducible dehydrin protein accumulation in wheat and barley revealed that the highly frost-tolerant winter wheat (Mironovskaya 808, Šárka) and barley (Odesskij 31, Dicktoo) genotypes started to accumulate detectable levels of cold-inducible dehydrin proteins at relatively higher temperatures with respect to the less tolerant winter ones (Vítámvás et al., 2010; Kosová et al., 2013b). Constitutively enhanced levels or more inducible levels of several stress-protective proteins in tolerant genotypes, then, protect the cellular proteomes of the tolerant genotypes from larger disturbances in cellular homeostasis caused by stress. As an example, a constitutively enhanced level of the detoxification enzyme lactoylglutathione lyase (glyoxalase I) involved in the glutathione-dependent detoxification of methylglyoxal was found in salt-tolerant sea barley grass (*Hordeum marinum*) as compared with salt-susceptible *H. vulgare* cv. Tadmor (Maršálová et al., 2016) as well as in salt-tolerant cultivated barley (*H. vulgare*) cv. Morex as compared with salt-susceptible cv. Steptoe (Witzel et al., 2009). As a consequence, tolerant genotypes usually suffer from a lower level of oxidative stress (ROS), which reveals positive impacts on crucial processes of energy metabolism, especially photosynthesis.

Lower levels of harmful compounds thus enable the tolerant plants to efficiently adjust their energy metabolism processes, especially photosynthesis and ATP biosynthesis, to the enhanced needs of the stress acclimation process. In susceptible plants, aerobic processes become downregulated under stress to eliminate ROS-induced damage, while tolerant plants show enhanced levels of aerobic processes such as photosynthesis to meet enhanced energy demands during stress acclimation (Askari et al., 2006; Pang et al., 2010). The plant metabolic response depends on the nature and severity of the given stress factor, with upregulation of energy metabolism under mild stress but downregulation of crucial fine-regulated aerobic processes (photosynthesis and aerobic respiration) under severe stress. There was an increase in proteins related to aerobic respiration under mild drought (35% soil water capacity [SWC]) and an increase in proteins related to anaerobic fermentation under severe drought (30% SWC) in spring barley Amulet (Vítámvás et al., 2015). For example, salt-susceptible *A. thaliana* showed an increase in several catabolic enzymes associated with aerobic respiration at 150 mM NaCl, indicating enhanced energy demands for stress acclimation, while salt-tolerant *T. halophila* showed an increase in some crucial photosynthesis-related proteins, indicating enhanced biosynthesis of new energy-rich compounds necessary for the stress response (Pang et al., 2010). Differential drought stress–coping strategies can be reflected also at proteome level. For example, Urban et al. (2017) studied proteome response to drought in oilseed rape (*Brassica napus*) and found that water-savers (a conservative strategy leading to stomatal closure under water deficit and exhibiting high water use efficiency) showed enhanced levels of proteins related to nitrogen assimilation, ATP biosynthesis, and redox homeostasis under drought, while water-spenders (open stomata under drought stress) showed an enhanced abundance of proteins involved in carbohydrate and energy metabolism and photosynthesis, and stress-related and tRNA processing proteins. Moreover, the differential ability to acclimate to stress conditions and to restore cellular metabolism can also be reflected by proteins involved in regulation of cell division and plant growth and development. For example, differential isoforms of eIF5A show opposite effects on cell division or death (apoptosis), with eIF5A-2 inducing cell division, while eIF5A-1 induces apoptosis (Thompson et al., 2004; Maršálová et al., 2016).

12.4 BRIEF CHARACTERISTICS OF MAJOR ENVIRONMENTAL STRESSES

The environmental stress factors reveal common as well as specific effects on plant cellular homeostasis leading to induction of the plant stress response. Brief characteristics of the major environmental stress factors, with a focus on their common as well as specific impacts on plant cell homeostasis, are given in the following subsections and in Table 12.1.

12.4.1 TEMPERATURE STRESS

Plants, as poikilothermic organisms, cannot regulate the temperature of their own bodies; thus, they adopt the temperature of the surrounding environment. Temperatures outside the optimum temperature range show adverse effects on biomolecular properties, such as phospholipid bilayers in cell membranes as well as the kinetics of enzymatically catalyzed reactions. Both low and high temperatures (cold, frost, and heat) lead to altered thermal movement of molecules, including phospholipids and proteins in plasma membranes. Altered

TABLE 12.1

Basic Physiological and Molecular Characteristics of the Impacts of the Major Abiotic Stress Factors on Plants

Temperature STress (cold, frost; heat)	Drought	Salinity	Waterlogging and Flooding	Heavy Metals
Primary signal:	Primary signal:	Primary signal:	Primary signal:	Primary signal:
Altered kinetics and physiochemical properties of biomolecules → altered fluidity of PM phospholipids (PM protein receptors) → signal transduction to nucleus	Cellular dehydration → ABA, K^+ → stomatal closure → Ca^{2+}-mediated signal transduction to nucleus	Enhanced Na^+ entering cell cytoplasm passively or via PM ion channels → SOS1/SOS2/SOS3 complex activation leading to Na^+ exclusion and signal transduction to nucleus	Enhanced water content in soil leading to root hypoxia (anoxia)	Enhanced metal ions in cell cytoplasm
Cellular environment:	Cellular environment:	Cellular environment:	Cellular environment (roots):	Cellular environment:
Frost: cellular dehydration	Cellular dehydration	Osmotic effect (decreased soil water potential leading to cellular dehydration)	Hypoxia/anoxia → fermentation → accumulation of organic acids → low pH → inhibition of enzymatic activities	Enhanced levels of free metal ions act as ROS catalysts; they also compete with essential metals in metalloenzyme active sites
Heat: protein misfolding		Ionic effect		
Energy metabolism:	Energy metabolism:	Energy metabolism:	Energy metabolism:	
Discrepancies in kinetics of electron transport and enzymatic reactions (aerobic electron transport processes in mitochondria and chloroplasts and the following enzymatic reactions) → enhanced risk of ROS, photoinhibition	In C3 plants, stomatal closure leads to a decrease in RubisCO carboxylation and an increase in RubisCO oxygenation activities → enhanced risk of ROS	Enhanced needs for ATP due to ATP-dependent mechanisms of toxic ion exclusion or intracellular (vacuolar) compartmentation	Hypoxia/anoxia	
Enhanced need for ROS scavenging, detoxification, enhanced chaperones (protein folding—heat)	Enhanced need for ROS scavenging, detoxification, enhanced chaperones (enhanced risk of protein misfolding under cellular dehydration)	Enhanced need for detoxification, toxic ion chelation	Acidification of cell cytoplasm → inhibition of novel protein biosynthesis and activities of several enzymes, PCD induction	Enhanced need for ROS scavenging, detoxification, metal ion chelation
Alterations in cellular structures: enhanced lignification (cold)	Alterations in cellular structures: enhanced lignification		Alterations in cellular structures: decreased lignification, aerenchyma formation as a result of PCD	Decreased growth, enhanced accumulation of heavy metals in vacuoles (hyperaccumulator plants)

Abbreviations: PCD: programmed cell death; M: plasma membrane; ROS: reactive oxygen species; SOS: salt-overly-sensitive (gene).

fluidity of plasma membrane phospholipids initiates conformational changes in transmembrane receptors, including two-component receptor-like histidine kinases (RLKs), leading to phosphorylation of their cytosolic kinase domains and thus, initiating a cascade of phosphorylation events leading to cold signal amplification and signal transduction to the nucleus. Two major types of cold signaling pathways can be distinguished, ABA-dependent and ABA-independent pathways, which coordinately act to transform the cold signal into alterations in gene expression. At the level of energy metabolism, cold leads to imbalances in photosynthesis, which are caused by decreased kinetics of enzymatically catalyzed secondary reactions with respect to primary electron transport processes,

resulting in chilling-enhanced photoinhibition. Regarding low-temperature tolerance, plants are generally divided into three groups: chilling-susceptible plants, which are seriously injured by low temperatures above zero (up to +15 °C), such as cowpea (*Vigna unguiculata*), chickpea (*Cicer arientinum*), sorghum (*Sorghum bicolor*), and maize (*Zea mays*); chilling-tolerant plants, which can grow under low above-zero temperatures but cannot survive below-zero temperatures, such as potato (*Solanum tuberosum*); and freezing-tolerant plants, which can survive below-zero temperatures, such as winter-type cereals (winter barley, wheat, triticale, and rye). Besides cold acclimation to low temperatures, winter-type plants growing in temperate climates with regular freezing periods during winter have adaptive epigenetic mechanisms preventing a premature transition to a cold-susceptible flowering stage, a phenomenon known as *vernalization* (Sung and Amasino, 2005).

In contrast, high temperatures lead to enhanced kinetics of chemical processes, resulting in an enhanced risk of protein misfolding. Therefore, enhanced levels of several heat shock proteins (HSPs) (HSP110, HSP90, HSP70, HSP60, and small HSPs) as well as heat shock factors (HSFs) (transcription factors regulating HSP expression) were found in heat-treated plants (Skylas et al., 2002; Majoul et al., 2004). Besides enhanced accumulation of heat-protective proteins from the HSP superfamily involved in preventing protein misfolding, the induction of specific heat-inducible thermostable protein isoforms was found in some proteins. One example is thermostable RubisCO activase B in Triticeae, which was found to substitute for heat-unstable RubisCO activase A in heat-treated cereals (Rollins et al., 2013).

12.4.2 DROUGHT

Drought, as an imbalance between water uptake by roots and water release by shoots, leads to cellular dehydration, which can be characterized by a decrease in cell water potential. Most higher plants, including major crops, are homoiohydric plants; i.e., they are susceptible to cellular dehydration; thus, they try to maintain well-hydrated cells by increasing water uptake and decreasing water release via both stomatal and cuticular transpiration. Active acclimation at root level relies on an accumulation of osmotically active compounds, so-called *osmotic adjustment* (OA), leading to a decrease in cell water potential. In leaves, the decreased water potential of cell cytoplasm in stomatal guard cells leads to stomatal closure, aimed at reducing stomatal transpiration. In C3 plants, stomatal closure leads to imbalances between carbon assimilation and photosynthetic primary electron transport processes, resulting in enhanced ROS formation (oxidative stress). As a response, multiple isoforms of small redox metabolism-related enzymes such as thioredoxins (Trx) are induced by drought to protect the photosynthetic electron transport chain (ETC) from overreduction (Hajheidari et al., 2007).

12.4.3 SALINITY

Salinity means enhanced levels of salt ions in soil solution. Salt ions decrease soil water potential and lead to cell dehydration (osmotic effect). Moreover, salt ions then penetrate plant cells, leading to enhanced levels of salt ions in the cell cytoplasm, including those that are toxic to plant cells, such as sodium ions (Na^+) (ionic effect). Plants actively respond to enhanced levels of salt ions in plant cells either by salt ion exclusion from the plant body or by intracellular salt ion compartmentation into vacuoles. The osmotic effect is rapid and common to all dehydrative stresses, while the ionic effect is specific to salinity stress and has a cumulative character; i.e., the level of salt ions in the cell cytoplasm increases with time (Munns, 2005; Munns and Tester, 2008). Similarly to heavy metal stress, free salt ions can act as inhibitors of several metalloenzymes; thus, plants induce enhanced levels of phytochelatins, phytosiderophores, and other chelating proteins in response to salinity stress (Witzel et al., 2009).

Regarding tolerance to enhanced salinity, plants are generally divided into glycophytes, which include the majority of plant species and which cannot tolerate enhanced salt levels (50 to 100 mM NaCl), and halophytes, which can complete their life cycle under enhanced salinity conditions, up to the salt levels approaching those in seawater (ca. 600 mM NaCl). Halophytes can grow under enhanced salinity due to efficient intracellular compartmentation mechanisms underlying so-called *tissue tolerance*. Halophytes comprise only ca. 1% of angiosperm plant species, but they are distributed in a wide range of plant families across both dicots and monocots. Although only a very few crops are halophytes, e.g., quinoa (*Chenopodium quinoa*), several important glycophytic crops have halophytic relatives, e.g., barley vs. halophytic sea barleygrass (*Hordeum marinum*), bread wheat vs. tall wheatgrass (*Thinopyrum ponticum*), or rice vs. tateoka *Porteresia coarctata* (for more details on glycophytes vs. halophytes comparisons, see Munns, 2005; Colmer et al., 2006; Kosová et al., 2013c).

12.4.4 WATERLOGGING AND FLOODING

Waterlogging means that soil is soaked with water approaching 100% soil water capacity, while flooding means water reaching over 100% soil water capacity and remaining on the soil surface. Due to the decreased air content in the soil, plant roots suffer from hypoxia, leading to the development of fermentation processes. As a consequence of enhanced accumulation of organic acids as products of fermentation processes, cellular pH values decrease, revealing an impact on various processes of cellular metabolism. Decreased pH also initiates processes of programmed cell death (PCD) leading to aerenchyma formation. At the molecular level, Komatsu et al. (2011) detected flooding-induced translocation of cytochrome c from mitochondria into cytoplasm via the voltage-dependent anion channel (VDAC), leading to the interaction of cytosolic cytochrome c with caspases and the induction of PCD.

12.4.5 HEAVY METAL STRESS

Free heavy metal ions act as competitors of some essential metals for their binding sites in enzyme catalytic centers or some regulatory motifs (e.g., the Zn-finger motif in DNA binding transcription factors and regulatory proteins). Moreover, free metal ions act as coactivators of ROS formation, thus enhancing the risk of excessive ROS production. Therefore, the plant strategy to cope with free metal ions lies in their chelation and thus, inactivation using plant-produced chelating compounds such as phytochelatins and phytosiderophores (glutathione oligomers), especially in hyperaccumulator plants such as *Alyssum lesbiacum* or *Thlaspi caerulescens* (reviewed in Ahsan et al., 2009).

12.4.6 EFFECTS OF MULTIPLE STRESS FACTORS

In the field, plants usually have to face simultaneous effects of multiple stress factors, such as heat and drought, freezing associated with either drought (so-called winter drought) or waterlogging, waterlogging associated with hypoxia (anoxia), and others. Proteomic studies revealed that the effects of combined stress treatments do not simply equal the sum of the individual stress factors; thus, they needed to be studied as novel stress conditions. A few examples of proteomic studies dealing with combinations of stress treatments are given in the following paragraphs.

A comparison of the effects of drought and salinity studied in two separate experiments on the susceptible common wheat cultivar Jinan 177 and its somatic hybrid with tall wheatgrass (*Thinopyrum ponticum*) cv. Shanrong 3 revealed more alterations in proteome composition in salinity-treated plants than in drought-treated ones due to the ionic effect of salinity stress (Peng et al., 2009).

In the field, summer seasons are often characterized by heat, which is frequently associated with drought, since high temperatures enhance water transpiration from plant shoots as well as water evaporation from the soil surface. The effects of drought (15% soil water content), heat (36 °C), and combined drought and heat treatment on proteomes of two drought-tolerant barley cultivars of different origin, a Syrian landrace Arta and an Australian cultivar Keel, were investigated by Rollins et al. (2013). Drought led to a reduction in plant growth while maintaining a relatively stable proteome composition. In contrast, heat led to enhanced protein damage, especially of PSII components, and thus, an enhanced need for energy due to an increased protein turnover. An enhanced abundance of ROS-scavenging enzymes and protective proteins (HSPs and a thermostable RubisCO activase B isoform) was found in heat-treated plants, indicating an imbalance between the rates of primary (light-dependent) and secondary (light-independent) photosynthetic reactions and an enhanced risk of protein damage under heat stress. The combined effects of drought and heat (32 °C) were also studied in the wheat grain proteome in the terminal spikelet and anthesis stages (Yang et al., 2011). The researchers found that several proteins showed a specific response to each stress treatment, while only a few common proteins involved in redox metabolism, defense, carbohydrate metabolism, and storage showed an analogous response to combined stress treatments. Exclusive responses of proteins to heat stress include increased cinnamoyl-CoA reductase, TCTP (translationally-controlled tumor protein), cell division control protein, and heat shock cognate 70 (HSC70), and decreased 14-3-3 protein. Significant impacts on the content of grain storage proteins (albumins, globulins, gliadins, and glutenins) were found.

Winter seasons represent crucial limitations to plant survival not only due to low temperatures but also due to several water stresses showing profound impacts on the plant water regime. Freezing temperatures act as a dehydration stress, while the thawing of snow cover can cause waterlogging. The combined effect of spring freezing with either drought or waterlogging was studied in winter wheat cv. Yannong leaves (Li et al., 2014). Differences between the individual treatments and the combined treatments were observed. For example, HSP70 decreased in response to a single freezing stress treatment, while the same protein increased in response to combined stress treatments. In contrast, decreased levels of chloroplast ATP synthase β subunit and mitochondrial ATP synthase α subunit in both single freezing and combined freezing and waterlogging treatments are consistent with a decrease in Ca^{2+} and Mg^{2+}-ATPase activities and the observed damage to PSII under waterlogging stress.

A comparison of contrasting water stresses—drought and flooding—was carried out in soybean seedlings by Oh and Komatsu (2015). The results showed differentially regulated stress responses—an increase in enzymes involved in the regulation of redox homeostasis was found in drought-stressed plants, while an increase in anaerobic metabolism–related (fermentation) enzymes was found in flooded plants. Moreover, an opposite pattern of changes in S-adenosylmethionine (SAM) synthetase (SAMS) was found in drought-treated plants vs. flooded ones, indicating an increase in SAMS under drought and a decrease in SAMS under flooding. Since SAM is a universal cell methylation agent involved in lignin biosynthesis, the differential pattern of SAMS may be related to changes observed in root cell wall lignification in soybean seedlings under the two treatments corresponding to increased cell wall lignification under drought and decreased cell wall lignification under flooding, respectively.

The effects of drought, cold, and paraquat herbicide treatments on the pea mitochondrial proteome were compared by Taylor et al. (2005). The strongest adverse effects on mitochondrial proteins resulting in oxidative damage were observed under paraquat treatment, followed by chilling, while drought showed the mildest effects. Mitochondria isolated from stressed pea plants maintained their ETC activity; complexes of the mitochondrial ETC were least damaged by oxidative stress, while the F_1Fo ATP synthase complex was more damaged, and the enzymes of carbon metabolism in the mitochondrial matrix were significantly modified by oxidation. Moreover, increased lipid peroxidation and a decrease in inner membrane import proteins were observed. Differential changes in abundance of several HSP proteins with the induction of HSP22 and the opposite patterns in several isoforms of HSP70 were also found under both chilling and drought treatments.

12.5 MAJOR BIOLOGICAL PROCESSES INVOLVED IN PLANT STRESS RESPONSE

Plant response to environmental stress signals includes a coordinated network of biological processes, starting with stress perception, signaling, and signal transduction to the nucleus and leading to alterations in gene expression, protein metabolism, and energy metabolism, and the synthesis of novel structural and regulatory proteins and novel stress-responsive metabolites, ultimately leading to stress-adapted phenotypes with enhanced stress resistance. The major proteins/protein

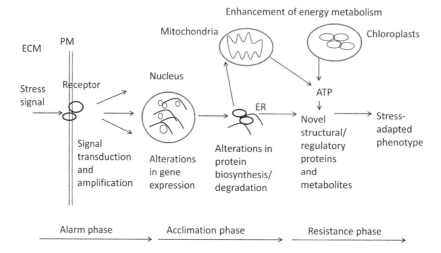

FIGURE 12.2 Schematic dynamics of plant abiotic stress response regarding major biological processes involved in plant stress response with respect to the major phases of plant stress response (alarm, acclimation, and resistance phases) leading to the acquisition of enhanced stress resistance. Abbreviations: ECM: extracellular matrix; ER: endoplasmic reticulum; PM: plasma membrane.

groups involved in plant stress response are discussed in the following subsections and summarized in Figure 12.2 and Table 12.2.

12.5.1 Stress Signaling and Signal Transduction

The plasma membrane represents a dynamic interface between the cell symplast and the outer environment. Thus, stress sensors are usually located in the plasma membrane. Altered environmental conditions lead to alterations in plasma membrane fluidity, which are sensed by plasma membrane–anchored proteins, including RLKs, heterotrimeric G proteins, and others (Zhu, 2016).

Some examples of stress sensors are as follows:

Cold: Two-component histidine kinases with a cytoplasmic kinase domain enabling signal transduction into the cytoplasm and signal amplification were identified in cyanobacteria (Murata and Los, 1997; Suzuki et al., 2000). COLD1, identified in rice cv. Nipponbare, is involved in chilling (0–15 °C) sensing and signal transduction via interaction with RGA1, the alpha subunit of heterotrimeric G protein, and via calcium signaling (Ma et al., 2015).

Salinity: Probably the most investigated stress-responsive receptor complex is the salt-responsive SOS1/SOS2/SOS3 complex in *Arabidopsis thaliana*, which is composed of SOS1, a transmembrane ATP-dependent Na^+/H^+ antiporter, and SOS2, a protein kinase interacting with SOS3, which is a calcium binding protein. Salinity stress is associated with passive Na^+ influx into cell cytoplasm, which leads to elevated Ca^{2+}, which is sensed by small protein SOS3 at the cytoplasmic side of the plasma membrane. SOS3/CBL4 (calcineurin B-like protein 4) binds cytoplasmic calcium, forms dimers, and interacts with SOS2/CIPK24, which is a protein kinase

inducing signal transduction from the plasma membrane to the nucleus. Moreover, the SOS3/SOS2 (CBL4/CIPK24) complex is targeted to the plasma membrane through a myristoyl fatty acid chain covalently bound to SOS3. Plasma membrane–targeted SOS3/SOS2 (CBL4/CIPK24) complex then phosphorylates SOS1, which is an ATP-dependent Na^+/H^+ antiporter encoded by the *NHX7* gene in *A. thaliana*. Phosphorylated SOS1 (NHX7) binds ATP and provides an active efflux of Na^+ ions from the cytoplasm to the apoplast (Shi et al., 2000, 2002; Zhu, 2000, 2002).

Signal transduction from the plasma membrane into the nucleus is ensured by cascades of signaling proteins using reversible PTMs: phosphorylation followed by signal transduction, modulation, and amplification. Crosstalk between differential signaling cascades integrates multiple primary signals, leading to their modulation and the formation of novel signals acting in the nucleus. During the evolution of the eukaryotes, signaling cascades such as the MAPK kinase cascade ensuring signal transduction via reversible phosphorylation have evolved. Differential phosphorylation of several components of signaling pathways, such as two MAPKs, components of Ca^{2+}-dependent signaling pathways, and components of ABA-dependent signaling pathways, SnRK2 and protein phosphatase 2C (PP2C), were found in wheat leaves under osmotic stress (Zhang et al., 2014). Besides multistep signaling cascades such as the MAPK(KK) cascade, small mobile Ca^{2+} binding proteins such as calmodulin, calreticulin, and calcineurin B-like proteins are involved in environmental signal transduction as secondary messengers interacting with kinases and phosphatases in signaling cascades. Increased levels of calmodulin and calreticulin were found under salinity in *Arabidopsis* roots (Jiang et al., 2007), in rice roots (Cheng et al., 2009), in grapevine shoots (Vincent et al., 2007), and elsewhere.

TABLE 12.2

Major Protein Functional Groups Involved in Plant Abiotic Stress Responses with a Focus on the Differential Proteins Regarding the Different Stresses and Differentially Tolerant Plant Genotypes

Major proteins (protein functional groups) affected by abiotic stress factors sorted according to the major biological processes involved in plant abiotic stress response

Receptors:

PM-associated proteins: RLKs, COLD1, heterotrimeric G proteins (chilling in rice); SOS1/SOS2/SOS3 complex (salinity)

Signaling pathways:

MAPK(KK), Ca^{2+} signaling (calmodulin, calreticulin, calcineurin B-like CBL); CIPK

ABA signaling: PP2C, SnRK

Transcription factors:

AREB/ABF (bZIP; ABA-dependent); MYB, MYC, NAC, NAM

Cold: ICE1 (bHLH), CBF/DREB1 (AP1 domain; ABA-independent)

Heat: HSF

Protein metabolism:

Protein biosynthesis: ribosomal proteins: translation initiation (eIF-3, eIF-5A and elongation factors (plastid eEF-G, eEF-Tu↑ (heat)

Protein degradation: proteases and protease inhibitors (serpins); components of ubiquitin-proteasome system

Energy metabolism:

Glycolysis, aerobic and anaerobic respiration (fermentation): FBP ALDO, GAPDH, TPI, ENO, PGM↑ (multiple stresses); ADH↑ (flooding and waterlogging, severe drought)

Aerobic respiration: AOX↑ (drought, oxidative stress)

ATP metabolism: NDPK, ATP synthase CF1α,β,γ↑ (cold, drought)

Photosynthesis:

Primary electron transport processes: OEE2 (OEC complex in PSII), D1, D2, CP47 (RC PSII), Rieske Fe-S (PSI)↑ tolerant, ↓ susceptible

Carbon assimilation:

RubisCO LSU and SSU↑ tolerant, ↓ susceptible; βCA, RCA, CPN60α,β↑; RCA B (heat in Triticeae)↑; Calvin cycle: PGK, PRK↑ tolerant, ↓ susceptible

Carbohydrate metabolism:

Carbohydrate catabolism: glycolysis enzymes↑ (several stresses)

Carbohydrate anabolism: SuSy; UDP-glucose pyrophosphorylase↑ (24 h cold-treated rice seedlings; Cui et al., 2005) ↓ (21-day cold-treated winter wheats; Vítámvás et al., 2012); XET↑ (drought)

Other metabolism:

SAM metabolism: SAMS1, SAMS2↑ (cold, drought), ↓ (flooding)

Lignin metabolism: PAL; CCOMT, COMT↑ (cold, drought), ↓ (flooding)

Stress and defense-related proteins:

Chaperons and chaperonins: CPN60α,β↑ (cold), HSP↑ (heat, drought), HSP90↓ (cold), HSP70/HSP90↑ (drought); PPI, UPR

LEA proteins: dehydrins (LEA-II), LEA-III↑ (cold, drought, salinity)

PR proteins: germin, germin-like (GLP), thaumatin-like↑ (cold)

ROS scavenging and redox homeostasis: APX, DHAR, MDAR, SOD, Trx isoforms (chloroplasts at drought, heat), PDI

Detoxification: GST↑, lactoylglutathione lyase↑

Ion-binding proteins: phytochelatins, phytosiderophores↑ (salinity, heavy metals)

Other: USP↑ (drought, salinity, senescence)

Structural proteins:

Cell wall: expansion

Cytoskeleton: actin, tubulin

Transport proteins: annexin, AQP↑ (drought), VDAC; ion transport: SOS1, NHX, V-ATPase, PPi-ase↑ (salinity); organellar transport: Ran G (nucleus; salinity); TOC75↑ (plastid; drought)

nsLTP

Regulatory proteins:

mRNA processing: sGRP

Protein modulation: 14-3-3

Cell cycle: Cdc48

PCD regulation: TCTP

Arrows: ↑ indicates an increase and ↓ indicates a decrease in protein relative abundance in stress-treated plants with respect to optimum (control) conditions. Abbreviations: ADH: alcohol dehydrogenase; AOX: alternative oxidase; APX: ascorbate peroxidase; AQP: aquaporin; CA: carbonic anhydrase; CBL: calcineurin B-like (protein); CCOMT: caffeoyl CoA O-methyltransferase; COMT: caffeic acid O-methyltransferase; DHAR: dehydroascorbate reductase; GLP: germin-like protein; GRP: glycine-rich protein; HSF: heat shock factor; HSP: heat shock protein; MDAR: monodehydroascorbate reductase; NDPK: nucleoside diphosphate kinase; nsLTP: non-specific lipid transfer proteins; OEC: oxygen-evolving complex; PAL: phenylalanine ammonia lyase; PCD: programmed cell death; PDI: protein disulfide isomerase; PGK: phosphoglycerokinase; PPI: peptidyl-prolyl cis-trans isomerase; PRK: phosphoribulokinase; PP2C: protein phosphatase 2C; RCA: RubisCO activase; RLK: receptor-like kinase; SnRK: sucrose nonfermenting related kinase; TCTP: translationally-controlled tumor protein; VDAC: voltage-dependent anion channel; XET: xyloglucan endo-transglycosylase.

Signal transduction to the nucleus leads to PTM modifications of transcription factors and alterations in gene expression.

12.5.2 REGULATION OF GENE EXPRESSION

As indicated earlier, the major stress-inducible signaling pathways can be distinguished into ABA-dependent and ABA-independent pathways, leading to the activation of ABA-dependent and ABA-independent transcription factors (TFs), respectively. However, it has to be kept in mind that there are multiple interactions (crosstalk) between both ABA-dependent and ABA-independent pathways and that the target TFs can contain regulatory elements related to both ABA-dependent and ABA-independent pathways in their promoters, thus revealing complex regulation of expression.

The major ABA-dependent pathway target TFs involve AREB/ABFs, which are bZIP motif-containing TFs that bind to ABRE regulatory elements in promoters of ABA-responsive genes, including stress-responsive proteins from the COR/LEA family (Rab proteins and others). In *A. thaliana*, CBF4/DREB1D and two DREB2 TFs (DREB2A and DREB2B) also belong to the ABA-responsive TFs, unlike CBF/DREB1 TFs. Other ABA-dependent TFs include MYB and MYC TFs, which bind to specific target motifs (MYBR and MYCR) in the promoters of stress-responsive genes (Yamaguchi-Shinozaki and Shinozaki, 2006).

ABA-independent pathway target TFs include CBF/DREB1 TFs with an AP2 domain, which bind to CRT/DRE/LTRE regulatory elements, which are present in the promoters of several cold-responsive genes (the Cor/Lea superfamily). NAC and NAM TFs are regulated by both ABA-dependent and ABA-independent pathways, and they bind as homooligomers or heterooligomers with zinc-finger TFs to specific target motifs (NACR and NAMR) in the promoters of stress-responsive genes from the Cor/Lea superfamily (Yamaguchi-Shinozaki and Shinozaki, 2006; Nuruzzaman et al., 2013).

Crosstalk between ABA-dependent and ABA-independent pathways enables the fine-tuning of the plant stress response. For example, ABA receptors of PYL proteins inhibit PP2C to relieve SnRK2 activity, which in turn phosphorylates bHLH TF ICE1 to activate the transcription of *CBFs* and *COR* (Gonzalez-Guzman et al., 2012; Zhan et al., 2015). Moreover, the resulting intensity of gene expression is determined by the number of copies of the individual stress-responsive promoter elements.

TFs are relatively rare among identified proteins in total proteome studies due to their relatively low abundance and nuclear localization. Enhanced levels of NAC α were found in salt-treated rice roots (Yan et al., 2005). Relatively higher proportions of TFs were identified in studies dealing with the nuclear proteome, such as a study on cold-responsive nuclear proteins in *A. thaliana* (Bae et al., 2003), where MYB, bZIP OBF4, and bHLH TFs were identified.

Recently, it is becoming obvious that stress-induced changes in gene expression also include epigenetic modifications leading to the remodeling of transcriptionally active and inactive DNA regions. Alterations in histone isoforms play a crucial part in the determination of transcriptionally active and inactive regions. For example, enhanced expression of the H2A.Z isoform in comparison to standard H2A was found in cold-treated winter barley Luxor (Janská et al., 2011). Besides different histone isoforms, histone PTMs such as methylation and acetylation are important epigenetic modifications involved in the regulation of transcriptionally active and inactive regions. For example, vernalization-induced inhibition of the expression of the major flowering repressor FLC in *A. thaliana* is associated with dimethylation of K9 and K27 lysine residues in histone H3 in the FLC coding region (Bastow et al., 2004).

12.5.3 PROTEIN METABOLISM

An active plant stress acclimation process requires profound alterations in both the biosynthesis and the biodegradation of several proteins. Ribosomes play a crucial role in protein biosynthesis. Thus, regarding protein biosynthesis, alterations in the biosynthesis of pre-ribosome forming proteins such as NOP1/NOP56 complex, which functions in 60S pre-ribosome biogenesis, were found in the nuclear proteome of flooded soybean root tips (Yin and Komatsu, 2016). Moreover, several alterations in the relative abundance of ribosomal proteins were found in stress-treated plants, such as 30S ribosomal protein S1, which is important for recognition and binding to mRNA upstream of the Shine-Dalgarno sequence, which is necessary for translation initiation (Boni et al., 1991; Fatehi et al., 2012; Maršálová et al., 2016). Moreover, alterations in 60S ribosomal proteins L12 and L13, which are known to bind to rRNA during ribosome assembly, were found in salt-treated barley (Fatehi et al., 2012; Maršálová et al., 2016) and drought-treated wheats (Faghani et al., 2015), respectively.

Translation initiation and elongation factors are involved in the interaction of aminoacyl-tRNA with mRNA on the ribosome (aminoacyl site, peptidyl site), enabling the formation of a peptide bond between a free amino acid and a nascent peptide. The translation initiation factors eIF-3 I-like and eIF-5A2 were transiently increased in cold-treated wheats, especially the spring cultivar (Kosová et al., 2013a). Moreover, some isoforms of translation initiation factors, such as eIF-5A isoforms, are involved in the regulation of cell division (Thompson et al., 2004); see Section 12.5.12 for more details.

Stress treatments also show significant impacts on organellar proteosynthesis, including alterations in protein components of organellar ribosomes. Increased chloroplast elongation factor eEF-G was reported in cold-treated *Thellungiella halophila* (Gao et al., 2009), while increased chloroplast elongation factor eEF-Tu was found in heat-treated winter wheat (Ristic et al., 2008). A differential effect of salinity stress was found on chloroplast 30S ribosomal protein S10, a protein crucial for binding of tRNA to the ribosomal surface and initiation of protein biosynthesis, which was found to be significantly decreased in salt-susceptible canola cultivar Sarigol compared with no significant change in salt-tolerant cultivar Hyola 308 when exposed to salt stress (Bandehagh et al., 2011).

Protein degradation is catalyzed by proteases, which are classified according to the amino acid residues at the cleavage site, e.g., serine proteases, cysteine proteases, etc. Proteases are often induced by pathogen infection to cleave pathogen-produced proteins. Moreover, plants induce expression of protease inhibitors against proteases produced by pathogens. As an example, inhibitors of serine proteases known as serpins were found in *Fusarium*-inoculated barley grain (Yang et al., 2010).

A plant's own cellular proteins are degraded via the ubiquitin–proteasome system. Proteins are first tagged by a polyubiquitin chain via a three-step cascade including ubiquitin-activating enzyme E1, ubiquitin-conjugating enzyme E2, and ubiquitin protein ligase E3. Polyubiquitin-tagged proteins are then targeted to the 26S proteasome, which is a supramolecular complex containing protein components with protease activity. The proteasome complex consists of a central barrel-shaped 20S core particle with protease function and 19S regulatory particles at each side of the central core, which regulate core particle proteolytic activity. AAA+-ATPases at the rim of 19S regulatory particles gate the entrance of ubiquitin-tagged proteins into the 20S core particle. The highest genetic variability was found for E3 ubiquitin protein ligases, which play an important role in the modulation of plant stress responses due to their specific interactions with stress-regulated TFs. For example, RING-type E3 ubiquitin ligase HOS1 interacts with ICE1 TF upstream of the CBF/COR pathway, leading to downregulation of the CBF/COR pathway and decreased low-temperature tolerance (Dong et al., 2006; Serrano et al., 2018). Enhanced levels of proteins involved in the ubiquitination–proteasomal pathway were found in several stress-treated plants, such as in salt-treated *Nitraria sphaerocarpa* (Chen et al., 2012) and drought-treated wheat (Cheng et al., 2016).

12.5.4 Energy Metabolism

The maintenance of sufficient energy production is crucial for stress acclimation. Aerobic energy-producing processes such as aerobic respiration and photosynthesis, which represent coordinated processes including both enzymatic and non-enzymatic electron transport reactions, are susceptible to stress. Stress leads to an imbalance between non-enzymatic electron transport processes and the following enzymatic carbon assimilation reactions leading to enhanced ROS formation (oxidative stress).

12.5.4.1 Photosynthesis

Photosynthesis represents the most crucial process of energy metabolism due to the conversion of solar energy to chemical energy, which is necessary for stress acclimation. However, photosynthesis is among the most susceptible processes affected by stress due to fine-tuning between cascades of ETC processes and chemical processes associated with carbon assimilation. Solar energy is primarily captured by light-harvesting protein complexes (LHCI, LHCII) binding chlorophylls (chl *a*, chl *b*) and carotenoids at peripheral sides of both photosystems. Stress usually leads to a decrease

in LHC proteins, especially Lhcb proteins in PSII, such as in drought-treated *A. thaliana* (Xu et al., 2012). Imbalances between light-dependent electron transport processes on thylakoid membranes and light-independent carbon assimilation processes in the chloroplast stroma lead to decreased capacity of ferredoxin-NADPH oxidoreductase at the acceptor side of the photosynthetic ETC, resulting in enhanced ROS formation and photoinhibition. ROS have damaging effects on components of both photosystems, especially on D1 protein in the reaction center of photosystem II (RC PSII) and on the components of the oxygen-evolving center (OEC) in PSII (Caruso et al., 2008). In contrast, an enhanced abundance of PSII components such as OEC components (OEE1, OEE2) and D2 protein in RC PSII was found in salt-tolerant halophytic plants such as *Hordeum marinum* (Maršálová et al., 2016), *Suaeda aegyptiaca* (Askari et al., 2006), and *Aeluropus lagopoides* (Sobhanian et al., 2010) under enhanced salinity, indicating the ability of halophytes to cope with salinity stress and to restore photosynthetic activity under salinity conditions. An increase in OEE1 protein was observed in drought-treated barley infected by *Piriformospora indica*, indicating an elicitor effect of the fungus on plant stress tolerance (Ghabooli et al., 2013). Wendelboe-Nelson and Morris (2012) have demonstrated a decrease of OEE1 in stressed leaves of the susceptible barley Golden Promise compared with the increase of OEE2 in tolerant Basrah. Kausar et al. (2013) showed an increase in OEE proteins under milder drought and in tolerant plant materials while also showing a decrease under severe drought or in sensitive plant materials. The fine tuning of redox homeostasis at the acceptor side of photosynthesis ETC is ensured by several isoforms of small redox proteins such as peroxiredoxins (Prx) and thioredoxins (Trx). The efficiency of photosynthetic ETC processes determines ATP synthesis via chloroplast CF1CF0-ATP synthase.

Carbon assimilation: Ribulose-1,5-bisphosphate carboxylase/oxygenase (RubisCO) is the most abundant enzyme on the earth and the major carbon assimilation enzyme, besides carbon prefixation via phosphoenolpyruvate carboxylase (PEPC), in C4 plants. In higher plants, the functional RubisCO holoenzyme consists of eight chloroplast-encoded RubisCO large subunits (RubisCO LSU) and eight nucleus-encoded RubisCO small subunits (RubisCO SSU). RubisCO shows dual enzymatic activities: carboxylation (carbon assimilation activity in photosynthesis) and oxygenation (photorespiration) activities, whose ratio depends on the CO_2/O_2 ratio. Therefore, other enzymes are involved in enhancing RubisCO carboxylation activity. The most important ones are RubisCO activase (RCA) and carbonic anhydrase (CA). RCA ensures binding of the substrate ribulose-1,5-bisphosphate (RuBP) to the enzyme active site. To ensure RubisCO activity under altered conditions of heat stress, cereals from tribe Triticeae express thermostable RubisCO activase B isoform, which was identified in heat-treated barley (Rollins et al., 2013). Increased RCA levels were also found in stress-treated plants that have only one RCA isoform, such as plants from the Brassicaceae family; enhanced RCA was reported in drought- and salt-treated rapeseeds (Bandehagh et al., 2011; Urban et al., 2017).

RubisCO requires CO_2 to catalyze RuBP carboxylation; however, in chloroplast stroma, CO_2 is mostly present in the form of hydrogencarbonate (HCO_3^-). CA catalyzes a reversible interconversion between CO_2 and HCO_3^-, thus ensuring novel CO_2 molecules in the vicinity of the RubisCO active site. Three CA families named α, β, and γ were reported in higher plants, each of them containing multiple protein isoforms. Opposite patterns of relative abundance changes between two βCA isoforms were found in drought-treated rapeseed (Urban et al., 2017).

Moreover, RubisCO carboxylation activity is modulated by PTMs, especially redox PTMs, which form spontaneously due to enhanced reactive molecular species (RMS) occurring as a result of imbalances in redox homeostasis during the initial phases of plant stress response. RubisCO S-nitrosylation (RubisCO SNO) leading to a decline in RubisCO carboxylation activity was reported in drought-treated *Brassica juncea* (Abat and Deswal, 2009).

Alterations in both RubisCO LSU and RubisCO SSU as well as Calvin cycle enzymes such as PGK and PRK, which catalyzes the regeneration of the RubisCO substrate ribulose-1,5-bisphosphate RuBP from ribulose-5-phosphate with the use of ATP, were found in stress-treated plants. In salt-susceptible durum wheat, a decrease in two RubisCO SSU isoforms, PGK and PRK, was found under 100 mM NaCl (Caruso et al., 2008). Enhanced carbonylation and SNO of RubisCO LSU along with other enzymes of energy metabolism (RCA, subunits of chloroplast and mitochondrial ATP synthase F1, enolase and other glycolytic enzymes, and alcohol dehydrogenase [ADH]) were found in salt-treated citrus (Tanou et al., 2009), indicating alterations in their catalytic activities.

12.5.4.2 Aerobic Respiration and Fermentation

The major catabolic pathways of plant energy metabolism include glycolysis followed by fermentation or aerobic respiration. Glycolysis is a 12-step enzymatic pathway of glucose conversion to two pyruvate molecules, yielding two ATP molecules. Pyruvate can be then either reduced to lactate (lactate fermentation), decarboxylated and reduced to ethanol (ethanolic fermentation), or activated via coenzyme A (CoA) to form acetyl-CoA, which enters the mitochondrial matrix to undergo a series of decarboxylation reactions known as the tricarboxylic acid cycle (TCA or Krebs cycle), which are connected with the respiratory ETC coupled with ATP biosynthesis in the inner mitochondrial membrane. Enhanced levels of components of the respiratory ETC, such as cytochrome c oxidase subunit 6b-1, were found in plants under stress, such as salt-treated *Arabidopsis thaliana* (Pang et al., 2010).

Alterations in Krebs cycle enzymes such as malate dehydrogenase (MDH) were found in cold-treated winter wheats, revealing genotypic differences in their acquired frost tolerance (Vítámvás et al., 2012). Apart from their enzymatic activity, these enzymes are also involved in buffering the pH of the mitochondrial matrix.

Stress generally leads to mobilization of catabolic processes due to enhanced energy requirements during the process of active stress acclimation. Therefore, enhanced levels of several components of the aerobic respiratory ETC were found in stress-treated plants. However, under severe stress, a shift from aerobic respiration to anaerobic fermentation accompanied by enhanced levels of ADH, formate dehydrogenase, or aldehyde dehydrogenase was found in susceptible plants, such as wheat seedlings under salinity (Fercha et al., 2013) or spring barley under drought (Vítámvás et al., 2015). Anaerobic fermentation processes show much lower efficiency in ATP production than aerobic respiration (only two ATP molecules per one glucose in fermentation vs. up to 38 ATP molecules per one glucose in aerobic respiration); however, the risk of cellular damage by ROS is significantly eliminated.

12.5.4.3 ATP Metabolism

Adenosine triphosphate (ATP) represents an immediate source of energy, which can be easily converted to other chemical forms. ATP is produced by both anaerobic and aerobic processes, with aerobic ones being more efficient. Anaerobic processes include glycolysis (two ATP molecules produced per one glucose molecule) and fermentation processes such as alcohol fermentation or lactate fermentation. Aerobic processes include electron transport processes at the inner membranes of chloroplasts and mitochondria. Both the photosynthetic ETC in thylakoid membranes and the respiratory ETC in inner mitochondrial membranes lead to the formation of H^+ gradients, which are used by ATP synthase complexes for ATP biosynthesis. Since plant stress acclimation reveals enhanced demands on immediate energy sources, enhanced relative abundance of several subunits of the ATP synthase complex, especially the CF1β and CF1γ subunits directly involved in ATP biosynthesis, was reported in several studies dealing with salinity (Wang et al., 2008; Pang et al., 2010) or low temperature (Han et al., 2013; Kosová et al., 2013a).

Phosphate transfer from ATP and other nucleoside triphosphates to nucleoside diphosphates is catalyzed by nucleoside diphosphate kinase (NDPK). Enhanced levels of NDPK isoforms were found in salt-treated *A. thaliana* (Pang et al., 2010) and pea (Kav et al., 2004) as well as in organellar proteomes such as in salt-treated wheat mitochondria (Jacoby et al., 2010) and maize chloroplasts (Zörb et al., 2009).

12.5.4.4 Crucial Role of Energy Metabolism in Plant Stress Tolerance

The ability to maintain the efficiency of the key processes of energy metabolism, especially photosynthesis, under stress conditions represents a crucial factor determining stress acclimation. Stress-tolerant (stress-adaptable) plants adjust their processes of energy metabolism, namely, photosynthesis and aerobic respiration, to altered environmental conditions without a significant decrease in energy (ATP) production while minimizing the harmful effects of oxidative stress. In contrast, susceptible plants show a significant decrease (downregulation) in aerobic processes involved in ATP production, especially in photosynthesis, thus leading to a lack of the energy necessary for active stress acclimation. As a response, plants activate catabolic processes leading to carbohydrate degradation. A downregulation of crucial aerobic processes, especially photosynthesis, has been repeatedly observed in susceptible plants; e.g., a decrease in the components of the photosynthetic ETC (OEE1) and carbon assimilation-related proteins (PRK, PGK, CA, and RubisCO SSU) was observed in drought-treated bread wheat (Ge et al., 2012) and salt-treated durum wheat (Caruso et al., 2008). In contrast, an increase in photosynthesis-related proteins was found in tolerant plants, e.g., the halophytic species *Suaeda aegyptiaca* (Askari et al., 2006) and *Aeluropus lagopoides* (Sobhanian et al., 2010) under increasing salinity up to 600 mM NaCl, which indicates that tolerant plants can efficiently cope with disbalances in their cellular homeostasis caused by stress and can enhance their energy metabolism to cover the enhanced energy requirements necessary for stress acclimation. However, when stress is too severe and the risk of oxidative damage very high, a shift from aerobic respiration to less efficient anaerobic fermentation is observed (Vítámvás et al., 2015).

12.5.5 STORAGE PROTEINS

Storage proteins typically accumulate in maturing seeds, where they represent an important source of energy and amino acids for the biosynthesis of novel compounds during plant embryo development. Seed storage proteins are known to be affected by stress factors, namely, heat and drought, resulting in adverse impacts on seed nutritional quality (Halford et al., 2015).

However, storage proteins can also accumulate in plant vegetative tissues or organs that are crucial for the survival of the whole plant during a period of harsh environmental conditions; for example, crown tissues in cereals, which are crucial for whole plant survival. An accumulation of legumin-like protein was found in cold-treated winter wheat crowns (Vítámvás et al., 2012).

12.5.6 CARBOHYDRATE METABOLISM

Carbohydrates do not only serve as the major storage energy source in plant cells; they are also important structural components of plant cells, namely, the cell wall, as well as showing important signaling and regulatory functions determining plant development due to their interactions with proteins and protein PTMs (N-glycosylation; O-glycosylation).

Carbohydrate catabolism: Saccharides as energy sources are tightly linked with energy metabolism. Under stress conditions, enhancement of catabolic processes such as glycolysis is often observed in plants to cover the enhanced energy requirements in the process of stress acclimation. Glycolysis and the following fermentation reactions are activated under severe stress to avoid enhanced risk of ROS as well as under stresses associated with hypoxia (anoxia), such as flooding. Enhanced levels of both cytoplasmic and plastidic isoforms of fructose bisphosphate aldolase (FBP ALDO) were found in salt-susceptible durum wheat, indicating an enhanced need for energy during stress acclimation (Caruso et al., 2008).

Carbohydrate anabolism: Monosaccharide (glucose) biosynthesis via the Calvin cycle carbon assimilation pathway is significantly affected by stress. As is reported for components of the photosynthetic ETC, decreased abundances of Calvin cycle enzymes such as PGK and PRK in salt-susceptible durum wheat correspond to decreased components of PSII OEC (OEE1) and ATP synthase subunit CF1α, indicating a downregulation of photosynthetic processes (Caruso et al., 2008).

Oligosaccharides (disaccharides and raffinose family oligosaccharides) are formed from nucleoside diphosphate (NDP)-activated monosaccharides and serve as saccharide transport forms as well as important cellular osmoprotectants due to their hydrophilic nature. Polysaccharides serve as storage energy sources (starch and inulin) as well as important structural components of plant cell walls (cellulose and hemicelluloses). To form glycosidic bonds, monosaccharide molecules have to be activated by binding to uracil diphosphate (UDP) or other nucleoside diphosphates (e.g., UDP-glucose, GDP-mannose). UDP-glucose can be directly used in oligo- and polysaccharide biosynthesis and can be synthesized by either sucrose synthase or UDP-pyrophosphorylase. In their study on cold-treated winter wheats, Vítámvás et al. (2012) found an increased level of sucrose synthase but a decreased level of UDP-pyrophosphorylase, which might indicate alterations in carbohydrate metabolism leading possibly to cold-induced alterations in cell wall polysaccharides. In contrast, an increased level of UDP-pyrophosphorylase was reported in cold-treated rice seedlings (Cui et al., 2005). Consistently with relatively lower growth inhibition in tolerant cultivars vs. sensitive ones under stress, a relatively increased level of enzymes involved in cell wall elongation, such as xyloglucan *endo*-transglycosylase (XET), was found in a drought-tolerant grapevine cultivar compared with a drought-susceptible one (Vincent et al., 2007).

Moreover, saccharides function as signaling molecules due to their interactions with specific proteins, such as proteins from the lectin superfamily. A jacalin-like lectin, VER2, is involved in shoot apical meristem development and transition to flowering in Triticeae. Increased levels of lectin VER2 were found in cold-treated winter wheat crowns, and the peak coincided with vernalization fulfilment followed by a decrease in vernalized winter wheat crowns (Rinalducci et al., 2011).

12.5.7 S-ADENOSYLMETHIONINE METABOLISM

S-adenosylmethionine (SAM) is a universal methylation agent in plant cells. Methylation reactions play important roles in a variety of biological processes, ranging from epigenetic modifications of both DNA bases and histones to several biosynthetic processes, such as lignin biosynthesis. Moreover, SAM is also a precursor in the biosynthesis of ethylene, which is an important stress-induced phytohormone, and polyamines, which are polycations involved in DNA protection in plant cells. SAM biosynthesis from methionine and ATP is ensured by S-adenosylmethionine synthase (SAMS), which is usually encoded by multiple genes in plant genomes, thus revealing multiple SAMS isoforms in proteomic analyses. Differential relative abundances of SAMS isoforms were found in roots of soybean seedlings exposed either to drought or to waterlogging stress, which may be associated with differences in lignin biosynthesis and remodeling, since lignin biosynthesis requires methylation of monomeric components (Wang et al., 2016).

12.5.8 PHYTOHORMONE METABOLISM

Plant metabolism under stress is redirected from active growth and development (flowering) to stress acclimation. These alterations represent the results of changes in phytohormone levels, because plant stress acclimation is associated with an increase in stress-responsive proteins such as ABA, jasmonic acid (JA), salicylic acid (SA), and ethylene and a decrease in growth-promoting phytohormones such as gibberellins. At the proteome level, significant alterations in several enzymes involved in phytohormone metabolism were found in stress-treated plants.

JA is a derivative of linolenic acid. Enzymes involved in the biosynthesis of JA from linolenic acid, including lipoxygenase 2 (LOX2), allene oxide cyclase (AOC), and allene oxide synthase (AOS), were found to be increased in *A. thaliana* in response to salinity of 150 mM NaCl (Pang et al., 2010).

DWARF3, an enzyme involved in gibberellin biosynthesis, showed a lower decrease in salt-tolerant *T. aestivum* × *Thinopyrum ponticum* hybrid with respect to its salt-susceptible *T. aestivum* parent, indicating lower inhibition of plant growth in the tolerant hybrid under high salinity of 200 mM NaCl (Peng et al., 2009).

12.5.9 SECONDARY METABOLISM (LIGNIN METABOLISM)

Under stress conditions, the metabolism of several secondary metabolites is also altered due to the roles of several secondary metabolites in plant stress acclimation.

The plant cell wall is a dynamic structure, whose major components are polysaccharides (cellulose, hemicelluloses, and pectins) and polyphenols (lignin). Alterations in enzymes involved in the phenylpropanoid pathway, such as phenylalanine ammonia lyase (PAL), ensuring the biosynthesis of lignin monomers were reported under various stress treatments, which affects cell wall mechanical properties. Dehydration stresses and low-temperature stress lead to enhanced cell wall lignification, which is indicated by enhanced levels of lignin biosynthetic enzymes (caffeyl O-methyltransferase [COMT]; caffeoyl-CoA methyl transferase [CCOMT]), including ROS-scavenging enzymes and methylation agents (SAM), which are necessary for polymerization of lignin monomers (Zhu et al., 2007; Li et al., 2018). In contrast, flooding leads to decreased lignification (Oh and Komatsu, 2015; Yu et al., 2015).

12.5.10 STRESS- AND DEFENSE-RELATED PROTEINS

Stress factors induce disturbances in cellular homeostasis, leading to an enhanced risk of protein misfolding, denaturation,

or damage by enhanced levels of RMS and toxic byproducts of cellular metabolism. As a response, stress-treated plants induce several proteins involved in the protection of biomolecules against damage caused by cellular dehydration or toxic compounds as well as proteins involved in detoxification.

An enhanced risk of protein misfolding leads to an enhanced accumulation of chaperones and chaperonins, among which heat shock proteins (HSPs) are the best characterized. The HSP superfamily consists of five groups sorted according to their molecular weight: HSP110, HSP90, HSP70, HSP60, and small HSPs. Several HSPs were identified in the grain protein fraction from heat-treated wheat (Skylas et al., 2002; Majoul et al., 2004). Recently, HSP70/HSP90 complexes were reported to play an important role in ABA-mediated stomatal closure (Clement et al., 2011). Enhanced biosynthesis of novel proteins at the endoplasmic reticulum (ER) during stress acclimation increases the risk of protein misfolding, leading to activation of the unfolded protein response (UPR; Ma and Hendershot, 2004). Other proteins involved in protein folding include peptidyl-prolyl cis-trans isomerases (PPI). They were reported to increase in wheat seedlings on drought priming (Wang et al., 2014a) as well as in drought-tolerant wheat cultivars (Ford et al., 2011; Cheng et al., 2016) but decreased in a drought-susceptible cultivar of common bean (Zadražnik et al., 2013). Proteins involved in photosynthesis are among the most susceptible in response to stress. Therefore, several stresses induce an accumulation of specific chaperonins, CPN60α and CPN60β, binding to the subunits of the major carbon assimilation protein RubisCO (Kosová et al., 2013a).

Another category of stress-inducible proteins with protective functions is the hydrophilic proteins (hydrophilins), including several COR/LEA protein groups (LEAII [dehydrins]; LEAIII). Hydrophilins, especially dehydrins from the LEAII family, accumulate in the cell cytoplasm and other intracellular compartments under dehydration stress, where they function as chaperones and molecular shields protecting other proteins from the loss of their hydration envelopes. Moreover, some dehydrin proteins also show cryoprotective, antifreeze, ROS scavenging, and ion-binding functions. For more information on dehydrins, see, e.g., the reviews of Rorat, (2006); Kosová et al., (2007, 2014a); Hara, (2010); Hanin et al., (2011).

Other stress-responsive proteins with largely unknown mitigating functions include universal stress proteins (USP), which were found to be constitutively enhanced in salt-tolerant *H. marinum* (Maršálová et al., 2016) as well as induced by drought and senescence in wheat (Bazargani et al., 2011). In rice, OsUSP1 was found to be involved in the ethylene-mediated stress response (Sauter et al., 2002).

ROS (RMS) scavenging and detoxification enzymes: Imbalances in cellular metabolism lead to enhanced formation of RMS, which are molecules (ions and radicals) showing enhanced chemical reactivity. To minimize the potential damaging effects of RMS, organisms evolved RMS-scavenging mechanisms, resulting in the conversion of RMS into less harmful compounds. RMS-scavenging systems usually consist of proteins using RMS-reactive compounds, such as metal ions or relatively small reactive molecules such as ascorbate or glutathione, as cofactors to decompose RMS. The ascorbate-glutathione cycle represents the best-known ROS scavenging pathway, which is aimed at hydrogen peroxide decomposition to water; it includes ascorbate peroxidase (APX), monodehydroascorbate reductase (MDAR), dehydroascorbate reductase (DHAR), and glutathione reductase (GR). Several isoforms of ascorbate-glutathione cycle enzymes were found to be altered under stress. Multiple isoforms of some redox-related proteins, such as thioredoxins, enable the plant to fine tune its redox metabolism in susceptible processes such as the photosynthetic ETC to minimize the risk of oxidative damage, as observed in drought-treated sugar beet (Hajheidari et al., 2007). Moreover, the activity of some ROS (RMS)-scavenging enzymes, such as thioredoxins, is regulated by other stress-responsive proteins such as cystathione-beta-synthase (CBS) (Bertoni, 2011). Enhanced CBS levels were found in drought-treated wheat during senescence (Bazargani et al., 2011). Besides their harmful effects RMS can activate other molecules to enable the synthesis of some biopolymers, such as lignin. Enhanced levels of proteins involved in the production of apoplastic ROS, namely H_2O_2, such as oxalate oxidase, germins, APX, Cu/Zn-SOD, and Trx, were found in the elongation zone of maize root cell walls (Zhu et al., 2007) as well as chickpea seedling shoot cell walls (Bhushan et al., 2007) subjected to dehydration stress. Apoplastic ROS are involved in the cross-linking of monolignols and other phenolics during the cell wall lignification process. RMS can also function as signaling molecules and can lead to protein PTMs such as carbonylation or nitrosylation, resulting in alterations of protein biological activity. For example, the S-nitrosylation of RubisCO subunits in cold-treated *Brassica juncea* leads to a decline in RubisCO carboxylation activity (Abat and Deswal, 2009).

Besides ROS and other RMS, plant metabolism under stress shows enhanced production of other toxic compounds, such as methylglyoxal as a toxic byproduct of carbohydrate and amino acid metabolism. Enhanced levels of enzymes with glyoxalase activities, such as lactoylglutathione lyase, were found in salt-tolerant barley plants when exposed to enhanced salinity (Witzel et al., 2009; Maršálová et al., 2016). Cyanide is produced as a toxic byproduct of ethylene biosynthesis, which is often enhanced under stress. Cyanide is an inhibitor of metalloenzymes, including cytochrome complexes in the mitochondrial ETC. Enhanced levels of cyanase, an enzyme involved in cyanide decomposition, were found in the halophytes *Sueada aegyptiaca* (Askari et al., 2006) and *H. marinum* (Maršálová et al., 2016). Another strategy of plant stress response to increased cyanide lies in enhanced levels of alternative oxidase (AOX), a crucial component of the cyanide-resistant respiratory pathway. Enhanced AOX was found in drought-treated pea mitochondria (Taylor et al., 2005).

The detoxification of xenobiotic or endobiotic metabolites starts with metabolite conjugation to glutathione (GSH), which is catalyzed by glutathione-S-transferases (GSTs). Several isoforms of GSTs are classified into groups marked by Greek letters, including the classes α, μ, π, ζ, φ, σ, τ, and θ. Several GST isoforms were found to be increased in salt-treated barley roots (Witzel et al., 2009) and drought-treated wheat leaves and roots (Faghani et al., 2015). Protein disulfide isomerases (PDI) catalyze a reversible interconversion between cysteine residues and inter- or intra-molecular disulfide bridges, thus acting as both redox enzymes and chaperones involved in protein folding. Enhanced PDI levels were found in copper- and PEG-treated wheat, respectively (Kang et al., 2012; Li et al., 2013), while decreased PDI levels were found in drought-treated wheat (Faghani et al., 2015).

Inhibitors of proteases are usually induced in response to proteases produced by plant pathogens. Among the most often induced protease inhibitors are serpins (serine protease inhibitors), which are induced in cereals in response to *Fusarium* infection (Yang et al., 2010; Eggert et al., 2011; Eggert and Pawelzik, 2011). Some protective and detoxification enzymes are classified into a large superfamily of pathogenesis-related (PR) proteins, which contains 16 groups (Edreva, 2005). These are primarily induced by plant pathogens (biotic stressors), but some of them can also be induced in response to abiotic stresses; for example, a thaumatin-like protein (PR5) and germin (PR15) and germin-like proteins (PR16) were found to be increased under cold and drought in winter and spring wheats, respectively (Kosová et al., 2013a; Faghani et al., 2015). Enhanced phosphorylation of several PR10 proteins was found in salt-treated peanut (Jain et al., 2006).

Ion-binding proteins: Free metal ions act as catalysts of ROS formation. Therefore, inactivation of metal ions by binding to specific proteins is necessary to eliminate the risk of ROS formation. Phytochelatins (polymers of glutathione) and phytosiderophores (IDS2,3) accumulate not only under heavy metal stress but also under other stresses associated with enhanced levels of free ions in cell cytoplasm, such as salinity (Witzel et al., 2009).

Differences in the relative abundance of several stress- and defense-related proteins have often been found in proteomic studies aimed at the comparison of genotypes showing differential stress tolerance. Stress-tolerant plants reveal constitutively enhanced levels of several defense-related proteins compared with susceptible plants. They can thus protect other proteins from harmful effects of stress (protein misfolding, oxidative stress, protein damage, etc.) more efficiently than susceptible plants.

12.5.11 STRUCTURAL PROTEINS

Structural proteins participate in the formation of cellular structures such as the cell wall, the plasma membrane and endomembranes, the cytoskeleton, and other intracellular structures. Stress factors also affect the composition and biological functions of cellular structures.

Plant responses to stress treatments are associated with an enhanced need for transport of several compounds. Transmembrane ion channels play a crucial role in salinity and heavy metal stress. They are involved in ATP-dependent salt ion exclusion (transport proteins in the plasma membrane, such as SOS1) or ion vacuolar compartmentation (transport proteins in the tonoplast, such as V-ATPases or PPi-ases, and endosomal Na⁺/H⁺ exchangers such as NHX). Enhanced levels of plasma membrane, vacuolar, and tonoplast ion transport channels were found in various salt-treated plants (Wang et al., 2008; Pang et al., 2010; Xu et al., 2010; Wakeel et al., 2011). Apart from membrane channels involved in salt or heavy metal ion transport, other membrane-associated transport proteins induced under dehydration stress include aquaporins (AQPs), which facilitate water transport across hydrophobic membranes, and monosaccharide sensing protein 2 (MSSP2), involved in monosaccharide transport; these proteins were found to be induced and phosphorylated in PEG-treated wheat roots (Zhang et al., 2014). Annexins are Ca^{2+}-binding proteins interacting with the MAPK signaling cascade. They can form oligomeric transmembrane channels facilitating vesicular trafficking. Increased annexins were found in salt-treated potato (Aghaei et al., 2008) and tomato (Manaa et al., 2011) as well as in drought-treated spring barley (Vítámvás et al., 2015). ABC transporters are known to participate in the transport of glutathione conjugates into vacuoles and were found to be elevated under Cd stress (Alvarez et al, 2009). Non-specific lipid transfer proteins (nsLTPs) are small proteins with a hydrophobic lipid binding site, which are involved in lipid transfer from donor membranes, such as endoplasmic reticulum, to acceptor membranes, such as chloroplasts, mitochondria, peroxisomes, and glyoxysomes. LTPs are also found in cell walls, where they are involved in cuticle formation. LTPs were reported to be induced by several pathogens and environmental stresses, including drought and salinity, in wheat (Jang et al., 2004).

Nucleus and semiautonomous organelles such as plastids and mitochondria are surrounded by two membranes, the inner ones representing an important selection barrier for molecular transport. Targeted organellar transport is ensured by specific transport proteins. Nuclear transport is provided by RanGTP-binding proteins (Ran GTPases) activated by specific proteins involved in GTP–GDP exchange. Increased levels of two RanGTPase activating proteins and one Ran-binding 1-c like protein were identified in salt-treated barleys (Maršálová et al., 2016), which corresponds to increased GTP-binding nuclear protein Ran1 in salt-treated shoots of *A. lagopoides* (Sobhanian et al., 2010). Semiautonomous organelles such as plastids and mitochondria have their own transmembrane transport proteins enabling molecular transport across both outer and inner membranes; these are named TOM and TIM in mitochondria and TOC and TIC in plastids, respectively. In drought-treated barley, a decrease in the chloroplast outer membrane transporter TOC75 was found (Vítámvás et al., 2015). Voltage-dependent anion channel (VDAC) is involved in metabolite transport across the inner mitochondrial membrane. Flooding stress–related translocation of cytochrome c from mitochondria into cytoplasm via VDAC leads to cytochrome c interaction with caspases and induction of PCD (Komatsu et al., 2011).

Proteins involved in the adhesion of several membrane-associated proteins and the interconnection of the plasma membrane–associated cytoskeleton with cell wall components such as ankyrins and fasciclins also reveal alterations in response to environmental stresses (Cheng et al., 2016; Maršálová et al., 2016). Alterations in these proteins may reveal relationships with alterations in cell wall proteins, including both structural proteins (expansins) and enzymes (XET), resulting in altered cell wall biomechanical properties in stress-treated plants (Budak et al., 2013).

While transmembrane protein channels enable or facilitate the movement of several compounds across cellular membranes, cytoskeleton components ensure cytoplasmic transport. Cytoskeletal proteins, including both actin (microfilaments) and tubulin (microtubules) cytoskeleton, are involved in molecular transport as well as in directing cell wall growth. β-tubulin was found to be increased in salt-treated *A. thaliana* leaves (Pang et al., 2010) and decreased in salt-treated *A. thaliana* roots (Jiang et al., 2007). A decline in β-tubulin and γ-tubulin-interacting protein was found in salt-treated *Nitraria sphaerocarpa* (Chen et al., 2012). Enhanced actin and β-tubulin levels were found in chicory roots exposed to cold (Degand et al., 2009), while a decrease in α- and β-tubulin was found in drought-stressed sunflower roots (Ghaffari et al., 2013) and copper-stressed wheat roots (Li et al., 2013).

12.5.12 REGULATORY PROTEINS

Proteins involved in the regulation of plant growth and development include proteins involved in cell cycle regulation and thus, affecting cell division. Stress usually leads to suppression of cell division due to the redirection of plant metabolism from active growth and development toward stress acclimation. A decrease in Cdc48 was found in salt-treated plants under high-salinity stress in both salt-susceptible plants such as *Hordeum vulgare* (Maršálová et al., 2016) and the halophyte *Tangut nitraria* (Cheng et al., 2015). Isoforms of translation initiation factor eIF-5A1, eIF-5A2 and eIF-5A3 differing in their hypusination level (PTM) were reported to be involved in affecting plant cell division (enhanced levels of eIF-5A2 and eIF-5A3) or plant cell senescence (eIF-5A1) leading to PCD (Thompson et al., 2004). Differential eIF-5A isoforms were reported in salt-treated barleys, with the eIF-5A1 isoform present in salt-susceptible *H. vulgare* and the eIF-5A2 isoform present in salt-tolerant *H. marinum* (Maršálová et al., 2016).

Several regulatory proteins are involved in modulation of the biological activity of other proteins or other compounds, such as nucleic acids, via mutual interactions. 14-3-3 proteins can interact with other proteins and modulate their activity. The interactions do not only include proteins produced by plant pathogens, which plays an important role in the biotic stress response, but were also reported to interact with vacuolar H⁺-ATPases, resulting in alterations of their activity. Therefore, alterations in several isoforms of 14-3-3 proteins were reported under salinity stress, including a decrease in *Puccinellia tenuiflora* (Yu et al., 2011) and an increase in *Kandelia candel* (Wang et al., 2014b).

Small glycine-rich proteins (GRPs) can bind to pre-mRNA, thus being involved in the regulation of transcription and mRNA processing. Several small GRPs in *A. thaliana*, such as AtGRP7 were reported to function as repressors of the major flowering repressor FLC and thus, as inducers of the autonomous flowering pathway (Quesada et al., 2005; Streitner et al., 2008). Increased levels of GRPs have been found under several stresses, such as in cadmium-treated poplar (Durand et al., 2010), transgenic *Arabidopsis* under salinity and dehydration stresses (Kwak et al., 2005), and in cold-treated spring wheat crowns (Kosová et al., 2013a).

The regulation of flowering in several temperate plants, such as *A. thaliana* and winter cereals, is ensured by epigenetic modification of genes encoding flowering inducers or repressors, respectively; a phenomenon known as *vernalization* (Sung and Amasino, 2005). In winter-type cereals, a dominant effect of the major vernalization gene *VRN1* on the expression of cold-regulated *Cor/Lea* genes was found (Danyluk et al., 2003). At the protein level, a dominant role of low temperatures under continuous cold treatment was found (Vítámvás et al., 2012); however, the effect of vernalization emerged as crucial to regulate novel accumulation of the cold-inducible dehydrin protein WCS120 under cold reacclimation treatment following previous plant deacclimation (Vítámvás and Prášil, 2008).

12.5.13 Programmed Cell Death (PCD) Regulation

Severe stress treatments can cause irreversible damage of plant cells, leading to programmed cell death (PCD). PCD initiation is associated with protein signaling events leading to alterations in chromatin structure and DNA fragmentation. Under severe stress treatments such as high salinity, enhanced levels of PCD-inducing proteins such as apoptotic chromatin condensation inducer in nucleus and FASassociated factor 2B were found (Maršálová et al., 2016). FAS associated factor 2B is involved in the interaction of FAS antigen with FAS ligand, leading to the initiation of apoptotic processes. The FAS associated factor binds to the FAS antigen, thus enhancing its interaction with FAS ligand and inducing signaling processes leading to apoptosis. Further PCD-related proteins identified under severe stress include plasminogen activator inhibitor 1 RNA-binding protein, involved in regulation of mRNA stability, and guanine-nucleotide binding protein (G protein), which is a phosphoprotein with a WD-repeat motif known to be induced by exogenous ABA. G proteins are known to be involved in a variety of cellular processes, including signal transduction, transcription regulation, cell cycle regulation, and apoptosis. Guanine-nucleotide binding protein subunit b was reported to be increased in salinity- and drought-treated rice (Dooki et al., 2006; Ke et al., 2009).

Translationally controlled tumor protein (TCTP) is a calcium-binding protein involved in PCD suppression. TCTP interaction with p53 counteracts PCD induction and promotes cell survival in tolerant barley genotypes exposed to salinity (Mostek et al.,2015; Maršálová et al., 2016).

12.6 ROLE OF PROTEIN ISOFORMS AND POSTTRANSLATIONAL MODIFICATIONS (PTMs)

One gene can give rise to multiple functional proteins as a result of postranscriptional and posttranslational modifications. Protein isoforms are either proteins that are products of a single gene that have arisen by differential posttranscriptional modifications, such as alternative splicing or primary transcript editing, or they represent products of homologous (orthologous or paralogous) genes. Protein isoforms differ in protein sequence. Protein PTMs differ in chemical modifications of amino acid residues, which can range from small molecules, such as NO, to whole peptides, such as ubiquitin or SUMO. Currently, more than 300 different kinds of PTMs have been described (Wu et al.,2016). In plant stress proteomics, the most-studied PTMs include phosphorylation, glycosylation, and PTMs caused by RMS, such as carbonylation, oxidation, or nitrosylation. Protein isoforms and PTMs differ in their MW or pI values; thus, they can be separated by 2-DE.

Protein isoforms can reveal the same, similar, or entirely different biological functions depending on their molecular constraints, i.e., cellular localization and interacting partners. Protein isoforms with the same cellular localization and biological function, such as various peroxiredoxin (Prx) and thioredoxin (Trx) isoforms, can reveal overlapping functions in the fine tuning of cellular redox metabolism, especially at the end of the photosynthetic ETC. Protein isoforms can differ in their cellular localization, such as SOD isoforms, which catalyze the dismutation of superoxide anion radical to hydrogen peroxide. Cu/Zn-SOD is localized in cytosol, while Mn-SOD is the mitochondrial and Fe-SOD the plastidic isoform.

Protein isoforms can differ in their abundance pattern under stress. For example, different isoforms of the SAM biosynthetic enzyme S-adenosylmethionine synthase (SAMS), named SAMS1 and SAMS2, respectively, were found in soybean under drought vs. flooding stress (Wang et al., 2016). SAMS catalyzes the biosynthesis of SAM, which is a universal methylation agent as well as a precursor of polyamines and ethylene. Environmental factors also affect protein PTM pattern; recently, a temperature-related interconversion of protein phosphoforms revealing an impact on their biological activity was reported (Vu et al., 2018).

Entirely different functions were found for nuclear isoforms of glycolytic enzymes with respect to their more abundant cytosolic or plastidial isoforms. For example, the nuclear isoform of enolase is encoded by the *LOS2* gene, and it has regulatory functions in crosstalk of cold-responsive signaling pathways (Lee et al., 2002). Nuclear enolase binds to the ZAT transcription factor, and it is involved in crosstalk with the CBF signaling pathway, thus affecting the expression of COR genes. The nuclear isoform of aldolase reveals DNA-binding function (Ronai et al., 1992), while the nuclear isoform of glyceraldehyde-3-phosphate dehydrogenase (GAPDH) may be involved in tRNA binding and its export from the nucleus into the cytoplasm (Singh and Green, 1993). Phosphorylation of FBP ALDO may lead to its translocation from the cytoplasm into the nucleus, where it may regulate the expression of its own gene (Gizak and Dzugaj, 2003). Several nuclear isoforms of glycolytic enzymes were found in drought-treated chickpea (Pandey et al., 2008). More details on differential biological functions of protein isoforms and PTMs are given in Kosová et al. (2018).

12.7 FUNCTIONAL STUDIES AND PROTEIN–PROTEIN INTERACTIONS

As demonstrated in the preceding text, the biological functions of a protein depend on its structure, including PTMs, as well as on its interactions with other protein and non-protein compounds (Figure 12.3). To study the function of a given protein in protein networks, mutants showing altered or null expression (knockout mutants) are created. The role of the given protein is then studied by a comparison of the knockout mutant with the same mutant with knockout gene heterologous complementation, i.e., the introduction of a homologous gene from another plant species to restore the original protein function.

Protein biological function represents a result of protein interactions with other protein and non-protein compounds. To precisely describe the protein–protein and protein–non-protein interactions of a given protein, interactomics studies are necessary to elucidate protein interactions determining their final biological functions in plants. For interactomics

studies, classical protein–protein interaction tools such as yeast-two-hybrid (Y2H) and split-GFP are used. An example of an interactomics study is the investigation of interactions between proteins involved in vernalization regulation in winter wheat by Tardif et al. (2007), who applied Y2H and split-GFP in *Nicotiana benthamiana* to study physical interactions of the vernalization-related proteins TaVRT1/VRN1, TaVRT2, VRN2, TaFT1/VRN3, and TaHd as well as between other protein groups, such as proteins involved in signaling (two phospholipases C, a receptor-like protein kinase, and G protein) and microtubule remodeling (α-tubulin and TaTil, a lipocalin known to be involved in microtubule polymerization).

Studies of the effect of a given protein on plant phenotype are based on either silencing, overexpression of a corresponding gene in the given organism, or heterologous expression of a corresponding gene in another organism. For heterologous expression, the target protein can be fused to GFP or another marker protein to facilitate the target protein visualization. An example of the protein knockout approach in proteomics is virus-induced gene silencing (VIGS). Zhang et al. (2016) employed the technique of VIGS for the study of plastid plastoglobule-located PAP6-like protein (also known as fibrillin) function in winter wheat under cold stress. Plants silenced for PAP6-like protein revealed decreased vigor under freezing temperature (−5 °C) compared with non-silenced (control) plants.

Recently, tools for site-directed mutagenesis, such as gene editing by CRISPR/Cas technology, transcription activator-like

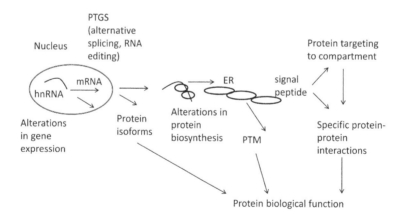

FIGURE 12.3 Major biological factors determining protein biological functions (posttranscriptional modifications such as alternative splicing or editing of primary transcript [hnRNA] leading to different protein isoforms; regarding the nascent protein, different posttranslational modifications and signal peptides determining protein targeting to cellular compartments lead to different cellular localization, protein–protein interactions, and protein interactions with other target compounds, determining the final biological function of the protein). ABA: abscisic acid; ADH: alcohol dehydrogenase; AOC: allene oxide cyclase; AOS: allene oxide synthase; AOX: alternative oxidase; APX: ascorbate peroxidase; AQP: aquaporin; ATP: adenosine triphosphate; CA: carbonic anhydrase; CBS: cystathione beta-synthase; CCOMT: caffeoyl-CoA methyl transferase; COMT: caffeyl O-methyltransferase; DHAR: dehydroascorbate reductase; ETC: electron transport chain; FBP ALDO: fructose bisphosphate aldolase; GAPDH: glyceraldehyde 3-phosphate dehydrogenase; GLP: germin-like protein; GR: glutathione reductase; GRP: glycine-rich protein; GST: glutathione-S-transferase; HSP: heat shock protein; LEA: late-embryogenesis abundant (protein); LOX: lipoxygenase; MDAR: monodehydroascorbate reductase; MDH: malate dehydrogenase; MSSP2: monosaccharide sensing protein 2; NDPK: nucleoside diphosphate kinase; nsLTP: non-specific lipid transfer protein; OA: osmotic adjustment; OEC: oxygen-evolving complex; PAL: phenylalanine ammonia lyase; PCD: programmed cell death; PDI: protein disulfide isomerase; PEG: polyethylene glycol; PGK: phosphoglycerokinase; PPI: peptidyl-prolyl cis-trans isomerase; PPi-ase: pyrophosphatase; PRK: phosphoribulokinase; Prx: peroxiredoxin; PTM: posttranslational modification; RCA: RubisCO activase; RMS: reactive molecular species; ROS: reactive oxygen species; SAM: S-adenosylmethionine; SAMS: S-adenosylmethionine synthase; SNO: S-nitrosylation; SOD: superoxide dismutase; SuSy: sucrose synthase; Trx: thioredoxin; UPR: unfolded protein response; USP: universal stress protein; VDAC: voltage-dependent anion channel; VIGS: virus-induced gene silencing; XET: xyloglucan endo-transglycosylase.

effector nucleases (TALEN), and other site-directed nucleases used for gene manipulation (reviewed in Sovová et al., 2017), have opened up new ways to study protein biological functions and their effects on the resulting plant phenotype.

12.8 USE OF PROTEOMIC STUDIES IN BREEDING FOR IMPROVED ABIOTIC STRESS TOLERANCE

Proteomic studies aimed at a comparison of the impacts of abiotic stress factors on plant proteome composition with respect to optimum growth conditions have generated large sets of data, mostly providing information on alterations in protein relative abundance between stress-treated and control plants and less often, information on alterations in other protein properties, such as protein cellular localization, protein isoforms and PTM variability, and alterations in potential protein interactions. Since proteins are "close" to the resulting plant phenotype, precise analysis of their quantitative (alterations in protein relative abundance) or qualitative (alterations in protein isoforms or PTMs) characteristics with respect to plant phenotypic characteristics associated with the acquired level of plant stress tolerance (e.g., lethal temperature of 50% sample (LT50) values as quantification of plant frost tolerance) can lead to the identification of potential protein markers of plant stress tolerance that reveal a genetic linkage to a given phenotypic trait associated with stress tolerance (e.g., LT50 in frost tolerance or ABA accumulation in drought tolerance). Examples include an observed relationship between the accumulation of some proteins (ABA45 and OSR40) and ABA accumulation in drought-treated maize leaves (Riccardi et al., 2004) or the potential use of cold-inducible K_n-type dehydrin protein relative abundance (WCS120 proteins in common wheat or DHN5 protein in barley) for the selection of winter wheat and barley cultivars with enhanced frost tolerance (Vítámvás et al., 2010; Kosová et al., 2013b; reviewed in the chapter on dehydrins by Kosová et al.,). Moreover, protein relative abundances determined by quantitative proteomics can be correlated with genetic markers, thus resulting in proteomic quantitative trait loci (pQTL) potentially of use in crop breeding (Burstin et al., 1994; Raorane et al., 2015); however, validation of laboratory data under field conditions is necessary (Zivy et al., 2015).

12.9 CONCLUSIONS AND FUTURE PERSPECTIVES

Proteins are direct effectors of an organism's interactions with the ambient environment; thus, they reflect the diversity and versatility of life. Unlike the genome, which can be described by publication of its full-length DNA sequence for a given plant genotype, it is practically impossible to fully describe the plant proteome corresponding to a single plant genotype, since the proteome dynamically responds to all developmental and environmental cues throughout the whole plant life cycle. Proteomic studies aimed at a comparison of proteome composition in plants under control and stress-treated conditions have

revealed profound impacts of environmental stress factors not only on stress- and defense-related proteins but also on a broad range of proteins involved in a diversity of biological processes: protein signaling and signal transduction; transcription factors and other proteins involved in the regulation of gene expression; proteins involved in protein metabolism, including both biosynthesis and biodegradation; enzymes involved in carbohydrate, lipid, phytohormone, S-adenosylmethionine, and secondary metabolism; enzymes involved in energy metabolism; and structural and regulatory proteins involved in the formation of the final plant cell shape and its future development. The plant proteome represents an open, dynamic system, which is a product of plant gene expression (genotypic determination) and the following posttranscriptional and posttranslational modifications; however, it actively shapes the plant phenotype in response to alterations in the plant's environment via several modes of plant–environment interaction, including posttranscriptional and posttranslational modifications (alternative splicing, RNA editing, folding and targeting of nascent proteins to different cellular compartments, and PTMs as chemical modifications of amino acid residues), which determine the final biochemical properties of a protein and its final biological function. It has to be kept in mind that to fully understand a protein's biological function in the given plant material, proteomic studies have to focus not only on the investigation of alterations in protein relative abundance but also on the investigation of protein isoforms and PTMs, as well as protein cellular localization and protein–protein interactions. Given all these aspects determining the final biological function of a protein in plant stress response, this enables the plant to finely tune its response to the given environmental stimuli to reach optimum adjustment of plant cellular structures and metabolism to the given ambient conditions.

Recently, a boom in novel technologies of genome editing has opened up new ways of designing protein structure and expression. However, the complexity of protein–protein interactions, as well as those between proteins and other molecules, in shaping the final protein biological function has to be kept in mind while designing altered proteins via genetic manipulation. Investigating the plant proteome response to abiotic stress thus presents a never-ending story, which can uncover to patient researchers the secrets of life, lying in immense adaptability to an ever-changing environment.

ACKNOWLEDGMENTS

The work was supported by Czech Ministry of Agriculture (MZe CR) institutional project MZe-RO0418 and project QK1710302.

REFERENCES

Abat, J. K. and R. Deswal. 2009. Differential modulation of S-nitrosoproteome of *Brassica juncea* by low temperature: change in S-nitrosylation of Rubisco is responsible for the inactivation of its carboxylase activity. *Proteomics* 9:4368–4380.

Aghaei, K., A.A. Ehsanpour, and S. Komatsu. 2008. Proteome analysis of potato under salt stress. *J. Proteome Res.* 7:4858–4868.

Ahsan, N., J. Renaut, and S. Komatsu. 2009. Recent developments in the application of proteomics to the analysis of plant responses to heavy metals. *Proteomics* 9:2602–2621.

Alvarez, S., B.M. Berla, J. Sheffield, R.E. Cahoon, J. M. Jez and L.M. Hicks.2009. Comprehensiveanalysis of the *Brassica juncea* root proteome in response to cadmium exposure bycomplementary proteomic approaches. *Proteomics* 9:2419–2431.

Askari, H., J.Edqvist, M.Hajheidari, M. Kafi, and G.H. Salekdeh. 2006. Effects of salinity levels on proteome of *Suaeda aegyptiaca* leaves. *Proteomics* 6:2542–2554.

Bae, M.S., E.J. Cho, E.Y. Choi, and O.K. Park. 2003. Analysis of the Arabidopsis nuclearproteome and its response to cold stress. *Plant J.* 36:652–663.

Bandehagh, A., G.H. Salekdeh, M. Toorchi, A. Mohammadi, and S. Komatsu. 2011. Comparativeproteomic analysis of canola leaves under salinity stress. *Proteomics*11:1965–1975.

Bastow, R., J.S. Mylne, C. Lister, Z. Lippman, R.A. Martienssen, C. Dean. 2004. Vernalization requires epigenetic silencing of FLC by histone methylation. *Nature* 427:164–167.

Bazargani, M.M.,E. Sarhadi, A.-A.S. Bushehri, A. Matros, H.-P. Mock, M.-R. Naghavi, V. Hajihoseini, M. Mardi, M.-R. Hajirezaei, F. Moradi, *et al.* 2011. A proteomics view on the role of drought-induced senescence and oxidative stress defense in enhanced stem reserves remobilization in wheat. *J. Proteom.* 74:1959–1973.

Bertoni, G. 2011.CBS domain proteins regulater edox homeostasis. *Plant Cell* 23:3562.

Bhushan, D., A. Pandey, M.K. Choudhary, A. Datta, S. Chakraborty, and N. Chakraborty. 2007. Comparative proteomics analysis of differentially expressed proteins in chickpea extracellular matrix during dehydration stress. *Mol. Cell. Proteomics* 6:1868–1884.

Boni, I.V., D.M. Isaeva, M.L. Musychenko, and N.V. Tzareva.1991. Ribosome-messenger recognition: mRNA target sites for ribosomal protein S1. *Nucleic Acids Res.* 19:155–162.

Budak, H., B.A. Akpinar, T. Unver, and M. Turktas. 2013. Proteome changes in wild and modern wheat leave supon drought stress by two-dimensional electrophoresis and nano LC-ESI-MS/MS. *Plant Mol. Biol.* 83:89–103.

Burstin, J., D. de Vienne, P. Dubreuil, and C. Damerval. 1994. Molecular markers and protein quantities as genetic descriptors in maize. I. Genetic diversity among 21 inbred lines. *Theor. Appl. Genet.* 89: 943–950.

Caruso, G., C. Cavaliere, C. Guarino, R. Gubbiotti, P. Foglia, and A. Lagana. 2008. Identification of changes in *Triticum durum* L. leaf proteome in response to salt stress by two-dimensionalelectrophoresis and MALDI-TOF mass spectrometry. *Anal. Bioanal. Chem.* 391:381–390.

Chen, J.H., T.L. Cheng, P.K. Wang, W.D. Liu, J. Xiao, Y.Q. Yang, *et al.* 2012. Salinity-induced changes in protein expression in the halophytic plant *Nitraria sphaerocarpa. J. Proteomics* 75:5226–5243.

Cheng, Y., Y. Qi, Q. Zhu, X. Chen, N. Wang, X. Zhao, *et al.* 2009. New changes in the plasma-membrane-associated proteomeofriceoots under salt stress. *Proteomics* 9:3100–3114.

Cheng, T., J. Chen, J. Zhang, S. Shi, Y. Zhou, L. Lu, *et al.* 2015. Physiological and proteomican alyses of leaves from the halophyte *Tangutnitraria* reveals diverse response pathway scritical for high salinity tolerance. *Front. Plant Sci.* 6:30.

Cheng, L., Y. Wang, Q. He, H. Li, X. Zhang, and F. Zhang. 2016. Comparative proteomics illustrates the complexity of drought resistance mechanisms in two wheat (*Triticum aestivum* L.) cultivars under dehydration and rehydration. *BMC Plant Biol.* 16: 188.

Clement, M., N. Leonhardt, M.J. Droillard, I. Reiter, J.L. Montillet, B. Genty, C. Lauriere, L. Nussaume, and L.D. Noel. 2011. The cytosolic/nuclear HSC70 and HSP90 molecular chaperones are important for stomatal closure and modulate abscisic acid-dependent physiological responses in Arabidopsis. *Plant Physiol.* 156:1481–1492.

Colmer, T.D., T.J. Flowers, and R. Munns. 2006. Use of wild relatives to improve salt tolerance in wheat. *J. Exp. Bot.* 57:1059–1078.

Cui, S., F. Huang, J. Wang, X. Ma, Y. Cheng, and J. Liu. 2005. A proteomic analysis of cold stress responses in rice seedlings. *Proteomics* 5: 3162–3172.

Danyluk, J., N.A. Kane, G. Breton, A.E. Limin, D.B. Fowler, and F. Sarhan. 2003. TaVRT-1, a putative transcription factor associated with vegetative to reproductive transition in cereals. *Plant Physiology* 132:1849–1860.

Degand, H., A.-M. Faber, N. Dauchot, D. Mingeot, B. Watillon, P. van Cutsem, P. Morsomme, and M. Boutry. 2009. Proteomic analysis of chicory root identifies proteins typically involved in cold acclimation. *Proteomics* 9:2903–2907.

Dong, C. H., M. Agarwal, Y. Zhang, Q. Xie, and J. K. Zhu. 2006. The negativeregulator of plant cold responses, HOS1, is a RING E3 ligase that mediatesthe ubiquitination and degradation of ICE1. *Proc. Natl. Acad. Sci. U.S.A.* 103:8281–8286.

Dooki, A.D., F.J. Mayer-Posner, H. Askari, A.A. Zaiee, and G.H. Salekdeh. 2006. Proteomic responsesofriceyoungpaniclestosalinity. *Proteomics* 6:6498–6507.

Durand, T.C., K. Sergeant, S. Planchon, S. Carpin, P. Label, D.Morabito, J.F.Hausman, and J. Renaut. 2010. Acute metal stress in *Populus tremula* x *P. alba* (717–1B4 genotype): Leaf and cambial proteome changes induced by cadmium. *Proteomics* 10:349–368.

Edreva, A. 2005. Pathogenesis-related proteins: research progress in the last 15 years. *Gen. Appl. Plant Physiol.* 31:105–124.

Eggert, K. andE. Pawelzik. 2011. Proteomeanalysis of *Fusarium* head blight in grains of naked barley (*Hordeumvulgare* subsp. *nudum*). *Proteomics* 11:972–985.

Eggert, K., C. Zörb, K.H. Mühling, and E. Pawelzik. 2011.Proteomeanalysis of *Fusarium* infection inemmergrain s(Triticumdicoccum). *Plant Pathol.* 60:918–928.

Faghani, E., J. Gharechahi, S. Komatsu, M. Mirzaei, R.A. Khavarinejad, F. Najafi, L.K. Farsad, and G.H. Salekdeh. 2015. Comparative physiology and proteomic analysis of two wheat genotypes contrasting in drought tolerance. *J. Proteomics* 114:1–15.

Fatehi, F., A. Hosseinzadeh, H. Alizadeh, T. Brimavandi, and P.C. Struik. 2012. The proteome response of salt-resistant and salt-sensitive barley genotypes to long-term salinity stress. *Mol. Biol. Rep.* 39:6387–6397.

Fercha, A., Capriotti, A.L., C aruso, G., Cavaliere, C., Gherroucha, H., Samperi, R., *et al.* 2013. Gel-free proteomics reveals potential biomarkers of priming-induced salt tolerance in durum wheat. *J. Proteomics* 91:486–499.

Ford, K.L., A. Cassin, and A. Bacic.2011. Quantitative proteomic analysis ofwheat cultivars with differing drought stress tolerance. *Front. Plant Sci.* 2:44.

Gao, F., Y. Zhou, W. Zhu, X. Li, L. Fan, and G. Zhang. 2009. Proteomic analysis of cold stress-responsive proteins in *Thellungiella* rosette leaves. *Planta* 230:1033–1046.

Ge, P., C. Ma, S. Wang, L. Gao, X. Li, G. Guo, W. Ma, and Y. Yan. 2012. Comparative proteomic analysis of grain development in two spring wheat varieties under drought stress. *Anal Bioanal Chem.* 402:1297–1313.

Gizak, A. and A. Dzugaj. 2003. FBPase is in the nuclei of cardiomyocytes. *FEBS Lett.* 539:51–55.

Ghabooli, M., B. Khatabi, F.S. Ahmadi, M. Sepehri, M. Mizraei, A. Amirkhani, *et al.* 2013. Proteomicsstudyrevealsthe molecular mechanismsunderlying water stress tolerance induced by *Piriformosporaindica* in barley. *J. Proteomics* 94:289–301.

Ghaffari, M., M. Toorchi, M. Valizadeh, and S. Komatsu. 2013. Differential response of root proteome todrought stress in drought sensitive and tolerant sunflower inbred lines. *Funct. Plant Biol.* 40:609–617.

Gonzalez-Guzman, M., G.A. Pizzio, R. Antoni, F. Vera-Sierera, E. Merilo, G.W. Bassel, M.A. Fernández, M.J. Holdsworth, M.A. Perez-Amador, H. Kollist, and P.L. Rodriguez. 2012. Arabidopsis PYR/PYL/RCAR receptors play a major role in quantitative regulation of stomatal aperture and transcriptional response to abscisic acid. *Plant Cell* 24:2483–2496.

Hajheidari, M., A. Eivazi, B.B. Buchanan, J.H. Wong, I. Majidi, and G.H. Salekdeh. 2007. Proteomics uncovers a role for redox in drought tolerance in wheat. J. Proteome Res. 6:1451–1460.

Halford, N.G., T.Y. Curtis, Z. Chen, and J. Huang. 2015. Effects of abiotic stress and crop management on cereal grain composition: implications for food quality and safety. *J. Exp. Bot.* 66:1145–1156.

Han, Q., Kang, G., and T. Guo. 2013. Proteomic analysis of spring freezestressresponsive proteins in leaves of bread wheat (*Triticum aestivum*L.). *Plant Physiol. Biochem.* 63:236–244.

Hanin, M., F. Brini, C. Ebel, Y. Toda , S. Takeda, and K. Masmoudi. 2011. Plant dehydrins and stress tolerance: Versatile proteins for complex mechanisms. *Plant Signalling Behaviour* 6:10.

Hara, M. 2010. The multifunctionality of dehydrins: An overview. *Plant Signalling Behaviour* 5: 503–508.

Hossain, Z. and S. Komatsu. 2013. Contribution of proteomic studies towards understanding plant heavy metal stress response. *Front. Plant Sci.* 3:310.

Hossain, Z., Nouri, M. Z., and S. Komatsu. 2012. Plant cell organelle proteomics in response to abiotic stress. *J. Proteome Res.* 11:37–48.

Jacoby, R. P., A.H. Millar, and N.L. Taylor. 2010. Wheat mitochondrial proteomes provide new links between antioxidant defense and plant salinity tolerance. *J. Proteome Res.* 9:6595–6604.

Jain, S., S. Srivastava, N.B. Sarin, and N.N.V. Kav. 2006. Proteomics reveals elevated levels of PR10 proteins in saline-tolerant peanut (*Arachis hypogaea*) calli. *Plant Physiol. Biochem.* 44:253–259.

Jang, C.S., H.J. Lee, S .J. Chang, and Y.W. Seo. 2004. Expression and promoter analysis of the TaLTP1gene induced by drought and salt stress in wheat (*Triticumaestivum* L.). *Plant Sci.* 167:995–1001.

Janmohammadi, M., L. Zolla, and S. Rinalducci. (2015). Low temperature tolerance in plants: changes at the protein level. *Phytochemistry* 117:76–89.

Janská, A., A. Aprile, J. Zámečník, L. Cattivelli, and J. Ovesná.2011.Transcriptional responses of winter barley to cold indicate nucleosomeremodelling as a specific feature of crown tissues. *Funct. Integr. Genomics* 11:307–325.

Jiang, Y., B. Yang, N.S. Harris, and M.K. Deyholos. 2007. Comparative proteomic analysis of NaClstress-responsive proteins in Arabidopsis roots. *J. Exp. Bot.* 58:3591–3607.

Johnová, P., J. Skalák, I. Saiz-Fernández, and B. Brzobohatý. 2016. Plant responses to ambient temperature fluctuations and water-limiting conditions: a proteome-wide perspective. *Biochim. Biophys. Acta Proteins Proteomics* 1864:916–931.

Jorrin-Novo, J.V., A.M.Maldonado, S. Echevarria-Zomeno, L. Valledor, M.A. Castillejo, M. Curto, J. Valero, B. Sghaeir, G. Donoso, and I. Redondo. 2009. Plant proteomics update

(20072008): Second-generation proteomic techniques, an appropriate experimental design, and data analysis to fulfill MIAPE standards, increase plant proteome coverage and expand biological knowledge. *J. Proteomics* 72:285–314.

Kang, G., G. Li, W. Xu, X. Peng, Q. Han, Y. Zhu, *et al.* 2012. Proteomics reveals the effects of salicylicacidon growth and tolerance to subsequent drought stress in wheat. *J. Proteome Res.* 11:6066–6079.

Kausar, R., M. Arshad, A. Shahzad, and S. Komatsu. 2013. Proteomicsanalysis of sensitive and tolerant barley genotypes under drought stress. *Amino Acids* 44:345–359.

Kav, N.N.V., S. Srivastava, L.Goonewardene, and S.F. Blade. 2004. Proteome level changes in the roots of *Pisum sativum* in response to salinity. *Annals of Applied Biology* 145:217–230.

Ke, Y.Q., G.Q. Han, H.Q. He, and J.X. Li. 2009. Differential regulation of proteins and phospho proteins in rice under drought stress. *Biochim. Biophys. Res. Commun.* 379:133–138.

Komatsu, S., A. Yamamoto, T. Nakamura, M.Z. Nouri, Y. Nanjo, K. Nishizawa, *et al.* 2011. Comprehensive analysis of mitochondria in roots and hypocotylsof soybean under flooding stress using proteomics and metabolomics techniques. *J. Proteome Res.* 10:3993–4004.

Komatsu, S., A.H.M. Kamal, and Z. Hossain. 2014. Wheat proteomics: proteome modulation and abiotic stress acclimation. *Front. PlantSci.* 5:684.

Kosová, K., P. Vítámvás, and I.T. Prášil. 2007. The role of dehydrins in plant response to cold. *Biol. Plant.* 51:601–617.

Kosová, K., P.Vítámvás, I.T. Prášil, and J. Renaut. 2011. Plant proteomechanges under abiotic stress—Contribution of proteomics studiesto understanding plant stress response. *J. Proteomics* 74:1301–1322.

Kosová, K., P. Vítámvás, S. Planchon, J. Renaut, R. Vanková, and I.T. Prášil. 2013a. Proteome analysis of cold response in spring and winterwheat (*Triticum aestivum*) crowns reveals similarities in stress adaptation and differences in regulatory processes between the growth habits. *J. Proteome Res.*12:4830–4845.

Kosová, K., P. Vítámvás, P. Prášilová, and I.T. Prášil. 2013b. Accumulation of WCS120 and DHN5 proteins in differently frost-tolerant wheat and barley cultivars grown under a broad temperature scale. *Biol. Plant.* 57:105–112.

Kosová, K., P. Vítámvás, M.O. Urban, and I.T. Prášil. 2013c. Plantproteome responses to salinity stress—Comparison of glycophytes and halophytes. *Funct. Plant Biol.* 40:775–786.

Kosová, K., P. Vítámvás, and I.T. Prášil. 2014a. Wheat and barley dehydrins under cold, drought, and salinity what can LEA-II proteins tell us about plant stress response? *Front. Plant Sci.* 5:343.

Kosová, K., P. Vítámvás, and I.T. Prášil. 2014b. Proteomics of stress responses in wheat and barley—search for potential protein markers of stress tolerance. *Front. Plant Sci.* 5:711.

Kosová K., P. Vítámvás, M.O. Urban, M. Klíma, A. Roy, and I.T. Prášil. 2015. Biological networks underlying abiotic stress tolerance in temperate crops—a proteomic perspective. *Int. J. Mol. Sci.* 16:20913–20942.

Kosová, K., M.O. Urban, P. Vítámvás, and I.T. Prášil. 2016. Drought stress response in common wheat, durum wheat, and barley: Transcriptomics, Proteomics, metabolomics, Physiology and breeding for an enhanced drought tolerance. In: Hossain, M.A. (ed.): *Drought Stress Tolerance in Plants*. Volume 2. pp. 277–314. Springer, Switzerland.

Kosová, K., P. Vítámvás, M.O. Urban, I.T. Prášil, and J. Renaut. 2018. Plant abiotic stress proteomics: The major factors determining alterations in cellular proteome. *Front. Plant Sci.* 9:122.

Kwak, K. J., Y. O. Kim, and H.S. Kang. 2005. Characterization of transgenic Arabidopsis plants overexpressing GR-RBP4 under high salinity, dehydration, or cold stress. *J. Exp. Bot.* 56:3007–3016.

Laloum, T., G. Martín, and P. Duque. 2018. Alternative splicing control of abiotic stress responses. *Trends Plant Sci.* 23: 140–150.

Lawlor, D.W. 2013. Genetic engineering to improve plant performanceunder drought: physiological evaluation of achievements, limitations, and possibilities. *J. Exp. Bot.* 64:83–108.

Lee, H., Y. Guo, M. Ohta, L.M. Xiong, B. Stevenson, and J.K. Zhu. 2002. *LOS2*, a genetic locus required for cold-responsive gene transcription encodes a bi-functional enolase. *EMBO J.* 21:2692–2702.

Levitt, J. 1980. Responses of Plants to Environmental Stresses: Chilling, Freezing and High Temperature Stresses. New York, NY: Academic Publisher.

Li, G., X. Peng, H. Xuan, L. Wei, Y. Yang, T. Guo, and G. Kang. 2013. Proteomic analysis of leaves and roots of common wheat (*Triticum aestivum* L.) under copper-stress conditions. *J. Proteome Res.12*: 4846–4861.

Li, X., J. Cai, F. Liu, T. Dai, W. Cao, and D. Jiang. 2014. Physiological, proteomic and transcriptional responses of wheat to combination of drought or waterlogging with late spring low temperature. *Funct. Plant Biol.* 41:690–703.

Li, Q., B. Byrns, M.A. Badawi, A.B. Diallo, J. Danyluk, F. Sarhan, D. Laudencia-Chingcuanco, J. Zou, and D.B. Fowler. 2018. Transcriptomic insights into phenological development and cold tolerance of wheat grown in the field. *Plant Physiol.* 176:2376–2394.

Ma, Y. and L.M. Hendershot. 2004. ER chaperone functions during normal and stress conditions. *J. Chem. Neuroanatomy* 28:51–65.

Ma, Y., X. Dai, Y. Xu, W. Luo, X. Zheng, D. Zeng, Y. Pan, X. Lin, H. Liu, D. Zhang, *et al.* 2015. COLD1 confers chilling tolerance in rice. *Cell* 160:1209–1221.

Majoul, T., E. Bancel, E. Triboï, J. Ben Hamida, and G. Branlard. 2004. Proteomic analysis of the effect of heat stress on hexaploid wheat grain: characterization of heat-responsive proteins from non-prolamins fraction. *Proteomics* 4:505–513.

Manaa, A., H. Ben Ahmed, B. Valot, J.-P. Bouchet, S. Aschi-Smiti, M. Causse, *et al.* 2011. Salt and genotype impact on plant physiology and root proteome variations in tomato. *J. Exp. Bot.* 62:2797–2813.

Maršálová, L.,P. Vítámvás, R.Hynek, I. T.Prášil, and K. Kosová. 2016. Proteomic response of *Hordeum vulgare* cv. Tadmor and *Hordeum marinum* to salinity stress: similarities and differences between a glycophyte and a halophyte. *Front. Plant Sci.* 7:1154.

Mostek, A., A. Börner, A. Badowiec, and S. Weidner. 2015. Alterations in rootproteome of salt-sensitive and tolerant barley lines under salt stress conditions.*J. Plant Physiol.* 174:166–176.

Munns, R.2005. Genes and salt tolerance: bringing them together. Tansley Review. *New Phytologist* 167:645–663.

Munns, R. and M. Tester. 2008. Mechanisms of salinity tolerance. *Annu. Rev. Plant Biol.*59:651–681.

Murata, N. and D.A. Los. 1997. Membrane fluidity and temperature perception. *Plant Physiol.* 115:875–879.

Nuruzzaman, M., A.M. Sharoni, and S. Kikuchi. 2013. Roles of NAC transcription factors in the regulation of biotic and abiotic stress responses in plants. *Front. Microbiol.* 4:248.

Oh, M., and S. Komatsu. 2015. Characterization of proteins in soybean roots under flooding and droughtstresses. *J. Proteomics*114:161–181.

Pandey, A., Chakraborty, S., Datta, A., and Chakraborty, N. 2008. Proteomics approach to identify dehydration responsive nuclear proteinsfrom chickpea (*Cicer arietinum* L.). *Mol. Cell. Proteomics* 7:88–107.

Pang, Q., Chen, S., Dai, S., Chen, Y., Wang, Y., and Yan, X. 2010. Comparativeproteomics of salt tolerance in *Arabidopsis thaliana* and *Thellungiella halophila*. *J. Proteome Res.* 9:2584–2599.

Peng, Z., M. Wang, F. Li, H. Lv, C. Li, and G. Xia. 2009. Aproteomic study of the response to salinity and drought stress in anintrogression strain of bread wheat. *Mol. Cell. Proteomics* 8:2676–2686.

Quesada, V., C. Dean, and G.G. Simpson. 2005. Regulated RNA processing in the control of Arabidopsis flowering. *Int. J. Dev. Biol.* 49:773–780.

Raorane, M.L., I.M. Pabuayon, B. Miro, R. Kalladan, M. Reza-Hajirezai, R.H. Oane, A. Kumar, N. Sreenivasulu, A. Henry, and A. Kohli. 2015. Variation in primary metabolites in parental and near-isogenic lines of the QTL qDTY 12.1: altered roots and flag leaves but similar spikelets of rice under drought. *Mol. Breeding* 35:138.

Riccardi, F., P. Gazeau, M.P. Jacquemot, D. Vincent, and M. Zivy. 2004. Deciphering genetic variations of proteome responses to water deficit in maize leaves. *Plant Physiol. Biochem.* 42:1003–1011.

Rinalducci, S., M.G. Egidi, S. Mahfoozi, S.J. Godehkahriz, and L. Zolla. 2011. The influence of temperature on plant development in a vernalization requiring winter wheat: a 2-DE based proteomic investigation. *J. Proteomics* 74:643–659.

Ristic, Z., U. Bukovnik, I. Momčilović, J. Fu, and P.V. Prasad. 2008. Heat-induced accumulation of chloroplast protein synthesis elongation factor, EF-Tu, in winter wheat. *J. Plant Physiol.* 165:192–202.

Rollins, J.A., E. Habte, S.E. Templer, T. Colby, J. Schmidt, and M. von Korff. 2013. Leaf proteome alterations in the context of physiological and morphological responses to drought and heat stress in barley (*Hordeum vulgare* L.). *J. Exp. Bot.* 64:3201–3212.

Ronai, Z., R. Robinson, S. Rutberg, P. Lazarus, and M. Sardana. 1992. Aldolase DNA interactions in a SEWA cell system. *Biochim. Biophys. Acta* 1130:20–28.

Rorat, T. 2006. Plant dehydrins tissue location, structure and function. *Cell. Mol. Biol. Letters* 11:536–556.

Sauter, M., G. Rewuski, T. Marwedel, and R. Lorbiecke. 2002. The novel ethylene-regulated gene *OsUsp1* from rice encodes a member of a plant protein family related to prokaryotic universal stress proteins. *J. Exp. Bot.* 53:2325–2331.

Serrano, I., L. Campos, and S. Rivas. 2018. Roles of E3 ubiquitin-ligases in nuclear protein homeostasis during plant stress responses. *Front. Plant Sci.* 9:139.

Shi, H., M. Ishitani, C. Kim, and J.K. Zhu. 2000. The *Arabidopsis thaliana* salt tolerance gene SOS1 encodes a positive Na+/H+ antiporter. *PNAS* 97:6896–6901.

Shi, H., F.J. Quintero, J.M. Pardo, and J.K. Zhu. 2002. The putative plasma membrane Na+/H+ antiporter SOS1 controls lon distance Na+ transport in plants. *Plant Cell* 14:465–477.

Singh, R., and M. R. Green.1993. Sequence-specific binding of transfer RNA by glyceraldehyde-3-phosphate dehydrogenase. *Science* 259:365–368.

Skylas, D.J., S.J. Cordwell, P.G. Hains, M.R. Larsen, D.J. Basseal, B.J. Walsh, *et al.* 2002. Heat shock of wheat during grain filling: proteins associated withheat-tolerance. *J. Cereal Sci.* 35:175–188.

Sobhanian, H., N. Motamed, F.R. Jazii, T. Nakamura, and S. Komatsu. 2010. Salt stress induceddifferential proteome and metabolome response in the shoots of *Aeluropus lagopoides* (*Poaceae*), a halophyte C4 plant. *J. Proteome Res.* 9:2882–2897.

Sovová, T., G. Kerrins, K. Demnerová, and J. Ovesná. 2017. Genome editing with engineered nucleases in economically important animals and plants: State of the art in the research pipeline. *Curr. Issues Mol. Biol.* 21:41–62.

Streitner, C.,S. Danisman, F. Wehrle, J.C. Schöning, J.R. Alfano, and D. Staiger. 2008. The small glycine-rich RNA binding protein AtGRP7 promotes floral transition in *Arabidopsis thaliana*. *Plant J.* 56:239–250.

Sung, S.and R. Amasino. 2005. Remembering winter. toward a molecular understanding ofvernalization. *Annu. Rev. Plant Biol.* 56:491–508.

Suzuki, I., D.A. Los, Y. Kanesaki, K. Mikami, and N. Murata. 2000. The pathway for perceptionand transduction of low-temperature signals in *Synechocystis*. *EMBO J.* 19:1327–1334.

Tamburino, R., M. Vitale, A. Ruggiero, M. Sassi, L. Sannino, S. Arena, *et al.* 2017. Chloroplast proteome response to drought stress andrecovery in tomato (*Solanum lycopersicum* L.). *BMC Plant Biol.* 17:40.

Tanou, G., C. Job, L. Rajjou, E. Arc, M. Belghazi, G. Diamantidis, A. Molassiotis, and D. Job 2009. Proteomics reveals the overlapping roles of hydrogen peroxide and nitric oxide in the acclimation of citrus plants to salinity. *Plant J.* 60:795–804.

Tardif, G., N.A. Kane, H. Adam, L. Labrie, G. Major, P. Gulick, *et al.* 2007. Interaction network of proteins associated with abiotic stress response and development in wheat. *Plant Mol. Biol.*63:703–718.

Taylor, N.L., J.L. Heazlewood, D.A. Day, and A.H. Millar. 2005. Differential impact of environmental stresses on the pea mitochondrial proteome. *Mol. Cell. Proteomics* 4:1122–1133.

Thompson, J.E., M.T. Hopkins, C. Taylor, and T.W. Wang. 2004. Regulation of senescence by eukaryotic translation initiation factor 5A: implications for plant growth and development. *Trends Plant Sci.* 9:174–179.

Urban, M.O., J. Vašek, M. Klíma, J. Krtková, K. Kosová, I.T. Prášil, and P. Vítámvás. 2017.Proteomic and physiological approach reveals drought-induced changes inrapeseeds: water-saver and water-spender strategy. *J. Proteomics* 152:188–205.

Vincent, D.,A. Ergul, M.C. Bohlman, E.A.R. Tattersall, R.L. Tillett, M.D. Wheatley, R. Woolsey, D.R. Quilici, J. Joets, K. Schlauch, *et al.* 2007. Proteomic analysis reveals differencesbetween *Vitis vinifera* L. cv. Chardonnay and cv. Cabernet Sauvignon and their responses towater deficit and salinity. *J. Exp. Bot.* 58:1873–1892.

Vítámvás, P. and I.T. Prášil. 2008. WCS120 protein family and frost tolerance during cold acclimation, deacclimation and reacclimation of winter wheat. *Plant Physiol. Biochem.*46:970–976.

Vítámvás P., Kosová K., Prášilová P., and Prášil I.T. 2010. Accumulation of WCS120 protein in wheat cultivars grown at 9°C or 17°C in relation to their winter survival. *Plant Breeding* 129:611–616.

Vítámvás, P., I.T. Prášil, K. Kosová, S. Planchon, and J. Renaut. 2012.Analysis of proteome and frost tolerance in chromosome 5A and 5Breciprocal substitution lines between two winter wheats during longtermcold acclimation. *Proteomics* 12:68–85.

Vítámvás, P., M.O. Urban, Z. Škodáček, K. Kosová, I. Pitelková, J. Vítámvás,J. Renaut,and I.T. Prášil. 2015. Quantitative analysis of proteome extracted from barleycrowns grown under different drought conditions. *Front. Plant Sci.* 6:479.

Vu, L.D., T. Zhu, I. Verstraeten, B. van de Cotte, IWGSCK. Gevaert, and I. De Smet. 2018. Temperature-induced changes in the wheat phosphoproteome reveal temperature-regulated interconversion of phosphoforms. *J. Exp. Bot.* 69:4609–4624.

Wakeel, A., A.R. Asif, B. Pitann, and S. Schubert. 2011. Proteome analysis of sugar beet (*Beta vulgaris* L.) elucidates constitutive adaptation during the first phase of salt stress. *J. Plant Physiol.* 168:519–526.

Wang, M. C., Z.Y. Peng, C.L. Li, F. Li, C. Liu, and G.M. Xia. 2008. Proteomic analysis on a high salt tolerance introgression strain of *Triticum aestivum/Thinopyrum ponticum*. *Proteomics* 8:1470–1489.

Wang, X., M. Vignjevic, D. Jiang, S. Jacobsen, and B. Wollenweber. 2014a. Improved tolerance to drought stress after anthesis due to priming before anthesis in wheat (*Triticum aestivum* L.) var. Vinjett. *J. Exp. Bot.* 65:6441–6456.

Wang, L., X. Liu, M. Liang, F. Tan, W. Liang, Y. Chen, YLin, L. Huang, J. Xing, and W. Chen. 2014b. Proteomic analysis of salt-responsive proteins in the leaves of mangrove *Kandelia candel* during short-term stress. *PLoS One*9:e83141.

Wang, X., M. W. Oh, and S. Komatsu. 2016. Characterization of S-adenosylmethionine synthetases in soybean under flooding and drought stresses. *Biol. Plant.* 60:269–278.

Wasinger, V.C., S.J. Cordwell, A. Cerpa-Poljak, J.X. Yan, A.A. Gooley, M.R. Wilkins, M.W. Duncan, R. Harris, K.L. Williams, and J. Humphrey-Smith. 1995. Progress with gene-product mapping of the Mollicutes: *Mycoplasma genitalium*. *Electrophoresis* 16:1090–1094.

Wendelboe-Nelson, C., and P.C. Morris. 2012. Proteins linked to drought tolerance revealed by DIGE analysis of drought resistant and susceptible barley varieties. *Proteomics* 12:3374–3385.

Witzel, K., A. Weidner, G.K. Surabhi, A. Börner, and H.P. Mock. 2009. Salt stress-induced alterations in the root proteome of barley genotypes with contrasting response towards salinity. *J. Exp. Bot.* 60: 3545–3557.

Wu, X., F. Gong, D. Cao, X. Hu, and W. Wang.2016. Advances in cropproteomics: PTMs of proteins under abiotic stress. *Proteomics* 16:847–865.

Xu, C.,T. Sibicky, and B. Huang. 2010. Protein profile analysis of salt-responsive proteins in leaves and roots in two cultivars of creeping bentgrass differing in salinity tolerance. *Plant Cell Rep.* 29:595–615.

Xu, Y.-H., R. Liu, L. Yan, Z.-Q. Liu, S.-C. Jiang, Y.-Y. Shen, X.-F. Wang, and D.-P. Zhang. 2012. Light-harvesting chlorophyll a/b-binding proteins are required for stomatal responseto abscisic acid in Arabidopsis, *J. Exp. Bot.* 63:1095–1106.

Yamaguchi-Shinozaki, K., and K. Shinozaki. 2006. Transcriptional regulatory networks in cellular responses and tolerance to dehydration and cold stresses. *Annu. Rev. Plant Biol.* 57:781–803.

Yan, S., Z. Tang, W. Su, and W. Sun. 2005. Proteomic analysis of salt stress responsive proteins in rice root. *Proteomics* 5:235–244.

Yang, F., J.D. Jensen, N.H. Spliid, B. Svensson, S. Jacobsen, L.N. Jørgensen, *et al.* 2010. Investigation of the effect of nitrogen on severity of *Fusarium* Head Blight in barley. *J. Proteomics* 73:743–752.

Yang, F., A.D. Jørgensen, H. Li, I. Søndergaard, C. Finnie, B. Svensson, D. Jiang, B. Wollenweber, and S. Jacobsen. 2011. Implications of high-temperature events and water deficits onprotein profiles in wheat (*Triticum aestivum* L. cv. Vinjett) grain. *Proteomics* 11:1684–1695.

Yin, X. and S. Komatsu. 2016. Nuclear proteomics reveals the role of protein synthesis and chromatin structure in root tip of soybean during the initial stage of flooding stress. *J. Proteome Res.* 15:2283–2298.

Yin, X.J. and S. Komatsu. 2017. Comprehensive analysis of response and tolerant mechanisms in early-stage soybean at initial-flooding stress. *J. Proteomics* 169:225–232.

Yu, J., S. Chen, Q. Zhao, T. Wang, C. Yang, C. Diaz, G. Sun, and S. Dai. 2011. Physiological andproteomic analysis of salinity tolerance in *Puccinellia tenuiflora*. *J. Proteome Res.* 10:3852–3870.

Yu, F., X. Han, C. Geng, Y. Zhao, Z. Zhang, and F. Qiu. 2015. Comparative proteomic analysisrevealing the complex network associated with waterlogging stress in maize (*Zea mays* L.)seedling root cells. *Proteomics* 15:135–147.

Zadražnik, T., K. Hollung, W. Egge-Jacobsen, V. Meglič, and J. Šuštar-Vozlič. 2013. Differential proteomic analysis of drought stress response in leaves of common bean (*Phaseolus vulgaris* L.). *J. Proteomics* 78: 254–272.

Zhan, X., J.K. Zhu, and L. Lang. 2015. Increasing freezing tolerance: kinase regulation of ICE1.*Developmental Cell* 32:257–258.

Zhang, M., D. Lv, P. Ge, Y. Bian, G. Chen, G. Zhu, X. Li, and Y. Yan. 2014. Phosphoproteome analysis reveals new drought response and defense mechanisms of seedling leaves in bread wheat (*Triticum aestivum* L.). *J. Proteomics* 109:290–308.

Zhang, N., W. Huo, L. Zhang, F. Chen, and D. Cui. 2016. Identification of winter-responsive proteins in bread wheat using proteomics analysis and virus-induced gene silencing (VIGS). *Mol. Cell. Proteomics* 15:2954–2969.

Zhu, J.K. 2000. Genetic analysis of plant salt tolerance using Arabidopsis. *Plant Physiol.* 124:941–948.

Zhu, J.K. 2002. Salt and drought stress signal transduction in plants. *Annu. Rev. Plant Biol.* 53:247–273.

Zhu, J.K. 2016. Abiotic stress signaling and responses in plants. *Cell* 167:313–324.

Zhu, J., S.Alvarez, E. L.Marsh, M. E. LeNoble, I.-J. Cho, M.Sivaguru, *et al.* 2007. Cell wall proteome in the maize primary root elongation zone. II. Region-specific changes in water soluble and lightly ionically bound proteins under water deficit. *Plant Physiol.* 145:1533–1548.

Zivy, M., S. Wienkoop, J. Renaut, C. Pinheiro, E. Goulas, and S. Carpentier. 2015. The quest for tolerant varieties: the importance of integrating "omics" techniques to phenotyping. *Front. Plant Sci.* 6:448.

Zörb, C., R. Herbst, C. Forreiter, and S. Schubert. 2009. Short-term effects of salt exposure on the maize chloroplast protein pattern. *Proteomics* 9:4209–4220.

Section III

Plants and Crops Responses: Physiology, Cellular and Molecular Biology, Microbiological Aspects, and Whole Plant Responses under Salt, Drought, Heat, Cold Temperature, Light, Nutrients, and Other Stressful Conditions

13 Responses of Photosynthetic Apparatus to Salt Stress
Structure, Function, and Protection

M. Stefanov, A.K. Biswal, M. Misra, A. N. Misra, and E.L. Apostolova

CONTENTS

13.1 INTRODUCTION

In recent years, extensive studies have been conducted to understand the effects and mechanism of tolerance of plants to environmental stress. The external abiotic and biotic factors that limit the rate of photosynthesis and reduce the ability of plants to convert energy to biomass, thereby reducing the growth development and ultimate yield, are defined as stress (Parihar et al., 2015). Plants are continuously exposed to a broad range of environmental stress factors, such as salinity (salt stress), drought, UV radiation, light, flooding, and temperature, which alter their physiological and biochemical processes (Misra et al., 2001a, 2001b, 2001c, 2002, 2012a, 2012b, 2014; Chaitanya et al., 2014; Suzuki et al., 2014; Misra and Misra, 2018a, 2018b). Decreased yields, as a result of the action of abiotic stress factors, the increasing human population, and the reduction in agricultural land, are leading to alarming predictions of depletion of food resources, and people are looking for new strategies that can guarantee the supply of food (Misra et al., 2002; Rasool et al., 2013; Shahbaz and Ashraf, 2013; Shrivastava and Kumar, 2015). The beginning of the twenty-first century is marked by global scarcity of water resources, environmental pollution, and increased salinization of soil and water. High salinity levels can lead to changes in soil properties, which negatively affect the environment, soil fertility, and agricultural production and result in serious harm to human health (Misra et al., 1995, 2001d; Brevik et al., 2015; Daliakopoulos et al., 2016; Hachicha et al., 2018). Furthermore, soil salinization leads to the alteration or even disruption of the earth's natural biological, biochemical, hydrological, and erosional cycles (Daliakopoulos et al., 2016).

Nearly 10% of the world's irrigated area (34.19×10^6 ha) (Mateo-Sagasta and Burke, 2011) and about 7–8% of the total land area (932.2×10^6 ha) (Daliakopoulos et al., 2016) is affected by salinization. It has been suggested that more than 50% of agricultural land will be affected by salinization by 2050, which will affect the food needs of the world's population (Rasool et al., 2013; Fahad et al., 2015; Shrivastava and Kumar, 2015). Salinity limits plant growth and development, causes physiological and metabolic imbalance in plants, and thus, decreases crop yields (Misra et al., 1990, 1995, 1996, 1997a, 2001a, 2001b; Krasensky and Jonak, 2012; Chaitanya et al., 2014). Salinization leads to destruction of enzyme structure, membrane integrity, and cell metabolism (Shokri-Garelo and Noparvar, 2018; Tanveer et al., 2018), which negatively affects germination, growth, flowering, photosynthesis, and respiration (Misra et al., 1990, 1995, 1996, 1997a, 2001a, 2001b; Ghosh et al., 2016; Stefanov et al., 2016a, 2016b, 2018a, 2018b, 2018c; Wungrampha et al., 2018).

The deleterious effects of salinization on plants are accompanied by osmotic stress, nutritional imbalance (Attia et al., 2011), ionic stress (Misra et al., 1990, 2001a, 2002; Munns and Tester, 2008), or a combination of these factors (Hapani and Marjadi, 2015). First, plants are subjected to a quick shock of osmotic stress, resulting in a physiological drought condition. Second, ions such as Na^+ and Cl^- cause ionic imbalance and further hinder the uptake of minerals such as K^+, Ca^{2+}, and Mn^{2+} (Misra et al., 2002; Ahmadizadeh et al., 2016; Nongpiur et al., 2016).

This chapter is focused on the effects of salt stress on the light reactions of photosynthesis and some defense mechanisms, such as osmoprotectants and the antioxidant defense system.

13.2 ORGANIZATION OF THE PHOTOSYNTHETIC APPARATUS

The light reactions of photosynthesis in oxygenic organisms such as higher plants, green algae, and cyanobacteria are mediated by four multi-subunit protein complexes: photosystem I (PSI), photosystem II (PSII), cytochrome b_6f complex, and ATPase (Joshi et al., 2013; Apostolova and Misra, 2014). These complexes are embedded in the thylakoid membrane of chloroplasts and carry out the conversion of light energy into usable chemical energy (Nevo et al., 2012). Both photosystems (PSII and PSI) are membrane-associated supercomplexes composed of a core complex and a peripheral antenna system, which is important for light harvesting. The stability of the photosynthetic apparatus is very important for the efficiency of photosynthesis (Karapetyan, 2004).

The light-driven reaction carried out in the PSII complex leads to the photolysis of water, which releases oxygen, electrons, and protons. The function of PSII is associated with charge separation across the thylakoid membranes (Misra, 2001b, 2001c; Govindjee, 2006; Joshi et al., 2013; Apostolova and Misra, 2014; Misra, 2018b), and this photosystem works together with the other main protein complexes, such as PSI and cytochrome b_6f complex, to perform the primary light reactions (Kouřil et al., 2012). The PSII supercomplex is composed of a PSII core and an outer light-harvesting complex system. The PSII core complex contains 20 different subunits, pigments, and lipids, many of which are evolutionarily conserved between cyanobacteria and plants, while the highly diverse antenna complex is found in the periphery of PSII (Watanade et al., 2009; Croce and Amerongen, 2011). The PSII reaction center is composed of the D1 and D2 polypeptide heterodimer, where the primary charge separation occurs. Closely associated with the reaction center are the inner antenna proteins CP43 and CP47 and a range of small hydrophobic polypeptides, which together comprise the PSII core complex (Büchel et al., 1999). To this complex belong the polypeptides of the oxygen-evolving complex (16, 23, and 33 kDa), which are luminal extrinsic subunits associated with oxygen evolution (Stoylova et al., 2000). The 33 kDa peripheral protein is called the *Mn-stabilizing protein*; it directly participates in the oxidation of water and also plays a role in the binding of Ca^{2+} and Cl^- ions to the Mn cluster (Semchonok, 2016). There is evidence that removal of

CP43 affects the photoreduction of Q_A; i.e., this protein stabilizes the Q_A binding site via the D2 subunit. In addition, CP_{47} interacts with three external PSII proteins (PsbO, P, and Q), which optimize the oxygen-evolving process, and thus, helps them to better bind to the PSII complex (Luciriski and Jackowski, 2006). Büchel et al. (1999) have shown that CP43 is not required for the water oxidase activity of PSII, although mutational studies demonstrate its involvement in normal PSII assembly and function *in vivo*. The authors suggest that CP47 plays an important role in the assembly and the function of the Mn clusters of the oxygen evolving complex. Recent investigations reveal that CP43 participates in the binding of the Mn_4Ca cluster and peripheral proteins of the oxygen evolving complex, and the detachment of this pigment–protein complex is responsible for the changes in the donor side of PSII (Laczkó-Dobos et al., 2008).

PSI (or plastocyanine:ferredoxin oxidoreductase) is a multisubunit pigment–protein complex, which catalyzes the light-dependent electron transport from the plastocyanine to the ferredoxin (Semchonok, 2016). This photosystem contains approximately 13 protein subunits, 96 chlorophyll molecules, 22 carotenoids, three [4Fe-4S] clusters, two philoquinones, and four lipid molecules (Ozakca, 2013; Mamedov et al., 2015). The core complex of PSI is composed of a heterodimer (PsaA and PsaB), to which a wide range of cofactors related to the light capture and the light harvesting as well as cofactors participating in the electron transport are linked (Jensen et al., 2007; Caffarri et al., 2014).

It has been found that PSII and PSI are heterogeneous in their structure and functions. On the basis of the antenna size, PSII has been classified into three forms (PSIIα, PSIIβ, and PSIIγ centers) (Mehta et al., 2010). The major part of the PSII complexes (PS2α) are located in the appressed grana regions of thylakoid membranes, while a smaller fraction of the PSII centers (PSIIβ and PSIIγ centers) is located in the stromal lamellae (Jajoo, 2014). PSIIγ has the smallest antenna system and at the same time has the longest lifetime in comparison with PSIIα and PSIIβ (Jajoo, 2014). The PSIIβ and PSIIγ centers are characterized by delayed photosynthetic electron transport due to the small and weak electron donor systems (Jajoo, 2014). The PSII centers possess heterogeneity depending on their ability to reduce the plastoquinone pool, and they are divided into so-called Q_B-reducing (active) or Q_B-nonreducing (inactive) centers (Jajoo, 2014). There are two types of PSI: PSIα and PSIβ. The population of PSIα is found in the unstacked areas of the grana structures, functionally associated with the PSIIα complexes, while the PSIβ is located in the stromal lamellae. Both types of PSI differ in the antenna size and the electron transport properties. According to the model presented by Albertsson (1995), PSIα is involved in the linear electron transport from water to $NADP^+$ and PSIβ in the cyclic electron transport.

13.3 EFFECTS OF SALT STRESS ON PLANTS

Salt stress strongly affects photosynthesis, which is the most fundamental and intricate physiological process in all green plants (Misra et al., 1990; Ashraf and Harris, 2013; Kalaji et al., 2014, 2018; Misra, 2018a). Having in mind that salt-induced changes in photosynthesis are directly related to total plant yield, it is very important to understand how this process is affected by high salt concentrations. Salt stress inflicts either short- or long-term effects on photosynthesis (Jajoo, 2014). The short-term effects (within minutes to hours) of salinity on this process are a result of stomatal limitation, which leads to a reduction in carbon assimilation, while the long-term effects result from the accumulation of salts in young leaves (Acosta-Motos et al., 2017). The short-term stress is partly related to the negative osmotic effect of salts on the capacity of roots to absorb soil water (Munns and Tester, 2008; Yan et al., 2015; Acosta-Motos et al., 2017). Abscisic acid, which accumulates in the stems and roots, is involved in the control and the regulation of stomatal opening/closing under salt stress (Chaves et al., 2009; Gupta and Huang, 2014; Wugrampha et al., 2018). Abscisic acid causes the activation of Ca^{2+} channels, the suppression of K^+ channels, and the modulation of the cytosolic pH of the guard cells (Wang and Song, 2008; Wugrampha et al., 2018). Stomatal closure minimizes the transpiration processes and the gas exchange and has a negative effect on the capture of light by the photosynthetic apparatus (Iyengar and Reddy, 1996). As a result of the stomatal closure, there is an imbalance between the light capture and the energy use, which leads to a disturbance in all photosynthetic processes (Silveira and Carvalho, 2016; Sun et al., 2016). A decrease in the carbon assimilation rate (Hniličkova et al., 2017) has been shown as a result of the inhibition of enzymes involved in carbon reduction (Misra et al., 1990; Ashraf and Harris, 2013; Kalaji et al., 2017, 2018).

In most cases, the negative effects of salinity are attributed to an increase in Na^+ and Cl^- ions in the plants, influencing different plant functions. A high salt concentration in the soil or in the irrigation water can also have a devastating effect on plant metabolism, disrupting the cellular homeostasis and uncoupling major physiological and biochemical processes (Misra et al., 1990, 2001a; Hasanuzzaman et al., 2012, 2013). The root uptake of Na^+ and Cl^- ions depends on the plant growth stage, the genetic characteristics, and environmental factors such as the temperature, the relative humidity, and the light intensity. In addition, the salt-induced changes in plants depend on the concentration and the duration of exposure to salts, the plant genotype, and environmental factors (Misra et al., 1990, 1997a, 1997b, 1999; Hasanuzzaman et al., 2013).

The entry of Na^+ and Cl^- ions into the transpiration stream and their accumulation in actively transpirational leaves induces ionic toxicity due to the nutritional and mineral imbalance in the cell cytosol (Munns, 2005; Parihar et al., 2015; Yan et al., 2015; Acosta-Motos et al., 2017). Excess Na^+ and Cl^- induces conformational changes in the protein structure and the membrane depolarization (Chinnusami and Zhu, 2006). Large amounts of Na^+ limit the uptake of K^+ from the plants, which is a reason for the disruption of the stomatal regulation and results in irreversible changes in the transpiration flow and a loss of cellular water content. The accumulation of Na^+ and Cl^- ions disrupts many physiological processes in plants. It is reported that the accumulation of Cl^- in plant cells is more dangerous than Na^+ (Tavakkoli et al., 2010; Hasanuzzaman et al., 2013), limiting chlorophyll biosynthesis and resulting in chlorine toxicity (Tavakkoli et al., 2010; Parihar et al., 2015). A high concentration of this ion is toxic to plants; critical levels for toxicity are 4–7 mg/g for chlorine-sensitive species and 15–50 mg/g for chlorine-tolerant species (Hasanuzzaman et al., 2013). Higher accumulation of Cl^- in plants leads to a significant reduction in the growth and the efficient use of water in plants. Therefore, salt stress occurs first in the form of osmotic stress; then, the effects of ionic stress appear with excessive uptake of Cl^- and Na^+ ions from the cells, resulting in Mn^{2+}, Ca^{2+}, and K^+ deficiencies (Ahmadizadeh et al., 2016; Acosta-Motos et al., 2017). The accumulation of high levels of Na^+ and Cl^- in the chloroplasts and the leakage of K^+ from the cells results in destruction of the thylakoid membranes and inhibition of photosynthetic electron transport (Percey et al., 2016; Li and Li, 2017). The effects of salt stress on the photosynthetic apparatus are summarized in Figure 13.1.

13.3.1 Plant Growth, Development, and Yields

A high salt concentration inhibits the growth of plants, reducing the fresh the dry weight of roots, stems, and leaves (Misra et al., 1996, 1997a, 1997b; Li and Li, 2017). It has been found that the stem diameter decreases due to the shrinking of the vascular tissues (Zhang et al., 2016; Wungrampha et al., 2018). Salt stress decreases the growth of plant tissues and organs, and stimulates the aging and falling of mature leaves (Li and Li, 2017). As a result of this stress, the leaf morphology changes, mainly increasing the thickness of leaf epidermal, mesophilic, and palisade cells (Li and Li, 2017). On the other hand, there is shrinkage of the intracellular space of the epidermal cells (Parida et al., 2004). Further, the salts slow down the processes leading to cell wall maturation (Taleisnik et al., 2009; Li and Li, 2017). Excess Na^+ in the leaves causes sodium toxicity, resulting in marginal yellowing with progressive necrosis. The symptoms are similar to those of micronutrient toxicity. Chloride ions cause premature yellowing or bronzing of the leaf tips and further, cause necrosis of the leaf (Wugrampha et al., 2018). Previous studies have revealed that the plant organs have different degrees of sensitivity to salt stress (Misra et al., 1996, 1997a, 1997b; Negrao et al., 2017; Mbarki et al., 2018). In fact, salinization limits the growth of aerial parts by decreasing the carbon allocation in the leaf tissues and increasing it in the root cells (Mbarki et al., 2018). Other studies have shown an opposite effect: the roots are affected more than the stems of plants (Hajlaoui et al., 2010). Genetic differences in the growth patterns of root, shoot, and leaf, as well as the metabolic differences between salt-tolerant and salt-sensitive plants, are reported by Misra et al. (1996, 1997a, 1997b). Low concentrations of salts stimulate the growth and the metabolism of tolerant plants, but susceptible plants have retarded growth at similar concentrations (Misra et al., 1996, 1997a, 1997b). The degree of growth retardation also varies among plant species, and at higher salt concentrations, the growth retardation of susceptible plants is

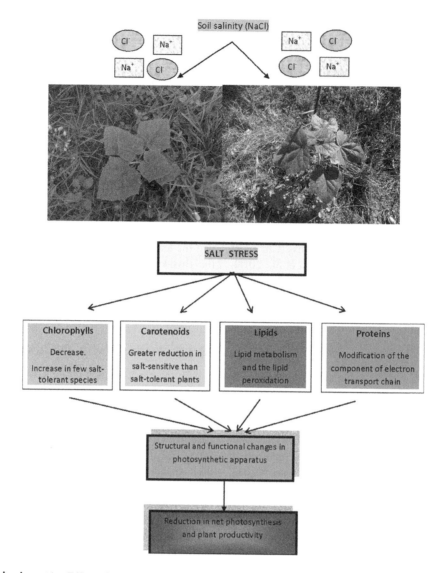

FIGURE 13.1 (See color insert.)   Effect of salt stress on the photosynthetic apparatus.

much faster compared with tolerant plants (Misra et al., 1996, 1997a, 1997b).

13.3.2   PHOTOSYNTHETIC PIGMENTS

Chlorophyll a, chlorophyll b and carotenoids are the main photosynthetic pigments, and they play an important role in photosynthesis, as the changes in their content affect the photosynthetic activity (Rahdari and Hoseini, 2011). The amount of chlorophylls is used as a sensitive indicator for the cellular metabolic state of plants (Parihar et al., 2015). Salt stress reduces the chlorophyll content, which is probably the result of an elevated level of the toxic cation Na$^+$ (Pinheiro et al., 2008; Yang et al., 2011). Previous investigations have revealed a decrease in the amounts of chlorophyll a and *b* in Indian mustard (Misra et al., 1995), rice (Misra et al., 1997b), Mung bean (Misra et al., 2001a), cabbage and kidney bean (Misra et al., 2006), maize (Deng et al., 2010), *Helianthus annuus* (Akram and Ashraf, 2011), *Medicago sativa* (Winicov and Seemann, 1990), *Triticum aestivum* (Perveen et al., 2010),

and *Ricinus communis* (Pinheiro et al., 2008). Salt-induced changes in the chlorophyll content may be due to impaired chlorophyll biosynthesis or accelerated chlorophyll degradation (Misra et al., 1995, 2006; Rahdari and Hoseini, 2011). It has also been shown that chlorophyll b can be converted into chlorophyll a (Eckardt, 2009). On the other hand, studies by Reddy and Vora (1986) revealed that the decrease in the total chlorophyll content in salt-stressed leaves is mainly due to degradation of chlorophyll a, which is much more sensitive to salinization than chlorophyll b (Misra et al., 1990). However, in rice cultivars differing in salt sensitivity, the chlorophyll a/b ratio increased under NaCl salinity stress (Misra et al., 1997b). Recent experiments with sunflower callus cultures indicate that salt stress affects chlorophyll biosynthesis more than chlorophyll degradation (Santos, 2004; Akram and Ashraf, 2011). It has been established that the changes in the chlorophyll content depend on the salt tolerance of the plant species, as in salt-tolerant species, chlorophyll content increases, while in salt-sensitive species, it decreases (Misra et al., 1995, 1997b, 1999, 2001a, 2006; Huang et al., 2005;

Khan and Shirazi, 2009; Akram and Ashraf, 2011; Qiu et al., 2014). According to some authors, the accumulation of chlorophylls may be used as a biochemical marker for salt tolerance in Indian mustard (Misra et al., 1995, 1999), rice (Misra et al., 1997b), mung bean (Misra et al., 1999, 2001a), wheat (Arfan et al., 2007), pea (Noreen et al., 2010), melon (Romero et al., 1997), and sunflower (Akram and Ashraf, 2011). However, the amount of pigments is not always related to the tolerance of plants to high salt concentrations. The research of Juan et al. (2005) shows a slight correlation between leaf accumulation of Na^+ and the amount of pigments in tomato cultivars with different salt tolerance. Misra et al. (1995, 1997b, 1999, 2001a) have reported that there is an enhancement in the chlorophyll content of leaves at low salt concentrations, but a decline in chlorophyll is reported with an increase in the concentration of salts and the duration of salt exposure of the plants. Also, the photosynthetic pigments of tolerant rice cultivars increase, whereas those of susceptible cultivar decrease at an early developmental stage (Misra et al., 1997b). The increase in the pigment content with a simultaneous decrease in leaf growth for different plant species under salt stress is interpreted as an increase in the number of chloroplasts per mesophyll cell in stressed leaves (Misra et al., 1995, 1997b, 1999). Our recent investigations with two lines of *Paulownia* with different sensitivity to a high salt concentration reveal an increase in chlorophyll content in the more tolerant line in comparison with the more sensitive line (Stefanov et al., 2016b, 2018c). It could be concluded that the influence of salt stress on the chlorophyll content depends on the plant species and the degree of genetic tolerance to salinity. The analysis of absorption spectra of *Paulownia* lines grown in soils with different salinity reveals variations in the amounts of chlorophyll a and chlorophyll b (Table 13.1). According to Percey et al. (2016b, 2018c), changes in the chlorophyll a/b ratio are an important indicator of the presence or the absence of abiotic stress. Our investigations reveal that there is no correlation between the degree of

sensitivity to salinization and changes in the chlorophyll a/b ratio (Stefanov et al., 2016b, 2018a, 2018c).

Carotenoids are an essential structural component of the photosynthetic apparatus; they participate in light-harvesting processes and in defense mechanisms against oxidative stress (Gomathi and Rakkiyapan, 2011; Shah, 2017). It has been established that the amount of carotenoids under salt stress varies depending on the salt tolerance of plants (Nagy et al., 2015). A higher reduction in the carotenoid level in salt-sensitive compared with salt-tolerant rice seedlings has been reported (Singh and Dubey, 1995). However, Misra et al. (1997b) have reported an interesting observation on the changes in the carotenoid content in salt susceptible and salt-tolerant rice cultivars. In the early stages of seedling development, only the carotenoid content of the susceptible rice cultivar decreased significantly compared with that of the tolerant genotype. However, older seedlings growing under saline conditions showed an increase in the carotenoid content, which was interpreted as an increase in the number of chloroplasts per leaf (Misra et al., 1997b). Salinization reduces the carotenoid content in beans, cotton (Brugnoli and Lauteri, 1991), rice (Lakra et al., 2017), *Arabidopsis thaliana* (Yu and Assmann, 2016), and tobacco (Shabala, 2017), which leads to a slowing of the photosynthetic process. Additionally, increased degradation of zeaxanthin and β-carotene, which are included in the plant defense system against damage to thylakoid membranes, has been shown (Sharma and Hall, 1992). On the other hand, the investigations of Gomathi and Rakkiyapan (2011) with sugarcane reveal that higher salt tolerance is related to higher membrane stability and significant higher chlorophyll and carotenoid content. Our recent study showed that the higher amount of carotenoids is one of the reasons for the long-term adaptation of plants to a high level of salts (Stefanov et al., 2018b, unpublished data). Investigations of salt-tolerant hot pepper suggest that the carotenoid content can be used as an indicator of salt tolerance in pepper (Ziaf et al., 2009).

TABLE 13.1
Absorption Spectral Ratios of Thylakoid Membranes Isolated from Leaves of *Paulownia tomentosa* × *fortunei* and *Paulownia elongata* x *elongata* Grown in Non-Saline (1.6 mS m⁻¹) and Saline (6.3 mS m⁻¹ and 14.0 mS m⁻¹) Soils*

Sample	A_{470}/A_{440}	A_{650}/A_{680}
P. tomentosa x *fortunei*		
1.6 mS m⁻¹	0.732 ± 0.004^{bc}	0.612 ± 0.025^{b}
6.3 mS m⁻¹	0.750 ± 0.007^{a}	0.475 ± 0.014^{c}
14.0 mS m⁻¹	0.765 ± 0.008^{a}	0.493 ± 0.011^{c}
P. elongata x *elongata*		
1.6 mS m⁻¹	0.716 ± 0.008^{c}	0.686 ± 0.004^{a}
6.3 mS m⁻¹	0.750 ± 0.006^{a}	0.498 ± 0.016
14.0 mS m⁻¹	0.746 ± 0.003^{a}	0.482 ± 0.006^{c}

Values in the same column followed by a different letter (a, b, c) are significantly different at $p<0.05$. The statistical differences between the means were determined using ANOVA.

13.3.3 LIPID COMPOSITION

The lipids in biological membranes, including the thylakoid membranes, are very important not only for the formation of the lipid bilayer but also for the structure and the function of membrane complexes. Duchene and Siegenthaler (2000) proposed that the current model of thylakoid membrane lipid composition is one that consists simultaneously of bulk and specific lipids. Various studies have indicated the influence of lipids in the assembly and the function of thylakoid membranes (Apostolova and Misra, 2014). The lipids of thylakoid membranes are characterized by high levels of unsaturation of their fatty acids, which determine their physical and biochemical characteristics and play an important role in the functions of the photosynthetic apparatus (Murata and Wada, 1995; Mullineaux and Kirchoff, 2009). The level of unsaturation is mediated by the activity of fatty acid desaturase. The expression of lipid desaturase genes is lightly regulated and thereby modulates the thylakoid membrane assembly (Kis et al., 1998).

Salt stress influences the total lipids and the fatty acid composition as a result of changes in the lipid metabolism. The influence of lipid biosynthesis and the lipid peroxidation with the increase of salinity have been shown in plants such as wheat, tomato, and purslane (Mehraban et al., 2017). The investigation of the lipid profile of *Chlamydomonas reinhardtii* showed a clear distinction between 0.1 M NaCl, which the algae could tolerate, and higher levels of NaCl (0.2 and 0.3 M), which caused cell death (Hounslow et al., 2016). The lipid content increases under salt conditions through either a long-term exposure to 0.1 M NaCl or a short-term exposure to 0.2–0.3 M NaCl. The study also reveals a significant increase in the amounts of palmitic acid and linolenic acid at the higher salinity. Various effects on fatty acid composition under salt stress have been found in six safflower genotypes using a salinity level of 8 dS/m (Javed et al., 2014). The authors show a higher ratio of unsaturated to saturated fatty acid in more tolerant genotypes. Moreover, the extent of unsaturation of fatty acids in these genotypes is correlated with the salinity tolerance and the potential of the photosynthetic apparatus to tolerate stress (Javed et al., 2014). Recent studies with *Nigella sativa* have shown an increase of total lipids under high salt conditions (Khlid, 2017). The author had suggested that the changes in the lipid metabolism of *Nigella sativa* under salinity stress may be due to the effect of salinity on enzyme activity and lipid biosynthesis. In addition, data reveal that the main fatty acids are linolenic and oleic acid. An increase of the lipid content under salt stress has been shown in some species of *Chlamydomonas* (Salama et al., 2013; Ahn et al., 2015).

On the other hand, investigation with a salt-tolerant yeast, *Zygosaccharomyces rouxii*, grown in a medium with 15% (w/v) NaCl revealed an increased amount of C16:1 and C18:1 fatty acids of acyl lipids, while C18:0 and C18:2 acids decreased in both whole cells and the plasma membrane, which led to a decrease of membrane fluidity in the presence of a high concentration of NaCl (Hosono, 1992). Bybordi et al. (2010) have shown a decrease of total lipid content and polyunsaturated fatty acids in canola (*Brassica napus* L.) and an increase of monounsaturated fatty acids in relation to increasing salinity. Investigations with seedlings of *Lotus corniculatus* demonstrate that the monogalactosyldiacylglycerol to digalactosyldiacylglycerol (MGDG/DGDG) ratio declined threefold, while the relative proportion of MGDG was 12-fold lower under drought, which is a component of salt stress (Yordanov et al., 2003). MGDG is the main lipid in thylakoid membranes. A significant decrease in the MGDG content can suggest a deleterious effect of salt stress on thylakoid membranes.

Lipid peroxidation in biological membranes is the most obvious symptom of oxidative damage in plants, and it is often used as a marker for oxidative stress (Taibi et al., 2016). Malondialdehyde is one of the final products of lipid oxidative modification and is responsible for membrane damage, including changes to the intrinsic properties of the membrane, such as fluidity, ion transport, enzyme activity, and protein cross-linking. These changes influence the structure and functions of the thylakoid membranes and can lead to cell death

(Sharma et al., 2012). It has been shown that the products of lipid degradation, as well as lipid peroxidation initiators (i.e., reactive oxygen species [ROS]), can mediate the signal transduction cascade (Labudda, 2013). Investigations with different genotypes of *Phaseolus vulgaris* have shown that NaCl treatments lead to variation in the increase of malondialdehyde depending on the salt tolerance of the genotypes; i.e., the increase is higher in the salt-tolerant genotype in comparison to the salt-sensitive genotypes (Taibi et al., 2016). In addition, it has been shown that after long-term salt treatment, the increase of lipid peroxidation in the salt-sensitive wheat cultivars is higher than that in the salt-sensitive cultivars (Parihar et al., 2015).

13.3.4 Chloroplasts and Thylakoid Membranes

13.3.4.1 Structure

It has been well established that salt stress causes ultrastructural changes to cellular organelles, such as chloroplasts and mitochondria (Bastías et al, 2013). Salinization leads to swelling of chloroplasts, increases in the number and size of plastoglobuli in these organelles, and subsequent destruction of their structure (Salama et al., 1994; Hernandez et al., 1999; Mitsuya et al., 2000; Li and Li, 2017). An electron microscopic study of *Cucumis sativus* revealed that salinity also causes damage to the chloroplast envelope (Shu et al., 2012). Salt stress accelerates leaf senescence, thereby accelerating the senescence-induced deterioration of chloroplasts and disorganization of thylakoid membranes, leading to the loss of photosynthetic function (Joshi et al., 2013). Studies of the effects of NaCl on the foliar ultrastucture of sweet potato suggest that the degradation of the cytoplasm and cell organelles, except the thylakoid membranes, occurs as a result of light-independent salt stress, while the degradation of the thylakoid membranes of chloroplasts in mesophyll cells occurs as a result of salt-induced oxidative damage (Mitsuya et al., 2000). This assumption is also confirmed in studies with rice (Mitsuya et al., 2003). In addition, it has been shown that a high salt level reduces the number and the thickness of stacked granal thylakoid membranes (Bruns and Hecht-Buchholz, 1990; Khavari-Nejad and Mostofi, 1998; Lee et al, 2013; Meng et al., 2016). Ultrastructural changes of chloroplasts under salt stress have been studied in *Eucalyptus microcorys*. These changes include the appearance of large starch grains, the swelling of thylakoid membranes and their disintegration, as well as the almost complete absence of grana (Keiper et al., 1998; Li and Li, 2017). In some plant species, the content of starch in chloroplasts decreases (Locy et al., 1996), while in others, it increases (Maiti et al., 2000). The degradation of Rubisco and formation of a Rubisco-containing body in salt-stressed *Glycine max* leaves have been shown, which correlates with the modifications of chloroplast ultrastructure and photosynthetic efficiency (He et al., 2014).

During photosynthesis, and on changes in environmental conditions, there are significant structural and functional changes in the photosynthetic machinery, which are connected with a variation of the composition and the thylakoid membrane ultrastructure (Joshi et al., 2013; Ünnep et al., 2014).

Earlier investigations by Sato et al. (1992) suggest that the 23 kDa protein of the oxygen evolving complex may be involved in the salt tolerance of plants. Later, differences were shown in the polypeptide composition of thylakoid membranes and PSII particles in rice cultivars that are sensitive or tolerant to salt stress (Wang et al., 2009). The authors reveal that the amounts of 43 kDa (the polypeptide of the inner antenna of PSII) and 33 kDa (the polypeptide of the oxygen evolving complex) are higher in the salt-tolerant than in the salt-sensitive cultivar. It has been shown that about 40% of D1 protein is lost under salinity, which leads to the inhibition of PSII activity (Sudhir et al., 2005).

It has been shown that salt stress in cyanobacteria leads to a loss of the 47 kDa chlorophyll protein and the 94 kDa core membrane linker protein that can attach the phycobilisome to the thylakoids (Garnier et al., 1994). The 23 kDa protein, which is extrinsically bound to PSII, is also dissociated under salt stress (Sudhir et al., 2005). Recent studies with two genotypes of wheat with different salt sensitivity have revealed that long-term treatment with 200 mM NaCl leads to a decrease of the PsaB protein (PSI-B core subunit of PSI) in the salt-sensitive genotype, while in the salt-tolerant genotype, it increases. It has also been shown that the amount of D1 protein of PSII in wheat is not changed under salt stress (Jusovich et al., 2018).

In an extensive radio-labelling and protein kinase study, Biswal et al. (2001) have shown that the variation in the salt susceptibility of plant photosynthetic apparatus could be due to the presence of 27.5, 33, 38, and 66 kDa polypeptides in salt-tolerant Brassica thylakoids. In contrast, the presence of 29, 40, 46, and 57.5 kDa polypeptides in salt-sensitive mung bean thylakoids was shown (Biswal et al., 2001). The photosynthetic inhibition of PSII in salt-susceptible mung bean thylakoids (Sahu et al., 1998; Misra et al., 1999, 2001a) is attributed to the absence of light-dependent protein kinase in the stressed chloroplast (Biswal et al., 2001). On the other hand, thylakoid redox-controlled protein kinase activity (Silverstein et al., 1993; Misra and Biswal, 2000) on LHCII (light-harvesting complex of PSII), D1, CP47, and LHCII kinase polypeptides could have played a major role in the adaptive strategy in the salt-tolerant mustard seedlings, regulating the energy dissipation process (Biswal et al., 2001). This plausible mechanism of energy dissipation by salt-tolerant plants is confirmed by analysis of the xanthophyll cycle (Misra et al., 2003, 2006), which is also a thylakoid redox-controlled process.

13.3.4.2 Function

The activity of photosynthesis is determined by the amount and the activity of light-harvesting complexes, electron transport components, and energy transduction processes as well as by carbon metabolism. The effects of salt stress on the photosynthetic components and their degree of damage depend on the concentration of the salt/ion and the duration of stress (Misra et al., 1997b, 2001a). Whereas some studies indicate an inhibition of photosynthesis under salt stress (Kao et al., 2001; Romero-Aranda et al., 2001), others show that this process is not affected by salinization or is stimulated by moderate salt concentrations (Kurban et al., 1998; Rajesh et

al., 1998). In the mulberry tree, the net assimilation of CO_2, the stomatal conductance, and the rate of transpiration are reduced under salt stress, whereas the intercellular concentration of CO_2 is increased (Agastian et al., 2000). In *Bruguiera parviflora*, the rate of photosynthesis increases at low salinity but decreases at high salinity (Parida et al., 2004; Mbarki et al., 2018). Our investigations with two salt-tolerance lines of *Paulownia* (*Paulownia tomentosa* × *fortunei* and *Paulownia elongata* × *elongata*) grown in soil with different salinity have revealed that the photosynthetic rate increases in the more salt-tolerant line in comparison to the line with lower tolerance to high salt concentration (Stefanov et al., 2016b, 2018c). Pulse-Amplitude-Modulation (PAM) chlorophyll fluorescence measurements in studied lines of *Paulownia* also revealed that salinity leads to an increase of the photochemical quenching coefficient (qP), stimulation of the linear electron transport rate (ETR), and improvement of the efficiency of the photochemical energy conversion (ΦP_{SII}) in both lines. On the basis of the experimental results, it was concluded that all these changes are the results of the impact of high salt concentration on Q_A reoxidation. The experimental data for the effect of salt stress on these *Paulownia* lines also revealed that high soil salinity led to a delay of the PSI-dependent cyclic electron transport in stroma lamellae (Stefanov et al., 2016b, 2018c). In addition, the low-temperature (77 K) chlorophyll fluorescence of *Paulownia* thylakoid membranes shows that the high salt concentrations have no influence on the energy redistribution between the two photosystems (Figure 13.2).

One of the most sensitive sites of the photosynthetic apparatus under salt stress is the PSII complex (Baker, 1991). The opinions of different authors on the effect of salt stress on this complex are contradictory. Some studies have shown that salt stress inhibits the activity of PSII (Loretto et al., 2003), while others show that it does not (Lu et al., 2003), suggesting that the sensitivity of this complex to a high salt concentration depends on the plant species. Studies on the influence of salt stress on sorghum have shown a significant decrease in the maximum quantum yield of PSII, the photochemical quenching coefficient (qp), and the electron transport rate (ETR) but an increase in the non-photochemical quenching (qN) (Netondo et al., 2004). However, studies of the effect of salinity on rice cultivars differing in salt tolerance have shown that the potential photochemical efficiency of PSII (maximal quantum yield Fv/Fm) is almost unaffected by salt stress, whereas the actual efficiency of photochemical energy conversion (actual quantum yield ΔF/Fm) declines with increasing salinity in salt-sensitive cultivars. An increase in qN in salt-sensitive cultivars under salt stress (Dionisio-Sese and Tobita, 2000) has been also shown. Recent investigations with wheat revealed that the changes in the function of the photosynthetic apparatus are accompanied by an increase of the energy transfer to PSI as a result of destacking of the thylakoid membranes and/or an alteration of PSI antenna size (Jusovich et al., 2018).

Misra's group studied extensively the mechanism of salt stress–induced damage to PSII using invasive and non-invasive techniques. They demonstrated that the PSII activity (measured as water to ferricyanide reduction) is severely affected by salt

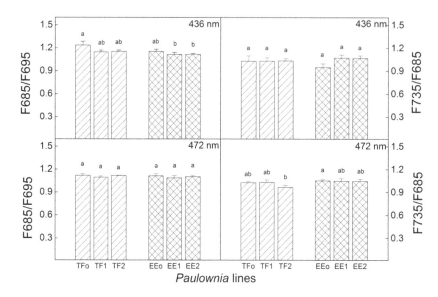

FIGURE 13.2 Low-temperature chlorophyll fluorescence emission ratio F_{685} and F_{695} of thylakoid membranes isolated from leaves of *Paulownia tomentosa × fortunei* (TF) and *Paulownia elongata × elongata* (EE) grown in non-saline (1.6 mS m^{-1} [TF$_0$ and EE$_0$]) and saline (6.3 mS m^{-1} [TF$_1$ and EE$_1$] and 14.0 mS m^{-1} [TF$_2$ and EE$_2$]) soils. Chlorophyll is excited at 472 and 436 nm. Values of respective ratios followed by different letters are significantly different at p < 0.05.

stress in the susceptible plant mungbean as compared with that in the Indian mustard plant (Misra et al., 1999) suggesting a stress-induced impairment of electron transfer damaging either the oxygen-evolving complex or the reaction center complex of PSII. These authors deciphered the lesion in PSII using a thermoluminescence technique and showed through several experiments that the charge recombination between $S_{2/3}Q_B^-$ and $S_2Q_A^-$ is differentially affected by the salt stress (Misra et al., 1999, 2001b, 2001c). The electron flow between Q_B and Q_A is dose and duration dependent and is gradual (Biswal et al., 2002). Through this experiment, it was illustrated that salt stress accelerates the aging process and decreases the size of both the PSII reaction center and the quinone pool, ultimately resulting in inhibition of the electron flow through PSII (Biswal et al., 2002). In a fast chlorophyll fluorescence study using Handy PEA et al. (2001a) show that this method is suitable for the assessment of salt/ion effects as well as for assessing the salt sensitivity of plants. These authors concluded that the differences in tolerance to salt stress are due to the abundance of active reaction centers and the capacity for energy dissipation. In further studies using PAM fluorimetry, it was observed that salt stress decreased NPQ in salt-tolerant plants but increased NPQ significantly in salt-susceptible plants (Misra et al., 2003, 2006). In these studies, the probable, but significant, role of the xanthophyll cycle is proposed by these authors (Misra et al., 2003, 2006). A significant increase in lutein de-epoxidation (Misra et al., 2003) and violaxanthin de-epoxidation simultaneously with the increase in NPQ in the salt-sensitive plant clearly indicates that salt stress induces photoinhibitory damage to PSII, and the xanthophyll cycle plays a major role in this process.

The effects of salt stress (the presence of 0.8 M NaCl) on the cyanobacterium *Spirulina platensis* have been shown to be a decrease in PSII-mediated oxygen-evolving activity as well as an increase in PSI activity and the amount of P700

(Sudhir et al., 2005). The authors concluded that salt-induced effects on the photosynthetic electron transport activities are due to alterations in the composition of thylakoid membrane proteins, which lead to the decreased energy transfer from the light-harvesting antenna to the PSII core.

In addition, the investigations of Lu and Vonshak (2002) with *Spirulina platensis* in the presence of 0.8 M NaCl have revealed that salt stress results in a decrease in the efficiency of electron transfer from Q_A to Q_B. Analyses of the polyphasic fluorescence transients (OJIP) have suggested that salt stress inhibits electron transport at both donor and acceptor sides of PSII, resulting in damage to the phycobilisome and shifting the distribution of excitation energy in favor of PSI (Lu and Vonshak, 2002). Salt-adapted cells can maintain a high conversion efficiency of excitation energy through the downregulation of the PSII reaction center (Lu and Vonshak, 2002). It has been reported in *Bruguiera parviflora* that at low salinity (100 mM), PSII-mediated electron transport activity increases, but this activity decreases at high salinity (Parida and AB, 2003). Recent studies on the influence of salt stress on *Desmostachya bipinnata* show that the photosynthetic rate is similar to that of the control at moderate salinity, but the rate is inhibited with a further increase in salinity (Asrar et al., 2017). The authors show that at moderate salinity, the photosynthetic pigments, electron transport rate, and photochemical quenching are not affected, but these parameters decline significantly at high salinity.

13.4 DEFENSE MECHANISM IN PLANTS AGAINST SALT STRESS

13.4.1 ANTIOXIDATIVE DEFENSE SYSTEM

Biochemical and molecular studies of the salt stress responses of plants have revealed significant increases of reactive oxygen species (ROS), including singlet oxygen ($^1O_2$), superoxide (O_2^-), hydroxyl radical (OH•), and hydrogen peroxide (H_2O_2) (Tunc-Ozdemir et al., 2009; Ahmad et al., 2010, 2012; Ahmad and Umar, 2011). ROS severely disrupt normal metabolism through oxidative damage to lipids, nucleic acids, and proteins, resulting in protein denaturation and degradation and the peroxidation of membrane lipids (Miller et al., 2010; Suo et al., 2017). The presence of ROS-producing centers, such as triplet chlorophylls and the electron transport chain in the photosynthetic apparatus, makes chloroplasts a major site of ROS production (Edreva, 2005), which leads to damage to chloroplasts. Impaired electron transport under salt stress leads to the transfer of electrons from excited photosynthetic pigment molecules to $^1O_2$ (Joshi et al., 2013; Anjum et al., 2017). Oxidative damage to the cell organelles is determined by the balance between ROS production and the activity of the antioxidant defense system. On the other hand, ROS influence the expression of a number of genes and signal transduction pathways (Apel and Hirt, 2004). Investigations with maize indicate a possible role of ROS in the systemic signaling from roots to leaves, allowing leaves to activate their defense mechanisms for better protection against the salt stress (AbdElgawad et al., 2016).

Plants cope with oxidative stress by the well-developed defense system, including enzymatic and non-enzymatic processes, which contribute to the detoxification of ROS (Gill and Tuteja, 2010; Miller et al., 2010; Baxter et al., 2014). Non-enzymatic antioxidants include ascorbic acid, glutathione, phenolic compounds, carotenoids, flavonoids, and tocopherol. Enzymatic antioxidants include superoxide dismutase (SOD), catalase (CAT), glutathione S-transferase, glutathione peroxidase (GP), ascorbate peroxidase (APX), peroxidase (POD), etc. (Tuncz-Ozdamir et al., 2009; Baxter et al., 2014; Anjum et al., 2017). The antioxidative enzyme activity changes significantly in salt-stressed plant cells (Das et al., 1992). Salt tolerance in some plants correlates with their ability to scavenge reactive oxygen (Azevedo Neto et al., 2006). The exposure of *Paulownia* plants to salt stress (150 and 200 mM NaCl) for 10 days led to an increase of the ferric reducing antioxidant power (FRAP) and the hydrogen donors capable of scavenging 2,2-diphenyl-1-picrylhydrazyl (DPPH) (Stefanov et al., 2018b, unpublished data.). A similar increase in the antioxidant activity under increasing salinity has been shown in other plant species. High levels of enzyme antioxidants have been shown in salt-tolerant cultivars of wheat (Bhutta, 2011), tomatoes (Dogan et al., 2010a, 2010b), barley (Jin et al., 2009), and canola (Ashraf and Ali, 2008). Recent studies with barley have revealed that salinity tolerance in barley correlates with its oxidative stress tolerance, but higher antioxidant activity at one particular time does not correlate with salinity tolerance and as such, cannot be used as a biochemical marker in barley (Maksimovič et al., 2013). The salt tolerance of radish cultivars is not also associated with higher antioxidant enzyme activity (Noreen and Ashraf, 2009). In addition, studies with safflower cultivars have shown that increases in the activity of SOD, APX, GP, and POD are not sufficient to prevent lipid peroxidation (Culha Erdal and Cakirlar, 2014). Transgenic

plants with high scavenging ability improve salt tolerance (Nagamiya et al., 2007).

The non-enzymatic antioxidants are generally small molecules, such as ascorbate (Asc) and glutathione, which are found in the thylakoid lumen and the stroma of chloroplasts, while lipophilic antioxidants such as α-tocopherol and β-carotene are in the lipid matrix of thylakoid membranes (Munne-Bosch, 2005). Ascorbate is a non-enzymatic scavenger, and its concentration can build up to the millimolar range, i.e., 20 mM in the cytosol and 20–300 mM in the chloroplast stroma (Smirnoff and Wheeler, 2000). This antioxidant directly reduces superoxide, singlet oxygen, and hydroxyl radicals, and it is a substrate for ascorbate peroxidase. Our investigation has suggested that one of the possible protective roles of ascorbate in photosynthetic membranes against oxidative damage under abiotic stress is modification of the oxygen-evolving complex (Dobrikova et al., 2013). The ascorbate content increases under oxidative stress (Noctor and Foyer, 1998) and influences the tocopherol content through the ascorbate-glutathione cycle (Munné-Bosch, 2005). Tocopherols are lipid-soluble antioxidants, considered as potential scavengers of ROS and as general antioxidants for the protection of membrane stability, including quenching or scavenging ROS such as singlet oxygen. In the case of protecting the membrane, tocopherols protect the lipids and other membrane components by physical quenching and react chemically with O_2 in the chloroplast (Suo et al., 2017). In *Arabidopsis*, γ-tocopherol is shown to protect membrane integrity (Sattler et al., 2004). Salt-stressed leaves develop plastoglobules (Ali-Reza et al., 2007), in which tocopherol biosynthesis also can occur (Austin et al., 2006; Vidi et al., 2006). Plastoglobules are thylakoid protrusions, composed of lipophilic constituents such as triacylglycerols, quinones, chlorophyll, carotenoids, and also MGDG and DGDG (Ghosh et al., 1994; Austin et al., 2006), which increase in number during senescence and in oxidative stress (Munné-Bosch and Alegre, 2004), when chlorophyll turnover is high. Free phytol from chlorophyll breakdown might be directly salvaged for tocopherol biosynthesis (Ischebeck et al., 2006; Dörmann, 2007), which is required for antioxidant protection in these conditions. As a second conclusion, γ-tocopherol cannot substitute for α-tocopherol to ensure better survival in salt stress, although markers for oxidative stress were decreased in transgenics compared with the wild type (Ali-Reza et al., 2007). Consequently, α-tocopherol appears to better mediate protection of macromolecules from denaturation by salt than γ-tocopherol. Alternatively, γ-tocopherol might not be able to influence cellular signaling as α-tocopherol does. Ali-Reza et al. (2007) concluded that γ-tocopherol exerts a specific function in osmoprotection during salt stress.

Phenolics are diverse secondary metabolites (flavonoids, tannins, hydroxycinnamate esters, and lignin) in plant tissues (Rice-Evans et al., 1997; Solecka, 1997), and they accumulate in salt-stressed plant cells (Das et al., 1990, 1992) Flavonoids act as antioxidants dependent on the reduction potential and the accessibility of their radicals (Harborne, 1964). They have high reactivity as electron donors and can alter the peroxidation kinetics by modifying the lipid packing order to decrease the fluidity of the membrane (Sakihama and Yamasaki, 2002). These changes could hinder the diffusion of free radicals and restrict peroxidation (Yadav and Sharma, 2016). Our recent study has shown that one of the reasons for the long-term adaptation of *Paulownia* is the significant increase of total flavonoids during the first days of NaCl treatment (Stefanov et al., 2018b, unpublished results).

13.4.2 XANTHOPHYLL CYCLE

The excess light absorbed by salt-stressed plants is dissipated through non-photochemical processes such as heat dissipation (Misra et al., 2001a, 2012a, 2018a) or non-photochemical quenching (NPQ) (Misra et al., 2006, 2012a, 2018a). The xanthophyll cycle is reported to play a significant role in the NPQ process (Muller et al., 2001; Demmig-Adams and Adams, 2002). Xanthophylls are a special group of carotenoids in the thylakoid membranes associated with the light-harvesting antenna pigment–protein complexes of all photosynthetic organisms (Pogson et al., 1998; Latowski et al., 2004). Besides their light-harvesting function, xanthophylls also regulate the assembly of antenna pigment–protein complexes and play a pivotal role in photoprotection, including membrane lipid peroxidation (Pogson et al., 1998; Davison et al., 2002; Latowski et al., 2004). The major xanthophyll pigments that are present in unstressed leaves are lutein (30–60%), neoxanthin (9–14%), and violaxanthin (Johnson et al., 1993; Kuhlbrandt et al., 1994). Under variable light conditions, violaxanthin is de-epoxidized enzymatically at low pH in the presence of ascorbate to form antheraxanthin and subsequently, zeaxanthin (Neubauer and Yamamoto, 1994; Bratt et al., 1995). Zeaxanthin decreases the fluidity of thylakoid membranes, thereby regulating the diffusion of ROS into the thylakoid membrane (Subczynski et al., 1991). Besides these functions, neoxanthin is considered as the precursor of abscisic acid (Milborrow and Lee, 1998), and the latter is reported to increase significantly under salinity stress and regulate photosynthesis through the stomatal pathway (Cramer and Quarrie, 2002; Makela et al., 2003). Lutein and violaxanthin de-epoxidation increases under salt stress and is assigned to the photoprotection of plants through NPQ (Misra et al., 2003, 2006). The inter-relationship and intricate biosynthetic pathways regulating xanthophyll cycle pigment dynamics, the redox regulation of this process, the need for ascorbate (which increases during salt stress), and the formation of abscisic acids through xanthophyll cycle pigments give an insight into the biochemical network that regulates the xanthophyll cycle under salt stress.

13.4.3 ACCUMULATION OF OSMOLYTES

Osmoprotectants (or osmolytes) are small, electrically neutral, non-toxic molecules at molar concentrations; they are highly soluble organic compounds that efficiently maintain osmotic balance and stabilize proteins and membranes under salt, drought, or other stresses. The accumulation of osmolytes (glycine betaine [GB], proline, gamma-aminobutyric acid, and sugars) is a common response observed in different plants under stress conditions (Suprasanna, 2016; Suo et al., 2017).

It has been shown that the level of accumulation of the major osmolyte (GB) in various plant species depends on their salt tolerance. Chen and Murata (2008) summarized plant species with the stress-induced accumulation of GB. It has been shown that highly tolerant species accumulate very high levels, low-tolerant species accumulate average levels, and sensitive species show low levels of this osmolyte (Rasool et al., 2013). The plants contain low levels of GB under non-stressed conditions, but its level increases under abiotic stress conditions (Rhodes and Hanson, 1993). Investigations with *Beta vulgaris, Spinacia oleracea, Hordeum vulgare, Triticum aestivum, and Sorghum bicolor* reveal accumulation of GB in the response to abiotic stress, as in salt-tolerant genotypes, the amounts are higher than in salt-sensitive genotypes (Suprasanna, 2016 and references therein). GB does not scavenge the ROS directly, but it alleviates the damaging effect of oxidative stress by activating ROS-scavenging enzymes and by repressing ROS production (Chen and Murata, 2008). Transgenic plants with an elevated level (overproduction) of GB exhibit an enhancement of tolerance to salt, cold, drought, or high-temperature stress (Rhodes and Hanson, 1993; Chen and Murata, 2011). The investigations of Chen and Murata (2008, 2011) suggest that GB decreases ROS accumulation under stress and protects the photosynthetic apparatus, activating stress-related genes and membrane protection. This osmolyte protects the quaternary structure of enzymes and proteins from the damaging effect of environmental stresses (Papageorgiou and Murata, 1995). An investigation with transgenic *Synechococcus* sp. PCC 7942 has revealed that GB counteracts salt-induced effects on the degradation and the synthesis of D1 protein (Ohnishi and Murata, 2006). This osmolyte is an antagonist of the inhibition of protein biosynthesis and thus, enhances PSII repair, which leads to increased stress tolerance (Chen and Murata, 2008). The investigations of Yang and Lu (2005) have suggested that photosynthesis is improved by the exogenous application of 10 mM GB to salt-stressed maize plants, which is associated with better stomatal conductance and PSII efficiency. The protective effects of GB observed in some plants (*Arabidopsis*, maize, and tomato) might be mediated by high GB levels that accumulate in reproductive organs and GB translocation to other organs such as leaves (Chen and Murata, 2008).

Many plants accumulate free proline in response to salt and osmotic stress (Das et al., 1990; Misra et al., 1996; Iyer and Caplan, 1998). Additionally, investigations with transgenic tobacco plants having an overproduction of proline reveal the increased tolerance to osmotic stress in these plants (Kavi Kishor et al., 1995). The enhancement of this amino acid under stress is a result of the stimulation of biosynthetic pathways in the plants, which plays a key role in osmoregulation in plants subjected to salinity stress and influences salt tolerance in different ways (Das et al., 1990; Misra et al., 1996). Proline acts like a molecular chaperone, protecting the proteins and the membrane integrity and increasing the activity of any enzyme (Rasool et al., 2013). On the other hand, proline can now be added to non-enzymatic antioxidants that plants need to counteract the inhibitory effects of ROS, having ROS-scavenging activity and singlet oxygen quenching ability (Gill and Tuteja, 2010). Furthermore, proline is known to induce the expression of salt stress–responsive genes, which possess proline-responsive elements in their promoters (Carillo et al., 2011). Our study with different lines of *Paulownia* (*Paulownia tomentosa* × *fortunei* and *Paulownia elongata* × *elongata*) exposure to salt stress (150 and 200 mM NaCl) for 10 days suggested that a much higher increase in the amount of proline in *Paulownia tomentosa* × *fortunei* (from 270% to 310%) in comparison to the *Paulownia elongata* × *elongata* line (146%) is one of the reasons for the survival of the *Paulownia tomentosa* × *fortunei* line at high NaCl concentration after long-term salt treatment (15–25 days) (Stefanov et al., 2018b, unpublished data).

Under salt stress, the accumulation of gamma-aminobutyric acid in plants is also reported (Suprasanna, 2016). Investigations with *Arabidopsis thaliana* have revealed that treatment with NaCl leads to upregulation of the level of this non-protein amino acid (Renalt et al., 2010). Additionally, the exogenous supply of gamma-aminobutyric acid to *Caragana intermedia* reveals its role as a signal molecule, which participates in the regulation of gene expression in plants under salt stress (Shi et al., 2010).

In previous investigations, it has been shown that salt stress increases the levels of soluble sugar, which plays a key role in the osmotic adjustment and in the stabilization of the membranes (Das et al., 1990; Misra et al., 1996; Suprasanna, 2016). It has been shown that survival of the resurrection plant *Craterostigma plantagineum* under almost complete tissue dehydration is associated with an accumulation of the sugar (Norwood et al., 2000). Sugar molecules stabilize the binding of the extrinsic membrane proteins associated with the oxygen-evolving complex of thylakoid membranes to the core complex of PSII (Tsvetkova et al., 1995). The disaccharide trehalose is produced in large quantities under stress, and it protects the biological membranes from desiccation-induced damage, which is a result of its unique feature of reversible water capacity. Trehalose also suppresses the aggregation of denatured proteins, maintaining them in a partially folded state from which they can be reactivated by molecular chaperones (Singer and Lindquist, 1998). The protection by sugar, as well as the better effect of trehalose in comparison to other disaccharides against dehydration of the plant cell membrane, has been shown in isolated thylakoid membranes and PSII particles (Busheva and Apostolova, 1989; Apostolova et al., 1992, 1994; Tsvetkova et al., 1995).

13.5 CONCLUSIONS

Salt stress is a major environmental stress factor, influencing photosynthesis, and thus, it has become a global research focus in recent years. Salinity causes inhibition of photosynthesis, influencing the main components of the photosynthetic apparatus (Figure 13.2). This stress modifies the components of the photosynthetic apparatus. It has been shown that the salt-induced variation in the amount of chlorophylls cannot be used as a marker for salt tolerance of plants, because in some species, the reduction of chlorophylls is associated with increased susceptibility to high salt content, whereas in other species, this trend is not observed. Alterations in chlorophyll content under salt stress are a result of influence on chlorophyll biosynthesis and/or degradation. On the other hand, the amount of carotenoids under salt stress varies depending on the salt tolerance of plants; i.e., there is a greater reduction in the carotenoid level in salt-sensitive than in salt-tolerant plants. Modification of the total lipids and the fatty acid composition in thylakoid membranes with increased salinity, as a result of lipid peroxidation and an influence on lipid biosynthesis, has also been shown. Moreover, investigations reveal that the extent of unsaturation of fatty acids correlates with the tolerance of plants to salinity. Studies on the influence of salt stress on plants also show modifications in the protein complexes of the photosynthetic apparatus. On the other hand, plants alleviate the damaging effect of oxidative stress, a secondary stress induced by salt stress, with a well-developed enzymatic and non-enzymatic defense system. The interplay between ROS and the antioxidative defense system plays a pivotal role in determining oxidative damage to membranes. It is concluded that the salt-induced alteration in the structure and function of thylakoid membranes depends on the plant species, the dose, and the duration of salt stress as well as the activity of the defense system of plants.

ACKNOWLEDGMENT

The salt stress studies of the Bulgarian group are supported by the project №17-135/01.08.2017 of Program for career development of young scientists, BAS.

BIBLIOGRAPHY

AbdElgawad, H. G., M. M. Zinta, R. Hegab, H. Pandey, W. Asard and W. Abuelsoud. 2016. High salinity induces different oxidative stress and antioxidant responses in maize seedlings organs. *Front Plant Sci.* 7: 276.

Acosta-Motos, J. R., M. F. Ortuño, A. Bernal-Vicente, P. Diaz-Vivancos, M. J. Sanchez-Blanco and J. A. Hernandez. 2017. Plant responses to salt stress: adaptive mechanisms. *Agronomy* 7: 18.

Agastian, P., S. J. Kingsley and M. Vivekanandan. 2000. Effect of salinity on photosynthesis and biochemical characteristics in mulberry genotypes. *Photosynthetica* 38: 287–290.

Ahmad, P. and S. Umar. 2011. *Oxidative Stress: Role of Antioxidants in Plants.* Studium Press, New Delhi.

Ahmad, P., Jeleel, CA, Azooz MM, Nabi G. 2009. Generation of ROS and non-enzymatic antioxidants during abiotic stress in Plants. *Bot. Res. Intern* 2: 11–20.

Ahmad, P., C. A. Jaleel and S. Sharma. 2010. Antioxidative defense system, lipid peroxidation, proline metabolizing enzymes and biochemical activity in two genotypes of Morus alba L. subjected to NaCl stress. *Russ. J. Plant Physiol.* 57: 509–517.

Ahmad, P., M. Ozturk and S. Gucel. 2012. Oxidative damage and antioxidants induced by heavy metal stress in two cultivars of mustard (*Brassica juncea* L.) plants. *Fresenius Environ. Bull.* 21: 2953–2961.

Ahmadizadeh, M., N. A. Vispo, C. Calapit-Palao and R. K. Singh. 2016. Reproductive stage salinity tolerance in rice: a complex trait to phenotype. *Ind. J. Plant Physiol.* (Springer) 21: 258.

Ahn, J. W., K. Hwangbo and C. J. Yin. 2015. Salinity-dependent changes in growth and fatty acid composition of new Arctic Chlamydomonas species, ArM0029A. *Plant Cell Tissue Organ Cult.* 120:1015–1021.

Akram, A. and M. Ashraf. 2011. Improvement in growth, chlorophyll pigments and photosynthetic performance in salt-stressed plants of sunflower (*Helianthus annuus* L.) by foliar application of 5-aminolevulinic acid. *Agrochimica* 55: 94–104.

Albertsson, P. A. 1995. The structure and function of the chloroplast photosynthetic membrane – a model for the domain organization. *Photosynth. Res.* 46: 141–149.

Ali-Reza, A., M. Hajirezaei, D. Hofius, U. Sonnewald and L. M. Voll. 2007. Specific roles of α- and γ-tocopherol in abiotic stress responses of transgenic tobacco. *Plant Physiol.* 143: 1720–1738.

Anjum, S.A., U. Ashraf, A. Zohaib, M. Tanveer, M. Naeem, I. Ali, T. Tabassum and U. Nazir. 2017. Growth and developmental responses of crop plants under drought stress: a review. *Zemdirbyste Agric.* 104: 267–276.

Apel, K. and H. Hirt 2004. Reactive oxygen species: metabolism, oxidative stress, and signal transduction. *Annu. Rev. Plant Biol.* 55: 373–399.

Apostolova, E. L. and A. N. Misra. 2014. Alterations in structural organization affect the functional ability of photosynthetic apparatus. In: *Handbook of Plant and Crop Physiology*, ed. M. Pessarakli, Marcel Dekker, New York, pp. 103–118.

Apostolova, E., M. Busheva and B. Tenchov. 1992. Freezing damage and protection of photosystem II by sucrose and trehalose. In: *Reserch in Photosynthesis*, Vol. IV, ed. N. Murata, Kluwer Academic Publishers, Netherlands, pp. 165–168.

Apostolova, E., M. Busheva and B. Tenchov. 1994. Inactivation and protection of photosystem 2 particles during freezing. *Photosynthetica* 30: 475–479.

Arfan, M., H. R. Athar and M. Ashraf. 2007. Does exogenous application of salicylic acid through the rooting medium modulate growth and photosynthetic capacity in two differently adapted spring wheat cultivars under salt stress? *J. Plant Physiol.* 164: 685–694.

Ashraf, M. and P. Harris. 2013. Photosynthesis under stressful environments: an overview. *Photosynthetica* 51: 163e190.

Ashraf, M. and Q. Ali. 2008. Relative membrane permeability and activities of some antioxidant enzymes as the key determinants of salt tolerance in Canola (*Brassica napus* L.). *Env. Experim. Bot.* 63: 266–273.

Asrar, H., T. Hussain and S. M. Hadi. 2017. Salinity induced changes in light harvesting and carbon assimilating complexes of *Desmostachya bipinnata* (L.) Staph. *Environ. Exp. Bot.* 135: 86–95.

Attia, H., N. Karray, N. Msilini and M. Lachaal. 2011. Effect of salt stress on gene expression of superoxide dismutases and copper chaperone in Arabidopsis thaliana. *Biol. Plant* 55: 159–163.

Austin, J. R., E. Frost, P. A. Vidi, F. Kessler and L. A. Staehelin. 2006. Plastoglobules are lipoprotein subcompartments of the chloroplast that are permanently coupled to thylakoid membranes and contain biosynthetic enzymes. *Plant Cell* 18: 1693–1703.

Azevedo Neto, A. D., J. T. Prico, J. Eneas-Filho, C. E. Braga de Abreu and E. Gomes-Filho. 2006. Effect of salt stress on antioxidative enzymes and lipid peroxidation in leaves and roots of salt-tolerant and salt-sensitive maize genotypes. *Env. Exp. Bot.* 56: 235–241.

Baker, N. R. 1991. Possible role of photosystem II in environmental perturbations of photosynthesis. *Physiol. Plant.* 81: 563–570.

Bastías, E., M. B. González-Moro and C. González-Murua. 2013. Interactive effects of excess boron and salinity on histological and ultrastructural leaves of Zea mays amylacea from the Lluta Valley (*Arica-Chile*). *Cienc. Inv. Agrar.* 40: 581–595.

Baxter, A., R. Mittler and N. Suzuki 2014. ROS as key players in plant stress signalling. *J. Exp. Bot.* 65:1229–1240.

Bhutta, W. M. 2011. Antioxidant activity of enzymatic system of two different wheat (*Triticum aestivum* L.) cultivars growing in under salt stress. *Plant Soil Env.* 57: 101–107.

Biswal, A. K., F. Dilnawaz, N. K. Ramaswamy, K. A. V. David and A. N. Misra. 2002. The termoluminescence characteristics of sodium chloride salt stressed Indian mustard seedlings. *Luminescence* 17: 135–140.

Biswal, A. K., N. K. Ramaswamy, M. Mathur and A. N. Misra. 2001. Light regulated protein kinase activity in thylakoid membranes of NaCl salt stressed seedlings. In: *Photosynthis*: PS2001. S24–006. CSIRO Publ., Melbourne, Australia.

Bratt, C. E., P. O. Arvidsson, M. Carlsson and H. E. Akerlund. 1995. Regulation of violaxanthin de-epoxidase activity by pH and ascorbate concentration. *Photosynth. Res.* 45: 169–175.

Brevik, E. C., A. Cerdà, J. Mataix-Solera, L. Pereg, J. N. Quinton, J. Six and K. Van Oost. 2015. The interdisciplinary nature of soil. *Soil* 1: 117–129.

Brugnoli, E. and M. Lauteri. 1991. Effects of salinity on stomatal conductance, photosynthetic capacity, and carbon isotop discrimination of salt-tolerant (*Gossypium hirsutum* L.) and salt-sensitive (*Phaseolus vulgaris* L.) C3 non-halophytes. *Plant Physiol.* 95: 628–635.

Bruns, S. and C. Hecht-Buchholz. 1990. Light and electrone microscope studies on the leaves of several potato cultivars after application of salt at various development stages. *Potato Research* 33: 33–41.

Büchel, C., J. Barber, G. Ananyev, S. Eshaghi, R. Watt and C. Dismukes. 1999. Photoassembly of the manganese cluster and oxygen evolution from monomeric and dimeric CP47-reaction center photosystem II complexes. *Proc. Natl. Acad. Sci. USA* 96: 14288–14293.

Busheva, M. and E. Apostolova. 1989. Cryoprotection of photosystem 2 particles by glycerol, saccharose and trehalose. *Photosynthetica* 23: 380–382.

Bybordi, A., M. B. Shako Zamani and K. T. Nezami. 2010. Effects of nacl salinity levels on lipids and proteins of canola (*Brassica napus* L.) cultivars. *Adv. Env. Biol.* 4: 397–403.

Caffarri, S., T. Tibiletti, R. Jennings and S. Santabrbara.2014. A comparison between plant photosystem I and photosystem II architecture and functioning. *Curr. Prot. Pep. Sci.* 15: 296–331.

Carillo, P., M. Grazia Annunziata, G. Pontecorvo, A. Fuggi and P. Woodrow. 2011. Salinity stress and salt tolerance. In: *Abiotic Stress in Plants—Mechanisms and Adaptations*, ed. A. K. Shanker and B. Venkateswarlu, INTECH, pp. 21–38.

Chaitanya, K. V., C. Rama Krishna, G. Venkata Ramana, and S.K. Khasim Beebi. 2014. Salinity stress and sustainable agriculture—a review. *Agric. Reviews* 35: 34–41.

Chaves, M. M., J. Flexas and C. Pinheiro. 2009. Photosynthesis under drought and salt stress: Regulation mechanisms from whole plant to cell. *Ann. Bot.* 103: 551–560.

Chen, T. H. and N. Murata. 2008. Glycinebetaine: an effective protectant against abiotic stress in plants. *Trends Plant Sci.* 13: 499–505.

Chen, T. H. and N. Murata. 2011. Glicinebetaine protects some plants against abiotic stress: mechanisms and biotechnological applications. *Plant Cell Environ.* 34: 1–20.

Chinnusamy, V. and J. K. Zhu. 2006. Salt stress signaling and mechanisms of plant salt tolerance. *Genetic Engineering* 27: 141–177.

Cramer, G. R. and S. A. Quarrie. 2002. Abscissic acid is correlated with leaf growth inhibition of four genotypes of maize differing in their response to salinity. *Funct. Plant Biol.* 29: 111–119.

Croce, R. and H. Amerongen. 2011. Light-harvesting and structural organization of Photosystem II: From individual complexes to thylakoid membrane. *J Photochem Photobiol B* 104:142–153.

Çulha Erdal, S. and H. Çakirlar. 2014. Impact of salt stress on photosystem II efficiency and antioxidant enzyme activities of safflower (*Carthamus tinctorius* L.) cultiva. *Turk. J. Biol.* 38: 549–560.

Daliakopoulos, L. N., I. K. Tsanis, A. Koutroulis, N. Kourgialas, A. E. Varouchakis, G. P. Karatzas and C. J. Ritsema. 2016. The threat of soil salinity: A European scale review. *Sci. Tot. Env.* 573: 727–739.

Das, N., M. Misra and A. N. Misra. 1990. Sodium chloride salt stress induced metabolic changes in pearl millet callus: Free solutes. *J. Plant Physiol.* 137: 244–246.

Das, N., M. Misra and A. N. Misra. 1992. Sodium chloride salt stress induced metabolic changes in pearl millet callus: oxidases. *Proc. Nat. Acad. Sci., India, Sect. B* 62: 263–268.

Davison, P. A., C. N. Hunter and P. Horton. 2002. Over expression of β-carotene hydroxylase enhances stress tolerance in *Arabidopsis*. *Nature* 418: 203–206.

Demmig-Adams, B. and W. W. Adams. 2002. Antioxidants in photosynthesis and human nutrition. *Sci.* 298: 2149–2153.

Deng, L., R. Fukuda, T. Kakihara, K. Narita and A. Ohta. 2010. Incorporation and remodeling of phosphatidylethanolamine containing short acyl residues in yeast. *Biochim. Biophys. Acta* 1801: 635–645.

Dionisio-Sese, M. and S. Tobita. 2000. Effect of salinity on sodium content and photosynthetic responses of rice seedlings differing in salt tolerance. *J. Plant Physiol.* 157: 54–58.

Dobrikova, A. G., L. Domonkos, O. Sözer, H. Laczkó-Dobos, M. Kis, A. Párducz and E. L. Apostolova. 2013. Effect of partial or complete elimination of light-harvesting complexes on the surface electric properties and the functions of cyanobacterial photosynthetic membranes. *Physiologia Plantarum* 147: 248–260.

Doğan, M, R. Tıpırdamaz and Y. Demir. 2010a. Effective salt criteria in callus—cultured tomato genotypes. *J. Biosci.* 65c: 613–618.

Doğan, M, R. Tıpırdamaz and Y. Demir. 2010b. Salt resistance of tomato species grown in sand culture. Plant Soil Env. 56: 499–507.

Dörmann, P. 2007. Functional diversity of tocochromanols in plants. *Planta* 225: 269–276.

Duchêne, S. and P.A. Siegenthaler. 2000. Cyclodextrins: a potential tool for studying the role of glycerolipids in photosynthetic membranes. *Lipids* 37: 201–208.

Eckardt, A. 2009. A new chlorophyll degradation pathway. *Plant Cell* 21: 700.

Edreva, A. 2005. Generation and scavenging of reactive oxygen species in chloroplasts: A submolecular approach. *Agr. Ecosyst. Env.* 106: 119–133.

Fahad, S., S. Hussain, A. Matloob, F. A. Khan, A. Khaliq, S. Saud, S. Hassan, D. Shan, F. Khan, N. Ullah, M. Faiq, M.R. Khan, A.K. Tareen, A. Khan, A. Ullah, N. Ullah and J. Huang. 2015. Phytohormones and plant responses to salinity stress: a review. *Plant Growth Reg.* 75: 391–404.

Garnier, F., J. P. Dubacq and J. C. Thomas. 1994. Evidence for a transient association of new proteins with the Spirulina maxima phycobilisomes in relation to light intensity. *Plant Physiol.* 106: 747–754.

Ghosh, B., M. N. Ali and S. Gantait. 2016. Response of rice under salinity stress: A review update. *J. Res. Rice* 4: 167.

Ghosh, S., K. A. Hudak, E. B. Dumbroff and J. E. Thompson. 1994. Release of photosynthetic protein catabolites by blebbing from thylakoids. *Plant Physiol.*106: 1547–1553.

Gill, S. S. and N. Tuteja. 2010. Reactive oxygen species and antioxidant machinery in abiotic stress tolerance in crop plants. *Plant Physiol Biochem.* 48: 909–930.

Gomathi, R. And P. Rakkiyapan. 2011. Comparative lipid peroxidation, leaf membrane thermostability, and antioxidant system in four sugarcane genotypes differing in salt tolerance. *Int. J. Plant Physiol. Biochem.* 3: 67–74.

Govindjee. 2006 Photosystem II: The light-driven water: Plastoquinone oxidoreductase. In: *Advances in Photosynthesis and Respiration*, ed. T. J. Wydrzynski and K. Satoh, Springer, Dordrecht, The Netherlands, *Photosynth. Res.* 87: 331–335.

Gupta, B. and B. Huang. 2014. Mechanism of salinity tolerance in plants: Physiological, biochemical, and molecular characterization. Hindawi Publishing Corporation, *Int. J. Genomics*, Article ID 701596, 18.

Hachicha, M., B. Kahlaoui, N. Khamassi, E. Misle and O. Jouzdan. 2018. Effect of electromagnetic treatment of saline water on soil and crops. *J. Saudi Soc. Agr. Sci.* 17: 154–162.

Hajlaoui, H., N. E. Ayeb, J. P. Garrec and M. Denden. 2010. Differential effects of salt stress osmotic adjustment and solutes allocation on the basis of root and leaf tissue senescence of two silage maize (*Zea mays* L.) varieties. *Ind. Crops Prod.* 31: 122–130.

Hapani, P. and D. Marjadi. 2015. Salt tolerance and biochemical responses as a stress indicator in plants to salinity: A review. CIB Tech. J. Biotechnol. 4: 33–46.

Harborne, J. B. 1964. Biochemistry of phenolic compounds. Academic, London, pp: 511–543.

Hasanuzzaman, M., M. A. Hossain, J. A. da Silva and M. Fujita. 2012. Plant responses and tolerance to abiotic oxidative stress: antioxidant defenses is a key factors. In: *Crop stress and its management: Perspectives and strategies*, ed. V. Bandi, A. K. Shanker, C. Shanker and M. Mandapaka, Springer, Berlin, pp. 261–316.

Hasanuzzaman, M., K. Nahar and M. Fujita. 2013. Plant response to salt stress and role of exogenous protectants to mitigate salt-induced damages. In: *Ecophysiology and Responses of Plants under Salt Stress*, ed. P. Ahmad, M. Azooz and M. Prasad, Springer, New York, NY, pp. 25–87.

He, Y., C. Yu, , L.Zhou, Y.Chen, A. Liu, J. Jin, J. Hong, Y.Qi and D. Jiang. 2014. Rubisco decrease is involved in chloroplast protrusion and Rubisco-containing body formation in soybean (Glycine max.) under salt stress. *Plant Physiol. Biochem.* 74: 118–124.

Hernandez, J, A. Campillo, A. Jimenez, J. Alarcon and F. Sevilla. 1999. Response of antioxidant systems and leaf water relations to NaCl stress in pea plants. *The New Phytologist* 141: 241–251.

Hnilickova, H., V. Hejnak, L. Nemcova, J. Martinkova, M. Skalicky, F. Hnilicka and P. Grieu. 2017. The effect of freezing temperature on physiological traitsin sunflower. *Plant Soil Env.* 63: 375–380.

Hosono, K. 1992. Effect of salt stress on lipid composition and membrane fluidity of the salt-tolerance yeast Zygosaccharomyces rouxii. *J. Gen. Microbiol.* 138: 91–96.

Hounslow, E., R. V. Kapoore, S. Vaidyanathan, D. J. Gilmour and P. C. Wright. 2016. The search for a lipid trigger: The effect of salt stress on the lipid profile of the model microalgal species *Chlamydomonas reinhardtii* for Biofuels Production. *Curr. Biotechnol.* 5: 305–313.

Huang, C., W. He, J. Guo, X. Chang, P. Su and L. Zhang. 2005. Increased sensitivity to salt stress in an ascorbate-deficient *Arabidopsis* mutant. *J. Exp. Bot.* 422: 3041–3049.

Ischebeck, T., A. M. Zbierzak, M. Kanwischer and P. Dörmann. 2006. A salvage pathway for phytol metabolism in *Arabidopsis*. *J. Biol. Chem.* 281: 2470–2477.

Iyengar, E. R. and M. P. Reddy. 1996. Photosynthesis in highly salt-tolerant plants. In: *Handbook of Photosynthesis*, ed. M. Pessarakli, Marcel Dekker; New York, p. 909.

Iyer, S. and A. Caplan. 1998. Products of proline catabolism can induce osmotically regulated genes in rice. *Plant Physiol.* 116: 203–211.

Jajoo, A. 2014. Changes in photosystem II heterogeneity in response to high salt stress. In: *Contemporary Problems of Photosynthesis*, ed. S. I. Allakhverdiev, A. B. Rubin and V. A. Shuvalov, Institute of Computer Science, Izhevsk, Moscow, pp. 397–413.

Javed, S., S. A. Bukhari, M. Y. Ashraf, S. Mahmood and T. Iftikhar. 2014. Effect of salinity on growth, biochemical parameters and fatty acid composition in safflower (*Carthamus tinctorius* L.). *Pak. J. Bot.* 46: 1153–158.

Jensen, P. E., R. Bassi, E. J. Boekema, J. P. Dekker, S. Jansson, D. Leister, C. Robinson and H. V. Scheller. 2007. Structure, function and regulation of plant photosystem I. Biochim. Biophys. Acta 1767: 335–352.

Jin, X., Y. Huang, F. Zeng, M. Zhou and G. Zang. 2009. Genotypic difference in response of peroxidase and superoxide dismutase isozymes and activities to salt stress in barley. *Acta Phys. Plant.* 31: 1103–1109.

Johnson, G. N., J. D. Scholes, P. Horton and A. J. Young. 1993. Relationship between carotenoid composition and growth habot in British plant species. *Plant Cell Environ.* 16: 681–686.

Joshi, P., A. N. Misra, L. Nayak and B. Biswal. 2013. Responce of mature, developing and senescing chloroplast to environmental stress. In: Plastid Development in Leaves during Growth and Senescence. (Eds. B. Biswal, K. Krupinska and U. C. Biswal), *Advances in Photosynthesis and Respiration* (Series Eds. Govindjee & TD Sharkey). Vol. 36, Chapter 28, Springer: Dordrecht, pp. 641–668.

Juan, M., M. Rivero, L. Romero and M. Ruiz. 2005. Evaluation of some nutritional and biochemical indicators in selecting salt-resistant tomato cultivars. *Environ. Exp. Bot.* 54: 193–201.

Jusovic, M., M. Y. Velitchkova, S. P. Misheva, A. Börner, E. L. Apostolova and A. G. Dobrikova. 2018. Photosynthetic responses of a wheat mutant (*Rht-B1c*) with altered DELLA proteins to salt stress. *J. Plant Growth Reg.* 37: 645–656.

Kalaji, H. M., A. Rastogi, M. Zivcak, M. Brestic, A. Daszkowska-Golec, K. Sitko, K. I. Alsharafa, R. Lotfi, P. Stypinski, L. A. Samborska and M. D. Cetner. 2018. Prompt chlorophyll fluorescence as a tool for phenotyping: an example of barley landcraces exposed to various abiotic stress factors. *Photosynthetica* 56: 953–961.

Kalaji, H. M., M. Schansker, M. Brestic, F. Bussotti, A. Calatayud, L. Ferroni, V. Goltsev, L. Guidi, A. Jajoo, P. Li, P. Losciale, V. K. Mishra, A. N. Misra, S. G. Nebauer, S. Pancaldi, S. Pancaldi, C. Penella, M. Pollastrini, K. Suresh, E. Tambussi, M. Yanniccari, M. Zivcak, M. D. Cetner, I. A. Samborska, A. Stirbet, K. Olsovska, K. Kunderlikova, H. Shelonzek, S. Rusinowski and W. Baba. 2017. Frequently asked questions about *in vivo* chlorophyll fluorescence, the sequel. *Photosynth Res.* 132: 13–66.

Kao, W. Y., H. C. Tsai, and T. T. Tsai. 2001. Effect of NaCl and nitrogen availability on growth and photosynthesis of seedlings of a mangrove species, *Kandelia candel* (L.) Druce. *J. Plant Physiol.* 158: 841–846.

Karapetyan, N. V. 2004. Interaction of pigment-protein complexes within aggregates stimulates dissipation of excess energy. Biochemistry (Mosc.) 69: 1299–1304.

Kavi Kishor, P. B., Z. Hong, C. H. Miao, C. A. Hu and D. S. Verma. 1995. Overexpression of A1-pyrroline-5 carboxylate synthetase increases proline production and confers osmotolerance in transgenic plants. *Plant Physiol.* 108: 1387–1394.

Keiper, F., D. Chen and L. De Filippis. 1998. Respiratory, photosynthetic and ultrastructural changes accompanying salt adaptation in culture of Eucalyptus microcorys. *J. Plant Physiol.* 152: 564–573.

Khalid, A. 2017. Khalid Changes in Lipids and Fatty Acids of Nigella sativa L. under Salinity Stress. *J. Mater. Environ. Sci.* 8: 3502–3507.

Khan, A. and U. Shirazi. 2009. Role of proline, K/Na ratio and chlorophyll content in salt tolerance of wheat (*Triticum aestivum* L.). *Pak. J. Bot.* 41: 633–638.

Khavari-Nejad, R. A. and Y. Mostofi. 1998. Effects of NaCl on photosynthetic pigments, saccharides and chloroplast ultrastructure in leaves of tomato cultivars. *Photosynthetica* 35: 151–154.

Kis, M., O. Zsiros, T. Farkas, H. Wada, F. Nagy and Z. Gombos. 1998. Light-induced expression of fatty acid desaturase genes. *Proc. Natl. Acad. Sci. USA* 95: 4209–4214.

Kouřil, R., J. P. Dekker and E. J. Boekema. 2012. Supramolecular organization of photosystem II in green plants. *Biochim. Biophys. Acta Bioenerg.* 1817: 2–12.

Krasensky, J. and C. Jonak. 2012. Drought, salt and temperature stress-induced metabolic rearrangements and regulatory networks. *J. Exp. Bot.* 63: 1593–1608.

Kuhlbrandt, W., D. N. Wang and Y. Fujiyoshi. 1994. Atomic model of plant light harvesting complex by electron crystallography. *Nature* 367: 614–621.

Kurban, H., H. Saneoka, K. Nehira, R. Adilla and K. Fujita. 1998. Effect of salinity on growth and accumulation of organic and inorganic solutes in the leguminous plants Alhagi pseudoalhagi and Vigna radiata. *Soil Sci. Plant Nutr.* 44: 589–597.

Labudda, M. 2013. Lipid peroxidation as a biochemical marker for oxidative stress during drought. An effective tool for plant breeding. E-wydawnictwo, Poland, http://www.e-wydawni ctwo.eu/Document/DocumentPreview/3342.

Laczko-Dobos, H., B. Ughy, S. Toth and Z. Gombos. 2008. Role of phosphatidylglycerol in the function and assembly of Photosystem II reaction center, studied in a cdsA-inactivated PAL mutant strain of *Synechocystis* sp. PCC6803 that lacks phycobilisomes. *Biochimica et Biophysica Acta* 1777: 1184–1194.

Lakra, N., C. Kaur, K. Anwar, S. Pareek and A. Pareek. 2017. Proteomics of contrasting rice genotypes: Identification of potential targets for raising crops for saline environment. *Plant. Cell Environ.* 41: 947–969.

Latowski, D., J. Grzyb and K. Strzalka. 2004. The xanthophyll cycle – molecular mechanism and physiological significance. *Acta Physiol. Plant.* 26: 197–212.

Lee, H. Y., J. S. Seo, J. H. Cho, H. Jung, J. K. Kim and J.S. Lee 2013. Oryza sativa COI homologues restore jasmonate signal transduction in Arabidopsis coi1-1 mutants. *PLoS One* 8: e52802.

Li, W. and Q. Li. 2017. Effect of environmental salt stress on plants and the molecular mechanism of salt stress tolerance. *Int. J. Environ. Sci. Nat. Res.* 7.

Locy, R. D., C. Ching-Chun , B. Nielsen and N. K. Singh. 1996. Photosynthesis in salt-adapted heterotrophic tobacco cells and regenerated plants. *Plant Physiol.* 110: 321–328.

Loreto, F., M. Centritto and K. Chartzoulakis. 2003. Photosynthetic limitations in olive cultivars with different sensitivity to salt stress. *Plant Cell Environ.* 26: 595e601.

Lu, C. and A. Vonshak. 2002, Effects of salinity on photosystem II function in cyanobacterial Spirulina platensis cells. *Physiol. Plant* 114: 405–413.

Lu, C., B. Jiang, B. Wang and T. Kuang. 2003. Photosystem II photochemistry and photosynthetic pigment composition in salt adapted chalophyte *Artemisia anethifolia* grown under outdoor conditions. *J. Plant Physiol.* 160: 403e408.

Luciński, R. and G. Jackowski. 2006. The structure, functions and degradation of pigment-binding proteins of photosystem II. *Acta Biochim. Polonica* 53: 693–708.

Maiti, R. K., S. Moreno-Limon and P. Wesche Ebeling. 2000. Responses of some crops to various abiotic stress factors and its physiological and biochemical basis of resistances. *Agric Rev.* 21: 155–167.

Makela, P., R. Munns, T. D. Colmer and P. Pelton-Sainio. 2003. Growth of tomato and an ABA deficient mutant (*sitiens*) under saline conditions. *Physiol. Plant.* 17: 58–63.

Maksimovič D. J. Zhang, F. Zeng, B. D. Živanovič, L. Shabala, M. Zhou and S. Shabala. 2013 Linking oxidative and salinity stress tolerance in barley: can root antioxidant enzyme activity be used as a measure of stress tolerance? *Plant Soil* 365:141–155

Mamedov, M. Govindjee, V. Nadtochenko and A. Semenov. 2015. Primary electron transfer processes in photosynthetic reaction centers from oxygenic organisms. *Photosynth Res.* 125: 51–63.

Mateo-Sagasta, J. and J. Burke. 2011. Agriculture and water quality interactions: a global overview SOLAW Background Thematic Report, FAO Rome.

Mbarki, S., O. Sytar, A. Cerda, M. Zivcak, A. Rastogi and X. He. 2018. Strategies to mitigate the salt stress effects on photosynthetic apparatus and productivity of crop plants. In: *Salinity Responses and Tolerance in Plants*, ed. V. Kumar et al., Vol. 1, pp. 85–136, Springer.

Mehraban, A., F. Kadali and M. Miri. 2017. Influence of salt stress on lipids metabolism, photorespiration, photosynthesis and chlorophyll fluorescence in crop plants. *Chem. Res. J.* 2: 127–132.

Mehta P., A. Jajoo, S. Mathur and S. Bharti. 2010. Chlorophyll a fluorescence study revealing effects of high salt stress on photosystem II in wheat leaves. *Plant Physiol. Biochem.* 48: 16–20.

Meng, F., Q. Luo, Q. Wang, X. Zhang, Z. Qi, F. Xu, X. Lei, Y. Cao, W.S. Chow and G. Sun. 2016. Physiological and proteomic responses to salt stress in chloroplasts of diploid and tetraploid black locust (*Robinia pseudoacacia* L.). *Sci. Rep.* 6: 23098.

Milborrow, B.V. and H. S. Lee. 1998. Endogenous biosynthetic precursor of (+)-Abscicic acid. VI.Carotenoids and ABA are formed by the 'non-mevalonate' trosepyruvate pathway in chloroplasts. *Aust. J. Plant Physiol.* 25: 507–512.

Miller, G., N. Suzuki, S. Ciftci-Yilmaz and R. Mittler. 2010. Reactive oxygen species homeostasis and signalling during drought and salinity stresses. *Plant Cell Environ.* 33: 453–467.

Misra, A. N. 2018a. Chlorophyll fluorescence: A practical approach to study ecophysiology of green plants. In: *Advances in Plant Ecophysiology Techniques.* ed. A. M. Sánchez-Moreiras and M. J. Reigosa, Springer International Publishing AG, part of Springer Nature, pp. 77–97.

Misra, A. N. 1995. Pearl millet seedling emergence and growth under soil crust. *J. Env. Biol.* 16: 113–117.

Misra, A. N. 2018b. Thermoluminescence: A tool to study ecophysiology of green plants. In: *Advances in Plant Ecophysiology Techniques.* ed. A. M. Sánchez-Moreiras and M. J. Reigosa, Springer International Publishing AG, part of Springer Nature, pp. 99–108.

Misra, A. N. and A. K. Biswal. 2000. Thylakoid membrane protein kinase activity as a signal transduction pathway in chloroplasts. *Photosynthetica* 38: 323–332.

Misra, A. N., A. K. Biswal and M. Misra. 2002. Physiological, biochemical and molecular aspects of water stress responses in plants, and the biotechnological applications. Proc. Nat. Acad. Sci.(India) 72B: 115–134.

Misra, A. N., A. Srivastava and R. J. Strasser. 2001a. Utilisation of fast chlorophyll a fluorescence technique in assessing the salt/ion sensitivity of mung bean and brassica seedlings. *J. Plant Physiol.* 158: 1173–1181.

Misra, A. N., B. Murmu, P. Singh and M. Misra. 1996. Growth and proline accumulation mung bean seedlings as affected by sodium chloride. *Biol. Plant.* 38: 531–536.

Misra, A. N., D. Latowski and K. Strzalka. 2003. De-epoxidation state of lutein and violaxanthin in the seedlings of salt sensitive and salt tolerant plant grown under NaCl salt stress. Data presented at Plant Biology 2003 Honolulu, Hawaii, USA, 2530 July.

Misra, A. N., D. Latowski and K. Strzalka. 2006. The xanthophylls cycle activity in kidney bean and cabbage leaves under salinity stress. *Russian J. Plant Physiol.* 53: 102–109.

Misra, A. N., F. Dilnawaz, M. Misra and A. K. Biswal. 2001b. Thermoluminescence in chloroplasts as an indicator of alterations in photosystem II reaction center by biotic and abiotic stress. *Photosynthetica* 39: 1–9.

Misra, A. N., F. Dilnawaz, M. Misra and A. K. Biswal. 2001c. Thermoluminescence in chloroplasts. In: *Biophysical Processes in Living Systems.* ed. B. Padha, pp. 303–311, Saradhi, Oxford/IBH, New Delhi/ Science Publ., Enfield (NH), USA, Plymouth, UK.

Misra, A. N., M. Misra and N. Das. 1990. Plant responses to salinity: Metabolic changes and the use of plant tissue culture—a perspective. In: *Environmental Concern and Tissue Injury*, ed. R. Prakash and S. M. Choubey, Jagmandir Books, New Delhi, pp. 77–84.

Misra, A. N., M. Misra and R. Singh. 2012a. Chlorophyll fluorescence in plant biology. In: *Biophysics.* ed. N. A. Misra Intech Open, pp. 171–192.

Misra, A. N., M. Misra and R. Singh. 2012b. Thermoluminescence in chloroplast thylakoid. In: *Biophysics.* ed. A. N. Misra, Intech Open, pp. 155–170.

Misra, A. N., S. M. Sahu and M. Misra. 1995. Soil salinity induced changes in pigment and protein content of cotyledons and leaves in Indian mustard *Brassica juncea* Coss. *Acta Physiol. Plant.* 17: 375–380.

Misra, A. N., S. M. Sahu, M. Misra, N. K. Ramaswamy and T. S. Desai. 1999. Sodium chloride salt stress induced changes in thylakoid pigment-protein complexes, PS II activity and TL glow peaks of chloroplasts from mungbean (*Vigna radiata* L.) and Indian mustard (*Brassica juncea* Coss.) seedlings. *Z. Naturforsch.* 54c: 640–644.

Misra, A. N., S. M. Sahu, M. Misra, P. Singh, I. Meera, N. Das, M. Kar and P. Sahu. 1997a. Root growth of a salt susceptible and a salt resistant rice (*Oryza sativa* L.) during seedling establishment under NaCl salinity. *J. Agron. Crop Sci.* 178: 9–14.

Misra, A., S. Sahu , M. Misra, P. Singh, I. Meera, N. Das, M. Kar and P. Sahu. 1997b. Sodium chloride induced changes in leaf growth, and pigment and protein contents in two rice cultivars. *Biol. Plant* 39: 257–262.

Misra, A. N., R. Vladkova, R. Singh, M. Misra, A. G. Dobrikova and E. L. Apostolova. 2014. Action and target sites of nitric oxide in chloroplasts. *Nitric Oxide* 39: 35–45.

Mitsuya, S., M. Kawasaki, M. Taniguchi and H. Miyake. 2003. Light dependency of salinity-induced chloroplast degradation. *Plant Prod. Sci.* 6: 219–223.

Mitsuya, S., Y. Takeoka and H. Miyake. 2000. Effects of sodium chloride on foliar ultrastructure of sweet potato (*Ipomoea batatas* Lam.) plantlets grown under light and dark conditions in vitro. *J. Plant Physiol.* 157: 661–667.

Müller, M., H. H. Kunz, J. I. Schroeder, G. Kemp, H. S. Young and H. E. Neuhaus. 2014. Decreased capacity for sodium export out of Arabidopsis chloroplasts impairs salt tolerance, photosynthesis and plant performance. *Plant J.* 78: 646–658.

Muller, P., X. P. Li and K. K. Niyogi. 2001. Non-photochemical quenching. A response to excess light energy. *Plant Physiol.* 125: 1558–1566.

Mullineaux, C. W. and H. Kirchhoff. 2009. Role of lipids in the dynamics of thylakoid membranes. In: *Lipids in Photosynthesis: Essential and Regulatory Functions*, ed. H Wada and N. Murata, Springer Science, London, pp. 283–294.

Munné-Bosch, S. 2005. The role of α-tocopherol in plant stress tolerance. *J. Plant Physiol.* 162: 743–748.

Munné-Bosch, S. and I. Alegre. 2004. Die and let live: leaf senescense contributes to plant survival under drought stress. *Funct. Plant Biol.* 31: 203–216.

Munns, R. 2005. Genes and salt tolerance: Bringing them together. *Plant Physiol.* 167: 645–663.

Munns, R. and M. Tester. 2008. Mechanisms of salinity tolerance. *Ann. Rev. Plant Biol.* 59: 651–681.

Murata, N. and H. Wada. 1995. Acyl-lipid desaturases and their importance in the tolerance and acclimatization to cold of cyanobacteria. *Biochem. J.* 308: 1–8.

Nagamiya, K., T. Motohashi, K. Nakao, S. H. Prodhan, E. Hattori, S. Hirose, K. Ozawa, Y. Ohkawa, T. Takabe, T. Takabe and A. Komamine. 2007. Enhancement of salt tolerance in transgenic rice expressing an *Escherchia coli* catalase gene, katE. *Plant Biotechnol. Rep.* 1: 49–55.

Nagy, Z., H. Daood, Z. Ambrozy and L. Helyes. 2015. Determination of polyphenols, capsaicinoids, and Vitamin C in new hybrids of chilli peppers. *J. Anal. Meth. Chem.* 33: 1–10.

Negrao, S., S. M. Schmockel and M. Tester. 2017. Evaluating physiological responses of plants to salinity stress. *Ann Bot.* 119: 1–11.

Netondo, G. W., J. C. Onyango and E. Beck. 2004. Sorghum and salinity: II. Gas exchange and chlorophyll fluorescence of sorghum under salt stress. *Crop Sci.* 44: 806–811.

Neubauer, C. and H. E. Yamamoto. 1994. Membrane barriers and Mehler peroxidase reaction limit the ascorbate availability for vialoxanthin de-epoxidase activity in intact chloroplasts. *Phtosynth. Res.* 39: 137–147.

Nevo, R., D. Charuvi, O. Tsabari and Z. Reich. 2012. Composition, architecture and dynamics of the photosynthetic apparatus in higher plants. *The Plant J.* 70: 157–176.

Noctor, G. and C. H. Foyer. 1998. Ascorbate and Glutathione: Keeping Active Oxygen under Control. *Annu. Rev. Plant Physiol. Plant Mol. Biol.* 49: 249–279.

Nongpiur, R. C., S. L. Singla-Pareek and A. Pareek. 2016. Genomics approaches for improving salinity stress tolerance in crop plants. *Curr. Genomics* 17: 343–357.

Noreen, Z. and M. Ashraf. 2009. Changes in antioxidant enzymes and some key metabolites in some genetically diverse cultivars of radish (*Raphanus sativus* L.). *Environ. Exp. Bot.* 67: 395–402.

Noreen, Z., M. Ashraf and N. A. Akram. 2010. Salt-induced modulation in some key gas exchange characteristics and ionic relations in pea (*Pisum sativum* L.) and their use as selection criteria. *Crop Pasture Sci.* 61: 369–378.

Norwood, M., M. R. Truesdale, A. Richterand and P. Scott. 2000. Photosynthetic carbohydrate metabolism in the resurrection plant Craterostigma plantagineum. *J. Exp. Bot.* 51: 159–165.

Ohnishi, N. and N. Murata. 2006. Glycinebetaine counteracts the inhibitory effects of salt stress on the degradation and synthesis of the D1 protein during photoinhibition in *Synechococcus* sp. PCC 7942. *Plant Physiol.* 141: 756–765.

Ozakca, D. 2013. Effect of abiotic stress on photosystem I-related gene transcription in photosynthetic organisms.

Papageorgiou, G. C. and N. Murata. 1995. The unusually strong stabilizing effects of glycinebetaine on the structure and function of the oxygen-evolving photosystem II complex. *Photosynth. Res.* 44: 243–252.

Parida, A. K., A. B. Das. and B. Mittra 2003. Effect of NaCl stress on the structure, pigment complex composition and photosynthetic activity of mangrove *Bruguiera parviflora* chlorplasts. *Photosynthesis* 41: 191–200.

Parida, A. K., A. B. Das and B. Mittra. 2004. Effects of salt on growth, ion accumulation, photosynthesis and leaf anatomy of the mangrove, *Bruguiera parviflora*. Trees-Structure Funct. 18: 167–174.

Parihar, P., S. Singh, R. Singh, V. P. Singh and S. M. Prasad. 2015. Effect of salinity stress on plants and its tolerance strategies: a review. *Environ. Sci. Pollut. Res.* 22: 4056–4075.

Percey, W. J., A. McMinn, J. Bose, C. Michael, C. Breadmore, M. Rosanne, E. Guijt and S. Shabala. 2016. Salinity effects on chloroplast PSII performance in glycophytes and halophytes. *Func. Plant Biol.* 43: 1003–1015.

Pinheiro, A., V. Silva and L. Endres. 2008. Leaf gas exchange, chloroplastic pigments and dry matter accumulation in castor bean (*Ricinus communis* L.) seedlings subjected to salt stress conditions. *Ind. Crop. Prod.* 27: 385–392.

Perveen, S., M. Shahbaz and A. Ashraf. 2010. Regulation in gas exchange and quantum yield of photosystem II in salt stressed and non-stressed wheat plantsraised from seed treated with triacontanol. *Pak. J. Bot.* 42: 3073–3081.

Pogson, B. J., K. K. Niyogi, O. Bjorkman and D. Della Penna. 1998. Altered xanthophylls compositions adversely affect chlorophyll accumulation and non-photchemical quenching in *Arabidopsis* mutants. *Proc. Natl. Acad. Sci. U. S. A.* 95: 13324–13329.

Qiu, Z., J. Guo, A. Zhu, L. Zhang, M. Zhang. 2014. Exogenous jasmonic acid can enhance tolerance of wheat seedlings to salt stress. *Ecotox. Environ. Safety* 104: 202–208.

Rahdari, P. and S. Hoseini. 2011. Salinity stress: A review. *Techn. J. Eng. Appl. Sci.* 1: 63–66.

Rahdari, P., S. Tavakoli and S. Hosseini. 2012. Studying of salinity stress effect on germination, proline, sugar, protein, lipid and chlorophyll content in Purslane (*Portulaca oleracea* L.)leaves. *J. Stress Physiol. Biochem.* 8: 182–193.

Rajesh, A., R. Arumugam and V. Venkatesalu. 1998. Growth and photosynthetic characteristics of Ceriops roxburghiana under NaCl stress. *Photosynthetica* 35: 285–287.

Rasool, S., A. Hameed, M. Azooz, M. Muneeb-u-Rehman, T.O. Siddlql and P. Ahmad. 2013. Salt stress: causes, types and responses. In: *Ecophysiology and Response of Plants under Salt Stress*, eds. P. Ahmad et al., Springer Sciences+Business media, LLC.

Reddy, M. and A. Vora. 1986. Changes in pigment composition, hill reaction activity and saccharide metabolism in Bajra (*Penniselum typhoids* H) leaves under NaCl salinity. *Photosynthetica* 20: 50–55.

Renault, H., V. Roussel, A. El Amrani, M. Arzel, D. Renault, A. Bouchereau and C. Deleu. 2010. The Arabidopsis pop2-1 mutant reveals the involvement of GABA transaminase in salt stress tolerance. *BMC Plant Biol.*10: 20.

Rhodes, D. and A. D. Hanson. 1993. Quaternary ammonium and tertiary sulfonium compounds in higher-plants. *Annu. Rev. Plant Physiol. Plant Mol. Biol.* 44: 357–384.

Rice-Evans, C., N. Miller and G. Paganga. 1997. Antioxidant properties of phenolic compounds. *Trends Plant Sci.* 2: 152–159.

Romero-Aranda, R., T. Soria and J. Cuartero. 2001. Tomato plant water uptake and plant water relationships under saline growth conditions. *Plant Sci.* 160: 265–272.

Romero, C., J. M. Belles, J. L. Vaya, R. Serrano and F. A. CulianezMacia. 1997. Expression of the yeast trehalose-6-phosphate synthase gene in transgenic tobacco plants: Pleiotropic phenotypes include drought tolerance. *Planta.* 201: 293–297.

Sahu, S. M., A. N. Misra, M. Misra, N. K. Ramaswamy and T. S. Desai. 1998. Sodium chloride salt stress induced changes in thylakoid pigment-protein complexes, PS II activity of mungbean (*Vigna radiata* L.) seedlings. In: *Photosynthesis: Mechanism and Effects*. Vol. IV, ed. G. Garab, Kluwer, Acad. Publ., The Netherlands, pp. 2625–2628.

Sakihama, Y. and H. Yamasaki. 2002. Lipid peroxidation induced by phenolics in conjunction with aluminum ions. *Biol. Plant.* 45: 249–254.

Salama, E. S., H. C. Kim, R. I. Abou-Shanab, M. K. Ji, Y. K. Oh and S. H. Kim. 2013. Biomass, lipid content, and fatty acid composition of freshwater Chlamydomonas mexicana and Scenedesmus obliquus grown under salt stress. *Bioprocess Biosyst. Eng.* 36: 827–833.

Salama, S., S. Trivedi, M. Busheva, A. A. Arafa, G. Garab, and L. Erdei. 1994. Effects of NaCl salinity on growth, cation accumulation, chloroplast structure and function in wheat cultivars differing in salt tolerance. *J. Plant Physiol.* 144: 241–247.

Santos, C., 2004, Regulation of chlorophyll biosynthesis and degradation by salt stress in sunflower leaves. *Sci. Hort.* 103: 93–99.

Sato, F., K. Murota, S.Aso and Y. Yamada 1992. Salt stress responses in cultured green tobacco cells. In: *Research in Photosynthesis*, Vol. IV, ed. N. Murata, Kluwer Academic Publishers, The Netherlands, pp. 259–262.

Sattler, S. E., L. U. Gilliland, M. Magallanes-Lundback, M. Pollard and D. DellaPenna. 2004. Vitamin E is essential for seed longevity, and for preventing lipid peroxidation during germination. *Plant Cell* 16: 1419–1432.

Semchonok, D. A. 2016. Structures of photosynthetic membrane complexes: Rijksuniversiteit Groningen.

Shabala, S. 2017. Signalling by potassium: another second messenger to add to the list? *J. Exp.Bot.* 68: 4003–4007.

Shah S. H., R. Houborg and M. F. McCabe. 2017. Response of chlorophyll, carotenoid and SPAD-502 measurement to salinity and nutrient stress in wheat (*Triticum aestivum* L.). *Agronomy* 7: 61.

Shahbaz, M. and M. Ashraf. 2013. Improving salinity tolerance in cereals. *Crit. Rev. Plant Sci.* 32: 237–249.

Sharma, P. K. and D. O. Hall. 1992. Changes in carotenoid composition and photosynthesis in sorghum under high light and salt stresses. *J. Plant Physiol.* 140: 661–666.

Sharma, P. K., A. B. Jha, R. S. Dubey and M. Pessarakli. 2012. Reactive oxygen species, oxidative damage and anti-oxidative defense mechanism in plants under stressful conditions. *J. Bot.* 2012: 1–26.

Shi, S. Q., Z. Shi, Z. P. Jiang, L. W. Qi X. M. Sun, C. X. Li, J. F. Liu, W. F. Xiao and S. G. Zhang. 2010. Effects of exogenous GABA on gene expression of Caragana intermedia roots under NaCl stress: regulatory roles for H2O2 and ethylene production. *Plant Cell Environ.* 33: 149–162.

Shokri-Gharelo, R. and P. M. Noparvar. 2018. Molecular response of canola to salt stress: insights on tolerance mechanisms. *Peer J.* 6:e4822.

Shrivastava, P. and R. Kumar. 2015. Soil salinity: A serious environmental issue and plant growth promoting bacteria as one of the tools for its alleviation. *Saudi J. Biol. Sci.* 22: 123–131.

Shu, S., R. Guo, J. Sun and Y. Yuan. 2012. Effects of salt stress on the structure and function of the photosynthetic apparatus in *Cucumis sativus* and its protection by exogenous putrescine. *Physiol. Plant.* 146: 285e296.

Silveira, J. A. G. And F. E. L. Carvalho. 2016. Proteomics, photosynthesis and salt resistance in crops: an integrative view. *J. Proteomics* 143: 24–35.

Silverstein, T., L. Cheng and J. F.Allen. 1993. Chloroplast thylakoid protein phosphatase reactions are redox-independent and kinetically heterogeneous. *FEBS Lett.* 334: 101–105.

Silverstein, T., L. Cheng and J. F. Allen. 1993a. Redox titration of multiple protein phosphorylations in pea chloroplast thylakoids. *Biochim. Biophys. Acta* 1183: 215–220.

Singer, M. A. and S. Lindquist. 1998. Multiple effects of trehalose on protein folding *in vitro* and *in vivo*. *Molecular Cell* 1: 639–648.

Singh, A. K. and R. S. Dubey. 1995. Changes in chlorophyll a and b contents and activity of photosystem I and II in rice seedlings induced by NaCl. *Photosynthetica* 31: 489–499.

Smirnoff, N. and G. L. Wheeler. 2000. Ascorbic acid in plants: biosynthesis and function. *Crit. Rev. Biochem. Mol. Biol.* 35: 291–314.

Solecka, D. 1997. Role of phenylpropanoid compounds in plant responses to different stress factors. *Acta Physiol. Plant* 19: 257.

Stefanov, M., E. Yotsova and E. L. Apostolova. 2018a. Salt- and light-induced changes in spectral characteristics of *Paulownia*. *Comp.* Rend. *Acad. Bulg. Sci.* 71: 1062–1069.

Stefanov, M., E. Yotsova, E. Gesheva, V. Dimitrova, Y. Markovska, S. Doncheva and E. L. Apostolova. 2018b. Effects of salt stress on the functions of the photosynthetic apparatus and the antooxydant activity of *Paulownia*. Data presented at: The 43rd FEBS Congress, Biochemistry for ever 2018, July 712, 2018, Prague, Czech Republic.

Stefanov, M., E. Yotsova, K. Ivanova, Y. Markovska and E. Apostolova. 2016a. Assessment of the impact of salinity on photosynthetic apparatus of *Paulownia elongata x kawakami*. *Comp.* Rend. *Acad. Bulg. Sci.* 69: 1325–1332.

Stefanov, M., E. Yotsova, K. Ivanova, Y. Markovska and E. Apostolova. 2018c. Effect of high light intensity on the photosynthetic apparatus of two hybrid lines of Paulownia grown on soil with different salinity.*Photosynthetica* 56: 832–840

Stefanov, M., K. Yotsova, G. Rashkov, K. Ivanova, Y. Markovska and E. Apostolova. 2016b. Effects of salinity on the photosynthetic apparatus of two Paulownia lines. *Plant Physiol. Biochem.* 101: 54–59.

Stoylova, S., T. D. Flint, R. C. Ford and A. Holzenburg. 2000. Structural analysis of photosystem II in far-red-light-adapted thylakoid membranes. *Eur. J. Biochem.* 267: 207–215.

Subczynski, W. K., E. Markowska and J. Sielewiesiuk. 1991. Effect of polar carotenoids on the oxygen diffusion oncentration product in lipid bilayers. *Biochim. Biophys. Acta* 1068: 68–72.

Sudhir, P. R., D. Pogoryelov, L. Kovacs, G. Garab and S. D. Murthy. 2005. The effects of salt stress on photosynthetic electron transport and thylakoid membrane proteins in the cyanobacterium spirulina platensis. *J. Biochem. Mol. Biol.* 38: 481–485.

Sun, Z. W., L. K. Ren, J. W. Fan, Q. Li, K. J. Wang, M. M. Guo, L. Wang, J. Li, G. X. Zhang, Z. Y. Yang, F. Chen and X. N. Li. 2016. Salt response of photosynthetic electron transport systemin wheat cultivars with contrasting tolerance. *Plant Soil Environ.* 62: 515–521.

Suo, J., Q. Zhao, L. David, S. Chen and S. Dai. 2017. Salinity response in chloroplasts: Insights from gene characterization. *Int. J. Mol. Sci.* 18: 1011

Suprasanna, P., Nikalje and A. N. Rai. 2016. Osmilyte accumulation and implication in abiotic stress tolerance. In: *Osmolytes and Plants Acclimation to Changing Environment: Emerging Omics Technologies*, ed. N. Iqbal, R. Nazar and N. Khan, Springer, New Delhi, pp. 1–12.

Suzuki, N., R. M. Rivero, V. Shulaev, E. Blumwald and R. Mittler. 2014. Abiotic and biotic stress combinations. *New Phytologist* 203: 32–43.

Taïbi, K., F. Taïbi, L. A. Abderrahim, A. Ennajah, M. Belkhodja and J. M. Mulet. 2016. Effect of salt stress on growth, chlorophyll content, lipid peroxidation and antioxidant defence systems in Phaseolus vulgaris L. *South Afr. J. Bot.* 105: 306–312.

Taleisnik, E, A. A. Rodríguez, D. Bustos, L. Erdei and L. Ortega. 2009. Leaf expansion in grasses under salt stress. *J. Plant Physiol.* 166: 1123–1140.

Tanveer, M., B. Shahzad, A. Sharma, S. Biju and R. Bhardwaj. 2018. 24- Epibrassinolide; an active brassinolide and its role in salt stress tolerance in plants: A review. *Plant Physiol. Biochem.* 130: 69–79.

Tavakkoli, E., F. Fatehi, S. Coventry, P. Rengasamy and G. K. McDonald. 2011. Additive effects of Na+ and Cl– ions on barley growth under salinity stress. *J. Exp. Bot.* 62: 2189–2203.

Tavakkoli, E., P. Rengasamy and G.K. McDonald. 2010. High concentrations of Na+ and Cl– ions in soil solution have simultaneous detrimental effects on growth of faba bean under salinity stress. *J. Exp. Bot.* 61: 4449–4459.

Tsvetkova, N., E. Apostolova, A. Brain, P. Quinn and P. Willams. 1995. Factors influencing PS II particle array formation in Arabidopsis thaliana chloroplasts and the relationship of such arrays to the thermostability of PS II. *Biochim Biophys. Acta* 1228: 201–210.

Tunc-Ozdemir, M., G. Miller, L. Song, J. Kim, A. Sodek, S. Koussevitzky, A. N. Misra, R. Mittler and D. Shintani. 2009. Thiamin confers enhanced tolerance to oxidative stress in *Arabidopsis. Plant Physiol.* 151: 421–432.

Unnep, R., O. Zsiros, K. Salymosi, L. Kovacs, P. H. Lambrev and T. Toth. 2014. The ultrastructure and flexibility of thylakoid membranes in leaves and isolated chloroplasts as revealed by small -angle neutron scattering. *Biochim. Biophys. Acta* 1837: 1572–1580.

Vidi, P. A., M. Kanwischer, S. Baginsky, J. R. Austin, G. Csucs, P. Dormann, F. Kessler and C. Brehelin. 2006. Tocopherol cyclase (VTE1) localization and vitamin E accumulation in chloroplast plastoglobule lipoprotein particles. *J. Biol. Chem.* 281: 11225–11234.

Wang, P. and C. P. Song. 2008. Guard cell signalling for hydrogen peroxide and abscisic acid. *New Phytol.* 178: 703–718.

Wang, R. L., C. Hua, F. Zhou and Q. C. Zhou. 2009. Effects of NaCl stress on photochemical activity and thylakoid membrane polypeptide composition of a salt-tolerant and a salt-sensitive rice cultivar. *Photosynthetica* 47: 125–127.

Watanabe, M., D. Watanabe, S. Nogami, S. Morishita and Y. Ohya. 2009. Comprehensive and quantitative analysis of yeast deletion mutants defective in apical and isotropic bud growth. *Curr. Genet.* 55: 365–380.

Winicov, I. and J. R. Seemann, 1990. Expression of genes for photosynthesis and the relationship to salt tolerance of Alfaalfa (*Medicago sativa*) cells. *Plant Cell Physiol.* 31: 1155–1161.

Wungrampha, S., R. Joshi, S. L. Singla-Pareek and A. Pareek. 2018. Photosynthesis and salinity: are these mutually exclusive? *Photosynthetica* 56: 366–381.

Yadav, N. and S. Sharma. 2016. Reactive oxygen species, oxidative stress and ROS scavenging system in plants. *J. Chem. Pharm. Res.* 8: 595–604.

Yan, K., C. Wu, L. Zhang and X. Chen. 2015. Contrasting photosynthesis and photoinhibition in tetraploid and its autodiploid honeysuckle (*Lonicera japonica* Thund.) under salt stress. *Front. Plant Sci.* 6: 227.

Yang, J. Y., W. Zheng, Y. Tian, Y. Wu and D. W. Zhou. 2011. Effects of various mixed salt-alkaline stresses on growth, photosynthesis and photosynthetic pigment concentrations of Medicago ruthenica seedlings. *Photosynthetica* 49: 275–284.

Yang, X. and C. Lu. 2005. Photosynthesis is improved by exogenous glycine betaine in salt-stressed maize plants. *Physiol. Plant.* 124: 343–352.

Yordanov, I., V. Velikova and T. Tsonev. 2003. Plant responses to drought and stress tolerance. *Bulg. J. Plant Physiol.* special issue: 187–206.

Yu, Y. and S. M. Assmann. 2016. The effect of NaCl on stomatal opening in Arabidopsis wild type and agb1 heterotrimeric G-protein mutant plants. *Plant Signal. Behav.* 11: e1085275.

Ziaf, K., M. Amjad, M. Aslam Pervez, Q. Iqbal, I. A. Rajwana and M. Ayyub. 2009. Evaluation of different growth and physiological traits as indices of salt tolerance in hot pepper (*Capsicum Annuum* L.). *Pak. J. Bot.* 41: 1797–1809.

14 Responses of Plants to Stresses of the Sonoran Desert

Thomas W Crawford, Jr

CONTENTS

14.1 SONORAN DESERT

14.1.1 GEOGRAPHY

The concept of a desert can be viewed differently depending upon the perspective of the beholder. Shreve considers the complexity of the concepts and definitions of "desert" by geographers, biologists, and economists. He presents an in-depth biologist's rationale for his definition of the boundaries of the North American Desert which he divides into the Great Basin, Mojave Desert, Sonoran Desert, and Chihuahuan Desert (Shreve 1936, 1942). Later, Shreve presented a detailed description of the Sonoran Desert and its subdivisions (Shreve 1951).

The boundaries of the Sonoran Desert and its subdivisions shown in Figure 14.1 were designated based upon criteria of its vegetation and flora, and it is treated as a desert in the sense that it is a region of biological unity. Vegetation refers to characteristics of a plant community such as biomass, cover, density, dominant species, and physiognomy; whereas, flora refers to the total species composition or all vascular plant species occurring in a particular plot, area, or region of interest (McLaughlin and Bowers 1999). The word, "Sonoran" is

used to describe the region of the Sonoran Desert, because more of the area lies within the Mexican state of Sonora and because of its brevity and convenience (Shreve 1951). The map of the Sonoran Desert presented by Shreve (1951) has approximately the same boundaries as those presented in a small-scale map by Harshberger (1911) and includes seven vegetational subdivisions (Figure 14.1). The boundaries of the Sonoran Desert are sharply defined wherever the topography is abrupt; whereas, in level or rolling regions, the boundary is poorly defined and there is a gradual transition from desert types to other types of vegetation (Shreve 1951).

While Shreve's boundaries and subdivisions of the Sonoran Desert are the most commonly accepted, there are at least five other major attempts to define these boundaries. One includes the Mojave Desert as part of the Sonoran Desert, while another excludes most of Baja California from the Sonoran Desert. Twelve years before Shreve's publication of a map of the Sonoran Desert (1951), Dice (1939) included an area defined by Shreve in 1951 as the Sonoran Desert and other areas such as the Mojave Desert and Imperial Valley of California in a larger area that he named the Sonoran Biotic Province. Shreve's and other delineations of the boundaries of the Sonoran Desert and its subdivisions are evidence of the

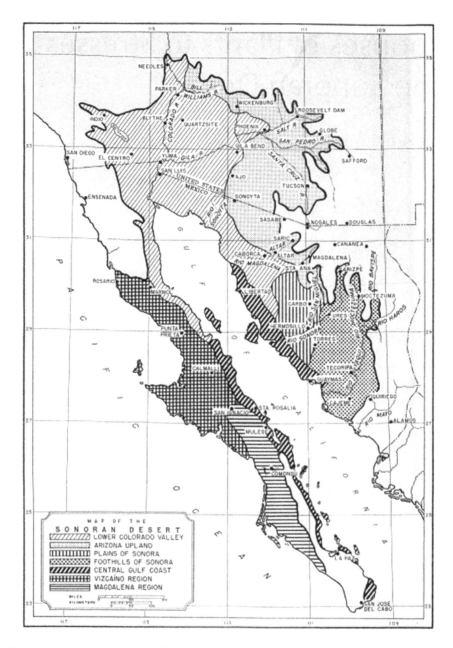

FIGURE 14.1 Map of the Sonoran Desert (Shreve 1951).

Courtesy of the Carnegie Institution for Science.

difficulty of defining precise boundaries in a geographic area with a great diversity of both geography and biota (Dimmitt 2000).

Another means of defining the boundaries of the Sonoran Desert is by using the abiotic "aridity index" of Emmanuel de Martonne (de Martonne 1926). This index is calculated for various locations, based upon annual values of precipitation in mm and temperature in °C. The general outline of the Sonoran Desert is shown in a map of the world composed in 1942 by de Martonne using the index of aridity (de Martonne 1942). Using annual data for precipitation and temperature that vary from year-to-year as a basis for calculating the index of aridity using de Martonne's formula would result in changes in the boundaries of the Sonoran Desert

from year-to-year. Boundaries of the Sonoran Desert determined using the aridity index would result in the area of the Sonoran Desert expanding, joining with the deserts of Nevada and Chihuahua, and occasionally with the Salt Lake Desert, whereas in wet periods the Sonoran Desert would contract (Ives 1949).

The area of the Sonoran Desert as defined by Shreve based on plant communities and flora is approximately 310,361 sq km (119,370 sq mi), and parts of the desert are within four states: Sonora (126,256 sq km, or 48,560 sq mi), Baja California (62,670 sq km, or 24,104 sq mi), Arizona (105,404 sq km, or 40,540 sq mi), and California (16,031 sq km, or 6166 sq mi). Elevations in the Sonoran Desert range from sea level to 915 m (3000 ft), except for a narrow band of land along its eastern edge in Arizona and in northern Sonora, where it is as high as 1050 m (3450 ft). Mountain ranges, so-called "Sky Islands," are found within 70% of the Sonoran Desert in what is called basin-and-range topography, and mountain ranges border the Sonoran Desert, rising gently or abruptly from its borders, except along 1) the line of separation from the Mojave Desert between Needles and Indio, California, 2) the line of separation from the Cape Region on the Pacific side of the tip of Baja California, and 3) the line between desert and thorn forest in southern Sonora (Shreve 1951).

At elevation increases, ascending above about 1000 m elevation in the mountains, or Sky Islands, within the basin-and-range Province of the Sonoran Desert, there is a transition from one vegetational community to another. A good example of these transitions from Sonoran Desert vegetation is in the Santa Catalina Mountains north of Tucson, Arizona. At the foothills of the Santa Catalina Mountains is the Sonoran Desert Scrub plant community. Ascending the mountain, the following plant communities are encountered in sequence from approximately 1000 to 3000 m above sea level: desert grassland, open oak woodland, pine – oak woodland, pine – oak forest, pine forest, montane fir forest, and subalpine forest (Whittaker and Niering 1965). The present chapter focuses mainly upon Sonoran Desert vegetation from sea level to approximately 1000 m above sea level.

14.1.2 CLIMATE

Of the four deserts in North America – Great Basin, Mojave, Chihuahua, and Sonoran – the Sonoran Desert is the wettest and hottest and may be the most botanically diverse desert in the world (Shreve and Wiggins 1964). Annual rainfall ranges from less than 50 mm in the Lower Colorado River Valley subdivision of the Sonoran Desert to 250 mm annually in the Plains of Sonora subdivision (Turner and Brown 1982). A major factor driving variations of rainfall from year-to-year is El Niño–Southern Oscillation (or ENSO, El Niño, and La Niña). In their study of ENSO using remote sensing and meteorological data, Zolotokrylin and others (2016) concluded that ENSO and late spring rainfall in the Sonoran Desert are related to different vegetation responses that depend on soil moisture induced by below- or above-normal rainfall. Based upon their data, they assert that a conspicuous characteristic of the Sonoran Desert is an immediate response of many plant species to an increase in moisture.

The climate in the Sonoran Desert includes droughts of various lengths of time, and recent research shows that since the mid- to late-1800s, there have been periods of the establishment of *Carnegiea gigantea* (saguaro) alternating with droughts during which few plants of this species germinated and became established (Conver et al. 2017). The authors see recent declines in establishment of stands of *C. gigantea* in the Sonoran Desert as being due to relatively recent, prolonged drought and higher temperatures, and they speculate that the destruction of nurse trees by widespread cutting in the Sonoran Desert in the early twentieth century has resulted in stress on small saguaro trees, diminishing establishment

of stands of the columnar cactus. Looking ahead to saguaros and rising temperatures in the twenty-first century, one bioclimatic model suggests that a 5°C increase in mean annual temperature could hypothetically shift the saguaro's cold limit approximately 500 km northward and approximately 600 m upward in the Southwestern United States, mainly outside of Arizona by 2090 (Rehfeldt et al. 2006).

Research on global warming indicates faster rates throughout the twenty-first century, and the predicted increase in climatic warming may result in contraction of the overall boundary of the Sonoran Desert in the south-east and expansion of its boundary northward, eastward, and upward in elevation. Changes may also occur in distributions of plant species and other characteristics of Sonoran Desert ecosystems (Weiss and Overpeck 2005). Estimates of an increase of 3°C, or an average, linear rate of increase of 0.03°C yr^{-1}, during the twenty-first century have been linked to the possible elimination of approximately 60% of species on the planet (Hansen et al. 2006).

Extreme climatic events such as unusual heat waves, regional freezes, floods, and droughts occur from time to time in the Sonoran Desert. Such stressors that reduce biotic resistance can enable non-native species to more easily compete with native vegetation (Diez et al. 2012).

14.1.2.1 Temperature

Temperatures fluctuate throughout the year throughout the Sonoran Desert, and extreme maximum and extreme minimum temperatures present stress to plants of the desert. Climate change affects the Sonoran Desert, and the Sonoran Desert Inventory and Monitoring Network (SODN) monitors climatic changes using its stations at different locations in the desert (National Park Service 2007). SODN characterizes the climate where it monitors as having little precipitation along with temperature extremes, winter temperatures being mild and summers being quite hot. Mean annual temperatures in the SODN range from about 10°C at Gila Cliff Dwellings National Monument at the headwaters of the Gila River in Southwest New Mexico to more than 20°C in western portions of the Organ Pipe National Monument in Arizona. In the summer within the SODN, maximum summer temperatures regularly exceed 40°C at lower elevations (National Park Service 2007). As do extremely hot temperatures, freezing temperatures in the Sonoran Desert affect plant distribution. For example, a cold wave of extraordinary duration and intensity began on January 18, 1937, and it produced temperatures generally colder than had been experienced in 24 years (Turnage and Hinckley 1938).

Ranges of temperature in which plants are stressed in the Sonoran Desert can be represented by annual measurements of air temperature, the extreme maximum (highest daily maximum temperature of the year) and extreme minimum (lowest daily minimum temperature of the year). At Tucson, Arizona, during the 50-year period from 1968 to 2017, the annual, extreme maximum temperature was within the range of approximately 40°C (104°F) to approximately 46°C (115°F), and the annual, extreme minimum temperature was within

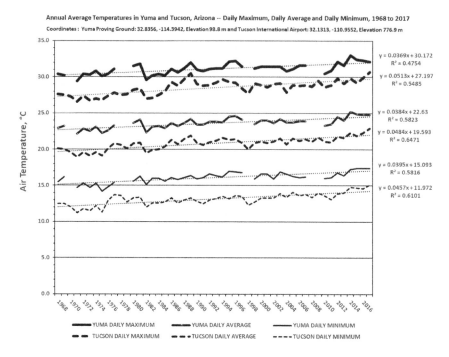

FIGURE 14.2   Annual average temperatures (°C) in Yuma and Tucson, Arizona, 1968–2017. (From NOAA/NCEI.)

the range of approximately 0°C (32°F) to −9°C (16°F). The measurement of temperature is a measurement of intensity, but both intensity and duration are critical determinants of stresses on plants of the Sonoran Desert and elsewhere.

The annual temperature data presented in Figures 14.2 through 14.4 and the monthly temperature data in Figures 14.5 and 14.6 are based upon climate data of the National Oceanic and Atmospheric Administration National Centers for Environmental Information (NOAA/NCEI) for Tucson International Airport at Tucson, Arizona and for Yuma Proving Ground located near Yuma, Arizona. The geographic coordinates of Tucson International Airport are 32.1313, −110.9552, elevation 776.9 m (2549 ft), and of the Yuma Proving Ground near Yuma, Arizona are 32.8356, −114.3942, elevation 98.8 m (325 ft).

Comparisons of annual average daily maximum, average (computed by adding the unrounded annual maximum and minimum temperatures and dividing by 2), and average minimum temperatures at Tucson, Arizona in the Arizona Upland subdivision and of Yuma, Arizona in the Lower Colorado Valley subdivision of the Sonoran Desert (Figure 14.1) show that, in general, heat stress on plants near Yuma has been greater than in the area around Tucson from 1968 to 2018 (Figure 14.2).

Also evident from the coefficients of the x term (x = year) of the linear regression equations of the annual temperature values of Figure 14.2 is a gradual warming trend of approximately 0.04 to 0.05°C yr^{-1}. This warming trend is evident for both locations and for all three temperature variables, the daily maximum, average, and minimum temperature. This warming trend indicates approximate rates of air-temperature increase greater than those in the Mojave Desert from 1904

to 2008 during which climate records indicate that annual air-temperature increased by approximately 2°C (36°F), or about 0.02°C yr^{-1} (Bai et al. 2011). The global surface temperature of the earth has increased at approximately 0.2°C per decade, or about 0.02°C yr^{-1} for three decades, as reported in 2006 (Hansen et al. 2006). Another estimate of global warming was reported in 2005 to be approximately 0.05°C yr^{-1} (Epstein and Bloom 2005). Mean annual air temperatures are reported to have increased 0.25°C per decade, or 0.025°C yr^{-1} Between 1949 and 2011 in the Tucson Basin (Brusca et al. 2013).

The very low values of the coefficients of determination, r^2, of Figure 14.2 suggest that factors affecting the general warming trend are not simply the passage of time, but others not included as variables in the equations. To determine, within the period of a year, the duration of the maximal, average and minimal temperatures shown in Figure 14.2, one can refer to the website of the National Centers for Environmental Information of the National Oceanic and Atmospheric Administration which is also a source of associated monthly and daily data (NOAA/NCEI at https://www.ncdc.noaa.gov/).

Deeper insight into thermal stress to plants of the Sonoran Desert is evident by considering the extreme annual maximum and annual extreme minimum temperatures in addition to the average daily maximum, daily average, and daily minimum temperatures. Comparing temperatures of Tucson, for example (Figure 14.3) with those of the Yuma Proving Ground near Yuma, Arizona (Figure 14.4), it is evident that high-temperature stress on plants is greater in the environments of Yuma, compared to Tucson, but that the stress from extreme, low temperatures tends to be less at Yuma, compared to Tucson. It is notable that positive slope of the linear regression least-squares fits of the extreme annual maximum and

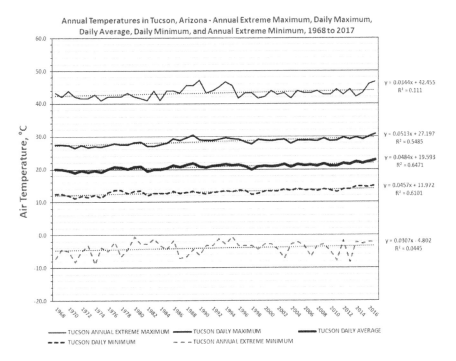

FIGURE 14.3   Annual temperatures (°C) in Tucson, Arizona, 1968–2017. (From NOAA/NCEI.)

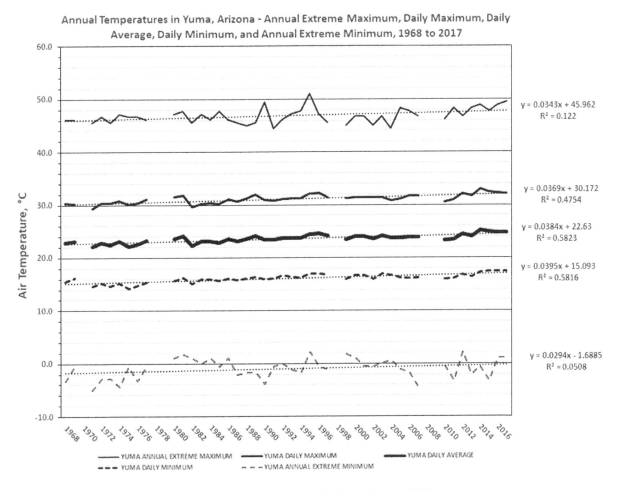

FIGURE 14.4   Annual temperatures (°C) in Yuma, Arizona, 1968–2017. (From NOAA/NCEI.)

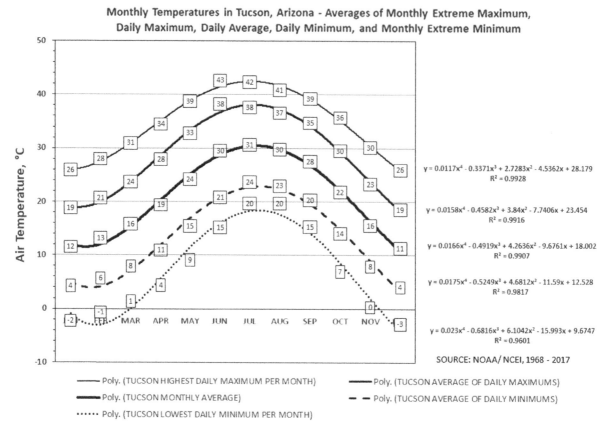

FIGURE 14.5   Monthly temperatures (°C) in Tucson, Arizona, 1968–2017. (From NOAA/NCEI.)

annual extreme minimum air temperatures data in Figures 14.3 and 14.4 are similar to those in Figure 14.2, in that they indicate an increase of both maximum and minimum air-temperature extremes. During the 50-year period of 1968 to 2017, the slight upward trend of annual extreme air temperatures at these two locations in the Arizona Upland and Colorado River Valley subdivisions of the Sonoran Desert is about 0.03°C for extreme maximum temperature and about 0.03 to 0.04°C for extreme minimum temperature. These estimated annual rates of increase are less than the estimated rates of increase shown for annual daily maximum, average and minimum air temperature (between 0.04 and 0.05°C yr^{-1} for the two locations during the same 50-year period. From the perspective of plant stress, it is evident from the data of Figures 14.2, 14.3, and 14.4 that climatic warming is occurring at the two sites in the Sonoran Desert and that warming trends are evident for all of the air-temperature variables depicted in the figures. There is a difference of approximately 3°C between Tucson and Yuma for the annual averages of daily maximum, average, and minimum air temperature (Figure 14.2) during the 50-year period of 1968–2017.

The gradual, annual increase in thermal stress during the summer on plants (and animals) at the two locations, Tucson and Yuma, Arizona, in the Sonoran Desert during this 50-year period (Figures 14.2, 14.3, and 14.4) presents a gradually selective trend of increasing pressure on plants

with differences almost imperceptible from year-to-year. Increasing temperatures in the mountains within the Sonoran Desert have been found to enable some plant species to expand their flowering range uphill. Of 93 species that shifted their range significantly during a 20-year period as temperatures increased, most were perennials (Cohn 2009). Thus, the trend of increasing air temperatures at higher elevations can diminish the cold stress on plants, and particularly perennial plants, in the Sky Islands of the Sonoran Desert.

Botanists have found large numbers of dead creosote (*Larrea tridentata*) plants along the northern edge of their range, probably indicating death from low temperatures; moreover, heavy snowfall can completely flatten creosote plants (Shreve 1940). Increases in extreme, minimum temperatures during the winter, as shown by the data for Tucson and Yuma (Figures 14.3 and 14.4), reduce stress for some plants and may enable them to increase northward of the geographic area in which they can grow. Laboratory simulations with *C. gigantea* and several other columnar cacti to study morphological changes with latitude indicate that the northern limit for *C. gigantea, Lemaireocerus thurberi,* and immature stems of *Lophocereus schottii* may be determined by the low temperatures causing freezing damage to the stem apex (Nobel 1980).

More evident than the subtle increases in temperature that occur year-to-year are changes in temperature that occur

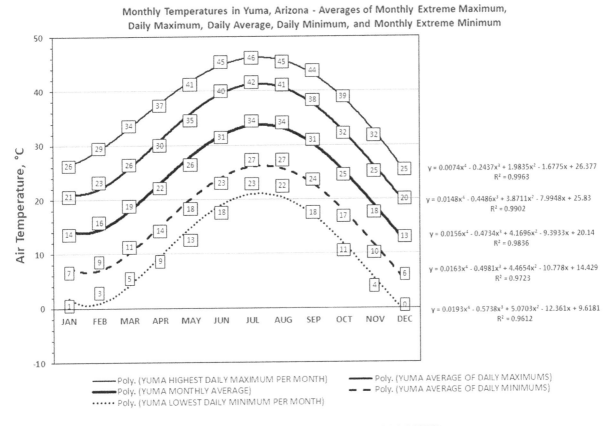

Monthly Temperatures in Yuma, Arizona - Averages of Monthly Extreme Maximum,
Daily Maximum, Daily Average, Daily Minimum, and Monthly Extreme Minimum

$y = 0.0074x^4 - 0.2437x^3 + 1.9835x^2 - 1.6775x + 26.377$
$R^2 = 0.9963$

$y = 0.0148x^4 - 0.4486x^3 + 3.8711x^2 - 7.9948x + 25.83$
$R^2 = 0.9902$

$y = 0.0156x^4 - 0.4734x^3 + 4.1696x^2 - 9.3933x + 20.14$
$R^2 = 0.9836$

$y = 0.0163x^4 - 0.4981x^3 + 4.4654x^2 - 10.778x + 14.429$
$R^2 = 0.9723$

$y = 0.0193x^4 - 0.5738x^3 + 5.0703x^2 - 12.361x + 9.6181$
$R^2 = 0.9612$

Poly. (YUMA HIGHEST DAILY MAXIMUM PER MONTH)   Poly. (YUMA AVERAGE OF DAILY MAXIMUMS)
Poly. (YUMA MONTHLY AVERAGE)   Poly. (YUMA AVERAGE OF DAILY MINIMUMS)
Poly. (YUMA LOWEST DAILY MINIMUM PER MONTH)

FIGURE 14.6   Monthly temperatures (°C) in Yuma, Arizona, 1968–2017. (From NOAA/NCEI.)

from month-to-month. Based on monthly temperature data of NOAA/NCEI for the 50-year period, 1968 to 2017, least-squares fits to the data are presented on a monthly basis using regression equations and mean values for Tucson (Figure 14.5) and Yuma (Figure 14.6) to show seasonal fluctuations in temperatures.

The most common temperatures experienced by plants of the Sonoran Desert in these two examples range between the average maximum and average minimum for each month, with average, exceptional extremes for each month indicated by the highest daily maximum and the lowest daily minimum for each month. Least-squares curve-fitting was done using mean values of each of the five variables for the period of 1968–2017. On a monthly basis, heat stress is greatest during June, July, and August in Tucson and Yuma and their environs, and, as is evident with the annual temperature data (Figures 14.2, 14.3, and 14.4), Yuma is warmer month-to-month than in Tucson.

Soil temperatures in the Sonoran Desert vary with depth. Temperature fluctuations at the soil surface are controlled by atmospheric conditions as well as by soil characteristics such as color and moisture-holding capacity.

14.1.2.2   Rainfall

Rainfall in the Sonoran Desert reduces the stress that arises from solar radiation that causes water to move to the atmosphere directly from soils by evaporation and indirectly

through plants by the process of transpiration. Between 1960 and 2010, the Sonoran Desert experienced a drought, with a 25 to 40% decrease in precipitation (Defenders of Wildlife 2010). Rainfall in the Sonoran Desert varies substantially in a two-dimensional sense, ranging from less than 100 mm per year to between 400 to 500 mm per year, depending upon location (Figure 14.5). Considering rainfall patterns at Yuma, Arizona, in the Colorado River Valley subdivision and Tucson, Arizona, in the Arizona Upland subdivision of the Sonora Desert, Figure 14.7 indicates annual rainfall within the range of 50 to 100 mm at Yuma and approximately 300 mm at Tucson. The isolines of "normal" temperatures of Figure 14.7 are based upon interpolated precipitation data from specific weather stations throughout the Sonoran Desert (Turner and Brown 1982) lack a temporal perspective other than the isolines being annual estimates of precipitation.

Precipitation data from the climate data of NOAA/NCEI referred to in Section 14.1.2.1 indicate substantial temporal variability of annual rainfall, during a 50-year period, in the context of the spatial variability of two locations, Tucson and Yuma, during the half-century from 1968 to 2017 (Figure 14.8). By virtue of Tucson's location to the east of Yuma and Tucson's higher elevation, namely 777 m (2549 ft), compared to 99 m (325 ft) at Yuma, Tucson receives more rain than Yuma. The differences in precipitation at Yuma cannot be calculated for 7 years lacking data for the Yuma Proving Ground (1978, 1979, 1980, 1982, 1998, 2008, and 2009). The small amount of

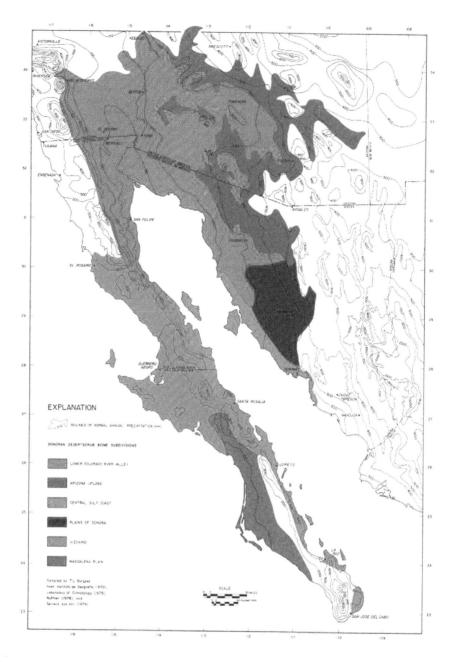

FIGURE 14.7 (See color insert.) Map of the Sonoran Desert with isolines of normal annual precipitation (mm) (Turner and Brown 1982). Courtesy of the Boyce Thompson Arboretum.

precipitation in the environs of Yuma is far more stressful to native vegetation, compared to the relatively abundant rainfall in Tucson. In the map with isolines for the Sonoran Desert (Figure 14.7), Tucson is located in between lines representing 200 and 300 mm of annual rainfall. Fifty years of NOAA/NCEI data for Tucson (1968–2017) show that there were 8 instances of annual rainfall less than 200 mm, 20 instances of annual rainfall greater than 200 mm, but less than 300 mm, and 22 years in which annual precipitation was greater than 300 mm (Figure 14.7). The NOAA/NCEI database with 43 years of data for the Yuma Proving Ground between 1968 and 2017 indicates that there were 9 years in which annual precipitation was less than 50 mm, 19 years during which annual rainfall was greater than 50 mm and less than 100 mm, and

15 years in which annual rainfall was greater than 100 mm in Yuma. The NOAA/NCEI data complement the map showing spatial aspects of rainfall distribution in the Sonoran Desert by providing insight into temporal variability of precipitation at various locations in the desert. Yuma, Arizona, is the location of the production of 90% of the leafy vegetables produced in the United States from November to March (http://arizona experience.org/land/yuma-county-americas-winter-vegetab le-capital), and this is because vegetable production is dependent upon irrigation using water from the Colorado River.

Comparing annual precipitation at two locations, Yuma in the Lower Colorado Valley subdivision and Tucson in the Arizona Upland subdivision, there are both annual spatial and temporal differences during the period from 1968 to

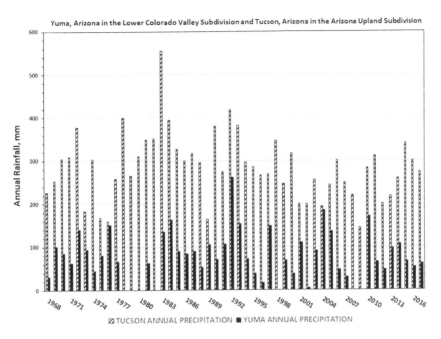

FIGURE 14.8 Annual rainfall (mm) at locations in two vegetational subdivisions of the Sonoran Desert, 1968–2017. (From NOAA/NCEI.)

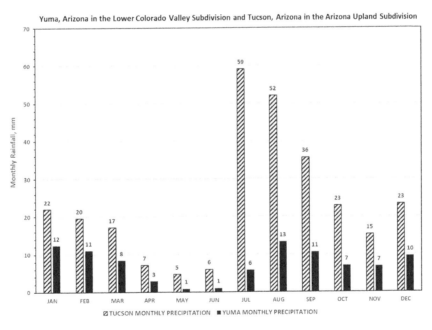

FIGURE 14.9 Average monthly rainfall (mm) at locations in two vegetational subdivisions of the Sonoran Desert, 1968–2017. (From NOAA/NCEI.)

2017. The stress on plants due to relatively little rainfall in the Lower Colorado Valley is evident (Figure 14.8), compared to the relatively abundant rainfall in the Arizona Upland subdivision of the Sonoran Desert.

A key factor in understanding moisture stress of plants in the Sonoran Desert is to understand the cyclic, temporal aspects of precipitation within the timeframe of a year. A prominent feature of many parts of the Sonoran Desert is bimodal rainfall, divided into two seasons. Regional, extratropical storm systems in the winter bring rain throughout the desert, but its distribution is geographically uneven, falling more in the northeastern part. In the summer, wind directions shift, bringing monsoon rains in localized, convective thunderstorms, mainly to the south-east part of the desert (Medeiros and Drezner 2012). Monthly precipitation data for Tucson and Yuma (NOAA/NCEI 50-year dataset) show clearly that there is intense, annual moisture stress in April, May, and June (Figure 14.9).

Although more rain falls every month in Tucson, compared to Yuma, there is an annual, bimodal precipitation pattern at both locations. In their "Sonoran Desert natural events" calendar, Hanson and Hanson (2000) define five, rather than four seasons:

- Winter – December, January, early February
- Spring – late February, March, April
- Foresummer Drought – May, June
- Summer Monsoon – July, August, Early September
- Fall – September, October, November

Clearly, during May and June, the annual foresummer drought, stress from minimal precipitation is a major factor in the Sonoran Desert, as indicated by rainfall data for Tucson, in the Arizona Upland subdivision, and Yuma in the Lower Colorado Valley subdivision of the desert.

14.1.3 Soils

Soils are derived from weathering of underlying rocks in the Sonoran Desert that results less from disintegration of the rocks by freezing temperatures and expansion of water, cracking the rock, but more from disintegration of rocks with surface temperatures of up to 65°C (150°F) to 70°C (160°F) that are weakened when rainwater of 15°C (60°F) to 21°C (70°F) contacts the very hot rocks heated by the sun. Wind and water cause erosion and deposition of soil materials, resulting in large areas of alluvial soils in the Sonoran Desert (Shreve 1951). As soils have developed in alluvial deposits of the Sonoran Desert, the downward movement of clay particles can result in argillic (clay-rich) horizons, and in many locations, calcium carbonate has been deposited within soils as water containing calcium and bicarbonate ions have moved downward (McAuliffe 1999). Deposition of calcium carbonate (lime) in soils diminishes the pore space available to hold water and air, creating an edaphic source of stress on plants, and formation of lime can entrap essential plant nutrients such as copper, iron, manganese, and zinc (Crawford et al. 2008). The extreme case of deposition of calcium carbonate in what is called a calcic horizon is called "caliche" when the pore space has been completely filled with calcium carbonate, making root growth impossible.

The water-holding capacity of soils of the Sonoran Desert varies greatly, depending upon the content and distribution of sand-, silt-, and clay-sized particles throughout the soil profile. The fine-earth fraction (National Soil Survey Center 2012) of coarse-textured soils that contain relatively little silt (2–50 μm diameter) and clay (<2 μm diameter) particles but predominantly sand (50 μm to 2 mm diameter) particles holds little water, and water moves relatively rapidly downward during rainfall events. Soils of this type can become "droughty" or lacking water, causing stress to plants with relatively shallow roots. On the other hand, deep-rooted plants have an advantage in such coarse-textured soils, being able to absorb water and mineral nutrients from depths at which such soils tend to maintain higher moisture content.

14.1.4 Vegetation

As noted in Section 14.1.1, the Sonoran Desert is defined by its plant communities. In a two-volume tome entitled *Vegetation and Flora of the Sonoran Desert*, Shreve and Wiggins (1964) present an exhaustive description, including maps, showing the distribution of plant species in the desert. It includes a smaller volume, *Part I Vegetation of the Sonoran Desert* originally published by the Carnegie Institution of Washington (Shreve 1951). Relationships between vegetation and the environment of the northern Sonoran Desert have been found in rocky uplands and alluvial flats to be influenced by slope angle, geologic substrate, and aspect. In the Organ Pipe Cactus National Monument, a nutrient gradient was also associated with differences in vegetation, with two soil characteristics, available magnesium and pH, and aspect being significant variables (Parker 1991).

14.2 STRESSES ON PLANTS OF THE SONORAN DESERT

Survival of plants in the Sonoran Desert, like the survival of plants in any ecosystem, depends upon their physiological reactions to stresses in their environment (Osmond et al. 1987). Abiotic stresses on plants in dryland conditions such as are found in the Sonoran Desert can include drought, nutrient deficiencies, high and low extremes of temperature, salinity, sodicity, and, in relatively rare cases, water-logging. Biotic stresses occur with competition for resources, such as mineral nutrients, including water, light, space and protection from herbivores, and protection from extremes of heat and cold among individual plants of the same species and plants of different species. Additional biotic stresses of plants in the Sonoran Desert are among species of plants and animals, involving such phenomena as allelopathy, parasitism, and herbivory.

14.2.1 Spatial and Temporal Aspects of Stress

14.2.1.1 The Horizontal Dimensions – X and Y

Stresses of all types, both abiotic and biotic can differ from place to place. The boundaries of the seven divisions of the Sonoran Desert were determined solely by "vegetation and flora" by Shreve (1951), and these plant communities result from interaction of plant genetic material (deoxyribonucleic acid, or DNA) with resources and stresses in various environments within the desert that enabled some plants to survive, while others have not. In the Sonoran Desert as a whole, geographic coordinates (latitude and longitude) can be associated with a wide range of variables associated with stresses on plants.

For example, Shreve (1951) presented a map showing the distribution of rainfall in the Sonoran Desert with 12.7-cm (5-inch) intervals based upon interpolation of rainfall data from 4 stations in California, 17 in Arizona, 16 in Sonora, and 3 in Baja California. These differences in rainfall show the stress of low rainfall to be greatest at greater longitudes at more westerly locations selected in the desert and the stress of low rainfall to be less at lesser longitudes to the east. Using more detailed isolines of precipitation presented in his study of the spatial characteristics of the distribution of two shrubs in the Sonoran Desert, Phillips (1979) showed that mean annual precipitation was greater for three locations (Ajo, Gila Bend, and Organ Pipe National Monument in Arizona) studied between longitudes of 112°W and 113°W, compared to mean annual precipitation at two study sites located between 114°W and 115°W (Blythe, California and Welton, Arizona). Phillips determined during his study of spatial relationships between individual plants and groups of plants of *L. tridentata* (*divaricata*), or creosote, and *Ambrosia* (or *Franseria*) *dumosa*, also known as burrow weed or white bursage, that climatic variables measured had no apparent effect on the spatial relationships between individual plants. Although he found that moisture was unimportant in the plant spacing of the two species studied, Phillips did find that different size classes of shrubs tended to change from aggregated to random to regular with increasing size of individual plants, leading him to conclude that some aspects of competition may account for changing spatial patterns of distribution of the two plant species.

14.2.1.2 The Vertical Dimension – Z

The vertical dimension can be viewed at many different scales in the Sonoran Desert. Much of the Sonoran Desert is part of the Basin and Range Province that includes slightly inclined plains on which are located many hills and mountains (Shreve 1951). Elevation above sea level is associated with climatic changes, and there are montane ecosystems within the Sonoran Desert that are popularly known as the Sky Islands. These mountain ranges include a graduated sequence of vegetation types, different from that of the valleys of the Sonoran Desert, with increasing elevation going from scrub to evergreen oaks to forests of pine, spruce, and fir at the higher elevations (Shreve 1915).

The vertical dimension of ecosystems in the Sonoran Desert has many facets. For example, although air, which is a fluid, permeates soils, it is part of a matrix of water, which is another fluid, and mineral and organic solids in soils. The interface at the surface of the soil between the soil and aboveground atmosphere is a point of division between stresses to plant roots and shoots. Solar radiation in the summer creates extreme stress which is manifest in high temperatures at the soil surface. The magnitude of fluctuations of soil temperature is greatest at the surface, and the magnitude of fluctuations of soil temperature diminishes with depth (Singer and Munns 1991). While temperature fluctuations at the surface of the soil are controlled by atmospheric forcing and are independent of soil texture, as heat moves through the soil profile, there are both amplitude damping and time lag differences associated both with depth and texture. At 48-cm (19-inch) depth, for example, comparisons of four different textural classes of soil show extremes of fluctuation of temperature during a 2-day period for sand (10 to 40°C, or 50 to 104°F),

loam (~23 to ~27°C, or ~73 to ~81°F), and silt and clay (~22 to ~23°C, or ~72 to ~73°F) (Clutter 2016). Temperature stress to soil microorganisms and to plant roots in the Sonoran Desert diminishes as depth increases below the soil surface.

Intense summer monsoon storms in the Sonoran Desert mainly impact only upper soil layers due to high runoff rates and high evapotranspiration rates (Crimmins et al. 2008). As Crimmins and others (2008) also point out, evapotranspiration rates are lower in the Sonoran Desert under cool winter temperatures, a factor that partially explains why more precipitation infiltrates soils in the Sonoran Desert to greater depths than in summer. Shreve (1934) measured rainfall, penetration, and runoff of precipitation for soil at Tucson, Arizona, in the Arizona Upland subdivision of the Sonoran Desert. From year-to-year, the amounts of these three variables varied, and from month-to-month, the data reveal that runoff occurred principally from intense rainfall events during the months of June through November, with almost no runoff during the gentler, more widespread rains from December through May. Roots of grasses and forbs in the Sonoran Desert are typically shallow, and these plants respond quickly to periodic rainfall events (Crimmins et al. 2008). However, some shrubs, such as the nurse plant, burroweed (*A. dumosa*) can have tap roots of mature plants that will vary from 1.5 to 5.5 m (5 to 18 ft) deep, depending on soil conditions. A few laterals may branch off the taproot of *A. dumosa* in the upper 46 cm (18 inches) or so, but seldom below that depth. Fluctuations in soil moisture become fewer and of smaller magnitude with increasing depth (Cable 1969).

14.2.1.3 The Fourth Dimension – Time

The temporal dimension of stresses upon plants of the Sonoran Desert ranges from long-term (years) to short-term (instantaneous to days). Long-term stresses include periodic stresses over the long-term such as recurring droughts, floods, periods of extremely high or low temperatures, and recurring stresses from herbivores, pests or pathogens. The seasonal changes in temperature and moisture due to the rotation of the Earth on its axis and around the sun subject plants of the Sonoran Desert to periodic extremes of temperature and moisture. The Drought Monitor (https://droughtmonitor.unl.edu/) provides maps and tabular data regarding various levels of drought that affect the United States and, in particular, Arizona. The data from the US Drought Monitor show that drought of varying severity affects parts of the Sonoran Desert on a frequent basis, in contrast to other parts of the United States that rarely, if ever, experience drought.

In the Arizona Upland subdivision of the Sonoran Desert, a bimodal rainfall pattern exists, and there are five seasons: winter, spring, dry foresummer, monsoon summer, and fall. In the dry foresummer which is generally from mid-May to June, the increasing temperature is conducive to plant growth, but a conflicting trend of less and less available water until the monsoon rains arrive severely stresses plants in the Sonoran Desert. To cope with the extreme stress due to lack of moisture between the spring and monsoon rains in July through

September, some trees of the Sonoran Desert such as *Salix* sp. (willows) and *Populus* sp. (cottonwood) survive near a constant water supply, such as a lake, pond, or perennial stream. Other tree species such as *Prosopis velutina* (mesquite) survive this periodic stress due to lack of water with deep root systems (Shreve 1915). Trees such as *Cercidium (palo verde)* sp. drop their leaves to diminish transpiration, as do shrubs such as *Fouquieria splendens* (ocotillo); these species grow more leaves when water becomes adequate during the monsoon summer and during the winter. Some columnar cacti such as *C. gigantea* change diameter during the year, having a pleat-like structure enabling them to expand to store water or to contract as water is transpired, depending upon the availability of soil moisture.

14.2.2 Abiotic Stresses

14.2.2.1 Radiant Energy Stress

Hot drylands, such as the Sonoran Desert, typically receive high solar radiation and as a result have high potential evaporation rates, high diurnal ranges in temperature, low precipitation and low atmospheric humidity (Verstraete and Schwartz 1991). There are seasonal changes in radiant energy that reaches the Sonoran Desert. Differences in insolation occur in the Sonoran Desert, depending upon aspect and slope at any particular location. Water availability tends to be greater on north-facing slopes because insolation and, therefore, evapotranspiration are lower than on south-facing slopes (Yeaton and Cody 1979). Although radiation is assumed by many not to be a limiting factor in deserts, it is important to understand that stomatal behavior of plants that restricts photosynthesis to periods of low evaporativity such as in the early morning when radiation is also low may restrict access to solar radiation necessary for photosynthesis (Noy-Meir 1973).

Modification of incoming and outgoing radiation flux occurs as a result of the growth of plants in the Sonoran Desert, for example, by two species of a common tree of the Sonoran Desert, palo verde (*Cercidium*) (Lowe and Hinds 1971). Ground temperatures on bare soil and under blue palo verde (*C. floridum*) and upland palo verde (*C. microphyllum*) trees that were 3 to 4 m (10 to 13 ft) high were observed. The effective (net) incoming radiation at ground level at midday in the open was more than twice that under the palo verde, and the effective (net) outgoing radiation in the open at night was more than twice as great as that under the tree. This mitigation of extremes of temperature and evapotranspiration due to intense radiant energy in the Sonoran Desert is associated with "nurse trees" such as palo verde (*Cercidium* sp.) and mesquite (*Prosopis* sp.) that enable other vegetation under their canopies to withstand extremes of heat and cold. Two species of plants in the Sonoran Desert that benefit from nurse trees are the saguaro (*C. gigantea*) and ocotillo (*F. splendens*) (Nobel and Zutta 2005).

Shading by native vegetation of the Sonoran vegetation creates microhabitats that can affect physiological processes such as carbon fixation by other plants. Near an equinox, shading by the nurse plant, *A. deltoidea* (triangle leaf bursage), reduced by 77% the amount of photosynthetically active radiation (PAR) received by a seedling of *C. gigantea*, reducing its predicted net CO_2 uptake by 90% compared with an unshaded seedling. In the same study, a seedling of the barrel cactus (*Ferocactus acanthodes*) located at the center of a plant of *Hilaria rigida* (a perennial bunchgrass) received 64% less total daily PAR and could fix 65% less CO_2 than could an unshaded seedling (Franco and Nobel 1989). In a comparative study of nurse plants of saguaro (*C. gigantea*), two relatively dense, leafy shrubs, triangleleaf bursage (*A. deltoidea*) and burroweed (*A. dumosa*), a shrub with a more open canopy, creosote bush (*L. tridentata*), and three large leguminous trees, upland palo verde (*C. microphyllum*), Ironwood (*Olneya tesota*) and mesquite (*Prosopis* sp.), were mapped and described in different parts of the Sonoran Desert (Drezner 2006). The study showed that saguaros are also disproportionately distributed beneath the warmer south side of their nurse in colder parts of their range. Opening the canopy structure (e.g., leaf size and density of the canopy) probably influences soil surface temperatures, amount of photosynthetically active radiation receipt, air temperature, and possibly humidity. Drezner's measurements (2006) show that few saguaros were established far from *L. tridentata*'s base, whereas saguaros were found under a larger portion of tree subcanopies. Means by which stresses to young plants of *C. gigantea* are reduced as a result of their nurse plant associations may include increasing shade, reducing the summertime radiation load, increasing night-time winter temperatures, increasing soil moisture, and reducing herbivory (Drezner 2006).

14.2.2.2 Low- and High-Temperature Stresses

Low temperatures during the winter months in the Sonoran Desert create stress on the biota of the desert and limit the areal extent of plant species, particularly in the northern parts of the desert. Investigating cold tolerance of columnar cacti of the Sonoran Desert, Shreve (1911) reported that *C. gigantea* was capable of withstanding 19 hours of continued freezing temperature as low as a minimum of −8.3°C (17°F). He concludes, based upon his research reported in 1911, which included controlled freezing of *C. gigantea*, that none of the individuals of the species *C. gigantea* living in the Sonoran Desert at that time has endured more than 20 hours of continuous freezing. Focusing on the possible causes of death of *C. gigantea* due to freezing temperatures, Shreve reasoned that even a single day without midday thawing during a period of freezing temperatures, coupled with cloudiness that would prevent the internal temperature of the cactus from exceeding that of the air could result in the death of *Carnegiea*. Nobel (1980) found that interspecific and intraspecific morphological differences of *C. gigantea* and other columnar cacti were factors affecting the northern extent of distribution of these cacti. He found that differences in apical minimum temperatures due to the presence or absence of pubescence and spines was related to the northernmost latitude at which various columnar cacti were found in the Sonoran Desert.

In January 1937, a polar mass of cold air spread over the western United States and caused disastrous freezes in the Sonoran Desert. The freezing effects of this large air mass were widespread, in contrast to annual, localized freezing associated with nocturnal ground inversions which are common in the Sonoran Desert. Ground inversions of air temperature are an atmospheric condition in which the temperature increases with elevation from the ground in a topographic basin up to as much as 300 m (about 1000 ft), above which the air temperature decreases with increasing elevation. These ground inversions occur on cold, clear nights and can be damaging to desert plants that grow in low areas where cold air collects. Damage to plants documented for specific plant species due to the freeze of 1937 in the Sonoran Desert includes leaves, branches, and flowers (Turnage and Hinckley 1938).

The range of the Sonoran Desert shrub *L. tridentata* is determined, in part by freezing temperatures that cause cavitation (air bubbles) to form in its vascular system. Pockman and Sperry (1997) found that minimum temperatures between −16°C (3°F) and −20°C (−4°F) completely eliminated hydraulic conductance. They also found that record (>20 years) minimum isotherms in this same range of temperatures corresponded closely with the northern limits of *L. tridentata* in both the Mojave and Sonoran deserts. More recent research indicates that with rising minimum temperatures in the Sonoran Desert with climate change, the range of *C. gigantea* may extend to higher elevations. However, while less frequent freeze events could allow saguaros to increase their range to include higher elevations, Springer and others (2015) see indications that increased fire activity related to the establishment and spread of invasive species such as buffelgrass could inhibit expansion of the range of saguaros.

Thermal limits for the survival of metabolically active tissues of vascular plants range from about −60°C (−76°F) to about +60°C (140°F) in different species, and some vascular plant tissues, such as seeds, can remain viable by tolerating an even wider range of temperature; such extreme tolerance to temperature extremes is usually greater when metabolic activity is minimal (Osmond et al. 1987). Non-succulent plants of the Sonoran Desert have adapted to extreme, high temperatures in the summer to facilitate photosynthesis. One adaption that enables trees such as *C. microphyllum* (foothill palo verde or little-leaved palo verde) to maintain leaf temperature close to the thermal optimum of photosynthetic enzymes, 25°C (77°F) to 35°C (95°F), when ambient air temperatures can exceed that range is microphylly, characteristic of small leaves. Small leaves and leaflets of desert plants that are smaller than those of plants growing in wetter habitats enable the desert plants to maintain leaf temperature below lethal temperatures, even without transpiration. Another adaption to cope with the stress of high temperatures in the Sonoran Desert is the thick, silvery hair that grows on the leaves of the *Encelia farinosa* (brittlebush) in response to drought. The highly reflective, silvery coating enables the shrub to avoid lethal leaf temperatures by diminishing heating of the leaves by infrared radiation while it reduces the level of photon flux to the leaves, lowering net photosynthesis by about a half of the rate when water is adequate (Gibson 1998).

Succulents that survive in the Sonoran Desert have adapted to extremely high temperatures and commonly high temperatures during the summer by tolerating relatively high tissue temperatures, the mean maximum tolerance being 64°C (147°F) (Smith et al. 1984). Two species of *Ferrocactus* (barrel cactus) tolerated an absolute maximum tissue temperature of 69°C (156°F). Morphological adaption of cylindropuntia cacti growing with a thin stem enables them to achieve lower maximum tissue temperatures, compared to more massive species of cacti, and the thinner-stemmed cacti appear to be less tolerant of high-temperature stress than are cacti with larger diameters. Acclimation of high-temperature tolerances in response to increasing day/night air temperatures was common to all 14 species of cacti studied by Smith and others (1984). For example, with the increase in air temperature from 40°C (104°F) day/30°C (86°F) night to 50°C (122°F) day/40°C (104°F) night, tolerable tissue temperatures increased an average of 6°C.

Approximately 140 species of cacti native to the Sonoran Desert are perennial stem succulents (Shreve and Wiggins 1964). To cope with high temperature during the day and drought, these succulents have adapted and survived by opening their stomata primarily at night to absorb CO_2 from the atmosphere when air and tissue temperatures are lower than in the day. The CO_2 is stored using Crassulacean Acid Metabolism (CAM) and is used in photosynthesis during the day when most of the stomata of the cacti are closed, resulting in higher water-use efficiency (mass of CO_2 absorbed per mass of water lost) (Nobel and Loik 1999). The CAM photosynthetic pathway is utilized by only about 6% of plant species, all of which have low productivity. By opening their stomates in their leaves and stems at night, plants such as cacti are able to absorb five to eight times more CO_2 per unit of water lost, compared to the 94% of plant species that open their stomates in the day (Nobel 2015).

Wildfires are catastrophic stress of extremely high temperature affecting ecosystems in the Sonoran Desert. Increases in invasive species, fires, and slow recovery by native species after a fire may have effects on desert biogeochemical cycles, since in many cases after a desert fire, soil inorganic nitrogen levels are higher, and soil carbon levels are lower. The intense heat that occurs at the soil surface penetrates and kills soil bacteria near the surface (Allen et al. 2011).

The temperature extremes and trends toward the higher high, average and low temperatures at Tucson and Yuma, Arizona (Figures 14.2 through 14.6) all indicate stress on plants of the Sonoran Desert. Temperature is a primary factor affecting plant growth and productivity, and pollination is one of the phenological stages most sensitive to extremely high temperatures. Referring to the findings of Hatfield and Prueger (2015) which focused on temperate-zone crops, increasingly high temperatures in the Sonoran Desert may select for plants that shed pollen during cooler periods of the day or are indeterminate, flowering over a longer period of the growing season.

14.2.2.3 Water Stress

Primary production of desert ecosystems is limited by, among other factors, the stresses of lack of precipitation and the lack of availability of mineral nutrients (Hadley and Szarek 1981). A general equation to express the amount of air-dry plant material produced per unit of water consumed is:

$$WUE = D / W$$

Where WUE is water-use efficiency, D is the mass (g) of dry plant matter produced, and W is the mass (g) of water used (Epstein and Bloom 2005). The water requirement of plants, expressed by Briggs and Shantz (1914), based on their experiments with many temperate-zone plant species, ranges from about 200 to 1000 g water transpired per gram of per gram of dry matter produced, corresponding to WUE of 5×10^{-3} to 1×10^{-3} or less (Epstein and Bloom 2005). Noy-Meir (1973) estimated that 25–75 mm (1–3 inches) of precipitation per year would be the minimum necessary to sustain vegetation in an arid ecosystem; he asserted that primary production above that minimum would occur with WUE of 0.5–2 mg dry weight^{-1} g water^{-1}, or WUE of 5×10^{-5} to 2×10^{-3} for above-ground desert primary production. The range of values of WUE based on experimental data of Briggs and Shantz for temperate-zone plants, approximately 5×10^{-3} to 1×10^{-3}, is mainly greater than the range of WUE for primary production of desert plants presented by Noy-Meir (1973), namely 5×10^{-5} to 2×10^{-3}, indicating generally greater productivity for temperate-zone crops than for plants growing in a desert.

Like all deserts, the Sonoran Desert has periods of drought stress, during which the growth rate of plants is suppressed. If the content of soil water is low enough, the matric suction (osmotic potential) of the soil is said to be at "permanent wilting point," indicating that the plant cannot regain turgidity after wilting. The internal water potential of plants in the Sonoran Desert changes both diurnally and seasonally. The shrub, *A. deltoidea* (*Franseria deltoidea*, or triangle leaf bursage) was measured to have developed water potential as low as −85 bar and demonstrated a large range of internal water potential both diurnally and annually to ensure that it has adequate water to carry out metabolic functions. The tree, *C. microphyllum* (palo verde), in contrast, has been shown to have only a slight response of changes in internal water potential both diurnally and annually, perhaps because of its deeper root system and ability to lose leaves during periods of extreme moisture stress (Halvorson and Patten 1974).

Succulents, among which are the cacti in the Sonoran Desert, survive periods of low soil moisture content by several physiological and morphological adaptions that have evolved over thousands of years. *Opuntia* sp. (prickly pear) cactus have evolved a number of characteristics to survive the paucity of water in the Sonoran Desert. Like all succulents, *Opuntia* sp. maintains low rates of transpiration and has waxy cuticles to retain large quantities of water, compared to nonsucculents. *O. laevis* (smooth prickly pear) and O. *polyacantha* (plains prickly pear), when compared to corn and soybean plants displayed little water loss, low nutrient requirements for growth, slight photosynthetic activity, slow translocation, and a low transpiration rate. The low number of stomata per unit of surface area of the prickly pear (41% of the stomata

of soybean per unit area) and the thick wax cuticle of prickly pears are important factors in reducing the diffusion of carbon dioxide necessary and penetration of light into the chlorophyll-containing grana of cells where photosynthesis occurs (Chow et al. 1966).

Cacti have an additional feature, areoles, which include spines (modified leaves) and glochids (tiny spines) that repel many herbivores and act as nodes for flowers and fruit to develop. Apical spines and pubescence of some cacti such as *C. gigantea* can help protect the growing points of the cactus from freezing (Nobel 1980). Soil moisture near the soil surface is important for desert succulents, since the roots of many desert succulent species may penetrate the soil no deeper than 0.1 m (Nobel 1976, 1977).

Seedling establishment of succulents in the Sonoran Desert is heavily dependent upon adequate soil moisture and, often, the presence of nurse plants or other shelters that can reduce the evaporative loss of water from the soil. Survival of a common succulent, *Agave deserti*, in the northwestern Sonoran Desert requires unusually wet years and the protection of sheltered microhabitats of rocks or nurse plants such as *A. dumosa* or the perennial bunchgrass *H. rigida*. *H. rigida* provided shelter for young plants of *A. deserti* that had substantially greater photosynthetic area and produced much more dry weight of living and dry leaves, compared to *A. deserti* plants under microhabitats on the north side of rocks (Jordan and Nobel 1979). In a study of the barrel cactus (*F. acanthodes*), Jordan and Nobel (1981) found that the cactus germinated in late summer and that a sufficiently long growing season was required relative to the length of subsequent droughts to allow establishment. Only 8 of the 18 years studied were found to be suitable for establishment of *F. acanthodes*. In another study, Jordan and Nobel (1982) found that *F. acanthodes* and *C. gigantea* occurred in places where at least 10% of the years were suitable for seedling establishment (Jordan and Nobel 1982). Drought stress in the Sonoran Desert presents severe selection pressure on young plants, resulting in cohorts of succulents such as *A. deserti* and cacti such as *F. acanthodes* and *C. gigantea* that can survive during prolonged droughts of more than 1 year when young. The ability of very small and young saguaros (*C. gigantea)* to survive in the Sonoran Desert is enhanced by larger plants called nurses that reduce environmental stresses to young saguaros where limitations such as freezing temperatures and recurring freezes in the north, as well as very limited rainfall in the west occur (Drezner 2004a). As saguaros grow, their water storage capacity increases, and as they grow, they are increasingly able to withstand periods of drought and high temperatures (Drezner 2004b).

Differences in slope are associated with greater or lesser stresses on plants in the Sonoran Desert, since clay-sized particles tend to accumulate lower on slopes, increasing water-holding capacity, whereas areas higher on slopes tend to have lower water-holding capacity and greater runoff. This difference in soil texture in soils at higher and lower elevations results in a gradient of increasing moisture content from higher to lower areas on a slope after rainfall (Yeaton and

Cody 1979). Moisture stress in riparian areas of the Sonoran Desert is seasonal, and during the dry foresummer, most herbaceous growth in the desert upland was found to be restricted above the floodplain of the San Pedro River (Arizona, USA). In the floodplain, a permanent source of water (shallow groundwater and associated capillary water) resulted in persistence of some herbaceous species during the dry summer season. Plant species diversity at both microscale and mesoscale increased along an elevational continuum from the drier uplands to the lowest floodplains. In the winter, however, the zone of maximum microscale plant diversity shifted to the upland, due to flood-related stresses such as flood scour in the floodplain (Stromberg 2007).

Production of seeds, interception of photosynthetically active radiation and uptake of CO_2 all increase as saguaro cacti (*C. gigantea*) develop branches, and branching increases with increasing moisture. By using stepwise regression to determine which climate, vegetation and soil variables best predict branching, Drezner (2003) found that, contrary to the literature, winter precipitation, particularly from January to April, was the best predictor of branching, not summer or annual rain. A study of the long-term demographics of *C. gigantea* in the Sonoran Desert, Pierson, Turner, and Betancourt (2013) determined that across the region, saguaro regeneration rates were highest from 1780 to 1860, which corresponded to a period of wet conditions and high *Pinus ponderosa* recruitment in the highlands. They speculate that milder and wetter winters and protection from livestock grazing may account for some surges of regeneration of *C. gigantea* at some sites within the region in the late twentieth century. Reduction of stress by increased rainfall enabled populations of *C. gigantea* to increase, but as Pierson and his colleagues point out, other stress factors such as grazing, extremely high or extremely low temperatures, and the aspect of the slope on which the saguaros grow must also be taken into account to understand demographic trends.

Long-term vegetation change at the Sierra del Pinacate Reserve in Sonora, Mexico, which is within the Sonoran Desert, was correlated with climatic data. Within the MacDougal Crater during the first half of the twentieth century, various populations of *L. tridentata* declined 50–90%, and the population of *Cercidium* spp. decreased 60%, while during the same period the population of *C. gigantea* increased fourfold. There were three major establishment peaks for *C. gigantea* during the periods, 1875–1890, 1905–1915, and 1945–1949. The high mortality of *Cercidium* and *L. tridentata* was probably a result of a prolonged drought during 1936–1964 (Turner 1990). Two-dimensional distribution patterns of *Larrea* (creosote) have been found to be associated with rainfall – more rainfall resulted in more dense populations of *Larrea*. At high rainfall levels, plants of *Larrea* were found to be grouped in clumps, and at low levels of rainfall, individual plants were regularly spaced, reflecting competition for rainwater absorbed in the soil (Woodell et al. 1969).

Drought stress of mesquite (*Prosopis glandulosa*) in the Sonoran Desert was mitigated by the ability of this phreatophytic tree to absorb water from 4 to 6 m (13 to 20 ft) deep

in the soil. Other adaptions of this species which enable it to withstand the stress of a lack of moisture are an osmotic adjustment and seasonally changing stomatal sensitivity to vapor pressure deficit (Nilsen et al. 1983). Another microphyllous plant, *L. tridentata* (creosote bush), is able to conserve water partially as a result of its leaves being covered by resin. This coating that is 2 to 4 μm (0.00008 to 0.00016 inches) in thickness is in epidermal cells, reduces epicuticular transpiration and also repels herbivores (Gibson 1998). The phenological characteristics of *L. tridentata* are adapted to the waxing and waning of drought stress, since active vegetative growth precedes renewal of reproductive activity. Flower buds of *L. tridentata* emerge and are sustained when soil moisture is adequate, and differences in phenological responses studied at four locations in the Sonoran Desert appeared to be caused mainly by local fluctuations in precipitation (Abe 1982).

The characteristic of being drought-deciduous is a way that many plants of the Sonoran Desert conserve water. Examples of plants that drop their leaves (aphylly) during periods of drought are the *Cercidium* sp. (palo verdes) and *F. splendens* (ocotillo). These two species of plants that drop their leaves during periods of drought have a photosynthetic pigment in their stems and can carry out photosynthesis without their leaves. *L. tridentata* has been found to allocate more carbon and nitrogen to reproductive, rather than to vegetative tissue under conditions of drought stress, and this allocation to reproductive tissue may confer a survival advantage to *L. tridentata* under conditions of drought and high-temperature stress (Sharifi et al. 1988).

Annual plants of the Sonoran Desert survive by evading the summer drought. Leaf water potentials are high, suggesting that photosynthesis and transpiration rates of many, if not most, desert annuals are higher than in desert evergreen and drought-deciduous perennials. Root:shoot ratios of both summer and winter desert annuals are quite low, indicating, perhaps, that a relatively small proportion of the carbon fixed is used for water acquisition. The greater proportion of the carbon fixed by these annual plants appears to be used for vegetative and reproductive growth of tissue above ground (Forseth et al. 1984). Avoiders of drought are not ephemerals with a short time to accomplish vegetative growth and produce seed. Drought avoiders in the Sonoran Desert have special structural modifications to avoid water loss. These include, but are not limited to, a thick epidermis, waxy coating of the epidermis, sunken stomata, rolling of leaves during drought, and dropping of leaves during drought. Some plants that have no leaves, such as cacti (e.g., *Opuntia* spp. and *C. gigantea*), and *Euphorbia* spp. and others that lose their leaves to conserve water during drought, such as *Cercidium* spp. and *Fouqueria splendens*, have the ability to carry out photosynthesis in their stems (Ram and Gupta 1997).

14.2.2.4 Edaphic Stresses

Large portions of the Sonoran Desert contain alkaline soils, the pH of which is greater than 7. Distribution of several Sonoran Desert plant species (*C. gigantea* and several of its nurse plant species, i.e., *A. deltoidea*, *A. dumosa*,

C. microphyllum, *L. tridentata*, *Prosopis* spp., and *Olneya tesota*) has been shown to be associated with a general east–west gradient where temperature increases and precipitation decreases westwards and calcium levels, total organic carbon, particle size, and soil pH increase westward (Medeiros and Drezner 2012). The researchers of that study concluded that a gradient in soil pH probably governs the range of several species (e.g., *A. deltoidea*, *A. dumosa*, and *L. tridentata*) that were found in calcic soils in western areas, in contrast to eastern regions with lower soil pH and higher precipitation. They concluded that the distribution of the dominant species reflects a temperature–precipitation–calcium–pH gradient.

Calcium carbonate, or lime, is common in many soils of the Sonoran Desert, and low availability of phosphorus (P) in soils can limit plant growth. Controlled experiments on the effects of calcium carbonate ($CaCO_3$) in soils upon uptake of P by two Sonoran Desert shrubs, *L. tridentata* (DC.) Cov. and *Parthenium incanum* H. B. K. show that increasing levels of $CaCO_3$ inhibited the uptake or availability of P to roots of *L. tridentata* seedlings that responded with increased root:shoot ratios, increased N:P ratios in tissues and decreased specific absorption rates of P (Lajtha and Schlesinger 1988). Declining P concentrations in the tissues of *L. tridentata*, particularly in the roots, and increased root growth are interpreted by Lajtha and Schlesinger as indications of reduced P availability. N concentrations of tissues of *Parthenium* seedlings were relatively unaffected by soil $CaCO_3$ levels, and although P concentrations in the tissue of *P. icanum* plants were greater with the $CaCO_3$-free treatment, compared to plants growing in soil with added $CaCO_3$, the differences were small, and the N:P ratios were similar. The stress of an edaphic factor, low soil P availability, results in responses of at least one Sonoran Desert plant. *L. tridentata*, such as increased root:shoot and lower N:P ratios in tissues.

In a study of 99 sites in the Organ Pipe Cactus National Monument, Arizona, slope angle and soil texture were key factors in the distribution of vegetation (Parker 1991). The patterns of monthly average temperature and rainfall at the Monument are similar to the corresponding 50-year averages for Tucson, Arizona, presented in Figures 14.5 and 14.9. Soils that developed from plutonic parent material, e.g., granite, tended to be lower in magnesium and tended to be of coarser texture, compared to soils of the volcanic parent material. Slope aspect was associated with stress. South-facing slopes with greater heat stress and lower moisture availability were more stressful than north-facing slopes with soils that were cooler and moister. The study pointed to soil chemical properties as being significant factors to modify the prominent influence of water availability on patterns of desert vegetation in the Sonoran Desert.

P. velutina (velvet mesquite) in the Sonoran Desert has been shown to reduce edaphic stress by creating islands of soil fertility with higher soil organic matter, net N mineralization, net nitrification rates, and microbial biomass under the trees' canopies (Schade and Hobbie 2005). In a study of soil characteristics under the canopy of *P. glandulosa* and a non-vegetated area between mesquite trees, it was found that under the mesquite canopy, total N, NO_3^--N, NH_4^+-N, organic C, $NaHCO_3$-extractable PO_4^{3-}-P, and saturation-extract K^+ were significantly higher under the mesquite (Virginia and Jarrell 1983). In the same study, Na^+ and Cl^- concentrations were significantly higher in the soil between the mesquite trees. Those conducting the study concluded that most of the N in the soil under the mesquite trees had been symbiotically fixed by bacteria in root nodules of the mesquite. Foliar analysis indicated that the mesquite was excluding Na^+, resulting in the leaf litter containing a lower concentration of Na+ than that of the soil between the mesquite trees. The accumulation of atmospheric N in mineral and organic forms as a result of the growth of the mesquite minimized the stress of lack of inorganic N in the soils under the mesquite trees.

Saline soils in the Sonoran Desert place stress both on plants and upon soil bacteria that fix atmospheric nitrogen which becomes available for uptake by the roots of plants in the desert. Rhizobial and bradyrhizobal symbiotic, nitrogen-fixing bacteria found in the Sonoran Desert have been found to have evolved to tolerate high salinity; this tolerance of salinity may be a key factor in the ability of a host of these nitrogen-fixing bacteria, mesquite (*Prosopis* sp.), to survive in marginal, saline soils (Jenkins 2003).

14.2.3 Biotic Stresses

14.2.3.1 Competition among Plants

In a field study of *L. tridentata*, Brisson and Reynolds (1994) found that competitive interactions among root systems of neighboring plants of the same species affected the arrangement of the root systems of creosote. The observed tendency of *L. tridentata* plants to reduce intraspecific root system overlap is in contrast to the extensive interspecific intermingling of root systems observed by Caldwell and Richards (1986). Roots of *Ambrosia dumosa* inhibited elongation of contacted roots of only other *A. dumosa* plants, and the apparent chemical communication among *A. dumosa* plants may be a detection-and-avoidance system to avoid the stress of competition for mineral nutrients, including water, in soil (Mahall and Callaway 1992).

Buffelgrass (*Pennisetum ciliare*) is an aggressive, non-native grass species introduced to the United States in the 1940s. This non-native grass is invasive in the Sonoran Desert and provides stress to native grass species in the Sonoran Desert by competing with and displacing the native grasses. Buffelgrass has been associated with reduced plant species diversity (Stevens and Falk 2009), and research has shown that buffelgrass may suppress germination of the seeds of some legume species by exuding allelopathic compounds (Fulbright and Fulbright 1990). The invasion success of buffelgrass is due, in part, to its ability to emerge following relatively low levels of rainfall (Ward and Smith 2006). Buffelgrass greatly increases fuel loads in the desert and increases fire which causes high mortality rates of *C. gigantea*, so buffelgrass is a threat not only to native grasses and legumes of the Sonoran Desert but also to the iconic saguaros (Drezner 2014).

14.2.3.2 Stress Relationships between Plants and Animals

Animals ranging in size from microscopic to large mammals interact with plants of the Sonoran Desert. Many interactions are mutually beneficial, such as pollination and seed dispersal accomplished by animals for the benefit of plants. Others such as herbivory of vegetative tissue of plants by animals tend to favor the animals more than the plant. Cacti are often low in protein, and mature cactus pads often contain less than 1% protein on a dry-weight basis, but their digestible-energy production per mm of rainfall is high, making them a nutritional complement to nitrogen-fixing plant species such as *Prosopis* (mesquite) (Russell and Felker 1987). *L. tridentata* experiences the stress of phytophagous arthropods that are seeking nutrients to survive. In experiments using fertilization with nitrogen, it was found that sap-sucking phytophagous insects were more responsive than leaf-chewing insects to the increased nitrogen content of *L. tridentata* bushes (Lightfoot and Whitford 1987). In the Sonoran Desert, trees of *Prosopis* spp. are stressed by *Prosopis* seed beetles, *Algarobius prosopis* and *Neltumus arizonensis*, that lay their eggs on the pods containing seeds. Females of *A. prosopis* can lay up to 300 eggs, and the females of *N. arizonensis* usually lay about 70 eggs. Once the larvae hatch from the eggs, they crawl on the pods, burrow through them and then chew their way into undamaged seeds. By damaging *Prosopis* seeds, these two pests reduce the number of viable seeds that drop from the tree to become part of the soil seed bank. In South Africa, these beetles are used for the biological control of weedy *Prosopis* trees (Klein 2002).

Birds are granivores of the Sonoran Desert and are important primary consumers as they harvest seeds. In the Sonoran Desert, resident, avian granivores include *Zenaidura macroura* (mourning dove), *Columbigallina passerina* (ground dove), *Carpodacus mexicanus* (house finch), *Lophortyx gambelii* (Gambel's quail), and *Amphispiza bilineata* (black-throated sparrow). Other resident species such as *Pipilo aberti* (*Melozone aberti* or Abert's towhee) and *Toxostoma curvirostre* (curve-billed thrasher), which, although primarily insectivorous, can consume large quantities of seeds. Migratory birds also contribute to seed-harvesting in the Sonoran Desert (Hadley and Szarek 1981). A curved-billed thrasher is shown in Figure 14.10 consuming seeds and pulp of the fruit of a saguaro tree at 10:02 a.m., July 5, 2018. The vegetative part of the tree is protected from herbivory by spines, water loss by transpiration is reduced by the waxy surface under the spines, and the pleats of the saguaro enable it to grow larger or smaller with net uptake or loss of water.

Most saguaro seeds fall to the ground when the fruit drops, so an animal that can walk can eat the pulp and seeds, and a portion of the seeds, once consumed, pass through the digestive tract and are deposited with feces of the seed-dispersing animal. In the Sonoran Desert, the viability of seeds of saguaros that have been dispersed in the feces of animals that have consumed them will depend upon such stress factors as moisture, temperature, and protection from herbivory at locations where the seeds are deposited in the animal feces. Nonetheless, while not measuring the viability of dispersed

FIGURE 14.10 (See color insert.) Curve-billed thrasher (*Toxostoma curvirostre*) consuming seeds and pulp of the fruit of saguaro (*C. gigantea*). (Source: T W Crawford, Jr.)

TABLE 14.1
Viability of Saguaro Seeds from Feces of Several Mammalian and Avian Species

Species	Common Name	% Germination
Canis latrans	Coyote	97.0
Pecari tajacu	Javelina	31.0
Spermophilus	Round-tailed ground squirrel	64.2
Citellus harrisi	Harris's antelope squirrel	53.5
Neotoma albigula	White-throated wood rat	50.0
Campylorhynchus brunneicapillus	Cactus wren	76.0
Toxostoma curvirostre	Curve-billed thrasher	54.7
Zenaida asiatica	White-winged dove	100.0

seeds in the wild, laboratory testing of the relative viability of saguaro seeds found in different animal feces in low-stress, closed petri dishes at 22°C with uniform moisture supply indicates (Table 14.1) the relative benefits of animal consumption and dispersion of saguaro seeds (National Park Service 2005):

Eddy (1959) reported finding scat of *P. tajacu* in which 85% of the saguaro seeds were undigested, but he apparently did not test their germination. The data of Table 14.1 and Eddy's findings indicate that mammalian and avian consumers and dispersers of viable saguaro seeds can contribute to the survival of *C. gigantea*. However, stress on the saguaro seeds due to passing through the digestive tract of the seed dispersers can occur as seeds transit from the fruit to where they are consumed to the locations where the seeds come to rest as feces are deposited on the landscape. Survival of saguaro seeds at locations where they are dropped in animal feces may be increased if the seeds are protected by nurse plants away from stressful areas of intensive animal foraging at the base of fruiting saguaro plants. Such protected locations reduce stresses of herbivory and extreme temperatures, increasing the probability of seed survival and offering more suitable conditions for seed germination and seedling establishment (National Park Service 2005).

Mammals which exert the stress of herbivory on stems (pads) of *Opuntia* spp. (prickly pear) include *Odocoileus virginianus* (white-tailed deer) and *Pecari tajacu* (javelina). Prickly pear fruits are eaten by a wide range of birds and various omnivores such as *Canis latrans* (coyote), *Procyon lotor* (raccoons), among others (Russell and Felker 1987). Although most *Opuntia* spp. are nontoxic, they reduce the stress of herbivory by being able to injure herbivores with spines and glochids (Delfelice 2004). *Opuntia* stems generally contain high concentrations of oxalates that bind calcium and can lead to calcium deficiency. *N. albigula* (white-throated woodrat) that consumes *Opuntia* is able to do so by having microflora in its gut that can ferment and degrade structural carbohydrates such as oxalate (Justice 1985). Stress on *Opuntia* by herbivory of *Tayassu tajacu* (javelina or collared peccary) is selective, since the javelina prefers to eat stems (pads) that are lower in oxalate content and stems that have relatively short spines (Theimer and Bateman 1992). Testing using confined javelina and two different morphological types of *Opuntia* pads with spines removed, it was found that inner pads of the two morphological types had lower neutral detergent fiber, lower levels of calcium oxalate crystals and a higher percentage of water than the outer pads; the javelina preferred the inner pads. Comparing consumption of pads with long or short spines, the javelina preferred to eat the pads with the shorter spines. In contrast to the white-throated woodrat that has gut microorganisms that can degrade oxalate, javelina have developed feeding behavior to avoid, or at least minimize, consumption of *Opuntia*'s stressful chemical defense, oxalate.

L. tridentata has extremely complex chemical defenses to ward off the stresses of competition with other plants for water and to combat the stresses of herbivory. Based on their experimental data, Mahall and Callaway (1992) reported that the roots of *L. tridentata* inhibited elongation of the roots of either *L. tridentata* or *A. dumosa* in their vicinity. The allelopathy of *Larrea* roots may be a factor to help explain the distribution of *Larrea* plants.

REFERENCES

Abe, Y. 1982. Phenology of tetraploid creosotebush, *Larrea tridentata* (DC.) COV. at the northeastern edge of the Sonoran Desert. PhD Dissertation, Tucson, AZ: University of Arizona.

Allen, E. B., R. J. Steers, and S. J. Dickens. 2011. Impacts of fire and invasive species on desert soil ecology. *Rangeland Ecol. Manage.* 64:450–462.

Bai, Y., T. A. Scott, W. Chen et al. 2011. Long-term variation in soil temperature of the Mojave Desert, southwestern USA. *Clim. Res.* 46:43–50.

Briggs, L. J., and H. L. Shantz. 1914. Relative water requirements of plants. J. Agric. Res. 3:1–63.

Brisson, J., and J. F. Reynolds. 1994. The effect of neighbors on root distribution in a creosotebush (*Larrea tridentata*) population. *Ecology.* 75(6):1693–1702.

Brusca, R. C., J. F. Wiens, W. M. Meyer et al. 2013. Dramatic response to climate change in the Southwest: Robert Whittaker's 1963 Arizona Mountain plant transect revisited. *Ecol. Evol.* 3(10):3307–3319.

Cable, D. R. 1969. Competition in the semidesert grass-shrub type as influenced by root systems, growth habits, and soil moisture extraction. *Ecology.* 50(1):27–38.

Caldwell, M. M., and J. H. Richards. 1986. Competing root systems: Morphology and models of absorption. In: *Economy of Plant Form and Function*, T. Givnish and R. Robichaud (eds.), 251–273. London, UK: Cambridge University Press.

Chow, P. N., O. C. Burnside, and T. L. Lavy. 1966. Physiological studies with prickly pear. *Weeds.* 14(1):58–62.

Clutter, M. 2016. The use of subsurface temperature fluctuations to estimate plant water use. M.S. Thesis, Tucson, AZ: University of Arizona.

Cohn, J. P. 2009. Sonoran Desert plants climb warming Santa Catalina Mountains. *BioScience.* 59(5):456.

Conver, J. L., T. Foley, D. E. Winkler et al. 2017. Demographic changes over >70 yr in a population of saguaro cacti (*C. gigantea*) in the northern Sonoran Desert. *J. Arid Environ.* 139:41–48.

Crawford, Jr., T., W. U. Singh, and H. Breman. 2008. Solving agricultural problems related to soil acidity in Central Africa's Great Lakes Region. Kigali, Rwanda: IFDC,

Crimmins, T. M., M. A. Crimmins, D. Bertelsen et al. 2008. Relationships between alpha diversity of plant species in bloom and climatic variables across an elevation gradient. Int. Biometeorol. 52:353–366.

de Martonne, Emmanuel. 1926. L'indice d'aridité. *Bulletin de l'Association de géographes français.* 3(9):3–5.

de Martonne, Emmanuel. 1942. Nouvelle carte mondial de l'indice d'aridité. *Annales de Géographie.* 51(288):241–250.

Defenders of Wildlife. 2010. Climate change and the Sonoran Desert region. https://defenders.org/sites/default/files/publications/climate_change_and_the_sonoran_desert_region.pdf

Delfelice, M. S. 2004. Prickly pear cactus. Opuntia spp: A spine-tingling tale. Weed Tech. 18(3):869–877.

Dice, L. R. 1939. The Sonoran biotic province. *Ecology.* 29(2):118–129.

Diez, J. M., C. M. D'Antonio, J. S. Dukes et al. 2012. Will extreme climatic events facilitate biological invasions? *Ecol. Environ.* 10(5):249–257.

Dimmitt, M. 2000. Biomes & communities of the Sonoran Desert region. In: *A Natural History of the Sonoran Desert*, S. J. Phillips and P. W. Comus (eds.), 3–18. Tucson, AZ: Sonora Desert Museum Press; Berkeley, CA, Los Angeles, CA, and London: University of California Press.

Drezner, T. D. 2003. A test of the relationship between seasonal rainfall and saguaro cacti branching patterns. *Ecogeography.* 26:393–404.

Drezner, T. D. 2004a. Saguaros and their nurses in the Sonoran Desert: A review. *Desert Plants.* 20(1):3–10.

Drezner, T. D. 2004b. Saguaro patterns and ecology over Arizona: A closer look at rainfall. *Desert Plants.* 20(1):24–32.

Drezner, T. D. 2006. Plant facilitation in extreme environments: The non-random distribution of saguaro cacti (*C. gigantea*) under their nurse associates and the relationship to nurse architecture. *J. Arid Environ.* 65:46–61.

Drezner, T. D. 2014. How long does the giant saguaro live? Life, death and reproduction in the desert. *J. Arid Environ.* 104:34–37.

Eddy, T. A. 1959. Foods of the collared peccary, *Pecari tajacu sonoriensis* (Mearns) in southern Arizona. MS Thesis, Tucson, AZ: University of Arizona, 102.

Epstein, E., and A. J. Bloom. 2005. *Mineral Nutrition of Plants: Principles and Perspectives.* Sunderland, MA: Sinauer Associates, Inc.

Forseth, I. N., J. R. Ehleringer, K. S. Werk, and C. S. Cook. 1984. Field water relations of Sonoran Desert annuals. *Ecology.* 65(5):1436–1444.

Franco, A. C., and P. S. Nobel. 1989. Effect of nurse plants on the microhabitat and growth of cacti. *J. Ecol.* 77(3):870–886.

Fulbright, N., and T. E. Fulbright. 1990. Germination of 2 legumes in leachate from introduced grasses. *J. Range Manage.* 43:466–467.

Gibson, A. C. 1998. Photosynthetic organs of desert plants. *BioScience.* 48(11):911–920.

Hadley, N. F., and S. R. Szarek. 1981. Productivity of desert ecosystems. *BioScience.* 31(10):747–753.

Halvorson, W. L., and D. T. Patten. 1974. Seasonal water potential changes in Sonoran Desert shrubs in relation to topography. *Ecology.* 55:173–177.

Hansen, J., M. Sato, R. Ruedy et al. 2006. Global temperature change. *PNAS.* 103(39):14288–14293.

Hanson, R. B., and J. Hanson. 2000. Sonoran Desert natural events. In: *A Natural History of the Sonoran Desert*, S. J. Phillips and P. W. Comus (eds.), 19–28. Tucson, AZ: Sonora Desert Museum Press; Berkeley, CA, Los Angeles, CA, and London: University of California Press.

Harshberger, J. W. 1911. Phytogeographic survey of North America. In: *Vegetation der Erde*, von A. Engler and O. Drude (eds.), Vol. 13, map.

Hatfield, J. L., and J. H. Prueger. 2015. Temperature extremes: Effect on plant growth and development. *Weather Clim. Extremes.* 10:4–10.

Ives, R. L. 1949. Climate of the Sonoran Desert region. *Ann. Assoc. Am. Geogr.* 39(3):143–187.

Jenkins, M. B. 2003. Rhizobial and bradyrhizobial symbionts of mesquite from the Sonoran Desert: Salt tolerance, facultative halophily and nitrate respiration. *Soil Biol. Biochem.* 35:1675–1682.

Jordan, P. W., and P. S. Nobel. 1979. Infrequent establishment of seedlings of *Agave deserti* (Agavaceae) in the northwestern Sonoran Desert. Am. J. Bot. 66(9):1079–1084.

Jordan, P. W., and P. S. Nobel. 1981. Seedling establishment of *Ferocactus acanthodes* in relation to drought. *Ecology.* 62(4):901–906.

Jordan, P. W., and P. S. Nobel. 1982. Height distributions of two species of cacti in relation to rainfall, seedling establishment and growth. Bot. Gaz. 143(4):511–517.

Justice, K. E. 1985. Oxalate digestibility of *Neotoma albigula* and *Neotoma mexicana*. *Oecologia.* 67(2):231–234.

Klein, H. 2002. Prosopis seed beetles (*Algarobius prosopis* and *Neltumius arizonensis*). PPRI Leaflet Series: Weeds Biocontrol, No. 3.2. ARC Plant Protection Research Institute.

Lajtha, K., and W. H. Schlesinger. 1988. The effect of CaCO3 on the uptake of phosphorus by two desert shrub species, *Larrea tridentata* (DC.) Cov. and *Parthenium incanum* H. B. K. *Bot. Gaz.* 149(3):328–334.

Lightfoot, D. C., and W. G. Whitford. 1987. Variation in insect densities on desert creosotebush: Is nitrogen a factor? *Ecology.* 68(3):547–557.

Lowe, C. H., and D. S. Hinds. 1971. Effect of paloverde (*Cercidium*) trees on the radiation flux at ground level in the Sonoran Desert in winter. *Ecology.* 52(5):916–922.

Mahall, B. E., and R. M. Callaway. 1992. Root communication mechanisms and intracommunity distributions of two Mojave Desert shrubs. *Ecology.* 73(6):2145–2151.

McAuliffe, J. R. 1999. The Sonoran Desert landscape complexity and ecological diversity. In: *Ecology of Sonoran Desert Plants and Plant Communities*, R. H. Robichaux (ed.), 68–114. Tucson, AZ: University of Arizona Press.

McLaughlin, S. P., and J. E. Bowers. 1999. Diversity and affinities of the flora of the Sonoran floristic province. In: *Ecology of Sonoran Desert Plants and Plant Communities*, R. H. Robichaux (ed.), 12–35. Tucson, AZ: University of Arizona Press.

Medeiros, A. S., and T. D. Drezner. 2012. Vegetation, climate, and soil relationships across the Sonoran Desert. *Ecoscience.* 19(2):148–160.

National Centers for Environmental Information of the National Oceanic and Atmospheric Administration (NOAA/NCEI). https://www.ncdc.noaa.gov/.

National Park Service. 2005. Chapter 3: The fate of the seed: Dispersal, attrition, and germination. In: *Ecology of the Saguaro: II* NPS Scientific Monograph No. 8. https://www.nps.gov/parkhistory/online_books/science/8/chap3.htm

National Park Service. 2007. Weather and Climate Inventory National Park Service Sonoran Desert Network Natural Resource Technical Report NPS/SODN/NRTR—2007/044.

National Soil Survey Center, Natural Resources Conservation Service, U. S. Department of Agriculture. 2012. Field book for describing and sampling soils. Version 3.0. September 2012.

Nilsen, E. T., M. R. Sharifi, P. W. Rudel et al. 1983. Diurnal and seasonal water relations of the desert phreatophyte *Prosopis glandulosa* (Honey mesquite) in the Sonoran Desert of California. *Ecology.* 64(6):1381–1393.

Nobel, P. S. 1976. Water relations and photosynthesis of a desert CAM plant, *Agave deserti*. Plant Physiol. 58:576–582.

Nobel, P. S. 1977. Water relations and photosynthesis of a barrel cactus, *Ferrocactus acanthodes*, in the Colorado Desert. *Oecologia.* 27:117–133.

Nobel, P. S. 1980. Morphology, surface temperatures and northern limits of columnar cacti in the Sonoran Desert. *Ecology.* 61(1):1–7.

Nobel, P. S. 2015. Letter to the editor by Dr. Park S. Nobel. http://jpacd.org/downloads/LE/Letter_to_Editor_by_Park_S.Nobel.pdf

Nobel, P. S., and M. E. Loik. 1999. Form and function of cacti. In: *Ecology of Sonoran Desert Plants and Plant Communities*, R. H. Robichaux (ed.), 143–163. Tucson, AZ: University of Arizona Press.

Nobel, P. S., and B. R. Zutta. 2005. Morphology, ecophysiology, and seedling establishment for *Fouquieria splendens* in the northwestern Sonoran Desert. *J Arid Environ.* 62:251–265.

Noy-Meir, I. 1973. Desert ecosystems: Environment and producers. *Ann. Rev. Ecol. Syst.* 4:25–51.

Osmond, C. B., M. P. Austin, J. A. Berry et al. 1987. Stress physiology and the distribution of plants. *BioScience* 37(1). *How Plants Cope: Plant Physiological Ecology*, 38–48.

Parker, K. C. 1991. Topography, substrate, and vegetation patterns in the Northern Sonora Desert. *J. Biogeogr.* 18(2):151–163.

Phillips, D. L. 1979. Competition and spacing patterns of shrubs in the Mojave and Sonoran deserts. Ph.D. Dissertation, Logan, UT: Utah State University.

Pierson, E. A., R. M. Turner, and J. L. Betancourt. 2013. Regional demographic trends from long-term studies of saguaro (*Carnegiea gigantea*) across the northern Sonoran Desert. *J. Arid Environ.* 88:57–69.

Pockman, W. T., and J. S. Sperry. 1997. Freezing-induced xylem cavitation and the northern limit of *Larrea tridentata*. *Oecologia.* 109:19–27.

Ram, H. Y. M., and P. Gupta. 1997. Plant life under extreme environments. *Curr. Sci.* 72(5):306–315.

Rehfeldt, G. E., N. L. Crookston, M. V. Warwell et al. 2006. Empirical analysis of plant-climate relationships for the Western United States. Int. J. Plant Sci. 167(6):1123–1150.

Russell, C. E., and P. Felker. 1987. The prickly-pears (*Opuntia* spp., Cactaceae): A source of human and animal food in semiarid regions. Econ. Bot. 41(3):433–445.

Schade, J. D., and S. E. Hobbie. 2005. Spatial and temporal variation in islands of fertility in the Sonoran Desert. *Biogeochemistry.* 73:541–553.

Sharifi, M. R., F. C. Meinzer, E. T. Nilsen et al. 1988. Effect of manipulation of water and nitrogen supplies on the quantitative phenology of *Larrea tridentata* (creosote bush) in the Sonoran Desert of California. Am. J. Bot. 75(8):1163–1174.

Shreve, F. 1911. The influence of low temperatures (*sic*) on the distribution of the giant cactus. *The Plant World*. 14(6):136–146.

Shreve, F. 1915. The vegetation of a desert mountain range as conditioned by climatic factors. Washington, DC: The Carnegie Institution of Washington, 112.

Shreve, F. 1934. Rainfall, runoff and soil moisture under desert conditions. *Ann. Assoc. Am. Geogr.* 24(3):131–156.

Shreve, F. 1936. The plant life of the Sonoran Desert. *Sci. Mon.* 42(3):195–213.

Shreve, F. 1940. The edge of the desert. *Yearb. Assoc. Pac. Coast Geogr.* 6:6–11.

Shreve, F. 1942. The desert vegetation of North America. Bot. Rev. 8(4):195–246.

Shreve, F. 1951. *Vegetation and Flora of the Sonoran Desert*. Volume I. Vegetation of the Sonoran Desert. Washington, DC: Carnegie Institution of Washington Publication, 591.

Shreve, F., and I. L. Wiggins. 1964. *Vegetation and Flora of the Sonoran Desert*. Vols. 1 and 2. Stanford, CA: Stanford University Press, 1740.

Singer, M. J., and D. N. Munns. 1991. *Soils – An Introduction*. New York, NY: Macmillan Publishing Company.

Smith, S. D., B. Didden-Zopfy, and P. S. Nobel. 1984. High-temperature responses of North American cacti. *Ecology.* 65(2):643–651.

Springer, A. C., D. E. Swann, and M. A. Crimmins. 2015. Climate change impacts on high elevation saguaro range expansion. *J. Arid Environ.* 116:57–62.

Stevens, J., and D. A. Falk. 2009. Can buffelgrass invasions be controlled in the American Southwest? Using invasion ecology theory to understand buffelgrass success and develop comprehensive restoration and management. *Ecol. Restor.* 27(4):417–427.

Stromberg, J. C. 2007. Seasonal reversals of upland-riparian diversity gradients in the Sonoran Desert. Diversity Distrib. 13:70–83.

Theimer, T. C., and G. C. Bateman. 1992. Patterns of prickly-pear herbivory by collared peccaries. J. Wildlife Manage. 56(2):234–240.

Turnage, W. V., and A. L. Hinckley. 1938. Freezing weather in relation to plant distribution in the Sonoran Desert. *Ecol. Monogr.* 8(4):529–550.

Turner, R. M. 1990. Long-term vegetation change at a fully protected Sonoran Desert site. *Ecology.* 7(12):464–477.

Turner, R. M., and D. E.Brown. 1982. 154.1 Sonoran Desert scrub. *Desert Plants.* 4(1–4):181–221.

Verstraete, M. M., and S. A. Schwartz. 1991. Desertification and global change. In: *Vegetation and Climate Interactions in Semi-arid Regions*, A. Henderson and A. J. Pitman (eds.). *Vegetation.* 91:3–13.

Virginia, R. A., and W. M. Jarrell. 1983. Soil properties in a mesquite-dominated Sonoran Desert ecosystem. *Soil Sci. Soc. Am. J.* 47:138–144.

Ward, J. P., S. E. Smith, and M. P. McClaran. 2006. Water requirements for emergence of buffelgrass (*Pennisetum ciliare*). Weed Sci. 54:720–725.

Weiss, J. L., and J. T. Overpeck. 2005. Is the Sonoran Desert losing its cool? *Global Change Biol.* 11:2065–2077.

Whittaker, R. H., and W. A. Niering. 1965. Vegetation of the Santa Catalina Mountains, Arizona: A gradient analysis of the south slope. *Ecology.* 46(4):429–452.

Woodell, S. R. J., H. A. Mooney, and A. J. Hill. 1969. The behaviour of *Larrea divaricata* (creosote bush) in response to rainfall in California. *J. Ecol.* 57(1):37–44.

Yeaton, R. I., and M. L. Cody. 1979. The distribution of cacti along environmental gradients in the Sonoran and Mojave deserts. *J. Ecol.* 67:529–541.

Zolotokrylin, A. N., T. B. Titkova, and L. Brito-Castillo. 2016. Wet and dry patterns associated with ENSO events in the Sonoran Desert from, 2000–2015. *J. Arid Environ.* 134:21–32.

15 Stresses in Pasture Areas in South-Central Apennines, Italy, and Evolution at Landscape Level

A. Fatica, L. Circelli, E. Di Iorio, C. Colombo, T. W. Crawford, Jr. and E. Salimei

CONTENTS

15.1 SOUTH-CENTRAL APENNINES OF ITALY: MOLISE REGION AND ITS PASTURES

The Molise region is located in the central Apennines of southern Italy, occupies an area of 443,758 ha, and has a range of elevations from sea level at the Adriatic Sea to 2,050 m a.s.l. The Molise region's borders are with Abruzzo region to the north, Lazio region to the west, Campania region to the southwest, Puglia region to the southeast, and the Adriatic Sea to the east (Figure 15.1a).

The population of the region is about 310,499 inhabitants, divided into 136 municipalities, with a density of about 70 inhabitants/km². The main regional town, Campobasso City, has 49,168 inhabitants and is located at the center of the region at about 700 m a.s.l. Another province of Molise is

represented by Isernia, a town with 21,685 inhabitants that is located at 423 m a.s.l.

The territory in the Molise region is predominantly mountainous, and forest, pasture, and natural meadows represent the most important land use of this territory, which needs to be protected from erosion and other hydrogeological phenomena such as flooding and landslides. In addition, the most important river of the Molise region is the Biferno, which has a torrential regime that is closely related to the intensity and duration of precipitation.

The presence of natural pastures in Molise has undergone a significant decrease over the past 27 years, from around 50,000 ha in 1990 (De Renzis et al. 1992a) to 36,627 ha in 2005 (INEA 2008) and to 37,690 ha in 2007 (ISTAT 2010). Considering the geographic and morphological characteristics

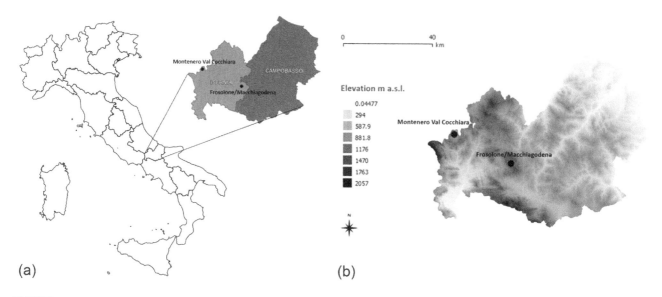

FIGURE 15.1   (a) Map of Molise region in Italy with the position of the two sample areas; (b) digital elevation model of Molise region.

of the Molise region, the need to manage the most fragile mountain areas exposed to erosive phenomena is evident.

This chapter focuses on one of the most important ecosystems of the region: natural pastures, which have always been used as a source of feed for animal husbandry, given Molise's long history of livestock production. However, the best practices to manage grazing can only be applied to livestock production if all the abiotic and biotic factors that stress production of forage are known in detail. These stresses are principally climatic (temperature and rainfall), edaphic, and biotic (competing vegetation in the pasture and stresses due to the presence of the animals).

To describe the best practices of sustainable management of the pastures of the central and southern Apennines, we first present a detailed description of the principal abiotic and biotic stress factors, and then, we report a case study carried out in two municipalities of the Molise region. The study areas are located in Isernia province near the municipalities of Montenero Val Cocchiara (41°43′N, 14°04′E) and Frosolone (41°36′N, 14°27′E), including the municipality of Macchiagodena (Figure 15.1a).

15.1.1   Geography

The Molise territory is nearly all mountainous (55% of the surface area) or hilly, with limited flat ground in the lower valleys and along the Adriatic coast (Figure 15.1b). The Apennines divide Molise into isolated mountains and a chaotic array of hills, which stretch to within a few kilometers of the coast, making communications difficult and creating a state of isolation. The highest mountains are located in part of the "Samnite Apennines" (the northern and eastern parts of the Campano Apennines, separate from the Abruzzese Apennines). The Samnite Apennines include the southern extreme of the Meta Mountains, culminating at 2185 m a.s.l., the northern slope of the calcareous Matese massif (Mt. Miletto, 2,050 m a.s.l.),

and the Mount Mutria group. In addition, the Molise border passes the Apennines watershed, including the upper valley of the Volturno River, between the Mainarde and Matese.

Toward the Adriatic, the mountainous landscape of the Apennines consists of a sequence of hills characterized by steeper slopes. Here, the land becomes increasingly lower as it approaches the Adriatic coastline. The Biferno river is entirely in the regional territory of Molise, while the Trigno and Fortore rivers cross the Abruzzo and Campania regions, respectively, and flow into the Adriatic Sea. These rivers, crossing the transverse valleys to the Apennines, lie semi-parallel to each other, flowing for a long distance at the limits of the regional territory onto the Tyrrhenian slope of the Volturno and the Tammaro rivers. Only the upper parts of the basins of these rivers lie in Molise. All the waterways are greatly affected by seasonal variations in precipitation and consequently, are torrential when precipitation is intense.

The studied area is characterized by human settlements that are organized into many sparsely urbanized small municipalities and rural areas. The latter, especially in the valleys, are devoted to marginal agriculture that takes advantage of the presence of water.

15.1.2   Climate

As a result of geographic and topographic differences between the coastline and the inland mountains, at varying distances from the sea, the climate of Molise has a wide range of characteristics that range from the typically maritime characteristics (modest variations in temperature, mild weather in all seasons, low precipitation in summer) to the continental cold-humid characteristics of the mountainous interior (Ludovico et al. 2018) (Figure 15.2).

The temperature shows marked differences in contrasting seasons and between day and night. Heavy precipitation, including snow, occurs up to over 2500 mm/year above 1000 m

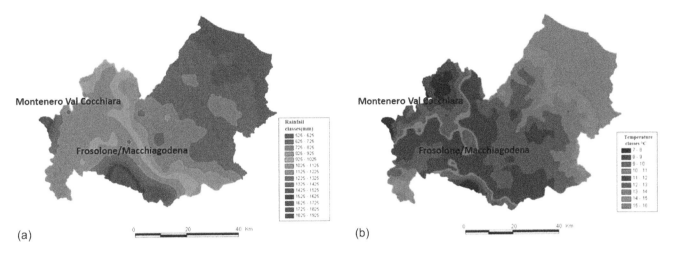

(a) (b)

FIGURE 15.2 Maps of climatic conditions in Molise region: rainfall classes on the left, temperature classes on the right.

a.s.l. Rainfall, although of short duration, is most frequently intense in autumn and spring, but rainfall lasts longer in winter, with peak levels in November. The lowest rainfall is in July.

The Montenero Val Cocchiara valley, an intermountain basin located at 950 m a.s.l. in the southern Apennines, was originally the bottom of an ancient lake. Currently, the valley is approximately 3 km long and 1 km wide and is partially covered by a peat bog. The local climate is characterized by cold-humid winters and hot-dry summers, as occurs in typical Mediterranean regions (Figure 15.3a).

The total mean annual rainfall in the Montenero Val Cocchiara valley is around 1000 mm, with an increasing trend during the 31 year period from 1987 to 2017 (Figure 15.3b). The annual rainfall, with a maximum in autumn and a minimum in summer, is usually enough to allow normal natural pasture vegetation. Although precipitation is most intense in autumn (more than 500 mm), droughts during the dry summers present stress, which favors the survival of drought-tolerant vegetation. The average temperature in this valley ranges from 4 °C in January to 21 °C in July, and the average annual temperature is about 12 °C, also in this case with an increasing trend in the same 31 year period (Figure 15.3c). During the winter, freezing temperatures occur, the temperature can drop to −8 °C, and when there is precipitation, it can snow.

In Frosolone/Macchiagodena, at 850 m a.s.l., the average temperature ranges from 8 °C in January to 20 °C in July and August (Figure 15.4a).

During winter and early spring in Frosolone, snowfall is frequent and plentiful, while in summer, the extreme high temperature does not exceed 35 °C. The variation of mean total annual precipitation is more pronounced than the variation of temperature, but with different trends (Figure 15.4b and c). During the period from 1987 to 2017, trend lines that estimate annual precipitation and annual temperature show that annual rainfall at Frosolone has tended to decrease (Figure 15.4b), whereas average temperature has tended to increase (Figure 15.4c). The variation of annual rainfall from

year to year at Frosolone indicates that drought stress occurs in the area in some years (Figure 15.4b).

15.1.3 Soil

The geological substrate in the central south Apennines is formed by Tertiary arenaceous, silty, marly sediment, namely, Miocene "Flysch" (Vezzani et al. 2010). The mountains are formed by hard calcareous rocks that result in a karstic landscape, while hilly and locally terraced morphology is dominated by marly limestone bedrock (Patacca and Scandone 2007). Based on mean annual precipitation (900–1400 mm), the soil moisture regime is udic and locally ustic in the valleys. The soil temperature regime is mesic and thermic. On flat land, soils are more developed, with pedogenic structure in depth. According to Food and Agriculture Organization World Reference Base (FAO-WRB) classification (FAO 2006), the weakly differentiated profiles are Eutric and Calcaric Cambisols (Eutrochrepts as reported by the United States Department of Agriculture [USDA] Soil Taxonomy; Soil Survey Staff 2010), soils with clay accumulation (Haplic and Gleyic Luvisols), and acid soils with organic matter accumulation (Humic Umbrisols). During spring and autumn, the soils are affected by intense erosion by water, resulting in the deposition of alluvial parent material, which results in soils (Eutric and Calcaric Regosols and Lithic Leptosols). In the sloping land along the steep sides of the hills, more intense erosion by water occurs at altitude 800–600 m a.s.l. on slopes >30%. This soil erosion results in changes associated with slope, stoniness, rockiness, and locally, changing morphology.

The Montenero Val Cocchiara area (289 ha; Figure 15.5) is located mainly on an alluvial plain of the Zittola river (Oligocene age), which is characterized by peaty layers with local outcropping. The boundary slopes of the plain consist of well-stratified limestone and alternating marly layers (Oligocene–Miocene age). The presence of a deep layer of peat (4 m) is due to ancient filling up and swamping of the plain. This area is subject to some environmental protection programs, such as the Sites of Community Importance

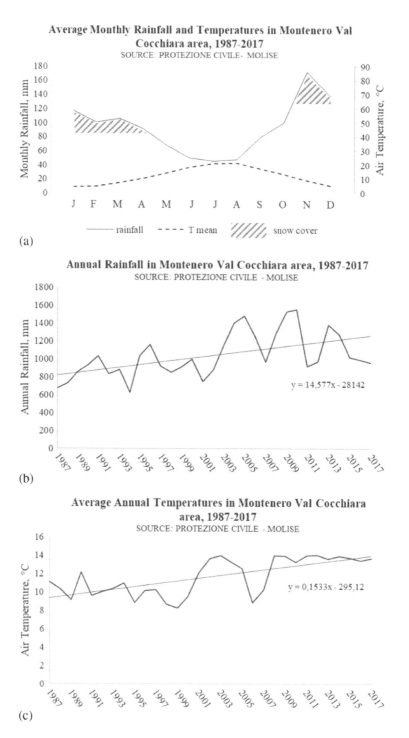

FIGURE 15.3 Climatic characterization of Montenero Val Cocchiara area: (a) average monthly rainfall and temperatures, (b) annual rainfall, (c) average annual temperatures.

(SIC) Program, set up by the European Union to safeguard the habitats and autochthonous animal and plant species. The soils sampled along the plain at locations highlighted in Figure 15.5 are very different in texture and profile development, but gley features, due to seasonal submersion, characterized them all. These soils, classified according to USDA Soil Taxonomy (Soil Survey Staff 2010) as Eutrochrepts, are characterized by the presence of gley features in B and C horizons (Montenero V.C.

3 and Montenero V.C. 4; Figure 15.6); their depth is adequate for root growth of pasture vegetation.

The Montenero soils sampled have clay–silty texture and are skeleton-free and calcareous, with a weakly to moderately alkaline reaction. They have very slow permeability and high available water holding capacity. The content of organic matter varies little throughout the profile sampled. Peaty horizons may be present below the depth of 1 m. The water table is

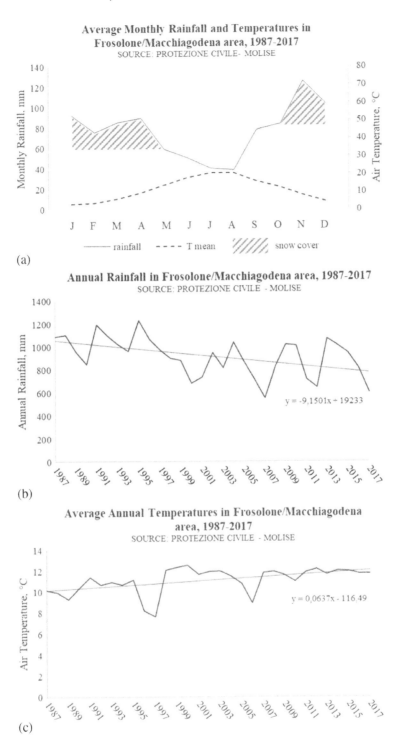

FIGURE 15.4 Climatic characterization of Frosolone/Macchiagodena area: (a) average monthly rainfall and temperatures, (b) annual rainfall, (c) average annual temperatures.

present for significant periods, starting from a depth of 20 cm from the surface during the wet season but present within 100 cm of depth at other times during the year.

The topography of the Frosolone/Macchiagodena area (249 ha; Figure 15.7) is terraced and flat, and the morphology of the soil surface is characterized by the absence of superficial stoniness and rockiness; the typical land use of this area is grazing. The degree of plant cover is always high (>90%), and

the vegetative composition of the Frosolone/Macchiagodena area is typical of Mediterranean grassland (Catorci et al. 2011). The vegetative cover diminishes the risk of stress to vegetation due to aridity and decreases the risk of potential erosion and compaction of the soil surface. Variations in the density or seasonality of grazing have significant impacts on plant community composition (Catorci and Gatti 2010). Calcaric Cambisols and Humic Humbrisols are sometimes present as lower soil

FIGURE 15.5 Aerial photo of Montenero Val Cocchiara area.

horizons. The soils of the Frosolone/Macchiagodena area (Frosolone 1 and Frosolone 5; Figure 15.6) are well drained, without skeleton, very deep, and with good depth for the roots. These soils, classified according to USDA Soil Taxonomy (Soil Survey Staff 2010) as Haplumbrepts and Hapludalfs, are characterized by a low calcium carbonate content, and their reaction is acid in the top soil and weakly acid in the subsoil. They have moderately slow permeability and very high water holding capacity. The soil organic matter content is high in the surface horizon. Based on geomorphic features, Aucelli et al. (2012) estimate spatially averaged water erosion rates for these basins as 0.13–0.23 mm/year, with valley (fluvial) incision rates of 0.46–0.71 mm/year for the same period.

These data suggest 200–400 m of downcutting since the mid-Pleistocene, which has set the boundary condition for hillslopes in the area. Hillslope gradients in the study area range from 10° along the top to more than 35° in the hillier northeast of the study area, with the majority of slopes being 8°.

15.1.4 VEGETATION

Pasture is a complex ecosystem whose productivity is influenced by natural factors, such as climate, soil, and vegetative characteristics, and anthropogenic factors, such as the mode and intensity of grazing (Di Rocco et al. 1992a).

Within the pasture ecosystem, the soil plays an important role, influencing the productivity of the vegetation as well as the most suitable management and conservation techniques.

First, the pedological characterization of the pasture areas allows assessment of the nutritional and habitability functions of the soil, which define its capacity to support the development of the herbaceous cover, influencing the productivity of the pasture from both a quantitative and a qualitative point of view.

Some soil parameters that directly influence the productivity of a pasture are

- Water holding capacity: it constitutes the reserve of water that the soil is able to retain for absorption by plants, in particular during the growing season.

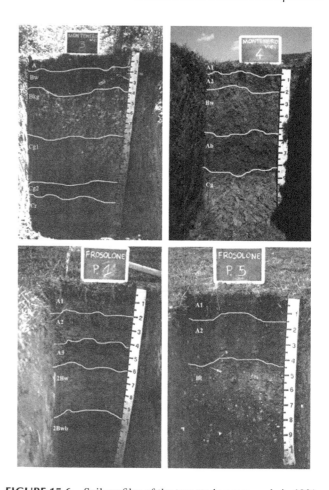

FIGURE 15.6 Soil profiles of the two study areas made in 1991–1993 (Di Rocco et al. 1991b), classified according to USDA Soil Taxonomy Montenero V.C. 3—*Eutrochrepts*: deep, well-drained soil with fine texture in all horizons. The top soil (A horizon), with a thickness of about 15 cm, has a very dark brown color (10YR 2/2) and a granular and subangular blocky structure; the exchange complex is saturated; the organic matter content is average (4.2%). The subsoil is characterized by the presence of a carbonate accumulation horizon (Bkg), with a coarse and very coarse subangular blocky structure toward massive. Montenero V.C. 4—*Eutrochrepts*: deep, well-drained soil with silty-clay texture, without skeleton. The topsoil is calcareous with a neutral reaction and a high content of organic matter (14%). During the wet season, the water table can also be found at a depth of 50 cm, the subsoil being characterized by a rather low permeability. This condition causes the presence of gley features in Cg horizon. Frosolone 1—*Haplumbrepts*: well-drained soil, without a skeleton and limestone, very deep, with high depth useful to the roots. The reaction is moderately acidic in the topsoil (pH 5.1) and weakly acidic in the subsoil (pH 6.1). The available water capacity is as high as the organic matter content (9.2%). In all horizons, there is a developed coarse subangular blocky structure. Frosolone 5—*Hapludalfs*: well-drained soil, very deep, with fine texture and high depth useful to the roots. The skeleton, of medium size, is frequent in the topsoil and abundant in the subsoil. The soil is without limestone, and it has a very strong acidic reaction in the topsoil (pH 4.9) and moderately acidic in the subsoil (pH 5.2). The available water capacity is moderate; the organic matter content is high (10.4%). In Bt horizon, there is a high content of clay (48.5%), and the granular structure is developed. (Soil Survey Staff 2010).

FIGURE 15.7 Aerial photo of Frosolone/Macchiagodena area.

- Nutritional status of the soil: it expresses the average level of both nutrients and organic substances in the soil, including their losses due to surface water erosion and leaching.
- Availability of oxygen for the roots: it is correlated with drainage of the soil. Indeed, limited availability of oxygen for the roots during the growing season influences the quality of forage produced by causing a decrease in the palatability of the forage.

However, as for all other ecosystems, even the pasture ecosystem is continually subjected to biotic and abiotic stress. According to Di Rocco et al. (1991b), the dynamics of degradation that can affect grazing areas are many.

Factors that can affect the soil erosion process are the intensity of the rainfall, the rate of surface water flow, the texture, structure, and permeability of the soil, the slope of the soil surface, and the presence or absence of cover crops. Furthermore, the action of humans on the soil, such as ploughing, deep ripping to improve drainage, installation of tile drains, or removal of vegetation, can strongly influence the intensity of the erosive process. Soil erosion is a process that takes place in two phases: detachment of the particles from the soil mass and transport of the soil particles. In conditions characterized by poor plant cover and soil with low infiltration capacity, the whole area of the slope contributes to generate surface runoff.

The Mediterranean basin is characterized by a climate that promotes the erosion process, with high intensity of rainfall and the alternation of strongly dry periods and very humid periods. The presence of dry periods is stressful to pastures, because the lack of water inhibits regular development of the vegetation and leaves the soil unprotected. Moreover, the topography of the Italian peninsula is very uneven. Two-thirds of the Italian peninsula is mountainous or hilly, and as elevation above sea level increases, the potential energy of the water flowing downhill by the force of gravity can cause erosion of the soil surface. According to Morgan (1986), it is estimated that one-sixth of the national surface (50,000 km²)

of Italy is subject to accelerated erosion and that soil losses range from 0.1 to 1.4 mm per year, with maximum values of 7 mm in particularly erosive years.

Another source of stress on pastures is degradation due to soil compaction caused by the passage of livestock and operating machines. The compaction of the soil involves the destruction of the structural aggregates and the reduction of both macroporosity and microporosity, which causes slower infiltration of the water in the soil. Overall, there is a worsening of the physical properties of the soil and its permeability as regards the movement of air, water, and plant roots of pasture species. Sometimes, reduction of the permeability of pasture soils results in the spread of competitive weed species.

Pasture includes herbaceous plants, bushes, and trees that can be directly used by the animals (Cavallero et al. 2002; Pardini 2005). However, there is a difference between pasture and pasture territory. Pasture territory indicates a vast surface covered by natural vegetation made by herbaceous, arboreal, and shrub species, used exclusively for animal grazing (Vallentine 1990). Pasture territory is not less than hundreds of hectares and excludes the possibility of intensive management.

Pasture presents a greater management intensity and less areal extension (Pardini 2005) of phytocenoses, whose biomass is used partly or wholly by herbivorous animals (Cavallero et al. 2002) that consume it directly.

As a result of stresses due to grazing by animals of different species and categories present on the pasture, the floral composition and the physical structure of the vegetation and the soil undergo modifications (Acciaioli and Esposito 2010).

However, meadows and pastures are not only a source of food for animals; they also have other functions, which are part of the cultural and social heritage of the territories (Cavallero et al. 2002). These functions are regulating the emission of gases in the atmosphere, controlling soil erosion phenomena (Gusmeroli 2004), and contributing to maintenance of the structure of the landscape (Sanderson et al. 2004).

According to the definition of Peeters et al. (2014), grasslands are represented by land devoted to the production of forage for harvest by grazing, cutting, or both, or used for other agricultural purposes such as renewable energy production. The vegetation can include grasses, grass-like plants, legumes, and other forbs. Woody species may also be present. Grasslands can be temporary or permanent, when they are not completely renewed or regenerated after destruction by ploughing or herbicide use, for 10 years or longer.

Two management categories of grasslands can be identified: meadows, i.e., grasslands that have been harvested predominantly by mowing over the last 5 years; and pastures, i.e., grasslands that have been harvested predominantly by grazing over the last 5 years (Peeters et al. 2014).

In both cases, the grassland types can be characterized by different tree associations. Especially in central Italy, along the Tyrrhenian coast, evergreen trees and shrubs (e.g., *Quercus ilex*, *Q. suber*, and *Q. coccifera*) can be found, while inland, in the hills, forests of deciduous trees (e.g., *Castanea sativa*) prevail, and in the mountains, forests of *Fagus sylvatica* prevail (Porqueddu et al. 2017). Finally, natural meadows are associated with various types of vegetative association (*Brachipodietum* spp., *Brometum* spp., *Arrhenatheretum* spp., *Festucetum* spp., and *Lolietocynosuretum* spp.) in high mountains. In southern Italy, grasslands are associated with thermoxerophytic shrubs along the coasts, together with forest and evergreen sclerophylls (Porqueddu et al. 2017). In the Apennine zone, above 500 m a.s.l., permanent grasslands are found. However, these areas were characterized by a gradual reduction from 1900 to 2000, while from 2000 to 2013, the area of permanent grasslands remained stable (EUROSTAT 2012).

Pastures have positive effects from different points of view (i.e., economic, ecological, managerial, and productive) linked, for example, to the increase of the number of palatable (edible) and not palatable species, the control of weed species, the reduction of soil losses due to erosive phenomena, the maintenance of biodiversity, and the decisions of the Natura 2000 network.

Natura 2000 is the main instrument of the European Union's policy for the conservation of biodiversity. This is an ecological network spread throughout the Union, established under Directive 92/43/EEC "Habitat" to ensure the long-term maintenance of natural habitats and endangered or rare species of flora and fauna at the Community level. The Natura 2000 network is constituted by the Sites of Community Interest (SIC), identified by the Member States as laid down in the Habitats Directive, which were subsequently designated as Special Conservation Areas (ZSC) and include the Area of Special Protection (ZPS) established under Directive 2009/147/EC "Birds" concerning the conservation of wild birds. The purpose of Article 2 of the Habitat Directive is to ensure the protection of nature by keeping and "taking account of economic, social and cultural needs, as well as regional and local particularities" (Directive 92/43/EEC); so, the aim of the directive is to preserve natural and semi-natural habitats (such as areas with traditional agriculture, woods that are used, or pastures).

In Italy, SIC, ZSC, and ZPS cover a total of about 19% of the national land and almost 4% of the marine territory. The Montenero Val Cocchiara area, an important SIC area, encompasses a wetland area of about 900 hectares, which represents one of the last wetlands existing in Europe (Tamburro et al. 2005).

However, the capacity of the pasture to sustain the requirements of the animals is determined by the quantity of biomass produced and the qualitative characteristics of the pasture, expressed as nutritional values.

In the following are reported the most important nutritional values of the pasture biomass collected in 2000–2003 in Montenero Val Cocchiara—Area 1 and Frosolone/Macchigodena—Area 2.

15.1.4.1 Montenero Val Cocchiara—Area 1

This area, included in the "Corine Biotope," "Bioitaly," and "Natura 2000" list (Lucchese 1995), is characterized by very rare plant populations such as *Salix pentandra* and *Dactyloriza*

incarnata, and most of the soil is a peat bog residue, which is very uncommon in the Apennine areas (Miraglia et al. 2001). The area is located between 800 and 900 m a.s.l. and is formed by a broad plain surrounded by wooded hills encompassing 2200 ha. The broad plain is used as grazing meadow (586 ha), while the remaining part is used as pasture. The broad plain (about 1000 ha) is divided into two sub-areas for the pasture area and three sub-areas for the grazing-meadow area, in which the animals cannot graze from the middle of April to the end of June (haymaking time).

In the following are reported some data from studies conducted in 2000 and 2001 and aimed at the preservation of the entire area (Miraglia et al. 2003), which is the habitat of the autochthonous Pentro horse (MiPAF 2003) as well as the habitat of the rare plant species already mentioned. The Pentro herd lives wild all year and is not fed using supplementary foods; moreover, cattle, sheep, and goats also graze in the area, resulting in a total of 1500 Adult Bovine Units (UBA, 24-month-old bovine) (Di Rocco et al. 1992a).

In the grazing-meadows system, there are a great number of plant species usually related to the pasturing and belonging to the Poaceae and Fabaceae families (80% of the plant species observed), identified according to Corral and Fenlon (1978). In the sub-areas used only as pasture, there is a different ratio between Gramineae (Poaceae) and Legume (Fabaceae) and a higher percentage of weeds such as *Juncus articulatus* and *Ranunculus acris*.

In their 2 year study of precipitation and temperature, Costantini et al. (2004) found abundant rainfall in fall and spring, while rainfall was scarce from June to September. During this period, when evapotranspiration is at the maximum level, there were differences in the rainfall: 85.2 and 150.4 mm of rain in the first and second year, respectively. In the first and second year of the study, the average temperature was 11.9 °C, and it ranged between 3.6 °C in January and 21.1 °C in July in both years.

Several variables measuring the chemical composition of forage from the grassland did not show differences between the two years; only the dry matter and protein percentage of the forage were different (Table 15.1), according to Costantini et al. (2004). The differences in dry matter and protein content of the forage are probably due to a greater drought stress in the first year, which could have caused a qualitative and quantitative deterioration of the grassland and resulted, at least temporarily, in proportionately more growth of the xerophytic (more drought-tolerant) species, which have a lower protein content.

A simulation model was used to evaluate the availability of forage in the investigated Area 1 (970 ha), assuming a 65% use of the herbaceous resources. In the year 2000, the production of forage in Montenero Val Cocchiara—Area 1 was estimated to have been 2500 t DM (dry matter), while in the second year (2001), it was estimated to have been 2900 t DM.

15.1.4.2 Frosolone/Macchiagodena—Area 2

This area is located between 1200 and 1400 m a.s.l. and is characterized by sloping areas with a slope between 10% and 20%, a high grassy covering, and flat areas destined for

TABLE 15.1

Montenero Val Cocchiara—Area 1: Chemical Composition of Pasture, 2000 and 2001 (Mean ± m.s.e.)

	2000	2001
Dry Matter (DM), %	32.68 ± 2.58	23.23 ± 1.8
Organic Matter, %DM	89.72 ± 0.33	90.22 ± 0.40
Crude Protein (CP), %DM	18.16 ± 0.93	21.45 ± 0.87
NDF, %DM	53.57 ± 0.98	52.37 ± 1.27
ADF, %DM	37.62 ± 1.24	35.17 ± 0.97
ADL, %DM	11.78 ± 0.58	10.63 ± 0.50

ADF: acid detergent fiber; ADL: acid detergent lignin; NDF: neutral detergent fiber.

permanent pastures that are characterized by a very high level of grasses and forbs, covering between 90% and 100% (Di Rocco et al. 1992a).

In the following are reported some data collected during 2002 and 2003 in a study conducted in the grassland managed by the "Comunità Montana Sannio," which was divided into three sub-areas in Frosolone municipality (Isernia) and one sub-area in Macchiagodena municipality (Isernia).

In these areas, the soil is acid, and the soil texture is light loam, with good organic matter content and a balanced C/N ratio. This soil can be used under permanent pasture by heavy animals, such as cattle and horses, without damaging the soil structure (Salimei et al. 2001). This area, used for bovine, equine, ovine, and caprine pasture, is nonetheless degraded, probably due to excessively high stocking rates (De Renzis et al. 1992b). The total cadastral area of the "fida pascolo" contract parcels is 1975 ha (Di Rocco et al. 1992a). Each sub-area has prefabricated structures used as animal shelters and for milk processing and storage of dairy products. These structures can all be reached through the regular road network, which has also promoted rural tourism, which, with the selling of typical products such as "Caciocavallo" and "Pecorino" cheese, represents additional income for the local farmers.

From the management authority, registered data in Area 2 indicate 3644 head at pasture (1800 UBA) in the first year (2002), and the grazing population increased to 4069 head in the following year (1950 UBA), due to an increase in grazing goats. The simulation model, described in Section 15.2.1, was applied to the data of the whole grazing area, permitting the estimation of the total available area, excluding spaces containing such features as streets, trees, buildings, and backwaters, at about 1555 ha, mainly composed by grass. The data of both Area 1 and Area 2 were analyzed similarly.

The grassland composition of Area 2, 52% annual grasses and 10% legumes, is typical of south-central Italy. In the heavily grazed and staging areas, there are ferns (*Pteridium aquilinum*) and species of the Carduaceae family (*Cardus* spp.,

Cirsium spp., and *Carlina acaulis*). There were also a single beech tree, an apple tree, and a pear tree in the grazing area, as well as some shrubs such as *Prunus spinosa* and *Crateagus* spp. In the second year of study, occurrences of rainfall were irregular, including the critical summer time period (July and August), during which biotic stress on pasture vegetation is greatest due to the presence of livestock.

In Area 2, the weather in the years 2002–2003 influenced the quality and quantity of grassland production: based on the experimental outcomes and considering the overlapping of the grazing zones (herbaceous, shrubby, and arboreal), the production of animal feed was estimated for the whole grazing area during the whole grazing time (180 d/year) to be approximately 6120 and 4300 t of dry matter in the first and second year, respectively.

Moreover, as also highlighted for the Montenero Val Cocchiara area (Table 15.1), many differences in the dry matter and chemical composition of pastured vegetables were observed for Frosolone/Macchiagodena (Table 15.2), comparing the two different pasture seasons.

Based on productivity and climatic conditions, pasture grass is generally available to animals from May to September, considering that its growth rate reaches a peak in autumn (15–25% of production) and another in spring (about 70%) (Pardini 2005). This trend is strongly influenced by relatively stable environmental factors, such as soil and altitude, and by changes in the weather (temperature and rainfall), changing seasons, and management practices (Pardini 2005). An important limiting factor of the growth of grass in temperate climates is the low winter temperature, while in tropical climates, variations in rainfall can limit the growth of grasses. In the Mediterranean climate, the principal climatic factors limiting the growth of pasture vegetation are low temperature in winter and high temperature in summer (Cavallero et al. 2002). Cavallero and his colleagues (2002) state that the average productivity of Italian pastures can range from 2–2.5 to 6.5 t dry matter/ha in the most productive pasture of the Apennine areas. However, as also shown by the two reported study cases, the qualitative and quantitative characteristics of forage can be affected by abiotic and biotic stresses.

TABLE 15.2

Frosolone/Macchiagodena—Area 2: Chemical Composition of Pasture, 2002–2003 (Mean ± m.s.e.)

	2002	2003
Dry Matter (DM), %	30.22 ± 1.02	34.10 ± 1.10
Organic Matter, %DM	89.54 ± 0.27	89.81 ± 0.29
Crude Protein (CP), %DM	16.55 ± 0.41	15.87 ± 0.44
NDF, %DM	49.16 ± 0.78	47.36 ± 0.85
ADF, %DM	30.81 ± 0.63	29.36 ± 0.68
ADL, %DM	7.87 ± 0.32	8.23 ± 0.35

ADF: acid detergent fiber; ADL: acid detergent lignin; NDF: neutral detergent fiber.

15.1.5 Abiotic Stresses of Plants

15.1.5.1 Temperature

The analysis of soil temperature data collected at the Montenero Val Cocchiara area shows that the average annual air temperature increased from 1987 to 2017 by about 2.3 °C. The coefficients of the x term (x = year) of the linear regression equations of the annual temperature values of Figure 15.3c indicate a gradual warming trend of approximately 0.15 °C/year. This warming trend is less evident in the Frosolone/Macchiagodena area, where the linear regression equations of the annual temperature values of Figure 15.4c is approximately 0.06 °C/year. This warming trend indicates that the approximate rate of air temperature increase was greater in Monetenero Val Cocchiara than in Frosolone, for which climate records from 1987 to 2017 indicate a mean annual air temperature increase of 0.95 °C. Extensive studies on the Alpine and Apennine climate show that the average annual air temperature over the Alps has increased in the last 100 years by approximately 1.5 °C (EEA 2009). The long-term climatic data available for the central Apennines indicate that average annual air temperature increased by approximately 0.027°C/year during the period 1950–2014 (Evangelista et al. 2016). In Figure 15.3c and Figure 15.4c, the low values of slope (the coefficient x, the independent variable, time) of the equations describing the trend of increasing temperature over the 31 year period indicate a relatively slow rate of increase in temperature for both the studied areas.

The increasing average annual temperatures recorded from 1987 to 2017 in the studied area may represent gradually increasing stress on the permanent pasture, which might affect the development of the stems and the aging of all the organs of the plant by gradually increasing lignin components and decreasing digestibility. Low temperatures have been shown to cause a decrease in nutritional value due to stunted growth of the aerial part of plants (Giardini 2012). As the temperature increases, there is generally an increase in the respiratory activity of plants, but the stress of very high temperatures has been shown to stop photosynthesis, thus depressing the accumulation of dry matter. In the same way, the stress of very low temperatures can stop the development of the plant by inhibiting its metabolic activities (Giardini 2012).

15.1.5.2 Rainfall

As regards mean annual rainfall series, data collected from 1987 to 2017 in the Montenero Val Cocchiara area show a trend for precipitation to increase at a rate of approximately 15 mm/year (Figure 15.3b). The opposite trend for mean annual rainfall during the same 31 year period was observed for the Frosolone/Macchiagodena area (Figure 15.4b), with an annual rate of decrease of mean annual rainfall of approximately 9 mm/year. Other climatic analyses conducted on other stations in Molise show a general tendency toward an increase in temperature, particularly in the minima (Costantini et al. 2013). However, the situation throughout Molise is by no means uniform, because besides stations that experienced substantial stability in the period examined, such as Frosolone/

Macchiagodena, there are others, such as Campobasso (the Region's capital) and Guardiaregia, that show positive trends, even more so than the Montenero Val Cocchiara area, in both mean temperature and annual rainfall. Precipitation trends in Molise are more temporally and spatially variable and can be asymmetric compared with trends in air temperature (Izzo et al. 2004). Climatic warming is predicted to cause changes in the seasonality of precipitation, with an expected increase in intra-annual variability, more intense precipitation extremes, and more potential for flooding and soil erosion by water (Gobiet et al. 2014). Projected changes in precipitation, snow cover patterns, and glacier storage in the Alps will also alter runoff regimes, leading to more droughts in the summer (EEA 2009).

The quantity and frequency of rainfall during the period of vegetative growth and the persistence of the snow cover that serves as thermal insulation are elements that can positively influence the productivity and the quality of grasses and forbs (Cavallero *et al.* 2002).

Rainfall is the main source of water supply of the pastures in the present study. However, plentiful rainfall, frequent and intense, can stress pastures by causing negative phenomena such as waterlogging, surface erosion by water, lack of flowering, or delay in the maturation of the various plants (Giardini 2012). Indeed, climatic warming effects and changes in rainfall seasonality and water availability have been proved to be important for ecosystem productivity in the forests of the Apennine regions, where changes in above-ground net primary productivity in response to a shift in the precipitation regime have been detected (Chelli et al. 2016).

15.1.5.3 Soils

The characteristics of soils, and in particular the pH and the calcium carbonate content, can influence the chemical composition, especially the mineral content, of the grasses and legumes that grow in Apennine pastures. At subalkaline pH, the lack of a microelement in soil often results in limited plant growth and a reduced concentration of the element in plant tissues (McDonald et al. 2011). In addition, micronutrient deficiencies of forage may affect both intake of the forage by animals and forage digestibility (McDonald et al. 2011). Some of the most common mineral nutrient deficiencies that result in stresses to pasture grasses include deficiencies of phosphorus, magnesium, copper, and cobalt. The reaction (pH) of soil is also an important factor that can affect the absorption of many elements; e.g., plants in limestone soils (McDonald et al. 2011) poorly absorb both manganese and cobalt. Stress due to drought caused by reduced precipitation or increased evapotranspiration can override the positive effects of higher temperatures on plant growth in hot, arid climatic zones.

15.1.5.4 Altitude

According to Zilotto et al. (2004), the temperature of the air decreases at an average of 0.6 °C per 100 m height, thus causing a shortening of the vegetative activity by about 5 days and a decrease in forage production by about 5%. In the context of recent changes in annual rainfall and average annual air temperature (Figure 15.3b and c and Figure 15.4b and c), an analysis of vegetative cover estimates that plant cover has increased in Apennine mountain sites (Chelli et al. 2017). For example, a general tendency toward increased forest vegetative cover was also observed on Matese mountain at 1400–1600 m a.s.l. The increase in plant cover over the last 30 years is most likely related to the expansion of the most thermophilic species or to the immigration of species from lower elevations. Indeed, in the case of the studied area, the increasing warming can affect ecosystems by increases in temperature, early snowmelt, and a prolonged growing season. These factors may have reduced climatic stresses on the plant communities studied by Chelli et al. (2017) and may have played a key role in the observed increase in plant cover. In fact, the air temperatures before snowmelt and after the meltdown (i.e., the May/June temperatures) are the main factor affecting plant growth in alpine ecosystems (Carbognani et al. 2014).

15.1.5.5 Solar Radiation

Solar radiation directly influences both the climate and the biological activity of plants and animals, and abundance or lack of solar radiation beyond the normal range of a plant can decrease the plant's growth. The energy necessary for plants to carry out photosynthesis comes from the sun, and the quantity of sun received by the plant can affect the concentration of sugars and the yield of grasses and forbs. In general, on a cloudy day, the soluble carbohydrate content of grass will be lower than on a sunny day (McDonald et al. 2011).

15.1.6 Biotic Stresses of Plants

15.1.6.1 Botanical Composition

The relationship between biotic stress and disturbance intensity not only affects the functional response of plant communities but can also lead to changes in plant community structure. The most common plant species for feeding animals belong to the families of Gramineae and Fabaceae (Acciaioli and Esposito 2010). Plants of the Gramineae are usually preponderant in spring at the beginning of the vegetative growth season and escape drought stress by seed production. Annual plants of the Gramineae family come quickly to maturity, and due to their superficial root system, they cannot withstand the stress of summer drought (Porqueddu et al. 2008). During the flowering phase, lignin begins to accumulate in the supporting tissues so that both the palatability and the digestibility of annual forages diminish: it follows that the ideal use of annual plants for forage is in the early vegetative phases before flowering. From the standpoint of forage quality for livestock, Gramineae have a higher carbohydrate content and energy value than Fabaceae, and Gramineae are richer in phosphorus, but poorer in protein and calcium, compared with Fabaceae (Cesaretti et al. 2009).

Because of their relatively deep root systems, Fabaceae, compared with Gramineae, are more resistant to drought stress, and Fabaceae are more present in the composition of the pastures when the vegetative growth stage season is advanced. In evaluating the composition of a pasture, it is also important to consider the presence of the less appetizing species, or weeds, which can strongly compromise the total value of the forage (Cesaretti et al. 2009). Depending on the geographical location, the composition of plant species of pastures may vary: in the alpine areas, perennial species predominate, while in the central-southern pastures, annual species prevail (Pardini 2005; Dumont et al. 2015).

15.1.6.2 Herbivory

Animals can stress forages in pastures in a number of ways, so proper management of grazing animals is essential to reduce stress on pastures (Colonna and Rosati 2015). The determination of the stocking rate, i.e., the crucial grazing management decision on the appropriate number of animals on a given surface in a certain period, allows the conservation of the forage's natural resources, as it ensures the balance between production and use of resources. Thus, appropriate stocking rates can help to prevent degradation of the forage, which can occur when a higher (overload) or lower (underload) load on the pasture is encountered (Pardini 2005). The stocking rate depends on the carrying capacity of the pasture, i.e., the quantity of forage produced per unit of the surface.

An excessive load (overload) of grazing animals leads to a breakdown of the forage plants due to plants not being able to build their reserves before the next access. Overstocking also causes morphological modifications of plants, which is manifest when they assume a creeping and prostrate habit with the roots arranged more and more superficially (Gusmeroli 2004). In the long term, therefore, excessive grazing pressure on forage plants can change the floristic composition of the pasture with a rapid decrease in palatable species and an increase in non-palatable species. Finally, due to excessive trampling, soil compaction and the formation of paths occur, which can increase erosion (Di Rocco et al. 1992a).

With an undersized load (underload), there is generally greater development of undesirable plant species, usually of poor palatable value, that do not require high fertility of the soil (Di Rocco et al. 1992a). Thus, the animals, having surplus food, concentrate their attention on the best forage, allowing the less appetizing plant species to produce seeds and resulting in a worsening of the floristic quality of the pasture. Finally, at the extreme, the effects of the underload can result in the abandonment of a pasture, which favors the return of forest ecosystems (Potenza and Fedele 2011).

15.1.7 Landscape Evolution in Abandoned Areas

Land degradation is more difficult to quantify in Italy than in other European countries because of widespread differences among Italy's landscapes and pedo-environmental conditions. This means that the application of soil and water conservation systems in Italy needs to be planned with a focus on local conditions. The most important driving forces of land degradation in the south-central Apennines are unfavorable climate conditions, poor soil management, improper land planning, and bad management of agriculture and animal husbandry.

Among the different drivers, poor management of soil in fragile environments (the responsibility of both public administrators and farmers) is the most important cause of soil degradation (Costantini and Lorenzetti 2013).

In the Apennines, significant land cover changes have occurred in the last 50 years, with expansion of forests and re-allocation of the urbanized areas (Borrelli et al. 2014). Changes in rural areas have taken place to promote more effective exploitation of agricultural and livestock areas in the most favorable sites, resulting in a gradual abandonment of hilly and mountainous areas and areas of increased hydrogeological instability to maximize food production and reduce the operational costs. This has led to a general degradation in cropland and livestock grazing quality in high-elevation rural areas (Bracchetti et al. 2012).

Despite rural soils of Molise being covered mainly by forests and permanent meadows, superficial soil erosion and landslides are frequent and are of concern to a large part of the Molise region. Landslides (Magliulo et al. 2008) have affected an area greater than 10% of the total regional surface of Molise (Figure 15.8).

Besides other factors, the anthropic influence on soil erosion by misuse of irrigation water and rainfall has been related to i) destruction of forest cover caused by fire, ii) abandonment of traditional hydraulic agrarian arrangements, especially of terracing, iii) diffusion of more intensively cultivated crops, iv) deep ploughing, and v) diffusion of excessive land levelling and slope reshaping during land preparation for specialized tree plantations. The importance of accelerated soil erosion by water in these soil regions is demonstrated by the fact that many agricultural soils have a low or very low organic matter content in the A horizon. In addition to increasing soil loss, the expansion of the aforementioned practices has caused the loss of the traditional landscape constituted by species diversity and in many cases, has damaged the land's capability and suitability for crops. The driving forces of soil degradation act on different levels: national, regional, municipal, and farm. Therefore, the most effective response to combat

land degradation will require integrated policy measures carried out in different spatial frames of reference (Colombo et al. 2011). The land degradation process that causes the most damage in Italy is certainly the irreversible loss of land caused by urbanization and other non-agricultural uses, which often affects the most fertile soils of the plains. Landscape that is more sensitive is mainly located in the upland areas of central Italy and to a lesser extent in the plains, where population density is generally higher. Actually, since the 1980s, the population has increased in moderately vulnerable areas, but it has decreased where there is highly vulnerable land affected by soil degradation and landslides (Torta 2005). At present, among the different ecosystem services that are lost with soil degradation, the diminished capability to produce food is particularly relevant, as it increases the gap between the Italian primary sectors and the primary sectors of other developed countries in terms of food self-sufficiency.

15.2 SOUTH-CENTRAL APENNINES STOCKING RATES EVOLUTION

In Europe, the flat land area allocated for cultivation is decreasing as a consequence of providing space for large and constantly expanding cities. The Directorate General of the European Commission recently issued an Annual Activity Report for Agriculture and Rural Development (2016). The report explains that this contraction of the cultivable area is proceeding in proportion to increasing the urban surface, resulting in a real need to intensify the production of primary goods and therefore, food. Furthermore, it is estimated that the demand for food and other agricultural products will increase by 50% between 2012 and 2050 due to factors such as population growth, urbanization, and per capita increases in income (FAO 2017).

In Italy, during the last 30 years, there has been a steady depopulation of the so-called "marginal" mountain areas in favor of the already heavily urbanized flat areas (MacDonald et al. 2000). This process, which led to a contraction of the number of small and medium-sized farms in the area, is also responsible for important economic, environmental, and social consequences. The progressive abandonment of mountain areas leads to a weakening of the entire economic network of the "mountain" (Ievoli et al. 2017) and therefore, of sectors such as agriculture, livestock, recreational activities, and tourism, resulting in an exponential increase in environmental stress. This phenomenon was mainly determined by the difficulties that often characterize agriculture and mountain economic activity because of difficulty in mechanizing agricultural practices, difficulty in finding manpower, difficulties in physically reaching the business centers due to roads often being completely absent, and the inability of small companies to join together to create cooperatives that have a greater resonance in the competitive market compared with small businesses.

Therefore, it is necessary to recover and protect all the "marginal" areas to enhance the touristic and cultural (agritourism) aspects of rural Molise, given the importance of the typical food products of the region (Xosé et al. 2006).

0 ▭▭▭ 2,75 km

FIGURE 15.8 Map of territories close to Frosolone/Macchiagodena area: different types of landslides, georeferenced by points, are shown in the figure.

The abandonment of transhumance and pastoralism in the central Apennines, with a consequent reduction in anthropogenic grasslands, took place around the 1950s, with major changes in grazing activities, such as a steep decrease of sheep units of about 30% (Falcucci et al. 2007).

Less Favoured Areas (LFAs), as marginal areas, cover 25% of the European surface, but in the main Euro-Mediterranean countries, namely, Greece, Italy, Portugal, and Spain, there are only 15.2 million ha of permanent grasslands (EUROSTAT 2012). In fact, these four countries show a considerably lower proportion of natural or agriculturally improved grasslands compared with the countries of northern and central Europe (EUROSTAT 2012).

Therefore, given the natural agricultural-livestock vocation of the marginal areas, meadows and natural pastures are increasingly important, because they represent the first source of food supplies for the animals. From the point of view of conservation and environmental protection, the grazing of pastures in mountainous areas of Molise appears to be one of the most effective forms of land use. Efficiently and effectively managing grazing in Molise while minimizing stresses to forages in meadows and pastures can control both degradation and abiotic stresses, such as erosion and hydrogeological instability, and stabilize the equilibrium among humans, their livestock, and their environment (Frattaroli et al. 2014). In fact, both meadows and natural pastures may be able to resist frequent, but moderate, disturbances (i.e., deforestation, fires, and overgrazing) by strategies to ensure sustainable animal production and sustainable ecosystems over several millennia (Plieninger et al. 2010; Zapata and Robledano 2014).

At present, consumers' demand for foodstuffs is mainly for products derived from agro-zootechnical activities whose marketing is directly linked to a positive image associated with the presence of meadows and pastures at high altitudes. On the other hand, as already mentioned, the growing interest in agri-tourism and local and traditional animal products also emphasizes the importance of management of grazing animals to preserve some indigenous plant and animal populations from extinction (Lucchese 1995; Miraglia et al. 2003; Iamarino et al. 2004).

So, pasture is an essential source of food for the welfare of native species and their traditionally obtained products, which offer the consumer unique aromas and dietary components with health-promoting properties (e.g., conjugated linoleic acid isomers). In this regard, different studies have demonstrated that pastures have an influence on the volatile chemical compounds of milk and its derivatives (Bendall 2001; Gaspardo et al. 2009; Coppa et al. 2011). A role of the chemical composition of forage in affecting the fatty acid composition of milk has been demonstrated (Lourenço et al. 2010), suggesting that unsustainable management of grassland not only leads to the onset of degradation of the vegetation and soil, but also contributes to malnutrition and stresses the grazers, with serious repercussions on the production of milk and meat and their compositional peculiarities. Both the nutrient requirements of grazing animals and the sustainability of the pasture are crucial factors in the microeconomics of the fragile LFAs. Different studies have been, and are continuing to be, carried out to assess systems for the evaluation of both pasture productivity and animal needs. According to Morenz et al. (2012), when a model for the prediction of animal feeding behavior, needs, and production is used to determine the stocking rate in a grazing area, it is necessary to verify the adaptation of the model to the characteristics of grazing animals (species and breeds), the climatic conditions of the area, and the type and quantity of plants available to animals. Furthermore, Baudracco et al. (2012) created an animal simulation model, named e-Cow, that predicts herbage intake, milk yield, and live weight in dairy cows grazing temperate pastures with and without supplementary feeding. The use of technology in many agricultural and pastoral practices has made possible increases in the efficiency of production processes while limiting environmental impacts (Zhang and Carter 2018).

In this general framework, it is worth considering that Molise is highly susceptible to landslides and hydrogeological instability due to its complex geological setting, characterized by several different structurally susceptible lithologies. More than 4000 mass movements and incipient erosional processes are known to affect the territory of Molise, despite its limited area, and as consequences of poor landscape management (Pisano et al. 2017). Soil erosion by water and mass movements of soil are still the most widespread forms of soil degradation in many regions of the central Apennines. Landslides and floods, soil organic matter decline, and loss of biodiversity are all linked to erosion by water. Besides reducing soil fertility and stressing plants, erosion by water impairs several other ecosystem services, e.g., quality of cultivable land, value of the landscape, and biodiversity. In this regard, in agreement with Cocca et al. (2012), it should be noted that a loss of cultivable land increases in areas with steeper slope, which are less productive and more difficult to manage but are richer in terms of landscape and biodiversity resources. Therefore, it is crucial to maintain a territorial network of traditional, extensive farms to avoid further landscape deterioration.

The current study aims to evaluate the contribution of pasture management to minimizing the environmental stresses in the two described areas of Molise in the Apennines, i.e., in the municipalities of Montenero Val Cocchiara (Area 1) and Frosolone/Macchiagodena (Area 2). The sustainable stocking rate was assessed and compared with the real stocking rates. The same areas were considered 30 years later to achieve suggestions for the future management of the land. To achieve this purpose, a series of environmental parameters have been used as inputs in a spreadsheet model for the assessment of a sustainable stocking rate with a nutritional approach (Pulina et al. 1999) adapted to the specific conditions of multispecies grazing animals in south-central Italy (Salimei et al. 2001, 2005).

15.2.1 SIMULATION MODEL WITH NUTRITIONAL APPROACH

With this model, considering the different grazing species, the best use of the pasture resource is hypothesized with respect to maintaining or increasing the biodiversity of plants and animals as a basis for supporting organizational decisions at the local level. For the construction of the model (Pulina et al. 1999), stocking rate, expressed as the number of animals per unit of territory and time, is calculated according to the type of animals and their diet, i.e., the dry matter (DM) and energy intake. For sustainable use of the pasture, the maximum availability of forage for each species among grass, shrubs, and trees is considered without compromising the long-term production of the pasture (Tothill and Gillies 1992). Specific areas were defined as portions of the land surface delimited by its intrinsic physical, biological, and socioeconomic characteristics (Sereni 1997), which has a minimum dimension of 100 km².

The original model (Pulina et al. 1999) assumes that

- The current state of the pastures is considered in its context through the application of intensity of grazing appropriate to the different vegetative dynamics.
- The feed intake of different animal species must not vary due to the different floristic composition of pastures.
- Feed supplements can be considered as possible inputs within the model.

The resulting stocking rates consider grazing species characteristics such as race, breeding techniques, and production levels (Pulina et al. 1999).

In more detail, the model simulates the livestock stocking rate based on the primary equation, which considers stocking rate (SR) as a function of the surface covered by tree (S_t), shrubs (S_s), and grass (S_g), and it is composed of three modules (Figure 15.9):

1. The forage module, estimating the quantity of forage available for animals
2. The animal module, evaluating the dry matter intake and the nutritive requirements of the animals
3. The stocking rate module, assessing the stocking rate on the basis of data elaborated from the previous two modules (Pulina et al. 1999)

The investigated areas of Molise have been divided into areas (named Grazing Unit Areas—GUAs) homogenous in terms of size, topography, soils, and vegetation cover, as described in the first paragraph of this chapter.

The GUAs represent sub-areas open to grazing by animals in the whole pasture area (Pulina et al. 1999). The module identifies the percentage of the surface covered by grass, shrubs, or trees and establishes the amount of total dry matter available to the animals. Therefore, grass, shrubs, and trees are divided into different heights according to the different animal feeding behaviors (Van Soest 1994; Papachristou 1997). Thus, sheep explore their feeding areas in two dimensions, grazing on grass and on the smaller bushes, while goats and cows move in a three-dimensional space (Figure 15.10).

The area covered by grass, shrubs, or trees is investigated using aerial photography, and inputs are represented by

- The area covered by vegetation (excluding rocks, bare soil, water sources, buildings, and roads)
- The surface occupied by trees and shrubs
- The grass production per hectare (DM/ha) assuming that the hectare is completely covered by vegetation
- The energy content of forages reported for the investigated years (2000–2003) from literature data (Baumont et al. 2000; Freer and Dove 2002; Martin-Rosset et al. 1994; National Research Council 2001)

To limit the problem of overestimation and/or underestimation of the consistency of the pasture, samples were collected using transects delimiting a precise area (Salimei et al. 2005).

15.2.1.1 Animal Module

The animal module calculates animal feed intake and requirements in terms of energy and DM. According to Van Soest (1994), given the different digestibility of feedstuffs and the different feeding behavior, in herbivorous animals, the use efficiencies of the various feeds can be defined as being constant.

In the animal module, the intake values have been calculated according to data in the literature (Baumont et al. 2000; Freer and Dove 2002; Martin-Rosset et al. 1994; National Research Council 2001). Inputs are represented by body weight (kg) of adult males, females, and young, average milk production (kg/d), and average body weight gain (g/d), as well as additional nutrient requirements in the case of late gestation.

15.2.1.2 Stocking Rate Simulations

All inputs described in the preceding paragraphs have been processed by the simulation model to calculate the stocking rate for 180 days of the grazing season per year (Salimei et al. 2005). In more detail, based on the diet and on dry matter intake detected on these two areas, the grassland use was estimated up to the limit beyond which all the vegetation would be compromised in the long term, considering three different stocking rates (minimum, medium, and maximum) up to the point of exhaustion of one of the three forage sources available, arboreal or shrubby or herbaceous, according to Pulina et al. (1999).

15.2.1.2.1 Montenero Val Cocchiara—Area 1

Based on chemical composition data of the pasture of Montenero Val Cocchiara—Area 1, reported in Section 15.1.4, the estimation of the maximum stocking rate (Table 15.3 B) is 1.14 UBA/ha in 2000 and 1.35 UBA/ha in 2001.

In the first year, the effective stocking rate (1.55 UBA/ha) was higher than the maximum stocking rate simulated by the model. Moreover, considering the estimated maximum stocking rate, a significant deficit of the arboreal and shrubby resources was calculated (Table 15.3 C); it was −1109 kg DM

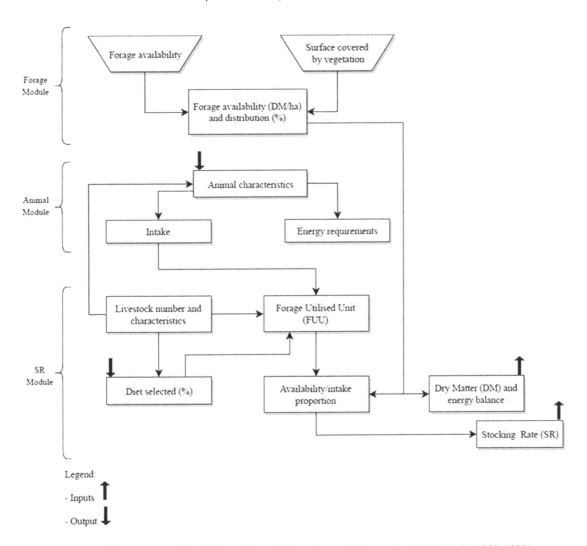

FIGURE 15.9 Flow chart of the three modules. (Adapted from Pulina, G., et al., *Livest. Prod. Sci.*, 61, 287–299, 1999.)

and −1329 kg DM in years 1 and 2, respectively. Due to the significant herbaceous percentage in the grassland (Table 15.3 A), it was not surprising that the minimum stocking rate (Table 15.3 B), estimated considering the complete shrubby (2000) and arboreal depletion (2001), results in an underuse of the herbaceous component (Table 15.3 C).

The simulation model used to evaluate the stocking rate suggests that to reduce stresses to soil and plants due to overgrazing, one must take into account all the management procedures, such as the reduction of stocking rate, the periodic exclusion of the more degraded areas from grazing, and the administration of complementary hay and concentrates.

15.2.1.2.2 Frosolone/Macchiagodena—Area 2

Based on the chemical composition of the pasture of Frosolone/Macchiagodena—Area 2, reported in Section 15.1.4, the estimated stocking rates in 2002 and 2003 (180 days/year) ranged between 0.006 UBA/ha (minimum level) and 1.66 UBA/ha (maximum level, observed in the first year) (Table 15.4 B).

The observed forage availability affected the determination of the maximum estimated stocking rate, 1.66 UBA/ha

(Table 15.4 A and B), which was higher than the stocking rate based on measurements (1.16 UBA/ha).

Considering that the data for Area 1 showed an opposite trend in year 1, it is important to highlight not only that overgrazing has negative, stressful effects on soil and grass but also that the reduction of the grazing pressure and the consequent abandonment of the mountain areas can result in dangerous environmental degradation (Nolan et al. 1998). Moreover, considering the maximum stocking rate estimated for the first year, a significant deficit of arboreal and shrubby forage is shown (−1708 kg DM), indicating the need for a careful evaluation and management of the grazing livestock so as not to compromise the arboreal and shrubby resources by excessive grazing stress. Balancing the stresses of foraging animal species on various species of grass, shrubs, and trees can be facilitated by the careful management of both the livestock and the forage vegetation.

In the second year, the arboreal and shrubby resources in Area 2 were scant (−1270 kg DM) (Table 15.4 C). The maximum estimated stocking rate (1.190 UBA/ha; Table 15.4 B), was similar although not identical to the measured one (1.25 UBA/ha).

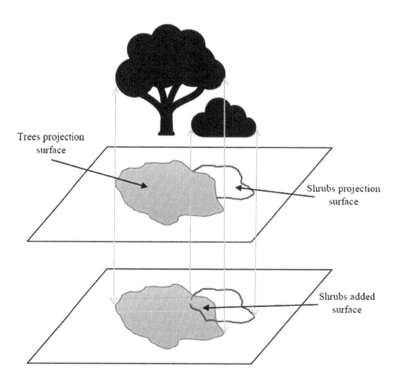

FIGURE 15.10 Volume of forage distribution in area covered by trees and shrubs. (Adapted from Pulina, G., et al., *Livest. Prod. Sci.*, 61, 287299, 1999.)

TABLE 15.3
Area 1: Pasture Stocking Rate and Feedstuffs Deficit Study in the Two Years

	2000			2001		
A	Forage availability:nutritional needs ratio					
	Trees	Shrubs	Grass	Trees	Shrubs	Grass
	0.01	0.01	1.44	0.01	0.01	1.46
B	Stocking rate, UBA/ha					
	max	med	min	max	med	min
	1.14	0.008	0.007	1.35	0.010	0.008
C	Feedstuffs deficit, kg DM					
Trees	−290.5	0.0	0.2	−349.1	-0.5	0.0
Shrubs	−818.5	−0.5	0.0	−980.8	0.0	1.4
Grass	0.0	2472.2	3473.7	0.0	2966.1	2970.4
Total	**−1109.0**	**2471.7**	**3473.9**	**−1329.9**	**2965.6**	**2971.8**

DM: dry matter; UBA: Adult Bovine Unit (24 months).

In times of climate change, the opportunity to have "real time" estimations of forage availability, consistent with the different animal feeding behaviors and adaptations, is of increasing importance for sustainable grassland management.

In conclusion, the stocking rate simulations, based on the field results of two case areas in south-central Italy, highlight the great importance of managing foraging stresses on arboreal and shrubby vegetation to sustainably maintain the forage resources. Climatic data on rainfall and temperature (Figure 15.3 and Figure 15.4) between 1987 and 2017 show a trend for temperature to increase in both Montenero Val

Cocchiara—Area 1 and Frosolone/Macchiagodena—Area 2. On the other hand, during the same period, annual rainfall tended to increase in Montenero Val Cocchiara—Area 1 but tended to decrease in Frosolone/Macchiagodena—Area 2. If these trends continue, forage plant species composition and growth in the two areas may differ, requiring differences in management of grazing livestock and the forage trees, shrubs, and grasses in the two areas. An estimation of the arboreal, shrubby, and herbaceous forage availability, together with knowledge of animal feeding behavior, can contribute to better management of grassland in the control of the negative

TABLE 15.4

Area 2: Pasture Stocking Rate and Feedstuffs Deficit Study in the Two Years

	2002			2003		
A	**Forage availability: nutritional needs ratio**					
	Trees	Shrubs	Grass	Trees	Shrubs	Grass
	0.01	0.01	1.43	0.01	0.01	1.46
B	**Stocking rate, UBA/ha**					
	max	med	min	max	med	min
	1.66	0.01	0.006	1.190	0.007	0.006
C	**Feedstuffs deficit, kg DM**					
Trees	−429.5	−0.4	0.0	−318.5	−0.1	0.0
Shrubs	−1278.5	0.0	1.1	−951.5	0.0	0.4
Grass	0.0	3899.9	3903.2	0.0	2741.6	2742.8
Total	**−1708.0**	**3899.5**	**3904.3**	**−1270**	**2741.5**	**2743.2**

effects of under- and overgrazing, enabling the preservation of biodiversity and edaphic resources, and vegetation while improving local microeconomies.

15.2.2 STRATEGIES OF LAND PROTECTION

The environment, climate, and intensity of land use can influence the floristic composition of permanent pasture (Pini et al. 2017). For example, a decrease in the presence of high-quality forage can result in the reduction of the nutritional value of the grassland, its palatability, and useful biomass production. These phenomena can result from alteration of the concentrations of nutrients within the soil that influence the development of one plant species rather than another.

Botanical degradation of the pastures in Molise can be associated mainly with the extensive presence of weeds which cause a decline in the forage quality of the grassland. The botanical degradation of pastures is determined by assessing the proportions of coverage of forage species and weed species. The problem of weeds is greater on permanent meadows and pastures where no rotation is practiced. There are two stages of the development of weeds: in the first phase, therophytes, or annual weeds, develop, while over the years, during a second phase, plants typical of set-aside lands proliferate, tending to reach a balance with the surrounding environment (Di Rocco et al. 1991a). In fact, in such environments, the dynamism is very limited because of the limited human interventions within the system. However, it is essential to clarify the term *weeds*, as a plant can be considered a weed in an annual pasture but not in a perennial one. Therefore, for a perennial pasture, it can be understood that weeds are all those plants that are poor-quality forage, that are not useful for livestock, or at worst, are toxic (Cantele et al. 1980).

Stress on forage can result from unbalanced distribution of a herd of cattle among different areas of a pasture due to factors related to traditional breeding activities or because of infrastructural problems associated with adminitrative and legal constraints. Due to these and other factors, some

pastures can go from being underused and abandoned areas to overgrazed areas. In both cases, the soil is not efficient in terms of production and environmental protection.

A significant problem in reducing biotic stress on forages is the challenge of managing grazing well, based on precise technical-agronomic requirements. The achievement of this aim depends upon moving from poorly controlled and indiscriminate exploitation of resources to unified, qualitative improvement of production to achieve productive and sustainable use of the potentialities of the same landscape (Malandra et al. 2018).

These problems can be overcome thanks to the adoption of a regional plan to enhance the rural resource, which includes a set of regulatory, organizational, technical, and training activities. The objectives of rural resource management plans will differ depending on whether the objective of the plan is to improve public or private pastures (Argenti et al. 2017). In either case, the priority will be to increase the productivity of the pastures both in primary (forage production) and in secondary terms (production of meat, milk, and wool) (Di Rocco et al. 1991c).

In addition, permanent pasture may have an important stabilization effect on soil on steep slopes. Cultivated areas have generally negative effects on slope stability. In the south-central Apennines, natural pasture areas, which are not cultivated, are less prone to landslides when compared with other cover types. Well-managed pastures can help conserve and efficiently utilize soil and water resources on slopes.

The objectives of a grazing management plan to provide economically sustainable livestock production while minimizing stresses on the forage resources may include

- Diffusion of income-management techniques through the application of agronomic techniques, including partial mechanization of cultivation operations, mowing of the pasture, and sizing of the grazing load
- Design of pasture management able to guarantee the stabilization of forage production during the pasturing season

- Forecasting of recovery of high-value areas and risk of degradation
- Monitoring of the rural resource on a regional scale

Another important strategy of permanent pasture protection is the creation of an inventory of the grazing resources through the study of the climatic, geomorphological, pedological, vegetational, agronomic, and socioeconomic aspects that characterize different types of pasture. Based on the data of such an inventory, one can proceed to the classification of areas of study, with the aim of presenting an inventory for quick consultation that highlights the potential, management, and type of conservation area. The purpose of this tool is to evaluate the management of the pasture, to be understood as productive activity, in qualitative and quantitative terms, sustainable over time, and contributing to the effective control of any irreversible degradation processes that occur in specific territories (Dibari et al. 2015).

Considering the characteristics of pastures that influence the qualitative and quantitative values of production (Martin-Rosset 2015; National Research Council 2007), management techniques should guarantee the achievement of the expected production results by meeting the requirements of economic profit, technical feasibility, and ecological sustainability. The management of livestock and forage should also reduce and, if necessary, reverse degradation processes of the resource by implementing specific conservation practices appropriate to the qualities of the territory examined (FAO 1977).

The qualities of LFAs are linked to:

- Climate, expressed in reference to the seasonal risk of drought responsible for the lack of development of the grass in the pastures
- Degree of deepening of the roots and therefore, the soil's ability to support forage plant species
- Vegetation, intended as productivity class and therefore, estimated on the basis of the floristic-vegetational surveys
- Fodder value (palatability) of the forage production, the expression of the extent to which fodder is appealing in the green state
- Management, possibly using mechanization
- Land conservation to counter widespread water erosion risks (rearrangement of surface horizons implies the degradation of the nutritional status of the soil)
- Landslide risks (i.e., pastures in the clayey hills characterized by limited productivity)

In conclusion, to preserve the characteristics of LFAs in marginal mountainous areas, a distinction can be made between ordinary interventions, such as works repeated annually to improve soil fertility, and those of an extraordinary nature, concerning works whose usefulness has a multi-year character.

Examples of ordinary improvements include:

1. Agronomic interventions, such as annual fertilization, weeding, spring mowing, spring rotary tillage, and seeding

2. Zootechnical interventions, such as correct proportioning of the livestock load based on the analysis of the quantitative elements (productivity of the pasture, surface, animals' feed needs, grazing time), qualitative elements (palatability, food value), climate (temperature and precipitation as rain and snow), use (animal species, grazing methods, availability of stocks, distribution of water points and admissions), and physical characteristics of areas being managed (e.g., slope, exposure, degree of coverage)

In contrast, examples of extraordinary improvements include:

1. Creation of production and service facilities (land improvement), such as shelters, drinking troughs, fences, and various infrastructure
2. Improvement of the ecosystem services that were decreased by land degradation; in particular, the capability to produce food, in terms of food self-sufficiency
3. Improvements of soil, such as drainage channels, and stone removal

An important factor for environmental protection and particularly for the protection of mountain pastures is the choice of livestock load on a given pasture.

Knowing the sustainable number of animals that can visit a grazing area enables the establishment of a balance between the productive potential of the forage and its use by the animals, guaranteeing protection against erosion. Therefore, to quantify the load of livestock, we must take into account the many functions that the forage performs.

It is widely demonstrated that in the long term, stresses of different types can cause negative effects on forage and grazing animals in the case of both grazing overload and underload (Di Rocco et al. 1991b). In the first case, there will be a direct negative effect on the animals, which will fail to satisfy their dietary needs, as well as on soil, because it will manifest rapid degradation of the forage and the appearance of erosive phenomena. In this situation, the plants are subject to the stress of high-use regimes that cause a decrease in total production and impede the development of a stable production system. With this is associated the innate feeding behavior of animals, which, when possible, prefer plants that are more palatable, leading to a drastic reduction of the more palatable plants, reducing competitive stress on infesting species of lower nutritional quality, and enabling them to thrive. Other floristic changes tend to occur in overgrazed pastures (e.g., ammonia flora increase: *Pteridium* spp. and Urticaceae family). Another result of overgrazing is the establishment of paths of compacted, denuded soil where grazing animals have created waterlogging of the soil that prevents plant growth (Di Rocco et al. 1991a).

In the second case, however, the negative effects of underload are more associated with the progressive deterioration of the floristic composition of the grasses and forbs. The excessive presence of available forage implies freedom of choice for grazers, resulting in disproportionate stress on high-quality forage and leading to almost exclusive consumption of plants with higher nutritional value. Such selective grazing behavior, resulting in higher stress on more palatable forage, will result in a qualitative decline of the pasture. Selective grazing of high-quality forage in underused pastures can become increasingly frequent if livestock are rotated between highly grazed areas and others infrequently used by livestock.

15.3 CONCLUSIONS

In conclusion, the balanced use of pastures to minimize stress on the forage, while achieving economically profitable and environmentally sustainable animal production, can be achieved by managing the grazing animals to ensure the optimal livestock load. Data on temperature and rainfall at Montenero Val Cocchiara and Frosolone in the south-central Apennines from 1987 to 2017 indicate a gradually increasing average annual air temperature for the two sub-montane grasslands. Gradually increasing rainfall at Montenero Val Cocchiara and gradually decreasing rainfall at Frosolone during the same period suggest that the species composition of forage grown in these two areas may require changes in management in the future. The modeling of stocking rates in the two areas of the present study indicates the importance of managing the stocking rates of cattle, horses, sheep, and goats to sustainably maintain the grasses, forbs, shrubs, and trees of pastures in the south-central Apennines. Our study provides in-depth knowledge that can be useful to manage the abiotic and biotic stresses of grazing under conditions of two gradually changing climatic variables, temperature and rainfall. Best management practices of managing pastures for sustainable production can result in both sustainable economic activity and sustainable ecosystems in Mediterranean regions such as the south-central Apennines.

ACKNOWLEDGMENTS

The authors are grateful to Protezione Civile—Molise for providing climate data included in the chapter. They are also grateful to Colonna M. and Rosati D. for the information on the permanent pasture documents of the ARSARP, providing vegetational and pedological information included in the experimental section.

This research was supported by the Education, Audiovisual and Culture Executive Agency (EACEA) funded by EU (561841-EPP-1-2015-1-IT-EPPKA2-CBHE-JP) - IUCLAND - International University Cooperation on Land Protection in European-Asiatic Countries (2015–2019).

REFERENCES

Acciaioli, A. and S. Esposito. 2010. Il funzionamento dell'ecosistema pastorale In La gestione e il recupero delle praterie dell'Appennino settentrionale, il pascolamento come strumento di tutela e salvaguardia della biodiversità. Ed. ARSIA ISBN 978- 88-8295- 121(1): 21–29.

Argenti, G., E. Ianchetto and F. Ferretti. 2017. Proposal of a simplified method for pastoral value assessment inside forest planning. Ann. Silv. *Res.* 41 (2): 67–73.

Aucelli, P.P.C., V. Amato, V. Scorpio, V. Bracone and C.M. Rosskopf. 2012. Paleo-landscape reconstruction and assessment of long-term erosion rates through DEM analysis: preliminary results from the Molise Apennine (Central-Southern Italy). *Rend. Online Soc. Geol. Ital.* 21: 1105–1107.

Baudracco, J., N. Lopez-Villalobos, C.W. Holmes, et al. 2012. E-Cow: An animal model that predicts herbage intake, milk yield and live weight change in dairy cows grazing temperate pastures, with and without supplementary feeding. *Animal* 6: 980–993.

Baumont, R., S. Prache, M. Meuret and P. Morand-Fehr. 2000. How forage characteristics influence behaviour and intake in small ruminants: a review. *Livest. Prod. Sci.* 64: 15–28.

Bendall, J.G. 2001. Aroma compounds of fresh milk from New Zealand cows fed different diets. *J Agric. Food Chem.* 49: 4825–4832.

Borrelli, P., S. Modugno, P. Panagos, M. Marchetti, B. Schütt and L. Montanarella. 2014. Detection of harvested forest areas in Italy using Landsat imagery. *Appl. Geogr.* 48: 102–111.

Bracchetti, L., L. Carotenuto and A. Catorci. 2012. Land-cover changes in a remote area of central Apennines (Italy) and management directions. *Landsc. Urban Plan.* 104 (2): 157–170.

Cantele, A., A. Cardinali, P. Catizone, G.G. Lorenzoni, C. Mallegni and P. Talamucci. 1980. Stato attuale della lotta alle malerbe nei prati e nei pascoli. Società italiana per lo studio della lotta alle malerbe, Firenze. pp. 9–124.

Carbognani, M., M. Tomaselli and A. Petraglia. 2014. Current vegetation changes in an alpine late snowbed community in the south-eastern Alps (N-Italy). *Alp. Bot.* 124(2): 105–113.

Catorci, A. and R. Gatti. 2010. Floristic composition and spatial distribution assessment of montane mesophilous grasslands in the Central Apennines (Italy): A multi-scale and diachronic approach. *Plant Biosyst* 144 (4): 793–804.

Catorci, A., G. Ottaviani, I. Vitasović Kosić and C. Cesaretti. 2011. Effect of spatial and temporal patterns of stress and disturbance intensities in a sub-Mediterranean grassland. Plant Biosyst. 1–16, iFirst article.

Cavallero, A.G., P. Rivoira, R. Talamucci, R. Baldoni and L. Giardini. 2002. Pascoli. In Coltivazioni erbacee, foraggere e tappeti erbosi. Patron Editore, Bologna.

Cesaretti, S., S. Castagna, B. Montenegro and A. Catorci. 2009. Zootechnical characterization of grassland vegetation in a pastoral system as a tool for biodiversity conservation: a case study of Umbria-Marches Apennines. *Inf Bot Ital.* 41(2): 247–258.

Chelli, S., R. Canullo, G. Campetella, A.O. Schmitt, S. Bartha, M. Cervellini, C. Wellstein and V. Vandvik. 2016. The response of sub-Mediterranean grasslands to rainfall variation is influenced by early season precipitation. *Appl. Veg. Sci.* 19(4): 611–619.

Chelli, S., C. Wellstein, G. Campetella, R. Canullo, R. Tonin, S. Zerbe and R. Gerdol. 2017. Climate change response of vegetation across climatic zones in Italy. *Clim. Res.* 71(3): 249–262.

Cocca, G., E. Sturaro, L. Gallo and M. Ramanzin. 2012. Is the abandonment of traditional livestock farming systems the main driver of mountain landscape change in Alpine areas? *Land Use Policy* 29: 878–886.

Colombo, C., G. Palumbo and A. Belliggiano. 2011. Il degrado della risorsa suolo, quale futuro per l'agricoltura e per l'ambiente? Italian J. Agronomy: 1–7. s2. e1.

Colonna, M. and D. Rosati. 2015. Aree pascolive e salvaguardia della biodiversità zootecnica molisana. ARSARP. ISBN: 978-88-94112-80-1.

Coppa, M., B. Martin, P. Pradel, B. Leotta, A. Piolo and V. Vasta. 2011. Effect of a hay-based diet or different upland grazing systems on milk volatile compounds. *J. Agric. Food Chem.* 59: 4947–4954.

Corral, A.J. and J.S. Fenlon. 1978. A comparative method for describing the seasonal distribution of production from grasses. *J. Agric. Sci.* 91: 61–67.

Costantini, M., N. Miraglia, M. Polidori and G. Meineri. 2004. 39 th Int. Symp. of Husbandry, Roma, 1016 June, 235.

Costantini, E.A.C., M. Fantappiè and G. L'Abate. 2013. Climate and pedoclimate of Italy. In: E.A.C. Costantini and C. Dazzi (eds.). *The Soils of* Italy, World Soils Book Series, Springer, pp. 19–38.

Costantini, E.A.C. and R. Lorenzetti. 2013. Soil degradation processes in the Italian agricultural and forest ecosystems. *Italian J. Agronomy* 8: e28.

De Renzis, M., M. Litterio, R. Motti and A. Di Gennaro. 1992a. I pascoli del Molise. Una sintesi metodologica.

De Renzis, M., M. Litterio, R. Motti and A. Di Gennaro. 1992b. *Molise Agricoltura.* 15 (10): 20.

Di Rocco, A., P. Raimondo, A. Tomaro, *et al.* 1991a. Analisi floristico-vegetazionale delle aree a pascolo della regione Molise finalizzata all'impostazione di un progetto di miglioramento a scala regionale. Rapporto tecnico finale. Regione Molise – Assessorato Agricoltura e Foreste. Associazione Provinciale Allevatori Campobasso.

Di Rocco, A., P. Raimondo, A. Tomaro, *et al.* 1991b. Indagine sui fenomeni degradativi in atto e sul rischio di degradazione potenziale delle aree a pascolo della regione Molise. Rapporto tecnico finale. Regione Molise – Assessorato Agricoltura e Foreste. Associazione Provinciale Allevatori Campobasso.

Di Rocco, A., P. Raimondo, A. Tomaro, *et al.* 1991c. Indagine sulle caratteristiche climatiche e pedo-agronomiche delle aree a pascolo della regione Molise, finalizzata all'impostazione di un programma di miglioramento della produzione foraggera a scala regionale. Rapporto tecnico finale. Regione Molise – Assessorato Agricoltura e Foreste. Associazione Provinciale Allevatori Campobasso.

Di Rocco, A., P. Raimondo, A. Tomaro, *et al.* 1992a. Inventario regionale della risorsa pascolo. Rapporto tecnico finale. Volume 1. Regione Molise – Assessorato Agricoltura e Foreste. Associazione Provinciale Allevatori Campobasso.

Di Rocco, A., P. Raimondo, A. Tomaro, *et al.* 1992b. Inventario regionale della risorsa pascolo. Rapporto tecnico finale. Volume 2. Regione Molise – Assessorato Agricoltura e Foreste. Associazione Provinciale Allevatori Campobasso.

Dibari, C., G. Argenti, F. Catolfi, M. Moriondo, N. Staglianò and M. Bindi. 2015. Pastoral suitability driven by future climate change along the Apennines. Italian J. Agronomy. 10:659, pp. 109–116.

Dumont, B., D. Andueza, V. Niderkorn, A. Lüscher, C. Porqueddu and C. PiconCochard. 2015. A meta-analysis of climate change effects on forage quality in grasslands: specificities of mountain and mediterranean areas. *Grass Forage Sci.* 70, 239–254.

EEA. 2009. Regional climate change and adaptation. The Alps facing the challenge of changing water resources. EEA Report number 8/2009 (ISSN 1725–9177).

European Commission. 2016. Annual Activity Report of the Directorate General for Agriculture and Rural Development. https://ec.euro pa.eu/info/publications/annual-activity-reports-2016.

EUROSTAT. 2012. https://ec.europa.eu/eurostat/data/database.

Evangelista, A., L. Frate, M.L. Carranza, F. Attorre, G. Pelino and A. Stanisci. 2016. Changes in composition, ecology and structure of high-mountain vegetation: a re-visitation study over 42 years. *AoB Plants* 8:1–11.

Falcucci, A., L. Maiorano and L. Boitani. 2007. Changes in land use/land-cover patterns in Italy and their implications for biodiversity conservation. *Landsc. Ecol.* 22: 617–631.

FAO. 1977. The state of food and agriculture 1976. FAO Agriculture Series number4. Rome

FAO. 2006. Guidelines for Soil Profile Description (Revised), 4th edn. Rome.

FAO. 2017. The future of food and agriculture. Trends and challenges. Rome.

Frattaroli, A.R., S. Ciabò, G. Pirone, D.M. Spera, A. Marucci and B. Romano. 2014. The disappearance of traditional agricultural landscapes in the Mediterranean basin. The case of almond orchards in Central Italy. *Plant Sociology* 51(2): 3–15.

Freer, M. and H. Dove. 2002. Sheep nutrition. CABI Publ., New York (NY, USA) p. 385.

Gaspardo, B., G. Procida, S. Volarič, S. Sgorlon and B. Stefanon. 2009. Determination of volatile fractions in raw milk and ripened cheese by means of GC-MS. Results of a survey performed in the marginal area between Italy and Slovenia. *Ital. J. Anim. Sci.* 8: 377–390.

Giardini, L. 2012. L'agronomia per conservare il futuro. Patron Editore, Bologna. Sesta edizione.

Gobiet, A., S. Kotlarski, M. Beniston, G. Heinrich, J. Rajczak and M. Stoffel. 2014. 21st centuryclimate change in the European Alps-a review. *Sci. Total Environ.* 493: 1138–1151.

Gusmeroli, F. 2004. Il piano di pascolamento: strumento fondamentale per una corretta gestione del pascolo. Quaderni SOZOOALP, 1. Il sistema delle malghe alpine, pp. 27–41.

Iamarino, D., M. Fidotti, N. Miraglia and F. Pilla. 2004. Genetic characterisation of Pentro young horses by microsatellite markers. Proc. 55th Annual Meeting European Association of Animal Production, 310.

Ievoli, C., R. Basile and A. Belliggiano. 2017. The spatial patterns of dairy farming in Molise. *Europ. Countrys* 9: 729–745.

INEA. 2008. L'agricoltura italiana Conta.

ISTAT. 2010. 6° Censimento Generale dell'Agricoltura.

Izzo M., P.P.C. Aucelli and A. Mazzarella. 2004. Recent changes in rainfall and air temperature at Agnone (Molise - Central Italy). *Annali di Geofisica. Roma.* 47(6): 1689–1698.

Lourenço, M., E. Ramos-Morales and R.J. Wallace. 2010. The role of microbes in rumen lipolysis and biohydrogenation and their manipulation. *Animal* 4: 1008–1023.

Lucchese, F. 1995. Elenco preliminare della flora spontanea del Molise. *Ann. Bot.* 53 (Suppl. 12): 1–386.

Ludovico, F., M.L. Carranza, A. Evangelista, A. Stinca, J.H.J. Schaminée and A. Stanisci. 2018. Climate and land use change impacts on Mediterranean high-mountain vegetation in the Apennines since the 1950s. *Plant Ecology Diversity* 1: 1–13.

MacDonald, D., J.R. Crabtree, G. Wiesinger, T. Dax, N. Stamou, P. Fleury, J. Gutierrez Lazpita and A. Gibon. 2000. Agricultural Abandonment in Mountain Areas of Europe: Environmental Consequences and Policy Response. *J. Environ. Mgmt.* 59 (1): 47–69.

Magliulo, P., A. Di Lisio, F. Russo and A. Zelano. 2008. Geomorphology and landslide susceptibility assessment using GIS and bivariate statistics: a case study in southern Italy. *Nat. Hazards* 47 (3): 411–435.

Malandra, F., A. Vitali, C. Urbinati and M. Garbarino. 2018. 70 Years of Land Use/Land Cover Changes in the Apennines (Italy): A Meta-Analysis. *Forests* 9, 551, pp. 1–15.

Martin-Rosset, W., M. Vermorel, M. Doreau, J.L. Tisserand and J. Andrieu 1994. The French horse feed evaluation systems and recommended allowances for energy and protein. *Livest. Prod. Sci.* 40: 37–56.

Martin-Rosset, W. 2015. Equine nutrition. INRA Nutrient requirements, recommended allowances and feed tables. Wageningen Academic Publisher, The Netherlands.

McDonald, P., R.A. Edwards, J.F.D. Greenhalgh, C.A. Morgan, L.A. Sinclair and R.G. Wilkinson. 2011. Animal Nutrition. Ed. 7th Pearson: pp. 481–498.

MiPAF, 2003. Disciplinare del registro anagrafico delle razze popolazioni equine riconducibili a gruppi etnici locali. D. MiPAF number 24347 del 5.11.

Miraglia, N., M. Polidori, F. Lucchese and E. Pietrolà. 2001. 52nd Congress EAAP, Budapest, 2629 August, 326.

Miraglia, N., M. Polidori and E. Salimei. 2003. Review of feeding strategies, feeds and management of equines in Central-Southern Italy. In Working animal in agriculture and transport, EAAP Technical Series n°6, ed. R.A. Pearson, P. Lhoste, M. Saastamoinen, W. Martin-Rosset, Wageningen Academic Publishers, (NL), pp. 103–112.

Morenz, M.J.F., J.F. Coelho da Silva, L.J. Magalhães Aroeira, et al. 2012. Evaluation of the Cornell Net Carbohydrate and Protein System model on the prediction of dry matter intake and milk production of grazing crossbred Cows. Rev. Bras. Zootec. 41: 398–406.

Morgan, R.P.C. 1986. Soil erosion and conservation. Longman Group UK Limited, Harlow.

National Research Council. 2001. Nutrient Requirements of Dairy Cattle 7th rev ed., National Academy Press, Washington D.C.

National Research Council. 2007. Nutrient Requirements of Small Ruminants Sheep, Goats, Cervids, and New World Camelids, National Academy Press, Washington D.C.

Nolan, T., J. Connolly, C. Sall, A.T. Diop, M.M. Vaz Lourenço, C. Zoumana, J. Cesar and N.G. Hegde. 1998. Natural Resource Development and Utilisation in the Sahel. Final Technical Report on EU Contract n8 TS3 CT-93-0202, Ireland/Senegal/France/Portugal/India. p. 380.

Papachristou, T.G. 1997. Foraging behaviour of goats and sheep on Mediterranean kermes oak shrublands. Small Ruminant Res. Mgmt. 54: 322–330.

Pardini, A. 2005. Gestione dei pascoli e dei territori pascolivi. Ed. Aracne.

Patacca, E. and P. Scandone. 2007. Geology of the Southern Apennines. Boll. Soc. Geol. Ital. Spec. Issue 7: 75–119.

Peeters, A., G. Beaufoy, R.M. Canals, et al. 2014. Grassland term definitions and classifications adapted to the diversity of European grassland-based systems. Grassland Sci. Eur. 19: 743–750.

Pini, R., C. Ravazzi, L. Raiteri, A. Guerreschi, L. Castellano and R. Comolli. 2017. From pristine forests to high-altitude pastures: An ecological approach to prehistoric human impact on vegetation and landscapes in the western Italian Alps. J. Ecology 105(6): 1580–1597.

Pisano, L., V. Zumpanoc, Z. Malek, C.M. Rosskopf and M. Parise. 2017. Variations in the susceptibility to landslides, as a consequence of land cover changes: A look to the past, and another towards the future. Sci. Total Environ. 601–602, 1147–1159.

Plieninger, T., V. Rolo and G. Moreno. 2010. Large-scale patterns of Quercus ilex, Quercus suber, and Quercus pyrenaica regeneration in Central-Western Spain. Ecosystems 13: 644–660.

Porqueddu, C., S. Nieddu and S. Maltoni. 2008. Drought survival of some perennial grasses in Mediterranean rainfed conditions: agronomic traits. In: Porqueddu, C. (ed.), M.M. Tavares de Sousa (ed.). Sustainable Mediterranean grasslands and their multi-functions. Zaragoza: CIHEAM / FAO / ENMP / SPPF pp. 231–235 (Options Méditerranéennes: Série A. Séminaires Méditerranéens; n. 79). Options Méditerranéennes, A-79, 231235.

Porqueddu C., R.A.M. Melis, A. Franca, F. Sanna, I. Hadjigeorgiou and I. Casasus. 2017. The role of grasslands in the less favoured areas of Mediterranean Europe. Grassland Sci. Eur. 22: 3–22.

Potenza, C. and V. Fedele. 2011. Effectiveness of the GAEC cross-compliance standard protection of permanent pasture in relation to grazing and pasture conservation management in marginal mountain areas. Italian J.Agronomy 6(s1): e11.

Pulina, G., E. Salimei, G. Masala and J.L.N. Sikosana. 1999. A spreadsheet model for the assessment of sustainable stocking rate in semi-arid and sub-humid regions of Southern Africa. Livest. Prod. Sci. 61: 287–299.

Salimei, E., A. Cappuccio, G. Porcaro and G. Pulina. 2001. 3° Convegno "Nuove acquisizioni in materia di alimentazione, allevamento e allenamento del cavallo sportivo", Campobasso, 1214 Luglio, 7986.

Salimei E., M. Costantini, N. Miraglia, M. Polidori and F. Pilla. 2005. Aree pascolive in Molise: gestione del carico animale quale mezzo di difesa idrogeologica e di valorizzazione delle risorse del territorio. X Convegno Interdisciplinare ISAPA/ISPALEM, Udine (UD) 3-4 Novembre.

Sanderson, M.A., R.H. Skinner, D.J. Barker, G.R. Edwards, B.F. Tracy and D.A. Weding. 2004. Plant species diversity and management of temperate forage and grazing land. Plant species diversity and management of temperate forage and grazing land ecosystems. Crop Sci. 44: 1132–1144.

Sereni, E. 1997. History of the Italian Agricultural Landscape. Princeton, N.J.: Princeton University Press. p. 390.

Soil Survey Staff. 2010. Keys to Soil Taxonomy. 11th ed. United States Department of Agriculture (USDA) and Natural Resources Conservation Service. U. S. Government Printing Office, Washington, DC.

Tamburro, R., E. Tolve, G. Sardella and A. Manuppella. 2005. La zona umida "Torbiera": individuazioni delle pressioni e del buffer. Caso studio: 'Pantano Zittola – Feudo Val Cocchiara' (Isernia – Molise). APAT & ARPA Molise. pp. 1–92.

Torta, G. 2005. Consequences of rural abandonment in a northern Apennines Landscape (Tuscany, Italy). Recent Dynamics of the Mediterranean Vegetation and Landscape: pp. 157–165.

Tothill, J.C. and C. Gillies. 1992. The Pasture Lands of Northern Australia: Their Condition, Productivity and Sustainability. Tropical Grassland Society of Australia, Brisbane, Australia.

Vallentine, J.F. 1990. Grazing management. San Diego: Academic Press Inc.

Van Soest, P.J. 1994. Nutritional Ecology of the Ruminant, second ed., Cornell University Press, Ithaca, New York State.

Vezzani, L., A. Festa and F. Ghisetti. 2010. Geology and tectonic evolution of the Central-Southern Apennines, Italy. Geol. Soc. Am. Spec. 469: 1–58.

Xosé, A., A. López and B.G. Martín. 2006. Tourism and quality agro-food products: an opportunity for the spanish countryside. Tijdschrift voor Economische en Sociale Geografie 97 (2): 166–177.

Zapata, V.M. and F. Robledano. 2014. Assessing biodiversity and conservation value of forest patches secondarily fragmented by urbanisation in semiarid southeastern Spain. J. Nat. Conserv. 22: 166–175.

Zhang, B. and J. Carter. 2018. FORAGE – An online system for generating and delivering property-scale decision support information for grazing land and environmental management. Computers Elect.Agri. 150: 302–311.

Zilotto U., M. Scotton and F. Da Ronch. 2004. Il sistema delle malghe alpine: aspetti agrozootecnici, paesaggistici e turistici. Quaderni della Società per lo Studio e la Valorizzazione dei Sistemi Zootecnici Alpi. 1: 1–220.

16 Turfgrass Nutrient Management under Stresses
A Part of Integrated Stress Management

Haibo Liu, Nick Menchyk, Frank Bethea, Christian Baldwin,
Jacob Taylor, and Caleb Patrick

CONTENTS

16.1 INTRODUCTION

'Turfgrass nutrient management under stresses' (TNMS) is not a new concept for turfgrass science and management (Beard, 1973; Carrow et al., 2001; Busey, 2003; McCarty and Tucker, 2005; Turgeon, 2012; Christians et al., 2016; McCarty, 2018). This concept has been used for turf since a fertilizer program was first implemented and investigated. However, there is lack of bibliographic reviews focusing on systematic approaches using nutrient management to reduce stresses.

Non-food and non-fiber grasses are important to human beings, including ornamental grasses and turfgrasses (Beard and Green, 1994). These grasses came from their origins by natural spreading or artificial collections requiring relatively less nutrient input in comparison with other agricultural crops with yield as their targeted goal. Turfgrass has been well accepted by human beings since the 1300s, perhaps even earlier (Beard, 2002). Today, home lawns exhibit the closest relationship between human life and grasses – and there are hundreds of millions of lawns surrounding homes on this earth; and often these managed turfgrass areas experience unfavorable stresses such as drought, low fertility, poor soil conditions, temperature extremes, salinity, shade, acidic or alkaline soils, traffic compaction, weed, disease, insect, mite pests, nematodes, and others at least once per season (Fry and Huang, 2004; Liu et al., 2008; Bell, 2011; Huang et al., 2014). Some may have to overcome multiple stresses, and some may have to face permanent stresses year-round particularly associated with lower mowing heights and intensive turf use. Intensive studies related to turfgrasses, particularly in the past 70 years, have focused on the mentioned stresses for enhanced turfgrass management, turfgrass improvement, integrated pest management (IPM), and environmental stewardship, with the findings being applied and adapted in turfgrass management.

The major turfgrass species are in the grass family, which includes major agricultural crops such as rice (*Oryza sativa* L.), wheat (*Triticum aestivum* L.), corn (*Zea mays* L.), sorghum (*Sorghum bicolor* (L.) Moench), barley (*Hordeum vulgare* L.), rye (*Secale cereale* L.), and other field and forage crops. By comparison with other crops, turfgrasses require relatively fewer nutrients, but fertilizers are still needed (Turner and Hammel, 1992; Carrow et al., 2001; Frank and Guertal, 2013a, 2013b; StJohn et al., 2013). Overall, turf use and quality can be enhanced by proper fertilizer application, systematic management programs, and improved turfgrasses to overcome stressful conditions. Reviews on turfgrass nutrient use, physiology, and stress resistance and management have been well documented in the last few decades (Dernoeden, 2000; McCarty and Miller, 2002; Fry and Huang, 2004; McCarty, 2011, 2018; Dernoeden, 2013; Christians et al., 2016).

16.2 TURF STRESSES

In order to discuss the topic thoroughly, turfgrass stresses are grouped into four sub-topics: biotic stresses, abiotic stresses, turf use and management stresses, and environmental pollution stresses (Table 16.2). Turfgrass nutrient enhancement under these stresses can minimize the severity and longevity of stresses and even can be a critical factor under certain conditions. Stresses to turfgrasses are generated from two major sources, natural stresses such as temperature extremes, poor soil conditions, and pests; and non-natural stresses such as traffic, cultivation, and mowing (Table 16.1). The natural stresses can be divided into biotic and abiotic stresses. Biotic stresses are caused by living organisms such as weeds, diseases, insects and mites, and other living pests. Abiotic stresses are unfavorable growth conditions such as heat, cold, drought, flooding, salinity, acidic soils, nutrient deficiency, nutrient imbalance and toxicity, and excessive organic matter content. Non-natural stresses are mainly related to turf use including lower (or higher) mowing heights, cultivation, traffic, over-use, and others caused by human activities. Environmental pollution stresses are associated with any pollutants deposited into turf areas including excessive fertilizer or. It is a challenge to enhance nutrient use under these stressful conditions, and general maintenance strategies can be complex and include multiple approaches. Since the basic practices of mowing, fertilizing, and watering can have significant impacts to a given stress or multiple stresses at a time, the definition of stress to turfgrasses is a concept covering all unfavorable conditions that cause negative impacts to turfgrass growth, performance, and function.

16.2.1 BIOTIC STRESSES

Turf biotic stresses include weeds, diseases, nematodes, earthworms, insects and mites, and other livings organisms (normally excluding anthropogenic impacts). Turfgrass nutrients strongly impact on these stresses including influences on turf recovery (Liu et al., 2008, 2010, 2019).

The severity of turf weed problems is often associated with low maintenance levels. A lower-maintenance-level turf is more vulnerable to weed invasion and proliferation. In general, weeds require fewer nutrients and water. With improved nutrient use efficiency, turfgrasses may have a better chance to be able to enhance turf vigor, compete, and even suppress weeds. Turfgrasses compete with weeds for nutrients, and different soil nutrient levels also favor one plant species over another. It is difficult to simply use soil nutrient levels and fertilizers as remedies to control weeds in turf and it will be unlikely to be the main tool in weed control in the future. However, the difference of nutrient uptake efficiency between a weed and turfgrass species or cultivar can be further enhanced (Busey, 2003; Turgeon et al., 2009; Turgeon, 2012; McCarty, 2018).

Turf and plant diseases have a complicated relationship with plant mineral nutrients (Datnoff et al., 2007). Excessive nitrogen is in favor of the most serious turf diseases in association with soil conditions, such as neutral to slightly alkaline soil conditions, which are favorable for some pathogens with the exception of dollar spots and a few of other diseases (Liu et al., 2008). Several nutrients, such as Mn and Si (Hill et al., 1999; Datnoff, 2005) have been found to suppress some turf diseases in addition to some mineral elements in pesticides, which is a topic beyond the scope of this chapter. A nutrient deficient turf will have a difficult time recovering from diseases. Schmid et al. (2017) recently reported that that anthracnose (*Colletotrichum cereale* Manns sensu lato Crouch, Clarke & Hillman) severity on annual bluegrass (*Poa annua* L.) putting-green turf was influenced by N sources (ammonium verse nitrate), and that low-rate applications of potassium nitrate every 7 days were most effective at reducing disease severity under moderately acidic soil conditions and moderate to low soil K levels, while the same rate of ammonium sulfate worsened the disease with the lowest soil pH.

Dollar spot (*Sclerotinia homoeocarpa* F.T. Bennett) is a common disease of creeping bentgrass (*Agrostis stolonifera* L.) on golf course putting greens, tee boxes and fairways. McCall et al. (2017) found that turf quality was improved with $FeSO_4$ in 2012, but not in 2015. Iron chelate suppressed dollar spot in 2012, but not in 2015. Sulfur had no impact on dollar spot or turf quality in either year. They concluded that $FeSO_4$ suppresses dollar spot by acting as a fungistat against the pathogen. In addition, Mattox et al. (2015) reported excellent *Microdochium nivale* (Fries) Samuels & Hallett disease control on annual bluegrass putting greens with 48.8 kg $FeSO_4$ ha^{-1} every 2 weeks from September to April in Oregon, although the mode of action still remains unclear.

TABLE 16.1

The Nutrient Elements Required by Higher Plants, Some Beneficial Elements, and Possible Microbial–Host Relationships to Enhance Turfgrass Stress Tolerances

Nutrients	Chemical Forms Available to Plants	Range of Concentration in Plants (Dry Weight Basis % or mg kg^{-1})	Main Functions in Plants	Potentials for Stress Reduction or Toxicity	Supportive Evidence from Other Crops or Plants from Grass Family or Comprehensive Review Articles	References
Nutrients That Plants Obtain from Water and Air without Normal Fertilizer Input						
Carbon (C)	CO_2, CO_3^{-2}, HCO_3^-	40–45%	Photosynthesis, carbohydrate metabolisms, photorespiration	Short and long term carbon pool; excessive thatch and organic matter negatively impact turf under stresses	Post and Kwon (2000); Halvorson et al. (2002); Lal (2004); Lal (2015)	Qian and Follett (2002); Qian et al. (2003); McCarty et al. (2007); Dai et al. (2009); Qian et al. (2010); Zirkle et al. (2011); Selhorst and Lal (2013); Cho et al. (2017); Law et al. (2017)
Hydrogen (H)	H^+, H_2O, OH^-, HCO_3^-, H_2O_2 and other forms associated with other nutrients	3–6%	Photosynthesis, carbohydrates, pH and proton gradient regulator	Reduce diseases by providing acidic soil conditions less favorable for pathogens	Zhu (2016)	Hill et al. (1999); Liu et al. (2008); Obear et al. (2016)
Oxygen (O)	O_2, CO_2, OH^-, H_2O, and other forms associated with other nutrients	40–45%	Photosynthesis, carbohydrates, respiration	Reduce root oxygen stress and water-logging stress	Choudhury et al. (2016)	Bertrand et al. (2003); Castonguay et al. (2009)
Macronutrients – Primary Nutrients with the Highest Potential for Deficiency and Required by Plants in a Range of 0.1 to 6% in Which N Is Often Most Needed in Quantity						
Nitrogen (N)	NH_4^+, NO_3^-, Urea, N_2, amino acids, and other N forms	2–6%	Amino acids, amides, proteins, nucleic acids, coenzymes, hormones, secondary metabolites	Complicated roles and sufficient N will speed recovery from stresses or worsen stresses	Lemaire (2015)	Dernoeden et al. (1991); Thompson et al. (1995); Hull and Liu (2005); Liu et al. (2008); Baldwin et al. (2009); Frank and Guertal (2013a); Obasa et al. (2013); Miller, et al. (2016); Koehler and Miller (2017)
Phosphorus (P)	$H_2PO_4^-$, HPO_4^{2-}	0.1–0.6%	Nucleotides, nucleic acids, phytic acid, actions of ATP	Germination and seedling requirement and association with mycorrhizae to overcome reduced P input	Pang et al. (2018)	Carroll, et al. (2005); Soldat and Petrovic (2008); Frank and Guertal (2013b); Chang et al. (2014)
Potassium (K)	K^+	0.5–4%	Cofactor of over 80 enzymes, regulator of cell turgor, competitive with Na^+	Heat, drought, cold, disease, and wear stresses	Jaiswal et al. (2016)	Dernoeden et al. (1991); Thompson et al. (1995); Miller and Dickens (1996); Trenholm et al. (2001); Cakmak (2005); Hoffman et al. (2010a, 2010b); Schmid et al. (2016)

(Continued)

TABLE 16.1 (CONTINUED)
The Nutrient Elements Required by Higher Plants, Some Beneficial Elements, and Possible Microbial–Host Relationships to Enhance Turfgrass Stress Tolerances

Nutrients	Chemical Forms Available to Plants	Range of Concentration in Plants (Dry Weight Basis % or mg kg⁻¹)	Main Functions in Plants	Potentials for Stress Reduction or Toxicity	Supportive Evidence from Other Crops or Plants from Grass Family or Comprehensive Review Articles	References
Secondary Nutrients with High Potential for Deficiency and Required by Plants in a Range of 0.1 to 2%						
Sulfur (S)	SO_4^{2-}, SO_2	0.1–1.5%	Amino acids including cystine and methionine, coenzyme A, proteins	Reduce soil pH, black layer	Capaldi et al. (2015)	Hodges (1992); Berndt and Vargas (2006, 2008)
Calcium (Ca)	Ca^{2+}	0.1–6%	Cell wall lamella, the second messenger in metabolism	Heat, cold, and stresses	Niu and Liao (2016)	Fu and Huang 2003; StJohn et al. (2003); StJohn et al. (2013)
Magnesium (Mg)	Mg^{2+}	0.05–1%	Chlorophyll, phosphate transfer	Heat, cold, and stresses	Guo et al. (2016)	Kamon (1973); Sartain (1993); Lee et al. (2007); Hua et al. (2008)
Micronutrients – With Variable Potentials for Deficiency and Normally Required by Plants in a Range of Less Than 0.1% (1,000 mg kg⁻¹) of the Dry Weight, and Toxicity Is Found for Some Nutrients						
Iron (Fe)	Fe^{2+}, Fe^{3+}	20–500	Cytochromes and photosynthesis	Black layer, reduction of P use	Anderson et al. (2018)	Hodges (1992); Xu and Mancino (2001); Berndt and Vargas (2006, 2008); Mattox et al. (2015); McCall et al. (2017)
Manganese (Mn)	Mn^{2+}	10–500	Dehygrogenases, photosynthesis and O_2 evolution	Disease reduction. Reduce photorespiration carbon losses?	Anderson et al. (2018)	Hill et al. (1999)
Zinc (Zn)	Zn^{2+}	10–250	Alcohol dehygronease, glutamic dehydrogenase, carbonic anhydrase	Zinc finger protein	Dietrich et al. (1997); Hänsch and Mendel (2009)	Hull (2001); Xu and Mancino (2001); Hua et al. (2008)
Copper (Cu)	Cu^{2+}	1–80	Conversion of amino acids to proteins, the formation of carbohydrates during photosynthesis and in the formation of lignin	Disease reduction	Hänsch and Mendel (2009); Lange (2017)	Curvetto and Rauser (1979); Hill et al. (1999); Faust and Christians (1999, 2000); Hua et al. (2008)
Boron (B)	H_3BO_3	0.2–800	Cell elongation and nucleic acid metabolism	Protein synthesis	Hänsch and Mendel (2009)	Guertal (2004)
Molybdenum (Mo)	MoO_4^{-2}, $HMoO_4^{-}$	0.1–10	Nitrogenase, nitrate reductase, xanthine dehydrogenase	Enhancement of symbiotic relationship	Gupta (1997)	Gupta (1997); Kaiser et al. (2005)
Chlorine (Cl)	Cl^{-}	10–80,000	Photosynthesis and O_2 evolution	Disease control, electrical charge balance, ATPase activity at the tonoplast	Hänsch and Mendel (2009).	Thompson et al. (1995); Mann et al. (2004)

(Continued)

TABLE 16.1 (CONTINUED)
The Nutrient Elements Required by Higher Plants, Some Beneficial Elements, and Possible Microbial–Host Relationships to Enhance Turfgrass Stress Tolerances

Nutrients	Chemical Forms Available to Plants	Range of Concentration in Plants (Dry Weight Basis % or mg kg⁻¹)	Main Functions in Plants	Potentials for Stress Reduction or Toxicity	Supportive Evidence from Other Crops or Plants from Grass Family or Comprehensive Review Articles	References
Nickel (Ni)	Ni^{2+}	0.05–5.0	Urease, antioxidant metabolism, glutathione biosynthesis	Foliar N absorption, biological N fixation (BNF)	Freeman et al. (2004); Fabiano et al. (2015); Macedo et al. (2016)	Chen et al. (2009); Menchyk (2012); Patrick. (2018)
Beneficial Nutrients – Required by Some Plants, and Except Si, the Concentration Range in Plants Is Similar to Micronutrients (mg kg⁻¹), and Toxicity Is Found for Some Nutrients						
Silicon (Si)	H_4SiO_4	0.1–100,000	Cell wall rigidity and elasticity	Disease, insect, wear, heat resistance, reduction of other element toxicity	Adrees et al. (2015)	Street et al. 1981; Trenholm et al. (2001); Datnoff et al. (2007); Brecht et al. (2007); Liang et al. (2009); Vaculika et al. (2009)
Sodium (Na)	Na^+	10–80,000	Regeneration of phosphoenolpyruvate in C_4 and CAM plants, substitute for K^+	Disease, weed control, nutrient imbalance, cold tolerance	Subbarao et al. (2003)	Munshaw et al. (2004)
Selenium	Se^{-2}, SeO_3^{-2}, SeO_4^{-2}	0.1–25	Selenoproteins, selenoenzymes	Disease resistance	Hartikainen et al. (2000)	Hopper and Parker (1999); Li et al. (2008); Zhu et al. (2009)

Microbes	Nutrients benefited	Type of association with host plants	Main benefits to host plants	Potentials for stress reduction		References
Microbial–host relationships						
Neotyphodium Endophytes	P, others	Endo infection above-ground tissues	Insect pest resistance and stress resistance	Al tolerance, drought stress, P use efficiency	Khan et al. (2013)	Funk et al. (1983); Latch (1993); Liu et al. (1996); Malinowski and Belesky (1999, 2000); Johnson-Cicalese et al. (2000); Zaurov et al. (2001)
Arbuscular-mycorrhizal (AM) symbiosis	P, N, others	Roots	P, water, and other nutrient uptake and stress resistance	Drought tolerance,	Rouphael et al. (2015)	Subramanian and Charest (1995); Pelletier and Dionne (2004)
Root zone and rhizosphere bacteria	N and C	Non-pathogenic and with habitats in root zone and rhizosphere	Reduction of excessive thatch and organic matter in root zones, enhancement of abiotic stress resistance	Drought and salinity stress reduction	Elliott et al. (2003)	Figueiredo et al.. (2008); Yang et al. (2009)
Rhizobia, diazotrophic bacteria	N_2	Root nodules	Nitrogen fixation	Promising	Sincik and Acikgoz (2007)	Raimam et al. (2007); Bi et al. (2009); Bonfante and Anca (2009); Heath and Tiffin (2009); Prakamhang et al. (2009)

Insects and mites in general like nutritious plants with plenty of protein and amino acid content; on the other hand, a sufficient nutrient supply can help turfgrasses recover from insect and mite damages (Davidson and Potter, 1995; Michalski and Cheng, 2018). Several nutrients or elements such as Si and Al may have effects on suppressing turf insects and mites by direct unattractive diets or enhanced lignin formation (Potter, 1998).

16.2.2 Abiotic Stresses

Nutrients themselves that become deficient or toxic are abiotic stresses to turfgrasses in addition to other abiotic stresses. These stresses include all physical or chemical unfavorable growth conditions (Table 16.2). A proper nutrient supply can minimize stress damage and negative consequences. Under stresses, the improvement of the unfavorable conditions is the primary step in combination with other approaches including nutrient enhancement that can save energy, costs, and resources. Such approaches require accurate and frequent soil and tissue nutrient analyses.

16.2.3 Turf Use and Management Stresses

Unlike other yield crop productions in agriculture and horticulture, turfgrasses are maintained for different uses as a functional crop requiring special management practices and skills. Frequent turf mowing is the most unique procedure that sets it apart from any other crops for maintenance requiring specialized mowers and equipment (Beard, 2002; McCarty, 2011; 2018). However, lower mowing heights are stressful for turfgrasses during unfavorable growth conditions such as hot summers for cool-season turfgrasses and cooler seasons for warm-season turfgrasses. Mowing with clipping removal consistently depletes turf nutrients through the loss of nutrients in the cut-off leaves and shoots. In addition, mowing equipment can cause soil compaction and traffic stress to the turf.

In order to improve the turf root system with a better oxygen supply and reduced organic matter such as excessive thatch and mat layers, turf cultivation is often applied during the best growing seasons of both cool- and warm-season turfgrasses. Cultivation includes hollow tine core aerations and core removal, solid tine aerations, verti-cutting, grooming, hydro-jet, dry-jet, and others. These physical cultivation methods can cause temporary stress to turfgrasses. Proper nutrient supply to supplement the losses by core removal is critical.

Turfgrass traffic, ball marks, divots, and wear are all caused by human use of the turf. Adequate fertilizer input can minimize the stresses to the turfgrass and speed up turf recovery from these stresses.

16.2.4 Pollution Stresses

Improper fertilizer input can cause negative consequences to the environment including environmental pollution potentials. Turfgrasses are sinks for pollutants from the air, water,

and soil, particularly in urban and industrial areas. Proper turfgrass nutrient management can enhance turfgrass's ability to reduce pollution and be a better sink for pollutants. On the other hand, improper and excessive nutrient and pesticide input can cause environmental pollution due to runoff and leaching, etc. Water pollution and poor water quality have been one of the major concerns for turfgrass and environmental relationships. Soil pollution includes mining which can create toxic levels of some minerals. Turfgrasses can effectively absorb Pb, Cd, Cr, Mn, Ni, Cu, Zn, and other heavy metals to be a part of phytoremediation (Qu et al., 2003; Kuo et al., 2005; Yesilonis et al., 2008; Duo et al., 2009) and function as the first vegetation on oil-shale mined land (Xia, 2004). Turfgrasses are effective absorbers of SO_2, NO, NO_2, O_3, and other greenhouse gases. As the dominant ground cover vegetation in urban areas, those functions are important for protecting the environment (Beard and Green, 1994; Carrow et al., 2001; Groffman et al., 2009). In addition, golf courses, sports fields, and sod farms have been sinks for recycling water, biosolid wastes, composts, and sewage wastes (Lockett et al., 2008; Groffman et al., 2009).

As sinks of the above-mentioned pollutants, turfgrass nutrient enhancement can strengthen the turf to overcome these stresses as environmental grasses. The general strategies for reducing these kinds of stress is to minimize the toxicity, enhance the uptake of deficient nutrients, and balance the overall nutrient pool as much as possible, in addition to the proper turfgrass selection suitable for the task. Furthermore, large-scale planning and multiple-dimension approaches using turfgrasses as plant sources for phytoremediation seem needed.

16.2.5 Definition of Integrated Stress Management (ISM)

Integrated stress management is a strategic approach for controling and minimizing crop stresses by applying multiple-stress relief methods. It has a similar definition as integrated pest management, but ISM includes IPM since a pest problem is a stress to a crop as well. Integrated stress management can be applied to any crops or plants which are under stresses, and ISM requires a thorough understanding of stress and the environmental conditions associated with the stress. Integrated stress management will be more intensively applied to crops and plants for multiple-stress resistance including nutrient use efficiency.

Integrated stress management can include multiple approaches to control a single stress and multiple approaches to control multiple stresses, with the later an ideal maintenance plan for ISM. For example, if a turfgrass can tolerate acidic soil conditions, it has the capability to resist Al toxicity, and it will also have resistance to most patch disease pathogens, which prefer neutral soil pH conditions. In addition, by keeping the above condition in favor of the turfgrass, if the turfgrass has an endophyte infection with natural resistance to insect pests, further enhancement of the turfgrass is a typical ISM approach through the direct improvement of Al toxicity

TABLE 16.2

General Relationships Between Nutrients and Stresses for Turfgrasses

Stresses	Relationship Between Management Level And Stresses	Nutrients and Elements That May Have More Significant Impacts And Influences for the Severity of Stresses Than Other Nutrients	Updated Evidence for Genetic Improvements	Supportive Evidence From Other Crops or Plants In Grass Family or Comprehensive Reviews	References
		Biotic Stresses			
Weeds	Lower maintenance turf suffers more	N, P, K, and Na	unknown	Ervin et al. (2017); McCarty (2018)	Busey, (2003); McCarty and Tucker (2005); Zhang et al. (2018)
Diseases	Turf mown lower suffers more with excessive moisture, shade, poor soil nutrient or excessive nutrient supplies, and other pathogenically favorable conditions	N, P, K, Mn, Al, Fe, Cu, Zn, and Si	Existed and promising	Walters and Bingham 2007; Datnoff et al. (2007); Amtmann et al. (2008); Dordas (2008)	Dernoeden et al. (1991); Fidanza and Dernoeden 1996; Hill et al. (1999); Datnoff (2005); Smiley et al. (2005); Obasa et al. (2013); Mattox et al. (2015); Miller, et al. (2016); Koehler and Miller 2017; McCall et al. (2017); Schmid et al. (2017)
Nematodes	Sandy, moist and warm soils	C, N, P, and Si	unknown	Youssef and Eissa (2014); Song et al. (2016); Song et al. (2017); Zhang et al. (2017)	Christie et al. (1954); Duo et al. (2017); Shaver et al. (2017)
Insect and mite pests	All types	N, P, and Si	Existed and promising		Reinert and Engelke (2001)
Earthworms	?	EDTA negatively affects earthworms	?		Zhao et al. (2017); Duo et al. (2019)
Other destructive living forms	?	?	?		Potter (1998)
		Abiotic Stresses			
Drought	Cool-season and lower mowing turfgrasses	N, K, Si, Ca, Cl, Fe, Na	Existed and promising	Huang et al. (2014); Zahoor et al. (2018)	Huang (2001); Karcher, et al. (2008); Richardson et al. (2008); Su et al. (2008); Li et al. (2010)
Water-logging	Carpetgrass, centipedegrass	Fe, Mn	Existed	Bush et al. (1999); Dennis et al. (2000)	Jiang and Wang (2006)
Salinity	Cool-season turfgrasses and warm-season turfgrasses	K, Ca, and N	Existed and very promising	Duncan and Carrow (1999); Grattan and Grieve (1999); Huang et al. (2014)	Marcum and Murdoch (1994); Marcum (2001); Qian et al., (2001); Lee et al. (2004a, 2004b); Baldwin et al. (2006); Marcum and Pessarakli (2006); Qian et al. (2007); Li et al. (2010)
Heat	Cool-season turfgrasses	N, K, Ca, Fe, and Si	Existed and promising	Huang et al. (2014); Liu et al. (2018)	Zhang and Ervin (2008); Brecht et al. (2009); Xu and Huang (2009); Zhang et al. (2010)
Low temperatures	Warm-season turfgrasses and cool-season turfgrasses	N, K, Ca, Na, and Si	Existed and promising	Duncan and Carrow (1999); Webster and Ebdon (2005); Huang et al. (2014)	Anderson and Taliaferro (2002); Ebdon et al. (2002); Anderson et al. (1996, 2002, 2003); Munshaw, et al. (2004); Webster and Ebdon (2005); Patton and Reicher (2007); Rukavina et al. (2007); Dionne et al. (2010)
Shade	Bermudagrasses and others	N, Fe	Existed	Valladares and Niinemets (2008)	Baldwin et al. (2008); Sarvis et al. (2009); Pease and Stier (2018)

(Continued)

TABLE 16.2 (CONTINUED)

General Relationships Between Nutrients and Stresses for Turfgrasses

Stresses	Relationship Between Management Level And Stresses	Nutrients and Elements That May Have More Significant Impacts And Influences for the Severity of Stresses Than Other Nutrients	Updated Evidence for Genetic Improvements	Supportive Evidence From Other Crops or Plants In Grass Family or Comprehensive Reviews	References
Acid Soils	All turfgrass species	N, Ca, Fe,	Existed	Kochian et al. (2015)	Murray and Foy (1978); Liu et al. (1995, 1996,1997); Foy and Murray (1998); Baldwin et al. (2005); Liu (2005); Yan et al. (2009)
Nutrient Imbalances	All types of turf	All nutrients	unknown	Fageria (2009)	Carrow et al. (2001)
		Turfgrass Use and Management Stresses			
Traffic and wear stresses	Sports turf and golf course putting greens and tees	N, K, Ca, and Si	Existed	Young et al. (2015); Dest and Ebdon (2017)	Trenholm et al. (2001); Hoffman et al. (2010a, 2010b)
Cultivations and soil organic matter management	Golf course putting greens, tee boxes, fairways, and sports turf	C, N, P, K	Unknown	Christians et al. (2016); McCarty (2018)	Fu and Dernoeden (2008); Fu et al. (2009); Rowland et al. (2009); Fidanza et al. (2017)
Mowing	Golf course putting greens and other fine turf areas requiring lower mowing heights	C, N, P, K, and all nutrients	Unknown	Christians et al. (2016); McCarty, (2018)	Heckman et al. (2000); Law et al., (2017)
		Environmental Pollution Stresses			
Water pollution and poor water quality	Arid and semi-arid regions, coastal areas	N, P, K, and Ca	Promising	Duncan et al. (2009)	Lockett et al. (2008); Groffman et al. (2009)
Air pollutions	Industrial and urban areas	S, N, O_3, and other greenhouse gases and dust	Promising	Beard and Green (1994)	Groffman et al. (2009); Braun and Bremer (2018a, 2018b)
Soil pollutions	Mine industry areas, urban soils, industrial and military sites,automobiles	Heavy metals	Promising	Beard and Green (1994)	Qu et al. (2003); Kuo et al. (2005); Yesilonis et al. (2008); Duo et al. (2009); Liu et al. (2019)

resistance and insect resistance and the indirect enhancement of patch disease resistance. Furthermore, the turfgrass may be even more drought resistant than other turfgrasses which provide new challenges and opportunities for further enhancements. Therefore, ISM also contains endless continuous efforts for crop and plant improvements for stress resistance.

16.3 TURFGRASS NUTRITION

Turfgrass nutrition has been well documented in turfgrass textbooks and related literature (Beard, 1973; Carrow et al., 2001; Turgeon, 2008; McCarty, 2011, 2018). However, this chapter focuses on a follow-up approach of the recent reviews of the enhancement of turfgrass nutrient use and the stresses that turfgrasses encounter during the growing season (Liu et al., 2008).

16.3.1 MACRO AND SECONDARY NUTRIENTS: N, P, K, S, CA, AND MG

Turfgrasses require supplemental nutrients from fertilizers. There are six macronutrients, nitrogen (N), phosphorus (P), potassium (K), sulfur (S), calcium (Ca), and magnesium (Mg), which are needed in the greatest quantity. Plants contain a concentration range between 0.5 to 6.0% of these six nutrients by dry weight in above-ground tissues. Among these nutrients, N, K, and P are the most critical for turfgrasses (Table 16.1) due to the high potential of deficiency (Carrow et al., 2001).

Nitrogen is the most needed nutrient in quantity for all green plants and crops including turfgrasses (Havlin et al., 2013; Barker and Pilbeam, 2015; Taiz et al., 2015; Fageria, 2016). It has been most actively investigated among all turf nutrients due to N's significant roles for turf growth, color, quality, and use. Nitrogen is a constituent of almost every compound in plants except carbohydrates including proteins, chlorophyll, hormones, nucleic acids, and secondary metabolites. Under N deficiency, turf stresses can be worsened, and N plays the most significant role among all nutrients to recover turf from these stresses (Liu et al., 2008). On the other hand, excessive N input can increase turf stress severity including reduced resistance to the extremes of environmental conditions for growth and higher potential damages from pests. Sometimes, excessive N may be even worse than a slight N deficiency for managing turf under stresses regardless of N loss potential from the turf-soil system.

Potassium is the second highest nutrient in tissue concentration among both cool-season and warm-season turfgrasses with a range of 1 to 4%. Unlike most nutrients, K does not form any compounds in plants, but it plays significant roles in plant metabolisms as a monovalent cation K^+. Potassium is required by more than 80 enzymes as a cofactor to function in plants (Marschner, 2012; Barker and Pilbeam, 2015). Potassium is the principal cation for establishing cell turgor and regulating plant hormones associated with water use, stress resistance, and nutrient balance in plants. Potassium deficiency is often found in sandy soils, and it competes with other cations for plant absorption and soil CEC sites. Negative impacts of K deficiency to turf stresses are much more significant than the excessiveness of K for turfgrasses. Potassium toxicity is rarely reported for turfgrasses. According to a recent survey, K input to golf course greens has already surpassed N input in the US (Throssell et al., 2009). In the future, K may be listed second next to N as NKP instead off NPK (a very important order for crop yield production) according to the deficient potential particularly associated with sandy soils for turf growth even with relatively high organic matter content (Carrow et al., 2001).

Phosphorus concentration in turfgrass tissues ranges from 0.2 to 0.6% by dry weight and P is a component of ATP, sugar phosphate, nucleic acids, phospholipids, phytic acids, and coenzymes. Phosphorus plays more significant roles in turf seedlings than mature turf in both cool-season and warm-season turfgrasses (Carrow et al., 2001). P is very often deficient due to its low concentrations (<10 mgL^{-1}) in most soil solutions. In acidic soils, high soluble aluminum and iron can react with P and form insoluble phosphate salts to cause severe P deficiency and metal toxicity. Excessive P applications can cause runoff, leaching, and contamination of P to the environment and water sources (Carrow et al., 2001; Sims and Sharpley, 2005; Soldat and Petrovic, 2008).

Sulfur is required by plants to synthesize the S-containing amino acids cystine, cysteine, and methionine. Grasses normally contain 0.1 to 0.5% S by dry weight. Sulfur deficiency is often observed in salt-affected sandy soils with high pH. Frequent removal of clippings can cause S depletion from some soils. However, frequent use of sulfur-coated urea and sulfate-containing fertilizers, including gypsum ($CaSO_4$), can add S to turfgrass-soil systems. Acidic soil conditions and high S contents in soils can suppress some patch diseases such as take-all patch, summer patch, and spring dead spot (Dernoeden and O'Neil, 1983; Carrow et al., 2001; Vargas, 2005). Sulfur can be found under anaerobic conditions associated with excessive Fe to form black layers in soils (Hodges, 1992; Carrow et al., 2001; Vargas, 2005). Black layers often occur in anaerobic soil conditions created by compaction and/or excessive irrigation. The anaerobic conditions cause sulfur-reducing bacteria to use sulfate or elemental sulfur instead of oxygen during the respiration process, which produces hydrogen sulfide (H_2S) gas, a poisonous compound to turfgrass roots. The hydrogen sulfide reacts with metal elements such as Fe, creating black deposits of FeS, which form black layers within the soil and inhibit turfgrass growth. The use of sulfuric acid to treat irrigation water is a common practice where carbonate (CO_3^{-2}) and bicarbonate (HCO_3^-) levels are high, and excess sodium (Na) exists. Bicarbonates can react with Ca and Mg in the water and can remove these cations from solution. This reduction of soluble Ca and Mg results in an increase in the sodium adsorption ration (SAR) of the water, which increases the likelihood of soil structure problems caused by Na deflocculating clay particles (Christians, 1999). The process of adding sulfuric acid to irrigation water requires a thorough evaluation of the soil conditions and water chemistry before it is initiated.

Primarily, calcium (Ca) is passively absorbed by plants at the root tip as Ca^{2+}. The concentration of Ca found in grasses range from 0.3–1.25% of dry matter (Carrow et al., 2001), which varies among turfgrass species and cultivars. Normally, irrigated turfgrass can receive appreciable amounts of Ca from the irrigation water. Calcium plays important roles that affect the susceptibility to stresses in four ways. First, Ca is important for the stability and function of plant membranes. When there is Ca deficiency, membrane leakage occurs with low-molecular-weight compounds, e.g., sugars and amino acids, from the cytoplasm to the apoplast, which may stimulate the infection by some pathogens (Marschner, 2012). Second, Ca is an important component of the cell wall structure required in the middle lamella for stability. When Ca concentration drops, there is an increased susceptibility to stress causing wilting symptoms. Third, Ca and K works together to activate ABA in stomata closure and opening in responding water deficit of cells. Fourth, Ca is a secondary messenger in plants and it involves signaling messages for heat, drought, and lower temperature stresses. In addition, Ca may be involved for seed production of crops to protect from diseases and stresses. For turfgrass management, Ca deficiency and toxicity are less frequently observed than the incidences of nutrient imbalanced of Ca associated with other cations (Carrow et al., 2001).

Magnesium (Mg) is absorbed from the soil as the Mg^{2+} ion. The Mg requirement for plant growth ranges from 0.15–1% of the dry weight. Magnesium is essential for numbers of fundamental biochemical processes in all living cells. For example, Mg^{2+} ions are involved in the interaction of the ribosome subunits, the counter-ions of ATP and the central ions in chlorophylls. Magnesium is a cofactor in numerous enzymes mainly involved in nucleotide metabolism (Marschner, 2012). Less than 0.15% dry leaf tissue Mg content is usually considered deficient and will vary depending upon species and cultivar

(Carrow et al., 2001). Turfgrass Mg deficiencies commonly occur in acid soils and soils that receive high application rates of basic cations. Saline or effluent waters with high concentrations of Ca, K, or Na can be negative for the soil Mg availability and Mg uptake. Sandy soils with low CEC receiving high irrigation rates are also susceptible to Mg loss through leaching. The direct involvement of stress relief of Mg in turfgrasses has not been reported, but the essential Mg role of photosynthesis helps turfgrasses in carbohydrate production to overcome stress particularly after cultivation during the most active growing seasons of both cool-season and warm-season turfgrasses.

16.3.2 Micronutrients: Fe, Cu, Zn, Mn, Cl, B, Mo, and Ni

Iron has received the greatest attention of all the micronutrients because of its effects on turfgrass appearance, color, and stress management (Carrow et al., 2001; Liu et al., another chapter of this book). For turfgrasses, micronutrient deficiencies do not always show clear symptoms due to relatively fine leaf textures and non-yield production in comparison with other crops. Actual deficiencies of Cu, Zn, Mn, B, Cl, Mo, and Ni are rarely reported in turfgrasses, but toxicities of Cu, Zn, Mn, Cl, B, Mo, and Ni can occur in areas contaminated by mining or other industrial operations. Soil amendments with high heavy metal contents or excessive use of micronutrient fertilizers can cause turfgrass stresses. With proper input or at normal ranges of concentrations, the most significant roles of micronutrients are to reduce plant stresses in addition to their essential physiological and biochemical functions to plants. Their direct and indirect roles in pest resistance and control particularly with diseases can be grouped as 1) pests are sensitive to some micronutrients; 2) some micronutrients are constituents of pesticides; 3) some micronutrients can function as protective or curative agents for host plants physically or chemically; and 4) some micronutrients can have indirect impacts on pests by forming symbiotic relationships with other microbes. In addition, certain ratios of micronutrients and interactions with each other can have impacts on plant stresses (Dordas, 2008).

16.3.3 Beneficial Nutrients: Si, Na, and Se

Silicon (Si) has been the most actively studied in recent years as a beneficial nutrient for plants, including turfgrasses, and numerous reports have identified its positive roles in stress relief and control and pest management. Evidence shows three major aspects of Si roles in plant protection and stress relief in addition to its physiological roles being identified in crops such as rice. First, Si may form a physical protection barrier as silicate on a plant leaf surface between the cuticle layer and epidermal cells. Second, Si stimulates antioxidant systems in plants. Third, Si interacts with other toxic metals to reduce toxicity (Liang et al., 2007).

Sodium (Na) is not an essential plant nutrient, and for cool-season turfgrasses excessive Na is toxic to C_3 plants. However, Na is an essential nutrient to warm-season turfgrasses

particularly for salinity tolerant species by enhancing PEP carboxylase, which deserves more attention (Munshaw et al., 2004; Marschner, 2012; Barker and Pilbeam, 2015). Sodium also plays a positive role in protection from freezing damages and suppression of weeds and pests. Although Na has never been reported being deficient for turfgrasses, it is often found that turfgrass tissue dry weight contains about 10 to 100 mg kg^{-1} of Na when turfgrasses are grown in sandy soils without salinity stress. Facing increased salinity stresses to turfgrasses (Carrow and Duncan, 1998; Qian and Mecham, 2005), the interactions of Na and other nutrients in alleviating stresses require further investigations such as K^+, Fe^{+2}, F^{+3}, Mn^{+2}, and Mg^{+2}.

Selenium (Se) is an essential micronutrient for many organisms, including plants, animals, and humans with a similar chemical function to sulfur. As a plant nutrient, Se ranges 0.1 to 25 ppm by dry weight in plants and its main function is to form selenoproteins and selenoenzymes (Barker and Pilbeam, 2015), which have protective functions against stresses. Although Se functions to plants have not been fully identified, Se is normally taken up by plants as selenate, SeO_4^{-2}, and Se is rarely reported deficient. Moat Se research focuses on its toxicity to plants and in soils particularly for biofortification and phytoremediation of Se-contaminated environments since some crops reported have the capability to accumulate Se (Hopper and Parker, 1999; Li et al., 2008; Zhu et al., 2009; White, 2018; Liu et al.).

16.3.4 Microbial–Host Relationships

The general interactions of plant mineral nutrients and soil microbes are not the focus of this chapter, but several relationships are important for turfgrass nutrient management under stresses related to nutrient use directly or indirectly.

Grasses infected with *Neotyphodium* spp. endophytes have an extraordinary impact on the ecology and economy of turfgrasses (Funk et al., 1983, 1994). Endophytes induce drought resistance, and mineral nutrient (N, P, Ca) use efficiency which affects the production of ergot alkaloids, which are toxic to livestock living on forage crops to but are beneficial for turfgrasses from insect pest attacks. It was reported that endophyte-infected tall fescue had an enhanced P uptake under P deficiency stress. The mechanisms initiated included an altered root morphology with reduced root diameters and longer root hairs, and a chemical modification of the rhizosphere resulting from exudation of phenolic-like compounds, which can form Al-chelating compounds in the rhizosphere to protect hosts from Al toxicity (Funk et al., 1983; Latch, 1993; Liu et al., 1996; Malinowski and Belesky, 1999, 2000; Johnson-Cicalese et al., 2000; Zaurov et al., 2001).

Pelletier and Dionne (2004) reported that two mycorrhizal species, *Glomus intraradices* Schenck & Smith and *G. etunicatum* Becker & Gerdemann, affected the establishment of a lawn mixture of Kentucky bluegrass (*Poa pratensis* L.), red fescue (*Festuca rubra* L.), and perennial ryegrass (*Lolium perenne* L.). Turfgrass inoculated with *G. intraradices* at rates between 40 and 60 mL m^{-2} established more quickly than turfgrass inoculated with *G. etunicatum* when inoculated at time of seeding, with no irrigation or fertilization inputs. These results confer arbuscular-mycorrhizal (AM) symbiosis' benefits to host plants, including improved tolerance to abiotic and biotic stresses. Although the majority of turfgrasses form an AM symbiosis in most soils, little is known of the mycorrhization of turfgrass species for stress control.

Plant-growth-promoting rhizobacteria (PGPR) are associated with plant roots and the thatch layer of turfgrasses. Recent work by several groups shows that PGPR may have functions for salinity and drought tolerance enhancement (Figueiredo et al., 2008; Yang et al., 2009). Although it has not been reported for turfgrasses, PGPR may have effects to reduce the need for fertilizers and preventing the accumulation of nitrates (denitrifies), sulfate, and phosphates (S and P efficient users) in turf soils.

Unlike legumes, grasses cannot fix nitrogen directly from N_2 as the major N supply, but some grasses including rice and maize as the major crops can fix certain amounts of nitrogen through different types of rhizobactria, and the future is promising (Raimam et al., 2007; Bi et al., 2009; Bonfante and Anca, 2009; Heath and Tiffin, 2009; Prakamhang et al., 2009) to further reduce nitrogen input to turf areas.

16.4 MAJOR MANAGEMENT STRATEGIES TO ENHANCE TURFGRASS NUTRIENT USE UNDER STRESSES

16.4.1 MOWING

Mowing removes foliar parts and shoots of turfgrasses and in general lower mowing heights have negative impacts to turf stresses except for physical removal of weeds and turf shoot density enhancement. However, in order to meet the turf use and function, low mowing heights for turf lower than 2.5 mm for golf putting greens are required. The major strategies for enhancing nutrient use for lower mowing height situations include 1) frequent and light fertilizer applications during the growing season; 2) increased utilization of liquid and foliar fertilizers rather than granular fertilizers; 3) recycling clippings to regain nutrients for maintaining turf use and quality; 4) frequent mowing to avoid the excessive removal of clippings at each single mowing event; 5) use of lower nutrient demand or nutrient use efficient turfgrass species or cultivars; 6) use of turf paint to maintain turf color to avoid winter overseeding, fertilizing, and mowing; 7) use of plant growth regulators to reduce growth and nutrient input without compromising turf quality; 8) use of turfgrass species and cultivars, which can tolerate lower mowing heights; and 9) use turfgrass species and cultivars with deeper root systems and better capability to take up nutrients to minimize the potential of nutrient losses.

Lower and more frequent mowing with clipping removal increases turfgrass nutrient demand and is why golf course putting greens require more fertilizer input than other turf areas during the growing season. Raising the mowing height will help turfgrass nutrient use under stress. The relationship between mowing and turf stress has been thoroughly and intensively investigated (Dernoeden, 2000; Beard, 2002; McCarty and Miller, 2002; McCarty, 2011, 2018). Putting greens in a range of 2.5 to 5 mm have received the most attention and need the most improvement in nutrient use enhancement and ISM to overcome stresses. It is never over-emphasized that slightly raising the mowing height of a putting green as allowed can be the key factor in stress relief particularly during unfavorable growing conditions for both cool- and warm-season turfgrasses.

Reduced mowing height associated with plant growth regulator application on turf can enhance turf nutrient use efficiency and even reduce fertilizer input without compromising turf quality and function. Lower mowing heights associated with proper water management and other management practices can also enhance turf nutrient use efficiency and overcome turf stresses.

16.4.2 IRRIGATION

Proper irrigation can enhance turfgrass nutrient use under stressful conditions. The general strategies may include two main aspects 1) the irrigation practice itself and 2) water sources applied to turfgrasses. Under stressful conditions, moist but not water-logging the root zone will assure water need and nutrients for the turfgrass.

Turf irrigation practices to enhance nutrient use under stresses may include several general approaches: 1) to water turf as needed and avoid excessive irrigation; 2) to avoid scheduled irrigation by using a sensing system to understand the exact turf water need; 3) to water the turf at early morning of a day to reduce stress; 4) to hand water the most needed area to avoid broadcasting irrigation; 5) to use wetting agents and plant growth regulators to save water and minimize the stress; 6) to select proper irrigation systems suitable for the turf without or minimize negative impacts to stresses; and 7) to use flexible fertigation system and effluent water. Irrigation design should be adapted to the site-specific for turf use and playability to avoid uneven irrigation causing or worsen stresses.

Poor water quality will worsen turf stresses. Using poor water quality tolerant turfgrasses can minimize the water quality problem impact on turf stresses in addition to the correction of the poor water quality. Pure portable water used for irrigation has minimum influence on available nutrient concentrations in the soil. However, poor water and effluent water can change nutrient concentrations and soil chemistry leading to negative even detrimental impacts on turfgrasses. Imbalanced nutrient supply and element toxicity are often encountered by using poor quality water and effluent water. Frequent soil nutrient, turfgrass tissue tests, and soil management can provide guidelines to correct poor quality water stress and enhance nutrient use for focusing on other stresses.

16.4.3 FERTILIZATION

Nutrient deficiency and toxicity cause stress on the turf. These stresses in conjunction with other abiotic and biotic stresses can be complicated problems to solve since there are more than 18 elements that a turfgrass needs. Each nutrient must be in the proper range of supply and interactions among the 18 elements must consistently exist. The concept of 'apply the exact fertilizer as the turf needs' makes perfect logical sense but it is difficult in practice due to the unique aspects of turfgrass management without the uniform measurement for quality such as yield. Turf quality and appearance evaluation can be subjective depending on locations, turfgrass species, turf use, climatic conditions, and individual evaluators. The following reasons may contribute to the complex nature of fertilization management for turf under stresses: (1) turfgrass management is not for production, and it is harder to measure the exact need for fertilizers; (2) the primary nutrients of N, P, and K do not play a significant role as yield production crops do; (3) greater differences in nutrient requirements exist among turfgrass species, cultivars, and turf use; (4) unique maintenance procedures are highly associated with nutrient status such as clipping removal from turf; and (5) as perennial functional crops, turfgrasses have thatch layers and receive traffic and soil compaction stresses, which are highly associated with nutrient use efficiency.

For professional turfgrass management, frequent soil and tissue testing with accurate interpretation of the turfgrass nutrient status under stresses may be the most important first step before any fertilization practices.

16.4.4 Pest Management

The turf pests include weeds, diseases, insects and mites, and others, which either use turfgrasses as their food sources or habitat on turfgrasses causing physical damages or aesthetic appearance disturbance. To enhance turfgrass nutrient use under these pest stresses is part of integrated pest management requiring a thorough understanding of turfgrass nutrient needs and pest control methods. The general strategies may include the following.

First, understanding of the major or dominant nutrient element associated with the current or potential pest problem. Lower soil pH will have a positive function to suppress patch disease pathogens such as take-all patch, brown patch, and spring dead spot, which are very serious diseases for creeping bentgrass and hybrid bermudagrasses. With a lower soil pH approach, other potential impacts to turfgrass nutrients must be considered such as reduced availability of soil P and increased soluble Fe, Al, and Mn. Soil Mn has a positive effect of suppressing take-all patch. Unlike the other two basic turfgrass practices of watering and mowing, turfgrass fertilizers include over 18 nutrient elements, and each nutrient has its unique way affecting turf quality and pest potential. Among the 18 turfgrass nutrients, they interact with each other, and it is often a combination of effects to turfgrasses each positive or negative.

Second, understanding nutrient ratios and nutrient interaction impacts to pest problems and potential. During the new turfgrass establishment, P input is encouraged. However, weed species are favored by high P supply such as carpetweed and other common summer annuals, which compete for the warm-season turfgrass during establishment.

Third, enhancing turfgrass with microbial association including endophytes, mycorrhizae, and rhizosphere microbial activities. These associations benefit turfgrasses. For example, mycorrhizae can enhance turfgrass P uptake and suppress Fe deficiency. Endophyte-infected turfgrasses have a better Al toxicity tolerance (Table 16.1).

Integrated stress management includes IPM since pests are biotic stresses and research is needed to enhance ISM related to turfgrass nutrients.

16.4.5 Root Zone Management

Turfgrasses belong to shallow-rooted crops, and all grasses have fibrous root systems. The root zone management of turfgrass includes cultivation, thatch control, root growth enhancement, soil modification and amendment, edaphic problem correction, and edaphic environment enhancement.

Turfgrass root zone management with modified sandy root zones either as a USGA green (85/15 v/v of sand and peat moss) and California green (100% sand) are commonly used for golf putting greens and sports fields. These modified root zones significantly improve water percolation and also face the challenge of low nutrient retention in the root zone.

The most significant turfgrass management practice that disturbs a golf course's playability is core cultivation. The speedy recovery from core cultivation requires proper nutrient supply. Other management techniques such as Sub-Air system for altering soil CO_2, O_2, soil solution levels, and microbial activities require research regarding abiotic and biotic stresses.

Fu et al. (2009) reported that both spring-only and especially spring-plus-summer coring caused substantial reductions in turf quality for a 2-week period. Spring-plus-summer coring resulted in increased chlorophyll levels as well as improved turf color and quality in late summer of 'Providence' creeping bentgrass putting green. During the study period of 2 years (2006 and 2007), the bentgrass was fertilized biweekly with 4.9 kg N ha^{-1} from urea between 1 May and 7 June and then weekly through 24 August for a total of 78.4 kg N ha^{-1} during the experimental period in 2006. In the autumn of 2006, 71 kg N ha^{-1} was applied between September and November. In 2007, the bentgrass was fertilized weekly at a rate of 4.9 kg N ha^{-1} with urea between 30 April and 27 August to provide a total of 88.2 kg N ha^{-1} during the experimental period. Although fertilizer rates were not compared for the study, the lower rates and frequent application of urea during summer months were important for the recovery from cultivation stresses.

16.4.6 Plant Growth Regulators (PGR), Surfactants (Wetting Agents), and Bio-Stimulants

For fine turfgrass management, such as golf course putting greens, tees, fairways, and sports fields, plant growth regulators, wetting agents, and bio-stimulants are commonly used. These products have been applied to turfgrass in the past several decades, and the use of these products will likely continue to grow. The major benefits of these products to turfgrasses include reduction of drought, heat, shade, and other stresses in addition to reduced mowing frequency and enhanced turf appearance. The association of these products with turfgrass nutrient and nutrient use efficiency needs more attention.

The combination of trinexapac-ethyl and reduced N input improve both cool-season and warm-season turfgrasses (McCullough et al., 2006; Baldwin et al., 2009; Menchyk et al., 2014). Summer applications of turf surfactants plus K, Ca, and Mg input improves creeping bentgrass summer stress (Fu and Huang, 2003; Sarvis, 2008). Bio-stimulants plus plant hormone regulations enhance turfgrass stresses to shade, drought, heat, and cold (Zhang et al., 2010; Zhang et al., 2017).

Applications of cytokinins have been shown to increase endogenous cytokinin content to improve plant stress tolerance. Goatley and Schmidt (1990) found that foliar application of 6-benzyladenine (6-BA) delayed senescence of Kentucky bluegrass (*Poa pratensis* L.). Liu et al. (2002) reported that root zone injection of zeatin riboside increased endogenous cytokinins and antioxidant activity and delayed senescence in heat-stressed creeping bentgrass. Wang et al. (2006) showed that the synthetic cytokinin, 6-BA, at 50 µg L^{-1} also increased creeping bentgrass heat resistance. Zhang et al. (2010) reported that seaweed (*Ascophyllum nodosum* Jol.) extract-based cytokinins (SWEC) led to improvements in turfgrass stress tolerance. Repeated foliar application of SWEC at 10 µM may be an effective approach for improving turfgrass performance under heat stress. Increased heat stress tolerance and repeated applications of SWEC might have an enhancement of N use in creeping bentgrass.

16.5 TURFGRASS IMPROVEMENT TO ENHANCE NUTRIENT USE UNDER STRESSES

16.5.1 Improved Nutrient Uptake and Use

Differences in nutrient uptake kinetics and enzyme activities exist in cool-season turfgrasses leading to different nutrient use efficiency (Cisar et al., 1989; Jiang and Hull, 1999; Bushoven and Hull, 2001, 2005; Hull and Liu, 2005; Liu and Hull, 2006). These differences can be further enhanced for future turfgrass improvement related to nutrient use, particularly under stresses. Guertal and Hicks (2009) reported that N source treatments of ammonium nitrate (AN) (34-0-0) or calcium nitrate (16-0-0; 39% Ca), both applied at 2.4, 4.9, 7.3, or 9.8 g N m^{-2} wk^{-1} to 'TifSport' and 'Tifway' bermudagrasses for 2 years and TifSport had a higher percentage of groundcover than Tifway. Shoot density of both turfgrasses increased as N rate increased, with shoot density maximized at N rates between 6.2 and 7.6 g N m^{-2} wk^{-1}. In both years, 2003 and 2004, it was found that the dry weight of stolons and rhizomes decreased as N rate increased. Guertal and Hicks (2009) also found that the highest rate of N used in the study (9.8 g N m^{-2} wk^{-1}) was not needed for satisfactory growth

of hybrid Bermuda grass. Lower rates of either NH$_4$NO$_3$ or Ca(NO$_3$)$_2$ fostered turfgrass coverage without a subsequent reduction in the dry weight of important turfgrass stems (stolons and rhizomes). Nitrogen source rarely affects the percentage of turf groundcover, shoot density, or dry weight of stolons and rhizomes, which was found by Stiglbauer et al. (2009) in 'Diamond' zoysiagrass establishment. These findings indicate genetic differences exist for both cool-season and warm-season turfgrasses in nutrient use, and the future of turfgrass nutrient use improvement is promising.

16.5.2 Stronger Competitiveness and Fitness

Garrison and Stier (2010) studied 10 turfgrass species that were planted into two anthropogenic prairies and monitored over a 2-year period. They found that colonies of most species, including Kentucky bluegrass (*Poa pratensis* L.), creeping bentgrass (*Agrostis stolonifera* L.), and tall fescue (*Schedonorus arundinaceus* (Schreb.) Dumort=*Lolium arundinaceum* (Schreb.) Darbysh, formerly *Festuca arundinacea* Schreb. var. arundinacea), decreased in size or died due to herbivores and environmental stresses. However, at one location, fine fescues (*Festuca* spp.) and colonial bentgrass (*A. capillaris* L.) colonies remained relatively constant or slightly increased in size, while native velvet bentgrass (*A. canina* L.) nearly tripled in colony diameter.

As mono-cultured crops, turfgrasses are rarely mixed with other grasses. The new cultivars of turfgrass species must have high-affinity transporters (HAT) for nutrient uptake allowing turfgrasses to be able to compete with weeds when external soil solutions with lower nutrient concentrations. HATs would require less active (energy cost) uptake leading to increased uptake efficiency under nutrient deficient stress (Hull and Liu, 2005; Liu et al., 2008).

16.5.3 Modified Growth Habit and Appearance

Turfgrasses can be genetically modified in growth habits and appearance to meet turf needs such as finer leaf textures, enhanced horizontal growth habits with more stolons and rhizomes, and dwarf types to tolerate lower mowing heights. Such changes may or may not enhance stress tolerance, but proper nutrient management will benefit these modified turfgrasses to overcome stress.

Genetically modified turfgrasses should have larger roots rather than smaller roots to overcome stressful conditions. Li et al. (2010) reported that the Arabidopsis vacuolar H$^+$-pyrophosphatase (AVP1), when over-expressed in creeping bentgrass transgenic (TG) plants, regulated root and shoot development via facilitation of auxin flux and enhanced plant resistance to salt and drought stresses. TG plants exhibited greater biomass production than WT controls under both normal and elevated salinity conditions. When subjected to salt stress, fresh (FW) and dry weights (DW) of both leaves and roots decreased more significantly in WT compared to TG plants. These results demonstrate the great potential of genetic manipulation of vacuolar H$^+$-pyrophosphatase expression in

TG perennial species for improvement of plant abiotic stress resistance with greater root growth.

Rhizomatous growth habit is the preferred horizontal growth to help turfgrass to form an even ground cover to meet the function. Tall fescues (Jernstedt and Bouton, 1985; De Battista and Bouton, 1990) have been identified and invested gated for potential enhancement in rhizomatous growth habit. In addition to the morphological roles of turfgrass rhizomes, nutrient related roles of rhizomes seem leading to opportunities for enhancement in practices. Kavanova and Gloser (2005) studied the nitrogen stores in a rhizomatous perennial grass (*Calamagrostis epigejos*) to investigate the regrowth after defoliation. They found that roots were the main net source of mobilized N and the root dominant N storage compounds were free amino acids. Free amino acids and soluble proteins in the roots decreased by 55 and 50%, respectively, and a substantial (38%) decrease in stubble protein was also observed after defoliation. Although the relative abundance of several soluble proteins in roots decreased during the initial recovery from defoliation, no evidence was found for vegetative storage protein (different from free amino acids and proteins) in rhizomes. They concluded that new leaf regrowth was entirely reliant on N stores present in the plant roots after defoliation. Mobilized N originated mainly from free amino acids and soluble proteins located in roots. Their data suggest that rhizomes play an important role in N transport but not in N storage. Such a finding may be important for winter overseeding practices related to N and other nutrients. Before overseeding, an aggressive vertical mowing can damage more rhizomes than roots for a warm-season turfgrass. This damage could result in the lost function of rhizomes for next spring transporting function of N and other nutrients from roots to new leaves when green upstarts.

16.5.4 INTEGRATED MANAGEMENT SCALES TO ENHANCE SUSTAINABILITY

With financial challenges and competition for golfers in the golf industry worldwide, new strategies must be implemented to enhance the management skills and economics of the turf industry to prepare for the future. Labor costs, energy costs, resource costs, and other costs can be creatively reduced by more integrated cooperation between golf courses beyond individual courses despite ownership and management structure. Recycling resources used for turfgrasses are also important for future turfgrass management. For turfgrass nutrient management, the recommendation of nutrient use is not only based on certain species or cultivars used but also the certain geographical locations and climates used with specific mowing heights and water use standards. Site-specific nutrient sensing and application will further enhance nutrient use efficiency (Xiong, 2007; Bell and Xiong, 2008; Krum et al., 2010). Finally, the discrepancy between essential management requirements and public perceptions for turfgrass management input will need to be further corrected and minimized by large-scale research (Groffman et al., 2009), extensive outreach, and public educational programs.

16.5.5 FUTURE TURFGRASS FERTILIZERS AND APPLICATIONS

It is unlikely that the costs for turfgrass fertilizers will go down because of several reasons: 1) costs of energy; 2) costs of labor; 3) more limited resources to produce fertilizers; 4) more expensive costs of fertilizer transportation; and 5) increased regulations implemented in fertilizer production.

The future turfgrass fertilizers may be required to meet the following to be 1) more responsive to turfgrass-soil and tissue test results; 2) increased liquid fertilizer production and application; 3) new formulations with pesticides, surfactants, bio-stimulants, colorants, bio-fertilizers and soil amendments (Acikgoz et al., 2016; Vaughn et al., 2018), and others; 4) increased production of natural and organic fertilizers with new technologies being implemented; 5) more microbial activity engaging and enhancing fertilizers; and 6) more turfgrass fertilizers designed to be applied to turf under stresses to speed the recovery and minimize stresses.

16.6 SUMMARY AND PROSPECTS

Turfgrass fertilization faces new challenges demanding new and exciting technologies and management strategies. Limited resources and restricted regulations can positively contribute to a reduction of excessive input potential of turf fertilizers, but sound fertilizer management programs as a part ISM are critical to maintain turfgrasses healthy and functional in nutrient aspects and benefit the environment. As functional crops, the economic, environmental, social, and political values of turfgrasses vary from different societies and locations. In spite of those differences, stressed turfgrasses are often seen and maintained than yield production crops. However, it is unnecessary to reach the maximum growth rate for turfgrasses to be the best functional crops and it is often that slightly deficient turfgrass nutrients are recommended.

Over the past 50 years, most studies have focused on complex turfgrass nutrition problems associated with N, P, K, S, Ca, Mg, Fe, Si have received the greatest attention because of their effects on turfgrass appearance, color, and stress management in comparison with other nutrients/micronutrients. Micronutrient deficiencies do not always show clear symptoms such as chlorosis and determining micronutrient toxicities or deficiencies can be difficult. Actual deficiencies of Cu, Zn, Mn B, Cl, Mo, and Ni are rarely reported in turfgrasses, but toxicities of Cu, Zn, Mn Mo, and Ni can occur in areas contaminated by mining or other industrial operations, soil amendments high in heavy metals, or excessive use of micronutrient fertilizers.

Sodium and Si can have effects on plant growth and development. Excess Na competes with other basic cations (K, Ca, and Mg mainly) for exchange sites and can cause leaching and ultimately deficiencies of these ions in addition to its salinity stress to plants. Silicon has one of the most actively researched non-essential plant nutrients in recent years, and some turf managers have adapted it as a turfgrass nutrient to apply to the turf as a fertilizer regularly.

New research on turfgrass stresses associated with nutrients, particularly with micronutrients, is needed. Integrated stress management combined with the importance of environmental stewardship will continue to drive the need for further research into turfgrass nutrition and physiology in addition to stress controls.

ACKNOWLEDGMENTS

The authors sincerely thank numerous organizations, companies, colleagues, former advisors, and former graduate and undergraduate students for the encouragement and support and being involved in the related researches of turfgrasses on N, P, K, Al, Fe, and Ni. Special thanks to both undergraduate and graduate students in four advanced turfgrass and soil science courses, HORT4200 Applied Turfgrass Stress Physiology, PES 4520/6520 Soil Fertility and Plant Nutrition (2012–2018), PES 8900 Plant Nitrogen Metabolism (2015–2018), and PES4850/6850 Environmental Soil Chemistry (2014–2018) at Clemson University for their spirits, encouragement, and support. Thanks to the PES 4850/6850 2018 class which included (in alphabetical order) graduate students: Ethan R. Barnett, Sarah K. Holladay, Jiwoo Park, Savannah R. Petrone, Coleman A. Scroggs, and Zolian S. Zoong-lwe; undergraduate students: Thomas C. Chapman, Adam W. Chastain, Payton Davis, Katrina R. Hale-Phillips, Kimberly J. Henning, Ryan M. Ponder, Logan T. Shelton, and Brandon T. Welch for their reviews of the chapter and contributions in composing the abstract.

BIBLIOGRAPHY

Acikgoz, E., U. Bilgili, F. Sahin, and K. Guillard. 2016. Effect of plant growth-promoting *Bacillus* sp. on color and clipping yield of three turfgrass species. *J. Plant Nutr.* 39 (10):1404–1411.

Adrees, M., S. Ali, M. Rizwan, M.Z. Rehman, M. Ibrahim, F. Abbas, M. Farid, M.K. Qayyum, and M.K. Irshad. 2015. Mechanisms of silicon-mediated alleviation of heavy metal toxicity in plants: a review. *Ecotoxicol. Environ. Saf.* 199:186–197.

Amtmann, A., S. Troufflard, and P. Armengaud. 2008. The effect of potassium nutrition on pest and disease resistance in plants. *Physiol. Plant.* 133:682–691.

Anderson, E., E. Peiter, and H. Kupper. 2018. Trace metal metabolism in plants. *J. Exp. Bot.* 69(5):909–954.

Anderson, J.A., and C.M. Taliaferro. 2002. Freeze tolerance of seed-producing turf bermudagrasses. *Crop Sci.* 42:190–192.

Anderson, J.A., C.M. Taliaferro, and D.L. Martin. 1993. Evaluating freeze tolerance of bermudagrass in a controlled environment. *HortScience.* 28:955.

Anderson, J.A., C.M. Taliaferro, and D.L. Martin. 2002. Freeze tolerance of bermudagrasses: vegetatively propagated cultivars intended for fairway and putting green use, and seed-propagated cultivars. *Crop Sci.* 42:975–977.

Anderson, J.A., C.M. Taliaferro, and D.L. Martin. 2003. Longer exposure durations increase freeze damage to turf bermudagrasses. *Crop Sci.* 43:973–977.

Baldwin, C.M., H. Liu, and L.B. McCarty. 2008. Diversity of 42 bermudagrass cultivars in a reduced light environment. II International Conference on Turfgrass Science and Management for Sports Fields. *Acta Hort. (ISHS).* 783:147–158.

Baldwin, C.M., H. Liu, L.B. McCarty, W.B. Bauerle, and J.E. Toler. 2005. Aluminum tolerance of warm-season turfgrasses. *Int. Turfgrass Soc. Res. J.* 10:811–817.

Baldwin, C.M., H. Liu, L.B. McCarty, W.L. Bauerle, and J.E. Toler. 2006. Effects of trinexapac-ethyl on the salinity tolerance of two ultradwarf bermudagrass cultivars. *HortScience.* 41:808–814.

Baldwin, C.M., H. Liu, L.B. McCarty, H. Luo, and J.E. Toler. 2009. Nitrogen and plant growth regulator influence on 'Champion' bermudagrass putting green under reduced sunlight. *Agron. J.* 101:75–81.

Barker, A.V., and D.J. Pilbeam. 2015. *Handbook of Plant Nutrition.* 2nd ed. CRC Press, Boca Raton, FL.

Beard, J.B. 1973. *Turfgrass: Science and Culture.* Prentice-Hall, Inc., Englewood Cliffs, NJ.

Beard, J.B. 2002. *Turf Management for Golf Courses.* 2nd ed. John Wiley & Sons, Chelsea, MI.

Beard, J.B., and R.L. Green. 1994. The role of turfgrasses in environmental protection and their benefits to humans. *J. Environ. Qual.* 23:452–460.

Bell, G.E. 2011. *Turfgrass Physiology and Ecology: Advanced Management Principles.* Wallingford, Oxfordshire; CABI, Cambridge, MA.

Bell, G.E., and X. Xiong. 2008. The history, role, and potential of optical sensing of practical turf management. pp. 641–660. In: M. Pessarakli (ed.) *Handbook of Turfgrass Management and Physiology.* CRC Press, New York, NY.

Berndt, W.L., and J.M. Vargas, Jr. 2006. Dissimilatory reduction of sulfate in black layer. *HortScience.* 41:815–817.

Berndt, W.L., and J.M. Vargas, Jr. 2008. Elemental sulfur reduces to sulfide in black layer soil. *HortScience.* 43:1615–1618.

Bertrand, A., Y. Castonguay, P. Nadeau, S. Laberge, R. Michaud, G. Bélanger, and P. Rochette. 2003. Oxygen deficiency affects carbohydrate reserves in overwintering forage crops. *J. Exp. Bot.* 54:1721–1730.

Bi, Y.M., S. Kant, J. Clark, S. Gidda, F. Ming, J. Xu, A. Rochon, B.J. Shelp, L. Hao, R. Zhao, R.T. Mullen, T. Zhu, and S. J. Rothstein. 2009. Increased nitrogen-use efficiency in transgenic rice plants over-expressing a nitrogen-responsive early nodulin gene identified from rice expression profiling. *Plant Cell Environ.* 32:1749–1760.

Bonfante, P., and I.A. Anca. 2009. Plants, mycorrhizal fungi, and bacteria: a network of interactions. *Annu. Rev. Microbiol.* 63:363–383.

Braun, R.C., and D.J. Bremer. 2018a. Nitrous oxide emissions in turfgrass systems: a review. *Agron. J.* 110:2222–2232.

Braun, R.C., and D.J. Bremer. 2018b. Nitrous oxide emissions from turfgrass receiving different irrigation amounts and nitrogen fertilizer forms. *Crop Sci.* 58:1762–1775.

Brecht, M., C. Stiles, and L. Datnoff. 2009. Effect of high temperature stress and silicon fertilization on pathogenicity of Bipolaris cynodontis and *Curvularia lunata* on Floradwarf bermudagrass. *Int. Turfgrass Soc. Res. J.* 11:165–180.

Busey, P. 2003. Cultural management of weeds in turfgrass: a review. *Crop Sci.* 43:1899–1911.

Bush, E.W., J.N. McCrimmon, D.P. Shepard, and P.W. Wilson. 1999. Carpetgrass and centipedegrass tissue iron and manganese accumulation in response to soil waterlogging. *J. Plant Nutr.* 22(3):435–444.

Bushoven, J.T., and R.J. Hull. 2001. Nitrogen use efficiency is linked to nitrate reductase activity and biomass partitioning between roots and shoots of perennial ryegrass and creeping bentgrass. *Int. Turfgrass Soc. Res. J.* 9:245–252.

Bushoven, J.T., and R.J. Hull. 2005. The role of nitrate in modulating growth and partitioning of nitrate assimilation between roots and leaves of perennial ryegrass (*Lolium perenne* L.). *Intl. Turfgrass Soc. Res. J.* 10: 834–840.

Cakmak, I. 2005. The role of potassium in alleviating detrimental effects of abiotic stresses in plants. *J. Plant Nutr. Soil Sci.* 168:521–530.

Capaldi, F.R., P.L. Gratão, A.R. Reis, L.W. Lima, and R.A. Azevedo. 2015. Sulfur metabolism and stress defense responses in plants. *Trop. Plant Biol.* 8:60–73.

Carroll, M.J., T.Q. Ngo, and J.M. Krouse. 2005. Tall fescue seedling growth in response to phosphorus fertilization. *Int. Turfgrass Soc. Res. J.* 10:310–317.

Carrow, R.N., and R.R. Duncan.1998. *Salt-Affected Turfgrass Sites: Assessment and Management.* John Wiley & Sons, Chelsea, MI.

Carrow, R.N., D.V. Waddington, and P.E. Rieke. 2001. *Turfgrass Soil fertility and Chemical Problems – Assessment and Management.* Ann Arbor Press, Chelsea, MI.

Castonguay, Y., G. Thibault, P. Rochette, A. Bertrand, S. Rochefort, and J. Dionne. 2009. Physiological responses of annual bluegrass and creeping bentgrass to contrasted levels of O2 and CO2 at low temperatures. *Crop Sci.* 49:671–689.

Chang, Z., X. Jin, and D. Li. 2014. Phosphorus responses vary among cool-season turfgrasses during establishment from seed. *Agron. J.* 106:1975–1980.

Chen, C., D. Huang, and J. Liu, 2009. Functions and toxicity of nickel in plants: recent advances and future prospects. *Clean Soil Air Water.* 37:304–313.

Cho, T., M. Kudo, and S. Kimura. 2017. Relationship of macronutrients and nonstructural carbohydrate concentration in clippings of creeping bentgrass putting greens in Japan. *Int. Turfgrass Soc. Res. J.* 13:404–409.

Choudhury, F.K., R.M. Rivero, E. Blumwald, and R. Mittler. 2016. Reactive oxygen species, abiotic stress and stress combination. *Plant J.* doi.

Christians, N.E., A.J. Patton, and Q.D. Law. 2016. *Fundamentals of Turfgrass Management.* 5th ed. John Wiley & Sons, Hoboken, NJ.

Christie, J.R., J.M. Good, and G.C. Walter. 1954. Nematodes associated with injury to turf. *Proc. Soil Sci. Soc. Florida.* 14:167–169.

Cisar, J.L., R.J. Hull, and D.T. Duff. 1989. Ion uptake kinetics of cool season turfgrasses. pp. 233–235. In: H. Takatoh (ed.) Proc. Int. Turfgrass Res. Conf., 6th, Tokyo, Japan. 31 July–5 August 1989. Int. Turfgrass Soc. and the Japanese Soc. of Turfgrass Sci., Tokyo.

Curvetto, N.R., and W.E. Rauser. 1979. Isolation and characterization of copper binding proteins from roots of Agrosits gigantea tolerant to excess copper. *Plant Physiol.* 63:59.

Dai, X., D.M. Vietor, F.M. Hons, T.L. Provin, R.H. White, T.W. Boutton, and C.L. Munster. 2009. Effect of composted biosolids on soil organic carbon storage during establishment of transplanted sod. *HortScience.* 44:503–507.

Danneberger, T.K. 1993. *Turfgrass Ecology and Management.* G.I.E. Media Inc., Cleveland, OH.

Datnoff, L.E., W.H. Elmer, and D. Huber. 2007. *Mineral Nutrition and Plant Disease.* American Phytopathological Society, St. Paul, MN.

Datnoff, L.E. 2005. Silicon in the life and performance of turfgrass. Online. *Appl. Turfgrass Sci.* Published 14 September 2005.

Davidson, A.W., and D.A. Potter. 1995. Response of plant-feeding, predatory, and soil-inhabiting invertebrates to Acremonium endophyte and nitrogen fertilization in tall fescue turf. *J. Econ. Entomol.* 88(2):367–379.

De Battista, J.P., and J.H. Bouton. 1990. Greenhouse evaluation of tall fescue genotypes for rhizome production. *Crop Sci.* 30:536–541.

Dennis, E.S., R. Dolferus, M. Ellis, M. Rahman, Y. Wu, F.U. Hoeren, A. Grover, K.P. Ismond, A.G. Good, and W.J. Peacock. 2000. *J. Exp. Bot.* 51:89–97.

Dernoeden, P.H., J.N. Crahay, and D.B. Davis. 1991. Spring dead spot and bermudagrass quality as influenced by nitrogen source and potassium. *Crop Sci.* 31:1674–1680.

Dernoeden, P.H. 2000. *Creeping Bentgrass Management: Summer Stresses, Weeds and Selected Maladies.* Ann Arbor Press, Chelsea, MI.

Dernoeden, P.H. 2013. *Creeping Bentgrass Management.* 2nd ed. CRC Press, Taylor & Francis Group, Boca Raton, FL.

Dest, W.M., and J.S. Ebdon. 2017. The effect of wear and soil compaction on kentucky bluegrass sod rooting and plant recovery. *Int. Turfgrass Soc. Res. J.* 13:338–345.

Dietrich, R.A, M.H. Richberg, R. Schmidt, C. Dean, and J.L. Dangl. 1997. A novel zinc finger protein is encoded by the Arabidopsis LSD1 gene and functions as a negative regulator of plant cell death. *Cell.* 88:685–694.

Dionne, J., S. Rochefort, D.R. Huff, Y. Desjardins, A. Bertrand, and Y. Castonguay. 2010. Variability for freezing tolerance among 42 ecotypes of green-type annual bluegrass. *Crop Sci.* 50:321–336.

Dordas, C. 2008. Role of nutrients in controlling plant diseases in sustainable agriculture: a review. *Agron. Sustain. Dev.* 28:33–46.

Duncan, R.R., and R.N. Carrow. 1999. Turfgrass – molecular genetic improvement for abiotic/edaphic stress environment. *Adv. Agron.* 67:233–306.

Duncan, R.R., R.N. Carrow, and M.T. Huck. 2009. *Turfgrass and Landscape Irrigation Water Quality: Assessment and Management.* CRC Press, Boca Raton, FL.

Duo, L., L. He, and S. Zhao. 2017. The impact of modified nanoscale carbon black on soil nematode assemblages under turfgrass growth conditions. *Eur. J. Soil Biol.* 80:53–58.

Duo, L., L. Yin, C. Zhang, and S. Zhao. 2019. Ecotoxicological responses of the earthworm *Eisenia fetida* to EDTA addition under turfgrass growing conditions. *Chemosphere.* 220:56–60.

Duo, L.A., F. Lian and S.L. Zhao. 2009. Enhanced uptake of heavy metals in municipal solid waste compost by turfgrass following the application of EDTA. *J. Environ. Monit. Assess.* 165(1–4):377–387.

Ebdon, J.S., R.A. Gagne, and R.C. Manley. 2002. Comparative cold tolerance in diverse turf quality genotypes of perennial ryegrass. *HortScience.* 37:826–830.

Elliott, M.L., E.A. Guertal, E.A. des Jardin, and H.D. Skipper. 2003. Effect of nitrogen rate and root-zone mix on rhizosphere bacterial populations and root mass in creeping bentgrass putting greens. *Biol. Fertil. Soils.* 37:348–354.

Ervin, E.H., N. Reams, X. Zhang, A. Boyd, and S. Askew. 2017. An integrated nutritional and chemical approach to *Poa annua* suppression in creeping bentgrass greens. *Crop Sci.* 57:567–572.

Fabiano, C.C., T. Tezotto, J.L. Favarin, J.C. Polacco, and P. Mazzafera. 2015. Essentiality of nickel in plants: a role in plant stresses. *Front. Plant Sci.* 6:754.

Fageria, N. 2009. *The Use of Nutrients in Crop Plants.* CRC Press, Boca Raton, FL.

Fageria, N.K. 2016. *Nitrogen Management in Crop Production.* CRC Press, Boca Raton, FL.

Faust, M.B., and N.C. Christians. 1999. AB-DTPA and Mehlich III soil tests unable to predict copper available to creeping bentgrass. *Commun. Soil Sci. Plant Anal.* 30:2475–2484.

Faust, M.B., and N.E. Christians. 2000. Copper reduces shoot growth and root development of creeping bentgrass. *Crop Sci.* 40:498–502.

Fidanza, M.A., and P.H. Dernoeden. 1996. Brown patch severity in perennial ryegrass as influenced by irrigation, fungicide, and fertilizers. *Crop Sci.* 36:1631–1638.

Fidanza, M.A., G.H. Snyder, and J.L. Cisar. 2017. Documenting changes in USGA specification rootzone properties in ultradwarf bermudagrass greens. *Int. Turfgrass Soc. Res. J.* 13:642–648.

Figueiredo, V.B., H.A. Burity, C.R. Martínez, and C.P. Chanway. 2008. Alleviation of drought stress in the common bean (*Phaseolus vulgaris* L.) by co-inoculation with *Paenibacillus polymyxa* and *Rhizobium tropici*. *Appl. Soil Ecol.* 40:182–188.

Foy, C. D., and J. J. Murray. 1998. Responses of Kentucky bluegrass cultivars to excess aluminum in nutrient solutions. *J. Plant Nutr.* 21:1967–1983.

Frank, K.W., and E.A. Guertal. 2013a. Nitrogen research in turfgrass, pp. 457–492. In: J.C. Stier, B.P. Horgan, and S.A. Bonos (eds.) *Turfgrass: Biology, Use, and Management*. ASA, SSSA, CSSA, Madison, WI.

Frank, K.W., and E.A. Guertal. 2013b. Potassium and phosphorus research in turfgrass, pp. 493–520. In: J.C. Stier, B.P. Horgan, and S.A. Bonos (eds.) *Turfgrass: Biology, Use, and Management*. Agron. Monogr. 56. ASA, Madison, WI.

Freeman, J.L., M.W. Persans, K. Nieman, C. Albrecht, W. Peer, I.J. Pickering, and D.E. Salt. 2004. Increased glutathione biosynthesis plays a role in nickel tolerance in Thlaspi nickel hyperaccumulators. *Plant Cell.* 16:2176–2191.

Fu, J. and B. Huang. 2003. Effects of foliar application of nutrients on heat tolerance of creeping bentgrass. *J. Plant Nutr.* 26:81–96.

Fu, J., and P.H. Dernoeden. 2008. Carbohydrate metabolism in creeping bentgrass as influenced by two summer irrigation practices. *J. Am. Soc. Hortic. Sci.* 133:678–683.

Fu, J., P.H. Dernoeden, and J.A. Murphy. 2009. Creeping bentgrass color and quality, chlorophyll content, and thatch-mat accumulation responses to summer coring. *Crop Sci.* 49:1079–1087.

Funk, C.R., F.C. Belanger, and J.A. Murphy. 1994. Role of endophytes in grasses used for turf and soil conservation, pp. 201–209. In: C.W. Bacon and J.F. White, Jr. (eds.) Biotechnology of Endophytic Fungi Conclusions of Grasses. CRC Press, Boca Raton, FL.

Funk, C.R., P.M. Halisky, M.C. Johnson, M.R. Siege, A.V. Stewart, S. Ahmad, R.H. Hurley, and I.C. Harvey. 1983. An endophytic fungus and resistance to sod webworms: association in *Lolium perenne* L. *Biotechnology*. 1:189–191.

Garrison, M.A., and J.C. Stier. 2010. Cool-season turfgrass colony and seed survival in a restored prairie. *Crop Sci.* 50:345–356.

Goatley, J.M., Jr., and R.E. Schmidt. 1990. Anti-senescence activity of chemicals applied to Kentucky bluegrass. *J. Am. Soc. Hortic. Sci.* 115:654–656.

Grattan, S.R., and C.M. Grieve. 1999. Salinity: mineral nutrient relations in horticultural crops. *Sci. Hort.* 78:127–157.

Groffman, P.M., C.O. Williams, R.V. Pouyat, L.E. Band, and I.D. Yesilonis. 2009. Nitrate leaching and nitrous oxide flux in urban forests and grasslands. *J. Environ. Qual.* 38:1848–1860.

Guertal, E.A. 2004. Boron fertilization of bentgrass. *Crop Sci.* 44:204–208.

Guertal, E.A. and C.A. Hicks. 2009. Nitrogen source and rate effects on the establishment of 'TifSport' and 'Tifway' hybrid bermudagrass. *Crop Sci.* 49: 690–695.

Guo, W., H. Nazim, Z. Liang, and D. Yang. 2016. Magnesium deficiency in plants: an urgent problem. *Crop. J.* 4:83–91.

Gupta, U.C. 1997. *Molybdenum in Agriculture*. Cambridge University Press, New York, NY.

Halvorson, A.D., B.J. Wienhold, and A.L. Black. 2002. Tillage, nitrogen, and cropping system effects on soil carbon sequestration contribution from USDA-ARS. *Soil Sci. Soc. Am. J.* 66:906–912.

Hänsch, R., and R.R. Mendel. 2009. Physiological functions of mineral micronutrients (Cu, Zn, Mn, Fe, Ni, Mo, B, Cl). *Curr. Opin. Plant Biol.* 12(3):259–266.

Hartikainen, H., T. Xue, and V. Piironen. 2000. Selenium as an anti-oxidant and pro-oxidant in ryegrass. *Plant Soil.* 225(1–2):193–200.

Havlin, J.L., S.L. Tisdale, W.L. Nelson, and J.D. Beaton. 2013. *Soil Fertility and Fertilizers – An Introduction to Nutrient Management*. 8th ed. Pearson, Upper Saddle River, NJ.

Heath, K.D., and P. Tiffin. 2009. Stabilizing mechanisms in legume-rhizobium mutualism. *Evolution.* 63:652–662.

Heckman, J. R., H. Liu, W. Hill, M. DeMilia, and W.L. Anastasia. 2000. Kentucky bluegrass responses to mowing and nitrogen fertility management. *J. Sustainable Agr.* 15:25–33.

Hill, W.J., J.R. Heckman, B.B. Clarke, and J.A. Murphy. 1999. Take-all patch suppression in creeping bentgrass with manganese and copper. *HortScience.* 34:891–892.

Hodges, C.F. 1992. Interaction of cyanobacteria and sulfate-reducing bacteria in sub-surface black-layer formation in high-sand content golf greens. *Soil Biol. Biochem.* 24:15–20.

Hoffman, L., J.S. Ebdon, W.M. Dest, and M. DaCosta. 2010a. Effects of nitrogen and potassium on wear mechanisms in perennial ryegrass: I. Wear tolerance and recovery. *Crop Sci.* 50:357–366.

Hoffman, L., J.S. Ebdon, W.M. Dest, and M. DaCosta. 2010b. Effects of nitrogen and potassium on wear mechanisms in perennial ryegrass: II. Anatomical, orphological, and physiological characteristics. *Crop Sci.* 50:367–379.

Hopper, J., and D.R. Parker. 1999. Plant availability of selenite and selenate as influenced by the competing ions phosphate and sulfate. *Plant Soil.* 210:199–207.

Hua, L., Y. Wang, W. Wu, M.B. McBride, and Y. Chen. 2008. Biomass and Cu and Zn uptake of two turfgrass species grown in sludge compost-soil mixtures. *Water Air Soil Pollut.* 188:225–234.

Huang, B. 2001. Nutrient accumulation and associated root characteristics in response to drought stress in tall fescue cultivars. *HortScience.* 36:148–152.

Huang, B., M. DaCosta, and Y. Jiang. 2014. Research advances in mechanisms of turfgrass tolerance to abiotic stress: from physiology to molecular biology. *Crit. Rev. Plant Sci.* 33:141–189.

Hull, R.J., and H. Liu. 2005. Turfgrass nitrogen: physiology and environmental impacts. *Int. Turfgrass Soc. Res. J.* 10:962–975.

Hull, R.J. 1997. Phosphorus usage by turfgrasses: the energy nutrient often neglected by turf managers. *Turfgrass Trends.* 6(5):1–12.

Hull, R.J. 2001. Zinc usage by turfgrasses. *Turfgrass Trends.* 10(7):7–11.

Jaiswal, D.K., J.P. Verma, S. Prakash, V.S. Meena, and R.S. Meena. 2016. Potassium as an important plant nutrient in sustainable agriculture: a state of the art, pp. 21–29. In: V.S. Meena, B.R. Maurya, J.P. Verma, and R.S. Meena (eds.) *Potassium Solubilizing microorganisms for Sustainable Agriculture*. Springer, India.

Jernstedt, J.A., and J.H. Bouton. 1985. Anatomy, morphology, and growth of tall fescue rhizomes. *Crop Sci.* 25:539–542.

Jiang, Y., and K. Wang. 2006. Growth, physiological, and anatomical responses of creeping bentgrass cultivars to different depths of waterlogging. *Crop Sci.* 46:2420–2426.

Jiang, Z., and R.J. Hull. 1998. Interrelationships of nitrate uptake, nitrate reductase, and nitrogen use efficiency in selected Kentucky bluegrass cultivars. *Crop Sci.* 38:1623–1632.

Johnson-Cicalese, J., M.E. Secks, C.K. Lam, W.A. Meyer, J.A. Murphy, and F.C. Belanger. 2000. Cross species inoculation of Chewings and strong creeping red fescues with fungal endophytes. *Crop Sci.* 40:1485–1489.

Kaiser, B.N., K.L. Gridle, J.N. Brady, T. Phillips, and S.D. Tyerman. 2005. The role of molybdenum in agricultural plant production. *Ann. Bot.* 965:745–54.

Kamon, Y. 1973. Magnesium deficiency in zoysiagrass. In: The 2nd Proc. Int. Turfgrass Res. Conf., pp. 145–148.

Karcher, D.E., M.D. Richardson, K. Hignight, and D. Rush. 2008. Drought tolerance of tall fescue populations selected for high root/shoot ratios and summer survival. *Crop Sci.* 48:771–777.

Kavanová, M., and V. Gloser. 2005. The use of internal nitrogen stores in the rhizomatous grass *Calamagrostis epigejos* during regrowth after defoliation. *Ann. Bot.* 95:457–463.

Khan, A.L., J. Hussain, A. Al-Harrasi, A. Al-Rawahi, and J.J. Lee. 2013. Endophytic fungi: resource for gibberellins and crop abiotic stress resistance. *Crit. Rev. Biotechnol.* 1–13

Koehler, J.F., and G.L. Miller. 2017. Impact of nitrogen source and a ph buffer on the in vitro growth and morphology of *Rhizoctonia solani* AG 2–2 LP. *Int. Turfgrass Soc. Res. J.* 13:198–202.

Krum, J.M., R.N. Carrow, and K. Karnok. 2010. Spatial mapping of complex turfgrass sites: site-specific management units and protocols. *Crop Sci.* 50:301–315.

Kuo, Y.J., Y.S. Chang, M.A. Lila, and H.Y. Chiu. 2005. Screening growth and root formation in cadmium-treated turfgrass using a whole-plant microculture system. *J. Plant Nutr.* 28:1041–1048.

Lal, R. 2004. Soil Carbon sequestration impacts on global climate change and food security. *Science.* 304 (5677):1623–1627.

Lal, R. 2015. Sequestering carbon and increasing productivity by conservation agriculture. *J. Soil Water Conserv.* 70(3):55A–62A.

Lange, B. 2017. Copper and cobalt accumulation in plants: a critical assessment of the current state of knowledge. *New Phytol.* 213:537–551.

Latch, G.C.M. 1993. Physiological interactions of endophytic fungi and their hosts. Biotic stress tolerance imparted to grasses by endophytes. *Agric. Ecosyst. Environ.* 44:143–156.

Law, Q.D., J.M. Trappe, Y. Jiang, R.F. Turco, and A.J. Patton. 2017. Turfgrass selection and grass clippings management influence soil carbon and nitrogen dynamics. *Agron. J.* 109:1719–1725.

Lee, G.J., R.N. Carrow, and R.R., Duncan. 2004a. Salinity tolerance of selected seashore paspalums and bermudagrasses: root and verdure responses and criteria. *HortScience.* 39:1143–1147.

Lee, G.J., R.R. Duncan, and R.N. Carrow. 2004b. Salinity tolerance of seashore paspalum ecotypes: shoot growth responses and criteria. *HortScience.* 39:1138–1142.

Lee, G.J., R.R. Duncan, and R.N. Carrow. 2007. Nutrient uptake responses and inorganic ion contribution to solute potential under salinity stress in halophytic seashore paspalums. *Crop Sci.* 47:2504–2512.

Lemaire, G. 2015. Crop responses to nitrogen, pp. 1–25. In: R.A. Meyers (ed.) Encyclopedia of Sustainability Science and Technology. Springer, New York, NY.

Li, H.F., S.P. McGrath, and F.J. Zhao. 2008. Selenium uptake, translocation and speciation in wheat supplied with selenate or selenite. *New Phytol.* 178:92–102.

Li, Z., C.M. Baldwin, Q. Qian, H. Liu, and H. Luo. 2010. Heterologous expression of Arabidopsis H+-pyrophosphatase enhances salt tolerance in transgenic creeping bentgrass (*Agrostis stolonifera* L.). *Plant Cell Environ.* 33:272–289.

Liang, Y., W. Sun, Y.G. Zhu, and P. Christie. 2007. Mechanisms of silicon-mediated alleviation of abiotic stresses in higher plants: a review. *Environ. Pollut.* 147:422–428.

Liu, H. 2005. Aluminum toxicity of seeded bermudagrass cultivars. *HortScience.* 40:221–223.

Liu, H., and R.J. Hull. 2006. Comparing cultivars of three cool-season turfgrasses for nitrogen recovery in clippings. *HortScience.* 41:827–831.

Liu, H., C.M. Baldwin, and H. Luo. 2008. Acid soil and aluminum tolerance in turfgrasses, pp. 373–386. In: M. Pessarakli (ed.) *Handbook of Turfgrass Management and Physiology.* CRC Press Taylor & Francis Group, New York, NY.

Liu, H., C.M. Baldwin, F.W. Totten, and L.B. McCarty. 2008. Foliar fertilization for turfgrasses. II International Conference on Turfgrass Science and Management for Sports Fields. *Acta Hort. (ISHS).* 783:323–332.

Liu, H., C.M. Baldwin, H. Luo, and M. Pessarakli. 2008. Enhancing turfgrass nitrogen use under stresses, pp. 555–599. In: M. Pessarakli (ed.) *Handbook of Turfgrass Management and Physiology.* CRC Press, Taylor & Francis Group, New York, NY.

Liu, H., J.R. Heckman, and J.A. Murphy. 1995. Screening Kentucky bluegrass for aluminum tolerance. *J. Plant Nutr.* 18:1797–1814.

Liu, H., J.R. Heckman, and J.A. Murphy. 1996. Screening fine fescues for aluminum tolerance. *J. Plant Nutr.* 19:677–688.

Liu, H., J.R. Heckman, and J.A. Murphy. 1997. Aluminum tolerance among genotypes of *Agrostis* species. *Int. Turfgrass Soc. Res. J.* 8:729–734.

Liu, H., N. Menchyk, F. Bethea, and C. Baldwin. 2011. Chapter 38 Turfgrass nutrient management under stresses: a part of integrated stress management, pp. 965–988. In: Mohammad Pessarakli (ed.) *Handbook of Plant and Crop Stress*, 3rd Ed. Taylor & Francis Group, Raton, FL.

Liu, R., L. Shi, T. Zhu, T. Yang, A. Ren, J. Zhu, and M.W. Zhao. 2018. Cross talk between nitric oxide and calcium-calmodulin regulates ganoderic acid biosynthesis in *Ganoderma lucidum* under heat stress. *Appl. Environ. Microbiol.* 84:e00043-18.

Liu, X., B. Huang, and G. Banowetz. 2002. Cytokinin effects on creeping bentgrass responses to heat stress: I. Shoot and root growth. *Crop Sci.* 42:457–465.

Lockett, A.M., D.A. Devitt, and R.L. Morris. 2008. Impact of reuse water on golf course soil and turfgrass parameters monitored over a 4.5-year period. *HortScience.* 43:2210–2218.

Macedo, F.G., J.D. Bresolin, E.F. Santos, F. Furlan, W.T. Silva, J.C. Polacco, and J. Lavres. 2016. Nickel availability in soil as influenced by liming and its role in soybean nitrogen metabolism. *Front. Plant Sci.* 7:1358.

Malinowski, D.P., and D.P. Belesky. 1999. *Neotyphodium coenophialum*-endophyte infection affects the ability of tall fescue to use sparingly available phosphorus. *J. Plant Nutr.* 22:835–853.

Malinowski, D.P., and D.P. Belesky. 2000. Adaptations of endophyte-infected cool-season grasses to environmental stresses: mechanisms of drought and mineral stress tolerance. *Crop Sci.* 40:923–940.

Mann, R.L., P.S. Kettlewell, and P. Jenkinson. 2004. Effect of foliar applied potassium chloride on septoria leaf blotch of winter wheat. *Plant Pathol.* 53:653–659.

Marcum, K.B. 2001. Salinity tolerance of 35 bentgrass cultivars. *HortScience.* 36:374–376.

Marcum, K.B., and C.L. Murdoch. 1994. Salinity tolerance mechanisms of six C4 turfgrasses. *J. Am. Soc. Hortic. Sci.* 119:779–784.

Marcum, K.B., and M. Pessarakli. 2006. Salinity tolerance and salt gland excretion efficiency of bermudagrass turf cultivars. *Crop Sci.* 46:2571–2574.

Marschner, H. 2012. *Marschner's Mineral Nutrition of Higher Plants*. 3rd ed. Academic Press, London.

Mattox, C., A.R. Kowalewski, and B.W. McDonald. 2015. The effects of different rates of nitrogen and iron sulfate on Microdochium patch development on trafficked annual bluegrass putting greens in the absence of fungicides. In: Annual Meeting Abstracts, ASA, CSSA, and SSSA, Madison, WI, p. 92685.

McCall, D.S., E.H. Ervin, C.D. Shelton, N. Reams, and S.D. Askew. 2017. Influence of ferrous sulfate and its elemental components on dollar spot suppression. *Crop Sci.* 57:581–586.

McCarty, L.B. 2011. *Best Golf Course Management Practices*. 3rd ed. Prentice-Hall Inc., Upper Saddle River, NJ.

McCarty, L.B. 2018. *Golf Turf Management*. CRC Press. Boca Raton, FL.

McCarty, L.B., M.F. Gregg, and J.E. Toler. 2007. Thatch and mat management in an established creeping bentgrass golf green. *Agron. J.* 99:1530–1537.

McCarty, L.B., and B.J. Tucker. 2005. Prospects for managing turf weeds without prospective chemicals. *Int. Turfgrass Soc. Res. J.* 10:34–41.

McCullough, P.E., H. Liu, L.B. McCarty, T. Whitwell, and J.E. Toler. 2006. Bermudagrass putting green growth, quality, and nutrient partitioning influenced by nitrogen and trinexapac-ethyl. *Crop Sci.* 46:1515–1525.

Menchyk, N. 2012. Foliar applied urea nitrogen metabolism in warm-season turfgrass under salinity stress. PhD Dissertation, Clemson University, Clemson, SC.

Menchyk, N., D.G. Bielenberg, S. Martin, C. Waltz, H. Luo, F. Bethea, Jr., and H. Liu. 2014. Nitrogen and trinexapac-ethyl applications for managing 'diamond' zoysiagrass putting greens in the transition zone, U.S. *HortScience*. 49:1076–1080.

Michalski, J., and Z. Cheng. 2018. Effects of 'Lights out' turfgrass renovation on plants, soil arthropod and nematode communities. *Appl. Soil Ecol.* 127:144–154.

Miller, G.L., D.T. Earlywine, R. Braun, J.D. Fry, and M.M. Kennelly. 2016. Influence of nitrogen source and application timing on large patch of zoysiagrass. *Crop Forage Turfgrass Manage.* 2:2015–0189.

Miller, G.L., and R. Dickens. 1996. Potassium fertilization related to cold resistance in bermudagrass. *Crop Sci.* 36:1290–1295.

Munshaw, G.C., X. Zhang, and E.H. Ervin. 2004. Effect of salinity on bermudagrass cold hardiness. *HortScience*. 39:420–423.

Murray, J.J., and C.D. Foy. 1978. Differential tolerances of turfgrass cultivars to an acid soil high in exchangeable aluminum. *Agron. J.* 70:769–774.

Niu, L., and W. Liao. 2016. Hydrogen peroxide signaling in plant development and abiotic responses: crosstalk with nitric oxide and calcium. *Front. Plant Sci.* 7:230.

Obasa, K., J. Fry, D. Bremer, R.S. John, and M. Kennelly. 2013. Effect of cultivation and timing of nitrogen fertilization on large patch disease of zoysiagrass. *Plant Dis.* 97:1075–1081.

Obear, G.R., P. Barak, and D.J. Soldat. 2016. Soil inorganic carbon accumulation in sand putting green soils II: acid–base relationships as affected by water chemistry and nitrogen source. *Crop Sci.* 56:851–861.

Pang, J., H.R. Megan, L. Hans, and K.H.M. Siddique. 2018. Phosphorus acquisition and utilization in crop legumes under global change. *Curr. Opin. Plant Biol.* 45:1–7.

Patrick, C.R. 2018. Nickel tolerance in cool-season grasses. MS Thesis. Clemson University, Clemson, SC.

Patton, A.J., and Z.J. Reicher. 2007. Zoysiagrass species and genotypes differ in their winter injury and freeze tolerance. *Crop Sci.* 47:1619–1627.

Pease, B.W., and J.C. Stier. 2018. Nitrogen rate and growth regulator effects on shaded velvet and creeping bentgrasses. *Agron. J.* 110:2151–2158.

Pelletier, S., and J. Dionne. 2004. Inoculation rate of Arbuscular-mycorrhizal fungi *Glomus intraradices* and *Glomus etunicatum* affects establishment of landscape turf with no irrigation or fertilizer inputs. *Crop Sci.* 44:335–338.

Post, W.M., and K.C. Kwon. 2000. Soil carbon sequestration and land-use change: processes and potential. *Glob. Change Biol.* 6:317–327.

Potter, D.A. 1998. *Destructive Turfgrass Insects: Biology, Diagnosis, and Control*. Ann Arbor Press, Inc., Chelsea, MI.

Prakamhang, J., K. Minamisawa, K. Teamtaisong, N. Boonkerd, and N. Teaumroong. 2009. The communities of endophytic diazotrophic bacteria in cultivated rice (*Oryza sativa* L.). *Appl. Soil Ecol.* 42:141–149.

Qian, Y., R.F. Follett, and J.M. Kimble. 2010. Soil organic carbon input from urban turfgrasses. *Soil Sci. Soc. Am. J.* 74:366–371.

Qian, Y.L., W. Bandaranayake, W.J. Parton, B. Mecham, M.A. Harivandi, and A.R. Mosier. 2003. Long-term effects of clipping and nitrogen management in turfgrass on soil organic carbon and nitrogen dynamics: The CENTURY model simulation. *J. Environ. Qual.* 32:1694–1700.

Qian, Y.L., and R.F. Follett. 2002. Assessing soil carbon sequestration in turfgrass systems using long-term soil testing data. *Agron. J.* 94:930–935.

Qian, Y.L., J.M. Fu, S.J. Wilhelm, D. Christensen, and A.J. Koski. 2007. Relative salinity tolerance of turf type saltgrass selections. *HortScience*. 42:205–209.

Qian, Y.L., S.J. Wilhelm, and K.B. Marcum. 2001. Comparative responses of two Kentucky bluegrass cultivars to salinity stress. *Crop Sci.* 41:1895–1900.

Qu, R.L., D. Li, R. Du, and R. Qu. 2003. Lead uptake by roots of four turfgrass species in hydroponic cultures. *HortScience*. 38:623–626.

Raimam, M.P., U. Albino, M.F. Cruz, G.M. Lovato, F. Spago, T.P. Ferracin, D.S. Lima, T. Goulart, C.M. Bernardi, M. Miyauchi, M.A. Nogueira, and G. Andrade. 2007. Interaction among free-living N-fixing bacteria isolated from Drosera villosa var. villosa and AM fungi (Glomus clarum) in rice (Oryza sativa). *Appl. Soil Ecol.* 35:25–34.

Reinert, J.A., and M.C. Engelke. 2001. Resistance in zoysiagrass, *Zoysia* spp., to the tropical sod webworm, *Herpetogramma phaeopteralis* Guenne. *Int. Turfgrass Soc. Res. J.* 9:798–801.

Richardson, M.D., D.E. Karcher, K. Hignight, and D. Rush. 2008. Drought tolerance and rooting capacity of Kentucky bluegrass cultivars. *Crop Sci.* 48:2429–2436.

Rouphael, Y., P. Franken, C. Schneider, D. Schwarz, M. Giovannetti, M. Agnolucci, S.D. Pascale, P. Bonini, and G. Colla. 2015. Arbuscular mycorrhizal fungi acts as biostimulants in horticultural crops. *Sci. Hortic.* 188:97–105.

Rowland, J.H., J.L. Cisar, G.H. Snyder, J.B. Sartain, and A.L. Wright. 2009. USGA ultradwarf bermudagrass putting green properties as affected by cultural practices. *Agron. J.* 101:1565–1572.

Rukavina, H., H.G. Hughes, and Y.L. Qian. 2007. Freezing tolerance of 27 saltgrass ecotypes from three cold hardiness zones. *HortScience*. 42:157–160.

Sartain, J.B. 1993. Interrelationships among turfgrasses, clipping recycling, thatch, and applied calcium, magnesium, and potassium. *Agron. J.* 85:40–43.

Sarvis, W.G. 2008. Creeping bentgrass summer stress management with K, Ca, and Mg associated with surfactant applications. MS Thesis, Clemson University, Clemson, SC.

Sarvis, W.G., H. Liu, L.B. McCarty, and J.E. Toler. 2009. Management of Pao trivialis as overseeded turf under shade conditions. *Int. Turfgrass Soc. Res. J.* 11:837–847.

Schmid, C.J., B.B. Clarke, and J.A. Murphy. 2017. Anthracnose severity and annual bluegrass quality as influenced by nitrogen source. *Crop Sci.* 57(Suppl 1):S-285–S-292.

Schmid, C.J., J.A. Murphy, B.B. Clarke, M. DaCosta, and J.S. Ebdon. 2016. Observations on the effect of potassium on winter injury of annual bluegrass in New Jersey in 2015. *Crops Forage Turfgrass Manage.* 2(1):1–4.

Selhorst, A., and R. Lal. 2013. Net carbon sequestration potential and emissions in home lawn turfgrasses of the United States. *Environ. Manage.* 51:198–208.

Shaver, B.R., P. Agudelo, and S.B. Martin. 2017. Damage functions for sting nematode (*Belonolaimus longicaudatus*) on bermudagrass turf. *Int. Turfgrass Soc. Res. J.* 13:517–523.

Sims, J.T., and A.N. Sharpley. 2005. *Phosphorus: Agriculture and the Environment.* Agron. Monogr. 46. ASA, CSSA, SSSA, Madison, WI.

Sincik, M., and E. Acikgoz. 2007. Effects of white clover inclusion on turf characteristics, nitrogen fixation, and nitrogen transfer from white clover to grass species in turf mixtures. *Commun. Soil Sci. Plant Anal.* 38:1861–1877.

Smiley, R.W., P.H. Dernoeden, and B.B. Clarke. 2005. *Compendium of Turfgrass Diseases.* 3rd ed. APS Press, St. Paul, MN.

Soldat, D.J., and A.M. Petrovic. 2008. The fate and transport of phosphorus in turfgrass ecosystems. *Crop Sci.* 48:2051–2065.

Song, D., K. Pan, A. Tariq, F. Sun, Z. Li, X. Sun, L. Zhang, O.A. Olusanya, and X. Wu. 2017. Large-scale patterns of distribution and diversity of terrestrial nematodes. *Appl. Soil Ecol.* 114:161–169.

Song, M., X. Li, S. Jing, L. Lei, J. Wang, and S. Wan. 2016. Responses of soil nematodes to water and nitrogen additions in an old-field grassland. *Appl. Soil Ecol.* 102:53–60.

Stiglbauer, B.J., H. Liu, L.B. McCarty, D.M. Park, J.E. Toler, and K.R. Kirk. 2009. Diamond zoysiagrass putting green establishment affected by sprigging rates, nitrogen sources, and rates in the southern transition zone. *HortScience.* 44:1757–1761.

StJohn, R.A., N.E. Christians, H. Liu, and N. Menchyk. 2013. *Secondary Nutrients and Micronutrient Fertilization in Turgrass Monograph 56.* Crop Science Society of America, Madison, WI, pp. 521–542.

StJohn, R.A., N.E. Christians, and H.G. Taber. 2003. Supplemental calcium applications to creeping bentgrass established on calcareous sand. *Crop Sci.* 43:967–972.

Su, K., D.J. Bremer, S.J. Keeley, and J.D. Fry. 2008. Rooting characteristics and canopy responses to drought of turfgrasses including hybrid bluegrasses. *Agron. J.* 100:949–956.

Subbarao, G.V., W.L. Ito, O. Berry, and R.M. Wheeler. 2003. Sodium: a functional plant nutrient. *Crit. Rev. Plant Sci.* 22:391–416.

Subramanian, K.S., and C. Charest. 1995. Influence of arbuscular mycorrhizae on the metabolism of maize under drought stress. *Mycorrhiza.* 5:273–278.

Taiz, L., E. Zeiger, I.M. Møller, and A.S. Murphy. 2015. *Plant Physiology and Development.* 6th ed. Sinauer Associates, Inc., Publishers, Sunderland, MA.

Tan, S., F. Dong, Y. Yang, Q. Zeng, B. Chen, and L. Jiang. 2017. Effects of Waterlogging and Cadmium on ecophysiological responses and metal bio-accumulation in bermudagrass (*Cynodon Dactylon*). *Environ. Earth Sci.* 719:1–11.

Thompson, D.C., B.B. Clarke, and J.R. Heckman. 1995. Nitrogen form and rate of nitrogen and chloride application for the control of summer patch in Kentucky bluegrass. *Plant Dis.* 79:51–56.

Throssell, C.S., G.T. Lyman, M.E. Johnson, and G.A. Stacey. 2009. Golf course environmental profile measures nutrient use and management and fertilizer restrictions, storage, and equipment calibration. Online. *Appl. Turfgrass Sci.*

Trenholm, L.E., R.N. Carrow, and R.R. Duncan. 2001. Wear tolerance, growth and quality of seashore paspalum in response to nitrogen and potassium. *HortScience.* 36:780–783.

Turgeon, A.J., 2012. *Turfgrass Management.* 9th ed. Prentice-Hall/Pearson Higher Education.

Turgeon, A.J., L.B. McCarty, and N.E. Christians. 2009. *Weed Control in Turf and Ornamentals.* Prentice-Hall Inc., Upper Saddle River, NJ.

Turner, R.S. and N.W. Hummel, Jr. 1992. Nutritional requirements and fertilization, pp. 385–439. In: D.V. Waddington, R.N. Carrow, and R.C. Shearman (eds.) *Turfgrass.* Agron. Monogr. 32. ASA, CSSA, and SSSA, Madison, WI.

Vaculíka, M., A. Luxa, M. Luxovác, E. Tanimotod, and I. Lichtscheidle. 2009. Silicon mitigates cadmium inhibitory effects in young maize plants. *Environ. Exp. Bot.* 67:52–58.

Valladares, F., and U. Niinemets. 2008. Shade tolerance, a key plant feature of complex nature and consequences. *Annu. Rev. Ecol. Evol. Syst.* 39:237–257.

Vaughn, S.F., F.D. Dinelli, M.A. Jackson, M.M. Vaughan, and S.C. Peterson. 2018. Biochar-organic amendment mixtures added to simulated golf greens under reduced chemical fertilization increase creeping bentgrass growth. *Ind. Crops Prod.* 111:667–672.

Walters, D.R., and I.J. Bingham. 2007. Influence of nutrition on disease development caused by fungal pathogens: implications for plant disease control. *Ann. Appl. Biol.* 151:307–324.

Wang, Z., J. Sun, J. Li, and Y. Zhu. 2006. Heat resistance enhanced by trinexapac-ethyl and benzyladenine combination in creeping bentgrass. *HortScience.* 41:1711–1714.

Webster, D.E., and J.S. Ebdon. 2005. Effects of nitrogen and potassium fertilization on perennial ryegrass cold tolerance during deacclimation in late winter and early spring. *HortScience.* 40:842–849.

White, P.J. 2018. Selenium metabolism in plants. *Biochim. Biophys. Acta (BBA) Gen. Subjects.* 1862(11):2333–2342.

Xia, H.P. 2004. Ecological rehabilitation and phytoremediation with four grasses in oil shale mined land. *Chemosphere.* 54:345–353.

Xiong, X., G.E. Bell, J.B. Solie, M.W. Smith, and B. Martin. 2007. Bermudagrass seasonal responses to nitrogen fertilization and irrigation detected using optical sensing. *Crop Sci.* 47:1603–1610.

Xu, X., and C.F. Mancino. 2001a. Annual bluegrass and creeping bentgrass response to varying levels of iron. *HortScience.* 36:371–373.

Xu, X., and C.F. Mancino. 2001b. Zinc requirements of annual bluegrass and creeping bentgrass. *HortScience.* 36:384–386.

Xu, Y., and B. Huang. 2009. Effects of foliar-applied ethylene inhibitor and synthetic cytokinin on creeping bentgrass to enhance heat tolerance. *Crop Sci.* 49:1876–1884.

Yan, J., J. Chen, T. Zhang, J. Liu, and H. Liu. 2009. Evaluation of aluminum tolerance and nutrient uptake of 50 centipedegrass accessions or cultivars. *HortScience.* 44:857–861.

Yang, J., J.W. Kloepper, and C.M. Ryu. 2009. Rhizosphere bacteria help plants tolerate abiotic stress. *Trends Plant Sci.* 14:1–4.

Yesilonis, I.D., B.R. James, R.V. Pouyat, and B. Momen. 2008. Lead forms in urban turfgrass and forest soils as related to organic matter content and pH. *J. Environ. Monit. Assess.* 146:1–17.

Young, J., M. Richardson, and D. Karcher. 2015. Creeping bentgrass putting green response to combined mowing, rolling, and foot traffic under environmental stress. *Agron. J.* 107:1959–1966.

Youssef, M.M.A., and M.F.M. Eissa. 2014. Biofertilizers and their role in management of plant parasitic nematodes: a review. *J. Biotechnol. Pharm. Res.* 5(1):1–6.

Zahoor, A., S. Anjum, E.A. Waraich, M.A. Ayub, T. Ahmad, R.M.S. Tariq, R. Ahmad, and M.A. Iqbal. 2018. Growth, physiology, and biochemical activities of plant responses with foliar potassium application under drought stress: a review. *J. Plant Nutr.* 41(13):1734–1743.

Zaurov, D.E., S. Bonos, J.A. Murphy, M. Richardson, and F.C. Belanger. 2001. Endophyte infection can contribute to aluminum tolerance in fine fescues. *Crop Sci.* 41:1981–1984.

Zhang, F., C. Gao, J. Wang, Y. Lu, Z. Shen, T. Liu, D. Chen, W. Ran, and Q. Shen. 2017. Coupling sugarcane yield to soil nematodes: implications from different fertilization regimes and growth stages. *Agric. Ecosyst. Environ.* 247:157–165.

Zhang, J., Y. Liu, J. Yu, W. Zhang, Y. Xie, and N. Ge. 2018. Key factors influencing weed infestation of cool-season turfgrass *Festuca arundinacea* Schreb. areas during early spring in the Tianjin Region, China. *HortScience.* 53(5):723–728.

Zhang, X., E.H. Ervin, W. Wu, N. Sharma, and A. Hamill. 2017. Auxin and trinexapac-ethyl impact on root viability and hormone metabolism in creeping bentgrass under water deficit. *Crop Sci.* 57(Suppl 1):S-130–S-137.

Zhang, X., K. Wang, and E.H. Ervin. 2010. Optimizing dosages of seaweed extract-based cytokinins and zeatin riboside for improving creeping bentgrass heat tolerance. *Crop Sci.* 50:316–320.

Zhang, X.Z., and E.H. Ervin. 2008. Impact of seaweed extract-based cytokinins and zeatin riboside on creeping bentgrass heat tolerance. *Crop Sci.* 48:364–370.

Zhao, S.L., L. He, Y.F. Lu, and L. Duo. 2017. The impact of modified nano-carbon black on the earthworm Eisenia fetida under turfgrass growing conditions: assessment of survival, biomass, and antioxidant enzymatic activities. *J. Hazard. Mater.* 338:218–223.

Zhu, J.K. 2016. Abiotic stress signaling and responses in plants. *Cell.* 167:313–324.

Zhu, Y.G., E.A.H. Pilon-Smits, F.J. Zhao, P.N. Williams, and A.A. Meharg. 2009. Selenium in higher plants: understanding mechanisms for biofortification and phytoremediation. *Trends Plant Sci.* 14:436–442.

Zirkle, G., R. Lal, and B. Augustin. 2011. Modeling carbon sequestration in home lawns. *HortScience.* 46:808–814.

17 Nutrient Management of Golf Course Putting Greens under Stress

Haibo Liu, Nick Menchyk, Frank Bethea, Christian Baldwin,
Jacob Taylor, and Caleb Patrick

CONTENTS

17.1 INTRODUCTION

Golf started between the 1300s and the 1400s and has become a much more popular game since World War II (Hurdzan, 2006), and the number of golf courses will continue to grow, even the world is currently facing various global challenges, with golf becoming an official game in the Olympics in 2016. This trend may be partially due to the fact that new golf courses have been rapidly developed within the last four to five decades in countries where golf is not traditionally played, such as Asian countries among others, along with the development of human civilization and the demand for higher living standards. Compared with other outside sports, golfing

suits the broadest age span, from childhood to senior age, and it has been a life-long hobby and exercise for many golfers (Murray et al., 2017). Currently, there are about 34,000 golf courses in the world, and it is estimated that the economic impact is about $50 to $100 billion each year just for turfgrass management. Focusing on turfgrass management aspects, the golf course industry employs more than a million part-time and full-time professionals to maintain the abovementioned 34,000 golf courses in the world by the routine maintenance practices of grasses, fertilizers, chemicals, irrigation systems, pest control, stress management, golf course design and construction, and renovations, among others (Beard, 2002; Haydu et al., 2008; Dernoeden, 2013; McCarty, 2018).

Although the total putting green area for an 18-hole golf course is only about 1 hectare (10,000 m^2), or about 1.6% of the whole golf course area on average (Beard, 2002; Hurdzan, 2006; McCarty, 2018), these putting greens absorb the most energy, cost, and labor for their higher maintenance demands than the other areas of a golf course (Beard, 2002; McCarty and Miller, 2002; Dernoeden, 2013; McCarty, 2018). Golf courses, sports fields, lawns, and utility turfs are specially mono-cultured functional crops. Among these functional turfgrass crops, golf course putting green management represents the highest level of turfgrass management requirements due to the lowest cutting heights (normally between 2.0 and 5 mm), the traffic, and the environmental stresses encountered. Well-maintained putting greens represent the quality of the golf course, with positive impacts on golfers, the game, the community, the environment, and the local economy. Golf putting green turfgrasses are normally mowed in the range of 2.0 to 5.0 mm to meet the ball rolling requirement for playability, and only a few turfgrass species from over 12,000 species (Soreng et al., 2015) of the grass family, one of the four largest higher plant families, have been adapted as putting green turfgrasses (Table 17.1). Obviously, there is great potential to explore new grass species and varieties, including from other plant families, for golf course and turf use and even for future putting greens.

Due to the low mowing height requirement, many fewer turfgrasses are suitable for putting greens than for other types of turf. The turfgrass species that have been adapted as putting green turfgrasses include both cool-season and warm-season turfgrasses (Beard, 2002; McCarty and Miller, 2002; Dernoeden, 2013; McCarty, 2018). Creeping bentgrass (*Agrostis stolonifera* L.), annual bluegrass (*Poa annua* L.), fine fescues (*Festuca* spp.), and velvet bentgrass (*Agrostis canina* L.) are the cool-season turfgrasses, plus the most commonly used winter overseeding turfgrass, rough bluegrass (*Poa trivialis* L.), and the hybrids in the *Agrostis* genus as future potential putting green turfgrasses. The dominant warm-season putting green turfgrass has been hybrid bermudagrass (*Cynodon dactylon* (L.) Pers. × *C. transvaalensis* Burtt-Davy). Seashore paspalum (*Paspalum vaginatum* Swartz.) and zoysiagrass cultivars (*Zoysia matrella* (L.) Merr.) (Engelke et al., 2002; Stiglbauer et al., 2009; Liu et al., 2013; Menchyk et al., 2014; Chandra et al., 2017) are the relatively newer members of the warm-season putting green turfgrasses, with Diamond zoysiagrass as a new putting green turfgrass, and have been adapted in recent years, plus the other upcoming zoysiagrass hybrids and cultivars (Liu et al., 2013; Menchyk et al., 2014; Chandra et al., 2017). In comparison with hybrid bermudagrasses, these two new warm-season putting green turfgrasses have advantages of improved salinity, poor water quality, and shade tolerance and are used in the warmer climatic zones in the world (Table 17.1).

Although there are only a few turfgrass species available as putting green turfgrasses, there are many more varieties or cultivars of these species developed for putting greens over many decades of breeding effort (McCarty, 2018). Creeping bentgrass has the largest number of cultivars among these species because of its long history of use as a putting green turfgrass (Dernoeden, 2013). To date, there have been several dozen creeping bentgrass cultivars available (Honig et al., 2016; McCarty, 2018), which are still the dominant putting green turfgrass for cool climatic zones and transition zones. With the efforts in bentgrass breeding and genetic research, the development of new hybrid bentgrass cultivars is promising to improve the shortcomings of current creeping bentgrass: poor summer heat tolerance and high disease potential (Belanger et al., 2003, 2004; Bonos et al., 2006; Tian et al., 2009; Meyer et al., 2017).

It is exciting that additional grass species, annual bluegrass, seashore paspalum, velvet bentgrass, Diamond zoysiagrass, and other zoysiagrass and bentgrass hybrids and their cultivars have become new members of the putting green turfgrasses or more popular than they used to be. In the past several decades, the turfgrass breeders of the world have successfully developed close to thousands of new turf cultivars and genetic lines to meet the turf industry's demand from over a dozen turfgrass species (www.ntep.org). Therefore, the future new putting green turfgrass cultivars or species are promising.

To maintain these putting green turfgrasses, fertilizers are needed (Turner and Hummel, 1992; Carrow et al., 2001; Throssell and Lyman, 2009; Frank and Guertal, 2013a, 2013b; StJohn et al., 2013). It is much more challenging to manage these turfgrass putting greens under a stress or multiple stresses than any other types of turf, mainly due to the low mowing and often intensive traffic. Also, it is very often the case that the putting greens of a golf course are managed under a single stress, multiple stresses, temporary stress, and/ or permanent stress to overcome unfavorable growth conditions as long as the golf course is in use, despite the location and climate. With new demands and increasing resource limitations, some of these stresses have become more critical than they used to be, such as water shortages and pesticide and fertilizer restrictions. Golf course management is a multi-dimensional approach, dealing with all aspects of plant growth and culture and overcoming all kinds of plant culture stress. This chapter has two goals: 1) to focus on and update putting green turfgrasses at the species level, with an overall review of the major stresses that putting green management has to encounter; and 2) to highlight possible approaches of integrated stress management (ISM) related to nutrient management to successfully overcome or reduce putting green stresses (Liu et al., 2019, and the previous chapters in this book).

17.2 GOLF COURSE PUTTING GREENS

The sizes of individual putting greens range from less than 100 m^2 to several hundred square meters, with the common sizes between 400 and 600 m^2. Putting greens are also the most attractive areas of a golf course from the appearance point of view than any other types of turf on the golf course because of the low mowing height and smooth surface maintained by frequent mowing. However, putting greens often encounter stresses (Beard, 2002; Dernoeden, 2013; McCarty,

TABLE 17.1

Major Stresses Encountered by Putting Green Turfgrasses and Improvement Potential

Putting Green Turfgrasses	Advantages	Main Challenges	Most Needed Improvement	References
		Cool-Season		
Creeping bentgrass	Has the largest number of cultivars as a putting green species available and serves the largest number of golf courses in the world	Summer heat, diseases, weed, thatch, and management in the transition zone from cool climate to sub-tropical climate	Summer heat resistance and disease resistance	Xu and Huang (2000); Chai et al. (2002); Bonos et al (2003); Wang and Luthe (2003); Zhang and Ervin (2004); Fu et al. (2005, 2006) ; McCarty et al. (2005, 2007); Bonos et al. (2006); Honig et al. (2016); Meyer et al. (2017)
Annual bluegrass (and as a weed in creeping bentgrass as well)	Fresh green color and low mowing height tolerance	Summer heat and diseases	Disease resistance	Dest and Guillard (1987); Hagley et al. (2002); Huff (2003); Vargas and Turgeon (2004); Bertrand et al. (2009); Dai et al. (2009); Inguagiato et al. (2009);Dionne et al. (2010); Inguagiato and Guillard (2016); Ervin et al. (2017); Han et al. (2017); Mattox et al. (2017); Schmid et al. (2017); Schmid et al. (2018); Guertal and McElroy (2018); McDonald et al. (2018)
Mixture of creeping bentgrass and annual bluegrass	Lower mowing height tolerance	Summer heat and diseases, differences in N use	Disease resistance and lack of weedy appearance in either grass	Frank et al. (2018); O'Connor et al. (2018);
Velvet bentgrass	Finest texture and green color	Summer heat and some diseases	Disease resistance	Briman and Meyer (2000); Brilman (2003); Koeritz and Stier (2009); Pease and Stier (2018)
Agrostis hybrids	Improved stress resistance to environment and diseases	Unknown	Further research and identification are needed	Belanger et al. (2003); Belanger et al. (2004)
Fine fescues, and fine fescue and creeping bentgrass mixture	Lower water and nutrient input and excellent cold tolerance	Poor heat tolerance and high thatch accumulation potential	Further research and identification are needed; competition with creeping bentgrass and annual bluegrass	Chen et al. (2018); Calvache et al. (2017); Grimshaw et al. (2018)
		Warm-Season		
Hybrid bermudagrasses	Excellent summer-month performance in the transition zone, sub-tropical and tropical zones	Shade, thatch layer accumulation, winter dormancy, spring dead sport disease, salinity tolerance	Shade tolerance, disease resistance, salinity tolerance with genetic variations, thatch control	Sartain (1985); Carrow et al. (1987); McCarty and Miller (2002); McCarty (2005); Baldwin et al. (2006); Zhang et al. (2006); Baldwin et al. (2009c); Bauer et al. (2009); Hodges et al. (2016); Park et al. (2017); McCarty (2018); Van Tran et al. (2018)
Seashore paspalum	Resistance to poor water quality and salinity	Cold stress and lower mowing height, ball rolling distance	Cold resistance, disease resistance, scalping reduction	Duncan and Carrow (2000); Duncan and Carrow (2005); Lee et al. (2005); Kopec et al. (2007); Stiglbauer et al. (2013); Canegallo et al. (2017)
Diamond zoysiagrass	Shade tolerance and salinity resistance	Lower mowing height, thatch control, ball rolling distance, and diseases	Zoysiagrass patch resistance, cold tolerance	Qian and Engelke (1999); Engelke et al. (2002); Baldwin et al. (2009c); Chen et al. (2009); Stiglbauer et al. (2009); Atkinson et al. (2012); Menchyk et al. (2014); Patton et al. (2017)
Other zoysiagrass species and cultivars	Shade tolerance and salinity resistance	Lower mowing height, thatch control, and diseases	Zoysiagrass patch resistance, insect resistance	Marcum (1999); Huang et al. (2016); Chandra et al. (2017); Ebina et al. (2017)

(Continued)

TABLE 17.1

Major Stresses Encountered by Putting Green Turfgrasses and Improvement Potential

Putting Green Turfgrasses	Advantages	Main Challenges	Most Needed Improvement	References
		Winter Overseeding Turfgrasses		
Rough bluegrass	Heat sensitive and fine texture	Diseases and salinity stress	Improvements in disease resistance and water quality tolerance	Liu et al. (2001); Camberato and Martin (2004); Rajasekar et al. (2006); Volterrani et al. (2009); Espinosa et al. (2013)
Perennial ryegrass	Fast germination	Diseases and salinity	Spring transition	Volterrani et al. (2009)
Annual ryegrass	Fast germination	Disease and salinity	Improvement in texture for putting green overseeding	Volterrani et al. (2009)
Fine fescues	Lower density	Heat sensitive	Mycorrhizal fungus enhancements	Volterrani et al. (2009); Calvache et al. (2017); Chen et al. (2018); Grimshaw et al. (2018)

2018). Due to the low mowing heights of putting greens, commonly seen stresses such as heat, cold, salinity, and drought (Huang et al., 2014), are more severe for putting greens than for other areas of golf courses, even with the same turfgrass species and cultivars used.

To provide rapid drainage, the putting green root zone is often modified with sand to avoid water-logging and ensure enough pore spaces to provide oxygen for root growth. There are two basic types of sandy root zone-mix modifications for putting greens: pure sand greens, and sand and peatmoss mix greens. The pure sand root zone was first tested in California, and it is also called *California green* (McCarty, 2018). A California green has a fast water percolation rate and is suitable for rapid water drainage designed for areas with more heavy storms within a short period of time, but it has poor water and nutrient retention capability. The other type of sandy green is called *USGA specification green*, because it was first invented by the United States Golf Association, and has been revised a few times by using sand and peatmoss in different ratios by volume (United States Golf Association Green Section Staff, 1993). Normally, sand occupies 80 to 90% by volume in the sand/peatmoss mix. USGA green overcomes the shortage of rapid water percolation and poor nutrient retention, and it is suitable for the majority of areas in the world. Native soils and partially modified putting greens found under poor subsurface water drainage conditions can be used as an economical way to build new putting greens (Beard, 2002; McCarty, 2018).

The putting green management includes the primary practices of mowing, fertilization, and watering with the addition of pest control, cultivation, and chemical applications of plant growth regulators, surfactant, and bio-stimulants. Other practices are often needed to maintain putting greens, such as dew and surface water removal, winter overseeding or painting, surface rolling, winter or summer covering, and microclimatic and growth condition modification, using fans to decrease air temperature and increase air circulation, Sub-Air systems to remove excessive water and improve temperature extremes of root zones, and hydronics (www.subairsystems.

com), a sub-surface tubing system, to increase or decrease root zone temperatures by running through hot and cool water during the winter or summer season to extend putting green turfgrass growth and color (Beard, 2002; McCarty, 2018).

17.3 GOLF COURSE PUTTING GREEN TURFGRASSES

Table 17.1 lists the current putting green turfgrasses, which belong to either cool-season or warm-season turfgrasses. Creeping bentgrass is the most dominant cool-season turfgrass used on putting greens because of its fine texture, stoloniferous growth habit, and low mowing-height tolerance (Beard, 1973, 2002; Warnke, 2003; Dernoeden, 2013; McCarty, 2018). Annual bluegrass serves areas without severe summer heat in cooler climate zones than creeping bentgrass (Huff, 2003; Vagas and Turgeon, 2004). Velvet bentgrass (Brilman and Meyer, 2000; Brilman, 2003) has the finest leaf texture among all these putting green turfgrasses and is the least popular cool-season putting green turfgrass so far. It has similar characteristics to annual bluegrass, adapted to a cool climate without severe summer heat, and it may have lower disease potential than creeping bentgrass (Koeritz and Stier, 2009). Rough bluegrass is often used as winter overseeding for warm-season putting green turfgrasses in warm climate zones (Hurley, 2003). These cool-season turfgrasses are suitable for cool and cold climatic zones, including the transition zone, with the challenge of summer heat stress but with the advantage of play in the winter months without requiring overseeding, as warm-season turfgrasses do.

Warm-season putting green turfgrasses include hybrid bermudagrasses, seashore paspalum, and Diamond zoysiagrass (Duncan and Carrow, 2000; Beard, 2002; Engelke et al., 2002; Kopec et al., 2007; Stiglbauer et al., 2009; McCarty, 2018). These warm-season turfgrasses are found on golf courses in hot tropical, sub-tropical, and warm climatic zones. Warm-season turfgrasses can be used in the transition zone with the challenge of winter kills but the advantage of no heat stress during the hot summer season. Hybrid bermudagrasses and seashore paspalum have several cultivars used for putting greens. To date, Diamond zoysiagrass has been the only zoysiagrass used for putting greens in the United States as the newest member of the putting green turfgrasses. However, some other putting green zoysiagrasses have been found in its countries of origin in Asia.

17.4 COMMON STRESSES ASSOCIATED WITH PUTTING GREEN TURFGRASSES

The most common stress of cool-season putting green turfgrasses is the summer heat and high disease potential, while among the warm-season putting green turfgrasses, the main stresses vary (McCarty and Miller, 2002; Dernoeden, 2013; Huang et al., 2014; McCarty, 2018). Shade and spring dead spot disease are the most limiting factors for hybrid bermudagrass putting greens in many locations. The most serious stresses for seashore paspalum and Diamond zoysiagrass as putting green turfgrasses have not been identified because of the relatively fewer number of golf courses that use them and their shorter history as putting green turfgrasses. However, seashore paspalum is relatively less cold hardy than zoysiagrass and bermudagrass, and that limits its distribution. Zoysiagrass often has a high brown patch disease potential, and brown patch resistant zoysiagrass is not yet available. All these putting green turfgrasses have to overcome one or multiple stresses during the growing seasons. For example, these turfgrasses have adapted to low mowing heights, and low mowing heights are directly or indirectly involved with the severity of many stresses. In other words, mowing heights of 2.5 to 3.2 mm are never favorable for any types of grasses but serve well for a golf game, and there is no stress-free putting green turfgrass.

17.4.1 SUMMER HEAT STRESS

High temperature limits the growth of cool-season turfgrasses during summer in many areas in the world, because cool-season turfgrasses are C_3 plants and grow most actively within a temperature range of 15 to 25 ° (Beard, 1973; DiPaola, 1984; DiPaola and Beard, 1992; Fry and Huang, 2004; Huang et al., 2014). All nutrients are important for cool-season turfgrasses during the summer; heat stress and high-temperature stress inhibit photosynthesis, limit carbohydrate accumulation, damage cell membranes, cause protein folding, and even lead to cell death (Fry and Huang, 2004). Reduction of fertilizer input, particularly with N, is highly recommended for cool-season putting green turfgrasses during the summer months to reduce the potential for fertilizer burn and root injury. Light and frequent foliar fertilizer input ensures the summer turf quality, and all cool-season turfgrasses have the ability to acquire heat tolerance to a certain degree by exposure to a gradual increase in temperature, which happens naturally as heat acclimation (Fry and Huang, 2004; Totten et al., 2008). Under heat stress, cool-season turfgrass root systems are more negatively affected than leaves (Liu et al., 2002), and liquid foliar nutrient application may be the only choice to provide

cool-season turfgrasses with nutrients to avoid or minimize any fertilizer salt burning and injury to roots (Liu et al., 2008).

The extremes of avoiding any N and other nutrient input during the heat-stressful summer tend to further increase heat stress, leading to further carbon depletion (Fry and Huang, 2004; Shen and Du, 2009; Zhang et al., 2010). Foliar and light nutrient applications are highly recommended for highly maintained turf, including golf course putting greens. The reasons for nutrient requirement under heat stress may include 1) to maintain basic metabolism to survive and 2) to activate mechanisms for heat-tolerant cultivars to overcome heat stresses, including the formation of heat-shock proteins (HSP), hormone regulation, and antioxidant enzyme production (Taiz and Zeiger, 2006). These formations and adjustments are all directly or indirectly associated with proper nutrient supply. However, the proper nutrient supply to enhance these heat-resistance mechanisms needs to be further investigated for turfgrasses, particularly for C_3 cool-season turfgrasses (Liu et al., 2008a, 2008b).

Wang and Luthe (2003) found that heat-sensitive creeping bentgrass variants fail to accumulate chloroplast HSP isoforms. The role of chloroplast-localized small HSPs (CP-sHSPs) in heat tolerance is to provide protection to photosystem II during heat stress. The accumulation of the additional CP-sHSP isoforms is genetically linked to heat tolerance, and the presence of the additional isoforms in the heat-tolerant creeping bentgrass variants indicates that the heat-tolerant creeping bentgrass can be enhanced at genetic and biochemical levels. Additional nutrient supply may not be significant in enhancing HSP formation and expression, but nutrient deficiency may negatively affect the heat-tolerant performance of heat-resistant turfgrasses during the hot summer months in addition to carbohydrate depletion losses.

With the situation of most putting green turfgrasses cultured on sandy soil, Ca rarely becomes deficient. However, Ca, as a signal plant nutrient, has been intensively studied in recent years for its role in reducing heat stress, particularly for C_3 plants, during the hot summer months. How practically to use Ca to enhance C_3 turfgrass heat tolerance is unclear and requires further investigation (Jiang and Huang, 2001; Fu and Huang, 2003; Saidi et al., 2009).

Although there is a lack of documented research, high soil moisture content and organic matter in root zones during hot summer months may negatively affect C_4 warm-season turfgrass nutrient uptake even though warm-season turfgrasses are not under heat stress (Brecht et al., 2009). However, with lower mowing heights and high potential for thatch accumulation, smaller root systems are often found in warm-season turfgrass putting greens during the summer months, and foliar nutrient applications also benefit the nutrient uptake of warm-season turfgrasses during hot summer months. For putting green turfgrasses, heat, drought, and salinity stresses are often associated with each other due to climatic conditions and poor soil and water supplies.

For both cool- and warm-season putting green turfgrasses, nutrient management of N, K, Ca, Si, Na, and Fe (Liu et al., 2008a, 2008b; Tables 17.2 and 17.3) can be significant in

TABLE 17.2

Major Turf Stresses Associated with Fe Status and Availability

Stress/Fe Status, Recommendations, and Future improvements	Fe[a] deficiency	Fe[a] toxicity	Management Recommendations to Reduce Stresses and Enhance Fe Turf Use	Evidence or Potential for Improvement	References
Sandy alkaline soils with soil pH >7.0	>/–, </+	>/+, </–	Light and frequent foliar Fe applications	Centipedegrass, creeping bentgrass, annual bluegrass	Carrow et al. (1988); Turner and Hummel (1992); Carrow et al. (2001); Xu and Mancino (2001)
High soil P content and water-logging conditions	>/–, </+	>/–, </+	Reduce P fertilizer input, more frequent Fe applications, and reduce Fe input during germination	*Arabidopsis*, rice	Ward et al. (2008); Fageria et al. (2008); Zheng et al. (2009)
Drought/heat stress	>/–, </+	>/+, </+	Iron forms plus plant growth regulators	Creeping bentgrass	Glinski et al. (1992); Zhang et al. (2002)
Disease	>/, </–	>/+, </–	FeSO₄ suppresses dollar spot	Creeping bentgrass	McCall et al. (2017)
Cold soil temperatures and fall and winter turf color	>/–, </+	>/?, </?	Light foliar applications	Bermudagrasses, Diamond zoysiagrass, creeping bentgrass, Kentucky bluegrass	White and Schmidt (1989); White and Schmidt (1990); Zhang et al. (2002); Devetter (2008); Taylor (2019)
Imbalanced soil nutrients including higher heavy metal contents (Zn, Cu, Mn, and Cd)	>/–, </+	>/+, </?	Reduce heavy metal nutrient element input, reduce heavy metal input sources, and use Fe-efficient cultivars	Wheat, rice	Takahashi (2003); Chen et al. (2004); Ghasemi-Fasaei and Ronaghi (2008)
Black layer	>/?, </+	>/–, </+	Reduce Fe and S input and improve poor soil drainage conditions	Cool-season turfgrass putting greens	Hodges (1992); Vargas (2005); (2006, 2008)

[a] For each pair of signs separated by /, the left <, >, or 0 sign indicates that the stress severity decreased or increased, or that there was no effect, respectively; the positive or negative sign on the right side indicates with or without Fe application or toxicity level respectively. The ? sign indicates currently unknown to the authors' knowledge.

reducing heat stress, which is often combined with drought and salinity stresses during hot summer months. Approaches with plant growth regulators (Xu and Huang, 2010) and bio-stimulant input for creeping bentgrass summer heat stresses can enhance heat resistance and indirectly favors nutrient use and metabolism under heat stress (Ookawa et al., 2004; Xu and Huang, 2009; Zhang et al., 2010).

17.4.2 SHADE STRESS

In general, C_3 plants do not need whole-day direct full sunshine, while C_4 plants need almost all the sunlight they can get, with a tolerance to high solar radiation during the growing season. However, today's putting green turfgrasses respond to shade stress differently, and they cannot be simply separated based on whether they are C_3 cool-season or C_4 warm-season turfgrasses. The newest member of the putting green turfgrasses, Diamond zoysiagrass, can tolerate low light intensity, as much as 70% shade, without losing its putting green quality. With the plant growth regulator trinexapac-ethyl applied regularly under shade, Diamond zoysiagrass shade tolerance can be improved even under 90% shade (Qian and Engelke, 1999; Baldwin et al., 2009d).

Except for hybrid bermudagrasses, in general, cool-season putting green turfgrasses plus seashore paspalum and Diamond zoysiagrass can tolerate partial shade. Reducing nutrients has been recommended for both warm- and cool-season turfgrasses under shade stress because of a low ratio of photosynthesis:respiration, which reduces turf growth and divot recovery (Beard, 1997; Carrow et al., 2001). Selecting shade-tolerant species and adjusting management techniques are the two approaches commonly recommended for maintaining turf in reduced light conditions. Raising mowing height to alleviate shade stress is commonly recommended as well.

Among all the nutrients, N reduction is critical for shade management. Goss et al. (2002) compared liquid applications of N to confirm that lower N rates (150-185 kg ha⁻¹ annually) resulted in better-quality turf than higher N rates (212–235 kg ha⁻¹). Granular forms of N must be dissolved in soil first; then, they are absorbed through roots and transported to shoots of a plant, a process that could be energy inefficient by forcing roots to use their carbohydrates for energy to assimilate and transport N to the shoots (Hull and Liu, 2005). In a shaded environment, turfgrass root development and energy budgets are stressed due to low photosynthetically active radiation.

TABLE 17.3

Major Putting Green Turfgrass Stresses and Potential Association with Si Applications for the Future

Putting Green Turfgrass Stresses	Stress Severity[a]/Si Application	Forms of Si Applied to Turfgrasses or Other Crops	Evidence or Potential for Improvement	References
Salinity	>/−, </+	$CaSiO_3$, $Na_2Si_3O_7$	Sugarcane, grape, Kentucky bluegrass, bermudagrass genetic variations	Ashraf et al. (2010); Soylemezoglu et al. (2009); Esmaeili et al. (2015); Esmaeili and Salehi. 2016; Sugiura et al. (2017)
Heavy metal or element toxicity	>/−, </+	K_2SiO_3, H_4SiO_4, $Na_2Si_3O_7$	Al, As, Mn, Cd, B	Liang et al. (2005); Guo et al. (2007); Doncheva et al. (2009); Exley (2009); Soylemezoglu et al. (2009); Vaculíka et al. (2009); Liu et al (2019) (this book)
Drought	>/−, </+	K_2SiO_3, $CaSiO_3$, Na_2SiO_3	Grasses, sorghum, cucumber, St. Augustinegrass, Kentucky bluegrass	Trenholm et al. (2004); Eneji et al. (2008); Hattori et al. (2008); Sonobe et al. (2009); Bae et al. (2017)
Cold	>/−, </+	K_2SiO_3	Wheat	Liang et al. (2008)
Heat	>/−, </+	$CaSiO_3$	Bermudagrass	Brecht et al. (2009)
Winter overseeding performance	0/−, >/+	CaSi	Rough bluegrass	Espinosa et al. (2013)
Diseases	>/−, </+	H_4SiO_4, $CaSiO_3$	*Bipolaris cynodontis*, brown patch, grey leaf spot, powdery mildew	Saigusa et al. (2000); Fauteux et al. (2006); Brecht et al. (2007); Datnoff et al. (2007); Nanayakkara et al. (2008); Brecht et al. (2009); Rahman et al. (2015); Wang et al. (2017)
Insects	0/−, </+?	K_2SiO_2	*Rusidrina depravata* Butler, aphids	Saigusa et al. (2000); Ranger et al. (2009)
Turf wear stress	>/−, </+	K_2SiO_2	Seashore paspalum	Trenholm et al. (2001)
Thatch layer accumulation	?/−, ?/+	?	?	Vaughn et al. (2018)

[a] For each pair of signs separated by /, the left <, >, or 0 signs indicate that the stress severity decreased or increased or that there was no effect, respectively; the positive or negative signs on the right indicate with or without Si application, respectively. The ? sign indicates currently unknown to the authors' knowledge.

Since the majority of N is used in the shoots, increased foliar absorption from liquid applications of N may increase N use efficiency and allow more photosynthate to be allocated to the roots of the plant, enabling the turfgrass to obtain more nutrients and water (Beard, 1997). Although research reports indicate that 100% liquid N and a combination of liquid N and granular N helped improve turf quality (Totten et al., 2008), comparisons between granular and liquid N applications and forms of N on shaded turf have been rarely reported.

In addition to N, Fe has been investigated. Baldwin et al. (2009c) reported that Champion bermudagrass putting green quality can be enhanced under 55% light reduction maintained at a 3.2 mm mowing height under reduced sunlight. Fe applications had minimum effect on reducing shade stress, but trinexapac-ethyl improved shade tolerance. Reduced N input benefited shade condition survival with reduced thatch accumulation. Baldwin et al. (2009a) also identified that winter month moderate shade (<60%) was not detrimental to a creeping bentgrass green, and the effects of Fe were not significant in enhancing shade tolerance during winter months, and a similar result was found for rough bluegrass overseeded on a Champion bermudagrass putting green (Baldwin et al., 2009c).

Another technique to improve turf in shaded conditions is the application of plant growth retardants (PGRs), particularly trinexapac-ethyl, which inhibits gibberellic acid biosynthesis. In addition to reducing clipping yields, multiple applications of PGRs increased turf density and color and the quality of both cool-season and warm-season turfgrasses. In addition, increased tillering and density of turf mowed at golf green height under shade were found for both warm- and cool-season turfgrasses, but turf quality was still unsatisfactory due to low irradiance for most warm-season turfgrasses, particularly hybrid bermudagrasses, when the light reduction was beyond 60% (Qian and Engelke, 1999; Baldwin et al., 2009c; Menchyk et al., 2014).

A 2-year study was conducted at Clemson University, Clemson, SC Turf Plots to determine the tolerance of a mature Diamond zoysiagrass (*Zoysia matrella* (L.) Merr.) golf green to reduced light environments (RLE) treated with trinexapac-ethyl (TE) at 0 and 0.013 kg ha⁻¹ wk⁻¹. Three levels of RLE, 0, 60, and 90%, were imposed to coincide with the maximum seasonal growth of zoysiagrass. Plant growth measurements included turf quality (TQ), chlorophyll concentration, clipping yield, and total nonstructural carbohydrates (TNC). Zoysiagrass grown in full sunlight and 60% RLE maintained commercially acceptable TQ (≥7) with and without TE throughout Year 1, while 90% RLE treatments fell below an acceptable level 2 months after study initiation. Sixty percent RLE treatments without TE demonstrated unacceptable TQ during Year 2. The application of TE sustained turf viability in 90% RLE through Year 2. Clipping yield increased in 60% RLE without TE compared with full sunlight treatments. The application of TE to 60% RLE treatments reduced clipping yield 30–76% in Year 1. Chlorophyll concentration in 60% RLE treatments was similar 10 weeks after RLE implementation compared with full sunlight treatments in Year 1 and higher in Year 2. The TNC concentration of roots, rhizomes, and stolons was similar in full sun and 60% RLE treatments at the end of both years and between 47 and 100% lower in 90% RLE treatments compared with full sun treatments. These results support the use of Diamond zoysiagrass in up to 60% reduced light putting green environments with concurrent use of a gibberellin biosynthesis inhibiting product such as TE (Atkinson et al., 2012).

Shade is still the number one limiting factor for hybrid bermudagrass putting greens close to trees under warm climatic conditions. Severe shade is even detrimental to hybrid bermudagrasses (Bunnell et al., 2005a, 2005b, 2005c; Baldwin et al., 2009a; Hodges et al., 2016). However, the recently released new shade-tolerant hybrid bermudagrass cultivar encourages further shade tolerance improvement for bermudagrasses (Hanna et al., 2010).

There is still a lack of research on seashore paspalum shade tolerance as a putting green turfgrass, but several other studies have demonstrated that seashore paspalum has a moderate shade tolerance in comparison with hybrid bermudagrasses (Jiang et al., 2004; Baldwin et al., 2009d). Golf course putting green shade is most likely caused by trees and shade intensity, penetrating light quality from tree canopies, and tree species and varieties affecting the outcome of light quality received on putting greens. Afternoon shade is more detrimental than morning shade for hybrid bermudagrass putting greens, and there is no difference for creeping bentgrass putting greens (Bell and Danneberger, 1999; Bunnell et al., 2005a, 2005b, 2005c; Baldwin et al., 2009a). A greenhouse study demonstrated that Diamond zoysiagrass and Seadwarf seashore paspalum had better growth and appearance than both hybrid and common bermudagrasses after treatment with blue, red, and yellow tarps for 8 weeks (Baldwin et al., 2009d).

17.4.3 Disease Stress

Nutrient enhancement of putting green turfgrasses under disease stresses can be complicated, but it is promising (Hull et al., 1979; Huber and Arny, 1985; Reuveni and Reuveni, 1998; Huber and Grahan, 1999; Datnoff, 2005; Dordas, 2008; Ghorbanim et al., 2008). Nutrient enhancement does not simply mean greater quantity of input; it means proper application with an understanding of the nature of stresses. There are many more research reports available on the effects of plant nutrients on disease severity than on nutrient enhancement for diseased plants. This may be partially due to the fact that the majority of the studies were based on annual yield-production crops, and those diseased crops had much less recovery potential from the yield losses than perennial function crops such as turfgrasses have to recover from diseases. Therefore, the concept of nutrient enhancement for stressed turfgrasses is much more important and practical as part of ISM (Liu et al., 2019, and the previous chapters in this book).

Dollar spot is the most common putting green disease, and fairy rings happen on all kind of putting greens. Cool-season putting green turfgrasses have several serious diseases, including pythium blight, brown patch, snow molds, and summer patch in addition to dollar spot. So far, seashore paspalum has not been found to have a unique disease, but hybrid bermudagrass and zoysiagrasses have been affected more often by spring dead spot and large brown patch, respectively.

Due to the lower mowing height, putting greens are the type of turfgrasses most vulnerable to diseases, and the expense of preventing and controlling putting green diseases can be a significant part of the budget for golf course management. Nitrogen has been most commonly studied as a nutrient element in relation to plant diseases. Excessive N applications encourage succulent growth and delayed dormancy of turfgrass, which may increase the susceptibility to pathogens and other stresses such as winter kill or summer decline. On the other hand, nitrogen deficiency weakens turf vigor and causes greater susceptibility to pathogens. Nitrogen form, source, rate, and time of application play important roles in turfgrass diseases. The relationships between N application practice and disease severity and occurrence among turfgrasses are rather complicated by the existence of host vertical and horizontal resistances to diseases, epidemiology of different diseases, co-existence of multiple diseases, and different host recovery mechanisms. In general, two groups of turf diseases are affected by N application rate: disease severity increased and disease severity reduced, plus a group of diseases not affected by N application rate or for which information is lacking (Liu et al., 2008a, 2008b).

Although limited information is available on the mechanisms of reduction of severity of diseases by N application, turfgrasses have been reported with either improved resistance to or speedy recovery from foliar diseases, including dollar spot (*Sclerotinia homoeocarpa* F. T. Bennett), leaf spot (*Drechslera* spp.), red thread (*Laetisaria fuciformis* (McAlpine) Burdsall), rusts (*Puccinia* spp.), and foliar anthracnose (*Colletotrichum graminicola* (Ces.) G. W. Wils.). Recent studies indicate that foliar N application can even more efficiently reduce or prevent these foliar diseases.

The worst group of turfgrass diseases are soil-borne crown diseases, including pythium blight (*Pythium aphanidermatum* (Edson Fitzp.)), brown patch (*Rhizoctonia solani* Kühn), gray snow molds (*Typhula incarnata* Fr.), gray leaf spot (*Pyricularia grisea* (Cooke) Sacc.), dead spot (*Ophiosphaerella agrostis*), and spring dead spot (*Ophiosphaerella* spp.), which are easily induced or worsened by heavy N rates, particularly when inorganic quick-release N fertilizers are used. Excessive application of N promotes thinner cell walls, more succulent tissues, and lower carbohydrate reserves, which encourage the penetration of the host plants by pathogens.

The influence of the two major N forms, NO_3^- and NH_4^+, on crop diseases have been intensively studied, and the two N forms normally have opposite effects on particular diseases. The influences of N form on disease severity include impacts on the host, the pathogen, and the soil environment. The effect of the specific N form on disease severity depends on many factors. Susceptible cultivars or species are easily affected by N forms, while resistant turfgrasses are not affected by diseases regardless of N forms.

The practices of N application to turf are rather more important than the N nutrient itself for the encouragement or discouragement of turf diseases. Improper application time, rate, and N form and sources will promote disease occurrence. For example, N fertilizers have effects on soil pH, and some turfgrass diseases are soil pH sensitive. Alkaline soil conditions promote take-all patch (*Gaeumannomyces graminis* (Sacc.) Arx & D. Oliver var. avenae), and slightly acidic soils suppress most patch diseases. There is a lack of evidence on the effect of N forms on the aggressiveness of turfgrass pathogens.

Limited information is available for the interaction between fertilizer application and non-fungal pathogenic turf diseases such as nematodes, viral diseases, and bacterial wilts for turfgrasses. However, recent findings suggest that the symbiotic relationship between the host and the endophyte may provide another possible dollar spot control method. Alumai et al. (2006) reported that entomopathogenic nematodes were correlated significantly with sand, silt, P, organic matter, and Mg content, but not with clay, pH, Ca, and K.

Although quantitative data are limited, golf course putting greens have the highest disease potential, whether using creeping bentgrass, bermudagrass, or zoysiagrass, because of the lower mowing height and more intensive traffic in comparison with any other type of turf. Nitrogen plays important roles in the recovery of turf from physical damage. Typically, golf course tees and sports fields receive more damage than golf putting greens, but tees and sports fields can recover more quickly than greens due to the greater mowing height and much lower pressure for diseases. The better reserves in roots, crowns, and other parts of higher-mowed turf benefit from quicker recovery from damage with a relatively lower N requirement compared with a lower and more frequently mowed turf. Due to the controversial responses of turf diseases to N, the zone for optimum N application is very restricted for most turfgrasses (Liu et al., 2008a, 2008b). In the middle zone, there is still a group of diseases independent of N status and supplies or for which information is lacking.

Nutrient management with other minerals has been applied to reduce crop and turfgrass diseases or as a remedy to cure crop or turfgrass diseases, including copper, manganese, silicon, and potassium, by either enhancing turf resistance to pathogens or suppressing pathogen infections. However, nutrients themselves may never be an efficient tool to cure turfgrass diseases directly, but the complicated roles of nutrients in different diseases deserve much more attention for management and future turfgrass improvement. For example, a selected genetic line of creeping bentgrass with natural resistance to dollar spot may also have greater capability to absorb N efficiently by roots and leaves. Nitrogen effects on non-pathogenic microbial activities exist, and these effects may have an influence on the turf quality.

17.4.4 WEED STRESS

Turfgrasses compete with weeds for nutrients, and different soil nutrient levels also favor one plant species over another. It is hard to use soil nutrient levels and fertilizers as remedies to control weeds in turf, and it is unlikely to be a main tool in weed control. The difference of nutrient uptake efficiency between a weed and a turfgrass species or cultivar can be further enhanced. There are generally three aspects to enhance nutrient use efficiency in weed suppression or control for turf: 1) implementing sound nutrient programs to maximize turf vigor to reduce weed potential; 2) applying different levels of nutrients to favor turfgrass and suppress weed species by understanding the nutrient requirement differences between turfgrasses and weed species; and 3) generating toxic nutrient or chemical levels (non-herbicide chemicals) to control weeds without damaging the turf species.

Weeds can take advantage of a turf if they have better resistance to stresses, including nutrient deficiencies. The combination of proper nutrient balance associated with other cultural practices ensures long-term turf vigor, which may minimize the potential for weeds. It is very common for weed invasion to start when a turf is under stress or during its early establishment. The documented research is limited on competition between turfgrass species and cultivars and weed species, and among multiple turfgrass and weed species regarding nutrient forms, supplies, and sources. However, under N deficiency, legumes that can fix N, such as clovers (*Trifolium* spp.), black medic (*Medicago lupulina* L.), and common lespedeza (*Kummerowia striata* (Thunb.)), along with grassy weeds such as bromesedge (*Andropogon virginicus* L.), quackgrass (*Agropyron repens* L.), bahiagrass (*Paspalum notatum* L.), and centipede grass (*Eremochloa ophiuroides* (Munro) Hack.), can be more troubling weeds for the turf. These weeds require very little N input and can survive well under N deficiency.

Under nutrient deficiency, turfgrasses lose their vigor to compete with weeds, and this becomes worse under stresses of drought, temperature extremes, nutrient deficiency, other pests, and traffic impacts. In the presence of weeds, the nutrients actually available to the turfgrass will be reduced depending on the degree of weed coverage. It is most likely that the worse the weed coverage, the worse will be the nutrient stresses to the turf, in addition to the differences between the nutrient uptake capabilities of the turfgrass and the weeds. Some broadleaf weeds may have further advantages due to deeper tap root systems for nutrient acquisition than grasses having relatively shorter fibrous root systems. Although there is a lack of research, it is not difficult to predict that some weeds may have the advantage of being able to absorb nutrients from soils at very low concentrations (<0.5 mM), at which some turfgrass species may already stop absorbing nutrients. Furthermore, recent studies by Hossain et al. (2004) and Sistani et al. (2003) showed aggressive N and phosphorus (P) uptake by the grassy weeds torpedograss (*Panicum repens* L.) and southern crabgrass (*Digitaria ciliaris* (Retz.) Koel.). The capability to accumulate N in the green tissues also further indicates the high N uptake efficiency of crabgrass species.

Optimal to excessive N supplies encourage the occurrence of several weed species, including annual bluegrass (*Poa annua* L.), bentgrasses (*Agrostis* spp.), bermudagrasses (*Cynodon* spp.), ryegrasses (*Lolium* spp.), and crabgrasses (*Digitaria* spp.) plus several annual broadleaf weeds. Encroachments between turfgrass species are also encouraged by a favorable nutrient supply for the more aggressive species in addition to other conditions and their own characteristics. For example, the more commonly observed encroachment is the invasion of creeping bentgrass putting greens by surrounding bermudagrass; however, the opposite invasion has been observed, partially due to an aggressive N fertilization program for the creeping bentgrass green versus the N-stressed bermudagrass surrounding it. Excessive N supply to turf promotes shoot, stolon, and rhizome growth with relatively smaller root systems, lower carbohydrate reserves, more succulent tissues, and greater vulnerability to some diseases and insects, which contribute to weakened vigor for the turfgrass to compete with weeds.

Turfgrasses also can lose their competitiveness with weeds under excessive nutrient supplies. At high external nutrient ion concentrations (normally >1.0 mM), the Vmax plateau of ion uptake kinetics is exceeded, as a low-affinity transport system (LATS) becomes functional. Therefore, the uptake rates increase in a linear function with ambient nutrient concentrations (between 5 and 100 mM) and exhibit no evidence of saturation kinetics. Some weeds have superior LATS systems to those of turfgrasses for absorbing nutrients when excessive nutrients are applied. Mowing practice adds complexity to nutrient effects on turfgrass and weed competition. Although both turfgrasses and weeds are normally encouraged by adequate nutrient input, proper mowing practices enhance turf quality and suppress weeds more significantly, particularly broadleaf weeds, depending on the turfgrass species and the weed species.

Turf thatch accumulation is encouraged by excessive N applications. Thatch normally provides a physical barrier to weed invasion, particularly from seed germination. Under proper nutrient supply, thatch functions positively to suppress weed invasion. A turf becomes more vulnerable to weed invasion after aerification. Weeds germinate much more easily on bare soils than on a turf with proper moisture and temperature. Some turfgrass species may naturally have allelopathic effects on weeds by producing toxins or having aggressive growth habits. However, there is a lack of evidence on nutrient roles in either the allelopathic effects of turfgrasses or the relationship between nutrient supply and enhancement of allelopathy.

Nutrient sources and forms also have ecological impacts on turf and weed competition. In general, annual summer grassy weeds will be encouraged by inorganic and quick-release fertilizers. Perennial grassy weeds with the closest similarity to turfgrasses are more difficult to control by cultural practices. Ammonium-type N fertilizers and acidic fertilizers have been encouraged for use with the potential to reduce several patch diseases in turfgrasses. Excessive application of ammonium-type and acidic fertilizers may lead to a lower soil pH with reduced available P, Ca, and Mg. Under such soil conditions, annual bluegrass as a weed is discouraged, since it needs more P and a more neutral soil pH than the desired creeping bentgrass.

Nutrients cannot serve as a remedy to control weeds effectively for turf, but a proper fertilization program plays an extremely important role, associated with other cultural practices to maintain strong turf vigor for healthy conditions to minimize weed invasion. Corn gluten meal has been used for both grassy and broadleaf weed control by suppressing weed seedling germination. Corn gluten meal on average contains 60% protein and 10% N by weight and also functions as an N fertilizer. The amino acids as dipeptide forms in corn gluten meal mainly inhibit new weed seedling root germination and growth, with negligible phototoxic effects on the existing turf. Brosnan et al. (2009) reported that a single granular application of fine salt (99% sodium chloride, 1% sodium silicoaluminate) at a rate of 1464 kg ha⁻¹ provided 84% and 23% control of sourgrass (*Paspalum conjugatum*) 6 weeks after initial treatment in seashore paspalum turf in Hawaii. Using sodium chloride (NaCl) as a weed control may only apply to salinity-tolerant turfgrasses such as seashore paspalum, which can tolerate salinity levels as high as 54 dS m⁻¹, a level at which most turfgrass species could not survive (Duncan and Carrow, 2000). Wiecko (2003) reported that applications of saline ocean water (electrical conductivity [EC] = 55 Ds m⁻¹) controlled large crabgrass (*Digitaria sanguinalis* L.) in seashore paspalum, but repeated applications of saline ocean water did not control yellow nutsedge (*Cyperus esculentus* L.).

Dandelion (*Taraxacum officinale* Weber), a very frequent broadleaf weed worldwide on turf, showed different responses to N and K. Johnson and Bowyer (1982) reported that after 4 years of fertilization treatments, dandelion cover was reduced when fertilized with a higher rate of N of 600 kg ha⁻¹ year⁻¹ than at a lower rate of N of 300 kg ha⁻¹ year⁻¹ in Kentucky turf, while Tilman reported that dandelion was a poorer competitor for potassium than cool-season turfgrasses such as *Festuca* species.

Fertilization time and methods affect weed and crop competition (Blackshaw et al., 2004a, 2004b; Blackshow et al., 2007; Blackshaw and Molnar, 2009). Dunn et al. (1993) reported that in the northern transition zone, N fertilizer applications later than September encouraged winter annual weed emergence during the time that zoysiagrass would normally become dormant. In addition, any fertilization during the slow growing season for the turfgrass may be more favorable for weeds (Busey, 2003).

Excessive nutrient supply can be toxic to plants by interrupting regular physiological metabolism. The sensitivity to nutrient toxicity between turfgrasses and weeds deserves more attention. There is a lack of literature on the synergetic effect between plant nutrients and herbicides for turfgrass weed control; even many herbicides contain nutrient elements. However, an antagonism may happen when nutrients and herbicides are co-applied. Scroggs et al. (2009) reported that in field studies, weed control was greatest when glyphosate was applied alone, with control of barnyardgrass, browntop millet, and Palmer amaranth ranging between 93% and 95%. When glyphosate was co-applied with formulations of zinc sulfate, the control of these three weed species was reduced to 39%, 39%, and 45%, respectively. These results indicate that glyphosate-based weed control is reduced when it is co-applied with zinc products at their current use rates. Antagonism of the herbicidal performance of glyphosate by other cations such as Fe, Ca, Mg, Na, and K has also been observed (Mueller et al., 2007).

Future nutrient-related research will illuminate areas such as turf–weed ecology with a better understanding of differences in nutrient uptake efficiency and nutrition physiology between turfgrasses and weeds; the impacts of nutrient application rates, fertilizer forms and sources, and time and method of application on the competition between turf and weeds; and helpful nutrient management programs for turfgrasses under weed stresses.

One of the challenges of creeping bentgrass putting greens is the invasion of annual bluegrass. O'Connor et al. (2018) recently reported on when both grasses were fertilized through either foliar or soil application with either urea or ammonium sulfate. Ammonium sulfate resulted in greater overall growth for both species. Foliar application resulted in greater shoot growth for annual bluegrass, and soil application resulted in greater root growth for creeping bentgrass. However, no statistical difference was observed in the stomatal number between the two species, but annual bluegrass possessed more aqueous pores. As an annual winter weed, annual bluegrass leaves may not have the same leaf cuticle layer characteristics as creeping bentgrass does, affected by foliar and granular fertilizers (Totten et al., 2008; Stiegler et al., 2011, 2013; Bethea et al., 2014; Xiao et al., 2018).

Quick-release nitrogen fertilizers have been applied to overseeded warm-season putting greens to speed the spring transition, and the fertilizers will speed up the spring green-up of the warm-season turfgrass and suppress the overseeded cool-season turfgrass with high N fertility in high spring temperatures.

17.4.5 SALINITY STRESS

Salinity is an important growth-limiting factor for most turfgrasses, which are non-halophytic plants. Excessive salts in soils inhibit turfgrass growth by osmotic stress, nutritional imbalance, and specific ion toxicity in addition to the structural damage to soils. Soil salinity can be progressively exacerbated by practices such as irrigation and fertilization, especially in arid regions. The proper use of N fertilization in saline soils is important to sustain the N supply for turfgrasses. On the other hand, over-fertilization with N may

contribute to soil salinization and increase the negative effects of soil salinity on turf performance.

The two major salinity sources for turfgrasses are the natural salinity conditions of soils and irrigation water with a high salt content applied to turfgrasses. The relationship between turfgrass salinity stress and N fertilization deserves much more attention because of the total areas of turf under salinity stress and the complicated soil chemical interactions among Na^+, Cl^-, NO_3^-, NH_4^+, other N forms, cations, and anions in soils. Turfgrasses demonstrate diversified salinity tolerance, and salinity-tolerant plants seem less affected in N uptake and metabolism than salinity-sensitive plants.

Putting green turfgrass responses to salinity may change with mowing height, turf use, seasons, and nutrient input. To date, there is a lack of proper specific fertilizer recommendations for each putting green turfgrass for either salinity-sensitive or tolerant turfgrass species.

Studies of turfgrass growth responses to N and soil salinity for different turfgrass species and turf use are important to reveal whether the amount of N applied alleviates or aggravates the detrimental effects of salinity. In addition, examining turfgrass growth during different seasons associated with other stresses may provide profound information to practitioners on how to enhance salt tolerance over time, such as winter overseeding and association with other pest stresses. Under salinity stresses, both NO_3^- and NH_4^+ uptakes were affected in tall fescues. Ammonium compared with $NO3^-$ nutrition is known to increase the salinity sensitivity in maize, wheat, and peas. Under both N and salinity stress, shoot growth is much more reduced than root growth.

Van Tran et al. (2018) investigated salinity tolerance among a large range of bermudagrasses (*Cynodon* spp.) relative to other halophytic and non-halophytic perennial C_4 grasses. They found that there was a wide range of genetic variation for salt tolerance in bermudagrasses, and the best salt tolerance was selected from saline natural environments. They also indicated that there was no correlation between salt tolerance and drought resistance. They used salinity levels above 20 dS m^{-1} for 8 weeks for detecting a large variation in salt tolerance in bermudagrasses.

17.4.6 LOW-TEMPERATURE STRESS

It has been well documented that increased cold acclimation could improve the freeze tolerance of turfgrasses and other crops, but the rate and level of cold acclimation primarily affected by temperature are also influenced by factors such as light intensity, day length, cultural practices, and other abiotic stresses such as drought and salinity. Turfgrasses possess various adaptive mechanisms for surviving freezing temperatures, such as increases in certain sugars or amino acids, the synthesis of novel proteins, and the rate of renewal of membrane lipid fatty acids. Cold stresses for turfgrasses include winter month N management for both cool- and warm-season turfgrasses. When soil is colder than the optimum temperature range for root activities, N uptake is significantly reduced, the demand for N falls to the minimum, and plant growth ceases. However, turfgrasses are perennial crops, and winter green color is ideal. Several reports stated that late-season N improved the fall and spring color of bermudagrass and had little effect on TNC levels, without any negative effects on post-dormancy recovery in the spring. On the other hand, aggressive late fall season N applications to turfgrasses may increase the potential for disease, including spring dead spot for bermudagrass, gray snow molds for creeping bentgrasses, and winter kill potential for warm-season turfgrasses.

Observations for two winters showed that foliar and lower-rate N enhanced the turf quality of overseeded rough bluegrass on bermudagrasses more than 100% granular N fertilizers. Warm-season turfgrasses possess various N-rich defensive metabolites and enzymes to cope with low-temperature stress, including some proteins. The protein content of Midiron bermudagrass during cold acclimation was correlated with its superior freezing tolerance compared with Tifway, and cultivars with greater stolon proline content exhibited greater freezing tolerance than those with less proline during the winter. Zhang et al. (2006) compared Riviera (cold-tolerant) and Princess-77 (cold-sensitive) bermudagrasses, which were either subjected to cold acclimation at 8/4°C (day/night) with a light intensity of 200 μmol m^{-2} s^{-1} over a 10 h photoperiod for 21 d or maintained at 25/23°C (day/night) with natural sunlight in a glasshouse. Cold acclimation induced accumulation of sugars, proline, and TNC in both cultivars, but protein accumulation was found only in Riviera and not in Princess-77. Superoxide dismutase (SOD) increased during the first 7 d and then declined, while catalase (CAT) and ascorbate peroxidase (APX) activity decreased in response to cold acclimation in both cultivars. Significant correlations of LT50 with sugars, proline, protein, CAT, and APX were obtained in Riviera, but only with proline and the antioxidant enzymes in Princess-77. These results suggest that screening cold-tolerant cultivars among bermudagrasses with rapid accumulation of C- and N-rich compounds during cold acclimation is possible. However, the direct N nutrient impacts were not included in these reports, and future research is needed. Numerous reports have stated that unsaturated membrane lipids play a significant role in both chilling and freezing tolerance, with little or no effect from saturated fatty acids. There is no direct evidence of N nutrient levels in relation to the saturation and turnover rate of membrane lipids, although transgenic plants have further demonstrated the importance of unsaturated membrane lipids.

17.4.7 Drought and Water-Logging Stresses

Due to the intensive management practices of putting greens, drought and water-logging situations are normally temporary stresses, which may last hours before the situation is improved. Any type of putting green may not survive a hot summer day without any irrigation, and daily irrigation is essential for putting greens during the hot summer months. However, water-logging situations may last longer, depending on the natural precipitation. The unhealthy relationships between water and nutrients for putting green turfgrasses due

to improper nutrient management or restrictions can be summarized as follows:

- Drought with excessive nutrient input
- Drought under nutrient deficiency
- Water-logging with excessive nutrient input
- Water-logging under nutrient deficiency

If these situations exist as well as temperature extremes and salinity stress, extended drought or water-logging periods can be detrimental to putting green turfgrasses. Nutrient availability and uptake by roots are affected under these unfavorable water conditions in combination with other factors. The general practice for enhancing nutrient use under water stress is to protect root systems and minimize the damage. Light and frequent fertilization is recommended under drought conditions.

When a putting green is under water stress, reduced fertilizer input can alleviate drought stress, and excessive nutrient input will worsen the situation due to increased osmotic stresses in soils. Turfgrasses under drought stress still need nutrient input, but at a reduced rate. The degree of reduction of the nutrient rate will depend on the severity of drought, the turfgrass used, and other environmental conditions. Drought periods can last from hours to months even in yearly or multiple-year cycles, but these longer periods of drought may affect other types of turf more than putting greens, and as mentioned earlier, daily irrigation is needed during the hot summer months. Any approaches to reducing drought stress to putting green turfgrasses will be favorable for nutrient use efficiency, and among the nutrients, N, K, and Ca have been reported. Bian et al. (2009) reported that with trinexapac-ethyl (TE) treatment of L-93 creeping bentgrass, during the later phase of 28 days of drought stress, TE-treated plants had increased accumulation of soluble sugars and inorganic ions (Ca and K) in the leaves. McCullough et al. (2007) reported that TE application can reduce N input for both L-92 creeping bentgrass and Tifeagle bermudagrass putting greens in a transition zone location, indicating the potential of controlled N use combined with plant growth regulators for both cool-season and warm-season putting green turfgrasses, although there is a shortage of reports on other cool-season and seashore paspalum and zoysiagrass putting greens.

Water-logging may cause double damage to putting green turfgrasses: they may lose root functions to effectively absorb nutrients, and nutrients can be lost by leaching from the soil. For nitrogen losses, denitrification can be more significant under water-logging conditions during warmer days than on cooler days. During the water-logging period, any nutrient application to soils is pointless, and the critical time is post-water-logging management with proper nutrient supply to replace the nutrients lost during the water-logging stress. Extended water-logging periods are more likely to result in nutrient deficiency, particularly N deficiency, in addition to the root damage, and this damage varies among creeping bentgrass cultivars (Jiang and Wang, 2006; Wang and Jiang, 2007; Tan et al., 2017). Due to the weak root systems immediately following a water-logging period, light and foliar N application is encouraged, with caution during high temperatures.

17.4.8 Traffic, Ball Mark, Divots, Thatch Layer, and Organic Matter Stresses

An excessive thatch layer and organic matter content in putting green soils have negative effects on the soil profile, such as reductions in hydraulic conductivity, decreased water infiltration, increased localized dry spots, reduced tolerance to temperature extremes, increased disease and pest problems, and reduced pesticide and fertilizer effectiveness (Beard, 1973, 2002; Turgeon, 2008; McCarty, 2018). Although there is a lack of documentation on the relationship between nutrients and thatch and organic matter control for putting greens, it is believed that excessive N input plays negative roles in thatch and organic matter accumulation for both cool-season and warm-season turfgrasses (Engel and Alderfer, 1967; Carrow et al., 1987, 2001; McCarty, 2005, 2007). In addition to N, excessive input of other nutrients, such as K, P, and Ca, to putting greens may have negative impacts on either thatch accumulation or thatch decomposition (Callahan et al., 1998). Ledeboer and Skogley (1967) reported that in a field study on 15-year-old velvet bentgrass putting turf, sucrose treatments decreased turf quality and modified the stimulating effects of fertilizer. Lime had no influence on turf quality. Sucrose tended to increase the incidence of dollar spot disease, while fertilizer decreased it. They found that the physical thatch structure showed that leaf sheaths were more resistant to decay than clippings or sloughed leaves. Nodes and crown tissues were most resistant to decomposition. To date, there are still mixed findings on thatch and organic matter content control of both cool-season and warm-season turfgrass putting greens, related to control methods such as coring, vertical mowing, and others; frequency of coring and others; topdressing practices; and chemical thatch control methods. However, it has been agreed that coring and vertical mowing are the two effective ways to remove thatch or organic matter from putting green root zones, and vertical mowing interferes less with playability (Samaranayake et al., 2008; Fu et al., 2009; Rowland et al., 2009; McCarty, 2018).

All types of turf require cultural and cultivation practices to improve soil compaction, reduce thatch accumulation, and enable recovery from wear, ball marks, and divots for sports and golf course turf use. The frequency, intensity, and type of cultivation heavily depend on the turf use, the turfgrass species and cultivar, and the environmental conditions. Golf course putting greens, tees, and fairways require hollow tine core cultivation at least once a year (in most cases, multiple cultivations per year) plus other types of cultivation to improve the soil and root zone conditions and reduce thatch accumulation (Beard, 2002; McCarty, 2018). Cultivation is also recommended once per year for sports fields and lawns. Nitrogen and other nutrients play a significant role in helping turfgrass recover from the temporary disruption caused by cultivation. Turfgrass cultivation includes hollow tines, solid tines, slicers, groomers, hydro-jet (high-pressure water beams vertically penetrating turf soil profile), and vertical cutting. These processes temporarily stress the turf, and proper nutrient supplies help the turf to recover. Soil cultivation with opening holes provides better contact between fertilizers and roots, and controlled-release N fertilizers plus other nutrients are normally recommended. However, if cultivation is conducted under stress, more caution is needed with the proper nutrient supply. In many such cases, light and frequent nutrient applications help the turf grow evenly. Proper nutrient application can minimize the time required to resume playability for golf courses and sports fields. Nutrient deficiencies, particularly of N, slow the recovery from cultivation disturbance or damage.

The positive effects of plant nutrients on turfgrass wear tolerance and recovery have been reported (Trenholm et al., 2001; Hoffman et al., 2010a, 2010b), and there is a lack of reports on putting green wear studies in general, particularly the wear associated with turf maintenance mechanical practices and traffic stresses.

17.4.9 Insect and Mite Pest Stresses

Insects and mites require high dietary N to meet their need for amino acids and proteins. Normally, high N contents in plants benefit insect and mite pests. High N fertility levels result in rapid growth and succulent tissues, which may also increase the chance of damage by insect and mite pests. Excessive N fertilization associated with improper mowing, watering, and fungicide application encourages the accumulation of a thatch layer in some turfgrasses. A thick layer of thatch is a good habitat for chinch bugs (*Blissus* spp.), billbugs (*Sphenophorus* spp.), sod webworms (*Crambus* spp.), two-lined spittlebugs (*Prosapia bicincta* (Say)), and several other insects and mites. Heavy N fertilization of turfgrasses stimulates rapid leaf and shoot growth at the expense of root growth and with relatively reduced root growth and development. In such a situation, the turfgrass will have to overcome more severe stress when the environmental conditions are not favorable for roots, such as the summer time for cool-season turfgrasses and the cool spring time for warm-season turfgrasses. In addition, turfgrasses with poor root systems will suffer more severe damages from subsurface feeders. On the other hand, a slightly higher N supply may encourage foliar growth and speedy recovery from damage by foliar feeders.

Endophyte infections with some turfgrasses, including *Fescue* and *Lolium* species, provide one of the most important natural means of resistance against surface feeders. It may be that subsurface feeders and N deficiency of the host turfgrasses can discourage the symbiotic relationship, since most of the alkaloids produced between the endophyte and the hosts are N-dependent secondary N metabolites. There is a lack of information on the influence of the N form on turf insect and mite pests, including earthworms as pests. However, recent reports indicate that black turfgrass ataenius (*Ataenius spretulus* (Haldeman)) density was encouraged by using N fertilizers as sewage sludge, but there was no effect when manure-rich N fertilizers were used. Turfgrass species with resistance to turf insects and mites vary at both the species and the cultivar level, and the resistance includes speedy recovery, which is closely related to the efficiency of nutrient acquisition and production of the N-related metabolites.

Multiple-factor ecosystems including the host N status, the pest, and its natural enemy are complicated and interesting, since plant N status can affect insect consumption rates and the population dynamics of herbivorous insects and mites. Changes in such ecosystems depend rather heavily on carbon deposit in addition to the plant's N status. For a turfgrass ecosystem, multiple factors are involved, including constant disturbance of insect and mite pests by mowing and traffic.

The development of insect-resistant turfgrass species or cultivars is promising, since diversified forms of resistance to insect and mite pests exist among turfgrass species. The resistant turfgrass species or cultivars need a sufficient N supply to produce enough toxins, which is most likely the case for insect and mite resistance, and unhealthy turfgrasses due to either N deficiency or excess will significantly lose resistance.

17.4.10 Nutrient Loss and Misbalancing Stresses

Golf course putting greens have been constructed with dominant sand and organic matter mixtures as the root zone materials, which are more favorable for proper drainage, root oxygen supply, and reduced traffic compaction (McCarty, 2018) than native soils. Such artificial putting green root zone media often face leaching, loss, and misbalancing of nutrients with excessive nutrient application and irrigation and precipitation inputs. A monitoring system for putting greens is the key to reducing such stresses.

Putting green granular fertilizer applications raise concern about the quantity of fertilizer granules collected from putting greens by mowing the first time after a granular fertilizer application (Mancino et al., 2001; Throssell et al., 2017). Immediate irrigation after granular fertilizer application to putting greens and a delayed next mowing event seem to be logical solutions for minimizing such nutrient losses from putting greens.

Turfgrass grow-in on sand-based putting greens usually incurs a high risk for nutrient leakage. Aamlid et al. (2017) evaluated how the substitution of a standard mineral fertilizer with an amino-acid-based fertilizer affects creeping bentgrass establishment rate and the concentration of nitrate and total N in drainage water using the United States Golf Association green field lysimeter facility at Landvik, Norway. The liquid fertilizers used were arGrow Turf (70% of N as arginine and 30% as lysine) and Wallco (60% of N as nitrate and 40% as ammonium) applied at 2-week intervals at the two rates of 1.5 and 3.0 g N m^{-2} for each application. They found significantly faster grow-in on plots receiving amino-acid-based fertilizer than on plots receiving mineral fertilizers.

17.5 SELECTED NUTRIENTS WITH IMPACTS ON PUTTING GREEN TURFGRASSES UNDER STRESS

Putting green management is different from traditional crop systems, and the relative importance of nutrient management of putting greens will depend on the turfgrass species and cultivars,

the location, and the maintenance levels. However, the general common characteristic associated with nutrient management is the challenge of sandy soils with relatively low cation exchange capacity (CEC) and frequent cycles of water saturation and dryness. Nutrient toxicities of putting greens are rarely reported unless recycling water is used continuously. The nutrient management of putting greens under stress has not been fully investigated, and many areas remain unclear. In the following sections, these nutrients are grouped according to the frequency of the associated stress problems observed by the authors. Exclusion of nutrients from this section is purely due to the length limitations of the chapter, and the authors have included an update on two beneficial nutrients, silicon and sodium.

17.5.1 The Group with Greatest Impact

17.5.1.1 Nitrogen

Nitrogen is the nutrient needed in the highest quantity for all green plants and crops, including turfgrasses, and it has been the most actively investigated of all turf nutrients due to its significant roles in turf growth, color, quality, and use. Nitrogen is a constituent of almost all compounds in plants except carbohydrates. These compounds include proteins, chlorophyll, hormones, nucleic acids, and secondary metabolites. Under N deficiency, turf stresses can be worsened, and N plays the most significant role among all nutrients in the recovery of turf under stress (Liu et al., 2008a, 2008b). On the other hand, excessive N input can increase turf stress severity, including lower resistance to extremes of environmental conditions for growth and higher potential damage from pests. Excessive N may very often be even worse than slight N deficiency for managing turf under stress, regardless of the potential N loss from turf-soil systems.

Putting greens require a higher N input than any other type of turfgrass, mainly due to the frequent removal of clippings. Despite the impossible solution of returning clippings to putting greens, the challenge to research is to find alternative solutions to save nutrient inputs to putting greens.

The relationship of putting green N input and ball rolling distance (BRD) still remains debatable with various putting green turfgrasses. Kopec et al. (2007) investigated SeaIsle 2000 seashore paspalum (*Paspalum vaginatum* Sw.); when it was maintained as a putting green surface, they found that root weights were not affected by mowing height or applied N fertilizer rates. Clipping dry weight generally increased with shorter mowing height and higher levels of applied N. There was a trend for leaf Fe levels to decrease with increasing mowing height. Leaf tissue N increased in response to increasing levels of N application. They also found that BRD was largely unaffected by N fertilization but was consistently affected by mowing height and rolling. With 2 years of repeated experiments, they found that mowing height significantly affected BRD values on seven of eight test dates in 2000 and on all 12 in 2001, with the greatest BRD occurring at the shortest mowing height. The maximum observed BRD was 277 cm. The mean BRD on rolled and unrolled turf surfaces mowed at the 0.3 cm height was 234 and 214 cm, respectively.

Pompeiano and Patton (2017) studied growth and root architecture responses of zoysiagrass to changes in fertilizer nitrate:urea ratio using a greenhouse experiment to see how zoysiagrass responds to nitrate and urea as N sources. *Z. japonica* cv. El Toro (fast-growing) and cv. Meyer (slow-growing) and *Z. matrella* cv. Zorro (fast-growing) and cv. Diamond (slow-growing) were chosen for this study. Plants were clonally propagated as phytomers and established in 21 cm deep cone-tainers filled with a sand-based growth medium. The treatment consisted of a modified, half-strength Hoagland's solution with five different nitrate:urea ratios (100:0, 75:25, 50:50, 25:75, 0:100). The cultivars had different above- and below-ground biomass and root architecture traits. El Toro had the highest total biomass production (2.083 g plant^{-1} dry weight [DW]), while Meyer and Zorro together averaged 0.734 g plant^{-1} DW (65% less than El Toro). Diamond was the least productive in terms of leaf, culm, and root biomass (0.278 g plant^{-1} DW; 87% less than El Toro). They also found that above- and below-ground DW production was greatest following treatment with 25:75 nitrate:urea, whereas 100% nitrate produced plants with the lowest DW, and zoysiagrass rooting traits were only minimally influenced by N source. However, as the concentration of urea increased, slight increases in root surface area and volume were observed, accompanied by a decline in the root volume ratio; these were also found by Menchyk (2012) when using urea as the sole N source in a greenhouse study.

Nitrogen sources and rates applied to turf have been reported controversially regarding turf diseases due to the complexity of interactions between the host plant and the plant pathogen with plant nutrients (Datnoff et al., 2007; Frank and Guertal, 2013a). Spring and fall ammonium sulfate applications have been shown to increase zoysiagrass green cover during spring large patch (*Rhizoctonia solani* AG 2-2 LP) active periods compared with calcium nitrate treatments and urea applications, although the results were inconsistent between trial locations and dates (Miller et al., 2016; Koehler and Miller, 2017). Ammonium sulfate had a negative impact on large patch pathogen growth when compared with calcium nitrate in both laboratory and field studies.

Cho et al. (2017) examined the relationships between fructan concentration and macronutrient (N, P, or K) concentration in clippings from creeping bentgrass putting greens across >200 golf courses in Japan over an 8 year period and found that the leaf N concentration was negatively correlated with fructan concentration in each month. Such findings may partially support and explain the hypothesis of the negative relationship between excessive N supplies and stress tolerance for cool-season turfgrasses.

17.5.1.2 Potassium

Potassium (K) is an essential macronutrient for turfgrasses, and the high demand for K when growing putting green turfgrasses is generally recognized by golf course superintendents. Potash (K_2O) fertilizers are often applied to putting greens to maintain the turf quality, and as a matter of fact, as much K as N has been applied to putting greens (Throssell

and Lyman, 2009). However, even for a well-fertilized putting green, K deficiency can frequently occur due to the predominantly sandy soils of putting greens with lower CEC and high potential for leaching. With increased costs of fertilizers, the nutrient use efficiency of putting greens has become more important for turfgrass nutrient management.

Under salinity stress, the efficiency of K use for putting green turfgrasses can be critical, and Na/K ratios play significant roles in salinity tolerance for many crops. However, there is a lack of information on specific recommendations for K for putting green turfgrasses with poor water quality and salinity stresses. An average of 200 to 300 kg ha^{-1} K has been applied to putting greens annually, and there is still room for further K efficiency improvement, although root zone K toxicities are rarely reported. Liquid and foliar K applications to putting greens require caution to avoid fertilizer burning, which is often associated with drought and heat stresses.

Under other stressful conditions, K supply is critical in many aspects for turfgrasses. Regardless of the positive impacts of proper K input on stresses such as cold, salinity, drought, and diseases, K input to a cool-season or warm-season turfgrass putting green differs according to rate and application method. Significant discrepancies in K effects related to stresses are found in the literature. Part of the reason may be the variable study conditions used and the different turfgrasses in addition to a lack of uniformity of control of stressful or favorable factors in the soils, media, and environment. However, as mentioned, direct K toxicities to plants have rarely been reported, and research on K interactions with other nutrients under stressful conditions for putting greens has not fully started yet.

Mirmow (2016) reported that traffic treatments, especially those applied in the morning at 08:00, had a significant impact on creeping bentgrass health and appearance. Supplemental K did not show any significant effect on turfgrass health. Even with the damage that occurred in the colder winter months, the turfgrass had recovered to acceptable quality by the spring-time months when temperatures warmed. After a long-term 6 year study on a creeping bentgrass putting green with no positive effects from K, Bier et al. (2018) questioned the accuracy of soil K analyses and additional input of K by regular topdressing, which also agreed with a hybrid Tifgreen bermudagrass putting green study by Goatley et al. (1994). Schmid et al. (2018) reported that K fertilization can reduce anthracnose severity, and that the severity of anthracnose (*Colletotrichum cereale* Manns sensu lato Crouch, Clarke & Hillman) can be correlated with mat and tissue K concentrations of annual bluegrass (AGB) turf based on 3 year field research; it is important to monitor K in ABG turf and to maintain sufficient levels to reduce disease severity and improve turfgrass quality. All K sources were effective at reducing anthracnose severity compared with N-alone fertilization; however, KCl was not as effective as KNO_3, K_2CO_3, or K_2SO_4. They also found an interaction between the effects of N and K on anthracnose disease severity in their previous study (Schmid et al., 2017).

17.5.1.3 Phosphorus

In plants, the most important function of P is to activate both enzymes and metabolic intermediates and provide reversible energy storage in the form of ATP. P is also a major structural component of nucleic acids and membrane lipids and takes part in regulatory pathways involving phospholipid-derived signaling molecules (e.g., phosphatidyl-inositol and inositol triphosphate) or phosphorylation reactions (e.g., MAP kinase cascades) (Marschner, 2012; Raghothama, 1999; Taiz and Zeiger, 2006; Amtmann and Armengaud, 2009).

Putting greens with sand as the dominant soil texture often require regular and routine P applications. Guertal (2006) stated that P fertilization of sand-based greens should not be neglected, and slightly higher rates or more frequent application than have been recommended by current soil-test recommendations may be warranted after a 2 year P nutrient use and uptake study on bermudagrass putting greens. Although there is a lack of specific research on the relationship between P and stress reduction, a recent report (Lyons et al., 2008) showed that the root distribution of creeping bentgrass could be manipulated by spatial localization of P supply in the root zone with buffered P sources.

Phosphite (PO_3^{3-}) is a reduced form of phosphorus (P) derived from the alkali metal salts of phosphorous acid (H_3PO_3). Phosphite is chemically similar to phosphate (PO_4^{3-}), but the different tetrahedral molecular structure of phosphite does not provide the same capability to react with P-related enzymes and the same functions as phosphate has in plants (Zhang et al., 2011). However, phosphite has significant properties as an inhibitor of turf pathogens and Cyanobacteria (syn. blue-green algae) (Burpee, 2005; Cook et al., 2005, 2009; Ervin et al., 2009; Dempsey, 2012; Chang, 2016; Inguagiato et al., 2017; Dempsey et al., 2018).

Putting green P application and the interaction of P with soil microbial activities, including symbiotic associations with mycorrhizae, are the important aspects to enhance P use efficiency, particularly under stress, although matured turf has been found to use less P than newly established turf (Frank and Guertal, 2013b).

17.5.1.4 Iron

The relationship of Fe with other turf stresses (Table 17.2) has not been intensively investigated yet, and the enhancement of Fe uptake and use by turfgrasses under stressful conditions can provide a bright future to meet the purpose of minimized input with maximized outcome, since Fe is one of the turf nutrients that can change turf color and appearance within hours. The relationship of Fe with other nutrients and heavy metals and its antagonism with P for turfgrass management deserve more attention. Recently, Fe interactions with herbicides used for turf weed control have been investigated, and contradictory results have been reported (Massey et al., 2006; McCullough and Hart, 2009).

Fe makes up about 5% by weight of the earth's crust, and Fe is the fourth most abundant element, next only to oxygen, silicon, and aluminum, in most soils. For turfgrasses, Fe is one of the micronutrients with dramatic effects on turf color

and quality, and it may be the most commonly used micronutrient for turf management, particularly for sand-based soils and golf putting greens, which often have a high potential for Fe deficiency with lower soil CEC. Soils are normally well provided with Fe. but in well-aerated and calcareous soils. Fe is found in oxide and hydroxide compounds with very low solubility. So, putting green turfgrasses often have to face an iron deficiency in sandy soils, and frequent Fe applications to putting greens are recommended. Most turfgrasses contain a normal range of Fe within 200 mg kg^{-1} with a normal range of 50 to 200 mg kg^{-1} in green tissues by dry weight. As a structural component of cytochrome, hemes, hematin, ferrichrome, and leghemoglobin, the essential compounds for oxidation-reduction reactions in both photosynthesis and respiration, Fe is essential for turf color, growth, and appearance.

Iron is absorbed by roots as inorganic forms of Fe^{2+} and Fe^{3+}, with Fe^{2+} as the dominant form being taken up by plants after it is reduced from Fe^{3+} in the rhizosphere. The activity of Fe^{3+} in the soil solution is highly pH dependent, and its activity decreases 1000-fold for each pH unit rise when the soil pH is above neutral. Under most soil conditions, grasses have the capability to excrete iron-chelators, called *phytosiderophores*, from the roots to solubilize the external insoluble Fe^{3+}. The amount of these phytosiderophores increases under Fe deficiency stress.

When the turfgrass tissue Fe content is lower than 50 mg kg^{-1}, Fe deficiency is likely to occur, and it may happen more often in cool-season turfgrasses, since most cool-season turfgrasses contain higher Fe in tissues. Iron deficiency symptoms appear in younger leaves first, because the mobility of Fe in plants is very limited. Iron deficiency may occur under several soil stressful conditions. Soil Fe availability is also affected by soil moisture, temperature, organic matter content, and the balance with other micronutrients or heavy metals. Excessive cation nutrients such as Cu, Mn, and Zn plus toxic elements such as Cd in soil solutions can inhibit Fe uptake and use by turfgrasses.

Fe applications to deficient grasses can improve the uptake of N and other elements (Marscher, 2012). Foliar Fe application of ferrous sulfate and of chelated Fe to turfgrass can correct Fe deficiency effectively. Detailed information on inorganic Fe and chelate Fe sources on turf is listed in Carrow et al. (2001). Summer heat–induced Fe chlorosis can occur during the high-temperature stress periods of midsummer for both cool-season and warm-season turfgrasses. This condition often occurs on a number of cool-season grasses grown both on sand and on clay-loam field soils. Devetter (2008) found that higher application rates in the range of 1.12 kg Fe ha^{-1} were required at the time of the onset of chlorosis to overcome the problem.

Under acidic and wet soil conditions, excessive Fe can become toxic, with consequences of Mn deficiency, and form black layers under the turf root zone when Fe reacts with excessive soil sulfur under anaerobic conditions. Black layers often occur in anaerobic soil conditions. Anaerobic bacteria produce hydrogen sulfide gas, which is poisonous to turfgrass roots. Hydrogen sulfide also reacts chemically with metal elements such as Fe, creating black deposits, which form black layers within the soil.

Fe use by turfgrass in shady conditions remained unknown until recently, when Taylor (2019) grew Diamond zoysiagrass (*Zoysia matrella* (L.) Merr) and Tifgrand bermudagrass (*Cynodon transvaalensis* Burt-Davy × *C. dactylon*) in pure sand under three light levels (Control, 40, and 60% shade) and treated them with three levels of iron, applied foliarly as $FeSO_4 \cdot H_2O$ (0, 3, and 5 kg ha^{-1} of Fe). Turfgrasses were evaluated based on the parameters of visual turfgrass quality (TQ), normalized difference vegetative index (NDVI), shoot chlorophyll content, clipping yield, plant tissue nutrient concentrations, and total root biomass. Turfgrass cultivar and shade level had significant impacts on TQ and NDVI for most weeks, whereas iron significantly impacted TQ only for most weeks. Diamond zoysiagrass produced consistently higher TQ ratings than Tifgrand bermudagrass. Both turfgrasses had acceptable TQ scores under light to moderate shade and only displayed unacceptable TQ under 60% shade. Applications of 3 and 5 kg ha^{-1} of Fe showed higher TQ scores compared with the control, illustrating that iron fertilization has beneficial impacts on shaded and non-shaded turfgrasses by improving visual turfgrass quality.

Dollar spot (caused by *Sclerotinia homoeocarpa* F.T. Bennett) is a common disease of *Agrostis stolonifera* L. and is especially devastating on putting greens (Dernoeden, 2013). Ervin et al. (2017) found that ferrous sulfate ($FeSO_4$) and its elemental components were capable of reducing dollar spot development when biweekly (May to September) foliar treatments were applied in four treatments: control, $FeSO_4$ (48.8 kg ha^{-1}), sulfur (10.3 kg ha^{-1}), and iron chelate (11.2 kg Fe ha^{-1}). However, the turf quality was improved with $FeSO_4$ in 2012 but not in 2015. Iron chelate suppressed dollar spot in 2012 but not in 2015, and sulfur had no impact on dollar spot or turf quality in either year. Without knowing the specific mechanism involved yet, there is also strong evidence that as rates of $FeSO_4$ applications increase, the severity of Microdochium patch (*Microdochium nivale* (Fries) Samuels & Hallett) decreases on annual bluegrass putting greens (Mattox et al., 2017).

17.5.2 THE IMPORTANT GROUP

17.5.2.1 Sulfur

Plants require sulfur (S) as a nutrient because S is required to synthesize the S-containing amino acids cystine, cysteine, and methionine. The main structural function of S in proteins is to form disulfide (-S-S-) bonds, which help proteins to form third- and fourth-level structures (Buchanan et al., 2000). Sulfur is also needed to form coenzyme A, which is important for oxidation of fatty acids in the citric acid cycle and for chlorophyll formation. In addition, S is involved in chlorophyll stabilization for green plants (Marschner, 2012).

Grasses normally contain 0.10 to 0.50% S by DW. Sulfur is actively absorbed as sulfate (SO_4^{-2}); however, very small amounts of SO_2 or other sulfur compounds can be absorbed by leaves or roots as gaseous forms (Hull, 1998; Carrow et al., 2001). It is common for sulfate to compete with other anions for plant uptake in the soil solution. The interactions of SO_4^{-2} with soil minerals are affected by the Fe/Al ratio, clay content, and soil pH.

The visual symptoms of S deficiencies resemble N, Fe, and Mg deficiencies with a yellowish, chlorotic condition of the leaves. Since S is not as mobile as N in plants, younger leaves can show S deficiency symptoms first. Relatively small amounts of S are needed by plants in relation to N. Although soil organic matter provides a better pool of available S than N, high N applications can cause turfgrass S deficiencies. This is particularly true in sandy soils subjected to frequent leaching (Hull, 1998; Carrow et al., 2001).

Sulfur deficiency is often observed with salt-affected sandy soils. Sulfur deficiencies for turfgrasses may not occur frequently because of the use of S-containing fertilizers and the capability of SO_2 absorption by turfgrass leaves. Turfgrass can also receive S from rain water. The use of high-sulfur coal in industry can add large amounts of S to the atmosphere. Killorn (1983) reported that the central part of the United States receives an estimated 13.5–16.8 kg ha^{-1} per year in rainfall.

However, frequent removal of clippings can cause S depletion from some soils, since leaves on average contain about 0.2 to 0.6% S by DW (Carrow et al., 2001).

High SO_2 concentrations in the atmosphere can be toxic to plants in some industrial areas due to production of sulfuric acid (H_2SO_4) on the turfgrass leaves. This acid is destructive to the cuticle layers of plant leaves. Direct S toxicity due to excessive S uptake by the plant is not a problem, and the excess S can be released to the atmosphere as H_2S from leaves (Carrow et al., 2001). Frequent use of sulfur-coated urea and sulfate-containing fertilizers, including gypsum ($CaSO_4$), can add S to turfgrasses. Acidic soil conditions and high S contents in soils can suppress some patch diseases, such as take-all patch, summer patch, and spring dead spots (Dernoeden and O'Neil, 1983; Carrow et al., 2001; Vargas, 2005).

Sulfur can be found under anaerobic conditions associated with excessive Fe to form black layers in soils (Hodges, 1992; Carrow et al., 2001). Black layers often occur in anaerobic soil conditions created by compaction and/or over-irrigation. The anaerobic conditions cause sulfur-reducing bacteria to use sulfate or elemental sulfur instead of oxygen during the respiration process, which produces hydrogen sulfide (H_2S) gas, which is poisonous to turfgrass roots. The hydrogen sulfide reacts with metal elements such as Fe, creating black deposits of FeS, which form black layers within the soil and inhibit turfgrass growth.

The use of sulfuric acid to treat irrigation water is a common practice where carbonate (CO_3^{2-}) and bicarbonate (HCO_3^-) levels are high and excess sodium (Na) exists. Bicarbonates can react with Ca and Mg in the water and can remove them from solution. This reduction of soluble Ca and Mg results in an increase in the sodium adsorption ratio (SAR) of the water, which increases the likelihood of soil structure problems caused by Na deflocculating clay particles (Christians, 1999). The process of adding sulfuric acid to irrigation water requires a thorough evaluation of the soil conditions and water chemistry before it is initiated.

17.5.2.2 Calcium

Calcium (Ca) is the fifth most abundant element in the earth's crust with an average soil calcium concentration of 3.6%. Calcium is passively absorbed as Ca^{2+}. After being absorbed, Ca^{2+} is preferentially attracted into the intercellular spaces and into the root cortex. Calcium deficiencies and toxicities are rarely reported for putting green turfgrasses. This is partially due to the relatively low requirement of turfgrasses for Ca and the abundance of Ca in the putting green root zones. The concentration of Ca in turfgrasses ranges from 0.3% to 1.25% of dry matter (Carrow et al., 2001), which varies among the various turfgrass species and even within cultivars of the same species. Irrigated turfgrass can receive certain amounts of Ca from the irrigation water, and rainfall typically has 8 mg L^{-1} Ca, whereas irrigation water frequently has 25–50 mg L^{-1} Ca (Carrow et al., 2001).

When deficiencies occur, symptoms are seen on newer leaves, leaf tips, and leaf margins, since Ca is relatively immobile in the plant. Since Ca is essential for cell elongation and division, the new leaf tips and margins are often distorted. The leaf tips and margins may have a reddish-brown color and may wither and die (Carrow et al., 2001). Calcium deficiencies can also reduce root elongation, mucilage production, and root cell division, creating shortened and stunted roots (Carrow et al., 2001).

Many soils that are low in Ca have high concentrations of soluble Al, Mg, Mn, H, and/or Na, which can out-compete Ca for the cation exchange sites in the soil and in the intercellular apoplasm of roots. Calcium deficiency symptoms are rarely seen on the shoots of plants grown on these types of soil. However, roots from plants grown in these types of soil are often stunted, thin, and spindly and are brownish/black in color. Adding calcium to soils with high levels of Al, Mn, H, or Na usually improves turfgrass growth and development due to the improvement of the damaged root system. In addition, proper Ca/K, Ca/Na, and Ca/Mn ratios also play a role in putting green soil nutrient balances and alleviate the potential for pests with physiological Ca signaling, salinity tolerance, and reduction functions of oxidative stresses of turfgrasses (Carrow et al., 2001; Larkindale and Huang, 2004; StJohn et al., 2013; Salahshoor and Kazemi, 2016; Wang and Huang, 2017).

True Ca phytotoxicity in turfgrasses has not been reported yet, and putting greens may be stressed by soluble salt fertilizer burn when high rates of $CaCl_2$ or $CaNO_3$ are applied, particularly under drought and heat stresses. Frequent applications of Ca can also reduce the levels of Mg and K found in turfgrass leaves or on cation exchange sites (StJohn et al., 2003, 2013). Putting green sandy soils often have a lower cation exchange capacity and become more easily affected by competition among cations, causing reductions in availability for uptake.

17.5.3 The Beneficial Group

This is an interesting group of nutrients, and there are more than two belonging to this group. The selected two, Si and Na, may be even more interesting or beneficial for putting green management currently or in the future, in addition to the active research into the positive roles of these two in plants and crops in recent years.

17.5.3.1 Silicon

Silicon has been studied for over 150 years for its roles in living things, and in the past two decades, it has rapidly gained interest for the management of crops, including turfgrass (Table 17.3). Some turf managers have routinely applied forms of silicon to their turf as a nutrient supply and for protection against abiotic and biotic stresses, regardless of whether or not it is defined as essential for plants (Hull, 2004; Datnoff, 2005). The significance and updated research of Si in turfgrasses have been recognized in a couple of relatively new reviews specifically focusing on turfgrasses.

Silicon is the second most abundant mineral element in soils next to oxygen, which makes up 38% of the earth's crust; Si comprises approximately 31% of the earth's crust by weight. Although the amount of Si is about four and six times that of the third and fourth most abundant elements, Al and Fe, which comprise 7% and 5% of the earth's crust by weight, respectively, neither mineral O nor Si, fortunately, is a major stressful factor to living things on the earth. In contrast, gaseous oxygen is essential for all living things except groups of anaerobic microorganisms, and to date, mineral forms of O have not been identified with serious direct impacts on plants.

Despite the abundance of Si in soils, Si deficiency can still occur due to the high demand and Si depletion from continuous planting of crops such as rice. Silicon deficiency mainly occurs in upland rice production fields with low soil pH and in highly weathered soils. Silicon has not yet been classified as an essential mineral nutrient for plants, but most grasses, including turfgrasses, on average contain 1% of their dry weight as Si. This is only just lower than N and K tissue concentrations, and some textbooks even list Si as a macronutrient. Although a standard or recommended Si concentration in any turfgrass species is not available yet, Si is applied to turfgrasses to increase turfgrass tissue Si concentrations with active uptake and to enhance wear tolerance and disease resistance in different turfgrass species.

Silicon often occurs in the form of insoluble minerals in the earth's crust, and soil solutions on average contain less than 20 mg soluble Si L^{-1}, which is still several times the concentration of soluble P. Soluble Si in soils is actively taken up by plant roots as silicic acid, $Si(OH)_4$. Some species in the grass family, such as wetland rice, can accumulate as much as 10–15% of their dry shoot weight as Si, and some broadleaf plants can reach 3% Si in dry shoot and leaf weight. This is much higher than the greatest demand for N, P, and K concentrations in these plants. Silicic acid is transported in plants from roots to leaves mainly through the xylem (Marschner, 2012), and Si either is deposited underneath the cuticle layer to protect plants from pathogens or stays in the cell walls to strengthen plants. Silicic acid concentrations inside plant roots can be many times the Si levels in the soil nutrient solution, and an active root Si uptake happens dominantly. Recently, root plasma membrane Si transporters have been identified in rice, which are responsible for the root membranes' Si influx and efflux. This type of membrane transporter is also found in corn and barley, with unique features being only localized at the root endodermis with no polarity.

As an essential (rice) or beneficial nutrient (many other plants), Si has been identified as playing physiological roles in strengthening cell wall structure, enhancing lignin biosynthesis, interacting with other essential nutrients, and reducing the toxicity of some heavy metals, including Al, Mn, and Fe. Silicon has been shown to interact with other macro- and micronutrients, in particular strengthening the compartmentation of micronutrients in plants and even the uptake. Silicon deficiency for rice mainly occurs in upland rice production fields with low soil pH and a long history of crop production and in highly weathered soils.

Enhanced disease resistance in response to Si applications in combination with other nutrients has been reported for both warm- and cool-season turfgrasses. Tremblay et al. (2002) reported that under both greenhouse and field conditions, Si applications controlled dollar spot on creeping bentgrass. Brecht et al. (2004) reported a 30% gray leaf spot control on St. Augustinegrass by Si applications. Recently, Brecht et al. (2009) identified the controlling effects of silicon fertilization on the pathogenicity of *Bipolaris cynodontis* and *Curvularia lunata* on Floradwarf bermudagrass. With further understanding of the turfgrass disease resistance enhancement by Si, nutrient management including Si may have the potential to target more serious putting green diseases caused by the pathogens *Pythium*, *Rhizotonia*, and *Typhula*.

Trenholm et al. (2001) found a 20% reduced wear injury following the foliar application of potassium silicate at 1.1 kg and 2.2 kg Si ha^{-1} and the application as a soil drench at 22.4 kg Si ha^{-1} to seashore paspalum putting greens. However, significant positive benefits of Si on turfgrass insect resistance have not been fully identified, even though it would be expected that high Si concentrations in plant tissue would play a role in suppression of insect feeding. With the effect of lignin biosynthesis enhancement, the application of Si may negatively influence forage crop quality, and the specific effects of Si on putting green thatch accumulation have not been identified yet.

Si has previously been identified as a potential inducer or modulator of plant defenses against different fungal pathogens. Rahman et al. (2015) inoculated perennial ryegrass with the causal agent of gray leaf spot (*Magnaporthe oryzae* (Catt.) B.C. Couch) with three different levels of calcium silicate (1, 5, and 10 metric tons ha^{-1}). When applied at a rate of 5 t ha^{-1}, calcium silicate was found to significantly suppress gray leaf spot in perennial ryegrass, including a significant reduction of disease incidence (39.5%) and disease severity (47.3%). They also found that following infection with *M. oryzae*, the grass levels of several phenolic acids, including chlorogenic acid and flavonoids, and relative expression levels of genes encoding phenylalanine ammonia lyase (PALa and PALb) and lipoxygenase (LOXa) significantly increased in Si-amended plants compared with those of non-amended control plants. These results suggest that Si-mediated increase of host defense

responses to fungal pathogens in perennial ryegrass has great potential to be part of an effective integrated disease management strategy against gray leaf spot development. On the other hand, Zhang et al. (2006), after 2 years of field studies, concluded that CaSiO$_3$ application to soil containing adequate Si should not be recommended for the control of brown patch on tall fescue, nor should CaSiO$_3$ be recommended to control brown patch on creeping bentgrass grown on low-Si soil or dollar spot on high-Si soil. CaSiO$_3$ application may exacerbate the incidence of brown patch disease, possibly because of nutrient imbalances, particularly in tall fescue.

Chen et al. (2016) found that sorghum (*Sorghum bicolor* (L.) Moench) leaf gas exchanges, whole-plant hydraulic conductance (Kplant), and root hydraulic conductance (Lpr) were reduced by K deficiency, but Si application moderated the K-deficiency-induced reductions, suggesting that Si alleviated the plant's hydraulic conductance. In addition, 29% of Si-alleviated transpiration was eliminated by HgCl$_2$ treatment, suggesting that aquaporin was not the primary cause for the reversal of plant hydraulic conductance. They also found that the K$^+$ concentration in xylem sap was significantly increased and the xylem sap osmotic potential was decreased by Si application, suggesting that the major cause of Si-induced improvement in hydraulic conductance could be ascribed to the enhanced xylem sap K$^+$ concentration, which increases the osmotic gradient and xylem hydraulic conductance. The results of this study indicate that Si is an element that is able to enhance resistance to multiple stresses.

There is very limited information on Si deficiency or toxicities in turfgrasses, and there is no information on symptoms of Si deficiency and toxicity among turfgrasses. The difficulty of accurate analysis of soil and plant tissue Si concentrations has not been fully resolved by using glass tubes for Si digestion from soil and plant tissues. Further investigations of the direct, indirect, physiological, pest resistance enhancement, and practical benefits of Si to turfgrass species and cultivars and golf putting green management are needed.

17.5.3.2 Sodium

The earth's crust has 2.8% Na and 2.6% K by weight, but the earth is a salty globe, with the majority of water as ocean water, containing on average 30 g NaCl L^{-1}. Sodium is an essential element for animals and humans, and Na must be present in relatively large amounts in their diet. Sodium is the principal electrolyte in animal and human systems to maintain the ionic balance of body tissues and fluids. The osmotic characteristics of Na are used in the bloodstream to regulate osmotic pressure within the cells and body fluids so as to avoid excessive loss of water. Plants use K as the principal electrolyte even with plenty of Na in the soil due to the totally different water use and regulation mechanisms in plants and animals. Sodium can substitute for K, particularly in C$_4$ plants, but plants still exhibit a strong preference for K (Subbarao et al., 2003; Marschner, 2012).

Sodium is not an essential plant nutrient element, but it may be worthy of respect from the plant nutritional point of view or as an interesting element for putting green turfgrass management for the future. For cool-season putting green turf-grass management, except to a certain degree for pest suppression by Na effects, excessive Na is toxic to C_3 plants. However, the possible role of Na in essential nutrition of warm-season turfgrasses, particularly salinity-tolerant species, deserves more attention (Munshaw et al., 2004). Although Na has never been reported as being deficient in putting green turfgrasses, it is often found that turfgrass tissue DW contains about 10 to 100 mg kg^{-1} of Na when turfgrasses are grown in sandy soils without salinity stress.

A halophyte is defined as any plant, especially a seed plant, that is able to grow in habitats excessively rich in salts. These plants have special physiological adaptations that enable them to absorb water from soils and from seawater, which have solute concentrations that non-halophytes could not tolerate. Some halophytes are actually succulent, with a high water-storage capacity.

17.6 SUMMARY AND PROSPECTS

With new challenges of limited resources, the options for turf-grass management are becoming more and more limited in terms of resources, but there are more options with integrated management and approaches. For putting greens, which are among the most intensively managed artificial mono-culture crop systems, the functions and economic values are unique and will evolve. ISM will not only bring multiple-dimensional approaches for future turf managers but also enhance the readiness for local, regional, and even international management challenges. The functions and demands of functional crops such as turfgrasses can meet the new demands, and this does not always mean the highest expense and the prettiest appearance; rather, the majority of the people using the turf are better served. Nutrient management for turfgrasses has been one of the most active research areas in turfgrass science and management since turfgrass research started more than a century ago, and it will continue with the ultimate goal: "use less and make the best out of it."

ACKNOWLEDGEMENTS

The authors are sincerely thankful for the encouragement and support of numerous organizations, companies, colleagues, former advisors, and former graduate and undergraduate students involved in the related research on N, P, K, Al, Fe, and Ni of turfgrasses. Special thanks go to both undergraduate and graduate students in four advanced turfgrass and soil science courses, HORT4200 Applied Turfgrass Stress Physiology, PES 4520/6520 Soil Fertility and Plant Nutrition (2012–2018), PES 8900 Plant Nitrogen Metabolism (2015–2018), and PES4850/6850 Environmental Soil Chemistry (2014–2018) at Clemson University for their spirit, encouragement, and support. Thanks to the PES 4850/6850 2018 class, which included (in alphabetical order) graduate students: Ethan R. Barnett, Sarah K. Holladay, Jiwoo Park, Savannah R. Petrone, Coleman A. Scroggs, and Zolian S. Zoong-lwe; undergraduate students: Thomas C. Chapman, Adam W. Chastain, Payton Davis, Katrina R. Hale-Phillips, Kimberly J. Henning, Ryan M. Ponder, Logan T. Shelton, and Brandon T. Welch for their reviews of the chapter and contributions to composing the abstract.

REFERENCES

Aamlid, T.S., A. Kvalbein, and T. Pettersen. 2017. Evaluation of an amino-acid-based fertilizer for grow-in of creeping bentgrass putting greens. *Crop Sci.* 57(Suppl1):S-357–S-360.

Ashraf, M., M. Rahmatullah, R. Afzal, F. Ahmed, A. Mujeeb, S.L. Ali, and L. Ali. 2010. Alleviation of detrimental effects of NaCl by silicon nutrition in salt-sensitive and salt-tolerant genotypes of sugarcane (*Saccharum officinarum* L.). *Plant Soil* 326(1–2):381–391.

Atkinson, J.L., L.B. McCarty, H. Liu, J. Faust, and J.E. Toler. 2012. Diamond zoysiagrass golf green response to reduced light environments with the use of trinexapac-ethyl. *Agron. J.* 104(4):847–852.

Bae, E., A. Hong, S. Choi, K. Lee, and Y. Park. 2017. Silicon pre-treatment alleviates drought stress and increases antioxidative activity in Kentucky bluegrass. *Int. Turfgrass Soc. Res. J.* 13(1):591–600.

Baldwin, C.M., H. Liu, L.B. McCarty, W.L. Bauerle, and J.E. Toler. 2006. Effects of trinexapac-ethyl on the salinity tolerance of two ultradwarf Bermudagrass cultivars. *Hort. Sci.* 41(3):808–814.

Baldwin, C.M., H. Liu, L.B. McCarty, H. Luo, and J.E. Toler. 2009a. "L-93" creeping bentgrass putting green responses to various winter light intensities in the southern transition zone. *Hort. Sci.* 44:1751–1756.

Baldwin, C.M., H. Liu, L.B. McCarty, H. Luo, and J.E. Toler. 2009b. Dormant Bermudagrass spring green-up influenced by shade. *Int. Turfgrass Soc. Res. J.* 11:711–721.

Baldwin, C.M., H. Liu, L.B. McCarty, H. Luo, and J.E. Toler. 2009c. Nitrogen and plant growth regulator influence on "Champion" Bermudagrass putting green under reduced sunlight. *Agron. J.* 101(1):75–81.

Baldwin, C.M., H. Liu, L.B. McCarty, H. Luo, J.E. Toler, and S.H. Long. 2008. Winter month foot and equipment traffic impacts on a "L-93" creeping bentgrass putting green. *HortScience* 43:922–926.

Baldwin, C.M., H. Liu, L.B. McCarty, H. Luo, C.E. Wells, and J.E. Toler. 2009d. Altered light spectral qualities impact on warm-season turfgrass growth and development. *Crop Sci.* 49:1444–1453.

Bauer, B.K., R.E. Poulter, A.D. Troughton, and D.S. Loch. 2009. Salinity tolerance of twelve hybrid Bermudagrass [*Cynodondactylon* (L.) Pers. x *C. transvaalensis* Burtt-Davy] genotypes. *Int. Turfgrass Soc. Res. J.* 11:313–326.

Bauer, S.J., B.P. Horgan, D.J. Soldat, D.T. Lloyd, and D.S. Gardner. 2017. Effects of low temperatures on nitrogen uptake, partitioning, and use in creeping bentgrass putting greens. *Crop Sci.* 57(2):1001–1009.

Beard, J.B. 1973. *Turfgrass: Science and Culture*. Prentice-Hall, Inc., Englewood Cliffs, NJ.

Beard, J.B. 1997. Shade stresses and adaptation mechanisms of turf-grasses. *Int. Turfgrass Soc. Res. J.* 8:1186–1195.

Beard, J.B. 2002. *Turf Management for Golf Courses*. 2nd ed. John Wiley & Sons, Chelsea, MI.

Beard, J.B., and H.J. Beard. 2005. *Beard's Turfgrass Encyclopedia for Golf Courses—Grounds—Lawns—Sports Fields*. Michigan State University Press, East Lansing, MI.

Belanger, F.C., S. Bonos, and W.A. Meyer. 2004. Dollar spot resistant hybrids between creeping bentgrass and colonial bentgrass. *Crop Sci.* 44(2):581–586.

Belanger, F.C., T.R. Meagher, P.R. Day, K. Plumley, and W.A. Meyer. 2003. Interspecific hybridization between *Agrostis stolonifera* and related *Agrostis* species under field conditions. *Crop Sci.* 43:240–246.

Bell, G., and T.K. Danneberger. 1999. Temporal shade on creeping bentgrass turf. *Crop Sci.* 39(4):1142–1146.

Berndt, W.L., and J.M. Vargas, Jr. 2006. Dissimilatory reduction of sulfate in black layer. *HortScience* 41(3):815–817.

Berndt, W.L., and J.M. Vargas, Jr. 2008. Elemental sulfur reduces to sulfide in black layer soil. *HortScience* 43(5):1615–1618.

Bertrand, A., Y. Castonguay, J. Cloutier, L. Couture, T. Hsiang, J. Dionne, and S. Laberge. 2009. Genetic diversity for pink snow mold resistance in green-type annual bluegrass (*Poa annua* var. reptans). *Crop Sci.* 49:589–599.

Bethea, Jr., F.G., D. Park, A. Mount, N. Menchyk, and H. Liu. 2014. Effects of acute moisture stress on creeping bentgrass cuticle morphology and associated effects on foliar nitrogen uptake. *HortScience* 49(12):1582–1587.

Bier, P.V., M. Persche, P. Koch, and D.J. Soldat. 2018. A long term evaluation of differential potassium fertilization of a creeping bentgrass putting green. *Plant Soil* 431(1–2):303–316.

Bonos, S.A., M.D. Casler, and W.A. Meyer. 2003. Inheritance of dollar spot resistance in creeping bentgrass. *Crop Sci.* 43(6):2189–2196.

Bonos, S.A., B.B. Clarke, and W.A. Meyer. 2006. Breeding for disease resistance in the major cool-season turfgrasses. *Annu. Rev. Phytopathol.* 44:213–234.

Brandenburg, R.L., and M.G. Villani. 1995. *Handbook of Turfgrass Insect Pests.* The Entomological Society of America, Hyattsville, MD.

Brecht, M., L. Datnoff, T. Kucharek, and R. Nagata. 2007. Effect of silicon on components of resistance to grey leaf spot in St. Augustinegrass. *J. Plant Nutr.* 30(7):1005–1021.

Brecht, M., C. Stiles, and L. Datnoff. 2009. Effect of high temperature stress and silicon fertilization on pathogenicity of *Bipolaris cynodontis* and *Curvularia lunata* on Floradwarf Bermudagrass. *Int. Turfgrass Soc. Res. J.* 11:165–180.

Brecht, M.O., L.E. Datnoff, T.A. Kucharek, and R.T. Nagata. 2004. Influence of silicon and chlorothalonil on the suppression of gray leaf spot and increase plant growth in St. Augustinegrass. *Plant Dis.* 88(4):338–344.

Brede, D. 2000. *Turfgrass Maintenance Reduction Handbook: Sports, Lawns, and Golf.* John Wiley & Sons, Chelsea, MI.

Brilman, L. 2003. Velvet bentgrass, pp. 201–205. *In*: M.D. Casler, and R.R. Duncan (eds.) *Turfgrass Biology, Genetics, and Breeding.* John Wiley & Sons, Hoboken, NJ.

Brilman, L.A., and W.A. Meyer. 2000. Velvet bentgrass: Rediscovering a misunderstood turfgrass: Past mistakes have damaged an excellent species reputation. *Golf Course Manage.* 68:70–75.

Broadley, M.R., and P.J. White. 2005. *Plant Nutritional Genomics.* Blackwell Publishing CRC Press, Boca Raton, FL.

Bunnell, B.T., L.B. McCarty, and W.C. Bridges, Jr. 2005a. "TifEagle" Bermudagrass response to growth factors and mowing height when grown at various hours of sunlight. *Crop Sci.* 45:575–581.

Bunnell, B.T., L.B. McCarty, and W.C. Bridges, Jr. 2005b. Evaluation of three Bermudagrass cultivars and Meyer Japanese zoysiagrass grown in shade. *Int. Turfgrass Soc. Res. J.* 10(2):826–833.

Bunnell, B.T., L.B. McCarty, J.E. Faust, W.C. Bridges, Jr., and N.C. Rajapakse. 2005c. Quantifying a daily light integral requirement of a "TifEagle" Bermudagrass golf green. *Crop Sci.* 45(2):569–574.

Burpee, L.L. 1994. *A Guide to Integrated Control of Turfgrass Diseases: Vol. 1—Cool Season Turfgrasses.* GCSAA Reference Materials, Lawrence, KS.

Burpee, L.L. 2005. Sensitivity of *Colletotrichium graminicola* to phosphonate fungicides. *Int. Turfgrass Soc. Res. J.* 10:163–169.

Callahan, L.L., W.L. Sanders, J.M. Parham, C.A. Harper, L.D. Lester, and E.R. McDonald. 1998. Cultural and chemical controls of thatch and their influence on rootzone nutrients in a bentgrass green. *Crop Sci.* 38(1):181–187.

Calvache, S., T. Espevig, T.E. Andersen, E.J. Joner, A. Kvalbein, T. Pettersen, and T.S. Aamlid. 2017. Nitrogen, phosphorus, mowing height, and arbuscular mycorrhiza effects on red fescue and mixed fescue–bentgrass putting greens. *Crop Sci.* 57(2):537–549.

Camberato, J.J., and S.B. Martin. 2004. Salinity slows germination of rough bluegrass. *HortScience* 39(2):394–397.

Campbell, C.L., and L.C. Madden. 1990. *Introduction to Plant Disease Epidemiology.* John Wiley and Sons, Inc., New York. NY.

Canegallo, A., S.B. Martin, J. Camberato, and S. Jeffers. 2017. Seashore paspalum cultivar susceptibility to large patch and fungicide evaluation for disease control in South Carolina. *Int. Turfgrass Soc. Res. J.* 13(1):185–190.

Carrow, R.N., and R.R. Duncan.1998. *Salt-affected Turfgrass Sites: Assessment and Management.* John Wiley & Sons, Chelsea, MI.

Carrow, R.N., B.J. Johnson, and R.E. Burns. 1987. Thatch and quality of Tifway Bermudagrass turf in relation to fertility and cultivation. *Agron. J.* 79(3):524–530.

Carrow, R.N., B.J. Johnson, and G.W. Landry, Jr. 1988. Centipedegrass response to foliar application of iron and nitrogen. *Agron. J.* 80(5):746–750.

Carrow, R.N., D.V. Waddington, and P.E. Rieke. 2001. *Turfgrass Soil Fertility and Chemical Problems—Assessment and Management.* Ann Arbor Press, Chelsea, MI.

Casler, M.D., and R.R. Duncan. 2003. *Turfgrass Biology, Genetics, and Cytotaxonomy.* Sleeping Bear Press, Chelsea, MI.

Chai, B., S.B. Maqbool, R.K. Hajela, D. Green, J.M. Vargas, Jr., D. Warkentin, R. Sabzikar, and M.B. Sticklen. 2002. Cloning of a chitinase-like cDNA (hs2), its transfer to creeping bentgrass (*Agrostis palustris* Huds.) and development of brown patch (*Rhizoctonia solani*) disease resistant transgenic lines. *Plant Sci.* 163(2):183–193.

Chandra, A., S. Milla-Lewis, and Q. Yu. 2017. An overview of molecular advances in zoysiagrass. *Crop Sci.* 57(Suppl1):S-73–S-81.

Chang, B. 2016. Phosphite in Soil and Turfgrass. M.S. Thesis. Auburn University, Auburn, Alabama, USA.

Chen, D., B. Cao, S. Wang, P. Liu, X. Deng, L. Yin, and S. Zhang. 2016. Silicon moderated the deficiency by improving the plant-water status in sorghum. *Sci. Rep.* 6:22882.

Chen, J., J. Yan, Y. Qian, Y. Jiang, T. Zhang, H. Guo, A. Guo, and J. Liu. 2009. Growth responses and ion regulation of four warm season turfgrasses to long-term salinity stress. *Sci. Hortic.* 122(4):620–625.

Chen, Y., T. Pettersen, A. Kvalbein, and T.S. Aamlid. 2018. Playing quality, growth rate, thatch accumulation and tolerance to moss and annual bluegrass invasion as influenced by irrigation strategies on red fescue putting greens. *J. Agron. Crop Sci.* 204(2):185–195.

Chen, Y., J. Shi, G. Tian, S. Zheng, and Q. Lin. 2004. Fe deficiency induces Cu uptake and accumulation in *Commelina communis*. *Plant Sci.* 166(5):1371–1377.

Chérif, M., J.G. Menzies, D.L. Ehret, C. Bogdanoff, and R.R. Bélanger. 1994. Yield of cucumber infected with *Pythium aphanidermatum* when grown with soluble silicon. *HortScience* 29(8):896–897.

Cho, T., M. Kudo, and S. Kimura. 2017. Relationship of macronutrients and nonstructural carbohydrate concentration in clippings of creeping bentgrass putting greens in Japan. *Int. Turfgrass Soc. Res. J.* 13(1):404–409.

Christians, N. 2007. *Fundamentals of Turfgrass Management.* 2nd ed. John Wiley & Sons, Hoboken, NJ.

Clarke, B.B., and A.B. Gould. 1993. *Turfgrass Patch Diseases Caused by Ectotrophic Root-infecting Fungi.* APS Press, St. Paul, MN.

Cook, P.J., P.J. Landschoot, and M.T. Schlossberg. 2005. Evaluation of phosphonate fungicides for control of anthracnose basal rot and putting green quality. http://plantscience.psu.edu/research/centers/turf/ research/annual-reports/2005/cook1.pdf.

Cook, P.J., P.J. Landschoot, and M.J. Schlossberg. 2009. Inhibition of *Pythium* spp. and suppression of pythium blight of turfgrasses with phosphonate fungicides. *Plant Dis.* 93(8):809–814.

Couch, H.B. 1995. *Diseases of Turfgrasses.* 3rd ed. Krieger Publishing Company, Malabar, FL.

Couch, H.B. 2000. *The Turfgrass Disease Handbook.* Krieger Publishing Co., Melbourne, FL.

Dai, J., D.R. Huff, and M.J. Schlossberg. 2009. Salinity effects on seed germination and vegetative growth of greens-type *Poa annua* relative to other cool-season turfgrass species. *Crop Sci.* 49:696–703.

Danneberger, T.K. 1993. *Turfgrass Ecology and Management.* G.I.E. Media Inc., Cleveland, OH.

Datnoff, L.E. 2005. Silicon in the life and performance of turfgrass. Online. *Appl. Turfgrass Sci.* Published 14 September 2005.

Datnoff, L.E., W.H. Elmer, and D.M. Huber. 2007. *Mineral Nutrition and Plant Disease.* The American Phytopathological Society, St. Paul, MN.

Datnoff, L.E., G.H. Snyder, and G.H. Korndorfer. 2001. *Silicon in Agriculture.* Elsevier Science, The Netherlands.

Dempsey, J.J., I. Wilson, P.T.N. Spencer-Phillips, and D. Arnold. 2018. Suppression of the in vitro growth and development of *Microdochium nivale* by phosphite. *Plant Pathol.* 67(6):1296–1306.

Dempsey, J.J., I.D. Wilson, P.T.N. Spencer-Phillips, and D.L. Arnold. 2012. Suppression of *Microdochium nivale* by potassium phosphite in cool-season turfgrasses. *Acta Agric. Scand. B* 62(suppl1):70–78.

Dernoeden, P.H. 2013. *Creeping Bentgrass Management.* 2nd ed. CRC Press, Taylor & Francis Group, Boca Raton, FL.

Dest, W.M., and K. Guillard. 1987. Nitrogen and phosphorus nutritional influence on bentgrass-annual bluegrass community composition. *J. Am. Soc. Hortic. Sci.* 112:769–773.

Devetter, D.N., N.E. Christians, and D. Minner. 2008. Dealing with summer induced chlorosis of turf. *Golf Course Manage.* 76(5):123–126.

Dionne, J., S. Rochefort, D.R. Huff, Y. Desjardins, A. Bertrand, and Y. Castonguay. 2010. Variability for freezing tolerance among 42 ecotypes of green-type annual bluegrass. *Crop Sci.* 50(1):321–336.

DiPaola, J.M. 1984. Syringing effects on the canopy temperatures of bentgrass greens. *Agron. J.* 76(6):951–953.

DiPaola, J.M., and J.B. Beard. 1992. Physiological effects of temperature stress. *In:* D.V. Waddington, R.N. Carrow, and R.C. Shearman (eds.) *Turfgrass.* Agronomy Monograph 32. American Society of Agronomy, Madison, WI.

Doncheva, S., C. Poschenrieder, Z.L. Stoyanova, K. Georgieva, M. Velichkova, and J. Barceló. 2009. Silicon amelioration of manganese toxicity in Mn-sensitive and Mn-tolerant maize varieties. *Environ. Exp. Bot.* 65(2–3):189–197.

Dordas, C. 2008. Role of nutrients in controlling plant diseases in sustainable agriculture: A review. *Agron. Sustain. Dev.* 28(1):33–46.

Duble, R. 1996. *Turfgrasses: Their Management and Use in the Southern Zone.* Texas A&M Univ. Press, College Station, TX.

Duncan, R.R. 2000. Plant tolerance to acid soil constraints: Genetic resources, breeding methodology, and plant improvement, pp. 1–38. *In:* R.E. Wilkinson (ed.) *Plant Environment Interaction.* 2nd ed. Marcel Dekker Inc., New York.

Duncan, R.R., and R.N. Carrow. 2000. *Seashore Paspalum: The Environmental Turfgrass.* Ann Arbor Press, Chelsea, MI.

Duncan, R.R., and R.N. Carrow. 2005. Managing seashore paspalum greens. *Golf Course Manage.* 73:114–118.

Ebina, M., M. Kobayashi, H. Tonogi, S. Tsuruta, H. Akamine, and S. Kasuga. 2017. Evaluation and breeding of zoysiagrass using Japan's natural genetic resources. *Int. Turfgrass Soc. Res. J.* 13(1):40–43.

Emmons, R.D., and F. Rossi. 2015. *Turfgrass Science and Management.* 5th ed. Cengage Learning, Clifton Park, NY.

Eneji, A.E., S. Inanaga, S. Muranaka, J. Li, T. Hattori, P. An, and W. Tsuji. 2008. Growth and nutrient use in four grasses under drought stress as mediated by silicon fertilizers. *J. Plant Nutr.* 31(2):355–365.

Engel, R.E., and R.B. Alderfer. 1967. The effect of cultivation, topdressing, lime, N, and wetting agent on thatch development on 1/4-inch bentgrass over a 10-year period. *N.J. Agric. Exp. Stn. Bull.* 818:32–45.

Engelke, M.C., and S. Anderson. 2003. Zoysiagrasses, pp. 271–286. *In:* M.D. Casler, and R.R. Duncan (eds.) *Turfgrass Biology, Genetics, and Breeding.* John Wiley & Sons, Hoboken, NJ.

Engelke, M.C., P.F. Colbaugh, J.A. Reinert, K.B. Marcum, R.H. White, B. Ruemmele, and S.J. Anderson. 2002. Registration of 'Diamond' zoysiagrass. *Crop Sci.* 42(1):304–305.

Epstein, E. 1999. Silicon. *Annu. Rev. Plant Physiol. Plant Mol. Biol.* 50(1):641–664.

Epstein, E. 2005. *Nutrition of Plants: Principles and Perspectives.* Sinauer Associates, Inc., Sunderland, MA.

Ervin, E.H., D.S. McCall, and B.J. Horvath. 2009. Efficacy of phosphite fungicides and fertilizers for control of Pythium blight on a perennial ryegrass fairway in Virginia. *Appl. Turfgrass Sci.* 6(1):1–3.

Ervin, E.H., N. Reams, X. Zhang, A. Boyd, and S. Askew. 2017. An integrated nutritional and chemical approach to *Poa annua* suppression in creeping bentgrass greens. *Crop Sci.* 57:567–572.

Esmaeili, S., and H. Salehi. 2016. Kentucky bluegrass (*Poa pratensis* L.) silicon-treated turfgrass tolerance to short- and long-term salinity condition. *Adv. Hortic. Sci.* 30(2):87–94.

Esmaeili, S., H. Salehi, and S. Eshghi. 2015. Silicon ameliorates the adverse effects of salinity on turfgrass growth and development. *J. Plant Nutr.* 38(12):1885–1901.

Espinosa, A., G.L. Miller, and L.E. Datnoff. 2013. Accumulation of silicon in *Cynodon dactylon* x *C. transvaalensis* and *Poa trivialis* used as an overseed grass. *J. Plant Nutr.* 36(11):1719–1732.

Exley, C. 2009. Darwin, natural selection and the biological essentiality of aluminum and silicon. *Trends Biochem. Sci.* 34(12):589–593.

Fageria, N.K., A.B. Santos, M.P. Barbosa Filho, and C.M. Guimarães. 2008. Iron toxicity in lowland rice. *J. Plant Nutr.* 31(9):1676–1697.

Fauteux, F., F. Chain, F. Belzile, J. Menzies, and R.R. Bélanger. 2006. The protective role of silicon in the *Arabidopsis*–powdery mildew pathosystem. *Proc. Natl. Acad. Sci. U.S.A.* 103(46):17554–17559.

Fermanian, T.W., M.C. Shurtleff, R. Randell, H.T. Wilkinson, and P.L. Nixon. 2002. *Controlling Turfgrass Pests.* 3rd ed. Prentice-Hall, Upper Saddle River, NJ.

Follett, R.F., and J.L. Hatfield. 2001. *Nitrogen and Environment: Sources, Problems, and Management.* Elsevier, New York, NY.

Frank, K.W., and E.A. Guertal. 2013a. Nitrogen research in turfgrass, pp. 457–491. *In*: B. Horgan, J. Stier, and S. Bonos (eds.) *Agronomy Monograph 56*. ASA, CSSA, and SSSA, Madison, WI.

Frank, K.W., and E.A. Guertal. 2013b. Potassium and phosphorus research in turfgrass, pp. 493–519. *In*: J.C. Stier, B.P. Horgan, and S.A. Bonos (eds.) *Turfgrass: Biology, Use, and Management, Agronomy Monograph 56*. ASA, CSSA, SSSA, Madison, WI.

Frank, K.W., T.A. Nikolai, A. Hathaway, and Z. Williams. 2018. The fate of late-fall-applied nitrogen in creeping bentgrass and annual bluegrass. *Crop Forage Turfgrass Manag.* 4(1):180029.

Fry, J.D., and B. Huang. 2004. *Applied Turfgrass Science and Physiology*. John Wiley & Sons, Hoboken, NJ.

Fu, D., P.C. St. Amand, Y. Xiao, S. Muthukrishnan, and G.H. Liang. 2006. Characterization of T-DNA integration in creeping bentgrass. *Plant Sci.* 170(2):225–237.

Fu, D., N.A. Tisserat, Y. Xiao, D. Settle, S. Muthukrishnan, and G.H. Liang. 2005. Overexpression of rice TLPD34 enhances dollar-spot resistance in transgenic bentgrass. *Plant Sci.* 168(3):671–680.

Fu, J., and B. Huang. 2003. Effects of foliar application of nutrients on heat tolerance of creeping bentgrass. *J. Plant Nutr.* 26(1):81–96.

Fu, J., P.H. Dernoeden, and J.A. Murphy. 2009. Creeping bentgrass color and quality, chlorophyll content, and thatch–mat accumulation responses to summer coring. *Crop Sci.* 49(3):1079–1087.

Ghasemi-Fasaei, R., and A. Ronaghi. 2008. Interaction of iron with copper, zinc, and manganese in wheat as affected by iron and manganese in a calcareous soil. *J. Plant Nutr.* 31(5):839–848.

Ghorbanim, A., S. Wilcockson, A. Koocheki, and C. Leifert. 2008. Soil management for sustainable crop disease control: A review. *Environ. Chem. Lett.* 6(3):149–162.

Glinski, D.S., R.N. Carrow, and K.J. Karnok. 1992. Iron fertilization effects on shoot/root growth, water use, and drought stress of creeping bentgrass. *Agron. J.* 84(3):496–503.

Goatley, J.M., V. Maddox, D.J. Lang, and K.K. Crouse. 1994. "Tifgreen" Bermudagrass response to late-season application of nitrogen and potassium. *Agron. J.* 86:7–10.

Goss, R.M., J.H. Baird, S.L. Kelm, and R.N. Calhoun. 2002. Trinexapac-ethyl and nitrogen effects on creeping bentgrass grown under reduced light conditions. *Crop Sci.* 42:472–479.

Goyal, S.S., R. Tischner, and A.S. Basra. 2005. *Enhancing the Efficiency of Nitrogen Utilization in Plants*. Food Products Press, New York, NY.

Grimshaw, A.L., Y. Qu, W.A. Meyer, E. Watkins, and S.A. Bonos. 2018. Heritability of simulated wear and traffic tolerance in three fine fescue species. *HortScience* 53(4):416–420.

Guertal, E.A., and J.S. McElroy. 2018. Soil type and phosphorus fertilization affect *Poa annua* growth and seedhead production. *Agron. J.* 110:2165–2170.

Guo, W., Y.G. Zhu, W.J. Liu, Y.C. Liang, C.N. Geng, and S.G. Wang. 2007. Is the effect of silicon on rice uptake of arsenate (AsV) related to internal silicon concentrations, iron plaque and phosphate nutrition? *Environ. Pollut.* 148(1):251–257.

Gussack, E., and F.S. Rossi. 2001. *Turfgrass Problems: Picture Clues and Management Options*. Cornell University Natural Resource, Agriculture, and Engineering Service (NRAES), Ithaca, NY.

Hagley, K.J., A.R. Miller, and A.C. Gange. 2002. Variation in life history characteristics of *Poa annua* L. in golf putting greens. *J. Turfgrass Sports Surf. Sci.* 78:16–24.

Han, K.M., J.E. Kaminski, and T.T. Lulis. 2017. Influence of nitrogen, plant growth regulators, and iron sulfate on annual bluegrass populations on a golf green. *Int. Turfgrass Soc. Res. J.* 13(1):661–669.

Hanna, W.W., S.K. Braman, and B.M. Schwartz. 2010. "ST-5", a shade-tolerant turf Bermudagrass. *HortScience* 45(1):132–134.

Hattori, T., K. Sonobe, S. Inanaga, P. An, and S. Morita. 2008. Effects of silicon on photosynthesis of young cucumber seedlings under osmotic stress. *J. Plant Nutr.* 31(6):1046–1058.

Hauck, R.D. 1984. *Nitrogen in Crop Production*. American Society of Agronomy, Crop Science Society of America, Soil Science Society of America, Madison, WI.

Havlin, J.L., J.D. Beaton, S.L. Tisdale, and W.L. Nelson. 2005. *Soil Fertility and Fertilizers*. 7th ed. Prentice Hall, Upper Saddle River, NJ.

Haydu, J.J., A.W. Hodges, and C.R. Hall. 2008. Estimating the economic impact of the U.S. golf course industry: Challenges and solutions. *HortScience* 43(3):759–763.

Hodges, B.P., C.M. Baldwin, B. Stewart, M. Tomaso-Peterson, J.D. McCurdy, E.K. Blythe, and H.W. Philley. 2016. Quantifying a daily light integral for establishment of warm-season cultivars on putting greens. *Crop Sci.* 56(5):2818–2826.

Hodges, C.F. 1992. Interaction of cyanobacteria and sulfate-reducing bacteria in sub-surface black-layer formation in high-sand content golf greens. *Soil Biol. Biochem.* 24(1):15–20.

Hoffman, L., J.S. Ebdon, W.M. Dest, and M. DaCosta. 2010a. Effects of nitrogen and potassium on wear mechanisms in perennial ryegrass: I. Wear tolerance and recovery. *Crop Sci.* 50(1):357–366.

Hoffman, L., J.S. Ebdon, W.M. Dest, and M. DaCosta. 2010b. Effects of nitrogen and potassium on wear mechanisms in perennial ryegrass: II. Anatomical, morphological, and physiological characteristics. *Crop Sci.* 50(1):367–379.

Hogendorp, B.K., R.A. Cloyd, and J.M. Swiader. 2009. Effect of silicon-based fertilizer applications on the reproduction and development of the citrus mealybug (Hemiptera: Pseudococcidae) feeding on green coleus. *J. Econ. Entomol.* 102(6):2198–2208.

Honig, J.A., C. Kubik, V. Averello, J. Vaiciunas, W.A. Meyer, and S.A. Bonos. 2016. Classification of bentgrass (*Agrostis*) cultivars and accessions based on microsatellite (SSR) markers. *Genet. Resour. Crop Evol.* 63(7):1139–1160.

Honig, J.A., C. Kubik, V. Averello, J. Vaiciunas, W.A. Meyer, and S.A. Bonos. 2016. Classification of bentgrass (*Agrostis*) cultivars and accessions based on microsatellite (SSR) markers. *Genet. Resour. Crop Evol.* 63(7):1139.

Hossain, M.A., Y. Ishimine, H. Akamine, and H. Kuramochi. 2004. Effect of nitrogen fertilizer application on growth, biomass production and N-uptake of torpedograss (*Panicum repens* L.). *Weed Biol. Manag.* 4(2):86–94.

Huang, B., M. DaCosta, and Y. Jiang. 2014. Research advances in mechanisms of turfgrass tolerance to abiotic stress: From physiology to molecular biology. *Crit. Rev. Plant Sci.* 33(2–3):141–189.

Huang, X., F. Wang, R. Singh, J.A. Reinert, M.C. Engelke, A.D. Genovesi, A. Chandra, and Q. Yu. 2016. Construction of high-resolution genetic maps of *Zoysia matrella* (L.) Merrill and applications to comparative genomic analysis and QTL mapping of resistance to fall armyworm. *BMC Genomics* 17:562.

Huber, D.M., and D.C. Arny. 1985. Interaction of potassium and with plant diseases, pp. 467–488. *In*: R.D. Munson (ed.) *Potassium in Agriculture*. American Society of Agronomy, Madison, WI.

Huff, D.R. 2003. Annual bluegrass, pp. 39–51. *In*: M.D. Casler, and R.R. Duncan (eds.) *Turfgrass Biology, Genetics, and Breeding*. John Wiley & Sons, Hoboken, NJ.

Hull, R.J. 2004. Scientists start to recognize silicon's beneficial effects. *Turfgrass Trends* 8:69–73.

Hull, R.J., N. Jackson, and C.R. Skogley. 1979. Influence of nutrition on stripe smut severity in Kentucky bluegrass turf. *Agron. J.* 71(4):553–555.

Hurdzan, M.J. 2006. *Golf Course Architecture: Evolutions in Design, Construction, and Restoration Technology.* 2nd ed. John Wiley & Sons, Hoboken, NJ.

Hurley, R. 2003. Rough bluegrass, pp. 67–73. *In*: M.D. Casler, and R.R. Duncan (eds.) *Turfgrass Biology, Genetics, and Breeding.* John Wiley & Sons, Hoboken, NJ.

Inguagiato, J.C., and K. Guillard. 2016. Foliar N concentration and reflectance meters to guide N fertilization for anthracnose management of annual bluegrass putting green turf. *Crop Sci.* 56(6):3328–3337.

Inguagiato, J.C., J.E. Kaminski, and T.T. Lulis. 2017. Effect of phosphite rate and source on cyanobacteria colonization of putting green turf. *Crop Sci.* 57(Suppl 1):S274–S284.

Inguagiato, J.C., J.A. Murphy, and B.B. Clarke. 2009. Anthracnose disease and annual bluegrass putting green performance affected by mowing practices and lightweight rolling. *Crop Sci.* 49(4):1454–1462.

Jiang, Y., and B. Huang. 2001. Effects of calcium and antioxidant metabolism and water relations associated with heat tolerance in two cool-season grasses. *J. Exp. Bot.* 355:341–349.

Jiang, Y., R.N. Carrow, and R.R. Duncan. 2005. Physiological acclimation of seashore paspalum and Bermudagrass to low light. *Sci. Hortic. Amsterdam* 105(1):101–115.

Jiang, Y., R.R. Duncan, and R.N. Carrow. 2004. Assessment of low light tolerance of seashore paspalum and Bermudagrass. *Crop Sci.* 44(2):587–594.

Koehler, J.F., and G.L. Miller. 2017. Impact of nitrogen source and a pH buffer on the in vitro growth and morphology of *Rhizoctonia solani* AG 2–2 LP. *Int. Turfgrass Soc. Res. J.* 13:198–202.

Koeritz, E.J., and J.C. Stier. 2009. Nitrogen rate and mowing height effects on velvet and creeping bentgrasses for low-input putting greens. *Crop Sci.* 49(4):1463–1472.

Kopec, D.M., J.L. Walworth, J.J. Gilbert, G.M. Sower, and M. Pessarakli. 2007. "SeaIsle 2000" paspalum putting surface response to mowing height and nitrogen fertilizer. *Agron. J.* 99:133–140.

Korndorfer, A.P., R. Cherry, and R. Nagata. 2004. Effect of calcium silicate on feeding and development of tropical sod webworms (Lepidoptera: Pyralidae). *Fla. Entomol.* 87(3):393–395.

Larkindale, J., and B. Huang. 2004. Thermotolerance and antioxidant systems in *Agrostis stolonifera*: Involvement of salicylic acid, abscisic acid, calcium, hydrogen peroxide, and ethylene. *J. Plant Physiol.* 161(4):405–413.

Lea, P.J., and J.F. Morot-Gaudry. 2001. *Plant Nitrogen.* Springer, New York.

Ledeboer, F.B., and C.R. Skogley. 1967. Investigations into the nature of thatch and methods for its decomposition. *Agron. J.* 59(4):320–323.

Lee, G., R.N. Carrow, and R.R. Duncan. 2005. Criteria for assessing salinity tolerance of the halophytic turfgrass seashore paspalum. *Crop Sci.* 45(1):251–258.

Leslie, A.R. 1994. *Handbook of Integrated Pest Management for Turf and Ornamentals.* Lewis Publishers, Ann Arbor, MI.

Liang, Y.C., J.W.C. Wong, and L. Wei. 2005. Silicon-mediated enhancement of cadmium tolerance in maize (*Zea mays* L.) grown in cadmium contaminated soil. *Chemosphere* 58(4):475–483.

Liang, Y.C., J. Zhu, Z. Li, G. Chu, Y. Ding, J. Zhang, and W. Sun. 2008. Role of silicon in enhancing resistance to freezing stress in two contrasting winter wheat cultivars. *Environ. Exp. Bot.* 64(3):286–294.

Liu, C., J.J. Camberato, S.B. Martin, and A.V. Turner. 2001. Rough bluegrass germination varies with temperature and cultivar/seed lot. *HortScience* 36(1):153–156.

Liu, H., C.M. Baldwin, H. Luo, and M. Pessarakli. 2008b. Enhancing turfgrass nitrogen use under stresses, pp. 555–599. *In*: M. Pessarakli (ed.) *Handbook of Turfgrass Management and Physiology.* CRC Press Taylor & Francis Group, New York.

Liu, H., C.M. Baldwin, F.W. Totten, and L.B. McCarty. 2008a. Foliar fertilization for turfgrasses II. International Conference on Turfgrass Science and Management for Sports Fields. *Acta Hort. (ISHS)* 783:323–332.

Liu, H., N. Menchyk, F.G. Bethea Jr., and H. Luo. 2013. "Diamond" zoysiagrass putting greens in the transition zone. *Int. Turfgrass Soc. Res. J.* 12:761–766.

Liu, X., B. Huang, and G. Banowetz. 2002. Cytokinin effects on creeping bentgrass responses to heat stress: I. Shoot and root growth. *Crop Sci.* 42:457–465.

Ma, J., and E. Takahashi. 1990. Effect of silicon on the growth and phosphorus uptake of rice. *Plant Soil* 126(1):115–119.

Ma, J.F., and N. Yamaji. 2008. Functions and transport of silicon in plants. *Cell. Mol. Life Sci.* 65(19):3049–3057.

Ma, J.F., K. Tamai, N. Yamaji, N. Mitani, S. Konishi, M. Katsuhara, M. Ishiguro, Y. Murata, and M. Yano. 2006. A silicon transporter in rice. *Nature* 440(7084):688–691.

Mancino, C.F., D. Petrunak, and D. Wilkinson. 2001. Loss of putting greens-grade fertilizer granules due to mowing. *HortScience* 36(6):1123–1126.

Marcum, K.B. 1999. Salinity tolerance mechanisms of grasses in the subfamily Chloridoideae. *Crop Sci.* 39(4):1153–1160.

Marschner, H. 2012. *Marschner's Mineral Nutrition of Higher Plants.* 3rd ed. Academic Press, London.

Massey, J.H., J.M. Tayler, N. Binbuga, K. Chambers, G.E. Coats, and W.P. Henry. 2006. Iron antagonism of MSMA herbicide applied to Bermudagrass: Characterization of the Fe^{2+}-MAA complexation reaction. *Weed Sci.* 54(1):23–30.

Mattox, C.M., A.R. Kowalewski, B.W. McDonald, J.G. Lambrinos, B.L. Daviscourt, and J.W. Pscheidt. 2017. Nitrogen and iron sulfate affect microdochium patch severity and turf quality on annual bluegrass putting greens. *Crop Sci.* 57(Suppl 1):S-293–S-300.

McCall, D.S., E.H. Ervin, C.D. Shelton, N. Reams, and S.D. Askew. 2017. Influence of ferrous sulfate and its elemental components on dollar spot suppression. *Crop Sci.* 57(2):581–586.

McCarty, L.B. 2018. *Golf Turf Management.* CRC Press, Taylor & Francis Group, Boca Raton, FL.

McCarty, L.B., and G.L. Miller. 2002. *Managing Bermudagrass Turf: Selection, Construction, Cultural Practices and Pest Management Strategies.* Sleeping Bear Press, Chelsea, MI.

McCarty, L.B., J.W. Everest, D.W. Hall, T.R. Murphy, and F. Yelverton. 2001. *Color Atlas of Turfgrass Weeds.* Ann Arbor Press, Chelsea, MI.

McCarty, L.B., M.F. Gregg, and J.E. Toler. 2007. Thatch and mat management in an established creeping bentgrass golf green. *Agron. J.* 99(6):1530–1537.

McCarty, L.B., M.F. Gregg, J.E. Toler, J.J. Camberato, and H.S. Hill. 2005. Minimizing thatch and mat development in a newly seeded creeping bentgrass golf green. *Crop Sci.* 45(4):1529–1535.

McCullough, P.E., and S.E. Hart. 2009. Chlelated iron and adjuvants influence bispyribac–sodium efficacy for annual bluegrass (*Poa annua*) control in cool-season turfgrasses. *Weed Technol.* 23(4):519–523.

McDonald, B., C. Mattox, M. Gould, and A. Kowalewski. 2018. Effects of sulfur and calcium source on pH, anthracnose severity, and *Microdochium* patch management on annual bluegrass in Western Oregon. *Crop Forage Turfgrass Manag.* 4(1):180018.

Menchyk, N. 2012. Foliar Applied Urea Nitrogen Metabolism in Warm-season turfgrass under Salinity Stress. Ph.D. Dissertation, Clemson University, Clemson, SC.

Menchyk, N., D.G. Bielenberg, S. Martin, C. Waltz, H. Luo, F. Bethea, Jr., and H. Liu. 2014. Nitrogen and trinexapac-ethyl applications for managing "diamond" zoysiagrass putting greens in the transition zone, U.S. *HortScience* 49(8):1076–1080.

Mengel, K., E.A. Kirby, H. Kosegarten, and T. Appel. 2001. *Principles of Plant Nutrition*. Kluwer Academic Publishers, Boston, MA.

Meyer, W.A., L. Hoffman, and S.A. Bonos. 2017. Breeding cool-season turfgrass cultivars for stress tolerance and sustainability in a changing environment. *Int. Turfgrass Soc. Res. J.* 13(1):3–10.

Miller, G.L., and R. Dickens. 1996. Potassium fertilization related to cold resistance in Bermudagrass. *Crop Sci.* 36(5):1290–1295.

Miller, G.L., D.T. Earlywine, R. Braun, J.D. Fry, and M.M. Kennelly. 2016. Influence of nitrogen source and application timing on large patch of zoysiagrass. *Crop Forage Turfgrass Manag.* 2(1). doi:10.2134/cftm2015.0189.

Miller, G.L., J.T. Edenfield, and R.T. Nagata. 2005. Growth parameters of Floradwarf and Tifdwarf bermudagrasses exposed to various light regimes. *Int. Turfgrass Soc. Res. J.* 10(2):879–884.

Mirmow, W.N. 2016. Fall Potassium Fertilization and Winter Traffic Effects on a Creeping Bentgrass Putting Green. M.S. Thesis, Clemson University, Clemson, SC.

Mitani, N., Y. Chiba, N. Yamaji, and J.F. Ma. 2009. Identification and characterization of maize and barley Lsi2-like silicon efflux transporters reveals a distinct silicon uptake system from that in rice. *Plant Cell* 21(7):2133–2142.

Morot-Gaudry, J.F. 2001. *Nitrogen Assimilation by Plants—Physiological, Biochemical and Molecular Aspects*. Science Publishers, Enfield, NH.

Mosier, A.R., J.K. Syers, and J.R. Freney. 2004. *Agriculture and the Nitrogen Cycle—Assessing the Impact of Fertilizer Use on Food Production and the Environment*. Island Press, Washington.

Munshaw, G.C., X. Zhang, and E.H. Ervin. 2004. Effect of salinity on Bermudagrass cold hardiness. *HortScience* 39(2):420–423.

Murray, A.D., L. Daines, D. Archibald, R.A. Hawkes, C. Schiphorst, P. Kelly, L. Grant, and N. Mutrie. 2017. The relationships between golf and health: A scoping review. *Br. J.Sports Med.* 51(1):12–19.

Nanayakkara, U.N., W. Uddin, and L.E. Datnoff. 2008. Effects of soil type, source of silicon, and rate of silicon source on development of gray leaf spot of perennial ryegrass turf. *Plant Dis.* 92(6):870–877.

Niemczyk, H.D. 2001. *Destructive Turf Insects*. 2nd ed. G.I.E. Media, Cleveland, OH.

O'Connor, K., F. Hébert, J.E. Powers, K.S. Jordan, and E.M. Lyons. 2018. Leaf morphology explains the disparity between annual bluegrass and creeping bentgrass growth under foliar fertilization. *J. Plant Nutr.* 41(5):596–608.

Ookawa, T., Y. Naruoka, A. Sayama, and T. Hirasawa. 2004. Cytokinin effects on ribulose-1,5-bisphosphate carboxylase/oxygenase and nitrogen partitioning in rice during ripening. *Crop Sci.* 44(6):2107–2115.

Park, D.M., J.L. Cisar, M.A. Fidanza, E.J. Nangle, G.H. Snyder, and K.E. Williams. 2017. Seasonal cultural management practices for aging ultradwarf Bermudagrass greens in the subtropics: I. Nitrogen and potassium fertilization. *Int. Turfgrass Soc. Res. J.* 13(1):280–290.

Patton, A.J., B.M. Schwartz, and K.E. Kenworthy. 2017. Zoysiagrass (*Zoysia* spp.) history, utilization, and improvement in the United States: A review. *Crop Sci.* 57:S1–S12.

Pease, B.W., and J.C. Stier. 2018. Nitrogen rate and growth regulator effects on shaded velvet and creeping bentgrasses. *Agron. J.* 0(6). doi:10.2134/agronj2018.01.0071.

Pira, E. 1997. *A Guide to Golf Course Irrigation System Design and Drainage*. Ann Arbor Press, Chelsea, MI.

Pompeiano, A., and A.J. Patton. 2017. Growth and root architecture responses of zoysiagrass to changes in fertilizer nitrate:urea ratio. *J. Plant Nutr. Soil Sci.* 180(5):528–534.

Potter, D.A. 1998. *Destructive Turfgrass Insects: Biology, Diagnosis, and Control*. Ann Arbor Press, Chelsea, MI.

Puhalla, J., J. Krans, and M. Goatley. 2002. *Sports Fields: A Manual for Construction and Maintenance*. John Wiley & Sons, Hoboken, NJ.

Puhalla, J., J. Krans, and M. Goatley. 2003. *Baseball and Softball Fields: Design, Construction, Renovation, and Maintenance*. John Wiley & Sons, Hoboken, NJ.

Qian, Y.L., and M.C. Engelke. 1999. "Diamond" zoysiagrass as affected by light intensity. *J. Turfgrass Manag.* 3(2):1–15.

Rahman, A., C. Wallis, and W. Uddin. 2015. Silicon induced systemic defense responses in perennial ryegrass against infection by *Magnaporthe oryzae*. *Phytopathology* 105(6):748–757.

Rajasekar, S., S. Fei, and N.E. Christians. 2006. Analysis of genetic diversity in rough bluegrass determined by RAPD markers. *Crop Sci.* 46(1):162–167.

Ranger, C.M., A.P. Singh, J.M. Frantz, L. Cañas, J.C. Locke, M.E. Reding, and N. Vorsa. 2009. Influence of silicon on resistance of *Zinnia elegans* to *Myzus persicae* (Hemiptera: Aphididae). *Environ. Entomol.* 38(1):129–136.

Redmond, C.T., and D.A. Potter. 2006. Silicon fertilization does not enhance creeping bentgrass resistance to cutworms and white grubs. Online. *Appl. Turfgrass Sci.* Published 10 November 2006.

Reuveni, M., D. Oppernheim, and R. Reuveni. 1998. Integrated control of powdery mildew on apple trees by foliar sprays of mono-potassium phosphate fertilizer and sterol inhibiting fungicides. *Crop Prot.* 17(7):563–568.

Reuveni, R., and M. Reuveni. 1998. Foliar-fertilizer therapy—a concept in integrated pest management. *Crop Prot.* 17(2):111–118.

Rodriguez, I.R., G.L. Miller, and L.B. McCarty. 2002. Bermudagrass establishment on high sand-content soils using various N–P–K ratios. *HortScience* 37:208–209.

Rowland, J.H., J.L. Cisar, G.H. Snyder, J.B. Sartain, and A.L. Wright. 2009. USGA ultradwarf Bermudagrass putting green properties as affected by cultural practices. *Agron. J.* 101(6):1565–1572.

Sachs, P.D. 2004. *Managing Healthy Sports Fields: A Guide to Using Organic Materials for Low-maintenance and Chemical-free Playing Fields*. John Wiley & Sons, Hoboken, NJ.

Sachs, P.D., and R.T. Luff. 2002. *Ecological Golf Course Management*. Chelsea, Ann Arbor, MI.

Saidi, Y., A. Finka, M. Muriset, Z. Bromberg, Y.G. Weiss, F.J.M. Maathuis, and P. Goloubinoff. 2009. The heat shock response in moss plants is regulated by specific calcium-permeable channels in the plasma membrane. *Plant Cell* 21(9):2829–2843.

Saigusa, M., K. Onozawa, H. Watanabe, and K. Shibuya. 2000. Effects of porous hydrate calcium silicate on the wear resistance, insect resistance, and disease tolerance of turf grass "Miyako". *Grassl. Sci.* 45:416–420.

Salahshoor, F., and F. Kazemi. 2016. Effect of calcium on reducing salt stress in seed germination and early growth stage of *Festuca ovina* L. *Plant Soil Environ.* 62(10):460–466.

Samaranayake, H., T.J. Lawson, and J.A. Murphy. 2008. Traffic stress effects on bentgrass putting green and fairway turf. *Crop Sci.* 48(3):1193–1202.

Sartain, J.B. 1985. Effect of acidity and N source on the growth and thatch accumulation of tifgreen Bermudagrass and on soil nutrient retention. *Agron. J.* 77(1):33–36.

Schmid, C.J., B.B. Clarke, and J.A. Murphy. 2017. Anthracnose severity and annual bluegrass quality as influenced by nitrogen source. *Crop Sci.* 57(Suppl 1):S-285–S-292.

Schmid, C.J., B.B. Clarke, and J.A. Murphy. 2018. Potassium nutrition affects anthracnose on annual bluegrass. *Agron. J.* 110(6):2171–2179.

Schumann, G.L., P.J. Vittum, M.L. Elliott, and P.P. Cobb. 1998. *IPM Handbook for Golf Courses*. Chelsea, Ann Arbor, MI.

Shen, H., H. Du, Z.Wang, and B. Huang. 2009. Differential responses of nutrients to heat stress in warm-season and cool-season turfgrasses. *HortScience* 44:2009–2014.

Shetlar, D., P. Heller, and P. Irish. 1990. *Turfgrass Insect and Mite Manual*. 3rd ed. Pennsylvania Turfgrass Council, Bellefonte, PA.

Sistani, K.R., G.A. Pederson, G.E. Brink, and D.E. Rowe. 2003. Nutrient uptake by ryegrass cultivars and crabgrass from a highly phosphorus-enriched soil. *J. Plant Nutr.* 26(12):2521–2535.

Smiley, R.W., P.H. Dernoeden, and B.B. Clarke. 2005. *Compendium of Turfgrass Diseases*. 3rd ed. APS Press, Pilot Knob, MN.

Smiley, R.W., P.H. Dernoeden, and B.B. Clarke. 2005. *Compendium of Turfgrass Diseases*. 3rd ed. The American Phytopathological Society, St. Paul, MN.

Sonobe, K., T. Hattori, P. An, W. Tsuji, E. Eneji, K. Tanaka, and S. Inanaga. 2009. Diurnal variations in photosynthesis, stomatal conductance and leaf water relation in sorghum grown with or without silicon under water stress. *J. Plant Nutr.* 32(3):433–442.

Soreng, R.J., P.M. Peterson, K. Romaschenko, G. Davidse, F.O. Zuloaga, E.J. Judziewicz, T.A. Filgueiras, J.I. Davis, and O. Morrone. 2015. A worldwide classification of the Poaceae (Gramineae). *J. Syst. Evol.* 2:117–137.

Soylemezoglu, G., K. Demir, A. Inal, and A. Gunes. 2009. Effect of silicon on antioxidant and stomatal response of two grapevine (*Vitis vinifera* L.) rootstocks grown in boron toxic, saline and boron toxic-saline soil. *Sci. Hortic.* 123(2):240–246.

Sprent, J.I. 1987. *The Ecology of the Nitrogen Cycle*. Cambridge University Press, New York.

Stiegler, J.C., M.D. Richardson, and D.E. Karcher. 2011. Foliar nitrogen uptake following urea application to putting green turfgrass species. *Crop Sci.* 51(3):1253–1260.

Stiegler, J.C., M.D. Richardson, D.E. Karcher, T.L. Roberts, and R.J. Norman. 2013. Foliar absorption of various inorganic and organic nitrogen sources by creeping bentgrass. *Crop Sci.* 53(3):1148–1152.

Stiglbauer, B.J., H. Liu, L.B. McCarty, D.M. Park, and J.E. Toler. 2013. Seashore paspalum green establishment affected by sprig rates, nitrogen sources and rates. *Int. Turfgrass Soc. Res. J.* 12:251–256.

Stiglbauer, B.J., H. Liu, L.B. McCarty, D.M. Park, J.E. Toler, and K.R. Kirk. 2009. Diamond zoysiagrass putting green establishment affected by sprigging rates, nitrogen sources, and rates in the southern transition zone. *HortScience* 44:1757–1761.

StJohn, R., N. Christians, H. Liu, and N. Menchyk. 2013. *Secondary Nutrients and Micronutrient Fertilization in Turfgrass Monograph 56*, pp. 521–542. Crop Science Society of America, Madison, WI.

StJohn, R., N.E. Christians, and H.G. Taber. 2003. Supplemental calcium applications to creeping bentgrass established on calcareous sand. *Crop Sci.* 43(3):967–972.

Street, J.R., P.R. Henderlong, and F.L. Himes. 1981. The effect of silica rates on the growth, silica deposition, and water absorption among three turfgrass species, pp. 259–268. *In*: R.W. Sheard (ed.) Proceedingsof the 4th International Turfgrass Research Conference, Guelph, Ontario, Canada, 19–23 July, Inter. Turfgrass Soc., Ontario Agric. College, Univ. of Guelph, Guelph, Ontario.

Subbarao, G.V., O. Ito, W.L. Berry, and R.M. Wheeler. 2003. Sodium—a functional plant nutrient. *Crit. Rev. Plant Sci.* 22(5):391–416.

Sugiura, S., S. Tanaka, C. Mizuniwa, and S. Takahashi. 2017. Sodium chloride transport in the stolons of *Zoysia matrella* in a heterogeneous saline environment. *Int. Turfgrass Soc. Res. J.* 13:610–613.

Taiz, L., and E. Zeiger. 2006. *Plant Physiology*. 4th ed. The Benjamin/Cummings Publishing Company, Redwood City, CA.

Takahashi, M. 2003. Overcoming Fe deficiency by a transgenic approach in rice. *Plant Cell Tissue Organ Cult.* 72(3):211–220.

Taliaferro, C.M. 2003. Bermudagrass, pp. 235–256. *In*: M.D. Casler, and R.R. Duncan (eds.) *Turfgrass Biology, Genetics, and Breeding*. John Wiley & Sons, Hoboken, NJ.

Tani, T., and J.B. Beard. 2002. *Color Atlas of Turfgrass Diseases*. John Wiley & Sons.

Taylor, J.W. 2019. Two Warm-season Turfgrass Responses to Different Levels of Shade and Iron. M.S. Thesis, Clemson University, Clemson, SC.

Throssell, C.S., J. Kruse, C. Bigelow, and J.A. Murphy. 2017. Fertilizer granule collection and nutrient removal from putting greens following mowing. *Int. Turfgrass Soc. Res. J.* 13(1):275–279.

Throssell, C.S., G.T. Lyman, M.E. Johnson, and G.A. Stacey. 2009. Golf course environmental profile measures nutrient use and management and fertilizer restrictions, storage, and equipment calibration. Online. *Appl. Turfgrass Sci.* 6. doi:10.1094/ATS-2009-1203-01-RS.

Tian, J., F.C. Belanger, and B. Huang. 2009. Identification of heat stress-responsive genes in heat-adapted thermal *Agrostis scabra* by suppression subtractive hybridization. *J. Plant Physiol.* 166(6):588–601.

Totten, F.W., H. Liu, L.B. McCarty, C.M. Baldwin, D.G. Bielenberg and J.E. Toler. 2008. Efficiency of foliar versus granular fertilization: A field study of creeping bentgrass performance. *J. Plant Nutr.* 31(5):972–982.

Tremblay, D., R. Belanger, and J. Dionne. 2002. Soluble silicon applications stimulate defense reactions and disease resistance of creeping bentgrass. Annual Meeting Abstract, ASA/CSSA/SSSA, Indianapolis, 10–14 November, 2002.

Trenholm, L.E., L.E. Datnoff, and R.T. Nagata. 2004. Influence of silicon on drought and shade tolerance of St. Augustinegrass. *HortTechnology* 14:487–490.

Trenholm, L.E., R.R. Duncan, R.N. Carrow, and G.H. Snyder. 2001. Influence of silica on growth, quality, and wear tolerance of seashore paspalum. *J. Plant Nutr.* 24(2):245–259.

Turgeon, A.J. 1994. *Turf Weeds and Their Control*. American Society of Agronomy, Crop Science Society of America, Madison, WI.

Turgeon, A.J. 2008. *Turfgrass Management*. 8th ed. Prentice-Hall, Upper Saddle River, NJ.

Turner, R.S., and N.W. Hummel, Jr. 1992. Nutritional requirements and fertilization, pp. 385–439. *In*: D.V. Waddington, R.N. Carrow, and R.C. Shearman (eds.) *Turfgrass Agronomy Monograph 32*. ASA, CSSA, and SSSA, Madison, WI.

United States Golf Association Green Section Staff. 1993. USGA Recommendations for a Method of Putting Green Construction. The 1993 Revision. *USGA Green Section Record* 31(2):1–3.

Uriarte, R.F., H.D. Shew, and D.C. Bowman. 2004. Effect of soluble silica on brown patch and dollar spot of creeping bentgrass. *J. Plant Nutr.* 27(2):325–339.

Vaculík, M., A. Lux, M. Luxová, E. Tanimoto, and I. Lichtscheidl. 2009. Silicon mitigates cadmium inhibitory effects in young maize plants. *Environ. Exp. Bot.* 67(1):52–58.

Van Tran, T., S. Fukai, H.E. Giles, and C.J. Lambrides. 2018. Salinity tolerance among a large range of bermudagrasses (*Cynodon* spp.) relative to other halophytic and non-halophytic perennial C4 grasses. *Environ. Exp. Bot.* 145:121–129.

Vargas, Jr., J.M. 2005. *Management of Turfgrass Disease*. 3rd ed. John Wiley & Sons, Hoboken, NJ.

Vargas, Jr., J.M., and A.J. Turgeon. 2004. *Poa annua—Physiology, Culture, and Control of Annual Bluegrass*. John Wiley & Sons, Hoboken, NJ.

Vaughn, S.F., F.D. Dinelli, J.A. Kenar, M.A. Jackson, A.J. Thomas, and S.C. Peterson. 2018. Physical and chemical properties of pyrolyzed biosolids for utilization in sand-based turfgrass rootzones. *Waste Manag.* 76:98–105.

Volterrani, M., N. Grossi, S. Magni, M. Gaetani, F. Lulli, P. Croce, A. De Luca, and M. Mocioni. 2009. Evaluation of seven cool-season turfgrasses for overseeding a Bermudagrass putting green. *Int. Turfgrass Soc. Res. J.* 11:511–518.

Wang, D., and D.S. Luthe. 2003. Heat sensitivity in a bentgrass variant: Failure to accumulate a chloroplast heat shock protein isoform implicated in heat tolerance. *Plant Physiol.* 133(1):319–327.

Wang, K., and Y. Jiang. 2007. Antioxidant responses of creeping bentgrass roots to waterlogging. *Crop Sci.* 47(1):232–238.

Wang, M., L. Gao, S. Dong, Y. Sun, Q. Shen, and S. Guo. 2017. Role of silicon on plant–pathogen interactions. *Front. Plant Sci.* 8:701.

Wang, X., and B. Huang. 2017. Lipid- and calcium-signaling regulation of HsfA2c-mediated heat tolerance in tall fescue. *Environ. Exp. Bot.* 136:59–67.

Ward, J.T., B. Lahner, E. Yakubova, D.E. Salt, and K.G. Raghothama. 2008. The effect of iron on the primary root elongation of *Arabidopsis* during phosphate deficiency. *Plant Physiol.* 147(3):1181–1191.

Warnke, S. 2003. Creeping bentgrass, pp. 175–185. *In*: M.D. Casler, and R.R. Duncan (eds.) *Turfgrass Biology, Genetics, and Breeding*. John Wiley & Sons, Hoboken, NJ.

White, R.H., and R.E. Schmidt. 1989. Bermudagrass response to chilling temperatures as influenced by iron and benzyladenine. *Crop Sci.* 29(3):768–773.

White, R.H., and R.E. Schmidt. 1990. Fall performance and post-dormancy growth of "Midiron" Bermudagrass in response to nitrogen, iron, and benzyladenine. *J. Am. Soc. Hortic. Sci.* 115(1):57–61.

Wiese, H., M. Nikolic, and V. Römheld. 2007. Silicon in plant nutrition: Effects on zinc, manganese and boron leaf concentrations and compartmentation, pp. 33–47. *In*: B. Sattelmacher, and W.J. Horst (eds.) *The Apoplast of Higher Plants: Compartment of Storage, Transport and Reactions*. Springer, Netherlands.

Xiao, M., K.W. Frank, and T.A. Nikolai. 2018. Foliar and granular fertilizer effects on creeping bentgrass and soil nutrient levels. *Crop Forage Turfgrass Manag.* 4(1):170539.

Xu, Q., and B. Huang. 2000. Effects of differential air and soil temperature on carbohydrate metabolism in creeping bentgrass. *Crop Sci.* 40(5):1368–1374.

Xu, X., and C.F. Mancino. 2001. Annual bluegrass and creeping bentgrass response to varying levels of iron. *HortScience* 36(2):371–373.

Xu, Y., and B. Huang. 2009. Effects of foliar-applied ethylene inhibitor and synthetic cytokinin on creeping bentgrass to enhance heat tolerance. *Crop Sci.* 49(5):1876–1884.

Xu, Y., and B. Huang. 2010. Responses of creeping bentgrass to trinexapac-ethyl and biostimulants under summer stress. *HortScience* 45(1):125–131.

Zhang, J., J. Geng, H. Ren, J. Luo, A. Zhang, and X. Wang. 2011. Physiological and biochemical responses of *Microcystis aeruginosa* to phosphite. *Chemosphere* 85(8):1325–1330.

Zhang, Q., J. Fry, K. Lowe, and N. Tisserat. 2006. Evaluation of calcium silicate for brown patch and dollar spot suppression on turfgrasses. *Crop Sci.* 46(4):1635–1643.

Zhang, X., and E.H. Ervin. 2004. Cytokinin-containing seaweed and humic acid extracts associated with creeping bentgrass leaf cytokinins and drought resistance. *Crop Sci.* 44(5):1737–1745.

Zhang, X., E.H. Ervin, and A.J. LaBranche. 2006. Metabolic defense responses of seeded Bermudagrass during acclimation to freezing stress. *Crop Sci.* 46(6):2598–2605.

Zhang, X., R.E. Schmidt, E.H. Ervin, and S. Doak. 2002. Creeping bentgrass response to natural plant growth regulators and iron under two regimes. *HortScience* 37:898–902.

Zhang, X., K. Wang, and E.H. Ervin. 2010. Optimizing dosages of seaweed extract-based cytokinins and zeatin riboside for improving creeping bentgrass heat tolerance. *Crop Sci.* 50(1):316–320.

Zhang, Y., A.C. Guenzi, M.P. Anderson, C.M. Taliaferro, and R.A. Gonzales. 2006. Enrichment of Bermudagrass genes associated with tolerance to the spring dead spot fungus *Ophiosphaerella herpotricha*. *Physiol. Mol. Plant Pathol.* 68(4–6):105–118.

Zheng, L., F. Huang, R. Narsai, J. Wu, E. Giraud, F. He, L. Cheng, et al. 2009. Physiological and transcriptome analysis of iron and phosphorus interaction in rice seedlings. *Plant Physiol.* 151(1):262–274.

18 Molecular Chaperones and Acquisition of Thermotolerance in Plants

Hitoshi Nakamoto and Tahmina Akter

CONTENTS

18.1 INTRODUCTION

Plants growing in natural environments encounter various biotic and abiotic stresses during all periods of their growth and development. Biotic stresses such as pests and pathogens and abiotic stresses such as high temperature, low temperature, high light (photo-oxidative), drought, nutrient and salt stresses, are the primary causes of crop yield and quality reductions. High temperature or heat stress can lead to retardation in plant growth and development, and even death. It causes protein dysfunction/damage and alters the cellular proteome, and it can lead to the accumulation of toxic protein aggregates. Plants, as well as other organisms, respond to heat stress by inducing heat-shock proteins (Hsps) for self-defense. Thermotolerance in plants is related to the expression of Hsps (Vierling, 1991). Among major Hsps, there are highly conserved proteins in living organisms, namely 'molecular chaperones.' They are expressed under normal and various stress

conditions. They play an important role in the maintenance of protein homeostasis in cells.

In this chapter, we summarize recent progress made toward understanding the roles of Hsps and molecular chaperones in terms of thermotolerance in plants.

18.2 HSPS AND MOLECULAR CHAPERONES

The unexpected discovery by Ritossa that a new puffing pattern of the polytene chromosomes in the salivary glands of the fruit fly *Drosophila* was induced upon a heat-shock treatment demonstrated a rapid new RNA synthesis at puff sites upon heat shock (Ritossa, 1962). This observation opened up a way to study cellular/biochemical responses to heat stress. When sodium dodecyl sulfate-polyacrylamide gel electrophoresis (SDS-PAGE) became available, newly synthesized proteins of the salivary glands upon heat shock, that is radiolabeled proteins due to the heat-induced incorporation of [35S] methionine, were separated. A set of heat-shock induced proteins was detected whereas the synthesis of most other proteins which had been synthesized before the heat shock was significantly suppressed (Tissières et al., 1974; Koninkx, 1976). The size of the most highly expressed protein was around 70 kDa. These heat-shock induced proteins, including the 70-kDa protein, were later called heat-shock proteins (Hsps). Initially, only several Hsps were detected, but now more than a hundred Hsps are shown to be induced in plants like Arabidopsis as the techniques have been improved to attain a higher resolution and sensitivity for the analysis of proteins (for example, see Palmblad et al., 2008). Not all Hsps, however, are molecular chaperones.

The term molecular chaperone was originally used by Laskey et al. (1978) in order to describe the properties of nucleoplasmin, an acidic nuclear protein necessary to correctly assemble nucleosome cores out of DNA (acidic) and histone (basic). The term was later defined and more generalized by Ellis (1996) as follows: Molecular chaperones are the class of proteins which assist the folding of certain other polypeptide chains so that their assembly into oligomeric structure form correctly. In this definition 'assembly' is used in a broad sense to include not only the folding of newly synthesized polypeptide chains and any association into oligomers that may happen subsequently, but also any changes that may occur when proteins are translocated across membranes, perform their normal functions, or are repaired/removed after damaged by stresses. Under heat stress, many proteins are prone to denaturation. Denatured proteins may form aggregates. Protein aggregation is potentially harmful to the cell. Molecular chaperones suppress protein aggregation, assist refolding of denatured proteins or promote their degradation after stress.

18.3 HEAT-SHOCK RESPONSE

An increase in temperature above the normal growth temperature of an organism induces a wide variety of perturbations in cellular structures and metabolic processes (see Nakamoto and Hiyama, 1999, and references therein). When the magnitude and duration of the heat stress exceeds a threshold, cells are irreversibly damaged and die. Plants, which cannot move, rely on transcriptomic regulation to remodel themselves to survive under stress. For example, genome-wide transcriptome analysis of rice leaves showed that 244 genes are expressed two-fold up-regulated upon heat shock at 42°C for 10–30 min and 238 genes at 42°C for 10 h (Jung et al., 2012). Transcriptional regulation depends on various transcription factors such as heat-shock factors (Hsfs) which we will describe in the next section (18.4). In addition, epigenetic regulations play important roles in transcriptional regulation.

The proteome is modified by the transcriptional changes. Proteome analysis of rice leaves showed that among 48 proteins differentially expressed upon heat shock at 42°C for 12 or 24 h, 18 up-regulated proteins were molecular chaperones (small Hsp, Hsp60/Cpn60, Hsp70, Hsp90) (Lee et al., 2007). Other differentially expressed proteins were categorized into classes related to energy metabolism, redox homeostasis and regulatory proteins. The time course of accumulation of different members of a certain molecular chaperone family (see Section 18.5 below) after heat shock is not the same. Arabidopsis was heat-treated at 38°C for 1, 2, and 4 h, and the dynamics of the heat-shock response was analyzed by quantitative proteomics (Palmblad et al., 2008). Some members of the Hsp70 family were constitutive, e.g., Hsc70-1 and Hsc70-3. Others were early induced, e.g., Hsp70b, and late or slowly induced, e.g., Hsc70-5 and Hsp70. The Hsp90 family also had constitutively expressed and heat-induced members. Many members of the small Hsp family, i.e., Hsp18.1-CI, Hsp17.6A-CI, Hsp17.6B-CI, Hsp17.7-CII, Hsp17.4-CI, and Hsp101 were not detected significantly at 20°C (the growth temperature) but were heat-induced.

The importance of the heat-induction of Hsps/molecular chaperones in cellular thermotolerance comes from the apparent correlation between the level of Hsps/molecular chaperones accumulation and cellular thermotolerance. When whole organisms or cultured cells are given short treatments at a moderately elevated temperature, their resistance to extreme heat increases dramatically (Parsell and Lindquist, 1994). For example, when yeast cells grown at 25°C are pretreated at 37°C, they can survive at 50°C 1000-fold better than non-pretreated cells. This increase in thermotolerance (i.e., acquired thermotolerance) is observed in virtually every organism studied (Parsell and Lindquist, 1994). Such tolerance inducing treatments generally also induce the synthesis of Hsps/molecular chaperones.

Besides a short and rapid increase in temperature, a treatment that has often been used in laboratory experiments but may not be relevant to plants in the field, a gradual increase in temperature also confers thermotolerance in plants (Altschuler and Mascarenhas, 1982; Howarth, 1991). This gradual increase in temperature effectively elicits the synthesis of Hsps/molecular chaperones as well. Arabidopsis thaliana was treated to severe heat stress (45°C) without acclimation or following two different acclimation treatments. Notably, a gradual increase to 45°C (22 to 45°C over 6 h) led to higher survival and higher-fold transcript changes than a step-wise acclimation (90 min at 38°C plus 120 min at 22°C before 45°C).

18.4 HEAT-SHOCK FACTORS

In general, the induction of Hsps, including molecular chaperones is rapid, intense and transient, suggesting that it is an important emergency response. Transcripts of Hsps/molecular chaperones can be detected as early as 5 min after heat shock (for example, see Sung et al., 2001). Their levels may reach a peak within 30 min of heat-shock exposure. In eukaryotic cells, the heat-induction of Hsps/molecular chaperones is mediated by Hsf1. The molecular structure of Hsf1 is highly conserved in eukaryotic species. The basic structure of Hsf1 consists of an N-terminal helix-turn-helix DNA-binding domain with an adjacent oligomerization domain that consists of hydrophobic heptad repeats (HR-A/B), and a C-terminal heptad repeat

(HR-C) (von Koskull-Doring, 2007; Vihervaara and Sistonen, 2014). Hsf1 binds to the heat-shock elements (HSEs) that are composed of inverted nGAAn pentamers. Hsf1 is functional as a trimer since each DNA-binding domain of the Hsf1 trimer recognizes a single nGAAn, and three nGAAn are necessary for an Hsf1 to bind the cis-acting element stably (Vihervaara and Sistonen, 2014). HSEs are conserved in promoters of heat-shock-inducible genes of all eukaryotes. Under normal conditions, molecular chaperones such as Hsp90 and Hsp70 suppress trimerization of Hsf1. Upon heat shock, molecular chaperones associate with denatured proteins and, thus, Hsf1 is freed from the molecular chaperones to form a trimer. The Hsf trimer binds to HSEs and induces Hsp/molecular chaperone gene expression. The basic working principle of Hsf is thought to be universal for all eukaryotic cells, but it is more complicated in plant cells because of a large number (19 to 52) of Hsfs in plants as compared with that in vertebrates, i.e., only ~4 (Scharf et al., 2012). On the basis of structural characteristics, plant Hsfs are categorized into three classes, A, B, and C. In *Arabidopsis thaliana*, there are 21 genes encoding Hsfs. Among them, 15 members are in class A, 5 members in class B and 1 in class C (Guo et al., 2008; Scharf et al., 2012).

18.4.1 HEAT-SHOCK FACTOR A1 (HSFA1S)

Heat shock factor A1s (HsfA1s) are categorized in class A. HsfA1s play major roles in the heat-shock response. In Arabidopsis, there are four HsfA1s. HsfA1a, HsfA1b, or HsfA1d alone, but not HsfA1e, can induce a heat-shock response. However, the triple knockout mutant of HsfA1a/HsfA1b/HsfA1d show drastic defects in heat-shock response, indicating that HsfA1a, HsfA1b, or HsfA1d are the master regulators of heat stress response in Arabidopsis (Yoshida et al., 2011). The four HsfA1-type proteins also play an important role in gene expression under normal conditions since the HsfA1a/HsfA1b/HsfA1d/HsfA1e quadruple mutant showed severe growth retardation and abnormal morphological phenotypes (Yoshida et al., 2011; Liu and Charng, 2013).

18.4.2 HEAT-SHOCK FACTOR A2 (HSFA2)

Expression of HsfA2 that is the most highly heat-induced among all Arabidopsis Hsfs is positively regulated by HsfA1 (Busch et al., 2005). HsfA2 is heat-induced whereas HsfA1s are expressed constitutively. HsfA2 sustains the expression of Hsp/molecular chaperone genes and extends the duration of acquired thermotolerance in Arabidopsis (Charng et al., 2007). Acquired thermotolerance of Arabidopsis decays faster in the absence of HsfA2. Salt and osmotic stresses also induce the HsfA2 gene expression, and HsfA2-overexpressing Arabidopsis transgenic lines show enhanced tolerance to salt and osmotic stresses as well as heat stress (Ogawa et al., 2007). HsfA1 not only regulates the transcription of HsfA2, but it is also required for the efficient nuclear transport of HsfA2 (Scharf et al., 1998). These Hsfs form hetero-oligomers. The physical interaction between HsfA1 and HsfA2 results in synergistic transcriptional activation of heat stress gene expression (Chan-Schaminet et al., 2009).

There are many reports regarding altered stress tolerance in mutant plants with altered Hsf expression (see Fragkostefanakis et al., 2015).

18.5 FIVE EVOLUTIONARILY CONSERVED MOLECULAR CHAPERONE FAMILIES

Five molecular chaperone families that are highly conserved among different species are classified on the basis of their approximate molecular masses (Vierling, 1991; Parsell and Lindquist, 1993; Boston et al., 1996; Nakamoto and Hiyama, 1999; Hartl et al., 2011): small Hsp (also called or abbreviated as low molecular weight/mass Hsp, small stress protein, sHsp, or α-Hsp), chaperonin/Hsp60/GroEL, Hsp70/DnaK, Hsp90/HtpG, and Hsp100/ClpB (Table 18.1).

A single plant species has multiple members for a chaperone family. Some, but not all, members are heat-induced. In the above-mentioned proteome analysis (Lee et al., 2007), three members of the Hsp100 family, seven of Hsp70 and seven of small Hsp were heat-induced. Some Hsps/molecular chaperones are produced at particular stages of the cell cycle or during development in the absence of stress (Vierling, 1991; Nakamoto and Hiyama, 1999; Wang et al., 2004).

As suggested from the fact that molecular chaperones are constitutively expressed and/or their expression is developmentally regulated, the importance of molecular chaperones extends beyond their role in protection from high temperature and other stresses. All the above-mentioned, representative molecular chaperones are involved in general and essential cellular functions; for example, protein folding and subunit assembly, protein translocation across membranes, signal transduction and intracellular protein breakdown. In fact, some molecular chaperones such as *Escherichia coli* GroEL (Hsp60), and yeast Hsp82 (Hsp90) have been shown to be essential for viability (Borkovich et al., 1989; Fayet et al., 1989).

TABLE 18.1

Molecular Chaperone Families That Are Highly Conserved, Ubiquitously Distributed

Chaperone Families	(Biochemical) Functions
Small Hsp (sHsp)	Prevents aggregation
Chaperonin/Hsp60/GroEL	Prevents aggregation
	Assists protein folding
Hsp70/DnaK	Prevents aggregation
	Assists protein folding
	Re-solubilizes protein aggregates
	Assists protein translocation
	Involved in protein degradation
Hsp90/HtpG	Prevents aggregation
	Binds and stabilizes/regulates transcriptions, protein kinases, etc
	Involved in protein degradation
Hsp104/ClpB	Re-solubilizes protein aggregates

18.5.1 Mechanism for Molecular Chaperones to Maintain Protein Homeostasis

Temperature increase can lead to denaturation of a protein or a protein complex, resulting in exposure of hydrophobic amino acid residues normally buried within the interior of the protein (Ellis, 1996). A denatured protein may misfold and/ or aggregate due to the interaction of the surface-exposed hydrophobic region with other such regions of itself or other denatured proteins. Molecular chaperone binds to the exposed hydrophobic surface of a non-native protein to prevent its misfolding and/or aggregation and subsequently facilitates refolding to its native conformation and assembly into a protein complex. Molecular chaperones can prevent improper interactions among surface-exposed hydrophobic regions of a protein(s). Their interaction with denatured proteins is transient, and they are not present in native proteins/protein complexes. Thus, they behave like enzymes.

Presently, three major functions which can be analyzed *in vitro* are described for molecular chaperones. First, '(un)foldase' activity that assists/facilitates protein folding. Second, 'holdase' activity which binds an unfolding protein to prevent its aggregation and holds it in a folding/assembly competent state. Third, 'disaggregase' activity that rescues proteins from the aggregated state. All molecular chaperones do not necessarily have all of these activities (Figure 18.1).

18.5.2 Small Hsps

The small Hsps are found in eubacteria, eukaryotes, and archaea (Narberhaus, 2002; Sun et al., 2002; Laksanalamai and Robb, 2004; Nakamoto and Vigh, 2006; Garrido et al., 2012; Waters, 2013; Haslbeck and Vierling, 2015). They are ubiquitous in terms of cellular localization as well as the biological world. They are present in the cytosol as well as in every membrane-bound compartment, e.g., chloroplast, mitochondrion, endoplasmic reticulum (ER), peroxisomes, and nucleus. Often, multiple members of the small Hsp family are

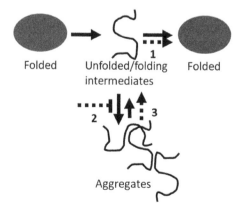

FIGURE 18.1 Various structures of proteins in the cell and functions of molecular chaperones to maintain protein homeostasis. Chaperone functions are indicated by broken lines/arrows. The #1 arrow indicates '(un)foldase' activity. The #2 line indicates 'holdase' activity. The #3 arrow indicates 'disaggregase' activity.

present in one cellular compartment. They are present even in the smallest microbial genome sequenced to date (e.g., *Nanoarchaeum equitans*) (Waters et al., 2003). The small Hsp family in plants is classified into cytosolic class I, cytosolic class II, cytosolic class III, chloroplast, mitochondrion, endoplasmic reticulum, and peroxisome small Hsps (Siddique et al., 2008; Waters, 2013).

The small Hsps are so named because their monomer is relatively small. Their monomeric molecular masses range from 12–42 kDa. Sometimes they are also referred to as the Hsp20 family because almost all members of this family have molecular masses ranging from 15–22 kDa (Waters, 2013). These are the most diverse in terms of primary structure among the major molecular chaperones. They are characterized by their secondary and tertiary structures. The primary amino acid sequence of small Hsps consists of a variable N-terminal region, the α-crystallin domain or Hsp20 domain of approximately 100 amino acids, and a C-terminal extension of variable length. The C-terminal extension may contain organelle specific retention of amino acid motifs (Waters, 2013). The α-crystallin domain which is conserved among small Hsps is built from an immunoglobulin fold that consists of a β-sandwich, comprising two anti-parallel β-sheets (Kim et al., 1998; van Montfort et al., 2001). The majority of small Hsps form oligomers of 12 to >32 subunits.

Small Hsps are ATP-independent holdases. They create the first line of defense against stress-induced cell damage by preventing aggregation of protein (Haslbeck and Vierling, 2015). They bind non-native proteins to form a soluble complex, preventing their aggregation. The N-terminal arm is likely critical for the ability of small Hsps to efficiently protect many different substrates although there may not be a discrete substrate-binding surface in small Hsps (Jaya et al., 2009). The N-terminal arm and/or C-terminal extension may also be involved in the oligomer formation of small Hsps, and thus dissociation of the small Hsp oligomers may be necessary to make them available for interaction with non-native proteins. The small Hsp/non-native protein complex serves as a transient reservoir of substrates for subsequent refolding by ATP-dependent chaperone systems such as the Hsp70/ DnaK chaperone system. Small Hsps even coaggregate with non-native proteins in order to mediate resolubilization of the aggregates and subsequent refolding by ATP-dependent chaperone systems such as Hsp100/ClpB and the Hsp70/DnaK chaperone system (see Nakamoto and Vigh, 2006, and references therein).

Angiosperms have 11 diverse small Hsp subfamilies (Waters, 2013). Among these subfamilies, six are localized in the cytoplasm/nucleus (CI–CVI), one is localized to the endoplasmic reticulum, one to peroxisome (PX), one to the chloroplast (CP) and two are present in the mitochondrion (MTI and MTII). From 11 subfamilies, one is present only in eudicots (CVI), and the rest (CP, MTI, MTII, ER, PX, CI, CII, CIII, CIV, and CV) are present in both monocots and eudicots. Arabidopsis contains 19 small Hsps (Scharf et al., 2001), rice 23 (Sarkar et al., 2009) and poplar 36 (Waters et al., 2008).

One notable structural feature of the chloroplast small Hsp (Hsp21) is that it contains a set of conserved methionines. Methionines M49, M52, M55, M59, M62, M67 are located on one side of an amphipathic helix, which may fold back over two other conserved methionines (M97 and M101), to form a binding groove lined with methionines, for recognition of proteins with an overall hydrophobic character (Sundby et al., 2005). As small Hsps protect other proteins from aggregation by binding to their hydrophobic surfaces, keeping the conserved methionines in a reduced form is a prerequisite for maintaining such binding. Their results suggest that Hsp21 kept active by the reducing power generated by photosynthetic electron transport protects proteins in chloroplast under heat stress.

Evidence for involvement of small Hsps in plant thermotolerance is found in literature; for example, enhanced thermotolerance of transformed tobacco plants with the introduction of the tomato mitochondrial small Hsp gene (Sanmiya et al., 2004), increased thermotolerance by overexpression of a small Hsp (sHsp17.7) in rice (Murakami et al., 2004), and increased thermotolerance observed in Arabidopsis that constitutively expresses a cytosolic class I small Hsp from *Rosa chinensis* (Jiang et al., 2009). Arabidopsis cytosolic small Hsps (CI and CII small Hsps) are also shown to be important for heat stress tolerance with RNA interference and overexpression lines (McLoughlin et al., 2016). Carrot transgenic cells and regenerated plants, which constitutively expressed the carrot small Hsp (Hsp17.7) gene, showed more thermotolerance than the vector controls (Malik et al., 1999). In contrast, heat-inducible Hsp17.7 antisense lines were less thermotolerant than the vector controls. Arabidopsis CI and CII small Hsps interact with eukaryotic translation elongation factor 1B and eukaryotic translation initiation factor 4A, suggesting that they protect protein translation factors during heat stress (McLoughlin et al., 2016). Constitutive overexpression of a small Hsp increases the thermal resistance of photosystem II (Nakamoto et al., 2000), indicating that small Hsp protects the photosynthetic apparatus under heat stress. Small Hsp also stabilizes membrane proteins such as the photosystems and soluble proteins such as phycobilisomes from oxidative damage (Sakthivel et al., 2009).

Small Hsp has not only the protein-protective activity but also an ability to stabilize lipid membranes (Torok et al., 2001). A mutant small Hsp with increased thylakoid association provided an elevated resistance against UV-B damage in the cyanobacterium *Synechocystis* sp. PCC 6803 (Balogi et al., 2008). In this cyanobacterium, it was shown that the wild-type small Hsp (Hsp17) is equally distributed between thylakoid and cytosolic fractions whereas a mutant small Hsp17 mutated in position L9P appeared primarily in the soluble, and another one mutated in position Q16R was found exclusively in the thylakoid membrane fraction. Compared with the wild type, the Q16R mutant had an enhanced lipid-mediated thylakoid membrane interaction, which directly affected the photosystem II complex and led to a greatly enhanced resistance to UV-induced photosystem II inactivation via facilitating photosystem II repair (Balogi et al., 2008). In this connection, it is interesting to note that immunocytochemical studies showed that cyanobacterial small Hsp has dynamic properties to change its localization between cytosol and thylakoid membranes during heat shock (Nitta et al., 2005), indicating that the only one small Hsp (HspA) in cyanobacteria takes care of both cytoplasmic proteins and membranes/membrane proteins. In fact, constitutive overexpression of HspA stabilizes both thylakoid membranes and the most abundant cytoplasmic proteins in cyanobacteria, phycobiliproteins (i.e., phycocyanins) under elevated temperatures (Nitta et al., 2005; Nakamoto and Honma, 2006).

18.5.3 Chaperonin/Hsp60/GroEL

Like small Hsp, chaperonin (Cpn60)/Hsp60/GroEL is found throughout all three domains. It is ubiquitous in terms of cellular localization as well as the biological world, existing in cytosol, chloroplast and mitochondrion. Cpn60/Hsp60/GroEL, at least one of them when multiple homologs are present in a cell, is essential (for example, see Fayet et al., 1989).

The two subfamilies of this family have been identified along recognizable evolutional lines (Horwich et al., 2007). Type I is found in bacteria (GroEL) and endosymbiotically related organelles, mitochondria (Hsp60), and chloroplasts (Cpn60 or Rubisco binding protein), whereas type II resides in the archaebacterial cytosol (thermosome)/eukaryotic cytosol (CCT/TriC). Type I and type II chaperonins are distantly related to each other. Both type I and type II chaperonins are megadalton-size double-ring assemblies that are composed of seven or eight, identical or similar ~60-kDa subunits. The assemblies provide an encapsulated cavity where a non-native polypeptide folds productively. What differentiates type I and type II chaperonins structurally is the presence and absence of a detachable 'lid' structure for encapsulation. Type I chaperonins require a co-chaperonin (Cpn10, Cpn20, Hsp10, or GroES) to close the cavity, while type II chaperonins have a built-in protrusion structure to perform this function.

Chaperonins suppress aggregation of denatured proteins. They provide essential assistance to the folding/refolding of newly translated/non-native and newly translocated proteins. Despite the different encapsulation mechanisms, the ATP-directed chaperone cycles of the two subfamilies appear to be similar (Horwich et al., 2007). We will discuss type I chaperonins, GroEL and chloroplast chaperonin below. We will not discuss type II chaperonins since little is known about the type II chaperonins in plants. Readers should consult recent reviews (Lopez et al., 2015; Skjærven et al., 2015) for detailed information and discussions concerning type II chaperonin mostly in non-plant organisms.

18.5.3.1 Type I Chaperonin

Among the type I chaperonins, the *E. coli* GroEL is most extensively studied. The chaperonin forms a large oligomer that is composed of 14, identical ~60-kDa GroEL subunit (Boisvert et al., 1996; Xu et al., 1997). The subunits are arranged in a barrel-like complex that is made up of two stacked heptameric rings, which enclose a 'cavity.' GroEL consists of three domains: an equatorial domain that forms the foundation of the assembly, an apical domain that forms the end of the ring, and an intermediate domain that connects the two (Braig et al., 1994; Xu et al., 1997). The equatorial domain includes most of the connections between monomers of the same ring and between rings and contains the MgATP/MgADP-binding pocket. The apical domain binds GroES or a substrate polypeptide. GroES also called co-chaperonin, forms a homo-heptamer ring and plays an important role in the GroEL's chaperone action as described below.

How does GroEL assist folding of a non-native protein in an ATP-dependent way? The principal mechanism is summarized as follows (Horwich et al., 2007; Saibil et al., 2013). An 'open' ring with or without ATP may accept a non-native polypeptide. Binding of ATP to the equatorial domains of the ring renders the apical domains competent to bind GroES, whose association is accompanied by a further large movement of the apical domains. This drives the release of polypeptide substrate protein from the cavity wall into the now encapsulated hydrophilic, so-called *cis* cavity, where folding then commences before ATP hydrolysis in the *cis* ring weakens the affinity of GroEL for GroES and leads the entry of ATP into the sites of the opposite, so-called *trans* ring. It takes ~10 seconds for the hydrolysis of ATP. The binding of ATP to the *trans* ring triggers the release of the *cis* ligands (GroES, the substrate protein, and ADP). At the same time, ATP binding in the *trans* ring also enables binding of GroES to that ring, triggering a further round of protein folding (Figure 18.2). The mechanism of protein folding in chloroplast Cpn60 is thought to be similar to that in GroEL.

18.5.3.2 Chloroplast Type I Chaperonin

Chloroplast Cpn60 was first discovered as the Rubisco binding protein (Barraclough and Ellis, 1980). There are two types of Cpn60, α, and β, whose nuclear-encoded precursors are synthesized outside the chloroplast and imported into it. These proteins are constitutively expressed, and their levels increase only slightly during heat shock. The two chaperonins are ~50% identical in terms of amino acid sequences, and the occurrence of the two types of Cpn60s in a higher plant plastid is general (Martel et al., 1990; Hill and Hemmingsen, 2001). To find out the significance of these isoforms, purified recombinant Cpn60α and Cpn60β from pea were subjected to *in vitro* reconstitution experiments (Dickson et al., 2000). In the presence of ATP and/or co-chaperonins, Cpn60β formed a homo-tetradecamer. In contrast, Cpn60α only assembled into an α/β hetero-tetradecamer in the presence of Cpn60β. Both Cpn60β- and Cpn60α/β-tetradecamer with co-chaperonins and ATP were active in folding a denatured protein *in vitro*. Similar observations have been reported with Cpn60α and Cpn60β from *Arabidopsis thaliana* and *Chlamydomonas reinhardtii* (Vitlin Gruber et al., 2013; Bai et al., 2015). The *A. thaliana* genome contains two genes encoding Cpn60α and four genes encoding Cpn60β (Hill and Hemmingsen, 2001). Hetero-tetradecamer consisting of Cpn60α, Cpn60β1, Cpn60β2 and Cpn60β3 is detected in *A. thaliana* chloroplast by an analysis of the stromal proteome in its oligomeric state extracted from highly purified chloroplasts (Peltier et al., 2006).

There are some reports suggesting the involvement of Cpn60 in the thermotolerance of photosynthetic organisms. Arabidopsis Cpn60β interacts with Rubisco activase which is heat-labile, indicating that it protects Rubisco activase (and hence photosynthesis) from thermal denaturation (Salvucci, 2008). Arabidopsis Cpn60β is involved in cell death since its mutation (len1 mutation) leads to accelerated cell death to heat-shock stress in comparison with wild-type plants (Ishikawa et al., 2003). One of the two GroELs of the thermophilic cyanobacterium *Thermosynechococcus elongates*, GroEL2, which is dispensable under normal growth conditions plays an essential role under stresses including heat and cold (Sato et al., 2008). Constitutive overexpression of two GroELs in the cyanobacterium *Synechocystis* sp. PCC 6803 displays improved cellular thermotolerance and also reduced photobleaching of

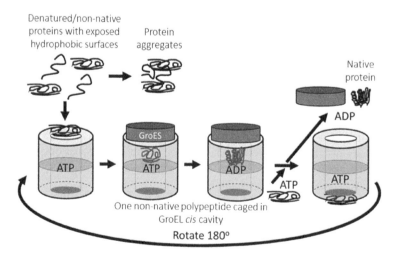

FIGURE 18.2 Chaperone mechanism of GroEL.

(Modified from Nakamoto, H. and K. Kojima, 2016).

phycocyanin, a major protein in a cyanobacterial cell under heat stress conditions (Nakamoto et al., 2003).

Cpn60 plays an important role under non-stress conditions. Arabidopsis Cpn60α1 is involved in chloroplast development and subsequently for the proper development of the plant embryo and seedling (Apuya et al., 2001). Arabidopsis Cpn60α and Cpn60β are suggested to be required for the formation of a normal plastid division apparatus since the reduction of the Cpn60 levels impaired normal FtsZ ring formation (Suzuki et al., 2009). Cpn60 is involved in protein import into chloroplasts since several proteins, e.g., small subunit of Rubisco, beta-subunit of ATP synthase, glutamine synthetase, Rieske iron-sulfur protein, and ferredoxin NADP$^+$ reductase, imported into chloroplasts form stable complexes with Cpn60 (Lubben et al., 1989; Madueno et al., 1993; Tsugeki and Nishimura, 1993). Cpn60 may assist the post-translocational folding of these proteins. Arabidopsis Cpn60β4 forms a hetero-oligomer with Cpn60α1 and other three Cpn60βs. This hetero-oligomer appears to be required to fold a subunit of the chloroplast NADH dehydrogenase-like complex (NDH) and/or for the assembly of NDH (Peng et al., 2011).

18.5.3.3 Chloroplast Co-Chaperonin, Cpn10, and Cpn20

In 1992, Bertsch et al. isolated a protein from pea chloroplast lysate that formed a stable complex with GroEL in the presence of ATP and assisted refolding of denatured bacterial Rubisco, where it replaced GroES (Bertsch et al., 1992). Remarkably, it had twice the molecular mass of the GroES co-chaperonin and termed as Cpn20 or Cpn21. Genes encoding this 'double' GroES-like co-chaperonin from spinach and Arabidopsis were cloned, and the predicted amino acid sequences revealed that the co-chaperonin consists of two GroES-like sequences fused head-to-tail to form a single protein (see Weiss et al., 2009, and references therein). Cpn20 shows 40–50% sequence identity between two GroES-like units (Hirohashi et al., 1999; Koumoto et al., 1999). Arabidopsis Cpn20 forms a tetramer (Koumoto et al., 1999). In addition to the unique co-chaperonin, Arabidopsis chloroplast has a Cpn10 that has only one GroES-like sequence (Hill and Hemmingsen, 2001; Koumoto et al., 2001). There are three co-chaperonin genes in *A. thaliana* which encode two isoforms of Cpn10 and one Cpn20 with predicted chloroplast transit peptides (Hill and Hemmingsen, 2001; Vitlin Gruber et al., 2014), whereas in *C. reinhardtii*, there are also three co-chaperonin genes, with one gene encoding Cpn10 (Cpn11) and the other two encoding Cpn20 and Cpn23 with tandem Cpn10 sequences (Tsai et al., 2012). *C. reinhardtii* Cpn10, Cpn20 and Cpn23 form hetero-oligomeric ring complexes consisting of seven ∼10-kDa domains. For example, a mixture of Cpn10 and Cpn23 yielded a species of 79,327 Da, corresponding to three Cpn10 in complex with two Cpn23. Although co-chaperonins alone are not able to assist GroEL-mediated refolding of bacterial Rubisco *in vitro* but can do it when Cpn20 and/or Cpn23 is combined with Cpn10 (Tsai et al., 2012). *A. thaliana* Cpn10 and Cpn20 also form functional homo-oligomers *in vitro* (Vitlin Gruber et al., 2014).

18.5.3.4 Evolution of Chloroplast Chaperonin Heterogeneity

In contrast to bacterial GroES/GroEL and mitochondrial Hsp10/Hsp60, chloroplast Cpn60 and Cpn10/20/23 show incredible heterogeneity. This heterogeneity may confer Cpn60s and their co-chaperonins great diversity in their structures and functions which may be important for chloroplast development and function.

The heterogeneity comes from the presence of multiple (co-)chaperonin paralogs in chloroplasts. What may be the evolutionary path for the paralogs? After gene duplications, Cpn60β may have retained a property of its ancestral Cpn60 that is involved in the tetradecamer assembly. On the other hand, Cpn60α may have lost it, but have retained other essential properties as a chaperone since Cpn60α deletion is lethal to plants as described above. In chloroplasts, Cpn60α and Cpn60β form a hetero-tetradecamer to function as a chaperone. Thus, we postulate that Cpn60α and Cpn60β partition the ancestral Cpn60 function, and may reflect an outcome of subfunctionalization (Nakamoto and Kojima, 2017).

Cyanobacterial chaperonin genes *groEL1* and *groEL2* appear to follow a different path from the chloroplast Cpn60s in the evolution after gene duplication since GroEL1 and GroEL2 do not form a hetero-oligomer (Nakamoto and Kojima, 2017). GroEL2 does not form a tetradecamer whereas GroEL1 does (Huq et al., 2010). We postulate that the cyanobacterial GroEL2 may reflect an outcome of neofunctionalization since the GroEL paralog appears to take on a totally new structure and function after a gene duplication event (Nakamoto and Kojima, 2017). Phylogenetic analysis indicates that chloroplast Cpn60s are derived from an ancestral cyanobacterial GroEL1 (Suzuki et al., 2009). Thus, we postulate evolutionary paths to chloroplast chaperonins and to cyanobacterial GroELs as shown in Figure 18.3.

18.5.4 Hsp70/DnaK

Hsp70 or DnaK (a prokaryotic homolog) is present in eukaryotic and eubacterial cells. Genes for Hsp70/DnaK homologs are found in a subset of archaea (Large et al., 2009). Hsp70s which are highly conserved proteins are present in the cytoplasm, mitochondrion, chloroplast, endoplasmic reticulum and nucleus of eukaryotic cells (Boston et al., 1996; Renner and Waters, 2007). The Arabidopsis genome contains at least

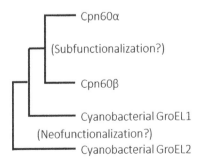

FIGURE 18.3 Evolution of chloroplast Cpn60 and cyanobacterial GroELparalogs.

14 genes encoding members of the Hsp70 family (Lin et al., 2001). Multiple members of the Hsp70 family can be present in one cellular compartment of higher plants. The Hsp70 family contains both heat-inducible and constitutively expressed members, the latter of which are sometimes called heat-shock cognate proteins (Hsc70). Many of higher plant Hsp70s show organ-specific expression pattern.

Hsp70/DnaK consists of an N-terminal ATPase domain of ~45 kDa and a C-terminal peptide-binding domain of ~25 kDa (Mayer and Bukau, 2005; Genevaux et al., 2007; Mayer, 2010). The C-terminal peptide-binding domain is further subdivided into a β-sandwich subdomain (a substrate-binding cavity) and an α-helical subdomain. The latter functions like a lid which closes the cavity, leading to an increase in the affinity of Hsp70/DnaK toward a substrate. Hsp70/DnaK interacts with extended, hydrophobic peptide segments in an ATP-controlled fashion (Mayer and Bukau, 2005; Genevaux et al., 2007; Mayer, 2010). ATP binding to the N-terminal ATPase domain triggers the transition to the low-affinity state (ATP-bound state) of the C-terminal peptide-binding domain, while ATP hydrolysis leads to the high-affinity state (ADP-bound state). In the ATP-bound state of DnaK, the β-sandwich subdomain and the α-helical subdomain are completely detached from each other, explaining the low affinity to a protein substrate (Kityk et al., 2012). The N-terminal ATPase domain and the C-terminal peptide-binding domain mutually communicate each other allosterically. Hsp40/DnaJ/J-protein and/or a substrate binding to Hsp70/DnaK stimulates ATP hydrolysis, resulting in stable binding of a substrate protein whereas nucleotide exchange factor (NEF) enhances the exchange of ADP to ATP, and thus releasing of the unfolded intermediate from Hsp70/DnaK. After releasing from the substrate-binding cavity of Hsp70/DnaK unfolded, it immediately and spontaneously refolds to its native state (Sharma et al., 2010). Hsp70 is thought to unfold stable, misfolded or aggregated proteins by the use of ATP and convert them into natively refoldable species (Sharma et al., 2010; Mattoo et al., 2013).

Hsp70 plays diverse roles in cells, including (re)folding of newly synthesized or unfolded polypeptides, assisting in transport of proteins into organelles, dissociation of macro-molecular complexes/aggregates, and targeting proteins to proteasomes for degradation (Mayer and Bukau, 2005; Bukau et al., 2006; Genevaux et al., 2007). Thus, Hsp70 assists protein homeostasis in cells.

Transgenic Arabidopsis whose level of Hsp70/Hsc70 was reduced by an Hsp70 antisense gene exhibited less thermotolerance than the wild type (Lee and Schoffl, 1996). Mutational studies indicated that a chloroplast member plays a role for thermotolerance. Two plastid Hsc70 (cpHsc70-1 and cpHsc70-2) T-DNA insertion knockout mutants, Δcphsc70-1 and Δcphsc70-2, of Arabidopsis were isolated. Although no visible phenotype was observed in the Δcphsc70-2 mutant under normal growth conditions, the Δcphsc70-1 mutant plants exhibited variegated cotyledons, malformed leaves, growth retardation and impaired root growth. After heat-shock treatment of germinating seeds, root growth from Δcphsc70-1 seeds was further impaired (Su and Li, 2008). Transgenic

tobacco plants that constitutively expressed elevated levels of nucleus-localized Hsp70 (NtHSP70-1) showed a higher level of thermotolerance than antisense transgenic seedlings or transgenic seedlings carrying only the vector (Cho and Choi, 2009). Hsp70 is involved in abiotic stress responses other than high-temperature response. Arabidopsis mutants which over-express cytosolic/nuclear Hsp70 (HSC70-1) show tolerance to salt, cadmium and arsenic (Cazale et al., 2009). Interestingly, Hsp70, as well as Cpn60, greatly increases in abundance in chromoplast differentiation during flower formation in *Narcissus pseudonarcissus* (Bonk et al., 1996). The plastid Hsp70 interacts with phytoene desaturase.

18.5.4.1 J-Protein

J-protein is characterized by the presence of J-domain with histidine, proline and aspartate (HPD) signature motif, which is essential for the interaction of the protein with Hsp70/DnaK (Cheetham and Caplan, 1998; Walsh et al., 2004; Kampinga and Craig, 2010). The J-domain consists of four helices. The turn/loop between the helix II and III has the sequence HPD. As described above, J-proteins such as Hsp40 and DnaK interact with Hsp70 to enhance its ATPase activity and controls its binding affinity for a substrate protein.

J-proteins are classified into three types, type I, type II and type III (Cheetham and Caplan, 1998; Walsh et al., 2004; Kampinga and Craig, 2010). Type I J-proteins have J-domain that is linked by a glycine/phenylalanine (G/F)-rich region to a zinc-finger domain followed by a carboxy-terminal domain/extension. The zinc-finger domain and the carboxy-terminal domain can interact with non-native protein substrates. Type II J-proteins lack a zinc-finger domain. Type III J-proteins have a J-domain but lack the other sequence features found in other types (Cheetham and Caplan, 1998; Walsh et al., 2004; Kampinga and Craig, 2010). In addition to these types, type IV J-proteins may be included (Walsh et al., 2004). Type IV J-proteins are classified as proteins having noticeable structural and sequence similarities with J-domain without HPD motif. J-proteins often far outnumber the number of Hsp70s in a cell or cellular compartment (Kampinga and Craig, 2010). Often, multiple J-proteins function with a single Hsp70, and functional diversity of Hsp70s is driven by a diverse class of J-proteins. Among 120 Arabidopsis J-proteins, 8 are type I, 16 are type II and 92 of them are type III and other 4 are in type IV (Rajan and D'Silva, 2009).

J-protein is located in various subcellular compartments such as chloroplast, mitochondrion, endoplasmic reticulum, vacuole, cytosol and nucleus of eukaryotes. Homologs are also present in prokaryotes. Based on Subcellular location database analysis for Arabidopsis proteins it has been predicted that among 120 J-proteins, 50 proteins present in cytosol, 19 in mitochondria, 12 in the chloroplast, 9 in the endoplasmic reticulum, 3 in the cytoskeleton, 1 in the plasma membrane, 24 in the nucleus and 2 in the vacuole.

J-proteins are involved in numerous and diverse functions including protein (re)folding, protein transport to intracellular organelles, and even protein degradation in the eukaryotic cells (Kampinga and Craig, 2010). During protein biosynthesis, J-proteins with Hsp70 provide the first line of defense against protein aggregation. Hsp70 and J-protein associated with ribosome bind to a nascent polypeptide chain that has just emerged from the ribosome tunnel and protect them from forming aggregates (Craig et al., 2003; Kampinga and Craig, 2010). In Arabidopsis mitochondria, J-proteins are present in both matrix and inner membrane. Mitochondrial membrane J- protein and J-like protein Pam18 and Pam16 are involved in transporting non-native protein into the matrix in cooperation with mitochondrial Hsp70 (D'Silva et al., 2003). Some J-proteins shunt substrate proteins toward degradative pathways besides facilitating protein folding. An Arabidopsis chloroplast J-protein (J20) is involved in the degradation of deoxyxylulose 5-phosphate synthase, the first enzyme of the plastidial isoprenoid pathway (Pulido et al., 2013).

In *E. coli*, DnaJ which is heat-induced is essential for its viability at high temperature (Saito and Uchida, 1977), indicating that J-proteins are involved in thermotolerance. A tomato chloroplast J-protein (SlCDJ2) which is also heat-induced contributes to thermotolerance since enhanced heat tolerance was observed in transgenic plants overexpressing SlCDJ2 (Wang et al., 2015). The J-protein is involved in the maintenance of CO_2 assimilation and Rubisco activity under heat stress. An Arabidopsis mitochondria J-protein (BIL2) participates in resistance against salinity and strong light stress (Bekh-Ochir et al., 2013). A soybean nuclear-localized type III J-protein (GmHsp40.1) plays a positive role in pathogen defense since silencing GmHsp40.1 enhanced the susceptibility of soybean plants to the mosaic virus (Liu and Whitham, 2013).

18.5.4.2 Nucleotide Exchange Factor

In addition to J-protein, NEF is also required as a co-chaperone in the Hsp70 chaperone system. NEF interacts with the N-terminal ATPase domain of Hsp70 (Bracher and Verghese, 2015). It accelerates nucleotide (ADP/ATP) exchange in Hsp70. In prokaryotes, GrpE which is an essential protein performs this exchange (Genevaux et al., 2007). It has a molecular mass of ~22 kDa and forms a stable dimer in solution. GrpE facilitates the replacement of Hsp70-bound ADP with ATP. Interestingly, the *E. coli* GrpE dimer acts as a thermosensor (Grimshaw et al., 2001). The GrpE shows a temperature-dependent reversible inactivation. Due to this inactivation, DnaK may be kept in the ADP-bound state, stabilizing denatured proteins against aggregation under thermal stress.

In eukaryotes, GrpE homologs are present in chloroplast and mitochondrion whereas a large diversity of Hsp70 NEFs has been discovered in the eukaryotic cytosols (Bracher and Verghese, 2015). These NEFs which are not homologous to GrpE belong to the Hsp110/Grp170, HspBP1/Sil1 and BAG domain protein families. BAG (Bcl-2 associated proteins athanogene) is a family of protein having a conserved region located near the C-terminal, named the BAG domain. The family members perform diverse functions including stress management. The BAG domain mediates direct interaction with the ATPase domain of Hsp70 (Brive et al., 2001). The *Arabidopsis thaliana* genome contains seven homologs of the

BAG family, including homologs with domain/motif organization similar and dissimilar to animal BAGs (Kabbage and Dickman, 2008). For example, *Arabidopsis thaliana* BAG5 is unique in that it has a calmodulin binding motif. It suggests a potential role in the plant calcium signaling pathway. Hsp110 and its ER-lumenal homolog Grp170 share their domain architecture with Hsp70, consisting of an N-terminal nucleotide binding domain followed by a β-sandwich and an α-helical domain, but have long insertions and C-terminal extensions compared to Hsp70. The Arabidopsis genome contains at least four members of the Hsp110/Grp170 family (Lin et al., 2001).

18.5.4.3 The Hsp70 Chaperone System

The chaperone cycle of the Hsp70 chaperone system is summarized as follows (Mayer and Bukau, 2005; Genevaux et al., 2007; Mayer, 2010). J-protein binds a non-native protein (and protects it from aggregation) and then interacts with the ATP-bound state of Hsp70. The interaction stimulates ATP hydrolysis and the closing of the substrate-binding cavity of Hsp70. Thus, the ADP-bound form of Hsp70 exhibits a high affinity for its substrate. NEF helps the release of ADP and binding of ATP to Hsp70 through its interaction with the Hsp70 ATPase domain. In the ATP-bound state of Hsp70, its substrate-binding cavity becomes opened. Thus, the ATP-bound Hsp70 exhibits a low affinity for its substrate, releasing the bound substrate from Hsp70. The chaperone cycle, that is, the binding and releasing a protein substrate, is thought to facilitate protein folding (Figure 18.4).

18.5.5 Hsp90/HtpG

Hsp90 is present in eukaryotic and eubacterial cells. The eubacterial homolog is called HtpG. Archaea generally lack genes for Hsp90/HtpG (Large et al., 2009). Eukaryotic members of the Hsp90 family are present in chloroplast, mitochondrion, endoplasmic reticulum and predominantly localized in the cytoplasm (Gupta, 1995; Johnson, 2012). The *Arabidopsis thaliana* genome contains seven members of the Hsp90/HtpG family. Four members which contain the highly conserved C-terminal pentapeptide MEEVD constitute the cytoplasmic subfamily, whereas the other three Hsp90s are predicted to be within the plastidial, mitochondrial and endoplasmic

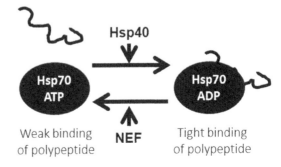

FIGURE 18.4 Chaperone mechanism of Hsp70/DnaK.

reticulum compartments, respectively (Krishna and Gloor, 2001).

The fundamental structure of Hsp90 is evolutionarily conserved. It forms a constitutive dimer (Pearl and Prodromou, 2006). Each monomer consists of three domains: N-terminal ATP binding domain, middle domain and C-terminal dimerization domain. Hsp90 shows very weak, intrinsic ATPase activity. Structural, biochemical and mutational analysis of Hsp90 showed that conformational/domain rearrangements of Hsp90 are coupled to the ATPase reaction (Pearl and Prodromou, 2006; Wandinger et al., 2008; Prodromou, 2012). Hsp90 is in an 'open' conformation in the absence of a nucleotide (i.e., ATP or ADP). The 'open' state is dynamic since nucleotide-free Hsp90 takes multiple conformations in equilibrium (Krukenberg et al., 2008). However, upon ATP binding to the N-terminal domain of Hsp90, the domain dimerizes transiently to form a 'closed' conformation. This conformational cycle driven by ATP binding is thought to drive structural changes of a substrate protein and its release (Pearl and Prodromou, 2006; Wandinger et al., 2008; Prodromou, 2012).

Substrate (or client) binding is thought to occur at multiple binding sites of the closed, ATP-bound conformation of Hsp90 (Krukenberg et al., 2009; Southworth and Agard, 2011). ATP binding and hydrolysis are essential to the biological function of Hsp90, which was proved by mutational analysis with yeast (Panaretou et al., 1998). When the ATP binding site of the Hsp90 N-terminal domain is blocked by geldanamycin, a benzoquinone ansamycin, the inhibitor of Hsp90 causes rapid degradation of Hsp90 clients such as various protein kinases (Whitesell et al., 1994). Hsp90 assists folding and conformational maturation/maintenance of a wide variety of client proteins including transcription factors and protein kinases under physiological conditions (for the list of clients and co-chaperones, see https://www.picard.ch/downloads/Hsp90interactors.pdf). At least the eukaryotic cytosol Hsp90 is not able to perform the chaperone functions without co-chaperones such as Cdc37, Aha1 and p23/Sba1 (Johnson, 2012), although Hsp90 can recognize and bind non-native proteins, thereby preventing their nonspecific aggregation (Wiech et al., 1992) like the other major classes of molecular chaperones. This anti-aggregation activity may be especially important under stress conditions. In plant cells, Hsp90-based chaperone heterocomplexes have been identified. For example, Hsp70, a p60/Sti1/Hop homolog, and high-molecular-weight immunophilins have been detected in these heterocomplexes (Owens-Grillo et al., 1996; Stancato et al., 1996; Reddy et al., 1998; Krishna and Gloor, 2001).

Surprisingly, under heat stress, the amount of Hsp90 in at least some eukaryotic cells increases by up to 6% of total cellular proteins under heat stress (Wegele et al., 2004), indicating the importance of Hsp90 under stress. However, as far as we know, there is no direct evidence that Hsp90 is involved in thermotolerance in higher plants. This may be due to the fact that Hsp90 is involved in diverse cellular functions as described below, and it is an essential protein (Borkovich et al., 1989). However, the following results with the TU8 mutant suggest that Hsp90 plays a role in thermotolerance

in plants (Ludwig-Muller et al., 2000). The TU8 mutant of Arabidopsis that is deficient in glucosinolate metabolism and pathogen-induced auxin accumulation was found to be less thermotolerant than in wild-type plants. Among different molecular chaperones/Hsps, the only expression level of cytoplasmic Hsp90 declined in the mutant at elevated temperatures. Transient expression of Hsp90 in mutant protoplasts increased their survival rate at higher temperatures to near equivalent that of wild-type protoplasts, suggesting that the reduced level of Hsp90 in the mutant may be the primary cause for the reduction in thermotolerance. However, the physiological function of Hsp90 to confer thermotolerance may be substituted by molecular chaperones/Hsps other than Hsp90. In Arabidopsis, pretreatment with Hsp90 inhibitors such as geldanamycin and radicicol induced the acquisition of high-temperature tolerance just like pre-heat treatment (Yamada et al., 2007). It is known that Hsp90 negatively regulates heat-inducible genes by suppressing Hsf function. In fact, many molecular chaperone genes including Hsp70 and Hsp101 are up-regulated by the Hsp90 inhibitors. Thus, the thermotolerance observed in the presence of the Hsp90 inhibitors is due to other molecular chaperones. In contrast to eukaryotes, HtpG (the bacterial Hsp90 homolog) is not essential under normal conditions, and thus it is possible to construct an *HtpG* knockout mutant to evaluate a role of Hsp90 under stress conditions. The *HtpG* knockout mutants of the cyanobacterium *Synechococcus elongatus* PCC 7942 exhibited a great loss of thermotolerance as compared with the wild-type strain, suggesting that Hsp90/HtpG plays a role for thermotolerance in photosynthetic organisms (Tanaka and Nakamoto, 1999). HtpG plays a role under oxidative stress and protects the photosystem II against photoinhibition (Hossain and Nakamoto, 2003), explaining part of the mechanism for the thermal protection by HtpG. Under heat stress, photosynthetic organisms are placed under oxidative stress due to inactivation of energy consumption.

The Hsp90 family contains constitutively expressed members as well as stress-inducible ones in eukaryotes. In fact, cytosolic Hsp90 in eukaryotes is essential at all temperatures (Borkovich et al., 1989). It has been assigned numerous and diverse functions, including protein folding, signal transduction, protein transport to intracellular organelles, and even protein degradation in the eukaryotic cells (Johnson, 2012). For example, chloroplast Hsp90 which is an essential protein function in membrane translocation into the organelle (Inoue et al., 2013). It interacts with import intermediates of nuclear-encoded preproteins and also with protein import components located in the outer and inner chloroplast envelope membranes. Radicicol inhibits the membrane translocation. One notable function of plant Hsp90 is that it is crucial for defense against pathogens. Hsp90 is involved in plant immunity signaling pathways (Shirasu, 2009). The SGT1–Hsp90 pair, a chaperone complex is required for maintenance of immune sensors. SGT1 is one of the Hsp90 co-chaperones.

In higher plants, the expression of Hsp90 is also regulated developmentally/tissue-specifically (e.g., Yabe et al., 1994). In wheat, there are at least three genes encoding cytosolic

Hsp90s (TaHsp90.1, TaHsp90.2, TaHsp90.3). Organ-specific transcription of these three genes are differentially controlled (Wang et al., 2011). Decreasing the expression of the TaHsp90.1 gene by virus-induced gene silencing (VIGS) showed impaired growth of seedling whereas the suppression of the TaHsp90.2 or TaHsp90.3 genes via VIGS resulted in a hypersensitive resistance response to fungus, indicating that these cytosolic Hsp90s are involved in normal growth, development, as well as biotic stress tolerance in plants.

Lastly, we should add that Hsp90 buffers genetic variation by keeping mutant proteins stable (Rutherford and Lindquist, 1998; Queitsch et al., 2002). When this buffering is compromised, for example, by heat stress that diverts Hsp90s from its normal, specific target proteins to denatured proteins, the variations are exposed, resulting in the production of an array of morphological phenotypes. The strength and breadth of Hsp90s effects on the buffering and release of genetic variation suggest it may have an impact on evolutionary processes.

18.5.6 Hsp100/ClpB

Hsp100/ClpB is present in eukaryotic and eubacterial cells. Archaea generally lack genes encoding members of this family (Large et al., 2009). Hsp100/ClpB has been identified in the chloroplast, mitochondrion and cytosol of plants (Hsp101) (Keeler et al., 2000; Katiyar-Agarwal et al., 2001; Lee et al., 2007). Intriguingly it is not found in the cytosol of animal cells (Mosser et al., 2004), while a homolog (Hsp104) is present in the yeast cytosol.

Hsp100/ClpB belongs to the class 1 family of Clp/Hsp100 AAA+ (ATPases associated with various cellular activities) proteins (Neuwald et al., 1999; Lee et al., 2004; Mogk et al., 2015). Hsp100/ClpB forms a hexameric ring-like structure with a central channel. It consists of two AAA domains, which are oriented head-to-tail. The AAA domain comprises conserved Walker A and Walker B motifs for nucleotide binding and hydrolysis. It is characterized in having a middle domain that is inserted in the first AAA domain. Unlike other Hsp100/Clp proteins, Hsp101, Hsp104 and ClpB are not involved in protein degradation. Rather, they function as a molecular chaperone. Yeast Hsp104 and E. coli ClpB neither promote protein folding nor suppress protein aggregation. It is an ATP-dependent disaggregase that mediates the dissolving of protein aggregates. The Hsp70/DnaK chaperone system collaborates with them for the disaggregation.

What is the mechanism for protein disaggregation? A probable mechanism may be as follows (Mogk et al., 2015). First, Hsp40/DnaJ/J-protein and then Hsp70/DnaK binds to protein aggregates. Hsp70/DnaK recruits Hsp100/ClpB by binding to the M-domain of Hsp100/ClpB. The M-domain binds to the ATPase domain of Hsp70/DnaK. The Hsp70 binding activates Hsp100/ClpB. The activated Hsp100/ClpB binds a substrate protein(s) that is subsequently translocated (mechanically pulled) through the central, narrow channel of the Hsp100/ClpB ring in response to ATP hydrolysis. The protein is unfolded during the translocation/pulling process. This pulling action is associated with unfolding because the substrate protein is forced to enter the channel that cannot otherwise be negotiated. Upon release from Hsp100/ClpB, the substrate protein refolds with or without the assistance of other molecular chaperones such as the Hsp70/DnaK system.

Pioneering works to decipher the role of Hsp100/ClpB for thermotolerance were done with yeast (Sanchez and Lindquist, 1990). Yeast Hsp104 plays a key role in promoting survival at extreme temperatures by mediating resolubilization of heat-denatured proteins from insoluble aggregates (Parsell and Lindquist, 1994). Successful complementation of yeast Hsp104 mutants with higher plant Hsp101 provided the first clue for their important role in imparting thermoprotection (Lee et al., 1994; Schirmer et al., 1994; Wells et al., 1998). While Hsp101 mutants such as hot-1 in Arabidopsis were found to be defective in the acquisition of thermotolerance against high temperature (Hong and Vierling, 2000), the constitutive expression of Hsp101 provided a significant growth advantage to the transgenic seedlings at high temperatures (Queitsch et al., 2000). When the expression of Hsp101 was reduced by antisense or co-suppression, the modified plants had impaired acquired thermotolerance (Queitsch et al., 2000). ClpB is also essential for thermotolerance in the cyanobacterium Synechococcus elongatus PCC 7942 (Eriksson and Clarke, 1996). In addition to the solubilization of protein aggregates, Arabidopsis Hsp101 is involved in the efficient release of mRNAs for ribosomal proteins from stress granules after heat shock, contributing rapid translation recovery after the stress (Merret et al., 2017).

Above studies with model plants or organisms may prompt someone to construct an improved thermotolerant crop plant by overexpression of Hsp101. High temperature is detrimental to both the vegetative and reproductive stages of rice (Pareek et al., 1998). Thus, the development of improved elite rice varieties with enhanced tolerance to high temperature can help in this regard. Arabidopsis AtHsp101 cDNA was employed for overexpression in an elite indica rice cultivar to obtain transgenic plants. Transgenic plants aged 40–45 days old were exposed to 45°C for 3 h and subsequently placed at 28°C for recovery. After 5 days, control plants totally collapsed whereas transgenic plants were green and healthy. Strong accumulation of AtHsp101 was observed in the transgenic lines at normal temperature as well as in response to high-temperature stress. On the other hand, native OsHsp100 was significantly induced to an almost similar extent in untransformed and transgenic plants at 47°C. Thus, it was speculated that heat-induced OsHsp100 alone was inadequate to impart a higher level of protection against heat stress and the constitutive overexpression of AtHsp101 provided the improved thermotolerance capacity to the transgenic plants (Katiyar-Agarwal et al., 2003).

In addition to its essential role in acquired thermal tolerance, Hsp101 provides a substantial fitness benefit under normal growth conditions (Tonsor et al., 2008). Hsp101 is developmentally regulated in plants (Young et al., 2001), indicating a role of Hsp101 under non-heat stress conditions. The T-DNA insertion mutants of chloroplast and mitochondrial homologs showed no evidence for heat stress phenotypes of seedling similar to those observed in *Hsp101* mutants (Lee et al., 2007). However, the chloroplast homolog was shown to be essential for chloroplast development.

18.6   A NOVEL HSP (HSA32) INVOLVED IN 'LONG-TERM' ACQUIRED THERMOTOLERANCE

A heat-stress-associated 32-kD protein (Hsa32), which is highly conserved in land plants but absent in most other organisms was identified (Charng et al., 2006). The gene responds to heat shock at the transcriptional level in moss, Arabidopsis and rice. Disruption of Arabidopsis Hsa32 by T-DNA insertion resulted in a great loss of 'long-term' acquired thermotolerance that was acquired at 37°C when recovery period at 22°C after the pre-heat-treatment was longer than 48 h, and severe heat shock was challenged at 44°C. This indicated that Hsa32 is essential for tolerance against a severe heat challenge after a long recovery following acclimation treatment, which is apparently due to a fast decay of acquired thermotolerance in the absence of Hsa32. Hsp101 is also involved with the long-term acquired thermotolerance, and there is an interplay between Hsp101 and Hsa32 for the long-term acquired tolerance (Wu et al., 2013; Lin et al., 2014). Hsp101 enhances the translation of Hsa32 during recovery after heat treatment whereas Hsa32 retards the decay/degradation of Hsp101.

18.7   ENGINEERING PLANTS TOLERANT TO STRESSES OTHER THAN HEAT

The roles of molecular chaperones in thermotolerance were already discussed throughout the above sections. Hence in this section, we will briefly introduce some publications that are related to the production of plants that are tolerant to stresses other than heat by introducing/inducing a molecular chaperone gene(s).

Temperature stress is not the only stress that leads to an elevated expression of molecular chaperones. The cold induction of some molecular chaperones (for example, see Neven et al., 1992) may be related to plant defense mechanisms against cold stress. Thus, it is reasonable to expect that the overexpression of molecular chaperones can improve cold stress tolerance in plants. Low-temperature storage is one of the most important methods of reducing post-harvest decay and maintaining the organoleptic and nutritional quality of fruits and vegetables. Exposing sensitive fruits to low temperatures induces chilling injury leading to significant changes in overall quality. There is a report showing a simple way to obtain a cold-tolerant tomato fruit by pre-heat-treatment to induce molecular chaperones. The Fortaleza tomato variety showed high sensitivity to cold storage (87% damage in untreated fruit after 21 days at 2°C). However, when they were heat-treated at 38°C in a chamber for 24 and 48 h, development of chilling-associated symptoms in a significant percentage of fruits (47 and 20% damage, respectively) were prevented (Polenta et al., 2007). They detected the increasing accumulation of a 17.6

kDa class I small Hsp in pericarp proteins of the tomato fruit by the heat treatment.

Expression of some Hsps/molecular chaperones in different organisms has been shown to be affected by a number of chemicals: arsenite, heavy metals such as cadmium, cobalt, copper, nickel and silver (Nakamoto and Hiyama, 1999; Wang et al., 2004). Cytosolic/nuclear Hsc70-1 overexpression in Arabidopsis specifically conferred gamma-ray hypersensitivity and tolerance to salt, cadmium and arsenite (Cazale et al., 2009). Plants overexpressing Hsc70-1 accumulated less cadmium, thus providing a possible molecular explanation for their tolerance phenotype. DnaK1 from the halotolerant cyanobacterium *Aphanothece halophytice* was overexpressed in the cytosol of transgenic tobacco plants and found to improve their salt tolerance (Sugino et al., 1999).

Interestingly, while evaluating researches aimed at developing transgenic crops and plants with enhanced tolerance to naturally occurring environmental conditions it was revealed that the response of plants to a combination of two different abiotic stresses is unique and cannot be directly extrapolated from the response of plants to each of the different stresses applied individually (reviewed in Ref. Mittler, 2006). Plant acclimation to a particular abiotic stress may require a specific response that is tailored to the precise environmental conditions the plant encounters. To illustrate this point, transcriptome profiling studies of plants subjected to different abiotic stress conditions prompted a somewhat unique response and little overlap in transcript expression could be found between the responses of plants to abiotic stress conditions such as heat, drought, cold, salt, high light or mechanical stress (reviewed in Mittler, 2006). Drought and heat stress can be an excellent example of this type, which may occur in the field simultaneously. It was observed that a combination of drought and heat stress had a significantly greater detrimental effect on the growth and productivity of maize, barley, sorghum as compared with each of the different stresses applied individually (Heyne and Brunson, 1940; Craufurd and Peacock, 1993; Savin and Nicolas, 1996; Jagtap et al., 1998). Transcriptome profiling studies of plants subjected to a combination of drought and heat stress suggest that the stress combination requires a unique acclimation response involving >770 transcripts that are not altered by drought or heat stress (Rizhsky et al., 2004). To develop transgenic crops with enhanced tolerance to field conditions, a combination of different molecular chaperones in transgenic plants may be able to improve the way.

18.8   CONCLUSIONS

Molecular chaperones are proteins that accompany and look after proteins that are in non-native/non-functional structural states. They facilitate the proper folding/assembly of proteins by binding to non-native proteins. They can disassemble and/or disaggregate protein complexes/aggregates. They even assist in the degradation of (denatured) proteins. Thus, they play an essential role in cellular protein homeostasis. They are involved in a broad array of cellular processes required for both normal cellular functions and survival under stress conditions. Many lines of evidence support that molecular chaperones are some of the most important entities to provide heat and other stress tolerance to plants since proteins with fragile structures are denatured/unfolded and aggregate under stresses. Genetic manipulation to introduce a molecular chaperone(s) and/or to overexpress it can result in the acquisition of tolerance to various stresses, including high temperature, high light, salt stresses. Understanding the structure and function of molecular chaperones are essential for development/cultivation of crop plants that can tolerate various environmental stresses.

ACKNOWLEDGMENTS

This work was supported in part by Grant-in-Aid for Scientific Research (C) [No. 18K05407] to H.N. from the Ministry of Education, Science, Sports and Culture of Japan. T.A. would like to thank the Japanese Government (MEXT) Scholarship for financial support to study in Japan.

REFERENCES

Altschuler, M. and J. P. Mascarenhas. 1982. Heat shock proteins and effect of heat shock in plants. *Plant Mol. Biol.* 1:103–115.

Apuya, N. R., R. Yadegari, R. L. Fischer, J. J. Harada, J. L. Zimmerman and R. B. Goldberg. 2001. The Arabidopsis embryo mutant schlepperless has a defect in the chaperonin-60alpha gene. *Plant Physiol.* 126:717–730.

Bai, C., P. Guo, Q. Zhao et al. 2015. Protomer roles in chloroplast chaperonin assembly and function. *Mol. Plant.* 8:1478–1492.

Balogi, Z., O. Cheregi, K. C. Giese, K. Juhász, E. Vierling, I. Vass, L. Vígh and I. Horváth. 2008. A mutant small heat shock protein with increased thylakoid association provides an elevated resistance against UV-B damage in *Synechocystis* 6803. *J. Biol. Chem.* 283:22983–22991.

Barraclough, R. and R. J. Ellis. 1980. Protein synthesis in chloroplasts. IX. Assembly of newly-synthesized large subunits into ribulose bisphosphate carboxylase in isolated intact pea chloroplasts. *Biochim. Biophys. Acta.* 608:19–31.

Bekh-Ochir, D., S. Shimada, A. Yamagami et al. 2013. A novel mitochondrial DnaJ/Hsp40 family protein BIL2 promotes plant growth and resistance against environmental stress in brassinosteroid signaling. *Planta.* 237:1509–1525.

Bertsch, U., J. Soll, R. Seetharam and P. V. Viitanen. 1992. Identification, characterization, and DNA sequence of a functional 'double' groES-like chaperonin from chloroplasts of higher plants. *Proc. Natl. Acad. Sci. USA.* 89:8696–8700.

Boisvert, D. C., J. Wang, Z. Otwinowski, A.L. Norwich and P. B. Sigler. 1996. The 2.4Å crystal structure of the bacterial chaperonin GroEL complexed with ATP gamma. *S. Nat. Struct. Biol.* 3:170–177.

Bonk, M., M. Tadros, J. Vandekerckhove et al. 1996. Purification and characterization of chaperonin 60 and heat-shock protein 70 from chromoplasts of *Narcissus pseudonarcissus*. *Plant Physiol.* 111:931–939.

Borkovich, K. A., F. W. Farrelly, D. B. Finkelstein et al. 1989. Hsp82 is an essential protein that is required in higher concentrations for growth of cells at higher temperatures. *Mol. Cell. Biol.* 9:3919–3930.

Boston, R. S., P. V. Viitanen and E. Vierling. 1996. Molecular chaperones and protein folding in plants. *Plant Mol. Biol.* 32:191–222.

Bracher, A. and J. Verghese. 2015. The nucleotide exchange factors of Hsp70 molecular chaperones. *Front. Mol. Biosci.* 2:10.

Braig, K., Z. Otwinowski, R. Hegde, D. C. Boisvert, A. Joachimiak, A.L. Horwich and P.B. Sigler. 1994. The crystal structure of the bacterial chaperonin GroEL at 2.8 Å. *Nature.* 371:578–586.

Brive, L., S. Takayama, K. Briknarová, S. Homma, S. K. Ishida, J. C. Reed and K.R. Ely. 2001. The carboxy-terminal lobe of hsc70 ATPase domain is sufficient for binding BAG1. *Biochem. Biophys. Res. Commun.* 289:1099–1105.

Bukau, B., J. Weissman and A. Horwich. 2006. Molecular chaperones and protein quality control. *Cell.* 125:443–451.

Busch, W., M. Wunderlich and F. Schöffl. 2005. Identification of novel heat shock factor-dependent genes and biochemical pathways in *Arabidopsis thaliana*. *Plant J.* 41:1–14.

Cazale, A. C., M. Clement, S. Chiarenza, M.A. Roncato, N. Pochon, A. Creff, E. Marin, N. Leonhardt and L. D. Noël. 2009. Altered expression of cytosolic/nuclear HSC70-1 molecular chaperone affects development and abiotic stress tolerance in *Arabidopsis thaliana*. *J. Exp. Bot.* 60:2653–2664.

Chan-Schaminet, K. Y., S. K. Baniwal, D. Bublak, L. Nover and K. D. Scharf. 2009. Specific interaction between tomato HsfA1 and HsfA2 creates hetero-oligomeric superactivator complexes for synergistic activation of heat stress gene expression. *J. Biol. Chem.* 284:20848–20857.

Charng, Y. Y., H. C. Liu, N. Y. Liu, F. C. Hsu and S. S. Ko.. 2006. Arabidopsis Hsa32, a novel heat shock protein, is essential for acquired thermotolerance during long recovery after acclimation. *Plant Physiol.* 140:1297–1305.

Charng, Y. Y., H. C. Liu, N. Y. Liu, W. T. Chi, C. N. Wang, S. H. Chang and T.T. Wang. 2007. A heat-inducible transcription factor, HsfA2, is required for extension of acquired thermotolerance in Arabidopsis. *Plant Physiol.* 143:251–262.

Cheetham, M. E. and A. J. Caplan. 1998. Structure, function and evolution of DnaJ: conservation and adaptation of chaperone function. *Cell Stress Chaperones.* 3:28–36.

Cho, E. K. and Y. J. Choi. 2009. A nuclear-localized HSP70 confers thermoprotective activity and drought-stress tolerance on plants. *Biotechnol. Lett.* 31:597–606.

Craig, E. A., H. C. Eisenman and H. A. Hundley. 2003. Ribosome-tethered molecular chaperones: the first line of defense against protein misfolding? *Curr. Opin.* 6:157–162.

Craufurd, P. Q. and J. S. Peacock. 1993. Effect of heat and drought stress on sorghum. *Exp. Agric.* 29:77–86.

Dickson, R., C. Weiss, R. J. Howard. 2000. Reconstitution of higher plant chloroplast chaperonin 60 tetradecamers active in protein folding. *J. Biol. Chem.* 275:11829–11835.

D'Silva, P. D., B. Schilke, W. Walter, A. Andrew and E. A. Craig. 2003. J protein cochaperone of the mitochondrial inner membrane required for protein import into the mitochondrial matrix. *Proc. Natl. Acad. Sci. USA.* 100:13839–13844.

Ellis, R. J. 1996. Chaperonins: introductory perspective. In: *The Chaperonins*, ed. R. J. Ellis, 1–25. Academic Press, San Diego.

Eriksson, M. J. and A. K. Clarke. 1996. The heat shock protein ClpB mediates the development of thermotolerance in the cyanobacterium *Synechococcus* sp. strain PCC 7942. *J. Bacteriol.* 178:4839–4846.

Fayet, O., T. Ziegelhoffer and C. Georgopoulos. 1989. The groES and groEL heat shock gene products of Escherichia coli are essential for bacterial growth at all temperatures. *J. Bacteriol.* 171:1379–1385.

Fragkostefanakis S., S. Röth, E. Schleiff and K. D. Scharf. 2015. Prospects of engineering thermotolerance in crops through modulation of heat stress transcription factor and heat shock protein networks. *Plant Cell. Environ.* 38:1881–1895.

Garrido, C., C. Paul, R. Seigneuric and H. H. Kampinga. 2012. The small heat shock proteins family: the long forgotten chaperones. *Int. J. Biochem. Cell Biol.* 44:1588–1592.

Genevaux, P., C. Georgopoulos and W. L. Kelley. 2007. The Hsp70 chaperone machines of Escherichia coli: a paradigm for the repartition of chaperone functions. *Mol. Microbiol.* 66:840–857.

Grimshaw, J. P., I. Jelesarov, H. J. Schönfeld and P. Christen. 2001. Reversible thermal transition in GrpE, the nucleotide exchange factor of the DnaK heat-shock system. *J. Biol. Chem.* 276:6098–6104.

Guo, J., J. Wu, Q. Ji, C. Wang, L. Luo, Y. Yuan, Y. Wang and J. Wang.. 2008. Genome-wide analysis of heat shock transcription factor families in rice and Arabidopsis. *J. Genet. Genomics.* 35:105–118.

Gupta, R. S. 1995. Phylogenetic analysis of the 90kD heat shock family of protein sequences and an examination of the relationship among animals, plants, and fungi species. *Mol. Biol. Evol.* 12:1063–1073.

Hartl, F. U., A. Bracher and M. Hayer-Hartl. 2011. Molecular chaperones in protein folding and proteostasis. *Nature.* 475:324–332.

Haslbeck, M. and E. Vierling. 2015. A first line of stress defense: small heat shock proteins and their function in protein homeostasis. *J. Mol. Biol.* 427:1537–1548.

Heyne, E. G. and A. M. Brunson. 1940. Genetic studies of heat and drought tolerance in maize. *J. Am. Soc Agron.* 32:803–814.

Hill, J. E. and S. M. Hemmingsen. 2001. Arabidopsis thaliana type I and II chaperonins. *Cell Stress Chaperones.* 6:190–200.

Hirohashi, T., K. Nishio and M. Nakai. 1999. cDNA sequence and overexpression of chloroplast chaperonin 21 from *Arabidopsis thaliana. Biochim. Biophys. Acta.* 1429:512–515.

Hong, S. W. and E. Vierling. 2000. Mutants of Arabidopsis thaliana defective in the acquisition of tolerance to high temperature stress. *Proc. Natl. Acad. Sci. USA.* 97:4392–4397.

Horwich, A. L., W. A. Fenton, E. Chapman and G. W. Farr. 2007. Two families of chaperonin: physiology and mechanism. *Annu. Rev. Cell. Dev. Biol.* 23:115–145.

Hossain, M. M. and H. Nakamoto. 2003. Role for the cyanobacterial HtpG in protection from oxidative stress. *Curr. Microbiol.* 46:70–76.

Howarth, C. J. 1991. Molecular responses of plants to an increased incidence of heat shock. *Plant Cell Environ.* 14:831–841.

Huq, S., K. Sueoka, S. Narumi, F. Arisaka and H. Nakamoto. 2010. Comparative biochemical characterization of two GroEL homologs from the cyanobacterium *Synechococcus elongatus* PCC 7942. *Biosci. Biotechnol. Biochem.* 74:2273–2280.

Inoue, H., M. Li and D. J. Schnell. 2013. An essential role for chloroplast heat shock protein 90 (Hsp90C) in protein import into chloroplasts. *Proc. Natl. Acad. Sci. USA.* 110: 3173–3178.

Ishikawa, A., H. Tanaka, M. Nakai and T. Asahi. 2003. Deletion of a chaperonin 60 beta gene leads to cell death in the Arabidopsis lesion initiation 1 mutant. *Plant Cell Physiol.* 44:255–261.

Jagtap, V., S. Bhargava, P. Streb and J. Feierabend. 1998. Comparative effect of water, heat and light stresses on photosynthetic reactions in *Sorghum bicolor* (L.) Moench. *J. Exp. Bot.* 49:1715–1721.

Jaya, N., V. Garcia and E. Vierling. 2009. Substrate binding site flexibility of the small heat shock protein molecular chaperones. *Proc. Natl. Acad. Sci. USA.* 106:15604–15609.

Jiang, C., J. Xu, H. Zhang, X. Zhang, J. Shi, M. Li and F. Ming. 2009. A cytosolic class I small heat shock protein, RcHSP17.8, of *Rosa chinensis* confers resistance to a variety of stresses to *Escherichia coli*, yeast and *Arabidopsis thaliana. Plant Cell. Environ.* 32:1046–1059.

Johnson, J. L. 2012. Evolution and function of diverse Hsp90 homologs and cochaperone proteins. *Biochim. Biophys. Acta.* 1823:607–613.

Jung, K. H., H. J. Ko, M. X. Nguyen, S.-R. Kim, P. Ronald and G. An. 2012. Genome-wide identification and analysis of early heat stress responsive genes in rice. *J. Plant Biol.* 55:458–468.

Kabbage, M. and M. B. Dickman. 2008. The BAG proteins: a ubiquitous family of chaperone regulators. *Cell. Mol. Life Sci.* 65:1390–1402.

Kampinga, H. H. and E. A. Craig. 2010. The HSP70 chaperone machinery: J proteins as drivers of functional specificity. *Nat. Rev. Mol. Cell. Biol.* 11:579–592.

Katiyar-Agarwal, S., M. Agarwal and A. Grover. 2003. Heat-tolerant basmati rice engineered by over-expression of Hsp101. *Plant Mol. Biol.* 51:677–686.

Katiyar-Agarwal, S., M. Agarwal, D. R. Gallie and A. Grover. 2001. Search for the cellular functions of plant Hsp100/Clp family proteins. *Crit. Rev. Plant. Sci.* 20:277–295.

Keeler, S. J., C. M. Boettger, J. G. Haynes, K. A. Kuches, M. M. Johnson, D. L. Thureen, C. L. Keeler Jr. and S. L. Kitto.. 2000. Acquired thermotolerance and expression of the HSP100/ClpB genes of lima bean. *Plant Physiol.* 123:1121–1132.

Kim, K. K., R. Kim and S. H. Kim. 1998. Crystal structure of a small heat-shock protein. *Nature.* 394:595–599.

Kityk, R., J. Kopp, I. Sinning and M. P. Mayer. 2012. Structure and dynamics of the ATP-bound open conformation of Hsp70 chaperones. *Mol. Cell.* 48: 863–874.

Koninkx, J. F. 1976. Protein synthesis in salivary glands of *Drosophila hydei* after experimental gene induction. *Biochem. J.* 158:623–628.

Koumoto, Y., T. Shimada, M. Kondo, T. Takao, Y. Shimonishi, I. Hara-Nishimura and M. Nishimura. 1999. Chloroplast Cpn20 forms a tetrameric structure in *Arabidopsis thaliana. Plant J.* 17:467–477.

Koumoto, Y., T. Shimada, M. Kondo, I. Hara-Nishimura and M. Nishimura. 2001. Chloroplasts have a novel Cpn10 in addition to Cpn20 as co-chaperonins in *Arabidopsis thaliana. J. Biol. Chem.* 276:29688–29694.

Krishna, P. and G. Gloor. 2001. The Hsp90 family of proteins in *Arabidopsis thaliana. Cell Stress Chaperones.* 6:238–246.

Krukenberg, K. A., F. Förster, L. M. Rice, A. Sali and D. A. Agard. 2008. Multiple conformations of *E. coli* Hsp90 in solution: insights into the conformational dynamics of Hsp90. *Structure.* 16:755–765.

Krukenberg, K. A., U. M. Bottcher and D. R. Southworth. et al. 2009. Grp94, the endoplasmic reticulum Hsp90, has a similar solution conformation to cytosolic Hsp90 in the absence of nucleotide. *Protein Sci.* 18:1815–1827.

Laksanalamai, P. and F.T. Robb. 2004. Small heat shock proteins from extremophiles: a review. *Extremophiles.* 8:1–11.

Large, A. T., M. D. Goldberg and P. A. Lund. 2009. Chaperones and protein folding in the archaea. *Biochem. Soc. Trans.* 37:46–51.

Laskey, R. A., B. M. Hond, A. D. Mills and J. T. Finch. 1978. Nucleosomes are assembled by an acidic protein which binds histones and transfers them to DNA. *Nature.* 275:416–420.

Lee, D. G., N. Ahsan, S. H. Lee et al. 2007. A proteomic approach in analyzing heat-responsive proteins in rice leaves. *Proteomics.*

Lee, J. H. and F. Schoffl. 1996. An Hsp70 antisense gene affects the expression of HSP70/HSC70, the regulation of HSF, and the acquisition of thermotolerance in transgenic *Arabidopsis thaliana. Mol. Gen. Genet.* 252:11–19.

Lee, S., M. E. Sowa, J. M. Choi and F. T. Tsai. 2004. The ClpB/Hsp104 molecular chaperone-a protein disaggregating machine. *J. Struct. Biol.* 146:99–105.

Lee, Y. J., R. T. Nagao and J. L. Key. 1994. A soybean 101-KD heat shock protein complements yeast HSP 104 deletion mutant in acquiring thermotolerance. *Plant Cell.* 6:1889–1897.

Lee, U., I. Rioflorido, S. W. Hong, J. Larkindale, E. R. Waters and E. Vierling. 2007. The Arabidopsis ClpB/Hsp100 family of proteins: chaperones for stress and chloroplast development. *Plant J.* 49:115–127.

Lin, B. L., J. S. Wang, H. C. Liu, R. W. Chen, Y. Meyer, A. Barakat and M. Delseny. 2001. Genomic analysis of the Hsp70 super-family in *Arabidopsis thaliana. Cell Stress Chaperones.* 6:201–208.

Lin, M., K. Chai, S. Ko, L. Y. Kuang, H. S. Lur and Y. Y. Charng.. 2014. A positive feedback loop between heat shock protein101 and heat stress-associated 32-kD protein modulates long-term acquired thermotolerance illustrating diverse heat stress responses in rice varieties. *Plant Physiol.* 164:2045–2053.

Liu J. Z. and S. A. Whitham. 2013. Over-expression of a nuclear-localized DnaJ domain-containing HSP40 from soybean reveals its roles in cell death and disease resistance. *Plant J.* 74:110–121.

Liu, H. C. and Y. Y. Charng. 2013. Common and distinct functions of Arabidopsis class A1 and A2 heat shock factors in diverse abiotic stress responses and development. *Plant Physiol.* 163:276–290.

Lopez, T., K. Dalton and J. Frydman. 2015. The mechanism and function of group II chaperonins. *J. Mol. Biol.* 427:2919–2930.

Lubben, T. H., G. K. Donaldson, P. V. Viitanen and A. A. Gatenby. 1989. Several proteins imported into chloroplasts form stable complexes with the GroEL-related chloroplast molecular chaperone. *Plant Cell.* 1:1223–1230.

Ludwig-Muller, J., P. Krishna and C. Forreiter. 2000. A glucosinolate mutant of Arabidopsis is thermosensitive and defective in cytosolic Hsp90 expression after heat stress. *Plant Physiol.* 123:949–958.

Madueno, F., J. A. Napier and J. C. Gray. 1993. Newly imported Rieske iron-sulfur protein associates with both Cpn60 and Hsp70 in the chloroplast stroma. *Plant Cell.* 5:1865–1876.

Malik, M. K., J. P. Slovin, C. H. Hwang and A. A. Gatenby. 1999. Modified expression of a carrot small heat shock protein gene, Hsp17.7, results in increased or decreased thermotolerance. *Plant J.* 20:89–99.

Martel, R., L. P. Cloney, L. E. Pelcher and S. M. Hemmingsen. 1990. Unique composition of plastid chaperonin-60: alpha and beta polypeptide-encoding genes are highly divergent. *Gene.* 94:181–187.

Mattoo, R. U. H., S. K. Sharma, S. Priya, A. Finka and P. Goloubinoff. 2013. Hsp110 is a bona fide chaperone using ATP to unfold stable misfolded polypeptides and reciprocally collaborate with Hsp70 to solubilize protein aggregates. *J. Biol Chem.* 288:21399–21411.

Mayer, M. P. 2010. Gymnastics of molecular chaperones. *Mol. Cell.* 39:321–331.

Mayer, M. P. and B. Bukau. 2005. Hsp70 chaperones: cellular functions and molecular mechanism. *Cell. Mol. Life Sci.* 62:670–684.

McLoughlin, F, E. Basha, M. E. Fowler, M. Kim, J. Bordowitz, S. Katiyar-Agarwal and E. Vierling. 2016. Class I and II small heat shock proteins together with HSP101 protect protein translation factors during heat stress. *Plant Physiol.* 172:1221–1236.

Merret, R., M. C. Carpentier, J. J. Favory et al. 2017. Heat shock protein HSP101 affects the release of ribosomal protein mRNAs for recovery after heat shock. *Plant Physiol.* 74: 1216–1225.

Mittler, R. 2006. Abiotic stress, the field environment and stress combination. *Trends Plant. Sci.* 11:15–19.

Mogk, A., E. Kummer and B. Bukau. 2015. Cooperation of Hsp70 and Hsp100 chaperone machines in protein disaggregation. *Front. Mol. Biosci.* 2:22.

Mosser, D. D., S. Ho and J. R. Glover. 2004. Saccharomyces cerevisiae Hsp104 enhances the chaperone capacity of human cells and inhibits heat stress-induced proapoptotic signaling. *Biochemistry.* 43:8107–8115.

Murakami, T., S. Matsuba, H. Funatsuki, K. Kawaguchi, H. Saruyama, M. Tanida and Y. Sato. 2004. Over-expression of a small heat shock protein, sHSP17.7, confers both heat tolerance and UV-B resistance of rice plants. *Mol. Breed.* 13:165–175.

Nakamoto, H. and D. Honma. 2006. Interaction of a small heat shock protein with light – harvesting cyanobacterial phycocyanins under stress conditions. *FEBS Lett.* 580: 3029–3034.

Nakamoto, H. and K. Kojima. 2017. Non-housekeeping, non-essential GroEL (chaperonin) has acquired novel structure and function beneficial under stress in cyanobacteria. *Physiol. Plant.* 161:296–310.

Nakamoto, H. and L. Vigh. 2006. The small heat shock proteins and their clients. *Cell Mol. Life Sci.* 64:294–306.

Nakamoto, H. and T. Hiyama. 1999. Heat shock proteins and temperature stress. In: *Handbook of Plant and Crop Stress*, ed. M. Pessarakli, 399–416. CRC Press, New York, NY.

Nakamoto, H., M. Suzuki and K. Kojima. 2003. Targeted inactivation of the hrcA repressor gene in cyanobacteria. *FEBS Lett.* 549:57–62.

Nakamoto, H., N. Suzuki and S. K. Roy. 2000. Constitutive expression of a small heat-shock protein confers cellular thermotolerance and thermal protection to the photosynthetic apparatus in cyanobacteria. *FEBS Lett.* 483:169–174.

Narberhaus, F. 2002. Alpha-crystallin-type heat shock proteins: socializing minichaperones in the context of a multichaperone network. *Microbiol. Mol. Biol. Rev.* 66:64–93.

Neuwald, A. F., L. Aravind, J. L. Spouge and E. V. Koonin. 1999. AAA+: A class of chaperone-like ATPases associated with the assembly, operation, and disassembly of protein complexes. *Genome Res.* 9:27–43.

Neven, L. G., D. W. Haskell, C. L. Guy, N. Denslow, P. A. Klein, L. G. Green and A. Silverman. 1992. Association of 70-Kilodalton heat-shock cognate proteins with acclimation to cold. *Plant Physiol.* 99:1362–1369.

Nitta, K., N. Suzuki, D. Honma, Y. Kaneko and H. Nakamoto. 2005. Ultrastructural stability under high temperature or intensive light stress conferred by a small heat shock protein in cyanobacteria. *FEBS Lett.* 579:1235–1242.

Ogawa, D., K. Yamaguchi and T. Nishiuchi. 2007. High-level overexpression of the Arabidopsis HsfA2 gene confers not only increased thermotolerance but also salt/osmotic stress tolerance and enhanced callus growth. *J. Exp. Bot.* 58:3373–3383.

Owens-Grillo, J. K., L. F. Stancato, K. Hoffmann, W.B. Pratt and P. Krishna. 1996. Binding of immunophilins to the 90 kDa heat shock protein (hsp90) via a tetratricopeptide repeat domain is a conserved protein interaction in plants. *Biochemistry.* 35:15249–15255.

Palmblad, M., D. J. Mills and L. V. Bindschedler. 2008. Heat-shock response in *Arabidopsis thaliana* explored by multiplexed quantitative proteomics using differential metabolic labeling. *J. Proteome Res.* 7:780–785.

Panaretou, B., C. Prodromou, S. M. Roe, R. O'Brien, J. E. Ladbury, P. W. Piper and L. H. Pearl. 1998. ATP binding and hydrolysis are essential to the function of the Hsp90 molecular chaperone in vivo. *EMBO J.* 17:4829–4836.

Pareek, A., S. L. Singla and A. Grover. 1998. Protein alterations associated with salinity, desiccation, high and low temperature stresses and abscisic acid application in seedlings of Pusa 169, a high-yielding rice (*Oryza sativa* L.) cultivar. *Curr. Sci.* 75:1023–1035.

Parsell, D. A. and S. Lindquist. 1993. The function of heat-shock proteins in stress tolerance: degradation and reactivation of damaged proteins. *Annu. Rev. Genet.* 27:437–496.

Parsell, D. A. and S. Lindquist. 1994. Heat shock proteins and stress tolerance. In: *The Biology of Heat Shock Proteins and Molecular Chaperones*, 457–494. Cold Spring Harbor Laboratory Press, New York, NY.

Pearl, L. H. and C. Prodromou. 2006. Structure and mechanism of the Hsp90 molecular chaperone machinery. *Annu. Rev. Biochem.* 75:271–294.

Peltier, J. B., Y. Cai, Q. Sun, V. Zabrouskov, L. Giacomelli, A. Rudella, A. J. Ytterberg, H. Rutschow and K. J. van Wijk. 2006. The oligomeric stromal proteome of *Arabidopsis thaliana* chloroplasts. *Mol. Cell. Proteomics.* 5:114–133.

Peng, L., Y. Fukao, F. Myouga, R. Motohashi, K. Shinozaki and T. Shikanai. 2011. A chaperonin subunit with unique structures is essential for folding of a specific substrate. *PLoS Biol.* 9(5) (April):1–13.

Polenta, G. A., J. J. Calvete and C. B. Gonzalez. 2007. Isolation and characterization of the main small heat shock proteins induced in tomato pericarp by thermal treatment. *FEBS J.* 274:6447–6455.

Prodromou, C. 2012. The 'active life' of Hsp90 complexes. *Biochim. Biophys. Acta.* 1823:614–623.

Pulido, P., G. Toledo-Ortiz, M. A. Phillips, L. P. Wright and M. Rodríguez-Concepción. 2013. Arabidopsis J-protein J20 delivers the first enzyme of the plastidial isoprenoid pathway to protein quality control. *Plant Cell.* 25:4183–4194.

Queitsch, C., S. W. Hong, E. Vierling and S. Lindquist. 2000. Heat shock protein 101 plays a crucial role in thermotolerance in Arabidopsis. *Plant Cell.* 12:479–492.

Queitsch, C., T. A. Sangster and S. Lindquist. 2002. Hsp90 as a capacitor of phenotypic variation. *Nature.* 417:618–624.

Rajan, V. B. and P. D'Silva. 2009. *Arabidopsis thaliana* J-class heat shock proteins: cellular stress sensors. *Funct. Integr. Genomics.* 9:433–446.

Reddy, R. K., I. Kurek, A. M. Silverstein, M. Chinkers, A. Breiman and P. Krishna. 1998. High molecular weight FK506-binding proteins are components of heat shock protein 90 heterocomplexes in wheat germ lysate. *Plant Physiol.* 118:1395–1401.

Renner, T. and E. R. Waters. 2007. Comparative genomic analysis of the Hsp70s from five diverse photosynthetic eukaryotes. *Cell Stress Chaperones.* 12:172–185.

Ritossa, F. 1962. A new puffing pattern induced by temperature shock and DNP in *Drosophila. Experientia.* 18:571–573.

Rizhsky, L., H. Liang, J. Shuman, V. Shulaev, S. Davletova and R. Mittler. 2004. When defense pathways collide. The response of Arabidopsis to a combination of drought and heat stress. *Plant Physiol.* 134:1683–1696.

Rutherford, S. L. and S. Lindquist. 1998. Hsp90 as a capacitor for morphological evolution. *Nature.* 396:336–342.

Saibil, H. R., W. A. Fenton, D. K. Clare and A. L. Horwich. 2013. Structure and allostery of the chaperonin GroEL. *J. Mol. Biol.* 425:1476–1487.

Saito, H. and H. Uchida. 1977. Initiation of the DNA replication of bacteriophage lambda in *Escherichia coli* K12. *J. Mol. Biol.* 113:1–25.

Sakthivel, K., T. Watanabe and H. Nakamoto. 2009. A small heat-shock protein confers stress tolerance and stabilizes thylakoid membrane proteins in cyanobacteria under oxidative stress. *Arch. Microbiol.* 191:319–328.

Salvucci, M. E. 2008. Association of Rubisco activase with chaperonin-60beta: a possible mechanism for protecting photosynthesis during heat stress. *J. Exp. Bot.* 59:1923–1933.

Sanchez, Y. and S. L. Lindquist. 1990. HSP104 required for induced thermotolerance. *Science.* 248:1112–1115.

Sanmiya, K., K. Suzuki, Y. Egawa et al. 2004. Mitochondrial small heat-shock protein enhances thermotolerance in tobacco plants. *FEBS Lett.* 557:265–268.

Sarkar, N. K., Y. K. Kim and A. Grover. 2009. Rice sHsp genes: genomic organization and expression profiling under stress and development. *BMC Genomics.* 10:393.

Sato, S., M. Ikeuchi and H. Nakamoto. 2008. Expression and function of a *groEL* paralog in the thermophilic cyanobacterium *Thermosynechococcus elongatus* under heat and cold stress. *FEBS Lett.* 582:3389–3395.

Savin, R. and M. E. Nicolas. 1996. Effects of short periods of drought and high temperature on grain growth and starch accumulation of two malting barley cultivars. *J. Plant Physiol.* 23:201–210.

Scharf, K. D., H. Heider, I. Höhfeld, R. Lyck, E. Schmidt and L. Nover. 1998. The tomato Hsf system: HsfA2 needs interaction with HsfA1 for efficient nuclear import and may be localized in cytoplasmic heat stress granules. *Mol. Cell Biol.* 18:2240–2251.

Scharf, K. D., M. Siddique and E. Vierling. 2001. The expanding family of *Arabidopsis thaliana* small heat stress proteins and a new family of proteins containing alpha-crystallin domains ACD proteins. *Cell Stress and Chaperones.* 6:225–237.

Scharf, K. D., T. Berberich, I. Ebersberger and L. Nover. 2012. The plant heat stress transcription factor (Hsf) family: structure, function and evolution. *Biochim. Biophys. Acta.* 1819:104–119.

Schirmer, E. C., S. Lindquist and E. Vierling. 1994. An Arabidopsis heat shock protein complements a thermotolerance defect in yeast. *Plant Cell.* 6:1899–1909.

Sharma, S. K., P. De Los Rios, P. Christen, A. Lustig and P. Goloubinoff. 2010. The kinetic parameters and energy cost of the Hsp70 chaperone as a polypeptide unfoldase. *Nat. Chem. Biol.* 6:914–920.

Shirasu, K. 2009. The HSP90-SGT1 chaperone complex for NLR immune sensors. *Annu. Rev. Plant Biol.* 60:139–164.

Siddique, M., S. Gernhard, P. von Koskull-Doring, E. Vierling and K. D. Scharf. 2008. The plant sHSP superfamily: five new members in *Arabidopsis thaliana* with unexpected properties. *Cell Stress Chaperones.* 13:183–197.

Skjærven, L., J. Cuellar, A. Martinez and M. Valpuesta. 2015. Dynamics, flexibility, and allostery in molecular chaperonins. *FEBS Lett.* 589:2522–2532.

Southworth, D. R. and D. A. Agard. 2011. Client-loading conformation of the Hsp90 molecular chaperone revealed in the Cryo-EM structure of the human Hsp90: hop complex. *Mol. Cell.* 42:771–781.

Stancato, L. F., K. A. Hutchison, P. Krishna and W. B. Pratt. 1996. Animal and plant cell lysates share a conserved chaperone system that assembles the glucocorticoid receptor into a functional heterocomplex with Hsp90. *Biochemistry.* 35:554–561.

Su, P. H. and H. M. Li. 2008. Arabidopsis stromal 70-kD heat shock proteins are essential for plant development and important for thermotolerance of germinating seeds. *Plant Physiol.* 146:1231–1241.

Sugino, M., T. Hibino, Y. Tanaka, N. Nii, T. Takabe and T. Takabe. 1999. Overexpression of DnaK from a halotolerant cyanobacterium *Aphanothece halophytice* acquires resistance to salt stress in transgenic tobacco plants. *Plant Sci.* 146:81–88.

Sun, W., M. Van Montagu and N. Verbruggen. 2002. Small heat shock proteins and stress tolerance in plants. *Biochim. Biophys. Acta.* 1577:1–9.

Sundby, C., U. Harndahl, N. Gustavsson, E. Åhrmana and D. J. Murphy. 2005. Conserved methionines in chloroplasts. *Biochim. Biophys. Acta.* 1703:191–202.

Sung, D. Y., E. Vierling, and C. L. Guy. 2001. Comprehensive expression profile analysis of the Arabidopsis Hsp70 gene family. *Plant Physiol.* 126:789–800.

Suzuki, K., H. Nakanishi, J. Bower, D. W. Yoder, K. W. Osteryoung and S. Y. Miyagishima. 2009. Plastid chaperonin proteins Cpn60 alpha and Cpn60 beta are required for plastid division in *Arabidopsis thaliana*. *BMC Plant Biol.* 9:38.

Tanaka, N. and H. Nakamoto. 1999. HtpG is essential for the thermal stress management in cyanobacteria. *FEBS Lett.* 458:117–123.

Tissières, A., H. K. Mitchell and U. M. Tracy. 1974. Protein synthesis in salivary glands of *Drosophila melanogaster*: relation to chromosome puffs. *J. Mol. Biol.* 84:389–398.

Tonsor, S. J., C. Scott, I. Boumaza, T. R. Liss, J. L. Brodsky and E. Vierling. 2008. Heat shock protein 101 effects in *A. thaliana*: genetic variation, fitness and pleiotropy in controlled temperature conditions. *Mol. Ecol.* 17:1614–1626.

Torok, Z., P. Goloubinoff, I. Horvath et al. 2001. *Synechocystis* HSP17 is an amphitropic protein that stabilizes heat-stressed membranes and binds denatured proteins for subsequent chaperone-mediated refolding. *Proc. Natl. Acad. Sci. USA.* 98:3098–3103.

Tsai, Y. C., O. Mueller-Cajar, S. Saschenbrecker, F. U. Hartl and M. Hayer-Hartl. 2012. Chaperonin cofactors, Cpn10 and Cpn20, of green algae and plants function as hetero-oligomeric ring complexes. *J. Biol. Chem.* 287:20471–20481.

Tsugeki, R. and M. Nishimura. 1993. Interaction of homologues of Hsp70 and Cpn60 with ferredoxin-NADP+ reductase upon its import into chloroplasts. *FEBS Lett.* 320:198–202.

van Montfort, R. L., E. Basha, K. L. Friedrich, C. Slingsby and E. Vierling. 2001. Crystal structure and assembly of a eukaryotic small heat shock protein. *Nat. Struct. Biol.* 8:1025–1030.

Vierling, E. 1991. The roles of heat shock proteins in plant. *Annu. Rev. Plant Physiol. Plant Mol. Biol.* 42:579–620.

Vihervaara, A. and L. Sistonen. 2014. HSF1 at a glance. *J. Cell Sci.* 127:261–266.

Vitlin Gruber, A., G. Zizelski, A. Azem and C. Weiss. 2014. The Cpn10(1) co-chaperonin of *A. thaliana* functions only as a hetero-oligomer with Cpn20. *PLoS One.* 9(11) (November):1–15. https://journals.plos.org/plosone/article?id=10.1371/journal.pone.0113835.pdf.

Vitlin Gruber, A., S. Nisemblat, A. Azem and C. Weiss. 2013. The complexity of chloroplast chaperonins. *Trends Plant Sci.* 18:688–694.

von Koskull-Doring, P., K. D. Scharf, and L. Nover. 2007. The diversity of plant heat stress transcription factors. *Trends Plant Sci.* 12:452–457.

Walsh, P., D. Bursać and Y. C. Law. 2004. The J-protein family: modulating protein assembly, disassembly and translocation. *EMBO R.* 5:567–571.

Wandinger, S. K., K. Richter and J. Buchner. 2008. The Hsp90 chaperone machinery. *J. Biol. Chem.* 283:18473–18477.

Wang, G., F. Kong, S. Zhang, X. Meng, Y. Wang and Q. Meng. 2015. A tomato chloroplast-targeted DnaJ protein protects Rubisco activity under heat stress. *J. Exp. Bot.* 66:3027–3040.

Wang, G. F., X. Wei, R. Fan et al. 2011. Molecular analysis of common wheat genes encoding three types of cytosolic heat shock protein 90 (Hsp90): functional involvement of cytosolic Hsp90s in the control of wheat seedling growth and disease resistance. *New Phytol.* 191(2):418–431.

Wang, W., B. Vinocur, O. Shoseyov and A. Altman. 2004. Role of plant heat-shock proteins and molecular chaperones in the abiotic stress response. *Trends Plant Sci.* 9:244–252.

Waters, E. R. 2013. The evolution, function, structure, and expression of the plant sHSPs. *J. Exp. Bot.* 64:391–403.

Waters, E. R., B. D. Aevermann and Z. Sanders-Reed. 2008. Comparative analysis of the small heat shock proteins in three angiosperm genomes identifies new subfamilies and reveals diverse evolutionary patterns. *Cell Stress Chaperones.* 13:127–142.

Waters, E., M. J. Hohn, I. Ahel et al. 2003. The genome of Nanoarchaeum equitans: insights into early archaeal evolution and derived parasitism. *Proc. Nat. Acad. Sci. USA.* 100:12984–12988.

Wegele, H., L. Müller and J. Buchner. 2004. Hsp70 and Hsp90 – a relay team for protein folding. *Rev. Physiol. Biochem. Pharmacol.* 151:1–44.

Weiss, C., A. Bonshtien, O. Farchi-Pisanty, A. Vitlin and A. Azem. 2009. Cpn20: siamese twins of the chaperonin world. *Plant Mol. Biol.* 69:227–238.

Wells, D. R., R. L. Tanguay, H. Le and D. R. Gallie. 1998. HSP101 functions as a specific translational regulatory protein whose activity is regulated by nutrient status. *Genes Dev.* 12:3236–3251.

Whitesell, L., E. G. Mimnaugh, B. De Costa, C. E. Myers and L. M. Neckers. 1994. Inhibition of heat shock protein HSP90-pp60v-src heteroprotein complex formation by benzoquinone ansamycins: essential role for stress proteins in oncogenic transformation. *Proc. Natl. Acad. Sci. USA.* 91:8324–8328.

Wiech, H., J. Buchner, R. Zimmermann and U. Jakob. 1992. Hsp90 chaperones protein folding in vitro. *Nature.* 358:169–170.

Wu, T. Y., Y. T. Juan, Y. H. Hsu, S. H. Wu, H. T. Liao, R. W. Fung and Y. Y. Charng. 2013. Interplay between heat shock proteins HSP101 and HSA32 prolongs heat acclimation memory posttranscriptionally in Arabidopsis. *Plant Physiol.* 161:2075–2084.

Xu, Z., A. L. Horwich and P. B. Sigler. 1997. The crystal structure of the asymmetric GroEL–GroES–(ADP)7 chaperonin complex. *Nature.* 388:741–750.

Yabe, N., T. Takahashi and Y. Komeda. 1994. Analysis of tissue-specific expression of *Arabidopsis thaliana* HSP90-family gene HSP81. *Plant Cell. Physiol.* 35:1207–1219.

Yamada, K., Y. Fukao, M. Hayashi, M. Fukazawa, I. Suzuki and M. Nishimura. 2007. Cytosolic HSP90 regulates the heat shock response that is responsible for heat acclimation in *Arabidopsis thaliana*. *J. Biol. Chem.* 282:37794–37804.

Yoshida, T., N. Ohama, J. Nakajima et al. 2011. Arabidopsis HsfA1 transcription factors function as the main positive regulators in heat shock-responsive gene expression. *Mol. Genet. Genomics.* 286:321–332.

Young, T. E., J. Ling, C. Geisler-Lee, R. L. Tanguay, C. Caldwell and D. R. Gallie. 2001. Development and thermal regulation of the maize heat shock protein, HSP101. *Plant Physiol.* 127:777–791.

19 Phytohormone Homeostasis and Crosstalk Effects in Response to Osmotic Stress

Omid Askari-Khorasgani and Mohammad Pessarakli

CONTENTS

ABBREVIATIONS

ABA	Abscisic acid
ABC	ATP-binding cassette
ABF	ABRE-binding factor
ABI	ABA-insensitive
ABRE	ABA-responsive element
AIB	2-aminoisobutyric acid; ethylene biosynthesis inhibitor
AMF	Arbuscular mycorrhizal fungi
AQPs	Aquaporins
AUX	Auxin
BG	β-glucosidase
BL	Brassinolide
BR	Brassinosteroid
CBF	C-repeat binding factor
CDT-1	Chromatin *licensing and DNA replication factor 1*
CK	Cytokinin
CKX3	Oxidase/dehydrogenase 3
CPK	Calcium-dependent protein kinases
CRT	Calreticulin
CS	Castasterone
DHAR	Dehydroascorbate reductase
DHN	Dehydrin
DXP	1-deoxy-D-xylulose-5-phosphate
ENT	*EQUILIBRATIVE NUCLEOSIDE TRANSPORT*
ET	Ethylene
GA	Gibberllin or gibberllic acid
GABA	Gamma-aminobutyric acid or γ-aminobutyric acid
GABA-T	Gamma-aminobutyric acid transaminase
GGPP	Geranylgeranyl pyrophosphate
GO	Gene ontology
IAA	Indole-3-acetic acid
IDP	Isopentenyl diphosphate
IPC	Intergovernmental Panel on Climate
IPP	Isopentenyl pyrophosphate
IPT	Isopentenyltransferase
JA	Jasmonic acid
JAZ	JASMONATE ZIM-DOMAIN
KAT1	*ARABIDOPSIS THALIANA* POTASSIUM CHANNEL 1
LEA	Late embryogenesis abundant
mDr	Mild drought
MeJA	Methyl-jasmonate
MEP	2-C-methyl-d-erythritol-4-phosphate
MVA	Mevalonate
MYB	Myeloblastosis
MYC	Myelocytomatosis
NCED	9-*cis*-epoxycarotenoid dioxygenase
NDGA	Nordihydroguaiaretic acid; ABA-biosynthesis inhibitor
NO	Nitric oxide
OST1	Open stomata 1

PAs	Polyamines
PEO-IAA	2-indol-3-yl-4-oxo-4-phenylbutanoic acid; antagonist of auxin receptor
PIP	Phosphatidylinositol 4,5-bisphosphate
PLCG	Phospholipase C gamma
PP2C	Protein phosphatase 2Cs
PPO	Polyphenol oxidase
PPP	Pentose phosphate pathway
ProDH	Proline dehydrogenase
PRT	Protease
P$_{SARK}$	Promoter senescence-associated receptor kinase
PUP	PURINE PERMEASE
PYL	PYRABACTIN RESISTANCE1/ REGULATORY-LIKE
PYR	Pyrabactin resistance
RAB	Responsive to ABA
RCAR	REGULATORY COMPONENTS OF ABA RECEPTORS
RD	RESPONSIVE TO DESICCATION
RFOs	Raffinose family oligosaccharides
ROS	Reactive oxygen species
RWC	Relative water content
SA	Salicylic acid
SLAC1	SLOW ANION CHANNEL-ASSOCIATED 1
SNF1	Sucrose non-fermenting-1
SnRK	Sucrose-non-fermenting-1-related protein kinase
TF	Transcription factor
UDP	Uridine diphosphate glucose
VP14	Viviparous14
WS	Water stress
WT	Wild-type
WUE	Water-use efficiency
ZEP	Zeaxanthin epoxidase

19.1 INTRODUCTION

Drought, which can be intensified by hydrological cycle imbalance, is one of the most severe stresses, limiting agricultural production and food security. Hydrological cycle imbalance is a major human concern, exhausting global food and water supply. It can be exacerbated by climate change, the imbalance in usage and source recharge exacerbated by the increasing trend of population growth mainly in limited areas worldwide. A change in climate can accelerate the hydrological cycle, which is affected by a change in precipitation, evapotranspiration, the magnitude and timing of run-off, and by any change in the intensity and frequency of floods and droughts (Eitzinger et al., 2003; Arnell, 2004; Siegel, 2008; Allaman et al., 2009; Ramos and Martínez-Casasnovas, 2010; Schiermeier, 2011; Wang et al., 2013; Hall, 2014; Pachauri et al., 2014; Intergovernmental Panel on Climate (IPC), 2015; Muscolo et al., 2015; Fant et al., 2016; Gonçalves, 2016; Tollefson, 2016). The increase in world population from the current 7.7 billion to more than 9.7 billion by 2050 intensifies human food security by causing more imbalance between food demand and supply (United Nation World Population Prospects (UNWPP),

2015; World Population Forecast (WPF), 2016). Based on the FAO World Soil Resources Report 2000, 64% of the global land area suffer from water deficit (Cramer et al., 2011), and it has been estimated that half of all agricultural lands will be abandoned by desertification by 2050 (Jamil et al., 2011). Water scarcity is a worldwide problem, seriously constraining global crop production and quality, causing a 50–70% yield loss (Gonçalves, 2016). Drought is a complex physical–chemical process, in which many biological macromolecules and small molecules are involved, such as nucleic acids (DNA, RNA, microRNA), proteins, carbohydrates, lipids, hormones, ions, free radicals, and mineral elements. Water stress (WS) threshold varies depending on plant species and genotypes and therefore plants employ different morphological, physiological, biochemical, cellular, and molecular adaptive mechanisms to deal with WS (Blum and Ebercon, 1981; Liu et al., 1998; Tripathy., 2000; HongBo et al., 2006; Lei et al., 2006). Physiological changes begin with gene expression and signal transduction, which are affected by genotype-by-environment interactions (Khodursky et al., 2000; Takeda and Matsuoka, 2008). Exploring drought resistance physiological mechanisms and signaling pathways associated with metabolic and hormone variations provide a scientific basis for developing tolerant species through conventional breeding and modern agricultural biotechnology.

On drought perception, many factors simultaneously act on molecular, biochemical, and cellular levels with complex interaction networks (Vasilev et al., 2016). Therefore, OMICs analytical procedures should be applied to avoid misleading conclusions by optimizing synchronized assessment of multiple factors under running an adequate number of experiments.

In the molecular levels, transcription factors (TFs) and kinases are involved in crosstalk between stress signaling pathways. Hormone signaling pathways regulated by abscisic acid (ABA), salicylic acid (SA), jasmonic acid (JA), and ethylene (ET), as well as reactive oxygen species (ROS) signaling pathways, play key roles in the crosstalk between biotic and abiotic stress signaling, leading to physiological and metabolic changes that increase the chance of plant survival (Fujita et al., 2006). Accordingly, this review is undertaken to elucidate the role of plant hormones and their crosstalk in response to water scarcity to provide insight into drought tolerance adaptation and to highlight the potential approaches for the future breeding programs.

19.2 PHYTOHORMONES

19.2.1 ABSCISIC ACID (ABA)

ABA is a 15-carbon plant growth regulator which belongs to a class of metabolites known as isoprenoids, also called terpenoids. They derive from a common five-carbon (C$_5$) precursor, isopentenyl pyrophosphate (IPP) is produced primarily in plastids via 1-deoxy-D-xylulose-5-phosphate (DXP) from pyruvate and glyceraldehydes-3-phosphate (Nambara and Marion-Poll, 2005; Danquah, et al., 2014). This leads to the sequential production of farnesyl pyrophosphate, geranylgeranyl pyrophosphate (GGPP), phytoene, ζ-carotene, lycopene, and β-carotene. β-carotene is converted to a xanthophyll, zeaxanthin, which is the first oxygenated carotenoid (Danquah et al., 2014). Although ABA contains 15 carbon atoms, in plants, it is not derived directly from the 15-carbon sesquiterpene precursor xanthoxin. It is rather synthesized via the oxidative cleavage of C_{40} carotenoid intermediates originating from the 2-C-methyl-d-erythritol-4-phosphate (MEP) pathway (cis-epoxycarotenoids violaxanthin and neoxanthin catalyzed by 9-cis-epoxycarotenoid dioxygenase, NCED, and zeaxanthin epoxidase, ZEP) to release the 15C compound xanthoxin, also known as xanthoxal, and the reaction is stimulated by drought conditions (Duckham et al., 1991; Nambara and Marion-Poll, 2005; Christmann et al., 2007; Khurana et al., 2008; Ton et al., 2009; Finkelstein, 2013; Zhang et al., 2014a). Xanthoxin is presumed to migrate from the plastid to the cytosol, where it is converted to ABA by different suggested possible pathways: via abscisic aldehyde, xanthoxic acid or abscisic alcohol (Danquah et al., 2014), isopentenyl diphosphate (IDP) derived from the mevalonate (MVA) pathway in fungi (Inomata et al., 2004), and IDP derived from the MVA-independent pathway in higher plants (Inomata et al., 2004; Nambara and Marion-Poll, 2005).

Neoxanthin synthesis was recently found to depend on the product of the *ABA4* locus (AT1G67080), a highly conserved unique plastid membrane-localized protein (Finkelstein, 2013). ABA2 (short-chain alcohol dehydrogenase) converts xanthoxin derived from cleavage of carotenoids into abscisic aldehyde, which is finally oxidized into ABA by abscisic aldehyde oxidase (Liu et al., 2015a). Liu et al. (2015) revealed that uridine diphosphate glucose (UDP)-glucosyltransferase (UGT), UGT71C5, plays an important role in ABA homeostasis by glucosylating ABA to an abscisic acid-glucose ester (GE) in *A. thaliana*. The transgenic plants with down-regulated UGT71C5 exhibit enhanced resistance to drought stress, consistent with the ABA-excessive phenotype in plants. In addition to the *de novo* ABA biosynthesis, the β-glucosidase (BG) homolog *Arabidopsis thaliana* β-glucosidase1/2 (*At*BG1/2) generates ABA from abscisic acid-glucosyl ester in endoplasmic reticulum and vacuole, respectively (Liu et al., 2015a). Xylem ABA can also be recycled from the phloem to the xylem without any contribution of roots. Glucose conjugated ABA represents one of these inactive pools, which may be hydrolyzed to active ABA by a stress-inducible β-glucosidase in the endoplasmic reticulum. Finally, ABA trapped in alkaline cell compartments may also be released in drought-stressed plants because of the alkalinization of their apoplast (Tardieu et al., 2010). At the site of action, ABA uptake into cells could be mediated by diffusion or by specific transporters. Therefore, both carrier-mediated and passive ABA transport mechanisms are suggested to exist. Recently, it was reported (Manzoor et al., 2016) that one of the ATP-binding cassette (ABC) transporter genes, *At*ABCG25, mediates ABA efflux from ABA biosynthesizing vascular cells and that ABA would diffuse into apoplastic areas. Another ABC transporter

*At*ABCG40 expressed in guard cells of Arabidopsis was found to be responsible for ABA influx into guard cells to facilitate stomatal closure (Manzoor et al., 2016).

It is now well acknowledged that WS is sensed not only in roots by releasing chemical signals such as Cl^-, NO_3^-, and $SO4_2^-$, ABA biosynthesis and redox signaling molecules (Marowa et al., 2016), but also in shoots (leaves and stems) by activation of phospholipase C gamma (PLCG) associated with ABA-induced Ca^{2+} (mainly cytosolic) elevation (Kamata and Hirata, 1999; Khurana et al., 2008; Reddy et al., 2011; Virdi et al., 2015), redox signaling and its associated regulation of transcription factors (García-Mata and Lamattina, 2001; Forman et al., 2004; Jakab et al., 2005; Scandalios, 2005; del Río et al., 2006; Kotchoni et al., 2006; Christmann et al., 2007; Hao and Zhang, 2009; Tanou et al., 2012; Caldeira et al., 2014; Diao et al., 2016; Li, 2016; Tavladoraki et al., 2016; Li et al., 2017; Nahar et al., 2017; Pourghayoumi et al., 2017), phosphatidic acid signaling, heterotrimeric G-protein-coupled signaling, and pH increases (Zhu et al., 2016), and an array of Ca^{2+}-binding proteins, also known as "EF-hand"-containing proteins, that function as Ca^{2+} sensors (Hashimoto et al., 2012; Virdi et al., 2015; Zheng et al., 2017) to allow timely adjustment of the plant's transpiration stream and hydraulic conductivity.

The role of ABA on stomatal closure in these conditions could be originated either from its biosynthesis in the roots or directly from guard cells. ABA can be synthesized in a wide range of tissues, including roots (root tips, vascular bundles of roots), and leaves and stem vascular systems (in the phloem companion cells and xylem parenchyma cells next to the phloem sieve cells and xylem vessels, respectively, in the leaf veins and florescence stems), as well as leaf mesophyll (Koiwai et al., 2004; Tardieu et al., 2010; Daszkowska-Golec and Szarejko, 2013; Malcheska et al., 2017), in leaf cuticle (Wang et al., 2011), and also in guard cells (Koiwai et al., 2004; Daszkowska-Golec and Szarejko, 2013). On receiving a stress signal, ABA formation starts in vascular tissues and is released from the biosynthesis site into other cells by means of particular ATP-dependent transporters. This permits the quick ABA distribution in neighboring tissues (Manzoor et al., 2016). Stomata respond to the concentration of ABA in the guard-cell apoplast, not symplast. Its physiological mechanisms remain poorly understood because firstly guard cells flanking stomata are small and contain only atomic amount of ABA (Zhang and Outlaw, 2001; Borel and Simonneau, 2002) and secondly the stomatal movement is under the complex regulatory networks of redox signaling, ion channels, and hormone crosstalks (Kamata and Hirata, 1999; Khurana et al., 2008; Reddy et al., 2011; Tanou et al., 2012; Virdi et al., 2015; Diao et al., 2016; Li, 2016; Marowa et al., 2016; Tavladoraki et al., 2016; Li et al., 2017; Nahar et al., 2017; Pourghayoumi et al., 2017). The stomatal movement is regulated by a complex hormone signaling network in which ABA, JA, salicylic acid, and brassinosteroids (BRs)-specific brassinolide (BL) are positive regulators of stomatal closure, while auxin (AUX) and cytokinins (CKs) are positive regulators of stomatal opening by reducing H_2O_2 and NO levels within guard cells, cytoplasmic alkalinization in guard cells, regulating the activity of G-proteins, slow anion channels, Ca permeable channels, and inwardly rectifying K channels that mediate K uptake within guard cells (Rai et al., 1986; Tanaka et al., 2006; Hao et al., 2011; Khokon et al., 2011; Daszkowska-Golec and Szarejko, 2013; Shi et al., 2015; Ha et al., 2016; Nazareno and Hernandez, 2017). However, different effects of these hormones, such as auxins and ethylene (ET), could mainly be due to their concentrations and genotype or species-specific effects, redox status, cytosolic pH, duration of exposure, activation of G-proteins, and their crosstalk effects as observed for ABA-dependent-induced stomatal closure by JA and BL, and stimulation of ABA biosynthesis by JA and SA (Tanaka et al., 2006; Acharya and Assmann, 2009; Hao et al., 2011; Khokon et al., 2011; Daszkowska-Golec and Szarejko, 2013; Shi et al., 2015; Ha et al., 2016). With dual function, ET may either induce stomatal closure by generating H_2O_2 in guard cells (Desikan et al., 2006) or delay stomatal closure by inhibiting ABA signaling (Tanaka et al., 2005). Regarding the concentration, in an ABA-dependent manner, BL may induce stomatal closure, while at higher concentrations inhibit ABA-induced stomatal closure (Ha et al., 2016).

With differential effects, plant phenolic compounds are involved in the stomatal movement (Rai et al., 1986). Because the stomatal movement is regulated by an integrative effect between phytohormones and redox signaling (Miura et al., 2013), the roles of redox, water, nutrient, and hormone levels, gene expression, circadian rhythm, and environmental conditions should be taken into account during the breeding programs.

ABA accumulates in shoot and root tissues subjected to water deficit to initiate processes involved in adaptation to drought and other environmental stresses (Koiwai et al., 2004; Tardieu et al., 2010; Wang et al., 2011; Daszkowska-Golec and Szarejko, 2013; Malcheska et al., 2017). Plants are able to respond rapidly to transient water shortage at midday by closing the stomata and causing midday suppression of photosynthesis. Such a short-term adjustment is difficult to reconcile with the long-distance transport of a chemical signal in trees. Transport of a chemical signal from roots to the canopy of tall trees may take more than a day based on the transpirational streaming in conifer xylem with transport velocities of up to 2 m h^{-1} (Christmann et al., 2007). The enzymes that convert xanthoxin into ABA are expressed in the vascular parenchyma cells of roots and shoots. Evidence for root-derived ABA as a long-distance signal has been obtained from split-root experiments with whole plants in which only one part of the root system experienced water deficit (Christmann et al., 2007). Likewise, maintaining the water status in leaves by pressurization of drought-exposed root systems did not prevent stomatal closure in herbaceous species (Wilkinson and Davies, 2010). However, the grafting of tomato shoots onto rootstocks with a deficiency in ABA biosynthesis indicated that root-generated ABA is not required, or, alternatively, that transported ABA from the shoot may compensate for ABA deficiency in roots (Holbrook et al., 2002). ABA can be stored as physiologically inactive ABA-glucose conjugates from which ABA is

released by a specific hydrolase that is activated under WS. However, the levels of ABA-glucose ester are too small for ABA release from this conjugate to contribute significantly to the overall increase in ABA during WS (Christmann et al., 2007). Christmann and co-workers (2007) found that the shoot response to the limited soil water supply is not affected by the capacity to generate ABA in the root; however, the response does require ABA biosynthesis and signaling in the shoot. Soil WS elicits a hydraulic response in the shoot, which precedes ABA signaling and stomatal closure. Based on their observations, Christmann et al. (2007) concluded that root-inflicted WS elicits shoot ABA responses in the absence of root-generated ABA, and shoot-derived ABA is necessary and sufficient for stomatal closure. These investigators' data did not support the model of root-derived ABA translocated from water-stressed roots to the shoot as the long-distance signal. Attenuation of root-to-shoot WS signaling by providing water directly to leaves of water-stressed plants revealed that the nature of the long-distance signal favored a hydraulic over an electric or chemical signal (Christmann et al., 2007). According to Christmann et al. (2007), the identification of a change in turgor pressure of mesophyll cells within minutes of root-evoked WS provided evidence for the involvement of a hydraulic signal, and stomatal closure in response to WS required both ABA and a hydraulic signal in the shoot. There was no indication of such a stomata-controlling signal when the hydraulic signal was attenuated. Instead, stomata remained open independent of the water potential of the feeding solution. Attenuation of the hydraulic response prevented ABA signaling and subsequent physiological responses. The attenuation was overcome by feeding ABA to the root system, compatible with a single response pathway of ABA and the hydraulic signal. Generation of the hydraulic signal neither depended on ABA biosynthesis nor on ABA signaling. In Christmann et al., (2007) study, in response to WS, the mutants *aba2* (ABA-deficient) or *abi1* (ABA-insensitive) revealed a hydraulic response comparable to that of wild-type (WT) plant; however, the stomata did not close in the mutants. Overall, the data revealed a signal pathway in which ABA acts downstream of the hydraulic signal in communicating WS between root and shoot (Christmann et al., 2007). ABA signal can also be modified by the pH of the xylem sap and by the water relations of the leaf. Additionally, the limitations in stomatal opening imposed by high concentrations of ABA can be rapidly and completely reversed by lowering the leaf temperature, and there is also clear interactions between the effect of ABA and CO_2 on stomata (apoplast of the guard cells), intercellular CO_2, and also evaporative demand. The influence of ABA on cell growth can also be modified with variation in tissue water status (Davies et al., 1994). It is now believed that ABA accumulation follows a bi-phasic pattern in which the early phase leads to stomatal closure, while the late phase activates signaling pathways to regulate gene expressions associated with stimulating protective proteins and metabolites associated with protection mechanisms (Jeong and Seo, 2017).

ABA, both endogenous synthesized and exogenous application, affect plant water status, water uptake, and growth in different ways such as water hydraulic conductivity both in roots and leaves, root system architecture, and aquaporins (AQPs) activity. In spite of contradictions, the main tendency is toward a positive effect of ABA on hydraulic conductivity, AQPs activity, maintenance of main root growth with less lateral roots, and plant recovery after rehydration, resulting in higher productivity. Thus, manipulation of ABA biosynthesis can be used as an efficient strategy for designing drought-resistant crops (Tardieu et al., 2010). A study on the grapevine showed that drought-induced ABA activated polyamine (PA) biosynthesis and exodus into the apoplast which, subsequently, contributed to the PA homeostasis and the secondary signal transduction, the later triggered by producing H_2O_2 as the result of PA oxidation by apoplastic amine oxidases, likely involved in stomata closure (Toumi et al., 2010).

The effect of ABA biosynthesis on biomass accumulation depends on multifarious factors, including plant species or genotype, specific organ, tissue sensitivity in an ABA dose-dependent manner (Liu et al., 2015a; Manzoor et al., 2016), circadian clock (Seung et al., 2012; Portolés and Zhang, 2014; Grundy et al., 2015; Jeong and Seo, 2017), developmental stage, severity, duration, plants' stress memory, and recurrence of different stress combinations, and environmental conditions (Manzoor et al., 2016; Avramova, 2017); hence the need for developing simulation and modeling systems in various environmental conditions. Stress signals and ABA are likely to share common elements in their respective signaling pathways (Jakab et al., 2005). ABA regulates many aspects of plant growth and development embracing embryo maturation, seed dormancy, seed germination, cell division and elongation (Khurana et al., 2008), stomatal closure, flowering time (Christmann et al., 2007), and also functions in response to environmental stresses such as drought, salinity, cold, pathogen attack and UV radiation (Khurana et al., 2008). Under osmotic stress, ABA regulates the expression of many genes during seed development and germination, some of them are related to the synthesis of osmotic protectants and late embryogenesis abundant (LEA) class proteins and thus contribute to drought tolerance. The ABA-responsive element (ABRE) is the major *cis*-element for ABA-responsive gene expression. ABRE-binding protein (AREB)/ABRE-binding factor (ABF) TFs regulate ABRE-dependent gene expression. Other TFs are also involved in ABA-responsive gene expression. Sucrose non-fermenting-1 (SNF1)-related protein kinases 2 (SnRK2) are the key regulators of ABA signaling, including the AREB/ABF regulon. Recently, ABA receptors and group A 2C-type protein phosphatases were shown to govern the ABA signaling pathway (Nakashima and Yamaguchi-Shinozaki, 2013). The ABA signaling crosstalk pathway with other signaling factors such as ROS and NO, and gamma-aminobutyric acid (GABA) biosynthesis and interactions with primary metabolism have been a topic of recent debates (Sequera-Mutiozabal et al., 2016). In addition to the implication of ABA on seed desiccation (Bartels et al., 1996) and protecting plants against near-lethal stresses (Borel et al., 2001), ABA has a key role in maintaining near homeostasis of leaf water status when plants are subjected to mild drought (mDr) or to changes

in evaporative demand (Borel et al., 2001), and also protection of dehydration-tolerant vegetative tissues (Bartels et al., 1996). In particular, it contributes to fine-tuning of stomatal conductance in such a way that day-time leaf water potential is largely maintained under water deficit, thereby reducing transpiration rates and depletion rates of soil water (Borel et al., 2001). It controls gene expression, either alone via ABA-responsive elements, or in interaction with other compounds such as sugars, AUXs, gibberellins (GAs), or brassinosteroids (BRs) (Tardieu et al., 2010). Interestingly, *AtbHLH68* over-expression in the vascular system of *Arabidopsis* enhanced drought tolerance by not only increasing sensitivity to ABA and/or increasing ABA level but also by improving water uptake as the result of developing lateral root elongation under WS (Le Hir et al., 2017).

ABA biosynthesis is known to up-regulate in response to osmotic stress; however, the signaling pathways by which ABA biosynthetic genes are up-regulated are still obscure. Nevertheless, both ABA-dependent and ABA-independent signal transduction cascades exist subsequent to the initial perception of the stress signal. According to Agarwal et al. (2006) the ABA biosynthesis is induced by dehydration and high salinity as a result of activating two main regulatory ABA-dependent gene expressions, including bZIP-type ABRE TF and myelocytomatosis/myeloblastosis (MYC/MYB) (Agarwal et al., 2006). AREB1/ABF2, AREB2/ABF4 and ABF3, which are activated by SnRK2s, such as, SRK2D/SnRK2.2, SRK2E/SnRK2.6 and SRK2I/SnRK2.3 (SRK2D/E/I), are transcription activators of ABRE-dependent ABA signaling in response to osmotic stress during vegetative growth (Fujita et al., 2005; Yoshida et al., 2015). The ABRE elements contain the DNA binding motif of the bZIP structure. The bZIP proteins are involved in UV light, salt, drought and SA defense signaling pathways. Most of the drought stress-inducible genes are also induced by ABA. However, in *aba* or *abi* *Arabidopsis* mutants, a number of other genes were induced by drought, salt, and cold. This suggests that some of the genes do not require ABA for their expression under drought, salt, and cold conditions. These genes include *rd29A/lti78/cor78*, *kin1*, *cor6.6/kin2* and *cor47/rd17* (Agarwal et al., 2006). ABA-independent stress-responsive gene expression has been thought to be regulated through dehydration-responsive element (DRE) *cis*-elements, while ABA-dependent pathways activate gene expression through abscisic acid responsive *cis*-elements (ABRE *cis*-elements). When the DRE/calreticulin (CRT)-binding protein DREB1/C-repeat binding factor (CBF) was overexpressed in transgenic *Arabidopsis* plants, changes in the expression of more than 40 stress-inducible genes were identified, and these changes led to increased freezing, salt, and drought tolerance. Overexpression of *ABF3* or *ABF4* genes conferred several ABA-associated phenotypes such as ABA hypersensitivity, sugar hypersensitivity, and enhanced drought tolerance, along with the altered expression of ABA- or stress-responsive genes such as *rd29B*, *ABA-insensitive 1* (*ABI 1*) and *ABI 2* (Khurana et al., 2008). Identification of the transcriptional activators of ABA signaling would be a promising tool to enhance drought tolerance (Khurana et al.,

2008). *Syntrichia caninervis DREB* genes were involved in ABA-dependent signal transduction, conferring tolerance to multiple abiotic stresses (Li et al., 2016a).

In *A. thaliana*, a responsive to ABA (RAB)-related gene, *RAB18*, has been shown to accumulate following drought stress and exogenous application of ABA in WT plants, but not in *aba* and *abi* mutants. Another gene, *RD29A* (*RESPONSIVE TO DESICCATION29A*), however, showed both ABA-independent and ABA-responsive expression because in its promoter region, which contains both drought-responsive element and ABA-responsive *cis*-acting elements (Jakab et al., 2005).

ABA induces stomatal closure and, therein, activates guard-cell anion channels in a calcium-dependent as well as-independent manner. ABA triggers the release of anions and K^+ from guard cells. The decrease in guard-cell osmotic pressure and volume results in stomatal closure, reducing transpirational loss of water from the leaf. Two key components of the ABA signaling pathway are the protein kinase OST1 (open stomata 1) and the protein phosphatase ABI1 (ABA-insensitive 1) (Geiger et al., 2009). The recently identified guard-cell anion channel SLAC1 (SLOW ANION CHANNEL-ASSOCIATED 1) appeared to be the key ion channel in this signaling pathway but remained electrically silent when expressed heterologously (Geiger et al., 2009). SLAC1 is directly activated by SNF1-related protein kinase 2 (SRK2E/OST1/SnRK2.6), which is involved in the ABA signaling complex of the ABA receptor pyrabactin resistance (PYR) family, and protein phosphatase 2Cs (PP2Cs) or by the calcium-dependent protein kinases, CPK21 and CPK23, and CPK3 and CPK6 (Geiger et al., 2009). SRK2E also inhibits KAT1 (*ARABIDOPSIS THALIANA* POTASSIUM CHANNEL 1) activity by phosphorylation. SLAC1 represents the slow, deactivating, weak voltage-dependent anion channel of guard cells controlled by phosphorylation/dephosphorylation. An ABA- and CO_2/O_3-insensitive mutant was shown to lack a gene encoding a putative guard-cell anion transporter named SLAC1 (Geiger et al., 2009). In guard cells of these mutant plants, anion currents appeared largely suppressed. When expressed heterologously, SLAC1, however, remained electrically silent. OST1 is identified as an interaction partner of SLAC1 and ABI1 (ABA-insensitive 1) (Geiger et al., 2009). The initial steps in ABA signal transduction have been shown to activate guard-cell anion channels, such as SLAC1, in a calcium-dependent as well as -independent manner, which causes depolarization of the plasma membrane of guard cells (Geiger et al., 2009; Osakabe et al., 2013). The depolarization of the plasma membrane decreases the activity of inward K^+ channels, such as KAT1/KAT2, and activates outward K^+ channels, such as the guard-cell outward rectifying K^+ channel, GORK, resulting in K^+ efflux from guard cells. Anion and K^+ efflux from guard cells leads to loss of guard-cell turgor and causes stomatal closure (Osakabe et al., 2013).

ABA-alternation pattern greatly varies depending on plant species and transgenic lines under different ranges of water potential. ABA concentration may greatly decline in some transgenic mutant plants under normal irrigation regime and thus rendering them incapable of growth, while it may remain unchanged in others both in roots in the xylem sap under normal condition and escalate under stress condition as a protection mechanism to cope with stresses. Plants may also respond differently to rapid and progressive stress, and alternation patterns in ABA accumulation may strikingly vary between detached parts of a plant and whole intact plant subjected to soil drying. The difference between ABA content in a detached part of a plant and the whole plant could be attributed to the differences in stress-induced expression of *ZEP* because the expression level of antisense mRNA (under the control of the 35S promoter) remained unchanged in all conditions (Borel et al., 2001). It was hypothesized that in response to rapid stress, root xanthophyll was rapidly metabolized, so it was necessary to produce them again. On the other hand, during a progressive stress, a slight increase in the flux of xanthophyll metabolized for ABA biosynthesis could be sufficient for withstanding during the whole stress period. Then, a slight increase of the *ZEP* mRNA level could be sufficient, possibly in combination with post-transcriptional regulation of this gene or of other genes implicated in ABA biosynthesis. For example, the induction of the gene encoding *9-cis-epoxycarotenoid dioxygenase* (*NCED*; which catalyzes the cleavage of 9-*cis*-epoxycarotenoids) (Borel et al., 2001) or *viviparous14* (*VP14*) (Caldeira et al., 2014) has been reported in leaves in response to WS. Plants under-expressing the *NCED/VP14* gene, which is involved in ABA biosynthesis, had a lower root hydraulic conductivity, *PIP* (*phosphatidylinositol 4,5-bisphosphate*) transcript abundance and protein content and consistently presented larger oscillations of leaf elongation rate than WT plants during continuous light, all of which were under the influence of genotype and circadian rhythms (Caldeira et al., 2014). According to Caldeira et al. (2014), in dehydrated cowpea (*Vigna unguiculata*) plants, the *VuNCED1* gene was markedly expressed in leaves and stems, but not in roots. Endogenous ABA levels escalated mainly in leaves and stems under drought conditions. The results suggested that ABA could be produced mainly in dehydrated cowpea leaves, but not in roots, and that leaves ABA mainly triggers stomata closure. In maize, strong expression of the *Vp14* gene was detected in the roots even under non-stress conditions (Iuchi et al., 2000). The mRNA of tomato *Vp14* homolog has been detected before and after drought stress (Iuchi et al., 2000). By contrast, the *VuNCED1* mRNA was not detected before stress treatment. Therefore, it seems that the VuNCED1 protein may mainly function under drought and high salt conditions, but not under normal growth condition. In cowpea, drought inducible *VuNCED1* gene encodes the 9-cis-epoxycarotenoid dioxygenase, and its product, which is localized in plastids, has a key role in ABA biosynthesis (Iuchi et al., 2000). Studies have so far suggested that overexpression of *NCED* genes may induce plant drought tolerance, lower the rate of transpiration, enhance metabolism of soluble sugars involved in osmotic adjustment, and eventually improve germination rate, relative water content (RWC), and plant water-use efficiency (WUE) (Zhang et al., 2014a; Manzoor et al., 2016). In addition to conferring drought tolerance, the higher ROS scavenging rate is associated with the overexpression of *Cr*NCED1 from *Citrus reshni* in transgenic tobacco (*Nicotiana nudicaulis*), which contained higher ABA levels than the WT under both normal and drought conditions (Xian et al., 2014). The transgenic lines displayed enhanced tolerance to multiple abiotic stresses such as dehydration, drought, salt, and oxidative stresses when compared with the WT plants (Xian et al., 2014). Under drought treatment, transgenic tobacco overexpressing wheat *TaNCED1* exhibited a higher germination rate, RWC, contents of soluble sugars and ABA when compared with the WT plants (Zhang et al., 2014b).

A land-plant-specific ABA signaling network is comprised of ABA receptors, negatively regulating group A PP2C, and transducing protein kinases. ABA receptors control the activity of a subfamily of three SNF1-related protein kinases (SnRK2 kinases) in response to environmental stress (Umezawa et al., 2009; Park et al., 2015; Li et al., 2016b). The SNF1-related kinase 2 (SnRK2) proteins from several plant species have been implicated in ABA signaling pathways (Geiger et al., 2009). These SnRK2 kinases autoactivate by *cis-* and *trans-*autophosphorylation on their activation loops but are continuously inactivated by clade A PP2Cs, which results in low basal kinase activity (Park et al., 2015). ABA binds to members of the PYL/RCAR (PYRABACTIN RESISTANCE1/REGULATORY-LIKE/REGULATORY COMPONENTS OF ABA RECEPTORS) ABA receptor family that initiate signal transduction by downregulating type 2CPP, which may occur even in the absence of ABA ligand known as "leaky" ABA signal transduction (Aleman et al., 2016; Li et al., 2016b). The PP2C-Rop-GEF-ROP/RAC is hypothesized as controlling the loop model, which is involved in shutting off ABA signal transduction, to counteract leaky ABA signal transduction (Li et al., 2016b). Small monomeric G-proteins of the plant ras (rat sarcoma viral oncogene homolog) related C_3 botulinum toxin substrate (RAC)/Rho of plants (ROP) family are molecular switches in signal transduction of many cellular processes (Nibau et al., 2006). RAC/ROP GTPases (GTPases: a large family of hydrolase enzymes that can bind and hydrolyze guanosine triphosphate) regulate polarized cell differentiation, division, and growth, cell morphogenesis, defense, and diverse signaling cascades, including hormones signaling and hormone-related gene expression, vesicle trafficking, subcellular gradients of Ca^{2+}, the organization of the actin cytoskeleton, and the production of reactive oxygen intermediates (Schultheiss et al., 2003; Lavy et al., 2007; Yang and Fu, 2007; Wu et al., 2011; Burkart et al., 2015). Small GTPases of the ROP/RAC family act as negative regulators of ABA signaling and maintain PP2C activity. Active PP2C protein phosphatases protect and stabilize RopGEF1 from ABA-mediated degradation, which contribute to ABA-mediated lateral root elongation (Li et al., 2016b).

The subgroup of ABA receptors is up- or down-regulated at the mRNA level by JA (Aleman et al., 2016). Aleman et al. (2016) reported a direct mechanism for transcriptional modulation mediated by an ABA receptor different from the core ABA signaling pathway, and a putative mechanistic link connecting ABA and JA signaling pathways. They found a link between ABA and JA signaling through a direct interaction of the ABA receptor PYL6 (RCAR9) with the basic helix-loop-helix TF MYC2. PYL6 and MYC2 interacted in yeast two-hybrid assays, and the interaction was enhanced in the presence of ABA. PYL6 and MYC2 interacted *in planta* based on bimolecular fluorescence complementation and co-immunoprecipitation of the proteins. Furthermore, PYL6 was able to modify transcription driven by MYC2 using JAZ6 (JASMONATE ZIM-DOMAIN 6) and JAZ8 DNA promoter elements in yeast one-hybrid assays. Finally, *pyl6* T-DNA mutant plants showed an increased sensitivity to the addition of JA along with ABA in cotyledon expansion experiments (Aleman et al., 2016). ABA is likewise able to sustainably induce dehydrin1 (DHN1) expression, whereas its expression could only transiently be enhanced by a low-temperature treatment or the application of JA (Beck et al., 2007). Park et al. (2015) findings provided a mechanistic basis for the activity levels of ABA, which improves water consumption and stress tolerance by controlling guard-cell aperture and other protective responses. One attractive strategy for controlling water use is to develop compounds that activate ABA receptors, but agonists approved for use have yet to be developed. In principle, an engineered ABA receptor that can be activated by an existing agrochemical could achieve this goal. Park et al. (2015) found a variant of the ABA receptor PYRABACTINRESISTANCE 1 (PYR1) that possesses nanomolar sensitivity to the agrochemical mandipropamid and demonstrated its efficacy for controlling ABA responses and drought tolerance in transgenic plants. Their results showed that when ABA levels rise during stress, the phytohormone binds to soluble ABA receptors and stabilizes their activated conformations, enabling them to bind to and inhibit PP2Cs. This, in turn, allows accumulation of activated SnRK2 kinases, whose direct targets include SLOW ANION CHANNEL 1, an anion channel that controls guard-cell aperture, and ABA RESPONSE-ELEMENT BINDING FACTORS, bZIP TFs that mediate ABA-regulated gene expression. Thus, ABA controls water-use and stress physiology by the receptor-mediated inhibition of PP2C activity and leads to SnRK2 kinase activation. They suggested that crystallographic studies provide a mechanistic basis for its activity and demonstrate the relative ease with which the PYR1 ligand-binding pocket can be altered to accommodate new ligands (Park et al., 2015).

Chromatin *licensing and DNA replication factor 1 (CDT-1) gene* isolated from the genome of resurrection plant *Craterostigma plantagineum* is a novel gene and no genes with similar sequences have been so far reported in other plant species by *in silico* analysis of available databases. The activation and constitutive expression of the *CDT-1* gene stimulate the osmotic tolerance of the callus which does not require ABA treatment to express desiccation tolerance and in which desiccation-responsive genes like *lea* genes are constitutively expressed. The mechanism by which *CDT-1* activates genes is not well understood. *CDT-1* may function as a regulatory non-coding RNA. In analogy, *CDT-1* RNAs could assume a different structure during water depletion and allow transcription of dehydration-induced mRNAs, e.g., by interacting with chromatin components (Bartels, 2005).

Cis-acting elements of drought stress and/or ABA-responsive promoters have mainly been identified by using transgenic plants such as tobacco or *Arabidopsis* to assay regulated gene expression (Bartels et al., 1996). In addition to the conservation of protein-coding sequences, desiccation tolerance is crucially linked to the correct functioning and recognition of *cis*-regulatory sequences. These regulatory sequences may even lie at a considerable distance from the protein-coding sequences whose expression they control (Bartels, 2005). As assay systems, promoter fragments were coupled with sensitive reporter genes such as those encoding β-glucuronidase or luciferase. In the context of drought stress, the best-characterized class of *cis*-elements are the ABA-responsive elements which contain the palindromic core motif CACGTG known as G-box. The specific function of ABA-regulated gene expression could be determined by the interaction between the G-box- and constitutive-exon(CE)-type-sequences (Bartels et al., 1996). Identifying the linkage between *cis*-regulatory sequences and protein-coding sequences will lead to a better understanding of how coordinated expression of dehydration-relevant genes has been achieved and how it can be exploited in manipulating desiccation tolerance (Bartels, 2005).

Plants regulate their transpiration rates by modifying stomatal aperture (Borel and Simonneau, 2002; Park et al., 2015) in the guard-cell (Borel and Simonneau, 2002), and consequently ABA receptors have emerged as attractive targets for water-use optimization; however, ABA agonists approved for this use have yet to be developed. ABA can be removed from or released into the transpiration stream before reaching guard cells, depending on the membrane pH gradients around the ABA transport pathway. Changes in transpiration rate are also thought to influence the local accumulation of ABA in guard-cell walls, while the metabolism of ABA can prevent its build up. An investigation showed that ABA concentration in sap extruded from ABA-fed leaves was markedly higher when leaves were sampled from the drought-stressed plants rather than the well-watered ones, but only slightly higher when leaves were pre-incubated in solution with elevated ABA concentration (Borel and Simonneau, 2002). Agrochemical control of plant water use could be accomplished in transgenic plants that express an engineered ABA receptor that responds to an existing agrochemical, a strategy based on orthogonal ligand–receptor systems, which have enabled selective chemical control of diverse targets. The recent findings demonstrate that it is possible to repurpose an existing agrochemical using receptor engineering. This strategy can be broadly applied to other plant receptors and agrochemicals and, therefore, opens new avenues for crop improvement (Park et al., 2015).

To investigate the response of *Vicia faba* to osmotic stress, Zhang and Outlaw (2001) studied the roots of the plants that were submerged in polyethylene glycol solution. The water potentials of root and leaf declined during 20 min of WS but recovered after stress relief. During stress, the ABA

concentration in the root apoplast increased, but in the leaf apoplast remained low. The ABA concentration in the guard-cell apoplast increased during stress, providing evidence for intra-leaf ABA redistribution and leaf apoplastic heterogeneity. Subsequently, the ABA concentration of the leaf apoplast increased, consistent with ABA import via the xylem. Throughout, the ABA contents of the guard-cell apoplast, but not the guard-cell symplast, were convincingly correlated with stomatal aperture size, identifying an external locus for ABA perception under these conditions. Apparently, ABA accumulates in the guard-cell apoplast by evaporation from the guard-cell-wall, so the ABA signal in the xylem is amplified maximally at high transpiration rates. Thus, stomata will display apparently higher sensitivity to leaf apoplastic ABA if stomata are wide open in a relatively dry atmosphere (Zhang and Outlaw, 2001). It has been suggested that both leaf ABA and ABA in the xylem stream entering the leaf should be considered to account for changes in stomatal conductance. This can be achieved by a substantial volume of sap (related to the leaf apoplastic water fraction) that contained extra-vascular leaf ABA together with xylem ABA. It can be of particular importance for predicting leaf ABA content, and xylem sap ABA is suspected (for example, in plants rewatered after a soil drying episode) (Borel and Simonneau, 2002). The comparative analysis of the leaf anatomical features of two olive cultivars showed that the leaf size of the epidermal and mesophyll cells with a parallel increase of the cell density decreased under WS. The anatomical strategies to reduce the transpiration rate referred to a significant reduction of the mesophyll intercellular space volume (internal blocking of water vapor movement) and to a greater increase of the density of non-glandular scales on the leaf surface (external blocking of water vapor movement). On the other hand, the rise in the density of stomata undoubtedly contributes to a better control of transpiration. Cell size reduction could prevent cell collapse in arid conditions; however, a greater number of epidermal cells per leaf surface reflects a better control of water loss through cuticular transpiration. Stomata became more numerous and smaller on the lower leaf surface, while the upper leaf surface was devoid of stomata in both cultivars concomitant with an increase of the cuticle thickness. On the other hand, non-glandular hairs (scales) greatly increased in number and covered both leaf surfaces (pubescence is thicker on the lower leaf surface). Reduction in transpiration by stomata closure parallels the reduction in photosynthesis rate and may be ascribed to the inactivation of Rubisco and the inhibition of photochemical reactions. In order for olive leaves to be able to carry out the necessary photosynthesis under drought stress conditions, the plants increase the number of mesophyll cells (consequently, the number of chloroplasts as well as the CO_2-uptaking cell surface) and also increase the number of the elongated sclereids, which are likely involved in light distribution within the mesophyll (Bosabalidis and Kofidis, 2002). Bosabalidis and Kofidis (2002) found that the non-glandular scales on the olive leaf surface play an insignificant role in the absorption of visible light so that photosynthesis can normally proceed (Bosabalidis and Kofidis,

2002). Application of exogenous ABA leads to the expression of most of the dehydration-induced proteins. Some of these gene products have been shown to accumulate in the cytosol, and three of them have been found in the chloroplast of dehydrated leaves. The stress-induced proteins are considered to form part of the protective mechanisms displayed by the plant under stress conditions. However, according to Alamillo and Bartels, (2001), the function of the desiccation-stress proteins remains elusive (Alamillo and Bartels, 2001). The underwater deficit leaves from all genotypes of cassava rapidly accumulated large amounts of ABA in both mature and young leaves. Correspondingly, young leaves halted leaf expansion growth, and transpiration rate decreased. Young leaves accumulated more ABA than the mature ones in both the control and the stressed treatments. The high ABA levels under water deficit were completely reversed to control levels after 1 day of rewatering. This rapid return to control ABA levels corresponded with a rapid recovery of leaf area growth rates. This rapid reduction in leaf area growth and stomatal closure could be due to cassava's ability to rapidly synthesize and accumulate ABA at an early phase of a water deficit episode. This drought response can be interpreted as an appropriate adaptation strategy for environments with cycles of intermittent or seasonal drought, followed by recovery of water status, when the bulk of the growth occurs (Alves and Setter, 2000). Overall, manipulation of ABA biosynthetic and transport-related genes and TFs, desiccation-stress proteins, Ca^{2+} sensor proteins and the genes responsible for anatomical modifications would be efficient strategies for improving WUE and engineering drought-tolerant crops.

19.2.2 BRASSINOSTEROID (BR)

Brassinosteroids (BRs) are a class of plant polyhydroxylated steroid phytohormones that regulate multiple morphophysiological and developmental processes, such as cell division, elongation, expansion, and cell death, vascular differentiation, pollen tube growth, seed germination, seed yield, flowering, proton pump activation to apoplast and into a vacuole by stimulation of transmembrane ATPases, as well as increasing the efficiency of photosynthesis by elevating the level of CO_2 assimilation and Rubisco activity, assimilate distribution, detoxification by antioxidant enzymes, cell membrane polarization, source/sink relationships, reproductive development, ions uptake into the plant cell, gene expression, nucleic acid and protein synthesis, signal perception and transduction and also confer plant resistance against various biotic and abiotic stresses (Zhang et al., 2008; Gruszka, 2013; Zhou et al., 2014; Gruszka et al., 2016; Todorova et al., 2016). BRs are likely involved in signal transduction pathways initiated by hormone perception by plasma membrane-associated receptor complex, transduced via phosphorylation/dephosphorylation cascade to regulate gene expression by a group of transcription factors (Gruszka, 2013). Recent studies disclosed that BR application may increase plant tolerance to drought stress probably by affecting regulation of gene expression and TFs, signal perception and transduction, as

well as hormonal crosstalk, biosynthesis, and homeostasis; however, plant responses may vary depending on plant tissue and organ type, genotype and BR concentration (Saini et al., 2015; Gruszka et al., 2016). The involvement of BRs in signal transduction, gene expression, and developmental biology such as cell expansion and organ growth, homeostasis, photomorphogenesis, regulatory effect on hormone pathways, as well as crosstalk effects with phytohormones such as AUX, CK, ABA, ET, GA, JA, and SA, and also PAs has been corroborated (Bouquin et al., 2001; Goda et al., 2004; Nemhauser et al., 2004; Gendron et al., 2008; Divi et al., 2010; Vercruyssen et al., 2011; Wang et al., 2012; Saini et al., 2015; Gruszka et al., 2016; Todorova et al., 2016). In barley, drought-induced GA_7 biosynthesis represented an enzymatic step which stimulated the effect of BRs on GA biosynthesis. Castasterone (CS) accumulation, which is inversely correlated with 28-homoCS accumulation, is suggested to end the BR biosynthetic pathway in monocots. Biosynthesis of JA and GA were also a BR- and species-dependent process. Drought stress significantly induced the accumulation of JA, ABA, and SA. ABA inhibits the action of growth-promoting hormones, including BR signaling, and masks BR effects in plant stress responses. It has been postulated that BR and ABA responses remain in a constant balance, regulating homeostasis and plant development. Application of excessive exogenous BR may stimulate ABA accumulation, thereby preventing enhanced BR responses within some feedback loop. The other way round, a deficiency in BR biosynthesis may result in hypersensitization to endogenous ABA, and consequently to enhanced drought tolerance, while BR-overproducing and constitutive BR signaling mutants are less tolerant to stresses, which suggests that BR negatively regulates stress responses. It is known that many of the physiological functions regulated by BRs are in fact modulated based on a complicated network of interactions with other hormones: AUX, GA, ABA, ET, and JA. This interaction is based on some crosstalk between BR biosynthesis and signal transduction pathways and signalosomes of other phytohormones, which has been mostly documented with regard to plant growth regulation under control rather than the stress condition. Moreover, BRs may promote some anti-stress mechanisms, that are independent of ABA, ET, JA, and SA at least to some extent (Gruszka et al., 2016). ABA-BRs crosstalk positively regulates cell elongation, stomatal movement, and photosynthesis. The increasing antioxidative activities of a BR-treated plant is ascribed to the BR-induced elevation of ABA content (Yuan et al., 2010). Rice plants overexpressing the genes associated with auxin response factor (ARF), such as, *OsARF19*, were sensitive to BR treatment, controlling leaf angles and, thus, photosynthesis and grain yield through AUX-BR crosstalk (Zhang et al., 2015). Overall, the goal is to improve drought tolerance by manipulating gene expression and signaling pathways in the specific cell, tissue, and organs to regulate hormone homeostasis.

Reactive oxygen species such as H_2O_2 are involved in the regulation of multiple plant responses to a variety of stresses and may function as a second messenger in phytohormone signaling and other important biological processes. BRs induce a transient increase in the *Respiratory burst oxidase homolog 1* (*RBOH1*) transcript, NADPH oxidase activity, H_2O_2 in the apoplast and nitric oxide (NO). BR-induced ROS production is important for BR-induced stress tolerance (Zhou et al., 2014) because ROS and NO signaling involve ABA crosstalk, GABA biosynthesis, and interactions with primary metabolism (Sequera-Mutiozabal et al., 2016). Zhou and co-workers (2014) concluded that BR-induced stress tolerance is associated with rapid and transient H_2O_2 production by NADPH oxidase. The process, in turn, triggers increased ABA biosynthesis, leading to further increases in H_2O_2 production and prolonged stress tolerance. Both endogenous and exogenous BRs increase ABA accumulation, the effect of which is more significant under stress conditions (Zhou et al., 2014). In maize, BR treatment increased the content of ABA and up-regulated the expression of the ABA biosynthetic *vp14* gene in leaves (Zhang et al., 2011). According to Zhang et al. (2011), BR treatment induced the generation of NO in mesophyll cells of leaves, and treatment with the NO donor sodium nitroprusside up-regulated the content of ABA and the *vp14* expression in the leaves (Zhang et al., 2011). With synergistic effects, BRs and ABA induce NADPH oxidase *RBOH* genes expression and increase apoplastic H_2O_2 production in both the apoplastic and chloroplastic compartments (Xia et al., 2009; Zhou et al., 2014). Zhang et al. (2011) results suggested that BR-induced NO production and NO-activated ABA biosynthesis are important mechanisms for BR-enhanced WS tolerance in leaves of maize plants (Zhang et al., 2011). Accordingly, the present review suggests that investigation on the effect of BRs on NO signaling pathway may provide deep insight into the role of BRs on signal transduction and stress tolerance.

Examining the effect of BR on the germination and seedling growth of different varieties of sorghum under osmotic stress showed that BRs could effectively increase the percentage of germination and seedling growth of all the varieties; the growth promotion being associated with enhanced levels of soluble proteins and free proline (Vardhini and Rao, 2003). Vardhini and Rao (2003) showed that BRs also affected plant cell detoxification by increasing catalase and superoxide dismutase and glutathione reductase, but reducing indole-3-acetic acid (IAA) oxidase, polyphenol oxidase, protease, peroxidase, ribonuclease, and ascorbic acid activities (Vardhini and Rao, 2003). Numerous studies indicate that the effects of exogenous BR treatments have been more noticeable under stressful conditions than normal conditions. In addition to promoting plant tolerance to biotic and abiotic stressors, the positive effect of BRs on improving both yield and quality traits such as mineral content and nutrient composition has also been reported (Vardhini et al., 2011; Kamthan et al., 2012; Serna et al., 2012; Freitas et al., 2014). Zhang et al. (2008) results showed that BR treatment prior to water deficit alleviated the negative impact of drought stress and promoted biomass accumulation and seed yield of both drought-stressed and control soybean plants due to improving plant water status, photosynthesis rate, accumulation of soluble sugars and proline, as well as antioxidant enzyme activity (Zhang et al., 2008).

19.2.3 Cytokinin (CK)

Cytokinins (CKs), which were named for their ability to promote cytokinesis, are a class of phytohormone derived from adenine. Naturally occurring cytokinins can be divided into two groups based on their side chain: those with isoprene-derived side chains, which are predominant in plants; and those with aromatic side chains (El-Showk et al., 2013). Natural CKs commonly contain an adenine moiety and a side chain modification at the adenine N6 position. Natural and artificial CKs are recognized by common CK receptors (Osugi and Sakakibara, 2015). CK regulates plant drought adaptation via a multistep component system consisting of histidine kinases, histidine phosphotransfer proteins, and type-A and type-B response regulators (type-A and type-B authentic response regulators, "ARRs") (Nguyen et al., 2016). Three histidine kinases and at least five type-B ARRs have been shown to act as positive regulators of CK signaling, while a number of type-A ARRs, and AHP6, act as negative regulators of the pathway (Hutchison and Kieber, 2007). CKs are involved in many different developmental and physiological processes in plants, including plant growth and development, cell proliferation and differentiation (Osugi and Sakakibara, 2015; Chang et al., 2016), embryogenesis, maintenance of root and shoot meristems, vascular development (Osugi and Sakakibara, 2015), delay leaf senescence (Chang et al., 2016), photosynthesis (Kimura et al., 2001; Dobra et al., 2010; Rivero et al., 2010), recovery (Dobra et al., 2010), WUE (Rivero et al., 2007), nutrient use efficiency and content, hormone homeostasis, chlorophyll content, root-to-shoot ratio (Werner et al., 2010; Kuppu et al., 2013; Reguera et al., 2013), assimilate partitioning, sink strength, and source/sink relationships (Reguera et al., 2013). CKs also modulate root elongation, lateral root number, root biomass, nodule formation, and apical dominance in response to environmental stimuli (Werner et al., 2010; Osugi and Sakakibara, 2015). While high CK concentration inhibits root growth and branching (Werner et al., 2010), the proper levels of ABA, ET, CK, and AUX can promote root system architecture and even tiller formation depending on their concentrations in specific tissues (Xu et al., 2008; Rowe et al., 2016). In broccoli, the effective use of water using regulated water stress enhanced CK biosynthesis in broccoli florets and, consequently, promoted its postharvest quality by delaying yellowing of their florets (Zaicovski et al., 2008). Hence, depending on the plant genotype and species, effective use of water by strategies like regulated deficit irrigation or partial root-zone drying could be employed to promote water, nutrient, and hormone homeostasis and, consequently, crop production and quality.

CK application lessens cell membrane lipid peroxidation (Chang et al., 2016), and can improve the plant tolerance to abiotic stresses such as drought by improving the antioxidant capacity against ROS damage (Reguera et al., 2013; Chang et al., 2016), cell membrane integrity, increased anthocyanin biosynthesis, ABA hypersensitivity, reduced stomatal aperture (but not to altered stomatal density) (Nguyen et al., 2016), and coordinated regulation of carbon and nitrogen metabolism (Reguera et al., 2013). The burst of excess ROS may lead to membrane lipids damage in plant cells, while the enzymatic systems are responsible for maintaining the membrane structure by improving the antioxidant capacity against ROS damage. Therefore, the stimulation of antioxidant enzymatic systems after the exogenous CK application could be one of the important protective mechanisms for the plants to avoid abiotic injuries (Chang et al., 2016). Application of CK can improve plant recovery after mDr (Novakova et al., 2007). CK signaling-mediated network controls plant adaptation to drought via many dehydration and/or ABA-responsive genes that can provide osmotic adjustment and protection to cellular and membrane structures. Repression of CK response, and thus CK signaling, e.g., type-B *ARR* genes involved in signaling, is one of the strategies plants use to cope with water deficit, providing novel insight for the design of drought-tolerant plants by genetic engineering. This stress-induced suppression of CK signaling appears to be a yin-yang strategy facilitating environmental acclimation/adaptation of plants. Detailed functional analysis information of all of the components of CK signaling and crosstalk network pathways with genes, TFs, and hormones in tobacco plant responses to drought are still elusive (Nguyen et al., 2016). CKs may improve ROS detoxification or cell differentiation by stimulating PA biosynthesis and homeostasis and subsequently PA-mediated CK-AUX function (Legocka and Żarnowska, 2000; Alabdallah et al., 2017). Reduction in CK levels under drought is attributed to the decrease in dihydrozeatin levels (Le et al., 2012). Synergistically, plants with higher proline-induced osmoregulatory and antioxidative effects, particularly in upper leaves, maintained higher CK levels associated with improved plant recovery during dehydration. Combined effects of drought and heat, however, exerted more severe effects by a further reduction in CK levels (Dobra et al., 2010). Consequently, maintenance of higher CK levels (Merewitz et al., 2010), particularly when CK is combined with potassium (Hu et al., 2013), melatonin (Zhang et al., 2017), and/or osmoprotectants such as PAs, glycine betaine, proline, other amino acids and polyols (Farooq et al., 2009; Wimalasekera et al., 2011), and a higher ratio of CK to ABA effectively improve plant water-use efficiency, photosynthetic capacity, stomatal closure, leaf senescence, chlorophyll and anthocyanin content, recovery, inflorescence meristem activity, higher root/shoot ratio, modified metabolic accumulation associated with stress signaling, osmoprotection, detoxification, and energy use efficiency, and eventually plant performance under multiple abiotic stresses, such as drought, salinity, and heat stress (Merewitz et al., 2010, 2011, 2012; Macková et al., 2013; Šmehilová et al., 2016; Joshi et al., 2017). Interestingly, CK/ABA/AUX homeostasis and their drought resistance effects, such as better osmoregulation by promoting osmoregulators (e.g., PAs and proline) and ROS detoxification, can be modulated by using plant growth-promoting rhizobacteria (Goicoechea et al., 1998; Liu et al., 2013; Kaushal and Wani, 2016) and beneficial microbiomes

(Marasco et al., 2012; Singh and Reddy, 2015; Boivin et al., 2016; Vurukonda et al., 2016). Arbuscular mycorrhizal fungi (AMF) may promote photosynthesis through better stomatal conductance by enhancing ABA content. AMF promotes root growth, which consequently improves plant nutrient and water uptake, osmotic adjustment (Wang et al., 2017), shoot growth (García et al., 2017), redox status (e.g., more H_2O_2 effluxes), G-protein signaling, soil structure by glomalin (Wang et al., 2017), nitrogen fixation, hormone homeostasis (e.g., AUX production), production of more PA, glutamate, glycine betaine, trehalose, and proline content (García et al., 2017). However, depending on the AMF strain and plant genotypes, their interaction effects on plant growth and stress tolerance greatly vary (Arzanesh et al., 2011; Shah et al., 2017). Like AMF, inoculation of rootstocks grapevine with bacterial microbiome improved hormone homeostasis, nutrient uptake, water status, root system, immune system to pathogens, antioxidative activities, and consequently plant performance (Rolli et al., 2015). Accordingly, to understand the differential responses, more investigations based on molecular, nutrients, and hormone variations of plant–microbiome or plant-fungal symbiosis associated with their tolerance and performance under different environmental conditions are yet required.

In a wide range of plant taxa, the endogenous CK levels and their associated CK responsive genes may depress under stress conditions (Zwack and Rashotte, 2015). Therefore, the application of beneficial fungus and bacteria in agriculture known as the second or evergreen revolution would promisingly allow farmers to use organic farming with the same benefits of modern biotechnological approaches.

Both rapid and developmental responses to drought are mediated primarily by the stress-related hormone ABA. However, growth-promoting hormones, such as GAs and CKs, are also involved in the response to drought. It seems that the balance between stress-related and growth-promoting hormones has an influence on plant performance during short episodes of drought and on their adaptation to prolonged drought conditions (Zhou et al., 2014; Farber et al., 2016). Thus, plant growth, yield, and quality improvement by hormone homeostasis, crosstalk, and signaling can be obtained through either application of hormones, manipulation of genes, enzymes, receptors, and TFs related to either growth promoter hormones or their inhibitors for reducing function or inactivation. The effects of CKs on morphological and physiological characteristics vary depending on plant species, and even in different leaves on the same plant (Farber et al., 2016). Thus, preliminary experiments by examining different genetic varieties, and different tissue, and organ types in different environmental conditions and stress combinations is required before conducting a large-scale commercial farming.

Li et al. (2015) findings demonstrated that down-regulation of the *UGT76C2* (*UDP-glycosyltransferase 76C2*) gene, a cytokinin glycosyltransferase and a member of family 1 UGTs in *Arabidopsis* by drought, osmotic stress, and ABA, plays an important role in drought stress adaptation. Compared to WT plants, transgenic lines ectopically expressing *UGT76C2* exhibited reduced tolerance to ABA and osmotic stress during postgermination growth, while enhanced adaptation to drought stress at the mature stage (Li et al., 2015); hence, showing the importance of assessing plant responses at different maturity stages. Several studies have reported that senescence could be delayed in transgenic plants expressing *isopentenyltransferase* (*IPT*, an enzyme that mediates CK synthesis), using different promoters. With similar effects on rice, tobacco, and cotton transgenic plants, expression of the *Agrobacterium tumefaciens IPT* gene under the control of a specific maturation- and stress-inducible promoter *senescence-associated receptor kinase* (P_{SARK}) (Hajouj et al., 2000) conferred a significant shift in the expression of hormone-associated genes, delayed senescence and flowering, and enhanced photosynthetic capacity and root-to-shoot ratio, consequently improved grain quality (higher mineral nutrients and starch content) and yield as well as drought tolerance. Under WS, rice transgenic plants overexpressing P_{SARK}::*IPT* exhibited an increase in CK level, which subsequently induced the expression of BR-related genes and repression of PP2C, *PCAR1*, and ABA-related genes (Rivero et al., 2010; Peleg et al., 2011). Nordihydroguaiaretic acid (NDGA; ABA-biosynthesis inhibitor) treatment was also useful in reducing JA and ET levels and, consequently, increasing photosynthesis and grain yield in rice (Tamaki et al., 2015). Overall, the findings indicate that the strong sink/source relationships, regulating the mobilization of stored carbohydrates from the vegetative tissues to the grains, and the increase in BR and reduction of ABA and JA levels are of key importance for improving rice yield under stress and control conditions (Rivero et al., 2010; Peleg et al., 2011; Kuppu et al., 2013; Tamaki et al., 2015). Changes in hormonal homeostasis and the associated crosstalks by overexpressing the P_{SARK}::*IPT* gene in tobacco resulted in the modification of source/sink relationships and a stronger sink capacity, photosynthetic apparatus, WUE, and biomass production of the plants under water (Rivero et al., 2010; Peleg et al., 2011) and nitrogen deficiency (Rubio-Wilhelmi et al., 2014). In another study, Tamaki et al. (2015) obtained similar results in transgenic rice plant using different plant hormone inhibitors of which some of them had the same homeostasis function and their effects were similar to the P_{SARK}::*IPT* plants in a Peleg et al. (2011). Among the tested chemical compounds, 2-indol-3-yl-4-oxo-4-phenylbutanoic acid (PEO-IAA; antagonist of AUX receptor), nordihydroguaiaretic acid (NDGA; ABA-biosynthesis inhibitor), and 2-aminoisobutyric acid (AIB; ethylene biosynthesis inhibitor) improved grain yield in a dose-dependent manner. Their effects were dependent on the plant developmental stage. Like CK effects, NDGA can inhibit ABA biosynthesis in some fruits during fruit maturation period, and in maize, soybean, and rice under drought stress conditions (Tamaki et al., 2015). NDGA and AIB treatments induced an increase in photosynthesis in rice flag leaves concomitant to the increments of starch content in flag leaves and grains (Tamaki et al., 2015). NDGA inhibited the expression of ABA-responsive gene, but did not significantly decrease ABA content. Instead, NDGA significantly decreased JA and JA-isoleucine. Thus, the specific inhibition of JA and ET biosynthesis resulted in grain yield increase in

rice (Tamaki et al., 2015). The *PURINE PERMEASE* (*PUP*) gene family encodes broad affinity transporters that are able to transport several CKs. The expression pattern of *PUP1* and *PUP2* suggests that they may play a role in the loading and unloading of CKs for long-distance transport. The *EQUILIBRATIVE NUCLEOSIDE TRANSPORT* (*ENT*) gene family has also been implicated in CK and nucleoside transport (El-Showk et al., 2013). An experiment showed that continuous moderate and severe drought stress, which stimulated the production of malondialdehyde and proline, could be alleviated by preliminary application of a synthetic phenylurea type CK, 4PU-30, promoting the endogenous biosynthesis of PAs (Todorova et al., 2016). According to Farber et al. (2016), transgenic tomato plants overexpressing the *A. thaliana* CK-degrading enzyme CK oxidase/dehydrogenase 3 (CKX3) maintained higher leaf water status under drought conditions due to reduced whole-plant transpiration. The reduced transpiration could be attributed to smaller leaf area and reduced stomatal density (Farber et al., 2016).

19.2.4 SALICYLIC ACID

Salicylic acid is a seven carbon (C) containing, naturally occurring phenolic compound and endogenously synthesized signaling molecule in plants that can function as a plant growth regulator. The shikimic acid pathway and the malonic acid pathway are the two main pathways known to be involved in the synthesis of plant phenolics. The shikimic acid pathway takes part in the biosynthesis of most plant phenolic compounds. It simply converts simple carbohydrate precursors derived from glycolysis and pentose phosphate pathway (PPP) to the aromatic amino acids, including SA precursor, phenylalanine. The most common pathway in plants for SA synthesis is the phenylalanine pathway; however, SA biosynthesis may also be accomplished via the isochorismate pathway (Khan et al., 2015). SA plays an important role in promoting signaling, seed germination, seedling establishment, cell growth, respiration, stomatal conductance and transpiration, senescence-associated gene expression, enhancement of antioxidative protection through both non-enzymatic (e.g., by increasing phenolic compounds such as flavonoids, anthocyanins, and total phenol contents) and enzymatic antioxidant activities, energy production, and photosynthesis under adverse environmental conditions. These studies suggest that SA may enhance tolerance to multiple biotic and abiotic stresses in plants by interactive effects and crosstalk signaling with other phytohormones on several functional molecules and transcriptional reprogramming. It can be hypothesized that plant redox status such as ROS, H_2O_2, and glyoxalase system play an important role in SA signaling and function, though adequate investigations are still missing to fully reveal their signaling pathways (Kang et al., 2012; Miura et al., 2013; Alam et al., 2014; Chen et al., 2014; Fayez and Bazaid, 2014; Khan et al., 2015; Liu et al., 2015b; Abbaspour and Ehsanpour, 2016; Lu et al., 2016; Sharma et al., 2017). Several studies suggest that SA may confer tolerance to several abiotic stresses by modulating the plant responses such as promoting basal thermotolerance, antioxidative activity, photosynthetic capacity, chlorophyll (Chl) contents, biomass production, RWC, osmoregulation (mediated by osmolytes such as proline, glycin betain, soluble sugars, amines, etc. and ion homeostasis such as Na/K balance and inhibition of Na and Cl accumulation), secondary metabolites, mineral nutrients, nitrogen metabolism, and hormone homeostasis, all of which may consequently enhance plant performance (Jakab et al., 2005; Bechtold et al., 2010; Chen et al., 2014; Fayez and Bazaid, 2014; Khan et al., 2015). SA may enhance drought tolerance indirectly by improving plant nutrient homeostasis. Under salinity, soil field or maize seeds SA pretreatment lowered Na and Cl accumulation and boron toxicity, while depending on plant genotype and experiments differentially promoted N, P, K, Mg, Fe, Cu, and Mn contents and, thus, plant growth and tolerance to multiple stresses (Gunes et al., 2005, 2007). Since the negative impact of SA on drought tolerance and photosynthesis, while increasing chilling resistance, for example in maize, has been documented (Németh et al., 2002), preliminary experiments are required to include the effect of plant genotype, hormone concentrations and their interactions, different stress combinations, environmental factors, and genotype-by-environmental interactions.

Interestingly, Liu et al. (2016) findings demonstrated that SA receptors activate JA signaling through a non-canonical pathway to promote effector-triggered immunity (Liu et al., 2016). Stimulated by stress-induced ROS signaling, ABA and SA signaling interactions and crosstalk effects can enhance drought tolerance in a dose-dependent manner by promoting antioxidative enzyme activities (Agarwal et al., 2005; Chojak-Koźniewska et al., 2017). In barley, SA enhanced the amount of ABA in the leaves of the stressed plants, conferring a protective role on the leaves cell membrane integrity (Bandurska and Stroi ski, 2005). Like SA, ascorbic acid or ascorbic acid combined with calcium chloride can enhance drought tolerance by increasing antioxidative enzyme activity and antioxidative non-enzyme activity by stimulating osmoregulators, proline, protein content, and photosynthetic pigments (Amin et al., 2009; Madany and Khalil, 2017). Several genes induced transcriptionally by osmotic stress are also part of plant defense responses to wounding and pathogen attack. Expression of peroxidase, *PR-1*, *PR-10*, and osmotin (*PR-5*) is increased by WS even though the role of these proteins in abiotic stress has not fully been clarified. During plant-pathogen interactions, the expression of the pathogenesis-related protein genes is induced by the plant hormone SA, an important regulator of systemic acquired resistance. SA treatment also increases the resistance of wheat (*Triticum aestivum*) seedlings to salinity and drought (Jakab et al., 2005). The nonprotein amino acid β-aminobutyric acid (BABA), a potent inducer of resistance against infection by various pathogens, in certain cases exerted its function via priming of SA-dependent defense mechanisms in Arabidopsis. In other cases, however, BABA acts through potentiation of ABA-dependent signaling pathways. Jakab and co-workers (2005) found that BABA priming protected Arabidopsis against abiotic stresses such as drought and high salinity based on ABA-dependent, but

SA-independent defense mechanisms. Due to the positive impact on hormone and ion regulation, detoxification and promoting protection mechanisms against biotic and abiotic stresses, application of SA and ascorbic acid decreased the harmful effect of stressors, such as drought and salinity, on germination and seedling growth of many plants (Motamedi et al., 2013; Batool et al., 2014; Farzane et al., 2014; Guan et al., 2015).

19.2.5 Jasmonate or Jasmonic Acid

Jasmonic acids are a class of lipidic plant hormones, synthesized from linolenic acid present in the chloroplast membrane. They are involved in plant development, biotic and abiotic stress responses, and plant-microbe interactions associated with defense and symbiosis systems. Derivatives such as methyl-jasmonate (MeJA) are volatile and participate in long-range signaling between plants (Aleman et al., 2016; Campos et al., 2016; Cao et al., 2016; Zheng et al., 2016; Chini et al., 2017). A central feature of JA signaling is the repression of JA responses by the JASMONATE ZIM-DOMAIN (JAZ) protein family. JAZ proteins function to inhibit the activity of transcription factors responsible for driving the expression of JA-responsive target genes (Roberts, 2016). JA biosynthesis and signaling were among the main enriched gene ontology (GO) categories in the down-regulated genes at the late stage of mDr in maize (*Zea mays*) and early stage in soybean (*Glycine max*), as they showed an early increase within 2 h of dehydration and a decrease in concentration afterward (Harb et al., 2010). Moreover, jasmonates were found to cause stomatal closure. This role was confirmed by the impaired stomatal response to exogenous jasmonates in jasmonate-insensitive mutants. There is a crosstalk between jasmonates and ABA as they utilize a similar cascade of events to stimulate stomatal closure, cell-wall modification (Harb et al., 2010), photosynthesis apparatus, and root elongation (Kazan and Manners, 2008). Both ABA and MeJA induce stomatal closure most likely by triggering the production of ROS in stomatal guard cells. ABA also appears to induce JA biosynthesis in *Arabidopsis*, and the increased JA levels found in ABA-treated plants could be due to the effects of ABA on suppression of SA gene expression by transcriptional repressors or producing specific proteins that mediate JA genes suppression (Kazan and Manners, 2008; Caarls et al., 2017) and/or SA precursor phenylpropanoids (Nugroho et al., 2002; Gupta et al., 2017). Nonetheless, as shown in *Arabidopsis*, without ABA variations, SA and JA biosynthesis and signaling were strongly up-regulated under a drought and heat combination (Gupta et al., 2017). To unravel their precise functions, their gene expression, metabolic changes, and signal transduction should be assessed during the whole stress episode in connection with circadian rhythm, stress severity, and combination of different stresses.

ABA-dependant genes *OST1* and *SLAC1* involved in stomatal closure and SA signaling were up-regulated under combined heat and drought stress (Gupta et al., 2017). Irreversibly, the expression of ABA genes may suppress the JA and ET signaling pathways and genes expression, which has been extrapolated as a preparatory response needed for drought acclimation (Harb et al., 2010; Gupta et al., 2017). JA is probably not required at a high concentration under drought stress, and an increase in its concentration might negatively affect plant growth response. Indeed, the interaction between cellulose synthesis and a high concentration of JA revealed a negative effect of JA on cell-wall modification and plant growth, which enhanced plant resistance to fungal pathogens. Moreover, JA induction in response to wounding and herbivory freezes the cell cycle, inhibits cell expansion, and results in stunted growth. To minimize the inhibitory effect of JA on plant growth under prolonged drought (late mDr), the downregulation of JA biosynthesis and signaling pathways can act in establishing new homeostasis in the acclimation process (Harb et al., 2010).

19.2.6 Auxin (AUX)

AUXs and CKs may antagonize the effect of ABA on stomatal movement (Tanaka et al., 2006). Besides, endogenous and exogenous AUX content regulates ROS detoxification/generation and influences antioxidative enzyme activity. More importantly, AUXs confer drought resistance through regulation of root system architecture (RSA) (Shi et al., 2014; Choudhury et al., 2017). Moreover, auxin significantly modulated some carbon metabolites, including amino acids, organic acids, sugars, sugar alcohols, and aromatic amines (Shi et al., 2014). Similar to AUX and with some overlapped gene expression patterns, BR positively regulates RSA (Teale et al., 2008). Along with synergistic effects with AUX, ET promotes root growth by regulating signaling, cell division, and expansion (Teale et al., 2008). GA is another important hormone that promotes root elongation by regulating cell elongation (Inada and Shimmen, 2000; Teale et al., 2008; Shani et al., 2013). ABA controls the AUX transport to root tip. Consequently, AUX activates the plasma membrane H^+-ATPase to release more protons along the root tip, encouraging cell-wall loosening and, consequently, root elongation and root hair development under mild water deficit (Xu et al., 2013). Like ABA, AUX may induce ROS and NO production, and both are involved in AUX-mediated signaling (Choudhury et al., 2017). Although high concentrations of CKs may antagonize the effect of AUX and pose negative effects on root growth, it is hypothesized that a low concentration of CK to AUX would be effective to regulate root gravitropism, and to stimulate protoxylem vessel or vascular differentiation and lateral root initiation (Aloni et al., 2006; Teale et al., 2008). In wheat, the proper ratio of kenitin:IAA alleviated the negative effects of NaCl-induced osmotic stress by improving the leaf area and photosynthetic pigments and reducing the proline content as the indicator of lower oxidative damage. The photosynthetic pigments were preserved at the highest level after the treatment with 20 ppm kinetin combined with 30 ppm IAA (Zoubida and Gherroucha, 2017). With crosstalk effects with AUX, sucrose, nitrogen, and phosphorus are also involved in improving RSA (Teale et al., 2008). Accordingly, to improve

RSA, the crosstalk effects and ratio of AUX:ET:ABA:GA:CK in relationship with redox status should be taken into account for designing drought-tolerant plants.

19.2.7 Gibberellins (GA)

The GAs are a large group of tetracyclic diterpenoid carboxylic acids, of which a very small number function as growth hormones in higher plants, the predominant bioactive forms being GA_1 and GA_4 (Colebrook et al., 2014). GA metabolism has shown to contribute to plant growth and development processes such as seed germination, leaf expansion, stem and root elongation, flowering, and fruit development, signal transduction, and tolerance to multiple biotic and abiotic stresses such as drought and salinity in different ways (Shimada et al., 2006; Mohammed, 2007; Colebrook et al., 2014). The ameliorative effects of GA could be due to the preservation of photosynthetic pigments, photosynthesis machinery, protein biosynthesis, reducing proline and PA accumulation (their accumulations are the indicators of more sensitivity), more effective antioxidative enzyme activity, nitrogen metabolism, ion uptake and nutrient homeostasis (lower Na and Cl uptake and higher K, Ca, and Mg levels), and hormone homeostasis associated with better water uptake (Mohammed, 2007).

In mungbean, foliar application of IAA, GA_3, and kinetin reduced the negative effects of salinity by improving plant water status and, consequently, photosynthesis and growth (Mohammed, 2007). Up-regulation of GA-responsive genes or down-regulation of a negative GA regulator may confer stress tolerance through maintaining GA and CK at high levels or up-regulating CK genes, GA-CK crosstalk, signaling pathways associated with transcriptional regulation, and consequently their ameliorative effects (Qin et al., 2011). Another way would be the expression of genes encoding regulatory proteins that regulate transcription factors and gene expression to maintain higher levels of a specific hormone(s). Apart from the related gene expression, the function of hormones depends on hormone metabolism, distribution, and degradation. In this context, the expression of genes encoding regulatory proteins, regulation of transcription factors, and hormone transporters associated with IAA, CK, and GA metabolism and in general the IAA:CK:GA:JA:SA:ABA homeostasis, particularly under the control of optimum osmoregulation by for example sufficient accumulation of osmolytes such as PA may enhance drought tolerance (Krishnan and Merewitz, 2017; Todaka et al., 2017).

19.2.8 Strigolactone (SLs) and Karrikins (KARs)

The plant hormones strigolactones (SLs) and smoke-derived karrikins (KARs) are butenolide compounds that have been identified as regulators of plant growth and development (Soundappan et al., 2015). SLs derive from β-carotenoid that convert to carlactone (Alder et al., 2012). SLs are signaling molecules that inhibit shoot branching and contribute to biological processes, including germination of parasitic plant seeds, stimulation of symbiotic fungi (Gomez-Roldan et al., 2008; Umehara et al., 2008; Tsuchiya et al., 2010; Alder et al., 2012;

Ha et al., 2014; Al-Babili and Bouwmeester, 2015), plants' aerial part growth, adventitious root formation, water and nutrient uptake reproductive development, leaf senescence, and stress responses (Ruiz-Lozano et al., 2016).

In the nonmycotrophic *Arabidopsis*, more *axillary growth (MAX)* genes, namely *MAX1/AT2G26170*, *MAX3/AT2G44990*, and *MAX4/AT4G32810*, have been identified to encode enzymes that are involved in the SL-biosynthetic pathway (Ha et al., 2014). In plants, SLs regulate a wide range of endogenous functions and developmental processes such as regulation of root and shoot system architecture, nutrient availability (Challis et al., 2013; Al-Babili and Bouwmeester, 2015) and tolerance to biotic stresses such as root parasitic weeds (López-Ráez et al., 2009) and abiotic stresses such as drought and salinity (Ha et al., 2014). SLs act as a root-derived hormone that moves upward in plant stems and can act as a long-distance messenger for auxin (Dun et al., 2009) and promote symbiosis between plants and AMF, possibly through its ability to induce AMF hyphal branching (Akiyama et al., 2005; Ha et al., 2014), while SLs may suppress shoot branching in plants (Kameoka and Kyozuka, 2015). Synergistically, AMF alters the plant genes to produce higher SL under drought condition, increasing the nutrient and water uptake, photosystem II efficiency, and plant growth and performance (Smith and Li, 2014; Ruiz-Lozano et al., 2016). The effect of AMF and SL on plant performance depends on the ABA level, which contributes not only on stress signaling and tolerance but also plant–fungus symbiosis (Ruiz-Lozano et al., 2016). Due to the hormone crosstalk effects, specific responses of specific plants and host diversity, more investigations on the effect of AMF–plant and also plant–microbiome interactions on plant hormone homeostasis are still required to suggest an effective procedure for plant cultivation.

Strigolactone esterase D14 encodes an α/β hydrolase, which is proposed to act in signaling or in the hydrolysis of SLs to an active compound and provides specificity to signaling via MAX2/RMS4/D3, an F-box protein that mediates both SL signaling and signaling of KARs. KARs are compounds structurally related to SLs that are found in smoke and act as germination stimulants for plants that colonize ground cleared by forest fires. Recent findings suggest that there may be a diversity of SL-like compounds, the response to which requires a D14/D14-like protein. The *MAX* genes can be broadly categorized into two groups based on the taxonomic distribution of their wider gene families. *MAX1*, *MAX3*, and *MAX4* were each found to be most similar to nonplant genes, while *MAX2*, *D14*, and *D27* are most similar to other plant-specific genes. The analysis of *MAX1* suggests that it may have played a significant role in this upstream SL signal diversity and its later refinement (Challis et al., 2013; Al-Babili and Bouwmeester, 2015). Although both KAR and SL may exhibit differential responses depending on the plant species, the symbiont strain, and the environmental condition, they have so far exhibited similar responses in promoting seed germination and development of plants and parasites, plants' root and shoot branching, RSA, water and nutrient use efficiency and, thus, plant performance under normal and stress conditions (Smith and Li, 2014).

19.3 CONCLUSIONS

Upon stress perception, in a close relationship, hormones interactively work together and with other regulatory networks and signaling pathways such as regulatory proteins as well as chemical and redox signaling pathways to control plant responses at all levels from molecular to the whole plant and morphological levels. Consequently, robust hormone homeostasis allows plants to improve root and shoot architecture systems, WUE, nutrient status, biochemistry, photosynthesis, and overall crop performance and yield. Breeding programs to obtain plants with superior hormone homeostasis require a deep understanding of hormone crosstalks and estimating their integrative effects with all plant regulatory systems and their impacts on plant responses under different environmental conditions. In this regard, the signaling pathways, the gene expression, the site of hormone biosynthesis, transport system, target cells, tissues, and organs, the hormones concentrations and their daily oscillations associated with the circadian clock, the plant growth stage, and the environmental changes should be taken into account. Accordingly, this review aimed to highlight the importance of finding advanced approaches that comprehensively evaluate the integrative effects of plant hormone homeostasis with other regulatory networks at all plant levels in relation to constant environmental changes during the whole growth period.

19.4 FUTURE REMARKS

Breeding programs for designing climate-resilient crops require a deep understanding of the plant regulatory networks, including stress perception and signaling pathways, plant redox, ion, nutrient, water, and hormone homeostasis and their integrative effects on crop performance. To reach this purpose, it is essential to consider more factors underlying plant molecular, cellular, biochemical, physiological, morphological, and anatomical aspects of different parts of the root-to-shoot structures during the whole growth period in regard to different environmental conditions. This would allow researchers to remove the limiting factors. Phytohormones function depends on their crosstalk effects, signaling pathways, daily oscillations associated with the circadian clock, concentrations, sources of biosynthesis, transport, and target cells, tissues, and organs, plant growth and developmental stages, plant genotypes with different sensitivity levels, and the environmental changes. Combination of different stresses, their severity, duration, variations, and recurrence at different plant growth stages trigger different unique signaling pathways and genes expressions that yield different responses. Thus, more advanced multidisciplinary approaches and statistical analysis software is required to consider the effects of hormone homeostasis in connection with the abovementioned factors. Application of plant symbiosis and precision agriculture as well as new technologies would also be safe and effective strategies to not only develop superior crops but also to combat climate change and global warming.

REFERENCES

Abbaspour, J., and Ehsanpour, A. (2016). The impact of salicylic acid on some physiological responses of *Artemisia aucheri* Boiss. under in vitro drought stress. *Acta Agriculturae Slovenica*, *107*(2), 287–298.

Acharya, B. R., and Assmann, S. M. (2009). Hormone interactions in stomatal function. *Plant Molecular Biology*, *69*(4), 451–462.

Agarwal, P. K., Agarwal, P., Reddy, M., and Sopory, S. K. (2006). Role of DREB transcription factors in abiotic and biotic stress tolerance in plants. *Plant Cell Reports*, *25*(12), 1263–1274

Agarwal, S., Sairam, R. K., Srivastava, G. C., and Meena, R. C. (2005). Changes in antioxidant enzymes activity and oxidative stress by abscisic acid and salicylic acid in wheat genotypes. *Biologia Plantarum*, *49*(4), 541–550.

Akiyama, K., Matsuzaki, K.-i., and Hayashi, H. (2005). Plant sesquiterpenes induce hyphal branching in arbuscular mycorrhizal fungi. *Nature*, *435*(7043), 824–827.

Alabdallah, O., Ahou, A., Mancuso, N., Pompili, V., Macone, A., Pashkoulov, D., Stano, P., Cona, A., Angelini, R., and Tavladoraki, P. (2017). The Arabidopsis polyamine oxidase/ dehydrogenase 5 interferes with cytokinin and auxin signaling pathways to control xylem differentiation. *Journal of Experimental Botany*, *68*(5), 997–1012.

Alam, M. M., Nahar, K., Hasanuzzaman, M., and Fujita, M. (2014). Exogenous jasmonic acid modulates the physiology, antioxidant defense and glyoxalase systems in imparting drought stress tolerance in different *Brassica* species. *Plant Biotechnology Reports*, *8*(3), 279–293.

Alamillo, J. M., and Bartels, D. (2001). Effects of desiccation on photosynthesis pigments and the ELIP-like dsp 22 protein complexes in the resurrection plant *Craterostigma plantagineum*. *Plant Science*, *160*(6), 1161–1170.

Al-Babili, S., and Bouwmeester, H. J. (2015). Strigolactones, a novel carotenoid-derived plant hormone. *Annual Review of Plant Biology*, *66*(1), 161–186.

Alder, A., Jamil, M., Marzorati, M., Bruno, M., Vermathen, M., Bigler, P., Ghisla, S., Bouwmeester, H., Beyer, P., and Al-Babili, S. (2012). The path from β-carotene to carlactone, a strigolactone-like plant hormone. *Science*, *335*(6074), 1348–1351.

Aleman, F., Yazaki, J., Lee, M., Takahashi, Y., Kim, A. Y., Li, Z., Kinoshita, T., Ecker, J. R., and Schroeder, J. I. (2016). An ABA-increased interaction of the PYL6 ABA receptor with MYC2 Transcription Factor: A putative link of ABA and JA signaling. *Scientific Reports*, *6*, 1–10.

Allamano, P., Claps, P., and Laio, F. (2009). Global warming increases flood risk in mountainous areas, *Geophysical Research Letters*, 36, L24404.

Aloni, R., Aloni, E., Langhans, M., and Ullrich, C. I. (2006). Role of cytokinin and auxin in shaping root architecture: Regulating vascular differentiation, lateral root initiation, root apical dominance and root gravitropism. *Annals of Botany*, *97*(5), 883–893.

Alves, A. A., and Setter, T. L. (2000). Response of cassava to water deficit: Leaf area growth and abscisic acid. *Crop Science*, *40*(1), 131–137.

Amin, B., Mahleghah, G., Mahmood, H. M. R., and Hossein, M. (2009). Evaluation of interaction effect of drought stress with ascorbate and salicylic acid on some of physiological and biochemical parameters in Okra (*Hibiscus esculentus* L.). *Research Journal of Biological Sciences*, 4, 380–387rjbsci.2009.380.387.

Arnell, N. W. (2004). Climate change and global water resources: SRES emissions and socio-economic scenarios. *Global Environmental Change*, *14*(1), 31–52.

Arzanesh, M. H., Alikhani, H. A., Khavazi, K., Rahimian, H. A., and Miransari, M. (2011). Wheat (*Triticum aestivum* L.) growth enhancement by *Azospirillum* sp. under drought stress. *World Journal of Microbiology and Biotechnology*, *27*(2), 197–205.

Avramova, Z. (2017). The JA- and ABA-signaling pathways crosstalk during one, but not repeated, dehydration stresses: A nonspecific "panicky," or a meaningful response? *Plant, Cell and Environment*, n/a–n/a.

Bandurska, H., and Stroi ski, A. (2005). The effect of salicylic acid on barley response to water deficit. *Acta Physiologiae Plantarum*, *27*(3), 379–386.

Bartels, D. (2005). Desiccation tolerance studied in the resurrection plant Craterostigma plantagineum. *Integrative and Comparative Biology*, *45*(5), 696–701.

Bartels, D., Furini, A., Ingram, J., and Salamini, F. (1996). Responses of plants to dehydration stress: A molecular analysis. *Plant Growth Regulation*, *20*(2), 111–118.

Batool, N., Ilyas, N., Noor, T., Saeed, M., Mazhar, R., Bibi, F., and Shahzad, A. (2014). Evaluation of drought stress effects on germination and seedling growth of *Zea mays* L. *International Journal of Biosciences*, *5*(4), 203–209.

Bechtold, U., Lawson, T., Mejia-Carranza, J., Meyer, R. C., Brown, I. R., Altmann, T., Ton, J., and Mullineaux, P. M. (2010). Constitutive salicylic acid defences do not compromise seed yield, drought tolerance and water productivity in the *Arabidopsis* accession C24. *Plant, Cell and Environment*, *33*(11), 1959–1973.

Beck, E. H., Fettig, S., Knake, C., Hartig, K., and Bhattarai, T. (2007). Specific and unspecific responses of plants to cold and drought stress. *Journal of Biosciences*, *32*(3), 501–510.

Blum, A., and Ebercon, A. (1981). Cell membrane stability as a measure of drought and heat tolerance in wheat. *Crop Science*, *21*(1), 43–47.

Boivin, S., Fonouni-Farde, C., and Frugier, F. (2016). How auxin and cytokinin phytohormones modulate root microbe interactions. *Frontiers in Plant Science*, *7*(1240).

Borel, C., Audran, C., Frey, A., Marion-Poll, A., Tardieu, F., and Simonneau, T. (2001). *N. plumbaginifolia* zeaxanthin epoxidase transgenic lines have unaltered baseline ABA accumulations in roots and xylem sap, but contrasting sensitivities of ABA accumulation to water deficit. *Journal of Experimental Botany*, *52*(suppl 1), 427–434.

Borel, C., and Simonneau, T. (2002). Is the ABA concentration in the sap collected by pressurizing leaves relevant for analysing drought effects on stomata? Evidence from ABA-fed leaves of transgenic plants with modified capacities to synthesize ABA. *Journal of Experimental Botany*, *53*(367), 287–296.

Bosabalidis, A. M., and Kofidis, G. (2002). Comparative effects of drought stress on leaf anatomy of two olive cultivars. *Plant Science*, *163*(2), 375–379.

Bouquin, T., Meier, C., Foster, R., Nielsen, M. E., and Mundy, J. (2001). Control of specific gene expression by gibberellin and brassinosteroid. *Plant Physiology*, *127*(2), 450–458. 010173.

Burkart, G. M., Baskin, T. I., and Bezanilla, M. (2015). A family of ROP proteins that suppresses actin dynamics, and is essential for polarized growth and cell adhesion. *Journal of Cell Science*, *128*(14), 2553–2564.

Caarls, L., Van der Does, D., Hickman, R., Jansen, W., Verk, M. C. V., Proietti, S., Lorenzo, O., Solano, R., Pieterse, C. M. J., and Van Wees, S. C. M. (2017). Assessing the role of ETHYLENE RESPONSE FACTOR transcriptional repressors in salicylic acid-mediated suppression of jasmonic acid-responsive genes. *Plant and Cell Physiology*, *58*(2), 266–278.

Caldeira, C. F., Jeanguenin, L., Chaumont, F., and Tardieu, F. (2014). Circadian rhythms of hydraulic conductance and growth are enhanced by drought and improve plant performance. *Nature Communications*, *5*(3565), 1–9.

Campos, M. L., Yoshida, Y., Major, I. T., de Oliveira Ferreira, D., Weraduwage, S. M., Froehlich, J. E., Johnson, B. F., Kramer, D. M., T. D., Sharkey, and Howe, G. A. (2016). Rewiring of jasmonate and phytochrome B signalling uncouples plant growth-defense tradeoffs. *Nature Communications*, *7*(12570), 1–10.

Cao, J., Li, M., Chen, J., Liu, P., and Li, Z. (2016). Effects of MeJA on *Arabidopsis* metabolome under endogenous JA deficiency. *Scientific Reports.*, *6*(1), 1–13.

Challis, R. J., Hepworth, J., Mouchel, C., Waites, R., and Leyser, O. (2013). A role for more AXILLARY GROWTH1 (MAX1) in evolutionary diversity in strigolactone signaling upstream of MAX2. *Plant Physiology*, *161*(4), 1885–1902.

Chang, Z., Liu, Y., Dong, H., Teng, K., Han, L., and Zhang, X. (2016). Effects of cytokinin and nitrogen on drought tolerance of creeping bentgrass. *PloS One*, *11*(4), 1–19.

Chen, Z. L., Li, X. M., and Zhang, L. H. (2014). Effect of salicylic acid pretreatment on drought stress responses of zoysiagrass (*Zoysia japonica*). *Russian Journal of Plant Physiology*, *61*(5), 619–625.

Chini, A., Ben-Romdhane, W., Hassairi, A., and Aboul-Soud, M. A. M. (2017). Identification of TIFY/JAZ family genes in *Solanum lycopersicum* and their regulation in response to abiotic stresses. *PloS One*, *12*(6), e0177381.

Chojak-Koźniewska, J., Linkiewicz, A., Sowa, S., Radzioch, M. A., and Kuźniak, E. (2017). Interactive effects of salt stress and *Pseudomonas syringae* pv. lachrymans infection in cucumber: Involvement of antioxidant enzymes, abscisic acid and salicylic acid. *Environmental and Experimental Botany*, *136*, 9–20.

Choudhury, F. K., Rivero, R. M., Blumwald, E., and Mittler, R. (2017). Reactive oxygen species, abiotic stress and stress combination. *The Plant Journal*, *90*(5), 856–867.

Christmann, A., Weiler, E. W., Steudle, E., and Grill, E. (2007). A hydraulic signal in root-to-shoot signalling of water shortage. *The Plant Journal*, *52*(1), 167–174.

Colebrook, E. H., Thomas, S. G., Phillips, A. L., and Hedden, P. (2014). The role of gibberellin signalling in plant responses to abiotic stress. *The Journal of Experimental Biology*, *217*(1), 67–75.

Cramer, G. R., Urano, K., Delrot, S., Pezzotti, M., and Shinozaki, K. (2011). Effects of abiotic stress on plants: A systems biology perspective. *BMC Plant Biology*, *11*(1), 163.

Danquah, A., de Zelicourt, A., Colcombet, J., and Hirt, H. (2014). The role of ABA and MAPK signaling pathways in plant abiotic stress responses. *Biotechnology Advances*, *32*(1), 40–52.

Daszkowska-Golec, A., and Szarejko, I. (2013). Open or close the gate – stomata action under the control of phytohormones in drought stress conditions. *Frontiers in Plant Science*, *4*(138), 1–16.

Davies, W. J., Tardieu, F., and Trejo, C. L. (1994). How do chemical signals work in plants that grow in drying soil? *Plant Physiology*, *104*(2), 309–314.

del Río, L. A., Sandalio, L. M., Corpas, F. J., Palma, J. M., and Barroso, J. B. (2006). Reactive oxygen species and reactive nitrogen species in peroxisomes. production, scavenging, and role in cell signaling. *Plant Physiology*, *141*(2), 330–335.

Desikan, R., Last, K., Harrett-Williams, R., Tagliavia, C., Harter, K., Hooley, R., Hancock, J. T., and Neill, S. J. (2006). Ethylene-induced stomatal closure in *Arabidopsis* occurs via AtrbohF-mediated hydrogen peroxide synthesis. *The Plant Journal*, *47*(6), 907–916.

Diao, Q.-N., Song, Y.-J., Shi, D.-M., and Qi, H.-Y. (2016). Nitric oxide induced by polyamines involves antioxidant systems against chilling stress in tomato (*Lycopersicon esculentum* Mill.) seedling. *Journal of Zhejiang University Science B*, *17*(12), 916–930.

Divi, U. K., Rahman, T., and Krishna, P. (2010). Brassinosteroid-mediated stress tolerance in *Arabidopsis* shows interactions with abscisic acid, ethylene and salicylic acid pathways. *BMC Plant Biology*, *10*(1), 151.

Dobra, J., Motyka, V., Dobrev, P., Malbeck, J., Prasil, I. T., Haisel, D., Gaudinova, A., Havlova, M., Gubis, J., and Vankova, R. (2010). Comparison of hormonal responses to heat, drought and combined stress in tobacco plants with elevated proline content. *Journal of Plant Physiology*, *167*(16), 1360–1370.

Duckham, S., Linforth, R., and Taylor, I. (1991). Abscisic-acid-deficient mutants at the *aba* gene locus of *Arabidopsis thaliana* are impaired in the epoxidation of zeaxanthin. *Plant, Cell and Environment*, *14*(6), 601–606.

Dun, E. A., Brewer, P. B., and Beveridge, C. A. (2009). Strigolactones: Discovery of the elusive shoot branching hormone. *Trends in Plant Science*, *14*(7), 364–372.

Eitzinger, J., Štastná, M., Žalud, Z., and Dubrovský, M. (2003). A simulation study of the effect of soil water balance and water stress on winter wheat production under different climate change scenarios. *Agricultural Water Management*, *61*(3), 195–217.

El-Showk, S., Ruonala, R., and Helariutta, Y. (2013). Crossing paths: Cytokinin signalling and crosstalk. *Development*, *140*(7), 1373–1383.

Fant, C., Schlosser, C. A., Gao, X., Strzepek, K., and Reilly, J. (2016). Projections of water stress based on an ensemble of socioeconomic growth and climate change scenarios: A case study in Asia. *PloS One*, *11*(3), e0150633.

Farber, M., Attia, Z., and Weiss, D. (2016). Cytokinin activity increases stomatal density and transpiration rate in tomato. *Journal of Experimental Botany*, *67*(22), 6351–6362.

Farooq, M., Wahid, A., Kobayashi, N., Fujita, D., and Basra, S. M. A. (2009). Plant drought stress: Effects, mechanisms and management. *Agronomy for Sustainable Development*, *29*(1), 185–212.

Farzane, M., Monem, R., Mirtaheri, S., and Kashani, S. (2014). Effect of salicylic acid on germination and growth seedling of 10 variety barley (Hordeum vulgare L.) under drought stress. *International Journal of Biosciences*, *5*(1), 445–448.

Fayez, K. A., and Bazaid, S. A. (2014). Improving drought and salinity tolerance in barley by application of salicylic acid and potassium nitrate. *Journal of the Saudi Society of Agricultural Sciences*, *13*(1), 45–55.

Finkelstein, R. (2013). Abscisic acid synthesis and response. *The Arabidopsis Book*, *11*, 1–36

Forman, H. J., Fukuto, J. M., and Torres, M. (2004). Redox signaling: Thiol chemistry defines which reactive oxygen and nitrogen species can act as second messengers. *The American Journal of Physiology – Cell Physiology*, *287*(2), C246–256.

Freitas, S. d. J., Santos, P., Berilli, S. d. S., Lopes, L., and de Carvalho, A. (2014). Sprouting, development and nutritional composition of pineapple plantlets from axillary buds subjected to brassinosteroid. *Revista Brasileira de Ciências Agrárias*, *9*(1), 19–24.

Fujita, M., Fujita, Y., Noutoshi, Y., Takahashi, F., Narusaka, Y., Yamaguchi-Shinozaki, K., and Shinozaki, K. (2006). Crosstalk between abiotic and biotic stress responses: A current view from the points of convergence in the stress signaling networks. *Current Opinion in Plant Biology*, *9*(4), 436–442.

Fujita, Y., Fujita, M., Satoh, R., Maruyama, K., Parvez, M. M., Seki, M., Hiratsu, K., Ohme-Takagi, M., Shinozaki, K., and Yamaguchi-Shinozaki, K. (2005). AREB1 is a transcription activator of novel ABRE-dependent ABA signaling that enhances drought stress tolerance in *Arabidopsis*. *The Plant Cell*, *17*(12), 3470–3488.

García, J. E., Maroniche, G., Creus, C., Suárez-Rodríguez, R., Ramirez-Trujillo, J. A., and Groppa, M. D. (2017). In vitro PGPR properties and osmotic tolerance of different *Azospirillum* native strains and their effects on growth of maize under drought stress. *Microbiological Research*, *202*, 21–29.

García-Mata, C., and Lamattina, L. (2001). Nitric oxide induces stomatal closure and enhances the adaptive plant responses against drought stress. *Plant Physiology*, *126*(3), 1196–1204.

Geiger, D., Scherzer, S., Mumm, P., Stange, A., Marten, I., Bauer, H., Ache, P., Matsci, S., Liese, A., Al-Rasheid, K. A. S., and Hedrich, R. (2009). Activity of guard cell anion channel SLAC1 is controlled by drought-stress signaling kinase-phosphatase pair. *Proceedings of the National Academy of Sciences*, *106*(50), 21425–21430.

Gendron, J. M., Haque, A., Gendron, N., Chang, T., Asami, T., and Wang, Z.-Y. (2008). Chemical genetic dissection of brassino-steroid–ethylene interaction. *Molecular Plant*, *1*(2), 368–379.

Goda, H., Sawa, S., Asami, T., Fujioka, S., Shimada, Y., and Yoshida, S. (2004). Comprehensive comparison of auxin-regulated and brassinosteroid-regulated genes in Arabidopsis. *Plant Physiology*, *134*(4), 1555–1573.

Goicoechea, N., Szalai, G., Antolín, M. C., Sánchez-Díaz, M., and Paldi, E. (1998). Influence of arbuscular mycorrhizae and Rhizobium on free polyamines and proline levels in water-stressed alfalfa. *Journal of Plant Physiology*, *153*(5), 706–711.

Gomez-Roldan, V., Fermas, S., Brewer, P. B., Puech-Pages, V., Dun, E. A., Pillot, J.-P., Letisse, F., Matusova, R., Danoun, S., Portais, J.-C., Bouwmeester, H., Bécard, G., Beveridge, C. A., Rameau, C., and Rochange, S. F. (2008). Strigolactone inhibition of shoot branching. *Nature*, *455*(7210), 189–194.

Gonçalves, S. (2016). MYB transcription factors for enhanced drought tolerance in plants. In P. Ahmad (Ed.), *Water Stress and Crop Plants: A Sustainable Approach* (Vol. 1, pp. 194–205). Chichester, UK: John Wiley and Sons, Ltd.

Grundy, J., Stoker, C., and Carré, I. A. (2015). Circadian regulation of abiotic stress tolerance in plants. *Frontiers in Plant Science*, *6*(648).

Gruszka, D. (2013). The brassinosteroid signaling pathway – New key players and interconnections with other signaling networks crucial for plant development and stress tolerance. *International Journal of Molecular Sciences*, *14*(5), 8740–8774.

Gruszka, D., Janeczko, A., Dziurka, M., Pociecha, E., Oklestkova, J., and Szarejko, I. (2016). Barley brassinosteroid mutants provide an insight into phytohormonal homeostasis in plant reaction to drought stress. *Frontiers in Plant Science*, *7*(1824), 1–14.

Guan, C., Ji, J., Jia, C., Guan, W., Li, X., Jin, C., and Wang, G. (2015). A GSHS-like gene from *Lycium chinense* maybe regulated by cadmium-induced endogenous salicylic acid and overexpression of this gene enhances tolerance to cadmium stress in Arabidopsis. *Plant Cell Reports*, *34*(5), 871–884.

Gunes, A., Inal, A., Alpaslan, M., Cicek, N., Guneri, E., Eraslan, F., and Guzelordu, T. (2005). Effects of exogenously applied salicylic acid on the induction of multiple stress tolerance and mineral nutrition in maize (*Zea mays* L.). *Archives of Agronomy and Soil Science*, *51*(6), 687–695.

Gunes, A., Inal, A., Alpaslan, M., Eraslan, F., Bagci, E. G., and Cicek, N. (2007). Salicylic acid induced changes on some physiological parameters symptomatic for oxidative stress and mineral nutrition in maize (*Zea mays* L.) grown under salinity. *Journal of Plant Physiology*, *164*(6), 728–736.

Gupta, A., Hisano, H., Hojo, Y., Matsuura, T., Ikeda, Y., Mori, I. C., and Senthil-Kumar, M. (2017). Global profiling of phytohormone dynamics during combined drought and pathogen stress in *Arabidopsis thaliana* reveals ABA and JA as major regulators. *Scientific Reports*, *7*, 4017.

Ha, C. V., Leyva-Gonzalez, M. A., Osakabe, Y., Tran, U. T., Nishiyama, R., Watanabe, Tanaka, M., Seki, M., Yamaguchi, S., Dong, N. V., Yamaguchi-Shinozaki, K., Shinozaki, K., Herrera-Estrella, L., and Tran, L.-S. P. (2014). Positive regulatory role of strigolactone in plant responses to drought and salt stress. *Proceedings of National Academy of Sciences of the United States*, *111*(2), 851–856.

Ha, Y., Shang, Y., and Nam, K. H. (2016). Brassinosteroids modulate ABA-induced stomatal closure in Arabidopsis. *Journal of Experimental Botany*, *67*(22), 6297–6308.

Hajouj, T., Michelis, R., and Gepstein, S. (2000). Cloning and characterization of a receptor-like protein kinase gene associated with senescence. *Plant Physiology*, *124*(3), 1305–1314.

Hall, J. W. (2014). Editorial: Steps towards global flood risk modelling. *Journal of Flood Risk Management*, *7*(3), 193–194.

Hao, G.-P., and Zhang, J.-H. (2009). The role of nitric oxide as a bioactive signaling molecule in plants under abiotic stress. In *Nitric Oxide in Plant Physiology* (pp. 115–138). Wiley-VCH Verlag GmbH and Co. KGaA.

Hao, J. H., Wang, X. L., Dong, C. J., Zhang, Z. G., and Shang, Q. M. (2011). Salicylic acid induces stomatal closure by modulating endogenous hormone levels in cucumber cotyledons. *Russian Journal of Plant Physiology*, *58*(5), 906–913.

Harb, A., Krishnan, A., Ambavaram, M. M., and Pereira, A. (2010). Molecular and physiological analysis of drought stress in Arabidopsis reveals early responses leading to acclimation in plant growth. *Plant Physiology*, *154*(3), 1254–1271.

Hashimoto, K., Eckert, C., Anschütz, U., Scholz, M., Held, K., Waadt, R., Reyer, A., Hippler, M., Becker, D., and Kudla, J. (2012). Phosphorylation of calcineurin B-like (CBL) calcium sensor proteins by their CBL-interacting protein kinases (CIPKs) is required for full activity of CBL-CIPK complexes toward their target proteins. *Journal of Biological Chemistry*, *287*(11), 7956–7968.

Holbrook, N. M., Shashidhar, V. R., James, R. A., and Munns, R. (2002). Stomatal control in tomato with ABA-deficient roots: Response of grafted plants to soil drying. *Journal of Experimental Botany*, *53*(373), 1503–1514.

HongBo, S., ZongSuo, L., and MingAn, S. (2006). Osmotic regulation of 10 wheat (*Triticum aestivum* L.) genotypes at soil water deficits. *Colloids and Surfaces B: Biointerfaces*, *47*(2), 132–139.

Hu, L., Wang, Z., and Huang, B. (2013). Effects of cytokinin and potassium on stomatal and photosynthetic recovery of *Kentucky bluegrass* from drought stress. *Crop Science*, *53*(1), 221–231.

Hutchison, C. E., and Kieber, J. J. (2007). Signaling via histidine-containing phosphotransfer proteins in Arabidopsis. *Plant Signaling and Behavior*, *2*(4), 287–289.

Inada, S., and Shimmen, T. (2000). Regulation of elongation growth by gibberellin in root segments of Lemna minor. *Plant Cell Physiol*, *41*(8), 932–939.

Inomata, M., Hirai, N., Yoshida, R., and Ohigashi, H. (2004). Biosynthesis of abscisic acid by the direct pathway via ionylideneethane in a fungus, *Cercospora cruenta*. *Bioscience, Biotechnology, and Biochemistry*, *68*(12), 2571–2580.

Intergovernmental Panel on Climate , IPC (2015). *Climate Change 2014: Mitigation of Climate Change: Working Group III Contribution to the IPCC Fifth Assessment Report.* Cambridge: Cambridge University Press.

Iuchi, S., Kobayashi, M., Yamaguchi-Shinozaki, K., and Shinozaki, K. (2000). A stress-inducible gene for 9-cis-epoxycarotenoid dioxygenase involved in abscisic acid biosynthesis under water stress in drought-tolerant cowpea. *Plant Physiology, 123*(2), 553–562.

Jakab, G., Ton, J., Flors, V., Zimmerli, L., Métraux, J.-P., and Mauch-Mani, B. (2005). Enhancing *Arabidopsis* salt and drought stress tolerance by chemical priming for its abscisic acid responses. *Plant Physiology, 139*(1), 267–274.

Jamil, A., Riaz, S., Ashraf, M., and Foolad, M. (2011). Gene expression profiling of plants under salt stress. *Critical Reviews in Plant Sciences, 30*(5), 435–458.

Jeong, Y. Y., and Seo, P. J. (2017). Bidirectional regulation between circadian clock and ABA signaling. *Communicative and Integrative Biology, 10*(2), e1296999.

Joshi, R., Sahoo, K. K., Tripathi, A. K., Kumar, R., Gupta, B. K., Pareek, A., and Singla-Pareek, S. L. (2017). Knockdown of an inflorescence meristem-specific cytokinin oxidase – OsCKX2 in rice reduces yield penalty under salinity stress condition. *Plant, Cell and Environment* [In press], n/a–n/a.

Kamata, H., and Hirata, H. (1999). Redox regulation of cellular signalling. *Cellular Signalling, 11*(1), 1–14.

Kameoka, H., and Kyozuka, J. (2015). Downregulation of rice DWARF 14 LIKE suppress mesocotyl elongation via a strigolactone independent pathway in the dark. *Journal of Genetics and Genomics, 42*(3), 119–124.

Kamthan, A., Kamthan, M., Azam, M., Chakraborty, N., Chakraborty, S., and Datta, A. (2012). Expression of a fungal sterol desaturase improves tomato drought tolerance, pathogen resistance and nutritional quality. *Scientific Reports, 2*(951), 1–10.

Kang, G., Li, G., Xu, W., Peng, X., Han, Q., Zhu, Y., and Guo, T. (2012). proteomics reveals the effects of salicylic acid on growth and tolerance to subsequent drought stress in Wheat. *Journal of Proteome Research, 11*(12), 6066–6079.

Kaushal, M., and Wani, S. P. (2016). Plant-growth-promoting rhizobacteria: Drought stress alleviators to ameliorate crop production in drylands. *Annals of Microbiology, 66*(1), 35–42.

Kazan, K., and Manners, J. M. (2008). Jasmonate signaling: Toward an integrated view. *Plant Physiology, 146*(4), 1459–1468.

Khan, M. I. R., Fatma, M., Per, T. S., Anjum, N. A., and Khan, N. A. (2015). Salicylic acid-induced abiotic stress tolerance and underlying mechanisms in plants. *Frontiers in Plant Science, 6*(462), 1–17.

Khodursky, A. B., Peter, B. J., Cozzarelli, N. R., Botstein, D., Brown, P. O., and Yanofsky, C. (2000). DNA microarray analysis of gene expression in response to physiological and genetic changes that affect tryptophan metabolism in *Escherichia coli. Proceedings of the National Academy of Sciences, 97*(22), 12170–12175.

Khokon, A. R., Okuma, E., Hossain, M. A., Munemasa, S., Uraji, M., Nakamura, Y., Mori, I., and Murata, Y. (2011). Involvement of extracellular oxidative burst in salicylic acid-induced stomatal closure in Arabidopsis. *Plant, Cell and Environment, 34*(3), 434–443.

Khurana, P., Vishnudasan, D., and Chhibbar, A. K. (2008). Genetic approaches towards overcoming water deficit in plants – special emphasis on LEAs. *Physiology and Molecular Biology of Plants, 14*(4), 277–298.

Kimura, T., Nakano, T., Taki, N., Ishikawa, M., Asami, T., and Yoshida, S. (2001). Cytokinin-induced gene expression in cultured green cells of *Nicotiana tabacum* identified by fluorescent differential display. *Biosci Biotechnol Biochem, 65*(6), 1275–1283.

Koiwai, H., Nakaminami, K., Seo, M., Mitsuhashi, W., Toyomasu, T., and Koshiba, T. (2004). Tissue-specific localization of an abscisic acid biosynthetic enzyme, AAO3, in *Arabidopsis. Plant Physiology, 134*(4), 1697–1707.

Kotchoni, S. O., Kuhns, C., Ditzer, A., KIRCH, H. H., and Bartels, D. (2006). Over-expression of different aldehyde dehydrogenase genes in *Arabidopsis thaliana* confers tolerance to abiotic stress and protects plants against lipid peroxidation and oxidative stress. *Plant, Cell and Environment, 29*(6), 1033–1048.

Krishnan, S., and Merewitz, E. B. (2017). Polyamine application effects on gibberellic acid content in creeping bentgrass during drought stress. *Journal of the American Society for Horticultural Science, 142*(2), 135–142.

Kuppu, S., Mishra, N., Hu, R., Sun, L., Zhu, X., Shen, G., Blumwald, E., Payton, P., and Zhang, H. (2013). Water-deficit inducible expression of a cytokinin biosynthetic gene ipt improves drought tolerance in Cotton. *PloS One, 8*(5), e64190.

Lavy, M., Bloch, D., Hazak, O., Gutman, I., Poraty, L., Sorek, N., Sternberg, H., and Yalovsky, S. (2007). A novel ROP/RAC effector links cell polarity, root-meristem maintenance, and vesicle trafficking. *Current Biology, 17*(11), 947–952.

Le, D. T., Nishiyama, R., Watanabe, Y., Vankova, R., Tanaka, M., Seki, M., Ham, L. H., Yamaguchi-Shinozaki, K., Shinozaki, K., and Tran, L.-S. P. (2012). Identification and expression analysis of cytokinin metabolic genes in soybean under normal and drought conditions in relation to cytokinin levels. *PloS One, 7*(8), e42411.

Le Hir, R., Castelain, M., Chakraborti, D., Moritz, T., Dinant, S., and Bellini, C. (2017). AtbHLH68 transcription factor contributes to the regulation of ABA homeostasis and drought stress tolerance in *Arabidopsis thaliana. Physiologia Plantarum, 160*(3), 312–327.

Legocka, J., and Żarnowska, A. (2000). Role of polyamines in the cytokinin-dependent physiological processes II. Modulation of polyamine levels during cytokinin-stimulated expansion of cucumber cotyledons. *Acta Physiologiae Plantarum, 22*(4), 395–401.

Lei, Y., Yin, C., and Li, C. (2006). Differences in some morphological, physiological, and biochemical responses to drought stress in two contrasting populations of *Populus przewalskii. Physiologia Plantarum, 127*(2), 182–191.

Li, H., Zhang, D., Li, X., Guan, K., and Yang, H. (2016). Novel DREB A-5 subgroup transcription factors from desert moss (*Syntrichia caninervis*) confers multiple abiotic stress tolerance to yeast. *Journal of Plant Physiology, 194*, 45–53.

Li, Y.-J., Wang, B., Dong, R.-R., and Hou, B.-K. (2015). AtUGT76C2, an *Arabidopsis* cytokinin glycosyltransferase is involved in drought stress adaptation. *Plant Science, 236*, 157–167.

Li, Z., Waadt, R., and Schroeder, J. I. (2016). Release of GTP exchange factor mediated down-regulation of abscisic acid signal transduction through ABA-induced rapid degradation of RopGEFs. *PLOS Biology, 14*(5), e1002461.

Li, Z.-G. (2016). Methylglyoxal and glyoxalase system in plants: Old players, new concepts. *The Botanical Review, 82*(2), 183–203.

Li, Z.-G., Duan, X.-Q., Min, X., and Zhou, Z.-H. (2017). Methylglyoxal as a novel signal molecule induces the salt tolerance of wheat by regulating the glyoxalase system, the antioxidant system, and osmolytes. *Protoplasma*, 1–12.

Liu, F., Xing, S., Ma, H., Du, Z., and Ma, B. (2013). Cytokinin-producing, plant growth-promoting rhizobacteria that confer resistance to drought stress in *Platycladus orientalis* container seedlings. *Applied Microbiology and Biotechnology, 97*(20), 9155–9164.

Liu, L., Sonbol, F.-M., Huot, B., Gu, Y., Withers, J., Mwimba, M., Yao, J., He, S. Y., and Dong, X. (2016). Salicylic acid receptors activate jasmonic acid signalling through a non-canonical pathway to promote effector-triggered immunity. *Nature Communications*, 7(13099), 1–10.

Liu, Q., Kasuga, M., Sakuma, Y., Abe, H., Miura, S., Yamaguchi-Shinozaki, K., and Shinozaki, K. (1998). Two transcription factors, DREB1 and DREB2, with an EREBP/AP2 DNA binding domain separate two cellular signal transduction pathways in drought- and low-temperature-responsive gene expression, respectively, in *Arabidopsis*. *The Plant Cell Online*, 10(8), 1391–1406.

Liu, X., Rockett, K. S., Korner, C. J., and Pajerowska-Mukhtar, K. M. (2015b). Salicylic acid signalling: New insights and prospects at a quarter-century milestone. *Essays in Biochemistry*, 58(0), 101–113.

Liu, Z., Yan, J.-P., Li, D.-K., Luo, Q., Yan, Q., Liu, Z.-B., Ye, L.-M., Wang, J.-M., Li, X.-F., and Yang, Y. (2015a). UDP-glucosyltransferase71c5, a major glucosyltransferase, mediates abscisic acid homeostasis in *Arabidopsis*. *Plant Physiology*, 167(4), 1659–1670.

López-Ráez, J. A., Matusova, R., Cardoso, C., Jamil, M., Charnikhova, T., Kohlen, W., Ruyter-Spira, C., Verstappen, F., and Bouwmeester, H. (2009). Strigolactones: Ecological significance and use as a target for parasitic plant control. *Pest Management Science*, 65(5), 471–477.

Lu, H., Greenberg, J. T., and Holuigue, L. (2016). Editorial: Salicylic acid signaling networks. *Frontiers in Plant Science*, 7(238), 1–3.

Macková, H., Hronková, M., Dobrá, J., Turečková, V., Novák, O., Lubovská, Z., Motyka, V., Haisel, D., Hájek, T., Prášil, I. T., A., Štorchová, H., Ge, E., Werner, T., Schmülling, and Vanková, R. (2013). Enhanced drought and heat stress tolerance of tobacco plants with ectopically enhanced cytokinin oxidase/dehydrogenase gene expression. *Journal of Experimental Botany*, 64(10), 2805–2815.

Madany, M., and Khalil, R. (2017). Seed priming with ascorbic acid or calcium chloride mitigates the adverse effects of drought stress in sunflower (*Helianthus annuus* L.) seedlings. *The Egyptian Journal of Experimental Botany*, 13(1), 119–133.

Malcheska, F., Ahmad, A., Batool, S., Müller, H. M., Ludwig-Müller, J., Kreuzwieser, J., Randewig, D., Hänsch, R,. Mendel, R. R., Hell, R., Wirtz, M., Geiger, D., Ache, P., Hedrich, R., Herschbach, C., and Rennenberg, H. (2017). Drought enhanced xylem sap sulfate closes stomata by affecting ALMT12 and guard cell ABA synthesis. *Plant Physiology*. 174(2), 798–814.

Manzoor, H., Athar, H. u. R., Rasul, S., Kanwal, T., Anjam, M. S., Qureshi, M. K., Bashir, N., Zafar, Z. U., Ali, M., and Ashraf, M. (2016). Avenues for improving drought tolerance in crops by ABA regulation. In P. Ahmad (Ed.), *Water Stress and Crop Plants: A Sustainable Approach* (pp. 177–193). Chichester, UK: John Wiley & Sons, Ltd.

Marasco, R., Rolli, E., Ettoumi, B., Vigani, G., Mapelli, F., Borin, S., Abou-Hadid, A. F., El-Behairy, U. A., Sorlini, C., Cherif, A., Zocchi, G., and Daffonchio, D. (2012). A drought resistance-promoting microbiome is selected by root system under desert farming. *PloS One*, 7(10), e48479.

Marowa, P., Ding, A., and Kong, Y. (2016). Expansins: Roles in plant growth and potential applications in crop improvement. *Plant Cell Reports*, 35(5), 949–965.

Merewitz, E. B., Du, H., Yu, W., Liu, Y., Gianfagna, T., and Huang, B. (2012). Elevated cytokinin content in ipt transgenic creeping bentgrass promotes drought tolerance through regulating metabolite accumulation. *Journal of Experimental Botany*, 63(3), 1315–1328.

Merewitz, E. B., Gianfagna, T., and Huang, B. (2010). Effects of SAG12-ipt and HSP18.2-ipt expression on cytokinin production, root growth, and leaf senescence in creeping bentgrass exposed to drought stress. *Journal of the American Society for Horticultural Science*, 135(3), 230–239.

Merewitz, E. B., Gianfagna, T., and Huang, B. (2011). Photosynthesis, water use, and root viability under water stress as affected by expression of SAG12-ipt controlling cytokinin synthesis in *Agrostis stolonifera*. *Journal of Experimental Botany*, 62(1), 383–395.

Miura, K., Okamoto, H., Okuma, E., Shiba, H., Kamada, H., Hasegawa, P. M., and Murata, Y. (2013). SIZ1 deficiency causes reduced stomatal aperture and enhanced drought tolerance via controlling salicylic acid-induced accumulation of reactive oxygen species in *Arabidopsis*. *The Plant Journal*, 73(1), 91–104.

Mohammed, A. (2007). Physiological aspects of mungbean plant (*Vigna radiata* L. Wilczek) in response to salt stress and gibberellic acid treatment. *Research Journal of Agriculture and Biological Sciences*, 3(4), 200–213.

Motamedi, M., Khodarahmpour, Z., and Ahakpaz, F. (2013). Influence of salicylic acid pretreatment on germination and seedling growth of wheat (*Triticum aestivum* L.) cultivars under salt stress. *International Journal of Biosciences*, 3(8), 226–233.

Muscolo, A., Junker, A., Klukas, C., Weigelt-Fischer, K., Riewe, D., and Altmann, T. (2015). Phenotypic and metabolic responses to drought and salinity of four contrasting lentil accessions. *Journal of Experimental Botany*, 66(18), 5467–5480.

Nahar, K., Hasanuzzaman, M., Alam, M. M., Rahman, A., Mahmud, J.-A., Suzuki, T., and Fujita, M. (2017). Insights into spermine-induced combined high temperature and drought tolerance in mung bean: Osmoregulation and roles of antioxidant and glyoxalase system. *Protoplasma*, 254(1), 445–460.

Nakashima, K., and Yamaguchi-Shinozaki, K. (2013). ABA signaling in stress-response and seed development. *Plant Cell Reports*, 32(7), 959–970.

Nambara, E., and Marion-Poll, A. (2005). Abscisic acid biosynthesis and catabolism. *Annual Review of Plant Biology*, 56(1), 165–185.

Nazareno, A. L., and Hernandez, B. S. (2017). A mathematical model of the interaction of abscisic acid, ethylene and methyl jasmonate on stomatal closure in plants. *PloS One*, 12(2), e0171065.

Németh, M., Janda, T., Horváth, E., Páldi, E., and Szalai, G. (2002). Exogenous salicylic acid increases polyamine content but may decrease drought tolerance in maize. *Plant Science*, 162(4), 569–574.

Nemhauser, J. L., Mockler, T. C., and Chory, J. (2004). Interdependency of brassinosteroid and auxin signaling in *Arabidopsis*. *PLoS Biology*, 2(9), e258.

Nguyen, K. H., Ha, C. V., Nishiyama, R., Watanabe, Y., Leyva-González, M. A., Fujita, Y., Tran, U. T., Li, W., Tanaka, M., Seki, M., Schaller, G. E., Luis Herrera-Estrella, L., and Tran, L.-S. P. (2016). *Arabidopsis* type B cytokinin response regulators ARR1, ARR10, and ARR12 negatively regulate plant responses to drought. *Proceedings of the National Academy of Sciences*, 113(11), 3090–3095.

Nibau, C., Wu, H. M., and Cheung, A. Y. (2006). RAC/ROP GTPases: "hubs" for signal integration and diversification in plants. *Trends in Plant Science*, 11(6), 309–315.

Novakova, M., Dobrev, P., Motyka, V., Gaudinova, A., Malbeck, J., Pospisilova, J., Haisel, D., Storchova, H., Dobra, J., Mok, M. C., Mok, D. W. S., Martin, R., and Vankova, R. (2007). Cytokinin Function in Drought Stress Response and Subsequent Recovery. In Z. Xu, J. Li, Y. Xue, and W. Yang

(Eds.), *Biotechnology and Sustainable Agriculture 2006 and Beyond: Proceedings of the 11th IAPTCandB Congress, August 18–31, 2006, Beijing, China* (pp. 171–174). Dordrecht: Springer Netherlands.

Nugroho, L. H., Verberne, M. C., and Verpoorte, R. (2002). Activities of enzymes involved in the phenylpropanoid pathway in constitutively salicylic acid-producing tobacco plants. *Plant Physiology and Biochemistry*, 40(9), 755–760.

Osakabe, Y., Arinaga, N., Umezawa, T., Katsura, S., Nagamachi, K., Tanaka, H., Ohiraki, H., Yamada, K., Seo, S.-U., Abo, M., Yohimura, E., Shinozaki, K., and Yamaguchi-Shinozaki, K. (2013). Osmotic stress responses and plant growth controlled by potassium transporters in *Arabidopsis*. *The Plant Cell*, 25(2), 609–624.

Osugi, A., and Sakakibara, H. (2015). QandA: How do plants respond to cytokinins and what is their importance? *BMC Biology*, 13(1), 102.

Pachauri, R. K., Allen, M. R., Barros, V. R., Broome, J., Cramer, W., Christ, R., Church, J. A., Clarke, L., Dahe, Q., Dasgupta, P., Dubash, N. K., Edenhofer, O., Elgizouli, I., Field, C. B., Forster, P., Friedlingstein, P., Fuglestvedt, J., Gomez-Echeverri, L., Hallegatte, S., Hegerl, G., Howden, M., Jiang, K., Jimenez Cisneroz, B., Kattsov, V., Lee, H., Mach, K. J., Marotzke, J., Mastrandrea, M. D., Meyer, L., Minx, J., Mulugetta, Y., O'Brien, K., Oppenheimer, M., Pereira, J. J., Pichs-Madruga, R., Plattner, G. K., Pörtner, H. O., Power, S. B., Preston, B., Ravindranath, N. H., Reisinger, A., Riahi, K., Rusticucci, M., Scholes, R., Seyboth, K., Sokona, Y., Stavins, R., Stocker, T. F., Tschakert, P., van Vuuren, D., and van Ypserle, J. P. (2014). *Climate Change 2014: Synthesis Report. Contribution of Working Groups I, II and III to the Fifth Assessment Report of the Intergovernmental Panel on Climate Change.* IPCC. 151 pp. ISBN: 978-92-9169-143-2.

Park, S.-Y., Peterson, F. C., Mosquna, A., Yao, J., Volkman, B. F., and Cutler, S. R. (2015). Agrochemical control of plant water use using engineered abscisic acid receptors. *Nature*, 520(7548), 545–548.

Peleg, Z., Reguera, M., Tumimbang, E., Walia, H., and Blumwald, E. (2011). Cytokinin-mediated source/sink modifications improve drought tolerance and increase grain yield in rice under water-stress. *Plant Biotechnology Journal*, 9(7), 747–758.

Portolés, S., and Zhang, D.-P. (2014). ABA Signaling and Circadian Clock. In D.-P. Zhang (Ed.), *Abscisic Acid: Metabolism, Transport and Signaling* (pp. 385–407). Dordrecht: Springer Netherlands.

Pourghayoumi, M., Rahemi, M., Bakhshi, D., Aalami, A., and Kamgar-Haghighi, A. A. (2017). Responses of pomegranate cultivars to severe water stress and recovery: Changes on antioxidant enzyme activities, gene expression patterns and water stress responsive metabolites. *Physiology and Molecular Biology of Plants*, 23(2), 321–330.

Qin, F., Kodaira, K. S., Maruyama, K., Mizoi, J., Tran, L. S., Fujita, Y., Morimot, K., Shinozaki, K., and Yamaguchi-Shinozaki, K. (2011). SPINDLY, a negative regulator of gibberellic acid signaling, is involved in the plant abiotic stress response. *Plant Physiology*, 157(4), 1900–1913.

Rai, V. K., Sharma, S. S., and Sharma, S. (1986). Reversal of ABA-induced stomatal closure by phenolic compounds. *Journal of Experimental Botany*, 37(1), 129–134.

Ramos, M. C., and Martínez-Casasnovas, J. A. (2010). Effects of precipitation patterns and temperature trends on soil water available for vineyards in a Mediterranean climate area. *Agricultural Water Management*, 97(10), 1495–1505.

Reddy, A. S., Ali, G. S., Celesnik, H., and Day, I. S. (2011). Coping with stresses: Roles of calcium- and calcium/calmodulin-regulated gene expression. *The Plant Cell*, 23(6), 2010–2032.

Reguera, M., Peleg, Z., Abdel-Tawab, Y. M., Tumimbang, E. B., Delatorre, C. A., and Blumwald, E. (2013). Stress-induced cytokinin synthesis increases drought tolerance through the coordinated regulation of carbon and nitrogen assimilation in rice. *Plant Physiology*, 163(4), 1609–1622.

Rivero, R. M., Gimeno, J., Van Deynze, A., Walia, H., and Blumwald, E. (2010). Enhanced cytokinin synthesis in tobacco plants expressing P_{SARK}::IPT prevents the degradation of photosynthetic protein complexes during drought. *Plant and Cell Physiology*, 51(11), 1929–1941.

Rivero, R. M., Kojima, M., Gepstein, A., Sakakibara, H., Mittler, R., Gepstein, S., and Blumwald, E. (2007). Delayed leaf senescence induces extreme drought tolerance in a flowering plant. *Proceedings of the National Academy of Sciences*, 104(49), 19631–19636.

Roberts, M. R. (2016). Jasmonic acid signalling. In *eLS*. John Wiley and Sons, Ltd., 1–9.

Rolli, E., Marasco, R., Vigani, G., Ettoumi, B., Mapelli, F., Deangelis, M. L., Gandolfi, C., Casati, E., Previtali, F., Gerbino, R., Pierotti Cei, F., Borin, S., Sorlini, C., Zocchi, G., and Daffonchio, D. (2015). Improved plant resistance to drought is promoted by the root-associated microbiome as a water stress-dependent trait. *Environmental Microbiology*, 17(2), 316–331.

Rowe, J. H., Topping, J. F., Liu, J., and Lindsey, K. (2016). Abscisic acid regulates root growth under osmotic stress conditions *via* an interacting hormonal network with cytokinin, ethylene and auxin. *The New Phytologist*, 211(1), 225–239.

Rubio-Wilhelmi, M. d. M., Reguera, M., Sanchez-Rodriguez, E., Romero, L., Blumwald, E., and Ruiz, J. M. (2014). P_{SARK}::IPT expression causes protection of photosynthesis in tobacco plants during N deficiency. *Environmental and Experimental Botany*, 98, 40–46.

Ruiz-Lozano, J. M., Aroca, R., Zamarreño, Á. M., Molina, S., Andreo-Jiménez, B., Porcel, R., García Mina, J. M., Ruyter-Spira, C., and López-Ráez, J. A. (2016). Arbuscular mycorrhizal symbiosis induces strigolactone biosynthesis under drought and improves drought tolerance in lettuce and tomato. *Plant, Cell and Environment*, 39(2), 441–452.

Saini, S., Sharma, I., and Pati, P. K. (2015). Versatile roles of brassinosteroid in plants in the context of its homoeostasis, signaling and crosstalks. *Frontiers in Plant Science*, 6, 1–17.

Scandalios, J. G. (2005). Oxidative stress: Molecular perception and transduction of signals triggering antioxidant gene defenses. *Brazilian Journal of Medical and Biological Research*, 38(7), 995–1014.

Schiermeier, Q. (2011). Increased flood risk linked to global warming: Likelihood of extreme rainfall may have been doubled by rising greenhouse-gas levels. *Nature*, 470(7334), 316–317.

Schultheiss, H., Dechert, C., Kogel, K.-H., and Hückelhoven, R. (2003). Functional analysis of barley RAC/ROP G-protein family members in susceptibility to the powdery mildew fungus. *The Plant Journal*, 36(5), 589–601.

Sequera-Mutiozabal, M., Tiburcio, A. F., and Alcázar, R. (2016). Drought stress tolerance in relation to polyamine metabolism in plants. *Drought Stress Tolerance in Plants* (Vol. 1, pp. 267–286). Springer.

Serna, M., Hernández, F., Coll, F., Coll, Y., and Amorós, A. (2012). Brassinosteroid analogues effects on the yield and quality parameters of greenhouse-grown pepper (*Capsicum annuum* L.). *Plant Growth Regulation*, 68(3), 333–342.

Seung, D., Risopatron, J. P. M., Jones, B. J., and Marc, J. (2012). Circadian clock-dependent gating in ABA signalling networks. *Protoplasma*, 249(3), 445–457.

Shah, D. A., Sen, S., D. , Ghosh, M., Grover, and Mohapatra, S. (2017). An auxin secreting Pseudomonas putida rhizobacterial strain that negatively impacts water-stress tolerance in *Arabidopsis thaliana*. *Rhizosphere*, *3*, Part 1, 16–19.

Shani, E., Weinstain, R., Zhang, Y., Castillejo, C., Kaiserli, E., Chory, J., Tsien, R. Y., and Estelle, M. (2013). Gibberellins accumulate in the elongating endodermal cells of *Arabidopsis* root. *Proceedings of the National Academy of Sciences of the United States of America*, *110*(12), 4834–4839.

Sharma, M., Gupta, S. K., Majumder, B., Maurya, V. K., Deeba, F., Alam, A., and Pandey, V. (2017). Salicylic acid mediated growth, physiological and proteomic responses in two wheat varieties under drought stress. *Journal of Proteomics*, *163*, 28–51.

Shi, C., Qi, C., Ren, H., Huang, A., Hei, S., and She, X. (2015). Ethylene mediates brassinosteroid-induced stomatal closure via Galpha protein-activated hydrogen peroxide and nitric oxide production in Arabidopsis. *The Plant Journal*, *82*(2), 280–301.

Shi, H., Chen, L., Ye, T., Liu, X., Ding, K., and Chan, Z. (2014). Modulation of auxin content in *Arabidopsis* confers improved drought stress resistance. *Plant Physiology and Biochemistry*, *82*, 209–217.

Shimada, A., Ueguchi-Tanaka, M., Sakamoto, T., Fujioka, S., Takatsuto, S., Yoshida, S., Sazuka, T., Ashikar, M., and Matsuoka, M. (2006). The rice SPINDLY gene functions as a negative regulator of gibberellin signaling by controlling the suppressive function of the DELLA protein, SLR1, and modulating brassinosteroid synthesis. *The Plant Journal*, *48*(3), 390–420.

Siegel, F. R. (2008). Water: An essential, limited, renewable resource. In *Demands of Expanding Populations and Development Planning: Clean Air, Safe Water, Fertile Soils* (pp. 83–112). Berlin, Heidelberg: Springer.

Singh, R. P., and Reddy, C. R. K. (2015). Unraveling the functions of the macroalgal microbiome. *Frontiers in Microbiology*, *6*(1488), 1–8.

Šmehilová, M., Dobrůšková, J., Novák, O., Takáč, T., and Galuszka, P. (2016). Cytokinin-specific glycosyltransferases possess different roles in cytokinin homeostasis maintenance. *Frontiers in Plant Science*, *7*(1264), 1–19.

Smith, S. M., and Li, J. (2014). Signalling and responses to strigolactones and karrikins. *Current Opinion in Plant Biology*, *21*, 23–29.

Soundappan, I., Bennett, T., Morffy, N., Liang, Y., Stanga, J. P., Abbas, A., Ottoline, L., and Nelson, D. C. (2015). SMAX1-LIKE/D53 family members enable distinct MAX2-dependent responses to Strigolactones and Karrikins in *Arabidopsis*. *The Plant Cell*, *161*(4), 1885–1902.

Takeda, S., and Matsuoka, M. (2008). Genetic approaches to crop improvement: Responding to environmental and population changes. *Nature Reviews Genetics*, *9*(6), 444–457.

Tamaki, H., Reguera, M., Abdel-Tawab, Y. M., Takebayashi, Y., Kasahara, H., and Blumwald, E. (2015). Targeting hormone-related pathways to improve grain yield in rice: A chemical approach. *PloS One*, *10*(6), e0131213.

Tanaka, Y., Sano, T., Tamaoki, M., Nakajima, N., Kondo, N., and Hasezawa, S. (2005). Ethylene inhibits abscisic acid-induced stomatal closure in *Arabidopsis*. *Plant Physiology*, *138*(4), 2337–2343.

Tanaka, Y., Sano, T., Tamaoki, M., Nakajima, N., Kondo, N., and Hasezawa, S. (2006). Cytokinin and auxin inhibit abscisic acid-induced stomatal closure by enhancing ethylene production in *Arabidopsis*. *Journal of Experimental Botany*, *57*(10), 2259–2266.

Tanou, G., Filippou, P., Belghazi, M., Job, D., Diamantidis, G., Fotopoulos, V., and Molassiotis, A. (2012). Oxidative and nitrosative-based signaling and associated post-translational modifications orchestrate the acclimation of citrus plants to salinity stress. *The Plant Journal*, *72*(4), 585–599.

Tardieu, F., Parent, B., and Simonneau, T. (2010). Control of leaf growth by abscisic acid: Hydraulic or non-hydraulic processes? *Plant, Cell and Environment*, *33*(4), 636–647.

Tavladoraki, P., Cona, A., and Angelini, R. (2016). Copper-containing amine oxidases and FAD-dependent polyamine oxidases are key players in plant tissue differentiation and organ development. *Frontiers in Plant Science*, *7*(824), 1–11.

Teale, W. D., Ditengou, F. A., Dovzhenko, A. D., Li, X., Molendijk, A. M., Ruperti, B., Paponov, I., and Palme, K. (2008). Auxin as a model for the integration of hormonal signal processing and transduction. *Molecular Plant*, *1*(2), 229–237.

Todaka, D., Zhao, Y., Yoshida, T., Kudo, M., Kidokoro, S., Mizoi, J., Kodaira, K.-S., Takebayashi, Y., Kojima, M., Sakakibara, H., Toyooka, K., Sato, M., Fernie, A. R., Shinozaki, K., and Yamaguchi-Shinozaki, K. (2017). Temporal and spatial changes in gene expression, metabolite accumulation and phytohormone content in rice seedlings grown under drought stress conditions. *The Plant Journal*, *90*(1), 61–78.

Todorova, D., Talaat, N. B., Katerova, Z., Alexieva, V., and Shawky, B. T. (2016). Polyamines and brassinosteroids in drought stress responses and tolerance in plants. In P. Ahmad (Ed.), *Water Stress and Crop Plants: A Sustainable Approach* (Vol. 2, pp. 608–627). Chichester, UK: John Wiley and Sons, Ltd.

Tollefson, J. (2016). Global warming already driving increases in rainfall extremes. *Nature*.

Ton, J., Flors, V., and Mauch-Mani, B. (2009). The multifaceted role of ABA in disease resistance. *Trends in Plant Science*, *14*(6), 310–317.

Toumi, I., Moschou, P. N., Paschalidis, K. A., Bouamama, B., Salem-Fnayou, A. B., Ghorbel, A. W., Milki, A., and Roubelakis-Angelakis, K. A. (2010). Abscisic acid signals reorientation of polyamine metabolism to orchestrate stress responses via the polyamine exodus pathway in grapevine. *Journal of Plant Physiology*, *167*(7), 519–525.

Tripathy, J. N., Zhang, J., Robin, S., Nguyen, T. T., and Nguyen, H. T. (2000). QTLs for cell-membrane stability mapped in rice (*Oryza sativa* L.) under drought stress. *Theoretical and Applied Genetics*, *100*(8), 1197–1202.

Tsuchiya, Y., Vidaurre, D., Toh, S., Hanada, A., Nambara, E., Kamiya, Y., Yamaguchi, S., and McCourt, P. (2010). A small-molecule screen identifies new functions for the plant hormone strigolactone. *Nature Chemical Biology*, *6*(10), 741–749.

Umehara, M., Hanada, A., Yoshida, S., Akiyama, K., Arite, T., Takeda-Kamiya, N., Magome, H., Kamiya, Y., Shirasu, K., Yoneyama, K., Kyozuka, J., and Yamaguchi, S. (2008). Inhibition of shoot branching by new terpenoid plant hormones. *Nature*, *455*(7210), 195–200.

Umezawa, T., Sugiyama, N., Mizoguchi, M., Hayashi, S., Myouga, F., Yamaguchi-Shinozaki, K., Ishihama, Y., Hirayama, T., and Shinozaki, K. (2009). Type 2C protein phosphatases directly regulate abscisic acid-activated protein kinases in Arabidopsis. *Proceedings of National Academy of Science of the United States of America*, *106*(41), 17588–17593.

United Nation World Population Prospects, UNWPP (2015). *United Nation World Population Prospects*. http://www.un.org/en/development/desa/publications/world-population-prospects-2015-revision.html.

Vardhini, B. V., and Rao, S. S. R. (2003). Amelioration of osmotic stress by brassinosteroids on seed germination and seedling growth of three varieties of sorghum. *Plant Growth Regulation*, 41(1), 25–31.

Vardhini, B. V., Sujatha, E., and Rao, S. S. R. (2011). Studies on the effect of brassinosteroids on the qualitative changes in the storage roots of radish. *Asian and Australasian Journal of Plant Science and Biotechnology*, 5(1), 27–30.

Vasilev, N., Boccard, J., Lang, G., Grömping, U., Fischer, R., Goepfert, S., Rudaz, S., and Schillberg, S. (2016). Structured plant metabolomics for the simultaneous exploration of multiple factors. *Scientific Reports*, 6(1), 1–15.

Vercruyssen, L., Gonzalez, N., Werner, T., Schmülling, T., and Inzé, D. (2011). Combining enhanced root and shoot growth reveals cross talk between pathways that control plant organ size in *Arabidopsis*. *Plant Physiology*, 155(3), 1339–1352.

Virdi, A. S., Singh, S., and Singh, P. (2015). Abiotic stress responses in plants: Roles of calmodulin-regulated proteins. *Frontiers in Plant Science*, 6, 1–19.

Vurukonda, S. S. K. P., Vardharajula, S., Shrivastava, M., and SkZ, A. (2016). Enhancement of drought stress tolerance in crops by plant growth promoting rhizobacteria. *Microbiological Research*, 184, 13–24.

Wang, G. Q., Zhang, J. Y., Xuan, Y. Q., Liu, J. F., Jin, J. L., Bao, Z. X., He, R. M., Liu, C. S., Liu, Y. l., and Yan, X. L. (2013). Simulating the impact of climate change on runoff in a typical river catchment of the Loess Plateau, China. *Journal of Hydrometeorology*, 14(5), 1553–1561.

Wang, W.-X., Zhang, F., Chen, Z.-L., Liu, J., Guo, C., He, J.-D., Zou, Y.-N., and Wu, Q.-S. (2017). Responses of phytohormones and gas exchange to mycorrhizal colonization in trifoliate orange subjected to drought stress. *Archives of Agronomy and Soil Science*, 63(1), 14–23.

Wang, Z.-Y., Bai, M.-Y., Oh, E., and Zhu, J.-Y. (2012). Brassinosteroid signaling network and regulation of photomorphogenesis. *Annual Review of Genetics*, 46, 701–724.

Wang, Z.-Y., Xiong, L., Li, W., Zhu, J.-K., and Zhu, J. (2011). The plant cuticle is required for osmotic stress regulation of abscisic acid biosynthesis and osmotic stress tolerance in *Arabidopsis*. *The Plant Cell*, 23(5), 1971–1984.

Werner, T., Nehnevajova, E., Köllmer, I., Novák, O., Strnad, M., Krämer, U., and Schmülling, T. (2010). Root-specific reduction of cytokinin causes enhanced root growth, drought tolerance, and leaf mineral enrichment in *Arabidopsis* and Tobacco. *The Plant Cell*, 22(12), 3905–3920.

Wilkinson, S., and Davies, W. J. (2010). Drought, ozone, ABA and ethylene: New insights from cell to plant to community. *Plant, Cell and Environment*, 33(4), 510–525.

Wimalasekera, R., Tebartz, F., and Scherer, G. F. E. (2011). Polyamines, polyamine oxidases and nitric oxide in development, abiotic and biotic stresses. *Plant Science*, 181(5), 593–603.

World Population Forecast, WPF (2016). *World Population Forecast*. http://www.worldometers.info/world-population/#table-fo recast.

Wu, H.-M., Hazak, O., Cheung, A. Y., and Yalovsky, S. (2011). RAC/ROP GTPases and auxin signaling. *The Plant Cell*, 23(4), 1208–1218.

Xia, X.-J., Wang, Y.-J., Zhou, Y.-H., Tao, Y., Mao, W.-H., Shi, K., Asami, T., Chen, Z., and Yu, J.-Q. (2009). Reactive oxygen species are involved in brassinosteroid-induced stress tolerance in cucumber. *Plant Physiology*, 150(2), 801–814.

Xian, L., Sun, P., Hu, S., Wu, J., and Liu, J.-H. (2014). Molecular cloning and characterization of CrNCED1, a gene encoding 9-cis-epoxycarotenoid dioxygenase in *Citrus reshni*, with functions in tolerance to multiple abiotic stresses. *Planta*, 239(1), 61–77.

Xu, W., Jia, L., Shi, W., Liang, J., Zhou, F., Li, Q., and Zhang, J. (2013). Abscisic acid accumulation modulates auxin transport in the root tip to enhance proton secretion for maintaining root growth under moderate water stress. *New Phytologist*, 197(1), 139–150.

Xu, Y., Tian, J., Gianfagna, T., and Huang, B. (2008). Effects of SAG12-ipt expression on cytokinin production, growth and senescence of creeping bentgrass (*Agrostis stolonifera* L.) under heat stress. *Plant Growth Regulation*, 57(3), 281.

Yang, Z., and Fu, Y. (2007). ROP/RAC GTPase signaling. *Current Opinion in Plant Biology*, 10(5), 490–494.

Yoshida, T., Fujita, Y., Maruyama, K., Mogami, J., Todaka, D., Shinozaki, K., and Yamaguchi-Shinozaki, K. (2015). Four arabidopsis AREB/ABF transcription factors function predominantly in gene expression downstream of SnRK2 kinases in abscisic acid signalling in response to osmotic stress. *Plant, Cell and Environment*, 38(1), 35–49.

Yuan, G.-F., Jia, C.-G., Li, Z., Sun, B., Zhang, L.-P., Liu, N., and Wang, Q.-M. (2010). Effect of brassinosteroids on drought resistance and abscisic acid concentration in tomato under water stress. *Scientia Horticulturae*, 126(2), 103–108.

Zaicovski, C. B., Zimmerman, T., Nora, L., Nora, F. R., Silva, J. A., and Rombaldi, C. V. (2008). Water stress increases cytokinin biosynthesis and delays postharvest yellowing of broccoli florets. *Postharvest Biology and Technology*, 49(3), 436–439.

Zhang, A., Zhang, J., Zhang, J., Ye, N., Zhang, H., Tan, M., and Jiang, M. (2011). Nitric oxide mediates brassinosteroid-induced ABA biosynthesis involved in oxidative stress tolerance in maize leaves. *Plant and Cell Physiology*, 52(1), 181–192.

Zhang, J., Shi, Y., Zhang, X., Du, H., Xu, B., and Huang, B. (2017). Melatonin suppression of heat-induced leaf senescence involves changes in abscisic acid and cytokinin biosynthesis and signaling pathways in perennial ryegrass (*Lolium perenne* L.). *Environmental and Experimental Botany*, 138, 36–45.

Zhang, M., Zhai, Z., Tian, X., Duan, L., and Li, Z. (2008). Brassinolide alleviated the adverse effect of water deficits on photosynthesis and the antioxidant of soybean (*Glycine max* L.). *Plant Growth Regulation*, 56(3), 257–264.

Zhang, S., Song, G., Li, Y., Gao, J., Liu, J., Fan, Q., Huang, C. Y., Sui, X., Chu, X. S., Guo, D., and Li, G. Y. (2014a). Cloning of 9-cis-epoxycarotenoid dioxygenase gene (TaNCED1) from wheat and its heterologous expression in tobacco. *Biologia Plantarum*, 58(1), 89–98.

Zhang, S., and Outlaw, W. (2001). The guard-cell apoplast as a site of abscisic acid accumulation in *Vicia faba* L. *Plant, Cell and Environment*, 24(3), 347–355.

Zhang, S., Wang, S., Xu, Y., Yu, C., Shen, C., Qian, Q., Geisler, M., Jian, D. A., and Qi, Y. (2015). The auxin response factor, OsARF19, controls rice leaf angles through positively regulating OsGH3-5 and OsBRI1. *Plant, Cell and Environment*, 38(4), 638–654.

Zhang, S. J., Song, G. Q., Li, Y. L., Gao, J., Liu, J. J., Fan, Q. Q., and Li, G. Y. (2014b). Cloning of 9-cis-epoxycarotenoid dioxygenase gene (TaNCED1) from wheat and its heterologous expression in tobacco. *Biologia Plantarum*, 58(1), 89–98.

Zheng, H., Pan, X., Deng, Y., Wu, H., Liu, P., and Li, X. (2016). *AtOPR3* specifically inhibits primary root growth in *Arabidopsis* under phosphate deficiency. *Scientific Reports*, 6(1), 1–11.

Zheng, Y., Liao, C., Zhao, S., Wang, C., and Guo, Y. (2017). The glycosyltransferase QUA1 regulates chloroplast-associated calcium signaling during salt and drought stress in *Arabidopsis*. *Plant and Cell Physiology*, 58(2), 329–341.

Zhou, J., Wang, J., Li, X., Xia, X.-J., Zhou, Y.-H., Shi, K., Chen, Z., and Yu, J.-Q. (2014). H_2O_2 mediates the crosstalk of brassinosteroid and abscisic acid in tomato responses to heat and oxidative stresses. *Journal of Experimental Botany*, *65*(15), 4371–4383.

Zhu, M., Monroe, J., Suhail, Y., Villiers, F., Mullen, J., Pater, D., Hauser, F., Jeon, B. W., Bader, J. S., Kwak, J. M., Schroeder, J. I., McKay, J. K., Assmann, S. M., and Kwak, J. M. (2016). Molecular and systems approaches towards drought-tolerant canola crops. *New Phytologist*, *210*(4), 1169–1189.

Zoubida, B., and Gherroucha, H. (2017). Improvement of salt tolerance in durum wheat (*Triticum durum* Desf.) by Auxin and Kenitin application. *European Scientific Journal*, *13*(9), 96–110.

Zwack, P. J., and Rashotte, A. M. (2015). Interactions between cytokinin signalling and abiotic stress responses. *Journal of Experimental Botany*, *66*(16), 4863–4871.

20 Heliotropism
Plants Follow the Sun

Yehouda Marcus

CONTENTS

20.1 INTRODUCTION

20.1.1 EARTH ROTATION AND PLANT MOVEMENT

Almost every environmental factor that affects plants has been changed during their evolution, besides the daily circling of Earth around its own axis and the yearly one around the sun that persist in a constant period ever since the formation of the Solar System. Of all the environmental factors that affect plants in terrestrial environments, sunlight undergoes the most prominent daily alterations. The oscillations in light intensity, duration, and direction prominently affect the local meteorological conditions (temperature, precipitation, and wind), and hence plant behavior and growth. Since light is the sole source of energy for carbon fixation and growth, it is an important morphogenetic agent that affects plant behavior. Absorbed light may also affect the energy balance of the plants as they are heated and so influence transpiration and water usage efficiency (Ehleringer and Forseth, 1980; Koller, 1990). The heating of flowers, particularly in cold regions, attracts pollinating insects and may enhance plant fertility (Kevan, 1975; Stanton and Galen, 1989; Totland, 1996; Atamian et al., 2016). On the other hand, excess light causes photoinhibitory damage to the photosynthetic machinery (Kok, 1956; Adir et al., 2003). Therefore, many plants use a mechanism that anticipates the regularly changing light conditions. They track the sun position by moving their aerial organs (leaves, buds, flowers, and inflorescences) toward or away from the sun's rays, and then during the night they resume their original position in preparation for the sunrise. This rhythmical movement has been defined as sun tracking or heliotropism (Koller, 1986;

Vandenbrink et al., 2014; Kutschera and Briggs, 2016; Serrano et al., 2018). The orientation of plant organs perpendicular to the plain of the sun's rays, which is called ortho- or dia-heliotropism, maximizes light absorbance, whereas the orientation parallel to the sun's rays, which is called para-heliotropism, minimizes light absorbance (Serrano et al., 2018). Plants track the azimuth (i.e., the horizontal angle from the north) of the sun from east to west or its elevation in the sky, or the tridimensional combination of both angles by their horizontal or vertical movements (Stanton and Galen, 1989; Totland, 1996; Zhang et al., 2010; Serrano et al., 2018). Not all heliotropic plants track the sun perfectly. By deviating from the accurate solar orbit, they control the quanta of light absorbed and consequently their temperature and thus optimize the photosynthesis and transpiration rate, which determines the water usage efficiency (Koller, 1990). Such oscillations may occur within 24 hours or once in a season (diurnal or seasonal heliotropism, respectively; Serrano et al., 2018). Diaheliotropic movements are, particularly, rewarding at suboptimal conditions, when light intensity limits the rate of photosynthesis (in the morning and evening, under a cloudy sky or shaded environment; Forseth and Ehleringer, 1980; Kutschera and Briggs, 2016). Conversely, paraheliotropic movement minimizes light absorbance and reduces photoinhibitory damage and transpiration. In extreme environmental conditions, diaheliotropic behavior may be changed to paraheliotropic, e.g., at midday when the sun is at its zenith, at a high light intensity, high ambient temperature, or under water deficiency. When the stress is relieved, or at dawn and twilight, the plants revert to diaheliotropic behavior (Forseth and Ehleringer, 1980; Koller, 1990).

20.2 MECHANISM

Heliotropic movements occur by asymmetric volume changes at both sides of the moving organ via reversible turgor-mediated alterations in the pulvinus (the motor organ of the leaf) shape (Koller, 1986) or cell elongation (Atamian et al., 2016).

20.2.1 Pulvinus-Driven Heliotropic Movement

The pulvinus found at the base of the leaf or the leaflet petiole is a vascular tissue surrounded by a cylinder-like cortex of thin-walled elastic parenchyma cells (Koller, 1986). Pulvinar bending results from stimulation of a membranal H^+ ATPase that drives ion movement primarily K^+ and Cl^- from one side to the opposite side of the pulvinus cortex accompanied by water movement that balances the alterations in the osmotic potential (Moran, 2007). This requires orchestrated opening and closure of a battery of ion and water channels on the vacuole and plasma membranes operating in inverse modes at opposite sides of the pulvinus cortex (Moshelion et al., 2002a,b; Uehlein and Kaldenhoff, 2008). Concomitantly, phosphorylation of the actin cytoskeleton and its rearrangement presumably facilitates alterations in the shape of the parenchyma cells leading to their expansion at one side and shrinkage at the opposite side (Kameyama et al., 2000; Kanzawa et al., 2006). As a result, the petiole bends in the direction of the contacting zone. Reversal of the ion movement stimulates bending in the opposite direction. Movement of the leaf vertically or horizontally is determined by the spatial location of the two antagonistic zones of the cortex and not by a directional signal (Koller, 1990).

20.2.2 Elongation-Driven Heliotropic Movements

Heliotropic movements also occur in plants and organs lacking pulvini by a differential and irreversible cell growth on two opposite sides of the organ (e.g., *Ranunculus adoneus* Stanton and Galen, 1993) or *Anemone rivularis* (Zhang et al., 2010)). One of the best-known examples is the diaheliotropic stem movement of young sunflower *(Helianthus annuus)* plants (Kutschera and Briggs, 2016). During their growth, sunflower buds track the azimuth of the sun throughout the day, whereas at night they reorient to face east in preparation for the dawn. Atamian et al. (2016) have shown that the heliotropic movements of sunflowers are the result of the slow-moving gradient of the growth-promoting hormone auxin from the western to the eastern side of the stem. This was demonstrated by the daily rhythm in the induction of auxin-stimulated genes at both sides of the stem, accompanied by the faster growth of cells on the shaded side, which resulted in bending toward the illuminated side. At night the auxin gradient, presumably regulated by a circadian clock, reverts and the stem reorients eastward in preparation for the dawn. As the plant matures, its elongation capability decreases along with the gradual decay in the heliotropic movement that eventually ceases upon anthesis. Then, the inflorescence permanently faces the east (Kutschera and Briggs, 2016).

While heliotropism in blooming sunflowers ceases, the flowers of other species, such as *Papaver* and *Ranunculus* spp., continue tracking the sun over the whole period of their flowering (Kevan, 1975; Totland, 1996).

20.2.3 Light Perception

Light, primarily blue, but also red in some plants, stimulates the heliotropic response in a mechanism that involves a circadian clock (Koller, 1986; Stanton and Galen, 1993). Red light is perceived by the phytochrome, whereas the identity of the blue-light receptor involved in the pulvinus-dependent movement is yet unknown. Still, phototropin has been suggested to be the blue-light receptor mediating the elongation-driven heliotropic movement (Serrano et al., 2018). The light is sensed in the pulvinus (e.g., *Leguminosae* and *Oxalidaceae* spp.), or in the leaf (*Malvaceae* spp., sunflower). However, the involvement of floral organs, such as petals and gynoecium, has also been shown to affect flower or inflorescence heliotropic movement (Koller, 1990; Serrano et al., 2018). The blue-light photoreceptor in the veins of *Lavatera cretica* (Malvaceae) has been shown to perceive the light direction and stimulate diaheliotropic responses (Koller, 1990). The daily heliotropic oscillations (e.g., sunflower) persisted in free-running conditions (constant light and temperature) but were severely impaired, when a period of light/dark cycles was extended to 30 hours. On the other hand, when the photoperiod was extended while the period of the cycle remained 24 hours, the stem of the sunflower still faced eastward at the end of the cycle. This was obtained by a faster west to east movement during the night (Atamian et al., 2016). The expression of circadian clock-related genes in sunflowers, although they altered rhythmically, was similar at both sides of the stem (Atamian et al., 2016). Covington and Harmer (2007) demonstrated that the response of *Arabidopsis* to auxin is clock-controlled. These phenomena indicate an interplay between light and the circadian clock that presumably dictates the timing of the responsiveness of auxin in the regulation of the heliotropic response in sunflowers. Several genes encoding channel proteins expressed in the pulvinus (e.g., Aquaporin; Moshelion et al., 2002a), K^+ and Cl^- channels (Moshelion et al., 2002b; Oikawa et al., 2018) have been shown to be regulated by diurnal and circadian rhythms. Although expression of these genes was correlated with alterations in water and ion permeability, as well as with leaf movement, it is still questionable whether expression of the channels and their gating or energization of ion transport determines the rhythmical movement of the plants. The mode of coordination of channel operation in the pulvinus during heliotropic movements is largely unknown. Still, progress has been made in understanding the regulation of pulvinus action in the non-directional movements of the mimosoide plants *Mimosa pudica* and *Samanea saman* in response to stimuli such as touch (seismonasty) or 'sleeping movements' in response to light/dark (nyctinasty). In these processes, cytosolic Ca^{+2} is a pivotal secondary messenger that modulates channel activity and phosphorylation of key factors in the pulvinar osmoregulation, affects membrane potential, and interacts with the phosphoinositide signal transduction cascade (Toriyama and Jaffe, 1972; Moran, 2007; Oikawa et al., 2018). It is plausible that calcium also plays a similar role in the pulvinar heliotropic movement. Calcium has been shown to be involved in the blue-light induced phototropin signaling (Harada and Shimazaki, 2007) and presumably also in the heliotropic movement mediated by this receptor, but its role in these processes remains obscure.

20.3 CONCLUSIONS AND OUTLOOK

The daily alterations in sun position and light intensity and on the other hand the absolute dependence of plants on light led to the evolution of heliotropic responses in plants to optimize sunlight utilization under unfavorable conditions. These responses reflect on plant organ movements driven by differential elongation or turgor changes on both sides of the moving organ. The heliotropic movements, which are induced by directional light regulated by the circadian rhythm and affected by the environmental conditions, enable the plant to prepare itself throughout the day for the forthcoming changes in light while allowing flexibility to respond to the actual environment. Despite recent studies of the regulation of elongation-dependent heliotropism (Atamian et al., 2016), the underlying molecular mechanisms and photoreceptors involved still remain to be unraveled.

ACKNOWLEDGMENTS

The author thanks Professor Michael Gurevitz and Dr. Nina Kamennaya who critically read the manuscript.

REFERENCES

Adir, N., Zer, H., Shochat, S., and Ohad, N. 2003. Photoinhibition—A historical perspective. *Phtosynth. Res.* 76: 343.

Atamian, H.S., Creux, N.M., Brown, E.A., Garner, A.G., Blackman, B.K., and Harmer, S.L. 2016. Circadian regulation of sunflower heliotropism, floral orientation, and pollinator visits. *Science.* 353: 587–590.

Covington, M.F., and Harmer, S.L. 2007. The circadian clock regulates auxin signaling and responses in *Arabidopsis. PLoS Biol.* 5: e222.

Ehleringer, J., and Forseth, I. 1980. Solar tracking by plants. *Science.* 210: 1094–1098.

Harada, A., and Shimazaki, K. 2007. Phototropins and blue light-dependent calcium signaling in higher plants. *Photochem. Photobiol.* 83: 102–111.

Kameyama, K., Kishi, Y., Yoshimura, M., Kanzawa, N., Sameshima, M., and Tsuchiya, T. 2000. Tyrosine phosphorylation in plant bending. *Nature.* 407: 37.

Kanzawa, N., Hoshino, Y., Chiba, M., Hoshino, D., Kobayashi, H., Kishi, N., Osumi, M., Sameshima, M., and Tsuchiya, T. 2006. Change in the actin cytoskeleton during seismonastic movement of *Mimosa pudica. Plant Cell Physiol.* 47: 531–539.

Kevan, P.G. 1975. Sun-tracking solar furnaces in high arctic flowers: Significance for pollination and insects. *Science.* 189: 723–726.

Kok, B. 1956. On the inhibition of photosynthesis by intense light. *Biochim. Biophys. Acta.* 21: 234–244.

Koller, D. 1986. The control of leaf orientation by light. *Photochem. Photobiol.* 44: 819–826.

Koller, D. 1990. Light-driven leaf movements. *Plant Cell Environ.* 13: 615–632.

Kutschera, U., and Briggs, W.R. 2016. Phototropic solar tracking in sunflower plants: An integrative perspective. *Ann. Bot.* 117: 1–8.

Moran, N. 2007. Osmoregulation of leaf motor cells. *FEBS Lett.* 581: 2337–2347.

Moshelion, M., Becker, D., Czempinski, K., Mueller-Roeber, B., Attali, B., Hedrich, R., and Moran, N. 2002a. Diurnal and circadian regulation of putative potassium channels in a leaf moving organ. *Plant Physiol.* 128: 634–642.

Moshelion, M., Becker, D., Biela, A., Uehlein, N., Hedrich, R., Otto, B., Levi, H., Moran, N., and Kaldenhoff, R. 2002b. Plasma membrane aquaporins in the motor cells of *Samanea saman*: Diurnal and circadian regulation. *Plant Cell.* 14: 727–739.

Oikawa, T., Ishimaru, Y., Munemasa, S., Takeuchi, Y., Washiyama, K., Hamamoto, S., Yoshikawa, N., Mutara, Y., Uozumi, N., and Ueda, M. 2018. Ion channels regulate nyctinastic leaf opening in *Samanea saman. Current Biol.* 28: 2230–2238.

Serrano, A.M., Arana, M.V., Vanhaelewyn, L., Ballare, C.L., Van Der Straeten, D., and Vandenbussche, F. 2018. Following the star: Inflorescence heliotropism. *Environ. Exp. Bot.* 147: 75–85.

Stanton, M.L., and Galen, C. 1989. Consequences of flower heliotropism for reproduction in an alpine buttercup (*Ranunculus adoneus*). *Oecologia.* 78: 477–485.

Stanton, M.L., and Galen, C. 1993. Blue light controls solar tracking by flowers of an alpine plant. *Plant Cell Environ.* 16: 983–989.

Toriyama, H., and Jaffe, M.J. 1972. Migration of calcium and its role in the regulation of seismonasty in the motor cells of *Mimosa pudica* L. *Plant Physiol.* 49: 72–81.

Totland, O. 1996. Flower heliotropism in an alpine population of *Ranunculus acris* (Ranunculaceae): Effects on flower temperature, insect visitation, and seed production. *Am. J. Bot.* 83: 452–458.

Uehlein, N., and Kaldenhoff, R. 2008. Aquaporins and plant leaf movements. *Ann. Bot.* 101: 1–4.

Vandenbrink, J.P., Brown, E.A., Harmer, S.L., and Blackman, B.K. 2014. Turning heads: The biology of solar tracking in sunflower. *Plant Sci.* 224: 20–26.

Zhang, S., Ai, H.L., Bin Yu, W., Wang, H., and Li, D.Z. 2010. Flower heliotropism of *Anemone rivularis* (Ranunculaceae) in the Himalayas: Effects on floral temperature and reproductive fitness. *Plant Ecol.* 209: 301–312.

21 Carbon Metabolic Pathways and Relationships with Plant Stress

Carlos M. Figueroa, Romina I. Minen, Florencio E. Podestá, and Alberto A. Iglesias

CONTENTS

21.1 INTRODUCTION

Carbohydrates are the most abundant organic compounds in the biosphere. However, their importance is due not only to their abundance but also to the multifaceted critical functions they play in all cells and organisms. In plants, carbohydrates contribute to cell operation by (i) being key intermediates in many metabolic routes; (ii) forming part of many structural components; and (iii) transporting carbon and energy between tissues (Figueroa et al. 2016). Concerning involvement in energetic metabolism, carbohydrates are the main intermediates in glycolysis and the oxidative pentose-P pathway. Furthermore, these routes have an important task in providing carbon skeletons for the synthesis of a large number of cellular components, including lipids, nucleic acids, organic acids, and proteins. Carbohydrates are the chief components of structural molecules such as cellulose, the most abundant biomolecule. Also, sugars appear in combination with other compounds, such as in glycolipids and glycoproteins, which decisively determine cell structure and function. Still, starch constitutes a major storage product, whereas sucrose and polyols are leader metabolites for carbon mobilization between photosynthetic and heterotrophic tissues in plants (Figueroa et al. 2016).

The sessile lifestyle enforces strict challenges to plants with respect to their capacity for coping with changing environments. Plants are affected by many stress conditions elicited, among other things, by (i) water shortage; (ii) nutrient deficiency; (iii) toxic compounds (natural or arising from contamination); and (iv) temperature extremes. To manage unfavorable conditions that may prevail in certain wild situations, plants have developed a series of mechanisms to accurately adapt their metabolic machinery. Many of the responses to stress engage changes in carbon metabolism. Examples of the latter are relocation of carbon skeletons to produce specific stress-related compounds, excretion of organic acids to deal with the presence of toxic metals, and modification of the main metabolic routes by using alternative (or secondary) steps. Adaptations include the transient fine-tuning of constitutive processes or the induction of latent ones to accommodate to specific milieus. More extensive changes can even include the adjustment of the whole photosynthetic machinery to endure long-term challenges.

This chapter deals with the main changes of primary carbon metabolism that endow plants to cope with the most common abiotic stresses and survive in ever-changing environments. The analysis will consider the different fates of carbohydrates produced by carbon photoassimilation, emphasizing on the flux of energy and reducing power within (and between) the different cells and tissues in a plant. The overall scenario will integrate relationships between carbohydrates and plant physiology with the ability to overcome different harmful conditions elicited under stress.

21.2 CARBON PARTITIONING IN PLANTS

Oxygenic photosynthesis is the process by which the energy from light is used to fix (reduce) inorganic CO_2. This is a central event on Earth, since most organisms obtain (directly or indirectly) energy from photoassimilated products. The global photosynthetic process can be considered as the integration of two phases: (i) a light phase, wherein the electromagnetic energy of the sun is transformed into chemical energy (ATP) and reducing power (NADPH); and (ii) a synthetic phase, wherein ATP and NADPH are biochemically used to fix atmospheric CO_2, with a primary production of carbohydrates (photosynthates). Carbon fixation takes place in the chloroplast of the photosynthetic cell via the Benson-Calvin cycle (BCC), which requires ATP and NADPH produced in the light phase (Figure 21.1). Triose-P and hexose-P are intermediates of the BCC that can be metabolized for the production of starch within the same cellular compartment. Alternatively, triose-P can be exported to the cytosol to feed

glycolytic routes or those synthesizing sucrose or polyols, the major compounds used for carbon translocation from source (photosynthetic) to sink (heterotrophic) tissues.

In this picture, photoassimilates are partitioned at two levels: intra- and inter-cellular. In the partition between the chloroplast and the cytosol, a molecule of triose-P is exchanged with Pi through a specific translocator located in the plastid envelope (TPT in Figure 21.1). In the cytosol, the transported metabolite can feed different pathways, mainly (i) catabolic glycolysis, resulting in the indirect export of photogenerated ATP and NADPH from the chloroplast, and (ii) anabolic synthesis of soluble sugars. Most plants mainly synthesize and accumulate sucrose, although many species also produce sugar-alcohols (for example, glucitol, abbreviated Gol; see Figure 21.1). The second level of partitioning involves the transport of carbohydrates to heterotrophic plant tissues, sucrose (or a sugar-alcohol, depending on the species) being the main carrying compound. Once in the non-photosynthetic cell, sucrose (or the sugar alcohol) can be metabolized by different routes, and once again, a partition of photoassimilates takes place between the cytosol and the plastid (Figure 21.2).

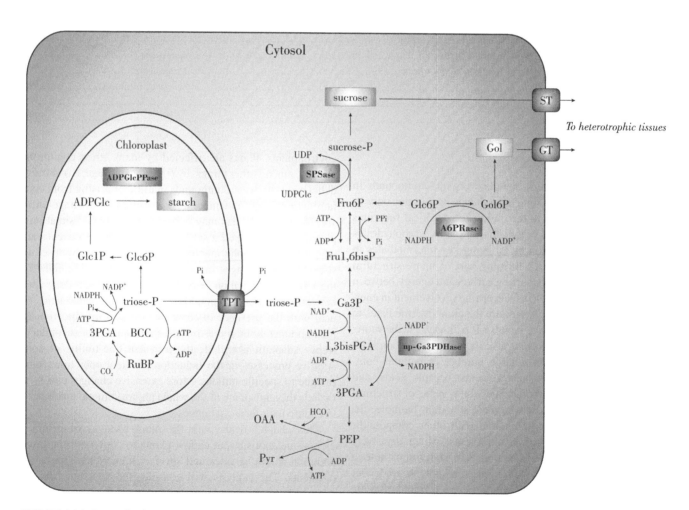

FIGURE 21.1 (See color insert.)   Schematic representation of carbon metabolism in plant photosynthetic cells. Key enzymes for carbon partitioning are colored in violet. The major photosynthetic products (starch, sucrose, and glucitol) are depicted in green. Transporters are highlighted in red (TPT: triose-P translocator; ST: sucrose transporter; GT: glucitol transporter). BCC is the Benson-Calvin cycle.

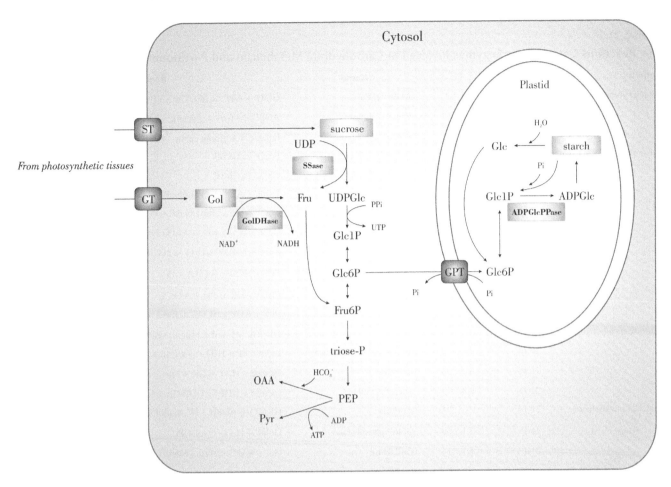

FIGURE 21.2 (See color insert.)   Schematic representation of carbon metabolism in plant heterotrophic cells. The enzymes involved in sucrose, starch, and glucitol metabolism are colored in violet. Photosynthates imported from leaves (sucrose and glucitol) are depicted in green. Starch, the major storage compound, is colored in cyan. The glucose-P translocator (GPT) is highlighted in red.

As detailed in Figures 21.1 and 21.2, starch and sucrose (or polyols) are the major carbohydrates produced by photosynthesis as a result of the partition of carbon between a relatively stationary (starch) and a mobile (sucrose, polyols) form. Understanding the processes of synthesis, partition (within one cell and between source and sink tissues), and storage of carbohydrates is important, because these molecules determine plant productivity. Different factors that affect the operation or control of these processes (at systemic or cellular levels) can modify the carbon and energy demands of different tissues (Figueroa et al. 2016). Starch biosynthesis takes place in plastids (Figures 21.1 and 22.2) with the involvement of three enzymatic steps sequentially catalyzed by ADP-glucose pyrophosphorylase (ADPGlcPPase, EC 2.7.7.27), starch synthase (EC 2.4.1.21), and branching enzyme (EC 2.4.1.18; Equations 21.1 through 21.3 in Table 21.1 (Ballicora et al. 2004). This polysaccharide constitutes a transitory storage glucan in source (green) tissues, and its level varies during the photoperiod, while long-term storage occurs in sinks (non-photosynthetic) tissues, such as fruits, roots, and tubers (Figure 21.2). Starch is one of the major storage compounds in higher plants in both photosynthetic and heterotrophic tissues.

21.3   CARBON METABOLISM IN THE CYTOSOL AND PLANT STRESS

Many carbohydrate-related metabolic routes take place in the cytosol, where they are connected to other carbon pathways. In this compartment, fructose 6-phosphate (Fru6P), glucose 6-phosphate (Glc6P), and glucose 1-phosphate (Glc1P) are part of a metabolic node that includes (i) the conversion between Fru6P and Glc6P, catalyzed by Glc6P isomerase (EC 5.3.1.9), and (ii) the reaction of phosphoglucomutase (EC 5.4.2.2), allowing the formation of Glc1P from Glc6P (Equations 21.4 and 21.5 in Table 21.1). In plants, the hexose-P pool in the cytosol feeds carbon skeletons to different metabolic fluxes to produce many metabolites, mainly glycolysis and the synthesis of soluble sugars. These central metabolic routes generate intermediate metabolites or final products that are critical for plant function under physiological as well as stress conditions (Figueroa et al. 2016).

21.3.1   ORGANIZATION OF PLANT GLYCOLYSIS

Glycolysis is a ubiquitous metabolic pathway by which most organisms convert glucose (Glc) into pyruvate (Pyr) with the

TABLE 21.1

Key Reactions Catalyzed by Enzymes Involved in Carbohydrate Metabolism and Partitioning in Plants

Pathway	Equation	EC	Name	Reaction
Starch synthesis	21.1	2.7.7.27	ADPGlcPPase	$Glc1P + ATP \rightleftharpoons ADPGlc + PPi$
	21.2	2.4.1.21	Starch synthase	$ADPGlc + (\alpha\text{-}1,4\text{-glucan})_n \rightarrow ADP + (\alpha\text{-}1,4\text{-glucan})_{n+1}$
	21.3	2.4.1.18	Branching enzyme	linear α-1,4-glucan chain $\rightarrow \alpha$-1,6-branched α-1,4-glucan chain
Glycolysis	21.4	5.3.1.9	Glc6P isomerase	$Fru6P \rightleftharpoons Glc6P$
	21.5	5.4.2.2	Phosphoglucomutase	$Glc6P \rightleftharpoons Glc1P$
	21.6	2.7.1.11	ATP-PFKase	$Fru6P + ATP \rightarrow Fru1,6bisP + ADP$
	21.7	2.7.1.90	PPi-PFKase	$Fru6P + PPi \rightleftharpoons Fru1,6bisP + Pi$
	21.8	1.2.1.12	Ga3PDHase	$Ga3P + NAD^+ + Pi \rightleftharpoons 1.3PGA + NADH + H^+$
	21.9	1.2.1.9	np-Ga3PDHase	$Ga3P + NADP^+ + H_2O \rightarrow 3PGA + NADPH + 2\ H^+$
	21.10	4.1.1.31	PEP carboxylase	$PEP + HCO_3^- \rightarrow OAA + Pi$
	21.11	2.7.1.40	PyrKase	$PEP + ADP \rightarrow Pyr + ATP$
	21.12	4.1.1.49	PEPCKase	$PEP + ADP + HCO_3^- \rightleftharpoons OAA + ATP$
Sucrose metabolism	21.13	2.4.1.14	SPSase	$UDPGlc + Fru6P \rightleftharpoons sucrose\text{-}6P + UDP$
	21.14	3.1.3.24	Sucrose-6P phosphatase	$sucrose\text{-}6P + H_2O \rightarrow sucrose + Pi$
	21.15	3.2.1.26	Invertase	$sucrose + H_2O \rightarrow Glc + Fru$
	21.16	2.4.1.13	Sucrose synthase	$sucrose + UDP \rightleftharpoons UDP\text{-}Glc + Fru$
Glucitol metabolism	21.17	1.1.1.200	A6PRase	$Glc6P + NADPH + H^+ \rightleftharpoons Gol6P + NADP^+$
	21.18	3.1.3.50	Gol6P phosphatase	$Gol6P + H_2O \rightarrow Gol + Pi$
	21.19	1.1.1.14	GolDHase	$Gol + NAD^+ \rightleftharpoons Fru + NADH + H^+$
Raffinose synthesis	21.20	2.7.7.64	USPPase	$Gal1P + UTP \rightleftharpoons UDP\text{-}Gal + PPi$
	21.21	2.4.1.123	Galactinol synthase	$UDP\text{-}Gal + myo\text{-inositol} \rightleftharpoons UDP + galactinol$
	21.22	2.4.1.82	Raf synthase	$galactinol + sucrose \rightleftharpoons Raf + myo\text{-inositol}$

production of two molecules of ATP (Givan 1999; Plaxton 1996; Podestá 2004). Glycolysis can be visualized as the first step of the fundamental process of intermediary metabolism known as respiration. Glycolysis develops in plants with unique characteristics (with respect to performance in other organisms), exhibiting a series of differential features and playing functions additional to the provision of ATP and Pyr for mitochondrial respiration (Plaxton and Podestá 2006). One singularity is the alternative use of PPi or ATP as phosphoryl donors (Plaxton and Podestá 2006; Stitt 1998). Also, the presence of subsidiary enzymes enlarges the classic 10-reaction pathway and lends metabolism a great degree of plasticity (Plaxton 1996). Most importantly, the final product of plant glycolysis is not necessarily Pyr, as the cytosolic pool of phosphoenolpyruvate (PEP) represents a metabolic branch point determining different fates for carbon skeletons (Plaxton 1996; Plaxton and Podestá 2006). Additionally, Glc is not the constrained starting metabolite for glycolysis, since the photosynthetically generated hexose-P or triose-P pools can both feed the route toward PEP or Pyr (Dennis and Blakeley 2000). Finally, it is worth noting that in plants, a second set of glycolytic enzymes (having distinctive characteristics) is localized in the plastid. All of these features provide plants

with useful tools to accurately cope with different types of stressful situations.

Starting from the top, down along the conventional route, the conversion of Fru6P to Fru1,6bisP is the first distinctive reaction found in cytosolic glycolysis performed in plant cells. As in most other organisms, it is a key regulatory step, but the phosphorylation can involve two different phosphofructokinases (PFKases) (Equations 21.6 and 21.7, respectively, in Table 21.1): a classical (but Fru2,6bisP-insensitive) ATP-dependent (ATP-PFKase, EC 2.7.1.11) or a PPi-dependent (potently activated by Fru2,6bisP; PPi-PFKase, EC 2.7.1.90) enzyme, which use ATP or PPi, respectively, as phosphoryl donors (Carnal and Black 1979; Dennis and Blakeley 2000; Figueroa et al. 2016; Givan 1999; Plaxton and Podestá 2006) (Figure 21.1). Plant PPi-PFKase is widespread in different species and tissues, and its subunit composition and activity respond to environmental and developmental cues (Nielsen 1995; Plaxton and Podestá 2006; Podestá and Plaxton 1994a; Theodorou and Plaxton 1996; Trípodi and Podestá 1997). The available information indicates that the enzyme most probably functions in the glycolytic direction (even when it catalyzes a reversible reaction), modulating the balance between triose-P and hexose-P pools and carbon partitioning among starch and sucrose (Hajirezaei et al. 1994; Plaxton 1996; Plaxton and Podestá 2006; Podestá and Plaxton 2003).

Following the conversion of Fru1,6bisP into triose-P, the first energy-conserving reaction takes place with the generation of ATP (energetic power) and NADH by the joint action of NAD-dependent, phosphorylating, glyceraldehyde 3-phosphate dehydrogenase (Ga3PDHase, EC 1.2.1.12) and phosphoglycerate kinase (EC 2.7.2.3). The first reaction (Equation 21.8 in Table 21.1) constitutes the only oxidative conversion during glycolysis, with oxidation coupled to phosphorylation (note that orthophosphate is a substrate) to conserve energy in the form of 1,3bisP-glyceric acid. In plants, the two enzymatic steps can be circumvented by a single one, catalyzed by NADP-dependent, non-phosphorylating, Ga3PDHase (np-Ga3PDHase, EC 1.2.1.9), which yields 3-phosphoglyceric acid (3PGA) and NADPH (reducing power; Equation 21.9 in Table 21.1) (Bustos and Iglesias 2002; Iglesias et al. 2002). The latter reaction is irreversible and renders a product that cannot be used to generate ATP, because oxidation is uncoupled from phosphorylation (note that water is the substrate instead of orthophosphate). The contribution of this enzyme thus leads to a nil ATP yield but produces NADPH, a key metabolic intermediate in carrying reducing equivalents.

This detailed branch point in plant cytosolic glycolysis, by which triose-P can be metabolized to produce ATP or NADPH, is clearly critical for cell energetics. It is thus expected that the enzymes involved in one and/or the other pathway should be under regulation. Many studies support such control of the activity of both Ga3PDHase and np-Ga3PDHase. Ga3PDHase is a target of post-translational modifications by redox (Piattoni et al. 2013; Schneider et al. 2018) and phosphorylating (Piattoni et al. 2017) mechanisms. Redox regulation elicits changes in activity (decreased by oxidation), protein interaction (mainly with thioredoxin), and subcellular localization (from cytosol to mitochondria and nucleus) (Schneider et al. 2018). Phosphorylation renders a less active enzyme and triggers mechanisms of its proteolysis, which are operative in particular tissues and developmental stages (as observed in seeds) (Piattoni et al. 2017). The current view states that Ga3PDHase has moonlighting activity, constituting a redox-switch with multiple functions that critically orchestrate energy metabolism (Schneider et al. 2018). It has been demonstrated that np-Ga3PDHase is phosphorylated in heterotrophic (not in photosynthetic) plant cells (Bustos and Iglesias 2002). The phosphorylated enzyme interacts with regulatory 14-3-3 proteins to arrange a structure with lower activity and enhanced sensitivity to inhibition by ATP and pyrophosphate (Bustos and Iglesias 2003). Studies performed in wheat endosperm revealed that phosphorylation of np-Ga3PDHase at serine-404 is mediated by an SNF1-related protein kinase (SnRK1), which is in turn allosterically inhibited by ribose 5-phosphate (Piattoni et al. 2011). It has been also reported (Bustos et al. 2008) that the np-enzyme is involved in the response to oxidative stress in wheat leaves, which is associated with the relative low reactivity of key cysteine residues in the protein with oxidant agents (Piattoni et al. 2013). In agreement with the latter, plants with a deficiency

in this enzyme were found to have a reduced glycolytic capacity and increased levels of oxidative stress (Rius et al. 2006).

From then on, two reactions drive to the production of PEP, a metabolite that, as stated earlier, plays a central function in plant carbohydrate metabolism as a regulatory molecule besides its significance as a branch-point intermediate. PEP is the substrate for a carboxylation reaction (Equation 21.10 in Table 21.1) mediated by PEP carboxylase (PEPCase, EC 4.1.1.31), a regulatory enzyme that generates oxaloacetate (OAA) and Pi (Figures 21.1 and 21.2) (Chollet et al. 1996; Iglesias et al. 1997). PEPCase exists in various isoforms distributed at different levels in plant tissues. It plays a major role in anaplerotic routes replenishing intermediates of the tricarboxylic acid cycle and thus, links cytosolic carbon metabolism with respiration and N assimilation taking place in other compartments. This enzyme is responsible for the primary CO_2 fixation in plants performing C_4 and crassulacean acid metabolism (CAM), initiating the CO_2-concentrating mechanism. PEPCase is tightly regulated by metabolites and by post-translational phosphorylation on a serine residue (Bakrim et al. 2001; Chollet et al. 1996; Trípodi and Plaxton 2005). Malate is a strong inhibitor, but its effect is dependent on pH and most importantly, on the phosphorylation status of the enzyme. When phosphorylated, PEPCase is usually more active and less sensitive to malate, mainly at pH values on the acidic side of the optimum (around pH 8.0) (Baur et al. 1992; Chollet et al. 1996; Hartwell et al. 1999; Moraes and Plaxton 2000; Trípodi and Plaxton 2005). Glc6P is a main allosteric effector, activating the enzyme and reducing its sensitivity to malate inhibition (Chollet et al. 1996; Plaxton 1996). The enzyme from C_3 plants (i.e., banana fruit and castor bean germinating endosperm) is highly sensitive to inhibition by aspartate and glutamate, which reinforces the link between N and C metabolism mediated by PEPCase (Law and Plaxton 1995; Trípodi and Plaxton 2005). Phosphorylation diminishes the effect of aspartate (Gregory et al. 2009; Law and Plaxton 1995; Trípodi and Plaxton 2005). The characterization of PEP carboxylase kinase has evidenced that its ability to phosphorylate the target enzyme is weakened in the presence of malate, while it is strengthened in the presence of PEP (Murmu and Plaxton 2007). Additionally, PEPCase is modulated by mono-ubiquitination, which adds another layer of complexity to the allosteric regulation of this enzyme. This post-translational modification reduces the apparent affinity for PEP and increases the sensitivity to malate (Ruiz-Ballesta et al. 2014; Uhrig et al. 2008). Therefore, the effect of mono-ubiquitination is opposite to that triggered by phosphorylation; indeed, it has been suggested that these processes are mutually exclusive (O'Leary et al. 2011).

PEP is also the substrate for the usual last reaction of glycolysis in most organisms. The step catalyzed by Pyr kinase (PyrKase, EC 2.7.1.40) produces Pyr and ATP (Equation 21.11 in Table 21.1) and constitutes the second energy-conserving reaction (Figures 21.1 and 21.2). PyrKase has been studied in many plant tissues, and its regulatory role on the whole glycolytic process has been recognized (Hu and Plaxton 1996; Lin et al. 1989; Moraes and Plaxton 2000; Plaxton 1989; Podestá and Plaxton 1991). In fact, the activity of this enzyme controls the flux of the upper part of glycolysis by regulating the levels of PEP (Givan 1999; Plaxton 1996) (see later for further analysis). PyrKase regulation is organ specific in plants, with the cytosolic form from germinating castor seed endosperm or cotyledons exhibiting a pH-dependent response to several metabolite inhibitors (Podestá and Plaxton 1991, 1994b). A concerted decrease in pH and inhibitor concentrations, as could be caused by anoxia, will provoke a rise in the enzyme activity. Glutamate is an important inhibitor of PyrKase from several sources (Hu and Plaxton 1996; Lin et al. 1989; Plaxton 1996; Plaxton and Podestá 2006; Podestá and Plaxton 1994b), while aspartate activates and relieves the inhibitory effect of the former. The reciprocal effects of aspartate on PyrKase and PEPCase grant an effective mechanism to balance both activities during N assimilation. In this scenario, aspartate accumulation trims down the flux through PEPCase, while it increases PyrKase activity in leaves, ripening banana fruit and castor bean endosperm (Plaxton and Podestá 2006; Smith et al. 2000; Turner et al. 2005; Turner and Plaxton 2000).

PEP carboxykinase (PEPCKase, EC 4.1.1.49) is another enzyme that uses PEP as a substrate, catalyzing its carboxylation coupled to phosphorylation of ADP in a reversible reaction (Chen et al. 2004; Daley et al. 1977; Leegood and Walker 2003; Rylott et al. 2003; Walker and Leegood 1996; Wingler et al. 1999) (Equation 21.12 in Table 21.1). Although it is generally implicated in gluconeogenesis, the activity of PEPCKase determines PEP concentration in the cytosol. The enzyme is particularly abundant in leaves of plants developing C_4 or CAM, where it is involved in the process of decarboxylation (Martin et al. 2011). *Arabidopsis thaliana* (*Arabidopsis*) has two genes coding for PEPCKase1 and PEPCKase2, respectively (Malone et al. 2007), the former having higher levels of expression in different tissues, according to the *Arabidopsis* eFP Browser (Winter et al. 2007).

The study of PEPCKase has been hindered by its extreme sensitivity to proteolysis on extraction (Martín et al. 2007), giving rise to an active form exhibiting few (if any) regulatory properties. In many (but not all) plants, the enzyme has been identified as a target of protein kinases (Leegood and Walker 2003). Phospho-proteomic (de la Fuente van Bentem et al. 2006) and degradomic (Tsiatsiani et al. 2013) studies indicate that post-translational modification of PEPCKase in *Arabidopsis* involves a protein kinase of the GSK3/Shaggy-like family and a metacaspase, respectively. The phosphorylation site would localize within the peptide excised by proteolysis, which is supported by the fact that changes in the enzyme properties induced by phosphorylation could not be evidenced on the proteolyzed protein (Leegood and Walker 2003). Studies with a preparation containing only the phospho or dephospho forms of PEPCKase are lacking, but the use of special reaction media designed to measure one form preferentially suggested that phosphorylated PEPCKase is the less active form (Leegood and Walker 2003). A strict coordination in the phosphorylation status of PEPCase and PEPCKase should operate *in vivo* to avoid a potential futile cycle produced by the simultaneous action of both enzymes.

This detailed scenario establishes that levels of PEP depend on the coordination of the activities of at least three regulatory enzymes. This is of relevance considering the key role of PEP as a regulatory metabolite. In plants, the conversion of Fru6P to Fru1,6bisP does not constitute the primary control point of glycolysis, in part due to the fact that Fru2,6bisP strongly activates PPi-PFKase but is ineffective to regulate ATP-PFKase (Kruger and Dennis 1987; Theodorou and Plaxton 1994; Trípodi and Podestá 1997). All plant cytosolic ATP-PFKases characterized so far are inhibited by PEP (Givan 1999; Lee and Copeland 1996), which also inhibits the synthesis of Fru2,6bisP (Plaxton and Podestá 2006). In this way, the flux through the upper part of glycolysis is controlled by PEP levels. This reinforces the role of PEPCase and PyrKase as pacemaker enzymes of the whole pathway, meaning that the primary control of glycolysis depends on them. ATP-PFKase and PPi-PFKase are still relevant for fine-tuning carbon flow in the cytosol, with activities linked to the levels of PEP and also to those of Pi, since Pi activates ATP-PFKase while strongly inhibiting the glycolytic reaction of PPi-PFKase (Plaxton and Podestá 2006; Podestá 2004).

21.3.2 Metabolism of Sucrose and Polyols

In leaves, during the light period, sucrose is produced from triose-P translocated from the chloroplast through the TPT in exchange for Pi (Figure 21.1). In the dark, starch mobilization allows the constant production of sucrose and the uninterrupted carbon flow to heterotrophic tissues (Stitt et al. 2010). Sucrose is synthesized by sucrose 6-phosphate synthase (SPSase) and sucrose 6-phosphatase (Equations 21.13 and 21.14 in Table 21.1 and Figure 21.1). The disaccharide is then transported to sink tissues, where it is degraded by two different types of enzymes (Equations 21.15 and 21.16 in Table 21.1 and Figure 21.2): invertases (EC 3.2.1.26, which catalyze the irreversible hydrolysis of sucrose to Glc and Fru) and sucrose synthase (EC 2.4.1.13, which catalyzes a reversible reaction) (Figueroa et al. 2016; Koch 2004; Winter and Huber 2000). The reaction catalyzed by sucrose synthase conserves the energy of the glycosidic bond in UDP-glucose (UDPGlc), which could be used for cellulose synthesis or provide carbon skeletons for glycolysis (Koch 2004). In heterotrophic tissues, Glc6P and Glc1P are the preferred molecules carried through a specific translocator (GPT; see Figure 21.2) inside plastids, where they can be used for starch synthesis (Figueroa et al. 2016).

In addition to sucrose and starch, certain plants synthesize Gol (also known as sorbitol) or mannitol as important photosynthetic products (Loescher and Everard 2000). In these organisms, the analysis of $[^{14}C]CO_2$ assimilation revealed the presence of two major soluble compounds: the polyol and sucrose. In apple leaves, about 70% of the newly photosynthetically fixed carbon was found as Gol and sucrose (Grant and Rees 1981). Likewise, almost 80% of the label was recovered as mannitol and sucrose in mature celery leaves, with similar amounts of each one on a molar basis (Loescher et al. 1992). Similar results were obtained in apricot, where Gol was found to be the compound with the highest label 30 minutes after the pulse (Bieleski and Redgwell 1977). Gol is a major photosynthetic product in many economically important fruit-bearing tree species of the Rosaceae family, including apple, peach, and pear. Because in many plants, sugar-alcohols are the main photosynthetic products, their metabolism is expected to be tightly regulated, probably at different levels. It is worth noting that sugar-alcohols could have important roles in plant tolerance to certain types of abiotic stresses (Loescher and Everard 2000). Indeed, recent studies showed that enzymes involved in Gol metabolism from peach leaves and fruits are redox-regulated (Hartman et al. 2014; Hartman et al. 2017).

Gol synthesis in mature leaves starts with the reaction catalyzed by the NADP-dependent aldose 6-phosphate reductase (A6PRase, EC 1.1.1.200; Equation 21.17 in Table 21.1 and Figure 21.1), which converts Glc6P into Gol6P (Figueroa et al. 2016; Loescher and Everard 2000). The phosphate group is then removed by a specific Gol6P phosphatase (EC 3.1.3.50; Equation 21.18 in Table 21.1 and Figure 21.1) (Zhou et al. 2003). It has been shown that these enzymes are mainly located in the cytosol (Loescher and Everard 2000). Authors have suggested that np-Ga3PDHase provides the NADPH necessary for the synthesis of the sugar-alcohol (Figure 21.1). Actually, the activity of np-Ga3PDHase in celery leaves is high enough to support mannitol production in this plant (Gao and Loescher 2000; Gomez Casati et al. 2000; Rumpho et al. 1983). The polyol produced in leaves is transported to sink and developing tissues, such as fruits or immature leaves, where it is converted to Fru by the activity of a NAD-dependent Gol dehydrogenase (GolDHase, EC 1.1.1.14; Equation 21.19 in Table 21.1 and Figure 21.2) (Figueroa et al. 2016; Loescher and Everard 2000).

21.3.3 BIOSYNTHESIS OF RAFFINOSE

The occurrence of raffinose (Raf; an α-1,6-galactosyl extension of sucrose) seems to be universal in angiosperms, whereas Raf-derived oligosaccharides (such as stachyose, verbascose, and ajugose) are only found in certain species (Peters and Keller 2009). Raf is usually found in seeds, where it plays a key role in membrane stabilization during the desiccation process (Downie et al. 2003). Plants from the Cucurbitaceae family (such as watermelon and cucumber) also use Raf to export carbon to heterotrophic tissues (Haritatos et al. 1996).

Additionally, plants accumulate Raf in source tissues under certain abiotic stress situations, including heat, cold, salinity, and drought (Bachmann et al. 1994; Panikulangara et al. 2004; Peters and Keller 2009; Taji et al. 2002; Zuther et al. 2004).

Raf levels are tightly related to the expression of the biosynthetic enzymes (Cunningham et al. 2003; Downie et al. 2003; Volk et al. 2003). The first step is the production of UDP-galactose (UDPGal), which is not specific to this pathway but provides one of the essential metabolites for Raf synthesis. UDPGal is an intermediate of the Leloir pathway, which requires three enzymes to interconvert Gal into Glc: galactokinase (EC 2.7.1.6), galactose 1-phosphate uridylyltransferase (EC 2.7.7.12), and UDPGal 4-epimerase (EC 5.1.3.2) (Allard et al. 2001; Frey 1996; Leloir 1951). Alternatively, UDPGal can be synthesized from Gal1P and UTP by UDP-sugar pyrophosphorylase (USPPase, EC 2.7.7.64; Equation 21.20 in Table 21.1) (Dai et al. 2006; Kotake et al. 2004, 2007). UDPGal is used for the synthesis of galactolipids (found in plastid membranes) (Dormann and Benning 2002; Kelly and Dormann 2002) and cell wall components (hemicellulose and pectin) (Bacic 2006; Loque et al. 2015; Pauly et al. 2013). UDPGal is also used for the synthesis of galactinol (composed by galactose and *myo*-inositol) in a reaction catalyzed by galactinol synthase (EC 2.4.1.123; Equation 21.21 in Table 21.1 and Figure 21.3) (Liu et al. 1995; Saravitz et al. 1987). Galactinol is then used for Raf synthesis by Raf synthase (EC 2.4.1.82; Equation 21.22 in Table 21.1 and Figure 21.3), which transfers the galactosyl moiety from galactinol to a preformed sucrose molecule (Gangl et al. 2015; Lehle and Tanner 1973; Peterbauer et al. 2002).

The pathway of Raf biosynthesis has been studied to some extent in the model plant *Arabidopsis* (Sengupta et al. 2015). It has been shown that expression of the genes encoding galactinol synthase is modulated by different types of abiotic stress. For instance, the transcripts *AtGolS1* and *AtGolS2* increase when plants are exposed to drought and salinity, whereas *AtGolS3* transcripts increase under cold stress (Taji et al. 2002). Similarly, *Arabidopsis* plants lacking Raf synthase 5 (*rs5-1* and *rs5-2*) fail to accumulate Raf in leaves when exposed to a plethora of abiotic stress conditions, including cold, drought, salinity, oxidation, and heat shock (Egert et al. 2013). Raf metabolism has gained increasing attention in the last decade; however, information on the

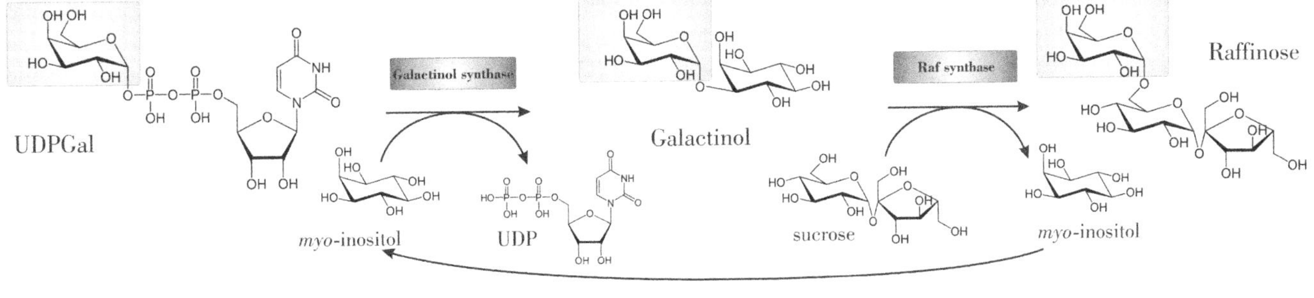

FIGURE 21.3 Raf biosynthesis in plants. Galactinol and Raf synthases are highlighted in violet. The Gal moiety from UDPGal, galactinol, and Raf is colored in green.

kinetic and structural properties of the enzymes involved in Raf biosynthesis is still scarce (Dai et al. 2006; Gangl et al. 2015; Gangl and Tenhaken 2016; Kotake et al. 2007; Li et al. 2007; Peterbauer et al. 2002). Detailed characterization of these enzymes should contribute to better understand the mechanisms underlying the regulation of Raf synthesis, not only in model plants but also in economically important crop species.

21.4 PLANT CARBOHYDRATE METABOLISM AND STRESS

21.4.1 RESPONSES OF GLYCOLYTIC CARBON METABOLISM ENZYMES TO STRESS

21.4.1.1 Hexose-P Metabolism

An early work reported that PPi-PFKase (but, not ATP-PFKase) is induced by anoxia in rice seedlings (Mertens et al. 1990). In this paper, the authors also found an increase in Fru2,6bisP levels. This metabolite helps to boost PPi-PFKase activity in a more acidic environment that could curtail this enzyme's activity. In seedlings, sucrose degradation takes place mainly via uridylates, thus requiring the sucrose synthase/nucleoside diphosphate kinase system and lowering the dependence on adenylates, which usually have lower levels in this condition (Ricard et al. 1991). Pi starvation increases PPi-PFKase levels in *Brassica nigra* and *B. napus* (Theodorou and Plaxton 1994, 1996). *B. nigra* seedlings respond to Pi starvation by increasing the activity of PPi-PFKase (raising the ratio PPi-PFKase:ATP-PFKase), because the amount of the alpha subunit becomes higher in relation to the beta subunit of the enzyme (Theodorou and Plaxton 1994). The same response in the α:β ratio was found in black mustard suspension cells exposed to Pi deficit (Theodorou et al. 1992). In both cases, the sensitivity of PPi-PFKase to Fru2,6bisP is increased. Thus, the response to Pi stress at the metabolic step involving hexose-Ps clearly avoids the use of adenylate (necessary for ATP-PFKase). This is a result of the acquired ability of plants to efficiently exploit the energy available in the anhydride bond of PPi, employing this as a phosphoryl donor (Stitt 1998). Assuming that PPi is a byproduct of anabolism, no ATP is required for the conversion of sucrose to hexose-P via the sucrose synthase pathway in heterotrophic tissues, while two ATPs are needed for the invertase pathway. Remarkable differences are found with respect to the stability and usage of Pi between animal and plant cells. In animals, PPi can be barely conceived as nothing more than a transient byproduct of several cell reactions that is readily hydrolyzed. Conversely, PPi levels in the cytosol of plant cells remain stable through a variety of conditions, including (but not limited to) severe Pi-deficiency or anoxia, which highlights the importance of PPi-dependent enzymes during stress (Plaxton and Podestá 2006; Stitt 1998). The weight of PPi in the plant cell economy has been demonstrated by the generation of transgenic plants expressing a bacterial pyrophosphatase in the cytosol (Jellito et al. 1992). These plants exhibited a threefold decrease in levels of PPi and also severely impaired growth. In contrast,

inhibition of the plastidial inorganic pyrophosphatase by RNAi produced potato tubers with increased ADPGlc levels and lower amounts of starch and amylose (Andersson et al. 2018), indicating the importance of this enzyme for proper synthesis of the polysaccharide.

The link between Pi nutritional status and the regulation of ATP-PFKase and PPi-PFKase is also evident considering that (i) ATP-PFKase is regulated by the Pi:PEP ratio, with Pi behaving as an activator and PEP as an inhibitor of the enzyme, and (ii) Pi is a potent inhibitor of PPi-PFKase in the glycolytic direction of catalysis (Dennis and Blakeley 2000). Under Pi deficiency, PPi-PFKase is active and ATP-PFKase inhibited, and thus, glycolysis proceeds via the adenylate-independent pathway. PPi-PFKase has also been implicated in response to cold stress (Falcone Ferreyra et al. 2006). These authors determined that the quaternary structure of PPi-PFKase in orange fruits changed (with a displacement of the α:β subunit ratio from 1.66 to 1.0) on exposure to frost. After such a modification in subunit composition, the enzyme almost doubled its activity and became more sensitive to Fru2,6bisP.

Cold stress primarily leads to a fall in sucrose synthesis, with accumulation of phosphorylated metabolites and a consequent Pi-limitation of photosynthesis (Stitt and Hurry 2002). Plants cope with this adversity by increasing synthesis of the disaccharide. *Arabidopsis* SPSase is activated by post-transcriptional phosphorylation within minutes after cold stress is initiated (Stitt and Hurry 2002). In the longer term, SPSase and cytosolic Fru1,6bisP phosphatase expression is boosted, releasing the Pi required to sustain photosynthesis at the expense of phosphorylated metabolites. The response is complemented by a movement of Pi from the vacuole to the cytosol, which facilitates the replenishment of phosphorylated metabolites without depleting free Pi (Hurry et al. 2000). Thus, as expressed by Stitt and Hurry (2002), "changes in Pi modulate and may even act as a signal in the regulation of photosynthetic/metabolic acclimation to low temperatures, and lead to major changes in the ability of the different genotypes to develop frost tolerance."

21.4.1.2 PEP metabolism

Levels of PEPCase vary in response to different abiotic stresses. It has been reported that frost-damaged orange fruits contain increased levels of PEPCase, with this enzyme also exhibiting lower sensitivity to its natural feedback inhibitor malate (Falcone Ferreyra et al. 2006). Since the fermentative pathway is higher in fruits, the authors proposed that PEPCase (in combination with malate dehydrogenase) could act as an ancillary fermentative enzyme, contributing to the provision of ATP when aerobic respiration is affected. Vu et al. (1995) showed that cold-hardy citrus varieties responded with an increase in extractable PEPCase activity on acclimation, whereas a sensitive cultivar showed a decrease in foliar levels of the enzyme. Similar trends have been identified in other cultivars exposed to cold, underscoring an as yet little-characterized function of PEPCase in plant primary metabolism. Abiotic stresses that affect water balance also caused an induction of PEPCase expression in wheat seedlings, which was root-specific under anoxia but also observed in shoots under cold stress (González et al. 2003). Overall, the induction of PEPCase has been linked to an enhanced requirement for the production of organic acids (for example, malic acid) in response to cytoplasmic alkalinization among other challenges.

PEPCase activity is also linked to Pi levels, with several studies showing an increase in the content of the enzyme on nutritional deprivation of Pi. The first report on the bypassing of adenylate-using enzymes due to Pi deprivation in *B. nigra* suspension cells detailed that PEPCase, np-Ga3PD-Hase, PPi-PFKase, and PEP phosphatase exhibited important increases in activity at the expense of ATP-PFKase, PyrKase, and Ga3PDHase (Duff et al. 1989). *B. napus* efficiently uses rock Pi by excreting organic acids to the environment with the concomitant solubilization of the former. Pi deprivation in *B. napus* and the related hedge mustard caused an increase in PEPCase activity and (probably as a result of this) in malate exudation (Hoffland et al. 1992). Moraes and Plaxton (2000) reported the purification and properties of the Pi-deficient *B. nigra* PEPCase, which increased by 2.5-fold after starvation and returned to near control levels after re-feeding. The study of the purified enzyme showed that the increased activity under Pi shortage cannot be attributed to an increase in its phosphorylation state (which would make it less susceptible to malate inhibition). It was proposed that at least in these cells, metabolite effects override the importance of phosphorylation in PEPCase control (Moraes and Plaxton 2000). Several reports (Andaluz et al. 2009; Thimm et al. 2001) note the induction of PEPCase in roots under iron deficiency, which follows the apoplastic acidification caused by iron starvation (Thimm et al. 2001). PEPCase induction in roots occurs in parallel to an increase of several glycolytic enzymes, namely, Ga3PDHase, phosphoglyceromutase, enolase, and PyrKase. These changes, together with an enhanced mitochondrial electron transport complement, support a respiratory surge as a response to iron stress.

21.4.2 Soluble Carbohydrates and Their Role against Stress

Many organisms accumulate low–molecular weight compounds such as disaccharides (sucrose and trehalose), Raf, sugar-alcohols, quaternary amines, or amino acids. It has been proposed that these molecules allow organisms to cope with certain abiotic stress conditions (such as salinity, cold, or drought). These compounds were defined as compatible solutes after studies performed with yeasts accumulating non-reducing sugars (such as trehalose) and sugar-alcohols (including glycerol) in response to osmotic stress (Brown and Simpson 1972). It is well known that different inorganic (K^+, Na^+, Cl^-, and SO_4^{2-}) and organic (reducing hexoses) molecules are critical for osmotic adjustments in aqueous systems. However, these compounds are different from compatible solutes, which may affect the properties of solutions in distinctive ways (Bohnert and Jensen 1996; Loescher and Everard

2000). Early works pointed out that high concentrations of compatible solutes have no effect on the *in vitro* activity of enzymes and that they can also protect the protein structure from the deleterious effects of salts or heat (Bohnert and Jensen 1996; Loescher and Everard 2000). For such a protective effect, compatible solutes have to reach concentration values as high as 500 mM, which, interestingly, are levels that can be achieved in cells (Nadwodnik and Lohaus 2008).

Different explanations have been proposed to account for the protective effect of compatible solutes on biological structures. One hypothesis is that these compounds substitute for water molecules in the hydration layer of proteins and membranes, thus allowing functionality of enzymes at extremely low water concentrations (Webb and Bhorjee 1968). Alternatively, it was suggested that compatible solutes might localize outside the hydration sphere of proteins, thus generating a particular rearrangement of the sphere that leads the biological structure to adopt a preferential hydration (Timasheff 1993). Studies performed in plants at the present time strongly suggest that cytosolic carbohydrates derived from different metabolic fates play a role as compatible solutes. The accumulation of these metabolites could enable the plant to overcome extreme conditions of low temperatures, high salt, and water deficit. The whole picture confirms that modifications in the balance of photosynthate partitioning would critically determine plant productivity and survival under physiological or stress conditions.

21.4.2.1 Osmotic Stress Adaptation

A plant's response to osmotic stress generally results in decreased starch and the use of the polysaccharide to synthesize and increase the levels of soluble sugars. This increase in carbohydrate concentration was associated with higher increased activity of the synthesizing enzymes. For example, SPSase activity noticeably increased in osmotically stressed spinach leaves and potato tubers as a consequence of the phosphorylation of a single residue (serine-424 in spinach leaf SPSase) that is different from that involved in dark-light modulation (Winter and Huber 2000). The site of modification is widely conserved among species, and its phosphorylation activates the enzyme, thus allowing sucrose synthesis in osmotically stressed leaves (which is restricted under normal conditions) (Toroser and Huber 1997). The accumulation of sucrose, cyclic or acyclic sugar-alcohols, proline, and quaternary amines (such as glycine-betaine) could potentially play a direct role in osmoregulation and could also quickly provide carbohydrates to be metabolized for energy production when carbon is diverted from growth to other functions (Hare et al. 1998).

It has been suggested that Raf is important for plant tolerance to osmotic stress. Over-expression of the *Thellungiella salsuginea* galactinol synthase gene in *Arabidopsis* significantly increased the content of galactinol and Raf. These plants showed increased germination rates and grew well during early development under osmotic stress (Sun et al. 2013). Comparative expression analyses carried out under various abiotic stresses suggest that several *galactinol synthase* and

Raf synthase genes from sesame (*Sesamum indicum*) are significantly regulated by osmotic, drought, salt, and waterlogging stresses but only slightly affected by cold stress (You et al. 2018). Consequently, galactinol and Raf significantly accumulated under osmotic stress in sesame.

21.4.2.2 Cold Stress

Carbohydrates such as sucrose, Raf, fructans, and sugar-alcohols can be important in tolerance and resistance to cold-induced damage (del Viso et al. 2009; Loescher and Everard 2000; Pontis 1989; Sengupta et al. 2015; Tognetti et al. 1990). Cold injury results by desiccation due to water demand from the protoplast as a consequence of the growing ice crystal (Loescher and Everard 2000). Even when results are not conclusive, many studies have linked cold stress with the accumulation of compatible solutes. For instance, it was found that Gol concentration in apple shoot xylem increased with leaf senescence and low temperatures (Williams and Raese 1974). Similar changes have been described in plum trees, where the highest amount of Gol in sap was found in plants exposed to temperatures below zero (Loescher et al. 1990). Further support for this view was obtained by (Hirai 1983), who reported an increase of Gol and A6PRase levels in loquat leaves during low-temperature seasons. Furthermore, accumulation of sucrose in photosynthetic and heterotrophic tissues has been linked with increased SPSase activity (and level of the protein) in spinach leaves (Guy et al. 1992) and potato tubers (Hill et al. 1996) exposed to non-freezing temperatures. In spinach leaves, the rate of SPSase protein synthesis seems to be responsible for the increase in activity, and the newly produced enzyme subunit appears to be identical to that found under normal conditions (Guy et al. 1992). On the other hand, cold-exposed potato tubers showed an increase in a particular subunit (1b) of SPSase, resulting in an enzyme form with distinct kinetic properties. Thus, the altered kinetics of SPSase may play an important role in the regulation of sucrose synthesis in cold-stored tubers (Hill et al. 1996).

It has been reported that Raf is present in chloroplasts (Schneider and Keller 2009) and that it plays a key role in stabilizing photosystem II in *Arabidopsis* during freeze-thaw cycles (Knaupp et al. 2011). Recently, authors characterized an *Arabidopsis* mutant lacking the gene encoding Raf synthase 5, which is important for plants to accumulate Raf under cold stress, among other types of abiotic stress conditions (Egert et al. 2013; Zuther et al. 2004). Seven genes coding for galactinol synthase were found in the *Arabidopsis* genome. Among these, the expression of three genes was induced by abiotic stresses, and interestingly, only *AtGolS3* was induced by cold stress but not by drought or high salinity (Taji et al. 2002). Saito and Yoshida (2011) found that the expression of two *galactinol synthase* and four *Raf synthase* genes was induced by low temperature in rice (*Oryza sativa*). These authors also showed that chilling treatment for a long period considerably increased the galactinol and Raf content in rice. In this study, Raf levels were higher in leaf blades than in leaf sheaths under cold stress. Similarly, galactinol and Raf levels were significantly higher in harvested, cold-treated peach fruits. Interestingly,

the levels of Raf were well correlated with the degree of mealiness (a symptom of chilling injury) resistance (Bustamante et al. 2016). The expression of two genes encoding galactinol synthase was induced by cold stress but not by drought, heat, or abscisic acid (ABA) treatments in wheat (*Triticum aestivum*). Transgenic lines of rice constitutively over-expressing each of these two *galactinol synthase* genes accumulated significantly higher levels of galactinol and Raf than wild-type plants and exhibited enhanced cold-stress tolerance (Shimosaka and Ozawa 2015).

21.4.2.3 Drought Stress

The relationship between drought stress and the accumulation of carbohydrates is complex, as a result of secondary effects, such as growth inhibition and dehydration (Loescher and Everard 2000). Still, it is important to draw attention to the accumulated evidence. For instance, Gol was the main soluble carbohydrate, and its content increased twofold, in plants from *Prunus* spp. exposed to drought stress (Ranney et al. 1991). In apple trees, drought triggered the preferential accumulation of Gol and Glc at the expense of sucrose and starch (Wang et al. 1995, 1996). Studies with genetically transformed plants demonstrated that the increased ability for synthesizing soluble sugar-alcohols upholds drought tolerance. Transformed tobacco plants synthesizing the non-reducing disaccharide trehalose exhibited enhanced ability to survive after a drought period, although their phenotypes were altered, and the growing rate was diminished up to 50% under normal growing conditions (Holmstrom et al. 1996). A similar work showed that transformed tobacco plants engineered to produce bacterial fructans grew better (giving higher levels of fresh and dry weight) than controls under polyethyleneglycol-induced water stress. On the other hand, compared with those producing trehalose, plants accumulating fructans did not show a difference from controls under normal conditions (Pilon-Smits et al. 1995).

Homologous over-expression of galactinol synthase 2 in *Arabidopsis* led to higher levels of galactinol and Raf with a concomitant reduction of foliar transpiration, which resulted in an increased tolerance to drought stress (Taji et al. 2002). Expression of this enzyme also conferred drought tolerance and increased grain yield in two different rice genotypes under dry field conditions. These transgenic lines displayed higher levels of galactinol than controls. The increased grain yield of transgenic rice under drought conditions was related to a larger number of panicles, grain fertility, and biomass. In this case, the amended drought tolerance was associated with higher relative water content of leaves, higher photosynthetic activity, lower reduction in plant growth, and faster recovering ability (Selvaraj et al. 2017). dos Santos et al. (2011) showed that there are at least three galactinol synthase isoforms related to abiotic stress in *Coffea arabica* trees. The three genes are differentially regulated at the mRNA level in leaves of coffee plants subjected to water deficit. *CaGolS1* was the only isoform highly up-regulated during the whole water deficit period and also after rehydration, while *CaGolS2* and *CaGolS3* were significantly expressed only under severe water deficit. In this study, stachyose levels in coffee leaves were much higher than Raf levels after the water deficit treatment. These changes were even more dramatic under severe stress conditions, which indicates that stachyose probably plays a major function as a protectant against damage caused by drought stress compared with Raf (dos Santos et al. 2011).

21.4.2.4 Salinity-Induced Stress

The correlation between salt stress and sugar accumulation is well established. Studies performed with celery exposed to stressing levels of NaCl (Everard et al. 1994) or macronutrient excess (Stoop and Pharr 1994) determined that changes in mannitol metabolism were the consequence of both carbon partitioning and use. Celery plants exposed to 300 mM NaCl showed similar rates of carbon flux into mannitol to those of controls, although carbon assimilation decreased by 70%. Mannitol synthesis was maintained at the expense of sucrose, thus increasing the ratio between labeled mannitol and sucrose fourfold. This change was linked to higher activity of mannose 6-phosphate reductase (EC 1.1.1.224). This increase in activity was not associated with an increase in the protein level, suggesting a possible post-translational modification of the enzyme (Everard et al. 1994). On the other hand, a study carried out with Japanese persimmon (*Diospyros kaki*, a plant that normally does not produce sugar-alcohols) transformed with the gene encoding for the A6PRase from apple leaves determined that lines capable of accumulating Gol exhibited an increase in tolerance to NaCl-induced stress (Gao et al. 2001).

Raf metabolism has also been linked to high salinity. It has been suggested that osmolytes such as Raf might facilitate water retention in the cytosol and allow Na^+ capture by the vacuole or the apoplast, contributing to the salt compartmentalization process during salt stress (dos Santos et al. 2011). The profiles of three *galactinol synthase* transcripts from *Coffea arabica* exposed to salt stress show a differential response, as observed in *Arabidopsis* plants (Taji et al. 2002). The high levels of *galactinol synthase* transcripts and the increased accumulation of Raf detected in coffee leaves subjected to salinity indicate that this metabolite has a major role in increasing plant tolerance to this type of abiotic stress. It should be noted that transgenic *Arabidopsis* plants that overexpressed a galactinol synthase from *Thellungiella salsuginea* show improved tolerance not only to osmotic but also to high-salinity stress. Compared with wild-type plants, salt-stressed transgenic *Arabidopsis* exhibited higher germination rate, photosynthetic capacity, and seedling growth (Sun et al. 2013).

21.4.2.5 Hydroxyl Radical Scavengers

Reduced stomatal conductance observed under salt and drought stress increases the production of free radicals. There is support for the idea that in fungi and other organisms, sugar-alcohols (such as Gol and mannitol) and cyclic polyols (such as *myo*-inositol) can operate as free radical scavengers (Jennings and Burke 1990; Smirnoff and Cumbes 1989). It was also shown that this effect could be found in stress caused by dehydration (Smirnoff 1993), although support from *in vivo* assays is scarce. Transformation of tobacco plants with a bacterial mannitol 1-phosphate dehydrogenase

(EC 1.1.1.17), directed to chloroplasts by the introduction of a transit peptide, generated a line capable of accumulating mannitol in chloroplasts reaching concentrations up to 100 mM with no alteration of the phenotype or the photosynthetic activity. The accumulation of mannitol in chloroplasts led to increased tolerance to oxidative stress generated by methylviologen. Importantly, the presence of the polyol did not reduce the abundance of reactive oxygen species, but it conferred an additional protection to that already present in non-transformed plants (Shen et al. 1997). Recent studies suggest that like other sugars, Raf and related oligosaccharides could act as antioxidants (ElSayed et al. 2014; Nishizawa et al. 2008). Indeed, transgenic *Arabidopsis* plants expressing galactinol synthases from chickpea (CaGolS1 and CaGolS2) showed significantly increased galactinol and Raf levels. These plants displayed better growth responses than control plants when exposed to heat and oxidative stress. The improved tolerance of transgenic lines was associated with lower accumulation of reactive oxygen species and lipid peroxidation (Salvi et al. 2018).

21.5 CONCLUDING REMARKS

As can be inferred from the many examples detailed in this chapter, carbon metabolism and partitioning critically affect plant productivity under physiological and stress conditions. Evaluation of the potential of plants to adapt in different environments requires understanding the function of key enzymes involved in different metabolic routes operating in plants. Many factors can affect the expression pattern of genes and the activity of enzymes involved in carbohydrate metabolism. In general, the modulation of enzyme activity by metabolites (allosterism) or post-translational modifications (redox or phosphorylation mechanisms) has been relatively well established in sucrose and starch biosynthetic pathways. Conversely, the enzymes implicated in the metabolism of sugar-alcohols and Raf have been scarcely characterized. In this context, efforts should be made to develop accurate protocols to express and purify these enzymes, and such work is currently the main objective of many laboratories. The development of molecular tools to identify the functionality of different enzymes to rationally modify plant metabolism may provide a practical way for handling plant behavior in diverse (and many times adverse) habitats.

ACKNOWLEDGMENTS

This work was supported by Agencia Nacional de Promoción Científica y Tecnológica (PICT 2015 1767 to AAI, PICT 2015 1074 to FEP and PICT 2015 0642 to CMF) and Universidad Nacional del Litoral (CAI+D 2016 to AAI).

REFERENCES

Allard, S. T., M. F. Giraud, and J. H. Naismith. 2001. Epimerases: Structure, function and mechanism. *Cell Mol Life Sci* 58:1650–1665.

Andaluz, S., J. Rodríguez-Celma, A. Abadía, J. Abadía, and A.-F. López-Millán. 2009. Time course induction of several key enzymes in *Medicago truncatula* roots in response to Fe deficiency. *Plant Physiol Biochem* 47:1082–1088.

Andersson, M., H. Turesson, S. Arrivault, et al. 2018. Inhibition of plastid PPase and NTT leads to major changes in starch and tuber formation in potato. *J Exp Bot* 69:1913–1924.

Bacic, A. 2006. Breaking an impasse in pectin biosynthesis. *Proc Natl Acad Sci U S A* 103:5639–5640.

Bachmann, M., P. Matile, and F. Keller. 1994. Metabolism of the raffinose family oligosaccharides in leaves of *Ajuga reptans* L. (Cold acclimation, translocation, and sink to source transition: Discovery of chain elongation enzyme). *Plant Physiol* 105:1335–1345.

Bakrim, N., J. Brulfert, J. Vidal, and R. Chollet. 2001. Phosphoenolpyruvate carboxylase kinase is controlled by a similar signaling cascade in CAM and C4 plants. Biochem Biophys Res Commun 286:1158–1162.

Ballicora, M. A., A. A. Iglesias, and J. Preiss. 2004. ADP-glucose pyrophosphorylase: A regulatory enzyme for plant starch synthesis. *Photosyn Res* 79:1–24.

Baur, B., K. J. Dietz, and K. Winter. 1992. Regulatory protein phosphorylation of phosphoenolpyruvate carboxylase in the facultative crassulacean-acid-metabolism plant *Mesembryanthemum crystallinum* L. *Eur J Biochem* 209:95–101.

Bieleski, R. L., and R. J. Redgwell. 1977. Synthesis of sorbitol in apricot leaves. *Funct Plant Biol* 4:1–10.

Bohnert, H. J., and R. G. Jensen. 1996. Strategies for engineering water-stress tolerance in plants. *Trends Biotechnol* 14:89–97.

Brown, A. D., and J. R. Simpson. 1972. Water relations of sugar-tolerant yeasts: The role of intracellular polyols. *J Gen Microbiol* 72:589–591.

Bustamante, C. A., L. L. Monti, J. Gabilondo, et al. 2016. Differential metabolic rearrangements after cold storage are correlated with chilling injury resistance of peach fruits. *Front Plant Sci* 7:1478.

Bustos, D. M., C. A. Bustamante, and A. A. Iglesias. 2008. Involvement of non-phosphorylating glyceraldehyde-3-phosphate dehydrogenase in response to oxidative stress. *J Plant Physiol* 165:456–461.

Bustos, D. M., and A. A. Iglesias. 2002. Non-phosphorylating glyceraldehyde-3-phosphate dehydrogenase is post- translationally phosphorylated in heterotrophic cells of wheat (*Triticum aestivum*). *FEBS Lett* 530:1–3.

Bustos, D. M., and A. A. Iglesias. 2003. Phosphorylated non-phosphorylating glyceraldehyde-3-phosphate dehydrogenase from heterotrophic cells of wheat interacts with 14-3-3 proteins. *Plant physiology* 133:2081–2088.

Carnal, N. W., and C. C. Black. 1979. Pyrophosphate-dependent 6-phosphofructokinase. A new glycolytic enzyme in pineapple leaves. *Biochem Biophys Res Commun* 86:20–26.

Cunningham, S. M., P. Nadeau, Y. Castonguay, S. Laberge, and J. J. Volenec. 2003. Raffinose and stachyose accumulation, galactinol synthase expression, and winter injury of contrasting alfalfa germplasms. *Crop Science* 43:562–570.

Chen, Z. H., R. P. Walker, L. I. Tecsi, P. J. Lea, and R. C. Leegood. 2004. Phosphoenolpyruvate carboxykinase in cucumber plants is increased both by ammonium and by acidification, and is present in the phloem. *Planta* 219:48–58.

Chollet, R., J. Vidal, and M. H. O'Leary. 1996. Phosphoenolpyruvate carboxylase: A ubiquitous, highly regulated enzyme in plants. *Annu Rev Plant Physiol Plant Mol Biol* 47:273–902.

Dai, N., M. Petreikov, V. Portnoy, et al. 2006. Cloning and expression analysis of a UDP-galactose/glucose pyrophosphorylase from melon fruit provides evidence for the major metabolic pathway of galactose metabolism in raffinose oligosaccharide metabolizing plants. *Plant Physiol* 142:294–304.

Daley, L. S., T. B. Ray, H. M. Vines, and C. C. Black. 1977. Characterization of phosphoenolpyruvate carboxykinase from pineapple leaves *Ananas comosus* (L.) Merr. *Plant Physiol* 59:618–622.

de la Fuente van Bentem, S., D. Anrather, E. Roitinger, et al. 2006. Phosphoproteomics reveals extensive in vivo phosphorylation of Arabidopsis proteins involved in RNA metabolism. *Nucleic Acids Res* 34:3267–3278.

del Viso, F., A. F. Puebla, C. M. Fusari, et al. 2009. Molecular characterization of a putative sucrose:fructan 6-fructosyltransferase (6-SFT) of the cold-resistant Patagonian grass *Bromus pictus* associated with fructan accumulation under low temperatures. *Plant Cell Physiol* 50:489–503.

del Viso, F., A. F. Puebla, H. E. Hopp, and R. A. Heinz. 2009. Cloning and functional characterization of a fructan 1-exohydrolase (1-FEH) in the cold tolerant Patagonian species *Bromus pictus*. *Planta* 231:13–25.

Dennis, D. T., and S. D. Blakeley. 2000. Carbohydrate metabolism. In *Biochemistry and Molecular Biology of Plants*, ed. B. B. Buchanan, W. Gruissem and R. L. Jones, 630–675. Rockville: American Society of Plant Physiologists.

Dormann, P., and C. Benning. 2002. Galactolipids rule in seed plants. *Trends Plant Sci* 7:112–118.

dos Santos, T. B., I. G. Budzinski, C. J. Marur, et al. 2011. Expression of three galactinol synthase isoforms in *Coffea arabica* L. and accumulation of raffinose and stachyose in response to abiotic stresses. *Plant Physiol Biochem* 49:441–448.

Downie, B., S. Gurusinghe, P. Dahal, et al. 2003. Expression of a GALACTINOL SYNTHASE gene in tomato seeds is up-regulated before maturation desiccation and again after imbibition whenever radicle protrusion is prevented. *Plant Physiol* 131:1347–1359.

Duff, S. M. G., G. B. G. Moorhead, D. D. Lefebvre, and W. C. Plaxton. 1989. Phosphate starvation inducible 'bypasses' of adenylate and phosphate dependent glycolytic enzymes in *Brassica nigra* suspension cells. *Plant Physiol* 90:1275–1278.

Egert, A., F. Keller, and S. Peters. 2013. Abiotic stress-induced accumulation of raffinose in Arabidopsis leaves is mediated by a single raffinose synthase (RS5, At5g40390). *BMC Plant Biol* 13:218.

ElSayed, A. I., M. S. Rafudeen, and D. Golldack. 2014. Physiological aspects of raffinose family oligosaccharides in plants: Protection against abiotic stress. *Plant Biol (Stuttg)* 16:1–8.

Everard, J. D., R. Gucci, S. C. Kann, J. A. Flore, and W. H. Loescher. 1994. Gas exchange and carbon partitioning in the leaves of celery (*Apium graveolens* L.) at various levels of root zone salinity. *Plant Physiol* 106:281–292.

Falcone Ferreyra, M. L., V. Perotti, C. M. Figueroa, et al. 2006. Carbohydrate metabolism and fruit quality are affected in frost-exposed Valencia orange fruit. *Physiol Plant* 128:224–236.

Figueroa, C. M., C. V. Piattoni, K. E. J. Trípodi, F. E. Podestá, and A. A. Iglesias. 2016. Carbon photoassimilation and photosynthate partitioning in plants. In *Handbook of Photosynthesis, Third Edition*, ed. M. Pessarakli, 509–535. Boca Raton: CRC Press.

Frey, P. A. 1996. The Leloir pathway: A mechanistic imperative for three enzymes to change the stereochemical configuration of a single carbon in galactose. *FASEB J* 10:461–470.

Gangl, R., R. Behmuller, and R. Tenhaken. 2015. Molecular cloning of AtRS4, a seed specific multifunctional RFO synthase/galactosylhydrolase in *Arabidopsis thaliana*. *Front Plant Sci* 6:789.

Gangl, R., and R. Tenhaken. 2016. Raffinose family oligosaccharides act as galactose stores in seeds and are required for rapid germination of Arabidopsis in the dark. *Front Plant Sci* 7:1115.

Gao, M., R. Tao, K. Miura, A. M. Dandekar, and A. Sugiura. 2001. Transformation of Japanese persimmon (*Diospyros kaki* Thunb.) with apple cDNA encoding NADP-dependent sorbitol-6-phosphate dehydrogenase. *Plant Sci* 160:837–845.

Gao, Z., and W. H. Loescher. 2000. NADPH supply and mannitol biosynthesis. Characterization, cloning, and regulation of the non-reversible glyceraldehyde-3-phosphate dehydrogenase in celery leaves. *Plant Physiol* 124:321–330.

Givan, C. V. 1999. Evolving concepts in plant glycolysis: Two centuries of progress. *Biol Rev Camb Philos Soc* 74:277–309.

Gomez Casati, D. F., J. I. Sesma, and A. A. Iglesias. 2000. Structural and kinetic characterization of NADP-dependent, non-phosphorylating glyceraldehyde-3-phosphate dehydrogenase from celery leaves. *Plant Sci* 154:107–115.

González, M. C., R. Sánchez, and F. J. Cejudo. 2003. Abiotic stresses affecting water balance induce phosphoenolpyruvate carboxylase expression in roots of wheat seedlings. *Planta* 216:985–992.

Grant, C. R., and T. a. Rees. 1981. Sorbitol metabolism by apple seedlings. *Phytochemistry* 20:1505–1511.

Gregory, A. L., B. A. Hurley, H. T. Tran, et al. 2009. *In vivo* regulatory phosphorylation of the phosphoenolpyruvate carboxylase AtPPC1 in phosphate-starved *Arabidopsis thaliana*. *Biochem J* 420:57–65.

Guy, C. L., J. L. Huber, and S. C. Huber. 1992. Sucrose phosphate synthase and sucrose accumulation at low temperature. *Plant Physiol* 100:502–508.

Hajirezaei, M., U. Sonnewald, R. Viola, et al. 1994. Transgenic potato plants with strongly decreased expression of pyrophosphate: Fructose-6 phosphate phosphotransferase show no visible phenotype and only minor changes in metabolic fluxes in their tubers. *Planta* 192:16–30.

Hare, P. D., W. A. Cress, and J. Van Staden. 1998. Dissecting the roles of osmolyte accumulation during stress. *Plant Cell Environ* 21:535–553.

Haritatos, E., F. Keller, and R. Turgeon. 1996. Raffinose oligosaccharide concentrations measured in individual cell and tissue types in *Cucumis melo* L. leaves: Implications for phloem loading. *Planta* 198:614–622.

Hartman, M. D., C. M. Figueroa, D. G. Arias, and A. A. Iglesias. 2017. Inhibition of recombinant aldose-6-phosphate reductase from peach leaves by hexose-phosphates, inorganic phosphate and oxidants. *Plant Cell Physiol* 58:145–155.

Hartman, M. D., C. M. Figueroa, C. V. Piattoni, and A. A. Iglesias. 2014. Glucitol dehydrogenase from peach (*Prunus persica*) fruits is regulated by thioredoxin *h*. *Plant Cell Physiol* 55:1157–1168.

Hartwell, J., A. Gill, G. A. Nimmo, et al. 1999. Phosphoenolpyruvate carboxylase kinase is a novel protein kinase regulated at the level of expression. *Plant J* 20:333–342.

Hill, L. M., R. Reimholz, R. Schröder, T. H. Nielsen, and M. Stitt. 1996. The onset of sucrose accumulation in cold-stored potato tubers is caused by an increased rate of sucrose synthesis and coincides with low levels of hexose-phosphates, an activation of sucrose phosphate synthase and the appearance of a new form of amylase. *Plant Cell Environ* 19:1223–1237.

Hirai, M. 1983. Seasonal changes in sorbitol-6-phosphate dehydropenase in loquat leaf. *Plant Cell Physiol* 24:925–931.

Hoffland, E., R. Van den Boogard, J. Nelemans, and G. Findenegg. 1992. Biosynthesis and root exudation of citric and malic acids in phosphate-starved rape plants. *New Phytol* 122:675–680.

Holmstrom, K.-O., E. Mantyla, B. Welin, et al. 1996. Drought tolerance in tobacco. *Nature* 379:683–684.

Hu, Z., and W. C. Plaxton. 1996. Purification and characterization of cytosolic pyruvate kinase from leaves of the castor oil plant. *Arch Biochem Biophys* 333:298–307.

Hurry, V., A. Strand, R. Furbank, and M. Stitt. 2000. The role of inorganic phosphate in the development of freezing tolerance and the acclimatization of photosynthesis to low temperature is revealed by the *pho* mutants of *Arabidopsis thaliana*. *Plant J* 24:383–396.

Iglesias, A. A., F. E. Podestá, and C. S. Andreo. 1997. Structural and regulatory properties of the enzymes involved in C3, C4 and CAM pathways for photosynthetic carbon assimilation. In *Handbook of Photosynthesis*, ed. M. Pessarakli, 481–503. New York: Marcel Dekker.

Iglesias, A. A., L. R. Vicario, D. F. Gómez-Casati, et al. 2002. On the interaction of substrate analogues with non-phosphorylating glyceraldehyde-3-phosphate dehydrogenase from celery leaves. *Plant Science* 162:689–696.

Jellito, T., U. Sonnewald, L. Willmitzer, M. R. Hajirezaei, and M. Stitt. 1992. Inorganic pyrophosphate content and metabolies in leaves and tubers of potato and tobacco plants expressing *E. coli* pyrophosphatase in their cytosol: Biochemical evidence that sucrose metabolism has been manipulated. *Planta* 188:238–244.

Jennings, D. H., and R. M. Burke. 1990. Compatible solutes - the mycological dimension and their role as physiological buffering agents. *New Phytol* 116:277–283.

Kelly, A. A., and P. Dormann. 2002. DGD2, an arabidopsis gene encoding a UDP-galactose-dependent digalactosyldiacylglycerol synthase is expressed during growth under phosphate-limiting conditions. *J Biol Chem* 277:1166–1173.

Knaupp, M., K. B. Mishra, L. Nedbal, and A. G. Heyer. 2011. Evidence for a role of raffinose in stabilizing photosystem II during freeze-thaw cycles. *Planta* 234:477–486.

Koch, K. 2004. Sucrose metabolism: Regulatory mechanisms and pivotal roles in sugar sensing and plant development. *Curr Opin Plant Biol* 7:235–246.

Kotake, T., S. Hojo, D. Yamaguchi, et al. 2007. Properties and physiological functions of UDP-sugar pyrophosphorylase in Arabidopsis. *Biosci Biotechnol Biochem* 71:761–771.

Kotake, T., D. Yamaguchi, H. Ohzono, et al. 2004. UDP-sugar pyrophosphorylase with broad substrate specificity toward various monosaccharide 1-phosphates from pea sprouts. *J Biol Chem* 279:45728–45736.

Kruger, N. J., and D. T. Dennis. 1987. Molecular properties of pyrophosphate: Fructose-6-phosphate phosphotransferase from potato tuber. *Arch Biochem Biophys* 256:273–279.

Law, R. D., and W. C. Plaxton. 1995. Purification and characterization of a novel phosphoenolpyruvate carboxylase from banana fruit. *Biochem J* 307:807–816.

H.-S. Lee, and L. Copeland. 1996. Phosphofructokinase from the host fraction 69. of chickpea nodules. Planta 96:607–614.

Leegood, R. C., and R. P. Walker. 2003. Regulation and roles of phosphoenolpyruvate carboxykinase in plants. *Arch Biochem Biophys* 414:204–210.

Lehle, L., and W. Tanner. 1973. The function of *myo*-inositol in the biosynthesis of raffinose. Purification and characterization of galactinol:sucrose 6-galactosyltransferase from *Vicia faba* seeds. *Eur J Biochem* 38:103–110.

Leloir, L. F. 1951. The enzymatic transformation of uridine diphosphate glucose into a galactose derivative. *Arch Biochem Biophys* 33:186–190.

Li, S., T. Li, W. D. Kim, et al. 2007. Characterization of raffinose synthase from rice (*Oryza sativa* L. var. Nipponbare). *Biotechnol Lett* 29:635–640.

Lin, M., D. H. Turpin, and W. C. Plaxton. 1989. Pyruvate kinase isozymes from the green alga, *Selenastrum minutum*. I. Purification and physical and immunological characterization. *Arch Biochem Biophys* 269:219–227.

Liu, J. J., W. Odegard, and B. O. de Lumen. 1995. Galactinol synthase from kidney bean cotyledon and zucchini leaf. Purification and N-terminal sequences. *Plant Physiol* 109:505–511.

Loescher, W., and J. Everard. 2000. Regulation of sugar alcohol biosynthesis. In *Photosynthesis: Physiology and metabolism*, ed. R. C. Leegood, T. D. Sharkey and S. von Caemmerer, 275–299. New York: Kluwer Academic Publishers.

Loescher, W. H., T. McCamant, and J. D. Keller. 1990. Carbohydrate reserves, translocation, and storage in woody plant roots. *HortScience* 25:274–281.

Loescher, W. H., R. H. Tyson, J. D. Everard, R. J. Redgwell, and R. L. Bieleski. 1992. Mannitol synthesis in higher plants: Evidence for the role and characterization of a NADPH-dependent mannose 6-phosphate reductase. *Plant Physiol* 98:1396–1402.

Loque, D., H. V. Scheller, and M. Pauly. 2015. Engineering of plant cell walls for enhanced biofuel production. *Curr Opin Plant Biol* 25:151–161.

Malone, S., Z. H. Chen, A. R. Bahrami, et al. 2007. Phosphoenolpyruvate carboxykinase in Arabidopsis: Changes in gene expression, protein and activity during vegetative and reproductive development. *Plant Cell Physiol* 48:441–450.

Martín, M., W. C. Plaxton, and F. E. Podestá. 2007. Activity and concentration of non-proteolyzed phosphoenolpyruvate carboxykinase in the endosperm of germinating castor oil seeds: Effects of anoxia on its activity. *Physiol Plant* 130:484–494.

Martin, M., S. P. Rius, and F. E. Podesta. 2011. Two phosphoenolpyruvate carboxykinases coexist in the Crassulacean Acid Metabolism plant *Ananas comosus*. Isolation and characterization of the smaller 65 kDa form. *Plant Physiol Biochem: PPB* 49:646–653.

Mertens, E., Y. Larondelle, and H. G. Hers. 1990. Induction of pyrophosphate:fructose 6-phosphate 1-phosphotransferase by anoxia in rice seedlings. *Plant Physiol* 93:584–587.

Moraes, T. F., and W. C. Plaxton. 2000. Purification and characterization of phosphoenolpyruvate carboxylase from *Brassica napus* (rapeseed) suspension cell cultures: Implications for phosphoenolpyruvate carboxylase regulation during phosphate starvation, and the integration of glycolysis with nitrogen assimilation. *Eur J Biochem* 267:4465–4476.

Murmu, J., and W. Plaxton. 2007. Phosphoenolpyruvate carboxylase protein kinase from developing castor oil seeds: Partial purification, characterization, and reversible control by photosynthate supply. *Planta* 226:1299–1310.

Nadwodnik, J., and G. Lohaus. 2008. Subcellular concentrations of sugar alcohols and sugars in relation to phloem translocation in *Plantago major, Plantago maritima, Prunus persica*, and *Apium graveolens*. *Planta* 227:1079–1089.

Nielsen, T. H. 1995. Fructose-1,6-bisphosphate is an allosteric activator of pyrophosphate:fructose-6-phosphate 1-phosphotransferase. *Plant Physiol* 108:69–620.

Nishizawa, A., Y. Yabuta, and S. Shigeoka. 2008. Galactinol and raffinose constitute a novel function to protect plants from oxidative damage. *Plant Physiol* 147:1251–1263.

O'Leary, B., J. Park, and W. C. Plaxton. 2011. The remarkable diversity of plant PEPC (phosphoenolpyruvate carboxylase): Recent insights into the physiological functions and post-translational controls of non-photosynthetic PEPCs. *Biochem J* 436:15–34.

Panikulangara, T. J., G. Eggers-Schumacher, M. Wunderlich, H. Stransky, and F. Schoffl. 2004. Galactinol synthase1. A novel heat shock factor target gene responsible for heat-induced synthesis of raffinose family oligosaccharides in Arabidopsis. *Plant Physiol* 136:3148–3158.

Pauly, M., S. Gille, L. Liu, et al. 2013. Hemicellulose biosynthesis. *Planta* 238:627–642.

Peterbauer, T., L. Mach, J. Mucha, and A. Richter. 2002. Functional expression of a cDNA encoding pea (*Pisum sativum* L.) raffinose synthase, partial purification of the enzyme from maturing seeds, and steady-state kinetic analysis of raffinose synthesis. *Planta* 215:839–846.

Peters, S., and F. Keller. 2009. Frost tolerance in excised leaves of the common bugle (*Ajuga reptans* L.) correlates positively with the concentrations of raffinose family oligosaccharides (RFOs). *Plant Cell Environ* 32:1099–1107.

Piattoni, C. V., D. M. Bustos, S. A. Guerrero, and A. A. Iglesias. 2011. Nonphosphorylating glyceraldehyde-3-phosphate dehydrogenase is phosphorylated in wheat endosperm at serine-404 by an SNF1-related protein kinase allosterically inhibited by ribose-5-phosphate. *Plant Physiol* 156:1337–1350.

Piattoni, C. V., D. M. L. Ferrero, I. Dellaferrera, A. Vegetti, and A. A. Iglesias. 2017. Cytosolic glyceraldehyde-3-phosphate dehydrogenase is phosphorylated during seed development. *Front Plant Sci* 8:522.

Piattoni, C. V., S. A. Guerrero, and A. A. Iglesias. 2013. A differential redox regulation of the pathways metabolizing glyceraldehyde-3-phosphate tunes the production of reducing power in the cytosol of plant cells. *Int J Mol Sci* 14:8073–8092.

Pilon-Smits, E., M. Ebskamp, M. J. Paul, et al. 1995. Improved performance of transgenic fructan-accumulating tobacco under drought stress. *Plant Physiol* 107:125–130.

Plaxton, W. C. 1989. Molecular and immunological characterization of plastid and cytosolic pyruvate kinase isozymes from castor-oil-plant endosperm and leaf. *Eur J Biochem* 181:443–451.

Plaxton, W. C. 1996. The organization and regulation of plant glycolysis. *Annu Rev Plant Physiol Plant Mol Biol* 47:185–214.

Plaxton, W. C., and F. E. Podestá. 2006. The functional organization and control of plant respiration. *Crit Rev Plant Sci* 25:159–198.

Podestá, F. E. 2004. Glycolysis. In *Encyclopedia of Plant and Crop Science*, ed. T. Goodman, 547–550. New York: Marcel Dekker.

Podestá, F. E., and W. C. Plaxton. 1991. Kinetic and regulatory properties of cytosolic pyruvate kinase from germinating castor oil seeds. *Biochem J* 279:495–501.

Podestá, F. E., and W. C. Plaxton. 1994a. Regulation of carbon metabolism in germinating *Ricinus communis* cotyledons. I. Developmental profiles for the activity, concentration, and molecular structure of pyrophosphate- and ATP-dependent phosphofructokinases, phosphoenolpyruvate carboxylase, and pyruvate kinase. *Planta* 194:374–380.

Podestá, F. E., and W. C. Plaxton. 1994b. Regulation of carbon metabolism in germinating *Ricinus communis* cotyledons. II. Properties of phosphoenolpyruvate carboxylase and cytosolic pyruvate kinase associated with the regulation of glycolysis and nitrogen assimilation. *Planta* 194:406–417.

Podestá, F. E., and W. C. Plaxton. 2003. Ligand binding to potato tuber pyrophosphate-dependent phosphofructokinase studied through intrinsic fluorescence quenching. Evidence of competitive binding among fructose-1,6-bisphosphate and fructose-2,6-bisphosphate. *Arch Biochem Biophys* 414:101–107.

Pontis, H. G. 1989. Fructans and cold stress. *J Plant Physiol* 134:148–150.

Ranney, T. G., N. L. Bassuk, and T. H. Whitlow. 1991. Osmotic adjustment and solute constituents in leaves and roots of water-stressed cherry (*Prunus*) trees. *J Amer Soc Hort Sci* 116:684–688.

Ricard, B., J. Rivoal, A. Spiteri, and A. Pradet. 1991. Anaerobic stress induces the transcription and translation of sucrose synthase in rice. *Plant Physiol* 95:669–674.

Rius, S. P., P. Casati, A. A. Iglesias, and D. F. Gomez-Casati. 2006. Characterization of an *Arabidopsis thaliana* mutant lacking a cytosolic non-phosphorylating glyceraldehyde-3-phosphate dehydrogenase. *Plant Mol Biol* 61:945–957.

Ruiz-Ballesta, I., A. B. Feria, H. Ni, et al. 2014. *In vivo* monoubiquitination of anaplerotic phosphoenolpyruvate carboxylase occurs at Lys624 in germinating sorghum seeds. *J Exp Bot* 65:443–451.

Rumpho, M. E., G. E. Edwards, and W. H. Loescher. 1983. A pathway for photosynthetic carbon flow to mannitol in celery leaves : Activity and localization of key enzymes. *Plant Physiol* 73:869–873.

Rylott, E. L., A. D. Gilday, and I. A. Graham. 2003. The gluconeogenic enzyme phosphoenolpyruvate carboxykinase in Arabidopsis is essential for seedling establishment. *Plant Physiol* 131:1834–1842.

Saito, M., and M. Yoshida. 2011. Expression analysis of the gene family associated with raffinose accumulation in rice seedlings under cold stress. *J Plant Physiol* 168:2268–2271.

Salvi, P., N. U. Kamble, and M. Majee. 2018. Stress-inducible galactinol synthase of chickpea (CaGolS) is implicated in heat and oxidative stress tolerance through reducing stress-induced excessive reactive oxygen species accumulation. *Plant Cell Physiol* 59:155–166.

Saravitz, D. M., D. M. Pharr, and T. E. Carter. 1987. Galactinol synthase activity and soluble sugars in developing seeds of four soybean genotypes. *Plant Physiol* 83:185–189.

Schneider, M., J. Knuesting, O. Birkholz, J. J. Heinisch, and R. Scheibe. 2018. Cytosolic GAPDH as a redox-dependent regulator of energy metabolism. *BMC Plant Biol* 18:184.

Schneider, T., and F. Keller. 2009. Raffinose in chloroplasts is synthesized in the cytosol and transported across the chloroplast envelope. *Plant Cell Physiol* 50:2174–2182.

Selvaraj, M. G., T. Ishizaki, M. Valencia, et al. 2017. Overexpression of an *Arabidopsis thaliana* galactinol synthase gene improves drought tolerance in transgenic rice and increased grain yield in the field. *Plant Biotechnol J* 15:1465–1477.

Sengupta, S., S. Mukherjee, P. Basak, and A. L. Majumder. 2015. Significance of galactinol and raffinose family oligosaccharide synthesis in plants. *Front Plant Sci* 6:656.

Shen, B., R. G. Jensen, and H. J. Bohnert. 1997. Increased resistance to oxidative stress in transgenic plants by targeting mannitol biosynthesis to chloroplasts. *Plant Physiol* 113:1177–1183.

Shimosaka, E., and K. Ozawa. 2015. Overexpression of cold-inducible wheat galactinol synthase confers tolerance to chilling stress in transgenic rice. *Breed Sci* 65:363–371.

Smirnoff, N. 1993. The role of active oxygen in the response of plants to water deficit and desiccation. *New Phytol* 125:27–58.

Smirnoff, N., and Q. J. Cumbes. 1989. Hydroxyl radical scavenging activity of compatible solutes. *Phytochemistry* 28:1057–1060.

Smith, C. R., V. L. Knowles, and W. C. Plaxton. 2000. Purification and characterization of cytosolic pyruvate kinase from *Brassica napus* (rapeseed) suspension cell cultures: Implications for the integration of glycolysis with nitrogen assimilation. *Eur J Biochem* 267:4477–4485.

Stitt, M. 1998. Pyrophosphate as an alternative energy donor in the cytosol of plant cells: An enigmatic alternative to ATP. *Bot Acta* 111:167–175.

Stitt, M., and V. Hurry. 2002. A plant for all seasons: Alterations in photosynthetic carbon metabolism during cold acclimation in Arabidopsis. *Curr Opin Plant Biol* 5:199–206.

Stitt, M., J. Lunn, and B. Usadel. 2010. Arabidopsis and primary photosynthetic metabolism - more than the icing on the cake. *Plant J* 61:1067–1091.

Stoop, J., and D. M. Pharr. 1994. Mannitol metabolism in celery stressed by excess macronutrients. *Plant Physiol* 106:503–511.

Sun, Z., X. Qi, Z. Wang, et al. 2013. Overexpression of TsGOLS2, a galactinol synthase, in *Arabidopsis thaliana* enhances tolerance to high salinity and osmotic stresses. *Plant Physiol Biochem* 69:82–89.

Taji, T., C. Ohsumi, S. Iuchi, et al. 2002. Important roles of drought- and cold-inducible genes for galactinol synthase in stress tolerance in *Arabidopsis thaliana*. *Plant J* 29:417–426.

Theodorou, M. E., F. A. Cornel, S. M. G. Duff, and W. C. Plaxton. 1992. Phosphate starvation inducible synthesis of the a-subunit of pyrophosphate-dependent phosphofructokinase in black mustard suspension cells. *J Biol Chem* 267:21901–21905.

Theodorou, M. E., and W. C. Plaxton. 1994. Induction of PPi-dependent phosphofructokinase by phosphate starvation in seedling of *Brassica nigra*. *Plant Cell Environ* 17:287–294.

Theodorou, M. E., and W. C. Plaxton. 1996. Purification and characterization of pyrophosphate-dependent phosphofructokinase from phosphate-starved *Brassica nigra* suspension cells. *Plant Physiol* 112:343–351.

Thimm, O., B. Essigmann, S. Kloska, T. Altmann, and T. J. Buckhout. 2001. Response of Arabidopsis to iron deficiency stress as revealed by microarray analysis. *Plant Physiol* 127:1030–1043.

Timasheff, S. N. 1993. The control of protein stability and association by weak interactions with water: How do solvents affect these processes? *Annu Rev Biophys Biomol Struct* 22:67–97.

Tognetti, J. A., C. L. Salerno, M. D. Crespi, and H. G. Pontis. 1990. Sucrose and fructan metabolism of different wheat cultivars at chilling temperatures. *Physiol Plant* 78:554–559.

Toroser, D., and S. C. Huber. 1997. Protein phosphorylation as a mechanism for osmotic-stress activation of sucrose-phosphate synthase in spinach leaves. *Plant Physiol* 114:947–955.

Trípodi, K. E. J., and W. C. Plaxton. 2005. *In vivo* regulatory phosphorylation of novel phosphoenolpyruvate carboxylase isoforms in endosperm of developing castor oil seeds. *Plant Physiol* 139:969–978.

Trípodi, K. E. J., and F. E. Podestá. 1997. Purification and structural and kinetic characterization of the pyrophosphate: Fructose-6-phosphate 1-phosphotransferase from the Crassulacean Acid Metabolism plant, pineapple. *Plant Physiol* 113:779–786.

Tsiatsiani, L., E. Timmerman, P. J. De Bock, et al. 2013. The Arabidopsis metacaspase9 degradome. *Plant Cell* 25:2831–2847.

Turner, W., V. Knowles, and W. Plaxton. 2005. Cytosolic pyruvate kinase: Subunit composition, activity, and amount in developing castor and soybean seeds, and biochemical characterization of the purified castor seed enzyme. *Planta* 222:1051–1062.

Turner, W. L., and W. C. Plaxton. 2000. Purification and characterization of cytosolic pyruvate kinase from banana fruit. *Biochem J* 352 Pt 3:875–882.

Uhrig, R. G., Y. M. She, C. A. Leach, and W. C. Plaxton. 2008. Regulatory monoubiquitination of phosphoenolpyruvate carboxylase in germinating castor oil seeds. *J Biol Chem* 283:29650–29657.

Volk, G. M., E. E. Haritatos, and R. Turgeon. 2003. Galactinol synthase gene expression in melon. *J Am Soc Hortic Sci* 128:8–15.

Vu, J. C. V., S. K. Gupta, G. Yelenosky, and M. S. B. Ku. 1995. Cold-induced changes in ribulose 1,5-bisphosphate carboxylase-oxygenase and phosphoenolpyruvate carboxylase in citrus. *Environ Exp Bot* 35:25–31.

Walker, R. P., and R. C. Leegood. 1996. Phosphorylation of phosphoenolpyruvate carboxykinase in plants. Studies in plants with C4 photosynthesis and Crassulacean acid metabolism and in germinating seeds. *Biochem J* 317:653–658.

Wang, Z., B. Quebedeaux, and G. W. Stutte. 1995. Osmotic adjustment: Effect of water stress on carbohydrates in leaves, stems and roots of apple. *Funct Plant Biol* 22:747–754.

Wang, Z., B. Quebedeaux, and G. W. Stutte. 1996. Partitioning of [14C]glucose into sorbitol and other carbohydrates in apple under water stress. *Funct Plant Biol* 23:245–251.

Webb, S. J., and J. S. Bhorjee. 1968. Infrared studies of DNA, water, and inositol associations. *Can J Biochem* 46:691–695.

Williams, M. W., and J. T. Raese. 1974. Sorbitol in tracheal Sap of apple as related to temperature. *Physiol Plant* 30:49–52.

Wingler, A., R. P. Walker, Z. H. Chen, and R. C. Leegood. 1999. Phosphoenolpyruvate carboxykinase is involved in the decarboxylation of aspartate in the bundle sheath of maize. *Plant Physiol* 120:539–546.

Winter, D., B. Vinegar, H. Nahal, et al. 2007. An "Electronic Fluorescent Pictograph" browser for exploring and analyzing large-scale biological data sets. *PLoS One* 2:e718.

Winter, H., and S. C. Huber. 2000. Regulation of sucrose metabolism in higher plants: Localization and regulation of activity of key enzymes. *CRC Crit Rev Plant Sci* 19:31–67.

You, J., Y. Wang, Y. Zhang, et al. 2018. Genome-wide identification and expression analyses of genes involved in raffinose accumulation in sesame. *Sci Rep* 8:4331.

Zhou, R., L. Cheng, and R. Wayne. 2003. Purification and characterization of sorbitol-6-phosphate phosphatase from apple leaves. *Plant Sci* 165:227–232.

Zuther, E., K. Buchel, M. Hundertmark, et al. 2004. The role of raffinose in the cold acclimation response of *Arabidopsis thaliana*. *FEBS Lett* 576:169–173.

22 Protein Synthesis by Plants Under Stressful Conditions

Pallavi Sharma and R. S. Dubey

CONTENTS

22.1 INTRODUCTION

Plants being sessile are continuously exposed to environmental stresses such as drought, salinity, cold, and hot temperatures which present major challenges in achieving sustainable food production. Plants are unable to express their full genetic potential for production in stressful environments (Zeigler, 1990; Gao et al., 2007; Blum, 2017). When subjected to stresses, a complex reaction is triggered in plants involving a wide variety of physiological and biochemical responses to overcome, avoid, or nullify the effects of stresses. Tolerance or sensitivity of plants toward a particular stressful condition depends on the genetic and biochemical make-up of the species. During environmental stresses, plants recognize the stress stimulus via various sensors which in turn activate a signal transduction cascade. Plant hormones, signal transducers, secondary messengers, and transcription regulators relay the signal, activating stress-responsive genes (Zhu, 2001; Kaur and Gupta, 2005; Sharma and Dubey, 2007; Kumar et al., 2008; Guo et al., 2009; Gupta et al., 2009; Cvikrová et al., 2013; Danquah et al., 2014; Gilroy et al., 2018). Several

signals thus converge to regulate stress-inducible genes that encode proteins and enzymes. The expression of these genes causes the accumulation or depletion of certain metabolites, alteration in the activity behaviors of many enzymes, overall changes in protein synthesis, and, of particular interest, synthesis of novel proteins specific to particular stress, which directly participate in stress metabolism and contribute to the specificity of the adaptive response under stress (Guo et al., 2009; Casaretto et al., 2016).

In plants, stress-induced alteration in protein synthesis has been suggested to provide evolutionary value to the plants for enhanced survival in adverse environmental situations. The synthesis of such stress-induced proteins and the enhanced levels of enzyme activity have been well documented under salinity stress (Singh et al., 1987; Naot et al., 1995; Aarati et al., 2003; Mahmoodzadeh, 2009; Cui et al., 2015; Xiong et al., 2017), osmotic stress or drought (Jiang and Huang, 2002; Demirevska et al., 2008; Khurana et al., 2008; Prinsi et al., 2018), heat shock (Heikkila et al., 1984; Mansfield and Key, 1988; Vierling, 1991; Cordewener et al., 1995; Waters et al., 1996; Lee et al., 2007; Yildiz and Terzi, 2008; Usman et al., 2014; Ungelenk et al., 2016), low-temperature treatment (Meza-Basso et al., 1986; Hahn and Walbot, 1989; Bruggemann et al., 1994; Griffith et al., 1997; Matsuba et al., 1997; John et al., 2009; Kikuchi and Masuda, 2009; Dalen et al., 2015; Wang et al., 2016), anaerobiosis (Ricard et al., 1991; Christopher and Good, 1996; Kato-Noguchi, 2000; Subbaiah and Sachs, 2003; Kamal et al., 2015; Loreti et al., 2016; Wang et al., 2017), infection with pathogens (Antoniw et al., 1980; Ohashi and Matsuoka, 1985; Abad et al., 1996; Herbers et al., 1996; Tornero et al., 1997; Almagro et al., 2009; Ebrahim et al., 2011; Jiang et al., 2015), wounding (Cabello et al., 1994; Jung et al., 1995; Schaller and Ryan, 1996; Jimenez et al., 2008; Dafoe et al., 2009; Sarkar et al., 2015; Feng et al., 2016), metal toxicity (Choi et al., 1995; Shah and Dubey, 1998a; Sharma and Dubey, 2007; Júnior et al., 2015; Zayneb et al., 2017), gaseous pollutants (Kirtikara and Talbot, 1996; Choudhary and Agrawal, 2014), and ultraviolet (UV) radiation (Jung et al., 1995; Rao et al., 1996; Xu et al., 2008; Pan et al., 2009; Pascual et al., 2017). Figure 22.1 shows the schematic diagram of the synthesis of proteins under different stresses in plants.

Depending on their severity and duration, different stresses cause differential expression of genes and synthesis of new proteins which in turn lead to alteration in protein contents and thus increased survival of the plants (Ben-Hayyim et al., 1989; Pandey et al., 2010). Under salinity stress, it is suggested that the newly synthesized proteins, together with amino acids and soluble nitrogenous compounds, act as components of the salt-tolerance mechanism. These might function as compatible cytoplasmic solutes in osmotic adjustment in order to equalize the osmotic potential of the cytoplasm with the vacuoles in adverse conditions of salinity (Greenway and Munns, 1980; Dubey and Rani, 1989). Under anaerobic stress, the polypeptides which are synthesized, have specific functions and belong to the enzymes of sugar phosphate metabolism (Ricard et al., 1991). Heat-shock proteins (HSPs), which are synthesized under heat stress, possibly assist in protein folding, protein–protein interactions, and the translocation of proteins across cellular compartments, and they have a possible role in protecting the organisms from heat stress (Cordewener et al., 1995; Wang et al., 2004; McLoughlin et al., 2016). Similarly, the pathogenesis-related proteins do act in the defense of the plants and have a putative role in pathogen resistance (Artlip and Funkhouser, 1995; Jiang et al., 2015; Han et al., 2017). Many novel proteins which are synthesized under specific stresses have been well characterized, and their physico-chemical parameters have been determined (Ben-Hayyim et al., 1989; Robertson and Chandler, 1994; Sachs et al., 1996; Waters et al., 1996; Efeoglu and Terzioglu, 2007; Huang and Xu, 2008; Oh and Komatsu, 2015), and data regarding association characteristics and amino acid sequences have been reported (Singh et al., 1987; Badur et al., 1994; Naot et al., 1995; Zhang et al., 1996).

Studies related to the stress-induced synthesis of proteins have been performed using cultured plant cells (Ericson and Alfinito, 1984; Singh et al., 1987; Ben-Hayyim et al., 1989; Vance et al., 1990; Sobkowiak and Deckert, 2006), seedlings (Mansfield and Key, 1988; Hahn and Walbot, 1989; Han and Kermode, 1996; Igarashi et al., 1997; Efeoglu and Terzioglu, 2007; Cui et al., 2015; Wan et al., 2017), excised plant organs (Hurkman and Tanaka, 1987; Stuiver et al., 1988), and intact plants (Burke et al., 1985; Ohashi and Matsuoka, 1985; Kee and Nobel, 1986; Popova et al., 1995). Among these systems, cultured plant cells have proven to be superior to other systems as they show a uniform response and are under better control of environmental parameters (Ben-Hayyim et al., 1989; Fadzilla et al., 1997). Cell cultures from tobacco, cowpea, potato, citrus, and many other plant species have been used to identify and characterize newly synthesized proteins under salinity, heat-shock, freezing, osmotic, and heavy-metal stresses (Singh et al., 1987; Ben-Hayyim et al., 1989; Vierling, 1991; Naot et al., 1995; Fadzilla et al., 1997).

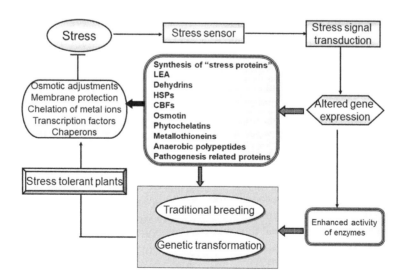

FIGURE 22.1 Stress-induced protein synthesis in plants. Stresses cause important modifications in gene expression in plants which lead to accumulation of stress-related proteins and enhanced activity of enzymes. These proteins provide enhanced survival to plants under adverse environmental situations and can be used to make abiotic stress-tolerant plants by plant breeding or genetic transformation.

Different methods such as electrophoresis, western analysis, enzyme kinetics studies, etc. are employed to measure quantitative and qualitative changes in proteins present in stressed plants. Currently, a high-throughput method for identification and characterization of stress-induced protein is possible using various proteomics tools. Improved extraction and protein purification protocols and sequence databases for peptide mass matches are available (Timperio et al., 2008). Recent progress in sensitivity and accuracy for proteome analysis has made it easier to identify and characterize novel proteins and to determine the alteration in protein types and their expression under stressful conditions (Qureshi et al., 2007; Abreu et al., 2013; Barkla et al., 2013; Ngara and Ndimba, 2014). Proteomic approaches are regarded as the best available tools for analyzing protein functions in stressed plants. Two-dimensional polyacrylamide gel electrophoresis (2DE) has been widely used for the separation and characterization of proteins from complex mixtures. Laser capture microdissection is a promising method for isolating particular cell types under microscope with the help of a laser beam (Longuespée et al., 2014), whereas free flow electrophoresis can be used to isolate pure membrane fractions and/or organelles including plasma membranes, tonoplast, mitochondria and Golgi apparatus (Barkla et al., 2013). A wealth of literature is available, dealing with environmental stress-induced proteome changes, based on whole tissue/organ analysis (Bae et al., 2003; Giacomelli et al., 2006; Goulas et al., 2006; Taylor et al., 2009; Komatsu and Hossain, 2013; Prinsi et al., 2018).

Environmental stresses adversely affect plant protein metabolism and induce the synthesis of many novel stress-specific proteins. These proteins might be involved in signal transduction, anti-freezing, metal-binding, heat shock, anti-pathogenesis, antioxidative defense, or osmolyte synthesis (Qureshi et al., 2007; Cui et al., 2015; Drira et al., 2016; Zayneb et al., 2017). Proteomic analysis has shown different stress responses in tolerant and sensitive genotypes (Ashoub et al., 2015; Hao et al., 2015; Prinsi et al., 2018). Many reports on the proteomics analysis of plants exposed to combined stresses are also available. These include humidity and high temperature in *Portulaca oleracea* (Yang et al., 2012), ozone and drought in poplar (Bohler et al., 2013), salinity and mercury in *Suaeda salsa* (Liu et al., 2013), heat and drought in barley, *Carissa spinarum* and *Arabidopsis thaliana* (Koussevitzky et al., 2008; Rollins et al., 2013), and salinity and drought in *Brassica oleracea* (Sahin et al., 2018). These analyses performed in various plant species exposed to one or different combinations of stresses show different protein synthetic responses (Koussevitzky et al., 2008; Suzuki et al., 2014). An overview of protein synthetic responses in plants and alterations in the levels of key enzymes under various stresses is presented in Figure 22.2. These proteins that have been discussed in detail in the following sections can be successfully used as

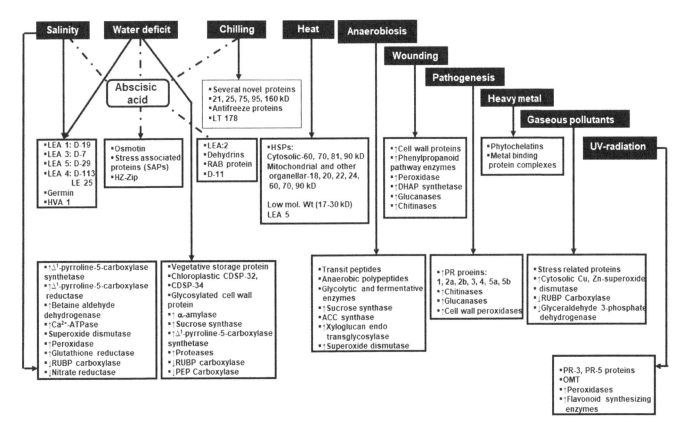

FIGURE 22.2 An overview of stress-induced protein synthetic responses in plants. Different stresses induce the synthesis of various groups of proteins and cause either elevation (↑) or decline (↓) in the levels of enzymes. Some of the responses of salinity, water deficit, and chilling are common and are mediated via elevated levels of abscisic acid. For details, see text.

attractive targets to produce stress-tolerant plants using bio-technological approaches (Table 22.1). The identification and characterization of novel stress-induced proteins provide new insights into plant stress responses. These proteins, which are stress-specific, present newer avenues for improving the stress tolerance of plants. It is well established that stress proteins are synthesized in the tissues of plants adapted to stresses; how-ever, specific metabolic functions for most of these proteins and mechanisms by which they confer adaptability toward stresses are still unknown (Ben-Hayyim et al., 1989; Artlip and Funkhouser, 1995; Khurana et al., 2008). Understanding the function and mechanism of their action will be helpful in the development of stress-tolerant plants using the traditional breeding or transgenic approaches. This chapter presents our current status of knowledge regarding the effect of vari-ous environmental stresses on the overall aspects of protein

TABLE 22.1

Transgenic Stress-Tolerant Plants Which Have Been Produced Using Stress-Related Genes/Proteins

Stress Protein	Gene	Stress	Plant	Reference
Late-embryogenesis abundant	OsLEA-3	Salt	Rice	Hu (2008)
proteins	AtLEA14	Salt	Arabidopsis	Jia et al. (2014)
	OsLEA4	Salt	Rice	Hu et al. (2016)
	SiLEA14	Osmotic	Arabidopsis, Foxtail	Wang et al. (2014b)
	HVA1	Drought	millet	Xu et al. (1996)
	OsEm1	Drought	Rice	Dalal et al. (2009)
	BnLEA4-1	Drought	Rice	Yu et al. (2016)
			Arabidopsis	
Dehydrins	OsDhn1	Salt, drought	Rice	Kumar et al. (2014)
	CsLEA	Salt	Alfalfa	Zhang et al. (2016)
	Dhn5	Salt	Arabidopsis	Drira et al. (2016)
	Dhn24	Chilling	Cucumber	Yin et al. (2006)
	Dhn5	Drought	Arabidopsis	Brini et al. (2007)
Heat-shock proteins	sHSP17.7	Drought	Rice	Sato and Yokoya, (2008)
	NtHSP70-1	Drought	Tobacco	Cho and Choi, (2009)
	CaHSP16.4	Drought	Arabidopsis	Huang et al., (2018)
	AtHsp90.2,	Salt, drought	Arabidopsis	Song et al., (2009)
	AtHsp90.5,	Heat stress	Tobacco	Cho and Choi (2009)
	AtHsp90.7	Heat stress	Arabidopsis	McLoughlin et al. (2016)
	NtHSP70-1			
	sHSPs			
CBF/DREB	DREB1A	Drought	Tobacco	Kasuga et al. (2004)
	HvCBF4	Low temperature	Rice	Oh et al. (2007)
	OsDREB1B	Drought, high-salt and low-temperature stress	Tobacco	Gutha and Reddy, (2008)
Osmotin	SindOLP	Salt	Sesame	Chowdhury et al., (2017)
	TlOsm	Salt	Rice	Le et al. (2018)
	NtOsm	Drought	Carrot	Annon et al. (2014)
	SindOLP	Drought	Sesame	Chowdhury et al. (2017)
	SnOLP	Drought	Soybean	Weber et al. (2014)
	NtOsm	Chilling	Tomato	Patade et al., (2013)
Phytochelatin	PdPCS1	Cd, Cr and Cu	Date palm	Zayneb et al. (2017)
	MnPCSs	Zn, Cd	Tobacco and	Fan et al. (2018)
			Arabidopsis	
Metallothionein	BjMT2	Cu, Cd	Arabidopsis	An et al. (2006)
	BcMT1 and	Cu, Cd	Arabidopsis	Lv et al. (2013)
	BcMT2 SpMTL	Cd	Sedum plumbizincicola	Peng et al. (2017)
Pathogenesis-related proteins	OgChitIVa	Botrytis cinerea	Arabidopsis	Pak et al. (2009)
	PR-5	Alternaria	Tobacco	Velazhahan and Muthukrishnan, (2003)
	SmPR10	alternate	Arabidopsis	Han et al. (2017)
	GmPRP	Salinity	Tobacco	Jiang et al. (2015)
		Phytophthora nicotianae		
Antifreeze protein	AFP	Freezing stress	Tobacco	Worrall et al. (1998); Fan et al. (2002)

synthesis in plants and the possible role of stress-specific proteins in conferring an enhanced survival value to the plants against various environmental stressful situations.

22.2 SALINITY

Soil salinity is a major environmental stress that significantly influences plant growth and productivity (Munns, 2002). It is a threat to crop cultivation in arid and semi-arid areas of the world. The build-up of salinity in soil has made millions of hectares of land unusable for cultivation. It is estimated that every year more than a million hectares of land is subjected to salinization. Soil salinity is hence threatening civilization by continuously decreasing the areas of crop cultivation. Salinity not only causes great losses in crop yield but also has an impact on other economic, environmental, social, and political problems in the affected countries. Salinity affects seed germination, nutrient uptake, metabolism, and plant growth owing to osmotic inhibition of water availability, ion imbalance, toxic effects of salt ions, and its effects on cellular gene expression and protein synthesis machinery (Dubey and Pessarakli, 1995; Abdelkader et al., 2007; Parihar et al., 2015). When exposed to salinity stress, plants employ strategies to combat the stress by overexpression and downexpression of ceratin proteins for maintaining proper osmotic and cellular ion homeostasis in the cell, for sustaining cellular metabolism, to meet antioxidative defense demand and the processes of detoxification and repair (Munns, 2002; Chinnusamy et al., 2005; Abdelkader et al., 2007). Salinity causes either a decrease (Popova et al., 1995) increases (Dubey, 1983; Elenany, 1997) in the level of total and/or soluble proteins. Depending on the plant parts studied, salinity leads to increased activity/synthesis of many enzymes (Dubey et al., 1987; Dubey and Rani, 1987; Dubey and Sharma, 1990; Mittal and Dubey, ; Mittal and Dubey, 1995; Igarashi et al., 1997; Mishra et al., 2013) and induces the synthesis of salt stress-specific proteins (Ben-Hayyim et al., 1989; Artlip and Funkhouser, 1995; Mahmoodzadeh, 2009).

22.2.1 SALT-INDUCED PROTEIN SYNTHESIS

Plants growing in a saline environment show distinct alterations in the pattern of synthesis and accumulation of proteins. Initially, most of the experiments on the salinity-induced synthesis of proteins have been conducted using plant cell cultures. Cell cultures rather than whole plant systems have proven to be more advantageous for such studies, because, in cell cultures, environmental parameters can be better controlled, and the stress-tolerant cell lines generated can be readily selected and assayed for newly synthesized proteins. Several investigators have shown salinity-induced production of novel proteins in cultured plant cells (Ericson and Alfinito, 1984; Singh et al., 1985, 1987; Ben-Hayyim et al., 1989; Artlip and Funkhouser, 1995; Elenany, 1997). In tobacco cells, enhancement in the level of certain proteins and a decline in the content of others is observed when cells are adapted to NaCl (Singh et al., 1987). Tomato cell cultures when grown in a medium with 25 mM NaCl and proline, synthesize extra

polypeptides of 190, 58, 45, and 26 kDa, and with the further increase of NaCl in the medium a new 67 kDa protein gets accumulated (Elenany, 1997). Yildiz (2007) observed the synthesis of two new low-molecular-weight (LMW) proteins (30.0 and 28.9 kDa) and one protein with an intermediate molecular weight (44.3 kDa) in response to NaCl treatment in wheat cultivar Ceyhan-99 (salt-sensitive), whereas six LMW proteins (18.6, 19.4, 25.7, 25.9, 26.0, and 27.6 kDa) were newly synthesized in wheat cultivar Firat-93 (salt-tolerant). The proteins that were newly synthesized were cultivar-specific and mostly acidic in nature with mol. wt. <31.6 kD. These newly synthesized proteins were suggested as important for salinity tolerance in plants (Yildiz, 2007). A novel salt-induced leucine-rich-repeat receptor-like protein kinase OsRPK1 was observed in root cells of rice exposed to salt treatment (Cheng et al., 2009). Newly synthesized protein with molecular weight (18, 20, and 30 kDa) were observed due to NaCl treatments in U-Taung-2, a salt-tolerant genotype, but not in Mut Pe Khaing, salt-sensitive. These proteins were suggested as participating in osmotic adjustment (Win and Oo, 2017).

The level of proteins differs in salt-sensitive and tolerant genotypes subjected to salinity stress (Dubey and Rani, 1989; Elsamad and Shaddad, 1997; Mahmoodzadeh, 2009; Tada and Kashimura, 2009; Cui et al., 2015; Xiong et al., 2017). Using germinating seeds of broad bean (*Vicia faba*) lines, Ahmed and coworkers (2001) showed that salinity stress caused an increase in the number of protein bands in salt-tolerant line 67 but not in salt-sensitive line 13. Electrophoretic analysis (SDS-PAGE) of total soluble protein profiles showed the salinity-induced synthesis of a few polypeptides in seeds of canola salt-tolerant cv. Okapi. In comparison to sensitive cultivar, the number of induced proteins was greater in tolerant cultivars which could be involved in biochemical adjustment to combat salinity stress (Mahmoodzadeh, 2009). Plant cells undergoing salt stress synthesize and accumulate osmotin which provides tolerance to multiple stresses (Ullah et al., 2017; Wan et al., 2017) (Figure 22.3). Osmotin is regarded as a unique 26 kDa protein associated with NaCl-adapted tobacco cells (Singh et al., 1987). Interestingly, the synthesis of osmotin is not induced by osmotic shock but starts only when cells are adapted to NaCl or polyethylene glycol (Singh et al., 1985). In salt-adapted cells, 10–12% of the total protein was contributed by osmotin (Singh et al., 1985). It is believed that osmotin plays a role in osmotic adjustment by facilitating solute accumulation or by causing certain structural or metabolic changes in the cells (Singh et al., 1987). It acts as a protectant of plasma membranes in plants during salt stress (Singh et al., 1989). Tada and Kashimura (2009) observed an enhanced content of osmotin in the lateral roots of B. *gymnorhiza*, a mangrove plant, due to salt treatment at early time points and concluded that osmotin may be associated with initial osmotic adaptation under salt stress. The protective abilities of osmotin against salinity stress have been confirmed by overexpression studies (Barthakur et al., 2001; Husaini and Abdin, 2008; Chowdhury et al., 2017; Le et al., 2018). Salinity led to a 1000-fold upregulation of a plasma membrane-localized osmotin (TlOsm) in the shoots and roots of

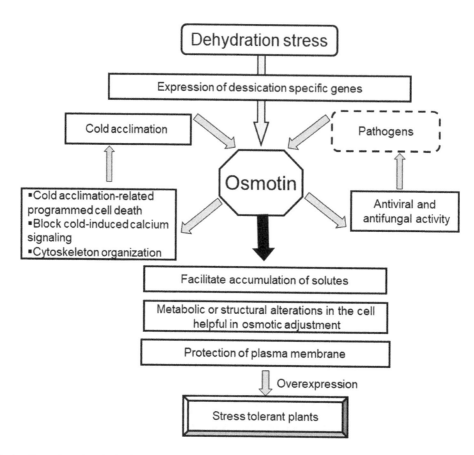

FIGURE 22.3   Plant cells undergoing desication stress synthesize and accumulate osmotin which provides osmotic adjustment to the cell. Osmotin also plays a role in cold acclimation and against viral and fungal pathogen infections.

resurrection plant *Tripogon loliiformis*. *TlOsm* overexpression in rice plants caused increased tolerance to salinity compared to wild-type (Le et al., 2018). Sesame plants transformed with an osmotin-like protein (OLP) gene (*SindOLP*) showed enhanced tolerance to salinity and increased expression of three genes encoding antioxidant enzymes and five defense-associated marker genes associated with JA/ET and SA pathways, namely Si-Ethylene-responsive factor, Si-Defensin, Si-Apetala2, Si-Thaumatin-like protein, and Si-Chitinase. It was suggested that transgenic lines exhibit enhanced survival under salinity probably by activating several components of the defense signaling cascade (Chowdhury et al., 2017). OLP and chitinase (Chi11) belonging to the pathogenesis-related (PR) class of proteins are induced due to abiotic and biotic stresses. Tomato plants expressing OLP and Chi11 showed enhanced tolerance for salt in comparison to untransformed plants. Co-immunoprecipitation showed that Chi11 co-expresses with phosphofructokinase2 (PFK2), which possibly plays a role in enhanced root biomass (Kumar et al., 2016).

In many plant species, certain hydrophilic proteins and their mRNAs are synthesized in response to salt stress. Some of these proteins and mRNAs are also induced by water deficit and treatment with abscisic acid (ABA) (Artlip and Funkhouser, 1995). Such proteins have been grouped in different classes based on DNA sequences of their genes and/or predicted functions of the proteins (Bray, 1995). In plants, late-embryogenesis abundant (LEA) proteins are involved in

various abiotic stresses including salinity stress (Ling et al., 2016). Salinity imposition during the period of seed development following maturation leads to the synthesis of LEA proteins in cotton, carrot, barley, and maize (Ramagopal, 1993). The LEA proteins contribute to salinity tolerance, desiccation tolerance and freezing tolerance in plants by playing roles in ion sequestration, water replacement, and in the stabilization of macromolecules and membranes. The LEA proteins can act *in vitro* as glassy state stabilizers, cryoprotectants, molecular sponges, prevent protein aggregation and bind to lipid vesicles or actin filaments (Figure 22.4). Thus, various groups of LEA proteins have been identified which have been classified based on notable structural domains predicted by amino acid sequences (Bray, 1995). LEA group 1 proteins include the Em family of proteins which are devoid of cysteine and tryptophan. In salt-treated finger millet seedlings, a 21 kDa protein belonging to the LEA 1 group of proteins appeared as the most prominent protein, whereas no measurable amount of this protein was observed in a cucumber species which was salinity sensitive. In comparison to salt-sensitive species setaria, salt-tolerant finger millet species proso revealed a higher expression of the 21 kDa protein showing its role in stress tolerance (Aarati et al., 2003).

LEA group 2 proteins include dehydrin, RAB (ABA-responsive), and D11 proteins which have characteristic lysine-rich regions and are also expressed owing to treatment with ABA (Bray, 1995). The most widely studied proteins

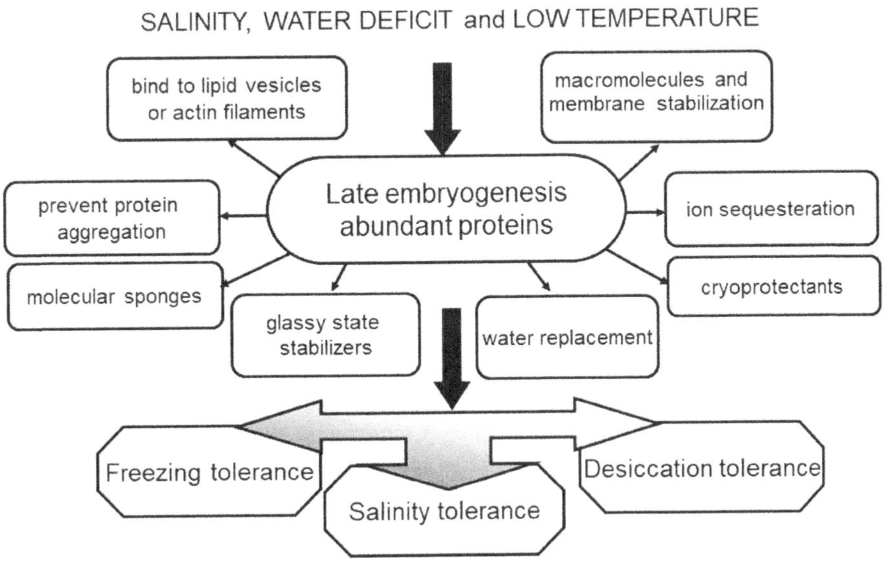

FIGURE 22.4 Roles of late-embryogenesis abundant proteins in freezing tolerance, salinity tolerance, and desiccation tolerance. The function of LEA proteins in water replacement, ion sequesteration and macromolecules, and membrane stabilization has been proposed. LEA proteins can act *in vitro* as glassy state stabilizers, cryoprotectants, molecular sponges, prevent protein aggregation, bind to lipid vesicles or actin filaments.

from many plant species that accumulate in response to dehydrative conditions like salinity, drought, and low temperature are dehydrins (group 2, LEA), which are hydrophilic Gly-rich proteins (Close, 1997). A. thaliana transformed with dehydrin *DHN-5* showed improved salinity tolerance (Brini et al., 2007). Alfalfa plant transformed with a *Cleistogenes songorica* dehydrin protein (CsLEA), exhibited a relatively lower content of Na+ and ability to maintain a higher relative water content, lesser photosystem changes, higher biomass of shoots, and reduced membrane injury due to osmotic stress (Zhang et al., 2016). *OsDhn1* gene encoding OsDhn1 protein is stress-inducible in rice. Increased salt tolerance and relatively low concentration of H_2O_2 was observed in rice plant overexpressing *OsDhn1* suggesting that OsDhn1 plays a key role in the stress tolerance via ROS detoxification (Kumar et al., 2014). Overexpression of conserved segments and full-length of wheat dehydrin DHN-5 in A. thaliana led to diverse responses and tolerance to environmental stresses (Drira et al., 2016). Although the specific biochemical function of dehydrins has not been demonstrated, they seemingly appear to play a role in protecting membranes from damage (Puhakainen et al., 2004) or act as surfactants which can inhibit the coagulation of a variety of macromolecules and thus maintain structural integrity (Close, 1997).

LEA group 3 and group 5 proteins are represented by D7 and D29 from cotton, respectively, and contain a repeated tract of 11 amino acids (Bray, 1995). Citrus cell suspensions grown in the presence of 0.2 M NaCl or leaves of citrus plants irrigated with NaCl accumulate a protein (C-LEA5) which has a high similarity with the cotton LEA5 protein (Naot et al., 1995). OsLEA3 is a LEA group 3 protein of rice (*Oryza sativa* L.), the expression of which is induced under salt stress. Transgenic rice plants which accumulated OsLEA3 in their vegetative tissues showed increased salt tolerance

(Hu, 2008). LEA group 4 protein represents the D113 protein, the synthesis of which is induced in drying cotton seeds, and this protein has a homolog in tomato LE25 (Bray, 1995). Transgenic rice plants overexpressing OsLEA4 conferred enhanced salt resistance in comparison to wild-type plants (Hu et al., 2016). AtLEA14 expressed ubiquitously in various tissues of Arabidopsis was induced significantly with an increased duration of salt treatment. Overexpression of *AtLEA14* in Arabidopsis showed increased salinity tolerance suggesting its important protective functions in Arabidopsis under salt stress (Jia et al., 2014). *SiLEA14*, a novel foxtail millet LEA gene localized in cytoplasm and nucleus, was remarkably induced by salinity and abscisic acid treatment. *SiLEA14* overexpression in transgenic Arabidopsis and foxtail millet plants showed higher salt tolerance than wild type (Wang et al., 2014b). Pathogenesis-related proteins are implicated in functional adaptations that help in tolerance against various environmental stresses. *Salix matsudana* PR protein (SmPR10) showed upregulation under salinity stress (100 mM NaCl). Arabidopsis plants transformed with SmPR10 exhibited enhanced salt tolerance (Han et al., 2017).

The proteomic approach has been applied to identify several novel proteins and differentially expressed proteins that were involved in increasing plant adaptation to salt stress. Yang and coworkers (2009) identified 38 salt-responsive proteins in leaves of *Populus cathayana* Rehder using peptide mass fingerprint analysis. Some of the novel salt-responsive proteins identified were suggested to be associated with the physiological and biochemical responses of P. *cathayana* to salt stress. Similarly, a novel salt-induced protein was identified and associated with salt tolerance in the mangrove plant *Bruguiera gymnorhiza* (Tada and Kashimura, 2009). In tomato, three novel proteins and six upregulated proteins including a salt-tolerance protein, EGF receptor-like protein, M2D3.3, enoyl-CoA hydratase, and phosphoglycerate mutase-like protein were identified which were associated with salt stress responses (Amin et al., 2007). Some novel salt-responsive proteins related to membrane transport (plasma membrane intrinsic proteins (PIPs) and tonoplast intrinsic proteins (TIPs)), cytoskeleton metabolism (fasciclin-like arabinogalactan proteins and actin-related protein2), signal transduction (leucine-rich repeat receptor-like kinase, LRR) and stress responses (universal stress protein, thaumatin-like protein, desiccation-related protein PCC13-62 and dirigent-like protein) were identified in salt-treated cotton roots (Li et al., 2015). In roots and shoots of the salt-treated seedlings of salt-tolerant alfalfa cultivar Zhongmu No 1, some novel salt-responsive proteins such as isoliquiritigenin 2'-O-methyltransferase, pathogenesis-related protein 2, CP12, harvest-induced protein were identified (Xiong et al., 2017).

In Arabidopsis, the enhanced relative level of enzymes is associated with glycolysis and the respiratory pathway, whereas in Thellungiella an enhanced level of protein is biosynthesis related (S7, S24, ribosomal protein S15A, 40S ribosomal proteins, eukaryotic translation initiation factor 3A), and photosynthesis-related (RUBISCO activase) proteins have been reported (Pang et al., 2010). An increased amount of β subunit of various ATP synthase involved in ATP biosynthesis has been observed in tolerant salt-treated plants while the reduced amount was observed in sensitive plants (Pang et al., 2010; Li et al., 2011). In salt-stressed rice panicles, some of these proteins had a higher constitutive expression level; some were associated with upregulation of antioxidants, upregulation of proteins involved in ATP generation, signal transduction, transcription, translation, etc. (Dooki et al., 2006). A subcellular proteomics approach was used to identify differentially expressed plasma membrane (PM)-associated proteins in a salt-tolerant rice variety, IR651, in response to salt stress. Most of the identified proteins including proteins involved in the regulation of PM channels and pumps, signal transduction, oxidative stress defense, membrane structure, protein folding, methyl cycle, etc. are associated with plant tolerance to salt stress (Nohzadeh et al., 2007). In the roots, 22 proteins associated mainly with signal processing, protein synthesis, water conservation, and biotic cross-tolerance were found to be regulated specifically in F63, salt-tolerant maize genotype but remained unchanged in F35, sensitive maize genotype suggesting that these could be the major contributors to the tolerance of F63 (Cui et al., 2015).

22.2.2 Protein Level in Salt-Stressed Plants

Protein synthesis in plants growing in saline environments is adversely affected and depends on the intensity of stress, the plant genotypes, and the plant parts studied. Salt stress results in an overall decrease in protein synthesis with a loss of polyribosomes (Artlip and Funkhouser, 1995). In germinating seeds, as well as during late growth stages of plants, salinity causes impairment in synthesis as well as degradation of proteins. To access the general impact of salt damage on plant growth and metabolism, various investigators have attempted to study the overall status of total proteins and soluble proteins and the pattern of protein synthesis in different parts of plants growing under salinity stress (Rahdari et al., 2012; Klados and Tzortzakis, 2014). In the majority of cases, salt stress lowers the amount of protein in salt-stressed plant parts due to reduced protein synthesis and increased protein-hydrolyzing enzyme activities. However, in some cases, an enhanced level of protein level is observed under salt stress, possibly because of the enhanced synthesis of novel salinity-induced proteins or reduced proteolytic enzyme activities (Kahlaoui et al., 2018).

High salinity significantly reduced total protein content in leaves of *Portulaca oleracea* (Rahdari et al., 2012) and *Cichorium spinosum* (Klados and Tzortzakis, 2014). The reduced protein level in salt-stressed plant parts is attributed to reduced protein synthesis, the reduced amino acid availability and the denaturation of the enzymes involved in amino acids and protein synthesis (Hall and Flowers, 1973; Popova et al., 1995). In chickpea (*Cicer arietinum* L.), one of the major legume crops for semi-arid tropics, a salinity treatment with 100 mM NaCl in nutrient solution caused a marked decrease in the level of proteins in developing seeds when plants were grown in sand cultures (Murumkar and Chavan, 1986). When rice seeds were germinated under increasing concentrations of NaCl, a decrease in total and soluble protein level was noted in the embryoaxes (Dubey and Rani, 1987). A greater decrease in protein level was obtained in the embryoaxes of salt-sensitive cultivars than tolerant cultivars under a similar level of salinization. Dubey and Rani (1987) observed that under 14 dS m^{-1} NaCl salinity, the soluble protein level of embryoaxes of salt-sensitive rice cultivars Ratna and Jaya reduced to almost one-third compared with non-salinized seeds at 120 hours of germination. A moderate salinity level of 7 dS m^{-1} NaCl had virtually no effect on the change in total and soluble protein levels in embryoaxes of germinating seeds of salt-tolerant rice cultivars CSR-1 and CSR-3, whereas higher salinity level caused a marked decrease in protein level in the embryoaxes of these cultivars (Dubey and Rani, 1987). In barley plants, the imposition of NaCl stress results in a decline in the leaf protein content and induces marked quantitative and qualitative changes in the polypeptide profiles (Popova et al., 1995).

Although salinity causes decreased protein synthesis and increased proteolysis in various plant species, in many cases increased protein levels are observed under salinity stress in germinating seeds (Dubey and Rani, 1987), growing seedlings (Dubey and Rani, 1989; Kahlaoui et al., 2018), and different plant parts (Elsamad and Shaddad, 1997). The NaCl salinity caused a delay in the breakdown of endospermic proteins and inhibition in translocation of hydrolyzed products from rice endosperms to growing embryoaxes (Dubey, 1983). In germinating seeds, endospermic protein hydrolysis is suppressed under salinization. When seeds of rice cultivars with a difference in salt tolerance are germinated under increasing levels of NaCl, it has been observed that salt treatment suppresses protein depletion from the endosperm of all cultivars, with greater suppression in salt-sensitive cultivars than in salt-tolerant cultivars (Dubey and Rani, 1987). In comparison to salt-sensitive tomato cultivar, Heinz-2274 irrigated with saline water, the level of total soluble protein increased significantly in the roots and leaves of Rio Grande, the salt-tolerant cultivar (Kahlaoui et al., 2018). Similar to rice, cowpea (*Vigna unuiculata* L.) seedlings, pea (*Pisum sativum* L.), Trigonella (*Trigonella foenum-graecum* L. and T. *aphanoneura* Rech. f.), beans (*Phaseolus vulgaris* L.), and pigeon pea (*Cajanus cajan*) plants, as well as in soybean (*Glycine max*) callus cultures, NaCl salinity treatment caused an increase in the protein content (Joshi, 1987; Mehta and Vora, 1987; Elenany, 1997; Niknam et al., 2006; Dantas et al., 2007). The increased protein level under salinization as noted in these cases appears to be due to the enhanced synthesis of pre-existing as well as certain new sets of proteins (Dubey and Rani, 1989). The enhanced synthesis of many novel proteins has been noticed in plants subjected to saline stress, but whether such increased synthesis is responsible for a net increase in the total and soluble protein level of stressed plants remains to be investigated.

To understand the mechanism of salt tolerance in crops, various investigators have studied the metabolic status of proteins and amino acids in germinating seeds and seedlings using cultivars varying in salt tolerance (Dubey and Rani, 1987; Dubey and Rani, 1989; Elsamad and Shaddad, 1997; Ahmed et al., 2001; Dantas et al., 2007). Especially in rice, a staple food crop for the majority of the world population, salt-tolerant cultivars are characterized by a higher value of protease-specific activity as well as a higher total and soluble protein content in germinating seed parts under control and salt treatments, when compared with sensitive seedlings (Dubey and Rani, 1989). This shows that the salt-tolerance ability is associated with a possible higher protein level in rice, seemingly endogenous proteins that are either not found or are very poorly expressed in sensitive cultivars. Soybean cultivars differing in salt tolerance show different levels of proteins and amino acids when grown in the presence of NaCl (Elsamad and Shaddad, 1997). Salt-tolerant soybean cultivars Clark and Forest accumulate higher levels of soluble proteins, whereas sensitive cultivar Kint shows a decrease in the soluble protein level when grown in saline soils (Elsamad and Shaddad, 1997). These observations indicate that, in rice and soybean plants, the salt-tolerance ability is associated with a higher level of proteins, which are seemingly endogenous proteins that are either not found or are very poorly expressed in sensitive cultivars on the imposition of salinity.

22.2.3 Enzyme Levels in Salt-Stressed Plants

Salinity induces alterations in the activities of proteolytic (Dubey, 1983; Dubey and Rani, 1987; Dubey and Rani, 1990), amylolytic (Dubey, 1983; Dreier et al., 1995), nucleolytic (Dubey, 1985), phosphorolytic (Dubey et al., 1987; Dubey and Sharma, 1989; Dubey and Sharma, 1990; Banuls et al., 1995; Lin et al., 1997; Nasri et al., 2016), oxidative (Mittal and Dubey, 1995), antioxidative (Piqueras et al., 1996; Azevedo-Neto et al., 2006; Khosravinejad et al., 2008; Shafi et al., 2009; Mishra et al., 2013), photosynthetic (Popova et al., 1995), and

nitrogen assimilatory enzymes (Dubey, 1997; Debouba et al., 2007; Kauser et al., 2014) in germinating seeds and in growing plants. Salinity causes either an increase or a decrease in the activity of enzymes, depending on the extent of stress, the nature of the enzymes, the genotypes of plant species differing in salt tolerance and the plant parts studied.

The level and activity of the key enzyme of photosynthesis in C3 plants, ribulose-1,5-bisphosphate carboxylase (RUBISCO), decrease in barley plants on the imposition of NaCl stress (Popova et al., 1995). In germinating rice seed endosperms, salt stress led to the reduced activity of hydrolytic enzymes α-amylase, ribonuclease, protease, phytase, and phosphatase (Dubey, 1983; Dubey, 1985; Dubey and Rani, 1987). The decrease was more in salt-sensitive than in salt-tolerant varieties. In growing seedlings of rice, salinity increases the activities of nucleases (Dubey, 1985), proteases (Dubey, 1985; Dubey and Rani, 1990), peptidases (Dubey and Rani, 1990), phosphatases (Dubey et al., 1987; Dubey and Sharma, 1989; Dubey and Sharma, 1990), and oxidases (Mittal and Dubey, 1991). In rice, NaCl treatment has been shown to reduce bioactive gibberellins content via an increase in bioactive GA inactivation, which in turn reduces germination of rice seeds by decreasing α-amylase activity via downregulation of α-amylase gene expression (Liu et al., 2018). Barley plants grown in the presence of 200 mM NaCl showed stimulation in β-amylase activity in leaves (Dreier et al., 1995). β-Amylases are regarded as stress-induced proteins in barley (Dreier et al., 1995). NaCl treatment resulted in a decrease in the activities of α-amylase, β-amylase, and α-glucosidase enzymes associated with a seed germination process of two lentil cultivar Castelluccio and Eston. In contrast, seeds of two bean cultivars Pantelleria and Ustica showed higher activities of enzymes associated with the germination process under salt stress (Sidari et al., 2008). Certain enzymes involved in the synthesis of osmolytes show a marked increase under salinity stress (Ramagopal, 1993). The enzymes of proline biosynthesis, Δ^1-pyrroline-5-carboxylate synthetase (P5CS) and Δ^1-pyrroline-5-carboxylate reductase (P5CR), the penultimate enzyme of betaine biosynthesis, betaine aldehyde dehydrogenase, and the enzyme of sorbitol biosynthesis, aldose reductase, show increased activity in many plants subjected to salt stress (Ramagopal, 1993; Igarashi et al., 1997).

Genes that code for key enzymes involved in osmolyte biosynthesis have been isolated in barley, spinach, sugar beet, soybean, and rice plants (Ramagopal, 1993). Salinity treatment of 7 and 14 dS m⁻¹ NaCl led to a significant increase in the levels of free amino acids with a substantially elevated level of proline in rice seedlings (Dubey and Rani, 1989). P5CS is the rate-limiting enzyme in proline biosynthesis in plants. It was observed that salt stress increased P5CS activity thus increasing the proline content in the roots of cucumber seedlings (Duan et al., 2006). Salt-tolerant varieties of rice show a higher level of expression of the enzyme P5CS and its mRNA compared with salt-sensitive varieties grown in a saline medium (Igarashi et al., 1997). Soybeans, transformed with a novel *Solanum torvum* P5CS (Zhang et al., 2015), tobacco plants overexpressing *p5cs* (Kishor et al., 1995) or

Arabidopsis P5CS (Yamchi et al., 2007) produced more proline and exhibited better performance under salt stress. The enzymes involved in membrane transport, such as plasma membrane-ATPase in cotton seedlings (Lin et al., 1997) and Ca^{2+}-ATPase in tomato plants (Ramagopal, 1993), show a higher level of activity under salinization. An increased turnover of the tonoplast H^+-ATPase was observed in leaves of *Citrus sinensis* plants under salinity stress (Banuls et al., 1995). Increased salt tolerance in the halophyte species, *Triplex lentiformis*, and *Chenopodium quinoa* has been related to the plant's ability to upregulate plasma membrane H^+-ATPase under salinity stress (Bose et al., 2015). The changes in the activity behavior of the enzymes, acid phosphatase, alkaline phosphatase, and ATPase isolated from the chloroplasts of two sets of rice seedlings varying in salt tolerance, when grown under increasing levels of NaCl salinity, are shown in Table 22.2. As it is evident from the table, acid phosphatase activity was more inhibited owing to salinity in salt-sensitive cultivars compared with the salt-tolerant ones, whereas the activity of alkaline phosphatase increased in the salt-sensitive seedlings but not in the salt-tolerant ones. Further, salinity caused enhancement of ATPase activity in both sets of rice seedlings, with greater enhancement in tolerant cultivars than in the sensitive ones. Acid phosphatase activity in roots increased in lettuce variety Romaine and decreased in Vista. In shoots, no difference in activity was noticed in the two varieties. However, in cotyledons, acid phosphatase activity was reduced in two varieties during the 24 hours after germination (Nasri et al., 2016).

Isoenzyme profiles of several enzymes are affected by salinity. In some cases, certain molecular forms of enzymes in non-salinized plants disappeared in stressed plants and new molecular forms of enzymes appeared under salinity stress. In shoots of 15-day-old non-salinized rice seedlings, four acid phosphatase isozymes were observed, whereas when seedlings were raised at a salinity level of 14dS m^{-1} NaCl, only one isoenzyme remained detectable (Dubey and Sharma, 1989). The reduced number of isoenzymes of acid phosphatase at a higher level of salt stress paralleled the reduced enzyme activity under such conditions (Dubey and Sharma, 1989). Phosphohydrolases show varying behavior under salinity in rice plants of differing salt tolerance. In the young embryo-axes of germinating seeds, certain new molecular forms of acid phosphatases appear under salinization. When acid phosphatase isoforms from the embryoaxes of germinating seeds of the salt-sensitive rice cultivar Jaya and the salt-tolerant cultivar CSR-1 were compared at 48 and 96 hours of germination under increasing levels of NaCl salinity, it was observed that certain new isoenzyme forms appeared in both sets of cultivars under salinization (Dubey and Sharma, 1990). Further, a great number of isoenzymes were observed in the embroaxes of salt-tolerant rice varieties than salt-sensitive varieties under both controls as well as salt treatments (Dubey and Sharma, 1990). Therefore, under salinization, certain isoforms of acid phosphatase are not synthesized, whereas the synthesis of certain new isoforms is induced depending on the plant parts and the genotypes studied. It has been shown that salt tolerance is associated with the presence of a large number of acid phosphatase isoenzymes (Dubey and Sharma, 1990). The prime enzyme of nitrate assimilation, nitrate reductase (NR), has been extensively studied for its behavior in different plant species under salinization (Dubey and Pessarakli, 1995; Dubey, 1997; Debouba et al., 2007). Salinity effects on NR activity are varied and depend on the type and extent of salinity as well as on genotypes of the plants studied. In intact tissues of wheat, lentil (*Lens esculanta*), mulberry (*Morus abla*), sorghum, and tobacco plants, NR activity decreases owing to NaCl salinity (Dubey, 1997), whereas in rice plants the behavior of NR varies in genotypes differing in salt tolerance when raised under NaCl salinity (Katiyar and Dubey, 1992). The salt-sensitive rice cultivars Ratna and Jaya show a lower NR activity in their shoots and roots compared to the salt-tolerant cultivars CSR-1 and CSR-3, and the activity in tolerant cultivars further increases under salinization (Katiyar and Dubey, 1992). However, in sorghum, nitrate reductase activity was lowered due to salinity in both tolerant (JS-2002 and Sandalbar) as well as sensitive genotypes (Noor and FJ-115) (Kauser et al., 2014).

In seedlings of salt-sensitive rice cultivar, Malviya-36 and salt-tolerant cultivar CSR-27, enhanced activity of ascorbate peroxidase (APX), superoxide dismutase (SOD), and its isoform Cu/Zn-SOD was observed under salt stress. The activity of catalase (CAT), guaiacol peroxidase (GPX), monodehydroascorbate reductase (MDHAR), dehydroascorbate reductase (DHAR), and glutathione reductase (GR) increased at a moderate salinity treatment of 7 dS m^{-1} NaCl in salt-sensitive seedlings, while decline in the activities of these enzymes was noticed at higher salinity level of 14 dS^{-1} NaCl (Mishra et al., 2013). Increased antioxidant enzyme activity was also

TABLE 22.2

Salinity-Induced Alteration in the Activity Behavior of Phosphorolytic Enzymes in Chloroplasts of 20-day Grown Rice Plants

Rice Cultivar	NaCl Treatment (dSm^{-1})	Acid Phosphatase	Alkaline Phosphatase	ATPase
CSR-1 (T)	0	4.00	0.68	0.12
	7	3.20	0.50	0.15
	14	3.00	0.45	0.22
CSR-3 (T)	0	4.30	0.70	0.12
	7	3.20	0.64	0.16
	14	3.00	0.48	0.26
Ratna (S)	0	3.20	0.62	0.06
	7	2.60	0.70	0.08
	14	1.40	0.76	0.11
Jaya (S)	0	2.80	0.59	0.05
	7	2.00	0.64	0.08
	14	1.20	0.68	0.14

Source: T and S in parentheses indicate tolerant and sensitive rice cultivars, respectively. Enzyme units are expressed as μmol substrate hydrolyzed h^{-1} mg^{-1} protein.

observed in halophyte *Limonium delicatulum* Kuntze to protect against salt-induced oxidative stress and to tolerate salt ion accumulation (Souid et al., 2016). The activities of the oxidative enzymes, polyphenol oxidase and indole-3-acetic acid oxidase increase in the seedlings of salt-tolerant as well as salt-sensitive rice cultivars under salinization, and the extent of the increase differs in the two sets of cultivars (Mittal and Dubey, 1995). The specific activities and patterns of peroxidase and SOD isoenzymes are altered significantly in salt-stressed plants. Rice seedlings, differing in salt tolerance possess a constitutively different number of peroxidase isoforms in non-salinized seedlings. When these seedlings were raised under NaCl salinity, certain new isoforms of peroxidases appeared, and the intensities of some of the pre-existing isoenzymes increased. In salt-tolerant rice cv. CSR-1, three and five peroxidase isoenzymes appeared in roots and shoots respectively, whereas in a salt-sensitive cv. Ratna, in roots as well as shoots, six isoenzymes were observed (Mittal and Dubey, 1991). An increase in peroxidase activity and induction of a new SOD isozyme was noticed in salt-tolerant embryonic callus cultures of lemon (*Citrus limon* L. Burn) (Piqueras et al., 1996). An increase in the activity of SOD, POD, and APX was observed in leaves of salt-tolerant 'Quickstart II' and salt-sensitive 'DP1' at 4 days after treatment with 250 mM NaCl but a lower activity was observed at 8 and 12 days after treatment. Five isoforms of SOD, POD, and two APX isoforms showed differential patterns between two genotypes (Hu et al., 2012). The presence of different isoenzyme patterns of phosphatases, peroxidases, SOD, and ribonucleases in salt-sensitive and salt-tolerant genotypes of crops strengthens the view that salt tolerance or sensitivity depends on the genetic and biochemical make-up of the species. Also, specific molecular forms of the isoenzymic proteins, which appear to be constitutive proteins, are possibly associated with the salt tolerance or sensitivity characteristics. However, the mechanism of the expression of intrinsic isoenzyme proteins related to sensitivity or tolerance and of those isoenzymic proteins that specifically appear under salinization remains to be investigated.

22.3 DROUGHT

Water deficit is one of the most common environmental factors that limits crop productivity. Drought is a natural calamity and has devastating effects on crop yields. Crop plants frequently encounter drought conditions during their lifetimes. However, certain stages, including germination, seedling, and flowering, are the most critical for water deficit damage. Stress encountered during these stages drastically affects crop yields. Drought reduces plant growth and causes many morphological and biochemical alterations in plants resulting in a drastic loss in yield. Studies conducted by various investigators related to the effects of drought stress on the protein synthesis in many important crops suggest that drought stress severely affects protein synthesis, alters gene expression and protein profiles in stressed tissues, and induces the synthesis of stress-induced-specific proteins. Many of these proteins

are hydrophilic and belong to specified families and have predicted functions in protecting the cells from drought stress.

Some of the stress-induced proteins appear to be tissue-specific, whereas others do not appear to be specific for any particular tissue or organ. A considerable amount of work has been done by various groups of investigators in the last few years to understand the mode of protein synthesis in plant parts under drought conditions (Bewley and Larsen, 1982; Dasgupta and Bewley, 1984; Heikkila et al., 1984; Baker et al., 1995; Zhang et al., 1996; Close, 1997; Pelah et al., 1997; Bibi et al., 2009; Ke et al., 2009), the level of proteins in stressed plants (Rai et al., 1983; Kumar and Singh, 1991; Mirzaei et al., 2012; Xie et al., 2016), and the activities of key enzymes influenced under drought (Lodh et al., 1977; Mali et al., 1980; Geigenberger et al., 1997; Igarashi et al., 1997; Sharma and Dubey, 2004; Sharma and Dubey, 2005a, 2005b; Hameed et al., 2013; Pyngrope et al., 2013a,b; Muscolo et al., 2014; Chmielewska et al., 2016).

22.3.1 DROUGHT-INDUCED PROTEIN SYNTHESIS

Drought causes alteration to gene expression in plants leading to an inhibition of protein synthesis as well as an enhanced synthesis of certain stress-specific proteins. Quantitative and qualitative changes occur in the synthesis of proteins in plants in response to water deficit. Drought causes tissue- and organ-specific differential genomic expression which leads to alterations in the patterns of protein synthesis in cells (Bewley and Larsen, 1982; Dasgupta and Bewley, 1984; Heikkila et al., 1984; Yoshimura et al., 2008; Ke et al., 2009; Si et al., 2009; Mirzaei et al., 2012; Mohammadi et al., 2012; Chmielewska et al., 2016). Studies of maize (Bewley and Larsen, 1982) and wheat (Scott et al., 1979) have indicated that increasing the level of drought causes a decrease in the polyribosome level. However, plant species which can survive under drought stress show a greater capacity for producing polyribosomes in the tissues. Various investigators have demonstrated the synthesis of drought-specific proteins in different crops (Dasgupta and Bewley, 1984; Heikkila et al., 1984; Singh et al., 1987; Claes et al., 1990; Baker et al., 1995; Zhang et al., 1996; Close, 1997; Pelah et al., 1997; Mohammadi et al., 2012). Many of proteins which are specifically synthesized under drought have been isolated and well characterized (Singh et al., 1987; Claes et al., 1990; Close, 1997; Pelah et al., 1997; Mohammadi et al., 2012). A relationship between the physiological adaptations and accumulation of drought-induced proteins has been suggested (Riccardi et al., 1998; Basu et al., 2010; Mirzaei et al., 2012; Mohammadi et al., 2012; Chmielewska et al., 2016). The induced synthesis of novel proteins, including enzymes associated with osmoprotectant biosynthesis, HSPs, LEA proteins, chaperones, enzymes associated with detoxification, transcription factors, kinases, and phosphatases in response to drought has been observed in many plant species (Jiang and Huang, 2002; Khurana et al., 2008; Ke et al., 2009; Chmielewska et al., 2016). The major families of drought-induced proteins are LEAs, RABs (responsive to ABA), dehydrins, and vegetative storage proteins (Artlip and Funkhouser, 1995).

LEA proteins have been further subdivided into several groups: group 1 (D19 protein from cotton), group 2 (D11 from cotton), group 3 (D7 from cotton), group 5 (D29 from cotton). Fifty-one genes encoding Arabidopsis LEA protein were identified and classified in nine groups based on sequence similarity (Hundertmark and Hincha, 2008). In *Citrus sinensis*, 72 genes were classified in seven groups (LEA 1, LEA 2, LEA 3, LEA 4, LEA 5, dehydrin, and seed maturation protein SMP) on the basis of a predicted sequence and phylogenetic relationship with LEA proteins of A. thaliana (Pedrosa et al., 2015). LEA group 1 (D19) proteins, which are devoid of cysteine and tryptophan, have been detected in cotton, barley, and carrot under water deficit (Ramagopal, 1993). It is suggested that LEA group 1 proteins function in a water-binding capacity creating a protective aqueous environment (Bray, 1995). LEA group 3 and LEA group 5 proteins, which are represented by D7 and D29 proteins, respectively, from cotton contain a repeated tract of 11 amino acids. These proteins have been isolated from desiccating mature cotton embryos, chloroplasts of *Craterostigma plantagineum*, and citrus seedlings exposed to drought (Bray, 1995; Naot et al., 1995). Citrus cell suspension in response to salt stress, leaves of citrus plants irrigated with NaCl or seedlings exposed to drought led to an osmotic stress-induced elevated level of LEA5 proteins and its mRNA (Naot et al., 1995). Another group of proteins, group 4, is represented by D113 protein, which has a homolog in tomato and is expressed in drying cotton seeds (Ramagopal, 1993; Bray, 1995). Many of these proteins are hydrophilic and are soluble on boiling and are therefore expected to be located in the cytosol. Substantial evidence suggests the involvement of LEAs in desiccation tolerance (Khurana et al., 2008). It is predicted that most of these proteins are associated with the protection of cellular structures and components from dehydration associated with water deficit. Using a transgenic approach, it has been shown that LEA proteins play a key role in protecting plants under drought stress. Overexpression of *OsEm1*, a drought-responsive gene which encodes group I LEA proteins, enhances plant survival ratio, ABA sensitivity, expression of other LEA genes, including *LEA3*, *RAB21*, *RAB16A*, *RAB16C*, and osmotic tolerance in rice (Yu et al., 2016). Expression of the barley (*Hordeum vulgare* L.) LEA protein gene, HVA1, in a rice cell suspension leads to a higher level of accumulation of this protein, and such plants show enhanced tolerance to water stress (Xu et al., 1996). Tobacco plant transformed with LEA 4 group genes from the resurrection plant *Boea hygrometrica* conferred dehydration tolerance (Liu et al., 2009). During dehydration, these proteins are probably involved in the general protection of the plant cells by affecting the stability of membranes and proteins. LEA4-1 cDNA cloned from *B. napus* was overexpressed in transgenic Arabidopsis plants. Arabidopsis plants transformed with *BnLEA4-1* showed enhanced tolerance to drought (Dalal et al., 2009). *SiLEA14*, a novel LEA gene significantly induced by osmotic stress in foxtail millet when overexpressed in Arabidopsis and foxtail millet led to higher tolerance to osmotic stress in these plants compared to untransformed plants (Wang et al., 2014b).

Dehydrins, a group of LEA II proteins, get induced in various plant species exposed to environmental stresses (Kosová et al., 2014). Among proteins in plants that accumulate in response to drought, dehydrins have been the most commonly observed. Rice, barley, maize, pea, Bermuda grass, and Arabidopsis plants show an increased synthesis of dehydrins under osmotic stress (Ramagopal, 1993; Su et al., 2013; Kosová et al., 2014). Dehydrins contain several typical domains combined together in a few specific patterns with several minor permutations. Dehydrin polypeptides are made up of less than 100 to nearly 600 amino acid residues (Close, 1997). Although the fundamental biochemical mode of action of dehydrins has not been established, it is believed that dehydrins are surfactants and thereby they reduce the coagulation of a variety of macromolecules and thus preserve the structural integrity of the cell (Close, 1997). The OsDhn1 gene has been reported to be stress inducible in rice. Overexpression of this gene led to increased drought tolerance via ROS scavenging in transgenic rice plants compared to control plants (Kumar et al., 2014). Among three Bermuda grass cultivars Latitude 36, Celebration, and Premier, Latitude 36 had the best visual quality and lower electrolyte leakage. Drought treatment led to the appearance of 16 and 23 kDa dehydrin proteins in 'Latitude 36' suggesting that these dehydrins could be associated with drought tolerance in Bermuda grass (Su et al., 2013). Genes encoding dehydrins are also ABA-regulated. In dehydrated leaves of tomato, maize, and Arabidopsis plants, endogenous ABA levels increase with the simultaneous increase in the levels of dehydrins and its mRNA (Bray, 1995). Dehydrins are localized primarily in the cytoplasm of root and shoot cells (Bray, 1995). In certain plant species, synthesis of dehydrin-like proteins has been observed under osmotic stress or under treatment with ABA. In Stellaria longipes, the synthesis of a dehydrin-like protein is induced as a result of treatment with ABA or under osmotic stress (Zhang et al., 1996). Sequence analysis of this protein indicates that it shares some similarity in structural features with the dehydrins of other plants and also exhibits certain unique characteristics (Zhang et al., 1996). In castor bean, dehydrin-like protein synthesis is tissue-specific and is dependent on the seed's physiological stage. Patterns of water deficit-induced dehydrin-related polypeptides in endosperms differ from those induced during late seed development (Han and Kermode, 1996). In drought-stressed roots and shoots of Lathyrus sativus, dehydrin-like transcripts accumulate, which are also expressed in unstressed seedlings owing to ABA treatment (Sinha et al., 1996). A novel protein with a 40 kDa molecular weight has been detected in pea plants under desiccation. Two lysine-rich blocks were evident in the deduced amino acid sequence of this protein, and the remaining sequence differed significantly from other pea dehydrins (Robertson and Chandler, 1994). By analogy with heat-shock cognate proteins, this protein has been designated as dehydrin cognate (Robertson and Chandler, 1994). Transgenic Arabidopsis plants overexpressing wheat dehydrin DHN-5 were more tolerant to water deprivation as compared to wild-type plants (Brini et al., 2007). It was proposed that DHN-5, contributes to an improved drought tolerance through osmotic adjustment (Brini et al., 2007).

Abscisic acid plays a major role in responses of the plant to environmental stresses including drought (Liu et al., 2013). Genetic expression studies reveal that among the stress-induced proteins which are well characterized, the majority are the product of ABA-responsive genes. How stress conditions signal an increased production of ABA, how ABA modulates the expression of these genes, and what is the functional role of stress-responsive proteins in dehydration tolerance, such as osmoprotectants, radical scavengers, protectants of subcellular organelles and macromolecules, or as regulatory proteins, remain to be investigated in detail. Many of these proteins also appear in response to the application of ABA, suggesting that ABA is a signal in the stress response. Synthesis of these proteins also occurs with ABA application. Similarly, a family of genes and their products, glycine-rich proteins, which are hydrophilic and ABA-responsive, have been identified in alfalfa plants under drought stress (Bray, 1995). However, no specific predictions about the functions of these Arabidopsis and alfalfa proteins have been made under stressful conditions. Like dehydrins, the RAB and D11 (group 2) family of proteins are also ABA-regulated and possess a characteristic lysine-rich region with consensus amino acid sequences repeated at least two times. Proteins of this family have been identified in many plant species, including maize, tomato, wheat, alfalfa, Arabidopsis, rice, and castor (Mundy and Chua, 1988; Ramagopal, 1993; Bray, 1995; Han and Kermode, 1996). Exposure of *A. thaliana* to drought stress results in the accumulation of the RAB 18 protein, and such plants develop enhanced freezing tolerance (Mantyla et al., 1995). Progressive water deficit in whole *S. tuberosum* plants leads to about a 2.5-fold increase in leaf ABA content and the synthesis of two chloroplastic proteins of 32 and 34 kDa, namely CDSP 32 and CDSP 34, which are synthesized in the stroma and in the thylakoids, respectively (Pruvot et al., 1996). Differences between the leaf chloroplast proteome of drought exposed ABA-deficient maize mutant Vp5 and wild-type Vp5 seedlings revealed 11 upregulated proteins and 5 downregulated proteins (associated with ATP synthesis and photosynthesis) indicating the key role of ABA in the regulation of the synthesis of drought-induced proteins in maize chloroplasts (Hu et al., 2012b).

Osmotin and osmotin-like proteins are PR-5 pathogenesis-related group of proteins which represent key dehydration-inducible proteins in plants. Osmotin, the 26 kDa protein which is synthesized and accumulates in cells experiencing osmotic adjustment to NaCl, also accumulates in cells experiencing osmotic adjustment to polyethylene glycol (Singh et al., 1987). ABA, which is known to induce osmotic adjustment in cells, also induces the synthesis of osmotin. Osmotin synthesis is regulated by ABA, but its accumulation is dependent on the extent of drought stress and the adjustability of the cells to stress. Like dehydrins, osmotin is also a much more extensively studied protein which accumulates under water and salinity stresses in various plant species like tobacco, triplex, tomato, and maize (Ramagopal, 1993). An

osmotically regulated glycine-and threonine-rich protein was identified in rice by Mundy and Chua (1988). This protein is a product of an ABA-responsive gene *small rab 21*. Osmotic stress imposed by polyethylene glycol or desiccation leads to a rise in the content of ABA and in turn the rab 21 gene is induced and expressed to synthesize these proteins in rice tissues. Constitutive expression of an osmotin gene in transgenic tobacco was shown to improve its drought tolerance (Barthakur et al., 2001). Parkhi and coworkers (2009) provided direct evidence for the protective role of osmotin in cotton plants experiencing drought stress by transforming cotton plants with apoplastically secreted tobacco osmotin. Under drought stress, the seedlings expressing osmotin showed reduced H_2O_2 content and lipid peroxidation. Overexpression of the tobacco osmotin gene in carrot led to slower wilting rate and faster recovery when drought stress was alleviated. Transgenic plants exposed to drought stress exhibited reduced H_2O_2 concentration, decreased electrolyte leakage and lipid peroxidation, and enhanced leaf water content suggesting that osmotin protects against drought stress (Annon et al., 2014). The induction of omotin like protein and chitinase (Chi11) which belong to the PR class of proteins has been observed during environmental stresses. Overexpression of OLP and Chi11 in tomato enhanced salt, drought, and fungal stress tolerance in comparison to the controls (Kumar et al., 2016). Sesame plants transformed with a SindOLP, an osmotin-like protein gene showed enhanced drought tolerance. SindOLP led to enhanced tolerance by inducing the activity of ROS scavenging enzymes and genes associated with SA and JA/ET pathways such as Si-Thaumatin-like, Si-Chitinase, Si-Apetala2, Si-Defensin, and Si-Ethylene-responsive factor (Chowdhury et al., 2017). Overexpression of *S. nigrum* OLP (SnOLP) in soybean lines led to better physiological responses and increased yield under water stress (Weber et al., 2014). A boiling-stable protein (BspA) has been shown to accumulate in the roots of *Populus popularis* plants under drought (Pelah et al., 1997). In addition to BspA, plants also show accumulation of the drought-related proteins dehydrin (DSP-16) and sucrose synthase under water deficit (Pelah et al., 1997). In a highly drought-tolerant legume, cowpea (*Vigna unguiculata*), which shows about 160-times higher accumulation of ABA in drought-stressed conditions compared to unstressed plants, two cDNA clones, CPRD 22, and CPRD 8, which encode putative proteins that are associated with group 2 LEA proteins and old yellow enzyme, respectively, were identified in drought-stressed plants (Iuchi et al., 1996b). However, in 10-hour dehydrated cowpea plants, two additional cDNA clones, CPRD 12 and CPRD 46, were identified which encode putative proteins related to nonmetallo-short-chain alcohol dehydrogenase (CPRD 12) and chloroplastic lipoxygenase (CPRD 46). These genes are also induced under salinity stress (Iuchi et al., 1996a). From rice cv. Tainchung native 1, a 15 kDa protein that accumulates in the sheaths and roots of mature plants and seedlings when subjected to either osmotic stress or after treatment with ABA has been isolated and characterized (Claes et al., 1990). Rice varieties show considerable differences in sensitivity to drought; however, in many of the varieties examined, water deficit created as a result of PEG (polyethylene glycol) led the induced synthesis of one 26 kDa protein with a pI of 6 (Perezmolphebalch et al., 1996). In certain varieties of rice, drought causes accumulation of 18 and 85 kDa proteins, called stress-associated proteins (SAPs), that also accumulate under salinity and high and low temperatures (Pareek et al., 1997).

Heat-shock proteins which are conserved polypeptides, respond to various stresses. HSPs are known to accumulate in plants under drought stress (Arora et al., 1998; Jiang and Huang, 2002; Chmielewska et al., 2016). These proteins are classified in five groups based on molecular weight: (I) Hsp60, (II) Hsp70, (III) Hsp90, (IV) Hsp100, and (V) small heat-shock proteins (sHSPs). HSPs are involved in signaling mechanisms, translation, carbohydrate metabolism, amino acid metabolism, and host defense interactions. They play a key role in controlling the genome (Augustine, 2016). One of the presumed functions of HSPs is related to prevention of protein denaturation during cellular dehydration. Overexpression of AtHsp90.2, AtHsp90.5, and AtHsp90.7 in *A. thaliana* enhanced plant sensitivity to drought stress (Song et al., 2009). When rice sHSP gene, sHSP17.7, was overexpressed in the rice cultivar Hoshinoyume, enhanced tolerance to drought stress was observed in transgenic lines (Sato and Yokoya, 2008). Only seedlings with higher sHSP17.7 expression regrew after rewatering. Transgenic tobacco plants constitutively expressing elevated levels of NtHSP70-1 showed enhanced tolerance to drought stress (Cho and Choi, 2009). CaHSP16.4 localized in nucleus and cytoplasm exhibited dynamic expression pattern when exposed to drought stress. The expression of CaHSP16.4 was initially downregulated and then upregulated. CaHSP16.4 silencing led to significant enhancement in malonaldehyde and decrease in total chlorophyll content, showing the importance of CaHSP16.4 in drought tolerance. Transformed Arabidopsis plants expressing CaHSP16.4 showed enhanced ROS scavenging ability and increased drought tolerance (Huang et al., 2018). In rice leaves, more downregulated proteins were observed in the early stages of drought, whereas more upregulated proteins were detected in severe drought. During a severe drought, proteins associated with transport and signaling such as aquaporins and G-proteins accumulated (Mirzaei et al., 2012).

The imposition of drought stress to two contrasting inbred lines of sunflower (*Helianthus annuus* L.) during the flowering stage led to an alteration in expression of 18 proteins in sensitive and 24 proteins in tolerant lines (Ghaffari et al., 2017). Soybean roots, exposed to drought stress showed an increased level of proteins associated with protein synthesis, redox-related proteins, glycolysis, and cell organization related proteins. Levels of three S-adenosylmethionine synthetase proteins were enhanced, and changes in S-adenosylmethionine synthetase mRNA expression exhibited a similar pattern to the alterations in protein abundance suggesting that S-adenosylmethionine synthetase proteins participate in the regulation of stress response (Oh and Komatsu, 2015). In nodulated alfalfa plants, drought stress caused reduced carboxylation which appeared to be as a result

of the lower content of protein RUBISCO, its lower activation state as well as low regeneration of RuBP. Aranjuelo et al. (2010) observed drought led to the reduction of nitrogenase activity in leaves of N2-fixing alfalfa. Protein expression profiling of rice revealed 100 differentially regulated proteins probably associated with various functions including cell signaling, cell wall modification, molecular chaperones, carbohydrate metabolism, and cell defense. Comparison of rice dehydration-responsive proteome with the proteome data of maize and chickpea revealed an evolutionary divergence in the organ specificity and dehydration response with few conserved proteins (Pandey et al., 2010). Ninety-six protein spots associated with defense, storage, nitrogen metabolism, photosynthesis, carbohydrate metabolism, and some other important functions were identified by comparing the developing grain proteomics of two wheat varieties, Chinese Spring (poor drought resistance) and Ningchun 4 (good drought resistance). During grain development, APX protein was reduced more in Chinese Spring compared to Ningchun 4, whereas tumor protein controlled translationally was upregulated in Ningchun 4 but was not present in Chinese Spring. The large subunit of RUBISCO, oxygen-evolving complex and triosephosphate isomerase were upregulated in Ningchun 4. These proteins were associated with higher drought resistance of Ningchun 4 in comparison to Chinese Spring (Ge et al., 2012). In barley roots and leaves, changes in drought-responsive proteomic revealed some biochemical mechanisms associated with drought tolerance (Chmielewska et al., 2016). Drought-induced enhanced accumulation of HSP70, carbohydrates, proline, ascorbic acid, and certain proteins and metabolites including Clp, APX, and GST appeared to be important for drought tolerance. The level of enzymes enolase and chloroplastic glyceraldehyde 3-phosphate dehydrogenase declined in the leaves of Maresi (drought-sensitive), but it increased in the leaves of Cam/B1/CI (drought tolerant) on the imposition of drought (Chmielewska et al., 2016). One isoform of fructokinase accumulated more in barley variety Cam/B1/CI compared to Maresi suggesting compromised glycolysis in Maresi under drought (Chmielewska et al., 2016). An immunoblot analysis performed in leaves of drought-sensitive (Sadovo and Miziya) and tolerant (Katya and Zlatitza) wheat varieties for detection of ATP dependent calpain protease, RUBISCO, RUBISCO binding protein, RUBISCO activase, some HSPs, dehydrins, and proteins under drought stress revealed that the drought-tolerant varieties contained higher levels of these proteins in comparison to the sensitive varieties (Demirevska et al., 2008). In drought-tolerant wheat cultivar 'Ningchun 47' some drought-stress-related proteins, such as enolase, 2-Cys peroxiredoxin BAS1, oxygen-evolving enhancer protein 2, Fibrillin-like protein,6-phosphogluconate dehydrogenase, and 70-kDa HSP were more upregulated at three leaf stage than those in drought-sensitive 'Chinese Spring' cultivar (Cheng et al., 2015). The comparative dehydration-responsive proteomic analysis of two genotypes 101.14 (sensitive) and M4 (tolerant) of grapevine rootstock revealed that many proteins associated with protein synthesis, secondary metabolism, hormone metabolism, energy metabolism, and stress exhibited considerable differences between the two genotypes (Prinsi et al., 2018).

22.3.2 Protein Level in Drought-Stressed Plants

The levels of the total, as well as soluble, proteins get altered in plants growing under drought conditions compared with plants growing under non-stressed conditions. Various workers have observed either a decline (Barnett and Naylor, 1966; Hsiao, 1973; Kumar and Singh, 1991; Gogorcena et al., 1995; Yu et al., 1996; Boo and Jung, 1999; Sharma and Dubey, 2005a; Bai et al., 2006; Akhzari and Pessarakli, 2016) or an increase (Rai et al., 1983; Kumar and Singh, 1991; Ashrafi Parchin and Shaban, 2014) in the levels of total or soluble proteins in different organs of plants subjected to drought stress. The increased or decreased levels of proteins depend on the plant species and organ studied as well as the severity and duration of the stress. An examination of sodium dodecyl sulfate-polyacrylamide gel electrophoresis (SDS-PAGE) analysis showed variation in the contents of leaf proteins in control and drought-stressed wheat plants revealing stress-induced regulation of protein content (Jahanbakhsh et al., 2017). Sharma and Dubey (2005a) observed a concomitant decrease in the content of total soluble proteins with increasing levels of water deficit in roots as well as shoots of growing rice seedlings. Shah and Loomis (1965) observed a decreased contents of soluble and total proteins in sugar beet leaves from data recorded on a per gram of dry weight basis when the plants were subjected to progressive drought stress. These investigators observed that the response to drought was quick and could be reversed by rewatering the plants. This indicates that the effects of drought are reversible to a certain extent. According to Hsiao (1973), the rapid response of plants undergoing drought stress and its quick reversibility by rewatering suggest that drought affects protein synthesis mainly at the translation level.

When Bermuda grass plants were grown under increasing levels of drought, a decline in the soluble protein level was observed (Barnett and Naylor, 1966). In whole chloroplasts and chloroplast membrane fractions, a decline in the protein content was observed in drought-resistant as well as drought-sensitive genotypes of drought-stressed wheat plants compared with non-stressed plants (Kulshrestha et al., 1987). Mung bean seedlings varying in drought stress tolerance, when raised under increasing levels of water deficit, showed a decline in the protein level in the axis (Kumar and Singh, 1991). A significant decline in the content of total soluble protein was reported in *Levistivum officinale* subjected to −1.0 MPa drought treatment (Akhzari and Pessarakli, 2016). A quantitative decrease in the content of some high molecular weight proteins was observed at 5th day of germination in chickpea seedlings under drought stress, and some new germination related proteins appeared on the 7th day. In cultivar CM-2000, a delayed expression of two proteins with mol. wt. 100 and 60.8 kDa was observed under drought stress, whereas earlier expression of these proteins was observed in cv. CM-94/99 (Bibi et al., 2009). In whole plants of *Lycopersicon*

chilense, as well as in cell suspensions, drought stress leads to decreased synthesis of a proline-rich 12.6 kDa protein in the cell walls (Yu et al., 1996). This is possibly attributed to the downregulation of its gene PTGRP under desiccation. In nodules of drought-stressed (−2.03 MPa) pea plants, about a 30% decline in soluble protein level was observed in comparison to the control plants (Gogorcena et al., 1995). When rice varieties varying in drought-stress tolerance were examined for changes in the protein profiles in different organs due to drought stress, it was observed that in the two cultivars, Sinaloa and IR 10120, the synthesis of many polypeptides decreased owing to PEG-induced drought (Perezmolphebalch et al., 1996). It is suggested that in rice, the extent of the reduction in the contents of proteins or changes in protein profiles in different organs due to water deficit are cultivar specific (Perezmolphebalch et al., 1996). A decreased level of the total, as well as the soluble proteins in drought-stressed tissues, appears to be due to more protein degradation and an overall reduction in protein synthesis under drought (Kumar and Singh, 1991).

It has been observed that drought-stressed plants show a high protease activity compared with non-stressed plants (Thakur and Thakur, 1987). The high protease activity in drought-stressed plants seems to be of adaptive significance, as it causes free amino acid accumulation due to the degradation of proteins. The increased free amino acid, organic acids, and quaternary ammonium compounds act as compatible solutes to maintain the osmotic balance between the vacuole and the cytoplasm under drought stress (Barnett and Naylor, 1966). Certain investigators have observed an increase in protein levels in plants due to drought stress (Rai et al., 1983; Kumar and Singh, 1991; Ashrafi Parchin and Shaban, 2014). Protein profiling of five wheat genotypes using SDS-polyacrylamide gels revealed increased seed storage proteins under drought compared to irrigated plants. However, no effect of drought stress on banding patterns of protein was observed (Ashrafi Parchin and Shaban, 2014). Genotypes of *C. arietinum* cultivars, differing in drought-stress tolerance, when raised under increasing osmotic potential levels, showed increased protein levels in shoots compared with non-stressed plants (Rai et al., 1983). A drought-resistant *C. arietinum* cv. C-214 showed an increase of 60% protein over the control at an osmotic potential of −3 atm, whereas a sensitive cultivar, G-130, showed a 15% increase over the control under similar conditions of stress (Rai et al., 1983). Similarly, when drought-resistant *Z. mays* cv. Ageti-76 plants were grown under increasing osmotic potentials in the range of 1 to 10 atm, an increase in protein content was noticed, reaching 190% of control (Rai et al., 1983). Similarly, in cotyledons of germinating mung beans under drought stress, an increased protein level was noticed when compared with cotyledons of non-stressed germinating seeds (Kumar and Singh, 1991).

Seedlings of drought-tolerant mung bean genotypes showed a higher protein content in embryoaxes as well as cotyledons compared with drought-sensitive genotypes when raised at a −10.0 bar moisture stress level (Kumar and Singh, 1991). Similarly, drought-resistant maize cultivars showed a high protease activity at higher levels of drought stress, whereas inhibition in protease activity is noticed under higher drought stress levels in sensitive cultivars (Thakur and Thakur, 1987). While comparing the total protein and free amino acid pool size in drought-resistant and drought-sensitive cultivars of *C. arietinum* and *Z. mays*, Rai et al. (1983) observed that resistant plants are characterized by an increase over non-stressed plants in total protein and free amino acid levels. These observations indicate that drought has varying effects on the level of proteins in different crop species, and the stress-induced response depends on the species of crop examined, and it may vary even in different organs within the same species.

22.3.3 Enzyme Levels in Drought-Stressed Plants

The normal metabolism of plants growing under drought conditions is adversely affected with a concomitant disturbance of the enzymatic constitution of the plants. Drought stress lowers the levels of many enzymes in the tissues (Mali et al., 1980; Thakur and Thakur, 1987; Gogorcena et al., 1995; Du et al., 1996; Reddy, 1996; Geigenberger et al., 1997; Sharma and Dubey, 2005b; Xu and Zhou, 2005; Muscolo et al., 2014). The activities of certain enzymes increase as a result of drought stress (Hsiao, 1973; Lodh et al., 1977; Thakur and Thakur, 1987; Bray, 1995; Igarashi et al., 1997; Sharma and Dubey, 2005a; Hameed et al., 2013; Pyngrope et al., 2013a,b; Hassan et al., 2015). Photosynthetic efficiency and nitrogen assimilation are decreased in drought-stressed plants mainly due to the reduced activities of the key enzymes associated with these processes. NR, the prime enzyme in the N-assimilation process, is markedly inhibited by drought (Hsiao, 1973; Xu

and Zhou, 2005). The effect of mild (−0.5 MPa) as well as moderate (−2 MPa) levels of drought imposed for 24 hours, on the level of protein and the activities of enzymes NR, glutamine synthetase (GS), alanine aminotransferase (AlaAT), and aspartate aminotransferase (AspAT) in roots and shoots of 20-day grown rice plants is shown in Table 22.3. As is evident from the table, drought stress causes a drastic decline in the protein levels as well as in the activities of the enzymes of NO_3^- assimilation, NR and GS, whereas the key enzymes of amino acid metabolism, AlaAT and AspAT, show increased activity under a mild level of drought; however, under a moderate drought stress of −2 MPa, a pronounced inhibition in the enzyme activity is noticed. The activity of NR is directly related to protein synthesis and plant growth, and both of these processes are adversely affected under drought (Sinha and Nicholas, 1981). Severe drought stress led to a significant reduction in both NRact (NR activity in the presence of Mg^{2+} representing the non-phosphorylated NR state) and NRmax (NR activity in the presence of EDTA representing maximum NR activity), whereas no significant change was observed in NR activation state (Sharma and Dubey, 2005b). Four genotypes (Eston, Castelluccio, Ustica, and Pantelleria) of lentil, when subjected to drought stress during germination, led to a reduced activity of the enzymes associated with the germination process. Among four genotypes, the α-amylase and β-glucosidase activities were most affected by drought in the drought susceptible Pantelleria and Ustica (Muscolo et al., 2014). The photosynthetic apparatus is sensitive to dehydration. Drought stress has a direct effect on carboxylating enzymes. The activities of enzymes RuBP carboxylase and PEP carboxylase decreased in the leaves of

TABLE 22.3
Drought Stress Induced a Decrease in the Level of Protein and Alteration in the Activity of Enzymes of N Metabolism in Roots (R) and Shoots (S) of 20-d Grown Rice Plants. Two Rice Cultivars Were Used for the Study

Rice Cultivar	Drought Stress		Protein (mg g⁻¹ fw)	Nitrate Reductase (nmol NO_2 min⁻¹ mg⁻¹ protein)	Glutamine Synthetase (µmol-γ-glutamyl Hydroxamate Formed min⁻¹ mg⁻¹ Protein)	Alanine Aminotransferase (nmol Pyruvate min⁻¹ mg⁻¹ Protein)	Aspartate Aminotransferase (nmol Pyruvate min⁻¹ mg⁻¹ Protein)
Ratna	0	(R)	13.00	7.80	0.56	18.00	42.00
		(S)	24.00	8.50	0.36	120.00	30.00
	0.5 MPa	(R)	6.00	6.00	0.48	32.00	54.00
		(S)	18.00	7.50	0.20	141.00	67.50
	−2.0 MPa	(R)	2.50	1.80	0.36	7.00	18.00
		(S)	10.80	2.00	0.08	15.00	17.50
Jaya	0	(R)	6.00	9.80	0.82	30.00	38.20
		(S)	14.60	11.0	0.65	62.00	45.00
	0.5 MPa	(R)	5.00	7.90	0.75	42.50	62.00
		(S)	9.20	8.00	0.38	112.50	75.00
	−2.0 MPa	(R)	3.50	1.05	0.30	5.40	22.50
		(S)	6.20	2.00	0.16	15.00	12.40

plants subjected to drought (Kaiser, 1987; Du et al., 1996). In sugarcane leaves, a decrease in the leaf water potential up to −0.37 and −0.85 MPa led to about a 2- to 9-times decrease in the activities of RuBP carboxylase, PEP carboxylase, fructose-1,6-bisphosphatase, NADP malic enzyme, and orthophosphate dikinase leading to an overall decreased rate of photosynthesis (Du et al., 1996). Drought stress alters carbon partitioning in plant parts owing to an alteration in the activities of sugar metabolizing enzymes. In leaves of sorghum plants, drought condition reduces sucrose formation owing to an inhibition in the activities of fructose-1,6-bisphosphatase and sucrose phosphate synthase (Reddy, 1996). However, in potato tubers, moderate drought stress leads to an activation of sucrose phosphate synthase and stimulation of sucrose synthesis (Geigenberger et al., 1997). More extreme drought stress reduces the activities of one or more of the enzymes associated with the terminal reactions of starch synthesis leading to a further alteration in carbon partitioning in potato tubers (Geigenberger et al., 1997). Levels of many enzymes increase under drought stress.

Many hydrolytic enzymes show an increased activity in drought-stressed tissues. The α-amylase activity increased under drought stress, which was responsible for increased starch hydrolysis *in vivo*, leading to increased levels of sugars and a decreased level of starch in drought-stressed tissues (Hsiao, 1973). Proteases have been shown to be induced under drought stress (Hameed et al., 2013). A thiol protease in pea and two cysteine proteinases in Arabidopsis have been identified which are induced under water deficit (Bray, 1995). Certain hydrolytic, as well as oxidative enzymes, show different behaviors in the crop cultivars differing in drought-stress tolerance. While investigating the behavior of drought-resistant and drought-sensitive maize cultivars for protease activity under drought stress, Thakur and Thakur (1987) observed an increasing trend in protease activity with an increasing osmotic potential in resistant cultivar Ageti-76, whereas in the sensitive cultivar Vijay, they observed a decreased protease activity under severe drought stress. In drought-sensitive rice cultivar Malviya-36, enhanced proteolysis was observed in shoots and roots under water stress in comparison to the tolerant cultivar Brown Gora (Pyngrope et al., 2013a). In-gel activity staining revealed more proteolytic bands and expression of oxidized proteins in roots compared to tolerant cultivar (Pyngrope et al., 2013a). In certain plant species, an increased synthesis of sucrose (Pelah et al., 1997) and proline (Dubey and Pessarakli, 1995) occurs under drought stress owing to a stress-induced increase in the activities of the enzymes synthesizing these metabolites. In *Populus popularis* plants, the accumulation of sucrose accompanied by the increased activity of its synthetic enzyme sucrose synthase occurs under water deficit (Pelah et al., 1997). The activity of the enzyme P5CS which is involved in the biosynthesis of proline, increases in rice seedlings under dehydration (Igarashi et al., 1997). In cotton plants, tolerant genotypes showed higher P5CS activities under drought than that of drought-sensitive genotypes (Parida et al., 2008). Petunia plants transformed by P5CS gene from *A. thaliana* (AtP5CS)

or from rice (OsP5CS) tolerated 14 days of drought stress (Yamada et al., 2005).

Drought induces oxidative damage by inducing the production of ROS and decreasing the activities of the antioxidant enzymes CAT, POD, and SOD (Mali et al., 1980; Gogorcena et al., 1995) in plants. De novo synthesis of monodehydroascorbate reductase (MDHAR), dehydroascorbate reductase (DHAR) and GR involved in ascorbate regeneration is one of the key drought-stress responses of plants to combat oxidative stress (Boo and Jung, 1999; Sharma and Dubey, 2005a). Ascorbate peroxidase (APX) serves as an important component of the antioxidative defense system under drought stress in plants (Sharma and Dubey, 2005a). In pea nodules, drought stress (−2.03 MPa) led to decreased activities of CAT (25%), APX (18%), DHAR (15%), GR (31%), and SOD (30%) with a simultaneous decrease in the contents of ascorbate (59%), reduced glutathione (57%), and oxidized glutathione (38%) (Gogorcena et al., 1995). The drought-sensitive cultivar, Malviya-36 showed a higher constitutive level of SOD and CAT in comparison to drought-tolerant cultivar Brown Gora. Underwater deficit SOD, CAT, and GPX, MDHAR, DHAR, APX, and GR showed a greater increase in tolerant cultivar compared to sensitive cultivar (Pyngrope et al., 2013b). Isozyme profiling revealed differential expression of APX isoforms in drought-tolerant cv. Brown Gora and drought-sensitive cv. Malviya-36. In comparison to Brown Gora, Malviya-36 showed a decrease in the content of protein thiol and an increase in the level of protein carbonyls (Pyngrope et al., 2013a). In rice, activities of CAT and SOD were decreased due to drought stress in salt-sensitive IR-29 and Pusa Basmati (PB) and remained unaltered in salt-tolerant Pokkali. Drought-induced GPX activity showed marked enhancement in both IR-29 and PB (Basu et al., 2010). Higher membrane stability index (MSI), lower accumulation of H_2O_2, and the enhanced activity of antioxidant enzymes, SOD, GPX, CAT, and APX, under water stress were observed in Sids, the more tolerant wheat cultivar to drought compared to Gmiza in the early vegetative stage (Hassan et al., 2015). Huseynova (2012) observed a significant increase in CAT, GR, and APX activities in drought-tolerant wheat cultivars Barakatli-95 (durum wheat) and Azamatli-95 (bread wheat) in comparison to sensitive cvs. Garagylchyg-2 (durum wheat) and Giymatli-2/17 (bread wheat) grown under severe water deficit compared to normal irrigation. The SOD activity was not affected in resistant cultivars but reduced significantly in sensitive cultivars (Huseynova, 2012). Underwater stress, higher activities of peroxidase, SOD, and CAT were observed in drought-tolerant wheat genotype FD-83 in comparison to sensitive genotype. In contrast, drought enhanced the activities of APX and proteases in sensitive genotype Nesser (Hameed et al., 2013). The overproduction of antioxidant enzymes in order to scavenge reactive oxygen species, which are greatly produced under drought, provides an elegant approach for engineering plant species for drought-stress tolerance. Transformed rice plants expressing pea manganese SOD have been shown to be more resistant to drought stress compared to normal plants (Wang et al., 2005).

22.4 HEAT STRESS

Due to global warming and changes in weather conditions, high temperatures or heat stress will probably become more widespread in the coming decades (Salomé, 2017). Heat stress adversely affects the growth and development of the plants and reduces crop yield in many areas of the world (Usman et al., 2014). Some plants can survive when the temperature exceeds 20°C above ambient, whereas in most field crops, temperatures above 40°C cause heat injury, severely limit photosynthesis, inhibit normal transcription and translation and alter protein metabolism by causing protein breakdown, protein denaturation, enzyme inactivation, induction in synthesis of certain novel proteins, etc. (He et al., 2005). Short-term severe heat stress can decrease the concentration of total protein and proteins associated with nutrient uptake and assimilation in tomato roots (Giri et al., 2017). Photosynthesis is very sensitive to high temperatures, and the primary sites of targets are PSII, RUBISCO, whereas Cytb559 and PQ are also affected. Production of HSPs, ROS and secondary metabolites are associated with high-temperature stress (Mathur et al., 2014). Proteome profiling of *Populus euphratica* subjected to heat stress revealed short-term upregulation of proteins associated with the restructuring of the cytoskeleton, destabilization of the membrane, assimilation of sulfur, biosynthesis of thiamine and hydrophobic amino acids, etc. Long-term upregulated proteins were associated with photosynthesis and redox homeostasis. Late downregulated proteins were mainly involved in carbon metabolism (Ferreira et al., 2006). In grapevine leaves, 23 upregulated proteins were identified, among which five were involved in photosynthesis, including three subunits of ATP synthase (α, β, γ) associated with photosystem electron-transfer reactions, PsaF, and a fructose-bisphosphate aldolase (FBA) involved in the Calvin cycle (Liu et al., 2014). Heat stress leads to protein dysfunction due to improper folding and protein aggregation. The ability of plants to respond to heat stress by keeping proteins in their proper conformation, inhibiting protein accumulation and inhibiting protein degradation by various proteases are very important for the survival of the cell. Exposure to heat stress causes alterations at the molecular level modulating the gene expression and transcript accumulation, causing the synthesis of stress-associated proteins as a stress-tolerance strategy (Wahid et al., 2007).

When plants are subjected to heat treatments, beyond optimum growth temperatures, normal protein synthesis declines owing to a coordinate loss of translational efficiency of most mRNAs, whereas synthesis of stress-related proteins increases (Iba, 2002). On exposure to heat stress, ribosomes fall off mRNAs but reassociate rapidly to start the synthesis of new proteins (Salomé, 2017). In *Arabidopsis thaliana*, a gene *hot3-1* encodes the translation initiation factor 5B (eIF5B) which is involved in the initiation of translation, including joining of a ribosomal subunit at the mRNA beginning to start the translation. In *hot3-1* mutants, which are sensitive to *hot* temperatures, this process was much slower leading to slower correction of the cellular damage induced

by heat stress as mutants were devoid of the capability to replenish the important proteins in a timely manner (Zhang et al., 2017a). Exposure to heat stress and signal perception leads to alteration at the molecular level modulating the gene expression and transcript accumulation, causing the synthesis of stress-associated proteins as a stress-tolerance strategy (Wahid et al., 2007). Plants increase the concentration of pre-existing molecular chaperones by expressing additional chaperones via a signaling mechanism (Reddy et al., 2016). A higher expression of HSPs has been associated with an induction of thermotolerance as an adaptive defense mechanism (He et al., 2005; Usman et al., 2014). The synthesis of HSPs occurs in diverse plant species when they are exposed to temperatures 10–15°C above ambient temperatures (Cooper and Ho, 1983; Heikkila et al., 1984; Kanabus et al., 1984; Kee and Nobel, 1986; Mansfield and Key, 1988; Singla and Grover, 1994; Cordewener et al., 1995; Waters et al., 1996; Wang et al., 2014a). Crop plants, such as maize, soybean, cowpea, and wheat, start synthesizing HSPs in the tissues with a rise in tissue temperatures beyond 32–33°C (Vierling, 1991; Efeoglu and Terzioglu, 2007). Ribosomal protein S1 and 14-3-3-like protein were upregulated by heat stress (Liu et al., 2014). HSP22 located in the endoplasmic reticulum was induced 3-fold under heat stress, whereas HSP21 located in chloroplast was induced 5.5-fold in grapevine leaves (Liu et al., 2014). The induction of HSP synthesis parallels the temperature increase. It has been observed that heat shock leads to stability as well as the rapid induction of specific mRNAs synthesis related to specific HSPs (Sullivan et al., 1990). The HSPs range in molecular mass from about 10 to 200 kDa (Schoffl et al., 1999). The phenomenon of heat-shock response is conserved among all biological organisms. Although HSPs provide the molecular basis for thermotolerance, but whether they act directly in signal transduction or induce the synthesis of secondary agents involved in protection is not yet clear. The HSPs are proposed to act as molecular chaperones (Gething, 1997). They stabilize proteins and membranes and can help in protein refolding under stresses (Wang et al., 2004). The possible different roles of HSPs in plants under heat stress are shown in Figure 22.5. The HSPs take part in protein disaggregation, stabilization of protein conformation, control of cell cycle, protein trafficking, and assist newly synthesized and translocated proteins to attain their native forms. HSPs have been shown to improve photosynthesis, partitioning of assimilates, membrane stability, water and nutrient use efficiency, etc. (Camejo et al., 2005; Ahn and Zimmerman, 2006; Momcilovic and Ristic, 2007).

22.4.1 SYNTHESIS OF HEAT-SHOCK PROTEINS

The synthesis of HSPs occurs in plant cell cultures undergoing thermoadaptation or intact plants subjected to heat stress (Vierling, 1991). It has been shown that not only heat stress but other conditions, such as treatment with arsenite, heavy metals, ethanol (Vierling, 1991), ABA, drought stress, and wounding (Heikkila et al., 1984), also induce expression of some of the HSPs mRNAs and synthesis of HSPs in plants.

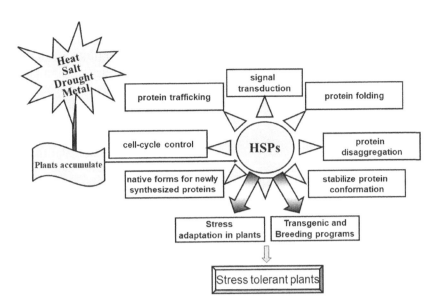

FIGURE 22.5 Roles of heat-shock proteins in plants under heat stress. HSPs take part in signal transduction, protein folding, protein disaggregation, stabilization of protein conformation, control of cell cycle, protein trafficking, and assist newly synthesized and newly translocated proteins to achieve their native forms and thus confer an enhanced survival value to the plants against heat stress.

This shows that the synthesis of HSPs can be induced even in the absence of heat stress and high-temperature protection can be provided without prior heat shock. He and coworkers (2005) investigated the effects of sudden heat stress and heat acclimation on the synthesis of proteins in creeping bentgrass (*Agrostis palustris* Huds.) and observed that sudden and gradual heat exposure induced the expression of HSPs of molecular weight 23, 36, and 66 kDa, whereas HSPs of mol. wt. 57 and 54 kDa were induced due to heat acclimation. New HSPs synthesized due to heat acclimation seem to be associated with increased thermotolerance of creeping bentgrass (He et al., 2005). A 2-D protein analysis revealed a higher and diverse response in wheat seedlings exposed to 37°C for 8 hours compared to the other treatments like 37°C for 4 hours, 45°C for 8 hours, 45°C for 4 hours, and 45°C for 2 hours. Five protein spots, ranging from 6–7.8 pI and 27–31.7 kDa molecular weight, were expressed when seedlings were exposed to 37°C for 2 hours and continued at 37 and 45°C for all exposure times. This suggests that these early proteins and other newly synthesized proteins may have protective effects at 37 and 45°C and appear to help in achieving healthy growth during the recovery period (Efeoglu and Terzioglu, 2007). Cytoplasmic distribution and subcellular localization of HSPs indicate that they remain either specifically associated with various organelles, such as nuclei, chloroplasts, mitochondria, plasma membrane, or as cytoplasmic aggregates distinct from ribosome granules (Mansfield and Key, 1988; Vierling, 1991). When the tissue temperature exceeds 32–33°C, HSPs are typically seen. The appearance of these proteins has been positively correlated with enhanced thermotolerance, and also provide a certain level of cross-protection to other kinds of stresses (Artlip and Funkhouser, 1995). In carrot cells, the heat-shock treatment causes inhibition of protein synthesis with the simultaneous appearance of new proteins (Pitto et al., 1983). In carrot cells, patterns of these newly synthesized

proteins become different depending on the growth stages of cells and culture conditions. It was shown by Kanabus et al. (1984) that tobacco cell suspensions synthesize HSPs in different phases of the growth cycle.

Different tissues of the same plant, as well as different developmental stages of the tissues, show a different pattern of HSPs (Burke et al., 1985). Among the wide range of HSPs which accumulate under heat shock, some are specifically associated with organelles, including the nucleus, nucleolus, chloroplast, mitochondria, and plasma membrane. Certain other HSPs are found to be associated with ribosome, or they remain as aggregates in the cytoplasm (Mansfield and Key, 1988). In pigeon pea (*Cajanus cajan*) plants, heat-shock proteins of 18, 20, 22, and 24 kDa are associated with mitochondrial and membrane fractions, whereas the 60, 70, and 81 kDa proteins are found in the soluble fraction (Kishore and Upadhyaya, 1994). In mitochondria of pea plants, a novel 22 kDa protein accumulates in the matrix when normal the growth temperature is shifted from 25 to 40°C (Lenne and Douce, 1994). Using 2-D gel electrophoresis, the effect of heat stress on soluble proteins from leaf tissues of wheat cultivars varying in sensitivity to heat stress was studied. Twenty-two of thirty-one differentially expressed proteins were newly synthesized LMW or small HSPs.

In heat-tolerant cultivars, the number of the sHSPs was higher compared to heat-sensitive cultivars. Some of the sHSPs were cultivar specific (Yildiz and Terzi, 2008). These proteins were suggested to play an important role in acquiring thermotolerance. Rice seedlings subjected to high temperatures showed 73 differentially expressed proteins, among which 48 proteins were identified, and it was observed that a group of sHSPs which constitute a diverse family of chaperones which co-aggregate with misfolded proteins were newly induced in rice by heat stress (Lee et al., 2007; Ungelenk et al., 2016). In a rice cultivar Nipponbare during anthesis and seedling stages, heat stress led to upregulation of five heat-shock proteins Hsp16.9A, Hsp17.4, Hsp17.9A, Hsp23.2, and Hsp26.7 (Chen et al., 2014). The level of all five HSPs was higher in a heat-tolerant rice cultivar Co39 in comparison to the heat-sensitive cultivar Azucena, indicating that the expression level of these five HSPs are positively related to heat tolerance (Chen et al., 2014).

22.4.2 TYPES OF HEAT-SHOCK PROTEINS

Heat stress induces the synthesis of a wide range of HSPs in plants. A general system of classification for these proteins is based on their molecular weights and their localization in the cell. The five major conserved families of HSPs/chaperones are (I) the HSP100/Clp family, (II) the HSP90 family, (III) the HSP70/DnaK family, (IV) the chaperonins (GroEL and HSP60), and (V) the sHSP (Huang and Xu, 2008). HSP100/Clp family chaperones are reported to be induced by different environmental stresses (Wang et al., 2004). Reduced thermotolerance was observed in tomato lines expressing an antisense construct of chloroplast-localized HSP100/ClpB (Yang et al., 2006). Many HSPs, like HSP60, HSP70, and HSP90, are present as constitutive proteins in the cytoplasm as well as other organelles like mitochondria and chloroplasts of plants in non-stressed conditions and their level increases under heat shock. Studies indicate that HSPs function in a fashion similar to molecular 'chaperons' and assist the self-assembly of nascent polypeptides into their correctly folded tertiary structures and also prevent the formation of an aggregation of non-functional proteins resulting from heat denaturation (Artlip and Funkhouser, 1995).

Small HSPs which range in size from 12 to 40 kDa and are encoded by six nuclear gene families accumulate to high levels in response to heat stress. These small HSPs bind denatured proteins partially, prevent irreversible inactivation and aggregation of proteins, and hence lead to the development of thermotolerance (Waters et al., 1996). Under extreme heat stress, A. thaliana RNAi and overexpression lines for plant cytosolic sHSPs showed decreased and increased tolerance, respectively. They have been proposed to be involved in the protection of specific translation factors in cytosol either directly or indirectly (McLoughlin et al., 2016). According to Harrington et al. (1994), HSPs have a possible function in signal transduction involving protein kinases and heat-shock-induced calmodulin-binding proteins. Transgenic tobacco plants expressing enhanced levels of NtHSP70-1 exhibited reduced degradation and fragmentation of nuclear DNA during heat stress. In addition, NtHSP70-1 overexpressing seedlings were healthy, whereas antisense seedlings died after heat stress (Cho and Choi, 2009).

22.5 LOW TEMPERATURE

Low temperature is one of the most severe constraints limiting crop productivity. Low environmental temperatures lead to chilling injury in plants and result in the loss of plasma membrane integrity, irreversible and proportional loss of

proteins from the cell, and ultimately the death of the cell (Levitt, 1980; Stuiver et al., 1988). According to Levitt (1980), freezing-induced dehydration within the cell leads to aggregation of proteins owing to the formation of disulfide bonds as well as the denaturation of soluble proteins. Synthesis of many key enzyme-proteins decreases when plants are exposed to low temperatures (Bruggemann et al., 1994; Matsuba et al., 1997; Janmohammadi et al., 2015), and synthesis of certain specific proteins is induced (Griffith et al., 1997). Many enzymic proteins, especially those of the carbon assimilation pathway, are extremely sensitive to chilling. Low-temperature stress reduces the levels of RUBISCO large subunit and D1 protein (Li et al., 2016). The key photosynthetic enzyme of C3 plants, RUBISCO, which constitutes up to 50–60% of the soluble proteins, undergoes changes in structure, conformation, and properties at low temperatures (Stuiver et al., 1988; Bruggemann et al., 1994). In *Zoysia japonica* plants, a drastic decline in the level of C4-cycle enzymes phosphoenol pyruvate carboxylase (PEPC) and phosphoenolpyruvate carboxykinase (PCK) occurs due to low-temperature stress. In tomato and maize plants, chilling stress results in an irreversible loss of RUBISCO and stromal fructose-1,6-bisphosphatase activities (Bruggemann et al., 1994). In the conifer *Pinus sylvestris*, the contents of the D1 protein of the photosystem II (PSII) reaction center and PSII light-harvesting complex (LHCII) proteins decline under low-temperature stress (Ottander et al., 1995). Similarly, the −SH-rich enzyme GR becomes partially inactivated by freezing (Guy and Carter, 1984).

It is suggested that the activity loss of many enzymes on chilling is as a result of modification of sulfhydryl groups or other side chains of the protein (Bruggemann et al., 1994). Substantial enhancement in protein content was observed in stress-tolerant plants exposed to short-term cold stress. This was suggested to be due to activation of the cold stress response mechanism (Kosakivska et al., 2008). Karimzadeh and coworkers (2006) observed an alteration in the electrophoretic pattern of the water-soluble proteins from winter and spring wheat cultivars and pointed out the accumulation of stress proteins in leaves on exposure to freezing temperature. In leaf cells of *Brassica campestris* var. olifera and *Amaranthus caudathus* L. (ruderals) and *Rumex patienta* L. x *Rumex tianshanicus* A. Los. (stress-tolerant), differences in protein patterns were observed in control and short-term low-temperature exposed plants (Kosakivska et al., 2008).

Due to low-temperature stress, expression and accumulation of many proteins were observed in three major cereal crops produced worldwide, wheat, rice, and barley. Low temperature leads to upregulated expression of proteins associated with carbohydrate metabolism, ROS detoxification, redox adjustment, cryoprotection, defense/detoxification, remodeling of the cell wall, cytoskeletal rearrangements, etc. In contrast, low temperature led to downregulation of many photosynthesis-associated proteins. These alterations are general adaptation mechanisms in plants, as observed in the model plant Arabidopsis (Janmohammadi et al., 2015). Protein expression patterns of rice roots undergoing chilling stress revealed 27 upregulated proteins including some novel proteins such as acetyl transferase, fructokinase, phosphogluconate dehydrogenase, NADP-specific isocitrate dehydrogenase, putative alpha-soluble N-ethylmaleimide-sensitive factor (NSF) attachment protein, PrMC3 and glyoxalase 1. These proteins were mainly involved in various processes, including detoxification, vesicular trafficking, energy production, and metabolism (Lee et al., 2009). Using 2-DE, Hashimoto and Komatsu (2007) detected the synthesis of four new proteins in rice plants exposed to cold stress. Comparative proteomic analysis of rice plants exposed to low temperature provided new insights into chilling stress responses. Yan et al. (2006) using mass spectrometry analysis identified 85 differentially expressed proteins, including RcbL, RcbA, APX, enolase, and HSPs which are well known low-temperature-responsive proteins and certain novel proteins such as 2-Cys peroxiredoxin, armadillo repeat-containing protein, and putative nascent polypeptide-associated complex. sHSPs have been suggested to play important roles in the acquisition of freezing tolerance in *Castanea sativa* (Lopez-Matas et al., 2004).

Subzero temperatures may be endured by cold-tolerant plants partially by reducing the ice crystals formation in the intercellular spaces by accumulating antifreeze proteins (AFP) and carbohydrates. Certain proteins associated with resistance to pathogens have the capability to bind and alter the ice crystal growth in plants. Seedlings of cold-treated *Picea abies* L. Karst. ecotypes showed thaumatin PR proteins and β-1,3-glucanase in the apoplastic fluid and extra accumulation of various PR chitinases isoforms (Dalen et al., 2015). The levels of antioxidant enzymes such as SOD, POD, and APX were enhanced in low temperature (20, 17°C) treated *Kappaphycus alvarezii* thalli (Li et al., 2016).

22.5.1 COLD ACCLIMATION

When plants are exposed to low non-freezing temperatures for a few hours or day, certain new sets of proteins are synthesized, and these plants develop the capacity to adapt to subsequent chilling or freezing temperatures. Such a mechanism of adaptation is known as cold acclimation (CA). Generally, temperatures from 4 to 15°C are considered to be chilling, whereas a temperature below 4°C, is considered to be freezing (Artlip and Funkhouser, 1995). CA results in altered gene expression leading to the synthesis of specific proteins and certain enzymes which are responsible for the development of freezing tolerance (Dhindsa and Mohapatra, 1990; Antikainen and Griffith, 1997; Griffith et al., 1997; Catalá et al., 2014).

Several pre-existing proteins abundant in the tissues of plants grown under normal temperatures, decline on exposure to low temperatures. However, many new transcripts and polypeptides are synthesized (Meza-Basso et al., 1986; Guy and Haskell, 1987; Griffith et al., 1997; Yan et al., 2006; Hashimoto and Komatsu, 2007; Janmohammadi et al., 2015), which appear to play a major role in acclimation of plants to freezing temperatures (Griffith et al., 1997; Janmohammadi et al., 2015). Plants show variation in low-temperature tolerance. In many crop plants, exposure to low non-freezing

temperatures induces freezing tolerance. Freezing-tolerant or cold-acclimated plants possess new proteins which are not present in normal or non-acclimatized plants. Hahn and Walbot (1989), while studying the effect of cold treatment on the pattern of protein synthesis in rice leaves, detected several novel proteins of 95, 75, 25, and 21 kDa that were synthesized during 1–7 days of 11 and 6°C cold treatment. These proteins were cold specific, other environmental stresses, such as salinity, drought stress, and acid treatment, could not induce the synthesis of such proteins. No uniform pattern was observed for protein synthesis among plant species during CA. This implies that CA-induced proteins are not highly conserved as heat-shock proteins. A characteristic feature of CA-induced proteins is that some of the synthesized proteins are transient, whereas others are stable, the synthesis of which continues for weeks (Guy and Haskell, 1987).

In herbaceous plants, overwintering and cold acclimation are expensive energetically and depend on a functional plastid metabolism. Cold shock (1 day) affected Arabidopsis plastid proteomes. Short-term (10 days) acclimation induced major alterations in the stromal proteins but few modulations in the lumen proteome (Goulas et al., 2006). Long-term acclimation (40 days) resulted in changes of the proteomes of both compartments, with the appearance of new proteins in the lumen and further alterations in protein level in the stroma. Forty-three differentially expressed proteins seemed to be involved in photosynthesis, hormone biosynthesis, other plastid metabolic functions, and stress perception and signal transduction (Goulas et al., 2006). In leaves of maize inbred line W9816 (a cold-resistance genotype) reduction of photodamage due to the over-energized state of the thylakoid membrane, enhanced production of stress-responsive proteins, energy generation via glycolysis and enhanced ability to detoxify ROS. Under chilling stress, post-transcriptional regulation and post-translational modifications also play key roles in the adaptation of maize (Wang et al., 2016). In rice seedlings, 32 early cold-regulated proteins were identified using 2-D. Within 5 minutes of cold treatment, 26 unique proteins exhibited an altered expression level. Among these proteins, the protein level of cellular phospholipase D (OsPLD-1) was enhanced as early as 1 minute after exposure to cold temperature (Huo et al., 2016). In freezing-tolerant cereal plants, such as rice, wheat, and barley, antifreeze proteins are synthesized during CA, which appear to play a significant role in increasing freezing tolerance (Antikainen and Griffith, 1997; Griffith et al., 1997). AFPs enable plants to survive under subzero temperatures and have properties to prevent ice recrystallization in the cells by modifying ice crystal morphology (Sharma and Deswal, 2014). Six AFPs with unique ability to absorb onto ice surface and ability to inhibit its growth have been isolated from protoplast of winter rye leaves where at subzero temperatures ice formation takes place (Griffith et al., 1997). Among the rye AFPs, two are chitinase-like, two endoglucanase-like, and two thaumatin-like proteins (Griffith et al., 1997). The accumulation of AFPs is not a common response to all plants, but it is a specific response playing an important role in the freezing tolerance for some plants. Transgenic tobacco plants expressing carrot AFP showed increased freeze resistance and inhibition of electrolyte leakage from cold-stressed cells (Worrall et al., 1998; Fan et al., 2002). The only grass indigenous to Antarctica, *Deschampsia antarctica* E. Desv. has well-developed freezing tolerance which is induced strongly by cold acclimation (John et al., 2009). *D. antarctica* showed expression of potential recrystallization inhibition activity that reduces the conversion of small ice crystals into large ones at low temperature. It is localized in the apoplasm and is proteinaceous in nature. In leaf tissues, *D. antarctica* ice recrystallization inhibition protein (DaIRIP) transcript levels are significantly increased after cold acclimation. Overexpression of a *DaIRIP* in *A. thaliana* demonstrated novel RI activity (John et al., 2009). In certain plants, such as citrus (Durham et al., 1991) and some herbaceous species (Guy et al., 1988), a very high molecular weight protein has been identified that is specifically synthesized under CA. Durham et al. (1991), while comparing polypeptide patterns resulting from *in vitro* translations of total RNA isolated from cold-acclimatized and non-cold-acclimatized leaf tissues of cold-sensitive *Citrus grandis* plants, observed a 160 kDa polypeptide in cold-acclimatized leaves that was not present in non-cold-acclimatized citrus leaves. This 160 kDa unique polypeptide has also been detected in cold-acclimatized spinach and sweet orange, *C. sinensis* (Guy et al., 1988). In *A. thaliana*, MSACIA and MSACIB, two glycine-rich proteins, accumulate during CA. The expression of these two proteins varies in timing and localization. Differential expression of these proteins involves both transcriptional regulation and post-transcriptional modifications (Ferullo et al., 1997). Comparisons among various cultivars of *A. thaliana* indicate that failure to accumulate these proteins is associated with low freezing tolerance.

Dehydrins are usually induced in response to stresses that cause cellular dehydration. Dehydrins are all highly hydrophilic proteins, and they differ in size, ranging from 188 to 475 amino acids (Xu et al., 2014). Extracellular freezing imposes cellular dehydration leading to accumulation of dehydrins and desiccation tolerance which are considered as a key component of the CA process. Accumulation of high levels of dehydrin transcripts has been observed in field-grown freezing-tolerant bromegrass (*Bromus inermis* L.) and rye plants (Robertson et al., 1994). When exposed to chilling stress, three dehydrin-like lysine-rich proteins with molecular weight 14, 60, and 65 kDa accumulated in floral buds of a woody perennial blueberry (*Vaccinium*) (Muthalif and Rowland, 1994). Field-acclimated plants accumulating high levels of dehydrin transcripts and proteins are considered more freeze tolerant (Robertson et al., 1994; Arora et al., 1997). Enhanced accumulation of dehydrin transcripts or proteins has been reported in plants during the coldest months (Welling et al., 2004; Wisniewski et al., 2006). Peng and coworkers (2008) observed that one of the dehydrins from leaf tissues of *Rhododendron catawbiense* RcDHN5 encodes an acidic, SK2 type dehydrin expression of which is upregulated and downregulated during seasonal CA and spring deacclimation (DA), respectively. In vitro partial water loss assays suggested that purified RcDHN5 provides protection to enzyme activity against a dehydration treatment.

Arabidopsis plants transformed with RcDHN5 exhibited improved 'constitutive' freezing tolerance compared with the control plants (Peng et al., 2008). Cucumber lines transformed with Dhn24 encoding a SK3-type DHN24 dehydrin from cold-acclimated species S. sogarandinum showed a significantly less chilling injury (Yin et al., 2006). In the perennial fruit crop Loquat (*Eriobotrya japonica*), the imposition of freezing led to upregulation of seven dehydrin proteins and this upregulation was much greater in freezing-tolerant cultivar than freezing sensitive cultivar (Xu et al., 2014). Osmotin also accumulates and plays an important role in cold acclimation of plants (D'Angeli and Altamura, 2007). Increased cold tolerance was observed in tomato plants transformed with tobacco osmotin (NtOsm) (Patade et al., 2013). Transgenic olive tree expressing tobacco osmotin gene demonstrated its association with cold acclimation-related programmed cell death, cold-induced blockage of calcium signaling and cold-induced alterations in the cytoskeleton (Parkhi et al., 2009). However, the underlying mechanism of osmotin induced cold tolerance is still unclear.

During cold stress, gene induction through activation of promoters containing C-repeat (CRT) cis-element by C-repeat-binding factors (CBF) is the best understood genetic pathway involved in cold tolerance. Low non-freezing temperatures increased freezing tolerance in *A. thaliana* via CBF regulatory pathway. Activation of CBF transcription factors indicates that cold stress perceived as a signal is transduced into the cells (Ruelland et al., 2009). CBFs overexpression in Arabidopsis caused increased freezing tolerance (Gilmour et al., 2000; Fowler and Thomashow, 2002). In Arabidopsis, low temperature but not NaCl, ABA, or drought stress induces expression of RARE COLD INDUCIBLE 1A (RCI1A; 14-3-3 psi) protein indicating a specific function for this protein in cold stress (Jarillo et al., 1994; Roberts et al., 2002). RCI1A negatively regulates cold acclimation and constitutive freezing tolerance by regulating cold-induced gene expression (Catalá et al., 2014). Overexpression of Arabidopsis transcriptional activator CBF1 induces cold-regulated (cor) genes and increases freezing tolerance in non-acclimated Arabidopsis plants (Jaglo-Ottosen et al., 1998). Rice plants expressing barley HvCBF4 showed enhanced abiotic stress tolerance (Oh et al., 2007). However, the CBF pathway has been suggested to be complex and highly interconnected, and enhancement in freezing tolerance accompanying cold acclimation is dependent partially on the CBF-CRT/DRE regulatory module (Park et al., 2015). A cis-acting promoter element, DRE (drought-responsive element) plays an important role in regulating gene expression under abiotic stresses. Plants transformed with cDNA encoding DREB1A induced the expression of several stress-tolerance genes under normal growing conditions and led to improved tolerance of plants toward drought, salt loading, and freezing (Liu et al., 1998; Kasuga et al., 1999; Kasuga et al., 2004; James et al., 2008). Similarly, Gutha and Reddy (2008) observed increased tolerance of transgenic tobacco plants overexpressing OsDREB1B toward the abiotic stresses osmotic stress and freezing. The inheritance of freezing tolerance appears to be a multigenic phenomenon, and the precise function of the proteins encoded by these genes is not fully known. Both transcriptional and post-transcriptional controls have been shown to be involved in the expression of these genes (Hughes and Dunn, 1996).

22.5.2 ABSCISIC ACID AND CA

An ABA-dependent cold signaling pathway is CBF-independent and has been studied in several plant species (Shi et al., 2015). Exogenous ABA induces freezing tolerance in many plant species (Dhindsa and Mohapatra, 1990; Tseng and Li, 1991), although the physiological basis of this phenomenon is poorly understood. In certain plants, an increase in the exogenous ABA level is observed following CA (Tseng and Li, 1991; Artlip and Funkhouser, 1995; Pagter et al., 2008; Kosová et al., 2012). Cold stress (4°C) responses in wheat varieties are associated with hormonal changes. When the contents of most abundant dehydrin WCS120 were compared in the crowns and leaves of the spring wheat cv. Sandra and winter wheat cv. Samanta, a rapid and stronger increase in the level of ABA and dehydrin WCS120 was observed within 1 day in winter wheat, accompanied with a decrease in bioactive auxin and cytokinins as well as increased deactivation of gibberellins, suggesting rapid suppression of growth in winter wheat as a result of cold exposure (Kosová et al., 2012), whereas during next 3–7 days of cold exposure winter wheat plants got adapted to cold with a decrease in the level of ABA and increase in cytokinins and auxins unlike spring wheat cv. Sandra. The inability of spring wheat cultivar to frost tolerance (FT) was correlated with the maintenance of a high level of cytokinins and auxins (Kosová et al., 2012). Plantlets of potato (*S. commersonii*) stem culture, when treated with ABA for 14 days, developed cold tolerance with the concomitant induction of 30 polypeptides (Tseng and Li, 1991). It is highly unlikely that ABA regulates all the genes associated with cold acclimation; however, it regulates many of the genes associated with an increase in freezing tolerance (Gusta et al., 2005).

Evidence suggests that there may be ABA-dependent and ABA-independent pathways involved in the cold acclimation process. Cold acclimation led to increased concentration of many amino acids such as aspartic acid, glutamic acid, glutamine, and proline in wheat plants, whereas ABA had a significant effect only on asparagine indicating that ABA does not mediate cold-induced alterations in free amino acid levels (Kovács et al., 2011). Several specific translatable mRNA populations and their *in vivo* translation products have been identified following ABA treatment of potato plantlets (Robertson et al., 1994). It is suggested that ABA alters gene expression leading to the development of cold hardiness (Mohapatra et al., 1988) by the synthesis of certain specific polypeptides that are similar to some of the polypeptides synthesized during CA (Tseng and Li, 1991; Mantyla et al., 1995). The ABA has been shown to induce the synthesis of certain polypeptides that are not synthesized in CA tissues (Tseng and Li, 1991). A comparative study of CA-induced proteins and ABA-induced proteins suggests that both CA

and ABA induce the synthesis of specific and certain common proteins. This also suggests that the full development of cold tolerance requires the synthesis of complete sets of CA-induced proteins, because certain genes, in addition to those responsive to ABA, are involved in the development of maximum freezing tolerance (Dhindsa and Mohapatra, 1990). Genetic studies indicate that ABA biosynthesis and signaling components are important for the expression of COR (cold-regulated) genes (Gilmour and Thomashow, 1991; Mantyla et al., 1995). ABA-insensitive-3 (ABI3) and ABA-responsive element (ABRE)-binding proteins were shown to function in the cold stress response. Expression of these proteins in seeds conferred an ability to express COR genes in vegetative tissues and enhanced freezing tolerance in Arabidopsis (Tamminen et al., 2001).

22.6 ANAEROBIC STRESS

Anaerobic stress is generally caused by excessively wet soil or flooding conditions. Anaerobiosis affects plant metabolism as a result of a low oxygen concentration in the rooting medium. Plants adapting to anaerobic stress switch from oxidative to fermentative carbohydrate metabolism (Ricard et al., 1991). Under anaerobic stress, normal protein synthesis is suppressed due to loss of polysomes, alteration in gene expression leads to the synthesis of specific sets of novel polypeptides commonly known as transition polypeptides (TPs) and anaerobic polypeptides (ANPs). Repression of pre-existing aerobic proteins and the synthesis of new proteins appear to be the immediate biochemical response of anaerobiosis (Mohapatra et al., 1988; Subbaiah and Sachs, 2003; Olgun et al., 2008; Alam et al., 2010; Loreti et al., 2016). In soybeans, a total of 365 nuclear proteins got altered in expression with flooding, and translation of most of the proteins was suppressed due to the continuous flooding (Yin and Komatsu, 2016). Most of the studies related to protein synthesis under anaerobic conditions have been performed in maize (Sachs et al., 1980), rice (Ricard et al., 1991), and Arabidopsis (Dolferus et al., 1997). Transient polypeptides are translated primarily during the first 5 hours of anoxia, and they are stable and last long after their synthesis declines, whereas anaerobic polypeptides appear after approximately 90 minutes of anoxia, and their synthesis continues for several days until cell death (Artlip and Funkhouser, 1995). In maize seedlings, anoxia led to an immediate termination of protein synthesis followed by synthesis of approximately 20 selective anaerobic proteins. These anaerobic proteins included enzymes associated with glycolytic-fermentative pathways required for rescuing the cell from the energy crisis, aerenchyma formation, and root tip death. Even partial depletion of oxygen induces 'anaerobic' genes (Subbaiah and Sachs, 2003). Because of anaerobic conditions in maize, initially the rapid synthesis of four 33 kDa transition polypeptides takes place, whereas after 90 minutes of anoxia, synthesis of 20 additional polypeptides occurs (Sachs et al., 1980; Sachs et al., 1996), which represent about 70% of the total proteins synthesized during anaerobiosis (Sachs et al., 1980).

Anaerobic stress-induced proteins are different from HSPs except for a few that are common to both types of stresses (Sachs et al., 1996). Proteome analysis of the coleoptile of an anoxia tolerant rice genotype revealed that one of the proteins synthesized during anoxia was ortho phosphate dikinase (PDK) that could help to produce pyrophosphate from ATP (Huang et al., 2005). Flooding led to a reduction in protein associated with protein synthesis and enhancement in fermentation, stress, wall-related proteins, and glycolysis-related proteins. Three S-adenosylmethionine synthetases generally declined under flooding conditions. Their mRNA expression levels also exhibited a similar tendency as the changes in the abundance of protein. These results indicated the involvement of S-adenosylmethionine synthetase in stress response regulation (Oh and Komatsu, 2015). Submergence tolerance of rice depends on SUB1A and/or SNORKEL genes, and germination of rice under anaerobic conditions requires trehalose-6-phosphate phosphatase (Loreti et al., 2016). Average amylolytic activity in rice embryos under anoxia was 2-fold higher compared to endosperms (Pompeiano and Guglielminetti, 2016). Various measures are taken by plants to limit resource consumption due to energy limitation during hypoxia. ATP reserves are spared by using pyrophosphate (PPi)-dependent enzymes which can utilize PPi as a substrate to catalyze reactions (Pucciariello et al., 2014). Most of the ANPs are apparently involved in maintaining ATP levels in the cells. Expression analyses revealed the importance of ATP and PPi-dependent phosphofructokinase gene family during anoxic stress (Mustroph et al., 2013).

Many of the enzymes associated with glycolysis or fermentative processes including alcohol dehydrogenase (ADH), lactate dehydrogenase (LDH), enolase, aldolase, pyruvate decarboxylase, glucose-phosphate isomerase, glyceraldehydes-3-phosphate dehydrogenase, and sucrose synthase were enhanced during anaerobiosis (Sachs et al., 1996). Among these enzymes, ADH is the best-characterized. In several tissues of maize examined, ADH gene expression is maximal with anoxia (Artlip and Funkhouser, 1995). Similarly, in *A. thaliana*, two ADH genes exist, one set is strongly induced by low oxygen stress mainly in roots, whereas the other set is constitutively expressed in roots and leaves (Dolferus et al., 1997). In maize seedlings during several days of hypoxic induction, LDH activity increases up to 3.5-fold. This enhanced activity resulted from increased protein levels, which can be correlated with the induction of 2 *ldh* transcripts of 1.3 and 1.7 kb (Christopher and Good, 1996). Ricard et al. (1991) observed a significantly enhanced level of sucrose synthase with a concomitant increase in its mRNA level in rice seedlings subjected to anaerobiosis. Unlike maize, only one sucrose synthase protein exists in rice. Its synthesis is enhanced with a concomitant increase in mRNA levels under anaerobiosis (Ricard et al., 1991) that indicates that its level of control is possibly transcriptional. Proteins involved in photosynthesis, tricarboxylic acid (TCA) cycle, RNA, DNA, and signaling were mainly affected in hypocotyl and leaf of seedlings facing flood. Proteins, including α-amylase and β-glucosidase, were enhanced in root and leaf under stresses indicating that α-amylase may be associated with mobilization of carbohydrate in soybean leaf under flooding stress (Wang et al., 2017).

Two enzymes have been identified which have functions different than ANPs, the level of which increases due to hypoxia. These are 1-aminocarboxylate-1-cyclopropane synthase (ACC synthase), which catalyzes the rate-limiting step in the synthesis of ethylene, and the other is xyloglucan endotransglycosylase, which is possibly involved in aerenchyma formation during flooding (Artlip and Funkhouser, 1995; Sachs et al., 1996). During hypoxia, pyruvate generated from glycolysis is efficiently utilized by the alanine aminotransferase/glutamate synthase cycle, causing accumulation of alanine and regeneration of NAD^+. Carbon is then kept in a nitrogen store to avoid its loss via ethanol fermentative pathway (Diab and Limami, 2016). In maize seedlings, it has been observed that treatment with ABA increases tolerance to anaerobic conditions (Hwang and Van Toai, 1991). Such an induction of tolerance is partly due to the synthesis of novel proteins. It was shown by Hwang and Van Toai (1991) that cycloheximide, when added together with ABA, reduced the survival rate of maize seedlings. However, ABA-induced tolerance appears to be specific, because results similar to those with maize are not observed in other crops.

Hypoxia causes oxidative stress in plants (Blokhina et al., 2003). In wheat roots under anoxia total SOD activity increased and additional SOD activity bands appeared in native gels. With increasing duration of anoxia, SOD mRNA levels declined, whereas protein content of different SOD isoforms was enhanced. Subcellular fractionation experiments showed that anoxia-responsive SOD isoforms are plastid-associated. SOD has been suggested to be a very stable enzyme which accumulates and remains active even under severe anoxia (Biemelt et al., 2000). In soybean roots, waterlogging stress led to upregulation of 14 proteins and induction of 5 new proteins. Several proteins such as enzymes involved in fermentation and glycolysis pathway were induced by anaerobiosis suggesting that, via the fermentation pathway, plants try to meet energy demand during anaerobiosis stress (Alam et al., 2010). The novel proteins are associated with many processes such as signal transduction, RNA processing, programmed cell death, energy metabolism and redox homeostasis (Alam et al., 2010). Flooding stress limits energy production and hence leads to growth inhibition and eventually death in most plants. Nineteen proteins that were observed under flooding stress in soybean were development and protein-related proteins. Ferritin was suggested to have a key role in plant cell protection against oxidative damage (Kamal et al., 2015).

22.7 PATHOGENESIS

When plants are infected with pathogens like bacteria, viruses, and fungi, certain novel proteins are synthesized (Fang et al., 2015; Jiang et al., 2015). These are host-coded proteins and are known as pathogenesis-related proteins which protect the plants from the invasion of the pathogen (Viktorova et al., 2012). Many of the PR proteins are also induced by altered levels of plant hormones, or the presence of various pollutants including heavy metals (Jung et al., 1995). Since these proteins are synthesized during infection, they appear to have a possible role in inducing resistance against further infection by pathogens (Tornero et al., 1997). The PR proteins form a heterogeneous family of plant proteins and have been divided into different classes depending on biological properties, enzyme activity, and coding sequence similarities. PR proteins are classified in 17 families including β-1,3-glucanases, chitinases, peroxidases, thaumatin-like proteins, ribosome-inactivating proteins (RIPs), lipid-transfer proteins, thionins, defensins, proteinase inhibitors, oxalate oxidase, and oxalate oxidase-like protein (Selitrennikoff, 2001; Ghosh, 2006).

β-1,3-glucanases and chitinases are the best-characterized PR proteins and belong to. PR-2 and PR-3 families of PR proteins respectively. Chitinases hydrolyze β-1,4-acetyl glucosamine linkages of chitin polymers, which are primary constituents of fungal cell walls, whereas glucanases hydrolyze β-1,3-glucan residues present in the fungal cell wall (Artlip and Funkhouser, 1995). Based on amino acid similarity from various organisms, chitinases have been grouped into five classes, Class I to V. Additionally, in class PR-4, some proteins showing low endochitinase activity are found among the chitin-binding proteins (Neuhaus et al., 1996). Chitinases show an increased level in many plant species, including rice (Kim et al., 2009), Arabidopsis (Samac and Shah, 1991), tobacco, wheat (Broekaert et al., 1988), pepper (Egea et al., 1996), tomato (Lawrence et al., 1996), and bean (Dann et al., 1996) due to infection by fungal pathogens or viruses. Chitinases have potent antifungal activity (Broekaert et al., 1988). In tomato plants, four chitinase isozymes with molecular weights of 26, 27, 30, and 32 kDa are induced on infection by the fungus *Alternaria solani* (Lawrence et al., 1996). A novel chitinase IVa gene (*OgChitIVa*) from *Oryza grandiglumis* was suggested to play an important role in the signal transduction process involved in resistance to *B. cinerea* in plants. Different β-1,3-glucanases isoforms have been characterized in the infected tissues. Groundnut leaves infection with *Cercospora arachidicola*, the early leaf spot pathogen, causes significant induction of the extracellular β-1,3-glucanase activity with the synthesis of its three isoforms (Roulin and Buchala, 1995). In pepper (*Capsicum annum*) plants, it has been shown that the glucanase activity is involved in the mechanism of resistance to the cucumber mosaic virus in tolerant cultivars (Egea et al., 1996).

It is suggested that a higher level of β-1,3-glucanase and chitinase and the induction of 30 kDa isoenzyme of a chitinase in early blight-resistant breeding lines is associated with genetically inherited resistance of tomato to *A. solani* (Lawrence et al., 1996). β-1,3-glucanases, chitinase and lignin content in plant leaves can be utilized as biochemical markers for identification of plant varieties which are resistant to fungal infection (Ebrahim et al., 2011). The overexpression of chitinase and/or β-1,3-glucanases in transgenic plants provides considerable protection against fungal pathogens. Overexpression of OgChitIVa from Arabidopsis caused mild resistance to *Botrytis cinerea*, the fungal pathogen, by decreasing the rate of disease and size of necrosis (Pak et al., 2009). Similarly,

strawberry plants transformed with β-1,3-glucanase gene *bgn13.1* from *T. harzianum* showed higher tolerance to crown rot disease (Mercado et al., 2015). Two class 1 PR proteins designated as acidic PR-1 protein (PR1a1) and basic PR-1 protein (PR1b1), which are low-molecular-weight proteins are encoded by two closely associated genes in tomato plants (Tornero et al., 1997). The expression of these two proteins is also induced by salicylic acid and ethylene. In transgenic tobacco plants infected with the tobacco mosaic virus (TMV), the *PR1b1* gene exhibited activation locally in tissues showing hypersensitive response (Tornero et al., 1997).

Thaumatin-like proteins represent one group (PR-5) of antifungal PR proteins which have been isolated from several plants. Tobacco plants transformed with PR-5 showed higher resistance to *Alternaria alternata* (Velazhahan and Muthukrishnan, 2003). Osmotin protein in plants is associated with defense responses to numerous pathogens (Viktorova et al., 2012). Osmotin, initially described in tobacco suspension cultures under salinity stress, belongs to the family of class 5a PR proteins, has antifungal activity (Artlip and Funkhouser, 1995). Osmotin also repressed the growth of various fungal pathogens including *Neurospora crassa*, *Candida albicans*, and *Trichoderma reesei* (Vigers et al., 1992; Abad et al., 1996). Osmotin level was increased because of fungal and viral pathogen infections in tobacco and tomato (Stintzi et al., 1991; Woloshuk et al., 1991). Sesame plants transformed with *OLP* gene (*SindOLP*), showed increased tolerance to charcoal rot pathogen (Chowdhury et al., 2017). Apoplastically secreted tobacco osmotin in cotton confers a moderate resistance to *Rhizoctonia solani* (Parkhi et al., 2009). Although some possible mechanisms of a osmotin led defense against biotic stresses are deciphered, its exact role in stress response is not clear (Parkhi et al., 2009). In the primary leaves of bean plants, 10 acidic and 8 basic PR proteins were identified following the southern bean mosaic virus infection, which included 4, 17 kDa serologically related, acidic proteins of unknown functions, 2 chitinases, 1 acidic (29 kDa), and 1 basic (32 kDa) possessing antifungal activities, and 4 (21, 28, 29, and 36 kDa) serologically related, acidic glucanases (Mohamed and Sehgal, 1997). Using SDS-PAGE, a stress-induced protein (17 kDa) designated as FISP17 was identified. Western blot analysis revealed reaction of antibody only with a 17 kDa protein in *F. solani* infected plant exudates, but no reaction was observed with culture filtrates of *F. solani* or exudates of healthy plants (Li et al., 2000).

Comparative 2-D gel analyses revealed 21 differentially expressed protein spots in secreted proteins due to *Magnaporthe grisea* and/or an elicitor in a cell suspension-culture of rice over the control. MALDI-TOF-MS and LC-ESI-MS/MS studies of these protein spots revealed that most of these assigned proteins were related to defense processes such as two germin A/oxalate oxidases, nine chitinases, secretory proteins, five domain unknown function 26, and β-expansin, (Kim et al., 2009). Some PR proteins are present in healthy tissues and are differentially expressed by signals involved in flowering and reproduction. This implies that they are associated with the normal physiological processes of the plants in addition to plant defense (Samac and Shah, 1991). According to certain investigators, salicylic acid is involved in the signal transduction pathway resulting in pathogen resistance and PR protein synthesis (Ferullo et al., 1997). The level of salicylic acid increases in plants following pathogen attack (Samac and Shah, 1991). In barley leaves, salicylic acid treatment induces the accumulation of two PR proteins and one salicylic acid-specific protein (Tamas and Huttova, 1996). In tobacco plants, salicylic acid acts as an endogenous signal for acidic PR-1 protein expression (Conrath et al., 1997). However, using transgenic tobacco plants accumulating high amounts of soluble sugars due to the cytosolic expression of an inorganic pyrophosphatase from *Escherichia coli*, the possible role of soluble sugars in the induction of PR protein has been suggested (Badur et al., 1994; Herbers et al., 1996). Such an induction appeared to be salicylic acid-independent in the source leaves of tobacco plants (Herbers et al., 1996).

According to Malamy et al. (1996), multiple pathways exist that are associated with defense response in plants, one of which appears to be independent of salicylic acid. More evidence is required to address the signal transduction pathways leading to the synthesis of PR proteins and the function of PR proteins in plant defense. A novel PR gene *GmPRP* showed upregulation in soybean leaves after infection with *P. sojae*. Tobacco plants transformed with *GmPRP* gene resulted in enhanced resistance to *P. sojae* race 1 and *Phytophthora nicotianae* (Jiang et al., 2015). Microbes or elicitor-induced signal transduction pathways cause increased production of reactive oxygen and nitrogen species in plants. Class III plant peroxidases are proteins known to be induced during the plant–pathogen interaction (Almagro et al., 2009). They are members of a large multigene family, and are involved in a wide variety of physiological processes, such as suberin and lignin formation, cell wall component cross-linking, and phytoalexins synthesis, or play a role in ROS and RNS metabolism, both starting the hypersensitive response, a type of programmed host cell death at the site of infection related to limited pathogen development (Almagro et al., 2009). The total protein content declined in leaves but enhanced up to 135% in the pumpkin fruits infected with yellow vein mosaic disease. In infected plant material, a significant increase in GPX, APX, SOD, and GR activities was noted in comparison to the control plants. The intensity of isozymic bands of GPX, CAT, APX, SOD, and glutamate dehydrogenase were higher, whereas the GR isozyme band decreased in the infected plants compared to controls (Jaiswal et al., 2013).

22.8 WOUNDING

Wounding induces new proteins, alters the levels of proteins and activities of many enzymes in plants (Mehta et al., 1991; Cabello et al., 1994; Schaller and Ryan, 1996; Chen et al., 2005; Dafoe et al., 2009; Turrà et al., 2009; Buron-Moles et al., 2014). In tobacco crown gall tumor tissues, a 16 kDa glycine-rich hydrophobic polypeptide was characterized, which is a cell wall protein and is induced by mechanical wounding (Yasuda et al., 1996). This polypeptide participates in the

process of wound-healing in tobacco plants by modifying the cell wall composition (Yasuda et al., 1996). In tomato plants, several systemic wound-response proteins (swarps) have been described (Mehta et al., 1991; Schaller and Ryan, 1996). Mehta et al. (1991) noticed the appearance of several novel proteins of 80.0, 63.0, 33.0, 29.0, 28.5, and 25.5 kDa and a decrease in the content of a 15 kDa protein due to wounding in tomato fruit tissues. Proteomic analysis of the root and leaf of wounded seedlings of *Fagus sylvatica* exhibited alteration in protein expression. Among the differentially expressed spots, identified proteins were associated with basic metabolism and typical stress-responsive mechanisms, including some involved in plant–pathogen interactions (Valcu et al., 2009).

In response to wounding, 'Golden Delicious' apples also showed an altered abundance of some proteins and synthesized a broad range of PR proteins (Buron-Moles et al., 2014). Both chitinase and glucanase activities were induced in the stems and roots of chickpea plants due to wounding (Cabello et al., 1994). In tomato, asparatic protease (LeAspP) mRNA was systemically induced by wounding (Schaller and Ryan, 1996). Chalcone synthase (CHS) is an important enzyme of the flavonoid/isoflavonoid biosynthetic pathway. Induced expression of this enzyme was observed in plants infected with bacteria or fungus (Dao et al., 2011). CHS expression leads to accumulation of flavonoid and isoflavonoid phytoalexins and participates in the salicylic acid defense pathway. One day after wounding, the appearance of ipamorelin (IPO) protein was observed in sweet potato leaves. This protein was lectin, a carbohydrate-binding protein (Chen et al., 2005). In potato, genotype-dependent expression of specific members of protease inhibitor gene families was observed in various tissues due to wounding (Turrà et al., 2009). A Kunitz trypsin inhibitor (TPI-2) (located in cell walls) was accumulated in wounded chickpea leaves (Jimenez et al., 2008). Shen and coworkers (2003) identified 10 proteins upregulated and 19 proteins downregulated in rice leaf sheath following wounding. Among all these proteins, Bowman-Birk trypsin inhibitor, calmodulin-related protein, and putative receptor-like protein kinase are confirmed to be wound-responsive proteins.

Using comparative 2-DE, it was observed that in poplar phloem exudate two proteins, a thaumatin-like protein, and pop3 were upregulated after 24 hours of wounding (Dafoe et al., 2009). In *F. sylvatica*, amino acid, energy, and carbohydrate metabolism were both systemically and locally regulated following leaf wounding. Several protein spots probably representing degradation products of RUBISCO, isocitrate dehydrogenase, triose-phosphate isomerase, formyltetrahydrofolate deformylase, flavodoxin-like quinone reductase decreased, whereas proteinase inhibitors, β-d-galactoside galactohydrolase isoform accumulated following wounding (Valcu et al., 2009). Enzymes of the phenylpropanoid pathway, peroxidase, dihydroxyacetonephosphate synthase (DHAP synthetase), glycine-rich and hydroxy-proline-rich cell wall proteins, protease inhibitors, and 1-aminocyclopropane-1-carboxylate synthase, chalcone synthase have been shown to increase following wounding (Davis et al., 1990; Mehta et al., 1991; Dao et al., 2011). It is believed that some of these enzymes and proteins are involved in lignification process and thus form a wound periderm to limit pathogenic attack (Davis et al., 1990). It has been shown that in wounded tissues of plants synthesis of specific ethylene response factor (ERF) occurs, which interact with transcription factors (TF) and coordinate with stress signaling process to take up repair mechanisms in the tissues (Heyman et al., 2018). The heterodimerized protein complexes of ERF-TF are regarded as potent cell division activators which help in repairing damaged tissues (Heyman et al., 2018).

22.9 METAL TOXICITY

An increase in industrialization has led to the increased introduction of several metals like Pb, Zn, Cd, Cu, and Hg in the soil environment. Elevated concentrations of heavy metals in the soil negatively affect plant growth, cause activation of signal transduction pathway, induction or inhibition of enzymes, and induce synthesis of metal-binding cysteine-rich polypeptides and metal transporter proteins (Maksymiec, 2007; Sinha et al., 2011; Peng and Gong, 2014; Singh et al., 2015a). Synthesis of certain novel proteins with a molecular weight higher than 14 kDa occurs when plants are treated with Cd (Choi et al., 1995; Shah and Dubey, 1998a). An 18 kDa Cd-binding protein complex was purified by Shah and Dubey (1998a) from rice plants. This complex had specific Cd content of 3.7 μmoles mg^{-1} peptide and had 4 –SH groups per protein molecule. It is suggested that in plants these protein complexes bind Cd with the help of –SH groups of the peptide in mercaptide bonds and help in sequestration of excess Cd ions (Shah and Dubey, 1998a). Isolation of a cDNA was reported by Choi and coworkers (1995), which was differentially expressed by 150 mM Cd in Arabidopsis plants and encoded an 18.3 kDa protein. Lupin roots treated with Pb, Cu, or nitrite ions led to enhanced synthesis of a 16 kDa polypeptide, which appeared to be a cytosolic Cu/n-SOD (Przymusinski et al., 1995). From HgCl$_2$-treated maize leaves, transcriptionally activated cDNA clones were isolated which represented several known proteins including glycine-rich proteins, PR proteins, chaperons, and membrane proteins (Didierjean et al., 1996). Phytochelatins have a primary structure of (γ-Glu-Cys)n-Gly or (γ-Glu-Cys) n-Ala, when n=2–11, and have an apparent function in the sequestration of metal ions within the plant. Phytochelatin synthesis occurs exclusively from glutathione (GSH) in a transpeptidation reaction catalyzed by enzyme phytochelatin synthase (PC synthase). This occurs in two steps: (I) cleavage of a glutathione molecule and generation of a γ-Glu-Cys unit, (II) transfer of γ-Glu-Cys unit to an acceptor molecule GSH or a phytochelatin peptide to produce PCn+1 (Grill et al., 1989; Chen et al., 1997; Pal and Rai, 2010; Singh et al., 2015b). Phytochelatins form complexes with metal ions resulting in sequestration of metal ions in plants and hence act as an important component in detoxification of metals in plants (Anjum et al., 2015). Apo-phytochelatin (non-metallated PCs) may get degraded by vacuolar hydrolases, and in turn phytochelatins may return to the cytosol where they could continue to carry out their shuttle role (Figure 22.6). When treated

with Cd, Cu, or As many plants showed an increased concentration of phytochelatins (Wójcik et al., 2005; Mishra and Dubey, 2011; Machado-Estrada et al., 2013). The best inducer observed was Cd, followed by the cations Ag, Bi, Pb, Zn, Cu, Hg, and Au.

In red spruce cell suspension cultures, the PC and its precursor, γ-glutamylcysteine (γ-EC) were enhanced by 2- to 4-fold with 12.5 to 200 mM Cd concentrations in comparison to the control. However, Zn-exposed cells revealed less than a 2-fold increase in γ-EC and PC concentration in comparison to control even at very high concentration of 800 mM Zn. In addition, higher chain PCs were also present in Zn as well as Cd-treated cells and their concentration enhanced significantly with increasing Zn and Cd concentrations (Thangavel et al., 2007). Negrin et al. (2017) suggested a complex relationship of PCs with concentrations of metal in three species of Portuguese salt marsh *Sarcocornia perennis*, *Spartina maritima* and *Halimione portulacoides*. In roots of all species, higher concentrations of metals were observed. However, a higher concentration of total PCs was present in stems or leaves of *Halimione portulacoides* and *Sarcocornia perennis* and large roots of *Spartina maritima*. Synthesis of PC$_2$ could be correlated positively with As, Zn, and Pb concentrations in the tissues (Negrin et al., 2017). In date palm, expression of type I PCS (PdPCS1) was enhanced in seedling hypocotyls treated with Cd, Cr, and Cu (Zayneb et al., 2017). However, PCS had contrasting effects on As and Cd content in rice grains (Uraguchi et al., 2017). In soils containing environmentally relevant As and Cd contents, reduced Cd accumulation but an enhanced As accumulation was observed in OsPCS1 mutant grains. Transgenic Indian mustard plants transformed with *A. thaliana*, AtPCS1 gene encoding PCS exhibited higher As and Cd tolerance in comparison to untransformed plants (Gasic and Korban, 2007). Overexpression of two mulberry PC synthase genes (*MnPCSs*) in transgenic tobacco and Arabidopsis resulted in Zn/Cd accumulation and tolerance (Fan et al., 2018).

Metallothioneins (MTs) are low-molecular-weight (6–7 kDa), 60–65 amino acid residue long, cysteine (20 molecules)-rich metal-binding (through mercaptide bonds) proteins (Liu et al., 2000). These are also proposed to act as antioxidants and play a role in plasma membrane repair (Salt et al., 1998; Dietz et al., 1999; Hassinen et al., 2011). Based on the arrangement and number of cysteine residues, plant MTs have been grouped in class II and can be further classified in four types on the basis of their amino acid sequences (Robinson et al., 1993; Cobbett and Goldsbrough, 2002). A novel metallothionein-like protein, the expression of which is regulated by metal ions, osmoticum, or ABA, has been isolated from Douglas fir trees during embryogenesis (Chatthai et al., 1997). In Cd/Zn hyperaccumulator *Sedum plumbizincicola*, a novel *SpMTL* gene encoding metallothionein-like protein was identified. SpMTL expression in roots demonstrated a positive correlation with Cd accumulation in leaves (Peng et al., 2017). It was proposed that enhanced expression and altered SpMTL protein sequences might contribute to hypertolerance and hyperaccumulation of Cd in *S. plumbizincicola*

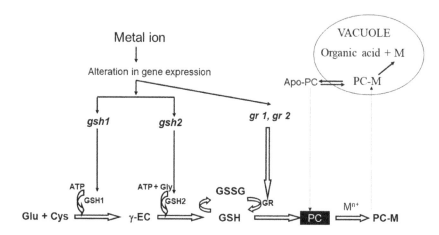

FIGURE 22.6 Mechanism involved in chelation and compartmentalization of metal ions in vacuoles. Apo-phytochelatin (non-metallated PCs) may get degraded by vacuolar hydrolases, and in turn phytochelatins (PCs) may return to the cytosol where they could continue to carry out their shuttle role. The synthesis of PCs accompanies with a decrease in cell glutathione pool and an increase in the activities of glytamyl cysteine synthetase (GSH 1), glutathione synthetase (GSH 2) as well as glutathione reductase. The elevated activities of GSH 1, GSH 2 and GR are correlated with enhanced expression of corresponding genes *gsh 1, gsh 2, gr1* and *gr 2*.

(Peng et al., 2017). *A. thaliana* transformation with a *Brassica juncea* metallothionein 2 (BjMT2), led to enhanced Cd and Cu tolerance (An et al., 2006). Similarly, overexpression of *Brassica campestris* metallothionein BcMT1 and BcMT2 led to enhanced Cu and Cd tolerance and reduced ROS production in *A. thaliana* (Lv et al., 2013).

Evidence suggests that enhanced HSP synthesis occurs in plants due to metal stress; however, the specific functions of HSPs and the structures that are protected by HSPs under metal stress remain unidentified (Heckathorn et al., 2004; Joseph et al., 2012). Cd exposure induces the synthesis of many stress proteins presumably HSPs with molecular weights ranging between 10 to 70 kDa (Sanita di Toppi and Gabbrielli, 1999; Júnior et al., 2015). In *Lycopersicon peruvinum* cell cultures treatment with 1 mM Cd, led to an accumulation of HSP70 bound to plasmalemma endoplasmic reticulum and mitochondrial membranes (Neumann et al., 1994). Enhanced SaHMA3 expression has been suggested to play an important role in Cd hyperaccumulation and hypertolerance in *Sedum alfredii*, a Cd hyperaccumulator (Zhang et al., 2016). HSP70 shows a high affinity for misfolded proteins and assists them in reaching their native conformation by reintegration in proper membrane complex (Sanita di Toppi and Gabbrielli, 1999). In maize plants, Ni, Cu, Zn, and Pb led to decline in net photosynthesis (Phn) and rise in the contents of chloroplastic small HSPs (smHSPs), with a time of exposure (Heckathorn et al., 2004). Under *in vitro* conditions, it was shown that addition of purified chloroplast smHSPs to the thylakoids protected photosynthetic electron transport (Phet) from Pb (not by Ni). Under *in vivo* conditions, Phn was protected from Ni and Pb in the presence of increased levels of smHSPs in a heat-tolerant *Agrostis stolonifera* genotype expressing additional chloroplast smHSPs compared to a near-isogenic heat-sensitive genotype (Heckathorn et al., 2004). These results suggest that HSPs protect photosynthesis from metals and are among the first to demonstrate that specific functions are protected by HSPs during metal

stress (Heckathorn et al., 2004). Proteomics study revealed 13 upregulated and 2 newly induced spots in chromium-treated *Miscanthus sinensis* roots.

Seedlings of *Miscanthus sinensis* exposed to Cr for 4 weeks showed elevated levels of proteins associated with nitrogen and carbohydrate metabolism, molecular chaperones and novel proteins such as nitrate reductase, inositol monophosphatase, formate dehydrogenase, a putative dihydrolipoamide dehydrogenase, and adenine phosphoribosyl transferase in their roots (Sharmin et al., 2012). In rice seeds germinating in the presence of 200 mM Cu for 6 days, 13 upregulated proteins were identified, including metallothionein-like protein, membrane-associated protein, putative wall-associated protein kinase, pathogenesis-related proteins and the putative small GTP-binding protein Rab2 (Zhang et al., 2009). *A. thaliana* plants treated with 10 mM Cd for 24 hours showed an increased synthesis of PCs and 41 spots indicated significant changes in protein abundance. Most of the identified proteins belonged to four different classes: metabolic enzymes such as ATP sulfurylase, glycine hydroxymethyltransferase, and trehalose-6-phosphate phosphatase, glutathione S-transferases, latex allergen-like proteins, and unknown proteins (Roth et al., 2006). The soybean cell suspension-culture treated with different concentrations of Cd showed the Cd-induced appearance of the SOD, chalcone synthase, histone H2B, and glutathione transferase proteins (Sobkowiak and Deckert, 2006). Excess copper treatment resulted in enhanced expression of three proteins in roots of 10-day old bean plants. The contents of an intracellular PR protein and a newly identified protein homologous to PvPR1, PvPR2, were enhanced with increasing concentration of Cu. At 50 mM Cu, PvPR1 and a homolog of *A. thaliana* thylakoid luminal 17.4 kDa protein appeared in leaves. Another protein which was enhanced slightly by Cu treatment had a homologous sequence to a mitochondrial precursor of glycine cleavage system H protein of *Flaveria pringlei* (Cuypers et al., 2005). There were seven proteins including oxygen-evolving

(OE), heavy-metal ATPase (HMA3), nonspecific lipid-transfer protein (nsLTP), SOUL heme-binding protein, pollen allergen-like protein, short-chain dehydrogenase/reductase that were enhanced by Cd treatment and accumulated at a constitutively higher level in hyperaccumulator *S. alfredii* in comparison to non-hyperaccumulator *S. alfredii* (Zhang et al., 2017b). In roots of Cd-treated *Amaranthus hybridus* L. plants, 10 proteins were specifically expressed under Cd treatment and these proteins related to signal transduction, protein, and energy metabolism, defense and stress indicating that redirecting root cell metabolism could be a key mechanism of survival in Cd-treated *A. hybridus* plants (Jin et al., 2016).

22.9.1 ENZYME LEVELS IN METAL-STRESSED PLANTS

Germinating seeds or plants growing under excess levels of metals show alteration in levels and activity behaviors of many enzymes. Suppression of proteolytic activity marked by decreased activities of protease and peptidase was observed due to Cd exposure in germinating rice seeds leading to altered contents of amino acids and proteins (Shah and Dubey, 1998b). Enhanced content of proteins and increased activity of carboxypeptidase accompanied with a decrease in the activity of protease and leucine aminopeptidase (LAP) and content of free amino acid pool was reported in rice seedlings subjected to As toxicity (Mishra and Dubey, 2006). Aluminum treatment caused a reduction in the activity of protease in roots of Al-sensitive rice cultivar HUR-105 and tolerant cultivar Vandana, whereas a reverse trend was obtained in shoots. Seedlings of rice cv. Vandana, when subjected to increased concentration of aluminum, exhibited new isozymes of protease in shoots (Bhoomika et al., 2014). Ni toxicity suppressed the protein and RNA hydrolysis in germinating rice seeds and seedlings by reducing the activity of protease and ribonuclease, respectively (Maheshwari and Dubey, 2007, 2008). Shah and Dubey (1995) observed an increase in ribonuclease activity in 100 μM Cd-treated rice seedlings, whereas higher Cd treatment of 500 μM was inhibitory to the enzyme. Similarly, rice seedlings were grown in medium containing 25 or 50 μM As showed a marked decline in ribonuclease activity (Mishra and Dubey, 2006). Tissue-specific inhibition of phosphatases activities both under *in situ* and *in vitro* conditions has been observed due to Cd and As in growing rice plants (Shah and Dubey, 1998c; Mishra and Dubey, 2008b).

Activities of inorganic pyrophosphatase, alkaline phosphatase, and acid phosphatase decreased in rice plants grown under high Cd, As, and Ni concentrations (Shah and Dubey, 1998c; Mishra and Dubey, 2008b; Maheshwari and Dubey, 2011). A significant reduction in the activities of nitrate assimilatory enzymes NR, nitrite reductase (NiR) and GS was observed in rice seedlings grown under high As concentration, whereas elevated aspartate and alanine aminotransferases activities were observed (Jha and Dubey, 2004b). Ni and Al toxicity is also reported to reduce functional NR in rice seedlings (Sharma and Dubey, 2005b; Mishra and Dubey, 2011). Rice seedlings subjected to 80 μM Al^{3+} demonstrated a reduced level of NR max (NR activity in the presence of EDTA) but led to higher NRact (NR activity in the presence of Mg^{2+} representing the non-phosphorylated NR state) and NR activation state. However, seedlings grown in medium with a higher level of 160 μM Al^{3+} showed a decrease in NRact and NRmax (Sharma and Dubey, 2005b). A high concentration of Cd reduced the carbonic anhydrase and NR activity in *Lepidium sativum* (Gill et al., 2012). A marked increase in aminating glutamate dehydrogenase (NADH-GDH) activity and deaminating GDH (NAD^+-GDH) and reduction in the activities of nitrate assimilatory enzymes GS and NR was reported in the roots and shoots of Al and Ni treated seedlings. The activities of AspAT and AlaAT increased due to Ni and Al exposure. It is suggested that GDH participates in $NH4^+$ assimilation under heavy-metal stress, whereas aminotransferases might assist in meeting increased amino acid demand under stressed conditions (Mishra and Dubey, 2011).

The effect of enhanced concentrations of Cd, As, Ni, and Al *in situ* on the activities of enzymes related to sugar and starch metabolism have been investigated in rice seedlings (Verma and Dubey, 2001; Jha and Dubey, 2004a; Mishra and Dubey, 2008a; Mishra and Dubey, 2013). The activity of enzymes associated with starch hydrolysis α-amylase, β-amylase declined, whereas sucrose hydrolyzing enzymes sucrose synthase (SS) and acid invertase were enhanced in the rice seedlings grown in the presence of these metals. Enzymes associated with sucrose synthesis, SPS showed declined activity in Cd, As, Al, and Ni treated rice seedlings in comparison to the controls (Verma and Dubey, 2001; Jha and Dubey, 2004a; Mishra and Dubey, 2008a; Mishra and Dubey, 2013). Enhanced activity of starch phosphorylase enzyme was reported in As and Ni treated rice seedlings (Jha and Dubey, 2004a; Mishra and Dubey, 2013). These results suggested that toxic concentrations of Cd, As, Ni, and Al in rice seedlings cause marked alterations in carbohydrate metabolism. In pea plants, Devi and coworkers (2007) also observed marked changes in the activity of enzymes of glycolysis, carbohydrate metabolism and pentose phosphate pathway in Cd-treated plants. They suggested that the increased activities of phospho glucoisomerase and hexokinase in shoots and roots of Cd-exposed seedlings might be due to preferable channeling of hexose toward glycolysis than pentose phosphate pathway. Enhanced S content and ATP-S activity were reported in *Lepidium sativum* plants exposed to Cd, suggesting activation of the S assimilation pathway in coordination with increased antioxidant enzyme activities and GSH for possible Cd tolerance (Gill et al., 2012).

High Cd concentration (100 mg Cd kg^{-1} soil) increased the activity of SOD, CAT, APX, and GR (Gill et al., 2012). Metals like Pb, Cd, Al, Ni, and Mn induce the activity of antioxidative enzymes viz., GPX, SOD, APX, GR, and enzymes of the ascorbate-glutathione cycle (MDHAR, DHAR, GR) in plants (Shah et al., 2001; Verma and Dubey, 2003; Qureshi et al., 2007; Sharma and Dubey, 2007; Maheshwari and Dubey, 2009; Mishra and Dubey, 2011; Srivastava and Dubey, 2011; Gill et al., 2012). However, Schützendübel and coworkers (2001) showed that metals like Cd could decrease some of the antioxidant enzymes leading to accumulation of H_2O_2 and hence such plants would suffer from oxidative stress. Ezaki and coworkers (2001) observed Al^{3+} led to the induction of tobacco peroxidase and glutathione S-transferase genes. In shoots of rice seedlings, high concentration of Pb (1000 µM) caused a decrease in the intensity of two pre-existing CAT isoforms (Verma and Dubey, 2003). Sharma and Dubey (2007) observed the appearance of two new isoenzymes of APX in roots of Al-treated seedlings compared to control. Using western blot analysis, Sharma and Dubey (2007) showed that alteration in the activity of APX due to Al^{3+} toxicity was as a result of changes in the amounts of enzyme protein. In both root and leaf tissues of water hyacinth, APX and POD activities were positively correlated with Pb concentration, while the activity of SOD and CAT increased with up to 800 mg/L Pb treatment and thereafter the activity declined at higher Pb concentrations (*Eichhornia crassipes* (Mart.)) (Malar et al., 2016). Cu, Pb, and Zn-treated bamboo showed a significant increase in SOD, CAT and POD activities (Emamverdian et al., 2018). APX and CAT activities were enhanced in *S. cheesmaniae* but declined in *S. lycopersicum* when subjected to Cu treatment (Branco-Neves et al., 2017). The remarkable enhancement in the activity of SOD and CAT was observed in Al-treated seedlings of tolerant rice cultivar Vandana compared to sensitive cultivar HUR-105 (Bhoomika et al., 2013). It was shown by Bhoomika and coworkers (2013) that the differential responses of antioxidant enzymes exist due to Al toxicity in rice cultivars differing in Al tolerance and that Al-tolerant rice cultivars are characterized by the marked presence and elevated activity of Fe-SOD under Al-excess. These studies suggest the important role of antioxidant enzymes in metal tolerance.

22.10 GASEOUS POLLUTANTS

Ozone (O_3), nitric oxide (NO), and sulfur oxide (SO_2) are considered as major air pollutants which affect the growth and metabolism of plants. These air pollutants when present in excess, cause increased production of ROS in the tissues and induce synthesis of novel proteins, inhibit the synthesis of many proteins, and induce activities of many enzymes (Conklin and Last, 1995; Glick et al., 1995; Carreras et al., 1996; Rao et al., 1996; Sharma and Davis, 1997; Agrawal et al., 2002a, 2002b; Torres et al., 2007; Sarkar et al., 2015; Feng et al., 2016). A 66 kDa leaf protein demonstrated a specific and strong cross-reaction with an antibody developed against anti-MAPKinase (ERK1). Its level was enhanced within 5 minutes of exposure to elevated O_3 in rice plants suggesting the potential involvement of this protein in the O_3-induced self-defense pathway(s) (Agrawal et al., 2002a). An O_3 induced transcript has been characterized in *A. thaliana* that encodes a basic protein of 8.6 kDa which represents a novel O_3 stress-related protein (Sharma and Davis, 1997). It is suggested that O_3-induced proteins include antioxidant enzymes and several proteins associated with responses to other environmental stresses and that O_3 led responses are partly due to activation of a salicylic acid-dependent signaling pathway (Sharma and Davis, 1997).

The leaf proteome of *Phaseolus vulgaris* and maize plants exposed to an elevated O_3 level revealed changes in RUBISCO protein levels (Torres et al., 2007). In bean leaves, the content of two SOD proteins (20 and 19 kDa) was reduced dramatically while expression of naringenin-7-O-methyltransferase (NOMT; 42 kDa), small HSP (33 kDa), and APX (25 kDa) increased (Torres et al., 2007). In maize depending on the developmental stages of the leaves, alteration in APX, SOD, and CAT expressions was observed due to O_3 exposure. In younger leaves, NOMT (41 kDa) and crossreacting HSPs (30 and 24 kDa) were strongly enhanced. The results revealed a clear alteration in the level of proteins related to secondary metabolism, heat shock and oxidative stress due to O_3 exposure. In bean, the appearance of a novel PR-2 protein was suggested as a potential marker for O_3 stress (Torres et al., 2007). Plants grown at places with enhanced industrial pollution showed significantly low contents of soluble proteins (Carreras et al., 1996). Excess O_3 shows adverse effects on many proteins including photosynthetic protein RUBISCO and enzymes related to energy metabolism (Feng et al., 2008; Sarkar and Agrawal, 2010). O_3 fumigation causes a decline in the level of mRNA corresponding to the small subunit of RUBISCO, 10 kDa protein of water-evolving complex of PSII and chlorophyll a/b-binding protein (Conklin and Last, 1995). Similarly, in potato plants, ozone accelerates senescence with a decline in RUBISCO small-subunit mRNA as well as a decline in the transcripts of glyceraldehyde-3-phosphate dehydrogenase (Glick et al., 1995). Agrawal et al. (2002b) reported 52 differentially expressed proteins; including RUBISCO and various defense/stress-related and PR proteins. Comparative proteomic analysis revealed 32 and 20 proteins differentially expressed in O_3-treated soybean chloroplast, and leaf, respectively. Proteins related to photosynthesis such as carbon assimilation and photosystem I/II were reduced following O_3 exposure. In contrast, proteins associated with carbon metabolism and antioxidant defense increased. The activity of enzymes associated with carbohydrate metabolism was enhanced following O_3 exposure, which was consistent with the decline in starch and rise in sucrose concentrations (Ahsan et al., 2010). Many antioxidant enzymes show an increased level when plants are exposed to ozone. Cytosolic Cu/Zn-SOD is the best-characterized enzyme, which shows increased activity with the simultaneous synthesis of new isoforms, in plants under O_3 exposure (Conklin and Last, 1995; Rao et al., 1996). In *A. thaliana*, O_3 exposure enhanced the activities of POD, SOD, GR, and ascorbate reductase and modified the substrate affinity of both GR and APX (Conklin and Last, 1995; Rao et al., 1996). However, in the chloroplasts, a decline in the levels of Fe-SOD and GR was observed (Conklin and Last, 1995). Under field conditions, two high yielding rice cultivars, Malviya dhan 36 and Shivani, showed induced SOD, CAT, POD, APX, and GR under elevated O_3 levels (Sarkar et al., 2015). Significant interactions between O_3 and SOD and CAT has been reported in winter wheat cultivar (Feng et al., 2016).

22.11 CONCLUSIONS

Environmental stresses such as salinity, drought, heat, chilling, anaerobiosis, heavy metals, gaseous pollutants, and UV radiation cause modulation in gene expression of plants causing induction of specific genes and enhanced abundance of their translatable mRNAs and proteins. As a result, the increased synthesis of certain novel proteins occurs in stressed plants with a concomitant decrease in the level of certain pre-existing proteins. The newly synthesized or overexpressed pre-existing proteins appear to increase the capacity of plants to adapt to a stressful environment through physiological and biochemical adjustments. Most of the stresses induce the synthesis of proteins specific to the particular stress. However, certain proteins are common and can be synthesized under more than one type of stress. For example, cold, drought, and salinity stresses cause the expression of some common proteins that can also be induced by treatment of normal tissues with abscisic acid. Most of the stress-induced proteins have been isolated and well characterized for their physicochemical properties, and their amino acid sequences have been determined. Although stress-specific proteins are believed to protect the plants against various stresses, the exact physiological functions and mechanisms of action of many of them are not very clear. Further experiments are necessary to determine the functions of these proteins. With the advent of the tools of genomics and proteomics, it has been possible to identify and characterize the whole set of stress-responsive genes, to determine their expression patterns under stresses and to identify and characterize the entire spectrum of proteins which are overexpressed and those which are repressed under stresses. The identification of novel stress-responsive proteins and especially the proteins associated with stress tolerance provides new insights into our understanding of the mechanisms of stress tolerance in plants. Functional analysis of novel stress-responsive proteins and understanding the role of specific proteins in tolerance toward a particular stress, has helped a long way in producing stress-tolerant crop plants using transgenic approaches. Many stress-tolerant crop plants have been produced which express genes/proteins associated with tolerance. Proteomic alterations in plants exposed to individual stresses have been studied in detail by various groups of workers. However, under natural field conditions plants are simultaneously exposed to many stresses which can interact antagonistically or synergistically. Understanding the roles and regulation of proteins accumulated under single and combinations of stresses would be helpful in the development of crop plants tolerant to multiple environmental stresses.

REFERENCES

Aarati, P., B. T. Krishnaprasad, G. M. Savitha, R. Gopalakrishna, G. Ramamohan, and M. Udayakumar. 2003. Expression of an ABA responsive 21 kDa protein in finger millet (*Eleusine coracana* Gaertn.) under stress and its relevance in stress tolerance. *Plant Sci.* 164:25–34.

Abad, L. R., M. P. D'Urzo, D. Liu et al. 1996. Antifungal activity of tobacco osmotin has specificity and involves plasma membrane permeabilization. *Plant Sci.* 118:11–23.

Abdelkader, A. F., H. Aronsson, and C. Sundqvist. 2007. High salt-stress in wheat leaves (*Triticum aestivum*) causes retardation of chlorophyll accumulation due to a limited rate of protochlorophyllide formation. *Physiol. Plant.* 130:157–166.

Abreu, I. A., A. P. Farinha, S. Negrão, N. Gonçalves et al. 2013. Coping with abiotic stress: proteome changes for crop improvement. *J. Proteomics* 93:145–168.

Agrawal, G. K., R. Rakwal, M. Yonekura, A. Kubo, and H. Saji. 2002a. Rapid induction of defense/stress-related proteins in leaves of rice (*Oryza sativa*) seedlings exposed to ozone is preceded by newly phosphorylated proteins and changes in a 66 kDa ERK-type MAPK. *J. Plant Physiol.* 159:361–369.

Agrawal, G. K., R. Rakwal, M. Yonekura, A. Kubo, and H. Saji. 2002b. Proteome analysis of differentially displayed proteins as a tool for investigating ozone stress in rice (*Oryza sativa* L.) seedlings. *Proteomics* 2:947–959.

Ahmed, A. M., A. M. Ismail, and M. M. Azooz. 2001. Protein patterns in germinating seeds of *Vicia faba* lines in response to interactive effects of salinity and vitamins treatments. *Phyton* 41:97–110.

Ahn, Y. J., and J. L. Zimmerman. 2006. Introduction of the carrot HSP 17.7 into potato (*Solanum tuberosum* L.) enhances cellular membrane stability and tuberization *in vitro*. *Plant Cell Environ.* 29:95–104.

Ahsan, N., Y. Nanjo, H. Sawada, Y. Kohno, and S. Komatsu. 2010. Ozone stress-induced proteomic changes in leaf total soluble and chloroplast proteins of soybean reveal that carbon allocation is involved in adaptation in the early developmental stage. *Proteomics* 10:2605–2619.

Akhzari, D., and M. Pessarakli. 2016. Effect of drought stress on total protein, essential oil content, and physiological traits of *Levisticum officinale* Koch. *J. Plant Nutr.* 39:1365–1371.

Alam, I., D. G. Lee, K. H. Kim et al. 2010. Proteome analysis of soybean roots under waterlogging stress at an early vegetative stage. *J. Biosci.* 35:49–62.

Almagro, L., L. V. G. Ros, S. Belchi-Navarro, R. Bru, A. R. Barcelo, and M. A. Pedreno. 2009. Class III peroxidases in plant defence reactions. *J. Exp. Bot.* 60:377–390.

Amini, F., A. A. Ehsanpour, Q. T. Hoang, and J. S. Shin. 2007. Protein pattern changes in tomato under in vitro salt stress. *Russ. J. Plant Physiol.* 54:464–471.

An, Z. G., C. J. Li, Y. G. Zu et al. 2006. Expression of BjMT2, a metallothionein 2 from *Brassica juncea*, increases copper and cadmium tolerance in *Escherichia coli* and *Arabidopsis thaliana*, but inhibits root elongation in *Arabidopsis thaliana* seedlings. *J. Exp. Bot.* 57:3575–3582.

Anjum, N. A., M. Hasanuzzaman, M. A. Hossain et al. 2015. Jacks of metal/metalloid chelation trade in plants – an overview. *Front. Plant Sci.* 6:192. doi:10.3389/fpls.2015.00192

Annon, A., K. Rathore, and K. Crosby. 2014. Overexpression of a tobacco osmotin gene in carrot (*Daucus carota* L.) enhances drought tolerance. *In Vitro Cell. Dev. Biol. Plant* 50:299–306.

Antikainen, M., and M. Griffith. 1997. Antifreeze protein accumulation in freezing-tolerant cereals. *Physiol. Plant.* 99:423–432.

Antoniw, J. F., C. E. Ritter, W. S. Pierpoint, and L. C. Van Loon. 1980. Comparison of three pathogenesis-related proteins from plants of two cultivars of tobacco infected with TMV. *J. Gen. Virol.* 47:79–87.

Aranjuelo, I., G. Molero, G. Erice, J. C. Avice, and S. Nogués. 2010. Plant physiology and proteomics reveals the leaf response to drought in alfalfa (*Medicago sativa* L.). *J. Exp. Bot.* 62:111–123.

Arora, R., D. S. Pitchay, and B. C. Bearce. 1998. Water-stress-induced heat tolerance in geranium leaf tissues: a possible linkage through stress proteins? *Physiol. Plant.* 103:24–34.

Arora, R., L. J. Rowland, and A. R. Panta. 1997. Chill-response dehydrins in blueberry: are they associated with cold hardiness or dormancy transitions? *Physiol. Plant.* 101:8–16.

Artlip, T. S., and E. A. Funkhouser. 1995. Protein synthetic responses to environmental stresses. In *Handbook of Plant and Crop Physiology*, ed. M. Pessarakli, pp. 627–644. New York: Marcel Dekker Inc.

Ashoub, A., M. Baeumlisberger, M. Neupaertl, M. Karas, and W. Brüggemann. 2015. Characterization of common and distinctive adjustments of wild barley leaf proteome under drought acclimation, heat stress and their combination. *Plant Mol. Biol.* 87:459–471.

Ashrafi Parchin, R., and M. Shaban. 2014. Study on protein changes in wheat under drought stress. *Int. J. Adv. Biol. Biomed. Res.* 2:317–320.

Augustine, S. M. 2016. Function of heat-shock proteins in drought tolerance regulation of plants. In *Drought Stress Tolerance in Plants*, Vol 1, pp. 163–185. Cham: Springer.

Azevedo-Neto, A. D., J. T. Prisco, J. Enéas-Filho, C. E. B. Abreu, and E. Gomes-Filho. 2006. Effect of salt stress on antioxidative enzymes and lipid peroxidation in leaves and roots of salt-tolerant and salt-sensitive maize genotypes. *Environ. Exp. Bot.* 56:87–94.

Badur, R., K. Herbers, G. Mönke, F. Ludewig, and U. Sonnewald. 1994. Induction of pathogenesis-related proteins in sugar accumulating tobacco leaves. *Photosynthetica* 30:575–582.

Bae, M. S., E. J. Cho, E. Y. Choi, and O. K. Park. 2003. Analysis of the *Arabidopsis* nuclear proteome and its response to cold stress. *Plant J.* 36:652–663.

Bai, L. P., F. G. Sui, T. D. Ge, Z. H. Sun, Y. Y. Lu, and G. S. Zhou. 2006. Effect of soil drought stress on leaf water status, membrane permeability and enzymatic antioxidant system of maize. *Pedosphere* 16:326–332.

Baker, E. H., K. J. Bradford, J. A. Bryant, and T. L. Rost. 1995. A comparison of desiccation-related proteins (dehydrin and QP47) in peas (*Pisum satium*). *Seed Sci. Res.* 5:185–193.

Banuls, J., R. Ratajczak, and U. Luttge. 1995. NaCl stress enhances proteolytic turnover of the tonoplast H+-ATPase of Citrus sinensis-appearance of a 35 kDa fragment of subunit A still exhibiting ATP-hydrolysis activity. *Plant Cell Environ.* 18:1341–1344.

Barkla, B. J., R. Vera-Estrella, and O. Pantoja. 2013. Progress and challenges for abiotic stress proteomics of crop plants. *Proteomics* 13:1801–1815.

Barnett, N. M., and A. W. Naylor. 1966. Amino acid and protein metabolism in Bermuda grass during water stress. *Plant Physiol.* 41:1222–1230.

Barthakur, S., V. Babu, and K. C. Bansal. 2001. Over-expression of osmotin induces proline accumulation and confers tolerance to osmotic stress in transgenic tobacco. *J. Plant Biochem. Biotechnol.* 10:31–37.

Basu, S., A. Roychoudhury, P. P. Saha, and D. N. Sengupta. 2010. Differential antioxidative responses of indica rice cultivars to drought stress. *Plant Growth Regul.* 60:51–59.

Ben-Hayyim, G., Y. Vaadia, and B. G. Williams. 1989. Proteins associated with salt adaptation in citrus and tomato cells: involvement of 26 kDa polypeptides. *Physiol. Plant.* 77:332–340.

Bewley, J. D., and K. M. Larsen. 1982. Differences in the responses to water stress of growing and non-growing regions of maize mesocotyls: protein synthesis on total, free and membrane bound polyribosome fractions. *J. Exp. Bot.* 33:406–415.

Bhoomika, K., S. Pyngrope, and R. S. Dubey. 2013. Differential responses of antioxidant enzymes to aluminum toxicity in two rice (*Oryza sativa* L.) cultivars with marked presence and elevated activity of Fe SOD and enhanced activities of Mn SOD and catalase in aluminum tolerant cultivar. *Plant Growth Regul.* 71:235–252.

Bhoomika, K., S. Pyngrope, and R. S. Dubey. 2014. Effect of aluminum on protein oxidation, non-protein thiols and protease activity in seedlings of rice cultivars differing in aluminum tolerance. *J. Plant Physiol.* 171:497–508.

Bibi, N., A. Hameed, H. Ali et al. 2009. Water stress induced variations in protein profiles of germinating cotyledons from seedlings of chickpea genotypes. *Pak. J. Bot.* 41:731–736.

Biemelt, S., U. Keetman, H. P. Mock, and B.Grimm. 2000. Expression and activity of isoenzymes of superoxide dismutase in wheat roots in response to hypoxia and anoxia. *Plant Cell Environ.* 23:135–144.

Blokhina, O. B., E. Virolainen, and K. V. Fagerstedt. 2003. Antioxidants, oxidative damage and oxygen deprivation stress: a review. *Ann. Bot.* 91:179–194.

Blum, A. 2017. Osmotic adjustment is a prime drought stress adaptive engine in support of plant production. *Plant Cell Environ.* 40:4–10.

Bohler, S., K. Sergeant, Y. Jolivet et al. 2013. A physiological and proteomic study of poplar leaves during ozone exposure combined with mild drought. *Proteomics* 13:1737–1754.

Boo, Y. C., and J. Jung. 1999. Water deficit induced oxidative stress and antioxidative defence in rice plants. *J. Plant Physiol.* 51:255–261.

Branco-Neves, S., C. Soares, A. de Sousa et al. 2017. An efficient antioxidant system and heavy metal exclusion from leaves make *Solanum cheesmaniae* more tolerant to Cu than its cultivated counterpart. *Food Energy Secure.* 6:123–133.

Bray, E. A. 1995. Regulation of gene expression during abiotic stresses, and the role of the plant hormone abscisic acid. In *Handbook of Plant and Crop Physiology*, ed. M. Pessarakli, pp. 733–752. New York: Marcel Dekker Inc.

Brini, F., M. Hanin, V. Lumbreras et al. 2007. Overexpression of wheat dehydrin DHN-5 enhances tolerance to salt and osmotic stress in *Arabidopsis thaliana*. *Plant Cell Rep.* 26:2017–2026.

Broekaert, W. F., J. V. Parijs, A. K. Allen, and W. J. Peumans. 1988. Comparison of some molecular, enzymatic and antifungal properties of chitinases from thorn-apple, tobacco and wheat. *Physiol. Mol. Plant. Pathol.* 33:319–331.

Bruggemann, W., S. Klaucke, and Maaskantel, K. 1994. Long-term chilling of young tomato plants under low-light V. Kinetic and molecular-properties of 2 key enzymes of the calvin cycle in *Lycopersicon esculentum* mill. and *L. peruvianum* mill. *Planta* 194:160–168.

Burke, J. J., J. L. Hatfield, R. R. Klein, and J. E. Mullet. 1985. Accumulation of heat shock proteins in field-grown cotton. *Plant Physiol.* 78:394–398.

Buron-Moles, G., R. Torres, F. Amoako-Andoh et al. 2014. Analysis of changes in protein abundance after wounding in 'Golden Delicious' apples. *Postharvest Biol. Technol.* 87:51–60.

Cabello, F., J. V. Jorrín, and M. Tena. 1994. Chitinase and beta-1,3-glucanase activities in chickpea (*Cicer arietinum*) – induction of different isoenzymes in response to wounding and ethephon. *Physiol. Plant.* 92:654–660.

Camejo, D., P. Rodrıguez, M. A. Morales, J. M. Dell'amico, A. Torrecillas, and J. J. Alarcon. 2005. High temperature effects on photosynthetic activity of two tomato cultivars with different heat susceptibility. *J. Plant Physiol.* 162:281–289.

Carreras, H. A., M. S. Canas, and M. L. Pignata. 1996. Differences in responses to urban air pollutants by *Ligustrum lucidum* AIT and *Ligustrum lucidum* AIT F tricolor (REHD). *Environ. Pollut.* 93:211–218.

Casaretto, J. A., A. El-kereamy, B. Zeng et al. 2016. Expression of OsMYB55 in maize activates stress-responsive genes and enhances heat and drought tolerance. *BMC Genomics* 17:312.

Catalá, R., R. López-Cobollo, M. M. Castellano et al. 2014. The Arabidopsis 14–3-3 protein RARE COLD INDUCIBLE 1A links low-temperature response and ethylene biosynthesis to regulate freezing tolerance and cold acclimation. *Plant Cell* 26:3326–3342.

Chatthai, M., K. H. Kaukinen, T. J. Tranbarger, P. K. Gupta, and S. Misra. 1997. The isolation of a novel metallothionein-related cDNA expressed in somatic and zygotic embryos of Douglas-fir: regulation by ABA, osmoticum and metal ions. *Plant Mol. Biol.* 34:243–254.

Chen, J., J. Zhou, and P. B. Goldsbrough. 1997. Characterization of phytochelatin synthase from tomato. *Physiol. Plant.* 101:165–172.

Chen, X., S. Lin, Q. Liu et al. 2014. Expression and interaction of small heat shock proteins (sHsps) in rice in response to heat stress. *Biochim. Biophys. Acta* 1844:818–828.

Cheng, Y., Y. Qi, Q. Zhu et al. 2009. New changes in the plasma-membrane-associated proteome of rice roots under salt stress. *Proteomics* 9:3100–3114.

Chen, Y. C., H. S. Chang, H. M. Lai, and S. T. Jeng. 2005. Characterization of the wound-inducible protein ipomoelin from sweet potato. *Plant Cell Environ.* 28:251–259.

Cheng, Z., K. Dong, P. Ge, Y. Bian, L. Dong, X. Deng, X. Li, and Y. Yan. 2015. Identification of leaf proteins differentially accumulated between wheat cultivars distinct in their levels of drought tolerance. *PLoS One.* 10:e0125302.

Chinnusamy, V., A. Jagendorf, and J. K. Zhu. 2005. Understanding and improving salt tolerance in plants. *Crop Sci.* 45:437–448.

Chmielewska, K., P. Rodziewicz, B. Swarcewicz et al. 2016. Analysis of drought-induced proteomic and metabolomic changes in barley (*Hordeum vulgare* L.) leaves and roots unravels some aspects of biochemical mechanisms involved in drought tolerance. *Front. Plant Sci.* 7:1108. doi:10.3389/fpls.2016.01108

Cho, E. K., and Y. J. Choi. 2009. A nuclear-localized HSP70 confers thermoprotective activity and drought-stress tolerance on plants. *Biotechnol. Lett.* 31:597–606.

Choi, S. Y., E. M. Baek, and S. Y. Lee. 1995. A cDNA differentially expressed by Cadmium stress in *Arabidopsis. Plant Physiol.* 101:699–700.

Chowdhury, S., A. Basu, and S. Kundu. 2017. Overexpression of a new osmotin-like protein gene (SindOLP) confers tolerance against biotic and abiotic stresses in sesame. *Front. Plant Sci.* 8:410. doi:10.3389/fpls.2017.00410

Christopher, M. E., and A. G. Good. 1996. Characterization of hypoxically inducible lactate dehydrogenase in maize. *Plant Physiol.* 112:1015–1022.

Claes, B., R. Dekeyser, R. Villarroel et al. 1990. Characterization of a rice gene showing organ-specific expression in response to salt stress and drought. *Plant Cell* 2:19–27.

Close, T. J. 1997. Dehydrins – a commonality in the response of plants to dehydration and low temperature. *Physiol. Plant.* 100:291–296.

Cobbett, C., and P. Goldsbrough. 2002. Phytochelatins and metallothioneins: roles in heavy metal detoxification and homeostasis. *Annu. Rev. Plant Biol.* 53:159–182.

Conklin, P. L., and R. L. Last. 1995. Differential accumulation of antioxidant mRNAs in *Arabidopsis thaliana* exposed to ozone. *Plant Physiol.* 109:203–212.

Conrath, U., H. Silva, and D. F. Klessig. 1997. Protein dephosphorylation mediates salicylic acid-induced expression of *PR-1* genes in tobacco. *Plant J.* 11:747–757.

Cooper, P., and T. H. D. Ho. 1983. Heat shock proteins in maize. *Plant Physiol.* 71:215–222.

Cordewener, J. H. G., G. Hause, E. Gorgen et al. 1995. Change in synthesis and localization of members of the 70 kDa class of heat-stress proteins accompany the induction of embryogenesis in *Brassica napus* L. microspores. *Planta* 196:747–755.

Cui, D., D. Wu, J. Liu et al. 2015. Proteomic analysis of seedling roots of two maize inbred lines that differ significantly in the salt stress response. *PLoS ONE* 10:e0116697.

Cuypers, A., K. M. Koistinen, H. Kokko, S. Karenlampi, S. Auriola, and J. Vangronsveld. 2005. Analysis of bean (*Phaseolus vulgaris* L.) proteins affected by copper stress. *J. Plant Physiol.* 162:383–392.

Cvikrová, M., L. Gemperlová, O. Martincová, and R. Vanková. 2013. Effect of drought and combined drought and heat stress on polyamine metabolism in proline-over-producing tobacco plants. *Plant Physiol. Biochem.* 73:7–15.

D'Angeli, S., and M. M. Altamura. 2007. Osmotin induces cold protection in olive trees by affecting programmed cell death and cytoskeleton organization. *Planta* 225:1147–1163.

Dafoe, N. J., A. Zamani, A. K. M. Ekramoddoullah, D. Lippert, J. Bohlmann, and C. P. Constabel. 2009. Analysis of the poplar phloem proteome and its response to leaf wounding. *J. Proteome Res.* 8:2341–2350.

Dalal, M., D. Tayal, V. Chinnusamy, and K. C. Bansal. 2009. Abiotic stress and ABA-inducible Group 4 LEA from *Brassica napus* plays a key role in salt and drought tolerance. *J. Biotechnol.* 139:137–145.

Dalen, L. S., Ø. Johnsen, A. Lönneborg, and M. W. Yaish. 2015. Freezing tolerance in Norway spruce, the potential role of pathogenesis-related proteins. *Acta Physiol. Plant.* 37:1–9.

Dann, E. K., P. Meuwly, J. P. Metraux, and B. J. Deverall. 1996. The effect of pathogen inoculation or chemical treatment on activities of chitinase and beta-1,3-glucanase and accumulation of salicylic acid in leaves of green bean, *Phaseolus vulgaris. Physiol. Mol. Plant Pathol.* 49:307–319.

Danquah, A., A. de Zelicourt, J. Colcombet, and H. Hirt. 2014. The role of ABA and MAPK signaling pathways in plant abiotic stress responses. *Biotechnol. Adv.* 32:40–52.

Dantas, B. F., L. S. Ribeiro, and C. A. Aragao. 2007. Germination, initial growth and cotyledon protein content of bean cultivars under salinity stress. *Rev. Bras. Sementes* 29:106–110.

Dao, T. T. H., H. J. M. Linthorst, and R. Verpoorte. 2011. Chalcone synthase and its functions in plant resistance. *Phytochem. Rev.* 10:397–412.

Dasgupta, J., and J. D. Bewley. 1984. Variations in protein synthesis in different regions of greening leaves of barley seedlings and effects of imposed water stress. *J. Exp. Bot.* 35:1450–1459.

Davis, M., W. Butler, and M. E. Vayda. 1990. Molecular responses to environmental stresses and their relationship to soft rot. In *Molecular and Cellular Biology of Potato*, eds. M. Vayda, and W. Park, pp. 71–87. Wallingford: CAB International.

Debouba, M., H. Maâroufi-Dghimi, A. Suzuki, M. H. Ghorbel, and H. Gouia. 2007. Changes in growth and activity of enzymes involved in nitrate reduction and ammonium assimilation in tomato seedlings in response to NaCl stress. *Ann. Bot.* 99:1143–1151.

Demirevska, K., L. Simova-Stoilova, V. Vassileva, I. Vaseva, B. Grigorova, and U. Feller. 2008. Drought-induced leaf protein alterations in sensitive and tolerant wheat varieties. *Gen. Appl. Plant Physiol.* 34:79–102.

Devi, R., N. Munjral, A. K. Gupta, and N. Kaur. 2007. Cadmium induced changes in carbohydrate status and enzymes of carbohydrate metabolism, glycolysis and pentose phosphate pathway in pea. *Environ. Exp. Bot.* 61:167–174.

Dhindsa, R. S., and S. S. Mohapatra. 1990. cDNA cloning, and expression of genes associated with freezing tolerance in alfalfa. In *Proceedings of International Congress of Plant Physiology*, New Delhi, pp. 908–915.

Diab, H., and A. M. Limami. 2016. Reconfiguration of N metabolism upon hypoxia stress and recovery: roles of alanine aminotransferase (AlaAT) and glutamate dehydrogenase (GDH). *Plants (Basel)* 5: E25. doi:10.3390/plants5020025

Didierjean, L., P. Frendo, W. Nasser, G. Genot, J. Marivet, and G. Burkard. 1996. Heavy-metal-responsive genes in maize – identification and comparison of their expression upon various forms of abiotic stress. *Planta* 199:1–8.

Dietz, K. J., M. Baier, and U. Krämer. 1999. Free radicals and reactive oxygen species as mediators of heavy metal toxicity in plants. In *Heavy Metal Stress in Plant*, eds. M. N. V. Prasad, and J. Hagemeyer, pp. 73–97. Heidelberg: Springer-Verlag.

Dolferus, R., M. Ellis, G. Debruxelles et al. 1997. Strategies of gene action in *Arabidopsis* during hypoxia. *Ann. Bot.* 79:21–31.

Dooki, A. D., F. J. Mayer-Posner, H. Askari, A. A. Zaiee, and G. H. Salekdeh. 2006. Proteomic responses of rice young panicles to salinity. *Proteomics* 6:6498–6507.

Dreier, W., C. Schnarrenberger, and T. Borner. 1995. Light and stress-dependent enhancement of amylolitic activities in white and green barley leaves-beta-amylases are stress-induced proteins. *J. Plant Physiol.* 145:342–348.

Drira, M., M. Hanin, K. Masmoudi, and F. Brini. 2016. Comparison of full-length and conserved segments of wheat dehydrin DHN-5 overexpressed in *Arabidopsis thaliana* showed different responses to abiotic and biotic stress. *Funct. Plant Biol.* 43:1048–1060.

Du, Y. C., Y. Kawamitsu, A. Nose et al. 1996. Effects of water stress on carbon exchange rate and activities of photosynthetic enzymes in leaves of sugarcane (*Saccharum* sp). *Aust. J. Plant Physiol.* 23:719–726.

Duan, J. J., S. R. Guo, H. F. Fan, S. P. Wang, and Y. Y. Kang. 2006. Effects of salt stress on proline and polyamine metabolisms in the roots of cucumber seedlings. *Acta Botanica Boreali-Occidentalia Sinica* 26:2486–2492.

Dubey, R. S. 1983. Biochemical changes in germinating rice seeds under saline stress. *Biochem. Physiol. Pflanzen.* 177:523–535.

Dubey, R. S. 1985. Effect of salinity on nucleic acid metabolism of germinating rice seeds differing in salt tolerance. *Plant Physiol. Biochem. (India)* 12:9–16.

Dubey, R. S. 1997. Nitrogen metabolism in plants under salt stress. In *Strategies for Improving Salt Tolerance in Higher Plants*, eds. P. K. Jaiwal, R. P. Singh, and A. Gulati, pp. 129–158. New Delhi: IBH Publn.

Dubey, R. S., and K. N. Sharma. 1989. Acid and alkaline phosphatases in rice seedlings growing under salinity stress. *Indian J. Plant Physiol.* 32:217–223.

Dubey, R. S., and K. N. Sharma. 1990. Behaviours of phosphatases in germinating rice in relation to salt tolerance. *Plant Physiol. Biochem. (Paris)* 28:17–26.

Dubey, R. S., and M. Pessarakli. 1995. Physiological mechanisms of nitrogen absorption, and assimilation in plants under stressful conditions. In *Handbook of Plant and Crop Physiology*, ed. M. Pessarakli, pp. 605–625. New York: Marcel Dekker Inc.

Dubey, R. S., and M. Rani. 1987. Proteases and proteins in germinating rice seeds in relation to salt tolerance. *Plant Physiol. Biochem. (India)* 14:174–182.

Dubey, R. S., and M. Rani. 1989. Influence of NaCl salinity on growth and metabolic status of proteins and amino acids in rice seedlings. *J. Agron. Crop Sci.* 162:97–106.

Dubey, R. S., and M. Rani. 1990. Influence of NaCl salinity on behaviours of protease, aminopeptidase and carboxypeptidase in rice seedling in relation to salt tolerance. *Aust. J. Plant Physiol.* 17:215–221.

Dubey, R. S., K. N. Sharma, and B. Singh. 1987. Salinity induced adenosine triphosphatase activity in germinating rice seeds. *Indian J. Plant Physiol.* 30:256–260.

Durham, R. E., G. A. Moore, D. Haskell, and C. L. Guy. 1991. Cold-acclimation induced changes in freezing tolerance and translatable RNA content in *Citrus grandis* and *Poncirus trifoliata*. *Physiol. Plant.* 82:519–522.

Ebrahim, S., K. Usha, and B. Singh. 2011. Pathogenesis related (PR) proteins in plant defense mechanism. *Sci. Against Microb. Pathog.* 2:1043–1054.

Efeoglu, B., and S. Terzioglu. 2007. Varying patterns of protein synthesis in bread wheat during heat shock. *Acta Biol. Hung.* 58:93–104.

Egea, C., M. D. Alcázar, and M. E. Candela. 1996. Beta-1,3-glucanase and chitinase as pathogenesis-related proteins in the defense reaction of two *Capsicum annuum* cultivars infected with cucumber mosaic virus. *Biol. Plant.* 38:437–443.

Elenany. A. E. 1997. Shoot regeneration and protein synthesis in tomato tissue cultures. *Biol. Plant.* 39:303–308.

Elsamad, H. M. A., and M. A. K. Shaddad. 1997. Salt tolerance of soybean cultivars. *Biol. Plant.* 39:263–269.

Emamverdian, A., Y. Ding, F. Mokhberdoran, and Y. Xie. 2018. Antioxidant response of bamboo (*Indocalamus latifolius*) as affected by heavy metal stress. *J. Elementol.* 23:341–352.

Ericson, M. E., and S. H. Alfinito. 1984. Protein produced during salt-stress in tobacco cell cultures. *Plant Physiol.* 74:506–509.

Ezaki, M. Katsuhara, M. Kawamura, and H. Matsumoto. 2001. Different mechanisms of four aluminum (Al)-resistant transgenes for Al toxicity in *Arabidopsis*. *Plant Physiol.* 127:918–927.

Fadzilla, N. M., R. P. Finch, and R. H. Burdon. 1997. Salinity, oxidative stress and antioxidant responses in shoot cultures of rice. *J. Exp. Bot.* 48:325–331.

Fan, W., Q. Guo, C. Liu et al. 2018. Two mulberry phytochelatin synthase genes confer zinc/cadmium tolerance and accumulation in transgenic Arabidopsis and tobacco. *Gene* 645:95–104.

Fan, Y., B. Liu, H. Wang, S. Wang, and J. Wang. 2002. Cloning of an antifreeze protein gene from carrot and its influence on cold tolerance in transgenic tobacco plants. *Plant Cell Rep.* 21:296–301.

Fang, X., J. Chen, L. Dai et al. 2015. Proteomic dissection of plant responses to various pathogens. *Proteomics* 15:1525–1543.

Feng, Y., S. Komatsu, T. Furukawa, T. Koshiba, and Y. Kohno. 2008. Proteome analysis of proteins responsive to ambient and elevated ozone in rice seedlings. *Agric. Ecosyst. Environ.* 125:255–265.

Feng, Z., L. Wang, H. Pleijel, J. Zhu, and K. Kobayashi. 2016. Differential effects of ozone on photosynthesis of winter wheat among cultivars depend on antioxidative enzymes rather than stomatal conductance. *Sci. Total Environ.* 572:404–411.

Ferreira, S., K. Hjerno, M. Larsen et al. 2006. Proteome profiling of *Populus euphratica* Oliv. upon heat stress. *Ann. Bot.* 98:361–377.

Ferullo, J. M., L. P. Vezina, J. Rail, S. Laberge, P. Nadeau, and Y. Castonguay. 1997. Differential accumulation of two glycine-rich proteins during cold-acclimation alfalfa. *Plant Mol. Biol.* 33:625–633.

Fowler, S., and M. F. Thomashow. 2002. Arabidopsis transcriptome profiling indicates that multiple regulatory pathways are activated during cold acclimation in addition to the CBF cold response pathway. *Plant Cell* 14:1675–1690.

Gao, J. P., D. Y. Chao, and H. X. Lin. 2007. Understanding abiotic stress tolerance mechanisms: recent studies on stress response in rice. *J. Integr. Plant Biol.* 49:742–750.

Gasic, K., and S. S. Korban. 2007. Transgenic Indian mustard (*Brassica juncea*) plants expressing an *Arabidopsis* phytochelatin synthase (AtPCS1) exhibit enhanced As and Cd tolerance. *Plant Mol. Biol.* 64:361–369.

Ge, P., C. Ma, S. Wang et al. 2012. Comparative proteomic analysis of grain development in two spring wheat varieties under drought stress. *Anal. Bioanal. Chem.* 402:1297–1313.

Geigenberger, P., R. Reimholz, M. Geiger, L. Merlo, V. Canale, and M. Stitt. 1997. Regulation of sucrose and starch metabolism in potato tubers in response to short-term water deficit. *Planta* 201:502–518.

Gething, M. J. 1997. *Guidebook to Molecular Chaperones, and Protein Folding Catalysts.* New York: Oxford University Press.

Ghaffari, M., M. Toorchi, M. Valizadeh, and M. Shakiba. 2017. Proteomic prospects for tolerance of sunflower (*Helianthus annuus*) to drought stress during the flowering stage. *Crop Pasture Sci.* 68:457–465.

Ghosh, M. 2006. Antifungal properties of haem peroxidase from *Acorus calamus. Ann. Bot.* 98:1145–1153.

Giacomelli, L., A. Rudella, and K. J. van Wijk. 2006. High light response of the thylakoid proteome in Arabidopsis wild type and the ascorbate-deficient mutant *vtc2–2.* A comparative proteomics study. *Plant Physiol.* 141:685–701.

Gill, S. S., N. A. Khan, and N. Tuteja. 2012. Cadmium at high dose perturbs growth, photosynthesis and nitrogen metabolism while at low dose it up regulates sulfur assimilation and antioxidant machinery in garden cress (*Lepidium sativum* L.). *Plant Sci.* 182:112–120.

Gilmour, S. J., A. M. Sebolt, M. P. Salazar, J. D. Everard, and M. F. Thomashow. 2000. Overexpression of the Arabidopsis CBF3transcriptional activator mimics multiple biochemical changes associated with cold acclimation. *Plant Physiol.* 124:1854–1865.

Gilmour, S. J., and M. F. Thomashow. 1991. Cold acclimation and cold-regulated gene expression in ABA mutants of *Arabidopsis thaliana. Plant Mol. Biol.* 17:1233–124.

Gilroy, S., K. Trebacz, and V. Salvador Recatala. 2018. Intercellular electrical signals in plant adaptation and communication. *Front. Plant Sci.* 9:643. doi:10.3389/fpls.2018.00643

Giri, A., S. Heckathorn, S. Mishra, and C. Krause. 2017. Heat stress decreases levels of nutrient-uptake and-assimilation proteins in tomato roots. *Plants (Basel)* 6:6. doi:10.3390/plants6010006

Glick, R. E., C. D. Schlagnhaufer, R. N. Arteca, and E. J. Pell. 1995. Ozone-induced ethylene emission accelerates the loss of ribulose-1,5-bisphosphate carboxylase/oxygenase and nuclear-encoded mRNAs in senescing potato leaves. *Plant Physiol.* 109:891–898.

Gogorcena, Y., I. Iturbeormaetxe, P. R. Escuredo, and M. Becana. 1995. Antioxidant defenses against activated oxygen in pea nodules subjected to water stress. *Plant Physiol.* 108:753–759.

Goulas, E., M. Schubert, T. Kieselbach et al. 2006. The chloroplast lumen and stromal proteomes of *Arabidopsis thaliana* show differential sensitivity to short and long-term exposure to low temperature. *Plant J.* 47:720–734.

Greenway, H., and R. Munns. 1980. Mechanisms of salt tolerance in nonhalophytes. *Ann. Rev. Plant Physiol.* 31:149–1190.

Griffith, M., M. Antikainen, W. C. Hon et al. 1997. Antifreeze proteins in winter rye. *Physiol. Plant.* 100:327–332.

Grill, E., S. Löffler, E. L. Winnacker, and M. H. Zenk. 1989. Phytochelatins, the heavy-metal-binding peptides of plants, are synthesized from glutathione by a specific γ-glutamylcysteine dipeptidyl transpeptidase (phytochelatin synthase). *Proc. Natl. Acad. Sci. USA* 86:6838–6842.

Guo, Y. H., Y. P. Yu, D. Wang et al. 2009. GhZFP1, a novel CCCH-type zinc finger protein from cotton, enhances salt stress tolerance and fungal disease resistance in transgenic tobacco by interacting with GZIRD21A and GZIPR5. *New Phytol.* 183:62–75.

Gupta, M., P. Sharma, N. B. Sarin, and A. K. Sinha. 2009. Differential response of arsenic stress in two varieties of *Brassica juncea* L. *Chemosphere* 74:1201–1208.

Gusta, L. V., R. Trischuk, and C. J. Weiser. 2005. Plant cold acclimation: the role of abscisic acid. *J. Plant Growth Regul.* 24:308–318.

Gutha, L. R., and A. R. Reddy. 2008. Rice *DREB1B* promoter shows distinct stress-specific responses and the overexpression of cDNA in tobacco confers improved abiotic and biotic stress tolerance. *Plant Mol. Biol.* 68:533–555.

Guy, C. L., and J. V. Carter. 1984. Characterization of partially purified glutathione reductase from cold-hardened and nonhardened spinach leaf tissue. *Cryobiology* 21:454–464.

Guy, C. L., and D. Haskell. 1987. Induction of freezing tolerance in spinach is associated with the synthesis of cold acclimation induced proteins. *Plant Physiol.* 84:872–878.

Guy, C. L., D. Haskell, and G. Yelenosky. 1988. Changes in freezing tolerance and polypeptide content of spinach and citrus at 5°C. *Cryobiology* 25:264–271.

Hahn, M., and V. Walbot. 1989. Effects of cold-treatment on protein-synthesis and mRNA levels in rice leaves. *Plant Physiol.* 91:930–938.

Hall, J. L., and T. J. Flowers. 1973. The effect of salt on protein synthesis in the halophyte *Suaeda maritinta. Planta* 110:361–368.

Hameed, A., M. Goher, and N. Iqbal. 2013. Drought induced programmed cell death and associated changes in antioxidants, proteases, and lipid peroxidation in wheat leaves. *Biol. Plant.* 57:370–374.

Han, B., and A. R. Kermode. 1996. Dehydrin-like proteins in castor bean seeds and seedlings are differentially produced in response to ABA and water-deficit-related stresses. *J. Exp. Bot.* 47:933–939.

Han, X., X. He, W. Qiu et al. 2017. Pathogenesis-related protein PR10 from *Salix matsudana* Koidz exhibits resistance to salt stress in transgenic *Arabidopsis thaliana. Environ. Exp. Bot.* 141:74–82.

Hao, P., J. Zhu, A. Gu et al. 2015. An integrative proteome analysis of different seedling organs in tolerant and sensitive wheat cultivars under drought stress and recovery. *Proteomics* 15:1544–1563.

Harrington, H. M., S. Dash, N. Dharmasiri, and S. Dharmasiri. 1994. Heat shock proteins – a search for functions. *Aust. J. Plant Physiol.* 21:843–855.

Hashimoto, M., and S. Komatsu. 2007. Proteomic analysis of rice seedlings during cold stress. *Proteomics* 7:1293–1302.

Hassan, N. M., Z. M. El-Bastawisy, A. K. El-Sayed, H. T. Ebeed, and M. M. N. Alla. 2015. Roles of dehydrin genes in wheat tolerance to drought stress. *J. Adv. Res.* 6:179–188.

Hassinen, V. H., A. I. Tervahauta, H. Schat, and S. O. Kärenlampi. 2011. Plant metallothioneins-metal chelators with ROS scavenging activity? *Plant Biol.* 13:225–232.

He, Y. L., X. Z. Liu, and B. R. Huang. 2005. Protein changes in response to heat stress in acclimated and nonacclimated creeping bentgrass. *J. Am. Soc. Hort. Sci.* 130:521–526.

Heckathorn, S. A., J. K. Mueller, S. LaGuidice et al. 2004. Chloroplast small heat-shock proteins protect photosynthesis during heavy metal stress. *Am. J. Bot.* 91:1312–1318.

Heikkila, J. J., J. E. T. Papp, J. D. Schultz, and J. D. Bewley. 1984. Induction of heat shock protein messenger RNA in maize mesocotyls by water stress, abscisic acid and wounding. *Plant Physiol.* 76:270–274.

Herbers, K., P. Meuwly, J. P. Métraux, and U. Sonnewald. 1996. Salicylic acid-independent induction of pathogenesis-related protein transcripts by sugars is dependent on leaf developmental stage. *FEBS Lett.* 397:239–244.

Heyman, J., B. Canher, A. Bisht, F. Christiaens, and L. D. Veylder. 2018. Emerging role of the plant ERF transcription factors in coordinating wound defense responses and repair. *J. Cell Sci.* 131. doi:10.1242/jcs.208215

Hsiao, T. C. Plant responses to water stress. 1973. *Ann. Rev. Plant Physiol.* 24:519–570.

Hu, L., H. Li, H. Pang, and J. Fu. 2012. Responses of antioxidant gene, protein and enzymes to salinity stress in two genotypes of perennial ryegrass (*Lolium perenne*) differing in salt tolerance. *J. Plant Physiol.* 169:146–156.

Hu, T., S. Zhu, L. Tan, W. Qi, S. He, and G. Wang. 2016. Overexpression of OsLEA4 enhances drought, high salt and heavy metal stress tolerance in transgenic rice (*Oryza sativa* L.). *Environ. Exp. Bot.* 123:68–77.

Hu, T. Z. 2008. OsLEA3, a late embryogenesis abundant protein gene from rice, confers tolerance to water deficit and salt stress to transgenic rice. *Russ. J. Plant Physiol.* 55:530–537.

Hu, X., X. Wu, C. Li et al. 2012b. Abscisic acid refines the synthesis of chloroplast proteins in maize (*Zea mays*) in response to drought and light. *PloS ONE* 7. doi:10.1371/journal.pone.0049500

Huang, B. R., and C. P. Xu. 2008. Identification and characterization of proteins associated with plant tolerance to heat stress. *J. Integr. Plant Biol.* 50:1230–1237.

Huang, L. J., G. X. Cheng, A. Khan et al. 2018. CaHSP16.4, a small heat shock protein gene in pepper, is involved in heat and drought tolerance. *Protoplasma* 1–13. doi:10.1007/s00709-018-1280-7

Huang, S., H. Greenway, T. D. Colmer, and A. H. Millar. 2005. Protein synthesis by rice coleoptiles during prolonged anoxia: implications for glycolysis, growth and energy utilization. *Ann. Bot.* 96:703–715.

Hughes, M. A., and M. A. Dunn. 1996. The molecular biology of plant acclimation to low temperature. *J. Exp. Bot.* 47:291–305.

Hundertmark, M., and D. K. Hincha. 2008. LEA (late embryogenesis abundant) proteins and their encoding genes in *Arabidopsis thaliana*. *BMC Genomics* 9:118. doi:10.1186/1471-2164-9-118

Huo, C., B. Zhang, H. Wang et al. 2016. Comparative study of early cold-regulated proteins by two-dimensional difference gel electrophoresis reveals a key role for phospholipase Dα1 in mediating cold acclimation signaling pathway in rice. *Mol. Cell Proteomics* 15:1397–1411.

Hurkman, W. J., and C. K. Tanaka. 1987. The effects of salt on the pattern of protein synthesis in barley roots. *Plant Physiol.* 83:517–524.

Husaini, A. M., and M. Z. Abdin. 2008. Overexpression of tobacco osmotin gene leads to salt stress tolerance in strawberry (*Fragaria* x *ananassa* Duch.) plants. *Indian J. Biotechnol.* 7:465–471.

Huseynova, I. M. 2012. Photosynthetic characteristics and enzymatic antioxidant capacity of leaves from wheat cultivars exposed to drought. *Biochim Biophys Acta.* 1817:1516–1523.

Hwang, S. Y., and T. T. Van Toai. 1991. Abscisic acid induces anaerobiosis tolerance in corn. *Plant Physiol.* 97:593–597.

Iba, K. 2002. Acclimative response to temperature stress in higher plants: approaches of gene engineering for temperature tolerance. *Annu. Rev. Plant. Biol.* 53:225–245.

Igarashi, Y., Y. Yoshiba, Y. Sanada, K. Wada, K. Yamaguchi-Shinozaki, and K. Shinozaki. 1997. Characterization of the gene for Δ1-pyrroline-5-carboxylate synthetase and correlation between the expression of the gene and salt tolerance in *Oryza sativa* L. *Plant Mol. Biol.* 33:857–865.

Iuchi, S., K. Yamaguchishinozaki, T. Urao, and K. Shinozaki. 1996a. Characterization of two cDNA for novel drought-inducible genes in the highly drought-tolerant cowpea. *J. Plant Res.* 109:415–424.

Iuchi, S., K. Yamaguchishinozaki, T. Urao, T. Terao, and K. Shinozaki. 1996b. Novel drought-inducible genes in the highly drought-tolerant cowpea: cloning of cDNAs and analysis of the expression of the corresponding genes. *Plant Cell Physiol.* 37:1073–1082.

Jaglo-Ottosen, K. R., S. J. Gilmour, D. G. Zarka, O. Schabenberger, and M. F. Thomashow. 1998. *Arabidopsis* CBF1 overexpression induces COR genes and enhances freezing tolerance. *Science* 280:104–106.

Jahanbakhshsh G. S., H. Chilan, and K. Razavi. 2017. Effects of water deficit on the physiological response, total protein, and gene expression of *Rab17* in wheat (*Triticum aestivum*). *J. Plant Proc. Func.* 5:35–42.

Jaiswal, N., M. Singh, R. S. Dubey, V. Venkataramanappa, and D. Datta. 2013. Phytochemicals and antioxidative enzymes defence mechanism on occurrence of yellow vein mosaic disease of pumpkin (*Cucurbita moschata*). *3 Biotech* 3:287–295.

James, V. A., I. Neibaur, and F. Altpeter. 2008. Stress inducible expression of the DREB1A transcription factor from xeric, *Hordeum spontaneum* L. in turf and forage grass (*Paspalum notatum* Flugge) enhances abiotic stress tolerance. *Transgenic Res.* 17:93–104.

Janmohammadi, M., L. Zolla, and S. Rinalducci. 2015. Low temperature tolerance in plants: changes at the protein level. *Phytochemistry* 117:76–89.

Jarillo, J. A., J. Capel, A. Leyva, J. M. Martínez-Zapater, and J. Salinas. 1994. Two related low-temperature-inducible genes of Arabidopsis encode proteins showing high homology to 14-3-3 proteins, a family of putative kinase regulators. *Plant Mol. Biol.* 25:693–704.

Jha, A. B., and R. S. Dubey. 2004a. Carbohydrate metabolism in growing rice seedlings under arsenic toxicity. *J. Plant Physiol.* 161:867–872.

Jha, A. B., and R. S. Dubey. 2004b. Arsenic exposure alters the activities of key nitrogen assimilatory enzymes in growing rice seedlings. *Plant Growth Regul.* 43:259–268.

Jia, F., S. Qi, H. Li et al. 2014. Overexpression of Late Embryogenesis Abundant 14 enhances Arabidopsis salt stress tolerance. *Biochem. Biophys. Res. Commun.* 454:505–511.

Jiang, L., J. Wu, S. Fan et al. 2015. Isolation and characterization of a novel pathogenesis-related protein gene (GmPRP) with induced expression in soybean (*Glycine max*) during infection with *Phytophthora sojae*. *PLoS ONE* 10. doi:10.1371/journal.pone.0129932

Jiang, Y., and B. Huang. 2002. Protein alterations in tall fescue in response to drought stress and abscisic acid. *Crop Sci.* 42:202–207.

Jimenez, T., I. Martin, J. Hernandez-Nistal, E. Labrador, and B. Dopico. 2008. The accumulation of a Kunitz trypsin inhibitor from chickpea (TPI-2) located in cell walls is increased in wounded leaves and elongating epicotyls. *Physiol. Plant.* 132:306–317.

Jin, H., M. Xu, H. Chen et al. 2016. Comparative proteomic analysis of differentially expressed proteins in *Amaranthus hybridus* L. roots under cadmium stress. *Water Air Soil Pollut.* 227:220. doi:10.1007/s11270-016-2914-z

John, U. P., R. M. Polotnianka, K. A. Sivakumaran et al. 2009. Ice recrystallization inhibition proteins (IRIPs) and freeze tolerance in the cryophilic Antarctic hair grass *Deschampsia antarctica* E. Desv. *Plant Cell Environ.* 32:336–348.

Joseph, B., J. George, and M. Jeevitha. 2012. Impact of heavy metals and Hsp response. *Int. J. Biosci* 2:51–64.

Joshi, S. 1987. Effect of soil salinity on nitrogen metabolism in *Cajanus cajan* L. *Indian J. Plant Physiol.* 30:223–225.

Jung, J. L., S. Maurel, B. Fritig, and G. Hahne. 1995. Different pathogenesis-related proteins are expressed in sunflower (*Helianthus annuus* L.) in response to physical, chemical and stress factors. *J. Plant Physiol.* 145:153–160.

Júnior, C. A. L., S. Barbosa Hde, R. Moretto Galazzi, H. H. Ferreira Koolen, F. C. Gozzo, and M. A. Arruda. 2015. Evaluation of proteome alterations induced by cadmium stress in sunflower (*Helianthus annuus* L.) cultures. *Ecotoxicol. Environ. Saf.* 119:170–177.

Kahlaoui, B., M. Hachicha, E. Misle, F. Fidalgo, and J. Teixeira. 2018. Physiological and biochemical responses to the exogenous application of proline of tomato plants irrigated with saline water. *J. Saudi Soc. Agric. Sci.* 17:17–23.

Kaiser, W. M. 1987. Effect of waters deficit on photosynthetic capacity. *Physiol. Plant.* 71:142–149.

Kamal, A. H., H. Rashid, K. Sakata, and S. Komatsu. 2015. Gel-free quantitative proteomic approach to identify cotyledon proteins in soybean under flooding stress. *J. Proteomics* 112:1–13.

Kanabus, J., C. S. Pikaard, and J. H. Cherry. 1984. Heat shock proteins in tobacco cell suspension during growth cycle. *Plant Physiol.* 75:639–644.

Karimzadeh, G., G. R. Sharifi-Sirchi, M. Jalali-Javaran, H. Dehghani, and D. Francis. 2006. Soluble proteins induced by low temperature treatment in the leaves of spring and winter wheat cultivars. *Pak. J. Bot.* 38:1015–1026.

Kasuga, M., Q. Liu, S. Miura, K. Yamaguchi-Shinozaki, and K. Shinozaki. 1999. Improving plant drought, salt and freezing tolerance by gene transfer of a single stress-inducible transcription factor. *Nature Biotechnol.* 17:287–291.

Kasuga, M., S. Miura, K. Shinozaki, and K. Yamaguchi-Shinozaki. 2004. A combination of the *Arabidopsis DREB1A* gene and stress-inducible *rd29A* promoter improved drought- and low-temperature stress tolerance in tobacco by gene transfer. *Plant Cell Physiol.* 45:346–350.

Katiyar, S., and R. S. Dubey. 1992. Influence of NaCl salinity on behaviour of nitrate reductase and nitrite reductase in rice seedlings differing in salt tolerance. *J. Agron. Crop Sci.* 169:289–297.

Kato-Noguchi, H. 2000. Anaerobically induced proteins in rice seedlings. *Plant Prod. Sci.* 3:225–228.

Kaur, N., and A. K. Gupta. 2005. Signal transduction pathways under abiotic stresses in plants. *Curr. Sci.* 88:1771–1780.

Kausar, A., M. Y. Ashraf, and M. Niaz. 2014. Some physiological and genetic determinants of salt tolerance in sorghum (*Sorghum bicolor* (L.) Moench): biomass production and nitrogen metabolism. *Pak. J. Bot.* 46:515–519.

Ke, Y., G. Han, H. He, and J. Li. 2009. Differential regulation of proteins and phosphoproteins in rice under drought stress. *Biochem. Biophys. Res. Commun.* 379:133–138.

Kee, S. C., and P. S. Nobel. 1986. Concomitant changes in high temperature tolerance and heat-shock proteins in desert succulents. *Plant Physiol.* 80:596–598.

Khosravinejad, F., R. Heydari, and T. Farboodnia. 2008. Antioxidant responses of two barley varieties to saline stress. *Res. J. Biol. Sci.* 3:486–490.

Khurana, P., D. Vishnudasan, and A. K. Chhibbar. 2008. Genetic approaches towards overcoming water deficit in plants – special emphasis on LEAs. *Physiol. Mol. Biol. Plants* 14:277–298.

Kikuchi, T., and K. Masuda. 2009. Class II chitinase accumulated in the bark tissue involves with the cold hardiness of shoot stems in highbush blueberry (*Vaccinium corymbosum* L.). *Sci. Hort.* 120:230–236.

Kim, S. T., Y. H. Kang, Y. Wang et al. 2009. Secretome analysis of differentially induced proteins in rice suspension-cultured cells triggered by rice blast fungus and elicitor. *Proteomics* 9:1302–1313.

Kirtikara, K., and D. Talbot. 1996. Alteration in protein accumulation, gene expression and ascorbate-glutathione pathway in tomato (*Lycopersicon esculentum*) under paraquat and ozone stress. *J. Plant Physiol.* 148:752–760.

Kishor, P. B. K., Z. Hong, G. H. Miao, C. A. Hu, and D. P. S Verma. 1995. Overexpression of Δ1-pyrroline-5-carboxylate synthetase increases proline production and confers osmotolerance in transgenic plants. *Plant Physiol.* 108:1387–1394.

Kishore, R., and K. C. Upadhyaya. 1994. Intracellular distribution of heat shock proteins in pigeon pea (*Cajanus cajan*). *J. Plant Biochem. Biotechnol.* 3:43–46.

Klados E., and N. Tzortzakis. 2014. Effects of substrate and salinity in hydroponically grown Cichorium spinosum. *J. Soil Sci. Plant Nutr.* 14:211–222.

Komatsu, S., and Z. Hossain. 2013. Organ-specific proteome analysis for identification of abiotic stress response mechanism in crop. *Front. Plant Sci.* 4:71. doi:10.3389/fpls.2013.00071

Kosakivska, I., D. Klymchuk, V. Negretzky, D. Bluma, and A. Ustinova. 2008. Stress proteins and ultrastructural characteristics of leaf cells of plants with different types of ecological strategies. *Gen. Appl. Plant Physiol.* 34:405–418.

Kosová, K., I. T. Prášil, P. Vítámvás et al. 2012. Complex phytohormone responses during the cold acclimation of two wheat cultivars differing in cold tolerance, winter Samanta and spring Sandra. *J. Plant Physiol.* 169:567–576.

Kosová, K., P. Vítámvás, and I. T. Prášil. 2014. Wheat and barley dehydrins under cold, drought, and salinity-what can LEA-II proteins tell us about plant stress response? *Front. Plant Sci.* 5:343. doi:10.3389/fpls.2014.00343

Koussevitzky, S., N. Suzuki, S. Huntington et al. 2008. Ascorbate peroxidase 1 plays a key role in the response of *Arabidopsis thaliana* to stress combination. *J. Biol. Chem.* 283:34197–34203.

Kovács, Z., L. Simon-Sarkadi, C. Sovány, K. Kirsch, G. Galiba, and G. Kocsy. 2011. Differential effects of cold acclimation and abscisic acid on free amino acid composition in wheat. *Plant Sci.* 180:61–68.

Kulshrestha, S., D. P. Mishra, and R. K. Gupta. 1987. Changes in contents of chlorophyll, proteins and lipids in whole chloroplast membrane fractions at different leaf water potentials in drought resistant and sensitive genotype of wheat. *Photosynthetica* 21:65–70.

Kumar, K., K. P. Rao, P. Sharma, and A. K. Sinha. 2008. Differential regulation of rice mitogen activated protein kinase kinase (MKK) by abiotic stress. *Plant Physiol. Biochem.* 46:891–897.

Kumar, M., S. C. Lee, J. Y. Kim, S. J. Kim, and S. R. Kim. 2014. Over-expression of dehydrin gene, *OsDhn1*, improves drought and salt stress tolerance through scavenging of reactive oxygen species in rice (*Oryza sativa* L.). *J. Plant Biol.* 57:383–393.

Kumar, P. K., and R. A. Singh. 1991. Germination and metabolism in susceptible and tolerant mung bean genotypes under moisture stress. *Indian J. Plant Physiol.* 34:267–270.

Kumar, S. A., P. H. Kumari, G. Jawahar et al. 2016. Beyond just being foot soldiers-osmotin like protein (OLP) and chitinase (Chi11) genes act as sentinels to confront salt, drought, and fungal stress tolerance in tomato. *Environ. Exp. Bot.* 132:53–65.

Lawrence, C. B., M. H. A. J. Joosten, and S. Tuzun. 1996. Differential induction of pathogenesis-related proteins in tomato by *Alternaria solani* and the association of a basic chitinase isozyme with resistance. *Physiol. Mol. Plant Pathol.* 48:361–377.

Le, T. T., B. Williams, and S. G. Mundree. 2018. An osmotin from the resurrection plant *Tripogon loliiformis* (TlOsm) confers tolerance to multiple abiotic stresses in transgenic rice. *Physiol. Plant.* 162:13–34.

Lee, D. G., N. Ahsan, S. H. Lee et al. 2007. A proteomic approach in analyzing heat-responsive proteins in rice leaves. *Proteomics* 7:3369–3383.

Lee, D. G., N. Ahsan, S. H. Lee et al. 2009. Chilling stress-induced proteomic changes in rice roots. *J. Plant Physiol.* 166:1–11.

Lenne, C., and R. Douce. 1994. A low molecular mass heat-shock protein is localized to higher plant mitochondria. *Plant Physiol.* 105:1255–1261.

Levitt, J. 1980. *Responses of Plants to Environmental Stresses.* New York: Academic Press.

Li, H., J. Liu, L. Zhang, and T. Pang. 2016. Effects of low temperature stress on the antioxidant system and photosynthetic apparatus of *Kappaphycus alvarezii* (Rhodophyta, Solieriaceae). *Marine Biol. Res.* 12:1064–1077.

Li, S. X., G. L. Hartman, B. S. Lee, and J. W. Widholm. 2000. Identification of a stress-induced protein in stem exudates of soybean seedlings root-infected with *Fusarium solani* f. sp glycines. *Plant Physiol. Biochem.* 38:803–809.

Li, W., C. Zhang, Q. Lu, X. Wen, and C. Lu. 2011. The combined effect of salt stress and heat shock on proteome profiling in *Suaeda salsa. J. Plant. Physiol.* 168:1743–1752.

Li, W., F. A. Zhao, W. Fang et al. 2015. Identification of early salt stress responsive proteins in seedling roots of upland cotton (*Gossypium hirsutum* L.) employing iTRAQ-based proteomic technique. *Front. Plant Sci.* 6:732. doi:10.3389/fpls.2015.00732

Lin, H., S. S. Salus, and K. S. Schumaker. 1997. Salt sensitivity and the activities of the H+-ATPases in cotton seedlings. *Crop Sci.* 37:190–197.

Ling, H., X. Zeng, and S. Guo. 2016. Functional insights into the late embryogenesis abundant (LEA) protein family from *Dendrobium officinale* (Orchidaceae) using an *Escherichia coli* system. *Sci. Rep.* 6:39693. doi:10.1038/srep39693

Liu, G. T., L. Ma, W. Duan, B. C. Wang, J. H. Li, H. G. Xu, X. Q. Yan, B. F. Yan, S. H. Li, and L. J. Wang. 2014. Differential proteomic analysis of grapevine leaves by iTRAQ reveals responses to heat stress and subsequent recovery. *BMC Plant Biol.* 14:110. doi:10.1186/1471-2229-14-110

Liu, J. Y., T. Lu, and N. M. Zhao. 2000. Classification and nomenclature of plant metallothionein-like proteins based on their cysteine arrangement patterns. *Acta Bot. Sin.* 42:649–652.

Liu, L., N. Li, C. Yao, S. Meng, and C. Song. 2013. Functional analysis of the ABA-responsive protein family in ABA and stress signal transduction in Arabidopsis. *Chin. Sci. Bull.* 58:3721–3730.

Liu, L., W. Xia, H. Li et al. 2018. Salinity inhibits rice seed germination by reducing α-amylase activity via decreased bioactive gibberellin content. *Front. Plant Sci.* 9:275.

Liu, Q., M. Kasuga, Y. Sakuma et al. 1998. Two transcription factors, DREB1 and DREB2, with an EREBP/AP2 DNA binding domain separate two cellular signal transduction pathways in drought- and low temperature- responsive gene expression, respectively, in *Arabidopsis. Plant Cell* 10:1391–1406.

Liu, X., Z. Wang, L. Wang, R. Wu, J. Phillips, and X. Deng. 2009. LEA 4 group genes from the resurrection plant *Boea hygrometrica* confer dehydration tolerance in transgenic tobacco. *Plant Sci.* 176:90–98.

Lodh, S. B., D. P. Bhattacharya, G. Ramkrishnaya, and S. P. Deb. 1977. Studies on the oxidative enzymes of rice in relation to soil moisture stress. *Indian Agric.* 21:181–186.

Longuespée, R., M. Fléron, C. Pottier et al. 2014. Tissue proteomics for the next decade? Towards a molecular dimension in histology. *OMICS* 18:539–552.

Lopez-Matas, M. A., P. Nuñez, A. Soto et al. 2004. Protein cryoprotective activity of a cytosolic small heat shock protein that accumulates constitutively in chestnut stems and is up-regulated by low and high temperatures. *Plant Physiol.* 134:1708–1717.

Loreti, E., H. van Veen, and P. Perata. 2016. Plant responses to flooding stress. *Curr. Opin. Plant Biol.* 33:64–71.

Lv, Y., X. Deng, L. Quan, Y. Xia, and Z. Shen. 2013. Metallothioneins BcMT1 and BcMT2 from *Brassica campestris* enhance tolerance to cadmium and copper and decrease production of reactive oxygen species in *Arabidopsis thaliana. Plant Soil* 367:507–519.

Machado-Estrada, B., J. Calderón, R. Moreno-Sánchez, and J. S. Rodríguez-Zavala. 2013. Accumulation of arsenic, lead, copper, and zinc, and synthesis of phytochelatins by indigenous plants of a mining impacted area. *Environ. Sci. Pollut. Res.* 20:3946–3955.

Maheshwari, R., and R. S. Dubey. 2007. Nickel toxicity inhibits ribonuclease and protease activities in rice seedlings: protective effects of proline. *Plant Growth Regul.* 51:231–243.

Maheshwari, R., and R. S. Dubey. 2008. Inhibition of ribonuclease and protease activities in germinating rice seeds exposed to nickel. *Acta Physiol. Plant.* 30:863–872.

Maheshwari, R., and R. S. Dubey. 2009. Nickel-induced oxidative stress and the role of antioxidative defense in rice seedlings. *Plant Growth Regul.* 59:37–49.

Maheshwari, R., and R. S. Dubey. 2011. Effect of nickel toxicity on the alteration of phosphate pool and the suppressing activity of phosphorolytic enzymes in germinating seeds and growing seedlings of rice. *Int. J. Plant Physiol. Biochem.* 3:50–59.

Mahmoodzadeh, H. 2009. Protein profiles in response to salt stress in seeds of *Brassica napus. Res. J. Environ. Sci.* 3:225–231.

Maksymiec, W. 2007. Signaling responses in plants to heavy metal stress. *Acta Physiol. Plant.* 29:177–187.

Malamy, J., P. Sanchezcasas, J. Hennig, A. Guo, and D. F. Klessig. 1996. Dissection of the salicylic acid signaling pathway in tobacco. *Mol. Plant Microbe Interact.* 9:474–482.

Malar, S., S. S. Vikram, P. J. Favas, and V. Perumal. 2016. Lead heavy metal toxicity induced changes on growth and antioxidative enzymes level in water hyacinths [*Eichhornia crassipes* (Mart.)]. *Bot. Stud.* 55:54. doi:10.1186/s40529-014-0054-6

Mali, P. C., B. B. Nanda, D. P. Bhattacharya, and S. B. Lodh. 1980. Changes in the activity of some enzymes and free proline in rice (*Oryza sativa* L.) during water stress. *Plant Biochem. J.* 7:126–132.

Mansfield, M. A., and J. L. Key. 1988. Cytoplasmic distribution of heat-shock proteins in soybean. *Plant Physiol.* 86:1240–1246.

Mantyla, E., V. Lang, and E. T. Palva. 1995. Role of abscisic acid in drought-induced freezing tolerance, cold acclimation and accumulation of LTI78 and RAB18 proteins in *Arabidopsis*. *Plant Physiol.* 107:141–148.

Mathur, S., D. Agrawal, and A. Jajoo. 2014. Photosynthesis: response to high temperature stress. *J. Photochem. Photobiol. B. Biol.* 137:116–126.

Matsuba, K., N. Imaizumi, S. Kaneko, M. Samejima, and R. Ohsugi. 1997. Photosynthetic responses to temperature of phosphoenolpyruvate carboxykinase type C-4 species differing in cold sensitivity. *Plant Cell Environ.* 20:268–274.

McLoughlin, F., E. Basha, M. E. Fowler et al. 2016. Class I and II small heat-shock proteins protect protein translation factors during heat stress. *Plant Physiol.* 172:1221–1236.

Mehta, N. J., and A. B. Vora. 1987. Metabolic changes induced by NaCl salinity in pea plants. In *International Conference of Plant Physiologists of SAARC Countries, Gorakhpur, India* (Abstr):47.

Mehta, R. A., B. L. Parsons, A. M. Mehta, H. L. Nakhasi, and A. K. Mattoo. 1991. Differential protein metabolism and gene expression in tomato fruit during wounding stress. *Plant Cell Physiol.* 32:1057–1065.

Mercado, J. A., M. Barceló, C. Pliego, M. Rey, J. L. Caballero, J. Muñoz-Blanco, D. Ruano-Rosa, C. López-Herrera, B. de los Santos, F. Romero-Muñoz, and F. Pliego-Alfaro. 2015. Expression of the β-1, 3-glucanase gene bgn13. 1 from *Trichoderma harzianum* in strawberry increases tolerance to crown rot diseases but interferes with plant growth. *Transgenic Res.* 24:979–989.

Meza-Basso, L., M. Alberdi, M. Raynal, M. L. Ferrero-Cadinanos, and M. Delseny. 1986. Changes in protein synthesis in rapeseed (Brassica napus) seedlings during a low temperature treatment. *Plant Physiol.* 82:733–738.

Mirzaei, M., D. Pascovici, B. J. Atwell, and P. A. Haynes. 2012. Differential regulation of aquaporins, small GTPases and V-ATPases proteins in rice leaves subjected to drought stress and recovery. *Proteomics* 12:864–877.

Mishra, P., K. Bhoomika, and R. S. Dubey. 2013. Differential responses of antioxidative defense system to prolonged salinity stress in salt-tolerant and salt-sensitive Indica rice (*Oryza sativa* L.) seedlings. *Protoplasma* 250:3–19.

Mishra, P., and R. S. Dubey. 2008a. Effect of aluminium on metabolism of starch and sugars in growing rice seedlings. *Acta Physiol. Plant.* 30:265–275.

Mishra, P., and R. S. Dubey. 2011. Nickel and Al-excess inhibit nitrate reductase but upregulate activities of aminating glutamate dehydrogenase and aminotransferases in growing rice seedlings. *Plant Growth Regul.* 64:251–261.

Mishra, P., and R. S. Dubey. 2013. Excess nickel modulates activities of carbohydrate metabolizing enzymes and induces accumulation of sugars by upregulating acid invertase and sucrose synthase in rice seedlings. *Biometals* 26:97–111.

Mishra, S., and R. S. Dubey. 2006. Inhibition of ribonuclease and protease activities in arsenic exposed rice seedlings: role of proline as enzyme protectant. *J. Plant Physiol.* 163:927–936.

Mishra, S., and R. S. Dubey. 2008b. Changes in phosphate content and phosphatase activities in rice seedlings exposed to arsenite. *Braz. J. Plant Physiol.* 20:19–28.

Mittal, R., and R. S. Dubey. 1991. Behaviour of peroxidases in rice: changes in enzyme activity and isoforms in relation to salt tolerance. *Plant Physiol. Biochem. (Paris)* 29:31–40.

Mittal, R., and R. S. Dubey. 1995. Influence of sodium chloride salinity on polyphenol oxidase, indole 3-acetic acid oxidase and catalase activities in rice seedlings differing in salt tolerance. *Trop. Sci.* 35:141–149.

Mohamed, F., and O. P. Sehgal. 1997. Characteristics of pathogenesis-related proteins induced in *Phaseolus vulgaris* cv. pinto following viral infection. *J. Phytopathol.* 145:49–58.

Mohammadi, P. P., A. Moieni, and S. Komatsu. 2012. Comparative proteome analysis of drought-sensitive and drought-tolerant rapeseed roots and their hybrid F1 line under drought stress. *Amino Acids* 43:2137–2152.

Mohapatra, S. S., R. J. Poole, and R. S. Dhindsa. 1988. Abscisic acid-regulated gene expression in relation to freezing tolerance in alfalfa. *Plant Physiol.* 87:468–473.

Momcilovic, I., and Z. Ristic. 2007. Expression of chloroplast protein synthesis elongation factor, EF-Tu, in two lines of maize with contrasting tolerance to heat stress during early stages of plant development. *J. Plant Physiol.* 164:90–99.

Mundy, J., and N. H. Chua. 1988. Abscisic acid and water-stress induce the expression of a novel rice gene. *EMBO J.* 7:2279–2286.

Munns, R. 2002. Comparative physiology of salt and water stress. *Plant Cell Environ.* 25:239–250.

Murumkar, C. V., and P. D. Chavan. 1986. Influence of salt stress on *Biochem*ical processes in chickpea, *Cicer arietinum* L. *Plant Soil* 96:439–443.

Muscolo, A., M. Sidari, U. Anastasi, C. Santonoceto, and A. Maggio. 2014. Effect of PEG-induced drought stress on seed germination of four lentil genotypes. *J. Plant Interact.* 9:354–363.

Mustroph, A., J. Stock, N. Hess, S. Aldous, A. Dreilich, and B. Grimm. 2013. Characterization of the phosphofructokinase gene family in rice and its expression under oxygen deficiency stress. *Front. Plant Sci.* 4:125. doi:10.3389/fpls.2013.00125

Muthalif, M. M., and L. J. Rowland. 1994. Identification of dehydrin-like proteins responsive to chilling in floral buds of blueberry (*Vaccinium*, Section *Cyanococcus*). *Plant Physiol.* 104:1439–1447.

Naot, D., G. Ben-Hayyim, Y. Eshdat, and D. Holland. 1995. Drought, heat and salt stress induce the expression of a citrus homologue of an atypical late-embryogenesis lea5 gene. *Plant Mol. Biol.* 27:619–622.

Nasri, N., S. Maatallah, R. Kaddour, and M. Lachâal. 2016. Effect of salinity on *Arabidopsis thaliana* seed germination and acid phosphatase activity. *Arch. Biol. Sci.* 68:17–23.

Negrin, V. L., B. Teixeira, R. M. Godinho, R. Mendes, and C. Vale. 2017. Phytochelatins and monothiols in salt marsh plants and their relation with metal tolerance. *Mar. Pollut. Bull.* 121:78–84.

Neuhaus, J. M., B. Fritig, H. J. M. Linthorst, F. Meins, J. D. Mikkelsen, and J. Ryals. 1996. A revised nomenclature of chitinase genes. *Plant Mol. Biol. Rep.* 14:102–104.

Neumann, D., O. Lichtenberger, D. Günther, K. Tschiersch, and L. Nover. 1994. Heat-shock proteins induce heavy-metal tolerance in higher plants. *Planta* 194:360–367.

Ngara, R., and B. K. Ndimba. 2014. Understanding the complex nature of salinity and drought-stress response in cereals using proteomics technologies. *Proteomics* 14:611–621.

Niknam, V., N. Razavi, H. Ebrahimzadeh, and B. Sharifizadeh. 2006. Effect of NaCl on biomass, protein and proline contents and antioxidant enzymes in seedlings and calli of two *Trigonella* species. *Biol. Plant.* 50:591–596.

Nohzadeh, M. S., R. M. Habibi, M. Heidari, and G. H. Salekdeh. 2007. Proteomics reveals new salt responsive proteins associated with rice plasma membrane. *Biosci. Biotechnol. Biochem.* 71:2144–2154.

Oh, M., and S. Komatsu. 2015. Characterization of proteins in soybean roots under flooding and drought stresses. *J. Proteomics* 114:161–181.

Oh, S. J., C. W. Kwon, D. W. Choi, S. I. Song, and J. K. Kim. 2007. Expression of barley *HvCBF4* enhances tolerance to abiotic stress in transgenic rice. *Plant Biotechnol. J.* 5:646–656.

Ohashi, Y., and M. Matsuoka. 1985. Synthesis of stress proteins in tobacco leaves. *Plant Cell Physiol.* 26:473–480.

Olgun, M., A. M. Kumlay, M. C. Adiguzel, and A. Caglar. 2008. The effect of waterlogging in wheat (*T. aestivum* L.). *Acta Agric. Scand. Sect. B* 58:193–198.

Ottander, C., D. Campbell, and G. Oquist. 1995. Seasonal changes in photosystem II organization and pigment composition in *Pinus sylvestris*. *Planta* 197:176–183.

Pagter, M., C. R. Jensen, K. K. Petersen, F. L. Liu, and R. Arora. 2008. Changes in carbohydrates, ABA and bark proteins during seasonal cold acclimation and deacclimation in *Hydrangea* species differing in cold hardiness. *Physiol. Plant.* 134:473–485.

Pak, J. H., E. S. Chung, S. H. Shin et al. 2009. Enhanced fungal resistance in *Arabidopsis* expressing wild rice PR-3 (*OgChitIVa*) encoding chitinase class IV. *Plant. Biotechnol. Rep.* 3:147–155.

Pal, R., and J. P. N. Rai. 2010. Phytochelatins: peptides involved in heavy metal detoxification. *Appl. Biochem. Biotechnol.* 160:945–963.

Pan, Q. H., L. Wang, and J. M. Li. 2009. Amounts and subcellular localization of stilbene synthase in response of grape berries to UV irradiation. *Plant Sci.* 176:360–366.

Pandey, A., U. Rajamani, J. Verma et al. 2010. Identification of extracellular matrix proteins of rice (*Oryza sativa* L.) involved in dehydration-responsive network: a proteomic approach. *J. Proteome Res.* 9:3443–3464.

Pang, Q., S. Chen, S. Dai, Y. Chen, Y. Wang, and X. Yan. 2010. Comparative proteomics of salt tolerance in *Arabidopsis thaliana* and *Thellungiella halophila*. *J. Proteome Res.* 9:2584–2599.

Pareek, A., S. L. Singla, A. K. Kush, and A. Grover. 1997. Distribution patterns of HSP 90 protein in rice. *Plant Sci.* 125:221–230.

Parida, A. K., V. S. Dagaonkar, M. S. Phalak, and L. P. Aurangabadkar. 2008. Differential responses of the enzymes involved in proline biosynthesis and degradation in drought tolerant and sensitive cotton genotypes during drought stress and recovery. *Acta Physiol. Plant.* 30:619–627.

Parihar, P., S. Singh, R. Singh, V. P. Singh, and S. M. Prasad. 2015. Effect of salinity stress on plants and its tolerance strategies: a review. *Environ. Sci. Pollut. Res.* 22:4056–4075.

Park, S., C. M. Lee, C. J. Doherty, S. J. Gilmour, Y. Kim, and M. F. Thomashow. 2015. Regulation of the Arabidopsis CBF regulon by a complex low-temperature regulatory network. *Plant J.* 82:193–207.

Parkhi, V., V. Kumar, G. S. Kumar, L. M. Campbell, N. K. Singh, and K. S. Rathore. 2009. Expression of apoplastically secreted tobacco osmotin in cotton confers drought tolerance. *Mol. Breed.* 23:625–639.

Pascual, J., M. J. Canal, M. Escandon, M. Meijon, W. Weckwerth, and L. Valledor. 2017. Integrated physiological, proteomic and metabolomic analysis of UV stress responses and adaptation mechanisms in *Pinus radiata*. *Mol. Cell Proteomics* 16:485–501.

Patade, V. Y., D. Khatri, M. Kumari, A. Grover, S. M. Gupta, and Z. Ahmed. 2013. Cold tolerance in Osmotin transgenic tomato (*Solanum lycopersicum* L.) is associated with modulation in transcript abundance of stress responsive genes. *Springerplus* 2:117. doi:10.1186/2193-1801-2-117

Pedrosa, A. M., C. D. P. S. Martins, L. P. Gonçalves, and M. G. C. Costa. 2015. Late-embryogenesis abundant (LEA) constitutes a large and diverse family of proteins involved in development and abiotic stress responses in sweet orange (*Citrus sinensis* L. Osb.). *PLoS ONE* 10:e0145785.

Pelah, D., W. Wang, A. Altman, O. Shoseyov, and D. Bartels. 1997. Differential accumulation of water stress-related proteins, sucrose synthase and soluble sugars in *Populus* species that differ in their water stress response. *Physiol. Plant.* 99:153–159.

Peng, J. S., G. Ding, S. Meng, H. Y. Yi, and J. M. Gong. 2017. Enhanced metal tolerance correlates with heterotypic variation in SpMTL, a metallothionein-like protein from the hyperaccumulator *Sedum plumbizincicola*. *Plant Cell Environ.* 40:1368–1378.

Peng, J. S., and J. M. Gong. 2014. Vacuolar sequestration capacity and long-distance metal transport in plants. *Front. Plant Sci.* 5:19. doi:10.3389/fpls.2014.00019

Peng, Y., J. L. Reyes, H. Wei et al. 2008. RcDhn5, a cold acclimation-responsive dehydrin from *Rhododendron catawbiense* rescues enzyme activity from dehydration effects in vitro and enhances freezing tolerance in RcDhn5-overexpressing Arabidopsis plants. *Physiol. Plant.* 134:583–597.

Y. H. J. L Reyes, H. Wei et al. 2008. RcDhn5, a cold acclimation-responsive dehydrin from *Rhododendron catawbiense* rescues enzyme activity from dehydration effects *in vitro* and enhances freezing tolerance in RcDhn5-overexpressing *Arabidopsis* plants. *Physiol. Plant.* 134:583–597.

Perezmolphebalch, E., M. Gidekel, M. Seguranieto, L. Herreraestrella, and N. Ochoaalejo. 1996. Effects of water stress on plant growth and root proteins in three cultivars of rice (*Oryza sativa*) with different levels of drought tolerance. *Physiol. Plant.* 96:284–290.

Piqueras, A., J. L. Hernandez, E. Olmos, F. Sevilla, and E. Hellin. 1996. Changes in antioxidant enzymes and organic solutes associated with adaptation of citrus cells to salt stress. *Plant Cell Tissue Organ Cult.* 45:53–60.

Pitto, L., F. Loschiavo, G. Giuliano, and M. Terzi. 1983. Analysis of heat-shock protein pattern during somatic embryogenesis of carrots. *Plant Mol. Biol.* 2:231–237.

Pompeiano, A., and L. Guglielminetti. 2016. Carbohydrate metabolism in germinating caryopses of *Oryza sativa* L. exposed to prolonged anoxia. *J. Plant Res.* 129:833–840.

Popova, L. P., Z. G. Stoinova, and L. T. Maslenkova. 1995. Involvement of abscisic acid in photosynthetic process in *Hordeum vulgare* L. during salinity stress. *J. Plant Growth Regul.* 14:211–218.

Prinsi, B., A. S. Negri, O. Failla, A. Scienza, and L. Espen. 2018. Root proteomic and metabolic analyses reveal specific responses to drought stress in differently tolerant grapevine rootstocks. *BMC Plant Biol.* 18:126. doi:10.1186/s12870-018-1343-0

Pruvot, G., J. Massimino, G. Peltier, and P. Rey. 1996. Effects of low temperature, high salinity and exogenous ABA on the synthesis of two chloroplastic drought-induced proteins in *Solanum tuberosum*. *Physiol. Plant.* 97:123–131.

Przymusinski, R., R. Rucinska, and E. A. Gwozdz. 1995. The stress-stimulated 16 kDa polypeptide from lupin roots has properties of cytosolic Cu–Zn-superoxide dismutase. *Environ. Exp. Bot.* 35:485–495.

Pucciariello, C., L. A. Voesenek, P. Perata, and R. Sasidharan. 2014. Plant responses to flooding. *Front. Plant Sci.* 5:226. doi:10.3389/fpls.2014.00226

Puhakainen, T., M. W. Hess, P. Makela, J. Svensson, P. Heino, and E. T. Palva. 2004. Overexpression of multiple dehydrin genes enhances tolerance to freezing stress in *Arabidopsis*. *Plant Mol. Biol.* 54:743–753.

Pyngrope, S., K. Bhoomika, and R. S. Dubey. 2013a. Oxidative stress, protein carbonylation, proteolysis and antioxidative defense system as a model for depicting water deficit tolerance in Indica rice seedlings. *Plant Growth Regul.* 69:149–165.

Pyngrope, S., K. Bhoomika, and R. S. Dubey. 2013b. Reactive oxygen species, ascorbate-glutathione pool, and enzymes of their metabolism in drought-sensitive and tolerant indica rice (*Oryza sativa* L.) seedlings subjected to progressing levels of water deficit. *Protoplasma* 250:585–600.

Qureshi, M. I., S. Qadir, and L. Zolla. 2007. Proteomics-based dissection of stress-responsive pathways in plants. *J. Plant Physiol.* 164:1239–1260.

Rahdari, P., S. Tavakoli, and S. M. Hosseini. 2012. Studying of salinity stress effect on germination, proline, sugar, protein, lipid and chlorophyll content in purslane (*Portulaca oleracea* L.) leaves. *J. Stress Physiol. Biochem.* 8:182–193.

Rai, V. K., G. Singh, P. K. Thakur, and S. Banyal. 1983. Protein and amino acid relationship during water stress in relation to drought resistance. *Plant Physiol. Biochem.* 10(s):161–167.

Ramagopal, S. 1993. Advances in understanding the molecular biology of drought, and salinity tolerance in plants-the first decade. In *Advances in Plant Biotechnology and Biochemistry*, eds. M. L. Lodha, S. L. Mehta, S. Ramagopal, and G. P. Srivastava, pp. 39–48. Kanpur: Indian Society of Agricultural Biochemists.

Rani, M. 1988. Influence of salinity on metabolic status of proteins, and amino acids during germination and early seedling stages of rice. PhD diss., Banaras Hindu University, India.

Rao, M. V., G. Paliyath, and D. P. Ormrod. 1996. Ultraviolet-B and ozone-induced biochemical changes in antioxidant enzymes of *Arabidopsis thaliana*. *Plant Physiol.* 110:125–136.

Reddy, A. R. 1996. Fructose 2,6-bisphosphate-modulated photosynthesis in sorghum leaves grown under low water regimes. *Phytochemistry* 43:319–322.

Reddy, P. S., T. Chakradhar, R. A. Reddy, R. B. Nitnavare, S. Mahanty, and M. K. Reddy. 2016. Role of heat shock proteins in improving heat stress tolerance in crop plants. In *Heat Shock Proteins and Plants*, pp. 283–307. Cham: Springer.

Ricard, B., J. Rivoal, A. Spiteri, and A. Pradet. 1991. Anaerobic stress induces the transcription and translation of sucrose synthase in rice. *Plant Physiol.* 95:669–674.

Riccardi, F., P. Gazeau, D. V. Vienne, and M. Zivy. 1998. Protein changes in responses to progressive water deficit in maize. *Plant Physiol.* 117:1253–1263.

Roberts, M. R., J. Salinas, and D. B. Collinge. 2002. 14-3-3 proteins and the response to abiotic and biotic stress. *Plant Mol. Biol.* 50:1031–1039.

Robertson, A. J., A. Weninger, R. W. Wilen, P. Fu, and L. V. Gusta. 1994. Comparison of dehydrin gene expression and freezing tolerance in *Bromus inermis* and *Secale cereale* grown in controlled environments, hydroponics and the field. *Plant Physiol.* 106:1213–1216.

Robertson, M., and P. M. Chandler. 1994. A dehydrin cognate protein from pea (*Pisum sativum* L.) with an atypical pattern of expression. *Plant Mol. Biol.* 26:805–816.

Robinson, N. J., A. M. Tommey, C. Kuske, and P. J. Jackson. 1993. Plant metallothioneins. *Biochem. J.* 295:1–10.

Rollins, J. A., E. Habte, S. E. Templer, T. Colby, J. Schmidt, and M. Von Korff. 2013. Leaf proteome alterations in the context of physiological and morphological responses to drought and heat stress in barley (*Hordeum vulgare* L.). *J. Exp. Bot.* 64:3201–3212.

Roth, U., E. von Roepenack-Lahaye, and S. Clemens. 2006. Proteome changes in *Arabidopsis thaliana* roots upon exposure to Cd2+. *J. Exp. Bot.* 57:4003–4013.

Roulin, S., and A. J. Buchala. 1995. The induction of 1,3-β-glucanases and other enzymes in groundnut leaves infected with *Cercospora arachidicola*. *Physiol Mol. Plant. Pathol.* 46:471–489.

Ruelland, E., M. N. Vaultier, A. Zachowski, and V. Hurry. 2009. Cold signalling and cold acclimation in plants. *Adv. Bot. Res.* 49:35–150.

Sachs, M. M., M. Freeling, and R. Okimoto. 1980. The anaerobic proteins of maize. *Cell* 20:761–767.

Sachs, M. M., C. C. Subbaiah, and I. N. Saab. 1996. Anaerobic gene expression and flooding tolerance in maize. *J. Exp. Bot.* 47:1–15.

Sahin, U., M. Ekinci, S. Ors, M. Turan, S. Yildiz, and E. Yildirim. 2018. Effects of individual and combined effects of salinity and drought on physiological, nutritional and biochemical properties of cabbage (*Brassica oleracea* var. capitata). *Sci. Hortic.* 240:196–204.

Salomé, P. A. 2017. Some like it HOT: protein translation and heat stress in plants. *Plant Cell* 29:2075.

Salt, D. E., R. D. Smith, and I. Raskin. 1998. Phytoremediation. *Annu. Rev. Plant Physiol. Plant Mol. Biol.* 49:643–668.

Samac, D. A., and D. M. Shah. 1991. Developmental and pathogen induced activation of the *Arabidopsis* acidic chitinase promoter. *Plant Cell* 3:1063–1072.

Sanita di Toppi, L., and R. Gabbrielli. 1999. Response to cadmium in higher plants. *Environ. Exp. Bot.* 41:105–130.

Sarkar, A., and S. B. Agrawal. 2010. Identification of ozone stress in Indian rice through foliar injury and differential protein profile. *Environ. Monit. Assess.* 161:205–215.

Sarkar, A., A. A. Singh, S. B. Agrawal, A. Ahmad, and S. P. Rai. 2015. Cultivar specific variations in antioxidative defense system, genome and proteome of two tropical rice cultivars against ambient and elevated ozone. *Ecotoxicol. Environ. Saf.* 115:101–111.

Sato, Y., and S. Yokoya. 2008. Enhanced tolerance to drought stress in transgenic rice plants overexpressing a small heat-shock protein, sHSP17.7. *Plant Cell Rep.* 27:329–334.

Schaller, A., and C. A. Ryan. 1996. Molecular cloning of a tomato leaf cDNA encoding an aspartic protease, a systemic wound response protein. *Plant Mol. Biol.* 31:1073–1077.

Schoffl, F., R. Prandl, and A. Reindl. 1999. Molecular responses to heat stress. In *Molecular Responses to Cold, Drought, Heat, and Salt Stress in Higher Plants*, eds. K. Shinozaki, and K. Yamaguchi-Shinozaki, pp. 81–98. Austin, TX: R. G. Landes Co.

Schützendübel, A., P. Schwanz, T. Teichmann et al. 2001. Cadmium-induced changes in antioxidative systems, hydrogen peroxide content, and differentiation in Scots pine roots. *Plant Physiol.* 127:887–898.

Scott, N. S., R. Munns, and E. W. R. Barlow. 1979. Polyribosome content in young and aged wheat leaves subjected to drought. *J. Exp. Bot.* 30:905–911.

Selitrennikoff, C. P. 2001. Antifungal proteins. *Appl. Environ. Microbiol.* 67:2883–2894.

Shafi, M., J. Bakht, M. J. Hassan, M. Raziuddin, and G. Zhang. 2009. Effect of cadmium and salinity stresses on growth and antioxidant enzyme activities of wheat (*Triticum aestivum* L.). *Bull. Environ. Contam. Toxicol.* 82:772–776.

Shah, C. B., and R. S. Loomis. 1965. Ribonucleic acid and protein metabolism in sugar beet during drought. *Physiol. Plant.* 18:240–254.

Shah, K., and R. S. Dubey. 1995. Effect of cadmium on RNA level as well as activity and molecular forms of ribonuclease in growing rice seedlings. *Plant Physiol. Biochem. (Paris)* 33:577–584.

Shah, K., and R. S. Dubey. 1998a. A 18 kDa Cd inducible protein complex: its isolation and characterization from rice (*Oryza sativa* L.) seedlings. *J. Plant Physiol.* 152:448–454.

Shah, K., and R. S. Dubey. 1998b. Cadmium elevates level of protein, amino acids and alters the activity of proteolytic enzymes in germinating rice seeds. *Acta Physiol. Plant.* 20:189–196.

Shah, K., and R. S. Dubey. 1998c. Cadmium suppresses phosphate level and inhibits the activity of phosphatases in growing rice seedlings. *J. Agron. Crop Sci.* 180:223–231.

Shah, K., R. G. Kumar, S. Verma, and R. S. Dubey. 2001. Effect of cadmium on lipid peroxidation, superoxide anion generation and activities of antioxidant enzymes in growing rice seedlings. *Plant Sci.* 161:1135–1144.

Sharma, B., and Deswal, R. 2014. Antifreeze Proteins in Plants: an overview with an insight into the detection techniques including nanobiotechnology. *J Proteins Proteomics* 5(2):89–107.

Sharma, P., and R. S. Dubey. 2004. Ascorbate peroxidase from rice seedlings: properties of enzyme isoforms, effects of stresses and protective roles of osmolytes. *Plant Sci.* 167:541–550.

Sharma, P., and R. S. Dubey. 2005a. Drought induces oxidative stress and enhances the activities of antioxidant enzymes in growing rice seedlings. *Plant Growth Regul.* 46:209–221.

Sharma, P., and R. S. Dubey. 2005b. Modulation of nitrate reductase activity in rice seedlings under aluminium toxicity and water stress: role of osmolytes as enzyme protectant. *J. Plant Physiol.* 162:854–864.

Sharma, P., and R. S. Dubey. 2007. Involvement of oxidative stress and role of antioxidative defense system in growing rice seedlings exposed to toxic concentrations of aluminum. *Plant Cell Rep.* 26:2027–2038.

Sharma, Y. K., and K. R. Davis. 1997. The effects of ozone on antioxidant responses in plants. *Free Rad. Biol. Med.* 23:480–488.

Sharmin, S. A., I. Alam, K. H. Kim et al. 2012. Chromium-induced physiological and proteomic alterations in roots of *Miscanthus sinensis*. *Plant Sci.* 187:113–126.

Shen, S., Y. Jing, and T. Kuang. 2003. Proteomics approach to identify wound-response related proteins from rice leaf sheath. *Proteomics* 3:527–535.

Shi, Y., Y. Ding, and S. Yang. 2015. Cold signal transduction and its interplay with phytohormones during cold acclimation. *Plant Cell Physiol.* 56:7–15.

Si, Y., C. Zhang, S. Meng, and F. Dane. 2009. Gene expression changes in response to drought stress in *Citrullus colocynthis*. *Plant Cell Rep.* 28:997–1009.

Sidari, M., C. Santonoceto, U. Anastasi, G. Preiti, and A. Muscolo. 2008. Variations in four genotypes of lentil under NaCl-salinity stress. *Am. J. Agric. Biol. Sci.* 3:410–416.

Singh, N. K., C. A. Bracker, P. M. Hasegawa et al. 1987. Characterization of osmotin, a thaumatin like protein associated with osmotic adaptation in plant cells. *Plant Physiol.* 85:529–536.

Singh, N. K., A. K. Handa, P. M. Hasegawa, and R. A. Bressan. 1985. Proteins associated with adaptation of cultured tobacco cells to NaCl. *Plant Physiol.* 79:126–137.

Singh, N. K., D. E. Nelson, D. Kuhn, P. M. Hasegawa, and R. A. Bressan. 1989. Molecular cloning of the osmotin and regulation of its expression by ABA and adaptation to low water potential. *Plant Physiol.* 90:1096–1101.

Singh, S., P. Parihar, R. Singh, V. P. Singh, and S. M. Prasad. 2015. Heavy metal tolerance in plants: role of transcriptomics, proteomics, metabolomics, and ionomics. *Front. Plant Sci.* 6:1143. doi:10.3389/fpls.2015.01143

Singh, S., D. K. Tripathi, D. K. Chauhan, and N. K. Dubey. 2015. Glutathione and phytochelatins mediated redox homeostasis and stress signal transduction in plants: an integrated overview. In *Plant Metal Interaction: Emerging Remediation Techniques*, ed. P. Ahmad, pp. 285–310. Elsevier.

Singla, S. L., and A. Grover. 1994. Detection and quantification of a rapidly accumulating and predominant 104 kDa heat stress polypeptide in rice. *Plant Sci.* 97:23–30.

Sinha, A. K., M. Jaggi, B. Raghuram, and N. Tuteja. 2011. Mitogen-activated protein kinase signaling in plants under abiotic stress. *Plant Signal. Behav.* 6:196–203.

Sinha, K. M., A. Sachdev, R. P. Johari, and S. L. Mehta. 1996. Lathyrus dehydrin – a drought inducible cDNA clone: isolation and characterization. *J. Plant Biochem. Biotechnol.* 5:97–101.

Sinha, S. K., and D. J. Nicholas. 1981. Nitrate reductase. In *The Physiology and Biochemistry of Drought Resistance in Plants*, eds. L. G. Paleg, and D. Aspinall, pp. 145–168. Sydney: Academic Press.

Sobkowiak, R., and J. Deckert. 2006. Proteins induced by cadmium in soybean cells. *J. Plant Physiol.* 163:1203–1206.

Song, H. M., R. M. Zhao, P. X. Fan, X. C. Wang, X. Y. Chen, and Y. X. Li. 2009. Overexpression of AtHsp90.2, AtHsp90.5 and AtHsp90.7 in *Arabidopsis thaliana* enhances plant sensitivity to salt and drought stresses. *Planta* 229:955–964.

Souid, A., M. Gabriele, V. Longo et al. 2016. Salt tolerance of the halophyte *Limonium delicatulum* is more associated with antioxidant enzyme activities than phenolic compounds. *Funct. Plant Biol.* 43:607–619.

Srivastava, S., and R. S. Dubey. 2011. Manganese-excess induces oxidative stress, lowers the pool of antioxidants and elevates activities of key antioxidative enzymes in rice seedlings. *Plant Growth Regul.* 64:1–16.

Stintzi, A., T. Heitz, S. Kauffmann, M. Legrand, and B. Fritig. 1991. Identification of a basic pathogenesis-regulated thaumatin-like protein of virus-infected tobacco as OSM. *Physiol. Mol. Plant. Pathol.* 38:137–146.

Stuiver, C. E. E., L. J. de Kok, and P. J. C. Kuiper. 1988. Freezing injury in spinach leaf tissue: effects on water-soluble proteins, protein-sulfhydryl and water-soluble non-protein-sulfhydryl groups. *Physiol. Plant.* 74:72–76.

Su, K., J. Q. Moss, G. Zhang, D. L. Martin, and Y. Wu. 2013. Bermudagrass drought tolerance associated with dehydrin protein expression during drought stress. *J. Am. Soc. Hortic. Sci.* 138:277–282.

Subbaiah, C. C., and M. M. Sachs. 2003. Calcium-mediated responses of maize to oxygen deprivation. *Russ. J. Plant Physiol.* 50:752–761.

Sullivan, C. Y., W. R. Jordan, A. Blum, and M. Traore. 1990. An overview of heat resistance. In *Proceedings of International Congress of Plant Physiology*, New Delhi, pp. 916–922.

Suzuki, N., R. M. Rivero, V. Shulaev, E. Blumwald, and R. Mittler. 2014. Abiotic and biotic stress combinations. *New Phytol.* 203:32–43.

Tada, Y., and T. Kashimura. 2009. Proteomic analysis of salt-responsive proteins in the mangrove plant, *Bruguiera gymnorhiza. Plant Cell Physiol.* 50:439–446.

Tamas, L., and J. Huttova. 1996. Accumulation of pathogenesis-related proteins in barley induced by phosphate and salicylic acid. *Biologia* 51:479–484.

Tamminen, I., P. MaÈkelaÈ, P. Heino, and E. T. Palva. 2001. Ectopic expression of ABI3 gene enhances freezing tolerance in response to abscisic acid and low temperature in *Arabidopsis thaliana. Plant J.* 25:1–8.

Taylor, N. L., Y. F. Tan, R. P. Jacoby, and A. H. Millar. 2009. Abiotic environmental stress induced changes in the *Arabidopsis thaliana* chloroplast, mitochondria and peroxisome proteomes. *J. Proteomics* 72:367–378.

Thakur, P. S., and A. Thakur. 1987. Protease activity in response to water stress in two differentially sensitive *Zea mays* L. cultivars. *Plant Physiol. Biochem. (India)* 14:136–139.

Thangavel, P., S. Long, and R. Minocha. 2007. Changes in phytochelatins and their biosynthetic intermediates in red spruce (*Picea rubens* Sarg.) cell suspension cultures under cadmium and zinc stress. *Plant Cell Tissue Organ Cult.* 88:201–216.

Timperio, A. M., M. G. Egidi, and L. Zolla. 2008. Proteomics applied on plant abiotic stresses: role of heat shock proteins (HSP). *J. Proteomics* 71:391–411.

Tornero, P., J. Gadea, V. Conejero, and P. Vera. 1997. Two PR-1 genes from tomato are differentially regulated and reveal a novel mode of expression for a pathogenesis-related gene during the hypersensitive response and development. *Mol. Plant Microbe Interact.* 10:624–634.

Torres, N. L., K. Cho, J. Shibato et al. 2007. Gel-based proteomics reveals potential novel protein markers of ozone stress in leaves of cultivated bean and maize species of Panama. *Electrophoresis* 28:4369–4381.

Tseng, M. J., and P. H. Li. 1991. Changes in protein synthesis and translatable messenger RNA populations associated with ABA-induced cold hardiness in potato (*Solatium commersonii*). *Physiol. Plant.* 81:349–358.

Turrà, D., D. Bellin, M. Lorito, and C. Gebhardt. 2009. Genotype-dependent expression of specific members of potato protease inhibitor gene families in different tissues and in response to wounding and nematode infection. *J. Plant Physiol.* 166:762–774.

Ullah, A., A. Hussain, M. Shaban et al. 2017. Osmotin: a plant defense tool against biotic and abiotic stresses. *Plant Physiol. Biochem.* 123:149–159.

Ungelenk, S., F. Moayed, C. T. Ho et al. 2016. Small heat shock proteins sequester misfolding proteins in near-native conformation for cellular protection and efficient refolding. *Nat. Commun.* 7:13673. doi:10.1038/ncomms13673

Uraguchi, S., N. Tanaka, C. Hofmann et al. 2017. Phytochelatin synthase has contrasting effects on cadmium and arsenic accumulation in rice grains. *Plant Cell Physiol.* 58:1730–1742.

Usman, M. G., M. Y. Rafii, M. R. Ismail, M. A. Malek, M. A. Latif, and Y. Oladosu. 2014. Heat shock proteins: functions and response against heat stress in plants. *Int. J. Sci. Technol. Res* 3:204–218.

Valcu, C. M., M. Junqueira, A. Shevchenko, and K. Schlink. 2009. Comparative proteomic analysis of responses to pathogen infection and wounding in *Fagus sylvatica. J. Proteome Res.* 8:4077–4091.

Vance, N. C., D. O. Copes, and J. B. Zaerr. 1990. Differences in proteins synthesized in needles of unshaded and shaded *Pinus ponderosa* var. Scopulorum seedlings during prolonged drought. *Plant Physiol.* 92:1244–1248.

Velazhahan, R., and S. Muthukrishnan. 2003. Transgenic tobacco plants constitutively overexpressing a rice thaumatin-like protein (PR-5) show enhanced resistance to *Alternaria alternata. Biol. Plant.* 47:347–354.

Verma, S., and R. S. Dubey. 2001. Effect of cadmium on soluble sugars and enzymes of their metabolism in rice. *Biol. Plant.* 44:117–123.

Verma, S., and R. S. Dubey. 2003. Lead toxicity induces lipid peroxidation and alters the activities of antioxidant enzymes in growing rice plants. *Plant Sci.* 164:645–655.

Vierling, E. 1991. The roles of heat shock proteins in plants. *Annu. Rev. Plant Physiol. Plant Mol. Biol.* 42:579–620.

Vigers, A. J., S. Wiedemann, W. K. Roberts, M. Legrand, C. P. Selitrennikoff, and B. Fritig. 1992. Thaumatin-like pathogenesis-related proteins are antifungal. *Plant Sci.* 83:155–161.

Viktorova, J., L. Krasny, M. Kamlar, M. Novakova, M. Mackova, and T. Macek. 2012. Osmotin, a pathogenesis-related protein. *Curr. Protein Pept. Sci.* 13:672–681.

Wahid, A., S. Gelani, M. Ashraf, and M. R. Foolad. 2007. Heat tolerance in plants: an overview. *Environ. Exp. Bot.* 61:199–223.

Wan, Q., S. Hongbo, X. Zhaolong, L. Jia, Z. Dayong, and H. Yihong. 2017. Salinity tolerance mechanism of osmotin and osmotin-like proteins: a promising candidate for enhancing plant salt tolerance. *Curr. Genomics* 18:553–556.

Wang, F. Z., Q. B. Wang, S. Y. Kwon, S. S. Kwak, and W. A. Su. 2005. Enhanced drought tolerance of transgenic rice plants expressing a pea manganese superoxide dismutase. *J. Plant Physiol.* 162:465–472.

Wang, K., X. Zhang, M. Goatley, and E. Ervin. 2014a. Heat shock proteins in relation to heat stress tolerance of creeping bentgrass at different N levels. *PloS ONE* 9:e102914.

Wang, M., P. Li, C. Li et al. 2014b. SiLEA14, a novel atypical LEA protein, confers abiotic stress resistance in foxtail millet. *BMC Plant Biol.* 14:290. doi:10.1186/s12870-014-0290-7

Wang, W., B. Vinocur, O. Shoseyov, and A. Altman. 2004. Role of plant heat-shock proteins and molecular chaperones in the abiotic stress response. *Trends Plant Sci.* 9:244–252.

Wang, X., E. Khodadadi, B. Fakheri, and S. Komatsu. 2017. Organ-specific proteomics of soybean seedlings under flooding and drought stresses. *J. Proteomics* 162:62–72.

Wang, X., X. Shan, Y. Wu et al. 2016. iTRAQ-based quantitative proteomic analysis reveals new metabolic pathways responding to chilling stress in maize seedlings. *J. Proteomics* 146:14–24.

Waters, E. R., G. J. Lee, and E. Vierling. 1996. Evolution, structure and function of the small heat shock proteins in plants. *J. Exp. Bot.* 47:325–338.

Weber, R. L. M., B. Wiebke-Strohm, C. Bredemeier et al. 2014. Expression of an osmotin-like protein from *Solanum nigrum* confers drought tolerance in transgenic soybean. *BMC Plant Biol.* 14:343. doi:10.1186/s12870-014-0343-y

Welling, A., P. Rinne, A. Vihera-Aarnio, S. Kontunen-Soppela, P. Heino, and E. T. Palva. 2004. Photoperiod and temperature differentially regulate the expression of two dehydrin genes during overwintering of birch (*Betula pubescens* Ehrh.). *J. Exp. Bot.* 55:507–516.

Win, K. T., and O. Zaw. 2017. Salt-stress-induced changes in protein profiles in two Blackgram (*Vigna mungo* L.) varieties differing salinity tolerance. *Adv. Plants Agric. Res.* 6:1–7.

Wisniewski, M. E., C. L. Bassett, J. Renaut, R. Farrell, T. Tworkoski, and T. S. Artlip. 2006. Differential regulation of two dehydrin genes from peach (*Prunus persica*) by photoperiod, low temperature and water deficit. *Tree Physiol.* 26:575–584.

Wójcik, M., J. Vangronsveld, and A. Tukiendorf. 2005. Cadmium tolerance in *Thlaspi caerulescens*: I. Growth parameters, metal accumulation and phytochelatin synthesis in response to cadmium. *Environ. Exp. Bot.* 53:151–161.

Woloshuk, C. P., J. S. Meulenhoff, M. Sela-Buurlage, P. J. M. van den Elzen, and B. J. C. Cornelissen. 1991. Pathogen-induced proteins with inhibitory activity toward *Phytophthora infestans*. *Plant Cell* 3:619–628.

Worrall, D., L. Elias, D. Ashford et al. 1998. A carrot leucine-rich-repeat protein that inhibits ice recrystallization. *Science* 282:115–117.

Xie, H., D. H. Yang, H. Yao, G. E. Bai, Y. H. Zhang, and B. G. Xiao. 2016. iTRAQ-based quantitative proteomic analysis reveals proteomic changes in leaves of cultivated tobacco (*Nicotiana tabacum*) in response to drought stress. *Biochem. Biophys. Res. Commun.* 469:768–775.

Xiong, J., Y. Sun, Q. Yang et al. 2017. Proteomic analysis of early salt stress responsive proteins in alfalfa roots and shoots. *Proteome Sci.* 15:19. doi:10.1186/s12953-017-0127-z

Xu, C. P., J. H. Sullivan, W. M. Garrett, T. J. Caperna, and S. Natarajan. 2008. Impact of solar Ultraviolet-B on the proteome in soybean lines differing in flavonoid contents. *Phytochemistry* 69:38–48.

Xu, D., X. Duan, B. Wang, B. Hong, T. H. D. Ho, and R. Wu. 1996. Expression of a late-embryogenesis abundant protein gene, *HVA1*, from barley confers tolerance to water deficit and salt stress in transgenic rice. *Plant Physiol.* 110:249–257.

Xu, Z. Z., and G. S. Zhou. 2005. Effects of water stress on photosynthesis and nitrogen metabolism in vegetative and reproductive shoots of *Leymus chinensis*. *Photosynthetica* 43:29–35.

Yamada, M., H. Morishita, K. Urano et al. 2005. Effects of free proline accumulation in petunias under drought stress. *J. Exp. Bot.* 56:1975–1981.

Yamchi, A., F. R. Jazii, A. Mousavi, and A. A. Karkhane. 2007. Proline accumulation in transgenic tobacco as a result of expression of Arabidopsis Delta(1)-pyrroline-5-carboxylate synthetase (P5CS) during osmotic stress. *J. Plant Biochem. Biotechnol.* 16:9–15.

Yan, S. P., Q. Y. Zhang, Z. C. Tang, W. A. Su, and W. N. Sun. 2006. Comparative proteomic analysis provides new insights into chilling stress responses in rice. *Mol. Cell Proteomics* 5:484–496.

Yang, F., X. W. Xiao, S. Zhang, H. Korpelainen, and C. Y. Li. 2009. Salt stress responses in *Populus cathayana* Rehder. *Plant Sci.* 176:669–677.

Yang, J., Y. Sun, A. Sun, S. Yi, J. Qin, M. Li, and J. Liu. 2006. The involvement of chloroplast HSP100/ClpB in the acquired thermotolerance in tomato. *Plant Mol. Biol.* 62:385 395.

Yang, Y., J. Chen, Q. Liu et al. 2012. Comparative proteomic analysis of the thermotolerant plant *Portulaca oleracea* acclimation to combined high temperature and humidity stress. *J. Proteome Res.* 11:3605–3623.

Yasuda, E., H. Ebinuma, and H. Wabiko. 1996. A novel glycine-rich/hydrophobic 16 kDa polypeptide gene from tobacco: similarity to proline-rich protein genes and its wound-inducible and developmentally regulated expression. *Plant Mol. Biol.* 33:667–678.

Yildiz, M. 2007. Two-dimensional electrophoretic analysis of soluble leaf proteins of a salt-sensitive (*Triticum aestivum*) and a salt-tolerant (*T. durum*) cultivar in response to NaCl stress. *J. Integr. Plant Biol.* 49:975–981.

Yildiz, M., and H. Terzi. 2008. Small heat shock protein responses in leaf tissues of wheat cultivars with different heat susceptibility. *Biologia* 63:521–525.

Yin, X., and S. Komatsu. 2016. Nuclear proteomics reveals the role of protein synthesis and chromatin structure in root tip of soybean during the initial stage of flooding stress. *J. Proteome Res.* 15:2283–2298.

Yin, Z., T. Rorat, B. M. Szabala, A. Ziolkowska, and S. Malepszy. 2006. Expression of a *Solanum sogarandinum* SK3-type dehydrin enhances cold tolerance in transgenic cucumber seedlings. *Plant Sci.* 170:1164–1172.

Yoshimura, K., A. Masuda, M. Kuwano, A. Yokota, and K. Akashi. 2008. Programmed proteome response for drought avoidance/tolerance in the root of a C3 xerophyte (wild watermelon) under water deficits. *Plant Cell Physiol.* 49:226–241.

Yu, J., Y. Lai, X. Wu, G. Wu, and C. Guo. 2016. Overexpression of OsEm1 encoding a group I LEA protein confers enhanced drought tolerance in rice. *Biochem. Biophys. Res. Commun.* 478:703–709

Yu, L. X., H. Chamberland, J. G. Lafontain, and Z. Tabaeizadeh. 1996. Negative regulation of gene expression of a novel proline-, threonine- and glycine-rich protein by water stress in *Lycopersicon chilense*. *Genome* 39:1185–1193.

Zayneb, C., R. H. Imen, K. Walid et al. 2017. The phytochelatin synthase gene in date palm (*Phoenix dactylifera* L.): phylogeny, evolution and expression. *Ecotoxicol. Environ. Saf.* 140:7–17.

Zeigler, H. 1990. Role of Plant physiology in assessing productivity potential under stress environment. In *Proceedings of the International Congress of Plant Physiology 88*, New Delhi, pp. 10–17.

Zhang, G. C., W. L. Zhu, J. Y. Gai, Y. L. Zhu, and L. F. Yang. 2015. Enhanced salt tolerance of transgenic vegetable soybeans resulting from overexpression of a novel Δ1-pyrroline-5-carboxylate synthetase gene from *Solanum torvum* Swartz. *Hortic. Environ. Biotechnol.* 56:94–104.

Zhang, H. X., C. L. Lian, and Z. G. Shen. 2009. Proteomic identification of small, copper-responsive proteins in germinating embryos of *Oryza sativa*. *Ann. Bot.* 103:923–930.

Zhang, J., Z. Duan, D. Zhang et al. 2016. Co-transforming bar and CsLEA enhanced tolerance to drought and salt stress in transgenic alfalfa (*Medicago sativa* L.). *Biochem. Biophys. Res. Commun.* 472:75–82.

Zhang, J., M. Zhang, M. J. I. Shohag, S. Tian, H. Song, Y. Feng, and X. Yang. 2016. Enhanced expression of SaHMA3 plays critical roles in Cd hyperaccumulation and hypertolerance in Cd hyperaccumulator *Sedum alfredii* Hance. *Planta* 243:577–589.

Zhang, L., X. Liu, K. Gaikwad, X. Kou, F. Wang, X. Tian, M. Xin, Z. Ni, Q. Sun, H. Peng, and E. Vierling. 2017a. Mutations in eIF5B confer thermosensitive and pleiotropic phenotypes via translation defects in *Arabidopsis thaliana*. *Plant Cell* 29:1952–1962.

Zhang, X. H., M. M. Moloney, and C. C. Chinnappa. 1996. Analysis of an ABA- and osmotic stress-inducible dehydrin from *Stellaria longipes. J. Plant Physiol.* 149:617–622.

Zhang, Z., H. Zhou, Q. Yu et al. 2017b. Quantitative proteomics analysis of leaves from two *Sedum alfredii* (Crassulaceae) populations that differ in cadmium accumulation. *Proteomics* 17.

Zhu, J. K. 2001. Cell signaling under salt, water and cold stresses. *Curr. Opin. Plant Biol.* 4:401–406.

23 Ultraviolet Radiation Effects on Plants
Harmful or Beneficial?

Arman Pazuki, Fatemeh Aflaki, and Mohammad Pessarakli

CONTENTS

23.1 INTRODUCTION

Human activities, particularly man-made chemicals, over recent decades have resulted in the depletion of stratospheric ozone and consequently, increased ultraviolet B (UVB) irradiation of the earth's surface. Each year, the ozone layer over the Antarctic pole reaches the thinnest thickness recorded for the year around early October, and the hole is usually largest in early September. The unit used in the measure of the column abundance of ozone in the atmosphere is Dobson Units (DU), corresponding to a layer of ozone 10 μm thick if the ozone was held at standard temperature and pressure (273 K, 1 atm pressure) (American Meteorological Society 2018). The average amount of ozone in the atmosphere is roughly 300 DU. Any recorded level thinner than 220 DU is called an *ozone hole*. The annual data for 21 September–16 October, 1979 to 2017, over Antarctica are below 220 DU,

which represents anthropogenic ozone losses from chlorine and bromine compounds (NASA Ozone Watch 2018a). As the observations show, Antarctica is experiencing an alarmingly thin ozone layer. However, the abundance of ozone in the Arctic has changed between 200 and 250 DU. The average minimum and mean ozone layer over the Arctic from 1 July 2017 to 30 June 2018 were 312 and 350 DU, respectively (NASA Ozone Watch 2018b). During the year 2010, measurements over much of the Northern Hemisphere unexpectedly indicated a high annual mean of the ozone columns. At the mid-latitudes, 45°N to 55°N, the annual mean exceeded 360 DU, which was one of the highest records over the last decades. This unusually high annual mean of the ozone layer over the Northern Hemisphere is surmised to be mostly due to the negative phase of the Arctic Oscillation (AO) and North Atlantic Oscillation (NAO) in 2010, coinciding with the easterly wind shear phase of the quasi-biennial oscillation of stratospheric winds at the equator (QBO), and partially, the recent decline of stratospheric chlorine and bromine (Steinbrecht et al. 2011). Therefore, it seems that the worst conditions are being experienced in Southern Hemisphere latitudes rather than their Northern counterparts. Apart from the irreversibly depleted ozone layer over the Antarctic continent due to anthropogenic activities, particular weather patterns also cause mini-holes in ozone. The differences between these two are that the mini-hole is a natural phenomenon, covering a smaller area, and reversible. For example, the highest and lowest minima over recent decades have fluctuated between 170 and 340 DU in the northern polar region (NASA Ozone Watch 2018c).

Photosynthesis originated when the sun shone on the earth. The majority of the solar radiation consists of visible and infrared radiation. Approximately 5% of the solar radiation reaching the earth's surface is ultraviolet (UV) radiation (100–400 nm). UV radiation comprises UVA (315–400–nm), UVB (280–315–nm), and UVC (100–280–nm). UV radiation causes damage to plant cells, which depends on the irradiating wavelength and the ability of cells to repair the damage. Stratospheric ozone completely absorbs UVC and most of the UVB, but it does not absorb any UVA. As a result, the radiation reaching the earth surface comprises about 95% UVA and 5% UVB. Therefore, in natural conditions, photosynthetic organisms are exposed to UVA and UVB. The UV radiation reaching a unit surface area is defined as a radiometric unit of energy ($J\ m^{-2}$) or power ($W\ m^{-2}$).

Throughout the evolution of photosynthesis, the earth's surface has been occupied by a myriad of light-using organisms. The genomic information determines the response of the organisms to environmental cues. However, it is the expression of genes in each cell of photosynthetic organisms that determines their traits and features. The expression of genes can be tuned by environmental changes. The quality and the intensity of sunlight reaching the earth can change depending on the solar cycle, the stratosphere, and the troposphere. Photosynthesizing organisms live in the troposphere, where unicellular or multicellular species of photosynthetic prokaryotes (cyanobacteria and anoxygenic photosynthetic bacteria), unicellular eukaryotes (euglena), multicellular algae, and plants exist. Many photosynthesizing species in fresh and salt water can move to find a favorable place. However, plants living out of water are sessile; thus, the irradiance they receive has undeniable effects on these species.

Many published articles suggest that enhancement of terrestrial UVB radiation due to the stratospheric depletion of ozone may result in tremendously adverse influences on photosynthesis and the productivity of crop plants. Based on observations, almost all forms of life can be negatively affected by UV radiation. Nevertheless, UV radiation can have benefits for many organisms. In this chapter, some of the many advantages that UV radiation may provide for living things are explained and discussed to balance out the notorious effects of UV radiation. To provide the evidence for beneficial effects of UV radiation, inevitably, some negative effects of UV radiation are also reviewed.

23.2 EFFECTS OF ULTRAVIOLET RADIATION ON THE GENOME

Photosynthesizing organisms have evolved to live under sunlight to convert light energy into chemical energy. Although this is a significant evolutionary advantage, they are also exposed to solar UV radiation, which induces stress and damage in the photosynthesizing apparatus, DNA, and tissues. Solar UVC radiation is almost entirely removed by the atmospheric ozone layer. However, some solar UVB and most UVA reach the earth's surface. The depletion of the ozone layer is leading to greater irradiation by UVB and a slight increase in UVA. Shorter wavelengths are more harmful. The damage that plants suffer directly due to UV irradiation can be classified into two major categories: damage to DNA and to physiological processes. The damage caused by UV radiation to DNA can be further classified into two sub-categories: mutations due to direct DNA alteration and mutations due to activation of mutagenic elements. However, the damage to physiological processes can affect many processes either directly or indirectly.

23.2.1 IMMEDIATE EFFECTS OF UV RADIATION ON GENOME

Most studies on the deleterious effects of increased UVB irradiation due to ozone depletion have been done in ecosystems of the South Pole. It was shown that the effect of photosynthetically active radiation (PAR) alone on the cell division rate of a diatom (*Chaetoceros socialis*) was similar to PAR + increased solar UVA, whereas the addition of UVB to the latter significantly decreased the division rate (Smith et al. 1992). In addition, it was indicated that UVB inhibition of photosynthesis increased linearly with increasing UVB dose (Smith et al. 1992). Moreover, it can also jeopardize terrestrial ecosystems. DNA absorbs UV radiation (mainly UVB and UVC), which can predispose plants to mutations. Researchers compared the effects of whole solar UV spectrum, selectively attenuated UVB radiation, and no UV treatment on plant DNA damage in southern South America, in the temperate latitudes (55° S). Higher UVB radiation passing through the ozone hole explained about half of the variation in plant DNA damage, as indicated by an increase in the steady-state level of the cyclobutane-type pyrimidine dimers (CPDs) at midday (Rousseaux et al. 1999). Another type of DNA damage products is pyrimidine (6–4) pyrimidinone dimer (6-4PP dimers), which are induced by UV radiation. By using *Arabidopsis* mutants defective in the repair of CPDs or 6–4 products, it was shown that both classes of pyrimidine dimers play a significant role in the inhibition of root elongation by UVB (Jiang et al. 1997). Photolyases are DNA repair enzymes that use the energy of blue light (400–500 nm) to repair 6-4PP dimers (Sancar 1994). *Arabidopsis* seedlings irradiated with UVB for 4 h and grown under light from CW1500 cool-white lamps (with blue light) or under F40GO gold fluorescent lights (without blue light) developed sensitivity symptoms, including withering and necrosis of expanded leaves, within 6 days of growing without blue light, whereas in the presence of blue light, the damage was alleviated, and the aforementioned symptoms were not observed (Landry et al. 1997). The vital role of blue light in UVB-induced DNA damage emphasizes the importance of light quality for plant survival (for a review, see Pazuki et al. 2017).

23.2.2 Delayed Effects of UV Radiation on Genome

As briefly discussed earlier, high-energy UV radiation induces immediate DNA damage, e.g., CPD and 6-4PP dimers. However, beyond the extent of immediate DNA damage, UV radiation also results in delayed damage to the genome, which may induce mutation. Transposable elements (TEs) make up a substantial proportion of the majority of organisms' genomes. They can change their position within a genome or may move from one genome to another; hence the name *transposon*. Hundreds of families of transposable elements comprise nearly 85% of the maize genome (Schnable et al. 2009). The maize nucleus, with a 2.3 Gb genome, contains one of the biggest genome sizes among crop plants (Michael and Jackson 2013), which can potentially result in more mutations. However, the majority of TEs are usually epigenetically silenced. Thus, plants are able to protect the genome integrity efficiently (Lisch 2009). The spontaneous reactivation of *Mutator* (*Mu*) elements (transposons in the maize genome) occurs with a frequency of about 10^{-4} (Walbot 1991). In stressful conditions, however, TEs can be activated. A study showed that in maize plants, transposon activity increases under UV radiation stress (Makarevitch et al. 2015). About 5% of the genome-wide transcriptome response to UV radiation stress could be attributed to the genes located near one of these TE families (Makarevitch et al. 2015). A study on allelic variation of maize showed a strong correlation between the presence of TE insertions and UV radiation stress–responsive upregulation of gene expression (Makarevitch et al. 2015), which could be due to facilitating transcription factors binding or influencing the local chromatin domain to be more accessible under the stress condition (Dixon et al. 2016). Maize pollen irradiated over 3 min with UVB simulating 33% ozone depletion activated immobile *Mu* transposons in maize sperm (Walbot 1999). Similarly, pollen irradiated with UVC (254 nm) showed an up to 40-fold increase in the reactivation rates of *Mu* element transcription in progeny kernels. Moreover, one of the observed consequences of the reactivation was an up to 40-fold increase in the copy number of *Mu9* elements (Walbot 1992), which can increase the chance of mutation. The expression levels of TEs in different plant species may also differ (Hirsch and Springer 2017). To keep the integrity of the genome and to minimize the detrimental effects of activation of TEs, plants may exploit silencing strategies to epigenetically control TE expression (Okamoto and Hirochika 2001). Short RNAs (sRNAs) with 24 nucleotides are a class of sRNAs that silence TEs through DNA methylation (Henderson and Jacobsen 2007). Researchers studying Norway spruce (*Picea abies*) TE transcription found that the expression of such 24-nucleotide sRNAs was highly tissue specific, observable only in the reproductive tissue samples, particularly in the male cone sample (Nystedt et al. 2013).

TEs' roles in the plant genome and transcriptome are being elucidated (Zhao et al. 2016). The insertion of a TE within the coding regions of a gene generally causes mutation, which may induce loss of function. Generally, it is thought that when the rate of mutation in a population is higher, thereby increasing the rate of variation, some of the individuals within the population may survive, adapt to the stressful conditions, and reproduce (Latta 2010). The benefits a plant under UV stress may receive as a result of activating TEs can arise from reshaping gene expression pattern or splicing in plants. The new changes are often tolerated by the plants if the insertions integrate within introns or untranslated regions of genes. The insertions may provide novel promoters, or they may modify chromatin in the regions near genes, which eventually can alter the gene expression level. TEs are suggested as key sources of mutation, variation, speciation, adaptation, domestication, breeding, and evolution (Biémont and Vieira 2006; Dubin et al. 2018). The variation regenerated as a result of controlled retrotransposition may help with breeding crop plants. Retrotransposition often occurs under particular stress conditions, such as UV radiation (Wessler 1996), which may be harnessed to provide beneficial traits for plant breeding through mutation and subsequent variation (Paszkowski 2015).

23.3 EFFECTS OF UV RADIATION ON ENVIRONMENT

23.3.1 Delayed Effects of UV Radiation on Biotic Environment Surrounding Organisms

In addition to the immediate and late effects of UVB radiation on organisms, UV radiation may affect organisms through delayed effects on their environment. For instance, UVA and UVB radiation inhibit algal consumers growing in benthic diatom communities in shallow freshwater because the herbivores are more sensitive to UV radiation than sympatric algae. As a result of the differential sensitivity to UV radiation, in habitats exposed to UVB, the algae number increased (Bothwell et al. 1994). The result of the experiment suggests that investigating the deleterious effects of short-term UVB irradiation on algal and plant photosynthesis, which has recently been the focus of numerous studies, may not provide a clear understanding of the indirect impact on an ecosystem of the sympatric organisms' response to increased UVB irradiance. The results of other studies suggested that the acute and chronic effects of UVB radiation on photosynthesizing species can differ. Generally, a higher incidence of UVB slows down the cell division rate. The effects of UVB + UVA + PAR or UVA + PAR on cell division rates of *Chaetoceros socialis*, a diatom, and *Phaeocystis* spp., prymnesiophytes, in Antarctic marine ecosystems were compared. While UVB irradiance decreased the cell division rate in *C. socialis*, it increased the division rate in *Phaeocystis* (Smith et al. 1992). Researchers studied the influence of solar UV radiation on the growth, biomass, and community composition of periphyton, complex communities of heterotrophic and photoautotrophic microorganisms. In comparison with only PAR irradiation, the addition of UV irradiation up to the third week inhibited the periphyton communities' growth indices (Chl a, mean cell volume, and cell number), whereas from the fourth to the fifth weeks the indices increased two- to fourfold (Bothwell et al. 1993). The results exemplify species-specific differences in the sensitivity of organisms to UV radiation.

23.3.2 Solar UV Radiation Effects on the Decomposition of Organic Material

The decomposition of organic material is important in global biogeochemical cycles, soil fertility, and nutrient availability. Solar UVB radiation is a high-energy emission from the sun that can influence the decomposition of organic compounds. However, UVB interacts with other influencing factors in this process. Studying the interactive effects of manipulated ambient solar UVB exposure of leaf litter under and away from the shrub canopies on decomposition rates indicated that during the first 50 days, about 60% of the litter was decomposed, whereas during the remaining days, over almost a year, it was partially decomposed (10% more). Moreover, in comparison, the initial C and N remaining in the litter were higher under the canopy, but the C remaining in the litter was lowest when it was away from the canopy and irradiated with UVB. The results of the experiment suggest that the decomposition rate of litter under UVB radiation can be changed or negated by other ambient factors, e.g., soil surface temperatures, precipitation, and solar emission outside the UVB region (Predick et al. 2018). A meta-analysis of a database containing 16 references and 50 experiments, largely from dryland ecosystems, showed that solar UV radiation increased mass loss by 23%, although a large variation in photo-degradation rates among and within ecosystems was reported (King et al. 2012). UV irradiation may alter plant exudates in the soil. Exposure to UV radiation may indirectly affect the productivity of plants as a result of changes in the exudates secreted by plant roots. The results of research suggest a possible mechanism for UV-dependent enhanced nitrogenase activity in fenugreek (*Trigonella foenum-graecum*). As compared with control (UVA + UVB + PAR), the exclusion of UVA and UVB increased nitrogenase activity by 80%, whereas the exclusion of UVB but irradiation with UVA + PAR increased the enzyme activity by 120% (Sharma and Guruprasad 2012). The result suggests that exclusion of UVA + UVB radiation might increase nitrogenase activity. However, irradiating UVA along with PAR may increase the enzyme activity more than when all UV radiation is excluded, hence showing the probable benefit of UVA (Sharma and Guruprasad 2012).

23.4 UVB RADIATION EFFECTS ON PLANT HORMONES

The spectral composition of irradiation enables plants to adjust their growth and development. Photoreceptors receive specific spectral bands and help the plant to perceive the seasons, the time of day, or the presence of competitors. One of the primary effects of UVB radiation is hormonal inactivation or activation to interact with the environment. The UVB-specific photoreceptor in plants is UV RESISTANCE LOCUS 8 (UVR8), which transduces a signaling cascade to regulate photo-morphogenesis in response to UVB radiation along with mediating photo-protective responses (Rizzini et al. 2011). The effects of UVB on the plant signal transduction pathway, hormones, morphology, and defense have previously been discussed (Brosché and Strid 2003; Vanhaelewyn et al. 2016). An overview of the beneficial effects of UV radiation mediated by some of the plant hormones involved in defense responses is provided here.

23.4.1 UV Radiation Effects on Auxin

Plant canopies are able to change the ratio of UV radiation to visible light reaching understory vegetation (Deckmyn et al. 2001). Far-red light transmitted through neighboring plants provides a cue for plants under dense canopy to detect the threat of shading and subsequently, to respond with rapid elongation (Franklin and Whitelam 2005). However, UVR8-mediated UVB perception can suppress shade avoidance symptoms by activating the pathways to inhibit the biosynthesis of auxin (IAA), a plant growth hormone (Hayes et al. 2014), and consequently reduces the content of IAA (Roro et al. 2017). In addition, some morphogenic responses to UVB radiation do not seem to need a UVB receptor system, but probably, they are the result of UVB-induced modification of the metabolism of secondary metabolites. Flavonoid aglycones can affect polar auxin transport, and phenol-oxidizing peroxidases are able to change auxin catabolism (Jansen 2001, 2002). Consequently, plants irradiated with UVB radiation exhibit a dwarf phenotype, short petioles, small and thick leaves, curling, short inflorescence, and a higher root to shoot ratio (Favory et al. 2009; Wargent et al. 2009; Hornitschek et al. 2012; Bernal et al. 2013; Hayes et al. 2014; Fierro et al. 2015; Innes et al. 2018). The observations suggest that UVB radiation, despite unwanted consequences for crop productivity, helps plants to understand their environment and respond accordingly.

23.4.2 UV Radiation Effects on Brassinosteroids

Brassinosteroids are steroidal hormones involved in the growth, development, and symbiosis of plants and have also been shown to enhance stress tolerance (Bajguz and Hayat 2009; Wei and Li 2016). Plants irradiated with UVB show a dwarf phenotype, which has been reported for other plants, e.g., *Arabidopsis*, tomato, and rice plants mutant for synthesizing brassinosteroids (Fujioka and Yokota 2003). Under long- and short-term exposure to UVB, researchers found a role for brassinosteroids in UVB signaling. The analysis of microarray data from *A. thaliana* plants exposed to UVB radiation indicated that numerous genes that are involved in hormone biosynthesis or metabolism, including brassinosteroids, were differentially expressed. The magnitudes of the observed changes were relatively small, but their effects on hormone homeostasis would be considerable (Hectors et al. 2007). Expression of the genes induced by UVB radiation requires a complete brassinosteroid synthesis pathway (Sävenstrand et al. 2004). Considering that brassinosteroids are involved in plant immune responses (De Bruyne et al. 2014), a low dose of UVB radiation may alter the defense gene expression regulated by the hormone transduction (Sävenstrand et al. 2004). Thus, UVB radiation is able to improve plant defense responses to pathogens through regulating brassinosteroid signaling.

23.4.3 UV RADIATION EFFECTS ON ABSCISIC ACID

Abscisic acid (ABA) is a plant hormone with diverse roles from stomatal closure and shoot growth to the induction of dormancy in seeds. As it appears, it is mostly involved in stress tolerance, particularly in drought stress. The ABA molecule has an absorption peak in the UVC region, but it has absorption up to short-wavelength UVB radiation (Milborrow 1974), which is able to isomerize a small fraction of the ABA to *cis*- or *trans*-ABA (Lindoo et al. 1979). It was indicated that mild UV irradiation of an aqueous solution of ABA reduces the half-life of the molecule to as little as 24 min (Cao et al. 2013). The results of some experiments suggest that drought cross-tolerance due to UV radiation is probably mediated through ABA hormone signal transduction. It was assumed that UVB and drought stress interaction induces an increase in salicylic acid (SA) accumulation, which enhances ABA concentration (Bandurska and Cieslak 2013), because empirical observation showed that in the same plant, SA application increased the ABA level (Bandurska and Stroinski 2005). UVB radiation enhanced the endogenous ABA level about 50–100% in *Artemisia annua*, depending on the treatment day in the plant's life (Pan et al. 2014), and 15–100% in rice (*Oryza sativa*), depending on the examined rice cultivars (Lin et al. 2002). It was suggested that the effects of UVB on plant responses cannot be solely explained by ABA signal transduction. However, it was observed that ABA is effective in increasing UVB-absorbing compounds (Duan et al. 2008). Therefore, ABA mediates the UVB-induced responses of plants, and the effects of UVB radiation on the signaling and accumulation of ABA seem to be species and growth stage dependent.

23.4.4 UV RADIATION EFFECTS ON JASMONATES

Jasmonates (JAs) are a class of plant hormones that play an important role in plant defense and some other roles in growth and development (Creelman and Mullet 1997). There are some studies indicating the effects of UVB radiation on modifying the quality of plant tissue, which consequently affects the interest of herbivorous insect in the tissue (Mewis et al. 2012). The damage resulting from an insect, *Plutella xylostella* L. (diamondback moth), feeding on Brassicaceae was more severe in the absence of UVB radiation in comparison with ambient UVB–irradiated *A. thaliana* leaves. In

addition, the insect preferred to deposit more eggs on plants grown under a lower level of UVB than on plants irradiated with ambient UVB. However, the important point is that this preference was also observed for a mutant with impaired jasmonic acid (JA) sensitivity (*jar1-1*); even when the mutant was irradiated with ambient UVB, the insect laid more eggs on it. Moreover, lower amounts of UV-absorbing compounds accumulated in the mutants (Caputo et al. 2006). The results of a study on *Nicotiana attenuata* plants demonstrated that UVB radiation activated JA-dependent and -independent pathways to increase the concentration of phenolic compounds, and through increasing sensitivity to JAs, it could upregulate the expression of wound-response genes (TPI), which led to a significant increase in resistance to herbivory (Demkura et al. 2010). The exposure of sprouting broccoli plants to moderate doses of UVB increased the concentration of phenolic compounds in the aerial part of the seedlings. Moreover, glucosinolates (defensive metabolites) accumulated in the tissue, and this was accompanied by increases in the expression of genes associated with SA and JA signaling defense pathways and increases in the transcription of genes responsive to fungal and bacterial pathogens (Mewis et al. 2012). It seems that UVB radiation, by changing JA signaling, is able to influence insect oviposition behavior and herbivory damage, and microbial pathogenesis.

23.5 STRATEGIES OF PLANTS TO EVADE UV RADIATION: REFLECTION, AVOIDANCE, AND ATTENUATION

Aquatic organisms can survive under solar UV radiation by living under water, which to some extent attenuates solar UV radiation depending on the concentration of dissolved organic compounds and particulate matter (Vincent et al. 2001; Huovinen et al. 2003). Despite the higher sensitivity of aquatic photosynthetic species to solar UV radiation in comparison with higher plants (Karentz et al. 1991), the photosynthetic species living in water can take shelter under water. In addition, some species of algae and cyanobacteria can avoid the high incidence of solar UV radiation by vertical migration in water (Eggersdorfer and Häder 1991; Qin et al. 2015). However, terrestrial photosynthetic organisms are sessile and subjected to higher solar UV radiation, which can damage DNA and membrane integrity. As a result, the rates of survival and energy conversion to organic substances can be decreased. Therefore, to thrive in such a high-stress environment, developing compounds to screen UV irradiance would be advantageous. Plant species, varieties, and even sexes are different in their sensitivity to UV radiation (Strømme et al. 2015; Strømme et al., 2018a). The differences are because of the defense mechanisms used by plants: DNA repair (Yang 2011), paraheliotropism (Grant 1999), leaf structure (Holmes and Keiller 2002), antioxidant content, and UV-absorbing compounds (Holmes and Keiller 2002).

Robberecht et al.'s (1980) study showed that the reflection of solar UVB radiation by the leaf surface of four species from exposed habitats in equatorial and tropical latitudes can vary from 5% for glabrous leaf to 40% for densely pubescent leaf. Of the unreflected radiation, 60–95% is absorbed. However, on average, less than 2% of the irradiated UVB may be transmitted into the mesophyll, which means that plant preventive mechanisms attenuate about 98% of UVB irradiation. The amount of wax deposited on the abaxial surface of tobacco plant (*N. tabacum*) leaf exposed to UVB radiation was comparable to that observed in non-UVB-treated plants, whereas the wax chemical composition differed significantly (Barnes et al. 1996). Plants with the lowest ability to reflect or attenuate irradiated UV using the cuticular membrane, the cell walls, and pubescence rely on absorption of UV radiation by flavonoids and phenolics (Caldwell et al. 1983). Therefore, the attenuating effect of absorbing mechanisms is greater than that of reflective mechanisms.

Some plant species avoid UV radiation by changing leaf orientation away from the sun. The results of a study indicated that soybean (*Glycine max*) cultivars tolerant to UV radiation, in fact, are able to change their leaves' inclination to relatively receive the lowest intensity of UV radiation (Grant 1999). Thus, although the radiation may not be changed, the irradiation is decreased. Most plant species owe their ability to absorb UV radiation mostly to flavonoids or phenolic derivatives and the cuticular membrane or the cell walls (Caldwell et al. 1983). A diverse array of these compounds has been reported from almost all terrestrial higher plant species (Caldwell et al. 1983). The accumulation of UV-screening pigments can be induced in plants irradiated with UVB (Fischbach et al. 1999). The screening compounds may efficiently protect the photosynthetic apparatus. However, this is at the expense of growth inhibition, perhaps due to the high metabolic cost of accumulating phenolic compounds.

23.6 THE PROBABLE BENEFITS OF UV RADIATION FOR CROP PLANTS

23.6.1 Effects of UV Radiation on Crop Productivity

Anthropogenic depletion of stratospheric ozone exposes agricultural plants to higher levels of solar UV radiation. Therefore, the studies on the effects of UV radiation are mainly focused on the potential detrimental influences of UV irradiation on plants and photosynthesizing organisms. However, recent evidence suggests that although some deleterious effects can be attributed to UV irradiation, it is also one of the main effectors that regulate plants' morphology, physiological and biochemical processes, and genetic information. UV radiation affects plants, including agricultural plants. Studying the effects of UV radiation on the morphology and productivity of pea (*Pisum sativum*) indicated that although morphological characteristics could be altered by UV radiation, in comparison with the UV-excluded condition, the impact of UV radiation on the number of pods was minor. It was observed that the influence of other climatic factors, such as irradiance, temperature, and vapor pressure deficit, on the productivity of pea was greater than that of UV radiation

(Roro et al. 2016). Moreover, the results of recent investigations show that UV radiation can increase crop plants' yield quality and quantity. For instance, by irradiating with UV for a short time, e.g., a couple of days before the final harvest, the quality and sometimes, the quantity of the harvest can be increased (Tsormpatsidis et al. 2010). Exposing woody Mediterranean plant species to UVA + UVB increased the leaf mass to area and the carotenoids to chlorophyll a + b ratio as compared with plants grown without UV or with UVA, respectively (Bernal et al. 2013). Some UV-irradiated plants can tolerate other stressors better than when UV is excluded. A meta-analysis by Newsham and Robinson (2009) in polar regions shows that UVB radiation did not result in any effects on several physiological and biochemical traits (e.g., carotenoid or chlorophyll concentrations, net photosynthesis, F_v/F_m or ΦPSII, below-ground or total biomass, leaf mass, leaf area, or specific leaf area). Irradiating lettuce (*Lactuca sativa* L.) with UVB produced higher harvestable yields and an increased photosynthetic rate as compared with UVB-excluded treatment. Plants pre-acclimatized to UVB radiation also showed higher relative growth and induced higher maximum photochemical efficiency of photosystem II after irradiation with high PAR (Wargent et al. 2011). However, in another experiment, lettuce plants irradiated with solar UVB yielded about 60% lower dry and fresh weights for the shoots. Similarly, solar UVA radiation decreased the dry weight of shoot and root to as low as 60% and 50% of that in UVA-excluded radiation-treated plants, respectively (Krizek et al. 1998). The results of a field experiment on *Trigonella foenum-graecum* showed that in comparison with control (UV + PAR), the exclusion of solar UV enhanced root biomass, number of nodules, nodule fresh weight, leghemoglobin, and hemechrome content in the nodules (Sharma and Guruprasad 2012). The meta-analysis indicated that UVB irradiation decreased aboveground biomass by 15%, reduced plant height up to 10%, and increased screening compounds in leaves or thalli (by 7% and 25%, respectively) and DNA damage by 90% (Newsham and Robinson 2009). Notwithstanding the damage to DNA, inducing plants with UV radiation to accumulate flavonoids in tissues can protect DNA from the damage resulting from the radiation (Stapleton and Walbot 1994).

23.6.2 Accumulation of Value-Added Compounds in Plant Tissues

From the point of view of a consumer, value-added in plants, particularly nutritionally improved traits, is very appealing. UV radiation as a stressor can induce photosynthesizing organisms to accumulate antioxidants and secondary metabolites (Stracke et al. 2010; Morales et al. 2013). The protective antioxidants consumed from plant sources can be categorized into various classes: vitamins, carotenoids, flavonoids, and polyphenols. UV radiation induces many plants to produce higher amounts of medicinal secondary metabolites. UVB radiation highly induces the activation of genes encoding phenylpropanoid biosynthetic enzymes, which leads to the accumulation of flavonol glycosides (Stracke et al. 2010).

The accumulated compounds protect the producer from UV radiation, and the same compounds may also protect the grazers and consumers from the detrimental effects of radiation (Jansen et al. 2008; Jiang et al. 2016; Petruk et al. 2018). Epidemiological studies suggest that the reason behind the increased longevity and decreased incidence of cardiovascular disease observed in people in the Mediterranean region is flavonoid-rich foods (Auger et al. 2004; Trichopoulou and Critselis 2004; Hooper et al. 2008). Many edible plants growing in tropical regions contain high amounts of flavonoids (Miean and Mohamed 2001; Sulaiman and Balachandran 2012). Quercetin, a plant flavonol, is one of the naturally occurring flavonoids, the largest group of naturally available plant phenolic compounds. In grapevine (*Vitis vinifera*), complete solar radiation resulted in the accumulation of the greatest amount of quercetin, and excluding UVB radiation caused greater accumulation in comparison with completely UV-excluded solar radiation (Kolb et al. 2003). This compound may have anti-prostanoid, anti-inflammatory, anti-atherosclerotic, anti-thrombic, anti-hypertensive, and anti-arrhythmic effects (Formica and Regelson 1995). In the same species (*V. vinifera*), elevated irradiation of UV leads to the production of higher levels of plant flavonols (Carbonell-Bejerano et al. 2014), which provides an efficient shield against UVA but an incomplete shielding against UVB radiation (Kolb et al. 2003). UVB and UVA irradiated from cool white fluorescent tubes increased the amounts of flavonoids in lettuce plants within a few days with no reduction in growth. Higher doses of the UVs radiated from UVB tubes induced a higher content of flavonoid in the tissues but with some reduction in growth (Rodriguez et al. 2014). Exposing black cumin (*Bunium persicum*), a medicinal plant, to short-term UVB radiation produced a higher seed yield and a higher content and antioxidative properties of essential oil (Saeidnejad et al. 2016). Pot rose (*Rosa × hybrida* "Toril") plants irradiated with UVB at 0.1 or 0.2 W m^{-2} for 1 or 2 h showed increasingly enhanced flavonol and anthocyanin contents, which were further increased if the plants were exposed for 15 days as compared with 10 or 5 days' exposure. However, UVB treatments did not significantly change the number of flower buds (Suthaparan et al. 2012). The results of the experiments suggest that regardless of the small negative outcomes, UV radiation can induce the accumulation of very important and economic compounds in the plants' tissues that we and other herbivores consume.

23.7 UV RADIATION MAY INDUCE CROSS-TOLERANCE TO OTHER TYPES OF STRESS

23.7.1 UV RADIATION-INDUCED CROSS-TOLERANCE TO MAJOR ABIOTIC STRESSES

Exposing plants to different stresses can cause the induction of similar morphogenic, physiologic, and metabolic responses (Potters et al. 2009). Enhanced exposure to UV radiation or drought stress is a trigger for plants to show a range of responses, including oxidative stress, causing damage to proteins, lipids, carbohydrates, and DNA (Bandurska et al.

2013). Cross-tolerance to UV radiation and water availability (drought stress) was observed in seedlings of six woody Mediterranean species (three mesophytes vs. three xerophytes) (Bernal et al. 2013). Irradiating *Pistacia lentiscus* (a xerophyte) with UVA in the well-watered condition decreased above-ground biomass, whereas in the low-watered condition, UVA radiation mitigated the reduction in growth that the plant might experience in an UV-free environment, which could be due to UVA-enhanced apparent electron transport rate values under low irrigation treatment. Similarly, under a low water supply, UVA increased the root biomass of the studied species (Bernal et al. 2013). Seedlings of another plant species (*Laurus nobilis*) also showed the same reaction to low water supply when exposed to above-ambient levels of UVA or UVA + UVB radiation. In reduced water supply conditions, while growing under higher than ambient UVA or UVA + UVB radiation, the plants produced more biomass and had a higher leaf relative water content. It was suggested that UVA radiation enhances the thermal dissipation of light energy, whereas the addition of UVB to UVA, by accumulating light-absorbing pigments or activating the antioxidant defense system, can mitigate UVA's negative effects (Bernal et al. 2015). Solar UV radiation also helped silver birch (*Betula pendula* Roth.) seedlings to cope with water scarcity. However, well-watered plants were slightly disadvantaged by the solar UV radiation (Robson et al. 2015). Similar results were also observed for *Populus yunnanensis* (Duan et al. 2008). Irradiating sunflower plants (*Helianthus annuus*) with UVB, while subjecting them to drought stress, implied a protective effect by reducing lipid peroxidation resulting from water stress (Cechin et al. 2008). In addition, a correlation between tolerance to drought and UVB radiation stresses was reported (Yang et al. 2005). Plant metabolic responses to drought stress and UVB radiation can be similar. The common responses of plants to UV or drought stresses are augmenting antioxidation capacity through the upregulation of enzymatic antioxidants, including superoxide dismutase, catalase, guaiacol peroxidase, ascorbate peroxidase, and glutathione reductase, or accumulating low–molecular weight antioxidants such as proline, glycine betaine, ascorbate, and glutathione, or increasing the amounts of secondary metabolites, e.g., polyamines, tocopherol, carotenoids, alkaloids, and flavonoids (Bandurska et al. 2013). UVB radiation may also increase plant tolerance to low-temperature stress (Chalker-Scott and Scott 2004). Moreover, the tolerance induced as a result of UV radiation can mitigate transplantation stress to increase plant survival and performance of the crop in the field after transplantation (Garner and Björkman 1999; Wargent et al. 2006). The reaction of stressed plants to elevated reactive oxygen species (ROS) is generally similar (Wrzaczek et al. 2013). Both UV radiation and heavy metal toxicity generate ROS (Babu et al. 2003). This similarity and overlap in response may explain the UV-induced increase in plant tolerance to heavy metal toxicity (Li et al. 2012). However, in higher plants, an optimum level of ROS can play beneficial roles in development, differentiation, redox levels, stress signaling, interactions with other organisms, systemic responses, and cell death (Mittler 2017).

Others suggested that the perception of stress from low doses of UVB radiation can even be good for plants, because it activates the pathways, making the plants ready to react efficiently to other stressors (Hideg et al. 2013). However, the cross-tolerances can also be species (Bernal et al. 2013) or genotype-specific (He et al. 2011). As the empirical results suggest, one advantage of UV radiation can be the induction of cross-tolerance to other types of stress.

23.7.2 UV Radiation May Indirectly Facilitate Initiation of Symbiotic Relationships to Mitigate Nutrient Deficiency Stress

The symbiotic relationship between plants from the Fabaceae family and bacteria of the genus *Rhizobium* provides reduced nitrogen for the host plants and carbohydrates for the nitrogen-fixing bacteria. The initiation of the symbiotic relationship requires signaling molecules to help both sides in recognizing each other. Flavonoids are plant signal molecules secreted by roots that activate the expression of nodulation genes in the bacteria (Liu and Murray 2016). In return, Rhizobia secrete lipo-chitooligosaccharides that act as nodulation factors, which under conditions of nitrogen limitation, induce division of the root cortex cells to develop into nodule primordia (Schultze and Kondorosi 1998).

The results of an experiment indicated that bean plants (*Phaseolus vulgaris*) could tolerate high irradiation of UVB if they also received high levels of PAR, because a lower incidence of PAR resulted in inhibition of photosystem II. The same research group found that UVB radiation can increase the amounts of flavonoids in the bean plants (Cen and Bornman 1990). UV radiation can alter the profile of gene expression to regulate the content of flavonoids (Casati and Walbot 2003). The presence of flavonoids in the seed coat can also efficiently protect seeds from high-energy, short-wavelength UV radiation at very high levels (Tepfer and Leach 2017). The flavonoids content of a type of bean (*Phaseolus lunatus*) was shown to be variable (Agostini-Costa et al. 2014). The flavonol content of Italian ecotypes of bean (*Phaseolus vulgaris*) varied from 0.19 to 0.84 g kg^{-1} of seed fresh weight (Dinelli et al. 2006). During germination, seedlings can use the stored compounds for communication with other organisms. Root exudates and the plant defense system are the key regulators of recognizing and initiating a symbiotic relationship with microbiotas, of which rhizobia are key members (Poole et al. 2018). In hairy root cultures of a plant from the mustard family (Brassicaceae), UVB radiation induced flavonoid accumulation. In comparison with the control, UVB increased the flavonoid concentration 16-fold (Jiao et al. 2018). Alfalfa (*Medicago sativa*) seed and root exudates were compared for flavonoid signals, e.g., chrysoeriol and luteolin, which upregulate the transcription of nodulation (nod) genes in *Rhizobium meliloti*. During the first 4 hours of imbibition, seeds released about 100-fold more flavonoids than roots of 72-hour-old seedlings (Hartwig et al. 1990). The literature suggests that plants generally accumulate flavonoids in response to UV radiation, and seeds' flavonoids come from the mother plant. In addition, it was shown that plants use flavonoids for communication with symbiotic microorganisms to support themselves. Therefore, it is reasonable to think that UV radiation is beneficial for plants, because inducing the accumulation of flavonoids indirectly helps the root to find more resources by initiating symbiosis with arbuscular mycorrhiza or rhizobacteria.

In addition to benefiting from symbiosis with rhizobacteria, similarly, plants can gain an advantage from fungi. Host plants can activate a signaling pathway called the *common symbiotic pathway*, which also controls symbiotic signals in arbuscular mycorrhiza (Gough and Cullimore 2011), a root endosymbiosis between plants and glomeromycete fungi. The fungi provide water and macro-nutrients, such as phosphate, nitrogen, potassium, and sulfur (Harrison et al. 2002; Guether et al. 2009; Sieh et al. 2013; Garcia and Zimmermann 2014), and even micro-nutrients, such as zinc, iron, manganese, and copper (Lehmann et al. 2014; Lehmann and Rillig 2015), for the roots of 70–90% of land plant species (Harley and Smith 1983).

23.7.3 UV Radiation and Its Consequences against Pathogens

Phenolics are known as important first-line compounds induced in plant defense against infection (Matern and Kneusel 1988). Conidia are asexual spores that help with the dispersal and environmental persistence of many fungi, which in pathogenic species, also contribute to host recognition and infection. Both solar UVA and UVB in sublethal doses result in reduced conidial germination speed and virulence, and after a few hours, they can kill conidia of most fungal species (reviewed in Braga et al. 2015). *Citrus aurantium* fruits were irradiated with UVC (254 nm) after inoculation with the fungus *Penicillium digitatum*. UVC radiation altered the flavonoid profile of the peel and significantly increased the resistance of the fruits to the fungus after 2 hours of irradiation (Arcas et al. 2000). Irradiating lettuce with UVB prior to inoculation with a suspension of *Bremia lactucae* (lettuce downy mildew) conidia significantly reduced spore formation. However, in addition to increasing resistance to the pathogen, leaf thickness and pigmentation were also increased, while leaf area was reduced (Wargent et al. 2006). Irradiating powdery mildew (*Podosphaera pannosa*), a fungal disease, on pot rose (*Rosa × hybrida* "Toril") plants with UVB for 2 min to 2 h during 15 days demonstrated that irradiating the plants before fungus inoculation was not effective in reducing the number of spores or disease severity. However, after inoculation of the fungus, UVB irradiation at 0.1 or 0.2 W m^{-2} for 1 or 2 h significantly decreased both observations (Suthaparan et al. 2012). Almost similar results were observed for cucumber (*Cucumis sativus*) plants inoculated with cucumber powdery mildew (*P. xanthii*) (Suthaparan et al. 2014). Elevating the intensity of UVB radiation up to 1.2 W m^{-2}, but exposing for a very short time (2 or 5 min), also reduced disease severity in the pot rose. The results indicated that the UVB efficiency against conidium production was mostly due to short-wavelength

UVB (below 290 nm) (Suthaparan et al. 2012). The effect of UVB on powdery mildew was suggested to be due to direct suppression of the pathogens rather than enhanced resistance of the host through the accumulation of secondary metabolites (Suthaparan et al. 2012, 2014). However, it was observed that daily light integral and daylight quality are influential in the efficiency of UV irradiation in controlling the disease (Suthaparan et al. 2017). These observations are not in agreement with Wargent et al.'s (2006) result reported for lettuce, for which irradiating lettuce with UVB prior to inoculation was effective against the pathogen. To sustainably manage pests and diseases in the agricultural intensification of crops, among many available management strategies, UV radiation is an alternative that can be used (Barrière et al. 2014).

23.7.4 UV Radiation May Enhance Plant Tolerance to Herbivorous Insects

The increase in global average temperature due to anthropogenic activities (Ekwurzel et al. 2017) may enhance the fitness of insect species at higher latitudes, where the temperature is currently cooler than the physiological optima of the species (Deutsch et al. 2008). The damage of herbivory can reduce plant growth, size, and seed set (Wilbur et al. 2013). It was demonstrated that some insect herbivores can alter the density of a plant population. However, the impacts of the herbivores on plant populations with high density are likely to be different from those on populations with low density or endangered species, for which the insects potentially are serious threats (Myers and Sarfraz 2017). ROS produced as a result of UV radiation may also defend plants against insects attacking them (Gatehouse 2002; Wrzaczek et al. 2013). Phenolics and flavonoids are constitutively produced chemicals in tissues of numerous species of bryophytes, lower and higher vascular plants (Stafford 1991). One of the strategies plants use to deter insects from grazing is concentrating phenolic compounds (Dreyer and Jones 1981; Agrawal et al. 2018). They are also accumulated as a result of exposing plants to other stresses, especially UV radiation (Winkel-Shirley 2002). Due to this accumulation, the plants may also be resistant to insects by avoiding attack or deterring the herbivore (Schaefer and Rolshausen 2006). Using a microarray analysis enriched in wound- and insect-responsive sequences (241 genes) in field-grown plants of *Nicotiana longiflora*, it was shown that the perception of stress from ambient UVB and caterpillar herbivory had correlation and similar effects on various functional groups of genes ($P < .001$; $R^2 = .22$) (Izaguirre et al. 2003). The effects of solar UVB radiation on the growth of the dominant plant species in a shrub-dominated ecosystem in southern Argentina, which has experienced high irradiation due to the Antarctic ozone hole, were studied. In the three studied plants, although leaf or frond length in the UVB attenuated condition increased by 12%, 25–75% of the leaf area was consumed by insect herbivory (Rousseaux et al. 2001). Since the efficiency of phenolic compounds in defense against insects can differ (Gatehouse 2002), some insects might even benefit from plants having higher levels of the compounds (Simmonds 2003).

23.8 NUTRIENT IMBALANCE MAY AGGRAVATE UV RADIATION STRESS

The negative effects of UV radiation on susceptible photosynthesizing species or ecosystems are undeniable. However, the damage that plants suffer from UV stress may not be merely due to UV radiation, considering millions of years of evolution and adaptation to sunlight. Murali and Teramura (1985) observed that when soybean plants (*Glycine max*) were provided with optimum amounts of phosphorus, as a result of UVB irradiation, the content of flavonoids decreased. In the phosphorus deficiency condition, however, UVB radiation increased flavonoids in soybean plants. Providing a sufficient nitrogen supply may alleviate the stress plants can experience under UV radiation (Guo et al. 2014). The application of nitrogen, phosphate, and potassium (NPK) fertilizers at 1.5 times more than the recommended amounts decreased nutrients in the edible part, growth, and economic yield of radish (*Raphanus sativus*) irradiated with UVB (7.2 kJ m^{-2} d^{-1}), which was likely due to allocating lower amounts of photosynthates toward roots as compared with shoots (Singh et al. 2011). On the other hand, nutrient deficiency may reduce plants' fitness under ambient UV radiation. Near-isogenic lines of maize varying in flavonoid biosynthesis genes while undergoing nutrient deficiency (the level of the treatments: sufficient nutrients, or 30 to 70% lower N, K, Mn, Fe, and Zn) were exposed to different irradiation regimes to examine changes in maximum net photosynthetic rate and chlorophyll fluorescence (F_v/F_m). The results indicated that flavonoid content is genotype dependent. In nutrient deficiency conditions, flavonoids, anthocyanins, chlorophyll a and b contents, photosynthetic rate, F_v/F_m, and stomatal conductance were significantly reduced. However, in the genotypes with higher levels of flavonoids, some of the negative effects were relatively mitigated (Lau et al. 2006). Photosynthetic organisms' lipid profile also may change due to the interaction of UV radiation with nutrient availability. Exposing seston to increasing phosphorus (P) concentrations in the absence and presence of UV radiation indicated that phosphorus enrichment lowered the content of highly unsaturated fatty acids and seston C:P ratio, hence decreasing food quality (Villar-Argaiz et al. 2009).

23.9 LONG-TERM EFFECTS OF UV RADIATION ON PLANTS

The majority of research articles report the detrimental effects of short-term exposure of plants to elevated levels of UV radiation. However, the long-term effects of the radiation have not been fully understood. Irradiating Douglas fir (*Pseudotsuga menziesii*) with enhanced levels of UVB radiation (two- or threefold) over 3 years in both glasshouse and field experiments indicated that although during the first year, photosynthetic rates and growth were slightly reduced, during the second and third years, they reached levels equal to, or even exceeding, those of trees under ambient UVB radiation. Moreover, quantum yield, chlorophyll, and UV screening compounds did not change under high levels of UV radiation

(Bassman et al. 2002). Similarly, the results of an experiment on the effects of a higher level of UVB radiation on Eurasian aspen (*Populus tremula*) indicated that the plant responses to UVB radiation decreased over three consecutive years (Strømme et al. 2018b). A meta-analysis indicated that ambient UVB radiation for woody or herbaceous plants, overall, has a negative effect on total biomass (7.0–14.6% reduction). Higher UVB irradiation did not result in significant changes in any variables for woody plants. However, the height and specific leaf area of herbaceous plants decreased, perhaps at the expense of producing UVB-absorbing compounds. The analysis suggested that in the case of higher UVB radiation, herbaceous plants can be more influenced than woody plants; thus, in the management of grassland ecosystems, they may need more attention (Li et al. 2010). In most of the experiments, to exclude or attenuate UV radiation, UV-transparent or opaque filters are used. These filters are satisfactorily used for artificial radiation sources (Casati and Walbot 2003). However, under solar radiation, it was observed that the accumulation of UVB-absorbing compounds in plants is more affected by filters *per se* rather than the filters' ability to attenuate UVB radiation (Comont et al. 2012). The results of the experiment strongly suggest including an unfiltered control in similar studies.

23.10 TERRESTRIAL UV RADIATION IN THE FUTURE

The depletion of stratospheric ozone has been a matter of concern in recent decades, and it is predicted that the consequent higher solar UV irradiation will continue for a few more decades (McKenzie et al. 2003). However, the fluctuation in ozone is not the major factor affecting UVB radiation reaching the earth's surface. It is believed that the factors altering UV radiation received at the earth's surface, in decreasing order, are zenith angle, cloud cover, the ozone layer, the distance between the sun and the earth, aerosols, altitude, and albedo (McKenzie et al. 2003). Cloud effects on UVB attenuation are about 15–30% (McKenzie et al. 2003). Man-made pollutants play important roles in attenuating UVB irradiance, typically reducing the UV irradiance by up to 20% (Wenny et al. 2001), but in some locations, anthropogenic aerosols can attenuate UVB irradiance by up to 54% (di Sarra et al. 2002). The altitude effect can be changed by surface albedo, aerosols, or tropospheric ozone (McKenzie et al. 2003). Snow and sea ice cover have influential effects on albedo. In addition, the anthropogenic contribution to increased global average temperature (Ekwurzel et al. 2017) may differentially mitigate some of the negative effects of increased transmission of UV radiation in different strains or species of photosynthetic organisms (Islam et al. 2018). Seasonal differences may also alleviate UVB impact on the exposed population. Phytoplankton species living in the spring season can be more sensitive to UVB radiation than the species living in summer, when total irradiance at the sea's surface is maximal (Hobson and Hartley 1983). The other environmental factors determining the magnitude of UVB's harmful effects are

the quantity of the irradiation, the quality of the irradiating spectra, and the duration of exposure. During recent years, advanced instruments for measuring spectral irradiance have become more available. Although they are cheaper, simpler, and accessible, one of the drawbacks of broadband measuring instruments is that they average all the spectral irradiances; hence, the radiation of a particular band cannot be obtained. Spectroradiometers, by measuring spectral irradiances and providing detailed information on each band, let the researcher investigate the specific effects of a band on plants and other photosynthesizing organisms (Pazuki et al. 2017). The complexity of studying variables affecting UV radiation reaching the earth's surface, differential distributions of spatial and temporal influencing factors, and possible interactions between fluctuating ozone, changing climate, and large-scale atmospheric circulation limit our confidence in the ability of models to predict future changes in UV irradiation on the global scale. The Montreal Protocol under the Vienna Convention has shown a beneficial effect on global ozone recovery. Further recovery of ozone abundances to those of the 1980s in the Arctic and Antarctic has been predicted as a result of the expected reduction of ozone-depleting substances (WMO 2014). A new United Nations report documented that implementation of the Montreal Protocol and its succeeding amendments are effectively phasing out the consumption and production of controlled ozone-depleting substances; thereby, the recovery of ozone in the stratosphere may have started (WMO 2018). The report highlights that since a severe ozone depletion in the polar regions has been avoided, the Antarctic ozone hole is also recovering, although it is expected to occur every year. Excluding the polar regions, since 2000, the ozone layer in the upper stratosphere has increased by 1–3% per decade. At projected rates, it is expected that the total column ozone in the mid-latitudes of the Northern and Southern Hemispheres will recover completely in the 2030s and the 2050s, followed by the Antarctic in the 2060s. Assuming full and sustained compliance with the Montreal Protocol and the Kigali Amendment, it is projected to avoid up to 0.4° C of the anticipated future global average warming by 2100 (WMO 2018).

23.11 CONCLUSIONS

Solar UV radiation affects almost all organisms and processes occurring in terrestrial ecosystems. Most of the research conducted so far has been focused on the possible negative effects of UV radiation on terrestrial life. However, mounting empirical evidence suggests that despite the numerous forms of damage attributed to UV radiation, the benefits that photosynthetic organisms may get from UV radiation, if harnessed optimally, have the potential to improve the yield of crop plants, both in quality and in quantity, and to support and bring advantages to the ecosystem. However, the detrimental effects of UV radiation in some exceptional cases are not denied. The exceptions, for example, are the effects of high UV irradiance on ecosystems under ozone holes, in high latitudes, or where the population is endangered. Apart from the aforementioned cases, UV radiation provides many advantages for plants and other photosynthetic organisms to adjust their behavior, growth, and development. UV radiation influences organisms surrounding plants and the processes they cause to take place, e.g., competition, herbivory, decomposition of litter, etc. In addition, UV radiation can significantly contribute to photo-degradation of organic and inorganic materials. Most often, to investigate the effects of UV radiation on terrestrial organisms, a limited number of species during a short time are examined, which does not seem to provide a reliable result that can be confidently extrapolated to an ecosystem. Climatological, ecological, physiological, biochemical, photobiological, genetic, and transcriptomic approaches must be integrated to comprehend the effects of UV radiation on an ecosystem and eventually, at a global scale. Therefore, it is suggested that in future research, more diverse species should be studied in long-term experiments to reach definitive conclusions about global changes as a result of enhanced UV radiation levels because of the ozone layer reduction or the consequences of atmospheric disturbances due to other anthropogenic activities. Moreover, as the discussed findings suggest, UV radiation can be efficiently and effectively used as an inducer for improving the protection, tolerance, and production of crop plants. Finally, from a broad physiological and biochemical point of view, the effects of UV radiation on plant processes in the long term do not seem seriously detrimental.

BIBLIOGRAPHY

Agostini-Costa, T.D.S., Teodoro, A.F.P., Alves, R.D.B.D.N., Braga, L.R., Ribeiro, I.F., Silva, J.P., Quintana, L.G. and Burle, M.L. 2015. Total phenolics, flavonoids, tannins and antioxidant activity of lima beans conserved in a Brazilian Genebank. *Ciência Rural*, 45(2): 335–341

Agrawal, A.A., Hastings, A.P., Fines, D.M., Bogdanowicz, S. and Huber, M. 2018. Insect herbivory and plant adaptation in an early successional community. *Evolution*, 72(5): 1020–1033

American Meteorological Society. 2018. Glossary of Meteorology. http://glossary.ametsoc.org/wiki/Dobson_unit

Arcas, M.C., Botía, J.M., Ortuño, A.M. and Del Río, J.A. 2000. UV irradiation alters the levels of flavonoids involved in the defence mechanism of *Citrus aurantium* fruits against *Penicillium digitatum*. *European Journal of Plant Pathology*, 106(7): 617–622

Auger, C., Al-Awwadi, N., Bornet, A., Rouanet, J.M., Gasc, F., Cros, G. and Teissedre, P.L. 2004. Catechins and procyanidins in Mediterranean diets. *Food Research International*, 37(3): 233–245

Babu, T.S., Akhtar, T.A., Lampi, M.A., Tripuranthakam, S., Dixon, D.G. and Greenberg, B.M. 2003. Similar stress responses are elicited by copper and ultraviolet radiation in the aquatic plant *Lemna gibba*: Implication of reactive oxygen species as common signals. *Plant and Cell Physiology*, 44(12): 1320–1329

Bajguz, A. and Hayat, S. 2009. Effects of brassinosteroids on the plant responses to environmental stresses. *Plant Physiology and Biochemistry*, 47(1): 1–8

Bandurska, H. and Cieślak, M. 2013. The interactive effect of water deficit and UV-B radiation on salicylic acid accumulation in barley roots and leaves. *Environmental and Experimental Botany*, 94: 9–18

Bandurska, H., Niedziela, J. and Chadzinikolau, T. 2013. Separate and combined responses to water deficit and UV-B radiation. *Plant Science*, 213: 98–105

Bandurska, H. and Stroiski, A. 2005. The effect of salicylic acid on barley response to water deficit. *Acta Physiologiae Plantarum*, 27(3): 379–386

Barnes, J.D., Percy, K.E., Paul, N.D., Jones, P., McLaughlin, C.K., Mullineaux, P.M., Creissen, G. and Wellburn, A.R. 1996. The influence of UV-B radiation on the physicochemical nature of tobacco (*Nicotiana tabacum* L.) leaf surfaces. *Journal of Experimental Botany*, 47(1): 99–109

Barrière, V., Lecompte, F., Nicot, P.C., Maisonneuve, B., Tchamitchian, M. and Lescourret, F. 2014. Lettuce cropping with less pesticides. A review. *Agronomy for Sustainable Development*, 34(1): 175–198

Bassman, J.H., Edwards, G.E. and Robberecht, R. 2002. Long-term exposure to enhanced UV-B radiation is not detrimental to growth and photosynthesis in Douglas-fir. *New Phytologist*, 154(1): 107–120

Bernal, M., Llorens, L., Badosa, J. and Verdaguer, D. 2013. Interactive effects of UV radiation and water availability on seedlings of six woody Mediterranean species. *Physiologia Plantarum*, 147(2): 234–247

Bernal, M., Verdaguer, D., Badosa, J., Abadía, A., Llusià, J., Peñuelas, J., Núñez-Olivera, E. and Llorens, L. 2015. Effects of enhanced UV radiation and water availability on performance, biomass production and photoprotective mechanisms of *Laurus nobilis* seedlings. Environmental and Experimental Botany, 109: 264–275

Biémont, C. and Vieira, C. 2006. Genetics: Junk DNA as an evolutionary force. *Nature*, 443(7111): 521–524

Bothwell, M.L., Sherbot, D., Roberge, A.C. and Daley, R.J. 1993. Influence of natural ultraviolet radiation on lotic periphytic diatom community growth, biomass accrual, and species composition: Short-term versus long-term effects 1, 2. *Journal of Phycology*, 29(1): 24–35

Bothwell, M.L., Sherbot, D.M. and Pollock, C.M. 1994. Ecosystem response to solar ultraviolet-B radiation: Influence of trophic-level interactions. *Science*, 265(5168): 97–100

Braga, G.U., Rangel, D.E., Fernandes, É.K., Flint, S.D. and Roberts, D.W. 2015. Molecular and physiological effects of environmental UV radiation on fungal conidia. *Current Genetics*, 61(3): 405–425

Brosché, M. and Strid, Å. 2003. Molecular events following perception of ultraviolet-B radiation by plants. *Physiologia Plantarum*, 117(1): 1–10

Caldwell, M.M., Robberecht, R. and Flint, S.D. 1983. Internal filters: Prospects for UV-acclimation in higher plants. *Physiologia Plantarum*, 58(3): 445–450

Cao, M., Liu, X., Zhang, Y., Xue, X., Zhou, X.E., Melcher, K., Gao, P., Wang, F., Zeng, L., Zhao, Y. and Deng, P. 2013. An ABA-mimicking ligand that reduces water loss and promotes drought resistance in plants. *Cell Research*, 23(8): 1043–1054

Caputo, C., Rutitzky, M. and Ballaré, C.L. 2006. Solar ultraviolet-B radiation alters the attractiveness of *Arabidopsis* plants to diamondback moths (*Plutella xylostella* L.): Impacts on oviposition and involvement of the jasmonic acid pathway. *Oecologia*, 149(1): 81–90

Carbonell-Bejerano, P., Diago, M.P., Martínez-Abaigar, J., Martínez-Zapater, J.M., Tardáguila, J. and Núñez-Olivera, E. 2014. Solar ultraviolet radiation is necessary to enhance grapevine fruit ripening transcriptional and phenolic responses. *BMC Plant Biology*, 14(1): 183

Casati, P. and Walbot, V. 2003. Gene expression profiling in response to ultraviolet radiation in maize genotypes with varying flavonoid content. *Plant Physiology*, 132(4): 1739–1754

Cechin, I., Corniani, N., de Fátima Fumis, T. and Cataneo, A.C. 2008. Ultraviolet-B and water stress effects on growth, gas exchange and oxidative stress in sunflower plants. *Radiation and Environmental Biophysics*, 47(3): 405–413

Cen, Y.P. and Bornman, J.F. 1990. The response of bean plants to UV-B radiation under different irradiances of background visible light. *Journal of Experimental Botany*, 41(11): 1489–1495

Chalker-Scott, L. and Scott, J.D. 2004. Elevated Ultraviolet-B Radiation induces cross-protection to cold in leaves of *Rhododendron* under field conditions. *Photochemistry and Photobiology*, 79(2): 199–204

Comont, D., Martinez Abaigar, J., Albert, A., Aphalo, P., Causton, D.R., Figueroa, F.L., Gaberscik, A., Llorens, L., Hauser, M.T., Jansen, M.A., Kardefelt, M., de la Coba Luque, P., Neubert, S., Núñez-Olivera, E., Olsen, J., Robson, M., Schreiner, M., Sommaruga, R., Strid, Å., Torre, S., Turunen, M., Veljovic-Jovanovic, S., Verdaguer, D., Vidovic, M., Wagner, J., Winkler, J.B., Zipoli G. and Gwynn-Jones, D. 2012. UV responses of *Lolium perenne* raised along a latitudinal gradient across Europe: A filtration study. *Physiologia Plantarum*, 145(4): 604–618

Creelman, R.A. and Mullet, J.E. 1997. Biosynthesis and action of jasmonates in plants. *Annual Review of Plant Biology*, 48(1): 355–381

De Bruyne, L., Höfte, M. and De Vleesschauwer, D. 2014. Connecting growth and defense: The emerging roles of brassinosteroids and gibberellins in plant innate immunity. *Molecular Plant*, 7(6): 943–959

Deckmyn, G., Cayenberghs, E. and Ceulemans, R. 2001. UV-B and PAR in single and mixed canopies grown under different UV-B exclusions in the field. *Plant Ecology*, 154(1–2): 123–133

Demkura, P.V., Abdala, G., Baldwin, I.T. and Ballaré, C.L. 2010. Jasmonate-dependent and-independent pathways mediate specific effects of solar ultraviolet B radiation on leaf phenolics and antiherbivore defense. *Plant Physiology*, 152(2): 1084–1095

Deutsch, C.A., Tewksbury, J.J., Huey, R.B., Sheldon, K.S., Ghalambor, C.K., Haak, D.C. and Martin, P.R. 2008. Impacts of climate warming on terrestrial ectotherms across latitude. *Proceedings of the National Academy of Sciences*, 105(18): 6668–6672

Di Sarra, A., Cacciani, M., Chamard, P., Cornwall, C., DeLuisi, J.J., Di Iorio, T., Disterhoft, P., Fiocco, G., Fuà, D. and Monteleone, F. 2002. Effects of desert dust and ozone on the ultraviolet irradiance at the Mediterranean island of Lampedusa during PAUR II. *Journal of Geophysical Research: Atmospheres*, 107(D18)

Dinelli, G., Bonetti, A., Minelli, M., Marotti, I., Catizone, P. and Mazzanti, A. 2006. Content of flavonols in Italian bean (*Phaseolus vulgaris* L.) ecotypes. *Food Chemistry*, 99(1): 105–114

Dixon, J.R., Gorkin, D.U. and Ren, B. 2016. Chromatin domains: The unit of chromosome organization. *Molecular Cell*, 62(5): 668–680

Dreyer, D.L. and Jones, K.C. 1981. Feeding deterrency of flavonoids and related phenolics towards *Schizaphis graminum* and *Myzus persicae*: Aphid feeding deterrents in wheat. *Phytochemistry*, 20(11): 2489–2493

Duan, B., Xuan, Z., Zhang, X., Korpelainen, H. and Li, C. 2008. Interactions between drought, ABA application and supplemental UV-B in *Populus yunnanensis*. *Physiologia Plantarum*, 134(2): 257–269

Dubin, M.J., Scheid, O.M. and Becker, C. 2018. Transposons: A blessing curse. *Current Opinion in Plant Biology*, 42: 23–29

Eggersdorfer, B., Häder, D.P. 1991. Phototaxis, gravitaxis and vertical migrations in the marine dinoflagellates, *Peridinium faeroense* and *Amphidinium caterea*. *Acta Protozool*, 30: 63–71.

Ekwurzel, B., Boneham, J., Dalton, M.W., Heede, R., Mera, R.J., Allen, M.R. and Frumhoff, P.C. 2017. The rise in global atmospheric CO2, surface temperature, and sea level from emissions traced to major carbon producers. *Climatic Change*, 144(4): 579–590

Favory, J.J., Stec, A., Gruber, H., Rizzini, L., Oravecz, A., Funk, M., Albert, A., Cloix, C., Jenkins, G.I., Oakeley, E.J. and Seidlitz, H.K. 2009. Interaction of COP1 and UVR8 regulates UV-B-induced photomorphogenesis and stress acclimation in *Arabidopsis*. *The EMBO Journal*, 28(5): 591–601

Fierro, A.C., Leroux, O., De Coninck, B., Cammue, B.P., Marchal, K., Prinsen, E., Van Der Straeten, D. and Vandenbussche, F. 2015. Ultraviolet-B radiation stimulates downward leaf curling in *Arabidopsis thaliana*. *Plant Physiology and Biochemistry*, 93: 9–17

Fischbach, R.J., Kossmann, B., Panten, H., Steinbrecher, R., Heller, W., Seidlitz, H.K., Sandermann, H., Hertkorn, N. and Schnitzler, J.P. 1999. Seasonal accumulation of ultraviolet-B screening pigments in needles of Norway spruce (*Picea abies* (L.) Karst.). *Plant, Cell & Environment*, 22(1): 27–37

Formica, J.V. and Regelson, W. 1995. Review of the biology of quercetin and related bioflavonoids. *Food and Chemical Toxicology*, 33(12): 1061–1080.

Franklin, K.A. and Whitelam, G.C. 2005. Phytochromes and shade-avoidance responses in plants. *Annals of Botany*, 96(2): 169–175

Fujioka, S. and Yokota, T. 2003. Biosynthesis and metabolism of brassinosteroids. *Annual Review of Plant Biology*, 54(1): 137–164

Garcia, K. and Zimmermann, S.D. 2014. The role of mycorrhizal associations in plant potassium nutrition. *Frontiers in Plant Science*, 5: 337

Garner, L.C. and Björkman, T. 1999. Mechanical conditioning of tomato seedlings improves transplant quality without deleterious effects on field performance. *HortScience*, 34(5): 848–851.

Gatehouse, J.A. 2002. Plant resistance towards insect herbivores: A dynamic interaction. *New Phytologist*, 156(2): 145–169

Gough, C. and Cullimore, J. 2011. Lipo-chitooligosaccharide signaling in endosymbiotic plant-microbe interactions. *Molecular Plant-Microbe Interactions*, 24(8): 867–878

Grant, R.H. 1999. Potential effect of soybean heliotropism on ultraviolet-B irradiance and dose. *Agronomy Journal*, 91(6): 1017–1023

Guether, M., Neuhäuser, B., Balestrini, R., Dynowski, M., Ludewig, U. and Bonfante, P. 2009. A mycorrhizal-specific ammonium transporter from *Lotus japonicus* acquires nitrogen released by arbuscular mycorrhizal fungi. *Plant Physiology*, 150(1): 73–83

Guo, X.R., Chang, B.W., Zu, Y.G. and Tang, Z.H. 2014. The impacts of increased nitrate supply on *Catharanthus roseus* growth and alkaloid accumulations under ultraviolet-B stress. *Journal of Plant Interactions*, 9(1): 640–646

Harley, J.L. and Smith, S.E. 1983. *Mycorrhizal Symbiosis*. Academic Press Inc., London.

Harrison, M.J., Dewbre, G.R. and Liu, J. 2002. A phosphate transporter from *Medicago truncatula* involved in the acquisition of phosphate released by arbuscular mycorrhizal fungi. *The Plant Cell*, 14(10): 2413–2429

Hartwig, U.A., Maxwell, C.A., Joseph, C.M. and Phillips, D.A. 1990. Chrysoeriol and luteolin released from alfalfa seeds induce nod genes in *Rhizobium meliloti*. *Plant Physiology*, 92(1): 116–122

Hayes, S., Velanis, C.N., Jenkins, G.I. and Franklin, K.A. 2014. UV-B detected by the UVR8 photoreceptor antagonizes auxin signaling and plant shade avoidance. *Proceedings of the National Academy of Sciences*, 111(32): 11894–11899

He, L., Jia, X., Gao, Z. and Li, R. 2011. Genotype-dependent responses of wheat (Triticum aestivum L.) seedlings to drought, UV-B radiation and their combined stresses. *African Journal of Biotechnology*, 10(20): 4046–4056

Hectors, K., Prinsen, E., De Coen, W., Jansen, M.A. and Guisez, Y. 2007. *Arabidopsis thaliana* plants acclimated to low dose rates of ultraviolet B radiation show specific changes in morphology and gene expression in the absence of stress symptoms. *New Phytologist*, 175(2): 255–270

Henderson, I.R. and Jacobsen, S.E. 2007. Epigenetic inheritance in plants. *Nature*, 447(7143): 418–424

Hideg, É., Jansen, M.A. and Strid, Å. 2013. UV-B exposure, ROS, and stress: Inseparable companions or loosely linked associates? *Trends in Plant Science*, 18(2): 107–115

Hirsch, C.D. and Springer, N.M. 2017. Transposable element influences on gene expression in plants. *Biochimica et Biophysica Acta (BBA)-Gene Regulatory Mechanisms*, 1860(1): 157–165

Hobson, L.A. and Hartley, F.A. 1983. Ultraviolet irradiance and primary production in a Vancouver Island fjord, British Columbia, Canada. *Journal of Plankton Research*, 5(3): 325–331

Holmes, M.G. and Keiller, D.R. 2002. Effects of pubescence and waxes on the reflectance of leaves in the ultraviolet and photosynthetic wavebands: A comparison of a range of species. *Plant, Cell & Environment*, 25(1): 85–93

Hooper, L., Kroon, P.A., Rimm, E.B., Cohn, J.S., Harvey, I., Le Cornu, K.A., Ryder, J.J., Hall, W.L. and Cassidy, A. 2008. Flavonoids, flavonoid-rich foods, and cardiovascular risk: A meta-analysis of randomized controlled trials. *The American Journal of Clinical Nutrition*, 88(1): 38–50

Hornitschek, P., Kohnen, M.V., Lorrain, S., Rougemont, J., Ljung, K., López-Vidriero, I., Franco-Zorrilla, J.M., Solano, R., Trevisan, M., Pradervand, S. and Xenarios, I. 2012. Phytochrome interacting factors 4 and 5 control seedling growth in changing light conditions by directly controlling auxin signaling. *The Plant Journal*, 71(5): 699–711

Huovinen, P.S., Penttilä, H. and Soimasuo, M.R. 2003. Spectral attenuation of solar ultraviolet radiation in humic lakes in Central Finland. *Chemosphere*, 51(3): 205–214

Innes, S.N., Solhaug, K.A., Arve, L.E. and Torre, S. 2018. UV radiation as a tool to control growth, morphology and transpiration of poinsettia (*Euphorbia pulcherrima*) in variable aerial environments. *Scientia Horticulturae*, 235: 160–168

Islam, M.A., Beardall, J. and Cook, P. 2018. Intra-strain variability in the effects of temperature on UV-B sensitivity of cyanobacteria. *Photochemistry and Photobiology*

Izaguirre, M.M., Scopel, A.L., Baldwin, I.T. and Ballaré, C.L. 2003. Convergent responses to stress. Solar ultraviolet-B radiation and *Manduca sexta* herbivory elicit overlapping transcriptional responses in field-grown plants of *Nicotiana longiflora*. *Plant Physiology*, 132(4): 1755–1767

Jansen, M.A. 2002. Ultraviolet-B radiation effects on plants: Induction of morphogenic responses. *Physiologia Plantarum*, 116(3): 423–429

Jansen, M.A., Hectors, K., O'Brien, N.M., Guisez, Y. and Potters, G. 2008. Plant stress and human health: Do human consumers benefit from UV-B acclimated crops? *Plant Science*, 175(4): 449–458

Jansen, M.A., van den Noort, R.E., Tan, M.A., Prinsen, E., Lagrimini, L.M. and Thorneley, R.N. 2001. Phenol-oxidizing peroxidases contribute to the protection of plants from ultraviolet radiation stress. *Plant Physiology*, 126(3): 1012–1023

Jiang, C.Z., Yee, J., Mitchell, D.L. and Britt, A.B. 1997. Photorepair mutants of *Arabidopsis*. *Proceedings of the National Academy of Sciences*, 94(14): 7441–7445

Jiang, N., Doseff, A.I. and Grotewold, E. 2016. Flavones: From biosynthesis to health benefits. *Plants*, 5(2): 27

Jiao, J., Gai, Q.Y., Yao, L.P., Niu, L.L., Zang, Y.P. and Fu, Y.J. 2018. Ultraviolet radiation for flavonoid augmentation in *Isatis tinctoria* L. hairy root cultures mediated by oxidative stress and biosynthetic gene expression. *Industrial Crops and Products*, 118: 347–354

Karentz, D., Cleaver, J. E., and Mitchell, D. L. 1991. Cell survival characteristics and molecular responses of Antarctic phytoplankton to ultraviolet-B radiation. Journal of Phycology, 27: 326–341

King, J.Y., Brandt, L.A. and Adair, E.C. 2012. Shedding light on plant litter decomposition: Advances, implications and new directions in understanding the role of photodegradation. *Biogeochemistry*, 111(1–3): 57–81

Kolb, C.A., Kopecký, J., Riederer, M. and Pfündel, E.E. 2003. UV screening by phenolics in berries of grapevine (*Vitis vinifera*). *Functional Plant Biology*, 30(12): 1177–1186

Krizek, D.T., Britz, S.J. and Mirecki, R.M. 1998. Inhibitory effects of ambient levels of solar UV-A and UV-B radiation on growth of cv. New Red Fire lettuce. *Physiologia Plantarum*, 103(1): 1–7

Landry, L.G., Stapleton, A.E., Lim, J., Hoffman, P., Hays, J.B., Walbot, V. and Last, R.L. 1997. An *Arabidopsis* photolyase mutant is hypersensitive to ultraviolet-B radiation. *Proceedings of the National Academy of Sciences*, 94(1): 328–332

Latta, R.G. 2010. Natural selection, variation, adaptation, and evolution: A primer of interrelated concepts. *International Journal of Plant Sciences*, 171(9): 930–944

Lau, T.S.L., Eno, E., Goldstein, G., Smith, C. and Christopher, D.A. 2006. Ambient levels of UV-B in Hawaii combined with nutrient deficiency decrease photosynthesis in near-isogenic maize lines varying in leaf flavonoids: Flavonoids decrease photoinhibition in plants exposed to UV-B. *Photosynthetica*, 44(3): 394–403

Lehmann, A. and Rillig, M.C. 2015. Arbuscular mycorrhizal contribution to copper, manganese and iron nutrient concentrations in crops–A meta-analysis. *Soil Biology and Biochemistry*, 81: 147–158

Lehmann, A., Veresoglou, S.D., Leifheit, E.F. and Rillig, M.C. 2014. Arbuscular mycorrhizal influence on zinc nutrition in crop plants–A meta-analysis. *Soil Biology and Biochemistry*, 69: 123–131

Li, F.R., Peng, S.L., Chen, B.M. and Hou, Y.P. 2010. A meta-analysis of the responses of woody and herbaceous plants to elevated ultraviolet-B radiation. *Acta Oecologica*, 36(1): 1–9

Li, X., Zhang, L., Li, Y., Ma, L., Bu, N. and Ma, C. 2012. Changes in photosynthesis, antioxidant enzymes and lipid peroxidation in soybean seedlings exposed to UV-B radiation and/or Cd. *Plant and Soil*, 352(1–2): 377–387

Lin, W., Wu, X., Linag, K., Guo, Y., He, H., Chen, F. and Liang, Y. 2002. Effect of enhanced UV-B radiation on polyamine metabolism and endogenous hormone contents in rice (*Oryza sativa* L.). *Ying Yong Sheng Tai Xue Bao (The Journal of Applied Ecology)*, 13(7): 807–813

Lindoo, S.J., Seeley, S.D. and Caldwell, M.M. 1979. Effects of ultraviolet-B radiation stress on the abscisic acid status of *Rumex patientia* leaves. *Physiologia Plantarum*, 45(1): 67–72

Lisch, D. 2009. Epigenetic regulation of transposable elements in plants. *Annual Review of Plant Biology*, 60: 43–66

Liu, C.W. and Murray, J.D. 2016. The role of flavonoids in nodulation host-range specificity: An update. *Plants*, 5(3): 33

Makarevitch, I., Waters, A.J., West, P.T., Stitzer, M., Hirsch, C.N., Ross-Ibarra, J. and Springer, N.M. 2015. Transposable elements contribute to activation of maize genes in response to abiotic stress. *PLoS Genetics*, 11(1): e1004915

Matern, U. and Kneusel, R.E. 1988. Phenolic compounds in plant disease resistance. *Phytoparasitica*, 16: 153–170

McKenzie, R.L., Björn, L.O., Bais, A. and Ilyasd, M. 2003. Changes in biologically active ultraviolet radiation reaching the Earth's surface. *Photochemical & Photobiological Sciences*, 2(1): 5–15

Mewis, I., Schreiner, M., Nguyen, C.N., Krumbein, A., Ulrichs, C., Lohse, M. and Zrenner, R. 2012. UV-B irradiation changes specifically the secondary metabolite profile in broccoli sprouts: Induced signaling overlaps with defense response to biotic stressors. *Plant and Cell Physiology*, 53(9): 1546–1560

Michael, T.P. and Jackson, S. 2013. The first 50 plant genomes. *The Plant Genome*, 6(2): 1–7

Miean, K.H. and Mohamed, S. 2001. Flavonoid (myricetin, quercetin, kaempferol, luteolin, and apigenin) content of edible tropical plants. *Journal of Agricultural and Food Chemistry*, 49(6): 3106–3112

Milborrow, B.V. 1974. The chemistry and physiology of abscisic acid. *Annual Review of Plant Physiology*, 25(1): 259–307

Mittler, R. 2017. ROS are good. *Trends in Plant Science*, 22(1): 11–19

Morales, L.O., Brosché, M., Vainonen, J., Jenkins, G.I., Wargent, J.J., Sipari, N., Strid, Å., Lindfors, A.V., Tegelberg, R. and Aphalo, P.J. 2013. Multiple roles for UV RESISTANCE LOCUS 8 in regulating gene expression and metabolite accumulation in *Arabidopsis* under solar UV radiation. *Plant Physiology*, 161: 744–759

Murali, N.S. and Teramura, A.H. 1985. Effects of ultraviolet-B irradiance on soybean. VI. Influence of phosphorus nutrition on growth and flavonoid content. *Physiologia Plantarum*, 63(4): 413–416

Myers, J.H. and Sarfraz, R.M. 2017. Impacts of insect herbivores on plant populations. *Annual Review of Entomology*, 62: 207–230

NASA Ozone Watch. 2018a. Southern Hemisphere Minimum Ozone. https://ozonewatch.gsfc.nasa.gov/statistics/o3_min_0921-1016.txt

NASA Ozone Watch. 2018b. Index of meteorology, figures, merra, ozone. https://ozonewatch.gsfc.nasa.gov/meteorology/figures/merra/ozone/toms_capn_2017_omps+merra2.txt

NASA Ozone Watch. 2018c. What are Ozone Mini-Holes? https://ozonewatch.gsfc.nasa.gov/facts/miniholes_NH.html

Newsham, K.K. and Robinson, S.A. 2009. Responses of plants in Polar Regions to UVB exposure: A meta-analysis. *Global Change Biology*, 15(11): 2574–2589

Nystedt, B., Street, N.R., Wetterbom, A., Zuccolo, A., Lin, Y.C., Scofield, D.G., Vezzi, F., Delhomme, N., Giacomello, S., Alexeyenko, A. and Vicedomini, R. 2013. The Norway spruce genome sequence and conifer genome evolution. *Nature*, 497(7451): 579

Okamoto, H. and Hirochika, H. 2001. Silencing of transposable elements in plants. *Trends in Plant Science*, 6(11): 527–534

Pan, W.S., Zheng, L.P., Tian, H., Li, W.Y. and Wang, J.W. 2014. Transcriptome responses involved in artemisinin production in *Artemisia annua* L. under UV-B radiation. *Journal of Photochemistry and Photobiology B: Biology*, 140: 292–300

Paszkowski, J. 2015. Controlled activation of retrotransposition for plant breeding. *Current Opinion in Biotechnology*, 32: 200–206

Pazuki, A., Aflaki, F., Pessarakli, M., Gurel, E., Gurel, S. 2017. Plant responses to Extended Photosynthetically Active Radiation (EPAR). *Advances in Plants & Agriculture Research*, 7(3): 00260

Petruk, G., Del Giudice, R., Rigano, M.M. and Monti, D.M. 2018. Antioxidants from plants protect against skin photoaging. *Oxidative Medicine and Cellular Longevity*, 2018

Poole, P., Ramachandran, V. and Terpolilli, J. 2018. Rhizobia: From saprophytes to endosymbionts. *Nature Reviews Microbiology*, 16(5): 291–303

Potters, G., Pasternak, T.P., Guisez, Y. and Jansen, M.A. 2009. Different stresses, similar morphogenic responses: Integrating a plethora of pathways. *Plant, Cell & Environment*, 32(2): 158–169

Predick, K.I., Archer, S.R., Aguillon, S.M., Keller, D.A., Throop, H.L. and Barnes, P.W. 2018. UV-B radiation and shrub canopy effects on surface litter decomposition in a shrub-invaded dry grassland. *Journal of Arid Environments*, 157: 13–21

Qin, H., Li, S., and Li, D. 2015. Differential responses of different phenotypes of Microcystis (Cyanophyceae) to UV-B radiation. *Phycologia*, 54(2): 118–129

Rizzini, L., Favory, J.J., Cloix, C., Faggionato, D., O'Hara, A., Kaiserli, E., Baumeister, R., Schäfer, E., Nagy, F., Jenkins, G.I. and Ulm, R. 2011. Perception of UV-B by the *Arabidopsis* UVR8 protein. *Science*, 332(6025): 103–106

Robberecht, R., Caldwell, M.M. and Billings, W.D. 1980. Leaf ultraviolet optical properties along a latitudinal gradient in the arctic-alpine life zone. *Ecology*, 61(3): 612–619

Robson, T.M., Hartikainen, S.M. and Aphalo, P.J. 2015. How does solar ultraviolet-B radiation improve drought tolerance of silver birch (*Betula pendula* Roth.) seedlings? *Plant, Cell & Environment*, 38(5): 953–967

Rodriguez, C., Torre, S. and Solhaug, K.A. 2014. Low levels of ultraviolet-B radiation from fluorescent tubes induce an efficient flavonoid synthesis in Lollo Rosso lettuce without negative impact on growth. *Acta Agriculturae Scandinavica, Section B–Soil & Plant Science*, 64(2): 178–184

Roro, A.G., Dukker, S.A., Melby, T.I., Solhaug, K.A., Torre, S., and Olsen, J.E. 2017. UV-B-induced inhibition of stem elongation and leaf expansion in pea depends on modulation of gibberellin metabolism and intact gibberellin signalling. *Journal of Plant Growth Regulation*, 36(3): 680–690

Roro, A.G., Terfa, M.T., Solhaug, K.A., Tsegaye, A., Olsen, J.E. and Torre, S., 2016. The impact of UV radiation at high altitudes close to the equator on morphology and productivity of pea (*Pisum sativum*) in different seasons. *South African Journal of Botany*, 106: 119–128

Rousseaux, M.C., Ballaré, C.L., Giordano, C.V., Scopel, A.L., Zima, A.M., Szwarcberg-Bracchitta, M., Searles, P.S., Caldwell, M.M. and Díaz, S.B. 1999. Ozone depletion and UVB radiation: Impact on plant DNA damage in southern South America. *Proceedings of the National Academy of Sciences of the United States of America*, 96(26): 15310–15315

Rousseaux, M.C., Scopel, A.L., Searles, P.S., Caldwell, M.M., Sala, O.E. and Ballaré, C.L. 2001. Responses to solar ultraviolet-B radiation in a shrub-dominated natural ecosystem of Tierra del Fuego (southern Argentina). *Global Change Biology*, 7(4): 467–478

Saeidnejad, A.H., Kafi, M., Khazaei, H.R. and Pessarakli, M. 2016. Combined effects of drought and UV stress on quantitative and qualitative properties of *Bunium persicum*. *Journal of Essential Oil Bearing Plants*, 19(7): 1729–1739

Sancar, A. 1994. Structure and function of DNA photolyase. *Biochemistry*, 33(1): 2–9

Sävenstrand, H., Brosché, M. and Strid, Å. 2004. Ultraviolet-B signalling: *Arabidopsis* brassinosteroid mutants are defective in UV-B regulated defence gene expression. *Plant Physiology and Biochemistry*, 42(9): 687–694

Schaefer, H.M. and Rolshausen, G. 2006. Plants on red alert: Do insects pay attention? *BioEssays*, 28(1): 65–71

Schnable, P.S., Ware, D., Fulton, R.S., Stein, J.C., Wei, F., Pasternak, S., Liang, C., Zhang, J., Fulton, L., Graves, T.A. and Minx, P. 2009. The B73 maize genome: Complexity, diversity, and dynamics. *Science*, 326(5956): 1112–1115

Schultze, M. and Kondorosi, A. 1998. Regulation of symbiotic root nodule development. *Annual Review of Genetics*, 32(1): 33–57

Sharma, S. and Guruprasad, K.N. 2012. Enhancement of root growth and nitrogen fixation in *Trigonella* by UV-exclusion from solar radiation. *Plant Physiology and Biochemistry*, 61: 97–102

Sieh, D., Watanabe, M., Devers, E.A., Brueckner, F., Hoefgen, R. and Krajinski, F. 2013. The arbuscular mycorrhizal symbiosis influences sulfur starvation responses of *Medicago truncatula*. *New Phytologist*, 197(2): 606–616

Simmonds, M.S. 2003. Flavonoid–insect interactions: Recent advances in our knowledge. *Phytochemistry*, 64(1): 21–30

Singh, S., Kumari, R., Agrawal, M. and Agrawal, S.B. 2011. Modification in growth, biomass and yield of radish under supplemental UV-B at different NPK levels. *Ecotoxicology and Environmental Safety*, 74(4): 897–903

Smith, R.C., Prezelin, B.B., Baker, K.E.A., Bidigare, R.R., Boucher, N.P., Coley, T., Karentz, D., MacIntyre, S., Matlick, H.A. and Menzies, D. 1992. Ozone depletion: Ultraviolet radiation and phytoplankton biology in Antarctic waters. *Science*, 255(5047): 952–959

Stafford, H.A. 1991. Flavonoid evolution: An enzymic approach. *Plant Physiology*, 96(3): 680–685

Stapleton, A.E. and Walbot, V. 1994. Flavonoids can protect maize DNA from the induction of ultraviolet radiation damage. *Plant Physiology*, 105(3): 881–889

Steinbrecht, W., U. Köhler, H. Claude, M. Weber, J. P. Burrows, and van der A. 2011. Very high ozone columns at northern mid-latitudes in 2010. *Geophysical Research Letters*, 38: L06803

Stracke, R., Favory, J.J., Gruber, H., Bartelniewoehner, L., Bartels, S., Binkert, M., Funk, M., Weisshaar, B. and Ulm, R. 2010. The *Arabidopsis* bZIP transcription factor HY5 regulates expression of the PFG1/MYB12 gene in response to light and ultraviolet-B radiation. *Plant, Cell & Environment*, 33(1): 88–103

Strømme, C.B., Julkunen-Tiitto, R., Krishna, U., Lavola, A., Olsen, J.E. and Nybakken, L. 2015. UV-B and temperature enhancement affect spring and autumn phenology in *Populus tremula*. *Plant, Cell & Environment*, 38(5): 867–877

Strømme, C.B., Julkunen-Tiitto, R., Olsen, J.E. and Nybakken, L. 2018a. The dioecious *Populus tremula* displays interactive effects of temperature and ultraviolet-B along a natural gradient. *Environmental and Experimental Botany*, 146: 13–26

Strømme, C., Sivadasan, U., Nissinen, K., Lavola, A., Randriamanana, T., Julkunen-Tiitto, R. and Nybakken, L. 2018b. Interannual variation in UV-B and temperature effects on bud phenology and growth in *Populus tremula*. *Plant Physiology and Biochemistry*

Sulaiman, C.T. and Balachandran, I. 2012. Total phenolics and total flavonoids in selected Indian medicinal plants. *Indian Journal of Pharmaceutical Sciences*, 74(3): 258–260

Suthaparan, A., Solhaug, K.A., Stensvand, A. and Gislerød, H.R. 2017. Daily light integral and day light quality: Potentials and pitfalls of nighttime UV treatments on cucumber powdery mildew. *Journal of Photochemistry and Photobiology B: Biology*, 175: 141–148

Suthaparan, A., Stensvand, A., Solhaug, K.A., Torre, S., Mortensen, L.M., Gadoury, D.M., Seem, R.C. and Gislerød, H.R. 2012. Suppression of powdery mildew (*Podosphaera pannosa*) in greenhouse roses by brief exposure to supplemental UV-B radiation. *Plant Disease*, 96(11): 1653–1660

Suthaparan, A., Stensvand, A., Solhaug, K.A., Torre, S., Telfer, K.H., Ruud, A.K., Mortensen, L.M., Gadoury, D.M., Seem, R.C. and Gislerød, H.R. 2014. Suppression of cucumber powdery mildew by supplemental UV-B radiation in greenhouses can be augmented or reduced by background radiation quality. *Plant Disease*, 98(10): 1349–1357

Tepfer, D. and Leach, S. 2017. Survival and DNA damage in plant seeds exposed for 558 and 682 Days outside the international space station. *Astrobiology*, 17(3): 205–215

Trichopoulou, A. and Critselis, E. 2004. Mediterranean diet and longevity. *European Journal of Cancer Prevention*, 13(5): 453–456

Tsormpatsidis, E., Henbest, R.G.C., Battey, N.H. and Hadley, P. 2010. The influence of ultraviolet radiation on growth, photosynthesis and phenolic levels of green and red lettuce: Potential for exploiting effects of ultraviolet radiation in a production system. *Annals of Applied Biology*, 156(3): 357–366

Vanhaelewyn, L., Prinsen, E., Van Der Straeten, D. and Vandenbussche, F. 2016. Hormone-controlled UV-B responses in plants. *Journal of Experimental Botany*, 67(15): 4469–4482

Villar-Argaiz, M., Medina-Sánchez, J.M., Bullejos, F.J., Delgado-Molina, J.A., Ruíz Pérez, O., Navarro, J. C. and Carrillo, P. 2009. UV radiation and phosphorus interact to influence the biochemical composition of phytoplankton. *Freshwater Biology*, 54(6): 1233–1245

Vincent, W.F., Kumagai, M., Belzile, C., Ishikawa, K. and Hayakawa, K. 2001. Effects of seston on ultraviolet attenuation in Lake Biwa. *Limnology*, 2(3): 179–184

Walbot, V. 1991. The mutator transposable element family of maize. In: Setlow J.K. (eds) *Genetic Engineering. Genetic Engineering* (Principles and Methods), Vol. 13. Springer, Boston, MA

Walbot, V. 1992. Reactivation of Mutator transposable elements of maize by ultraviolet light. *Molecular and General Genetics MGG*, 234(3): 353–360

Walbot, V. 1999. UV-B damage amplified by transposons in maize. *Nature*, 397(6718): 398–399

Wargent, J.J., Elfadly, E.M., Moore, J.P. and Paul, N.D. 2011. Increased exposure to UV-B radiation during early development leads to enhanced photoprotection and improved long-term performance in *Lactuca sativa*. *Plant, Cell & Environment*, 34(8): 1401–1413

Wargent, J.J., Gegas, V.C., Jenkins, G.I., Doonan, J.H. and Paul, N.D. 2009. UVR8 in *Arabidopsis thaliana* regulates multiple aspects of cellular differentiation during leaf development in response to ultraviolet B radiation. *New Phytologist*, 183(2): 315–326

Wargent, J.J., Taylor, A. and Paul, N.D. 2006. UV supplementation for growth regulation and disease control. *In V International Symposium on Artificial Lighting in Horticulture*, 711: 333–338

Wei, Z. and Li, J. 2016. Brassinosteroids regulate root growth, development, and symbiosis. *Molecular Plant*, 9(1): 86–100

Wenny, B.N., Saxena, V.K. and Frederick, J.E. 2001. Aerosol optical depth measurements and their impact on surface levels of ultraviolet-B radiation. *Journal of Geophysical Research: Atmospheres*, 106(D15): 17311–17319

Wessler, S.R. 1996. Plant retrotransposons: Turned on by stress. *Current Biology*, 6(8): 959–961

Wilbur, H., Alba, C., Norton, A. and Hufbauer, R. 2013. The effect of insect herbivory on the growth and fitness of introduced *Verbascum thapsus* L. *Neobiota*, 19: 21

Winkel-Shirley, B. 2002. Biosynthesis of flavonoids and effects of stress. *Current Opinion in Plant Biology*, 5(3): 218–223

WMO (World Meteorological Organization) Assessment for Decision-Makers: Scientific Assessment of Ozone Depletion: 2014, 88 pp., Global Ozone Research and Monitoring Project-Report No. 56, Geneva, Switzerland.

WMO (World Meteorological Organization), Executive Summary: Scientific Assessment of Ozone Depletion: 2018, 67 pp., World Meteorological Organization, Global Ozone Research and Monitoring Project – Report number 58, Geneva, Switzerland, 2018

Wrzaczek, M., Vainonen, J.P., Gauthier, A., Overmyer, K. and Kangasjärvi, J. 2011. Reactive oxygen in abiotic stress perception-from genes to proteins. In: *Abiotic Stress Response in Plants-Physiological, Biochemical and Genetic Perspectives*. InTech. Doi

Yang, W. 2011. Surviving the sun: Repair and bypass of DNA UV lesions. *Protein Science*, 20(11): 1781–1789

Yang, Y., Yao, Y., Xu, G. and Li, C. 2005. Growth and physiological responses to drought and elevated ultraviolet-B in two contrasting populations of *Hippophae rhamnoides*. *Physiologia Plantarum*, 124(4): 431–440

Zhao, D., Ferguson, A.A. and Jiang, N. 2016. What makes up plant genomes: The vanishing line between transposable elements and genes. *Biochimica et Biophysica Acta (BBA)-Gene Regulatory Mechanisms*, 1859(2): 366–380

Section IV

Plants and Crops Responses under
Pollution and Heavy Metal Stresses

24 Plant Heavy Metal Interactions and Pollution Stress

Mojtaba Kordrostami, Ali Akbar Ebadi, Babak Rabiei, and Mohammad Mafakheri

CONTENTS

24.1 INTRODUCTION

Generally, pollution refers to every possible factor that is capable of disturbing natural conditions and generating detrimental impacts on the environment (Argüello, Raimunda and González-Guerrero 2012, Krishna et al. 2017). In other words, undesirable changes in the environment that disrupt substantial biological processes and threaten vital resources and organisms are so-called *pollution* (Martinez 2009). Pollution in natural ecosystems is contamination in a place where it naturally should not be in the first place or has reached a concentration higher than the natural level and negatively affects organisms (Rai 2015).

24.2 POLLUTANTS

There are different views on pollution and pollutants. In one view, pollutants are categorized into two groups: pointwise (in a specific area, usually with high concentrations of pollutants) and non-pointwise (with a lower concentration and wider dispersion) (Slater 2001). Currently, non-pointwise sources of pollutants are considered to be the main and highly adverse contaminating agents of water and soil on a global scale, and agriculture and transportation have the largest share in the

generation of these pollutants (Edwards 2002). In another view, pollutants are defined as two groups: mineral and organic pollutants (Dixit et al. 2015):

1. Mineral pollutants: metals (Fe, Pb, Cd, Cr, Ni, etc.), pseudo-metals (Se, As, etc.), nutrient elements (N, S, K, P, etc.), and radioactive particles (U, Cs, etc.).
2. Organic pollutants: organic compounds are a major source of contaminants for soil and water. They include polychlorinated biphenyls, petroleum hydrocarbons, polycyclic hydrocarbons, organic pesticides, etc.

A group of mineral pollutants is composed of heavy metals and trace elements (Tchounwou et al. 2012). Trace elements are those that have a low concentration in soil and plants (Tyler 2004). These elements may or may not be necessary for the growth and development of plants (Zhang et al. 2013).

The term *heavy metal* in various scientific fields has definite definitions. For example, in petrology, this term refers to metals that react with dithizone (Duffus 2002). But, in most common applications, the term refers to metals whose specific gravity exceeds 4.5 g/cm^3 (sometimes, it is defined as 4.5 or 6 g/cm^3) (Tchounwou et al. 2012).

The sources of water and air pollution are as follows (Kjellstrom et al. 2006):

1. Industrial pollution
2. Wastewaters and garbage
3. Contamination through agricultural waste and the use of chemical inputs

Soil contamination is very important because of its direct impact on agricultural production and nutrition (Blum 2013). Most pollutants are found in industrial areas around big cities and mines (Goel 2006). One of the main problems in agriculture in these areas is soil contamination with heavy metals (Saha et al. 2017). Heavy metals are first transferred to the plants and then, enter the food chain. They are also significantly stable and non-degradable (Jan et al. 2015). Although many of them are required in small amounts in biological cycles, they are often toxic in high concentrations (Abollino et al. 2003). Unlike other pollutants, it is very difficult to dispose of these metals from the environment, because they are not chemically or biologically degradable and destructible, but they can be oxidized, reduced, or complexed by organic materials (Ali, Khan and Sajad 2013).

24.3   DEFINITION OF HEAVY METALS

Metal elements with a specific gravity of more than 5 g/cm^3 are referred to as *heavy metals* (Jaishankar et al. 2014). These metals exist in nature as cations and oxidized anions. Elements such as nickel, chromium, cobalt, mercury, zinc, cadmium, copper, and manganese are cationic in the soil, while elements such as molybdenum, selenium, arsenic, and boron are combined with oxygen in the soil and have a negative charge (White and Brown 2010, Maret 2016).

Heavy metals are also defined as elements with particular properties (flexibility, conductivity, and stability, such as cations, specific ligands, etc.) and an atomic number greater than 20 (Marques, Rangel and Castro 2009, Jin et al. 2018). Generally, heavy metals are divided into two distinct groups based on their Lewis acidity and their tendency toward different ligands (Duffus 2002, Jan et al. 2015). The first group consists of soft metal ions, such as mercury, cadmium, silver, copper, and platinum, which are preferably bonded to polar ligands by means of covalent bonds (Hall 2002, Appenroth 2010). The second group includes metal ions such as iron, zinc, nickel, cobalt, and lead, which are more likely to be linked to intermediate ligands such as amines, amides, and imines (Diomides 2005, Oves et al. 2016). All metal ions are toxic to plants at concentrations above their critical levels in the soil, regardless of which groups they belong to (Chibuike and Obiora 2014, Xie et al. 2016).

Nowadays, in most industrialized countries, a specific limitation on the concentration of heavy metals has been determined (Singh et al. 2011, Tchounwou et al. 2012). However, due to the fact that the concentration of these elements varies from one country to another, and the range between the minimum and maximum concentrations sometimes reaches

100-fold, this problem cannot be generalized (Olaniran, Balgobind and Pillay 2013, Ayangbenro and Babalola 2017).

24.4   HEAVY METAL SOURCES

These elements enter the soil in two ways (Tchounwou et al. 2012, Oves et al. 2016):

a. Weathering of minerals
b. Human activities

Pollutants caused by human activities include (Duruibe, Ogwuegbu and Egwurugwu 2007)

a. Industrial activities
b. Atmospheric sediments
c. Agricultural activities, such as applying pesticides containing heavy metals, fertilizers, and soil amendments and insecticides

24.5   TOXICITY OF HEAVY METALS IN SOIL

Many heavy metals, such as iron, zinc, and copper, have physiological functions in plants and animals, but they are toxic at high concentrations. Some other metals, such as cadmium and lead, are poisonous even at low concentrations (Tangahu et al. 2011, Jaishankar et al. 2014).

It is necessary to point out that the concentration of an element in the soil may be higher than its toxic level, but its toxicity does not appear due to the absence of plant absorption (Rout and Sahoo 2015). Heavy metals in very low concentrations have been proved to have extraordinary effects.

Their consistency and stability, as well as natural accumulation of heavy metals in the body of organisms and transfer to the other levels of the food chain, have led scientists to investigate various patterns of aggregation and diffusion of these heavy metals in edible and non-edible plant tissues from different perspectives, particularly in consumer health, which is very important (Khan et al. 2015, ul Islam et al. 2007). The mechanism of the heavy metals' effects from a biochemical viewpoint has emerged from the extraordinary tendency of their cations to interact with sulfur. Heavy metal cations contained in organic compounds can get into the body through eating and readily attach to the widespread hydrosulfide groups (-SH) in the human body and other organisms (Catenacci 2014, Jan et al. 2015). The sulfur–metal bonds commonly affect enzymes that control the fundamental metabolic and catabolic reactions. Therefore, these enzymes cannot carry out their common duties, which results in threatening the health of organisms and sometimes causing their death (Rahal et al. 2014).

The characteristics of heavy metal toxicity include the following (Ercal, Gurer-Orhan and Aykin-Burns 2001, Tchounwou et al. 2012): a) long-term toxicity effects on soil; b) some of the heavy metals in the environment may change from a low-toxicity compound to another one with an augmented toxic effect, of which mercury is a superb instance; c. natural accumulation and the increased bioavailability of heavy metals in food chains can harm very basic physical activities and hence, jeopardize human life; d) heavy metals can only be substituted in their chemical capacity, but not reduced or decreased, even by biological processes (Barakat 2011, Jaishankar et al. 2014, Jan et al. 2015).

Sources of heavy metals include soil materials, chemical fertilizers, sewage sludge, irrigation water, coal-burning residues, the metal smelting industry, materials released from automobiles, etc. (Chopra, Pathak and Prasad 2009, Wuana and Okieimen 2011). These metals are continuously introduced into the biosphere by volcanoes, natural weathering of rocks, and human activities such as combustion of fossil fuels and sewage (Violante et al. 2010).

Among the heavy metal elements, cadmium, chromium, mercury, manganese, lead, bismuth, tungsten, and zinc can be mentioned as often problematic elements (Bhat and Khan 2011, Singh et al. 2011). Recently, high concentrations of heavy metals such as arsenic, cadmium, copper, iron, zinc, and lead have been reported in soils of various countries (Singh and Prasad 2015). The metal elements are absorbed only through the roots of plants in *a* and *b* forms (Nouri et al. 2009).

24.6 PLANT–METAL INTERACTIONS

Overall, among the elements that exist in a plant, only 1% or fewer are heavy metals. Some of these metals, such as iron, zinc, copper, manganese, and nickel, are required for plant growth and metabolism, while so far, no biological role for other metals, such as cadmium, lead, arsenic, and mercury, has been identified (Nagajyoti, Lee and Sreekanth 2010, Tangahu et al. 2011). Based on the roles of metals in plants, they can be grouped into one or more of the following groups (Emamverdian et al. 2015, Singh et al. 2016):

1. Metals that enter the composition and structure of the plant, such as calcium in calcium pectates of the middle lamella
2. Metals that are required for enzyme activity, such as zinc for the activity of the aldolase enzymes
3. Metals that are a major component of certain enzymes and proteins, such as copper in cytochrome oxidase and zinc in alcohol dehydrogenase

While some heavy metals are essential for life, it is common that their accumulation has a poisonous effect on living organisms. Increased concentrations of essential and unnecessary heavy metals in the soil usually result in poisoning symptoms and halting further plant growth. Toxicity symptoms appear to be due to a range of interactions at the cellular/molecular level (Kim, Kim and Seo 2015). In general, the toxic effects of heavy metals on plants can be the results of the following reasons (Hall 2002, Singh et al. 2011):

a. Heavy metals such as cadmium, lead, copper, silver, and gold cause changes in the permeability of the cell membrane, which results in ion absorption and hemostasis.

b. Certain heavy metal cations, such as lead, silver, and zinc, form bonds with protein sulfhydryl groups (including enzymes) that damage their structure and function.

c. Some heavy metals compete with essential ions because of their similarity to them, especially to the cations, to occupy their binding sites, thus reducing the absorption of essential cations.

d. Several heavy metals, including molybdenum, selenium, arsenic, and boron, which are found in soil as anions, occupy essential group positions such as phosphate and nitrate groups and, thus, damage the plant.

e. Reactions with metals, for example, zinc and boron, cause plant disturbance through a link with phosphate bonds and active adenosine triphosphate (ATP) and adenosine diphosphate (ADP) groups.

Their capability to bond tightly with oxygen, nitrogen, and sulfur atoms is an important reason for their toxicity. These atoms are particularly prevalent in the structure of proteins, and the effect of these metals on the protein structures, especially enzymes, is generally high (Osredkar and Sustar 2011, Jan et al. 2015). In addition, high concentrations of heavy metals stimulate the formation of free radicals and ROS (Abdal Dayem et al. 2017).

The effects of heavy metal toxicity, in general, include prevention of seed germination, pollen germination, and growth of the pollen tube as well as plant growth and development. They can cause cytogenetic abnormalities in plants; destroy many biochemical and physiological processes, including damage to cell membranes, decreasing cell respiration, protein degradation, damage to the photosynthetic apparatus, and decreasing photosynthesis rate; affect the functionality of some enzymes; and cause lipid peroxidation (Nagajyoti, Lee and Sreekanth 2010, Emamverdian et al. 2015).

24.6.1 FACTORS AFFECTING HEAVY METAL ABSORPTION BY PLANTS

Heavy metals in contaminated soils are mainly introduced into the plant through root absorption. Many factors have been reported to be involved in the root absorption of these elements, which include (Tangahu et al. 2011, Wuana and Okieimen 2011)

a. Physiochemical properties of the soil such as pH, temperature, humidity, oxidation-reduction potential, cation exchange capacity, phosphate content, organic matter, and clay content of the soil

b. The type of elements and their amounts in the soil as well as their availability to plants

c. Plant type, size, and root status

d. Competition between this element and other elements

An increase in factors such as clay, organic matter, iron oxide, manganese oxide, and ion exchange capacity (CEC) reduces plants' access to heavy metals. Also, the increase of cations in the soil reduces the absorption of heavy metals as a result of their competition in binding to the cell surfaces, such as membranes, as well as competition in the transmission mechanisms (Mishra and Dubey 2006, Jan et al. 2015).

It has been generally stated that at neutral pH, plants have maximum access to essential elements, and the toxic effects of metals are minimal. Also, heavy metal cations have the highest mobility under acidic conditions. Improving the condition of soil ventilation (or, in other words, increasing the oxidation state of the environment) increases the absorption of the elements by the plant (Pezeshki and DeLaune 2012).

24.6.2　Absorption of Metals by Roots and Their Transfer to the Shoots

The entry of heavy metals into the plant is usually carried out by the roots from the soil or maybe, through the leaves. The effective factors in the absorption of heavy metals by the roots are the growth conditions, plant type, and amount and solubility of the elements in the soil (Shah et al. 2010, Li, Yu and Luan 2015).

Only a fraction of the total amount of ions bound to the root is absorbed into the cells, since these ions are absorbed physically in the negatively charged extracellular regions (COO^-) of the root (Lasat 1999). It should be noted that these particles bound to the cell wall are not transmitted to the aerial parts of the plant; therefore, the plant mainly presents significant accumulation of the metals in the root. For example, many plants accumulate lead in the roots, but the transfer of lead to the aerial parts is infinitesimal. In support of this theory, Pourrut et al. (2011) pointed out that the limiting step in the phytoextraction of lead is the long transition distance from the roots to the aerial parts of the plant.

Due to being charged, metal ions are not able to move freely through cell membranes with lipophilic structure. Thus, the transfer of ions into cells is carried out by membrane proteins called *transporters*. These transporters have a transmembrane structure and also have an extracellular binding domain for ion binding prior to transfer. The binding domain is ion specific, and the transmembrane structure provides the ability to transfer bonded ions from the extracellular space of the membrane's hydrophobic environment to the cell (Argüello, Raimunda and González-Guerrero 2012). The absorption of metals into the cells of the root is the entry point of metals into the living tissues. The metals are transferred to the plant's organs through the vessels after entering the plant, initially controlled by two processes: root pressure and leaf transpiration (Conn and Gilliham 2010). For transfer of the elements to the aerial parts, metal ions must first be passed through a Casparian strip, which is a limiting step in the transfer of ions to the shoots (Chen et al. 2011). The cell wall of xylem also has a high CEC, which slows down the movement of metal cations. Nonetheless, chelating compounds such as citrate, malate, phytochelatins, and metallothionein facilitate the movement of certain metals in the transpiration pathway. Therefore, the mobility of metal elements in the soil and in the plant is different; in other words, they have different transmission coefficients (Mirecki et al. 2015). The transfer coefficient

of each metal can be calculated according to the amount of metal in the aerial parts of the plant in relation to the total amount of metal in the soil. Considering this case, cadmium, titanium, and zinc have the highest transfer coefficients in the plant, while this ratio for lead is about 100 to 1000 times lower (Tuna et al. 2002).

24.7　MECHANISMS OF HEAVY METAL RESISTANCE AND DETOXIFICATION

In general, ion homeostasis is the basis of heavy metal resistance (Meharg, Cumbes and Macnair 1993). Plants that are not aggregating metals have general mechanisms for controlling the homeostasis of metal ions within the cell (Emamverdian et al. 2015). The adjustment of ion entry by stimulating the transporters at low metal concentrations, their inhibition in the presence of high amounts of metal, and secretion of ions from the cell to the external solution are a few of these mechanisms (Emamverdian et al. 2015, Singh et al. 2016). However, ultra-aggregating species, which accumulate metals at very high levels, have a range of mechanisms at the cellular and intracellular levels that are involved in toxicity and thus, tolerance to heavy metal stress. These mechanisms include the key roles of mycorrhizal fungi (due to limited metal uptake by the root); reduced absorption and accumulation of heavy metals by the cell wall and plasma membrane; extracellular secretions; cellular binding of proteins such as metallothioneins and phytochelatins, which is caused by low–molecular weight ligands such as organic acids, amino acids, and peptides; induction of thermal shock proteins; and metallic compartmentation in the vacuole (Lasat 1999).

24.7.1　Transfer and Accumulation of Metals in the Vacuole

The structural properties of the cell wall and its role as a metal tolerance mechanism have been controversial. Cell walls are in direct contact with the metal ions in the soil solution. Therefore, ions must pass through the cell wall before being absorbed by the cell membrane. Some compounds of the cell wall with negative charges are capable of binding with metal ions, thereby controlling the metal entry into the symplast (Pourrut et al. 2011). On the other hand, the gradual accumulation of heavy metals in the cell wall rather than in cytosol reduces the toxic effects of metal on the active metabolic sites (cytosol). In fact, the first (binding of metal ions to the cell wall constituents) is an avoidant mechanism, and the second (accumulation of metals in the cell wall) is a tolerance mechanism. The most important compounds of the cell wall that bind to the heavy metals are polygalacturonic acid, proteins, and silica substances (Vandenbossche et al. 2015).

24.7.2　Amino Acids

Some of the amino acids and their derivatives are among the most important chelates or ligands of heavy metals, and they are critical to heavy metal tolerance (Flora and Pachauri 2010).

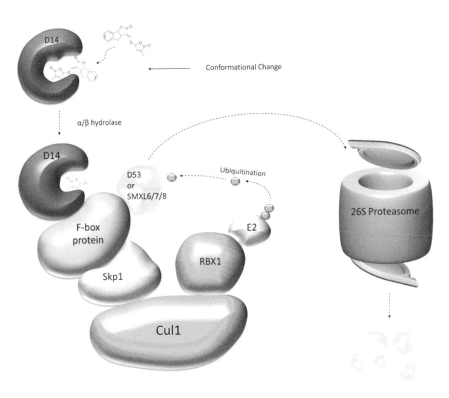

FIGURE 11.3 SL's signal transduction pathway. The picture shows D14, a non-canonical hormone receptor for SL, and F-box protein as a part of SCF-ubiquitin ligase protein complex (Skp1-Cullin-F-box) which results in ubiquitination and subsequent degradation of SMXL6–8 by the 26S proteasome.

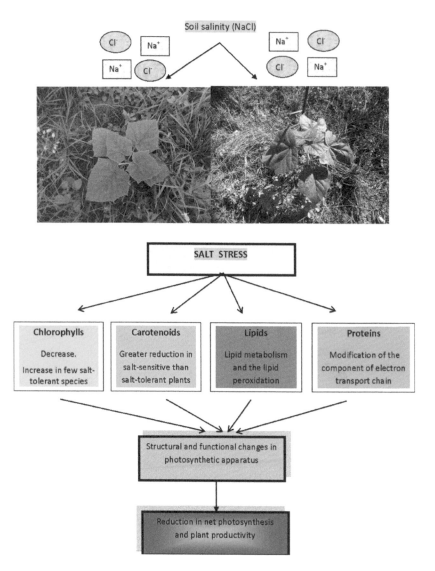

FIGURE 13.1 Effect of salt stress on the photosynthetic apparatus.

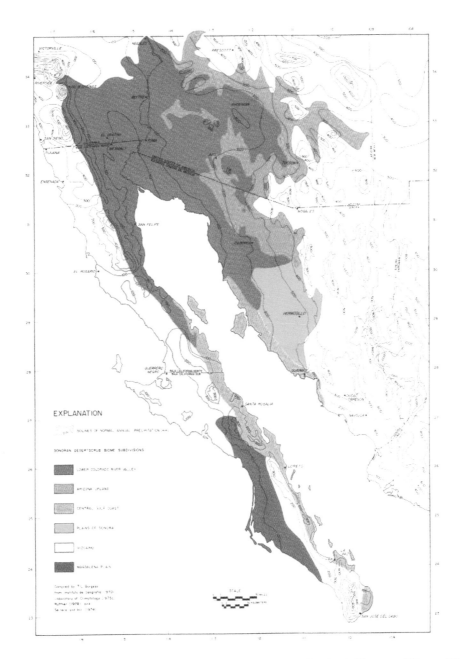

FIGURE 14.7 Map of the Sonoran Desert with isolines of normal annual precipitation (mm) (Turner and Brown 1982).

Courtesy the Boyce Thompson Arboretum.

FIGURE 14.10 Curve-billed thrasher (*Toxostoma curvirostre*) consuming seeds and pulp of the fruit of saguaro (*C. gigantea*). (From T W Crawford, Jr.)

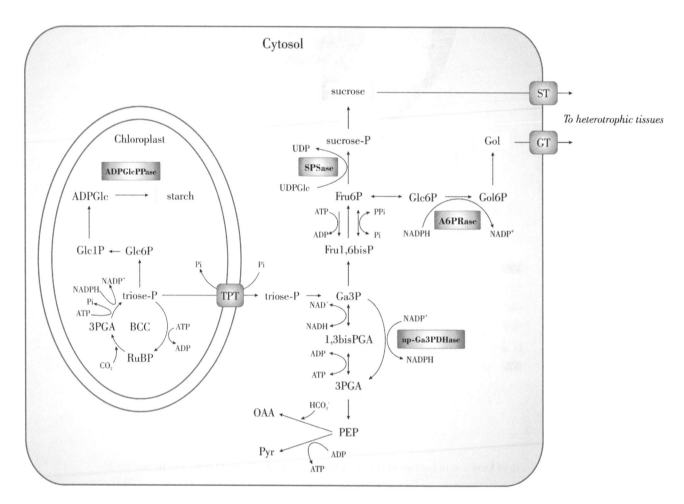

FIGURE 21.1 Schematic representation of carbon metabolism in plant photosynthetic cells. Key enzymes for carbon partitioning are colored in violet. The major photosynthetic products (starch, sucrose, and glucitol) are depicted in green. Transporters are highlighted in red (TPT: triose-P translocator; ST: sucrose transporter; GT: glucitol transporter). BCC is the Benson-Calvin cycle.

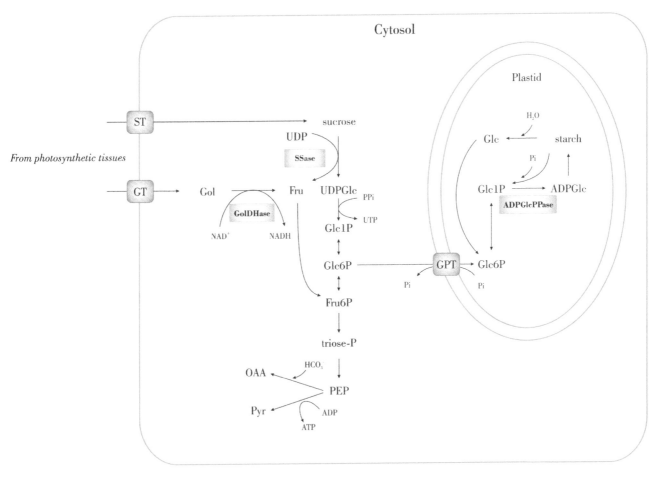

FIGURE 21.2 Schematic representation of carbon metabolism in plant heterotrophic cells. The enzymes involved in sucrose, starch, and glucitol metabolism are colored in violet. Photosynthates imported from leaves (sucrose and glucitol) are depicted in green. Starch, the major storage compound, is colored in cyan. The glucose-P translocator (GPT) is highlighted in red.

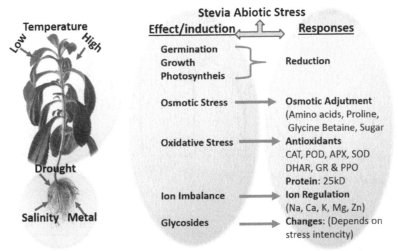

Effect and response of stevia plant to various components of abiotic stress

FIGURE 33.1 Abiotic stress tolerance of stevia.

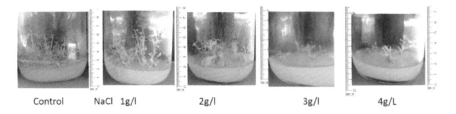

Control   NaCl 1g/l   2g/l   3g/l   4g/L

FIGURE 33.2 NaCl treatment on *in vitro* plants of stevia (unpublished data).

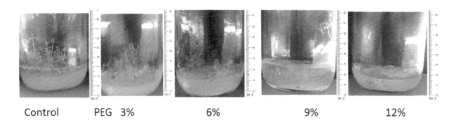

Control   PEG 3%   6%   9%   12%

FIGURE 33.3 PEG treatment on *in vitro* plants of stevia (unpublished data).

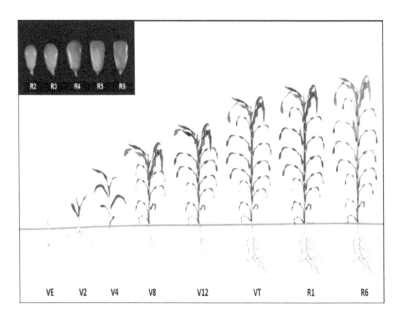

FIGURE 37.1 Vegetative and reproductive growth stages of a corn plant. Vegetative growth stages (VE-VT): VE, seed emergence; tassel or anthesis (VT); reproductive stages (R1–R6): silking, R1; blister, R2; milk, R3; dough, R4; dent, R5; physiological maturity or black layer, R6.

FIGURE 37.3 Vegetative growth of corn at the University of Wyoming Powell Research and Extension Center. Top: under subsurface drip irrigation; bottom: under on-surface drip irrigation. Plants clearly show more vigorous growth under subsurface than on-surface drip irrigation.

FIGURE 40.1 A declined, unproductive citrus tree (a 20-year-old 'Hamlin' orange tree on Swingle Citrumelo rootstock (*Citrus paradisi* Macfad. × *Poncirus trifoliata* (L.) Raf.) showing its fruits at harvest time in 2003 in Mollisol in a low-lying, poorly-drained citrus orchard in Osceola County, central Florida.

FIGURE 40.2 The layout of the declined citrus trees resulting from water stress and pest infestation in Mollisol in a low-lying, poorly-drained orchard in Osceola County, central Florida. The trees were 20-year-old 'Hamlin' orange trees on Swingle Citrumelo rootstock and under the declined tree canopy was a pyramidal Tedders trap for the root weevil monitoring in summer 2013.

FIGURE 40.3 Testing the flooding effects on the physiological development of citrus rootstock seedlings of different varieties ('Swingle Citrumelo' and 'Carrizo') in the greenhouse.

(A)   (B)

FIGURE 40.4   Comparison of the plant roots and canopies of the citrus rootstock (Swingle Citrumelo) seedlings grown under non-flooding (A) and 30-day flooding (B) conditions.

FIGURE 40.8   Citrus trees ('Hamlin' orange trees on Swingle Citrumelo rootstock) under flooding occurred after a heavy rain in the fall (2003). Flooded anoxic soil had a negative soil redox potential E_h that caused tree decline.

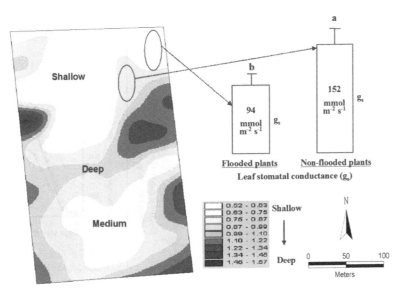

FIGURE 40.9   Comparison of citrus tree water stress status with leaf stomatal conductance (g_s) measurements in a wetter (shallow water table 0.52–0.63 m, g_s 94 mmol m^{-1}s^{-1}) area and in a medium water table area (medium water table 0.75–1.00 m, g_s 152 mmol m^{-1}s^{-1}). Citrus trees in the wetter area were more water stressed than in the drier area in Osceola orchard in Florida. (From Li et al., 2004.)

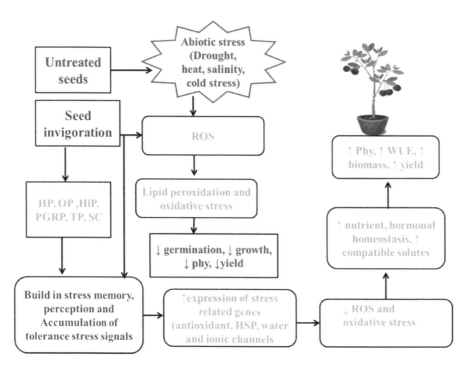

FIGURE 41.2 Mechanism of seed invigoration induced stress tolerance. ↑ = increase, ↓ = decrease, HP = hydropriming, OP = osmopriming, HlP = halopriming, PGRP = plant growth regulator priming, TP = thermal priming, SC = seed coating, ROS = reactive oxygen species, LP = lipid peroxidation, CS = compatible solutes, Phy = photosynthesis, WUE = water use efficiency.

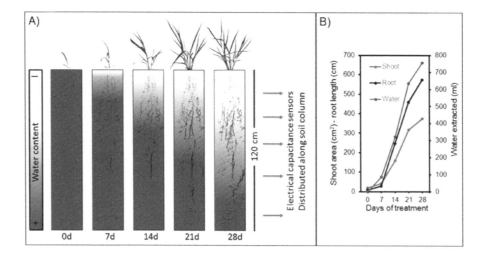

FIGURE 42.1 Representation of the method to record responses of tropical forages grasses under drought conditions. (A) The method allows simultaneous recording of shoot and root growth and water uptake over a period of time. The electrical capacitance sensors are distributed down the soil profile to estimate water content at different soil profiles, while a weighing scale records the whole weight of the soil column to calculate gravimetric soil water content. (B) Actual data from 1A.

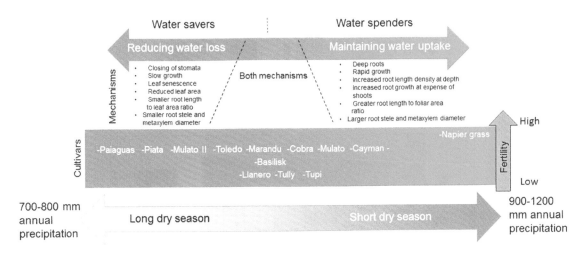

FIGURE 42.2 Different strategies of water use and associated mechanisms among a number of tropical forage grass cultivars that are being used by farmers. A fine interplay exists between the acquisition of water by roots in drying soil and water loss through transpiration. These two components tend to act simultaneously. Nonetheless, a certain strategy of water use tends to supersede over the other one. The mode of water use was evaluated in conjunction with contrasting soil fertility levels for targeting of forage grasses to different patterns of precipitation and soil fertility levels.

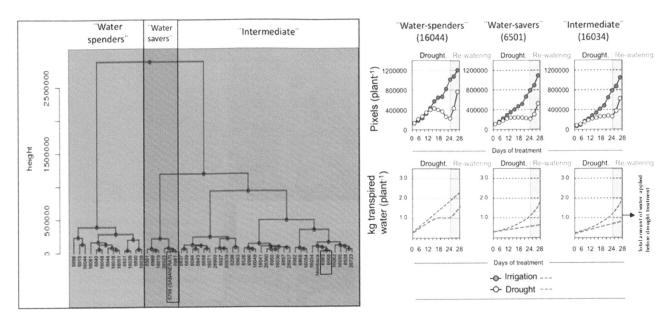

FIGURE 42.3 An example of hierarchical clustering of models of water use in 50 genotypes of *Panicum maximum*. Examples of three different genotypes—with their accession numbers from CIAT Genetic Resources Program—showing differences in their water use strategies are shown on the right-hand side of the image. The responses of these genotypes based on accumulated growth (pixel-based) and transpired water over a drought period and recovery period are also shown.

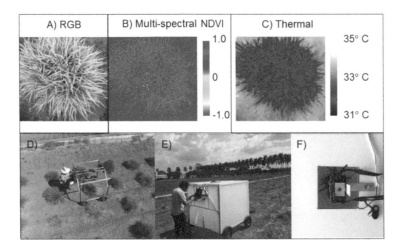

FIGURE 42.4 A hand-propelled vehicle where a number of sensors can be mounted (e.g., cameras). A) an image recorded with an off-the-shelf camera; B) a processed image showing NDVI values; C) a thermal image; D) the propelled vehicle showing just the frame and gathering spectroradiometric data from different genotypes of *Brachiaria*; E) the propelled vehicle covered with a fabric to diffuse light for even illumination and no shadows when capturing images; a laptop computer is connected to the sensors for viewing, triggering and capturing data; F) a top view of the covered vehicle, with an open space to mount an off-the-shelf camera. NDVI, Normalized Difference Vegetation Index; Green healthy plants show values of NDVI close to 1, whereas chlorotic ones and bare soil, show NDVI values close to 0.

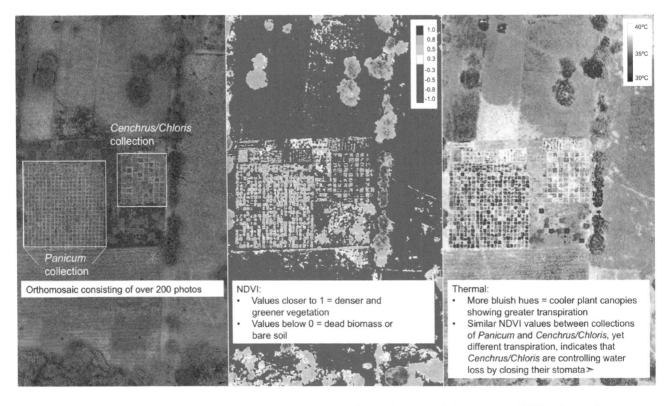

FIGURE 42.5 An example of images collected from an UAV showing from left to right: RGB (normal), NDVI values and canopy temperatures of a collection of *Panicum maximum*, *Cenchrus ciliaris*, and *Chloris gayana* germplasm accessions (a total of 153 germplasm accessions from CIAT's Genetic Resources Program and the International Livestock Research Institute, ILRI). The use of UAVs allows simultaneous assessment of a large number of genotypes.

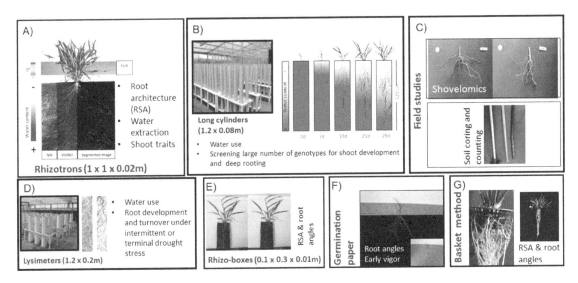

FIGURE 42.6 Methods available at CIAT to study roots. A) Large rhizotrons that are used to estimate soil water content and root distribution and root angles of roots. This is also used to estimate canopy temperature (using a FLIR system). B) system as the one described in Figure 1; C) shovelomics and soil coring; D) large lysimeters to perform long term studies; E, F, and G) methods using rhizoboxes, germination paper, and basket method to study root angles.

FIGURE 44.1 Images of morphological and taxonomic features of nutritionally useful parts of moringa, tamarind, and fenugreek.

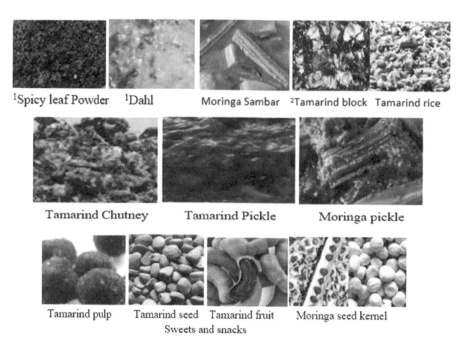

FIGURE 44.2 Images of food or food products made from moringa and tamarind. ([1]Dahls and spicy powders of leaves from all crops look alike; [2]tamarind block used for soups and rice.)

FIGURE 44.3 Images of products used in livestock and other sectors of agriculture.

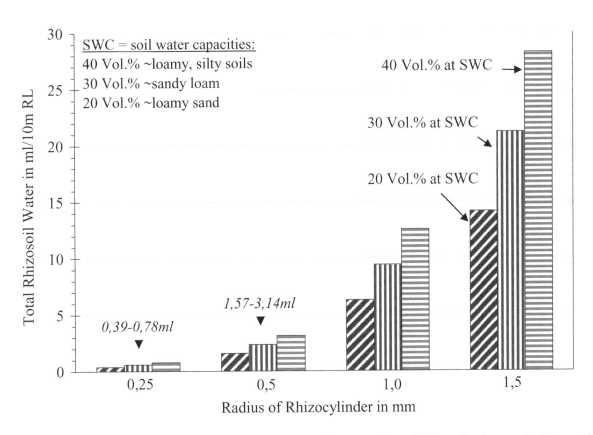

FIGURE 48.4 Total amount of rhizosoil water as related to soil texture and root morphology (SWC = soil water capacity, RL = root length).

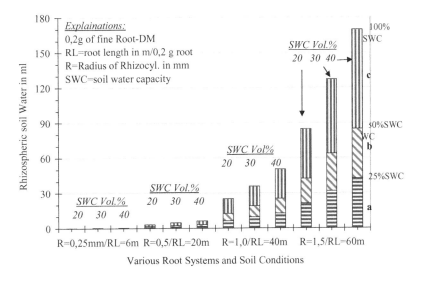

FIGURE 48.5 Rhizospheric soil water fractions (ml) in various soils around roots of different morphology.

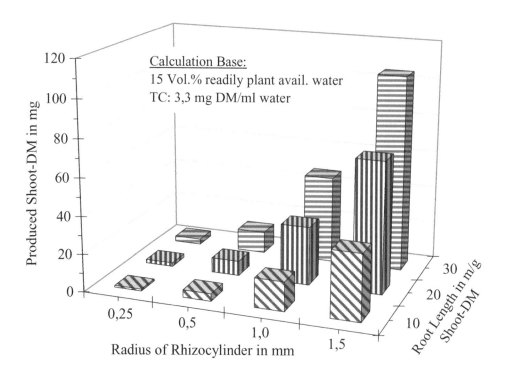

FIGURE 48.6 Shoot-DM produced from the readily plant available rhizospheric soil water fraction of a medium textured soil (15% Vol at 50% SWC) by roots of different morphology and length (transpiration coeff. TC 300ml/g DM shoot).

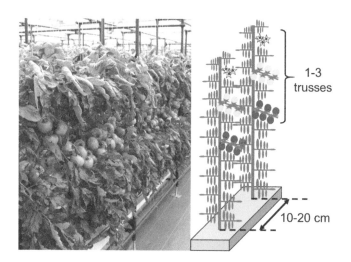

FIGURE 50.1 Low truss with a high-density tomato production system.

FIGURE 50.5 Commercial production system for spinach with low NO_3^- concentration.

The most important amino acid in this regard is histidine, which was observed to increase in some nickel-aggregating and non-aggregating plants. For example, the increase of histidine content in the xylem sap of *Alyssum lesbiacum* was more than 36% (Kerkeb and Krämer 2003). The amount of histidine in non-aggregating species enhances improvement of the resistance level to nickel and ultimately, elevates its transfer to aerial parts (Nasibi et al. 2013). However, it is noteworthy that the increase of histidine is not a general mechanism of resistance to nickel in all plants. For example, no histidine increase was observed in plants that do not accumulate nickel, such as *Thlaspi goesingense* (Persans et al. 1999).

24.7.3 METALLOTHIONEINS (MTs)

Metallothioneins are small, 6 to 7 kDa proteins, which are highly conserved during evolution (Pula et al. 2015). These compounds were first identified as cadmium-binding proteins in mammalian tissues. So far, more than 50 pseudo-metallothionein sequences have been identified in different plants. These peptides are rich in cysteine amino acids and lacking aromatic amino acids, which are bound by cysteine amino acids to cadmium, zinc, and mercury and reduce their toxic effects (Isani and Carpenè 2014). By 1997, the only protein that was introduced as a metallothionein was the EC protein in wheat (Giordani et al. 2000).

24.7.4 ORGANIC ACIDS

Organic acids such as citrate, malate, and oxalate are important ligands for heavy metals in the detoxification of metal ions (Osmolovskaya, Vu and Kuchaeva 2018). They also play a role in processes such as the transport of metals through xylem and accumulation of metals in the vacuole. In all the plants of the Brassicaceae family, malate has a significant role in zinc chelation in leaves (Mourato et al. 2015). Harmens et al. (1994) reported that citrate has a high potential for zinc complexation; therefore, cytosolic citrate is more suitable than malate for transferring zinc to the vacuole. It has been observed that extracellular chelation is important in aluminum resistance mechanisms by organic acids (Wenzl et al. 2001). It is believed that due to the high content of malate and citrate in some non-aggregating plants, these materials play a substantial role in the formation of a complex with metals for their transfer to the xylem.

24.7.5 PLASMA MEMBRANE

The plasma membrane of plants is considered to be the first living structure exposed to heavy metal damage (Yadav et al. 2011). The function and structure of the plasma membrane are rapidly impacted by heavy metals. Plasma membrane function seems to be affected in two ways: first, the plasma membrane is permeable in the presence of high amounts of heavy metals, which results in the excretion of protons and potassium ions; second, the H^+-ATPase activity of the plasma membrane is inhibited in these conditions, which affects the absorption of nutrients (Janicka-Russak et al. 2008).

Therefore, the resistance mechanism is to maintain the strength of the plasma membrane against damage by heavy metals. There is, of course, little evidence of how this process works. One of the important factors in maintaining the strength of the plasma membrane in the presence of heavy metals is the increased repair of the membrane after injury, which is carried out by heat shock proteins or metallothioneins (Emamverdian et al. 2015).

24.7.6 PHYTOCHELATINS (PCs)

Higher plants have two important types of cysteine-rich peptides that are capable of binding with heavy metals: phytochelatins and metallothioneins (Cobbett 2000). Metal chelation in the cytosol by high-affinity ligands is an important mechanism in the detoxification of metals and tolerance to heavy metals (Rauser 1995). Phytochelatins are a family of complex metal peptides that are found in all plants and some microorganisms. Their synthesis in plants is rapidly induced by heavy metals, especially cadmium and arsenic (Liu et al. 2015). In this case, the phytochelatin synthase enzyme is activated by metallic ions, and it produces phytochelatin from glutathione (Sunitha et al. 2013). Phytochelatins, in addition to metallic detoxification (especially for cadmium), play a key role in the homeostasis of heavy metals, iron metabolism, and sulfur, and as an antioxidant in the cell (Yadav 2010).

24.7.7 ROOT DISCHARGE

Certain plant root secretions, such as organic acids and amino acids, are metal-chelating agents that increase the absorption of specific elements (Ahmad 2016). One of the clearest examples is the role of root-end secretions in relation to organic and toxic alcohols. For example, buckwheat, in response to aluminum stress, secretes oxalic acid from the roots and accumulates inorganic aluminum oxalate in the leaves (Zheng et al. 2005). Therefore, detoxification occurs both internally and externally. There is evidence that such secretion occurs in wheat and corn from the roots through anionic aluminum-activated channels.

24.7.8 VACUOLE COMPARTMENTATION

Ion removal from the cell membrane and transfer to vacuoles are ways that reduce the levels of toxic metals in the cytosol and therefore, are important mechanisms for heavy metal resistance in plants (Yadav 2010). For instance, the accumulation of phytochelatin–cadmium complexes in the vacuole occurs through the ABC transporter in the tonoplast, which also plays an important role in the accumulation of other metal ions, such as zinc (Song et al. 2014).

24.8 METHODS OF ELIMINATING POLLUTION

Although soils have various mechanisms, including sedimentation, absorption, and reduction reactions, as a natural capability to reduce the accessibility and mobility of metals, with increasing concentrations of heavy metals, these pollutants can be mobile and thus, lead to developing contamination in agricultural products and groundwater (Shi et al. 2009). Therefore, there are many methods for the remediation of heavy metals in water and soil, including electrochemical methods, reverse osmosis, ion exchange method, plant intermediates, chemical stabilization, soil leaching, oxidation, reduction, adsorption, and biological methods (Gunatilake 2015, Ayangbenro and Babalola 2017).

In most cases, more than one method is needed to optimize the remediation process. Often, these methods are faced with problems such as high cost, being time-consuming, low performance on a large scale, or inefficient performance at high concentrations of pollutants (Rajasulochana and Preethy 2016). Generally, one of the ways of improving the environment and preventing the entry of heavy metals such as cadmium into the human food chain and livestock is to reduce the availability of these metals in soil for plants and micro-organisms (Tchounwou et al. 2012, Ayangbenro and Babalola 2017). There are several ways to remove metal from contaminated soil: soil washing, chemical methods, and biological methods. In chemical stabilization, the mobility of heavy metals is reduced by addition of non-mobile materials (Wu et al. 2010, Wuana and Okieimen 2011). Reducing the mobility of metals in soil using organic and inorganic materials has been attributed to various mechanisms, including adsorption (Adriano 2001), precipitation with sulfides, phosphates, hydroxides, or carbonates, and the formation of insoluble metal complexes (Naidu et al. 1997). Although there are several methods for removing metals, most of the processes mentioned have significant disadvantages, such as high energy requirements, resulting in a costly process, low efficiency, and consequently, low cost-effectiveness of the remediation process (Barakat 2011, Ayangbenro and Babalola 2017). Furthermore, the physical or physicochemical methods used to improve contaminated soils do not only damage the physical building of the soil and stop its biological activity but also cause a secondary contamination in the soil that need to be refined (Singh and Prasad 2015).

The chemical stabilization of heavy metals is superior to other methods due to its low cost and high efficiency. The use of instruments, through absorption mechanisms, acid-free reactions, deposition, oxidation-reduction, complexation, cation exchange, and humidification, can lead to instability and the stabilization of heavy metals in the soil (Cho, Oh and Kim 2005). Choosing the appropriate additive reduces the mobility and the availability of contaminants and prevents leaching and absorption by plants and other living organisms (Pinto et al. 2014). The selection of modifiers depends on the type of pollutant, the soil characteristics, and the expected efficiency to be considered. Modifiers should also be easily accessible, relatively inexpensive, and easy to use and should not cause further damage to the environment.

24.9 CONCLUSIONS

Many heavy metals have physiological functions in plants and animals, but they are toxic at high concentrations. Plants have different mechanisms to cope with heavy metals. There are also many methods for the remediation of heavy metals in water and soil, including electrochemical methods, reverse osmosis, ion exchange method, plant intermediates, chemical stabilization, soil leaching, oxidation, reduction, adsorption, and biological methods. Often, these methods are faced with problems such as high cost, being time-consuming, low performance on a large scale, and inefficient performance at high concentrations of pollutants. The chemical stabilization of heavy metals is superior to other methods due to its low cost and high efficiency. So, the selection of an easy and inexpensive method can help farmers to better control heavy metals on their farms.

REFERENCES

Abdal Dayem, A, MK Hossain, SB Lee, K Kim, SK Saha, G-M Yang, HY Choi, and S-G Cho. 2017. "The role of reactive oxygen species (ROS) in the biological activities of metallic nanoparticles." *International Journal of Molecular Sciences* 18 (1):120.

Abollino, O, M Aceto, M Malandrino, C Sarzanini, and E Mentasti. 2003. "Adsorption of heavy metals on Na-montmorillonite. Effect of pH and organic substances." *Water Research* 37 (7):1619–1627.

Adriano, DC. 2001. "Arsenic." In *Trace Elements in Terrestrial Environments*, 219–261. Springer.

Ahmad, P. 2016. *Plant Metal Interaction: Emerging Remediation Techniques*: Elsevier.

Ali, H, E Khan, and MA Sajad. 2013. "Phytoremediation of heavy metals—concepts and applications." *Chemosphere* 91 (7):869–881.

Appenroth, K-J. 2010. "Definition of "heavy metals" and their role in biological systems." In *Soil Heavy Metals*, 19–29. Springer.

Argüello, JM, D Raimunda, and M González-Guerrero. 2012. "Metal transport across biomembranes: Emerging models for a distinct chemistry." *Journal of Biological Chemistry* 287 (17):13510–13517.

Ayangbenro, AS, and OO Babalola. 2017. "A new strategy for heavy metal polluted environments: A review of microbial biosorbents." *International Journal of Environmental Research and Public Health* 14 (1):94.

Barakat, MA. 2011. "New trends in removing heavy metals from industrial wastewater." *Arabian Journal of Chemistry* 4 (4):361–377.

Bhat, UN, and AB Khan. 2011. "Heavy metals: An ambiguous category of inorganic contaminants, nutrients and toxins." *Research Journal of Environmental Sciences* 5 (8):682–690.

Blum, WEH. 2013. "Soil and land resources for agricultural production: General trends and future scenarios-a worldwide perspective." *International Soil and Water Conservation Research* 1 (3):1–14.

Catenacci, A. 2014. "Heavy metal removal from water: Characterization and applicability of unconventional media." PhD School of Politecnico di Milano.

Chen, T, X Cai, X Wu, I Karahara, L Schreiber, and J Lin. 2011. "Casparian strip development and its potential function in salt tolerance." *Plant Signaling & Behavior* 6 (10):1499–1502.

Chibuike, GU, and SC Obiora. 2014. "Heavy metal polluted soils: Effect on plants and bioremediation methods." *Applied and Environmental Soil Science* 2014.

Cho, H, D Oh, and K Kim. 2005. "A study on removal characteristics of heavy metals from aqueous solution by fly ash." *Journal of Hazardous Materials* 127 (1–3):187–195.

Chopra, AK, C Pathak, and G Prasad. 2009. "Scenario of heavy metal contamination in agricultural soil and its management." *Journal of Applied and Natural Science* 1 (1):99–108.

Cobbett, CS. 2000. "Phytochelatin biosynthesis and function in heavy-metal detoxification." *Current Opinion in Plant Biology* 3 (3):211–216.

Conn, S, and M Gilliham. 2010. "Comparative physiology of elemental distributions in plants." *Annals of Botany* 105 (7):1081–1102.

Diomides, CJ. 2005. *An Investigation of Inorganic Background Soil Constituents with a Focus on Arsenic Species*: Victoria University (Melbourne, Vic.).

Dixit, R, D Malaviya, K Pandiyan, UB Singh, A Sahu, R Shukla, P Singh, JP Rai, PK Sharma, and H Lade. 2015. "Bioremediation of heavy metals from soil and aquatic environment: An overview of principles and criteria of fundamental processes." *Sustainability* 7 (2):2189–2212.

Duffus, JH. 2002. "Heavy metals" a meaningless term? (IUPAC Technical Report)." *Pure and Applied Chemistry* 74 (5):793–807.

Duruibe, JO, MOC Ogwuegbu, and JN Egwurugwu. 2007. "Heavy metal pollution and human biotoxic effects." *International Journal of Physical Sciences* 2 (5):112–118.

Edwards, CA. 2002. "Assessing the effects of environmental pollutants on soil organisms, communities, processes and ecosystems." *European Journal of Soil Biology* 38 (3–4):225–231.

Emamverdian, A, Y Ding, F Mokhberdoran, and Y Xie. 2015. "Heavy metal stress and some mechanisms of plant defense response." *The Scientific World Journal* 2015.

Ercal, N, H Gurer-Orhan, and N Aykin-Burns. 2001. "Toxic metals and oxidative stress part I: Mechanisms involved in metal-induced oxidative damage." *Current Topics in Medicinal Chemistry* 1 (6):529–539.

Flora, SJS, and V Pachauri. 2010. "Chelation in metal intoxication." *International Journal of Environmental Research and Public Health* 7 (7):2745–2788.

Giordani, T, L Natali, BE Maserti, S Taddei, and A Cavallini. 2000. "Characterization and expression of DNA sequences encoding putative type-II metallothioneins in the seagrass *Posidonia oceanica*." *Plant Physiology* 123 (4):1571–1582.

Goel, PK. 2006. *Water Pollution: Causes, Effects and Control*: New Age International.

Gunatilake, SK. 2015. "Methods of removing heavy metals from industrial wastewater." *Methods* 1 (1).

Hall, JL. 2002. "Cellular mechanisms for heavy metal detoxification and tolerance." *Journal of Experimental Botany* 53 (366):1–11.

Harmens, H, PLM Koevoets, JAC Verkleij, and WHO Ernst. 1994. "The role of low molecular weight organic acids in the mechanism of increased zinc tolerance in Silene vulgaris (Moench) Garcke." *New Phytologist* 126 (4):615–621.

Isani, G, and E Carpenè. 2014. "Metallothioneins, unconventional proteins from unconventional animals: A long journey from nematodes to mammals." *Biomolecules* 4 (2):435–457.

Jaishankar, M, T Tseten, N Anbalagan, BB Mathew, and KN Beeregowda. 2014. "Toxicity, mechanism and health effects of some heavy metals." *Interdisciplinary Toxicology* 7 (2):60–72.

Jan, AT, M Azam, K Siddiqui, A Ali, I Choi, and QMR Haq. 2015. "Heavy metals and human health: Mechanistic insight into toxicity and counter defense system of antioxidants." *International Journal of Molecular Sciences* 16 (12):29592–29630.

Janicka-Russak, M, K Kabała, M Burzyński, and G Kłobus. 2008. "Response of plasma membrane H+-ATPase to heavy metal stress in Cucumis sativu s roots." *Journal of Experimental Botany* 59 (13):3721–3728.

Jin, Y, Y Luan, Y Ning, and L Wang. 2018. "Effects and mechanisms of microbial remediation of heavy metals in soil: A critical review." *Applied Sciences* 8 (8):1336.

Kerkeb, L, and U Krämer. 2003. "The role of free histidine in xylem loading of nickel in *Alyssum lesbiacum* and *Brassica juncea*." *Plant Physiology* 131 (2):716–724.

Khan, A, S Khan, MA Khan, Z Qamar, and M Waqas. 2015. "The uptake and bioaccumulation of heavy metals by food plants, their effects on plants nutrients, and associated health risk: A review." *Environmental Science and Pollution Research* 22 (18):13772–13799.

Kim, HS, YJ Kim, and YR Seo. 2015. "An overview of carcinogenic heavy metal: Molecular toxicity mechanism and prevention." *Journal of Cancer Prevention* 20 (4):232.

Kjellstrom, T, M Lodh, T McMichael, G Ranmuthugala, R Shrestha, and S Kingsland. 2006. "Air and water pollution: Burden and strategies for control." In *Disease Control Priorities in Developing Countries*. World Bank: Washington, DC.

Krishna, IVM, V Manickam, A Shah, and N Davergave. 2017. *Environmental Management: Science and Engineering for Industry*: Butterworth-Heinemann.

Lasat, MM. 1999. "Phytoextraction of metals from contaminated soil: A review of plant/soil/metal interaction and assessment of pertinent agronomic issues." *Journal of Hazardous Substance Research* 2 (1):5.

Li, J, H Yu, and Y Luan. 2015. "Meta-analysis of the copper, zinc, and cadmium absorption capacities of aquatic plants in heavy metal-polluted water." *International Journal of Environmental Research and Public Health* 12 (12):14958–14973.

Liu, W, X Zhang, L Liang, C Chen, S Wei, and Q Zhou. 2015. "Phytochelatin and oxidative stress under heavy metal stress tolerance in plants." In *Reactive Oxygen Species and Oxidative Damage in Plants Under Stress*, 191–217. Springer.

Maret, W. 2016. "The metals in the biological periodic system of the elements: Concepts and conjectures." *International Journal of Molecular Sciences* 17 (1):66.

Marques, APGC, AOSS Rangel, and PML Castro. 2009. "Remediation of heavy metal contaminated soils: Phytoremediation as a potentially promising clean-up technology." *Critical Reviews in Environmental Science and Technology* 39 (8):622–654.

Martinez, JL. 2009. "Environmental pollution by antibiotics and by antibiotic resistance determinants." *Environmental Pollution* 157 (11):2893–2902.

Meharg, AA, QJ Cumbes, and MR Macnair. 1993. "Pre-adaptation of yorkshire fog, *Holcus lanatus* L. (Poaceae) to arsenate tolerance." *Evolution* 47 (1):313–316.

Mirecki, N, R Agic, L Sunic, L Milenkovic, and S Ilic. 2015. "Transfer factor as indicator of heavy metals content in plants." *Fresenius Environmental Bulletin* 24 (11c):4212–4219.

Mishra, S, and RS Dubey. 2006. "Heavy metal uptake and detoxification mechanisms in plants." *International Journal of Agricultural Research* 1 (2):122–141.

Mourato, MP, IN Moreira, I Leitão, FR Pinto, JR Sales, and LL Martins. 2015. "Effect of heavy metals in plants of the genus Brassica." *International Journal of Molecular Sciences* 16 (8):17975–17998.

Nagajyoti, PC, KD Lee, and TVM Sreekanth. 2010. "Heavy metals, occurrence and toxicity for plants: A review." *Environmental Chemistry Letters* 8 (3):199–216.

Naidu, R, RS Kookana, ME Sumner, RD Harter, and KG Tiller. 1997. "Cadmium sorption and transport in variable charge soils: A review." *Journal of Environmental Quality* 26 (3):602–617.

Nasibi, F, T Heidari, Z Asrar, and H Mansoori. 2013. "Effect of arginine pre-treatment on nickel accumulation and alleviation of the oxidative stress in Hyoscyamus niger." *Journal of Soil Science and Plant Nutrition* 13 (3):680–689.

Nouri, J, N Khorasani, B Lorestani, M Karami, AH Hassani, and N Yousefi. 2009. "Accumulation of heavy metals in soil and uptake by plant species with phytoremediation potential." *Environmental Earth Sciences* 59 (2):315–323.

Olaniran, AO, A Balgobind, and B Pillay. 2013. "Bioavailability of heavy metals in soil: Impact on microbial biodegradation of organic compounds and possible improvement strategies." *International Journal of Molecular Sciences* 14 (5):10197–10228.

Osmolovskaya, N, DV Vu, and L Kuchaeva. 2018. "The role of organic acids in heavy metal tolerance in plants." *Biological Communications* 63 (1):9–16–9–16.

Osredkar, J, and N Sustar. 2011. "Copper and zinc, biological role and significance of copper/zinc imbalance." *Journal of Clinical Toxicology* 3:2161–0495.

Oves, M, M Saghir Khan, A Huda Qari, M Nadeen Felemban, and T Almeelbi. 2016. "Heavy metals: Biological importance and detoxification strategies." *Journal of Bioremediation and Biodegradation* 7 (334):2.

Persans, MW, X Yan, J-MML Patnoe, U Krämer, and DE Salt. 1999. "Molecular dissection of the role of histidine in nickel hyperaccumulation in *Thlaspi goesingense* (Hálácsy)." *Plant Physiology* 121 (4):1117–1126.

Pezeshki, SR, and RD DeLaune. 2012. "Soil oxidation-reduction in wetlands and its impact on plant functioning." *Biology* 1 (2):196–221.

Pinto, AP, A de Varennes, R Fonseca, and D Martins Teixeira. 2014. "Phytoremediation of soils contaminated with heavy metals." *Phytoremediation: Management of Environmental Contaminants* 1:133.

Pourrut, B, M Shahid, C Dumat, P Winterton, and E Pinelli. 2011. "Lead uptake, toxicity, and detoxification in plants." In *Reviews of Environmental Contamination and Toxicology Volume 213*, 113–136. Springer.

Pula, B, T Tazbierski, A Zamirska, B Werynska, A Bieniek, J Szepietowski, J Rys, P Dziegiel, and M Podhorska-Okolow. 2015. "Metallothionein 3 expression in normal skin and malignant skin lesions." *Pathology & Oncology Research* 21 (1):187–193.

Rahal, A, A Kumar, V Singh, B Yadav, R Tiwari, S Chakraborty, and K Dhama. 2014. "Oxidative stress, prooxidants, and antioxidants: The interplay." *BioMed Research International* 2014.

Rai, Prabhat. 2015. *Biomagnetic Monitoring of Particulate Matter: In the Indo-Burma Hotspot Region*: Elsevier.

Rajasulochana, P, and V Preethy. 2016. "Comparison on efficiency of various techniques in treatment of waste and sewage water– A comprehensive review." *Resource-Efficient Technologies* 2 (4):175–184.

Rout, GR, and S Sahoo. 2015. "Role of iron in plant growth and metabolism." *Reviews in Agricultural Science* 3:1–24.

Saha, JK, R Selladurai, MV Coumar, ML Dotaniya, S Kundu, and AK Patra. 2017. *Soil Pollution-An Emerging Threat to Agriculture*. Vol. 10: Springer.

Shah, FUR, N Ahmad, KR Masood, and JR Peralta-Videa. 2010. "Heavy metal toxicity in plants." In *Plant adaptation and phytoremediation*, 71–97. Springer.

Shi, W-Y, H-B Shao, H Li, M-A Shao, and S Du. 2009. "Progress in the remediation of hazardous heavy metal-polluted soils by natural zeolite." *Journal of Hazardous Materials* 170 (1):1–6.

Singh, A, and SM Prasad. 2015. "Remediation of heavy metal contaminated ecosystem: An overview on technology advancement." *International Journal of Environmental Science and Technology* 12 (1):353–366.

Singh, R, N Gautam, A Mishra, and R Gupta. 2011. "Heavy metals and living systems: An overview." *Indian Journal of Pharmacology* 43 (3):246.

Singh, S, P Parihar, R Singh, VP Singh, and SM Prasad. 2016. "Heavy metal tolerance in plants: Role of transcriptomics, proteomics, metabolomics, and ionomics." *Frontiers in Plant Science* 6:1143.

Slater, DH. 2001. *Pollution: Causes, Effects and Control*: Royal Society of Chemistry.

Song, W-Y, DG Mendoza-Cozatl, Y Lee, JI Schroeder, S-N Ahn, H-S Lee, T Wicker, and E Martinoia. 2014. "Phytochelatin– metal (loid) transport into vacuoles shows different substrate preferences in barley and A rabidopsis." *Plant, Cell & Environment* 37 (5):1192–1201.

Sunitha, MS, S Prashant, SAKumar, S Rao, ML Narasu, and PBK Kishor. 2013. "Cellular and molecular mechanisms of heavy metal tolerance in plants: A brief overview of transgenic plants overexpressing phytochelatin synthase and metallothionein genes." *Plant Cell Biotechnology and Molecular Biology* 14 (1–2):33–48.

Tangahu, BV, S Abdullah, S Rozaimah, H Basri, M Idris, N Anuar, and M Mukhlisin. 2011. "A review on heavy metals (As, Pb, and Hg) uptake by plants through phytoremediation." *International Journal of Chemical Engineering* 2011.

Tchounwou, PB, CG Yedjou, AK Patlolla, and DJ Sutton. 2012. "Heavy metal toxicity and the environment." In *Molecular, Clinical and Environmental Toxicology*, 133–164. Springer.

Tuna, AL, B Burun, İ Yokas, and E Coban. 2002. "The effects of heavy metals on pollen germination and pollen tube length in the tobacco plant." *Turkish Journal of Biology* 26 (2):109–113.

Tyler, G. 2004. "Rare earth elements in soil and plant systems-A review." *Plant and Soil* 267 (1–2):191–206.

ul Islam, E, X Yang, Z He, and Q Mahmood. 2007. "Assessing potential dietary toxicity of heavy metals in selected vegetables and food crops." *Journal of Zhejiang University Science B* 8 (1):1–13.

Vandenbossche, M, M Jimenez, M Casetta, and M Traisnel. 2015. "Remediation of heavy metals by biomolecules: A review." *Critical Reviews in Environmental Science and Technology* 45 (15):1644–1704.

Violante, A, V Cozzolino, L Perelomov, AG Caporale, and M Pigna. 2010. "Mobility and bioavailability of heavy metals and metalloids in soil environments." *Journal of Soil Science and Plant Nutrition* 10 (3):268–292.

Wenzl, P, GM Patino, AL Chaves, JE Mayer, and IM Rao. 2001. "The high level of aluminum resistance in signalgrass is not associated with known mechanisms of external aluminum detoxification in root apices." *Plant Physiology* 125 (3):1473–1484.

White, PJ, and PH Brown. 2010. "Plant nutrition for sustainable development and global health." *Annals of Botany* 105 (7):1073–1080.

Wu, G, H Kang, X Zhang, H Shao, L Chu, and C Ruan. 2010. "A critical review on the bio-removal of hazardous heavy metals from contaminated soils: Issues, progress, eco-environmental concerns and opportunities." *Journal of Hazardous Materials* 174 (1–3):1–8.

Wuana, RA, and FE Okieimen. 2011. "Heavy metals in contaminated soils: A review of sources, chemistry, risks and best available strategies for remediation." *Isrn Ecology* 2011.

Xie, Y, J Fan, W Zhu, E Amombo, Y Lou, L Chen, and J Fu. 2016. "Effect of heavy metals pollution on soil microbial diversity and bermudagrass genetic variation." *Frontiers in Plant Science* 7:755.

Yadav, S, M Irfan, A Ahmad, and S Hayat. 2011. "Causes of salinity and plant manifestations to salt stress: A review." *Journal of Environmental Biology* 32 (5):667.

Yadav, SK. 2010. "Heavy metals toxicity in plants: An overview on the role of glutathione and phytochelatins in heavy metal stress tolerance of plants." *South African Journal of Botany* 76 (2):167–179.

Zhang, C, Q Li, M Zhang, N Zhang, and M Li. 2013. "Effects of rare earth elements on growth and metabolism of medicinal plants." *Acta Pharmaceutica Sinica B* 3 (1):20–24.

Zheng, SJ, JL Yang, YF He, XH Yu, L Zhang, JF You, RF Shen, and H Matsumoto. 2005. "Immobilization of aluminum with phosphorus in roots is associated with high aluminum resistance in buckwheat." *Plant Physiology* 138 (1):297–303.

25 Plant Responses to Stress Induced by Toxic Metals and Their Nanoforms

Kráľová Katarína, Masarovičová Elena, and Jampílek Josef

CONTENTS

25.1 INTRODUCTION

There are 350,000 plant species in the world, and about 80,000 are edible for humans. However, at present only about 150 species are actively cultivated, directly for human food or as feed for animals, and from these 30 produce 95% of human calories and proteins (Füleky 2009), whereby only three of them, namely rice, maize, and wheat, contribute nearly 60% of calories and proteins obtained by humans from plants (FAO 2004). Besides the adverse impacts of toxic metals on the morphological and biochemical processes of crops, manifested in undesirably reduced yield, accumulation of toxic metals in edible parts of crops could also seriously threaten human health. Consequently, the gradual increase in the contamination of the environment with toxic metals or their nanoforms due to anthropogenic activities represents a risk when providing safe, sustainable, and healthy food for a growing world population, because metals present in bioavailable forms and at excessive levels are phytotoxic (Nasreddine and Parent-Massin 2002; Reeves and Chaney 2008; Antisari et al. 2015; Khan et al. 2015; Clemens and Ma 2016; Zhou et al. 2016a). Thus, in the interest of better food safety, it is important to reduce toxic element accumulation in crops. This could be achieved by the remediation of polluted metal soils using metal tolerant plants, including some fast-growing woody plants that are able to uptake metal contaminants

479

from soil and translocate them to harvestable plant parts (e.g., (Masarovičová, Kráľová, 2017, 2018)).

Recently several monographs were devoted to heavy metal stress in plants, heavy metal tolerance in plants, plant–metal interactions, and detoxification of heavy metals in plants (Sherameti and Varma 2011; Furini 2012; Gupta et al. 2013; Parvaiz 2015; Jan and Parray 2016). To avoid the deleterious effects caused by heavy metal excess, such as impaired photosynthesis, oxidative stress, and impaired nutrient uptake resulting in reduced growth, plants respond at the structural, physiological, and molecular levels and the defense strategies of plants related to detoxification of excess toxic metals that may result at cell wall, cell membrane or protoplasm level include, for example, transport of the metals especially to apparent free spaces, intercellular or intracellular vacuoles and bodies like lysosomes to sequester the metals, increasing metal tolerance by protecting the integrity of the plasma membrane, or increased synthesis of phytochelatins, metallothioneins, and enzymes with antioxidant activity (Aery 2012; Ovečka and Takáč 2014; Panda et al. 2016).

The need to reduce applied doses of metal-based pesticides or fertilizers in order to reduce contamination of the environment, as well as financial costs, resulted in increased effort to use nanoscale metals and metal oxides with the same chemical composition as their microscale/bulk counterparts, while showing different physical and chemical properties resulting in a distinct impact on the plants. The toxicity of such nanoparticles (NPs) is affected by their morphology and size, small particles being usually more effective due to their high specific surface area and ability to penetrate organisms (Nair et al. 2010; Masarovičová et al. 2013, 2014; Jampílek and Kráľová 2015, 2017, 2019; Ruttkay-Nedecky et al. 2017).

This chapter is focused on the classification of metal ions according to their toxicity and physico-chemical characteristics, and comprehensively summarizes the adverse effects of toxic metals on Rubisco and on photosynthetic electron transport with specification of their sites of action in the photosynthetic apparatus; it also presents an up-to-date review of the recent findings related to the impact of toxic metals on woody plants, some important agronomic crops and medicinal plants, including defense reactions, and detoxification mechanisms induced by toxic metals in plants. Moreover, the phytotoxic effects of metal/metal oxide nanoparticles on algae and vascular plants are briefly outlined. The potential of different plant species to be used in phytoremediation technologies is highlighted as well.

25.2 CLASSIFICATION OF METAL IONS ACCORDING TO THEIR TOXICITY AND PHYSICO-CHEMICAL CHARACTERISTICS SUITABLE TO PREDICT METAL TOXICITY

From a physiological point of view, metals can be classified into three main groups: I) essential and basically non-toxic, II) essential, but harmful at higher concentrations, and III)

toxic. The toxicity of metal ions and metal-based compounds is closely connected with bioavailability that depends on biological parameters and on the physico-chemical properties of metals, their ions, and their compounds. The physico-chemical properties depend on the atomic structure of the metals that are systematically described by the periodic table (Duffus 2002).

Nieboer and Richardson (1980) classified metals based on their ionic and covalent bonding tendencies and donor-atom preference. In this classification, two indices were used: I) covalent index, $(_{\chi m})^2 r$, which reflects the degree of covalent interactions in the metal–ligand complex relative to ionic interactions and II) ionic index, z^2/r, expressing an effective measure of ionic interactions (where z is the ion charge, $_{\chi m}$ is the Pauling electronegativity and r is the ionic radius corresponding to the most common coordination number). The dependence of the covalent index vs. the ionic index was used as a base for classification of metal and metalloid ions into three classes. Metals of class A are oxygen donor-atom seekers, whereas those of class B are nitrogen and sulfur seekers, and metals ranged to borderline metals are characterized by an ambivalent affinity for all three donor-atoms. In general, for a fixed value of the ionic index, toxicity increases with increasing magnitude of covalent index; conversely, for a fixed value of the covalent index, toxicity increases with increasing magnitude of the ionic index. Donor preference is an important reactivity principle that has implications for environmental and biological monitoring (Nieboer et al. 1999). Thus, it can be stated that in the toxicology of the metals, besides the atomic property determinants of reactivity such as ionic and covalent bonding tendencies and donor-atom preference, metal-ion complex formation and stability, kinetic aspects, solubility, metal speciation, radical formation as well as particle size and shape are also crucial, whereby the ultimate toxic outcomes are unambiguously modulated by anatomical structures, biological processes and genetic determinants as well (Nieboer et al. 1999).

In Pearson's classification (Pearson 1963) metal acceptors and the ligand donors are divided into 'hard and soft categories.' A 'hard acceptor' is characterized by low polarizability, low electronegativity, and large positive charge density, whereas a hard donor by low electron mobility or polarizability but high electronegativity and high negative charge density. The characteristics of 'soft donors and acceptors' are the opposite. According to HSAB (Hard Soft Acids and Bases), the hard acceptors prefer to bind hard donors, and soft acceptors prefer to bind soft donors to form stable compounds (Shaw et al. 2004). Class A (hard) metals are Lewis acids (electron acceptors) of small size and low polarizability (deformability of the electron sheath or hardness), e.g., Na, Mg, Al, K. On the other hand, class B (soft) metals are Lewis acids (electron acceptors) of large size and high polarizability (softness): Cu(I), Pd, Ag, Cd, Ir, Pt, Au, Hg, Ti, Pb(II), and to the borderline (intermediate) metals belong V, Cr, Mn, Fe(II), Co, Ni, Cu(II), Zn, Rh, Pb(IV), and Sn (Duffus 2002).

The toxicity of metal ions was also correlated with some other physico-chemical characteristics such as, for example, the stability constant reflecting the relative affinity of the metal ion for the ligand that could be used as quantitative measure of relative stabilities of various metal complexes or the first hydrolysis constant representing a measure of the ability of the metal ion to form a metal hydroxide and reflecting the metal-ion affinity to intermediate ligands with oxygen donor-atoms (Enache et al. 2003). The softness index ($_{op}$) quantifies the ability of a metal ion to accept an electron during interaction with a ligand and reflects the importance of covalent interactions relative to electrostatic interactions in determining intermetal trends in bioactivity.

Wolterbeek and Verburg (2001) focused attention on the possibilities for predicting metal toxicity on the basis of general metal properties including electrochemical potential, the ionization potential, the ratio between atomic radius and atomic weight, and the electronegativity χ_m and found that metal properties contribute to the observed toxicities in relative importance depending on the specific effect, effect level, exposure time, and selected organisms. Improved scales for metal-ion softness and toxicity were analyzed by Kiniaide (2009). For predicting the effects of untested metals during risk assessment activities, the quantitative ion character-activity relationships (QICARs) could be useful, too (Newman et al. 1998; Walker et al. 2003).

25.3 EFFECTS OF TOXIC METALS ON PHOTOSYNTHETIC ELECTRON TRANSPORT

Natural photosynthesis is the process by which sunlight is captured and converted into the energy of chemical bonds of organic molecules. The 'light reactions' of photosynthesis include light absorption, transfer of excitation energy to reaction centers (RCs), followed by electron and proton transfer reactions that produce NADPH, ATP, and O_2. Subsequent electron transfer steps prevent the primary charge from recombining by transferring the electron through the photosynthetic electron transport (PET) chain by a system of suitable electron acceptors and electron donors. Photosystem (PS) II is a protein complex that can oxidize water, which results in the release of O_2, whereby the catalytic cleaving of water occurs at a cluster consisting of four manganese atoms and one calcium atom that is situated at the luminal side of two key polypeptides of PS II, D_1, and D_2. Electrons formed during the photochemical cleavage of water are then transferred to P_{680}, the reaction center chlorophyll (or the primary electron donor) of PS II, via a redox-active tyrosine residue on D_1 protein (Tyr_Z), and thereafter they are transported to a mobile pool of plastoquinone (PQ) molecules by subsequent redox reactions via pheophytin and redox-active quinones, Q_A and Q_B. The electrons are further transmitted to the cytochrome complex and to the PS I, whereby the final electron acceptor of PS I is $NADP^+$ (e.g., (Whithmarsh 1998; Govindjee 2000; Barber and Tran 2013)).

PET can be investigated by several methods in intact preparations (plant leaves, algae, cyanophytes), in isolated plant chloroplasts (intact chloroplasts or chloroplasts without outer membrane), or in experimentally prepared fractions of the photosynthetic apparatus (PS I or PS II). Estimation of O_2 or CO_2 concentration in the atmosphere or in the water in the closed space surrounding the sample or measurements of chlorophyll (Chl) fluorescence are suitable methods for the study of PET in intact samples (e.g., (Masarovičová and Kráľová 2005; Kalaji et al. 2017)). Convenient methods to study PET in isolated chloroplasts are electrochemical measurements of oxygen concentration using the Clark electrode (Clark 1956; Masarovičová and Kráľová 2005) enabling estimation of PET through the whole photosynthetic apparatus or spectrophotometric methods that can be applied in monitoring of PET through individual sections of PET chain.

The Hill reaction is defined as the photoreduction of an artificial electron acceptor, e.g., 2,6-dichlorophenol-indophenol (DCPIP), by the hydrogens of water occurring in a chloroplast suspension, which is connected with O_2 evolution (Hill 1937) and decoloration of the initially purple-colored DCPIP to colorless $DCPIPH_2$. Consequently, by measuring absorbance at 600 nm, the rate of DCPIP photoreduction and so the rate of water decomposition can be estimated. This method is suitable for PET monitoring through PS II because the site of DCPIP action is the PQ pool on the acceptor side of PS II (Izawa 1980). On the other hand, the rate of PET through PS I can be followed spectrophotometrically by $DCPIPH_2$ photooxidation (Xiao et al. 1997); however, it is necessary to secure PET interruption between PS II and PS I of chloroplasts (for this purpose usually DCMU is used because its site of action is situated in Q_B on the acceptor side of PS II), whereby $DCPIPH_2$ prepared using ascorbate serves as an artificial electron donor in the site of plastocyanin on the donor side of PS I (Marsho and Kok 1980). The PET can be inhibited by a wide scale of PET inhibitors, including transition metals, a site of action of which can be more exactly specified using Chl$a$ fluorescence (e.g., (Joshi and Mohanty 2004; Kalaji et al. 2016, 2017), or by electron paramagnetic resonance (EPR) (Hoff 1979).

The EPR method is suitable for detection of compounds which contain unpaired spins, including radicals formed during electron transport through the photosynthetic apparatus. EPR spectroscopy is a useful tool to study photosynthesizing organisms exposed to abiotic stresses (Šeršeň and Kráľová 2013a), and it is applicable not only to study the site of action of metal inhibitors of PET in the photosynthetic apparatus of isolated chloroplasts, but it also enables detection of the interactions of metal ions with some amino acid residues of photosynthetic proteins.

Intact chloroplasts of both algae and vascular plants are known to exhibit EPR signals in the region of free radicals (g = 2.00), which are stable during several hours (Hoff 1979) and could be registered at laboratory temperature by conventional continual wave EPR apparatus. These signals were first observed by Commoner et al. (1957), and they were later denoted as signal I (g=2.0026, $\Delta B_{pp} = 0.8$ mT) and signal II (g = 2.0046, $\Delta B_{pp} = 2$ mT) indicating their connection with PS I and PS II, respectively (Weaver 1968). Signal II consists of two components, namely signal II_{slow} that is observable in the dark and signal $II_{very\ fast}$ that occurs at the irradiation of chloroplasts by visible light and represents an intensity increase of signal II at the irradiation of chloroplasts by the visible light. It was found that signal II_{slow} belongs to the intermediate D• and signal $II_{very\ fast}$ belongs to the intermediate Z•. Intermediates Z• and D• are tyrosine radicals (Tyr_Z• and TyrD•, respectively) situated at the 161st position in D_1 and D_2 proteins located on the donor side of PS II (Svensson et al. 1991). The EPR signal I is associated with a cation-radical of a Chl dimer situated in the core of PS I (Hoff 1979). Some ions of transition metals (Cu^{2+}, Fe^{2+}, Fe^{3+}, Ni^{2+}, Mn^{2+}, etc.) having unpaired spins are paramagnetic and exhibit characteristic EPR spectra depending on the properties of the ligand. Addition of such metal ions/compounds to chloroplasts results in interaction with some amino acid residues of photosynthetic proteins, which is reflected in the changes in EPR spectra of these ions (Šeršeň et al. 1997a).

25.3.1 COPPER

Although copper is an essential metal for plants, higher concentrations of Cu^{2+} ions adversely affect the photosynthetic apparatus of plants acting at several sites in the PET chain (Barón et al. 1995; Yruela 2005; Masarovičová et al. 2010a).

PS I as the site of Cu^{2+} action was indeed reported by Baszyński et al. (1988), Samuelsson and Öquist (1980), and Bohner et al. (1980); however, Shoi et al. (1978a) situated it directly in the ferredoxine at the acceptor side of PS I, while Singh and Singh (1987) estimated the site of Cu^{2+} action as being the cytochrome b on the donor side of PS I.

The sites of Cu^{2+} action situated on both sides of PS II were reported as well (Baszyński et al. 1988; Maksymiec et al. 1994; Ouzounidou et al. 1995). Some researchers did not offer precise specification of the site of Cu^{2+} action on the donor side of PS II (Cedeno-Maldonado et al. 1972; Bohner et al. 1980; Samuelsson and Öquist 1980; Samson et al. 1988), while others reported the precise site for the action of Cu^{2+} as directly in protein D_1 (Vavilin et al. 1995), between Tyr_Z and P_{680} (Shoi et al. 1978a; Shoi et al. 1978b; Hsu and Lee 1988; Renganathan and Bose 1989; Lidon and Henriques 1991), directly in Tyr_Z (Schröder et al. 1994), in both intermediates (Tyr_Z and Tyr_D) (Kráľová et al. 1994; Šeršeň et al. 1996), in the manganese cluster on the donor side of PS II (Kráľová et al. 1994; Šeršeň et al. 1996) as well as in Q_A and Q_B on the acceptor side of PS II (Mohanty et al. 1989a; Yruela et al. 1991, 1992, 1993; Smirnova et al. 1998; Utchig et al. 2000), on both sides of PS II, namely in Tyr_Z and in Q_A and Q_B, respectively (Renger et al. 1993; Jegerschöld et al. 1995), in nonheme Fe^{2+} in PS II (Jegerschöld et al. 1999), in cytochrome b_{559} in photosystem II thylakoids (Burda et al. 2002, 2006) or in the antenna Chl of PS II (Lidon et al. 1993). Moreover, the direct interaction of Cu^{2+} ions with the thylakoid membranes of the photosynthetic apparatus (Szalontai et al. 1999) and the alteration in the lipid composition caused by these metal ions, which was manifested in the decreased molar ratio of monogalactosyldiacylglycerol to digalactosyldiacylglycerol in PS II, were reported as well (Quartacci et al. 2000).

Investigation of the effects of some Cu(II) compounds with different ligands on photosynthetic apparatus using EPR spectroscopy showed that Cu^{2+} ions interacted with Z/D intermediates or with amino acids constituting photosynthetic peptides D_1 and D_2 which prevented the oxidation of Z/D intermediates (Kráľová et al. 1994, 1998; Šeršeň et al. 1997a) because in Cu^{2+}-treated chloroplasts a decrease of the intensity of both EPR signals, i.e., signal II_{slow} and $II_{very\ fast}$ was observed. Figure 25.1 shows the EPR spectra of an untreated chloroplast suspension as well as those in the presence of diaqua-(N-pyruvidene-β-alaninatocopper(II) monohydrate (Cu(pyr-β-ala)) in the dark (full lines) and in the light (dotted lines). The EPR signal at g = 2.0046 and line width ΔB_{pp} ~2 mT belongs almost completely to signal II_{slow} (Figure 25.1A, full line), while the EPR signal induced by light corresponds approximately to signal $II_{very\ fast}$ (Figure 25.1A, the difference between spectra recorded in the light and in the dark). The EPR signals II_{slow} and $II_{very\ fast}$ completely disappeared at treatment of chloroplasts with 50 μmol dm^{-3} of Cu(pyr-β-ala)

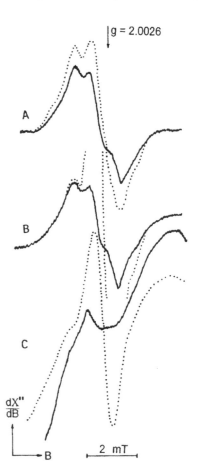

FIGURE 25.1 The EPR spectra of untreated spinach chloroplasts (A) and chloroplasts with 10 mol m^{-3} (B) and 50 mol m^{-3} (C) of Cu(pyr-β-ala). The full lines were recorded in the dark and the dotted ones were recorded in the light. The dotted line in C was recorded at 0.5 amplification compared to other lines. B is magnetic induction given in militesla (mT) and dχ''/dB is the first derivative of the imaginary part of magnetic susceptibility χ with respect to B. (From Šeršeň F., Kráľová K., Bumbálová A, Švajlenová O. *J. Plant Physiol.* 1997; 151:299–305. With permission.).

(Figure 25.1C, full line), suggesting interaction of Cu^{2+} ions of Cu(pyr-β-ala) not only with the intermediate $Tyr_Z^{\bullet}$, but also with $TyrD^{\bullet}$ situated on the donor side of PS II. Moreover, a great increase of signal I of chloroplasts at g = 2.0026 and width $\Delta B_{pp} = 0.7$ mT was observed (Figure 25.1C, dotted line). EPR signal I in the light belongs to the cation-radical of a Chl dimer situated in the core of PS I (P_{700}) (Hoff 1979). While in undamaged chloroplasts the reduction of P_{700} that is excited in the light and easily oxidizes due to the interaction with the primary acceptor of PS I is secured by the supply of electrons from PS II, in Cu(pyr-β-ala)-treated chloroplasts due to disrupted PET this reduction does not occur, P_{700} remains in the oxidized state, which is reflected by the considerable increase of signal I in the light. From this finding, it could be concluded that P_{700} was not impaired by Cu(pyr-β-ala). Schröder et al. (1994) and Jegerschöld et al. (1995) who treated the chloroplasts only with very low Cu^{2+} concentrations reported that Cu^{2+} ions exclusively damaged intermediate Z (Tyr_Z).

In the EPR spectrum of Cu(pyr-β-ala)-treated chloroplasts, the appearance of characteristic lines belonging to Mn^{2+} ions can also be observed; however, in the spectrum shown in Figure 25.2 only 3 from 6 lines corresponding to free Mn^{2+} ions are resolved (marked with arrows I_1, I_2, I_3 on line C), the other three lines remain unresolved due to high signal intensity of Cu^{2+} ions. The appearance of characteristic lines belonging to free Mn^{2+} ions also confirmed the interaction of Cu^{2+} ions with a manganese cluster that is a component of an oxygen-evolving complex (OEC) situated on the donor side of PS II, resulting in the release of Mn^{2+} ions into the interior of the thylakoid membrane (Blankenship and Sauer 1974; Kráľová et al. 1994). The loss of manganese ions from PS II was also observed by Jegerschöld et al. (1995).

Based on estimated copper amount in the sediment and supernatant after centrifugation of Cu(pyr-β-ala)-treated chloroplasts using radionuclide fluorescence analysis (RFA), it was found that the number of copper ions bound to one set of PS I and PS II exceeded the sum of bivalent ions (Cu^{2+}, Ca^{2+}, Fe^{2+}, and Mn^{2+}) that are present in untreated chloroplasts. This indicates that Cu^{2+} ions from Cu(pyr-β-ala) are also bound to the proteins of photosynthetic centers, whereby Cu(pyr-β-ala) ligands will be substituted by amino acids of these proteins. Based on the RFA spectra of untreated and Cu(pyr-β-ala)-treated chloroplasts it could be assumed that Cu^{2+} can replace other bivalent ions (Ca^{2+}, Mn^{2+}) located in the photosynthetic centers (Figure 25.3).

The interaction of Cu(pyr-β-ala) with aromatic amino acids was also confirmed by the quenching of the fluorescence band at 332 nm in Cu(pyr-β-ala)-treated chloroplasts (Figure 25.4).

The changed shape of EPR signals of Cu^{2+} (Figure 25.2, compare lines B and C) in Cu(pyr-β-ala)-treated chloroplasts also supports the formation of chelate-bound Cu in chloroplasts and it could be assumed that due the binding of the Cu^{2+} ions to the voluminous protein molecules the motion of the

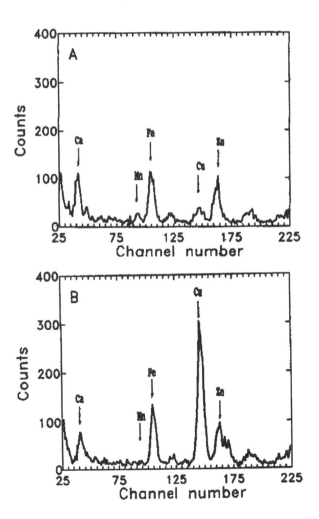

FIGURE 25.3 The RFA spectrum of untreated chloroplast (A) and that of chloroplasts treated with Cu(pyr-β-ala). (From Šeršeň F., Kráľová K., Bumbálová A, Švajlenová O. *J. Plant Physiol.* 1997; 151:299–305. With permission.).

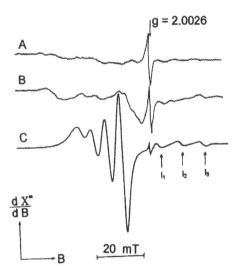

FIGURE 25.2 The EPR spectra of untreated chloroplasts (A) and that of copper (II) present in the re-suspended sediment of centrifuged (10 min at 15 000 g at 4°C) treated with 10 mol m⁻³ Cu(pyr-β-ala) (B) and that of Cu(pyr-β-ala) in chloroplasts suspension before centrifugation (C). (From Šeršeň F., Kráľová K., Bumbálová A, Švajlenová O. *J. Plant Physiol.* 1997; 151:299–305. With permission.).

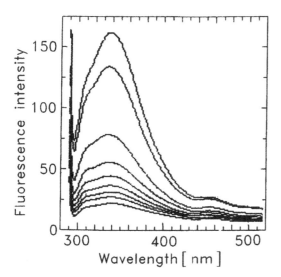

FIGURE 25.4 The fluorescence emission spectra of aromatic amino acids contained in untreated chloroplasts and in chloroplasts treated with 0.04, 0.138, 0.253, 0.48, 0.539, 0.619 and 0.791 mol m^{-3} of Cu(pyr-β-ala) (the curves from top to bottom). The excitation wavelength was 275 nm. (From Šeršeň F., Kráľová K., Bumbálová A, Švajlenová O. *J. Plant Physiol.* 1997; 151:299–305. With permission.).

spin of the Cu^{2+} electron will be restricted. This is reflected in the diminution of the hyperfine structure in the EPR spectrum of Cu^{2+} present in the re-suspended sediment of centrifuged spinach chloroplasts treated with 10 μmol dm^{-3} of Cu(pyr-β-ala) (Figure 25.2, line B).

The above-discussed effects of Cu(pyr-β-ala) on the photosynthetic apparatus of spinach chloroplasts were also found for a set of copper compounds with biologically active ligands using freshwater algae *Chlorella vulgaris* (Šeršeň et al. 1996; Kráľová et al. 1998). However, it could be stressed that the manganese cluster in algae was damaged only by diaqua(4-chloro-2-methylphenoxyacetato)copper(II) complex (Šeršeň et al. 1996) and not by copper(II) complexes with ligands showing anti-inflammatory activity (e.g., flufenamate, niflumate, naproxenate) (Kráľová et al. 1998).

The inhibitory action of Cu^{2+} ions significantly correlated with the stability constants of Cu(II) compounds and different coordinating modes of acidoligands were found to notably affect the PET-inhibiting activity of Cu(II) compounds in spinach chloroplasts (Šeršeň et al. 1997b; Kráľová et al. 2000a). The PET-inhibiting activities of the simple carboxylatocopper (II) complexes of the general formula $Cu(RCOO)_2.n$ H_2O were found to be by one or two orders higher than that of complex cuprates with Cu-NCO(S)-Cu' bridges within dimeric $(Cu_2(TSB)_2X_2)$ unit containing tridentate Schiff base (TSB^{2-}) of *N*-salicylideneaminoacidato type and pseudohalogenide (X) ligands (Kráľová et al. 2000a).

25.3.2 MERCURY

Mercury belongs to the group of toxic metals acting at several sites of the photosynthetic apparatus (e.g., (Šeršeň et al.

1998a; Masarovičová et al. 2010a)). Singh et al. (1989) situated the site of action of Hg^{2+} ions at the donor side of PS I without precise specification, other researchers situated it at the site of the plastocyanin (Kimura and Katoh 1972; Radmer and Kok 1974; Rai et al. 1991), at the acceptor side of PS I either at ferredoxin action site (Honeycutt and Krogmann 1972; De Filippis et al. 1981) or in the F_B iron-sulfur cluster (Kojima et al. 1987; Jung et al. 1995). Some researchers reported that Hg^{2+} ions target PS II, however, without exact determination of the site of action (Honeycutt and Krogmann 1972; Kimura and Katoh 1972; Rai et al. 1991), the others situated the site of the action of Hg^{2+} ions on the donor side of PS II, namely in the OEC (De Filippis et al. 1981; Singh and Singh 1987; Samson et al. 1990; Bernier et al. 1993; Bernier and Carpentier 1995), directly in the core of PS II (Murthy and Mohanty 1995) or on the acceptor side of PS II between the Q_A and Q_B quinones (Miles et al. 1973; Prokowski 1993; Kukarskikh et al. 2003).

Interactions of Hg^{2+} ions with phycobilisomes constituting a part of the light-harvesting complex (LHC) of *Spirulina platensis* (Murthy et al. 1989; Murthy and Mohanty 1991, 1995), with the LHC of PS II in spinach chloroplasts (Nahar and Tajmir-Riahi 1994, 1995) and inhibition of the ATP formation by Hg^{2+} (Honeycutt and Krogmann 1972; De Filippis et al. 1981; Singh and Singh 1987) were also described.

Due to a strong affinity of Hg^{2+} ions to C = O, C–N, C–S and C–SH groups, the mechanisms of Hg^{2+} action could be connected with the formation of complexes with amino acids in the proteins of photosynthetic centers (Nahar and Tajmir-Riahi 1994, 1995; Bernier and Carpentier 1995).

In the EPR spectra of Hg^{2+}-treated spinach chloroplasts considerable changes were observed compared to the untreated ones (Šeršeň et al. 1998a, 1998b) such as decreased signal intensities of both signals II_{slow} (line II in Figures 25.5B, C) and $II_{very fast}$ (abstraction of lines II in the light and in the dark, Figures 25.5B, C) suggesting the interaction of Hg^{2+} ions with $Z^•/D^•$ intermediates or their near vicinity. As the oxidation of Z/D tyrosines cannot occur, PET between PS II and PS I will be interrupted (which was also manifested by the increase of EPR signal I intensity in the light) (lines I in Figures 25.5B, 25.5C). On the other hand, the damage of PS I in Hg-treated chloroplasts (line III in Figures 25.5B, C, full lines) due to oxidation of Chl dimer in the core of PS I was confirmed by the increase of signal I intensity in the dark. Moreover, Hg^{2+} induced the release of Mn^{2+} ions from the manganese cluster into the interior of the thylakoid membranes (Figure 25.6). The binding of Hg^{2+} ions to photosynthetic proteins was confirmed by RFA method and by the quenching of the fluorescence of aromatic amino acids (Šeršeň et al. 1998a).

Among PET activities of the cyanobacterium *Nostoc muscorum* exposed for 72h to Hg, the most sensitive response was observed in the whole chain activity compared to that of PS II and PS I and the donor side of PS II was estimated as the site of Hg action (Singh et al. 2012). In purple photosynthetic bacterium *Rhodobacter sphaeroides* wild type, the treatment with Hg^{2+} (<100 μM) damaged PET at donor and acceptor sites of PS II; the concentration of the photoactive RCs and the connectivity of photosynthetic units were reduced; cells

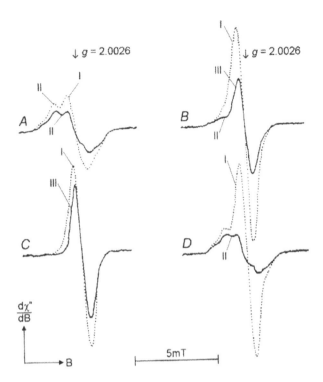

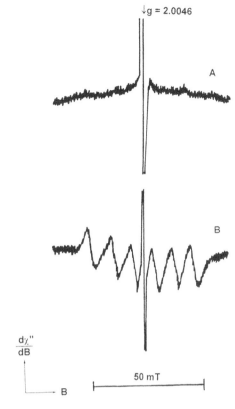

FIGURE 25.5 EPR spectra of untreated spinach chloroplasts (A) and chloroplasts treated with 8 (B) or 40 (C) mol m^{-3} of HgCl$_2$ or with 5 mol m^{-3} DCMU (D). The full line spectra were recorded in the dark and the dotted ones were recorded in the light. (From Šeršeň F., Kráľová K., Bumbálová A. *Photosynthetica* 1998; 35:551–559. With permission.).

FIGURE 25.6 EPR spectra of Mn^{2+} ions in untreated chloroplasts and in chloroplasts treated with 50 mol m^{-3} HgCl$_2$ (B). (From Šeršeň F., Kráľová K., Bumbálová A. *Photosynthetica* 1998; 35:551–559. With permission.).

that were kept in the light showed a higher extent of inhibition of the photosynthetic apparatus than those kept in the dark; and the light-harvesting system and the bc1 complex demonstrated notably higher resistance against Hg^{2+} than the RCs (Asztalos et al. 2012).

25.3.3 Organometallic Compounds

Organometallic compounds are agents containing at least one metal-to-carbon bond, in which the carbon is part of an organic group. In the presence of MeHg$^+$, impairment of PET from Q$_A$ to Q$_B$ was observed in *Chlamydomonas reinhardtii* suggesting damage on the donor side of PS II, whereby treatment with MeHg$^+$ resulted in considerably higher inhibition of the photosynthetic processes of algae than that with HgCl$_2$ (Kukarskikh et al. 2003). Decreased photochemical activity of the PS II RCs of diatom *Thalassiosira weissflogii* due to treatment with MeHg$^+$ (10^{-6}–10^{-7} mol dm^{-3}) was reported, whereby a similar effect was observed only at higher HgCl$_2$ concentrations (Antal et al. 2004). Graevskaya et al. (2003) confirmed these results and also reported that both methylmercury chloride and HgCl$_2$ decreased the rate of PS II reparation and increased a heat pathway of excitation dissipation in a PS II antennae complex of *T. weissflogii*.

Organomercuric compounds (methylmercuric chloride, phenylmerucic acetate, and phenylmerucic borate) were found to inhibit PET in spinach chloroplasts with lower effectiveness than HgCl$_2$ (Šeršeň and Kráľová 2013b), which could be explained with the fact that the stability constants of organomercury complexes with ligands such as amino acid residues in peptides are consistently lower than those of the corresponding Hg^{2+} complexes (Jackson 1998). Using EPR spectroscopy it was shown that these organomercuric compounds, in contrast to HgCl$_2$ action, do not interact with intermediates Z$^•$/D$^•$, and with the manganese cluster and as their site of action in photosynthetic apparatus ferredoxin on the acceptor side of PS I, the quinone electron acceptors Q$_A$ or Q$_B$ on the reducing side of PS II could be suggested (Šeršeň and Kráľová 2013b). Honeycutt and Krogmann (1972) previously proposed that ferredoxine situated on the acceptor side of PS I was the site of action of phenylmercuric acetate.

Organotin compounds are known to inhibit the growth of algae (Fargašová and Kizlink 1996; Fargašová 1998). Exposure of marine diatom *Thalassiosira pseudonana* to 0.5 µg dm^{-3} triphenyltin chloride resulted in upregulation of photosynthesis-related genes, while in the presence of 1.0 µg dm^{-3} triphenyltin chloride these genes were down-regulated (Yi et al. 2014). Inhibition of PET as well as the oxygen-evolving system in PS II of *Chlorella* green algae by (C$_6$H$_5$)$_3$PbCl and (C$_6$H$_5$)$_3$SnCl was reported by Murkowski and Skorska (2010), whereby (C$_6$H$_5$)$_3$PbCl caused stronger inhibition than (C$_6$H$_5$)$_3$SnCl, particularly at higher concentrations. The model tributyltin compound, tributyltin naphthenate, inhibited PET in spinach chloroplasts. The sites of its inhibitory action were found to be the Z•/D• intermediates or their vicinity as well as the manganese cluster in the OEC and PS I. The mechanism of inhibitory action is probably connected with the interaction between tributyltin naphthenate and amino acids in photosynthetic proteins which was confirmed by fluorescence measurements (Šeršeň et al. 1997c).

25.3.4 CADMIUM

Cd is one of the major environmental pollutants adversely affecting the photosynthetic apparatus of vascular plants by direct and indirect mechanisms of action (e.g., (Krupa 1999)).

Several researchers reported PS II as the site of Cd^{2+} action, however, without precise localization (Hampp et al. 1976; Nedunchezhian and Kulandaivelu 1995). Many researchers situated the site of Cd^{2+} on the donor side of PS II, namely in the OEC or in its vicinity (Bazzaz and Govindjee 1974; Van Duijvendijk-Matteoli and Desmet 1975; Baszyński et al. 1980; De Filippis et al. 1981; Atal et al. 1991; Šeršeň and Kráľová 2001; Sigfridson et al. 2004; Pagliano et al. 2006; Bartlett et al. 2008) and in Z/D intermediates (Šeršeň and Kráľová 2001; Sigfridson et al. 2004), or at the site of Q_A or Q_B on the acceptor side of PS II (Singh and Singh 1987; Fodor et al. 1996; Sigfridson et al. 2004). On the other hand, PS I activity was reported to be affected by Cd^{2+} only slightly (Bazzaz and Govindjee 1974; Baszyński et al. 1980; Krupa et al. 1987; Atal et al. 1991), although Fagioni et al. (2009) estimated high sensitivity of PS I to Cd in hydroponically cultivated *Spinacia oleracea* L. plants, while the effects on PS II were not significant. Further toxic effects of Cd^{2+} ions were reflected in the inhibition of ATP (De Filippis et al. 1981) and Chl synthesis (Stobart et al. 1985; Pagliano et al. 2006; Fagioni et al. 2009), disorganization of thylakoid membranes (Fernandez-Pinas et al. 1995; Ouzounidou et al. 1997), changes in the lipid composition of thylakoid membranes (Skórzyńska-Polit and Baszyński 1997), and in replacing Ca^{2+} ions by Cd^{2+} from the Ca/Mn cluster constituting the oxygen-evolving center (Ono and Inoue 1989; Matysik et al. 1998, 2000; Sigfridson et al. 2004; Bartlett et al. 2008). Loss of the electron transfer from intermediates Z/D to P_{680}•+, slow-down of electron transfer between Q_A•− to Q_B, and inhibition of steady-state O_2 evolution as a result of Cd^{2+}-treatment was estimated by Sigfridson et al. (2004) using flash-induced variable fluorescence and EPR spectroscopy.

The interaction of Cd^{2+} ions with Z•/D• intermediates (or their near vicinity) and with the manganese cluster in OEC resulting in the release of Mn^{2+} ions into the interior of thylakoid membranes was confirmed by EPR spectroscopy (Šeršeň and Kráľová (2001). Moreover, Chla dimer in the core of PS I was also oxidized in the dark suggesting interaction of Cd^{2+} ions with the primary donor of PS I (P_{700}) and the decrease of EPR signal I after switching off the light, which differed from that observed with the control sample, suggested the deleterious effect of Cd^{2+} ions on direct, cyclic, and noncyclic electron flow through PS I (Figure 25.7). In the

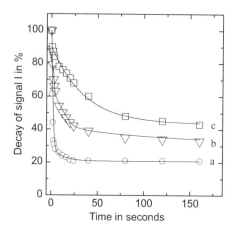

FIGURE 25.7 The time dependence of EPR signal I intensity after switching off the light in chloroplasts treated with 5 mM DCMU (a) 0.05 M $CdCl_2$ and 0.05 M $HgCl_2$ (c). (From Šeršeň F., Kráľová K. *Photosynthetica* 2001; 39:575–580. With permission.).

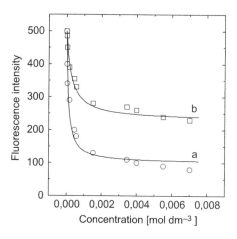

FIGURE 25.8 The dependencies of fluorescence intensity of aromatic amino acid residues in chloroplast peptides on concentrations of $HgCl_2$ (a) and $CdCl_2$ (b). (From Šeršeň F., Kráľová K. *Photosynthetica* 2001; 39:575–580. With permission.).

control sample, in which all mechanisms (direct, cyclic and noncyclic, respectively) of the reduction of $P700^+$ were not damaged, after switch-off the decrease of signal I intensity to the original value in the range of 5 seconds was observed (Figure 25.7, line a). Application of DCMU to control chloroplasts resulted in complete inhibition of the electron flow from PS II to PS I, however, without damaging PS I, while signal I intensity was restricted due to the addition of $CdCl_2$ or $HgCl_2$ (Figure 25.7 lines b, c) suggesting that these metal compounds impaired all reduction mechanisms of P_{700}^+, $HgCl_2$ being a more effective inhibitor. Moreover, the decrease of the intensity of the fluorescence emission band at 334 nm in $CdCl_2$-treated chloroplasts (Figure 25.8) confirmed the interaction of $CdCl_2$ with the aromatic amino acid residues in photosynthetic proteins and such fluorescence quenching was also estimated in chloroplasts treated with $HgCl_2$ (Šeršeň et al. 1998a, 1998b). Reduced intensity of this fluorescence emission band belonging mainly to the tryptophan residues (Chen 1986) is connected with the formation of complexes between Cd^{2+} ions and the aromatic amino acids contained in the peptides of the photosynthetic apparatus. The equilibrium constants (K) estimated according to Tominaga et al. (1995) by computer fitting of the fluorescence quenching curves *a* and *b* (Figure 25.8) were 10,200 dm^3 mol^{-1} ($r^2 = 0.97$) for $HgCl_2$ and 3700 dm^3 mol^{-1} ($r^2 = 0.97$) for $CdCl_2$ indicating the increased ability of Hg^{2+} ions to form complexes with aromatic amino acids compared to Cd^{2+} ions.

In another study the stability constants of the formation of complexes between tryptophan and metal ions determined by the above-mentioned method were found to be: 3.60×10^8 dm^3 mol^{-1} for Cu^{2+}, 1.82×10^5 dm^3 mol^{-1} for Hg^{2+}, 1.70×10^4 dm^3 mol^{-1} for Cd^{2+}, and 3.72×10^3 dm^3 mol^{-1} for Ni^{2+} and showed a close correlation with the PET-inhibiting effectiveness of corresponding metal ions (Cigáň et al. 2003).

25.3.5 Nickel

Ni^{2+} ions inhibited PS II activity in pea chloroplasts, and they were found to inhibit the $S_2Q_B^-$ charge recombination acting at the level of the secondary quinone electron acceptor, Q_B, possibly by directly modifying the Q_B site (Mohanty et al. 1989b). Ni(II) complexes with *N*-donor ligands affected PET through photosynthetic apparatus due to an interaction with $Z^•/D^•$ intermediates and the manganese cluster in the OEC (Šeršeň et al. 1997d).

In the PS II submembrane fraction, Ni^{2+} treatment resulted in a strong inhibition of O_2 evolution and depletion of the extrinsic polypeptides of 16 and 24 kDa associated with the OEC in PS II, suggesting that the interaction of Ni^{2+} with these polypeptides caused a conformational change inducing their release together with Ca^{2+} ions from the OEC, which was reflected in PET inhibition (Boisvert et al. 2007). El-Naggar (1998) reported Ni^{2+} ions as inhibiting O_2 evolution in *Chlamydomonas reinhardtii* by inactivation only the PS II activity, and while the addition of an artificial electron donor 1,5-diphenylcarbazide to broken *C. reinhardtii* cells was not restored, the donor side of PS II was suggested as their site of action. Treatment of *Chlorella vulgaris* with NiO NPs also resulted in considerable inhibition in photochemical reactions of photosynthesis (Oukarroum et al. 2017). The inhibitory effects of Ni^{2+} ions could also be connected with their ability to form complexes with amino acid residues in photosynthetic proteins (Cigáň et al. 2003).

The investigation of *in vivo* toxicity of Ni^{2+} ions on the photosynthetic system of primary bean leaves confirmed an indirect effect of Ni^{2+} ions on photosystems related to the disturbances caused by the metal in the Calvin cycle reactions and down-regulation, or even feedback inhibition, of electron transport by the excessive amounts of ATP and NADPH accumulated due to non-efficient dark reactions (Krupa et al. 1993).

25.3.6 Lead

The PS II of *Nostoc muscorum* responded sensitively to both low and high concentrations of Pb^{2+} ions and an increase in

Chl*a* fluorescence at high Pb concentrations suggested that the electron flow was inhibited at the reducing side of the PS II RC. On the other hand, notable inhibition of PS I was estimated only in the presence of high concentrations of Pb^{2+} ions (Prasad et al. 1991). Besides reduction of the efficiency of PET in PS II, high concentrations of Pb^{2+} ions pronouncedly affected non-photochemical fluorescence quenching suggesting an increase in the proton gradient across the thylakoid membrane and a decrease of photophosphorylation (Kastori et al. 1998). Reduction of the rate of whole chain electron transport, photochemical activities of PS II, and oxygen evolution due to Pb^{2+} treatments was also confirmed by Wu et al. (2008), while only minor changes in the photoreduction activities of PS I were observed. However, in the presence of Pb^{2+} ions, the absorption of light in spinach chloroplasts decreased, and excitation energy to be absorbed by LHC II and transferred to PS II was inhibited, which resulted in decelerated electron transport, water photolysis, and oxygen evolution. Pb^{2+} induced inhibition of PS II activity mediated via the water-splitting system by displacement of native inorganic cofactors (Cl^-, Ca^{2+}, and Mn^{2+}) from the OEC was reported by Rashid and Popovic (1990). Pb^{2+} ions notably dissociated extrinsic 23 and 17 kDa polypeptides and partially dissociated the 33 kDa polypeptide which protects the binding sites of these cofactors with a shield (Rashid et al. 1994).

Pb^{2+} ions accumulated in PS II of *Spirodela polyrrhiza* damaged its secondary structure, decreased the absorbance of visible light, inhibited energy transfer among amino acids within the PS II protein-pigment complex, and reduced energy transport from tyrosine residue to Chl*a* (Ling and Hong 2009). Using fluorescence and thermoluminescence techniques, Belatik et al. (2013) did not estimate direct interaction of Pb^{2+} ions with Chl molecules in PS II, however, they interacted with the OEC, which resulted in the perturbation of charge recombination between the quinone acceptors of PS II and the S_2 state of the Mn_4Ca cluster, whereby significant retardation of the electron transfer between Q_A and Q_B can be attributed to a transmembrane modification of the acceptor side of PS II.

Dao and Beardall (2016) investigated effects of Pb on PS II activity and heterogeneity of two green microalgae *Chlorella* and *Scenedesmus* using various non-invasive Chl fluorescence measurements and a decrease in maximum quantum yield, and the performance index indicated inhibition of PS II performance. Pb treatment also resulted in considerable alteration in antenna heterogeneity manifested by the decrease of PS II energetic connectivity and an increase of antenna size; however, an increase in Q_B non-reducing and inactive (non-O_2 evolving) RCs was observed only in *Chlorella*. Comparison of rice seedlings inoculated with an endophytic fungus and non-inoculated seedlings showed that the inhibitory effects of Pb^{2+} on PS II connectivity, the OEC, and on the PET from the Q_A to Q_B were reduced in inoculated seedlings suggesting that the photosynthetic energy conservation in inoculated seedlings was more effective than in inoculated ones (Li and Zhang 2015).

Both $Pb(NO_3)_2$ and $Pb(CH_3CHOO)_2$ were found to inhibit PET through PS I and PS II in spinach chloroplasts, $Pb(NO_3)_2$

with IC_{50} of 337 µmol dm^{-3} being a more effective PET inhibitor than lead acetate (IC_{50} = 643 µmol dm^{-3}), and as sites of Pb^{2+} action in the photosynthetic apparatus, the OEC and the $Z^\bullet/D^\bullet$ intermediates on the donor side of PS II were estimated. However, treatment with Pb^{2+} ions also resulted in the impairment of the cyclic electron flow in chloroplasts suggesting that the acceptor side of PS I was also damaged, likely at the site of ferredoxin. Based on the fluorescence emission spectra of Chl$a$ of untreated chloroplasts as well as of chloroplasts treated with the lead compounds, which were recorded at 25°C and at the temperature of liquid nitrogen, it could be concluded that Pb^{2+} interacted mainly with pigment–protein complexes in PS II. Using fluorescence spectroscopy formation of complexes between Pb^{2+} and amino acid residues in photosynthetic proteins was also confirmed and calculated constants of complex formation among Pb^{2+} and aromatic amino acids were as follows: 6942 $\pm$ 1843 dm^3 mol^{-1} for $Pb(NO_3)_2$ and 6067$\pm$1666 dm^3 mol^{-1} for $Pb(CH_3COO)_2$, respectively (Seršeň et al. 2014).

25.3.7 CHROMIUM

Dichromate application was found to decrease the overall photosynthetic efficiency of PS II and PS I of the green fronds of *Spirodela polyrhiza* (Susplugas et al. 2000). The experiments using *Chlamydomonas reinhardtii* showed that $Cr_2O_7^{2-}$ has different sites of inhibition that are associated with PS II, PS I and electron transport sink beyond photosystems (Perreault et al. 2009). In Cr-treated *Spirodela polyrrhiza*, a decrease in the number of active PS II RCs and damage of the OEC were reported (Appenroth et al. 2001) and PS II (the core complex as well as connecting antenna) responded to chromate treatment more sensitively than PS I and the peripheral LHC of PS II (Appenroth et al. 2003). On the other hand, Bishnoi et al. (1993) estimated that addition of Cr^{6+} notably inhibited the activity of PS I in isolated pea (*Pisum sativum* L.) chloroplasts, while that of PS II was not much affected. In Cr-treated *Ocimum tenuiflorum* L., the induction of lipid peroxidation coupled with potassium leakage and a reduction of photosynthetic pigments, protein, cysteine, ascorbic acid, and non-protein thiol contents was observed (Rai et al. 2004). In *Lemna gibba* exposed to Cr for 96 h, PET inhibition at both donor and acceptor sides of PS II was estimated and the inhibitory effects of Cr that were situated at the OEC and Q_A reduction correlated with the decreased content of the D_1 protein and the amount of 24 and 33 kDa proteins of OEC (Ali et al. 2006).

It was found that Cr^{6+} reduced O_2 evolution and inhibited the water-splitting system in PS II of *Synechocystis* sp., increased the proportion of inactivated PS II (PS II_X) and PS II_β RCs resulting in increased fluxes of dissipated energy, inhibited PET from Q_A^- to Q_B/Q_B^- and induced accumulation of P_{680}^+ (Pan et al. 2009). Cr^{6+} inhibited the efficiency of PS II to use energy under high light to a greater extent than of PS I and also showed lower light adaptability than PS I, suggesting less sensitivity of PS I to high light and Cr^{6+}. When increasing Cr^{6+} concentration, an increase of energy dissipation through non-light-induced non-photochemical fluorescence quenching

was observed and the stimulated cyclic electron flow notably contributed to quantum yield and PET of PS I, securing the protection of PS I from abiotic stresses caused by Cr^{6+} and high light (Wang et al. 2013).

25.3.8 ZINC

Based on the investigation of PET inhibition by Zn^{2+} ions using PS II submembrane preparations an active Zn^{2+} inhibitory site on the donor side of PS II was estimated and it was assumed that at higher concentrations the Zn^{2+} ions strongly perturb the conformation of the PS II core complex and could also affect the reducing side of PS II (Rashid et al. 1991). In Mn-depleted PS II, it was observed that Zn^{2+} ions caused the displacement of the non-heme Fe^{2+} (Jegerschöld et al. 1999). In an experiment using *Synechocystis aquatilis* f. *aquatilis* Sauvageau treated with Zn^{2+}, rapid inactivation of electron transport through PS II was recorded, and the site of PET inhibition was situated on the reducing side of PS II (Chaloub et al. 2005). However, according to Mohanty et al. (1989b), Zn^{2+} does not inhibit the PET between Q_A and Q_B on the acceptor side of PS II, but possibly directly modifies the Q_B site, which results in the loss of PS II activity.

Tajmirriahi and Ahmed (1993) reported that the Zn^{2+} ions applied at higher concentrations interacted with the light-harvesting proteins LHC II of chloroplast thylakoid membranes, the C = O and C–N groups being the main coordination sites of Zn^{2+}, while at low metal concentration binding with the protein carbonyl groups occurs. Zn^{2+} cations irreversibly suppressed the oxygen-evolving function and produced disaggregation of the dimer structure of the OEC. In isolated submembrane fragments enriched by PS II, the oxygen-evolving activity was found to decrease in following order: Zn < Cd < Pb < Hg (Shutilova 2006).

25.3.9 IRON

Although iron is an essential bio-element, its excess adversely affects PET in photosynthesizing organisms (e.g., (Mallic and Rai 1992; Kampfenkel et al. 1995)). Excess Fe caused PET inhibition in *Anabaena doliolum* and *Chlorella vulgaris*, whereby PS II showed greater sensitivity to iron in both tested species (Mallic and Rai 1992). In *Nicotiana plumbaginifolia* plants exposed to higher Fe concentrations for 12 h, a 40% decrease of the photosynthetic rate and increased reduction of PS II were observed (Kampfenkel et al. 1995).

Iron excess caused a notable increase in cyt b_6/f content of thylakoids of pea plants resulting in the enhanced susceptibility of PS II to photoinhibition and a higher rate of singlet oxygen photoproduction in thylakoids (Suh et al. 2002). In *Pisum sativum* L. seedlings cultivated hydroponically under higher Fe^{2+} concentration, reduction of Chl content and PET rate, deactivated RC of PS II, declined plant net photosynthetic rate, and increased fluidity of the chloroplast membranes were estimated (Xu et al. 2015). An excess of Fe caused an increased level of stress in broccoli plants which resulted in a decrease of the maximum photochemical quantum yield of

PS II (F_v/F_m), PS II operating efficiency $_{\Phi II}$, coefficient of photochemical quenching (q_P) and electron transport rate (Peña-Olmos and Casierra-Posada 2013).

Investigation of the effects of different Fe^{3+} compounds on PET in spinach chloroplasts using EPR spectroscopy showed that Fe^{3+} ions interacted with $Z^{\bullet}/D^{\bullet}$ intermediates (or their near vicinity) situated at the donor side of PS II resulting in PET interruption between the photosynthetic centers PS II and PS I. Moreover, treatment with $[Fe(nia)_3(H_2O)_2](ClO_4)_3$ in which ClO_4^- anions are not bound to the Fe atom by a coordination bond and water molecules in the coordination sphere of $(Fe(nia)_3(H_2O)_2)(ClO_4)_3$ can easily be substituted by another ligand (e.g., residues of amino acids in proteins) resulting in a release of Mn^{2+} from the OEC (Kráľová et al. 2008a).

25.3.10 ALUMINUM

Treatment of tobacco plants with Al caused PET inhibition and a decrease of PS II photochemical activity *in vivo* connected with interacting or replacing the non-heme iron between Q_A and Q_B by Al, and resulted in restricted electron transfer between Q_A and Q_B (Li et al. 2012). In Al-sensitive maize cultivar B-73 hydroponically cultivated in the presence of 200 µM Al for 13 days, the PS II activity was severely affected (Mihailovic et al. 2008). In seedlings of salt-tolerant grass *Thinopyrum bessarabicum* exposed to 1 mM Al in nutrient solution at a pH of 9.0 for 24 h, the destruction of PS II RCs and increased reduction of Q_A was observed suggesting that a high proportion of PS II centers were unable to reduce Q_B and it was supposed that PET inhibition was connected with changes in the chloroplast ultrastructure due to the altered membrane permeability (Moustakas, et al. 1997). Al was found to decrease the efficiency of excitation energy capture by open PS II RCs, the photochemical quenching coefficient of PS II fluorescence, and the *in vivo* quantum yield of PS II photochemistry in seedlings of the salt-tolerant grass *Thinopyrum bessarabicum* that were exposed to 1 mM Al (pH 9.0; 48 h). This metal also increased the non-photochemical dissipation of excitation energy suggesting that Al stress slowed down the photosynthetic metabolism and closed the stomata (Moustakas et al. 1996).

Al-induced photoinhibition in seedlings of *Citrus grandis* (L.) Osbeck cv. Tuyou irrigated daily for 5 months with a nutrient solution containing 0.2–1.6 mM Al was reported to occur at both the donor (i.e., OEC) and the acceptor sides of PS II, and it may be associated with growth inhibition (Jiang et al. 2008).

Treatment with 3 mM Al^{3+} caused a destabilization of the Mn_4O_5Ca cluster of the OEC in isolated thylakoid membranes of spinach, and it was supposed that the retarded electron transfer between Q_A and Q_B was connected to a transmembrane structural modification of PS II polypeptides due to the interaction of Al^{3+} at the OEC. A notably retarded fluorescence induction and considerably reduced F_v/F_m together with the maximal amplitude of Chl fluorescence induced by a single turnover flash observed at treatment with >3 mM Al^{3+} could be explained with the formation of P_{680}^+ due to PET

inhibition between tyrosine 161 of the D_1 subunit and P_{680} (Hasni et al. 2013).

Gaziyev et al. (2011) reported that the site of action of Al^{3+} is situated between Q_A and Q_B on the acceptor side of PS II and can be observed only under acidic conditions.

25.3.11 LANTHANIDES

Lanthanides were found to compete with Ca^{2+} for binding sites on the donor side of PS II, whereby the OEC, in which Ca^{2+} ions were replaced by trivalent lanthanide, was unable to proceed to the higher oxidation states. Using a low-temperature EPR study, a close interaction between the lanthanides and the tyrosine radical $TyrD^{\bullet}$ situated on the donor side of PS II was confirmed (Bakou et al. 1992). A lanthanide-substituted OEC was found to display a thermoluminescence band arising from a $S_2Q_A^-$ charge recombination, indicating that the Mn cluster was oxidized to the S_2 state (Ono 2000).

On the other hand, study of PS II from *Arabidopsis thaliana* treated by La^{3+}, Ce^{3+}, and Nd^{3+} resulted in induction of a generous expression of LHC IIb, which notably increased the content of LHC II and its trimer on the thylakoid membrane and improved O_2 evolution rate suggesting that the investigated lanthanides could improve the efficiency of light absorption and regulate excitation energy distribution from PS I to PS II, whereby their efficiency decreased in the following order: $Ce^{3+}>Nd^{3+}>La^{3+}>$control (Liu et al. 2009). In PS II isolated from spinach, the enhancement of the photochemical activity and O_2 evolution rate by Ce^{3+} treatment was previously observed by Liu et al. (2007).

25.3.12 ANTIMONY

Sb damaged PS II reaction cores of Chinese cork oak (*Quercus variabilis* Bl.) trees but elicited a defense mechanism at the donor side of PS II, and it was found to affect the electron transport flow after Q_A^- more strongly than the OEC and light-harvesting pigment–protein complex LHC II (Zhao et al. 2015). Sb^{3+} complexes were reported to exhibit good inhibitory efficiency with the photosynthetic and carbonic anhydrase activity of PS II (Karacan et al. 2016). Zhang et al. (2010) estimated that exposure of *Synechocystis* sp. to Sb resulted in reduced O_2 evolution suggesting the OEC as a target site of Sb, which was also supported with the progressive and drastic increase in F_0/F_v with Sb concentration, and it could be assumed that Sb replaced Mn from the water-splitting apparatus situated on the donor side of PS II. Using Chl fluorescence measurements, *in vivo* inhibition of electron transport from Q_A^- to Q_B/Q_B^- and accumulation of P_{680}^+ as a result of Sb treatment was observed, whereby once electron transfer from Q_A^- to Q_B/Q_B^- was inhibited by Sb, more Q_A^- was reoxidized via an $S_2(Q_AQ_B)^-$ charge recombination and oxidation by PQ molecules, and Sb exposure steadily increased the proportion of inactive centers PS II_X, and increased the effective antenna size per RC and the portion of PS II_β.

The high concentration of Sb^{5+} (100 mg dm^{-3}) significantly inhibited PS II, the quantum yield for photochemistry, the density of RCs and photosynthesis performance index decreased, while the dissipated energy increased, the sites of Sb^{5+} action being situated mainly on the donor side of PS II (Wang and Pan 2012). In *Microcystis aeruginosa* treated with Sb^{5+}, a more serious inhibition of PS II compared to PS I was observed, whereby the cyclic electron transport rate notably contributed to the rate of PET of PS I, especially at Sb^{5+} concentrations >100 mg dm^{-3} and high illumination (>555 μmol photons m^{-2} s^{-1}) (Wang et al. 2015a).

25.3.13 ARSENIC

The treatment of *Microcystis aeruginosa* with 10 mg dm^{-3} As^{3+} for a period ≥24 h resulted in flattening of the fluorescence transient and a drastic decrease of amplitude of the fast phase of Q_A^- reoxidation kinetics, while a decrease in the quantum yield for primary photochemistry, density of RCs, and photosynthesis performance index, and an increase of the dissipated energy were observed. It was found that following PET inhibition from Q_A^- to Q_B under As^{3+} exposure more Q_A^- was reoxidized via a $S_2(Q_AQ_B)$–charge recombination and As^{3+} stress may result in increased stability of the $S_2Q_B^-$ and $S_2Q_A^-$ recombination (Wang et al. 2012a).

25.4 RUBISCO AND RUBISCO ACTIVASE UNDER METAL STRESS

Rubisco (ribulose-1,5-bisphosphate carboxylase/oxygenase), is the most abundant enzyme on earth representing 50% of soluble leaf protein and 20–30% of total leaf nitrogen in C_3 plants, and 30% of soluble leaf protein and 5–9% of total leaf nitrogen in C_4 plants (Feller et al. 2008), catalyzes CO_2 assimilation and also functions as an oxygenase in the plants (see Buchanan et al. 2001; Pessarakli 2005; Andersson and Backlund 2008). For the catalysis, an activation process including selective carbamylation of the active site, lysine is indispensable. The carbamyl group is then stabilized by an Mg^{2+} ion under *in vivo* conditions. *In vitro* the Mg^{2+} ion can be replaced by other divalent metal ions acting as activators; however, this substitution has not only a considerable effect on the catalytic activity of the enzyme, but the individual metal ions also exhibit a different effect on the ratio of the carboxylation reaction and the competing oxygenation reaction (Karkehabadi et al. 2003). The deactivation of the enzyme after substitution of Mg^{2+} in the ternary Rubisco complex by metal cations can result in a loss of carboxylation capacity (Van Assche and Clijsters 1990). Rubisco is a hexaclecameric enzyme composed of two subunits: a small subunit encoded by a nuclear gene (rbcS), and a large subunit encoded by a plastid gene (rbcL). Due to its high abundance, Rubisco represents an interesting target to express peptides or small proteins as fusion products at high levels (Rumeau et al. 2004). Using *in situ* cryo-electron tomography, Engel et al. (2015) found that in *Chlamydomonas* chloroplasts the Rubisco complexes are hexagonally packed within the pyrenoid, with ~15 nm between their centers. Oxygenase activity of Rubisco causing crop losses in plants can be reduced either

by concentrating CO_2 around Rubisco or by modifying its kinetic properties, whereby potential enlarging of the range of kinetic properties (for example, by introduction of novel amino acid sequences) that would be impossible to reach via normal evolution, could result in improved photosynthesis (Evans 2013).

While a decrease in the activity of Rubisco in response to drought or high temperature is associated with down-regulation of the activation state of the enzyme (e.g., by de-carbamylation and/or binding of inhibitory sugar phosphates), the adverse effects of low temperature, heavy metal stress, ozone, and UV-B irradiation are associated with changes in the concentration of Rubisco (Galmes et al. 2013). The plants regulate Mn and Mg activities in the chloroplast to alleviate deleterious changes in their nitrogen/carbon balance in response to changes in atmospheric CO_2, and Rubisco contributes to improved energy transfers between photorespiration and nitrate assimilation (Bloom and Kameritsch 2017).

Rubisco activase (RCA) is an enzyme activating Rubisco in an ATP-dependent reaction that is supposed to power conformational changes in Rubisco that support the release of inhibitory sugar phosphates from the catalytic site (Khairy et al. 2016), and it plays a vital role in the response of photosynthesis to temperature (Portis 2003). Heating results in the aggregation of RCA, which is connected with the disruption of secondary structure content and formation of insoluble protein (Barta et al. 2010). The RCA gene expression in early *Brachypodium distachyon* seedlings was found to be considerably reduced at increased drought stress (Bayramov and Guliyev 2014). RCA has been reported to play an important role in the regulation of non-steady-state photosynthesis at any leaf temperature, and to a lesser extent of steady-state photosynthesis at high temperatures (Yamori et al. 2012). The reduced content and activity of Rubisco and RCA in tobacco plants cultured under Cd-treatment *in vitro* was found to be recovered by ethylsalicylic acid (Roh and Cui 2014).

Inactivation of −SH groups was suggested to explain the inhibitory effects of Pb^{2+}, Cd^{2+}, Zn^{2+}, and Cu^{2+} ions on the activity of chloroplast enzyme Rubisco and phosphoribulokinase *in vitro* (Stiborová 1986), and inhibition of Rubisco activity by toxic metal ions was confirmed in several plant species (Mysliwa-Kurdziel et al. 2004).

Lee and Roh (2003) observed considerable reduction of Rubisco activity in *Canavalia ensiformis* L. leaves following treatment with 5.0 μM Cd; the activation and induction of Rubisco were inhibited by Cd^{2+} ions, and the change in the levels of RCA resulted in subsequent alteration of Rubisco levels. The evaluation of the protein profile of saffron in response to Cd stress showed that Rubisco and a further 10 proteins including ferredoxin-NADP reductase, a 70 kDa heat shock-related protein, and three synthesis-associated proteins were down-regulated following Cd treatment (Rao et al. 2017). The Rubisco small subunit of *Spinacia oleracea* L. was particularly sensitive to Cd stress showing a 49% decrease in the relative abundance (compared to a 17% decrease estimated for the Rubisco large subunit) (Bagheri et al. 2017).

Pérez-Romero et al. (2016) found that the maximum carboxylation rate of the Cd-accumulator *Salicornia ramosissima* was not affected until 0.20 mM Cd suggesting high tolerance of Rubisco carboxylation capacity to Cd stress which could be connected with many factors associated with the presence of an enzyme tolerant isoform, the reduction of protein degradation, or with improved protein expression and upregulation under Cd stress conditions.

Treatment of hydroponically cultivated *Oryza sativa* L. seedlings with 50 and 100 μM Cd resulted in a significant reduction of the levels of the Rubisco large subunit compared to the control, whereby mainly application of higher concentrations caused extremely strong reduction of this subunit (Wang et al. 2014). A 74 and 95% decrease of Rubisco activity in lettuce leaves treated with 10 and 50 μM Cd, respectively, compared to the control plants was observed, whereby plants exposed to 1 μM Cd showed similar Rubisco activities as the control plants (Dias et al. 2013).

On the other hand, in Cd-treated *Cynara cardunculus* L. plants showing prompt translocation of Cd to the shoot ultrastructural injuries of chloroplasts, but no decline in photochemistry was observed; however, the levels of Rubisco and D_1 protein increased 7- and 4.5-fold, respectively, suggesting that Cd in shoot tissue probably triggers the defense mechanism as a compensatory response to neutralize chloroplast damage (Arena et al. 2017).

Exposure of tomato seedlings to 50 μM Cd that resulted in H_2O_2 accumulation in chloroplasts due to inactivation of ascorbic peroxidase (APX) was associated with a 10% loss of Rubisco content and 60% loss of Rubisco activity, whereby the level of the carbonylated Rubisco large subunit increased 5-fold and −SH groups in the Rubisco large subunit were oxidized (Liu et al. 2008). Inhibition of Rubisco activity under Cd stress could be connected with an irreversible dissociation of the Rubisco large and small subunits (Malik et al. 1992), replacing Mg^{2+} in its catalytic sites with Cd^{2+} (Dias et al. 2013), reducing Chl synthesis at the Cd treatment and shifting Rubisco activity toward oxygenation reactions (Pietrini et al. 2003) or due to the binding of Cd^{2+} to thiol residues of Rubisco activase (Portis 2003).

The study of the mechanism of molecular interactions between Pb^{2+} and Rubisco using spectral methods showed that increasing concentrations of Pb^{2+} ions caused the gradual decrease of the carboxylase activity of Rubisco, whereby Pb^{2+} ions were directly bound to Rubisco. The replacement of Mg^{2+} in the Rubisco catalytic center by Pb^{2+} resulted in a completely altered primary conformation of Rubisco, and it could be suggested that the Pb^{2+} coordination created a new metal-ion-active site form of Rubisco, which was reflected in the reduced carboxylase activity of the enzyme (Wang et al. 2009). Following the spraying of plants cultivated in a field experiment with different concentrations of $PbCl_2$ solution, a notable reduction of the activities of the key enzymes of carbon assimilation, especially RCA, were estimated (Xiao et al. 2008). In a Pb-tolerant *Pisum sativum* cultivar, the application of 1000 mg kg^{-1} Pb resulted in a 2.6-fold decrease of Rubisco activity with respect to the control plants (Rodriguez et al. 2015).

Rubisco activity in rice (*Oryza sativa* L.) plants cultivated in nutrient solution showed a gradual decrease with increasing Cu levels in hydroponic solution and an adverse effect of excess Cu on the activity of Rubisco reduced the overall photosynthetic activity (Lidon and Henriques 1991). In field-grown *Avena sativa* plants, the metal stress due to combined application of Cu^{2+} and Pb^{2+} resulted in a notable reduction (47%) of Chl(*a*+*b*) content, accompanied by proportional changes in Rubisco activity (Moustakas et al. 1994). According to Lidon and Henriques (1991), the prevailing toxic effect of these metal ions was on Rubisco activity, which in turn limited the overall photosynthetic activity. The nitric oxide (NO) donor sodium nitroprusside was found to play a protective role in the regulation of *in vitro* grown tobacco (*Nicotiana tabacum*) plant responses to Cd and Cu stress by enhancing contents and activities of Rubisco and RCA reduced by toxic metal application (Khairy et al. 2016).

Treatment of wheat plants cultivated in Hoagland solution with 50 and 100 μM Ni for 7 days resulted in the reduction of Rubisco protein concentration by 18 and 29%, respectively, compared to the control that may result from a stress-induced proteolysis evoked by oxidative stress and Ni-induced proteolytic degradation of Rubisco (Gajewska et al. 2013). Adverse effects of applied Ni concentrations became evident starting from the 4th day of growth on the metal-containing media and reduced Rubisco level was accompanied with declines in net photosynthetic rate and stomatal conductance, while Ni-induced stimulation of phosphoenolpyruvate carboxylase was detected.

Zinc predominantly inhibits the carboxylase activity of this enzyme without affecting its oxygenase function (Van Assche and Clijsters 1986a, 1986b). Treatment of *Landoltia punctata* with Co^{2+} and Ni^{2+} at a dose of 5 mg dm^{-3} resulted in a 48.8 and 51.1% reduction of Rubisco activity, respectively, and was reflected in a decrease of photosynthesis (Guo et al. 2017).

Harmful effects on the Rubisco activity of pea plants exposed to 200–2000 mg dm^{-3} Cr^{6+} were estimated by Rodriguez et al. (2012), whereby at the highest applied Cr^{6+} concentration a decrease in activity of 90% compared to the control plants was observed, which could be related to the local substitution of Mg by Cr resulting in the reduced affinity of the enzyme to CO_2. A decrease of Rubisco activity estimated in Al-sensitive rye (*Secale cereale*) genotype Riodeva treated with 1.11 and 1.85 mM Al during 3 weeks might be related to the reduced net photosynthetic rate (Silva et al. 2012). In leaves of 2-week-old rice seedlings exposed to two arsenate doses (50 or 100 μM) for 4 days notably reduced the Rubisco large subunit was estimated as resulting in a decreased photosynthesis rate (Ahsan et al. 2010). Partial disruption of the photosynthetic processes with prominent fragmentation of the Rubisco was observed in the adaptative response of hydroponically cultivated *Agrostis tenuis* plants treated with arsenate and arsenite (Duquesnoy et al. 2009). Activation of the Rubisco small subunit, Rubisco large subunit, glutamine synthetase, and glutamate synthase genes significantly contributed to the increase of growth and seed yield of *Brassica juncea* plants following foliar treatment with FeS NPs (2–10 ppm) that were able to activate carbon and nitrogen assimilatory pathways at specific growth stages (Rawat et al. 2017).

The changes in Rubisco activity and protein content in the leaves of trees can also be used as a sensitive diagnostic parameter for ascertaining the negative effects of abiotic and biotic factors of the environment. The results of a study focused on the effects of internal (ontogenetical development) and external factors (meteorological conditions and pollutants) on specific Rubisco activity and protein content in three Slovak autochtonous oak species: *Quercus cerris* L., *Q. robur* L., and *Q. dalechampii* L. stemming from forest stands with different degrees of pollution damage showed that both biochemical characteristics in the oak leaves were significantly reduced (Table 25.1), protein content being the most sensitive parameter for the damage of oak leaves. *Q. robur* or *Q. dalechampii* exhibited higher sensitivity of both biochemical characteristics to adverse effects of the environment than *Q. cerris*, whereby reduction of the protein content in damaged oak leaves was also manifested in inhibition of *de novo* Rubisco synthesis. Higher content of toxic metals (Al, Cu, Zn, and Pb) together with lower content of essential bio-elements (K, Na, Ca, Mg, and S) estimated in damaged leaves of all three oak species suggested that acid rain as well as other sources of pollution increase the amount of toxic metals (especially Al) and flush out the bio-elements (especially Ca and K) from the soil and leaves (Table 25.2) (Konečná et al. 1989). The life functions of woody plants grown under different stress conditions are negatively affected, and the adverse effects of the environment reflected especially in biochemical (enzymes and chloroplasts activities, protein, and Chl contents), and physiological processes, e.g., photosynthesis is finally manifested in a lower biomass production (see Masarovičová et al. 1999a; Masarovičová and Kráľová 2017, 2018).

TABLE 25.1
Values of the Specific Rubisco Activity and Protein Content in the *Quercus dalechampii, Q. robur,* and *Q. cerris* Leaves

	Specific Activity (Bq μg⁻¹ (Protein))			Protein Content (μg mm⁻³)		
	Healthy Samples	Damaged Samples		Healthy Samples	Damaged Samples	
		A	B		A	B
Q. dalechampii:						
	58.55	37.42	29.96	7.20	1.92	0.29
	107.42	31.13	30.41	5.56	1.88	0.32
	97.22	27.78	20.01	2.38	1.88	0.45
	104.52	–	19.34	3.01	–	0.49
				2.82		
x	91.93	32.11	24.93	4.19	1.89	0.39
s_x	11.33	2.83	3.04	0.93	0.01	0.05
Q. robur:						
	41.98	25.41	20.10	6.90	1.16	0.45
	69.95	33.60	31.15	4.86	1.76	0.33
	83.77	17.78	28.87	4.38	1.16	0.39
	–	–	31.76	3.42	–	0.49
			25.39			0.57
x	65.23	25.60	27.45	4.89	1.36	0.45
s_x	12.29	4.57	2.15	0.73	0.20	0.04
Q. cerris:						
	24.58	27.72	27.81	6.04	1.28	0.45
	34.63	20.78	30.53	4.47	1.36	0.73
	27.74	17.38	31.67	3.12	1.36	0.84
	26.57	–	23.50	2.84	–	0.77
	–	–	19.55	1.92	–	0.77
x	28.38	21.96	26.61	3.68	1.33	0.71
s_x	2.18	3.04	2.26	0.72	0.03	0.07

Source: Konečná, B. et al., *Photosynthetica* 23:566, 1989. With permission.

Note: A – leaves of the seedlings transferred in the spring 1987 from damaged forest stand and transplanted into the garden; B – leaves of the seedlings that were processed immediately after sampling (July 1987); x – mean, s_x – S.E.

25.5 IMPACT OF TOXIC METALS ON WOODY PLANTS USED AS METAL PHYTOREMEDIATORS

The EU has close to 182 million hectares of forests and other wooded lands, corresponding to 43% of the EU land area, and the overall level of EU-28 roundwood production reached an estimated 425 million m^3 in 2014, whereby resources from forestry and forest industries are the main contributor to bioenergy production (cf. Masarovičová et al. 2009; Bentsen and Felby 2012; Eurostat 2016). The economic weight of the wood-based industries in the EU-28 as measured by gross value added was equivalent to EUR 129 billion or 7.9% of the manufacturing total in 2013 (Eurostat 2016). Moreover, some woody plants such as poplar, willow, black locust, ash, or alder could also be successfully used for remediation of substrates contaminated by organic and inorganic pollutants, including toxic metals. These species accumulate only moderate toxic metal concentrations in their shoots (in contrast to metal hyperaccumulators characterized generally by low biomass production), but their large biomass predestines them to be successfully used in the phytoremediation of metal-contaminated soil. For example, the fast-growing tree poplar (3–5m per year) characterized by a high transpiration rate (100 liters per day) producing a large biomass that is not a part of the food chain can be considered as an excellent species for phytoremediation purposes (cf. Chapell 1997; Masarovičová et al. 2010b; Masarovičová and Kráľová 2017, 2018).

An extensive and massive root system of fast-growing trees (e.g., willows or poplars) that penetrate deeply into the soil ensure efficient uptake of water contaminated with pollutants from the substrate. The fast-growing trees, due to their perennial characteristics, long life span, high transpiration rate, quick regeneration of their removed aboveground parts, and easy vegetative reproduction, are advantageous in phytoremediation processes (see Stomp et al. 1993). For example, poplars could

TABLE 25.2
Contents of the Metallic and Non-Metallic Elements in the Healthy and Damaged Oak Trees

Species	Content of Elements (mg kg⁻¹ dm)								
	K	Na	Ca	Mg	S	Al	Cu	Zn	Pb
Healthy samples									
Q.cerris	8 750	75	12 820	1 671	800	3 727	55	36	6
Q.robur	8 470	100	24 120	2 023	810	1 090	78	62	3
Q.dalechampii	6 750	50	15 580	2 673	836	1 878	463	63	3
Damaged samples									
Q.cerris	8 000	80	8 086	4 376	632	6 738	212	26	7
Q.robur	6 620	100	8 288	851	824	1 697	68	23	2
Q.dalechampii	3 720	60	12 330	1 823	989	2 109	97	22	3

Source: Konečná, B. et al. *Photosynthetica* 23:566, 1989. With permission.

take up some inorganic contaminants including toxic metals, such as Cd (Koprivova et al. 2002), Hg (Rugh et al. 1998), and Zn (Di Baccio et al. 2003), even though their tolerance to toxic metals is limited (Dietz and Schnoor 2001), are suitable for several cycles of decontamination, while their leaves could be easily collected and subsequently incinerated (Dietz and Schnoor 2001; Bittsánszky et al. 2005). However, it should be stressed that the precondition for utilization of woody plants in phytoremediation technologies is their sufficient toxic metal tolerance, and it is necessary to thoroughly investigate the toxic effects of metals on the structure and function of trees.

Since plants growing on contaminated substrates absorb contaminants through the roots and store them in root tissues and/or transport them into the stems and/or leaves, from the aspect of phytoremediation effectiveness it is important to estimate the bioconcentration factor (BCF) that is defined as the ratio of metal concentrations in plant dry mass (µg g⁻¹ dm) to those in the soil (µg g⁻¹ soil) (Tu and Ma 2002; Masarovičová et al. 2010b) and translocation factor (TF) defined as the total metal content in plant shoots to that in the roots, reflecting the effectiveness of plants in translocating metal from the root to the shoot (Stoltz and Greger 2002; Tu and Ma 2002; Deng et al. 2004; Masarovičová et al. 2010b).

25.5.1 Willows

Cadmium is one of the most dangerous environmental pollutants showing adverse effects on plants and has been found to decrease root elongation, cause root tip damage, collapse root hairs or decrease of their number, reduce root biomass, and increase or decrease lateral root formation (see Barceló and Poschenrieder 1999). Šottníková et al. (2003) found that Cd-treatment suppressed rooting and root growth (length and biomass production) as well as its development in willow (*Salix viminalis* L., *S. alba* L., *S. purpurea* L., and *S. cinerea* L.) and poplar (*Populus* × *euroamerica* cv. Gigant and *Populus* × *euroamerica* cv. Robusta) species. The roots of tested species responded to Cd treatment more sensitively than the shoots; the highest tolerance of root systems to Cd

stress was observed in *S. cinerea*, *S. alba*, and *Populus* cv. Robusta, and Cd was predominantly accumulated in the roots.

The comparison of above-mentioned willow and poplar species, rooted directly in 12 µM Cd(NO₃)₂ (direct treatment) with plants firstly rooted in Knop nutrient solution and transferred to Cd(NO₃)₂ afterwards (indirect treatment), demonstrated that rooting in Knop nutrient solution showed beneficial effects on root cumulative length, number and biomass production of *S. alba* roots as well as assimilation pigment and starch contents, net photosynthetic rate and specific leaf mass of *S. alba* leaves (Lunáčková et al. 2003a). On the other hand, roots and shoots of *P. robusta* rooted in Knop nutrient solution were more sensitive to the toxic effect of Cd than plants cultivated directly in Cd treatment. Root apices of both species in both experimental variants were negatively affected by Cd treatment, while central cylinders of more distant root parts were not seriously influenced. The cambial activity started, and lateral root primordia were formed close to the root apex. Structural changes induced by Cd indicated a better adaptation of the roots of directly Cd-treated plants of both species than of the roots of indirectly Cd-treated plants. Cd application resulted in enhanced values of specific leaf mass in both species and caused the xeromorphic character of leaves – increased stomata density but reduced stomata sizes. Indirect treatment positively affected assimilation pigment and starch contents, net photosynthetic rate, and specific leaf mass, and it lowered root Cd uptake in willow, Cd accumulation in cuttings of both species, and Cd accumulation in poplar shoot. It could be noted that directly Cd-treated poplar roots exhibited unusual defense activity of the root apical meristem. According to researchers, *S. alba* could be recommended for phytoextraction of toxic metals from contaminated substrates. Cd treatment was found to increase root respiration rate of willow and poplar plants suggesting that the toxic effect of Cd-induced energy required for increased metal ions uptake into the roots and for repairing mechanisms as a consequence of metabolism damages (Lunáčková et al. 2003b).

The performance and toxic metal uptake by *S. viminalis* in the field were found to be 2–10-fold higher than by plants in the growth chamber. However, the growth of willow grown on strongly polluted soil with up to 18 mg Cd kg⁻¹, 1400 mg Cu kg⁻¹, 500 mg Pb kg⁻¹, and 3300 mg Zn kg⁻¹ was notably reduced, and therefore despite high Cd and Zn concentrations in the leaves (≥80 mg Cd kg⁻¹ and ≥3000 mg Zn kg⁻¹) *S. viminalis* was found to be unsuitable for phytoremediation purposes at the given soil pollution (Jensen et al. 2009). On the other hand, willows grown on moderately polluted soils (2.5 mg Cd kg⁻¹ and 400 mg Zn kg⁻¹) extracted 0.13% of total Cd and 0.29% of the total Zn per year, probably representing the most mobile fraction, suggesting that they are promising candidate for metal extraction if pollution with metals is not too high.

Between two willow clones cultivated in a greenhouse pot experiment on six sediment-derived soils supplemented with Cd (0.9−41.4 mg kg⁻¹) dry weight root biomass and total shoot length of *S. viminalis* were significantly lower compared to *S. fragilis* for all treatments; however, both clones accumulated high Cd and Zn levels in aboveground plant parts, BCFs of Cd and Zn in the leaves being highest for the treatments with the lowest soil Cd and Zn concentration (Vandecasteele et al. 2005). The investigation of the effect of Ca^{2+}/Mg^{2+} ratio on Cu accumulation, photosynthetic activity and growth of *S. viminalis* L. 'Cannabina' treated with Cu^{2+} at doses 1, 2, and 3 mM showed that at 1/10 Ca^{2+}/Mg^{2+} ions ratio the plants were adapted to high Cu^{2+} concentrations and they accumulated not only the highest Cu amounts in the roots but also showed effective Cu translocation to aboveground plant organs (Borowiak et al. 2012). Among 20 different clones of willow and poplar species cultivated hydroponically in the presence of 4.45 µmol dm⁻³ Cd or 76.5 µmol dm⁻³ Zn, the largest metal concentrations in the leaves were detected in *S. dasyclados* (315 mg Cd kg⁻¹ dm) and a *S. smithiana* clone (3180 mg Zn kg⁻¹ dm) showing low metal tolerance. On the other hand, the clones of *S. matsudana*, *S. fragilis* L., and *S. purpurea* L. produced large biomass, even though there was a lower metal contents in leaves than in the above-mentioned two willow species, and could be considered as suitable candidates for phytoextraction (Dos Santos Utmazian et al. 2007).

A multiple-level study of metal (Zn, Cu, Cd, Ni) tolerance in *S. fragilis* and *S. aurita* clones was performed by Evlard et al. (2014) using rooted cuttings planted in contaminated soil and cultivated for 100 days. Based on estimated growth reduction and the proteomic changes in *S. fragilis*, it was concluded that this clone adjusted its metabolism to maintain cellular homeostasis. On the other hand, *S. aurita* maintained growth at the cost of cellular deregulation as indicated by the physiological and proteomics data, and consequently it can be stated that the high biomass is not linked with a good tolerance strategy.

In cuttings of *S. viminalis* L cv. Cannabina cultivated in Knop's medium in the presence of 3 mM Ni for 14 days, the estimated Ni content in shoots and roots was about 15 mg kg⁻¹ d.w. and inhibition of root biomass, shoot, root and leaf elongation, and the photosynthetic area were observed. On the other hand, a notable increase in soluble carbohydrates in leaves (340%) and a 4-fold increase of phenolics content

compared to the control was estimated, and treatment with 2.5 mM Ni resulted in approximately a 68-fold higher salicylic acid content than in untreated plants. The tested willow cultivar exhibited sufficient resistance to Ni to be cultivated at heavily contaminated sites, and the observed powerful plant growth inhibition probably resulted from Ni-induced disturbances in nutrient uptake accompanied by oxidative stress (Drzewiecka et al. 2012).

Mg and Fe deficiency in mature plants of hydroponically cultivated *Salix viminalis* in the presence of 5 µg cm⁻³ Cd resulted in a notable increase in Cd accumulation in roots, stems, and leaves. It can be supposed that Cd utilizes the same uptake and transport pathways as Mg and Fe and the uptake and translocation of Mg and Fe could be further facilitated by plants as an adaptive response to the deficiency of these essential metals, whereby such physiological reaction could additionally stimulate Cd accumulation. Based on the high Cd content in aerial plant parts (51.5−130.6 µg g⁻¹), the tested *S. viminalis* genotype could be considered as suitable for phytoextraction (Borisev et al. 2016).

In three willow species (*S. alba*, *S. matsudana*, and *S. nigra*) exposed to very high Cd concentrations in hydroponics, the F_v/F_m ratio was found to be a less sensitive indicator of Cd stress than the rate of photosynthesis and pigment concentration and based on biomass production of *S. nigra* and *S. matsudana* that showed a high tolerance to Cd, independently on the applied Cd concentration; these two willow species could be utilized in the sanitation of Cd-contaminated sites (Nikolic et al. 2015).

In five *Salix* spp. exposed to high Cd and Zn concentrations, the concentrations of both metals in plant organs decreased as follows: root>leaf>stem, regardless of the species. In a combined Cd and Zn treatment, more effective translocation of Cd from root to aboveground tissue was observed, while the same treatment suppressed Zn translocation from root to leaf and stem. In general, a higher reduction of photosynthetic parameters in Zn alone and combined Cd and Zn treatments was estimated compared to the treatment using only Cd and *S. caprea*, which showed the lowest reduction of photosynthesis relative to the control could be considered as a suitable candidate for phytoremediation of Cd- and Zn-contaminated sites (Han et al. 2013).

Symbiotic association between *S. purpurea* L. and *Rhizophagus irregularis* under Cu stress reduced gas exchange associated with changes in Chl content, affected the sequestration of Cu into the cell walls by modifying vessels anatomy, and upheld leaf specific conductivity and root hydraulic conductance, which may be related to a dynamic aquaporin gene regulation of plasma membrane intrinsic proteins $PIP_{1;2}$ along with an upregulation of tonoplast intrinsic protein $TIP_{2;2}$ in the roots of inoculated willow (Almeida-Rodriguez et al. 2016).

25.5.2 Poplars

The toxic effects of Cd on hybrid poplar plants were manifested by stunted growth (plant height and biomass), decreased root length, and chlorosis of young leaves, whereby stem and leaf growth was more affected than root growth, and the plants showed reduced photosynthetic activity compared to the control plants as a consequence of a decline in Chl synthesis. Plants exposed to 10 µM Cd accumulated approximately 40-fold higher Cd concentration in the roots than in the leaves and stems (Nikolic et al. 2008). In 13 poplar clones, the lowest concentration of metals was found in wood, while the highest was generally estimated in senescing leaves, making removal and treatment of fallen leaves necessary (Laureysens et al. 2004). Gu et al. (2007) reported that 100 µM Cd notably inhibited root growth of four poplar cultivars; Cd accumulation in all poplar organs increased significantly with increasing Cd concentration (10–100 µmol dm^{-3}) and with time; Cd was accumulated predominantly in the roots and Cd concentrations in bark were pronouncedly higher than in wood.

Acclimation to the Cd-induced oxidative stress following a period of functional decline was estimated in hydroponically cultivated poplar (*Populus jacquemontiana* subsp. *glauca* cv. 'Kopeczkii') plants treated with 10 µM Cd for 4 weeks, and increased activity of peroxidases dominantly contributed to H$_2$O$_2$ elimination and thus the recovery of photosynthesis. The function of superoxide dismutase (SOD) isoforms was found to be less important, and the biosynthesis of the photosynthetic apparatus and some antioxidative enzymes was supported by redistribution of the Fe content of leaf mesophyll cells into the chloroplasts. Since a decline in the level of lipid peroxidation was observed before the increase in photosynthetic activity, it could be suggested that for the restoration of the photosynthetic apparatus during long-term Cd-treatment elimination of oxidative stress damage by acclimation mechanisms is required (Solti et al. 2016). A nitrogen supplement effectively alleviated Cd toxicity in poplar plants through mitigation of the Cd-induced inhibition of Chl synthesis causing an unblocking of the corresponding pathway (Zhang et al. 2014).

Nikolic et al. (2017) cultivated *P. deltoides* (clone B-81) and *P. × euramericana* (clone Pannonia) in soil containing 8.14 mg Cd kg^{-1} soil under semi-controlled conditions for 6 weeks and found that biomass production and morphological characteristics in *P. × euramericana* were adversely affected to a greater extent than in *P. deltoides* plants and tolerance index of *P. deltoides* was also higher than that of *P. × euramericana* (82.2 vs. 66.5, respectively). The researchers related the higher tolerance to Cd toxicity observed in *P. deltoides* to unchanged proline content and undisturbed nitrogen metabolism. Differences were also estimated among the metal amounts accumulated in the individual plant organs of both poplars of the total Cd content uptaken by plants, which were 58.0 and 46.7% in the roots, 18.2 and 39.9% in the stems, and 23.8 and 13.4% in leaves of *P. × euramericana* and *P. deltoids*, respectively, and it could be stated that *P. deltoides* is suitable for Cd phytostabilization and phytoextraction as well.

Treatment of *P. × euramericana* cv. Neva seedlings with Cd (50–500 µM) for 40 days resulted in notable ultrastructural changes in the leaves such as swollen chloroplast thylakoids, dissolved thylakoid grana, disintegrated chloroplasts, numerous plastoglobuli in chloroplasts, and the content of soluble protein increased with an increasing Cd concentration and declined with the prolonged duration of the treatment; this caused a significant decrease of Chl*a*, Chl*b*, and Chl(*a*+*b*) reaching 47.69, 37.10, and 45.49% of the control at treatment with 500 µM Cd (Ge et al. 2015). A chemical, morphological, and proteomics study focused on the response of *Populus* spp. to Cd stress showed that Cd toxicity was mainly influenced by proteins related to general defense, stress response, and carbohydrate metabolism (Marmiroli et al. 2013). Gaudet et al. (2011) investigated the intraspecific variation of physiological and molecular responses to Cd stress in *Populus nigra* L. using two *P. nigra* genotypes, originating from contrasting environments in the northern (genotype 58-861) and southern (genotype Poli) regions of Italy that were treated with Cd in hydroponics for 3 weeks. Although Cd stress affected biomass production and photosynthesis in both clones, the southern clone was found to be notably more tolerant. The northern clone accumulated thiols but down-regulated the glutathione S-transferase (GST) gene, whereas the southern clone accumulated phytochelatins and up-regulated the GST gene that could be advantageous to complex and detoxify Cd.

In *P. × euramericana* (Dade) cv. Nanlin-95 and *S. jiangsuensis* CL. '172' cultivated in elevated CO$_2$ (EC) in Cd-contaminated soil, plant growth stimulation by increasing leaf photosynthesis and higher efficiency of phytoremediation was observed, mainly at high levels of Cd exposure, which was connected to decreased oxidative damage due to increased antioxidant enzyme activities and because of lower reactive oxygen species (ROS) levels membrane lipid peroxidation was avoided (Guo et al. 2015). Similar results related to an EC effect on plant growth, gas exchange, root morphology, and Cd uptake of *P. euramericana* cv. '74/76' and two willow genotypes (*Salix jiangsuensis* CL. '799' and *S. jiangsuensis* CL. '172') were published by Wang et al. (2012b), who observed an elevated total Cd uptake by tested species due to an increased biomass under EC suggesting an alternative way to improve the efficiency of the phytoremediation of contaminated soil.

Exogenous abscisic acid was found to decrease Zn concentrations in *P. × canescens* under excess Zn for 7 days, likely by modulating the transcript levels of key genes involved in Zn uptake and detoxification (Shi et al. 2015). Investigation of the Ni phytoremediation potential of *P. alba* clones grown *in vitro* using tissue culture showed that a higher Ni concentration notably inhibited accumulation of fresh mass and especially contents of photosynthetic pigments, whereby at 1 mM Ni serious disturbance of growth in the clones was estimated and application of 0.1 mM Ni was reflected in a considerable increase of Ni accumulation, while the increase of biomass accumulation was observed in a treatment with 10 µM Ni (Katanic et al. 2015). The percentage of linolenic acid, which is mainly associated with thylakoid lipids, was markedly lower in leaves of *P. nigra* trees grown on the metal-contaminated landfill compared to trees cultivated on the control area and Pb and Cr contents were significantly negatively correlated to C$_{18/3}$/(C$_{18/0}$+C$_{18/1}$+C$_{18/2}$) fatty acid ratios, whereby the decline of C$_{18/3}$/(C$_{18/0}$+C$_{18/1}$+C$_{18/2}$) fatty acid ratio was already observed at a low leaf metal content (Le Guedard et al. 2012).

25.5.3 *Robinia pseudoacacia* L.

Among tree species cultivated on a mound constructed of dredged sediment slightly polluted with toxic metals, *Robinia pseudoacacia* and white poplar had the highest growth rates, while ash, maple, and alder showed the highest survival rates (>90%); however, they exhibited stunted growth. In the foliage of ash, alder, maple, and *Robinia pseudoacacia* normal concentrations of Cd, Cu, Pb, and Zn were estimated, and these species could be considered as suitable candidates for phytostabilization under the given environmental conditions. On the other hand, white poplar that accumulated 8.0 mg Cd kg^{-1} and 465 mg Zn kg^{-1} in its leaves might cause a risk of Cd and Zn input into the ecosystem because of autumn litterfall and therefore it is unsuitable for phytostabilization (Mertens et al. 2004). Based on the strong correlation between the degree of metal contamination and metal concentrations in leaves of *R. pseudoacacia* estimated at sites with different degrees of metal pollution (industry, urban roadside, suburban), this species could be used as an effective biomonitor (bioindicator) of environmental quality in areas subjected to both industrial and traffic pollutions (Celik et al. 2005). The metal accumulation index values of trees growing in urban biotopes with different levels of toxic metal contamination in the city of Dabrowa Gornicza (southern region of Poland, Silesian Voivodeship) were 7.3–20.6 for *R. pseudoacacia* and 10.5–27.2 for *Betula pendula*. The increasing tendency of proline content in the leaves of *B. pendula* grown on biotopes characterized with high traffic was estimated, while in the leaves of *R pseudoacacia* increased non-protein thiols content was observed (Nadgorska-Socha et al. 2017).

Photosynthetic sensitivity to Cd and Pb of the 1-year-old seedlings of *R. pseudoacacia* sprayed on the leaves twice over a period of 10 days with Cd and Pb (250–2000 mg dm^{-3}) dissolved in saline was found to be weak, the PS II chlorophyll pigments were not damaged, and a slight impact on Chl fluorescence was observed only in treatments with the two highest concentrations (Dezhban et al. 2015). In *R. pseudoacacia* seedlings exposed to Pb stress, the chloroplasts swelled owing to local plasmolysis and lost most of their starch content; their thylakoid lamellae gradually became disordered and loosely packed and with an application of Pb concentration exceeding 1000 mg kg^{-1} the chloroplast envelope was lost and lipid globules were released into the surrounding mesophyll cells, whereby the photosynthetic function was most significantly affected by stomatal conductance and inactivity of the PS II reaction center, while PET inhibition exhibited only minute effects (Zhou et al. 2017a).

Four black locust families exposed to Cd (6 ppm), Ni (100 ppm), and Pb (40 ppm) in hydroponics were found to be highly tolerant to metal stress with tolerance index >60, although their photosynthetic activities were highly sensitive to applied metals, net photosynthetic rate of Ni-treated plants being the most affected. While Cd and Ni concentrations in stems and leaves of *R. pseudoacacia* families exceeded 100 mg Cd kg^{-1} and 1000 mg Ni kg^{-1}, in both single metal treatments as well as the simultaneous application of all three metals, Pb

accumulation in aboveground biomass was greatly affected by a multi-pollution treatment (Zupunski et al. 2016).

25.5.4 Eucalypt Species

Iori et al. (2017) tested the Cd phytoremediation ability of two clones of eucalypt plants cultivated in hydroponics and found that Cd was accumulated mainly in the roots, with low metal level reaching the aerial parts; however, the plants exhibited a notable tolerance to submersion and showed large environmental adaptability suggesting that they could be used for the phytoremediation of Cd-polluted wastewaters.

Following exposure of two eucalypt genotypes (hybrid clones of *Eucalyptus camaldulensis* × *E. globulus* ssp. *bicostata* J.B. Kirkp named Velino ex 7 and Viglio ex 358) for 3 weeks to 50 µM CdSO$_4$ under hydroponics, the metal binding and particularly the content of thiols like cysteine and glutathione (GSH), organic acids like oxalate and citrate, and polyamines in the clones showed differences suggesting different defense responses to Cd between the clones, although low TF was estimated in both clones (Pietrini et al. 2015).

The acid and Al stress adversely affected the photosynthetic activity of seedlings of four vegetatively propagated *Eucalyptus* clones. Al reduced Chl content and inhibited net photosynthetic rate, transpiration rate, and water use efficiency, whereby at the same Al concentration (4.4 mM) the highest impairment of photosynthetic parameters was estimated at pH 3. Moreover, Al stress-induced changes in leaf morphology result in a reduced thickness of leaf epidermis and palisade tissue, the descendant palisade tissue/spongy tissue ratio and leaf tissue looseness. On the other hand, differences in tolerance to Al toxicity between the clones were estimated to improve *Eucalyptus* plantation productivity by selecting Al tolerant clones (Yang et al. 2015). Among seedlings of six clones of the *Eucalyptus* species (*E. globulus* Labill., *E. urophylla* S.T. Blake, *E. dunnii* Maiden, *E. saligna* Sm., *E. cloeziana* F. J. Muell., and *E. grandis* w. Hill ex Maiden), cultivated for 10 days in hydroponics containing 0–2.5 µM Al, *E. globulus* and *E. urophylla* were found to be more tolerant to Al toxicity and *E. grandis* and *E. cloeziana* were more susceptible to Al-induced damage. As *E. globulus* seedlings were highly sensitive to lanthanum, it could be supposed that they have the tolerance mechanism that is specific for Al. The uptaken Al was predominantly retained in the root system, and metal amount translocated to the shoots was found particularly in older leaves. In root tips of seedlings exposed to Al, an increase in malic acid concentration 200% was estimated, whereby the rise in malate concentration correlated with the degree of Al tolerance of the species, and it could be suggested that tolerance is mainly derived from the internal detoxification of Al by complexation with malic acid (Silva et al. 2004).

25.5.5 Some Other Broadleaved Species

Theriault et al. (2016) performed a comprehensive transcriptome analysis of responses to Ni stress in white birch (*Betula papyrifera*) using Ni resistant and Ni susceptible genotypes. A total of 208,058 trinity genes were identified that were assembled to 275,545 total trinity transcripts. Expression profiles were generated, and 62,587 genes were found to be significantly differentially expressed among the Ni resistant, susceptible, and untreated libraries. A downregulation of genes associated with translation (in ribosome), binding, and transporter activities was estimated as the main Ni resistance mechanism in white birch and five identified candidate genes associated with Ni resistance including GST, thioredoxin family protein, putative transmembrane protein, and two Nramp transporters were reported to be suitable for the genetic engineering of birch trees.

B. pendula trees grown on highly contaminated mounds of metallurgical waste accumulated high amounts of Zn in its leaves and their effect on the cleansing of the soil in the heavily polluted environment was reflected in considerably lower Zn content in the soil, and in the leaves of *Solidago virgaurea* grown near the trees, compared to plants grown at 7 m away, suggesting the phytoremediation potential of *B. pendula* to clean soils contaminated with Zn. Moreover, the researchers proposed the inclusion of *B. pendula* in the group of hyperaccumulators of Zn (Dmuchowski et al. 2014). Pajak et al. (2017) evaluated accumulation capacities of Pb, Zn, Cd, Cu, and Cr in washed and unwashed needles of *Acer platanoides* L. and leaves of *B. pendula* Roth growing in a contaminated area and found that the metal accumulation in plant tissues correlated with soil concentrations only in the case of Zn and Pb, respectively. The levels of Fe, Mn, and Zn in the leaves and bark of *B. pendula* L. and *Acer platanoides* L. grown in an urbanized environment exceeded that of other elements, and in the leaves of Norway maple the highest mean content of Fe and Mn was found, while in *B. pendula* leaves the highest mean content of Zn was observed. The impact of the sampling sites on the content of trace elements in tree bark was less regular than the analogous effect in leaves, and significant correlations between the content of trace elements in soil and their accumulation in the leaves and bark of investigated trees was confirmed (Kosiorek et al. 2016).

Van Nevel et al. (2011) investigated the effects of *B. pendula*, *Quercus robur*, and *Q. petraea*, *R. pseudoacacia* and *P. tremula* on the soil pH, organic carbon content and cation exchange capacity and on the redistribution of Cd and Zn on a polluted sandy soil, whereby soil and biomass were sampled in 10-year-old stands growing on former agricultural land. Increased total as well as the NH_4OAc-EDTA-extractable Cd and Zn concentrations in the topsoil compared to deeper soil layers and to other tree species were observed with *P. tremula* because of Cd and Zn accumulation in its leaf litter; aspen cultivation also resulted in notable higher topsoil pH, content of organic carbon and cation exchange capacity compared to the majority of the other species. Based on these results, aspen should be avoided when afforesting Cd and Zn-contaminated lands.

Seedlings of Holm oak (*Q. ilex* subsp. *ballota*) and mastic shrub (*Pistacia lentiscus*), frequently used in the afforestation of degraded soils in mining areas that were treated with Cd (20–200 mg dm^{-3}) and Tl (2–20 mg dm^{-3}) in a sand culture exhibited a strong decrease of net assimilation rates and stomatal conductance compared to the control plants. Seedlings treated with Cd showed an increase in the root mass ratio and a decrease in the specific leaf area in oak seedlings. On the other hand, application of Tl did not cause such a response; however, Tl affected the antennae complexes and RCs of the PS II in plants more than Cd, which resulted in stronger phytotoxic effects, *P. lentiscus* plants being more sensitive to both tested metals than oak plants. The tolerance of Holm oak to Cd was found to be higher than to Tl, which is connected with its high capacity for Cd retention at the root level, while Tl was effectively translocated into the leaves and this species also showed higher resistance to both metals than the mastic shrub (Dominguez et al. 2011).

Seedlings of three *Quercus* spp. (*Q. shumardii*, *Q. phellos*, and *Q. virginiana*) and rooted cuttings of two *Salix* spp. (*S. matsudana* and *S. integra*) that were transplanted into pots containing 50 and 100% Pb/Zn mine tailings were evaluated for their tolerance to heavy metals. *Q. virginiana* was found to be highly tolerant to heavy metals, it was able to grow normally in the Pb/Zn tailings, and it showed low values of metal TFs and therefore could be considered as a candidate for the phytostabilization of mine tailings. In general, the TFs estimated for all species were low, notably higher TF values for Zn (1.42–2.18) and Cd (1.03–1.45) compared to other studied species were observed only with *S. integra* (Shi et al. 2017). *In vitro* treatment of *Q. pubescens* plantlets with high Cd or Cu concentrations (50 and 250 μM, respectively) resulted in reduced dry biomass of shoots and roots and the plant tolerance index (Di Santo et al. 2017). *Q. variabilis* grown in soils containing high concentrations of Sb was found to tolerate and accumulate high Sb levels in roots (1623.39 mg kg^{-1} d.w.) and while the Sb concentration in roots and stems gradually increased, in leaves it remained at a certain level for several weeks. Sb showed a higher adverse effect on PET after Q_A^- than on the OEC and light-harvesting pigment–protein complex II, which reduced leaf Chl concentrations and inhibited plant growth; however, it elicited defense mechanisms on the donor side of PS II and acclimation to Sb toxicity was observed as well (Zhao et al. 2015).

Seedlings of *Alnus incana* and *A. glutinosa* which were grown for 158 days in forest soil (control) and in soil from a Cu smelter area, with low nutrient content and high concentrations of Cu (1510 mg kg^{-1}) and Pb (490 mg kg^{-1}) accumulated in their roots Cu and stored Pb, Zn (*A. incana* approximate to *A. glutinosa*) and Cd (*A. glutinosa*>*A. incana*), concentrations of Cu, Pb, Zn, and Cd being lower in shoots than in roots. Nodules stored Cu, Cd (*A. incana*>*A. glutinosa*), Pb, and Zn, while concentrations of P and Ca were lower in nodules than in the control and reduction in the rate of nitrogen fixation in nodules when alders were grown in polluted soil was observed (Lorenc-Plucinska et al. 2013).

Study of the accumulation and translocation of Cu in the organs of two apple tree cultivars (Fuji and Rails) showed that Cu accumulated mostly within apple tree roots (especially in the fibrous roots); however, cv. Rails showed considerably higher Cu concentrations in fibrous roots and lower concentrations in the aboveground parts. Cu accumulated in the fibrous root of cv. Fuji showed higher *in vivo* mobility, while higher Cu amounts were sequestrated in the cell wall and vacuole with cv. Rails (Wang et al. 2016a).

The concentration and accumulation of oxalic acid and malic acid in the roots of *Malus hupehensis* (Pamp.) Rehd seedlings exposed to low and toxic Zn concentrations showed a 20–60% increase after a short-time metal treatment compared to the control plants and contributed to the resistance of Zn stress through a balancing of Zn homeostasis (Liu et al. 2012). As the primary strategy to reduce Cd mobility and toxicity in four apple rootstocks exposed to 300 µM $CdCl_2$ immobilization of Cd in the cell wall and soluble fraction (most likely in the vacuole) and conversion into pectate- or protein-integrated forms and undissolved Cd phosphate forms was suggested. Among tested apple rootstocks *Malus baccata* showed not only lower concentrations of ROS in roots and bark, H_2O_2 in roots and leaves and malonedialdehyde (MDA) in roots, wood and bark, but also higher concentrations of soluble sugars in the bark and starch in the roots and leaves as well as enhanced levels of antioxidants which were found to be highly tolerant to Cd stress (Zhou et al. 2017b). *M. baccata* exposed to Cd stress also showed the smallest BCF and TF and the lowest Cd concentrations in the tissues as well as lower transcript levels of genes related to Cd uptake and translocation, but stronger gene expression for Cd detoxification compared to the other three tested apple rootstocks suggesting that it could be used as rootstock in Cd-contaminated soil.

Potted pear trees grafted on quince (*Cydonia oblonga* Mill.) Sydo stock treated with Cu in sandy and clay-loam soils showed reduced shoot growth and leaf area only for pear grown in sandy soils containing >400 mg Cu kg^{-1} and Cu accumulation in roots increased linearly with increasing Cu levels in sandy soil, while in clay-loam soil a quasi-parabolic course was observed, whereby with increasing soil Cu concentration a decrease in root Zn concentration was estimated (Toselli et al. 2008). *Prosopis juliflora* seedlings exposed to short-term (48 h) Cu stress at doses of 50 and 100 mM accumulated notable Cu amounts in their roots and showed an evident reduction in the maximum photochemical efficiency of PS II (F_v/F_m) and the activity of PS II (F_v/F_0) in leaves treated with 100 mM Cu^{2+} with respect to the control after 4 h of exposure suggesting the photosynthetic apparatus of *P. juliflora* as the primary target of the Cu^{2+} action (Michel-Lopez et al. 2016a). In a similar experiment with *P. juliflora* seedlings exposed to short-term (48 h) Cd stresses the F_v/F_m and F_v/F_0 ratios in *P. juliflora* leaf did not show significant changes during the experiment indicating that the photosynthetic apparatus was not the primary target of the Cd^{2+} action (Michel-Lopez 2016b).

25.5.6 Coniferous Trees

Pinus sylvestris grown on an abandoned historic Cu deposit in Ľubietová (Slovakia) exhibited the strategy of an accumulator of Pb and Zn, whereby Zn translocated into leaves

(TF = 2.95), and Pb remained fixed in the roots (TF = 0.18) (Andras et al. 2016). In mycorrhized pine seedlings grown in the presence of lead, Pb was primarily accumulated in the roots, and a smaller metal amount was estimated in the pine needles. Pb was concentrated in nanocrystalline aggregates attached to the cell wall, it was associated with P and Cl and in both roots and needles pyromorphite, $Pb_5(PO_4)_3Cl$, was present. Such biomineralization could be interpreted as a defense mechanism of the plant against Pb pollution (Bizo et al. 2017).

Arencibia et al. (2016) investigated the tolerance of 1-year-old seedlings of *Pinus* spp. (*P. radiata*, *P. pinaster*, and *P. canariensis*) to abiotic stress induced by $CuSO_4$ (300 mM) or $AlCl_3$ (100 mM) in *ex vitro* conditions and found that a higher number of tolerant seedlings originated from seeds collected from mother trees growing in a more severe environment. *P. pinaster* showed the highest activity level for the enzymes peroxidase (POX), SOD, and catalase (CAT) estimated after 60 days of metal treatment, while the intermediate and lowest values were observed in *P. canariensis* and *P. radiata*. The most tolerant species to Cu-induced stress was found to be *P. pinaster*, and it was shown that abiotic stress in the maternal environment could induce 'transgenerational plasticity,' which could affect progeny performances. Under Cu excess, notable inhibition of growth and development of 6-week-old *P. sylvestris* seedlings cultivated in hydroculture was observed along with the loss of dominance of the main root and strong inhibition of lateral root development suggesting a lack of adaptive reorganization of the root system architecture under metal excess. Moreover, application of 10 µM $CuSO_4$ completely suppressed the absorption of Mn and Fe from the nutrient solution. Scots pine could be considered as a Cu excluder because only minor metal translocation into aerial organs was estimated (Ivanov et al. 2016a). The toxic effect of Zn on 6-week-old seedlings of *P. sylvestris* cultivated in hydroculture was manifested mainly in changes in the architecture of the root system and the suppression of its growth. At application of 300 µM $ZnSO_4$ roots accumulated up to 35-fold higher Zn amounts than shoots and up to a 3.5-fold reduction in Mn content in the roots and needles was estimated, while an increase in Fe content of up to 23.7% in the roots and up to 42.3% in the needles was observed. Low molecular weight antioxidants were suggested to play the determinant role in the prevention of oxidative stress in the organs of *P. sylvestris* seedlings (Ivanov et al. 2016b).

Dordevic et al. (2017a) cultivated embryogenic cell masses of *Abies alba* and *Picea abies* on media enriched with 50–500 µM Cd^{2+} or Pb^{2+} and observed a notable reduction of the proliferation rate of *A. alba* and *P. abies* embryogenic cell masses. Application of 500 µM Cd^{2+} significantly slowed or stopped the growth of embryogenic cell masses in both species, and during the maturation stage the total number of somatic embryos was reduced under Cd exposure. On the other hand, the proliferation ratio at application of all Pb concentrations remained relatively high for the *P. abies* cell lines, while the formation of early precotyledonary and cotyledonary somatic embryos in both species was reduced, Cd being more toxic than Pb. In *A. alba* and *P. abies* embryogenic cell masses treated with Cd^{2+} and Pb^{2+} at doses 50, 250, and 500 µM during the 21-day proliferation period the metal treatment induced synthesis of phytochelatins suitable to bind heavy metal ions and thereby avoid the production of reactive oxygen species. Similar increases of all antioxidative thiol compounds were observed following treatment with Pb^{2+} in *A. alba* and Cd^{2+} in both species, whereby in *P. abies* the highest antioxidative response was obtained using 50 µM Pb and an extremely low uptake of Pb suggested that the ability of *P. abies* to reduce Pb uptake after longer exposure times results in its higher resistance (Dordevic et al. 2017b).

Mg^{2+} ions were found to alleviate the negative effects of Pb on growth, photosynthesis, and ultrastructural alterations of *Torreya grandis* seedlings; they improved the growth of the Pb-stressed seedlings, increased Chl content and the net photosynthetic rate, enhanced chloroplasts development, increased root growth and oxidative activity, and also protected the root ultrastructure. Moreover, in the presence of Mg^{2+}, a 5-fold higher Pb concentration in the roots and a 4-fold higher efficiency of Mg^{2+} translocation to the shoots was observed compared to plants without a Mg^{2+} treatment (Shen et al. 2016).

25.6 IMPACT OF TOXIC METALS ON CROPS

The topic of the effect of toxic substances including heavy metals on the physiological and production characteristics of the crops is in general extraordinarily important, because from a global aspect crops have a dominant use as food and fodder, although in advanced and highly-developed countries they are also used as technical plants for alternative sources of energy or for environment protection. Consequently, a constantly increasing number of research papers are focused on the investigation of the impact of toxic metals on important agronomic crops, including the adverse effects manifested on the molecular level up to the outcomes resulting in impairment of the physiological and production characteristics of the crops estimated in field experiments. Considering all the above-mentioned aspects, the impact of toxic metals on crops would require a separate chapter in this book. Therefore, besides providing an overview of recent review papers dealing with the adverse effects of individual toxic metals on plants and their corresponding detoxification mechanisms, only brief selection related to the effects of metals on the most important crops will be presented.

In general, the effects of abiotic stresses, including heavy (or toxic) metal stress, are usually dependent on the plant species, the genotype, the age of the plant, the timing of the stress application, and the intensity of this stress (Le Gall et al. 2015). Toxic metals could be transferred from the soil, or the atmosphere, to the crops and subsequently could be accumulated in their edible parts. However, metals can only exert toxic effects on crops when they are present in bioavailable forms and at excessive levels. For plants cultivated in soil, the phytoavailability of metals that are affected by soil associated factors, such as pH, redox potential, cation exchange capacity, soil type, and soil texture, and by plant-associated factors,

such as root exudates and root rhizosphere processes (microorganisms), is pivotal for their uptake by plants (Sheoran et al. 2016). Some agricultural practices, including excessive use of metal-based pesticides and management of post-harvest residues or irrigation by water contaminated with toxic metals, could adversely impact the environment and human health (Udeigwe et al. 2015). Consequently, appropriate combinations of soil types, plant species/cultivars, and agronomic practices could result in a restricted transfer of toxic metals to the food chain and/or extraction of energy and metals for industrial use (Tang et al. 2012). As the adverse effects of the ions of toxic metals are reflected in damage to the structural, enzymatic, and non-enzymatic components of plant cells, often leading to the loss of cell viability resulting in decreased plant growth (Dutta et al. 2018), better food safety could be achieved by the reduction of toxic metal accumulation in crops using the cultivation of crop varieties with strongly reduced concentrations of toxic elements in their edible parts (Clemens and Ma 2016). However, not only metal excess but also metal deficiency results in reduced growth and decreased productivity of crops, and consequently in human malnutrition (Vatansever et al. 2017).

Adverse effects of toxic metals on plants are manifested mainly as impaired seed germination, reduced nutrient uptake through roots, inhibition of root metabolism, disruption of chloroplast ultrastructure and cell membrane permeability, inhibition of the photosynthetic apparatus connected with impairment of PET and reduction of Rubisco activity, reduced levels of photosynthetic pigments, induction of alterations in leaves respiratory activities, ROS production, genotoxic damages, and triggering of enzyme and non-enzymatic antioxidants (as a defense to oxidative damage) finally causing plant growth reduction and lower biomass production.

Several review papers were published dealing with the adverse effects of individual toxic metals on plants and their corresponding detoxification and tolerance mechanisms, including Cd (Andresen and Kuepper 2013; Rizwan et al. 2016a, 2016b), Zn (Rout and Das 2009), Cu (Yruela 2009; Adrees et al. 2015), Fe (Krohling et al. 2016), Ni (Yusuf et al. 2011; Sreekanth et al. 2013), Pb (Pourrut et al. 2011; Ashraf et al. 2015), Cr (Ertani et al. 2017; Shahid et al. 2017), As (Chandrakar et al. 2016), Sb (Feng et al. 2013) or rare toxic metals and metalloids and rare earth elements (Babula et al. 2008).

A critical review dealing with heavy metal induced oxidative stress on seed germination and seedling development was presented by Seneviratne et al. (2017). The long-distance transport of heavy metals via the xylem and phloem and the interactions between the two transport systems were overviewed by Page and Feller (2015). The effect of stress on cell wall metabolism at the physiological, morphogenic, transcriptomic, proteomic, and biochemical levels was described by Le Gall et al. (2015). Findings related to the responses of plant roots to heavy metal ions in the rhizosphere, metal-ion-induced reactions at the cell wall–plasma membrane interface, and different aspects of heavy metal-ion uptake and transport in plants via membrane transporters at the molecular and proteomic levels were summarized by Ovečka and Takáč (2014), and the aspects related to the physiology of heavy metal stress, ROS, and several molecular events that are associated with metal tolerance were discussed from a genomics perspective by Panda et al. (2015).

In order to detoxify the excessive levels of heavy metals, plants have adopted different defense strategies, whereby the detoxification may result at the cell wall, cell membrane or protoplasm level. Metal excess in plants can be transported and sequestered primarily in apparent free spaces; intercellular or intracellular vacuoles and bodies like lysosomes and heat shock proteins or metallothioneins chelating heavy metals contribute to the protection of plasma membrane integrity resulting in metal tolerance (Aery 2012). While plants that do not belong to the hyperaccumulators of heavy metals largely sequester toxic metals in root vacuoles, the hyperaccumulator plants usually sequester them in leaf cell vacuoles following efficient long-distance translocation (Sharma et al. 2016). High tolerance to heavy metal toxicity could rely either on reduced uptake or increased internal sequestration of metal that is manifested by an interaction between a genotype and its environment (Solanki and Dhankhar 2011).

As the main adaptive strategies of wheat under Cd stress, the promotion of antioxidant defense system, osmoregulation, ion homeostasis and overproduction of signaling molecules could be considered (Rizwan et al. 2016b). Sarwar et al. (2010) overviewed the role of mineral nutrition in minimizing cadmium accumulation by plants, and it was reported that the Cd^{2+} moves across the plasmalemma of the wheat root via Ca^{2+} channels and pretreatment with the Ca^{2+} channel blockers such as La^{3+} or Gd^{3+} and verapamil results in a notable decrease of Cd content in the whole plant (Li et al. 2017a; Yue et al. 2017). The Rht-B1c mutation in wheat plants was found to improve plant tolerance to Cd stress connected with structural changes in the mutant photosynthetic membranes ensuring better protection of the Mn cluster of the OEC and increased capacity for PS I cyclic electron transport, which results in the protection of photochemical activity of the photosynthetic apparatus under Cd stress (Dobrikova et al. 2017).

Soybean plants were reported to be able to respond rapidly to Cd-induced oxidative stress by improving the availability of the NADPH necessary for the ascorbate-glutathione cycle (Perez-Chaca et al. 2014). Pb treatment notably increased ROS and MDA levels in *Brassica napus* plant organs and antioxidant activities under Pb stress could be increased by application of H_2S also resulting in enhanced contents of total GSH and GSH/GSSG in leaves and roots (Ali et al. 2014). Maize seedlings were reported to have strong capacities for adapting to low Cd concentrations by consuming GSH and for developing an antioxidative enzyme system to defend against high Cd stress (Xu et al. 2014). A comparison of two maize cultivars showed that cultivars that accumulated more Cd in their roots exhibited higher antioxidant activities and thus were better at alleviating Cd-induced oxidative damage (Anjum et al. 2015). The high exudation of dicarboxylic organic acids of maize plants exposed to Cd caused an improvement in nutrient uptake and the activities of antioxidants (Javed et al. 2017).

Besides the direct protective cellular mechanisms against Cd toxicity, such as synthesis of phytochelatins and metallothioneins, metal compartmentalization in vacuoles, and the increased activity of antioxidant enzymes, indirect mechanisms such as microelements (Zn, Fe, Mn, and Se) interfering with Cd uptake contribute to a reduction of Cd concentration in plants resulting finally in the alleviation of Cd-induced toxicity (Choppala et al. 2014). The phytotoxic effect of Cd on tomato plants grown in soil was reflected in a nutritional imbalance, particularly in the metabolisms of N, P, and Mn and enhanced levels of MDA and H_2O_2 as well as increased activity of CAT, APX, and guaiacol peroxidase in plant tissues were estimated (Nogueirol et al. 2016). The treatment of tomato plants grown in soil supplemented with Cd (160–1280 µM) reduced the fruit biomass, moisture content, total fiber, protein, glucose, and fructose contents, and decreased Ca^{2+} and K^+ levels in the fruit, suggesting that fruit quality was strongly deteriorated, whereby a cultivar showing greater Cat2 gene expression and low H_2O_2 and MDA contents was found to effectively tolerate Cd stress (Hussain et al. 2017).

The application of a magnetic field or laser radiation on wheat seed was found to improve the growth of young seedlings treated with Pb or Cd (Chen at al. 2017), and Cd stress in germinating wheat grains could be ameliorated by hydrogen sulfide causing modulation of the antioxidant system (Huang et al. 2016a). Cd uptake by plants can be reduced by phosphorus addition (Arshad et al. 2016), by pretreatment of wheat seedlings with salicylic acid (Shakirova et al. 2016), or by silicon supply causing an increase of root oxalate exudation (Rizwan et al. 2016c; Wu et al. 2016) and enhanced phytochelatin formation in wheat plants (Greger et al. 2016). The altered lipid profile of developing mustard (*Brassica juncea* L.) seeds under Cd stress was manifested by an increase in the saturated/unsaturated ratio indicating that either the synthesis or activity of the olelyl-CoA desaturase enzyme was considerably affected (Beniwal et al. 2015).

As leafy vegetable consumption was identified as a dominant exposure pathway of Cd in the human body, it is important to minimize Cd accumulation in these crops, which can be achieved by breeding Cd pollution-safe cultivars (Huang et al. 2017). The Cd amount in rice can be controlled by gene families that code for putative transition metal transporters or metal chaperones and quantitative trait loci, and reduction of Cd content in rice could also be obtained by generation of Cd excluder rice cultivars by transgenics (Sebastian and Prasad 2014). For Cd minimization in food, crops could also contribute Cd-resistant plant growth promoting rhizobacteria (Sharma and Archana 2016). On the other hand, the application of microorganisms including bacteria and mycorrhiza may facilitate the application of metal phytoextraction at a commercially large scale (Sheoran et al. 2016).

Treatment of wheat seedlings grown in hydroponics with 25 and 50 µM Ni resulted in pronouncedly reduced plant heights, Chl content that was significantly affected in a dose-dependent manner, and notably increased proline content (Parlak 2016). Excess Ni in wheat plants could inhibit the transfer of Cu, Mn, and Mg to grains resulting in the accumulation of Ca, Mg, and Mn in the shoots and shells of wheat plants, and ready accumulation of Ni in the grains at low Ni concentrations could represent health risks (Wang et al. 2015b). Ni content in the shoots and roots of wheat notably increased in the presence of histidine and it was assumed that the Ni(His) was most likely to be taken up as a complex or that receptors at the membrane were able to enhance Ni uptake from the Ni(His) complex (Dalir and Khoshgoftarmanesh 2015), while application of glutamine and glycine enhanced the symplastic to apoplastic Ni ratio in the roots of wheat plants more effectively than histidine; however, more Ni was translocated from the roots to the shoots in the presence of histidine (Dalir and Khoshgoftarmanesh 2014). Similarly, histidine, but not glycine, enhanced Zn uptake and translocation into the aboveground parts of triticale, probably due to the formation of strong complexes with Zn (Khodamoradi et al. 2015). Application of Ca alleviated the negative effects of Ni on rice plants and decreased the translocation of Ni into the shoots (Aziz et al. 2015). Ni^{2+} was found to impair sugar metabolism, which was reflected in the decline in the activity of sucrose and starch hydrolyzing enzymes resulting in the inhibition of radicle growth in germinating wheat seeds (Negi et al. 2014). A review paper focused on the specific aspects of nickel's effects on the growth, morphology, photosynthesis, mineral nutrition, and enzyme activity of plants was presented by Sreekanth et al. (2013). Ni stress resulted in the accumulation of glucose and decline in concentration of sucrose as well as considerable increases in the concentrations of malic and citric acids in wheat seedlings suggesting that excess Ni redirected the carbon metabolism of developing wheat shoots to provide carbon skeletons for the synthesis of amino acids and organic acids as well as to supply reducing power to sustain normal metabolic processes and to support defense mechanisms against oxidative stress (Gajewska et al. 2013). Ni also altered the total fatty acid composition in wheat seedlings resulting in a significant decrease in linolenic acid (C18:3) and increased linoleic (C18:2) and oleic (C18:1) acid contents in the plant organs, whereby in the shoot, a parallel increase in palmitic acid (C16:0) and a decrease in palmitoleic acid (C16:1) were estimated and these alterations could be partly related to lipid peroxidation and increased electrolyte leakage reflecting the damage of cells caused by Ni (Gajewska et al. 2012).

Treatment of *B. napus* plants with 400 µM Cr resulted in a broken cell wall, immature nucleus, a number of mitochondria, ruptured thylakoid membranes and enlarge size of the vacuole and starch grains in leaf ultrastructures, and disruption of the golgibodies and diffused cell wall in the root cells (Gill et al. 2015). An investigation of Hg-induced cyto- and genotoxicity in *Pisum sativum* L. seedlings showed that root cell cycle impairment occurred at lower concentrations (≥1 µM) than structural DNA damages (≥10 µM). Hg concentration affected the cytostatic effects with delays during the S-phase at lower doses and arrests at G_1 at higher ones and a decreased mitotic index (MI) and cell proliferation index (CPI) (Azevedo et al. 2018). Doses exceeding 5 mg dm^{-3} caused significant DNA damage in *Lactuca sativa* plants,

DNA fragmentation was detected at ≥5 mg dm^{-3}, while the presence of micronuclei at 20 mg dm^{-3} and cell cycle impairment at doses of 0.05 mg dm^{-3} was revealed; at 0.5 mg dm^{-3} mostly G_2 arrest was estimated (Silva et al. 2017). The Pb-treatment of *Allium cepa* induced a dose-dependent decrease in MI in the root tips, caused mitotic abnormalities such as distorted metaphase, fragments, sticky chromosomes, laggards, vagrant chromosomes and bridges, and treatment with higher concentrations arrested cell cycle resulting in cell death, while at lower Pb doses some repair responses were generated (Kaur et al. 2014). The mineral nutrient imbalance was found to be involved in changes of antioxidant levels, and DNA damages of the *Phaseolus vulgaris* L. seedlings treated with toxic metals (Gjorgieva et al. 2013) and Zn-induced genotoxic effects in root meristems of barley seedlings was described by Truta et al. (2013).

Wang et al. (2017) studied the accumulation of seven heavy metals in soil-maize and soil-wheat systems and found that wheat was more likely to accumulate heavy metals than maize and the BCF of individual metals in grains of both crops decreased exponentially with their average concentrations in soil. The accumulation of tested metals in maize grains increased as follows: Pb < Cr < Zn < As < Cu < Cd < Hg, while for the metal accumulation in wheat grains the following rank was estimated: Zn < Pb < Cr < Cu < As < Hg < Cd.

Findings related to uptake capacity, oxidative damage, and biochemical and physiological tolerance and defense mechanisms, and the heavy metal toxicity of six economically important species from the *Brassica genus* (*B. juncea*, *B. napus*, *B. oleracea*, *B. carinata*, *B. rapa*, and *B. nigra*) belonging to heavy metal accumulators were summarized by Mourato et al. (2015). *B. juncea* and *B. napus* plants were considered to be promising species for long-term phytoextraction of Cd-contaminated farmlands for bioenergy production (Shi et al. 2016). According to Thao et al. (2015), the development of transgenic crops with optimized endogenous ethylene levels under heavy metal stress will result in their improved tolerance against heavy metals. Compared with hyperaccumulators, the phytoremediation efficiency of field crops such as *Helianthus annuus* L., *Zea mays* L., and *Brassica napus* L. is rarely high, but they may compensate for lower accumulated metal concentrations with a greater biomass yield (Vamerali et al. 2010).

25.7 RESPONSE OF MEDICINAL PLANTS TO METAL STRESS

Christenhusz and Byng (2016) reported that the currently known, described, and accepted number of plant species is ca. 374,000, of which approximately 308,312 are vascular plants (with 295,383 flowering plants) whereby phytotaxa currently contributes more than a quarter of the ca. 2000 species that are described every year. In Europe, with its long tradition in the use of botanicals, about 2000 medicinal and aromatic plant (MAP) species are used on a commercial basis, two-thirds of them being native to this region (Lange 1998). The majority of the MAP species are still harvested from the wild;

however, in terms of volumes: >50% of medicinal plant material used worldwide may already be sourced from cultivation (Kathe 2006). Barbieri (2013) analyzed particular regulations on the cultivation and collection of MAPs in the European Union.

Medicinal plants could be regarded as potential plant factories for new natural drugs. Therefore, it is necessary to check and monitor the herbs used as phytopharmaceuticals for the content of harmful substances including toxic metals (see Kráľová and Masarovičová 2006). Moreover, medicinal plants also have great potential for their exploitation in modern phytotechnologies; such as phytoremediation and phytofortification (e.g., (Masarovičová and Kráľová 2007, 2017; Masarovičová et al. 2010b)). In the next section of this chapter, we would like to present only a brief overview focused on the impact of toxic metals on medicinal plants, thus for more detail information see Chapter 39, entitled 'Responses of Medicinal Plants to Abiotic Stresses', in this *Handbook*.

It was found that some medicinal plants produce specific secondary metabolites that can detoxify some of the toxic metals. For example, 5 month-old-plants of *Hypericum perforatum* that accumulated 290 g Cd g^{-1} dm in their leaves following exposure to 12 µM Cd in hydroponics for 7 days (Kráľová et al. 2000b) can be considered as Cd hyperaccumulators, while it pronouncedly exceeded the value of 100 µg Cd g^{-1} dm generally defined for Cd hyperaccumulators (Baker 1995; Masarovičová et al. 2010b). The high tolerance of *H. perforatum* for metal ions such as Cd^{2+} could be connected with an additive mechanism of tolerance, in which chelation of metal ions by some of its specific secondary metabolites (e.g., naphthodianthrones hypericin and pseudohypericin) results in the change of the toxic ionic form of metal into non-toxic metal chelate (Masarovičová and Kráľová 2007). Previously, Palivan et al. (2001) also reported that hypericin forms a four-coordinated copper species where the solvent participates to the coordination sphere of the metal and Falk and Mayr (1997) described that the fringelites are able to form *peri* chelate complexes with transition metal ions. Consequently, the chelatation of Cd^{2+} ions with hypericin could contribute to the enhanced tolerance of *H. perforatum* plants to Cd toxicity and to their Cd hyperaccumulating ability, which was already reported previously (Marquard and Schneider 1998; Masarovičová et al. 1999b; Masarovičová and Kráľová et al. 2004).

Metal contamination resulted in multiple reductions of the concentration of pseudohypericin and hypericin in *H. perforatum* plants (Murch et al. 2003) and exposure to Cd caused a decrease in total alkaloid levels in the roots of *Catharanthus roseus* L. (Pandey et al. 2007) as well as a significant decrease of essential metals such as Mn, Fe, and Cu in the roots and stems of Cd-treated St John's wort plants compared to the control plants (Kráľová et al. 2000b).

In hydroponically cultivated *H. perforatum* plants exposed to CdSO$_4$, CdSeO$_4$, CdSeO$_3$, Cd(NCS)$_2$(nia)$_2$, and Cd(NCSe)$_2$(nia)$_2$, a strong effect of the Se oxidation state as well as the Cd−Se interaction on the uptake and translocation of both elements within the plant were estimated. The portion of Cd allocated to *H. perforatum* shoots, related to the total amount accumulated by the plant, was the highest for CdSO$_4$ and Cd(NCS)$_2$(nia)$_2$-treated plants and the lowest for a plant exposed to CdSeO$_3$ showing approximately half of the corresponding value estimated in treatments with Cd(NCSe)$_2$(nia)$_2$ and CdSeO$_4$. Notable differences were also observed in the Se amount accumulated by plant shoots that was the lowest for Cd(NCSe)$_2$(nia)$_2$ treatment, and the value estimated for treatment with CdSeO$_4$ was approximately two-fold higher than that for treatment with CdSeO$_3$ (Peško et al. 2010).

Pb treatment (75–1500 mg Pb kg^{-1} soil) significantly influenced the morphology, anatomy, and hypericin content in *H. perforatum* plants grown in soil, and an increase in the size of the epidermal cells, mesophyll tissue, and diameter of the stems and roots, as well as an amplified vascular bundle and pith area was observed. Moreover, Pb concentration in the roots exceeded that in the shoots, and maximum hypericin content was observed at a dose of 600 mg Pb kg^{-1} (Zarinkamar et al. 2013).

Medicinal plant *Matricaria recutita* L., which is often mentioned as the 'star among medicinal species' (Singh et al. 2011), can tolerate Cd concentrations corresponding to middle-strong contaminated soils (Grejtovský and et al. 2001; Masarovičová et al. 2003). The potential of *M. recutita* plants to accumulate high Cd levels from the soil was firstly reported by Marquard and Schneider (1998). The content of the important secondary metabolites of *M. recutita* cultivars, cultivated in Slovakia and belonging to chemical type A of chamomile, decreases as follows: α-bisabolool>α-bisabolool oxide B>α-bisabolool oxide A (Šalamon et al. 2007). Correlation between accumulated Cd in chamomile anthodium dry mass cultivated in the field during the period of 1999–2001 in Eastern Slovakia and actual climatic relations reflecting the total amount of precipitations and the mean daily temperatures exceeding 10°C for the investigated period was estimated (Šalamon et al. 2007).

Among two tetraploid cultivars of *M. recutita* L. cv. Goral and cv. Lutea exposed to 12 µM Cd in hydroponics, production characteristics, photosynthetic pigments, and photosynthesis were inhibited to a greater extent in cv. Lutea compared to cv. Goral, although it accumulated less Cd in its shoots than cv. Goral. However, in a treatment with 120 µM Cd showing higher accumulated Cd amounts in shoots than plants exposed to 12 µM Cd, higher levels of photosynthetic pigments were observed compared to a 12 µM Cd treatment suggesting that secondary chamomile metabolites also contribute to detoxification mechanisms induced by Cd (Pavlovič et al. 2006). Based on high Cd content in chamomile shoots (>300 µg g^{-1} at 12 µmol dm^{-3} Cd in hydroponium), *M. recutita* could be classified as a Cd hyperaccumulator species, and this medicinal plant was also reported as a Cd-tolerant species by Chizzola and Mitteregger (2005) and Kováčik (2013).

Tetraploid chamomile cultivar Goral exposed to 120 µmol dm^{-3} Cd in hydroponics accumulated 2-fold more Cd in both shoots and roots compared to more tolerant diploid cultivar Novbona, whereby Cd amounts accumulated in roots of both cultivars which was more than 16-fold higher than those accumulated in shoots (Kráľová and Masarovičová 2003).

In hydroponically cultivated chamomile seedlings, $CaCl_2$ amendment was found to ameliorate the $CdCl_2$-induced stress what was reflected in increased plant growth, reduction of ROS levels and oxidative injury, and a reduction of Cd uptake and total phenolics in plants (Farzadfar et al. 2013).

Exposure of chamomile plants to Ni excess resulted in elevated levels of root-soluble proteins, caused a higher accumulation of soluble phenolics in both the rosettes and the roots and H_2O_2 in the roots, increased proline content, and elevated levels of histidine in the roots, suggesting its involvement in Ni retention. In a treatment with 120 µmol dm^{-3} Ni the leaf rosettes accumulated 174.1 µg Ni g^{-1} d.w., while Ni accumulation in the roots was 6.1-fold higher (Kováčik et al. 2009a). In combined treatments of Cd or Ni with salicylic acid (SA), SA promoted an enhancement in soluble proteins of *M. recutita* roots and reduced their water content, and enhanced phenylalanine ammonia-lyase activity and accumulation of total soluble phenols, particularly in the roots, and increased accumulated Ni amount in the shoots, while reduced Cd levels in the shoots were detected (Kováčik et al. 2009b). Tetraploid chamomile cultivar Lutea and diploid cultivar Novbona responded differently to a treatment with 60 µmol dm^{-3} Cu for 7 days. Exposure to Cu resulted in lower root tissue water content and dry biomass, higher lignin accumulation and cinnamyl alcohol dehydrogenase activity, a higher amount of Cu and soluble phenols in both shoots and roots, higher reduction of potassium content, and faster phenolic metabolism in the diploid cultivar compared to tetrapoid cultivar. On the other hand, phenylalanine ammonia-lyase activity was stimulated in tetraploid but reduced in diploid roots, while reduction of soluble proteins and promotion of sulfur-containing amino acids in the shoots of Cu-treated plants was observed independently on the ploidy level (Kováčik et al. 2011).

In *M. recutita* shoots grown in Cd-contaminated soils supplemented with an appropriate Zn concentration, a reduced Cd accumulation was estimated (Chizzola and Mitteregger 2005). Co-treatment of hydroponically cultivated *M. recutita* plants with 12 µM Cd and 120 or 180 µmol dm^{-3} Zn, respectively, resulted in increased Zn and reduced Cd accumulation in the roots but increased Cd content in the shoots, whereby plant dry mass, photosynthetic pigments, maximum efficiency of PS II, and root respiration rate were negatively affected (Kummerová et al. 2010). In cv. Novbona plants exposed to Pb in a pot experiment (50 mg Pb kg^{-1} dry soil), the content of Pb in chamomile inflorescences was 3-fold higher compared to the control plants; however, it was 5-fold lower than the limit set by the WHO for the highest acceptable level of Pb in the chamomile drug (Grejtovský et al. 2008).

Intrinsic factors (ontogenetic development, redistribution of assimilates between root and shoot) and external factors such as date of outplanting or cultivation conditions (especially temperature affecting the transpiration flow) pronouncedly affected Cd translocation from roots to shoots in hydroponically cultivated chamomile plants, cv. Bona, whereby 3-month-old chamomile plants originating from sowing in agrotechnical date exhibited higher physiological activity and accumulated higher Cd amounts in plant tissues than 2-month-old plants that were sown later, thus after the agrotechnical date (Peško et al. 2011).

The results related to Cd and Se accumulation in hydroponically cultivated *M. recutita* plants, cv. Lutea (Kráľová et al. 2007a) and more tolerant cv. Goral (Lešíková et al. 2007) exposed to 12, 24, and 60 µmol dm^{-3} of $CdSeO_4$, $CdSeO_3$, $Cd(NCSe)_2(nia)_2$ (in which the selenocyanate anions are bonded to Cd atoms as *N*-donor monodentate ligands; nia = nicotinamide) and $Cd(NCS)_2(nia)_2$, correlated with those observed in an experiment with *H. perforatum* (Peško et al. 2010): the TF values related to Cd estimated in the experiments with $CdSeO_4$ and $Cd(NCS)_2(nia)_2$ were more than 2-fold higher than those found for $CdSeO_3$ and $Cd(NCSe)_2(nia)_2$ and the highest fraction of Cd allocated to the shoots related to the total amount of Cd accumulated by the plant was observed for $CdSeO_4$, while the lowest was for $Cd(NCSe)_2(nia)_2$ (Kráľová et al. 2007a; Lešíková et al. 2007). In a treatment of cv. Goral plants with $CdSeO_4$ and $Cd(NCS)_2(nia)_2$ approximately 40% from the total amount of Cd accumulated by the plant were allocated to shoots, while approximately ≥80% of the total amount of metal accumulated by the plant remained in the roots after the treatment with $CdSeO_3$ and $Cd(NCSe)_2(nia)_2$ (Lešíková et al. 2007). Se translocation into the shoots also notably depended on the oxidation state of Se and at the application of $CdSeO_4$ the BCF values related to shoots were approximately 2.5-fold higher than those determined for roots, and the corresponding BCF values for $Cd(NCSe)_2(nia)_2$ were similar to those of $CdSeO_3$. With the application of $CdSeO_4$, 90% of uptaken Se and 50% of uptaken Cd was situated in the shoots (Lešíková et al. 2007). Similar results were also reported by Shanker et al. (1996). Lower Cd translocation into the shoots in a treatment with $CdSeO_3$ could be connected with the reduction of less mobile SeO_3^{2-} to selenide, which tends to form Cd–Se complex which is unavailable to the plants. Previously it was found that foliar spraying of Na_2SeO_4 on *M. recutita* plants affected the contents of secondary metabolites in the chamomile essential oil and significantly increased Se content in the chamomile flowers (Jakovljevic et al. 2000).

The addition of chelators to soils results in increased total dissolved metal concentration as well as in the change of the primary route of plant metal uptake from the symplastic to the apoplastic pathway. Consequently, depending on the metal, plant species, and chelating agent concentration, a notable enhancement of metal uptake could be achieved (Nowack et al. 2006).

For example, the addition of citric acid enhanced Zn and Cd accumulation, mostly at the root level (Mugnai et al. 2006), and the application of copper in the form of Cu(II) chelates resulted in more effective Cu translocation into the shoots in comparison with a $CuSO_4.5H_2O$ treatment (Kráľová et al. 2007b). Using the artificial chelator ethylenediaminetetraacetic acid (EDTA), Cu translocation into the shoots in hydroponically cultivated *M. recutita* plants in the presence of 60 µM Cu significantly increased, and about 59% Cu accumulated by plants was allocated to the shoots, while in the absence of chelator this portion represented only 4.8% (Kráľová and

Masarovičová 2008b). *Calendula officinalis* exhibited high Cu tolerance up to 400 mg Cu kg⁻¹, which is far above the phytotoxic range estimated for non-hyperaccumulators and in plants treated with 300 mg Cu kg⁻¹ soil leaf and root accumulations of 4675 and 3995 µg g⁻¹ d.w. were estimated, suggesting that *Calendula officinalis* is a Cu hyperaccumulator with TF>1 (Goswami and Das 2016). Mycorrhizal fungi improved growth and yield of potted marigold cultivated under Pb and Cd stress as well as phytoremediation performance of this medicinal plant by increasing accumulation of metals in the plant organs, and treatment with a dose 80 mg Cd kg⁻¹ soil resulted in accumulation of 833.3 and 1585.8 mg Cd in the shoots and roots, respectively (Tabrizi et al. 2015). Moustakas et al. (2011) observed that co-treatment of marigold plants cultivated under glasshouse conditions with Cd and Zn resulted in a notable reduction of Cd concentration in the leaves as Zn concentration in soil exceeded 5 mg Zn kg⁻¹, while Zn concentrations in the leaves showed a decrease and those in the petals increased with increasing rates of applied Cd. The Cd/Zn ratio of plant tissue was enhanced by Cd application, while it decreased with Zn application. The application of ethylenegluatarotriacetic acid increased the total Cd in *Calendula* up to 217% and humic acid, ethylenediamine-*N*,*N*'-disuccinic acid, and EDTA also enhanced the ability of this plant to clean-up the Cd-contaminated soil (Mani and Patel 2014).

The reduced seed germination in *S. officinalis* observed upon exposure to Cd represented 81% (10 ppm), 59.66% (50 ppm), and 21.33% (100 ppm) compared to the control (93.33%), and treatment with 10 ppm Cd stimulated relative root elongation, while both higher Cd concentrations showed an inhibitory effect (Petrescu et al. 2014). *S. officinalis* cultivated in soil contaminated with toxic metals was able to accumulate Cd, Pb, and Zn, which resulted in a plant biomass reduction, but essential oil yield and quality was not reduced, despite a decline in α- and β-thujones and elevated camphor content in the sage leaves (Stancheva et al. 2010). Marquard and Schneider (1998) classified this medicinal plant species as excluder of Cd.

The fresh biomass, root yield, and essential oil yield of *Mentha piperita* notably decreased with increasing Cr and Pb applications (30.0 and 60.0 mg kg⁻¹ soil), and the metals notably affected the levels of α-pinene, β-pinene, sabinene, β-myrcene, limonene, menthone, and isomenthone in the essential oil. The metals were accumulated more in roots than in shoots, whereby shoot Cr and Pb concentrations increased with increasing concentration of the applied metal, and similarly *M. piperita* was found to be a suitable crop for cultivation in Cr- and Pb-contaminated soils (Prasad, et al. 2010). Akountianaki-Ioannidou et al. (2015) investigated the effects of Cd and Zn interactions on the concentration of both metals in *M. piperita* L. using metal concentrations of 1, 5, 10 mg kg⁻¹ Cd and Zn, respectively. Cd concentration in leaves of peppermint increased significantly at each Cd level with increasing Zn additions, but Zn concentration was not affected by Cd addition to the growth medium. Although *M. piperita* was found to remove only moderate amounts of toxic metals from the soils, according to Zheljazkov et al. (1999) in the long-term it could be used for phytoremediation of metal polluted

soils. Exposure of 1-month-old *Melissa officinalis* L. plants to 500 µmol dm⁻³ Ni resulted in a notable loss in chlorophylls and carotenoids content, Ni stress caused H₂O₂ accumulation, while lipid peroxidation pronouncedly decreased and proline accumulation was not affected. Ni was accumulated predominantly in *M. officinalis* roots and TF< 1 was estimated (Maivan et al. 2017).

25.8 IMPACT OF NANOSCALE METAL AND METAL OXIDES ON ALGAE AND VASCULAR PLANTS

The US National Nanotechnology Initiative (2004) defined nanotechnology in 2004 as 'the understanding and control of matter at dimensions of roughly 1 to 100 nanometers, where unique phenomena enable novel applications.' According to the recommendation on the definition of a nanomaterials adopted by the European Commission in 2011, the term 'nanomaterial' means 'a natural, incidental or manufactured material containing particles, in an unbound state or as an aggregate or as an agglomerate and where, for 50% or more of the particles in the number size distribution, one or more external dimensions is in the size range 1–100 nm. In specific cases and where warranted by concerns for the environment, health, safety or competitiveness the number size distribution threshold of 50% may be replaced by a threshold between 1 and 50%.' Also, a material having the specific surface area by volume of the material >60 m² cm⁻³ could be considered as nanomaterial (European Commission 2011). Nanomaterials with the same chemical composition as microscale/bulk materials may demonstrate different physical and chemical properties and have a distinct impact on the environment, including living organisms.

The metal/metal oxide NPs, similarly to their counterparts, showing larger dimensions could exhibit beneficial effects on physiological and biochemical processes in non-vascular plants (including algae) and vascular plants when applied at low doses and toxic effects at higher concentrations (e.g., (Nair et al. 2010; Masarovičová and Kráľová 2013; Masarovičová et al. 2014; Jampílek and Kráľová 2015, 2017; Ruttkay-Nedecky et al. 2017)) and therefore it is necessary to reduce their harmful content in environmental matrices. By application of nanoscale metal fertilizers not only stimulation of plant growth could be achieved using a lower effective dose compared to bulk metal form but also the contamination of environment with excess metal could be reduced (e.g., (Jampílek and Kráľová 2017)). The toxic effects of metal NPs notably contribute to the dissolution of metal ions (e.g., (Ji et al. 2011; Turner et al. 2012; Liu et al. 2016; Zouzelka et al. 2016; El-Kassas and Okbah 2017)) and oxidative stress induced is by NPs resulting in the impairment of cellular components (e.g., (Liu et al. 2016; Reyes et al. 2016; Li et al. 2017b)), whereby the toxicity of NPs is also affected by their morphology and particle size, small particles being more effective due to their high specific surface area and ability to penetrate organisms. To the toxicity of metal NPs could also contribute dissolved metal ions (e.g., (Xiang et al. 2015; Lei 2016)).

It should be noted that due to dynamic processes, such as physical, chemical, and biological processes, occurring in the environment the metal oxide NPs undergo transformations affecting their fate, transport, persistence, bioavailability, and toxic effects (Amde et al. 2017).

Such environmental transformation could be inhibited or limited by the surface functionalization of NPs resulting in modified particle aggregation, mobility, dissolution, and eco-toxic potential (Shevlin et al. 2018; Zhang et al. 2018). Metal-based NPs are usually coated with natural organic compounds including chitosan, dextran, alginate, or citric acid, whereby the coating agents, besides the reduction of aggregation, could provide negative or positive charges to the particles and consequently modify their interaction with the environment (Lopez-Moreno et al. 2018). Like this, coating with natural organic matter generally lowers the toxicity of zero-valent iron NPs (Lei et al. 2018); humic acid coatings on the surfaces of CuO NPs were found to increase electrostatic and steric repulsion between the NPs and the plant cell wall/membrane resulting in reduced contact between NPs, and in this manner mitigating oxidative stress in rice seedlings induced by CuO NPs (Peng et al. 2015). Polymer-coated CuO NPs were reported to be more toxic against *Chlamydomonas reinhardtii* than the uncoated ones, which could be connected with their increased capacity to penetrate the algal cell and intracellular interactions between NPs and the cellular system (Perreault et al. 2012). Exposure of *Lemna minor* to surface-coated AgNPs showed that AgNPs coated with polyvinylpyrrolidone were more harmful on growth rate and fronds per colony, while the citrate-coated AgNPs had a greater impact on chlorosis and the activities of guaiacol peroxidase and GST (Pereira et al. 2018). The physiological responses of *Wolffia globosa* to AgNPs coated with adenosine triphosphate (ATP-nAg) and AgNPs coated with citrate (cit-nAg) also differed from each other suggesting that in a treatment with cit-nAg, citrate could possibly serve as the substrate for the tricarboxylic acid cycle and accumulated sugar may stimulate pentose phosphate pathways, while in an exposure of *W. globosa* to ATP-nAg, ATP would act as an exogenous energy source for the plant metabolisms (Zou et al. 2017). In *Ocimum basilicum* plants exposed to 500 mg kg^{-1} of unmodified TiO$_2$ NPs, hydrophobic TiO$_2$ NPs coated with aluminum oxide and dimethicone and hydrophilic TiO$_2$ NPs coated with aluminum oxide and glycerol differences in the homeostasis of essential elements was estimated: unmodified TiO$_2$ NPs increased Cu (104%) and Fe (90%); hydrophilic TiO$_2$ NPs increased Fe (90%), while hydrophobic TiO$_2$ NPs pronouncedly increased Mn (339%) but reduced Ca (71%), Cu (58%), and P (40%). Reduction of total sugar content by hydrophilic NPs (66%) exceeded that achieved of a treatment with unmodified and hydrophobic TiO$_2$ NPs (39 and 39%, respectively), whereby unmodified NPs diminished mainly content of reducing sugar (34%) and hydrophobic NPs considerably reduced the amount of starch (35%). Consequently, coated TiO$_2$ NPs significantly affected the nutritional quality of basil by altering some essential elements, starch, and reducing sugars (Tan et al. 2017). Quigg et al. (2013) reported that exopolymeric substances produced by algae could also reduce the bioavailability and toxic effects of metal NPs and/or their ions on cellular activities of algae, and large aggregates of metal oxide NPs entrapping and wrapping the algal cells results in shading and could affect the nanotoxicity as well (Ji et al. 2011; Aravantinou et al. 2017; Deng et al. 2017).

25.8.1 Effects of Metal Nanoparticles on Algae

Algae in freshwater and marine ecosystems are of great importance, because they produce approx. 50% of all oxygen on earth and provide the basis of the aquatic food chain (Chapman 2013). Consequently, because in fact animals at an upper level of the food chain, including man, depend on algae, it is necessary to estimate the impact of engineered NPs occurring in the environment on these photosynthesizing organisms, and gradually increasing metal nanopollution of the marine and freshwater environment represents a serious risk.

The biphasic dose response of marine diatom *Phacodactylum tricornutum* to CeO$_2$ NPs manifested as algal growth stimulation for a treatment with 5 mg dm^{-3} CeO$_2$ NPs and inhibition at doses of $\geq$10 mg dm^{-3} CeO$_2$ NPs was reported by Deng et al. (2017). Application of Fe$_3$O$_4$ NPs with a particle size of 20 nm also stimulated growth of *Picochlorum* sp during late growth stages (stationary and decline phases) at different concentrations; however, treatment with a dose of 200 mg dm^{-3} resulted in a notable reduction of the viability of algae during the exponential growth phase (Hazeem et al. 2015).

The toxic effect of metal/metal oxide NPs on physiological and biochemical processes in algae is usually manifested by inhibition of PET, reduced the content of photosynthetic pigments, oxidative stress by generating ROS, and lipid peroxidation resulting in deterioration of membranes leads ultimately to reduced algal growth.

Impairment of photosynthesis in algae due to PET inhibition in PS II was observed in CuNPs treated *Phacodactylum tricornutum* (Zhu et al. 2017) and *Chlamydomonas reinhardtii* exposed to core-shell CuO NPs (Saison et al. 2010), NiO NPs treated *C. vulgaris* (Oukarroum et al. 2017), Cr$_2$O$_3$ NPs treated *C. reinhardtii* (da Costa et al. 2016a), AgNPs treated *Apatococcus lobatus* (Nowicka-Krawczyk et al. 2017), *Skeletonema costatum* (Huang et al. 2016b) or *C. reinhardtii* Matorin et al. (2013), while in *S. obliquus* exposed to Ag nanoclusters, beside PET inhibition, reduced content of Rubisco was also estimated (Zhang et al. 2017).

The reduced content of chlorophylls, mainly of Chl$a$ following exposure to metal/metal oxide NPs was reported for *Lyngbya majuscula* exposed to biosynthesized CuNPs (El-Kassas and Okbah 2017), CuNPs treated *C. reinhardtii* (Reyes et al. 2016, ZnO NPs and TiO$_2$ treated *C. vulgaris* (Liu et al. 2016) and *Picochlorum* sp. (Hazeem et al. 2016), *C. vulgaris* exposed to NiNPs (Morgalev et al. 2017) and NiO NPs (Oukarroum et al. 2017), AgNPs treated *Skeletonema costatum* (Huang et al. 2016b), *Chlamydomonas acidophila* (Oukarroum et al. 2014), *C. vulgaris*, and *Dunaliella tertiolecta* (Oukarroum et al. 2012) or filamentous green algae

Pithophora oedogonia and *Chara vulgaris* (Dash et al. 2012), Fe_3O_4 NPs treated *Picochlorum* sp. (Hazeem et al. 2015), Cr_2O_3 NPs treated *C. reinhardtii* (da Costa et al. 2016a) or *Chlorella* and *Scenedesmus* sp. exposed to TiO_2 NPs (Sadiq et al. 2011), and *Chlorella* sp. exposed to anatase and rutile NPs (Iswarya et al. 2015).

Generation of harmful ROS was observed in *C. reinhardti* treated with CuO NPs (Perreault et al. 2012; Melegari et al. 2013), core-shell CuO NPs (Saison et al. 2010) and Cr_2O_3 NPs (da Costa et al. 2016a), ZnO NPs treated *Scenedesmus obliquus* (Bhuvaneshwari et al. 2015), *Phaeodactylum tricornutum* and *Alexandrium minutum* (Castro-Bugallo et al. 2014), *C. vulgaris* exposed to NiO NPs (Oukarroum et al. 2017), AgNPs treated *C. vulgaris* and *Dunaliella tertiolecta* (Oukarroum et al. 2012) as well as TiO_2 NPs treated *Karenia brevis* and *Skeletonema costatum* (Li et al. 2015), and nitrogen-fixing cyanobacteria *Anabaena variabilis* (Cherchi et al. 2011).

Increased lipid peroxidation was estimated in a treatment of *C. reinhardtii* with CuO NPs (Melegari et al. 2013), *C. vulgaris* and *Dunaliella tertiolecta* with AgNPs (Oukarroum et al. 2012) and *P. subcapitata* with CeO_2 NPs (Rogers et al. 2010), and membrane damage was observed in *C. reinhardtii* exposed to CuNPs (Reyes et al. 2016), *C. vulgaris* exposed to ZnO NPs, TiO_2 NPs (Liu et al. 2016) and NiO NPs (Li et al. 2017b), CeO_2 NPs treated *Phaeodactylum tricornutum*, *Nannochloris atomus*, *C. reinhardtii* (Sendra et al. 2017) and *P. subcapitata* (Rogers et al. 2010), and TiO_2 NPs treated *Anabaena variabilis* (Cherchi et al. 2011) and *Karenia brevis* (Li et al. 2015).

The toxic effect of metal oxide NPs on growth of *C. vulgaris* was estimated as based on algal Chl content, and cell count decreased in the order: $ZnO>NiO>CuO>TiO_2>Fe_2O_3$ (Ko et al. 2018). Using iron/iron oxide NPs of 20–30 nm, the growth of *Chlorella pyrenoidosa* declined as follows: zero-valent iron $NPs>Fe_3O_4$ $NPs>Fe_2O_3$ NPs suggesting a reduced inhibitory activity with the oxidation of NPs (Lei et al. 2016).

The growth of *Chlorella* sp. and *Scenedesmus* sp was inhibited more effectively in a treatment with TiO_2 NPs (72 h EC_{50}: 16.12 and 21.2 mg dm^{-3}, respectively) compared to bulk micro-sized TiO_2 (72 h EC_{50}: 35.50 and 44.40 mg dm^{-3}, respectively) (Sadiq et al. 2011), and the 72 h IC_{50} value estimated for CeO_2 NPs related to the growth inhibition of *P. subcapitata* was found to be about 6-fold lower than that of bulk CeO_2 (10.3 ± 1.7 and 66 ± 22 mg dm^{-3}, respectively) (Rogers et al. 2010). Dissolved Zn^{2+} ions from ZnO NPs were not estimated as being the dominant mechanism for the growth inhibition of *Chlorella* sp., because at concentrations of >50 mg dm^{-3} ZnO NPs showed a higher toxicity to algae than the Zn^{2+} ions, released less Zn^{2+} ions into the culture media than bulk ZnO and an experiment in the presence and absence of illumination excluded the shading effects of NPs on the nanotoxicity (Ji et al. 2011).

25.8.2 Effects of Metal Nanoparticles on Vascular Plants

Although some metal-based NPs applied at low concentrations could exhibit beneficial effects on plants (e.g., improved germination and photosynthesis), as mentioned above, at higher concentrations they could negatively affect the growth and development of plants and decrease the yield and quality of important agricultural crops by altering mineral nutrition, impairing photosynthesis, generating ROS which cause damage to the cellular membranes, proteins and nucleic acids, and inducting genotoxicity (e.g., (Du et al. 2017a; Jampílek and Kráľová 2017, 2019; Rizwan et al. 2017; Ruttkay-Nedecky et al. 2017)). However, besides the size, shape, crystal structure, and physico-chemical properties of NPs, their impact on crops is also affected by plant species, growth stages, growth conditions, dose, and duration of the NPs exposure. At the exposure of crop plants to metal-based NPs, the increasing amount of NPs in different plant parts including fruits and grains represents a threat to human health.

Plants as sessile organisms are increasingly exposed to metal-based NPs that can enter plant tissues through either the root tissues or the aboveground organs and tissues due to deposition of atmospheric particles. For uptake and translocation these NPs must overcome a series of chemical and physiological barriers, which control the size exclusion limits representing the size of the largest particle that can pass through a pore. However, some metal-based NPs, such as AgNPs, TiO_2 NPs, and ZnO NPs, were found to induce the formation of new and larger pores in cell walls and cuticles, and ruptures and disruption of microfilaments enabling internalization of larger NPs (Wang et al. 2016c). The physiological and biochemical responses of crops following exposure to metal NPs and mechanisms of interaction between plants and NPs were overviewed by Zuverza-Mena et al. (2017). As the coding regions of the genome are responsible for plant adaptation to adverse conditions, protein signatures provide insights into the phytotoxicity of NPs at the proteome level. A review paper summarizing the recent contributions of plant proteomic research to elaborate on the complex molecular pathways of plant responses to NP stress was presented by Hossain et al. (2015), and the changes in the protein profiles of plants exposed to metal-containing NPs estimated by the use of proteomic techniques were discussed by Mustafa and Komatsu (2016).

Oxidative stress generated by overproduction of ROS results in cells failing to maintain normal physiological redox-regulated functions leading to subsequent DNA damage (Fu et al. 2014). Possible detoxification pathways that might enable plant resistance to oxidative stress and facilitate metal-based NP detoxification were reviewed by Ma et al. (2015). While at low NP concentrations an increase of the activities of anti-oxidant enzymes could neutralize the toxic effects, treatment with high NP concentrations results in a decrease of these activities, and consequently in the reduction of crop yield (Du et al. 2017a; Rizwan et al. 2017). The detoxification mechanisms of nanomaterials described at physiological, proteomic, transcriptomic, and metabolomic levels were summarized by Dev et al. (2018). A critical review focused on barriers, pathways, and processes for uptake, translocation and accumulation of nanomaterials in plants was presented by Schwab et al. (2016).

With respect to the great rise in the number of papers dealing with the adverse impact of metal-based NPs on crops, we will present only a selection of recent findings related to this research topic.

Although Cu is an essential metal, CuO NPs applied at higher concentrations are phytotoxic For example, CuO NPs adversely affected rice germination rate, root and shoot length, and biomass; they accumulated in *Oryza sativa* plant organs and negatively affected a number of thylakoids per granum, photosynthetic rate, transpiration rate, stomatal conductance, and photosynthetic pigment content, increased MDA and proline contents, and caused elevated expression of APX and SOD (Da Costa et al. 2016b). The toxic effects of CuO NPs on cultivated crop plants by inhibiting seed germination were observed by Rajput et al. (2018); they were shown to cause a decrease in the shoot and root lengths, reduction in photosynthesis and respiration rate, and morphological as well enzymatic changes. In *Brassica oleracea* var. *botrytis* and *Solanum lycopersicum* plants treated with CuO NPs (20–40 nm) at concentrations of 100 and 500 mg dm^{-3}, respectively, a significant reduction of Chl content and a concentration-dependent increase of lipid peroxidation, electrolyte leakage, and antioxidant enzyme activity in tomato accompanied with an enhancement in SOD and H$_2$O$_2$ formation in leaves was observed and at exposure to 500 mg dm^{-3} CuO NPs deposition of lignin in the roots of both plants was estimated (Singh et al. 2017). An enhancement of pectin content in wheat plants treated with CuO NPs (500–800 mg dm^{-3}) due to changes in the content and conformation of cell wall polysaccharides as well as an increase of lipid peroxidation, amide content and lignin content in the leaves was reported by Sharm and Uttam (2017). A considerable reduction of Chls and to a lesser extent also a reduction of carotenoids and an increased MDA accumulation in the roots compared to the control at a 2-day exposure of wheat seedlings to Cu NPs, Ni NPs, and CuO NPs at concentrations of 0.0125–1.0 mol dm^{-3} was estimated as well (Korotkova et al. 2017). Using a metabolomics approach, Zhao et al (2016) observed pronounced metabolic changes in cucumber root exudates under exposure to CuNPs (10 and 20 mg dm^{-3}) suggesting an active defense mechanism against stress induced by nanoscale copper particles. Wang et al. (2012c) reported direct evidence for the bioaccumulation and biotransformation of CuO NPs with sizes 20–40 nm in maize suggesting the potential risk of these NPs from the aspect of food safety.

Application of 400 mg dm^{-3} ZnO NPs with a size of 24 ± 3 nm considerably reduced germination and root growth of maize plants (Lopez-Moreno et al. 2017), while a dose of 500 mg dm^{-3} caused a pronounced increase in cell-wall bound peroxidase activity and lignification of root cells resulting in reduced root growth of wheat plants (Nair and Chung 2016). ZnO NP-induced anatomical and ultrastructural modifications of the different types of cell in the root and leaf of *Brassica napus* plants were attributed, at least partly, to the Zn^{2+} dissolution by ZnO NPs rather than their absorption by the root and their subsequent effects (Kouhi et al. (2015). ZnNPs and ZnO NPs exhibited superior negative effects on

seed germination, growth characteristics, and photosynthetic pigment (Chl and carotenoid) contents of tomato and wheat plants compared to their bulk counterparts, ZnO NPs being more phytotoxic, whereby increased H$_2$O$_2$ accumulation and MDA levels following exposure to both NPs were estimated as being higher in tomato plants. Zn accumulation in plant organs decreased in following order: Zn^{2+} ions>ZnNPs>ZnO NPs>bulk ZnO>control (Amooaghaie et al. 2017).

Application of phytogenic AgNPs to rice, maize, and peanut resulted in size dependent effects on their germination, growth, and biochemical parameters such as, Chl content, carotenoid, and protein content (Prasad et al. 2017). Following foliar exposure of *Lactuca sativa* to AgNPs, Ag was effectively entrapped by the cuticle, penetrated in the leaf tissue through the stomata, it diffused in leaf tissues, and AgNPs were oxidized and Ag$^+$ ions formed complexes with thiol-containing molecules (Larue et al. 2014). In contrast to foliar exposure, root exposure to AgNPs pronouncedly decreased biomass of soybean and rice plants and also a higher increase in the MDA and H$_2$O$_2$ contents of leaves was observed at root exposure. At similar AgNPs application levels, foliar exposure resulted in 17–200-fold higher Ag bioaccumulation than root exposure and NPs were stored in the cell wall and plasmalemma of the leaves. Detection of AgNPs in plants following treatment with Ag$^+$ ions, indicates their *in vivo* formation. Due to translocation processes in plants exposed to AgNPs, fruits, seeds, and other edible parts may become contaminated with this toxic metal (Li et al. 2017c).

The rice root proteome study revealed that at exposure to AgNPs the AgNPs-responsive proteins were primarily associated with the oxidative stress response pathway, Ca^{2+} regulation and signaling, transcription, protein degradation, cell-wall synthesis, cell division, and apoptosis, and based on estimated increased abundance of defense-related proteins, such as SOD, APX, and GST, accelerated production of ROS could be assumed (Mirzajani et al. 2014). In wheat plants treated with pristine AgNPs, sulfidized Ag$_2$S NPs, and Ag$^+$, various chemical transformations were observed on the epidermis and inside the roots, even for Ag$_2$S NPs, leading to an exposure to multiple Ag forms that likely evolve over time, and the Ag source impacted in different ways also the genes involved in defense against oxidative stress, defense against pathogens, and metal homeostasis (Pradas del Real et al. 2017).

Exposure to 50 and 100 mg dm^{-3} γ-Fe$_2$O$_3$ NPs pronouncedly reduced the root length of maize plants, decreased Chl contents, and roots treated with 20–100 mg dm^{-3} γ-Fe$_2$O$_3$ NPs showed 5–7-fold higher MDA levels compared to the control (Li et al. 2016). Wang et al. (2016b) reported that the reduced toxicity of zero-valent iron due to 'aging' that lasted 2 or 4 weeks was connected with the incomplete oxidation of these NPs. Exposure of 55-day-old *Helianthus annuus* plants to 50 and 100 mg dm^{-3} nanomaghemite in hydroponic solution for 5 days resulted in a significant reduction of the root hydraulic conductivity and reduced concentrations of the macronutrients Ca, K, Mg, and S in the shoots compared to the control plants resulting in lower contents of Chl pigments (Martinez-Fernandez et al. 2016). On the other hand, magnetic Fe$_3$O$_4$

NPs applied at a dose of 2000 mg dm^{-3} were found to reduce uptake of heavy metals applied at concentrations of 1 mmol dm^{-3} and alleviate their toxicity in wheat seedlings by activation of protective mechanisms against oxidative stress resulting in an increase in SOD and POD activities and reduction of MDA content, whereby the beneficial effect of Fe$_3$O$_4$ NPs was associated with their adsorption capacity for heavy metals (Konate et al. 2017).

A prolonged treatment of hydroponically grown kidney bean plants with CeO$_2$ NPs at a dose of 500 mg dm^{-3} resulted in pronouncedly reduced root antioxidant enzyme activities and caused an increase in the root-soluble protein (by 204%) and guaiacol peroxidase activity in the leaves (Majumdar et al. 2014). In the fruit of cucumber plants grown in soil treated with CeO$_2$ NPs at 400 mg kg^{-1}, increased starch content, altered non-reducing sugar content, and decreased globulin and glutelin, while at a dose of 800 mg kg^{-1} a notably decreased phenolic content was estimated but it did not impact flavonoid content (Zhao et al. 2014.). Treatment with 800 mg kg^{-1} CeO$_2$ NPs reduced the yield of cucumber plants by 31.6% compared to the control (Zhao et al. 2013). CeO$_2$ NPs caused destruction of chloroplasts and vascular bundles and alteration of nutrient absorption on conventional and Bt-transgenic cotton plants, toxic effects being more pronounced in Bt-transgenic cotton plants (Nhan et al. 2015). Genotoxic effects from the exposure of soybean plants to CeO$_2$ NPs were reported by Lopez-Moreno et al. (2010).

Although at ambient CO$_2$ level (370 μmol dm^{-3}), TiO$_2$ NPs (50 and 200 mg kg^{-1}) were not phytotoxic to rice plants, under high CO$_2$ concentration (570 μmol dm^{-3}) they pronouncedly reduced rice biomass and grain yield and at higher applied concentration they increased accumulation of Ca, Mg, Mn, P, Zn, and Ti but reduced fat and total sugar in grains (Du et al. 2017b). TiO$_2$ NPs with sizes <100 nm (tetragonal crystals) or spherical shaped 10 nm TiO$_2$ NPs exerted specific effects at different levels of toxicity to *Vicia faba* L., depending on their size and shape and they were internalized in root cells; however, application of bulk TiO$_2$ caused stronger adverse effects in *V. faba* roots (Castiglione et al. 2016).

Treatment of wheat plants with 13 nm Al$_2$O$_3$ resulted in lignin accumulation, callose deposition, and cellular damage in root cortex cells correlating with reduced root elongation, decreased proline content, increased POD activity, and induced DNA fragmentation (Yanik and Vardar 2015).

BIBLIOGRAPHY

Adrees, M., S. Ali, M. Rizwan et al. 2015. The effect of excess copper on growth and physiology of important food crops: A review. *Environ. Sci. Pollut. Res.* 22(11):8148–8162.

Aery, N. C. 2012. Plant defence against heavy metal stress. In *Plant Defence: Biological Control (Progress in Biological Control)* Vol. 12, eds. J. M. Merillon, and K. G. Ramawat, pp. 241–269. Dordrecht: Springer Science & Business Media.

Ahsan, N., D. G. Lee, K. H. Kim et al. 2010. Analysis of arsenic stress-induced differentially expressed proteins in rice leaves by two-dimensional gel electrophoresis coupled with mass spectrometry. *Chemosphere* 78(3):224–231.

Akountianaki-Ioannidou, A., K. Papadimitriou, P. Barouchas, and N. Moustakas. 2015. The effects of Cd and Zn interactions on the concentration of Cd and Zn in sweet bush basil (*Ocimum basilicum* L.) and peppermint (*Mentha piperita* L.). *Fresen. Environ. Bull.* 24(1):77–83.

Ali, B., T. M. Mwamba, R. A. Gill et al. 2014. Improvement of element uptake and antioxidative defense in *Brassica napus* under lead stress by application of hydrogen sulfide. *Plant Growth Regul.* 74(3):261–273.

Ali, N. A., D. Dewez, O. Didur, and R. Popovic. 2006. Inhibition of photosystem II photochemistry by Cr is caused by the alteration of D1 protein and oxygen evolving complex. *Photosynth. Res.* 89(2–3):81–87.

Almeida-Rodriguez, A. M., M. P. Gomes, A. Loubert-Hudon, S. Joly, and M. Labrecque. 2016. Symbiotic association between *Salix purpurea* L. and *Rhizophagus irregularis*: Modulation of plant responses under copper stress. *Tree Physiol.* 36(4):407–420.

Amooaghaie, R., M. Norouzi and M. Saeri. 2017. Impact of zinc and zinc oxide nanoparticles on the physiological and biochemical processes in tomato and wheat. *Botany* 95(5):441–455.

Andersson, I., and A. Backlund. 2008. Structure and function of Rubisco. *Plant Physiol. Biochem.* 46(3):275–291.

Andras, P., I. Turisova, G. Buccheri, J. M. X. de Matos, and V. Dirner. 2016. Comparison of heavy-metal bioaccumulation properties in *Pinus* sp and *Quercus* sp in selected European Cu deposits. *Web Ecol.* 16:81–87.

Andresen, E., and H. Kuepper. 2013. Cadmium toxicity in plants. *Met. Ions Life Sci.* 11:395–413.

Anjum, S. A., M. Tanveer, S. Hussain et al. 2015. Cadmium toxicity in maize (*Zea mays* L.): Consequences on antioxidative systems, reactive oxygen species and cadmium accumulation. *Environ. Sci. Pollut. Res.* 22(21):17022–17030.

Antal, T. K., E. E. Graevskaya, D. N. Matorin et al. 2004. Study of chloride mercury and methylmercury effects on the photosynthetic activity of diatom *Thalassiosira weissflogii* by fluorescence methods. *Biofizika* 49:72–78.

Antisari, L. V., F. Orsini, L. Marchetti, G. Vianello, and G. Gianquinto. 2015. Heavy metal accumulation in vegetables grown in urban gardens. *Agron. Sustain. Dev.* 34:1139–1147.

Appenroth, K. J., A. Keresztes, E. Sarvari, A. Jaglarz, and W. Fischer. 2003. Multiple effects of chromate on *Spirodela polyrhiza*: Electron microscopy and biochemical investigations. *Plant Biol.* 5(3):315–323.

Appenroth, K. J., J. Stöckel, A. Srivastava, and R. J. Strasser. 2001. Multiple effects of chromate on the photosynthetic apparatus of *Spirodela polyrhiza* as probed by OJIP chlorophyll a fluorescence measurements. *Environ. Pollut.* 115(1):49–64.

Aravantinou A. F., F. Andreou, and I. D. Manariotis. 2017. Long-term toxicity of ZnO nanoparticles to *Scenedesmus rubescens* cultivated in different media. *Sci. Rep.* 7:13454.

Arena, C., F. Figlioli, M. C. Sorrentino et al. 2017. Ultrastructural, protein and photosynthetic alterations induced by Pb and Cd in *Cynara cardunculus* L., and its potential for phytoremediation. *Ecotox. Environ. Saf.* 145:83–89.

Arencibia, A. D., C. Rodriguez, L. Roco, C. Vergara, N. Gonzalez-Soto, and R. Garcia-Gonzalez. 2016. Tolerance to heavy metal stress in seedlings of three pine species from contrasting environmental conditions in Chile. *IFOREST* 9(6):937–945.

Arshad, M., S. Ali, A. Noman et al. 2016. Phosphorus amendment decreased cadmium (Cd) uptake and ameliorates chlorophyll contents, gas exchange attributes, antioxidants, and mineral nutrients in wheat (*Triticum aestivum* L.) under Cd stress. *Arch. Agron. Soil Sci.* 62(4):533–546.

Ashraf, U., A. S. Kanu, Z. W. Mo et al. 2015. Lead toxicity in rice: Effects, mechanisms, and mitigation strategies-a mini review. *Environ. Sci. Pollut. Res.* 22(23):18318–18332.

Asztalos, E., G. Sipka, M. Kis, M. Trotta, and P. Maroti. 2012. The reaction center is the sensitive target of the mercury(II) ion in intact cells of photosynthetic bacteria. *Photosynth. Res.* 112(2):129–140.

Atal, N., P. P. Saradhi, and P. Mohanty. 1991. Inhibition of the chloroplast photochemical reactions by treatment of wheat seedlings with low concentrations of cadmium: Analysis of electron transport activities and changes in fluorescence yield. *Plant Cell Physiol.* 32(7):943–951.

Azevedo, R., E. Rodriguez, R. J. Mendes et al. 2018. Inorganic Hg toxicity in plants: A comparison of different genotoxic parameters. *Plant Physiol. Biochem.* 125:247–254.

Aziz, H., M. Sabir, H. R. Ahmad et al. 2015. Alleviating effect of calcium on nickel toxicity in rice. *Clean-Soil Air Water* 43(6):901–909.

Babula, P., V. Adam, R. Opatrilova, J. Zehnalek, L. Havel, and R. Kizek. 2008. Uncommon heavy metals, metalloids and their plant toxicity: A review. *Environ. Chem. Lett.* 6(4):189–213.

Bagheri, R., J. Ahmad, H. Bashir, M. Iqbal, and M. I. Qureshi. 2017. Changes in rubisco, cysteine-rich proteins and antioxidant system of spinach (*Spinacia oleracea* L.) due to sulphur deficiency, cadmium stress and their combination. *Protoplasma* 254(2):1031–1043.

Baker, A. J. M. 1995. *Metal hyperaccumulation by plants: Our present knowledge of the ecophysiological phenomenon. In Will Plants Have a Role in Bioremediation?*, eds. D. Randall, I. Raskin, A. J. M. Baker, D. Blevins, *and* R. Smith, *pp.* 7–8. Columbia, MO: *University of Missouri.*

Bakou, A., C. Buser, G. Dandulakis, G. Brudvig, and D. F. Ghanotakis. 1992. Calcium binding site(s) of photosystem II as probed by lanthanides. *BBA-Bioenerg.* 1099(2):131–136.

Barber, J., and Tran P. D. 2013. From natural to artificial photosynthesis. *J. R. Soc. Interface* 10(81):20120984.

Barbieri, C. 2013. Medicinal and aromatic plants legislation in the European Union, Italy and several of its regions. *Nat. Prod. Res.* 27(17):1576–1582.

Barceló, J., and C. Poschenrieder. 1999. Structural and ultrastructural changes in heavy metal exposed plants. In *Heavy Metal Stress in Plants*, eds. M. N. V. Prasad and J. Hagemeyer, pp. 183–205. Berlin Heidelberg: Springer-Verlag.

Barón, M., J. B. Arellano, and J. L. Gorgé. 1995. Copper and photosystem II: A controversial relationship. *Physiol. Plant.* 94(1):174–180.

Barta, C., A. M. Dunkle, R. M. Wachter, and M. E. Salvucci. 2010. Structural changes associated with the acute thermal instability of Rubisco activase. *Arch. Biochem. Biophys.* 499(1–2):17–25.

Bartlett, J. E., S. V. Baranov, G. M. Ananyev, and G. C. Dismukes. 2008. Calcium controls the assembly of the photosynthetic water-oxidizing complex: A cadmium(II) inorganic mutant of the Mn4Ca core. *Phil. Trans. R. Soc. B – Biol. Sci.* 363(1494):1253–1261.

Baszyński, T., A. Tukendorf, M. Ruszakowska, E. Skorzyńska, and W. Maksymiec. 1988. Characteristics of the photosynthetic apparatus of copper non-tolerant spinach exposed to excess copper. *J. Plant Physiol.* 132(6):708–713.

Baszyński, T., L. Wajda, M. Król, D. Wolińska, Z. Krupa, and A. Tukendorf. 1980. Photosynthetic activities of cadmium-treated tomato plants. *Physiol. Plant.* 48(3):365–370.

Bayramov, S., and N. Guliyev. 2014. Changes in Rubisco activase gene expression and polypeptide content in *Brachypodium distachyon. Plant Physiol. Biochem.* 81:61–66.

Bazzaz, N. B., and Govindjee. 1974. Effects of cadmium nitrate on spectral characteristics and light reaction of chloroplasts. *Environ. Lett.* 6(1):1–12.

Belatik, A., S. Hotchandani, and R. Carpentier. 2013. Inhibition of the water oxidizing complex of photosystem II and the reoxidation of the quinone acceptor QA by Pb2+. *PLoS ONE* 8(7):e68142.

Beniwal, V., K. S. Nehra, and V. Chhokar. 2015. Cadmium induced alteration in lipid profile of developing mustard (*Brassica juncea* L.) seed. *Biocatal. Agric. Biotechnol.* 4(3):416–422.

Bentsen N. S., and C. Felby. 2012. Biomass for energy in the European Union – A review of bioenergy resource assessments. *Biotechnol. Biofuels* 5(1):25.

Bernier, M., and R. Carpentier. 1995. The action of mercury on the binding of the extrinsic polypeptides associated with the water oxidizing complex of photosystem II. *FEBS Lett.* 360(3):251–254.

Bernier, M., R. Popovic, and R. Carepentier. 1993. Mercury inhibition at the donor side of photosystem II is reversed by chloride. *FEBS Lett.* 321(1):19–23.

Bhuvaneshwari M., V. Iswarya, S. Archanaa et al. 2015. Cytotoxicity of ZnO NPs towards freshwater algae *Scenedesmus obliquus* at low exposure concentrations in UV-C, visible and dark conditions. *Aquat. Toxicol.* 162:29–38.

Bishnoi, N. R., L. K. Chugh, and S. K. Sawhney. 1993. Effect of chromium on photosynthesis, respiration and nitrogen fixation in pea (*Pisum sativum* L.) seedlings. *J. Plant Physiol.* 142(1):25–30.

Bittsánszky, A., T. Kömives, G. Gullner et al. 2005. Ability of transgenic poplars with elevated glutathione content to tolerate zinc(2+) stress. *Environ. Int.* 31(2): 251–254.

Bizo, M. L., S. Nietzsche, U. Mansfeld et al. 2017. Response to lead pollution: Mycorrhizal *Pinus sylvestris* forms the biomineral pyromorphite in roots and needles. *Environ. Sci. Pollut. Res.* 24(16):14455–14462.

Blankenship, R. E., and K. Sauer. 1974. Manganese in oxygen evolution. I. Electron paramagnetic resonance study of the environment of manganese in Tris-washed chloroplasts. *Biochim. Biophys. Acta* 357(2):252–266.

Bloom, A. J., and P. Kameritsch. 2017. Relative association of Rubisco with manganese and magnesium as a regulatory mechanism in plants. *Physiol. Plant.* 161(4):545–559.

Bohner, H., H. Böhme, and P. Böger. 1980. Reciprocal formation of plastocyanin and cytochrome c-553 and the influence of cupric ions on photosynthetic electron transport. *Biochim. Biophys. Acta* 592(1):103–112.

Boisvert, S., D. Joly, S. Leclerc, S. Govindachary, J. Harnois, and R. Carpentier. 2007. Inhibition of the oxygen-evolving complex of photosystem II and depletion of extrinsic polypeptides by nickel. *Biometals* 20(6):879–889.

Borisev, M., S. Pajevic, N. Nikolic et al. 2016. Magnesium and iron deficiencies alter Cd accumulation in *Salix viminalis* L. *Int. J. Phytorem.* 18(2):164–170.

Borowiak, K., K. Drzewiecka, Z. Magdziak, M. Gasecka, and M. Mleczek. 2012. Effect of Ca/Mg ions ratio on copper accumulation, photosynthetic activity and growth of Cu2+-treated *Salix viminalis* L. 'Cannabina.' *Photosynthetica* 50(3):353–361.

Buchanan, B. B., W. Gruissem, and R. L. Jones. 2001. *Biochemistry and Molecular Biology of Plants.* Rockville, MD: America Society of Plant Physiologist, Courier Companies.

Burda, K., J. Kruk, K. Strzalka, and G. H. Schmid. 2002. Stimulation of oxygen evolution in photosystem II by copper (II) ions. *Z. Naturforsch c – J. Biosci.* 57(9–10):853–857.

Burda, K., J. Kruk, K. Strzalka, J. Stanek, G. H. Schmid, and O. Kruse. 2006. Mössbauer studies of Cu(II) ions interaction with the non-heme iron and cytochrome b559 in a *Chlamydomonas reinhardtii* PS I minus mutant. *Acta Phys. Pol. A* 109:237–247.

Castiglione, M. R., L. Giorgetti, L. Bellani, S. Muccifora, S. Bottega, and C. Spano. 2016. Root responses to different types of TiO2 nanoparticles and bulk counterpart in plant model system *Vicia faba* L. *Environ. Exp. Bot.* 130:11–21.

Castro-Bugallo, A., A. Gonzalez-Fernandez, C. Guisande, and A. Barreiro. 2014. Comparative responses to metal oxide nanoparticles in marine phytoplankton. *Arch. Environ. Contam. Toxicol.* 67(4):483–493.

Cedeno-Maldonado, A., J. A. Swader, and R. L. Heath. 1972. The cupric ion as an inhibitor of photosynthetic electron transport in isolated chloroplasts. *Plant Physiol.* 50(6):698701.

Celik, A., A. A. Kartal, A. Akdogan, and Y. Kaska. 2005. Determining the heavy metal pollution in Denizli (Turkey) by using *Robinia pseudo-acacia* L. *Environ. Int.* 31(1):105–112.

Chaloub, R. M., C. C. P. de Magalhaes, and C. P. dos Santos. 2005. Early toxic effects of zinc on PS II of *Synechocystis aquatilis* f. aquatilis (Cyanophyceae). *J. Phycol.* 41(6):1162–1168.

Chandrakar, V., S. C. Naithani, and S. Keshavkant. 2016. Arsenic-induced metabolic disturbances and their mitigation mechanisms in crop plants: A review. *Biologia* 71(4):367–377.

Chapell, J. 1997. Phytoremediation of TCE using *Populus*, Status report prepared for the U.S. EPA Technology Innovation Office under a National Network of Environmental Management Studies Fellowship. http://www.clu-in.org/products/intern/phytotce.htm

Chapman, R. L. 2013. Algae: The world's most important 'plants'– an introduction. *Mitig. Adapt. Strateg. Glob. Change* 18(1):5–12.

Chen, R. F. 1986. Fluorescence quenching as a parameter for measuring complex formation between metal ions and aromatic amino acids and peptides. *Anal. Lett.* 22(9–10):963–967.

Chen, Y. P., D. Chen, and Q. Liu. 2017. Exposure to a magnetic field or laser radiation ameliorates effects of Pb and Cd on physiology and growth of young wheat seedlings. *J. Photochem. Photobiol. B* 169:171–177.

Cherchi, C., T. Chernenko, M. Diem, and A. Z. Gu. 2011. Impact of nano titanium dioxide exposure on cellular structure of *Anabaena variabilis* and evidence of internalization. *Environ. Toxicol.* 30(4):861–869.

Chizzola, R., and U. S. Mitteregger. 2005. Cadmium and zinc interactions in trace element accumulation in chamomile. *J. Plant Nutr.* 28(8):1383–1396.

Choppala, G., Saifullah, N. Bolan et al. 2014. Cellular mechanisms in higher plants governing tolerance to cadmium toxicity. *Crit. Rev. Plant Sci.* 33(5):374–391.

Christenhusz, M., and K. W. Byng. 2016. The number of known plant species in the world and its annual increase. *Phytotaxa* 261(3):201–217.

Cigáň, M., F. Šeršeň, and K. Kráľová. 2003. Relationship between the ability of heavy metals to form complexes with tryptophan and their photosynthesis-inhibiting activity. In *Ksienga Konferencyjna/Proceedings. ECOpole'03*, Jamrozowa Polana, Poland, October 16–18, 2003, eds. M. Waclawek and W. Waclawek, pp. 35–38. Opole: Towarzystwo Chemii a Inzynierii Ekologicznej.

Clark, L. C. 1956. Monitor and control of blood and tissue oxygen tension. *Trans. Am. Soc. Artif. Intern. Organs* 2: 41.

Clemens, S., and J. F. Ma. 2016. Toxic heavy metal and metalloid accumulation in crop plants and foods. *Ann. Rev. Plant Biol.* 67:489–512.

Commoner, B., J. J. Heise, B. B. Lippincot, R. E. Norberg, J. V. Pssoneau, and J. Townsend. 1957. Biological activity of free radicals. *Science* 126(3263):57–63.

da Costa, C. H., F. Perreault, A. Oukarroum, S. P. Melegari, R. Popovic, and W. G. Matias. 2016a. Effect of chromium oxide (III) nanoparticles on the production of reactive oxygen species and photosystem II activity in the green alga *Chlamydomonas reinhardtii*. *Sci. Total Environ.* 565:951–960.

Da Costa, M. V. J., and P. K. Sharma. 2016b. Effect of copper oxide nanoparticles on growth, morphology, photosynthesis, and antioxidant response in *Oryza sativa*. *Photosynthetica* 54(1):110–119.

Dalir, N., and A. H. Khoshgoftarmanesh. 2014. Symplastic and apoplastic uptake and root to shoot translocation of nickel in wheat as affected by exogenous amino acids. *J. Plant Physiol.* 171(7):531–536.

Dalir, N., and A. H. Khoshgoftarmanesh. 2015. Root uptake and translocation of nickel in wheat as affected by histidine. *J. Plant Physiol.* 184:8–14.

Dao, L. H. T., and J. Beardall. 2016. Effects of lead on two green microalgae *Chlorella* and *Scenedesmus*: Photosystem II activity and heterogeneity. *Algal Res.* 16:150–159.

Dash, A., A. P. Singh, B. R. Chaudhary, S. K. Singh, and D. Dash. 2012. Effect of silver nanoparticles on growth of eukaryotic green algae. *Nano Micro Lett.* 4(3):158–165.

De Filippis, L. F., R. Hampp, and H. Ziegler. 1981. The effects of sublethal concentrations of zinc, cadmium and mercury on *Euglena*. Adenylates and energy charge. *Z. Pflanzenphysiol.* 103(1):1–7.

Deng, H., Z. H. Ye, and M. H. Wong. 2004. Accumulation of lead, zinc, copper and cadmium by 12 wetland plant species thriving in metal contaminated sites in China. *Environ. Pollut.* 132(1):29–40.

Deng, X. Y., J. Cheng, X. L. Hu, L. Wang, D. Li, and K. Gao. 2017. Biological effects of TiO2 and CeO2 nanoparticles on the growth, photosynthetic activity, and cellular components of a marine diatom *Phaeodactylum tricornutum*. *Sci. Total Environ.* 575:87–96.

Dev, A., A. K. Srivastava, and S. Karmakar. 2018. Nanomaterial toxicity for plants. *Environ. Chem. Lett.* 16(1):85–100.

Dezhban, A., A. Shirvany, P. Attarod, M. Delshad, M. Matinizadeh, and M. Khoshnevis. 2015. Cadmium and lead effects on chlorophyll fluorescence, chlorophyll pigments and proline of *Robinia pseudoacacia*. *J. For. Res.* 26(2):323–329.

Dias, M. C., C. Monteiro, J. Mouthino-Pereira, C. Correia, B. Goncalves, and C. Santos. 2013. Cadmium toxicity affects photosynthesis and plant growth at different levels. *Acta Physiol. Plant.* 35(4):1281–1289.

Di Baccio, D., R. Tognetti, L. Sebastiani, and C. Vitagliano. 2003. Responses of *Populus deltoids* × *Populus nigra* (*Populus_euramericana*) clone I-214 to high zinc concentrations. *New Phytol.* 159(2):443–452.

Dietz, A. C., and J. L. Schnoor. 2001. Advances in phytoremediation. *Environ. Health Perspect.* 109(Suppl. 1):163–168.

Di Santo, P., C. Cocozza, R. Tognetti, G. Palumbo, E. Di Iorio, and B. Paura. 2017. A quick screening to assess the phytoextraction potential of cadmium and copper in *Quercus pubescens* plantlets. *IFOREST* 10:93–98.

Dmuchowski, W., D. Gozdowski, P. Bragoszewska, A. H. Baczewska, and I. Suwara. 2014. Phytoremediation of zinc contaminated soils using silver birch (*Betula pendula* Roth). *Ecol. Eng.* 71(2014):32–35.

Dobrikova, A. G., E. K. Yotsova, A. Boerner, S. P. Landjeva, and E. L. Apostolova. 2017. The wheat mutant DELLA-encoding gene (Rht-B1c) affects plant photosynthetic responses to cadmium stress. *Plant Physiol. Biochem.* 114:10–18.

Dominguez, M. T., T. Maranon, J. M. Manuel, and S. Redondo-Gomez. 2011. Response of Holm oak (*Quercus ilex* subsp *ballota*) and mastic shrub (*Pistacia lentiscus* L.) seedlings to high concentrations of Cd and Tl in the rhizosphere. *Chemosphere* 83(8):1166–1174.

Dordevic, B., J. Krajnakova, D. Hampel, D. Gomory, and L. Havel. 2017a. Effects of cadmium and lead stress on somatic embryogenesis of coniferous species. Part I: Evaluation of the genotype-dependent response. *Acta Physiol. Plant.* 39:140.

Dordevic, B., M. Praskova, D. Hampel, and L. Havel. 2017b. Effects of cadmium and lead stress on somatic embryogenesis of coniferous species. Part II: Changes of thiol substances. *Acta Physiol. Plant.* 39:141.

Dos Santos Utmazian, M. N., G. Wieshammer, R. Vega, and W. W. Wenzel. 2007. Hydroponic screening for metal resistance and accumulation of cadmium and zinc in twenty clones of willows and poplars. *Environ. Pollut.* 148(1):155–165.

Drzewiecka, K., M. Mleczek, M. Gasecka, Z. Magdziak, and P. Golinski. 2012. Changes in *Salix viminalis* L. cv. 'Cannabina' morphology and physiology in response to nickel ions – Hydroponic investigations. *J. Hazard. Mater.* 217:429–438.

Du, W. C., J. L. Gardea-Torresdey, Y. W. Xie et al. 2017b. Elevated CO2 levels modify TiO2 nanoparticle effects on rice and soil microbial communities. *Sci. Total Environ.* 578:408–416.

Du, W. C., W. J. Tan, J. R. Peralta-Videa et al. 2017a. Interaction of metal oxide nanoparticles with higher terrestrial plants: Physiological and biochemical aspects. *Plant Physiol. Biochem.* 110:210–225.

Duffus, J. H. 2002. 'Heavy metals' – A meaningless term? *Pure Appl. Chem.* 74(5):793–807.

Duquesnoy, I., P. Goupil, I. Nadaud, G. Branlard, A. Piquet-Pissaloux, and G. Ledoigt. 2009. Identification of *Agrostis tenuis* leaf proteins in response to As(V) and As(III) induced stress using a proteomics approach. *Plant Sci.* 176(2):206–213.

Dutta, S., M. Mitra, P. Agarwal et al. 2018. Oxidative and genotoxic damages in plants in response to heavy metal stress and maintenance of genome stability. *Plant Signal. Behav.* 5:1–49.

El-Kassas, H. Y., and M. A. Okbah. 2017. Phytotoxic effects of seaweed mediated copper nanoparticles against the harmful alga: *Lyngbya majuscula*. *J. Genet. Eng. Biotechnol.* 15(1):41–48.

El-Naggar, A. H. 1998. Toxic effects of nickel on photosystem II of *Chlamydomonas reinhardtii*. *Cytobios* 93:93–101.

Enache, M., J. C. Dearden, and J. D. Walker. 2003. QSAR analysis of metal ion toxicity data in sunflower cultures (*Helianthus annuus* 'Sunspot'). *QSAR Comb. Sci.* 22(2):234–240.

Engel, B. D., M. Schaffer, L. K. Cuellar, E. Villa, J. M. Plitzko, and W. Baumeister. 2015. Native architecture of the *Chlamydomonas* chloroplast revealed by in situ cryo-electron tomography. *eLife* 4:e04889.

Ertani, A., A. Mietto, M. Borin, and S. Nardi. 2017. Chromium in agricultural soils and crops: A review. *Water Air Soil Pollut.* 228(5):190.

European Commission. 2011. Recommendation of 18 October 2011on the definition of nanomaterial (2011/696/EU). Official Journal of European Union L 275/38. https://eur-lex.europa.eu/legal-content/EN/TXT/?uri=CELEX%3A32011H0696

Eurostat. 2016. *Agriculture, Forestry and Fishery Statistics*, 2016 edn. Luxembourg: Publications Office of the European Union.

Evans, J. R. 2013. Improving photosynthesis. *Plant Physiol.* 162(4):1780–1793.

Evlard, A., K. Sergeant, B. Printz et al. 2014. A multiple-level study of metal tolerance in *Salix fragilis* and *Salix aurita* clones. *J. Proteomics* 101:113–129.

Fagioni, M., G. M. D'Amici, A. M. Timperio, and L. Zolla. 2009. Proteomic analysis of multiprotein complexes in the thylakoid membrane upon cadmium treatment. *J. Proteome Res.* 8(1):310–326.

Falk, H., and E. Mayr. 1997. Concerning *bay* salt and *peri* chelate formation of hydroxyphenanthroperylene quinones (fringelites). *Monatsh. Chem.* 128(4):353–360.

FAO. 2004. Building on Gender, Agrobiodiversity and Local Knowledge. http://www.fao.org/docrep/007/y5609e/y5609e02.htm

Fargašová, A. 1998. Comparison of effects of tributyl-, triphenyl-, and tribenzyltin compounds on freshwater benthos and alga *Scenedesmus quadricauda*. *Bull. Environ. Contam. Toxicol.* 60(1):9–15.

Fargašová, A., and J. Kizlink. 1996. Effect of organotin compounds on the growth of the freshwater alga *Scenedesmus quadricauda*. *Ecotox. Environ. Saf.* 34(2):156–159.

Farzadfar, S., F. Zarinkamar, S. A. Modarres-Sanavy, and M. Hojati. 2013. Exogenously applied calcium alleviates cadmium toxicity in *Matricaria chamomilla* L. plants. *Environ. Sci. Pollut. Res. Int.* 20(3):1413–1422.

Feller, U., I. Anders, and T. Mae. 2008. Rubiscolytics: Fate of Rubisco after its enzymatic function in a cell is terminated. *J. Exp. Bot.* 59(7):1615–1624.

Feng, R. W., C. Y. Wei, S. X. Tu, Y. Z. Ding, R. G. Wang, and J. K. Guo. 2013. The uptake and detoxification of antimony by plants: A review. *Environ. Exp. Bot.* 96:28–34.

Fernandez-Pinas, F., P. Mateo, and I. Bonila. 1995. Ultrastructural changes induced by selected cadmium concentrations in the cyanobacterium *Nostock* UAM 208. *J. Plant Physiol.* 147:452–456.

Fodor, F., É. Sárvári, F. Láng, Z. Szigeti, and E. Cseh. 1996. Effects of Pb and Cd on cucumber depending on the Fe-complex in culture solution. *J. Plant. Physiol.* 148(3–4):434–439.

Fu, P. P., Q. S. Xia, H. M. Hwang, P. C. Ray, and H. Yu. 2014. Mechanisms of nanotoxicity: Generation of reactive oxygen species. *J. Food Drug Anal.* 22(1):64–75.

Füleky, G. 2009. Cultivated plants, primarily as food sources. *Vol. I. Encyclopedia of Life Support Systems (EOLSS)*. https://www.eolss.net/sample-chapters/C10/E5-02.pdf

Furini, A. 2012. *Plants and Heavy Metals*. Netherlands: Springer Science & Business Media.

Gajewska, E., P. Bernat, J. Dlugonski, and M. Sklodowska. 2012. Effect of nickel on membrane integrity, lipid peroxidation and fatty acid composition in wheat seedlings. *J. Agron. Crop Sci.* 198(4):286–294.

Gajewska, E., E. Niewiadomska, K. Tokarz, M. Slaba, and M. Sklodowska. 2013. Nickel-induced changes in carbon metabolism in wheat shoots. *J. Plant Physiol.* 170(4):369–377.

Galmes, J., I. Aranjuelo, H. Medrano, and J. Flexas. 2013. Variation in Rubisco content and activity under variable climatic factors. *Photosynth. Res.* 117(13):73–90.

Gaudet, M., F. Pietrini, I. Beritognolo et al. 2011. Intraspecific variation of physiological and molecular response to cadmium stress in *Populus nigra* L. *Tree Physiol.* 31(12):1309–1318.

Gaziyev, A., S. Aliyeva, I. Kurbanova, R. Ganiyeva, S. Bayramova, and R. Gasanov. 2011. Molecular operation of metals into the function and state of photosystem II. *Metallomics* 3(12):1362–1367.

Ge, W., Y. Q. Jiao, J. H. Zou, W. S. Jiang, and D. H. Liu. 2015. Ultrastructural and photosynthetic response of *Populus 107* leaves to cadmium stress. *Pol. J. Environ Stud.* 24(2):519–527.

Gill, R. A., L. L. Zang, B. Ali et al. 2015. Chromium-induced physico-chemical and ultrastructural changes in four cultivars of *Brassica napus* L. *Chemosphere* 120:154–164.

Gjorgieva, D., T. K. Panovska, T. Ruskovska, K. Baceva, and T. Stafilov. Mineral nutrient imbalance, total antioxidants level and DNA damage in common bean (*Phaseolus vulgaris* L.) exposed to heavy metals. *Physiol. Mol. Biol. Plants* 19(4):499–507.

Goswami, S., and S. Das. 2016. Copper phytoremediation potential of *Calandula officinalis* L. and the role of antioxidant enzymes in metal tolerance. *Ecotox. Environ. Saf.* 126:211–218.

Govindjee. 2000. Milestones in photosynthesis research. In *Probing Photosynthesis: Mechanisms, Regulation and Adaptation*, eds. M. Yunus, U. Pathre, and P. Mohanty, pp. 9–39. London: Taylor & Francis.

Graevskaya, E. E., T. K. Antal, D. N. Matorin, E. N. Voronova, S. I. Pogosyan, and A. B. Rubin. 2003. Evaluation of diatomea algae *Thalassiosira weissflogii* sensitivity to chloride mercury and methylmercury by chlorophyll fluorescence analysis. *J. Phys. IV France* 107:569–572.

Greger, M., A. H. Kabir, T. Landberg, P. J. Maity, and S. Lindberg. 2016. Silicate reduces cadmium uptake into cells of wheat. *Environ. Pollut.* 211:90–97.

Grejtovský, A., K. Markušová, and L. Nováková. 2008. Lead uptake by *Matricaria chamomilla*. *Plant Soil Environ.* 54(2):47–54.

Grejtovský, A., and R. Prič. 2000. The effect of high cadmium concentration in soil on growth, uptake of nutrient and some heavy metals on *Chamomilla recutita* (L.) Rauschert. *J. Appl. Bot. Angew. Bot.* 74(5–6):169–174.

Grejtovský, A., M. Repčák, A. Eliášová, and K. Markušová. 2001. Effect of cadmium on active principle contents of *Matricaria recutita* L. *Herba Pol.* 48:203–208.

Gu, J. G., L. W. Qi, W. S. Jiang, and D. H. Liu. 2007. Cadmium accumulation and its effects on growth and gas exchange in four *Populus* cultivars. *Acta Biol. Cracoviensia.-Ser. Bot.* 49:7–14.

Guo, B. H., S. X. Dai, R. G. Wang et al. 2015. Combined effects of elevated CO2 and Cd-contaminated soil on the growth, gas exchange, antioxidant defense, and Cd accumulation of poplars and willows. *Environ. Exp. Bot.* 115:1–10.

Guo, L., Y. Q. Ding, Y. K. Xu et al. 2017. Responses of *Landoltia punctata* to cobalt and nickel: Removal, growth, photosynthesis, antioxidant system and starch metabolism. *Aquat. Toxicol.* 190:87–93.

Gupta, D. K., F. J. Corpas, and J. M. Palma. 2013. *Heavy Metal Stress in Plants*. Heidelberg: Springer-Verlag Berlin.

Hampp, R., K. Beulich, and H. Ziegler. 1976. Effects of zinc and cadmium on photosynthetic CO2-fixation and Hill activity of isolated spinach chloroplasts. *Z. Pflanzenphysiol.* 77(4):336–344.

Han, S. H., D. H. Kim, and S. J. Shin. 2013. Bioaccumulation and physiological response of five willows to toxic levels of cadmium and zinc. *Soil Sediment Contam.* 22(3):241–255.

Hasni, I., S. Hamdani, and R. Carpentier. 2013. Destabilization of the oxygen evolving complex of photosystem II by Al3+. *Photochem. Photobiol.* 89(5):1135–1142.

Hazeem, L. J., M. Bououdina, S. Rashdan, L. Brunet, C. Slomianny, and R. Boukherroub. 2016. Cumulative effect of zinc oxide and titanium oxide nanoparticles on growth and chlorophyll a content of *Picochlorum* sp. *Environ. Sci. Pollut. Res.* 23(3):2821–2830.

Hazeem, L. J., F. A. Waheed, S. Rashdan et al. 2015. Effect of magnetic iron oxide (Fe3O4) nanoparticles on the growth and photosynthetic pigment content of *Picochlorum* sp. *Environ. Sci. Pollut. Res.* 22(15):11728–11739.

Hill, R. 1937. Oxygen evolution by isolated chloroplasts. *Nature* 139:881–882.

Hoff, A. J. 1979. Application of ESR in photosynthesis. *Phys. Rep.* 54:75–200.

Honeycutt, R. C., and D. W. Krogmann. 1972. Inhibition of chloroplast reactions with phenylmercuric acetate. *Plant Physiol.* 49(3):376–380.

Hossain, Z., G. Mustafa, and S. Komatsu. 2015. Plant responses to nanoparticle stress. *Int. J. Mol. Sci.* 16(11):26644–26653.

Hsu, B. D., and J. Y. Lee. 1988. Toxic effects of copper on photosystem II of spinach chloroplasts. *Plant Physiol.* 87(1):116–119.

Huang, Z. Q., S. C. Ye, L. Y. Hu et al. 2016a. Hydrogen sulfide promotes wheat grain germination under cadmium stress. *Proc. Indian Natl. Sci. Acad. B Biol. Sci.* 86(4):887–895.

Huang, J., J. P. Cheng, and J. Yi. 2016b. Impact of silver nanoparticles on marine diatom *Skeletonema costatum*. *J. Appl. Toxicol.* 36(10):1343–1354.

Huang, Y. Y., C. T. He, C. Shen et al. 2017. Toxicity of cadmium and its health risks from leafy vegetable consumption. *Food Funct.* 8(4):1373–1401.

Hussain, I., M. A. Ashraf, R. Rasheed et al. 2017. Cadmium-induced perturbations in growth, oxidative defense system, catalase gene expression and fruit quality in tomato. *Int. J. Agric. Biol.* 19(1):61–68.

Iori, V., F. Pietrini, D. Bianconi, G. Mughini, A. Massacci, and M. Zacchini. 2017. Cadmium phytoremediation ability of eucalypt plants under hydroponics. *IFOREST* 10(2):416–421.

Iswarya, V., M. Bhuvaneshwari, S. A. Alex et al. 2015. Combined toxicity of two crystalline phases (anatase and rutile) of titania nanoparticles towards freshwater microalgae: *Chlorella* sp. *Aquat. Toxicol.* 161:154–169.

Ivanov, Y. V., A. V. Kartashov, A. I. Ivanova, Y. V. Savochkin, and V. V. Kuznetsov. 2016a. Effects of copper deficiency and copper toxicity on organogenesis and some physiological and biochemical responses of Scots pine (*Pinus sylvestris* L.) seedlings grown in hydroculture. *Environ. Sci. Pollut. Res.* 23(17):17332–17344.

Ivanov, Y. V., A. V. Kartashov, A. I. Ivanova, Y. V. Savochkin, and V. V. Kuznetsov. 2016b. Effects of zinc on Scots pine (*Pinus sylvestris* L.) seedlings grown in hydroculture. *Plant Physiol. Biochem.* 102:1–9.

Izawa, S. 1980. Acceptors and donors for chloroplast electron transport. *Methods Enzymol.* 69:413–434.

Jackson, T. A. 1998. Mercury in aquatic ecosystems. In *Metal Metabolism in Aquatic Environments*, eds. W. J. Langston, and M. J. Bebianno, pp. 77–158. London: Chapman & Hall.

Jakovljevic, M., S. Antic-Mladenovic, M. Ristic, S. Maksimovic, and S. Blagojevic. 2000. Influence of selenium on the yield and quality of chamomile (*Chamomilla* recutita (L.) Rausch.). *Rostlinná VYROBA-Plant Production* 46:123–126.

Jampílek, J., and K. Kráľová. 2015. Application of nanotechnology in agriculture and food industry, its prospects and risks. *Ecol. Chem. Eng. S* 22(3):321–361.

Jampílek, J., and K. Kráľová. 2017. Nanomaterials for delivery of nutrients and growth-promoting compounds to plants. In *Nanotechnology: An Agricultural Paradigm*, eds. R. Prasad, M. Kumar, and V. Kumar, pp. 177–226. Berlin Heidelberg: Springer-Verlag.

Jampílek, J., and K. Kráľová. 2019. Impact of nanoparticles on photosynthesizing organisms and their use in hybrid structures with some components of the photosynthetic apparatus. In *Plant Nanobionics*, ed. R. Prasad. 255–331, Cham, Switzerland: Springer Nature.

Jan, S., and J. A. Parray. 2016. *Approaches to Heavy Metal Tolerance in Plants*. Singapore: Springer-Verlag/Sci-Tech/Trade.

Javed, M. T., M. S. Akram, K. Tanwir et al. 2017. Cadmium spiked soil modulates root organic acids exudation and ionic contents of two differentially Cd tolerant maize (*Zea mays* L.) cultivars. *Ecotox. Environ. Saf.* 141:216–225.

Jegerschöld, C., J. B. Arellano, W. P. Schröder, P. J. M. van Kan, M. Barón, and S. Styring. 1995. Cu(II) inhibition of the electron transfer through photosystem II studied by EPR spectroscopy. *Biochemistry* 34:12747–12754.

Jegerschöld, C., F. MacMillan, W. Lubitz, and A. W. Rutherford. 1999. Effects of copper and zinc ions on photosystem II studied by EPR spectroscopy. *Biochemistry* 38:12439–12445.

Jensen, J. K., P. E. Holm, J. Nejrup, M. B. Larsen, and O. K. Borggaard. 2009. The potential of willow for remediation of heavy metal polluted calcareous urban soils. *Environ. Pollut.* 157(3):931–937.

Ji, J., Z. F. Long, and D. H. Lin. 2011. Toxicity of oxide nanoparticles to the green algae *Chlorella* sp. *Chem. Eng. J.* 170(2–3):525–530.

Jiang, H. X., L. S. Chen, J. G. Zheng, S. Han, N. Tang, and B. R. Smith. 2008. Aluminum-induced effects on Photosystem II photochemistry in *Citrus* leaves assessed by the chlorophyll a fluorescence transient. *Tree Physiol.* 28(12):1863–1871.

Joshi, M. K., and P. Mohanty. 2004. Chlorophyll a fluorescence as a probe of heavy metal ion toxicity in plants. In *Chlorophyll a Fluorescence: A Signature of Photosynthesis. Advances in Photosynthesis and Respiration*, Vol. 19, eds. G. C. Papagorgiou, and Govindjee, pp. 637–661. Dordrecht: Springer.

Jung, Y. S., I. Yu, and J. H. Golbeck. 1995. Reconstitution of iron-sulfur center FB results in complete restoration of NADP+ photoreduction in Hg-treated photosystem I complexes from *Synechococus* sp. PCC 6301. *Photosynth. Res.* 46(1–2):249–255.

Kalaji, H. M., V. N. Goltsev, K. Zuk-Golaszewska, M. Živčák, and M. Brestič. 2017. *Chlorophyll Fluorescence: Understanding Crop Performance – Basics and Applications*. CRC Press.

Kalaji, H. M., A. Jajoo, A. Oukarroum et al. 2016. Chlorophyll a fluorescence as a tool to monitor physiological status of plants under abiotic stress conditions. *Acta Physiol. Plant.* 38:102.

Kampfenkel, K., M. Van Montagu, and D. Inze. 1995. Effects of iron excess on *Nicotiana plumbaginifolia* plants. Implications to oxidative stress. *Plant Physiol.* 107(3):725–735.

Karacan, M. S., M. V. Rodionova, T. Tunc et al. 2016. Characterization of nineteen antimony(III) complexes as potent inhibitors of photosystem II, carbonic anhydrase, and glutathione reductase. *Photosynth. Res.* 130(1–3):167–182.

Karkehabadi, S., T. C. Taylor, and I. Andersson. 2003. Calcium supports loop closure but not catalysis in Rubisco. *J. Molec. Biol.* 334(1):65–73.

Kastori, R., M. Plesnicar, Z. Sakac, D. Pankovic, and I. Arsenijevic-Maksimovic. 1998. Effect of excess lead on sunflower growth and photosynthesis. *J. Plant Nutr.* 21(1):75–85.

Katanic, M., B. Kovacevic, B. Dordevic et al. 2015. Nickel phytoremediation potential of white poplar clones grown *in vitro*. *Rom. Biotech. Lett.* 20(1):10085–10096.

Kathe, W. 2006. Revision of the 'Guidelines on the conservation of medicinal plants' by WHO, IUCN, WWF and TRAFFIC. In *Medicinal and Aromatic Plants*, eds. R. J. Bogers, L. E. Craker, and D. Lange, pp. 109–120. Netherlands: Springer.

Kaur, G., H. P. Singh, D. R. Batish, and R. K. Kohli. 2014. Pb-inhibited mitotic activity in onion roots involves DNA damage and disruption of oxidative metabolism. *Ecotoxicology* 23(7):1292–1304.

Khairy, A. I. H., M. J. Oh, S. M. Lee, D. S. Kim, and K. S. Roh. 2016. Nitric oxide overcomes Cd and Cu toxicity in *in vitro*-grown tobacco plants through increasing contents and activities of rubisco and rubisco activase. *Biochimie Open* 2:41–51.

Khan, A., S. Khan, M. A. Khan, Z. Qamar, and M. Waqas. 2015. The uptake and bioaccumulation of heavy metals by food plants, their effects on plants nutrients, and associated health risk: A review. *Environ. Sci. Pollut. Res.* 22(18):13772–13799.

Khodamoradi, K., A. H. Khoshgoftarmanesh, N. Dalir, M. Afyuni, and R. Schulin. 2015. How do glycine and histidine in nutrient solution affect zinc uptake and root-to-shoot translocation by wheat and triticale? *Crop Pasture Sci.* 66(11):1105–1110.

Kimura, M., and S. Katoh. 1972. Studies on electron transport associated with photosystem I. I. Functional site of plastocyanin: Inhibitory effects of HgCl2 on electron transport and plastocyanin in chloroplasts. *Biochim. Biophys. Acta* 283(2):279–292.

Kiniaide, T. B. 2009. Improved scale for metal ion softness and toxicity. *Environ. Toxicol. Chem.* 28(3):525–533.

Ko, K. S., D. C. Koh, II. C. Kong. 2018. Toxicity evaluation of individual and mixtures of nanoparticles based on algal chlorophyll content and cell count. *Materials* 11(1):121.

Kojima, Y., Y. Niinomi, S. Tsuboi, T. Hiyama, and H. Sakurai. 1987. Destruction of photosystem I iron-sulfur centers of spinach and *Anacystis nidulans* by mercurials. *Bot. Mag.* 100(3):243–253.

Konate, A., X. He, Z. Y. Zhang et al. 2017. Magnetic (Fe3O4) nanoparticles reduce heavy metals uptake and mitigate their toxicity the in wheat seedling. *Sustainability* 9(5):790.

Konečná, B., F. Frič, and E. Masarovičová. 1989. Ribulose-1,5-bisphosphate carboxylase activity and protein content in pollution damaged leaves of three oaks species. *Photosynthetica* 23(4):566–574.

Koprivova, A., S. Kopriva, D. Jäger, B. Will, L. Jouanin, and H. Rennenberg. 2002. Evaluation of transgenic poplar lines overexpressing enzymes of glutathione synthesis for phytoremediation of cadmium. *Plant Biol.* 4(6):664–670.

Korotkova, A. N., S. V. Lebedev, F. G. Kayumov, and E. A. Sizova. 2017. Biological effects of wheat (*Triticum vulgare* L.) under the influence metal nanoparticles (Fe, Cu, Ni) and their oxides (Fe3O4, CuO, NiO). *Sel'skokhozyaistvennaya Biologiya* 1:172–182.

Kosiorek, M., B. Modrzewska, and M. Wyszkowski. 2016. Levels of selected trace elements in Scots pine (*Pinus sylvestris* L.), silver birch (*Betula pendula* L.), and Norway maple (Acer platanoides L.) in an urbanized environment. *Environ. Monit. Assess.* 188(10):598.

Kouhi, S. M. M., M. Lahouti, A. Ganjeali, and M. H. Entezari. 2015. Long-term exposure of rapeseed (*Brassica napus* L.) to ZnO nanoparticles: Anatomical and ultrastructural responses. *Environ. Sci. Pollut. Res.* 22(14):10733–10743.

Kováčik, J. 2013. Hyperaccumulation of cadmium in *Matricaria chamomilla*: A never-ending story? *Acta Physiol. Plant.* 35(5):1721–1725.

Kováčik, J., J. Gruz, J. Hedbavny, B. Klejdus, and M. Strnad. 2009b. Cadmium and nickel uptake are differentially modulated by salicylic acid in *Matricaria chamomilla* plants. *J. Agric. Food Chem.* 57(20):9848–9855.

Kováčik, J., B. Klejdus, J. Hedbavny, P. Martonfi, F. Stork, and L. Martonfiova. 2011. Copper uptake, physiology and cytogenetic characteristics in three *Matricaria chamomilla* cultivars. *Water Air Soil Pollut.* 218(1–4):681–691.

Kováčik, J., B. Klejdus, J. Kaduková, and M. Bačkor. 2009a. Physiology of *Matricaria chamomilla* exposed to nickel excess. *Ecotox. Environ. Saf.* 72(2):603–609.

Kráľová, K., F. Šeršeň, and M. Blahová. 1994. Effects of Cu(II) complexes on photosynthesis in spinach chloroplasts. Aqua(aryloxyacetato)copper(II) complexes. *Gen. Physiol. Biophys.* 13: 483–491.

Kráľová, K., F. Šeršeň, and M. Melník. 1998. Inhibition of photosynthesis in *Chlorella vulgaris* by Cu(II) complexes with biologically active ligands. *J. Trace Microprobe Techn.* 16:491–500.

Kráľová, K., K. Kissová, and O. Švajlenová. 2000a. Effects of carboxylatocopper (II) complexes on photosynthesizing organisms. *Chem. Inz. Ekol.* 7(10):1077–1083.

Kráľová, K., E. Masarovičová, and A. Bumbálová. 2000b. Toxic effect of cadmium on *Hypericum perforatum* plants and green alga *Chlorella vulgaris*. *Chem. Inz. Ekol.* 7(11):1200–1205.

Kráľová, K., and E. Masarovičová. 2003. *Hypericum perforatum* L. and *Chamomilla recutita* (L.) Rausch. – accumulators of some toxic metals. *Pharmazie* 58(5):359–360.

Kráľová, K., and E. Masarovičová. 2004. Could complexes of heavy metals with secondary metabolites induce enhanced metal tolerance of *Hypericum perforatum*? In *Macro and Trace Elements. Mengen- und Spurenelemente. 22.* Workshop, Jena, Germany, September 24–25, 2004, eds. M. Anke et al., pp. 411–416. Jena: Friedrich Schiller Universität.

Kráľová, K., and E. Masarovičová. 2006. Plants for the future. *Ecol. Chem. Eng.* 13(11):1179–1207.

Kráľová, K., E. Masarovičová, I. Ondrejkovičová, and M. Bujdoš. 2007a. Effect of selenium oxidation state on cadmium translocation in chamomile plants. *Chem. Pap.* 61(3):171–175.

Kráľová, K., E. Masarovičová, J. Kubová, and O. Švajlenová. 2007b. Response of *Matricaria recutita* plants to some copper(II) chelates. *Acta Hort. (ISHS)* 749:237–243.

Kráľová, K., E. Masarovičová, F. Šeršeň, and I. Ondrejkovičová. 2008a. Effect of different Fe(III) compounds on photosynthetic electron transport in spinach chloroplasts and on iron accumulation in maize plants. *Chem. Pap.* 62(4): 358–363.

Kráľová, K., and E. Masarovičová. 2008b. EDTA-assisted phytoextraction of copper, cadmium and zinc using chamomile plants. *Ecol. Chem. Engin.* 15(3):213–220.

Krohling, C. A., F. J. Eutropio, A. A. Bertolazi et al. 2016. Ecophysiology of iron homeostasis in plants. *Soil Sci. Plant Nutr.* 62(1):39–47.

Krupa, Z. 1999. Cadmium against higher plant photosynthesis – A variety of effects and where do they possible come from? *Z. Naturforsch.* 54c:723–729.

Krupa, Z., A. Siedlecka, W. Maksymiec, and T. Baszynski. 1993. *In vivo* response of photosynthetic apparatus of *Phaseolus vulgaris* to nickel toxicity. *J. Plant Physiol.* 142:664–668.

Krupa, Z., E. Skorzyńska, W. Maksymiec, and T. Baszyński. 1987. Effect of cadmium treatment on photosynthetic apparatus and its photochemical activities in greening radish seedlings. *Photosynthetica* 21:156–154.

Kukarskikh, G. P., E. E. Grayevskaya, T. E. Krendeleva, K. N. Timofeev, and A. B. Rubin. 2003. Effect of methylmercury on the primary photosynthetic activity of green microalgae *Chlamydomonas reinhardtii*. *Biofizika* 48:853–859.

Kummerová, M., Š. Zezulka, K. Kráľová, and E. Masarovičová. 2010. Effect of zinc and cadmium on physiological and production characteristics in *Matricaria recutita*. *Biol. Plant.* 54(2):308–314.

Lange, D. 1998. *Europe's Medicinal and Aromatic Plants: Their Use, Trade and Conservation*. Cambridge: TRAFFIC International.

Larue, C., H. Castillo-Michel, S. Sobanska et al. 2014. Foliar exposure of the crop *Lactuca sativa* to silver nanoparticles: Evidence for internalization and changes in Ag speciation. *J. Hazard. Mater.* 264:98–106.

Laureysens, I., R. Blust, L. De Temmerman, C. Lemmens, and R. Ceulemans. 2004. Clonal variation in heavy metal accumulation and biomass production in poplar coppice culture: I. Seasonal variation in leaf, wood and bark concentrations. *Environ. Pollut.* 131(3):485–494.

Lee, K. R., and K. S. Roh. 2003. Influence of cadmium on Rubisco activation in *Canavalia ensiformis* L. leaves. *Biotech. Bioprocess Engin.* 8:94–100.

Le Gall, H., F. Philippe, J. M. Domon, F. Gillet, J. Pelloux, and C. Rayon. 2015. Cell wall metabolism in response to abiotic stress. *Plants-Basel* 4(1):112–166.

Le Guedard, M., O. Faure, and J. J. Bessoule. 2012. *Populus nigra* grown on a metallurgical landfill. *Chemosphere* 88(6):693–698.

Lei, C., Y. Q. Sun, D. C. W. Tsang, and D. H. Lin. 2018. Environmental transformations and ecological effects of iron-based nanoparticles. *Environ. Pollut.* 232:10–30.

Lei, C., L. Q. Zhang, K. Yang, L. Z. Zhu, and D. H. Lin. 2016. Toxicity of iron-based nanoparticles to green algae: Effects of particle size, crystal phase, oxidation state and environmental aging. *Environ. Pollut.* 218:505–512.

Lešíková, J., K. Kráľová, E. Masarovičová, J. Kubová, and I. Ondrejkovičová. 2007. Effect of different cadmium compounds on chamomile plants. *Acta Hort. (ISHS)* 749:223–229.

Li, C. C., F. Dang, M. Li et al. 2017c. Effects of exposure pathways on the accumulation and phytotoxicity of silver nanoparticles in soybean and rice. *Nanotoxicology* 11(5):699–709.

Li, F. M., Z. Liang, X. Zheng, W. Zhao, M. Wu, and Z. Y. Wang. 2015. Toxicity of nano-TiO2 on algae and the site of reactive oxygen species production. *Aquat. Toxicol.* 158:1–13.

Li, J., J. Hu, C. X. Ma et al. 2016. Uptake, translocation and physiological effects of magnetic iron oxide (γ-Fe2O3) nanoparticles in corn (*Zea mays* L.). *Chemosphere* 159:326–334.

Li, L. Z., C. Tu, W. J. G. M. Peijnenburg, and Y. M. Luo. 2017a. Characteristics of cadmium uptake and membrane transport in roots of intact wheat (*Triticum aestivum* L.) seedlings. *Environ. Pollut.* 221:351–358.

Li, X. M., and L. H. Zhang. 2015. Endophytic infection alleviates Pb2+ stress effects on photosystem II functioning of *Oryza sativa* leaves. *J. Hazard. Mater.* 295:79–85.

Li, Y. Q., R. Xiao, Z. L. Liu, X. J. Liang, and W. Feng. 2017b. Cytotoxicity of NiO nanoparticles and its conversion inside *Chlorella vulgaris*. *Chem. Res. Chinese U* 33(1):107–111.

Li, Z., F. Q. Xing, and D. Xing. 2012. *Characterization of target site of aluminum phytotoxicity in photosynthetic electron transport by fluorescence techniques in tobacco leaves*. Plant Cell Physiol. *53(7)*:1295–1309.

Lidon, F. C., and F. S. Henriques. 1991. Limiting step on photosynthesis of rice plants treated with varying copper levels. *J. Plant Physiol.* 138(1):115–118.

Lidon, F. C., J. Ramalho, and F. S. Henriques. 1993. Copper inhibition of rice photosynthesis. *J. Plant Physiol.* 142(1):12–17.

Ling, Q. F., and F. H. Hong. 2009. Effects of Pb2+ on the structure and function of photosystem II of *Spirodela polyrrhiza*. *Biol. Trace Elem. Res.* 129(1–3):251–260.

Liu, D., A. H. Liu, C. He, J. H. Wang, and Y. A. Wang. 2012. Response of organic acids to zinc homeostasis in zinc-deficient and zinc-toxic apple rootstock roots. *Pedosphere* 22(6):803–814.

Liu, K. L., L. Shen, J. Q. Wang, and J. P. Sheng. 2008. Rapid inactivation of chloroplastic ascorbate peroxidase is responsible for oxidative modification to Rubisco in tomato (*Lycopersicon esculentum*) under cadmium stress. *J. Integr. Plant Biol.* 50(4):415–426.

Liu, X. Q., H. Huang, C. Liu, M. Zhou, and F. S. Hong. 2009. Physico-chemical property of rare earths-effects on the energy regulation of photosystem II in *Arabidopsis thaliana*. *Biol. Trace Elem. Res.* 130(2):141–151.

Liu, X. Q., M. G. Su, C. Liu, L. Zhang, W. H. Si, and F. S. Hong. 2007. Effects of CeCl3 on energy transfer and oxygen evolution in spinach photosystem II. *J. Rare Earth* 25(5):624–630.

Liu, X. Y., H. Z. Yao, F. Ahmad, and Y. Zhou. 2016. Photosynthetic toxicity of ZnO and TiO2 nanoparticles to *Chlorella vulgaris*. *Nanomedicine* 12(2):538–538.

Lopez-Moreno, M. L., Y. Cedeno-Mattei, S. J. Bailon-Ruiz et al. 2018. Environmental behavior of coated NMs: Physicochemical aspects and plant interactions. *J. Hazard Mater.* 347:196–217.

Lopez-Moreno, M. L., G. de la Rosa, G. Cruz-Jimenez, L. Castellano, J. R. Peralta-Videa, and J. L. Gardea-Torresdey. 2017. Effect of ZnO nanoparticles on corn seedlings at different temperatures; X-ray absorption spectroscopy and ICP/OES studies. *Microchem. J.* 134:54–61.

Lopez-Moreno, M. L., G. de la Rosa, J. A. Hernandez-Viezcas, et al. 2010. Evidence of the differential biotransformation and genotoxicity of ZnO and CeO2 nanoparticles on soybean (*Glycine max*) plants. *Environ. Sci. Technol.* 44(19):7315–7320.

Lorenc-Plucinska, G., M. Walentynowicz, and A. Niewiadomska. 2013. Capabilities of alders (*Alnus incana* and *A. glutinosa*) to grow in metal-contaminated soil. *Ecol. Eng.* 58:214–227.

Lunáčková, L, E. Masarovičová, K. Kráľová, and V. Streško. 2003b. Response of fast growing woody plants from family Salicaceae to cadmium treatment. *Bull. Environ. Contam. Toxicol.* 70(3):576–585.

Lunáčková, L., A. Šottníková, E. Masarovičová, A. Lux, and V. Streško. 2003a. Comparison of cadmium effect on willow and poplar in response to different cultivation conditions. *Biol. Plant.* 47(3):403–411.

Ma, C. X., J. C. White, O. P. Dhankher, and B. S. Xing. 2015. Metal-based nanotoxicity and detoxification pathways in higher plants. *Environ. Sci. Technol.* 49(12):7109–7122.

Maivan, E. S., T. Radjabian, P. Abrishamchi, and D. Talei. 2017. Physiological and biochemical responses of *Melissa officinalis* L. to nickel stress and the protective role of salicylic acid. *Arch. Agron. Soil Sci.* 63(3):330–343.

Majumdar, S., J. R. Peralta-Videa, S. Bandyopadhyay, J. A. Hernandez-Viezcas, S. Sahi, and J. L. Gardea-Torresdey. 2014. Exposure of cerium oxide nanoparticles to kidney bean shows disturbance in the plant defense mechanisms. *J. Hazard. Mater.* 278:279–287.

Maksymiec, W., R. Russa, T. Urbanik-Sypniewska, and T. Baszyński. 1994. Effect of excess Cu on the photosynthetic apparatus of runner bean leaves treated at two different growth stages. *Physiol. Plant.* 91(4):715–721.

Malik, D., I. S. Sheoran, and R. Singh. 1992. Carbon metabolism in leaves of cadmium-treated wheat seedlings. *Plant Physiol. Biochem.* 30(2):223–229.

Mallick, N., and L. C. Rai. 1992. Metal induced inhibition of photosynthesis, photosynthetic electron transport chain and ATP content of *Anabaena doliolum* and *Chlorella vulgaris*: Interaction with exogenous ATP. *Biomed. Environ. Sci.* 5(3):241–250.

Mani, D., and N. K. Patel. 2014. Humic acid, EDDS and EDTA induced phytoremediation of cadmium contaminated alluvial soil by *Calendula officinalis* L. *J. Ind. Chem. Soc.* 91(11):2073–2082.

Marmiroli, M., D. Imperiale, E. Maestri, and N. Marmiroli. 2013. The response of *Populus* spp. to cadmium stress: Chemical, morphological and proteomics study. *Chemosphere* 93(7):1333–1344.

Marquard, R., and M. Schneider. 1998. Zur Cadmiumproblematik im Arzneipflanzenbau. In *Fachtagung Arznei- und Gewürzpflanzen, Giessen, Germany*, October 1–2, 1998, eds. R. Marquard and E. Schubert, pp. 9–15.

Marsho, T. V., and B. Kok. 1980. P700 detection. *Methods Enzymol.* 69: 280–289.

Martinez-Fernandez, D., D. Barroso, and M. Komarek. 2016. Root water transport of *Helianthus annuus* L. under iron oxide nanoparticle exposure. *Environ. Sci. Pollut. Res.* 23(2):1732–1741.

Masarovičová, E., A. Cicák, and I. Štefančík. 1999a. Plant responses to air pollution and heavy metal stress. In *Handbook of Plant and Crop Stress*, ed. M. Pessarakli, pp. 569–598. New York: Marcel Dekker.

Masarovičová, E., and K. Kráľová. 2005. Approaches to measuring plant photosynthetic activity. In *Handbook of Photosynthesis*, 2nd edn, ed. M. Pessarakli, pp. 617–656. Boca Raton, FL: Taylor & Francis Group.

Masarovičová, E., and K. Kráľová. 2007. Medicinal plants – Past, nowadays, future. *Acta Hort. (ISHS)* 749:19–27.

Masarovičová, E., and K. Kráľová. 2013. Metal nanoparticles and plants. *Ecol. Chem. Eng. S* 20(1):9–22.

Masarovičová E., and K. Kráľová. 2017. Essential elements and toxic metals in some crops, medicinal plants, and trees. In *Phytoremediation: Management of Environmental Contaminants*, Vol. 5, eds. A. A. Ansari, S. S. Gill, R. Gill, G. R. Lanza, and L. Newman, pp. 183–255. Cham: Springer International Publishing AG.

Masarovičová, E., and K. Kráľová. 2018. Woody species in phytoremediation applications for contaminated soils. In *Phytoremediation: Management of Environmental Contaminants*, Vol. 6, ed. A. A. Ansari. S.S. Gill, R. Gill, G. Lanza, L. Newman, 319–373, Cham, Switzerland: Springer Nature.

Masarovičová, E., K. Kráľová, and M. Kummerová. 2010b. Principles of classification of medicinal plants as hyperaccumulators or excluders. *Acta Physiol. Plant.* 32(5):823–829.

Masarovičová, E., K. Kráľová, and M. Peško. 2009. Energetic plants – Cost and benefit. *Ecol. Chem. Engin. S* 16(3):263–276.

Masarovičová, E., K. Kráľová, and F. Šeršeň. 2010a. Plant responses to toxic metal stress. In *Handbook of Plant and Crop Stress*, 3rd edn, pp. 595–634. Boca Raton, FL: CRC Press.

Masarovičová, E., K. Kráľová, F. Šeršeň, A. Bumbálová, and A. Lux. 1999b. Effect of toxic metals on medicinal plants. In *Mengen- und Spurelemente. 19. Arbeitstagung, Jena, February 3–4, 1999*, eds. M. Anke et al., pp. 189–196. Leipzig: Verlag Harald Schubert.

Masarovičová, E., K. Kráľová, and V. Streško. 2003. Effect of metal ions on some medicinal plants. *Chem. Inz. Ekol.* 10(3–4):275–279.

Masarovičová, E., K. Kráľová, and S. S. Zinjarde. 2014. Metal nanoparticles in plants. Formation and action. In *Handbook of Plant and Crop Physiology*, 3rd edn, ed. M. Pessarakli, pp. 683–731. Boca Raton, FL: CRC, Taylor and Francis.

Matorin, D. N., D. A. Todorenko, N. Kh. Seifullina, B. K. Zayadan, and A. B. Rubin. 2013. Effect of silver nanoparticles on the parameters of chlorophyll fluorescence and P-700 reaction in the green alga *Chlamydomonas reinhardtii*. *Microbiology* 82(6):809–814.

Matysik, J., A. Alia, H. J. van Gorkom, and H. J. M. de Groot. 1998. Substitution of calcium by cadmium in Photosystem II complex. In *Photosynthesis: Mechanisms and Effects*, Vol. II, ed. G. Garab, pp. 1423–1426. Dordrecht: Kluwer Academic Publishers.

Matysik, J., A. Alia, G. Nachtegaal, H. J. van Gorkom, A. J. Hoff, and H. J. M. de Groot. 2000. Exploring the calcium-binding site in photosystem II membranes by solid-state Cd-113 NMR. *Biochemistry* 39(23):6751–6755.

Melegari, S. P., F. Perreault, R. H. R. Costa, R. Popovic, and W. G. Matias. 2013. Evaluation of toxicity and oxidative stress induced by copper oxide nanoparticles in the green alga *Chlamydomonas reinhardtii*. *Aquat. Toxicol.* 142:431–440.

Mertens, J., P. Vervaeke, A. De Schrijver, and S. Luyssaert. 2004. Metal uptake by young trees from dredged brackish sediment: Limitations and possibilities for phytoextraction and phytostabilisation. *Sci. Total Environ.* 326(1–3):209–215.

Michel-Lopez, C. Y., F. E. Y. Gil, G. F. Ortiz et al. 2016b. Bioaccumulation and effect of cadmium in the photosynthetic apparatus of *Prosopis juliflora*. *Chem. Spec. Bioavailab.* 28:1–6.

Michel-Lopez, C. Y., F. E. Y. Gil, G. F. Ortiz, J. M. Santamaria, and D. Gonzalez-Mendoza. 2016a. Bioaccumulation and changes in the photosynthetic apparatus of *Prosopis juliflora* exposed to copper. *Bot. Sci.* 94(2):323–330.

Mihailovic, N., G. Drazic, and Z. Vucinic. 2008. Effects of aluminium on photosynthetic performance in Al-sensitive and Al-tolerant maize inbred lines. *Photosynthetica* 46:476–480.

Miles, D., P. Bolen, S. Farag et al. 1973. Hg++ – A DCMU independent electron acceptor of photosystem II. *Biochim. Biophys. Res. Commun.* 50:1113–1119.

Mirzajani, F., H. Askari, S. Hamzelou et al. 2014. Proteomics study of silver nanoparticles toxicity on *Oryza sativa* L. *Ecotox. Environ. Saf.* 108:335–339.

Mohanty, N., I. Vass, and S. Demeter. 1989a. Copper toxicity affects photosystem II electron transport at the secondary quinone acceptor QB. *Plant Physiol.* 90(1):175–179.

Mohanty, N., I. Vas, and S. Demeter. 1989b. Impairment of photosystem 2 activity at the level of secondary quinone electron acceptor in chloroplasts treated with cobalt, nickel and zinc ions. *Physiol. Plant.* 76(3):386–390.

Morgalev, N. Y., V. A. Kurovsky, A. I. Gosteva, G. T. Morgaleva, S. Y. Morgalev, and A. A. Burenina. 2017. Influence of metal-containing nanoparticles on the content of photosynthetic pigments of unicellular alga *Chlorella vulgaris* Baijer. *NHC* 13:255–262.

Mourato, M. P., I. N. Moreira, I. Leitao, F. R. Pinto, J. R. Sales, and L. L. Martins. 2015. Effect of heavy metals in plants of the genus *Brassica*. *Int. J. Mol. Sci.* 16(8):17975–17998.

Moustakas, M., E. P. Eleftheriou, and G. Ouzounidou. 1997. Short-term effects of aluminium at alkaline pH on the structure and function of the photosynthetic apparatus. *Photosynthetica* 34(2):169–177.

Moustakas, M., T. Lanaras, L. Symeonidis, and S. Karataglis. 1994. Growth and some photosynthetic characteristics of field-grown *Avena sativa* under copper and lead stress. *Photosynthetica* 30(3):389–396.

Moustakas, M., G. Ouzounidou, E. P. Eleftheriou, and R. Lannoye. 1996. Indirect effects of aluminium stress on the function of the photosynthetic apparatus. *Plant Physiol. Biochem.* 34:553–560.

Moustakas, N. K., A. Akoumianaki-Ioannidou, and P. E. Barouchas. 2011. The effects of cadmium and zinc interactions on the concentration of cadmium and zinc in pot marigold (*Calendula officinalis* L.). *Aust. J. Crop. Sci.* 5(3):274–279.

Mugnai, S., E. Azzarello, C. Pandolfi, and S. Mancuso. 2006. Zinc and cadmium accumulation in *Hyssopus officinalis* L. and *Satureja montana* L. *Acta Hort. (ISHS)* 723:361–366.

Murch, S. J., K. Haq, H. P. V. Rupasinghe, and P. K. Saxena. 2003. Nickel contamination affects growth and secondary metabolite composition of St. John's wort (*Hypericum perforatum* L.). *Environ. Exp. Bot.* 49(3):251–257.

Murkowski, A., and E. Skorska. 2010. Effect of (C6H5)3PbCl and (C6H5)3SnCl on delayed luminescence intensity, evolving oxygen and electron transport rate in photosystem II of *Chlorella vulgaris*. *Bull. Environ. Contam. Toxicol.* 84(2):157–160.

Murthy, S. D. S., and P. Mohanty. 1991. Mercury induces alteration of energy transfer in phycobilisome by selectively affecting the pigment protein, phycocyanin, in the cyanobacterium *Spirulina platensis*. *Plant Cell Physiol.* 32(2):231–237.

Murthy, S. D. S., and P. Mohanty. 1995. Action of selected heavy metal ions on the photosystem 2 activity of the cyanobacterium *Spirulina platensis*. *Biol. Plant.* 37:79–84.

Murthy, S. D. S., S. C. Sabat, and P. Mohanty. 1989. Mercury-induced inhibition of photosystem II activity changes in the emission of fluorescence from phycobilisomes in intact cells of the cyanobacterium *Spirulina platensis*. *Plant Cell Physiol.* 30(8):1153–1157.

Mustafa, G., and S. Komatsu. 2016. Toxicity of heavy metals and metal-containing nanoparticles on plants. *BBA−Proteins Proteom.* 1864(8):932–944.

Mysliwa-Kurdziel, B., M. N. V. Prasad, and K. Strzalka. 2004. Photosynthesis in heavy metal stressed plants. In *Heavy Metal Stress in Plants: From Biomolecules to Ecosystems*, 2nd edn, ed. M. N. V. Prasad, pp. 146–181. Berlin: Springer.

Nadgorska-Socha, A., M. Kandziora-Ciupa, M. Trzesicki, and G. Barczyk. 2017. Air pollution tolerance index and heavy metal bioaccumulation in selected plant species from urban biotopes. *Chemosphere* 183:471–482.

Nahar, S., and H. A. Tajmir-Riahi. 1994. A comparative study of Fe(II) and Fe(III) ion complexation with proteins of the light-harvesting complex of chloroplast thylakoid membranes. *J. Inorg. Biochem.* 54(2):79–90.

Nahar, S., and H. A. Tajmir-Riahi. 1995. Do metal ions the protein secondary structure of light-harvesting complex of thylakoid membranes. *J. Inorg. Biochem.* 58(3):112–234.

Nair, P. M. G., and I. M. Chung. 2014. A mechanistic study on the toxic effect of copper oxide nanoparticles in soybean (*Glycine max* L.) root development and lignification of root cells. *Biol. Trace Elem. Res.* 162(1-3):342–352.

Nair, R., S. H. Varghese, B. G. Nair, T. Maekawa, Y. Yoshida, and D. S. Kumar. 2010. Nanoparticulate material delivery to plants. *Plant Sci.* 179(3):154–163.

Nasreddine, L., and D. Parent-Massin. 2002. Food contamination by metals and pesticides in the European Union. Should we worry? *Toxicol. Lett.* 127(1–3):29–41.

Nedunchezhian, N., and G. Kulandaivelu. 1995. Effect of Cd and UV-B radiation on polypeptide composition and photosystem activities of *Vigna unguiculata* chloroplasts. *Biol. Plant.* 37:437–441.

Negi, A., H. P. Singh, D. R. Batish, and R. K. Kohli. 2014. Ni+2-inhibited radicle growth in germinating wheat seeds involves alterations in sugar metabolism. *Acta Physiol. Plant.* 36(4):923–929.

Newman, M. C., J. T. McCloskey, and C. R Tatara. 1998. Using metal-ligand binding characteristics to predict metal toxicity: Quantitative ion character-activity relationships (QICARs). *Environ. Health Perspect.* 106(Suppl. 6):1419–1425.

Nhan, J. V., C. X. Ma, and Y. K. Rui. 2015. Phytotoxic mechanism of nanoparticles: Destruction of chloroplasts and vascular bundles and alteration of nutrient absorption. *Sci. Rep.* 5:11618.

Nieboer, E., G. G. Fletchera, and Y. Thomassen. 1999. Relevance of reactivity determinants to exposure assessment and biological monitoring of the elements. *J. Environ. Monit.* 1(1):1–14.

Nieboer, E., and D. Richardson. 1980. The replacement of the nondescript term 'heavy metals' by a biologically and chemically significant classification of metal ions. *Environ. Pollut. Ser. B* 1(1):3–26.

Nikolic, N., D. Kojic, A. Pilipovic et al. 2008. Responses of hybrid poplar to cadmium stress: Photosynthetic characteristics, cadmium and proline accumulation, and antioxidant enzyme activity. *Acta Biol. Cracov.-Ser. Bot.* 50(2):95–103.

Nikolic, N., L. Zoric, I. Cvetkovic et al. 2017. Assessment of cadmium tolerance and phytoextraction ability in young *Populus deltoides* L. and *Populus × euramericana* plants through morpho-anatomical and physiological responses to growth in cadmium enriched soil. *IFOREST* 10:635–644.

Nikolic, N. P., M. K. Borisev, S. P. Pajevic et al. 2015. Photosynthetic response and tolerance of three willow species to cadmium exposure in hydroponic culture. *Arch. Biol. Sci.* 67(4):1411–1420.

Nogueirol, R. C., F. A. Monteiro, P. L. Gratao, B. K. D. da Silva, and R. A. Azevedo. 2016. Cadmium application in tomato: Nutritional imbalance and oxidative stress. *Water Air Soil Pollut.* 227(6):210.

Nowack, B., R. Schulin, and B. H. Robinson. 2006. A critical assessment of chelant- enhanced metal phytoextraction. *Environ. Sci. Technol.* 40(17):5225–5232.

Nowicka-Krawczyk, P., J. Zelazna-Wieczorek, and T. Kozlecki. 2017. Silver nanoparticles as a control agent against facades coated by aerial algae – A model study of *Apatococcus lobatus* (green algae). *PLoS ONE* 12(8):e0183276.

Ono, T. 2000. Effects of lanthanide substitution at Ca2+-site on the properties of the oxygen evolving center of photosystem II. *J. Inorg. Biochem.* 82(1–4):85–91.

Ono, T. A., and Y. Inoue. 1989. Roles of Ca2+ in O2 evolution in higher-plants photosystem II-effects of replacement of Ca2+ site by other cations. *Arch. Biochem. Biophys.* 275:440–448.

Oukarroum, A., S. Bras, F. Perreault, and R. Popovic. 2012. Inhibitory effects of silver nanoparticles in two green algae, *Chlorella vulgaris* and *Dunaliella tertiolecta*. *Ecotox. Environ. Saf.* 78:80–85.

Oukarroum, A., M. Samadani, and D. Dewez. 2014. Influence of pH on the toxicity of silver nanoparticles in the green alga *Chlamydomonas acidophila*. *Water Air Soil Pollut.* 225(8):2038.

Oukarroum, A., W. Zaidi, M. Samadani, and D. Dewez. 2017. Toxicity of nickel oxide nanoparticles on a freshwater green algal strain of *Chlorella vulgaris*. *Biomed. Res. Int.* 2017:9528180.

Ouzounidou, G., M. Moustakas, and E. P. Eleftheriou. 1997. Physiological and ultrastructural effects of cadmium on wheat (*Triticum sativum* L.) leaves. *Arch. Environ. Contam. Toxicol.* 32(2):154–160.

Ouzounidou, G., M. Moustakas, and R. Lannoye. 1995. Chlorophyll fluorescence and photoacoustic characteristics in relation to changes in chlorophyll and Ca2+ content of a Cu-tolerant *Silene compacta* ecotype under Cu treatment. *Physiol. Plant.* 93(3):551–557.

Ovečka, M., and T. Takáč. 2014. Managing heavy metal toxicity stress in plants: Biological and biotechnological tools. *Biotechnol. Adv.* 32(1):73–86.

Page, V., and U. Feller. 2015. Heavy metals in crop plants: Transport and redistribution processes on the whole plant level. *Agronomy-Basel* 5(3):447–463.

Pagliano, C., M. Raviolo, F. Dalla Vecchia et al. 2006. Evidence for PS II donor side damage and photoinhibition induced by cadmium treatment on rice (*Oryza sativa* L.). *J. Photochem. Photobiol. B – Biol.* 84(1):70–78.

Pajak, M., W. Halecki, and M. Gasiorek. 2017. Accumulative response of Scots pine (*Pinus sylvestris* L.) and silver birch (*Betula pendula* Roth) to heavy metals enhanced by Pb-Zn ore mining and processing plants: Explicitly spatial considerations of ordinary kriging based on a GIS approach. *Chemosphere* 168:851–859.

Palivan, C. G., G. Gescheidt, and L. Weiner. 2001. The formation of copper complexes with hypericin, in solutions: An EPR study. *J. Inorg. Biochem.* 86:369–369.

Pan, X. L., X. Chen, D. Z. Zhang et al. 2009. Effect of chromium (VI) on photosystem II activity and heterogenity of *Synechocystis* sp (Cyanophyta): Studied with *in vivo* chlorophyll fluorescence tests. *J. Phycol.* 45(2):386–394.

Panda, P., L. Sahoo, and S. K. Panda. 2015. Heavy metal and metalloid stress in plants: The genomics perspective. In *Abiotic Stresses in Crop Plants*, eds. U. Chakraborty, and B. Chakraborty, pp. 164–177. Boston: CAB International.

Panda, S. K., S. Choudhary, and H. K. Patra. 2016. Heavy-metal-induced oxidative stress in plants: Physiological and molecular perspectives. In *Abiotic Stress Response in Plants 1*, eds. N. Tuteja, and S. S. Gill, pp. 219–232. Wiley VCH Verlag.

Pandey, S., K. Gupta, and A. K. Mukherjee. 2007. Impact of cadmium and lead on *Catharanthus roseus* – A phytoremediation study. *J. Environ. Biol.* 28(3):655–662.

Parlak, K. U. 2016. Effect of nickel on growth and biochemical characteristics of wheat (*Triticum aestivum* L.) seedlings. *NJAS-Wagen J. Life Sci.* 76:1–5.

Parvaiz, A. 2015. *Plant Metal Interaction*. Elsevier.

Pavlovič, A., E. Masarovičová, K. Kráľová, and J. Kubová. 2006. Response of chamomile plants (*Matricaria recutita* L.) to cadmium treatment. *Bull. Environ. Contam. Toxicol.* 77(5):763–771.

Pearson, R. G. 1963. Hard and soft acids and bases. *J. Am. Chem. Soc.* 85(22):3533–3539.

Peña-Olmos, J. E., and F. Casierra-Posada. 2013. Photochemical efficiency of photosystem II (PSII) in broccoli plants (*Brassica oleracea* var Italica) affected by excess iron. *Orinoquia* 17(1):15–22.

Peng, C., H. Zhang, H. X. Fang et al. 2015. Natural organic matter-induced alleviation of the phytotoxicity to rice (*Oryza sativa* L.) caused by copper oxide nanoparticles. *Environ. Toxicol. Chem.* 34(9):1996–2003.

Pereira, S. P. P., F. Jesus, S. Aguiar et al. 2018. Phytotoxicity of silver nanoparticles to *Lemna minor*: Surface coating and exposure period-related effects. *Sci. Total Environ.* 618:1389–1399.

Perez-Chaca, M. V., M. Rodriguez-Serrano, A. S. Molina et al. 2014. Cadmium induces two waves of reactive oxygen species in *Glycine max* (L.) roots. *Plant Cell Environ.* 37(7):1672–1687.

Pérez-Romero, J. A., S. Redondo-Gomez, and E. Mateos-Naranjo. 2016. Growth and photosynthetic limitation analysis of the Cd-accumulator *Salicornia ramosissima* under excessive cadmium concentrations and optimum salinity conditions. *Plant Physiol. Biochem.* 109:103–113.

Perreault, F., N. A. Ali, C. Saison, R. Popovic, and P. Juneau. 2009. Dichromate effect on energy dissipation of photosystem II and photosystem I in *Chlamydomonas reinhardtii*. *J. Photochem. Photobiol. B* 96(1):24–29.

Perreault F., A. Oukarroum, S. P. Melegari, W. G. Matias, and R. Popovic. 2012. Polymer coating of copper oxide nanoparticles increases nanoparticles uptake and toxicity in the green alga *Chlamydomonas reinhardtii*. *Chemosphere* 87(11):1388–1394.

Peško, M., K. Kráľová, and E. Masarovičová. 2010. Response of *Hypericum perforatum* plants to supply of cadmium compounds containing different forms of selenium. *Ecol. Chem. Eng. S* 17(3):279–287.

Peško, M., K. Kráľová, and E. Masarovičová. 2011. Growth of *Matricaria recutita* L. and accumulation of Cd and Cu within plant organs in different stage of ontogenetical development (In Slovak). In *Effect of Abiotic and Biotic Stressors on Plant Features*, eds. V. Bláha, and F. Hnilička, pp 208–211. Praha: Research Institute of Plant Production.

Pessarakli, M. 2005. *Handbook of Photosynthesis*, 2nd edn. Boca Raton, FL: Taylor & Francis Group.

Petrescu, I., I. Dobosan, E. Madosa et al. 2014. Effect of cadmium on seed germination and seedling development at *Salvia officinalis* L. *J. Biotechnol.* 185:S115–S115.

Pietrini, F., M. Iannelli, S. Pasqualini, and A. Massacci. 2003. Interaction of cadmium with glutathione and photosynthesis in developing leaves and chloroplasts of *Phragmites australis* (Cav.) Trin. ex Steudel. *Plant Physiol.* 133(2):829–837.

Pietrini, F., V. Iori, D. Bianconi, G. Mughini, A. Massacci, and M. Zacchini. 2015. Assessment of physiological and biochemical responses, metal tolerance and accumulation in two eucalypt hybrid clones for phytoremediation of cadmium-contaminated waters. *J. Environ. Manag.* 162:221–231.

Portis, A. R. 2003. Rubisco activase–Rubisco's catalytic chaperone. *Photosynth. Res.* 75(1):11–27.

Pourrut, B., M. Shahid, C. Dumat, P. Winterton, and E. Pinelli. 2011. Lead uptake, toxicity, and detoxification in plants. *Rev. Environ. Contam. Toxicol.* 213:113–136.

Prasad, A., A. K. Singh, S. Chand, C. S. Chanotiya, and D. D. Patra. 2010. Effect of chromium and lead on yield, chemical composition of essential oil, and accumulation of heavy metals of mint species. *Commun. Soil Sci. Plant Anal.* 41(18):2170–2186.

Prasad, S. M., J. B. Singh, L. C. Rai, and H. D. Kumar. 1991. Metal-induced inhibition of photosynthetic electron transport chain of the cyanobacterium *Nostoc muscorum*. *FEMS Microbiol. Lett.* 82(1):95–100.

Prasad, T. N. V. K. V., S. Adam, P. V. Rao, R. R. Reddy, and T. G. Krishna. 2017. Size dependent effects of antifungal phytogenic silver nanoparticles on germination, growth and biochemical parameters of rice (*Oryza sativa* L), maize (*Zea mays* L) and peanut (*Arachis hypogaea* L). *IET Nanobiotechnol.* 11(3):277–285.

Pradas del Real, A. E., V. Vidal, M. Carriere et al. 2017. Silver nanoparticles and wheat roots: A complex interplay. *Environ. Sci. Technol.* 51(10):5774–5782.

Prokowski, Z. 1993. Effects of HgCl2 on long-lived delayed luminescence in *Scenedesmus quadricauda*. *Photosynthetica* 28:563–566.

Quartacci, M. F., C. Pinzino, C. L. M. Sgherri, F. Dalla Vecchia, and F. Navari-Izzo. 2000. Growth in excess copper induces changes in the lipid composition and fluidity of PSII-enriched membranes in wheat. *Physiol. Plant.* 108(1):87–93.

Quigg, A., W. C. Chin, C. S. Chen et al. 2013. Direct and indirect toxic effects of engineered nanoparticles on algae: Role of natural organic mater. *ACS Sustain. Chem. Eng.* 1(7):686–702.

Radmer, R., and B. Kok. 1974. Kinetic observation of the system II electron acceptor pool isolated by mercuric ion. *Biochim. Biophys. Acta* 357(2):177–180.

Rai, L. C., A. K. Singhet, and N. Mallik. 1991. Studies on photosynthesis, the associated electron transport system and some physiological variables of *Chlorella vulgaris* under heavy metal stress. *J. Plant Physiol.* 137(4):419–424.

Rai, V., P. Vajpayee, S. N. Singh, and S. Mehrotra. 2004. Effect of chromium accumulation on photosynthetic pigments, oxidative stress defense system, nitrate reduction, proline level and eugenol content of *Ocimum tenuiflorum* L. *Plant Sci.* 167(5):1159–1169.

Rajput, V. D., T. Minkina, S. Suskova et al. 2018. Effects of copper nanoparticles (CuO NPs) on crop plants: A mini review. *Bionanoscience* 8(1):36–42.

Rao, J. F., W. D. Lv, and J. M. Yang. 2017. Proteomic analysis of saffron (*Crocus sativus* L.) grown under conditions of cadmium toxicity. *Biosci. J.* 33(3):713–720.

Rashid, A., and R. Popovic. 1990. Protective role of CaCl2 against Pb2+ inhibition in photosystem II. *FEBS Lett.* 271(1–2):181–184.

Rashid, A., M. Bernier, L. Pazdernick, and R. Carpentier. 1991. Interaction of Zn2+ with the donor side of photosystem II. *Photosynth. Res.* 30(2–3):123–130.

Rashid, A., E. L. Camm, and A. K. Ekramoddoullah. 1994. Molecular mechanism of action of Pb2+ and Zn2+ on water oxidizing complex of photosystem II. *FEBS Lett.* 350(2–3):296–298.

Rawat, M., R. Nayan, B. Negi, M. G. H. Zaidi, and S. Arora. 2017. Physio-biochemical basis of iron-sulfide nanoparticle induced growth and seed yield enhancement in *B. juncea*. *Plant Physiol. Biochem.* 118:274–284.

Reeves, P. G., and R. L. Chaney. 2008. Bioavailability as an issue in risk assessment and management of food cadmium: A review. *Sci. Total Environ.* 398(1–3):13–19.

Renganathan, M., and S. Bose. 1989. Inhibition of primary photochemistry of photosystem II by copper in isolated pea chloroplasts. *Biochim. Biophys. Acta* 974(3):247–253.

Renger, G., H. M. Gleiter, E. Haag, and F. Reifarth. 1993. Photosystem II: Thermodynamics and kinetics of electron transport from QA to QB(QB*) and deleterious effects of copper (II). *Z. Naturforsch.* 48c(3–4):234–240.

Reyes, V. C., M. R. Spitzmiller, A. Hong-Hermesdorf et al. 2016. Copper status of exposed microorganisms influences susceptibility to metallic nanoparticles. *Environ. Toxicol. Chem.* 35(5):1148–1158.

Rizwan, M., S. Ali, T. Abbas et al. 2016b. Cadmium minimization in wheat: A critical review. *Ecotox. Environ. Saf.* 130:43–53.

Rizwan, M., S. Ali, M. Adrees et al. 2016a. Cadmium stress in rice: Toxic effects, tolerance mechanisms, and management: A critical review. *Environ. Sci. Pollut. Res.* 23(18):17859–17879.

Rizwan, M., S. Ali, M. F. Qayyum et al. 2017. Effect of metal and metal oxide nanoparticles on growth and physiology of globally important food crops: A critical review. *J. Hazard. Mater.* 322A:2–16.

Rizwan, M., J. D. Meunier, J. C. Davidian, O. S. Pokrovsky, N. Bovet, and C. Keller. 2016c. Silicon alleviates Cd stress of wheat seedlings (*Triticum turgidum* L. cv. *Claudio*) grown in hydroponics. *Environ. Sci. Pollut. Res.* 23(2):1414–1427.

Rodriguez, E., C. Santos, R. Azevedo, J. Moutinho-Pereira, C. Correia, and M. C. Dias. 2012. Chromium (VI) induces toxicity at different photosynthetic levels in pea. *Plant Physiol. Biochem.* 53:94–100.

Rodriguez, E., M. D. Santos, R. Azevedo et al. 2015. Photosynthesis light-independent reactions are sensitive biomarkers to monitor lead phytotoxicity in a Pb-tolerant *Pisum sativum* cultivar. *Environ. Sci. Pollut. Res.* 22(1):574–585.

Rogers, N. J., N. M. Franklin, S. C. Apte et al. 2010. Physico-chemical behaviour and algal toxicity of nanoparticulate CeO2 in freshwater. *Environ. Chem.* 7(1):50–60.

Roh, K. S., and Q. J. Cui. 2014. Effects of ethylsalicylic acid on growth and Rubisco/Rubisco activase in tobacco plant cultured under cadmium treatment *in vitro*. *J. Life Sci.* 24(5):558–566.

Rout, G. R., and P. Das. 2009. Effect of metal toxicity on plant growth and metabolism: I. Zinc. In *Sustainable Agriculture*, eds. E. Lichtfouse, M. Navarette, P. Debaeke, V. Souchere, and C. Alberola, pp. 873–884. Springer Science+Business Media B.V.

Rugh, C. L., J. F. Senecoff, R. B. Meagher, and S. A. Merkle. 1998. Development of transgenic yellow poplar for mercury phytoremediation. *Nat. Biotechnol.* 16(10):925–928.

Rumeau, D., N. Becuwe-Linka, A. Beyly et al. 2004. Increased zinc content in transplastomic tobacco plants expressing a polyhistidine-tagged Rubisco large subunit. *Plant Biotech. J.* 2(5):389–399.

Ruttkay-Nedecky, B., O. Krystofova, L. Nejdl, and V. Adam. 2017. Nanoparticles based on essential metals and their phytotoxicity. *J. Nanobiotechnol.* 15(1):33.

Sadiq, I. M., S. Dalai, N. Chandrasekaran, and A. Mukherjee. 2011. Ecotoxicity study of titania (TiO2) NPs on two microalgae species: *Scenedesmus* sp and *Chlorella* sp. *Ecotox. Environ. Saf.* 74(5):1180–1187.

Saison, C., F. Perreault, J. C. Daigle et al. 2010. Effect of core-shell copper oxide nanoparticles on cell culture morphology and photosynthesis (photosystem II energy distribution) in the green alga, *Chlamydomonas reinhardtii. Aquat. Toxicol.* 96(2):109–114.

Šalamon, I., K. Kráľová, and E. Masarovičová. 2007. Accumulation of cadmium in chamomile plants cultivated in Eastern Slovakia regions. *Acta Hort. (ISHS)* 749:217–222.

Samson, G., J. C. Morissette, and R. Popovic. 1988. Copper quenching of the variable fluorescence in *Dunaliella tertiolecta*. New evidence for a copper inhibition effect on PS II photochemistry. *Photochem. Photobiol.* 48(3):329–352.

Samson, G., J. C. Morisette, and R. Popovic. 1990. Determination of four apparent mercury interaction sites in photosystem II by using a new modification of the Stern-Volmer analysis. *Biochem. Biophys. Res. Commun.* 166(2):873–878.

Samuelsson, G., and G. Öquist. 1980. Effects of copper chloride on photosynthetic electron transport and chlorophyll-protein complexes of *Spinacia oleracea. Plant Cell Physiol.* 21(3):445–454.

Sarwar, N., Saifullah, S. S. Malhi et al. 2010. Role of mineral nutrition in minimizing cadmium accumulation by plants. *J. Sci. Food Agric.* 90(6):925–937.

Schröder, W. P., J. B. Arellano, T. Bittner, M. Barón, H. J. Eckert, and G. Renger. 1994. Flash-induced absorption spectroscopy studies of copper interaction with photosystem II in higher plants. *J. Biol. Chem.* 269(52):32865–32870.

Schwab, F., G. S. Zhai, M. Kern, A. Turner, J. L. Schnoor, and M. R. Wiesner. 2016. Barriers, pathways and processes for uptake, translocation and accumulation of nanomaterials in plants – Critical review. *Nanotoxicology* 10(3):257–278.

Sebastian, A., and M. N. V. Prasad. 2014. Cadmium minimization in rice. A review. *Agron. Sustain. Dev.* 34(1):155–173.

Sendra, M., P. M. Yeste, I. Moreno-Garrido, J. M. Gatica, and J. Blasco. 2017. CeO2 NPs, toxic or protective to phytoplankton? Charge of nanoparticles and cell wall as factors which cause changes in cell complexity. *Sci. Total Environ.* 590:304–315.

Seneviratne, M., N. Rajakaruna, M. Rizwan, H. M. S. P. Madawala, Y. S. Ok, and M. Vithange. 2017. Heavy metal-induced oxidative stress on seed germination and seedling development: A critical review. *Environ. Geochem. Health.* doi:10.1007/s10653-017-0005-8

Šeršeň, F., and K. Kráľová. 2001. New facts about CdCl2 action on the photosynthetic apparatus of spinach chloroplasts and its comparison with HgCl2 action. *Photosynthetica* 39(4):575–580.

Šeršeň, F., and K. Kráľová. 2013a. EPR spectroscopy – A valuable tool to study photosynthesizing organisms exposed to abiotic stresses. In *Photosynthesis*, ed. Z. Dubinsky, pp. 248–283. Rijeka: InTech.

Šeršeň, F., and K. Kráľová. 2013b. Action of some organomercury compounds on photosynthesis in spinach chloroplasts. *Ecol. Chem. Eng. S* 20(3):489–498.

Šeršeň, F., K. Kráľová, and M. Blahová. 1996. Photosynthesis of *Chlorella vulgaris* as affected by diaqua(4-chloro-2-methyl-phenoxyacetato)copper(II) complex. *Biol. Plant.* 38:71–75.

Šeršeň, F., K. Kráľová, and A. Bumbálová. 1998a. Action of mercury on the photosynthetic apparatus of spinach chloroplasts. *Photosynthetica* 35(4): 551–559.

Šeršeň, F., K. Kráľová, and A. Bumbálová. 1998b. EPR study of mercury action on the photosynthetic apparatus of spinach chloroplasts. In *Photosynthesis: Mechanisms and effects*, Vol. IV, ed. G. Garab, pp. 2697–2700. Dordrecht: Kluwer Academic Publishers.

Šeršeň, F., K. Kráľová, A. Bumbálová, and O. Švajlenová. 1997a. The effect of Cu(II) ions bound with tridentate Schiff base ligands upon the photosynthetic apparatus. *J. Plant Physiol.* 151(3):299–305.

Šeršeň, F., K. Kráľová, and A. Fargašová. 1997c. Effect of tributyltin compounds on photosynthetic processes. In *Progress in Coordination and Organometallic Chemistry, Monograph Series of the International Conferences on Coordination Chemistry Held Periodically at Smolenice in Slovakia*, Vol. 3, eds. G. Ondrejovič, and A. Sirota, pp. 227–232. Bratislava: Slovak Technical University Press.

Šeršeň, F., K. Kráľová, E. Jóna, and A. Sirota. 1997d. Effects of some Ni(II) complexes with *N*-donor ligands on photosynthetic electron transport in spinach chloroplasts. *Chem. Listy* 91:685.

Šeršeň, F., K. Kráľová, M. Peško, and M. Cigáň. 2014. Effect of Pb2+ ions on photosynthetic apparatus. *Gen. Physiol. Biophys.* 33:131–136.

Šeršeň, F., K. Kráľová, and J. Sokolik. 1997b. Effect of two structural types of carboxylatocopper(II) complexes on photosynthesis in spinach chloroplasts. *Chem. Listy* 91:684.

Shahid, M., S. Shamshad, M. Rafiq et al. 2017. Chromium speciation, bioavailability, uptake, toxicity and detoxification in soil-plant system: A review. *Chemosphere* 178:513–533.

Shakirova, F. M., C. R. Allagulova, D. R. Maslennikova, E. O. Klyuchnikova, A. M. Avalbaev, and M. V. Bezrukova. 2016. Salicylic acid-induced protection against cadmium toxicity in wheat plants. *Environ. Exp. Bot.* 122:19–28.

Shanker, K., S. Mishra, S. Srivastava et al. 1996. Effect of selenite and selenate on plant uptake of cadmium by maize (*Zea mays*). *Bull. Environ. Contam. Toxicol.* 56(3):419–424.

Sharm, S., and K. N. Uttam. 2017. Rapid analyses of stress of copper oxide nanoparticles on wheat plants at an early stage by laser induced fluorescence and attenuated total reflectance Fourier transform infrared spectroscopy. *Vib. Spectrosc.* 92:135–150.

Sharma, R. K., and G. Archana. 2016. Cadmium minimization in food crops by cadmium resistant plant growth promoting rhizobacteria. *Appl. Soil Ecol.* 107:66–78.

Sharma, S. S., K. J. Dietz, and T. Mimura. 2016. Vacuolar compartmentalization as indispensable component of heavy metal detoxification in plants. *Plant Cell Environ.* 39(5):1112–26.

Shaw, B. P., S. K. Sahu, and R. K. Mushra. 2004. Heavy metal induced oxidative damage in terrestrial plants. In *Heavy Metal Stress in Plants: From Biomolecules to Ecosystems*, 2nd edn, ed. M. N. V. Prasad, pp. 84–126. Berlin: Springer.

Shen, J., L. L. Song, K. Muller et al. 2016. Magnesium alleviates adverse effects of lead on growth, photosynthesis, and ultrastructural alterations of *Torreya grandis* seedlings. *Front Plant Sci.* 7:1819.

Sheoran, V., A. S. Sheoran, and P. Poonia. 2016. Factors affecting phytoextraction: A review. *Pedosphere* 26(2):148–166.

Sherameti, I., and A. Varma. 2011. *Detoxification of Heavy Metals*. Berlin: Springer-Verlag.

Shevlin, D., N. O'Brien, and E. Cummins. 2018. Silver engineered nanoparticles in freshwater systems – Likely fate and behaviour through natural attenuation processes. *Sci. Total Environ.* 621:1033–1046.

Shi, G. R., S. L. Xia, C. F. Liu, and Z. Zhang. 2016. Cadmium accumulation and growth response to cadmium stress of eighteen plant species. *Environ. Sci. Pollut. Res.* 23(22):23071–23080.

Shi, W. G., H. Li, T. X. Liu, A. Polle, C. H. Peng, and Z. B. Luo. 2015. Exogenous abscisic acid alleviates zinc uptake and accumulation in *Populus × canescens* exposed to excess zinc. *Plant Cell Environ.* 38(1):207–223.

Shi, X., S. F. Wang, H. J. Sun et al. 2017. Comparative of *Quercus* spp. and *Salix* spp. for phytoremediation of Pb/Zn mine tailings. *Environ. Sci. Pollut. Res.* 24(4):3400–3411.

Shoi, Y., H. Tamai, and T. Sasa 1978a. Inhibition of photosystem II in green alga, *Ankistrodesmus falcatus* by copper. *Physiol Plant* 44(4):434–438.

Shoi, Y., H. Tamai, and T. Sasa. 1978b. Effects of copper on photosynthetic electron transport systems in spinach chloroplasts. *Plant Cell Physiol.* 19(2):203–209.

Shutilova, N. I. 2006. Molecular mechanisms of the inhibitory action exerted by heavy metals on oxygen-evolving pigment-lipoprotein complex of chloroplast membranes. *Biologicheskie Membrany* 23:355–363.

Sigfridson, K. G. V., G. Bernat, F. Mamedov, and S. Styring. 2004. Molecular interference of Cd2+ with Photosystem II. *Biochim. Biophys. Acta – Bioenerg.* 1659(1):19–31.

Silva, I. R., R. F. Novais, G. N. Jham et al. 2004. Responses of eucalypt species to aluminum: The possible involvement of low molecular weight organic acids in the Al tolerance mechanism. *Tree Physiol.* 24:1267–1277.

Silva, S., G. Pinto, M. C. Dias et al. 2012. Aluminium long-term stress differently affects photosynthesis in rye genotypes. *Plant Physiol. Biochem.* 54:105–112.

Silva, S., P. Silva, H. Oliveira et al. 2017. Pb low doses induced genotoxicity in *Lactuca sativa* plants. *Plant Physiol. Biochem.* 112:109–116.

Singh, A., N. B. Singh, I. Hussain, and H. Singh. 2017. Effect of biologically synthesized copper oxide nanoparticles on metabolism and antioxidant activity to the crop plants *Solanum lycopersicum* and *Brassica oleracea* var. *botrytis*. *J. Biotechnol.* 262:11–27.

Singh, D. P., P. Khare, and P. S. Bisen. 1989. Effect of Ni2+, Hg2+ and Cu2+ on growth, oxygen evolution and photosynthetic electron transport in *Cylindospermum* IU 942. *J. Plant. Physiol.* 134:406–412.

Singh, D. P., and S. P. Singh. 1987. Action of heavy metals on Hill activity and O2 evolution in *Anacystis nidulans*. *Plant Physiol.* 83:12–14.

Singh, O., Z. Khanam, N. Misra, and M. K. Srivastava. 2011. Chamomile (*Matricaria chamomilla* L.): An overview. *Pharmacognosy Rev.* 5(9):82–95.

Singh, R., P. K. Srivastava, V. P. Singh, G. Dubey, and S. M. Prasad. 2012. Light intensity determines the extent of mercury toxicity in the cyanobacterium *Nostoc muscorum*. *Acta Physiol. Plant.* 34(3):1119–1131.

Skórzyńska-Polit, E., and T. Baszyński. 1997. Differences in sensitivity of the photosynthetic apparatus in Cd-stressed runner bean plants in relation to their age. *Plant Sci.* 128:11–21.

Smirnova, I. A., A. Blomberg, L. E. Andreasson, and P. Brzezinski. 1998. Localization of light-induced structural changes in bacterial photosynthetic reaction centers. *Photosynth. Res.* 56:45–55.

Solanki, R., and R. Dhankhar. 2011. Biochemical changes and adaptive strategies of plants under heavy metal stress. *Biologia* 66(2):195–204.

Solti, A., E. Sárvári, E. Szöllősi et al. 2016. Stress hardening under long-term cadmium treatment is correlated with the activation of antioxidative defence and iron acquisition of chloroplasts in *Populus*. *Z. Naturforsch. c* 71(9–10):323–334.

Šottníková, A., L. Lunáčková, E. Masarovičová, A. Lux, and V. Streško. 2003. Changes in the rooting and growth of willows and poplars induced by cadmium. *Biol. Plant,* 46(1):129–131.

Sreekanth, T. V. M., P. C. Nagajyothi, K. D. Lee, and T. N. V. K. V. Prasad. 2013. Occurrence, physiological responses and toxicity of nickel in plants. *Int. J. Environ. Sci. Technol.* 10(5):1129–1140.

Stancheva, I., M. Geneva, M. Hristozkova, Y. Markovska, and I. Šalamon. 2010. Antioxidant capacity of sage grown on heavy metal-polluted soil. *Russ. J. Plant Physiol.* 57(6):799–805.

Stiborová, M., M. Doubravová, and S. Leblová. 1986. A comparative study of the effect of heavy metal ions on ribulose-1,5-bisphosphate carboxylase and phosphoenol pyruvate carboxylase. *Biochem. Physiol. Pflanz.* 181(6):373–379.

Stobart, A. K., W. T. Griffiths, I. Ameen-Bukhari, and R. P. Sherwood. 1985. The effect of Cd2+ on the biosynthesis of chlorophyll in leaves of barley. *Physiol. Plant.* 63:293–298.

Stoltz, E., and M. Greger. 2002. Accumulation properties of As, Cd, Cu, Pb and Zn by four wetland plant species growing on submerged mine tailings. *Environ. Exp. Bot.* 47(3):271–280.

Stomp, A. M., K. H. Han, S. Wilbert, and M. P. Gordon. 1993. Genetic improvement of tree species for remediation of hazardous wastes. *In Vitro Cell Develop. Biol.- Plant* 29(4):227–232.

Suh, H. J., C. S. Kim, J. Y. Lee, and J. Jung. 2002. Photodynamic effect of iron excess on photosystem II function in pea plants. *Photochem. Photobiol.* 75:513–518.

Susplugas, S., A. Srivastava, and R. J. Strasser. 2000. Changes in the photosynthetic activities during several stages of vegetative growth of *Spirodela polyrhiza*: Effect of chromate. *J. Plant Physiol.* 157:503–512.

Svensson, B., I. Vass, and S. Styring. 1991. Sequence analysis of D1 and D2 reaction center proteins of photosystem II. *Naturforsch.* 46c:765–776.

Szalontai, B., L. I. Horvath, M. Debreczeny, M. Droppa, and G. Horvath. 1999. Molecular rearrangements of thylakoids after heavy metal poisoning, as seen by Fourier transform infrared (FTIR) and electron spin resonance (ESR) spectroscopy. *Photosynth. Res.* 61:241–252.

Tabrizi, L., S. Mohammadi, M. Delshad, and B. M. Zadeh. 2015. Effect of arbuscular mycorrhizal fungi on yield and phytoremediation performance of pot marigold (*Calendula officinalis* L.) under heavy metals stress. *Int. J. Phytorem.* 17(12):1244–1252.

Tajmirriahi, H. A., and A. Ahmed. 1993. Complexation of copper and zinc ions with proteins of a light-harvesting complex (LHC II) of chloroplast thylakoid membranes studied by FT-IR spectroscopy. *J. Mol. Struct.* 297:103–108.

Tan, W. J., W. C. Du, A. C. Barrios et al. 2017. Surface coating changes the physiological and biochemical impacts of nano-TiO2 in basil (*Ocimum basilicum*) plants. *Environ. Pollut.* 222:64–72.

Tang Y. T., T. H. B. Deng, Q. H. Wu et al. 2012. Designing cropping systems for metal-contaminated sites: A review. *Pedosphere* 22(4):470–488.

Thao, N. P., M. I. R. Khan, N. B. A. Thu et al. 2015. Role of ethylene and its cross talk with other signaling molecules in plant responses to heavy metal stress. *Plant Physiol.* 169(1):73–84.

Theriault, G., P. Michael, and K. Nkongolo. 2016. Comprehensive transcriptome analysis of response to nickel stress in white birch (*Betula papyrifera*). *PLoS ONE* 11(4):e0153762.

Tominaga, T. T., H. Imasato, O. R. Nascimento, and M. Tabak. 1995. Interaction of tyrosine dipeptides with Cu2+ ions: A fluorescence study. *Anal. Acta* 315:217–224.

Toselli, M., E. Baldi, G. Marcolini et al. 2008. Response of potted pear trees to increasing copper concentration in sandy and clay-loam soils. *J. Plant Nutr.* 31(12):2089–2104.

Truta, E. C., D. N. Gherghel, I. C. I. Bara, and G. V. Vochita. 2013. Zinc-induced genotoxic effects in root meristems of barley seedlings. *Not. Bot. Horti. Agrobot. Cluj Napoca* 41(1):150–156.

Tu, C., and L. Q. Ma. 2002. Effects of arsenic concentrations and forms on arsenic uptake by the hyperaccumulator ladder brake. *J. Environ. Qual.* 31(2):641–647.

Turner, A., D. Brice, and M. T. Brown. 2012. Interactions of silver nanoparticles with the marine macroalga, *Ulva lactuca*. *Ecotoxicology* 21(1):148–154.

Udeigwe, T. K., J. M. Teboh, P. N. Eze et al. 2015. Implications of leading crop production practices on environmental quality and human health. *J. Environ. Manag.* 151:267–279.

U.S. National Nanotechnology Initiative. 2004. The National Nanotechnology Initiative Strategic Plan. December 2004. https://www.nano.gov/sites/default/files/pub_resource/nni_st rategic_plan_2004.pdf

Utchig, L. M., O. Poluetkov, D. M. Tiede, and M. C. Thurnauer. 2000. EPR investigation of Cu2+-substituted photosynthetic bacterial reaction centers: Evidence for histidine ligation at the surface metal site. *Biochemistry* 39:2961–2969.

Vamerali, T., M. Bandiera, and G. Mosca. 2010. Field crops for phytoremediation of metal-contaminated land. A review. *Environ. Chem. Lett.* 8(1):1–17.

Van Assche, F., and H. Clijsters. 1986a. Inhibition of photosynthesis in *Phaseolus vulgaris* by treatment with toxic concentration of Zn: Effects on electron transport and photophosphorylation. *Physiol. Plant.* 66:717–721.

Van Assche, F., and H. Clijsters. 1986b. Inhibition of photosynthesis in *Phaseolus vulgaris* by treatment with toxic concentration of Zn: Effect on ribulose-1,5-biphosphate carboxylase/oxygenase. *J. Plant Physiol.* 125:355–360.

Van Assche, F., and H. Clijsters. 1990. Effects of metals on enzyme activity in plants. *Plant Cell Environ.* 13(3):195–206.

Vandecasteele, B., E. Meers, P. Vervaeke, B. De Vos, P. Quataert, and F. M. G. Tack. 2005. Growth and trace metal accumulation of two *Salix* clones on sediment-derived soils with increasing contamination levels. *Chemosphere* 58(8):995–1002.

Van Duijvendijk-Matteoli, M. A., and G. M. Desmet. 1975. On the inhibitory action of cadmium on the donor side of photosystem II in isolated chloroplasts. *Biochim. Biophys. Acta* 408:164–169.

Van Nevel, L., J. Mertens, J. Staelens et al. 2011. Elevated Cd and Zn uptake by aspen limits the phytostabilization potential compared to five other tree species. *Ecol. Eng.* 37(7):1072–1080.

Vatansever, R., I. I. Ozyigit, and E. Filiz. 2017. Essential and beneficial trace elements in plants, and their transport in roots: A review. *Appl. Biochem. Biotechnol.* 181(1):464–482.

Vavilin, D. V., V. A. Polynov, D. N. Matorin, and P. S. Venediktov. 1995. The subletal concentrations of copper stimulate photosystem II photoinhibition in *Chlorella pyrenoidosa*. *J. Plant. Physiol.* 146:609–614.

Walker, J. D., M. Enache, and J. C. Dearden. 2003. Quantitative cationic–activity relationships for predicting toxicity of metals. *Environ. Toxicol. Chem.* 22(8):1916–1935.

Wang, J., Z. Q. Fang, W. Cheng, P. E. Tsang, and D. Y. Zhao. 2016b. Ageing decreases the phytotoxicity of zero-valent iron nanoparticles in soil cultivated with *Oryza sativa*. *Ecotoxicology* 25(6):1202–1210.

Wang, P., E. Lombi, F. J. Zhao, and P. M. Kopittke. 2016c. Nanotechnology: A new opportunity in plant sciences. *Trends Plant Sci.* 21(8):699–712.

Wang, Q. Y., J. S. Liu, and B. Hu. 2016a. Integration of copper subcellular distribution and chemical forms to understand copper toxicity in apple trees. *Environ. Exp. Bot.* 123:125–131.

Wang, R. G., S. X. Dai, S. R. Tang et al. 2012b. Growth, gas exchange, root morphology and cadmium uptake responses of poplars and willows grown on cadmium-contaminated soil to elevated CO2. *Environ. Earth Sci.* 67(1):1–13.

Wang, S. Y., W. Y. Wu, F. Liu, R. K. Liao, and Y. Q. Hu. 2017. Accumulation of heavy metals in soil-crop systems: A review for wheat and corn. *Environ. Sci. Pollut. Res.* 24(18):15209–15225.

Wang, S. Z., F. L. Chen, S. Y. Mu, D. Y. Zhang, X. L. Pan, and D. J. Lee. 2013. Simultaneous analysis of photosystem responses of *Microcystis aeruginoga* under chromium stress. *Ecotox. Environ. Saf.* 88:163–168.

Wang, S. Z., and X. L. Pan. 2012a. Effects of Sb(V) on growth and chlorophyll fluorescence of *Microcystis aeruginosa* (FACHB-905). *Curr. Microbiol.* 65:733–741.

Wang, S. Z., X. L. Pan, and D. Y. Zhang. 2015a. PSI showed higher tolerance to Sb(V) than PSII due to stimulation of cyclic electron flow around PSI. *Curr. Microbiol.* 70(1):27–34.

Wang, X. M., Y. G. Ze, X. Wu et al. 2009. Effect of Pb2+ on the kinetic and spectral characterization of ribulose-1,5-bisphosphate carboxylase/oxygenase. *Chin. J. Chem.* 27(4): 727–731.

Wang, Y., S. L. Wang, Z. R. Nan et al. 2015b. Effects of Ni stress on the uptake and translocation of Ni and other mineral nutrition elements in mature wheat grown in sierozems from northwest of China. *Environ. Sci. Pollut. Res.* 22(24):19756–19763.

Wang, Y. W., X. H. Jiang, K. Li et al. 2014. Photosynthetic responses of *Oryza sativa* L. seedlings to cadmium stress: Physiological, biochemical and ultrastructural analyses. *Biometals* 27(2):389–401.

Wang, Z. Y., X. Y. Xie, J. Zhao et al. 2012c. Xylem- and phloembased transport of CuO nanoparticles in maize (*Zea mays* L.). *Environ. Sci. Technol.* 46(8):4434–4441.

Weaver, E. C. 1968. EPR studies of free radicals in photosynthetic systems. *Annu. Rev. Plant. Physiol.* 19:283–294.

Whitmarsh, J. 1998. Electron transport and energy transduction. In *Photosynthesis: A Comprehensive Treatise*, ed. A. S. Raghavendra, pp. 87–110. Cambridge: University Press.

Wolterbeek, H. T., and T. G. Verburg. 2001. Predicting metal toxicity revisited: General properties vs. specific effects. *Sci. Total Environ.* 279:87–115.

Wu, J. W., C. M. Geilfus, B. Pitann, and K. H. Muehling. 2016. Silicon-enhanced oxalate exudation contributes to alleviation of cadmium toxicity in wheat. *Environ. Exp. Bot.* 131:10–18.

Wu, X., F. Hong, C. Liu et al. 2008. Effects of Pb2+ on energy distribution and photochemical activity of spinach chloroplast. *Spectrochim. Acta A Mol. Biomol. Spectrosc.* 69:738–742.

Xiang, L., H. M. Zhao, and Y. W. Li. 2015. Effects of the size and morphology of zinc oxide nanoparticles on the germination of Chinese cabbage seeds. *Environ. Sci. Pollut. Res.* 22(14):10452–10462.

Xiao, R., S. Ghosh, A. R. Tanaka, B. M. Greenberg, and E. B. Dumbroff. 1997. A rapid spectrophotometric method for measuring Photosystem I and Photosystem II activities in a single sample. *Plant Physiol. Biochem.* 35:411–417.

Xiao, W., L. Chao, C. X. Qu et al. 2008. Effects of lead on activities of photochemical reaction and key enzymes of carbon assimilation in spinach chloroplast. *Biol. Trace Element Res.* 126(1–3):269–279. doi:10.1007/s12011-008-8196-6

Xu, S. J., D. M. Lin, H. L. Sun, X. M. Yang, and X. F. Zhang. 2015. Excess iron alters the fatty acid composition of chloroplast membrane and decreases the photosynthesis rate: A study in hydroponic pea seedlings. *Acta Physiol. Plant.* 37:212.

Xu, X. H., C. Y. Liu, X. Y. Zhao, R. Y. Li, and W. J. Deng. 2014. Involvement of an antioxidant defense system in the adaptive response to cadmium in maize seedlings (*Zea mays* L.). *Bull. Environ. Contam. Toxicol.* 93(5):618–624.

Yamori, W., C. Masumoto, H. Fukayama, and A. Makino. 2012. Rubisco activase is a key regulator of non-steady-state photosynthesis at any leaf temperature and, to a lesser extent, of steady-state photosynthesis at high temperature. *Plant J.* 71(6):871880.

Yang, M., L. Tan, Y. Y. Xu et al. 2015. Effect of low pH and aluminum toxicity on the photosynthetic characteristics of different fast-growing *Eucalyptus* vegetatively propagated clones. *PLoS One* 10(6):e0130963.

Yanik, F., and F. Vardar. 2015. Toxic effects of aluminum oxide (Al2O3) nanoparticles on root growth and development in *Triticum aestivum*. *Water Air Soil Pollut.* 226(9):296.

Yi, A. X., P. T. Y. Leung, and K. M. Y. Leung. 2014. Photosynthetic and molecular responses of the marine diatom *Thalassiosira pseudonana* to triphenyltin exposure. *Aquat. Toxicol.* 154, 48–57.

Yruela, I. 2005. Copper in plants. *Braz. J. Plant Physiol.* 17(1):145–156.

Yruela, I. 2009. Copper in plants: Acquisition, transport and interactions. *Funct. Plant Biol.* 36(5):409–430.

Yruela, I., P. J. Alonso, I. O. De Zarate, G. Montoya, and R. Picorel. 1993. Precise location of the Cu(II) inhibitory binding site in higher plant and bacterial photosynthetic reaction centers as probed by light-induced absorption changes. *J. Biol, Chem.* 268(3):1684–1689.

Yruela, I., G. Montoya, P. J. Alonso, and R. Picorel. 1991. Identification of the pheophytin-QA-Fe domain of the reducing side of the photosystem II as the Cu(II) inhibitory binding site. *J. Biol. Chem.* 266(34):22847–2285.

Yruela, I., G. Montoya, and R. Picorel. 1992. The inhibitory mechanism of Cu(II) on the photosystem II electron transport from higher plants. *Photosynth. Res.* 33(3):227–233.

Yue, J. Y., X. Zhang, and N. Liu. 2017. Cadmium permeates through calcium channels and activates transcriptomic complexity in wheat roots in response to cadmium stress. *Genes Genomics* 39(2):183–196.

Yusuf, M., Q. Fariduddin, S. Hayat, and A. Ahmad. 2011. Nickel: An overview of uptake, essentiality and toxicity in plants. *Bull. Environ. Contam. Toxicol.* 86(1):1–17.

Zarinkamar, F., S. Ghelich, and S. Soleimanpour. 2013. Toxic effects of Pb on anatomy and hypericin content in *Hypericum perforatum* L. *Bioremediat. J.* 17(1):40–51.

Zhang, D. Y., X. L. Pan, G. J. Mu, and J. L. Wang. 2010. Toxic effects of antimony on photosystem II of *Synechocystis* sp as probed by *in vivo* chlorophyll fluorescence. *J. Appl. Phycol.* 22(4):479–488.

Zhang, F., X. Q. Wan, and Y. Zhong. 2014. Nitrogen as an important detoxification factor to cadmium stress in poplar plants. *J. Plant Interact.* 9(1):249–258.

Zhang, L., N. Goswami, J. P. Xie, B. Zhang, and Y. L. He. 2017. Unraveling the molecular mechanism of photosynthetic toxicity of highly fluorescent silver nanoclusters to *Scenedesmus obliquus*. *Sci. Rep.* 7:16432.

Zhang, W. C., B. D. Xiao, and T. Fang. 2018. Chemical transformation of silver nanoparticles in aquatic environments: Mechanism, morphology and toxicity. *Chemosphere* 191:324–334.

Zhao, K. J., Y. X. Huang, J. Hu, H. J. Zhou, A. S. Adeleye, and A. A. Keller. 2016. 1H NMR and GC-MS based metabolomics reveal defense and detoxification mechanism of cucumber plant under nano-Cu Stress. *Environ. Sci. Technol.* 50(4):2000–2010.

Zhao, L. J., J. R. Peralta-Videa, C. M. Rico et al. 2014. CeO2 and ZnO nanoparticles change the nutritional qualities of cucumber (*Cucumis sativus*). *J. Agric. Food Chem.* 62(13):2752–2759.

Zhao, L. J., Y. P. Sun, J. A. Hernandez-Viezcas et al. 2013. Influence of CeO2 and ZnO nanoparticles on cucumber physiological markers and bioaccumulation of Ce and Zn: A life cycle study. *J. Agric. Food Chem.* 61(49):11945–11951.

Zhao, X. L., L. Y. Zheng, X. L. Xia et al. 2015. Responses and acclimation of Chinese cork oak (*Quercus variabilis* Bl.) to metal stress: The inducible antimony tolerance in oak trees. *Environ. Sci. Pollut.* 22:11456–11466.

Zheljazkov, V. D., E. A. Jeliazkova, L. E. Craker et al. 1999. Heavy metal uptake by mint. *Acta Hort. (ISHS)* 500:111–117.

Zhou, H., W. T. Yang, X. Zhou et al. 2016a. Accumulation of heavy metals in vegetable species planted in contaminated soils and the health risk assessment. *Int. J. Environ. Res. Public Health* 13:289.

Zhou, J., Z. P. Jiang, J. Ma, L. F. Yang, and Y. Wei. 2017a. The effects of lead stress on photosynthetic function and chloroplast ultrastructure of *Robinia pseudoacacia* seedlings. *Environ. Sci. Pollut. Res.* 24:10718–10726.

Zhou, J. T., H. X. Wan, J. L. He, D. G. Lyu, and H. F. Li. 2017b. Responses to understand cadmium tolerance in apple rootstocks. *Front. Plant Sci.* 8:966.

Zhou, J. T., H. X. Wan, S. J. Qin et al. 2016b. Net cadmium flux and gene expression in relation to differences in cadmium accumulation and translocation in four apple rootstocks. *Environ. Exp. Bot.* 130:95–105.

Zhu, Y., J. Xu, T. Lu et al. 2017. A comparison of the effects of copper nanoparticles and copper sulfate on *Phaeodactylum tricornutum* physiology and transcription. *Environ. Toxicol. Pharmacol.* 56:43–49.

Zou, X. Y., P. H. Li, J. Lou, and H. W. Zhang. 2017. Surface coating-modulated toxic responses to silver nanoparticles in *Wolffia globosa*. *Aquat. Toxicol.* 189:150–158.

Zouzelka, R., P. Cihakova, J. R. Ambrozova, and J. Rathousky. 2016. Combined biocidal action of silver nanoparticles and ions against Chlorococcales (*Scenedesmus quadricauda*, *Chlorella vulgaris*) and filamentous algae (*Klebsormidium* sp.). *Environ. Sci. Pollut. Res.* 23(9):8317–8326.

Zupunski, M., M. Borisev, S. Orlovic et al. 2016. Hydroponic screening of black locust families for heavy metal tolerance and accumulation. *Int. J. Phytorem.* 18(6):583–591.

Zuverza-Mena, N., D. Martinez-Fernandez, and W. C. Du. 2017. Exposure of engineered nanomaterials to plants: Insights into the physiological and biochemical responses-A review. *Plant Physiol. Biochem.* 110:236–264.

26 Turfgrass Hyper-Accumulative Characteristics to Alleviate Heavy and Toxic Metal Stresses

Haibo Liu, Nick Menchyk, Frank Bethea, Christian Baldwin,
Jacob Taylor, and Caleb Patrick

CONTENTS

26.1 INTRODUCTION

Abundant elements (Na, Al, Cl, and Fe) and heavy metals occur naturally in soils, waters, and the environment either as natural or anthropogenic where industrial activities have caused environmental pollutions (Cd, Pb, Cu, Mn, Zn, Ni, Hg, As, Cr, Se, and Co, Table 26.1) (McGrath, 1994; Salt et al., 1998). Most turfgrass species are able to tolerate abundant elements and heavy metals in variable degrees due to their nature without yield harvesting unlike other agricultural crops as well as perennial plants with fibrous and dense root systems and dense canopies. However, understanding of these turfgrasses' different capabilities of tolerating excessive levels of these elements, which allow these grasses to hyper-accumulate these often toxic elements from the environment, will further help enhance turfgrass environmental protection potential as well as possessing a network of environmental plants with elemental homeostasis in soils. Therefore, this chapter aims for a review that has not been documented before, focusing on turfgrass responses to turfgrass hyper-accumulative characteristics in soils and the environment.

The turfgrass industry in the US is a significant agricultural sector and on average it counts for about 1/5 to 1/6 of employment in the US amounting to 5 to 6 million, about 1 million of which is farming population, and its annual economic values are more than $100 billion with over 15,000 golf courses, 80 million home lawns, over 47,000 km² of utility turf areas, and over 750,000 sports fields covered with turfgrasses. Turfgrasses occupy about 181,300 km² in the US, which is about the same size of the whole state of Oklahoma,

1.87% of total US land or 18% of the total land used for corn (*Zea mays* L.), soybeans (*Glycine max* (L.) Merr.), wheat (*Triticum* spp. L.), and cotton (*Gossypium* spp.) in the US (personal communications). With such a significant land area covered with turfgrasses, especially closer to urban communities, these turfgrasses can serve as recycling sites with different sources as well as sinks of excessive elements as phytoremediation crops to reduce soil and water pollution (McGrath, 1994; Gaskin et al., 2003; Palazzo et al., 2003; Cheng, et al., 2007; Li et al., 2010; Zhao et al., 2011; Carrow and Duncan, 2012; Zhao et al., 2016; Meana et al., 2019). Earlier researches of heavy metals and other elements associated with turfgrass species are documented as early as the 1950s (Bradshaw, 1952; Gregory and Bradshaw, 1964). To date, almost all turfgrasses have been investigated for their tolerance to heavy metals and other elements as well as their impacts on the environment (Tables 26.2 and 26.3). However, thorough reviews on turfgrass hyper-accumulative characteristics in soils, waters, air, and the environment are still lacking. This review excludes the primary and secondary plant nutrients of turfgrasses, N, P, K, Ca, Mg, and S.

26.2 EXCESSIVE METAL AND OTHER ELEMENTS IN TURFGRASSES

Turfgrasses are a group of functional crops and are often perennials (Beard, 1973; Turgeon, 2012; McCarty, 2018). Contaminated soils and waters pose a major environmental and human health problem, which may be partially solved

TABLE 26.1

Critical Levels, Critical Toxic Levels, and Hyper-Accumulation of Elements in Land Plants

Elements	Critical Levels for Plants mg kg^{-1}	Critically Toxic Levels in Plants mg kg^{-1}	Most Toxic Forms if Excessiveness Exists	Hyper-accumulation Levels mg kg^{-1}	References
			Plant Nutrients		
Iron (Fe)	100–200	>500	Fe^{+3}	>14,697	Beard et al. (1977); Yust et al. (1984); Cooper and Spokas (1991); Lee et al. (1996); Hull (1999); Yang et al. (2000); Nagajyoti et al. (2010); Taylor, (2019)
Chloride (Cl)	100–200	>500	Cl^{-1}	>3,923	Cordukes and Parups (1971); Hull (2003c); Munns and Tester (2008); Cox et al. (2018)
Manganese (Mn)	20–500	>500	Mn^{+2} Mn^{+3} Mn^{+4}	>14,900	Lee et al. (1996); Hull (2001a); Heckman et al. (2003); Beard and Beard (2005); Mora et al. (2009); Kramer (2010); Reeves et al. (2018);
Boron (B)	20–100	>100	B(OH)$_3$	>1,970	Oertli et al. (1969); Lee et al. (1996); Nable et al. (1997); Hull (2002a)
Zinc (Zn)	50–150	>150	Zn^{+2}	>10,000	Lee et al. (1996); Xu and Mancino (2001); Hull (2001b); Kramer (2010)
Copper (Cu)	5–100	>100	Cu^{+3}	>1,298	Wong, (1982); Nus et al. (1993); Hull (2002b); Kramer (2010); Reeves et al. (2018)
Molybdenum (Mo)	2–20	>100	MoO$_4^{-2}$	>3,640	Lee et al. (1996); Gaskin et al. (2003); Hull (2003b); Mashhadi et al. (2011)
Nickel (Ni)	0.01–10	>100	Ni^{+2}	>10,000	Yang et al. (2000); Hull (2003a); Kramer (2010); Menchyk (2012); Patrick, (2018); Reeves et al. (2018)
			Non-Plant Nutrients		
Sodium (Na)	–	>1,000	Na$^+$	>10,000	Harivandi et al. (2008); Munns and Tester (2008); Yamamoto et al. (2016); Schiavon and Baird (2018)
Aluminum (Al)	–	>1,000	Al(H$_2$O)$_6^{+3}$	>1,000	Foy and Murray (1998); Baldwin et al. (2005); Cartes et al. (2012); Kochian et al. (2015); Singh et al. (2017)
Cadmium (Cd)	–	>2.5	Cd^{+1}, Cd^{+2}	>1,326	Jarvis et al. (1976); Jarvis and Jones (1978); King (1981); Lasat (2002); Nagajyoti et al. (2010); Hu et al. (2013); Fard et al. (2016b); Wang et al. (2019)
Lead (Pb)	–	>13	Pb$^+$, Pb^{+2}, Pb^{+3}, Pb^{+4}	>5,310	Jones et al. (1973a); Wong (1982); Yesilonis et al. (2008); Kramer (2010); Nagajyoti et al. (2010); Wang et al. (2010); Luo et al. (2013); Fard et al. (2016a)
Arsenic (As)	–	>7	Elemental arsenic and arsenic compounds (particularly inorganic)	>4,980	Wang et al. (1973); Ma et al. (2001); Gadepalle et al. (2008); Hartley and Lepp (2008); Kramer (2010); Nagajyoti et al. (2010); Ansah, (2012)
Mercury (Hg)	–	>0.11	Hg^{+1}, Hg^{+2} and its compounds	>28,000	Wang et al. (1973); McCartney et al. (2001); Temmerman et al. (2007); Comino et al. (2009); Shiyab et al. (2009); Nagajyoti et al. (2010)
Chromium (Cr)	–	>50	Cr$^+$, Cr^{2+}, Cr^{3+}, Cr^{4+}, Cr^{6+}	>18,656	Zhang et al. (2007); Wu et al. (2014); Huang et al., (2018)
Selenium (Se)	–	>50	H$_2$Se	>1,191	Smith and Watkinson (1984); Wu and Huang (1991); Valle et al. (1993); Wu (1994); Hartikainen et al. (2000); Valle et al. (2002); Dhillon and Dhillon (2003). Fordyce (2005)
Cobalt (Co)	–	>50	Co^{2+}, Co^{3+}	>1,971	Faucon et al. (2007); Taghizadeh and Solgi (2017); Kabeya et al. (2018)

TABLE 26.2

Some Related References on Turfgrass Responses to Heavy Metals and Other Elements

Turfgrass Species	Evidence of Toxic Effects and Tolerant Mechanisms	Toxic Elements and Metals	References Cited
		C₃ **Cool-Season Turfgrasses**	
Tall fescue (*Festuca arundinacea* Schreb.)	EDTA added, Nitric oxide, hyper-accumulative, genetic variations, exclusion mechanisms, hyper-tolerant transformants	Al, As, Cd, Cl, Cr, Cu, Na, Ni, Pb, Se, and Zn	Carlson and Rolfe (1979); King (1981); Wu and Huang (1991); Wu (1994); Foy and Murray (1998); Begonia et al. (2001); Zhao and Duo (2002); Palazzo et al. (2003); Qu et al. (2003); Begonia et al. (2005); Duo et al. (2005); Soleimani et al. (2009); Jin et al. (2010); Zhao et al. (2010); Gao et al. (2012); Zhao et al. (2014); Fard et al. (2016a, 2016b); Chen et al. (2018a); Fei et al. (2018); Huang et al. (2017); Zhuo et al. (2017); Zhu et al. (2018); Wang et al. (2019)
Perennial ryegrass (*Lolium perenne* L.)	EDTA added, nitric oxide, nitrilotriacetic acid (NTA) added, hyper-accumulative, air-absorptive, antioxidative enzymes, and root exudation of carboxylates	Al, As, Cd, Co, Cu, Hg, Mn, Ni, Pb, Se, and Zn	Jones et al. (1973); Carlson and Rolfe (1979); Wong and Bradshaw (1982); Smith and Watkinson (1984); Hartikainen et al. (2000); Duo et al. (2005); Temmerman et al. (2007); Gadepalle et al. (2008); Hartley and Lepp (2008); Mora et al. (2009); Gunawardana et al. (2010); Cartes et al. (2012); Liu et al. (2012); He et al. (2013); Hu et al. (2013); Fard et al. (2016a, 2016b); Zhao et al. (2016); Taghizadeh and Solgi. 2017; Chen et al. (2018b); Chi et al. (2018); Patrick, (2018)
Annual ryegrass (*Lolium multiflorum* L.)	Hyper-accumulative	As, Cd, Pb, Zn	Daniel and Daniel (1955a, 1955b); Carrow et al. (1975); Shamima and Sugiyama (2008); Zhu et al. (2015)
Kentucky bluegrass (*Poa pratensis* L.)	Hyper-accumulative, 7,511 mg Al kg⁻¹, rhizosphere pH adjustment	Al, B, Cd, Cl, Cu, Fe, Mn, Mo, Pb, and Zn	Deal and Engel (1965); Cooper and Spokas (1991); Nus et al. (1993); Lee et al. (1996); Foy and Murray (1998b); de Lespinay et al. (2010); Putra et al. (2013); Xu and Wang (2013); Fard et al. (2016a, 2016b); Fei et al. (2018)
Creeping bentgrass (*Agrostis stolonifera* L.)	Disease suppression, Hyper-accumulative to 1,298 mg Cu kg⁻¹, 1,500 mg Zn kg⁻¹, 2,166 mg Pb kg⁻¹	Cd, Cu, Hg, Mn, Ni, Pb, and Zn	Jowett (1958); Peterson (1969); Gilmour and Miller 1973; Spear and Christians (1991); Heckman et al. (2003); Gladkov et al. (2011); McCarty et al. (2014); Patrick (2018); Yuan et al. (2018)
Red top (*Agrostis alba* L.)	Hyper-accumulative	Cd	Jowett (1958); Shamima and Sugiyama (2008).
Colonial bentgrass (*Agrostis tenuis* Sibth.) (*Agrostis capillaris* L)	Hyper-accumulative, mitochondrial adaptations	Cu, and Zn	Jowett (1958); Gregory and Bradshaw (1964); Peterson (1969); Turner and Marshall (1972); Wainwright and Woolhouse (1977).
Blue sheep fescue (*Festuca ovina* L.),	Hyper-accumulative, chelation	Cu, and Pb	Garland and Wilkins (1981); Ebrahimi and Díaz (2014); Fard et al. (2016a, 2016b)
Red fescue (*Festuca rubra* L.)	Hyper-accumulative	Co, Cu, Pb, and Zn	Wang, (1982); Fard et al. (2016); Taghizadeh and Solgi (2017);
Annual bluegrass (*Poa annua* L.)	Hyper-accumulative	As, Hg, and Zn	Daniel and Daniel (1955a, 1955b); Xu and Mancino (2001); Comino et al. (2009)
Velvetgrass (*Holcus lanatus* L.)	Phosphate enhancement of uptake	As	Meharg et al. (1994)
Canada bluegrass (*Poa compressa* L.)	Hyper-accumulative	Zn	Palazzo et al. (2003)
Hard fescue (*Festuca brevipila* R. Tracey)	Exclusive with uptake	Zn	Palazzo et al. (2003)
		C₄ **warm-season turfgrasses**	
Bermudagrass (*Cynodon dactylon* (L.) Pers.)	Salt glands, hyper-accumulative, exclusions, organic acid exclusion, nitric oxide	Al, Cd, Cu, Hg, Mo, Na, Ni, Pb, Se, and Zn	Wang et al. (1973); Weaver et al. (1984); Wu et al. (1988); Valle et al. (1993); Shu et al. (2002); Gaskin et al. (2003); Feng et al. (2005); Liu (2005); Marcum and Pessarakli (2006); Soleimani et al. (2009); Matteson et al. (2014); Shi et al. (2014); Xie et al. (2014); Wu et al. (2014); Mahoney et al. (2015a, 2015b)
Buffalograss (*Buchlöe dactyloides* (Nutt.) Engelm.)	Hyper-accumulative	B, Cl, Cu, Fe, Mn, Mo, Pb, Se, and Zn	Wu et al. (1988); Schumacher et al. (1993); Jackson et al. (1995); Qu et al. (2003);

(Continued)

TABLE 26.2 (CONTINUED)

Some Related References on Turfgrass Responses to Heavy Metals and Other Elements

Turfgrass Species	Evidence of Toxic Effects and Tolerant Mechanisms	Toxic Elements and Metals	References Cited
Centipedegrass (*Eremochloa ophiuroides* (Munro) Hack.)	Hyper-accumulative, superoxide dismutase and catalase	Cu, and Pb	Qu et al. (2003); Chen et al. (2007); Liu et al. (2012); Li et al. (2016)
Saltmeadow cordgrass (groundcover grass) (*Spartina patens* (Aiton) Muhl.)	Hyper-accumulative	Pb	Qu et al. (2003);
Seashore paspalum (*Paspalum vaginatum*)	Hyper-accumulative, halophytic effects	Cd	Wang (2010); Chen et al. (2016)
Zoysiagrass (*Zoysia japonica* Steud.)	Hyper-accumulative, Melatonin	Al	Baldwin et al. (2005); Huang et al. (2017); Luo et al. (2018)
Matrella Zoysiagrass – (*Zoysia matrella* (L.) Merr.)	Hyper-accumulative	Al, and Cu	Baldwin et al. (2005); Chen et al. (2007)
Carpetgrass (*Axonopus* Beauv.)	Hyper-accumulative, Hyper-tolerant	Al, Cd, Cu, Ni, Pb, and Zn	Baldwin et al. (2005); Matthews-Amune and Kakulu (2012).
St. Augustinegrass (*Stenotaphrum secundatum* (Walt.)) Kuntze]	Hyper-accumulative	Al, Cd, Cu, Ni, Pb, and Zn	Dudeck (1985); Baldwin et al. (2005)

by emerging phytoremediation technology (Salt et al., 1998). Those excessive soil and environmental elements can be divided into two groups for turfgrasses either as essential micronutrients such as Fe, Cl, B, Mn, Cu, Zn, Mo, and Ni (Carrow et al., 2001; Marschner, 2012; StJohn and Christians, 2013) or toxic elements not needed by turfgrasses such as Na, Al, Cd, Pb, As, Hg, Cr, Se, and Co toxic to plants, humans, animals, and other living forms (Table 26.1).

Turfgrasses do not necessarily belong to the "food chain" with regular mowing as cut-off clipping removal (Figure 26.1) or cycling in the turf-soil system. The fibrous root systems, dense canopies of groundcover plants such as turfgrasses, which have been used for centuries by humans, are not hyper-accumulators, but all have hyper-accumulative characteristics (Tables 26.1, 26.2, and 26.3).

26.2.1 Natural Sources

Soils are major sources of these elements, and these elements' contents vary from soil to soil, and generally these elements exist in most soils as carbonates, sulfides, oxides, or salts (Sarwar et al., 2017). Heavy metals and metalloids in soils are derived from the soil parent materials. The local landscape, water movements, and interactions, as well as soil weathering processes all have impacts on the availability of these elements either as plant micronutrients or toxic elements since nearly the total earth crust, 95%, is made up of ingenious rocks and about 5% sedimentary rocks (Thornton, 1981). Other natural factors such as soil pH, cation exchange capacity (CEC),

organic matter, and vegetation also have impacts on these elements causing them either to become available or toxic.

26.2.2 Anthropogenic Sources

Agricultural practices and industry advancements, fertilizer and pesticide applications to soils, are major anthropogenic sources of these elements, although abundant soil elements such as aluminum and iron often count for 8% and 5% of the soil by weight. Cheng et al. (2007) studied the growth of perennial ryegrass (*Lolium perenne* L.) in soils amended with 5–100% composted sewage sludge (CSS), and the impacts of a CSS amendment on the soil's physical and chemical properties. They found that soils amended with ≤20% CSS did not significantly affect seedling emergence, while the contents of chlorophyll, nitrogen, phosphorous, and potassium of perennial ryegrass grown in such soils were greatly improved. Bulk density, water retention, and nutrient content of the soil were also improved with the amendment of CSS. However, high CSS contents introduced excessive amounts of heavy metals and soluble salts. They found that Cu, Zn, and Pb accumulated slightly (up to ~2.3 times) in clippings of perennial ryegrass grown in CSS-amended soils compared to those grown in the base and reference soils, while no significant Cd absorption in the shoots of perennial ryegrass occurred (Cheng et al., 2007).

Consistently, applications of phosphate rock phosphorous (P) fertilizers such as triple superphosphate and calcium phosphate contain heavy metals (Zn and Cd) in varying concentrations depending on rock phosphate sources (Sarwar

TABLE 26.3

Some Related References on Toxic Heavy Metals and Other Elements Found in Turfgrass Species

Elements and Metals	Evidence of Turfgrass Species with Tolerances	Mechanisms Associated with Tolerances	References Cited
		Plant Essential Metals	
Fe	Kentucky bluegrass Matrella zoysiagrass bermudagrass	Hyper-accumulative	Deal and Engel (1965); Beard et al. (1977); Yust et al. (1984); Wong and Lau (1985); Taylor (2019)
Cl	Seashore paspalum, bermudagrass zoysiagrass, tall fescue, creeping bentgrass, Kentucky bluegrass, Tall fescue,	Hyper-accumulative	Cordukes and Parups (1971); Tran et al. (2018)
B	Seashore paspalum, Tall wheatgrass	Activated related enzymes to enhance cold tolerance	Oertli et al. (1969); Schuman (1969); Yu et al. (2005)
Mn	Kentucky bluegrass, tall fescue	Hyper-accumulative Organic matter enhanced exclusion	Deal and Engel (1965); Brye and Pirani 2005.
Cu	Perennial ryegrass, red fescue, bermudagrass, blue sheep fescue, colonial bentgrass	Biochar added, hyper-accumulative	Wainwright and Woolhouse (1977); Karami et al. (2011)
Zn	Colonial bentgrass, creeping bentgrass, Tall fescue, hard fescue, Canada bluegrass	Hyper-accumulative >1,553 mg kg^{-1}	Peterson (1969); Turner and Marshall (1972); Wainwright and Woolhouse (1977); Spear and Christians (1991); Palazzo et al. (2003)
Mo	Seashore paspalum	Activated related enzymes to enhance cold tolerance	Yu et al. (2005)
Ni	Tall fescue, Bermudagrass, creeping bentgrass, zoysiagrass, bermudagrass, perennial ryegrass	Hyper-accumulative >800 mg kg^{-1}	Soleimani et al. (2009); Menchyk (2012); Liu, (2018); Patrick (2018)
		Non-Plant Essential Nutrient Elements and Metals	
Na	Seashore paspalum, bermudagrass zoysiagrass, tall fescue, creeping bentgrass	Plant tissue tolerance, root uptake avoidance, sequestration to older leaves or vacuoles, salt glands, and compartmentalization	Marcum (2001); Carrow and Duncan (2012); Huang et al., (2014); Tran et al. (2018)
Al	Tall fescue, bermudagrass, fine fescues, Kentucky bluegrass, creeping bentgrass, zoysiagrass, centipedegrass	Genetic variation, endophyte infection, Al accumulations, Al tolerant genes	Murray and Foy (1978); Liu et al. (1995); Liu et al. (1996); Liu et al. (1997a, 1997b); Liu (2001); Zaurov et al. (2001); Baldwin et al. (2005); Liu (2005); Yan et al. (2009); Huang et al. (2017); Chen et al. (2018a); Luo et al. (2018)
Cd	Tall fescue, perennial ryegrass, seashore paspalum, centipedegrass, creeping bentgrass, Kentucky bluegrass	Increased uptake by adding organic acids,	Liu et al. (2012); Zhao et al. (2011); Hu et al. (2013); Zhao et al. (2014); Zhao et al. (2016); Zhuo et al. (2017); Chi et al. (2018); Yuan et al. (2018); Zhu et al. (2018)
Pb	Tall fescue, perennial ryegrass, bermudagrass, Buffalograss, centipedegrass, Kentucky bluegrass	Increased uptake by adding organic acids, Sulfur added, biochar added, hyper-accumulative >6,000 mg Pb kg^{-1}	Jones et al. (1973b); Qu et al. (2003); Begonia et al. (2005); Yesilonis et al. (2008); Karami et al. (2011); Zhao et al. (2011); Liu et al. (2012); Luo et al. (2013); Putra et al. (2013); Zhao et al. (2016)
As	Annual bluegrass, creeping bentgrass, Kentucky bluegrass, perennial ryegrass, tall fescue, bermudagrass, velvetgrass,	Herbicide degrading, hyper-accumulative,	Daniel (1955a, 1955b); Carrow et al. (1975); Weaver et al. (1984); Meharg et al. (1994); Feng et al. (2005); Dong et al. (2008); Comino et al. (2009); Jin et al. (2010); Matteson et al. (2014); Mahoney et al. (2015a, 2015b)
Hg	Annual bluegrass, bermudagrass, creeping bentgrass	Hg$_2$Cl and HgCl$_2$ were used as fungicides	Gilmour and Miller 1973; Weaver et al. (1984); Comino et al. (2009)
Cr	Tall fescue	Nitric oxide	Brye and Pirani (2005); Zhang et al. (2007); Wu et al. (2014); Huang et al. (2018)
Se	Tall fescue, buffalograss, creeping bentgrass, bermudagrass, perennial ryegrass	Chloride and sulfate added, genetic variations	Wu et al. (1988); Wu and Huang (1991); Hartikainen et al. (2000)
Co	Perennial ryegrass, red fescue	Hyper-accumulative	Taghizadeh and Solgi (2017)

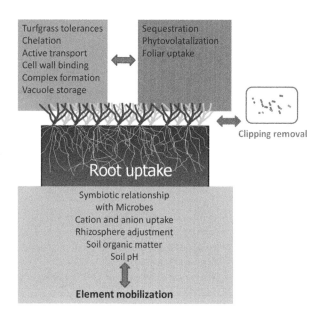

FIGURE 26.1 A turfgrass ecosystem with hyper-accumulative potentials of excessive elements. A partial adaptation of Sarwar et al. (2017).

et al., 2017). The extent of the contamination of soils with heavy metals through irrigation varies from location to location depending upon the level of contamination in the irrigation water and recycling water element contents (Qian and Mecham, 2005).

Dry and wet deposits of emissions from different point and non-point sources, including the steel industry, metal smelters, metal refineries, foundries, and cement industries, also contribute to these elements accumulating in soils. Emissions from automobiles using lead-enriched fuel are responsible for significant Pb accumulation in soils near highways. Mining activities have also been reported as contaminating the soils with heavy metals in localized areas (Sarwar et al., 2017).

Increasing water shortage in the arid and semiarid agricultural areas requires the use of recycled wastewater (RWW) when possible. As recycled wastewater has become a common water source for irrigating golf courses, lawns, and urban landscapes, the effects of RWW irrigation on soil chemical properties have been a major concern for the environment particularly with heavy metal and salty material deposits with multiple years of RWW usage (Mancino, 1992; Pepper and Mancino,1993; United States Golf Association, and (American Society of Golf Course Architects, 1994; Pessarakli, 2014)). Qian and Mecham (2005) compiled soil test data from fairways of 10 golf courses that were near metropolitan Denver and Fort Collins, CO, they found the golf course sites with RWW irrigation, exhibited 0.3 units of higher pH and 200, 40, and 30% higher concentrations of extractable Na, B, and P, respectively. The sites irrigated with RWW exhibited 187% higher EC and a 481% higher sodium adsorption ratio (SAR). Comparison of soil chemical properties before and 4 or 5 years after RWW irrigation on the two golf courses studied also found soils conatained 1) an 89 to 95% increase in Na

content; 2) a 28 to 50% increase in B content; and 3) an 89 to 117% increase in P content at the surface depth.

26.3 HARMFUL EFFECTS OF EXCESSIVE PLANT MICRONUTRIENTS AND HEAVY METALS TO TURFGRASSES

Although eight elements (Fe, Cl, B, Mn, Cu, Zn, Mo, and Ni) are essential micronutrients for turfgrasses grown in low concentrations, excessive amounts in the soil above threshold values can result in toxicity. This toxicity varies with the nature of an element as well as turfgrass species (Tables 26.1 through 26.3). Excessive micronutrient toxicity in plants depends on soil characteristics such as soil pH, CEC, organic matter, parent materials, and others, as well as the bioavailability of these elements in soil solutions, which is a function of water uptake and element uptake capabilities. Nonessential metals/metalloids such as Cd, Pb, As, Hg, Chr, Se, and Co, discussed in this chapter (Tables 26.1 through 26.3), are toxic both in their chemically combined or elemental forms, and turfgrass responses to these elements vary across a broad spectrum from the tolerance of hyper-accumulation to toxicity. Excessive concentrations of the mentioned elements may replace essential metals in pigments or enzymes disrupting their function and causing oxidative stress. The toxicity hinders the growth process of the underground and aboveground turfgrass parts and the activity of turfgrass growth and development. To avoid and tolerate the toxicity, turfgrasses have developed mechanisms by which toxic elements are excluded, retained at the root level, or transformed into physiologically tolerant forms as hyper-accumulative characteristics in general.

Elements including Na, Al, As, Cd, Pb, Hg, Se, Cr, and Co are not plant nutrients and are toxic to plants although some are human and animal micronutrients (Table 26.1). Other metals such as Fe, Mn, Cu, Zn, and Ni, plus Cl, B, and Mo, are plant micronutrients in a concentration range of 0.1 to 200 mg kg^{-1} in most turfgrasses (StJohn and Christians, 2013) and excessive amounts of these metals are also toxic to both plants and humans, particularly in their ionic forms (Clemens and Ma, 2016). Cd and inorganic As are classified as carcinogens, As and Hg are also toxic in their methylated forms (Clemens and Ma, 2016). Although the primary targets of toxicity for these elements are still not known for turfgrasses, these elements can damage different cellular structures and a variety of tissues and organs, causing stress and even death to the turfgrasses (Menchyk, 2012; Patrick, 2018). One major cause of toxicity is the strong interactions between different protein and enzyme groups which interferes with the homeostasis of essential nutrient elements. Both Cd and Pb can replace zinc (Zn) in proteins. Arsenate disturbs phosphate metabolism because of the chemical similarity of the two anions (Sarwar et al., 2017).

Most turfgrass species are not halophytes, which are the flora of saline soils (Marcum, 1999). Halophytes, including seashore paspalum (*Paspalum vaginatum* Sw.) (Chen et al., 2016), can adjust osmotically to soil salinity by accumulating

Na^+ and Cl^- ions and sequestering the vast majority of ions in vacuoles, often associating cytoplasm organic solute accumulation to prevent adverse effects on metabolism. When non-halophytes encounter high salinities, growth is inhibited. This is caused by toxicities to the metabolism of Na^+ and Cl^- in the cytoplasm associated with insufficient osmotic adjustment, resulting in reduced net photosynthesis because of stomatal closure, with drought stresses without optimal levels of K^+ required for maintaining 60 to 80 enzyme activities; also, with possible damage from reactive oxygen species or changes in hormonal concentrations. Turfgrasses with salt glands on their leaves (Marcum and Murdoch, 1990; Marcum and Pessarakli, 2006) may provide opportunities not only to remove Na^+ and Cl^- but also other toxic elements, which may deserve further investigations. On the other hand, NaCl addition was found to reduce Cd toxicity in *Carpobrotus rossii* (Haw.) Schwantes (Cheng et al., 2018). It was found that the addition of NaCl to Cd-containing solutions improved plant growth along with 70–87% less shoot Cd accumulation, resulting from decreases in Cd root uptake and root-to-shoot translocation irrespective of Cd^{2+} activity in solutions. They also reported that Cd increased the concentrations of amino acids in plant shoots; the effect of NaCl on the synthesis of amino acids was inconsistent. Chen et al. (2016) studied 32 salinity-tolerant clones harboring 18 salinity-tolerance genes and 20 Cd-tolerant clones, including five Cd-tolerance genes using seashore paspalum, and they concluded that the 18 salinity-tolerance genes and five Cd-tolerance genes could potentially be used as candidate genes for the genetic modification of glycophytic grass species to improve salinity and Cd-tolerance and for further analysis of molecular mechanisms regulating salinity and Cd-tolerance. The interactions of these cations and anions associated with plant uptake and avoidance are not the focus of this chapter, but obviously it is a major factor affecting each in their plant metabolism and toxicity resistances.

Aluminum (Al) toxicity is a major constraint for crop production in acidic soil worldwide (Kochian et al., 2015). When the soil pH is lower than 5.5, Al^{3+} as a cation is released to the soil solution, and it can be taken up by plants. Al toxicity mainly happens at the turfgrass root tip, in which Al inhibits cell elongation and cell division, leading to root stunting accompanied by reduced water and nutrient uptake (Liu et al., 1995, 1996, 1997a, 1997b; Liu, 2001, 2005; Baldwin et al., 2005; Liu, 2005; Yan et al., 2013). To date, a variety of genes have been identified that are induced or repressed upon Al exposure (Kochian et al., 2015; Hufnagel et al., 2018; Silva et al., 2018). Baldwin et al. (2005) identified several warm-season turfgrass species with capabilities to absorb Al^{3+} in a range of 800 to 1,400 mg kg^{-1} in root and leaves.

Since soils contain about 8% of Al by weight and Al is released into the soil, solutions become available to plants when soil pH is below 5.5 (Foy and Murray, 1998a, 1998b; Kochian et al., 2015). Phytoremediation normally does not apply to Al, Na, Fe, and Cl as the abundant elements in soils have rather effective ways for controlling excessive levels of toxicity which focuses on plant resistance and avoidance. The rest of the other elements discussed in this chapter are

excellent target elements for turfgrass phytoremediation to protect the environment, particularly urban communities and the polluted urban soils and air and water sources.

26.4 FACTORS AFFECTING EXCESSIVE METAL AND OTHER ELEMENT BIOAVAILABILITY IN TURFGRASSES

A number of factors control bioavailability of an element in the soil, including soil organic matter, soil pH (Chlopecka et al., 1996; Yesilonis et al., 2008), soil cation exchange capacity, competitive ion concentration, root exudates, and species of plants present in the soil and plant age. Turfgrasses are closer to social communities than other row crops and are dominant plant covers in urban areas. The factors that affect excessive metal and other elements bioavailability in turfgrasses interact with the soil, water, and air sources as well as all fertilizers, soil amendments, and bio-solids input into these turf systems. In general, soils serve as the most important factor in determining the bioavailability levels of these excessive metal and other elements in turfgrasses.

Acidic soils (pH <5.5) contain toxic ionic forms of aluminum as $Al(H_2O)_6^{+3}$ with six water molecules surrounding it, although it becomes precipitated when pH is above 5.5 (Kochain et al., 2015). Phytoremediation practices normally do not apply to abundant soil elements such as Al, Na, and Cl but some turfgrasses have the super capability to absorb these elements from soils and water (Liu, 2005).

Obviously, pollution sources are critical to all turf sites. For example, in the US, utility turf, counting for about 26% of the total turf area of 181,300 km^2 covering airports, highways, and bank areas along rivers and lakes, can absorb these excessive elements and maintain them as less toxic to the environment serving as sinks. The dense turf canopies can also absorb air pollutions of these elements and a total of 12 million tons of dust have been trapped annually from the air by the total turf area in US (personal communications). Although there is a lack of literature with other examples, the trees in the central part of Beijing removed 1,261.4 tons of pollutants from the air in 2002 (Yang et al., 2005).

Another unique situation with managed turf areas is mowing. The cut-off clippings can either be returned back to the turf or removed (Beard, 1973; McCarty, 2018). Both practices have benefits in phytoremediation of the turf either to keep the toxic elements in the system or remove them with further treatment of the clippings. The advantages of recycling clippings are that they will return turf nutrients back to the soil and turf creating a more sustainable system with minimum fertilizer input with additional nutrient protection management (Frank and Guertal, 2013a, 2013b).

26.5 MECHANISMS OF HYPER-ACCUMULATIVE CHARACTERISTICS AMONG TURFGRASSES

Mechanisms of hyper-accumulative characteristics among turfgrasses are listed in Table 26.4. Hyper-accumulation is

TABLE 26.4

Mechanisms of Hyper-Accumulative Characteristics among Turfgrasses

Mechanisms of Hyper-accumulative Characteristics	Elements	Turfgrasses	References
Hyper-accumulative with resistant genes, hyper-tolerant transformants	Al, As, B, Cd, Cl, Co, Cr, Cu Fe, Hg, Mn, Mo, Na, Ni, Pb, and Zn	All turfgrass species discussed in the chapter	Marcum (1999); Baldwin et al, (2005); Clemens and Ma (2016); Chen et al, (2018a); Yuan et al, (2018)
Avoidance and exclusion (salt glands), rhizosphere pH and organic acid efflux	Al, Cl, Cd, and Na	Warm-season turfgrasses and cool-season turfgrass	Marcum (1999); Liao and Huang (2002); Baldwin et al, (2005); Chen et al, (2018a)
Root accumulation, phosphate precipitation	Cd, and Pb	Kentucky bluegrass, Tall fescue	Begonia et al, (2001; 2005); Xu and Wang, (2013)
Compartmentation of heavy metal in a vacuole	All toxic elements	Highly possible but unknown yet with specific turfgrasses	Zenk (1996); Hall (2002); He et al. (2013); Clemens and Ma (2016)
Cell wall compartmentation	All toxic elements	Highly possible but unknown yet with specific turfgrasses	Zenk (1996); Hall, (2002); He et al, (2013); Clemens and Ma (2016)
Endophyte and mycorrhizae	Al, As, Cd, Ni, and Zn	Fine fescues, tall fescues, perennial ryegrass, bermudagrass	Liu et al. (1995); Takacs et al. (2001); Zaurov et al. (2001); Dong et al. (2008); Ren et al. (2011); Hu et al. (2013); Wu et al. (2014)
Nitric oxide regulation	Cd, and Zn	Perennial ryegrass, bermudagrass, tall fescue	Shi et al. (2014); Chen et al. (2018b); Chi et al. (2018)
Chelation agents and biological chelators	Cd	Tall fescue, blue sheep fescue	Dou (2005); Zhao et al. (2011; 2013); Ebrahimi and Díaz (2014);
Enzymatic adjustments, superoxide dismutase, and catalase	Cu, and Pb	Tall fescue, perennial ryegrass, centipedegrass	Lee et al. (2007); Zhao et al. (2010); Li et al. (2016); Hasan et al. (2017)
Metallothionein (MT) A family of cysteine-rich, low molecular weight (MW ranging from 500 to 14,000 Da) proteins with the capacity to bind both metals.	All metals	Highly possible but unknown yet with specific turfgrasses	Hall (2002); Freisinger (2011); Clemens and Ma (2016)
Glutathione	Cl, and Na, Cl	Creeping bentgrass	Xu et al. (2010)
Melatonin	Al	zoysiagrass	Luo et al. (2018)
Silicon-mediated alleviation	Al, As, Cl, Co, Cr, Cu, Fe, Mn, Pb, and Se	Highly possible but unknown yet with specific turfgrasses	Nanayakkara et al. (2008); Adrees et al. (2015); Tripathi et al. (2017)
Hormone regulations	Pb	Increased compartmentation to cell walls	He et al. 2013

a dominant mechanism in turfgrasses (Tables 26.1, 26.2, 26.3, 26.4) in addition to exclusion and avoidance and other regulations.

The effects of lower molecular weight organic acids such as oxalic acid, citric acid, and acetic acid, as well as larger molecular weight organic acids such as humic acid, on the toxicity of Cd during perennial ryegrass growth were studied (Liao and Huang, 2002). They found that Cd toxicity was enhanced gradually with the concentration increase of low molecular weight organic acids resulting in the decrease of chlorophyll concentrations in ryegrass plants and the biomass affecting by such an order of oxalic acid < acetic acid < citric acid. As expected, Cd toxicity was reduced as a result of the addition of a larger molecular weight of humic acid, and consequently the perennial ryegrass concentration of chlorophyll and the biomass increased. Lower molecular weight organic acids unlikely reduced Cd uptake as effectively as humic acid by the perennial ryegrass roots (Liao and Huang, 2002).

Under copper stress using a hydroponic culture, the root growth and protective enzymes of tall fescue and perennial

ryegrass were found to be different (Zhao et al., 2010), although Cu stress significantly inhibited root growth (root elongation and dry biomass) of both turfgrasses. The authors indicated that Malondialdehyde (MDA) content in the roots of both turfgrasses significantly increased under copper stress. In tall fescue root, superoxide dismutase (SOD) activities increased with an increasing Cu concentration; while peroxidase (POD) activity increased at low Cu levels (<60 mg L^{-1}) and decreased at high Cu levels (>60 mg L^{-1}). Increased MDA content indicated the formation of free radicals under Cu stress, while increased SOD activity pointed to the operation of a scavenging mechanism. In roots of perennial ryegrass, however, SOD and POD were not activated by copper. The authors concluded that these results demonstrated that turfgrass cultivars clearly differ in tolerance to Cu stress, and tolerance depends on the enhanced activities of its antioxidant system.

The phytoextraction efficiency of cadmium (Cd) contaminated soil mainly depends upon the mechanisms of plants in absorption, translocation, distribution, and detoxification of Cd. A pot experiment was designed to investigate Cd distribution and accumulation among the different leaves of tall fescue and Kentucky bluegrass and its regulation by nitrilotriacetic acid (NTA), a biodegradable chelating agent (Fei et al., 2018). The results indicated that a young leaf protection mechanism might be involved in their Cd hypertolerance. The Cd preferential accumulation could lead to a novel phytoextraction strategy by continuously harvesting the senescent and dead leaves of tall fescue and Kentucky bluegrass. Nitrilotriacetic acid promoted the young leaf protection in tall fescue, but not in Kentucky bluegrass, indicating different mechanisms associated with two turfgrasses in their Cd hypertolerance (Fei et al., 2018). Yuan et al. (2018) investigated differentially expressed genes (DEGs) under lower (BT2_5) and higher (BT43) Cd concentration treatments of 2.5 and 43 mM $CdCl_2$, respectively, in creeping bentgrass. A total of 463,184 unigenes were obtained from creeping bentgrass leaves using RNA sequencing technology. They identified four key transcription factor (TF) families, WRKY, bZIP, ERF, and MYB, that were associated with Cd stress in creeping bentgrass, and these four TFs play crucial roles during the creeping bentgrass response to Cd stress. These results indicate the regulatory mechanisms of responding to the Cd stress of creeping bentgrass which is enriching information for future potential phytoremediation using turfgrasses.

Xu and Wang (2013, 2014) compared Kentucky bluegrass and tall fescue in Cd hyper-accumulations. They concluded that both turfgrasses could tolerate and accumulate high levels of Cd. They also found that the Cd hypertolerance of Kentucky bluegrass and tall fescue involved detoxification by forming the undissolved Cd phosphates in the leaves with some similarity to the Al rhizosphere adjustments (Kochian et al., 2015). The superior Cd-tolerance of tall fescue to Kentucky bluegrass is related to less Cd being uploaded into the stele of roots and less Cd being transported to the leaves.

Mercury uptake by annual bluegrass indicates that it (Comino et al., 2009) can serve as a plant species in phytoremediation of Hg, although other turfgrass species should be studied in comparison with another extreme Hg hyper-accumulator; Indian mustard (*Brassica juncea* L.) can accumulate as high as 2.8% of Hg as dry weight in its roots (Shiyab et al., 2009).

Turfgrasses differ in their Pb uptakes, hyper-accumulative, and hyper-tolerant characteristics (Qu et al., 2003; Wang et al., 2010; Liu et al., 2012; Yang et al., 2014; Fard et al., 2016). The effects of growth-promoting hormone diethyl aminoethyl hexanoate (DA-6) and EDTA, either alone or in combination, when applied to original soil or lead (Pb) spiked soil on Pb phytoextraction, subcellular distribution and chemical forms in perennial ryegrass were studied (He et al., 2013). EDTA addition alone significantly reduced plant biomass though it increased Pb accumulation. Foliar spray of DA-6 alone increased both plant biomass and Pb accumulation, with 10 μM DA-6 being the most effective with the maximum Pb accumulation. At the subcellular level, about 35–66% of Pb was distributed in the cell wall and 21–42% in a soluble fraction, with a minority present in a cellular organelles fraction treated with 10 μM DA-6. These results suggest that Pb fixation by pectates and proteins in cell wall and compartmentalization

by vacuole might be responsible for Pb detoxification in plants, and the combined use of EDTA and 10 μM DA-6 appears to be optimal for improving the remediation efficiency of perennial ryegrass for Pb contaminated soil (He et al., 2013).

Arsenic (As) is the 20th most abundant element in the geosphere, in the earth's crust As is known to be approximately 1.8 mg kg^{-1}, ranked as the 14th in seawater, and has been recognized as an extremely toxic metalloid for humans as well as for fauna and flora (Hasanuzzaman et al., 2018).

Monosodium methyl arsenate (MSMA) is a commonly used herbicide for weed control in turfgrass and other crop systems with high association to soil anions such as phosphate (Daniel, 1955a, 1955b; Carrow et al., 1975; Mattson et al., 2014; Mahoney et al., 2015a, 2015b). In addition to natural and water source arsenic pollutions (Ravenscroft et al., 2009), there are concerns that arsenic from applied MSMA could leach to groundwater or runoff into other water sources threatening human and ecosystem health. The US-EPA has proposed a phase-out of the herbicide. After MSMA application to turfgrass-covered and bare ground lysimeters, the majority of arsenic was retained in turfgrass foliage and soils throughout year-long experiments, with 50 to 101% of the applied arsenic recovered in turfgrass systems and 55 to 66% recovered in bareground systems (Mattson et al., 2014; Mahoney et al., 2015a, 2015b). It was suggested that under aerobic conditions, minimal arsenic leaching into groundwater would occur after a typical application of MSMA to turfgrass. However, repeated MSMA application may pose environmental risks.

Carrow et al. (1975) found that arsenic had no effect on the seed germination of creeping bentgrass, Kentucky bluegrass, and annual bluegrass when the As was applied and incubated 7 weeks prior to seeding. However, with increasing As levels, the growth of all three grasses decreased regardless of P level. Annual bluegrass was most affected by As. Phosphorus had little or no effect on As toxicity as exhibited by turfgrass growth even at very high P levels. However, Niazi et al. (2017) reported a phosphate-assisted increase in As uptake was substantially (up to two times) greater for *Brassica napus*, notably due to higher shoot As concentration and dry biomass yield, compared to *B. juncea* at the P 100 mg kg^{-1} level when examining the potential role of phosphate (P; 0, 50, 100 mg kg^{-1}) on growth, gas exchange attributes, and photosynthetic pigments of two *Brassica* species under arsenic stress (0, 25, 50, 75 mg kg^{-1}) in a pot experiment. The findings demonstrate that phosphate supplementation has the potential to improve As phytoextraction efficiency, predominantly for *B. napus*, by minimizing As-induced damage to plant growth, as well as by improving the physiological and photosynthetic attributes. Mandal et al. (2017) further confirmed that Chinese brake fern (*Pteris vittata* L.) As accumulations were enhanced by applying phosphatic fertilizers of di-ammonium phosphate (DAP) and single superphosphate (SSP) with a high As accumulation as high as 4,504 mg kg^{-1}.

Biochar (BC) exhibits great potential for being an applicable material for the water and soil remediation of toxic metal and metalloids due to its physio-chemical surface characteristics (Vithanage et al., 2017). Phytostabilization can be applied in the use of turfgrasses as perennial groundcover crops to reduce the bioavailability of pollutants in the environment and keep turfgrasses away from the food chain. Ozdemir at al. (2018) concluded the applicability of poultry abattoir sludge compost (PASC) and biochar on the establishment of turfgrass as promising agro-industrial-based fertilizers in turfgrass sod production for sustainable soil and nutrient management.

Among all the turfgrass species, tall fescue, perennial ryegrass, and bermudagrass have received the most attention in their hyper-accumulative characteristics (Tables 26.1, 26.2, 26.3, 26.4). As cool-season turfgrasses, adapted in the northern states and the transition zone as well, tall fescue and perennial ryegrass will provide more solutions to phytoremediation aspects due to their strong stress hypertolerance of excessive elements in the environment in general when comparing with other cool-season turfgrasses. However, different approaches and goals exist between turfgrass tall fescue and forage tall fescue with a similarity with bermudagrasses, which is the most popular warm-season turfgrass in the world (Christians et al., 2016; McCarty, 2018) as the most intensively studied warm-season turfgrass with hyper-accumulative characteristics (Tables 26.1, 26.2, 26.3, 26.4). However, unlike the more thorough reaches of salinity and acid soil tolerances among turfgrass cultivars, the turfgrass cultivar level and genetic differences in hyper-accumulation and hyper-toleration characteristics deserve more investigation and data collection.

Plant signaling molecules such as nitric oxide, metallothionein (MT), glutathione, melatonin, and silicon-mediated alleviation have been studied intensively in the past decade or more and have provided new insights into the hyper-accumulative characteristics associated with turfgrasses and their hypertolerance to those elements (Table 26.4). In general, these signaling molecules alleviate oxidative stresses caused by the excessive amounts absorbed from soils and in the soil environment.

Integrated approaches to further enhance turfgrass hyper-accumulation characteristics may open many new research opportunities as well as solutions to very critical practical needs in future societies with globe-wide increasing turf areas.

26.6 CONCLUSIONS AND FUTURE PERSPECTIVES

Based on the literature and knowledge pool to date on turfgrass hyper-accumulative and hyper-tolerant characteristics of excessive elements, particularly with heavy metals and metalloids, turfgrasses, as a unique group of perennial and groundcover plants with great potentials to protect the environment, can play significant roles in phytoremediation for the future. The future is promising, and the following methods are unlimited and practical for future turfgrass management:

- Phytoextraction, the use of pollutant-accumulating turfgrasses to remove excessive elements or organics from soils by concentrating them in harvestable clippings.

- Phytodegradation, the use of turfgrasses associated with other groundcover plants as well as the associated microorganisms to degrade organic pollutants and reduce the severity of heavy metal toxicity in the environment.
- Phytostabilization, the use of turfgrasses as perennial groundcover crops to reduce the bioavailability of pollutants in the environment and keep turfgrasses away from the food chain.
- Phytovolatilization, the use of turfgrasses to volatilize and remove pollutants from the air.

Based on all the hyper-accumulative and hyper-tolerant characteristics of turfgrasses discussed in this chapter, the authors hope the excessive element stresses in the environment provide future opportunities for managing turf in multiple dimensions and with integrated approaches.

ACKNOWLEDGMENTS

The authors sincerely thank numerous organizations, companies, colleagues, former advisors, and former graduate and undergraduate students for the encouragement and support given in the related researches of turfgrasses on N, P, K, Al, Fe, and Ni. Special thanks to both undergraduate and graduate students in four advanced turfgrass and soil science courses, HORT4200 Applied Turfgrass Stress Physiology, PES 4520/6520 Soil Fertility and Plant Nutrition (2012–2018), PES 8900 Plant Nitrogen Metabolism (2015–2018), and PES4850/6850 Environmental Soil Chemistry (2014–2018) at Clemson University for their spirit, encouragement, and support. Thanks to the PES 4850/6850 2018 class which included (in alphabetical order) graduate students: Ethan R. Barnett, Sarah K. Holladay, Jiwoo Park, Savannah R. Petrone, Coleman A. Scroggs, and Zolian S. Zoong-lwe; undergraduate students: Thomas C. Chapman, Adam W. Chastain, Payton Davis, Katrina R. Hale-Phillips, Kimberly J. Henning, Ryan M. Ponder, Logan T. Shelton, and Brandon T. Welch for their reviews of the chapter and contributions in composing the abstract.

BIBLIOGRAPHY

Adrees, M., S. Ali, M. Rizwan, M.Z. Rehman, M. Ibrahim, F. Abbas, M. Farid, M.K. Qayyum, and M.K. Irshad. 2015. Mechanisms of silicon-mediated alleviation of heavy metal toxicity in plants: a review. *Ecotoxicol. Environ. Saf.* 119:186–197.

Ansah, K.O. 2012. Warm season turfgrasses as potential candidates to phytoremediate arsenic pollutants at Obuasi Goldmine in Ghana. MS Thesis. Colorado State University, Fort Collins, CO.

Baldwin, C.M., H. Liu, L.B. McCarty, W.B. Bauerle, and J.E. Toler. 2005. Aluminum tolerance of warm-season turfgrasses. *Int. Turfgrass Soc. Res. J.* 10:811–817.

Beard, J.B., and H. Beard. 2005. *Beard's Turfgrass Encyclopedia for Golf Courses, Grounds, Lawns, Sports Fields*, p. 466. East Lansing, MI. Michigan State University Press.

Beard, J.B., J.H. Eckhardt, and G. Horst. 1977. Iron application – rates, carriers, and toxicity comparisons. Turfgrass Res. Rep. Tex. A&M Turfgrass Field Day, pp. 36–37.

Beard, J.B. 1973. *Turfgrass: Science and Culture*. Upper Saddle River, NJ: Prentice Hall.

Begonia, M.T., G.B. Begonia, M. Ighoavodha, and D. Gilliard. 2005. Lead accumulation by tall fescue (*Festuca arundinacea* Schreb.) grown on a lead-contaminated soil. *Int. J. Environ. Res. Public Health*. 2(2):228–233.

Begonia, M.T., G.B. Begonia, M. Ighavodha, O. Okuyiga-Ezem, and B. Crudup. 2001. Chelate-induced phytoextraction of lead from contaminated soils using tall fescue (*Festuca arundinacea*). *J. Mississippi Acad. Sci*. 46(1): 15.

Bradshaw, A.D. 1952. Populations of *Agrostis tenuis* resistant to lead and zinc poisoning. *Nature*. 169:1098.

Brye, K.R., and A.L. Pirani. 2005. Metal uptake by tall fescue (*Festuca arundinacea*) as affected by poultry litter application. *Grass For. Sci*. 61:192–199.

Carlson, R.W., and G.L. Rolfe. 1979. Growth of ryegrass and fescue as affected by lead-cadmium-fertilizer interaction. *J. Environ. Q*. 8(3):348.

Carrow, R.N., and R.R. Duncan. 2012. *Best Management Practices for Saline and Sodic Turfgrass Soils: Assessment and Reclamation*. Boca Raton, FL. CRC Press.

Carrow, R.N., D.V. Waddington, and P.E. Rieke. 2001. *Turfgrass Soil fertility and Chemical Problems – Assessment and Management*. Chelsea, MI. Ann Arbor Press.

Carrow, R.N., P.E. Rieke, and B.G. Ellis. 1975. Growth of Turfgrasses as affected by soil phosphorus and arsenic. *Soil Sci. Soc. Am. J*. 39:1121–1124.

Cartes, P., M. McManus, C. Wulff-Zottele, S. Leung, A. Gutiérrez-Moraga, and M. Mora. 2012. Differential superoxide dismutase expression in ryegrass cultivars in response to short term aluminum stress. *Plant Soil*. 350 (1–2):353–363.

Chen, F., H. Ai, M. Wei, C. Qin, Y. Feng, S. Ran, Z. Wei, H. Niu, Q. Zhu, H. Zhu, L. Chen, J. Sun, H. Hou, K. Chen, and H. Ye. 2018a. Distribution and phytotoxicity of soil labile aluminum fractions and aluminum species in soil water extracts and their effects on tall fescue. *Ecotoxicol. Environ. Saf*. 163:180–187.

Chen, L.L., X.T. Bao, and Z.L. Wang. 2007. Effects of soil copper contamination on the growth of five turf grasses. *Prat. Sci*. 24(7):101–104.

Chen, W., Y. Dong, G. Hu, and X. Bai. 2018b. Effects of exogenous nitric oxide on cadmium toxicity and antioxidative system in perennial ryegrass. *J. Soil Sci. Plant Nut*. 18(1):129–143.

Chen, Y., C. Chen, Z. Tan, J. Liu, L. Zhuang, Z. Yang, and B. Huang. 2016. Functional identification and characterization of genes cloned from halophyte seashore paspalum conferring salinity and cadmium tolerance. *Front. Plant Sci*. 7:102.

Cheng, H., W. Xu, J. Liu, Q. Zhao, Y. He, and G. Chen. 2007. Application of composted sewage sludge (CSS) as a soil amendment for turfgrass growth. *Ecol. Eng*. 29:96–104.

Cheng, M.M., A. Wang, Z.Q. Liu, A.R. Gendall, S. Rochfort, and C.X. Tang. 2018. Sodium chloride decreases cadmium accumulation and changes the response of metabolites to cadmium stress in halophyte Carpobrotus rossii. *Ann. Bot*. doi.org/10.1093/aob/mcy077.

Chi, S., Y. Qin, W. Xu, Y. Chai, D. Feng, Y. Li, T. Li, M. Yang, and Z. He. 2018. Differences of Cd uptake and expression of OAS and IRT genes in two varieties of ryegrasses. *Environ. Sci. Poll. Res*. doi.org/10.1007/s11356-018-2509-x.

Chlopecka, A., J.R. Bacon, M.J. Wilson, and J. Kay. 1996. Forms of cadmium, lead, and zinc in contaminated soils from southwest Poland. *J. Environ. Qual*. 25:69–79.

Clemens, S., and J.F. Ma. 2016. Toxic heavy metal and metalloid accumulation in crop plants and foods. *Ann. Rev. Plant Biol*. 67:489–512.

Comino, E., A. Fiorucci, S. Menegatti, and C. Marocco. 2009. Preliminary test of arsenic and mercury uptake by *Poa annua*. *Ecol. Eng*. 35:343–350.

Cooper, R.J., and L.A. Spokas. 1991. Growth, quality and foliar iron concentration of Kentucky blue grass treated with chelated iron source. *J. Am. Soc. Hort. Sci*. 116:798–801.

Cordukes, W.E., and E.V. Parups. 1971. Chloride uptake by various turfgrass species and cultivars. *Canad. J. Plant Sci*. 51(6):485–490.

Cox, D.D., N.A. Slaton, W.J. Ross, and T.L. Roberts. 2018. Trifoliolate leaflet chloride concentrations for characterizing soybean yield loss from chloride toxicity. *Agron. J*. 110:1589–1599.

Daniel, W.H. 1955a. *Poa annua* control with arsenic materials. Golf Course Rep. 23(1):5–8.

Daniel, W.H. 1955b. Arsenic control of *Poa annua* points to fertilizing study. *Golfdom*. 29(4):70–76.

de Lespinay, A., H. Lequeux, B. Lambillotte, and S. Lutts. 2010. Protein synthesis is differentially required for germination in *Poa pratensis* and *Trifolium repens* in the absence or in the presence of cadmium. *Plant Growth Regul*. 61:205–214.

Deal, E.E., and R.E. Engel. 1965. Iron, manganese, boron and zinc: effects on growth of Merion Kentucky bluegrass1. *Agron. J*. 57:553–555.

Dhillon, K.S., and S.K. Dhillon. 2003. Distribution and management of seleniferous soils. Adv. Agron. 79:119–184.

Dong, Y., Y.G. Zhu, F.A. Smith, Y. Wang, and B.D. Chen. 2008. Arbuscular mycorrhiza enhance arsenic resistance of both white clover (*Trifolium repens*, Linn) and ryegrass (*Lolium perenne* L.) plants in an arseniccontaining soil. *Environ. Pollut*. 155:174–181.

Dudeck, A.E. 1985. Heavy metal effects on bermudagrass and St. Augustinegrass. Ann. *Meet. Abstr*. 115.

Duo, L., Y. Gao, and S. Zhao. 2005. Heavy metal accumulation and ecological responses of turfgrass to rubbish compost with EDTA addition. *J. Integr. Plant Biol*. 47(9):1047–1054.

Ebrahimi, M., and F.M. Díaz. 2014. Use *of Festuca ovina* L. in chelate assisted phytoextraction of copper contaminated soils. *J. Rangel. Sci*. 4(3):171–182.

Fard, K.G., M. Ghasemnezhad, H. Zakizadeh, M. Kafi, and F. Rejali. 2016a. Evaluation of six cold-season turfgrasses responses to lead phytotoxicity for screening tolerant species. *Caspian J. Environ. Sci*. 14(3):215–226.

Fard, K.G., H. Zakizadeh, M. Ghasemnezhad, M. Kafi, and F. Rejali. 2016b. Response of six turf species to cadmium stress: displaying the most tolerant cultivar. *Vegetos*. 29:3.

Faucon, M.P., M.N. Shutcha, and P. Meerts. 2007. Revisiting copper and cobalt concentrations in supposed hyperaccumulators from SC Africa: influence of washing and metal concentrations in soil. *Plant Soil*. 301:29–36.

Fei, L., P. Xu, Q. Dong, Q. Mo, and Z. Wang. 2018. Young leaf protection from cadmium accumulation and regulation of nitrilotriacetic acid in tall fescue (*Festuca arundinacea*) and Kentucky bluegrass (*Poa pratensis*). *Chemosphere*. 212:124–132.

Feng, M., J.E. Schrlau, R. Snyder, G.H. Snyder, M. Chen, J.L. Cisar, and Y. Cai. 2005. Arsenic transport and transformation associated with MSMA application on a golf course green. *J. Agric. Food Chem*. 53:3556–3562.

Fordyce, F. 2005. Selenium deficiency and toxicity in the environment. *In*: [O Selinus, B Alloway, J Centeno, R Finkelman, R Fuge, U Lindh and P Smedley, editors], pp. 373–415, *Essentials of Medical Geology*. London, Elsevier.

Foy C.D., and J.J. Murray. 1998a. Developing aluminum-tolerant strains of tall fescue for acid soils. *J. Plant Nutr*. 21:1301–1325.

Foy, C.D., and J.J. Murray. 1998b. Responses of Kentucky bluegrass cultivars to excess aluminum in nutrient solutions. *J. Plant Nutr.* 21:1967–1983.

Frank, K.W., and E.A. Guertal. 2013a. Nitrogen research in turfgrass. In: [JC Stier, BP Horgan, and SA Bonos, editors], pp. 457–492, *Turfgrass: Biology, Use, and Management.* ASA, SSSA, CSSA, Madison, WI.

Frank, K.W., and E.A. Guertal. 2013b. Potassium and phosphorus research in turfgrass. In: [JC Stier, BP Horgan, and SA Bonos, editors], pp. 493–520, *Turfgrass: Biology, Use, and Management.* Agron. Monogr. 56. ASA, Madison, WI.

Freisinger, E. 2011. Structural features specific to plant metallothioneins. *J. Biol. Inorg. Chem.* 16:1035–1045.

Gadepalle, V.P., S.K. Ouki, R. Van Herwijnen, and T. Hutchings. 2008. Effects of amended compost on mobility and uptake of arsenic by rye grass in contaminated soil. *Chemosphere.* 72:1056–1061.

Gao, Y., D. Li, and Y. Chen. 2012. Differentiation of carbonate, chloride, and sulfate salinity responses in tall fescue. *Scientia Horticulturae.* 139:1–7.

Garland, C.J., and D.A. Wilkins. 1981. Effect of calcium on the uptake and toxicity of lead in *Hordeum vulgare* L. and *Festuca ovina* L. *New Phytologist.* 87(3):581–593.

Gaskin, J.W., R.B. Brobst, W.P. Miller, and E.W. Tollner. 2003. Long-term biosolids application effects on metal concentrations in soil and bermudagrass forage. *J. Environ. Qual.* 32:146–152.

Gilmour, J.T., and M.S. Miller. 1973. Fate of a mercuric-mercurous chloride fungicide added to turfgrass. *J. Environ. Qual.* 2:145–148.

Gladkov, E.A., O.N. Gladkova, and L.S. Glushetskaya. 2011. Estimation of heavy metal resistance in the second generation of creeping bentgrass (*Agrostis solonifera*) obtained by cell selection for resistance to these contaminants and the ability of this plant to accumulate heavy metals. *Appl. Biochem. Microbiol.* 47(8):776–779.

Gregory, R.P.G., and A.D. Bradshaw. 1964. Heavy metal tolerance in populations of *Agrostis tenuis* sibth. and other grasses. *New Phytol.* 64(1):131–143.

Gunawardana, B., N. Singhal, and A. Johnson. 2010. Amendments and their combined application for enhanced copper, cadmium, lead uptake by *Lolium perenne*. *Plant Soil*. 329(1–2):283–294.

Hall, J.L. 2002. Cellular mechanisms for heavy metal detoxification and tolerance. *J. Exp. Bot.* 53(366):1–11.

Harivandi, M.A., K.B. Marcum, and Y. Qian. 2008. Recycled, gray, and saline water irrigation for turfgrass. In: [JB Beard and MP Kenna, editors], pp. 243–257, *Water Quality and Quantity Issues for Turfgrasses in Urbanlandscapes.* Council for Agricultural Science and Technology, Ames, IA.

Hartikainen, H., T. Xue, and V. Piironen. 2000. Selenium as an anti-oxidant and pro-oxidant in ryegrass. *Plant Soil.* 225(1–2):193–200.

Hartley, W., and N.W. Lepp. 2008. Effect of in situ soil amendments on arsenic uptake in successive harvests of ryegrass (*Lolium perenne* cv Elka) grown in amended As-polluted soils. *Environ. Pollut.* 156:1030–1040.

Hasan, M.K., Y. Cheng, M.K. Kanwar, X.Y. Chu, G.J. Ahammed, and Z.Y. Qi. 2017. Responses of plant proteins to heavy metal stress-a review. *Front. Plant Sci.* 8:1492.

Hasanuzzaman, M., K. Nahar, and M. Fujita. 2018. *Mechanisms of Arsenic Toxicity and Tolerance in Plants.* Singapore: Springer Singapore.

He, S.Y. Q.L. Wu, and Z.L. He. 2013. Effect of DA-6 and EDTA alone or in combination on uptake, subcellular distribution and chemical form of Pb in *Lolium perenne*. *Chemosphere.* 93:2782–2788.

Heckman, J.R., B.B. Clarke, and J.A. Murphy. 2003. Optimizing manganese fertilization for the suppression of take-all patch disease on creeping bentgrass. *Crop Sci.* 43:1395–1398.

Hu, J.L., S.C. Wu, F.Y. Wu, H.M. Leung, X.G. Lin, and M.H. Wong. 2013. Arbuscular mycorrhizal fungi enhance both absorption and stabilization of Cd by Alfred stonecrop (*Sedum alfredii* Hance) and perennial ryegrass (*Lolium perenne* L.) in a Cd-contaminated acidic soil. *Chemosphere.* 93:1359–1365.

Huang, B., M. DaCosta, and Y. Jiang. 2014. Research advances in mechanisms of turfgrass tolerance to abiotic stress: from physiology to molecular biology. *Crit. Rev. Plant Sci.* 33:141–189.

Huang, C., G. Liu, and C. Bai. 2017. Evaluation of aluminum resistance among Zoysia Willd. accessions from China. *Hortsci.* 52(2):225–229.

Huang, M., H. Zhu, J. Zhang, D. Tang, X. Han, L. Chen, D. Du, J. Yao, K. Chen, and J. Sun. 2017. Toxic effects of cadmium on tall fescue and different responses of the photosynthetic activities in the photosystem electron donor and acceptor sides. *Sci. Rep.* 7(14387):1–10.

Huang, M., H. Ai, X. Xu, K. Chen, H. Niu, H. Zhu, J. Sun, D. Du, and L. Chen. 2018. Nitric oxide alleviates toxicity of hexavalent chromium on tall fescue and improves performance of photosystem II. *Ecotoxicol. Environ. Saf.* 164:32–40.

Hufnagel, B., C.T. Guimaraes, E.J. Craft, J.E. Shaff, R.E. Schaffert, L.V. Kochian, and J.V. Magalhaes. 2018. Exploiting sorghum genetic diversity for enhanced aluminum tolerance: allele mining based on the AltSB locus. *Sci. Rep.* 8:10094.

Hull, R J. 2002a. Recent research offers clues to boron's purpose. *TurfGrass Trends.* 11(3):11–16.

Hull, R.J. 1999. Iron usage by turfgrasses. *TurfGrass Trends.* 8(2):1–10.

Hull, R.J. 2001a. Manganese usage by turfgrasses. *TurfGrass Trends.* 10(9):6–13.

Hull, R.J. 2001b. Zinc usage by turfgrasses. *TurfGrass Trends.* 10(7):7–11.

Hull, R.J. 2002b. Copper is essential bridge for many plant functions. *TurfGrass Trends.* 11(9):14–16.

Hull, R.J. 2003a. How do turfgrasses use nickel? *TurfGrass Trends.* 52:54–55.

Hull, R.J. 2003b. Micronutrient management: Small amounts of molybdenum streamlines nitrogen uptake. *TurfGrass Trends.* 42–44.

Hull, R.J. 2003c. Ubiquitous chlorine performs vital tasks in turf. *TurfGrass Trends.* 58–62.

Jackson, M.B., C.W. Lee, M.A. Schumacher, M.E. Duysen, J.R. Self, and R.C. Smith. 1995. Micronutrient toxicity in buffalograss. *J. Plant Nutr.* 18(6):1337–1349.

Jarvis, S.C., and L.H.P. Jones. 1978. Uptake and transport of cadmium by perennial ryegrass from flowing solution culture with a constant concentration of cadmium. *Plant Soil.* 49(2):333–342.

Jarvis, S.C., L.H.P. Jones, and M.J. Hopper. 1976. Cadmium uptake from solution by plants and its transport from roots to shoots. *Plant Soil.* 44(1):179–191.

Jin, J., Y. Xu, and Y. Huang. 2010. Protective effect of nitric oxide against arsenic-induced oxidative damage in tall fescue leaves. *Afr. J. Biotechnol.* 9(11):1619–1627.

Jones, L.H.P., C.R. Clement, and M.J. Hopper. 1973a. Lead uptake from solution by perennial ryegrass and its transport from roots to shoots. *Plant Soil.* 38:403–414.

Jones, L.H.P., S.C. Jarvis, and D.W. Cowling. 1973b. Lead uptake from soils by perennial ryegrass and its relation to the supply of an essential element (sulphur). *Plant Soil.* 38:605–619.

Jowett, D. 1958. Populations of *Agrostis* spp. tolerant of heavy metals. *Nature.* 182:816–817.

Kabeya, F.I., P. Pongra, B. Lange, M.P. Faucon, J.T. van Elteren, M. Šala, V.S. Šelih, E.V. Eeckhoudt, and. N. Verbruggen. 2018. Tolerance and accumulation of cobalt in three species of Haumaniastrum and the influence of copper. *Environ. Exp. Bot.* 149:27–33.

Karami, N., R. Clemente, E. Moreno-Jiménez, N.W. Lepp, and L. Beesley. 2011. Efficiency of green waste compost and biochar soil amendments for reducing lead and copper mobility and uptake to ryegrass. *J. Hazardous Mater.* 191(1–3):41–48.

King L.D. 1981. Effects of swine manure lagoon sludge and municipal sewage sludge on growth, nitrogen recovery and heavy metal content of fescue grass. *J. Environ. Qual.* 10:465–472.

Kochian, L.V., M.A. Piñeros, L. Jiping, and J.V. Magalhaes. 2015. Plant adaptation to acid soils: the molecular basis for crop aluminum resistance. *Ann. Rev. Plant Biol.* 66:571–598.

Kramer, U. 2010. Metal hyperaccumulation in plants. *Ann. Rev. Plant Biol.* 61:517–534.

Lasat, M.M. 2002. Phytoextraction of toxic metals. *J. Environ. Qual.* 31:109–120.

Lee, C.W., M.B. Jackson, M.E. Duysen, T.P. Freeman, and J.R. Self. 1996. Induced micronutrient toxicity in "Touchdown" Kentucky bluegrass. *Crop Sci.* 36:705–712.

Lee, S.H., N. Ahsana, K.W. Lee, D.H. Kim, D.G. Lee, S.S. Kwak, S.Y. Kwon, T.H. Kim, and B.H. Lee. 2007. Simultaneous overexpression of both CuZn superoxide dismutase and ascorbate peroxidase in transgenic tall fescue plants confers increased tolerance to a wide range of abiotic stresses. *J. Plant Physiol.* 164:1626–1638.

Li, T., W.H. Xu, Y.R. Chai, X.B. Zhou, Z.Y. Wang, and D.T. Xie. 2017. Differences of Cd uptake and expression of Cd-tolerance related genes in two varieties of ryegrasses. *Bulg. Chem. Commun.* 49(3):697–705.

Li, X., H. Cen, Y. Chen, S. Xu, L. Peng, H. Zhu, and Y. Li. 2016. physiological analyses indicate superoxide dismutase, catalase, and phytochelatins play important roles in Pb tolerance in *Eremochloa Ophiuroides*. *Int. J. Phytoremediation.* 18(3):251–260.

Li, Y., R.L. Chaney, G. Siebielec, and B.A. Kerschner. 2010. Response of four turfgrass cultivars to limestone and biosolids-compost amendment of a zinc and cadmium contaminated soil at Palmerton, Pennsylvania. *J. Environ. Qual.* 29:1440–1447.

Liao, M., and C. Huang. 2002. Effects of organic acids on the toxicity of cadmium during ryegrass growth. *J. Appl. Ecol.* 13:109–112.

Liu, H. 2001. Soil acidity and aluminum toxicity response in turfgrass. *Int. Turfgrass Soc. Res. J.* 9:180–188.

Liu, H. 2005. Aluminum toxicity of seeded bermudagrass cultivars. *HortSci.* 40: 221–223.

Liu, H. 2018. Nickel and turfgrasses. ASA, CSSA, and SSSA International Annual Meetings. p. 112253. Baltimore, MD.

Liu, H., J. Han, H. Liu, and X. Kang. 2012. Influence of lead gradient stress on the physiological and biochemical characteristics of perennial ryegrass (*Lolium perenne*) seedlings. *Acta Pratacult. Sin.* 21:57–63.

Liu, H., J.R. Heckman, and J.A. Murphy. 1995. Screening Kentucky bluegrass for aluminum tolerance. *J. Plant Nutri.* 18:1797–1814.

Liu, H., J.R. Heckman, and J.A. Murphy. 1996. Screening fine fescues for aluminum tolerance. *J. Plant Nutri.* 19:677–688.

Liu, H., J.R. Heckman, and J.A. Murphy. 1997a. Greenhouse screening of turfgrasses for aluminum tolerance. *Int. Turfgrass Soc. Res. J.* 8:719–728.

Liu, H., J.R. Heckman, and J.A. Murphy. 1997b. Aluminum tolerance among genotypes of Agrostis species. *Int. Turfgrass Soc. Res. J.* 8:729–734.

Liu, Y., K. Wang, P. Xu, and Z. Wang. 2012. Physiological responses and tolerance threshold to cadmium contamination in *Eremochloa ophiuroides*. *Int. J. Phytoremediation.* 14:(5) 467–480.

Lou, Y., H. Luo, T. Hu, H. Li, and J. Fu. 2013. Toxic effects, uptake, and translocation of Cd and Pb in perennial ryegrass. *Ecotoxicology* 22:207–214.

Luo, H., C. He, and L. Han. 2018. Heterologous expression of ZjOMT from *Zoysia japonica* in Escherichia coli confers aluminum resistance through melatonin production. *PLoS ONE.* 13(5): e0196952.

Ma, L.Q., K.M. Komar, C. Tu, W. Zhang, and Y. Cai. 2001. A fern that hyperaccumulates arsenic. *Nature.* 409:579.

Mahoney, D.J., T.W. Gannon, M.D. Jeffries, A.R. Matteson, and M.L. Polizzotto. 2015a. Management considerations to minimize environmental impacts of arsenic following monosodium methylarsenate (MSMA) applications to turfgrass. *J. Environ. Manage.* 150:444–450.

Mahoney, D.J., T.W. Gannon, M.D. Jeffries, and M.L. Polizzotto. 2015b. Arsenic distribution and speciation in a managed turfgrass system following a monosodium methylarsenate (MSMA) application. *Crop Sci.* 55:2877–2885.

Malinowski, D.P., and D.P. Belesky. 1999. Tall fescue aluminum tolerance is affected by neotyphodium coenophialum endophyte. *J. Plant Nutr.* 22:(8)1335–1349.

Mancino, C.F. 1992. Irrigation of turfgrass with secondary sewage effluent: soil quality. *Agron. J.* 84:650–654.

Mandal, A., T.J. Purakayastha, A.K. Patra, and B. Sarkar. 2017. Arsenic phytoextraction *by Pteris vittata* improves microbial properties in contaminated soil under various phosphate fertilizations. *Appl. Geochem.* 10.1016/j.apgeochem.2017.04.008.

Marcum, K.B., and C.L. Murdoch. 1994. Salinity tolerance mechanisms of six C4 turfgrasses. *J. Am. Soc. Hortic. Sci.* 119:779–784.

Marcum, K.B. 1999. Salinity tolerance mechanisms of grasses in the subfamily Chloridoideae. *Crop Sci.* 39:1153–1160.

Marcum, K.B. 2001. Salinity tolerance of 35 bentgrass cultivars. *HortSci.* 36(2):374–376.

Marcum, K.B., and C.L. Murdoch. 1990. Salt glands in the Zoysieae. *Ann. Bot.* 66:1–7.

Marcum, K.B., and M. Pessarakli. 2006. Salinity tolerance and salt gland excretion efficiency of bermudagrass turf cultivars. *Crop Sci.* 46:2571–2574.

Marcum, K.B., S.J. Anderson, and M.C. Engelke. 1998. Salt gland ion secretion: a salinity tolerance mechanism among five zoysiagrass species. *Crop Sci.* 38:806–810.

Marschner, H. 2012. *Marschner's Mineral Nutrition of Higher Plants.* 3rd ed. Academic Press, London.

Mashhadi, M., A. Boojar, and Z. Tavakkoli. 2011. New molybdenum-hyperaccumulator among plant species growing on molybdenum mine- a biochemical study on tolerance mechanism against metal toxicity. *J. Plant Nutr.* 34:10:1532–1557.

Matteson, A.R., T.W. Gannon, M.D. Jeffries, S. Haines, D.F. Lewis, and M.L. Polizzotto. 2014. Arsenic retention in foliage and soil after monosodium methyl arsenate (msma) application to turfgrass. *J. Environ. Qual.* 43:379–388.

Matthews-Amune, O.C., and S. Kakulu. 2012. Determination of heavy metals in forage grasses (carpet grass (*Axonopus Ompressus*), guinea grass (*Panicum Maximum*) and elephant grass (*Pennisetum Purpureum*)) in the Vicinity of Itakpe Iron Ore Mine, Nigeria. *Int. J. Pure Appl. Sci. Technol.* 13(2):16–25.

McCarty, L.B. 2018. *Golf Turf Management*. CRC Press. Taylor & Francis Group, Boca Raton, FL.

McCarty, L.B., J.R. Gann, C.E. Wells, and P.D. Gerard. 2014. Creeping bentgrass field response to pigment-containing products. *Agron. J.* 106(4):1533–1539.

McCartney, D., Y. Zhang, and C. Grant. 2001. Characterization of compost produced at a golf course: Impact of historic mercury accumulations in putting green soil. *Compost Sci. Util.* 9(1):73–91.

McGrath, S.P. 1994. Effects of heavy metals from sewage sludge on soil microbes in agricultural ecosystems. In: [SM Ross, editor], pp. 242–274, *Toxic Metals in Soil-Plant Systems*. John Wiley, Chichester.

Meena, M.D., R.K. Yadav, B. Narjary, G. Yadav, H.S. Jat, P. Sheoran, M.K. Meena, R.S. Antil, B.L. Meena, H.V. Singh, V.S. Meena, P.K. Rai, A. Ghosh, and P.C. Moharana. 2019. Municipal solid waste (msw): strategies to improve salt affected soil sustainability: a review. *Waste Manage.* 84:38–53.

Meharg, A.A., J. Naylor, and M.R. Macnair. 1994. Phosphorus nutrition of arsenate-tolerant and nontolerant phenotypes of velvet-grass. *J. Environ. Qual.* 23:234–238.

Menchyk, N. 2012. Foliar applied urea nitrogen metabolism in warm-season turfgrass under salinity stress. PhD Dissertation, Clemson University, Clemson, SC.

Mora, M., A. Rosas, A. Ribera, and Z. Rengel. 2009. Differential tolerance to Mn toxicity in perennial ryegrass genotypes: involvement of antioxidative enzymes and root exudation of carboxylates. *Plant Soil.* 320 (1–2):79–89.

Munns, R., and M. Tester. 2008. Mechanisms of salinity tolerance. *Ann. Rev. Plant Biol.* 59:651–681.

Murray, J.J., and C.D. Foy. 1978. Differential tolerances of turfgrass cultivars to an acid soil high in exchangeable aluminum. *Agron. J.* 70:769–774.

Nable, R.O., G.S. Banuelos, and J.G. Paull. 1997. Boron toxicity. *Plant Soil.* 193:181–198,

Nagajyoti, P.C., K.D. Lee, and T.V.M. Sreekanth. 2010. Heavy metals, occurrence and toxicity for plants: a review. *Environ. Chem. Lett.* 8:199–216.

Nanayakkara, U.N., W. Uddin, and L.E. Datnoff. 2008. Application of silicon sources increases silicon accumulation in perennial ryegrass turf on two soil types. *Plant Soil.* 303(1–2):83–94.

Niazi, N.K., I. Bibi, A. Fatimah, M. Shahid, M.T. Javed, H. Wang, Y.S. Ok, S. Bashir, B. Murtaza, Z.A. Saqib, and M.B. Shakoor. 2017. Phosphate-assisted phytoremediation of arsenic by *Brassica napus* and *Brassica juncea*: morphological and physiological response *Int. J. Phytoremediation*.

Nus, J.L., N.E. Christians, and K.L. Diesburg. 1993. High phosphorus applications influence soil-available potassium and Kentucky bluegrass copper content. *HortSci.* 28:639–641.

Oertli, J.J., O.R. Lunt, and V.B. Youngner. 1961. Boron toxicity in several turfgrass species. *Agron. J.* 53(4):262–265.

Ozdemir, S., N.N. Nuhoglu, O.H. Dede, and K. Yetilmezsoy. 2018. Mitigation of soil loss from turfgrass cultivation by utilizing poultry abattoir sludge compost and biochar on low-organic matter soil. *Environ. Technol.* DOI.

Palazzo, A.J., T.J. Cary, S.E. Hardy, and C.R. Lee. 2003. Root growth and metal uptake in four grasses grown on zinc-contaminated soils. *J. Environ. Qual.* 32:834–840.

Patrick, C.R. 2018. Nickel Tolerance in Cool-Season Grasses. MS Thesis. Clemson University. Clemson, SC.

Pepper, I.L., and C.F. Mancino. 1993. Irrigation of turf with effluent water. In: [M. Pessarakli, editor], *Handbook of Plant and Crop Stress*. pp. 623–641, New York, NY: Marcel Dekker.

Pessarakli, M. 2014. *Handbook of Plant and Crop Physiology*, 3rd ed. Edited by M. Pessarakli (Revised and Expanded). Boca Raton, FL: CRC Press; Taylor & Francis Publishing Group.

Peterson, P.J. 1969. The distribution of zinc-65 in *Agrostis tenuis* Sibth. and *A. stolonifera* L. tissues. *J. Exp. Bot.* 20(4):863–875.

Putra, R.S., Y. Ohkawa, and S. Tanak. 2013. Application of EAPR system on the removal of lead from sandy soil and uptake by Kentucky bluegrass (*Poa pratensis* L.). *Sep. Purif. Technol.* 102:34–42.

Qian, Y.L., and B. Mecham. 2005. Long-term effects of recycled wastewater irrigation on soil chemical properties on golf course fairways. *Agron. J.* 97:717–721.

Qu, R.L., D. Li, R. Du, and R. Qu. 2003. Lead uptake by roots of four turfgrass species in hydroponic cultures. *HortSci.* 38(4):623–626.

Ravenscroft, P., H. Brammer, K. Richards. 2009. *Arsenic Pollution: A Global Synthesis*. RGS-IBG Book Series, Chichester, UK: Wiley-Blackwell.

Reeves, R.D., A. Van Der Ent, and A.J. Baker. 2018. Global distribution and ecology of hyperaccumulator plants. In pp. 75–92, *Agromining: Farming for Metals*. Springer.

Ren, A.Z., C. Li, and Y.B. Gao. 2011. Endophytic fungus improves growth and metal uptake of *Lolium arundinaceum* Darbyshire Ex. Schreb. *Int. J. Phytoremediation.* 13:233–243.

Salt, D.E., R.D. Smith, and I. Raskin. 1998. Phytoremediation. *Ann. Rev. Plant Physiol. Plant Mol. Biol.* 49:643–668.

Sarwar, N., M. Imran, M.R. Shaheen, W. Ishaque, M.A. Kamran, A. Matloob, A. Rehim, and S. Hussain. 2017. Phytoremediation strategies for soils contaminated with heavy metals: modifications and future perspectives. *Chemosphere.* 171:710–721.

Schiavon, M., and J.H. Baird. 2018. Evaluation of products to alleviate irrigation salinity stress on bermudagrass turf. *Agron. J.* 110:2136–2141.

Schuman, G. E. 1969. Boron tolerance of tall wheatgrass. *Agron. J.* 61(3):445–447.

Schumacher, M.A., M.B. Jackson, M.E. Duysen, R.C. Smith, and C.W. Lee. 1993. Micronutrient toxicity symptoms in buffalograss. *HortSci.* 28(5):551.

Shahid, M., C. Dumat, S. Khalid, E. Schreck, T. Xiong, and N.K. Niazie. 2017. Foliar heavy metal uptake, toxicity and detoxification in plants: a comparison of foliar and root metal uptake. *J. Hazardous Mater.* 325:36–58.

Shamima, S., and S. Sugiyama. 2008. Cadmium phytoextraction capacity in eight C3 herbage grass species. *Grassland Sci.* 54:27–32.

Shi, H., T. Ye, and Z. Chan. 2014. Nitric oxide-activated hydrogen sulfide is essential for cadmium stress response in bermudagrass (*Cynodon dactylon* (L) Pers). *Plant Physiol. Biochem.* 74:99–107.

Shiyab, S., J. Chen, F.X. Han, D.L. Monts, F.B. Matta, M. Gu, and Y. Su. 2009. Phytotoxicity of mercury in Indian mustard (*Brassica juncea* L.) *Ecotoxicol. Environ. Saf.* 72:619–625.

Shu, W.S., Z.H. Ye, C.Y. Lan, Z.Q. Zhang, and M.H. Wong. 2002. Lead, zinc and copper accumulation and tolerance in populations of *Paspalum distichum* and *Cynodon dactylon*. *Environ. Pollut.* 120(2):445–453.

Silva, C.M.S., C. Zhang, G. Habermann, E. Delhaize, and P.R. Ryan. 2018. Does the major aluminium-resistance gene in wheat, TaALMT1, also confer tolerance to alkaline soils? *Plant Soil.* 424:451–462.

Singh, S., D.K. Tripathi, S. Singh, S. Sharma, N.K. Dubey, D.K. Chauhan, D. K. Chauhan, and M. Vaculík. 2017. Toxicity of aluminium on various levels of plant cells and organism: a review. *Environ. Exp. Bot.* 137:177–193.

Smith, G.S., and J.H. Watkinson. 1984. Selenium toxicity in perennial ryegrass and white clover. *New Phytol.* 97(4):557–564.

Soleimani, M., M.A. Hajabbasi, M. Afyuni, A.H. Charkhabi, and H. Shariatmadari. 2009. Bioaccumulation of nickel and lead by bermuda grass (*Cynodon dactylon*) and tall fescue (*Festuca arundinacea*) from two contaminated soils. *Caspian J. Environ. Sci.* 7(2):59–70.

Spear, G.T., and N.E. Christians. 1991. Creeping bentgrass response to zinc in modified soil. *Commun. Soil Sci. Plant Anal.* 22:19–20.

StJohn, R. N. Christians, H. Liu, and N. Menchyk. 2013. Secondary nutrients and micronutrient fertilization. In: *Turfgrass Monograph 56 Crop Science Society of America.* pp. 521–542, Madison, WI: Crop Science Society of America.

Taghizadeh, M., and E. Solgi. 2017. Impact of heavy metal stress on in vitro seed germination and seedling growth indices of two turfgrass species. *J. Rangeland Sci.* 7(3):220–231.

Takacs, T., B. Biro, and I. Voros. 2001. Arbuscular mycorrhizal effect on heavy metal uptake of ryegrass (*Lolium perenne* L.) in pot culture with polluted soil. In: [WJ Horst, editor], *Plant Nutrition, Food Security and Sustainability of Agro-ecosystems*, pp. 480–481. The Netherlands: Kluwer Academic Publishers.

Taylor, J.W. 2019. Two warm-season turfgrass responses to different levels of shade and iron. MS Thesis. Clemosn University, Clemson, SC.

Temmerman, L.D. N. Claeys, E.Roekens, and M. Guns. 2007. Biomonitoring of airborne mercury with perennial ryegrass cultures. *Environ. Pollut.* 146:458–462.

Thornton, I. 1981. Geochemical aspects of the distribution and forms of heavy metals in soils. In [NW Lepp, editor], *Effect of Heavy Hetal Pollution on Plants: Metals in the Environment*, pp. 1–34, vol. II. London and New Jersey: Applied Science Publishers.

Tran, T.V., S. Fukai, H.E. Giles, and C.J. Lambrides. 2018. Salinity tolerance among a large range of bermudagrasses (*Cynodon* spp.) relative to other halophytic and non-halophytic perennial C4 grasses. *Environ. Exp. Bot.* 145:121–129.

Tripathi, D.K., S. Singh, V.P. Singh, S.M. Prasad, N.K. Dubey, and D.K. Chauhan. 2017. Silicon nanoparticles more effectively alleviated UV-B stress than silicon in wheat (*Triticum aestivum*) seedlings. *Plant Physiol. Biochem.* 110:70–81.

Turgeon, A.J. 2012. *Turfgrass Management.* 9th ed. Upper Saddle River, NJ: Prentice-Hall Inc.

Turner, R.G., and C. Marshall. 1972. The accumulation of zinc by subcellular fractions of roots of *Agrostis tenuis* Sibth. in relation to zinc tolerance. *New Phytol.* 71:671–676.

United States Golf Association, and American Society of Golf Course Architects. 1994. *Wastewater Reuse for Golf Course Irrigation.* Boca Raton, FL: Lewis Publishers.

Valle, G., L.R. McDowell, N.S. Wilkinson, and D. Wright. 1993. Selenium concentration of bermudagrass after spraying with sodium selenate. *Commun. Soil. Sci. Plant Anal.* 24(13–14):1763–1768.

Valle, G., L.R. McDowell, D.L. Prichard, P.J. Chenoweth, D.L. Wright, F.G. Martin, W.E. Kunkle, and N.S. Wilkinson. 2002. Selenium concentration of fescue and bahiagrasses after applying a selenium fertilizer. *Commun. Soil. Sci. Plant Anal.* 33(9–10):1461–1472.

Verkleij, J., and H. Schat. 1990. Mechanisms of metal tolerance in higher plants. In: [AJ Shaw, editor], pp. 179–194, *Heavy Metal Tolerance in Plants: Evolutionary Aspects.* Boca Raton, FL: CRC Press.

Vithanage, W., I. Herath, S. Joseph, J. Bundschuh, N. Bolan, Y.S. Ok, M.B. Kirkham, and J. Rinklebe. 2017. Interaction of arsenic with biochar in soil and water: a critical review. *Carbon.* 113:219–230.

Wainwright, S.J., and H.W. Woolhouse. 1977. Some Physiological aspects of copper and zinc tolerance in *Agrostis tenuis* Sibth.: cell elongation and membrane damage. *J. Exp. Bot.* 28(4):1029–1036.

Wang, D.S., R.W. Weaver, and J.R. Melton. 1973. As and Hg concentrations in common bermuda grass grown on soils amended with As and Hg. Ann. *Meet. Abstr.* 65:189.

Wang, K. 2010. The stress responses and tolerance thresholds to soil lead, cadmium and zinc contamination in centipedegrass and seashore paspalum. Master's Degree Thesis, Shanghai Jiaotong University, Shanghai, China.

Wang, Y., D. Meng, L. Fei, Q. Dong, and Z. Wang. 2019. A novel phytoextraction strategy based on harvesting the dead leaves: cadmium distribution and chelator regulations among leaves of tall fescue. *Sci. Total Environ.* 650:3041–3047.

Wang, Z., L. Li, and Y. Xuejun. 2010. Research progress on lead poisoning of turfgrasses. *Grassland Turf.* 30(2):8–15.

Weaver, R.W., J.R. Melton, D. Wang, and R.L. Duble. 1984. Uptake of arsenic and mercury from soil by bermudagrass *Cynodon dactylon. Environ. Pollut. Series A Ecolo. Biol.* 33(2):133–142.

Wong, M.H. 1982. Metal cotolerance to copper, lead and zinc in *Festuca rubra. Environ. Res.* 29:42–47.

Wong, M.H., and A.D. Bradshaw. 1982. A comparison of the toxicity of heavy metals, using root elongation of rye grass, Lolium perenne. *New Phytol.* 91:255–261.

Wong, M.H., and W.M. Lau. 1985. Toxic effects of iron tailings and the response of watercress from tailings at high concentrations of Fe, Zn and Mn. *Environ. Pollut. A Ecol. Biol.* 38(2):129–140.

Wong, M.H., W.M. Lau, S.W. Li, and C.K. Tang. 1983. Root growth of two grass species on iron ore tailings at elevated levels of manganese, iron and copper. *Environ. Rese.* 30:26–33.

Wu, L. 1994. Selenium accumulation and colonization of plants in soils with elevated selenium and salinity. In: [WT Frankenberger and S Benson, editors], *Selenium in the Environment*, pp. 279–326. New York, NY: Marcel-Dekker.

Wu, L., and Z. Huang. 1991. Chloride and sulfate salinity effects on selenium accumulation by tall fescue. *Crop Sci.* 31:114–118.

Wu, L., Z. Huang, and R.G. Burau. 1988. Selenium accumulation and selenium-salt cotolerance in five grass species. *Crop Sci.* 28:517–522.

Wu, S.L., B.D. Chen, Y.Q. Sun, B.H. Ren, X. Zhang, and Y.S. Wang. 2014. Chromium resistance of dandelion (*Taraxacum platypecidum* Diels.) and bermudagrass (*Cynodon dactylon* (Linn.) Pers.) is enhanced by arbuscular mycorrhiza in Cr(VI) contaminated soils Environ. *Toxicol. Chem.* 33:2105–2113.

Xie, Y., L. Hu, Z. Du, X. Sun, E. Amombo, J. Fan, and J. Fu. 2014. Effects of cadmium exposure on growth and metabolic profile of bermudagrass [*Cynodon dactylon* (L.) Pers.]. *PLoS ONE.* 9(12): e115279.

Xu, C., T. Sibicky, and B. Huang. 2010. Protein profile analysis of salt-responsive proteins in leaves and roots in two cultivars of creeping bentgrass differing in salinity tolerance. *Plant Cell Rep.* 29:595–615.

Xu, P.X., and Z.L. Wang. 2013. Physiological mechanism of hypertolerance of cadmium in Kentucky bluegrass and tall fescue: chemical forms and tissue distribution. *Environ. Exp. Bot.* 96:35–42.

Xu, P.X., and Z.L. Wang. 2014. A comparison study in cadmium tolerance and accumulation in two cool-season turfgrasses and *Solanum nigrum* L. *Water Air Soil Pollut.* 225:1938.

Xu, X., and C.F. Mancino. 2001. Zinc requirements of annual bluegrass and creeping bentgrass. *HortSci.* 36:784–786.

Yamamoto, A., M. Hashiguchi, R. Akune, T. Masumoto, M. Muguerza, Y. Saeki and R. Akashi. 2016. The relationship between salt gland density and sodium accumulation/secretion in a wide selection from three zoysia species. *Aust. J. Bot.* 64(4):277–284.

Yan, J., J. Chen, T. Zhang, J. Liu, and H. Liu. 2009. Evaluation of aluminum tolerance and nutrient uptake of 50 centipedegrass accessions or cultivars. *HortSci.* 44(3):857–861.

Yang, J., J. McBride, J.X. Zhou, and Z.Y. Sun. 2005. The urban forest in Beijing and its role in air pollution reduction. *Urban For. Urban Green.* 3:65–78.

Yang, M.Y., Y.Y. Liang, D.B. Zen, C. Tao, and C. Hua. 2014. Effects of lead stress on accumulation capacity and physiological metabolism of ryegrass. *J. Northwest A & F Univ.* 42(12):97–101.

Yang, X., G. Wang, C. Zhang, and V. Römheld. 2000. Characteristics of iron uptake in Ni-tolerant mutant and wild-type of ryegrass. *J. Plant Nutr.* 23:1867–1876.

Yesilonis, I.D., B.R. James, R.V. Pouyat, and B. Momen. 2008. Lead forms in urban turfgrass and forest soils as related to organic matter content and pH. *Environ. Monit. Assess.* 146:1–17.

Yu, M., Y.J. Cheng, H.D. Xiao, H.F. Wang, F. Wang, and X.H. Sun. 2005. Influences of boron and molybdenum on active oxygen species in turfgrass seashore paspalum under low temperature. *Acta Agron. Sin.* 6:755–759.

Yuan, J., Y. Bai, Y. Chao, X. Sun, C. He, X. Liang, L. Xie, and L. Han. 2018. Genome-wide analysis reveals four key transcription factors associated with cadmium stress in creeping bentgrass (*Agrostis stolonifera* L.). *Peer J.* 6:e5191 https://doi.org/10.7717/peerj.5191.

Yust, A.K., D.J. Wehner, and T.W. Fermanian. 1984. Foliar application of N and Fe to Kentucky bluegrass. *Agron. J.* 76:934–938.

Zaurov, D.E., S. Bonos, J.A. Murphy, M. Richardson, and F.C. Belanger. 2001. Endophyte infection can contribute to aluminum tolerance in fine fescues. *Crop Sci.* 41:1981–1984.

Zenk, M.H. 1996. Heavy metal detoxification in higher plants-a review. *Gene.* 179:21–30.

Zhang, X.H., J. Liu, H.T. Huang, J. Chen, Y. Zhu, and D.Q. Wang. 2007. Chromium accumulation by the hyperaccumulator plant *Leersia hexandra* Swartz. *Chemosphere.* 67:1138–1143.

Zhao, C., J. Xu, Q. Li, S. Li, P. Wang, and F. Xiang. 2014. Cloning and characterization of a *Phragmites australis* phytochelatin synthase (PaPCS) and achieving Cd tolerance in tall fescue. *PLoS ONE.* 9(8):e103771.

Zhao, S., Q. Liu, Y. Qi, and L. Duo. 2010. Responses of root growth and protective enzymes to copper stress in turfgrass. *Acta Biologica Cracoviensia Series Botanica.* 52(2):7–11.

Zhao, S.L., F. Lian, and L. Duo. 2011. EDTA-assisted phytoextraction of heavy metals by turfgrass from municipal solid waste compost using permeable barriers and associated potential leaching risk. *Bioresour. Technol.* 102:621–626.

Zhao, S.L., and L.A. Duo. 2002. Initial growth effect and ecological threshold of *Festuca arundinacea* L., under progressive stress of Cu2+ and Zn2+. *Acta Ecol. Sin.* 22:1099–1105.

Zhao, S.L., L.N. Jia, and L. Duo. 2013. The use of a biodegradable chelator for enhanced phytoextraction of heavy metals by *Festuca arundinacea* from municipal solid waste compost and associated heavy metal leaching. *Bioresour. Technol.* 129:249–255.

Zhao, S.L., L.N. Jia, and L. Duo. 2016. Combining nitrilotriacetic acid and permeable barriers for enhanced phytoextraction of heavy metals from municipal solid waste compost by *Lolium perenne* and reduced metal leaching. *J. Environ. Qual.* 45:933–939.

Zhu, H., H. Ai, L. Cao, R. Sui, H. Ye, D. Du, J. Sun, J. Yao, K. Chen, and L. Chen. 2018. Transcriptome analysis providing novel insights for Cd-resistant tall fescue responses to Cd stress. *Ecotoxicol. Environ. Saf.* 160:349–356.

Zhu, T., D. Fu, and F. Yang. 2015. Effect of saponin on the phytoextraction of Pb, Cd and Zn from soil using Italian ryegrass. *Bull. Environ. Contam. Toxicol.* 94(1):129–133.

Zhuo, Y.Q., S. Qiu, E. Amombo, Q. Zhu, D.Tang, M. Huang, X. Han, L. Chen, S. Wang, K. Chen, and J. Sun. 2017. Nitric oxide alleviates cadmium toxicity in tall fescue photosystem II on the electron donor side. *Environ. Exp. Bot.* 137:110–118.

Section V

Plant and Crop Responses under Biotic Stress

27 How Crops Stress Weeds

Jack Dekker

CONTENTS

27.1 INTRODUCTION

There are numerous publications available on the effects of biotic and abiotic stresses on plants/crops. Some recent sample publications include Maghsoudi et al. (2018), Valizadeh-Kamran et al. (2018), Askari-Khorasgani et al. (2017), Gheidary et al. (2017), Ghouchani et al. (2017), Mardani et al. (2017), Marzban et al. (2017), Pessarakli 2015, 2017), Pessarakli et al. 2015, 2017), Sheikh Mohamadi et al. (2017), Khorsandy et al. (2016), Parker et al. (2016), Saeidnejad et al. (2016), Akhzari et al. (2015a, 2015b), Ashrafi et al. (2015), Lotfi et al. (2015), Izadi et al. (2014), Kadkhodaie et al. (2014), Nikbakht et al. (2014), Sanchez et al. (2014), Pirasteh-Anosheh et al. (2013), Emam et al. (2012), Shekoofa et al. (2012), Shobbar et al. (2012), Pessarakli and Kopec (2011), Heydari and Pessarakli (2010), and Marcum and Pessarakli (2010). While these stresses are imposed on plants, there are some counter-stresses by plants that impose stress on the stressors. Weeds are among the most common biotic stressors, which often impose stress on plants/crops; however, they are also among the most attacked stressors, both by the plants/crops and by humans, who fight either mechanically or by using agrochemicals (e.g., herbicides) to eradicate them and reduce the stresses on plants. Although numerous sources are available on the effects of the biotic and abiotic stresses on plants/crops and the fact that plants/crops also defend themselves by exerting stress on the stressors, the literature available on the counter-stress by plants/crops on weeds is rare or non-existent. Therefore, this chapter aims to provide information on this subject and fill this gap.

27.2 CROPS AND WEEDS INTERACT

Weed infestation and control are primary concerns in crop production management. Growers focus on weed stress to their crops to minimize yield losses and reduce future infestations. Crop management uses tactics as ways and means to mitigate these stresses. There exists an opportunity to view crop–weed interactions from an alternate point of view (Dekker, 2016b). The natural tendency of growers to see agronomic practice from the perspective of weed stress on crops can obscure the importance of stresses caused by crops on their neighboring weeds. A more complete understanding of the stresses crops exert on weeds could provide the basis for improved crop management tactics and strategies.

What are these opportunities?

27.3 WHAT IS CROP–WEED STRESS?

The nature of biology is mutual stress. These stresses take many forms. Physical stress occurs when strong physical pressure is applied to an object, deforming its shape or producing a strain. This physical action is accompanied by an opposing, resisting reaction by the object to the stress. Biological stress occurs with external aggression by a physical or chemical force, or infective agent, causing internal stress to an organism and resulting in a response attempting to restore the previous conditions. Stress to an organism is any factor that restricts growth and reproduction or causes a potentially adverse change or disturbance to its equilibrium. A stressor is a chemical or biological agent, environmental condition, external stimulus, or event that triggers the stress response in an organism. Disturbance is an interruption with direct or indirect spatial, temporal, biological, or abiological effects that alters or destroys a biological individual or community.

27.4 CROP–WEED INTERACTIONS

Crops and weeds can facilitate (cooperate, synergize) or interfere with each other. These interactions can lead to positive, negative, and neutral consequences to crops, their weedy neighbors, or both.

Antagonistic interactions occur when one species or plant benefits and the neighbor is harmed. Examples include the growth of a vine growing on top of a support plant (kudzu smothering a sorghum plant), parasitism (dodder (*Cuscuta* spp.) or *Orabanche* feeding on maize) or predation (grazing). Amensalistic interactions see one species unaffected while a neighbor is harmed. For example, allelopathic toxins produced by the unaffected species can poison or inhibit a neighbor plant to its detriment. Competition occurs when several interacting species are all harmed. These harmful interactions arise when limited opportunity space-time (resources, conditions, disturbance, neighbors) is shared.

Mutualism occurs when both species or plants benefit from the presence of each other. This can occur when neighbors exploit different opportunities at different times. Examples include legume–grass pastures and plant–pollinator interactions. Plants whose growth and development occur at different times can provide shelter and protection to each other over time: for example, a hairy vetch cover crop underlying a maize field. Commensalism arises when one species or plant is benefited by the other, which is unaffected. Plants growing in the same locality, but at different times of the year, provide an example of commensalism, as do epiphytes growing on a supporting tree (e.g., orchids). Neutralism is observed when two neighbors have little or no effect on each other.

27.5   CROPS AND WEEDS ARE A FAMILY

The origin of weeds and crops is the same: wild progenitor species are selected by humans for their use as food, fiber, fuel, and drugs. The most common pattern for the origins of agricultural plants is the inter-fertile wild-crop–weed complex, in which both crop and weed were derived from the same wild plant relative. Extensive genetic interaction between wild, weed, and cultivated phenotypes provides an extensive genetic reservoir metapopulation available to all inter-fertile species of the complex. This genetic condition of pervasive promiscuity exists whether human managers like it or not. Continuous introgression between members of these plant complexes often completely masks the wild forms; only crop–weed complexes remain.

27.6   LIFE HISTORY OF CROP-WEED INTERACTIONS

Interactions between crops and weeds in agriculture occur throughout their life histories. A plant's life history is the significant features of the life cycle through which an organism passes, with particular reference to phenotypic traits and strategies influencing survival and reproduction. Life history is how long a plant typically lives, how long it usually takes to reach reproductive size, how often it reproduces, and a number of other attributes that have demographic and fitness consequences.

Crop and weed plant life histories correspond to the seasonal events of crop production. The establishment of a new crop in a field often begins with various tillage operations (e.g., plowing, harrowing, and hand weeding), which eliminate weeds that might subsequently interfere with the crop being grown. Tillage is a stressor that sets ecological succession in that field back to its starting point. Herbicides applied just before or after planting are weed stressors. Planting crop seeds establishes a crop in a locality. The choice of crop species, including the use of herbicide-resistant crop varieties, also defines the type and amount of interaction with local weeds that will follow during the growing season. Important phenotypic traits of both crop and weed will define that mutual struggle. The time of planting will have considerable influence on the subsequent interactions and relative success of both types of plants: seedling emergence time, vigor, and number. Plant growth and development during the season determine the relative success of both crops and weeds: shoot and leaf growth for shading and solar exploitation, root infiltration, and exploitation of soil resources and conditions. Crop harvesting eliminates crop plants, giving a useful yield, and destroys lingering weed infestations for future cropping. Postharvest tillage is a stressor setting ecological succession back to zero.

Cooperative or interfering interactions between crops and weeds are revealed over time as the growing season progresses. Observing these interactions as crops develop in a field highlights the stresses crops can induce in weeds to their advantage.

27.7   HOW DO WEEDS STRESS CROPS?

Weeds seize locally available opportunity space-time created by crop production. Weeds react to crop stress by gradual evolution of phenotypic traits in aid of improved competitive ability (e.g., quick and numerous seedling emergence), avoidance (e.g., weed seed dormancy), synergy (e.g., exploiting opportunities unused or neglected by crops), and the introgression of new favorable traits drawn from the wild-crop–weed gene pool.

Weeds have evolved over millennia due to the stresses imposed on them by crops and human management practices. Weed evolution is the natural selection of phenotypic traits that allow individual plants to exploit opportunity space-time: un- or under-used resources (e.g., light, water, nutrients, and gases), pervasive conditions (e.g., heat and terroir), disturbance (e.g., herbicides, tillage, and frozen soil), and neighbor organisms (e.g., crops, other weeds, diseases, and insects) in a locality at a particular time. Weeds evolve clever traits to avoid death and ensure reproduction when confronted by crops and their human sponsors. The challenge to agriculture is to embrace natural selection realistically and exploit it to the detriment of weed populations. Unlike weeds, which must find their way through nature on their own, crops are guided by human intervention. It is therefore through human intervention that stresses on weeds are created and enhanced.

27.8   HOW DO CROPS STRESS WEEDS?

Crops seize locally available opportunity space-time created by human managers and inherent conditions and resources in the cropping field. Crops stress weeds by direct interactions (e.g., shading), by human activity modulating the environment (e.g., tillage), by cropping practices exploiting weed life-cycle weaknesses (e.g., crop variety selection), and by other sophisticated agricultural technologies (e.g., herbicides).

Weeds stress crops in innumerable ways. The outcome for growers is always the same: loss of crop productivity and continuing weed infestations in the future. Crop–weed interactions are not all one way, though. Crops and crop production practices stress weeds, reducing their population size (mortality) and productivity (reproduction). Weeds evolve naturally, slowly, on their own without direct human help. Crops can evolve quickly; a new cultivar can potentially become available every growing season with improved plant breeding. And those crop cultivars can be aided and abetted by their human managers with improved cropping practices. Weeds are, and always have been, at a severe disadvantage when confronted by enlightened and technologically gifted humans. Crop cultivars and farming practices today are designed for high productivity, resistance to stress, and the ability to stress neighboring plants.

But opportunity space-time never rests.

27.9 STRESS, GENETICS, AND PLANT IMPROVEMENT

Some benefit may be gained from looking at agricultural improvement as a dialogue between crops and humans and weeds from the weeds' point of view. How can we maximize crop stress on weeds? How can we limit the opportunity space-time available to local weed populations?

When crop improvement is viewed from the weeds' point of view, the opportunities and vulnerabilities of plants in the field may be more readily apparent. Two possible approaches are acceleration of crop improvement and incorporation of weedy traits into crop varieties.

Weed evolution occurs at an inherently slower pace than is possible with human genetic transformation technology. The Red Queen principle operates in agricultural evolutionary systems. Continuing crop improvement and enhanced weed competitive development are needed to maintain the relative fitness of both co-evolving crops and weeds. An accelerated rate of traditional and transgenic improvement may allow crops to stay ahead of weed evolution. An excellent contemporary example of crop enhancement of weed and soil insect stress can be found with transgenic maize and cotton that have resistance to glyphosate (and other herbicides) for weed control, coupled with BT for stress on soil insect infestations.

Successful weed species can be directly exploited by incorporation of their useful traits into elite crop germplasm. Conceptually, what are the weed traits that cause the greatest stress on crops? These traits are a direct indication of a crop's vulnerabilities to specific weed species infesting them. Breeding crops with these weedy traits by conventional and transgenic techniques may provide both relief from weed stress and enhanced crop competitiveness.

What weedy traits appear most advantageous for crop improvement?

Weed species often succeed in avoiding stress from vigorous crop species by means of seed dormancy: optimal timing of seedling emergence for local conditions and subsequent growth and development (e.g., Dekker, 2014). The incorporation of variable crop seedling emergence timing might be an approach to enhancing weed stress. Related approaches may be useful in more precise cover crop and double-crop emergence timing of germination in the field.

Crop canopy closure occurs when leaves completely shade the ground, preventing or inhibiting weed germination and development. Leaf growth is a function of photosynthesis in crop leaves. Rapid leaf development is already one of the most useful weed stressor traits in current plant breeding. Another unexploited approach may be to incorporate more efficient photosynthetic capabilities in crops to take advantage of sub-optimal light conditions. The s-triazine resistant *Brassica napus* (R) has been shown to be at a distinct photosynthetic advantage over susceptible (S) phenotypes in stressful conditions (Dekker, 2016a). A mixture of R and S individual plants or individual leaves on the same plant may allow this enhancement of weed stress through improved light use by crop cultivars.

These approaches are only two among many other possibilities. The key in all cases should be recognition of the power of natural and human-mediated evolution at work in agriculture.

REFERENCES

Akhzari, D., M. Pessarakli, F. Mahmoodi, and B. Farokhzadeh. 2015b. Effects of Grazing and Fire on Soil and Vegetation Properties in a Semi-arid Rangeland. *The Journal of Ecopersia*, 3(1): 901–916.

Akhzari, D., M. Pessarakli, and S. Eftekhari Ahandani. 2015a. Effects of Grazing Intensity on Soil and Vegetation Properties in a Mediterranean Rangeland. *Communications in Soil Science and Plant Analysis*, 46(22): 2798–2806.

Ashrafi, E., J. Razmjoo, M. Zahedi, and M. Pessarakli. 2015. Screening Alfalfa for Salt Tolerance Based on Lipid Peroxidation and Antioxidant Enzymes. *Agronomy Journal*, 107(1): 167–173.

Askari-Khorasgani, O., S. Emadi, F. Mortazaienezhad, and M. Pessarakli. 2017. Differential Responses of Three Chamomile Genotypes to Salinity Stress with Respect to Physiological, Morphological, and Phytochemical Characteristics. *Journal of Plant Nutrition*, 40(18): 2619–2630.

Dekker, J. 2016b. *Evolutionary Ecology of Weeds*, 2nd edition, p. 552. CreateSpace Publishing, Weeds-R-Us Press, Ames, IA.

Dekker, J. 2014. Seed Dormancy, Germination and Seedling Recruitment in Weedy *Setaria*. In: *Handbook of Plant and Crop Physiology*, 3rd edition, Chapter 2, pp. 33–102; Pessarakli, M., Editor. Taylor and Francis Group, CRC Press, Boca Raton, FL.

Dekker, J. 2016a. Evolutionary Ecology of s-Triazine Resistant Plants: Pleiotropic Photosynthetic Reorganization in the Chloroplast Chronomutant. In: *Handbook of Photosynthesis*, 3rd Edition, Chapter 39, pp. 705–722; Pessarakli, M., Editor. CRC Press, Taylor and Francis Group, Boca Raton, FL.

Emam, Y., A. Shekoofa, F. Salehi, A.H. Jalali, and M. Pessarakli. 2012. Drought Stress Effects on Two Common Bean Cultivars with Various Growth Habits. *Archives of Agronomy and Soil Science*, 58(5): 527–534.

Gheidary, S., D. Akhzari, and M. Pessarakli. 2017. Effects of Salinity, Drought, and Priming Treatments on Seed Germination and Growth Parameters of *Lathyrus sativa* L. *Journal of Plant Nutrition*, 40(10): 1507–1514.

Ghouchani, R., H. Abbaspour, A. Saed-Moucheshi, and M. Pessarakli. 2017. Colonization with Endo-Mycorrhiza Affects the Resistance of Safflower in Response to Salinity Condition. *Journal of Plant Nutrition*, 40(13): 1856–1867.

Hossein Saeidnejad, A., M. Kafi, H. Reza Khazaei, and M. Pessarakli. 2016. Combined Effects of Drought and UV Stress on Quantitative and Qualitative Properties of *Bunium persicum*. *Journal of Essential Oil Bearing Plants*, 19(7): 1729–1739.

Izadi, M.H., J. Rabbani, Y. Emam, M. Pessarakli, and A. Tahmasebi. 2014. Effects of Salinity Stress on Physiological Performance of Various Wheat and Barley Cultivars. *Journal of Plant Nutrition*, 37(4): 520–531.

Kadkhodaie, A., J. Razmjoo, M. Zahedi, and M. Pessarakli. 2014. Selecting Sesame Genotypes (*Sesamum indicum* L.) for Drought Tolerance Based on Some Physiochemical Traits. *Agronomy Journal*, 106(1): 111–118.

Khorsandy, S., A. Nikbakht, M.R. Sabzalian, and M. Pessarakli. 2016. Effect of Fungal Endophytes on Morphological Characteristics, Nutrients Content and Longevity of Plane Trees (*Platanus orientalis* L.). *Journal of Plant Nutrition*, 39(8): 1156–1166.

Lotfi, R., M. Pessarakli, P. Gharavi-Kouchebagh, and H. Khoshvaghti. 2015. Physiological Responses of *Brassica napus* to Fulvic Acid under Water Stress: Chlorophyll A Fluorescence and Antioxidant Enzymes Activity. *The Crop Journal*, 3(5): 434–439.

Maghsoudi, K., Y. Emam, A. Niazi, M. Pessarakli, and M.J. Arvin. 2018. P5CS Expression Level and Proline Accumulation in the Sensitive and Tolerant Wheat Cultivars under Control and Drought Stress Conditions in the Presence/Absence of Silicon and Salicylic Acid. *Journal of Plant Interactions*, 13(1): 461–471.

Marcum, K.B., and M. Pessarakli. 2010. Salinity Tolerance of Ryegrass Turf Cultivars. *HortScience*, 45(12): 1882–1884.

Mardani, S., S.-H. Tabatabaei, M. Pessarakli, and H. Zareabyaneh. 2017. Physiological Responses of Pepper Plant (*Capsicum annuum* L.) to Drought Stress. *Journal of Plant Nutrition*, 40(10): 1453–1464.

Marzban, L., D. Akhzari, A. Ariapour, B. Mohammadparast, and M. Pessarakli. 2017. Effects of Cadmium Stress on Seedlings of Various Rangeland Plant Species (*Avena fatula* L., *Lathyrus sativus* L. and *Lolium temulentum* L.): Growth, Physiological Traits and Cadmium Accumulation. *Journal of Plant Nutrition*, 40(15): 2127–2137.

Mohamadi, M.H.S., N. Etemadi, A. Nikbakht, and M. Pessarakli. 2017. Physiological Responses of Two Cool-season Grass Species to Trinexapac-ethyl under Traffic Stress. *HortScience*, 52(1): 99–109.

Nikbakht, A., M. Pessarakli, N. Daneshvar-Hakimi-Maibodi, and M. Kafi. 2014. *Agronomy Journal*, 106(2): 585–595.

Pakar, N., H. Pirasteh-Anosheh, Y. Emam, and M. Pessarakli. 2016. Barley Growth, Yield, Antioxidant Enzymes and Ions Accumulation Affected by PGRs under Salinity Stress. *Journal of Plant Nutrition*, 39(10): 1372–1379.

Pessarakli, M. 2015. Effects of a Bio-Stimulant and Salinity Stress on Growth and Quality of Ryegrass (*Lolium prenne* L.), an Urban Desert Landscape and Forage Crop, for Sustainable Agriculture in Arid Regions. *International Journal of Water Resources and Arid Environments*, 4(2): 94–104.

Pessarakli, M. 2017. Growth Responses of Sacaton Grass (*Sporobolus airoides* Torr.) and Seashore Paspalum (*Paspalum vaginatum* Swartz) under Prolonged Drought Stress Condition. *Advances in Plants and Agriculture Research*, 7(4).

Pessarakli, M., and D.M. Kopec. 2011. Responses of Various Saltgrass (*Distichlis spicata*) Clones to Drought Stress at Different Mowing Heights. *Journal of Food, Agriculture, and Environment (JFAE)*, 9(3 and 4): 665–668.

Pessarakli, M., D.D. Breshears, J. Walworth, J.P. Field, and D.J. Law. 2017. Candidate Halophytic Grasses for Addressing Land Degradation: Shoot Responses of *Sporobolus airoides* and *Paspalum vaginatum* to Weekly Increasing NaCl Concentration. *Arid Land Research & Management Journal*, 31(2): 169–181.

Pessarakli, M., M. Haghighi, and A. Sheibanirad. 2015. Plant Responses under Environmental Stress Conditions. *Advances in Plants and Agriculture Research*, 2(6).

Pirasteh-Anosheh, H., Y. Emam, and M. Pessarakli. 2013. Changes in Endogenous Hormonal Status in Corn (*Zea mays*) Hybrids under Drought Stress. *Journal of Plant Nutrition*, 36(11): 1695–1707.

Sanchez, P.L., M.-K. Chen, M. Pessarakli, H.J. Hill, M.A. Gore, and M.A. Jenks. 2014. Effects of Temperature and Salinity on Germination of Non-pelleted and Pelleted Guayule (*Parthenium argentatum* A. Gray) Seeds. *Industrial Crops and Products*, 55: 90–96.

Shekoofa, A., Y. Emam, and M. Pessarakli. 2012. Effect of Partial Defoliation after Silking Stage on Yield Components of Three Grain Maize Hybrids under Semi-arid Conditions. *Archives of Agronomy and Soil Science*, 58(7): 777–788.

Shobbar, M.S., O. Azhari, Z.S. Shobbar, V. Niknam, H. Askari, M. Pessarakli, and H. Ebrahimzadeh. 2012. Comparative Analysis of Some Physiological Responses of Rice Seedlings to Cold, Salt and Drought Stresses. *Journal of Plant Nutrition*, 35(7): 1037–1052.

Valizadeh-Kamran, R., M. Toorchi, M. Mogadam, H. Mohammadi, and M. Pessarakli. 2018. Effects of Freeze and Cold Stress on Certain Physiological and Biochemical Traits in Sensitive and Tolerant Barley (*Hordeum vulgare*) Genotypes. *Journal of Plant Nutrition*, 41(1): 102–111.

Section VI

Genetic Factors and Plant/Crop
Genomics under Stress Conditions

28 Candidate Gene Expression Involved in Plant Osmotic Tolerance

Mojtaba Kordrostami and Ali Akbar Ebadi

CONTENTS

28.1 INTRODUCTION

Drought and salinity are the most important abiotic stresses limiting the production of staple food in the world (Turkan, 2011). More than 6% of the world's land is affected by salinity or by the conditions associated with sodium (Munns, 2005; Shrivastava and Kumar, 2015; Machado and Serralheiro, 2017). Under salinity stress conditions, plant growth and development are affected by osmotic stress due to the salt accumulation in the root environment and ionic toxicity due to the accumulation of sodium and chloride in the leaves (Kordrostami, Rabiei and Kumleh, 2016). The results of many types of research have shown that under salinity conditions, the early decline in growth is due to factors associated with osmotic stress (Fricke, 2004; Qados, 2011; Rahnama et al., 2011). However, only after a long time and as a result of increasing concentration of sodium in older leaves will the damage caused by ionic toxicity, especially in the old leaves, be visible (Munns and Tester, 2008; Flowers, Munns and Colmer, 2014).

Increasing salinity of soil or water causes a disturbance in plant physiological and biochemical processes and causes problems such as 1) ionic imbalance, 2) mineral deficiency, 3) osmotic stress, 4) ionic toxicity, and 5) oxidative stress (Machado and Serralheiro, 2017). These conditions interact with cellular components such as DNA, lipids, and pigments, and reduce the growth and development of most crops (Kasote et al., 2015). Salinity tolerance is a polygenic trait that includes (a) the division of large amounts of salt within the

plant; (b) osmotic regulation; and (c) morphological changes (Kordrostami, Rabiei and Kumleh, 2017a). Studies in the field of salt stress tolerance include (a) classical breeding programs that, despite some successes, have been limited due to the polygenic nature of this trait; (b) the use of mutations that lead to the elimination of gene function for the identification and study of genes responsible for stress. For example, in *Arabidopsis*, due to the ease of genetic manipulation of this plant, several types of research have been carried out in this field; and (c) cultivation methods under laboratory conditions in plants such as alfalfa, rice, and potatoes (Hanin et al., 2016).

Plants have many physiological and biochemical strategies to overcome unsuitable conditions (Shinozaki and Yamaguchi-Shinozaki, 2000; Hasanuzzaman et al., 2013). In response to stress, they produce different compounds that need to be explored and fully understood (Wu et al., 2005; Rejeb, Pastor and Mauch-Mani, 2014). The molecular mechanisms of abiotic stress resistance are based on the activity and regulation of certain genes (Rodríguez, Canales and Borrás-Hidalgo, 2005; Bechtold and Field, 2018). The mechanisms of salinity tolerance in resistant cultivars may be more rapid and stable than in susceptible ones (Hanin et al., 2016). The gene products activated in stress are classified into two groups: (a) functional proteins for stress resistance (e.g., scavenging enzymes and proteins associated with the biosynthesis of osmotic regulating agents and carbohydrates) and (b) signal transduction regulator proteins and the genes associated with stress (such as transcription factors) (Rabbani et al., 2003; Wu et al., 2005; Bechtold and Field, 2018). In other words, the reception and transduction of stress messages through the presence of messenger molecules forms an important part of plant response to unfavorable conditions resulting from the activation of genes associated with stress, and following the transduction of this message (and with the activity of another group of genes), various proteins are synthesized that involve different metabolic and physiological responses (Tuteja and Sopory, 2008; Lee and Luan, 2012). Munns (2005) classified the salinity resistance genes into three groups: genes involved in the salinity signal transduction, osmotic and supportive genes, and genes that affect the faster growth of the plant in saline soils.

Improving the performance of crops in saline soils and areas with water limitation seems to be necessary; hence, the study of physiological traits and candidate genes involved in improving the crop yield in dry and saline areas is very important.

28.2 SALINITY

According to the United States Department of Agriculture (USDA) definition, a saline soil is a soil with electrical conductivity of 4 dS/m in its saturated extract. This amount is approximately 40 mM NaCl, which produces an osmotic pressure of 0.2 MPa and will significantly reduce crop yields (Munns and Tester, 2008; Machado and Serralheiro, 2017). High salinity is an environmental limiting factor that affects a large range of crops and disturbs the growth, physiological, and metabolic processes of plants (Shrivastava and Kumar, 2015). More than 800 million hectares of land are affected by salinity around the world. This amount includes more than 6% of the total available land (Yadav et al., 2011). Most of these lands are located in arid and semi-arid regions. The weathering of bedrock causes the release of sodium chloride, calcium, and magnesium, in higher amounts, and sulfates and carbonates, in lower amounts; among them, the most frequent is sodium chloride (NaCl). Another factor in the accumulation of these salts is their transport by wind and rain. Of the 1500 million hectares of rainfed lands, 32 million hectares (2%) are affected by secondary salinity, and of 230 million hectares of irrigated lands, 45 million hectares (20%) are under salinity (Munns and Tester, 2008).

28.3 CANDIDATE GENES FOR SALT TOLERANCE

28.3.1 Salt Uptake and Transport

The root is the first part of the plant that absorbs sodium ions. Due to the potential difference in the plasma membrane (120–200 mV), the increase in sodium outside the cell creates a large electrochemical gradient that causes the passive transport of this ion into the cell (Volkov, 2015). The exact system of sodium adsorption is not fully identified, but the results of various studies indicate that Na^+ and K^+ are transported by common proteins (Alberts et al., 2002). Probably, the similarity between the radii of these two ions makes it difficult to differentiate them from each other and creates the basis

of Na^+ toxicity (Tester and Davenport, 2003). Two adsorption systems have been identified, with high and low affinity for potassium, which can also play a role in sodium absorption. The two systems are different in terms of maximal velocity (Vmax) and half-saturation concentration (Km). Absorption is usually carried out by the transporters in the first system and by the channels in the second system (Britto and Kronzucker, 2008).

According to Britto and Kronzucker's (2008) findings, high-affinity K^+ transporters (Km = 10–30 μm) make it possible to absorb very low concentrations of K^+. These transporters are symporters (K^+-H^+ symporters) that bring K^+ into the cell along with protons. In this case, the transfer of potassium depends on the activity of the H^+-ATPase proton pump for the proton slope production. The K^+ uptake transporter-high-affinity K^+ transporter (KUP-HAK) family, which are highly associated with potassium transporters in bacteria and fungi, and the high-affinity K^+ transporter (HKT) family are two important groups in this category that have been identified in *Arabidopsis*, barley, and rice (Rodríguez-Navarro and Rubio, 2006). In the high-affinity K^+ (HAK) transporter, the motion of K is coupled to the proton slope and is highly specific for potassium ion = 1000 (K^+/Na^+), but sodium can block these transporters in millimolar concentrations. The low K^+/Na^+ ratio can lead to a significant reduction in intracellular K^+ levels. The second family (HKT) has a lower affinity for potassium than HAK transporters. It contains a specific binding site for sodium and a binding site for two potassium and sodium ions. Therefore, the binding sites of this transporter, especially in high concentrations of NaCl, are saturated with sodium and can easily absorb this ion (Blumwald, 2000; Munns and Tester, 2008). The results of several studies have shown that the expression of *HAK* genes is upregulated under salinity stress conditions in *Mesembryanthemum crystallinum* (Su et al., 2002) and *Arabidopsis* (Maathuis, 2005). The important role of *HAK* genes in K deficiency is indicated in rice (Banuelos et al., 2002), *Arabidopsis* (Ahn, Shin and Schachtman, 2004; Qi et al., 2008), barley (Santa-María et al., 1997), and tomato (Wang, Garvin and Kochian, 2002).

Potassium inward rectifying channels (KIRC) such as *AKT1*, in *Arabidopsis*, which activate the potassium flow by a negative difference in the plasma membrane potential and have a high selectivity for K^+/Na^+, in K^+ physiological concentrations and external Na^+ concentration, are the first group of low-affinity potassium transporters. In a mutant of *Arabidopsis* (*akt1-1*) in which the gene is silenced, salinity sensitivity is similar to that of wild plants, which suggests that these channels are not likely to play a role in Na^+ absorption (Blumwald, 2000).

The second group of low-affinity transporters is potassium outward rectifying channels (KORC), which can play a role in sodium entry into the plant cells. These channels open at the plasma membrane (i.e., by varying the potential difference of the plasma membrane to more positive values), causing K^+ exit and Na^+ entrance (Blumwald, 2000).

The third group is voltage independent cation channels (VICs) or non-selective cation channels (NSCCs). Generally, these channels are non-selective for mono-atomic ions. It is likely that the role of these channels is less important than that of KIRC in the absorption and entry of this ion, because they are less abundant compared with KIRC. But, these channels have a relatively high selectivity for Na^+/K^+ and, unlike the voltage-dependent channels (KIRC, KORC), are not opened by changing the voltage. In various studies, it has been shown that in high salt concentrations, VICs play a significant role in the absorption and entry of Na^+ into the cell (Maathuis and Amtmann, 1999; Munns and Tester, 2008).

28.3.2 SODIUM TRANSFER

After sodium entry into the root epidermal cells, its lateral movement begins towards the xylem. This lateral movement is done through the symplastic or apoplastic pathways up to the cortex. Subsequently, due to the presence of a Casparian strip in the endoderm area of the root, the apoplastic flow is stopped to some extent, and the symplastic movement continues up to the vascular cylinder cells. Then, it enters the apoplast of the vascular cells, and the sodium is then introduced into the xylem by Na^+/H^+ antiporters in the vascular cell membrane and moves to the shoot through the transpirational flow. Sodium is discharged from the xylem to the root by transporters such as HKT (Apse and Blumwald, 2007). The *HKT* gene was first isolated from wheat root (Schachtman and Schroeder, 1994). The return of sodium from the xylem to the root is one of the first steps in controlling the Na concentration in the shoots (Munns and Tester, 2008), which is handled by this gene. The sodium movement is carried out from the shoots to the roots through the phloem, but its loading and discharge in the phloem are not well understood (Apse and Blumwald, 2007).

28.4 GENES INVOLVED IN ION HOMEOSTASIS UNDER SALINITY STRESS

Homeostasis can be defined as the tendency of a cell or living organism to maintain a stable internal state against disturbing environmental stimuli. One of these stimuli is an increase in the concentration of ions and salinity from salts, including NaCl (Rodríguez, Canales and Borrás-Hidalgo, 2005). As the ion flows inside and outside the plant and its cells, it is controlled, and an ion balance is established as needed. But when plant cells are exposed to NaCl, the ion status, such as Na^+, K^+, Ca^{2+}, and Cl^-, is impaired. Therefore, it is essential for a plant to restore the ion balance in its cells to maintain its metabolic function and growth (Kordrostami, Rabiei and Kumleh, 2016). The result of this action is salinity adaptation (Niu et al., 1995; Ahmad, Azooz and Prasad, 2013). Some salinity-tolerant plants have complex and specific mechanisms for salinity tolerance (Kordrostami, Rabiei and Kumleh, 2017b). These unique mechanisms, such as the presence of salt glands, are seen in a limited number of plant species. Therefore, it is impossible to consider these methods as a basic and general strategy for all species. But mechanisms that have a cellular basis are commonly seen in all species and contribute to salinity tolerance. In the meantime, processes that cause more harmful ions to escape from the root or to be sequestered within the cell are more important (Niu et al., 1995). The ionic balance in saline environments depends on the plasma membrane and vacuole protein transporters, including H^+-ATPase, pyrophosphatase, Ca^{2+}-ATPase, secondary transporters, and channels responsible for ionic flows (Hasegawa et al., 2000).

28.4.1 SODIUM HOMEOSTASIS

Unlike in animal cells, Na^+ is not essential for most plants (Wu, 2018). Although sometimes this ion is referred to as an osmotic regulating agent in the cell, excessive accumulation of Na has harmful effects on plants (Shokri-Gharelo and Noparvar, 2018). One of the harmful effects of sodium is its competition with potassium to enter the cell. This is due to their structural similarities and the competition between these two ions to connect to potassium-binding sites in the cell, which inhibits potassium-dependent metabolic processes. In most plants, the amount of sodium ion in the cytosol is balanced by the restriction of its entry into the cell or excretion to vacuoles and apoplasts (Maathuis and Amtmann, 1999).

In vitro, protein synthesis requires a concentration of 100 to 150 mM potassium and is inhibited by a concentration of 100 mM Na^+. This sensitivity of enzymes to salinity is observed in both halophytes and glycophytes. Therefore, halophytes have more effective strategies for reducing the Na concentration and increasing the K/Na ratio. In most plants, excretion of Na from the cytosol and its fractionation in vacuoles are two important solutions that reduce the cytosolic concentration of this ion (Blumenthal, Goldberg and Brinckmann, 2000; Tester and Davenport, 2003).

The transfer of sodium ions into vacuoles, in addition to reducing their toxic effects in the cytosol, is physiologically important under salinity or potassium deficiency conditions for osmotic regulation and maintaining cell volume and growth (Hasegawa et al., 2000). Of course, it should be noted that with increasing sodium concentration in vacuoles under high-salinity conditions, their osmotic pressure should be increased to maintain the volume of the cell and cellular components such as cytoplasm. This can be done by increasing the K^+ concentration to below its toxic levels and the accumulation of compatible solutes (Maathuis, Ahmad and Patishtan, 2014).

Intracellular Na^+ concentrations in plant cells are regulated by Na^+/H^+ antiporters in the plasma membrane and the vacuole membrane. These transporters in the plasma membrane cause Na^+ excretion from the cytosol into apoplasts, and in the tonoplast, they cause Na^+ accumulation within the vacuole. The force required for this antiporter activity is due to the proton pump electrochemical gradient generated by the activity of the proton pumps, including P- and V-H^+-ATPase, and vacuolar H^+-pyrophosphatase (Niu et al., 1995). This proton slope helps reduce the cytosolic Na^+ concentration by 100 to 1000 times compared with the apoplast and vacuole. This will prevent the formation of toxic Na and Cl concentrations in the cytosol and maintain the K^+ concentration at 80 mM under salinity stress conditions (Sze, Li and Palmgren, 1999).

Full recognition of the activity and molecular properties of these transporters was possible after the sequencing of the *Arabidopsis* genome and during the study of the *Arabidopsis* salt overly sensitive (*SOS*) mutants (Zhu, 2002). *AtSOS1* encodes the Na^+/H^+ antiporter in the *Arabidopsis* plasma membrane, and with NaCl treatment, its transcript levels increase in *Arabidopsis* (Yang et al., 2009). Also, the overexpression of *SOS1* increases salinity tolerance in this plant, possibly due to a decrease in the Na^+ accumulation in its cells (Shi et al., 2002). In addition, the expression of the *nhaA* gene (which encodes the plasma membrane Na^+/H^+ antiporter in *Escherichia coli*) also increases salinity and drought resistance in rice and *Arabidopsis* (Wu et al., 2005).

Salinity causes overexpression of the genes encoding proton pumps, H^+-pyrophosphatase, and vacuolar Na^+/H^+ antiporters in various plants such as *Arabidopsis*, rice, and tobacco, and transgenic plants with overexpression of these genes, in comparison with wild-type plants, show more salinity tolerance (Gaxiola et al., 2001; Shi and Zhu, 2002). Overexpression of H^+-pyrophosphatase and H^+-ATPase allows the accumulation of more Na^+ in the vacuole (Gaxiola et al., 2001; Shi and Zhu, 2002). Similar results were observed in wheat. The overexpression of vacuolar Na^+/H^+ antiporter in the roots and shoots of tolerant wheat genotypes facilitates Na^+ excretion from the cytosol and its entrance into the vacuole. In contrast to these results, Yang et al. (2009) compared the various types of transgenic *Arabidopsis* and concluded that the plants with *AtNHX1* overexpression showed no more salinity tolerance than the wild types. But, simultaneous overexpression of *AtNHX1* and *SOS3* increased their salinity tolerance. They also stated that the overexpression of *SOS1* in transgenic plants increases the plant's salinity tolerance alone, but its activity depends on the presence of two other members of this pathway, *SOS2* and *SOS3*.

28.4.2 Potassium Homeostasis

Potassium is usually the most abundant ion available in the cytosol, and its concentration is estimated to be in the range of 40 to 200 mM. This ion plays an important role in enzyme homeostasis, cellular turgor pressure, protein synthesis, and osmotic regulation (Mitsuya et al., 2002). Potassium is absorbed by the plant through two different absorption systems. If the external concentration of K^+ is low (lower than 1 mM), absorption is carried out by high-affinity systems, and at higher concentrations (greater than 1 mM), it is carried out by low-affinity systems (Britto and Kronzucker, 2008; Munns and Tester, 2008). The presence of this ion in the cytoplasm can help salinity tolerance (Munns and Tester, 2008). For example, it has been observed that the presence of more potassium in the culture medium of *SOS Arabidopsis* mutants reduces their sensitivity and leads to a better phenotype. This may be due to an increase in the concentration of cytosolic potassium (Zhu, 2002). In addition, the ability of plants to increase the K^+/Na^+ ratio, especially in the shoots, is a sign of better salinity stress tolerance (Maathuis and Amtmann, 1999). The most important way to maintain this ratio is to

control and reduce the presence of sodium in the cytosol. For this reason, except for the entry of sodium into the plant, it is important to transfer it from roots to shoots. For example, during the investigation of sensitive and tolerant rice genotypes, it was observed that after salinity treatment in both cultivars, potassium concentration increased in the shoots and decreased in the roots. This increase was far higher in the tolerant genotype (Diédhiou, 2006).

28.4.3 Calcium Homeostasis

Calcium is an essential element for plants, which plays an important role in maintaining the structure of the cell wall and membrane and regulates the growth and development of the plant. However, to prevent toxic effects of calcium, its concentration should be maintained at a low and appropriate level through the activity of calcium pumps and the maintenance of excess calcium in the organelles and intracellular components such as vacuoles and the endoplasmic reticulum (Hepler, 2005). The calcium resting potential in the cytosol is established with Ca^{2+}-ATPases and a Ca^{2+}/H^+ antiporter (Boudsocq and Sheen, 2009). The first attempt to identify the Ca^{2+}/H^+ antiporter genes was carried out using a yeast suppression assay (Hirschi et al., 1996). In this study, the strain lacking Ca^{2+}/H^+ antiporter genes (*VCX1*) had slower growth in a medium with high calcium content (Cunningham and Fink, 1996). Lately, two genes, *CAX1* and *CAX2*, were isolated from *Arabidopsis* that allowed the yeast to grow normally in a medium containing high Ca (Hirschi et al., 1996). These results showed that *CAX1* is a Ca^{2+}/H^+ antiporter, which plays an important role in Ca transport into the vacuole (Hirschi et al., 2000). Subsequently, many of the genes encoding the Ca^{2+}/H^+ antiporter were identified in different plants (Kamiya et al., 2006; Shigaki and Hirschi, 2006).

Under salinity conditions, calcium is able to reduce sodium absorption by disabling the ionic channels (Maathuis and Amtmann, 1999). Also, in the presence of calcium, the concentration of sodium in the shoots of resistant and susceptible cultivars of plants decreases, which is highly correlated with the decrease of sodium leakage in the apoplastic pathway and its diversion (Anil et al., 2005). Therefore, Ca has a negative effect on the apoplastic absorption of sodium and reduces Na^+ absorption and transfer (Maathuis and Amtmann, 1999). Of course, the effect of calcium depends on the plant species and the amount of salinity. Wu and Wang (2012), for example, observed a positive effect of calcium on the reduction of sodium accumulation, increased potassium, and K^+/Na^+ ratio in rice shoots only in low salinity (25 mM NaCl). However, calcium did not have a significant effect in a medium containing 125 mM NaCl. Different environmental stimuli increase intracellular calcium concentrations in plants. This increase is due to the entry of calcium by calcium channels from intracellular and apoplastic stores and according to the electrochemical potential. These channels open after hyperpolarization, depolarization, and the binding of ligands such as glutamate, inositol triphosphate (IP3), and cyclic nucleotide monophosphates (cNMPs). Calcium excretion also occurs from the cytosol by Ca^{2+}-ATPases and transporters (Boudsocq and Sheen, 2009).

28.5 SIGNAL TRANSDUCTION UNDER SALINITY STRESS

In living cells, changes in the extracellular environment must be transmitted to the intracellular medium and reach the nucleus in some way to express the related genes. In other words, when plants are exposed to biotic and abiotic stresses, they activate the signal transduction pathways so that they can adapt to even the slightest environmental changes (Chinnusamy, Jagendorf and Zhu, 2005). In this way, first, the ligand must be received by a membrane receptor or intracellular receptor and then, the message transmitted by changes within the cytosol (Zhu, 2002). One of the most important changes that enable the message to be understood is the production of secondary messengers. With the activity of these secondary messengers, the concentration of intracellular calcium and consequently, the cascade reactions of protein phosphorylation increase. Finally, the expression of the stress-reactive genes begins with phosphorylation of the transcription factors (Xiong, Schumaker and Zhu, 2002).

Adaptive responses to stress can be examined in a number of ways. One of them is homeostasis or balance, which includes ionic balance or osmotic balance and is mainly related to salinity and osmotic stress (Assaha et al., 2017). Another aspect of controlling stress-induced damage is restoration, scavenging, and lastly, controlling plant growth. Accordingly, the signaling of stresses such as drought and salinity is divided into three functional groups (Zhu, 2002): a) osmotic and ionic stress signals to restore cellular balance after stress, b) scavenging signals for controlling and repairing stress-induced damage, and c) cell division and development coordinator signals to restore a more favorable position than stress conditions. Because calcium is an essential element in signal transduction, this chapter will further discuss calcium signaling and the genes encoding this path.

28.5.1 Ca^{2+} SIGNALING

Salinity, drought, and cold stress cause a temporary increase in calcium entry from the apoplastic space or release from internal storage sources into the cell cytoplasm. Increasing the calcium concentration, as one of the secondary messengers, can be a source of biological information that is specifically identified and activates the next path (Wan, Lin and Mou, 2007). The oscillation of cell calcium is due to the activity of ion channels, Ca^{2+}-ATPase, and Ca^{2+}/H^+ antiporters (Sanders, Brownlee and Harper, 1999). DeWald et al. (2001) concluded that after the osmotic stress, in addition to IP3, the synthesis of its precursor, phosphatidylinositol-4,5-bisphosphate (PIP2), and phospholipase C also increased. Meanwhile, the amount of intercellular calcium in the root increases with increasing IP3, and this process continues for 30 minutes. Phospholipase C converts phosphatidylinositol,4,5-bisphosphate to diacylglycerol (DAG) and IP3, which activate protein kinases and

calcium channels, respectively. The temporary increase in IP3 is similar to the increase in Ca^{2+} under stress conditions, which expresses its role in Ca^{2+} signal transduction (DeWald et al., 2001; Tuteja and Sopory, 2008). Also, the role of IP3 in salinity stress signal transduction was determined during *FRY1* locus examination in *Arabidopsis*. This gene encodes inositol polyphosphate-1-phosphatase, which is involved in IP3 catabolism (Berdy et al., 2001). In fact, IP3 and DAG (derived from phosphatidylinositol-4,5-bisphosphate hydrolysis) are two secondary messengers that increase as a result of environmental stresses such as salinity. Due to these, especially IP3, the expression of the stress-reactive genes is increased, and the plant's defense responses are activated (Xiong, Schumaker and Zhu, 2002).

28.5.2 Ca^{2+} SENSORS

Any changes in calcium concentration should be understood and identified by the target cells to provide an appropriate response to the stimulant. Calcium sensors can be divided into two groups. The first group is those that do not inherently have enzyme activity and only transfer calcium-induced changes to target proteins. Calmodulins (CaMs) and Calcineurin B-like proteins (CBLs) are in this group. The second group also includes calcium-dependent protein kinases (CDPKs) and calcium and calmodulin-dependent protein kinases (CCaMKs), whose enzymatic activity begins directly on calcium binding.

The first group of calcium sensors are protein kinases that are directly activated by binding to calcium and then exert their effect (Boudsocq and Sheen, 2009). Calcium-dependent protein kinase (CDPK) enzymes have kinase activity domains adjacent to the amine terminals. These enzymes are coded by a gene family. There are 34 genes in *Arabidopsis* and 29 genes in rice for CDPK enzymes. The presence of a large number of genes in this gene family has also been observed in corn, tomato, and soybean (Harmon, Gribskov and Harper, 2000; Cheng et al., 2002; Asano et al., 2005). Similarly, CDPKs differ in the number of EF-hands of the calmodulin-like domains. Most of them have four and some have one to three EF-hands, and each EF-hand also binds to Ca^{2+} (Cheng et al., 2002). Asano et al. (2005) found that of 29 CDPKs in rice, *OsCPK5* and *OsCPK25* have three EF-hands, and *OsCPK6* has two EF-hands. This difference can lead to variability in the threshold of stimulating these sensors against different concentrations of calcium. That is, the different levels of calcium in each stress can activate certain types of enzymes (Cheng et al., 2002). The members of this enzyme family are active in the cytosol, chloroplasts, peroxisomes, and nucleus (Boudsocq and Sheen, 2009). To connect to the membrane, most of these enzymes require myristic acid, and some also require palmitic acid; these are bound to glycine and cysteine amino acids, respectively, at the N terminal (Cheng et al., 2002). In plants, CDPK enzymes have several roles, including participation in hormonal signaling pathways, the growth of various parts of plants, the changing of the stomatal guard cells, and carbon metabolism. But, one of their most important tasks is to collaborate on abiotic stress signaling pathways, so that in various experiments, increased transcription and enzymatic activity of these protein kinases due to abiotic stresses have been clearly observed (Cheng et al., 2002). The expression of *OsCDPK7* in salinity and cold stress was reported in 10 days seedlings of rice. Also, the overexpression of this gene increases the salt tolerance of plants (Saijo et al., 2000).

Calmodulins have four EF-hand sites for Ca^{2+} binding. Calcium binding to these sites transforms the calmodulin spherical structure into an open structure and allows the binding of its sub-targets (i.e., the factors involved in signal transduction, such as protein kinases, transcription factors, and protein effectors such as enzymes and the transporters participating in physiological responses) as well as their adjustment (Guo et al., 2001; Boudsocq and Sheen, 2009). In addition, a series of calmodulin-like proteins have been identified in plants. Plant calmodulins have an amino acid sequence similarity to animal calmodulins. For example, in *Arabidopsis*, this similarity is more than 95% for calmodulins and 75–50% in the calmodulin-like proteins. Accordingly, calmodulin-like proteins such as *CaM8* are limited in binding to the target proteins (Luan et al., 2002; Zielinski, 2002). Although the role of these sensors has not yet been fully elucidated in plants, it has been observed that the expression of the gene encoding *OsMSR2* calmodulin-like protein increases under a wide range of stress conditions and at different developmental stages of rice plants. Also, the expression of this gene in *Arabidopsis* increases its salinity and drought resistance (Xu et al., 2011). Stress-activated calmodulins have different protein targets that are involved in various processes, including transcription, signal transduction, ion transport, and metabolism. These sensors positively or negatively affect the activity of the transcription factors involved in the expression of genes (Boudsocq and Sheen, 2009).

Calcineurin B-like proteins (CBLs) are another type of Ca^{2+} sensor. Calcineurin is a serine/threonine calmodulin-dependent protein phosphatase and a eukaryotic calcium. This heterodimer protein has two subunits, A and B. The A catalytic subunit contains the active site, and the B regulating subunit has a Ca^{2+} binding site (Rusnak and Mertz, 2000). CBLs, which are likely to be present only in signal transduction in the plant (Luan et al., 2002), are small proteins consisting of two spherical domains connected by a short section; each domain also consists of two EF-hand motifs. These sensors are largely conserved in terms of size and sequence and have the ability to bind to a different number of Ca^{2+} ions. For example, *CBL4/SOS3* binds to three and *CBL2* binds to two calcium ions. Binding to calcium by changing the receptor's structure provides the ability to communicate with other proteins (Kolukisaoglu et al., 2004; Boudsocq and Sheen, 2009). CBLs and calmodulins lack significant similarity with each other in terms of the amino acid sequence, except in the EF-hand motif. The EF-hand is a motif with a helix and loop structure that contains 12 amino acids in its loop-like part. Amino acids at positions 1, 3, 5, 7, 9, and 12 can be bonded to calcium. Positions 1, 6, and 12 are highly conserved and are aspartate, glycine, and glutamate (Luan et al., 2002). The target proteins with which CBLs communicate are members of the serine/threonine family of protein kinases, also called calcineurin B-like protein-interacting protein kinases (CIPKs). These enzymes are highly conserved in the N-terminal with kinase activity and are less conserved in the C-terminal with regulatory activity. However, in their regulatory unit, the highly conserved NAF domain with 24 amino acids is essential for binding this protein to the CBLs and activating it (Albrecht et al., 2001). Different members of this family of enzymes can be expressed in response to various environmental stresses, such as salinity, drought, cold, abscisic acid (ABA), and polyethylene glycol (PEG) treatments. In other words, any CIPK enzyme may be expressed in such a way as to respond to several stresses, although the expression or overexpression of some of these enzymes, such as *OsCIPK03* in cold stress, has been observed only in response to a specific stimulus (Xiang, Huang and Xiong, 2007). In addition to the multiplicity of CIPK enzymes, there are also different types of CBL sensors. For example, in *Arabidopsis* and rice (as two plant models in genetic studies), 10 genes encoding *CBL* and 25 and 30 genes encoding different types of *CIPK* have been identified (Kolukisaoglu et al., 2004; Xiang, Huang and Xiong, 2007). Due to the diversity of these sensors and enzymes, the formation of different types of communication between these two components is possible (Batistic and Kudla, 2004). For example, *CIPK1* acts in response to environmental stresses in interaction with *CBL9* and *CBL1* in an ABA-dependent pathway and a non-ABA pathway, respectively (Albrecht et al., 2001; Boudsocq and Sheen, 2009). *CBL4–CIPK24* (or the SOS3–SOS2 complex action in response to ion stress) is the famous complex of these two components, which have proved their activity in different plants (Halfter, Ishitani and Zhu, 2000; Boudsocq and Sheen, 2009). Thus, the formation of various CBL–CIPK complexes forms a highly orderly network in which extremely delicate and accurate reactions to abiotic stresses are provided for the plant (Boudsocq and Sheen, 2009).

After the stress perception and activation of calcium sensors, the environmental stress signal transduction to the nucleus is performed through various pathways. Under salinity stress due to osmotic stress, two major pathways for signal transduction to the nucleus are MAPK and SOS.

28.5.3 SOS Pathway

In response to ion effects caused by salinity stress, a signaling pathway based on the activity of *SOS* genes has been identified. This pathway is activated by high concentrations of Na^+ in the plant (and after an increase in the amount of cytosolic Ca^{2+}) and ultimately changes the expression and activity of the Na^+, K^+, and H^+ transporters. Under salinity stress, NaCl is detected before and after entering the cell. Extracellular sodium is detected by membrane receptors, and intracellular sodium is detected by membrane proteins and the enzymes sensitive to this ion; then, the level of intracellular calcium is increased (Shi et al., 2000). The CBL protein, encoded by *SOS3*, senses an increase in calcium levels. The calcium binding is necessary for SOS3 dimerization, leading to the final deformation of the protein. This causes the calcium signal transduction induced by salt stress (Liu et al., 2000). Then, SOS3 communicates with a serine/threonine protein kinase (CIPK), also known as SOS2, and activates it (Liu et al., 2000). SOS2 has regulatory activity with 446 amino acids in the C-terminal and has a catalytic activity similar to that of the yeast SNF1 protein kinase and mammalian AMPK in the N-terminal. These two regions interact with one another, and possibly, in normal cellular conditions, the association between these two parts blocks the catalytic portion and inhibits the activity of this enzyme. The regulatory terminal also interacts with the sequence of 21 amino acids or FISL motifs with SOS3. It has been observed that FISL removal is effective in activating SOS2 without the need for SOS3 (Guo et al., 2001). Finally, the SOS2–SOS3 complex increases the expression level of *SOS1*. The result of the expression of this gene is the production of the Na^+/H^+ antiporter, which is also probably one of the sodium sensors (Liu et al., 2000). This transporter excretes sodium from the cell's cytoplasm. In this way, using this mechanism, plants can control the movement of Na^+ throughout the plasma membrane by adjusting the entry and exit of this cation. Na^+ excretion from the cytoplasm is the hardest process for coping with salinity stress in plants.

The overexpression of *SOS1* in transgenic plants improves salinity tolerance. *AtSOS1* encodes the Na^+/H^+ antiporter in the *Arabidopsis* plasma membrane. Increasing the expression of *SOS1* under 50–200 mM salinity increases the tolerance in *Arabidopsis*, possibly due to decreased Na^+ accumulation in root cells, sap, and plant shoots (Shi et al., 2002). Also, the expression of the *nhaA* gene (which is the plasma membrane Na^+/H^+ antiporter in *Escherichia coli*) increases salinity tolerance in rice and *Arabidopsis* (Wu et al., 2005). The expression of *SOS1* occurs in different parts of the plant, but its overexpression in the epidermal cells of the root tip and the cells around the xylem indicates that *SOS1* as a plasma membrane antiporter plays a vital role in Na excretion from the root cells (Shi et al., 2000). Of course, this antiporter is also involved in the transfer of sodium from roots to shoots. But, there are different outcomes and perspectives on how it contributes to this transfer. In an experiment to determine the role of *SOS1* in the transfer of Na^+ to shoots, after the concentration of this ion was measured in the raw sap of *Arabidopsis*, it was found that the Na^+ concentration of raw sap increased in both maternal and *sos1* mutants under 100 mM salinity conditions but that the value was always higher in *sos1* seedlings. This conclusion suggests that the *SOS1* antiporter in the wild-type plants transfers Na^+ from the transpirational flow of xylem (Shi and Zhu, 2002).

In addition to the activity of SOS1 antiporter, the SOS2–SOS3 complex also activates *NHX1* genes, which help to accumulate Na^+ within the vacuole. This protein complex also inhibits the activity of *HKT1* plasma membrane transporters and sodium entry into the cytoplasm (Liu et al., 2000; Shi et al., 2000).

In one of the SOS signal transduction pathways, a series of sensors, calcium-dependent kinases, and antiporters are operating. One of the members of this complex is the

serine/threonine kinase enzyme called CIPK. Twenty-five and 30 CIPKs have been identified in *Arabidopsis* and rice, respectively, and these interact with different types of CBLs (Boudsocq and Sheen, 2009). By studying CBL1 and CBL9 proteins in *Arabidopsis*, it was found that CBLs that are similar in amino acid sequence do not necessarily interact with the same CIPKs. For example, in *Arabidopsis*, four CIPK products, *AtCIPK1*, *AtCIPK8*, *AtCIPK18*, and *AtCIPK24*, interacted with two, CBL1 and CBL9. Other CIPKs may also interact with one of these two sensors. For example, the protein derived from the expression of *AtCIPK21* interacts with AtCBL9 and AtCBL7, and the protein derived from the expression of *AtCBL17* interacts only with AtCBL1 (Shi et al., 2000; Albrecht et al., 2001).

Also, members of this enzyme family have different expressions and functions in terms of changing environmental conditions. For example, *AtCIPK3* transcripts have been shown to increase in conditions of frost, drought, salinity, injury, and ABA. While *AtCIPK8* is active in nitrate (NO^{-3}) signaling, *AtCIPK9* and *AtCIPK23* are active in response to low potassium levels, and *AtCIPK11* is active under ABA treatment and salinity and drought stresses (Das and Pandey, 2010). In rice affected by ABA, cold, drought, PEG, and salinity, at least one of the 30 known *OsCIPK* genes is activated. *OsCIPK03* and *OsCIPK12*, for example, cause cold and drought resistance, respectively. *OsCIPK15* and *OsCIPK24* also contribute to salinity tolerance (Xiang, Huang and Xiong, 2007; Das and Pandey, 2010).

CIPK24, one of the kinases of this family, commonly known as *SOS2*, in interaction with CBL10 and CBL4 in tonoplast and plasma membrane, plays an important role in salt stress tolerance (Zhu, 2002). Determining the cDNA sequence of this gene and comparing it with the genomic sequence determined that the *SOS2* gene or *CIPK24* has a total of 13 exons and 12 introns, and it produces a protein containing 446 amino acids with a molecular weight of 51 kDa. *SOS2* is a protein kinase with two catalytic and regulatory domains at the N and C terminals, respectively (Liu et al., 2000; Guo et al., 2001). The sequencing of *sos2* mutants has shown that both catalytic and regulatory sections are required for the activity of this protein. *SOS2* is produced in roots and shoots of plants, and its mRNA levels increase due to salt stress. It has also been shown that a mutation that eliminates the autophosphorylation activity of this enzyme plays an important role in the creation of super-susceptible saline plants. This mutation causes the Lys-40 amino acid to be altered in the subunit II of the catalytic domain and consequently, inhibits the autophosphorylation of the enzyme. Therefore, this enzyme is one of the essential elements for creating the balance of Na^+ and K^+ ions inside cells and inducing more resistance to salinity stress in plants (Liu et al., 2000).

The partitioning of Na ions into vacuole cells is also done by specific cationic antiporters. Proton/cation antiporters are grouped into two families: CPA1 and CPA. The CPA1 family has evolved from prokaryotic *NhaP* genes. One of the genes associated with the *NhaP* genes is the Na^+/H^+ antiporter present in the *Arabidopsis* plasma membrane (*AtSOS1*), which has been shown to have similar sequences (SOS) in the plant series.

Among the most important family members of the CPA1, many of which have been studied, are plasma membrane NHE antiporters (PM-NHE), which are found only in vertebrates. Recently, *NHE/NHX* sequences have also been identified in plants, animals, and fungi, which are the result of the evolution of the NHE sequences of the plasma membrane. This family is divided into two sub-groups, I and II, with 20–25% identity. The members of the first sub-group are very diverse compared with other NHE/NHX sequences and are found in the angiosperms, gymnosperms, monocots, and dicotyledons. By examination of the isoforms of sub-group I, their placement on the vacuolar membrane was proved to be a unique feature of the group members. In all the plants, there are different isoforms of NHX proteins, most of which are found under normal conditions, while some are found following salinity stress in leaves and roots, and others are found following ABA treatment, osmotic stress, etc. (Rodríguez-Rosales et al., 2009).

28.6 CONCLUSIONS

Plants have many physiological and biochemical approaches to overcome unsuitable conditions. There are several genes that cooperate in osmotic stress tolerance. The genes for salt uptake and transport, ion homeostasis (sodium, potassium, and calcium), and signal transduction are the main genes for salinity tolerance. The coordination between the expression of these genes and the antioxidant system of the plant can allow a plant to tolerate osmotic stress. Therefore, the difference between a susceptible and a tolerant plant is the type of action of the transporters and its antioxidant system when facing osmotic stress.

REFERENCES

Ahmad, P, MM Azooz, and MNV Prasad. 2013. *Salt Stress in Plants.* Springer.

Ahn, SJ, R Shin, and DP Schachtman. 2004. "Expression of *KT/KUP* genes in Arabidopsis and the role of root hairs in K+ uptake." *Plant Physiology* 134 (3):1135–1145.

Alberts, B, A Johnson, J Lewis, M Raff, K Roberts, and P Walter. 2002. "Electron-transport chains and their proton pumps." *Molecular Biology of the Cell* 4.

Albrecht, V, O Ritz, S Linder, K Harter, and J Kudla. 2001. "The *NAF* domain defines a novel protein–protein interaction module conserved in Ca^{2+}-regulated kinases." *The EMBO Journal* 20 (5):1051–1063.

Anil, VS, P Krishnamurthy, S Kuruvilla, K Sucharitha, G Thomas, and MK Mathew. 2005. "Regulation of the uptake and distribution of Na+ in shoots of rice (*Oryza sativa*) variety Pokkali: role of Ca^{2+} in salt tolerance response." *Physiologia Plantarum* 124 (4):451–464.

Apse, MP and E Blumwald. 2007. "Na+ transport in plants." *FEBS Letters* 581 (12):2247–2254.

Asano, T, N Tanaka, G Yang, N Hayashi, and SKomatsu.2005. "Genome-wide identification of the rice calcium-dependent protein kinase and its closely related kinase gene families: comprehensive analysis of the *CDPKs* gene family in rice." *Plant and Cell Physiology* 46 (2):356–366.

Assaha, DVM, A Ueda, H Saneoka, R Al-Yahyai, and MW Yaish. 2017. "The role of Na+ and K+ transporters in salt stress adaptation in glycophytes." *Frontiers in Physiology* 8:509.

Banuelos, MA, B Garciadeblas, B Cubero, and A Rodriguez-Navarro. 2002. "Inventory and functional characterization of the HAK potassium transporters of rice." *Plant Physiology* 130 (2):784–795.

Batistic, O and J Kudla. 2004. "Integration and channeling of calcium signaling through the CBL calcium sensor/CIPK protein kinase network." *Planta* 219 (6):915–924.

Bechtold, U and B Field. 2018. *Molecular Mechanisms Controlling Plant Growth during Abiotic Stress.* UK: Oxford University Press.

Berdy, SE, J Kudla, W Gruissem, and GE Gillaspy. 2001. "Molecular characterization of *At5PTase1*, an inositol phosphatase capable of terminating inositol trisphosphate signaling." *Plant Physiology* 126 (2):801–810.

Blumenthal, M, A Goldberg, and J Brinckmann. 2000. "Integrative medicine communications." *Herbal Medicines, Austin*: 419–423.

Blumwald, E. 2000. "Sodium transport and salt tolerance in plants." *Current Opinion in Cell Biology* 12 (4):431–434.

Boudsocq, M and J Sheen. 2009. "Stress signaling II: calcium sensing and signaling." In *Abiotic Stress Adaptation in Plants*, 75–90. Springer.

Britto, DT and HJ Kronzucker. 2008. "Cellular mechanisms of potassium transport in plants." *Physiologia Plantarum* 133 (4):637–650.

Cheng, SH, MR Willmann, H-C Chen, and J Sheen. 2002. "Calcium signaling through protein kinases. The Arabidopsis calcium-dependent protein kinase gene family." *Plant Physiology* 129 (2):469–485.

Chinnusamy, V, A Jagendorf, and J-K Zhu. 2005. "Understanding and improving salt tolerance in plants." *Crop Science* 45 (2):437–448.

Cunningham, KW and GR Fink. 1996. "Calcineurin inhibits *VCX1*-dependent H+/Ca^{2+} exchange and induces Ca^{2+} ATPases in *Saccharomyces cerevisiae.*" *Molecular and Cellular Biology* 16 (5):2226–2237.

Das, R and GK Pandey. 2010. "Expressional analysis and role of calcium regulated kinases in abiotic stress signaling." *Current Genomics* 11 (1):2–13.

DeWald, DB, J Torabinejad, CA Jones, JC Shope, AR Cangelosi, JE Thompson, GD Prestwich, and H Hama. 2001. "Rapid accumulation of phosphatidylinositol 4, 5-bisphosphate and inositol 1, 4, 5-trisphosphate correlates with calcium mobilization in salt-stressed *Arabidopsis.*" *Plant Physiology* 126 (2):759–769.

Diédhiou, CJ. 2006. *Mechanisms of Salt Tolerance: Sodium, Chloride and Potassium Homeostasis in Two Rice Lines with Different Tolerance to Salinity Stress.* Bielefeld (Germany): Bielefeld University.

Flowers, TJ, R Munns, and TD Colmer. 2014. "Sodium chloride toxicity and the cellular basis of salt tolerance in halophytes." *Annals of Botany* 115 (3):419–431.

Fricke, W. 2004. "Rapid and tissue-specific accumulation of solutes in the growth zone of barley leaves in response to salinity." *Planta* 219 (3):515–525.

Gaxiola, RA, J Li, S Undurraga, LM Dang, GJ Allen, SL Alper, and GR Fink. 2001. "Drought-and salt-tolerant plants result from overexpression of the *AVP1* H+-pump." *Proceedings of the National Academy of Sciences* 98 (20):11444–11449.

Guo, Y, U Halfter, M Ishitani, and J-K Zhu. 2001. "Molecular characterization of functional domains in the protein kinase *SOS2* that is required for plant salt tolerance." *The Plant Cell* 13 (6):1383–1400.

Halfter, U, M Ishitani, and J-K Zhu. 2000. "The *Arabidopsis SOS2* protein kinase physically interacts with and is activated by the calcium-binding protein *SOS3*." *Proceedings of the National Academy of Sciences* 97 (7):3735–3740.

Hanin, M, C Ebel, M Ngom, L Laplaze, and K Masmoudi. 2016. "New insights on plant salt tolerance mechanisms and their potential use for breeding." *Frontiers in Plant Science* 7:1787.

Harmon, AC, M Gribskov, and JF Harper. 2000. "*CDPKs*—a kinase for every Ca²⁺ signal?" *Trends in Plant Science* 5 (4):154–159.

Hasanuzzaman, M, K Nahar, MdM Alam, R Roychowdhury, and M Fujita. 2013. "Physiological, biochemical, and molecular mechanisms of heat stress tolerance in plants." *International Journal of Molecular Sciences* 14 (5):9643–9684.

Hasegawa, PM, RA Bressan, J-K Zhu, and HJ Bohnert. 2000. "Plant cellular and molecular responses to high salinity." *Annual Review of Plant Biology* 51 (1):463–499.

Hepler, PK. 2005. "Calcium: a central regulator of plant growth and development." *The Plant Cell* 17 (8):2142–2155.

Hirschi, KD, VD Korenkov, NL Wilganowski, and GJ Wagner. 2000. "Expression of *Arabidopsis CAX2* in tobacco. Altered metal accumulation and increased manganese tolerance." *Plant Physiology* 124 (1):125–134.

Hirschi, KD, R-G Zhen, KW Cunningham, PA Rea, and GR Fink. 1996. "*CAX1*, an H⁺/Ca²⁺ antiporter from *Arabidopsis*." *Proceedings of the National Academy of Sciences* 93 (16):8782–8786.

Kamiya, T, T Akahori, M Ashikari, and M Maeshima. 2006. "Expression of the vacuolar Ca²⁺/H⁺ exchanger, *OsCAX1a*, in rice: cell and age specificity of expression, and enhancement by Ca²⁺." *Plant and Cell Physiology* 47 (1):96–106.

Kasote, DM, SS Katyare, MV Hegde, and H Bae. 2015. "Significance of antioxidant potential of plants and its relevance to therapeutic applications." *International Journal of Biological Sciences* 11 (8):982.

Kolukisaoglu, Ü, S Weinl, D Blazevic, O Batistic, and J Kudla. 2004. "Calcium sensors and their interacting protein kinases: genomics of the *Arabidopsis* and rice CBL-CIPK signaling networks." *Plant Physiology* 134 (1):43–58.

Kordrostami, M, B Rabiei, and HH Kumleh. 2016. "Association analysis, genetic diversity and haplotyping of rice plants under salt stress using SSR markers linked to *SalTol* and morpho-physiological characteristics." *Plant Systematics and Evolution* 302 (7):871–890.

Kordrostami, M, B Rabiei, and HH Kumleh. 2017a. "Biochemical, physiological and molecular evaluation of rice cultivars differing in salt tolerance at the seedling stage." *Physiology and Molecular Biology of Plants* 23 (3):529–544.

Kordrostami, M, B Rabiei, and HH Kumleh. 2017b. "Different physiobiochemical and transcriptomic reactions of rice (*Oryza sativa* L.) cultivars differing in terms of salt sensitivity under salinity stress." *Environmental Science and Pollution Research* 24 (8):7184–7196.

Lee, SC and S Luan. 2012. "ABA signal transduction at the crossroad of biotic and abiotic stress responses." *Plant, Cell and Environment* 35 (1):53–60.

Liu, J, M Ishitani, U Halfter, C-S Kim, and J-K Zhu. 2000. "The *Arabidopsis thaliana SOS2* gene encodes a protein kinase that is required for salt tolerance." *Proceedings of the National Academy of Sciences* 97 (7):3730–3734.

Luan, S, J Kudla, M Rodriguez-Concepcion, S Yalovsky, and W Gruissem. 2002. "Calmodulins and calcineurin B–like proteins: calcium sensors for specific signal response coupling in plants." *The Plant Cell* 14 (suppl 1):S389–S400.

Maathuis, FJM. 2005. "The role of monovalent cation transporters in plant responses to salinity." *Journal of Experimental Botany* 57 (5):1137–1147.

Maathuis, FJM and A Amtmann. 1999. "K⁺ nutrition and Na⁺ toxicity: the basis of cellular K⁺/Na⁺ ratios." *Annals of Botany* 84 (2):123–133.

Maathuis, FJM, I Ahmad, and J Patishtan. 2014. "Regulation of Na⁺ fluxes in plants." *Frontiers in Plant Science* 5:467.

Machado, RMA and RP Serralheiro. 2017. "Soil salinity: effect on vegetable crop growth. Management practices to prevent and mitigate soil salinization." *Horticulturae* 3 (2):30.

Mitsuya, S, K Yano, M Kawasaki, M Taniguchi, and H Miyake. 2002. "Relationship between the distribution of Na and the damages caused by salinity in the leaves of rice seedlings grown under a saline condition." *Plant Production Science* 5 (4):269–274.

Munns, R. 2005. "Genes and salt tolerance: bringing them together." *New Phytologist* 167 (3):645–663.

Munns, R and M Tester. 2008. "Mechanisms of salinity tolerance." *Annual Review of Plant Biology* 59:651–681.

Niu, X, RA Bressan, PM Hasegawa, and JM Pardo. 1995. "Ion homeostasis in NaCl stress environments." *Plant Physiology* 109 (3):735.

Qados, AMSA. 2011. "Effect of salt stress on plant growth and metabolism of bean plant *Vicia faba* (L.)." *Journal of the Saudi Society of Agricultural Sciences* 10 (1):7–15.

Qi, Z, CR Hampton, R Shin, BJ Barkla, PJ White, and DP Schachtman. 2008. "The high affinity K⁺ transporter *AtHAK5* plays a physiological role in planta at very low K⁺ concentrations and provides a caesium uptake pathway in *Arabidopsis*." *Journal of Experimental Botany* 59 (3):595–607.

Rabbani, MA, K Maruyama, H Abe, MA Khan, K Katsura, Y Ito, K Yoshiwara, M Seki, K Shinozaki, and K Yamaguchi-Shinozaki. 2003. "Monitoring expression profiles of rice genes under cold, drought, and high-salinity stresses and abscisic acid application using cDNA microarray and RNA gel-blot analyses." *Plant Physiology* 133 (4):1755–1767.

Rahnama, A, R Munns, K Poustini, and M Watt. 2011. "A screening method to identify genetic variation in root growth response to a salinity gradient." *Journal of Experimental Botany* 62 (1):69–77.

Rejeb, IB, V Pastor, and B Mauch-Mani. 2014. "Plant responses to simultaneous biotic and abiotic stress: molecular mechanisms." *Plants* 3 (4):458–475.

Rodríguez, M, E Canales, and O Borrás-Hidalgo. 2005. "Molecular aspects of abiotic stress in plants." *Biotecnología Aplicada* 22 (1):1–10.

Rodríguez-Navarro, A, and F Rubio. 2006. "High-affinity potassium and sodium transport systems in plants." *Journal of Experimental Botany* 57 (5):1149–1160.

Rodríguez-Rosales, MP, FJ Gálvez, R Huertas, MN Aranda, M Baghour, O Cagnac, and K Venema. 2009. "Plant NHX cation/proton antiporters." *Plant Signaling and Behavior* 4 (4):265–276.

Rusnak, F and P Mertz. 2000. "Calcineurin: form and function." *Physiological Reviews* 80 (4):1483–1521.

Saijo, Y, S Hata, J Kyozuka, K Shimamoto, and K Izui. 2000. "Over-expression of a single Ca²⁺-dependent protein kinase confers both cold and salt/drought tolerance on rice plants." *The Plant Journal* 23 (3):319–327.

Sanders, D, C Brownlee, and JF Harper. 1999. "Communicating with calcium." *The Plant Cell* 11 (4):691–706.

Santa-María, GE, F Rubio, J Dubcovsky, and A Rodríguez-Navarro. 1997. "The *HAK1* gene of barley is a member of a large gene family and encodes a high-affinity potassium transporter." *The Plant Cell* 9 (12):2281–2289.

Schachtman, DP and JI Schroeder. 1994. "Structure and transport mechanism of a high-affinity potassium uptake transporter from higher plants." *Nature* 370 (6491):655.

Shi, H and J-K Zhu. 2002. "Regulation of expression of the vacuolar Na+/H+ antiporter gene *AtNHX1* by salt stress and abscisic acid." *Plant Molecular Biology* 50 (3):543–550.

Shi, H, M Ishitani, C Kim, and J-K Zhu. 2000. "The *Arabidopsis thaliana* salt tolerance gene *SOS1* encodes a putative Na+/H+ antiporter." *Proceedings of the National Academy of Sciences* 97 (12):6896–6901.

Shi, H, FJ Quintero, JM Pardo, and J-K Zhu. 2002. "The putative plasma membrane Na+/H+ antiporter *SOS1* controls long-distance Na+ transport in plants." *The Plant Cell* 14 (2):465–477.

Shigaki, T and KD Hirschi. 2006. "Diverse functions and molecular properties emerging for *CAX* cation/H+ exchangers in plants." *Plant Biology* 8 (4):419–429.

Shinozaki, K and K Yamaguchi-Shinozaki. 2000. "Molecular responses to dehydration and low temperature: differences and cross-talk between two stress signaling pathways." *Current Opinion in Plant Biology* 3 (3):217–223.

Shokri-Gharelo, R and PM Noparvar. 2018. "Molecular response of canola to salt stress: insights on tolerance mechanisms." *PeerJ* 6:e4822.

Shrivastava, P and R Kumar. 2015. "Soil salinity: a serious environmental issue and plant growth promoting bacteria as one of the tools for its alleviation." *Saudi Journal of Biological Sciences* 22 (2):123–131.

Su, H, D Golldack, C Zhao, and HJ Bohnert. 2002. "The expression of HAK-type K+ transporters is regulated in response to salinity stress in common ice plant." *Plant Physiology* 129 (4):1482–1493.

Sze, H, X Li, and M G Palmgren. 1999. "Energization of plant cell membranes by H+-pumping ATPases: regulation and biosynthesis." *The Plant Cell* 11 (4):677–689.

Tester, M and R Davenport. 2003. "Na+ tolerance and Na+ transport in higher plants." *Annals of Botany* 91 (5):503–527.

Turkan, I. 2011. *Plant Responses to Drought and Salinity Stress: Developments in a Post-genomic Era*. Vol. 57. Academic Press.

Tuteja, N and SK Sopory. 2008. "Chemical signaling under abiotic stress environment in plants." *Plant Signaling and Behavior* 3 (8):525–536.

Volkov, V. 2015. "Salinity tolerance in plants. Quantitative approach to ion transport starting from halophytes and stepping to genetic and protein engineering for manipulating ion fluxes." *Frontiers in Plant Science* 6:873.

Wan, B, Y Lin, and T Mou. 2007. "Expression of rice Ca2+-dependent protein kinases (CDPKs) genes under different environmental stresses." *FEBS Letters* 581 (6):1179–1189.

Wang, Y-H, DF Garvin, and LV Kochian. 2002. "Rapid induction of regulatory and transporter genes in response to phosphorus, potassium, and iron deficiencies in tomato roots. Evidence for cross talk and root/rhizosphere-mediated signals." *Plant Physiology* 130 (3):1361–1370.

Wu, GQ and SM Wang. 2012. "Calcium regulates K+/Na+ homeostasis in rice (*Oryza sativa* L.) under saline conditions." *Plant Soil Environment* 58 (3):121–127.

Wu, H. 2018. "Plant salt tolerance and Na+ sensing and transport." *The Crop Journal*.

Wu, L, Z Fan, L Guo, Y Li, Z-L Chen, and L-J Qu. 2005. "Overexpression of the bacterial nhaA gene in rice enhances salt and drought tolerance." *Plant Science* 168 (2):297–302.

Xiang, Y, Y Huang, and L Xiong. 2007. "Characterization of stress-responsive *CIPK* genes in rice for stress tolerance improvement." *Plant Physiology* 144 (3):1416–1428.

Xiong, L, KS Schumaker, and J-K Zhu. 2002. "Cell signaling during cold, drought, and salt stress." *The Plant Cell* 14 (suppl 1):S165–S183.

Xu, G-Y, PSCF Rocha, M-L Wang, M-L Xu, Y-C Cui, L-Y Li, Y-X Zhu, and X Xia. 2011. "A novel rice calmodulin-like gene, *OsMSR2*, enhances drought and salt tolerance and increases ABA sensitivity in *Arabidopsis*." *Planta* 234 (1):47–59.

Yadav, S, M Irfan, A Ahmad, and S Hayat. 2011. "Causes of salinity and plant manifestations to salt stress: a review." *Journal of Environmental Biology* 32 (5):667.

Yang, Q, Z-Z Chen, X-F Zhou, H-B Yin, X Li, X-F Xin, X-H Hong, J-K Zhu, and Z Gong. 2009. "Overexpression of *SOS* (Salt Overly Sensitive) genes increases salt tolerance in transgenic *Arabidopsis*." *Molecular Plant* 2 (1):22–31.

Zhu, J-K. 2002. "Salt and drought stress signal transduction in plants." *Annual Review of Plant Biology* 53 (1):247–273.

Zielinski, RE. 2002. "Characterization of three new members of the *Arabidopsis thaliana* calmodulin gene family: conserved and highly diverged members of the gene family functionally complement a yeast calmodulin null." *Planta* 214 (3):446–455.

29 Drought-Induced Gene Expression Reprogramming Associated with Plant Metabolic Alterations and Adaptation

Omid Askari-Khorasgani and Mohammad Pessarakli

CONTENTS

ABBREVIATIONS

$[Ca^{2+}]_{cyt}$	Cytosolic free Ca^{2+}
$[Ca^{2+}]_{nuc}$	Nuclear Ca^{2+}
$[Ca^{2+}]_{org}$	Organellar Ca^{2+}
ABA	Abscisic acid
ABF	Abscisic acid-binding factor
ABI	*ABSCISIC ACID-INSENSITIVE*
ABRE	ABA-responsive promoter elements
ADF	*Actin depolymerizing factor*
AREB	ABA RESPONSIVE ELEMENTS-BINDING PROTEIN
AtHB	*Arabidopsis thaliana* homeobox
bHLH	Basic helix-loop-helix
BPMs	BTB/POZ AND MATH DOMAIN proteins
bZIP	Basic leucine zipper
CaM	Calmodulin
CBF	C-repeat binding factor
Chl	Chlorophyll
DRE	Dehydration responsive element
DREB	Dehydration responsive elements binding protein
ER	Endoplasmic reticulum
ERD	EARLY RESPONSIVE TO DEHYDRATION
ERFs	Ethylene response factors
H3K4me3	Histone H3 trimethyl lysine 4
HD-Zip	Homeodomain-leucine zipper
HECT	Homologous to the E6-AP Carboxy
HSPs	Heat shock proteins
LEA	Late embryogenesis abundant
miRNA	MicroRNA
mRNA	Messenger RNA
MYB	Myeloblastosis

MYC	Myelocytomatosis
NAC	Nitrogen assimilation control protein
NADP-ME	NADP-malic enzyme
NCED	9-*cis*-epoxycarotenoid dioxygenase
NF-Y	Nuclear factor Y
PARP	Poly(ADP-ribose) polymerase
PDH	*Pyruvate dehydrogenase*
PIP	Plasma membrane intrinsic protein
piRNA	Piwi-interacting RNA
ProDH	Proline dehydrogenase
PSII	Photosystem II
RAB	Ras-associated GTP-binding protein
RBPs	RNA-binding proteins
RD	RESPONSIVE TO DESICCATION
Rma1H1	RING membrane-anchor 1 homolog 1
RNP	Ribonucleoprotein
ROS	Reactive oxygen species
SA	Salicylic acid
SCF	S-phase kinase-associated protein 1-Cullin-F-Box protein
Ser5P Pol II	Serine 5 phosphorylated polymerase II
siRNA	Small interfering RNA
SKP1	S-phase kinase-associated protein 1
SNF	Sucrose non-fermenting
sp1	*Specificity protein 1*
TCA	Tricarboxylic acid
TFs	Transcription factors
Ub	Ubiquitin
WRKY	W-boxes receptor protein kinase
WT	Wild-type
ZFP	Cys2His2 zinc finger protein

29.1 INTRODUCTION

All organisms require the ability to sense their surroundings and adapt. Such capabilities allow them to thrive and survive in a wide range of habitats and to withstand environmental changes (Woodson, 2016). This can be achieved by using different types of adaptations such as structural, physiological, biochemical, molecular, and behavioral adaptations, all of which are under the control of gene expression regulation and signal transduction (De Boer and Volkov, 2003; Zhang et al., 2004; Vasquez-Robinet et al., 2008; Ahuja et al., 2010; Hornung, 2011). Water stress sensing triggers signaling pathways followed by altering large-scale transcriptional reprogramming that mitigates the deleterious effect of drought stress by maintaining and/or restoring cellular processes through improved homeostasis (Zhu, 2002; Berkhout et al., 2013; Han and Wagner, 2014). The adaptation of all living organisms requires the capacity for transcriptional regulation of genes and metabolic pathways as two fundamental biological processes necessary for homeostasis, growth, development, and differentiation. A growing body of evidence indicates that cofactors or cosubstrates of chromatin-modifying enzymes are the key metabolic regulators, linking the metabolic programming, gene expression, and chromatin states, with high selectivity depending on time and dose (Caldana et al., 2012; van der Knaap and Verrijzer, 2016). Plants and other eukaryotes use their energy-producing organelles (i.e., mitochondria and chloroplasts) to sense their surroundings and adapt through a detoxification and defense system (Berry, 2003; Woodson, 2016). In response to a changing cellular or external environment, these organelles can emit "retrograde" signals to the nucleus that alter gene expression and cell physiology, regulating growth, development, and responses to environmental cues. This signaling is important in plants, fungi, and animals and controls transcription factors (TFs) that act as master regulators of gene expression programs involved in diverse cellular functions, including photosynthesis, energy production/storage, stress responses, growth, cell death, aging, and tumor progression. Chloroplast retrograde signals, singlet oxygen, ubiquitination, the 26S proteasome, and the cellular degradation machinery in plants are known to lead to the gene transcriptional and post-transcriptional regulatory mechanisms (Caldana et al., 2012; van der Knaap and Verrijzer, 2016; Woodson, 2016). Classic metabolic enzymes may directly regulate gene expression at both nuclear and mitochondrial levels or via interplay between metabolites, enzymes, and regulatory proteins. These dual-function proteins may provide a direct link between metabolic programming and the control of gene expression (Hall et al., 2004; Chubukov et al., 2012; Castello et al., 2015; van der Knaap and Verrijzer, 2016). In all organisms, RNA-binding proteins (RBPs), which form ribonucleoprotein (RNP) complexes, RNA–protein interactions, and protein–protein interactions have critical roles in post-transcriptional modification and RNA processing, such as microRNA biogenesis, pre-mRNA splicing, polyadenylation, stability, function, transport (from the nucleus to the cytoplasm), localization, translation, and degradation, as well as various cellular processes, such as growth, development, and stress management (Glisovic et al., 2008; Hogan et al., 2008; Kang et al., 2013; Klass et al., 2013; Muppirala et al., 2013; Gerstberger et al., 2014; Kechavarzi and Janga, 2014; Castello et al., 2015; Popova et al., 2015; Stoiber et al., 2015; Francisco-Velilla et al., 2016; Treiber et al., 2017). In addition, post-translational modifications of chromatin play a major role in the activation or repression of gene transcription and the modulation of cellular processes such as genome stability, and DNA replication and repair. These include covalent modifications of histones and DNA, such as acetylation, methylation, phosphorylation of the histones, and DNA methylation, and non-covalent alteration of the nucleosome occupancy, positioning (nucleosome destabilization or substitution of canonical histones by histone variants), or composition. Some of these chromatin modifications are involved in first, the maintenance of stable patterns of gene expression and second, enzyme modification, usually referred to as *epigenetic regulation* and known as *stress memory*, the molecular mechanisms of which are still elusive (Ding et al., 2012; Han and Wagner, 2014; Jarillo et al., 2014; van der Knaap and Verrijzer, 2016). Therefore, due to the importance of the regulation of gene expression in plant responses to environmental stimuli and, thus, plant performance, this review discusses the regulatory pathways of gene expression and the associated metabolic modifications during plant adaptation processes to stress, particularly drought, to highlight the importance of metabolites and regulatory pathways that require future studies in breeding programs for plant/crop stress tolerance.

29.2 REGULATION OF GENE EXPRESSION

29.2.1 CALCIUM/HORMONE/REDOX SIGNALING

Several abiotic challenges result in immediate, transient oscillations or spikes in Ca^{2+} concentrations, mostly cytosolic free Ca^{2+} ($[Ca^{2+}]_{cyt}$) but in some cases, nuclear Ca^{2+} ($[Ca^{2+}]_{nuc}$) or organellar Ca^{2+} ($[Ca^{2+}]_{org}$) concentrations too, which are restored to basal levels within minutes. Gene expression regulation in both plants and animals is initiated by Ca^{2+} oscillations, sensed and signaled by Ca^{2+} sensor proteins or specific protein kinases, such as calmodulin (CaM), that interact with TFs to regulate gene expression (Khurana et al., 2008; Reddy et al., 2011; Hashimoto et al., 2012; Scholz et al., 2015; Virdi et al., 2015). The signalization generated by the interplay of Ca^{2+} oscillations, the phytohormone abscisic acid (ABA), and redox status regulates plant physiological processes and thus, plant growth, development, and defense mechanisms (Kamata and Hirata, 1999; Trachootham et al., 2008; Vickers et al., 2009; Spoel and Loake, 2011; Tanou et al., 2012; Alam et al., 2014; Dietz, 2014; Diao et al., 2016; Li, 2016; Tavladoraki et al., 2016; Nahar et al., 2017; Pourghayoumi et al., 2017). Phytohormones and metabolic and protein alterations caused by environmental stimuli affect cellular processes, signalization, growth, and development. Pioneer efforts in identifying the earliest molecular events controlling cell division in response to osmotic stress have evidenced that ethylene signaling, initiated by ethylene response factors (ERFs), interacts with gibberellin and DELLA and consequently, acts upstream on cell cycle arrest and on cyclin-dependent kinase A activity, independently of transcriptional control. ERFs interact with gibberellin, DELLA (a negative regulator of gibberellin and a regulator of ABA metabolism and GA-ABA pathways), ABA, and redox signaling to control redox and hormone homeostasis, various cellular processes, signaling, gene regulation, and stress responses. Consequently, the ethylene/ABA/gibberellin/DELLA signaling pathway affects cell growth, development, tolerance, root and shoot system architecture, and hence, plant adaptation and performance under optimal and stress conditions (Eckardt, 2007; Ubeda-Tomas et al., 2008; Bonhomme et al., 2012; Hauvermale et al., 2012; Müller and Munné-Bosch, 2015; Zhang et al., 2016a; Phukan et al., 2017; Van De Velde et al., 2017).

29.2.2 Regulation of Signal Transduction and Associated Cellular Processes Induced by Protein Processing and Homeostasis

Depending on early or late responses to stress, gene expression is regulated by three protein groups. With a profound impact on all aspects of cell processes, these proteins are involved in growth and development, environmental sensing, signaling, defense responses, intercellular communication, and selective exchange interfaces. In the first group, functional proteins with known enzymatic or structural functions can be induced by genes encoding protective proteins (e.g., hydrophilins and detoxifying enzymes), enzymes related to carbohydrate metabolism, and proteins involved in the transport of water and other molecules. The second group contains regulatory proteins (proteins and RNAs) and the third, proteins with as yet unknown functions (Showalter, 1993; Bartels, 2005; Umezawa et al., 2006; Augustine, 2016). Functional proteins operate in direct protection from stress, avoiding damage to membranes and proteins, and include chaperones, heat shock proteins (HSPs) as molecular chaperones, late embryogenesis abundant (LEA) proteins, antifreeze proteins, ion transporters, messenger RNA (mRNA)-binding proteins, key enzymes for osmolyte biosynthesis, water channel proteins, sugar and proline (Pro) transporters, detoxification enzymes, osmotin, and various proteases. The second group of stress-inducible genes codes for regulatory proteins involved in the control of the downstream signaling pathways and gene expression. These include TFs, protein kinases, phosphatases, protein phosphatases, enzymes involved in phospholipid metabolism, and other signaling molecules such as CaM-binding protein, hormones, RNA molecules, etc. (Shinozaki and Yamaguchi-Shinozaki, 2007; Khurana et al., 2008; Augustine, 2016; Gonçalves, 2016). Interestingly, a growing body of evidence indicates that some proteins, such as RBPs and chaperones, classified as functional proteins may have both functional and regulatory roles, covering growth, development, and stress responses. For example, chaperons are involved in signaling, protein processing, and defense responses as well as cellular

processes and homeostasis (Panter et al., 2000; Wang et al., 2004; Sakurai and Ota, 2011; Kang et al., 2013; Jacob, 2017). Engineering RBPs and chaperones is critical in improving tolerance not only in plants but also in bacteria, as well as their symbiotic relationships (Balsiger et al., 2004; Asadi Rahmani et al., 2009; Alexandre and Oliveira, 2011). The phosphorylation of proteins is of considerable importance for regulating plant growth. As a rapid and transient post-translational modification, protein phosphorylation achieves a fine-tuned regulation of protein function in a wide array of cellular processes during development or in response to environmental cues, from signaling cascades to gene expression (Bonhomme et al., 2012). In all living organisms, protein phosphorylation/dephosphorylation plays a critical role in the evolution of the genome; signaling and gene expression; metabolism (mainly carbohydrate and nitrogen); assimilate compartmentalization and availability, and, thus, symbiosis; protein processes (e.g., degradation, localization, and functional interactions); DNA-related processes and metabolism; and cell processes, such as the cell cycle, cell function, signaling, growth, division, differentiation, motility, organelle trafficking, membrane transport, and immunity (also muscle contraction, learning, and memory in humans) (Cohen, 2002; Yoshida and Parniske, 2005; Huber, 2007; Ciesla et al., 2011; Swaney et al., 2013; Tripodi et al., 2015; Garcia-Garcia et al., 2016; Jin et al., 2016; Lee and Yaffe, 2016; Jayaraman et al., 2017; Vlastaridis et al., 2017; Zipfel and Oldroyd, 2017). Particularly, early protein phosphorylation/dephosphorylation events are essential for rapid growth adjustment in dehydrated plants (Bonhomme et al., 2012). Similar to phosphorylation/dephosphorylation, DELLA proteins positively regulate signaling pathways and transcriptional regulation associated with nodulation, bacterial colonization, symbiosis, and, thus, stress tolerance (Jin et al., 2016). Protein phosphorylation and stress responses can be regulated by the status of plant hormones such as ABA (Kline et al., 2010) and salicylic acid (SA) (Wu et al., 2015). Advanced analytical procedures such as OMICs, and more specifically phosphoproteomics, are required to decipher the relationships between early stress signaling; protein processing, such as phosphorylation/dephosphorylation, particularly for photosynthetic enzymes, photosystem II (PSII) core proteins, regulatory proteins, protein kinases such as sucrose non-fermenting (SNF)-related protein kinases, and protein phosphatases; carbohydrate (e.g., sucrose and trehalose) and osmolyte metabolism; ABA- and SA- induced responses and their integrative roles in gene ontology; cellular processes, metabolism, and photosynthesis under normal and different stress conditions (Shin et al., 2007; Kline et al., 2010; Bonhomme et al., 2012; Chen and Hoehenwarter, 2015; Ilhan et al., 2015; Wu et al., 2015; Alqurashi et al., 2017; Arias-Baldrich et al., 2017; Betterle et al., 2017; Chen et al., 2017; Uhrig et al., 2017). *EARLY RESPONSIVE TO DEHYDRATION (ERD)* genes are a group of plant genes having functional (e.g., a nucleic localization function and a transcriptional activation function) and probably regulatory roles (i.e., regulation of ABA and SA defense pathways; regulation of drought and salinity responses before ABA accumulation, and thus presumably in an ABA-independent manner, by encoding a Clp (caseinolytic protease) protease regulatory subunit; regulation of C-repeat binding factor (CBF); and regulation of TFs and post-transcriptional genes (the rehydration-inducible *ERD5* gene encodes the proline dehydrogenase [*ProDH*] gene, which is involved in degradation of accumulated proline during drought stress and dehydration and regulation of rehydration-inducible genes) in various plant cellular processes (e.g., growth; development; accumulation of soluble sugars, proteins, and antioxidant enzymes; and programmed or stress-induced cell death), as well as signaling, hormones (e.g., ABA and SA), multiple stress responses (mainly drought, high salinity, ABA, and cold stress), and thus, rapid adaptation (Satoh et al., 2002; Nakashima and Yamaguchi-Shinozaki, 2005; Alves et al., 2011; Rai et al., 2012, 2016; Shao et al., 2014; Yu et al., 2017). In *Arabidopsis*, the overexpression of the *Brassica juncea ERD4* (*BjERD4*) chloroplast gene, which increases under stress conditions (dehydration, sodium chloride, low temperature, heat, or ABA and SA treatments), encodes a novel RBP, which is subsequently involved in functional and regulatory pathways including plant growth and development, stress responses, and adaptation, as described earlier (Rai et al., 2012, 2016; Kang et al., 2013).

The addition of ubiquitin (Ub) to a substrate protein, so-called *ubiquitination*, which is catalyzed in three consecutive steps by E1, E2, and E3 enzymes, is a reversible post-translational modification of proteins, affecting many aspects of cellular and protein processes (Lee et al., 2009; Jackson and Durocher, 2013; Sharma et al., 2016; Han et al., 2017; Wang et al., 2017). In higher plants, E3 Ub ligases are encoded by a large gene family comprised of diverse isoforms. Based on the subunit composition, E3s can be classified into two main groups. The homologous to the E6-AP Carboxy (HECT) and RING/U-box E3 classes consist of a single subunit, forming different covalent bonds and thus binding affinity and target specificity, while the S-phase kinase-associated protein 1 (SKP1)-Cullin-F-Box protein (SCF) and anaphase-promoting complex E3 ligases consist of multiple polypeptides (Ciechanover et al., 1982; Mazzucotelli et al., 2006; Lee et al., 2009; McDowell and Philpott, 2016). Ubiquitination regulates a myriad of cellular processes, including cell division; differentiation; hormone perception, biosynthesis, and signaling; degradation of hormone-specific transcription factors; improving stress responses, probably through hormone (e.g., ABA) signaling; protein processes (e.g.., biosynthesis, stability, trafficking, degradation, modification, distribution, activity, and interactions: for example, expression of HSPs and modulating the amount, activity, and interactions of regulatory proteins and stress-responsive TF proteins), and thus, stress responses and tolerance (Santner and Estelle, 2010; Lyzenga and Stone, 2011; Schwarz and Patrick, 2012; Chen and Hellmann, 2013; Sharma et al., 2016; Uckelmann and Sixma, 2017). Ubiquitination is also critical in the regulation of DNA replication and repair, genome stability, gene silencing, and multi-protein interactions (Lyzenga and Stone, 2011; Schwertman et al., 2016; Sharma et al., 2016; Uckelmann and

Sixma, 2017; Zhang et al., 2017). Ubiquitination reacts with sumoylation, phosphorylation, and acetylation to regulate plant cell signaling and protein processing and thus, growth, development, and immunity or defense responses against biotic and abiotic stresses (He et al., 2017; Wang et al., 2017). Plant E3 Ub ligases are involved in a myriad of cellular and physiological processes in plants. Their activity and functions vary depending on their expression levels, which are under the influence of stress conditions, such as drought stress severity, and also plant tissue and organ types (e.g., roots and leaves). To take an example, E3 Ub ligase RING membrane-anchor 1 homolog 1 (Rma1H1), a hot pepper (*Capsicum annuum*) homolog of a human RING membrane-anchor 1 E3 Ub ligase, and Rma1, an *Arabidopsis* homolog of Rma1H1, can rapidly be induced by various abiotic stresses and confer drought tolerance by lowering the levels of plasma membrane–localized water channel aquaporin (AQP) in leaf protoplasts via inhibition of AQP trafficking from the endoplasmic reticulum (ER) membrane to the plasma membrane and subsequent proteasomal degradation (Lee et al., 2009). This Rma1H1-induced reduction of PIP2;1 was inhibited by MG132, an inhibitor of the 26S proteasome. Rma1H1 interacted with PIP2;1 *in vitro* and ubiquitinated it *in vivo*. Reversibly, suppression of Rma homologs elevated the levels of PIP2;1 in *Arabidopsis* leaf protoplasts (Lee et al., 2009). In one study, antisense PIP2–transformed lines resulted in reduced hydraulic conductance and alterations in plant morphology, while the overexpression of a plasma membrane *AQP PIP1b* in transgenic tobacco improved plant vigor under favorable growth conditions but not under drought or salt stress (Khurana et al., 2008). Plant responses mediated by AQPs are dependent on plant genotype; plant tolerance and defense mechanisms; protein processing (e.g., phosphorylation, glycosylation, proteolytic, methylation, ubiquitination, deamidation, heteromerization, disulfide bond formation, and protonation processes); mRNA levels; AQP regulatory mechanisms, which can be modified at both transcriptional and post-translational levels; AQP inhibitors; AQP type and expression pattern; the localization and stability of AQPs and their interactions; cell, tissue, and organ type; cell metabolism; cell pH; cell expansion; cell CO_2 diffusion pathway; permeability; intracellular and cell-to-cell transport pathways; cell signaling triggered by calcium, redox, and hormones; symbiosis; plant development stage; transpiration; hydraulic conductivity, light (intensity, quality, and duration), relative humidity, circadian rhythms, and environmental biotic and abiotic stimuli (Johansson et al., 2000; Chaumont and Tyerman, 2014; Ruiz-Lozano and Aroca, 2017; Santoni, 2017; Uehlein et al., 2017). Therefore, approaches involving alterations in plant plasma membrane AQPs should be viewed with prudence. A decreased transcript level of AQPs in roots can also be seen as enabling the construction of a barrier against water efflux from roots to dry soil due to reduced membrane water permeability (Bogeat-Triboulot et al., 2007). Taken together, depending on the early or late responses to dehydration and stress conditions, advanced analytical procedures such as OMICs are of critical importance to unraveling the relationships between gene expression and metabolic pathways, such as protein, hormone, carbohydrate, and osmolyte metabolism and also homeostasis, as well as cell cycles, stress signaling, and ubiquitination under different stress conditions.

29.2.3 Gene Expression Modifies Plant Metabolic Pathways Associated with the Activity of Enzymes and Proteins as a Plant Adaptation Mechanism to Cope with Stress Conditions

Different stress conditions, such as different combinations of stresses, severity, duration, and repetition, induce unique signaling pathways with differential stress responses, all of which are under the influence of plant genetic and epigenetic processes, growth and development stage, tolerance, defense mechanisms, and circadian rhythms (Mittler, 2006; Harb et al., 2010; Strozycki et al., 2010; Atkinson and Urwin, 2012; Ding et al., 2012; Atkinson et al., 2013; Caldeira et al., 2014a,b; Kissoudis et al., 2014; Suzuki et al., 2014; Pandey et al., 2015; Trerotola et al., 2015; van der Knaap and Verrijzer, 2016). Stress-induced signaling targets TF proteins, the regulation of which is under the influence of protein processing, mainly phosphorylation/dephosphorylation, mitogen-activated protein kinase signaling, and calcium/hormone/redox signaling, to regulate a large variety of gene expression patterns, and there exists an overlap between the expression patterns, leading to metabolic change and thus, stress responses (Barolo and Posakony, 2002; Agarwal et al., 2006; Whitmarsh, 2007; Scholz et al., 2015). Studying *Arabidopsis* plants, Harb and coworkers (2010) described that plant sensing and responding processes to mild drought stress were analogous to the more drastic progressive wilting or dehydration causing lethality (Harb et al., 2010). According to Harb and coworkers (2010), the early stage of mild drought stress induced the expression of some characteristic genes, such as *Dehydration Responsive Elements Binding 2A* (*DREB2A*), *Abscisic Acid-Binding Factor 3* (*ABF3*), and *9-Cis-Epoxycarotenoid Dioxygenase 3* (*NCED3*), and downstream gene regulation (*RESPONSIVE TO DESICCATION* [*RD22, RD29A, RD29B*] and *Ras-associated GTP-binding protein 18* [*RAB18*]). They concluded that ABA is needed for normal drought response and that any perturbation in ABA biosynthesis or signaling will negatively affect plant growth under drought. They found that stress perception and signaling occurred at an early stage of mild drought treatment and consequently, facilitated a molecular genetic and physiological dissection of subsequent responses to the stress (Harb et al., 2010). Mild stress treatment, therefore, could be a good model system to dissect the response and resistance of plants to drought (Harb et al., 2010).

As an acclimation strategy, the *A. thaliana homeobox domain leucine-zipper 7* and *12* (*AtHB7* and *AtHB12*) genes act as negative growth regulators of plants to allow plants to survive under prolonged drought stress (Olsson et al., 2004). The transcription of *AtHB7* and *AtHB12* depends on water deficit conditions and ABA level, which affects the activity of the serine/threonine phosphatases *ABSCISIC ACID-INSENSITIVE 1* and *2* (*ABI1* and *ABI2*) (Olsson et al., 2004). Depending on growth and development stages, expression levels, ABA levels, and water stress conditions, *AtHB7* and *AtHB12* differentially and divergently regulate plant growth, development, and water stress responses. The functional differences could be due to their dissimilar effects on protein processes, such as phosphorylation motifs, amino acid production pathways, protein–protein interactions, and target-binding in ABA-dependent and -independent manners (Olsson et al., 2004; Re et al., 2014). The expression of *AtHB12*, which could be induced by either ABA or water deficit, improved root vasculature differentiation; root elongation; inflorescence stem, leaf, and flower development in young plants under standard growth conditions; and seed production in water-stressed plants. *AtHB7* promoted leaf development, chlorophyll levels, and photosynthesis and reduced stomatal conductance and delayed senescence with more reduced seed yield in mature plants. The high-level expression of either gene deferred inflorescence stem growth and elongation, causing plants to develop more branched inflorescence stems, forming rosette leaves with a more rounded shape and with shorter petioles (Olsson et al., 2004; Re et al., 2014).

Depending on plant, environmental, and stress conditions, plant cell processes can be modulated by ABA-dependent and -independent signaling pathways. ABA biosynthesis and levels, ABA-induced gene expression and the associated stress responses, such as AQP levels, activity, and function; signal transduction; and hydraulic conductance, can be affected by stress conditions; plant genotype; cell, tissue and organ type; as well as circadian rhythms (Ito et al., 2009; Draye et al., 2010; Seung et al., 2012; Caldeira et al., 2014a,b; Lobet et al., 2014; Portolés and Zhang, 2014; Grundy et al., 2015; Singh and Laxmi, 2015; Lee et al., 2016; Manzoor et al., 2016). Environmental stimuli are initially transmitted into plant cells by the first round of signal transduction generated by crosstalk effects between pH/calcium/redox/hormone/chemical signaling (Trachootham et al., 2008; Pérez-Alfocea et al., 2011; Spoel and Loake, 2011; Tanou et al., 2012; Dietz, 2014; Li et al., 2017). Stress-induced ABA biosynthesis and subsequently, the signaling pathways generated by ABA lead to cell calcium elevation as well as the activation of TFs and calcium binding proteins (calcium sensors or EF-hands) that regulate a second round of signal transduction, gene expression, metabolic reprogramming (e.g., antioxidant activities), morphological and physiological responses, and thus, growth, development, and adaptation (Kamata and Hirata, 1999; Agarwal et al., 2005, 2006; Seki et al., 2007; Khurana et al., 2008; Hashimoto et al., 2012; Caldeira et al., 2014b; Das et al., 2014; Virdi et al., 2015; Ranty et al., 2016; Zheng et al., 2017). TFs are regulatory proteins critical in ABA-dependent or -independent regulation of specific sets of genes. Subsequently, these genes act as either functional genes that maintain cell structure and function or regulatory genes that regulate a third round of signal transduction and activation of regulatory proteins (Singh et al., 2002; Fujita et al., 2006; Ambawat et al., 2013; Nuruzzaman et al., 2013). The crosstalk regulatory system between regulatory genes and regulatory proteins controls all aspects of a plant's life.

Phosphorylation of the ABA receptor (by the PYR/PYL/RCAR [PYL] protein family, the negative regulator type 2C protein phosphatase [PP2C], and the positive regulator class III SNF1-related protein kinase 2 [SnRK2]) activates ABA-dependent TFs (Wasilewska et al., 2008; Cutler et al., 2010; Zhang et al., 2015; Guo et al., 2017). Recently, a greater number of TFs have been recognized to function in an ABA-dependent manner, including responsive elements-binding factor (ABF)/ABA responsive elements-binding protein (AREB)-type basic leucine zipper (bZIP), homeodomain-leucine zipper (HD-Zip), myeloblastosis (MYB), CBF, myelo-cytomatosis (MYC), ABA-binding factor (ABF), Cys2His2 zinc finger protein (ZFP), W-boxes receptor protein kinase (WRKY), responsive to dehydration (RD), nuclear factor Y (NF-Y), and nitrogen assimilation control protein (NAC) TFs. Some of them can be regulated via both ABA-dependent and -independent signaling pathways and have different sensitivity levels to ABA signals (Yamaguchi-Shinozaki and Shinozaki, 1993; Singh et al., 2002; Anderson et al., 2004; Knight et al., 2004; Agarwal et al., 2006; Fujita et al., 2006; Nishiyama et al., 2008; Valdes et al., 2012; Ambawat et al., 2013; Jaradat et al., 2013; Nuruzzaman et al., 2013; Joseph et al., 2014; Xu et al., 2014; Roy, 2016; Zang et al., 2016; Zhang et al., 2016b,c; Guo et al., 2017; Mao et al., 2017; Nazareno and Hernandez, 2017; Shen et al., 2017). The regulatory signals induced by phosphorylation/dephosphorylation and ubiquitination interact with ABA-induced signals and perhaps other regulatory signals to activate TFs; the molecular pathways by which this occurs are still unknown (Takahashi et al., 2016; Banerjee and Roychoudhury, 2017; Sanyal et al., 2017; Yang et al., 2017). ABA-induced signals interact with other signaling pathways (e.g., induced by hormone crosstalk between ethylene, cytokinin, SA, and jasmonic acid, and redox and chemical signaling such as calcium, cyclic nucleotides, polyphosphoinositides, nitric oxide, and sugars) to translate a code into changes in cellular activities and thus, metabolic reprogramming (Anderson et al., 2004; Tuteja and Sopory, 2008; Zhang et al., 2014; Yin et al., 2016; Guo et al., 2017; Nazareno and Hernandez, 2017). Finally, the activated TFs bind to (i) ABA-responsive promoter binding element (AREB) or (ii) DNA binding proteins (also referred to as *cis-* or *trans*-acting elements, dehydration responsive element [DRE], or coupling elements) to regulate the expression patterns of ABA-responsive or ABA-independent genes and consequently, signal transduction and plant adaptation mechanisms (Kucho et al., 2003; Hernandez-Garcia and Finer, 2014; Joshi et al., 2016; Zagorchev et al., 2016; Agarwal et al., 2017; Senavirathne et al., 2017).

A successful gene expression pattern depends on the composition of different copies of TFs and a specific set of coupling elements (Joshi et al., 2016), the numbers of TFs phosphorylation/dephosphorylation pathways and sites, TF interaction, target degradation (e.g., by 26S proteasome–mediated proteolysis assisted by DRIP1 [DREB2A-interacting protein1] and DRIP2 proteins) (Qin et al., 2008; Joshi et al., 2016), natural antisense transcripts and their targets (NATs) (Lapidot and Pilpel, 2006; Wight and Werner, 2013), post-translational modification and stability of TFs, ubiquitination (Qin et al., 2008), and the TF recognition binding sequence motif on the promoter region of the exon within a gene (Hobo et al., 1999; Agarwal et al., 2006). To take an example for the last item, the base substitution analysis of the promoter region of the *rd29A* gene revealed that a nine-base-pair conserved sequence, TACCGACAT (DRE), is essential for the regulation of *rd29A* induction by dehydration or cold. Overexpression of the active form of *AtDREB2A* (without the negative regulatory domain, i.e., with deletion of a region between residues 136 and 165) up-regulates the downstream drought-inducible genes and improves the drought stress tolerance of *Arabidopsis* (Agarwal et al., 2006). Transgenic peanut plants simultaneously co-expressing *AtDREB2A, AtHB7*, and *AtABF3* showed increased tolerance to drought, salinity, and oxidative stresses compared with wild-type (WT) plants, with an increase in total plant biomass (Pruthvi et al., 2014). According to Pruthvi et al. (2014), transgenic plants exhibited improved membrane and chlorophyll (Chl) stability due to the promotion of redox status, regulation of the expression of genes linked with protein processing and ubiquitination, i.e., *AhRbx1, AhProline amino peptidase, AhHSP70, AhDIP*, and *AhLea4*, and osmotic adjustment by proline synthesis under stress conditions. The improvements in stress tolerance in transgenic lines were associated with induced downstream gene expression of various cellular tolerance-related genes (Pruthvi et al., 2014). Mediated by the negative regulatory domain, BTB/POZ AND MATH DOMAIN proteins (BPMs) substrate adaptors for Cullin3-based E3 ubiquitin ligase ubiquitinate and degrade *AtDREB2A* through proteolysis processing (Morimoto et al., 2017). However, more studies are yet required to uncover the molecular basis for AtDREB2A stabilization and activation. The regulation of osmotin and osmotin-like protein is highly important for improving crop performance and protection against multiple biotic and abiotic stresses, improving signaling (e.g., hormone and protein kinase signaling to regulate redox status), osmoprotection (e.g., by increasing proline content), detoxification, membrane integrity, the root and shoot architecture system, stomatal conductance, water status, Chl content, photosynthesis, secondary metabolite production, recovery, survival rate, and yield components (Annon et al., 2014; Weber et al., 2014; Khan et al., 2015; Chowdhury et al., 2017).

Pre-exposure to mild stimuli such as "priming" and "acclimation" are examples of stress memory, improving crop tolerance and performance to recurring stress exposure (van der Knaap and Verrijzer, 2016). Epigenetic systems may act as the conduit for environmental cues initiating short-term, long-term, and translational changes in stress memory without altering the gene sequences (Ding et al., 2012; Trerotola et al., 2015). Epigenetic mechanisms have been shown to include incomplete erasure of DNA methylation, parental effects, transmission of distinct RNA types (e.g., mRNA, noncoding RNA, microRNAs [miRNAs], small interfering RNAs [siRNA], and piwi-interacting RNA [piRNA]), and persistence of subsets of histone marks (Trerotola et al., 2015). In the first category, short-term "storage" of the stress memory, also termed *transcriptional memory*, is a non-heritable, but persistent, change in the chromatin organization, methylome, small RNAs, histone variants, and modifications through different protein processing. Long-term stress memory relies on heritable post-translational modification of chromatin. The last of these, true transgenerational stress memory, leads to mitotically and meiotically heritable changes in the chromatin organization (Chinnusamy and Zhu, 2009; Han and Wagner, 2014; Müller-Xing et al., 2014; Yelina et al., 2015; Martinez and Köhler, 2017; Santos et al., 2017). Accordingly, priming stress in "somatic memory" modulates plant responses to environmental changes either by epigenetic "stress memory" within the life span of an organism or by its transmission to the progeny through "transgenerational inheritance", depending on plant genotype, plant growth and development stage, and environmental conditions (Han and Wagner, 2014). Chromatin modification by enzyme activity is critically dependent on central metabolites as cofactors or cosubstrates. Thus, the availability of metabolites that are required for the activity of histone-modifying enzymes may connect metabolism to chromatin structure and gene expression. Finally, selective metabolic enzymes act in the nucleus to adjust gene transcription in response to changes in metabolic state (van der Knaap and Verrijzer, 2016).

Altered responses to consecutive stresses imply that plants exercise a form of "stress memory." For example, tobacco plants pre-exposed to methyl jasmonate increased nicotine pools 2 days earlier when exposed again as compared with plants without previous exposure (Baldwin and Schmelz, 1996). Pre-treatment of plants with SA or its synthetic analogue, benzothiadiazole S-methylester, resulted in increased transcription from a subset of genes on a subsequent stress. Some stress memory effects could be perpetuated to the next generation, as observed with flagellin protein or ultraviolet radiation treatments. Therefore, epigenetic systems may act as the conduit for environmental cues initiating both short- and long-term changes in gene expression in response to stress. During recurring dehydration stresses, *Arabidopsis* plants display transcriptional stress memory, demonstrated by an increase in the rate of transcription and elevated transcript levels of a subset of the stress-response genes (trainable genes). The transcriptional memory is associated with two distinct marks found only at the trainable genes during their recovery from stress-induced transcription: high levels of histone H3 trimethyl lysine 4 (H3K4me3) nucleosomes and stalled serine 5-phosphorylated polymerase II (Ser5P Pol II). In contrast, H3K4me3 and Ser5P dynamically increase when non-trainable genes are induced and then decrease to basal levels during recovery. During recovery (watered) states, trainable genes produce transcripts at basal (pre-induced) levels but remain associated with atypically high H3K4me3 and Ser5P polymerase II levels as "memory marks" for as long as the transcriptional memory responses last, indicating that RNA polymerase II is stalled. It is tempting to suggest that under natural conditions, stress memory is activated by the previous day's dehydration stress, persists through the recovery period

at night (when transpiration is lower and leaf water potential recovers), and then facilitates the plant's response to dehydration stress encountered during the next day, when adequate soil water is unavailable to maintain leaf water potential at the higher transpiration rates occurring during the day. The duration of dehydration stress memory, persisting for up to 5 days in trained *Arabidopsis* plants recovering in watered soil, is sufficiently long and may be a mechanism used by plants during soil-water deficit conditions (Ding et al., 2012).

Stress-induced signaling pathways activate the defense mechanism to change protein processing and the gene expression pattern to modify cellular processes such as metabolic reprogramming; for example, antioxidant and osmolyte biosynthesis. This defense mechanism consequently reduces the negative effects of stressors to maintain cell structure and function. For example, the *Medicago sativa aldose–aldehyde reductase* (*MsALR*) gene is responsible for encoding an aldose–aldehyde reductase from alfalfa (Bartels, 2001). The ectopic expression of the *aldose–aldehyde reductase* (*ALR*) gene in tobacco plants has been effective in regulating redox status by scavenging methylglyoxal and reactive oxygen species (ROS) (e.g., by reducing the level of reactive aldehydes and H_2O_2); proline and sorbitol accumulation; ion homeostasis; photosynthetic pigment degradation and accordingly, improved cell membrane integrity; water status; photosynthesis; and tolerance to multiple stresses such as cold, heavy metals, drought, salinity, and ultraviolet stresses (Oberschall et al., 2000; Hideg et al., 2003; Hegedüs et al., 2004; Kumar et al., 2013). Bartels (2001) reported that the transcript of the *MsALR* gene was present in all tissues of an alfalfa plant, and the recombinant alfalfa aldose reductase protein exhibited specific enzyme activities and reduced aldose and aldehyde substrates by means of NADPH as a cofactor, but not NADH. The transgenic tobacco plants overexpressing the alfalfa aldose reductase showed reduced damage (measured by Chl fluorescence) when exposed to oxidative stress agents such as paraquat (a herbicide that produces ROS) or H_2O_2 (Bartels, 2001). The transgenic tobacco plants recovered better from damage caused by water deficit than the untransformed WT control plants. The overexpression of genes encoding antioxidant enzymes or ROS-scavenging enzymes improves plant redox status and thereby, plant tolerance and stress responses. For example, transgenic tobacco or alfalfa plants expressing Mn-superoxide dismutase tend to have reduced injury from water deficit stress (Bartels, 2001). In a similar way, the application of SA (Tari et al., 2010) and the overexpression of glyoxalase enzymes (Yadav et al., 2005) and enzymes that reduce aldehyde and increase aldose reductase activity, such as succinic semialdehyde, γ-aminobutyrate (Allan et al., 2008), can modify plant responses to different stressors by improving redox homeostasis.

The actin cytoskeleton plays an essential role in various cellular processes, including growth, development, viability (e.g., cytoplasmic organization, establishment of cell polarity, cell elongation, polar tip growth, intracellular trafficking, and cytokinesis), and responses to hormones and biotic and abiotic stresses. Intracellular actin filament activity can be regulated by the actin depolymerizing factor (ADF) proteins (cofilin, myosin, fibrin, and villin) and the small actin-binding protein profilin. Also, pH changes, N-terminal phosphorylation, and specific polyphosphoinositides reversibly affect the activity of ADF and the actin cytoskeleton (Staiger et al., 1997; Kovar and Staiger, 2000; Ruzicka et al., 2007), suggesting that it may play a pivotal role in affecting the organization of the plant actin cytoskeleton in response to extrinsic and intrinsic stimuli (Kovar and Staiger, 2000; Augustine et al., 2008) and drought stress (Tuberosa and Salvi, 2006). Drought stress and ABA affect regulatory pathways related to the protein processing of phosphoproteins, such as phosphorylation, calcium signaling, glycolytic metabolism, G protein beta subunit-like protein, ascorbate peroxidase, manganese superoxide dismutase, and triosephosphate isomerase (He and Li, 2008; Ke et al., 2009; Que et al., 2012). Therefore, regulation of phosphoproteins, phosphorylation of the small actin-binding protein profilin, and expression of *ADF* genes would be effective strategies to improve plant responses to drought stress.

Another strategy to improve plant responses to drought stress and also their performance is to regulate genes, enzymes, proteins, and pigments related to plant photosynthetic machinery. Observations on *Populus euphratica* revealed that transcriptional changes of photosynthesis-related genes were negligible under moderate water deficit, despite low internal CO_2 concentrations. However, this was accompanied by an increased abundance of photosynthesis-related proteins, such as oxygen-evolving complex 33-kD PSII, Rubisco activase, carbonate dehydratase (or carbonic anhydrase), chloroplast glyceraldehyde-3-P dehydrogenase, and phosphoglycerate kinase. Oxygen-evolving complex 33-kD PSII protein, an extrinsic subunit of PSII probably involved in the stabilization of the PS components, was also affected under mild drought stress in spruce (*Picea abies*). Rubisco activase, which regulates the activity of Rubisco in response to changes in light or temperature via ADP-to-ATP ratio and redox potential, also accumulated in rice under drought stress (Bogeat-Triboulot et al., 2007). Carbonic anhydrases may be candidates for the coregulation of mesophyll conductance to CO_2 in the liquid phase and leaf net CO_2 assimilation rate (photosynthetic rate) and play an important role during drought and salinity stress. Glyceraldehyde-3-phosphate dehydrogenase and phosphoglycerate kinase are enzymes involved in the pentose phosphate cycle but could be impaired under drought stress. This early occurrence of an increased abundance of photosynthesis-related proteins during the stress treatment may have partly counterbalanced the decreased internal CO_2 concentration and contributed to the partial maintenance of photosynthesis during the first stages of water deficit. On the other hand, a putative pheophorbide A oxygenase displayed an increased transcript level, which was consistent with the observed increase of the Chl *a* to Chl *b* ratio, a known indication of Chl catabolism when Chl *b* is converted to Chl *a* during senescence (Bogeat-Triboulot et al., 2007). At peak stress intensity, the repression of photosynthesis-related genes (Rubisco small subunit and PSI reaction center subunits VI and X) may be due to stress severity and could indicate the beginning of

senescence (Bogeat-Triboulot et al., 2007). Among the identified regulated proteins were HSPs and chaperonins, involved in protein repair and protection against denaturation, which are normally synthesized on abiotic stress exposure. Like HSP function, the *specificity protein 1* (*sp1*) gene in *Populus tremula* plants was found to be up-regulated shortly after the application of different abiotic stresses, such as salt, cold, heat, and mannitol, but *sp1* was severely down-regulated after 24 h of exposure to mannitol (Bogeat-Triboulot et al., 2007). Similarly, the abundance of SP1 proteins was slightly reduced in water stress–acclimatized *P. euphratica* plants (Bogeat-Triboulot et al., 2007).

Metallothioneins (small, cysteine-rich, low–molecular weight metal-binding proteins) and protein and non-protein thiols are found across most taxonomic groups. They are intracellular, low–molecular weight, cysteine-rich proteins belonging to a small multigene family and are involved in post-translational protein modification; heavy metal and ROS detoxification; cell homeostasis; redox sequestration and regulation, and/or metabolic regulation via Zn donation homeostasis; signalization; and regulation of responses to various stresses (Coyle et al., 2002; Domènech et al., 2006; Bogeat-Triboulot et al., 2007; Cadenas and Packer, 2010; Grennan, 2011; Hassinen et al., 2011; Zagorchev et al., 2013; Cejudo et al., 2014; Pivato et al., 2014). Improving sulfur metabolism, e.g., by sulfur-containing amino acids (Wulff-Zottele et al., 2010; Khan et al., 2013) or methionine (Zagorchev et al., 2013), and antioxidant enzymes such as glutathione (Cheng et al., 2015), peroxidase, superoxide dismutase, and catalase (Askari-Khorasgani et al., 2017) also promotes drought tolerance and reduces yield loss through improved redox status and photosynthesis.

Plant polyamine content has been modulated by the overexpression or down-regulation of arginine decarboxylase (*adc*), ornithine decarboxylase (*odc*), and *S*-adenosylmethionine decarboxylase (*samdc*). Putrescine biosynthesis can be stimulated by the overexpression of heterologous *adc* or *odc* cDNAs in plants, most often in parallel with a small increase in spermidine and spermine concentrations with strict homeostatic regulation. The manipulation of polyamine biosynthesis, which can also be mediated by nitric oxide and ethylene regulatory signals, can produce drought-tolerant germplasm (Capell et al., 2004; Montilla-Bascón et al., 2017). To take an example, generated transgenic rice plants expressing the *Datura stramonium adc* gene produced much higher levels of putrescine than WT plants under stress, promoting spermidine and spermine synthesis and ultimately protecting the plants from drought stress (Capell et al., 2004).

Microarray analysis of barley genotypes during the reproductive stage showed that one gene expressed in a drought-tolerant genotype under drought stress encoded malate dehydrogenase (oxaloacetate-decarboxylating) (NADP+) (EC 1.1.1.40) or NADP-malic enzyme (NADP-ME) located in guard cell complexes of a C_3 plant (Alexandersson et al., 2005). According to Alexandersson et al. (2005), NADP-ME facilitates lignin biosynthesis by providing NADPH and regulates cytosolic pH through balancing the synthesis and degradation of malate, which can also influence the AQP activity and the water permeability of plasma membrane intrinsic proteins (PIPs). Another possible role of NADP-ME is to control stomatal closure by degrading malate during the day under water-deficit conditions, because NADP-ME expression leads to decreased stomatal aperture and increased fresh mass gained per unit water used. Therefore, manipulation of organic anion metabolism in guard cells through regulating NADP-ME expression has been proposed as an approach for drought avoidance and water conservation. The pyruvate produced by NADP-ME could be further degraded and used in other pathways. The *Pyruvate dehydrogenase* (*PDH*) gene, which is known to be involved in the oxidative decarboxylation of pyruvate to generate acetyl coenzyme A for the tricarboxylic acid (TCA) cycle, was also expressed in response to water stress (Guo et al., 2009). Since the TCA cycle provides carbon skeletons for many biosynthetic pathways, a high level of *PDH* provides rich carbon sources for diverse uses within a plant. Therefore, up-regulation of NADP-ME and *PDH* in drought-tolerant genotypes suggests that carbon metabolism is critical for acclimation to water deficit (Guo et al., 2009).

Plants contain two genes that code for poly(ADP-ribose) polymerase (PARP): *parp1* and *parp*. Both PARPs are activated by DNA damage caused by, for example, ROS overproduction. On activation, polymers of ADP-ribose are synthesized on a range of nuclear enzymes using NAD$^+$ as substrate (Block et al., 2005). PARPs catalyze the transfer of multiple poly(ADP-ribose) units onto target proteins. Poly(ADP-ribosyl)ation plays a crucial role in a variety of cellular processes, including, most prominently, auto-activation of PARP at sites of DNA breaks to activate DNA repair processes (Song et al., 2015). Plant stressors, such as drought, high light, and heat, activate PARP, causing NAD$^+$ breakdown and ATP consumption. When the PARP activity is reduced by means of chemical inhibitors or by gene silencing, cell death is inhibited, and plants become tolerant to a broad range of abiotic stresses, such as high light, drought, heat (Block et al., 2005; Zhu et al., 2016), and pathogens (by improving the immune system) (Song et al., 2015). Under stress conditions, plant lines with low poly(ADP-ribosyl)ation activity maintain their energy homeostasis by reducing NAD$^+$ breakdown and, consequently, energy consumption. The higher energy-use efficiency avoids the need for too intense mitochondrial respiration and, consequently, reduces the formation of ROS (Block et al., 2005), thereby improving plant tolerance to multiple stresses.

29.3 CONCLUDING REMARKS

At gene expression and metabolic levels, a review of the pertinent literature demonstrated that redox, carbohydrate, osmolyte, and protein homeostasis are essential to maintain plant cell processes under drought stress conditions. Since plant responses greatly vary depending on stress conditions, such as different combinations of stresses, severity, duration, repetition of stress, plant and cell growth and development stage, circadian clock, growth season, plant genotype, organ, cell and tissue types, and plant–environment interactions, future studies will be more successful if these factors are included in the experiments. Moreover, the co-expression of genes encoding antioxidant and photosynthetic enzymes, carbohydrates, osmolytes, AQPs, and proteins is among the most promising tools to both minimize drought stress effects and optimize plant performance under drought conditions.

REFERENCES

Agarwal, P. K., Agarwal, P., Reddy, M., and Sopory, S. K. 2006. Role of DREB transcription factors in abiotic and biotic stress tolerance in plants. *Plant Cell Reports*, 25(12), 1263–1274.

Agarwal, P. K., Gupta, K., Lopato, S., and Agarwal, P. 2017. Dehydration responsive element binding transcription factors and their applications for the engineering of stress tolerance. *Journal of Experimental Botany*, 68(9), 2135–2148.

Agarwal, S., Sairam, R. K., Srivastava, G. C., and Meena, R. C. 2005. Changes in antioxidant enzymes activity and oxidative stress by abscisic acid and salicylic acid in wheat genotypes. *Biologia Plantarum*, 49(4), 541–550.

Ahuja, I., de Vos, R. C. H., Bones, A. M., and Hall, R. D. 2010. Plant molecular stress responses face climate change. *Trends in Plant Science*, 15(12), 664–674.

Alam, M. M., Nahar, K., Hasanuzzaman, M., and Fujita, M. 2014. Exogenous jasmonic acid modulates the physiology, antioxidant defense and glyoxalase systems in imparting drought stress tolerance in different *Brassica* species. *Plant Biotechnology Reports*, 8(3), 279–293.

Alexandersson, E., Fraysse, L., Sjövall-Larsen, S., Gustavsson, S., Fellert, M., Karlsson, M., Johanson, U., and Kjellbom, P. 2005. Whole gene family expression and drought stress regulation of aquaporins. *Plant Molecular Biology*, 59(3), 469–484.

Alexandre, A. and Oliveira, S. 2011. Most heat-tolerant rhizobia show high induction of major chaperone genes upon stress. *FEMS Microbiology Ecology*, 75(1), 28–36.

Allan, W. L., Simpson, J. P., Clark, S. M., and Shelp, B. J. 2008. γ-Hydroxybutyrate accumulation in *Arabidopsis* and tobacco plants is a general response to abiotic stress: putative regulation by redox balance and glyoxylate reductase isoforms. *Journal of Experimental Botany*, 59(9), 2555–2564.

Alqurashi, M., Thomas, L., Gehring, C., and Marondedze, C. 2017. A microsomal proteomics view of H$_2$O$_2$ and ABA-dependent responses. *Proteomes*, 5(3), 22.

Alves, M. S., Fontes, E. P. B., and Fietto, L. G. 2011. Early responsive to dehydration 15, a new transcription factor that integrates stress signaling pathways. *Plant Signaling and Behavior*, 6(12), 1993–1996.

Ambawat, S., Sharma, P., Yadav, N. R., and Yadav, R. C. 2013. MYB transcription factor genes as regulators for plant responses: an overview. *Physiology and Molecular Biology of Plants*, 19(3), 307–321.

Anderson, J. P., Badruzsaufari, E., Schenk, P. M., Manners, J. M., Desmond, O. J., Ehlert, C., Maclean, D. J., Ebert, P. R., and Kazan, K. 2004. Antagonistic interaction between abscisic acid and jasmonate-ethylene signaling pathways modulates defense gene expression and disease resistance in *Arabidopsis*. *The Plant Cell*, 16(12), 3460–3479.

Annon, A., Rathore, K., and Crosby, K. 2014. Overexpression of a tobacco osmotin gene in carrot (*Daucus carota* L.) enhances drought tolerance. *In Vitro Cellular and Developmental Biology—Plant*, 50(3), 299–306.

Arias-Baldrich, C., de la Osa, C., Bosch, N., Ruiz-Ballesta, I., Monreal, J. A., and García-Mauriño, S. 2017. Enzymatic activity, gene expression and posttranslational modifications of photosynthetic and non-photosynthetic phosphoenolpyruvate carboxylase in ammonium-stressed sorghum plants. *Journal of Plant Physiology*, 214(Supplement C), 39–47.

Asadi Rahmani, H., Saleh-rastin, N., Khavazi, K., Asgharzadeh, A., Fewer, D., Kiani, S., and Lindström, K. 2009. Selection of thermotolerant bradyrhizobial strains for nodulation of soybean (*Glycine max* L.) in semi-arid regions of Iran. *World Journal of Microbiology and Biotechnology*, 25(4), 591–600.

Askari-Khorasgani, O., Emadi, S., Mortazaienezhad, F., and Pessarakli, M. 2017. Differential responses of three chamomile genotypes to salinity stress with respect to physiological, morphological, and phytochemical characteristics. *Journal of Plant Nutrition*, 1–12.

Atkinson, N. J., Lilley, C. J., and Urwin, P. E. 2013. Identification of genes involved in the response of *Arabidopsis* to simultaneous biotic and abiotic stresses. *Plant Physiology*, 162(4), 2028–2041.

Atkinson, N. J. and Urwin, P. E. 2012. The interaction of plant biotic and abiotic stresses: from genes to the field. *Journal of Experimental Botany*, 63(10), 3523–3543.

Augustine, R. C., Vidali, L., Kleinman, K. P., and Bezanilla, M. 2008. Actin depolymerizing factor is essential for viability in plants, and its phosphoregulation is important for tip growth. *The Plant Journal*, 54(5), 863–875.

Augustine, S. M. 2016. Function of heat-shock proteins in drought tolerance regulation of plants, In: *Drought Stress Tolerance in Plants*, Hossain M., Wani S., Bhattacharjee, S., Burritt, D., and Tran, L. S. (eds.), Volume 1. Springer, Cham, pp. 163–185.

Baldwin, I. T. and Schmelz, E. A. 1996. Immunological "memory" in the induced accumulation of nicotine in wild tobacco. *Ecology*, 77(1), 236–246.

Balsiger, S., Ragaz, C., Baron, C., and Narberhaus, F. 2004. Replicon-specific regulation of *small heat shock* genes in *Agrobacterium tumefaciens*. *Journal of Bacteriology*, 186(20), 6824–6829.

Banerjee, A. and Roychoudhury, A. 2017. Abscisic-acid-dependent basic leucine zipper (bZIP) transcription factors in plant abiotic stress. *Protoplasma*, 254(1), 3–16.

Barolo, S. and Posakony, J. W. 2002. Three habits of highly effective signaling pathways: principles of transcriptional control by developmental cell signaling. *Genes and Development*, 16(10), 1167–1181.

Bartels, D. 2001. Targeting detoxification pathways: an efficient approach to obtain plants with multiple stress tolerance? *Trends in Plant Science*, 6(7), 284–286.

Bartels, D. 2005. Desiccation tolerance studied in the resurrection plant *Craterostigma plantagineum*. *Integrative and Comparative Biology*, 45(5), 696–701.

Berkhout, J., Teusink, B., and Bruggeman, F. J. 2013. Gene network requirements for regulation of metabolic gene expression to a desired state. , *Scientific Reports*, 3(1)1417.

Berry, S. 2003. Endosymbiosis and the design of eukaryotic electron transport. *Biochimica et Biophysica Acta (BBA) Bioenergetics*, 1606(1), 57–72.

Betterle, N., Poudyal, R. S., Rosa, A., Wu, G., Bassi, R., and Lee, C.-H. 2017. The STN8 kinase-PBCP phosphatase system is responsible for high-light-induced reversible phosphorylation of the PSII inner antenna subunit CP29 in rice. *The Plant Journal*, 89(4), 681–691.

Block, M. D., Verduyn, C., Brouwer, D. D., and Cornelissen, M. 2005. Poly (ADP-ribose) polymerase in plants affects energy homeostasis, cell death and stress tolerance. *The Plant Journal*, 41(1), 95–106.

Bogeat-Triboulot, M.-B., Brosché, M., Renaut, J., Jouve, L., Le Thiec, D., Fayyaz, P., Vinocur, B., Witters, E., Laukens, K., and Teichmann, T. 2007. Gradual soil water depletion results in reversible changes of gene expression, protein profiles, ecophysiology, and growth performance in *Populus euphratica*, a poplar growing in arid regions. *Plant Physiology*, 143(2), 876–892.

Bonhomme, L., Valot, B., Tardieu, F., and Zivy, M. 2012. Phosphoproteome dynamics upon changes in plant water status reveal early events associated with rapid growth adjustment in maize leaves. *Molecular and Cellular Proteomics*, 11(10), 957–972.

Cadenas, E. and Packer, L. 2010. Thiol redox transitions in cell signaling, Part A: chemistry and biochemistry of low molecular weight and protein thiols. In: *Methods in Enzymology*, Volume 473, 1st ed. Academic Press, pp. 1–359, p. 416.

Caldana, C., Fernie, A. R., Willmitzer, L., and Steinhauser, D. 2012. Unraveling retrograde signaling pathways: finding candidate signaling molecules via metabolomics and systems biology driven approaches. *Frontiers in Plant Science*, 3, 267.

Caldeira, C. F., Bosio, M., Parent, B., Jeanguenin, L., Chaumont, F., and Tardieu, F. 2014a. A hydraulic model is compatible with rapid changes in leaf elongation under fluctuating evaporative demand and soil water status. *Plant Physiology*, 164(4), 1718–1730.

Caldeira, C. F., Jeanguenin, L., Chaumont, F., and Tardieu, F. 2014b. Circadian rhythms of hydraulic conductance and growth are enhanced by drought and improve plant performance. *Nature Communications*, 5(5365), 1–9.

Capell, T., Bassie, L., and Christou, P. 2004. Modulation of the polyamine biosynthetic pathway in transgenic rice confers tolerance to drought stress. *Proceedings of the National Academy of Sciences of the United States of America*, 101(26), 9909–9914.

Castello, A., Hentze, M. W., and Preiss, T. 2015. Metabolic enzymes enjoying new partnerships as RNA-binding proteins. *Trends in Endocrinology and Metabolism*, 26(12), 746–757.

Cejudo, F. J., Meyer, A. J., Reichheld, J.-P., Rouhier, N., and Traverso, J.A. 2014. Thiol-based redox homeostasis and signaling. *Frontiers in Plant Science*, 5, 266.

Chaumont, F. and Tyerman, S. D. 2014. Aquaporins: highly regulated channels controlling plant water relations. *Plant Physiology*, 164(4), 1600–1618.

Chen, L. and Hellmann, H. 2013. Plant E.3 ligases: flexible enzymes in a sessile world. *Molecular Plant*, 6(5), 1388–1404.

Chen, Y. and Hoehenwarter, W. 2015. Changes in the phosphoproteome and metabolome link early signaling events to rearrangement of photosynthesis and central metabolism in salinity and oxidative stress response in *Arabidopsis*. *Plant Physiology*, 169(4), 3021–3033.

Chen, Y.-E., Zhang, C.-M., Su, Y.-Q., Ma, J., Zhang, Z.-W., Yuan, M., Zhang, H.-Y., and Yuan, S. 2017. Responses of photosystem II and antioxidative systems to high light and high temperature co-stress in wheat. *Environmental and Experimental Botany*, 135(Supplement C), 45–55.

Cheng, M.-C., Ko, K., Chang, W.-L., Kuo, W.-C., Chen, G.-H., and Lin, T.-P. 2015. Increased glutathione contributes to stress tolerance and global translational changes in *Arabidopsis*. *The Plant Journal*, 83(5), 926–939.

Chinnusamy, V. and Zhu, J.-K. 2009. Epigenetic regulation of stress responses in plants. *Current Opinion in Plant Biology*, 12(2), 133–139.

Chowdhury, S., Basu, A., and Kundu, S. 2017. Overexpression of a new osmotin-like protein gene (*SindOLP*) confers tolerance against biotic and abiotic stresses in sesame. *Frontiers in Plant Science*, 8, 410.

Chubukov, V., Zuleta, I. A., and Li, H. 2012. Regulatory architecture determines optimal regulation of gene expression in metabolic pathways. *Proceedings of the National Academy of Sciences of the United States of America*, 109(13), 5127–5132.

Ciechanover, A., Elias, S., Heller, H., and Hershko, A. 1982. "Covalent affinity" purification of ubiquitin-activating enzyme. *Journal of Biological Chemistry*, 257(5), 2537–2542.

Ciesla, J., Fraczyk, T., and Rode, W. 2011. Phosphorylation of basic amino acid residues in proteins: important but easily missed. *Acta Biochimica Polonica*, 58(2), 137–148.

Cohen, P. 2002. The origins of protein phosphorylation. *Nature Cell Biology*, 4, E127.

Coyle, P., Philcox, J., Carey, L., and Rofe, A. 2002. Metallothionein: the multipurpose protein. *Cellular and Molecular Life Sciences*, 59(4), 627–647.

Cutler, S. R., Rodriguez, P. L., Finkelstein, R. R., and Abrams, S. R. 2010. Abscisic acid: emergence of a core signaling network. *Annual Review of Plant Biology*, 61(1), 651–679.

Das, R., Pandey, A., and Pandey, G. K. 2014. Role of calcium/calmodulin in plant stress response and signaling. In: *Approaches to Plant Stress and Their Management*, Gaur, R. K. and Sharma, P. (eds.). Springer, New Delhi, pp. 53–84, p. 403.

De Boer, A. H. and Volkov, V. 2003. Logistics of water and salt transport through the plant: structure and functioning of the xylem. *Plant, Cell and Environment*, 26(1), 87–101.

Diao, Q.-N., Song, Y.-J., Shi, D.-M., and Qi, H.-Y. 2016. Nitric oxide induced by polyamines involves antioxidant systems against chilling stress in tomato (*Lycopersicon esculentum* Mill.) seedling. *Journal of Zhejiang University Science B*, 17(12), 916–930.

Dietz, K.-J. 2014. Redox regulation of transcription factors in plant stress acclimation and development. *Antioxidants and Redox Signaling*, 21(9), 1356–1372.

Ding, Y., Fromm, M., and Avramova, Z. 2012. Multiple exposures to drought "train" transcriptional responses in *Arabidopsis*. *Nature Communications*, 3(740), 1–9.

Domènech, J., Mir, G., Huguet, G., Capdevila, M., Molinas, M., and Atrian, S. 2006. Plant metallothionein domains: functional insight into physiological metal binding and protein folding. *Biochimie*, 88(6), 583–593.

Draye, X., Kim, Y., Lobet, G., and Javaux, M. 2010. Model-assisted integration of physiological and environmental constraints affecting the dynamic and spatial patterns of root water uptake from soils. *Journal of Experimental Botany*, 61(8), 2145–2155.

Eckardt, N. A. 2007. GA signaling: direct targets of DELLA proteins. *The Plant Cell*, 19(10), 2970–2970.

Francisco-Velilla, R., Fernandez-Chamorro, J., Ramajo, J., and Martinez-Salas, E. 2016. The RNA-binding protein Gemin5 binds directly to the ribosome and regulates global translation. *Nucleic Acids Research*, 44(17), 8335–8351.

Fujita, M., Fujita, Y., Noutoshi, Y., Takahashi, F., Narusaka, Y., Yamaguchi-Shinozaki, K., and Shinozaki, K. 2006. Crosstalk between abiotic and biotic stress responses: a current view from the points of convergence in the stress signaling networks. *Current Opinion in Plant Biology*, 9(4), 436–442.

Garcia-Garcia, T., Poncet, S., Derouiche, A., Shi, L., Mijakovic, I., and Noirot-Gros, M.-F. 2016. Role of protein phosphorylation in the regulation of cell cycle and DNA-related processes in bacteria. *Frontiers in Microbiology*, 7, 184.

Gerstberger, S., Hafner, M., and Tuschl, T. 2014. A census of human RNA-binding proteins. *Nature Reviews Genetics*, 15(12), 829–845.

Glisovic, T., Bachorik, J. L., Yong, J., and Dreyfuss, G. 2008. RNA-binding proteins and post-transcriptional gene regulation. *FEBS Letters*, 582(14), 1977–1986.

Gonçalves, S. 2016. MYB transcription factors for enhanced drought tolerance in plants. In: *Water Stress and Crop Plants: A Sustainable Approach*, Ahmad, P. (ed.), John Wiley & Sons, Chichester, UK, pp. 194–205.

Grennan, A. K. 2011. Metallothioneins, a diverse protein family. *Plant Physiology*, 155(4), 1750–1751.

Grundy, J., Stoker, C., and Carré, I. A. 2015. Circadian regulation of abiotic stress tolerance in plants. *Frontiers in Plant Science*, 6(648).

Guo, D., Zhou, Y., Li, H.-L., Zhu, J.-H., Wang, Y., Chen, X.-T., and Peng, S.-Q. 2017. Identification and characterization of the abscisic acid (ABA) receptor gene family and its expression in response to hormones in the rubber tree. 7, 45157.

Guo, P., Baum, M., Grando, S., Ceccarelli, S., Bai, G., Li, R., Von Korff, M., Varshney, R. K., Graner, A., and Valkoun, J. 2009. Differentially expressed genes between drought-tolerant and drought-sensitive barley genotypes in response to drought stress during the reproductive stage. *Journal of Experimental Botany*, 60(12), 3531–3544.

Hall, D. A., Zhu, H., Zhu, X., Royce, T., Gerstein, M., and Snyder, M. 2004. Regulation of gene expression by a metabolic enzyme. *Science*, 306(5695), 482–484.

Han, S.-K. and Wagner, D. 2014. Role of chromatin in water stress responses in plants. *Journal of Experimental Botany*, 65(10), 2785–2799.

Han, Y., Sun, J., Yang, J., Tan, Z., Luo, J., and Lu, D. 2017. Reconstitution of the plant ubiquitination cascade in bacteria using a synthetic biology approach. *The Plant Journal*, 91(4), 766–776.

Harb, A., Krishnan, A., Ambavaram, M. M., and Pereira, A. 2010. Molecular and physiological analysis of drought stress in *Arabidopsis* reveals early responses leading to acclimation in plant growth. *Plant Physiology*, 154(3), 1254–1271.

Hashimoto, K., Eckert, C., Anschütz, U., Scholz, M., Held, K., Waadt, R., Reyer, A., Hippler, M., Becker, D., and Kudla, J. 2012. Phosphorylation of calcineurin B-like (CBL) calcium sensor proteins by their CBL-interacting protein kinases (CIPKs) is required for full activity of CBL-CIPK complexes toward their target proteins. *Journal of Biological Chemistry*, 287(11), 7956–7968.

Hassinen, V. H., Tervahauta, A. I., Schat, H., and Karenlampi, S. O. 2011. Plant metallothioneins-metal chelators with ROS scavenging activity? *Plant Biol (Stuttg)*, 13(2), 225–232.

Hauvermale, A. L., Ariizumi, T., and Steber, C. M. 2012. Gibberellin signaling: a theme and variations on DELLA repression. *Plant Physiology*, 160(1), 83–92.

He, H. and Li, J. 2008. Proteomic analysis of phosphoproteins regulated by abscisic acid in rice leaves. *Biochemical Biophysical Research Communications*, 371(4), 883–888.

He, Z., Huang, T., Ao, K., Yan, X., and Huang, Y. 2017. Sumoylation, phosphorylation, and acetylation fine-tune the turnover of plant immunity components mediated by ubiquitination. *Frontiers in Plant Science*, 8(1682).

Hegedüs, A., Erdei, S., Janda, T., Tóth, E., Horváth, G., and Dudits, D. 2004. Transgenic tobacco plants overproducing alfalfa aldose/aldehyde reductase show higher tolerance to low temperature and cadmium stress. *Plant Science*, 166(5), 1329–1333.

Hernandez-Garcia, C. M. and Finer, J. J. 2014. Identification and validation of promoters and cis-acting regulatory elements. *Plant Science*, 217218(Supplement C), 109–119.

Hideg, É., Nagy, T., Oberschall, A., Dudits, D., and Vass, I. 2003. Detoxification function of aldose/aldehyde reductase during drought and ultraviolet-B (280–320 nm) stresses. *Plant, Cell and Environment*, 26(4), 513–522.

Hobo, T., Asada, M., Kowyama, Y., and Hattori, T. 1999. ACGT-containing abscisic acid response element (ABRE) and coupling element 3 (CE3) are functionally equivalent. *The Plant Journal*, 19(6), 679–689.

Hogan, D. J., Riordan, D. P., Gerber, A. P., Herschlag, D., and Brown, P. O. 2008. Diverse RNA-binding proteins interact with functionally related sets of RNAs, suggesting an extensive regulatory system. *PLOS Biology*, 6(10), e255.

Hornung, E. 2011. Evolutionary adaptation of oniscidean isopods to terrestrial life: structure, physiology and behavior. *Terrestrial Arthropod Reviews*, 4(2), 95–130.

Huber, S. C. 2007. Exploring the role of protein phosphorylation in plants: from signalling to metabolism. *Biochemical Society Transactions*, 35(1), 28–32.

Ilhan, S., Ozdemir, F., and Bor, M. 2015. Contribution of trehalose biosynthetic pathway to drought stress tolerance of *Capparis ovata* Desf. *Plant Biology*, 17(2), 402–407.

Ito, S., Kawamura, H., Niwa, Y., Nakamichi, N., Yamashino, T., and Mizuno, T. 2009. A genetic study of the *Arabidopsis* circadian clock with reference to the *TIMING OF CAB EXPRESSION 1 (TOC1)* gene. *Plant and Cell Physiology*, 50(2), 290–303.

Jackson, S. P. and Durocher, D. 2013. Regulation of DNA damage responses by ubiquitin and SUMO. *Molecular Cell*, 49(5), 795–807.

Jacob, P., Hirt, H., and Bendahmane, A. 2017. The heat-shock protein/chaperone network and multiple stress resistance. *Plant Biotechnology Journal*, 15(4), 405–414.

Jaradat, M. R., Feurtado, J. A., Huang, D., Lu, Y., and Cutler, A. J. 2013. Multiple roles of the transcription factor AtMYBR1/AtMYB44 in ABA signaling, stress responses, and leaf senescence. *BMC Plant Biology*, 13(1), 192.

Jarillo, J. A., Gaudin, V., Hennig, L., Köhler, C., and Piñeiro, M. 2014. Plant chromatin warms up in Madrid: Meeting summary of the 3rd European Workshop on Plant Chromatin 2013, Madrid, Spain. *Epigenetics* 9(4), 644–652.

Jayaraman, D., Richards, A. L., Westphall, M. S., Coon, J. J., and Ane, J. M. 2017. Identification of the phosphorylation targets of symbiotic receptor-like kinases using a high-throughput multiplexed assay for kinase specificity. *The Plant Journal*, 90(6), 1196–1207.

Jin, Y., Liu, H., Luo, D., Yu, N., Dong, W., Wang, C., Zhang, X., Dai, H., Yang, J., and Wang, E. 2016. DELLA proteins are common components of symbiotic rhizobial and mycorrhizal signalling pathways. *Nature Communications*, 7, 12433.

Johansson, I., Karlsson, M., Johanson, U., Larsson, C., and Kjellbom, P. 2000. The role of aquaporins in cellular and whole plant water balance. *Biochimica et Biophysica Acta Biomembranes*, 1465(1–2), 324–342.

Joseph, M. P., Papdi, C., Kozma-Bognár, L., Nagy, I., López-Carbonell, M., Rigó, G., Koncz, C., and Szabados, L. 2014. The *Arabidopsis* zinc finger protein3 interferes with abscisic acid and light signaling in seed germination and plant development. *Plant Physiology*, 165(3), 1203–1220.

Joshi, R., Wani, S. H., Singh, B., Bohra, A., Dar, Z. A., Lone, A. A., Pareek, A., and Singla-Pareek, S. L. 2016. Transcription factors and plants response to drought stress: current understanding and future directions. *Frontiers in Plant Science*, 7, 1029.

Kamata, H. and Hirata, H. 1999. Redox regulation of cellular signalling. *Cellular Signalling*, 11(1), 1–14.

Kang, H., Park, S. J., and Kwak, K. J. 2013. Plant RNA chaperones in stress response. *Trends in Plant Science*, 18(2), 100–106.

Ke, Y., Han, G., He, H., and Li, J. 2009. Differential regulation of proteins and phosphoproteins in rice under drought stress. *Biochemical and Biophysical Research Communications*, 379(1), 133–138.

Kechavarzi, B. and Janga, S. C. 2014. Dissecting the expression landscape of RNA-binding proteins in human cancers. *Genome Biology*, 15(1), R14.

Khan, M. I. R., Asgher, M., Iqbal, N., and Khan, N. A. 2013. Potentiality of sulphur-containing compounds in salt stress tolerance. In: *Ecophysiology and Responses of Plants under Salt Stress*, Ahmad, P., Azooz, M. M., and Prasad, M. N. V. (eds.), Springer New York, New York, NY, pp. 443–472.

Khan, M. S., Ahmad, D., and Khan, M. A. 2015. Utilization of genes encoding osmoprotectants in transgenic plants for enhanced abiotic stress tolerance. *Electronic Journal of Biotechnology*, 18(4), 257–266.

Khurana, P., Vishnudasan, D., and Chhibbar, A. K. 2008. Genetic approaches towards overcoming water deficit in plants—special emphasis on LEAs. *Physiology and Molecular Biology of Plants*, 14(4), 277–298.

Kissoudis, C., van de Wiel, C., Visser, R. G., and van der Linden, G. 2014. Enhancing crop resilience to combined abiotic and biotic stress through the dissection of physiological and molecular crosstalk. *Frontiers in Plant Science*, 5, 207.

Klass, D. M., Scheibe, M., Butter, F., Hogan, G. J., Mann, M., and Brown, P. O. 2013. Quantitative proteomic analysis reveals concurrent RNA–protein interactions and identifies new RNA-binding proteins in *Saccharomyces cerevisiae*. *Genome Research*, 23(6), 1028–1038.

Kline, K. G., Barrett-Wilt, G. A., and Sussman, M. R. 2010. In planta changes in protein phosphorylation induced by the plant hormone abscisic acid. *Proceedings of the National Academy of Sciences*, 107(36), 15986–15991.

Knight, H., Zarka, D. G., Okamoto, H., Thomashow, M. F., and Knight, M. R. 2004. Abscisic acid induces CBF gene transcription and subsequent induction of cold-regulated genes via the CRT promoter element. *Plant Physiology*, 135(3), 1710–1717.

Kovar, D. R. and Staiger, C. J. 2000. Developments in plant and soil sciences. Chapter: actin depolymerizing factor. In: *Actin: A Dynamic Framework for Multiple Plant Cell Functions*, Staiger, C. J., Baluška, F., Volkmann, D., and Barlow, P. W. (eds.), Volume 89. Springer Netherlands, Dordrecht, pp. 67–85, p.669.

Kucho, K.-i., Yoshioka, S., Taniguchi, F., Ohyama, K., and Fukuzawa, H. 2003. *Cis*-acting elements and DNA-binding proteins involved in CO_2-responsive transcriptional activation of CAH1 encoding a periplasmic carbonic anhydrase in *Chlamydomonas reinhardtii*. *Plant Physiology*, 133(2), 783–793.

Kumar, D., Singh, P., Yusuf, M. A., Upadhyaya, C. P., Roy, S. D., Hohn, T., and Sarin, N. B. 2013. The *Xerophyta viscosa* aldose reductase (ALDRXV4) confers enhanced drought and salinity tolerance to transgenic tobacco plants by scavenging methylglyoxal and reducing the membrane damage. *Molecular Biotechnology*, 54(2), 292–303.

Lapidot, M. and Pilpel, Y. 2006. Genome-wide natural antisense transcription: coupling its regulation to its different regulatory mechanisms. *EMBO Reports*, 7(12), 1216–1222.

Lee, H. G., Mas, P., and Seo, P. J. 2016. MYB96 shapes the circadian gating of ABA signaling in *Arabidopsis*. *Scientific Reports*, 6(1), 17754.

Lee, H. K., Cho, S. K., Son, O., Xu, Z., Hwang, I., and Kim, W. T. 2009. Drought stress-induced Rma1H1, a RING membrane-anchor E3 ubiquitin ligase homolog, regulates aquaporin levels via ubiquitination in transgenic *Arabidopsis* plants. *The Plant Cell*, 21(2), 622–641.

Lee, M. J. and Yaffe, M. B. 2016. Protein regulation in signal transduction. *Cold Spring Harbor Perspectives in Biology*, 8(6).

Li, W., Jia, L., and Wang, L. 2017. Chemical signals and their regulations on the plant growth and water use efficiency of cotton seedlings under partial root-zone drying and different nitrogen applications. *Saudi Journal of Biological Sciences*, 24(3), 477–487.

Li, Z.-G. 2016. Methylglyoxal and glyoxalase system in plants: old players, new concepts. *The Botanical Review*, 82(2), 183–203.

Lobet, G., Couvreur, V., Meunier, F., Javaux, M., and Draye, X. 2014. Plant water uptake in drying soils. *Plant Physiology*, 164(4), 1619–1627.

Lyzenga, W. J. and Stone, S. L. 2011. Abiotic stress tolerance mediated by protein ubiquitination. *Journal of Experimental Botany*, 63(2), 599–616.

Manzoor, H., Athar, H. U. R., Rasul, S., Kanwal, T., Anjam, M. S., Qureshi, M. K., Bashir, N., Zafar, Z. U., Ali, M., and Ashraf, M. 2016. Avenues for improving drought tolerance in crops by ABA regulation. In: *Water Stress and Crop Plants: A Sustainable Approach*, Ahmad, P. (ed.), John Wiley & Sons, Chichester, UK, pp. 177–193.

Mao, C., Lu, S., Lv, B., Zhang, B., Shen, J., He, J., Luo, L., Xi, D., Chen, X., and Ming, F. 2017. A rice NAC transcription factor promotes leaf senescence via ABA biosynthesis. *Plant Physiology*, 174(3), 1747–1763.

Martinez, G. andKöhler, C. 2017. Role of small RNAs in epigenetic reprogramming during plant sexual reproduction. *Current Opinion in Plant Biology*, 36(Supplement C), 22–28.

Mazzucotelli, E., Belloni, S., Marone, D., De Leonardis, A. M., Guerra, D., Di Fonzo, N., Cattivelli, L., and Mastrangelo, A. M. 2006. The E3 ubiquitin ligase gene family in plants: regulation by degradation. *Current Genomics*, 7(8), 509–522.

McDowell, G. S. and Philpott, A. 2016. Chapter Two: New insights into the role of ubiquitylation of proteins, In: Jeon, K.W. (ed.), *International Review of Cell and Molecular Biology*, Academic Press, pp. 35–88.

Mittler, R. 2006. Abiotic stress, the field environment and stress combination. *Trends in Plant Science*, 11(1), 15–19.

Montilla-Bascón, G., Rubiales, D., Hebelstrup, K. H., Mandon, J., Harren, F. J. M., Cristescu, S. M., Mur, L. A. J., and Prats, E. 2017. Reduced nitric oxide levels during drought stress promote drought tolerance in barley and is associated with elevated polyamine biosynthesis. *Scientific Reports*, 7(1), 13311.

Morimoto, K., Ohama, N., Kidokoro, S., Mizoi, J., Takahashi, F., Todaka, D., Mogami, J., et al. 2017. BPM-CUL3 E3 ligase modulates thermotolerance by facilitating negative regulatory domain-mediated degradation of DREB2A in *Arabidopsis*. *Proceedings of the National Academy of Sciences*, 114(40), E8528–E8536.

Müller, M. and Munné-Bosch, S. 2015. Ethylene response factors: a key regulatory hub in hormone and stress signaling. *Plant Physiology*, 169(1), 32–41.

Müller-Xing, R., Xing, Q., and Goodrich, J. 2014. Footprints of the sun: memory of UV and light stress in plants. *Frontiers in Plant Science*, 5, 474.

Muppirala, U. K., Lewis, B. A., and Dobbs, D. L. 2013. Computational tools for investigating RNA-protein interaction partners. *Journal of Computer Science and Computational Biology*, 6(4), 182.

Nahar, K., Hasanuzzaman, M., Alam, M. M., Rahman, A., Mahmud, J.-A., Suzuki, T., and Fujita, M. 2017. Insights into spermine-induced combined high temperature and drought tolerance in mung bean: osmoregulation and roles of antioxidant and glyoxalase system. *Protoplasma*, 254(1), 445–460.

Nakashima, K. and Yamaguchi-Shinozaki, K. 2005. Molecular studies on stress-responsive gene expression in *Arabidopsis* and improvement of stress tolerance in crop plants by regulon biotechnology. *Japan Agricultural Research Quarterly*, 39(4), 221–229.

Nazareno, A. L. and Hernandez, B. S. 2017. A mathematical model of the interaction of abscisic acid, ethylene and methyl jasmonate on stomatal closure in plants. *PLoS One*, 12(2), e0171065.

Nishiyama, R., Fujita, M., Seki, M., Kim, J.-M., Oono, Y., and Shinozaki, K. 2008. Posttranslational regulation of the NAC transcription factors in an ABA-dependent stress-signaling pathway. *Plant and Cell Physiology Supplement*, 2008, 0767.

Nuruzzaman, M., Sharoni, A. M., and Kikuchi, S. 2013. Roles of NAC transcription factors in the regulation of biotic and abiotic stress responses in plants. *Frontiers in Microbiology*, 4(248).

Oberschall, A., Deák, M., Török, K., Sass, L., Vass, I., Kovács, I., Fehér, A., Dudits, D., and Horváth, G. V. 2000. A novel aldose/aldehyde reductase protects transgenic plants against lipid peroxidation under chemical and drought stresses. *The Plant Journal*, 24(4), 437–446.

Olsson, A., Engström, P., and Söderman, E. 2004. The homeobox genes *ATHB12* and *ATHB7* encode potential regulators of growth in response to water deficit in *Arabidopsis*. *Plant Molecular Biology*, 55(5), 663–677.

Pandey, P., Ramegowda, V., and Senthil-Kumar, M. 2015. Shared and unique responses of plants to multiple individual stresses and stress combinations: physiological and molecular mechanisms. *Frontiers in Plant Science*, 6, 723.

Panter, S., Thomson, R., de Bruxelles, G., Laver, D., Trevaskis, B., and Udvardi, M. 2000. Identification with proteomics of novel proteins associated with the peribacteroid membrane of soybean root nodules. *Molecular Plant-Microbe Interactions*, 13(3), 325–333.

Pérez-Alfocea, F., Ghanem, M. E., Gómez-Cadenas, A., and Dodd, I. C. 2011. Omics of root-to-shoot signaling under salt stress and water deficit. *OMICS: A Journal of Integrative Biology*, 15(12), 893–901.

Phukan, U. J., Jeena, G. S., Tripathi, V., and Shukla, R. K. 2017. Regulation of Apetala2/ethylene response factors in plants. *Frontiers in Plant Science*, 8, 150.

Pivato, M., Fabrega-Prats, M., and Masi, A. 2014. Low-molecular-weight thiols in plants: functional and analytical implications. *Archives of Biochemistry and Biophysics*, 560, 83–99.

Popova, V. V., Kurshakova, M. M., and Kopytova, D. V. 2015. Methods to study the RNA-protein interactions. *Molecular Biology*, 49(3), 418–426.

Portolés, S. and Zhang, D.-P. 2014. ABA signaling and circadian clock. In: *Abscisic Acid: Metabolism, Transport and Signaling*, Zhang, D.-P. (ed.), Springer Netherlands, Dordrecht, pp. 385–407.

Pourghayoumi, M., Rahemi, M., Bakhshi, D., Aalami, A., and Kamgar-Haghighi, A. A. 2017. Responses of pomegranate cultivars to severe water stress and recovery: changes on antioxidant enzyme activities, gene expression patterns and water stress responsive metabolites. *Physiology and Molecular Biology of Plants*, 23(2), 321–330.

Pruthvi, V., Narasimhan, R., and Nataraja, K. N. 2014. Simultaneous expression of abiotic stress responsive transcription factors, AtDREB2A, AtHB7 and AtABF3 improves salinity and drought tolerance in peanut (*Arachis hypogaea* L.). *PLoS One*, 9(12), e111152.

Qin, F., Sakuma, Y., Tran, L. S., Maruyama, K., Kidokoro, S., Fujita, Y., Fujita, M., et al. 2008. *Arabidopsis* DREB2A-interacting proteins function as RING E3 ligases and negatively regulate plant drought stress-responsive gene expression. *Plant Cell*, 20(6), 1693–1707.

Que, S., Li, K., Chen, M., Wang, Y., Yang, Q., Zhang, W., Zhang, B., Xiong, B., and He, H. 2012. PhosphoRice: a meta-predictor of rice-specific phosphorylation sites. *Plant Methods*, 8(1), 5.

Rai, A., Suprasanna, P., D'Souza, S. F., and Kumar, V. 2012. Membrane topology and predicted RNA-binding function of the "early responsive to dehydration (ERD4)" plant protein. *PLoS One*, 7(3), e32658.

Rai, A.N., Tamirisa, S., Rao, K., Kumar, V., and Suprasanna, P. 2016. Brassica RNA binding protein ERD4 is involved in conferring salt, drought tolerance and enhancing plant growth in *Arabidopsis*. *Plant Molecular Biology*, 90(45), 375–387.

Ranty, B., Aldon, D., Cotelle, V., Galaud, J.-P., Thuleau, P., and Mazars, C. 2016. Calcium sensors as key hubs in plant responses to biotic and abiotic stresses. *Frontiers in Plant Science*, 7(327).

Re, D. A., Capella, M., Bonaventure, G., and Chan, R. L. 2014. Arabidopsis AtHB7 and AtHB12 evolved divergently to fine tune processes associated with growth and responses to water stress. *BMC Plant Biology*, 14(1), 150.

Reddy, A. S., Ali, G. S., Celesnik, H., and Day, I. S. 2011. Coping with stresses: roles of calcium- and calcium/calmodulin-regulated gene expression. *Plant Cell*, 23(6), 2010–2032.

Roy, S. 2016. Function of MYB domain transcription factors in abiotic stress and epigenetic control of stress response in plant genome. *Plant Signaling and Behavior*, 11(1), e1117723.

Ruiz-Lozano, J. M. and Aroca, R. 2017. Plant aquaporins and mycorrhizae: their regulation and involvement in plant physiology and performance. In: *Plant Aquaporins: From Transport to Signaling*, Chaumont, F. and Tyerman, S.D. (eds.), Springer International Publishing, Cham, pp. 333–353, p. 353.

Ruzicka, D. R., Kandasamy, M. K., McKinney, E. C., Burgos-Rivera, B., and Meagher, R. B. 2007. The ancient subclasses of *Arabidopsis* actin depolymerizing factor genes exhibit novel and differential expression. *The Plant Journal*, 52(3), 460–472.

Sakurai, H. and Ota, A. 2011. Regulation of chaperone gene expression by heat shock transcription factor in *Saccharomyces cerevisiae*: importance in normal cell growth, stress resistance, and longevity. *FEBS Letters*, 585(17), 2744–2748.

Santner, A. and Estelle, M. 2010. The ubiquitin-proteasome system regulates plant hormone signaling. *The Plant Journal*, 61(6), 1029–1040.

Santoni, V. 2017. Plant aquaporin posttranslational regulation. In: *Plant Aquaporins: From Transport to Signaling*, Chaumont, F. and Tyerman, S. D. (eds.), Springer International Publishing, Cham, pp. 83–105, p. 353.

Santos, A. P., Ferreira, L. J., and Oliveira, M. M. 2017. Concerted flexibility of chromatin structure, methylome, and histone modifications along with plant stress responses. *Biology*, 6(1), 3.

Sanyal, S. K., Kanwar, P., Yadav, A. K., Sharma, C., Kumar, A., and Pandey, G. K. 2017. *Arabidopsis* CBL interacting protein kinase 3 interacts with ABR1, an APETALA2 domain transcription factor, to regulate ABA responses. *Plant Science*, 254(Supplement C), 48–59.

Satoh, R., Nakashima, K., Seki, M., Shinozaki, K., and Yamaguchi-Shinozaki, K. 2002. ACTCAT, a novel *cis*-acting element for proline- and hypoosmolarity-responsive expression of the *ProDH* gene encoding proline dehydrogenase in *Arabidopsis*. *Plant Physiology*, 130(2), 709–719.

Scholz, S. S., Reichelt, M., Vadassery, J., and Mithofer, A. 2015. Calmodulin-like protein CML37 is a positive regulator of ABA during drought stress in *Arabidopsis*. *Plant Signal and Behavior*, 10(6), e1011951.

Schwarz, L. A. and Patrick, G. N. 2012. Ubiquitin-dependent endocytosis, trafficking and turnover of neuronal membrane proteins. *Molecular and Cellular Neurosciences*, 49(3), 387–393.

Schwertman, P., Bekker-Jensen, S., and Mailand, N. 2016. Regulation of DNA double-strand break repair by ubiquitin and ubiquitin-like modifiers. *Nature Reviews Molecular Cell Biology*, 17(6), 379–394.

Seki, M., Umezawa, T., Urano, K., and Shinozaki, K. 2007. Regulatory metabolic networks in drought stress responses. *Current Opinion in Plant Biology*, 10(3), 296–302.

Senavirathne, W., Jayatilake, D., Herath, V., and Wickramasinghe, H. 2017. Evaluation of genetic diversity of *cis*-acting elements of abscisic acid responsive element binding protein (ABRE-BP) in selected Sri Lankan rice varieties. *Tropical Agricultural Research*, 28(2), 120–132.

Seung, D., Risopatron, J. P. M., Jones, B. J., and Marc, J. 2012. Circadian clock-dependent gating in ABA signalling networks. *Protoplasma*, 249(3), 445–457.

Shao, H. H., Chen, S. D., Zhang, K., Cao, Q. H., Zhou, H., Ma, Q. Q., He, B., et al. 2014. Isolation and expression studies of the ERD15 gene involved in drought-stressed responses. *Genetics and Molecular Research*, 13(4), 10852–10862.

Sharma, B., Joshi, D., Yadav, P. K., Gupta, A. K., and Bhatt, T. K. 2016. Role of ubiquitin-mediated degradation system in plant biology. *Frontiers in Plant Science*, 7, 806.

Shen, J., Lv, B., Luo, L., He, J., Mao, C., Xi, D., and Ming, F. 2017. The NAC-type transcription factor OsNAC2 regulates ABA-dependent genes and abiotic stress tolerance in rice. *Scientific Reports*, 7, 40641.

Shin, R., Alvarez, S., Burch, A. Y., Jez, J. M., and Schachtman, D. P. 2007. Phosphoproteomic identification of targets of the *Arabidopsis* sucrose nonfermenting-like kinase SnRK2.8 reveals a connection to metabolic processes. *Proceedings of the National Academy of Sciences*, 104(15), 6460–6465.

Shinozaki, K.and Yamaguchi-Shinozaki, K. 2007. Gene networks involved in drought stress response and tolerance. *Journal of Experimental Botany*, 58(2), 221–227.

Showalter, A. M. 1993. Structure and function of plant cell wall proteins. *The Plant Cell*, 5(1), 9–23.

Singh, D. and Laxmi, A. 2015. Transcriptional regulation of drought response: a tortuous network of transcriptional factors. *Frontiers in Plant Science*, 6(895).

Singh, K. B., Foley, R. C., and Oñate-Sánchez, L. 2002. Transcription factors in plant defense and stress responses. *Current Opinion in Plant Biology*, 5(5), 430–436.

Song, J., Keppler, B. D., Wise, R. R., and Bent, A. F. 2015. PARP2 is the predominant poly (ADP-ribose) polymerase in *Arabidopsis* DNA damage and immune responses. *PLOS Genetics*, 11(5), e1005200.

Spoel, S. H.and Loake, G. J. 2011. Redox-based protein modifications: the missing link in plant immune signalling. *Current Opinion in Plant Biology*, 14(4), 358–364.

Staiger, C. J., Gibbon, B. C., Kovar, D. R., and Zonia, L. E. 1997. Profilin and actin-depolymerizing factor: modulators of actin organization in plants. *Trends in Plant Science*, 2(7), 275–281.

Stoiber, M. H., Olson, S., May, G. E., Duff, M. O., Manent, J., Obar, R., Guruharsha, K. G., et al. 2015. Extensive cross-regulation of post-transcriptional regulatory networks in *Drosophila*. *Genome Research*, 25(11), 1692–1702.

Strozycki, P. M., Szymanski, M., Szczurek, A., Barciszewski, J., and Figlerowicz, M. 2010. A new family of ferritin genes from *Lupinus luteus*—comparative analysis of plant ferritins, their gene structure, and evolution. *Molecular Biology and Evolution*, 27(1), 91–101.

Suzuki, N., Rivero, R. M., Shulaev, V., Blumwald, E., and Mittler, R. 2014. Abiotic and biotic stress combinations. *New Phytologist*, 203(1), 32–43.

Swaney, D. L., Beltrao, P., Starita, L., Guo, A., Rush, J., Fields, S., Krogan, N. J., and Villen, J. 2013. Global analysis of phosphorylation and ubiquitylation cross-talk in protein degradation. *Nature Methods*, 10(7), 676–682.

Takahashi, Y., Kinoshita, T., Matsumoto, M., and Shimazaki, K. 2016. Inhibition of the *Arabidopsis* bHLH transcription factor by monomerization through abscisic acid-induced phosphorylation. *The Plant Journal*, 87(6), 559–567.

Tanou, G., Filippou, P., Belghazi, M., Job, D., Diamantidis, G., Fotopoulos, V., and Molassiotis, A. 2012. Oxidative and nitrosative-based signaling and associated post-translational modifications orchestrate the acclimation of citrus plants to salinity stress. *The Plant Journal*, 72(4), 585–599.

Tari, I., Kiss, G., Deér, A. K., Csiszár, J., Erdei, L., Gallé, Á., Gémes, K., et al. 2010. Salicylic acid increased aldose reductase activity and sorbitol accumulation in tomato plants under salt stress. *Biologia Plantarum*, 54(4), 677–683.

Tavladoraki, P., Cona, A., and Angelini, R. 2016. Copper-containing amine oxidases and FAD-dependent polyamine oxidases are key players in plant tissue differentiation and organ development. *Frontiers in Plant Science*, 7(824), 1–11.

Trachootham, D., Lu, W., Ogasawara, M. A., Valle, N. R.-D., and Huang, P. 2008. Redox regulation of cell survival. *Antioxidants and Redox Signaling*, 10(8), 1343–1374.

Treiber, T., Treiber, N., Plessmann, U., Harlander, S., Daiss, J. L., Eichner, N., Lehmann, G., Schall, K., Urlaub, H., and Meister, G. 2017. A compendium of RNA-binding proteins that regulate microRNA biogenesis. *Molecular Cell*, 66(2), 270284. e213.

Trerotola, M., Relli, V., Simeone, P., and Alberti, S. 2015. Epigenetic inheritance and the missing heritability. *Human Genomics*, 9(1), 17.

Tripodi, F., Nicastro, R., Reghellin, V., and Coccetti, P. 2015. Post-translational modifications on yeast carbon metabolism: regulatory mechanisms beyond transcriptional control. *Biochimica et Biophysica Acta General Subjects*, 1850(4), 620–627.

Tuberosa, R. and Salvi, S. 2006. Genomics-based approaches to improve drought tolerance of crops. *Trends in Plant Science*, 11(8), 405–412.

Tuteja, N. and Sopory, S. K. 2008. Chemical signaling under abiotic stress environment in plants. *Plant Signaling and Behavior*, 3(8), 525–536.

Ubeda-Tomas, S., Swarup, R., Coates, J., Swarup, K., Laplaze, L., Beemster, G. T., Hedden, P., Bhalerao, R., and Bennett, M. J. 2008. Root growth in *Arabidopsis* requires gibberellin/DELLA signalling in the endodermis. *Nature Cell Biology*, 10(5), 625–628.

Uckelmann, M. and Sixma, T. K. 2017. Histone ubiquitination in the DNA damage response. *DNA Repair*, 56, 92–101.

Uehlein, N., Kai, L., and Kaldenhoff, R. 2017. Plant aquaporins and CO_2. In: *Plant Aquaporins: From Transport to Signaling*, Chaumont, F. and Tyerman, S. D. (eds.), Springer International Publishing, Cham, pp. 255–265, p. 353.

Uhrig, R. G., Schläpfer, P., Mehta, D., Hirsch-Hoffmann, M., and Gruissem, W. 2017. Genome-scale analysis of regulatory protein acetylation enzymes from photosynthetic eukaryotes. *BMC Genomics*, 18(1), 514.

Umezawa, T., Fujita, M., Fujita, Y., Yamaguchi-Shinozaki, K., and Shinozaki, K. 2006. Engineering drought tolerance in plants: discovering and tailoring genes to unlock the future. *Current Opinion in Biotechnology*, 17(2), 113–122.

Valdes, A. E., Overnas, E., Johansson, H., Rada-Iglesias, A., and Engstrom, P. 2012. The homeodomain-leucine zipper (HD-Zip) class I transcription factors ATHB7 and ATHB12

modulate abscisic acid signalling by regulating protein phosphatase 2C and abscisic acid receptor gene activities. *Plant Mol Biology*, 80(45), 405–418.

Van De Velde, K., Ruelens, P., Geuten, K., Rohde, A., and Van Der Straeten, D. 2017. Exploiting DELLA signaling in cereals. *Trends in Plant Science*, 22(10), 880–893.

van der Knaap, J. A., and Verrijzer, C. P. 2016. Undercover: gene control by metabolites and metabolic enzymes. *Genes and Development*, 30(21), 2345–2369.

Vasquez-Robinet, C., Mane, S. P., Ulanov, A. V., Watkinson, J. I., Stromberg, V. K., De Koeyer, D., Schafleitner, R., et al. 2008. Physiological and molecular adaptations to drought in Andean potato genotypes. *Journal of Experimental Botany*, 59(8), 2109–2123.

Vickers, C. E., Gershenzon, J., Lerdau, M. T., and Loreto, F. 2009. A unified mechanism of action for volatile isoprenoids in plant abiotic stress. *Nature Chemical Biology*, 5(5), 283–291.

Virdi, A. S., Singh, S., and Singh, P. 2015. Abiotic stress responses in plants: roles of calmodulin-regulated proteins. *Frontiers in Plant Science*, 6, 1–19.

Vlastaridis, P., Papakyriakou, A., Chaliotis, A., Stratikos, E., Oliver, S. G., and Amoutzias, G. D. 2017. The pivotal role of protein phosphorylation in the control of yeast central metabolism. *G3: Genes Genomes Genetics*, 7(4), 1239–1249.

Wang, J., Yu, H., Xiong, G., Lu, Z., Jiao, Y., Meng, X., Liu, G., Chen, X., Wang, Y., and Li, J. 2017. Tissue-specific ubiquitination by IPA1 interacting protein1 modulates IPA1 protein levels to regulate plant architecture in rice. *The Plant Cell*, 29(4), 697–707.

Wang, W., Vinocur, B., Shoseyov, O., and Altman, A. 2004. Role of plant heat-shock proteins and molecular chaperones in the abiotic stress response. *Trends in Plant Science*, 9(5), 244–252.

Wasilewska, A., Vlad, F., Sirichandra, C., Redko, Y., Jammes, F., Valon, C., Frei dit Frey, N., and Leung, J. 2008. An update on abscisic acid signaling in plants and more. *Molecular Plant*, 1(2), 198–217.

Weber, R. L. M., Wiebke-Strohm, B., Bredemeier, C., Margis-Pinheiro, M., de Brito, G. G., Rechenmacher, C., Bertagnolli, P. F., et al. 2014. Expression of an osmotin-like protein from *Solanum nigrum* confers drought tolerance in transgenic soybean. *BMC Plant Biology*, 14(1), 343.

Whitmarsh, A. J. 2007. Regulation of gene transcription by mitogen-activated protein kinase signaling pathways. *Biochimica et Biophysica Acta (BBA). Molecular Cell Research*, 1773(8), 1285–1298.

Wight, M. and Werner, A. 2013. The functions of natural antisense transcripts. *Essays in Biochemistry*, 54, 91–101.

Woodson, J. D. 2016. Chloroplast quality control—balancing energy production and stress. *New Phytologist*, 212(1), 36–41.

Wu, L., Hu, X., Wang, S., Tian, L., Pang, Y., Han, Z., Wu, L., and Chen, Y. 2015. Quantitative analysis of changes in the phosphoproteome of maize induced by the plant hormone salicylic acid. *Scientific Reports*, 5(1), e18155.

Wulff-Zottele, C., Gatzke, N., Kopka, J., Orellana, A., Hoefgen, R., Fisahn, J., and Hesse, H. 2010. Photosynthesis and metabolism interact during acclimation of *Arabidopsis thaliana* to high irradiance and sulphur depletion. *Plant, Cell and Environment*, 33(11), 1974–1988.

Xu, L., Lin, Z., Tao, Q., Liang, M., Zhao, G., Yin, X., and Fu, R. 2014. Multiple nuclear factor Y transcription factors respond to abiotic stress in *Brassica napus* L. *PloS One*, 9(10), e111354.

Yadav, S. K., Singla-Pareek, S. L., Reddy, M. K., and Sopory, S. K. 2005. Transgenic tobacco plants overexpressing glyoxalase enzymes resist an increase in methylglyoxal and maintain higher reduced glutathione levels under salinity stress. *FEBS Letters*, 579(27), 6265–6271.

Yamaguchi-Shinozaki, K. and Shinozaki, K. 1993. The plant hormone abscisic acid mediates the drought-induced expression but not the seed-specific expression of rd22, a gene responsive to dehydration stress in *Arabidopsis thaliana. Molecular and General Genetics MGG*, 238(1), 17–25.

Yang, W., Zhang, W., and Wang, X. 2017. Post-translational control of ABA signalling: the roles of protein phosphorylation and ubiquitination. *Plant Biotechnology Journal*, 15(1), 4–14.

Yelina, N., Diaz, P., Lambing, C., and Henderson, I. R. 2015. Epigenetic control of meiotic recombination in plants. *Science China. Life Sciences*, 58(3), 223–231.

Yin, Y., Adachi, Y., Nakamura, Y., Munemasa, S., Mori, I. C., and Murata, Y. 2016. Involvement of OST1 protein kinase and PYR/PYL/RCAR receptors in methyl jasmonate-induced stomatal closure in *Arabidopsis* guard cells. *Plant Cell Physiology*, 57(8), 1779–1790.

Yoshida, S. and Parniske, M. 2005. Regulation of plant symbiosis receptor kinase through serine and threonine phosphorylation. *Journal of Biological Chemistry*, 280(10), 9203–9209.

Yu, D., Zhang, L., Zhao, K., Niu, R., Zhai, H., and Zhang, J. 2017. VaERD15, a transcription factor gene associated with cold-tolerance in Chinese wild *Vitis amurensis. Frontiers in Plant Science*, 8(297).

Zagorchev, L., Seal, E. C., Kranner, I., and Odjakova, M. 2013. A central role for thiols in plant tolerance to abiotic stress. *International Journal of Molecular Sciences*, 14(4).

Zagorchev, L., Teofanova, D., and Odjakova, M. 2016. Ascorbate-glutathione cycle: controlling the redox environment for drought tolerance. In: *Drought Stress Tolerance in Plants*, Hossain, M. A., Wani, S. H., Bhattacharjee, S., Burritt, D. J., and Tran, L. -S. P. (eds.), Volume 1: *Physiology and Biochemistry*. Springer International Publishing, Cham, Switzerland, p. 187–226, p. 526.

Zang, D., Li, H., Xu, H., Zhang, W., Zhang, Y., Shi, X., and Wang, Y. 2016. An *Arabidopsis* zinc finger protein increases abiotic stress tolerance by regulating sodium and potassium homeostasis, reactive oxygen species scavenging and osmotic potential. *Frontiers in Plant Science*, 7(1272).

Zhang, H., Li, A., Zhang, Z., Huang, Z., Lu, P., Zhang, D., Liu, X., Zhang, Z.-F., and Huang, R. 2016a. Ethylene response factor TERF1, regulated by ethylene-insensitive3-like factors, functions in reactive oxygen species (ROS) scavenging in tobacco (*Nicotiana tabacum* L.). *Scientific Reports*, 6, 29948.

Zhang, D., Tong, J., Xu, Z., Wei, P., Xu, L., Wan, Q., Huang, Y., He, X., Yang, J., Shao, H., and Ma, H. 2016b. Soybean C2H2-type zinc finger protein GmZFP3 with conserved QALGGH motif negatively regulates drought responses in transgenic *Arabidopsis. Frontiers in Plant Science*, 7, 325.

Zhang, G.-B., Yi, H.-Y., and Gong, J.-M. 2014. The *Arabidopsis* ethylene/jasmonic acid-NRT signaling module coordinates nitrate reallocation and the trade-off between growth and environmental adaptation. *The Plant Cell*, 26(10), 3984–3998.

Zhang, X., Zhang, B., Li, M. J., Yin, X. M., Huang, L. F., Cui, Y. C., Wang, M. L., and Xia, X. 2016c. OsMSR15 encoding a rice C2H2-type zinc finger protein confers enhanced drought tolerance in transgenic *Arabidopsis. Journal of Plant Biology*, 59(3), 271–281.

Zhang, X. L., Jiang, L., Xin, Q., Liu, Y., Tan, J. X., and Chen, Z. Z. 2015. Structural basis and functions of abscisic acid receptors PYLs. *Frontiers in Plant Science*, 6, 88.

Zhang, X., Zang, R., and Li, C. 2004. Population differences in physiological and morphological adaptations of *Populus davidiana* seedlings in response to progressive drought stress. *Plant Science*, 166(3), 791–797.

Zhang, Z., Jones, A. E., Wu, W., Kim, J., Kang, Y., Bi, X., Gu, Y., et al. 2017. Role of remodeling and spacing factor 1 in histone H2A ubiquitination-mediated gene silencing. *Proceedings of the National Academy of Sciences*, 114(38), E7949–E7958.

Zheng, Y., Liao, C., Zhao, S., Wang, C., and Guo, Y. 2017. The glycosyltransferase QUA1 regulates chloroplast-associated calcium signaling during salt and drought stress in *Arabidopsis. Plant and Cell Physiology*, 58(2), 329–341.

Zhu, J.-K. 2002. Salt and drought stress signal transduction in plants. *Annual Review of Plant Biology*, 53(1), 247–273.

Zhu, M., Monroe, J., Suhail, Y., Villiers, F., Mullen, J., Pater, D., Hauser, F., Jeon, B. W., Bader, J. S., and Kwak, J. M. 2016. Molecular and systems approaches towards drought-tolerant canola crops. *New Phytologist*, 210(4), 1169–1189.

Zipfel, C. and Oldroyd, G. E. D. 2017. Plant signalling in symbiosis and immunity. *Nature*, 543(7645), 328–336.

Section VII

Plant/Crop Breeding under Stress Conditions

30 Marker-Assisted Breeding for Disease Resistance in Legume Vegetable Crops

Bhallan Singh Sekhon, Akhilesh Sharma, and Rakesh Kumar Chahota

CONTENTS

30.1 OVERVIEW

This chapter provides an overview of marker-assisted breeding (MAB) for disease resistance in legume vegetable crops. Basic concepts of resistance breeding have been elaborated for the easy understanding of the readers. Moreover, this chapter will shed light on the role of MAB to supplement the conventional breeding approaches. Finally, comprehensive information has been provided to the readers about the progress made in MAB along with the identified QTLs for disease resistance

in economically important leguminous vegetable crops: pea, French bean, cowpea, and cluster bean.

30.2 IMPORTANCE OF LEGUME VEGETABLES

Malnutrition, hidden hunger, and climate change are major concerns to agricultural scientists in view of the increasing global population. Legume vegetables, on account of their versatile advantages, could play a significant role in meeting these challenges. They are rich in dietary protein, calories, dietary fiber, and minerals (Singh, 2000) and also contribute to reducing the emission of greenhouse gases (Stagnari et al., 2017). Legumes also play an important role in cropping systems by enhancing crop diversification and soil fertility through atmospheric nitrogen fixation. However, the cultivation of vegetable legumes is highly affected by several biotic and abiotic factors, such as diseases, insect pests, drought, low soil fertility, and poor crop management (Mwang'ombe et al., 2007). Of these, losses caused by diseases are very significant and devastating in legume vegetables.

30.3 IMPORTANT DISEASES OF LEGUMINOUS VEGETABLES

Legume vegetables are highly prone to biotic stresses and are generally affected by a wide range of pathogens including fungi, bacteria, and viruses (Figures 30.1 and 30.2).

30.4 WHAT IS THE NEED FOR RESISTANCE BREEDING?

General principles of disease prevention and control include the basic cultural methods such as crop rotation, burning of infected residues of earlier crops, roguing of affected plants, adjusting the date of sowing, and the use of certified planting material and well-rotted manures. Additionally, chemical control through sprays is the most widely adopted method by vegetable growers, but it has certain limitations. Its effectiveness in controlling fungal and bacterial pathogens is limited to prevention and not cure, since chemicals are unable to penetrate the plant tissues to the extent that pathogens penetrate. Furthermore, the high cost of pesticides and their residual toxicity in the consumer chain pose serious limitations,

especially in legume vegetables. The alternative method is to raise crop varieties with acquired or inherited resistance. Cultivars with improved stress resistance can reduce reliance on pesticides in high-input systems, avert the risk of yield loss from pests in low- and high-input systems, and enable more stable production across diverse and adverse environments (low precipitation, high humidity, etc.) and poor soil conditions (Mooney, 2007). Based on the reaction/behavior of different cultivars to the pathogens, they were classified as either susceptible (rate of reproduction of pathogen, $r = 1$) or immune ($r = 0$). Other terms include resistance, where the rate of reproduction of the pathogen is always more than 0 but less than 1. Another phenomenon, tolerance (the disease is there but symptoms are not visible), is used by breeders to select cultivars for yield improvement. A spectrum of pathogens and a number of genes governing their expression are used as the basis for describing the genetics of disease resistance in crop plants. A classification of resistance on a genetic and epidemiological basis is given in Figure 30.3.

30.5 CONVENTIONAL BREEDING VERSUS MARKER-ASSISTED BREEDING

The development of cultivars with improved resistance to biotic and abiotic stresses is a primary goal of crop breeding programs throughout the world. Conventional breeding methods for disease resistance are based primarily on the principles of Mendelian genetics. Selection and hybridization are the methods commonly followed for developing disease resistant cultivars. Generally, the pedigree method is followed for the transfer of many genes, while backcross methods are considered efficient for single gene (major) transfer. However, classical breeding methods have certain

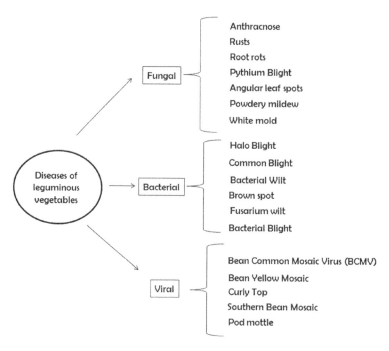

FIGURE 30.1 Important diseases of leguminous vegetables.

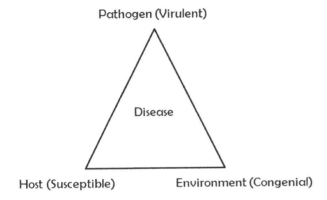

FIGURE 30.2 Host pathogen interaction.

limitations. As well as genotype–environment interactions (Taran et al., 2002), the quantitative nature of disease-related traits makes it difficult to achieve rapid progress through classical breeding (Mwang'ombe et al., 2007). In addition, breeding is complicated by pathogen variability and the different genes conditioning resistance. The identification of plants carrying two or more resistance alleles of different genes using standard inoculation test is impractical, because several races would be needed to screen for specific alleles (Yu et al., 2000). Thus, classical breeding is limited by the length of screening procedures and reliance on environmental factors. Hence, deployment of the molecular markers linked to resistance genes could be an alternative. This is the most reliable screening procedure to increase the efficiency of isolating disease resistance lines using marker-assisted selection (MAS). An efficient resistance breeding program in the present-day context must focus on the use of the emerging technologies of genomics, MAS, and bioinformatics. Basically, marker-assisted technologies are used as a supplement to classical breeding approaches, particularly in disease resistance breeding programs, for efficient selection, to reduce the time of the breeding procedure, and for introgression of useful traits with minimum linkage drag in existing cultivars (Figure 30.4).

30.6 CONCEPTS OF MARKER-ASSISTED BREEDING

The concept of MAS was suggested by Soller and Beckmann in 1983 and by Smith and Simpson in 1986 (Ben-Ari and Lavi, 2012). The selection of plants carrying genomic regions responsible for the expression of target traits through the use of molecular markers is termed *marker-assisted selection* (Lema, 2018). Molecular technology is considered a potent and trustworthy tool on account of its multiple applications, such as germplasm evaluation, genetic mapping, map-based gene discovery, and characterization of traits (Xu, 2010; Jiang, 2013). MAS has become possible for traits governed by both major genes as well as quantitative trait loci (QTLs) with the advent of an array of molecular markers and genetic maps in crop plants.

Important molecular markers, along with the year of their development, are depicted in Figure 30.5.

Molecular markers are generally classified as monomorphic or polymorphic on the basis of their capacity to reveal the genetic differences between individuals of the same or different species of crop plants. Polymorphic markers may also be described as codominant or dominant. Codominant markers may have many different alleles and are able to differentiate between homozygotes and heterozygotes, allowing the determination of genotypes and allele frequencies at loci. In contrast, a dominant marker only has two alleles and does not determine heterozygosity. Further, markers have been classified into polymerase chain reaction (PCR)- and non-PCR-based markers. The former requires a smaller quantity of genetic material in comparison to the latter and is generally preferred in MAS. The commonly used molecular markers along with their nature are listed in Table 30.2. Plant breeders have an interest in MAS mainly with respect to tolerance to biotic and abiotic stresses as well as for the improvement of horticultural traits. Molecular markers are generally exploited for the recognition of resistance genes and for their introgression and pyramiding into new cultivars in a shorter time and a more efficient manner.

30.7 MAS FOR DISEASE RESISTANCE IN LEGUME VEGETABLES

MAS, as well as genomics-assisted breeding (GAB), has been explored at a significant level in cereals, but it has not achieved much in legume crops on account of the lack of available genomic resources and optimized bioinformatics tools (Collard and Mackill, 2008; Bohra et al., 2014; Kulaeva et al., 2017). Different research workers reported several reasons for the lack of research in legumes with respect to MAB. Large genome size (Ellis et al., 2011), complicated genome assembly, and the presence of repetitive elements (Reid and Ross, 2011) are the major hindrances reported in pea. However, the most comprehensive source of pea gene sequence was made available by Ambrose (1995), Zohary and Hopf (2000), Moose and Mumm (2008), and Smykal et al. (2011). In French bean, inaccurate estimation of the main QTLs and epistatic and QTL × environmental interaction effects are the reported concerns in the use of MAS (Gonzalez et al., 2017). The use of genome-wide association studies (GWAS) is suggested for improving French bean via MAS through direct gene tagging for QTL mapping of biotic stress related traits. In the case of cowpea, MAS is limited on account of the paucity of trait-linked markers and lack of research related to gene structure and organization. Muchero et al. (2009) advocated the use of transcript derived SNPs for automated genotyping and revealing the genome structure through synteny analysis. Timko et al. (2008) called cowpea genome an ideal target for reduced representation sequencing. In cluster bean, limited availability of genomic resources restricts the use of MAS in breeding programs (Tanwar et al., 2017). Despite such barriers, scientists were able to reveal the genome size of these leguminous

TABLE 30.1

List of Important Diseases and Losses (%) Reported in Leguminous Vegetables

Crop	Common Name	Causal Organism	Losses (%)	Reference
Pea	Powdery mildew	*Erysiphe pisi* var. *pisi* D.C.	25–50	Janila and Sharma (2004)
	Fusarium wilt	*F. solani* f. sp. *pisi* and *F. avenaceum*	35–57	Chittem et al. (2015)
	Ascochyta blight or Black Spot disease	Complex of *Ascochyta pinodes, A. pinodella, A. pisi,* and/or *Phoma koolunga*	10–60	Liu et al. (2016)
	Pea rust	*Uromyces pisi* and *U. viciae-fabae*	More than 30	Barilli et al. (2014)
	Aphanomyces root rot	*Aphanomyces euteiches*	40–85	Pfander and Hagedorn (1983)
	Pea bacterial blight	*Pseudomonas syringae* pv. *Pisi*	10–80	Provvidenti and Hampton 1991
	Pea seed-borne mosaic virus (PSbMV)	Aphid transmitted	25–36	Coutts et al. (2009); Congdon et al. (2017)
	Pea leafroll virus	Aphid transmitted	19–81	Kaiser (1972)
	Tomato spotted wilt virus	Transmitted through young larval thrips	57.1–84.4	Fajardo et al. (1998)
	Pea early browning virus	Transmitted by soil-inhabiting trichodorid nematodes and through seeds	27–77	Fiederow (1983)
French bean	Angular leaf spot	*Phaeoisariopsis griseola*	As high as 80	Schwartz et al. (1981); Mwang'ombe et al. (2007)
	Anthracnose	*Colletotrichum lindemuthianum*	40–100	Mohammed (2013)
	Rust	*Uromyces appendiculatus*	85–100	Azmeraw and Hussien (2017)
	Fusarium root rot	*Fusarium solani* f. sp. *phaseoli*	42–52	Steadman et al. (1975)
	Fusarium wilt or yellows	*Fusarium oxysporum* f. sp. *phaseoli*	Up to 100	Maina et al. (2017)
	Rhizoctonia root rot	*Rhizoctonia solani*	8.5–64.7	Sharma and Sohi (1980 and 1981)
	Wilt and seed rot	*Pythium* spp. *Pythium*	Up to 70	Otsyula et al. (2003)
	Charcoal rot or ashy stem blight	*Macrophomina phaseolina*	65	Zaumeyer and Thomas (1957)
	Aphanomyces root rot	*Aphanomyces eufeches* f. sp. *phaseoli*	40–85%	Pfander and Hagedorn (1983)
	Bean common mosaic virus (BCMV)	Transmitted through seed or spread by aphids	53–68	Hampton (1975)
	Bean common mosaic necrosis virus (BCMNV)	Transmitted through seed or spread by aphids	100	Singh and Schwartz (2010)
	Bean yellow stipple virus	Transmitted through *Olpidium brassicae*	44–75	Sanchez and Bencomo (1979)
	Pea leafroll virus	Transmitted through aphids (*Myzus persicae*)	96–100	Kaiser (1972)
	Bean golden mosaic virus (BGMV)	Transmitted through whitefly (*Bemesia tabaci*)	37.7	Rodas (1975)
Cowpea	Anthracnose	*Colletotrichum lindemuthianum*	74	Masangwa et al. (2013)
	Rust	*Macrophomina phaseolina*	50	Chandrasekhar et al. (1989)
	Bean golden yellow mosaic virus	Transmitted through whitefly (*Bemesia tabaci*)	10 and 100	Shoyinka (1974)
	BCMV	Transmitted through seeds, pollens, and aphid insect vector	35–98	Prasad et al. (2007)
	BCMNV	Transmitted through seed or spread by aphids	100	Singh and Schwartz (2010)
	Common bacterial blight	*Xanthomonas axonopodis* pv. *vignicola*	2.7–92.2	Kishun (1989)
Cluster bean	Bacterial blight	*Xanthomonas axonopodis* pv. *cyamopsidis*	50–70	Amin et al. (2017)
	BCMV	Transmitted through seeds, pollens, and aphid insect vector	35–98	Prasad et al. (2007)
	BCMNV	Transmitted through seed or spread by aphids	100	Singh and Schwartz (2010)
	Anthracnose	*Colletotrichum capsici* f. sp. *Cyamopsicola*	84 (chilli)	Thind and Jhooty (1985)

vegetable crops (Table 30.3). Further, the important research studies, along with comprehensive reviews related to the genomics of these crops, have been listed in Table 30.4.

Several QTLs have been identified and mapped, and through their use, the role of MAS with respect to disease resistance in legume vegetables is discussed in the following subheadings.

30.7.1 PEA

30.7.1.1 Powdery Mildew

An obligate ecto-parasitic biotrophic fungus, *Erysiphe pisi* var. *pisi* D.C., is reported to be the causal agent of pea powdery mildew, which mainly damages the crop in late sowings or in late-maturing varieties (Fondevilla and Rubiales, 2011),

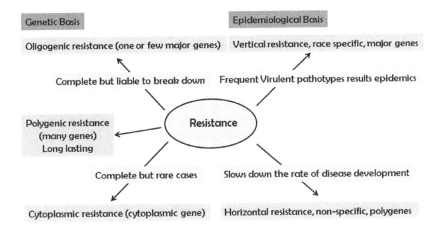

FIGURE 30.3 Concepts of resistance breeding.

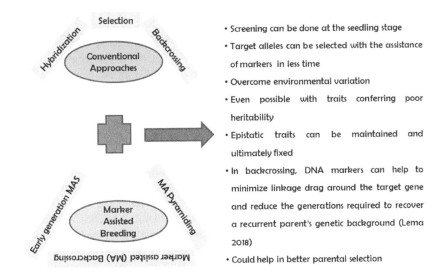

FIGURE 30.4 Marker-assisted breeding as an aid to conventional breeding.

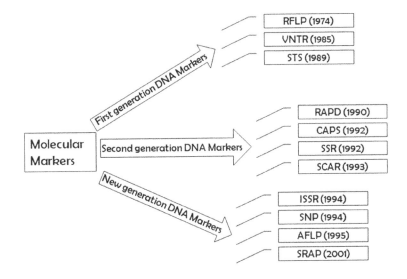

FIGURE 30.5 Advent of commonly used molecular markers.

TABLE 30.2

Nature of Molecular Markers Commonly Used in Crop Improvement

DNA Marker	Full name	Nature	Reference
RFLP	Restriction fragment length polymorphism	Codominant, non-PCR based	Botstein et al. (1980)
STS	Sequence tagged sites	STS created from RFLPs codominant, otherwise dominant	Olsen et al. (1989)
RAPD	Random amplified polymorphic DNA	Dominant, PCR based	Williams et al. (1990)
CAPS	Cleaved amplified polymorphic sequence	Codominant, PCR based	Akopyanz et al. (1992)
SSR	Simple sequence repeats	Codominant, PCR based	Akkaya et al. (1992).
SCAR	Sequence characterized amplified region	Codominant/dominant, PCR based	Paran and Michelmore (1993)
ISSR	Inter simple sequence repeats	Dominant, PCR based	Ziekiewicz et al. (1994)
SNP	Single nucleotide polymorphism	Codominant/dominant, PCR based	Jordan and Humphries (1994)
AFLP	Amplified fragment length polymorphism	Dominant, PCR based	Vos et al. (1995).
SRAP	Sequence-related amplified polymorphism	Codominant/dominant, PCR based	Li and Quiros (2001)

TABLE 30.3

Progress in the Genomics of Legume Vegetable Crops

Crop	Genome Size	Reference
Garden pea	4.45 Gbp	Dolezel and Greilhuber (2010)
French bean	587 Mbp	Schmutz et al. (2014)
Cowpea	620 Mbp	Boukar et al. (2018)
Cluster bean	152,530 bp (chloroplast)	Kaila et al. (2017)

TABLE 30.4

Studies Related to the Genomics of Legume Vegetable Crops

Crop	Reference
Pea	Macas et al. (2007); Smykal et al. (2012); Zhukov et al. (2015)
French bean	Schmutz et al. (2014); Resende et al. (2018)
Cowpea	Boukar et al. (2016); Boukar et al. (2018)
Cluster bean	Rawal et al. (2017); Kaila et al. (2017); Tyagi et al. (2018)

and yield losses of 25–50% have been reported (Janila and Sharma, 2004; Katoch et al., 2010). The most effective method of combating this disease is genetically controlled resistance (Fondevilla and Rubiales, 2012). Hammarlund (1925) reported four recessive genes for powdery mildew, which was not supported in further studies. These genes were designated as *er* by Harland (1948), and the resistance was reported to be under the control of a single gene in the homozygous recessive condition. This was later supported by Mishra and Shukla (1984), Timmerman-Vaughan et al. (1994), Janila and Sharma (2004), and Srivastava and Mishra (2004). Since then, genes *er2* and *Er3* have also been reported, but the *er1* gene emerged as the resistant source in most of the cultivars. Afterwards, digenic segregation (resistance under the control of both *er1*

and *er2*) was agreed by several workers (Heringa et al., 1969; Sokhi et al., 1979; Tiwari et al., 1997). Several markers have been developed for the use of MAS in pea improvement.

The availability of a highly saturated consensus map and linked markers for key biotic and abiotic stress tolerances provides important resources for pea molecular breeding programs (Sudeesh et al., 2014). First, the random amplification of polymorphic DNA (RAPD)-derived sequence characterized amplified region (SCAR) marker ScOPD 10_{650} was developed in Canadian germplasm (Timmerman-Vaughan et al., 1994). Later on, restriction fragment length polymorphism (RFLP), RAPD/SCAR, and simple sequence repeat (SSR) markers were developed and associated to *er1* on linkage group (LG) VI (Dirlewanger et al., 1994; Timmerman-Vaughan et al., 1994; Tiwari et al., 1998; Tiwari et al., 1999; Ek et al., 2005; Pereira et al., 2010). Closely linked SCAR markers, Sc-OPO-18_{1200} and Sc-OPE-16_{1600}, were also developed in relation to powdery mildew disease (Tiwari et al., 1998). This specific primer pair Sc-OPO-18_{1200} is expected to be useful for the identification of homozygous resistant individuals in F_2 and subsequent segregating generations in the near future. Furthermore, the polymorphic nature was not observed by Janila and Sharma (2004), as they were unable to differentiate resistant and susceptible lines of Indian origin. A substantial increase in the efficiency of MAS in pea breeding for powdery mildew was reported when the markers ScOPD-10_{650} and OPU-17 were coupled with the allele causing resistance (Janila and Sharma, 2004). *Er3* was mapped between the SCAR marker Scw4637 and the RAPD marker OPAG05_1240, located on an uncertain pea LG (Fondevilla et al., 2008). Mapping of the *er2* gene to LG III has been done recently (Katoch et al., 2010). To further facilitate the MAS, Srivastava et al. (2012) cloned and sequenced the 880 bp polymorphic band of the tightly linked RAPD marker OPX 04_{880} and developed a SCAR marker ScOPX 04_{880}. This marker was linked with the *er1* gene at 0.6 cM in coupling phase as ScOPX 04_{880}–*er1*–ScOPD 10_{650}. They also reported that ScOPX 04_{880} will correctly differentiate homozygous resistant plants from susceptible accessions with more than 99% accuracy. ScOPX 04$_{880}$ can help in an error-free MAS in combination with the repulsion phase marker ScOPD 10$_{650}$. Recently, different types of functional markers corresponding to five *er1* alleles were developed by Pavan et al. (2013), including an STS dominant marker er1-2/MGB targeting the *er1-2* allele, according to the sequence difference. To date, er1-2/MGB has not been tested in other resistance cultivars or in derived populations carrying *er1-2*, except Franklin. Specific SCAR molecular markers and a sequence-tagged site (STS) for the mutation er1mut1 were developed (Santo et al., 2013). Sun et al. (2015a) reported that the resistance in Xucai 1 was conferred by *er1-2* and suggested that its linked markers will be useful in pea breeding programs. The resistance gene was close to markers AD60 and c5DNAmet on LG VI in both the populations, suggesting that the resistance gene was an *er1* allele. An STS marker specific for *er1-2* was invalidated when tested in Chinese pea germplasms (Sun et al., 2015b; Wang et al., 2015). Fortunately, another functional marker specific for *er1-2*, PsMLO1-650, a coupling-phase marker, was developed and validated in the F$_2$ population derived from Bawan 6×X9002 (Wang et al., 2015). The functional marker for *er1-6* could distinguish pea germplasm harboring *er1-6* from pea germplasm with a non-*er1-6* genotype (Sun et al., 2015a). Moreover, the functional marker specific for *er1-6* was validated in pea germplasms. Sun et al. (2016) characterized a novel *er1* allele, "*er1-7*," conferring pea powdery mildew resistance by a 10 bp deletion in wild type PsMLO1 cDNA and also developed an ideal and co-dominant functional marker specific for *er1-7*, InDel111–120, which can accurately detect the *er1-7* gene solely based on PCR product size. The resistance gene *er1-7* and its functional marker reported in this study are expected to be of great importance for powdery mildew resistance in pea breeding programs. To date, four SCAR markers, five SSR markers, and a gene marker have been identified as being useful for detecting *er1* in genetic populations. Thus, these 10 *er1*-linked markers were used to identify polymorphisms between the crossed parents (Sun et al., 2016). A summary of various markers linked to the resistance gene *er1* located at different genetic distances in different mapping populations as reported by different researchers is provided in Table 30.5.

30.7.1.1.1 Parallel Assessment of Boron (B) Toxicity Tolerance and Powdery Mildew Resistance

B toxicity is a major problem in several production zones, including southern Australia, India, Pakistan, Iraq, Peru, and the United States (Yau and Ryan, 2008). B toxicity is difficult to manage by manipulation of agricultural systems, creating an incentive for solutions based on genetic improvement. Two major genes for tolerance to high B concentrations were detected, which interact with each other with incomplete dominance at each locus (Bagheri et al., 1996). B tolerance was reported to be tightly linked with the *er1* locus on chromosome Ps VI (Timmerman et al., 1994; Sudheesh et al., 2014).

A major implication arising from the parallel assessment of B toxicity tolerance and powdery mildew resistance is that both traits in the Kaspa×ps1771 population are controlled by single QTLs of large magnitude within the same interval (Sudheesh et al., 2014). The ps1771 genotype can hence function as a common donor for the two traits. Fortunately, both the linked traits are favorable in nature, as otherwise, recombination events within the QTL containing region would be required to separate the determinants. A nearby diagnostic marker, PsMlo, which is reported to be associated with powdery mildew (PM) resistance and boron (B) tolerance as well as its linked markers being associated with salinity tolerance across a diverse set of pea germplasm, was evaluated (Javid et al., 2015). The PsMlo1 marker predicted the PM and B phenotypic responses with high levels of accuracy (>80%) across a wide range of field pea genotypes; therefore, it has the potential to be widely adopted to identify PM resistant and B tolerant germplasm, which makes it suitable for implementation within the pea breeding program. The association of a single marker with a pair of traits of interest would be like hitting two birds with a single stone.

30.7.1.2 *Fusarium* species cause root rot (*F. solani* f.sp. *pisi* and *F. avenaceum*) or wilt (*F. oxysporum* f.sp. *pisi*)

Fusarium root rot is a major disease of pea worldwide (Kraft and Pfleger, 2001). To develop cultivars with superior *Fusarium* root rot resistance, the microsatellite markers

---

TABLE 30.5
Summary of Various Markers Linked to the Resistance Gene *er1* Located at Different Genetic Distances in Different Mapping Populations as Reported by Different Researchers

Gene Associated	Marker	Reported Distance	References
er1 gene	RFLP marker p236	9.8 cM	Dirlewanger et al. (1994)
er1 gene	RAPD marker OPD10650	2.1 cM	Timmerman et al. (1994)
er1 gene	SCAR ScOPD10650	3.4 cM	Janila and Sharma (2004)
er1 gene	PSMPSAD60 (named AD60 in this study)	10.4 cM	Ek et al. (2005)
er1 gene	Repulsion-phase marker ScOPD10650	2.2 cM	Srivastava et al. (2012)
er1 gene	Coupling-phase marker ScOPX04880	0.6 cM	Srivastava et al. (2012)
er1 gene	AD60 and c5DNAmet (Xucai 1×Bawan 6 F$_2$)	9.9 and 15.4 cM	Sun et al. (2015a)
er1 gene	AD60 and c5DNAmet (Qizhen 76×Xucai 1 F$_2$)	8.7 and 8.1 cM	Sun et al. (2015a)

that are closely linked to this QTL may be useful for stacking QTLs from "Carman" and other resistance sources. SSR markers flanking resistance loci useful for MAS were reported recently (Feng et al., 2011), including one QTL on LG VII and three QTLs on LG II, III, and VI. Single genes are available for *Fusarium* wilt resistance, such as *Fw* for resistance to race 1 and *Fwf* for resistance to race 5 (Coyne et al., 2000; Grajal-Martin and Muehlbauer, 2002; McClendon et al., 2002; Loridon et al., 2005; Okubara et al., 2005). Three QTLs with flanking SSR markers including the hypothesized single gene *Fnw* were identified (McPhee et al., 2012), controlling resistance to *Fusarium wilt* race 2. The authors also concluded that results of this research provide a basis for MAS of the major *Fnw* loci in both green and dry pea breeding programs, but additional research is necessary to fully characterize the complementary gene action governing resistance of the two minor loci identified.

Resistance to *Fusarium wilt* race 1 is governed by a single gene, *Fw*, located on LG III (Kwon et al., 2013). Using 80 F_8 recombinant inbred lines (RILs) derived from the cross "Green Arrow × PI 179449," they amplified 72 polymorphic markers between resistant and susceptible lines with the target region amplified polymorphism (TRAP) technique. They developed three user-friendly dominant SCAR markers, Fw_Trap_480, Fw_Trap_340, and Fw_Trap_220, which were found to be tightly linked to and only 1.2 cM away from the Fw locus and are therefore ideal for MAS. These newly identified markers are useful to assist in the isolation of the *Fusarium wilt* race 1 resistance gene in pea.

30.7.1.3 *Ascochyta* Blight of Pea

Both a single gene (*Rap2*) and QTLs have been reported to confer resistance to Ascochyta blight caused by *Mycosphaerella pinodes* (Tar'an et al., 2003; Prioul et al., 2004; Timmerman-Vaughan et al., 2004; Fondevilla et al., 2011). Single genes (*Rmp1*, *Rmp2*, *Rmp3*, and *Rmp4*) for *M. pinodes* resistance were identified when studies were conducted with very few markers per LG (Clulow et al., 1991). Ten and six QTLs associated with *Ascochyta* blight resistance at the seedling and adult plant stages, respectively, and four QTLs independent of developmental stage were identified using a linkage map based on RAPD, SSR, and STS marker polymorphism (Prioul et al., 2004; Timmerman-Vaughan et al., 2004). One may understand the quantitative inheritance of resistance to *M. pinodes* in pea through a series of QTL studies. Fondevilla et al. (2011) confirmed three of these QTLs in a second RIL population.

30.7.1.4 Pea Rust

In temperate regions, pea rust is caused by *Uromyces pisi*, while *U. viciae-fabae* is the causal organism in tropical and subtropical regions (Barilli et al., 2010). One major (Qruf) and one minor (Qruf1) QTL for rust resistance on LG VII were revealed by using composite interval mapping (CIM) (Rai et al., 2011). They also found two flanking SSR markers, AA505 and AA446 (10.8 cM), for the major QTL. The minor QTL was environment specific, and it was detected only in

the polyhouse. It was flanked by SSR markers, AD146 and AA416 (7.3 cM). The major QTL Qruf was consistently identified across all the four environments. Therefore, the SSR markers flanking Qruf would be useful for MAS for pea rust (*U. fabae*) resistance. They also explain up to 58% and 12% of the phenotypic variation for resistance to *U. viciae-fabae* based on two QTLs on LG VII, flanked by SSRs for MAS. Phenotypic variation in a *P. fulvum* cross for resistance to *U. pisi* based on one QTL on LG III was also explained to the extent of 63% (Barilli et al., 2010).

30.7.1.5 *Aphanomyces* Root Rot

Genetic resistance to *Aphanomyces* root rot (*Aphanomyces euteiches*) in pea is quantitative with moderate heritability (Pilet-Nayel et al., 2005; Hamon et al., 2010). Five consistent QTLs with codominant SSRs for MAS in France and the United States over multiple environments were identified, and these confer high levels of partial resistance (Hamon et al., 2010). Meta-analysis conducted using 244 individual QTLs reported previously in three mapping populations (Puget × 90–2079, Baccara × PI180693, and Baccara × 552) and in a fourth mapping population in this study (DSP × 90–2131) resulted in the identification of 27 meta-QTLs for resistance to *A. euteiches* (Hamon et al., 2013). Seven resistance meta-QTLs, including six of the highly consistent genomic regions, co-localized with six of the meta-QTLs were identified in this study for earliness and plant height along with three morphological genes (*Af*, *A*, and *R*). The authors concluded that QTL meta-analysis provided an overview of the moderately low diversity of loci controlling partial resistance to *A. euteiches* in four main sources of resistance in pea, and further work will be required to identify the best combinations of QTLs for durably increasing partial resistance to *A. euteiches*.

30.7.1.6 Pea Bacterial Blight

Resistance to pea bacterial blight (*Pseudomonas syringae* pv. *pisi*) is reported to be controlled by the single dominant genes *Ppi1* (for race R2), *Ppi3* (for race R3), and *Ppi4* (for race R4) (Hunter et al., 2001). *Pisum abyssinicum* accessions (16 originated from Ethiopia and one from Yemen) are resistant or partially resistant to all races, including race 6, for which there are no known resistant cultivars (Fondevilla et al., 2012). This resistance is controlled by a major recessive gene together with a number of modifiers (Elvira-Recuenco and Taylor, 2001).

30.7.1.7 Viruses

Pea seed-borne mosaic virus (PSbMV), a member of the genus *Potyvirus*, family *Potyviridae*, is a serious pathogen resulting in yield losses from 10% to 80%. Two clusters of recessive resistance genes to various potyviruses are present in pea as shown by genetic studies (Provvidenti and Hampton, 1991). The genes *bcm*, *cyv-1*, *mo*, *pmv*, and *sbm-2* are included in a cluster on LG II and confer resistance to *bean common mosaic virus* (BCMV), *clover yellow vein virus* (CYVV), *pea mosaic virus* (PMV), and the L1 (P2) pathotype of PSbMV. The second cluster on LG VI includes *cyv-2*, *wlv*, and *sbm-1*, conferring resistance to *clover yellow vein virus* (ClYVV), white lupin strain of *bean yellow mosaic virus* (BYMV-W), and the P1 pathotype of PSbMV.

Recently, two homologous *eIF4E* and *eIF(iso)4E* genes were identified by a candidate gene approach based on the *Medicago truncatula* genome to be responsible for PSbMV and BYMV-W resistance at the *sbm-1* and *sbm-2* loci (Gao et al., 2004; Bruun-Rasmussen et al., 2007). *eIF4E* in pea was shown to control resistance to ClYVV at the *cyv-2* locus and to BYMV-W virus at the *wlv* locus (Andrade et al., 2009).

Based on these studies, reliable and allele-specific testing of single nucleotide polymorphism (SNP) and co-dominant amplicon length polymorphism was developed, which proved to be 100% reliable and faster and more cost efficient compared with classical virological testing (Smykal et al., 2010). Naturally occurring diversity can be extended by mutagenesis and refined using a TILLING approach, from which mutations were identified in pea *eIF4E* gene (Dalmais et al., 2008). For resistance to *pea enation mosaic virus*, markers closely linked to the *En* gene were reported (Yu et al., 1995; Randhawa and Weeden, 2011).

30.7.1.8 Pea Early Browning Virus

Virus-induced gene silencing (VIGS) has become an important reverse genetics tool for functional genomics. VIGS vectors based on pea early browning virus (PEBV, genus Tobravirus) are available for *Pisum sativum* and were successfully used to silence pea genes involved in the symbiosis with nitrogen-fixing *Rhizobium* as well as development (Constantin et al., 2004; Gronlund et al., 2010).

30.7.2 French Bean

30.7.2.1 Angular Leaf Spot

Resistance against angular leaf spot (ALS) caused by *Phaeoisariopsis griseola* is controlled by major genes that are either dominant or recessive with single or duplicate inheritance and may interact in an additive manner with or without epistasis (Mahuku et al., 2003). Diverse sources of resistance to ALS in bean genotypes have been reported (Beebe and Pastor-Corrales, 1991; Mahuku et al., 2003). The resistant cultivars include A 75, A 140, A 152, A 175, A 229, BAT 76, BAT 431, BAT 1432, BAT 1458, and G5686, MAR 1 and MAR 2 (Ferreira et al., 2000). The sources of resistance reported from Africa include GLP 24, GLP X-92, GLP – 806, and GLP 77. Ragagnin et al. (2005) found that the resistance to race 63:23 of ALS in genotype "AND 277" is controlled by a single dominant gene (*Pgh-1*). Genotype "Cornell 49-242" carries gene *Pgh-2*, which confers resistance to pathotype 31:17. Genotypes "G5686" and "Mexico 54" display fairly good levels of resistance to nearly all races (Ferreira et al., 2000). These cultivars are not only good sources of resistance to ALS but could also serve as reliable indicators of new races of the pathogen in the future (Ngulu, 1999). "Mexico 54" showed resistance to all isolates of ALS characterized in Africa (Mahuku et al., 2002). The simple inheritance of resistance to specific isolates of ALS has been reported, with identification of molecular

markers for some of these resistance genes (Carvalho et al., 1998). SCAR markers for the identification of genes for resistance to ALS include SH13 for the *phg-1* gene in LG VI and SNO2 for the *phg-2* gene in LG VIII (Miklas et al., 2002). Other markers include SAA19, SBA16, and SMO_2 (Queiroz et al., 2004), which is the Ouro Negro dominant gene.

30.7.2.2 Anthracnose

Anthracnose, caused by *Colletotrichum lindemuthianum*, is a highly variable seed-borne fungal pathogen of common bean that is found on all continents where beans are grown (Melotto et al., 2000). Several resistant lines to ALS have been developed by traditional methods (Singh et al., 2003), although the development of new cultivars must continue because of the high pathogenic variability of the fungus (Mahuku et al., 2011). The resistance to ALS is controlled by several independent genes, which possess one or more alleles resistant to several races of the fungus. A genotype with only one resistance gene or allele can manage the disease for a few years only, until the appearance of new races of the fungus (Pastor-Corrales et al.,1998).

Resistance to anthracnose is conditioned primarily by nine major independent genes, *Co-1* to *Co-10*. Mendez-Vigo et al. (2005) recently showed that the *Co-3* and *Co-9* genes are allelic. The majority of the nine genes are dominant, except *co-8* with recessive gene inheritance, while multiple alleles exist at the *Co-1*, *Co-3*, and *Co-4* loci (reviewed by Kelly and Vallejo, 2004). The nine resistance genes *Co-2* to *Co-10* are of Middle American origin, whereas *Co-1* is the only locus from the Andean gene pool. An order of dominance exists among the four alleles at the *Co-1* locus. Eight resistance loci have been mapped to the integrated bean linkage map (Freyre et al., 1998), and the three *Co*-genes mapped to the LGs B1, B4, and B11 cluster with the *Ur*-genes for rust resistance (Kelly et al., 2003). The *Co-1* gene resides on B1, *Co-2* on B11, *Co-3/ Co-9* on B4 (Mendez-Vigo et al., 2005), *Co-4* on B8 (Melotto et al., 2004), *Co-5* (Campa et al., 2005) and *Co-6* on B7, and *Co-10* on B4 (Kelly and Vallejo, 2004). With the exception of the *Co-3/Co-9* gene cluster on B4, none of the other major *Co*-genes appear to be linked. In addition, there is co-localization with major resistance genes and QTLs that condition partial resistance to anthracnose (Geffroy et al., 2000).

The 10 *Co*-genes are represented in the anthracnose differential cultivars (Melotto et al., 2000) but are present as part of a multi-allelic series or in combination with other *Co*-genes, making it difficult to characterize more complex races of *C. lindemuthianum*. Although the *Co*-genes behave as major Mendelian factors, they most likely exist as resistance gene clusters, as has been demonstrated at the molecular level for the B4 *R*-gene cluster (Ferrier-Cana et al., 2003). Molecular markers linked to the majority of major *Co*-genes have been widely reported, and these provide the opportunity to enhance disease resistance through MAS (reviewed by Kelly and Vallejo, 2004; Kelly et al., 2003). Pyramiding genetically diverse resistance genes using MAS and deploying different gene combinations in different geographic regions is proposed as the most practical and realistic approach to provide efficient long-term control of bean anthracnose (Balardin and Kelly, 1998). MAS has been used successfully to breed for enhanced resistance to anthracnose in the cultivar Perola in Brazil (Ragagnin et al., 2003) and in pinto beans in the United States, but there is a need for caution based on the unsuccessful attempts to introgress the *Co-42* gene using marker-assisted backcrossing in two landrace bean cultivars from Ecuador (Ernest and Kelly, 2004). Indirect selection should be periodically verified by direct selection to ensure that the resistance gene is being transferred. Bean breeders have a unique opportunity to improve on natural gene pyramids in landrace cultivars (Young et al., 1998) by combining resistance genes from the two major gene pools to develop complementary resistance to a wide range of pathogenic races. To design effective gene pyramids, breeders need information on the pathogenic variability of *C. lindemuthianum* present in production areas. For example, 16 races of *C. lindemuthianum* from Guatemala (Muhuka, personal communication, 2004) defeated the *Co-2*, *Co-5*, and *Co-6* genes and supported the potential value of the *Co-1* and *Co-4* genes, both of which have suffered a major breakdown of resistance in Ecuador and Mexico, respectively.

Gonzalez et al. (2015) identified 10 and 16 main effect QTLs for resistance to anthracnose races 23 and 1545, respectively, using a multi-environment QTL mapping approach. The homologous genomic regions corresponding to 17 of the 26 main effect QTLs detected were positive for the presence of a resistance-associated gene cluster encoding nucleotide-binding and leucine-rich repeat (NL) proteins. The main effect QTLs detected on LG 05 for resistance to race 1545 in stem, petiole, and leaf were located within a 1.2 Mb region. The NL gene *Phvul.005G117900*, which can be considered an important candidate gene for the non-organ-specific QTL identified, is located in this region.

In North America, combining the *Co-42*, *Co-5*, and *Co-6* genes would be effective, whereas for areas of Central America, the most suitable gene pair would be the *Co-12* and *Co-42* gene combination. Since *Co-42* is recognized as the most broadly based resistance gene (Balardin and Kelly, 1998), it would be invaluable to include in gene pyramids with other Middle American genes in countries such as the Dominican Republic and Ecuador, where Andean races prevail. Since the genes differ in their effectiveness in controlling the highly variable races of the anthracnose pathogen, continued evaluation of resistance sources suggests that better tailored gene pyramids can be developed, provided information is available on the race diversity in specific regions. When selecting for anthracnose resistance in a particular region, bean breeders should carefully choose a gene pair that, if deployed singly, would confer resistance to all known races in that region. Sousa et al. (2014) found the linkage between the g12333250 marker and the *Co-52* allele identified in the MSU 7-1 French bean breeding line, which was found to be of great importance for marker-assisted introgression of this gene into commercial cultivars and elite cultivars, which could expand the resistance spectrum of common bean cultivars. Thus, the g12333250 marker may become an efficient tool for use in bean breeding programs.

30.7.2.3 Rust

The highly variable nature of the rust pathogen, caused by *Uromyces appendiculatus*, and the rapid breakdown of major gene resistance present in bean cultivars has challenged bean breeders to develop durable resistance to bean rust. Pyramiding different resistance genes and mechanisms (specific, adult plant, slow rusting, reduced pustule size, and pubescence) should prolong the life of a bean cultivar by creating a more durable resistance complex (Mmbaga et al., 1996). The importance of such resistance gene pyramids was observed in Honduras. Bean lines carrying the broadly effective *Ur-11* resistance gene of Middle American origin were infected by a newly identified "race 108" of rust pathotype (Stavely et al., 1997), whereas lines possessing the hypostatic *Ur-4* resistance gene of Andean origin in addition to *Ur-11* were not infected. Nine major rust resistance genes, *Ur-3*, *Ur-4*, *Ur-5*, *Ur-6* (Park et al., 2004), *Ur-7* (Park et al., 2003), *Ur-9* (Jung et al., 1998), *Ur-11*, *Ur-12* (Jung et al., 1998), and *Ur-13* (Mienie et al., 2005), and four unnamed genes, one each in breeding line BAC 6 (Jung et al., 1996) and Ouro Negro (Correa et al., 2000; Faleiro et al., 2000) and two in Dorado (Miklas et al., 2000a), have been characterized, tagged, and mapped (reviewed by Kelly et al., 2003 and Miklas et al., 2002). Based on map location and previous inheritance studies, the existence of gene clusters appears to be more common for rust than for anthracnose resistance genes in bean. For instance, Stavely (1984) showed that resistance to individual rust races in the bean line B-190 (*Ur-5* gene) is conditioned by single dominant genes linked in coupling that appear to be inherited as a complex linkage block. The apparent linkage of an additional unnamed gene from the cultivar Dorado suggests that the *Ur-5* genomic region on B4 may contain an even greater complex of linked genes than was previously considered. The tight linkage between *Ur-3* and *Ur-11* and the apparent linkage of a different unnamed gene from Dorado are indicative of another rust resistance gene block on B11. More effort on pyramid resistance in new cultivars is needed, as too much emphasis is being placed on the use of the single *Ur-3* gene in North America, despite the multigene pyramids available in breeding lines in three major U.S. commercial bean seed classes (Pastor-Corrales, 2003). The inheritance of rust resistance was studied by Divya (2014) in an F_2 generation using molecular markers. Resistance was confirmed in the parent genotype IC525236 by using SCAR (SK 14), which is controlled by the single dominant gene *Ur*. Moreover, genes for resistance to angular leaf spot, common bacterial blight, *Fusarium* root rot, and white mold have been located at the bottom of Chr1 (Miklas and Singh, 2007), as well as the *Phg-1* and *Ur-9* genes, which confer resistance against *Pseudocercospora griseola* and *U. appendiculatus*, respectively (Kelly and Vallejo, 2004; Gonçalves-Vidigal et al., 2011).

30.7.2.4 Root Rot

Root rot is caused by a complex of soil-borne pathogens that includes *Fusarium solani* f. sp. *phaseoli* (*Fusarium* root rot), *F. oxysporum* f. sp. *phaseoli* (*Fusarium* wilt or yellows), *Rhizoctonia solani* (*Rhizoctonia* root rot), *Pythium* spp. (*Pythium* wilt and seed rot), *Macrophomina phaseolina* (charcoal rot or ashy stem blight), *Thielaviopsis basicola* (black root rot), and *Aphanomyces eufeches* f. sp. *phaseoli* (*Aphanomyces* root rot), and it is the major limiting factor of common bean (Abawi and Pastor-Corrales, 1990). Root rots are economically important in most bean production areas (Snapp et al., 2003) but are particularly problematic in regions characterized by low soil fertility, limited crop rotation, and intensive seasonal bean production.

Quantitative resistance conditioned by four QTLs with relatively minor effect (13–19%) was reported in the Dorado/XAN 176 mapping population (Miklas et al., 1998). Two of the larger-effect QTLs that expressed across environments were located within resistance gene clusters on linkage groups B4 and B7 (Miklas et al.,, 2000c). Mayek-Perez et al. (2001) reported a similar inheritance of resistance in BAT 477. However, lack of map integration and validation studies of the *Mp* gene- and QTL-linked markers in additional populations has restricted the use of the markers in breeding for resistance to charcoal rot. The widespread nature and importance of *F. solani* as the predominant root rot pathogen in common bean emphasizes the need for effective control through the development of resistant cultivars (Boomstra and Bliss, 1977; Schneider et al., 2001; Chowdbury et al., 2002; Navarro et al., 2003).

Over 30 QTLs associated with root rot resistance have been reported in RIL populations derived from four resistance sources, many with minor effect. Sixteen QTLs for *Fusarium* root rot resistance were identified in an RIL population derived from a cross between susceptible cultivar Montcalm and resistant line FR266 (Schneider et al., 2001). Chowdbury et al. (2002) identified two QTLs in an RIL population derived from susceptible cultivar AC Compass crossed to resistant line NY2114-12. Similarly, Navarro et al. (2003) identified six QTLs in an RIL population derived from susceptible snap bean cultivar "Eagle" crossed with resistant line "Puebla 152," while nine QTLs were identified in two inbred backcross line populations derived from the susceptible cultivars "Red Hawk" and "C97407" crossed to resistant line "Negro San Luis" by Roman-Aviles and Kelly (2005). A single large-effect QTL was detected by Roman-Aviles and Kelly (2005) on B5 that accounted for up to 53% of the phenotypic variation for *Fusarium* root rot resistance that could be backcrossed into susceptible germplasm using MAS. A second QTL on B5 that explained up to 30% of the variation for resistance was linked to one of the markers, previously identified as associated to root rot resistance (Schneider et al., 2001). Most of the QTLs located on LGs B2 and B3 of the integrated bean map (Freyre et al., 1998) were close to a region where defense response genes, *Pgip* and *ChS*, and pathogenesis-related proteins, *PvPR-1* and *PvPR-2*, have been identified (Schneider et al., 2001). The detection of QTLs in the same genomic regions as previously reported QTLs for root rot resistance would suggest that different resistance sources might possess similar genes or resistance mechanisms associated with known defense response genes in *Phaseolus vulgaris*. Other than the dominant monogenic

resistance to *Fusarium* wilt (Cross et al., 2000) mapped to B10 (Fall et al., 2001), the inheritance of resistance to other root rot pathogens (*Rhizoctonia*, *Pythium*, *Aphanomyces*, etc.) have not been studied extensively, nor have resistance genes or QTLs been identified.

30.7.3 Cowpea

30.7.3.1 Anthracnose

Major work in beans with regard to disease resistance has provided the opportunity for bean breeders to pyramid up to four major anthracnose resistance genes, namely, *Co-1*, *Co-5*, *Co-6*, and *Co-2*, using RAPD markers for MAS. Young and Kelly (1996) reported four RAPD markers linked to alleles at the *Co-1*, *Co-5*, and *Co-6* loci and stated that $OF10_{530}$ (RAPD) would facilitate the introgression of the *Co-1* gene across *Phaseolus* gene pools for resistance to anthracnose. Linkage drag is a source of cost for introgression of genes. MAS reduces linkage drag and helps easy selection. Young et al. (1998) identified two RAPD markers, $OAS13_{950}$ and $OAL9_{740}$, linked to the *Co-4²* allele. Miklas and Kelly (2002) used the SAS13 marker to backcross the *Co-4²* allele into highly susceptible Durango race pinto beans grown widely in North America.

30.7.3.2 Rust Resistance

Miklas et al. (1993) identified an RAPD $OA14_{1100}$ marker tightly linked to the Up 2 locus. Development of the rust resistant navy bean lines BelMiDak-RR1 used $OA14_{1100}$ to verify its presence in combination with the *Ur-11* gene (Stavely, 1998). Li et al. (2007) developed an amplified fragment length polymorphism (AFLP) marker (E-AAG/M-CTG) and converted it to a SCAR marker, named ABRSAAG/CTG98, for rust resistance, which is controlled by a single dominant gene designated *Rr1*. Muchero et al. (2010) identified the QTL for *Macrophomina phaseolina* in cowpea with SNP markers, which may serve as an important tool for efficient introgression of resistance.

30.7.3.3 Common Bacterial Blight

Five QTLs conferring resistance to Xap were identified by Nodari et al. (1993). The BC420 marker was identified by Yu et al. (2000). Effective use of MAS for the QTLs has been repeatedly demonstrated by researchers (Jung et al.,, 1999; Park et al., 1999; Yu et al., 2000; Fourie and Herselman, 2002; Mutlu et al., 2002). Agbicodo et al. (2010) identified three SNP-based QTLs: CoBB-1, CoBB-2, and CoBB-3, on LG3, LG5, and LG9, respectively.

30.7.3.4 Bean Golden Yellow Mosaic Virus

Host resistance to the virus is the most practical means of control. A major gene, *bgm-1*, carries the resistance inherited by a recessive gene. A codominant RAPD marker was identified for the *bgm-1* gene (Urrea et al., 1996) that was subsequently converted to a SCAR marker named SR2 at CIAT (Centro Internacional de Agricultura Tropical). To counter the loss of pod quality and yield, the recessive *bgm-1* gene should be combined with other resistance genes conditioning non-deformed pods (Molina Castenado and Beaver, 1998) and the *bgm-2* for reduced mosaic symptoms (Velez et al., 1998). The pole garden bean cultivar Genuine is a result of MAS with moderate resistance to BGMV carrying the *bgm-1* gene (Kelly et al., 2003).

30.7.3.5 BCMV and BCMNV

Two genes, BCMV and BCMNV, the dominant *I* genes, located on B2 (Gepts, 1999), independent of three recessive *bc* loci have been currently mapped (Kelly et al., 2003). The independence of the BCMV resistance genes provides opportunities to use gene pyramiding as a strategy in breeding for durable resistance. The combination of the dominant *I* gene with recessive *bc* resistance genes offers durability over single gene resistance to BCMV and BCMNV, since the two types of genes have distinctly different mechanisms of resistance (Kelly, 1997). Miklas et al. (2002) and Miklas and Kelly (2002) have used markers linked to the *I* gene to develop enhanced germplasm with the *I + bc-3* gene combination. Markers linked to the other resistance genes, such as the *bc-u*, *bc-2²*, and *bc-3* genes, are needed to facilitate their introgression and gene pyramiding into germplasm possessing the *I* gene (Kelly et al., 2003).

30.7.4 CLUSTER BEAN

30.7.4.1 Bacterial Blight

Bacterial blight is caused by *Xanthomonas campestris*. Several SCAR markers have been found linked to resistance QTLs from different sources: UBC420 linked to the QTL on chromosome 1 with resistant alleles from "XAN 159," SU91 linked to the QTL on chromosome 3 from the same sources, and SAP6 linked to QTL on chromosome 8 with resistance from great northern "Nebraska No. Sel. 27" (Jung et al., 1997; Yu et al., 2000; Miklas et al., 2000; Pedrosa et al., 2003). These sources have been used in practical breeding to pyramid resistance genes. More tightly linked markers to the common bacterial blight resistance QTL on chromosome 1 of XAN 159 have been designed and validated to be useful in different genetic backgrounds for MAS (Liu et al., 2007).

30.7.4.2 Bean Common Mosaic Virus

SW13 is linked to the *I* gene for resistance to *bean common mosaic virus* (BCMV) and has proved to be very reliable in different genetic backgrounds (Melotto et al., 1996; Miklas et al., 2006). SCAR marker SBD5.1300 is tightly linked to *bc-1²*, which confers resistance to a specific strain of BCMV and *bean mosaic necrosis virus* (BCMNV). However, its resistance is masked by *bc-2²* and *bc-3* (Miklas et al., 2006). A recessive gene, *bgm-1*, confers bean gold mosaic virus resistance. Its tightly linked marker SR2 is also close to *bc-1* (Blair et al., 2007). The linkage between two loci may facilitate MAS.

30.7.4.3 Anthracnose

Breeding for anthracnose (caused by *Colletotrichum* spp.) resistance from different sources uses MAS to combine different genes (*Co-1* to *Co-10*) conferring resistance to various predominant races based on geographic regions in a practical and realistic manner (Balardin and Kelly, 1998). SAS13 is

linked to the *co-4²* gene, which has the broadest resistance to fungal races (Melotto and Kelly, 2001). However, application in MAS using this marker is not very consistent or reliable.

30.8 CONCLUSIONS

Plant breeders have traditionally and routinely used various recurrent selection methods to cumulate favorable alleles for yield and other polygenic traits. These methods of selection provide the population or breeding lines with diverse genetic recombination. The selection methods using classical breeding should be supplemented with MAS. To make it more useful to the breeder, gains made from MAS must be cost-effective compared with gains achieved through classical breeding. It is anticipated that the applications and technology improvements will reduce the cost of markers, which will subsequently pave the way for greater adoption of the use of molecular markers in plant breeding. The obstacles to using MAS are costly equipment, infrastructure, and consumables.

BIBLIOGRAPHY

Abawi GS and Pastor-Corrales MA. 1990. Root rots of beans in Latin America and Africa: Diagnosis, research methodologies, and management strategies. *Centro Internacional de AgriculturaTropical* (CIAT), Cali, Colombia, p. 114

Agbicodo EM, Fatokun CA, Bandyopadhyay R, Wydra K, Diop NN, Muchero W, Ehlers JD, Roberts PA, Close TJ, Visser RGF and van der Linden CG. 2010. Identification of markers associated with bacterial blight resistance loci in cowpea [*Vigna unguiculata* (L.) Walp.]. *Euphytica* 175:215–226.

Akkaya MS, Bhagwat AA and Cregan PB. 1992. Length polymorphism of simple sequences repeats DNA in soybean. *Genetics* 132:1131–1139.

Akopyanz N, Bukanov NO, Westblom TU and Berg DE. 1992. PCR-based RFLP analysis of DNA sequence diversity in the gastric pathogen Helicobacter pylori. *Nucleic Acid Research* 20:6221–6225.

Ambrose MJ. 1995. From near east centre of origin the prized pea migrates throughout world. *Diversity* 11:118–119.

Amin AM, Patel NR, Jaiman RK, Prajapati DB and Amin AU. 2017. Effect of date of sowing on the development of bacterial blight of clusterbean. *Environment and Ecology* 35(2A):967–970.

Andrade M, Abe Y, Nakahara KS and Uyeda I. 2009. The cyv-2 resistance to Clover yellow vein virus in pea is controlled by the eukaryotic initiation factor 4E. Journal of General Plant Pathology 75:241–249.

Azmeraw Y and Hussien T. 2017. Management of common bean rust (*Uromyces appendiculatus*) through host resistance and fungicide sprays in Hirna district, eastern Ethiopia. *Advances in Crop Science and Technology* 5:314.

Bagheri A, Paull JG and Rathjen AJ. 1996. Genetics of tolerance to high concentrations of soil boron in peas (*Pisum sativum* L.). *Euphytica* 87:69–75.

Balardin RS and Kelly JD. 1998. Interaction among races of *colletotrichum lindemuthianum* and diversity in *phaseolus vulgaris*. *Journal of the American Society for Horticultural Science* 123:1038–1047.

Barilli E, Satovic Z, Rubiales D and Torres AM. 2010. Mapping of quantitative trait loci controlling partial resistance against rust incited by *Uromyces pisi* (Pers.) Wint. in a *Pisum fulvum* L. intraspecific cross. *Euphytica* 175:151–159.

Barilli E, Sillero JC, Prats E and Rubiales D. 2014. Resistance to rusts (*Uromyces pisi* and *U. viciae-fabae*) in pea. *Czech Journal of Genetics and Plant Breeding* 50:135–143.

Beebe SE and Pastor-Corrales MA. 1991. Breeding for disease resistance. In: van Schoonhoven A and Voysest O (eds.), Common beans: Research for crop improvement, C.A.B. Intl., Wallingford, UK and CIAT, Cali, Colombia, pp. 561–617.

Ben-Ari G and Lavi U. 2012. Marker assisted selection in plant breeding. In: Altman A, Michae P and Lasegawa (eds), Plant biotechnology and agriculture: Prospects for the twenty-firstst century. pp. 163–184.

Blair MW, Rodriguez LM, Pedraza F, Morales F, Beebe S. 2007. Genetic mapping of common bean golden mosaic gemivirus resistance gene Bgm-1 and linkage with potyvirus resistance in common bean. *Theoretical and Applied Genetics* 114:261–271.

Bohra A, Pandey MK, Jha UC, Singh B, Singh IP and Datta D. 2014. Genomics-assisted breeding in four major pulse crops of developing countries: present status and prospects. Springer Berlin Heidelberg, *Theoretical and Applied Genetics* 127:1263–1291.

Boomstra AG and Bliss FA. 1977. Inheritance of resistance to *Fusarium solani* f.sp. *phaseoli* in beans (*Phaseolus vulgaris* L.) and breeding strategy to transfer resistance. *Journal of the American Society for Horticultural Sciences* 102:186–188.

Botstein D, White RL, Skolnick M and Davis RW. 1980. Construction of a genetic linkage map in man using restriction fragment length polymorphisms. *American Journal of Human Genetics* 32(3):314–331.

Boukar O, Belko N, Chamarthi S et al. 2018. Cowpea (*Vigna unguiculata*): Genetics, genomics and breeding. *Plant Breeding* 00:1–10.

Boukar O, Fatokun CA, Huynh BL, Roberts PA and Close TJ. 2016. Genomic tools in cowpea breeding programs: Status and perspectives. *Frontiers in Plant Science* 7:757.

Bruun-Rasmussen M, Moller IS, Tulinius G, Hansen KR, Lund OS and Johansen IE. 2007. The same allele of translation initiation factor 4E mediates resistance against two potyvirus spp. in *pisum sativum*. Molecular Plant Microbe Interactions 20:1075–1082.

Campa A, Rodr´ıguez-Su´arez C, Pa˜neda A, Giradles R and Ferreira JJ. 2005. The bean anthracnose res´ıstance gene Co-5 is located in linkage group B7. *Annual Report of Bean Improvement Cooperation* 48:68–69.

Carvalho GA, Paular TJ, Alzate-Marin AL, Nietsche S, Barros EG and Moreira MA. 1998. Inheritance of line resistance AND277 of common bean plants of the race 63-23 of *Phaeoisariopsis griseola* and identification of RAPD markers linked to the resistant gene). *Fitopatologia Brasia* 23:482–485.

Chandrashekar M, Kumar TBA, Saifulla M and Yadahally YH. 1989. Effect of different dates of sowing cowpea on the severity of leaf rust caused by *Uromyces phaseoli* var. *vignea*. *Tropical Agriculture* 66 (2):149–152.

Chittem K, Mathew FM, Gregoire M, Lamppa RS, Chang YW, Markell SG et al. 2015. Identification and characterization of Fusarium spp. associated with root rots of field pea in North Dakota. *European Journal of Plant Pathology* 143(4):641–649.

Chowdbury MA, Yu K and Park SJ. 2002. Molecular mapping of root rot resistance in common bean. *Annual Report of Bean Improvement Cooperation* 45:96–97.

Clulow SA, Matthews P and Lewis BG. 1991. Genetical analysis of resistance to Mycosphaerella pinodes in pea seedlings. *Euphytica* 58:183–189.

Collard BCY and Mackill DJ. 2008. Marker-assisted selection: an approach for precision plant breeding in the twenty-first century. *Philosophical Transactions of the Royal Society B* 363(1491):577–572.

Congdon BS, Coutts BA, Renton M, Banovic M and Jones RAC. 2016. Pea seed-borne mosaic virus in field pea : Widespread infection, genetic diversity, and resistance gene effectiveness. *Plant Disease* 100:2475–2482.

Constantin GD, Krath BN, MacFarlane SA, Nicolaisen M, Johansen IE and Lund OS. 2004. Virus induced gene silencing as a tool for functional genomics in a legume species. *Plant Journal* 40:622–631.

Correa RX, Costa MR, Good-God PI, Ragagnin VA, Faleiro FG, Moreira MA and Barros EG 2000. Sequence characterized amplified regions linked to rust resistance genes in the common bean. *Crop Science* 40:804–807.

Coutts BA, Prince RT and Jones RAC. 2009. Quantifying effects of seedborne inoculums on virus spread, yield losses and seed infection in Pea seed borne mosaic virus in field pea pathosystem. *Phytopathology* 99:1156–1167.

Coyne CJ, Inglis DA, Whitehead SJ and Muehlbauer FJ. 2000. Chromosomal location of Fwf in pea. Pisum Genetics 32:20–22.

Cross H, Brick MA, Schwartz HF, Panella LW and Byrne PF. 2000. Inheritance of resistance to *Fusarium* wilt in two common bean races. *Crop Science* 40:954–958.

Dalmais M, Schmidt J, Le Signor C, Moussy F, Burstin J, Savois V, Aubert G, Brunaud V, de Oliveira Z and Guichard C. 2008. UTILLdb, a *Pisum sativum* in silico forward and reverse genetics tool Genome Biol 9.

Dirlewanger E, Isaac P, Ranade S, Belajouza M, Cousin R and Devienne D. 1994. Restriction fragment length polymorphism analysis of loci associated with disease resistance genes and developmental traits in *Pisum sativum* L. *Theoretical and Applied Genetics* 88:17–27.

Divya B, Aghora TS, Mohan N and Sudeep HP. 2014. Breeding Pole Type French Bean (*Phaseolus vulgaris*) for Rust (*Uromyces phaseoli* Reben Wint) Resistance using Molecular Markers. *International Journal of Horticulture* 4(10):50–52.

Dolezel J and Greilhuber J. 2010. Nuclear genome size. Are we getting closer? *Cytometry* 77:35–642.

Ek M, Eklund M, von Post R, Dayteg C, Henriksson T, Weibull P, Ceptilis A, Issac P and Tuvesson S. 2005. Microsatellite markers for powdery mildew resistance in pea (*Pisum sativum* L.). *Hereditas* 142:86–91.

Ellis THN, Hofer JI, Timmerman-Vaughan GM, Coyne CJ and Hellens RP. 2011. Mendel, 150 years on. *Trends in Plant Sciences* 16:590–596.

Elvira-Recuenco M and Taylor JD. 2001. Resistance to bacterial blight (Pseudomonas syringae pv. pisi) in Spanish pea (*Pisum sativum*) landraces. *Euphytica* 118:305–311.

Ernest EG and Kelly JD. 2004. The Mesoamerican anthracnose resistance gene Co-42 does not confer resistance in certain Andean backgrounds. *Annual Report of the Bean Improvement Cooperative* 47:245–246.

Fajardo TVM, Lopes CA, Silva WCC and De Avila AC. 1998. Dispersal of disease and yield reduction in processing tomato infected with Tospovirus in the federal distinct. *Brazilian Review of Plant Pathology* 77:302.

Faleiro FG, Vinhadelli WS, Ragagnin VA, Corrˆea RX, Moreira MA and Barros EG. 2000. RAPD markers linked to a block of genes confering rust resistance to the common bean. *Genetics and Molecular Biology* 23:399–402.

Fall AL, Byrne PF, Jung J, Coyne DP, Brick MA and Schwartz HF. 2001. Detection and mapping of a major locus for fusarium wilt resistance in common bean. *Crop Science* 41:1494–1498.

Feng J, Hwang R, Chang KF, Conner RL, Hwang SF, Strelkov SE, Gossen BD, McLaren DL and Xue AG. 2011. Identification of microsatellite markers linked to quantitative trait loci controlling resistance to Fusarium root rot in field pea. *Canadian Journal of Plant Sciences* 91:199–204.

Ferreira CF, Borem A, Caravalho GA, Neitsche S, Paula TJ and de Barros EG. 2000. Inheritance of angular leaf spot resistance in common bean and identification of a RAPD marker linked to a resistance gene. *Crop Science* 40:1130–1133.

Ferrier Cana E, Geffroy V, Macadre C, Creusot F, Imbert Bollore P, Sevignac M and Langin T. 2003. Characterization of expressed NBS-LRR resistance gene candidates from common bean. *Theoretical and Applied Genetics* 106:251–261.

Fiederow ZG. 1983. Pea early browning virus on horse bean (*Vicia faba* L. ssp. *minor*). Zeszyty Problemowe Postepo´w Nauk Rolniczych 291:97–109.

Fondevilla S, Almeida NF, Satovic Z, Rubiales D, Vaz Patto MC, Cubero JI and Torres AM. 2011. Identification of common genomic regions controlling resistance to *Mycosphaerella pinodes*, earliness and architectural traits in different pea genetic backgrounds. *Euphytica* 18:43–52.

Fondevilla S, Martın-Sanz A, Satovic Z, Fernández-Romero MD, Rubiales D and Caminero C. 2012. Identification of quantitative trait loci involved in resistance to *Pseudomonas syringae* pv. *syringae* in pea (*Pisum sativum* L.). *Euphytica* 186:805–812 doi.

Fondevilla S and Rubiales D. 2011. Powdery mildew control in pea. A review. *Agronomy for Sustainable Development* 32(2):401-409 doi.

Fondevilla S and Rubiales D. 2012. Powdery mildew control in pea. *A review Agronomy for Sustainable Development* 32:401–409.

Fondevilla S, Rubiales D, Moreno MT and Torres AM. 2008. Identification and validation of RAPD and SCAR markers linked to the gene Er3 conferring resistance to *Erysiphe pisi* DC in pea. *Molecular Breeding* 22:193–200.

Fourie D and Herselman L. 2002. Breeding for common blight resistance in dry beans in South Africa. *Annual report of the Bean Improvement Cooperative* 45:50–51.

Freyre R, Skroch P, Geffroy V, Adam-Blondon AF, Shirmohamadali A, Johnson W, Llaca V, Nodari R et al. 1998. Towards an integrated linkage map of common bean. 4. Development of a core map and alignment of RFLP maps. *Theoretical and Applied Genetics* 97:847–856.

Gao Z, Eyers S, Thomas C, Ellis N and Maule A. 2004. Identification of markers tightly linked to sbm recessive genes for resistance to Pea seed-borne mosaic virus. *Theoretical and Applied Genetics* 109:488–494.

Geffroy V, S´evignac M, De Oliveira J, Fouilloux G, Skroch P, Thoquet P, Gepts P, Langin T and Dron M. 2000. In: heritance of partial resistance against *Colletotrichum lindemuthianum* in *Phaseolus vulgaris* and co-localization of QTL with genes involved in specific resistance. *Molecular Plant Microbe Interactions* 13:287–296.

Gepts P. 1999. Development of an integrated linkage map. In: Singh SP (ed.), Developments in plant breeding. Common bean improvement in the Twenty-First century. Kluwer Academic Publishers, Dordrecht, The Netherlands, pp. 53–91.

Gonzalez AM, Yuste-Lisbona FJ, Godoy L, Fernández-Lozano A, Rodiño AP, De Ron AM et al. 2016. Exploring the quantitative resistance to *pseudomonas syringae* pv. *phaseolicola* in common bean (*Phaseolus vulgaris* L.). *Molecular Breeding* 36:166 doi.

Graham PH and Vance CP. 2003. Legumes: importance and constraints to greater use. *Plant Physiology* 131(3):872–877.

Grajal-Martin MJ and Muehlbauer FJ. 2002. Genomic location of the Fw gene for resistance to fusarium wilt race 1 in peas. Journal of Heredity 93:291–293.

Gronlund M, Olsen A, Johansen IE and Jakobsen I. 2010. Protocol: using virus-induced gene silencing to study the arbuscular mycorrhizal symbiosis in *Pisum sativum*. *Plant Methods* 6:28.

Hammarlund C. 1925. Zur Genetik, biologic und physiologic einiger Erysiphaeceen. *Hereditas* 6:1–126.

Hamon C, Baranger A, Coyne CJ, McGee RJ, Le Goff I, L'Anthoene V, Esnault R, Rivière JP, Klein A and Mangin P et al. 2010. New consistent QTL in pea associated with partial resistance to *Aphanomyces euteiches* in multiple field and controlled environments from France and the United States. *Theoretical and Applied Genetics* 123:261–281.

Hamon C, Baranger A, Coyne CJ, McGee RJ, Lesne A, Esnault R, Riviere JP, Klein A, Mangin P et al. 2013. QTL meta-analysis provides a comprehensive view of loci controlling partial resistance to *Aphanomyces euteiches* in four sources of resistance in pea. *Céline BMC Plant Biology* 13(1):45.

Hampton RO. 1975. The nature of bean yield reduction by bean yellow and bean common mosaic viruses. *Phytopathology* 65:1342–1346.

Harland SC. 1948. Inheritance of immunity to mildew in Peruvian forms of *Pisum sativum*. *Heredity* 2:263–269.

Heringa RJ, Van Norel A and Tazelaar MF. 1969. Resistance to powdery mildew (*Erysiphe polygoni* D.C.) in peas (*Pisum sativum* L.). *Euphytica* 18:163–169.

Hunter PJ, Ellis N and Taylor JD. 2001. Association of dominant loci for resistance to *Pseudomonas syringae* pv. *pisi* with linkage groups II, VI and VII of *Pisum sativum*. *Theoretical and Applied Genetics* 103:129–135.

Janila P and Sharma B. 2004. RAPD and SCAR markers for powdery mildew resistance gene er in pea. *Plant Breeding* 123:271–274.

Javid M, Rosewarne GM, Sudheesh S, Kant P, Leonforte A, Lombardi M, Kennedy PR, Cogan OIN, Slater AT and Kaur S. 2015. Validation of molecular markers associated with boron tolerance, powdery mildew resistance and salinity tolerance in field peas. Frontiers in Plant Sciences 6:917.

Jiang GL. 2013. Molecular markers and marker-assisted breeding in plants. In: Anderson SB (ed.), Plant breeding from laboratories to fields, InTech, Croatia, pp. 45–83.

Jordan SA and Humphries P. 1994. Single nucleotide polymorphism in exon 2 of the BCP gene on 7q31-q35. *Human Molecular Genetics* 3(10):1915.

Jung G, Coyne DP, Bokosi JM, Steadman JR and Nienhuis J. 1998. Mapping genes for specific and adult plant resistance to rust and abaxial leaf pubescence and their genetic relationship using Random Amplified Polymorphic DNA (RAPD) markers in common bean. *Journal of American Society for Horticultural Sciences* 123:859–863.

Jung G, Coyne DP, Scroch PW, Nienhuis J, Arnaud-Santana E, Bokosi J, Ariyarathne HM, Steadman, Beaver JS and Kaeppler SM. 1996. Molecular markers associated with plant architecture and resistance to common blight, web blight, and rust in common beans. *Journal of the American Society for Horticultural Science* 121:794–803.

Jung G, Skroch PW, Coyne DP, Nienhuis J, Arnaud- Santana E, Ariyarathne HM, Kaeppler SM and Bassett MJ. 1997. Molecular-marker-based genetic analysis of tepary bean derived common bacterial blight resistance in different developmental stage of common bean. *Journal of the American Society for Horticultural Sciences* 122:329–337.

Jung G, Skroch PW, Nienhuis J, Coyne DP, Arnaud-Santana E, Ariyarathne HM and Marita JM. 1999. Confirmation of QTL associated with common bacterial blight resistance in four different genetic backgrounds in common bean. *Crop Sciences* 39:1448–1455.

Kaila T, Chaduvla PK, Rawal HC, Saxena S, Tyagi A, Mithra SVA et al. 2017. Chloroplast Genome Sequence of Clusterbean (*Cyamopsis tetragonoloba* L.): Genome Structure and Comparative Analysis. *Genes* 8(9):212.

Kaiser WJ. 1972. Diseases of food legumes caused by pea leaf roll virus in Iran. FAO, Plant Protection Bulletin 20:127–133.

Katoch V, Sharma S, Pathania S, Banayal K, Sharma SK and Rathour R. 2010. Molecular mapping of pea powdery mildew resistance gene er2 to pea linkage group III. *Molecular Breeding* 25:229–237.

Kelly JD. 1997. A review of varietal response to bean common mosaic potyvirus in *Phaseolus vulgaris*. *Plant Varieties and Seeds* 10:1–6.

Kelly JD, Gepts P, Miklas PN and Coyne DP. 2003. Tagging and mapping of genes and QTL and molecular-marker assisted selection for traits of economic importance in bean and cowpea. *Field Crops Research* 82:135–154.

Kelly JD and Vallejo VA. 2004. A comprehensive review of the major genes conditioning resistance to anthracnose in common bean. *Hortscience* 39:1196–1207.

Kishun R. 1989. Appraisal of loss in yield of cowpea due to *Xanthomonas campestris* pv. *vignicola*. *Indian Phytopathology* 42:241–246.

Kraft JM and Pfleger FL. 2001. Compendium of Pea Diseases and Pests; APS Press: St. Paul, MN, USA pp. 14–17.

Kulaeva OA, Zhernakov AI, Afonin AM, Boikov SS, Sulima AS, Tikhonovich IA et al. 2017. Pea Marker Database (PMD) – A new online database combining known pea (*Pisum sativum* L.) gene-based markers. *PLoS One* 12(10): e0186713.

Kwon SJ, Smykal P, Hu J, Wang M, Kim SJ, McGee RJ, McPhee K and Coyne CJ. 2013. User-friendly markers linked to Fusarium wilt race 1 resistance Fwgene formarker-assisted selection in pea. *Plant Breeding* 132:642–648.

Lema M. 2018. Marker Assisted Selection in Comparison to Conventional Plant Breeding: Review Article. *Agricultural Research and Technology* 14(2).

Li G and Quiros C. 2001. Sequence-related amplified polymorphism (SRAP), a new marker system based on a simple PCR reaction: its application to mapping and gene tagging in Brassica. *Theoretical and Applied Genetics* 103:455.

Li GJ, Liu YH, Ehlers JD, Zhu ZJ, Wu XH and Wang BG. 2007. Identification of an AFLP fragment linked to rust resistance in asparagus bean and its conversion to a SCAR marker, *HortScience* 42 (5):1153–1156.

Liu N, Xu S, Yao X, Zhang G, Mao W, Hu Q, Feng Z and Gong Y. 2016. Studies on the Control of Ascochyta Blight in Field Peas (*Pisum sativum* L.) caused by *Ascochyta pinodes* in Zhejiang Province, China. *Frontiers in Microbiology* 7:481 doi.

Liu S, Yu K and Park SJ. 2007. Development of STS markers and QTL validation for common bacterial blight resistance in common bean. *Plant Breeding* 127(1):62–68.

Loridon K, McPhee KE, Morin J, Dubreuil P, Pilet-Nayel ML, Aubert G, Rameau C, Baranger A, Coyne CJ and Lejeune-Hénault I et al. 2005. Microsatellite marker polymorphism and mapping in pea (*Pisum sativum* L.). *Theoretical and Applied Genetics* 111:1022–1031.

Macas J, Neumann P and Navrátilová A. 2007. Repetitive DNA in the pea (*Pisum sativum* L.) genome: comprehensive characterization using 454 sequencing and comparison to soybean and *Medicago truncatula*. *BMC Genomics* 8:427.

Mahuku GS, Henriquez MA, Munoz J and Buruchara RA. 2002. Molecular markers dispute the existence of the Afro-Andean Group of the bean angular leaf spot pathogen, *Phaeoisariopsis griseola*. *Phytopathology* 92:580–589.

Mahuku GS, Jara C, Cajiao C and Beebe S. 2003. Sources of resistance to angular leaf spot (*Phaeoisariopsis griseola*) in common bean core collection, wild *Phaseolus vulgaris* and secondary gene pool. *Euphytica* 130:303–313.

Maina PK, Wachira PM, Okoth SA and Kimenju JW. 2017. Cultural, Morphological and Pathogenic Variability among *Fusarium oxysporum* f. sp. *phaseoli* Causing Wilt in French Bean (*Phaseolus vulgaris* L.). *Journal of Advances in Microbiology* 2(4):1–9.

Masangwa JIG, Aveling TAS and Kritziger Q. 2013. Screening of plant extracts for antifungal activities against Colletotrichum species of common bean (*Phaseolus vulgaris*) and cowpea (*Vigna unguiculata* L.). *Journal of Agricultural Science* 151:482–491.

Mayek-Perez N, Lopez-Castaneda C, Lopez-Salinas E and Acosta-Gallegos JA. 2001. Inheritance of genetic resistence to *Macrophomina phaseolina* (Tassi) Goid in common bean. *Agrociencia* 35:637–648.

McClendon MT, Inglis DA, McPhee KE and Coyne CJ. 2002. DNA markers for Fusarium wilt race 1 resistance gene in pea. *Journal of American Society for Horticultural Sciences* 127:602–607.

McPhee KE, Inglis DA, Gunderson B and Coyne CJ. 2012. Mapping a resistance gene for Fusarium wilt Race 2 on LG IV of pea (*Pisum sativum* L.). *Plant Breeding* pp. 1–7 doi.

Melotto M, Afanador L and Kell JD. 1996. Development of a SCAR marker linked to the I gene in common bean. *Genome* 39:1216–1219.

Melotto M, Balardin RS and Kelly JD. 2000. Host–pathogen interaction and variability of *Colletotrichum lindemuthianum*. In: Prusky D, Freeman S and Dickman MB, (eds.), *Colletotrichum host specificity, pathology, and host–pathogen interaction*, APS Press, St. Paul, MN, pp. 346–361.

Melotto M and Kelly JD. 2001. Fine mapping of the Co-4 locus of common bean reveal a resistance gene candidate, COK-4 that encodes for a protein kinase. *Theoretical and Applied Genetics* 103:503–517.

Mendez-Vigo B, Rodrıguez-Suarez C, Paneda A, Ferreira JJ and Giraldez R. 2005. Molecular markers and allelic relationships of anthracnose resistance gene cluster B4 in common bean. *Euphytica* 141:237–245.

Mienie CMS, Liebenberg MM, Pretorius ZA and Miklas PN. 2005. SCAR markers linked to the *Phaseolus vulgaris* rust resistance gene *Ur-13*. *Theoretical and Applied Genetics* 94:87–94.

Miklas PN, Hang AN, Kelly JD, Strausbaugh CA and Forster RL. 2002b. Registration of three kidney bean germplasm lines resistant to bean common mosaic and necrosis potyviruses: USLK-2 light red kidney, USDK-4 dark red kidney, and USWK-6 white kidney. *Crop Science* 42:674–675.

Miklas PN and Kelly JD. 2002. The use of MAS to develop pinto bean germplasm possessing Co-42 gene for anthracnose resistance. *Annual Report of Bean Improvement and Cooperative* 45:68–69.

Miklas PN, Kelly JD, Beebe SE and Blair MW. 2006a. Common bean breeding for resistance against biotic and abiotic stresses: from classical to MAS breeding. *Euphytica* 147:105–131

Miklas PN, Pastor-Corrales MA, Jung G, Coyne DP, Kelly JD, McClean PE and Gepts P. 2002a. Comprehensive linkage map of bean rust resistance genes. *Bean Improvement Cooperative* 45:125–129.

Miklas PN, Smith JR, Riley R, Grafton KF, Singh SP, Jung G and Coyne DP. 2000b. Marker-assisted breeding for pyramided resistance to common bacterial blight in common bean. *Annual Report of the Bean Improvement Cooperative* 43:39–40.

Miklas PN, Smith JR and Singh SP. 2006b. Registration of common bacterial blight resistant dark red kidney bean germplasms line USKD-CBB-15. *Crop Science* 46:1005–1006.

Miklas PN, Stavely JR and Kelly JD. 1993. Identification and potential use of a molecular marker for rust resistance in common bean. *Theoretical and Applied Genetics* 85:745–749.

Miklas PN, Stone V, Daly MJ, Stavely JR, Steadman JR, Bassett MJ, Delorme R and Beaver JS. 2000a. Bacterial, fungal, and viral disease resistance loci mapped in a recombinant inbred common bean population ('Dorado'/XAN 176). *Journal of American Society and Horticultural Sciences* 125:476–481.

Miklas PN, Stone V, Urrea CA, Johnson E and Beaver JS. 1998. Inheritance and QTL analysis of field resistance to ashy stem blight. *Crop Science* 38:916–921.

Mishra SP and Shukla P. 1984. Inheritance of powdery mildew resistance in pea. *Z Pflanzenzuchtg* 93:251–254.

Mmbaga M, Steadman JR and Stavely JR. 1996. The use of host resistance in disease management of rust in common bean. *Integrated Pest Management Review* 1:191–200.

Mohammed A. 2013. An Overview of Distribution, Biology and the Management of Common Bean Anthracnose. *Journal of Plant Pathology and Microbiology* 4:193 doi.

Molina CA and Beaver JS. 1998. Inheritance of normal pod development in bean golden mosaic resistant common beans. *Annual Report of Bean Improvement and Cooperative* 41:3–4.

Mooney and Daniel F. 2007. The economic impact of disease-resistant bean breeding research in Northern ecuador. M.S. thesis, Michigan State University, p. 153.

Moose SP and Mumm RH. 2008. Molecular plant breeding as the foundation for twenty-first century crop improvement. *Plant Physiology* 147:969–977.

Muchero W, Ehlers JD and Roberts PA. 2010. QTL analysis for resistance to foliar damage caused by *Thrips tabaci* and *Frankliniella schultzei* (Thysanoptera: Thripidae) feeding in cowpea [*Vigna unguiculata* (L.) Walp.]. *Molecular Breeding* 25:47–56.

Muchero WJD, Ehlers TJ and Close PA Roberts. 2009. Mapping QTL for drought stress-induced premature senescence and maturity in cowpea [*Vigna unguiculata* (L.) Walp.], *Theoretical and Applied Genetics* 118 (5):849–863.

Mutlu N, Coyne DP, Steadman JR, Reiser J and Sutton L. 2002. Progress in backcross breeding with RAPD (SCAR) molecular markers to pyramid QTLs for resistance to common bacterial blight in pinto and great northern beans. *Annual Report of Bean Improvement and Cooperative* 45:70–71.

Mwangombe AW, Wagara IN, Kimenju JW and Buruchara R. 2007. Occurance and severity of Angular leaf spot of common bean in Kenya as influenced by geographical location, altitude and agroecological zones. *Plant Pathology Journal* 6 (3):235–241.

Navarro F, Sass M and Nienhuis J. 2003. Identification and mapping bean root rot resistance in a population of Mesoamerican×Andean origin. *Annual Report of Bean Improvement and Cooperative* 46:213–214.

Ngulu FS. 1999. Final report on pathogenic variation in phaeoisariopsis griseola in Tanzania. Selian Agricultural Research Institute, Arusha, Tanzania, p.13.

Nodari RO, Tsai SM, Gilbertson RL and Gepts P. 1993. Towards an integrated linkage map of common bean. II. Development of an RFLP-based linkage map. *Theoretical and Applied Genetics* 85:513–520.

Okubara PA, Keller KE, McClendon MT, McPhee KE, Inglis DA and Coyne CJ. 2005. Y15_999Fw, a dominant SCAR marker linked to the Fusarium wilt race 1 (Fw) resistance gene in pea. *Pisum Genetics* 37:32–35.

Olson M, Hood L, Cantor C and Botstein D. 1989. A common language for physical mapping of the human genome. *Science* 245(4925):1434–1435.

Otsyula RM, Buruchara RA, Mahuku G and Rubaihayo P. 2003. Inheritance and transfer of root rots (*Pythium*) resistance to bean genotypes. *African Crop Science Society* 6:295–298.

Paran I and Michelmore RW. 1993. Development of reliable PCR-based markers linked to downy mildew resistance genes in lettuce. *Theoretical and Applied Genetics* 85:985 https://doi.org/10.1007/BF00215038.

Park SO, Coyne DP and Steadman JR. 1999. Molecular markers linked to the Ur-7 gene conferring specific resistance to rust in common bean. *Annual Report of Bean Improvement and Cooperative* 42:31–32.

Park SO, Coyne DP, Steadman JR, Crosby KM, Brick MA. 2004. RAPD and SCAR markers linked to the Ur-6 Andean gene controlling specific rust resistance in common bean. *Crop Science* 44:1799–1807.

Park SO, Coyne DP, Steadman JR and Skroch PW. 2003. Mapping of the Ur-7 gene for specific resistance to rust in common bean. *Crop Science* 43:1470–1476.

Pastor-Corrales JR, Steadman and Kelly JD. 2003. Common bean gene pool information provides guidance for effective deployment of disease resistance genes. *Phytopathology* 93(suppl.):S70.

Pavan S, Schiavulli A, Appiano M, Miacola C, Visser RGF, Bai Y, Lotti C and Ricciardi L. 2013. Identification of a complete set of functional markers for the selection of er1 powdery mildew resistance in *Pisum sativum* L. *Molecular Breeding* 31:247–253.

Pedrosa A, Vallejos CE and Bachmair A. 2003. Integration of common bean (*Phageolus vulgaris* L.) linkage and chromosomal maps. *Theoretical and Applied Genetics* 106:205–212.

Pereira G, Marques C, Ribeiro R, Formiga S, Damaso M, Tavares-Sousa M, Farinho M and Leitao JM. 2010. Identification of DNA markers linked to an induced mutated gene conferring resistance to powdery mildew in pea (*Pisum sativum* L.). *Euphytica* 171:327–335.

Pfander WF and Hagedorn DJ. 1983. Disease progress and yield loss in Aphanomyces root rot of peas. *Phytopathology* 73:1109–1113.

Pilet-Nayel ML, Muehlbauer FJ, McGee RJ, Kraft JM, Baranger A and Coyne CJ. 2005. Consistent QTL in pea for partial resistance to Aphanomyces euteiches isolates from United States and France. *Phytopathology* 95:1287–1293.

Prasad HP, Uday A, Shankar C, Bhuvanendra H, Kumar SH and Prakash HS. 2007. Management of Bean common mosaic Virus strain blackeye cowpea mosaic (BCMV-BICM) in cowpea using plant extracts. *Achieves of Phytopathology and Plant Protection* 40(2):139–147.

Prioul S, Frankewitz A, Deniot G, Morin G and Baranger A. 2004. Mapping of quantitative trait loci for partial resistance to Mycosphaerella pinodes in pea (*Pisum sativum* L.), at the seedling and adult plant stages. *Theoretical and Applied Genetics* 108:1322–1334.

Provvidenti R and Hampton RO. 1991. Chromosomal distribution of genes for resistance to seven potyviruses in Pisum sativum. *Pisum Genetics* 23:26–28.

Queiroz VT, Sousa CS, Costa MR, Sanglad DA, Arruda KMA, Souza TLPO, Regagnin VA Barros EG and Moreira MA. 2004. Development of SCAR markers linked to common bean anthracnose resistance genes Co-4 and Co-6. *Annual Report of the Bean Improvement* Cooperative 47:249–250.

Ragagnin V, Sanglard D, de Souza TL, Costa M, Moreira M and Barros EA. 2005. New inoculation procedure to evaluate angular leaf spot disease in bean plants (*Phaseolus vulgaris* L.) for breeding purposes. *Annual Report of the Bean Improvement* Cooperative 48:90–91.

Ragagnin VA, Sanglard DA, de Souza TLPO, Moreira MA and de Barros EG. 2003. Simultaneous transfer of resistance genes for rust, anthracnose, and angular leaf spot to cultivar Perola assisted by molecular markers. *Annual Report of the Bean Improvement Cooperative* 46:159–160.

Rai R, Singh AK, Singh BD, Joshi AK, Chand R and Srivastava CP. 2011. Molecular mapping for resistance to pea rust caused by *Uromyces fabae* (Pers.) de-Bary. *Theoretical and Applied Genetics* 123:803–813.

Randhawa H and Weeden NF. 2011. Refinement of the position of En on LG III and identification of closely linked DNA markers. *Pisum Genetics*.

Rawal HC, Kumar S, Mithra SVA, Solanke AU, Nigam D, Saxena S, Tyagi A et al. 2017. High quality unigenes and microsatellite markers from tissue specific transcriptome and development of a database in clusterbean (*Cyamopsis tetragonoloba*, L. Taub). *Genes* 8:313.

Reid JB and Ross JJ. 2011. Mendel's genes: Toward a full molecular characterization. *Genetics* 189:3–10.

Resende RT, Resende MDV, Azevedo CF, Silva FF, Melo LC, Pereira HS, Souza TLPO, Valdisser PAMR, Brondani C and Vianello RP. 2018. Genome-Wide Association and Regional Heritability Mapping of Plant Architecture, Lodging and Productivity in *Phaseolus vulgaris*. *G3: Genes, Genomes, Genetics* 8(8):2841–2854.

Rodas RSC. 1975. Evaluation of systematic insecticides to control white fly vector of yellow mosaic virus. Annual Meeting of the Central American Cooperative Programme for Improving food Crops, Santa Tecla, Elsalvador, 7th April 1975, pp. 327–335.

Roman-Aviles B and Kelly JD. 2005. Identification of quantitative trait loci conditioning resistance to *Fusarium* root rot in common beans. *Crop Science* 45:92–95.

Sanchez BN and Bencomo P. 1979. Damage due to bean yellow stipple virus in different bean cultivars. *Ciencias de la Agricultura* 4:135–143.

Santo T, Rashkova M, Alabaca C and Leitao J. 2013. The ENU-induced powdery mildew resistant mutant pea (*Pisum sativum* L.) lines S (erl1mut1) and F (erl1mut2) harbour early stop codons in the PsMLO1 gene. *Molecular Breeding* 32:723–727.

Schmutz J, McClean PE, Mamidi S, Wu GA, Cannon SB, Grimwood J et al. 2014. A reference genome for common bean and genome-wide analysis of dual domestications. *Nature Genetics* 46(7):707–713.

Schneider KA, Grafton KF and Kelly JD. 2001. QTL analysis of resistance to *Fusarium* root rot in bean. *Crop Science* 41:535–542.

Sharma SR and Sohi HS. 1980. Assessment of losses in French bean caused by *Rhizoctonia solani* Kuhn. *Indian Phytopathology* 33:366–369.

Sharma SR and Sohi HS. 1981. Effect of different fungicides against Rhizoctonia root rot of French beans (*Phaseolus vulgaris* L.). *Indian Journal of Mycology and Plant Pathology* 11:216–220.

Shoyinka SA. 1974. Status of virus diseases of cowpea in Nigeria. 1974. In Proceedings of the First IITA grain Legumes Improvement Workshop IITA, Ibadan, Nigeria,pp. 270–273.

Singh SP. 2000. Broadening the genetic base of common bean cultivars: A Review. *Crop Science* 41:1659–1675.

Singh SP and Schwartz HF. 2010. Breeding common bean for resistance to diseases: A review. *Crop Sciences* 50:2199–2223.

Smykal P, Aubert G, Burstin J, Coyne CJ, Ellis NTH et al. 2012. Pea (*Pisum sativum* L.) in the Genomic Era. *Agronomy* 2(2):74–115.

Smykal P, Kenicer G, Flavell AJ, Kosterin O, Redden RJ, Ford R, Zong X, Coyne CJ, Maxted N, Ambrose MJ and Ellis THN. 2011. Phylogeny, phylogeography and genetic diversity of the Pisum genus. *Plant Genet Resources* 9:4–18.

Smykal P, Safarova D, Navratil M and Dostalova R. 2010. Marker assisted pea breeding: eIF4E allele specific markers to pea seed-borne mosaic virus (PSbMV) resistance. *Molecular Breeding* 26:425–438.

Snap S, Kirk W, Roman-Aviles B and Kelly JD. 2003. Root traits play a role in integrated management of Fusarium root rot in snap beans. *Hortscience* 38:187–191.

Sokhi SS, Jhooty JS and Bains SS. 1979. Resistance in pea against powdery mildew. *Indian Phytopathology* 32:571–574.

Sousa LL, Cruz AS, Filho PSV, Vallejo VA, Kelly JD and Gonçalves-Vidiga MC. 2014. Genetic mapping of the resistance allele Co-52 to Colletotrichum lindemuthianum in the common bean MSU 7-1 line. *Australian Journal of Crop Science* 8:317–323.

Srivastava RK and Mishra SK. 2004. Inheritance of powdery mildew resistance using near-isogenic lines (NILs) in pea (*Pisum sativum* L.). *Indian Journal of Genetics* 64(4):303–305.

Srivastava RK, Mishra SK, Singh AK and Mohapatra T. 2012. Development of a coupling-phase SCAR marker linked to the powdery mildew resistance gene 'erl' in pea (*Pisum sativum* L.). *Euphytica* 186:855–866.

Stagnari F, Maggio A, Galieni A and Michele Pisante. 2017. Multiple benefits of legumes for agriculture sustainability: An overview. *Chemical and Biological Technologies in Agriculture* 4:2.

Stavely JR. 1984. Genetics of resistance to *Uromyces phaseoli* in a *Phaseolus vulgaris* line resistant to most races of the pathogen. *Phytopathology* 74:339–344.

Stavely JR. 1998. Recombination of two major dominant rust resistance genes that are tightly linked in repulsion. *Annual Report Bean Improvement Cooperative* 41:17–18.

Stavely JR, Kelly JD, Grafton KF, Mullins CA, Straw A, McMillan RT, Beaver JS, Miklas PN, Steinke J, Steadman JR, Coyne DP, Lindgren DT and Silbernagel MJ. 1997. Rust resistant bean germplasm releases, 1994–1996. *Annual Report Bean Improvement Cooperative* 40:120–121.

Steadman JR, Kerr ED and Mumm RF. 1975. Root rot of bean in Nebraska: Primary patho-gen and yield loss appraisal. *Plant Disease Reporter* 59:305–308.

Sudheesh S, Lombardi M, Leonforte A, Cogan N, Materne M, Forster JW and Kaur S. 2014. Consensus Genetic Map Construction for Field Pea (*Pisum sativum* L.), Trait Dissection of Biotic and Abiotic Stress Tolerance and Development of a Diagnostic Marker for the erl Powdery Mildew Resistance Gene. *Plant Molecular Biology Reporter* 33:1391–1403.

Sun S, Deng D, Wang Z, Duan C, Wu X, Wang X, Zong X and Zhu Z. 2016. A novel erl allele and the development and validation of its functional marker for breeding pea (*Pisum sativum* L.) resistance to powdery mildew. *Theoretical and Applied Genetics* doi 10.1007/s00122-016-2671-9.

Sun S, Fu H, Wang Z, Duan C, Zong X and Zhu Z. 2015a. Discovery of a novel erl allele conferring powdery mildew resistance in Chinese pea (*Pisum sativum* L.) landraces. PLos ONE.

Sun S, Wang Z, Fu H, Duan C, Wang X and Zhu Z. 2015b. Resistance to powdery mildew in the pea cultivar Xucai 1 is conferred by the gene erl. *The Crop Journal* 3:489–499.

Tanwar UK, Pruthi V and Randhawa GS. 2017. RNA-Seq of Guar (*Cyamopsis tetragonoloba*, L. Taub.) Leaves: De novo Transcriptome Assembly, Functional Annotation and Development of Genomic Resources. *Frontiers in Plant Science* 8:91 http://doi.org/10.3389/fpls.2017.00091.

Taran B, Michaels TE and Pauls KP. 2002. Genetic mapping of agronomic traits in common bean. *Crop Science* 42:544–556.

Taran, B, Warkentin, T, Somers DJ, Miranda D, Vandenberg A, Blade S, Woods S, Bing D, Xue A and DeKoeyer D et al. 2003. Quantitative trait loci for lodging resistance, plant height and partial resistance to mycosphaerella blight in field pea (*Pisum sativum* L.). *Theoretical and Applied Genetics* 107:1482–1491.

Thind TS and Jhooty JS. 1985. Relative prevalence of fungal diseases of chilli fruits in Punjab. *Indian Journal of Mycology and Plant Pathology* 15:305–307.

Timko MP, Rushton PJ, Laudeman TW, Bokowiec MT, Chipumuro E, Cheung F, Town CD and Chen X. 2008. Sequencing and analysis of the gene-rich space of cowpea. *BMC Genomics* 27(9):103.

Timmerman-Vaughan GM, Frew TJ, Butler R, Murray S, Gilpin M, Falloon K, Johnston P, Lakeman MB, Russell A and Khan T. 2004. Validation of quantitative trait loci for Ascochyta blight resistance in pea (*Pisum sativum* L.), using populations from two crosses. *Theoretical and Applied Genetics* 109:1620–1631.

Timmerman-Vaughan GM, Frew TJ and Weeden NF. 1994. Linkage analysis of er1, a recessive Pisum sativum gene for resistance to powdery mildew fungus (Erysiphe pisi D.C). *Theoretical and Applied Genetics* 88:1050–1055.

Tiwari KR, Penner GA and Warkentin TD. 1997. Inheritance of powdery mildew resistance in pea. Canadian Journal of Plant Science 77:307–310.

Tiwari KR, Penner GA and Warkentin TD. 1998. Identification of coupling and repulsion phase RAPD markers for powdery mildew resistance gene er-1 in pea. *Genome* 41:440–444.

Tiwari KR, Penner GA and Warkentin TD. 1999. Identification of AFLP markers for powdery mildew resistance gene er2 in pea. *Pisum Genetics* 31:27–29.

Tyagi A, Nigam D, Mithra SVA, Solanke AU, Singh NK, Sharma TR and Gaikwad K. 2018. Genome-wide discovery of tissue specific miRNAs in clusterbean (*Cyamopsis tetragonoloba*) indicates their association with galactomannan biosynthesis. *Plant Biotechnology Journal* 16(6):1241–1257.

Urrea CA, Miklas PN, Beaver JS and Riley RH. 1996. A codominant RAPD marker useful for indirect selection of BGMV resistance in common bean. *Journal of American Society for Horticultural Sciences* 121:1035–1039.

Velez JJ, Bassett MJ, Beaver JS and Molina A. 1998. Inheritance of resistance to bean golden mosaic virus in common bean. *Journal of American Society for Horticultural Sciences* 123:628–631.

Vos P, Hogers R, Bleeker M, Reijans M, van de Lee T, Hornes M et al. 1995. AFLP: A new technique for DNA fingerprinting. *Nucleic Acids Research* 23(21):4407–4414.

Wang Z, Fu H, Sun S, Duan C, Wu X, Yang X and Zhu Z. 2015. Identification of powdery mildew resistance gene in pea line X9002. *ACTA Agronomica Sinica* 41:515–523.

Williams JGK, Kubelik AR, Livak KJ, Rafalski JA and Tingey SV. 1990. DNA polymorphisms amplified by arbitrary primers are useful asgenetic markers. *Nucleic Acids Research* 18:6531–6535.

Xu Y. 2010. Molecular plant breeding. *AGRIS*. CAB International. Wallingford, UK.

Yau SK and Ryan J. 2008. Boron toxicity tolerance in crops: a viable alternative to soil amelioration. *Crop Sciences* 48:854–865.

Young R and Kelly JD. 1996. RAPD Markers flanking the Are gene for anthracnose resistance in common bean. *Journal of American Society of Horticultural Science* 121:37–41.

Young RA, Melotto M, Nodari RO and Kelly JD. 1998. Marker assisted dissection of the oligogenic anthracnose resistance in common bean cultivar, G 2333. *Theoretical and Applied Genetics* 96:87–94.

Yu J, Gu WK, Provvidenti Rand Weeden N. 1995. Identifying and mapping two DNA markers linked to the gene conferring resistance to Pea Enation Mosaic Virus. *Journal of American Society of Horticultural Science* 120:730–733.

Yu K, Park SJ and Poysa V. 2000. Marker-assisted selection of common beans for resistance to common bacterial blight: Efficacy and economics. *Plant Breeding* 119:411–415.

Zaumeyer WJ and Thomas C. 1957. A monographic study of bean diseases and methods for their control, Technical Bulletin No. 868, United States Department of Agriculture.

Zhukov VA, Zhernakov AI, Kulaeva OA, Ershov NI, Borisov AY and Tikhonovich IA. 2015. "De Novo Assembly of the Pea (*Pisum sativum* L.) Nodule Transcriptome. *International Journal of Genomics* Article ID 695947, p. 11.

Zietkiewicz E, Rafalski A and Labuda D. 1994. Genome fingerprinting by simple sequence repeat (SSR)-anchored polymerase chain reaction amplification. *Genomics* 20:176–183.

Zohary D and Hopf M. 2000. Domestication of Plants in the Old World. Oxford University Press, New York, pp. 151–159.

31 Breeding for Improved Crop Resistance to Osmotic Stress

Mojtaba Kordrostami and Babak Rabiei

CONTENTS

31.1 INTRODUCTION

Arid and semi-arid climatic conditions have limited water resources and thereby crop production is also limited (Prinz and Singh 2003). The rapid growth of the population and the reduction of resources makes it necessary to study methods for the optimal use of food production potential for greater productivity (Godfray et al. 2010). Because in most parts of the world water is the main limiting factor for agricultural production, its optimal use is very important (Mancosu et al. 2015). This is more serious for countries with a dry climate, most of which are classed as arid and semi-arid. Paying attention to the economy and management of water resources is important not only in arid and semi-arid regions but also in areas with good rainfall as it influences production costs (Heyns 2009). In this regard, knowing the reaction of plants and determining the sensitivity of different stages of their growth to osmotic stress is of great importance (Osakabe et al. 2014).

Unfavorable environmental factors, such as drought stress, which is one of the most common environmental stresses, limits the production of crop plants. Drought has destructive and harmful effects on most plant growth stages, such as the germination stage and seedling establishment as well as organ structure and activity. Since it is almost impossible to create and maintain a pure water potential in the soil, establishment of drought stress conditions using various osmotic substances to create osmotic potentials is considered as one of the most important methods of studying the effects of drought stress on plants (Fahad et al. 2017). Any attempt at the genetic improvement of drought tolerance in plants using existing genetic diversity requires an efficient evaluation method that should be fast and able to select a large population (Chakhchar et al. 2017). Drought tolerance is the result of various morphological, physiological, and biochemical traits. Therefore, we can use these different components as selection indices for selecting the ideal plant type (Lamaoui et al. 2018).

31.2 THE ROLE AND IMPORTANCE OF WATER IN THE PLANT

The most important water–plant relationship is the water balance during dehydration because this balance controls the physiological processes and the conditions in which the quality and quantity of plant growth depend (Ings et al. 2013). In addition to the role of water in the growth of the whole plant, it is clear that water shortages in various ways affect the growth of the various organs of the plant (Chaves et al. 2002). These effects include:

A. Reduction of the shoot:root ratio
B. Reduction of the leaf:stem ratio
C. Decrease in the growth ratio of lateral roots to the total length of the root

It should be noted that sometimes due to the low relative humidity of the air, the high heat and wind speed, although there is a lot of available moisture in the root environment, the transpiration rate is more than the rate of water absorption (Seyednasrollah 2017). Under such conditions, the effect of drought stress is severe. It is worth mentioning that metabolism will only be disturbed during the daily periods with a maximum water deficit, so if these periods gradually become longer, the stomata closure during these periods will result in reduced transpiration, photosynthesis, and CO_2 absorption, and leaf temperature rises (Amedie 2013).

Increasing stress reduces photosynthesis. By reducing cellular turgor pressure, the rate of cell elongation decreases and the leaf area extends slowly, resulting in a decreased growth rate (Baird et al. 2017). When the moisture stress in the soil approaches the permanent wilting point, the turgor pressure of the plant also drops to zero, cell elongation stops, and the cell division rate drops dramatically, the stomata remain closed for most of the day and transpiration continues only through the cuticle, the leaf temperature rises, and most metabolic processes, such as transpiration increases and apparent photosynthesis and dry matter production reach zero. Also, the disruption of the normal metabolism of the cell is accompanied by the denaturation of proteins and carbohydrates, P and N in the leaves are transmitted from the older leaves to the stems, each cell and tissue will then be destroyed, the roots and root hairs disappear, and finally the older roots become corky (Chaves et al. 2002; Osakabe et al. 2014). It should be noted that the harmonization of plant phenology by supplying moisture is always an important breeding goal and phenological traits have a direct or indirect effect on grain yield. Therefore, it is important that the critical stage of plant growth, which is most susceptible to drought, coincides with the usual dry period in the region (Wilhite 2000).

31.3 THE CONCEPTS OF STRESS

From a biological point of view, any kind of change in environmental conditions that causes a significant drop in plant growth and development is called stress (Cramer et al. 2011). Physiologically, stress is the result of abnormal physiological processes that result from the effects of one or a combination of biological and environmental factors (Fathi and Tari 2016). Overall, stress refers to any condition that reduces the dry matter production in the plant, under its genetic potential (Negrão, Schmöckel, and Tester 2017).

31.3.1 Types of Environmental Stresses

In general, stresses are divided into two categories: biotic and abiotic stresses. Biotic stresses include pathogens such as fungi and bacteria, and abiotic stresses include heat stress, cold, salinity, drought, and heavy metals. Stresses are limiting factors in crop production which reduce growth and ultimately yield by disrupting metabolism (Jaleel et al. 2009).

31.4 DROUGHT STRESS

Drought stress is one of the most important global threats to food production. In addition, climate change and population growth in the world are broadening the scope of this problem. Generally, drought is the absence of rainfall for a given period, so that this period is sufficient to cause soil moisture to evaporate and damage the plant (Anderegg, Anderegg, and Berry 2013). The term "drought stress" is used to reduce rainfall shortages. If the plant is artificially exposed to water scarcity, the term water deficit stress will be used (Obidiegwu et al. 2015). One of the solutions to this problem is to create new cultivars with more tolerance to drought stress (Fahad et al. 2017). Environmental stresses are very important factors in the distribution of plant species on the surface of the planet and reduce the potential of plant production. As a result of stresses, the average yield of plants is less than 10–20% of their actual potential (Shrivastava and Kumar 2015). Decreasing growth due to drought stress is far more than other environmental stresses (Farooq et al. 2010). Drought stress refers to conditions where cells and tissues are placed in a position where their inflammation is not complete. This can range from a slight reduction in water potential to a permanent wilting point of the plant. Simply, water scarcity occurs when transpiration is more than water absorption (Tardieu 2013). The reaction of plants to drought stress is associated with the metabolic activity, morphological, growth, and potential yield of the crop, and the yield of plants under water deficit conditions depends on the total available water and water use efficiency (WUE) of the plant (Jaleel et al. 2009).

31.4.1 Drought Classification Based on the Time of Occurrence

Drought stress is divided into three categories according to the time of occurrence:

A. Primary season drought stress
B. Periodic drought stress
C. Terminal season drought stress

Primary season drought stress occurs at the beginning of the growing cycle, while there is enough moisture at the end of the growing season, and this is seen in the southern part of Latin America. In the second type, periodic drought stress, cultivating is carried out using the remaining moisture stored in the soil. Terminal season drought stress is also dominant in the Mediterranean regions, in areas where there is usually enough moisture in the early stages of the growth cycle, but the plants at the end of their growth stage are facing water shortages (Fahad et al. 2017).

31.5 PHYSIO-BIOCHEMICAL TRAITS ASSOCIATED WITH DROUGHT/ OSMOTIC STRESS

31.5.1 Effects of Osmotic Stress on Proline Content

Plants tolerate drought, salinity, and heat stresses with the storage of osmotic regulators. Osmotic regulators contain amino acids, sugars, and some mineral ions, hormones, and proteins. Proline is one of the active amino acids in the osmotic regulation phenomenon that plays a significant role in the formation and maintenance of osmotic pressure inside the plant (Hayat et al. 2012). Usually, the amount of free proline in irrigated plants is very low and about 0.2–0.6 mg/g dry matter. The amount of this substance increases by up to 40–50 mg/g dry matter after the reduction of the tissue water content (Lalelou et al. 2010). In some plants, in the early stages of drought stress, several amino acids increase, which, with prolonged dehydration, only proline amino acids accumulate and store (Kovács et al. 2012). Proline is accumulated in all parts of the whole plant during osmotic stress, but its fastest accumulation occurs in the leaves (Mattioli, Costantino, and Trovato 2009). To escape from stress, proline rapidly decomposes and produces potent reducing agents to produce phosphorylated oxidative mitochondria and ATP to recover and repair stress-related injuries (Mattioli, Costantino, and Trovato 2009). In relation to proline, different views have been suggested as a cause for drought tolerance. Some researches have reported that there is a positive correlation between proline accumulation and drought tolerance (Nayyar and Walia 2003; Forlani, Bertazzini, and Cagnano 2018). However, the results of others indicate that proline is a positive factor in adaptation under stress conditions in plants (Nayyar and Walia 2003; Pakniyat and Armion 2007; Khan et al. 2009; Kordrostami, Rabiei, and Kumleh 2017).

31.5.2 Effects of Osmotic Stress on Chlorophyll Content and Chlorophyll Fluorescence

Chlorophyll is the main pigment of the chloroplasts of the leaf cells, which by absorbing the energy of photons induces the light reactions of photosynthesis (Berg, Tymoczko, and Stryer 2002). Due to the particular importance of these pigments, the cells try to apply certain mechanisms to protect them (Yang et al. 2006). Osmotic stress is one of the environmental factors limiting plant photosynthesis (Shrivastava and Kumar 2015; Fahad et al. 2017). Increasing the efficiency of mechanisms that scavenge the reactive oxygen species (ROS), increasing carotenoids as mechanisms for preventing them from occurring, and increasing the amount of alternative oxidase (AOX) in chloroplasts of plant cells are among the methods for preventing damage to chlorophylls (Tripathy and Oelmüller 2012; Hossain and Dietz 2016).

Most studies show that under osmotic stress conditions, photosynthesis is reduced by interruptions in biochemical reactions (Lamaoui et al. 2018). The PSII system is very sensitive to environmental factors, and osmotic stress causes damage to the PSII reaction centers (Jajoo 2013). PSII is heavily influenced by water. Drought and high light stress effects on PSII can damage its photochemical reactions and increase the amount of electron transfer inhibitory activity (Lu et al. 2002). When the stomata are closed due to osmotic stress or high temperature, carbon dioxide is reduced and, as a result, building power is limited (Chaves, Flexas, and Pinheiro 2009). The study of photosynthesis status is a reliable criterion for evaluating the adaptability of plants to their environment. This criterion is non-destructive in both laboratory and field studies with the use of a fluorometer. In fact, what the fluorometer device represents is the ratio of Fv:Fm and its curve (Van Kooten and Snel 1990). The Fv:Fm value represents the maximum quantum efficiency of photosystem II and a measure of the mode of action of plant photosynthesis. So, this parameter is about 0.83 for most plant species under normal environmental conditions. Values less than this amount are observed when the plant is exposed to stress, which indicates the photo-inhibition phenomenon (Fracheboud 2004). Therefore, in different genotypes, quantum yield reduction or fluorescence variation (Fv = Fm-Fo) has been used as a measure of stress tolerance or resistance in a time interval (Bibi, Oosterhuis, and Gonias 2008; Eshghizade and Ehsanzadeh 2009).

31.5.3 Effects of Osmotic Stress on Soluble Sugars

Soluble sugars are also compatible osmolytes that accumulate under drought stress and may act as an osmotic agent or Osmo-protector (Gangola and Ramadoss 2018). In the first case, the increase in sugar content due to stress is related to osmotic regulation and maintenance of turgor pressure. Increasing soluble sugars in response to drought stress are attributed to their slower transfer from leaf to stem and slower consumption due to reduced growth and other changes such as starch hydrolysis (Wang et al. 1999). Low levels of soluble sugars under non-stress conditions are largely related to the continued use of produced sugars at the growing points of the plant (Sharma and Kuhad 2006). The increase in soluble sugars due to osmotic stress is due to starch hydrolysis. Increasing soluble sugars is usually accompanied by a reduction in starch, and in some cases reduced sucrose. At the same time, increasing soluble sugars also increase the activity of hydrolyzing enzymes such as alpha-amylase and invertase, which degrade starch and sucrose, respectively (Devi et al. 2013). Increasing the soluble sugar content is one of the reasons for increasing the internal osmotic pressure of the plant, which attempts to neutralize the osmotic pressure of the environment and even absorb the water in the soil as much as possible (Williams 1963).

31.6 MORPHOLOGICAL TRAITS ASSOCIATED WITH DROUGHT/OSMOTIC STRESS

Root characteristics are one of the most important traits in determining the reaction of plants to drought stress (Farooq et al. 2010; Comas et al. 2013). Some experiments show that drought stress reduces root weight as well as decreasing the shoot weight (Ferreira et al. 2013). Roots are important for keeping the plant in the soil, absorbing water and nutrients. Root control is the underlying part of the development and evolution of the plant, enabling the plant to respond to changes in the environmental conditions and preserve its life (Osakabe et al. 2014). Increasing root growth improves the ability to extract soil moisture and acts as a basic mechanism of drought tolerance (Lopes et al. 2011). When the plant is under drought stress conditions, its transpiration rate increases. In this case, an efficient rooting system has the ability to absorb water to compensate for transpiration and also stored water in the leaves (Xu, Zhou, and Shimizu 2010). Some researchers believe that the existence of surface roots and shallow roots is one of the ways to improve plant performance (Haling et al. 2013). Because, under rainy conditions after temporary drought stress, these roots quickly absorb water and help to improve the plant conditions (Xu, Zhou, and Shimizu 2010). In addition, the ultrasensitivity of some plant tissues causes them to react as soon as water deficiency occurs. Roots have different levels of susceptibility to drought stress. As soil moisture decreases, the death of hairy roots occurs much earlier than the thick roots. Under these conditions, the root and shoot reactions are different in response to drought stress. With a decrease in soil moisture, the roots send signals to the organs to increase the drought resistance. In these conditions, the aerial parts of the plants first deal with the drought by decreasing transpiration and maintaining the moisture content of the leaf, and subsequently reduce or stop their growth. This is while the roots continue to grow. The result will be a root development to provide more water and improve the root:shoot ratio (Chiatante et al. 2006). Increasing root growth increases the ability to extract soil moisture and acts as a basic mechanism of drought resistance (Farooq et al. 2012).

Investigating traits such as root area and volume and their relationships is necessary to determine the water absorption capacity through the root (Judd, Jackson, and Fonteno 2015). Kordrostami, Rabiei, and Kumleh (2016) stated that rice plants with higher root length, more root length density, and root:shoot ratio showed more tolerance to osmotic stress than plants that do not have this feature. The results of Kao (1981) showed that water absorption by root is an important characteristic of drought resistance. Therefore, plants that have a higher root:shoot ratio at the beginning of the growing season have a greater ability to maintain turgor pressure, and, consequently to improving their photosynthesis rate in the later periods of stress. In addition, the yield stability depends on the ability of the roots to absorb the water and nutrients in the soil, and this property will only be achieved through the mechanisms of adaptation associated with the roots and the shoots of the plants (Basu et al. 2016).

Root length is an index of the plant's ability to absorb water from deeper layers of the soil and better root penetration in the soil (Kulkarni and Phalke 2009). Root depth and ability to continuously absorb water are one of the most important factors in rainfed conditions (Pierret et al. 2016). Many researchers observed that osmotic stress decreased root length. Kordrostami, Rabiei, and Kumleh (2016) stated that osmotic stress reduces the root length in rice plants, but plants with higher root length showed more tolerance to osmotic stress than plants that do not have this feature. Increasing root dry weight in tolerant genotypes could be due to increased root growth and lateral root growth in drought conditions, which is considered as an acceptable trait for drought tolerance (Fang et al. 2017). Under stress conditions, the roots of the plants penetrate deeper into the soil where more water is accessible. This is one of the reasons for increasing the growth and dry weight of the roots and one of the mechanisms of drought tolerance (Liu, Li, and Xu 2004).

31.7 THE CONCEPTS OF DROUGHT RESISTANCE

Drought resistance is the ability of the plant to obtain and maintain water and continue the metabolic activities in tissues that are under a low water potential. The result is a large number of morphological and physiological characteristics but their interaction is not well defined. We can also express drought resistance as the survival capability of a species under water stress, from generation to generation (da Silva et al. 2013). Beyond that, drought resistance is the ability to survive under drought stress conditions without damage to the plant (Basu et al. 2016). Drought tolerance can also be determined by considering a set of plant capabilities such as survival under drought stress conditions, drought stress tolerance and water use efficiency (Rukundo et al. 2014). In fact, drought resistance is a generic term that includes a range of different mechanisms by which crop plants can tolerate drought conditions in arid areas (Ashraf 2010). Generally, stress resistance is caused by a gene family and not by a single gene. The production of transgenic plants with a desirable performance against stresses is more difficult due to the genetic complexity of the involved processes, and it is expected that only one gene would increase the growth rate of the plant in response to these stresses (Lawlor 2012). There are exceptional cases, such as genes responsible for osmotic regulation under water stress conditions, or the genes that form a complex with heavy metal contamination of the soil. Therefore, the production of abiotic stress resistant plants is one way to cope with each of the environmental stresses, and genetic engineering has effectively been instrumental in this field (Sreenivasulu, Sopory, and Kishor 2007). The stress-induced genes are classified into two main groups. The first group is the genes that directly protect the plant and the second group is the genes that regulate gene expression and signal transduction under stress conditions (Cattivelli et al. 2008).

31.8 DIFFERENT MECHANISMS OF DROUGHT RESISTANCE

Drought stress is one of the basic problems of agriculture in the world and is an important factor in reducing plant yields. According to FAO-Stat (2012), one-third of the world's arable land suffers from a lack of adequate water for irrigation. Application of agronomic methods and the use of drought-tolerant cultivars allows for optimal production in areas with water shortage. However, it can be said that in plants there are three general mechanisms for drought resistance (Basu et al. 2016) as follows:

A. Drought avoidance
B. Drought tolerance
C. Drought escape

31.8.1 DROUGHT AVOIDANCE

Drought avoidance is the ability of the plant to maintain water and turgor pressure; even under stress conditions; therefore, the plant avoids stress and its consequences. Drought avoidance is mainly due to the morphological and anatomical characteristics of the plant (Basu et al. 2016). These characteristics, in turn, are the result of physiological processes that have been caused by drought. It should be noted that the drought avoidance mechanism is a major mechanism of survival in herbaceous plants (De Micco and Aronne 2012). In order to maintain the function and metabolism of the crops, to produce satisfactory products, there should be enough water in the cells (Lawlor 2012). Whenever the root system is more active and developed, it will increase the amount of water available to the plant and its organs. Therefore, when the plants are exposed to drought stress, root:shoot dry weight ratio is increased, which is due to the higher allocation of dry matter to the root, in order to keep the plant away from drought (Xu et al. 2015). During research to find drought-resistant cultivars at the seedling stage, it was observed that resistant cultivars had more root length and root dry weight than the sensitive cultivars under stress conditions (Riaz et al. 2013). In fact, plants can resist drought by maintaining a small amount of water inside their cells (Xu, Zhou, and Shimizu 2010).

In dry regions, many plants have thick cuticles. A thick cuticle acts as insulation against incoming radiation, resulting in reduced transpiration and, consequently, reduced cuticle evaporation (Jones and Rotenberg 2001). The waxiness of the cuticle in wheat is controlled by a major effect gene, and some small effect genes modify its intensity (Yeats and Rose 2013). The heritability of waxiness should be high. Leaf wax is affected by external factors such as water stress, high temperature, and high radiation (Flowers 1989). Stomata are also one of the most important factors in water loss in the plant. So, when the plant is exposed to drought stress, the plant decreases the loss of water by closing its stomata (Pirasteh-Anosheh et al. 2016). Since the heritability of stomatal conductance is unknown, it is difficult to measure it, even its amount changes a lot during the day, and it is unlikely to be used in breeding

programs (Araus et al. 2008). Stomata closure is controlled by various factors in the plant; one of the most important of them is Abscisic acid (ABA) (Hunt et al. 2003). This plant growth regulator increases due to drought stress. Abscisic acid is produced through the Mevalonic acid pathway in the leaves (chloroplasts and other plastids), and this process is stimulated by environmental stresses, especially drought stress (Finkelstein 2013). Abscisic acid causes drought, salinity, and cold stress resistance. Increasing Abscisic acid is a process that is commonly observed after abiotic stresses such as drought, salinity, and cold (Sah, Reddy, and Li 2016). So, this increase probably has a relationship with drought tolerance and interferes with the prevention of cellular acquisition desiccation in plants. Many morphological and physiological adaptations to osmotic stress are controlled by the ABA hormone. It seems that reducing the cell's turgor pressure induces ABA production, and subsequently, ABA causes the expression of various genes (Vishwakarma et al. 2017).

31.8.2 DROUGHT TOLERANCE

In this case, the plant may survive despite the low internal water potential. This mechanism of drought resistance is called tolerance (hardness), and a plant that has this mechanism is able to supply moisture to regrow (Fang and Xiong 2015). The mechanism of drought tolerance is the basic method for the survival of herbaceous plants as well as pollen grains and other parts of higher plants which are dormant (Shah, Husain, and Raja 2017). The mechanism of drought tolerance is also more primitive than the drought avoidance mechanism because it is the only mechanism that exists in lower plants (Van Oosten et al. 2016). Plants with this type of resistance can tolerate drought by the following mechanisms (Jaleel et al. 2009):

A. Resist against water loss.
B. By using the dormancy mechanism, they can pass through the dry period and have a good ability to return from drought stress.

Osmotic pressure regulation or osmotic adjustment is one of the mechanisms of drought tolerance (Wang et al. 2016). Osmotic regulation means increasing the amount of intracellular soluble molecules in response to the reduction of external water potential. The effect of this action is to reduce the water flow from the cell and, thereby, reduce the loss of turgor pressure (Finan and Guilak 2010). The regulation of osmotic pressure is one of the important characteristics of plant drought tolerance, which over the last few years has received a lot of attention (Dolferus 2014). The concept of osmotic pressure regulation is the reduction of osmotic potential in response to water stress (Osakabe et al. 2014). Free proline accumulation also occurs in response to water stress in cells of many crops (Akıncı and Lösel 2012). Bajji, Lutts, and Kinet (2001) reported that under drought stress, wheat plants accumulated glycine, saccharose, betaine, valine, proline, asparagine, and glutamine in their tissues, which the proline accumulation was higher than the rest.

Drought tolerance depends on the cell's ability to maintain cell membranes under normal conditions and to prevent changes in the nature of the protein (Wang et al. 2016). The plant may survive despite low internal water potential (Basu et al. 2016). This mechanism is called tolerance, and a plant that has this mechanism is able to supply moisture to regrow (Fang and Xiong 2015).

31.8.3 Drought Escape

Drought escape is an important strategy for improving plant phenology in areas where the growing season is dominated by drought (Araus et al. 2002). The plant ends its life cycle before the start of stress-induced environmental conditions to avoid drought in its tissues (Basu et al. 2016). Probably, prematurity is the most common and simplest trait to breed drought resistance (Hossain et al. 2016). Prematurity, makes the plants provide their peak performance before the start of drought stress (Lobos et al. 2017). The occurrence of drought stress causes a relative prematurity. Generally, local genotypes are later mature than the improved cultivars. However, this difference decreases with increasing stress. Tillering is also another strategy for drought escape, which is under the genetic control and it has been observed that most of the genotypes of cereals have a great variety for this trait (Basu et al. 2016). In wheat and barley, higher and more crop yields are obtained from high-tillering genotypes. Tillering seems to reduce the yield sensitivity to plant densities. In general, drought escape is the most useful form of resistance because genotypes can reach yields before severe water constraints (Cattivelli et al. 2008).

31.9 PLANT BREEDING FOR OSMOTIC STRESS RESISTANCE

Drought is the most important factor limiting the growth and yield of plants, especially in arid and semi-arid areas, which affects 40–60% of cultivated lands (Fahad et al. 2017). One of the most effective ways to reduce dehydration damage is to breed the plants to tolerate water stress (Ullah et al. 2017). Hence, the breeders are working with various breeding methods to achieve varieties that tolerate drought stress. Genetic modification of crops to increase drought tolerance is a very important issue in plant breeding, and the detection and evaluation of drought-tolerant germplasms is a very difficult and time-consuming process (Lawlor 2012). Breeding the cultivars for drought resistance is one of the most important solutions to combat the drought problem. Drought tolerance is a quantitative trait, the genetic and physiological mechanisms associated with it are not well recognized, and there is no direct measurement method for drought tolerance (Fritsche-Neto and Borém 2012). This makes it difficult to identify drought-tolerant genotypes. Also, the polygenic nature of this trait, the control of drought resistance by a large number of genes, its dependence on time and the stress intensity, and the genotype × environment interactions, make it one of the most difficult issues in plant breeding studies (Abarshahr, Rabiei, and Lahigi 2011). Throughout the centuries, plant scientists

have endeavored endlessly to create drought-tolerant plants through the crosses of different species or varieties (Fahad et al. 2017; Lamaoui et al. 2018). These studies, which are based on the rules of classic plant breeding, rely on the selection and screening of plant populations to achieve drought or other adverse conditions such as cold and salinity resistance.

According to Rischkowsky and Pilling (2007), in each breeding program there are mainly three stages:

A. Creating genetic diversity
B. Selection within the genetic diversity to create a variety of desirable plant species
C. Evaluation of selected lines for commercial production

Genetic diversity can be achieved as follows (Govindaraj, Vetriventhan, and Srinivasan 2015):

A. Insertion of varieties or segregated generations from the other countries (introduction) or production within the country
B. Hybridization
C. Mutation

31.9.1 Breeding Methods for the Production of High Yielding Varieties under Drought Stress

Breeding methods for the production of high yielding varieties under drought stress include (Fritsche-Neto and Borém 2012; da Silva et al. 2013):

A. Provide varieties that are highly compatible with the drought stress

This method is more suitable for environments where the plant completes its life cycle using soil moisture reserves from the previous season. Varieties selected for high performance under normal conditions do not necessarily have high performance under stress conditions, and a stress tolerant variety should be evaluated and, then, selected under stress conditions (Rauf et al. 2016). One of the problems with this method is the variability of celestial precipitation from year to year in many semi-arid regions. Because the fluctuation of rainfall in different years changes the environmental selection index of the breeders from one year to another, the breeder needs to consider the size of the test area and the size of the population larger (Basu et al. 2016).

B. Provide varieties that are compatible with a wide range of environmental conditions

This breeding method is a method for the preparation of varieties that are compatible with a wide range of environmental conditions (Khan, Sovero, and Gemenet 2016). This procedure is most appropriate when plants are receiving rainfall during their growing season or in an optimal growth climate where periodic droughts occur.

In fact, plant breeders try to develop varieties that have the lowest yields in moderate stress conditions. The adaptation of the plant to environmental conditions is carried out in four general ways (Ahanger et al. 2017):

A. Phenological
B. Morphological
C. Physiological
D. Metabolic

Evaluation of plant traits in these four levels for adaptation to environmental stress is one of the objectives of the research related to stress resistance, and recognizing these adaptive traits is critical to the stress in plant breeding. Selection of proper genotypes for stress tolerance is performed in two ways (Cattivelli et al. 2008):

A. Direct, or observational method: which applies to absolute performance under controlled stress conditions and many improvements have been made in this way.
B. Indirect method: which is the screening and selection for morphological characteristics correlated with environmental stress tolerance. Indirect selection criteria for breeding the stress resistance and high performance are determined by physiologists and breeders. The breeding method used for drought resistance is the same as for other breeding purposes. In general, pedigree and bulk selection methods can be used to modify self-pollinated plants, and a recurrent selection method can be used to modify cross-pollinated crops.

The first trait for selecting drought stress tolerant varieties in breeding programs is grain yield, but sometimes the selection of resistant cultivars based on other traits is also important. Under stress conditions, other important traits are also related to grain yield. Accordingly, traits that play an important role in drought resistance are divided into different groups (Kamoshita et al. 2008).

One of the major goals of rice breeding programs is increasing yields per unit area. Achieving this goal is possible by precisely understanding the genetic control of yield and its components (Acquaah 2009). Due to the quantitative nature of the traits that contribute to increased performance, as well as their control by several genes and the severe effects of environmental factors, their exact genetic study is very difficult (Collard et al. 2005). It is important for a breeder to know the genetic basis of the trait and its quantitative changes as a fundamental genetic potential, and mapping quantitative trait loci (QTLs) not only provides information about the position of genes, but also allows the complex genetic model to decompose into single genetic components, and, thus, it is possible to find a better understanding of the genetic structure of these traits (Kordrostami and Rahimi 2015). In this case, quantitative traits will also be studied with the efficiency of single-gene traits and the results can be well utilized in the future

breeding programs (Collard et al. 2005). On the other hand, one of the limitations of plant cultivation, including rice, especially in recent years is water shortage and the inappropriate distribution of precipitation during the plant growth period, especially in flowering and pollination stages, which are the most sensitive stages to drought stress in rice (Bernier et al. 2007). Therefore, identification of QTLs controlling the yield-related traits and their tightly linked markers under drought stress conditions will allow better use of genetic resources and selection of drought-tolerant genotypes to increase production (Mardani et al. 2013).

31.9.2 Developing Tolerant Cultivars and Molecular Markers

Molecular genetics is another breeding method that identifies and maps many of the QTL genes and sequences affecting yield under drought stress conditions (Kordrostami and Rahimi 2015). There are two ways to develop tolerant cultivars through molecular markers. The first approach is to use grain yield as a criterion for stress tolerance, so a quantitative response to stress is required to identify the resistance gene QTLs based on the comparison of yield in both stressed and non-stressed environments. The second approach is to identify genes associated with stress tolerance and tolerance mechanisms. Therefore, in this approach, the candidate gene sequence is used as a marker. Before using these strategies, physiological studies should be conducted to confirm the positive effects of traits or genes related to yield under stress conditions and in different environments (Miklas et al. 2006). Also, molecular biology has identified useful genes with effective performance under stress conditions so that they can be used in gene transfer and genetic engineering (Deinlein et al. 2014).

31.10 CONCLUSIONS

One of the most effective ways to reduce dehydration damage is to breed the plants to tolerate water stress. Hence, the breeders are working with various breeding methods to achieve varieties that tolerate drought stress. Molecular genetics is a useful breeding method that identifies and maps many of the QTL genes and sequences affecting yield under drought stress conditions. Today, the use of molecular biology to enhance plant osmotic stress tolerance has increased. This method, because of being inexpensive and not time-consuming, is highly sought by plant breeders. Also, molecular biology has identified useful genes with effective performance under stress conditions so that they can be used in gene transfer and genetic engineering.

REFERENCES

Abarshahr, M, B Rabiei, and HS Lahigi. 2011. "Assessing genetic diversity of rice varieties under drought stress conditions." *Notulae Scientia Biologicae* 3 (1):114–123.

Acquaah, G. 2009. *Principles of Plant Genetics and Breeding.* London: John Wiley & Sons.

Ahanger, MA, NA Akram, M Ashraf, MN Alyemeni, L Wijaya, and P Ahmad. 2017. "Plant responses to environmental stresses—from gene to biotechnology." *AoB Plants* 9 (4).

Akıncı, Ş, and DM Lösel. 2012. "Plant water-stress response mechanisms." In *Water Stress*, London: InTech.

Amedie, FA. 2013. "Impacts of Climate Change on Plant Growth, Ecosystem Services, Biodiversity, and Potential Adaptation Measure." MasterThesis. Program Study of Biological and Environmental Science, University of Gothenburg, Sweden.

Anderegg, LDL, WRL Anderegg, and JA Berry. 2013. "Not all droughts are created equal: Translating meteorological drought into woody plant mortality." *Tree Physiology* 33 (7).

Araus, JL, GA Slafer, MP Reynolds, and C Royo. 2002. "Plant breeding and drought in C3 cereals: What should we breed for?" *Annals of Botany* 89 (7):925–940.

Araus, JL, GA Slafer, C Royo, and MD Serret. 2008. "Breeding for yield potential and stress adaptation in cereals." *Critical Reviews in Plant Science* 27 (6):377–412.

Ashraf, M. 2010. "Inducing drought tolerance in plants: Recent advances." *Biotechnology Advances* 28 (1):169–183.

Baird, AS, LDL Anderegg, ME Lacey, J HilleRis Lambers, and E Van Volkenburgh. 2017. "Comparative leaf growth strategies in response to low-water and low-light availability: Variation in leaf physiology underlies variation in leaf mass per area in Populus tremuloides." *Tree Physiology* 37 (9):1140–1150.

Bajji, M, St Lutts, and J-M Kinet. 2001. "Water deficit effects on solute contribution to osmotic adjustment as a function of leaf ageing in three durum wheat (Triticum durum Desf.) cultivars performing differently in arid conditions." *Plant Science* 160 (4):669–681.

Basu, S, V Ramegowda, A Kumar, and A Pereira. 2016. "Plant adaptation to drought stress." *F1000Research* 5:1–10.

Berg, JM, JL Tymoczko, and L Stryer. 2002. "Light absorption by chlorophyll induces electron transfer." 5 ed, In *Biochemistry*.

Bernier, J, A Kumar, V Ramaiah, D Spaner, and G Atlin. 2007. "A large-effect QTL for grain yield under reproductive-stage drought stress in upland rice." *Crop Science* 47 (2):507–516.

Bibi, AC, DM Oosterhuis, and ED Gonias. 2008. "Photosynthesis, quantum yield of photosystem II and membrane leakage as affected by high temperatures in cotton genotypes." *Journal of Cotton Science*.

Cattivelli, L, F Rizza, F-W Badeck, E Mazzucotelli, AM Mastrangelo, E Francia, C Mare, A Tondelli, and AM Stanca. 2008. "Drought tolerance improvement in crop plants: An integrated view from breeding to genomics." *Field Crops Research* 105 (1–2):1–14.

Chakhchar, A, M Haworth, C El Modafar, M Lauteri, C Mattioni, S Wahbi, and M Centritto. 2017. "An assessment of genetic diversity and drought tolerance in argan tree (*Argania spinosa*) populations: Potential for the development of improved drought tolerance." *Frontiers in Plant Science* 8:276.

Chaves, MM, J Flexas, and C Pinheiro. 2009. "Photosynthesis under drought and salt stress: Regulation mechanisms from whole plant to cell." *Annals of Botany* 103 (4):551–560.

Chaves, MM, JS Pereira, J Maroco, ML Rodrigues, Cândido PP Ricardo, ML Osório, I Carvalho, T Faria, and C Pinheiro. 2002. "How plants cope with water stress in the field? Photosynthesis and growth." *Annals of Botany* 89 (7):907–916.

Chiatante, D, A Di Iorio, S Sciandra, GS Scippa, and S Mazzoleni. 2006. "Effect of drought and fire on root development in Quercus pubescens Willd. and Fraxinus ornus L. seedlings." *Environmental and Experimental Botany* 56 (2):190–197.

Collard, BCY, MZZ Jahufer, JB Brouwer, and ECK Pang. 2005. "An introduction to markers, quantitative trait loci (QTL) mapping and marker-assisted selection for crop improvement: The basic concepts." *Euphytica* 142 (1–2):169–196.

Comas, L, S Becker, VMV Cruz, PF Byrne, and DA Dierig. 2013. "Root traits contributing to plant productivity under drought." *Frontiers in Plant Science* 4:442.

Cramer, GR, K Urano, S Delrot, M Pezzotti, and K Shinozaki. 2011. "Effects of abiotic stress on plants: A systems biology perspective." *BMC Plant Biology* 11 (1):163.

da Silva, EC, MB de Albuquerque, AD de Azevedo Neto, and CD da Silva Junior. 2013. "Drought and its consequences to plants—from individual to ecosystem." In *Responses of Organisms to Water Stress*. London: InTech.

De Micco, V, and G Aronne. 2012. "Morpho-anatomical traits for plant adaptation to drought." In *Plant Responses to Drought Stress*, 37–61. Berlin: Springer.

Deinlein, U, AB Stephan, T Horie, W Luo, G Xu, and JI Schroeder. 2014. "Plant salt-tolerance mechanisms." *Trends in Plant Science* 19 (6):371–379.

Devi, R, N Munjral, AK Gupta, and N Kaur. 2013. "Effect of exogenous lead on growth and carbon metabolism of pea (Pisum sativum L) seedlings." *Physiology and Molecular Biology of Plants* 19 (1):81–89.

Dolferus, R. 2014. "To grow or not to grow: A stressful decision for plants." *Plant Science* 229:247–261.

Eshghizade, HR, and P Ehsanzadeh. 2009. "Maize hybrids performance under differing irrigation regimes: II. Grain yield, components and water use efficiency." *Iranian Journal of Field Crop Science* 40 (2):145–153.

Fahad, S, AA Bajwa, U Nazir, SA Anjum, A Farooq, A Zohaib, S Sadia, W Nasim, S Adkins, and S Saud. 2017. "Crop production under drought and heat stress: Plant responses and management options." *Frontiers in Plant Science* 8:1147.

Fang, Y, and L Xiong. 2015. "General mechanisms of drought response and their application in drought resistance improvement in plants." *Cellular and Molecular Life Sciences* 72 (4):673–689.

Fang, Y, Y Du, J Wang, A Wu, S Qiao, B Xu, S Zhang, KHM Siddique, and Y Chen. 2017. "Moderate drought stress affected root growth and grain yield in old, modern and newly released cultivars of winter wheat." *Frontiers in Plant Science* 8:672.

FAO-Stat. 2012. *Food and Agriculture Organization of the United Nations. FAOSTAT database.*

Farooq, M, M Hussain, Abdul Wahid, and KHM Siddique. 2012. "Drought stress in plants: An overview." In *Plant responses to drought stress*, 1–33. Berlin: Springer.

Farooq, M, A Wahid, D-J Lee, SA Cheema, and T Aziz. 2010. "Drought stress: Comparative time course action of the foliar applied glycinebetaine, salicylic acid, nitrous oxide, brassinosteroids and spermine in improving drought resistance of rice." *Journal of Agronomy and Crop Science* 196 (5):336–345.

Fathi, A, and DB Tari. 2016. "Effect of drought stress and its mechanism in plants." *International Journal of Life Sciences* 10 (1):1–6.

Ferreira, ACdB, FM Lamas, GGd Brito, and ALDC Borin. 2013. "Water deficit in cotton plant originated from seeds treated with growth regulator." *Pesquisa Agropecuária Tropical* 43 (4):417–423.

Finan, JD, and F Guilak. 2010. "The effects of osmotic stress on the structure and function of the cell nucleus." *Journal of Cellular Biochemistry* 109 (3):460–467.

Finkelstein, R. 2013. "Abscisic acid synthesis and response." *The Arabidopsis Book/American Society of Plant Biologists* 11.

Flowers, TJ. 1989. *Plants under Stress: Biochemistry, Physiology and Ecology and Their Application to Plant Improvement.* Vol. 39. Cambridge University Press.

Forlani, G, M Bertazzini, and G Cagnano. 2018. "Stress-driven increase in proline levels, and not proline levels themselves, correlates with the ability to withstand excess salt in a group of 17 Italian rice genotypes." *Plant Biology* Accepted Author Manuscript.

Fracheboud, Y. 2004. "Using chlorophyll fluorescence to study photosynthesis." Presentation from the Institute of Plant Sciences ETH, Universitätstrasse 2.

Fritsche-Neto, R, and A Borém. 2012. *Plant Breeding for Abiotic Stress Tolerance.* Berlin: Springer.

Gangola, MP, and BR Ramadoss. 2018. "Sugars Play a Critical Role in Abiotic Stress Tolerance in Plants." In *Biochemical, Physiological and Molecular Avenues for Combating Abiotic Stress Tolerance in Plants*, 17–38. Elsevier.

Godfray, HCJ, IR Crute, L Haddad, D Lawrence, JF Muir, N Nisbett, J Pretty, S Robinson, C Toulmin, and R Whiteley. 2010. "The future of the global food system." The Royal Society.

Govindaraj, M, M Vetriventhan, and M Srinivasan. 2015. "Importance of genetic diversity assessment in crop plants and its recent advances: An overview of its analytical perspectives." *Genetics Research International* 2015.

Haling, RE, LK Brown, AG Bengough, IM Young, PD Hallett, PJ White, and TS George. 2013. "Root hairs improve root penetration, root–soil contact, and phosphorus acquisition in soils of different strength." *Journal of Experimental Botany* 64 (12):3711–3721.

Hanson, AD, CE Nelsen, AR Pedersen, and EH Everson. 1979. "Capacity for proline accumulation during water stress in barley and its implications for breeding for drought resistance 1." *Crop Science* 19 (4):489–493.

Hayat, S, Q Hayat, MN Alyemeni, AS Wani, J Pichtel, and A Ahmad. 2012. "Role of proline under changing environments: A review." *Plant Signaling & Behavior* 7 (11):1456–1466.

Heyns, P. 2009. "Water conservation in Arid and Semi-Arid Regions." eds. Hubert HG Savenije and Arjen Y. Hoekstra, *Water Resources Management* 1:113–150.

Hossain, MA, SH Wani, S Bhattacharjee, DJ Burritt, and L-SP Tran. 2016. *Drought Stress Tolerance in Plants, Vol 2: Molecular and Genetic Perspectives.* Berlin: Springer.

Hossain, MS, and K-J Dietz. 2016. "Tuning of redox regulatory mechanisms, reactive oxygen species and redox homeostasis under salinity stress." *Frontiers in Plant Science* 7:548.

Hunt, L, LN Mills, C Pical, CP Leckie, FL Aitken, J Kopka, B Mueller-Roeber, MR McAinsh, AM Hetherington, and JE Gray. 2003. "Phospholipase C is required for the control of stomatal aperture by ABA." *The Plant Journal* 34 (1):47–55.

Ings, J, LAJ Mur, PRH Robson, and M Bosch. 2013. "Physiological and growth responses to water deficit in the bioenergy crop Miscanthus x giganteus." *Frontiers in Plant Science* 4:468.

Jajoo, A. 2013. "Changes in photosystem II in response to salt stress." In *Ecophysiology and Responses of Plants under Salt Stress*, 149–168. Berlin: Springer.

Jaleel, CA, P Manivannan, A Wahid, M Farooq, HJ Al-Juburi, R Somasundaram, and R Panneerselvam. 2009. "Drought stress in plants: A review on morphological characteristics and pigments composition." *International Journal of Agriculture and Biology* 11 (1):100–105.

Jones, HG, and E Rotenberg. 2001. "Energy, radiation and temperature regulation in plants." *eLS: Essential for Life Science.*

Judd, LA, BE Jackson, and WC Fonteno. 2015. "Advancements in root growth measurement technologies and observation capabilities for container-grown plants." *Plants* 4 (3):369–392.

Kamoshita, A, RC Babu, NM Boopathi, and S Fukai. 2008. "Phenotypic and genotypic analysis of drought-resistance traits for development of rice cultivars adapted to rainfed environments." *Field Crops Research* 109 (1–3):1–23.

Kao, CH. 1981. "Senescence of rice leaves VI. Comparative study of the metabolic changes of senescing turgid and water-stressed excised leaves." *Plant and Cell Physiology* 22 (4):683–688.

Khan, A, V Sovero, and D Gemenet. 2016. "Genome-assisted breeding for drought resistance." *Current Genomics* 17 (4):330–342.

Khan, MA, MU Shirazi, Muhammad Ali Khan, SM Mujtaba, E Islam, S Mumtaz, A Shereen, RU Ansari, and M Yasin Ashraf. 2009. "Role of proline, K/Na ratio and chlorophyll content in salt tolerance of wheat (Triticum aestivum L.)." *Pakistan Journal of Botany* 41 (2):633–638.

Kordrostami, M, B Rabiei, and HH Kumleh. 2016. "Association analysis, genetic diversity and haplotyping of rice plants under salt stress using SSR markers linked to SalTol and morpho-physiological characteristics." *Plant Systematics and Evolution* 302 (7):871–890.

Kordrostami, M, B Rabiei, and HH Kumleh. 2017. "Biochemical, physiological and molecular evaluation of rice cultivars differing in salt tolerance at the seedling stage." *Physiology and Molecular Biology of Plants* 23 (3):529–544.

Kordrostami, M, and M Rahimi. 2015. "Molecular markers in plants: Concepts and applications." *Genetics in The third Millennium* 13:4024–4031.

Kovács, Z, L Simon-Sarkadi, I Vashegyi, and G Kocsy. 2012. "Different accumulation of free amino acids during short-and long-term osmotic stress in wheat." *The Scientific World Journal* 2012.

Kulkarni, M, and S Phalke. 2009. "Evaluating variability of root size system and its constitutive traits in hot pepper (Capsicum annum L.) under water stress." *Scientia Horticulturae* 120 (2):159–166.

Lalelou, FS, MR Shakiba, AD Mohammadi-Nassab, and SA Mohammadi. 2010. "Effects of drought stress and nitrogen nutrition on seed yield and proline content in bread and durum wheat genotypes." *Journal of Food, Agriculture & Environment* 8 (3–4):857–860.

Lamaoui, M, M Jemo, R Datla, and F Bekkaoui. 2018. "Heat and drought stresses in crops and approaches for their mitigation." *Frontiers in Chemistry* 6:26.

Lawlor, DW. 2012. "Genetic engineering to improve plant performance under drought: Physiological evaluation of achievements, limitations, and possibilities." *Journal of Experimental Botany* 64 (1):83–108.

Liu, H-S, F-M Li, and H Xu. 2004. "Deficiency of water can enhance root respiration rate of drought-sensitive but not drought-tolerant spring wheat." *Agricultural Water Management* 64 (1):41–48.

Lobos, GA, AV Camargo, A del Pozo, JL Araus, R Ortiz, and JH Doonan. 2017. "Plant phenotyping and phenomics for plant breeding." *Frontiers in Plant Science* 8:2181.

Lopes, MS, JL Araus, PDR Van Heerden, and CH Foyer. 2011. "Enhancing drought tolerance in C4 crops." *Journal of Experimental Botany* 62 (9):3135–3153.

Lu, Q, C Lu, J Zhang, and T Kuang. 2002. "Photosynthesis and chlorophyll a fluorescence during flag leaf senescence of field-grown wheat plants." *Journal of Plant Physiology* 159 (11):1173.

Mancosu, N, RL Snyder, G Kyriakakis, and D Spano. 2015. "Water scarcity and future challenges for food production." *Water* 7 (3):975–992.

Mardani, Z, B Rabiei, H Sabouri, and A Sabouri. 2013. "Mapping of QTLs for germination characteristics under non-stress and drought stress in rice." *Rice Science* 20 (6):391–399.

Mattioli, R, P Costantino, and M Trovato. 2009. "Proline accumulation in plants: Not only stress." *Plant Signaling & Behavior* 4 (11):1016–1018.

Miklas, PN, JD Kelly, SE Beebe, and MW Blair. 2006. "Common bean breeding for resistance against biotic and abiotic stresses: From classical to MAS breeding." *Euphytica* 147 (1–2):105–131.

Nayyar, H, and DP Walia. 2003. "Water stress induced proline accumulation in contrasting wheat genotypes as affected by calcium and abscisic acid." *Biologia Plantarum* 46 (2):275–279.

Negrão, S, SM Schmöckel, and M Tester. 2017. "Evaluating physiological responses of plants to salinity stress." *Annals of Botany* 119 (1):1–11.

Obidiegwu, JE, GJ Bryan, HG Jones, and A Prashar. 2015. "Coping with drought: Stress and adaptive responses in potato and perspectives for improvement." *Frontiers in Plant Science* 6:542.

Osakabe, Y, K Osakabe, K Shinozaki, and L-SP Tran. 2014. "Response of plants to water stress." *Frontiers in Plant Science* 5:86.

Pakniyat, H, and M Armion. 2007. "Sodium and proline accumulation as osmoregulators in tolerance of sugar beet genotypes to salinity." *Pakistan Journal of Biological Sciences* 10 (22):4081–4086.

Pierret, A, J-L Maeght, C Clément, J-P Montoroi, C Hartmann, and S Gonkhamdee. 2016. "Understanding deep roots and their functions in ecosystems: An advocacy for more unconventional research." *Annals of Botany* 118 (4):621–635.

Pirasteh-Anosheh, H, A Saed-Moucheshi, H Pakniyat, and M Pessarakli. 2016. Stomatal responses to drought stress. Vol. 2, *Water Stress and Crop Plants: A Sustainable Approach.* London: Wiley.

Prinz, D, and AK Singh. 2003. "Water resources in arid regions and their sustainable management." In *Water Resources in Arid Regions and Their Sustainable Management.* UNEP.

Rauf, S, JM Al-Khayri, M Zaharieva, P Monneveux, and F Khalil. 2016. "Breeding strategies to enhance drought tolerance in crops." In *Advances in Plant Breeding Strategies: Agronomic, Abiotic and Biotic Stress Traits,* 397–445. Berlin: Springer.

Riaz, A, A Younis, AR Taj, A Karim, U Tariq, S Munir, and S Riaz. 2013. "Effect of drought stress on growth and flowering of marigold (Tagetes erecta L.)." *Pakistan Journal of Botany* 45 (S1):123–131.

Rischkowsky, B, and D Pilling. 2007. "Genetic improvement methods to support sustainable utilization. Section D." *The state of the world's genetic resources for food and agriculture,* FAO:380–422.

Rukundo, P, HG Betaw, S Ngailo, and F Balcha. 2014. "Assessment of drought stress tolerance in root and tuber crops." *African Journal of Plant Science* 8 (4):214–224.

Sah, SK, KR Reddy, and J Li. 2016. "Abscisic acid and abiotic stress tolerance in crop plants." *Frontiers in Plant Science* 7:571.

Seyednasrollah, B. 2017. "Ecosystem response to a changing climate: Vulnerability, impacts and monitoring." Disseraton, Duke University.

Shah, LR, M Husain, and A Raja. 2017. "Abiotic stress mechanism in herbaceous crops: An overview." *The Pharma Innovation Journal* 6 (11).

Sharma, KD, and MSMS Kuhad. 2006. "Influence of Potassium level and soil moisture regime on biochemical metabolites of Brassica Species." *Brassica Journal* 8:71–74.

Shrivastava, P, and R Kumar. 2015. "Soil salinity: A serious environmental issue and plant growth promoting bacteria as one of the tools for its alleviation." *Saudi Journal of Biological Sciences* 22 (2):123–131.

Sreenivasulu, N, SK Sopory, and PB Kavi Kishor. 2007. "Deciphering the regulatory mechanisms of abiotic stress tolerance in plants by genomic approaches." *Gene* 388 (1):1–13.

Tardieu, F. 2013. "Plant response to environmental conditions: Assessing potential production, water demand, and negative effects of water deficit." *Frontiers in Physiology* 4:17.

Tripathy, BC, and R Oelmüller. 2012. "Reactive oxygen species generation and signaling in plants." *Plant Signaling & Behavior* 7 (12):1621–1633.

Ullah, A, H Sun, X Yang, and X Zhang. 2017. "Drought coping strategies in cotton: Increased crop per drop." *Plant Biotechnology Journal* 15 (3):271–284.

Van Kooten, O, and JFH Snel. 1990. "The use of chlorophyll fluorescence nomenclature in plant stress physiology." *Photosynthesis Research* 25 (3):147–150.

Van Oosten, MJ, A Costa, P Punzo, S Landi, A Ruggiero, G Batelli, and St Grillo. 2016. "Genetics of drought stress tolerance in crop plants." In *Drought Stress Tolerance in Plants,* Vol 2, 39–70. Berlin: Springer.

Vishwakarma, K, N Upadhyay, N Kumar, G Yadav, J Singh, RK Mishra, V Kumar, R Verma, RG Upadhyay, and M Pandey. 2017. "Abscisic acid signaling and abiotic stress tolerance in plants: A review on current knowledge and future prospects." *Frontiers in Plant Science* 8:161.

Wang, H-L, P-D Lee, L-F Liu, and J-C Su. 1999. "Effect of sorbitol induced osmotic stress on the changes of carbohydrate and free amino acid pools in sweet potato cell suspension cultures." *Botanical Bulletin of Academia Sinica* 40.

Wang, X, X Cai, C Xu, Q Wang, and S Dai. 2016. "Drought-responsive mechanisms in plant leaves revealed by proteomics." *International Journal of Molecular Sciences* 17 (10):1706.

Wilhite, DA. 2000. Drought as a Natural Hazard: Concepts and Definitions. In *Drought, A Golbal Assessment.*

Williams, WT. 1963. Plant–Water Relationships in Arid and Semi-Arid Conditions (1961). *JSTOR.*

Xu, W, K Cui, A Xu, L Nie, J Huang, and S Peng. 2015. "Drought stress condition increases root to shoot ratio via alteration of carbohydrate partitioning and enzymatic activity in rice seedlings." *Acta Physiologiae Plantarum* 37 (2):9.

Xu, Z, G Zhou, and H Shimizu. 2010. "Plant responses to drought and rewatering." *Plant Signaling & Behavior* 5 (6):649–654.

Yang, X, X Chen, Q Ge, B Li, Y Tong, A Zhang, Z Li, T Kuang, and C Lu. 2006. "Tolerance of photosynthesis to photoinhibition, high temperature and drought stress in flag leaves of wheat: A comparison between a hybridization line and its parents grown under field conditions." *Plant Science* 171 (3):389–397.

Yeats, TH, and JKC Rose. 2013. "The formation and function of plant cuticles." *Plant Physiology* pp. 113.222737.

32 Breeding for Improved Plant–Symbiont Thermotolerance and Symbiotic Performance by Regulating Heat Shock Proteins, RNA Binding Proteins, and Chaperones

Omid Askari-Khorasgani and Mohammad Pessarakli

CONTENTS

GLOSSARY

ABA: Abscisic acid
AP2/ERF: APETALA2/Ethylene Response Factor
CBF: C-repeat binding factor
Clp: Casein lytic proteinase
Cpn: Chaperonin
CSD: Cold shock domain
CSDP: Cold shock domain protein
CSP: Cold shock protein
DRBPs: DNA- and RNA-binding proteins
DRE: Dehydration-responsive element
ds: Double-stranded
ENOD8: Early nodulin 8
ER: Endoplasmic reticulum
FKBP: FK506-binding proteins
GA: Gibberellin
GB: Glycine-betaine
GR-RBPs: Glycine-rich RNA-binding proteins
Grxs: Glutaredoxins
H$_2$O$_2$: Hydrogen peroxide
Hfq: Host factor Qβ RNA phage

HSF: Heat shock transcription factor
HSP: Heat shock protein
kDa: Kilodalton
LOS: Low expression of osmotically responsive genes
MAPK: Mitogen-activated protein kinases
miRNA: MicroRNA
MON: Monocillin I
mtHSP: Mitochondrial HSP
NTRs: NADPH-thioredoxin reductases
PBM: Peribacteroid membrane
PDI: Protein disulfide isomerase
PLDα: Phospholipase Dα
PPI: Peptidyl-prolyl cis/trans isomerase
PPI: Peptidyl-prolyl isomerases
pre-miRNA: Precursor microRNA
Prx: Peroxiredoxin
RBP: RNA binding protein
RCF: Regulator of CBF gene expression
RH: RNA helicase
RNA: Ribonucleic acid
ROS: Reactive oxygen species
r-proteins: Ribosomal proteins

RRMs: RNA-recognition motifs
rRNA: Ribosomal ribonucleic acids
Rubisco: Ribulose-1,5-bisphosphate carboxylase/oxygenase
SA: Salicylic acid
SD: Shine-Dalgarno
sHSP: Small HSP
siRNA: Small interfering RNA
sRNA: Small RNA
ss: Single-stranded
STRS: STRESS RESPONSE SUPPRESSOR
TDX: TRX-like protein
TFs: Transcription factors
tiRNA: Transcription initiation RNA
TOGR: Thermotolerant Growth Required
TPR: Tetra-trico-peptide repeat
TRX: Thioredoxin
wHTH: Winged helix–turn–helix

32.1 INTRODUCTION

Since all organisms are constantly exposed to different stresses, they have evolved a variety of adaptation mechanisms through molecular, anatomical, morphological, physiological, biochemical, and behavioral alterations to cope with stress conditions. The adaptation mechanisms are under the control of gene expression regulation and signal transduction (Cho and Hong 2006; Klopf et al. 2009; de Nadal et al. 2011; Basu et al. 2016; Woodson 2016). The adaptation mechanisms mitigate the adverse effects of stress conditions by protecting the cell cycle, biological processes, energy status, structures, homeostasis, and functions (Lopez-Maury et al. 2008; Ni et al. 2009; Baena-González 2010; van der Knaap and Verrijzer 2016). Stress responses for the adaptation of organisms require the ability to regulate and translate stress-induced signal circuits into extensive gene expression and metabolic reprogramming, in which ribonucleic acid (RNA) and protein metabolism is of critical importance (Zhu et al. 2007; Kim et al. 2010a). RNA metabolism includes post-transcriptional regulation of RNA processes, such as pre-mRNA splicing, mRNA export, localization, turnover, and translational control (Kang et al. 2013). RNA chaperones are the types of RNA-binding proteins (RBPs) that prevent RNA misfolding or destabilize misfolded RNA species and, subsequently, provide assistance with their correct folding into their functional native conformation during RNA metabolism (Herschlag 1995; Kang et al. 2013). RNA chaperone activities consequently promote RNA stabilization, protein translation and protection, as well as RNA–RNA and protein–RNA interactions to regulate cellular processes under normal and stressful conditions, particularly extreme temperatures (Hammond and Helenius 1995; Cho and Hong 2006; Kang et al. 2013; Ishiguro et al. 2017; Jagodnik et al. 2017). The regulation of gene expression and transcription factors (TFs), such as heat shock proteins (HSPs), primarily induced by heat shock transcription factors (HSFs), upregulates the transcription of molecular chaperones to protect proteins and maintain cellular processes under extreme

temperatures (Lee and Schöffl 1996; Cho and Hong 2006; Sakurai and Ota 2011). HSPs and chaperones are found in most prokaryotes and eukaryotes and even some viruses (Littlefield and Nelson 1999; Maaroufi and Tanguay 2013; Gomez-Pastor et al. 2017; Jacob et al. 2017) and are involved in regulating diverse signaling pathways as well as cellular protein quality control and homeostasis. They play critical roles in protein folding, assembly, translocation, and degradation in many normal cellular processes, stabilize proteins and membranes, and can assist in protein refolding under stressful conditions. Therefore, HSP/chaperones protect plants against various stresses by re-establishing normal protein conformation and cellular homeostasis under normal and stressful conditions (Wang et al. 2004; Gonçalves et al. 2011; Sakurai and Ota 2011; Jacob et al. 2017). Regulation of the HSPs, RBPs, and other molecular chaperones has the potential to be used in plant and symbiont breeding programs to improve plant–symbiont symbiosis, performance, and tolerance to different stresses and more importantly, extreme temperatures (Verma et al. 1986; Fischer et al. 1993; Panter et al. 2000; Smykal et al. 2000; Minder et al. 2001; Kevei et al. 2002; Balsiger et al. 2004; Catalano et al. 2004; Räsänen et al. 2004; Lee et al. 2005a; Liu et al. 2006; Castro-Sowinski et al. 2007; Alkhalfioui et al. 2008; Castiglioni et al. 2008; Asadi Rahmani et al. 2009; Iturriaga et al. 2009; Morsy et al. 2010; Torres-Quesada et al. 2010; Alexandre and Oliveira 2011; Zhu et al. 2011a; Barnett et al. 2012; Bazin et al. 2012; Oger et al. 2012; Chae et al. 2013; Fu 2014; Jiménez-Zurdo and Robledo 2015; Paço et al. 2016; Sasaki et al. 2016; Rennella et al. 2017; Robledo et al. 2017). Thus, the present review aims to evaluate the most effective approaches regarding the regulation of HSPs and chaperones to improve stress responses, particularly thermotolerance, performance, and symbiosis.

32.2 STRUCTURAL AND FUNCTIONAL DIVERSITY OF HSPS/CHAPERONES

HSPs are classified into five major families based on their kilodalton (kDa) molecular weight and two groups according to their functions (Table 32.1). In the first group, the "foldases group," the HSP70 (DnaK/Ssa), HSP60 (GroEL or chaperonins [Cpn]), HSP90 (HtpG), and HSP100 (ClpA/B/C; casein lytic proteinase) families are ATP-dependent molecular chaperones involved in folding nascent polypeptides and refolding or correct folding stress-induced misfolded or unfolded proteins (Table 32.1). The second group, the ATP-independent "holdases group," includes the small HSP (sHSP) chaperone family, comprised of a low subunit molecular mass of 12–43 kDa; these sequesters unfolded or partially folded proteins, which are subsequently processed by the foldases (Table 32.1). The sHSP chaperones bind to the non-native substrate proteins and destabilize misfolded conformations and their intermediates, thereby preventing or reversing protein aggregations under stressful conditions such as extreme temperatures (Table 32.1) (Koyasu et al. 1989; Hoffmann et al. 2004; Wang et al.

TABLE 32.1

Functional Diversity and Influence of Five Major Classes of HSPs and Their Subfamilies on Plant–Symbiont Growth, Development, Performance, Stress Responses, and Symbiosis

Classes	Representative Members	Isolated from/ expressed in	Intracellular Localization	Major Functions	Refs
HSP70				HSP70 prevents protein aggregation and assists in protein refolding, degradation of damaged proteins, protein import and translocation, signal transduction, and transcriptional activation. In addition, HSP70 interacts with other chaperones such as HSP100, HSP90, HSP60, and sHSPs, regulating their activities, such as protein solubilization, disaggregation, refolding, and homeostasis.	(Mayer and Bukau 2005; Cho and Hong 2006; Kao et al. 2015; Reeg et al. 2016; Hasan et al. 2017; Zwirowski et al. 2017)
	HSP/HSC70	MsHSP70-1 isolated from *Medicago sativa*/expressed in *Arabidopsis*; tHSC70-1 isolated from *Arabidopsis*/expressed in *Arabidopsis*	Cytosol, nucleus, plastid, ER, mitochondria, peroxisome chloroplasts, symbiosome membrane, peribacteroid membrane, glyoxysome, and protein bodies	Some of the HSP/HSC70s are also upregulated under normal conditions and are involved in plant growth and development and tolerance to biotic and abiotic stresses, such as extreme temperatures, and assist the folding of *de novo* synthesized polypeptides and the import/translocation of precursor proteins. HSP70s are involved in protein homeostasis through ATP-dependent cycles and essentially bind non-native polypeptides and folding intermediates that are either newly synthesized or stress induced. They can be affected by development processes such as seed maturation and root growth. *MsHSP70-1* (*Medicago sativa* L. cv. Algonquin HSP70-1) is involved in alfalfa nodule development. Because *MsHSP70-1* expresses in various alfalfa tissues under normal conditions, it can be categorized as HSP/HSC70. MsHSP70-1 and AtHSC70-1 have similar functions in plant growth and development.	(Panter et al. 2000; Zhang et al. 2004; Mayer and Bukau 2005; Ma et al. 2006; He et al. 2008; Lee et al. 2009c; Li et al. 2017b)
	AtHSC70-1	*Arabidopsis*/*Arabidopsis*	Cytosol, nuclear	Overexpression of cytosolic/nuclear HSC70-1 caused heat tolerance and reduced the activity of root and shoot meristems.	(Cazalé et al. 2009)
	Thhsc70	*Thellungiella halophile*/ *Arabidopsis*	Cytosol	Increased tolerance to heat and cold stress. *Thhsp70* expression was high in leaves, low in stems, and almost absent in roots.	(Zhang et al. 2004)
	CnHSP70	*Chrysanthemum nankingense*	Cytoplasm	*CnHSP70* is induced under heat stress and thus, is believed to impart heat tolerance.	(Yang et al. 2016)
	NtHSP70-1	Tobacco/Tobacco	Cytoplasm, nucleus	*NtHSP70-1* (*Nicotiana tabacum HSP70-1*) overexpression induced drought tolerance in tobacco.	(Cho and Hong 2006)
	BiPHSP70		ER	BiP is directly involved in ER-associated protein degradation by recognizing and targeting abnormally folded or unfolded proteins for degradation. BiP proteins bind to and subsequently activate specific TFs to regulate gene expression, cell death signaling, water status, and cell homeostasis.	(Carvalho et al. 2014a., b)
HSP60/chaperonin			Chloroplast, mitochondria, cytosol, symbiosome membrane, plastid, peribacteroid membrane	HSP60 prevents protein denaturation and assists in correct refolding. HSP60-HSP70 complexes are involved in co- or post-translational import of cytoplasmic proteins and their translocation into symbiosomes, avoid protein aggregation and denaturation, and assist in the correct assembly of unfolded polypeptides generated under stress conditions. Protein refolding in *E. coli* requires chaperonin 60, chaperonin 10, and ATPase activity.	(Neupert et al. 1990; Martin et al. 1992; Truscott et al. 1994; Panter et al. 2000; Kim et al. 2013; Kao et al. 2015; Koldewey et al. 2017)

(Continued)

TABLE 32.1 (CONTINUED)

Functional Diversity and Influence of Five Major Classes of HSPs and Their Subfamilies on Plant–Symbiont Growth, Development, Performance, Stress Responses, and Symbiosis

Classes	Representative Members	Isolated from/expressed in	Intracellular Localization	Major Functions	Refs
HSP60 Subfamilies: Group I: Cpn60; Group II: CCT					
Group I	Cpn60	*Medicago truncatula, Sinorhizobium meliloti, Bradyrhizobium japonicum*	Chloroplast, mitochondria, symbiosome	*GroESL* (Cpn60 + Cpn10) and *dnaJ* (HSP40) improve symbiotic performance (e.g., in several rhizobia species) and tolerance to extreme temperatures (e.g., in rhizobia and *Escherichia coli*) and enhance mRNA stability. *GroESL* genes affect symbiotic effectiveness by co-regulating symbiotic genes such as nitrogen fixation genes. TCP-1 (mammalian)/cpn60 chaperonin, referred to as *molecular chaperone*, is involved in correct folding and subsequent assembly into oligomers and is required for normal cell growth.	(Milos and Roy 1984; Cannon et al. 1986; Hemmingsen et al. 1988; Fischer et al. 1993; Catalano et al. 2004; Balczun et al. 2006; Rajaram and Apte 2008; Alexandre and Oliveira 2011)
	OsCpn60α	Rice	Plastid	Cpn60 is involved in the maturation precursor RNAs, Rubisco assembly, nitrogen metabolism, and photosynthesis. Since nitrogen and C are essential in symbiosis, Cpn60 overexpression might benefit symbiosis processes. Unlike *Cpn60* with β subunit, the expression of *OsCpn60α1* and *OsCpn60α2* was strongly induced by heat stress and thus, might be effective to improve thermotolerance. *OsCpn60α1* is also required for the folding of Rubisco large subunit.	(Kim et al. 2013)
Group II	*Arabidopsis, Crocus sativus* (plant). *Caenorhabditis elegans* (animal), *Delia antiqua* (insect)	CCT	Cytosol	Expression of mRNA for *DaTCP-1* (TCP-1 refers to chaperonin containing tailless polypeptide-1, or TCP-1 ring complex, TRiC) improves cold adaptation; thus, overexpression of the *Arabidopsis* CCT gene might confer cold tolerance, which needs to be verified.	(Hill and Hemmingsen 2001; Kayukawa et al. 2005)
HSP90		*Medicago truncatula, Sinorhizobium meliloti, Arabidopsis,* etc.	Cytosol, nucleoplasm, ER, mitochondria and chloroplasts, symbiosome	Involved in maturation, structural folding, and maintenance of the conformational integrity of various proteins, collectively referred to as *clients*, and several cellular signal transduction proteins, including many steroid hormone receptors and kinases.	(Catalano et al. 2004; McLellan et al. 2007; Yamada et al. 2007; Wayne et al. 2011; Guo and Ki 2012)
	HSP90	*Brassica napus, Arabidopsis,* and *Spinacea oleracea*	Cytosol	Improves tolerance to both cold and heat stresses. *Arabidopsis'* symbiotic relationship with the rhizosphere fungus *Paraphaeosphaeria quadriseptata* upregulates the HSP90, leading to improved thermotolerance.	(Krishna et al. 1995; Ludwig-Müller et al. 2000; McLellan et al. 2007)
HSP100/Clp Subfamilies: Class I: ClpA, ClpB, ClpC, ClpD, ClpE (68–110 kDa); Class II: ClpM, ClpN, ClpX, ClpY (40–50 kDa)					
			Primarily located in stroma	HSP100/Clp proteins are involved in protein disaggregation and refolding.	(Katiyar-Agarwal et al. 2001; Sjögren et al. 2004; Kumar et al. 2016) *(Continued)*

TABLE 32.1 (CONTINUED)

Functional Diversity and Influence of Five Major Classes of HSPs and Their Subfamilies on Plant–Symbiont Growth, Development, Performance, Stress Responses, and Symbiosis

Classes	Representative Members	Isolated from/expressed in	Intracellular Localization	Major Functions	Refs
	AtHSP93	Arabidopsis	Cytoplasm	Involved in normal chloroplast development.	(Constan et al. 2004)
	AtHSP101/ClpB	Arabidopsis/rice, Saccharomyces cerevisiae, wheat	Cytoplasm, nucleoid/nucleus	Improves growth, development, seed germination, chloroplast development, and performance under normal and stress conditions. Enhances tolerance to high temperatures and other stresses (e.g., cold, salt, drought, ethanol, heavy metal stresses), required for the normal translation of certain mRNAs, maintenance and propagation of prion-like factors and thus, epigenetic effects. AtHSP101 interacts with sHSPs to promote heat tolerance. Chloroplast localized Arabidopsis pale-green 6 (APG6), as an AtHSP101 homolog, is involved in normal growth, chloroplast development, and heat tolerance.	(Garwal et al. 2001; Hong and Vierling 2001; Agarwal et al. 2003; Katiyar-Agarwal et al. 2003; Lee et al. 2005b; Myouga et al. 2006; Lee et al. 2007)
S. cerevisiae, Synechocystis, cyanobacteria, Synechococcus	HSP104		Cytoplasm, nucleus, mitochondria	Involved in resolubilization of aggregated proteins, conformational repair of heat-damaged ER glycoproteins, and proper protein refolding through its reactions with other chaperones, particularly HSP70. Enhances tolerance to high temperatures. Zinc finger proteins 2 and 4 (Msn2/4p) expressed in cytoplasm under normal and nucleus under stress conditions, and Hsfp can regulate the expression of HSP104 and HSP70.	(Amorós and Estruch 2001; Katiyar-Agarwal et al. 2001; Lum et al. 2004; Rikhvanov et al. 2004; Tkach and Glover 2008)
	ClpB	Ca36WT and Ca36pPHUclpB/Mesorhizobium mediterraneum (chickpea)	Not reported	ClpB overexpression in Mesorhizobium ciceri LMS-1 and Mesorhizobium mediterraneum UPM-Ca36T improves chickpea root hair curling, root nodule formation, development and nodule numbers, plant growth, symbiosis, and tolerance to heat and acid shocks. ClpB requires the collaboration of DnaK proteins to promote solubilization and reactivation of proteins. DnaKJ–GrpE–ClpB interactions catalytically cooperate and release heat-inactivated non-native protein to provide assistance in protein folding. In addition, ClpB protects proteins from denaturation and aggregation.	(Watanabe et al. 2000; Brígido et al. 2012; Paço et al. 2016)
	Clp protease (ClpP)	Wheat, Arabidopsis, tobacco	Chloroplast	Rubisco binding protein cpn60, ClpP, Rubisco activase. Rubisco large and small subunits improve chloroplast development, photosynthesis, recovery, and tolerance to heat and drought. ClpP and HSP70 protect plants against stress-induced oxidative stress.	(Shikanai et al. 2001; Demirevska-Kepova et al. 2005; Demirevska et al. 2008; Pulido et al. 2017)
Arabidopsis	ClpC (ClpC1 and ClpC2)		Chloroplast	ClpC chaperones are essential for chloroplast function, leaf development, and thus, photosynthesis.	(Sjögren et al. 2004)

(Continued)

TABLE 32.1 (CONTINUED)

Functional Diversity and Influence of Five Major Classes of HSPs and Their Subfamilies on Plant–Symbiont Growth, Development, Performance, Stress Responses, and Symbiosis

Classes	Representative Members	Isolated from/ expressed in	Intracellular Localization	Major Functions	Refs
	Wheat	ClpB, C	Chloroplast (*ClpB-p*), mitochondria (*Clp-m*), cytoplasm (*Clp-cyt*)	In wheat, TaClpB1 subgroups TaClpC and TaClpD are in chloroplast, TaClpB5 is in mitochondria, and TaClpB2, TaClpB3, and TaClpB4 are in cytoplasm. *ClpB-p (TaClpB1)*, *ClpB-cyt (TaClpB2* and *TaClpB3)*, and *CpB-m (TaClpB5)* in both sensitive and tolerant wheat genotypes were upregulated under heat stress, and thus their overexpression might improve tolerance to heat stress. Under cold stress, *TaClpC1* and *TaClpD1* were upregulated, while *TaClpB2* and *TaClpB3* were downregulated. Hence, *TaClpC1* and *TaClpD1* overexpression might improve cold adaptation. The abundance of *TaClpB1* and *TaClpD2* indicates that their overexpression might improve plant growth, while *TaClpB2* and *TaClpB3* in embryo, endosperm, caryopsis, and root indicates that they might improve seed development and germination.	(Muthusamy et al. 2016)
Class II			Chloroplast	Compared with the class I, few studies have been conducted on the class II HSP100. Thus, more experiments are required to show their roles in thermotolerance and symbiosis.	(Wang et al. 2004; Muthusamy et al. 2016)
sHSP 12–43 kDa				In an ATP-independent manner, sHSPs bind to HSP40, HSP70, HSP100 (ClpB), and misfolded substrate proteins to subsequently, cooperatively in an ATP-dependent manner, facilitate protein solubilization, disaggregation, and refolding, improving cell viability, recovery, and tolerance to both heat and cold stresses.	(Smykal et al. 2000; Liberek et al. 2008; Laskowska et al. 2010; Sun et al. 2010; Muthusamy et al. 2017; Zwirowski et al. 2017)
	OsHSP16.9	Rice/*E. coli*, rice	Cytoplasm	Improves thermotolerance in *E. coli* and drought and salt tolerance in rice. Its influence and cooperative regulation with other genes influential on symbiosis will provide valuable information in future.	(Yu Jin et al. 2014; Muthusamy et al. 2017)
	HSP17	*Synechocystis* (bacteria)	Thylakoid membranes	Prevents lipid membrane destabilization and assists in refolding of misfolded proteins under heat stress. AtHSP17.8 increases the efficiency of chloroplasts under heat stress. Hence, upregulation of AtHSP17.8 and HSP17 from *Synechocystis* might increase the photosynthetic efficiency under heat stress.	(Török et al. 2001; Kim et al. 2011a)
	JrsHSP17.3 HsfA3, sHSP17.4-CII, 23.8-M, sHsp 21-P	Walnut/*S. cerevisiae* (yeast) Tomato	Nucleoid sHSP17.4-CII in cytoplasm, 23.8-M in mitochondria, sHsp 21-P in chloroplast	Enhances tolerance to heat, cold, freezing, oxidative, and salt stresses. Prevent postharvest chilling injury in tomato.	(Zhai et al. 2016) (Ré et al. 2017)
	HSP17.7	Carrot/*S. cerevisiae* (yeast)	Nucleoid	Enhances growth, viability, and tolerance to heat, cold, acid, and osmotic stresses.	(Ko et al. 2017)

(Continued)

TABLE 32.1 (CONTINUED)

Functional Diversity and Influence of Five Major Classes of HSPs and Their Subfamilies on Plant–Symbiont Growth, Development, Performance, Stress Responses, and Symbiosis

Classes	Representative Members	Isolated from/expressed in	Intracellular Localization	Major Functions	Refs
	CsHSP17.7, CsHSP18.1, and CsHSP21.8	*Camellia sinensis*/yeast and *Arabidopsis*	CsHSP17.7 and CsHSP18.1 in cytosol, CsHSP21.8 in ER	*CsHSP17.7, CsHSP18.1,* and *CsHSP21.8* were upregulated in yeast under cold and heat stress and might promote thermotolerance. Overexpression of *Rosa chinensis RcHSP17.8* enhanced yeast viability under cold stress; *LimHSP16.45* enhanced *Arabidopsis* viability under abiotic stresses; rice *OsHSP18.2* protected proteins from denaturation and oxidative damage, leading to improved seed vigor, germination, longevity, and seedling establishment.	(Wang et al. 2017)
	NtHSP18	Tobacco pollen grain/*E. coli*	Cytosol	In a dose-dependent manner, ATP nucleotides suppressed the activity of HSP18 and substrate citrate synthase. Unlike HSP90 and HSP60 function, NtHSP18 protein did not prevent citrate synthase inactivation. However, NtHSP18 existence prior to heat stress protected citrate synthase from irreversible aggregation. It can be hypothesized that under stress conditions, where ATP is reduced, HSP18 functions to mitigate the adverse effects of cold and heat stresses. Reversibly, under normal conditions, normal ATP concentration inhibits the unnecessary activity of HSP18.	(Smykal et al. 2000)
	ThHSP18.3	*Tamarix hispida*/S. cerevisiae (yeast)	Nucleus	Enhances tolerance to multiple stresses, such as, heat, cold, and salinity.	(Gao et al. 2012)
	PtsHSP19.3	*Pyropia tenera*/*Chlamydomonas reinhardtii* and tobacco	Cytoplasm	Improves tolerance to heat, salt, and oxidative stresses.	(Yujin et al. 2017)
	HSP20	*Tigriopus japonicus*/*E. coli*	Mitochondria, but most other compartments in other plants	HSP20, HSP21, and HSP22 are upregulated by heat stress (30 °C), and HSP20 is downregulated by cold stress (4–10 °C); thus, they may impart heat tolerance. However, there is a paucity of knowledge regarding their influence on cold adaptation and symbiosis. HSP20 genes are localized in mitochondria (*T. japonicus*), nucleo-cytoplasm, ER, peroxisome, and mitochondria (tomato and pepper), and plastid (pepper). Wheat: 14 *TaHSP20* family genes are under the regulation of 8 *TamiRNA* genes and are good candidates for improving tolerance to multiple abiotic stresses. Wheat *TaHSP23.7-MTI* is particularly important to enhance heat tolerance. In rice, *OsHSP26.7, OsHSP23.2, OsHSP17.9A, OsHSP17.4,* and *OsHSP16.9A* are promising candidates to improve thermotolerance.	(Lund et al. 1998; Seo et al. 2006; Zhong et al. 2013; Guo et al. 2015; Yu et al. 2016; Muthusamy et al. 2017; Muthusamy et al. 2017)
	HSP21	*Arabidopsis*	Plastid, nucleus	AtHSP21 interacts with the chloroplast thylakoid membrane to prevent cessation of its development under heat stress.	(Waters 2013; Zhong et al. 2013)

2004; Sharma et al. 2013; Kapoor and Roy 2015; Zhang et al. 2015a). The sHSPs interact with HSP40, HSP70, HSP100, and misfolded substrate proteins to facilitate protein solubilization, disaggregation, refolding, and consequently, protein availability and homeostasis, improving stress responses such as heat and cold tolerance, cell viability, and recovery (Smykal et al. 2000; Liberek et al. 2008; Laskowska et al. 2010; Sun et al. 2010; Zwirowski et al. 2017). The chaperone "holdases," although incapable of folding polypeptides *per se*, impart a vital force to prevent protein aggregation by binding to unfolded or misfolded conformations and stabilizing their solubility until they are delivered to the "foldase" complexes (Table 32.1) (Çetinbaş and Shakhnovich 2015; Kapoor and Roy 2015). The HSP90 assists in the maturation of a limited set of substrate proteins, which are collectively referred to as *client proteins* (Wayne et al. 2011). Client proteins will then be released for refolding with the help of other ATP-dependent chaperones when the conditions become optimal (Zhang et al. 2015a). The sHSPs have been observed to exist as large homo-oligomers, and the reversible dissociation of their oligomers has been reported to be important for their enhanced chaperone activity at high temperatures. Such oligomeric dissociation presumably increases the exposure of their hydrophobic regions, which are buried in the oligomers in normal physiological conditions but are needed for binding non-native substrate proteins (Zhang et al. 2015a). A comparison between cold shock proteins (CSPs) and HSPs in arctic and temperate strains of rhizobia showed that heat (up to 46.4 °C, at which protein synthesis was inhibited) induced HSP synthesis with a distinctly higher molecular weight than the CSPs induced at temperatures at or below 0 °C. The CSPs are low molecular weight, ranging from 11.1 to 56.1 kDa in rhizobia (Cloutier et al. 1992) and from 7 to 10 kDa in bacteria (Jiang et al. 1997), and contain nucleic acid binding activity sufficient for their function as RNA chaperones. The cold shock domain (CSD) contains a polynucleotide binding surface consisting of essentially five nucleotides and three aromatic residues (F18, F31, and H33), which allows them to bind selectively to single-stranded (ss)DNA and ssRNA molecules with low sequence specificity and moderate binding affinity so as to subsequently act as an RNA chaperone (Castiglioni et al. 2008; Rennella et al. 2017). Therefore, the proper chaperone function and stress management require balance between intracellular levels of holdases, foldases, and client proteins with particular structural (molecular weight and sequences) and functional attributes, though some sHSP might be able to perform a dual function as both holdases and foldases (Li et al. 2012; Saibil 2013; Kapoor and Roy 2015; Preissler et al. 2015; Zhang et al. 2015a; Sekhar et al. 2016). In addition, protein processing such as phosphorylation, acetylation, and other post-translational modifications may regulate chaperone machinery and affect their functions (Khan et al. 1998; Wang et al. 2012a; Cloutier and Coulombe 2013; Saibil 2013; Nitika and Truman 2017). The functional diversity of such proteins also depends on their cellular localization. Molecular HSPs/chaperones are located in both the cytoplasm and organelles, such as nucleus, mitochondria, chloroplasts, symbiosome,

endoplasmic reticulum (ER) (Table 32.1) (Panter et al. 2000; Wang et al. 2004; Reddy et al. 2016), and peroxisome (Ma et al. 2006). Symbiosis leads to the formation of a new compartment, the so-called *symbiosome*, in the plant cell when bacteroids enter the plant cell by endocytosis and are then are surrounded by the peribacteroid membrane (PBM), originating from the plant plasma membrane (Kudryavtseva et al. 2014). The regulation of specific chaperones has the potential to improve symbiosis and, thereby, plant tolerance and performance (Tables 32.1 through 32.3) (Verma et al. 1986; Panter et al. 2000; Catalano et al. 2004; Castiglioni et al. 2008; Paço et al. 2016; Rennella et al. 2017). For example, the regulation of the symbiosome's protein chaperones, such as HSP60, HSP70, HSP90, and BiPHSP70 proteins (Table 32.1), protein disulfide isomerase (PDI) (Table 32.3), proteases (a serine and a thiol protease), H+-ATPase, and nodulins (e.g., nodulin 23, 24, 25, early nodulin (ENOD)16, and ENOD8) (Table 32.3), plays critical roles in nodule development and symbiosis. However, not all of their physiological roles have been discovered so far (Verma et al. 1986; Panter et al. 2000; Catalano et al. 2004). The additional expression of specific chaperones has been effective in promoting plant and bacterial growth and symbiotic performance (e.g., by ClpB, CspA, and CspB in Tables 32.1 and 32.2) (Castiglioni et al. 2008; Paço et al. 2016; Rennella et al. 2017) and stress responses (e.g., by HSP70, HSP90, and HSP101 in Table 32.1) (McLellan et al. 2007). HSPs can be regulated by genetic manipulation, by the application of HSP inhibitors, or by establishing effective symbiotic relationships. For instance, in *Arabidopsis*, the application of the HSP90 inhibitor monocillin I (MON) isolated from the Sonoran Desert in a dose-dependent manner and symbiosis with the rhizosphere fungus *Paraphaeosphaeria quadriseptata* can suppress HSP90 and transiently upregulate the mRNAs of HSP101 and HSP70, allowing the plant to survive under normally lethal high temperatures (McLellan et al. 2007).

In Rhizobiaceae, plants, and yeast, membrane phospholipids such as phosphatidylcholine, phosphatidylcholine-hydrolyzing phospholipase C, phosphocholine cytidylyltransferase, and the genes and the precursors involved in phosphatidylcholine formation are essential for root and nodule development, increasing the number of bacteroids, hormone (e.g., brassinolide) signaling, and stress adaptation (including heat stress) and establishing effective symbiosis (Minder et al. 2001; Tasseva et al. 2004; Kiewietdejonge et al. 2006; Kocourková et al. 2011). However, not enough information is available to show their influence on cold adaptation. The 14-3-3 proteins, phospholipase Dα (PLDα), and H+-ATPase are implicated in improving cold adaptation (Muzi et al. 2016; Barrero-Sicilia et al. 2017). Jasmonate and salicylate are believed to improve chilling tolerance by inducing sHSPs (Ré et al. 2017). Cold stress inducible PLDα and H+-ATPase mediate the conversion of phosphatidylcholine to phosphatidylethanolamine and, subsequently, phosphatidic acid and lysophospholipid to participate in cold adaptation (Muzi et al. 2016; Barrero-Sicilia et al. 2017). Thus, understanding the relationships between symbiosome proteins, 14-3-3 proteins,

TABLE 32.2
Functional Diversity and Effects of Molecular Chaperones Classified as RBPs on Plant–Symbiont Growth, Development, Performance, Stress Responses, and Symbiosis

Classes	Representative Members	Isolated from/ expressed in	Intracellular Localization	Major Functions	Refs
GR-RBPs	GRP7 and CSDP1	*Arabidopsis/E. coli*	GRP7 is localized in nucleus and cytoplasm; CSDP1 is not reported.	Nodulins, nodulin-like proteins, and GR-RBPs play critical roles in establishing symbiosis even in non-nodulating plant species such as *Arabidopsis*, rice, maize, and poplar. In addition to improving stress responses, nodule specific *GRP* genes are believed to improve infection, nodule development and thereby, symbiosis. Heterologous expression of *Arabidopsis CSDP1* or *GRP7* enhanced *E. coli* tolerance to cold stress (15 °C for 12–24 h) by promoting DNA melting and RNAase activities. While both the C- and N-terminal regions of *CSDP1* were necessary for cold adaptation, the N-terminal region of *GRP7* was sufficient for survival under cold stress. *CSDP1* and *GRP7* are involved in mRNA export and consequently, regulation of proteins involved in stress adaptation.	(Schröder et al. 1997; Kevei et al. 2002; Kim et al. 2007; Kim et al. 2008a; Denance et al. 2014)
	AtRZ-1a, b, c	*Arabidopsis/ Arabidopsis* and *E. coli*	AtRZ-1a is in nucleus and cytoplasm; AtRZ-1b and AtRZ-1c are in nucleus.	GRPs named RZ-1, which contain N-terminal RRM and C-terminal GR domain, have been reported in different plant species including tobacco, *Arabidopsis*, and rice. AtRZ-1a enhances tolerance to cold stress (15 °C for 10 h) in *Arabidopsis* and cold-sensitive BX04 quadruple *E. coli* mutant lacking four CSPs by modulating expression of stress-responsive genes and chaperone activity by affecting the translation of the target genes and not transcription level. AtRZ-1a preferentially and selectively binds to G- or U-rich RNA sequences and also to ssDNA more strongly than dsDNA, while AtRZ-1b and AtRZ-1c proteins non-specifically bind to RNA and DNA sequences. AtRZ-1b enhanced cold tolerance in CspA-deficient *E. coli* mutant cells but not in *Arabidopsis* during seed germination and seedling growth, and its CCHC-zinc finger motif of the C-terminal domain was necessary for complementation ability. However, the N-terminal domain of atRZ-1a was necessary to allow BX04 cells to survive under cold stress. Full-length AtRZ-1c had no complementation ability, while its N- or C-terminal regions partially induced cold adaptation.	(Kim et al. 2005; Kim and Kang 2006b; Kim et al. 2010b)
CSDPs	OsCSP1 and OsCSP2	*Oryza sativa*	Not reported.	OsCSP1 and OsCSP2 encode putative proteins consisting of an N-terminal CSD and GR regions interspersed by four (OsCSP1) and two (OsCSP2) CCHC zinc fingers. However, because of the sensitivity of rice to low temperatures, their roles in cold adaptation remain unknown.	(Chaikam and Karlson 2008)
CSD	CSD	*Arabidopsis/ Arabidopsis* or corn	CSD1 in cytoplasm; CSD2 in chloroplasts; CSD3 in peroxisome.	In *Arabidopsis*, heat stress (37 °C) moderately downregulates the expression of *miR393* while upregulating the expression of *miR169* and *miR398* precursors, particularly *miR398b*, more than *miR398a* and *miR398c*. Expression of heat-inducible *miR398* downregulates the expression of *CSD1, CSD2,* and *CCS* while upregulating four HSF genes (HSFA1e, HSFA2, HSFA3, and HSFA7b) and HSPs (HSP17.6, HSP70B, and HSP90.1), conferring heat tolerance. In addition to heat stress, 5 μM CuSO$_4$ in MS medium upregulates the expression of *miR398b* and *miR398*, leading to heat tolerance.	(Guan et al. 2013a)

(Continued)

TABLE 32.2 (CONTINUED)

Functional Diversity and Effects of Molecular Chaperones Classified as RBPs on Plant–Symbiont Growth, Development, Performance, Stress Responses, and Symbiosis

Classes	Representative Members	Isolated from/expressed in	Intracellular Localization	Major Functions	Refs
	AtCSP2 and AtCSP4	Arabidopsis/Arabidopsis	AtCSP2 is in nucleus and cytoplasm.	Reduced AtCSP2 and no AtCSP4 expression upregulated CBF TFs and enhanced freezing tolerance.	(Sasaki et al. 2013)
	ArCspA	Arthrobacter globiformis S155/Saccharomyces cerevisiae (yeast) or tobacco	Not reported.	Arthrobacter globiformis S155 is a psychrotropic bacterium capable of growing at −5 to 32 °C. ArCspA expression is induced under cold stress. ArCspA overexpression enhances tolerance to cold (long-term exposure to 15 °C) and freezing (−25 °C for 12–24 h) stresses.	Lee et al. (2014)
	AtCSP3	Arabidopsis	Nucleolus and nucleoplasm.	AtRH7-AtCSP3 is mainly in the nucleolus, with a smaller portion in the nucleoplasm, enhancing tolerance to cold and freezing stresses by unwinding duplex RNA and regulating RNA secondary structure, pre-rRNA processing, ribosome biogenesis, and rRNA metabolism.	(Kim et al. 2009a; Liu et al. 2016)
	CspA from E. coli and CspB from Bacillus subtilis	CSPs are 7–10 kDa in size in bacteria.	Probably in nucleolus, nucleoplasm, and cytoplasm, since ribosome biogenesis take place in these organelles.	Expression of CSPs in maize improved plant tolerance to multiple stresses without negative pleiotropic effects and yield drag. CspA (cold shock DEAD-box protein A) acts as both DEAD-box RHs and as a CSP. CsdA is involved in ribosome assembly at low temperature, possibly rearranging the RNA structure of a 50S precursor. CspA-overexpressed plants tested so far include: maize, rice, and Arabidopsis. CspA and CspB are RNA chaperones that are able to bind ssRNA and ssDNA without apparent sequence specificity, thus preventing the formation of secondary structures that prohibit translation at low temperatures.	(Charollais et al. 2004; Castiglioni et al. 2008; Stampfl et al. 2013; Liu et al. 2016; Rennella et al. 2017)
RHs	DEAD-box AtRH7/PRH75	Arabidopsis/Arabidopsis or E. coli	Mainly in nucleolus and to a lesser extent in nucleoplasm.	RHs enhance tolerance to both heat and cold stresses in addition to other stresses and concomitantly promote growth and development processes. Improves tolerance to cold stress (4 °C for 6 weeks). AtCSP3–AtRH7 interaction regulates mRNA and pre-rRNA accumulation, correct refolding of misfolded pre-rRNA, rRNA biogenesis in the nucleus, and suppression of CBF1, CBF2, and CBF3 genes.	(Hunger et al. 2006; Owttrim 2013) (Huang et al. 2016; Liu et al. 2016)
	AtRH25	Arabidopsis/Arabidopsis	Nucleolus and cytoplasm.	Improves tolerance to salt and cold (long-term exposure to 4 °C) and freezing stresses more than AtRH9.	(Kim et al. 2008b)
	DEAD-box RH CshA and CshB	Bacillus subtilis	Around the nucleoid.	RH CshA and CshB and CspB interactions are implicated in B. subtilis cold adaptation by modifying (destabilizing and correctly folding) misfolded mRNAs to initiate translation.	(Hunger et al. 2006)
	STRS1 and STRS2	Arabidopsis	Nucleolus, nucleoplasm, and chromocenters.	STRS1 and STRS2 are negative regulators to multiple stresses by epigenetic silencing of gene expression, RNA-dependent ATPase, and RNA unwinding activities to prevent over-activation of stress responses and thereby, related growth arrest under stress conditions.	(Kant et al. 2007; Khan et al. 2014)
	TOGR1	Rice	Nucleolus.	Heat-inducible Thermotolerant Growth Required1 (TOGR1) improves RNA metabolism, rRNA homeostasis, growth, and tolerance to heat stress.	(Wang et al. 2016a)

(Continued)

TABLE 32.2 (CONTINUED)

Functional Diversity and Effects of Molecular Chaperones Classified as RBPs on Plant–Symbiont Growth, Development, Performance, Stress Responses, and Symbiosis

Classes	Representative Members	Isolated from/ expressed in	Intracellular Localization	Major Functions	Refs
Hfq			Cytoplasm, nucleoid	Hfq binds hundreds of small noncoding RNAs and facilitates their interactions with mRNAs, regulating post-transcriptional gene regulation, translation, and thereby, bacterial stress responses, virulence, nutrient use, metabolism, and symbiosis. Hence, overexpression of Hfq can be used as an effective strategy to promote symbiosis, stress responses, and performance. Deletion of *Hfq* genes can protect plants by preventing the symbiosis between plants and pathogenic bacteria.	(Lee and Feig 2008; Diestra et al. 2009; Barra-Bily et al. 2010; Santiago-Frangos et al. 2016)
Hfq	*Sinorhizobium meliloti* (bacterium)		Cytoplasm, nucleoid.	Hfq and RNase E post-transcriptionally regulate the expression of the bacterial *NifA* gene, thereby regulating bacterial growth, carbon metabolism, colonization, and symbiosis. Hfq is involved in establishing the symbiosis between *Sinorhizobium meliloti* and *Medicago sativa*, nodulation, and promoting redox status.	(An et al. 2007; Zhang and Hong 2009; Barra-Bily et al. 2010; Torres-Quesada et al. 2010)

TABLE 32.3

Functional Diversity and Effects of Other Molecular Chaperones with RNA Chaperone Activities on Plant–Symbiont Growth, Development, Performance, Stress Responses, and Symbiosis

Other Types of RNA Chaperones	Representative Members	Isolated from/ expressed in	Intracellular Localization	Major Functions	Refs
Ribosomal RNA (rRNA)		In most organisms except viruses		rRNAs act as both proteins and RNA chaperones, improving growth, development, and stress responses.	(Ting et al. 2016; Ji et al. 2017)
	AtP3B	Arabidopsis/sweet potato	Nucleolus	AtP3B overexpression enhanced antioxidant activity, photosynthesis, yield, and tolerance to both heat and cold temperatures in sweet potato.	(Ji et al. 2017)
Thioredoxin (TRX)		In all organisms	Cytosol, nucleus, chloroplast, mitochondria, ER, plastid, plasma membrane, peroxisome, apoplast, extracellular matrices, phloem sap	TRX is a 12 kDa redox-active protein, containing a dithiol-disulfide active site that regulates redox status and thereby, many aspects of cellular processes. The chloroplast NADPH-dependent TRX NTRC activates peroxiredoxin (Prx), a new family of peroxidases, to reduce oxidative damage (caused by methyl viologen and *tert*-butyl hydroperoxide) by catalyzing the reduction of 2-Cys Prx, protecting the photosynthetic apparatus. Overexpression of genes encoding TRX proteins enhances tolerance to oxidative, cold, freezing, heat, and salt stresses. TRXs are typically reduced by NADPH via NADPH-thioredoxin reductases (NTRs).	(Knoblach et al. 2003; Machida et al. 2009; Chibani et al. 2011; Chae et al. 2013; Lee et al. 2013; Moon et al. 2015)
	AtNTRC	Arabidopsis, rice, cyanoacteria/Arabidopsis, rice, and E. coli	NTRC is in plastids of photosynthetic tissue (chloroplast), while NTRA and NTRB are mainly in cytosol and mitochondria, respectively	NADPH-dependent thioredoxin reductase type C (NTRC) contains an N-terminal TR domain and a TRX domain at the C-terminus. The TRX C-terminal is essential for the nucleic acid–protein complex formation. NTRC can directly reduce a TRX peroxidase and Prx, whereas NTRA and NTRB require specific TRX proteins to reduce Prx. NTRC overexpression improves tolerance to cold (4 d at 4 °C), freezing (−5 °C for 5 h), and heat stresses due to holdase activities, although it also acts as a disulfide reductase and a foldase chaperone. Cold stress stimulates expression of AtNTRC genes at mRNA level. AtNTRC overexpression protects RNA and DNA from RNase A and metal-catalyzed oxidation damage, respectively. NTRCs promote redox homeostasis by alleviating oxidative damage, protect proteins from aggregation, and promote root growth and lateral root formation.	(Reichheld et al. (2005); Kirchsteiger et al. (2012); Machida et al. (2012); Chae et al. (2013); (Moon et al. 2015)
	CvNTRC	Chlorella vulgaris (alga)/C. vulgaris and E. coli	Chloroplast	Expression of mature CvNTRC protein enhances C. vulgaris and E. coli tolerance to cold and freezing stresses.	(Machida et al. 2009)
	OsTRXh1	Rice/rice	Mainly in cytoplasm, and to a lesser extent in mitochondria, nucleus, extracellular matrix of Nicotiana alata, and phloem sap of rice	OsTRXh1 overexpression enhances tolerance to multiple stresses by promoting redox homeostasis. In mature seeds, TRXh in the nucleus of both aleurone and scutellum cells is implicated in seed development and germination by improving redox status.	(Serrato and Cejudo 2003; Zhang et al. 2011a)

(Continued)

TABLE 32.3 (CONTINUED)

Functional Diversity and Effects of Other Molecular Chaperones with RNA Chaperone Activities on Plant–Symbiont Growth, Development, Performance, Stress Responses, and Symbiosis

Other Types of RNA Chaperones	Representative Members	Isolated from/ expressed in	Intracellular Localization	Major Functions	Refs
	TRXs	Medicago truncatula	ER	TRXs genes are expressed in Medicago truncatula nodules, roots, and leaves, particularly those that are in a symbiotic relationship with Sinorhizobium meliloti and arbuscular Glomus species, regulating nodule development. Besides, an experiment on bacteria verified that TRX is linked with melanin biosynthesis, which subsequently regulates redox homeostasis and protects plants and symbionts from different biotic and abiotic stresses.	(Castro-Sowinski et al. 2007; Alkhalfioui et al. 2008)
	GmTRX	Glycine max/S. cerevisiae (yeast)	Primarily exist in cytosol, but might be transported to the peroxisome to interact with peroxisomal proteins	GmTRXs are expressed in the pericycle of soybean nodules and promote ROS scavenging and nodule development. The functions and regulation of GmTRX protein are mediated by its interaction with nodulin-35 and other peroxisomal proteins, including catalase (AtCAT), transthyretin-like protein 1 (AtTTL1), and acyl-coenzyme A oxidase 4 (AtACX4).	(Lee et al. 2005a; Du et al. 2015)
				Similarly to TRX, glutaredoxins (Grxs) are thiol disulfide oxidoreductase enzymes that act as reductants and disulfide isomerases, regulating redox status and protecting proteins against oxidative stress; however, not enough information is available to show their chaperone activities and their roles in thermotolerance and symbiosis.	(Rodríguez-Manzaneque et al. 1999; Meyer et al. 2009)
TRX-like protein (TDX)				Similarly to TRX, TDX functions as a disulfide reductase, foldase chaperone, and holdase chaperone; however, the role of holdase is predominant in inducing stress tolerance.	(Lee et al. 2009a)
	AtTDX	Arabidopsis/Arabidopsis	Cytosol	AtTDX contains three tetratricopeptide repeat (TPR) domains and a TRX motif. AtTDX function depends on oligomerization status and the activity of TPR domains. TPR domains of AtTDX block the active site of TRX and are necessary for holdase activity. Heat shock shifts the oligomerization status of AtTDX from low– to high–molecular weight complex, thereby switching disulfide reductase and foldase activities to holdase activity to confer thermotolerance and promote poststress recovery.	(Lee et al. 2009a)
	PDI	PDI is abundant in eukaryotes, including plants, animals, yeast, archaea, mammals, and bacteria.	Chloroplast, symbiosome, ER, nucleus, cytosol, mitochondria, Golgi, vacuole, cell wall, plasma membrane, peribacteroid membrane, and extracellular milieu	Protein disulfide isomerase (PDI) is a 57 kDa protein that contains TRX domains. A mammalian study showed that PDIs act as a hormone reservoir and are involved in apoptotic signaling and regulating cell fate. In plants, high PDI/substrate ratios promote holdase chaperone activity of TDX proteins.	(Panter et al. 2000; Catalano et al. 2004; Ding et al. 2008; Lee et al. 2009a; Santos et al. 2009; Cho et al. 2011; Grek and Townsend 2014; Pan et al. 2014)

(Continued)

TABLE 32.3 (CONTINUED)
Functional Diversity and Effects of Other Molecular Chaperones with RNA Chaperone Activities on Plant–Symbiont Growth, Development, Performance, Stress Responses, and Symbiosis

Other Types of RNA Chaperones	Representative Members	Isolated from/ expressed in	Intracellular Localization	Major Functions	Refs
	PDI protein MTH1745	Thermophilic archaea, *Methanothermobacter thermoautotrophicum* ΔH/*E. coli*	Need to be tested.	MTH1745 is an 18 kDa PDI protein containing a single TRX domain isolated from *Methanothermobacter thermoautotrophicum* ΔH, which grows at an optimal temperature of 65 °C, and is upregulated at cold shock (4 °C). MTH1745 allow the organism to survive under extreme temperatures by disulfide reductase and chaperone activities such as folding the citrate synthase after thermodenaturation.	(Ding et al. 2008)
	BdPDI and BdPDIL	*Brachypodium distachyon*	ER	Differentially regulate growth, development, and responses to multiple stresses.	(Zhu et al. 2014)
	Dsb	Gram-negative bacteria	Periplasm, inner membrane, ER	Disulfide bonds (Dsb) are structural motifs (a single TRX-like active site consisting of a pair of cysteines in a CXXC motif) in many bioactive peptides and proteins that act as chaperones, aiding in correct folding and stability of nascent proteins. Due to their chaperone activities, PDI and Dsb proteins have the potential to promote cold adaptation. Future experiments are recommended to evaluate the effectiveness of PDI and Dsb in conferring thermotolerance and responses to stress conditions, particularly oxidative stress, and also establishing symbiosis under different environmental conditions.	(Chen et al. 1999; Kurokawa et al. 2000; Ding et al. 2008; Patil et al. 2015)
Peroxiredoxins (Prxs)	2-Cys Prxs, types I and II and 1Cyc PrxVI are located in cytoplasm; 2Cys PrxIII is in mitochondria; 2Cys peroxiredoxin is in cytoplasm, mitochondria, and peroxisomes; and 2Cys PrxIV is in ER.	Plant, bacteria, yeast, fungi		Prxs are thiol-specific peroxidases that act as both antioxidant enzymes and chaperones. Stimulated by heat shock and oxidative stresses, Prxs protect proteins from denaturation and DNA from damage. Prxs are involved in H_2O_2 sensing–signaling pathways by interacting with PDI and oxidizing TRX proteins, thereby regulating photosynthesis and respiration, metabolism, seed dormancy, and development processes. Although they are good candidates for promoting cold adaptation and symbiosis, not enough studies have been so far conducted to demonstrate their effectiveness in this respect.	(Kim et al. 2009b; Dietz 2011; Kim et al. 2011b; Brown et al. 2013; Breitenbach et al. 2015; Lee et al. 2015; Mir et al. 2015; Netto and Antunes 2016)

(Continued)

TABLE 32.3 (CONTINUED)

Functional Diversity and Effects of Other Molecular Chaperones with RNA Chaperone Activities on Plant–Symbiont Growth, Development, Performance, Stress Responses, and Symbiosis

Other Types of RNA Chaperones	Representative Members	Isolated from/ expressed in	Intracellular Localization	Major Functions	Refs
Immunophilins		In all organisms	Chloroplast, ER, mitochondria, nucleus, nucleolus, periplasm	Immunophilins are molecular chaperones including cyclophilins and intercellular proteins termed FK506-binding proteins (FKBPs), some of which possess peptidyl-prolyl *cis/trans* isomerase (PPI) activities that inhibit protein aggregation, promote protein folding, regulate protein–protein and DNA/RNA–protein interactions and mRNA remodeling, and thereby, regulate gene expression, growth, development, and stress responses. PPI can be induced by heat stress and return to its normal level after the recovery. Hence, PPI has the potential to protect photosynthetic machinery under stress conditions and confer tolerance to both cold and heat stresses. Nonetheless, more studies are required to demonstrate the roles of immunophilins and PPI in thermotolerance and symbiosis.	(Kurek et al. 1999; Ou et al. 2001; He et al. 2004; Ingelsson et al. 2009; Monneau et al. 2013; Thapar 2015)
Calnexin and calreticulin		Eukaryotes and some types of yeast	ER	Calnexin and calreticulin are glucose binding lectins that act together as co-chaperones. Calnexin and calreticulin recognize unfolded, misfolded, or incompletely assembled polypeptide portions of glycoproteins that pass through the ER to prevent aggregation and provide assistance with degradation of misfolded proteins, correct folding, and subunit assembly. Calnexin and calreticulin might be induced or suppressed under stress conditions and subsequently interact with other HSPs, particularly HSP70 and HSP90, to regulate growth and stress responses; however, more studies are required to demonstrate their effectiveness in thermotolerance and symbiosis.	(Basu et al. 2001; Leach and Williams 2003; Rizvi et al. 2004; Gupta and Tuteja 2011; Nouri et al. 2012)
CCS	AtCCS (CSD1 and CSD2))	*Arabidopsis*	Cytosol (CSD1), chloroplasts (CSD2)	Copper/zinc chaperone for superoxide dismutase gene (CCS) regulated by miR398; allows the optimal use of Cu for photosynthesis by encoding both CSD1 and CSD2.	(Abdel-Ghany et al. 2005; Beauclair et al. 2010; Guan et al. 2013a)
Histone chaperones	*OsNAPL6* (H3/ H4 histone)	Rice/rice	Nucleus	*OsNAPL6* histone chaperone is involved in assembling the nucleosome-like structure and regulating gene expression, DNA repair, and abrogation of programmed cell death, thereby protecting plants against multiple stresses. OsNAPL6-Ox transgenic plants exhibited improved seed yield under multiple stress conditions.	(Tripathi et al. 2016)
	AtASF1A/B	*Arabidopsis/Arabidopsis*	Nucleus, cytoplasm	Expression of *AtASF1A/B* (*A. thaliana* anti-silencing function1 A/B) genes facilitates H3K56 acetylation and regulates the expression of some *HSF* and *HSP* genes, promoting growth, development, and tolerance to heat stress. Their influence on symbiosis and nodule development has not been studied yet.	(Zhu et al. 2011b; Weng et al. 2014)

PLDα, and phospholipids with chaperones provide valuable information to improve their influence on symbiosis, growth, development, and thermotolerance.

Different classes of molecular chaperones appear to bind to specific non-native substrates to determine their conformational states. Chaperone proteins do not covalently bind to their targets and do not form part of the final product (Wang et al. 2004). Chaperonins, including HSP60 (Chaperonin [Cpn] 60 or GroEL), HSP10 (Cpn10 or GroES), and TCP-1 (the chaperonin containing t-complex polypeptide 1), are ATP-dependent chaperones involved in protein translocation (e.g., between endosymbiotic bacteroids into peribacteroid space), protection against denaturation, correct folding, refolding, and subsequent assembly and stabilization of some proteins (Viitanen et al. 1990; Walters et al. 2002). Accordingly, due to the structural and functional diversity of such proteins (i.e., RBPs, HSPs, and chaperones), it is highly recommended to evaluate their optimal performance under particular protein balance and symbiotic relationship, plant, and environmental conditions.

After the perception of environmental cues, calcium sensor proteins regulate the expression of HSFs and, thereby, HSPs (Liu et al. 2005) and nodulin genes (Liu et al. 2006). Hence, regarding the influence of stress-induced redox, calcium, and hormone signaling pathways, which can be also mediated by calcium sensor proteins (Zeng et al. 2015), understanding their interactions and regulatory effects on the expression of HSPs, chaperones, and symbiotic genes provides valuable information to improve thermotolerance and symbiosis. Regulation of calcium sensor proteins, HSFs, and HSPs can be applied to promote symbiosis, e.g., due to the co-regulation of HSPs and symbiotic genes (Fischer et al. 1993; McLellan et al. 2007), protecting plants against not only abiotic but also biotic stresses by exerting their toxicity effects, e.g., by HSP70 inhibitors, on plant pathogens (McLellan et al. 2007). More investigations regarding the influence of HSPs, particularly in the symbiosome, on symbiosis and stress responses are required to provide further information to improve plant–symbiont thermotolerance and performance.

32.3 CLASSIFICATIONS, FUNCTIONAL DIVERSITY, AND REGULATORY ROLES OF RBPS AND OTHER MOLECULAR CHAPERONES DURING ADAPTATION TO STRESSES

32.3.1 RBPs

Proteins that bind to DNA or RNA—termed—*DNA-* and *RNA-binding proteins* (DRBPs)—play critical regulatory roles in cellular processes, including transcription, translation, gene silencing, microRNA (miRNA) biogenesis, and telomere maintenance (Hudson and Ortlund 2014). RNA-binding proteins (RBPs) are characterized by the presence of several conserved motifs and domains, including the RNA-recognition motif (RRM), GR domain, K homology domain, RGG-box, and zinc-finger motif (Kim et al. 2010a). The RBPs function as the key post-transcriptional modulators of gene expression and also, post-translational modifiers of proteins to control a wide range of cellular processes under normal and stressful conditions, such as heat and cold stresses. RBPs bind to different types of RNAs encoding the components of the mitogen-activated protein kinases (MAPK) and negatively regulate MAPK signaling by stabilizing MAPK phosphatase mRNA in the cytosol to regulate gene expression and diverse biological processes, which in turn, regulate RBPs and their target RNAs (Kondoh et al. 2007; Glisovic et al. 2008; Sugiura et al. 2011; Kang et al. 2013; Brown et al. 2015; Satoh et al. 2017; Tsanov and Daley 2017). RBPs' regulatory pathways modify protein processes, for example, by regulating the methylation, glycosylation, and phosphorylation of RNA chaperones, and RNA processing and metabolism, and through the interaction with precursor microRNA (pre-miRNA) proteins as well as the interaction with chaperone proteins (e.g., HSP70–HSP40, mitochondrial HSP70 [mtHSP70]–HSP10, HSP60–mtHSP70, and HSP70–client proteins), ubiquitin ligase CHIP (carboxy-terminus of Hsc70 interacting protein), and other proteins (Glisovic et al. 2008; He et al. 2008; Rajan and D'Silva 2009; Schlecht et al. 2011; Lee 2012; Kang et al. 2013; Nawrot et al. 2013; Bottinger et al. 2015; Kao et al. 2015; Zhang et al. 2015b; Alves and Goldenberg 2016; Treiber et al. 2017).

The RBPs are highly induced under cold stress and display RNA chaperone activity by their nucleic acid–binding property for disrupting the over-stabilized secondary structures in mRNAs and efficient translation, post-transcriptional regulation, and consequently, cold adaptation (Burd and Dreyfuss 1994; Gendra et al. 2004; Nomata et al. 2004; Bocca et al. 2005; Kim et al. 2005; Arluison et al. 2007; Castiglioni et al. 2008; Lee et al. 2009b; Park et al. 2010; Kang et al. 2013; Nawrot et al. 2013; Ciuzan et al. 2015). Among the plant RBPs, glycine-rich RNA-binding proteins (GR-RBPs), cold shock domain proteins (CSDPs), and RNA helicases (RHs) have been determined to function as RNA chaperones under various stress conditions, such as drought, salinity, and cold stresses. However, unlike RBPs (Table 32.2), which often remain bound to diverse types of RNA molecules to preserve their structures and functions, RNA chaperones (Table 32.3) are no longer needed once RNA molecules have been folded into their native conformations. RNA chaperones are proteins that bind transiently and non-specifically to RNA to resolve kinetically trapped, misfolded conformers by disrupting RNA–RNA interactions and by the loosening of RNA structures (Rajkowitsch et al. 2007; Castiglioni et al. 2008; Kang et al. 2013; Marchese et al. 2016). Consequently, RBPs regulate cellular processes by protein–RNA interactions and RNA chaperone activities (Castiglioni et al. 2008; Kang et al. 2013; Marchese et al. 2016), and thereafter, RNA chaperones regulate gene expression at transcription, post-transcription, and post-translation levels to control plant cell processes, including growth, development, and stress responses (Arluison et al. 2007; Cloutier and Coulombe 2013; Sasaki et al. 2015a).

32.3.1.1 Glycine-Rich RNA-Binding Proteins (GR-RBPs)

GR-RBPs are a highly conserved family of proteins involved in seed germination, growth, development, and stress responses, particularly cold adaptation, by post-transcriptional regulation of gene expression (Kim et al. 2005), DNA- and RNA-melting abilities (Kim et al. 2010a), preventing the formation of secondary structures in mRNAs (Lee et al. 2009b), and aiding mRNA export from the nucleus to the cytoplasm under various stress conditions (Kim et al. 2010a). RBPs can be classified based on the presence of additional motifs and a glycine-rich domain arranged in (Gly)n-X repeats, which are highly flexible and may act in protein–protein interactions (Nawrot et al. 2013). GR-RBPs have the ability to bind to RNA domains by using one or more RRMs, dsRNA or ssRNA binding domains, zinc finger, and other domains; however, CSDPs can bind to ssDNA and ssRNA (Lunde et al. 2007; Castiglioni et al. 2008; Masliah et al. 2013; Ciuzan et al. 2015; Rennella et al. 2017). The CSDPs do not recognize specific RNAs but bind preferentially to GR-RNA sequences to assist in DNA melting activity and enhance RNase activity under cold stress conditions (Kim et al. 2007). In plants and bacteria, CSPs that contain CSDs require functional (i) RNA-binding motifs, which facilitate nucleic acid recognition and binding, e.g., in *Escherichia coli* and maize, or (ii) RRM, e.g., in cyanobacteria lacking any CSDPs, for cold adaptation processes (Kim et al. 2005; Castiglioni et al. 2008; Kang et al. 2013). Similar to the Y-box proteins (Wolffe 1994), CSDPs are involved in the regulation of gene expression at the transcription or translation level (Kim et al. 2007). However, unlike plant CDPs that contain a single N-terminal CSD and a large glycine-rich region interspersed with CCHC zinc fingers (two or four in monocots; one, two, four, five, or seven in dicots) at the C-terminus (Kingsley and Palis 1994; Karlson et al. 2002; Karlson and Imai 2003; Nakaminami et al. 2006; Sasaki et al. 2007), all vertebrate Y-box proteins contain various auxiliary C- and N-terminal domains and a highly conserved CSD (Wistow 1990; Wolffe et al. 1992; Wolffe 1994; Graumann and Marahiel 1998), the structure of which affects their roles.

32.3.1.2 Cold Shock Domain Proteins (CSDPs)

The CSPs have been identified in a variety of organisms, from bacteria to mammals, and function as RNA chaperones (e.g., in plants, bacteria, yeasts, fungi, etc.). Ectopic expression of the CSPs has been successful in improving tolerance to different stresses, particularly cold stress, and modulating an organism's growth and development processes by altering gene expression and metabolic pathways under normal and stressful conditions (Table 32.2) (Sasaki et al. 2007; Kim et al. 2009a; Fang and St Leger 2010; Kang et al. 2013; Lee et al. 2014; Sasaki et al. 2015a; Sasaki et al. 2015b). The CSDPs contribute to the regulation of stress responses in different ways. For example, *Arabidopsis AtCSP2* is upregulated under stress conditions, such as cold and salinity, and negatively regulates seed germination and cold and salt stress responses by downregulating *C-repeat binding factor* (*CBF*) and *cold-regulated* (*COR*) genes (Table 32.2) (Sasaki et al. 2013; Sasaki

et al. 2015a; Sasaki et al. 2015b). The CSPs may confer tolerance to cold stress by destabilizing the over-stabilized secondary structures in mRNAs and, thereby, facilitating translation at low temperatures (Kim et al. 2010b). In *Arabidopsis*, *AtCSP2*-overexpressed seeds exhibited delayed germination as an escape strategy under salt stress and germinated after stress removal, while more than 60% of the wild-type seeds germinated in the same conditions. The delayed germination of *AtCSP2* seeds occurred by increasing the expression of an abscisic acid (ABA) catabolic gene (*CYP707A2*) and reducing gibberellin (GA) biosynthesis genes (*GA20ox* and *GA3ox*) (Sasaki et al. 2015a). In a different way, the *Arabidopsis* double mutant with reduced *AtCSP2* and no *AtCSP4* expression upregulated CBF TFs and, thereby, enhanced tolerance to freezing temperature (Sasaki et al. 2013). The downregulation of *Arabidopsis AtCSP2* (Sasaki et al. 2007) and the overexpression of *Arthrobacter* sp. *ArCspA* (Lee et al. 2014) and *Arabidopsis AtCSP3* (Kim et al. 2009a) genes positively improved tolerance to freezing temperatures. Kim et al.'s (2009) experiment suggested that *AtCSP3* is involved in constitutive and cold-induced tolerance against freezing temperature independently of the CBF/DREB1 pathway (Table 32.2) (Kim et al. 2009a).

The GR-RBPs contain one or more RRMs as binding domains at the conserved protein amino (N or NH$_2$)-terminal region and a variety of auxiliary motifs at the carboxy (C)-terminal region in the GR domain. The CSDPs, however, contain only a single RRM at the N-terminus GR CSD. Other RBPs may have different C-terminus domains, such as arginine-rich, cysteine-rich, short repeat (SR), repeat domain alone (RD), and acidic domains. These domains are important for the RNA chaperone activity of RBPs to regulate pre-mRNA processing (i.e., pre-mRNA capping, polyadenylation, splicing, mRNA transport, mRNA stability and degradation, and translation of the functional mRNA) and protein–protein interactions to regulate stress responses such as cold adaptation processes (Burd and Dreyfuss 1994; Gendra et al. 2004; Nomata et al. 2004; Bocca et al. 2005; Kim et al. 2005, 2010a; Arluison et al. 2007; Castiglioni et al. 2008; Lee et al. 2009b; Park et al. 2010; Kang et al. 2013; Nawrot et al. 2013; Ciuzan et al. 2015).

The GR-RBPs with C-terminal domains are comprised of CCHC-type zinc fingers interspersed in the glycine-rich domain interrupted mostly by arginine or aromatic amino acid residues and contain arginine–glycine-rich motifs (Bocca et al. 2005; Kim et al. 2010a; Mangeon et al. 2010). The N-terminal RRM is important for nucleic acid binding, while the overall folding and disordered flexible C-terminal GR region of RBPs and CSDPs are required for full RNA chaperone activity by sequence-specific binding as well as interaction with other proteins or ligand molecules (Kim et al. 2007; Kang et al. 2013).

Overexpression of the *AtRZ-1a* gene (*A. thaliana zinc finger-containing GR-RBP*), which can be upregulated by cold stress and downregulated by drought stress or ABA, had a positive role in improving *Arabidopsis* seed germination, seedling growth, and freezing tolerance (Kim et al. 2005).

It is suggested that the RBPs confer cold stress adaptation by regulating genes and proteins involved in adaptation processes, e.g., malate oxidoreductase (At2g13560), betaine aldehyde dehydrogenase (At1g74920), the key proline biosynthetic enzyme deltapyrroline-5-carboxylate synthase (At3g55610), HSPs (At1g79930), poly(A)-binding protein (At3g16830), and 20S proteasome subunit (At1g21720) modulated by *atRZ-1a* in *Arabidopsis* (Kim et al. 2005).

32.3.1.3 RNA Helicase Chaperones (RHs)

Helicases are enzymes involved in almost all aspects of RNA and DNA metabolism and gene regulation and, thus, play critical roles in improving plant growth, development, and responses to various stresses (Table 32.2) (Fairman-Williams et al. 2010; Huang et al. 2016; Liu et al. 2016; Shivakumara et al. 2017). Helicases are grouped into four superfamilies based on their amino acid sequences and functions: helicase superfamily (SF)1 and SF2 are prevalent, while SF3 and SF4 have been reported in viruses (Fairman-Williams et al. 2010). The RNA helicases (RHs) are a diverse group of RNA-dependent ATPases that contribute to numerous RNA metabolic processes, such as ribosomal RNA (rRNA) biogenesis, RNA stability, sensing dsRNA and cytoplasmic DNA, RNA unwinding, pre-mRNA splicing, RNA remodeling, export, translation, and degradation, and miRNA expression and, thus, regulate all aspects of RNA metabolism (Table 32.2) (Jarmoskaite and Russell 2011; Zhang et al. 2011b, 2015c; Wang et al. 2012b; König et al. 2013; Huang et al. 2016). Most RHs belong to the SF2, characterized by ATP-dependent translocation on the nucleic acid substrate and/or induction of RNA conformational changes and further sub-grouped as DEAD-, DEAH- and DExH-box helicases (Table 32.2) (Jarmoskaite and Russell 2011; König et al. 2013). Many DEAD-box RHs are multifunctional ATPase proteins involved in diverse RNA metabolism, ranging from RNA synthesis to RNA degradation. DEAD-box RHs use the energy from ATP hydrolysis to unwind short duplex RNA in an unusual process that involves little or no translocation and remodel RNA–protein complexes, but they can also function as RNA clamps to provide nucleation centers that establish larger RNA–protein complexes or even form tight compaction of ssRNA. This activity, coupled with mechanisms to direct different DEAD-box proteins to their physiological substrates, allows them to promote RNA folding steps and rearrangements and to accelerate the remodeling of RNA–protein or protein–protein complexes in ATP-dependent or ATP-independent reactions (Table 32.2) (Jarmoskaite and Russell 2011; Linder and Jankowsky 2011; Zhao and Jain 2011; Kang et al. 2013). The DEAD-box RHs regulate the expression of stress-responsive transcriptional activators and function in both ABA-dependent and ABA-independent manners, targeting different genes depending on the stress conditions (Kant et al. 2007). The *STRESS RESPONSE SUPPRESSOR1* (*STRS1*; At1g31970) and *STRS2* (At5g08620) genes encode proteins that are members of the large family of approximately 50 *Arabidopsis* DEAD-box RHs, acting as negative regulators of the stress-responsive genes by RNA-directed DNA methylation–mediated epigenetic silencing of gene expression under various stress conditions (Table 32.2) (Kant et al. 2007; Barak et al. 2014). Thus, subversion of *STRS1* and *STRS2* genes by means of T-DNA insertion enhanced the tolerance of *strs* mutant seedlings to salt and osmotic stresses and basal and acquired heat stresses in ABA-dependent and ABA-independent manners (Table 32.2) (Kant et al. 2007). Studies on the DEAD-box RHs demonstrated that the C-terminal domain is important in non-specific binding of the RNA chaperones to large RNA substrates, while the intrinsically disordered flexible region of the Hfq (host factor Qβ RNA phage) and virus-encoded RNA chaperones are important for the RNA chaperone activity (Table 32.2) (Kang et al. 2013).

According to Huang et al. (2016), expression of the DEAD-box *AtRH7/PRH75* gene in *Arabidopsis* plants enhanced mRNA and pre-rRNA accumulation and rRNA biogenesis, suppressed the expression of *C-repeat binding factor (CBF)1*, *CBF2*, and *CBF3* under cold stress conditions, and improved cell division, plant development, and cold adaptation (Table 32.2). Since *AtRH7* was asymmetrically expressed in the heart, torpedo, and cotyledon stages during *Arabidopsis* embryogenesis and was essential for pre-18S rRNA processing during globular stage development, Huang and coworkers proposed that AtRH7 might be involved in the regulation of cell differentiation during early embryogenesis or mitotic cell division during globular stage development, which requires future investigation (Table 32.2) (Huang et al. 2016). Kim et al.'s (2008) experiment demonstrated that expression of *AtRH25* enhanced *Arabidopsis* seed germination and tolerance to both salt and freezing stresses more than *AtRH9*, probably due to the differences in their binding properties (Kim et al. 2008b). Another study showed that expression of the *regulator CBF gene expression1* (*RCF1*) was essential for encoding DEAD (Asp-Glu-Ala-Asp)-box RH and regulation of the cold-responsive genes and pre-mRNA splicing and cold tolerance (Table 32.2) (Guan et al. 2013b). Thus, different RHs might have different regulatory effects on gene expression and consequently, stress responses. CsdA is one of the five *E. coli* DEAD-box RHs and as a CSP, provides assistance in RNA remodeling and chaperone activities to promote resistance against low-temperature stresses (Table 32.2) (Stampfl et al. 2013). At 12 °C, *Arabidopsis* growth, development, and tolerance to cold stress were dependent on AtRH7–AtCSP3 interactions (Liu et al. 2016). Similar to AtRH7–AtCSP3 interaction in *Arabidopsis* (Table 32.2) (Liu et al. 2016), the interactions between cold shock helicase-like protein (Csh)A, CshB, and CshC with CSPs are essential for regulating growth, development, and tolerance to cold stress, as reported in *Bacillus cereus* (Hunger et al. 2006; Owttrim 2013). The RHs can also enhance tolerance to hot temperatures. For instance, *TOGR1* (*Thermotolerant Growth Required1*) RH expression effectively improved RNA metabolism, rRNA homeostasis, and rice growth under high temperatures (Table 32.2) (Wang et al. 2016a). In addition to chaperone activities, the RHs such as DEAD helicase are involved in mRNA translation and stabilization under low temperatures and small RNA (sRNA)-mRNA duplex degradation, which is under the influence of Hfq interaction with RNase E, and therefore, affect growth, development, and tolerance to extreme temperatures (Owttrim 2013). Besides the interactions between positive and negative regulators, the expression patterns of stress-responsive genes encoding metabolic reprogramming involved in the regulation of stress responses can be transiently affected by circadian rhythms and different stress conditions due to differential interaction responses of sub-regulatory networks of different stresses (Kant et al. 2007). Overall, RHs can be used as a promising tool for improving plant tolerance to multiple stresses accompanied by optimized productivity (Chen et al. 2015; Wang et al. 2016a). Since the functions of nucleic binding proteins such as RHs and CSPs depend on their interactions (Huang et al. 2016; Liu et al. 2016) and binding properties (Kim et al. 2008b), understanding their signaling and regulatory networks, sequences, binding properties, and co-expression effects related to their gene expression and metabolic reprogramming will provide further insights into developing tolerant crops with improved performance under both normal and stress conditions.

32.3.2 Other Molecular Chaperones

Eukaryotic ribosomes are composed of rRNAs and ribosomal proteins (r-proteins). The rRNAs and r-proteins play fundamental roles in cellular processes (i.e., growth, development, and stress responses), acting as both proteins and RNA chaperones (Table 32.3) (Ting et al. 2016; Ji et al. 2017). The r-proteins are translated in the cytoplasm and then imported into the nucleus for assembly with the rRNAs. The r-proteins interact with RNA, DNA, or under particular conditions, with other proteins to regulate cellular processes (Kovacs et al. 2009; Kothe 2010). Recently, Ji and coworkers showed that overexpression of r-proteins can be used as an effective tool to improve both plant performance and tolerance to extreme temperatures (Ji et al. 2017). They showed that sweet potato (*Ipomoea batatas* [L.] Lam) overexpressing the *Arabidopsis* ribosomal P3 (*AtP3B*) gene had greater antioxidant activity, photosynthesis, and tolerance to both heat and cold temperatures and, consequently, improved yield and storage ability under stress conditions (Table 32.3) (Ji et al. 2017).

Thioredoxin (TRX) (Jee et al. 2005; Vieira Dos Santos and Rey 2006; Svensson and Larsson 2007; Lee et al. 2013; Zhu et al. 2014), TRX-like protein (TDX) (Lee et al. 2009a), NADPH-TRX reductase (Chae et al. 2013), glucose-regulated protein (GRP94), peptidyl-prolyl isomerases (PPI) or immunophilins (Gupta and Tuteja 2011), histone chaperones (Klopf et al. 2009; Weng et al. 2014; Probst and Mittelsten Scheid 2015; Tripathi et al. 2016), PDI (Gupta and Tuteja 2011; Wu et al. 2012; Zhu et al. 2014), and molecular chaperone binding proteins, such as calnexin and calreticulin proteins (Wang et al. 2004; Ma et al. 2016), are other molecular chaperones that participate in cellular processes, such as growth, development, and stress responses (Table 32.3). In this context, their transcriptional regulation, for example, by NADPH-TRX reductase-type C overexpression, can be regarded as a promising tool to improve plant and symbiont growth, development,

yield, tolerance, and symbiosis (Lee et al. 2005a; Alkhalfioui et al. 2008; Chae et al. 2013) (Table 32.3). Histone chaperones' activities induce the expression of genes at transcription levels and are thus involved in the regulation of protein homeostasis and responses to multiple stresses. The regulation of histone chaperones has the potential to promote plant stress responses by regulating heritable and non-heritable chromatin and epigenetic-based stress memory. Histone chaperones are grouped into seven major families: CAF1 (Chromatin assembly factor I), FACT (Facilitates chromatin transcription), ASF1 (Antisilencing factor 1), HIRA (Histone regulatory homolog A), NASP (Nuclear Autoantigenic Sperm Protein), SPT6 (Suppressor of Ty element 6), and NAP (Nucleosome Assembly Protein) (Klopf et al. 2009; Weng et al. 2014; Probst and Mittelsten Scheid 2015; Tripathi et al. 2016) (Table 32.3). However, knowledge about the structural, functional diversity, and regulatory roles of these RNA chaperones, particularly in symbiosis, is still limited and requires further investigations for crop improvement.

32.4 CONTRIBUTING FACTORS RESPONSIBLE FOR THE FUNCTIONAL DIVERSITY OF RBPS AND MOLECULAR CHAPERONES DURING ADAPTATION TO STRESSES

The RBPs control the cellular RNA chaperone activities and are differentially regulated at transcriptional and posttranscriptional levels depending on the cell, tissue and organ type, developmental stage, and environmental stimuli (Kang et al. 2013) via transcription initiation RNA (tiRNA) upregulation, the differential regulation of miRNA expression, and a fine-tuned layer of post-transcriptional miRNA modifications (Mesitov et al. 2017). The roles of RNA chaperones are more prominent when cells are exposed to low temperatures, because misfolded RNA molecules become over-stabilized and cannot assume a native conformation without the help of RNA chaperones (Kang et al. 2013). Additionally, the functional diversity and stress responses of molecular chaperones are highly dependent on their isoforms and intracellular locations (Song et al. 2009).

Depending on sequences and structures, the C- and N-terminal domains of chaperones differentially affect their functions and, consequently, growth and stress responses. The number and length of the zinc finger C-terminal GR domains in CSPs and GR-RBPs are crucial to the full activity of the RNA chaperones; ssDNA, double-stranded (ds)DNA, or mRNA binding activities; as well as RNA and DNA melting activities and, consequently, regulation of gene expression and cold adaptation (Nomata et al. 2004; Kim et al. 2007; Lee et al. 2009b; Kang et al. 2013). For example, GRP4 with a shorter C-terminus GR region compared with GRP7 had no RNA and DNA melting activities and had no influence on cold resistance (Kim et al. 2007). Furthermore, the C-terminal region of CSDP1 containing seven zinc fingers was more effective in stimulating growth and improving cold adaptation as compared with the N-terminal region of CSDP1 lacking zinc fingers or the C-terminal region of CSDP2, which contains two zinc fingers (Table 32.3) (Kim et al. 2007). In another study, overexpression of the N-terminal half domain (from amino acids 1 to 87) of atRZ-1a in the cold-sensitive BX04 quadruple *E. coli* mutant, in which four CSP genes (*cspA*, *cspB*, *cspE*, and *cspG*) were deleted, was more effective in improving cold adaptation as compared with the BX04 cells overexpressing the C-terminal half (from amino acids 88 to 246) (Kim and Kang 2006a). Ectopic expressions of bacterial CSPs have been successful in improving tolerance to cold stresses. *Arabidopsis AtCSP2* is an RNA chaperone involved in cold adaptation as well as the developmental process (Kang et al. 2013).

32.5 REGULATION OF HSPs AND MOLECULAR CHAPERONES

As mentioned earlier, many plants have evolved different defense mechanisms by regulating TFs that target specific sets of genes to subsequently control cellular processes at a wide range from molecular to anatomical levels (Cho and Hong 2006; Klopf et al. 2009; de Nadal et al. 2011; Basu et al. 2016; Woodson 2016). Among these TFs, HSFs function as the main regulators of the gene expression involved in HSP/chaperone induction. The HSFs are a family of winged helix–turn–helix (wHTH) DNA-binding proteins, grouped into three classes: A, B, and C (Littlefield and Nelson 1999; Maaroufi and Tanguay 2013; Gomez-Pastor et al. 2017; Jacob et al. 2017). The HSFs can be found in inactive form in the cytoplasm of stressed and unstressed cells (Morimoto 1998; Scharf et al. 1998; Kotak et al. 2004) and translocate into the nucleus on activation by stress-induced signaling pathways such as redox signaling (Giesguth et al. 2015). Regulation of HSF gene expression occurs at transcriptional and post-translational levels to control HSP/chaperone expression (Hietakangas et al. 2006; Swindell et al. 2007; Cohen-Peer et al. 2010; Gomez-Pastor et al. 2017). The expression of genes encoding HSFs is regulated by thermal and non-thermal stress-induced signaling pathways (Swindell et al. 2007) such as redox signaling (Jacquier-Sarlin and Polla 1996; Miller and Mittler 2006; Giesguth et al. 2015), calcium signaling (Mosser et al. 1990), and protein processing regulatory pathways such as ubiquitination (Parsell and Lindquist 1993; Mathew et al. 1998; Morimoto 1998), phosphorylation (Hietakangas et al. 2003), the sumoylation pathway (Cohen-Peer et al. 2010), and interactions between proteins such as TFs (Scharf et al. 1998). After activation, HSFs subsequently bind to heat shock promoter elements (HSEs) in the promoters of target genes and translocate from the cytosol to the nucleus to control the stress response (Morimoto 1998; Giesguth et al. 2015). The binding of HSFs to DNA can be promoted by a number of different factors, such as salicylic acid (SA) (Snyman and Cronjé 2008). The application of exogenous SA to both control and heat-stressed 3-week-old tomato seedlings enhanced HSF–DNA binding and markedly potentiated the upregulation of HsfA1 and HsfB1. Subsequently, in SA-treated heat-shock seedlings, the upregulation of HsfA1 and HsfB1 was followed by the upregulation of HSP70; however, the induction of

HSP70 levels by SA was insignificant in the control seedlings (Snyman and Cronjé 2008).

Intercellular redox status is another critical factor that regulates cell signaling, physiological processes, stress programs (cell death, growth, development, growth inhibition, and stress defense mechanisms) (García-Mata and Lamattina 2001; Forman et al. 2004; Jakab et al. 2005; Scandalios 2005; del Río et al. 2006; Kotchoni et al. 2006; Hao and Zhang 2009; Tanou et al. 2012; Diao et al. 2016; Li 2016; Tavladoraki et al. 2016; Woodson 2016; Li et al. 2017a; Nahar et al. 2017; Pourghayoumi et al. 2017), protein functions, and thereby, symbiotic performance (Oger et al. 2012). For example, cellular hydrogen peroxide (H_2O_2)-induced oxidative stress interacts with protein cysteine residues, forming a sulfenic acid (-SOH), in a process known as *sulfenylation*. Consequently, sulfenylation regulates various cell biological processes, including enzyme activity, metal binding, protein turnover, interactions and functions, sensing and signaling pathways, the regulation of expression of some TFs and related genes, and metabolic pathways (Oger et al. 2012). In addition to the influence of reactive oxygen species (ROS) on symbiosis at early symbiotic steps, the optimal H_2O_2-mediated sulfenylation promotes symbiotic performance by modulating amino-acid and carbohydrate metabolism in nodules and, thereby, growth and development of the symbiotic root organ and nodule (Oger et al. 2012). For instance, *Medicago truncatula– Sinorhizobium meliloti* symbiosis was affected by the sulfenylation of the nitrogenase iron protein, component II (nifH), and the nitrogenase molybdenum-iron protein beta chain (nifK), antioxidant enzymes regulating redox status (i.e., peroxiredoxins, class III peroxidases), proteins involved in defense mechanisms (e.g., lipoxygenase, chitinase, allene oxide synthase, and pathogenesis-related protein), cell wall enzymes (acidic glucanase, glucan-endo 1,3-beta-glucosidase, pectin methylesterase), and the bacterial HtpG (high-temperature protein G from HSP90 protein family), proteins involved in carbohydrate (glycolytic pathway and the tricarboxylic acid cycle, such as glyceraldehyde 3-phosphate dehydrogenase, malate dehydrogenase, and sucrose synthase) and amino-acid metabolism (glutamine synthetase and NADH-dependent glutamate synthase, aspartate amino transferase, and asparagine synthetase), early nodulin 8 (ENOD8), and HSP70 (Oger et al. 2012). The sulfenylation of proteins involved in calcium chelation, such as calreticulin and calnexin, may affect plant development and responses to stresses as well as symbiosis via control of the cell calcium signaling (Oger et al. 2012).

Another important group of TFs that is involved in the regulation of gene expression during cold acclimation is the CBF TFs, which control the expression patterns of *COR* genes. The promoters of many of these genes contain one or several copies of the dehydration-responsive element (DRE)/C-repeat *cis*-element, which has the core sequence CCGAC. The CBFs, also known as DRE binding proteins, are upstream transcription factors in the APETALA2/Ethylene Response Factor (AP2/ERF) family that bind to the promoter *cis*-element and activate the expression of these cold-responsive genes (Jaglo-Ottosen et al. 1998). Regulation of AP2/ERF-type TFs has been an effective tool for improving plant metabolism (the primary and secondary metabolite and phytohormone homeostasis), growth, developmental processes, and tolerance to multiple stresses, such as cold, salinity, alkalinity, heavy metals, and drought stresses (Licausi et al. 2013; Jisha et al. 2015; Abiri et al. 2017; Gu et al. 2017; Phukan et al. 2017; Yu et al. 2017). For example, heterologous *Arabidopsis CBF1* expression in transgenic tomato (Hsieh et al. 2002) and over-expression of *DREB1A* and *DREB2A* genes in *Arabidopsis* (Liu et al. 1998) have been effective in improving tolerance to cold stress. The interaction effects between RBPs, chaperone proteins, and CBF TF proteins on plant growth, symbiosis, and stress responses are complex and require more investigation. Their reactions vary depending on many factors, such as RBP localization; cell, tissue, and organ; environmental conditions; RBP isomers; and the circadian clock, as well as the growth and developmental stage. For example, as described earlier, overexpression of *AtCSP2* negatively regulated tolerance to cold and salinity in *Arabidopsis* by downregulating CBFs, while overexpression of *AtCSP3* positively enhanced plant tolerance to freezing temperature (Sasaki et al. 2013, 2015a,b). In addition to plant growth and development stage, stress responses and symbiosis are under the influence of environmental conditions and developmental stages of symbiosis. For example, early events in the symbiosis process, such as molecular signaling, rhizobia attachment, root-hair curling, infection thread formation, and nodule initiation, are particularly sensitive to high temperatures, salinity, acidity, and other environmental stresses. Furthermore, during the infection process, rhizobia have to deal with adverse conditions within the host cells and with the plant's innate immunity that induce physiological stress responses and may interfere with the symbiosis (Brígido et al. 2012). In this regard, it is proposed to evaluate the transcriptional regulation of genes during the whole of the plant and symbiont growth and developmental stages and symbiosis stages. Furthermore, the influence of circadian rhythms and environmental changes should also be taken into account to understand the effectiveness of gene regulation on performance, stress responses, and symbiosis.

AtRH7/PRH75 suppressed the expression of CBF1, CBF2, and CBF3 (Huang et al. 2016), while *regulator of CBF gene expression1* (*RCF1*) was essential for encoding cold-inducible DEAD (Asp-Glu-Ala-Asp)-box RH and negatively regulated the expression of the CBF (e.g., RD29A and COR15A) genes and their downstream target genes under cold stress. The CBF isomers also interact with each other, so that CBF2 negatively regulates CBF1 and CBF3, and Zn TRANSPORTER OF *Arabidopsis thaliana12* (ZAT12; a C2H2 zinc finger protein) (Guan et al. 2013b). Therefore, this review suggests that understanding the interactions of RBPs and other chaperones with CBFs will provide valuable information for further improvements in breeding programs in the future.

In bacteria, at least three different mechanisms are known to confer temperature regulation on the sHSP-encoding genes: sigma factors such as RpoH (RNA polymerase sigma factor), RheA (repressor of HSP18), and ROSE (repressor of heat shock gene expression) (Balsiger et al. 2004). In many organisms, alternative sigma factors, such as RpoH, mediate regulation of gene expression and protein metabolism, affecting growth, development, and symbiosis under both normal and stressful conditions. The RpoH-type sigma factors mediate transcription of the heat shock gene, chaperones, proteases, and symbiotic genes by selectivity of RNA polymerase under normal and stressful conditions (Ono et al. 2001; Balsiger et al. 2004; Barnett et al. 2012; Alexandre et al. 2014; Sasaki et al. 2016). RheA inhibits transcription of heat-inducible sHsp18 (Balsiger et al. 2004), which naturally is upregulated above 30 °C (Servant and Mazodier 1996), in *Streptomyces albus*. Thus, it is thought that sHsp18 regulation affects organism stress responses such as thermotolerance (Servant and Mazodier 1996) and also symbiosis via its chaperone activity and interaction with other HSPs (Smykal et al. 2000; Fu 2014). Similarly, ROSE post-transcriptionally regulates sHSP expression in rhizobia (Balsiger et al. 2004). The sequence is predicted to fold into a secondary structure that represses translation at low temperatures by masking the Shine-Dalgarno (SD) sequence at low temperatures. Local melting of that structure at higher temperatures liberates the SD sequence and allows translation (Balsiger et al. 2004). However, knowledge about their influence on stress responses is still limited, and their effects on symbiosis remain elusive.

In enterobacteria, most small RNAs (sRNAs) require Hfq, which acts as an RNA chaperone and as a riboregulator. The Hfq modulates the sensitivity of RNA to ribonucleases and facilitates pairing of sRNA with their messenger RNAs (mRNAs) and recognition and attachment to a polyadenylated fragment of specific *rpsO* (*ribosomal protein S15*) mRNAs (Folichon et al. 2003; Barra-Bily et al. 2010; Becker 2016). Hence, Hfq affects sRNA–mRNA interaction and, thereby, RNA translation and turnover rates of specific transcripts and contributes to complex post-transcriptional regulation of gene expression (Vogel and Luisi 2011). Hfq controls translation either directly, by stimulating the polyadenylation of an mRNA via coupling with poly(A) polymerase, or indirectly, via its action on sRNAs by modulating the translation of sigma factors and the addition of poly(A) tails on certain sRNAs (Barra-Bily et al. 2010; Vogel and Luisi 2011). In this context, regulation of sRNA expression by Hfq together with sigma factors is a promising strategy to control bacterial stress responses, e.g., by determining stress-responsive miRNAs or small interfering RNAs (siRNAs) as well as the bacterial population and symbiotic effectiveness under stress and normal conditions (Gottesman et al. 2006; Soper et al. 2010; De Lay et al. 2013; Jiménez-Zurdo and Robledo 2015; Chao and Vogel 2016; Wang et al. 2016b; Robledo et al. 2017).

In *E. coli*, where salt and heat stresses and stress-induced polyamines inhibited chaperone activity by destabilizing the active ClpB oligomer, stress-induced K-glutamate and glycine-betaine (GB) had positive effects on chaperone-mediated protein disaggregation and refolding and, thus, stress responses (Diamant et al. 2003). GB antagonizes the inhibition of protein biosynthesis and, thus, enhances photosystem II (PSII) repair, which leads to both improved photosynthesis and tolerance to different stresses by stabilizing the activity of repaired proteins. GB improves tolerance to multiple abiotic stresses by improving cell water status and osmotic adjustment, protecting the thylakoid membrane system, membrane integrity, ROS detoxification, photosystem and photosynthetic electron transport chain protection, and thereby, performance (Giri 2011; Liu et al. 2011; Kumar et al. 2017). In addition to protein homeostasis, breeding for improved enzymatic and non-enzymatic antioxidants, osmolytes, and polyamines is believed to maintain chaperone activity, ROS homeostasis, and symbiosis. Among these compounds, trehalose and GB are the most promising, which have strong potential to improve plant and symbiont growth, development, performance, tolerance, and symbiotic effectiveness (Räsänen et al. 2004; Iturriaga et al. 2009; Morsy et al. 2010).

MicroRNAs (miRNAs) are a class of small, non-coding regulatory RNAs that post-transcriptionally interact with specific target mRNAs and coordinately silence genes by targeting cognate mRNAs for cleavage, degradation, or translation repression, regulating all aspects of cellular processes, such as growth, development, symbiosis, and stress responses (Zhu et al. 2011a; Bazin et al. 2012). For example, heat-inducible microRNA398 (miR398) downregulates the expression of CCS (copper/zinc chaperones for superoxide dismutase), namely, CSD1 (copper/zinc superoxide dismutase1) and CSD2, and coordinately upregulates the transcription of HSF and HSP genes to confer tolerance to heat and oxidative stresses (Guan et al. 2013a). At the same time, miR398 controls copper use efficiency for optimal photosynthesis under stress and normal conditions and, thereby, affects growth and recovery (Table 32.3) (Beauclair et al. 2010).

32.6 CONCLUDING REMARKS AND FUTURE PERSPECTIVES

Establishing efficient symbiosis plays an essential role in improving plant cellular processes and, thereby, tolerance and performance to meet growing global demands. In this context, breeding tolerant symbionts would be a promising tool to both promote crop production and escape from the unwanted changes in non-target genes and molecules during plant tissue culture and transformation in food products. However, more studies are required to understand the influence of biotech symbionts in agriculture. Thus, this review was undertaken to incentivize the important roles of breeding symbiont and to emphasize the important roles of proteins (HSPs, RBPs, and molecular chaperones) in promoting thermotolerance, symbiosis, and crop production. More studies are also required to understand the relationships and the effects of protein–protein and RNA–protein interactions; the circadian clock; plant growth and development stages; symbiotic relationships and stages; stress conditions (e.g., severity, duration, repetition, stress combinations, etc.), gene sequencing of resistant and sensitive species and the diversity of their proteins, with different functions and structures in both host plants and symbionts; transcriptome and metabolite analysis in different cell, tissue, and organ types; intracellular locations; protein structure, function, and isomers; interactions between plant, multi-symbionts, and environmental conditions; and gene co-expression and silencing in agriculture. Long-term safety assessments of different plants exposed to biotech symbionts and their impact on the ecosystem are still suggested to verify the value of symbiont breeding in agriculture.

REFERENCES

Abdel-Ghany SE et al. (2005) AtCCS is a functional homolog of the yeast copper chaperone Ccs1/Lys7. *FEBS Lett* 579:2307–2312.

Abiri R et al. (2017) Role of ethylene and the APETALA 2/ethylene response factor superfamily in rice under various abiotic and biotic stress conditions. *Environ Exper Bot* 134:33–44.

Agarwal M, Katiyar-Agarwal S, Sahi C, Gallie DR, Grover A (2001) Arabidopsis thaliana Hsp100 proteins: Kith and kin. *Cell Stress Chaperones* 6:219–224.

Agarwal M et al. (2003) Molecular characterization of rice hsp101: Complementation of yeast hsp104 mutation by disaggregation of protein granules and differential expression in indica and japonica rice types. *Plant Mol Biol* 51:543–553.

Alexandre A, Laranjo M, Oliveira S (2014) Global transcriptional response to heat shock of the legume symbiont *Mesorhizobium loti* maff303099 comprises extensive gene downregulation. *DNA Res* 21:195–206.

Alexandre A, Oliveira S (2011) Most heat-tolerant rhizobia show high induction of major chaperone genes upon stress. *FEMS Microbiol Ecol* 75:28–36.

Alkhalfioui F et al. (2008) A novel type of thioredoxin dedicated to symbiosis in legumes. *Plant Physiol* 148:424–435.

Alves LR, Goldenberg S (2016) RNA-binding proteins related to stress response and differentiation in protozoa. *World J Biol Chem* 7:78–87.

Amorós M, Estruch F (2001) Hsf1p and Msn2/4p cooperate in the expression of Saccharomyces cerevisiae genes HSP26 and HSP104 in a gene- and stress type-dependent manner. *Mol Microbiol* 39:1523–1532.

An Q, Dong Y, Wang W, Li Y, Li J (2007) Constitutive expression of the *nifA* gene activates associative nitrogen fixation of *Enterobacter gergoviae* 57–7, an opportunistic endophytic diazotroph. *J App Microbiol* 103:613–620.

Arluison V, Hohng S, Roy R, Pellegrini O, Régnier P, Ha T (2007) Spectroscopic observation of RNA chaperone activities of Hfq in post-transcriptional regulation by a small non-coding RNA. *Nucleic Acids Res* 35:999–1006.

Asadi Rahmani H, Saleh-rastin N, Khavazi K, Asgharzadeh A, Fewer D, Kiani S, Lindström K (2009) Selection of thermotolerant bradyrhizobial strains for nodulation of soybean (*Glycine max* L.) in semi-arid regions of Iran. *World J Microbiol Biotechnol* 25:591–600.

Baena-González E (2010) Energy signaling in the regulation of gene expression during stress. *Mol Plant* 3:300–313.

Balczun C, Bunse A, Schwarz C, Piotrowski M, Kück U (2006) Chloroplast heat shock protein Cpn60 from *Chlamydomonas reinhardtii* exhibits a novel function as a group II intron-specific RNA-binding protein. *FEBS Lett* 580:4527–4532.

Balsiger S, Ragaz C, Baron C, Narberhaus F (2004) Replicon-specific regulation of small heat shock genes in agrobacterium tumefaciens. *J Bacteriol* 186:6824–6829.

Barak S, Singh Yadav N, Khan A (2014) DEAD-box RNA helicases and epigenetic control of abiotic stress-responsive gene expression. *Plant Signal Behav* 9:e977729.

Barnett MJ, Bittner AN, Toman CJ, Oke V, Long SR (2012) Dual RpoH sigma factors and transcriptional plasticity in a symbiotic bacterium. *J Bacteriol* 194:4983–4994.

Barra-Bily L et al. (2010) Proteomic alterations explain phenotypic changes in *Sinorhizobium meliloti* lacking the RNA chaperone Hfq. *J Bacteriol* 192:1719–1729.

Barrero-Sicilia C, Silvestre S, Haslam RP, Michaelson LV (2017) Lipid remodelling: Unravelling the response to cold stress in *Arabidopsis* and its extremophile relative *Eutrema salsugineum*. *Plant Sci* 263:194–200.

Basu S, Binder RJ, Ramalingam T, Srivastava PK (2001) CD91 is a common receptor for heat shock proteins gp96, hsp90, hsp70, and calreticulin. *Immun* 14:303–313.

Basu S, Ramegowda V, Kumar A, Pereira A (2016) Plant adaptation to drought stress. *F1000Res* 5:F1000 Faculty Rev-1554.

Bazin J, Bustos-Sanmamed P, Hartmann C, Lelandais-Brière C, Crespi M (2012) Complexity of miRNA-dependent regulation in root symbiosis. *Philos Trans R Soc Lond B Biol Sci* 367:1570–1579.

Beauclair L, Yu A, Bouché N (2010) microRNA-directed cleavage and translational repression of the copper chaperone for superoxide dismutase mRNA in Arabidopsis. *Plant J* 62:454–462.

Becker A (2016) Classic spotlight: Hfq, from a specific host factor for phage replication to a global player in riboregulation. *J Bacteriol* 198:2279–2280.

Bocca SN, Magioli C, Mangeon A, Junqueira RM, Cardeal V, Margis R, Sachetto-Martins G (2005) Survey of glycine-rich proteins (GRPs) in the *Eucalyptus* expressed sequence tag database (ForEST). *Genet Mol Biol* 28:608–624.

Bottinger L, Oeljeklaus S, Guiard B, Rospert S, Warscheid B, Becker T (2015) Mitochondrial heat shock protein (Hsp) 70 and Hsp10 cooperate in the formation of Hsp60 complexes. *J Biol Chem* 290:11611–11622.

Breitenbach M, Weber M, Rinnerthaler M, Karl T, Breitenbach-Koller L (2015) Oxidative stress in fungi: Its function in signal transduction, interaction with plant hosts, and lignocellulose degradation. *Biomolecules* 5:318–342.

Brígido C, Robledo M, Menéndez E, Mateos PF, Oliveira S (2012) A ClpB chaperone knockout mutant of *Mesorhizobium ciceri* shows a delay in the root nodulation of chickpea plants. *Mol Plant Microbe Interact* 25:1594–1604.

Brown AS, Mohanty BK, Howe PH (2015) Computational identification of post translational modification regulated RNA binding protein motifs. *PLOS ONE* 10:e0137696.

Brown Jonathon D, Day Alison M, Taylor Sarah R, Tomalin Lewis E, Morgan Brian A, Veal Elizabeth A (2013) A peroxiredoxin promotes H_2O_2 signaling and oxidative stress resistance by oxidizing a thioredoxin family protein. *Cell Rep* 5:1425–1435.

Burd CG, Dreyfuss G (1994) Conserved structures and diversity of functions of RNA-binding proteins. *Science* 265:615–621.

Cannon S, Wang P, Roy H (1986) Inhibition of ribulose bisphosphate carboxylase assembly by antibody to a binding protein. *J Cell Biol* 103:1327–1335.

Carvalho HH et al. (2014a) The molecular chaperone binding protein bip prevents leaf dehydration-induced cellular homeostasis disruption. *PLOS ONE* 9:e86661.

Carvalho HH et al. (2014b) The endoplasmic reticulum binding protein bip displays dual function in modulating cell death events. *Plant Physiol* 164:654–670.

Castiglioni P et al. (2008) Bacterial RNA chaperones confer abiotic stress tolerance in plants and improved grain yield in maize under water-limited conditions. *Plant Physiol* 147:446–455.

Castro-Sowinski S, Matan O, Bonafede P, Okon Y (2007) A thioredoxin of *Sinorhizobium meliloti* CE52G is required for melanin production and symbiotic nitrogen fixation. *Mol Plant Microbe Interact* 20:986–993.

Catalano CM, Lane WS, Sherrier DJ (2004) Biochemical characterization of symbiosome membrane proteins from *Medicago truncatula* root nodules. *Electrophoresis* 25:519–531.

Cazalé A-C et al. (2009) Altered expression of cytosolic/nuclear HSC70-1 molecular chaperone affects development and abiotic stress tolerance in *Arabidopsis thaliana. J Exp Bot* 60:2653–2664.

Çetinbaş M, Shakhnovich Eugene I (2015) Is catalytic activity of chaperones a selectable trait for the emergence of heat shock response? *Biophys J* 108:438–448.

Chae HB et al. (2013) Thioredoxin reductase type c (NTRC) orchestrates enhanced thermotolerance to Arabidopsis by its redox-dependent holdase chaperone function. *Mol Plant* 6:323–336.

Chaikam V, Karlson D (2008) Functional characterization of two cold shock domain proteins from *Oryza sativa. Plant Cell Environ* 31:995–1006.

Chao Y, Vogel J (2016) A 3′ UTR-derived small RNA provides the regulatory noncoding arm of the inner membrane stress response. *Mol Cell* 61:352–363.

Charollais J, Dreyfus M, Iost I (2004) CsdA, a cold-shock RNA helicase from *Escherichia coli*, is involved in the biogenesis of 50S ribosomal subunit. *Nucleic Acids Res* 32:2751–2759.

Chen J, Song J-l, Zhang S, Wang Y, Cui D-f, Wang C-c (1999) Chaperone activity of DsbC. *J Biol Chem* 274:19601–19605.

Chen J et al. (2015) Overexpression of an *Apocynum venetum* DEAD-Box helicase gene (*AvDH1*) in cotton confers salinity tolerance and increases yield in a saline field. *Front Plant Sci* 6:1227.

Chibani K, Tarrago L, Schürmann P, Jacquot J-P, Rouhier N (2011) Biochemical properties of poplar thioredoxin z. *FEBS Lett* 585:1077–1081.

Cho EJ, Yuen CYL, Kang B-H, Ondzighi CA, Staehelin LA, Christopher DA (2011) Protein disulfide isomerase-2 of *Arabidopsis* mediates protein folding and localizes to both the secretory pathway and nucleus, where it interacts with maternal effect embryo arrest factor. *Mol Cells* 32:459–475.

Cho EK, Hong CB (2006) Over-expression of tobacco NtHSP70-1 contributes to drought-stress tolerance in plants. *Plant Cell Rep* 25:349–358.

Ciuzan O, Hancock J, Pamfil D, Wilson I, Ladomery M (2015) The evolutionarily conserved multifunctional glycine-rich RNA-binding proteins play key roles in development and stress adaptation. *Physiol Plant* 153:1–11.

Cloutier J, Prevost D, Nadeau P, Antoun H (1992) Heat and cold shock protein synthesis in arctic and temperate strains of rhizobia. *Appl Environ Microbiol* 58:2846–2853.

Cloutier P, Coulombe B (2013) Regulation of molecular chaperones through post-translational modifications: Decrypting the chaperone code. *Biochim Biophys Acta Gene Regul Mech* 1829:443–454.

Cohen-Peer R, Schuster S, Meiri D, Breiman A, Avni A (2010) Sumoylation of *Arabidopsis* heat shock factor A2 (HsfA2) modifies its activity during acquired thermotholerance. *Plant Mol Biol* 74:33–45.

Constan D, Froehlich JE, Rangarajan S, Keegstra K (2004) A stromal Hsp100 protein is required for normal chloroplast development and function in *Arabidopsis. Plant Physiol* 136:3605–3615.

De Lay N, Schu DJ, Gottesman S (2013) Bacterial small RNA-based negative regulation: Hfq and its accomplices. *J Biol Chem* 288:7996–8003.

de Nadal E, Ammerer G, Posas F (2011) Controlling gene expression in response to stress. *Nat Rev Genet* 12:833.

del Río LA, Sandalio LM, Corpas FJ, Palma JM, Barroso JB (2006) Reactive oxygen species and reactive nitrogen species in peroxisomes. production, scavenging, and role in cell signaling. *Plant Physiol* 141:330–335.

Demirevska-Kepova K, Holzer R, Simova-Stoilova L, Feller U (2005) Heat stress effects on ribulose-1,5-bisphosphate carboxylase/oxygenase, Rubisco binding protein and Rubisco activase in wheat leaves. *Biol Plantarum* 49:521–525.

Demirevska K, Simova-Stoilova L, Vassileva V, Feller U (2008) Rubisco and some chaperone protein responses to water stress and rewatering at early seedling growth of drought sensitive and tolerant wheat varieties. *Plant Growth Regul* 56:97.

Denance N, Szurek B, Noel LD (2014) Emerging functions of nodulin-like proteins in non-nodulating plant species. *Plant Cell Physiol* 55:469–474.

Diamant S, Rosenthal D, Azem A, Eliahu N, Ben-Zvi AP, Goloubinoff P (2003) Dicarboxylic amino acids and glycinebetaine regulate chaperone-mediated protein-disaggregation under stress. *Mol Microbiol* 49:401–410.

Diao Q-N, Song Y-J, Shi D-M, Qi H-Y (2016) Nitric oxide induced by polyamines involves antioxidant systems against chilling stress in tomato (*Lycopersicon esculentum* Mill.) seedling. *J Zhejiang Univ Sci B* 17:916–930.

Diestra E, Cayrol B, Arluison V, Risco C (2009) Cellular electron microscopy imaging reveals the localization of the Hfq protein close to the bacterial membrane. *PLOS ONE* 4:e8301.

Dietz K-J (2011) Peroxiredoxins in plants and cyanobacteria. *Antioxid Redox Signal* 15:1129–1159.

Ding X, Lv Z-M, Zhao Y, Min H, Yang W-J (2008) MTH1745, a protein disulfide isomerase-like protein from thermophilic archaea, *Methanothermobacter thermoautotrophicum* involving in stress response. *Cell Stress Chaperones* 13:239–246.

Du H, Kim S, Hur YS, Lee MS, Lee SH, Cheon CI (2015) A cytosolic thioredoxin acts as a molecular chaperone for peroxisome matrix proteins as well as antioxidant in peroxisome. *Mol Cells* 38:187–194.

Fairman-Williams ME, Guenther UP, Jankowsky E (2010) SF1 and SF2 helicases: Family matters. *Curr Opin Struct Biol* 20:313–324.

Fang W, St Leger RJ (2010) RNA binding proteins mediate the ability of a fungus to adapt to the cold. *Environ Microbiol* 12:810–820.

Fischer HM, Babst M, Kaspar T, Acuña G, Arigoni F, Hennecke H (1993) One member of a gro-ESL-like chaperonin multigene family in Bradyrhizobium japonicum is co-regulated with symbiotic nitrogen fixation genes. *EMBO J* 12:2901–2912.

Folichon M, Arluison V, Pellegrini O, Huntzinger E, Régnier P, Hajnsdorf E (2003) The poly(A) binding protein Hfq protects RNA from RNase E and exoribonucleolytic degradation. *Nucleic Acids Res* 31:7302–7310.

Forman HJ, Fukuto JM, Torres M (2004) Redox signaling: Thiol chemistry defines which reactive oxygen and nitrogen species can act as second messengers. *Am J Physiol Cell Physiol* 287:C246–256.

Fu X (2014) Chaperone function and mechanism of small heat-shock proteins. *Acta Biochim Biophys Sin* 46:347–356.

Gao C, Jiang B, Wang Y, Liu G, Yang C (2012) Overexpression of a heat shock protein (ThHSP18.3) from *Tamarix hispida* confers stress tolerance to yeast. *Mol Biol Rep* 39:4889–4897.

García-Mata C, Lamattina L (2001) Nitric oxide induces stomatal closure and enhances the adaptive plant responses against drought stress. *Plant Physiol* 126:1196–1204.

Gendra E, Moreno A, Albà MM, Pages M (2004) Interaction of the plant glycine-rich RNA-binding protein MA16 with a novel nucleolar DEAD box RNA helicase protein from *Zea mays*. *Plant J* 38:875–886.

Giesguth M, Sahm A, Simon S, Dietz K-J (2015) Redox-dependent translocation of the heat shock transcription factor AtHSFA8 from the cytosol to the nucleus in *Arabidopsis thaliana*. *FEBS Lett* 589:718–725.

Giri J (2011) Glycinebetaine and abiotic stress tolerance in plants. *Plant Signal Behav* 6:1746–1751.

Glisovic T, Bachorik JL, Yong J, Dreyfuss G (2008) RNA-binding proteins and post-transcriptional gene regulation. *FEBS Lett* 582:1977–1986.

Gomez-Pastor R, Burchfiel ET, Thiele DJ (2017) Regulation of heat shock transcription factors and their roles in physiology and disease. *Nat Rev Mol Cell Biol* advance online publication.

Gonçalves D et al. (2011) Eucalyptustranscriptome analysis revealed molecular chaperones highly expressed in xylem. *BMC Proc* 5:P109.

Gottesman S et al. (2006) Small RNA regulators and the bacterial response to stress. *Cold Spring Harb Symp Quant Biol* 71:1–11.

Graumann PL, Marahiel MA (1998) A superfamily of proteins that contain the cold-shock domain. *Trends Biochem Sci* 23:286–290.

Grek C, Townsend DM (2014) Protein disulfide isomerase superfamily in disease and the regulation of apoptosis. *Endoplasmic Reticulum Stress Dis* 1:4–17.

Gu C, Guo Z-H, Hao P-P, Wang G-M, Jin Z-M, Zhang S-L (2017) Multiple regulatory roles of AP2/ERF transcription factor in angiosperm. *Bot Stud* 58:6.

Guan Q, Lu X, Zeng H, Zhang Y, Zhu J (2013a) Heat stress induction of miR398 triggers a regulatory loop that is critical for thermotolerance in Arabidopsis. *Plant J* 74:840–851.

Guan Q, Wu J, Zhang Y, Jiang C, Liu R, Chai C, Zhu J (2013b) A DEAD box RNA helicase is critical for pre-mRNA splicing, cold-responsive gene regulation, and cold tolerance in Arabidopsis. *Plant Cell* 25:342–356.

Guo M et al. (2015) Genome-wide analysis of the *CaHsp20* gene family in pepper: Comprehensive sequence and expression profile analysis under heat stress. *Front Plant Sci* 6.

Guo R, Ki J-S (2012) Differential transcription of heat shock protein 90 (HSP90) in the dinoflagellate *Prorocentrum minimum* by copper and endocrine-disrupting chemicals. *Ecotoxicol* 21:1448–1457.

Gupta D, Tuteja N (2011) Chaperones and foldases in endoplasmic reticulum stress signaling in plants. *Plant Signal Behav* 6:232–236.

Hammond C, Helenius A (1995) Quality control in the secretory pathway. *Curr Opin Cell Biol* 7:523–529.

Hao G-P, Zhang J-H (2009) The role of nitric oxide as a bioactive signaling molecule in plants under abiotic stress. In: Hayat S, Mori M, Pichtel J, Ahmad A, eds. *Nitric Oxide Plant Physiol*. Wiley-VCH Verlag GmbH & Co. KGaA, Weinheim, pp. 115–138.

Hasan MK, Cheng Y, Kanwar MK, Chu X-Y, Ahammed GJ, Qi Z-Y (2017) Responses of plant proteins to heavy metal stress—A review. *Front Plant Sci* 8:1492.

He Z, Li L, Luan S (2004) Immunophilins and parvulins. Superfamily of peptidyl prolyl isomerases in Arabidopsis. *Plant Physiol* 134:1248–1267.

He Z, Xie R, Wang Y, Zou H, Zhu J, Yu G (2008) Cloning and characterization of a heat shock protein 70 gene, *MsHSP70-1*, in *Medicago sativa*. *Acta Biochim Biophys Sin* 40:209–216.

Hemmingsen SM et al. (1988) Homologous plant and bacterial proteins chaperone oligomeric protein assembly. *Nature* 333:330–334.

Herschlag D (1995) RNA chaperones and the RNA folding problem. *J Biol Chem* 270:20871–20874.

Hietakangas V et al. (2003) Phosphorylation of serine 303 is a prerequisite for the stress-inducible SUMO modification of heat shock factor 1. *Mol Cell Biol* 23:2953–2968.

Hietakangas V, Anckar J, Blomster HA, Fujimoto M, Palvimo JJ, Nakai A, Sistonen L (2006) PDSM, a motif for phosphorylation-dependent SUMO modification. *Proc Natl Acad Sci USA* 103:45–50.

Hill JE, Hemmingsen SM (2001) *Arabidopsis thaliana* type I and II chaperonins. *Cell Stress Chaperones* 6:190–200.

Hoffmann JH, Linke K, Graf PCF, Lilie H, Jakob U (2004) Identification of a redox-regulated chaperone network. *EMBO J* 23:160–168.

Hong S-W, Vierling E (2001) Hsp101 is necessary for heat tolerance but dispensable for development and germination in the absence of stress. *Plant J* 27:25–35.

Hsieh T-H, Lee J-T, Yang P-T, Chiu L-H, Charng Y-y, Wang Y-C, Chan M-T (2002) Heterology expression of the *arabidopsis C-repeat/dehydration response element binding factor 1* gene confers elevated tolerance to chilling and oxidative stresses in transgenic tomato. *Plant Physiol* 129:1086–1094.

Huang CK, Shen YL, Huang LF, Wu SJ, Yeh CH, Lu CA (2016) The DEAD-Box RNA helicase AtRH7/PRH75 participates in Pre-rRNA processing, plant development and cold tolerance in Arabidopsis. *Plant Cell Physiol* 57:174–191.

Hudson WH, Ortlund EA (2014) The structure, function and evolution of proteins that bind DNA and RNA nature reviews. *Mol Cell Biology* 15:749–760.

Hunger K, Beckering CL, Wiegeshoff F, Graumann PL, Marahiel MA (2006) Cold-induced putative DEAD box RNA helicases CshA and CshB are essential for cold adaptation and interact with cold shock protein B in *Bacillus subtilis*. *J Bacteriol* 188:240–248.

Ingelsson B, Shapiguzov A, Kieselbach T, Vener AV (2009) Peptidyl–prolyl isomerase activity in chloroplast thylakoid lumen is a dispensable function of immunophilins in *Arabidopsis thaliana*. *Plant Cell Physiol* 50:1801–1814.

Ishiguro T et al. (2017) Regulatory role of RNA chaperone TDP-43 for RNA misfolding and repeat-associated translation in SCA31. *Neuron* 94:108–124.e107.

Iturriaga G, Suárez R, Nova-Franco B (2009) Trehalose metabolism: from osmoprotection to signaling. *Int J Mol Sci* 10:3793–3810.

Jacob P, Hirt H, Bendahmane A (2017) The heat-shock protein/chaperone network and multiple stress resistance. *Plant Biotechnol J* 15:405–414.

Jacquier-Sarlin MR, Polla BS (1996) Dual regulation of heat-shock transcription factor (HSF) activation and DNA-binding activity by H2O2: Role of thioredoxin. *Biochem J* 318 (Pt 1):187–193.

Jaglo-Ottosen KR, Gilmour SJ, Zarka DG, Schabenberger O, Thomashow MF (1998) *Arabidopsis CBF1* overexpression induces *COR* genes and enhances freezing tolerance. *Sci (New York, NY)* 280:104–106.

Jagodnik J, Brosse A, Le Lam TN, Chiaruttini C, Guillier M (2017) Mechanistic study of base-pairing small regulatory RNAs in bacteria. *Methods* 117:67–76.

Jakab G, Ton J, Flors V, Zimmerli L, Métraux J-P, Mauch-Mani B (2005) Enhancing Arabidopsis salt and drought stress tolerance by chemical priming for its abscisic acid responses. *Plant Physiol* 139:267–274.

Jarmoskaite I, Russell R (2011) DEAD-box proteins as RNA helicases and chaperones. *Wiley Interdiscip Rev RNA* 2:135–152.

Jee C, Vanoaica L, Lee J, Park BJ, Ahnn J (2005) Thioredoxin is related to life span regulation and oxidative stress response in *Caenorhabditis elegans*. *Genes cells* 10:1203–1210.

Ji CY et al. (2017) Overexpression of Arabidopsis P3B increases heat and low temperature stress tolerance in transgenic sweetpotato. *BMC Plant Biol* 17:139.

Jiang W, Hou Y, Inouye M (1997) CspA, the major cold-shock protein of *Escherichia coli*, is an RNA chaperone. *J Biol Chem* 272:196–202.

Jiménez-Zurdo JI, Robledo M (2015) Unraveling the universe of small RNA regulators in the legume symbiont *Sinorhizobium meliloti*. *Symbiosis* 67:43–54.

Jisha V, Dampanaboina L, Vadassery J, Mithöfer A, Kappara S, Ramanan R (2015) Overexpression of an AP2/ERF type transcription factor oserebp1 confers biotic and abiotic stress tolerance in rice. *PLOS ONE* 10:e0127831.

Kang H, Park SJ, Kwak KJ (2013) Plant RNA chaperones in stress response. *Trends in Plant Sci* 18:100–106.

Kant P, Kant S, Gordon M, Shaked R, Barak S (2007) STRESS RESPONSE SUPPRESSOR1 and STRESS RESPONSE SUPPRESSOR2, two DEAD-box RNA helicases that attenuate Arabidopsis responses to multiple abiotic stresses. *Plant Physiol* 145:814–830.

Kao TY et al. (2015) Mitochondrial Lon regulates apoptosis through the association with Hsp60-mtHsp70 complex. *Cell Death Dis* 6:e1642.

Kapoor M, Roy SS (2015) Chapter 1: Heat-shock proteins and molecular chaperones: Role in regulation of cellular proteostasis and stress management. In *Abiotic Stresses in Crop Plants*. CABI, Wallingford, UK.

Karlson D, Imai R (2003) Conservation of the cold shock domain protein family in plants. *Plant Physiol* 131:12–15.

Karlson D, Nakaminami K, Toyomasu T, Imai R (2002) A cold-regulated nucleic acid-binding protein of winter wheat shares a domain with bacterial cold shock proteins. *J Biol Chem* 277:35248–35256.

Katiyar-Agarwal S, Agarwal M, Gallie DR, Grover A (2001) Search for the cellular functions of plant Hsp100/Clp family proteins. *Crit Rev Plant Sci* 20:277–295.

Katiyar-Agarwal S, Agarwal M, Grover A (2003) Heat-tolerant basmati rice engineered by over-expression of *hsp101*. *Plant Mol Biol* 51:677–686.

Kayukawa T, Chen B, Miyazaki S, Itoyama K, Shinoda T, Ishikawa Y (2005) Expression of mRNA for the t-complex polypeptide–1, a subunit of chaperonin CCT, is upregulated in association with increased cold hardiness in *Delia antiqua*. *Cell Stress Chaperones* 10:204–210.

Kevei Z, Vinardell JM, Kiss GB, Kondorosi A, Kondorosi E (2002) Glycine-rich proteins encoded by a nodule-specific gene family are implicated in different stages of symbiotic nodule development in *Medicago* Spp. *Mol Plant Microbe Interact* 15:922–931.

Khan A et al. (2014) The Arabidopsis STRESS RESPONSE SUPPRESSOR DEAD-box RNA helicases are nucleolar- and chromocenter-localized proteins that undergo stress-mediated relocalization and are involved in epigenetic gene silencing. *Plant J* 79:28–43.

Khan IU, Wallin R, Gupta RS, Kammer GM (1998) Protein kinase A-catalyzed phosphorylation of heat shock protein 60 chaperone regulates its attachment to histone 2B in the T lymphocyte plasma membrane. *Proc Natl Acad Sci USA* 95:10425–10430.

Kiewietdejonge A et al. (2006) Hypersaline stress induces the turnover of phosphatidylcholine and results in the synthesis of the renal osmoprotectant glycerophosphocholine in *Saccharomyces cerevisiae*. *FEMS Yeast Res* 6:205–217.

Kim DH, Xu Z-Y, Na YJ, Yoo Y-J, Lee J, Sohn E-J, Hwang I (2011a) Small heat shock protein Hsp17.8 functions as an AKR2A cofactor in the targeting of chloroplast outer membrane proteins in Arabidopsis. *Plant Physiol* 157:132–146.

Kim JS et al. (2008a) Glycine-rich RNA-binding protein 7 affects abiotic stress responses by regulating stomata opening and closing in *Arabidopsis thaliana*. *Plant J* 55:455–466.

Kim JS, Kim KA, Oh TR, Park CM, Kang H (2008b) Functional characterization of DEAD-box RNA helicases in *Arabidopsis thaliana* under abiotic stress conditions. *Plant cell physiol* 49:1563–1571.

Kim JS et al. (2007) Cold shock domain proteins and glycine-rich RNA-binding proteins from *Arabidopsis thaliana* can promote the cold adaptation process in *Escherichia coli*. *Nucleic Acids Res* 35:506–516.

Kim JY, Kim WY, Kwak KJ, Oh SH, Han YS, Kang H (2010a) Glycine-rich RNA-binding proteins are functionally conserved in *Arabidopsis thaliana* and *Oryza sativa* during cold adaptation process. *J Exp Bot* 61:2317–2325.

Kim MH, Sasaki K, Imai R (2009a) Cold shock domain protein 3 regulates freezing tolerance in *Arabidopsis thaliana*. *J Biol Chem* 284:23454–23460.

Kim SY et al. (2009b) Oligomerization and chaperone activity of a plant 2-Cys peroxiredoxin in response to oxidative stress. *Plant Sci* 177:227–232.

Kim S-R, Yang J-I, An G (2013) OsCpn60α1, encoding the plastid chaperonin 60α subunit, is essential for folding of *rbcL*. *Mol Cells* 35:402–409.

Kim SY et al. (2011b) The 1-Cys peroxiredoxin, a regulator of seed dormancy, functions as a molecular chaperone under oxidative stress conditions. *Plant Sci* 181:119–124.

Kim WY, Kim JY, Jung HJ, Oh SH, Han YS, Kang H (2010b) Comparative analysis of Arabidopsis zinc finger-containing glycine-rich RNA-binding proteins during cold adaptation. *Plant Physiol Biochem* 48:866–872.

Kim Y-O, Kang H (2006a) The role of a zinc finger-containing glycine-rich rna-binding protein during the cold adaptation Process in *Arabidopsis thaliana*. *Plant Cell Physiol* 47:793–798.

Kim Y-O, Kim JS, Kang H (2005) Cold-inducible zinc finger-containing glycine-rich RNA-binding protein contributes to the enhancement of freezing tolerance in *Arabidopsis thaliana*. *Plant J* 42:890–900.

Kim YO, Kang H (2006b) The role of a zinc finger-containing glycine-rich RNA-binding protein during the cold adaptation process in *Arabidopsis thaliana*. *Plant Cell Physiol* 47:793–798.

Kingsley PD, Palis J (1994) GRP2 proteins contain both CCHC zinc fingers and a cold shock domain. *Plant Cell* 6:1522–1523.

Kirchsteiger K, Ferrández J, Pascual MB, González M, Cejudo FJ (2012) NADPH thioredoxin reductase c is localized in plastids of photosynthetic and nonphotosynthetic tissues and is involved in lateral root formation in Arabidopsis. *Plant Cell* 24:1534–1548.

Klopf E et al. (2009) Cooperation between the INO80 complex and histone chaperones determines adaptation of stress gene transcription in the yeast *Saccharomyces cerevisiae*. *Mol Cell Biol* 29:4994–5007.

Knoblach B et al. (2003) ERp19 and ERp46, new members of the thioredoxin family of endoplasmic reticulum proteins. *Mol Cell Proteomics* 2:1104–1119.

Ko E, Kim M, Park Y, Ahn Y-J (2017) Heterologous expression of the carrot *Hsp17.7* gene increased growth, cell viability, and protein solubility in transformed yeast (*Saccharomyces cerevisiae*) under heat, cold, acid, and osmotic stress conditions. *Curr Microbiol* 74:952–960.

Kocourková D et al. (2011) The phosphatidylcholine-hydrolysing phospholipase C NPC4 plays a role in response of Arabidopsis roots to salt stress. *J Exp Bot* 62:3753–3763.

Koldewey P, Horowitz S, Bardwell JCA (2017) Chaperone-client interactions: Non-specificity engenders multi-functionality. *J Biol Chem* 292:12010–12017.

Kondoh K, Sunadome K, Nishida E (2007) Notch signaling suppresses p38 MAPK activity via induction of MKP-1 in myogenesis. *J Biol Chem* 282:3058–3065.

König SLB, Liyanage PS, Sigel RKO, Rueda D (2013) Helicase-mediated changes in RNA structure at the single-molecule level. *RNA Biol* 10:133–148.

Kotak S, Port M, Ganguli A, Bicker F, von Koskull-Doring P (2004) Characterization of C-terminal domains of Arabidopsis heat stress transcription factors (Hsfs) and identification of a new signature combination of plant class A Hsfs with AHA and NES motifs essential for activator function and intracellular localization. *Plant J* 39:98–112.

Kotchoni SO, Kuhns C, Ditzer A, KIRCH HH, Bartels D (2006) Over-expression of different aldehyde dehydrogenase genes in *Arabidopsis thaliana* confers tolerance to abiotic stress and protects plants against lipid peroxidation and oxidative stress. *Plant Cell Environ* 29:1033–1048.

Kothe U (2010) 5.13 - Recent progress on understanding ribosomal protein synthesis. In: *Comprehensive Natural Products* II. Elsevier, Oxford, pp. 353–382.

Kovacs D, Rakacs M, Agoston B, Lenkey K, Semrad K, Schroeder R, Tompa P (2009) Janus chaperones: Assistance of both RNA- and protein-folding by ribosomal proteins. *FEBS Lett* 583:88–92.

Koyasu S, Nishida E, Miyata Y, Sakai H, Yahara I (1989) HSP100, a 100-kDa heat shock protein, is a Ca^{2+}-calmodulin-regulated actin-binding protein. *J Biol Chem* 264:15083–15087

Krishna P, Sacco M, Cherutti JF, Hill S (1995) Cold-induced accumulation of *hsp90* transcripts in *Brassica napus*. *Plant Physiol* 107:915–923.

Kudryavtseva NN, Sof'in AV, Bobylev GS, Sorokin EM (2014) A comparative study of phase states of the peribacteroid membrane from yellow lupin and broad bean nodules. *Biochem Res Int* 2014:6.

Kumar R, Singh AK, Lavania D, Siddiqui MH, Al-Whaibi MH, Grover A (2016) Expression analysis of ClpB/Hsp100 gene in faba bean (*Vicia faba* L.) plants in response to heat stress. *Saudi J Biol Sci* 23:243–247.

Kumar V, Shriram V, Hoque TS, Hasan MM, Burritt DJ, Hossain MA (2017) Glycinebetaine-mediated abiotic oxidative-stress tolerance in plants: Physiological and biochemical mechanisms. In: Sarwat M, Ahmad A, Abdin MZ, Ibrahim MM (eds) *Stress Signaling in Plants: Genomics and Proteomics Perspective*, Volume 2. Springer International Publishing, Cham, pp. 111–133.

Kurek I, Aviezer K, Erel N, Herman E, Breiman A (1999) The wheat peptidyl prolyl cis-trans-isomerase FKBP77 is heat induced and developmentally regulated. *Plant Physiol* 119:693–704.

Kurokawa Y, Yanagi H, Yura T (2000) Overexpression of protein disulfide isomerase DsbC stabilizes multiple-disulfide-bonded recombinant protein produced and transported to the periplasm in *Escherichia coli*. *Appl Environ Microbiol* 66:3960–3965.

Laskowska E, Matuszewska E, Kuczynska-Wisnik D (2010) Small heat shock proteins and protein-misfolding diseases. *Curr Pharm Biotechnol* 11:146–157.

Leach MR, Williams DB (2003) Calnexin and calreticulin, molecular chaperones of the endoplasmic reticulum. In: Eggleton P, Michalak M (eds) *Calreticulin*: Second Edition. Springer US, Boston, pp. 49–62.

Lee EK (2012) Post-translational modifications of RNA-binding proteins and their roles in RNA granules. *Curr Protein Pept Sci* 13:331–336.

Lee JH, Schöffl F (1996) An.Hsp70 antisense gene affects the expression of HSP70/HSC70, the regulation of HSF, and the acquisition of thermotolerance in transgenic *Arabidopsis thaliana*. *Mol Gen Genet* 252:11–19.

Lee JR et al. (2009a) Heat-shock dependent oligomeric status alters the function of a plant-specific thioredoxin-like protein, AtTDX. *Proc Natl Acad Sci USA* 106:5978–5983.

Lee M-O, Kim KP, Kim B-g, Hahn J-S, Hong CB (2009b) Flooding stress-induced glycine-rich RNA-binding protein from *Nicotiana tabacum*. *Mol Cells* 27:47–54.

Lee M-Y et al. (2005a) Induction of thioredoxin is required for nodule development to reduce reactive oxygen species levels in soybean roots. *Plant Physiol* 139:1881–1889.

Lee S-K, Park S-H, Lee J-W, Lim H-M, Jung S-Y, Park I-C, Park S-C (2014) A putative cold shock protein-encoding gene isolated from *Arthrobacter* sp. A2-5 confers cold stress tolerance in yeast and plants. *J Korean Soc Appl Biol Chem* 57:775–782.

Lee S, Jia B, Liu J, Pham BP, Kwak JM, Xuan YH, Cheong G-W (2015) A 1-Cys peroxiredoxin from a thermophilic archaeon moonlights as a molecular chaperone to protect protein and DNA against stress-induced damage. *PLOS ONE* 10:e0125325.

Lee S, Kim SM, Lee RT (2013) Thioredoxin and thioredoxin target proteins: From molecular mechanisms to functional significance. *Antioxid Redox Signal* 18:1165–1207.

Lee S et al. (2009c) Heat shock protein cognate 70–4 and an E3 ubiquitin ligase, CHIP, mediate plastid-destined precursor degradation through the ubiquitin-26s proteasome system in Arabidopsis. *Plant Cell* 21:3984–4001.

Lee T, Feig AL (2008) The RNA binding protein Hfq interacts specifically with tRNAs. *RNA* 14:514–523.

Lee U, Rioflorido I, Hong S-W, Larkindale J, Waters ER, Vierling E (2007) The Arabidopsis ClpB/Hsp100 family of proteins: Chaperones for stress and chloroplast development. *Plant J* 49:115–127.

Lee U, Wie C, Escobar M, Williams B, Hong S-W, Vierling E (2005b) Genetic analysis reveals domain interactions of Arabidopsis Hsp100/ClpB and cooperation with the small heat shock protein chaperone system. *Plant Cell* 17:559–571.

Li J, Soroka J, Buchner J (2012) The Hsp90 chaperone machinery: Conformational dynamics and regulation by co-chaperones. *Biochim Biophys Acta - Mol Cell Res* 1823:624–635.

Li Z-G (2016) Methylglyoxal and glyoxalase system in plants: Old players, new concepts. *Bot Rev* 82:183–203.

Li Z-G, Duan X-Q, Min X, Zhou Z-H (2017a) Methylglyoxal as a novel signal molecule induces the salt tolerance of wheat by regulating the glyoxalase system, the antioxidant system, and osmolytes. *Protoplasma* 254:1–12.

Li Z et al. (2017b) Molecular cloning and functional analysis of the drought tolerance gene MsHSP70 from alfalfa (*Medicago sativa* L.). *J Plant Res* 130:387–396.

Liberek K, Lewandowska A, Ziętkiewicz S (2008) Chaperones in control of protein disaggregation. *EMBO J* 27:328–335.

Licausi F, Ohme-Takagi M, Perata P (2013) APETALA2/Ethylene Responsive Factor (AP2/ERF) transcription factors: Mediators of stress responses and developmental programs. *New Phytol* 199:639–649.

Linder P, Jankowsky E (2011) From unwinding to clamping — the DEAD box RNA helicase family. *Nat Rev Mol Cell Biol* 12:505.

Littlefield O, Nelson HCM (1999) A new use for the 'wing' of the 'winged' helix-turn-helix motif in the HSF–DNA cocrystal. *Nat Struct Biol* 6:464.:10.1038/8269

Liu H-T, Sun D-Y, Zhou R-G (2005) Ca²⁺ and AtCaM3 are involved in the expression of heat shock protein gene in Arabidopsis. *Plant Cell Environ* 28:1276–1284.

Liu J et al. (2006) Recruitment of novel calcium-binding proteins for root nodule symbiosis in *Medicago truncatula*. *Plant Physiol* 141:167–177.

Liu J, Wisniewski M, Droby S, Vero S, Tian S, Hershkovitz V (2011) Glycine betaine improves oxidative stress tolerance and biocontrol efficacy of the antagonistic yeast *Cystofilobasidium infirmominiatum*. *Int J Food Microbiol* 146:76–83.

Liu Q, Kasuga M, Sakuma Y, Abe H, Miura S, Yamaguchi-Shinozaki K, Shinozaki K (1998) Two transcription factors, DREB1 and DREB2, with an EREBP/AP2 DNA binding domain separate two cellular signal transduction pathways in drought- and low-temperature-responsive gene expression, respectively, in Arabidopsis. *Plant Cell* 10:1391.

Liu Y, Tabata D, Imai R (2016) A Cold-Inducible DEAD-Box RNA helicase from *Arabidopsis thaliana* regulates plant growth and development under low temperature. *PLOS ONE* 11:e0154040.

Lopez-Maury L, Marguerat S, Bahler J (2008) Tuning gene expression to changing environments: From rapid responses to evolutionary adaptation. *Nat Rev Genet* 9:583–593.

Ludwig-Müller J, Krishna P, Forreiter C (2000) A glucosinolate mutant of Arabidopsis is thermosensitive and defective in cytosolic Hsp90 expression after heat stress. *Plant Physiol* 123:949–958.

Lum R, Tkach JM, Vierling E, Glover JR (2004) Evidence for an unfolding/threading mechanism for protein disaggregation by *Saccharomyces cerevisiae* Hsp104. *J Biol Chem* 279:29139–29146.

Lund AA, Blum PH, Bhattramakki D, Elthon TE (1998) Heat-stress response of maize mitochondria. *Plant Physiol* 116:1097–1110.

Lunde BM, Moore C, Varani G (2007) RNA-binding proteins: Modular design for efficient function. *Nat Rev Mol Cell Biol* 8:479.

Ma C, Haslbeck M, Babujee L, Jahn O, Reumann S (2006) Identification and characterization of a stress-inducible and a constitutive small heat-shock protein targeted to the matrix of plant peroxisomes. *Plant Physiol* 141:47–60.

Ma J, Wang D, She J, Li J, Zhu J-K, She Y-M (2016) Endoplasmic reticulum-associated N-glycan degradation of cold-upregulated glycoproteins in response to chilling stress in Arabidopsis. *New Phytol* 212:282–296.

Maaroufi H, Tanguay RM (2013) Analysis and phylogeny of small heat shock proteins from marine viruses and their cyanobacteria Host. *PLOS ONE* 8:e81207.:10.1371/journal.pone.0081207

Machida T et al. (2009) Expression pattern of a chloroplast NADPH-dependent thioredoxin reductase in *Chlorella vulgaris* during hardening and its interaction with 2-Cys peroxiredoxin bioscience. *Biotechnol Biochem* 73:695–701.

Machida T et al. (2012) Chloroplast NADPH-dependent thioredoxin reductase from chlorella vulgaris alleviates environmental stresses in yeast together with 2-Cys peroxiredoxin. *PLOS ONE* 7:e45988.

Mangeon A, Junqueira RM, Sachetto-Martins G (2010) Functional diversity of the plant glycine-rich proteins superfamily. *Plant Signal Behav* 5:99–104.

Marchese D, de Groot NS, Lorenzo Gotor N, Livi CM, Tartaglia GG (2016) Advances in the characterization of RNA-binding proteins. *Wiley Interdiscip Rev RNA* 7:793–810.

Martin J, Horwich AL, Hartl FU (1992) Prevention of protein denaturation under heat stress by the chaperonin Hsp60. *Sci (New York, NY)* 258:995–999.

Masliah G, Barraud P, Allain FH (2013) RNA recognition by double-stranded RNA binding domains: A matter of shape and sequence. *Cell Mol Life Sci* 70:1875–1895.

Mathew A, Mathur SK, Morimoto RI (1998) Heat shock response and protein degradation: Regulation of HSF2 by the ubiquitin-proteasome pathway. *Mol Cell Biol* 18:5091–5098.

Mayer MP, Bukau B (2005) Hsp70 chaperones: Cellular functions and molecular mechanism. *Cell Mol Life Sci* 62:670–684.

McLellan CA et al. (2007) A rhizosphere fungus enhances *Arabidopsis* thermotolerance through production of an HSP90 inhibitor. *Plant Physiol* 145:174–182.

Mesitov MV et al. (2017) Differential processing of small RNAs during endoplasmic reticulum stress. *Sci Rep* 7:46080.

Meyer Y, Buchanan BB, Vignols F, Reichheld J-P (2009) Thioredoxins and glutaredoxins: Unifying elements in redox biology. *Annu Rev Genet* 43:335–367.

Miller G, Mittler R (2006) Could heat shock transcription factors function as hydrogen peroxide sensors in plants?. *Ann Bot* 98:279–288.

Milos P, Roy H (1984) ATP-released large subunits participate in the assembly of RuBP carboxylase. *J Cell Biochem* 24:153–162.

Minder AC, De Rudder KEE, Narberhaus F, Fischer H-M, Hennecke H, Geiger O (2001) Phosphatidylcholine levels in *Bradyrhizobium japonicum* membranes are critical for an efficient symbiosis with the soybean host plant. *Mol Microbiol* 39:1186–1198.

Mir AA, Park S-Y, Sadat MA, Kim S, Choi J, Jeon J, Lee Y-H (2015) Systematic characterization of the peroxidase gene family provides new insights into fungal pathogenicity in *Magnaporthe oryzae*. *Sci Rep* 5:11831.

Monneau YR, Soufari H, Nelson CJ, Mackereth CD (2013) Structure and activity of the peptidyl-prolyl isomerase domain from the histone chaperone Fpr4 toward histone H3 proline isomerization. *J Biol Chem* 288:25826–25837.

Moon JC et al. (2015) Overexpression of Arabidopsis NADPH-dependent thioredoxin reductase C (AtNTRC) confers freezing and cold shock tolerance to plants. *Biochem Biophys Res Commun* 463:1225–1229.

Morimoto RI (1998) Regulation of the heat shock transcriptional response: Cross talk between a family of heat shock factors, molecular chaperones, and negative regulators. *Genes Dev* 12:3788–3796.

Morsy MR, Oswald J, He J, Tang Y, Roossinck MJ (2010) Teasing apart a three-way symbiosis: Transcriptome analyses of *Curvularia protuberata* in response to viral infection and heat stress. *Biochem Biophys Res Commun* 401:225–230.

Mosser DD, Kotzbauer PT, Sarge KD, Morimoto RI (1990) In vitro activation of heat shock transcription factor DNA-binding by calcium and biochemical conditions that affect protein conformation. *Proc Natl Acad Sci USA* 87:3748–3752.

Muthusamy SK, Dalal M, Chinnusamy V, Bansal KC (2016) Differential regulation of genes coding for organelle and cytosolic ClpATPases under biotic and abiotic stresses in wheat. *Front Plant Sci* 7:929.

Muthusamy SK, Dalal M, Chinnusamy V, Bansal KC (2017) Genome-wide identification and analysis of biotic and abiotic stress regulation of small heat shock protein (HSP20) family genes in bread wheat. *J Plant Physiol* 211:100–113.

Muzi C, Camoni L, Visconti S, Aducci P (2016) Cold stress affects H⁺-ATPase and phospholipase D activity in Arabidopsis. *Plant Physiol Biochem* 108:328–336.

Myouga F, Motohashi R, Kuromori T, Nagata N, Shinozaki K (2006) An Arabidopsis chloroplast-targeted Hsp101 homologue, APG6, has an essential role in chloroplast development as well as heat-stress response. *Plant J* 48:249–260.

Nahar K, Hasanuzzaman M, Alam MM, Rahman A, Mahmud J-A, Suzuki T, Fujita M (2017) Insights into spermine-induced combined high temperature and drought tolerance in mung bean: Osmoregulation and roles of antioxidant and glyoxalase system. *Protoplasma* 254:445–460.

Nakaminami K, Karlson DT, Imai R (2006) Functional conservation of cold shock domains in bacteria and higher plants. *Proc Natl Acad Sci USA* 103:10122–10127.

Nawrot R, Tomaszewski L, Czerwoniec A, Gozdzicka-Jozefiak A (2013) Identification of a coding sequence and structure modeling of a glycine-rich RNA-binding protein (CmGRP1) from *Chelidonium majus* L. *Plant Mol Biol Rep* 31:470–476.

Netto LE, Antunes F (2016) The roles of peroxiredoxin and thioredoxin in hydrogen peroxide sensing and in signal transduction. *Mol Cells* 39:65–71.

Neupert W, Hartl F-U, Craig EA, Pfanner N (1990) How do polypeptides cross the mitochondrial membranes? *Cell* 63:447–450.

Ni F-T, Chu L-Y, Shao H-B, Liu Z-H (2009) Gene expression and regulation of higher plants under soil water stress. *Curr Genomics* 10:269–280.

Nitika, Truman AW (2017) Cracking the chaperone code: Cellular roles for Hsp70 phosphorylation. *Trends Biochem Sci* 42:932–935.

Nomata T, Kabeya Y, Sato N (2004) Cloning and characterization of glycine-rich RNA-binding protein cDNAs in the moss physcomitrella patens. *Plant Cell Physiol* 45:48–56.

Nouri MZ, Hiraga S, Yanagawa Y, Sunohara Y, Matsumoto H, Komatsu S (2012) Characterization of calnexin in soybean roots and hypocotyls under osmotic stress. *Phytochem* 74:20–29.

Oger E, Marino D, Guigonis J-M, Pauly N, Puppo A (2012) Sulfenylated proteins in the *Medicago truncatula–Sinorhizobium meliloti* symbiosis. *J Proteom* 75:4102–4113.

Ono Y, Mitsui H, Sato T, Minamisawa K (2001) Two RpoH homologs responsible for the expression of heat shock protein genes in *Sinorhizobium meliloti*. *Mol Gen Genet* 264:902–912.

Ou W-B, Luo W, Park Y-D, Zhou H-M (2001) Chaperone-like activity of peptidyl-prolyl cis-trans isomerase during creatine kinase refolding. *Protein Sci* 10:2346–2353.

Owttrim GW (2013) RNA helicases: Diverse roles in prokaryotic response to abiotic stress. *RNA Biol* 10:96–110.

Paço A, Brígido C, Alexandre A, Mateos PF, Oliveira S (2016) The symbiotic performance of chickpea rhizobia can be improved by additional copies of the clpB chaperone gene. *PLOS ONE* 11:e0148221.

Pan S et al. (2014) Cell surface protein disulfide isomerase regulates natriuretic peptide generation of cyclic guanosine monophosphate. *PLOS ONE* 9:e112986.

Panter S, Thomson R, de Bruxelles G, Laver D, Trevaskis B, Udvardi M (2000) Identification with proteomics of novel proteins associated with the peribacteroid membrane of soybean root nodules. *Mol Plant Microbe Interact* 13:325–333.

Park SJ, Kwak KJ, Jung HJ, Lee HJ, Kang H (2010) The C-terminal zinc finger domain of Arabidopsis cold shock domain proteins is important for RNA chaperone activity during cold adaptation. *Phytochem* 71:543–547.

Parsell DA, Lindquist S (1993) The function of heat-shock proteins in stress tolerance: Degradation and reactivation of damaged proteins. *Annu Rev Genet* 27:437–496.

Patil AN, Tailhades J, Hughes AR, Separovic F, Wade DJ, Hossain AM (2015) Cellular disulfide bond formation in bioactive peptides and proteins. *Int J Mol Sci* 16:1791–1805.

Phukan UJ, Jeena GS, Tripathi V, Shukla RK (2017) Regulation of Apetala2/Ethylene response factors in plants. *Front Plant Sci* 8:150.

Pourghayoumi M, Rahemi M, Bakhshi D, Aalami A, Kamgar-Haghighi AA (2017) Responses of pomegranate cultivars to severe water stress and recovery: Changes on antioxidant enzyme activities, gene expression patterns and water stress responsive metabolites. *Physiol Mol Biol Plants* 23:321–330.

Preissler S et al. (2015) Physiological modulation of BiP activity by trans-protomer engagement of the interdomain linker. *eLife* 4:e08961.

Probst AV, Mittelsten Scheid O (2015) Stress-induced structural changes in plant chromatin. *Curr Opin Plant Biol* 27:8–16.

Pulido P, Llamas E, Rodriguez-Concepcion M (2017) Both Hsp70 chaperone and Clp protease plastidial systems are required for protection against oxidative stress. *Plant Signal Behav* 12:e1290039.

Rajan VBV, D'Silva P (2009) *Arabidopsis thaliana* J-class heat shock proteins: Cellular stress sensors. *Funct Integr Genomics* 9:433.

Rajaram H, Apte SK (2008) Nitrogen status and heat-stress-dependent differential expression of the cpn60 chaperonin gene influences thermotolerance in the cyanobacterium *Anabaena*. *Microbiol (Reading, England)* 154:317–325.

Rajkowitsch L et al. (2007) RNA chaperones, RNA annealers and RNA helicases. *RNA Biol* 4:118–130.

Räsänen LA, Saijets S, Jokinen K, Lindström K (2004) Evaluation of the roles of two compatible solutes, glycine betaine and trehalose, for the *Acacia senegal–Sinorhizobium* symbiosis exposed to drought stress. *Plant Soil* 260:237–251.

Ré MD, Gonzalez C, Escobar MR, Sossi ML, Valle EM, Boggio SB (2017) Small heat shock proteins and the postharvest chilling tolerance of tomato fruit. *Physiol Plant* 159:148–160.

Reddy PS, Chakradhar T, Reddy RA, Nitnavare RB, Mahanty S, Reddy MK (2016) Role of heat shock proteins in improving heat stress tolerance in crop plants. In: Asea AAA, Kaur P, Calderwood SK (eds) *Heat Shock Proteins and Plants*. Springer International Publishing, Cham, pp. 283–307.

Reeg S, Jung T, Castro JP, Davies KJA, Henze A, Grune T (2016) The molecular chaperone Hsp70 promotes the proteolytic removal of oxidatively damaged proteins by the proteasome. *Free Radic Biol Med* 99:153–166.

Reichheld JP, Meyer E, Khafif M, Bonnard G, Meyer Y (2005) AtNTRB is the major mitochondrial thioredoxin reductase in *Arabidopsis thaliana*. *FEBS Lett* 579:337–342.

Rennella E et al. (2017) RNA binding and chaperone activity of the E. coli cold-shock protein CspA. *Nucleic Acids Res* 45:4255–4268.

Rikhvanov EG, Rachenko EI, Varakina NN, Rusaleva TM, Borovskii GB, Voĭnikov VK (2004) Induction of synthesis of Hsp104 of *Saccharomyces cerevisiae* in heat shock is controlled by mitochondria. *Genetika* 40:437–444

Rizvi SM, Mancino L, Thammavongsa V, Cantley RL, Raghavan M (2004) A polypeptide binding conformation of calreticulin is induced by heat shock, calcium depletion, or by deletion of the C-terminal acidic region. *Mol Cell* 15:913–923.

Robledo M et al. (2017) A conserved alpha-proteobacterial small RNA contributes to osmoadaptation and symbiotic efficiency of rhizobia on legume roots. *Environ Microbiol* 19:2661–2680.

Rodríguez-Manzaneque MT, Ros J, Cabiscol E, Sorribas A, Herrero E (1999) Grx5 glutaredoxin plays a central role in protection against protein oxidative damage in *Saccharomyces cerevisiae*. *Mol Cell Biol* 19:8180–8190.

Saibil H (2013) Chaperone machines for protein folding, unfolding and disaggregation. *Nat Rev Mol Cell biology* 14:630–642.

Sakurai H, Ota A (2011) Regulation of chaperone gene expression by heat shock transcription factor in Saccharomyces cerevisiae: Importance in normal cell growth, stress resistance, and longevity. *FEBS Lett* 585:2744–2748.

Santiago-Frangos A, Kavita K, Schu DJ, Gottesman S, Woodson SA (2016) C-terminal domain of the RNA chaperone Hfq drives sRNA competition and release of target RNA. *Proc Natl Acad Sci USA* 113:E6089–E6096.

Santos CX et al. (2009) Protein disulfide isomerase (PDI) associates with NADPH oxidase and is required for phagocytosis of *Leishmania chagasi* promastigotes by macrophages. *J Leukoc Biol* 86:989–998.

Sasaki K, Kim MH, Imai R (2007) Arabidopsis COLD SHOCK DOMAIN PROTEIN2 is a RNA chaperone that is regulated by cold and developmental signals. *Biochem Biophys Res Commun* 364:633–638.

Sasaki K, Kim M-H, Imai R (2013) Arabidopsis COLD SHOCK DOMAIN PROTEIN 2 is a negative regulator of cold acclimation. *New Phytol* 198:95–102.

Sasaki K, Kim MH, Kanno Y, Seo M, Kamiya Y, Imai R (2015a) Arabidopsis cold shock domain protein 2 influences ABA accumulation in seed and negatively regulates germination. *Biochem Biophys Res Commun* 456:380–384.

Sasaki K, Liu Y, Kim MH, Imai R (2015b) An RNA chaperone, AtCSP2, negatively regulates salt stress tolerance. *Plant Signal Behav* 10:e1042637.

Sasaki S, Minamisawa K, Mitsui H (2016) A *Sinorhizobium meliloti* RpoH-regulated gene is involved in iron-sulfur protein metabolism and effective plant symbiosis under intrinsic iron limitation. *J Bacteriol* 198:2297–2306.

Satoh R, Hagihara K, Sugiura R (2017) Rae1-mediated nuclear export of Rnc1 is an important determinant in controlling MAPK signaling. *Curr Genet* 63:103–108.

Scandalios JG (2005) Oxidative stress: Molecular perception and transduction of signals triggering antioxidant gene defenses. *Braz J Med Biol Res* 38:995–1014.

Scharf K-D, Heider H, Höhfeld I, Lyck R, Schmidt E, Nover L (1998a) The tomato Hsf system: HsfA2 needs interaction with HsfA1 for efficient nuclear import and may be localized in cytoplasmic heat stress granules. *Mol Cell Biol* 18:2240–2251.

Schlecht R, Erbse AH, Bukau B, Mayer MP (2011) Mechanics of Hsp70 chaperones enables differential interaction with client proteins. *Nat Struct Mol Biol* 18:345–351.

Schröder G, Frühling M, Pühler A, Perlick AM (1997) The temporal and spatial transcription pattern in root nodules of *Vicia faba* nodulin genes encoding glycine-rich proteins. *Plant Mol Biol* 33:113–123.

Sekhar A, Rosenzweig R, Bouvignies G, Kay LE (2016) Hsp70 biases the folding pathways of client proteins. *Proc Natl Acad Sci USA* 113:E2794–E2801.

Seo JS, Lee Y-M, Park HG, Lee J-S (2006) The intertidal copepod *Tigriopus japonicus* small heat shock protein 20 gene (Hsp20) enhances thermotolerance of transformed *Escherichia coli*. *Biochem Biophys Res Commun* 340:901–908.

Serrato AJ, Cejudo FJ (2003) Type-h thioredoxins accumulate in the nucleus of developing wheat seed tissues suffering oxidative stress. *Planta* 217:392–399.

Servant P, Mazodier P (1996) Heat induction of hsp18 gene expression in Streptomyces albus G: Transcriptional and posttranscriptional regulation *J Bacteriol* 178:7031–7036.

Sharma S, Sarkar S, Paul SS, Roy S, Chattopadhyay K (2013) A small molecule chemical chaperone optimizes its unfolded state contraction and denaturant like properties. *Sci Rep* 3:3525.

Shikanai T, Shimizu K, Ueda K, Nishimura Y, Kuroiwa T, Hashimoto T (2001) The chloroplast *clpP* gene, encoding a proteolytic subunit of ATP-dependent protease, is indispensable for chloroplast development in tobacco. *Plant Cell Physiol* 42:264–273.

Shivakumara TN et al. (2017) Overexpression of Pea DNA Helicase 45 (PDH45) imparts tolerance to multiple abiotic stresses in chili (*Capsicum annuum* L.). *Sci Rep* 7:2760.

Sjögren LLE, MacDonald TM, Sutinen S, Clarke AK (2004) Inactivation of the *clpC1* gene encoding a chloroplast Hsp100 molecular chaperone causes growth retardation, leaf chlorosis, lower photosynthetic activity, and a specific reduction in photosystem content. *Content Plant Physiol* 136:4114–4126.

Smykal P, Masin J, Hrdy I, Konopasek I, Zarsky V (2000) Chaperone activity of tobacco HSP18, a small heat-shock protein, is inhibited by ATP. *Plant J* 23:703–713.

Snyman M, Cronjé MJ (2008) Modulation of heat shock factors accompanies salicylic acid-mediated potentiation of Hsp70 in tomato seedlings. *J Exp Bot* 59:2125–2132.

Song H, Zhao R, Fan P, Wang X, Chen X, Li Y (2009) Overexpression of AtHsp90.2, AtHsp90.5 and AtHsp90.7 in *Arabidopsis thaliana* enhances plant sensitivity to salt and drought stresses. *Planta* 229:955–964.

Soper T, Mandin P, Majdalani N, Gottesman S, Woodson SA (2010) Positive regulation by small RNAs and the role of Hfq. *Proc Natl Acad Sci USA* 107:9602–9607.

Stampfl S, Doetsch M, Beich-Frandsen M, Schroeder R (2013) Characterization of the kinetics of RNA annealing and strand displacement activities of the *E. coli* DEAD-box helicase CsdA. *RNA Biol* 10:149–156.

Sugiura R, Satoh R, Ishiwata S, Umeda N, Kita A (2011) Role of RNA-binding proteins in MAPK signal transduction pathway. *J Signal Transduct* 2011:8.

Sun J-h, Chen J-y, Kuang J-f, Chen W-x, Lu W-j (2010) Expression of *sHSP* genes as affected by heat shock and cold acclimation in relation to chilling tolerance in plum fruit. *Postharvest Biol Technol* 55:91–96.

Svensson MJ, Larsson J (2007) Thioredoxin-2 affects lifespan and oxidative stress in *Drosophila*. *Hereditas* 144:25–32.

Swindell WR, Huebner M, Weber AP (2007) Transcriptional profiling of Arabidopsis heat shock proteins and transcription factors reveals extensive overlap between heat and non-heat stress response pathways. *BMC Genomics* 8:125.

Tanou G, Filippou P, Belghazi M, Job D, Diamantidis G, Fotopoulos V, Molassiotis A (2012) Oxidative and nitrosative-based signaling and associated post-translational modifications orchestrate the acclimation of citrus plants to salinity stress. *Plant J* 72:585–599.

Tasseva G, Richard L, Zachowski A (2004) Regulation of phosphatidylcholine biosynthesis under salt stress involves choline kinases in *Arabidopsis thaliana*. *FEBS Lett* 566:115–120.

Tavladoraki P, Cona A, Angelini R (2016) Copper-containing amine oxidases and FAD-dependent polyamine oxidases are key players in plant tissue differentiation and organ development. *Front Plant Sci* 7:1–11.

Thapar R (2015) Roles of prolyl isomerases in RNA-mediated gene expression. *Biomolecules* 5:974–999.

Ting Y-H, Lu T-J, Johnson AW, Shie J-T, Chen B-R, Kumar.S S, Lo K-Y (2016) Bcp1 is the nuclear chaperone of the 60S ribosomal protein Rpl23 in *Saccharomyces cerevisiae*. *J Biol Chem* 292:585–596.

Tkach JM, Glover JR (2008) Nucleocytoplasmic trafficking of the molecular chaperone Hsp104 in unstressed and heat-shocked cells. *Traffic (Copenhagen, Denmark)* 9:39–56.

Török Z et al. (2001) Synechocystis HSP17 is an amphitropic protein that stabilizes heat-stressed membranes and binds denatured proteins for subsequent chaperone-mediated refolding. *Proc Natl Acad Sci USA* 98:3098–3103.

Torres-Quesada O et al. (2010) The *Sinorhizobium meliloti* RNA chaperone Hfq influences central carbon metabolism and the symbiotic interaction with alfalfa. *BMC Microbiology* 10:71.

Treiber T et al. (2017) A compendium of RNA-binding proteins that regulate MicroRNA. *Biogenesis Mol Cell* 66:270–284.e213.

Tripathi AK, Pareek A, Singla-Pareek SL (2016) A NAP-family histone chaperone functions in abiotic stress response and adaptation. *Plant Physiol* 171:2854–2868.

Truscott KN, HØJ PB, Scopes RK (1994) Purification and characterization of chaperonin 60 and chaperonin 10 from the anaerobic thermophile *Thermoanaerobacter brockii*. *Eur J Biochem* 222:277–284.

Tsanov KM, Daley GQ (2017) Signaling through RNA-binding proteins as a cell fate regulatory mechanism. *Cell Cycle* 16:723–724.

van der Knaap JA, Verrijzer CP (2016) Undercover: Gene control by metabolites and metabolic enzymes. *Genes Dev* 30:2345–2369.

Verma DPS, Fortin MG, Stanley J, Mauro VP, Purohit S, Morrison N (1986) Nodulins and nodulin genes of *Glycine max*. *Plant Mol Biol* 7:51–61.

Vieira Dos Santos C, Rey P (2006) Plant thioredoxins are key actors in the oxidative stress response. *Trends Plant Sci* 11:329–334.

Viitanen PV, Lubben TH, Reed J, Goloubinoff P, O'Keefe DP, Lorimer GH (1990) Chaperonin-facilitated refolding of ribulose bisphosphate carboxylase and ATP hydrolysis by chaperonin 60 (groEL) are potassium dependent. *Biochem* 29:5665–5671.

Vogel J, Luisi BF (2011) Hfq and its constellation of RNA. *Nat Rev Microbiol* 9:578.

Walters C, Errington N, Rowe AJ, Harding SE (2002) Hydrolysable ATP is a requirement for the correct interaction of molecular chaperonins cpn60 and cpn10. *Biochem J* 364:849–855.

Wang D, Huang J, Hu Z (2012b) RNA helicase DDX5 regulates microRNA expression and contributes to cytoskeletal reorganization in basal breast cancer cells. *Mol Cell Proteomics* 11:M111.011932.

Wang D, Qin B, Li X, Tang D, Zhang Ye, Cheng Z, Xue Y (2016a) Nucleolar DEAD-box RNA helicase TOGR1 regulates thermotolerant growth as a pre-rRNA chaperone in rice. *PLOS Genet* 12:e1005844.

Wang M, Zou Z, Li Q, Xin H, Zhu X, Chen X, Li X (2017) Heterologous expression of three *Camellia sinensis* small heat shock protein genes confers temperature stress tolerance in yeast and *Arabidopsis thaliana*. *Plant Cell Rep* 36:1125–1135.

Wang Q, Liu N, Yang X, Tu L, Zhang X (2016b) Small RNA-mediated responses to low- and high-temperature stresses in cotton. *Sci Rep* 6:35558.

Wang W, Vinocur B, Shoseyov O, Altman A (2004) Role of plant heat-shock proteins and molecular chaperones in the abiotic stress response. *Trends Plant Sci* 9:244–252.

Wang X, Lu X-a, Song X, Zhuo W, Jia L, Jiang Y, Luo Y (2012a) Thr90 phosphorylation of Hsp90α by protein kinase A regulates its chaperone machinery. *Biochem J* 441:387–397.:10.1042/bj20110855

Watanabe Y-h, Motohashi K, Taguchi H, Yoshida M (2000) Heat-inactivated proteins managed by DnaKJ-GrpE-ClpB chaperones are released as a chaperonin-recognizable non-native. *Form J Biol Chem* 275:12388–12392.

Waters ER (2013) The evolution, function, structure, and expression of the plant sHSPs. *J Exp Bot* 64:391–403.

Wayne N, Mishra P, Bolon DN (2011) Hsp90 and client protein maturation. *Methods Mol Biol (Clifton, NJ)* 787:33–44.

Weng M, Yang YUE, Feng H, Pan Z, Shen W-H, Zhu YAN, Dong A (2014) Histone chaperone ASF1 is involved in gene transcription activation in response to heat stress in *Arabidopsis thaliana*. *Plant Cell Environ* 37:2128–2138.

Wistow G (1990) Cold shock and DNA binding. *Nature* 344:823–824.

Wolffe AP (1994) Structural and functional properties of the evolutionarily ancient Y-box family of nucleic acid binding proteins. *BioEssays* 16:245–251.

Wolffe AP, Tafuri S, Ranjan M, Familari M (1992) The Y-box factors: A family of nucleic acid binding proteins conserved from *Escherichia coli* to man. *New Biol* 4:290–298

Woodson JD (2016) Chloroplast quality control - balancing energy production and stress. *New Phytol* 212:36–41.

Wu H, Dorse S, Bhave M (2012) In silico identification and analysis of the protein disulphide isomerases in wheat and rice. *Biologia* 67:48–60.

Yamada K, Fukao Y, Hayashi M, Fukazawa M, Suzuki I, Nishimura M (2007) Cytosolic HSP90 regulates the heat shock response that is responsible for heat acclimation in *Arabidopsis thaliana*. *J Biol Chem* 282:37794–37804.

Yang HY, Zhang Y, Sun M (2016) Molecular cloning and expression analysis of heat shock protein 70 from *Chrysanthemum nankingense* under heat shock. In: Acta Hortic, Leuven, Belgium, pp. 147–154.

Yu J et al. (2016) Genome-wide identification and expression profiling of tomato hsp20 gene family in response to biotic and abiotic stresses. *Front Plant Sci* 7:1215.

Yu Jin J et al. (2014) Overexpression of *Oshsp16.9* gene encoding small heat shock protein enhances tolerance to abiotic stresses in rice plant. *Breed Biotechnol* 2:370–379.

Yu Y et al. (2017) A novel AP2/ERF family transcription factor from *Glycine soja*, GsERF71, is a DNA binding protein that positively regulates alkaline stress tolerance in Arabidopsis. *Plant Mol Biol* 94:509–530.

Yujin J, Sungwhan Y, Sungoh I, Won-Joong J, EunJeong P, Dong-Woog C (2017) Overexpression of the small heat shock protein, PtsHSP19.3 from marine red algae, *Pyropiatenera* (Bangiales, Rhodophyta) enhances abiotic stress tolerance in *Chlamydomonas*. *J Plant Biotechnol* 44:287–295.

Zeng H, Xu L, Singh A, Wang H, Du L, Poovaiah BW (2015) Involvement of calmodulin and calmodulin-like proteins in plant responses to abiotic stresses. *Front Plant Sci* 6:600.

Zhai M, Sun Y, Jia C, Peng S, Liu Z, Yang G (2016) Over-expression of *JrsHSP17.3* gene from *Juglans regia* confer the tolerance to abnormal temperature and NaCl stresses. *J Plant Biol* 59:549–558.

Zhang C-J, Zhao B-C, Ge W-N, Zhang Y-F, Song Y, Sun D-Y, Guo Y (2011a) An apoplastic H-type thioredoxin is involved in the stress response through regulation of the apoplastic reactive oxygen species in rice plant. *Physiol* 157:1884–1899.

Zhang H et al. (2015b) A bipartite interaction between Hsp70 and CHIP regulates ubiquitination of chaperoned client proteins. *Structure (London, England: 1993)* 23:472–482.

Zhang K et al. (2015a) A novel mechanism for small heat shock proteins to function as molecular chaperones. *Sci Rep* 5:8811.

Zhang X, Guo S-L, Yin H-B, Xiong D-J, Zhang H, Zhao Y-X (2004) Molecular cloning and identification of a heat shock cognate protein 70 gene, *Thhsc70*, in *Thellungiella halophila*. *Acta Bot Sin* 46:1212–1219.

Zhang Y, Hong G (2009) Post-transcriptional regulation of *NifA* expression by Hfq and RNase E complex in *Rhizobium leguminosarum* bv. viciae. *Acta Biochim Biophys Sin* 41:719–730.

Zhang Y, You J, Wang X, Weber J (2015c) The DHX33 RNA helicase promotes mRNA translation initiation. *Mol Cell Biol* 35:2918–2931.

Zhang Z et al. (2011b) DDX1, DDX21, and DHX36 helicases form a complex with the adaptor molecule TRIF to sense dsRNA in Dendritic. *Cells Immun* 34:866–878.

Zhao X, Jain C (2011) DEAD-box proteins from *Escherichia coli* exhibit multiple ATP-independent activities. *J Bacteriol* 193:2236–2241.

Zhong L et al. (2013) Chloroplast small heat shock protein HSP21 interacts with plastid nucleoid protein pTAC5 and is essential for chloroplast development in *Arabidopsis* under heat stress. *Plant Cell* 25:2925–2943.

Zhu C, Ding Y, Liu H (2011a) MiR398 and plant stress responses. *Physiol Plant* 143:1–9.

Zhu C et al. (2014) Molecular characterization and expression profiling of the protein disulfide isomerase gene family in *Brachypodium distachyon*. *L PLoS ONE* 9:e94704.

Zhu J, Dong C-H, Zhu J-K (2007) Interplay between cold-responsive gene regulation, metabolism and RNA processing during plant cold acclimation. *Curr Opin Plant Biol* 10:290–295.

Zhu Y, Weng M, Yang Y, Zhang C, Li Z, Shen W-H, Dong A (2011b) Arabidopsis homologues of the histone chaperone ASF1 are crucial for chromatin replication and cell proliferation in plant development. *Plant J* 66:443–455.

Zwirowski S et al. (2017) Hsp70 displaces small heat shock proteins from aggregates to initiate protein refolding. *EMBO J* 36:783–796.

Section VIII

Examples of Empirical Investigations of Specific Plants and Crops Grown under Salt, Drought, and Other Environmental Stress Conditions

33 Abiotic Stress Impact and Tolerance of Natural Sweetener Plant Stevia

Rout Nutan Prasad, Rodríguez-Garay Benjamín, Barranco-Guzmán Angel Martín, Gómez-Entzin Veronica, and Rincón-Hernández Manuel

CONTENTS

33.1 INTRODUCTION

Stevia rebuaidana Bertoni is an important crop because it produces the high potency sweetener rebaudioside-A and other compounds with many health benefits. Paraguay is the country of origin of *Stevia rebuaidana* Bertoni. It is a perennial plant, of the *Asteraceae* family and its foliar parts contain diterpenoid steviol glycosides (SGs), a non-toxic, high potency sweetener being nearly 300 times sweeter than sucrose but non-calorific, and it is used in many countries of the world, including many developed countries (Singh and Rao, 2005; Yadav et al., 2011). Geuns (2003) reported that nearly 30% of stevia's dry weight of leaves contains SGs group of diterpene glycosides and its major part is stevioside whose yield is 0.6–7.9%, followed by rebaudioside-A (reb-A) whose yield is 0.3–6.5% (w/w) (Vouillamoz et al., 2016), and other SGs present in smaller amounts are: steviolbioside, rebaudioside-B, -C, -D, -E, -F, and dulcoside-A (Starratt et al., 2002; Ceunen and Geuns, 2013; Well et al., 2013). Apart from the non-calorific sweetener, it has been reported as having many medicinal uses such as anticariogenic, antineoplastic, antihypertensive, anti-inflammatory, and antihyperglycemic by many scientists (Mohd-Radzman et al., 2013; Aranda-González et al., 2014). The stevia market was estimated to value at US$ 347 million in 2014 and it is expected to reach US$ 565.2 million by 2020 (https://www.futuremarketinsights.com/reports/global-stevia-market).

Stress or biological stress is defined as disruption or modulation of hemostasis of the cell or organism, and plant stress as any external factor that negatively influences growth, productivity, reproductive capacity, or survival. This can be broadly divided into two main categories: abiotic stress and biotic stress (Rhodes and Nadolska-Orczyk, 2001; Ördög and Molnár, 2011). Abiotic stress is defined as non-living factors which disrupt or modulate the hemostasis of the cell negatively on living organisms in a specific environment. The main abiotic stresses are deficiencies or excesses of water (drought and flooding), low or high temperature, deficiencies or excesses of nutrients, high salinity, heavy metals, and other environmental extremes, which are the major causes of poor plant growth and reduced agricultural crop yields around the world (Acquaah, 2012).

Stevia is believed to have a moderate tolerance to abiotic stress; however, it affects the plant in all stages of its growth and development (Fallah et al., 2017), influences growth and physiological processes (Aggarwal et al., 2013; Karimi et al., 2014; Bidabadi and Masoumian, 2018; Hajihashemi et al., 2018b), and osmotic stress of stevia (Hajihashemi and Ehsanpour, 2013; Cantabella et al., 2017; Moradi-Peynevandi et al., 2018) to a great extent. Abiotic stress also induces oxidative stress, and stevia has a well advanced protective mechanism to cope with this stress condition (Hendawey and Reda, 2015; Afshar and Ansari, 2017; Cantabella et al., 2017; Bidabadi and Masoumian, 2018; Hajihashemi and Sofo, 2018). Effects and responses of the stevia plant to various components of abiotic stress are shown in Figure 33.1.

Considering the importance of this plant, there are scanty reports on an abiotic stress-related study on it. This may be because of its commercial use, or that researchers are not publishing the information or are not interested. In any case, scientists need to step up and investigate the abiotic tolerance mechanism and improvement of steviol glycoside production in stevia.

33.2 ABIOTIC STRESS CONCEPT IN PLANTS

The three main menaces of the twenty-first century are climate change, poverty, and food insecurity (Singh et al., 2018), and there exists evidence that shows that earth has undergone warming since the middle of the nineteenth century (Lobell et al., 2011; Wheeler and von Braun, 2013). The abiotic stress caused by climate change is currently a major threat to food production. To cope with abiotic stress, plants can respond with physiological changes and cellular processes to cope with this danger. It is important for man to study and understand the mechanisms that plants naturally have to defend themselves from all types of stresses in order to genetically improve crops or to develop management strategies to ensure the production of food. The use of the plant genetic resources (germplasm) of crops, wild plants, and landraces is one of the best alternatives to alleviate the stress problem suffered by plants. Wild relatives with a good tolerance to abiotic stress are important for plant breeders, especially when a cultivated species has genetic diversity limitations (Huang et al., 2013).

Wild biodiversity as well as crops, are negatively affected by climate change, and on the other hand biodiversity contributes to mitigating diverse kinds of stress; thus, knowledge, conservation and sustainable use of biodiversity are important elements to solve many problems imposed by climate change, mainly for the production of food (Fei et al., 2017; Singh et al.,

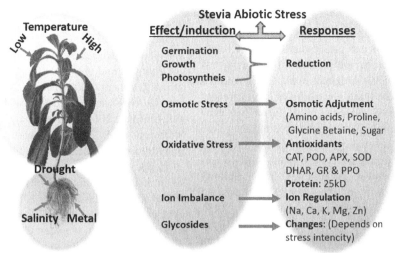

Effect and response of stevia plant to various components of abiotic stress

FIGURE 33.1 (See color insert.) Abiotic stress tolerance of stevia.

2018). In this regard, Cramer et al. (2011) reviewed the most recent progress on the systematic analyses (systems biology) of plant complex responses to abiotic stress that include transcriptomics, metabolomics, proteomics, and other integrated approaches, from the physiological to the molecular level.

Lichtenthaler (1996, 1998) defined plant stress as "any unfavorable condition or substance that affects or blocks a plant's metabolism, growth or development." Stress and strain are terms taken from physics that mean limits, "after which the material deforms elastically (it can bend back, strain), then plastically (it cannot bend back, stress) until it ruptures"; in other words, an "elastic response" implies a reversible damage that can be repaired, so that the plant viability is maintained. A "plastic response" entails irreversible damage meaning that the repair mechanisms do not work (Kranner et al., 2010). Following Lichtenthaler (1996) the term "eu-stress" is a kind of stress that is a positive element for plant development, while "dis-stress" is a negative stress that may lead the plant to death.

In this context, the types of short-term and long-term stresses must be well defined and understood. The first ones, if they are mild, can be repaired, while the second ones can cause considerable damage, producing little or no yield or finally death (Lichtenthaler, 1996).

33.3 ABIOTIC STRESS IN STEVIA

33.3.1 SALINITY STRESS

Excess salt in the soil and water termed as salinity is the most important environmental stress factor which adversely affects agriculture production worldwide (Munns and Tester, 2008). The whole landmass is also contaminated with different kinds of salts (Pasternak, 1987; Szabolcs, 1994). The most abundant salt is Na^+ constituting nearly two-thirds followed by Mg^{2+}, Ca^{2+}, and K^+ (Curtion et al., 1993; Egan and Unger, 1998). The estimated degraded soil by salinity is about 77 million

hectares (mha) around the world (Ghassemi et al., 1995; Munns and Rechards, 1996) and is a growing problem worldwide; as many as 1.5 million hectares are taken out of production each year as a result of it (Munns and Tester, 2008). Salinity affects plants through a series of stress events such as water imbalance, ion homeostasis, nutritional disorders, osmotic adjustment, generation of oxidative radicals, membrane disorder, and genotoxicity, etc. (Hasegawa et al., 2000; Munns, 2002; Zhu, 2007; Carillo et al., 2011). Almost all the physiological processes are affected by excess salt; it regulates pigmentation, reduces photosynthesis, and affects respiration in plants (Brown et al., 1987; Belkhodja, 1999; Munns et al., 2006; Hodges et al., 2016). Salinity reduces the uptake of the essential ions, imbalances the ionic concentration and osmotic potential within the cell (Reddy and Iyenger, 1999; Munns, 2005; Tang et al., 2015). Plants try to adjust the osmotic process by the accumulation of organic molecules (Horie et al., 2011) and changes the concentration of total organic acids (Perez-Alfocea et al., 1994; Acosta-Motos et al., 2017), proline (Zhang et al., 1995; Szabados and Savoure, 2010; Tang et al, 2015), soluble sugars (Borrelli et al., 2018), glycine betaine (Santa-Cruz et al., 1997; Ashraf and Fooland, 2007), nitrogen compounds, and carbohydrates (Flower et al., 1977).

Oxidative stress is the secondary effect of salinity leading to irreversible damage and metabolic dysfunction, and cell death (Vaidyanathan et al., 2003; Fenollosa and Munné-Bosch, 2018). Plants have well defined antioxidative defense systems and contain superoxide dismutase (SOD) that can remove excess superoxide anions converting H_2O_2 from O_2 and then further reducing it to H_2O by catalase and peroxidase (Wang et al., 2016). Another level of defense system in plants is through induction of protein genes during salinity tolerance by making ionic adjustment (He et al., 2005) enhancing vacuolar H^+ transport (Park et al., 2009), improving ABA-dependent regulation (Liu et al., 2014), and increasing the activity of antioxidant enzymes level (Tanaka et al., 1999).

33.3.2 Salinity Stress in Stevia

Stevia rebaudiana is a herbaceous perennial shrub which contains non-caloric sweeteners and is cultivated for its sweetening steviol glycosides, as a sugar substitute for diabetic patients and its medicinal use (Mohd-Radzman et al., 2013; Aranda-González et al., 2014). The main glycosides of stevia are stevioside and rebaudioside-A (Kalpana et al., 2009; Srivastav and Srivastav, 2014). Despite stevia plants accumulating high numbers of steviosides, it is believed to have moderate tolerance to salinity; salinity still affects the plant in all stages of its growth and development (Fallah et al., 2017). Salinity greatly influences the growth and physiological processes of stevia (Akandi et al., 2017; Bidabadi and Masoumian, 2018), including seed production (Liopa-Tsakalidi et al., 2012), growth and net photosynthesis (Bidabadi and Masoumian, 2018), and ionic imbalance (Cantabella et al., 2017). Salinity also induces oxidative stress, lipid peroxidation in stevia (Cantabella et al., 2017; Shahverdi et al., 2017a, 2017b; Bidabadi and Masoumian, 2018), and enhanced rebaudioside-A accumulation (Debnath et al., 2018). It seems to be of high importance to know the effect and mechanism of salt effect and tolerance of this plant for better growth to obtain high biomass and the commercial sweetener rebaudioside-A. Some reports available on salt stress in stevia are summarized in Table 33.1.

33.3.2.1 Effect on Germination and Growth

High salinity concentration reduced the germination percentage and seed germination time in plants (Jamil et al., 2005; Rouhi et al., 2011; Yadav et al., 2011). In stevia, a gradual decrease in the percentage of germination with increasing salt concentration and seed germination and velocity in NaCl was higher than the corresponding ones in NaHCO$_3$ (Liopa-Tsakalidi et al., 2012). It was also reported that salinity reduces germination uniformity, seedling strength, and vigor (Shahverdi et al., 2017b).

On the other hand, growth is a morphological trait with visible symptoms that are not prominent under low concentration treatments. The growth parameters such as biomass, stem diameter, plant height, number of green leaves per plant, leaf fresh weight, root dry weight, and total dry weight linearly decreased with increasing NaCl treatment (Zeng et al., 2013; Pandey and Chikara, 2014, 2015; Reis et al., 2015; Akandi et al., 2017; Hussain et al., 2017; Rameeh et al., 2017; Sarami et al., 2017; Shahverdi et al., 2017c). It has also been observed that leaves turned yellow with a reduction of the diameter of stem and shoot and root development (González et al., 2017). Figure 33.2 shows salinity treatments on *in vitro* plants of stevia (unpublished data).

33.3.2.2 Effect on Photosynthesis

Photosynthesis is the biochemical process of CO$_2$ assimilation, and it is known to be sensitive to many environmental stresses (Dubey, 1997). The effect of abiotic stress on the stevia plant showed that the physiological traits were more sensitive to salinity stress than the morphological traits (Akandi

et al., 2017). Growth and final yield are generally expected to be first reflected by some alterations in the photosynthetic pigment composition and/or their levels. Pigments are primary components for photosynthesis. In stevia, the levels of chlorophylls and chlorophyll a/b ratio have been affected by saline treatment (Akandi et al., 2017). The level of pigmentation decrease was reported in NaCl treatment *in vitro* (Pandey and Chikara, 2014; Sharuti et al., 2014; Rameeh et al., 2017; Shahverdi et al., 2017b; Bidabadi and Masoumian, 2018) and in hydroponic culture of stevia plants (Debnath et al., 2018).

In plants, including stevia, the reduction of photosynthesis in response to high salinity is a considerable stress factor (Bidabadi and Masoumian, 2018). A reduction of the photosynthesis rate in stevia by 47% was detected, and it was linked to a disequilibrium between CO$_2$•$^-$ assimilation and electron transport rates, and due to a decrease in stomatal conductance (Hussain et al., 2017). Further, Hussain et al. (2017) demonstrated that at elevated CO$_2$ levels, enhanced CO$_2$ assimilation in control and high-salt-stressed plants was likely due to significant increases in intercellular CO$_2$ concentration that could be a key factor for salinity tolerance in stevia.

33.3.2.3 Induction of Osmotic Stress and Ion Imbalance

Salinity induces osmotic stress and ion imbalance in the plants. High accumulation of salt in the cells changes ion hemostatic, reduces essential ion uptake which can damage proteins and membranes (Garcia et al., 1997; Rout and Shaw, 2001a; Iqbal et al., 2015; Abdallah et al., 2016), closes stomata opening, and decreases intracellular CO$_2$ concentration (Munns and Tester, 2008; Hussain et al., 2017).

In stevia, salinity enhanced the electrolyte leakage (Bidabadi and Masoumian, 2018), partially changed ionic imbalance, and reports are also available that show an increase in levels of Na, Ca, K, Mg, Zn, and Mn, and a decrease in the concentration of Cu, B, and Mo, neutral in the concentration of P, S, and Fe in *in vitro* experiments (Cantabella et al., 2017); whereas, Debnath et al. (2018) reported an increase of essential ions in lower NaCl concentrations and a decrease in higher concentrations in hydroponic culture. The Na$^+$/K$^+$ ratio in root and shoot also showed a decrease in the *ex vitro* culture of stevia (Shahverdi et al., 2017a).

33.3.2.4 Induction of Oxidative Stress

The production of oxygen radicals during salinity stress is a well-established phenomenon (Rout and Shaw, 2001b; He and Zhu, 2008; Aghaleh et al., 2009). Hydrogen peroxide (H$_2$O$_2$) is an indicator of oxidative stress, as it is generated from superoxide radical (O$_2$•$^-$) which leads to the formation of hydroxyl radical (•OH), causing peroxidative damage to almost all the biomolecules (Shaw et al., 2004; Kim et al., 2005). *In vitro* treatment with NaCl increased lipid peroxidation (MDA), H$_2$O$_2$ and O$_2$•$^-$ was observed in stevia plant (Cantabella et al., 2017; Bidabadi and Masoumian, 2018); a similar report was also confirmed in hydroponic and *ex vitro* stevia cultures (Sarami et al., 2017; Shahverdi et al., 2017a).

TABLE 33.1

Effect of Salt Stress on Different Parameters in Stevia

Culture Condition	Salinity range	Parameter Studied	Remarks	References
Controlled Condition	30–270 mM NaCl	S, R, L, PL, WC, FW, DW, Pig, EL,	Physiological traits were more sensitive to salinity stress than morphological traits.	9Akandi et al. (2017).
In vitro	2.5 g/L NaCl	Pig, P, EL, MDA, H_2O_2, SOD, CAT, POD	AMF inoculation was capable of alleviating the damage caused by salinity on stevia plants by reducing oxidative stress and improving photosynthesis efficiency.	Bidabadi and Masoumian (2018).
In vitro	2–5 g/L NaCl	G, K, Ca, proline, pigment, APX MDHAR DHAR, GR, SOD, CAT, POX, LP, St, Reb-A	NaCl-induced oxidative stress induces tolerance mechanisms in order to minimize the deleterious effects of salt stress.	Cantabella et al. (2017).
Hydroponics	50–300 mM NaCl	Pig, R, S, OP, Na, K, Mg, Ca, St, Reb-A, S, AA	Increase in ions, amino acids, amines, and sugars notably increased the rebaudioside-A concentration and helped the plant to tolerate salinity.	Debnath et al. (2018).
In vitro	20–80 mM NaCl	R, S, UGT, St, Reb-A	Morphological traits decreased, increase of both stevioside and rebaudioside-A along with its biosynthesis enzymes during salinity stress and the plant survived under salt stress, but it has the best performance under lower salinity.	Fallah et al. (2017).
Ex vitro	Conductivity 0.3–7.5 dS m^{-1}	L, R, S, DW	Salinity stress results in yellow leaves, reduce stem diameter, root and production of leaves.	González et al. (2017).
In vitro	NaCl, 0.05–2% Na_2CO_3 (0.25–1%)	G, S, R, Pa	In spite of causing some growth reduction, application of chemical stress can enhance the production of SGs up to three-fold compared to control plants.	Gupta et al. (2016).
Ex vitro	50–100 mM NaCl	G, S, R, P, FW	The results imply that elevated CO_2 level could ameliorate some of the detrimental effects of salinity, conferring higher tolerance and survival of plants.	Hussain et al. (2017).
Ex vitro	2–5% NaCl and $NaHCO_3$ 0–300 mM	SG	Seed germination reduced compared to water and Seed germination velocity in NaCl was higher than the corresponding ones in $NaHCO_3$.	Liopa-Tsakalidi et al. (2012).
In vitro	25–100 mM NaCl	S, R, L, FW, DW, Pa Pr, Pig	Salinity and drought stress significantly reduce the growth and yield components by affecting endogenous growth hormones.	Pandey and Chikara (2014).
In vitro	25–100 mM NaCl	L, S, R, FW, DW, UGT, St, Reb-A	NaCl is acting as an enhancer and manitol acting as a repressor of transcription of genes of steviol glycoside biosynthesis pathway that could alter the production of steviol glycosides.	Pandey and Chikara (2015).
In vitro	NaCl 50–100 mM with Glycine 50 mM	G, R, S, L, FW, Pig, St, Reb-A, GB	Owing to the amending effect of glycine betaine, its high concentrations made less hazarding effects of salinity on the studied traits.	Rameeh et al. (2017).
In vitro	NaCl 25–125 mM	Pig, S, Pa, Phe	Chlorophyll amount was observed to be decreased as compared to sugars, proline, and phenols with increased salt concentrations.	Sharuti et al. (2014).
Field Condition	Conductivity 2–0.3 dS m–1.	G, Dev, Y	Stevia crop tolerance to salinity was greater than the one of the sugar cane.	Reis et al. (2015).
Hydroponic	40–120 mM NaCl	B, R, S, EL, H_2O_2, Su, Pr, Pa	Using appropriate concentrations of auxin and cytokinin hormones (especially 0.1 mg/L IAA treatment) in stevia can be effective in increasing plant resistance to salinity.	Sarami et al. (2017).
Ex vitro	30–150 mM NaCl	G, R, S, L, FW, DW, Pig, MDA, Na	The result showed stevia as a moderate NaCl tolerant plant.	Shahverdi et al. (2017a).
Ex vitro	30–90 mM NaCl	SG, Pig, Pa, CAT, SOD	Nutri-priming with nutrients, especially Se and Fe increased the antioxidant capacity of the plant to improve germination and seedling growth under salinity stress.	Shahverdi et al. (2017b).

(Continued)

TABLE 33.1 (CONTINUED)

Effect of Salt Stress on Different Parameters in Stevia

Culture Condition	Salinity range	Parameter Studied	Remarks	References
Ex vitro	30–90 mm NaCl	G, L, B, Y, St, Reb-A, SVglys	The foliar application of selenium either alone or in combination with iron could alleviate the adverse effects of NaCl stress on the studied traits.	Shahverdi et al. (2018).
Ex vitro	30–150 NaCl	G, R, FW, DW, CAT, POX, PPO, SVglys	The high level of SVglys at lower salinity levels is one of the reasons for salinity tolerance in stevia.	Shahverdi et al. (2017c).
Growth Chamber	60–90 mM NaCl	G, S, R, L. DW, FW, Pig, Na, K, SVglys, SOD, CAT, POX Reb-A, St	The plant is moderately tolerant to salt stress. Hyposaline soil can be utilized in the plantations and may be profitable for optimizing the steviol glycoside composition.	Zeng et al. (2013).

G: Growth, B: Biomass, Pig: Pigment, P: Photosynthesis, S: Shoot, R: Root, PL: Plant Height; L: Leaves, DW: Dry Weight, FW: Fresh Weight, EL: Electrolyte Leakage, Dev: Development, Y: Yield, SG: Seed Germination, NaCl: Sodium Chloride, Ca: Calcium, K: Potassium, LP/MDA: Lipid Peroxidation, ROS: Reactive Oxygen Species, PA: Proline Accumulation, PPO: Polyphenol Oxidase, St: Steviolbioside, Rab A: Rebaudioside-A, SVglys: Steviol Glycosides, OP: Osmotic Potential, Su: Sugar, Pr: Protein, UGT: UDP: Dependent Glycosyltransferases, GB: Glycine Betaine, Phe: Phenols, CB: Carbohydrate Metabolism, Na: Sodium Accumulation., FAA: Free Amino Acids, WC: Water Content.

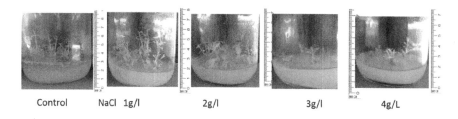

| Control | NaCl 1g/l | 2g/l | 3g/l | 4g/L |

FIGURE 33.2 (See color insert.)   NaCl treatment on *in vitro* plants of stevia (unpublished data).

The continuous decrease in protein content *in vitro* (Pandey and Chikara, 2014) and hydroponic culture was observed in all treatments explaining the toxicity of NaCl-induced stress effects in stevia (Sarami et al., 2017; Debnath et al., 2018).

33.3.3 Salinity Tolerance Mechanism in Stevia

Plants are of two types based on their salinity tolerance such as glycophytes or halophytes (Acosta-Motos et al., 2017). The complete mechanism of tolerance depends on ion compartmentation and exclusion, osmotic adjustment, and antioxidant defense ability (Munns, 2005; Abraham and Dhar, 2010; van Kempen et al., 2013; Tang et al., 2015). Stevia can tolerate salinity to a moderate level (Shahverdi et al., 2018).

33.3.3.1 The Ionic Regulation and Interaction

Plants having a higher capability of ion compartmentation and exclusion are called halophytes (Munns et al., 2006); thus, a high electrolyte leakage during NaCl treatment in stevia has been reported (Bidabadi and Masoumian, 2018); also it was observed that this plant accumulates many essential ions (Cantabella et al., 2017; Debnath et al., 2018). Stevia has a high capability for ion compartmentation by increasing the level of Na, Ca, K, Mg, Zn, and Mn when subject to NaCl treatments (Cantabella et al., 2017; Debnath et al., 2018) and the decrease in Na$^+$:K$^+$ ratio in root and shoot in *ex vitro*

cultures of stevia (Shahverdi et al., 2017a) make it a species of moderate tolerance to salinity.

33.3.3.2 Osmotic Adjustment or Regulations

Osmoregulation in the plant is the phenomenon of maintenance of osmotic potential with the immediate environment (Heuer and Feigin, 1993; Hasegawa et al., 2000). Apart from salts, there are many compounds that help with osmotic adjustment: carbohydrates (sucrose, sorbitol, mannitol, glycerol, arabinitol, pinitol) (Bohnert and Jenson, 1996; Hare et al., 1998); nitrogenous compounds (proteins, betaine, glutamate, aspartate, glycine, proline, 4-gamma aminobutiric acid), and organic acids (malate and oxalate) (Lamosa et al., 1998; Kinnersley and Turano, 2000). Like other plants, in stevia *in vitro* cultures, there was a high accumulation of carbohydrates such as sugars and high levels of hexose phosphatase involved in cellular protection (Sharuti et al., 2014; Debnath et al., 2018). The changes in the nitrogenous compounds are more prominent with NaCl stress in stevia. High accumulations of amino acids, amines, and proline was observed *in vitro* (Pandey and Chikara, 2014; Sharuti et al., 2014; Cantabella et al., 2017), *ex vitro* (Shahverdi et al., 2017b), and in hydroponic cultures and indicated the roles of these compounds in osmoregulation in stevia. Unlike proline, the increasing of glycine betaine in stevia supports its amending effect during salinity stress (Rameeh et al., 2017). Many organic acids, including tricarboxylic acid (TCA) cycle intermediates decrease during salinity stress; however, their role in stevia is not clear (Debnath et al., 2018).

33.3.3.3 Salinity Tolerance and Glycosides Contents of Stevia

Stevia is the plant for compound glycosides (SGs), the secondary metabolites responsible for the sweetness. It contains stevioside (St) and rabaudioside-A (RA) which are the most abundant with about 4–13% under normal conditions (Tavarini and Angelini, 2013; Urban et al., 2015). All the reports available indicate a positive correlation between SGs accumulation and salinity tolerance. Mild salt treatment did not affect stevioside and rebaudioside-A contents (Cantabella et al., 2017), whereas moderate treatment notably increased the concentration of both compounds (Shahverdi et al., 2017a, 2017c, 2018; Debnath et al., 2018), while higher salinity adversely affected them (Zeng et al., 2013). In addition, higher expression of both stevioside and rebaudioside-A genes are reported under salt stress (Pandey and Chikara, 2015; Fallah et al., 2017). The increase in the steviol glycoside contents and its gene expression in moderate salinity treatment and notable decrease of the content of rebaudioside-A (RA) and stevioside (ST) in higher saline treatments, indicate the roles of these compounds in salinity tolerance in stevia (Zeng et al., 2013; Pandey and Chikara, 2015; Shahverdi et al., 2017a, 2017c; Shahverdi et al., 2018).

33.3.3.4 Antioxidant Defenses in Stevia

Salinity induces reactive oxygen species (ROS) that lead to oxidative damage of plants (Rout and Shaw, 2001b; He and Zhu, 2008; Aghaleh et al., 2009). Plants have well-armed enzymatic and non-enzymatic antioxidant systems to scavenge the free oxygen radicals generated during stress conditions (Corpas et al., 2015). The ROS superoxide radicals ($O2\bullet^-$), hydrogen peroxide (H_2O_2), hydroxyl radicals (^1OH), and singlet oxygen ($^1O_2$) are produced primarily in organelles and the antioxidant enzymes (superoxide dismutase and catalase (CAT)), the ascorbate-glutathione (ASC-GSH) cycle enzymes (ascorbate peroxidase (APX), monodehydroascorbate reductase (MDHAR), dehydroascorbate reductase (DHAR) and glutathione reductase) scavenge the ROS radicals (Asada, 2006; Acosta-Motos et al., 2015; Corpas et al., 2015; Gupta et al., 2018).

It is evident that salinity induced oxidative stress in stevia through lipid peroxidation (MDA), and production of H_2O_2 and $O2\bullet^-$ (Cantabella et al., 2017; Sarami et al., 2017; Shahverdi et al., 2017a; Bidabadi and Masoumian, 2018) and a result of which is the continuous decrease in protein content (Pandey and Chikara, 2014; Sarami et al., 2017; Debnath et al., 2018).

Despite little differences, stevia has a well-developed antioxidant protection enzyme system under salinity with enhanced activities of SOD, CAT, APX, MDHAR, DHA, and DHAR. Bidabadi and Masoumian (2018) reported that SOD and POX activities were concentration dependent, and on the other hand CAT activity increased under low salinity followed by a decrease under high salinity conditions. However, observations by Cantabella et al. (2017) were a little contradictory, and they demonstrated that APX, GR, CAT activities were of a concentration-dependent manner and SOD, POX, MDHAR, and DHAR activities decreased under high salinity in in vitro conditions. In this contest, Shahverdi et al. (2017b) observed enhanced activity values for both enzymes (CAT and SOD) in germination stage treatments. Overall, all authors concluded that well-developed enzyme systems increase the antioxidant capacity, reduce oxidative stress, and induce tolerance mechanisms in order to minimize the deleterious effects of salt stress in stevia (Cantabella et al., 2017; Shahverdi et al., 2017b; Bidabadi and Masoumian, 2018).

33.4 DROUGHT STRESS TOLERANCE OF STEVIA

Drought is the most significant environmental stress in agriculture, and this physiological condition is caused by factors that tend to disrupt plant equilibrium. Drought affects growth, yield, membrane integrity, pigments content, osmotic adjustment, water relations, and photosynthetic activity. Under drought, plants accumulate proline, sucrose, soluble carbohydrates, glycine betaine, and other solutes in the cytoplasm to maintain cell turgor. Desiccation tolerance is the result of a complex cascade of molecular events, which can be divided into signal perception, signal transduction, gene activation, and biochemical alterations, leading to the acquisition of tolerance.

33.4.1 Introduction

Plants are often exposed to many stresses such as drought, low temperature, salt, floods, and heat, which severely affect their growth (Srivastava and Srivastava, 2014). Water comprises 80 to 90% of the biomass of plants and is the central molecule in all physiological processes of transport for metabolites and nutrients (Rohbakhsh, 2013). Drought is the stress that occurs when the demand for water exceeds the amount of water available during a given period or when poor quality restricts its use. It often occurs in areas with little rainfall and high population density or in areas where agricultural or industrial activities are intense (Yousfi et al., 2016). Water stress causes changes in the morphological, physiological, and biochemical responses of plants. As a result, plant growth and crop production are adversely affected (Benhmimou et al., 2017).

Drought stress reduces the size of the leaves, the extension of the stem and the proliferation of roots, disturbs the water relations of the plants, and reduces the efficiency in the use of water. Plants show a variety of physiological and biochemical responses at the cellular level and throughout the body toward the prevailing drought stress, which makes it a complex phenomenon (Anjum et al., 2011). The assimilation of CO_2 by the leaves is mainly reduced by the closure of the stomata; damage to the membrane and altered activity of several enzymes, especially those of CO_2 fixation and synthesis of adenosine triphosphate (Farooq et al., 2009). Plants exhibit a variety of mechanisms to resist drought stress. The main mechanisms include reduced water loss, enhanced water uptake with profound and prolific root systems, and smaller leaves to reduce transpiration loss (Farooq et al., 2012). The leaves of the plants change their angle of inclination, and the root maintains its rate of growth, while the aerial part decreases (Anjum et al., 2011). Among the physiological and metabolic changes that occur is a decrease in protein synthesis and, therefore, in the rate of growth, the wax increases leaf coverage, changes in perspiration, respiration, photosynthesis, and the distribution of nutrients (Anjum et al., 2011).

Among the nutrients, potassium ions help in the osmotic adjustment. Low molecular weight osmolytes, such as glycine betaine, proline, and other amino acids, organic acids, and polyols, are crucial for maintaining cellular functions during drought. Plant growth substances, such as salicylic acid, auxins, gibberellins, cytokinin, and abscisic acid, modulate the plant's responses to drought (Farooq et al., 2009).

The generation of reactive oxygen species is one of the earliest biochemicals produced under drought stress. The production of ROS in plants is a known plant defense response to water stress and acts as a secondary messenger in plants. ROS include oxygen ions, free radicals, and peroxides. During drought, the ROS levels increase, cause degradation in the proteins, DNA fragmentation, and cell death (Apel and Hirt, 2004). A method of physiological improvement is seed priming; it is reported that it improves the germination uniformity and higher tolerance against environmental stress, activating enzymes such as peroxidase (POD), catalase (CAT), and the accumulation of osmoprotectants (proline) that eliminate free radicals (Rouhi et al., 2012). Gorzi et al. (2018) reported that the priming of seeds of stevia with salicylic acid (SA), zinc (Zn), and iron (Fe) subjected to drought stress with polyethylene glycol (PEG) 6000 showed an increase in proline accumulation and the enzymatic activity of catalase (CAT), peroxidase (POD), and superoxide dismutase in all priming treatments.

Stevia rebaudiana is a perennial plant of the *Asteraceae* family and is a plant of great commercial interest for its production of sweetening compounds (steviol glycosides). The two main stevia glycosides are stevioside (5–10% dry leaves) and rebaudioside-A (2–4%). Due to their non-caloric properties and sweeteners, stevioside has gained attention with an increasing demand for low carbohydrate and low sugar dietary alternatives (Kalpana et al., 2009). Notable alterations have been reported in the physiology and biochemistry of stevia plants grown under water stress conditions (Srivastava and Srivastava, 2014).

Understanding how this plant responds to water stress can play an essential role in improving crop management and yield, especially because climate change scenarios suggest an increase in aridity in many areas of the world (Chaves et al., 2009).

33.4.2 Effect of Drought Stress in Stevia

S. rebaudiana is susceptible to water stress; it limits the production and growth of this crop, which requires an adequate level of soil moisture to ensure its development due to the shallowness of the root system. Therefore, in deficit water conditions, severe damage occurs at the cellular level (Villalba and Nakashima, 2016). Drought imposes serious effects on the germination and post/germination of seeds (Gozdi et al., 2017), growth (Karimi et al., 2014), photosynthesis (Ren and Shi, 2012), transpiration (Benhmimou et al., 2017), cellular architecture (Zhu et al., 2010; Golldack et al., 2014), induces osmotic stress (Hajihashemi and Ehsanpour, 2013), and increases production of oxygen radicals and lipid peroxidation in stevia (Torres et al., 2008; Hajihashemi and Ehsanpour, 2013; Hajihashemi and Sofo, 2018).

33.4.2.1 Germination

Factors adversely affecting seed germination may include sensitivity to drought stress and salt tolerance (Okçu et al., 2005). The first physiological disorder, which takes place during germination, is the reduction in imbibition of water by seeds which leads to a series of metabolic changes, including changed enzyme activities and a general reduction in hydrolysis and utilization of the seed reserve (Sidari et al., 2008). These effects may be the consequence of the reduced enzymatic activities of the glyoxylate cycle, which operates in the conversion of fats to carbohydrates during the germination of oil-rich seeds (Chia et al., 2005).

The quite negative water potentials that occur in soils during periods of drought impede the absorption of water, affecting the germination process of seeds and the growth of seedlings (Fanti and Andrade, 2004). The simulated drought with polyethylene glycol (PEG) affected the germination and growth of wheat seedlings and more markedly the increase in osmotic pressure (González et al., 2005). In stevia, poor germination capacity is a major problem in its cultivation, where drought stress influences the germination of seeds and the early growth of seedlings of many crops. It has been observed that germination (germination percentage, germination rate, mean germination time, germination value, length of the seedling, and seedling vigor index) is negatively affected by drought stress (Gorzi et al., 2018).

33.4.2.2 Growth

Drought stress progressively decreases CO_2 assimilation rates due to reduced stomatal conductance in plants (Anjum et al., 2011). It reduces leaf size, stems, and root proliferation, disturbs plant water relations and reduces water use efficiency, and produces a reduction in plant growth and productivity (Anjum et al., 2011). Direct inhibition of shoot growth by water deficit was also reported in plants (Pandey and Chikara, 2014). Under drought stress conditions, the roots induce a signal cascade to the shoots via xylem causing physiological changes and eventually determining the level of adaptation to the stress. Abscisic acid (ABA), cytokinins, ethylene, malate, and other unidentified factors have been implicated in the root–shoot signaling. This drought-induced root-to-leaf signaling through the transpiration stream results in stomatal closure (Anjum et al., 2011). One effect of drought stress on plants is the reduction in the production of fresh and dry

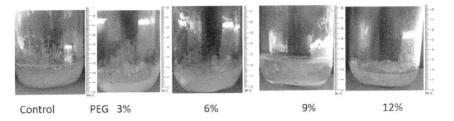

Control   PEG 3%   6%   9%   12%

FIGURE 33.3 (See color insert.)   PEG treatment on *in vitro* plants of stevia (unpublished data).

biomass due to the decrease in the number of leaves per plant and the size of the leaves (Anjum et al., 2011).

Stevia dry leaf yield exhibited decreasing trends with continuous drought stress, and there were differences among genotypes (Ren and Shi, 2012). Also, it is suggested that the soil moisture depletion has different effects on stevia plant organs. In fact, it was found that the stevia leaves were more sensitive to water stress than stems. The same results were observed in a field experiment carried out in south Italy (Lavini et al., 2008), where more leaf senescence was observed through decreasing irrigation volumes. Karimi et al. (2014) have reported significant reductions in leaf dry yield, and they recommended that stevia should not experience serious water stress during its vegetative growth.

33.4.2.3 Photosynthesis

Photosynthesis, together with cell growth, is among the primary processes to be affected by drought (Chaves, 1991). The effects can be direct, as the decreased CO_2 availability caused by diffusion limitations through the stomata and the mesophyll (Chaves et al., 2009). On the other hand, the alterations of photosynthetic metabolism are mostly present under multiple stress conditions and can seriously affect leaf photosynthetic machinery (Ort, 2001).

Photosynthetic response to drought stress is highly complex. It involves the interplay of limitations taking place at different sites of the cell/leaf and at different time scales in relation to plant development (Chaves et al., 2009).

Ren and Shi (2012) studied the effects of drought stress on photosynthesis and dry leaf yield in different germplasms of *Stevia rebaudiana*. They mentioned that net photosynthesis (Pn), transpiration rate (Tr), and dry leaf production of stevia showed different changes according to the amount of time under the stress condition.

33.4.2.4 Osmotic Stress

Under salt and drought conditions responses are triggered by primary osmotic stress signals (Chaves et al., 2003) or by secondary signal metabolites that generally increase or decrease in a transient mode. Osmotic stress is referred to in situations where water is insufficient, and its availability limits plant growth and development. These results can be induced by drought or excessive salt in water. The osmotic stress limits plant growth, affecting various metabolic processes such as photosynthesis, protein synthesis, respiration, nitrogen assimilation, and phytohormone turnover (Muscolo et al., 2003). Akula and Ravishankar (2011) mentioned that drought, high salinity, and

freezing impose osmotic stress and affect the growth, the productivity of crops, and secondary metabolites in plants. Few reports are available with relation to osmotic stress in stevia; there are some studies about the stress effect rather than direct involvement, except electrolyte leakage (Hajihashemi and Ehsanpour, 2013; Gupta et al., 2016; Hajihashemi and Sofo, 2018). Hajihashemi and Ehsanpour (2013) observed that in stevia plants stressed by drought, MDA increased in a positive correlation with the increase in electrolyte leakage. Figure 33.3 shows the effects of different PEG concentrations in *in vitro* cultures of stevia (unpublished results).

33.4.2.5 Oxidative Stress

As a consequence of the many environmental stresses, including dehydration, there is oxidative stress, i.e., the accumulation of reactive oxygen species, which damage cellular structures (Srivastava and Srivastava, 2014). When plants are subjected to stresses, a variety of ROS are generated, such as superoxide radical (O_2), hydrogen peroxide (H_2O_2), and hydroxyl radical (OH) (Choudhury and Panda, 2013). Different plants and genotypes within species respond differently to low water availability (Hong-Bo et al., 2006). Oxidative stress (ROS) is created in plants exposed to drought and has been shown to cause oxidation and damage to the lipid structure of the membrane that manifests itself due to the release of malondialdehyde (MDA), as well as the regulation of antioxidant enzymes (Zhang et al., 2007).

As in other stress inductions, direct evidence of the oxidative stress effect and radical measurements in stevia are scarce. In stevia, ROS production, H_2O_2 accumulation, and lipid peroxidation in response to PEG treatment have been reported in *in vitro* as well as in seeding stage treatment studies (Hajihashemi and Ehsanpour, 2013; Hajihashemi and Ehsanpour, 2014). In addition, under these stresses, stevia accumulated higher levels of phenolic compounds to eliminate free radicals. Paclobutrazol (PBZ) reduced the negative effect of PEG treatment by limiting SOD activity and H_2O_2 accumulation and increasing CAT activity. PBZ also decreased the effect of PEG on polyphenol oxidase (PPO) and phenols (Hajihashemi and Ehsanpour, 2014).

The decrease in chlorophyll content under drought stress has been considered as a typical symptom of oxidative stress and may result from the photo-oxidation of pigments and the degradation of chlorophylls (Farooq et al., 2009). Under conditions of *in vitro* culture and using polyethylene glycol to simulate drought stress in stevia, weights of fresh and dried leaves, water content, and chlorophylls were reported to be adversely affected by drought stress (Hajihashemi and Ehsanpour, 2013). Oxidative stress (ROS) is created in plants exposed to drought and has been shown to cause oxidation and damage to the lipid structure of the membrane (Ashraf, 2009) that manifests itself due to the release of malondialdehyde (MDA), as well as the regulation of antioxidant enzymes (Hajihashemi and Ehsanpour, 2013).

33.4.3 Drought Stress Tolerance of Stevia

S. rebaudiana is susceptible to water stress; water stress limits the production and growth of this crop, which requires an adequate level of soil moisture to ensure its development due to the shallowness of the root system. Therefore, under deficit water conditions, severe damage occurs at the cellular level (Villalba and Nakashima, 2016). Stevia induces tolerance by inducing osmotic adjustment in different ways (Karimi et al., 2015), accumulates proline, glycine betaine, and sugars (Hajihashemi and Ehsanpour, 2013; Pandey and Chikara, 2015). The antioxidative system is also well organized against drought stress, induces POD and CAT and accumulates ascorbate and glutathione to minimize the toxic effect of oxygen radicals at the cellular level (Gorzi et al., 2018; Hajihashemi and Sofo, 2018). Stevia also induces protein and steviosides as part of the tolerance mechanism against drought stress (Hajihashemi and Ehsanpour, 2013; Benhmimou et al., 2017).

33.4.3.1 Osmotic Adjustment

Osmotic adjustment is usually considered as a physiological process that helps to preserve water in plant tissues under soil water depletion (Chaves et al., 2003). In general, enzymes are sensitive to high amounts of ions, such as Na^+. The accumulation of ions during osmotic adjustment occurs in the vacuole, while in the cytoplasm solutes accumulate and are not related to the functionality of cellular macromolecules (Buchanan et al., 2000). These solutes are organic molecules of low molecular weight (osmolytes) such as polyols (sugars), methylamines, free amino acids, and amino acid derivatives. Osmotic adjustment is usually considered as a physiological process that helps to preserve water in plant tissues under soil water depletion, and it is considered a primary acclimation response through the increase of soluble cellular solutes such as soluble sugars, in the cytosol. It seems that osmolyte accumulation and osmotic adjustment have occurred in stevia in response to drought stress. Stevia induces tolerance by inducing osmotic adjustment in different ways (Hajihashemi and Ehsanpour, 2013; Karimi et al., 2015; Pandey and Chikara, 2015; Hajihashemi and Sofo, 2018). The sugar deficiency could be a cause of SVglys reduction in stevia under severe drought stress, and it can be assumed that glucose units were used for osmotic adjustment in order to create an acclimation process in the stressed tissues. The accumulation of amino acids is the first response of plants exposed to water deficit to reduce damage to cells and improve the tolerance of the plant (Anjum et al., 2011). Stevia induced accumulation of proline and increased the activity of its biosynthesis enzymes during *in vitro* PEG treatments (Hajihashemi and Ehsanpour, 2013;

Pandey and Chikara, 2015; Gupta et al., 2016). Other osmo-protectants also increase in stevia, and the accumulation of glycine betaine improves the ability of the plant to withstand the stress and can enable the plant to maintain water potential (Hajihashemi and Ehsanpour, 2013).

33.4.3.2 Oxidative Protection

Plants that grow under conditions of osmotic stress are severely affected by oxidative stress. Therefore, it is important to understand the antioxidant mechanisms by which *S. rebaudiana* responds to water deficit. Drought stress promotes the production of reactive oxygen species, including superoxide (O_2), singlet oxygen ($^1O_2$), hydroxyl (OH) and hydrogen peroxide (H_2O_2). ROS react with lipids, proteins, and DNA, resulting in peroxidation of lipids, denaturation of proteins, and DNA damage (Torres et al., 2008).

Plants have evolved in both enzymatic and non-enzymatic defense systems to eliminate and detoxify ROS. In enzyme systems, superoxide dismutase removes $O_2\bullet^-$ to H_2O_2, while peroxidase (POD), catalase (CAT), peroxidase ascorbate (APX), glutathione reductase (GR), monohydroascorbate reductase and dehydroascabasa reduce the volume in different cellular locations (Noctor and Foyer, 1998). The leaf extract of stevia has great potential to be used as a natural antioxidant agent (Hajihashemi and Geuns, 2013).

In stevia leaves, some compounds other than steviol glycosides, e.g., folic acid, pyrogallol, phenolics, and flavonoids have ROS uptake activity, a high antioxidant activity due to the presence of phenolic compounds. Kim et al. (2011) conducted a study and reported that the antioxidant response of stevia to polyethylene glycol in *in vitro* culture mobilizes the enzymes APX, CAT, PPO, and POD. This was reported by several other investigators (Hajihashemi and Ehsanpour, 2014; Gorzi et al., 2018). Also, under these stresses, stevia accumulated higher levels of phenolic compounds to eliminate free radicals. The positive role of ascorbate (Hajihashemi and Ehsanpour, 2014) glutathione (Gorzi et al., 2018; Hajihashemi and Sofo, 2018), anthocyanin (Hajihashemi and Ehsanpour, 2013) and steviol glycosides have been postulated together with antioxidant activity. Drought stress also increased a 25 kD protein with a critical function in plant development under these stresses.

33.4.4 Drought Tolerance and Glycosides Contents of Stevia

Stevia is a source of a great number of sweet ent-kaurene diterpenoid glycosides called steviol glycosides (SVglys), a group of no-calorie and sweet compounds (Karimi et al., 2015). Generally, the total SVglys content ranges from 4–20% of leaf dry weight, depending on many factors (Starratt et al., 2002; Karimi et al., 2015). Drought stress induces accumulation of high numbers of steviol glycosides *in vitro* as well as in field conditions, and higher accumulation is positively correlated with stress level (Lavini et al., 2008; Hajihashemi and Ehsanpour, 2013; Gupta et al., 2016; Benhmimou et al., 2017) and implies the role of these compounds in abiotic stress tolerance, including drought, in stevia.

33.4.5 Conclusions

The complexity of drought tolerance mechanisms explains the slow progress in yield improvement under drought-prone environments. In recent years, crop physiology and genomics have led to new insights in drought tolerance. It is necessary to carry out more studies about the mechanism of action of the *Stevia rebaudiana* plant to tolerate the stress caused by drought since currently there is very little information available and it is necessary to perform this type of work for the proper development of stevia plants in areas with water deficit. The integration of molecular genetics and physiology is leading to the identification of the most relevant loci controlling drought tolerance and drought-related traits. Finally, the effects of drought stress on stevia are summarized in Table 33.2.

33.5 TEMPERATURE STRESSES

The extreme temperatures (low temperature or high temperature) have an effect on the development of plants, causing an alteration in physiology, metabolic mechanism, gene expression, and development (Chinnusamy et al., 2007; Pereira, 2016). Plants exposed to stress with extreme temperatures (heat or cold) can result in greater tolerance and reduced impacts on physiology (Antoniou et al., 2016). Several studies have induced tolerance to heat or cold in a large number of species (Xing and Rajashekar, 2001; Zhang et al., 2013; Li et al., 2014; Xu et al., 2015). The effect that extreme temperatures have on plants is diverse; cold stress in tomato plants damaged the functions of photosystems and the assimilation of carbon dioxide (Zhou et al., 2012), while heat shock increased the synthesis of putrescine and proline (Zhang et al., 2013). In *Oryza sativa*, heat stress caused an increase in the content of ascorbic acid that functions as an antioxidant to oxidative stress (Chao and Kao, 2010), similar results were reported in *Vitis vinifera* in cold conditions (Wan et al., 2009). The increase of other hormones such as jasmonic acid, abscisic acid, and hydrogen peroxide has also been reported by heat or cold stress (Wan et al., 2009; Hu et al., 2013). Stress by temperature in plants is classified in three types: heat, cold, and frost (Raju et al., 2018), and these affect physiological, biochemical, and molecular factors (Kumar et al., 2014). Often, biotic or abiotic stress in plants causes the accumulation of secondary metabolites, which help plants adapt quickly to the environment and overcome stress (Akula and Ravishankar, 2011). Temperature influences growth, development, and function of plants through metabolic activity (Morison and Lawlor, 1999; Legris et al., 2017). Under conditions of high or low temperatures, plants undergo abiotic stress (Wang et al., 2003; Kotak et al., 2007). In addition, stress due to temperatures can increase or decrease the yield of a crop.

33.5.1 Temperature Stress on Stevia

Different responses have been observed in *Stevia rebaudiana* plants in response to stress by temperature; for example, increases in the production of carbohydrates, leaf growth, development of axillary shoots, reduction of photosynthesis, generation of shoots, etc. The effect of temperature on the growth and development of *Stevia rebaudiana* has been the subject of study, considering that the optimal temperature of the crop is 25/20°C (day/night) (Ceunen and Geuns, 2013). Other studies have shown that stevia plants can grow at temperatures ranging from −6 to +46°C (Singh and Rao, 2005). Research and published reports related to temperature stress tolerance are limited and are cited in Table 33.3.

33.5.1.1 Effect of Temperature on Growth

The growth of plants involves a large number of biochemical reactions that are highly sensitive to temperature, heat stress or low-temperature stress induce structural changes in tissues and cellular organelles, disorganization of cell membranes, alteration of water relations of leaves, and impedance of photosynthesis (Źróbek-Sokolnik, 2012).

The application of low temperatures to *Stevia rebaudiana in vitro*, favored growth, increasing the thickness of stems and leaves as a form of protection to the photosynthetic system (Hajihashemi et al., 2018a); in addition, the application of compounds such as polyamines helps overcome cold stress, improving plant performance (Moradi-Peynevandi et al., 2018). The stimulation of cold stress with previous treatment of endogenous signaling components (calcium chloride, hydrogen peroxide, 6-benzylaminopurine, salicylic acid) was used to improve the development of stevia plants under cold stress conditions (Soufi et al., 2016). The addition of these endogenous compounds induced greater biomass in the leaves of stevia seedlings, and out of these compounds BA was the one that increased the highest biomass rate (Soufi et al., 2015).

Other studies showed that during the acclimation of stevia seedlings produced under *in vitro* conditions, gradual increases in temperatures (28, 30, 32 and 36°C) are viable to achieve a high survival rate (Rafiq et al., 2007), as well as a direct temperature increase from *in vitro* to greenhouse of 30±2°C (Ibrahim et al., 2008). Increases in temperature help the hardening of the seedlings (Hossain et al., 2008). On the other hand, Carneiro et al. (1997) reported good growth and production of stevia in greenhouses in temperature ranges of 9 to 36°C.

33.5.1.2 Effect of Temperature on Root and Buds

The environmental temperature is a factor that controls the development of roots and shoots in plants, affecting the temperature of the soil (Lopushinsky and Max, 1990), the capacity of absorption of nutrients, and cellular respiration of the root (Aroca et al., 2011; Nieves-Cordones et al., 2012).

Studies showed that the increase of temperature in the soil favors the development of roots and the growth of leaves in stevia, and the use of mulches in stevia cultures cause a greenhouse effect (Kumar et al., 2014), allowing the passage of the sun's rays to reach the ground; these then heat the air and soil under the mulch and trapped heat increases the soil temperature (Hu, 1995). A higher soil temperature accelerates the development of leaves, allowing the crop to quickly

TABLE 33.2

The Effect of Water Stress on *Stevia rebaudiana*

Level of Study	Types of Drought Stress	Parameter Studied	Result and Remark	Reference
In vitro	PEG 6000 1–10,000 ppm	PH, LR, FW, DW	Concentrations higher than 10 ppm caused a reduction in growth.	Villalba and Nakashima (2016).
In vitro	PEG up to 20%	G, P, An, WUE, FW, DW, Chl, P5C5, Pig, Su, Pr, EL, MDA, GB, protein	Treatments with PEG and PBZ significantly reduced the accumulation (SG). Drought stress increased a 25 kD protein with a critical function in plant development under stresses.	Hajihashemi and Ehsanpour (2013).
In vivo/ field	CON$_2$H$_4$ 2.60 g/pot, KH$_2$PO$_4$ 4.30 g/pot, KCl 5.5 g/pot	P, Tr, and DW	The varying degrees decrease of dry leaf yield in different materials with the same drought stress showed different materials with different tolerance to drought stress.	Ren and Shi (2012).
Field	Irrigation regimes (100, 80 and 50%)	PH, NS, FW, DW, ST, Reb-A, Reb-B, Reb-C, Reb-D, Reb-F, Svgys	Stevia is sensitive to water deficit but adopts adaptive strategies that maintain its yield and increase the content of steviol glycosides.	Benhmimou et al. (2017).
In vitro	PEG 10,000–30,000ppm	PS, LS, NS, FW, ST	The selection based on survival % would be more effective in improving stevioside content of stevia plants under drought stress conditions.	Badran et al. (2015).
In vitro/ Greenhouse	10 mg L-1 of PBZ	SVglys (Reb-A, ST, Reb-B, Reb-F, Dul A, SB, Rub)	PBZ increased the ratio of Reb-A:ST.	Hajihashemi and Geuns (2017).
Field	Temperature 20, 30, and 35°C.	ETc, ETos, and Kc	Evapotranspiration was similar in all treatments. ETo is higher than ETc because of the initial growing phase.	Fronza and Folegatti (2003).
In vitro	Mannitol 1–5%	LR, NL, DW, FW, FW, and G	Osmotic stress significantly reduces the growth and yield components.	Ghaheri et al. (2015).
In vitro	PEG 6000 3–9 bar	Gr, P, Chl, POD and CAT	Seed priming with SA, Fe, Zn and particularly the integrated application of these three agents at a suitable concentration can promote the poor germination performance and improve the seedling growth by increasing the antioxidant capacity under drought conditions.	Gorzi, et al. (2018).
In vitro	Proline 2.5–10% and PEG 2.5–10%	SVglys, ST, Reb-A, BB, S, Pr	Despite causing some growth reduction, application of chemical stress can enhance the production of SGs up to three-fold compared to control plants.	Gupta et al. (2016).
In vitro	PEG 6000 5–155	SVglys, ST, Reb-A, Reb-B, Reb-C, Reb-F, Dul A, Rub, and SB	PEG 6000 affected the accumulation of SVglys in leaves.	Hajihashemi and Geuns (2016).
Field	Irrigation Control (T100) T66 and T33	ETc, PH, ETo, DW, DW, WUE, SVglys, and HI	The harvest index and water use efficiency showed no differences between the two cuts for the same treatments, while the values of both indices decreased with the increase in irrigation regime. Stevioside, rebaudioside-A and cation content in the leaves was unaffected by irrigation regime.	Lavini et al. (2008).
In vitro	Mannitol 25–100 mM	P, Chl, BB, FBE, MNS, MSL, MNL, and MRL	Drought stress significantly reduces the growth and yield components by affecting endogenous growth hormones.	Pandey and Chikara (2014).
In vitro	Mannitol at 5–10 bar	NS, LS, N, NL, NR, LR and SR	Mannitol produced negative effects and was more severe on plantlet growth than salinity stress.	Mubarak et al. (2012)
In vitro	NaCl (0–100 mM) Mannitol (0–100 mM)	MNS, MSL, MNR, MRL, Pr	The NaCl or mannitol produced negative effects on all parameters of growth.	Pandey, and Chikara (2015).
Seedling stage	PEG 5–15%	H$_2$O$_2$, P, Pig, Su, ROS, LP, MDA, Asb, Glutathione, GB	PEG-induced oxidative stress, due to insufficient antioxidant mechanisms, provoked damages to the cell membrane and photosynthetic apparatus, with consequently reduced carbohydrates and plant growth.	Hajihashemi and Sofo (2018).

(Continued)

TABLE 33.2 (CONTINUED)
The Effect of Water Stress on *Stevia rebaudiana*

Level of Study	Types of Drought Stress	Parameter Studied	Result and Remark	Reference
Greenhouse	Irrigation 90–45% reduction	L, G, DW, Su, TAC	The improvement of antioxidant capacity and soluble sugar content by soil water stress conditions could be considered as physiological and biochemical responses to progressive drought stress.	Karimi et al. (2018).

AG₃: Gibberellic Acid, An: Anthocyanin, Asb: Ascorbate, BB: Budbreak, BE: Bud elongation, CAT: Catalase, Chl: Chlorophyll, Dul A: Dulcoside-A, DW: Dry Weight, EL: Electrolyte Leakage, Etc: Crop Evapotranspiration, Eto: Evapotranspiration, Eto: Reference Evapotranspiration, FBE: Frequency of Bud Elongation, FW: Fresh Weight, GB: Glycine betaine, G: Germination, GHS: Glutathione, Gr: Growth, HI: Harvest Index, HI: Harvest Index, KC: Coefficient Crop, L: Leaf, L: Leaves, LP: Lipid Peroxidation, LR: Root Length, LS: Length of Shoots, MDA: Malondialdehyde, MNL: Mean Number of Leaves, MNR: Mean Number of Root, MNS: Mean Number of Shoots, MRL: Mean Root Length, MSL: Mean Shoots Length, NL: Number of Leaves, N: Nodes, NR: Roots Number, NS: Number of Shoots, P5CS-D1: Pyrroline-5-Carboxylate Synthetase, PBZ: Paclobutrazol, PEG: Polyethylene Glycol, PH: Height of the Plant, Pig: Pigment, POD: Peroxidase, P: Photosynthesis, Pr: Proline, PS: Survival percentage, Reb-A: Rebaudioside-A, ROS: Reactive Oxygen Species, R: Root, SB: Steviolbioside, SR: Survival Rate., S: Shoot, ST: Steviosides, Su: Sugar, SVglys: Steviol Glycosides, SV: Seedling Vigor, TAC: Total Antioxidant Capacity, Tr: Transpiration Rate, WUE: Water Use Effacing.

TABLE 33.3
Effect of Temperature Stress on Different Parameters in Stevia

Culture Condition	Temperature Range	Parameter Studied	Remarks	References
In vitro	5°C	QYPII, WUE, ChA, ChB, Car, NP, ICO₂, EPSI/II, WSC, Glu, StN, SD, StI, StL, StW, ECI, Stdi, LM	Cold stress reduces the activity of some photosynthetic traits, but the biomass content in the leaf and carbohydrates increase significantly.	Hajihashemi and Sofo (2018a).
Chamber	6–10°C	QYPII, MCI, Reb-A, Reb-B, Reb-C, Reb-C, Dulc A, St	Plants pretreatment with salicylic acid, hydrogen peroxide, 6-benzylaminopurine, and calcium chloride could induce tolerance to chilling.	Soufi et al. (2016).
In vitro	5°C	SG	The treatment of stress by pre-chilling the stevia seeds increased the germination rate of the seeds.	Macchia et al. (2007).
Chamber	4°C	RCh, ChF, ChA, ChB, Car, HP, Mal, SVglys, Pr, GB, CAT, Asc, PPO, Prol, FAA, ProA	The application of polyamines increases the adaptation to cold stress, inducing antioxidant defense system, membrane stability, and upregulating many regulatory and structural proteins biosynthesis.	Moradi-Peynevandi et al. (2018).
Chamber	0°C	Sto	The optimum seed storage temperature is 0°C.	Shuping and Shizhen (1995).
Chamber	20 and 25°C	SG	The highest seed germination rate was observed at 20°C.	Simlat et al. (2016).
Chamber	20, 25 and 30°C	SG	The highest seed germination rate was observed at 25°C.	Takahashi et al. (1996).
In vitro	5±1°C	SG	The samples were stored at 5±1 and 20% relative humidity before every experiment favored germination.	Hossa et al. (2017).
Greenhouse	30°C	BG	Warm temperatures activated the growth of axillary buds.	Smitha and Umesha (2012).
In vitro	20°C	ASG	At a temperature of 20°C, the growth was reduced without damaging the seed.	Nower (2014).
Field	29°C	NP	The application of a potassium fertilizer improves the tolerance of stevia to high temperatures and the relief of photosynthesis at noon.	Ma et al. (2012).
Chamber	5 and 15°C	Pr	Cold stress did not cause significant changes in the total protein content.	Afshar andAnsari (2017).
Chamber	10°C/6°C (day/night)	EPSI/II	Treatment with BA and hydrogen peroxide increased the efficiency of PSII.	Soufi et al. (2015).

FAA: Free Amino Acids, AP: Ascorbate Peroxidase, ASG: Artificial Seed Growth, Pr: Protein, BG: Bud Growth, Car: Carotenoids, CAT: Catalase, Cha: Chlorophyll-A, Chb: Chlorophyll-B, Chf: Chlorophyll Fluorescence, Dulc-A: Dulcoside-A, ECI: Epidermal Cell Index, EPS I/II: Efficiency Of Photosystems I and II, GB: Glycine Betaine, Glu: Glucose, HP: Hydrogen Peroxide, ICO₂: Intercellular CO₂, LM: Leaf Mass, Mal: Malondialdehyde, MCI: Membrane Cells Integrity, NP: Net Photosynthesis, PPO: Polyphenol Oxidase, Prol: Proline, Pra: Protease Activity, QYPII: Quantum Yield Of Photosystem II, Reb-A: Rebaudioside-A, Reb-B: Rebaudioside-B, Reb-C: Rebaudioside C, Reb-D: Rebaudioside D, Rch: Relative Chlorophyll, SG: Seed Germination, SD: Stomatal Density, Std: Stem Diameter, SVglys: Steviol Glycosides, St: Stevioside, Sti: Stomatal Index, Stl: Stomatal Length, Stn: Stomatal Number, Sto: Storage, Stw: Stomatal Width, WUE: Water Use Efficiency, WSC: Water-Soluble Carbohydrates.

reach the maximum leaf area index. Warm temperatures favor the activation of buds in the greenhouse; stevia propagation experiments carried out on cuttings showed that a quality temperature around 30°C and an abundant relative humidity favor the activation and the fast growth of the latent buds (Smitha and Umesha, 2012).

33.5.1.3 Effect of Temperature on Germination of Seeds

Stress by heat or low temperature induces physiological changes in seeds, altering the synthesis of auxins, cytokinins, and gibberellins that favor their germination (Weidner, 2005).

The germination rate of stevia seeds is low; however, few studies have been carried out in order to increase the germination rate (Simlat et al., 2016). With respect to the optimal temperature of germination, some studies have differed, for example, Carneiro (1990) reported an optimum temperature of 25°C, while Sakaguchi and Kan (1982) found a range of 15 to 30°C. Other studies like that of Macchia et al. (2007) evaluated the effect of temperature on the germination rate of seeds; the results showed that alternation between temperatures and a previous cooling generates a higher germination rate. On the other hand, in a study of temperature and LED light, it was observed that at a temperature of 20°C the highest rate of germination of seeds was obtained, while LED light was not a significant factor. Possibly cold prevents the latency of the seed; a similar method was used by Hossa et al. (2017), incubating stevia seeds at a temperature of 5±1°C before germination. Furthermore, the optimum storage temperature of stevia seeds is 0°C (Shuping and Shizhen, 1995). It is possible that cold prevents the dormancy of the seed, a similar method was used by Hossa et al. (2017), placing the seeds at a temperature of 5±1°C before sowing them for germination. The optimal storage temperature of stevia seeds was found to be 0°C (Shuping and Shizhen, 1995). Stevia seeds are sensitive to temperature; therefore, temperature plays an important role during storage and germination.

33.5.1.4 Effect of Temperature on Photosynthesis

Temperature stress causes biochemical changes that reduce the photosynthetic rate (Djanaguiraman et al., 2018), causing damage to the photosynthetic apparatus that can be reversible or irreversible depending on the time of exposure to stress (Yan et al., 2011; Son et al., 2014).

The growth of *Stevia rebaudiana* plants subjected to stress at low temperatures (5°C) causes a decrease in the maximum quantum yield of photosystems I and II, in addition to decreasing other photosynthetic characteristics (net photosynthesis, chlorophyll a, chlorophyll b, and carotenoids), but an increase in water-soluble carbohydrates, leaf size, and stomatal index and density, among others (Hajihashemi et al., 2018a). On the other hand, due to the high sensitivity of stevia plants to low temperatures, research has been developed to induce cold tolerance. Soufi et al. (2016) reported that previous treatment with endogenous signaling components such as salicylic acid, 6-benzylaminopurine, calcium chloride, and hydrogen peroxide induce tolerance of stevia to low temperatures increasing photosynthesis, in addition to increasing metabolites such as

rebaudiosides and steviosides. Stevia seedlings subjected to stress by cooling and treated with BA and hydrogen peroxide increased the photochemical reaction efficiency, suggesting that these compounds favor the reaction centers of photosystem II under cold conditions (Soufi et al., 2015). Another similar study with the pretreatment of polyamines increased the tolerance of stevia leaves to cold stress, and increased levels of chlorophyll a and chlorophyll b (Moradi-Peynevandi et al., 2018). Furthermore, the addition of potassium fertilizers increases tolerance to high temperatures and improves photosynthesis in stevia plants (Ma et al., 2012).

33.5.1.5 Effect of Temperature on Protein

Oxidative stress caused by extreme temperatures generates changes in the protein level of plants (Shulaev and Oliver, 2006), as well as modifications of protein activity (Irar et al., 2010). When subjected to cold stress, the stevia plants did not show an increase of total proteins (Afshar and Ansari, 2017); likewise, Moradi-Peynevandi et al. (2018) subjected the stevia plants to cooling during the four days, the total protein content did not increase. Both studies agree that apparently cold stress does not increase the level of total protein in stevia. However, the use of polyamines favored the total protein content during stress.

33.5.1.6 Effect of Temperature on Osmotic Stress

Osmotic stress is a water imbalance in the cell, which can be caused by the dehydration of the cell, and in order to maintain their osmotic balance the cells synthesize a variety of organic compounds (osmolytes) of low molecular weight such as proline, sucrose, glycine betaine, alaninebetaine, prolinebetaine (Rhodes and Hanson, 1993; Glenn and Brown, 1999).

Stevia shoots obtained from *in vitro* culture were encapsulated with 4% calcium alginate and subjected to an osmotic stress with mannitol and sorbitol to maintain a reduced growth, the results indicated that with a concentration of 0.5 M of mannitol and at a temperature of 20°C, the plants had better growth without damaging the artificial seeds (Nower, 2014). On the other hand, when leaves of adult plants were subjected to low temperatures, their glycine betaine contents decreased; glycine betaine is considered a compound that helps to maintain osmotic potential, while the level of proline remained the same during the stress period. In contrast, in stevia plants subjected to pretreatments with polyamines, the levels of osmolytes such as proline, glycine betaine, and free amino acids increased (Moradi-Peynevandi et al., 2018).

33.5.1.7 Effect of Temperature on Oxidative Stress

The plant cells suffer physiological injuries due to temperature stress (Mazorra et al., 2002), which cause the generation of oxidative stress through the synthesis of reactive oxygen substances such as hydrogen peroxide, hydroxyl radical, and superoxide radical (Mittler, 2002) that ends up damaging the cell membrane, proteins, and lipids (Sairam and Tyagi, 2004; Xu et al., 2006).

Stevia plants subjected to stress at low temperatures had a high content of hydrogen peroxide, as well as a significantly increased level of malondialdehyde between 42 and 50%, both reactive oxygen substances. Malondialdehyde causes lipid peroxidation (lesions of the membrane system) and cellular metabolic deterioration (Liu et al., 2013). However, the synthesis of these compounds was reduced by treating the plants with polyamines during the stress period. The use of polyamines in *Stevia rebaudiana* plants helps to reduce the levels of hydrogen peroxide and malondialdehyde by an increase in the production of glycine betaine (Moradi-Peynevandi et al., 2018). Hydrogen peroxide and malondialdehyde are considered oxidative stress agents, which have been shown to be reduced by the use of compounds such as glycine betaine (Megha et al., 2014). Glycine betaine also acts by protecting proteins and membranes of cellular organelles against damage at low temperatures.

33.5.2 TEMPERATURE STRESS TOLERANCE OF STEVIA

Stevia rebaudiana plants are very sensitive to low temperatures (9°C); however, the addition of compounds of endogenous signaling such as salicylic acid, hydrogen peroxide, calcium chloride and 6-benzylaminopuria, helps stevia plants to be tolerant to low-temperature conditions and favor their development (Soufi et al., 2016). Treatments with polyamines also increase the thermotolerance of stevia. Polyamines are molecules with a protective effect and regulatory functions in plants subjected to abiotic stress conditions. On the other hand, at low temperatures, some stevia plants have been shown to have a degree of tolerance (Moradi-Peynevandi et al., 2018). This tolerance has also been increased with the help of polyamines.

33.5.2.1 Avoidance Mechanism

Under conditions of stress due to temperature, plants activate various mechanisms for overcoming stress (Hasanuzzaman et al., 2013), among which the most important ones are photosynthetic acclimation, accumulation of metabolites, and synthesis of stress adaptation proteins (Wahid et al., 2012).

Chilling stress in *Stevia rebaudiana* causes lesions in the cell membrane, and these lesions induce the leakage of cellular electrolytes and cause oxidative stress. The treatment of stevia plants with endogenous signaling molecules (salicylic acid, hydrogen peroxide, chloride of calcium, and 6-benzylaminopuria) showed a high tolerance to cold stress and increased the production of rebaudioside-A and stevioside (Soufi et al., 2016). Although the mechanism of action of signaling molecules has not been described in stevia, it is known that these molecules favor thermotolerance by reducing oxidative stress. These endogenous molecules activate antioxidant systems that increase the enzymatic activity of peroxidases and catalases (Larkindale and Huang, 2004). Likewise, polyamines are compounds with a high capacity to induce tolerance and defense to temperature stress (Alcázar et al., 2010), increasing up to 29% the activity of hydrolytic enzymes or antioxidants such as catalase, ascorbate peroxidase, polyphenol oxidase, and protease. Cold stress in stevia reduces chlorophyll content (Moradi-Peynevandi et al., 2018),

and this decrease is related to the inhibition of some enzymes involved in chlorophyll biosynthesis such as Mg-chelatase, Fe-chelatase, and protoporphyrin-oxide IX oxidase. On the other hand, treatment with polyamines increased the level of chlorophyll in stevia plants due to its ability to attenuate free radicals that prevent chlorophyll biosynthesis (Von Deutsch et al., 2005).

33.5.2.2 Osmotic Adjustment

As a mechanism to overcome osmotic stress, plant cells tend to accumulate organic solutes such as; free amino acids, proline, and glycine betaine (Belkheiri and Mulas, 2013), these compounds help reduce membrane damage, improving osmotic potential, and cellular turgor (Yildiztugay et al., 2014).

The use of endogenous polyamines helps plants to tolerate abiotic stress (Groppa and Benavides, 2008), in *Stevia rebaudiana* the use of polyamines has been favorable to cope with cold stress, helping to maintain the efficiency of photosystem II and increasing compounds such as proline, glycine betaine, and the maintenance (in some cases increase) of photosynthetic pigments (chlorophyll a, chlorophyll b, carotenoids). In addition, inducing the synthesis of other amino acids such as ornithine and arginine. The osmolytes considered responsible for helping plants to be tolerant to cold stress are proline and glycine betaine. The protein content also increased in stevia plants treated with polyamines (Moradi-Peynevandi et al., 2018). Afshar and Ansari (2017) reported that cold stress in stevia plants without previous treatments did not increase total protein and proline levels.

33.5.2.3 Antioxidant Defense of Stevia

The activation of enzymes such as catalase, ascorbate peroxidase, superoxide dismutase, glutathione reductase, and the synthesis of compounds such as nonprotein amino acids, ascorbic acid, and phenolic compounds are the defense machinery against oxidative damage caused by the high production of reactive oxygen species in plants (Gill and Tuteja, 2010).

The application of polyamines reduces and prevents the levels of hydrogen peroxide and malonyl aldehyde, and this is attributed to their membrane stabilizing, antioxidant and acid neutralizing effects (Alcázar et al., 2010). Stevia's response to cold stress and the ability to activate catalase enzymes were probably activated by pretreatments with polyamines or endogenous signaling agents to which they were subjected during stress (Moradi-Peynevandi et al., 2018), however, some stevia plants did not show oxidative stress damage due to low temperatures for short periods, attributing this fact to a reduced capacity of stevia to tolerate temperature stress (Afshar and Ansari, 2017). Usually, the described mechanism of stevia to counteract stress by temperature is through the activation of the antioxidant mechanisms of enzymes catalase and peroxidases (Moradi-Peynevandi et al., 2018). Likewise, in *S. rebaudiana* cultivars subjected to cold stress, a significant reduction in photosynthetic activity has been reported, as a measure to increase cold photoprotection (Hajihashemi et al., 2018a).

33.5.2.4 Protein and Genes

As a measure of the response of plants to stress conditions, they change their protein content and gene expression; these changes generate information that helps to know the response the plant will have to overcome the stress condition. In addition, proteins function as antioxidants (Hakeem et al., 2012) and stevia's response to low-temperature stress does not seem to have a significant effect on total protein production.

33.5.3 CONCLUSIONS

There is little knowledge about the mechanisms of *Stevia rebaudiana* responses to temperature stress, despite being a plant of great commercial value. The few studies that exist are based on the application of compounds that favor tolerance to cold stress, but there is no experimental evidence on the effect that these compounds can have on stevia plants subjected to high temperatures stress. However, a response favoring stress tolerance as observed in other species would be expected. Also, it is well known that stress due to temperature directly affects the accumulation of metabolites, and this is why the discovery of stevia plants with greater tolerance to temperature stress is necessary for stable production of metabolites.

33.6 METAL STRESS

Metals are elements which at high concentration in the medium may be toxic for life and development of plants. Some can be essential as iron, copper, or zinc for important physiological functions, for example, photosynthesis or as cofactors for some enzymatic reactions in the cell. Other metals are highly toxic for plants, such as silver or cadmium that even in low concentrations can produce plant death. Metal stress tolerance in stevia is not well studied, and only a few research reports are available (Table 33.4). Impact of metal stress has positive or negative effects depending upon concentration as well as the type of metals or heavy metals.

33.6.1 METAL STRESS AND IMPACT ON PLANTS

Metals with molecular weight over 20 g/mol and/or density higher than 5 g/mL are termed as heavy metals (Duffus, 2003; Stankovic et al., 2014; Kim et al., 2015). They are

TABLE 33.4

Effect of Metals on Different Parameters of Stevia Plants

Level of study	Metal evaluated	Parameter studied	Remarks	References
In vitro	Cd, Co, Cu, Pb, Zn, Hg and Ag	G, R, S	Zinc and cobalt improved the formation of shoots and length. Silver is apparently the most phytotoxic for stevia.	Aggarwal et al. (2013).
In vitro	Cd	G, L, S	Cd had a negative impact on growth in stevia.	Kumar et al. (2012).
Greenhouse	Zn and P	G, L, PL, B, FW	Higher biomass and yield with Zn without P.	Das et al. (2005).
In vitro	Cu	G, B, Chl, S, R	Pretreatment enhanced biomass and regeneration capacity of *in vitro* plants of stevia.	Jain et al. (2009).
Laboratory	As, Cd, Cr, Cu, Fe, Mg, Pb, Se, Zn, Al, Ag, Co, Ca, Mn, Ni.	MC	Metal accumulation is not at threshold level.	Hajar et al. (2014).
Greenhouse	Nanoparticles of Si, Fe, Cu	DW, FW, SOD, CAT, Glu, MC, Cal, Sgly	Nanoparticles of Si, Fe, Cu induced glutathione, catalase and SOD in stevia.	Hendawey and Reda (2015).
In vitro	CuSO$_4$	PPO, POX, IB, CP, Vac, CGA, P, FC, S	Copper induced antioxidant enzymes and organic compounds in stevia.	Jain et al. (2014).
In vitro	Zn,	Sgly, G, FL, P, DPPHs, ROS, Reb-A, RP, AA	Increase of steviosides and antioxidant activity in stevia response to zinc.	Javed et al. (2016).
In vitro	Cu	L, G, S, FC, Fl, TAC, RP, DPPHs, FW, Sgly	Copper oxide nanoparticles functioned as stimulators of the production of bioactive components and can be employed in *in vitro* batch cultures.	Javed et al. (2017a).
In vitro	Zn, Cu	Sgly, Cal, AA, TAC, Cal, Fl, FC, RP, DPPHs, G, FW, DW	CuO nanoparticles were more toxic to stevia callus as compared to ZnO nanoparticles.	Javed et al. (2017b).
In vitro	Cu, Au	B, FC	Application of nanoparticles showed a positive effect in enhancing biomass and secondary metabolites production in adventitious root cultures of stevia.	Ghazal et al. (2018).

G: Growth, R: Root, S: Shoots, L: Leaves, PL: Plant Length, B: Biomass, FW: Fresh Weight, DW: Dry Weight, Chl: Chlorophyll, Cal: Callus, MC: Mineral Content, SOD: Super Oxide Dismutase, CAT: Catalase, Glu: Glutathione, POX: Peroxidase, PPO: Polyphenol Oxidase, IB: Inclusion Bodies, Vac: Vacuoles, CGA: Chlorogenic Acid, CP: Chloroplast, DPPHs: 2,2-Diphenyl-1-Picrylhydrazyl Scavenging Activity, ROS: Reactive Oxygen Species, Reb-A: Rebaudioside-A, RP: Reductive Power, Fl: Flavonoids, P: Photosynthesis, AA: Antioxidant Activity, TAC: Total Antioxidant Capacity, FC: Phenolic Compound, Fl: Flavonoids, Sgly: Steviol Glycosides serious threats to the environment and are non-essential for biological systems; some examples of these heavy metals are Cd, Hg, Pb, and Tl. But, some of the essential metals such as Co, Cu, Fe, Mn, and Zn are toxic after a certain threshold concentration (Van Bussel et al., 2014). Some metals can reduce vital functions or even kill the plants, even in very low concentrations, such metals are Al, Ga, In, Tl, Sn, Pb, Sb, Po, Bi, Ge, and Te (Järup, 2003; Navarro-Aviñó et al., 2007; Appenroth, 2010; Kohzadi, 2018). Each metal has different action in different systems of the plant, for example, the beneficial effect of cobalt is in the fixation of nitrogen, but its presence at more than 10×10^{-3} mg/g is toxic. An excess of metals in the environment can inhibit plant development and reduce crop productivity (Emamverdian et al., 2015; White and Pongrac, 2017).

Metals are not static elements in the environment; these are transported commonly from soil to roots and later to leaves by xylem (Navarro-Aviñó et al., 2007). Metal stress has effects on different developmental phases and plant functions such as germination, growth, photosynthesis, and interference with the activity of enzymes and isozymes, producing enzymatic failures related to protein and polypeptide hydrolysis. Also, reducing the availability of amino acids and nitrogen, interrupting nutrient supply, restricting carbohydrates movement forcing the cell to look for another pathway to balance failures in enzymes, like malate dehydrogenase, succinate dehydrogenase, succinate cytochrome c reductase, among others (Smiri et al., 2009; Jaouani et al., 2018).

33.6.2 Effect of Metal Stress on Stevia

The tolerance mechanisms of stevia to metal stress impact are not well understood. There are a few research reports limited to physiological assessments (Table 33.4). The available reports suggest some positive as well as some negative impacts on stevia growth and development. Metal stress decreases growth and induces shoot formation (Aggarwal et al., 2013), enhances number and size of shoots (Kumar et al., 2012), reduces biomass and yield (Das et al., 2005; Jain et al., 2009), and induces oxidative stress (Jain et al., 2014; Hendawey and Reda, 2015).

33.6.2.1 Growth and Development

Growth is the most evident effect of metal pollution consequences observed in plants. Each element results in different effects in different parts of the plants, but the majority of metals inhibit or reduce growth (Cakmak and Marschner, 1993; Choi et al., 1996; Fontes and Cox, 1998), cause chlorosis (Ebbsard and Kochian, 1997), and result in morphological modifications and death of the plants (Di Toppi and Gabrieli, 1999).

The effects of metal in stevia depends on the metal element. The application of cadmium, lead, mercury, and silver reduced the size significantly, the number of primordia shoots and leaves, and differentiation in a concentration dependence manner (Kumar et al., 2012; Aggarwal et al., 2013). Cobalt increases the number and length of shoots at low

concentrations but decreases them under higher concentrations of more than 50 mg/L (Kumar et al., 2012; Aggarwal et al., 2013). The same type of observation was reported for copper treatment under *in vitro* culture conditions, lower concentrations showed positive shoots development, and biomass, but greater than 5 μM caused adverse effects, leading the plant to death (Jain et al., 2009; Hendawey and Reda, 2015). The study of iron, copper, and silicon nanoparticles reported a positive impact on stevia under *in vitro* culture; these elements enhanced the increase of dry and fresh weights of callus at concentrations of 0.5, 1, and 8 ppm (Hendawey and Reda, 2015). Zinc had a positive effect at low levels which induced more shoot proliferation and leaves size; however, at concentrations greater than 100 nM, it was phytotoxic (Desai et al., 2015). The most toxic effect was observed in stevia plants in response to silver and lead treatments in *in vitro* as well as in *ex vitro* conditions (Das et al., 2005; Aggarwal et al., 2013).

33.6.2.2 Photosynthesis

Photosynthesis is one of the most important processes for plant life; however, some plants are not affected by metals due to their high tolerance (Papazoglou et al., 2005). Naturally, the presence of metal ions in the environment decreases chlorophyll contents in leaves (Clijsters and Assche, 1985), inhibits photosynthetic rates (Marin et al., 1993) and transpiration by altering stomatal activity (Bazzaz et al., 1974). The studies concerning stevia's photosynthetic response to heavy metals have focused mainly on the role of copper within the photosynthetic system. In stevia, the presence of copper in media increased chlorophyll production as well as number and size of electron inclusion bodies, vacuole size, and photosynthesis efficiency (Jain et al., 2009, 2014).

33.6.2.3 Induction of Osmotic Stress

Osmotic stress can be defined as limiting access to water for plants because of environmental conditions or components of the medium in which they are located, restrictions that can lead to different responses of the plant at all levels (Jian et al., 1997). Metals can generate damage in the cell, mainly in the membrane and the electron transport chain, changes in the permeability capacities, decrease in the production of the necessary elements for the cell wall, and greater flow of ions between the inside and the outside of the cell. There is not really a direct investigation on osmotic impact and tolerance mechanism in stevia; however, the application of polyamines reduced toxicity in stevia (Jain et al., 2014) and interaction of metals with essential elements showed some osmotic stress and ion imbalance in stevia during metal stress (Das et al., 2005).

33.6.2.4 Induction of Oxidative Stress

Metal stress generally leads to damage to cells and tissues (Shaw et al., 2004), produces ROS such as superoxide radicals ($O_2 \bullet^-$), free hydroxyl radicals (OH^-), singlet oxygen (O_2^*), and hydrogen peroxide (H_2O_2) (Mano et al., 2002; Temple et al., 2005), and further induces lipid peroxides and alters membrane integrity (Prasad and Freitas, 2005; Hao et al., 2006;

Gajewska and Skłodowska, 2007). In stevia, it is also reported that metals like Zn (Javed et al., 2017a), $CuSO_4$ (Jain et al., 2014), Si, and Fe (Hendawey and Reda, 2015) generate reactive oxygen species in *in vitro* as well as under greenhouse conditions.

33.6.3 Metal Stress Tolerance of Stevia

Tolerance to heavy metals is dependent on plant species, concentration, exposure time, and chemical form of the elements (Emamverdian et al., 2015; White and Pongrac, 2017); it is what determines their toxicity, and the consequences on plants hinge on the ability of the plant to eliminate them from the organism. All plants uptake metals from their medium and plant defense strategies against metals have been registered, such as sequestration by symbiotic organisms in the roots and antioxidant substances, among others (Emamverdian et al., 2015).

Plants produce compounds to protect cells against heavy metal derived damage from ROS for example. When a plant cell is stressed, there are induction signals to produce protective substances such as metallothionein, thiols, phytochelatines, phenolics, flavonoids, etc. One of the most important of these compounds is glutathione which is synthesized from cysteine, glutamate, and glycine (Meister and Anderson, 1983), where the expression of the glutathione induction gene is overexpressed under heavy metal stress (Xiang and Oliver, 1998).

It has been found that stevia plants can accumulate metals in different organs, such as leaves, flowers, and stems. Elements like arsenic, cadmium, chromium, copper, iron, magnesium, selenium, zinc, aluminum, silver, cobalt, calcium, manganese, and nickel were found in plants, but levels in samples were not phytotoxic (Hajar et al., 2014). It was observed that stevia accumulates secondary metabolites; for example, steviosides, probably used for resistance to heavy metals, which promote the activity of ascorbate peroxidase and the production of proline in plants under metal stress conditions (Nevmerzhitskaya et al., 2014) and altering the metal transport from roots making plants less susceptible to metal derived damage (Mohan and Robert, 2009; Nevmerzhitskaya et al., 2013).

33.6.3.1 Osmotic and Oxidative Stress Tolerance Mechanism

The effect of heavy metals in plants induces osmotic stress that, in turn, may initiate synthesis of metabolites that play an important role in metal binding, antioxidant defense and signaling (Sharma and Dietz, 2006; Joseph and Jini, 2010; Bhat and Khan, 2011). These osmolytes are sugars, sugar alcohols, proline, glycine betaine, proline, and polyamines (Mudgal et al., 2010; Dhir et al., 2012). Investigations in stevia do not indicate a direct role of these compounds during metal stress; however, the application of polyamines reduced the metal's toxicity (Jain et al., 2014); interaction of metals with essential elements revealed some osmotic stress and ion imbalance in stevia during metal stress (Das et al., 2005).

Plants need to balance between generation and removal of ROS; they possess an efficient antioxidant system for survival under stress conditions (Schutzendubel and Polle, 2002; Kovalchuk et al., 2003; Gratão et al., 2005; Pitzschke et al., 2006). Plants have a well-developed antioxidant protection system that comprises enzymatic antioxidants such as superoxide dismutase, ascorbate peroxidase (APX), glutathione reductase (GR), peroxidase (POX), catalase (CAT), and nonenzymatic antioxidants like ascorbic acid (AsA) and glutathione (GSH) (Chen et al., 2000; Schützendübel et al., 2001).

Stevia exposed to Cu nanoparticles in the medium showed beneficial effects in glycosides production, phenolic compounds, flavonoids, and antioxidant content (Javed et al., 2017a). The addition of nanoparticles of gold and copper to root cell submerged culture of stevia increased the generation of phenolic compounds and flavonoids (Ghazal et al., 2018). Rise of glutathione in response to Fe and Si nanoparticles was also reported, as well as the total antioxidant content was also augmented and increased in steviosides concentration under high concentration treatments of Fe, Cu, and Si nanoparticles (Hendawey and Reda, 2015). The maximum 1,1-diphenyl-2-picrylhydrazyl (DPPH) activity production in response to zinc, gold, and copper, and secondary metabolites production in adventitious root and shoot culture demonstrated a metal induced oxidative scavenging mechanism of stevia (Javed et al., 2017b; Ghazal et al., 2018). Stevia also demonstrated induced enzymatic antioxidative protection against Fe, Cu, and Si stress (Jain et al., 2014; Hendawey and Reda, 2015; Javed et al., 2017a). The enhanced activity of superoxide dismutase and catalase enzymes has been reported in *ex vitro* experiments in stevia plants in response to nanoparticles of Si, Fe, and Cu (Hendawey and Reda, 2015), and regenerated plants with higher levels of copper sulfate treatments *in vitro* exhibited higher PPO enzyme activities (Jain et al., 2014).

33.6.4 Steviol Glycosides and its Relationship with Metal Stress in Stevia

Abiotic stress caused by metals and/or heavy metals is known as an inhibitor or inductor of secondary metabolites in plants (Akula and Ravishankar, 2011). Cu in culture medium enhanced the steviol productivity as well as rebaudioside-A; however, more than 1000 mgL^{-1} acted as an inhibitor (Javed et al., 2017a, 2017b). In contrast, Desai et al. (2015) reported a decrement in the percentage of glycosides of steviol as the concentration of Zn increased; also nanoparticles of Fe increased steviosides concentrations at a minimum as 8 ppm Fe nanoparticles (Hendawey and Reda, 2015).

33.6.5 Conclusions

Despite scarce information about the effects and mechanisms of metals on the stevia plant, it can be concluded, in general terms, that most of the elements known as heavy metals are toxic at certain concentrations; however, at low doses, they can be used to improve the productivity of crops as promoters of growth or compounds of interest, related to abiotic stress.

It is still necessary to carry out research on the response of stevia in genetic and molecular aspects to better clarify what happens inside the plant during the stress induced by metals.

33.7 OVERALL CONCLUSIONS AND FUTURE PERSPECTIVES

Stevia rebaudiana is a species of great importance in modern agriculture around the world due to its positive implications on human health as a sweetener for food. As it has been seen before, stevia has the same or similar agronomic problems as other crops. Abiotic stress imposes important restrictions on the cultivation of stevia that decrease its productivity. In addition, global climate change is imposing more and more drastic types of stress in very short times. Thus, it is necessary to abound in the study of physiological and molecular defense mechanisms against abiotic stress in order to count abundantly with the metabolites produced by this plant species for human health.

REFERENCES

Abdallah, S. B., Aung, B., Amyot, L., Lalin, I., Lachâal, M., Karray-Bouraoui, N., and Hannoufa, A. (2016). Salt stress (NaCl) affects plant growth and branch pathways of carotenoid and flavonoid biosyntheses in *Solanum nigrum*. *Acta Physiologiae Plantarum*, 38(3): 72.

Abraham, G., and Dhar, D. W. (2010). Induction of salt tolerance in *Azolla microphylla* Kaulf through modulation of antioxidant enzymes and ion transport. *Protoplasma*, 245(1–4): 105–111.

Acosta-Motos, J. R., Diaz-Vivancos, P., Álvarez, S., Fernández-García, N., Sanchez-Blanco, M. J., and Hernández, J. A. (2015). Physiological and biochemical mechanisms of the ornamental *Eugenia myrtifolia* L. plants for coping with NaCl stress and recovery. *Planta*, 242(4): 829–846.

Acosta-Motos, J. R., Ortuño, M. F., Bernal-Vicente, A., Diaz-Vivancos, P., Sanchez-Blanco, M. J., and Hernández, J. A. (2017). Plant responses to salt stress: Adaptive mechanisms. *Agronomy*, 7(1): 18.

Acquaah, G. (2012). *Principles of Plant Genetics and Breeding*. 2nd ed. Wiley,:Oxford, UK, p 20.

Aggarwal, V., Shankar, V., and Agrawal, V. (2013). In vitro mass production of Stevia *rebaudiana* bertoni (an antidiabetic endangered herb) and assessment of phytotoxicity of different heavy metals on stevioside synthesis. *Acta Horticulturae*, 972: 121–133.

Aghaleh, M., Niknam, V., Ebrahimzadeh, H., and Razavi, K. (2009). Salt stress effects on growth, pigments, proteins and lipid peroxidation in *Salicornia persica* and *S. europaea*. *Biologia Plantarum*, 53(2): 243–248.

Akandi, Z. N., Pirdashti, H., Yaghoubian, Y., and Omran, V. G. (2017). Quantifying the response of growth and physiological parameters of stevia (*Stevia rebaudiana* Bertoni) medicinal plant to salinity stress under controlled conditions. *Journal of Science and Technology of Greenhouse Culture*, 8(1): 9–20.

Akula, R., and Ravishankar, G. A. (2011). Influence of abiotic stress signals on secondary metabolites in plants. *Plant Signaling & Behavior*, 6(11): 1720–1731.

Alcázar, R., Altabella, T., Marco, F., Bortolotti, C., Reymond, M., Koncz, C., and Tiburcio, A. F. (2010). Polyamines: Molecules with regulatory functions in plant abiotic stress tolerance. *Planta*, 231: 1237–1249.

Anjum, S. A., Xie, X. Y., Wang, L. C., Saleem, M. F., Man, C., and Lei, W. (2011). Morphological, physiological and biochemical responses of plants to drought stress. *African Journal of Agricultural Research*, 6(9): 2026–2032.

Antoniou, C., Savvides, A., Christou, A., and Fotopoulos, V. (2016). Unravelling chemical priming machinery in plants: The role of reactive oxygen–nitrogen–sulfur species in abiotic stress tolerance enhancement. *Current Opinion in Plant Biology*, 33: 101–107.

Apel, K., and Hirt, H. (2004). Reactive oxygen species: Metabolism, oxidative stress, and signal transduction. *Annual Review of Plant Biology*, 55: 373–399.

Appenroth, K. J. (2010). Definition of "heavy metals" and their role in biological systems. *In: Soil Heavy Metals*, Sherameti, I., and Verma, A. (Eds.), pp. 19–29. Springer, Berlin, Heidelberg.

Aranda-González, I., Segura-Campos, M., Moguel-Ordoñez, Y., and Betancur-Ancona, D. (2014). *Stevia rebaudiana* Bertoni. Un potencial adyuvante en el tratamiento de la diabetes mellitus. *CyTA-Journal of Food*, 12(3): 218–226.

Aroca, R., Porcel, R., and Ruiz-Lozano, J. M. (2011). Regulation of root water uptake under abiotic stress conditions. *Journal of Experimental Botany*, 63: 43–57.

Asada, K. (2006). Production and scavenging of reactive oxygen species in chloroplasts and their functions. *Plant Physiology*, 141(2): 391–396.

Ashraf, M. (2009). Biotechnological approach of improving plant salt tolerance using antioxidants as markers. *Biotechnology Advances*, 27(1): 84–93.

Ashraf, M., and Foolad, M. R. (2007). Roles of glycine betaine and proline in improving plant abiotic stress resistance. *Environmental and Experimental Botany*, 59(2): 206–216.

Badran, A. E., Alhady, M. R. A., and Hassan, W. A. (2015). In vitro evaluation of some traits in *Stevia rebaudiana* (Bertoni) under drought stress and their relationship on stevioside content. *American Journal of Plant Sciences*, 6(05): 746.

Bazzaz, F. A., Carlson, R. W., and Rolfe, G. L. (1974). The effect of heavy metals on plants: Part I. Inhibition of gas exchange in sunflower by Pb, Cd, Ni and Tl. *Environmental Pollution*, 7(4): 241–246.

Belkheiri, O., and Mulas, M. (2013). The effects of salt stress on growth, water relations and ion accumulation in two halophyte *Atriplex* species. *Environmental and Experimental Botany*, 86: 17–28.

Belkhodja, R., Morales, F., Abadia, A., Medrano, H., and Abadia, J. (1999). Effect of salinity on chlorophyll fluorescence and photosynthesis of barley (*Hordeum vulgare* L) grown under a triple-line-source sprinkler system in the field. *Photosynthetica*, 36: 375–387.

Benhmimou, A., Ibriz, M., Al Faïz, C., Douaik, A., Khiraoui, A., Amchra, F. Z., and Lage, M. (2017). Productivity of new sweet plant in Morocco (*Stevia rebaudiana* Bertoni) under water stress. *Journal of Medicinal Plants*, 5(5): 126–131.

Bhat, U. N., and Khan, A. B. (2011). Heavy metals: An ambiguous category of inorganic contaminants, nutrients and toxins. *Research Journal of Environmental Sciences*, 5(8): 682–690.

Bidabadi, S. S., and Masoumian, M. (2018). Arbuscular mycorrhizal symbiosis improves growth and antioxidative response of *Stevia rebaudiana* (Bert.) under salt stress. *Trends in Horticulture*, 1: 1–10.

Bohnert, H. J., and Jensen, R. G. (1996). Metabolic engineering for increased salt tolerance—the next step. *Functional Plant Biology*, 23(5): 661–667.

Borrelli, G. M., Fragasso, M., Nigro, F., Platani, C., Papa, R., Beleggia, R., and Trono, D. (2018). Analysis of metabolic and mineral changes in response to salt stress in durum wheat

(*Triticum turgidum* ssp. durum) genotypes, which differ in salinity tolerance. *Plant Physiology and Biochemistry*, 133: 57–70.

Brown, S., Day, D. A., and Critchley, C. (1987). Salt tolerance-does leaf respiration have a contribution to make? *In: Plant Mitochondria: Structure, Functional and Physiological Aspects*, Moore, AL, and Beechey, RB (Eds.), p. 393. Plenum Press, New York, NY.

Buchanan, C. D., Lim, S., Salzman, R. A., Kagiampakis, I., Morishige, D. T., Weers, B. D., and Mullet, J. E. (2005). *Sorghum bicolor*'s transcriptome response to dehydration, high salinity and ABA. *Plant Molecular Biology*, 58(5): 699–720.

Cakmak, I., and Marschner, H. (1993). Effect of zinc nutritional status on activities of superoxide radical and hydrogen peroxide scavenging enzymes in bean leaves. *In: Plant Nutrition, Developments in Plant and Soil Sciences*, Barrow, N. J. (Eds.), 54: 133–136. Springer, Dordrecht.

Cantabella, D., Piqueras, A., Acosta-Motos, J. R., Bernal-Vicente, A., Hernández, J. A., and Diaz-Vivancos, P. (2017). Salt-tolerance mechanisms induced in *Stevia rebaudiana* Bertoni: Effects on mineral nutrition, antioxidative metabolism and steviol glycoside content. *Plant Physiology and Biochemistry*, 115: 484–496.

Carillo, P., Annunziata, M. G., Pontecorvo, G., Fuggi, A., and Woodrow, P. (2011). Salinity stress and salt tolerance. *In: Abiotic Stress in Plants-mechanisms and Adaptations*, Shanker, A. (Eds.). InTech, Rijeka, Croatia.

Carniero, J. W. (1990). *Stevia rebaudiana* (Bert.) Bertoni. MSc Thesis, State University of Maringa, Brazil.

Carneiro, J. W. P., Muniz, A. S., and Guedes, T. A. (1997). Greenhouse bedding plant production of *Stevia rebaudiana* (Bert) Bertoni. *Canadian Journal of Plant Science*, 77: 473–474.

Ceunen, S., and Geuns, J. M. (2013). Steviol glycosides: Chemical diversity, metabolism, and function. *Journal of Natural Products*, 76: 1201–1228.

Chao, Y. Y., and Kao, C. H. (2010). Heat shock-induced ascorbic acid accumulation in leaves increases cadmium tolerance of rice (*Oryza sativa* L.) seedlings. *Plant and Soil*, 336: 39–48.

Chaves, M. M. (1991). Effects of water deficits on carbon assimilation. *Journal of Experimental Botany*, 42(1): 1–16.

Chaves, M. M., and Oliveira, M. M. (2004). Mechanisms underlying plant resilience to water deficits: Prospects for water-saving agriculture. *Journal of Experimental Botany*, 55: 2365–2384.

Chaves, M. M., Flexas, J., and Pinheiro, C. (2009). Photosynthesis under drought and salt stress: Regulation mechanisms from whole plant to cell. *Annals of Botany*, 103(4): 551–560.

Chen, L. M., Lin, C. C., and Kao, C. H. (2000). Copper toxicity in rice seedlings: Changes in antioxidative enzyme activities, H_2O_2 level, and cell wall peroxidase activity in roots. *Botanical Bulletin of Academia Sinica*, 41: 99–103.

Chia, T. Y., Pike, M. J., and Rawsthorne, S. (2005). Storage oil breakdown during embryo development of *Brassica napus* (L.). *Journal of Experimental Botany*, 56(415): 1285–1296.

Chinnusamy, V., Zhu, J., and Zhu, J. K. (2007). Cold stress regulation of gene expression in plants. *Trends in Plant Science*, 12: 444–451.

Choi, J. M., Pak, C. H., and Lee, C. W. (1996). Micro nutrient toxicity in French marigold. *Journal of Plant Nutrition*, 19(6): 901–916.

Choudhury, S., Panda, P., Sahoo, L., and Panda, S. K. (2013). Reactive oxygen species signaling in plants under abiotic stress. *Plant Signaling & Behavior*, 8(4): e23681–e23685.

Clijsters, H. V., and Van Assche, F. (1985). Inhibition of photosynthesis by heavy metals. *Photosynthesis Research*, 7(1): 31–40.

Corpas, F. J., Gupta, D. K., and Palma, J. M. (2015). Production sites of reactive oxygen species (ROS) in organelles from plant cells. *In: Reactive Oxygen Species and Oxidative Damage in Plants Under Stress*, Corpas, F. J., Gupta, D. K., and Palma, J. M. (Eds.), pp. 1–22. Springer, Cham.

Cramer, G. R., Urano, K., Delrot, S., Pezzotti, M., and Shinozaki, K. (2011). Effects of abiotic stress on plants: A systems biology perspective. *BMC Plant Biology*, 11:163.

Curtion, D., Steppuhn, H., and Selles, F. (1993). Plant response to sulphate and chloride salinity: Growth and ionic relations. *Soil Science Society of America Journal*, 57(5): 1304–1310.

Das, K., Dang, R., Shivananda, T. N., and Sur, P. (2005). Interaction between phosphorus and zinc on the biomass yield and yield attributes of the medicinal plant stevia (*Stevia rebaudiana*). *The Scientific World Journal*, 5: 390–395.

Debnath, M., Ashwath, N., Hill, C. B., Callahan, D. L., Dias, D. A., Jayasinghe, N. S., Roessner, U. et al. (2018). Comparative metabolic and ionomic profiling of two cultivars of *Stevia rebaudiana* Bert. (Bertoni) grown under salinity stress. *Plant Physiology and Biochemistry*, 129: 56–70.

Desai, C. V., Desai, H. B., Suthar, K. P., Singh, D., Patel, R. M., and Taslim, A. (2015). Phytotoxicity of zinc-nanoparticles and its influence on stevioside production in *Stevia rebaudiana* Bertoni. *Applied Biological Research*, 17(1): 1–7.

Dhir, B., Nasim, S. A., Samantary, S., and Srivastava, S. (2012). Assessment of osmolyte accumulation in heavy metal exposed *Salvinia natans*. *International Journal of Botany*, 8: 153–158.

Di Toppi, L. S., and Gabrieli, R. (1999). Response to cadmium in higher plants. *Environmental and Experimental Botany*, 41(2): 105–130.

Djanaguiraman, M., Boyle, D. L., Welti, R., Jagadish, S. V. K., and Prasad, P. V. V. (2018). Decreased photosynthetic rate under high temperature in wheat is due to lipid desaturation, oxidation, acylation, and damage of organelles. *BMC Plant Biology*, 18(1): 55.

Dubey, R. S. (1997). Photosynthesis in plants under stressful conditions. *In: Hand Book of Photosynthesis*, Pessarakli, M. (Ed.), p. 859. Marcel Dekker, New York, NY.

Duffus, J. H. (2003). "Heavy metals"—A meaningless term? (IUPAC Technical Report). *Pure and Applied Chemistry*, 75(9): 1357.

Ebbsard, S. D., and Kochian, L. V. (1997). Toxicity of zinc and copper to *Brassica* species: Implications for phytoremediation. *Journal of Environmental Quality*, 26(3): 776–781.

Egan, T. P., and Ungar, I. A. (1998). Effect of different salts of sodium and potassium on the growth of *Atriplex prostrata* (Chenopodiaceae). *Journal of Plant Nutrition*, 21(10): 2193–2205.

Emamverdian, A., Ding, Y., Mokhberdoran, F., and Xie, Y. (2015). Heavy metal stress and some mechanisms of plant defense response. *The Scientific World Journal*, 2015:1–18.

Fallah, F., Nokhasi, F., Ghaheri, M., Kahrizi, D., Agha, A. B. A., Ghorbani, T., Ansarypour, Z. et al. (2017). Effect of salinity on gene expression, morphological and biochemical characteristics of *Stevia rebaudiana* Bertoni under in vitro conditions. *Cellular and Molecular Biology (Noisy le Grand)*, 63(7):102–106.

Fanti, S. C., and de Andrade, S. C. J. G. (2004). Process germinativo de sementes de paineira sob estresses hídrico e salino. *Pesquisa Agropecuária Brasileira*, 39(9): 903–909.

Farooq, M., Hussain, M., Wahid, A., and Siddique, K. H. M. (2012). Drought stress in plants: An overview. *In: In Plant Responses to Drought Stress*, Lichtfouse, E., Navarrete, M., Debaeke, P., Véronique, S., and Alberola, C. (Eds.), pp. 1–33. Springer, Berlin, Heidelberg.

Farooq, M., Wahid, A., Kobayashi, N., Fujita, D. and Basra, S. M. A. (2009). Plant drought stress: Effects, mechanisms and management. *In: Sustainable Agriculture*, Lichtfouse, E., Navarrete, M., Debaeke, P., Véronique, S., and Alberola, C. (Eds.), pp. 153–188. Springer, Dordrecht.

Fei, S., Desprez, J. M., Potter, K. M., Jo I, Knott, J. A., and Oswalt, C. M. (2017). Divergence of species responses to climate change. *Science Advances*, 3 (5), e1603055.

Fenollosa, E., and Munné-Bosch, S. (2018). Photoprotection and photo-oxidative stress markers as useful tools to unravel plant invasion success. *In: Advances in Plant Ecophysiology Techniques*, Sánchez-Moreiras, D. M., and Reigosa, M. J. (Eds.), pp. 153–175. Springer, Cham.

Flowers, T. J., Troke, P. F., and Yeo, A. R. (1977). The mechanism of salt tolerance in halophytes. *Annual Review of Plant Physiology*, 28(1): 89–121.

Fontes, R. L. F., and Cox, F. R. (1998). Zinc toxicity in soybean grown at high iron concentration in nutrient solution. *Journal of Plant Nutrition*, 21(8): 1723–1730.

Fronza, D., and Folegatti, M. V. (2003). Water consumption of the stevia (*Stevia rebaudiana* Bert.) Bertoni crop estimated through microlysimeter. *Scientia Agricola*, 60(3): 595–599.

Gajewska, E., and Skłodowska, M. (2007). Effect of nickel on ROS content and antioxidative enzyme activities in wheat leaves. *Biometals*, 20(1): 27–36.

Garcia, A. B., Engler, J. D. A., Iyer, S., Gerats, T., Van Montagu, M., and Caplan, A. B. (1997). Effects of osmoprotectants upon NaCl stress in rice. *Plant Physiology*, 115(1): 159–169.

Geuns, J. M. (2003). Stevioside. *Phytochemistry*, 64(5): 913–921.

Ghaheri, M., Kahrizi, D., and Bahrami, G. (2015). Mannitol effect of on some morphological characteristics of in vitro *Stevia rebaudiana* Bertoni. *Biharian Biology*, 11(2): 94–97.

Ghassemi, F., Jakeman, A. J., and Nix, H. A. (1995). *Salinisation of Land and Water Resources: Human Causes, Extent, Management and Case Studies*. UNSW Press, Sydney, Australia, and CAB International, Wallingford, UK.

Ghazal, B., Saif, S., Farid, K., Khan, A., Rehman, S., Reshma, A., and Ahmad, N. (2018). Stimulation of secondary metabolites by copper and gold nanoparticles in submerge adventitious root cultures of *Stevia rebaudiana* (Bert.). *IET Nanobiotechnology*, 12(5): 569–573.

Gill, S. S., and Tuteja, N. (2010). Reactive oxygen species and antioxidant machinery in abiotic stress tolerance in crop plants. *Plant Physiology and Biochemistry*, 48: 909–930.

Glenn, E. P., Brown, J. J., and Blumwald, E. (1999). Salt tolerance and crop potential of halophytes. *Critical Reviews in Plant Sciences*, 18: 227–255.

Golldack, D., Li, C., Mohan, H., and Probst, N. (2014). Tolerance to drought and salt stress in plants: Unraveling the signaling networks. *Frontiers in Plant Science*, 5: 151–156.

González, L. M., Argentel, L., Zaldívar, N., and Ramírez, R. (2005). Efecto de la sequía simulada con PEG-6000 sobre la germinación y el crecimiento de las plántulas de dos variedades de trigo. *Cultivos Tropicales*, 26(4): 49–52.

González, Y., Villafañe, R., Basso, C., Trujillo, A., and Pérez, D. (2017). Tolerancia de la stevia (*Stevia rebaudiana* (Bertoni) Bertoni) a la salinidad. *Acta Científica Venezolana*, 67(2): 35–47.

Gorzi, A., Omidi, H., and Bostani, A. B. (2018). Morphophysiological responses of stevia (*Stevia rebaudiana* bertoni) to various priming treatments under drought stress. *Applied Ecology and Environmental Research*, 16(4): 4753–4771.

Gratão, P. L., Polle, A., Lea, P. J., and Azevedo, R. A. (2005). Making the life of heavy metal-stressed plants a little easier. *Functional Plant Biology*, 32(6): 481–494.

Groppa, M. D., and Benavides, M. P. (2008). Polyamines and abiotic stress: Recent advances. *Amino Acids*, 34(1): 35–45.

Gupta, P., Sharma, S., and Saxena, S. (2016). Effect of abiotic stress on growth parameters and steviol glycoside content in *Stevia rebaudiana* (Bertoni) raised in vitro. *Journal of Applied Research on Medicinal and Aromatic Plants*, 3(4): 160–167.

Gupta, D. K., Palma, J. M., and Corpas, F. J. (Eds.). (2018). *Antioxidants and Antioxidant Enzymes in Higher Plants*. Springer , Cham.

Hajar, E. W. I., Sulaiman, A. Z. B., and Sakinah, A. M. (2014). Assessment of heavy metals tolerance in leaves, stems and flowers of *Stevia rebaudiana* plant. *Procedia Environmental Sciences*, 20: 386–393.

Hajihashemi, S., and Ehsanpour, A. A. (2013). Influence of exogenously applied paclobutrazol on some physiological traits and growth of *Stevia rebaudiana* under in vitro drought stress. *Biologia*, 68(3): 414–420.

Hajihashemi, S., and Ehsanpour, A. A. (2014). Antioxidant response of *Stevia rebaudiana* B. to polyethylene glycol and paclobutrazol treatments under in vitro culture. *Applied Biochemistry and Biotechnology*, 172(8): 4038–4052.

Hajihashemi, S., and Geuns, J. M. (2013). Free radical scavenging activity of steviol glycosides, steviol glucuronide, hydroxytyrosol, metformin, aspirin and leaf extract of Stevia rebaudiana. *Free Radicals and Antioxidants*, 3: S34–S41.

Hajihashemi, S., and Geuns, J. (2016). Gene transcription and steviol glycoside accumulation in *Stevia rebaudiana* under polyethylene glycol-induced drought stress in greenhouse cultivation. *FEBS Open Bio*, 6(9): 937–944.

Hajihashemi, S., and Geuns, J. M. (2017). Steviol glycosides correlation to genes transcription revealed in gibberellin and paclobutrazol-treated *Stevia rebaudiana*. *Journal of Plant Biochemistry and Biotechnology*, 26(4): 387–394.

Hajihashemi, S., and Sofo, A. (2018). The effect of polyethylene glycol-induced drought stress on photosynthesis, carbohydrates and cell membrane in *Stevia rebaudiana* grown in greenhouse. *Acta Physiologiae Plantarum*, 40(8): 142.

Hajihashemi, S., Noedoost, F., Geuns, J. M., Djalovic, I., and Siddique, K. H. (2018a). Effect of cold stress on photosynthetic traits, carbohydrates, morphology, and anatomy in nine cultivars of *Stevia rebaudiana*. *Frontiers in Plant Science*, 9: 1–12.

Hajihashemi, S., Rajabpoor, S., and Djalovic, I. (2018b). Antioxidant potential in *Stevia rebaudiana* callus in response to polyethylene glycol, paclobutrazol and gibberellin treatments. *Physiology and Molecular Biology of Plants*, 24(2): 335–341.

Hakeem, K. R., Chandna, R., Ahmad, A., and Iqbal, M. (2012). Reactive nitrogen inflows and nitrogen use efficiency in agriculture: An environment perspective. *In: Environmental Adaptations and Stress Tolerance of Plants in the Era of Climate Change*, Ahmad, P., and Prasad, M. N. V. (Eds.), pp. 217–232. Springer, New York, NY.

Hao, F., Wang, X., and Chen, J. (2006). Involvement of plasmamembrane NADPH oxidase in nickel-induced oxidative stress in roots of wheat seedlings. *Plant Science*, 170(1): 151–158.

Hare, P. D., Cress, W. A., and Van Staden, J. (1998). Dissecting the roles of osmolyte accumulation during stress. *Plant, Cell & Environment*, 21(6): 535–553.

Hasanuzzaman, M., Nahar, K., Alam, M., Roychowdhury, R., and Fujita, M. (2013). Physiological, biochemical, and molecular mechanisms of heat stress tolerance in plants. *International Journal of Molecular Sciences*, 14: 9643–9684.

Hasegawa, P. M., Bressan, R. A., Zhu, J. K., and Bohnert, H. J. (2000). Plant cellular and molecular responses to high salinity. *Annual Review of Plant Biology*, 51(1): 463–499.

He, C., Yan, J., Shen, G., Fu, L., Holaday, A. S., Auld, D.,...., and Zhang, H. (2005). Expression of an Arabidopsis vacuolar sodium/proton antiporter gene in cotton improves photosynthetic performance under salt conditions and increases fiber yield in the field. *Plant and Cell Physiology*, 46(11): 1848–1854.

He, Y., and Zhu, Z. J. (2008). Exogenous salicylic acid alleviates NaCl toxicity and increases antioxidative enzyme activity in *Lycopersicon esculentum*. *Biologia Plantarum*, 52(4): 792.

Hendawey, M. H., Reda, E., and Fadl, A. E. (2015). Biochemical role of some nanoparticles in the production of active constituents in *Stevia rebaudiana* L. callus. *Life Science Journal*, 12(7): 144–156.

Heuer, B., and Feigin, A. (1993). Interactive effects of choloride and nitrate on photosynthesis and related growth parameters in tomatoes. *Photosynthetica*, 28: 549.

Hodges, M., Dellero, Y., Keech, O., Betti, M., Raghavendra, A. S., Sage, R., and Weber, A. P. (2016). Perspectives for a better understanding of the metabolic integration of photorespiration within a complex plant primary metabolism network. *Journal of Experimental Botany*, 67(10): 3015–3026.

Hong-Bo, S., Xiao-Yan, C., Li-Ye, C., Xi-Ning, Z., Gang, W., Yong-Bing, Y., and Zan, M. H. (2006). Investigation on the relationship of proline with wheat anti-drought under soil water deficits. *Colloids and Surfaces B: Biointerfaces*, 53(1): 113–119.

Horie, T., Kaneko, T., Sugimoto, G., Sasano, S., Panda, S. K., Shibasaka, M., and Katsuhara, M. (2011). Mechanisms of water transport mediated by PIP aquaporins and their regulation via phosphorylation events under salinity stress in barley roots. *Plant and Cell Physiology*, 52(4): 663–675.

Hossa, K. R., Carneiro, J. W. P., Guedes, T. A., and Braccini, A. L. (2017). *Stevia rebaudiana* (Bert) Bertoni: Influence of osmotic stress and seed priming on seed germination under laboratory conditions. *Acta Scientiarum. Agronomy*, 39: 379–384.

Hossain, M. A., Shamim Kabir, A. H. M., Jahan, T. A., and Hasan, M. N. (2008). Micropropagation of stevia. *International Journal of Sustainable Crop Production*, 3: 1–9.

Hu, W. (1995). High yield technology for groundnut. *International Arachis Newsletter*, 15: 1–22.

Hu, Y., Jiang, L., Wang, F., and Yu, D. (2013). Jasmonate regulates the inducer of CBF expression-C-repeat binding factor/DRE binding factor1 cascade and freezing tolerance in Arabidopsis. *Plant Cell*, 25: 2907–2924.

Huang, J., Levine, A., and Wang, Z. (2013). Plant abiotic stress. *The Scientific World Journal*, 2013.

Hussain, S., Geissler, N., El-Far, M. M., and Koyro, H. W. (2017). Effects of salinity and short-term elevated atmospheric CO_2 on the chemical equilibrium between CO_2 fixation and photosynthetic electron transport of *Stevia rebaudiana* Bertoni. *Plant Physiology and Biochemistry*, 118: 178–186.

Ibrahim, I. A., Nasr, M. I., Mohammed, B. R., and El-Zefzafi, M. M. (2008). Plant growth regulators affecting *in vitro* cultivation of *Stevia rebaudiana*. *Sugar Technology*, 10: 254–259.

Iqbal, N., Umar, S., and Khan, N. A. (2015). Nitrogen availability regulates proline and ethylene production and alleviates salinity stress in mustard (*Brassica juncea*). *Journal of Plant Physiology*, 178: 84–91.

Irar, S., Brini, F., Goday, A., Masmoudi, K., and Pagès, M. (2010). Proteomic analysis of wheat embryos with 2-DE and liquid-phase chromatography (ProteomeLab PF–2D) a wider perspective of the proteome. *Journal of Proteomics*, 73: 1707–1721.

Jain, P., Kachhwaha, S., and Kothari, S. L. (2009). Improved micropropagation protocol and enhancement in biomass and chlorophyll content in *Stevia rebaudiana* (Bert.) Bertoni by

using high copper levels in the culture medium. *Scientia Horticulturae*, 119(3): 315–319.

Jain, P., Kachhwaha, S., and Kothari, S. L. (2014). Chloroplast ultra-structure, photosynthesis and enzyme activities in regenerated plants of *Stevia rebaudiana* (Bert.) Bertoni as influenced by copper sulphate in the medium. *Scientia Horticulturae*, 52: 898–904.

Jamil, M., Lee, C. C., Rehman, S. U., Lee, D. B., Ashraf, M., and Rha, E. S. (2005). Salinity (NaCl) tolerance of *Brassica* species at germination and early seedling growth. *Electronic Journal of Environmental, Agricultural and Food Chemistry*, 4(4): 970–976.

Jaouani, K., Karmous, I., Ostrowski, M., El Ferjani, E., Jakubowska, A., and Chaoui, A. (2018). Cadmium effects on embryo growth of pea seeds during germination: Investigation of the mechanisms of interference of the heavy metal with protein mobilization-related factors. *Journal of Plant Physiology*, 226: 64–76.

Järup, L. (2003). Hazards of heavy metal contamination. *British Medical Bulletin*, 68:167–182.

Javed, R., Mohamed, A., Yücesan, B., Gürel, E., Kausar, R., and Zia, M. (2017a). CuO nanoparticles significantly influence in vitro culture, steviol glycosides, and antioxidant activities of *Stevia rebaudiana* Bertoni. *Plant Cell Tissue and Organ Culture*, 131: 611–620.

Javed, R., Usman, M., Yücesan, B., Zia, M., and Gürel, E. (2016). Effect of zinc oxide (ZnO) nanoparticles on physiology and steviol glycosides production in micropropagated shoots of *Stevia rebaudiana* Bertoni. *Plant Physiology and Biochemistry*, 110: 94–99.

Javed, R., Yücesan, B., Zia, M., and Gürel, E. (2017b). Elicitation of secondary metabolites in callus cultures of *Stevia rebaudiana* bertoni grown under ZnO and CuO nanoparticles stress. *Sugar Technology*, 20(2): 194–201.

Joseph, B., and Jini, D. (2010). Insight into the role of antioxidant enzymes for salt tolerance in plants. *International Journal of Botany*, 6(4): 456–464.

Kalpana, M., Anbazhagan, M., and Natarajan, V. (2009). Utilization of liquid medium for rapid micropropagation of *Stevia rebaudiana* Bertoni. *Journal of Ecobiotechnology*, 1: 16–20.

Karimi, M., Ahmadi, A., Hashemi, J., Abbasi, A., and Angelini, L. G. (2014). Effect of two plant growth retardants on steviol glycosides content and antioxidant capacity in Stevia (*Stevia rebaudiana* Bertoni). *Acta Physiologiae Plantarum*, 36(5): 1211–1219.

Karimi, M., Ahmadi, A., Hashemi, J., Abbasi, A., Tavarini, S., Guglielminetti, L., and Angelini, L. G. (2015). The effect of soil moisture depletion on stevia (*Stevia rebaudiana* Bertoni) grown in greenhouse conditions: Growth, steviol glycosides content, soluble sugars and total antioxidant capacity. *Scientia Horticulturae*, 183: 93–99.

Kim, R. H., Smith, P. D., Aleyasin, H., Hayley, S., Mount, M. P., Pownall, S., and Westaway, D. (2005). Hypersensitivity of DJ-1-deficient mice to 1-methyl-4-phenyl-1, 2, 3, 6-tetrahydropyrindine (MPTP) and oxidative stress. *Proceedings of the National Academy of Sciences*, 102(14): 5215–5220.

Kim, I. S., Yang, M., Lee, O. H., and Kang, S. N. (2011). The anti-oxidant activity and the bioactive compound content of *Stevia rebaudiana* water extracts. *LWT-Food Science and Technology*, 44(5): 1328–1332.

Kim, R. Y., Yoon, J. K., Kim, T. S., Yang, J. E., Owens, G., and Kim, K. R. (2015). Bioavailability of heavy metals in soils: Definitions and practical implementation—a critical review. *Environmental Geochemistry and Health*, 37: 1041–1061.

Kinnersley, A. M., and Turano, F. J. (2000). Gamma aminobutyric acid (GABA) and plant responses to stress. *Critical Reviews in Plant Sciences*, 19(6): 479–509.

Kohzadi, S., Shahmoradi, B., Ghaderi, E., Loqmani, H., and Maleki, A. (2018). Concentration, source, and potential human health risk of heavy metals in the commonly consumed medicinal plants. *Biological Trace Element Research*, 1–10.

Kotak, S., Larkindale, J., Lee, U., von Koskull-Döring, P., Vierling, E., and Scharf, K. D. (2007). Complexity of the heat stress response in plants. *Current Opinion in Plant Biology*, 10(3): 310–316.

Kovalchuk, I., Filkowski, J., Smith, K., and Kovalchuk, O. (2003). Reactive oxygen species stimulate homologous recombination in plants. *Plant, Cell & Environment*, 26(9): 1531–1539.

Kranner, I., Minibayeva, F. V., Beckett, R. P., and Seal, C. E. (2010). What is stress? Concepts, definitions and applications in seed science. *New Phytologist*, 188: 655–673.

Kumar, P., Dwivedi, P., and Singh, P. (2012). Role of polyamine in combating heavy metal stress in *Stevia rebaudiana* Bertoni under in vitro conditions. *International Journal of Agriculture, Environment and Biotechnology*, 5: 193–198.

Kumar, R., Sood, S., Sharma, S., Kasana, R. C., Pathania, V. L., Singh, B., and Singh, R. D. (2014). Effect of plant spacing and organic mulch on growth, yield and quality of natural sweetener plant Stevia and soil fertility in western Himalayas. *International Journal of Plant Production*, 8: 311–334.

Lamosa, P., Martins, L. O., Da Costa, M. S., and Santos, H. (1998). Effects of temperature, salinity, and medium composition on compatible solute accumulation by *Thermococcus spp. Applied and Environmental Microbiology*, 64(10): 3591–3598.

Larkindale, J., and Huang, B. (2004). Thermotolerance and antioxidant systems in Agrostis stolonifera: Involvement of salicylic acid, abscisic acid, calcium, hydrogen peroxide, and ethylene. *Journal of Plant Physiology*, 161: 405–413.

Lavini, A., Riccardi, M., Pulvento, C., De Luca, S., Scamosci, M., and d'Andria, R. (2008). Yield, quality and water consumption of *Stevia rebaudiana* Bertoni grown under different irrigation regimes in Southern Italy. *Italian Journal of Agronomy*, 3(2): 135–143.

Legris, M., Nieto, C., Sellaro, R., Prat, S., and Casal, J. J. (2017). Perception and signaling of light and temperature cues in plants. *The Plant Journal*, 90: 683–697.

Li, S. L., Xia, Y. Z., Liu, J., Shi, X. D., and Sun, Z. Q. (2014). Effects of cold-shock on tomato seedlings under high temperature stress. *The Journal of Applied Ecology*, 25: 2927–2934.

Lichtenthaler, H. K. (1996). Vegetation stress: An introduction to the stress concept in plants. *Journal of Plant Physiology*, 148: 4–14.

Lichtenthaler, H. K. (1998). The Stress Concept in Plants: An Introduction. *Annals of the New York Academy of Sciences*, 851(1): 187–198.

Liopa-Tsakalidi, A., Kaspiris, G., Salahas, G., and Barouchas, P. (2012). Effect of salicylic acid (SA) and gibberellic acid (GA3) pre-soaking on seed germination of stevia (*Stevia rebaudiana*) under salt stress. *Journal of Medicinal Plants Research*, 6(3): 416–423.

Liu, C., Mao, B., Ou, S., Wang, W., Liu, L., Wu, Y., Wang, X. et al. (2014). OsbZIP71, a bZIP transcription factor, confers salinity and drought tolerance in rice. *Plant Molecular Biology*, 84(1–2): 19–36.

Liu, W., Yu, K., He, T., Li, F., Zhang, D., and Liu, J. (2013). The low temperature induced physiological responses of *Avena nuda* L., a cold-tolerant plant species. *The Scientific World Journal*, 2013: 1–7.

Lobell, D. B., Wolfram Schlenker, W., and Costa-Roberts, J. (2011). Climate trends and global crop production since 1980. *Science*, 333(6042): 616–620.

Lopushinsky, W., and Max, T. A. (1990). Effect of soil temperature on root and shoot growth and on budburst timing in conifer seedling transplants. *New Forests*, 4: 107–124.

Ma, L., Ren, G. X., and Shi, Y. (2012). Effects of potassium fertilizer on diurnal change of photosynthesis in *Stevia rebaudiana* Bertoni. *Advanced Materials Research*, 343: 1087–1091.

Macchia, M., Andolfi, L., Ceccarini, L., and Angelini, L. G. (2007). Effects of temperature, light and pre-chilling on seed germination of *Stevia rebaudiana* (Bertoni) Bertoni accessions. *Italian Journal of Agronomy*, 2: 55–62.

Mano, J. I., Torii, Y., Hayashi, S. I., Takimoto, K., Matsui, K., Nakamura, K., Asada, K. et al. (2002). The NADPH: Quinone oxidoreductase P1-ζ-crystallin in Arabidopsis catalyzes the α, β-hydrogenation of 2-alkenals: Detoxication of the lipid peroxide-derived reactive aldehydes. *Plant and Cell Physiology*, 43(12): 1445–1455.

Marin, A. R., Pezeshki, S. R., Masschelen, P. H., and Choi, H. S. (1993). Effect of dimethylarsenic acid (DMAA) on growth, tissue arsenic, and photosynthesis of rice plants. *Journal of Plant Nutrition*, 16:865–880.

Mazorra, L. M., Nunez, M., Hechavarria, M., Coll, F., and Sanchez-Blanco, M. J. (2002). Influence of brassinosteroids on antioxidant enzymes activity in tomato under different temperatures. *Biologia Plantarum*, 45: 593–596.

Megha, S., Basu, U., and Kav, N. N. (2014). Metabolic engineering of cold tolerance in plants. *Biocatalysis and Agricultural Biotechnology*, 3: 88–95.

Meister, A., and Anderson, M. E. (1983). Glutathione. *Annual Review of Biochemistry*, 52: 711–760.

Mittler, R. (2002). Oxidative stress, antioxidants and stress tolerance. *Trends in Plant Science*, 7: 405–410.

Mohd-Radzman, N. H., Ismail, W. I. W., Adam, Z., Jaapar, S. S., and Adam, A. (2013). Potential roles of *Stevia rebaudiana* Bertoni in abrogating insulin resistance and diabetes: A review. *Evidence-Based Complementary and Alternative Medicine*, 2013: 718049.

Moradi-Peynevandi, K., Razavi, S. M., and Zahri, S. (2018). The ameliorating effects of polyamine supplement on physiological and biochemical parameters of *Stevia rebaudiana* Bertoni under cold stress. *Plant Production Science*, 21: 123–131.

Morison, J. I. L., and Lawlor, D. W. (1999). Interactions between increasing CO_2 concentration and temperature on plant growth. *Plant Cell & Environment*, 22: 659–682.

Mubarak, M. H., El Dein, A. B. T., and Sarag, E. L. (2012). *In vitro* response *stevia rebaudiana* growth under salinity and drought stress. *El- Minia, Egypt*, 1: 1369–1371.

Mudgal, V., Madaan, N., and Mudgal, A. (2010). Biochemical mechanisms of salt tolerance in plants: A review. *International Journal of Botany*, 6(2): 136–143.

Munns, R. (2002). Salinity, growth and phytohormones. *In: Salinity: Environment-plants-molecules*, Läuchli, A., and Lüttge, U. (Eds.), pp. 271–290. Springer, Dordrecht.

Munns, R. (2005). Genes and salt tolerance: Bringing them together. *New Phytologist*, 167(3): 645–663.

Munns, R., James, R. A., and Läuchli, A. (2006). Approaches to increasing the salt tolerance of wheat and other cereals. *Journal of Experimental Botany*, 57(5): 1025–1043.

Munns, R., and Richards, R. A. (1996). Improving crop productivity in saline soils, *In: Proceedings of the 2nd International Crop Science Congress*, 453. Enfield, N.H. Science Publishers, New Delhi.

Munns, R., and Tester, M. (2008). Mechanisms of salinity tolerance. *Annual Review of Plant Biology*, 59: 651–681.

Muscolo, A., Panuccio, M. R., and Sidari, M. (2003). Effects of salinity on growth, carbohydrate metabolism and nutritive properties of kikuyu grass (*Pennisetum clandestinum* Hochst). *Plant Science*, 164(6): 1103–1110.

Navarro-Aviñó, J. P., Alonso, I. A., and López-Moya, J. R. (2007). Aspectos bioquímicos y genéticos de la tolerancia y acumulación de metales pesados en plantas. *Revista Ecosistemas*, 16(2):10–25.

Nevmerzhitskaya, J. Y., Mikhailov, A. L., Strobykina, A. S., and Timofeeva, O. A. (2014). Influence of stevioside and heavy metals on physiological and biochemical parameters of winter wheat. *Biology and Medicine*, 6(3): 1.

Nevmerzhitskaya, Y. Y., Timofeeva, O. A., Mikhaylov, A. L., Strobykina, A. S., Strobykina, I. Y., and Mironov, V. F. (2013). Stevioside increases the resistance of winter wheat to low temperatures and heavy metals. *Biological Sciences*, 452: 287–290.

Nieves-Cordones, M., Alemán, F., Fon, M., Martínez, V., and Rubio, F. (2012). K+ nutrition, uptake, and its role in environmental stress in plants. *In: Environmental Adaptations and Stress Tolerance of Plants in the Era of Climate Change*, pp. 85–112. Springer, New York, NY.

Noctor, G., and Foyer, C. H. (1998). Ascorbate and glutathione: Keeping active oxygen under control. *Annual Review of Plant Biology*, 49(1): 249–279.

Nower, A. A. (2014). *In vitro* propagation and synthetic seeds production: An efficient method for *Stevia rebaudiana* Bertoni. *Sugar Technology*, 16: 100–108.

Okcu, G., Kaya, M. D., and Atak, M. (2005). Effects of salt and drought stresses on germination and seedling growth of pea (*Pisum sativum* L.). *Turkish Journal of Agriculture and Forestry*, 29(4): 237–242.

Ördög, V., and Molnár, Z. (2011). Plant Physiology Debreceni Egyetem, Nyugat-Magyarországi Egyetem, Pannon Egyetem. http://www.tankonyvtar.hu/en/tartalom/tamop425/0010_1A_Book_angol_01_novenyelettan/ch03s05.html Accessed on November 22, 2018.

Ort, D. R. (2001). When there is too much light. *Plant Physiology*, 125(1): 29–32.

Pandey, M., and Chikara, S. K. (2014). *In vitro* regeneration and effect of abiotic stress on physiology and biochemical content of *Stevia Rebaudiana* "Bertoni." *Journal of Plant Science & Research*, 1(3): 113–121.

Pandey, M., and Chikara, S. K. (2015). Effect of salinity and drought stress on growth parameters, glycoside content and expression level of vital genes in steviol glycosides biosynthesis pathway of *Stevia rebaudiana* (Bertoni). *International Journal of Genetics*, 7(1): 153–160.

Papazoglou, E. G., Karantounias, G. A., Vemmos, S. N., and Bouranis, D. L. (2005). Photosynthesis and growth responses of giant reed (*Arundo donax* L.) to the heavy metals Cd and Ni. *Environment International*, 31: 243–249.

Park, M. Y., Chung, M. S., Koh, H. S., Lee, D. J., Ahn, S. J., and Kim, C. S. (2009). Isolation and functional characterization of the Arabidopsis salt-tolerance 32 (AtSAT32) gene associated with salt tolerance and ABA signaling. *Physiologia Plantarum*, 135(4): 426–435.

Pasternak, D. (1987). Salt tolerance and crop production-a comprehensive approach. *Annual Review of Phytopathology*, 25(1): 271–291.

Pereira, A. (2016). Plant abiotic stress challenges from the changing environment. *Frontiers in Plant Science*, 7: 1123.

Pérez-Alfocea, F., Santa-Cruz, A., Guerrier, G., and Bolarin, M. C. (1994). NaCl stress-induced organic solute changes on leaves and calli of *Lycopersicon esculentum*, *L. pennellii* and their interspecific hybrid. *Journal of Plant Physiology*, 143(1): 106–111.

Pitzschke, A., Fornazi, C., and Hirt, H. (2006). Reactive oxygen species signalling in plants. *Antioxidants and Redox Signalling*, 8: 1757–1764.

Prasad, M. N. V., and Freitas, H. (2005). Metal-tolerant plants: Biodiversity prospecting for phytoremediation technology. *In: Trace Elements in the Environment*, Prasad, M. N. V., Sajwan, K. S., and Naidu, R. (Eds.), pp. 488–511. CRC Press.

Rafiq, M., Dahot, M. U., Mangrio, S. M., Naqvi, H. A., and Qarshi, I. A. (2007). *In vitro* clonal propagation and biochemical analysis of field established *Stevia rebaudiana Bertoni*. *Pakistan Journal of Botany*, 39: 2467–2474.

Raju, S. K. K., Barnes, A. C., Schnable, J. C., and Roston, R. L. (2018). Low-temperature tolerance in land plants: Are transcript and membrane responses conserved?. *Plant Science*, 276: 73–86.

Rameeh, V., Gerami, M., Omran, V. G., and Ghavampour, S. (2017). Impact of glycine betaine on salinity tolerance of stevia (*Stevia rebaudiana* Bertoni) under in vitro condition. *Cercetari Agronomice in Moldova*, 50(3): 95–105.

Reddy, M. P., and Iyenger, E. R. R. (1999). Crop responses to salt stress: Seawater application and prospects. *In: Handbook of Plant and Crop Stress*, Pessarakli, M. (Ed.), p. 1041. Marcel Dekker Inc, New York, NY, Basel.

Reis, M., Coelho, L., Santos, G., Kienle, U., and Beltrão, J. (2015). Yield response of stevia (*Stevia rebaudiana* Bertoni) to the salinity of irrigation water. *Agricultural Water Management*, 152: 217–221.

Ren, G. X., and Shi, Y. (2012). The effects of drought stress on the photosynthetic parameters and dry leaf yield of *Stevia rebaudina* Bertoni. *Advanced Materials Research*, 518: 4786–4789.

Rhodes, D., and Hanson, A. D. (1993). Quaternary ammonium and tertiary sulfonium compounds in higher plants. *Annual Review of Plant Biology*, 44: 357–384.

Rhodes, D., and Nadolska-Orczyk, A. (2001). *Plant Stress Physiology*. Chichester: eLS, John Wiley & Sons, Ltd.

Rohbakhsh, H. (2013). Alleviating adverse effects of water stress on growth and yield of forage sorghum by potassium application. *Advances in Environmental Biology*, 7(1): 40–46.

Rouhi, H. R., Aboutalebian, M. A., and Sharif-Zadeh, F. (2011). Effects of hydro and osmopriming on drought stress tolerance during germination in four grass species. *International Journal of Agro Science*, 1(2): 107–114.

Rouhi, H. R., Aboutalebian, M. A., Moosavi, S. A., Karimi, F. A., Karimi, F., Saman, M., and Samadi, M. (2012). Change in several antioxidant enzymes activity of Berseem clover (*Trifolium alexandrinum* L.) by priming. *International Journal of Agro Science*, 2(3): 237–243.

Rout, N. P., and Shaw, B. P. (2001a). Salt tolerance in aquatic macrophytes: Ionic relation and interaction. *Biologia Plantarum*, 44(1): 95–99.

Rout, N. P., and Shaw, B. P. (2001b). Salt tolerance in aquatic macrophytes: Possible involvement of the antioxidative enzymes. *Plant Science*, 160(3): 415–423.

Sairam, R. K., and Tyagi, A. (2004). Physiology and molecular biology of salinity stress tolerance in plants. *Current Science*, 86: 407–421.

Sakaguchi, M., and Kan, T. (1982). As pesquisas japonesas com *Stevia rebaudiana* (Bert.) Bertoni eo esteviosideo. (Japanese researches on *Stevia rebaudiana* (Bert.) Bertoni and stevioside.). *Ciência e Cultura*, 34: 235–248.

Santa-Cruz, A., Estan, M. T., Rus, A., Bolarin, M. C., and Acosta, M. (1997). Effects of NaCl and mannitol iso-osmotic stresses on the free polyamine levels in leaf discs of tomato species differing in salt tolerance. *Journal of Plant Physiology*, 151(6): 754–758.

Sarami, R., Omidi, H., and Bostani, A. A. (2017). The effect of auxin and cytokinin on the biochemical parameters and peroxidase activity (H_2O_2) of stevia (*Stevia rebaudiana* Bertoni) under salinity stress. *Journal of Science and Technology of Greenhouse Culture—Isfahan University of Technology*, 8(3): 91–105.

Schützendübel, A., and Polle, A. (2002). Plant responses to abiotic stresses: Heavy metal-induced oxidative stress and protection by mycorrhization. *Journal of Experimental Botany*, 53(372): 1351–1365.

Schützendübel, A., Schwanz, P., Teichmann, T., Gross, K., Langenfeld-Heyser, R. et al. (2001). Cadmium-induced changes in antioxidative systems, hydrogen peroxide content, and differentiation in Scots pine roots. *Plant Physiol.*, 127: 887–898.

Shahverdi, M. A., Omidi, H., and Tabatabaei, S. J. (2017a). Stevia (*Stevia rebaudiana* Bertoni) responses to NaCl stress: Growth, photosynthetic pigments, diterpene glycosides and ion content in root and shoot. *Journal of the Saudi Society of Agricultural Sciences*. In Press.

Shahverdi, M. A., Omidi, H., and Tabatabaei, S. J. (2017b). Effect of nutri-priming on germination indices and physiological characteristics of stevia seedling under salinity stress. *Journal of Seed Science*, 39(4): 353–362.

Shahverdi, M. A., Omidi, H., and Tabatabaei, S. J. (2017c). Morpho-physiological response of stevia (*Stevia rebaudiana* bertoni) to salinity under hydroponic culture condition (a case study in iran). *Applied Ecology and Environmental Research*, 16(1): 17–28.

Shahverdi, M. A., Omidi, H., and Tabatabaei, S. J. (2018). Plant growth and steviol glycosides as affected by foliar application of selenium, boron, and iron under NaCl stress in *Stevia rebaudiana* Bertoni. *Industrial Crops & Products*, 125: 408–415.

Sharma, S. S., and Dietz, K. J. (2006). The significance of amino acids and amino acid-derived molecules in plant responses and adaptation to heavy metal stress. *Journal of Experimental Botany*, 57(4): 711–726.

Sharuti, R., Narender, S., and Singh, S. K. (2014). Influence of NaCl on biochemical parameters of two cultivars of *Stevia rebaudiana* regenerated in vitro. *Journal of Stress Physiology & Biochemistry*, 10(2): 288–296.

Shaw, B. P., Sahu, S. K, and Mishra, R. K. (2004). Heavy metal induced oxidative damage in terrestrial plants. *In: Heavy Metal Stress in Plants from Biomolecules to Ecosystems*, MNV Prasad (Ed.), pp. 84–126. Springer-Verlag, New York, NY.

Shulaev, V., and Oliver, D. J. (2006). Metabolic and proteomic markers for oxidative stress. New tools for reactive oxygen species research. *Plant Physiology*, 141: 367–372.

Shuping, C., and Shizhen, S. (1995). Study on storage technique of *Stevia rebaudiana* seed. *Acta Agronomica Sinica*, 21: 102–105.

Sidari, M., Mallamaci, C., and Muscolo, A. (2008). Drought, salinity and heat differently affect seed germination of Pinus pinea. *Journal of Forest Research*, 13(5): 326–330.

Simlat, M., Ślęzak, P., Moś, M., Warchoł, M., Skrzypek, E., and Ptak, A. (2016). The effect of light quality on seed germination, seedling growth and selected biochemical properties of *Stevia rebaudiana* Bertoni. *Scientia Horticulturae*, 211: 295–304.

Singh, S. D., and Rao, G. P. (2005). Stevia: The herbal sugar of 21st century. *Sugar Technology*, 7: 17–24.

Singh, A. K., Singh, R. M., Velmurugan, A., Kumar, R. R., and Biswas, U. (2018). Harnessing genetic resources in field crops for developing resilience to climate change. *In: Biodiversity and Climate Change Adaptation in Tropical Islands*, Sivaperuman, C., Velmurugan, A., Singh, A. K., Jaisankar, I. (Eds.). pp. 597–621. Elsevier Inc.

Smiri, M., Chaoui, A., and El Ferjani, E. (2009). Respiratory metabolism in the embryonic axis of germinating pea seed exposed to cadmium. *Journal of Plant Physiology*, 166: 259–269.

Smitha, G. R., and Umesha, K. (2012). Vegetative propagation of Stevia (*Stevia rebaudiana* (Bertoni) Hemsl.) through stem cuttings. *Journal of Tropical Agriculture*, 50: 72–75.

Song, Y., Chen, Q., Ci, D., Shao, X., and Zhang, D. (2014). Effects of high temperature on photosynthesis and related gene expression in poplar. *BMC Plant Biology*, 14: 111.

Soufi, S., D'Urso, G., Pizza, C., Rezgui, S., Bettaieb, T., and Montoro, P. (2016). Steviol glycosides targeted analysis in leaves of *Stevia rebaudiana* (Bertoni) from plants cultivated under chilling stress conditions. *Food Chemistry*, 190: 572–580.

Soufi, S., Rezgui, S., and Bettaeib, T. (2015). Early effects of chilling stress on the morphological and physiological status of pretreated *Stevia rebaudiana* Bert. seedlings. *Journal of New Sciences*, 14: 467–472.

Srivastava, S., and Srivastava, M. (2014). Morphological changes and antioxidant activity of *Stevia rebaudiana* under water stress. *American Journal of Plant Sciences*, 5(22): 3417–3422.

Stankovic, S., Kalaba, P., and Stankovic, A. R. (2014). Biota as toxic metal indicators. *Environmental Chemistry Letters*, 12: 63–84.

Starratt, A. N., Kirby, C. W., Pocs, R., and Brandle, J. E. (2002). Rebaudioside F, a diterpene glycoside from *Stevia rebaudiana*. *Phytochemistry*, 59(4): 367–370.

Szabados, L., and Savoure, A. (2010). Proline: A multifunctional amino acid. *Trends in Plant Science*, 15(2): 89–97.

Szabolcs, I. (1994). Soils and salinisation. *In: Handbook of Plant and Crop Stress*, M. Pessarakli (Ed.), pp. 1–3. Marcel Dekker, New York, NY.

Takahashi, L., Melges, E., and Carneiro, J. W. P. (1996). Germination performance of seeds of *Stevia rebaudiana* (Bert.) Bertoni under different temperatures. *Revista Brasileira de Sementes*, 18: 6–9.

Tanaka, Y., Hibino, T., Hayashi, Y., Tanaka, A., Kishitani, S., Takabe, T., and Yokota, S. (1999). Salt tolerance of transgenic rice overexpressing yeast mitochondrial Mn-SOD in chloroplasts. *Plant Science*, 148(2): 131–138.

Tang, X., Mu, X., Shao, H., Wang, H., and Brestic, M. (2015). Global plant-responding mechanisms to salt stress: Physiological and molecular levels and implications in biotechnology. *Critical Reviews in Biotechnology*, 35(4): 425–437.

Tavarini, S., and Angelini, L. G. (2013). *Stevia rebaudiana* Bertoni as a source of bioactive compounds: The effect of harvest time, experimental site and crop age on steviol glycoside content and antioxidant properties. *Journal of the Science of Food and Agriculture*, 93(9): 2121–2129.

Temple, M. D., Perrone, G. G., and Dawes, I. W. (2005). Complex cellular responses to reactive oxygen species. *Trends in Cell Biology*, 15(6): 319–326.

Torres, F. M. L., Contour, A. D., Zuily, F. Y., and Pham T. A. T. (2008). Molecular cloning of glutathione reductase cDNAs and analysis of GR gene expression in cowpea and common bean leaves during recovery from moderate drought stress. *Journal of Plant Physiology*, 165(5): 514–521.

Urban, J. D., Carakostas, M. C., and Taylor, S. L. (2015). Steviol glycoside safety: Are highly purified steviol glycoside sweeteners food allergens? *Food and Chemical Toxicology*, 75: 71–78.

Vaidyanathan, H., Sivakumar, P., Chakrabarty, R., and Thomas, G. (2003). Scavenging of reactive oxygen species in NaCl-stressed rice (*Oryza sativa* L.) differential response in salt-tolerant and sensitive varieties. *Plant Science*, 165: 1411–1418.

van Bussel, C. G., Schroeder, J. P., Mahlmann, L., and Schulz, C. (2014). Aquatic accumulation of dietary metals (Fe, Zn, Cu, Co, Mn) in recirculating aquaculture systems (RAS) changes body composition but not performance and health of juvenile turbot (*Psetta maxima*). *Aquacultural Engineering*, 61:35–42.

van Kempen, M. M., Smolders, A. J., Bögemann, G. M., Lamers, L. L., Visser, E. J., and Roelofs, J. G. (2013). Responses of the *Azolla filiculoides* Stras. *Anabaena azollae* Lam. association to elevated sodium chloride concentrations: Amino acids as indicators for salt stress and tipping point. *Aquatic Botany*, 106: 20–28.

Villalba, L. B., and Nakashima, H. N. (2016). Efecto del estrés hídrico inducido con PEG 6000 sobre el crecimiento *in vitro* de plantas de *Stevia rebaudiana* cv. "KH-IAN/VC-142." *Biotecnología Vegetal*, 16(3): 189–192.

Von Deutsch, A. W., Mitchell, C. D., Williams, C. E., Dutt, K., Silvestrov, N. A., Klement, B. J., and Von-Deutsch, D. A. (2005). Polyamines protect against radiation-induced oxidative stress. *Gravitation and Space Biology Bulletin*, 18: 109–110.

Vouillamoz, J. F., Wolfram-Schilling, E., Carron, C. A., and Baroffio, C. A. (2016). Agronomical and phytochemical evaluation of *Stevia rebaudiana* genotypes. *Julius-Kühn-Archives*, 86(453): 86–88.

Wahid, A., Farooq, M., Hussain, I., Rasheed, R., and Galani, S. (2012). Responses and management of heat stress in plants. *In: Environmental Adaptations and Stress Tolerance of Plants in the Era of Climate Change*, Ahmad, P., and Prasad, M. N. V. (Eds.), pp. 135–157. Springer, New York, NY.

Wan, S. B., Tian, L., Tian, R. R., Pan, Q. H., Zhan, J. C., Wen, P. F., and Huang, W. D. (2009). Involvement of phospholipase D in the low temperature acclimation-induced thermotolerance in grape berry. *Plant Physiology and Biochemistry*, 47: 504–510.

Wang, M., Zhao, X., Xiao, Z., Yin, X., Xing, T., and Xia, G. (2016). A wheat superoxide dismutase gene TaSOD2 enhances salt resistance through modulating redox homeostasis by promoting NADPH oxidase activity. *Plant Molecular Biology*, 91(1–2): 115–130.

Wang, W., Vinocur, B., and Altman, A. (2003). Plant responses to drought, salinity and extreme temperatures: Towards genetic engineering for stress tolerance. *Planta*, 218: 1–14.

Weidner, S. (2005). Metabolism of phenolic compounds in *Vitis riparia* seeds during stratification and during germination under optimal and low temperature stress conditions. *Acta Physiologiae Plantarum*, 27: 313–320.

Well, C., Frank, O., and Hofmann, T. (2013). Quantitation of sweet steviol glycosides by means of a HILIC-MS/MS-SIDA approach. *Journal of Agricultural and Food Chemistry*, 61(47), 11312–11320.

Wheeler, T., and von Braun, J. (2013). Climate change impacts on global food security. *Science*, 341(6145): 508–513.

White, P. J., and Pongrac, P. (2017). 12 heavy-metal toxicity in plants. *Plant Stress Physiology*, 2(5): 300.

Xiang, C., and Oliver, D. J. (1998). Glutathione metabolic genes coordinately respond to heavy metals and jasmonic acid in *Arabidopsis*. *The Plant Cell*, 10: 1539–1550.

Xing, W., and Rajashekar, C. B. (2001). Glycine betaine involvement in freezing tolerance and water stress in *Arabidopsis thaliana*. *Environmental and Experimental Botany*, 46: 21–28.

Xu, L., Zhang, M., Zhang, X., and Han, L. B. (2015). Cold acclimation treatment–induced changes in abscisic acid, cytokinin, and antioxidant metabolism in zoysiagrass (*Zoysia japonica*). *Hortscience*, 50: 1075–1080.

Xu, S., Li, J., Zhang, X., Wei, H., and Cui, L. (2006). Effects of heat acclimation pretreatment on changes of membrane lipid peroxidation, antioxidant metabolites, and ultrastructure of chloroplasts in two cool-season turfgrass species under heat stress. *Environmental and Experimental Botany*, 56: 274–285.

Yadav, P. V., Kumari, M., and Ahmed, Z. (2011). Seed priming mediated germination improvement and tolerance to subsequent exposure to cold and salt stress in capsicum. *Research Journal of Seed Science*, 4(3): 125–136.

Yan, K., Chen, P., Shao, H., Zhang, L., and Xu, G. (2011). Effects of short-term high temperature on photosynthesis and photosystem II performance in sorghum. *Journal of Agronomy and Crop Science*, 197: 400–408.

Yildiztugay, E., Ozfidan-Konakci, C., Kucukoduk, M., and Duran, Y. (2014). Variations in osmotic adjustment and water relations of *Sphaerophysa kotschyana*: Glycine betaine, proline and choline accumulation in response to salinity. *Botanical Studies*, 55: 6.

Yousfi, S., Márquez, A. J., Betti, M., Araus, J. L., and Serret, M. D. (2016). Gene expression and physiological responses to salinity and water stress of contrasting durum wheat genotypes. *Journal of Integrative Plant Biology*, 58(1): 48–66.

Zeng, J., Chen, A., Li, D., Yi, B., and Wu, W. (2013). Effects of salt stress on the growth, physiological responses, and glycoside contents of *Stevia rebaudiana* Bertoni. *Journal of Agricultural and Food Chemistry*, 61(24): 5720–5726.

Zhang, C. S., Lu, Q., and Verma, D. P. S. (1995). Removal of feedback inhibition of Δ1-pyrroline-5-carboxylate synthetase, a bifunctional enzyme catalyzing the first two steps of proline biosynthesis in plants. *Journal of Biological Chemistry*, 270(35): 20491–20496.

Zhang, L. X., Li, S. X., Zhang, H., and Liang, Z. S. (2007). Nitrogen rates and water stress effects on production, lipid peroxidation and antioxidative enzyme activities in two maize (*Zea mays* L.) genotypes. *Journal of Agronomy and Crop Science*, 193(6): 387–397.

Zhang, X., Shen, L., Li, F., Meng, D., and Sheng, J. (2013). Arginase induction by heat treatment contributes to amelioration of chilling injury and activation of antioxidant enzymes in tomato fruit. *Postharvest Biology and Technology*, 79: 1–8.

Zhou, J., Wang, J., Shi, K., Xia, X. J., Zhou, Y. H., and Yu, J. Q. (2012). Hydrogen peroxide is involved in the cold acclimation-induced chilling tolerance of tomato plants. *Plant Physiology and Biochemistry*, 60: 141–149.

Zhu, J. K. (2007). Plant salt stress. *In eLS*, (Ed.).

Zhu, J. K., Hasegawa, P. M., Bressan, R. A. and Bohnert, H. J. (1997). Molecular aspects of osmotic stress in plants. *Critical Reviews in Plant Sciences*, 16(3): 253–277.

Zhu, J., Lee, B. H., Dellinger, M., Cui, X., Zhang, C., Wu, S., Zhu, J. K. et al. (2010). A cellulose synthase-like protein is required for osmotic stress tolerance in Arabidopsis. *The Plant Journal*, 63(1): 128–140.

Źróbek-Sokolnik, A. (2012). Temperature stress and responses of plants. *In: Environmental Adaptations and Stress Tolerance of Plants in the Era of Climate Change*, Ahemad, P., and Prasad, M. N. V. (Eds.), pp. 85–112. Springer, New York, NY.

34 Responses of Green Beans (*Phaseolus vulgaris* L.) in Terms of Dry Matter Production, Nitrogen Uptake, and Water Absorption under Salt Stress Conditions

Mohammad Pessarakli

CONTENTS

34.1 INTRODUCTION

The gradual progress of desertification due to the detrimental effects of natural stress factors such as low precipitation, long-term drought, heat, and erosion, coupled with improper human activities as a result of overgrazing, over-use of land, and the application of insufficient management decisions and improper agricultural practices, urbanization, and industrial activities, has left extensive arable lands at potential risk of conversion to unusable soils. These problems are more severe in arid and semi-arid regions, where soils already encounter salinity and sodicity problems and are more vulnerable to stress conditions.

Salinity stress is one of the major abiotic stresses limiting crop productivity and the geographical distribution of many important plants/crops worldwide. The accumulation of high soluble salts in a soil can significantly decrease the value and productivity of agricultural land. Salt and water stress have been recognized as major agricultural problems, especially in arid and semi-arid regions, for a long time. The retardation of crop yield by salinization has also been known for a long time. Since the early 1900s, various investigations of the effects of salts on plant/crop growth have been undertaken, covering a range of aspects, from plant response to salinity to salt behavior in soils [3, 4, 6–8, 16, 17, 23, 32, 34, 38, 46, 50, 52, 62, 63, 79, 85, 90, 101, 116, 117, 121, 125, 128, 135, 162, 164, 167, 169, 176, 178, 179, 202, 207, 215, 223, 226, 227, 231, 234–236, 243, 246, 251]. Physiological studies have revealed that the major effects of salinity on plant growth retardation are osmotic and specific ion effects [5, 20, 36, 40, 41, 55–57, 60, 68, 111, 119, 126, 127, 133, 135, 156, 196, 198, 199, 209, 228]. Furthermore, reduced nutrient uptake by plants grown in saline environments has been observed in several species of plants [11, 12, 23, 33, 36, 44, 52, 63, 80, 81, 85, 91, 103, 111, 119, 133, 164, 172–174, 178, 180–183, 194, 197, 209, 220]. Differences in salt tolerance among plant species also have been long recognized [20, 29, 35, 58, 93–97, 111, 112, 115, 122, 123, 127, 132, 139–143, 146, 155, 198, 210, 216, 240, 241, 247]. Although scientists in agriculture started to work on the salinity tolerance of plants over 50 years ago, in the early 1950s, there is still a great deal of interest by researchers in working on this subject. For some sample reports on this subject in the last few decades, see references [2, 4–28, 31–33, 36–38, 45–54, 59–62, 64–67, 72–78, 82, 84, 85, 87, 90–92, 94, 98, 99, 101–109, 114, 116, 117, 120–125, 128–131, 134–138, 143–145, 147, 149–154, 161, 163–171, 175–179, 184–195, 199, 201–205, 207, 208, 211–215, 217–246, 248–253]. According to Qadir et al. [186], cultivation of salt-tolerant grasses in a saline or saline-sodic soil may mobilize the native lime ($CaCO_3$) in these soils through root action. This may substitute for the chemical approach to the reclamation of such soils. Apte and Thomas [14] reported that the simultaneous application of halotolerant nitrogen-fixing cyanobacteria during crop growth seems to be an attractive possibility for the reclamation and improvement of saline soils, especially since it can also supplement the nitrogen requirement of the crop. De Villiers et al. [61], also assessing the salinity tolerance of different plant species, found that perennials seemed to be better suited for rehabilitation purposes under saline soil conditions. However, the role that salt tolerance plays in causing differences in growth and development, nutrient uptake, and metabolism between various plants, among plant species, and at different stages of growth is still a major concern among investigators and is not yet fully understood. The discovery of the physiological basis of salt tolerance in crops and the use of this knowledge to obtain more tolerant cultivars by modern plant breeding procedures should result in substantial increases in world food production.

The effect of salt stress on nutrient element use and nutrition as well as metabolism in plants has been studied for various plants using different methods. The results are still inconclusive. However, the change in nutrient metabolism induced by excess salt is commonly accepted among scientists as one of the most important factors responsible for abnormal plant metabolism and reduced growth. Bernstein et al. [42] found that despite the decrease in total N uptake, leaf N concentration of some grain and vegetable crops increased with increasing salinity at all N fertilization levels. Increased N concentration of corn (*Zea mays* L.) and cotton (*Gossypium hirsutum* L.) plants under salinity stress was reported by Khalil et al. [118]. The uptake and metabolism of ^{15}NHeq \O()$_4^+$ and ^{15}NOeq \O()$_3^-$ in red kidney beans (*Phaseolus vulgaris* L.) was adversely affected by both salt and water stress at -0.4 MPa osmotic potential [80, 81, 197]. Reduced ^{15}N uptake and metabolism as well as impaired protein synthesis under stress conditions by various crops have also been reported by several other investigators: Helal and Mengel [97] (barley, *Hordeum vulgare* L.); Pessarakli and Tucker [182] and Al-Rawahy et al. [11, 12] (tomato, *Lycopersicon esculentum* Mill.); Pessarakli and Tucker [183] (eggplant, *Solanum melongena* L.); Pessarakli et al. [174] (corn, *Zea mays* L.); Pessarakli [164] and Pessarakli et al. [172, 173], Assimakopoulou et al. [25] (green beans, *Phaseolus vulgaris* L.). However, Pessarakli and Tucker [180, 181] found that ^{15}N uptake and protein synthesis by cotton (*Gossypium hirsutum* L.) plants increased under low levels (-0.4 MPa osmotic potential) of NaCl salinity. Pessarakli and co-workers [170, 175, 178] have also found increased ^{15}N uptake in saltgrass (*Distichlis spicata*), a true halophytic plant species, under salinity stress. An increased total N concentration of plants grown in saline substrate was also reported by Bernstein and Pearson [43].

To explain these different results, a dilution or concentration effect (depending on the relative severity of salt stress on growth or nutrient, i.e., N, uptake) was reported as a cause of the fluctuations in N content or concentration in plants [80, 164, 180, 182, 183].

Among the various environmental stress factors, salinity appears to have been given more attention than any other factor both in the past and at the present time. This is clearly seen from the continuous investigations and the voluminous reports that are continuously being generated on this subject. Hundreds of publications are annually added to the literature on this subject, dealing with plant and crop stress caused by salinity. For some recent reports during the last few decades, see references [2, 4–28, 31–33, 36–38, 45–54, 59–62, 64–67, 72–78, 82, 84, 85, 87, 90–92, 94, 98, 99, 101–109, 114, 116, 117, 120–125, 128–131, 134–138, 143–145, 147, 149–154, 161, 163–171, 175–179, 184–195, 199, 201–205, 207, 208, 211–215, 217–246, 248–253].

Despite the voluminous publications dealing with the effects of salt stress on plant growth and nutrient (i.e., N) nutrition, the literature concerning this issue in green beans is scarce. The reports of Alislail and Bartels [10], Ashraf and Rasul [24], Bahmaniar and Sepanlou [33], Balasubramanian and Sinha [34], Bhivare and Nimbalkar [44], Coons and Pratt [55], Csizinsky [57], Frota and Tucker [80, 81], Gauch and Wadleigh [83], Harbir-Singh et al. [95], Hoffman et al. [100], Kant et al. [113], Lopez et al. [130], Maliwal and Paliwal [140], Papiernik et al. [160], Saad [197], Salim and Pitman [200], Savvas et al. [204], van Hoorn et al. [222], Velagaleti et al. [224], Wang and Shannon [230], Wignarajah [234], and recently those of Abdul Qados [2], Al Hassan et al. [9], Asfaw [18], Aydin et al. [28], Dong et al. [65], Elzaawely et al. [69], Emongor et al. [70], Erdal et al. [72], Farhangi-Abriz and Torabian [74], Ge et al. [84], Hanafy et al. [94], Hernansez et al. [98], Hernández-lucero et al. [99], Janmohammadi et al. [106], Karami Chame et al. [114], Khataar et al. [120], Liu et al. [129], Mahmood et al. [136], Mahmut et al. [137], Maqshoof et al. [143], Miransari et al. [147], Nahar et al. [150, 151], Radi et al. [188], Rady et al. [189,191], Rady and Mohamed [190], Sun et al. [217], Tavakkoli et al. [218, 219], Weisany et al. [233], Yin [244], You et al. [245], and Zhang et al. [248] deal primarily with the effects of salinity on the growth and/or chemical composition of other types of beans and are not concerned with green beans. Among the cited references in this chapter, in addition to the author's own research work, only that of Bernstein and Pearson [43] reported the influence of exchangeable sodium ions on the yield and chemical composition of green beans. However, Bernstein and Pearson's report [36] was not concerned with N (labeled or non-labeled) uptake and metabolism by the reported plant species. Thus, green beans were selected to be covered in this chapter primarily because they are classified as salt-sensitive plants [30] or according to Assimakopoulou et al. [25], an extremely salt-sensitive plant species. Also, the effects of salinity on the growth and nutrient uptake and use by these plants have not been studied and documented sufficiently. In addition, these salt-sensitive plant species were selected for this chapter to compile information regarding the responses of different plant species that are discussed by different authors in this book. This information is being compiled in a volume to assist readers in comparing all these various salt-tolerant plant types under stressful environmental conditions.

Thus, this chapter is concerned with growth, nitrogen (total and ^{15}N) uptake, protein synthesis, and water absorption by three cultivars (Tender Improved, Slim Green, and Kentucky Wonder) of green beans at the vegetative stage of growth under normal and NaCl stress conditions with the following objectives:

(1) To compare the growth of these cultivars by evaluating their dry matter yield under normal and NaCl stress conditions
(2) To compare total N and ^{15}N uptake and distribution in plant roots and shoots by these cultivars as affected by salinity
(3) To evaluate protein synthesis by these plant species under normal and salt stress conditions
(4) To study the water absorption by these cultivars as influenced by sodium chloride (NaCl) stress

34.2 FACTORS EVALUATED REGARDING THE RESPONSES OF GREEN BEANS TO SALT STRESS

34.2.1 DRY MATTER PRODUCTION

The effects of NaCl salinity on dry matter production of the three cultivars of green beans have been examined in several studies [164, 172, 173, 185]. All these studies reported that the NaCl stress significantly reduced total dry matter yield for all three cultivars, but Tender Improved was the least severely affected at all salinity levels (Table 34.1, data from Pessarakli et al. [172]). The degree of reduction in dry matter yield increased with increasing salt stress level and over time. Other investigators have also reported reductions in growth, dry matter production, and yields of other bean cultivars [2, 18, 44, 55, 57, 69, 70, 80, 95, 100, 106, 114, 120, 140, 190, 191, 197, 218, 219, 248] and a number of other plants/crops

and trees [3, 6–8, 12, 16, 22, 23, 27, 32, 36, 38, 42, 45, 46, 50, 60–63, 66–68, 78, 79, 85–93, 96–98, 101–104, 109, 112, 116, 119–122, 126–129, 135, 152–164, 168–171, 174–184, 186, 198, 199, 202, 207, 209, 215, 220–224, 226–228, 231, 235, 236, 240–244, 248, 249, 251].

Under NaCl stress, shoot and root growth were substantially lower for the Slim Green and Kentucky Wonder cultivars as compared with the Tender Improved [164, 172, 173, 185]. This phenomenon indicates the presence of significant interaction effects between salinity and cultivars. Roots appear to be affected less than shoots by salt stress for all cultivars.

Dry matter production and growth period were linearly correlated (r^2 values ranged from .89 to .99 for different treatments) [164]. Several studies conducted on these cultivars of green beans [164, 172, 173, 185] reported that for all cultivars, dry matter yield increased as growth period progressed.

34.2.2 TOTAL N UPTAKE BY PLANTS

According to Pessarakli [164] and Pessarakli et al. [172, 173], total N uptake by green bean plants was significantly decreased with increasing salinity of the nutrient solutions for all cultivars, at all three harvests. The results of Pessarakli's study [164] are presented here (Table 34.2). Slim Green contained substantially lower total N than the other two cultivars at each harvest for all corresponding treatments, except for the control shoots at the third harvest. The uptake values were markedly lower at the first harvest for this cultivar, indicating slower initial N uptake and slower early growth rate for the Slim Green cultivar. Several nitrogen uptake studies conducted on these cultivars of green beans [164, 172, 173] reported that for all cultivars, shoots contained substantially

---

TABLE 34.1

Dry Matter Yield of Three Green Bean Cultivars under Various NaCl Stress Levels at Different Harvest Times

Cultivar	Salt Stress (Osmotic potential)	Dry Weight of Plant Parts[a]					
		Shoots			Roots		
		Harvest[b]					
		1	2	3	1	2	3
	Mpa			g			
Tender	Control (−0.03)	3.12	5.18	7.25	0.84	0.98	1.48
Improved	−0.25	2.73	4.45	6.53	0.68	0.93	1.45
	−0.50	1.92	3.64	4.56	0.46	0.81	1.14
Slim	Control (−0.03)	1.76	4.24	7.22	0.36	0.78	1.51
Green	−0.25	1.32	2.16	3.51	0.34	0.49	0.88
	−0.50	0.76	0.92	1.34	0.21	0.32	0.41
Kentucky	Control (−0.03)	3.12	4.33	7.54	0.67	0.85	1.53
Wonder	−0.25	1.67	2.45	3.18	0.55	0.77	1.28
	−0.50	0.95	1.24	2.41	0.43	0.51	0.72
LSD (0.05)[c]		0.42	0.76	0.96	0.18	0.24	0.35

Source: Pessarakli, M., et al., *J. Plant Nutr.*, 12(10), 1105–1121, 1989.

[a] Represents the means for pots containing two plants with three replications.

[b] Harvests 1, 2, and 3 are for 5, 10, and 15 day ^{15}N uptake periods, respectively.

[c] Represents the least significant difference between the treatment means at the 0.05 level of confidence.

TABLE 34.2

Total N Uptake of Plant Parts of Three Green Bean Cultivars as Affected by Three Levels of NaCl Stress at Three Harvest Times

		Total-N Content of Plant Parts					
		Shoots			Roots		
	Salt Stress	Harvest[a]					
Cultivar	(Osmotic potential)	1	2	3	1	2	3
	Mpa	mg N pot[b]					
Tender	Control (−0.03)	104.8	146.2	210.4	24.4	27.6	40.6
Improved	−0.25	68.6	97.4	184.2	18.8	26.8	38.9
	−0.50	42.8	82.8	113.4	13.1	24.1	28.8
Slim	Control (−0.03)	46.6	118.8	215.3	10.4	23.5	44.5
Green	−0.25	33.2	58.7	89.2	9.3	14.4	22.6
	−0.50	18.2	24.5	35.7	6.0	8.7	10.2
Kentucky	Control (−0.03)	99.2	126.2	204.5	21.6	27.1	48.3
Wonder	−0.25	48.6	71.1	86.4	16.8	21.1	38.1
	−0.50	26.6	32.6	65.6	12.5	14.7	20.4
LSD(0.05) salinity x cultivar		3.4	5.3	13.1	1.6	2.3	3.1
Summary of the significance of variance sources							
Cultivar (C)		**	*	*	**	*	*
Salinity (S)		**	**	**	**	**	**
C X S		**	**	**	**	**	**

Source: Pessarakli, M., *Crop Science*, 31(6), 1633–1640, 1991.

*, ** Significant at P = .05 and .01, respectively.

[a] Harvests 1, 2, and 3 are for 5, 10, and 15 d 15N-uptake periods, respectively.

[b] Represents the means for pots containing two plants with three replicates.

more total N than roots, probably due to the larger dry weights of shoots than roots (larger sink size).

The reduction in total N uptake was similar to the reduction pattern for total dry matter yield by plants under NaCl stress. The similar reduction pattern for total N uptake and dry matter yield indicates that the major portion of the absorbed N was incorporated into protein and contributed to plant growth and development. As N uptake decreased, dry matter yield also decreased under the NaCl stress condition. This is supported by reports of several investigators [80, 81, 83, 88, 97, 171–174, 178, 180–183, 197], which indicated that changes in N metabolism caused by salinity stress are among the most important factors responsible for abnormal plant metabolism, reduced growth, and decreased crop yield.

34.2.3 Total N Concentration in Plant Tissues

All three studies conducted by Pessarakli [164] and Pessarakli et al. [172, 173] reported that total N concentrations in all three cultivars generally were lower in plants subjected to salinity, especially at the highest NaCl stress levels, as compared with controls. Table 34.3, obtained from Pessarakli [164], indicates this finding. However, for a salt-tolerant cotton plant, the N concentration was significantly higher in NaCl-stressed plants even at a higher level of salinity (−0.8 MPa osmotic potential) as observed by Pessarakli and Tucker [180, 181]. An increased N concentration of corn and cotton plants under salt stress conditions was also reported by Khalil et al. [118]. Therefore,

differences in N concentrations of these different crops (cotton as compared with green beans) under salt stress are probably due to differences in their salt tolerance. At each stress level, the total N concentration of Tender Improved generally tended to be lower than those of the other cultivars [164, 172, 173]. This is probably due to a dilution effect, since Tender Improved produced significantly higher dry matter than the other cultivars at each stress level for each harvest. The total N concentration of roots was generally higher than that of the shoots at each harvest, for each cultivar, for any corresponding treatment, except for the control Tender Improved plants [164].

34.2.4 Nitrogen-15 Uptake by Plants and Distribution of 15N in Shoots and Roots

The results of several studies [164, 172, 173] on different cultivars of green beans showed that total 15N uptake by plants was decreased with increasing salinity of nutrient solutions at all three harvests, for all three cultivars. The 15N results of an experiment completed by Pessarakli [164] are presented here in Table 34.4. The reduction in 15N uptake followed the same reduction patterns as total N and dry matter yield under stress conditions. This is an indication that the absorbed 15N was incorporated into protein and contributed to plant growth and development as reflected in dry matter production. The Slim Green cultivar absorbed the smallest amount of 15N under NaCl stress conditions. The absorbed 15N values were higher for Kentucky Wonder and generally highest for Tender

TABLE 34.3

Nitrogen Concentration of Plant Parts of Three Green Bean Cultivars as Affected by Three Levels of NaCl Stress at Three Harvest Times

Cultivar	Salt Stress (Osmotic potential)	Shoots Harvest[a] 1	2	3	Roots Harvest[a] 1	2	3
	MPa	mg N g⁻¹ dry wt.[b]					
Tender	Control (−0.03)	33.6	28.2	29.0	29.0	28.2	27.4
Improved	−0.25	25.1	21.9	28.2	27.6	28.8	26.8
	−0.50	22.3	22.7	24.9	28.4	29.7	25.3
Slim	Control (−0.03)	26.5	28.0	29.8	28.9	30.1	29.5
Green	−0.25	25.2	27.2	25.4	27.4	29.4	25.7
	−0.50	23.9	26.6	26.6	28.6	27.2	24.9
Kentucky	Control (−0.03)	31.8	29.1	27.1	32.2	31.9	31.6
Wonder	−0.25	29.1	29.0	27.2	30.5	27.4	29.8
	−0.50	28.0	26.3	27.2	29.1	28.8	28.3
LSD(0.05) salinity x cultivar		1.4	1.5	1.3	1.3	1.3	1.3

Summary of the significance of variance sources

Cultivar (C)		**	NS	*	*	*	**
Salinity (S)		*	*	*	*	*	**
C X S		*	*	*	*	*	**

Source: Pessarakli, M., *Crop Science*, 31(6), 1633–1640, 1991.

*, ** Significant at *P* = .05 and .01, respectively.

[a] Harvests 1, 2, and 3 are for 5, 10–, and 15 d ¹⁵N-uptake periods, respectively.

[b] Represents the means for pots containing two plants with three replicates.

TABLE 34.4

Nitrogen (¹⁵N) Content of Plant Parts and Shoot to Root ¹⁵N Ratios of Three Green Bean Cultivars as Affected by Three Levels of NaCl Stress at Three Harvest Times

Cultivar	Salt Stress (Osmotic potential)	Shoots Harvest[a] 1	2	3	Roots Harvest[a] 1	2	3	Shoot to Root ¹⁵N Ratio 1	2	3
	MPa	mg ¹⁵N pot⁻¹[b]								
Tender	Control (−0.03)	3.23	5.58	8.57	0.98	1.19	1.91	3.30	4.69	4.49
Improved	−0.25	1.56	3.37	7.22	0.69	1.24	1.88	2.12	2.72	3.84
	−0.50	1.08	2.94	4.15	0.38	1.06	1.30	2.84	2.77	3.19
Slim	Control (−0.03)	1.29	4.19	9.46	0.40	1.06	2.13	3.23	3.95	4.44
Green	−0.25	0.74	2.04	3.51	0.34	0.63	1.09	2.18	3.24	3.22
	−0.50	0.31	0.71	1.06	0.19	0.36	0.39	1.63	1.97	2.72
Kentucky	Control (−0.03)	3.06	4.94	9.44	0.87	1.23	2.40	3.52	4.02	3.93
Wonder	−0.25	1.22	2.34	3.33	0.66	0.89	1.76	1.85	2.63	1.89
	−0.50	0.55	0.98	2.37	0.44	0.63	0.82	1.25	1.56	2.89
LSD(0.05) salinity x cultivar		0.14	0.22	0.32	0.03	0.06	0.10	0.17	0.21	0.20

Summary of the significance of variance sources

Cultivar (C)		**	**	*	**	*	**	*	**	*
Salinity (S)		**	**	**	**	**	**	**	**	**
C X S		**	**	**	**	**	**	**	**	**

Source: Pessarakli, M., *Crop Science*, 31(6), 1633–1640, 1991.

*, ** Significant at *P* = .05 and .01, respectively.

[a] Harvests 1, 2, and 3 are for 5, 10, and 15 d ¹⁵N-uptake periods, respectively.

[b] Represents the means for pots containing two plants with three replicates.

TABLE 34.5

Nitrogen (^{15}N) Concentration of Plant Parts of Three Green Bean Cultivars as Affected by Three Levels of NaCl Stress at Three Harvest Times

		^{15}N Concentration of Plant Parts					
		Shoots			Roots		
	Salt Stress (Osmotic potential)	Harvest[a]					
Cultivar		1	2	3	1	2	3
	MPa	mg ^{15}N kg^{-1} dry wt.[b]					
Tender	Control (−0.03)	1035	1077	1182	1167	1214	1291
Improved	−0.25	571	757	1106	1015	1333	1297
	−0.50	563	808	910	826	1309	1140
Slim	Control (−0.03)	733	988	1310	1111	1359	1411
Green	−0.25	561	944	1000	1000	1286	1239
	−0.50	408	772	791	905	1125	951
Kentucky	Control (−0.03)	981	1141	1252	1299	1447	1569
Wonder	−0.25	731	955	1047	1200	1156	1375
	−0.50	579	790	983	1023	1235	1139
LSD(0.05) salinity × cultivar		21	25	27	16	18	20
Summary of the significance of variance sources							
Cultivar (C)		**	**	*	*	**	**
Salinity (S)		**	**	**	**	**	**
C X S		**	**	**	**	**	**

Source: Pessarakli, M., *Crop Science*, 31(6), 1633–1640, 1991.

*, ** Significant at P = .05 and .01, respectively.

[a] Harvests 1, 2, and 3 are for 5, 10, and 15 d ^{15}N-uptake periods, respectively.

[b] Represents the means for pots containing two plants with three replicates.

Improved under stress conditions. However, Tender Improved contained significantly lower ^{15}N in both shoots and roots, at the third harvest, under normal (non-saline) conditions as compared with the other two cultivars [164]. Substantial differences between the ^{15}N uptakes by the cultivars at each salinity level (Table 34.4) imply a significant interaction effect between salinity and cultivars at each harvest, for each plant part. Significant decreases in ^{15}N uptake by these plants under a high salinity level is in agreement with experimental data obtained with red kidney beans [80, 197], cotton [180], barley [97], tomato [11, 12, 182], and eggplant [183]. However, a low level of NaCl salinity (−0.4 MPa osmotic potential) in cotton slightly enhanced ^{15}N uptake [180]. Similar results were reported for saltgrass (*Distichlis spicata*), a true halophyte, by Pessarakli et al. [170, 175, 178]. This phenomenon is probably due to the difference in the salt tolerance of these different plant types (cotton as compared with green beans).

Nitrogen-15 contents of green bean shoots were reported [164] as being of higher magnitude than those in roots, for all three cultivars, at all salinity levels (Table 34.4). These differences are considered to be due to the larger dry weights of shoots than those of roots for all cultivars (larger sink size). The shoot/root ratios of ^{15}N content tended to increase with time and decrease with increasing salinity for all cultivars, except for Kentucky Wonder at the third harvest with −0.25 MPa stress. This may have been due to the retarded translocation of ^{15}N from roots to shoots caused by salt stress and accumulation of ^{15}N in shoots as the growth period

progressed. This observation is also clearly seen by comparing the ^{15}N concentration values between shoots and roots (Table 34.5, data from Pessarakli [164]). The concentration of ^{15}N in both shoots and roots increased as the growth period progressed and decreased as the salinity level increased for all cultivars. This pattern was similar to the shoot/root ratios of ^{15}N content of plants. The ^{15}N concentration in roots was far greater than that in shoots, for all cultivars. This higher concentration of ^{15}N in roots can be explained, in part, as the absorption of NHeq \O()$_4$$^+$ onto the root surface or the infusion of ammonium and nitrate ions into the root apparent free space, as suggested by Pessarakli and Tucker [180, 182, 183] and Pessarakli [164].

34.2.5 NITROGEN-15 UPTAKE RATES

The ^{15}N-uptake rate, expressed as mg ^{15}N absorbed per kg dry matter produced by plants per day, is presented in Table 34.6 (data from Pessarakli [164]). At each salinity level, ^{15}N uptake rates peaked at the earliest harvest and decreased as the growth period progressed for both the shoots and roots in each cultivar. This finding indicates that the younger plants absorbed ^{15}N at a faster rate than the older ones, regardless of stress level. Nevertheless, ^{15}N-uptake rates significantly decreased under NaCl stress as compared with the controls, at each harvest, for all cultivars, in both plant parts, except for the roots of the Tender Improved at the second and third harvests. Slim Green shoots, at the earliest harvest, had substantially lower

TABLE 34.6

Nitrogen (^{15}N) Uptake Rate of Plant Parts of Three Green Bean Cultivars as Affected by Three Levels of NaCl Stress at Three Harvest Times

		^{15}N Uptake Rate of Plant Parts					
	Salt Stress	Shoots			Roots		
	(Osmotic	Harvest[a]					
Cultivar	potential)	1	2	3	1	2	3
	MPa	mg ^{15}N kg^{-1} dry wt. d^{-1b}					
Tender	Control (−0.03)	207	108	79	233	121	86
Improved	−0.25	114	76	74	203	133	87
	−0.50	113	81	61	165	131	76
Slim	Control (−0.03)	147	99	87	222	136	94
Green	−0.25	112	94	67	200	129	83
	−0.50	82	77	53	181	113	63
Kentucky	Control (−0.03)	196	114	84	260	145	105
Wonder	−0.25	146	96	70	240	116	92
	−0.50	116	79	66	205	124	76
LSD(0.05) salinity x cultivar		11	10	7	8	6	5
Summary of the significance of variance sources							
Cultivar (C)		**	*	NS	*	**	*
Salinity (S)		**	**	**	**	**	**
C X S		**	**	*	**	**	**

Source: Ref. [164].

*,** Significant at P = 0.05 and 0.01, respectively.

[a] Harvests 1, 2, and 3 are for 5-, 10-, and 15-d ^{15}N-uptake periods, respectively.

[b] Represents the means for pots containing two plants with three replicates.

^{15}N-uptake rate than the other two cultivars under normal condition.

34.2.6 PROTEIN SYNTHESIS BY PLANTS

The crude protein contents of both shoots and roots of the three green bean cultivars were markedly lower under stress conditions as compared with the controls (Table 34.7, data from Pessarakli et al. [172]). Under stress conditions, the Tender Improved cultivar produced significantly more protein than the other two cultivars. Protein synthesis in shoots was substantially higher than that in roots for all the three cultivars. This significant difference appears to be due to the higher dry matter production of shoots than roots for any treatment for any of the three cultivars. Pessarakli et al. [173] used two sources of N (ammonium and nitrate) for evaluating protein synthesis in green beans and found that under normal (non-saline) conditions, the nitrate-treated plants synthesized appreciably more protein than the ammonium-treated ones, at each harvest, for all three cultivars. This phenomenon was more noticeable in roots than in shoots for each cultivar. However, except for the Tender Improved cultivar at the first harvest, the crude protein content of plants was substantially lower under stress as compared with the controls for either source of N. Salt stress had the most severe effect on protein synthesis in Slim Green, among the three cultivars, for both NH$_4$-N and NO$_3$-N sources of N.

Impaired protein synthesis under stress conditions by other bean cultivars, such as red kidney beans [81,197], and other types of plants, such as barley [97], cotton [181], alfalfa (*Medicago sativa* L.) [171], peas [110], wheat (*Triticum aestivum* L.) [1], tobacco (*Nicotiana tabacum* L.) [39], corn [148], and soybean [33], has been reported previously by many investigators. In these studies, either decreased amino acid incorporation into protein or the reduction in polyribosome levels due to the salt stress was reported as the reason for the depressed protein synthesis by plants. This may be a reason for the reduction in protein synthesis in green beans.

34.2.7 WATER UPTAKE BY PLANTS

For all three cultivars of green beans, total water uptake decreased with increased salinity (Table 34.8, data from Pessarakli [164]), and the decrease patterns were similar to those of dry matter production [164]. Tender Improved absorbed more water than Kentucky Wonder and Slim Green cultivars under NaCl stress conditions. However, under normal conditions, Kentucky Wonder absorbed significantly more water than the other two cultivars at the second and third harvests [164]. The absorbed water values for Slim Green were the lowest among the three cultivars, at each harvest, for any corresponding treatment. Reduction in water uptake by other plants, or other bean cultivars, due to salt stress has been reported by many investigators [80, 124, 130, 153, 158, 159,

TABLE 34.7

Crude Protein Content of Three Green Bean Cultivars Under Various NaCl Stress Levels at Different Harvest Times

		Crude Protein Content of Plant Parts[a]					
		Shoots			Roots		
		Harvest[b]					
Cultivar	Salt Stress (Osmotic Potential)	1	2	3	4	5	6
	MPa			mg			
Tender Improved	Control (−0.03)	361	521	766	92	108	149
	−0.25	239	453	653	56	113	165
	−0.50	181	314	416	47	101	108
Slim Green	Control (−0.03)	172	379	866	50	71	180
	−0.25	156	260	465	43	51	93
	−0.50	63	91	138	27	32	41
Kentucky Wonder	Control (−0.03)	259	350	683	79	89	256
	−0.25	142	187	332	61	68	157
	0.50	103	161	295	46	51	66
LSD (0.05)[c]		43.2	61.5	101.4	16.8	18.3	35.1

Source: Pessarakli, M., *Crop Science*, 31(6), 1633–1640, 1991.

[a] Represents the means for pots containing two plants with three replicates.

[b] Harvests 1, 2, and 3 are for 5, 10, and 15 d ^{15}N-uptake periods, respectively.

[c] Represents the least significant difference between the treatment means at the 0.05 level of confidence.

174, 180, 182–184, 197, 206]. These investigators generally agreed that plant root permeability (expressed as hydraulic conductivity of the root system) decreased significantly under salt-stress conditions. This may explain the reduction in the water uptake rate and may contribute to a similar reduction in nutrient absorption, resulting in retarded plant growth and decreased dry matter production under salt stress conditions.

34.2.8 WATER USE EFFICIENCY OF PLANTS

Water use efficiency, expressed as milliliters of water absorbed per gram dry matter produced by plants, is exhibited in Table 34.8 (data from Pessarakli [164]). These data indicate that all three cultivars tended to use water more efficiently at the earliest harvest than at later harvests under either normal or stress conditions. This appears to be due to the faster rate of growth and higher dry matter production rate (grams dry matter produced per day, which can be calculated from the dry matter data; Table 34.1) at the earliest harvest than at later harvests (a dilution effect). Nevertheless, all cultivars at each harvest (except for Slim Green and Kentucky Wonder at the first harvest) used substantially less water for each unit of dry matter produced under stress conditions as compared with the controls. Omami et al. [158] also reported that plants with different salinity tolerance showed different water use efficiency. In Pessarakli's [164] study, at each harvest, for any corresponding treatment (except control plants at the first harvest), Tender Improved used substantially less water for each unit of dry matter produced (used water more efficiently) than the other two cultivars. Reduced water use efficiency under saline conditions was also reported by Zhang et al. [250]. However, these

investigators found that the application of mulch was significantly effective in saving water and improving water use efficiency and crop yield under saline irrigation conditions. Their results showed that the use of mulches significantly reduced evapotranspiration (ET) (improved water use efficiency) of Swiss chard under salinity stress condition imposed by saline irrigation. Recently, Khataar et al. [120] studied the interaction of soil salinity and matric potential on the water use (WU), water use efficiency (WUE), and yield response factor (Ky) of bean and wheat plants. These investigators reported that aeration porosity is the predominant factor controlling WU, WUE, Ky, and shoot biomass (Bs) at high soil water potentials; as matric potential decreased, soil aeration improved, with Bs, WU, and Ky reaching maximum values at −6 to −10 kPa under all salinity levels. The WUE of wheat remained almost unchanged by reduction in matric potential under low salinity levels (conductivity [EC] ≤8 dSm^{-1}) but increased under higher salinity levels (EC ≥8 dSm^{-1}), as did bean WUE at all salinity levels, as matric potential decreased to −33 kPa. The WUE of wheat exceeded that of bean in both sandy loam and clay loam (different textures) soils [120]. Overall, their results showed that salinity caused all the studied parameters (WU, WUE, and Ky) of bean and wheat plants to decrease, particularly at high salinity levels.

34.3 SUMMARY AND CONCLUSIONS

The effects of NaCl stress on physiological responses in terms of dry matter production, total-N, ^{15}N, crude protein, and water uptake by three green bean cultivars were discussed in this chapter.

TABLE 34.8

Total Water Absorption and Water Use Efficiency by Three Green Bean Cultivars as Affected by Three Levels of NaCl Stress at Three Harvest Times

		Water Uptake			Water Use Efficiency		
	Salt Stress	Harvest[a]					
Cultivar	(Osmotic potential)	1	2	3	1	2	3
	MPa	mL H_2O pot^{-1b}			mL H_2O g^{-1} dry wt.$^{b-}$		
Tender	Control (−0.03)	2035	4160	6010	514	676	689
Improved	−0.25	1310	3275	4800	384	610	602
	−0.50	800	2125	3325	336	478	584
Slim	Control (−0.03)	1085	2985	5685	513	1057	1150
Green	−0.25	840	2005	3555	506	757	815
	−0.50	485	1310	2010	500	596	652
Kentucky	Control (−0.03)	1800	4315	7840	500	1019	905
Wonder	−0.25	1085	2225	3525	490	834	865
	−0.50	690	1780	2830	475	692	792
LSD (0.05) salinity x cultivar		86	118	143	42	47	61
Summary of the significance of variance sources							
Cultivar (C)		**	**	**	NS	*	**
Salinity (S)		**	**	**	*	**	**
C X S		**	**	**	*	**	**

Source: Pessarakli, M., *Crop Science*, 31(6), 1633–1640, 1991.

*, ** Significant at P = .05 and .01, respectively.

[a] Harvests 1, 2, and 3 are for 5, 10, and 15 d ^{15}N-uptake periods, respectively.

[b] Represents the means for pots containing two plants with three replicates.

Total dry matter production was greater for Tender Improved than for Kentucky Wonder and Slim Green cultivars, for any corresponding treatment, at each harvest. For all three cultivars, total dry weight decreased significantly with increasing salinity. Reduction in dry weight due to NaCl stress was less for Tender Improved than for the other two cultivars. Total-N and ^{15}N uptake, by all three cultivars, substantially decreased under NaCl stress conditions. Nitrogen-15 concentration and shoot/root ratios of ^{15}N decreased with increasing salinity. Nitrogen-15 concentrations of shoots were less than those of roots for all plants. Sodium chloride stress severely reduced the crude protein content of plant parts for all three cultivars, at all three harvests. However, the Tender Improved appeared to be less affected by salinity than the other two cultivars. Shoots of all plants contained substantially higher total crude protein than roots for all treatments. This appears to be due to the higher biomass of shoots than roots for any corresponding treatment. Nevertheless, shoots were more severely affected than roots by salinity when salinized plants were compared with the controls for each plant part. Sodium chloride stress severely decreased the crude protein content of all three cultivars at each harvest for both sources of ^{15}N. However, the Tender Improved appeared to be the least and the Slim Green the most severely affected by salinity among the three cultivars.

Under normal (non-saline) conditions, green beans appear to absorb and use more NO_3-N than NH_4-N for protein synthesis. In contrast, under salt stress, NO_3-N seems more severely affected than NH_4-N in terms of being incorporated into protein. Furthermore, any level of salt stress will likely cause a drastic reduction in protein content and N metabolism in salt-sensitive bean plants.

For all cultivars, water uptake also was substantially decreased under stress conditions, particularly at the highest level of stress. Among the three cultivars, Tender Improved was the least and Slim Green the most severely affected by salinity in all aspects of stress. This is an indication of the difference in the salt tolerance of these cultivars. Therefore, among the three cultivars discussed here, the Tender Improved cultivar of green beans appears the most suitable for growing under field conditions. Furthermore, since there are numerous cultivars of green beans, additional testing of their response under saline conditions could detect a wider range of tolerance and susceptibility to soil salinity. This will enable researchers to select the most salt-tolerant cultivars to be recommended to growers.

REFERENCES

1. Abdul-Kadir, S.M. and G.M. Paulsen. 1982. Effect of salinity on nitrogen metabolism in wheat. *Journal of Plant Nutrition*, 5:1141–1151.
2. Abdul Qados, A.M.S. 2011. Effect of salt stress on plant growth and metabolism of bean plant Vicia faba (L.). *Journal of the Saudi Society of Agricultural Sciences*, 10(1):7–15.

3. Aceves, N.E., L.H. Stolzy, and G.R. Methuys. 1975. Effect of soil osmotic potential produced with two salt species on plant water potential, growth and grain yield of wheat. *Plant Soil*, 42:619–627.

4. Adcock, D., A.M. McNeill, G.K. McDonald, and R.D. Armstrong. 2007. Subsoil constraints to crop production on neutral and alkaline soils in South-eastern Australia: A review of current knowledge and management strategies. *Australian Journal of Experimental Agriculture*, 47(11):1245–1261.

5. Afzal, I., S.M.A. Basra, A. Hameed, and M. Farooq. 2006. Physiological enhancements for alleviation of salt stress in wheat. *Pakistan Journal of Botany*, 38(5):1649–1659.

6. Ahmadi, A., Y. Emam, and M. Pessarakli. 2009. Response of various cultivars of wheat and maize to salinity stress. *Journal of Agriculture, Food, and Environment (JAFE)*, 7(1):123–128.

7. Akinci, S., K. Yilmaz, and I.E. Akinci. 2004. Response of tomato (*Lycopersicon esculentum* Mill.) to salinity in the early growth stages for agricultural cultivation in saline environments. *Journal of Environmental Biology*, 25(3):351–357.

8. Al-Busaidi, A., T. Yamamoto, M. Inoue, M. Irshad, Y. Mori, and S. Tanaka. 2007. Effects of seawater salinity on salt accumulation and barley (*Hordeum vulgare* L.) growth under different meteorological conditions. *Journal of Food Agriculture and Environment (JAFE)*, 5(2):270–279.

9. Al Hassan, M., M. Morosan, M. del P. López-Gresa, J. Prohens, O. Vicente., and M. Boscaiu. 2016. Salinity-induced variation in biochemical markers provides insight into the mechanisms of salt tolerance in common (*Phaseolus vulgaris*) and runner (*P. coccineus*) beans. *International Journal of Molecular Sciences*, 17(9), 1582

10. Alislail, N.Y. and P.G. Bartels. 1990. Effects of sodium chloride on Tepary bean. *In*: Vegetable Report (N.F. Oebker and M. Bantlin, Eds.), University of Arizona Agriculture Experiment Station pp. 110–111.

11. Al-Rawahy, S.A., J.L. Stroehlein, and M. Pessarakli. 1990. Effect of salt stress on dry-matter production and nitrogen uptake by tomatoes. *Journal of Plant Nutrition*, 13:567–577.

12. Al-Rawahy, S.A., J.L. Stroehlein, and M. Pessarakli. 1992. Dry-matter yield and nitrogen-15, Na , Cl–, and K+ content of tomatoes under sodium chloride stress. *Journal of Plant Nutrition*, 15(3):341–358.

13. Altman, A. 2003. From plant tissue culture to biotechnology: Scientific revolutions, abiotic stress tolerance, and forestry. *In-Vitro Cellular and Developmental Biology-Plant*, 39(2):75–84.

14. Apte, S.K. and J. Thomas. 1997. Possible amelioration of coastal soil salinity using halotolerant nitrogen-fixing *Cyanobacteria*. *Plant Soil*, 189(2):205–211.

15. Arzani, A. 2008. Improving salinity tolerance in crop plants: A biotechnological view. *In-Vitro Cellular and Developmental Biology-Plant*, 44(5):373–383.

16. Asch, F. and M.C.S. Wopereis. 2001. Responses of field-grown irrigated rice cultivars to varying levels of floodwater salinity in a semi-arid environment. *Field Crops Research*, 70(2):127–137.

17. Asch, F., M. Dingkuhn, and K. Dorffling. 2000. Salinity increases CO2 assimilation, but reduces growth in field-grown, irrigated rice. *Plant and Soil*, 218(1–2):1–10.

18. Asfaw, K.G. 2011. The response of some Haricot Bean (*Phaseolus vulgaris*) varieties for salt stress during germination and seedling stage. *Current Research Journal of Biological Sciences*, 3(4):282–288.

19. Ashraf, M. 1994. Breeding for salinity tolerance in plants. Critical Review, Plant Sciences, 13:17–42.

20. Ashraf, M. 2004. Some important physiological selection criteria for salt tolerance in plants. *Flora*, 199:361–376.

21. Ashraf, M., H.R. Athar, P.J.C. Harris, and T.R. Kwon. 2008. Some prospective strategies for improving crop salt tolerance. *Advanced Agronomy*, 97:45–110.

22. Ashraf, M. and M.R. Foolad. 2005. Pre-sowing seed treatment – A shotgun approach to improve germination, plant growth, and crop yield under saline and non-saline conditions. *Advances in Agronomy*, 88(Special Issue):223–271.

23. Ashraf, M. and A. Orooj. 2006. Salt stress effects on growth, ion accumulation and seed oil concentration in an arid zone traditional medicinal plant Ajwain (*Trachyspermum ammi* L.) sprague. *Journal of Arid Environments*, 64(2):209–220.

24. Ashraf, M. and E. Rasul. 1988. Salt tolerance of mung bean (*Vigna radiata* L.) at two growth stages. *Plant Soil*, 110(1):63–67.

25. Assimakopoulou, Anna, et al. 2015. Effect of salt stress on three green bean (*Phaseolus vulgaris* L.) cultivars. *Notulae Botanicae Horti Agrobotanici Cluj-Napoca*, 43(1). (online).

26. Athar, H.R., A. Khan, and M. Ashraf. 2008. Exogenously applied ascorbic acid alleviates salt induced oxidative stress in wheat. *Environmental Experimental Botany*, 63:224–231.

27. Awasthi, P., H. Karki, K. Bargali, and S.S. Bargali. 2016. Germination and seedling growth of pulse crop (*Vigna* spp.) as affected by soil salt stress. *Current Agriculture Research Journal*, 4(2), 159–170.

28. Aydin, A., C. Kant, and M. Turan. 2012. Humic acid application alleviate salinity stress of bean (*Phaseolus vulgaris* L.) plants decreasing membrane leakage. *African Journal of Agricultural Research*, 7(7):1073–1086.

29. Ayers, A.D., J.W. Brown, and C.H. Wadleigh. 1952. Salt tolerance of barley and wheat in soil plots receiving several salinization regimes. *Agronomy Journal*, 44:307–310.

30. Ayers, R.S. and D.W. Westcot. 1985. Water quality for agriculture. Food and Agriculture Organization (FAO) Irrigation and Drainage Paper 29 (Rev. 1), Food and Agriculture Organization, United Nations, Rome Italy, 174 p.

31. Babitha, K.C., R.S. Vemanna, K.N. Nataraja, and M. Udayakumar. 2015. Overexpression of *EcbHLH57* transcription factor from *Eleusine coracana* L. in tobacco confers tolerance to salt, oxidative and drought stress. *PLoS ONE*, 10(9), e0137098

32. Bahaji, A., I. Mateu, A. Sanz, and M.J. Cornejo. 2002. Common and distinctive responses of rice seedlings to saline- and osmotically-generated stress. *Plant Growth Regulation*, 38(1):83–94.

33. Bahmaniar, M.A. and M.G. Sepanlou. 2008. Influence of saline irrigation water and gypsum on leaf nutrient accumulation, protein, and oil seed in soybean cultivars. *Journal of Plant Nutrition*, 31(3):485–495.

34. Balasubramanian, V. and S.K. Sinha. 1976. Effects of salt stress on growth, nodulation and nitrogen fixation in cow-pea and mungbeans. *Physiologia Plantarum*, 36(2):197–200.

35. Ballantyne, A.J. 1962. Tolerance of cereal crops to saline soils in Saskatchewan. *Canadian Journal of Soil Science*, 42:307–310.

36. Banuls, J., F. Legaz, and E. Primo-Millo. 1991. Salinity-calcium interactions on growth and ionic concentration of Citrus plants. *Plant Soil*, 133(1):39–46.

37. Bao, A.K., S.M. Wang, G.Q. Wu, J.J. Xi, J.L. Zhang, and C.M. Wang. 2009. Over-expression of the Arabidopsis H+-PPase enhanced resistance to salt and drought stress in transgenic alfalfa (*Medicago sativa* L.). *Plant Science*, 176(2):232–240.

38. Ben-Gal, A. and U. Shani. 2002. Yield, transpiration and growth of tomatoes under combined excess boron and salinity stress. *Plant and Soil*, 247(2):211–221.

39. Ben-Zioni, A., C. Itai, and Y. Vaadia. 1967. Water and salt stress, Kinetin and protein synthesis in tobacco leaves. *Plant Physiology*, 42:361–365.

40. Bernstein, L. 1961. Osmotic adjustment of plants to saline media I. Steady state. *American Journal of Botany*, 48:909–918.

41. Bernstein, L. 1963. Osmotic adjustment of plants to saline media II. Dynamic phase. *American Journal of Botany*, 40:360–370.

42. Bernstein, L., L.E. Francois, and R.A. Clark. 1974. Interactive effects of salinity and fertility on yields of grains and vegetables. *Agronomy Journal*, 66:412–421.

43. Bernstein, L. and G.A. Pearson. 1956. Influence of exchangeable sodium ions on the yield and chemical composition of plants: I. Green beans, garden beans, clover, and alfalfa. *Soil Science*, 82:247–258.

44. Bhivare, N.V. and J.D. Nimbalkar. 1984. Salt stress effects on growth and mineral nutrition of French beans. *Plant Soil*, 80(1):91–98.

45. Bhuiyan, M., A. Raman, D. Hodgkins, D. Mitchell, and H. Nicol. 2015. Physiological response and ion accumulation in two grasses, one legume, and one saltbush under soil water and salinity stress. *Ecohydrology*, 8:1547–1559.

46. Blanco, F.F., Folegatti, M.V., Gheyi, H.R., and P.D. Fernandes. 2008. Growth and yield of corn irrigated with saline water. *Scientia Agricola*, 65(6):574–580.

47. Bochow, H., S.F. El-Sayed, H. Junge, A. Stavropoulou, and G. Schmiedeknecht. 2001. Use of *Bacillus subtilis* as biocontrol agent. IV. Salt-stress tolerance induction by *Bacillus subtilis* FZB24 seed treatment in tropical vegetable field crops, and its mode of action. *Zeitschrift Fur Pflanzenkrankheiten Und Pflanzenschutz - Journal of Plant Diseases and Protection*, 108(1):21–30.

48. Bonilla, I., A. El-Hamdaoui, and L. Bolanos. 2004. Boron and calcium increase *Pisum sativum* seed germination and seedling development under salt stress. *Plant and Soil*, 267(1–2):97–107.

49. Borsani, O., J. Cuartero, J.A. Fernandez, V. Valpuesta, and M.A. Botella. 2001. Identification of two loci in tomato reveals distinct mechanisms for salt tolerance. *Plant Cell*, 13(4):873–887.

50. Campos, C.A.B., P.D. Fernandes, H.R. Gheyi, F.F. Blanco, and S.A.F. Campos. 2006. Yield and fruit quality of industrial tomato under saline irrigation. *Scientia Agricola*, 63(2):146–152.

51. Cantrell, I.C. and R.G. Linderman. 2001. Preinoculation of lettuce and onion with VA mycorrhizal fungi reduces deleterious effects of soil salinity. *Plant and Soil*, 233(2):269–281.

52. Carter, C.T., C.M. Grieve, and J.A. Poss. 2005. Salinity effects on emergence, survival, and ion accumulation of *Limonium perezii*. *Journal of Plant Nutrition*, 28(7):1243–1257.

53. Chandrasekaran, M., K. Kim, R. Krishnamoorthy, D. Walitang, S. Sundaram, M.M. Joe, and T. Sa. 2016. Mycorrhizal symbiotic efficiency on C3 and C4 plants under salinity stress – A meta-analysis. *Frontiers in Microbiology*, 7, 1246

54. Chaparzadeh, Nader, Younes Aftabi, Meysam Dolati, Faramarz Mehrnejad, and Mohammad Pessarakli. 2014. Salinity tolerance ranking of various wheat landraces from west of Uremia Lake in Iran by using physiological parameters. *Journal of Plant Nutrition*, 37:1025–1039.

55. Coons, J.M. and R.C. Pratt. 1988. Physiological and growth responses of *Phaseolus vulgaris* and *Phaseolus acutifolius* when grown in fields at two levels of salinity. Bean Improvement Cooperative Annual Report, Geneva, NY 31:88–89.

56. Cramer, G.R. 1986. Na+–Ca2+ interactions in roots of salt-stressed cotton (*Gossypium hirsutum* L.). Dissertation Abstracts International, B Science and Engineering, 46(11):3667B.

57. Csizinsky, A.A. 1986. Influence of total soluble salt concentration on growth and elemental concentration of winged bean seedlings (*Psophocarpus tetragonolobus* L.). *Communications in Soil Science and Plant Analysis*, 17:1009–1018.

58. Das, S.K. and C.L. Mehrotra. 1971. Salt tolerance of some agricultural crops during early growth stages. *Indian Journal of Agricultural Science*, 41(10):882–888.

59. Dasgan, H.Y., H. Aktas, K. Abak, and I. Cakmak. 2002. Determination of screening techniques to salinity tolerance in tomatoes and investigation of genotype responses. *Plant Science*, 163(4):695–703.

60. De Pascale, S., C. Ruggiero, G. Barbieri, and A. Maggio. 2003. Physiological responses of pepper to salinity and drought. *Journal of the American Society for Horticultural Science*, 128(1):48–54.

61. De Villiers, A.J., M.W. Van Rooyen, G.K. Theron, and A.S. Claassens. 1997. Tolerance of six namaqualand pioneer species to saline soil conditions. *South African Journal of Plant and Soil*, 14(1):38–42.

62. di Caterina, R., M.M. Giuliani, T. Rotunno, A. de Caro, and Z. Flagella . 2007. Influence of salt stress on seed yield and oil quality of two sunflower hybrids. *Analysis of Applied Biology*, 151(2):145–154.

63. Dilley, D.R., A.L., Kenworthy, E.J. Benne, and S.T. Bass. 1958. Growth and nutrient absorption of apple, cherry, peach, and grape plants as influenced by various levels of chloride and sulfate. *Proceeding of the American Society of Horticultural Science*, 72:64–73.

64. Djilianov, D., E. Prinsen, S. Oden, H. van Onckelen, and J. Muller. 2003. Nodulation under salt stress of alfalfa lines obtained after *in-vitro* selection for osmotic tolerance. *Plant Science*, 165(4):887–894.

65. Dong, Z., L. Shi Y. Wang L. Chen, Z. Cai, J. Jin, and X. Li. 2013. Identification and dynamic regulation of microRNAs involved in salt stress responses in functional soybean nodules by high-throughput sequencing. *International Journal of Molecular Sciences*, 14(2):2717–2738.

66. Egamberdieva, D., K. Davranov, S. Wirth, A. Hashem, and F.F. Abd_Allah. 2017. Impact of soil salinity on the plant-growth – promoting and biological control abilities of root associated bacteria. *Saudi Journal of Biological Sciences*, 24(7), 1601–1608

67. Elangumaran, G. and D.L. Smith. 2017. Plant growth promoting rhizobacteria in amelioration of salinity stress: a systems biology perspective. *Frontiers in Plant Science*, 8, 1768

68. Elsheikh, E.A.E. and M. Wood. 1989. Response of chickpea and soybean rhizobia to salt: Osmotic and specific ion effects of salts. *Soil Biology and Biochemistry*, 21(7):889–895.

69. Elzaawely, A.A., M.E. Ahmed, H.F. Maswada, and T.D. Xuan. 2017. Enhancing growth, yield, biochemical, and hormonal contents of snap bean (*Phaseolus vulgaris* L.) sprayed with moringa leaf extract. *Archives of Agronomy and Soil Science*, 63(5):687–699.

70. Emongor, V.E. 2015. Effects of moringa (*Moringa oleifera*) leaf extract on growth, yield and yield components of snap beans (*Phaseolus vulgaris*). *Brazilian Journal of Science and Technology*, 6:114–122.

71. Epstein, E., J.D. Norlyn, D.W. Rush, R.K. Kingsbury, D.B. Kelley, and A.F. Warna. 1980. Saline culture of crops: A genetic approach. *Science*, 210:339–404.

72. Erdal, S., M.Genisel, H.Turk, and Z. Gorcek. 2012. Effects of progesterone application on antioxidant enzyme activities and K+/Na+ ratio in bean seeds exposed to salt stress. *Toxicology and Industrial Health*, 28(10):942–946.

73. Fahad, S., M. Adnan, M. Noor, M. Arif, M. Alam, I.A. Khan, H. Ullah, F. Wahid, I.A. Mian, Y. Jamal, A. Basir, S. Hassan, S.SAmanullah, M.Riaz, C.Wu, M.A. Khan and D. Wang. 2019. Major constraints for global rice production. *In: Advances in Rice Research for Abiotic Stress Tolerance*, pp.1–22, M. Hasanuzzaman, M. Fujita, K. Naha, and J.K. Biswas (eds.), Elsevier Inc., Wikipedia, London, UK.

74. Farhangi-Abriz, S. and S. Torabian. 2017. Antioxidant enzyme and osmotic adjustment changes in bean seedlings as affected by biochar under salt stress. *Journal of Ecotoxicology and Environmental Safety*, 137:64–70.

75. Flowers, T.J. 2004. Improving crop salt tolerance. *Journal of Experimental Botany*, 55(96):307–319.

76. Food and Agriculture Organization (FAO) of the United Nations. 2005. Global network on integrated soil management for sustainable use of salt-affected soils. Rome, Italy: FAO Land and Plant Nutrition Management Service. http://www.fao.org/ag/agl/agll/spush.

77. Foolad, M.R. 2004. Recent advances in genetics of salt tolerance in tomato. *Plant Cell Tissue and Organ Culture*, 76(2):101–119.

78. Forni, C., D. Duca, and B.R. Glick. 2017. Mechanisms of plant response to salt and drought stress and their alteration by rhizobacteria. *Plant Soil*, 410:335.

79. Francois, L.E., E.V. Maas, T.J. Donovan, and V.L. Youngs. 1986. Effect of salinity on grain yield and quality, vegetative growth, and germination of semi-dwarf and durum wheat. *Agronomy Journal*, 78(6):1053–1058.

80. Frota, J.N.E. and T.C. Tucker. 1978a. Absorption rates of ammonium and nitrate by red kidney beans under salt and water stress. *Soil Science Society of America Journal*, 42:753–756.

81. Frota, J.N.E. and T.C. Tucker. 1978b. Salt and water stress influence nitrogen metabolism in red kidney beans. *Soil Science Society of America Journal*, 42:743–746.

82. Gao, S.M., H.W. Zhang, Y. Tian, F. Li, Z.J. Zhang, X.Y. Lu, X.L. Chen, and R.F. Huang. 2008. Expression of TERF1 in rice regulates expression of stress-responsive genes and enhances tolerance to drought and high-salinity. *Plant Cell Reports*, 27(11):1787–1795.

83. Gauch, H.G. and C.H. Wadleigh. 1944. The influence of high salt concentrations on the growth of bean plants. *Botanical Gazet*, 105:379–387.

84. Ge, H., N. Zhao, Y. Miao, M. Chen, and X. Wang. 2016. Research progress on salt tolerance of faba bean. *Agricultural Science & Technology*, 17(3):569–572.

85. Gibberd, M.R., N.C. Turner, and R. Storey. 2002. Influence of saline irrigation on growth, ion accumulation and partitioning, and leaf gas exchange of carrot (*Daucus carota* L.). *Analysis of Botany*, 90(6):715–724.

86. Gill, K.S. 1987. Effect of salinity on dry matter production and some physiological parameters at the vegetative growth stage in bajra genotypes. *Plant Physiology and Biochemistry, India*, 14(1):82–86.

87. Goodman, J., J. Maschinski, P. Hughes, J. McAuliffe, J. Roncal, D. Powell, and L.O. Sternberg. 2012. Differential response to soil salinity in endangered key tree cactus: Implications for survival in a changing climate. *PLoS ONE*, 7(3):e32528

88. Greenway, H. 1973. Salinity, plant growth, and metabolism. *Journal of Australian Institute of Agricultural Sciences*, 39:24–34.

89. Greenway, H. and R. Munns. 1980. Mechanisms of salt tolerance in non-halophytes. *Annual Review of Plant Physiology*, 31:149–190.

90. Grieve, C.M., L.E. Francois, and J.A. Poss. 2001. Effect of salt stress during early seedling growth on phenology and yield of spring wheat. *Cereal Research Communications*, 29(1–2):167–174.

91. Grieve, C.M., J.A. Poss, S.R. Grattan, P.J. Shouse, J.H. Lieth, and L. Zeng. 2005. Productivity and mineral nutrition of *Limonium* species irrigated with saline wastewaters. *HortScience*, 40(3):654–658.

92. Gulnaz, A., J. Iqbal, S. Farooq, and F. Azam. 1999. Seed treatment with growth regulators and crop productivity. I. 2,4-D as an inducer of salinity-tolerance in wheat (*Triticum aestivum* L.). *Plant and Soil*, 210(2):209–217.

93. Gupta, G.M., S. Mohan, and K.G. Prasad. 1987. Salt tolerance of selected tree seedlings. *Journal of Tropical Forestry*, 3(3):217–227.

94. Hanafy, M.S., El-banna, A., Schumacher, H.M. Jacobsen, F.M., and Hassan, F.S. 2013. Enhanced tolerance to drought and salt stresses in transgenic faba bean (*Vicia faba* L.) plants by heterologous expression of the PR10a gene from potato. *Plant Cell Reports*, 32(5):663–674.

95. Harbir-Singh, H., J. Prakash, and H. Singh. 1986. Salinity tolerance of field bean (*Vicia faba* L.) genotypes during germination. *Seed Research*, 14(1):127–129.

96. Heenan, D.P., L.G. Lewin, and D.W. McCaffery. 1988. Salinity tolerance in rice varieties at different growth stages. *Australian Journal of Experimental Agriculture*, 28(3):343–349.

97. Helal, H.M. and K. Mengel. 1979. Nitrogen metabolism of young barley plants as affected by NaCl-salinity and potassium. *Plant Soil*, 51:457–462.

98. Hernansez, L., O. Loyola-Gonzalez, B. Valle, J. Martinez, L. Diaz-Lopez, C. Aragon, and J.C. LORENZO. 2015. Identification of discriminant factors after exposure of maize and common bean plantlets to abiotic stresses. *Notulae Botanicae Horti Agrobotanici Cluj-Napoca*, 43(2):589.

99. Hernández-lucero, E., A.A. Rodríguez-hernández, M. Ortega-amaro, and J.F. Jiménez-bremont. 2014. Differential expression of genes for tolerance to salt stress in common bean (*Phaseolus vulgaris* L.). *Plant Molecular Biology Reporter*, 32(2):318–327.

100. Hoffman, G.J., J.A. Jobes, Z. Hanscom, and E.V. Maas. 1978. Timing of environmental stress affects growth, water relations and salt tolerance of pinto bean. *Transactions of the American Society of Agricultural Engineering (ASAE)*, July/Aug 1978, 21(4):713–718.

101. Hokmabadi, H., K. Arzani, and P.F. Grierson. 2005. Growth, chemical composition, and carbon isotope discrimination of pistachio (*Pistacia vera* L.) rootstock seedlings in response to salinity. *Australian Journal of Agricultural Research*, 56(2):135–144.

102. Hou, M., L. Zhu, and Q. Jin. 2016. Surface drainage and mulching drip-irrigated tomatoes reduces soil salinity and improves fruit yield. *PLoS ONE*, 11(5): e0154799

103. Irshad, M., T. Honna, A.E. Eneji, and S. Yamamoto. 2002. Wheat response to nitrogen source under saline conditions. *Journal of Plant Nutrition*, 25(12):2603–2612.

104. Islam, F., J. Wang, M.A. Farooq, C. Yang, M. Jan, T.M. Mwamba, F. Hannan, L. Xu, and W. Zhou. 2019. Rice responses and tolerance to salt stress: Deciphering the physiological and molecular mechanisms of salinity adaptation. *In: Advances in Rice Research for Abiotic Stress Tolerance*, pp. 791–819, M. Hasanuzzaman, M. Fujita, K. Naha, and J.K. Biswas (eds.), Elsevier Inc., Wikipedia, London, UK.

105. Izadi, M.H., J. Rabbani, Y. Emam, M. Pessarakli, and A. Tahmasebi. 2014. Effects of salinity stress on physiological performance of various wheat and barley cultivars. *Journal of Plant Nutrition*, 37:520–531.

106. Janmohammadi, M., A. Abbasi, and N. Sabaghnia. 2012. Influence of NaCl treatments on growth and biochemical parameters of castor bean (*Ricinus communis* L.). *Acta Agriculturae Slovenica*, 99(1):31.

107. Jiang, C., E.J. Belfield, Y. Cao, J.A.C. Smith, and N.P. Harberd. 2013. An *Arabidopsis* soil-salinity–tolerance mutation confers ethylene-mediated enhancement of sodium/potassium homeostasis. *The Plant Cell*, 25(9), 3535–3552

108. Jithesh, M.N., S.R. Prashanth, K.R. Sivaprakash, and A.K. Parida. 2006. Antioxidative response mechanisms in halophytes: Their role in stress defense. *Journal of Genetics*, 85(3):237–254.

109. Jouyban, Z. 2012. The effects of salt stress on plant growth. *Technical Journal of Engineering and Applied Sciences*, 7(10): Online.

110. Kahane, I. and A. Poljakoff-Mayber. 1968. Effect of substrate salinity on the ability for protein synthesis in pea roots. *Plant Physiology*, 43:1115–1119.

111. Kannan, S. 1987. Differential uptake of sodium and chloride in crop cultivars and their relationship to salt tolerance. *Journal of Indian Society of Coastal Agricultural Research*, 5(1):139–143.

112. Kannan, S. and S. Ramani. 1988. Evaluation of salt tolerance in cowpea and tobacco: Effects of NaCl on growth, relative turgidity and photosynthesis. *Journal of Plant Nutrition*, 11(4):435–448.

113. Kant, C., A. Aydin, and M. Turan. 2008. Ameliorative effect of hydro gel substrate on growth, inorganic ions, proline, and nitrate contents of bean under salinity stress. *Journal of Plant Nutrition*, 31(8):1420–1439.

114. Karami Chame, S., B. Khalil-Tahmasbi, P. ShahMahmoodi, A. Abdollahi, A. Fathi, S.J. Seyed Mousavi, M. Hossein Abadi, S. Ghoreishi, and S. Bahamin. 2016. Effects of salinity stress, salicylic acid and Pseudomonas on the physiological characteristics and yield of seed beans (*Phaseolus vulgaris* L.). *Scientia Agriculturae*, 14(2):234–238.

115. Karim, N.H. and M.Z. Haque. 1986. Salinity tolerance at the reproductive stage of five rice varieties. *Bangladesh Journal of Agriculture*, 11(4):73–76.

116. Katerji, N. J.W. van Hoorn, A. Hamdy, and M. Mastrorilli. 2003. Salinity effect on crop development and yield, analysis of salt tolerance according to several classification methods. *Agricultural Water Management*, 62(1):37–66.

117. Kaushik, A., N. Saini, S. Jain, P. Rana, R.K. Singh, and R.K. Jain. 2003. Genetic analysis of a CSR10 (Indica) x Taraori Basmati F-3 population segregating for salt tolerance using ISSR markers. *Euphytica*, 134(2):231–238.

118. Khalil, M.A., A. Fathi, and M.M. Elgabaly. 1967. A salinity fertility interaction study on corn and cotton. *Soil Science Society of America Proceeding*, 31:683–686.

119. Khan, A.H. and M.Y. Ashraf. 1988. Effect of sodium chloride on growth and mineral composition of sorghum. *Acta Physiologica Plantarum*, 10(3):257–264.

120. Khataar, M., Mohhamadi, M. H., & Shabani, F. 2018. Soil salinity and matric potential interaction on water use, water use efficiency and yield response factor of bean and wheat. *Scientific Reports*, 8:2679

121. Khosh-Kholgh SimaN.A., H. Askari, H. Hadavand Mirzaei, and M. Pessarakli. 2009. Genotype-dependent differential responses of three forage species to Ca supplement in saline conditions. *Journal of Plant Nutrition*, 32(4):579–597.

122. Kilic, C.C., Y.S. Kukul, and D. Anac. 2008. Performance of purslane (*Portulaca oleracea* L.) as a salt-removing crop. *Agricultural Water Management*, 95(7):854–858.

123. Kim, K.S., Y.K. Yoo, and G.J. Lee. 1991. Comparative salt tolerance study in Korean lawngrasses: I. Comparison with western turfgrasses via in vitro salt tolerance test. *Journal of Korean Society for Horticultural Science*, 32(1):117–123.

124. Koyro, H.W. 2006. Effect of salinity on growth, photosynthesis, water relations and solute composition of the potential cash crop halophyte *Plantago coronopus* (L.). *Environmental and Experimental Botany*, 56(2):136–146.

125. Koyro, H.W. and S.S. Eisa. 2008. Effect of salinity on composition, viability and germination of seeds of *Chenopodium quinoa* Willd. *Plant and Soil*, 302(1–2):79–90.

126. Kuhad, M.S., I.S. Sheoran, and S. Kumari. 1987. Alleviation and separation of osmotic and ionic effect during germination and early seedling growth in pearl millet by pre-soaking the seeds with growth regulators. *Indian Journal of Plant Physiology*, 30(2):139–143.

127. Lagerwerff, J.V. 1969. Osmotic growth inhibition and electrometric salt-tolerance evaluation of plants; a review and experimental assessment. *Plant Soil*, 31(1):77–96.

128. Lee, M.K. and M.W. van Lersel. 2008. Sodium chloride effects on growth, morphology, and physiology of chrysanthemum (*Chrysanthemum xmorifolium*). *Hortscience*, 43(6):1888–1891.

129. Liu, H.R., G.W. Sun, L.J. Dong, L.Q. Yang, S.N. Yu, S.L. Zhang, and J.F. Liu. 2017. Physiological and molecular responses to drought and salinity in soybean. *Biologia Plantarum*, 1(3):557–564.

130. Lopez, C.M.L., H. Takahashi, and S. Yamazaki. 2002. Plant-water relations of Kidney bean plants treated with NaCl and foliarly applied glycinebetaine. *Journal of Agronomy and Crop Science*, 188(2):73–80.

131. Luo, J.-Y., S. Zhang, J. Peng, X.-Z. Zhu, L.-M. Lv, C.-Y. Wang, and J.-J. Cui. 2017. Effects of soil salinity on the expression of Bt Toxin (Cry1Ac) and the control efficiency of *Helicoverpa armigera* in field-grown transgenic Bt cotton. *PLoS ONE*, 12(1): e0170379

132. Maas, E.V. and G.J. Hoffman. 1977. Crop salt tolerance--Current assessment. *American Society of Civil Engineers (ASCE) Journal of Irrigation and Drainage Division*, 103:1115–134.

133. Maas, E.V., G. Ogata, and M.J. Caber. 1972. Influence of salinity on uptake of Fe, Mn, and Zn by plants. *Agronomy Journal*, 64:793–795.

134. Maggio, A., P.M. Hasegawa, R.A. Bressan, M.F. Consiglio, and R.J. Joly. 2001. Unravelling the functional relationship between root anatomy and stress tolerance. *Australian Journal of Plant Physiology*, 28(10):999–1004.

135. Maggio, A., S. de Pascale, G. Angelino, C. Ruggiero, and G. Barbieri. 2004. Physiological response of tomato to saline irrigation in long-term salinized soils. *European Journal of Agronomy*, 21(2):149–159.

136. Mahmood, S., I. Daur, S.G. Al-Solaimani, S. Ahmad, M.H. Madkour, M. Yasir, and Z. Ali. 2016. Plant growth promoting rhizobacteria and silicon synergistically enhance salinity tolerance of mung bean. *Frontiers in Plant Science*, 7:876

137. Mahmut, C.H., B. Canher, H. Niron, and M. Turet. 2014. Transcriptome analysis of salt tolerant common bean (*Phaseolus vulgaris* L.) under saline conditions. *PLoS One*, 9(3)http://dx.doi.org.ezproxy1.library.arizona.edu/10.1371/journal.pone.0092598

138. Maliro, M.F.A., D. McNeil, B. Redden, J.F. Kollmorgen, and C. Pittock. 2008. Sampling strategies and screening of chickpea (*Cicer arietinum* L.) germplasm for salt tolerance. *Genetic Resources and Crop Evolution*, 55(1):53–63.

139. Maliwal, G.L. 1973. Salt tolerance of vegetable crops. *Farmer Parliament*, 8(5):17–22.

140. Maliwal, G.L. and K.V. Paliwal. 1982. Salt tolerance of some mungbean (*Vigna radiata* L.), urdbean (*Vigna mungo* L.) and guar (*Cyamopsis tetragonoloba* L.) varieties at germination and early growth stages. *Legume Research*, 5(1):23–30.

141. Maliwal, G.L. and K.V. Paliwal. 1984. Salt tolerance of some paddy, maize, sorghum, cotton and tobacco varieties at germination and early growth stage. *Agricultural Science Digest, India*, 4(3):147–149.

142. Mangal, J.L., S. Lal, and P. S. Hooda. 1989. Salt tolerance of the onion seed crop. *Journal of Horticultural Science*, 64(4):475–477.

143. Maqshoof, A., A. Zahir, H. Naeem, and M. Asghar. 2011. Inducing salt tolerance in mung bean through coinoculation with rhizobia and plant-growth-promoting rhizobacteria containing 1-aminocyclopropane-1-carboxylate deaminase. *Canadian Journal of Microbiology*, 57(7):578–589.

144. Marcum, K.B. and M. Pessarakli. 2006. Salinity tolerance and salt gland excretion activity of bermudagrass turf cultivars. *Crop Science Society of America Journal*, 46(6):2571–2574.

145. Marcum, K.B., M. Pessarakli, and D.M. Kopec. 2005. Relative salinity tolerance of 21 turf-type desert saltgrasses compared to bermudagrass. *HortScience*, 40(3):827–829.

146. Mehta, P.K., A. Kachroo, M.K. Kaul, and R. Yamdagni. 1988. Salt tolerance in fruit crops - A review. *Agricultural Reviews*, 9(2):57–68.

147. Miransari, M., H. Riahi, F. Eftekhar, A. Minaie, and D.L. Smith. 2013. Improving soybean (*Glycine max* L.) N2 fixation under stress. *Journal of Plant Growth Regulation*, 32(4):909–921.

148. Morilla, C.A., J.S. Boyer, and R.H. Hageman. 1973. Nitrate reductase activity and polyribosomal content of corn (*Zea Mays* L.) having low leaf water potentials. *Plant Physiology*, 51:817–824.

149. Munns, R., R.A. James, and A. Lauchli. 2006. Approaches to increasing the salt tolerance of wheat and other cereals. *Journal of Experimental Botany*. 57(5):1025–1043.

150. Nahar, K., M. Hasanuzzaman, M. Alam, and M. Fujita. 2015. Roles of exogenous glutathione in antioxidant defense system and methylglyoxal detoxification during salt stress in mung bean. *Biologia Plantarum*, 59(4):745–756.

151. Nahar, K., M. Hasanuzzaman, A. Rahman, M. Alam, J. Al-Mahmud, T. Suzuki, and M. Fujita. 2016. Polyamines confer salt tolerance in mung bean (*Vigna radiata* L.) by reducing sodium uptake, improving nutrient homeostasis, antioxidant defense, and methylglyoxal detoxification systems. *Frontiers in Plant Science*, 7:1104.

152. Negrão, S., S.M. Schmöckel, and M. Tester. 2017. Evaluating physiological responses of plants to salinity stress. *Annals of Botany*, 119(1):1–11.

153. Netondo, G.W., J.C. Onyango, and E. Beck. 2004. Sorghum and salinity: I. Response of growth, water relations, and ion accumulation to NaCl salinity. *Crop Science*, 44(3):797–805.

154. Nguyen, P.D., C.L. Ho, J.A. Harikrishna, M.C.V.L. Wong, and R.A. Rahim. 2007. Functional screening for salinity tolerant genes from *Acanthus ebracteatus* Vahl using *Escherichia coli* as a host. *Trees-Structure and Function*, 21(5):515–520.

155. Nukaya, A., M. Masui, and A. Ishida. 1983. Salt tolerance of muskmelons at different growth stages as affected by diluted sea water. *Journal of Japanese Society for Horticultural Science*, 52(3):286–293.

156. O'Leary, J.W. 1971. Physiological basis for plant growth inhibition due to salinity. *In: Food, Fiber and the Arid Lands*, pp. 331–336, W.G. McGinnies, B.J. Goldman, and P. Paylore (eds.), University of Arizona Press, Tucson, AZ.

157. O'Leary, J.W. 1974. Salinity-induced changes in hydraulic conductivity of roots. pp. 309–314. *In: Structure and function of primary root tissues*, J. Kolek (ed.), Veda Publishing House of the Slovak Academy of Science, Bratislava, Czechoslovakia.

158. Omami, E.N., P.S. Hammes, and P.J. Robbertse. 2006. Differences in salinity tolerance for growth and water-use efficiency in some amaranth (*Amaranthus* spp.) genotypes. *New Zealand Journal of Crop and Horticultural Science*, 34(1):11–22.

159. Papadopoulos, I., V.V. Rendig, and F.E. Broadbent. 1985. Growth, nutrition, and water uptake of tomato plants with divided roots growing in differentially salinized soil. *Agronomy Journal*, 77:21–26.

160. Papiernik, S.K., C.M. Grieve, S.M. Lesch, and S.R. Yates. 2005. Effects of salinity, imazethapyr, and chlorimuron application on soybean growth and yield. *Communications in Soil Science and Plant Analysis*, 36(7–8):951–967.

161. Paul, D. and S. Nair. 2008. Stress adaptations in a plant growth promoting *rhizobacterium* (PGPR) with increasing salinity in the coastal agricultural soils. *Journal of Basic Microbiology*, 48(5):378–384.

162. Pearson, G.A. and L. Bernstein. 1959. Influence of exchangeable sodium on yield and chemical composition of plants. I. Wheat barley, oats, rice, tall fescue, and tall wheatgrass. *Soil Science*, 86:254–261.

163. Peng, J., J. Liu, L. Zhang, J. Luo, H. Dong, Y. Ma, and Y. Meng. 2016. Effects of soil salinity on sucrose metabolism in cotton leaves. *PLoS ONE*, 11(5): e0156241

164. Pessarakli, M. 1991. Dry-matter yield, nitrogen-15 absorption, and water uptake by green bean under sodium chloride stress. *Crop Science*, 31(6):1633–1640.

165. Pessarakli, M. 2005a. Supergrass: Drought-tolerant turf might be adaptable for golf course use. Golfweek's SuperNews Magazine, November 16, 2005, p 21 and cover page. http://www.supernewsmag.com/news/golfweek/supernews/20051116/p21.asp?st=p21_s1.htm

166. Pessarakli, M. 2005b. Gardener's delight: Low-maintenance grass. Tucson Citizen, Arizona, Newspaper Article, September 15, 2005, Tucson, AZ, U.S.A. Gardener's delight: Low-maintenance grass http://www.tucsoncitizen.com/

167. Pessarakli, M. 2007. Saltgrass (*Distichlis spicata*), a potential future turfgrass species with minimum maintenance/management cultural practices. *In: Handbook of Turfgrass Management and Physiology*, pp. 603–615, M. Pessarakli (Ed.), CRC Press, Taylor and Francis Group, Florida.

168. Pessarakli, M., D.D. Breshears, J. Walworth, J.P. Field, and D.J. Law. 2017. Candidate halophytic grasses for addressing land degradation: Shoot responses of *Sporobolus airoides* and *Paspalum vaginatum* to weekly increasing NaCl concentration. *Arid Land Research and Management Journal*, 31(2):169–181.

169. Pessarakli, M., N. Gessler, and D.M. Kopec. 2008. Growth responses of saltgrass (*Distichlis spicata*) under sodium chloride (NaCl) salinity stress. United States Golf Association (USGA) Turfgrass and Environmental Research Online (TERO), October 15, 2008, 7(20):1–7. http://turf.lib.msu.edu/tero/v02/n14.pdf

170. Pessarakli, M., M.A. Harivandi, D.M. Kopec, and D.T. Ray. 2012. Growth responses and nitrogen uptake by saltgrass (*Distichlis spicata* L.), a halophytic plant species, under salt stress, using the 15N technique. *International Journal of Agronomy*, 2012: Article ID 896971, 9 pages, doi.

171. Pessarakli, M. and J.T. Huber. 1991. Biomass production and protein synthesis by alfalfa under salt stress. *Journal of Plant Nutrition*, 14(3):283–293.

172. Pessarakli, M., J.T. Huber, and T.C. Tucker. 1989a. Protein synthesis in green beans under salt stress conditions. *Journal of Plant Nutrition*, 12(10):1105–1121.

173. Pessarakli, M., J.T. Huber, and T.C. Tucker. 1989b. Protein synthesis in green beans under salt stress with two nitrogen sources. *Journal of Plant Nutrition*, 12(11):1361–1377.

174. Pessarakli, M., J.T. Huber, and T.C. Tucker. 1989c. Dry matter yield, nitrogen absorption, and water uptake by sweet corn under salt stress. *Journal of Plant Nutrition*, 12(3):279–290.

175. Pessarakli, M. and D.M. Kopec. 2010. Growth Responses and Nitrogen Uptake of Saltgrass (*Distichlis spicata* L.), a True Halophyte, under Salinity Stress Conditions using 15N Technique. *In: Proceedings of the International Conference on Management of Soils and Ground Water Salinization in Arid Regions*, Vol. 2, 1–11, Muscat, Sultanate of Oman.

176. Pessarakli, M. and D.M. Kopec. 2005. Responses of twelve inland saltgrass accessions to salt stress. United States Golf Association (USGA) Turfgrass and Environmental Research Online (TERO) 4(20):1–5. http://turf.lib.msu.edu/tero/v02/n14.pdf

177. Pessarakli, M. and D.M. Kopec. 2008. Establishment of three warm-season grasses under salinity stress. *Acta HortScience*, ISHS, 783:29–37.

178. Pessarakli, M., K.B. Marcum, and D.M. Kopec. 2005. Growth responses and nitrogen-15 absorption of desert saltgrass (*Distichlis spicata*) to salinity stress. *Journal of Plant Nutrition*, 28(8):1441–1452.

179. Pessarakli, M. and Hayat Touchane. 2006. Growth responses of bermudagrass and seashore paspalum under various levels of sodium chloride stress. *Journal of Agriculture, Food, and Environment (JAFE)*, 4(3&4):240–243.

180. Pessarakli, M. and T.C. Tucker. 1985a. Uptake of nitrogen-15 by cotton under salt stress. *Soil Science Society of America Journal*, 49:149–152.

181. Pessarakli, M. and T.C. Tucker. 1985b. Ammonium (15N) metabolism in cotton under salt stress. *Journal of Plant Nutrition*, 8:1025–1045.

182. Pessarakli, M. and T.C. Tucker. 1988a. Dry matter yield and nitrogen-15 uptake by tomatoes under sodium chloride stress. *Soil Science Society of America Journal*, 52:698–700.

183. Pessarakli, M. and T. C. Tucker. 1988b. Nitrogen-15 uptake by eggplant under sodium chloride stress. *Soil Science Society of America Journal*, 52(6):1673–1676.

184. Pessarakli, M., T. C. Tucker, and K. Nakabayashi. 1991. Growth response of barley and wheat to salt stress. *Journal of Plant Nutrition*, 14(4):331–340.

185. Pessarakli, M. and M. Zhou. 1990. Effect of salt stress on nitrogen fixation by different cultivars of green beans. *Journal of Plant Nutrition*, 13(5):611–629.

186. Qadir, M.A., R.H. Qureshi, N. Ahmad, and M. Ilyas. 1996. Salt-tolerant forage cultivation on a saline-sodic field for biomass production and soil reclamation. *Land Degradation and Development*, 7(2):11–18.

187. Quesada, V., S. Garcia-Martinez, P. Piqueras, M.R. Ponce, and J.L. Micol. 2002. Genetic architecture of NaCl tolerance in Arabidopsis. *Plant Physiology*, 130(2):951–963.

188. Radi, A.A., F.A. Farghaly, and A.M. Hamada. 2013. Physiological and biochemical responses of salt-tolerant and salt-sensitive wheat and bean cultivars to salinity. *Journal of Biology and Earth Sciences*, 3(1): Online.

189. Rady, M.M., C. Bhavya Varma, and S.M. Howladar. 2013. Common bean (*Phaseolus vulgaris* L.) seedlings overcome NaCl stress as a result of presoaking in *Moringa oleifera* leaf extract. *Scientia Horticulturae*, 162:63–70.

190. Rady, M.M. and G.F. Mohamed. 2015. Modulation of salt stress effects on the growth, physio-chemical attributes and yields of *Phaseolus vulgaris* L. plants by the combined application of salicylic acid and *Moringa oleifera* leaf extract. *Scientia Horticulturae*, 193:105–113.

191. Rady, M.M., W.M. Semida, Kh.A. Hemida, and M.T. Abdelhamid. 2016. The effect of compost on growth and yield of *Phaseolus vulgaris* plants grown under saline soil. *International Journal of Recycling of Organic Waste in Agriculture* 5:311–321.

192. Rahman, A., M.S. Hossain, J.-A. Mahmud, K. Nahar, M. Hasanuzzaman, and M. Fujita. 2016. Manganese-induced salt stress tolerance in rice seedlings: Regulation of ion homeostasis, antioxidant defense and glyoxalase systems. *Physiology and Molecular Biology of Plants*, 22(3), 291–306

193. Reynolds, M.P., A. Mujeeb-Kazi, and M. Sawkins. 2005. Prospects for utilizing plant-adaptive mechanisms to improve wheat and other crops in drought- and salinity-prone environments. *Analysis of Applied Biology*, 146(2):239–259.

194. Rezaei, H., N.A. Khosh-Kholgh Sima, M.J. Malakouti, and M. Pessarakli. 2006. Salt tolerance of canola in relation to accumulation and xylem transportation of cations. *Journal of Plant Nutrition*, 29(11):1903–1917.

195. Rogers, M.E., A.D. Craig, R. Munns, T.D. Colmer, P.G.H. Nichols, C.V. Malcolm, E.G. Barrett-Lennard, A.J. Brown, W.S. Semple, P.M. Evans, K. Cowley, S.J. Hughes, R. Snowball, S.J. Bennett, G.C. Sweeney, B.S. Dear, and M.A. Ewing. 2005. The potential for developing fodder plants for the salt-affected areas of Southern and Eastern Australia: An overview. *Australian Journal of Experimental Agriculture*, 45:301–329.

196. Romo, J.T. and M.R. Haferkamp. 1987. Forage kochia germination response to temperature, water stress, and specific ions. *Agronomy Journal*, 79(1):27–30.

197. Saad, R. 1979. Effect of atmospheric carbon dioxide levels on nitrogen uptake and metabolism in red kidney beans (*Phaseolus vulgaris* L.) under salt and water stress. Ph.D. Dissertation, University of Arizona. University Microfiche, Ann Arbor, Mich. (Dissertation Abstracts B., 40:4057).

198. Salim, M. 1989. Salinity effects on growth and ionic relations of two triticale varieties differing in salt tolerance. *Journal of Agronomy and Crop Science*, 162(1):35–42.

199. Salim, M. 1991. Comparative growth responses and ionic relations of four cereals during salt stress. *Journal of Agronomy and Crop Science*, 66(3):204–209.

200. Salim, M. and M.G. Pitman. 1987. Salinity tolerance of mung bean (*Vigna radiata* L.): Seed production. *Biologia Plantarum*, 30(1):53–57.

201. Saqib, M., C. Zorb, and S. Schubert. 2008. Silicon-mediated improvement in the salt resistance of wheat (*Triticum aestivum*) results from increased sodium exclusion and resistance to oxidative stress. *Functional Plant Biology*, 35(7):633–639.

202. Savvas, D. and F. Lenz. 2000. Response of eggplants grown in recirculating nutrient solution to salinity imposed prior to the start of harvesting. *Journal of Horticultural Science and Biotechnology*, 75(3):262–267.

203. Savvas, D., D. Giotis, E. Chatzieustratiou, M. Bakea, and G. Patakioutas. 2009. Silicon supply in soilless cultivations of zucchini alleviates stress induced by salinity and powdery mildew infections. *Environmental and Experimental Botany*, 65(1):11–17.

204. Savvas, D., N. Mantzos, R.E. Barouchas, I.L. Tsirogiannis, C. Olympios, and H.C. Passam. 2007. Modelling salt accumulation by a bean crop grown in a closed hydroponic system in relation to water uptake. *Scientia Horticulturae*, 111(4):311–318.

205. Schwabe, K.A., I. Kan, and K.C. Knapp. 2006. Drain water management for salinity mitigation in irrigated agriculture. *American Journal of Agricultural Economy*, 88(1):133–149.

206. Shalhevet, J. and L. Bernstein. 1968. Effects of vertically heterogenous soil salinity on plant growth and water uptake. *Soil Science*, 106:85–93.

207. Shani, U. and L.M. Dudley. 2001. Field studies of crop response to water and salt stress. *Soil Science Society of America Journal*, 65(5):1522–1528.

208. Shannon, M.C. 1998. Adaptation of plants to salinity. *Advances in Agronomy*, 60:75–119.

209. Shannon, M.C., J.W. Gronwald, and M. Tal. 1987. Effects of salinity on growth and accumulation of organic and inorganic ions in cultivated and wild tomato species. *Journal of American Society for Horticultural Science*, 112(3):416–423.

210. Shimose, N. 1972. Physiology of salt injury in crops, A. Salt tolerance of barley, wheat, and asparagus. *Scientific Reports of the Faculty of Agriciculture*, Okayma University, 40:57–68.

211. Singh, D., C.K. Singh, S. Kumari, et al. 2017. Discerning morpho-anatomical, physiological and molecular multiformity in cultivated and wild genotypes of lentil with reconciliation to salinity stress. Aroca, R., ed. *PLoS ONE*. 12(5):e0177465.

212. Singla-Pareek, S.L., M.K. Reddy, and S.K. Sopory. 2003. Genetic engineering of the glyoxalase pathway in tobacco leads to enhanced salinity tolerance. *Proceedings of the National Academy of Sciences of the United States of America*, 100(25):14672–14677.

213. Singla-Pareek, S.L., S.K. Yadav, A. Pareek, M.K. Reddy, and S.K. Sopory. 2008. Enhancing salt tolerance in a crop plant by overexpression of glyoxalase II. *Transgenic Research*, 17(2):171–180.

214. Srivastava, S., B. Fristensky, and N.N.V. Kav. 2004. Constitutive expression of a PR10 protein enhances the germination of *Brassica napus* under saline conditions. *Plant and Cell Physiology*, 45(9):1320–1324.

215. Steppuhn, H., M.T. van Genuchten, and C.M. Grieve. 2005. Root-zone salinity: I. Selecting a product-yield index and response function for crop tolerance. *Crop Science*, 45(1):209–220.

216. Strogonov, B.P. 1964. *Physiological basis of salt tolerance of plants*. Academic Science, Union Soviet society of Russia (USSR) (Translated from Russian.), Israel Program for Scientific Translations, Jerusalem, 279 p.

217. Sun, Z., Y. Wang, F. Mou, Y. Tian, L. Chen, S. Zhang, Q. Jiang, X. Li. 2015. Genome-wide small RNA analysis of soybean reveals auxin-responsive microRNAs that are differentially expressed in response to salt stress in root apex. *Frontiers in Plant Science*, 6(1273), 40p. https://www.ncbi.nlm.nih.gov/pmc/articles/PMC4716665/

218. Tavakkoli, E., Rengasamy, P., & McDonald, G. K. 2010. High concentrations of Na+ and Cl– ions in soil solution have simultaneous detrimental effects on growth of faba bean under salinity stress. *Journal of Experimental Botany*, 61(15):4449–4459

219. Tavakkoli, E., J. Paull, P. Rengasamy, and G.K. McDonald. 2012. Comparing genotypic variation in faba bean (*Vicia faba* L.) in response to salinity in hydroponic and field experiments. *Field Crops Research*, 127:99–108.

220. Tuna, A.L., C. Kaya, M. Ashraf, H. Altunlu, I. Yokas, and B. Yagmur. 2007. The effects of calcium sulphate on growth, membrane stability and nutrient uptake of tomato plants grown under salt stress. *Environmental and Experimental Botany*, 59(2):173–178.

221. Tunçtürk, M., R. Tunçtürk, B. Yildirim, and V. Çiftçi. 2011. Effect of salinity stress on plant fresh weight and nutrient composition of some Canola (*Brassica napus* L.) cultivars. *African Journal of Biotechnology*, 10(10): Online.

222. van Hoorn, J.W., N. Katerji, A. Hamdy, and M. Mastrorilli. 2001. Effect of salinity on yield and nitrogen uptake of four grain legumes and on biological nitrogen contribution from the soil. *Agricultural Water Management*, 51(2):87–98.

223. Veatch, M.E., S.E. Smith, and G. Vandemark. 2004. Shoot biomass production among accessions of *Medicago truncatula* exposed to NaCl. *Crop Science*, 44(3):1008–1013.

224. Velagaleti, R.R., S. Marsh, D. Kramer, D. Fleischman, and J. Corbin. 1990. Genotypic differences in growth and nitrogen fixation among soybean (*Glycine max* L. Merr) cultivars grown under salt stress. *Tropical Agriculture*, 67(2):169–177.

225. Verma, D., S.L. Singla-Pareek, D. Rajagopal, M.K. Reddy, and S.K. Sopory. 2007. Functional validation of a novel isoform of Na+/H+ antiporter from *Pennisetum glaucum* for enhancing salinity tolerance in rice. *Journal of Biosciences*, 32(3):621–628.

226. Villora, G., D.A. Moreno, G. Pulgar, and L.M. Romero. 1999. Zucchini growth, yield, and fruit quality in response to sodium chloride stress. *Journal of Plant Nutrition*, 22(6):855–861.

227. Waheed, A., I.A. Hafiz, G. Qadir, G. Murtaza, T. Mahmood, and M. Ashraf. 2006. Effect of salinity on germination, growth, yield, ionic balance and solute composition of pigeon pea (*Cajanus cajan* L., Millsp). *Pakistan Journal of Botany*, 38(4):1103–1117.

228. Wahid, A. 2004. Analysis of toxic and osmotic effects of sodium chloride on leaf growth and economic yield of sugarcane. *Botanical Bulletin of Academia Sinica*, 45(2):133–141.

229. Wahid, A., M. Perveen, S. Gelani, and S.M.A. Basra. 2007. Pretreatment of seed with H2O2 improves salt tolerance of wheat seedlings by alleviation of oxidative damage and expression of stress proteins. *Journal of Plant Physiology*, 164(3):283–294.

230. Wang, D. and M.C. Shannon. 1999. Emergence and seedling growth of soybean cultivars and maturity groups under salinity. *Plant and Soil*, 214(1–2):117–124.

231. Wang, D., J.A. Poss, T.J. Donovan, M.C. Shannon, and S.M. Lesch. 2002. Biophysical properties and biomass production of elephantgrass under saline conditions. *Journal of Arid Environments*, 52(4):447–456.

232. Wang, Y., W. Shen, Z. Chan, and Y. Wu. 2015. Endogenous cytokinin overproduction modulates ROS homeostasis and decreases salt stress resistance in *Arabidopsis Thaliana*. *Frontiers in Plant Science*, 6:1004

233. Weisany, W., Y. Sohrabi, G. Heidari, A. Siosemardeh, and K. Ghassemi-Golezani. 2012. Changes in antioxidant enzymes activity and plant performance by salinity stress and zinc application in soybean ('*Glycine max*' L.) [online]. *Plant Omics*, 5(2):60–67.

234. Wignarajah, K. 1990. Growth response of *Phaseolus vulgaris* to varying salinity regimes. *Environmental and Experimental Botany*, 30(2):141–147.

235. Wilson, C., X. Liu, S.M. Lesch, and D.L. Suarez. 2006. Growth response of major US cowpea cultivars. I. Biomass accumulation and salt tolerance. *HortScience*, 41(1):225–230.

236. Wilson, C. and J.J. Read. 2006. Effect of mixed-salt salinity on growth and ion relations of a barnyardgrass species. *Journal of Plant Nutrition*, 29(10):1741–1753.

237. Winicov, I. and D.R. Bastola. 1999. Transgenic overexpression of the transcription factor alfin1 enhances expression of the endogenous MsPRP2 gene in alfalfa and improves salinity tolerance of the plants. *Plant Physiology*, 120(2):473–480.

238. Yaish, M.W., A. Al-Lawati, G.A. Jana, H. Vishwas Patankar, and B.R. Glick. 2016. Impact of soil salinity on the structure of the bacterial endophytic community identified from the roots of caliph medic (*Medicago truncatula*). *PLoS ONE*, 11(7), e0159007

239. Yang, H., J. Hu, X. Long, Z. Liu, and Z. Rengel. 2016. Salinity altered root distribution and increased diversity of bacterial communities in the rhizosphere soil of Jerusalem artichoke. *Scientific Reports*, 6:20687

240. Yang, Y.W., R.J. Newton, and F.R. Miller. 1990a. Salinity tolerance in Sorghum: I. Whole plant response to sodium chloride in Sorghum bicolor and Sorghum halepense. *Crop Science*, 30(4):775–781.

241. Yang, Y.W., R.J. Newton, and F.R. Miller. 1990b. Salinity tolerance in Sorghum: II. Cell culture response to sodium chloride in Sorghum bicolor and Sorghum halepense. *Crop Science*, 30 (4):781–785.

242. Yildirim, E., A.G. Taylor, and T.D. Spittler. 2006. Ameliorative effects of biological treatments on growth of squash plants under salt stress. *Scientia Horticulturae*, 111(1):1–6.

243. Yilmaz, D.D. 2007. Effects of salinity on growth and nickel accumulation capacity of *Lemna gibba* (Lemnaceae). *Journal of Hazardous Materials*, 147(1–2):74–77.

244. Yin, Y. 2015. Comparative proteomic and physiological analyses reveal the protective effect of exogenous calcium on the germinating soybean response to salt stress. *Journal of Proteomics*, 113:110–126.

245. You, M.P., T.D. Colmer, and M.J. Barbetti. 2011. *Salinity drives host reaction in Phaseolus vulgaris (common bean) to Macrophomina phaseolina. Functional Plant Biology*, 38(12):984–992.

246. Zehra, A. and M.A. Khan. 2007. Comparative effect of NaCl and sea-salt on germination of halophytic grass *Phragmites karka* at different temperature regimes. *Pakistan Journal of Botany*, 39(5):1681–1694.

247. Zeroni, M. 1988. Plant tolerance of salinity in greenhouses - physiological and practical considerations. Acta HortScience, No. 229, 55–72.

248. Zhang, H., J. Cui, T. Cao, J. Zhang, Q. Liu, and H. Liu. 2011. Response to salt stress and assessment of salt tolerability of soybean varieties in emergence and seedling stages. *Shengtai Xuebao/Acta Ecologica Sinica*, 31(10):2805–2812.

249. Zhang, H.X., J.N. Hodson, J.P. Williams, and E. Blumwald. 2001. *Engineering salt-tolerant Brassica plants*: Characterization of yield and seed oil quality in transgenic plants with increased vacuolar sodium accumulation. *Proceedings of the National Academy of Sciences of the United States of America*, 98(22):12832–12836.

250. Zhang, Q.T., M. Inoue, K. Inosako, M. Irshad, K. Kondo, G.Y. Qiu, and S.P. Wang. 2008. Ameliorative effect of mulching on water use efficiency of Swiss chard and salt accumulation under saline irrigation. *Journal of Food Agriculture and Environment*, 6(3–4):480–485.

251. Zhao, G.Q., B.L. Ma, and C.Z. Ren. 2007. Growth, gas exchange, chlorophyll fluorescence, and ion content of naked oat in response to salinity. *Crop Science*, 47(1):123–131.

252. Zhu, H., G.H. Ding, K. Fang, F.G. Zhao, and P. Qin. 2006. New perspective on the mechanism of alleviating salt stress by spermidine in barley seedlings. *Plant Growth Regulation*, 49(2–3):147–156.

253. Zhu, J.K. 2001. Plant salt tolerance. *Trends in Plant Sciences*, 6:66–71.

35 Growth Responses of Pepper Plant (*Capsicum annuum* L.) in Terms of Biomass Production and Water Uptake under Deficit Irrigation System, Mild Water Stress Conditions

Sara Mardani, Mohammad Pessarakli, and Rachel McDaniel

CONTENTS

35.1 INTRODUCTION

The agricultural sector is a major consumer of water and produces more than 40% of the annual global food production by 70% of all freshwater withdrawals (Food and Organization, 2008). As drought becomes a great concern all around the world, improving the efficiency of water resources used in the agricultural sector is a priority for enhanced food security. It is suggested that food demand will increase 50% by 2030 as the world population grows beyond 7.5 billion (Winfield, 2012). A major challenge for agriculture is to provide the world's growing population with a sustainable and secure supply of safe and sufficient food that meets food preferences for an active and healthy life. Drought and flooding caused by climate change is a great threat to the sustainability of water resources and directly affect global economies, local agriculture, and the environment (Gutiérrez et al., 2014). No single definition exists for drought; however, the prevailing definition of drought is defined as low moisture conditions over a period of time (McDaniel et al., 2017). Drought is considered as the main constraint on the extension of cultivated lands and the increase of crop production in many parts of the world (Dubois, 2011). Therefore, in a scenario of population growth, climate change, and global water scarcity, a more efficient management of the water resources is a must to conserve water as much as possible and assure crop production (Raza

et al., 2012). Deficit irrigation (DI) is a strategy for farmers facing water availability challenges. It allows a crop to tolerate some degree of water deficit to reduce cost and potentially increase net income where water supplies are limited, or water costs are high (Kirda et al., 1999). In this irrigation strategy, the amount of applied water is kept below the actual irrigation requirement, and slight stress that is developed has a minimal effect on yield. It may range from moderate to extremely severe (Heng, 2002). The physiological responses of plants to water stress and their relative importance for crop productivity vary with species, cultivars, soil type, nutrients, and climate (Akıncı and Lösel, 2012; Forni et al., 2017). Some crops are relatively resistant to water stress, or they use some strategies to cope with the soil moisture limitations (López-López et al., 2018). They may avoid stress by deep rooting, allowing access to soil moisture lower in the soil profile (Chaves et al., 2002; Lisar et al., 2012). DI can lead to many advantageous if it is properly applied (Galindo et al., 2018). An improvement of fruit quality was found in several studies with little or no yield decline (George and Nissen, 1992; Gheysari et al., 2017; Patanè et al., 2011; Yang et al., 2017); however, an undesirable yield penalty was reported in some studies in parallel with improvement of water use efficiency (WUE) and quality of some crops (Chen et al., 2013; She et al., 2018; Zhang et al., 2017; Zheng et al., 2013). The decline in water availability for irrigation and the positive results obtained by DI in some fruit crops have improved the interest in developing information on DI for a variety of crops (Chai et al., 2016; Consoli et al., 2017; Yang et al., 2017).

Pepper (*Capsicum annum* L.) is an important commercial crop, cultivated for vegetable, spice, and value-added processed products (Kumar and Rai, 2005). Its production and consumption are increasing worldwide. Total world production of hot pepper has been estimated to be 14–15 million tons of fresh fruit per year (Weiss, 2002). Irrigation is essential for pepper production because pepper is considered one of the most susceptible crops to water stress in horticulture (Ferrara et al., 2011; Showemimo and Olarewaju, 2007). DI of pepper plant would be difficult to manage because reductions in crop yield, quality, and quantity can result from even brief periods of water stress on this crop (Dalla Costa and Gianquinto, 2002; Fernandez et al., 2005; Jaimez et al., 2000; Katerji et al., 1992). To optimize pepper production and profitability in drought conditions, and to ensure the most efficient use of limited water resources, there is a clear need for comprehensive information regarding crop water use of pepper grown under DI systems. Therefore, the main objective of this study was to evaluate growth responses of the pepper plant (*Capsicum annuum* L.) in terms of biomass production and root water uptake under DI system. In this experiment, a full irrigation level and three deficit irrigation levels, including 80, 60, and 40% of the plant's water requirement were applied in a completely randomized design. All the physiological responses of pepper plant to different irrigation levels were measured. Evapotranspiration and root water uptake were also estimated. Eventually, water use efficiency and crop yield were also calculated.

For the effects of deficit irrigation on pepper plant, various irrigation treatments including control (full irrigation level, fully irrigated (FI)) and three DI levels, 80, 60, and 40% of the plant's water requirement called DI_{80}, DI_{60}, and DI_{40}, respectively, were studied. For details see Mardaninejad et al. (2017) and Mardani et al. (2017).

35.2 FACTORS EVALUATED REGARDING THE GROWTH RESPONSES OF PEPPER PLANT UNDER DEFICIT IRRIGATION SYSTEM

35.2.1 Soil Moisture Changes in Different Layers

By daily measurement of the volumetric soil moisture (VSM), the changes in the VSM content in different layers of the soil were identified during the determined period. These changes were more noticeable in the first and the second layers of the soil in all treatments, especially in the DI_{60} and DI_{40} treatments. Figure 35.1 shows the differences in the soil moisture contents in all soil layers in different irrigation treatments. It also illustrates that in each treatment, the soil moisture contents in the first and the second layers of the soil sharply decreased in the day before the next irrigation. By considering these changes, it was possible to determine from which layers the root water uptake (RWU) was higher. Figure 35.1 also shows when the roots reached the fifth layer in different

irrigation treatments, as the FI, DI_{80}, and DI_{60} treatments reached the fifth layer at 108, 114, and 132 days after planting, respectively; however, the roots in the DI_{40} treatment could not penetrate into this layer.

35.2.2 Average Daily Root Water Uptake

The results showed that the average root water uptake (ARWU) was significantly different among the various treatments. Figure 35.2 shows the fluctuation patterns of the daily water uptakes in all treatments for 64 days. According to Figure 35.2, the measured ARWU rate in the DI_{80}, DI_{60}, and DI_{40} treatments was less than that in the control treatment. Receiving less irrigation water in these treatments compared with the control was probably the main reason for this approach.

As shown in Figure 35.2, the maximum and the minimum ARWU was found in the FI and DI_{40} treatments as 8.5 and 2.7 mm/day, respectively. The ARWU in the DI_{80}, DI_{60}, and DI_{40} treatments were reduced by 17.08, 48.72, and 68.25%, respectively. Furthermore, only in the DI_{80} treatment, the reduced rate of water uptake was less than that of the plant's applied water. This is probably due to the plant regulatory mechanism that it uses under water stress, where it adjusts its roots' RWU with environmental conditions (Aliyari, 2010). The results also showed that the ARWU was significantly different in each layer in different treatments (Table 35.1). The results also illustrated that in all treatments, the upper layers of the soil could play a big role in RWU. However, the role of the lower layers of the soil in RWU was not noticeable in the DI_{60} and DI_{40} treatments; therefore, it indicated that although applied water in each treatment was different, the highest RWU rates were found in the upper layers of the soil in all treatments.

Figure 35.3 also shows that the proportion of daily ARWU in each layer to the total ARWU in all layers was dependent on the root length densities. The maximum RWU proportions of the upper layers of the soil were found in the DI_{40} and DI_{60} treatments as 78.9 and 74.7%, respectively. However, the minimum of the RWU values were found in the FI and DI_{80} treatments as 65.6 and 67.2%, respectively. Therefore, the results indicated that although applied water in the DI_{40} and DI_{60} treatments were less than that in the FI and DI_{80} treatments, the pepper plant could use its maximum capacity to uptake the soil moisture from the two upper layers of the soil in deficit irrigation. It means when the moisture was sufficient in the upper layers of the soil, the plant roots could cope with deficit irrigation via more water uptake from the surface layers of the soil.

35.2.3 Average Daily Evapotranspiration (ET$_C$)

The daily values of ET$_C$ in different treatments were identified during the 64 days by daily measurement of ET$_C$ (Figure 35.4). The results showed that the maximum and the minimum ET$_C$ were found in the FI and DI_{40} treatments by 6.8 and 2.3 mm/day, respectively. The summations of ET$_C$ across the determined period in the FI, DI_{80}, DI_{60}, and DI_{40} treatments

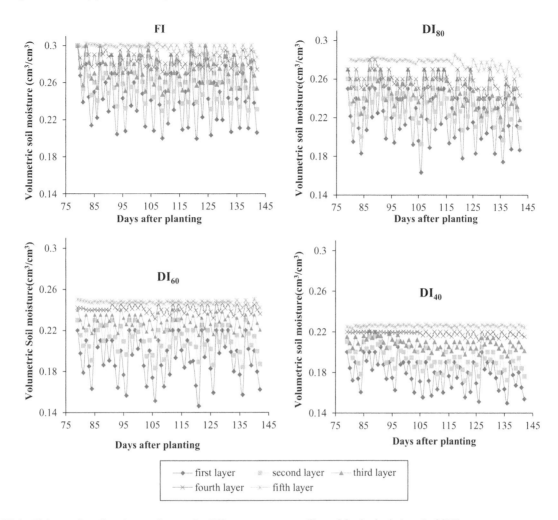

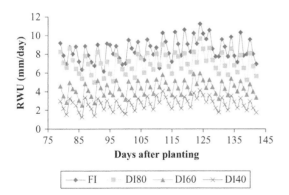

FIGURE 35.1 Volumetric soil moisture changes in different treatments. (From Mardaninejad et al., 2017.)

FIGURE 35.2 The daily RWU changes in different treatments. (From Mardaninejad et al., 2017.)

35.2.4 Physiological Responses of Plant to Deficit Irrigation

The details of the statistical analysis of the measured parameters such as the number of fruits per plant, fruit fresh/dry weight, root fresh/dry weight, shoot fresh/dry weight, and root zone volume in different irrigation treatments are shown in Table 35.2. Figure 35.5 and Table 35.2 show that the measured parameters were significantly different in different treatments. The results indicated that the maximum and the minimum of all measured parameters were found in the FI and DI_{40} treatments, respectively. The shoot dry weight in the DI_{80}, DI_{60}, and DI_{40} treatments also decreased by 16.9, 33.9, and 53.1%, respectively. Furthermore, the root dry weight decreased by 31, 47.8, and 62.6%, respectively in the DI_{80}, DI_{60}, and DI_{40} treatments. The physiological responses of pepper plant to deficit irrigation showed the sensitivity of pepper plant to water scarcity.

The details of the statistical analysis of the root length and density (RLD) are shown in Table 35.3. The results showed that the RLD in the first and the second depths of the soil column had significant differences among all treatments. The

were 401.1, 344.2, and 191.7 mm, respectively. The results showed that the ET_C values in the DI_{80}, DI_{60}, and DI_{40} treatments decreased by 14.2, 37.4, and 52.2%, respectively. It was noticeable that in the deficit-irrigated treatments, the reduced rates of ET_C were less than those of the plant's applied water, especially in the DI_{80} treatment.

TABLE 35.1

Variance Analysis of the Daily ARWU in All Layers

Variance Sources	df	First Layer	Second Layer	Third Layer	Fourth Layer	Fifth Layer
Treatments	3	0.037*	0.012*	0.0033*	0.0077*	0.0055*
Prediction error	8	0.00007	0.00006	0.00007	0.00002	0.00004
CV	–	0.89	0.95	1.65	0.9	1.2
P	–	<0.001	<0. 001	<0. 001	<0.001	<0.001

Source: Mardaninejad et al. (2017).
*Significant at the 0.05 probability level.

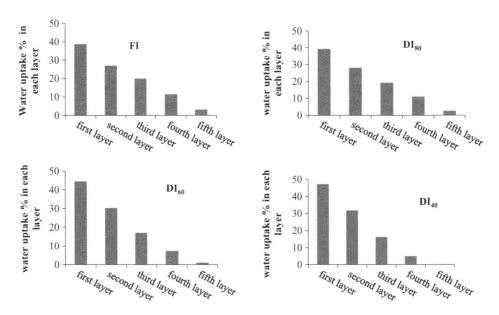

FIGURE 35.3 The ratio of daily RWU in each layer to the total RWU under each treatment. (From Mardaninejad et al., 2017.)

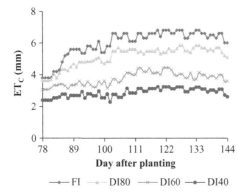

FIGURE 35.4 The daily values of ET_C in different treatments. (From Mardani et al., 2017.)

results also showed that in all treatments, the highest RLD was found in the upper depths of the soil column, and that can be the main reason why the highest ARWU rate occurred in the surface layers of the soil (Figure 35.3).

35.2.5 EFFECT OF WATER STRESS ON WATER USE EFFICIENCY AND CROP YIELD

The details of the statistical analysis of water use efficiency and crop yield in different irrigation treatments are shown in Table 35.2. The results showed that WUE and crop yield were not significantly different in the FI and DI_{80} treatments; however, they were significantly different between the other irrigation treatments (Figure 35.6). Increase in irrigation deficit resulted in a decrease in WUE. The highest values of WUE were found in the FI and DI_{80} treatments, while the lowest values were recorded for the DI_{60} and DI_{40} treatments. These results showed that the total dry mass of the fruit was significantly affected by DI. The results also showed that the maximum and the minimum WUE and crop yield were found in the FI and DI_{40} treatments by 0.50 and 0.38 L/g for WUE, and 16.7 and 5 tons/ha for crop yield, respectively. Furthermore, the results showed that WUE and crop yield reduced in the DI_{80}, DI_{60}, and DI_{40} treatments by 1, 11.2, and 23.3%, respectively, for WUE, and 29.4, 52.7, and 69.5%, respectively, for crop yield.

TABLE 35.2

Variance Analysis of the Plant Physiological Responses to Different Irrigation Treatments

Variance Sources	Treatments	Prediction Error	CV	P
df	3	8	–	–
No. of fruits/plant	5830*	393.433	21.81	< 0.001
Fruit fresh weigh/plant (g)	27973.86*	749.16	11.70	< 0.001
Fruit dry weigh/plant (g)	164.85*	18.57	19.17	< 0.001
Shoot fresh weigh/plant (g)	62542*	656	4.37	< 0.001
Shoot dry weigh/plant (g)	3935.2*	92	8.13	< 0. 001
Root fresh weigh/plant (g)	1250**	7.78	1.69	< 0.001
Root dry weigh/plant (g)	322**	0.512	2.86	< 0.001
Root zone volume (cm^3)	1089*	5.19	1.52	< 0.001
Water use efficiency (L/g)	0.0026*	0.0015	1.21	< 0.001
Crop yield (tons/ha)	24.55*	1.90	19.91	< 0.001

Source: Mardani et al. (2017).

*, **Significant at the 0.05 and 0.01 probability level, respectively.

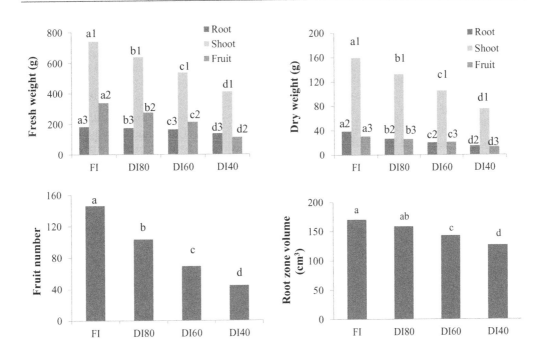

FIGURE 35.5   Physiological responses of plant to different irrigation treatments. (From Mardani et al., 2017.)

35.3 OVERALL EVALUATION OF THE FACTORS AFFECTING THE GROWTH RESPONSES OF PEPPER PLANT UNDER DEFICIT IRRIGATION SYSTEM

Asseng et al. (1998), Fereres and Soriano (2006), and Khan et al. (2009) found that total root growth reduced under deficit irrigation. The results of the present research also indicated that in the FI and DI_{80} treatments, the roots could reach the fifth layer, while in the DI_{40} treatment the roots could not penetrate as far as this layer; therefore, the RWU in this layer was zero. Asseng et al. (1998), Aliyari (2010), and Dathe et al.

(2014) reported that the maximum and the minimum percentages of water uptake were found in the full irrigation treatment and the treatment that experienced the highest range of deficit irrigation. In the present research, the highest and the lowest ranges of water uptakes were also found in the FI and DI_{40} treatments as 8.5 and 2.7 mm/day, respectively. Additionally, Aliyari (2010) found that in all treatments, the reduced rates of water uptake were less than the reduced rate of the plant's applied water; however, in the present research only in the DI_{80} treatment, the reduced rate was less than that of the plant's applied water. Luo et al. (2003), Dathe et al. (2014), and Verma et al. (2014) reported that the maximum percentages of water

TABLE 35.3

Variance Analysis of The Root Length And Density (RLD) In Different Depths

Variance Sources	df	RLD (cm/cm³)		
		0–17	17–34	34–51
Treatments	3	0.043*	0.0089*	0.0068*
Prediction error	8	0.0011	0.00021	0.00061
CV	–	3.87	2.86	12.52
P	–	<0.001	<0.001	<0.001

Source: Mardani et al. (2017).
*Significant at the 0.05 probability level.

(Fereres and Soriano, 2006; Fernandez et al., 2005; Moreno et al., 2003). The reduction in crop yield of pepper under deficit irrigation might be due to the reduction in fruit size and numbers (Fernandez et al., 2005). Reduction in pepper fruit size and numbers appears as the controlling factor for fruit yield. In the present study, deficit irrigation significantly reduced crop yield in terms of the fresh mass of fruit per plant. The fully irrigated treatment resulted in the highest yield because increasing soil water content led to increasing plant height and number of branches, resulting in an increase in the number of fruits and total yield. Similar results were reported by Antony and Singandhupe (2004), Dorji et al. (2005), Ismail (2010), Owusu-Sekyere et al. (2010), and Adu et al. (2018). It is indisputable that for high yields, adequate water supply and relatively moist soils are required during the entire growing season. Reduction in water supply during the growing period, in general, has an adverse effect on yield. The greatest reduction in yield occurs when there is a continuous water shortage until the time of the first picking (Dalla Costa and Gianquinto, 2002; Kang et al., 2001; Owusu-Sekyere et al., 2010; Sezen et al., 2006). In the present study, by considering the fact that the total dry mass of pepper plant was markedly affected by DI, DI led to decreased water use efficiency and crop yield. This indicates that water movement into fruit may have decreased with the progressive development of water deficit together with affecting the translocation of dry matter into the fruit and resulted in a decrease in mass production per unit of water, which led to a lower WUE and crop yield. Interestingly, in regards to WUE and crop yield, there were no significant differences found between the FI and DI_{80} treatments. The results showed that WUE and crop yield were not affected by moderate water stress (DI_{80}) imposed on pepper plant investigated in the present study. Whereas, under severe water stress (DI_{60} and DI_{40}), WUE and crop yield significantly decreased. Similar findings were reported using a variety of crop plants such as millet, barley, sorghum, and wheat (Boutraa et al., 2010; Ibrahim, 1997).

uptake were found in the top layers of the soil since most of the root systems were distributed in these sections. Zakerinia et al. (2008) and Besharat et al. (2010) reported that not only the maximum root water uptake was found in the top layers of the soil, but also the maximum available water was observed in these layers. The present research also showed although the volume of applied water in each treatment was different, in all treatments the highest rate of the RWU was found in the upper layers of the soil. Additionally, in all treatments, the maximum RWU was found in the first and the second layers of the soil. Furthermore, the maximum RWU proportions of the upper layers of the soil were found in the DI_{40}, DI_{60}, DI_{80}, and FI treatments as 78.9, 74.7, 67.2, and 65.6%, respectively.

Vegetative growth of pepper was closely linked to the amount of water applied. DI throughout the determined period decreased the fruit number, fruit fresh/dry weight, root fresh/dry weight, shoot fresh/dry weight, and root zone volume in different irrigation treatments. The results indicated that the pepper plant's physiological responses to DI were completely negative. Similar results were obtained by Kashiwagi et al. (2006), Gonzalez-Dugo et al. (2007), and Ismail (2010), whom all reported decreases in yield under water stress conditions. Additionally, in the present study, the evapotranspiration values (ET_C) in the DI_{80}, DI_{60}, and DI_{40} treatments substantially decreased by DI, indicating that the effect of DI on the ET_C values in different irrigation treatments was noticeable. The present findings on the ET_C values in DI conditions are in agreement with other studies

35.4 CONCLUSIONS

The results showed a great sensitivity of this horticultural crop to deficit irrigation. Water stress negatively affected most of the studied physiological parameters. The maximum

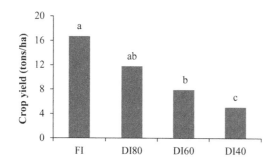
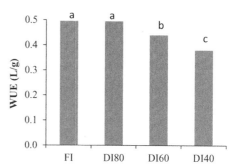

FIGURE 35.6 Water use efficiency and crop yield in different irrigation treatments. (From Mardani et al., 2017.)

and the minimum ARWU, evapotranspiration (ET_C), fruit number, root zone volume, root length density, and crop yield were found in the FI and DI_{40} treatments. Also, the ARWU, ET_C, and crop yield in the DI_{80}, DI_{60}, and DI_{40} treatments were reduced compared with those of the control treatment (FI), indicating that in the DI_{80} treatment, the reduced rate of uptake was less than the reduced rate of the plant's applied water. The maximum RWU was also found in the first and second layers of the soil. WUE showed no significant difference in the FI and DI_{80} treatments. In addition, there was no meaningful difference among the FI and DI_{80} treatments in crop yield. Based on the results of this study, it can be concluded that 20% DI had no significant reduction on the yield of pepper plant, but above this threshold, there was an adverse effect on the plant's growth and yield. Therefore, in the regions with limited water resources, the rate of plant's applied water can be decreased by 20%.

REFERENCES

Adu, M. O., D. O. Yawson, F. A. Armah, P. A. Asare, and K. A. Frimpong. 2018. Meta-analysis of crop yields of full, deficit, and partial root-zone drying irrigation. *Agricultural Water Management* 197:79–90.

Akıncı, Ş., and D. M. Lösel. 2012. Plant water-stress response mechanisms. In *Water stress*. InTech.

Aliyari, H. 2010. Effects of water stress on root distribution and soil water uptake by beans plant. Unpublished MSc Thesis, Faculty of Agriculture, Shahre-Kord University, Shahre-Kord, Iran, p. 76.

Antony, E., and R. Singandhupe. 2004. Impact of drip and surface irrigation on growth, yield and WUE of capsicum (Capsicum annum L.). *Agricultural Water Management* 65(2):121–132.

Asseng, S., J. Ritchie, A. Smucker, and M. Robertson. 1998. Root growth and water uptake during water deficit and recovering in wheat. *Plant and Soil* 201(2):265–273.

Besharat, S., A. H. Nazemi, and A. A. Sadraddini. 2010. Parametric modeling of root length density and root water uptake in unsaturated soil. *Turkish Journal of Agriculture and Forestry* 34(5):439–449.

Boutraa, T., A. Akhkha, A. A. Al-Shoaibi, and A. M. Alhejeli. 2010. Effect of water stress on growth and water use efficiency (WUE) of some wheat cultivars (Triticum durum) grown in Saudi Arabia. *Journal of Taibah University for Science* 3(1):39–48.

Chai, Q., Y. Gan, C. Zhao, H.-L. Xu, R. M. Waskom, Y. Niu, and K. H. Siddique. 2016. Regulated deficit irrigation for crop production under drought stress. A review. *Agronomy for Sustainable Development* 36(1):3.

Chaves, M. M., J. S. Pereira, J. Maroco, M. L. Rodrigues, C. P. P. Ricardo, M. L. Osório, I. Carvalho, T. Faria, and C. Pinheiro. 2002. How plants cope with water stress in the field? Photosynthesis and growth. *Annals of Botany* 89(7):907–916.

Chen, J., S. Kang, T. Du, R. Qiu, P. Guo, and R. Chen. 2013. Quantitative response of greenhouse tomato yield and quality to water deficit at different growth stages. *Agricultural Water Management* 129:152–162.

Consoli, S., F. Stagno, D. Vanella, J. Boaga, G. Cassiani, and G. Roccuzzo. 2017. Partial root-zone drying irrigation in orange orchards: Effects on water use and crop production characteristics. *European Journal of Agronomy* 82:190–202.

Dalla Costa, L., and G. Gianquinto. 2002. Water stress and watertable depth influence yield, water use efficiency, and nitrogen recovery in bell pepper: Lysimeter studies. *Australian Journal of Agricultural Research* 53(2):201–210.

Dathe, A., D. Fleisher, D. Timlin, J. Fisher, and V. Reddy. 2014. Modeling potato root growth and water uptake under water stress conditions. *Agricultural and Forest Meteorology* 194:37–49.

Dorji, K., M. Behboudian, and J. Zegbe-Dominguez. 2005. Water relations, growth, yield, and fruit quality of hot pepper under deficit irrigation and partial rootzone drying. *Scientia Horticulturae* 104(2):137–149.

Dubois, O. 2011. *The State of the World's Land and Water Resources for Food and Agriculture: Managing Systems at Risk*. London: Earthscan.

Fereres, E., and M. A. Soriano. 2006. Deficit irrigation for reducing agricultural water use. *Journal of Experimental Botany* 58(2):147–159.

Fernandez, M., M. Gallardo, S. Bonachela, F. Orgaz, R. Thompson, and E. Fereres. 2005. Water use and production of a greenhouse pepper crop under optimum and limited water supply. *The Journal of Horticultural Science and Biotechnology* 80(1):87–96.

Ferrara, A., S. Lovelli, T. Di Tommaso, and M. Perniola. 2011. Flowering, growth and fruit setting in greenhouse bell pepper under water stress. *Journal of Agronomy* 10(1):12–19.

Food and A. Organization. 2008. *Climate Change and Food Security: A Framework Document*. Food and Agriculture Organization of the United Nations Rome.

Forni, C., D. Duca, and B. R. Glick. 2017. Mechanisms of plant response to salt and drought stress and their alteration by rhizobacteria. *Plant and Soil* 410(1–2):335–356.

Galindo, A., J. Collado-González, I. Griñán, M. Corell, A. Centeno, M. Martín-Palomo, I. Girón, P. Rodríguez, Z. Cruz, and H. Memmi. 2018. Deficit irrigation and emerging fruit crops as a strategy to save water in Mediterranean semiarid agrosystems. *Agricultural Water Management* 202:311–324.

George, A., and R. Nissen. 1992. Effects of water stress, nitrogen and paclobutrazol on flowering, yield and fruit quality of the low-chill peach cultivar,'Flordaprince.' *Scientia Horticulturae* 49(3–4):197–209.

Gheysari, M., S.-H. Sadeghi, H. W. Loescher, S. Amiri, M. J. Zareian, M. M. Majidi, P. Asgarinia, and J. O. Payero. 2017. Comparison of deficit irrigation management strategies on root, plant growth and biomass productivity of silage maize. *Agricultural Water Management* 182:126–138.

Gonzalez-Dugo, V., F. Orgaz, and E. Fereres. 2007. Responses of pepper to deficit irrigation for paprika production. *Scientia Horticulturae* 114(2):77–82.

Gutiérrez, A. P. A., N. L. Engle, E. De Nys, C. Molejón, and E. S. Martins. 2014. Drought preparedness in Brazil. *Weather and Climate Extremes* 3:95–106.

Heng, L. K. 2002. *Deficit irrigation practices*. number 22. Food & Agriculture Org.

Ibrahim, Y. M. 1997. A practical approach for water use efficiency in arid areas.*Qatar University Science Journal*.

Ismail, S. M. 2010. Influence of deficit irrigation on water use efficiency and bird pepper production (*Capsicum annuum* L.) *Meteorology, Environment and Arid Land Agriculture Sciences* 21(2).

Jaimez, R., O. Vielma, F. Rada, and C. García-Núñez. 2000. Effects of water deficit on the dynamics of flowering and fruit production in Capsicum chinense Jacq in a tropical semiarid region of Venezuela. *Journal of Agronomy and Crop Science* 185(2):113–119.

Kang, S., L. Zhang, X. Hu, Z. Li, and P. Jerie. 2001. An improved water use efficiency for hot pepper grown under controlled alternate drip irrigation on partial roots. *Scientia Horticulturae* 89(4):257–267.

Kashiwagi, J., L. Krishnamurthy, J. Crouch, and R. Serraj. 2006. Variability of root length density and its contributions to seed yield in chickpea (*Cicer arietinum* L.) under terminal drought stress. *Field Crops Research* 95(2–3):171–181.

Katerji, N., M. Mastrorilli, and A. Hamdy. 1992. Effects of water stress at different growth stages on pepper yield. In *International Symposium on Irrigation of Horticultural Crops 335*.

Khan, M., A. Farooque, M. Haque, M. Rahim, and M. Hoque. 2009. Effects of water stress at various growth stages on the physio-morphological characters and yield in Chili pepper. *Bangladesh Journal of Agriculture* 33:353–362.

Kirda, C., C. Hera, P. Moutonnet, and D. Nielsen. 1999. *Crop Yield Response to Deficit Irrigation: Report of an FAO/IAEA Co-ordinated Research Program by Using Nuclear Techniques: Executed by the Soil and Water Management & Crop Nutrition Section of the Joint FAO/IAEA Division of Nuclear Techniques in Food and Agriculture*. Springer Science & Business Media.

Kumar, S., and M. Rai. 2005. Chile in India. *Chile Pepper Institute Newsletter* 22:1–3.

Lisar, S. Y., R. Motafakkerazad, M. M. Hossain, and I. M. Rahman. 2012. Water stress in plants: Causes, effects and responses. In *Water stress*. London: InTech.

López-López, M., M. Espadador, L. Testi, I. J. Lorite, F. Orgaz, and E. Fereres. 2018. Water use of irrigated almond trees when subjected to water deficits. *Agricultural Water Management* 195:84–93.

Luo, Y., Z. Ouyang, G. Yuan, D. Tang, and X. Xie. 2003. Evaluation of macroscopic root water uptake models using lysimeter data. *Transactions of the ASAE* 46(3):625.

Mardani, S., S. H. Tabatabaei, M. Pessarakli, and H. Zareabyaneh. 2017. Physiological responses of pepper plant (*Capsicum annuum* L.) to drought stress. *Journal of Plant Nutrition* 40(10):1453–1464.

Mardaninejad, S., H. Zareabyaneh, S. H. Tabatabaei, M. Pessarakli, and A. Mohamadkhani. 2017. Root water uptake of pepper plants (*Capsicum annuum* L.) under deficit irrigation system. *Journal of Plant Nutrition* 40(11):1569–1579.

McDaniel, R. L., C. Munster, and J. T. Cothren. 2017. Crop and location specific agricultural drought quantification: Part I. Method development. *Transactions of the ASABE* 60(3):721–728.

Moreno, M., F. Ribas, A. Moreno, and M. Cabello. 2003. Physiological response of a pepper (Capsicum annuum L.) crop to different trickle irrigation rates. *Spanish Journal of Agricultural Research* 1(2):65–74.

Owusu-Sekyere, J., P. Asante, and P. Osei-Bonsu. 2010. Water requirement, deficit irrigation and crop coefficient of hot pepper (*Capsicum frutescens*) using irrigation interval of four (4) days. *Journal of Agricultural and Biological Science* 5(5):72–78.

Patanè, C., S. Tringali, and O. Sortino. 2011. Effects of deficit irrigation on biomass, yield, water productivity and fruit quality of processing tomato under semi-arid Mediterranean climate conditions. *Scientia Horticulturae* 129(4):590–596.

Raza, A., J. Friedel, and G. Bodner. 2012. Improving water use efficiency for sustainable agriculture. In *Agroecology and strategies for climate change*, 167–211. Springer.

Sezen, S. M., A. Yazar, and S. Eker. 2006. Effect of drip irrigation regimes on yield and quality of field grown bell pepper. *Agricultural Water Management* 81(1–2):115–131.

She, D., M. Wang, N. Liu, G. Shao, and S. Deng. 2018. Fruit quality and yield of tomato as influenced by rain shelters and deficit irrigation.

Showemimo, F., and J. Olarewaju. 2007. Drought tolerance indices in sweet pepper (*Capsicum annuum* L.). *Int J Plant Breed Genet* 1(1):29–33.

Verma, P., S. P. Loheide II, D. Eamus, and E. Daly. 2014. Root water compensation sustains transpiration rates in an Australian woodland. *Advances in Water Resources* 74:91–101.

Weiss, E. 2002. *World Production and Trade*. Wallingford, UK: CABI Publishing, CAB International.

Winfield, I. J. 2012. FAO STATISTICAL YEARBOOK 2012: WORLD FOOD AND AGRICULTURE-Edited by A. Prakash and M. Stigler. *Journal of Fish Biology* 81(6):2095–2096.

Yang, H., T. Du, R. Qiu, J. Chen, F. Wang, Y. Li, C. Wang, L. Gao, and S. Kang. 2017. Improved water use efficiency and fruit quality of greenhouse crops under regulated deficit irrigation in northwest China. *Agricultural Water Management* 179:193–204.

Zakerinia, M., T. Sohrabi, M. Shahabifar, F. Abbasi, and M. Neishabouri. 2008. Effects of water stress on water adsorption by different parts of the root. *Journal of Agriculture and Natural Resources* 15:166–178.

Zhang, H., Y. Xiong, G. Huang, X. Xu, and Q. Huang. 2017. Effects of water stress on processing tomatoes yield, quality and water use efficiency with plastic mulched drip irrigation in sandy soil of the Hetao Irrigation District. *Agricultural Water Management* 179:205–214.

Zheng, J., G. Huang, D. Jia, J. Wang, M. Mota, L. S. Pereira, Q. Huang, X. Xu, and H. Liu. 2013. Responses of drip irrigated tomato (*Solanum lycopersicum* L.) yield, quality and water productivity to various soil matric potential thresholds in an arid region of Northwest China. *Agricultural Water Management* 129:181–193.

36 Effects of Salinity Stress on Tomato Plants and the Possibility of Its Mitigation

Maryam Mozafarian Meimandi, Noémi Kappel, and Mohammad Pessarakli

CONTENTS

36.1 INTRODUCTION

Tomato (*Solanum lycopersicum* L.) belongs to the *Solanaceae* family, and it is one of the most widely produced vegetable crops in the world. Tomato fruit is used in fresh consumption, cooked, and for processing (canning, juice, pulp, paste, and sauces). Several studies have shown strong inverse correlations between tomato consumption and the risk of certain types of cancer (e.g., prostate cancer/enlargement), cardiovascular diseases, and age-related muscular degeneration (reviewed (by Dorais et al., 2008)). This horticulture crop is a perennial diploid dicotyledon ($2n = 24$), but it is grown as an annual plant in temperate regions as plants and fruit are injured when they are exposed to non-freezing temperatures below 12°C. There are large numbers of tomato cultivars for open field and greenhouse cultivations. These include five major fruit types: classic round tomatoes (the most popular varieties), cherry and cocktail tomatoes (smaller than the classic tomatoes), plum and baby plum tomatoes, beefsteak tomatoes, and vine or truss tomatoes. The global production of tomato totaled 170.8 million tons in 2017. China, India, and the United States are the countries with the highest production of tomatoes in the world. Because they are easily grown, this vegetable crop is used as a model plant in many types of research, including biochemical, molecular, and genetic studies, as well as physiological and seed germination experiments.

The irrational use of fertilizers and agrochemicals, overpumping of groundwater for irrigation, consequent sea-water intrusion into fresh aquifers, and crop intensification cause excessive sodium chloride (NaCl) concentration in both soil and water. Increasing salinity in agricultural lands reduced up to 50% of cultivable lands (Mahajan and Tuteja, 2005).

Tomato plants are considered as moderately sensitive to salt stress with a maximum soil electrical conductivity level of 2.5 dS m^{-1} without a significant yield loss (Jones, 1999). In general, salinity can reduce plant growth and yield by reduction of leaf area, stomata closure, and through its impact on plant water relations, while the level of fructose, glucose, total soluble solids, amino acids, and organic acids in fruits can increase (Wu and Kubota, 2008).

36.2 PHYSIOLOGICAL AND BIOCHEMICAL EFFECTS OF SALT STRESS ON TOMATOES

Seed germination is the most sensitive developmental stage to salt stress. Exposure to even relatively low concentrations of NaCl can cause a reduction in tomato seed germination depending on the varieties. Singh et al. (2012) tried to find the response of 55 tomato genotypes to increased salinity during seed germination. Under increasing NaCl concentrations (1–3%), the germination of tomato seeds was significantly reduced and the time necessary to complete germination was lengthened. Moreover, root/shoot dry weight ratio and Na$^+$ content increased, while K$^+$ bioconcentration decreased.

It was shown that the growth (leaf area, biomass), as well as tomato fruit yield (cv. Floradade), was reduced after using saline drainage water (electrical conductivity of 4.2–4.8 dS m^{-1}) to irrigate field-grown plants (Malash et al., 2008). Salinity also caused a reduction in leaf growth rate due to an adverse effect on the cellular turgor pressure, diminished photosynthetic activity, and activation of metabolic signaling between stress perception and adaptation (Munns, 2002). Increasing NaCl concentration reduced yield, fruit number, and dry weight in tomato plants grown in rockwool (Adams, 1991). The number of trusses per plant and the number of flowers per truss were influenced by salinity depending on temperature, tomato cultivar, the severity of salinity, and duration of salinization (Heuvelink, 2018). Also, small fruit cultivars would be less affected by salinity than larger fruit size cultivars. So, under highly or moderately saline conditions, smaller fruit cultivars are suggested for cultivation.

Salinity in spring, summer, or during the day can cause a higher yield reduction than during the night or in fall. This is because of the higher temperature and illumination, lower humidity, lower water potential in plant, and higher transpiration rate in spring, summer, and day time compared to fall and night (reviewed by Cuartero and Fernández-Muñoz, 1998). Moreover, the salt sensitivity of tomato is different under soil and hydroponic conditions because there is a delay in the build-up in salinity in the soil. The sensitivity of tomato plants cultivated in a greenhouse is lower than in the open field due to a higher atmospheric humidity under controlled environment and protected cultivation (Huevelink and Dorais, 2005).

36.2.1 Effect of Salinity on Fruit Quality

Blossom end rot (BER), as a physiological disorder related to environmental factors including salinity, causes yield losses of up to 50% (Saure, 2001). Salinity results in the increased occurrence of BER in tomatoes due to decreasing absorption and translocation of Ca^{2+} to fruit (Johnson et al., 2003). However, in some reports, salinity had no influence on BER and Ca^{2+} concentration in tomatoes and there was not a linear relationship between salinity stress and BER occurrence (Chrétien et al., 2000).

The quality of vegetables is defined by multiple factors such as taste, visual appearance, and aroma as well as factors that affect quality indirectly such as chemical composition, the content of bioactive compounds, absence of agrochemicals, and anti-nutrient compounds (Kyriacou and Rouphael, 2018). Flavor (defined as bitterness, sourness (acidity), sweetness, and volatile compounds contents) is related to genotype, pre-harvesting, and post-harvesting factors. Salinity has an unfavorable effect on the vegetable quality and yield, while application of a low level of NaCl during a specific phenological stage may have beneficial effects on the nutritional value and chemical composition of tomatoes (Dorais et al., 2008; Schwarz et al., 2009; Segura et al., 2009).

The most common flavor index in tomato is total soluble solids (TSS) associated directly with sugar and organic acid concentrations in the juice. About 95% of sugar content in tomatoes is glucose and fructose, and a few percent is sucrose. Fructose is almost twice as sweet as glucose (Wu and Kubota, 2008). Increasing sugar content in vegetable fruits under saline conditions is a very common phenomenon (Slama et al., 2015). Wu and Kubota (2008) reported that the TSS of fruit increased with increasing EC values. Similarly, Claussen et al. (2006) observed that glucose and fructose contents in tomato fruit were elevated when nutrient concentration increased. Moreover, it was found that salinity improved earliness, fruit color, and sugar content due to upregulation genes related to ripening (1-aminocyclopropane-1-carboxylate oxidase), carotenoid biosynthesis (phytoene synthase 1), and gluconeogenesis (phosphoenolpyruvate carboxykinase) (Saito et al., 2008). Increasing vitamin C, total phenols, and carotenoids (Krauss et al., 2006); TSS and lycopene content (Wu et

al., 2004) were reported in several pieces of research depending on the growth conditions and fruit development stage. On the other hand, negative or no effects of salinity stress on the tomato fruit quality have been reported by Dorais et al. (2008) and Choudhary et al. (2010).

Lycopene level, as a powerful antioxidant, was enhanced under raising EC (Wu et al., 2004; Krauss et al., 2006; Fanasca et al., 2007). High EC treatment increased lycopene concentration in tomatoes at different stages (pink, light red, and red) due to an upregulation of genes encoding enzymes involved in the key steps of lycopene biosynthesis (Dorais et al., 2008). Similarly, Wu et al. (2004) reported that increasing salinity from 2.4 to 4.5 dS m^{-1} increased lycopene content (by 34–85%). Meanwhile, Dorais et al. (2000) found that an increase in the salinity level did not significantly change lycopene content, but caused a decrease in β-carotene and vitamin C contents and an increase in lutein concentration. Rouphael and Kyriacou (2018) concluded that management of nutrient solution EC is an effective tool in the improvement of tomato fruit quality in terms of chemical composition and bioactive compounds content depending on cultivars and growing conditions. However, the effect of salt stress on plants depends on the concentration and time of exposure to salt, plant genotypes, and environmental factors.

36.3 AMELIORATION OF SALT STRESS IN TOMATOES

The attempts to improve tomato's salt tolerance through plant breeding have been unsuccessful due to the multigenic origin of the adaptive responses. Therefore, some researchers focus on physiological, biochemical, and metabolic aspects to increase tomato resistance against salt stress.

36.3.1 Modified Mineral Nutrition for Increasing Salt Tolerance in Tomatoes

Several studies reported that intensive nutrition with selected elements could decrease the deleterious effects of salt stress in tomatoes. Parvin et al. (2015) investigated the mitigation of salt stress in tomatoes using exogenous Ca^{2+} application and reported that Ca^{2+} supplementation significantly reduced adverse effects of salinity on plant biomass, morphology, physiology, and fruit production. The Ca^{2+} supply can restrict the entry of Na^+ ions into the plant cell (Hussain et al., 2010); control ion homeostasis pathways (Yokoi et al., 2002), and cause amelioration of Na^+ toxicity in plants. Moreover, Ca^{2+} supplementation improved tomato growth and fruit yield by the maintenance of the membrane integrity and increased foliar K^+, Ca^{2+}, and N^+ level at high NaCl concentrations (Tuna et al., 2007).

Potassium (K) as a macronutrient is an inorganic osmolyte that plays an important ameliorative role under stress conditions through enzyme activation, regulation of osmotic pressure, maintenance of the membrane potential and turgor, opening and closing of stomata, and tropism (Lester et al., 2005; Abogadallah et al., 2010). According to Amjad et al. (2016), K$^+$ is an effective ameliorating agent against salt-induced oxidative damage. Its elevated concentration in the salt-exposed plants can prevent lipid peroxidation, modulate the activity of antioxidant enzymes, and thus exert a beneficial effect on plant growth. A high NaCl concentration (60 mM) significantly decreased leaf chlorophyll, plant growth, plant water use, and membrane permeability and induced P and K deficiencies in tomatoes, while foliar application of P and K improved these parameters (Kaya et al., 2001).

Beneficial effects of silicon (Si) in enhancing the tolerance of plants to salt stress have been described by numerous researchers. Silicon (1 mM) alleviated the harmful effects of salinity on tomato seed germination in terms of germination percentage, germination rate, and mean germination time (Haghighi et al., 2012). According to the results of Yunus and Zari (2017), application of Si significantly decreased Na$^+$ content and increased K$^+$, Ca^{2+}, and Mg^{2+} contents in the leaves of tomatoes growing under salt stress conditions. Silicon also increased the fruit yield and pH of tomato juice at high concentrations of NaCl (Korkmaz et al., 2018). Improving the photochemical efficiency of PSII and enhanced activities of antioxidant enzymes in tomato leaves after application of Si under salt stress was observed (Al-Aghabary et al., 2004). Recently, Li et al. (2015) reported that Si supplementation decreased Na$^+$ and Cl$^-$ concentrations and enhanced antioxidant defense system in tomato roots, improved the root growth, hydraulic conductance and water status in the leaves, photosynthetic rate and shoot growth and thus salt tolerance.

The mechanism by which Si plays a role in the mitigation of salt damages in plants, suggested by Zhu and Gong (2014), include improvement of photosynthesis and reduced transpiration rate, reduced ion toxicity and salt-induced oxidative damage, maintenance of the water content and regulation of the biosynthesis of compatible solutes and plant hormones. A similar mechanism of the beneficial effect of Si has also been reported under drought stress conditions (Maghsoudi et al., 2016).

Selenium (Se) is an essential element for humans and animals and beneficial even for plants. It can mitigate the harmful effects of biotic and abiotic stresses in plants (Haghighi et al., 2016a; Mozafariyan et al., 2017). Haghighi et al. (2016b) reported that selenium had a beneficial effect on growth and the photosynthetic attributes of cucumbers grown under greenhouse conditions. Mozafariyan et al. (2017) studied the effects of selenium on some morphological and physiological traits of tomato plants grown under hydroponic conditions. These investigators found Si alleviated the adverse effects of salinity on the plants by showing beneficial effects on all the studied parameters under salt stress. Hawrylak-Nowak (2018) reviewed mechanisms of Se in plant tolerance against stress: improvement of photosynthesis, antioxidant defense, and higher uptake of water and nutrients. Diao et al. (2014) found that Se applications reduced hydrogen peroxide (H$_2$O$_2$) accumulation and the level of lipid peroxidation, increased the activity of antioxidant enzymes, and enhanced photosynthetic capacity which are important mechanisms to inhibit damages under salinity stress in tomatoes. Moreover, the enrichment of tomato growth medium with Se under salt stress increased biomass and improved cell membrane stability (Mozafariyan et al., 2016).

For more information about the role of nutrients on abiotic stresses see Chapter 46.

36.3.2 Enhanced Tomato Salt Tolerance by Grafting

The improvement of salinity tolerance or resistance of tomatoes via breeding programs has been limited due to its genetic and physiological complexity (Cuartero et al., 2006). Vegetable grafting has been considered as a rapid alternative to the slow breeding method for increasing plant resistance against abiotic stresses (Flores et al., 2010; Haghighi et al., 2016a; Penella et al., 2017). Grafting vegetables onto compatible rootstocks offers a number of advantages; for instance, help to compete against biotic and abiotic stresses. Grafting is a beneficial way of reducing production losses in tomatoes under NaCl stress conditions. Santa-Cruz et al. (2002) grafted tomato cv. Moneymaker and UC-82B onto a commercial tomato hybrid (cv. Kyndia) at different NaCl concentrations (0, 50, and 100 mM). They showed that the shoot biomass reduction and accumulation of Na$^+$ and Cl$^-$ induced by salinity was lower in UC-82B when grafted onto cv. Kyndia, which did not occur when Moneymaker was used as scion (Santa-Cruz et al., 2002). The effect of grafting may vary according to the rootstock, scion, and growing conditions.

For more details on this subject see Chapter 47, entitled "The Role of Grafting by Vegetable Crops for Reducing Biotic and Abiotic Stresses" in this *Handbook*.

36.3.3 Other Methods for Minimizing the Effects of Salt Stress on Tomatoes

The use of mycorrhiza inoculations and plant growth promoting bacteria are a safe and eco-friendly choice for mitigation of salt stress in tomatoes. Currently, arbuscular mycorrhizal fungi (AMF) inoculations are used for alleviating salt stress in plants (Haghighi et al., 2017). Haghighi et al. (2017) studied the effects of mycorrhiza inoculation on cucumber growth irrigated with saline water. These investigators found a substantial improvement in the growth of the plants treated with AMF compared to the controls (untreated plants). Al-Karaki (2000), Al-Karaki et al. (2001), He et al. (2007), and Hajiboland et al. (2010) reported that inoculation with AMF improved plant growth, nutrient uptake and decreased tomato fruit yield losses under salt stress conditions. Huang et al. (2010) inoculated tomato cultivar Zhongza 9 with AMF under different NaCl levels and observed that AMF could alleviate the growth limitation by increasing SOD, POD, and ASA-POD activities in leaves and roots. Abdel Latef and Chaoxing (2011) reported that AMF could protect tomato plants against salt stress in terms of plant growth, leaf area, chlorophyll content, and fruit yield. AMF inoculations increased P and K concentrations and reduced the amount of Na$^+$, enhanced the activity of SOD, POD, CAT, and APX under both saline and non-saline conditions (Abdel Latef and (Chaoxing, 2011)).

Plant growth promoting bacteria (PGPB) containing 1-aminocyclopropane 1 (ACC) deaminase is one of the novel biological approaches for alleviating the negative impact of salt stress on tomato plants. *Achromobacter piechaudii* ARV8 can alleviate the adverse effects of salinity on tomato seedlings via a reduction in the amount of ethylene, increased uptake of P and K, increased uptake capacity of water and improved photosynthesis (Mayak et al., 2004). Similar effects were reported by Tank and Saraf (2010), Yan et al. (2014), Hassan et al. (2014), and Egamberdieva et al. (2017). Three PGPR strains (*Pseudomonas fluorescens*, *Pseudomonas aeruginosa*, and *Pseudomonas stutzeri*) stimulated the production of phytohormones and ACC deaminase enzyme and increased the salt resistance in tomatoes (Tank and Saraf, 2010). Ali et al. (2014) treated tomato plants with ACC deaminase, and the results showed a better growth and an increased number of flowers and buds in plants growing under salt stress conditions in comparison to the control plants.

Salicylic acid (SA) participates in the regulation of physiological processes in plants; it is an endogenous growth regulator of a phenolic nature (Shakirova et al., 2003). Application of SA increased plant growth, photosynthesis rate, and reduced electron leakage under salt stress conditions and it caused a better salinity stress tolerance in tomatoes (Stevens et al., 2006).

ACKNOWLEDGMENTS

This research was supported by the Higher Education Institutional Excellence Program (1783-3/2018/FEKUTSRAT). The financial support was awarded by the Ministry of Human Capacities within the framework of water-related researches of Szent István University.

REFERENCES

Abdel Latef, A.A.H., and H. Chaoxing. 2011. Effect of arbuscular mycorrhizal fungi on growth, mineral nutrition, antioxidant enzymes activity and fruit yield of tomato grown under salinity stress. *Sci. Hortic.*, 12(3): 228–233.

Abogadallah, G.M., Serag, M., and W.P. Quick. 2010. Fine and coarse regulation of reactive oxygen species in the salt tolerant mutants of barnyard grass and their wild type parents under salt stress. *Plant Physiol.*, 138: 60–73.

Adams, P. 1991. Effects of increasing the salinity of the nutrient solution with major nutrients or sodium chloride on the yield, quality and composition of tomatoes grown in rockwool. *J. Hortic. Sci.*, 66(2): 201–207.

Al-Aghabary, K., Zhu, Z., and Q. Shi. 2004. Influence of silicon supply on chlorophyll content, chlorophyll fluorescence, and antioxidative enzyme activities in tomato plants under salt stress. *J. Plant Nutr.*, 12: 2101–2115.

Ali, S., Charles, T.C., and B.R. Glick. 2014. Amelioration of high salinity stress damage by plant growth-promoting bacterial endophytes that contain ACC deaminase. *Plant Physiol. Biochem.*, 80: 160–167.

Al-Karaki, G.N. 2000. Growth of mycorrhizal tomato and mineral acquisition under salt stress. *Mycorrhiza*, 10: 51–54.

Al-Karaki, G.N., Hammad, R., and M. Rusan. 2001. Response of two tomato cultivars differing in salt tolerance to inoculation with mycorrhizal fungi under salt stress. *Mycorrhiza*, 11: 43–47.

Amjad, M., Akhtar, J., Anwar-ul-Haq, M., Atif Riaz, M., Ahmad Saqib, Z., Murtaza, B., and M. Asif Naeem. 2016. Effectiveness of potassium in mitigating the salt-induced oxidative stress in contrasting tomato genotypes. *J. Plant Nutr.*, 39(13): 1926–1935.

Choudhary, O.P., Ghuman, B.S., Dhaliwal, M.S., and N. Chawla. 2010. Yield and quality of two tomato (*Solanum lycopersicum* L.) cultivars as influenced by drip and furrow irrigation using waters having high residual sodium carbonate. *Irrig. Sci.*, 28: 513–523.

Chrétien, S., Gosselin, A., and M. Dorais. 2000. High electrical conductivity and radiation-based water management improve fruit quality of greenhouse tomatoes grown in rockwool. *HortScience*, 35: 627–631.

Claussen, W., Brückner, B., Krumbein, A., and F. Lenz. 2006. Long-term response of tomato plants to changing nutrient concentration in the root environment – the role of proline as an indicator of sensory fruit quality. *Plant Sci.*, 171: 323–331.

Cuartero, J., and R. Fernández-Muñoz. 1998. Tomato and salinity. *Sci. Hortic.*, 78(1–4): 83–125.

Cuartero, J., Bolarin, M.C., Asins, M.J., and V. Moreno. 2006. Increasing salt tolerance in tomato. *J. Exp. Bot.*, 57: 1045–1058.

JesúsCuartero, J., and R. Fernández-Muñoz. 1988. Tomato and salinity. *Sci. Hortic.*, 78(1–4): 83–125.

Diao, M., Ma, L., Wang, J., Cui, J., Fu, A., and H.-Y. Liu. 2014. Selenium promotes the growth and photosynthesis of tomato seedlings under salt stress by enhancing chloroplast antioxidant defense system. *J. Plant Growth Regul.*, 33: 671.

Dorais, M., Ehret, D.L., and A.P. Papadopoulos. 2008. Tomato (*Solanum lycopersicum*) health components: from the seed to the consumer. *Phytochem. Rev.*, 7: 231.

Dorais, M., Turcotte, G., Papadopoulos, A.P., Hao, X., and A. Gosselin. 2000. Control of tomato fruit quality and flavour by EC and water management. Agriculture and Agri-Food Canada Report, pp. 18–21.

Egamberdieva, D., Davranov, K., Wirth, S., Hashem, A., and E.F. Abd_Allah. 2017. Impact of soil salinity on the plant-growth–promoting and biological control abilities of root associated bacteria. *Saud. J. Biol. Sci.*, 24: 1601.

Fanasca, S., Martino, A., Heuvelink, E., and C. Stanghellini. 2007. Effect of electrical conductivity, fruit pruning, and truss position on quality in greenhouse tomato fruit. *J. Hort. Sci. Biotechnol.*, 82: 488–494.

Flores, F.B., Sanchez-Bel, P., Estañ, M.T., Martinez-Rodriguez, M.M., Moyano, E., Morales, B., Campos, J.F., Garcia-Abellán, J.O., Egea, M.I., Fernández-Garcia, N., Romojaro, F., and M.C. Bolarín. 2010. The effectiveness of grafting to improve tomato fruit quality. *Sci. Hortic.*, 125(3): 211–217.

Haghighi, M., Afifipour, Z., and M. Mozafarian. 2012. The alleviation effect of silicon on seed germination and seedling growth of tomato under salinity stress. *Veg. Crop. Res. Bull.*, 76(1): 119–126.

Haghighi, M., Mohammadnia, S., Attai, Z., and M. Pessarakli. 2017. Effects of mycorrhiza inoculation on cucumber growth irrigated with saline water. *J. Plant Nutr.*, 40(1): 128–137. http://www.tandfonline.com/doi/pdf/10.1080/01904167.2016.1201499

Haghighi, M., Sheibanirad, A., and M. Pessarakli. 2016a. Cucurbits grafting; methods, physiology, and responses to stresses. In: Pessarakli, M. (Ed.), *Handbook of Cucurbits, Growth, Cultural Practices, and Physiology*, pp. 255–272. CRC Press, Taylor & Francis Publishing Group, Boca Raton, FL.

Haghighi, M., Sheibanirad, A., and M. Pessarakli. 2016b. Effects of selenium as a beneficial element on growth and photosynthetic attributes of greenhouse cucumber. *J. Plant Nutr.*, 39(10):

1493–1498. http://www.tandfonline.com/doi/pdf/10.1080/01904167.2015.1109116

Hajiboland, R., Aliasgharzadeh, A., Laiegh, S.F., and C. Poschenrieder. 2010. Colonization with arbuscular mycorrhizal fungi improves salinity tolerance of tomato (*Solanum lycopersicum* L.) plants. *Plant Soil*, 331: 313–327.

Hassan, W., David, J., and F. Bashir. 2014. ACC-deaminase and/or nitrogen-fixing rhizobacteria and growth response of tomato (*Lycopersicon pimpinellfolium* Mill.). *J. Plant Interact.*, 9: 869–882.

Hawrylak-Nowak, B., Hasanuzzaman, M., and M. Renata. 2018. Mechanisms of selenium-induced enhancement of abiotic stress tolerance in plants. In: Hasanuzzaman, M., Fujita, M., Oku, H., Nahar, K., and Hawrylak-Nowak, B. (Eds.), *Plant Nutrients and Abiotic Stress Tolerance*, pp. 269–295. Springer, Singapore.

He, Z., He, C., Zhang, Z., Zou, Z., and H. Wang. 2007. Changes of antioxidative enzymes and cell membrane osmosis in tomato colonized by arbuscular mycorrhizae under NaCl stress. *Colloids Surf. B: Biointerfaces*, 59: 128–133.

Heuvelink, E. 2018. *Tomatoes*. Crop Production Science in Horticulture Series, 2018, CABI Publishing.

Heuvelink, E., and M. Dorais. 2005. Crop growth and yield. In: Heuvelink, E. (Ed.), *Tomatoes*, pp. 85–144. CABI Publishing, Cambridge, MA.

Huang, Z., Hez, Q., Zou, Z.-R., and Z.-B.Zhang 2010. The Effects of arbuscular mycorrhizal fungi on reactive oxyradical scavenging system of tomato under salt tolerance. *Agric. Sci. China*, 9(8): 1150–1159.

Hussain, K., Nisar, M.F., Majeed, A., Nawaz, K., Bhatti, K.H., Afghan, S., Shahazad, A., and S. Zia-ul-Hussnian. 2010. What molecular mechanism is adapted by plants during salt stress tolerance? *Afr. J. Biotechnol.*, 9: 416–422.

Johnson, H.E., Broadhurst, D., Goodacre, R., and A.R. Smith. 2003. Metabolic fingerprinting of salt-stressed tomatoes. *Phytochemistry*, 62(6): 919–928.

Jones, J. Benton. 1999. Tomato Plant Culture: In the Field, Greenhouse, and Home Garden. CRC Press Boca Raton, FL.

Jones, Jr., B. 2007. *Tomato Plant Culture: In the Field, Greenhouse, and Home Garden*, Second Edition. CRC Press.

Kaya, C., Kirnak, H., and D. Higgs. 2001. Enhancement of growth and normal growth parameters by foliar application of potassium and phosphorus in tomato cultivars grow at high (NaCl) salinity. *J. Plant Nutr.*, 24(2): 357–367.

Korkmaz, A., Karagöl, A., Akınoğlu, G., and H. Korkmaz. 2018. The effects of silicon on nutrient levels and yields of tomatoes under saline stress in artificial medium culture. *J. Plant Nutr.*, 41(1): 123–135.

Krauss, S., Schnitzler, W.H., Grassmann, J., and M. Woitke. 2006. The influence of different electrical conductivity values in a simplified recirculating soilless system on inner and outer fruit quality characteristics of tomato. *J. Agric. Food Chem.*, 54: 441–448.

Kyriacou, M.C., and Y. Rouphael. 2018. Towards a new definition of quality for fresh fruits and vegetables. *Sci. Hortic.*, 234: 463–469.

Lester, G.E., Jifon, J.L., and G. Rogers. 2005. Supplemental foliar potassium applications during muskmelon fruit development can improve fruit quality, ascorbic acid, and beta-carotene contents. *J. Am. Soc. Hortic. Sci.*, 130: 649–653.

Levent, T.A., Kaya, C., Ashraf, M., Altunlu, H., Yokas, I., and B.Yagmur 2007. The effects of calcium sulphate on growth, membrane stability and nutrient uptake of tomato plants grown under salt stress. *Environ. Exper. Bot.*, 59(2): 173–178.

Li, H., Zhu, Y., Hu, Y., Han, W., and H. Gong. 2015. Beneficial effects of silicon in alleviating salinity stress of tomato seedlings grown under sand culture. *Acta Physiol. Plant.*, 37: 71.

Maghsoudi, K., Emam, Y., and M. Pessarakli. 2016. Effect of silicon on photosynthetic gas exchange, photosynthetic pigments, cell membrane stability and relative water content of different wheat cultivars under drought stress conditions. *J. Plant Nutr.*, 39(7): 1001–1015. http://www.tandfonline.com/doi/pdf/10.1080/01904167.2015.1109108

Mahajan, S., and N. Tuteja. 2005. Cold, salinity and drought stresses: an overview. *Arch. Biochem. Biophys.*, 444(2):139–158.

Malash, N.M., Flowers, T.J., and R. Ragab. 2008. Effect of irrigation methods, management and salinity of irrigation water on tomato yield, soil moisture and salinity distribution. *Irrig. Sci.*, 26: 313.

Mayak, S., Tirosh, T., and B.R. Glick. 2004. Plant growth-promoting bacteria that confer resistance in tomato to salt stress. *Plant Physiol. Biochem.*, 42: 565–572.

Mozafariyan, M., Kamelmanesh, M.M., and B. Hawrylak-Nowak. 2016. Ameliorative effect of selenium on tomato plants grown under salinity stress. *Arch. Agron. Soil Sci.*, 62: 1368–1380.

Mozafariyan, M., Pessarakli, M., and K. Saghafi. 2017. Effects of selenium on some morphological and physiological traits of tomato plants grown under hydroponic condition. *J. Plant Nutr.*, 40(2): 139–144. http://www.tandfonline.com/doi/pdf/10.1080/01904167.2016.1201500

Munns, R. 2002. Comparative physiology of salt and water stress. *Plant Cell Environ.*, 25: 239–250.

Parvin, K., Ahamed, K.U., Islam, M.M., and Md. Nazmul Haque. 2015. Response of tomato plant under salt stress: role of exogenous calcium. *J. Plant Sci.*, 10: 222–233.

Penella, C., Nebauer, S.G., López-Galarza, S., Quiñones, A., San Bautista, A., and Á. Calatayud. 2017. Grafting pepper onto tolerant rootstocks: an environmental-friendly technique overcome water and salt stress. *Sci. Hortic.*, 226: 33–41.

Rouphael, Y., and M. Kyriacou. 2018. Enhancing quality of fresh vegetables through salinity eustress and biofortification applications facilitated by soilless cultivation. *Front. Plant Sci.*, 9: 1254.

Saito, T., Matsukura, C., Ban, Y., Shoji, K., Sugiyama, M., Fukuda, N., and S. Nishimura. 2008. Salinity stress affects assimilate metabolism at the gene-expression level during fruit development and improves fruit quality in tomato (*Solanum lycopersicum* L.). *J. Japanese Soc. Hortic. Sci.*, 77: 61–68.

Santa-Cruz, A., Martinez-Rodriguez, M.M., Perez-Alfocea, F., Romero-Aranda, R., and M.C. Bolarin. 2002. The rootstock effect on the tomato salinity response depends on the shoot genotype. *Plant Sci.*, 162(5): 825–831.

Saure, M.C. 2001. Blossom-end rot of tomato (*Lycopersicon esculentum* Mill.) – A calcium- or a stress-related disorder? *Sci. Hortic.*, 90(3–4): 193–208.

Schwarz, D., Franken, P., Krumbein, A., Kläring, H.P., and B. Bar-Yosef. 2009. Nutrient management in soilless culture in the conflict of plant, microorganism, consumer and environmental demands. *Acta Hortic.*, 843: 27–34.

Segura, M.L., Contreras, J.I., Salinas, R., and M.T. Lao. 2009. Influence of salinity and fertilization level on greenhouse tomato yield and quality. *Commun. Soil Sci. Plant Anal.*, 40: 485–497.

Shakirova, F., Sakhabutdinova, A., Bezrukova, M., Fatkhutdinova, R., and D. Fatkhutdinova. 2003. Changes in the hormonal status of wheat seedlings induced by salicylic acid and salinity. *Plant Sci.*, 164(3): 317–322.

Singh, J., Divakar Sastry, E.V., and V. Singh. 2012. Effect of salinity on tomato (*Lycopersicon esculentum* Mill.) during seed germination stage. *Physiol. Mol. Biol. Plants.*, 18(1): 45–50.

Slama, C.A., Bouchereau, A., Flowers, T., and A. Savouré. 2015. Diversity, distribution and roles of osmoprotective compounds accumulated in halophytes under abiotic stress. *Ann. Bot.*, 115: 433–447.

Stevens, J., Senaratna, T., and K. Sivasithamparam. 2006. Salicylic acid induces salinity tolerance in tomato (*Lycopersicon esculentum* cv. Roma): associated changes in gas exchange, water relations and membrane stabilisation. *Plant Growth Regul.*, 49: 77.

Tank, N., and M. Saraf. 2010. Salinity-resistant plant growth promoting rhizobacteria ameliorates sodium chloride stress on tomato plants. *J. Plant Interact.*, 5: 51–58.

Tuna A.L., C. Kaya, M. Ashraf, H. Altunlu, I. Yokas, and B. Yagmur. 2007. The effects of calcium sulphate on growth, membrane stability and nutrient uptake of tomato plants grown under salt stress. *Environmental and Experimental Botany*, 59:173–178.

Wu, M., and C. Kubota. 2008. Effects of high electrical conductivity of nutrient solution and its application timing on lycopene, chlorophyll and sugar concentrations of hydroponic tomatoes during ripening. *Sci. Hortic.*, 116(2): 122–129.

Wu, M., Buck, J.S., and C. Kubota. 2004. Effects of nutrient solution EC, plant microclimate and cultivars on fruit quality and yield of hydroponic tomatoes (*Lycopersicum esculentum* L.). *Acta Hortic.*, 659: 541–547.

Yan, J., Smith, M.D., Glick, B.R., and Y. Ling. 2014. Effects of ACC deaminase containing rhizobacteria on plant growth and expression of Toc GTPases in tomato (*Solanum lycopersicum*) under salt stress. *Botany*, 92: 775–781.

Yokoi, S., Bressan, R.A., and P.M. Hasegawa. 2002. Salt stress tolerance of plants. JIRCA Working Report, pp. 25–33.

Yunus, Q., and M. Zari. 2017. Effect of exogenous silicon on ion distribution of tomato plants under salt stress. *Commun. Soil Sci. Plant Anal.*, 48(16): 1843–1851.

Zhu, Y.X., and H.J. Gong. 2014. Beneficial effects of silicon on salt and drought tolerance in plants. *Agron. Sustain Dev.*, 34: 455–472.

37 Water Stress Effects on Growth and Physiology of Corn

M. Anowarul Islam and Abdelaziz Nilahyane

CONTENTS

37.1 INTRODUCTION

Stressful environments such as drought, salinity, and high temperature (heat) cause alterations in a wide range of physiological, biochemical, and molecular processes in plants. These changes, compounded by climate change, threaten the sustainability of crops and challenge their growth and production. Drought or water stress is considered one of the most serious problems, which drastically affects the growth and physiology of corn (*Zea mays* L.) as well as many other crops. Understanding corn's response to water stress at different levels will help to take the right actions and the appropriate strategies to limit the negative impact of water stress. Corn requires adequate amounts of water during all growth stages to achieve optimum growth and high yield. It is therefore important to match the water needs for each specific growth stage and avoid water stress during the sensitive growth periods of the growing season. In this chapter, we will focus on water stress and its effects on the growth and physiology of corn.

37.2 CORN: THE MOST IMPORTANT CEREAL CROP IN THE WORLD

37.2.1 IMPORTANCE

Corn, or maize, is one of the most important cereal crops cultivated in the world. Corn is the most widely distributed crop, because it can be grown in a wide range of edapho-climatic environment under both dryland and irrigated conditions. Due to its relatively low price and global distribution, corn is a source of high protein and energy and consequently, has a major role in human nutrition and animal feed (Memon et al., 2011). As a C4 plant, corn has the ability to efficiently use sunlight radiation, which leads to a high yield per hectare compared with other grain crops (Du Plessis, 2003).

Corn production is about 1038 million tons worldwide, in which the United States is the leader with 385 million tons, followed by China (220 million tons) and Brazil (87 million tons) (USDA-FAS, 2017). In 2016, the total acreage harvested in the United States was about 35 million hectares for corn for grain and 2.5 million hectares for corn for silage. The total acreage of corn production in the United States is 37.5 million hectares, which is higher than soybean (33 million hectares), hay (22 million hectares), and wheat (18 million hectares) (USDA-NASS, 2017). The total use of corn in the United States is about 370 million tons; 29% is used for fuel ethanol production, 38% for animal feed and residual, 15% for export, 8% for distillers dried grains with solubles (DDGs), and 9% for food, seed, and industrial (FSI), including high-fructose corn syrup, sweeteners, starch, cereal/other, beverage/alcohol, and seed (USDA-ERS, 2017).

Corn is an incredible crop. All parts of the corn plant are useable. The kernels are used for food. The stalks and leaves are usually used as animal feed, while the corn silk is used for medicine (Sailer, 2012). This clearly shows the importance of corn and indicates its significant role in human and animal feed.

37.2.2 Growth and Development

It is important to understand the growth and development of the corn plant so as to take the right actions in the field at the right time during the growing season. The growth and development of the corn plant are divided into vegetative and reproductive stages. The vegetative growth stages are determined based on the leaf tip (Nielsen, 2014) or leaf collar (Ritchie et al., 1993) appearance scales and undergo several sub-stages, during which the plants develop up to 21 leaves in total, depending on the cultivar and the environmental conditions. The vegetative growth stages begin with seed emergence (VE) and end with full emergence of the tassel or anthesis (VT). At the VT stage, the tassel is entirely visible, and the corn plant is at its full height. The pollen also starts to shed during this stage. The reproductive stages then start with silking (R1), followed by blister (R2), milk (R3), dough (R4), dent (R5), and physiological maturity or black layer stage (R6). At the silking stage, the silk is visible outside the husk and starts to capture the shed pollen. The blister stage is characterized by clear liquid inside the kernels and yellow outside (Figure 37.1). The milk stage occurs when the kernels develop milky white liquid inside and yellow outside. The dough stage is characterized by pasty and thick fluid inside the kernels. The dent stage is reached when the majority of kernels are partially dented. The last reproductive stage, R6, occurs when the milk line disappears and a black layer is formed on the kernels. At this stage, the maximum dry matter is reached, the whole plant is no longer green, and the crop is ready to harvest (Figure 37.1).

37.2.3 Physiology

Corn is a C4 plant, and it has the common Kranz leaf anatomy (Agepati and Rowan, 2011). As with other C4 plants, the photosynthetic pathway of corn requires the presence of two types of cells, the mesophyll and bundle sheath cells (Figure 37.2). The leaves of corn plants have two to four mesophyll cells between adjacent bundle sheaths (Morgan and Brown, 1979). Therefore, leaf thickness is constrained in corn plants to a much narrower range than in C3 plants (Dengler et al., 1994), which helps in identifying and grouping them. Corn plants have a higher net photosynthesis rate and higher water and sunlight use efficiency as compared with C3 plants. Corn plants also have a lower photorespiration rate and a lower light compensation point (0 to 10 ppm) than C3 plants (50 to 150 ppm).

37.2.4 Factors Affecting Growth and Physiology

As with many crops, corn is subject to many biotic and abiotic stresses. These stresses adversely affect the crop performance (e.g., growth and yield). For instance, biotic stresses, such as pests and diseases, reduce plant growth by damaging the plant tissues, affecting the photosynthetic rate and accelerating leaf senescence (Maheswari et al., 2012). Also, biotic stresses, such as weeds and undesirable species, can cause a significant decline in crop productivity by competing with the crops for light, nutrients, and water. It has been reported that biotic stresses may cause up to 16% yield loss in cereals such as wheat, barley, rice, and corn as well as potato, cotton, and soybean (Chakraborty et al., 2000).

On the other hand, drought, heat, flood, hail, frost damage, salinity, and nutrient deficiency are among the major abiotic stresses that greatly affect corn productivity. For example,

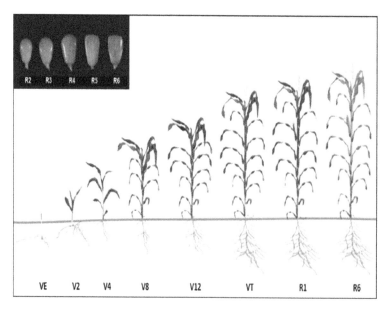

FIGURE 37.1 (See color insert.) Vegetative and reproductive growth stages of a corn plant. Vegetative growth stages (VE-VT): VE, seed emergence; tassel or anthesis (VT); reproductive stages (R1–R6): silking, R1; blister, R2; milk, R3; dough, R4; dent, R5; physiological maturity or black layer, R6.

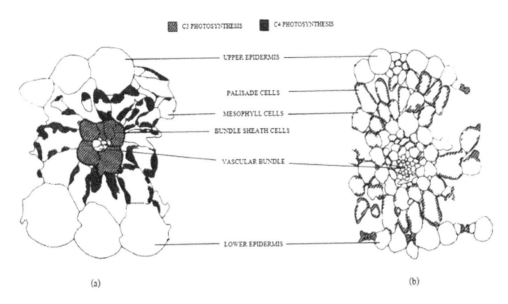

FIGURE 37.2 Comparison of leaf anatomy between C4 photosynthesis (a) and C3 photosynthesis (b). (From Black, C.C., in *Advances in Ecological Research*, Academic Press, New York, 1971. With permission.)

corn is very sensitive to temperatures below 10 °C, which limits the ability of the crop to emerge and accumulate dry matter. Cool temperatures combined with long-time chilling exposure may cause irreversible injuries as a result of physical and physiological changes within the plant (Maheswari et al., 2012). At the opposite extreme, high temperatures cause imbalance in CO_2 assimilation and reduce the time for photosynthesis as a result of increasing rate of reproductive development and limitation of vegetative development. In addition, high temperature causes high transpiration and evaporative demands as a result of plant water deficit.

Salinity is also one of the most important stresses that affect corn growth and physiology. High salt concentration causes osmotic and ionic stresses. Osmotic stress, commonly known as physiological drought, is a result of a decrease in the osmotic potential of the soil solution, which causes water stress in corn plants (Munns, 2009). However, ionic stress is due to the accumulation of high concentrations of sodium ion in plant tissues. Corn plants lack the ability to sequester sodium into vacuoles, unlike halophytes. In addition, high salt levels affect mineral nutrition and cause imbalance of the nutrients in the soil, which lead, to severe deficiencies.

37.3 WATER STRESS EFFECTS ON CORN GROWTH

Water stress is the result of low water supply and high water demand from the plant. The water supply is determined by the amount of water present in the soil layers as a result of water inputs into the system, such as irrigation water and rainfall. The water demand is mainly determined by soil evaporation and plant transpiration, which in turn, are driven by environmental conditions such as vapor pressure deficit, wind, and temperature. Transpiration and soil evaporation are commonly combined in the term *evapotranspiration*. Once the crop evapotranspiration surpasses the water supply,

water stress occurs. It has been reported that plant height, leaf area, stem diameter, and dry matter accumulation are greatly reduced under water stress (Khan et al., 2001; Zhao et al., 2006). The magnitude and severity of water stress in corn vary depending on weather conditions, soil water deficit, and crop growth stage.

37.3.1 EFFECT OF WATER STRESS AT EARLY VEGETATIVE GROWTH STAGES

Water stress has a great impact on seed germination, imbibition, and the germination potential of corn. Adequate moisture content is essential to maintain osmotic potential, convert stored food into consumable forms, and break down dormancy. Water stress reduces the germination rate of corn seeds by reducing their viability (Aslam et al., 2015). Corn seedlings have low daily water requirements during the first 3 to 4 weeks after planting, which makes the plant less water stressed during this period. However, as the season advances, corn plants show more water stress symptoms when rainfall or water supply is omitted. Çakir (2004) conducted a study of water stress on corn at different growth stages and concluded that the plant height was greatly reduced as a result of water stress during the rapid vegetative growth stage. Similarly, water stress during vegetative stages reduces the root and shoot elongation and decreases the size of the leaves, which leads to a reduced leaf area index. Extended water stress results in leaf burning and great yield loss (Aslam et al., 2015).

37.3.2 EFFECT OF WATER STRESS AT ANTHESIS (FLOWERING)

Water stress at pre-anthesis delays leaf tip emergence and reduces the leaf area, which causes up to 25% yield loss (NeSmith and Ritchie, 1992). Water stress at anthesis causes an imbalance in pollen shedding and silk emergence. Pollen sterility occurs as a result of dehydration of pollen and lack of moisture (Aylor, 2004). Saini and Westgate (2000) reported that an increase in abscisic acid (ABA) accumulation and decrease of invertase activity are the main causes of pollen sterility under water stress. Aslam et al. (2015) concluded that pollen viability, pollen size and weight, and pollen moisture content are all affected by water stress. Setter et al. (2001) reported that under 5 days of water stress, the kernel set was severely affected during pre- and early post-anthesis. Similarly, the harvest index and number of kernels per cob and total grain yield are severely reduced as a result of water stress at anthesis (Anjum et al., 2011). Çakir (2004) concluded that water stress during flowering significantly affected the final yield and reported a 30% to 40% grain yield loss if an irrigation event is missed during anthesis. This clearly indicates that this is one of the most critical periods of water stress, suggesting that moisture should not be omitted during this period to avoid severe yield loss.

37.3.3 EFFECT OF WATER STRESS AT REPRODUCTIVE GROWTH STAGES

Corn roots reach their maximum depth 50 to 60 days after planting and require more water supply, especially during reproductive stages (Kirkpatrick et al., 2006). Reproductive stages, including tasseling, silking, pollination, and early seed filling, are the most sensitive stages for water stress (Benham, 1998). Water stress during the first 3 weeks of reproductive stages can significantly reduce yield (Oktem, 2008). Prolonged water stress during tasseling and the ear formation stage may cause 66% to 93% grain yield loss (Çakir, 2004). Silk emerges 2 to 4 days after tassel emergence in normal conditions. However, silk appearance is delayed and the anthesis–silking interval increased under water stress conditions (Dass et al., 2001). This increase in the interval drastically affects the pollination process, and the silk becomes non-receptive to pollen, which in turn, can cause 40% to 50% yield loss (Nielsen, 2005). Water stress after silking affects kernel weight, shortens the grain-filling period, and can cause 20% to 30% yield loss during the kernel development phase (Heinigre, 2000). Kernels are very susceptible to abortion during the first 2 weeks after the silking stage if there is a lack of moisture. Lauer (2003) reported a 2.5% to 5.8% daily yield loss when corn is subject to water stress during the kernel development stage. Water stress during the dent stage is less devastating but can hasten maturity and may still cause significant yield and quality loss.

37.4 WATER STRESS EFFECTS ON CORN PHYSIOLOGY

Corn plant growth is controlled by several physiological, biochemical, and molecular processes. However, photosynthesis is a key process that contributes significantly to plant growth and development (Ashraf and Harris, 2013). In this section, we will focus on the effect of water stress on gas exchange, photosystems, and photosynthetic enzyme activity.

37.4.1 EFFECT OF WATER STRESS ON GAS EXCHANGE

Water stress has a great impact on physiological parameters such as photosynthesis, stomatal conductance, transpiration, and relative water content (RWC) (Farooq et al., 2009). There are two key events occurring during the process of photosynthesis: light reactions, in which light energy is converted into ATP and NADPH and O_2 is released, and dark reactions, in which CO_2 is fixed into carbohydrates using the products of light reactions (Taiz and Zeiger, 2010). Gas exchange such as CO_2 assimilation (photosynthesis) and water exiting the leaf surface (transpiration) are commonly used by crop physiologists to define water use efficiency (WUE) as carbon assimilated per unit of transpiration (Viets, 1962). It has been reported that both stomatal and non-stomatal limitations are the main causes of decreasing photosynthesis as a result of water stress (Shangguan et al., 1999). Atteya (2003) reported that stomata open quickly when water stress is relieved or controlled under stress to allow some carbon fixation, thus enhancing WUE.

Water stress reduces the translocation of photosynthetic assimilates as a result of a decrease in photosynthesis (Yadav et al., 2004). Similarly, drought stress increases the leaf temperature, which inhibits the enzymatic and photosynthetic activity (Chaves et al., 2002). Anjum et al. (2011) reported reduced activity of starch and sucrose synthesizing enzymes, which affect the grain filling of corn under water stress. Water stress causes the deterioration of thylakoid membranes as well as a decrease in chlorophyll (Chl) content (Anjum et al., 2011; Din et al., 2011). It has been also reported that net CO_2 assimilation rate decreased greatly as the RWC and leaf water potential decreased (Ashraf and Harris, 2013). It has been well documented that water stress can inhibit leaf photosynthesis and stomatal conductance, with a greater inhibitory effect on transpiration than that of CO_2 diffusion into the leaf tissue (Chaves et al., 2009; Sikuku et al., 2010). Water stress also reduces the efficiency of mesophyll cells to use the available CO_2 (Karaba et al., 2007). Thus, the decrease of plant photosynthesis exposed to water stress is normally caused by a reduction in the mesophyll conductance and stomatal closure.

When corn plants are under stress, stomata are closed to avoid water losses. This regulation of transpiration by stomatal closure is necessary for the plant to prevent dehydration (Sperry, 2000). However, it is well known that limiting gas exchange can lead to yield reduction and carbon starvation (Chae and Lee, 2001). Thus, an understanding of corn plant physiology and gas exchange response to water stress is required to minimize water supply with less impact on yield reduction. Stomatal closure is the early plant response to water stress to control water loss (Harb et al., 2010). It is generally caused by a decrease in leaf turgor and atmospheric vapor pressure (Chaves et al., 2009) as well as the action of the ABA hormone. It has been reported that high amounts of ABA lead to an increase in cytosolic Ca^{2+} and activation of plasma membrane–localized anion channels (Kohler and Blatt, 2002). This causes guard cell depolarization, loss of turgor in guard cells, high H_2O_2 production, and thus, stomatal closure (Wang et al., 2012).

Stomatal limitation is considered to be the key factor of photosynthesis reduction under water stress (Bousba et al., 2009). This was attributed to a decrease in both CO_2 assimilation rate and intracellular CO_2 concentration (C_i), which in turn, inhibits overall photosynthesis (Ashraf and Harris, 2013). This inhibition of photosynthesis is primarily compounded by damage to the photosystem machinery and deactivation of the enzymatic activity under water stress.

37.4.2 EFFECT OF WATER STRESS ON PHOTOSYSTEMS

It has been widely reported that photosynthesis is affected by several biotic and abiotic factors, which considerably impact fluorescence emission characteristics (Baker, 2008). The fluorescence parameters, such as Fv, Fm, Fo, Fi, and especially their ratios, are commonly used to determine metabolic disorders in leaves exposed to various stresses (Baker, 2008). For instance, the Fv/Fm ratio is used to determine the maximum quantum efficiency of photosystem II (PSII) and thus, the overall photosynthetic capacity (Balouchi, 2010). In general, Fv/Fm is usually around 0.8 for healthy leaves, while low values commonly suggest that photoinhibition occurs when the plant is under stress (Ashraf and Harris, 2013).

Several studies have shown that water stress decreases the electron transport through the PSII and PSI reaction systems (Liu et al., 2006; Zlatev, 2009) and alters the Chl a fluorescence (Zhang et al., 2011). In addition, *in vivo* studies have shown that water stress significantly affects the oxygen-evolving center (OEC) related to the PSII system as well as inactivating the PSII reaction center as a result of degradation of the D1, D2, and LHCII-PSII proteins (Duan et al., 2006; Kawakami et al., 2009; Zlatev, 2009). This leads to the generation of ROS, which eventually causes photoinhibition (Anjum et al., 2011). Liu et al. (2009) have reported accelerated dephosphorylation of PSII proteins as a result of water stress. This fast dephosphorylation is caused by both intrinsic and extrinsic membrane protein phosphatases, which comprise a key molecular mechanism in the process of repair of PSII proteins (Los et al., 2010). It has also been reported that water stress is behind the release of TLP40, a potential inhibitor of the membrane phosphatases (Liu et al., 2009).

37.4.3 EFFECT OF WATER STRESS ON ACTIVITY OF PHOTOSYNTHETIC ENZYMES

As a result of water stress, the lower C_i concentration leads to the deactivation of Rubisco and many other enzymes, such as nitrate reductase (NR) and sucrose phosphate synthase (SPS) (Mumm et al., 2011). This impairs the synthesis of carbohydrates as well as the WUE of corn, which in turn, negatively affects the overall plant yield (Gill et al., 2011). Jeanneau et al. (2002) used a transgenic approach to develop a corn line with improved expression of C_4-PEPC. They found a 30% increase in WUE and a 20% increase in dry biomass for this transgenic line exposed to moderate water stress. This suggests that enhancing the gene expression of endogenous enzymes involved in photosynthesis is a useful tool in developing water

stress–tolerant cultivars. In another study, it has been reported that the activities of several C_4 photosynthetic enzymes, such as Rubisco, PEPcase, and MADP-malic enzyme (NADP-ME), decreased two to four times under water stress, while the activity of phosphopyruvate dikinase (PPDK) decreased nine times, suggesting that PPDK is potentially the limiting enzyme to photosynthesis under water stress (Ashraf and Harris, 2013).

37.5 MANAGEMENT OPTIONS TO AVOID WATER STRESS ON CORN

Achieving high yield and quality of corn requires an understanding of the most sensitive growth stages and the peak periods of water demand by the crop. Corn growers can adopt integrated soil and water management strategies, including best agronomic practices to conserve soil moisture and avoid periods of high temperature and water shortage during the growing season. Farmers can select hybrids and cultivars with considerable drought tolerance, especially in areas with limited water resources and drastic environmental conditions.

Research should be oriented toward the selection of improved growth and drought-tolerant cultivars by transcriptional activation/inactivation of specific genes, accumulation of compatible solutes and protective enzymes, increase in ABA and antioxidant levels, and suppression of energy-consuming pathways (Waseem et al., 2011). Future research must combine new genetic, genomic, and bioinformatics tools to well depict and understand the ecophysiological response of corn plants to drought. Elucidating the genetic mechanisms underlying drought stress, as well as breeding programs for developing adequate cultivars for overcoming drought, may enable significant progress in choosing drought-tolerant genotypes. Farmers can also minimize soil tillage to retain the soil moisture and avoid loss through evaporation. In addition, appropriate fertilizers and irrigation scheduling programs are key (Nilahyane et al., 2018) to obtain vigorous and healthy plants (Figure 37.3). Crops are subjected to a combination of

FIGURE 37.3 (See color insert.) Vegetative growth of corn at the University of Wyoming Powell Research and Extension Center. Top: under subsurface drip irrigation; bottom: under on-surface drip irrigation. Plants clearly show more vigorous growth under subsurface than on-surface drip irrigation.

FIGURE 37.4 Corn cobs at blister (R2) stage affected by irrigation water. Left: 60% crop evapotranspiration (60ETc); middle: 80ETc; right: 100ETc.

multiple stresses in field conditions. It is thus essential to test newly developed cultivars to multiple stresses (Figure 37.4) and to carry out extensive field studies in a large spectrum of conditions to assess tolerance as absolute yield increases (ISAAA, 2008; Waseem et al., 2011; Nilahyane et al., 2018).

37.6 CONCLUSIONS

Corn is moderately tolerant to limiting water during the early growth stages. However, reducing the water supply during the late vegetative stages and early reproductive stages reduced the growth components, the yield, and the efficient use of the water applied. More specifically, silking is the most critical stage for water stress. Drought coupled with high air temperature during the silking stage can lead to total yield loss. The yield reduction under water stress conditions is coupled to a decrease in leaf area, disturbed light interception, stomatal closure (accumulation of ABA in leaves), reduction of CO_2 diffusion, and consequently, reduction in photosynthesis. To date, there is still a lack of knowledge regarding the mechanisms of cellular responses to water stress and how it is signaled and transcribed at the cell level to different symptoms at the plant level.

REFERENCES

Agepati, S.R., and F.S. Rowan. 2011. *C4 Photosynthesis and Related CO₂ Concentrating Mechanisms*, pp. 161–195. Springer Science and Business Media B.V.

Anjum, S.A., X.Y. Xie, L.C. Wang, M.F. Saleem, C. Man, and W. Lei. 2011. Morphological, physiological and biochemical responses of plants to drought stress. *Afr. J. Agron. Res.* 6:2026–2032.

Ashraf, M., and P.J.C. Harris. 2013. Photosynthesis under stressful environments: an overview. *Photosynthetica* 51:163–190.

Aslam, M., M.A. Maqbool, and R. Cengiz. 2015. *Drought Stress in Maize (Zea mays L.). Effects, Resistance, Mechanisms, Global Achievements and Biological Strategies for Improvement*. Springer Briefs in Agriculture. Springer Cham, Heidelberg, New York and London.

Atteya, A.M. 2003. Alteration of water relations and yield of corn genotypes in response to drought stress. *Bulg. J. Plant Physiol.* 29:63–76.

Aylor, D.E. 2004. Survival of maize (*Zea mays*) pollen exposed in the atmosphere. *Agric. Meteorol.* 123:125–133.

Baker, N.R. 2008. Chlorophyll fluorescence: a probe of photosynthesis in vivo. *Ann. Rev. Plant Biol.* 59:89–113.

Balouchi, H.R. 2010. Screening wheat parents of mapping population for heat and drought tolerance, detection of wheat genetic variation. *Int. J. Biol. Life Sci.* 6:56–66.

Benham, B.L. 1998. *NebGuide: Irrigating Corn*. Cooperative Extension, Institute of Agriculture and Natural Resources, University of Nebraska-Lincoln.

Black, C.C. 1971. Ecological implications of dividing plants into groups with distinct photosynthetic production capacities. In: Cragg, J.B. (ed.), *Advances in Ecological Research*, pp. 87–114. Academic Press, New York.

Bousba, R., N. Ykhlef, and A. Djekoun. 2009. Water use efficiency and flag leaf photosynthetic in response to water deficit of durum wheat (*Triticum durum* Desf.). *World J. Agric. Sci.* 5:609–616.

Çakir, R. 2004. Effect of water stress at different development stages on vegetative and reproductive growth of corn. *Field Crops Res.* 89:1–16.

Chae, H.C., and W.L. Lee. 2001. Ethylene and enzyme-mediated superoxide production and cell death in carrot cells grown under carbon starvation. *Plant Cell Reports* 20:256–261.

Chakraborty S., A.V. Tiedemann, and P.S. Teng. 2000. Climate change: potential impact on plant diseases. *Environ. Pollut.* 108:317–326.

Chaves, M.M., J. Flexas, and C. Pinheiro. 2009. Photosynthesis under drought and salt stress: regulation mechanisms from whole plant to cell. *Ann. Bot.* 103:551–560.

Chaves, M.M., J.S. Pereira, J. Maroco, M.L. Rodrigues, C.P.P. Ricardo, M.L. Osorio, I. Carvalho, T. Faria, and C. Pinheiro. 2002. How plants cope with stress in the field: photosynthesis and growth. *Ann. Bot.* 89:907–916.

Dass, S., P. Arora, M. Kumari, and D. Pal. 2001. Morphological traits determining drought tolerance in maize (*Zea mays* L.). *Indian. J. Agric. Res.* 35:190–193.

Dengler, N.G., R.E. Dengler, P.M. Donnelly, and P.W. Hattersley. 1994. Quantitative leaf anatomy of C3 and C4 grasses (Poaceae): bundle sheath and mesophyll surface area relationships. *Ann. Bot.* 73:241–255.

Din, J., S.U. Khan, I. Ali, and A.R. Gurmani. 2011. Physiological and agronomic response of canola varieties to drought stress. *J. Anim. Plant Sci.* 21:78–82.

Duan, H.G., S. Yuan, and W.J. Liu. 2006. Effects of exogenous spermidine on photosystem II of wheat seedlings under water stress. *J. Integr. Plant Biol.* 48:920–927.

Du Plessis, J. 2003. Maize production. Department of Agriculture. Pretoria, South Africa. Online at: http://www.arc.agric.za/arc-gci/Fact%20Sheets%20Library/Maize%20Production.pdf

Farooq, M., A. Wahid, N. Kobayashi, D. Fujita, and S.M.A. Basra. 2009. Plant drought stress: effects, mechanisms and management. *Agron. Sustain. Dev.* 29:185–212.

Gill, S.S., N.A. Khan, and N. Tuteja. 2011. Differential cadmium stress tolerance in five Indian mustard (*Brassica juncea* L.) cultivars: an evaluation of the role of antioxidant machinery. *Plant Signal Behav.* 6:293–300.

Harb, A., A. Krishnan, and M.R. Madana. 2010. Molecular and physiological analysis of drought stress in Arabidopsis reveals early responses leading to acclimation in plant growth. *Plant Physiol.* 154:1254–1271.

Heinigre, R.W. 2000. Irrigation and drought management. Crop Science Department. Online at: http://www.ces.ncsu.edu/p lymouth/cropsci/cornguide/Chapter4.tml

ISAAA. 2008. Biotechnology for the development of drought tolerant crops. Pocket K No. 32. Online at: https://isaaa.org/re sources/publications/pocketk/32/default.asp

Jeanneau, M., D. Gerentes, and X. Foueillassar. 2002. Improvement of drought tolerance in maize: towards the functional validation of the Zm-Asr1 gene and increase of water use efficiency by over-expressing C4–PEPC. *Biochimie* 84:1127–1135.

Karaba, A., S. Dixit, and R. Greco. 2007. Improvement of water use efficiency in rice by expression of HARDY, an Arabidopsis drought and salt tolerance gene. *Proc. Natl. Acad. Sci. USA.* 104:15270–15275.

Kawakami, K., Y. Umenab, N. Kamiyab, and J. Shen. 2009. Location of chloride and its possible functions in oxygen-evolving photosystem II revealed by X-ray crystallography. *Proc. Natl. Acad. Sci. USA* 106:8567–8572.

Khan, M.B., N. Hussain, and M. Iqbal. 2001. Effect of water stress on growth and yield components of maize variety YHS 202. *J. Res. Sci.* 12:15–18.

Kirkpatrick, A., L. Browning, J.W. Bauder, R. Waskom, M. Neibauer, and G. Cardon. 2006. *Irrigating with Limited Water Supplies: A Practical Guide to Choosing Crops Well-Suited to Limited Irrigation.* Montana State University, Bozeman, MT.

Kohler, B., and M.R. Blatt. 2002. Protein phosphorylation activates the guard cell Ca2+ channel and is a prerequisite for gating by abscisic acid. *Plant J.* 32:185–194.

Lauer, J., 2003. What happens within the corn plant when drought occurs? University of Wisconsin Extension. Online at: http://www.uwex.edu/ces/ag/issues/drought2003/corneffec t.html

Liu, N., S. Ko, K.C. Yeh, and Y. Charng. 2006. Isolation and characterization of tomato Hsa32 encoding a novel heat-shock protein. *Plant Sci.* 170:976–985.

Liu, X., Z. Wang, and L. Wang. 2009. LEA 4 group genes from the resurrection plant *Boea hygrometrica* confer dehydration tolerance in transgenic tobacco. *Plant Sci.* 176:90–98.

Los, D.A., A. Zorina, and M. Sinetova. 2010. Stress sensors and signal transducers in cyanobacteria. *Sensors* 10:2386–2415.

Maheswari, M., S.K. Yadav, A.K. Shanker, M.A. Kumar, and B. Venkateswarlu. 2012. Overview of plant stresses: mechanisms, adaptations and research pursuit. In: Venkateswarlu, B., A. Shanker, C. Shanker, and M. Maheswari (eds), *Crop Stress and Its Management: Perspectives and Strategies*, pp. 1–18. Springer, Dordrecht.

Memon, S.Q., M. Zakria, G.R. Mari, M.H. Nawaz, and M.Z. Khan. 2011. Effect of tillage methods and fertilizer levels on maize production. *Pak. J. Agric. Sci.* 48:115–117.

Morgan, J.A., and R.H. Brown. 1979. Photosynthesis in grass species differing in CO2 fixation pathways. II. Search for species with intermediate gas exchange and anatomical characteristics. *Plant Physiol.* 64:257–262.

Mumm, P., T. Wolf, and J. Fromm. 2011. Cell type-specific regulation of ion channels within the maize stomatal complex. *Plant Cell Physiol.* 52:1365–1375.

Munns, R. 2009. Strategies for crop improvement in saline soils. In: Ashraf, M., M. Ozturk, H.R. Athar (eds), *Salinity and Water Stress*, pp. 99–110. Springer, Dordrecht.

NeSmith, D.S., and J.T. Ritchie. 1992. Effects of soil water deficits during tassel emergence on development and yield component of maize (*Zea mays*). *Field Crops Res.* 28:251–256.

Nielsen, R.L. (Bob). 2005. Tassel emergence and pollen shed. Corny News Network, Purdue University. Online at: http://www.agry.purdue.edu.

Nielsen, R.L. (Bob). 2014. Determining leaf stages. Corny News Network, Purdue University. Online at: http://www.kingcorn.org/news/timeless/VStageMethods.html

Nilahyane, A., M.A. Islam, A.O. Mesbah, and A. Garcia y Garcia. 2018. Evaluation of silage corn yield gap: an approach for sustainable production in the semi-arid region of USA. *Sustainability* 10:2523. doi:10.3390/su10072523

Oktem, A. 2008. Effect of water shortage on yield, and protein and mineral compositions of drip irrigated sweet corn in sustainable agricultural systems. *Agric. Water Manage.* 95:1003–1010.

Ritchie, S.W., J.J. Hanway, and G.O. Benson. 1993. How a corn plant develops. Spec. Rep. 48 (revised). Iowa State University of Science and Technology, Cooperative Extension Service, Ames, IA.

Sailer, L. 2012. The importance of corn. The field position. Online at: http://www.thefieldposition.com/2012/06/the-importance-of-corn/

Saini, H.S., and M.E. Westgate. 2000. Reproductive development in grain crops during drought. *Adv. Agron.* 68:59–96.

Setter, T.L., B.A. Flannigan, and J. Melkonian. 2001. Loss of kernel set due to water deficit and shade in maize: carbohydrate supplies, abscisic acid, and cytokinins. *Crop Sci.* 41:1530–1540.

Shangguan, Z., M. Shao, and J. Dyckmans. 1999. Interaction of osmotic adjustment and photosynthesis in winter wheat under soil drought. *J. Plant Physiol.* 154:753–758.

Sikuku, P.A., G.W. Netondo, J.C. Onyango, and D.M. Musyimi. 2010. Chlorophyll fluorescence, protein and chlorophyll content of three NERICA rainfed rice varieties under varying irrigation regimes. *ARPN J. Agric. Biol. Sci.* 5:19–25.

Sperry, J.S. 2000. Hydraulic constraints on plant gas exchange. *Agric. Forest Meteorol.* 104:13–23.

Taiz, L., and E. Zeiger. 2010. *Plant Physiology*, 5th Ed. Sinauer Associates, Sunderland.

USDA-ERS. 2017. United States Department of Agriculture–Economic Research Service. Feed Outlook. Online at: http://usda.mannlib.cornell.edu/usda/current/FDS/FDS-10-16-2017.pdf

USDA-FAS. 2017. United States Department of Agriculture–Foreign Agricultural Service. Grain: World markets and trade. Online at: https://apps.fas.usda.gov/psdonline/circulars/grain.pdf

USDA-NASS. 2017. United States Department of Agriculture–National Agricultural Statistics Service. Crop production 2016 summary. Online at: http://usda.mannlib.cornell.edu/usd a/current/CropProdSu/CropProdSu-01-12-2017.pdf

Viets Jr., F.G. 1962. Fertilizers and the efficient use of water. *Adv. Agron.* 14:223–264.

Wang, W.H., X.Q. Yi, and A.D. Han. 2012. Calcium-sensing receptor regulates stomatal closure through hydrogen peroxide and nitric oxide in response to extracellular calcium in Arabidopsis. *J. Exp. Bot.* 63:177–190.

Waseem, M., A. Asghar, M. Tahir, M.A. Nadeem, M. Ayub, A. Tanveer, R. Ahmad, and M. Hussain. 2011. Mechanism of drought tolerance in plant and its management through different methods. *Continental J. Agric. Sci.* 5:10–25.

Yadav, R.S., C.T. Hash, F.R. Bidinger, K.M. Devos, and C.J. Howarth. 2004. Genomic regions associated with grain yield and aspects of post-flowering drought tolerance in pearl millet across environments and tester background. *Euphytica* 136:265–277.

Zhang, L., Z. Zhang, and H. Gao. 2011. Mitochondrial alternative oxidase pathway protects plants against photoinhibition by alleviating inhibition of the repair of photodamaged PSII through preventing formation of reactive oxygen species in Rumex K-1 leaves. *Physiol. Plant.* 143:396–407.

Zhao, T.J., S. Sun, Y. Liu, J.M. Liu, Q. Liu, Y.B. Yan, and H.M. Zhou. 2006. Regulating the drought responsive element (DRE)-mediated signaling pathway by synergic functions of trans active and trans inactive DRE binding factors in *Brassica napus. J. Biol. Chem.* 281:10752–10759.

Zlatev, Z. 2009. Drought-induced changes in chlorophyll fluorescence of young wheat plant. *Biotechnology* 23:437–441.

38 Moisture Stress and Its Effects on Forage Production Systems

M. Anowarul Islam and Albert T. Adjesiwor

CONTENTS

38.1 INTRODUCTION

Moisture stress is one of the most important factors limiting crop yields worldwide. This is because water is an absolute necessity for plant growth and development. As such, water availability is one of the environmental factors with a profound effect on plant growth. In fact, water is the most used substance among all the substances absorbed by plants (Chapman and Carter, 1976). Water performs a myriad of functions in plant growth. These functions can be categorized as follows: an electron source for carbon fixation; a cell constituent; a universal solvent allowing chemical reactions and transport of essential nutrients; and the maintenance of cell turgidity and elongation. These important and critical functions of water in plants suggest that inadequate water (i.e., moisture stress) can greatly affect plant growth and development. Plant moisture stress occurs when plant water status is reduced sufficiently to affect normal plant functioning (Gimenez et al., 2005). Plant moisture stress is a complex function of interaction among plant, soil, and atmospheric factors. Thus, an absolute value of plant moisture stress cannot be defined unless these complex interactions are taken into account (Gimenez et al., 2005). Water requirement may differ among crop species or even cultivars of the same crop species (Ibrahim, 1995). Among the major food crops, rice (*Oryza sativa* L.) requires the greatest amount of water, and sorghum (*Sorghum bicolor* L.) requires the least (it is the most drought tolerant) (Chapman and Carter,

1976). However, moisture stress affects plant growth, development, and productivity regardless of crop type. Understanding plant response to moisture stress is thus critical for increasing crop productivity while ensuring efficient use of water. This is particularly important in non-food crops grown for biomass, such as forages. This is because non-food crops are often grown under water-limited conditions (Barbanti et al., 2015). This chapter is therefore focused mainly on moisture stress in forage production systems.

Grazed and harvested forages are undoubtedly the mainstay of livestock production worldwide. Climate change projections and simulations have shown that most continents will experience severe, frequent, and prolonged droughts, coupled with other climate-related extremes such as heat waves and floods (IPCC, 2014). These projections suggest that forage crops are likely to be exposed to more frequent moisture stress conditions. Moisture stress affects plants all the way from germination to reproduction. Forages respond to moisture stress in one of three ways: escape or evasion, avoidance (internal adjustment, e.g., water accumulation), and tolerance (reduced growth). Forage crops that respond through an escape or evasion strategy hasten their growth cycle (from seed to seed) as moisture becomes limiting (Chapman and Carter, 1976). No matter the response strategy, moisture stress reduces germination and results in weak seedlings and poor crop stand. In established forage crops, nodulation in legumes, forage accumulation, and nutritive value are affected by moisture stress

(Abdel-Wahab et al., 2002; Friedericks et al., 1991; Jahanzad et al., 2013; Küchenmeister et al., 2013; Staniak and Harasim, 2018). Understanding how forage crops respond to moisture stress is important for increasing productivity under future climate scenarios.

38.2 WATER USE AND WATER USE EFFICIENCY OF FORAGE CROPS

Water use and water use efficiency in forages vary by crop species, growth form, and crop diversity, among other things (Table 38.1). Alfalfa (*Medicago sativa* L.) water use ranged from as low as 30 mm to more than 200 mm, depending on the time of harvest and irrigation amount. Water use was often greater under adequate moisture supply than in moisture-stressed plants (Collino et al., 2005). The water use efficiency of forage sorghum (*Sorghum bicolor* (L.) Moench) grown for grazing and hay was 5 and 7 MT mL^{-1}, respectively (Monjardino et al., 2015). The water use efficiency of berseem clover (*Trifolium alexandrinum* L.) increased from 1.41 kg m^{-3} to 2.48 kg m^{-3} when irrigation was reduced by 50% (Daneshnia et al., 2015). Similarly, the seasonal crop evapotranspiration of birdsfoot trefoil (*Lotus corniculatus* L.) grown for seed under adequate moisture was 437 mm compared with 246 to 306 mm for moisture-stressed

TABLE 38.1

Soil Water Use (WU) and Water Use Efficiency (WUE) in Response to Forage Mixture Diversity and Nitrogen Fertilizer Application, Sheridan, Wyoming, 2015

Forage Mixture	WU (mm)	WUE (kg mm^{-1})
Alfalfa (100%)	317	50
Alfalfa + meadow brome (50:50%)	306	60
Alfalfa + meadow brome (30:70%)	301	74
Alfalfa + sainfoin + meadow brome (25:25:50%)	305	52
Sainfoin[a] (100%)	310	44
Sainfoin + meadow brome[b] (50:50%)	308	44
Sainfoin + meadow brome (30:70%)	307	57
Alfalfa + birdsfoot trefoil + meadow brome (25:25:50%)	305	67
Birdsfoot trefoil (100%)	318	49
Birdsfoot trefoil + meadow brome (50:50%)	306	54
Birdsfoot trefoil + meadow brome (30:70%)	305	57
Alfalfa + sainfoin + birdsfoot trefoil + meadow brome (16.7: 16.7: 16.7:50%)	312	63
Meadow brome (100%) + 0 kg N ha^{-1}	306	57
Meadow brome (100%) + 56 kg N ha^{-1}	301	67
Meadow brome (100%) + 112 kg N ha^{-1}	314	69
P value	0.013	0.005
LSD[§] (0.05)	16.7	26.6

Adapted from Adjesiwor and Islam (2015).

[a] *Onobrychis viciifolia* Scop.

[b] *Bromus biebersteinii* Roem. & Schult.

plants (Garcia and Steiner, 1999). Total seasonal water application has a strong positive linear relationship with seasonal evapotranspiration (Garcia-Diaz and Steiner, 2000). Garcia-Diaz and Steiner (2000) demonstrated that seasonal birdsfoot trefoil evapotranspiration has a strong linear relationship with total aboveground biomass. Perennial and annual forage crops have different moisture requirements. For example, water use efficiency was greater for annual pastures (30–37 kg ha.mm^{-1}) compared with double-cropped or perennial pastures (21–27 kg ha.mm^{-1}) (Lawson et al., 2009). Also, C_4 crops have greater water use efficiency compared with C_3 forages (Jacobs et al., 2006). This might explain differences in water use between annual and perennial cool-season grass. The water requirement for non-irrigated birdsfoot trefoil ranged from 240 to 255 mm (Garcia and Steiner, 1999). The water use efficiency of orchardgrass (*Dactylis glomerata* L.) cultivars was greater than that of ryegrass (*Lolium perenne* L.) at varying water levels (Jensen et al., 2002). Even water use and water use efficiency vary among cultivars. When soil moisture was not limiting, "Tridan-8" tall fescue (*Schedonorus arundinaceus* (Schreb.) Dumort.) was more efficient in using soil moisture compared with "JS-2002." Conversely, JS-2002 was more efficient in using soil moisture under moisture stress (Barbanti et al., 2015). Soil moisture regime influenced water use efficiency in small-leaflet-size alfalfa populations compared with large-leaflet-size populations (Estill et al., 1993). Similarly, water use efficiency was greater in the pale alfalfa leaf chlorophyll variant (1.47 g kg^{-1}) than in the dark alfalfa leaf chlorophyll variant (1.22 g kg^{-1}) (Estill et al., 1991). Water use efficiency in five African clover species (*Trifolium* spp.) increased by more than 100% under moisture stress conditions (Friedericks et al., 1991).

38.3 DEFICIT IRRIGATION AND DRYLAND FORAGE PRODUCTION

Forage crops are low-value crops mostly grown under rainfed or dryland conditions. Thus, the availability of water throughout the growing season is one of the most limiting factors for improving productivity. Deficit irrigation, defined as the "the application of water below full crop-water requirements (evapotranspiration)" (Fereres and Soriano, 2006), can be a useful management strategy for efficient water use under limited water supply (Jacobs et al., 2006). Jacobs et al. (2006) demonstrated that the most efficient use of irrigation water was a weekly application of 50% of the crop water requirement in perennial pastures. Forage accumulation in three alfalfa cultivars increased by 62.3 to 76.6% under irrigation compared with the rainfed condition. Similarly, forage accumulation in wildrye (*Elymus nutans* Griseb.) increased by 81.5% under irrigation compared with the rainfed condition (Yan et al., 2018). Animal productivity from perennial grass and legume pastures irrigated at 75% of total water requirement was similar to that from fully irrigated pastures. However, reducing irrigation to <50% of total water requirement resulted in substantial reductions in forage accumulation and animal performance (Ates et al., 2013).

Forage accumulation in non-irrigated birdsfoot trefoil was 53% of the least stressed treatment (Garcia-Diaz and Steiner, 2000). Reducing irrigation to 50% of crop water requirement reduced forage accumulation in berseem clover by almost 20% compared with fully irrigated crops (Daneshnia et al., 2015). However, deficit irrigation increased the forage crude protein, water-soluble carbohydrate, and neutral detergent fiber composition of berseem clover (Daneshnia et al., 2016). Water requirement differs among forage species or cultivars. The choice of species or cultivar is critical in forage production under dryland or limited irrigation. For example, when water was not limiting, forage accumulation was greater in orchardgrass compared with meadow bromegrass. However, meadow bromegrass outyielded orchardgrass under limited irrigation (Jensen et al., 2001). Among eight cool-season grasses (orchardgrass, tall fescue, meadow brome, smooth brome [*Bromus inermis* Leyss.], intermediate wheatgrass [*Thinopyrum intermedium* (Host) Barkworth & D.R. Dewey], crested wheatgrass [*Agropyron cristatum* (L.) Gaertn.], and perennial ryegrass), forage accumulation and stability were greatest in meadow brome, orchardgrass, and tall fescue under full irrigation. Under limited irrigation, intermediate wheatgrass produced the greatest forage accumulation (Smeal et al., 2005). In a related study, tall fescue, meadow brome, and orchardgrass produced stable yields even under limited irrigation and thus, are species of choice for limited irrigation conditions (Waldron et al., 2002). Under non-limiting water conditions, tall fescue, prairie grass (*Bromus catharticus* M. Vahl), and kikuyugrass (*Pennisetum clandestinum* Hochst. ex Chiov.) had the greatest mean annual forage accumulation among 15 perennial forages evaluated. When irrigation was reduced to 33% of full irrigation, kikuyugrass accumulated the greatest amount of forage (Neal et al., 2009). Under irrigated conditions, berseem clover and alfalfa were the highest-yielding species. However, forage accumulation was greatest in hairy and winter vetch (*Vicia*, *Lathyrus* spp.) under dryland conditions (Fraser et al., 2004). In alfalfa, 35 to 105 days of irrigation water withdrawal resulted in 16% to 72% reduction in forage accumulation (Takele and Kallenbach, 2001). Similarly, reducing irrigation from 783 to 83 mm (89% reduction) reduced annual forage accumulation from 11.6 to 8.9 Mg ha^{-1} (23% reduction) and from 20.7 to 16.3 Mg ha^{-1} (21% reduction) in sainfoin and alfalfa, respectively (Peel et al., 2003). Neal et al. (2009) reported that among 15 perennial forages, alfalfa was the most tolerant to deficit irrigation. Reducing irrigation to 33% of full irrigation resulted in a 22% reduction in mean annual forage accumulation. Tall fescue cultivars generally did not respond differently to an irrigation gradient. However, the *Neotyphodium* endophyte-infected cultivar ("KY 31") accumulated more forage compared with its endophyte-free counterpart (Asay et al., 2001). Forage accumulation among sainfoin cultivars was different when watering levels were changed. Reducing irrigation from 783 to 83 mm reduced forage accumulation in "Pola" (the highest-yielding sainfoin cultivar) from 20.7 to 16.3 Mg ha^{-1} and in "Artemovsk" (the lowest-yielding sainfoin cultivar) from 9.0 to 7.5 Mg ha^{-1}. This represented a 21% and a 17% reduction in forage accumulation in Pola and Artemovsk, respectively (Peel et al., 2003).

38.4 RESPONSE OF FORAGES TO MOISTURE STRESS

Probably the most obvious response to moisture stress is change in plant morphology. Plants can modify both root and shoot growth in response to moisture stress. One of the most critical response strategies to moisture stress is root growth. Plants growing under moisture stress need to maintain access to soil water through either deeper root growth or larger root volume. This is evident in the greater root mass and greater root to shoot ratio of moisture-stressed plants compared with well-watered plants (Kyriazopoulos et al., 2014). Forage crops such as fenugreek (*Trigonnella foenum-graecum* L.) are also able to extend their roots into deeper layers of the soil to extract water under moisture stress (Dadrasan et al., 2015). Some plants form symbiotic associations with mycorrhizal fungi to increase the root surface area for moisture uptake. Under moisture stress, plants also modify leaf and canopy growth to reduce radiation load and water use (Barker and Caradus, 2001). This is accomplished through reduced leaf size, leaf rolling or curling, slow rate of leaf appearance and expansion, and reduced plant height (Barbanti et al., 2015; Lefi et al., 2004). In forage grasses, moisture stress slows down the appearance of tillers (Bade et al., 1985). Since forage accumulation is a function of the number of tillers and weight per tiller, reduction in the number of tillers reduces forage accumulation under moisture stress (Bade et al., 1985).

Moisture deficit reduces relative plant water content and leaf water potential. Since one of the principal functions of water is maintaining cell turgidity and elongation, a reduction in plant water content slows down the rate of cell expansion. Moisture-stressed plants have a reduced cell water potential. Just as with many crops, moisture stress limits forage crop growth and productivity through interference with photosynthetic processes at the cellular or canopy level. This can occur directly or indirectly through feedback inhibition due to limited photosynthate transport to sink organs (Jones and Corlett, 1992). Moisture deficit affects photosynthesis mainly through reduction in CO_2 diffusion to the chloroplast and metabolic constraints. However, the relative impact of these limitations varies with the intensity, duration, and rate of progression of the moisture stress (Pinheiro and Chaves, 2011). Transpiration rate, stomatal conductance, and net photosynthesis in alfalfa and *Festulolium* hybrid were lower in drought-stressed plants compared with well-watered plants (Staniak et al., 2018). The rate of photosynthesis at the vegetative and reproductive growth stages of orchardgrass, meadow fescue (*Schedonorus pratensis* (Huds.) P. Beauv.), *Festulolium braunii*, and Italian ryegrass (*Lolium multiflorium* Lam.) was reduced under moisture stress (Staniak and Kocon, 2015). Reduction in photosynthesis is related to decreased stomatal conductance, reduction in CO_2 diffusion, and decreased RuBisCo activity. Plants adjust to moisture stress through osmotic adjustment and compatible solute accumulation. These are dehydration adaptations that help attract water into plant cells and maintain turgor.

Thus, forages respond to moisture stress through morphological, anatomical, physiological, and cellular adjustments (Farooq et al., 2009). These responses can be grouped into three categories: escape or evasion, avoidance (or dehydration postponement), and tolerance (protoplasmic tolerance) (Chapman and Carter, 1976; Turner, 1986).

38.4.1 Drought Escape

The drought escape strategy involves plants completing their life cycle before soil moisture becomes limiting (Chapman and Carter, 1976; Islam and Obour, 2014). This involves rapid phenological development and developmental plasticity (Turner, 1986). This is common in annual forage legumes. Annual crops such as subclovers (*Trifolium* spp.) and medics (*Medicago* spp.) complete their life cycle before soil moisture becomes limiting and naturally self-reseed when there is adequate moisture. These crops are therefore important in mitigating the effects of drought on forage production (Ates et al., 2014; Dumont et al., 2015). Similarly, underground vetch (*Vicia sativa*) can naturally reseed through aerial and underground pods, thereby enhancing its drought tolerance (Ates et al., 2014). In perennial forage grasses, summer dormancy and dehydration tolerance are primary adaptations to drought (Norton et al., 2012; Norton and Volaire, 2012).

38.4.2 Dehydration Avoidance

Dehydration avoidance involves mechanisms that maintain water uptake and reduce water loss from plant canopies under water-limiting conditions. Turner (1986) elucidated that plants can postpone dehydration through maintenance of water uptake, maintenance of volume, osmotic adjustment, and reduction in water loss. The reduced water loss is accomplished through stomatal control of respiration. Some plants develop extensive and prolific root systems to maintain water update under limiting soil moisture. This is an important drought avoidance strategy in forage crops such as fenugreek (Dadrasan et al., 2015). Also, symbiotic root association between plant roots and mycorrhizal fungi increases the root surface area for water uptake. Some crops also have waxy leaves that maintain high tissue water potential under moisture stress conditions.

38.4.3 Dehydration Tolerance

Dehydration tolerance involves the ability of plants to maintain physiological functions at very low cell water potentials (Islam and Obour, 2014). This usually involves osmoprotection, antioxidation, and scavenging defense systems. Antioxidant defense under moisture stress requires high antioxidant activities of enzymes (e.g., peroxidase, glutathione reductase, catalase, ascorbate peroxidase, and superoxide dismutase) and high concentrations of non-enzymatic constituents (e.g., cysteine) (Farooq et al., 2009).

38.5 MOISTURE STRESS AND FORAGE ACCUMULATION

It is well known that moisture stress reduces the growth and dry matter accumulation of forages (Carter and Sheaffer, 1983; Grant et al., 2014). This is because crop water use is directly related to growth and forage accumulation (Rogers et al., 2016). Chapman and Carter (1976) elucidated that as plants grow, the roots need to be extended to penetrate deeper layers of the soil to forage for moisture and nutrients to support aboveground growth. Thus, if moisture stress limits root growth, total plant growth and development are reduced. As a rule of thumb, "a plant does not develop more vegetative parts than its roots can support" (Chapman and Carter, 1976). Despite this direct relationship between crop water use and forage accumulation, forage species differ in drought sensitivity. Drought stress reduced forage accumulation in red clover, birdsfoot trefoil, alfalfa, and cicer milkvetch (*Astragalus cicer* L.) by 13%, 21%, 33%, and 19%, respectively (Peterson et al., 1992).

Drought sensitivity can be different among cultivars of the same species, but this depends on the selection of cultivars (Liu et al., 2018; Yan et al., 2018). For example, "Gold Queen" alfalfa is more drought tolerant than the "Suntory" cultivar (Liu et al., 2018). In a related study, Yan et al. (2018) reported that three cultivars of alfalfa evaluated had similar drought sensitivity.

The varying degree of drought sensitivity among forage species suggests that the magnitude of reduction in forage accumulation will differ between monocrops and forage mixtures. Under drought stress, forage accumulation in alfalfa, *Festulolium*, and alfalfa + *Festulolium* mixture was reduced by 32–44%, 10–17%, and 21–28%, respectively (Staniak et al., 2018; Staniak and Harasim, 2018). Similarly, forage accumulation of birdsfoot trefoil, yellow-flowered alfalfa (*Medicago sativa* subsp. *falcata* (L.) Arcang), sainfoin, and white clover (*Trifolium repens* L.) monocultures under drought stress was reduced by 40–47%, 28–42%, 29–49%, and 36–56%, respectively. However, in mixtures with perennial ryegrass, forage accumulation of birdsfoot trefoil, yellow-flowered alfalfa, sainfoin, and white clover under drought stress was reduced by 23–30%, 19–21%, 18–33%, and 44–53%, respectively (Küchenmeister et al., 2013).

The relationship between drought stress and forage accumulation is mostly linear (Figure 38.1), and this has been observed in various forage crops (Rogers et al., 2016). However, this relationship can be quadratic, depending on the forage species (Asay et al., 2001).

38.6 MOISTURE STRESS AND FORAGE NUTRITIVE VALUE

Generally, moisture stress increases forage nutritive value (Grant et al., 2014). In safflower (*Carthanus tinctorius* L.), forage nutritive value decreased with increasing amount of irrigation water (Bar-Tal et al., 2008). The increase in nutritive value under moisture stress is mostly achieved through

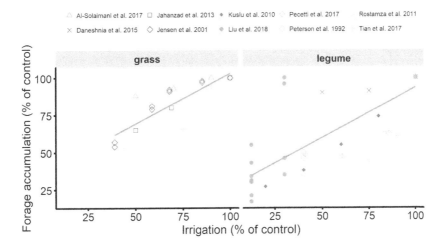

FIGURE 38.1 Relationship between irrigation and forage accumulation in grass and legumes. For irrigation, the control (greatest level of water) in each study was taken as 100%. For forage accumulation, the forage accumulation in the control treatment (greatest level of water) in each study was taken as 100%. References in the legend are the sources of data. (Al-Solaimani et al., 2017; Daneshnia et al., 2015; Jahanzad et al., 2013; Jensen et al., 2001; Kuslu et al., 2010; Liu et al., 2018; Pecetti et al., 2017; Peterson et al., 1992; Rostamza et al., 2011; Tian et al., 2017.)

reduction in fiber concentration, increased crude protein, and increased digestibility (Dumont et al., 2015; Jahanzad et al., 2013; Küchenmeister et al., 2013; Peterson et al., 1992). In addition, an increase in water-soluble carbohydrate concentration in forages has been reported under moisture stress (Jahanzad et al., 2013; Küchenmeister et al., 2013). Peterson et al. (1992) elucidated that improved nutritive value under drought stress conditions could be due to delayed maturity and greater leaf to stem ratio. The effects of moisture stress on forage crude protein concentration, fiber concentration, and forage digestibility are discussed in the following subsections.

38.6.1 Crude Protein Concentration

Crude protein concentration often increases with deficit irrigation or moisture stress (Figures 38.2 and 38.3). A linear relationship between moisture level and nutritive value has been reported in forage crops (Robins, 2016). The relationship is stronger in grasses (Figure 38.2) than in legumes (Figure 38.3) possibly because of the number of studies considered. Studies have reported increased crude protein under moisture stress. This spans grasses, legumes, and brassicas (Jahanzad et al., 2013; Keshavarz Afshar et al., 2012). The crude protein concentration of alfalfa and Elymus (*Elymus nutans* Griseb) was greater under irrigation compared with rainfed conditions (Yan et al., 2018).

38.6.2 Fiber Concentration and Digestibility

Forage digestibility is determined by fiber concentration. Thus, reduction in fiber may improve digestibility. Neutral detergent fiber (NDF) increased with increasing irrigation level, and digestibility decreased with increasing irrigation level (Figures 38.2 and 38.3), but these linear relationships were not strong. In alfalfa, the NDF concentration decreased about 12% under moisture stress (Pecetti et al., 2017).

Moisture stress increased the leaf to stem ratio of alfalfa, and this resulted in increased digestibility (Carter and Sheaffer, 1983). In three old world bluestems (*Bothriochloa* spp.), fiber concentration decreased while digestibility increased under moisture stress conditions (Philipp et al., 2005). In safflower, acid detergent fiber (ADF) and NDF increased with increasing irrigation level (Bar-Tal et al., 2008). Similarly, safflower digestibility decreased with increasing irrigation level (Bar-Tal et al., 2008). Forage digestibility increased by about 7% under drought conditions (Dumont et al., 2015). Highly stressed forage legumes had a reduced fiber concentration (Küchenmeister et al., 2013; Peterson et al., 1992). This is because moderate moisture stress often delays plant maturity, and this maintains the forage nutritive value at high levels (Buxton, 1996). In a related study, sorghum digestibility increased with deficit irrigation (Carmi et al., 2006; Jahanzad et al., 2013). In contrast, berseem clover digestibility decreased with deficit irrigation (Daneshnia et al., 2016). Thus, the effect of moisture stress on forage nutritive value might be species-dependent.

38.7 MANAGEMENT OPTIONS UNDER MOISTURE STRESS

The best management for moisture stress is irrigation. However, moisture stress arises because of lack of irrigation facilities or a limited supply of irrigation water to supplement precipitation. While there might be limited options for forage yield increase under moisture stress conditions in the short term, adopting best management practices could minimize stand losses and maintain forage yields until soil moisture improves (Lemus, 2011). Moisture stress in forages can be managed through breeding for drought tolerance, selection of appropriate species and cultivars, and adoption of best management practices. The objective of adjusting agronomic practices is to ensure that the most sensitive crop stages do not coincide with moisture stress (Farooq et al., 2009). Similarly,

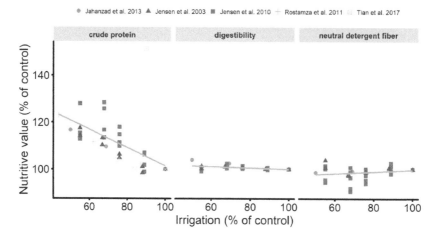

FIGURE 38.2 Relationship between irrigation and forage nutritive value (crude protein, dry matter digestibility, and neutral detergent fiber) of grasses. The control (greatest level of water) in each study was taken as 100%. Nutritive value in the control treatment (greatest level of water) in each study was taken as 100%. References in the legend are the sources of data. (Jahanzad et al., 2013; Jensen et al., 2003; Jensen et al., 2010; Rostamza et al., 2011; Tian et al., 2017.)

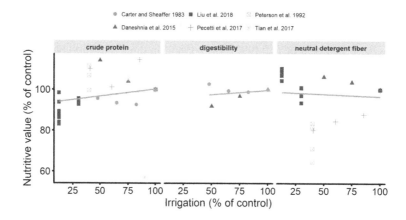

FIGURE 38.3 Relationship between irrigation and forage nutritive value (crude protein, dry matter digestibility, and neutral detergent fiber) of legumes. The control (greatest level of water) in each study was taken as 100%. Nutritive value in the control treatment (greatest level of water) in each study was taken as 100%. References in the legend are the sources of data. (Carter and Sheaffer, 1983; Daneshnia et al., 2015; Liu et al., 2018; Pecetti et al., 2017; Peterson et al., 1992; Tian et al., 2017.)

breeding for drought tolerance entails producing plant cultivars that are high-yielding and can adapt to a wide range of growing environments. These management practices are discussed in the following subsections.

38.7.1 Species and Cultivar Selection

Forage species and cultivars have varied sensitivity to moisture stress. Generally, warm-season grasses such as Bermudagrass (*Cynodon dactylon* (L.) Pers.), bahiagrass (*Paspalum notatum* Flueggé), and blue grama (*Bouteloua gracilis* (Willd. ex Kunth) Lag. ex Griffiths) are more drought tolerant (Islam and Obour, 2014). However, where cool-season grasses are desirable, forages species such as tall fescue, meadow brome, and orchardgrass can produce stable yields under limited irrigation and moisture stress (Waldron et al., 2002). In addition, forage legumes such as sainfoin and alfalfa have deep

taproots that can forage for soil moisture under moisture stress conditions. Thus, sainfoin is a drought-tolerant species that performs well under moisture stress (Carbonero et al., 2011; Sintim et al., 2016). Under dryland conditions in a semi-arid climate, "Remont," "Shoshone," and "Rocky Mountain" sainfoin accumulated more forage than "Delaney" (Sintim et al., 2016). Similar differences in moisture stress sensitivity have been reported in "Pola" and "Artemovsk" sainfoin cultivars (Peel et al., 2003). In tall fescue, the *Neotyphodium* endophyte-infected cultivar ("KY 31") accumulated more forage compared with its endophyte-free counterpart under limited irrigation (Asay et al., 2001). Thus, cultivar selection is important for increasing forage accumulation under moisture stress. Grass–legume mixtures comprising compatible drought-tolerant grasses and legumes can improve forage accumulation and seasonal distribution during moisture stress (Islam and Obour, 2014).

38.7.2 Grazing and Haying Management

Soil moisture is critical for the regrowth of perennial forages after grazing or clipping. Thus, grazing or clipping must be avoided during moisture stress conditions to prolong persistence and forage productivity. Overgrazing or clipping forages before moisture stress is not advisable, as this reduces carbohydrate storage in the roots. These carbohydrate reserves are important for plant regrowth and survival under moisture stress conditions. Thus, avoiding clipping or grazing too close to ground level (i.e. leaving high stubbles) will reduce stand losses. Also, high stubble height ensures that there are enough auxiliary buds for regrowth. This is difficult to achieve, because forage growth is slow and limited under moisture stress. Thus, the natural response will be to graze any green material available. While this might provide short-term relief, it will have long-term economic implications.

38.7.3 Fertility Management

Good soil fertility management improves forage productivity under an optimum soil moisture regime. However, application of fertilizers under moisture stress conditions is not advisable. Fertilizer application when soil moisture is adequate ensures good plant growth and the accumulation of carbohydrates essential for stand survival under moisture stress. Nitrogen volatilization increases under moisture stress conditions. This is especially true for urea fertilizers. Fertilizer volatilization under moisture stress can increase plant tissue damage. This will slow down the rate of plant recovery from drought stress (Lemus, 2011). It is not recommended to apply phosphorus and potassium fertilizers under moisture stress. However, in situations when these fertilizers are applied before moisture stress becomes severe, a routine soil test after the drought period will be necessary to determine the amount of residual phosphorus and potassium. This is because of reduced nutrient uptake due to less forage accumulation during moisture stress. Forage crops such as oats (*Avena* spp.), triticale (*Triticosecale rimpaui* C. Yen & J.L. Yang [*Secale cereale* × *Triticum aestivum*]), bromegrass (*Bromus* spp.), orchardgrass, millet, sorghum, fescue, Sudangrass (*Sorghum bicolor* (L.) Moench), corn (*Zea mays* L.), alfalfa, and sweetclover (*Melilotus* spp.) can accumulate toxic amounts of nitrate under drought stress. In addition, weeds such as quackgrass (*Elymus repens* (L.) Gould), common lambsquarters (*Chenopodium album* L.), kochia (*Kochia scoparia* (L.) Schrad.), and pigweeds (*Amaranthus* spp.) accumulate nitrate at toxic levels under moisture stress (Cash et al., 2006; Glunk et al., 2015). Split application of nitrogen could prevent excessive nitrate accumulation in forage crops and weeds. Also, since weeds tend to thrive more under moisture stress and are, therefore, more likely to be grazed by livestock, weed control could help prevent nitrate poisoning.

38.8 CONCLUSIONS

Moisture stress affects forage crops at different stages of growth. Regardless of the forage growth stage, forage crops initiate morphological, physiological, and cellular coping strategies to survive and grow. These strategies can be in the form of escape, avoidance, or tolerance. These processes, when initiated, affect forage accumulation and nutritional value. Although moisture stress often increases forage nutritive value, the reduction in forage accumulation outweighs any benefits in improved nutritive value. Moisture stress is unavoidable in forage production systems. However, producers can adopt best management practices (e.g. weed control, nutrient management, select appropriate cultivars and drought-tolerant species), and adjust management practices (e.g. planting date, grazing and harvest timing, grazing management, stubble height) to reduce the impact of moisture stress on forage crops. In addition, breeding crops with combined tolerance to drought and heat will improve forage crop productivity and sustainability.

REFERENCES

Abdel-Wahab AM, Shabeb MSA, Younis MAM (2002) Studies on the effect of salinity, drought stress and soil type on nodule activities of *Lablab purpureus* (L.) sweet (Kashrangeeg). *J Arid Environ* 51:587–602.

Adjesiwor AT, Islam MA (2015) Grass, legumes, and grass-legume mixtures: yield, nutritive value, and soil water use. *Proceedings of the Western Alfalfa and Forage Symposium Reno*, Nevada, USA.

Al-Solaimani SG, Alghabari F, Ihsan MZ, Fahad S (2017) Water deficit irrigation and nitrogen response of Sudan grass under arid land drip irrigation conditions. *Irrig Drain* 66:365–376.

Asay KH, Jensen KB, Waldron BL (2001) Responses of tall fescue cultivars to an irrigation gradient. *Crop Sci* 41:350–357.

Ates S, Feindel D, El Moneim A, Ryan J (2014) Annual forage legumes in dryland agricultural systems of the West Asia and North Africa Regions: research achievements and future perspective. *Grass Forage Sci* 69:17–31.

Ates S, Isik S, Keles G, Aktas AH, Louhaichi M, Nangia V (2013) Evaluation of deficit irrigation for efficient sheep production from permanent sown pastures in a dry continental climate. *Agric Water Manage* 119:135–143.

Bade D, Conrad B, Holt E (1985) Temperature and water stress effects on growth of tropical grasses. *J Range Manage* 38: 321–324.

Barbanti L, Sher A, Di Girolamo G, Cirillo E, Ansar M (2015) Growth and physiological response of two biomass sorghum (*Sorghum bicolor* (L.) Moench) genotypes bred for different environments, to contrasting levels of soil moisture. *Ital J Agron* 10:208–214.

Barker DJ, Caradus JR (2001) Adaptation of forage species to drought. In Proceedings of the Xix International Grassland Congress, pp. 241–246.

Bar-Tal A, Landau S, Li-Xin Z, Markovitz T, Keinan M, Dvash L, Brener S, Weinberg ZG (2008) Fodder quality of safflower across an irrigation gradient and with varied nitrogen rates. *Agron J* 100:1499–1505.

Buxton DR (1996) Quality-related characteristics of forages as influenced by plant environment and agronomic factors. *Anim Feed Sci Tech* 59:37–49.

Carbonero CH, Mueller-Harvey I, Brown TA, Smith L (2011) Sainfoin (*Onobrychis viciifolia*): a beneficial forage legume. *Plant Genet Resour* 9:70–85.

Carmi A, Aharoni Y, Edelstein M, Umiel N, Hagiladi A, Yosef E, Nikbachat M, Zenou A, Miron J (2006) Effects of irrigation and plant density on yield, composition and in vitro digestibility of a new forage sorghum variety, Tal, at two maturity stages. *Anim Feed Sci Tech* 131:120–132.

Carter P, Sheaffer C (1983) Alfalfa response to soil water deficits. I. Growth, forage quality, yield, water use, and water-use efficiency 1. *Crop Sci* 23:669–675.

Cash D, Funston R, King M, Wichman D (2006) Nitrate Toxicity of Montana Forages. Montana State University Extension. MontGuide, MT 200205 AG. Available online: http://ani malrangeextension.montana.edu/range/grazing-management/ gm-nitrate-toxicity.html. Accessed September 8, 2018.

Chapman SR, Carter LP (1976) *Crop Production: Principles and Practices.* San Francisco, CA: W.H. Freeman.

Collino DJ, Dardanelli JL, De Luca MJ, Racca RW (2005) Temperature and water availability effects on radiation and water use efficiencies in alfalfa (*Medicago sativa* L.). *Aust J Exp Agric* 45:383–390.

Dadrasan M, Chaichi MR, Pourbabaee AA, Yazdani D, Keshavarz-Afshar R (2015) Deficit irrigation and biological fertilizer influence on yield and trigonelline production of fenugreek. *Ind Crop Prod* 77:156–162.

Daneshnia F, Amini A, Chaichi MR (2015) Surfactant effect on forage yield and water use efficiency for berseem clover and basil in intercropping and limited irrigation treatments. *Agric Water Manage* 160:57–63.

Daneshnia F, Amini A, Chaichi MR (2016) Berseem clover quality and basil essential oil yield in intercropping system under limited irrigation treatments with surfactant. *Agric Water Manage* 164:331–339.

Dumont B, Andueza D, Niderkorn V, Luscher A, Porqueddu C, Picon-Cochard C (2015) A meta-analysis of climate change effects on forage quality in grasslands: specificities of mountain and Mediterranean areas. *Grass Forage Sci* 70:239–254.

Estill K, Delaney RH, Ditterline RL, Smith WK (1993) Water relations and productivity of alfalfa populations divergently selected for leaflet size. *Field Crop Res* 33:423–434.

Estill K, Delaney RH, Smith WK, Ditterline RL (1991) Water relations and productivity of alfalfa leaf chlorophyll variants. *Crop Sci* 31:1229–1233.

Farooq M, Wahid A, Kobayashi N, Fujita D, Basra S (2009) *Plant Drought Stress: Effects, Mechanisms and Management,* pp. 153–188. Sustainable agriculture: Springer.

Fereres E, Soriano MA (2006) Deficit irrigation for reducing agricultural water use. *J Exp Bot* 58:147–159.

Fraser J, McCartney D, Najda H, Mir Z (2004) Yield potential and forage quality of annual forage legumes in southern Alberta and northeast Saskatchewan. *Can J Plant Sci* 84:143–155.

Friedericks JB, Hagedorn C, Reneau RB (1991) Evaluation of African annual clovers to moisture stress in 2 Ethiopian highland soils. *Plant Soil* 133:271–279.

Garcia C, Steiner J (1999) Birdsfoot trefoil seed production: I. Crop-water requirements and response to irrigation. *Crop Sci* 39:775–783.

Garcia-Diaz CA, Steiner JJ (2000) Birdsfoot trefoil seed production: II. Plant-water status on reproductive development and seed yield. *Crop Sci* 40:449–456.

Gimenez C, Gallardo M, Thompson RB (2005) Plant–water relations. In Hillel D, ed. *Encyclopedia of Soils in the Environment,* pp. 231–238. Oxford: Elsevier.

Glunk E, Olson-Rutz Kathrin , King M, Wichman D, Jones C (2015) Nitrate Toxicity of Montana Forages. Available online: http: //animalrangeextension.montana.edu/forage/nitrate_toxic ity.html. Montana State University Extension, MontGuide MT200205AG. 8 p. Accessed September 8, 2018.

Grant K, Kreyling J, Dienstbach LFH, Beierkuhnlein C, Jentsch A (2014) Water stress due to increased intra-annual precipitation variability reduced forage yield but raised forage quality of a temperate grassland. *Agric Ecosyst Environ* 186:11–22.

Ibrahim YM (1995) Response of sorghum genotypes to different water levels created by Sprinkler irrigation. *Ann Arid Zone* 34:283–287.

IPCC (2014) Climate Change 2014: Synthesis Report. Contribution of Working Groups I, II and III to the Fifth Assessment Report of the Intergovernmental Panel on Climate Change [Core Writing Team, R.K. Pachauri and L.A. Meyer (eds.)]. IPCC, Geneva, Switzerland, 151 p.

Islam MA, Obour AK (2014) Drought physiology of forage crops. In *Handbook of Plant and Crop Physiology*, pp. 456–469. CRC Press.

Jacobs JL, Ward GN, McKenzie FR, Kearney G (2006) Irrigation and nitrogen fertiliser effects on dry matter yield, water use efficiency and nutritive characteristics of summer forage crops in south-west Victoria. *Aust J Exp Agric* 46:1139–1149.

Jahanzad E, Jorat M, Moghadam H, Sadeghpour A, Chaichi MR, Dashtaki M (2013) Response of a new and a commonly grown forage sorghum cultivar to limited irrigation and planting density. *Agric Water Manage* 117:62–69.

Jensen KB, Asay KH, Johnson DA, Waldron BL (2002) Carbon isotope discrimination in orchardgrass and ryegrasses at four irrigation levels. *Crop Sci* 42:1498–1503.

Jensen KB, Asay KH, Waldron BL (2001) Dry matter production of orchardgrass and perennial ryegrass at five irrigation levels. *Crop Sci* 41:479–487.

Jensen KB, Waldron BL, Asay KH, Johnson DA, Monaco TA (2003) Forage nutritional characteristics of orchardgrass and perennial ryegrass at five irrigation levels. *Agron J* 95:668–675.

Jensen KB, Waldron BL, Peel MD, Robins JG (2010) Nutritive value of herbage of five semi-irrigated pasture species across an irrigation gradient. *Grass Forage Sci* 65:92–101.

Jones H, Corlett J (1992) Current topics in drought physiology. *J Agric Sci* 119:291–296.

Keshavarz Afshar R, Chaichi M, Moghadam H, Ehteshami SM (2012) Irrigation, phosphorus fertilizer and phosphorus solubilizing microorganism effects on yield and forage quality of turnip (*Brassica rapa* L.) in an arid region of Iran. *Agric Res*: 1:370–378.

Küchenmeister K, Küchenmeister F, Kayser M, Wrage-Mönnig N, Isselstein J (2013) Influence of drought stress on nutritive value of perennial forage legumes. *Int J Plant Prod* 7:693–710.

Kuslu Y, Sahin U, Tunc T, Kiziloglu FM (2010) Determining water-yield relationship, water use efficiency, seasonal crop and pan coefficients for alfalfa in a semiarid region with high altitude. *Bulg J Agric Sci* 16:482–492.

Kyriazopoulos AP, Orfanoudakis M, Abraham EM, Parissi ZM, Serafidou N (2014) Effects of arbuscular mycorrhiza fungi on growth characteristics of *Dactylis glomerata* L. under drought stress conditions. *Not Bot Horti Agrobo* 42:132–137.

Lawson AR, Greenwood KL, Kelly KB (2009) Irrigation water productivity of winter-growing annuals is higher than perennial forages in northern Victoria. *Crop Pasture Sci* 60:407–419.

Lefi E, Gulias J, Cifre J, Ben Younes M, Medrano H (2004) Drought effects on the dynamics of leaf production and senescence in field-grown *Medicago arborea* and *Medicago citrina*. *Ann Appl Biol* 144:169–176.

Lemus R (2011) Pasture and Grazing Management under Drought Conditions. Mississippi State University Extension Service. Forage News 4: 1–3. Available online: https://www.nrcs.usd

a.gov/Internet/FSE_DOCUMENTS/nrcs142p2_016821.pdf. Accessed September 7, 2018.

Liu Y, Wu Q, Ge G, Han G, Jia Y (2018) Influence of drought stress on alfalfa yields and nutritional composition. *BMC Plant Biol* 18:13.

Monjardino M, MacLeod N, McKellar L, Prestwidge D (2015) Economic evaluation of irrigated forage production in a beef cattle operation in the semi-arid tropics of northern Australia. *Agric Syst* 139:122–143.

Neal JS, Fulkerson WJ, Lawrie R, Barchia IM (2009) Difference in yield and persistence among perennial forages used by the dairy industry under optimum and deficit irrigation. *Crop Pasture Sci* 60:1071–1087.

Norton M, Volaire F (2012) Selection of pasture and forage species adapted to changing environmental conditions in Mediterranean climates. In Acar Z, López-Francos A, Porqueddu C eds. *New Approaches for Grassland Research in a Context of Climate and Socio-Economic Changes.* CIHEAM Zaragoza. *Opt Méditerr A* 102:119–127.

Norton MR, Lelievre F, Volaire F (2012) Summer dormancy in *Phalaris aquatica* L., the influence of season of sowing and summer moisture regime on two contrasting cultivars. *J Agron Crop Sci* 198:1–13.

Pecetti L, Annicchiarico P, Scotti C, Paolini M, Nanni V, Palmonari A (2017) Effects of plant architecture and drought stress level on lucerne forage quality. *Grass Forage Sci* 72:714–722.

Peel MD, Asay KH, Johnson DA, Waldron BL (2003) Forage production of sainfoin across an irrigation gradient. *Crop Sci* 44:614–619.

Peterson P, Sheaffer C, Hall M (1992) Drought effects on perennial forage legume yield and quality. *Agron J* 84:774–779.

Philipp D, Allen V, Mitchell R, Brown C, Wester D (2005) Forage nutritive value and morphology of three old world bluestems under a range of irrigation levels. *Crop Sci* 45:2258–2268.

Pinheiro C, Chaves M (2011) Photosynthesis and drought: can we make metabolic connections from available data? *J Exp Bot* 62:869–882.

Robins JG (2016) Evaluation of warm-season grasses nutritive value as alternatives to cool-season grasses under limited irrigation. *Grassl Sci* 62:144–150.

Rogers ME, Lawson AR, Kelly KB (2016) Lucerne yield, water productivity and persistence under variable and restricted irrigation strategies. *Crop Pasture Sci* 67:563–573.

Rostamza M, Chaichi MR, Jahansouz MR, Alimadadi A (2011) Forage quality, water use and nitrogen utilization efficiencies of pearl millet (*Pennisetum americanum* L.) grown under different soil moisture and nitrogen levels. *Agric Water Manage* 98:1607–1614.

Sintim HY, Adjesiwor AT, Zheljazkov VD, Islam MA, Obour AK (2016) Nitrogen application in sainfoin under rain-fed conditions in Wyoming: productivity and cost implications. *Agron J* 108:294–300.

Smeal D, O'Neill MK, Arnold RN (2005) Forage production of cool-season pasture grasses as related to irrigation. *Agric Water Manage* 76:224–236.

Staniak M, Bojarszczuk J, Księżak J (2018) Changes in yield and gas exchange parameters in Festulolium and alfalfa grown in pure sowing and in mixture under drought stress. *Acta Agric Scand B Soil Plant Sci* 68:255–263.

Staniak M, Harasim E (2018) Changes in nutritive value of alfalfa (*Medicago × varia* T. Martyn) and Festulolium (*Festulolium braunii* (K. Richt) A. Camus) under drought stress. *J Agron Crop Sci* 204:456–466.

Staniak M, Kocon A (2015) Forage grasses under drought stress in conditions of Poland. *Acta Physiol Plant* 37:1–10.

Takele E, Kallenbach R (2001) Analysis of the impact of alfalfa forage production under summer water-limiting circumstances on productivity, agricultural and growers returns and plant stand. *J Agron Crop Sci-Zeitschrift Fur Acker Und Pflanzenbau* 187:41–46.

Tian Y, Liu Y, Jin J (2017) Effect of irrigation schemes on forage yield, water use efficiency, and nutrients in artificial grassland under arid conditions. *Sustainability* 9:2035.

Turner NC (1986) Crop water deficits: a decade of progress. In *Advances in Agronomy*, pp. 1–51, ed. N.C. Brady. Elsevier.

Waldron BL, Asay KH, Jensen KB (2002) Stability and yield of cool-season pasture grass species grown at five irrigation levels. *Crop Sci* 42:890–896.

Yan YL, Wan ZQ, Chao R, Ge YQ, Chen YL, Gu R, Gao QZ, Yang J (2018) A comprehensive appraisal of four kinds of forage under irrigation in Xilingol, Inner Mongolia, China. *Rangeland J* 40:171–178.

39 Responses of Medicinal Plants to Abiotic Stresses

Masarovičová Elena, Kráľová Katarína, Vykouková Ivana, and Kriššáková Zuzana

CONTENTS

39.1 INTRODUCTION

The survival of plants (including medicinal plants) is considerably dependent on their ability to adapt to the environmental conditions under which they live. Because of their immobility, plants need to recognize and respond to external stimuli sensitively and properly to ensure their survival. Abiotic stress responses in plants occur at various organ levels among which the root-specific processes are of particular importance. Under normal growth conditions, the root absorbs water and nutrients from the soil and supplies them throughout the plant body, thereby playing a pivotal role in maintaining cellular homeostasis. However, this balanced system is altered during the stress period when roots are forced to adopt several structural and functional modifications. Examples of these modifications include phenotypic, cellular, and molecular changes such as alteration of metabolism and membrane characteristics, hardening of the cell wall, and reduction of root length. These changes are often caused by a single or combined effect of several abiotic stress responsive pathways that can be best explored at the global level using high-throughput approaches such as proteomics (Ghosh and Xu 2014). Plant neurobiology, as a recently focused field of plant biology research, aims to understand how plants process the information they obtain from their environment to optimally develop, prosper, and reproduce. The behavior plants exhibit is coordinated across the whole organism by some form of integrated signaling, communication, and response system. This system includes long-distance electrical signals, vesicle-mediated transport of auxin in specialized vascular tissue, and the production of chemicals known to be neuronal in animals (Brenner et al. 2006; Baluška and Mancuso 2009). Plant neurobiology emphasizes the transit section of root apices (Baluška et al. 2010) considered as the "brain-like" organ of the plant where the synchronization of electrical activity was found (Baluška et al. 2009). It should be emphasized that these findings are extraordinarily important for plants that are under stress conditions.

39.2 HISTORY OF MEDICINAL PLANTS APPLICATION

Medicinal plants (often called medicinal herbs) have been discovered and used in traditional medicine practices (phytotherapy) since prehistoric times. Moreover, these plants synthesize hundreds of biologically active chemical compounds for functions including the defense against insects, fungi, diseases, and herbivorous animals (especially mammals). Wild medicinal plant species as a substantial component of natural phytocoenoses are also responsible for the biodiversity and stability of natural ecosystems.

39.2.1 INTRODUCTION

The earliest historical records of herbs originated from the Sumerian civilization (c. 4500–c. 1900 BC), where hundreds of medicinal plants (MP) were listed on clay tablets. The *Ebers Papyrus* from ancient Egypt describes over 850 plant medicines, while *Dioscorides* documented over 1000 recipes for medicines using over 600 MPs forming the basis of pharmacopoeias for some 1500 years (Ahn 2017).

Plants, including many now used culinary herbs and spices, have been used as medicines (not necessarily effectively) from prehistoric times. Spices have been used partly to counter food spoilage bacteria, especially in hot climates (Tapsell et al. 2006; Billing and Sherman, 1998) and also in meat dishes which spoiled more readily (Sherman and Hash 2001). In ancient Sumeria, hundreds of medicinal plants including myrrh and opium are listed on clay tablets. The ancient Egyptian *Ebers Papyrus* lists over 800 plant medicines such as aloe, cannabis, castor bean, garlic, juniper, and mandrake (Sumner 2008). In the Early Middle Ages, Benedictine monasteries preserved medicinal knowledge in Europe, translating and copying Classical texts and maintaining herb gardens (van Arsdall 2002; Mills 2000). The Early Modern period saw the flourishing of illustrated herbals across Europe, starting with the 1526 *Grete Herball*. John Gerard wrote his famous *The Herball or General History of Plants* in 1597, based on a publication named *The English Physician Enlarged* (Singer 1923). Many new plant medicines arrived in Europe as products of Early Modern exploration and resulting Columbian Exchange, in which livestock, crops, and technologies were transferred between the Old World and the Americas in the fifteenth and sixteenth centuries (cf. Gimmel 2008).

Collins (2000) published comprehensive monograph *Medieval Herbals: The Illustrative Traditions* representing a new, wide-ranging, and generously illustrated study of manuscript herbals produced between 601–1450 (seventh to fifteenth century). This book examines the two principal herbal traditions of Classical descent: the Dioscorides manuscripts in Greek, Arabic, and Latin and the *Latin Herbarius* of Apulcius Platonicus. It shows, from 1300, the illustrations of the *de herbis Traetatus* treatises, the first of which is in the British Library. The herbal manuscript, Egerton MS 747 in the British Library, showed a new observation of nature, paving the way in the fifteenth century for the French *Livres des Simples* and the magnificent plant paintings of later Italian herbals. Medieval herbals provide some of the few syntheses in English of existing research on the subject along with the issues of dating, location, production, and ownership of the individual codices. The above-mentioned author, Minta Collins, demonstrates how many herbals were not only codices for medical scholars but expensively illustrated books for bibliophiles, of equal interest to students of manuscripts, to historians of medicine and botany, and to art historians. Some years later, Tobyn et al. (2010) provided, in the illustrated color guide *Western Herbal Tradition: 2000 Years of Medicinal Plant Knowledge*, a comprehensive exploration of 27 plants that are central to the herbalist's repertoire and they offer analysis of these herbs through the examination of historical texts and discussion of current applications and research.

Drug research makes use of ethnobotany to search for pharmacologically active substances in nature and has in this way discovered hundreds of useful compounds. MPs are widely used in non-industrialized societies, mainly because they are readily available and cheaper than modern medicines. In many countries, there is little regulation of traditional

medicine, the World Health Organization (WHO) coordinates a network to encourage safe and rational usage. MPs face both general threats, such as climatic change and habitat destruction, and the specific threat of over-collection to meet market demand (Ahn 2017).

Medicinal plants may provide three main kinds of benefit: health benefits to the people who consume them as medicines; financial benefits to people who harvest, process, and distribute them for sale; and society-wide benefits, such as job opportunities, taxation income, and a healthier labor force (Smith-Hall et al. 2012). However, the development of plants or extracts having potential medicinal uses is blunted by weak scientific evidence, poor practices in the process of drug development, and insufficient financing (Ahn 2017).

The place of plants in medicine was radically altered in the nineteenth century by the application of chemical analysis. As chemistry progressed, additional classes of pharmacologically active substances were discovered in MPs (Petrovska 2012; Atanasov et al. 2015). Around the end of the nineteenth century, the mood of pharmacy turned against medicinal plants, as enzymes often modified the active ingredients when whole plants were dried, and alkaloids and glycosides purified from plant material started to be preferred (Petrovska 2012). Drug discovery from plants continued to be important through the twentieth and into twenty-first century, with important anti-cancer drugs from yew-tree (*Taxus baccata*) and a plant species named Madagaskar periwinkle (*Catharanthus roseus*) (Atanasov et al. 2015).

At present, a medicinal plant is a plant that is used in an attempt to maintain health, to be administered for a specific condition, or both, whether in modern medicine or in traditional medicine (Smith-Hall et al. 2012; Ahn 2017). The "Food and Agriculture Organization" estimated in 2002 that over 50,000 medicinal plants are used across the world (Schippmann et al. 2002). The "Royal Botanic Gardens" (KEW, United Kingdom) more conservatively estimated in 2016 that 17,810 plant species have medicinal use, out of some 30,000 plants for which use of any kind is documented (Anonymus 2016). In modern medicine, around a quarter of the drugs prescribed to patients are derived from medicinal plants, and they are rigorously tested (Farnsworth et al. 1985; Smith-Hall et al. 2012). In other systems of medicine, medicinal plants may constitute the majority of what are often informal attempted treatments, not tested scientifically (Tilburt and Kaptchuk 2008). The use of plant-based materials including herbal or natural health products with supposed health benefits is increasing in developed countries (Ekor 2014).

It could be concluded, that plant medicines are still in wide use around the world. In most of the developing world, especially in rural areas, local traditional medicine, including herbalism, is the only source of health care for people, while in the developed world, alternative medicine including the use of dietary supplements is marketed aggressively using the claims of traditional medicine (Dharmananda 1979). WHO has set out a strategy for traditional medicines with four objectives: (I) to integrate then as policy into national health care system, (II) to provide knowledge and guidance on their safety, efficacy, and quality, (III) to increase their availability and affordability, (IV) and to promote their rational, therapeutically sound usage (Saslis-Lagoudakis et al. 2012).

39.3 DROUGHT

Throughout evolution, plants have faced extreme variations in the environment, and they have survived and adapted themselves under different ecological conditions. Drought is one of the most serious threats to plant production all over the world and is likely to worsen with anticipated changes in climate conditions. Drought impairs normal growth, disturbs water relations, reduces water-use efficiency and negatively affects the other physiological processes in plants, as well (Kumar et al. 2018). Drought stress in plants is caused by the disturbance of water flow through the xylem. Interrupting the flow of water causes a decrease in cell turgor pressure. The decrease in turgor affects the process of mitosis, cell elongation, and expansion, as a result of which plant growth and development are severely affected and production decreases (Chhapekar et al. 2018). Plants, however, have a variety of physiological and biochemical responses at organism and cellular levels, making it a more complex phenomenon. Researchers have been trying to understand and dissect the mechanisms of plant tolerance to drought stress using various approaches (Kumar et al. 2018).

A drought is a physiological form of water deficit where the soil water available to the plants is inadequate, which adversely affects the plant's metabolism. However, plants possess multiple morphological (reduced leaf area, reduced stem length, leaf molding, wax content, efficient rooting system, number of branches), physiological (transpiration (E) water-use efficiency (WUE), osmotic adjustment, stomatal size, and activity, respectively), and biochemical responses (accumulation of proline, polyamine, trehalose, increasing of nitrate reductase activity, and storage of carbohydrate) under drought stress, making it a more complex phenomenon (Haworth et al. 2013). Of the various plant responses to water deficiency, enhanced abscisic acid (ABA) accumulation is one of the key mechanisms of adaptation to water stress (Bano et al. 2012).

Stomatal response, reactive oxygen species (ROS) scavenging, metabolic changes, and photosynthesis are mainly affected when plants are exposed to water stress. Moreover, plants accumulate biomolecules that are harmless and do not interfere with other biochemical processes. They can include protective proteins such as dehydrins, heat shock proteins (HSPs), late embryogenesis abundant (LEA) proteins (Lipiec et al. 2013), osmolytes like proline, trehalose, and sugars (Ilhan et al. 2015), glycine and betaine (Wang et al. 2010). Some signaling molecules include polyamines (Rangan et al. 2014), inositol (Sengupta et al. 2008), and hormones like ABA (Saradhi et al. 2000), ethylene and methyl jasmonate (Bartels and Sunkar 2005). In future years, scientists will have to develop new varieties not only for crops but also for medicinal plants through classical breeding tools or using new genomic approaches like molecular-assisted breeding, transgenic or genome editing technologies. To achieve success in this

direction it is essential to understand the mechanisms of plant perception to abiotic stress, the signaling pathways, and the identification of genes that are responsible for plant stress tolerance (cf. Chaves et al. 2003; Kulkarni et al. 2017). However, in nature exists not only water deficit in the soil but also in the air. Water content in the atmosphere (atmospheric humidity) is expressed as vapor pressure deficit (VPD). Many studies have also reported a high VPD reduced plant growth because of depressed photosynthesis. Namely, during photosynthesis, the pathway of CO_2 diffusion from the atmosphere to the sites of carboxylation is dominated by stomatal and mesophyll conductance (e.g., Du et al. 2018).

Leaf pigments (chlorophyll (Chl) a, b; carotenoids (Cars)) of *Digitalis lanata* Ehrh were followed after water stress (leaf water potential (Ψ) decreased from -0.7 to -2.5 MPa) (Stuhlfauth et al. 1990). It was found that the Chl a and b contents expressed per leaf area were not significantly altered, indicating that the photosynthetic apparatus remained basically intact. The unchanged Chl contents correlated with the constancy of dark and maximal fluorescence. However, the content of β-carotene increased by about 25%, whereas the amounts of the xanthophylls lutein, neoxanthin, and violaxanthin were reduced by 15–30%. Such a contrary course of oxidized and non-oxidized carotenes indicated that under water stress reaction center protection was favored against the maximum antenna radiant energy trapping and xanthophyll cycle activity.

In a later published paper by Fock et al. (1992), results of further experiments with the same medicinal plant species (*Digitalis lanata* Ehrh) and water stress were presented. After withholding irrigation, the Ψ decreased to -2.5 MPa, but leaves remained turgid even at a severe water stress. However, in water-stressed leaves, reduction of the leaf conductance, net photosynthetic rate (P_N), and E were observed, whereas gross CO_2 uptake and photorespiration were less affected. The stomatal closure induced an increase in reassimilation of internally released CO_2. The increased CO_2 recycling in relation to the water stress was high in 350 cm^3 $CO_2 m^{-3}$ and still substantial in 1000 cm^3 $CO_2 m^{-3}$, whereby consummation of a substantial amount of radiant energy in the form of adenosine triphosphate (ATP) and reduction equivalents were observed. In the severely stressed plants, metabolic demand for radiant energy was reduced by less than 40%, P_N was diminished by more than 70%, and the quantum efficiency of photosystem PSII was reduced, too. It was concluded, that this (and possibly other mechanisms) enabled the stressed plants to avoid overreduction of the photosynthetic electron transport chain.

Gray et al. (2003) attended to the effects of acute periods of drought stress on dry weight, and alkamide and phenolic acid content in the roots of purple coneflower (*Echinacea purpurea* (L.) Moench). Plants grown under brief drought stress during two seasons, in the initial flowering stage, produced fall-harvested roots with significantly greater cichoric acid concentration than the well-watered controls. Total alkamide (including the tetraenoic acid isomers and chlorogenic acid) from fall-harvested roots were unaffected by drought stress, regardless of when the stress occurred developmentally. The alkamide concentration in 3-year roots was significantly less (an average decrease of 50.5%) than that in 2-year roots, but total phenolic acids increased 67.1% for all variants. Moreover, root dry weight increased significantly by an average of 70.0% for all drought-stressed plants from 2 to 3 years of age, compared to an increase of 35.2% for well-watered controls. These results suggested that controlled drought stress can stimulate both an increase of root dry weight and root cichoric acid content, and thus root age is the predominant factor determining overall phytochemical content variation.

Drought stress was induced in six herbaceous ornamental landscape perennials (*E. purpurea* (L.) Moench, *Gaillardia aristata* Pursh., *Lavandula angustifolia* P. Mill., *Leucanthemum × superbum* (J.W. Ingram) Berg. ex Kent. "Alaska," *Penstenton barbatus* Roth var. praecox nanus rondo, and *Penstemon × mexicali* Mitch. "Red Rocks") over a 2-year period, whereby the first year was an establishment period (Zollinger et al. 2006). Plants were irrigated at frequencies of 1 week (control) or 4 weeks between June and September, simulating well-watered conditions, moderate drought or severe drought and osmotic potential, gas exchange, visual quality, leaf area, and dry weight were estimated. It was shown that *P. barbatus* has the greatest tolerance to all levels of drought avoiding desiccation by increasing root/shoot ratio and decreasing stomatal conductance, *L. angustifolia* and *P. × mexicali* possessed tolerance to moderate drought conditions but died after exposure to the first period of severe drought, and both *G. aristata* and *L. superbum* were not able to regulate shoot water loss effectively. Compared to control plants, *G. aristata* shoot dry weight was reduced by 50 and 84% and *L. × superbum* by 47 and 99%, whereby root dry weights were affected similarly for both species. *E. purpurea* exhibited poor visual quality at all irrigation intervals; in particular, wilting severely in both drought treatments. Based on the results, it was concluded that *P. barbatus* is recommended for ornamental landscapes that receive little or no supplemental irrigation and *E. purpurea* is not recommended for low water landscapes because of a low visual quality under even mild drought.

Similar physiological characteristics were measured in the leaves of four species (*E. purpurea* (L.) Moench, *Rudbeckia fulgida var. Sullivantii* (Beadle and Boynt.) Cronq., *Monarda didyma* L., and *Helianthus angustifolius* L.) during the lethal drying period (Chapman and Auge 1994). *E. purpurea* had low Ψ and relative water content (RWC) at stomatal closure, and low lethal Ψ and RWC indicated high dehydration tolerance, relative to the other three species. *R. fulgida* var. sullivantii had a similar low Ψ at stomatal closure and low lethal Ψ and displayed relatively large osmotic adjustment. *M. didyma* possessed the highest Ψ and RWC at stomatal closure and an intermediate lethal Ψ. *H. angustifolius* became desiccated more rapidly than the other species, despite having a high Ψ at stomatal closure. Moreover, this species had a high lethal Ψ, and displayed very little osmotic adjustment, indicating relatively low dehydration tolerance. Despite differences in stomatal sensitivity, dehydration tolerance, and osmotic adjustment, all four perennials fell predominantly in the drought-avoidance category.

Biomass and total phenolic content (TPC) were estimated in the plants of lemon balm (*Mellisa officinalis* L.) and thyme (*Thymus vulgaris* L.) cultivated under a different water supply (25, 40, and 70% saturation of soil water capacity (SWC)) (Németh-Zámbori et al. 2016). Differences between studied species were determined in both biomass and TPC, whereby lower SWC reduced biomass production only of lemon balm plants. The total content of flavonoids as well as the rosmarinic acid was not affected by the water stress. Correlation was confirmed between TPC and antioxidant capacity (r=0.766−0.883). The rosmarinic acid content of lemon balm plants thus contributed to the antioxidant capacity (r=0.679−0.869) of this species.

Similar experiments by Radácsi et al. (2016) performed with lemon balm plants (*M. officinalis* L. cv. "Soroksar") affected by water stress 1 and 2 (the soil water capacity decreased stepwise from 70 to 30% and sharply from 70 to 30%, respectively), and stress 3 caused the strongest and permanent water deficit when only 30% soil water capacity was applied. Water deficit reduced the plant height, shrub diameter, fresh and dry mass of the plants, but in the root length fresh and dry mass no significant differences were detected. Water deficit also influenced the phenolic compounds, whereby under water stress 1 and stress 2 a decreased accumulation of flavonoids (by 14−22%) was observed, while the drought conditions resulted in an enhancement of the rosmarinic acid concentration (1−22% compared to the control). The highest antioxidant capacity (201 mg ascorbic acid g^{-1} dry mass) and total phenolic content (178 mg gallic acid g^{-1} dry mass) were measured under stress 3. The results showed that not only the level of water deficit but its timing and dynamics might have significant effects on the quantity and quality of lemon balm production.

Farahani et al. (2009) used four levels of water deficit (field capacity of 100% (T_1), 80% (T_2), 60% (T_3), 40% (T_4), and 20% (T_5), respectively) to compare differences in biological yield, essential oil yield, essential oil %, leaf yield, height of plant, tiller number, stem diameter, stem yield, and internode length of *Melissa officinalis* L. plants. It was shown that water deficit significantly affected all studied parameters, whereby the highest biological yield and height of plant were achieved under T_1, highest essential oil yield was achieved under T_3, and highest stem diameter and essential oil % were determined under T_5. Based on these results, it was concluded that essential oil yield was reduced, but essential oil % increased under this condition.

Endogenous concentrations of some antioxidants (α-tocopherol: α-T and ascorbate: Asc), their redox states, and other indicators of oxidative stress in chloroplasts of three *Lamiaceae* species (differing in their carnosic acid (CA) contents), exposed to drought stress in the field were determined. Rosemary (*Rosmarinus officinalis*) and sage (*Salvia officinalis*) species contained CA; *M. officinalis* was CA-free species (Munne-Bosch and Alegre 2003). The authors found that the photosynthetic apparatus of all studied species were not damaged with leaf water contents between 86 and 58%, as indicated by constant maximum efficiency of PSII photochemistry ratios and malondialdehyde (MDA) levels in the chloroplasts. Moreover, these species showed a significant increase in α-T, a shift of the redox state of α-T toward its reduced state, and increased Asc levels in chloroplasts under stress conditions. Lemon balm showed the highest increase in α-T and Asc in chloroplasts under stress, which might compensate for the lack of CA. The authors also emphasized that CA, in combination with other low molecular weight antioxidants, helps to prevent oxidative damage in chloroplasts of water-stressed plants.

In another paper, also dealing with *M. officinalis* (Munne-Bosch and Alegre 1999), the effect of drought and simulated daily dew on the water relations and photosynthetic capacity of lemon balm plants grown in Mediterranean field conditions were evaluated. Water stress during the summer period caused a large decrease in Ψ to −3 MPa and RWC to 34%. These plants showed a one-peaked photosynthetic daily pattern with a maximum peak during the morning. However, as drought stress progressed the one-peaked photosynthetic daily pattern was maintained, but the maximum photosynthetic daily peak observed earlier in the morning decreased by ca. 50%. Simulated daily dew applied to water-stressed plants caused not only maintenance or even an improvement in photosynthetic rate depending on the time of day but also led to a complete recovery of plant water status and leaf pigment content. Based on these results, the authors concluded that dew may have an important role in rehydrating and reactivating the metabolism of water-stressed *M. officinalis* plants during the summer.

Munne-Bosch and Alegre (2000) also measured diurnal variations in pigment composition (Chls; Cars), α-tocopherol, and photosynthesis during the summer drought and recovery after autumn rainfalls in *M. officinalis* plants. Summer drought caused an extreme decrease of leaf RWC to ca. 35% and Ψ to −3 MPa, that indicated severe water stress. Under drought conditions, Chls content decreased progressively (associated with the inability to increase total Cars), but α-tocopherol increased by ca. 60% and showed a significant decrease at midday in severely stressed plants. Although *M. officinalis* plants under water stress showed a significant decrease in the endogenous content of photosynthetic pigments, the antioxidant function of α-tocopherol and the dissipation of excess excitation energy by the xanthophyll cycle may help to avoid irreversible damage to the photosynthetic apparatus.

Korkmaz et al. (2015) sprayed pepper seedlings with three concentrations (0.5, 5, or 25 mM) of glycinebetaine (GB) and afterward subjected them to water stress for 10 days. Water stress caused substantial reductions of shoot dry weight, leaf area, Chl content, Ψ, gas-exchange characteristics, and efficiency of PSII, but increased membrane permeability and lipid peroxidation. After foliar application of GB improved Ψ, RWC and antioxidant enzymatic activity were observed. Among the GB concentrations applied, enhanced water stress tolerance was obtained with 5 mM GB pre-treatment. These results indicated that GB, applied as a foliar spray, could be used as an ameliorative agent for pepper seedlings against the harmful effects of water stress.

Two pepper (*Capsicum annuum* L.) cultivars (Shanshu-2001 and Nongchengjiao-2) were subjected to four water supply regimes (80, 60, 40, and 20% of field capacity) representing control, mild, moderate, and severe water stress, respectively (Anjum et al. 2012). Shanshu-2001 showed a higher leaf RWC, protein, and proline accumulation than Nongchengjiao-2 in all water regimes. The total soluble proteins and proline continually increased with water stress in cultivar Shanshu-2001, whereas in cultivar Nongchengjiao-2 only proline raised. The activities of superoxide dismutase (SOD), peroxidase (POD), and catalase (CAT) were higher in Shanshu-2001 than in Nongchengjiao-2, which resulted in improved growth and yield of the first cultivar. Overall the cultivar Shanshu-2001 was better able to resist drought stress as indicated by better growth and yield due to higher activities of antioxidant enzymes, reduced lipid peroxidation, better accumulation of osmolytes, and maintenance of tissue water contents.

A field experiment was carried out to investigate the physiological changes following water stress in bell pepper (*C. annuum* L.) plants and their consequences on fruit yield. Plants were grown under irrigated and rainfed conditions, and leaf water content and Ψ were established (Delfine et al. 2002). Aboveground biomass, gas-exchange characteristics, Ψ, and intercepted solar radiation were lower in rainfed plants than in the irrigated control. Leaf ion and protein concentrations were higher in rainfed plants, but ion accumulation in fruits did not change between treatments. This experiment showed that photosynthetic limitations of plants growing in the field and subjected to moderate stress conditions were mainly due to decreasing stomatal conductance. In rainfed conditions, the photosynthetic rate was also decreased by the reduction of mesophyll conductance. However, the photochemical efficiency of the leaves was resistant to water stress. The photosynthetic limitation was sufficient to reduce plant growth, fruit yield, and fruit marketability. The results indicated that irrigation is necessary not only to optimize bell pepper production but may delay the decline of gas exchange and plant growth as the soil is dried.

Gas exchange, Chl fluorescence, and antioxidant enzyme activity in the leaves of 3-year-old saplings of *Hippophae rhamnoides* grown in pots were tested using different soil water supplies (Liu et al. 2017). When SWC was between 38.9 and 70.5%, stomatal limitation was responsible for the reduced P_N; when SWC was lower than 38.9%, nonstomatal limitation was the main factor restricting P_N. Moderate water stress improved the WUE of the leaf. Water stress significantly influenced both fluorescence parameters and the antioxidant enzyme system. When SWC was between 38.9 and 70.5%, nonphotochemical quenching increased and then decreased, indicating that thermal energy dissipation was a significant photoprotection mechanism. Antioxidant enzymes were activated when SWC ranged from 48.3 to 70.5%; under severe water stress (SWC < 38.9%), the antioxidant enzyme system was damaged, the activity of the antioxidant enzymes declined, and membranes were damaged. In the semiarid loess hilly region, the maximum SWC of 38.9% could sustain *H. rhamnoides* under natural conditions.

Growth and physiological parameters in response to drought were compared between two sea buckthorn (*H. rhamnoides* L.) populations coming from the southeast part of the Qinghai-Tibetan Plateau of China (Yang et al. 2010). The experimental design consisted of two water regimes (100 and 25% of field capacity) and two populations from the low and high altitude zone. It was found that drought tolerance was highly related to the plant antioxidant capacity and WUE as well as leaf nutrient status. In response to the drought, the highland population compared with lowland population showed greater inhibition of growth and leaf expansion, lower leaf nitrogen and phosphorus content, lower root nodule biomass and root mass/leaf area ratio, and higher leaf water content loss being parallel with a higher content of ABA.

According to the results obtained in the study of Keshavarz Afshar et al. (2015), drought stress can enhance the accumulation of silymarin in milk thistle (*Silybum marianum* L.) seeds. Under drought stress, the content of the silybin, which possesses the greatest portion of biological activity among the silymarin components increased. Plants grown under moderate and severe drought stress had 4 and 17% higher content of silymarin than those grown in well-watered conditions, which was attributed to a higher content of silybin, isosilybin, and silychristin, while silydianin content was lower under drought condition. It was concluded that drought stress enhanced the accumulation of silymarin in milk thistle seeds and improved its quality by increasing the portion of silybin.

Sedghi et al. (2012) evaluated the effect of phytohormones on the changes of antioxidant enzymes and Cars in petals of pot marigold (*Calendula officinalis* L.) under drought stress. It was shown that the activities of SOD and CAT increased by 47 and 73%, respectively in petals under water deficit compared with the control plants. Spraying with gibberellic acid (GA_3) and benzyl amino purine (BAP) alleviated drought effects, but the application of ABA, jasmonic acid (JA), salicylic acid (SA), and brassinolide induced the activity of these enzymes. The concentration of Cars was affected by drought and hormone treatments as well. The concentration of Cars increased under water deficit but GA_3, BAP, and JA had inhibitory effects on lycopene and carotene synthesis. Spraying with GA_3 increased luteoxanthin concentration in petals by 35 and 20% in comparison with the non-stressed as well as stressed environments. A decrease in POD activity under water stress confirmed that other mechanisms might be involved for ROS scavenging in petals of pot marigold.

Water deficit and ABA application on leaf gas-exchange and flavonoid content in marigold (*C. officinalis* L.) were observed by Pacheco et al. (2011). Water deficit was tested by withholding water (control-irrigation, 3, 6, and 9 days without irrigation) followed by 3 ABA concentrations (0, 10, and 100 µM) applied at the beginning of blooming. The RWC and the leaf gas exchange (P_N and E) as well as stomatal conductance, intercellular CO_2 concentration, and WUE were evaluated. At the end of 9 days of water deficit, there were significant decreases in all observed characteristics, independent of ABA application that suggested that the main effect of ABA was to cause a reduction in stomatal conductance which was accompanied by a reduction of P_N only when the plants suffered from water deficit. ABA restricted the flavonoids biosynthesis in control and water-stressed plants, too.

It could be concluded that it is vital to integrate crop physiology, genomics, and breeding approaches to resolve complex traits, understand the molecular basis of drought tolerance, and develop new generation plants not only for the crops but also for medicinal plants under a changing climate.

39.4 FLOODING AND WATERLOGGING

A flood is an overflow of water that submerges land that is usually dry. Flooding may occur as an overflow of water from water bodies (rivers, lakes, or ocean), due to an accumulation of rainwater on saturated ground in an areal flood or when the flow rate exceeds the capacity of the river channel. Global warming is shifting rainfall patterns, making heavy rain more frequent in many areas of the country. Moreover, human alteration of the land – like the engineering of rivers, the destruction of natural protective systems, and increased construction on floodplains – contributes to a greater risk of experiencing destructive and costly floods. Waterlogging occurs when roots cannot respire due to excess water in the soil profile; i.e., the soil is so wet that there is insufficient oxygen in the pore space for plant roots to be able to adequately respire. Waterlogging is becoming a critical threat to plants growing in areas prone to flooding (Parent et al. 2008; Kundzewicz et al. 2010; Ashraf 2012).

Maintenance of a functional root system upon flooding is essential because roots are vital for plant function. Flooding-resistant plants possess a number of adaptations that help maintain oxygen supply to the root (e.g., initiating organogenesis to replace their original root system with adventitious roots if oxygen supply becomes impossible (Sauter 2013)). Dryland species could affect response to flooding of the soil by developmental modifications in the shoot (e.g., epinastic leaf curvature, stomatal closure, and slowing of leaf expansion) in order to compensate the diminished input of resources from the roots (Jackson 2002).

Anoxia in plant tissues results in an energy crisis (Gibbs and Greenway 2003); therefore, plant cells need to reduce their energy requirements for maintenance, whereby during anoxia lasting several days a proportion of the produced energy is used to mitigate the acidifying effect of cytosolic pH. The anoxia tolerance of plants could be based on reduced rates of anaerobic carbohydrate catabolism or on accelerated rates ("Pasteur effect") (Greenway and Gibbs 2003).

It was reported that a 2-week flooding treatment had a negative effect on seedling performance of four meadow species younger than 6 weeks, whereby summer floods with high floodwater temperatures had especially detrimental effects on seedlings (Gattringer et al. 2017). Increased flooding duration and flooding depth negatively affected the performance of seedlings, and in the flooding tolerance of meadow plant species the interactive effect of these two factors was assumed to play a crucial role (Gattringer et al. 2018).

The seedlings of *Urtica dioica* L. cultivated in pot from seeds sown in April and flooded during spring (May–June) showed increased mortality and none of the plants survived more than 4 weeks of summer flooding (June–August), while all plants survived the whole period of autumn flooding (August–October), although their dry mass decreased as a result of leaf and root mortality (Klimesova 1994).

An investigation of the rooting-capacity of cuttings or crowns taken from *Rosmarinus officinalis* mother plants waterlogged for different periods showed that rooting percentage increased and dry weight ratio decreased when cuttings were obtained from mother plants waterlogged for 48 h or longer (Fernandez et al. 2000). *Mentha arvensis* showed a better response toward waterlogging in comparison to *M. piperita* and under stress conditions it maintained a healthy posture, showed a flourished vegetative growth, a lower degree of lipid peroxidation, and vigorous outgrowth of adventitious roots to assist waterlogging tolerance (Phukan et al. 2014).

Pot experiments on the flooding of sage (*S. officinalis* L.) plants showed that the effects of waterlogging were severe only when the temperature was high and flooding prolonged and as such all plants survived the flooding in winter, but 30–40% of these plants died during a 17-day flood in summer. In another experiment focused on the investigation of sage roots, anoxia strongly reduced root development, while a hypoxia condition increased the production of lateral roots close to the surface enabling them to acclimate to the subsequent anoxia, which markedly enhanced their chances of survival (Cameron et al. 2010).

Waterlogging stress during the seedling stage of medicinal plant *Chrysanthemum morifolium* resulted in reduced Chl content and relative leaf Ψ, while MDA, glutathione (GSH), and soluble sugar increased. After waterlogging treatments lasting for 4 days, the physiology and biochemistry characteristics could not be restored (Zhang et al. 2009).

In *Garcinia brasiliensis* plants, the fruit of which contains metabolite with anticancer activity, flooding caused a reduction in the dry mass of roots and shoots as well as in the length, surface area, and volume of roots, and roots presented thicker exodermis and greater xylem number, thicker phloem, and fewer xylem fibers. Considerably higher activities of SOD, ascorbate peroxidase (APX), and CAT in roots of flooded plants was observed as well, and after 70 and 90 days of flooding a notable increase in soluble sugars, and a reduction in starch content (at 90 days) was estimated (de Souza et al. 2013).

Baczek-Kwinta et al. (2008) reported that stomata of *M. officinalis* plants did not respond to flooding treatment and plant growth, water content, gas exchange, and fluorescence were not affected, while plants had enhanced anthocyan content in their leaves. On the other hand, basil (*Ocimum basilicum* L.) plants exhibited less resistance to flooding, whereby this stress induced a decrease of gas exchange to 1/3 of the control.

Upland and coastal ecotypes of tea tree (*Melaleuca alternifolia*) exhibited morphological adaptations for flood tolerance, with both ecotypes possessing a similar maximal capacity for adventitious roots and aerenchyma; however, under a prolonged flood, reduced plant growth was observed compared

to the well-watered control. Coastal plants were found to respond more rapidly to flood indicating that upland plants may delay morphological modifications until the flooding is more protracted (Shepherd et al. 2015). *M. alternifolia* seedlings survived and grew through 180 days of flooding with a subsequent 60-day recovery period in drained conditions suggesting that they would be suitable for afforestation in areas exposed to intermittent flooding (Jing et al. 2009).

Salvia officinalis plants survived 17 days flooding in winter, while *Lavandula angustifolia* also survived an equivalent flooding duration during summer. Sage, which is considered as a species showing "intermediate" tolerance to flooding, definitely demonstrated adaptations to waterlogging, e.g., acclimation against anoxia when pre-treated with hypoxia (King et al. 2012).

In *Momordica charantia* L., a flood-intolerant plant, after 4 days of flooding the hypocotyl cross-sectional area and numbers of adventitious roots notably increased and the aerobic respiratory rate of flooded roots dropped to 42% of the control level on the first day of flooding. A prolonged accumulation of ethanol in flood-stressed *M. charantia* roots resulted in the death of the seminal root system (Liao and Lin 1995). Both stomatal and metabolic factors were found to be responsible for the reduction in leaf CO_2 exchange rate during flooding stress, and the internal CO_2 concentration of leaf tissues of flooding-treated plants was assumed to be involved in lowering the activation of ribulose-1,5-bisphosphate carboxylase/oxygenase (Rubisco) during flooding stress (Liao and Lin 1994).

39.5 HIGH AND LOW TEMPERATURES

The response of plants to high-temperature (or low-temperature including frost) stress depends on its level, its duration, and the plant species. It should be emphasized that high temperatures (usually $\geq 30°C$) under natural conditions are also connected with high light because sunlight is a source not only of light but also heat energy. Heat stress is also one of the major limiting factors for plants all over the world. As a consequence of climate change and global warming, heat stress has an increasingly negative impact on plant growth, survival, and overall productivity (cf. Taiz and Zeiger 2010; Taiz et al. 2014; Szymańska et al. 2017). Heat stress and drought stress often occur simultaneously and can induce cellular damage due to the accumulation of ROS. Heat shock proteins, also known as molecular chaperones, are present in all plants and are important for maintaining and restoring the homeostasis of proteins. Apart from HSPs, the role of several other heat-responsive genes which are mainly transcription factors have been investigated (Wahid et al. 2012). With rising atmospheric CO_2 concentration, heat stress has become a more prevalent problem for checking plant productivity. Generally, a transitory phase when the temperature exceeds ambient temperature by 10–15°C is referred to as heat shock or heat stress (Wahid et al. 2007). Transient or persistent high temperature negatively impacts plant growth and development thus limiting productivity (Song et al. 2014). Heat waves particularly during the anthesis and grain-filling stages are detrimental to the photosynthetic system and significantly reduce yields (Feng et al. 2014). PSII complex is the most heat-intolerant part of the photosynthetic apparatus on the light reaction side of photosynthesis. Oxygen-evolving complex (OEC), PSII reaction center, and the light-capturing complexes are the primary components that are damaged by high temperatures (Mathur et al. 2014). Thermal stress leads to the release of two of the four Mn ions present per PSII (Tyystjärvi 2008). It is known that heat stress to chloroplasts liberates 33, 23, and 17 kDa proteins, which are extrinsically attached to the PSII complex. Discharge of the 33 kDa protein is responsible for the release of Mn ions and destabilization of the OEC (Allakhverdiev et al. 2008). On the cellular and molecular level, membranes and the cytoskeleton, proteins, enzymes, and RNA are the most affected (Hasanuzzaman et al. 2013).

On the plant organ level, heat stress markedly affects the water status in leaves and intracellular CO_2 concentration which is an effect of stomata closure (Greer and Weedon 2012). All of these changes lead to a reduction in the photosynthetic rate. However, the mechanism of photosynthetic inhibition under heat stress remains unclear. One mechanism suggested is the inhibition of the ribulose 1,5-bisphosphate (RuBP) carboxylation rate (Kurek et al. 2007). Within chloroplasts, under elevated temperatures, the most affected are the PSII, Rubisco, and ATP synthase (Ashir 2015). Moreover, apart from determining changes at the molecular level, a response at the whole plant or even natural plant population level which includes, e.g., photosynthetic rate and seeds production, is required to properly understand plant reactions to stress situations (Dhondt et al. 2013).

Low temperatures (0–15°C) are common in natural conditions and leads to cellular damage in many plant species, limiting plant productivity and distribution (Theocharis et al. 2012). In non-tolerant plants, this damage occurs in the first hours of stress. While short exposure to low temperatures may only trigger transitory changes, prolonged exposure to stress causes plant necrosis or death. On the other hand, tolerant plants have the ability to reorganize their molecular and physiological parameters to counteract the destructive effects of the cold. This phenomenon, known as cold acclimation, is the central factor for plant cold tolerance (Theocharis et al. 2012).

It should be mentioned that abiotic stress including cold, freezing damage, drought, and high salinity damage have been shown to be induced by similar mechanisms, most notably dehydration or water stress. Plants lose water when they find themselves in a freezing situation by osmosis. Cold stress is a major and serious threat to plant sustainability which leads to a loss in plant production. Many physiological and production changes occur in response to cold stress (Yadav 2010).

There are various phenotypic symptoms in response to cold stress including poor germination, stunted seedlings, yellowing of leaves (chlorosis), reduced leaf expansion and wilting, and may lead to the death of tissue (necrosis). Cold stress also severely hampers the reproductive development of plants. However, the major negative effect of cold stress is that it induces severe membrane damage. This damage is largely due to the acute dehydration associated with freezing during cold stress. Cold stress is perceived by the receptor in the cell membrane. Then a signal is transduced to switch on the cold-responsive genes and transcription factors for mediating stress tolerance. Understanding the mechanism of cold stress tolerance and genes involved in the cold stress signaling network is important for crop improvement (Yadav 2011). In this contribution, the major points discussed are the following: (I) the physiological effects of cold stress, (II) the sensing of cold temperatures and signal transduction, and (III) the role of various cold-responsive genes and transcription factors in the mechanism of cold stress tolerance. Based on the results, it can be concluded that cold stress is sensed by a yet unknown receptor. The cold stress signal is transduced through several components of the signal transduction pathways. Major components are calcium, ROS, protein kinase, protein phosphatase, and lipid signaling cascades. ABA also mediates the response of cold stress. The cold stress signal leads to the regulation of transcription factors and effector genes, collectively called cold-regulated genes. The effector genes encoding proteins under this category include chaperones, late embryogenesis abundant proteins, osmotin, antifreeze proteins, key enzymes for osmolyte biosynthesis such are proline, water channel proteins, sugar and proline transporters, detoxification enzymes, enzymes for fatty acid metabolism, and lipid-transfer proteins. Use of these genes and transcription factors in the genetic modification of plants can improve their cold tolerance and production.

When plants suffer freeze-induced damage, the plasma membrane is thought to be the primary site of injury because of its central role in the regulation of various cellular processes. Cold tolerant species, however, adapt to such freezing conditions by modifying cellular components and functions (cold acclimation). One of the most important adaptation mechanisms to freezing is the alteration of plasma membrane compositions and functions. Advanced proteomic technologies have succeeded in the identification of many candidates that may play roles in the adaptation of the plasma membrane to freezing stress. Proteomics results suggest that the adaptations of the plasma membrane functions to low temperature are associated with alterations of protein composition during cold acclimation. Some of the proteins identified by proteomic approaches have been verified in their functional roles in freezing tolerance mechanisms further. Thus, accumulation of proteomic results in the plasma membrane is of importance for application to molecular breeding efforts for increasing cold tolerance in plants (Takahashi et al. 2013).

Low-temperature stress also inhibits the repair of PSII. There are many reports confirming that low-temperature stress inhibits the repair of PSII but does not affect the photodamage to PSII (Allakhverdiev and Murata 2004). One of the major consequences of temperature stress is the excess generation of ROS, leading to oxidative stress. Investigations on cold-resistant plants have shown the high activity of antioxidant enzymes. Thus, an improvement of antioxidant enzyme activity is one of the effective ways to acquire cold resistance in plants (see Szymańska et al. 2017).

The effect of climate change on the composition of antioxidant and biochemical metabolites in five medicinal herbs (*Hypericum perforatum* L., *Matricaria chamomilla* L., *Thymus vulgaris* L., *Cynara cardunculus* L., and *E. purpurea* L.) that were selected and grown at two altitudes in the Western Himalayas (Jammu 305 m and Srinagar 1730 m above sea level, respectively) were observed (Kaur et al. 2016). At the Srinagar site, it was determined that there was a variable (up to 13-fold) increase in the content of phenols and flavonoids as well as in the content of proteins (1.3–1.8 times), sugars (2.8–4.1 times), and free amino acid (1.04–1.22 times). *H. perforatum* and *M. chamomilla* had a higher content of phenols, xanthophylls, and proline even at the subtropical environment at the Jammu site, suggesting the potential for increasing their geographical area. The results demonstrated that under changing environmental conditions the content of bioactive substances in studied medicinal plants is significantly affected as a defense strategy of these temperate species.

The freezing hardness (expressed as LT_{50}, i.e., temperature at which 50% of the plants are killed by cold stress), changes of content of total protein, phenolic, flavonoid, lipid peroxidation, MDA, respectively, activities of SOD, APX, CAT, as well as Chl fluorescence (F_v/F_m, where F_v is variable fluorescence and F_m maximum fluorescence) in *Echinacea purpurea* (L.) Moench plants were studied by Asadi-Sanam et al. (2015). Five-month-old seedlings were kept at 4°C for 2 weeks to induce cold acclimation. Afterward, the acclimated seedlings were treated with freezing temperatures (0, –4, –8, –12, –16, and –20°C) for 6 h, but unfrozen (control) seedlings remained at 4°C. Exposing seedlings to freezing temperatures was accompanied by decreasing dark-adapted Chl fluorescence (F_v/F_m) and the reduction of SOD, APX, and CAT activity except with the seedlings exposed to 0°C. Moreover, total protein and antioxidant capacity of these leaves declined significantly after exposure to freezing temperatures and thereafter reached the highest level at –8°C. Total phenolic content of freezing-treated seedlings was also significantly lower in comparison to control seedlings. However, total flavonoid content significantly increased with lowering freezing temperatures. It was ascertained that LT_{50} of *Echinacea* seedlings under artificially simulated freezing stress was –7°C.

Kochankov et al. (1998) performed experiments with purple coneflower (*E. purpurea* (L.) Moench) affected by temperature (from –5 to 35°C) together with growth regulators and other chemicals (Ethephon 1445 mg dm^{-3} and gibberellic acid GA$_3$ 60 mg dm^{-3}). The greatest final germination percentage and the shortest germination time were observed at 25°C, osmotic potential –0.5 MPa decreased germination at 20 and 25°C. Osmotic potential below –0.5 MPa completely inhibited germination at both temperatures. Application of above-mentioned concentrations of Ethephon and GA$_3$ on seeds with enhanced germination rates increased final germination percentage of water-stressed seeds and also enhanced seedling development. The seedlings survived 17 h at 0 and –2°C, but 17 h exposure to –5°C killed all seedlings.

The optimal temperature range (when the highest seed germination in the shortest time period is achieved) were determined in four medicinal plants (*E. purpurea*, *Echinacea pallida*, *Tanacetum parthenium*, and *Valeriana officinalis*) exposed to temperatures from 16 to 36°C for 7 days. A germination index or rate were calculated to determine the critical temperature for the maximum speed of germination. As an optimal temperature range for seed germination was found 19–23°C for *E. pallida* and *E. purpurea*, 17–21°C for *T. parthenium*, and 19–27°C for *V. officinalis*. Maximal germination rate occurred after 3 days for *E. purpurea* and *V. officinalis*, and after 5 days for *E. pallida* and *T. parthenium* (Hassell et al. 2001).

Parmenter et al. (1996) observed that seeds of *E. angustifolia* stratified for more than 2 weeks achieved maximal germination rates of 65–80%, but this parameter estimated for *E. purpurea* (84%) was unaffected by periods of stratification. The time required for maximal germination rate declined as the length of the stratification period increased. It was confirmed that trimming the seed coat of *E. angustifolia* had no significant effect on the germination rate of the unstratified seed.

Since quantitative information concerning the temperature and Ψ effects on seed germination in lemon balm (*Melisa officinalis* L.) were scarce, Atashi et al. (2014) quantified seed germination responses of this medicinal plant to different levels of temperature and Ψ under laboratory conditions using a segmented model. From 0 to –0.76 MPa, when Ψ increased, the base temperature was a constant 7.2°C to –0.38 MPa and increased linearly to 20.1°C as Ψ decreased. Based on the obtained results it was concluded that the segmented model and its parameters could be used in lemon balm germination simulation models.

It is known that low temperature is one of the major abiotic factors limiting pepper (*Capsicum annuum* L.) production during winter and early spring in non-tropical regions. However, the application of exogenous ABA effectively alleviates the symptoms of chilling injury, such as wilting and the formation of necrotic lesions on pepper leaves. Thus, Guo et al. (2013) identified genes that were differentially up- or downregulated in ABA-pretreated pepper seedlings incubated at 6°C for 48 h, using a suppression subtractive hybridization method. The expression profiles of 18 selected genes were analyzed, whereby the expression levels of 10 of these genes were at least two-fold higher in the ABA-pretreated seedlings under chilling stress than in water-pretreated (control) plants under chilling stress. These results suggested that ABA can regulate genes in pepper plants under chilling stress.

Photosynthetic and photorespiratory rates, Chl fluorescence, and activities of antioxidant enzymes (SOD and APX) were studied in two pepper (*C. annuum* L.) cultivars (Zhengjiao 13 and Longkouzaojiao) grown under drought and heat, as well as under drought and heat in combination (Hu et al. 2010). The drought-tolerant cv. Zhengjiao 13 exhibited greater photosynthetic rate and cytochrome respiratory pathway activity, and lower contents of superoxide radical and H_2O_2, as compared to the drought-sensitive cv. Longkouzaojiao. In both of these cultivars, photosynthetic rate and cytochrome respiratory pathway activity decreased and ROS production increased under drought and combined drought and heat stress. Drought as a lone stress factor also increased SOD and APX activities in the cytosol, chloroplasts, and mitochondria in both cultivars. The results indicated that drought and heat in combination were more harmful than either stress alone.

Using the same pepper cultivar Zhengjiao No. 13 and another cultivar Jizhao × Jilin, Hu et al. (2008) investigated the effects of heat and drought separately as well as in combination on growth and respiration activity. Heat stress resulted in a growth reduction in a cultivar of Jizhao × Jilin, but not in Zhengjiao No. 13. Drought, especially drought under heat, significantly inhibited the growth of peppers. Drought also increased the ion leakage from leaves more than heat. Heat-induced increases in both the cytochrome and alternative pathway respirations, while drought strongly decreased cytochrome respiration, but increased alternative respiration. These results suggest that higher heat or/and drought tolerance of pepper is associated with the capacity to keep high total and alternative respirations under stress.

A sensitive commercial hybrid (Canon) and a new hybrid bred to enhance abiotic stress tolerance (S103) of *C. annuum* L. differing in their response to chilling were subjected to three root-zone temperatures 7, 17, and 27°C (Aidoo et al. 2017). Plant height, shoot dry mass, root maximum length, root projected area, number of root tips, and the root dry mass of hybrid S103 were less affected by the treatment. Low root-zone temperatures significantly inhibit gaseous exchange, with a greater effect on the hybrid Canon. Metabolite profiling showed a greater increase in the root than in the leaves, whereby leaf response between the two cultivars differed significantly, too. The roots of both cultivars accumulated stress-related metabolites including γ-aminobutyric acid, proline, galactinol, and raffinose and at 7°C an increase of sugars was observed. The presented results suggested that the enhanced tolerance of S103 to root cold stress is likely linked to a more effective regulation of photosynthesis facilitated by the induction of stress-related metabolism.

C. annum L. plants (Airaki et al. 2012) after 24 h of exposure at 8°C exhibited visible symptoms of the wilting of stems and leaves, which was accompanied by significant changes in reactive nitrogen species and ROS with an increase of both protein tyrosine nitration and lipid peroxidation, indicating that low temperatures induce nitrosative and oxidative stress. These results confirmed that the changes have a role in the process of plant cold acclimation through the effect on the redox state of the cell.

Similar results were published by Gulen et al. (2016) for two cultivars (Amazon and Kekova) of *C. annuum* L. Pepper seedlings were grown for 4 weeks at 25/10°C day/night temperatures in a greenhouse and watered on a needs basis avoiding any additional stress factors. Gradual and shock heat stress were applied (from 35 up to 50°C) to the plant in a growth chamber and then heat stress tolerance was estimated. Leaf RWC decreased gradually from control to the highest temperature, while the loss of turgidity increased in both heat stress types. Found changes in the activities of ROS producing and ROS scavenging enzymes confirmed that the cultivar Amazon is superior with respect to its antioxidant defense systems and should be more tolerant than Kekova due to higher ROS scavenging systems.

In an experiment by Zobayed et al. (2005), 70-day-old St. John's wort plants were exposed for 15 days to 15, 20, 25, 30, and 35°C. The relatively high (35°C) or low (15°C) temperatures reduced the photosynthetic efficiency that resulted in low CO_2 assimilation and low maximal quantum efficiency of PSII, too. High temperatures (35°C) increased both leaf total peroxidase activity and the hypericin, pseudohypericin, and hyperforin concentrations in the shoot. These results confirmed that temperature is an important environmental factor in optimizing the secondary metabolite production in St. John's wort.

Chen et al. (2009) found significant differences among 15 hawthorns cultivars (*Crataegus* sp.) grown under natural high-temperature stress during the summer period in Shanghai in the net photosynthetic and transpiration rates, stomatal conductance, and WUE. Adaptation mechanism under high temperature can be caused by increasing heat dissipation, reducing the temperature of the leaf surface and relatively high electron transport or photochemical efficiency to minimize the damage of photosynthetic organism resulting in stable photosynthesis. The adaptation mechanism under high temperatures of the heat-tolerant hawthorn species can be summarized with increasing heat dissipation, reducing the temperature of the leaf surface and holding relatively high electron transport or photochemical efficiency, so as to minimize the damage of the photosynthetic organism and achieve stable photosynthesis. It indicated that the efficiency of the primary conversion of light energy of PSII and its potential activities differed significantly among the cultivars. F_v/F_m was suggested to be an index for heat tolerance of hawthorns species.

When 2-year-old saplings of *Ginkgo biloba* L. Were exposed to high temperatures it decreased the photosynthetic rate, leaf Ψ, and root/shoot ratio, but increased the stomatal conductance and light compensation point, while changes in the light saturation point, flavonol glycoside, and terpene lactone were not significant (Zhang et al. 2005). Saplings subjected to drought had a lower net photosynthetic rate, stomatal conductance, and leaf Ψ, but a higher point of light compensation, while it remained almost unchanged in root/shoot ratio, apparent quantum yield, and light saturation point. Under combined exposure to high temperatures and drought, the saplings had higher values of light compensation point and dark respiration rate, but lower values of light saturation point as well as contents of flavonol glycoside and terpene lactone. Although there was no significant change in root/shoot ratio, the total dry weight per plant decreased. The findings showed that drought and high temperature not only decreased the adaptability to light, light use efficiency, and stomatal conductance of the saplings but increased dark respiration rate, thus resulting in inhibited growth; however, the high temperature had a greater impact on secondary metabolites than the drought.

39.6 HIGHLIGHT

Plants perform a series of complex reactions that convert light into carbohydrates, a process known as photosynthesis. Photosynthesis is one of the most fundamental components of plant growth and productivity. Many of the photosynthetic components are severely affected by different abiotic stresses and greatly reduce yield components. Light is fundamental to the process of photosynthesis. However, on the other hand, light intensities (or irradiance) above the light saturation point of photosynthesis, termed as high light stress, are harmful to plants (e.g., Pascual et al. 2017; Szymańska et al. 2017). A light-induced drop in photosynthetic rate is generally referred to as photoinhibition. A high light level mainly impairs the PSII complex, which is the initiation site for linear electron flow and oxygen evolution from water. The degree of PSII photoinhibition is a measure of the difference between its rate of photodamage and repair (Takahashi and Badger 2011). Multiple mechanisms of PSII photoinhibition occur simultaneously under high light stress. The process, which dominates in photoinhibition, depends on the intensity and quality of light. The primary step in the UV-induced photodamage of PSII is the release of functional Mn^{2+} ions from the OEC by high light, rendering the OEC inactive. The dysfunctional OEC is unable to deliver electrons from water to the primary donor of PSII, $P680^+$. Since $P680^+$ is a powerful oxidant, it impairs the reaction center by oxidizing proteins in its vicinity, especially the D_1 polypeptide. In addition, the direct damaging effects of UV light also extend to redox-active quinones Q_A and Q_B situated on the acceptor side of PSII and the D_1 polypeptide. In the visible region of light, ROS generation is mainly responsible for the PSII photodamage. Additionally, inactivation of the Mn cluster also occurs due to the weak Mn^{2+} absorption in the red part of the visible spectrum. When incident light energy absorbed by Chl molecules in the PSII antennae complex is not utilized efficiently, conversion of singlet chlorophyll to harmful triplet chlorophyll, which transfers energy to O_2 forming singlet oxygen ($^1O_2$), occurs (Vass 2012). Furthermore, limitation in electron transport on the PSII acceptor side is accompanied by full reduction of the plastoquinone (PQ) pool, when the Q_B site becomes unoccupied by PQ and the forward electron from Q_A to Q_B is blocked and back electron transport from $Q_A{}^{\bullet-}$ to pheophytin (Pheo) and consequent recombination of $Pheo^{\bullet-}$ with $P680^{\bullet+}$ forms deleterious triplet chlorophyll which transfers to O_2, forming $^1O_2$. The primary radical pair $^1(P680^{\bullet+}Pheo^{\bullet-})$ formed by the reverse electron transport from $Q_A{}^{\bullet-}$ to Pheo either recombines to ground-state P680 or converts into a triplet radical pair $^3(P680^{\bullet+}Pheo^{\bullet-})$ by a change in the spin orientation, whereby a recombination of the triplet radical pair $^3(P680^{\bullet+}Pheo^{\bullet-})$ forms triplet excited-state chlorophyll, $^3P680^*$. The reaction of triplet chlorophyll molecules formed due to excess light energy and $^3P680^*$ formed due to a restricted electron flow to Q_A from P680, with O_2, produces the highly reactive $^1O_2$, which damages the proteins surrounding it, particularly the D_1 protein, thereby impairing the PSII complex (Pospíšil 2016). Besides PSII, PSI is also deactivated at high light intensities. PSI photoinhibition is caused due to the ROS-induced specific degradation of one (PSI-B) of the two large subunits of the PSI reaction center. This PSI photoinhibition is more pronounced at low temperatures, wherein both the polypeptides, PSI-A and PSI-B, of the reaction center are degraded (Cheng et al. 2016).

Plant responses to high light stress include photoinhibition, thermal energy dissipation of absorbed light energy and state transitions, cyclic electron flow around the PSI, photorespiratory pathways, redox state antioxidants, and redox signaling (Szymańska et al. 2017). Indeed, there were many published papers dealing with the relationship between photosynthesis and high light; this topic was recently comprehensively compiled by Singh and Thakur (2018) in their review that provides a huge repertoire of information.

The authors Jan et al. (2016) evaluated physiological adjustment of three medicinal herbs (*Atropa acuminata* Royle ex Lindl, *Lupinus polyphyllus* Lind, and *Hyoscyamus niger* L.) to the winter period characterized by intensive UV flux in the Kashmir valley (the North-Western Himalaya). The study verified the hypothesis that UV reflectance and absorbance at low temperatures are directly related to differences in alkaloid accumulation. In a field experiment, concentrations of the alkaloids quinolizidine (QA) and tropane (TA) were analyzed in these herbs at two different altitudes via gas chromatography-mass spectrometry (GC-MS) and determined by principal component analysis (PCA). Among the QA in *L. polyphyllus*, ammodendrine and lupanine accumulated at higher concentrations (186.36 and 95.91% in ammodendrine and lupanine, respectively) at the studied sites. TA together with hyoscyamine were recorded as the most abundant constituents irrespective of the plant and site while apotropine was accumulated in lesser quantities in *A. acuminata* than *H. niger*. However, the determination of apotropine concentrations demonstrated a significant variation of 175% between both sites. The final concentration of QA and TA reflects not only the relationship between the reflectance and absorbance of UV radiation but also confirmed that the spectral response of UV light contributes directly to alkaloid biosynthesis.

It is known that both photosynthetically active radiation (PAR) and Ultraviolet B (UV-B) radiation are among the main environmental factors acting on herbal yield and the biosynthesis of bioactive compounds in medicinal plants. Therefore, the aim of this study (Manukyan 2013) was to evaluate the influence of UV-B light (250–315 nm) and PAR (400–700 nm) on herbal yield, content, composition, as well as antioxidant capacity of essential oils and polyphenols in lemon catmint (*Nepeta cataria* L. f. citriodora), *M. officinalis* L., and *Salvia officinalis* L. under controlled greenhouse cultivation. The authors found that intensive UV-B radiation (2.5 kJ m^{-2} d^{-1}) positively influenced the herbal yield, whereby content and composition of the essential oil in the studied herbs were affected by both studied abiotic factors. In general, additional low-dose UV-B radiation (1 kJ m^{-2} d^{-1}) was most effective for the biosynthesis of polyphenols in medicinal herbs. Essential oils and polyphenol-rich extracts of radiated herbs showed significant differences in antioxidant capacity, too.

To understand the effect of UV-B irradiation on the metabolism of *M. officinalis*, metabolomics based on GC-MS was used in the paper published by Kim et al. (2012). Thirty-seven metabolites were identified in *M. officinalis* L. plants from different chemical classes, including alcohols, amino acids, inorganic acids, organic acids, and sugars. Metabolite profiles of the groups of *M. officinalis* irradiated with UV-B were separated and differentiated according to their irradiation times (0, 1, and 2 h). It was found that metabolic patterns differed among the three groups, and the 1 h-irradiated group was more similar to the control group (0 h irradiation) than the 2 h-irradiated group, whereby the UV-B irradiation of the plants led to a decrease in sugars such as fructose, galactose, sucrose, and trehalose and an increase in metabolites in the tricarboxylic acid cycle, the proline-linked pentose phosphate pathway, and the phenylpropanoid pathway. This study demonstrated that metabolite profiling is useful for gaining a holistic understanding of UV-induced changes in plant metabolism.

In general, the modulation of light is important during cultivation of medicinal plants to obtain desirable morphological and physiological changes associated with the maximum production of biologically active substances. A study by Oliveira et al. (2016) focused on evaluating the effect of the light spectrum transmitted by colored shade nets on growth, essential oil production, and photosynthetic characteristics in plants of lemon balm. Plants were cultivated in pots under black, red, and blue nets with 50% shading, and full sunlight exposure. The results showed that despite being considered a partial shade plant, this species is able to adapt to full sunlight conditions without increasing biomass production, whereby any noticeable changes in leaf anatomy were observed. However, the use of blue net resulted in increments of 116% in plant height, 168% in leaf area, 42% in leaf Chl content, and 30% in yield of essential oil.

Hoffmann et al. (2015) investigated the influence of light quality on the sensibility of pepper plants to water deficit. The plants were cultivated either under compact fluorescence lamps (CFL) or light-emitting diodes (LED) providing similar photon fluence rates (total number of photons incident from all directions on a small sphere divided by the cross-sectional area of the sphere and per time interval) but different light quality. CFL emitted a wide-band spectrum with dominant peaks in the green and red spectral region, whereas LEDs had narrow band spectra with dominant peaks at blue and red regions. After 1-week acclimation to light conditions plants were exposed to a water deficit by withholding irrigation. In general, plants grown under CFL suffered more from water deficit than plants grown under LEDs resulting in less biomass production compared to the control plants. As affected by water deficit, plants grown under CFL had a stronger decrease in electron transport rate and a more pronounced increase in heat dissipation. The elevated amount of blue light suppressed plant growth and biomass formation and consequently reduced the water demand of plants grown under LEDs. The results explained the potential of the target-use of light quality to induce structural and functional acclimations improving plant performance under stress conditions.

In another experiment with two cultivars of pepper (Fu et al. 2010), different light conditions on the relationship between leaf growth, stomatal parameters, and photosynthetic characteristics were analyzed. Photosynthetic active radiation was used of either 450–500 µmol m^{-2} s^{-1} (high light, HL) or 80–100 µmol m^{-2} s^{-1} (low light, LL). It was shown that stomatal density (SD) and stomatal index (SI; the percentage which the number of stomata forms to the total number of epidermal cells, each stoma being counted as one cell) increased with leaf area expansion, and then decreased. After 3 days of leaf expansion, SD and SI were lower under the LL condition. For both cultivars, regulation of photosynthesis and electron transport components were observed in LL-grown plants as indicated by lower light- and CO_2-saturated photosynthetic rate. Inhibition of the photosynthesis could be explained by the decrease of SD, SI, and Rubisco content and by the changes of the chloroplast structure, too. The LL condition also resulted in lower values of total biomass, root/shoot ratio, and smaller leaf thickness. However, specific leaf area and leaf pigment contents were higher in LL-treatment. It was concluded that variations in the photosynthetic characteristics of pepper plants grown under different light conditions reflected the physiological adaptations to the changing light conditions.

There is evidence of Chls and Cars degradation by low temperature (LT) and UV-B radiation, and production of phenolic compounds as defense components against these stress factors, but there is scarce information concerning the interactions between LT and UV-B radiation. In a recent paper by Leon-Chan et al. (2017), the contents of Chl, Cars, and phenolic compounds were analyzed in response to LT and UV-B radiation and the combination of both (LT+UV-B) abiotic factors in the leaves of *C. annuum* L. The authors observed a greater degradation of Chl, higher accumulation of Cars, chlorogenic acid, and some of the flavonoids under LT+UV-B conditions compared to the LT and UV-B conditions applied separately. UV-B radiation induced a higher total flavonoid concentration than LT, but the highest flavonoid concentration was observed in the leaves exposed to LT+UV-B. The higher accumulation of chlorogenic acid and apigenin-7-*O*-glucoside biosynthesis in leaves exposed to LT than in control indicated a higher resistance of the plants to UV-B radiation damage.

Since shading nets help to alleviate the heat stress of the plants, the effect of shade level (0, 30%, 47, 62, and 80%) on the microenvironment, plant growth, leaf gas exchange, and mineral nutrient content of field-grown *C. annum* L. plants (cultivars Camelot, Lafayette, Sirius, and Stiletto) was investigated (Diaz-Perez 2013). Photosynthetically active radiation and air, leaf, and root-zone temperatures decreased as shade level increased. With increased shade conditions, total plant leaf area, individual leaf area, and individual leaf weight increased, whereas leaf number per plant and specific leaf weight decreased. With increased shade values, net photosynthetic rate and stomatal conductance decreased, but concentrations of N, K, Ca, Mg, Mn, S, Al, and B increased. Based on the results, it could be concluded that morphological changes such as taller plants and thinner and larger leaves likely enhanced light capture of the leaves under shaded conditions compared with unshaded plants. High shade conditions reduced leaf temperature, excessive leaf transpiration rate, and photosynthetic rate, too. Thus, moderate shade levels (30 and 47%) were the most favorable for pepper plant growth.

To elucidate the relationships between environmental conditions and the amount of carbohydrates, the effect of night temperature and irradiance in changes of carbohydrates concentrations in the pepper leaves was studied (Kang et al. 2008). An experiment was conducted with two levels of irradiance (30 and 180 Wm^{-2}) and two levels of night temperature (13±1 and 18±2°C). In general, the amount of carbohydrates increased with increasing irradiance. However, changes of carbohydrate concentrations in the leaves of the studied pepper plants were constantly maintained regardless of irradiance levels during night time, in comparison with carbohydrates concentrations in flowers. This finding suggested that carbohydrates formed in the leaves were preferentially used to maintain leaves and then carbohydrates remaining in leaves might be expected to translocate to sink organs, such as flowers.

In a study by Yang et al. (2005) two contrasting populations of *H. rhamnoides* from the low and high altitudinal regions in the southeast of the Qinghai-Tibetan Plateau of China were used to investigate the effects of drought, UV-B light, and their combination on this plant species. The experimental design included two watering regimes (well-watered and drought-stressed) and two levels of UV-B (with and without UV-B supplementation). It was found that drought significantly decreased total biomass, total leaf area, and specific leaf area, but increased root/shoot ratio, fine root/thick root ratio, and content of ABA in both plant populations. However, the high altitudinal population was more sensitive to drought than the low altitudinal population. Elevated UV-B induced the increase of anthocyanin concentration in both populations, whereas the accumulation of UV-absorbing compounds occurred only in the low altitudinal population. These results demonstrated that there were different adaptive responses between two contrasting populations, whereby the high altitudinal population exhibited higher tolerance to drought and UV-B than the low altitudinal population.

Four levels of glasshouse light intensities (310, 460, 630, and 790 μmol m^{-2} s^{-1}) were used in order to analyze the effect of light intensity on the production, accumulation, and partitioning of total phenolics (TP), total flavonoids (TFl), and antioxidant activities in two varieties (Halia Bara and Halia Bentong) of Malaysian young ginger (*Zingiber officinale* Roscoe) (Ghasemzadeh et al. 2010). The authors recorded that TFl biosynthesis was highest in the Halia Bara variety under 310 μmol m^{-2} s^{-1} and TP was high in this variety under a light intensity of 790 μmol m^{-2} s^{-1}. The highest amount of these components was accumulated in the leaves and after that in the rhizomes. Antioxidant activities in both of varieties increased significantly with increasing TFl concentration, whereby high antioxidant activity was observed in the leaves of Halia Bara variety grown under 310 μmol m^{-2} s^{-1}. These results indicated the ability of different light intensities to enhance the medicinal compounds and antioxidant activities of the leaves and young rhizomes of studied ginger varieties.

Effect of different γ-irradiation doses (0, 5, 10, 15, and 20 G) on the enhancement of secondary metabolites production and antioxidant properties of rosemary callus culture was investigated (El-Beltagi et al. 2011). The recorded data confirmed a highly metabolic modification of chemical compounds and different antioxidant defense enzymes (APX, CAT, SOD, and glutathione reductase (GR)), which gradually increased in response to radiation doses, while reduced GSH, ascorbic acid contents, total soluble protein, total soluble amino acids, total soluble sugars, and phenylalanine ammonia lyase activity positively correlated with the increased doses. Meanwhile, higher doses of γ-irradiation positively enhanced the accumulation of TP and TFl in rosemary callus culture.

In a pot experiment with marigold grown under low and high light conditions (35 and 100% of full daylight), the content of Chls and Cars, Chl fluorescence parameters and electron transport rate were evaluated (Casierra-Posada and Ávila-León 2015). Compared to full sun plants, shaded plants had higher values of Chl a to Chl b ratio and lower values of Cars to Chls ratio. In shaded plants values for F_m, F_v and maximum quantum yield of PSII (F_v/F_m) increased, but the electron transport rate decreased. These results suggested that marigold plants are very sensitive to low-light conditions.

Since responses of growth and secondary metabolites to light intensity are useful for the cultivation of medicinal plants, Xu et al. (2014) studied the growth, flavonols content, flavonols yield per plant, and expression of flavonoid biosynthesis-related genes in 2-year old ginkgo (*Ginkgo biloba* L.) seedlings at four different light intensities (100, 76, 40, and 25% of full sunlight). Plants of 76% sunlight treatment had the highest values of plant biomass (root, stem, and leaf) indicating the negative effects of either full light or heavy shading on ginkgo seedling development. Flavonols (total flavonol, quercetin, kaempferol, and isorhamnetin) content and expression of flavonoid biosynthesis-related genes in the leaves were the highest under 100% sunlight, suggesting that full sunlight increases flavonoid biosynthesis and promotes the expression of these genes.

39.7 SALINITY

In the past few decades, salinity has become one of the major limiting factors for the growth and productivity of plants since most of the plants are glycophytes which cannot withstand salinity. High soil salinity is due to the increasing use of poor quality water for irrigation and soil salinization. Recently, a review paper was published (Daliakopoulos et al. 2016) dealing with soil salinification as one of the major soil degradation threats occurring in Europe. It was emphasized that the effects of salinification could be observed in numerous vital ecological and non-ecological soil functions. Drivers of salinization can be detected both in the natural and man-made environment, with climate and the foreseen climate change also playing an important role.

Plant adaptation or tolerance to salinity stress involves complex physiological traits, metabolic pathways, and molecular or gene networks (in detail see Mäkelä et al. 2000; Flowers 2004;

Rozema and Flowers 2008; Gupta et al. 2013). Salt tolerance is an important economic trait for plants growing in both irrigated fields and marginal lands. The plant kingdom contains plant species that possess highly distinctive capacities for salt tolerance as a result of evolutionary adaptation to their environments. Most of the plants (except halophytes) are sensitive to high salinity. Damage to plants due to high salinity includes reduction of leaf expansion, stomata closure, reduced photosynthesis, and biomass loss due to water deficit caused by osmotic imbalance. Thus, salt tolerance is a complex trait that is controlled by multiple genes and involves various biochemical and physiological mechanisms. The functions of the distinct sets of genes involved in specific biochemical and physiological mechanisms must be combined to achieve a substantial increase in salt tolerance (Zhang and Shi 2013). The physiological and molecular mechanisms of tolerance to osmotic and ionic components of salinity stress are reviewed at the cellular, organ, and whole-plant level. Plant growth responds to salinity in phases: a rapid, osmotic phase that inhibits the growth of young leaves and a slower, ionic phase that accelerates senescence of mature leaves. Plant adaptation to salinity is of three distinct types: osmotic stress tolerance, Na^+ or Cl^- exclusion, and the tolerance of tissue to accumulated Na^+ or Cl^- (see Munns and Tester 2008).

According to the Gupta and Huang (2014), a comprehensive understanding on how plants respond to salinity stress at different levels and an integrated approach of combining molecular tools with physiological and biochemical techniques are imperative for the development of salt-tolerant varieties of plants in salt-affected areas. Recent research has identified various adaptive responses to salinity stress at physiological, metabolic, cellular, and molecular levels, although mechanisms underlying salinity tolerance are far from being completely understood. These authors emphasized that among various salinity responses, mechanisms or strategies controlling ion uptake, transport balance, osmotic regulation, hormone metabolism, antioxidant metabolism, and stress signaling play critical roles in plant adaptation to salinity stress. However, there is a lack of integration of information from studies using genomics, transcriptomics, proteomics, and metabolomics, and the combined approach is essential for the determination of the key pathways or processes controlling salinity tolerance. In addition, in spite of the significant progress in the understanding of plant stress responses, there is still a large gap in our knowledge of transmembrane ion transport, sensors and receptors in the signaling transduction, molecules in long-distance signaling, and metabolites in energy supply. The future focus should be on the study of intercellular and intracellular molecular interaction involved in salinity stress responses. Genetic engineering has proven to be an efficient approach for the development of salinity-tolerant plants, and this approach will become more powerful as more candidate genes associated with salinity tolerance are identified and widely utilized (Barkla et al. 2013; Gupta and Huang 2014). The complex molecular system involved in stress tolerance and adaptation in plants can be easily deciphered with the help of the different above-mentioned "omics" studies (Gupta et al. 2013).

Growth is reduced by salinity via several quite distinct processes, which are related either to the accumulation of salt in the shoot, or which are independent of shoot salt accumulation. These can be experimentally distinguished by measuring effects immediately (within minutes to a few days) upon addition of salt (before there has been time for salt to accumulate in the shoot) or measured after much longer time (several days to weeks) after there has been time for salt to accumulate in the shoot and affect shoot growth. Within minutes of the application of salt in an experimental system, there are several rapid responses occurring. Because of their rapid onset, these effects are clearly independent of the accumulation of salts in the shoot. The two best-documented effects are stomatal closure, with concomitant increases in leaf temperature, and inhibition of shoot elongation. Hence, the primary consequence is the overall reduction in the production of new leaves and a significant reduction in shoot growth (Roy et al. 2014). It should also be emphasized that the interaction of salts with other mineral nutrients may result in nutrient imbalances and deficiencies that can ultimately lead to plant death (cf. Sairam and Tyagi 2004). In a comprehensive publication by Läuchli and Lüttge (2004), processes were described that were performed in the "environment–plants–molecules" system.

The latest papers deal with the application of nanomaterials also on medicinal plants grown under saline soils. For example, Safikhan et al. (2018) found that graphene nanomaterials (graphene oxide) could ameliorate the salt stress in milk thistle (*Silybum marianum* L.) plants and thus it could be commercially and economically beneficial for the production of this medicinal plant under saline conditions.

Plant growth is often limited by low levels of soil micronutrients such as Cu, mainly in calcareous salt-affected soils of arid and semiarid regions. Mehrizi et al. (2012) followed the individual and combined effects of salinity (0, 50, and 100 mM NaCl) and Cu (0, 0.5, and 1.0 µM $CuSO_4$) on growth, leaf RWC, cell membrane permeability, lipid peroxidation, and TPC of *Rosmarinus officinalis* L. in hydroponic conditions. The salt stress caused a significant decrease in leaf RWC (especially without $CuSO_4$). Salt treatments with 50 and 100 mM NaCl increased electrolyte leakage and MDA content of rosemary, whereby this increase was greater in the control (0 µM $CuSO_4$) than at 1.0 µM $CuSO_4$. In the presence of 50 and 100 mM NaCl, significant increases in TPC by 13 and 29%, respectively, were estimated. After application of 1.0 µM Cu at the 50 mM NaCl, an increase of TPC by 19% compared to control was observed. Since Cu nutrition resulted in greater accumulation of phenolic compounds in plant roots, a decrease of lipid peroxidation under salt stress conditions occurred.

The effect of exogenous Si (0–2.5 mM) on salt-stressed (120 mM NaCl) borage (*Borago officinalis* L.) plants was investigated by Torabi et al. (2015). Salt stress reduced the fresh and dry weights of the plants, protein contents, and CAT activity, but increased proline, glycine betaine, MDA, activity of SOD, and APX. Moreover, salt stress caused not only an increase of both thickness of the leaf blade and palisade parenchyma but induced changes in the structure and numbers of trichome and stomata. It was confirmed that the treatment of the plants with Si could moderate the negative effect of NaCl on the studied anatomical features.

Similarly, Jaffel et al. (2011) attended the effect of increasing salt concentrations on the growth, electrolyte leakage, lipid peroxidation, and major antioxidant enzyme activities (SOD, CAT, APX, and GR) of borage plants grown in a half-strength Hoagland nutrient solution in the presence of 0, 25, 50, and 75 mM NaCl. Increasing salt levels caused a significant reduction in leaf area, stem length, stem diameter, flower number, and dry masses of different organs. As a consequence of salinity stress, lipid peroxidation and membrane permeability were increased. The antioxidant activity showed an increase in the activity of SOD, a non-induced activity of CAT and APX, and a slight increase in GR. Based on the results, it could be indicated that borage plants appear to be sensitive to salt stress since enzymes related to the antioxidant enzymatic defense system in treated leaves should be highly active.

Jaffel-Hamza et al. (2013), in another experiment, cultivated borage in a hydroponic medium and studied the effect of salinity on growth, seed yield, and fatty acid composition of this medicinal plant using three NaCl concentrations (25, 50, and 75 mM). The results showed that salinity significantly reduced plant growth by 56.5% at 75 mM compared with the control, suppressed seed yield at 50 and 75 mM, and increased lipid peroxidation. Raising NaCl concentration caused a decrease in total fatty acid content by 77% at 75 mM NaCl. Moreover, the polyunsaturated fatty acid content decreased, and the saturated fatty acids increased with respect to increasing salinity. However, the lowest concentration of NaCl (25 mM) did not influence the fatty acid composition of the seeds.

In the research of Torabi et al. (2013), the effect of salinity on the development of anther in hydroponically-grown borage (*Borago officinalis* L.) was studied, whereby the formation, development, and structure of anthers were observed. In control plants, young anther consisted of four pollen sacs. Anther wall development followed the typical dicotyledonous pattern and was composed of an epidermal layer, an endothecium layer, and the tapetum. Microspore tetrads were tetrahedral. Salinity caused abnormalities during pollen development, such as the destruction of the anther wall and both the degeneration and production of abnormal pollen grains. It was concluded that possibly a general response to the harmful effects of salinity included a decrease in plant fertility, which involved aborting pollen, followed by a change in resource from reproductive activities to metabolic reactions.

Atropa belladonna L. (family *Solanaceae*) commonly known as belladonna or deadly nightshade is an invasive species (containing tropane alkaloids) belonging to the most toxic plant species in Middle Europe. Toxins produced by belladonna include atropine, scopolamine, and hyoscyamine, which are used as pharmaceutical anticholinergics with antisialagogue effects (decreasing saliva production), and produce some level of sedation, both being advantageous in surgical procedures. Ali (2000), in an experiment with the plants of *Atropa belladonna* L., studied the effects of putrescine (10^{-2} mM)

on germination and seedling growth under the influence of NaCl. Germination, growth, endogenous putrescine, and alkaloids content decreased when seeds were exposed to salt stress. Pre-soaking seeds in putrescine and treatment with NaCl reduced the accumulation of Na^+ and Cl^- ions in different organs of the plant. It could be concluded that pre-soaking seeds in putrescine can mitigate the harmful effect of NaCl during germination and early seedling growth of belladonna and increase the concentration of alkaloids and endogenous putrescine, too.

Growth and composition of *Atropa belladonna* L. plants were observed after separate and combined additions of NaCl and $NiCl_2$ to the nutrient medium (modified Johnson solution) for 8 weeks until the formation of the fifth leaf pair (Stetsenko et al. 2017). In a 7-day treatment, $NiCl_2$ was introduced at concentrations of 100 and 150 μM into the medium either separately or in combination with NaCl (100 mM). The presence of $NiCl_2$ in growth media reduced the increments in a fresh mass of shoots and roots, and the content of water, pigments, and Fe in the leaves, initiated leaf chlorosis, but the leaves of Ni-treated plants accumulated larger amounts of atropine, putrescine, proline, and MDA. In comparison with the action of Ni alone, the combined application of NaCl and $NiCl_2$ caused increases of both leaf water content as well as pigment concentration. It was confirmed that the presence of NaCl in the medium restricted the entry of Ni into the roots and reduced the levels of MDA and proline in leaves. In plants cultivated in the presence of 100 and 150 μM $NiCl_2$, nickel was located in the root outer cortex and the rhizoderm. In plants treated with 150 μM $NiCl_2$, nickel was also observed in tissues of the central cylinder, mostly in the pericycle, phloem, and xylem. In plants grown in the presence of 150 μM $NiCl_2$ and 100 mM NaCl, the decreased accumulation of nickel was noted in the tissues of the central cylinder in the root hair zone. The authors emphasized that the combined action of Ni and moderate salinity reduced Ni accumulation in the roots and shoots. The toxic effect of Ni on the plants was thus mitigated which was manifested in the stabilization of leaf water status, in an increase of photosynthetic pigment content, and in the alleviation of oxidative stress.

Sabra et al. (2012) evaluated the physiological and biochemical responses of the three Echinacea species (*Echinacea purpurea*, *E. pallida*, and *E. angustifolia*) to NaCl salinity (0, 50, 75, and 100 mM) after a 2-week cultivation in hydroponics. Salinity did not alter the root or shoot biomass of the three Echinacea species. *E. angustifolia* plants showed a decrease in stomatal conductance, photosynthetic and transpiration rates at all salt concentrations, whereas in *E. purpurea* and *E. pallida*, the photosynthetic rate was reduced only at 75 and 100 mM NaCl. The decline in the leaf gas exchange was highly correlated with Na^+ and Cl^- contents in all Echinacea species. *E. angustifolia* plants showed high injury index and electrolyte leakage values compared to the other two species, which indicates its sensitivity to salinity-induced membrane damage even at the lowest salt concentration tested. *E. purpurea* showed increased SOD and APX activities at all salt concentrations. On the other hand, salinity induced a reduction in CAT activity with no change in GR activity in any of the species. Summarizing the results, it could be indicated that the Echinacea species studied showed a limited salt tolerance.

Ozturk et al. (2004) carried out an experiment to ascertain the effects of salt stress (measured as electrical conductivity (EC)) of 0.25, 1.00, 2.00, 4.00, and 6.00 dS m^{-1} (deciSiemens per meter) and water deficiency (0, 12.5, 25.0, 37.5, and 50.0%) on yield components and essential oil contents of *Melissa officinalis* L. The results showed moderate tolerance of this medicinal plant against salt with the death of all plants when irrigation water had a salinity level of 6 dS m^{-1}. Dry yield per plant under salt stress varied between 4.33–14.87 g with the maximum yield at 1 dS m^{-1} salt concentration. Compared to salt stress, lemon balm was found highly tolerant against water stress. Dry yield under water deficit varied from 13.05 to 19.20 g per plant. Reduction in yield was not statistically significant until a 25% water deficiency. It was also found that each increase of salt stress was accompanied by a reduction in essential oil.

Recently, Manivannan et al. (2016) demonstrated the salinity stress resistance induced by Si when seedlings of *Capsicum annuum* L. were hydroponically treated with NaCl (50 mM) with or without Si (1.8 mM) for 15 days. It was confirmed that saline stress significantly reduced photosynthetic parameters, growth, and biomass of the plants, but increased the electrolyte leakage potential, lipid peroxidation, and H$_2$O$_2$ level. Addition of Si into the hydroponics improved photosynthesis of the plants and prevented oxidative damage by increasing the activities of antioxidant enzymes. Proteomic analysis by two-dimensional gel electrophoresis showed that Si treatment upregulated the accumulation of proteins involved in several metabolic processes, particularly those associated with nucleotide binding and transferase activity. Based on the results, it could be concluded that Si application induced resistance against salinity stress in *C. annuum* plants by regulating the physiological processes, antioxidant metabolism, and protein expression.

The influence of saline stress was also studied on the leaf anatomy of sweet pepper (*Capsicum annuum* L.) plants that were watered with NaCl (50 mM) or water (Valera and Garcia 2014). Leaves under saline stress showed a higher width of lamina and reduction of stomatal density, but stomata index was similar in both experimental variants. The increase of leaf succulence and unchanged Si could contribute to the reduction of the harmful effect of saline stress on sweet pepper plants.

A paper recently published by de Melo et al. (2017) provided an examination of photosynthetic and transpiration rates, photosynthetic pigments (Chl a, Chl b, and Cars), stomatal conductances, instantaneous carboxylation efficiency, and WUE in bell pepper irrigated with saline solutions. These solutions (0, 1, 3, 5, 7, and 9 dS m^{-1}) were prepared using two sources: NaCl and a mixture of Ca, Mg, K, Na and Cl salts, in randomized blocks with a 6 × 2 and 4 replicates, in total 48 experimental plots. It should be mentioned that bell pepper is a cultivar group of the species *Capsicum annuum* L. During experiments, EC of saline solutions that were expressed in

dS m^{-1} units were measured. All studied photosynthetic pigments decreased with increasing EC, whereby the Chl a was the most sensitive to salinity stress. Reversely, WUE increased with the increment of EC. No differences between the solutions of NaCl and mixture of salts were found for gas exchanges in bell pepper plants subjected to salinity. However, plants irrigated using water with EC of 1 dS m^{-1} showed an increase in net photosynthetic rate (48.86%), transpiration rate (38.10%), stomatal conductance (105.55%), internal CO$_2$ concentration (6.59%), and instantaneous carboxylation efficiency (39.45%), in comparison to the control plants.

Similarly, Aktas et al. (2012) studied the effects of 75 and 150 mM NaCl (EC was 8.50 and 15.35 dS m^{-1}, respectively) on a salt-tolerant and salt-sensitive pepper (*C. annuum* L.) genotypes. The salt-tolerant genotype showed lower declines in RWC, no change in Chl content, lower increases in lipid peroxidation, and greater increases in SOD activity, total protein content, and GSH content. The salt-sensitive genotype showed greater decreases in RWC, Chl content, SOD activity, and in guaiacol peroxidase activity, and higher increases in lipid peroxidation and the amount of proline, with a negligible increase in GSH content. These results showed that increases occurred in some antioxidative stress enzymes in the salt-tolerant pepper genotype, as well as increases in GSH content under salinity stress. The responses by these biochemical substances provide better protection against the influence of ROS.

The growth and photosynthetic characteristics of 2-year-old seedlings of *H. rhamnoides* L. and *Shepherdia argentea* (Pursh) Nutt. were followed after application of 200, 400, and 600 mmol dm^{-3} NaCl (Qin et al. 2009). The results confirmed that the biomass and total leaf area per plant of both studied species showed a considerable decrease with increasing NaCl concentration. When increasing NaCl concentration and stress time prolongation, the P$_N$ and E rates and stomatal conductance of *H. rhamnoides* and *S. argentea* seedlings declined markedly. However, the intercellular CO$_2$ concentration increased after an initial decrease, whereas the WUE and stomatal limiting value decreased after an initial increase. The P$_N$ and E rates and stomatal conductance of both studied species declined markedly, the intercellular CO$_2$ concentration increased after an initial decrease, whereas the WUE and stomatal limiting value decreased after an initial increase. The found results indicated that the decline of P$_N$ was mainly caused by the stomatal limitation in short-term stress, and by non-stomatal limitation in long-term stress. Morphological symptoms of salt injury in *H. rhamnoides* appeared on the 10th day, and all of its seedlings died on the 22nd day under 600 mmol dm^{-3} NaCl stress. In contrast, *S. argentea* could tolerate 600 mmol dm^{-3} NaCl stress for above 30 days, illustrating that this introduced tree species had higher salt tolerance than *H. rhamnoides*, and could be planted widely in saline regions of China.

Ruan and Xie (2002) investigated annual seedlings of *Hippophae rhamnoides* L. that were treated with Hoagland solution with 0, 100, 200, and 300 mmol dm^{-3} NaCl. After 30 days of saline stress. the highest water content, fresh weight, and dry weight of *H. rhamnoides* seedlings was observed under the treatment of 100 mmol dm^{-3} NaCl. However, with increasing NaCl concentration, total fresh and dry weight and osmotic potential gradually decreased, whereas the content of Na$^+$ and Cl$^-$, as well as soluble sugar contents progressively increased. These results provided the basis for forestation of *H. rhamnoides* in saline-alkali soil.

Experiments with two *Aloe vera* cultivars (170 and F50) were conducted to study physiological processes and ecological demands of this species under salt stress. The researchers (Jin et al. 2007) observed decreases in tissue water, total soluble sugars, and glucose, increases in dry matter and membrane injury occurred in the studied cultivars irrigated with 60% seawater. Less cell membrane injury and higher leaf K$^+$ and Ca^{2+} contents in the stems and roots were observed in an F50 cultivar. This cultivar had a relative superiority in growth under salinity conditions due to higher K$^+$/Na$^+$ ratio and lower Na$^+$/Ca^{2+} ratio than F50.

In another experiment with the aloe species (Lu et al. 2006), 1-year-old seedlings were placed under three types of environmental stress, which included salt (NaCl at 1.8%), low temperature (10°C), and drought (simulated with PEG-6000 at 25% (w/v)) for 7 days. It was found that all the three kinds of environmental stress could damage the membrane system, organelle structure stabilities, mitochondria structure, and nucleus structure of aloe in various degrees. Many vesicles also appeared around the chloroplasts, which resulted in the cellular endomembrane system falling into disorder and the organelles becoming destabilized. Under salt stress, the Golgi bodies disintegrated in the cytoplasm; under both salt stress and low-temperature stress there appeared a phenomenon that the mitochondria and chloroplast membranes fused and as a result the mitochondria embedded into the chloroplasts. Moreover, salt stress and low-temperature stress damaged the cellular membrane more severely than drought stress which indicated aloe performed better in drought resistance than in salt resistance.

Similarly, some of the cellular signaling events including cell death, ROS generation, and the behaviors of organelles were analyzed by Li et al. (2015) in a salt-tolerant cultivar (Keyuan-1) of peppermint (*Mentha × piperita* L.) under a 200 mM NaCl treatment. This NaCl concentration elicited cell death and a dramatic burst of ROS, whereby the major ROS accumulation occurred in the mitochondria and chloroplasts. Moreover, under saline stress, the mitochondrial activity and photosynthetic capacity exhibited the obvious decrease in the ROS-dependent manner, while the activities of APX, GR, and dehydroascorbate reductase changed in a concentration-dependent manner. The presented data can be used for peppermint tolerance screening, and contribute to the understanding of the cellular responses and molecular mechanisms of this species to salinity stress.

The ripe seeds of *Silybum marianum* L. Gaertn contain active substances of flavonoides (silymarin and silybin) that are important not only in the pharmaceutical industry but also for the resistance of this medicinal plant against saline stress (Ghavami et al. 2008). A pot experiment was conducted under seven levels of salinity in the concentration range 0.35–15 dS m^{-1} for two cultivars, wild-type Ahwaz and cultivated German cultivar Royston. Growth parameters such as plant height, the number of leaves per plant, the number of capitula per plant, the main shoot capitulum's diameter, and seed yield were reduced with salinity exceeding 9 dS m^{-1} in both genotypes. However, concentrations of silymarin and silybin in seeds significantly increased compared to the control plants. Amount of silybin was approximately 4-times greater in the plants cultivated under a salinity of 15 dS m^{-1} than in the plants grown under nonsaline conditions. Based on the results, it was concluded that both *S. marianum* cultivars survived under salinity as high as 15 dS m^{-1} levels and produced seeds with higher silymarin and silybin concentrations.

Iqbal et al. (2015) compared the influence of a semiarid environment (saline and saline-alkaline soils, respectively) on nutritional (total proteins, phenolics and riboflavin contents, and CAT activity) and antinutritional (hydrogen peroxide and MDA contents) properties of the leaves, flowers, and root of marigold (*Calendula officinalis* L.). It was found that salinity decreased plant biomass, increased H$_2$O$_2$ concentration in the flowers and roots, but it did not affect total protein and phenolics contents in the plant's organ. The high content of riboflavin was observed in all parts of the plants grown in saline soil, while plants grown under saline-alkali soil had relatively more riboflavin content only in the flowers. Plants exposed to both saline and saline-alkali soil had greater CAT activity in the flowers and leaves, but plants exposed to salinity alone had a higher content of MDA. It was concluded that calendula plants could grow in mild saline-alkali (EC ≤ 7; pH=8.5) soils without affecting its nutraceutical properties.

Bayat et al. (2012) also conducted pot experiments with marigold plants grown in greenhouse conditions and studied the effect of NaCl (0, 100, and 200 mM) in combination with exogenous application of phytohormone salicylic acid used at three concentrations (0 as a control and 1, 2 mM SA) on growth and ornamental characteristics. At the flowering stage, SA was applied by spraying 2-times in 2-week intervals, and NaCl was also applied as a drench (200 ml per pot) in 2-day intervals. The results showed that salinity decreased the growth and concentration of Chls. Foliar applications of SA on calendula plants grown under salt stress resulted in higher values of the root, shoot and total dry weight, plant height, and leaf area, too. Based on the results, it was concluded that foliar application of SA could ameliorate the negative effects of salinity on the growth and ornamental characteristics of the studied plant species.

In another experiment (Chaparzadeh et al. 2003) with marigold plants under salinity stress, it was observed that high salinity (100 mM NaCl) caused a reduction in the relative growth rate and leaf area ratio, a decrease of osmotic potential and increase of turgor potential. Salinity thus caused an alteration of nutrient uptake, and under salinity conditions, marigold maintained osmotic pressure mainly by accumulation of inorganic solutes.

Male and female seedlings of *Ginkgo biloba* L. were used to compare changes in gas exchange, WUE, free proline content, and POD and SOD activities under 40 mmol dm^{-3} NaCl soil salt stress (Jiang et al. 2009). The authors recorded that the P_N, stomatal conductance, and E rate in salt-stressed female seedlings decreased by 45.87, 25.00, and 16.47%, respectively, while the stomatal conductance, intercellular CO_2 concentration, and transpiration rate in salt-stressed male seedlings increased by 10.00, 8.10, and 22.95%, respectively, compared to the control. Considerably lower values of the WUE were observed in both salt treated female and male seedlings which decreased by 30.47 and 46.38%, respectively. The SOD activity in both treated seedlings decreased by 22.96 and 23.18%, respectively, without significant variations. It was concluded that under 40 mmol dm^{-3} NaCl stress female seedlings maintained a higher photosynthetic rate to accumulate energy, lower transpiration rate, and higher WUE to reduce the variance of water content, and higher antioxidant enzyme activities to relieve oxidative stress resulting in reduced adverse effects under salt stress condition compared to male seedlings.

39.8 ACIDITY

The acidification of soils (under outdoor conditions) is the consequence of a natural process occurring in the ecosystem. Usually, soil acidification is a slow process. Acid soils are therefore common in areas where ecosystem and soil development continued for long, geological periods of time. It should be emphasized that the driving force for the processes involved in soil acidification as well as the rate of this process remain unclear. However, with the advent of industrialization within the last two centuries, enormous quantities of acids have been mobilized and released into the atmosphere, much of which has been deposited in terrestrial ecosystems. In order to evaluate the importance of the acid deposition for soil acidification, the rate of natural and anthropogenically caused acid load in soils had to be compared. In solving this problem, a new ecosystem-oriented perspective on soil acidification was developed (cf. Ulrich and Sumner 1991).

The principal anthropogenic sources of acid deposition are H_2SO_4, HNO_3, and NH^+_4 derived from SO_2, NO_x, and NH_3, respectively. These compounds are emitted primarily by the burning of fossil fuels, industrial activities, and agricultural and livestock production. Ammonia interacts in the atmosphere and on the surface of vegetation to form NH^+_4, which may subsequently undergo nitrification in the soil to produce HNO_3. The impact of acid deposition on terrestrial and aquatic ecosystems is mediated primarily through interactions with soil biogeochemical processes. However, the direct impact of acid deposition on biological processes is often difficult to determine, because, for example, high H^+ concentrations lead to elevated levels of soluble Al^{3+}, which have been shown to produce severe ramifications on terrestrial and aquatic species. High Al^{3+} concentration can be directly toxic to plants resulting in the death of fine roots and mycorrhizae symbionts and can interfere with the acquisition of base cations and other nutrients from the soil solution inducing nutrient deficiencies and imbalances in plants (Chesworth 2008). The topic of soil

acidification with respect to the pH value (soil reaction) as the most important and meaningful soil parameter was recently analyzed in detail by Blume et al. (2016).

Besides acid soils, acidic deposition in the atmosphere is one of the more highly publicized aspects of atmospheric pollution. This acid deposition includes rainfall (acid rain), acidic fogs, mists, snowmelt, gases, and dry particulate matter. The primary origin of acidic deposition is the emission of SO_2 and nitrogen oxides (NO_x) from fossil fuel combustion. When acidic substances are deposited in natural ecosystems, a number of adverse environmental effects occur, including damage of vegetation and changes in soil and surface water chemistry. Damage from acidification is often not directly due to the presence of excessive H^+ but is caused by changes in other elements (Vance 2006).

In many areas, where the quantities of acid substances released and deposited, serious perturbations in the environment were observed. Thus, many soils throughout the world are or have become considerably acidic to restrict the growth and/or performance of roots of sensitive plant species in the soil profile. Therefore, the productivity and longevity of both natural and agricultural or other man-made ecosystems have been adversely affected. Moreover, soil acidity is one of the major factors regulating the species composition of ecosystems. While the process involved in soil acidification have been relatively well understood for many years, there has been much debate concerning the ecophysiological mechanisms by which soil acidity influences the plants (Ulrich and Sumner 1991).

One of the physiological processes greatly affected by the abiotic stresses (including acidity) is photosynthesis. The decline in the photosynthetic capacity of plants due to these stresses is directly associated with a reduction in the yield. Therefore, it is essential to examine the effects of acidity on each of the components of photosynthetic machinery for the in-depth understanding of the causes responsible for the decrease in P_N. There is an urgent need to develop a new stress-resilient photosynthetic apparatus in order to sustain plant productivity. Recent developments in the field of acidity with the help of "omics" tools have led to the accumulation of vast knowledge about plant adaptation methods to stress conditions vis-a-vis photosynthesis (Singh and Thakur 2018).

According to Radanovic et al. (2002), *H. perforatum* L. and *Achillea millefolium* L. are plants which can be classified as toxic metal accumulators. Including relating soils samples differing in pH reaction, 14 samples of St. John's wort and 9 samples of yarrow were collected from different localities in Yugoslavia and Republic Srpska. The total metal content (Mn, Zn, Cu, Pb, Ni, and Cd) was determined in the collected herb material by standard analytical method. The concentration of Mn and Zn in the herb decreased linearly with the increase of soil pH, while Cd concentration decreased exponentially. Cd content in St. John's wort was mostly above the limit of 0.5 ppm when soil pH (in nKCl) was lower than 5.9, and 4.5 for yarrow. Higher Cd content in the yarrow herb, and especially St. John's wort herb from acidic soils, points to the significance of both required control of Cd content in the raw materials collected in the wild and necessary avoidance of such soils for cultivated production of studied medicinal plant. However, the relationship of Ni content in the herb with soil pH was found only for *H. perforatum* (r=−0.80***).

Dagar et al. (2015) reported that alkali lands in India occupy about 3.8 million ha. Due to poor physical properties, excessive exchangeable sodium and high pH ranging from 8.39 to 9.84, most of these lands are with low biodiversity. In recent times, many of the medicinal plants are in great demand for both internal requirements and export. However, as these crops are non-conventional in nature, farmers are not convinced to cultivate them on fertile lands. The marginal lands (especially those affected by salinity, sodicity, and waterlogging problems) could be successfully utilized for the cultivation of some high-value stress-tolerant medicinal crops. Results presented in this study indicated that licorice (*Glycyrrhiza glabra* Linn.), which is quite remunerative and high in demand, could successfully be grown on alkali soils. Besides getting 2.4–6.1 tons ha^{-1} forage per year, a root biomass of 6.0–7.9 tons ha^{-1} year^{-1} could be obtained in 3 years of growth fetching about 600,000 to 800,000 rupees (Rs) ha^{-1} (approx. 7024 to 9366 EUR), i.e., 200,000 to 265,000 Rs year^{-1} ha^{-1} (approx. 2341 to 3102 EUR). Besides, other than growing licorice plants, the sodic lands could also be reclaimed substantially in terms of reducing soil pH and exchangeable sodium percentage.

The content of potentially toxic heavy metals (Pb, Ni, Cr, Cd, and Co) in peppermint herb (*Mentha piperita* L.) cultivated on seven soil types from Serbia (Humic Gleysols: pH 6.6, Chernozems: pH 8.0, Eutric Cambisols: pH 6.1, Distric Cambisols: pH 4.8, Fluvisols: pH 8.1, Solonetz: pH 7.2 and Vertisols: pH 6.4) in both field and vegetation trials were analyzed by Radanovic et al. (2001). The investigated heavy metals in peppermint were within their normal content in the plant material. The exception was found only on Vertisols (Glogovac), where contamination of soil and plants by an anthropogenic input (ferronickel smelter) occurred. Contents of Pb and Cr were higher in the plants under field conditions, contents of Cd and Ni were higher in the plants grown in pots, and Co contents were mainly equal. Amounts of Pb content in peppermint varied more in field conditions (2.9–5.9 ppm), in comparison with ones from the pot trial (1.2–2.6 ppm). The same findings were obtained for Cr (1.7–6.3 ppm, in comparison with 0.5–1.6 ppm). The content of Ni in peppermint plants varied more in field conditions (2.5–4.9 ppm), in comparison with its content from the pot trial (1.5–6.5 ppm), with the exception of the sample with a very high Ni concentration (19.2 ppm), which originated from the anthropogenic contamination of leaves (Glogovac). The concentration of Cd in the plants was significantly lower under field conditions (0.1–0.2 ppm), in comparison with the pot trial (0.2–0.3 ppm). Content of Co in peppermint was equal in both types of experiments (1.6–2.3 ppm). The available amounts of the listed heavy metals in the investigated soils (DTPA extraction and 2M HNO$_3$ extraction for Cr) did not have a significant correlation with their contents in peppermint (except for Cd). Significant correlation coefficients were obtained only for Ni and Cd

content in peppermint plants and soil pH value. According to the authors, this finding is understood since the investigated soils have quite different chemical properties, where a metal uptake is under the simultaneous influence of numerous different factors (available content, pH, cation exchange capacity (CEC), humus, etc.).

It is known that uptake by plants is a potential pathway of metal transfer to the human food chain. Using data for soil and *Urtica dioica* L. plants from 382 locations north-east of Ghent, the relation between soil characteristics (clay content, organic carbon content, CEC, soil pH, and total metal content) and concentration of metals in plants was studied (Tack and Verloo 1996). Contents of Zn and Mn in the plants varied widely as a consequence of the low values of clay (< 10%), CEC (< 15 cmol$_c$kg^{-1}) and organic carbon (< 3%) in the soil, and they were lower and less variable for higher values of these soil parameters. However, no systematic trend was observed as a function of soil pH, whereby the mean value of pH was 5.8. For other metals (Cd, Co, Cr, Cu, Fe, Ni, and Pb) ranges in plant metal concentrations were not affected by the studied soil characteristics. Based on the results, the authors emphasized that bulk soil parameters do not constitute sufficient information for a precise prediction of plant metal contents in the field.

The aim of the study published by Adamczyk-Szabela et al. (2015) was to estimate the influence of soil pH on the uptake of Cu, Zn, and Mn by *Valeriana officinalis* L. Preliminary studies involved soil analyses in determining acidity, organic matter content, and total Cu, Zn, and Mn being in bioavailable forms. The study involved atomic absorption spectrometry (AAS) in determining the concentration of the elements; soil pH=5.1 was used as being typical for central Poland. It was found that the intensity of germination was strongly pH dependent with the highest yield obtained in original, unmodified soil. According to the authors, it was surprising that high soil alkalinity stimulated Cu and Mn uptake while at the same time decreased Zn content was observed.

Heavy metals in four perennial medicinal plants (*M. recutita*, *Taraxacum officinale*, *Achillea millefolium*, and *H. perforatum*) collected from the Navodari region (Romania) were quantitatively analyzed using Flame Atomic Absorption Spectrometry (FAAS). Seven heavy metals (Cd, Cr, Co, Cu, Pb, Ni, and Zn) in plant samples and nine heavy metals (Cd, Cr, Co, Cu, Pb, Fe, Mn, Sn, and Zn) in the soil were selected on the basis of their effects upon health (Radulescu et al. 2013). Absorption of heavy metal in a medicinal plant is governed by soil characteristics such as pH, salinity, conductivity, and organic matter content. The pH of the examined soil samples was moderately basic. The result showed that the mean pH values were between 7.44 and 7.82. The smaller values of pH were observed in samples collected from the Rompetrol Refinery area suggesting a weak basic character of these soils. It was found that Zn concentration ranged between 12.180 mg kg^{-1} in *T. officinale* collected from the Ovidiu zone and 48.976 mg kg^{-1} in *A. millefolium* collected from the Rompetrol Refinery location. The above-mentioned authors found that the physiological activities of the medicinal plant influence Zn absorption and the interactions with many elements such as Fe and Cu. *T. officinale* accumulated more Cd (e.g., 0.783 mg kg⁻¹ DW in the Channel of Rompetrol Refinery location) than *A. millefolium*, *M. recutita*, and *H. perforatum*. These plants absorbed Pb from the soil and from the diethyl lead traffic-induced contaminated atmosphere because the collected points were near roads.

To assess the growth of *Rosmarinus officinalis*, the contents of Cd, Cu, Fe,Mn, Pb, Zn, Ca, Mg, K, N, and P were measured in this plant species cultivated on two substrates: pine bark (PB; pH 4.0, 80.5% organic matter) and pruning wastes-biosolids (BS; pH 6.9, 47.5% organic matter). The plants (initially 3.5±0.5 g dry weight and 31.1±6.9 cm high) were maintained under greenhouse conditions for 7 months. Nutrient solution samples were taken from each substrate in situ by rhizon probes, indicating that the concentrations of soluble Mn and Zn in PB were significantly greater than in the nutritive solution BS. At the end of the assay, the dry weight of leaves and height was significantly greater in plants cultivated in BS (40.0±2.2 g and 75.9±14.3 cm) than in PB (27.5±4.0 g and 62.4±10.2 cm). Plants cultivated in PB showed symptoms of slight chlorosis attributed to the high concentration of Mn in the leaves (106.6±7.8 mg kg⁻¹), which was much greater than in plants cultivated in BS (8.2±0.9 mg kg⁻¹). The concentration of toxic metals Cd and Pb in plants cultivated on both substrates did not exceed the recommended levels for the consumption of the leaves as the spice. In a case where *R. officinalis* is cultivated on the substrate of pine bark with an acid pH for food or medicinal use, the accumulation of Mn must also be considered (Tapia et al. 2014).

It is interesting that *R. officinalis* L. was suggested as useful in the restoration of the soil and vegetation. This potential use of studied species for the restoration of contaminated soils was performed with soils divided into two groups: pH < 5 and pH > 5. In an experiment by Moreno-Jimenez et al. (2011), native shrubs *R. officinalis*, *Myrtus communis*, *Retama sphaerocarpa*, and *Tamarix gallica* were planted and left to grow without intervention. Metal concentrations in the soils and plants, their extractability in soils, transfer factors, and plant survival were used to identify the most-interesting species for the above-mentioned biotechnological application. Total As was higher in soils with pH < 5. Ammonium sulfate-extractable zinc, Cu, Cd, and Al concentrations were higher in very acidic soils, but As was extracted more efficiently when soil pH was > 5. Unlike As, which was either fixed by Fe oxides or retained as sulfide, the extractable metals showed significant relationships with the corresponding total soil metal concentration and inverse relationships with soil pH. The species *T. gallica*, *R. officinalis*, and *R. sphaerocarpa* survived better in soils with pH > 5, while *M. communis* had better survival at pH < 5. It was found that *R. sphaerocarpa* showed the highest survival (30%) in all soils. Metal transfer from soil to harvestable parts was low for all plant species, whereby some species may be able to decrease metal availability in the soil. Found results suggested that *R. sphaerocarpa* is an adequate plant species for phytostabilizing these soils, although more research is needed to address the self-sustainability of both

this remediation technique and the associated environmental changes.

The growth of many plant species is limited in acid soils, which are characterized by high levels of potentially toxic elements as well as by low nutrient availability. Although plant–soil interactions are traditionally studied during the growing season, the highest concentrations of toxic elements in the soil may occur during the winter months. Falkengren-Grerup (1998) investigated the effects of a 3-month exposure to either an acid or a reference soil, at temperatures fluctuating around freezing point, on subsequent survival and growth of seven natural herbs (*Brachypodium sylvaticum*, *Carex pilulifera*, *Geum urbanum*, *Luzula pilosa*, *Mycelis muralis*, *Silene dioica*, and *Stellaria nemorum*) and medicinal plant species *Veronica officinalis*. All studied plants were exposed to ambient weather conditions from December to March; afterward, they were replanted in fresh reference soil and transferred into a glasshouse. Biomass of the plants was measured 5 weeks later. It was found that the studied plant species differed in their responses to the soils, in a manner reflecting their natural field distributions. All plants of the most acid-tolerant species (*C. pilulifera* and *L. pilosa*) survived in both treatments, whereas the more sensitive species showed lower survival rates after growth in the acid than in the reference soil. Similar results were found for the regrowth: *C. pilulifera* and *L. pilosa* (the most acid-tolerant species) were unaffected by the soil treatments (ratios between biomass in acid compared to reference soils were 0.8 and 1.1, respectively), whereas *G. urbanum*, *M. muralis*, *S. nemorum*, and *V. officinalis* were negatively affected (the above-mentioned ratios ranged from 0.3 to 0.5). Effects of the studied factors on above- and belowground biomass were similar. The results indicated that soil chemistry during the winter could be important for both survival and growth during the vegetative period that follows.

The seed extract of *Silybum marianum* contains seven flavonolignans known collectively as silymarin. These metabolites were formed in hairy root cultures of this medicinal plant. The effect of temperature (30/25, 25/25, and 15/20°C in 16/8 h cycle) and pH (5, 5.7, 6, and 7) were evaluated with respect to the root biomass and silymarin production. Incubation temperature 25/25°C promoted the silymarin production in 4-week old hairy roots (0.18 mg g⁻¹ DW) as compared with the cultures treated with 15/20 and 30/25°C (0.13 and 0.12 mg g⁻¹ DW, respectively). Maximal increases in biomass and silymarin accumulation occurred in the root cultures grown in pH 5 and 25/25°C (0.45 g and 0.26 mg g⁻¹ DW). The content of silybin, isosilybin, silychristin, silydianin were 0.025, 0.024, 0.061, and 0.095 mg g⁻¹ DW, respectively, which were higher than those grown in higher pH. These results suggested that 25/25°C and the acidic environment of the medium are beneficial for silymarin production using hairy root cultures. Furthermore, lipoxygenase (LOX) activity was strongly affected by pH which suggested that acidic environment may act as inducing signal for LOX activity and subsequently greater silymarin production (Rahimi and Hasanloo 2016).

de Campos Nóia et al. (2014) evaluated the effects of soil liming on dry matter production of alfalfa (*Medicago sativa* L.)

plants in a dystrophic red latosol. The applied six doses of lime, 0, 1.23, 2.34, 3.42, 4.53, and 5.63 t ha^{-1} were expected to raise bases saturation to 13, 30, 45, 60, 75, and 90%. Soil analyses indicated that at zero concentration of lime the pH$_{CaCl2}$ was 4.05, after application of 5.63 t lime ha^{-1}, pH$_{CaCl2}$ increased to 5.78. Forty days after soil liming, the plants were cut at the soil surface level to determine their dry matter content, and soil samples were analyzed. The results showed a quadratic regression effect of doses of lime on dry matter production and the dose of 4.83 t ha^{-1} of lime was found to be adequate for maximum productivity of dry matter (1.452 g vase^{-1}). It was concluded that liming increased alfalfa dry matter production and had a positive effect on all soil chemical variables.

In this part of the chapter we only briefly outlined the general aspects of plant–soil interactions associated with acid soils because it is such a comprehensive topic that it should be analyzed in a special chapter of this *Handbook*.

39.9   OZONE AND CARBON DIOXIDE

Troposphere ozone (O$_3$), which is known to be highly phytotoxic, is predominantly produced by photochemical reactions involving precursors generated by natural processes and to a much larger extent by anthropogenic activities (Krupa and Manning 1988). Global annual mean CO$_2$ concentration has increased by more than 45% since the start of the Industrial Revolution (around 1750) from 280 ppm (Eggleton 2012) to 410 ppm as of mid-2018 (Dlugokencky 2016; ESRL Global Monitoring Division 2018). In the late 1950s and early 1960s, yearly average ozone values were in the 15–20 ppb range, which doubled by the end of the twentieth century, in agreement with the high elevation European ozone trends compiled by Marenco et al. (2014) and Staehelin et al. (1994). These ozone increases coincided with rising European NO$_x$ emissions that increased by a factor of 4.5 between 1955 and 1985 (Staehelin et al. 1994). Ozone triggers an in planta generation and accumulation of H$_2$O$_2$ and/or O$_2^{\bullet-}$ depending on the species, accession, and cultivar, and both these ROS are involved in the induction of cell death in sensitive crop and native plants (Wohlgemuth et al. 2002; Langebartels et al. 2002).

Medicinal plants *Achillea millefolium* and *Rumex acetosa* were reported to be tolerant, while *Matricaria recutita* and *Tanacetum vulgare* showed sensitivity to ozone (Dohrmann and Tebbe 2005). In *Salvia officinalis* plants exposed to 120 ppb of ozone for 90 consecutive days (5 h day^{-1}), a strong reduction of photosynthetic activity, loss of Chls, and cellular water deficit at the end of the fumigation was observed. Oxidative stress in the chloroplasts was manifested by strong degradation of β-carotene despite the photoprotection conferred by the xanthophyll cycle, increase in antioxidant compounds (in particular caffeic acid and rosmarinic acid), and water-soluble carbohydrates (especially monosaccharides). However, based on these results, it can be stated that *S. officinalis* was able to activate an adaptive survival mechanism allowing the plant to complete its life cycle even under stressful oxidative conditions (Pellegrini et al. 2015).

Ozone exposure induced the flavonol accumulation and NO generation of *Ginkgo biloba* cells and significantly enhanced the nitrate reductase activity of the cells suggesting that nitrate reductase-mediated NO signaling is involved in ozone-induced flavonol production of *G. biloba* cells (Xu et al. 2012). Results of an experiment investigating *G. biloba* leaves at different developmental stages after a long-term O$_3$ exposure (O$_3$ concentration approx. 80 nmol mol^{-1}) showed that activities of SOD, APX, and CAT were induced to a higher level in younger leaves with a 50% expansion size under elevated O$_3$ exposure, while an ozone-induced decrease in APX and SOD activities as well as significant increase of MDA and H$_2$O levels was observed in the 80 and 100% expanded leaves. Higher resistance to O$_3$ in the younger leaves is connected with the positive responses of antioxidant systems (Yan et al. 2010).

In the leaves of 4-years-old *G. biloba* L. trees exposed to air with twice the ambient O$_3$ concentration in Shenyang in 2006, increased concentrations of terpenes and decreased concentrations of phenolics were estimated. Whereas reduced synthesis of phenolics could result in the decreased resistance of *G. biloba* to O$_3$-induced stress, increased synthesis of terpenes may improve the antioxidant abilities in its leaves at the end of ozone fumigation (He et al. 2009). After 50 days exposure to 80±8 nmol mol^{-1} elevated ozone, the net photosynthetic rate of *G. biloba* leaf decreased significantly, and this decline had no correlation with the changes of Chl content but correlated markedly with membrane lipid peroxidation. However, the decline of photosynthesis resulted in a reduction of leaf soluble protein and starch contents and in the slow growth of the tree (Zhang et al. 2007).

While the antioxidant system in *G. biloba* exposed to elevated ozone concentration in an urban area did respond by acclimating in the early season, late in the season the constant higher level of ROS and declining enzyme activities suggested that the system could not withstand long-term exposure, although no visible injury was observed (He et al. 2006).

Hypericin contents of the *Hypericum perforatum* cells in a suspension culture treated with 60–180 nanoliters per liter ozone were found to be considerably higher than those of the control, showing that ozone exposure may stimulate hypericin synthesis, whereby hypericin production by the cells exposed to 90 nanoliter O$_3$ at late exponential phase for 3 h was approx. four-fold higher than in the control (Xu et al. 2011).

After *in vitro* exposure of *M. officinalis* shoots to 200 ppb ozone for 3 h in controlled environmental conditions, an evident increase in lipid peroxidation, pronounced drop in the levels of total Cars, and an increase in non-terpenoid compounds was observed suggesting the breakdown of cells, an association between volatile products of the lipoxygenase pathway, and membrane degradation (D'Angiolillo et al. 2015). In ozone-treated (200 ppb, 3 h) *M. officinalis*, aseptic shoots, an activation of enzymes involved in phenolic metabolism, a development of cellular barriers with a higher degree of polymerization of monolignols, an accumulation of phenolic compounds, in particular rosmarinic acid (about four-fold compared to control plants cultivated in filtered air), and an increase of antioxidant capacity was observed. Cell death and H_2O_2 deposition concomitant with a prolonged $O_2^{\bullet-}$ generation estimated after the end of the treatment suggested that a transient oxidative burst occurred (Tonelli et al. 2015).

In *M. officinalis* plants exposed to background ozone dosages (80 ppb for 5 h), 24 h from the beginning of exposure rosmarinic acid and phenylalanine ammonia lyase (PAL) were upregulated, whereby the specific activity of rosmarinic acid synthase was closely correlated with a decrease of rosmarinic acid concentration, while the specific activity of PAL increased at 12 h from beginning of exposure to 163% in comparison to control levels (Döring et al. 2014a). During the same exposure of *M. officinalis* plants to ozone (80 ppb for 5 h), impairment of carboxylation efficiency and membrane damage resulting in enhanced solute leakage as well as rearrangement of the pigment composition of the photosynthetic apparatus and considerable activation of photoprotective mechanisms was observed (Döring et al. 2014a,b).

Ozone-stressed (200 ppb, 5 h) *M. officinalis* plants exhibited foliar injury and cell death induced by the biphasic production of H_2O_2 and $O_2^{\bullet-}$, significant production of ethylene, salicylic, jasmonic, and abscisic acid, O_3 degradation to ROS and their detoxification by some enzymatic (e.g., SOD), and non-enzymatic antioxidant systems (e.g., ascorbic acid, GSH and Cars), that worked in cooperation without providing a defense against free radicals. Consequently, it can be stated that ROS interact with hormonal signaling pathways regulating cell death and the sensitivity of *M. officinalis* to O_3 (Pellegrini et al. 2013).

Carbon dioxide is a greenhouse gas that absorbs and emits radiant energy within the thermal infrared range. The majority of anthropogenic CO_2 emissions originating from the combustion of fossil fuels (coal, oil, and natural gas) significantly contributes to greenhouse effects and could adversely affect global climate. On the other hand, CO_2 is indispensable for photosynthesis and, being one of its most limiting factors, has a great impact on plant biomass and secondary plant metabolites.

The ongoing rise in the air's CO_2 content was reported to increase not only plant biomass but also concentrations of health-promoting substances in several medicinal plants belonging to C_3 plants (Center for the Study of Carbon Dioxide and Global Change 2014). Enhanced plant biomass, as well as enhanced levels of hypericin and pseudohypericin, were estimated in *H. perforatum* plants grown under exposure to 1000 ppm (Zobayed and Saxena 2004) and 1500 ppm CO_2, respectively (Mosaleeyanon et al. 2005).

The exposure of Malaysian medicinal herb *Labisia pumila* Blume to higher than ambient CO_2 levels (400, 800, and 1,200 µmol mol^{-1}) for 15 weeks enhanced the production of total phenolics, flavonoids in the plant attributed to enhanced antioxidative properties, which might be due to the increased sucrose levels improving the phytomedicinal properties of the plants (Ibrahim and Jaafar 2011).

In *Zingiber officianale* cultivated under high levels of CO_2 (800 µmol mol^{-1}), an increased content of total carbohydrate, sucrose, glucose, and fructose in the leaf and rhizome (Ghasemzadeh et al. 2014), as well as elevated contents of flavonoids and phenolic compounds and antioxidant activities, were estimated (Ghasemzadeh et al. 2010). Elevated levels of CO_2 (600 and 900 ppm) enhanced phenol, flavonoid, carbohydrate, tannin, and alkaloid in *Catharanthus roseus* as well as fresh weight, shoot length, and number of leaves (Saravanan and Karthi 2014). In *Cannabis sativa* plants elevated CO_2 concentration (700 µmol mol^{-1}) considerably promoted photosynthesis, WUE, and internal CO_2 concentration, and suppressed transpiration and stomatal conductance as compared to the ambient CO_2 concentration (390 µmol mol^{-1}) (Chandra et al. 2011). In general, elevated CO_2 is beneficial for the growth of medicinal plants and the production of their valuable secondary metabolites showing healing effects.

39.10 RESPONSES OF MEDICINAL PLANTS TO TOXIC METALS PRESENCE

Medicinal plants have increasing economic importance, and the food, pharmaceutical, and cosmetic industry produce important goods with these plants. In general, products based on natural substances enjoy an increasing value (e.g., Choffnes 2016). However, anthropogenic activity and its effects on the environment showed that medicinal plants have also responded to changing environmental conditions. In the last few years, the practical use of alternative medicine in healing processes has showed a continually increasing tendency. Several species of medicinal plants can also be used as supplementary nutrition due to their ability to accumulate some essential nutrition elements (e.g., Se, Zn, and Fe) in the edible parts of these plants. Such fortification of plants with essential nutrients (phytofortification) in an easily assimilated form can help to feed the rapidly increasing world population and improve human health through balanced mineral nutrition. Therefore, data related to toxic metal contents in pharmaceutically utilized parts of the medicinal plants are also considered from the aspect of "food safety." According to the World Health Organization global survey related to the national policy on traditional medicine and regulation of herbal medicines (World Health Organization 2005), in 2003 the number of member states which had some form of regulation on herbal medicines was 83. On the other hand, some medicinal plants belong to toxic metal hyperaccumulators that can be used in the future in phytoremediation technologies (see Masarovičová and Kráľová 2007).

In the former 3rd edition of the *Handbook of the Plant and Crop stress*, we presented the results from three medicinal species: *Hypericum perfolatum* L., *Matricaria recutiia* L., and *Salvia officinalis* L. which are in general the most frequent medicinal plants used in phytotherapy (Masarovičová et al. 2011). Some of the most important results also presented in this chapter are extended by further findings.

H. perforatum (St. John's wort) is a plant which has been used as a medicinal herb since ancient times. *H. perforatum* is native to Europe, Asia, and North Africa and in temperate regions (e.g., North and South America, South Africa, Australia, and New Zealand) where it was introduced, and it has often become an aggressive weed (e.g., Čellárová et al. 1995). The chemical composition of St. John's wort has been well-studied. Documented pharmacological activities, including antidepressant, antiviral, and antibacterial effects, provide supporting evidence for several of the traditional uses stated for this medicinal plant. Many pharmacological activities appear to be attributable to hypericin and to the flavonoid constituents, whereby hypericin is also reported to be responsible for photosensitive reactions (Barnes et al. 2001). Major constituents of these plant extracts include several classes of compound exemplified by flavonols, flavonol glycosides, biflavones, naphthodianthrones, phloroglucinols, tannins, coumarins, essential oils, xanthophylls, and others (Nahrstedt and Butterweck 1997; Tatsis et al. 2007). The content of naphthodianthrone derivatives hypericin and pseudohypericin is approximately 0.05–0.15%, that of flavonoid quercetin and biflavone biapigenin is 0.3 and 0.26%, respectively. From the other secondary metabolites, the highest contents belong to phlorogluclinole derivatives hyperforin and adhyperforin (up to 4%) (Bilia et al. 2002). These compounds are very important for the medicinal plants to preserve them against environmental stress. Thus we studied tolerance of *H. perforatum* to the toxic effect of copper and cadmium with respect to metal accumulation in individual plant organs (Kráľová and Masarovičová 2004). The 6-week- (Cu experiment) or 5-month-old plants (Cd experiment) were exposed in hydroponics for 7 days to the following metal concentrations: 15, 30, 60, 90, and 120 μmol dm^{-3} CuCl$_2$.2H$_2$O and 12 μmol dm^{-3} Cd(NO$_3$)$_2$.4H$_2$O. Thereafter, the length and dry mass of shoots and roots were measured and metal content in plant organs determined. It was confirmed that the most sensitive parameter to Cu treatment was the root dry mass, the length of the shoot, as well as shoot dry mass, was not significantly affected. Lower values of root dry mass could be explained with a significant reduction of lateral roots and root hairs by Cu treatment. The roots of *H. perforatum* accumulated markedly higher concentrations of Cu than the shoots. The metal accumulated in both plant organs showed an increase with increasing metal concentration. Bioaccumulation factors (BAF), i.e., quotients obtained by dividing the concentration of the metal in dry mass of individual plant tissues (root and shoot, respectively) by its concentration in the external exposure medium, were also calculated. Taking into account the actual dry mass of individual plant organs (root and shoot), The Cu portion in the shoot was in the investigated concentration range approximately 20% from the total uptaken metal content by the whole plant. With respect to a relatively high Cd content in the shoot dry mass (1087 μg g^{-1}), *H. perforatum* could be classified as a Cd hyperaccumulator. In this paper, the possible contribution of the formation of metal complexes with the secondary metabolites of *H. perforatum* to plant metal tolerance was discussed. Later, we stated (Masarovičová and Kráľová 2007) that for plants producing specific secondary metabolites (medicinal plants) the further, additive mechanism of tolerance arose. This additive mechanism is connected with chelation of metal ions by some specific secondary metabolites, such as, for example,

hypericin and pseudohypericin produced by *H. perforatum*. The toxic ionic form of metal is thus changed into non-toxic metal chelate. This assumption is based on the findings of Falk and Schmitzberger (1992) and Falk and Mayr (1997) who stated that the pronounced acidity of the bay-region hydroxyl groups of hypericin makes salt formation a definite possibility: hypericin is present in the plant material mainly as potassium salt. In addition, in structurally similar bay-hydroxylated fringelites salt formation with divalent ions, such as Ca^{2+}, yields polymeric systems, which, because of their extreme insolubility, are highly stable in fossils.

hypericin   pseudohypericin

The peri-hydroxyl groups situated in the neighborhood of the carbonyl groups display the best prerequisites for forming chelates with transition metal ions. Such coordination complexes could be characterized in the case of fringelite D and Zn^{2+} (Falk and Mayr 1997).

Afterward, Palivan et al. (2001) investigated the formation of copper complexes with hypericin, in solutions using EPR spectroscopy, and found that hypericin forms a four-coordinated copper species where the solvent participates in coordination with the sphere of the metal. An excess of metal was not compulsory for the formation of the hypericin-copper complex; however, a higher aggregate (chain structure) could not be ruled out.

Taking into account the above-mentioned results concerning a complex formation between hypericin and copper, it could be assumed that this secondary metabolite of *H. perforatum*, as well as structurally similar pseudohypericin, will form similar complexes with further transition metal ions. Due to such an interaction, the concentration of free metal ions will decrease, and their toxic effect will be diminished. Thus, the formation of complexes between heavy metals and the above-mentioned secondary metabolites could be regarded as further mechanisms contributing to the enhanced tolerance of *H. perforatum* against such divalent metals as copper and cadmium.

The above-described and illustrated chelatation of cadmium ions with hypericin could contribute to the enhanced tolerance of *H. perforatum* plants to cadmium stress and to their cadmium hyperaccumulating ability. It should be mentioned that classification of this medicinal plant species as a Cd hyperaccumulator was first confirmed by Marquard and Schneider (1998) and consequently verified in our research (Masarovičová et al. 1999; Kráľová et al. 2000).

In *H. perforatum* plants grown on serpentine and nonserpentine sites, with concentrations of essential (Fe, Mn, Cr, Co, Cu, Zn) and toxic (Ni, Cd, Pb) metal elements varied across sites, the plants were found to be Ni tolerant, but they were able to accumulate trace elements at toxic levels and the amounts of Cd, Ni, and Cr were above the permissible limits in a dry plant (Pavlova et al. 2015). Similarly, *H. perforatum* plants from wild populations of Northern Turkey included an enormous amount of Cu, Mn, Fe, Zn, and Pb suggesting that this plant may be a metal hyperaccumulator (Ayan et al. 2006).

According to Murch et al. (2003), metal contamination can change the chemical composition of *H. perforatum*, thereby seriously impacting the quality, safety, and efficacy of natural seedlings that completely lose the capacity to produce or accumulate hyperforin and demonstrate a 15–20-fold decrease in the concentration of pseudohypericin and hypericin. Similarly, total alkaloid content was also found to be decreased in the roots of *Catharanthus roseus* (L.) plants treated with $CdCl_2$ (Pandey et al. 2007).

In our further paper (Kráľová et al. 2000), the effect of 12 µM $Cd(NO_3)_2$ (pH=5.5) on growth, plant biomass (root and shoot), and root dark respiration rate of *H. perforatum* (cultivated hydroponically) as well as cadmium accumulation in all plant organs was investigated. The highest Cd concentration was found in the root (1792 µg g^{-1} DW) which was 8-times higher than the concentration in the stem and 6-times higher than in the leaves. On the basis of the results, it could be concluded that Cd supported the release (permeability) of cell membranes in both roots and shoots. Consequently, the above-mentioned metal ions were transported into the leaves where their higher content was estimated. The effect of cadmium treatment on the content of iron, manganese, and copper in individual plant organs has also been determined. Cd administration did not affect the growth and dry biomass of the shoot and root and the root/shoot ratio. However, the root dark respiration rate of the Cd-treated plants was faster than those of the control plants (Kráľová et al. 2000).

It has already been mentioned that *Matricaria recutita* L., showing a wide scale of pharmacological effects (McKay and Blumberg 2006; Murti et al. 2012), is the most favored medicinal plant not only in Slovakia but also all over the world (see Masarovičová and Kráľová 2007; Singh et al. 2011) and the chamomile drug is included in the pharmacopoeia of 26 countries (Pamukov and Achtardžiev 1986). From a historical aspect, the native areas of the chamomile species were Asia, Northern Africa, Southern, and Eastern Europe. The notions about chamomile's healing power came from the antique scientific works of Hippocrates, Pliny, Dioscorides, and Galen into the old herbals and then current phytomedicine. Chamomile was already introduced into the register of the synanthropic plants of vegetation of settlements of Bratislava and Trnava (Slovakia, formerly Austro-Hungarian monarchy) in 1774 and 1791, respectively (Eliáš 1994). In the mythology, chamomile belongs to the nine holy herbs which Odin (also called Woden or Wodan), the main god of North-European German folks, donated to mortals for the improvement of their life (Val 2015). Nowadays, chamomile (*M. recutita* L., earlier synonyms also *Chamomilla recutita* (L.) Rausch. or *Matricaria chamomilla* L.) is the most favored and most frequently used medicinal plant all over the world (Singh et al. 2011). In the former Czechoslovakia, *M. recutita* represented the base of more than 32 official mass-produced phytotherapeutic preparations (Karmazín et al. 1984).

M. recutita produces a variety of volatile secondary metabolites, e.g., chamomillol, gossonorol, cubenol, α-cadinol, chamazulene, β-farnesene, (-)-α-bisabolol, (-)-α-bisabololoxide A, (-)-α-bisabololoxide B, 1-azulenethanol acetate, and (-)-α-bisabolol acetate (Magiatis et al. 2001), herniarin (Repčák et al. 1998), etc. In 1973, Schlicher identified four main chemical types of the intraspecific variability of chamomile based on the composition of essential oil. For the chamomile varieties cultivated in Slovakia, it is characteristic that the content of important secondary metabolites decreases in the following order: α-bisabolole > α-bisabolole oxide B > α-bisabolole oxide A >, i.e., these cultivars belong to chemical type A of *M. recutita* (see Šalamon et al. 2007).

Traditionally, in Eastern Slovakia, large regions are used for commercial chamomile cultivation. As chamomile species are long-term cultivated in field conditions, it is important to know how much Cd is taken up from the soil, transported, and accumulated in individual parts of plants. Therefore, Cd content in a pharmaceutical important plant part anthodium was also estimated by Šalamon et al. (2007). In this paper, we also evaluated the relationship between Cd content in chamomile anthodium dry mass and the mean hydrothermic coefficient of Seljaninov (HC) as an integrated index of hydrothermic parameters. This coefficient could be calculated according to the following formula: $HC = \sum R/(0.1 \times TS_m)$ where $\sum R$ is the total amount of precipitation and TS_m is the thermal sum of the mean daily temperatures exceeding 10°C in the investigated period. Correlation between accumulated Cd and HC value supported the enormous significance of actual climatic relations on the metal uptake and accumulation, while in the investigated period (1999–2002) the accumulated Cd in chamomile anthodium dry mass increased approximately linearly with an increasing HC value.

Marquard and Schneider (1998) were the first authors who confirmed that chamomile plants had the potential to accumulate high levels of cadmium from the soil. In our paper (Pavlovič et al. 2006), two tetraploid cultivars of *Matricaria recutita* L. (cv. Goral and cv. Lutea) were investigated in response to Cd application. Treated plants were cultivated in Hoagland solution with the following Cd concentrations: 3, 6, 12, 24, and 60 µmol dm^{-3} $Cd(NO_3)_2$. The plants growing in Hoagland solution without Cd served as a control. Primary root length, root increment (root length after treatment – root length before treatment) was calculated ((root increment of Cd-treated plants/root increment of control plants) × 100) in percentage, root and shoot dry mass, as well as Cd content in plant organs, were determined after 7 days of treatment. In other experiments, the older plants were grown in greenhouse conditions in the soil for 7 weeks after germination. Plants used for photosynthetic and respiration measurements as well as for analysis of assimilation pigment concentrations were grown under greenhouse conditions for 9 weeks after germination. Then their roots were washed, transferred to hydroponic Hoagland solutions (control) and Hoagland solution with 12 µM $Cd(NO_3)_2$, and placed in the growth chamber for 10 days. The concentration gradient that was used in our experiments reflects Cd content in the soil from noncontaminated to highly contaminated sites (Linkeš et al. 1997). For estimation of Cd toxicity to roots, we used primary root length and root increment, which are considered to be reliable parameters for heavy-metal tolerance (Murphy and Taiz 1995). At the beginning of Cd treatment, a non-significant variance in root length was observed. Significant inhibition of root growth was observed in both chamomile cultivars after Cd treatment. We did not find any differences in Cd accumulation in the root between cultivars, but cv. Lutea accumulated a slightly higher amount of Cd in the shoot. No differences between cultivars were recorded in the growth parameters after Cd treatment. In the root test, we observed the fragility, browning, and twisting of roots. In shoots, leaf roll, chlorosis, and leaf growth inhibition occurred. During the root test, chamomile plants cv. Goral formed the anthodia in all concentrations except the control, despite the fact that the plants were only 3 weeks old. According to our observation, the plants started blossoming when they were 8–12 weeks old; however, Cd treatment resulted in reduced flower size. From 4 to 5 weeks earlier, blossoming under Cd administration was also recorded for Cd hyperaccumulator *Arabidopsis halleri* (Küpper et al. 2000). Cd concentration of 12 µmol dm⁻³ in hydroponic solution (used in our other experiment) represents strong contaminated soil (Linkeš e al. 1997); however, the Cd effect on plant was stronger in comparison to the soil, because Cd^{2+} ions are not bound to the soil particles and so all ions are available for plant uptake. Cv. Lutea seemed to be more sensitive to Cd treatment, e.g., it exhibited greater leaf chlorosis. The measurements confirmed a higher inhibition of photosynthesis in cv. Lutea. A similar decrease in shoot dry weight in both cultivars was also detected. A decrease of the net photosynthetic rate could be due to structural and functional disorders in many different levels. Shoot and root respiration rates were not changed significantly in both chamomile cultivars. We confirmed that chamomile belongs to the group of Cd accumulator species. If we take into account the high content of Cd in the chamomile shoot (over 300 µg g⁻¹ at 12 µmol dm⁻³ Cd in solution), only a small amount of damage occurred in Cd-treated plants. Therefore this medicinal plant species exhibited high tolerance to Cd treatment. This fact was also confirmed by Masarovičová et al. (2003) when the effect of cadmium and zinc separately (10 µmol dm⁻³ for

Cd and 50 µmol dm⁻³ for Zn), as well as combined with the application of these ions on physiological processes (photosynthetic rate and dark respiration rates of leaves and roots, Chl concentration) and production parameters (shoot and root biomass, shoot/root ratio, length of shoots and roots) of young plants of *H. perforatum* and *M. recutita* was investigated.

Several authors agreed that *M. recutita* species tolerates Cd concentrations corresponding to the middle-strong contaminated soils (Grejtovský and Pirč 2000; Masarovičová et al. 2003) and high Cd concentration in shoots also assigns this medicinal plant as a Cd accumulators (Kráľová and Masarovičová 2003) or to Cd hyperaccumulators (Masarovičová et al. 2010). Similarly, Chizzola et al. (2003) confirmed that chamomile has a tendency to accumulate Cd. On the other hand, Kováčik et al. (2006) declared that chamomile is only tolerant to this toxic metal. These authors studied the influence of low (3 µM) and high (60 and 120 µM) Cd concentrations on 4-week-old chamomile plants. After 10 days exposure, dry mass accumulation and nitrogen content were not significantly altered under any of the levels of Cd. However, there was a significant decline in Chl and water content in the leaves. Among coumarin-related compounds, herniarin was not affected by Cd, while its precursors significantly increased at all tested Cd concentrations. Cd did not have any effect on umbelliferone, a stress metabolite of chamomile. Lipid peroxidation was also not even affected by 120 µM Cd. Cadmium accumulation was approx. 7-(60 µM) to 11-(120 µM)fold higher in the roots than in the leaves. The results indicated that chamomile is tolerant to this toxic metal. Existence of the earlier discussed additive mechanism of tolerance due to chelate formation between some chamomile secondary metabolites and cadmium thus has to be experimentally confirmed. Later Gjorgieva et al. (2013) published a paper in which different plant organs (leaves, flowers, stems, or roots) from three medicinal plant species: *Urtica dioica* L. (*Urticaceae*), *Taraxacum officinale* L. (*Asteraceae*), and *M. recutita* L. (*Asteraceae*) were evaluated as possible bioindicators of heavy-metal pollution in the Republic of Macedonia. Concentrations of Pb, Cu, Cd, Mn, Ni, and Zn were determined in unwashed plant parts collected from areas with different degrees of metal pollution by inductively coupled plasma atomic emission spectroscopy (ICP-AES). Summarizing the results, it was concluded that *T. officinale* and *U. dioica* were better metal accumulators and *M. recutita*, as opposed to our findings, was classified as an avoider of these metals.

In another paper, Kováčik et al. (2008) observed the physiological responses of *Matricaria chamomilla* plants exposed to Cd and copper excess (3, 60, and, 120 µM for 7 days) with special emphasis on phenolic metabolism. Cu at 120 µM reduced chamomile growth, especially in the roots where it was more abundant than Cd. Notwithstanding the low leaf Cu amount (37.5 µg g⁻¹ DW) in comparison with Cd (237.8 µg g⁻¹ DW) at 120 µM, it caused reduction of biomass accumulation, F_v/F_m ratio, and soluble proteins. In combination with the high accumulation of phenolics, a strong reduction of proteins and high guaiacol peroxidize (GPX) activity in the roots, supports the severe redox properties of Cu. In terms of leaf PAL activity, it seems that Cd had a stimulatory effect during the course of the experiment, whereas Cu was found to stimulate it after 7-day exposure. The opposite trend was visible in the roots, where Cd had a stimulatory effect at high doses but Cu mainly at the highest dose. This supports the assumption of different PAL time dynamics under Cd and Cu excess. A dose of 60 and 120 μM Cu led to 2- and 3-times higher root lignin accumulation while the same Cd doses increased it by 33 and 68%, respectively. A Cu dose of 120 μM can be considered as limiting for chamomile growth under conditions of the present research, while resistance to high Cd doses was confirmed. However, PAL and phenolics seemed to play an important role in detoxification of Cd- and Cu-induced oxidative stress.

Besides Cd and Cu, Kováčik et al. (2009) also examined the response of chamomile plants to Ni effects with respect to phenolic acids, total soluble phenolics and flavonoids, and activities of phenolic metabolism-related enzymes (shikimate dehydrogenase (SKDH), PAL, cinnamyl alcohol dehydrogenase (CAD), polyphenol oxidase (PPO)). Tested plants were exposed to 3, 60, and 120 μM of Ni for 10 days. Ni showed low toxicity as indicated by the unaltered content of total soluble phenolics in the leaf rosettes. In the roots, the effects of Ni were more visible, including increased levels of total phenolics and PAL activity, but a decrease in PPO activity was observed. CAD activity was not affected by any of the tested Ni concentrations. Cinnamic acid derivatives were affected by more than benzoic acid derivatives. Accumulation of chlorogenic acid, an important antioxidant compound, was enhanced by Ni treatment (ca. four-fold in 120 μM Ni), while the accumulation of protocatechuic acid, a phenol with high chelating strength, decreased in the leaf rosettes. The authors discussed these observations in connection with the antioxidative properties of phenolic metabolites as well as with previously tested metals (Cd and Cu).

In a paper by Kováčik et al. (2012), the phenolic metabolism of Al-exposed chamomile plants was modulated with four regulators: 2-aminoindane-2-phosphonic acid (AIP), salicylic acid, sodium nitroprusside (SNP), and dithiothreitol (DTT). Physiological parameters (tissue water content, soluble proteins, reducing sugars, K^+ content), root lignin content, and free amino acids (increase in root proline and alanine) were the most affected in the SA + Al variant, indicating the negative impact of SA on Al-induced changes. SNP showed the least visible impact, suggesting the protective effect of nitric oxide. A complex comparison between Al alone and combined treatments revealed that SA and DTT stimulated an increase in shoot phenolic acids (mainly vanillic acid), the sum of flavonols and soluble phenols, but decreased the levels of coumarin-related compounds ((Z)- and (E)-2-β-D-glucopyranosyloxy-4-methoxycinnamic acids), leading to elevation of shoot Al. A positive correlation between phenolic acids (mainly ferulic and chlorogenic acids), soluble phenols and total Al was found in the roots of SA and DTT variants. These events were not observed in AIP and SNP treatments. These data, for the first time, exactly confirm that phenolic metabolites may affect shoot Al uptake and this relation is rather

positive in terms of simple phenols (and negative in terms of coumarin-related compounds).

Jakovljevic et al. (2000) investigated the influence of the different doses of sodium selenite (0, 100, and 500 g Se per hectare) applied by foliar spraying on the yield and quality of chamomile. The applied doses of Se did not influence the formation of dry chamomile flowers yield and the content of essential oil. However, the applied Se caused the significant increase of the content of bisabolol oxide A and B, followed by the decrease of the chamazulene content in the chamomile essential oil. Significant increase of Se content in the chamomile flowers (12.9 to 53.6 ppm) was also observed. In our experiments with hydroponically cultivated chamomile plants, cv. Lutea (Kráľová et al. 2007) and cv. Goral (Lešíková et al. 2007) treated with $CdSeO_4$, $CdSeO_3$, and $Cd(NCSe)_2(nia)$ we investigated Cd accumulation in the roots and shoots of plants. The highest applied $CdSeO_3$ concentration (60 μmol dm^{-3}) caused higher Cd content in roots than the amount observed employing $CdSeO_4$, whereas already in the case of treatment with 12 and 24 μmol dm^{-3} solutions of Cd(II) selenite and selenite adverse effects were observed. This could be connected with the damage of the root cell membrane system due to the high concentration of cadmium and selenite ions. Immobilization of Cd ions in root tissue was manifested by a large amount of bioaccumulated Cd in this plant organ. The highest Cd content in the shoots was observed after application of $CdSeO_4$. The change of S to Se in the complex $Cd(NCX)_2(nia)_2$ led to the evident decrease of Cd content in the shoots. In general, the content of Cd accumulated in plant organs after application of $Cd(NCSe)_2(nia)_2$ was comparable with that observed after application of $CdSeO_3$. The values of the translocation factor for Cd estimated for the experiments with $CdSeO_4$ and $Cd(NCS)_2(nia)_2$ were more than 2-times higher than those found for $CdSeO_3$ and $Cd(NCSe)_2(nia)_2$. The highest fraction of Cd accumulated in shoots was observed for $CdSeO_4$, while the lowest was for $Cd(NCSe)_2(nia)_2$. The obtained results correspond with those obtained for the chamomile cultivar Goral (Lešíková et al. 2007) that was found to be more tolerant to the cadmium exposure compared to the cultivar Lutea (Kráľová et al. 2007a). The treatment with $CdSeO_4$ and $Cd(NCS)_2(nia)_2$ caused that approximately 40% of the total amount of Cd accumulated by the plant were allocated to the shoots. On the other hand, approximately 80% (or more) from the total amount of metal accumulated by the plant remained in the roots after the treatment with $CdSeO_3$ and $Cd(NCSe)_2(nia)_2$. These data correlate well with the results of Shanker et al. (1996). This finding could be explained by taking into account the fact that less mobile selenite, after being reduced to selenide, tends to form a Cd–Se complex, which appears to be unavailable to the plants. On the other hand, the more mobile anion selenate is available for Cd–Se formation only after following a more complicated redox process involving Se(VI) in SeO_4^{2-}, Se(IV) in SeO_3^{2-}, and Se(0) species. According to Whanger (1992), the presumed protective effect of Se against cadmium and mercury toxicity is through the diversion in their binding from low-molecular-mass proteins to higher-molecular-mass ones. The experiments with chamomile cv. Goral also showed that the BAF values related to Se accumulation in plant organs were significantly influenced by the oxidation state of Se – the application of selenate resulted in intensive translocation of Se into the shoots, and for this compound the BAF values determined for shoots were approximately 2.5-times higher than those determined for roots. The corresponding BAF values for $Cd(NCSe)_2(nia)_2$ were similar to those of $CdSeO_3$. At application of $CdSeO_4$, 90% of uptaken Se and 50% of uptaken Cd was situated in the shoots (Lešíková et al. 2007).

The presence of chelators can alter the mobility and transport of Zn, Cd, and Ni in soils because of the formation of water-soluble chelates, thus increasing the potential for metal pollution of natural waters. However, chelators could also increase the bioavailability and uptake of toxic metals. Chelated metals are taken up via the apoplastic pathway. Disruption of the Casparian band is required to achieve the high shoot concentrations needed for phytoextraction. Therefore, adding chelators to soil increases not only the total dissolved metal concentration but also changes the primary route of plant metal uptake from the symplastic to the apoplastic pathway, and depending on metal, plant species, and chelant concentration, significant increases in metal uptake are likely (Nowack et al. 2006). The addition of an organic chelator (citric acid) enhanced zinc and cadmium accumulation, mostly at the root level (Mugnai et al. 2006). Our experiments with a set of Cu(II) chelates confirmed that application of copper in the form of chelate led to more effective Cu translocation into the shoots in comparison to $CuSO_4.5H_2O$ treatment (Kráľová et al. 2007b). Artificial chelator ethylenediaminetetraacetic acid (EDTA) promoted Cu translocation very effectively into the shoots of hydroponically cultivated chamomile plants which was reflected in the fact that at the treatment with higher metal concentrations (24 and 60 µmol dm^{-3}), even 45 and 59% Cu accumulated by the plants, was allocated to the shoots (in the absence of a chelator this portion reached only 5.2 and 4.8%, respectively). The very efficient translocation of copper into the shoots observed in the presence of EDTA could be connected with the largest value of the Cu-EDTA stability constant (log K_1=18.8). Whereas chelate formation between EDTA and Zn or Cu resulted in significantly decreased metal uptake into chamomile roots; the decrease of Cd uptake due to chelate formation was very low (Kráľová and Masarovičová 2008).

Salvia officinalis L. is in general also one of the most important medicinal and aromatic plants with a great spectrum of application in the phytotherapy (e.g., Ghorbani and Esmaeilizadeh 2017), cosmetic and food industries. The genus *Salvia* includes more than 400 species. *S. officinalis* as a perennial plant originates from the Mediterranean region. In regard to the analysis of sage essential oil, the major compounds are thujone, cineole, camphor, and caryophyllene. These secondary metabolites are biologically active compounds in *herba salviae* having application in phytotherapy. In the food industry, this aromatic plant species is recommended as a spice or additive substance (cf. Langer et al. 1996; Perry et al. 1999). From all the above-mentioned aspects, it is important to have information on toxic metal effects on growth and metal accumulation into the different plant organs of this species. Since Marquard and Schneider (1998) characterized S. *officinalis* as an excluder of cadmium, we studied the effect of a large external concentration range of cadmium (30–480 µmol dm^{-3} $Cd(NO_3)_2$) on production characteristics (the length of roots and shoots as well as dry mass of roots and shoots) of this species. We tested two cultivars: cv. Krajova (Slovakian provenance) and cv. Primorska (Yugoslavian provenance). Two-month-old plants were cultivated hydroponically for 7 days under controlled conditions in Hoagland solution both with and without the presence of $Cd(NO_3)_2$ (Masarovičová et al. 2004). The plants were exposed to hydroponics for 7 days in controlled conditions: the control variant in Hoagland solution and metal treated variants in Hoagland solution with 30, 60, 120, 240, 360, and 480 µmol dm^{-3} $Cd(NO_3)_2$. Then the length and dry mass of shoots and roots were estimated. Metal content in the aboveground and underground parts of studied species were determined using FAAS. There were differences found in the phenology and production parameters between the two tested cultivars of different provenance. Cv. Krajova was already sensitive to the concentration of 60 µmol dm^{-3} $Cd(NO_3)_2$ when the oldest leaves dried. At a concentration of 120 µmol dm^{-3} $Cd(NO_3)_2$, all the older leaves dried out and the younger leaves wilted. At a concentration 240 µmol dm^{-3} $Cd(NO_3)_2$, brown spots were observed on the leaves, and at the highest applied metal concentrations to both (360 and 480 µmol dm^{-3} $Cd(NO_3)_2$, respectively) all leaves were dried and on the apical side of the leaves depigmentation was observed. Cultivar Primorska seems to be more tolerant to metal treatment. Visual changes occurred until the concentration of 120 µmol dm^{-3} $Cd(NO_3)_2$ when only some of older leaves of the plant dried out. At a concentration of 240 µmol dm^{-3} $Cd(NO_3)_2$, the leaves dried out but they were still green colored. This fact confirmed the disturbance of the water regime and indicated strong water stress. At the highest tested Cd concentrations (360 and 480 µmol dm^{-3} $Cd(NO_3)_2$, respectively) all the leaves were already dried, and there was damage to the leaf pigmentation as brown colored spots were observed. In spite of the high concentration of Cd (30–480 µmol dm^{-3}), the length of the roots in both cultivars were almost not affected. For the shoots of both cultivars, only a slight reduction of length was found, too. On the other hand, dry mass of the shoots decreased at all applied Cd concentrations more expressively than the dry mass of the roots. The negative effect of the high Cd concentrations on the shoot dry mass was manifested mainly in cv. Krajova. Both cultivars took up the greatest portion of Cd into their roots, but cv. Primorska accumulated approx. 2-times more Cd in the shoot than cv. Krajova. However, differences were found in the translocation of cadmium from the roots into the shoots. Cv. Krajova did not allocate Cd from the roots into the shoots at 240 µmol dm^{-3} $Cd(NO_3)_2$, which confirms the existence of barriers in the roots. BAF for root, depending on Cd concentration, ranged from 565–357, and BAF for the shoot was 47–22; 27.8–29.0% of Cd was allocated in the shoots. On the other hand, for cv. Primorska, with increasingly applied Cd concentrations, the

Cd translocation from the roots into the shoots increased in the whole applied Cd concentration range. BAF for roots ranged from 817 to 419 and for the shoot from 30 to 81, and 9.4–41.0% of Cd was allocated to the shoots suggesting that cv. Primorska is more effective mainly for the translocation of cadmium into aboveground parts of the plant. Our findings for both studied cultivars of *Salvia officinalis*, cv. Primorska and cv. Krajova did not support the earlier results of Marquard and Schneider (1998) that characterized this medicinal plant species as an excluder of Cd.

Medicinal, aromatic, and spice plants grown in different regions of Austria were monitored by Chizzola et al. (2003) as to their Cd, Cu, Fe, Mn, Pb, and Zn contents. Since the plants were grown under common field conditions, the essential elements were within the usual ranges for plant material. The contamination level of the toxic heavy metals, Pb and Cd, can be classified as normally low. Most samples contained less than 0.2 mg kg^{-1} Cd and less than 1.5 mg kg^{-1} Pb on a dry weight basis. Comparison with previous investigations suggests that contaminations with Pb occur rather by chance, whereas enhanced Cd values are restricted to some species having a tendency to accumulate this heavy metal. Some such species are St. John's wort, poppy, yarrow, chamomile, and absinthe. Careful choice of growing site and appropriate soil management can reduce the Cd uptake of these critical species. The above-mentioned authors emphasized that these precautions are important when larger amounts of the product are consumed.

The results presented in a paper published by Schulz et al. (2008) provided evidence for the role of phytochelatins (PCs) in the detoxification of arsenic in six nonhyperaccumulating plant species, *Agropyron repens*, *Leonurus marrubiastrum*, *Lolium perenne*, and *Zea mays* as well as two medicinal plants, *Glecoma hederacea* and *Urtica dioica*, in a pot experiment with high phosphate treatment. These plants differed in their As sensitivities and were selected to investigate whether PCs with longer chains in the roots of As-tolerant species are synthesized. Raised concentrations of total PCs were measured in plant species with a range of sensitivities to As at equivalent levels of arsenic exposure, determined as the inhibition of root biomass. In addition, the production of PCs as a function of accumulated As was studied. Long-term PC synthesis (over a 5-week period) was positively, but non-linearly, correlated with arsenic, suggesting that probably not all As is bound by PCs. Moreover, it could be shown that the synthesis of different chain lengths of PCs is associated with differences in As tolerance. In the more tolerant grasses, *A. repens* and *L. Perenne*, it was chiefly the dithiol PC2 which was estimated. In contrast, the dominant PC species in the less tolerant plants *U. dioica*, *G. hederacea*, *L. marrubiastrum*, and *Z. mays* was PC3, while PC2 and PC3 were detected as well.

To investigate the antioxidative response of GSH metabolism in *Urtica dioica* to a Cd-induced oxidative stress, activities of GR, glutathione-S-transferase (GST), and glutathione peroxidase (GSH-Px), content of reduced (GSH) and oxidized (GSSG) glutathione, lipid peroxidation (LPO), and also accumulation of Fe, Zn, Mn, Cu, and Cd were determined in the roots, stems, and leaves of plants exposed to 0 (control), 0.045, and 0.09 mM CdCl$_2$ for 58 h; whereas the Cd content continuously increased in all organs, the Fe, Zn, Mn, and Cu content decreased with dependence on the applied Cd concentration and incubation time. The Cd treatment resulted in increased GR and GST activities in all plant organs; however, GSH-Px activity was dependent on Cd concentration and plant organ. The GSH/GSSG ratio maintained above the control level in the stems at both Cd concentrations. The LPO was generally close to the control values in the roots and stems, but it increased in the leaves, especially at 0.09 mM Cd (Tarhan and Kavakcioglu 2016).

Ginger (*Zingiber officinale* Rosc.) is an important ingredient of spices and herbals. The monitoring of toxic heavy metals in the rhizome of ginger is important for protecting public health against the hazards of metal toxicity. The concentration of volatile and non-volatile metals (As, Hg, Pb, and Cd) in the soil and rhizome of ginger were thus analyzed using atomic absorption spectrometry (Gupta et al. 2010). Soil analysis profile showed uniformity in the metal contents, in the active root-zone and subsoil, except Hg, which was present in higher quantities in one out of the four sectors of the field. The metal content in the soil in increasing order was Cd < As < Pb < Hg. In a ginger rhizome, the volatile toxic heavy metals As and Hg varied from not detected to 0.13 µg g^{-1} and 0.01 to 0.42 µg g^{-1}, respectively. The non-volatile metals Pb and Cd ranged from 0.06 to 0.64 µg g^{-1} and 0.002 to 0.03 µg g^{-1}, respectively. The results illustrated the findings that soil is the major but not the only source of metal accumulation in the plants. In the mentioned study, the volatile metal content (As, Hg) was found more in rhizomes collected from Himachal Pradesh while the non-volatile metals were predominant in samples from Uttarakhand.

In the study (Ali et al. 2006) of tissue cultured *Panax ginseng* C.A. Meyer roots in a bioreactor, growth responses, carbonyl content, phenolics, flavonoids, diphenyl-2-picryl-hydrazyl (DPPH) free radical scavenging activity, cysteine, non-protein thiol (NPSH) content, phenolics related enzymes, and lignin content were analyzed after exposure to CuSO$_4$ at various concentrations. Cu^{2+} significantly inhibited root growth and caused an increase in protein oxidation (21%) (measured as carbonyl content) at 50 µM. Cu^{2+} stress resulted in the increased activities of glucose-6-phosphate dehydrogenase, SKDH, PAL, and CAD. The ensuing effect was an accumulation of phenolics and lignin. The phenolics, flavonoids, cysteine and NPSH contents as well as the DPPH activity, increased by 26, 83, 30, 40, and 76%, respectively, at a 50 µM Cu^{2+} treatment compared to the control after 40 days. The induced activities of substrate-specific peroxidases (caffeic acid peroxidase), chlorogenic acid peroxidase, PPO, and β-glucosidase indicated that these enzymes played an important role in the synthesis of phenolics compounds. The results also provided evidence that the application of Cu^{2+} at 25 and 50 µM was accompanied by a substantial increase in oxidative stress as indicated by protein oxidation and reduced root growth. The increase in phenolic compounds, related enzymes, and lignin clearly reflects the protective response to cellular damage induced by higher levels of Cu^{2+}.

Accumulation of Pb, Cd, and Ni in plants growing in different moisture sections around three in-field ponds was monitored by Malinowska et al. (2017). The following plants were analyzed: *Potentilla anserina* L., *Mentha arvensis* L., *Achillea millefolium* L., *Comarum palustre* L., *Lysimachia vulgaris* L., and *Lycopus europaeus* L. Plant and soil samples were collected from three moisture sections: wet, periodically wet, and dry. After dry mineralization, ICP-AES was used to measure the concentration of monitored metals in plant material, in soil and in the bottom sediment. The findings show that of all the analyzed herb plants only a few of them exceeded the limits of Pb and Cd concentration while the concentration of Ni was always below the limit. The content of heavy metals in the soil was much lower than their natural content in noncontaminated soil. The highest concentration of these metals was noted in the bottom sediment from the pond with surrounding arable land. Herb plants growing in soil with low water content accumulated more Pb, Cd, and Zn than plants growing in soil with higher moisture. The highest accumulation of Pb was observed in the weed *P. anserina* L., while the highest Cd and Ni was in the medicinal herb *C. palustre* L.

In 2011, a study was carried out (later published by Malinowska et al. 2015) analyzing the effects of road traffic on the bioaccumulation of Zn and Cu in selected species of dicotyledonous plants growing on adjacent grasslands. The plants were sampled from the 9-km-long Siedlce bypass (Masovian Voivodeship, Poland), a part of the international route E-30, and collected during the flowering stage at following distances from the road: 1, 5, 10, and 15 m. The content of Zn and Cu was determined using the AAS method, with dry mineralization done before. The highest concentration of the elements, regardless of the distance from the road, was found in *Taraxacum* spec. Among the tested plants, the lowest Zn content was in *Vicia cracca* and the lowest Cu content in *Rumex acetosa*. The limit for copper content was exceeded in *Taraxacum* spec. and, slightly, in *Achillea millefolium* growing at the roadside, closest to the road.

Heavy metals in four perennial medicinal plants such as *M. recutita* L., *Taraxacum officinale* L., *Achillea millefolium* L., and *H. perforatum* L., collected from the Navodari region (Constanţa County, Romania), were quantitatively analyzed by using FAAS. Seven heavy metals (Cd, Cr, Co, Cu, Pb, Ni, and Zn) in plant samples and nine heavy metals (Cd, Cr, Co, Cu, Pb, Fe, Mn, Sn, and Zn) in the soil were selected on the basis of their effects upon health (Radulescu et al. 2013). The absorption of heavy metals in medicinal plants is governed by soil characteristics such as pH, salinity, conductivity, and organic matter content. In this respect, Zn concentration ranged between 12.180 mg kg⁻¹ in *Taraxacum officinale* collected from the Ovidiu zone, and 48.976 mg kg⁻¹ in *Achillea millefolium* L. collected from the Rompetrol Refinery location. The researchers found that the physiological activities of the medicinal plants influenced Zn absorption and the interactions with many elements such as Fe and Cu. *T. officinale* L. was able to accumulate more Cd (e.g., 0.783 mg kg⁻¹ DW in the Channel of Rompetrol Refinery location) than *A. millefolium* L., *M. recutita*, and *H. perforatum* L. These plants can absorb Pb from the soil and from the diethyl lead traffic-induced contaminated atmosphere because the collected points were near roads. The analysis of Cr in medicinal plants indicated that Cr concentration ranged between 0.127 and 0.970 mg kg⁻¹ on the Channel of Rompetrol Refinery and from 0.059 to 0.704 mg kg⁻¹ on the Rompetrol Refinery, respectively.

According to Radanovic et al. (2002), *H. perforatum* L. and *A. millefolium* L. are plants which can be characterized as heavy-metal accumulators. Including relating soils samples differing in pH reaction, 14 samples of St. John's wort and 9 samples of yarrow were collected from different localities in Yugoslavia and Republic Srpska. The total heavy-metal content (Mn, Zn, Cu, Pb, Ni, and Cd) was determined in the collected herb material (for St. John's wort – 25 cm from the top of the plant and for yarrow – 30 cm) by standard analytical methods. Heavy-metal content in the herb mostly depended on soil composition, for both species. The concentration of Mn and Zn in the herb decreased linearly with the increase of soil pH, while Cd concentration decreased exponentially. Cd content in St. John's wort was mostly above the limit of 0.5 ppm when soil pH (in nKCl) was lower than 5.9, and 4.5 for yarrow. Higher Cd content in the yarrow herb, and especially the St. John's wort herb, from acidic soils, points to the significance of I) required control of Cd content in the raw materials collected in the wild and II) necessary avoidance of such soils for cultivated production of these species. A relationship of Ni content in the herb with soil pH was found only for St. John's wort (r=−0.80***). Correlation between heavy-metal content in the herb and humus content in the soil was analyzed, but statistically significant regression coefficients were not obtained.

The variation of the antioxidant activity of *Aloe vera* (L.) Burm. f. under metal stress with different concentrations of cadmium was studied by Cai et al. (2013). The scavenging capacity of potted *A. vera* plants to OH•⁻ and O₂•⁻ was determined, and the effects of Cd pollution on the antioxidant activity of *A. vera* were evaluated using flow injection chemiluminescence methods. The results showed that at a low Cd pollution level the antioxidant activity of *A. vera* increased with the increase of Cd content in the *A.vera* body; while at a high Cd pollution level, the antioxidant activity of *A. vera* decreased with the increase of Cd content. This indicated that low Cd levels could stimulate the self-protection system of *A. vera*, resulting in an increase in the antioxidant activity, high Cd concentrations would damage the aloe self-defended system, leading to a decrease in antioxidant activity.

Methyl jasmonate (MeJA) is known to elicit protective effects as a form of plant response to abiotic stress; however, related studies on plant response to metal stress are insufficient. Yan et al. (2013) examined the effects of MeJA on the growth and physiological responses of *Capsicum frutescens* L. seedlings exposed to Cd stress and the results showed that low exogenous MeJA concentrations exhibited protective effects on the growth and physiology of *C. frutescens* seedlings under Cd stress.

Selenium (Se) is an essential element for humans but is not considered as essential for plants. However, its beneficial role in improving plant growth and stress tolerances is well established. In order to study the role of Se in Cd toxicity in pepper (*Capsicum frutescens* L. cv. Suryankhi Cluster), Shekari et al. (2017) carried out an experiment in greenhouse conditions. Treatments comprised Cd (0, 0.25, and 0.5 mM $CdCl_2$) and Se (0, 3, and 7 μM Na_2SeO_3) with three replications. The results showed that Cd decreased Chl a, Chl b, and Cars, whereas Se supplementation diminished Cd toxicity on photosynthetic pigments. Selenium significantly enhanced the antioxidant activity of leaves, which was diminished by Cd toxicity. In general, Se has a beneficial effect on plant growth and acted as an antioxidant enzyme of pepper under Cd stress and non-stress conditions as well.

Badea (2015) determined the concentration and dispersion of heavy metals Pb, Cd, and Hg, in medicinal herbs *Chelidonium majus* L., *Crataegus monogyna* L., *Artemisia absinthium* L., and *H. perforatum* L., from spontaneous flora bordering coal power plants using AAS. The metals detected in the coal used as fuel were found in all samples analyzed: ash, slag, soil, vegetation, medicinal plants of spontaneous flora. The concentration in the soil exceeded the normal national value for Pb: 55.10 mg kg^{-1}. The metal bioaccumulation in medicinal plants was reflected in the values of transfer coefficient (TF): Cd (0.47) > Hg (0.15) > Pb (0.13). The concentration level related to the maximum extent permitted under the force rules for herbs: FAO/WHO (2011), Kelp monograph 1426 Ph. Eur. (2007), Commission Regulation EC (2006) were exceeded for Pb in *H. perforatum* (7.21 mg kg^{-1}), *A. absinthium* (5.37 mg kg^{-1}). The trophic transfer of toxic heavy metals in plants used as medicines may represent a potential health hazard to consumers, a fact which imposes the need for systematic control.

Florev et al. (2008) studied the accumulation of gold and some of the toxic metals (Cu, Zn) in horsetail (*Equisetum arvense*) in comparison with milfoil (*A. millefolium* L.). Plant samples were collected from the Rosia Montana district in the Carpathian Mountains (Romania), i.e., from an area in which gold had already been extracted for a long time during the Roman Empire. Presently, the quantity of gold is diminished and the actual tendency to obtain Au by cyanide technique is very dangerous for the environment. Using horsetail and milfoil as biological prospectors, the preferential accumulation of Au by horsetail can be observed. The aim of this study was to demonstrate the accumulation of Au in the plants of the Rosia Montana area in important quantities compared with other metal concentrations induced by the thousands of years of exploitation. As the authors underlined, the reserves of gold remaining in the soil are not important enough to develop industrial production, but they are very dangerous for the environment.

It was found that some of the medicinal plants accumulated increased amounts of toxic elements like Cd or Pb. Less is known about the accumulation of other hazardous elements like As and Sb in these species. Vaculík et al. (2013) investigated selected medicinal plants naturally growing on old mining sites in Slovakia, Central Europe, contaminated by As and Sb. Both these elements are nonessential for plants and, at higher levels, might be phytotoxic. The soil concentration of As and Sb at three different localities extensively used for the mining of Sb ores in former times highly exceed values characteristic for noncontaminated substrates and ranged between 146 and 540 mg kg^{-1} for As and 525 and 4,463 mg kg^{-1} for Sb. Extraction experiments of soils show differences between As and Sb leaching, as the highest amount of mobile As was released in acetic acid while Sb was predominantly released in distilled water. In total, seven different plant species were investigated (*Fragaria vesca*, *Taraxacum officinale*, *Tussilago farfara*, *Plantago major*, *Veronica officinalis*, *Plantago media*, and *Primula elatior*), and the concentration of the investigated elements in shoot ranged between 1 and 519 mg kg^{-1} for As and 10 and 920 mg kg^{-1} for Sb. Differences in the bioaccumulation of As and Sb as well as in the translocation of these elements from root to shoot within the same species growing on different localities have been found. This indicates that the efficiency of As and Sb uptake might vary between individual plants of the same species on different sites. Increased bioaccumulation of As and Sb in biomass of investigated plants might be dangerous for human when used for traditional medicinal purposes.

Liu et al. (2013) attended to the relationship between tree leaf micro-morphology and features in adsorbing air suspended particulate matter and accumulating heavy metals. Seven tree species, including *Ginkgo biloba* L., grown at heavy traffic density sites in Huainan (China) were selected to analyze the frequency of air particulate matter retained by leaves, the particle amount of different sizes per unit leaf area retained by leaves and its related micro-morphology structure, and the relationship between particle amount of different sizes per unit leaf area retained by leaves and its related accumulation of heavy metals. They found that the species was characterized by a small leaf area, special epidermis with abundant wax, and highly uneven cell wall, as well as big and dense stomata and without trichomes which mainly absorbed fine particulate matter; while those species with many trichomes mainly retained coarse particulate matter. The accumulation of heavy metals in leaves of the seven species was significantly different except for Pb. Tree species with high capacities in heavy-metal accumulation were *Ginkgo biloba*, *Ligustrum lucidum*, and *Cinnamomum camphora*. Accumulation of Cd, Cr, Ni, Zn, Cu, and total heavy-metal concentration for seven tree species was positively related to the amount of particulate matter absorbed (mainly fine particulates).

Rare earth elements, a type of abiotic elicitor, were used to investigate the shift in the mechanism of flavonoid accumulation from primary metabolites in a *Ginkgo* suspension cell culture. The changes in flavonoids, mineral ions, photosynthetic pigments, Chl fluorescence, and antioxidant enzymes with varying $Ce(NO_3)_3$ doses (0.01–5.0 mM) were studied by Chen et al. (2015). Low doses of Ce^{3+} (0.01–0.1 mM) improved cell growth; the highest increased dry weight (2.67 g dm^{-3} medium) was estimated at a 0.1 mM Ce^{3+} dose. Moderate doses (0.5–1.0 mM) limited cell growth and initiated a self-protective mechanism through modulation of the pigments, ions, and flavonoid content, Chl fluorescence and antioxidant enzymes, while high doses (2.0–5.0 mM) inhibited the cell growth and even caused cell mortality. The Chl content, K^+, Zn^{2+}, and growth mass peaked at 0.1 mM Ce^{3+}. The F_v/F_m, and F_v'/F_m' values, and the amounts of P, Ca^{2+}, Fe^{2+}, and Zn^{2+} reached a maximum at 0.5 mM Ce^{3+} while the flavonoids content was highest at 1.0 mM Ce^{3+} (11.1 mg dm^{-3} medium). The antioxidant enzyme activities were high from 0.05 to 0.5 mM Ce^{3+}. Moreover, the changes of the Chl fluorescence images of the suspension cells were consistent with the cell growth, relative electrical conductivity, and MDA content with increasing Ce^{3+} doses. The findings suggested that doses of 0.1–0.5 mM Ce^{3+} shifted the metabolism in *Ginkgo* cells from primary to secondary processes, diverting the metabolism from growth to defense-related pathways as a result of the flavonoids accumulation. The high ratio of Car/Chl, high Car content, and 0.5–1.0 mM Ce^{3+} were beneficial for the accumulation of flavonol glycosides in *Ginkgo* suspension cells.

Yu et al. (2012) discussed the suitable spraying amount of Mg, Zn, Mo for high yield and good quality of leaf-use ginkgo. The effects of spraying different levels of Mg, Zn, Mo fertilizers on leaf yield and medicinal quality of 2-year-old ginkgo seedlings was tested using an indoor stimulation experiment and a pot experiment. The main results were submitted as follows: I) combining application of Mg, Zn, Mo fertilizers increased leaf thickness, leaf area, single plant leaf area, and leaf biomass; II) positive effect of combining Mg with Zn, Mo fertilizers on Chl, soluble sugar, and soluble protein contents decreased with the increasing of Mg concentration. The same was also the effect of combining Zn with Mg, Mo fertilizers and combining Mo with Mg, Zn fertilizers on Chl, and soluble sugar contents. In contrast, the effects on soluble protein content correlated positively with Zn or Mo concentration; iii) combining the application of Mg, Zn, Mo fertilizers increased total flavonoid and terpene lactones contents, and total flavonoid and terpene lactones yields per plant. The results indicate that appropriate combined fertilization management can increase Chl, soluble sugar, soluble protein contents and improve ginkgo growth and medicinal quality which increases component factors of leaf yield and medicinal quality.

Mentha piperita L. is an aromatic and medicinal species of the family Lamiaceae, known as mint or peppermint, and its leaves and branches produce essential oil rich in menthol. A study of David et al. (2014) was aimed at evaluating physiological indexes, macro- and micronutrients in the shoots and essential oil of *Mentha piperita* L. grown in nutrient solution number 2 of Hoagland and Arnon (1950) with different N, P, K, and Mg levels. Shoot length, dry mass of the different organs, total dry mass, leaf area, essential oil yield and composition, and macronutrient (N, P, K, Mg, Ca, S) and micronutrient (Mn, Cu, Fe, Zn) contents in the shoot were evaluated. Plants treated with 65% N, 50% P, 25% K, 100% Mg had a tendency toward longer shoots, greater root and leaf blade dry masses, higher essential oil yield, higher menthol levels, and lower menthone levels. The results showed that *Mentha* could be grown in nutrient solution by reducing 65% N, 50% P, 25% K, and 100% Mg. This solution ensured better plant development compared to the other tested treatments. Therefore, the researches recommend *Mentha piperita* L. to be grown with such nutrient levels.

Several physiological responses of *M. piperita* L. leaves at different positions along the stem were investigated by Candan and Tarhan (2010) under Cu^{2+} deficiency conditions and compared with excess and control. Content of Chls and Cars and the Chls/Cars ratio in all leaf positions were significantly lower than those of the control under Cu^{2+} deficiency. The highest decreases were determined under excess condition. SOD, CAT, ascorbate-dependent POD, and guaiacol-dependent POD activities in all leaf positions in the absence of Cu^{2+} were higher than in the control and showed a positive correlation with each other. SOD activities under excess conditions were higher than at deficiency, while all of the other enzyme activities were significantly lower than the control. Co-operative functions of all these enzymes in the absence of Cu^{2+} resulted in lower lipid peroxidation levels (LPO) than in the control and excess conditions. Except for guaiacol-dependent POD, all other antioxidant enzyme activities reached their maximum, while LPO level had its minimum at leaf position 6. Although when enhancing the antioxidant enzyme activities, LPO levels also increased approximately two-fold compared to the control in the absence of Cu^{2+} on the 12th day; its levels significantly increased with decreasing of Cu^{2+} afterward.

Heavy-metal extraction from soils is one of the functions of plants which is widely studied and applied worldwide. However, little is known about the extent medicinal plants can accumulate these metals and cause problems to human health. Boechat et al. (2016) evaluated the accumulation of heavy metal/metalloid in plant tissues, nutritional imbalance, and the effect of heavy-metal concentrations in the soil on the medicinal plants. The experiment was conducted in a factorial scheme with three contaminated soil samples and a soil sample from an uncontaminated field and three medicinal species: *Cynara scolymus*, *Ocimum basilicum*, and *Rosmarinus officinalis*. The heavy-metal content in the biomass increased with increasing soil samples concentration. Biomass production and nutritional imbalance by nutrients did not show consistent results according to soil contamination criteria and are not good indicators of heavy metals presence in plant tissues, since they did not allow for the prediction of the presence of metal in the plants, due to the different behavior of the elements and plant species. There was a high concentration of Cd, Cr, Pb, and As and micronutrients Fe, Zn, and Cu in the plant tissues, above the limits recommended by the World Health Organization. Therefore, as the components of *C. scolymus*, *O. basilicum*, and *R. officinaliss* are used to prepare teas, condiments, or consumed raw, coupled with the ability of such species to concentrate toxic metals, the continued use of these plant products containing these metals can pose a potential health concern.

Arceusz and Wesolowski (2015) studied the relationships between the levels of essential metals (Zn, Fe, Na, Mg, Ca, and K) and phenolic acids (caffeic, chlorogenic, ferulic, gallic, rosmarinic, and syringic) in commercial herbs (lemon balm, thyme, rosemary, mint, sage, and angelica) and spices (caraway, lovage, hyssop, and oregano). In the herbs, higher quantities of metals and phenolic compounds were found than in spices. All plants contained high levels of Ca, K, and rosmarinic acid, but low levels of Zn and gallic acid. By using principal component and hierarchical cluster analyses, several clusters were identified, grouping samples originating from a plant of a particular botanical species. Multivariate analysis has also shown that the contents of phenolic acids had a stronger impact on the scattering of herbs and spices than on the metal levels. Furthermore, statistically significant correlations were found between calcium and ferulic, gallic, rosmarinic, and syringic acids as well as between zinc and sodium and caffeic acid. This suggests cooperation between these biologically active constituents in metabolic processes occurring in plants.

In recent years, nanotechnology has become increasingly important in almost every field. The new and improved physical, chemical, and biological properties of the material at the nanoscale have far-reaching implications in the fields of science and technology. Nanoparticles' (NPs) effect on various plant species must be investigated to develop a comprehensive toxicity profile for nanoparticles. The current study (Salman Khan et al. 2016) strives to evaluate the effects of nine types of metal NPs including monometallic and bimetallic alloy NPs (Ag, Au, Cu, AgCu (1/3), AgCu (3/1), AuCu (1/3), AuCu (3/1), AgAu (1/3), AgAu (3/1)) on seed germination, root and shoot growth, and biochemical profile of *Silybum marianum* plants. Seed germination was greatly affected and increased significantly upon treatment with NP suspensions and was recorded highest for the AgNPs suspension. Metal NPs also had a significant effect on the biochemical profile of *S. marianum*. For the first week, the effect on DPPH, total phenolics content, total flavonoids content, total protein content, peroxidase activity, and SOD activity was enhanced but declined as time progressed. Among the nanoparticles being used, the effect of AgNPs was mostly enhancing. The results obtained are significant in mapping the effects of different monometallic and bimetallic NPs on medicinal plant species.

Heavy metals have the potential to interact and induce several stress responses in the plants; thus, effects of heavy-metal stress on DNA damage and total antioxidants level in *Urtica dioica* L. leaves and stems was investigated (Gjorgieva et al. 2013). The samples were taken from areas with different metal exposition. Metal content was analyzed by ICP-AES, for a total antioxidants level assessment the Ferric-Reducing Antioxidant Power (FRAP) assay was used, and genomic DNA isolation from frozen plant samples was performed to obtain DNA fingerprints of an investigated plant. It was found that heavy-metal contents in stems generally changed synchronously with those in the leaves of the plant, and extraneous metals led to an imbalance of the mineral nutrient elements. DNA damages were investigated using the Random Amplified Polymorphic DNA (RAPD) technique, and the results demonstrated that the samples exposed to metals yielded a large number of new fragments (total 12) in comparison with the control sample. This study showed that DNA stability is highly affected by metal pollution which was identified by RAPD markers. Results suggested that heavy-metal stress influences the antioxidant status and also induces DNA damage in *U. dioica* which may help to understand the mechanisms of metals genotoxicity.

The aim of a study carried out by Jedynak et al. (2012) was to assess the strategy developed by terrestrial plants growing in an area contaminated by arsenic to avoid or minimize the toxic effects caused by this element. Eight plant species from two As-contaminated areas were selected for the investigation. Arsenic uptake by different plants was discussed. The speciation analysis of arsenic in plant leaves was performed as well, and both inorganic As(III) and As(V) were detected. Moreover, the concentration of phytochelatins in the investigated terrestrial plants was determined. It was noted that the highest concentration of As was found in the herb *Geranium robertianum*, 21 mg kg^{-1}, and common nettle (*Urtica dioica*), 5.3 mg kg^{-1}, growing in Zloty Stok (Lower Silesian Voivodeship, Poland) and Lomianki (Masovian Voivodeship), respectively. Phytochelatins were present in all investigated plant species: PC3 was present in the highest concentration in plants from Zloty Stok (compared to other phytochelatins) while none of the phytochelatins dominated in plants from Lomianki. A correlation between concentrations of phytochelatin and As was found in one of the nettle samples from Lomianki.

Summarizing the above-mentioned responses of medicinal plants to toxic metals it should be emphasized that specific secondary metabolites formed in these plants play an extraordinarily important role in their defense mechanism against the harmful effect of these metals (mainly in natural conditions). Specific secondary metabolites are released from these plants into the soil or atmosphere and thus (directly or indirectly through allelopathy) influence not only the biological part of the soil but also the competitiveness of the other plant species in the plant community. This fact is consequently manifested in the structure, biodiversity as well as stability of the ecosystem, too.

We want to mention that in Chapter 24 of this *Handbook* entitled "Plant responses to stress induced by toxic metals and their nanoforms," further findings related to the different medicinal plant species and metals, metal complexes as well as metal nanoparticles are briefly reported.

39.11 CONCLUDING REMARKS

Since medicinal plants are frequently exposed to various environmental stresses in their natural conditions, they have evolved physiological, biochemical, and molecular mechanisms to counteract the deleterious effects of these stresses. Abiotic stresses such as drought, flooding and waterlogging, heat, temperature, salinity, acidity, ozone, or toxic metals adversely affect the production of medicinal plants as well as the number of genes encoding for biosynthesis of stress protecting the compound. These stresses in isolation and/or in combination with control plant growth, affect development and productivity by causing physiological disorders, ion toxicity, and hormonal and nutritional imbalances. At present, genetic engineering techniques offer many applications in the improvement of medicinal plants for abiotic stress tolerance. However, to predict the functioning of the medicinal plant population in the future, a mechanistic understanding is required of how plants cope with different stressors under the coming climate conditions with elevated CO_2 concentrations, warmer temperatures, and expressive drought. Researchers thus face a challenge for better understanding the abiotic stress tolerance of the plants as well as improvement of various traits related to the quality of different medicinal plants. Moreover, both scientists and politicians will have to accept fundamental bioethical principles to ensure the sustainable development of human society as well as the essential protection of the environment and nature.

ACKNOWLEDGMENT

The financial support from the Slovak Scientific Grant Agency VEGA projects (1/0614/17 and 2/0118/18) is gratefully acknowledged.

BIBLIOGRAPHY

Ahn, K. 2017. The worldwide trend of using botanical drugs and strategies for developing global drugs. *BMB Rep.* 50:111–116.

Aidoo, M. K., T. Sherman, N. Lazarovitch, A. Fait, and S. Rachmilevitch. 2017. A bell pepper cultivar tolerant to chilling enhanced nitrogen allocation and stress-related metabolite accumulation in the roots in response to low root-zone temperature. *Physiol. Plant.* 161:196–210.

Airaki, M., M. Leterrier, R. M. Mateos, et al. 2012. Metabolism of reactive oxygen species and reactive nitrogen species in pepper (*Capsicum annuum* L.) plants under low temperature stress. *Plant Cell Environ.* 35:281–295.

Aktas, H., K. Abak, and S. Eker. 2012. Anti-oxidative responses of salt-tolerant and salt-sensitive pepper (*Capsicum annuum* L.) genotypes grown under salt stress. *J. Hort. Sci. Biotech.* 87:360–366.

Ali, R. M. 2000. Role of putrescine in salt tolerance of *Atropa belladonna* plant. *Plant Sci.* 152:173–179.

Ali, M. B., N. Singh, A. M. Shohael, E. J. Hahn, and K. Y. Paek. 2006. Phenolics metabolism and lignin synthesis in root suspension cultures of *Panax ginseng* in response to copper stress. *Plant Sci.* 171:147–154.

Allakhverdiev, S. I., and N. Murata. 2004. Environmental stress inhibits the synthesis de novo of proteins involved in the photodamage-repair cycle of photosystem II in *Synechocystis* sp. PCC 6803. *Biochim. Biophys. Acta.* 1657:23–32.

Allakhverdiev, S. I., V. D. Kreslavski, V. V. Klimov, D. A. Los, R. Carpentier, and P. Mohanty. 2008. Heat stress: An overview of molecular responses in photosynthesis. *Photosynth. Res.* 98:541–550.

Anjum, S. A., M. Farooq, X. Y. Xie, X. J. Liu, and M. F. Ijaz. 2012. Antioxidant defense system and proline accumulation enable hot pepper to perform better under drought. *Sci. Hortic.* 140:66–73.

Anonymous. 2016. State of the World´s Plants, Report 2016. KEW, Royal Botanic Gardens, Retrieved September 25, 2017.

Arceusz, A., and M. Wesolowski. 2015. Essential metals and phenolic acids in commercial herbs and spices. Multivariate analysis of correlations among them. *Open Chem.* 13:1196–1208.

van Arsdall, A. 2002. *Medieval Herbal Remedies. The Old English Herbarium and Anglo-Saxon Medicine.* Psychology Press. 259.

Asadi-Sanam, S., H. Pirdashti, A. Hashempour, M. Zavareh, and G. A. Nematzadeh, and Y. Yaghubian. 2015. The physiological and biochemical responses of eastern purple coneflower to freezing stress. *Fiziol. Rast.* 62:515–523.

Ashraf, M. A. 2012. Waterlogging stress in plants: A review. *Afr. J. Agric. Res.* 7:1976–1981.

Asthir, B. 2015. Protective mechanisms of heat tolerance in crop plants. Review article. *J. Plant Int.* 10:202–210.

Atanasov, A. G., B. Waltenberger, E. M. Pferschy-Wenzig, et al. 2015. Discovery and resupply of pharmacologically active plant-derived natural products: A review. *Biotechnol. Adv.* 33:1582–1614.

Atashi, S., E. Bakhshandeh, Z. Zeinali, E. Yassari, and J. A. Teixeira da Silva. 2014. Modeling seed germination in *Melisa officinalis* L. in response to temperature and water potential. *Acta Physiol. Plant.* 36:605–611.

Ayan, A. K., R. Kizilkaya, C. Cirak, and K. Kevseroglu. 2006. Heavy metal contents of St. John's wort (*Hypericum perforatum* L.) growing in Northern Turkey. *J. Plant Sci.* 1(3):182–186.

Baczek-Kwinta, R., K. Tokarz, and I. Czyczylo-Mysza. 2008. Different response of lemon balm (*Melissa officinalis* L.) and basil (*Ocimum basilicum* L.) to the impact of growth and root submergence. *Zes. Prob. Post. Nauk Rol.* 524:127–135.

Badea, D. N. 2015. Determination of potentially toxic heavy metals (Pb, Hg, Cd) in popular medicinal herbs in the coal power plant area. *Rev. Chim.* 66:1132–1136.

Baluška, F., and S. Mancuso. 2009. Plant neurobiology. From stimulus perception to adaptive behaviour of plants, via integrated chemical and electrical signaling. *Plant Signal. Behav.* 4:475–476.

Baluška, F., S. Mancuso, D. Volkmann, and P. W. Barlow. 2009. The "root-brain" hypothesis of Charles and Francis Darwin: Revival after more than 125 years. *Plant Signal. Behav.* 4:1121–1127.

Baluška, F., S. Mancuso, D. Volkmann, and P. W. Barlow. 2010. Root apex transition zone: A signaling-response nexus in the root. *Trends Plant Sci.* 15:402–408.

Bano, A., F. Ullah, and A. Nosheen. 2012. Role of abscisic acid and drought stress on the activities of antioxidant enzymes in wheat. *Plant Soil Environ.* 58:181–185.

Barkla, B. J., T. Castellanos-Cervantes, J. L. de León, et al. 2013. Elucidation of salt stress defense and tolerance mechanisms of crop plants using proteomics – Current achievements and perspectives. *Proteomics.* 13:1885–1900.

Barnes, J., L. A. Anderson, and J. D. Phillipson. 2001. St John's wort (*Hypericum perforatum* L.): a review of its chemistry, pharmacology and clinical properties. *J. Pharm. Pharmacol.* 53(5):583–600.

Bartels, D., and R. Sunkar. 2005. Drought and salt tolerance in plants. *Crit. Rev. Plant Sci.* 24:23–58.

Bayat, H., M. Alirezaie, and H. Neamati. 2012. Impact of exogenous salicylic acid on growth and ornamental characteristics of calendula (*Calendula officinalis* L.) under salinity stress. *J. Stress Physiol. Biochem.* 8:258–267.

Bilia, A. R., S. Gallori, and F. F. Vincieri. 2002. St. John´s wort and depression. Efficacy safety and tolerability – An update. *Life Sci.* 70:3077–3096.

Billing, J., and P. W. Sherman. 1998. Antimicrobial functions of spices: Why some like it hot. *Q. Rev. Biol.* 73:3–49.

Blume, H. P., G. W. Brümmer, H. Fleige, et al. 2016. Chemical properties and processes. In *Scheffer/Schachtschabel Soil Science*, 123–174. Berlin, Heidelberg, Springer-Verlag.

Boechat, C. L., F. S. Carlos, C. Gianello, and F. A. de Oliveira Camargo. 2016. Heavy metals and nutrients uptake by medicinal plants cultivated on multi-metal contaminated soil samples from an abandoned gold ore processing site. *Water Air Soil Pollut.* 227:1–11.

Brenner, E. D., R. Stahlberg, S. Mancuso, J. Vivanco, F. Baluška, and E. Van Volkenburgh. 2006. Plant neurobiology: An integrated view of plant signaling. *Trends Plant Sci.* 11:413–419.

Cai, Z., X. L. He, D. F. Lu, X. L. Qiu, and W. C. Yue. 2013. Distribution of cadmium in *Aloe vera* and its hazard impact on the antioxidant activity. *Adv. Mater. Res.* 610–613:306–310, Part: 1–4.

Cameron, R. W., C. M. King, and S. Robinson. 2010. Ornamental Mediterranean plants in the UK: Root adaptations to hypoxia and anoxia. *Acta Hortic.* 881:469–474.

Candan, N., and L. Tarhan. 2010. Alterations of the antioxidative enzyme activities, lipid peroxidation levels, chlorophyll and carotenoid contents along the peppermint (*Mentha piperita* L.) leaves exposed to copper deficiency and excess stress conditions. *J. Appl. Bot. Food Qual.* 83:103–109.

Casierra-Posada, F., and O. F. Ávila-León. 2015. Shade tolerance of marigold plants (*Calendula officinalis*). *Rev. U.D.C.A Actual. Divulg. Cient.* 18:119–126.

Center for the Study of Carbon Dioxide and Global Change. 2014. Health-promoting effects of elevated CO_2 on medicinal plants. http://www.co2science.org/subject/h/summaries/co2health medicinal.php.

Chandra, S., H. Lata, I. A. Khan, and M. A. Elsohly. 2011. Photosynthetic response of *Cannabis sativa* L., an important medicinal plant, to elevated levels of CO_2. *Physiol. Mol. Biol. Plants.* 17:291–295.

Chaparzadeh, N., R. A. Khavari-Nejad, F. Navari-Izzo, and R. Izzo. 2003. Water relations and ionic balance in *Calendula officinalis* L. under salinity conditions. *Agrochimica.* 47:1–2.

Chapman, D., and R. M. Auge. 1994. Physiological-mechanisms of drought resistance in 4 native ornamental perennials. *J. Am. Soc. Hortic. Sci.* 119:299–306.

Chaves, M. M., J. P. Maroco, and J. S. Pereira. 2003. Understanding plant responses to drought – From genes to the whole plant. *Funct. Plant Biol.* 30:239–264.

Chen, X. B., S. J. Li, Y. Li, and D. S. Zhang. 2009. Photosynthetic and chlorophyll fluorescence characteristic of ornamental hawthorns under summer heat stress. *Xibei Zhiwu Xuebao.* 29:2294–230.

Chen, Y., Y. Y. Luo, N. F. Qiu, F. Hu, L. Sheng, R. Wang, and F. Cao. 2015. Ce3+ induces flavonoids accumulation by regulation of pigments, ions, chlorophyll fluorescence and antioxidant enzymes in suspension cells of *Ginkgo biloba* L. *Plant Cell Tissue Organ Cult.* 123:283–296.

Cheng, D. D., Z. S. Zhang, X. B. Sun, M. Zhao, G. Y. Sun, and W. S. Chow. 2016. Photoinhibition and photoinhibition-like damage to the photosynthetic apparatus in tobacco leaves induced by *Pseudomonas syringae* pv. Tabaci under light and dark conditions. *BMC Plant Biol.* 16:29.

Chesworth, W. 2008. Acid deposition effects on soils. In *Encyclopedia of Soil Science*, ed. W. Chesworth, 2–5. Dordrecht, the Netherlands, Springer.

Chhapekar, S. S., V. Jaiswal, I. Ahmad, and R. Gau. 2018. Progress and prospects in *Capsicum* breeding for biotic and abiotic stresses. In *Biotic and Abiotic Stress Tolerance in Plants*, ed. S. Vats, 279–322. Springer Nature Singapore Pte Ltd.

Chizzola, R., H. Michitsch, and C. Franz. 2003. Monitoring of metallic micronutrients and heavy metals in herbs, spices and medicinal plants from Austria. *Eur. Food Res. Technol.* 216:407–411.

Choffnes, D. 2016. *Nature´s Pharmacopeia. A World of Medicinal Plants.* New York, NY: Columbia University Press.

Collins, M. 2000. Studies in medieval culture. London, British Library and University of Toronto Press, 334.

Commission Regulation EC. 2006. Setting maximum levels for certain contaminants in foodstuffs. Commission Regulation (EC) No 1881/2006. *Official J. Europ. Union.* L, 364:5–24.

Čellárová, E., K. Kimáková, Z. Daxnerová, and P. Mártonfi. 1995. *Hypericum perforatum* (St. John's wort): *In vitro* culture and the production of hypericin and other secondary metabolites. In *Biotechnology in Agriculture and Forestry 33: Medicinal and Aromatic Plants*, ed. Y. P. S. Bajaj, 261–275. Berlin, Germany, Springer-Verlag.

Dagar, J. C., R. K. Yadav, S. R. Dar, and S. Ahamad. 2015. Liquorice (*Glycyrrhiza glabra*): A potential salt-tolerant, highly remunerative medicinal crop for remediation of alkali soils. *Curr. Sci.* 108:1683–1688.

Daliakopoulos, I. N., I. K. Tsanis, A. Koutroulis, N. N. Kourgialas, A. E. Varouchakis, G. P. Karatzas, and C. J. Ritsema. 2016. The threat of soil salinity: A European scale review. *Sci. Total Environ.* 573:727–739.

D'Angiolillo, F., M. Tonelli, E. Pellegrini, C. Nali, G. Lorenzini, L. Pistelli, and L. Pistelli. 2015. Can ozone alter the terpenoid composition and membrane integrity of *in vitro Melissa officinalis* shoots? *Nat. Prod. Commun.* 10(6):1055–1058.

David, E. F. S., M. M. Mischan, M. O. M. Marques, and C. S. F. Boarao. 2014. Physiological indexes macro- and micronutrients in plant tissue and essential oil of *Mentha piperita* L. grown in nutrient solution with variation in N, P, K and Mg levels. *Rev. Bras. Plant. Med.* 16:106.

de Campos Nóia, N. R., M. C. P. da Cruz, M. J. Davalo, M. C. Oliveira, and C. K. Fujita. 2014. Produção de matéria seca de alfafa (*Medicago sativa* L.) em função de doses de calcário. (Dry matter production by alfalfa plants (*Medicago sativa* L.) as influenced by lime doses). *Cient. Jaboticabal.* 42:310–315.

de Melo, H. F., E. R. de Souza, H. H. F. Duarte, J. C. Cunha, and H. R. B. Santos. 2017. Gas exchange and photosynthetic pigments in bell pepper irrigated with saline water. *Rev. Bras. Eng. Agríc. Ambient.* 21:38–43.

Delfine, S., R. Tognetti, F. Loreto, and A. Alvino. 2002. Physiological and growth responses to water stress in field-grown bell pepper (*Capsicum annuum* L.). *J. Hortic. Sci. Biotech.* 77:697–704.

de Souza, T. C., E. D. Souza, S. Dousseau, E. M. de Castro, and P. C. Magalhães. 2013. Seedlings of *Garcinia brasiliensis* (Clusiaceae) subjected to root flooding: Physiological, morphoanatomical, and antioxidant responses to the stress. *Aquat. Bot.* 111:43–49.

Dharmananda, S., and J. F. Feldman. 1979. The methods of preparation of herb formulas: Decoctions, dried decoctions, powders, pills, tablets, and Tinctures. Institute of Traditional Medicine, Portland, Oregon. ITM Articles. Essay by Subhuti Dharmananda, Ph.D., Director, Institute for Traditional Medicine.

Dhondt, S., N. Wuyts, and D. Inzé. 2013. Cell to whole-plant phenotyping the best is yet to come. *Trends Plant Sci.* 18:428–439.

Díaz-Pérez, J. C. 2013. Bell pepper (*Capsicum annum* L.) crop as affected by shade level: Microenvironment, plant growth, leaf gas exchange, and leaf mineral nutrient concentration. *Hortscience.* 48:175–182.

Dlugokencky, E. 2016. Annual mean carbon dioxide data. Earth System Research Laboratory. National Oceanic & Atmospheric Administration. ftp://aftp.cmdl.noaa.gov/produ cts/trends/co2/co2_annmean_gl.txt, Accessed October 17, 2018.

Dohrmann, A. B., and C. C. Tebbe. 2005. Effect of elevated tropospheric ozone on the structure of bacterial communities inhabiting the rhizosphere of herbaceous plants native to Germany. *Appl. Environ. Microbiol.* 71(12):7750–7758.

Döring, A. S., E. Pellegrini, A. Campanella, et al. 2014a. How sensitive is *Melissa officinalis* to realistic ozone concentrations? *Plant Physiol. Biochem.* 74:156–164.

Döring, A. S., E. Pellegrini, M. Della Batola, C. Nali, G. Lorenzini, and M. Petersen. 2014b. How do background ozone concentrations affect the biosynthesis of rosmarinic acid in *Melissa officinalis*? *J. Plant Physiol.* 171(5):35–41.

Du, Q., G. Xing, X. Jiao, X. Song, and J. Li. 2018. Stomatal responses to long-term high vapour pressure deficits mediated most limitation of photosynthesis in tomatoes. *Acta Physiol. Plant.* 40:1–12.

Eggleton, T. 2012. *A Short Introduction to Climate Change.* Cambridge University Press, 52.

Ekor, M. 2014. The growing use of herbal medicines: Issues relating to adverse reactions and challenges in monitoring safety. *Front. Pharmacol.* 4:177.

El-Beltagi, H. S., O. K. Ahmed, and W. El-Desouky. 2011. Effect of low doses γ-irradiation on oxidative stress and secondary metabolites production of rosemary (*Rosmarinus officinalis* L.) callus culture. *Radiat. Phys. Chem.* 80:968–976.

Eliáš, P. 1994. Research of flora and vegetation of settlements (towns, villages, castle ruins) in Slovakia [In Slovak]. *Zpravodaj Čes. Bot. Spol.* 29:45–75.

ESRL Global Monitoring Division. 2018. Trends in atmospheric carbon dioxide. https://www.esrl.noaa.gov/gmd/ccgg/trends/, Accessed October 17, 2018.

Falk, H., and E. Mayr. 1997. Concerning *bay* salt and *peri* chelate formation of hydroxyphenanthroperylene quinones (fringelites). *Monatsh. Chem.* 128:353–360.

Falk, H., and W. Schmitzberger. 1992. On the nature of soluble hypericin in *Hypericum* species. *Monatsh. Chem.* 123:731–739.

FAO/WHO. 2011. Joint FAO/WHO, Food standards programme codex committee on contaminants in foods, fifth session, 2011, 8–25.

Farahani, H. A., S. A. Valadabadi, J. Daneshian, and M. A. Khalvati. 2009. Evaluation changing of essential oil of balm (*Melissa officinalis* L.) under water deficit stress conditions. *J. Med. Plant. Res.* 3:329–333.

Farnsworth, N. R., O. Akerele, A. S. Bingel, D. D. Soejarto, and Z. Guo. 1985. Medicinal plants in therapy. *Bull. World Health Organ.* 63:965–981.

Feng, B., P. Liu, G. Li, S. T. Dong, F. H. Wang, L. A. Kong, and J. W. Zhang. 2014. Effect of heat stress on the photosynthetic characteristics in flag leaves at the grain-filling stage of different heat-resistant winter wheat varieties. *J. Agron. Crop Sci.* 200:143–155.

Florev, R. M., A. I. Stoica, M. Ionica, and G. E. Baiulescu. 2008. Differential accumulation of gold and other metals by plants. *Rev. Chim.* 59:1019–1021.

Flowers, T. J. 2004. Improving crop salt tolerance. *J. Exp. Bot.* 55:307–319.

Fock, H. P., K. Biehler, and T. Stuhlfauth. 1992. Use and degradation of light energy in water-stressed *Digitalis-lanata*. *Photosynthetica*. 27:571–577.

Fu, Q. S., B. Zhao, Y. J. Wang, S. Ren, and Y. D. Guo. 2010. Stomatal development and associated photosynthetic performance of capsicum in response to differential light availabilities. *Photosynthetica*. 48:189–198.

Gattringer, J. P., K. Ludewig, S. Harvolk-Schöning, T. W. Donath, and A. Otte. 2018. Interaction between depth and duration matters: Flooding tolerance of 12 floodplain meadow species. *Plant Ecol.* 219(8):973–984.

Gattringer, J. P., T. W. Donath, R. L. Eckstein, K. Ludewig, A. Otte, and S. Harvolk-Schöning. 2017. Flooding tolerance of four floodplain meadow species depends on age. *PLOS ONE*. 12(5):e0176869.

Ghasemzadeh, A., H. Z. Jaafar, and A. Rahmat. 2010. Elevated carbon dioxide increases contents of flavonoids and phenolic compounds, and antioxidant activities in Malaysian young ginger (*Zingiber officinale* Roscoe.) varieties. *Molecules*. 15:7907–7922.

Ghasemzadeh, A., H. Z. Jaafar, A. Rahmat, P. E. M. Wahab, and M. R. A. Halim. 2010. Effect of different light intensities on total phenolics and flavonoids synthesis and anti-oxidant activities in young ginger varieties (*Zingiber officinale* Roscoe). *Int. J. Mol. Sci.* 11:3885–3897.

Ghasemzadeh, A., H. Z. Jaafar, E. Karimi, and S. Ashkani. 2014. Changes in nutritional metabolites of young ginger (*Zingiber officinale* Roscoe) in response to elevated carbon dioxide. *Molecules*. 19(10):16693–16706.

Ghavami, N., and A. A. Ramin. 2008. Grain yield and active substances of milk thistle as affected by soil salinity. *Commun. Soil Sci. Plant Anal.* 39:2608–2618.

Ghorbani, A., and M. Esmaeilizadeh. 2017. Pharmacological properties of *Salvia officinalis* and its components. *J. Tradit. Complement. Med.* 7(4):433–440.

Ghosh, D., and J. Xu. 2014. Abiotic stress responses in plant roots: A proteomics perspective. *Front. Plant Sci.* 5:6.

Gibbs, J., and H. Greenway. 2003. Mechanisms of anoxia tolerance in plants. I. Growth, survival and anaerobic catabolism. *Funct. Plant Biol.* 30(1):1–47.

Gimmel, M. 2008. Reading medicine in the Codex De La Cruz Badiano. *J. Hist. Ideas.* 69:169–192.

Gjorgieva, D., T. Kadifkova Panovska, T. Ruskovska, K. Baceva, and T. Stafilov. 2013. Influence of heavy metal stress on antioxidant status and DNA damage in Urtica dioica. *BioMed Res. Int.* 2013:6.

Gray, D. E., S. G. Pallardy, H. E. Garrett, and G. E. Rottinghaus. 2003. Acute drought stress and plant age effects on alkamide and phenolic acid content in purple coneflower roots. *Planta Med.* 69:50–55.

Greenway, H., and J. Gibbs. 2003. Mechanisms of anoxia tolerance in plants. II. Energy requirements for maintenance and energy distribution to essential processes. *Funct. Plant Biol.* 30(10):999–1036.

Grejtovský, A., and R. Pirč. 2000. The effect of high cadmium concentration in soil on growth, uptake of nutrient and some heavy metals on *Chamomilla recutita* (L.) Rauschert. *J. Appl. Bot. Angew. Bot.* 74:169–174.

Gulen, H., A. Ipek, and E. Turhan. 2016. Effects of antioxidant enzymes on heat stress tolerance of pepper (*Capsicum annuum* L.) seedlings. *ISHS Acta Hortic.* 1145:43–49.

Guo, W. L., R. G. Chen, Z. H. Gong, Y. X. Yin, and D. W. Li. 2013. Suppression subtractive hybridization analysis of genes regulated by application of exogenous abscisic acid in pepper plant (*Capsicum annuum* L.) leaves under chilling stress. *PLOS ONE.* 8:e66667.

Gupta, B., and B. Huang. 2014. Mechanism of salinity tolerance in plants: Physiological, biochemical and molecular characterization. *Int. J. Genonomics.* 2014:701596.

Gupta, B., A. Sengupta, J. Saha, and K. Gupta. 2013. Plant abiotic stress: "Omics" approach. *Plant Biochem. Physiol.* 01:1–3.

Gupta, S., P. Pandotra, A. P. Gupta, et al. 2010. Volatile (As and Hg) and non-volatile (Pb and Cd) toxic heavy metals analysis in rhizome of *Zingiber officinale* collected from different locations of North Western Himalayas by Atomic Absorption Spectroscopy. *Food Chem. Toxicol.* 48:2966–2971.

Hasanuzzaman, M., K. Nahar, M. M. Alam, R. Roychowdhury, and M. Fujita. 2013. Physiological, biochemical, and molecular mechanisms of heat stress tolerance in plants. *Int. J. Mol. Sci.* 14:9643–9684.

Hassell, R. L., R. J. Dufault, J. W. Rushing, and B. M. Shepard. 2001. Influence of temperature gradients on medicinal plant seed germination. *Hortscience.* 36:448.

Haworth, M., C. Elliott-Kingston, and J. C. McElwain. 2013. Co-ordination of physiological and morphological responses of stomata to elevated CO_2 in vascular plants. *Oecologia.* 171:71–82.

He, X. Y., W. Huang, W. Chen, et al. 2009. Changes of main secondary metabolites in leaves of *Ginkgo biloba* in response to ozone fumigation. *J. Environ. Sci.* 21:199–203.

He, X. Y., Y. N. Ruan, W. Chen, and T. Lu. 2006. Responses of anti-oxidative system in leaves of *Ginkgo biloba* to elevated ozone concentration in urban area. *Bot. Stud.* 47(4):409–416.

Hoffmann, A. M., G. Noga, and M. Hunsche. 2015. Acclimations to light quality on plant and leaf level affect the vulnerability of pepper (*Capsicum annuum* L.) to water deficit. *J. Plant Res.* 128:295–306.

Hu, W. H., Y. A. Xiao, J. J. Zeng, and X. H. Hu. 2010. Photosynthesis, respiration and antioxidant enzymes in pepper leaves under drought and heat stresses. *Biol. Plant.* 54:761–765.

Hu, W. H., Y. L. Cao, J. J. Zeng, K. Shi, and J. Q. Yu. 2008. Effects of heat and drought stresses on growth and respiration in different cultivar of pepper (*Capsicum annuum* L.). *Bull. Bot. Res.* 28:199–204.

Ibrahim, M. H., and H. Z. E. Jaafar. 2011. Increased carbon dioxide concentration improves the antioxidative properties of the Malaysian herb Kacip Fatimah (*Labisia pumila* Blume). *Molecules.* 16:6068–6081.

Ilhan, S., F. Ozdemir, and M. Bor. 2015. Contribution of trehalose biosynthetic pathway to drought stress tolerance of *Capparis ovate* Desf. *Plant Biol.* 17:402–407.

Iqbal, M., I. Hussain, A. Habib, M. A. Ashraf, and R. Rasheed. 2015. Effect of semiarid environment on some nutritional and antinutritional attributes of Calendula (*Calendula officinalis* L.). *J. Chem.* 2015:1–8.

Jackson, M. B. 2002. Long-distance signalling from roots to shoots assessed: The flooding story. *J. Exp. Bot.* 53(367):175–181.

Jaffel, K., S. Sai, N. K. Bouraoui, R. B. Ammar, L. Legendre, M. Lachâal, and B. Marzouk. 2011. Influence of salt stress on growth, lipid peroxidation and antioxidative enzyme activity in borage (*Borago officinalis* L.). *Plant Biosyst.* 145:362–369.

Jaffel-Hamza, K., S. Sai-Kachout, J. Harrathi, M. Lachaâl, and B. Marzouk. 2013. Growth and fatty acid composition of borage (*Borago officinalis* L.) leaves and seeds cultivated in saline medium. *J. Plant Growth Regul.* 32:200–207.

Jakovljevic, M., S. Antic-Mladenovic, M. Ristic, S. Maksimovic, and S. Blagojevic. 2000. Influence of selenium on the yield and quality of chamomile (*Chamomilla* recutita (L.) Rausch.). *Rost. Výroba.* 46:123–126.

Jan, S., A. N. Kamili, J. A. Parray, Y. S. Bedi, and P. Ahmad. 2016. Microclimatic variation in UV perception and related disparity in tropane and quinolizidine alkaloid composition of *Atropa acuminata*, *Lupinus polyphyllus* and *Hyoscyamus niger*. *J. Photochem. Photobiol. B Biol.* 161:230–235.

Jedynak, L., J. Kowalska, and A. Leporowska. 2012. Arsenic uptake and phytochelatin synthesis by plants from two arsenic-contaminated sites in Poland. *Pol. J. Environ. Stud.* 21:1629–1633.

Jiang, X. M., J. Y. Hu, W. H. Qi, G. D. Chen, and X. Xu. 2010. Different physiological responses of male and female *Ginkgo biloba* (Ginkgoaceae) seedlings to salt stress. *Acta Bot. Yunnanica* 31:447–453.

Jin, Z. M., C. H. Wang, Z. P. Liu, and W. J. Gong. 2007. Physiological and ecological characters studies on *Aloe vera* under soil salinity and seawater irrigation. *Process Biochem.* 42:710–714.

Jing, Y. X., G. L. Li, B. H. Gu, D. J. Yang, L. Xiao, R. X. Liu, and C. L. Peng. 2009. Leaf gas exchange, chlorophyll fluorescence and growth responses of *Melaleuca alternifolia* seedlings to flooding and subsequent recovery. *Photosynthetica.* 47(4):595–601.

Kang, Y. I., H. J. Kim, S. Y. Lee, H. Chun, N. J. Kang, and B. R. Jeong. 2008. Changes of carbohydrates in pepper (*Capsicum annuum* L.) leaves and flowers under different irradiance and night temperature regimes. *Hort. Environ. Biotech.* 49:397–402.

Karmazín, M., J. Hubík, and J. Dušek. 1984. *Catalog of Medicaments of Plant Origin (In Czech)*, 5th Edition. Praha, VHJ Spofa.

Kaur, T., R. Bhat, and D. Vyas. 2016. Effect of contrasting climates on antioxidant and bioactive constituents in five medicinal herbs in Western Himalayas. *J. Mt. Sci.* 13:484–492.

Kelp monograph 1426. 2007. *Ph. Eur.*, 6th Edition. Strasbourg, France, Council of Europe, (Vol. 2.4.27).

Keshavarz Afshar, R., M. R. Chaichi, M. Ansari Jovini, E. Jahanzad, and M. Hashemi. 2015. Accumulation of silymarin in milk thistle seeds under drought stress. *Planta.* 242:539–543.

Kim, S., E. J. Yun, M. A. Hossain, H. Lee, and K. H. Kim. 2012. Global profiling of ultraviolet-induced metabolic disruption in *Melissa officinalis* by using gas chromatography-mass spectrometry. *Anal. Bioanal. Chem.* 404:553–562.

King, C. M., J. S. Robinson, and R. W. Cameron. 2012. Flooding tolerance in four "Garrigue" landscape plants: Implications for their future use in the urban landscapes of north-west Europe? *Landsc. Urban Plan.* 107:100–110.

Klímešova, J. 1994. The effects of timing and duration of floods on growth of young plants of *Phalaris arundinacea* L. and *Urtica dioica* L. – An experimental study. *Aquat. Bot.* 48:21–29.

Kochankov, V. G., M. Grzesik, M. Chojnowski, and J. Nowak. 1998. Effect of temperature, growth regulators and other chemicals on *Echinacea purpurea* (L.) Moench seed germination and seedling survival. *Seed Sci. Technol.* 26:547–554.

Korkmaz, A., Ö Değer, and F. Kocaçınar. 2015. Alleviation of water stress effects on pepper seedlings by foliar application of glycinebetaine. *New Zeal. J. Crop Hort. Sci.* 43:18–31.

Kováčik, J., B. Klejdus, and M. Bačkor. 2009. Phenolic metabolis of *Matricaria chamomilla* plants exposed to nickel. *J. Plant Physiol.* 166:1460–1464.

Kováčik, J., F. Stork, B. Klejdus, J. Grúz, and J. Hedbavny. 2012. Effect of metabolic regulators on aluminium uptake and toxicity in *Matricaria chamomilla* plants. *Plant Physiol. Biochem.* 54:140–148.

Kováčik, J., M. Bačkor, and J. Kaduková. 2008. Physiological response of *Matricaria chamomilla* to cadmium and copper excess. *Environ. Toxicol.* 23:123–130.

Kováčik, J., J. Tomko, M. Bačkor, and M. Repčák. 2006. *Matricaria chamomilla* is not a hyperaccumulator, but tolerant to cadmium stress. *Plant Growth Regul.* 50:239–247.

Kráľová, K., and E. Masarovičová. 2003. *Hypericum perforatum* L. and *Chamomilla recutica* (L.) Rausch – Accumulators of some toxic metals. *Pharmazie.* 58:359–360.

Kráľová, K., and E. Masarovičová. 2004. Could complexes of heavy metals with secondary metabolites induce enhanced metal tolerance of *Hypericum perforatum*? In *Macro and Trace Elements. Mengen- und Spurenelemente*, eds M. Anke, et al., 411–416. Jena, Germany, Friedrich Schiller Universität.

Kráľová, K., and E. Masarovičová. 2008. EDTA-assisted phytoextraction of copper, cadmium and zinc using chamomile plants. *Ecol. Chem. Engin.* 15:213–220.

Kráľová, K., E. Masarovičová, and A. Bumbálová. 2000. Toxic effect of cadmium on *Hypericum perforatum* plants and green alga *Chlorella vulgaris. Chem. Inz. Ekol.* 7:1200–1205.

Kráľová, K., E. Masarovičová, I. Ondrejkovičová, and M. Bujdoš. 2007a. Effect of selenium oxidation state on cadmium translocation in chamomile plants. *Chem. Pap.* 61:171–175.

Kráľová, K., E. Masarovičová, and J. Jampílek. 2020. Plant responses to stress induced by toxic metals and their nanoforms. In *Handbook of Plant and Crop Stress.* Expanded edition, 4th Edition, ed. M. Pessarakli. Boca Raton, FL, Taylor & Francis Group. 479–522.

Kráľová, K., E. Masarovičová, J. Kubová, and O. Švajlenová. 2007b. Response of *Matricaria recutita* plants to some copper(II) chelates. *Acta Hortic.* 749:237–243.

Krupa, S. V., and W. J. Manning. 1988. Atmospheric ozone: Formation and effects on vegetation. *Environ. Pollut.* 50(1–2):101–137.

Kulkarni, M., R. Soolanayakanahally, S. Ogawa, Y. Uga, M. G. Selvaraj, and S. Kagale. 2017. Drought response in wheat: Key genes and regulatory mechanisms controlling root system architecture and transpiration efficiency. *Front.Chem.* 5:106.

Kumar, S., S. Sachdeva, K. V. Bhat, and S. Vats. 2018. Plant responses to drought stress: Physiological, biochemical and molecular basis. In *Biotic and Abiotic Stress Tolerance in Plants*, ed. S. Vats, 1–25. Springer Nature Singapore Pte Ltd.

Kundzewicz, Z. W., N. Lugeri, R. Dankers, et al. 2010. Assessing river flood risk and adaptation in Europe – Review of projections for the future. *Mitig Adapt. Strat. Glob. Change.* 15:641–656.

Küpper, H., E. Lombi, F. J. Zhao, and S. P. McGrath. 2000. Cellular compartmentation of cadmium and zinc in relation to other elements in the hyperaccumulator *Arabidopsis halleri. Planta.* 212:75–84.

Kurek, I., T. K. Chang, S. M. Bertain, A. Madrigal, L. Liu, M. W. Lassner, and G. Zhu. 2007. Enhanced thermostability of Arabidopsis Rubisco activase improves photosynthesis and growth rates under moderate heat stress. *Plant Cell.* 19:3230–3241.

Langebartels, C., H. Wohlgemuth, S. Kschieschan, S. Grün, and H. Sandermann. 2002. Oxidative burst and cell death in ozone-exposed plants. *Plant Physiol. Biochem.* 40(6–8):567–575.

Länger, R., C. Mechtler, and J. Jurenitsch. 1996. Composition of essential oils of commercial samples of *Salvia officinalis* L. and *S. fruticosa* Mill.: A comparison of oils obtained by extraction and steam distillation. *Phytochem. Anal.* 7:289–293.

Läuchli, A., and U. Lütthe. 2004. *Salinity: Environment – Plants – Molecules.* New York, NY, Boston, MA, Dordrecht, London, Moscow, Kluwer Academic Publishers.

León-Chan, R. G., M. López-Meyer, T. Osuna-Enciso, J. A. Sañudo-Barajas, J. B. Heredia, and J. León-Félix. 2017. Low temperature and ultraviolet-B radiation affect chlorophyll content and induce the accumulation of UV-B-absorbing and antioxidant compounds in bell pepper (*Capsicum annuum*) plants. *Environ. Exp. Bot.* 139:143–151.

Lešíková, J., K. Kráľová, E. Masarovičová, J. Kubová, and I. Ondrejkovičová. 2007. Effect of different cadmium compounds on chamomile plants. *Acta Hortic.* 749:223–229.

Li, L., C. F. He, Y. M. Dong, H. Zhao, and J. X. Liu. 2006. Effects of environmental stress on ultra-structure of aloe. *Xibei Zhiwu Xuebao.* 26:1940–1945.

Li, Z., H. T. Yang, X. Q. Wu, K. Guo, and J. S. Li. 2015. Some aspects of salinity responses in peppermint (*Mentha* x *piperita* L.) to NaCl treatment. *Protoplasma.* 252:885–899.

Liao, C. T., and C. H. Lin. 1994. Effect of flooding stress on photosynthetic activities of *Momordica charantia. Plant Physiol. Biochem.* 32:479–485.

Liao, C. T., and C. H. Lin. 1995. Effect of flood stress on morphology and anaerobic metabolism of *Momordica charantia. Environ. Exp. Bot.* 35:105–113.

Linkeš, V., J. Kobza, M. Švec, P. Ilka, P. Pavlenda, G. Barančíková, and L. Matúšková. 1997. *Soil Monitoring in Slovakia. Actual State of the Soils In 1992–1996.* [In Slovak]. Research Institute of the Soil Fertility, Bratislava.

Lipiec, J., C. Doussan, A. Nosalewicz, and K. Kondracka. 2013. Effect of drought and heat stresses on plant growth and yield: A review. *Int. Agrophys.* 27:463–477.

Liu, J., R. Zhang, G. Zhang, J. Guo, and Z. Dong. 2017. Effects of soil drought on photosynthetic traits and antioxidant enzyme activities in *Hippophae rhamnoides* seedlings. *J. Forest. Res.* 28:255–263.

Liu, L., Y. M. Fang, S. C. Wang, Y. Xie, and D. D. Yang. 2013. Leaf micro-morphology and features in adsorbing air suspended particulate matter and accumulating heavy metals in seven stress species. *HuanJing Kexue.* 34:2361–2367.

López, D., N. Carazo, and M. C. Rodrigo. 2001. Waterlogging effects on rooting capacity and quality of cuttings: *Rosmarinus officinalis* and *Limonium pectinatum. Acta Hortic.* 559:391–394.

Magiatis, P., A. Michaelakis, A. L. Skaltsounis., and S. A. Haroutounian. 2001. Volatile secondary metabolite pattern of callus cultures of *Chamomilla recutita. Nat. Prod. Lett.* 15:125–130.

Malinowska, E., J. Sosnowski, and K. Jankowski. 2017. Accumulation of some heavy metals by herb plants growing in fields with ponds. *Fresen. Environ. Bull.* 26:2156–2162.

Malinowska, E., K. Jankowski, B. W. K. Wiśniewska-Kadżajan, J. Sosnowski, R. Kolczarek, J. Jankowska, and G. A. Ciepiela. 2015. Content of zinc and copper in selected plants growing along a motorway. *Bull. Environ. Contam. Toxicol.* 95:638–643.

Manivannan, A., P. Soundararajan, S. Muneer, C. H. Ko, and B. R. Jeong. 2016. Silicon mitigates salinity stress by regulating the physiology, antioxidant enzyme activities, and protein expression in *Capsicum annuum* "Bugwang." *BioMed Res. Int.* 2016:1–14.

Manukyan, A. 2013. Effects of PAR and UV-B radiation on herbal yield, bioactive compounds and their antioxidant capacity of some medicinal plants under controlled environmental conditions. *Photochem. Photobiol.* 89:406–414.

Marenco, A., H. Gouget, P. Nédélec, J. P. Pagés, and F. Karcher. 1994. Evidence of a long-term increase in tropospheric ozone from Pic du Midi series: Consequences: Positive radiative forcing. *J. Geophys. Res.* 99:16617–16632.

Marquard, R., and M. Schneider. 1998. Zur Cadmiumproblematik im Arzneipflanzenbau. In *Fachtagung Arznei- und Gewűrzpflanzen*, eds R. Marquard, and E. Schubert, 9–15. Giessen, Germany.

Masarovičová, E., and K. Kráľová. 2007. Medicinal plants – Past, nowadays, future. *Acta Hortic.* 749:19–27.

Masarovičová, E., K. Kráľ'ová, and F. Šeršen. 2011. Plant responses to toxic metal stress. In *Handbook of Plant and Crop Stress*, 3rd Edition, ed. M. Pessarakli, 595–634. Boca Raton, FL, London, New York, NY, CRC, Taylor and Francis Group.

Masarovičová, E., K. Kráľ'ová, F. Šeršeň, A. Bumbálová, and A. Lux. 1999. Effect of toxic effects on medicinal plants. In *Mengen- und Spurelemente. 19. Arbeitstagung*, ed. M. Anke et al., 189–196. Leipzig, Germany: Verlag Harald Schubert.

Masarovičová, E., K. Kráľová, and M. Kummerová. 2010. Principles of classification of medicinal plants as hyperaccumulators or excluders. *Acta Physiol. Plant.* 32:823–829.

Masarovičová, E., K. Kráľová, and V. Streško. 2003. Effect of metal ions on some medicinal plants. *Chem. Inz. Ekol.* 10:275–279.

Masarovičová, E., K. Kráľová, and V. Streško. 2004. Comparative study of uptake, accumulation and some effects of cadmium in two cultivars of *Salvia officinalis* L. *Chem. Inz. Ekol.* 11:209–214.

Mathur, S., D. Agrawal, and A. Jajoo. 2014. Photosynthesis: Response to high temperature stress. *J. Photochem. Photobiol. B Biol.* 137:116–126.

Mäkelä, P., J. Karkkainen, and S. Somersalo. 2000. Effect of glycinebetaine on chloroplast ultrastructure, chlorophyll and protein content, and RuBPCO activities in tomato grown under drought or salinity. *Biol. Plant.* 43:471–475.

McKay, D. L., and J. B. Blumberg. 2006. A review of the bioactivity and potential health benefits of chamomile tea (*Matricaria recutita* L.). *Phytother. Res.* 20:519–530.

Mehrizi, M. H., H. Shariatmadari, H. Khoshgoftarmanesh, and F. Dehghani. 2012. Copper effects on growth, lipid peroxidation, and total phenolic content of rosemary leaves under salinity stress. *J. Agric. Sci. Tech.* 14:205–212.

Mills, F. A. 2000. Botany. In *Encyclopedia of Monasticism: M–Z*, ed. W. M. Johnston, 179. Taylor and Francis.

Mosaleeyanon, K., S. M. A. Zobayed, F. Afreen, and T. Kozai. 2005. Relationships between net photosynthetic rate and secondary metabolite contents in St. John's wort. *Plant Sci.* 169:523–531.

Mugnai, S., E. Azzarello, C. Pandolfi, and S. Mancuso. 2006. Zinc and cadmium accumulation in *Hyssopus officinalis* L. and *Satureja montana* L. *Acta Hortic.* 723:361–366.

Munné-Bosch, S., and L. Alegre. 1999. Role of dew on the recovery of water-stressed *Melissa officinalis* L. plants. *J. Plant Physiol.* 154:759–766.

Munné-Bosch, S., and L. Alegre. 2000. The significance of β-carotene, α-tocopherol and the xanthophyll cycle in droughted *Melissa officinalis* plants. *Functional Plant Biol.* 27:139–146.

Munné-Bosch, S., and L. Alegre. 2003. Drought-induced changes in the redox state of α-tocopherol, ascorbate, and the diterpene carnosic acid in chloroplasts of *Labiatae* species differing in carnosic acid contents. *Plant Physiol.* 131:1816–1825.

Munns, R., and M. Tester. 2008. Mechanisms of salinity tolerance. *Ann. Rev. Plant Biol.* 59:651–681.

Murch, S. J., K. Haq, H. P. V. Rupasinghe, and P. K. Saxena. 2003. Nickel contamination affects growth and secondary metabolite composition of St. John's wort (*Hypericum perforatum* L.). *Environ. Exp. Bot.* 49:251–257.

Murphy, A., and L. Taiz. 1995. A new vertical mesh transfer technique for metal tolerance studies in *Arabidopsis* ecotypic variation and copper-sensitive mutants. *Plant Physiol.* 108:29–38.

Murti, K., M. A. Panchal, V. Gajera, and J. Solanki. 2012. Pharmacological properties of *Matricaria recutita*: A review. *Pharmacologia.* 3(8):348–351.

Nahrstedt, A., and V. Butterweck. 1997. Biologically active and other chemical constituents of the herb of *Hypericum perforatum* L. *Pharmacopsychiatry.* 30:129–134.

Németh-Zámbori, É., Z. Pluhár, K. Szabó, et al. 2016. Effect of water supply on growth and polyphenols of lemon balm (*Melissa officinalis* L.) and thyme (*Thymus vulgaris* L.). *Acta Biol. Hung.* 67:64–74.

Nowack, B., R. Schulin, and B. H. Robinson. 2006. Critical assessment of chelant-enhanced metal phytoextraction. *Environ. Sci. Technol.* 40:5225–5232.

Oliveira, G. C., W. L. Vieira, S. C. Bertolli, and A. C. Pacheco. 2016. Photosynthetic behavior, growth and essential oil production of *Melissa officinalis* L. cultivated under colored shade nets. *Chilean J. Agric. Res.* 76:123–128.

Ozturk, A., A. Unlukara, A. Ipek, and B. Gurbuz. 2004. Effects of salt stress and water deficit on plant growth and essential oil content of lemon balm (*Melissa officinalis* L.). *Pak. J. Bot.* 36:787–792.

Pacheco, A. C., P. R. Camargo, and C. G. M. Souza. 2011. Water deficit and ABA application on leaf gas exchange and flavonoid content in marigold (*Calendula officinalis* L.). *Acta Sci. Agron.* 33:275–281.

Palivan, C. G., G. Gescheidt, and L. Weiner. 2001. The formation of copper complexes with hypericin, in solutions: An EPR Study. *J. Inorg. Biochem.* 86:369–369.

Pandey, S., K. Gupta, and A. K. Mukherjee. 2007. Impact of cadmium and lead on *Catharanthus roseus* – A phytoremediation study. *J. Environ. Biol.* 28:655–662.

Parent, C., N. Capelli, A. Berger, M. Crèvecoeur, and J. F. Dat. 2008. An overview of plant responses to soil waterlogging. *Plant Stress.* 2:20–27.

Parmenter, G. A., L. C. Burton, and R. P. Littlejohn. 1996. Chilling requirement of commercial *Echinacea* seed. *New Zeal. J. Crop Hort. Sci.* 24:109–114.

Pascual, J., M. Rahikainen, and S. Kangasjärvi. 2017. Plant light stress. In *eLS*. Chichester, UK, John Wiley & Sons Ltd. https://doi.org/10.1002/9780470015902.a0001319.pub3.

Pavlova, D., I. Karadjova, and I. Krasteva. 2015. Essential and toxic element concentrations in *Hypericum perforatum*. *Aust. J. Bot.* 63:152–158.

Pavlovič, A., E. Masarovičová, K. Kráľová, and J. Kubová. 2006. Response of chamomile plants (*Matricaria recutita* L.) to cadmium treatment. *Bull. Environ. Contam. Toxicol.* 77:763–771.

Pellegrini, E., A. Francini, G. Lorenzini, and C. Nali. 2015. Ecophysiological and antioxidant traits of *Salvia officinalis* under ozone stress. *Environ. Sci. Pollut. Res.* 22:13083–13093.

Pellegrini, E., A. Trivellini, A. Campanella, A. Francini, G. Lorenzini, C. Nali, and P. Vernieri. 2013. Signaling molecules and cell death in *Melissa officinalis* plants exposed to ozone. *Plant Cell Rep.* 32:1965–1980.

Perry, N. B., R. E. Anderson, N. J. Brennan, M. H. Douglas, A. J. Heaney, J. A. McGimpsey, and B. M. Smallfield. 1999. Essential oils from dalmatian sage (*Salvia officinalis* L.): Variations among individuals, plant parts, seasons, and sites. *J. Agric. Food Chem.* 47:2048–2054.

Petrovska, B. B. 2012. Historical review of medicinal plants´ usage. *Pharmacogn. Rev.* 6:1–5.

Phukan, U. J., S. Mishra, K. Timbre, S. Luqman, and R. K. Shukla. 2014. *Mentha arvensis* exhibit better adaptive characters in contrast to *Mentha piperita* when subjugated to sustained waterlogging stress. *Protoplasma.* 251:603–614.

Pospíšil, P. 2016. Production of reactive oxygen species by photosystem II as a response to light and temperature stress. *Front. Plant Sci.* 7:1–7.

Qin, J., K. N. He, G. D. Tan, Z. L. Wang, and J. Chen. 2009. Effects of NaCl stress on *Hippophae rhamnoides* and *Shepherdia argentea* seedlings growth and photosynthetic characteristics. *Yin Yong Shengtai Xuebao.* 20:791–797.

Radácsi, P., K. Szabó, D. Szabó, E. Trócsányi, and É Németh-Zámbori. 2016. Effect of water deficit on yield and quality of lemon balm (*Melissa officinalis* L.). *Zemdirbyste Agric.* 103:385–390.

Radanovic, D., S. Antic-Mladenovic, and M. Jakovljevic. 2002. Influence of some soil characteristics on heavy metal content in *Hypericum perforatum* L. and *Achillea millefolium* L. *Acta Horticult.* 576:295–301.

Radanovic, D., S. Antic-Mladenovic, M. Jakovljevic, and S. Maksimovic. 2001. Content of Pb, Ni, Cr, Cd and Co in peppermint (Mentha piperita L.) cultivated on different soil types from Serbia. *Rost. Výroba.* 47:111–116.

Radulescu, C., C. Stihi, I. V. Popescu, et al. 2013. Assessment of heavy metals level in some perennial medicinal plants by flame atomic absorption spectrometry. *Rom. Rep. Phys.* 65:246–260.

Rahimi, S., and T. Hasanloo. 2016. The effect of temperature and pH on biomass and bioactive compounds production in *Silybum marianum* hairy root cultures. *Res. J. Pharmacogn.* 3:53–59.

Rangan, P., R. Subramani, R. Kumar, A. K. Singh, and R. Sing. 2014. Recent advances in polyamine metabolism and abiotic stress tolerance. *Biomed. Res. Int.* 214:1–9.

Repčák, M., A. Eliášová, and A. Ruscančinová. 1998. Production of herniarin by diploid and tetraploid *Chamomilla recutita*. *Pharmazie.* 53:278–279.

Roy, S. J., S. Negrão, and M. Tester. 2014. Salt resistant crop plants. *Curr. Opin. Biotechnol.* 26:115–124.

Rozema, J., and T. Flowers. 2008. Ecology: Crops for a salinized world. *Science.* 322:1478–1480.

Ruan, C. J., and Q. L. Xie. 2002. Osmotic adjustment effect of *Hippophae rhamnoides* L. under salt-stress. *J. Plant Resour. Environ.* 11:45–47.

Sabra, A., F. Daayf, and S. Renault. 2012. Differential physiological and biochemical responses of three *Echinacea* species to salinity stress. *Sci. Hortic.* 135:23–31.

Safikhan, S., M. R. Chaichi, K. Khoshbakht, A. Amini, and B. Motesharezadeh. 2018. Application of nanomaterial graphene oxide on biochemical traits of Milk thistle (*Silybum marianum* L.) under salinity stress. *Aust. J. Crop Sci.* 12:931–936.

Sairam, R. K., and A. Tyagi. 2004. Physiology and molecular biology of salinity stress tolerance in plants. *Curr. Sci.* 86:407–421.

Salman Khan, M., M. Zaka, B. Haider Abbasi, L. Rahman, and A. Shah. 2016. Seed germination and biochemical profile of *Silybum marianum* exposed to monometallic and bimetallic alloy nanoparticles. *IET Nanobiotechnol.* 10:359–366.

Saradhi, P. P., I. Suzuki, A. Katoh, A. Sakamoto, P. Sharmila, D. -J. Shi, and N. Murata. 2000. Protection against the photoinduced inactivation of the photosystem II complex by abscisic acid. *Plant Cell Environ.* 23:711–718.

Saravanan, S., and S. Karthi. 2014. Effect of elevated CO_2 on growth and biochemical changes in *Catharanthus roseus* – A valuable medicinal herb. *J. Pharm. Pharm. Sci.* 3:411–422.

Saslis-Lagoudakis, C. H., V. Savolainen, E. M. Williamson, et al. 2012. Phylogenies reveal predictive power of traditional medicine in bioprospecting. *Proc. Natl Acad. Sci. USA.* 109:15835–15840.

Sauter, M. 2013. Root responses to flooding. *Curr. Opin. Plant Biol.* 16:282–286.

Sedghi, M., R. S. Sharifi, A. R. Pirzad, and B. Amanpour-Balaneji. 2012. Phytohormonal regulation of antioxidant systems in petals of drought stressed pot marigold (*Calendula officinalis* L.). *J. Agric. Sci. Technol.* 14:869–878.

Sengupta, S., B. Patra, S. Ray, and A. L. Majumder. 2008. Inositol methyl transferase from a halophytic wild rice, *Porteresia coarctata* Roxb. (Tateoka): Regulation of pinitol synthesis under abiotic stress. *Plant Cell Environ.* 31:1442–1459.

Shanker, K., S. Mishra, S. Srivastava, R. Srivastava, S. Dass, S. Prakash, and M. M. Srivastava. 1996. Effect of selenite and selenate on plant uptake of cadmium by maize (*Zea mays*). *Bull. Environ. Contam. Toxicol.* 56:419–424.

Shekari, L., M. M. Kamelmanesh, M. Mozafariyan, M. Hasanuzzaman, and F. Sadeghi. 2017. Role of selenium in mitigation of cadmium toxicity in pepper grown in hydroponic condition. *J. Plant. Nutr.* 40:761–772.

Shepherd, M., R. Wood, C. Bloomfield, and C. Raymond. 2015. Ecotypic responses to flood and drought in tea tree (*Melaleuca alternifolia*). *Crop Pasture Sci.* 66:864–876.

Sherman, P. W., and G. A. Hash. 2001. Why vegetable recipes are not very spicy. *Evol. Hum. Behav.* 22:147–163.

Schippmann, U., D. J. Leaman, and A. B. Cunningham. 2002. Impact of cultivation and gathering of medicinal plants on biodiversity: Global trends and issues. In *FAO. 2002. Biodiversity and the Ecosystem Approach in Agriculture, Forestry and Fisheries. Satellite event on the occasion of the Ninth Regular Session of the Commission on Genetic Resources for Food and Agriculture*, Rome, October 12–13, 2002, 1–21.

Schlicher, H. 1973. Neuere Erkenntnisse bei der Qualitätsbeurteilung von Kamillenbluten bzw. Kamillenöl. Teil 2: Qualitative Beurteilung de Ätherischen Öles *in Flores Chamomillae*. Aufteilung der Handelskamillen in vier, bzw. fünf chemischen Typen. *Planta Med.* 28:133–144.

Schulz, H., S. Härtling, and H. Tanneberg. 2008. The identification and quantification of arsenic-induced phytochelatins – Comparison between plants with varying As sensitivities. *Plant Soil.* 303:275–287.

Sing, J., and K. Thakur. 2018. Photosynthesis and abiotic stress in plants. In *Biotic and Abiotic Stress Tolerance in Plants*, ed. S. Vats, 27–46. Springer Nature Singapore Pte Ltd.

Singer, C. 1923. Herbals. *Edinburg Rev.* 237:95–112.

Singh, O., Z. Khanam, N. Misra, and M. K. Srivastava. 2011. Chamomile (*Matricaria chamomilla* L.): An overview. *Pharmacogn. Rev.* 5(9):82–95.

Smith-Hall, C., H. O. Larsen, and M. Pouliot. 2012. People, plants and health: A conceptual framework for assessing changes in medicinal plant consumption. *J. Ethnobiol. Ethnomed.* 8:43.

Song, Y., Q. Chen, D. Ci, X. Shao, and D. Zhang. 2014. Effects of high temperature on photosynthesis and related gene expression in poplar. *BMC Plant Biol.* 14:111.

Staehelin, J., J. Thudium, R. Buehler, A. Volz-Thomas, and W. Graber. 1994. Trends in surface ozone concentrations at Arosa (Switzerland). *Atmos. Environ.* 28:75–87.

Stetsenko, L. A., A. D. Kozhevnikova, and A. V. Kartashov. 2017. Salinity attenuates nickel-accumulating capacity of *Atropa belladonna* L. plants. *Russ. J. Plant Physiol.* 64:486–496.

Stuhlfauth, T., B. Steuer, and H. Fock. 1990. Chlorophylls and carotenoids under water-stress and their relation to primary metabolism. *Photosynthetica.* 24:412–418.

Sumner, J. 2008. *The Natural History of Medicinal Plants*. Timber Press, 252, ISBN 978-088-192-95-77.

Szymańska, R., I. Ślesak, A. Orzechowska, and J. Kruk. 2017. Physiological and biochemical responses to high light and temperature stress in plants. Review. *Environ. Exp. Bot.* 139:165–177.

Šalamon, I., K. Kráľová, and E. Masarovičová. 2007. Accumulation of cadmium in chamomile plants cultivated in Eastern Slovakia regions. *Acta Hortic.* 749:217–222.

Tack, F. M., and M. G. Verloo. 1996. Metal contents in stinging nettle (*Urtica dioica* L) as affected by soil characteristics. *Sci. Total Environ.* 192:31–39.

Taiz, L., and E. Zeiger. 2010. *Plant Physiology*, 5th Edition. Sunderland, MA, Sinauer Associates Inc., 782.

Taiz, L., E. Zeiger, I. M. Moller, and A. Murphy. 2014. *Plant Physiology and Development*, 6th edition. New York, NY, Oxford University Press Inc., 761.

Takahashi, D., B. Li, T. Nakayama, Y. Kawamura, and M. Uemura. 2013. Plant plasma membrane proteomics for improving cold tolerance. *Front. Plant Sci.* 4:1–5.

Takahashi, S., and M. R. Badger. 2011. Photoprotection in plants: A new light on photosystem II damage. *Trends Plant Sci.* 16:53–60.

Tapia, Y., O. Salazar, F. Nájera, A. Gárate, E. Eymar, and A. Masaguer. 2014. Accumulation of Mn in leaves of *Rosmarinus officinalis* cultivated in substrates of pine bark. *Commun. Soil Sci. Plant Anal.* 45:1961–1973.

Tapsell, L. C., I. Hemphill, L. Cobiac, et al. 2006. Health benefits of herbs and spices: The past, the present, the future. *Med. J. Aust.* 185:S4–24.

Tarhan, L., and B. Kavakcioglu. 2016. Glutathione metabolism in *Urtica dioica* in response to cadmium based oxidative stress. *Biol. Plant.* 60:163–172.

Tatsis, E. C., S. Boeren, V. Exarchou, A. N. Troganis, J. Vervoort, and I. P. Gerothanassis. 2007Identification of the major constituents of Hypericum perforatum by LC/SPE/NMR and/or LC/MS. *Phytochemistry.* 68:383–393.

Theocharis, A., C. Clément, and E. A. Barka. 2012. Physiological and molecular changes in plants grown at low temperatures. *Planta.* 235:1091–1105.

Tilburt, J. C., and T. J. Kaptchuk. 2008. Herbal medicine research and global health: An ethical analysis. *Bull. World Health Organ.* 86:577–656.

Tobyn, G., A. Denham, and M. Whitelegg. 2010. *The Western Herbal Tradition: 2000 Years of Medicinal Plant Knowledge.* London, Churchill Livingstone.

Tonelli, M., E. Pellegrini, F. D'Angiolillo, M. Petersen, C. Nali, L. Pistelli, and G. Lorenzini. 2015. Ozone-elicited secondary metabolites in shoot cultures of *Melissa officinalis* L. *Plant Cell Tissue Organ Cult.* 120:617–629.

Torabi, F., A. Majd, and S. Enteshari. 2015. The effect of silicon on alleviation of salt stress in borage (*Borago officinalis* L.). *Soil Sci. Plant Nutr.* 61:788–798.

Torabi, F., A. Majd, S. Enteshari, S. Irian, and S. Nabiuni. 2013. Effects of salinity on the development of hydroponically grown borage (*Borago officinalis* L.) male gametophyte. *Not. Bot. Horti Agrobot. Cluj Napoca.* 41:65–72.

Tyystjärvi, E. 2008. Photoinhibition of photosystem II and photodamage of the oxygen evolving manganese cluster. *Coord. Chem. Rev.* 252:361–376.

Ulrich, B., and M. E. Sumner. 1991. *Soil Acidity.* Berlin, Heidelberg, Springer Verlag, 5–6.

Vaculík, M., Ľ. Jurkovič, P. Matejkovič, M. Molnárová, and A. Lux. 2013. Potential risk of arsenic and antimony accumulation by medicinal plants naturally growing on old mining sites. *Water Air Soil Poll.* 224:1546.

Val, T. 2015. Medical and magical treasures in Anglo-Saxon herbals. http://www.norwichsphere.org.uk/essays/anglosaxonherbtreasures.pdf.

Valera, R. E., and G. R. Garcia. 2014. Leaf anatomy in sweet pepper (*Capsicum annuum* L.) plants under saline stress. *Rev. Fac. Agron. Univ. Zulia.* 31:281–287.

Vance, G. F. 2006. Acid rain and N deposition. In *Encyclopedia of Soil Science*, Vol. 1, 2nd edition. ed. R. Lal, 5–10. Taylor and Francis.

Vass, I. 2012. Molecular mechanisms of photodamage in the photosystem II complex. *Biochim. Biophys. Acta.* 1817:209–217.

Wahid, A., M. Farooq, I. Hussain, R. Rasheed, and S. Galani. 2012. Responses and management of heat stress in plants. In *Environmental Adaptations and Stress Tolerance of Plants in the Era of Climate Change*, eds P. Ahmad, and M. N. V. Prasad, 183–205.New York, NY, Springer.

Wahid, A., S. Gelani, M. Ashraf, and M. R. Foolad. 2007. Heat tolerance in plants: An overview. *Environ. Exp. Bot.* 61:199–223.

Wang, G. P., X. Y. Zhang, F. Li, Y. Luo, and W. Wang. 2010. Over accumulation of glycine betaine enhances tolerance to drought and heat stress in wheat leaves in the protection of photosynthesis. *Photosynthetica.* 48:117–126.

Whanger, P. D. 1992. Selenium in the treatment of heavy-metal poisoning and chemical carcinogenesis. *J. Trace Elem. Electrolytes Health Dis.* 6:209–221.

Wohlgemuth, H., K. Mittelstrass, S. Kschieschan, et al. 2002. Activation of an oxidative burst is a general feature of sensitive plants exposed to the air pollutant ozone. *Plant Cell Environ.* 25(6):717–726.

World Health Organization. 2005. *National Policy on Traditional Medicine and Regulation of Herbal Medicines. Report of a WHO Global Survey 2005.*

Xu, M., Y. Zhu, J. F. Dong, et al. 2012. Ozone induces flavonol production of *Ginkgo biloba* cells dependently on nitrate reductase-mediated nitric oxide signaling. *Environ. Exp. Bot.* 75:114–119.

Xu, M. J., B. Yang, J. F. Dong, et al. 2011. Enhancing hypericin production of *Hypericum perforatum* cell suspension culture by ozone exposure. *Biotechnol. Prog.* 27:1101–1106.

Xu, Y., G. Wang, F. Cao, C. Zhu, G. Wang, and Y. A. El-Kassaby. 2014. Light intensity affects the growth and flavonol biosynthesis of ginkgo (*Ginkgo biloba* L.). *New Forest.* 45:765–776.

Yadav, S. K. 2010. Cold stress tolerance mechanisms in plants. A review. *Agron. Sustain. Dev.* 30:515–527.

Yadav, S. K. 2011. Cold stress tolerance mechanisms in plants. In *Sustainable Agriculture*, eds E. Lichtfouse, M. Hamelin, M. Navarrete, and P. Debaeke, 605–620. Springer Science+Business Media B.V., Dordrecht, Heidelberg, London, New York, NY.

Yan, K., X. Y. He, W. Chen, T. Lu, and K. Yan. 2010. Ozone-induced changes in antioxidant systems of *Ginkgo biloba* in relation to the developmental stage of the leaves. 2010 4TH International Conference on Bioinformatics and Biomedical Engineering (ICBBE 2010), Chengdu, Peoples R China, Jun 18–20, 2010. Book Series: International Conference on Bioinformatics and Biomedical Engineering.

Yan, Z. Z., J. Chen, and X. Z. Li. 2013. Methyl jasmonate as modulator of Cd toxicity in *Capsicum frutescens var. fasciculatum* seedlings. *Ecotoxicol. Environ. Saf.* 98:203–209.

Yang, B. Z., Z. B. Liu, S. D. Zhou, et al. 2016. Exogenous Ca^{2+} alleviates waterlogging-caused damages to pepper. *Photosynthetica.* 54:620–629.

Yang, Y., Y. Yao, and X. Zhang. 2010. Comparison of growth and physiological responses to severe drought between two altitudinal *Hippophae rhamnoides* populations. *Silva Fenn.* 44:603–614.

Yang, Y. Q., Y. Yao, G. Xu, and C. Y. Li. 2005. Growth and physiological responses to drought and elevated ultraviolet-B in two contrasting populations of *Hippophae rhamnoides*. *Physiol. Plant.* 124:431–440.

Yu, W. W., F. L. Cao, and G. L. Wu. 2012. Leaf growth and medicinal quality of ginkgo seedling with spraying Mg, Zn, Mo fertilizers. *Xibei Zhiwu Xuebao.* 32:1214–1221.

Zhang, C. J., J. Q. Guo, G. X. Chen, and H. C. Xie. 2005. Effects of high temperature and/or drought on growth and secondary metabolites in *Ginkgo biloba* leaves. *Rural Eco Environ.* 21(3):11–15.

Zhang, J. L., and H. Shi. 2013. Physiological and molecular mechanisms of plant salt tolerance. *Photosynth. Res.* 115:1–22.

Zhang, W. W., T. H. Zhao, M. Y. Wang, X. Y. He, and S. L. Fu. 2007. Effects of elevated ozone concentration on *Ginkgo biloba* photosynthesis. *Shengtaixue Zazhi.* 26(5):645–649.

Zhang, Z. Z., Q. S. Guo, and Q. S. Shao. 2009. Effects of waterlogging stress on physiological and biochemical characteristics of medicinal *Chrysanthemum morifolium* during seedling stage. *Zhongguo Zhongyao zazhi. China. J. Chin. Mater. Med.* 34(18):2285–2299.

Zobayed, S., and P. K. Saxena. 2004. Production of St. John's Wort plants under controlled environment for maximizing biomass and secondary metabolites. *In Vitro Cell. Dev. Biol. Plant.* 40:108–114.

Zobayed, S. M. A., F. Afreen, and T. Kozai. 2005. Temperature stress can alter the photosynthetic efficiency and secondary metabolite concentrations in St. John's wort. *Plant Physiol. Biochem.* 43:977–984.

Zollinger, N., R. Kjelgren, T. Cerny-Koenig, K. Kopp, and R. Koenig. 2006. Drought responses of six ornamental herbaceous perennials. *Sci. Hortic.* 109:267–274.

40 Citrus Plant Botanic Characteristics and Its Abiotic and Biotic Stress

Hong Li

CONTENTS

ABBREVIATIONS

NF   non-flooding
NL   non-larval feeding
SFS   Smooth Flat Seville
SWI   Swingle Citrumelo
THR   tree health rating.

40.1 CITRUS PLANT BOTANIC CHARACTERISTICS

40.1.1 CITRUS TREE BOTANY

Citrus (*Citrus sinensis* (L.) Osb.), a warm climate fruit tree in the *Rutaceae* family, sub-family *Aurantoideae*, is the most economically important fruit tree crop in the world (Rieger, 2005; Syvertsen and Hanlin, 2008; Agusti et al., 2013; Albrecht et al., 2017). The four most important fruit crops in order of economic importance are citrus, grape, banana, and apple. Citrus production can be found mainly in warm subtropical areas. Citrus is the main fruit tree crop in the world with production of 104 million tons grown on 7.1 million ha (Agusti et al., 2013). In 2015, the total world production of citrus had increased by 16.4% to 121 million tons (FAO, 2016). Citrus is largely cultivated in tropical and subtropical regions in the world, and sweet oranges are the most important citrus production. Geographically, the largest producers of citrus are China (24.4%), Brazil (15.6%), and the USA (7.7%) based on the FAO statistics (2016). In North America,

citrus production can be found in Florida (from Orlando to the south), California, and Texas. Florida produces 82% of the total citrus production in North America (Rieger, 2005).

While *Citrus* is by far the most economically important genus, the *Citrus* genus originated in tropical and subtropical Southeast Asia (Rieger, 2005). There are six citrus groups of economic interest: sweet orange (*C. sinensis* (L.) Osb.), common mandarin (*C. reticulata* Blanco), Satsuma mandarin (*C. unshiu* Marc.), grapefruit (*C. paradisi* Macf.), lemon (*C. limon* Burm.), and lime (*C. aurantifolia* L.) in addition to a group of hybrids (Agusti et al., 2013). Two genera contain species important in citriculture. They are *Fortunella* spp. (kumquats): evergreen small trees (2–4 m), native to southern China; and *Poncirus trifoliata* L. Raf. of trifoliate orange, which was used as a male parent in production of citrange (sweet orange × trifoliate orange) rootstocks (Rieger, 2005).

Major hybrids of citrus are five categories that are commercially important. Among them, they are orange, tangerine, grapefruit, lemon, and limes. Orange can be divided into two categories: sweet orange (*Citrus* × *sinensis*) and sour orange (*Citrus* × *aurantium*); grapefruit (*Citrus* × *paradisi*); lemon (*Citrus* × *limon*), and lime can also be divided into Rangpur lime (*Citrus* × limonia), and Persian lime (*Citrus* × *latifolia*). Large sweet juicy hybrids of tangerine and grapefruit have a thick wrinkled skin (Gmitter, 2004; Rieger, 2005).

Most cultivars of citrus are self-pollinated. Cross-pollination is necessary only for some tangerines and tangerine hybrids (Agusti et al., 2013). The ovary position is superior, and flowers are white and have five petals and sepals fused at the base to form a small cup (Rieger, 2005). Flower volatiles vary qualitatively and quantitatively among citrus types during blooming. It is reported that flower volatiles can be suitable markers for revealing the genetic relationships between citrus cultivars (Azam et al., 2013). Up to 110 volatiles can be detected in citrus flowers, and the highest amounts of volatiles are present in fully opened flowers of most citrus, except for pomelos. All cultivars were characterized by a high percentage of either oxygenated monoterpenes or monoterpene hydrocarbons, and the presence of a high percentage of nitrogen-containing compounds was also observed (Azam et al., 2013). The citrus ovary is compound with 10–14 locules. In the fruit, the endocarp is divided into 10–14 segments separated by thin septa. Each segment is composed of juice and seeds with long stalks attached to the outer wall (Rieger, 2005).

As a warm climate tree crop, the citrus plant is adapted to grow in a wide variety of natural conditions (Falivene et al., 2006; Ashok et al., 2008; Albrecht et al., 2017). Most of the citrus varieties are subject to damage in cool temperatures and citrus trees perform best in subtropical climates, with slight changes of season, but little or no chance of freezing weather (Grosser and Chandler, 2004). Flowers, leaves, and stems can be lost to a freezing temperature, depending on the stage of acclimation, species, and age of tissue of the trees (Rieger, 2005).

Citrus plants can grow well in a wide variety of soil types, and the tree fruit crop is tolerant more or less to high (or low) soil pH or salinity (Fernandez-Ballester, et al., 2003; Alva et al., 2004; Li et al., 2007a, 2007b). The trees can be found growing in many soil types from pure sand (i.e., soils containing as many as 96% of sand in central Florida), organic muck (i.e., near the Everglades in Florida), and loamy, heavy, high pH soils (i.e., in Texas and in the San Joaquin Valley of California) (Obeza and Collins, 2002; Alva et al., 2004; Rieger, 2005; Syvertsen and Hanlin, 2008). Yet, recent studies have found that soil condition based rootstock alternatives, site-specific nutrient management, and integrated soil management systems are the viable means of combating an untimely decline in citrus orchards' productivity (Li et al., 2005; Li, 2009; Srivastava and Singh, 2009; Agusti et al., 2013).

All citrus are non-climacteric fruits, meaning that the fruits have already done most of their ripening on the plant and will slowly begin to abscise from the trees after picking. The external color changes during ripening. The color of citrus fruits only develops in climates with a (diurnal) cool winter (Rieger, 2005; Agusti et al., 2013). In tropical regions with no winter (i.e., in Hainan Island, citrus fruits remain green even until maturity. The lime plant is extremely sensitive to cool conditions; thus it is usually never exposed to cool enough conditions to develop color (Rieger, 2005; Syvertsen and Hanlon, 2008).

The major diseases in citrus include the largely reported citrus canker (*Xanthomonas axonopodis*), Huanglongbing (HLB, *Candidatus Liberibacter asiaticus*), phytophthora, bacterial spot (*Xanthomonas campestris* pv. *citrumelo*), and

black pit (*Pseudomonas syringae*) (Graham et al., 2003; Bove, 2006; De Souza et al., 2009; Castle and Futch, 2015). The major insects in citrus are western leaf-footed bugs, citrus aphids, citrus blackfly, root weevils, etc. (Jones and Schroeder, 1983; Lapointe, 2000; Rieger, 2005). *Diaprepes abbreviatus* (L.) (Coleoptera: Curculionidae) root weevils were originally from the Caribbean. Yet, the root weevil has become a major pest of citrus in recent years (Nigg et al., 2001, 2003; McCoy et al., 2003; Li et al., 2004, 2007a; Stuart et al., 2004). Phytophthora and *Diaprepes* root weevil complex have been reported in Florida (Graham et al., 2003).

Cultivars producing fruit with improved quality over an extended harvesting season has been the new technology development for increasing demand and consumption of citrus products (Grosser and Chandler, 2004). Also, the development of seedless fresh fruit cultivars of citrus is the new research effort to enable the expansion of the domestic and international marketplace for fresh fruit (Gmitter, 2004). Greater adaptation of rootstock cultivars, coupled with higher levels of potential productivity, late maturity, and cold-hardiness, has enabled growers to minimize inputs and maximize profit margins. Resistance to disease, pests, and other stresses can result in similar benefits (Gmitter, 2004).

Citrus fruits obtain their highest internal quality (juice content, sugar, and acid levels) in Florida subtropical humid climates, while irrigated citrus fruits achieve the best external quality in California Mediterranean climate (Rieger, 2005). However, citrus species generally do not tolerate high soil acidity, suboptimum nutrient levels, or flooding for more than a few days without injury (Blazquez, 1991; Ashok et al., 2008; Syvertsen and Hanlon, 2008; Agusti et al., 2013). Florida citrus soils range from well-drained Entisols on relatively high, rolling landscapes to poorly-drained Alfisoils and Spodosols on low-lying flatwoods (Obreza and Collins, 2002). Most flatwood soils contain high levels of active hydrogen because of high rainfall, and aluminum from the soil reacting with water to give free hydrogen (Li et al., 2007a, 2007b).

Although citrus plants are adapted to grow in a wide variety of soil and water conditions, the trees can be injured from anaerobic disturbance, oxygen deprivation, and root injury that can lead to tree decline (Figure 40.1).

When environmental waterlogging, soil acidity, and pest infestation occur simultaneously, trees in the entire orchard can severely decline within a short time (Figure 40.2).

40.1.2 CITRUS ROOTSTOCKS

In citrus production, an important factor to consider is the choice of rootstock (Gmitter, 2004; Castle and Futch, 2015; Albrecht et al., 2017). One plant is selected for its roots, and this is called the stock or rootstock. The other plant is selected for its stems, leaves, flowers, or fruits and is called the scion (Rieger, 2005). Citrus rootstocks are the roots and lower trunk of a citrus tree used to grow a variety that cannot be propagated by other seedlings. A citrus plant grown from a seed is called a rootstock. To produce productive citrus trees, the tree materials (a bud or a shoot) must be budded or grafted onto

FIGURE 40.1 (See color insert.) A declined, unproductive citrus tree (a 20-year-old 'Hamlin' orange tree on Swingle Citrumelo rootstock (*Citrus paradisi* Macfad. × *Poncirus trifoliata* (L.) Raf.) showing its fruits at harvest time in 2003 in Mollisol in a low-lying, poorly-drained citrus orchard in Osceola County, central Florida.

FIGURE 40.2 (See color insert.) The layout of the declined citrus trees resulting from water stress and pest infestation in Mollisol in a low-lying, poorly-drained orchard in Osceola County, central Florida. The trees were 20-year-old 'Hamlin' orange trees on Swingle Citrumelo rootstock and under the declined tree canopy was a pyramidal Tedders trap for the root weevil monitoring in summer 2013.

a citrus rootstock. Certain citrus varieties have been selected as seedling rootstocks (Grosser and Chandler, 2004; Gmitter, 2004). Choosing the right rootstock and scion combination can result in higher economic returns (Syvertsen et al., 2000; Castle and Futch, 2015; Albrecht et al., 2017).

Budding and grafting are effective methods of plant propagation widely used in tree fruit production. The only difference between budding and grafting is the size (or form) of the scion. Grafting is using a shoot, and in budding only one single bud is used. The tissues of one variety of plant are encouraged to fuse with those of another variety of plant. In citrus, within days after budding or grafting, new, continuous cambium forms, and new vascular tissue forms from cambiums. It is an easy and fruitful method of production because cutting the scion of an easily rooted plant can be easily propagated (Rieger, 2005; Agusti et al., 2013).

Rootstocks affect scion vigor, yield, fruit size, juice quality, and pest tolerance (Albrecht et al., 2017). Reasons for budding or grafting citrus trees include inducing resistance to disease, virus, and insect attacks, and grafting onto hardier plants for cold tolerance (Grosser and Chandler, 2004; Castle and Futch, 2015). Budded/grafted citrus trees are of high yield, and better quality as grafted trees accelerate flowering and fruiting to produce more high-quality fruits per unit area (Syvertsen et al., 2000; Castle and Futch, 2015). Other reasons for budding and grafting rootstocks are for tree adaptation to environmental limitations such as excessive soil acidity and salinity (Fernandez-Ballester, 2003; Albrecht et al., 2017).

The development of new citrus scion and rootstock cultivars is essential for the viability and profitability of citrus industry (Gmitter, 2004; Castle and Futch, 2015). New rootstock hybrids can be produced by sexual and somatic hybridization, screened for pest and disease resistance (Grosser and Chandler, 2004). Scion improvements and rootstock developments are determined using new scion and rootstock materials for nursery characteristics, tree size control, and field performance for primary selection and advanced testing (Gmitter, 2004). Citrus seedlings were produced from successful rootstock crosses, and interploid scion crosses, and several field trials of both rootstocks and new scions have been conducted in several growing areas (Gmitter, 2004; Grosser and Chandler, 2004; Castle and Futch, 2015).

It is well known that citrus rootstocks such as Carrizo, Swingle Citrumelo, Cleopatra, Volkamer lemon, C-35, US-802, US-897, US-942, UFR-1, and UFR-6 are developed and commercialized well in Florida (Castle and Futch, 2015; Albrecht et al., 2017). Yet, choosing a rootstock and a scion variety is an important decision because it can have an important long-term impact on citrus performance, tree survival, and financial outcomes (Castle and Futch, 2015). Rootstock selection should be based on soil adaptability, soil pH, pest pressure, and desired tree spacing and size control. Rootstocks mostly perform satisfactorily on the well-drained sandy soils. Except for Cleopatra, which is able to tolerate conditions of higher salinity, most rootstocks such as C-35, Carrizo, and Swingle perform poorly in high salinity soils (Albrecht et al., 2017).

Rootstocks have effects on pests and diseases and the choice of scion to be used in combination with the selected rootstock is very important (Castle and Futch, 2015). Rootstocks US-802, US-897, US-942, and UFR-1 are more tolerant to phytophthora infection than other rootstocks, but the damages in these rootstocks from phytophthora are still exacerbated when compromised by HLB infection (Castle and Futch, 2015; Albrecht et al., 2017). Novel varieties of citrus rootstocks are expected to have better field performance (disease tolerance) and better fruit quality for fresh markets;

however, there is still lack of information on their long-term performance under different environmental conditions (Castle and Futch, 2015; Albrecht et al., 2017).

Citrus rootstock and citrus tree combined abiotic/biotic stresses have been assessed through a series of multi-year environmental studies on different scales (greenhouse simulation studies and citrus production orchards across central and southern counties in Florida) during 2001 and 2004 (Li et al., 2003, 2004, 2005, 2006a, 2007a, 2007c, 2007d). These specific studies have examined the water and insect stresses of citrus rootstocks and spatial overlay patterns of naturally flooded and insect infected citrus trees in production orchards. The measurements also include the determination of citrus rootstock susceptibility to *Diaprepes* larval root feeding injury, leaf stomatal conductance, soil redox potential, shoot growth, tree health rating, and leaf areas related to the soil's physical and chemical conditions and *Diaprepes* adult population development. Together the information obtained from the studies has given further insights for developing management strategies to reduce biotic and abiotic stresses of citrus plants.

Most stresses faced by plants are either abiotic, e.g., anaerobe, flooding, acidity, salinity, heat, drought, or nutrient deficits, or biotic, e.g., disease infection and insect attacks (Blum, 1996; Gutschick, 1999; Hatch and Blaustein, 2000; Jones et al., 2003; Blanke and Cooke, 2004; Laughlin and Abella, 2007; Li et al., 2008). This chapter attempts to summarize the abiotic and biotic stresses of citrus rootstocks in the environmentally controlled greenhouses and citrus trees in the production orchards, associated with soil anaerobe, soil waterlogging, environmental acidity, and *Diaprepes abbreviatus* root weevil infestation, and to give an overview of the strategies for reducing plant abiotic and biotic stresses in citrus.

40.2 CITRUS PLANT ABIOTIC STRESS

40.2.1 Citrus Rootstock Abiotic Stress from Simulated Flooding

Plants demand oxygen, water, and nutrients for growth and plants show some degree of stress when exposed to unfavorable environments and these stresses can collectively contribute to affect plant growth (Mattson and Haack, 1987; Lichenthaler, 1996; Lower et al., 2003; Mittler, 2006; Li, 2009, 2015, 2016). Flooding can result in oxygen deprivation in the root system, leading to disturb plant–soil system equilibrium and arrest plant growth (Ruiz-Sanchez et al., 1996; Pezeshki and Delaune, 1998; Oren et al., 2001; Li et al., 2003, 2004, 2006, 2007b; Saqib et al., 2004). Plant physiological responses to flooding stress are reflected by reduced leaf stomatal conductance (g_s), leaf gas exchange, and leaf water potential (Oren et al., 2001; Li et al., 2003, 2004, 2006a; Mielke et al., 2003; Blanke and Cooke, 2004). Flooding induced plant leaf stomatal closure and, therefore, leaf conductance and leaf turgor potential were significantly reduced in flooded 2-year-old sour orange plants (Ruiz-Sanchez et al., 1996). Also, soil oxidation-reduction (redox) potential (E_h), can be reduced by

flooding causing increased plant water stress (Pezeshki and Delaune, 1998; Li et al., 2003, 2004, 2006a).

In citrus, flooding and its related soil waterlogging can be critical stress factors for rootstock growth. Two environmentally controlled studies were conducted in the greenhouse during 2002 and 2003 to investigate changes in leaf stomatal conductance, soil redox potential, and the plant growth of different citrus rootstocks under flooded conditions. Because flooding prohibits gas exchange in the plant–soil system and flooded soil is compacted, these environmentally controlled studies were attempting to quantify how strongly rootstock seedling plants could provide a response to flood stress.

The greenhouse study was conducted at the Citrus Research and Education Center, University of Florida in 2002 (Li et al., 2003). Three-month-old seedlings of two commercial citrus rootstock varieties, Swingle Citrumelo (*Citrus paradisi* Raf., SWI), and Smooth Flat Seville (*Citrus aurantium* L., SFS) were used in the study. The treatment consisted of four levels of flood duration (0, 10, 20, or 30 days), and two levels of *Diaprepes* larvae (0 or 5 larvae per seedling), feeding for 40 days, arranged with eight replicates in a completely randomized design. A total of 80 rootstock seedlings were selected for uniformity of root density and canopy size for each variety; each was transplanted into a 130-cm^3 plastic pot for the submergence test. The soil was a Candler fine sand, pH 6, containing 965 g kg^{-1} (or 97%) sand, similar to the average sand content of citrus soils in Florida (Li et al., 2004, 2007b).

The flooding treatments were applied by submerging seedlings up to 2 cm above the tops of seedling pots, and the shoots remained in the atmosphere (Figure 40.3). To compare the flooding effects within the treatments, the 30-day treatment seedlings were submerged for 10 days; then the 20-day treatments were flooded. After flooding for 10 more days, the 10-day treatments were submerged (Li et al., 2003, 2006a). After being flooded for another 10 days, all the flooded seedlings were removed from the water then put for drainage.

Another flooding greenhouse experiment was conducted using citrus rootstock seedlings of the same varieties as the above experiment (SWI and SFS) to verify the rootstock water stress with a longer flood duration (0, 20, or 40 days). The seedlings were selected for their age (5 months old) and, also, for their uniformity. Also, the methods for submergence and plant–soil measurements were the same as described in Li et al. (2007b). For both greenhouse studies, the control treatment (non-flooding) seedlings received the same amounts of fertilizers and waters as the other treatment's seedlings. Temperatures in the greenhouse varied between 28 ± 4°C throughout the experiment (Li et al., 2003, 2006; Li, 2007b).

The results in the first, shorter flooding duration (30-day) study showed that plant roots were damaged, and canopy growth was prevented, as shown by the differences in the roots and canopies of citrus rootstock (Swingle Citrumelo) seedlings grown under non-flooding (A) and 30-day flooding (B) conditions (Figure 40.4). In all rootstock seedlings, leaf stomatal conductance (g_s), measured using the Delta-T A4 porometer, was high (350 mmol m^{-2} s^{-1} on average) before flooding (Figure 40.5). The g_s values in non-flooded treatment

FIGURE 40.3 (See color insert.) Testing the flooding effects on the physiological development of citrus rootstock seedlings of different varieties ('Swingle Citrumelo' and 'Carrizo') in the greenhouse.

(A)   (B)

FIGURE 40.4 (See color insert.) Comparison of the plant roots and canopies of the citrus rootstock (Swingle Citrumelo) seedlings grown under non-flooding (A) and 30-day flooding (B) conditions.

rootstocks remained similarly high. Yet in flooded rootstock seedlings, the g_s values decreased significantly within increasing flooding duration (Figure 40.5). Their g_s values started to decrease after 2 days of flooding and then decreased consistently with the duration of flooding (Li et al., 2006a). The g_s values were as low as 40 mmol m^{-2} s^{-1} by the end of 30 days of flooding (Figure 40.5).

Soil redox potential (E_h), measured using the Orion oxidation-reduction probe, decreased abruptly following submergence of each treatment measured in the 30-day flooding study (Figure 40.5). The soil E_h values became negative within 1–3 days of flooding, and the E_h varied within –100 and –180 mV (Figure 40.5). The results showed a complete lack of oxygen was attained as quickly as 1 day after submergence (Li et al., 2006a). The analysis of variance showed

that flooding treatment and interactions between variety and flooding had significant effects on soil E_h and citrus leaf g_s (Table 40.1). The contrast test also showed that rootstock leaf g_s and soil E_h were significantly different between flooding treatments (Table 40.2).

The effects of flooding duration were significant on shoot growth and root rating for the two rootstock varieties (Li et al., 2006a). With initial shoot lengths of 39.4 ± 3.9 cm in SWI and 27.6 ± 6.4 cm in SFS, shoot lengths grew faster in SWI (2.7 ± 1.5, 1.5 ± 1.2, and 0.3 ± 0.6 cm) than in SFS (0.3 ± 0.5, 0.4 ± 0.7, and 0.3 ± 0.4 cm) for the F10, F20, and F30 treatments, respectively. The root volumes were significantly higher in the 10-day flooding than in the 30-day flooding treatments.

Rootstock leaf g_s, soil redox potential E_h, shoot growth, leaf areas, and root dry weights were significantly higher in the control (non-flooded) than in the flooded treatments for both rootstocks measured in the 40-day flooding experiment (Table 40.3). The rootstock plant leaf g_s and soil E_h in the flooded treatments decreased dramatically with flooding duration, showing similar patterns as shown in the first study. Shoot growth of citrus tree seedlings arrested during the flooding period. Root injury, estimated by classifying the whole seedling root system by percentage damage as 0% (control), 0–25%, 25–50%, 50–75%, and >75% damage, was mainly attributed to flooding anaerobic (Li et al., 2006a).

Compared to the shorter flooding duration (30 days), the different results determined in the longer flood (40 days) study were that flooded soil pH was found to increase significantly at the longest flood duration (40 days) level. Also, the pH value of 40-day floodwater increased by 0.7 units by the end of a 40-day flooding period compared to 0.3 units for the 20-day flooding treatments (Li et al., 2007b).

The responses of citrus rootstocks to flooded soil conditions could vary in citrus shoots, leaf hydraulic conductivity, and leaf stomatal conductance, but their tendency in these variables was the sudden reduction when the seedlings were

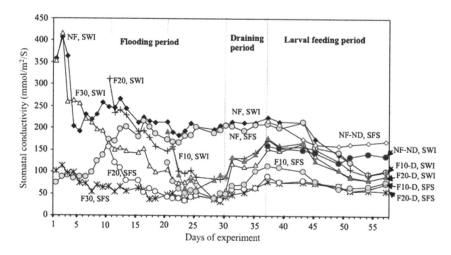

FIGURE 40.5 Citrus rootstock seedling temporal patterns (60 days) of leaf stomatal conductance (g_s) for Swingle Citrumelo (SWI) and Smooth Flat Seville (SFS) during the 30-day flooding and 30-day experiment. The treatments were as follows: NF, non-flooded; F30, 30-day flooded; F20, 20-day flooded; F10, 10-day flooded; NF-ND, non-flooded and non-larvae; F30, 30-day flooded with larvae; F20, 20-day flooded with larvae; F10, 10-day flooded with larvae. Each point represents the mean of $n = 8$ measurements. (From Li et al., 2006a.)

TABLE 40.1

Significance of Effects and Interactions of the Experimental Treatments (Citrus Rootstock Variety V and Flooding Duration F) on Rootstock Plant Abiotic Stress Variables (Leaf Stomatal Conductance g_s and Soil Redox Potential E_h) during the Submerging Period. The Two Rootstock Varieties Were Swingle Citrumelo and Smooth Flat Seville

Sources	df	Soil E_h[a]	Leaf g_s[a]
Model	42	236**	7.25**
Rep	7	0.55 ns	2.69*
Variety (V)	1	0.47 ns	89.5**
Flood (F)	3	3286**	7.30**
Rep × V	7	0.44 ns	1.9 ns
V × F	3	8.0**	43.1**
Rep × F	21	0.56 ns	1.5 ns
R^2		0.99	0.94

Source: Li et al. (2003).
[a] *F* values. ns, not significant.
*$P < 0.05$.
**$P < 0.01$.

under the anaerobic conditions (Li et al., 2003, 2006a, 2007b; Syvertsen and Hanlin, 2008). The degree of plant physiological stress from soil flooding events reflected by g_s, E_h, and shoot growth showed the cause of citrus plant decline from soil waterlogging. The tested citrus rootstocks were likely to have a maximum capacity of 20 days to resist to flooding damage, yet 10 days of flooding the negative effect had already caused a significantly low leaf stomatal conductance that could slow the shoot growth of citrus rootstocks (Li et al., 2006, 2007b).

40.2.2 CITRUS TREE ABIOTIC STRESS FROM NATURAL FLOODING, SOIL ACIDITY, AND EXCESSIVE IRON

From citrus trees' perspective, shoot development is the first priority, followed by root growth to support shoot growth, and any changes in water and soil characteristics will affect shoot and root development (Syvertsen et al., 2000; Syvertsen and Hanlin, 2008). Flooding is linked to soil acidity and nutrient suboptimum (Ruiz-Sanchez et al., 1996; Lower et al., 2003; Mielke et al., 2003; Yoo and James, 2003; Blanke and Cooke, 2004; Ramirez-Rodriguez et al., 2005). It is reported that flooding increased soil pH, decreased nutrient availability and leaf dry matter yield (Li et al., 2003; Yoo and James, 2003). In citrus, as its rootstock seedlings are susceptible to flood damage, citrus trees of hybrid varieties can also decline in relation to soil flooding from rainfall and high soil metal concentrations such as iron (Fe) ions that are related to water saturation conditions (Li et al., 2007a) (Figure 40.6).

In a low-lying, poorly-drained citrus orchard in Osceola County, central Florida, the plants consisting predominantly of 20-year-old 'Hamlin' orange trees on Swingle Citrumelo rootstock (*Citrus paradisi* Macfad. × *Poncirus trifoliata* (L.) Raf.), were planted in raised two-row beds with drainage furrows. The trees were generally in decline and had been infested by *Diaprepes* root weevil over the last 10 years (McCoy et al., 2003). The trees received regular liming, irrigation, fertilization and weeding but no chemical treatments for pest control during the study period (Li et al., 2004).

In the study area, there were a total of 1409 mature trees and 758 young trees replanted in 2000 for research purposes (Figure 40.7). The mature trees were weak, shown by a reduced fruit yield (24–28 Mg ha^{-1}). The soil, classified as a loamy Mollisol, was formed in the flatwoods sediments (Obreza and Collins, 2002). Tree flooding occurrence was dependent on local rain pattern. Total rainfall was 1340 mm with 60% falling in April through September during the study

TABLE 40.2

Contrast Analysis of Significant Differences in Citrus Rootstock Plant Abiotic Stress Variables (Soil Redox Potential E_h, Leaf Stomatal Conductance g_s, and Shoot Length), and Biotic Stress Variables (Root Weevil Larval Survival and Root Rating) between the Experimental Treatments (Flooding Duration F, Non-Flooding NF, *Diaprepes* Larval Feeding D, and Non-Larval Feeding ND) for Mean Values of Two Citrus Rootstocks (Swingle Citrumelo and Smooth Flat Seville). Each Treatment Replicates Are n = 8 per Flooding and Non-Flooding Treatment per Rootstock.

Contrasts	df	E_h[a]	g_s[a]	Shoot Length[a]	Larval Survival[a]	Root Rating[a]
Flooding period						
NF vs. F[b]	1	1018**	8.22**	15.6**		
F10 vs. F20[b]	1	7.96**	4.6* ns	4.7*		
F10 vs. F30[b]	1	6395**	17.1**	7.6**		
Larval feeding period						
ND vs. D[c]	1	45.7**	3.7 ns	8.6**		6.8**
NF-ND vs. NF-D[c]	1	2.5 ns	0.1 ns	3.2 ns		6.0*
F10-D vs. F20-D[c]	1	1.4 ns	5.7*	1.2 ns	26.4**	27.1**
F10-D vs. F30-D[c]	1	3.5 ns	14.0**	2.1 ns	461.7**	41.3**

Source: Li et al. (2007b).

[a] *F* values. ns, nonsignificant.

[b] F10, 10-day flooded; F20, 20-day flooded; F30, 30-day flooded; F40, 40-day flooded treatments.

[c] ND, no-*Diaprepes* larvae; NF, non-flooded treatments. Source: Li et al. (2006)

* & **significant at $P < 0.05$ & $P < 0.01$.

TABLE 40.3

Analysis of Variance on the Effects and Interactions of the Experimental Treatments (Citrus Rootstock Variety V, Flooding Duration F, and *Diaprepes* Larval Infestation D) on Soil Redox Potential (E_h), Leaf Stomatal Conductance (g_s), Seedling Shoot Length Growth, Larval Survival, Root Damage Rating, Root Dry Weight, and Leaf Area in Florida Loam in Longer Flooding Duration (40 Days) and Longer *Diaprepes* Root Weevil Larval Feeding Period (40 Days) Experiment

Sources	df	E_h[a]		Shoot Length[a]	Leaf Area	Root Dry Weight[a]	Larval Survival[a]	Root Rating
Variety (V)	1	ns	ns	13.8**	6.57**	14.6**	7.5**	6.33**
Flooding (F)	2	100.6**	3.67**	21.5**	20.5**	51.0**	9.1**	25.3**
Diaprepes larvae (D)	1	ns	ns	4.77*	ns	22.1**	168**	66.2**
V × F	2	ns	ns	ns	3.77*	ns	7.34**	ns
V × D	1	ns	ns	ns	ns	ns	3.75*	ns
F × D	2	ns	ns	ns	ns	ns	4.55*	6.54"
V × F × D	2	ns	ns	ns	ns	ns	3167*	ns
Model R^2		0.80**	0.22 ns	0.54**	0.55**	0.70**	0.78**	0.74**
CV		4.94	1.05	62.3	36.1	43.3	51.8	32.5
Mean[b]		357	46.3	3.85	78.1	0.81	1.94	2.97
RMSE[c]		17.6	48.9	2.39	28.2	0.35	1.00	0.96

Source: Li et al. (2007b).

[a] *F* values. ns, nonsignificant.

[b] Mean E_h in mV; g_s in mmol m^{-2} s^{-1}; shoot length in cm; root dry weight in g; and leaf area in cm^2.

[c] RMSE: root mean square error.

* and **significant at $P < 0.05$ and $P < 0.01$, respectively.

period, and flooding could happen by heavy rainfall and last for a few weeks (Figure 40.7). Flooding resulted in soil redox potential (E_h) falling far below zero (–260 mV) in the flooded areas that could be linked to tree decline (Figure 40.8).

Water table depth, a component for distinguishing the soil's unsaturated and saturated zones, varied between 0.6–1.4 m across the orchard during the dry period (Figure 40.9). After 3 weeks of flooding, plant water stress of the flooded trees

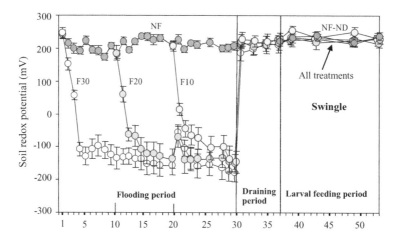

FIGURE 40.6 Comparison of reduced soil redox potential under different flooding durations that could contribute to citrus rootstock abiotic stress in Swingle Citrumelo throughout the 30-day flooding (F) and 30-day *Diaprepes* larval feeding (D) experiment. The flooding treatments were: NF, non-flooded; F30, 30-day flooded; F20, 20-day flooded; F10, 10-day flooded treatments. Each point represents the mean and standard error of n = 8 measurements. (From Li et al., 2006a.)

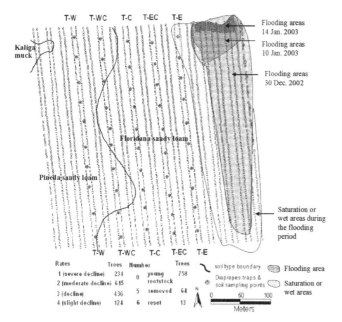

FIGURE 40.7 Citrus trees ('Hamlin' orange trees on Swingle rootstock) and its decline rating along with the soil flooding area lines in the Osceola orchard. The map also shows the soil type boundary, tree beds, and Tedders trap locations. T, transect (tree bed); T-W, west transect; T-WC, west-center transect; T-C, center transect; T-EC, east-center transect; T-E, east transect. (From Li et al., 2007a.)

FIGURE 40.8 (See color insert.) Citrus trees ('Hamlin' orange trees on Swingle Citrumelo rootstock) under flooding occurred after a heavy rain in the fall (2003). Flooded anoxic soil had a negative soil redox potential E_h that caused tree decline.

in the shallow water table areas (0.52–0.63 m) were more water stressed than the trees in the medium water table areas (0.75–1.00 m) because of difference in water saturation (Li et al., 2004).

Citrus tree decline, assessed by visual quantification of the tree decline symptoms using the characteristics referred to Blazquez (1991) using a numerical 1–4 ranking system, showed that a total of 16.6% of the mature trees were in decline and were rated as severely declined trees (rating 1). Moderately declined (rating 2) trees represented 43.6% of the

total trees, declined (rating 3) 30.9%, and slightly declined (rating 4) 8.8% (Li et al., 2004). The rating tree characteristics were: 1 = severe decline: canopy is easily seen through, flush on major limbs only or on less than half of the tree, leaves small; 2 = moderate decline: canopy easily seen through, flush on secondary and higher limbs scattered around the entire canopy, leaves small; 3 = decline: well-defined canopy, more than half of which cannot be seen through, flush on secondary and higher limbs, leaves large; and 4 = slight decline: well-shaped and well-defined canopy that cannot be seen through, flush on secondary and higher limbs, leaves large and green (L2 et al., 2004).

Overall, severe citrus tree decline was found in shallow water table areas and flooded areas, where the stomatal conductance (g_s) of newly matured leaves of the citrus rootstock seedlings decreased from 152 mmol m^{-2}s^{-1} to 94 mmol m^{-2}s^{-1} on average under the flooded conditions (Figure 40.9). However, leaf water potential P_a did not differ between the

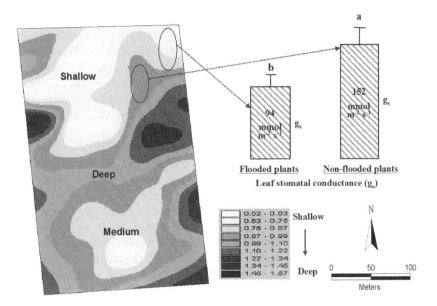

FIGURE 40.9 (See color insert.) Comparison of citrus tree water stress status with leaf stomatal conductance (g_s) measurements in a wetter (shallow water table 0.52–0.63 m, g_s 94 mmol m^{-1}s^{-1}) area and in a medium water table area (medium water table 0.75–1.00 m, g_s 152 mmol m^{-1}s^{-1}). Citrus trees in the wetter area were more water stressed than in the drier area in Osceola orchard in Florida. (From Li et al., 2004.)

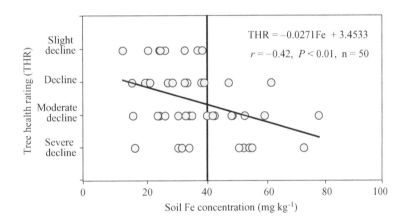

FIGURE 40.10 Regression relationship of soil Fe concentration vs. tree decline (rating class: 1 = severe decline, 2 = moderate decline, 3 = decline, and 4 = slight decline) on the Mollisol in Osceola orchard in Florida. (From Li et al., 2007a.)

trees in the flooded areas (0.66 ± 0.18 MPa) vs. the non-flooded areas (0.66 ± 0.10 MPa) (Li et al., 2004).

As a result of waterlogging, the soil was highly acidic (pH 4.9 ± 0.4) and high in Fe ions (36 ± 14 mg kg^{-1}). The soil Fe was negatively correlated with water table depth ($r = -0.38$, $P < 0.01$). Soil Fe is more soluble under low pH, anaerobic conditions (Kidd and Proctor, 2001) and high concentrations of Fe^{2+} in the soil solution could form a plaque to affect the absorption of nutrients by roots (Liu et al., 2007a). The increase of soil Fe concentrations with increasing soil water holding capacity and reducing soil pH levels are consistent with the elemental chemistry (Kidd and Proctor, 2001; Ramirez-Rodriguez et al., 2005; Li, 2015, 2016).

In addition, Fe is a micronutrient for plant growth, but too much Fe ions in the soil can become toxic to the plants (Ramirez-Rodriguez et al., 2005; Li, 2015, 2016). The regression plot of the tree health rating against the soil Fe

concentrations showed that tree decline was linearly correlated to soil Fe concentrations (Figure 40.10). More than 50% of the severe decline (rating 1) and moderately decline (rating 2) trees were situated within a high soil Fe concentration of 40–80 mg kg^{-1}. Severely decline, and moderately decline trees were 60.2% of the total infested trees. Only two trees of rating 3 (decline trees, or 30.9% of total infested trees) were within a soil Fe concentration >40 mg kg^{-1}. All the healthier trees (rating 4, slight decline) were in areas low in Fe concentrations between 13 and 39 mg kg^{-1} (Li et al., 2007a).

Stepwise multiple linear models for tree health rating related to soil water content (SWC) (g kg^{-1}), soil pH, and soil Fe (mg kg^{-1}) showed the trends as follows:

$$THR = 4.6598 - 0.00304SWC - 0.03503pH$$

$$-0.0326Fe \ \left(R^2 = 0.26, P < 0.0087 \right) \qquad (40.1)$$

The model Eq. (40.1) was also significant (F = 4.37, df = 3, 46), and the estimated parameters were significant for the intercept (*P* <0.0343), and Fe (*P* <0.0017) (Li et al., 2007a). Severely declined trees were associated with areas high in soil Fe concentrations >40 mg kg^{-1} (Figure 40.10), meaning that excessive Fe ions in the soils could cause citrus plant stress to affect citrus growth. Increasing soil pH levels with lime applications and improving soil drainage could be strategies for correcting soil Fe and water levels for the management of tree decline (Li et al., 2007a).

40.3 CITRUS PLANT BIOTIC STRESS

40.3.1 CITRUS ROOTSTOCK BIOTIC STRESS FROM ROOT WEEVIL LARVAL FEEDING

Plant biotic stress from insect and disease pathogen infestations can cause the loss of crops (Mattson and Haack, 1987; Lower et al., 2003; Mittler, 2006). In citrus, plant environmental stress from flooding, soil acidity, and nutritional elements can occur simultaneously with root weevil (Li et al., 2004, 2006a, 2007a), or *phytophthora* (Graham et al., 2003), or Alternaria (De Souza et al., 2009) or Huanglongbing infestations (Bove, 2006; Castle and Futch, 2015). Citrus root weevil *Diaprepes abbreviatus* (L.) infections in Florida orchards were originally from the Caribbean (Rogers et al., 2000; Nigg et al., 2001; Futch, 2003; McCoy et al., 2003; Stuart et al., 2004). The root weevil has become a major pest of citrus and other agricultural crops in Florida and it is the subject of intensive research in the state (Quintela and McCoy, 1997; Lapointe, 2000; Li et al., 2003, 2004, 2006, 2007a, 2007b; Nigg et al., 2003; Stuart et al., 2004).

The root weevil adults are citrus leaf feeders of all tree varieties (McCoy et al., 2003; Stuart et al., 2004). Females deposit eggs in masses glued between leaves in the citrus canopy. Hatching neonates fall to the soil surface and move into the soil where they feed on roots and subsequently pupate (McCoy et al., 2003). *Diaprepes* larval growth is consistently fast at an ambient temperature of 22–26°C. The time for a single generation from *Diaprepes* oviposition to adult emergence is estimated to be about 22 weeks (or 150 days) at 26°C under laboratory conditions (Lapointe, 2000).

Because of small sizes, the neonate larvae are virtually impossible to detect in the soil, and the initial injury to roots are difficult to quantify (Jones and Schroeder, 1983; Quintela and McCoy, 1997; Rogers et al., 2000). Citrus rootstock tissue injury in different rootstocks by *Diaprepes* neonate larval feeding was visual after 79 days of infection and larvae can consume 20–80% of the citrus seedling roots within 6 weeks of infestation in well-drained soil conditions (Rogers et al., 2000). However, in another study, leaf beetle larval-pupal weight was not influenced by the nutrient or flooding conditions of the plant (Lower et al., 2003).

Citrus rootstocks are vulnerable to attack by the *Diaprepes* weevils. Citrus rootstock plant environmental stress from flooding can occur simultaneously with root weevil infestations (Li et al., 2003, 2004, 2006). Is the combination of flooding and larval

feeding complicated for weevil control? In the two environmentally controlled flooding studies in the greenhouses described above (Section 40.2.1), the seedlings survival rates were put to the test by *Diaprepes* weevil larval feedings after the flooding test in the greenhouse. After completing the flooding duration treatments in the first greenhouse study (0, 10, 20, or 30 days), all seedlings were infested by five *Diaprepes* neonate larvae per seedling for 30 days. In the second greenhouse study, after completing the flooding duration treatments (0, 20, or 40 days), all seedlings were infested by five *Diaprepes* neonate larvae per seedling for 40 days. The larvae were selected for the active, 1-day-old ones, then five larvae were inoculated on to the soil for each of the seedlings a week after the flooding procedure was completed. There was a control for the non-flooding and non-larval feeding treatment (NF-ND), consisting of a total of 16 seedlings (2 varieties × 8 replicates). These seedlings were not flooded during the larval feeding period, and all treatments received the same rates of fertilizers and waters. The temperatures were kept between 28 ± 4°C throughout the larval feeding testing period (Li et al., 2006a).

The results showed that after being inoculated into the soil, larval survival rates and root injury ratings were significantly different form the previous flood treatments (Table 40.2). With an initial infestation of five-neonate larvae per rootstock seedling, the lowest larval survival in the 30-day feeding study was found in the non-flooded treatment (60 ± 22%), and the survival rate was significantly higher in the 30-day flooded treatments (88 ± 10%) for both varieties Swingle Citrumelo (SWI) and Smooth Flat Seville (SFS). In addition, differences in larval survival rates were significant between the two citrus rootstock varieties (SWI 78 ± 22%, and SFS 82 ± 14%).

In the 40-day feeding experiment, among the two varieties, larval survival was significantly lower, 66 ± 19% for SWI and 50 ± 22% for SFS seedlings flooded for 20–40 days. Also, there was an increase of 0.5 unit pH in the 40-day flooded soil. The root weevil larval survival and its growth reduced from 80 to 60% when soil pH increased from 4.8 to 5.6 in acidic soil (Li et al., 2007b).

In the larval feeding experiment with shorter duration (30 days), the growths of *Diaprepes* larval weights were within 370 times of its initial weights after 30 days of root feeding. Total weight of survival larvae varied between 25 and 182 mg per seedling. The initial neonate larval weight was on average 0.5 mg (mean of five 1-day-old neonates). The larval weights could be an increase of 36–375 times of their weights as 1-day-old neonates within the 30 days of feeding in the greenhouse (Li et al., 2003, 2006). The larval growths were increased by 50–364 times within 40 days of feeding on the rootstock seedlings in the second greenhouse study (Li et al., 2007b). In both larval feeding studies, larval growth was the highest in the non-flooded treatments. These results indicate the high potential of neonate larval growth in both flooded and non-flooded citrus orchards.

The relative vulnerability of citrus rootstock seedlings to the root weevil larval feeding injury was shown by root injury from larval feeding that was increased with the duration of the previous flooding. In the shorter feeding (30 days) study, the root injury ranged between 0–3% of the root areas for the non-flooded rootstock seedlings, yet the root injury was up to 6 and 12% of the root areas for the 10, 20, and 30 days of previously flooded seedlings in both rootstock varieties Swingle Citrumelo (SWI) and Smooth Flat Seville (Li et al., 2006).

In the 30-day feeding experiment, the rootstock seedling growth data measured during the larval infestation period showed that shoot growth was greater in SFS (2.1 cm on average) than in SWI (0.9 cm on average), compared to the 2.4 cm in SFS and 1.3 cm in SWI in the non-larval feeding controls. There was a significant reduction in shoot growth for the larval feeding treatments compared to the non-larval feeding treatments (Table 40.2).

These measurements showed that citrus seedling root injury was attributed to both flooding and larval infestation. In the 40-day larval feeding experiment, larval root feeding damage increased with the previous flooding duration (Figure 40.11). As compared with the non-flooded control, about 25–50% of seedlings with root damage were found in the rootstock seedlings previously flooded for 20 days, and root larval feeding damage increased significantly, found in 75% of the rootstock seedlings that were previously flooded for 40 days (Li et al., 2007b).

The effects of larval feeding were significant on shoot growth, leaf area, and root dry weight in the 40-day feeding experiment II (Table 40.3). Shoot growth, leaf area, and root dry weight of the two rootstock varieties were significantly different within the previously 20-day flooded seedlings with or without the larval feeding (Figure 40.12).

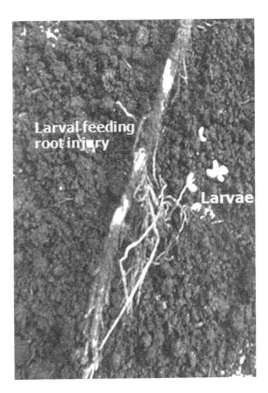

FIGURE 40.11 Citrus rootstock 'Carrizo' 40-day flooded roots injured by larval feeding after infested with five 1-day-old *Diaprepes* root weevil neonates for 40 days. The mean five-neonate weight was 0.5 mg and per rootstock they grew into big larvae of 120 mg on average.

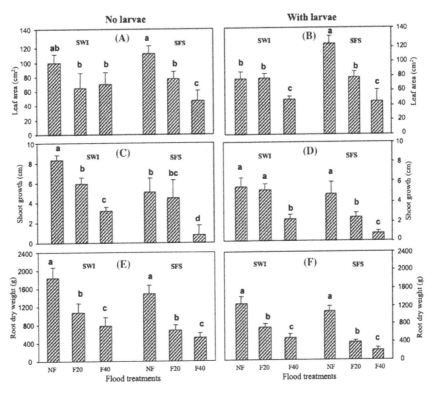

FIGURE 40.12 Comparisons of leaf area (A-B), shoot growth (C-D), and root dry weight (E-F) with and without *Diaprepes* larval feeding in a 40-day larval feeding experiment. The treatments were as follows: SWI, Swingle Citrumelo; SFS, Smooth Flat Seville; NF, non-flooded; F20, 20-day flooded; F40, 40-day flooded. Each bar represents the mean and standard error of n = 15 measurements. (From Li et al., 2006a.)

It is suggested that a 20-day flooding duration is the threshold for citrus rootstock seedling plant water stress and root injury from flooding damage (Li et al., 2006a). A negative soil redox potential and a decrease in leaf stomatal conductance could be an early indicator of plant water stress and root damage from flooding, and soil type and soil texture could also be factors affecting larval survival in the field. The causes related to soil-inhabiting *Diaprepes* larval survival and rootstock injury were complex. In the case of citrus rootstock, its seedling roots could be injured by waterlogging and larval feeding, but higher soil pH levels could be unfavorable to larval survival for rootstock protection.

40.3.2 CITRUS TREE COMBINED STRESS FROM ROOT WEEVIL FEEDING AND WATER SATURATION

Soil environment and plant quality influences the abundance of most herbivorous insects (White, 1984; Mattson and Haack, 1987; Orians and Fritz, 1996; Lower et al., 2003), which has been linked to plant biotic stress (Hatch and Blaustein, 2000; Laughlin and Abella, 2007; Li et al., 2008). *Diaprepes* root weevils have been dispersed not only in the rootstock nursery, but also the citrus orchards and root injury inflicted by the larvae resulted in plant decline and tree death (Nigg et al., 2001, 2003; Futch, 2003; McCoy et al., 2003; Stuart et al., 2004). Citrus tree biotic stress from root weevil infestations can occur simultaneously with environmental flooding, and excessive soil iron and high acidity (Li et al., 2004, 2007a, 2007c, 2007d).

In the Osceola citrus orchard where the tree water stress from flooding and citrus tree decline rating were determined (Section 40.2.2), the hatching *Diaprepes* root weevil neonates dropped from the mature tree canopy at a rate varying between 370 and 940 neonates per m² of soil surface under the trees (McCoy et al., 2003; Nigg et al., 2003). Certain neonates were able to survive after dropping onto the ground; then the survival neonates could move into the soils where they fed on tree roots and subsequently pupated. By removing a total of 60 mature trees and sieving soils sampled around the central root areas, the numbers of pupating larvae found in the soils

under the tree canopy was an average of 50 larvae per m³ of fresh soil per tree (n = 60 trees) (McCoy et al., 2003).

The larvae grow into adults in the soils; then the adults emerge from soils into the tree canopy. In the same Osceola citrus orchard, adult weevils emerging were monitored weekly all year long using modified pyramidal Tedders traps placed near tree trunks (Figure 40.7) in five 10-trap transects (five mature tree beds) in a 34 m × 25 m grid pattern. The adults were emerging from the soils in April when the temperatures became warm for tree growth. The weekly peak (128 weevils) appeared in mid-June when the young tree leaves were grown. A smaller weekly peak (91 weevils) occurred in mid-September, showing an interval of 3 months for a new generation of adults that were emerging from the soil for tree canopy feeding. The highest monthly catches occurred in June (monthly total 450 weevils, 9 adult weevils per trap), and the remaining months, except for April, had a similar level of 3.1–4.7 weevils per trap (Li et al., 2004).

A total of 1400 adults were captured in the 50 traps along the five transects over the year period. Males totaled 946 (19 ± 14 weevils per trap) and females 454 (9 ± 9 weevils per trap). An increase or a decrease in males was generally proportional to an increase or a decrease of females among monitoring traps (Li et al., 2004). The weevil peaks appeared in the south where there was less water saturation than in the north (Figure 40.13).

The difference in adult weevil among the transect areas was significant (ANOVA, F = 3.74, df = 4, 36, P = 0.0121). The adult weevil density was high in the western and center of the orchard (per transect; T-W, 31 ± 21; T-WC, 35 ± 27; and T-C, 38.8 ± 26 weevils weekly), compared to the eastern areas (per transect; T-EC, 16 ± 12; and TE, 19 ± 11 weevils) where there was soil flooding. Overall, severely and moderately declined trees were situated mainly in the southwest area of the orchard, extending from the north for 120 m to the south in all transects, especially T-W, T-WC, and T-C (Figure 40.13).

After emerging into the tree canopy, the root weevils might be attractive to local citrus trees because the adults move relatively little (McCoy et al., 2003; Nigg et al., 2003). In this

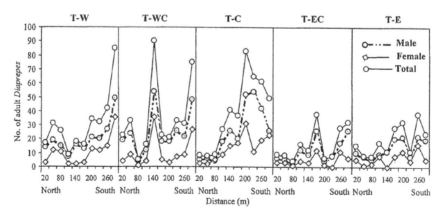

FIGURE 40.13 Spatial patterns of weekly *Diaprepes* adult weevils emerging from the soils along the west transect (T-W), west-center transect (T-WC), center transect (T-C), east-center transect (T-EC), and east transect (T-E). Each point is the mean of n = 32 monitored across the Osceola orchard in 2002. (From Li et al., 2004.)

Osceola citrus orchard, only 25% (or 146 weevils) of the captured and released weevils were recaptured over a 10-week period after release, with a total of 580 adults that were captured and marked (Nigg et al., 2003). Among the recaptured, marked adult weevils, 40% of them moved within 0–24 m, and 41% of them moved within 25–72 m from the release points. The movement of all recovered adult weevils was within 72 m from the release point during this release period (Nigg et al., 2003).

In another *Diaprepes* root weevil infested flatwoods citrus orchard in Hendry County, South Florida, the 10-year-old 'Hamlin' orange trees on 'Swingle Citrumelo' rootstocks were also damaged, and the orange fruit yield declined from 54 to 37 Mg ha^{-1} in this orchard. The *Diaprepes* adult weevils emerging from the soils were monitored in a 30 × 12 m grid under the citrus tree canopy using 100 pyramidal Tedders traps across the orchard over a period of 3 years (2001–2003). The data showed that 962, 945, and 549 adult weevils were trapped in 2001, 2002, and 2003, respectively (Li et al., 2007d). The weekly adult density varied between 5.5–9.6 weevils per trap. The weevil population also peaked in the spring (Figure 40.14), when the temperatures were warm, up to about 25°C each year. Difference in weevil density between the 3 years was significant (ANOVA, $F = 8.90$, $df = 2$, 297, $P < 0.0002$) (Li et al., 2007d). The weevil density varied between 0.010.42 weevils per m^2 per week under the citrus tree canopy over the 3 years.

The relationship between the root weevil adults and soil chemical characteristics are shown by regression trends (Figure 40.15). The relationship between citrus trees, soils, and root weevils can be explained by the adult weevil movement in the orchard. Because the adult weevils move relatively little and tend to stay close to where they emerge from the soils in the orchard (Nigg, 2001, 2003), tree decline, soil analysis, and root weevil relationships could be more or less correlated (Table 40.4). The correlation analysis showed that adult weevil development was positively correlated to soil H$^+$ ion concentrations ($r = 0.26^*$, n = 50) and negatively correlated with Mg ion concentrations ($r = -0.31^*$, n = 50) at the Oscerla orchard (Li et al., 2004), and negatively correlated with soil pH and Ca and My concentrations ($0.0121 < P < 0.0284$, n = 100) at the DeSoto orchard (Li et al., 2007a).

Most life stages, including larvae, pupae, and general adults, occur in the soil (McCoy et al., 2003; Nigg et al., 2003), and characteristics of the soil and the citrus trees on which root weevil feed can directly or indirectly influence adult weevil development (Li et al., 2004, 2007a, 2007c, 2007d). Water and soil nutrients influence plant performance (Ruiz-Sanchez et al., 1996; Yoo and James, 2003; Ramirez-Rodriguez et al., 2005; Li et al., 2008; Syvertsen and Hanlin, 2008; Srivastava and Singh, 2009; Li, 2015, 2016) and plant quality affects pest development (White, 1984; Mattson and Haack, 1987; Orians and Fritz, 1996; Lower et al., 2003); therefore, the citrus root

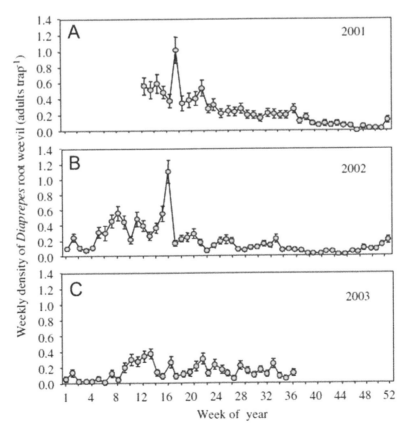

FIGURE 40.14 Temporal patterns of weekly *Diaprepes* adult weevils emerging from the soils across the DeSoto orchard. Each point is the mean and standard error value of n = 100 traps, monitored in 2001 (A), 2002 (B), and 2003 (C). (From Li et al., 2007d.)

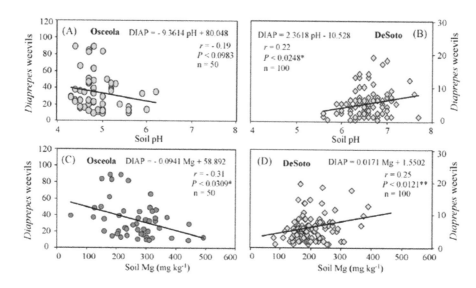

FIGURE 40.15 Regression relationship of *Diaprepes* adult weevil emerging from the soils vs. soil pH and Mg concentrations on the Mollisol in the Osceola orchard (A and C) and on the Spodosol in the DeSoto orchard (B and D). Data were collected in 2002 from the Osceola orchard and in 2003 from the DeSoto orchard. (From Li et al., 2007a.)

TABLE 40.4

Significance (*P*-Values) of the Correlation Coefficients for Citrus Tree Biotic Stress Variables (*Diaprepes* Root Weevil Adults and Tree Decline Rating) and Soil Characteristics (Soil Ph, Soil Water Content, Exchangeable Nutrients Mg and Fe Concentrations, and Soil Structural Sand Content) in the Mollisol (Osceola Orchard) in Central Florida and Spodosol (Desoto Orchard) in South Florida

	Sites									
	Mollisol (Osceola site)[a]					Spodosol (DeSoto site)[a]				
Variables	pH	SWC[b]	Mg	Fe	Sand	pH	SWC[b]	Mg	Fe	Sand
Diaprepes	ns	ns	0.0309*	ns	ns	0.0248*	ns	0.0121**	ns	0.0282*
Tree rating	ns	ns	ns	<0.0001***	ns	–	–	–	–	–
Clay	0.0544*	0.0278*	ns	ns	<0.0001***	ns	ns	ns	ns	<0.0001***
Ca	0.0306*	<0.0001***	<0.0001***	0.0217*	ns	<0.0001***	0.0053**	<0.0001***	<0.0001***	ns
SOM[b]	0.0008***	<0.0001***	<0.0001***	0.0413*	ns	ns	<0.0001**	ns	ns	ns
K	ns	0.0088**	ns	<0.0001***	ns	ns	0.0506*	ns	ns	0.0215*
P	ns	ns	<0.0001***	0.0193*	ns	ns	ns	0.0546*	ns	ns
Cu	ns	ns	ns	0.0067**	ns	0.0015**	0.0145**	0.0002***	<0.0001***	ns
CEC[b]	ns	<0.0001***	<0.0001***	0.0448*	ns	0.0146**	0.0068**	<0.0001***	0.0435*	ns

*, **, and *** significant at *P* <0.05, *P* <0.01 & *P* <0.001, respectively.

[a] Data measured in 2002 in the Osceola orchard (*n* = 50), and in 2003 in the DeSoto orchard (*n* = 100).

[b] SWC, soil gravimetric water content; SOM, soil organic matter; CEC, cation exchange capacity. Soils sampled in the 0–0.3 m depth. Source: Li et al. (2007a).

weevils might be more attracted to healthier, fuller trees for feeding and egg laying for more abundant adult occurrence (Li et al., 2004, 2007a).

Citrus tree decline symptoms are not apparent until the *Diaprepes* larvae are well-established on the roots and extensive damage has occurred (Graham et al., 2003; McCoy et al., 2003). Long periods of feeding of the larvae can break the resistance of citrus roots to infection by phytophthora, and

citrus trees were vulnerable to attack by the *Phytophthora-Diaprepes* weevil complex (Graham et al., 2003). A variety of approaches of chemical and biological controls have been tested, but there was still a lack of effective and safe control methods, and this species is still spreading. Flooding/water-related high soil acidity and excessive irons and root weevil infestation have led to tree decline to an unproductive state and tree death.

40.4 STRATEGIES FOR REDUCING CITRUS COMBINED ABIOTIC AND BIOTIC STRESSES

40.4.1 Synthesis of Citrus Combined Abiotic/Biotic Stresses

Plant stress is the state of a plant under excessive unfavorable pressure resulting in damages that can no longer be compensated for by the plants (Lichtenthaler, 1996; Mittler, 2006; Li, 2009). Plant abiotic and biotic stresses are related to unfavorable environmental and biological constraints (Gutschick, 1999; Laughlin and Abella, 2007; Li, 2009). In the case of citrus plants, simulation studies in greenhouses have shown citrus rootstock root injury under both flooding anaerobic effects and *Diaprepes* root weevil larval feeding pressure (Li al., 2003, 2006a, 2007b). Field-scale studies of *Diaprepes* root weevil life cycling (neonate drooping from tree canopy to soil, adult weevil emergence from soil, active adult weevil movement) and its biological controls have been the research focus for management of this pest (Lapointe, 2000; Rogers et al., 2000; Nigg et al., 2001, 2003; Graham et al., 2003; McCoy et al., 2003; Stuart et al., 2004). Orchard scale studies of citrus tree biology, soil chemical characteristics, and *Diaprepes* root weevil distributions done primarily in different counties across central and southern Florida have provided useful information for understanding citrus plant–soil–pest environmental relations for further management purposes (Li et al., 2004, 2005, 2006a, 2007a, 2007c, 2007d).

Citrus tree abiotic and biotic stress in this specific area is the combination of unfavorable constraints from soil anoxia from flooding, waterlogging, suboptimum nutrient levels, and root weevil feeding injury. Flooding may be beneficial by reducing larval survival (Li et al., 2006a) and adult weevil emergence (Li et al., 2007a), but it can be critical for citrus plant survival, based on the notion that flooding could significantly reduce soil redox potential, plant leaf stomatal conductance, and shoot growth (Li et al., 2004, 2006a). A negative soil redox potential and a reduced leaf stomatal conductance could be used as early indicators of plant water stress from waterlogging. The combined effects of these stress factors and *Diaprepes* root weevil development with time have given insight into how strongly the trees, soils, and insects can be directly and indirectly related to each other.

Abiotic and biotic stress conditions may result in an extensive loss in citrus production (Fernandez-Ballester et al., 2003; Li et al., 2006a; Syvertsen and Hanlon, 2008). It is estimated that economic costs from citrus tree infestation of *Diaprepes* root weevil and the associated root disease in citrus could be as much as $600 per hectare (Graham et al., 2003). It can be concluded that the most important factors influencing citrus tree health status in these specific areas are the anaerobic flooding associated water saturation, high soil acidity, excessive iron, and root weevil attacks, soil pH level, soil Fe, Mg, and Ca concentrations, and *Diaprepes* root weevil infestations.

The combination of flooding and larval feeding has complicated any treatments for drainage and root weevil control in reducing citrus plant abiotic and biotic stresses. Flooded rootstocks became weaker which made them more susceptible to injury by weevil larvae than the non-flooded rootstocks. A 20-day flooding period seems to be a critical duration for plant tolerance to water stress because leaf stomatal conductance could be significantly reduced, and roots could be damaged during a 20-day (or longer) flooding period. Although the information about citrus susceptibility to flooding stress and root weevil larval feeding has been provided, the time range that rootstock plants can tolerate flood stress and weevil attacks is still not clear enough.

Overall, the citrus tree decline was due to the combined biotic and abiotic stresses from *Diaprepes* root weevil infestation and water saturation related to high soil acidity and excessive iron ions. The simultaneous occurrence of soil flooding, excessive acidity and elements, and root weevil infestation has been the challenge to the sustainable production of citrus. Citrus growers need regular soil testing to determine soil chemical conditions for soil liming and tree fertilization purposes; therefore, soil characteristics have been useful for planning liming and nutrient applications for adequate soil pH and nutrient levels for citrus tree management and root weevil control.

40.4.2 Management Tools for Reducing Citrus Plant Abiotic/Biotic Stresses

Understanding plant, soil and insect relations can lead to the development of management tools for reducing plant abiotic/biotic stresses. The final goals to summarize citrus plant stress are to use citrus tree decline data and the information about leaf stomatal conductance, soil redox potential, soil water and chemical element characteristics, and root weevil survival for development of management tools for reducing citrus tree decline from abiotic and biotic stress. Time series models, multi-regression and exponential growth models, have been reported for reducing the cost of field monitoring and for a less frequent spray for integrated citrus tree management (Li et al., 2007a, 2007c, 2007d).

Mathematical models were useful to define problems, understand the systems, and make predictions for purposes of insect and soil management to reduce costs from field measurements and monitoring tasks (Worner, 1991; Tobin et al., 2001; Byers and Castle, 2005; Crowder and Onstad, 2005; Li et al., 2006b). Mathematical equations, derived from the field measurement data, would be useful for predicting its development patterns in space and time (Klironomos et al., 1999; Li et al., 2006b; Li, 2009) and for reducing the area to treat and management costs for control of the weevils (Li et al., 2007a, 2007d).

The traditional way to manage citrus root weevil pests is often by repeated foliar applications of chemical treatments when pest densities exceed an economic threshold that requires treatment (Futch, 2003; Graham et al., 2003). Biological control of the root weevils was reported by

Stuart et al. (2004). Yet, typically in the absence of early detections and effective management tools, citrus growers control the *Diaprepes* weevil population using four applications of insecticides with each application per season each year, and usually uniform rates are applied over the orchards (McCoy et al., 2003). However, the excessive use of costly pesticides can harm the environment (Byers and Castle, 2005).

The implications of a time series model for reducing citrus biotic stress are based on the spatial and temporal patterns of weevil development, obtained from the weekly monitoring data from 100 traps under the same citrus canopies in the citrus orchards over 3 years in the DeSoto orchard (Li et al., 2007d). Using the moving average forecast model with PROC EXPAND procedure (SAS Institute, 1993), future *Diaprepes* population patterns can be predicted against time (t) based on the average of the last *N* (a week as the moving average interval) monitoring data of the underlying time series. The new time series of weekly *Diaprepes* weevil development is generated by computation of the moving average of the original series, the 3-year mean of 7800, 10,400, and 7200 weekly monitoring data shown in Li et al. (2007d). The forecasted pattern exhibits a smoother change with time steps than the 3-year field dataset (Li et al., 2007d). The regression analysis shows that the forecast data are strongly related to the multi-year field monitoring data ($R^2 = 0.88$), shown in Li et al. (2007d).

Models probably would be adequate for a specific situation in predicting pest population, their damage, and control costs that can be incorporated in a decision-making framework (Worner, 1991; Tobin et al., 2001; Byers and Castle, 2005; Li et al., 2007a, 2007d). For practical management, the forecast trend suggests that equal timing of insecticide applications would not be efficient based on the weevil temporal pattern, and insecticides should be applied in the spring, in the summer, and early in the fall. Also, higher rates of insecticides should be applied in the spring, and lower rates should be applied in the summer and in the fall (Li et al., 2007d). Using this approach, time series model predictions will help reduce the frequency of the spraying of costly pesticides that can harm the environment.

As abiotic and biotic stresses of citrus trees are associated with time and environmental soil variables, the time series model and mathematical equations derived from these tree, soil, and weevil variable correlations in subsequent years can be useful for integration with multivariate linear stepwise models, soil unit management zones, and environmental mapping for improving the management of the orchards and the root weevil. It is also suggested that regulating drainage systems and practicing lime applications are useful management tools for reducing citrus crop abiotic/biotic stress.

ACKNOWLEDGMENT

I thank China Tobacco, Hainan Cigar Research Institute and the Chinese Academy of Tropical Agricultural Science, Environment and Plant Protection Institute for support (2018hzsJ008, ZDYF2016052, HNGDg12015). Also, I thank the Florida Citrus Production Research Advisory Council and Drs. Syvertsen, Futch, and McCoy at the University of Florida for support in conducting the greenhouse and the field studies in Florida.

REFERENCES

Agusti, M., C. Mesejo, C. Reig, and A. Martinez-Fuentes. 2013. Citrus production. In: *Tropical and Subtropical Crops*. Eds: Dixon G., and Aldous D.E. Springer, New York. p. 159–95.

Albrecht, U., F. Alferez, and M. Zekri. 2017. Citrus production guide: rootstock and scion selection. Publication HS-1308. Horticultural Sciences Department, IFAS Extension, University of Florida, Florida.

Alva, A.K., T.J. Baugh, K.S. Sajwan, and S. Paramasivam. 2004. Soil pH and anion abundance effects on copper adsorption. *Journal of Environmental Science and Health* 39:903–910.

Ashok, K., A.K. Alva, D. Mattos Jr., S. Paramasivam, B. Patil, H.T. Dou, and K.S. Sajwan. 2008. Potassium management for optimizing citrus production and quality. *International Journal of Fruit Sciences* 6:3–43.

Azam, M., M. Song, F. Fan, B. Zhang, Y. Xu, C. Xu, and K. Chen. 2013. Comparative analysis of flower volatiles from nine citrus at three blooming stages. *International Journal of Molecular Science* 14:46–67.

Blanke, M.M., and D.T. Cooke. 2004. Effects of flooding and drought on stomatal activity, transpiration, photosynthesis, water potential and water channel activity in strawberry stolons and leaves. *Plant Growth Regulation* 42:153–160.

Blazquez, C.H. 1991. Measurements of citrus tree health with a scanning densitometer from aerial color infrared photographs. *Plant Disease* 75:370–372.

Blum, A. 1996. Crop responses to drought and the interpretation of adaptation. *Plant Growth Regulation* 20:135–148.

Bove, J.M. 2006. Huanglongbing: a destructive, newly-emerging, century-old disease of citrus. *Journal of Plant Pathology* 88:7–37.

Byers, J.A., and S.J. Castle. 2005. Areawide models comparing synchronous versus asynchronous treatments for control of dispersing insect pests. *Journal of Economic Entomology* 98:1763–1773.

Castle, W.S., and S.H. Futch. 2015. Choose the right citrus rootstock. University of Florida, Horticultural Sciences Department, UF/IFAS Extension. Publication #HS1260.

Crowder, D.W., and D.W. Onstad. 2005. Using a generational time-step model to simulate dynamics of adaptation to transgenic corn and crop rotation by western corn rootworm (Coleoptera: Chrysomelidae). *Journal Economic Entomology* 98:518–533.

De Souza, M.C., E.S. Stuchi, A. de Goes. 2009. Evaluation of tangerine hybrid resistance to Alternaria alternate. *Scientia Horticulturae* 123:1–4.

Falivene, S., J. Giddings, S. Hardy, and G. Sanderson. 2006. Managing citrus orchards with less water. Primefact 427. NSW Department of Primary Industries, State of New South Wales, Australia.

FAO. 2016. Citrus fruit statistics 2015. http://www.fao.org/3/a-i5558e.pdf.

Fernandez-Ballester, G., F. Garrcia-Sanchez, A. Cerda, and V. Martinez. 2003. Tolerance of citrus rootstock seedlings to saline stress based on their ability to regulate ion uptake and transport. *Tree Physiology* 23:265–271.

Futch, S.H. 2003. Where the *Diaprepes* root weevils are. *Citrus Industry* 84:22–25.

Garrett, K.A., S.P. Dendy, E.E. Frank, M.N. Rouse, and S.E. Trave. 2006. Climate change effects on plant disease: genomes to ecosystems. *Annual Review of Phytopathology* 44:489–509.

Gmitter, F.G. 2004. Citrus rootstock and scion cultivar development. https://portal.nifa.usda.gov/web/crisprojectpages/0178451-ci trus-rootstock-and-scion-cultivar-development.html

Graham, J.H., D.B. Bright, and C.W. McCoy. 2003. *Phytophthora-Diaprepes* complex: *Phytophthora* spp. relationship with citrus rootstocks. *Plant Disease* 87:85–90.

Grosser, J.W., and J.L. Chandler. 2004. Production of twelve new allotetraploid somatic hybrid citrus breeding parents with emphasis on late maturity and cold-hardiness. *Journal of the American Pomological Society* 58:21–28.

Gutschick, V.P. 1999. Biotic and abiotic consequences of differences in leaf structure. *New Phytology* 143:3–18.

Hatch, A.C., and A.R. Blaustein. 2000. Combined effects of UV-B, nitrate, and low pH reduce the survival and activity level of larval *Cascades Frogs* (*Rana cascadae*). *Archives of Environmental Contamination and Toxicology* 39:494–499.

Jones, H.G, N. Archer, E. Rotenberg, and R. Casa. 2003. Radiation measurement for plant ecophysiology. *Journal of Experimental Botany* 54:879–889.

Jones, I.F., and W.J. Schroeder. 1983. Study of first-instar *Diaprepes abbreviatus* (Coleoptera: Curculionidae) activity for control purposes. *Journal of Economical Entomology* 76:567–569.

Kidd, P.S., and J. Proctor. 2001. Why plants grow poorly on very acid soils: are ecologists missing the obvious? *Journal of Experimental Botany* 52:791–799.

Klironomos, J.N., M.C. Rillig, and M.F. Allen. 1999. Designing belowground field experiments with the help of semivariance and power analyses. *Applied Soil Ecology* 12:227–238.

Lapointe, S.L. 2000. Thermal requirements for development of *Diaprepes abbreviatus* (Coleopera: Curculionidae). *Environmental Entomology* 29:150–156.

Laughlin, D.C., and S.R. Abella. 2007. Abiotic and biotic factors explain independent gradients of plant community composition in ponderosa pine forests. *Ecological Modeling* 205:231–240.

Li, H. 2009. Citrus tree abiotic and biotic stress and implication of simulation and modeling tools in tree management. *Tree and Forestry Science and Biotechnology* 3:66–78.

Li, H. 2015. Co-limitation of soil iron in root nodulation and chlorophyll-bean formation of yardlong-bean plants in tropical, humid environment. *International Journal of Bioscience, Biochemistry and Bioinformatics* 5:232–240.

Li, H. 2016. Botany and crop rotation management of new specialty Japanese melon in humid, tropical climate zones. Chapter 8. In: M. Pessarakli (ed.), *Handbook of Cucurbits*. Taylor & Francis Publishers, CRC Press, New York, pp. 139–148.

Li, H., S.H. Futch, R.J. Stuart, J.P. Syvertsen, and C.W. McCoy. 2007a. Association of soil iron with citrus tree decline and variability of soil pH, water, magnesium and *Diaprepes* Root weevil: two-site study. *Environmental and Experimental Botany* 59:321–333.

Li, H., S.H. Futch, and J.P. Syvertsen. 2007c. Cross-correlation patterns of air and soil temperatures, rainfall and citrus *Diaprepes abbreviatus* (L.) root weevil. *Pest Management Science* 63:1116–1123.

Li, H., S.H. Futch, J.P. Syvertsen, and C.W. McCoy. 2007d. Time series forecast and soil characteristics-based simple and multivariate linear models for management of *Diaprepes abbreviatus* (L.) root weevil in citrus. *Soil Biology & Biochemistry* 39:2436–2447.

Li, H., C.W. McCoy, and J.P. Syvertsen. 2007b. Controlling factors of environmental flooding, soil pH and *Diaprepes abbreviatus* (L.) root weevil feeding in citrus: larval survival and larval growth. *Applied Soil Ecology* 35:553–565.

Li, H., L.E. Parent, and A. Karam. 2006b. Simulation modeling of soil and plant nitrogen use in a potato cropping system in the humid and cool environment. *Agriculture, Ecosystem and Environment* 115:248–260.

Li, H., W.A. Payne, G.J. Michels and C.M. Rush. 2008. Reducing plant abiotic and biotic stress: drought and attacks of greenbugs, corn leaf aphids and virus disease in dryland sorghum. *Environmental Experimental Botany* 63:305–316.

Li, H., J.P. Syvertsen, C.W. McCoy, and A. Schumann. 2003. Soil redox potential and leaf stomatal conductance of two citrus rootstocks subjected to flooding and *Diaprepes* root weevil feeding. *Proceedings of Florida State Horticultural Society* 116:252–256.

Li, H., J.P. Syvertsen, R.J. Stuart, C.W. McCoy, and A. Schumann. 2005. Delineating management zones in citrus using overlay patterns of soil organic matter content, electrical conductivity and *Diaprepes* root weevil. In: D. Mulla et al. (eds.), *Proceedings of 7th International Conference of Precision Agriculture, ASA, CSSA, and SSSA*, Madison, WI, pp. 1962–1971.

Li, H., J.P. Syvertsen, R.J. Stuart, C.W. McCoy, and A. Schumann. 2006a. Water stress and root injury from simulated flooding and *Diaprepes* root weevil feeding in citrus. *Soil Science* 171:138–151.

Li, H., J.P. Syvertsen, R.J. Stuart, C.W. McCoy, A. Schumann, and W.S. Castle. 2004. Soil and *Diaprepes* root weevil spatial variability in a poorly drained citrus grove. *Soil Science* 169:650–662.

Lichenthaler, H.K. 1996. Vegetation stress: an introduction to the stress concept in plants. Journal of Plant Physiology (Germany) 148:4–14.

Liu, W.J., Y.G. Zhu, F.A. Smith, and S.E. Smith. 2004. Do iron plaque and genotypes affect arsenate uptake and translocation by rice seedlings (*Oryza sativa* L.) grown in solution culture? *Journal of Experimental Botany* 55:1707–1723.

Lower, S.S., S. Kirshenbaum, and C.M. Orians. 2003. Preference and performance of a willow-feeding leaf beetle: soil nutrient and flooding effects on host quality. *Oecologia* 136:402–422.

Mattson, W.J., and R.A. Haack. 1987. The role of drought in outbreaks of plant-eating insects. *BioScience* 37:110–118.

McCoy, C.W., R.J. Stuart, and N.N. Nigg. 2003. Seasonal life stage abundance of *Diaprepes abbreviatus* (L.) in irrigated and non-irrigated citrus plantings in central Florida. *Florida Entomology* 86:34–42.

Mielke, M.S., A.A.F. de Almeida, F.P. Gomes, M.A.G. Aguilar, and P.A.O. Mangabeira. 2003. Leaf gas exchange, chlorophyll fluorescence and growth responses of *Genipa americana* seedlings to soil flooding. *Environmental Experimental Botany* 50:221–231.

Mittler, R. 2006. Abiotic stress, the field environment and stress combination. *Trends in Plant Science* 11:15–19.

Nigg, N.N., S.E. Simpson, L.E. Ramos, T. Tomerlin, J.M. Harrison, and N. Cuyler. 2001. Distribution and movement of adult *Diaprepes abbreviatus* (Coleoptera: Curculionidae) in a Florida citrus grove. *Florida Entomology* 84:641–651.

Nigg, N.N., S.E. Simpson, R.J. Stuart, L.W. Duncan, C.W. McCoy, F.G. Jr. Gmitter. 2003. Abundance of *Diaprepes abbreviatus* (L.) (Coleoptera: Curculionidae) neonates falling to the soil under tree canopies in Florida citrus groves. *Horticultural Entomology* 96:835–843.

Obreza, T.A., and M.E. Collins. 2002. Common soils used for citrus production in Florida. Document No. SL 193. University of Florida, IFAS. Gainesville, Florida.

Oren, R., J.S. Sperry, B.E. Ewers, D.E. Pataki, N. Phillips, and J.P. Megonigal. 2001. Sensitivity of mean canopy stomatal conductance to vapor pressure deficit in a flooded *Taxodium distichum* L. forest: hydraulic and non-hydraulic effects. *Oecologia* 126:21–29.

Orians, C.M., and R.S. Fritz. 1996. Genetic and soil-nutrient effects on the abundance of herbivores on willow. *Oecologia* 105:388–396.

Pezeshki, S.R., and R.D. Delaune. 1998. Responses of seedlings of selected woody species to soil oxidation-reduction conditions. *Environmental and Experimental Botany* 40:123–133.

Ramirez-Rodriguez, V., J. Lopez-Bucio, and L. Herrera-Estrella. 2005. Adaptive responses in plants to nonoptimal soil pH. In: Jenks MA and Hasegawa PM (eds.), *Plant Abiotic Stress*. Blackwell Publishing, New York, pp. 145–170.

Ruiz-Sanchez, M.C., R.D. Domingo, D. Morales, and A. Torrecillas. 1996. Water relations of Fino lemon plants on two rootstocks under flooded conditions. *Plant Science* 120:119–125.

Quintela, E., and C.W. McCoy. 1997. Effects of imidacloprid on development, mobility, and survival of first instars of *Diaprepes abbreviatus* (Coleoptera: Curculionidae). *Journal of Economical Entomology* 90:988–995.

Rieger, M. 2005. *Introduction to Fruit Crops*. Haworth Food & Agricultural Products Press, New York, 462 p.

Rogers, S., J.H. Graham, and C.W. McCoy. 2000. Larval growth of *Diaprepes abbreviatus* (L.) and resulting injury to three citrus varieties in two soil types. *Journal of Economical Entomology* 93:380–387.

Saqib, M., J. Akhtar, and R.H. Qureshi. 2004. Pot study on wheat growth in saline and waterlogged compacted soil II. Root growth and leaf ionic relations. *Soil and Tillage Research* 77:179–187.

SAS Institute.1993. SAS/ETS User's guide. Version 6, 2nd ed. SAS Inst. Cary, NC.

Srivastava, A., and K.S. Singh. 2009. Citrus decline: soil fertility and plant nutrition. *Journal of Plant Nutrition* 32:197–245.

Stuart, R.J., D.I. Shapiro-Ilan, R.R. James, K.B Nguyen, and C.W. McCoy. 2004. Virulence of new and mixed strains of the entomopathogenic nematode *Steinernema riobrave* to larvae of the citrus root weevil *Diaprepes abbreviatus*. *Biological Control* 30:439–445.

Syvertsen, J.P., J.W. Grosser, and L.S. Lee. 2000. Growth and physiological characteristics of diploid and tetraploid citrus rootstock seedlings grown at elevated CO2. *Journal of American Horticultural Science Society* 125:228–234.

Syvertsen, J.P., and E.A. Hanlin. 2008. Citrus tree stresses: effects on growth and yield. Publication HS-1138. Horticultural Science Department, University of Florida, Granville, FL.

Tobin, P.C., S. Nagarkatti, and M.C. Saunders. 2001. Modeling development in Grape Berry Moth (Lepidoptera: Tortricidae). *Environmental Entomology* 30:692–699.

White, T.C.R. 1984. The abundance of invertebrate herbivores in relation to the availability of nitrogen in stressed food plants. *Oecologia* 63:90–105.

Worner, S.P. 1991. Use of models in applied entomology: the need for perspective. *Environmental Entomology* 20:768–773.

Yoo, M.S., and B.R. James. 2003. Zinc extractability and plant uptake in flooded, organic waste-amended soils. *Soil Science* 168:686–698.

Section IX

*Future Promises: Improving Plant
and Crop Adaptation/Tolerance and
Cultivation under Stressful Conditions*

41 Improving Crop Resistance to Abiotic Stresses Through Seed Invigoration

M. Farooq, A. Wahid, S.M.A. Basra, Abdul Rehman, and Kadambot H.M. Siddique

CONTENTS

41.1 INTRODUCTION

The global average surface temperature will rise by 1.5–2°C by the end of this century (Field 2014), which is likely to result in adverse episodes of extreme weather (cold or hot temperature) and changes in rainfall pattern (Fischer and Knutti 2015). This expected climate change will increase the incidence of abiotic stresses (heat, cold, drought, salinity, and waterlogging), which may result in huge losses in crop yield (Deryng et al. 2014; Loreti et al. 2016). Globally, ~34 Mha of irrigated land is affected by salinity, while salt and waterlogging combined affect 60 to 80 Mha (FAO 2011). High salinity is displacing 1.5 Mha of productive land annually (Munns and Tester 2008). Likewise, around 63.5 Mha of arable land is affected by drought (Narciso and Hossain 2002). Furthermore, heat stress, cold stress, salinity, and drought cause estimated yield losses of 40%, 15%, 17%, and 20%, respectively, with 8% from other factors (Ashraf and Harris 2005).

Temperature fluctuations strongly influence the growth, yield, and quality of crops (Wahid et al. 2007a). Chilling/cold stress can have an effect similar to drought stress, as water freezing concentrates the solutes in cells to create a state of water deficit for the plant. Due to the intricate interactions between stress factors and plant development mechanisms (Zhu 2002), tolerance to environmental stresses, at both the cellular and the whole plant level, is a complex phenomenon (Foolad et al. 2003; Farooq et al. 2009a, 2009b, 2009c, 2009d). The development of plant genotypes with high abiotic stress tolerance is a promising strategy for overcoming low crop yields in ever-changing climatic conditions and issues with food security in developed and developing countries. However, understanding the mechanisms of stress-related damage and tolerance is crucial. In the past three decades,

significant advances in biotechnological research have unraveled tolerance mechanisms to environmental stresses at the molecular level (Prabhavathi et al. 2002; Farooq et al. 2009a, 2009b, 2009d; Chen and Arora 2013; Kubala et al. 2015). The basic cellular responses of plants to environmental stresses are similar; depending on the stress, there may be inter- and intra-specific variations for tolerance mechanisms at different developmental stages (Zhu 2001a, 2001b, 2002; Farooq et al. 2009a, 2009b, 2009d). Likewise, oxidative and osmotic stresses and protein denaturation may occur in plants on exposure to environmental stresses, which can lead to adaptive responses, e.g., stress protein expression, accumulation of compatible solutes, and the antioxidant system at the cellular level (Zhu 2002). However, the expression of these adaptive responses is usually not enough to cope with abiotic stresses. Therefore, adaptation strategies need to be developed to improve a plant's ability to tolerate stress and sustain crop yield under the more variable climate predicted for the future.

Seed invigoration is a broad term involving several techniques aimed at improving seed delivery and crop performance under a wide range of environmental conditions (Farooq et al. 2009c). Different seed invigoration tools have been used for better seedling establishment or plant development/yield in stressful environments (Table 41.1; Farooq et al. 2008a, 2008b, 2008c, 2008d, 2009a, 2009b, 2009c, 2009d, 2009e, 2009f, 2009g; Fercha et al. 2014; Kubala et al. 2015; Zhang et al. 2015; Zheng et al. 2016; Farooq et al. 2017; Abid et al. 2018; Tabassum et al. 2018). Several review articles have discussed strategies to improve tolerance to abiotic stresses (e.g., Farooq et al. 2009a, 2009b, 2009c, 2009d; Jisha et al. 2013; Paparella et al. 2015; Savvides et al. 2016). There are also reviews on the role of pre-sowing seed treatments in

TABLE 41.1

Mechanism of Abiotic Stress Tolerance in Plants through Seed Invigoration Techniques

Plant	Seed Treatment	Stress	Mechanism	Reference
Indian mustard	Hydropriming	Iso-osmotic salt (150 mM NaCl) or PEG-8000 (20%) stress	↑ biomass, ↑ chlorophyll, ↑ SOD, ↑ GR, ↓ MDA	Srivastava et al. (2010)
Napa cabbage	Hydropriming	50, 100, 150, 200, 250 mM NaCl	↑ seed germination, ↑ seedling growth, ↑ POD, ↑ CAT, ↑ proline	Yan (2016)
Sunflower	Hydropriming	4, 8, 16 ds m⁻¹	↑ germination rate, ↑ seedling vigor, ↑ chlorophyll, ↑ TSS, ↑ proline	Matias et al. (2018)
Wheat	Osmopriming (−0.9 MPa PEG at 18 °C for 30 h)	35–40% FC at tillering and jointing	↑ phy, ↓ MDA, ↓ H_2O_2, ↑ CAT, ↑ APX, ↑ GR, ↑ GY	Abid et al. (2018)
Wheat	$CaCl_2$ (50 mg L⁻¹); osmopriming	10.65 dS m⁻¹	↑ stand establishment, ↑ TSP, ↑ phenolics, ↑ α-amylase, ↑ protease, ↑ K⁺, ↓ Na⁺, ↑ yield components, ↑ grain yield	Jafar et al. (2012)
Rice	Osmopriming (PEG 5%)	5% PEG	↑ seed germination, ↑ seedling growth, ↓ ROS, ↓ MDA, ↑ GPX activity	Goswami et al. (2013)
Rice	30% PEG	Nano-ZnO (NPs ≤100 nm), i.e. 0, 250, 500, and 750 mg L⁻¹	↑ photosynthetic pigments, ↓ oxidative MDA, ↓ CAT, ↓ POD, ↓ SOD	Salah et al. (2015)
Sorghum	Osmoprimed sorghum seeds (PEG 8000 20% solution)	15% soil moisture content	↑ emergence, ↑ chlorophyll, ↑ free amino acid, ↑ TSP, ↑ proline, ↑ reducing sugars, ↑ TSS, ↑ APX, ↑ SOD, ↑ CAT, ↑ POD, ↓ MDA	Zhang et al. (2015)
Trifolium repens L.	Osmopriming with PEG (−6.7 MPa)	Heavy metal stress (100 μM $CdCl_2$ or 1 mM $ZnSO_4$)	↑ emergence, ↑ seedling growth, ↑ light photosynthetic phase, ↓ MDA, ↓ Zn, ↓ Cd, ↓ alteration in root/leaf mesophyll cell ultrastructure	Galhaut et al. (2014)
Rice	Halopriming KNO_3 (0.5%)	15% PEG solution (pot); 15–18% MC (field)	↑ seedling growth, ↑ TSS, ↑ antioxidant activities, ↓ MDA	Zheng et al. (2016)
Wheat	Halopriming KCl and $CaCl_2$ (100 mM)	100 mM NaCl	↑ CAT, ↑ POD, ↑ APX, ↑ K⁺/Na⁺ ratio, ↓ Na⁺, ↑ proline, ↓ H_2O_2	Islam et al. (2015)
Cotton	Halopriming KNO_3 (2%)	Cold 18 °C	↑ seed germination, ↑ seedling growth, ↑ dry weight	Cokkizgin and Bolek (2015)
Silybum marianum L.	Halopriming KNO_3 (0.25 mM)	Salinity $CaCl_2$ (150 and 250 mM)	↑ seed germination, ↑ seedling growth, ↑ seed proteins, ↓ peroxidase	Zavariyan et al. (2015)
Trifolium repens L.	Hormonal priming with GA_3 (0.1 mM)	Heavy metal stress (100 μM $CdCl_2$ or 1 mM $ZnSO_4$)	↑ emergence, ↑ seedling growth, ↑ light photosynthetic phase, ↓ MDA, ↓ Zn, ↓ alteration in root/leaf mesophyll cell ultrastructure	Galhaut et al. (2014)
Rice	Seed priming with spd (0.5 mmol L⁻¹)	15% PEG solution (pot), 15–18% MC (field)	↑ seedling growth, ↑ TSS, ↑ antioxidant activities, ↓ MDA	Zheng et al. (2016)
Radish	Seed priming with 28-homobrassinolide (10⁻⁷ M)	Cr (0.5, 1.0, 1.5 mM)	↑ APX, ↑ CAT, ↑ GR, ↓ POD, ↓ MDA, ↑ photosynthetic pigments, ↑ TSP, ↑ proline	Sharma et al. (2011)
Tomato	0.5 mM salicylic acid	Heat stress 32/26 °C	↑ seed germination, ↑ TSS, ↑ lycopene, ↑ vitamin C, ↑ fruit yield	Singh and Singh (2016)
Faba bean	Seed priming with proline (5, 10 mM)	Salinity (3.3 and 6.25 dS m⁻¹)	↑ plant height, ↑ seedling dry weight, ↑ photosynthetic pigments, ↑ FAA, ↑ phenolics, ↑ proline, ↑ TSC, ↑ vessel diameter, ↑ fiber tissue thickness	Taie et al. (2013)
Wheat	0.1 mmol L⁻¹ SNP solution	Cold stress 12/8 °C	↑ seed respiration, ↑ seedling dry weight, ↑ α-amylase, ↑ SOD, ↓ MDA, ↓ H_2O_2	Li et al. (2013)
Strawberry	SNP (100 μM) or H_2O_2 (10 mM)	100 mM NaCl	↑ CAT, ↑ GR, ↑ MnSOD, ↑ MDHAR, ↑ DHAR	Christou et al. (2014)
Soybean	Seed coating with melatonin (300 μL per 100-seed reagent)	Salt (1% (w/v) NaCl) and drought (20% field capacity)	↑ growth, ↑ grain yield, ↑ fatty acids, ↑ gene expression of antioxidants	Wei et al. (2014)
Maize	Seed coating with SA (5% w/v)-loaded thermo-sensitive hydrogel (seeds:coating agents = 5:4 [w/w])	Chilling stress (5 °C)	↑ seed germination, ↑ seedling growth, ↑ seedling dry weight, ↑ antioxidant activities, ↓ MDA	Guan et al. (2015)

↑ = increase; ↓ = decrease; FAA = free amino acid; TSC = total soluble carbohydrates; FC = field capacity; MC = moisture content; GY = grain yield.

abiotic stress tolerance (Ashraf and Foolad 2005; Jisha et al. 2013; Savvides et al. 2016) and rice seed invigoration (Farooq et al. 2009c). This comprehensive review encompasses previous work on improving abiotic stress tolerance using seed invigoration approaches.

41.2 SEED INVIGORATION FOR STRESS RESISTANCE

Seed invigoration includes seed treatment techniques (seed priming, coating, and pelleting) that are helpful for enhancing crop performance (Farooq et al. 2009c). Seed priming is generally used interchangeably with seed invigoration (Farooq et al. 2006c) but includes several pre-sowing seed treatments. The four most common seed invigoration techniques are pre-sowing hydration methods (Basra et al. 2006; Farooq et al. 2006a, 2006b, 2006c, 2006d, 2007a, 2007b), low–molecular weight osmoprotectant seed treatments (Taylor et al. 1998), coating technologies (Song et al. 2005; Farooq et al. 2012; Rehman and Farooq 2016), and pre-sowing thermal treatments (Farooq et al. 2004b, 2005). These treatments focus on shortening the time to seedling emergence and protecting seeds from biotic and abiotic factors during critical phases of seedling establishment. Such treatments synchronize emergence, resulting in uniform and vigorous stands and improved yields. The role of seed invigoration in crop resistance to abiotic stresses is discussed in the following sections.

41.2.1 SEED HYDRATION TREATMENTS

Seeds require moisture, optimum temperatures, oxygen, and sometimes light for germination. Water uptake follows a triphasic pattern (Figure 41.1; Bewley 1997; Rajjou et al. 2012).

Phase I is imbibition, which commences with the physical uptake of water by seeds. Imbibition usually occurs rapidly due to the large difference in water potential between dry seeds and water. In living seeds, little metabolic activity occurs during this phase. Dead seeds imbibe water at the same rate as do viable seeds. Phase II is the lag period, when water uptake is minimal, but considerable metabolic activity is evident. The seed converts stored reserves (proteins, fats, and lipids) into compounds needed for germination. Phase III is radicle protrusion, which usually coincides with radicle emergence and is characterized by a rapid water uptake (rapid increase in fresh weight). Phases I and III are controlled by the amount of water available to seeds (Taylor et al. 1998). Seed priming improves stress tolerance in plants by (a) accelerating germination metabolism (e.g., endosperm weakening, respiration, and gene expression), which helps in the transition of quiescent dry seeds to the germination state under stressful conditions, and (b) the imposition of abiotic stress (exposure of plants to stratification, scarification, low–osmotic potential solutions of polyethylene glycol [PEG], salts, etc.), which stifles radicle protrusion but incites stress responses (e.g., accumulation of late embryogenesis abundant proteins [LEAs]), which possibly induce cross-tolerance. Both mechanisms build a priming memory (buildup of dormant stress tolerance signals) inside primed seed, which is conscripted (perception and systematic transduction of signals) when the plants are grown from primed seed exposed to stress and helps in stress tolerance by stimulating stress responses (Figure 41.2; Chen and Arora 2013). Patade et al. (2012) studied gene expression in priming-induced cold-tolerant *Capsicum annuum*; seedlings raised after thiourea priming (TU; 1.3 mM) and subjected to cold stress (4 °C for 24 h) showed either higher or earlier induced expression of genes involved in transcript

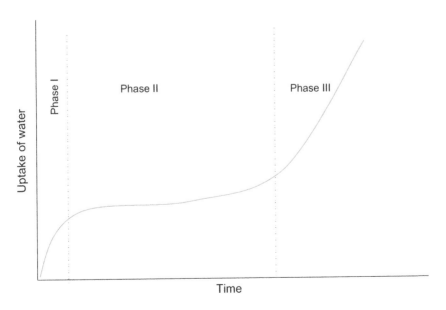

FIGURE 41.1 Tri-phasic curve of seed germination. During seed priming, seeds are placed in a solution with low water potential, which prevents them from imbibing sufficient water to germinate. As a result, the duration of Phase II is extended, which increases the hydrolysis of reserved food material. (Adapted from Farooq, M. et al., 2010.)

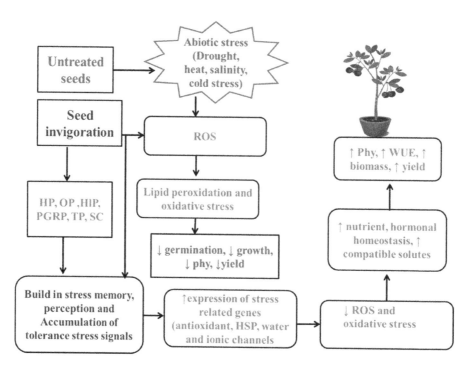

FIGURE 41.2 (See color insert.) Mechanism of seed invigoration induced stress tolerance. ↑ = increase, ↓ = decrease, HP = hydropriming, OP = osmopriming, HlP = halopriming, PGRP = plant growth regulator priming, TP = thermal priming, SC = seed coating, ROS = reactive oxygen species, LP = lipid peroxidation, CS = compatible solutes, Phy = photosynthesis, WUE = water use efficiency.

regulation (*CaWRKY30*), osmotic adjustment (*PROX1*, Osmotin), antioxidant defense (*Cu/Zn SOD*), and metabolite biosynthesis through the phenylpropanoid pathway (cinnamic acid hydroxylase [CAH]) than hydroprimed or untreated seeds (Table 41.2). Depending on the nature of the hydration method, pre-sowing hydration techniques can be classified into two groups: pre-soaking and seed priming. A brief account of each, with special reference to abiotic stress tolerance, is given in the following.

41.2.1.1 Pre-Soaking

For pre-soaking methods, seeds are soaked in water with or without aeration (Thornton and Powell 1992). As no osmoticum or salt is added, water imbibition is not controlled. There are several pre-soaking hydration techniques with potential for seed vigor enhancement: hydropriming (Farooq et al. 2006e), hardening (Farooq et al. 2004a), and on-farm priming (Harris et al. 2000). Of these, hydropriming has shown promise in improving crop resistance to abiotic stresses.

For hydropriming, seeds are soaked, misted in water, and then re-dried before completing germination, which minimizes chemical use and avoids discarding materials (Soon et al. 2000). In many studies, hydropriming substantially improved crop performance under stress conditions, e.g., improving the resistance to salinity in wheat (Basra et al. 2005; Fercha et al. 2013), rice (Jisha and Puthur 2014), barley (Kibite and Harker 1991), maize (Ashraf and Rauf 2001), sunflower (*Helianthus annuus* L.) (Kaya et al. 2006; Matias et al. 2018), pigeon pea (Jyotsna and Srivastava 1998), *Acacia* (Rehman et al. 1998), Indian mustard (*Brassica juncea* L.) (Srivastava et al. 2010), and Napa cabbage (*Brassica rapa*

subsp. pekinensis) (Yan 2016) and to drought in rice (Jisha and Puthur 2014). Hydroprimed Napa cabbage seed had improved resistance to salt stress through better seed germination and seedling growth, which was associated with higher catalase (CAT) and peroxidase (POD) activities and proline accumulation, resulting in lower malondialdehyde (MDA) levels (Table 41.1; Yan 2016). Srivastava et al. (2010) found that Indian mustard plants raised from hydroprimed seeds, when subjected to iso-osmotic salt (150 mM NaCl) or PEG-8000 (20%) stress, produced more biomass with higher chlorophyll content, superoxide reductase (SOD) and glutathione reductase (GR) activities, and lower MDA content than unprimed seeds. In the same study, seedlings from osmoprimed (CaCl₂) seeds accumulated lower levels of glycine betaine (GB), proline, total soluble sugars (TSS), and phenolics. Hydropriming performed better than osmopriming and hormonal priming with abscisic acid (ABA). In another study, hydropriming of sunflower seeds resulted in better seedling emergence, seedling growth, chlorophyll content, and TSS and proline accumulation under salt stress (4, 8, and 16 dS m⁻¹) than untreated seeds (Matias et al., 2018). Recently, Zheng et al. (2016) found that priming rice seeds with water improved seedling root and shoot growth and TSS with enhanced activities of α-amylase, SOD, POD, and CAT and reduced MDA levels under drought stress relative to untreated seeds. Fercha et al. (2013) studied the proteomics of salt tolerance induced by seed priming in wheat; hydroprimed seeds of wheat exhibited significant alterations in 72 proteins that are mostly involved in protein synthesis, proteolysis, metabolism, and disease response.

In summary, seed pre-soaking enhances plant tolerance to abiotic stresses by improving the plant's antioxidant system

TABLE 41.2

Changes in the Expression Pattern of Genes, Protein, and Transcription by Seed Treatment under Abiotic Stresses

Plant	Seed Treatment	Stress	Mechanism	Reference
Capsicum annuum	Thiourea priming (TU; 1.3 mM)	Cold stress	↑ *CaWRKY30, PROX1*	Patade et al. (2012)
Rice	Osmopriming (PEG 5%)	Drought	↑ *MnSOD, HSP70*	Goswami et al. (2013)
Alfalfa	Osmopriming (PEG −1 MPa)	Salinity	↑ 84 stress-related proteins	Yacoubi et al. (2013)
Rice	Osmopriming (PEG 30%)	Nano-Zn stress	↓ *APXa*, ↓ *APXb*, ↓ *CATa*, ↓ *CATb*, ↓ *CATc*, ↓ *SOD1*, ↓ *SOD2*, ↓ *SOD3*	Salah et al. (2015)
Rice (indica type)	Seed priming (spr [5 mM] and spd [5 mM])	Salinity	↑ *CAT*, ↑ *SOD*, ↑ *GR*, ↑ *ANS*, ↑ *BADH1*, ↑ *P5CS*, ↓ *PDH*, ↑ *SPDS*, ↑ *SPMS*, ↑ *SAMDC*, ↓ *PAO*, ↑ *NHX1*, ↑ *TRAB-1*, ↑ *WRKY-71*, ↑ *RbcS*	Paul and Roychoudhary (2017a)
Rice (aromatic type)	Seed priming (spr [5 mM] and spd [5 mM])	Salinity	↑ *SOD*, ↑ *APX*, ↑ *ANS*, ↑ *CAT*, ↑ *GR*, ↑ *P5CS*, ↑ *BADH1*, ↓ *PDH*, ↑ *HvProT*, ↑*TRAB-1*, ↑*WRKY-71*, ↑ *NCED3*, ↑*Osem*, ↓ *NHX1*, ↑ *RbcS*, ↑ *SAMDC*, ↑ *SPDS*, ↑ *SPMS*	Paul and Roychoudhary (2017b)
Rice (indica type)	Seed priming (spr [2.5 mM])	Salinity	↑ *CAT2*, ↑ *SOD1*, ↑ *SOD2*, ↑ *SOD3*	Paul and Roychoudhury (2016a)
Spinach	Osmopriming (PEG −0.6 MPa)	Drought, chilling	↑ *SoδTIP*, ↑ *SoPIP1;1*, ↑ *SoPIP2;1*	Chen et al. (2013)
Wheat	Ascorbic acid priming (0.5 mmol L⁻¹)	Salinity	↑ 35 stress-related proteins	Fercha et al. (2014)
Wheat	Ascorbic acid priming (0.5 mmol L⁻¹)	Salinity	Change in 83 protein expressions related to protein metabolism, antioxidant protection, repair processes, and methionine-related metabolism	Fercha et al. (2013)
Rapeseed	Osmopriming (PEG 15%)	Salinity	↑ *P5CS*, ↓ *PDH*	Kubala et al. (2015)
Soybean	Seed coating with melatonin (300 µL per 100-seed reagent)	Salt, drought	↑ *SUS6-1*, ↑ *SUS6-2*, ↑ *GroES-2*, ↑ *ADH1*, ↑ *ALDH3F*, ↑ *PFK, GAPC1*, ↑ *GAPCP-2*	Wei et al. (2014)

↑ = increase; ↓ = decrease

capacity and enhancing protein expression. However, priming duration is critical in this regard; soaked seeds must be re-dried to safe moisture levels before sowing.

41.2.1.2 Seed Priming

Seed priming is a controlled hydration technique that allows partial seed hydration to initiate germination-related metabolism without actual germination and then re-dries seed to facilitate routine handling (Bradford 1986). To control hydration, seeds are placed in solutions with high osmotic potential to prevent seeds from entering Phase III of hydration by extending and holding seeds within the lag phase (Phase II) (Figure 41.1; Rajjou et al. 2012). As seeds are metabolically active during this period, they convert stored reserves for germination, such that membrane and genetic repair is better than under normal imbibition. Seeds are then removed from the priming solution, rinsed with water, and dried. These seeds germinate faster than unprimed seeds (Harris et al. 2002; Farooq et al. 2006b).

Primed seeds usually have better and more synchronized germination than unprimed seeds (Farooq et al. 2009c) due to the reduced lag time to imbibition (Taylor et al. 1998; Welbaum et al. 1998; McDonald 2000; Brocklehurst and Dearman 2008; Rajjou et al. 2012), build-up of germination-enhancing metabolites (Farooq et al. 2006d; Rajjou et al.

2012), metabolic repair during imbibition (Burgass and Powell 1984; Bray et al. 1989), and osmotic adjustment (Bradford 1986). Seed priming can be accomplished by different means, as detailed in the following.

41.2.1.2.1 Osmopriming

Seeds are soaked in low–water potential solution during osmoconditioning or osmopriming (Jisha et al. 2013) to control water imbibition. Seeds are dipped in PEG, glycerol, sorbitol, mannitol, $CaCl_2$, KNO_3, KH_2PO_4, $MgSO_4$, etc. solutions to create a low water potential and prevent radicle protrusion (Jisha et al. 2013). Osmopriming has improved stand establishment in various crops under normal and suboptimal conditions (Table 41.1; Brocklehurst and Dearman 2008; Kubala et al. 2015; Zhang et al. 2015; Zheng et al. 2016; Farooq et al. 2017; Abid et al. 2018).

When planted in the field, primed seeds usually germinate quickly and evenly under stressful conditions. Osmoprimed seed improves crop performance under suboptimal growth conditions by enhancing germination, antioxidant activities, and osmolyte accumulation, reducing lipid peroxidation and damage by reactive oxygen species (ROS) (Li and Zhang, 2012; Goswami et al. 2013), and enhancing gene expression/transcription factors involved in stress signaling (Yacoubi et al. 2013; Salah et al. 2015). For instance, seeds osmoprimed with PEG improved drought tolerance in rice plants through better seedling germination, photosynthetic pigment, higher enzymatic antioxidant activities (Li and Zhang 2012; Goswami et al. 2013), osmolyte accumulation (free amino acids, TSP, proline, reducing sugars, and TSS contents), and enhanced expression of genes responsible for antioxidant activities (e.g., SOD) and stress-related proteins (HSP70) (Goswami et al. 2013). Recently, Abid et al. (2018) demonstrated that osmoprimed seeds under drought stress performed better than untreated seeds, as PEG priming invoked the stress memory (enhanced signal perception and transduction to initiate stress tolerance responses) in seeds on exposure to drought stress. Plants grown from osmoprimed seeds had lower ROS and MDA levels and higher CAT, APX, and GR activities, which resulted in better growth rates and higher biomass production and grain yields under water deficit (Abid et al. 2018). Likewise, osmoprimed sorghum seeds (PEG 8000 20% solution) improved stand establishment, chlorophyll content, compatible solutes (free amino acids, total soluble phosphorus [TSP], proline, reducing sugars, and TSS) and antioxidant activities (ascorbate peroxidase [APX], SOD, CAT, and POD) with lower MDA content under drought stress (Zhang et al. 2015). Osmopriming (with mannitol) and hydropriming substantially improved chickpea seedling growth under drought stress (Kaur et al. 2002). Sucrose phosphate synthase (SPS), amylase, sucrose synthase (SS), and acid and alkaline invertase activities in shoots improved substantially with seed priming. In addition, the specific activity of SPS in cotyledons of primed seedlings increased twofold (Kaur et al. 2002). Similarly, osmopriming (−1.0 MPa PEG) improved drought tolerance in asparagus (Bittencourt et al. 2004) and sugarcane (Patade et al. 2009).

Osmopriming with mannitol and hydropriming improved seedling growth and antioxidant activities and reduced lipid peroxidation in alfalfa under salt stress (Amooaghaie 2011). However, osmopriming was more effective at reducing salinity damage in alfalfa (Amooaghaie 2011). Improved salinity tolerance in osmoprimed (−1.0 MPa PEG 8000) alfalfa seedlings under salt stress was attributed to changes in the accumulation pattern of 84 proteins in osmoprimed seeds under salt stress compared with untreated seeds, which resulted in greater seedling vigor (Yacoubi et al. 2013). Osmopriming (PEG 6000) of rape (*Brassica napus* L. cv Libomir) seeds for 7 days at 25 °C before exposure to salt stress (100 mM NaCl) improved germination, proline accumulation, and up-regulation of the pyrroline-5-carboxylate synthetase (P5CS) gene, which augmented P5CS activity and down-regulated the proline dehydrogenase (*PDH*) gene (Kubala et al. 2015). Osmopriming has induced salinity tolerance in tomato (*Lycopersicon esculentum*) and asparagus (*Asparagus officinalis*) (Pill et al. 1991), stand establishment in cucumber (*Cucumis sativus*) primed with mannitol (Passam and Kakouriotis 1994) and sugarcane (Patade et al. 2009), and germination and stand establishment in Bermudagrass (*Cynodon dactylon*) (Al-Humaid 2002) through enhanced accumulation of osmolytes, expression of stress-tolerant protein, and

antioxidant activities (Zhu 2002; Yacoubi et al. 2013; Kubala et al. 2015).

The available evidence suggests that osmopriming improves crop resistance to chilling stress (Farooq et al. 2008c, 2008d). He et al. (2002) reported better and quicker germination of osmoprimed rice seeds under low-temperature (5 °C) stress. Likewise, Posmyk et al. (2001) observed that osmopriming could have mitigated chilling-induced injuries in soybean (*Glycine max* (L.) Merr.). Recently, Farooq et al. (2017) demonstrated that osmopriming chickpea seeds with CaCl₂ (−1.25 MPa) and hydropriming substantially improved chickpea performance under chilling stress. Seedlings raised from osmoprimed seeds had higher CO₂ assimilation, photochemical efficiency, α-amylase activity, trehalose content, and water relations under chilling stress. Likewise, osmopriming (−0.6 MPa PEG 8000) in *Spinacia oleracea* L. improved germination and stress tolerance through the enhanced accumulation of 19, 26, and 30 kDa dehydrin-like proteins (DLPs) on exposure to cold (5 °C) and drought stress (−0.8 MPa PEG 8000 at 10 °C), suggesting that cross-tolerance was initiated during priming and recruited on exposure to abiotic stress post-priming (Figure 41.2). Moreover, enhanced expression of *CAP85* and DLPs 30, 26, and 19 kDa was observed in osmoprimed seeds during germination (Chen et al. 2012). In another study, Chen et al. (2013) studied the expression of aquaporin (AQP) coding genes in spinach seeds during germination and post-germination under optimal and abiotic stresses (drought [−0.8 MPa PEG 8000 at 10 °C] and chilling [5 °C] after osmopriming [−0.6 MPa PEG 8000 at 15 °C for 1, 2, 4, and 8 d]). Generally, all AQPs were down-regulated under drought and chilling stress, but *SoδTIP*, *SoPIP1;1*, and *SoPIP2;1* expression was higher in osmoprimed seeds (2–4 d) that also had higher drought tolerance. Likewise, *SoPIP2;1* expression was higher in the osmoprimed seed with higher chilling tolerance (Table 41.2).

Osmopriming was effective in improving resistance to heavy metals in crop plants. For instance, Galhaut et al. (2014) reported that osmopriming of *Trifolium repens* L. with PEG (−6.7 MPa) and GA (0.1 mM) improved the performance under heavy metal stress (100 μM CdCl₂ or 1 mM ZnSO₄) by improving emergence and seedling growth and photosystem II (PSII) photochemical efficiency and reducing MDA levels and Zn and Cd accumulation in osmoprimed seedlings. Salah et al. (2015) found that rice plants grown from seeds primed with 30% PEG and subjected to Zn-nanoparticle stress had enhanced photosynthetic pigments and reduced oxidative damage due to lower MDA levels and antioxidant activities (CAT, POD, and SOD) through down-regulation of the CATa, CATb, CATc, SOD1, SOD2, SOD3, APXa, APXb, and CATa genes (Table 41.2; Salah et al. 2015). Moreover, osmopriming reduced the adverse effects of nano-ZnO stress on the ultrastructure of root and leaf mesophyll cells.

In summary, osmopriming improves seedling establishment and crop performance under stress conditions by reducing oxidative stress and augmenting the activities and expression of enzymes and proteins related to abiotic stress tolerance.

41.2.1.2.2 Halopriming

Halopriming involves soaking seeds in an aerated solution of inorganic salts with a low water potential to control water imbibition and prevent radicle protrusion; the seeds are then dried to closer to their original weight. Some studies have shown that halopriming significantly improves seed germination, seedling emergence and establishment, and final crop yield under stress (Table 41.1). For instance, halopriming improved salinity tolerance in *Festuca arundinacea* and *Festuca ovina* (Shakarami et al. 2011), pansy (Dorna et al. 2014), *Solanum lycopersicum* (Nakaune et al. 2012), *Cucumis melo* (Farhoudi et al. 2011), *Helianthus annuus* (Bajehbaj 2010), *Silybum marianum* (Zavariyan et al. 2015), and canola (Hassanpouraghdam et al. 2009).

Rice direct seeding is a common practice in upland rainfed areas. Du and Tuong (2002) demonstrated that seed priming with saturated $CaHPO_4$ and KCl was best for improving the stand establishment, phenology, and grain yield of direct-seeded rice. Several other studies found that halopriming with mixed salts improved the germination, stand establishment, and early seedling growth of rice under saline conditions (He et al. 2002; Ruan et al. 2003; Afzal et al. 2012; Jisha and Puthur 2014). Likewise, seed priming with KCl and $CaCl_2$ induced salinity tolerance in rice cultivars through better germination, seedling growth, and biomass production under salt stress (Afzal et al. 2012).

Seed priming with urea and KNO_3 improved germination, seedling growth, and proline accumulation in maize hybrids subjected to drought and salt stress (Anosheh et al. 2011). Ashraf and Ruaf (2001) observed that priming maize seeds with solutions containing NaCl, KCl, or $CaCl_2.2H_2O$ mitigated the adverse effects of salinity during germination. Halopriming with $CaCl_2.2H_2O$ was the most effective at improving germination and seedling growth. Recently, Zheng et al. (2016) demonstrated that halopriming (KNO_3; 0.5%) improved seedling growth and TSS, reduced MDA levels, and increased antioxidant activities relative to untreated seeds in rice under drought stress. Likewise, osmopriming ($CaCl_2$) and hormonal priming (ABA) of mustard seeds enhanced germination, seedling growth, and chlorophyll content under salinity and drought stress (Srivastava et al. 2010). Patade et al. (2009) observed significant improvements in the percentage and rate of sprouting in tolerant and moderately tolerant varieties of sugarcane by salt priming with NaCl (100 mM). The haloprimed seeds also reduced leaf senescence in sugarcane under salinity and drought stress (Patade et al. 2009).

Low-temperature stress can drastically affect normal plant function and metabolism, which reduces growth and yield. However, halopriming with KCl and $CaCl_2$ induced chilling tolerance in maize (Farooq et al. 2008a, 2008d, 2008e). Seeds primed in solutions of either 2.5% K_2HPO_4 or 2.5% K_2HPO_4 + KNO_3 improved chilling tolerance compared with untreated seeds, with the effect largely retained after the seeds were dried. Embryo phospholipid fractions and sterols increased during salt priming, and the proportion of phospholipid as diphosphatidyl glycerol also increased. An increase in diphosphatidyl glycerol content is likely due to the enhanced internal organization of mitochondrial membranes in primed seedlings (Basra et al. 1988).

Seed priming with $CaCl_2$, followed by KCl and NaCl, resulted in significant improvements in dry matter accumulation, net photosynthesis, and grain yield under salinity stress (Iqbal and Ashraf 2007b). Moreover, seed priming with these salts also reduced shoot Na^+, more so with KCl (Iqbal and Ashraf 2007b). Seed priming with NaCl alleviated the adverse effects of salinity during germination and later vegetative growth stages of pearl millet (Ashraf et al. 2002). Seed priming with $CaCl_2$, kinetin, salicylic acid (SA), or ascorbate improved yield-related traits and grain yield of wheat on salt-affected soil (Jafar et al. 2012). Seeds primed with $CaCl_2$ had better emergence, α-amylase and protease activities, and TSP and TS proteins than the other priming agents, but this was closely followed by ascorbate priming (Jafar et al. 2012). Seed priming with KNO_3 in soybean substantially increased germination, seedling growth, and biomass production under salt stress relative to unprimed seeds (Ahmadvand et al. 2012). Similarly, priming cotton seeds with KNO_3 accelerated the germination rate and induced chilling tolerance in cotton seedlings (Cokkizgin and Bolek 2015). Canola plants derived from primed seeds had relatively lower Na^+ content and higher K^+ content than those from unprimed seeds (Hassanpouraghdam et al. 2009). Germinating maize seeds primed with NaCl, KCl, or $CaCl_2$ had substantially higher Na^+, K^+, and Ca^{2+} concentrations, respectively, than unprimed seeds. The highest Cl^- contents were reported in germinating seeds primed with $CaCl_2$, followed by seeds primed with NaCl and KCl.

Resistance to salinity stress in germinating wheat was tested by osmopriming with different inorganic salts ($CaCl_2$, NaCl, and $CaSO_4$), with $CaCl_2$ being the most effective priming agent (Iqbal et al. 2006a). Likewise, wheat seeds primed with $CaCl_2$ germinated earlier than unprimed seeds. $CaCl_2$ priming also improved total sugars and non-reducing sugars in wheat plants under salinity stress (Afzal et al. 2008; Jafar et al. 2012). Ruan et al. (2002) reported improved seedling emergence, stand establishment, and seedling vigor index in flooded soil by priming with $CaCl_2$ and $CaCl_2$ + NaCl. Halopriming with $CaCl_2$ (100 mM) or KCl (100 mM) improved the resistance to salt stress (100 mM NaCl) in wheat by improving germination, proline and phenolic accumulation, and antioxidant activities, and reducing MDA and H_2O_2 levels, relative to untreated seeds (Islam et al., 2015). Seed priming with silicon (Na_2SiO_3) improved salt tolerance in wheat through the enhanced accumulation of K^+ and Ca^{2+} ions (Azeem et al. 2015). Seed priming with $CaCl_2$ (1.5%) improved salt tolerance (100 mM NaCl) in wheat by improving water relations, leaf expansion, osmolyte accumulation (proline and GB), and grain yield and reducing Na concentration and MDA level (Tabassum et al., 2017). Likewise, seed priming with 1.5% $CaCl_2$ improved leaf area, water relations, osmolyte accumulation, and grain yield in wheat under drought stress (Tabassum et al., 2018). Priming wheat seeds with $CaCl_2$ (50 mg) substantially improved seedling establishment and grain yield in saline soil by improving protease and α-amylase activities, leaf K⁺ concentration, total soluble phenolics, and total soluble proteins and reducing Na uptake (Jafar et al. 2012).

Several reports have indicated that along with improved stand establishment, halopriming enhanced the subsequent growth and ultimately, the final crop yield in broad bean (*Vicia faba*; Sallam 1999), rice (Ruan et al. 2002), and maize (Farooq et al. 2008a–c), grain yield in wheat (Jafar et al. 2012) and soybean (Eleiwa 1989), and both growth and seed yield in *Pennisetum americanum* and sorghum (Kadiri and Hussaini 1999) and rice (Du and Toung 2002) under stress conditions.

To summarize, halopriming is a cost-effective tool to improve crop stand establishment and subsequent growth and yield in stress-prone environments by improving antioxidant activities, membrane stability, and ionic homeostasis.

41.2.1.2.3 Matripriming

For matripriming, seeds are mixed with moist solid matrix carriers that create matrix forces to hold water and facilitate its slow absorption by the seed (Beckman et al. 1993; Taylor et al. 1998), such as granulated clay particles, sand, or vermiculite (Gray et al. 1990; Hardegree and Emmerich 1992a, 1992b). After treatment, the seed is separated from the solid carrier, washed, and allowed to dry.

Matripriming can improve crop performance under suboptimal field conditions. For instance, matripriming (sand priming) improved cold stress resistance in direct-seeded rice (Zhang et al. 2006), salinity resistance in maize (Zhang et al. 2007b), the germination percentage of alfalfa in saline soil (Hu et al. 2006), and the germination rate in wheat cultivars (Basra et al. 2003). Pandita et al. (2010) found that solid matripriming along with *Trichoderma viride* improved emergence and yield in okra under chilling stress. Likewise, matripriming enhanced seed germination in soybean (Mercado and Fernandez 2002). In celery, matripriming substantially reduced the adverse effects of heat stress on seed germination (Parera et al. 1993). Matripriming hastened seed germination and increased seedling growth and dry weights compared with unprimed seeds in maize (Zhang et al. 2007b) and improved germination and stand establishment in bitter gourd (*Momordica charantia*) under chilling stress (Lin and Sung 2001). QianQian et al. (2009) demonstrated that seed priming with vermiculite enhanced hot pepper vigor and salinity tolerance.

In conclusion, although research reports are limited, matripriming has potential, and is an inexpensive alternative, for improving crop stress resistance.

41.2.1.2.4 Priming with Hormones and Other Organic Sources

Seed soaking in a solution containing plant growth regulators (PGRs) can improve crop performance under stressful conditions (Jisha et al. 2013). Several growth regulators are commonly used for seed priming, including auxins (indole acetic acid [IAA], indole-3-butyric acid [IBA], and 1-naphthalene acetic acid [NAA]), ABA, gibberellins (GAs), ascorbic acid (AA), polyamines (PAs), kinetin, brassinolides, jasmonic acid (JA), SA, triacontanol, and benzyl aminopurine (BAP). Seed treatment with PGRs improved crop germination under suboptimal conditions. The use of PGRs in the priming solution and other pre-sowing treatments substantially improved resistance to abiotic stresses in many vegetable and field crops (Table 41.1; Farooq et al. 2008c, 2009e, 2009f, 2013; Afzal et al. 2013; Galhaut et al. 2014; Razaji et al. 2014; Bajwa et al. 2018).

Priming wheat seeds with GA induced salt tolerance by modulating ionic uptake and hormonal homeostasis and improving grain yield in salt-affected soil (Iqbal et al. 2011). In wheat, seed priming with AA, SA, and kinetin improved crop productivity under salt stress by improving K⁺ accumulation and limiting Na⁺ and Cl⁻ uptake (Figure 41.2; Afzal et al. 2013). Similarly, seed priming with kinetin (150 mg L⁻¹) enhanced photosynthesis, water use efficiency (WUE), and grain yield under salt stress (Iqbal and Ashraf 2005a). Priming wheat seed in a solution containing tryptophan (Trp) improved grain yields on saline soil and induced salt resistance through ionic homeostasis, while seed primed with IAA and Trp accumulated more SA than unprimed seed (Iqbal and Ashraf 2007a).

Kim et al. (2006) found that IAA and GA both induced salt tolerance in rice, but IAA was better than GA. Priming rice seeds with SA (1 mmol L⁻¹) substantially improved seedling growth, WUE, photosynthesis, photosynthetic pigments, proline contents, and SOD and POD activities, and reduced CAT activity under drought stress (Li and Zhang 2012). Hormonal priming of wheat seeds with AA improved germination, plant growth, water relations, and grain yield under low-moisture stress (Farooq et al. 2013). Seed priming with GA and ABA-induced drought tolerance in tall wheatgrass (*Agropyron elongatum* Host) by enhancing SOD and CAT activities (Eisvand et al. 2010). Likewise, seed priming with AA improved germination, seedling growth, and antioxidant activities in rapeseed under water deficit (Razaji et al. 2014).

Hormonal priming can help plants to tolerate heavy metals. Hormonal priming of *Trifolium repens* with GA₃ (0.1 mM) improved its performance under heavy metal stress (Cd and Zn) through better stand establishment and photosynthesis and lower MDA levels (Galhaut et al. 2014). Priming wheat seeds with AA (0.5 mmol L⁻¹) induced salt tolerance, and the primed seeds exhibited significant changes in expression of 83 proteins involved in antioxidant protection, repair, protein metabolism, and methionine-linked metabolism (Fercha et al. 2013). In another study, Fercha et al. (2014) reported that AA priming improved salt tolerance (NaCl 250 mmol L⁻¹) during early seedling growth in wheat by increasing/decreasing changing protein abundance in the embryo and surrounding tissues through changes in embryo gene expression involved in metabolic processes and protein synthesis.

SA is an endogenous phytohormone of phenolic nature with high antioxidant activity that regulates several physiological and biochemical processes in plants (Hayat et al. 2010), such as photosynthesis (Hayat et al. 2005), water relations (Rehman et al. 2012), and membrane stability (Farooq et al. 2008c). In maize, hormonal priming with SA enhanced thermal-stress resistance by hastening emergence with better seedling growth, biomass production, water relations, and membrane stability (Rehman et al. 2012). Likewise, seed priming with JA or SA enhanced growth and carbohydrate accumulation and induced chilling tolerance in sunflower (Gornik and Lahuta 2017). Seed priming with acetyl SA or JA in KNO₃ solution increased chilling resistance in sweet pepper (Korkmaz 2005). Recently, Singh and Singh (2016) demonstrated that seed priming with SA (0.25, 0.5, and 0.75 mM) improved vegetative and reproductive growth with better fruit quality and high yields under heat stress. Seed priming with SA induced chilling tolerance in maize (Farooq et al. 2008c), whereas SA and polyamines induced drought tolerance in rice (Farooq et al. 2009e, 2009f). Likewise, SA and AA priming augmented seedling emergence, growth, and dry weight under salt stress (Afzal et al. 2005, 2006).

Several other compounds have been tested for their potential use as priming agents, including BAP alone or in combination with sorghum water extract to induce resistance to salinity stress in wheat through the enhanced accumulation of TSS, TSP, phenolics, chlorophyll content, leaf K⁺ concentration, and α-amylase activity (Bajwa et al. 2018). The use of 5-aminolevulenic acid (25 and 50 ppm ALA) in the priming medium (KNO₃ solution) induced low-temperature stress tolerance in red pepper (*Capsicum annuum* cv. Sena) and enhanced the seedling germination even after storage at 4 °C and 25 °C for 1 month (Korkmaz and Korkmaz 2009). Seed priming with Se (75 µM) improved drought tolerance in wheat by improving seedling emergence and root development. Moreover, Se-priming seed enhanced soluble sugars and free amino acid accumulation (Nawaz et al. 2013). Seed priming with kinetin (150 mg L⁻¹) improved CO₂ fixation, WUE, and grain yield of wheat under salt stress (Iqbal and Ashraf 2005a). Seed priming of *Vigna radiata* (L.) Wilczek with β-aminobutyric acid (BABA) (1 mM) improved seedling growth and biomass production, PSI and PSII activity, chlorophyll fluorescence, mitochondrial activity, SOD, POD, and nitrate reductase (NR) activities, and proline, total protein and carbohydrate accumulation, and reduced MDA levels (Jisha and Puthur 2016).

PAs are low–molecular weight aliphatic amines that play a crucial role in several physiological and biochemical processes of plants related to growth and development (Alcázar et al. 2010; Roychoudhury et al. 2011) and stress tolerance (Alcázar et al. 2010). Seed priming with PAs can reduce the adverse effects of abiotic stresses in plants (Hussain et al. 2013; Chunthaburee et al. 2014). Seed priming with putrescine (0.1 mM) improved resistance to drought stress in maize by improving seedling growth and plant water relations (Hussain et al. 2013). Likewise, the incorporation of putrescine in the priming medium enhanced chilling tolerance in tobacco (*Nicotiana tabacum*) by improving germination, seedling growth, and the antioxidant defense system in plant cells (Xu et al. 2011). Rice seed priming with spermidine (spd; 0.5 mmol L⁻¹) substantially improved seedling emergence, growth, TSS, and starch metabolism, due to higher α-amylase activity, and reduced ROS damage due to

higher CAT, POD, and SOD activities and lower MDA levels (Zheng et al. 2016). In a laboratory study, Khan et al. (2012) demonstrated that seed priming of hot pepper in putrescine, spermidine. or spermine solution improved germination and seedling growth relative to untreated seeds. Similarly, Naidu and Williams (2004) reported progressive improvements in germination and seedling vigor in rice seedlings grown from seeds primed with nitrogenous compounds, including PAs (betaine, putrescine, spermine, spermidine, and proline) under low temperature. Iqbal and Ashraf (2005b) opined that PAs (putrescine, spermine, or spermidine) are effective at reducing the adverse effects of salinity stress in wheat, but the effect of PAs was genotype specific. Moreover, seed priming with PAs (spermidine and spermine) enhanced the expression of genes related to enzymatic antioxidants, PA synthesis, and RuBisCO activity in indica and aromatic rice under salt stress (Paul and Roychoudhury 2016; Paul et al. 2017a, 2017b). Seed priming with spermidine and GA induced salt tolerance in rice by lowering the salinity-induced decline in photosynthetic pigments, Na uptake, ROS levels with higher phenolic accumulation, and antioxidant activities. Furthermore, spd was more effective than GA for improving salt tolerance in black glutinous rice (Chunthaburee et al. 2014). Recently, Zheng et al. (2016) demonstrated that spd priming (0.5 mmol L⁻¹) enhanced drought tolerance in direct-seeded rice through better seedling emergence, growth, dry weight, and enzyme activities.

The brassinosteroids (BRs) are a group of phytohormones that perform various roles in physiological and biochemical processes of plants related to seed germination, senescence, and abiotic resistance (Rao et al., 2002). Seed priming with BR substantially enhanced the activities of antioxidant enzymes (SOD, POD, and CAT) in alfalfa genotypes (Hu et al. 2006). Furthermore, during seedling growth, primed seeds had significantly lower MDA accumulation than unprimed seeds. Sand priming enhanced the activities of CAT, POD, and SOD and soluble sugar content and reduced MDA accumulation under salt stress (Hu et al. 2006). Similarly, Zhang et al. (2007a) found that alfalfa seed priming with BR increased seedling growth and biomass production. In radish (*Raphanus sativus* L.), seed priming with 28-homoBL ameliorated the adverse effects of Cr stress by increasing antioxidant activities and proline and protein accumulation and reducing lipid peroxidation (Sharma et al. 2011).

In summary, pre-soaking seed treatments with PGR and other similar organic substances improved crop resistance to abiotic stresses by regulating starch metabolism, plant water relations, the antioxidant defense system, and hormonal and nutrient homeostasis. Since PGRs are costly, field application is unlikely unless commercial formulations at affordable prices are developed.

41.2.1.2.5 Priming with Low–Molecular Weight Osmolytes and Stress Signaling Molecules

Many plants synthesize different organic compounds called *osmolytes/osmoprotectants* that can help them tolerate stresses. These compounds, including proline, GB, and TSS

(Hoque et al. 2007), maintain cell turgor pressure and help to stabilize the structure and function of key macromolecules under stressful conditions. Exogenous application of these compounds can improve abiotic stress tolerance (Table 41.1; Posmyk and Janas 2007; Taie et al. 2013). For instance, seed priming with proline enhanced salt tolerance in faba bean through increased seedling growth and photosynthesis and reduced Na^+ and Cl^- uptake (Taie et al. 2013).

Seed priming of mungbean with proline improved seedling germination and growth at low temperature (Posmyk and Janas 2007). Wahid and Shabbir (2005) demonstrated that barley plants grown from GB-primed seeds performed better under high-temperature stress than unprimed seeds, as GB priming improved photosynthesis, water relations, and membrane stability under heat stress. A pre-sowing seed treatment of IAA and GB for rice effectively ameliorated the adverse effects of chilling stress (Chen et al. 2005). Seed priming with GB induced drought tolerance in rice by enhancing antioxidant capacity and maintaining cell membrane stability under low water stress (Farooq et al. 2008f). Similarly, GB-primed maize seeds had better germination and stand establishment under low-temperature stress than unprimed seeds (Farooq et al. 2008b).

Stress signaling molecules in plants, such as ROS, have received attention mostly for their damaging and protective effects. However, seed priming with these molecules influences the cellular signaling function related to the regulation of transcriptional and posttranscriptional processes. Many reports show the effects of seed priming with ROS against abiotic stress, especially H_2O_2, when used in low concentration as priming agents (Fotopoulos et al. 2015; Savvides et al., 2016). Reactive N species and nitrous oxide (NO) are also among the most studied compounds of this group (Savvides et al., 2016). Exogenous application of stress signaling molecules as seed priming or a foliar spray improved abiotic stress tolerance in several plant species (reviewed by Savvides et al., 2016), as seed primed with ROS had higher antioxidant capacities and α- and β-amylase activities, which led to enhanced sugar accumulation and starch degradation and hence, to higher respiration, seed viability, and germination under stressful conditions (Zheng et al. 2009; Christou et al. 2013, 2014).

Seed priming with NO or H_2O_2 enhanced photosynthetic pigments, chlorophyll fluorescence, and relative water content, and reduced electrolyte leakage and lipid peroxidation, under salt stress compared with unprimed seeds (Christou et al. 2014). Furthermore, plants raised from primed seed mitigated oxidative stress with better redox homeostasis and expression of antioxidant transcription factors (cytosolic APX [cAPX], CAT, GR, monodehydroascorbate reductase [MDHAR], dehydroascorbate reductase [DHAR], and MnSOD) along with ascorbate and glutathione biosynthesis in strawberry leaves (Christou et al. 2014). Seed priming with NO in wheat improved seed germination and crop establishment under salt stress through enhanced seed respiration rate and ATP synthesis with lower starch and higher soluble sugar content. Further, NO-primed seedlings showed reduced MDA content and H_2O_2 and superoxide anions ($O_2^{\bullet-}$) and increased CAT and SOD activities (Zheng et al. 2009). Similar observations were reported by Duan et al. (2007) in wheat under salt stress. Similarly, wheat seed priming with H_2O_2 enhanced stand establishment and dry matter accumulation under salt stress with enhanced nutrient uptake and heat-stable proteins (Wahid et al. 2007b). A pre-sowing seed treatment with H_2O_2 improved seedling growth in rice under low-temperature stress (Sasaki et al. 2005). Likewise, seed priming with SNP as an NO donor in wheat improved germination and seedling growth under chilling stress (Li et al. 2013). Seed priming with SNP improved seedling growth, plant water relations, and general metabolism under water deficit conditions in rice (Farooq et al. 2009g).

In summary, seed treatments with low–molecular weight osmolytes and stress signaling molecules may improve crop resistance to abiotic stresses; hence, their application on a commercial scale may be beneficial.

41.2.2 THERMAL AND RADIATION TREATMENTS

Exposure of seeds to high or low temperature for a certain period helps improve germination and seedling growth under suboptimal conditions. Generally, dry heat treatments are used to control seed-borne pathogens (Fourest et al. 1990) and break seed dormancy (Zhang 1990; Dadlani and Seshu 1990). Quite often, high-temperature exposure of seeds in dry heat results in the loss of viability and seed vigor, but it has proved advantageous in breaking rice seed dormancy (Lee et al. 2002).

Stratification of seeds before sowing is common practice to prevent precocious germination and improve germination and tolerance to abiotic stresses (Bewley and Black 1994). For instance, a pre-sowing chilling treatment improved salt tolerance in wheat through better seedling germination, photosynthesis, biomass production, and grain yield (Iqbal and Ashraf 2010). Furthermore, a pre-sowing chilling treatment improved the levels of IAA, ABA, SA, and spd with higher K^+ and Ca^+ uptake and lower Na accumulation (Iqbal and Ashraf 2010). Likewise, Wang et al. (2011) demonstrated that cold stratification improved wheat seedling growth under saline conditions. Similarly, Zhang et al. (2018) reported improved wheat seedling growth under salt stress after seed stratification alone or with halopriming. Sharma and Kumar (1999) demonstrated that pre-sowing chilling treatments of Indian mustard seeds improved salinity tolerance by accelerating K^+ and Ca^{2+} accumulation and decreasing Na^+ uptake. Likewise, incubation of pearl millet (*Pennisetum glaucum*) seed for 2 days at 5 °C enhanced the germination percentage under salt stress (Ashraf et al. 2003). Similar stress tolerance responses have been observed in parsnip (*Pastinaca sativa*) (Finch-Savage and Cox 1982) and Indian grass (*Sorghastrum nutans*) (Watkinson and Pill 1998).

In summary, while limited information is available regarding the potential of thermal treatments to improve crop resistance to abiotic stresses, such treatments may improve seedling establishment and crop performance under stress conditions, and their potential needs to be explored.

41.2.3 SEED COATING AND PELLETING

Seed coating is the practice of covering seed coats with external materials (chemicals, nutrients, or PGRs) to improve protection and handling and to some extent, germination and stand establishment (Pedrini et al. 2016). In seed coating, a finely ground solid or liquid form of the desired material is stuck to the seed surface using inert materials to form a continuous layer covering the seed coat (Scott 1989).

Seed pelleting with lime improved stand establishment in grain and pasture legumes in acidic soils (Pijnenborg and Lie 1990). A decade ago, Murata et al. (2008) established that seed pelleting with chloride, nitrate, sulfate, and carbonate of calcium, or Calcimax, decreases seedling mortality in low-pH soils. Seed pelleting of groundnut with Ca improved plant growth (Murata et al. 2008). Among the tested pelleting materials, $CaCO_3$ was the most effective at reducing seedling mortality (Murata et al. 2008). Coating seeds of native plant species for ecological restoration has received little attention (Madsen et al. 2012, 2014; Williams et al. 2016). So far, this field remains overlooked by many seed scientists and seed technology companies.

Guan et al. (2015) demonstrated that a thermoresponsive coating agent containing poly (N-isopropylacrylamide-co-butylmethacrylate [P(NIPAm-co-BMA)] hydrogel and SA substantially improved germination rate, seedling growth, and antioxidant activities, and reduced MDA content under chilling stress. They further reported that the thermoresponsive coating was more effective than the SA coating alone. Likewise, seed coating with SA (0.5, 1, and 1.5 g kg^{-1} seed) using hydrogel on pelleted tobacco seed improved final emergence and seedling growth under drought stress (Guan et al. 2014). Wei et al. (2014) demonstrated that coating soybean seed with melatonin enhanced growth, grain yield, and fatty acid contents under drought and salt stress (Table 41.1). Transcriptomic analysis showed that melatonin enhanced the expression of genes involved in cell division, fatty acid synthesis, carbohydrate metabolism, and photosynthesis. Furthermore, it also enhanced the expression of genes involved in secondary metabolite synthesis and oxidoreductase activity. Zhang et al. (2007c) inferred that rice seedlings grown from cold-tolerant seed-coating agents were improved under chilling stress by maintaining higher root vigor, chlorophyll content, soluble sugars, free proline, and CAT, POD, and SOD activities and had lower MDA content, electrolyte leakage, and plant injury rates than the control.

In conclusion, seed coating and pelleting treatments, with the help of private enterprise, may have the potential to formulate and market seeds for a particular environment.

41.3 ON-FARM PRACTICAL APPLICATIONS OF INVIGORATION TECHNIQUES

The findings of any research endeavor remain unproductive until adopted by the end user. Recent research efforts in many crop species have shown that soaking seeds in water followed by surface drying before sowing (on-farm priming) speeds up germination and emergence and results in more vigorous seedling growth, resulting in higher-yielding crops (Harris et al. 1999, 2000; Musa et al. 1999). On-farm seed priming is a simple, low-cost, low-risk method for promoting rapid seedling establishment and vigorous and faster seedling growth.

In Pakistan, on-farm seed priming of mung bean cv. NM 92 substantially enhanced grain yield compared with unprimed seeds (Rashid et al. 2004). Similarly, on-farm priming (water) of maize seed improved germination at low temperature in the field and enhanced growth and grain yield (Finch-Savage et al. 2004). Harris et al. (2002) reported that upland rainfed areas use primed rice seed to improve stand establishment, phenology, and grain yield in rice. Several farmers now use seed invigoration techniques to boost crop performance under adverse soil conditions. Of these, on-farm seed priming is the most popular in farming communities (personal observation).

41.4 CONCLUSIONS

Seed invigoration tools have the potential to improve emergence and stand establishment under a range of field conditions. Of interest are hydropriming, osmopriming, halopriming, hormonal priming, and the use of highly soluble and low–molecular weight chemicals. These techniques may enhance crop performance under saline, submerged, and drought-prone conditions and on marginal land. During the past few years, the molecular mechanisms of seed priming–induced abiotic stress tolerance have received attention, and the expression patterns of many genes and proteins in primed seeds have been identified. Yet, the molecular mechanisms of seed priming–induced stress tolerance are not clear. Variation exists within crops and varieties/genotypes/hybrids in their responses to various priming treatments, which will enable researchers to identify useful accessions for further work, such as

- More precise invigoration techniques using a range of salts, PGRs, jasmonates, osmolytes, and stress signaling compounds at varying concentrations and durations
- Optimal water potential, temperature range, and requirements for oxygenation
- Commercial fertilizers as priming and seed-coating agents
- Performance of invigorated seeds under a wide range of field conditions
- Thermal treatments with alternate cycles of low and high temperatures
- Storage potential of primed seeds—prolonged storage of primed and hardened seeds may be critical for technology transfer and the marketing of primed seeds
- Exploration of mechanism(s) of combined stress tolerance through seed invigoration techniques

REFERENCES

Abid, M., A. Hakeem, Y. Shao, Y. Liu, R. Zahoor, Y. Fan, J. Suyu, S.T.Ata-Ul-Karim, Z. Tian, D. Jiang and J.L. Snider. 2018. Seed osmopriming invokes stress memory against postgerminative drought stress in wheat (*Triticum aestivum* L.). *Environmental and Experimental Botany* 145:12–20.

Afzal, I., A. Butt, H.U. Rehman, S.M.A. Basra and A. Afzal. 2012. Alleviation of salt stress in fine aromatic rice by seed priming. *Australian Journal of Crop Science* 6:1401–1407.

Afzal, I., S. Rauf, S.M.A. Basra and G. Murtaza. 2008. Halopriming improves vigor, metabolism of reserves and ionic content in wheat seedling under salt stress. *Plant Soil and Environment* 549:382–388.

Afzal, I., S.M.A. Basra, M. Farooq and A. Nawaz. 2006. Alleviation of salinity stress in spring wheat by hormonal priming with ABA, salicylic acid and ascorbic acid. *International Journal of Agriculture and Biology* 8:23–28.

Afzal, I., S.M.A. Basra, M.A. Cheema, M. Farooq, M.Z. Jafar, M. Shahid and A. Yasmeen. 2013. Seed priming: a shotgun approach for alleviation of salt stress in wheat. *International Journal of Agriculture and Biology* 15:1199–1203.

Afzal, I., S.M.A. Basra, N. Ahmad and M. Farooq. 2005. Optimization of hormonal priming techniques for alleviation of salinity stress in wheat (*Triticum aestivum* L.). *Caderno de Pesquisa série Biologia* 7:95–108.

Ahmadvand, G., F. Soleimani, B. Saadatian and M. Pouya. 2012. Effects of seed priming on germination and emergence traits of two soybean cultivars under salinity stress. *Journal of Basic and Applied Science Research* 3:234–241.

Alcázar, R., T. Altabella, F. Marco, C. Bortolotti, M. Reymond, C. Koncz, P. Carrasco and A.F. Tiburcio. 2010. Polyamines: molecules with regulatory functions in plant abiotic stress tolerance. *Planta* 231(6):1237–1249.

Al-Humaid, A.I. 2002. Effects of osmotic priming on seed germination and seedling growth of bermudagrass (*Cynodon dactylon* L.) under saline conditions. *Bulletin of the Faculty of Agriculture*, Cairo University 53:265–274.

Amooaghaie, R. 2011. The effect of hydro and osmopriming on alfalfa seed germination and antioxidant defenses under salt stress. *African Journal of Biotechnology* 10:6269–6275.

Anosheh, H.P., H. Sadeghi and Y. Emam. 2011. Chemical priming with urea and KNO3 enhances maize hybrids (*Zea mays* L.) seed viability under abiotic stress. *Journal of Crop Science and Biotechnology* 14(4):289–295.

Ashraf, M. and H. Rauf. 2001. Inducing salt tolerance in maize (*Zea mays* L.) through seed priming with chloride salts: Growth and ion transport at early growth stages. *Acta Physiologia Plantarum* 23:407–414.

Ashraf, M. and P.J.C. Harris. 2005. Abiotic stresses. *In*: *Plant Resistance Through Breeding and Molecular Approaches*. New York: Haworth Press.

Ashraf, M., and M.R. Foolad. 2005. Pre-sowing seed treatment – a shotgun approach to improve germination, plant growth, and crop yield under saline and non-saline conditions. *Advances in Agronomy* 88:223–271.

Ashraf, M., Kausar, A. and Ashraf, M.Y. 2003. Alleviation of salt stress in pearl millet (*Pennisetum glaucum* (L.) R. Br.) through seed treatments. *Agronomie* 23:227–234.

Ashraf, M.Y., G. Sarwar, M. Ashraf, R. Afaf and A. Sattar. 2002. Salinity-induced changes in amylase activity during germination and early cotton seedling growth. *Biologia Plantarum* 45:589–591.

Azeem, M., N. Iqbal, S. Kausar, M.T. Javed, M.S. Akram and M.A. Sajid. 2015. Efficacy of silicon priming and fertigation to modulate seedling's vigor and ion homeostasis of wheat (*Triticum*

aestivum L.) under saline environment. *Environmental Science and Pollution Research* 22:14367–14371.

Bajehbaj, A.A. 2010. The effects of NaCl priming on salt tolerance in sunflower germination and seedling grown under salinity conditions. *African Journal of Biotechnology* 9(12):1764–1770.

Bajwa, A.A., M. Farooq and A. Nawaz. 2018. Seed priming with sorghum extracts and benzyl aminopurine improves the tolerance against salt stress in wheat (*Triticum aestivum* L.). *Physiology and Molecular Biology of Plants* 24(2):239–249.

Basra, A.S., S. Bedi and C.P. Malik. 1988. Accelerated germination of maize seeds under chilling stress by osmotic priming and associated changes in embryo phospholipids. *Annals of Botany* 61:635–639.

Basra, S.M.A, I.A. Pannu and I. Afzal. 2003. Evaluation of seedling vigor of hydro and matriprimed wheat (*Triticum aestivum* L.) seeds. *International Journal of Agriculture and Biology* 5(2):121–123.

Basra, S.M.A, Afzal, I., Rashid, R.A. and A. Hameed. 2005. Inducing salt tolerance in wheat by seed vigor enhancement techniques. *International Journal of Biology and Biotechnology* 2:173–179.

Basra, S.M.A., Farooq, M., Tabassum, R. and N. Ahmed. 2006. Evaluation of seed vigor enhancement techniques on physiological and biochemical basis in coarse rice. *Seed Science and Technology* 34:741–750.

Beckman, J.J., Moser, L.E., Kubik, K. and S.S. Waller. 1993. Big bluestem and switchgrass establishment as influenced by seed priming. *Agronomy Journal* 85:199–202.

Bewley J.D. 1997. Seed germination and dormancy. *The Plant Cell* 9:1055–1066.

Bewley, J.D. and M. Black. 1994. Seeds. *In*: *Physiology of Development and Germination*. New York: Plenum Press.

Bittencourt, M.L.C., D.C.F.S. Dias, L.A.S. Dias and E.F. Araújo. 2004. Effects of priming on asparagus seed germination and vigour under water and temperature stress. *Seed Science and Technology* 32:607–616.

Bradford, K.J. 1986. Manipulation of seed water relations via osmotic priming to improve germination under stress conditions. *Horticultural Science* 21:1105–1112.

Bray C.M., Davison, P.A., Ashraf, M. and R.M. Taylor. 1989. Biochemical changes during osmopriming of leek seeds. *Annals of Botany* 63:185–193.

Brocklehurst, P.A. and J. Dearman. 2008. Interaction between seed priming treatments and nine seed lots of carrot, celery and onion. II. Seedling emergence and plant growth. *Annals of Applied Biology* 102:583–593.

Burgass, R.W. and A.A. Powell. 1984. Evidence for repair processes in the invigoration of seeds by hydration. *Annals of Applied Biology* 53:753–757.

Chen, D., T.A. Gunawardena, B.P. Naidu, S. Fukai and J. Basnayake. 2005. Seed treatment with gibberellic acid and glycinebetaine improves seedling emergence and seedling vigour of rice under low temperature. *Seed Science and Technology* 33:471–479.

Chen, K. and R. Arora. 2013. Priming memory invokes seed stress-tolerance. *Environmental and Experimental Botany* 94:33–45.

Chen, K., A. Fessehaie and R. Arora. 2012. Dehydrin metabolism is altered during seed osmopriming and subsequent germination under chilling and desiccation in *Spinacia oleracea* L. cv. Bloomsdale: possible role in stress tolerance. *Plant Science* 183:27–36.

Chen, K., A. Fessehaie and R. Arora. 2013. Aquaporin expression during seed osmopriming and post-priming germination in spinach. *Biologia Plantarum* 57:193–198.

Christou, A., G.A. Manganaris and V. Fotopoulos. 2014. Systemic mitigation of salt stress by hydrogen peroxide and sodium nitroprusside in strawberry plants via transcriptional regulation of enzymatic and non-enzymatic antioxidants. *Environmental and Experimental Botany* 107:46–54.

Christou, A., G.A. Manganaris, I. Papadopoulos and V. Fotopoulos. 2013. Hydrogen sulfide induces systemic tolerance to salinity and non-ionic osmotic stress in strawberry plants through modification of reactive species biosynthesis and transcriptional regulation of multiple defence pathways. *Journal of Experimental Botany* 64(7):1953–1966.

Chunthaburee, S., J. Sanitchon, W. Pattanagul and P. Theerakulpisut. 2014. Alleviation of salt stress in seedlings of black glutinous rice by seed priming with spermidine and gibberellic acid. *Notulae Botanicae Horti Agrobotanici Cluj-Napoca* 42(2):405–413.

Cokkizgin, H. and Y. Bolek. 2015. Priming treatments for improvement of germination and emergence of cotton seeds at low temperature. *Plant Breeding Seed Science* 71:121–134.

Dadlani, M. and D.V. Seshu. 1990. Effect of wet and dry heat treatment on rice seed germination and seedling vigor. *International Rice Research Newsletter* 15:21–22.

Deryng, D., D. Conway, N. Ramankutty, J. Price and R. Warren. 2014. Global crop yield response to extreme heat stress under multiple climate change futures. *Environmental Research Letters* 9

Dorna, H., W. Li and D. Szopinska. 2014. The effect of priming on germination and vigour of pansy (Viola× Wittrockiana Gams.) seeds. *Acta Scientiarum Polonorum Hortorum Cultus* 13(6):15–29.

Du L.V. and T.P. Tuong. 2002. Enhancing the performance of dry-seeded rice: effects of seed priming, seedling rate, and time of seedling. *In: Direct seeding: Research Strategies and Opportunities*, ed. S. Pandey, M. Mortimer, L. Wade, T.P. Tuong, K. Lopes, and B. Hardy, pp. 241–256, Manila, Philippines: International Rice Research Institute.

Duan, P., F. Ding, F. Wang and B.-S. Wang. 2007. Priming of seeds with nitric oxide donor sodium nitroprusside (SNP) alleviates the inhibition on wheat seed germination by salt stress. *Journal of Plant Physiology and Molecular Biology* 33:244–250.

Eisvand, H.R., R. Tavakkol-Afshari, F. Sharifzadeh, H. Maddah-Arefi and S.M. Hesamzadeh-Hejaz. 2010. Effects of hormonal priming and drought stress on activity and isozyme profiles of antioxidant enzymes in deteriorated seed of tall wheatgrass (*Agropyron elongatum* host). *Seed Science and Technology* 38:280–297.

Eleiwa, M.E. 1989. Effect of prolonged seed soaking on the organic and mineral components of immature pods of soybeans. *Egyptian Journal of Botany* 32:149–160.

FAO. 2011. The state of the world's land and water resources for food and agriculture (SOLAW)—managing systems at risk. In: Food and Agriculture.Organization of the United Nations, Rome and Earthscan, London.

Farhoudi, R., S. Saeedipour and D. Mohammadreza. 2011. The effect of NaCl seed priming on salt tolerance, antioxidant enzyme activity, proline and carbohydrate accumulation of muskmelon (*Cucumis melo* L.) under saline condition. *African Journal of Agricultural Research* 6:1363–1370.

Farooq, M. S.M.A. Basra, A. Wahid, A. Khaliq and N. Kobayashi. 2009c. Rice seed invigoration. *In: Sustainable Agriculture Reviews. Book Series*, ed. E. Lichtfouse. Springer.

Farooq, M., A. Wahid and D.-J. Lee. 2009e. Exogenously applied polyamines increase drought tolerance of rice by improving leaf water status, photosynthesis and membrane properties. *Acta Physiologia Plantarum* 31:937–945.

Farooq, M., A. Wahid and K.H. Siddique. 2012. Micronutrient application through seed treatments: a review. *Journal of Soil Science and Plant Nutrition* 12:125–142.

Farooq, M., A. Wahid, N. Kobayashi, D. Fujita and S.M.A. Basra. 2009d. Plant drought stress: effects, mechanisms and management. *Agronomy for Sustainable Development* 29:185–212.

Farooq, M., A. Wahid, O. Ito, D.J. Lee and K.H.M. Siddique. 2009a. Advances in drought resistance of rice. *Critical Reviews in Plant Sciences* 28:199–217.

Farooq, M., A. Wahid, S.M.A. Basra and K.H.M. Siddique. 2010. Improving Crop Resistance to Abiotic Stresses through Seed Invigoration. *In: Handbook of Plant and Crop Stress*, M. Pessarakli (ed.), CRC Press, USA.

Farooq, M., A. Wahid, T. Aziz, D.J. Lee and K.H.M. Siddique. 2009b. Chilling tolerance in maize: physiological and agronomic implications. *Crop and Pasteur Science* 60:501–516.

Farooq, M., M. Hussain, A. Nawaz, D.J. Lee, S.S. Alghamdi and K.H. Siddique. 2017. Seed priming improves chilling tolerance in chickpea by modulating germination metabolism, trehalose accumulation and carbon assimilation. *Plant Physiology and Biochemistry* 111:274–283.

Farooq, M., M. Irfan, T. Aziz, I. Ahmad and S.A. Cheema. 2013. Seed priming with ascorbic acid improves drought resistance of wheat. *Journal of Agronomy and Crop Science* 199:12–22.

Farooq, M., S.M.A. Basra and A. Wahid. 2006b. Priming of field-sown rice seed enhances germination, seedling establishment, allometry and yield. *Plant Growth Regulation* 49:285–294.

Farooq, M., S.M.A. Basra and K. Hafeez. 2006c. Seed invigoration by osmohardening in fine and coarse rice. *Seed Science and Technology* 34:181–187.

Farooq, M., S.M.A. Basra and M.B. Khan. 2007b. Seed priming improves growth of nursery seedlings and yield of transplanted rice. *Archives of Agronomy and Soil Science* 53:311–322.

Farooq, M., S.M.A. Basra and N. Ahmad. 2007a. Improving the performance of transplanted rice by seed priming. *Plant Growth Regulation* 51:129–137.

Farooq, M., S.M.A. Basra, A. Wahid and H. Rehman. 2009g. Exogenously applied nitric oxide enhances the drought tolerance in fine grain aromatic rice (*Oryza sativa* L.). *Journal of Agronomy and Crop Science* 195:254–261.

Farooq, M., S.M.A. Basra, A. Wahid, N. Ahmad and B.A. Saleem. 2009f. Improving the drought tolerance in rice (*Oryza sativa* L.) by exogenous application of salicylic acid. *Journal of Agronomy and Crop Science* 195:237–246.

Farooq, M., S.M.A. Basra, A. Wahid, Z.A. Cheema, M.A. Cheema and A. Khaliq. 2008e. Physiological role of exogenously applied glycinebetaine in improving drought tolerance of fine grain aromatic rice (*Oryza sativa* L.). *Journal of Agronomy and Crop Science* 194:325–333.

Farooq, M., S.M.A. Basra, H. Rehman and B.A. Saleem. 2008a. Seed priming enhances the performance of late sown wheat (*Triticum aestivum* L.) by improving the chilling tolerance. *Journal Agronomy and Crop Science* 194:55–60.

Farooq, M., S.M.A. Basra, H.A. Karim and I. Afzal. 2004a. Optimization of seed hardening techniques for rice seed invigoration. *Emirates Journal of Agricultural Sciences* 16:48–57.

Farooq, M., S.M.A. Basra, I. Afzal and A. Khaliq. 2006e. Optimization of hydropriming techniques for rice seed invigoration. *Seed Science and Technology* 34:507–512.

Farooq, M., S.M.A. Basra, K. Hafeez and E.A. Warriach. 2004b. Influence of high and low temperature treatments on the seed germination and seedling vigor of coarse and fine rice. *International Rice Research Notes* 29:75–77.

Farooq, M., S.M.A. Basra, K. Hafeez and N. Ahmad. 2005. Thermal hardening: a new seed vigor enhancement tool in rice. *Journal of Integrative Plant Biology* 47:187–193.

Farooq, M., S.M.A. Basra, M. Khalid, R. Tabassum and T. Mehmood. 2006d. Nutrient homeostasis, reserves metabolism and seedling vigor as affected by seed priming in coarse rice. *Canadian Journal of Botany* 84:1196–1202.

Farooq, M., S.M.A. Basra, R. Tabassum and I. Afzal. 2006a. Enhancing the performance of direct seeded fine rice by seed priming. *Plant Production Science* 9:446–456.

Farooq, M., T. Aziz, M. Hussain, H. Rehman, K. Jabran and M.B. Khan. 2008b. Glycinebetaine improves chilling tolerance in hybrid maize. *Journal of Agronomy and Crop Science* 194:152–160.

Farooq, M., T. Aziz, S.M.A. Basra, A. Wahid, A. Khaliq and M.A. Cheema. 2008d. Exploring the role of calcium to improve the chilling tolerance in hybrid maize. *Journal of Agronomy and Crop Science* 194:350–359.

Farooq, M., T. Aziz, S.M.A. Basra, M.A. Cheema and H. Rehman. 2008c. Chilling tolerance in hybrid maize induced by seed priming with salicylic acid. *Journal of Agronomy and Crop Science* 194:161–168.

Farooq, M., T. Aziz, Z.A. Cheema, A. Khaliq and M. Hussain. 2008f. Activation of antioxidant system by KCl treatments improves the chilling tolerance in hybrid maize. *Journal of Agronomy and Crop Science* 194:438–448.

Fercha, A., A.L. Capriotti, G. Caruso, C. Cavaliere, H. Gherroucha, R. Samperi and A. Lagana. 2013. Gel-free proteomics reveal potential biomarkers of priming-induced salt tolerance in durum wheat. *Journal of Proteomics* 91:486–499.

Fercha, A., A.L. Capriotti, G. Caruso, C. Cavaliere, R. Samperi, S. Stampachiacchiere and A. Laganà. 2014. Comparative analysis of metabolic proteome variation in ascorbate-primed and unprimed wheat seeds during germination under salt stress. *Journal of Proteomics* 108:238–257.

Field, C.B. ed. 2014. *Climate Change 2014–Impacts, Adaptation and Vulnerability: Regional Aspects.* Cambridge University Press, Cambridge.

Finch-Savage, W.E. and C.J. Cox. 1982. A cold treatment technique to improve the germination of vegetable seeds prior to fluid drilling. *Scientia Horticulturae* 16:301–311.

Finch-Savage, W.E., K.C. Dent and L.J. Clark. 2004. Soak conditions and temperature following sowing influence the response of maize (*Zea mays* L.) seeds to on-farm priming (pre-sowing seed soak). *Field Crops Research* 90:361–374.

Fischer, E.M. and R. Knutti. 2015. Anthropogenic contribution to global occurrence of heavy-precipitation and high-temperature extremes. *Nature Climate Change* 5(6):560.

Foolad, M.R., P. Subbiah, C. Kramer, G. Hargrave and G.Y. Lin. 2003. Genetic relationships among cold, salt and drought tolerance during seed germination in an interspecific cross of tomato. *Euphytica* 130:199–206.

Fotopoulos, V., A. Christou, C. Antoniou and G.A. Manganaris. 2015. REVIEW ARTICLE Hydrogen sulphide: a versatile tool for the regulation of growth and defence responses in horticultural crops. *The Journal of Horticultural Science and Biotechnology* 90(3):227–234.

Fourest, E., L.D. Rehms, D.C. Sands, M. Bjarko and R.E. Lund. 1990. Eradication of *Xanthomonas campestris* pv translucens from barley seed with dry heat treatment. *Plant Disease* 74:816–818.

Galhaut, L., A. de Lespinay, D.J. Walker, M.P. Bernal, E. Correal and S. Lutts. 2014. Seed priming of *Trifolium repens* L.: improved germination and early seedling growth on heavy metal-contaminated soil. *Water, Air, & Soil Pollution* 225:1905.

Gornik, K. and B.L. Lahuta. 2017. Application of phytohormones during seed hydropriming and heat shock treatment on sunflower (*Helianthus annuus* L.) chilling resistance and changes in soluble carbohydrates. *Acta Physiologiae Plantarum* 39:118.

Goswami, A., R. Banerjee, S. Raha. 2013. Drought resistance in rice seedlings conferred by seed priming: role of the anti-oxidant defense mechanisms. *Protoplasma* 250:1115–1129.

Gray D., J.R.A. Steckel and L.J. Hands. 1990. Responses of vegetable seeds to controlled hydration. *Annals of Botany* 66:227–235.

Guan, Y., H. Cui, W. Ma, Y. Zheng, Y. Tian and J. Hu. 2014. An enhanced drought-tolerant method using SA-loaded PAMPS polymer materials applied on tobacco pelleted seeds. *The Scientific World Journal* 2014:9.

Guan, Y., Z. Li, F. He, Y., Huang, W. Song and J. Hu. 2015. "On-off" thermoresponsive coating agent containing salicylic acid applied to maize seeds for chilling tolerance. *PLoS ONE* 10(3):e0120695.

Hardegree S.P. and W.E. Emmerich. 1992a. Effect of matric-priming duration and priming water potential on germination of four grasses. *Journal of Experimental Botany* 43:233–238.

Hardegree, S.P. and W.E. Emmerich. 1992b. Seed germination response of four south-western range grasses to equilibration at sub-germination matric-potentials. *Agronomy Journal* 84:994–998.

Harris, D., A. Joshi, P.A. Khan, P. Gothkar and P.S. Sodhi. 1999. On-farm seed priming in semi-arid agriculture: Development and evaluation in maize, rice and chickpea in India using participatory methods. *Experimental Agriculture* 35:5–29.

Harris, D., R.S. Tripathi and A. Joshi. 2000. On-farm seed priming to improve crop establishment and yield in dry direct-seeded rice. *In: Proceedings of the International Workshop on Dry-Seeded Rice Technology*, pp. 25–28. Bangkok, Thailand.

Harris, D., R.S. Tripathi and A. Joshi. 2002. On-farm seed priming to improve crop establishment and yield in dry direct-seeded rice. *In: Direct Seeding: Research Strategies and Opportunities*, ed. S. Pandey, M. Mortimer, L. Wade, T.P. Tuong, K. Lopes, and B. Hardy, pp. 231–240. Manila, Philippines: International Rice Research Institute.

Hassanpouraghdam, M.B., J.E. Pardaz and N.F. Akhtar. 2009. The effect of osmopriming on germination and seedling growth of *Brassica napus* L. under salinity conditions. *Journal of Food Agriculture and Environment* 7:620–622.

Hayat, Q., S. Hayat, M. Irfan and A. Ahmad. 2010. Effect of exogenous salicylic acid under changing environment: a review. *Environment and Experimental Botany* 68:14–25.

Hayat, S., Q. Fariduddin, B. Ali and A. Ahmad. 2005. Effect of salicylic acid on growth and enzyme activities of wheat seedlings. *Acta Agronomica Hungarica* 53:433–437.

He, C.Z., J. Hu, Z.Y. Zhu, S.L. Ruan and W.J. Song. 2002. Effect of seed priming with mixed- salt solution on germination and physiological characteristics of seedling in rice (*Oryza sativa* L.) under stress conditions. *Journal of Zhejiang University (Agricultural and Life Sciences)* 28:175–178.

Hoque, M.A., E. Okuma, M.N.A. Banu, Y. Nakamura, Y. Shimoishi and Y. Murata. 2007. Exogenous proline mitigates the detrimental effects of salt stress more than the betaine by increasing antioxidant enzyme activities. *Journal of Plant Physiology* 164:553–561.

Hu, J., X.J. Xie, Z.F. Wang and W.J. Song. 2006. Sand priming improves alfalfa germination under high-salt concentration stress. *Seed Science and Technology* 34:199–204.

Hussain, S., M. Farooq, M.A. Wahid and A. Wahid. 2013. Seed priming with putrescine improves the drought resistance of maize hybrids. *International Journal of Agriculture and Biology* 15(6):1349–1353.

Iqbal, M. and M. Ashraf. 2007a. Seed treatment with auxins modulates growth and ion partitioning in salt-stressed wheat plants. *Journal of Integrative Plant Biology* 49:1003–1015.

Iqbal, M. and M. Ashraf. 2007b. Seed preconditioning modulates growth, ionic relations, and photosynthetic capacity in adult plants of hexaploid wheat under salt stress. *Journal of Plant Nutrition* 30:381–396.

Iqbal, M. and M. Ashraf. 2010. Changes in hormonal balance: a possible mechanism of pre-sowing chilling-induced salt tolerance in spring wheat. *Journal of Agronomy and Crop Science* 196(6):440–454.

Iqbal, M., and Ashraf, M. 2005b. Changes in growth, photosynthetic capacity and ionic relations in spring wheat (*Triticum aestivum* L.) due to pre-sowing seed treatment with polyamines. *Plant Growth Regulation* 46:19–30.

Iqbal, M., and M. Ashraf. 2005a. Presowing seed treatment with cytokinins and its effect on growth, photosynthetic rate, ionic levels and yield of two wheat cultivars differing in salt tolerance. *Journal of Integrative Plant Biology* 47:1315–1325.

Iqbal, M., M. Ashraf, A. Jamil and S.U. Rehman. 2006a. Does seed priming induce changes in the levels of some endogenous plant hormones in hexaploid wheat plants under salt stress? *Journal of Integrative Plant Biology* 48:181–189.

Iqbal, N., R. Nazar, M.R.K. Iqbal, A. Masood and A.K. Nafees. 2011. Role of gibberellins in regulation of source sink relations under optimal and limiting environmental conditions. *Current Science* 100:998–1007.

Islam, F., T. Yasmeen, S. Ali, B. Ali, M.A. Farooq and R.A. Gill. 2015. Priming-induced antioxidative responses in two wheat cultivars under saline stress. *Acta Physiologiae Plantarum* 37(8).

Jafar, M.Z., M. Farooq, M.A. Cheema, I. Afzal, S.M.A. Basra, M.A. Wahid, T. Aziz and M. Shahid. 2012. Improving the performance of wheat by seed priming under saline conditions. *Journal of Agronomy and Crop Science* 198(1):38–45.

Jisha, K.C. and J.T. Puthur. 2014. Seed halopriming outdo hydropriming in enhancing seedling vigor and osmotic stress tolerance potential of rice varieties. *Journal of Crop Science and Biotechnology* 17(4):209–219.

Jisha, K.C. J.T. Puthur. 2016. Seed priming with BABA (β-amino butyric acid): a cost-effective method of abiotic stress tolerance in *Vigna radiata* (L.) Wilczek. *Protoplasma* 253(2):277–289.

Jisha, K.C., K. Vijayakumari and J.T. Puthur. 2013. Seed priming for abiotic stress tolerance: an overview. *Acta Physiologiae Plantarum* 35(5):1381–1396.

Jyotsna, V. and A.K. Srivastava. 1998. Physiological basis of salt stress resistance in pigeonpea (*Cajanus cajan* L.) – II. Pre-sowing seed soaking treatment in regulating early seedling metabolism during seed germination. *Plant Physiology and Biochemistry (New Delhi)* 25:89–94.

Kadiri, M. and M.A. Hussaini. 1999. Effect of hardening pretreatments on vegetative growth, enzyme activities and yield of *Pennisetum americanum* and *Sorghum bicolor*. *Global Journal of Pure Applied Science* 5:179–183.

Kaur, S., A.K. Gupta and N. Kaur. 2002. Effect of osmo- and hydro-priming of chickpea seeds on seedling growth and carbohydrate metabolism under water deficit stress. *Plant Growth Regulation* 37:17–22.

Kaya, M.D., G. Okcu, M. Atak, Y. Cıkılı and O. Kolsarıcı. 2006. Seed treatments to overcome salt and drought stress. *European Journal of Agronomy* 24:291–295.

Khan, H.A., K. Ziaf, M. Amjad and Q. Iqbal. 2012. Polyamines improves germination and early seedling growth of hot pepper. *Chilean Journal of Agricultural Research* 72(3):429–433.

Kibite, S. and K.N. Harker. 1991. Effect of seed hydration on agronomic performance of wheat, barley and oats in central Alberta. *Canadian Journal of Plant Science* 71:515–518.

Kim, S.K., T.K. Son, S.Y. Park et al. 2006. Influences of gibberellin and auxin on endogenous plant hormone and starch mobilization during rice seed germination under salt stress. *Journal Environmental Biology* 27:181–186.

Korkmaz, A. 2005. Inclusion of acetyl salicylic acid and methyl jasmonate into the priming solution improves low-temperature germination and emergence of sweet pepper. *HortScience* 40:197–200.

Korkmaz, A. and Y. Korkmaz. 2009. Promotion by 5-aminolevulenic acid of pepper seed germination and seedling emergence under low-temperature stress. *Scientia Horticulturae* 119:98–102.

Kubala, S., Ł. Wojtyla, M. Quinet, K. Lechowska, S. Lutts and M. Garnczarska. 2015. Enhanced expression of the proline synthesis gene P5CSA in relation to seed osmopriming improvement of *Brassica napus* germination under salinity stress. *Journal of Plant Physiology* 183:1–12.

Lee, S.Y., J.H. Lee and T.O. Kwon. 2002. Varietal differences in seed germination and seedling vigor of Korean rice varieties following dry heat treatments. *Seed Science and Technology* 30:311–321.

Li, X. and L. Zhang. 2012. SA and PEG-induced priming for water stress tolerance in rice seedling. *Information Technology and Agricultural Engineering* 134:881–887.

Li, X., H. Jiang, F. Liu, J. Cai, T. Dai, W. Cao and D. Jiang 2013 Induction of chilling tolerance in wheat during germination by pre-soaking seed with nitric oxide and gibberellin. *Plant Growth Regulation* 71(1):31–40.

Lin, J.M. and J.M. Sung. 2001. Pre-sowing treatments for improving emergence of bitter gourd seedlings under optimal and sub-optimal temperatures. *Seed Science and Technology* 29:39–50.

Loreti, E., H. van Veen and P. Perata. 2016. Plant responses to flooding stress. *Current Opinion in Plant Biology* 33:64–71.

Madsen, M.D., K.W. Davies, C.J. Williams and T.J. Svejcar. 2012. Agglomerating seeds to enhance native seedling emergence and growth. *Journal of Applied Ecology* 49:431–438.

Madsen, M.D., K.W. Davies, D.L. Mummey and T.J. Svejcar. 2014. Improving restoration of exotic annual grass-invaded rangelands through activated carbon seed enhancement technologies. *Rangeland Ecology & Management* 67:61–67.

Matias, J.R., S.B. Torres, C.C. Leal, M.D.S. Leite and S. Carvalho. 2018. Hydropriming as inducer of salinity tolerance in sunflower seeds. *Revista Brasileira de Engenharia Agrícola e Ambiental* 22(4):255–260.

McDonald, M.B. 2000. Seed priming. *In: Seed Technology and Its Biological Basis*, ed. M. Black, and J.D. Bewley, pp. 287–325. Sheffield, UK: Sheffield Academic Press.

Mercado, M.F.O. and P.G. Fernandez. 2002. Solid matrix priming of soybean seeds. *Philippine Journal of Crop Science* 27:27–35.

Munns, R. and M. Tester. 2008. Mechanisms of salinity tolerance. *Annual Review of Plant Biology* 59:651–681.

Murata, M.R., G.E. Zharare and P.S. Hammes. 2008. Pelleting or priming seed with calcium improves groundnut seedling survival in acid soils. *Journal of Plant Nutrition* 31:1736–1745.

Musa, A.M., C. John, J. Kumar and D. Harris. 1999. Response of chickpea seeds to seed priming in the brain tract of Bangladesh. ICPN 6:20–22.

Naidu, B.P. and R. Williams. 2004. *Seed treatment and foliar application of osmoprotectants to increase crop establishment and cold tolerance at flowering in rice*. Report for the Rural Industries Research and Development Corporation. RIRDC Publication No. 04/004.

Nakaune, M., A. Hanada, Y.G. Yin, C. Matsukura and S. Yamaguchi. 2012. Molecular and physiological dissection of enhanced seed germination using short-term low-concentration salt seed priming in tomato. *Plant Physiology and Biochemistry* 52:28–37.

Narciso, J. and M. Hossain. 2002. *World Rice Statistics*. Los Baños, Philippines: International Rice Research Institute.

Nawaz, F., M.Y. Ashraf, R. Ahmad and E.A. Waraich. 2013. Selenium (Se) seed priming induced growth and biochemical changes in wheat under water deficit conditions. *Biological Trace Element Research* 151(2):284–293.

Pandita, V.K., A. Anand, S. Nagarajan, R. Seth and S.N. Sinha. 2010. Solid matrix priming improves seed emergence and crop performance in okra. *Seed Science and Technology* 38:665–674.

Paparella, S., S.S. Araújo, G. Rossi, M. Wijayasinghe, D. Carbonera and A. Balestrazzi. 2015. Seed priming: state of the art and new perspectives. *Plant Cell Reports* 34:1281–1293.

Parera, C.A., P. Qiao and D.J. Cantliffe. 1993. Enhanced celery germination at stress temperature via solid matrix priming. *Horticulture Science* 28:20–22.

Passam, H. C. and D. Kakouriotis. 1994. The effects of osmoconditioning on the germination, emergence and early plant growth of cucumber under saline conditions. *Scientia Horticulturae* 57:233–240.

Patade, V.Y., Bhargava, S. and P. Suprasanna. 2009. Halopriming imparts tolerance to salt and PEG induced drought stress in sugarcane. *Agriculture, Ecosystems and Environment* 134:24–28.

Patade, V.Y., D. Khatri, K. Manoj, M. Kumari and Z. Ahmed. 2012. Cold tolerance in thiourea primed capsicum seedlings is associated with transcript regulation of stress responsive genes. *Molecular Biology Reports* 39:10603–10613.

Paul, S. and A. Roychoudhury. 2016. Seed priming with spermine ameliorates salinity stress in the germinated seedlings of two rice cultivars differing in their level of salt tolerance. *Tropical Plant Research* 3:616–633.

Paul, S. and A. Roychoudhury. 2017b. Effect of seed priming with spermine/spermidine on transcriptional regulation of stress-responsive genes in salt-stressed seedlings of an aromatic rice cultivar. *Plant Gene* 11:133–142.

Paul, S. and A. Roychoudhury. 2017a. Seed priming with spermine and spermidine regulates the expression of diverse groups of abiotic stress-responsive genes during salinity stress in the seedlings of indica rice varieties. *Plant Gene* 11:124–132.

Pedrini, S., D.J. Merritt, J. Stevens and K. Dixon. 2017. Seed coating: science or marketing spin? *Trends in Plant Science* 22:106–116.

Pijnenborg, J.W.M. and T.A. Lie. 1990. Effect of lime pelleting on the nodulation of Lucerne (*Medicago sativa* L.) in acid soil: a comparative study carried out in the field, in pots, and in rhizotrons. *Plant and Soil* 121:225–234.

Pill, W.G., J.J. Frett and D.C. Morneau. 1991. Germination and seedling emergence of primed tomato and asparagus seeds under adverse conditions. *HortScience* 26:1160–1162.

Posmyk, M.M. and K.M. Janas. 2007. Effects of seed hydropriming in presence of exogenous proline on chilling injury limitation in *Vigna radiata* L. seedlings. *Acta Physiologiae Plantarum* 29:509–517.

Posmyk, M.M., F. Corbineau, D. Vinel, C. Bailly and D. Côme. 2001. Osmoconditioning reduces physiological and biochemical damage induced by chilling in soybean seeds. *Physiologia Plantarum* 111:473–482.

Prabhavathi, V., J.S. Yadav, P.A. Kumar and M.V. Rajam. 2002. Abiotic stress tolerance in transgenic eggplant (*Solanum melongena* L.) by introduction of bacterial mannitol phophodehydrogenase gene. *Molecular Breeding* 9:137–147.

QianQian, Q., L. Ming, Y. DonWei, C. Lei and Z. YuLin. 2009. Effects of priming treatment with vermiculite on the seed germination and seedlings' antioxidant characteristics of hot pepper under NaCl stress. *Acta Agriculturae Shanghai* 25:47–50.

Rajjou, L., M. Duval, K. Gallardo, J. Catusse, J. Bally, C. Job and D. Job. 2012. Seed germination and vigor. *Annual Reviews of Plant Biology* 63:507–533.

Rao, S.S.R., B.V. Vardhini, E. Sujatha and S. Anuradha. 2002. Brassinosteroids – new class of phytohormones. *Current Science* 82:1239–1245.

Rashid, A., D. Harris, P. Hollington and M. Rafiq. 2004. Improving the yield of mungbean *Vigna radiata* in the North West Frontiers province of Pakistan using on farm seed priming. *Experimental Agriculture* 40:233–244.

Razaji, A., M. Farzanian and S. Sayfzadeh. 2014. The effects of seed priming by ascorbic acid on some morphological and biochemical aspects of rapeseed (*Brassica napus* L.) under drought stress condition. *International Journal of Bioscience* 4(1):432–442.

Rehman, A. and M. Farooq. 2016. Zinc seed coating improves the growth, grain yield and grain biofortification of bread wheat. *Acta Physiologiae Plantarum* 38:238.

Rehman, H., I. Afzal, M. Farooq, T. Aziz and S.M. Ahmad. 2012. Improving temperature stress resistance in spring maize by seed priming. Proceedings of 3rd International Conference 'Frontiers in Agriculture'. Dankook International Cooperation on Agriculture, Dankook University,

Rehman, S., P.J.C. Harris and W.F. Bourne. 1998. The effect of hardening on the salinity tolerance of *Acacia* seeds. *Seed Science Technology* 26:743–754.

Roychoudhury, A., S. Basu and D.N. Sengupta. 2011. Amelioration of salinity stress by exogenously applied spermidine or spermine in three varieties of indica rice differing in their level of salt tolerance. *Journal of Plant Physiology* 168:317–328.

Ruan S., Q. Xue and K. Tylkowska. 2002. The influence of priming on germination of rice (*Oryza sativa* L.) seeds and seedling emergence and performance in flooded soils. *Seed Science and Technology* 30:61–67.

Ruan S.L., X.Q. Zhong, W.Q. Hua, S.L. Ruan Z.Q. Xue and Q.H. Wang. 2003. Physiological effects of seed priming on salt-tolerance of seedlings in hybrid rice (*Oryza sativa* L.). *Scientia Agriculturae Sinica* 36:463–468.

Salah, S.M., G. Yajing, C. Dongdong, L. Jie, N. Aamir, H. Qijuan, H. Weimin, N. Mingyu and H. Jin. 2015. Seed priming with polyethylene glycol regulating the physiological and molecular mechanism in rice (*Oryza sativa* L.) under nano-ZnO stress. *Scientific Reports* 5:14278.

Sallam, H.A. 1999. Effect of some seed-soaking treatments on growth and chemical components on faba bean plants under saline conditions. *Annals of Agricultural Sciences (Cairo)* 44:159–171.

Sasaki, K., S. Kishitani, F. Abe and T. Sato. 2005. Promotion of seedling growth of seeds of rice (*Oryza sativa* L. cv. Hitomebore) by treatment with H2O2 before sowing. *Plant Production Science* 8:509–514.

Savvides, A., S. Ali, M. Tester and V. Fotopoulos. 2016. Chemical priming of plants against multiple abiotic stresses: mission possible? *Trends in Plant Science* 21(4):329–340.

Scott, J.M. 1989. Seed coatings and treatments and their effects on plant establishment. *Advances in Agronomy* 42:43–83.

Shakarami, B., T.G. Dianati, M. Tabari and B. Behtari. 2011. The effect of priming treatments on salinity tolerance of *Festuca arundinacea* Schreb and *Festuca ovina* L. seeds during germination and early growth. *Iranian Journal of Rangelands Forest Plant Breeding and Genetic Research* 18(2):318–328.

Sharma, I., P.K. Pati and R. Bhardwaj. 2011. Effect of 28-homo-brassinolide on antioxidant defence system in *Raphanus sativus* L. under chromium toxicity. *Ecotoxicology* 20:862–874.

Sharma, P.C. and P. Kumar. 1999. Alleviation of salinity stress during germination in *Brassica juncea* by pre-sowing chilling treatments to seeds. *Biologia Plantarum* 42:451–455.

Singh, S.K. and P.K. Singh. 2016. Effect of seed priming of tomato with salicylic acid on growth, flowering, yield and fruit quality under high temperature stress conditions. *International Journal of Advanced Research* 4(2):723–727.

Song, W.J., J. Hu, J. Qiu, H.Y. Geng and R.M. Wang. 2005. Primary study on the development of special seed coating agents and their application in rice (*Oryza sativa* L.) cultivated by direct seeding. *Journal of Zhejiang University (Agriculture and Life Sciences)* 31:368–373.

Soon, K.J., C.Y. Whan, S.B. Gu, A.C. Kil and C.J. Lai. 2000. Effect of hydropriming to enhance the germination of gourd seeds. *Journal of Korean Society of Horticultural Sciences* 41:559–564.

Srivastava, A.K., V.H. Lokhande, V.Y. Patade, P. Suprasanna, R. Sjahril and S.F. D'souza. 2010. Comparative evaluation of hydro-, chemo-, and hormonal priming methods for imparting salt and PEG stress tolerance in Indian mustard (*Brassica juncea* L.). *Acta Physiologiae Plantarum* 32:1135–1144.

Tabassum, T., M. Farooq, R. Ahmad, A. Zohaib, A. Wahid and M. Shahid. 2018. Terminal drought and seed priming improve drought tolerance in wheat. *Physiology and Molecular Biology of Plants* 24(5):845–856.

Tabassum, T., M. Farooq, R. Ahmad, A. Zohaib and A. Wahid. 2017. Seed priming and transgenerational drought memory improve tolerance against salt stress in bread wheat. *Plant Physiology and Biochemistry* 118:362–369.

Taie, H.A.A., MT. Abdelhamid, M.G. Dawood and M.G. Nassar. 2013. Pre-sowing seed treatment with proline improves some physiological, biochemical and anatomical attributes of faba bean plants under sea water stress. *Journal of Applied Science and Research* 9(4):2853–2867.

Taylor, A.G., P.S. Allen, M.A. Bennett, J.K. Bradford J.S. Burris and M.K. Misra. 1998. Seed enhancements. *Seed Science Research* 8:245–256.

Thornton, J.M. and A.A. Powell. 1992. Short-term aerated hydration for the improvement of seed quality in *Brassica oleracea*. *Seed Science Research* 2:41–49.

Wahid, A. and A. Shabbir. 2005. Induction of heat stress tolerance in barley seedlings by pre-sowing seed treatment with glycine betaine. *Plant Growth Regulation* 46:133–141.

Wahid, A., M. Perveen, S. Gelani and S.M.A. Basra. 2007b. Pretreatment of seed with H2O2 improves salt tolerance of wheat seedlings by alleviation of oxidative damage and expression of stress proteins. *Journal of Plant Physiology* 164:283–294.

Wahid, A., S. Gelani, M. Ashraf and M.R. Foolad. 2007a. Heat tolerance in plants: an overview. *Environmental and Experimental Botany* 61:199–223.

Wang, L., W.L. Wang, C.H. Yin and C.Y. Tian 2011. Cold stratification, but not stratification in salinity, enhances seedling growth of wheat under salt treatment. *African Journal of Biotechnology* 10(66):14888–14890.

Watkinson, J.I. and W.G. Pill. 1998. Gibberellic acid and pre-sowing chilling increase seed germination of Indian grass [(*Sorghastrum nutans* L.) Nash.]. *HortScience* 33:849–851.

Wei, W., Q.T. Li, Y.N. Chu, R.J. Reiter, X.M. Yu, D.H. Zhu, W.K. Zhang, B. Ma, Q. Lin, J.S. Zhang and S.Y. Chen. 2014. Melatonin enhances plant growth and abiotic stress tolerance in soybean plants. *Journal of Experimental Botany* 66(3):695–707.

Welbaum, G.E., V. Shen, O.M. Oluoch and L.W. Jett. 1998. The evolution and effects of priming vegetable seeds. *Seed Technology* 20:209–235.

Williams, M.I., R.K. Dumroese, D.S. Page-Dumroese and S.P. Hardegree. 2016. Can biochar be used as a seed coating to improve native plant germination and growth in arid conditions? *Journal of Arid Environments* 125:8–15.

Xu, S., J. Hu, Y. Li, W. Ma, Y. Zheng and S. Zhu. 2011. Chilling tolerance in *Nicotiana tabacum* induced by seed priming with putrescine. *Plant Growth Regulation* 63:279–290.

Yacoubi, R., C. Job, M. Belghazi, W. Chaibi and D. Job. 2013. Proteomic analysis of the enhancement of seed vigour in osmoprimed alfalfa seeds germinated under salinity stress. *Seed Science Research* 23:99–110.

Yan, M. 2016. Hydro-priming increases seed germination and early seedling growth in two cultivars of Napa cabbage (*Brassica rapa* subsp. *pekinensis*) grown under salt stress. *The Journal of Horticultural Science and Biotechnology* 91(4):421–426.

Zavariyan, A., M. Rad and M. Asghari. 2015. Effect of seed priming by potassium nitrate on germination and biochemical indices in *Silybum marianum* L. under salinity stress. *International Journal of Life Sciences* 9:23–29.

Zhang, C.F., J. Hu, J. Lou, Y. Zhang and W.M. Hu. 2007b. Sand priming in relation to physiological changes in seed germination and seedling growth of waxy maize under high-salt stress. *Seed Science and Technology*. 35:733–738.

Zhang, F., J. Yu, C.R. Johnston, Y. Wang, K. Zhu, F. Lu, Z. Zhang and J. Zou. 2015. Seed Priming with polyethylene glycol induces physiological changes in sorghum (*Sorghum bicolor* L. Moench) seedlings under suboptimal soil moisture environments. *PLoS One* 10:e0140620.

Zhang, H.-Q., Y.-B. Zou, G.-C. Xiao and Y.-F. Xiong. 2007c. Effect and mechanism of cold tolerant seed-coating agents on the cold tolerance of early Indica rice seedlings. *Agricultural Sciences in China* 6:792–801.

Zhang, L., C. Tian and L. Wang. 2018. Cold stratification pretreatment improves salinity tolerance in two wheat varieties during germination. *Seed Science and Technology* 46(1):87–92.

Zhang, S., J. Hu, N. Liu and Z. Zhu. 2006. Pre-sowing seed hydration treatment enhances the cold tolerance of direct-sown rice. *Seed Science and Technology* 34:593–601.

Zhang, S., J. Hu, Y. Zhang, X.J. Xie and Allen Knapp. 2007a. Seed priming with brassinolide improves lucerne (*Medicago sativa* L.) seed germination and seedling growth in relation to physiological changes under salinity stress. *Australian Journal of Agricultural Research* 58:811–815.

Zhang, X.G. 1990. Physiochemical treatments to break dormancy in rice. *International Rice Research Newsletter* 15:22.

Zheng, C., D. Jiang, F. Liu, T. Dai, W. Liu, Q. Jing and W. Cao. 2009. Exogenous nitric oxide improves seed germination in wheat against mitochondrial oxidative damage induced by high salinity. *Environmental and Experimental Botany* 67(1):222–227.

Zheng, M., Y. Tao, S. Hussain, Q. Jiang, S. Peng, J. Huang, K. Cui and L. Nie. 2016. Seed priming in dry direct-seeded rice: consequences for emergence, seedling growth and associated metabolic events under drought stress. *Plant Growth Regulation* 78(2):167–178.

Zhu, J.K. 2001a. Cell signaling under salt, water and cold stresses. *Current Opinion in Plant Biology* 4:401–406.

Zhu, J.K. 2001b. Plant salt tolerance. *Trends in Plant Science* 2:66–71.

Zhu, J.K. 2002. Salt and drought stress signal transduction in plants. *Annual Review of Plant Physiology and Plant Molecular Biology* 53:247–273.

42 Drought Resistance of Tropical Forage Grasses
Opening a Fertile Ground for Innovative Research

Juan Andrés Cardoso and Idupulapati M. Rao

CONTENTS

42.1 INTRODUCTION

Drought negatively affects the productivity of plants globally. Anthropogenic climate change is expected to create great uncertainty about the timing and frequency of drought events in the future across the globe (IPCC, 2014). However, there is no consensual definition of drought. Here we refer to drought as insufficient soil moisture that results in inhibition of normal plant growth and function. In other words, drought stress to plants occurs when the supply of water does not meet the water requirements.

This chapter relates to tropical forage grasses and drought, mostly from an eco-physiological point of view. However, this chapter is not meant to be an exhaustive review of the peculiarities of tropical forage grasses under drought stress.

This by itself is a herculean task due to the vast number of tropical forage grasses (for information purposes, check http://tropicalforages.info; an interactive selection tool with most of the accumulated information on the adaptation, use, and management of tropical forages).

Studies on tropical forage grasses and their mechanisms to cope with drought are few. Furthermore, a great deal of the information regarding drought responses is full of contradictions, and with the democratization of internet access, misleading information spreads like wildfire. To the best of our knowledge, studies from the last 10 years regarding tropical forage grasses and drought are mostly published in gray literature and predatory journals. Thus, the aim of this chapter is to provide an overview of four aspects, and these include 1) drought stress, 2) tropical forage grasses, 3) the International

Center for Tropical Agriculture (CIAT) strategy for evaluating drought stress responses and mechanisms, and 4) future outlook.

42.1.1 Drought Stress

Levitt (1980) clearly dissected the mechanisms of drought resistance into three components: escape, dehydration tolerance, and dehydration avoidance. Our view is that Levitt's definition makes the most compelling case for dissecting plant's capacity to perform under drought stress. However, tropical forage grasses are "unique" in the sense that most of them are used and treated as perennial crops. That is, forage grasses experience water deficits throughout several periods or years. As such, maintaining a constant supply of forage to feed livestock over different drought periods is of great economic interest and is mostly achieved by forage grasses using *dehydration avoidance* strategy.

42.1.2 Tropical Forage Grasses

We define tropical forage grasses, their importance, and their common traits. We briefly mention reasons for the misconception of drought resistance among tropical forage grasses. We reviewed credible information on mechanisms of drought resistance in tropical forage grasses.

42.1.3 CIAT Strategy for Evaluating Drought Stress Responses and Mechanisms

CIAT aims to identify drought resistant tropical forage grasses by characterizing an "ideotype" for different water use strategies (water spending vs. water saving). CIAT is currently implementing a phenomics approach for addressing such an important task.

42.1.4 Future Outlook

We briefly mention some relevant issues that ought to be sorted out in the short-term to further increase the understanding of responses and coping mechanisms of tropical forage grasses under drought stress. This, in turn, will contribute toward their better utilization in different drought-prone target environments.

We provide below more detailed information on each one of the above four aspects.

42.2 DROUGHT STRESS

Drought stress is a major abiotic factor restricting plant growth and productivity around the world (Boyer, 1982). There are many definitions and considerable disagreements about the concept of drought. At least seven definitions of drought are available depending on certain disciplinary perspective. As such, perspectives on drought stress exist in terms of meteorology, climate, atmosphere, hydrological cycle, water management, agriculture (Subrahmanyam, 1967; Passioura,

2007), physiology (Hsiao, 1973), and ecology (Crausbay et al., 2017). Although it is useful to view drought under different perspectives, the boundaries separating them are often too vague (Wilhite and Glantz, 1985). Considering the aforementioned, we consider the term drought as almost meaningless. In the present chapter, we simply refer to drought stress as an imbalance between water loss through transpiration and water uptake through the root system. Such imbalance leads to a wide range of responses from genes to the whole plant level (Chaves et al., 2003). Drought stress is determined by water supply, in turn, influenced by type, texture, and hydraulic conductivity of the soil, but also to the atmospheric evaporative demand (Loka et al., 2018). Furthermore, the drought stress depends not only on its duration, type, and intensity but also on the developmental stage of the plant.

Levitt (1980) also clearly dissected the main strategies that plants have developed in order to continue growth or ensure survival under drought stress conditions. Such strategies are drought escape, dehydration tolerance, and dehydration avoidance. Drought escape is associated with phenology, by entering a state of dormancy or by ensuring plant survival by the production of seeds. Dehydration tolerance is mostly associated with the ability to sustain low tissue water potential and maintain turgor under drought conditions. Dehydration avoidance is achieved by maintaining high tissue water potentials by reducing water loss from the shoot and/or maintaining water uptake from the roots. Osmotic adjustment is considered as a component of dehydration avoidance because it facilitates water uptake from drying soil. For further reading, Blum (2010) provides a comprehensive description of different drought resistance strategies. Dehydration avoidance in tropical forage grasses is the prevalent strategy to cope with drought stress.

42.2.1 Dehydration Avoidance

The mechanisms by which tropical forage grasses can avoid dehydration can be divided into two: 1) by reducing water loss and 2) by maintaining water uptake.

42.2.1.1 Reduction of Water Loss

Reduction of water loss can be achieved mostly by 1) reducing transpirable area and 2) stomatal control.

42.2.1.1.1 Reducing the Transpirable Area

One of the most obvious responses to the limited water supply is to restrict water lost through the shoot tissue. As such, one of the earliest responses of tropical forage grasses is the reduction of shoot growth under limited water supply. In grasses, shoot growth is the outcome of tillering and leaf growth that result from processes occurring at intercalary, apical, and axillary meristems. The leaf growth zone is the main site of shoot growth. Cell expansion and division are components of leaf growth, with cell expansion particularly sensitive to water deficit (Hsiao, 1973). Reductions in leaf size result in smaller plants with lower total leaf area, thereby lowering transpirational water loss.

Leaf folding or rolling is a symptom of turgor loss. In grasses, folding is the outcome of loss of turgor in bulliform cells. Rolling decreases the load of photosynthetically active radiation (PAR) to the leaf surface, particularly at midday, and thereby reduces transpiration while protecting photosystem II (PSII) from damage (Blum, 2010). Leaf posture is also considered important, and erect leaves receive lesser amounts of direct light at midday. Additionally, leaves of grasses might reflect greater amounts of radiation in the presence of leaf epicuticular waxes or pubescence. Leaf epicuticular waxes can also reduce cuticular transpiration (Blum, 2010). To our knowledge, all of the aforementioned have not been systematically recorded in tropical forage grasses under drought stress.

42.2.2.1.1.2 Stomatal Control

One of the earliest responses to the limited water supply is stomatal closure. Stomata are small pores that control the flow of gases in and out of the leaf. Stomatal density and size influences the amount of gas exchange in and out of the leaf (Lawson and Blatt, 2014). In this sense, stomata not only control the amount of water lost, but also that of CO_2 entering the leaf. This is a trade-off under limited water conditions: to save up water, carbon assimilation and thereby growth is hindered. There are several interacting factors controlling stomatal opening–closure including CO_2, light, pH, ion status, aquaporins, hydraulic and hormonal signals (Blum, 2010). Among hormonal signals affecting stomatal control, abscisic acid (ABA) is probably the most studied (Cai et al., 2017).

42.2.1.2 Maintaining Water Uptake

Maintaining water uptake is an obvious function of the root system. Under drought conditions, maintaining water uptake is facilitated by osmotic adjustment.

42.2.1.2.1 Root System

The rate of plant dehydration is shaped by rates of water loss and uptake. Water uptake is a crucial function of the root system, which in turn determined by factors such as its size and distribution down the soil profile, anatomy (e.g., the diameter of xylem vessels) and senescence. The root system varies greatly according to soil properties (physical, chemical, and biological). Water uptake is considered to be proportional to root length density (root length per unit of soil volume, cm root cm^{-3} soil, RLD). For some authors, RLD values of over 1.0 to 1.5 cm/cm^3 are required to extract all available water in a drying soil (Passioura, 1983; Wasson et al., 2012). Plants with greater RLD values in deep soil layers are better able to maintain water status and stomatal conductance during soil drying than those with lower values of RLD. However, this might also not be always true as shown by some researchers (Passioura, 1991; Kirkegaard et al., 2007). Root hair development contributes to increasing root surface area to explore the water resource (Hofer, 1991). Xylem vessels with larger diameter have a greater capacity for conducting water than with smaller diameter (Richards and Passioura, 1989). Yet, large xylem vessels are prone to cavitation.

42.2.1.2.2 Osmotic Adjustment

Maintaining water uptake is facilitated by osmotic adjustment. Osmotic adjustment is a process where solutes accumulate in the cell to maintain cell hydration and turgor in leaf tissue and also in other metabolically active cells (Sanders and Arndt, 2012).

42.3 TROPICAL FORAGE GRASSES

The term forage refers to plant material that is used to feed livestock. In the present chapter, we refer to tropical forages as those often found in lower latitudes (tropics to subtropics) and low elevations (below to 1,850 meters above sea level). The definition of tropical forages is then mostly based on their ability to grow well under warm temperatures (25–35°C), and their lack of freezing tolerance.

42.3.1 IMPORTANCE OF TROPICAL FORAGE GRASSES

Over 600 million people of low-income depend on livestock production to sustain their livelihoods (Perry and Sones, 2007; Herrero et al., 2009). Most of them live in the developing countries in the tropics (Thornton et al., 2002) where livestock is mainly fed from forage grasses (Herrero et al., 2010). Tropical forages grasses are mostly located or grown on marginal lands that are not suitable for growing crops (Humphreys, 1981, 1994; Rao, 2014; Schultze-Kraft et al., 2018).

The most abundant tropical forages are grasses, and no doubt, the most economically important ones have originated from Africa. Tropical forage grasses are primarily from the tribes *Andropogoneae*, *Paniceae*, *Chlorideae*, and *Eragosteae* (Kretschmer and Pitman, 2001). Nowadays, the most economically important tropical forage grasses are probably *Brachiaria* spp. (syn. *Urochola*), *Panicum maximum* (syn. *Megathyrus maximus*), *Cenchrus ciliaris* and *C. purpureus* (syn. *Pennisetum purpureum*) grasses. African grasses have been introduced widely into the global tropics, and since then have been either naturalized or actively cultivated. The first departure of African grasses (e.g., *B. mutica*, *P. maximum*, *Hyparrhenia rufa*) to different parts of the tropics likely occurred during the slave trade to the Americas from the seventeenth century (Parsons, 1972; Carney, 2010). The second departure of African grasses elsewhere was the outcome of several germplasm explorations by different organizations in the 1970s and early 1980s (Attere, 1992). Such explorations aimed to assemble diverse collections of tropical forage species with the obvious objective of conservation, but also that of rapid cultivar development through agronomic evaluation or plant breeding. Since their introduction, African grasses—at least in tropical America—have outcompeted the native grasses. This success is associated with a number of plant characteristics that make these grasses valuable forages, these include high biomass production; rapid establishment; greater nutritive value and palatability to grazing animals; and persistence when there is little husbandry, yet high productivity under grazing pressure (Baruch and Jackson, 2005; Miles et al., 2004; Overholt and Franck, 2017).

African grasses are remarkable in their combination of several features, but a large part of their success can be attributed to the presence of the C_4 photosynthetic pathway.

42.3.2 C_3 and C_4 Photosynthesis

C_4 photosynthesis is a set of anatomical and physiological traits that concentrate CO_2 around the carbon-fixing enzyme Rubisco (Edwards et al., 2010). As a consequence, C_4 plants have the potential to achieve higher rates of photosynthesis than C_3 plants under high irradiation. This explains the success and dominance of C_4 grasses in low latitudes and altitudes (Edwards et al., 2010). Furthermore, the C_4 photosynthetic pathway present in grasses offers a lower expenditure of water (i.e., high water use efficiency) than that of C_3 grasses (Osborne and Freckleton, 2008). However, it has long been debunked that C_4 grasses are more resistant to drought than C_3 species. The concept of water use efficiency has been somehow mixed up with drought resistance in tropical forage grasses, and this can be misleading.

42.3.3 Studies on Responses of Tropical Forage Grasses to Drought

A literature search in Google Scholar with the terms tropical forage grasses and drought found over 174,000 records. Unsurprisingly, search results were mostly restricted to economically important forage grasses (*Panicum maximum*, *Brachiaria* spp., *Cenchrus ciliaris*). Interestingly, most of the work performed on such tropical forage grasses and their coping mechanisms under water stress were mostly performed/published in the late 1970s and back again in the late 1990s to early–mid-2000s. More interestingly is that most information on such topics from the mid-2000s is obtained from gray literature (above 90%), or poorly reputed journals (https://predatoryjournals.com/journals/) that do not appear on the Web of Science (https://clarivate.com/products/web-of-science/). Unfortunately, information from gray literature provides mostly preliminary results and does not provide full details of the methods used. Of great concern is that available information from poorly reputed journals is full of contradictions and fantastical claims, and with the democratization of internet access misleading information spreads like wildfire. For a newcomer to the topic who reads such literature, it would seem that tropical forage grasses are drought resistant mostly by virtue of their deep root system and high water use efficiency. Albeit this might be true, in most gray literature and in publications from poorly reputed journals, none of the aforementioned traits were measured. Another problem identified from such sources of information has been the disregard to plant size. This is a prevalent pitfall in recent studies in tropical forage grasses. There is hardly any information where plant phenology, shoot and root development and function, is studied in terms of water relations and drought stress. The topic is important since plant size and phenotypic plasticity plays a major role in improving the adaptation of genotypes to any given environment and certainly to the water regime.

The excellent work performed in the late 1970s to the early 1980s by Australia based researchers (Ludlow, Ng, and Wilson) and in the late 90s to mid-2000s by Venezuela based researchers (Baruch and Guenni) was mostly focused on growth, water relations, and gas exchange of a few tropical forage grasses (*Panicum maximum*; *Brachiaria mutica*; *Brachiaria humidicola*; *Brachiaria brizantha*; *Brachiaria decumbens*; *Cenchrus ciliaris*). Common ground for those studies was the reduction in water status traits in response to drought. However, the magnitudes of decrease differed greatly among these studies. These differences were mainly the result of interactions between drought stress and the high variability of various plant types. Thus, because of the very limited information available and the difficulty to compare the results of such individual studies (even if the studies were performed by a very small community), the outcome of this review is unsurprisingly short.

Work performed using several *Brachiaria* spp. showed that wilting symptoms developed at predawn water potentials ranging from –1.5 to –2.0 MPa (Guenni et al., 2002). For *Cenchrus ciliaris*, leaf water potential measured at 14:00 h decreased to a minimum of –6.9 MPa (Wilson and Ludlow, 1976). The same authors showed that critical leaf water potentials at which stomata closed and the photosynthetic rate was reduced, occurred over a range of predawn water potentials of –2.0 to –5.0 MPa. Ludlow and Ng (1976) showed that for *Panicum maximum*, stomata began to close at –0.6 MPa of leaf water potential. Wilson and Ludlow (1983a, 1983b) showed that water stress induced an osmotic adjustment of up to 1.0 MPa in *Panicum maximum* and *Cenchrus ciliaris*. These authors showed that osmotic adjustment delayed the loss of turgor in water-stressed leaves by about 4 days. Guenni et al. (2004) showed that for *Brachiaria* species, osmotic adjustment ranged between 0.38 (*B. decumbens*) to 0.87 MPa (*B. humidicola*), and it was absent in *B. brizantha* and *B. mutica*. In *Panicum maximum*, extreme levels of water stress down to –9.0 MPa inhibited photosynthesis, but the latter was recovered after re-watering (Ludlow and Ng, 1983a). Work with water-stressed *Panicum maximum*, however, showed no evidence of stimulation of growth following relief of stress (Ng et al., 1975).

In comparative studies by Guenni et al. (2002, 2004, 2006), they showed that differences in tolerance to mild water stress among four *Brachiaria* species were associated with contrasting patterns of root growth, soil water use, biomass allocation, and leaf osmotic adjustment leading to differences in time of wilting, carbon assimilation, and tolerance to water stress. These authors showed that the more tolerant species (*B. humidicola*) showed a deeper root system, smaller leaf area, and greater osmotic adjustment than *B. brizantha*, *B. mutica*, and *B. decumbens*. Guenni et al. (2004, 2006) suggested that the different combination of traits in *Brachiaria* lead to certain "ideotypes" of water use and thereby could be targeted to different conditions of precipitation. For further explanation of the concept of "ideotypes" for different water use strategies (i.e., water spending vs. water saving), see Chapter 43 of this book.

Following the framework described by Guenni et al. (2004, 2006) for targeting tropical forage grasses for different conditions of precipitation based on several traits, more recent work by Cardoso et al. (2015) recorded responses of Napier grass (*Pennisetum purpureum*, syn. *Cenchrus purpureus*) and a *Brachiaria* hybrid (cv. Mulato II) under drought conditions. These authors showed that Napier grass and Mulato II were similar in drought resistance (in terms of the absolute shoot dry mass production and the relative reduction of shoot dry mass over a period of 21 days). However, each grass showed different strategies for coping with water deficit conditions. Napier grass exhibited a larger root system than Mulato II and attempted to maximize carbon assimilation while there was the availability of soil water. Whereas, Mulato II showed a root system hypothetically large enough to extract most water under drying soil, yet it restricted water loss by early stomatal closure. The authors suggested that Napier grass showed a "*water-spending*" behavior that might be targeted to areas with a short dry season, whereas Mulato II showed a "*water-saving*" behavior that could be directed to areas with longer dry periods.

Humphreys (1981) stated that mechanistic studies of drought responses were key to assist targeting of several tropical forages to specific environments, yet very few studies regarding that aspect existed at that time. The aforementioned still holds true almost 40 years later despite the importance of tropical forage grasses to livestock production. The Tropical Forages Program of the International Center for Tropical Agriculture (http://www.ciat.cgiar.org) recognizes this drawback and thereby has embarked on implementing innovative approaches to describe complex plant phenotypes.

42.4 CIAT STRATEGY FOR EVALUATING DROUGHT STRESS RESPONSES AND MECHANISMS

CIAT aims to identify drought resistant tropical forage grasses by characterizing an "ideotype" for different water use strategies (water spending vs. water saving). Furthermore, CIAT is currently implementing innovative approaches for tackling such an important task.

42.4.1 INNOVATIVE APPROACHES FOR EVALUATING DROUGHT STRESS RESPONSES AND MECHANISMS

Studies on plant responses to drought have mostly relied upon conventional phenotyping procedures which are labor-intensive, time-consuming, and thereby of lower throughput. On the other hand, the development of electronics, robotics, computer science, and sensor systems is revolutionizing how plants are studied and understood under different environmental stresses. Such technologies are aimed to capture multiple phenotypic values at high temporal and spatial resolutions in a high throughput manner (i.e., phenomics). Phenomic approaches have opened a great avenue for improvement on how tropical forage grasses are studied under drought stress.

Phenomics aim for the systematic characterization of phenotypes through the acquisition of high-dimensional phenotypic data at different organizational scales of the plant (Holue et al., 2010). In particular, phenomics can contribute to filling the gap regarding 1) comparative and quantitative assessment of responses of tropical forage grasses under drought conditions, and 2) associate such responses to different shoot and root traits. Responses of tropical forage grasses during drought such as growth, morphological, and architectural features of both shoot and roots in 2D or 3D, greenness and senescence, can be recorded at high spatial and temporal resolution using digital cameras (either in the visible or near infrared spectrum of light). Thermal cameras can be used to determine canopy temperature. A relatively lower canopy temperature indicates the capacity to take up water from drying soil or to maintain plant water status. A comprehensive review of imaging and phenotyping was provided by Perez-Sanz et al. (2017) with details on image acquisition technologies and image analysis algorithms. Spectroradiometric apparatus allows fast and non-destructive evaluation of different traits such as total green biomass, canopy architecture, plant water status, nitrogen concentration, and photosynthetic efficiency (Araus et al., 2008; Araus and Kefauver, 2018). Currently, most of the aforementioned traits measured indirectly by spectroradiometry are based on indexes obtained from measurements of bands in the visible and infrared regions of the spectrum. A number of vegetation indices that are commonly used are listed in Table 42.1.

42.4.2 CURRENT PLATFORMS AVAILABLE AT CIAT TO STUDY DROUGHT RESPONSES

CIAT currently uses two platforms, one for studies under controlled conditions (greenhouse) and another one for field studies. The data generated from both systems are being made available for other researchers through open access.

TABLE 42.1
Vegetation Indices That Are Commonly Used

Physiological Parameter	Index
Leaf area, chlorophyll, green biomass, etc.	$NDVI = (R_{NIR} - R_{RED})/R_{NIR} + R_{RED})$
	$MCARI = (R_{700} - R_{670}) - 0.2 (R_{700} - R_{550})$
	$\times (R_{700}/R_{670})$
	$G = R_{554}/R_{677}$
Chlorophyll degradation	$NPQI = (R_{415} - R_{435})/R_{415} + R_{435})$
Carotenoid/Chlorophyll ratio	$SIPI = (R_{800} - R_{435})/R_{415} + R_{435})$
Photosynthetic radiation use efficiency	$PRI = (R_{531} - R_{570})/(R_{531} + R_{570})$
Water content	$WI = R_{900}/R_{970}$

R refers to the reflectance values; number–or text–in subscript next to R, refers to specific wavelengths–or ranges–of the electromagnetic spectrum. NDVI: Normalized Difference Vegetation Index; MCARI: Modified Chlorophyll Absorption Ratio Index: G: Greenness index: NPQI: Normalized Phaeophytinization Index; SIPI: Structure Insensitive Pigment Index; PRI: Photochemical Reflectance Index: WI: Water Band Index.

42.4.2.1 Controlled Conditions

Several methods have been applied for the last few years with the aim to record responses to different tropical forage grasses under drought conditions. Nowadays studies are mostly performed on plants growing in containers (long tubes of at least 80 cm length, with different diameters filled with soils collected from target field sites) under greenhouse conditions. Among these is the simultaneous assessment of shoot and root growth over a period of time using digital cameras recording information, either in the visible and/or near infrared spectrum of light. Such a method also allows determination of change of leaf greenness and senescence over time. Besides the aforementioned, soil water content is recorded (gravimetrically and by electrical conductivity) in parallel over the course of the experiment (Figure 42.1) and can be applied to

screen hundreds of plants per day. One of the advantages of this method is that it can detect whether roots with enough root length density (<1.0 cm/cm^3), that have grown deep in the moist soil layers are able—or not—to take up water. The use of this method—in conjunction with more intensive phenotyping—has also given insights into different strategies of water use and associated mechanisms among a number of tropical forage grasses (Figure 42.2). The same method is also used to make comparisons among a larger number of genotypes, and based on their phenotypic similarity or dissimilarity, grouped them together. Interestingly, resulting groups coincided to water use strategies supporting the notion that similar phenotypes lead to certain ideotypes of water use (Figure 42.3). This phenotyping method is yielding very useful information, and it needs to be further refined.

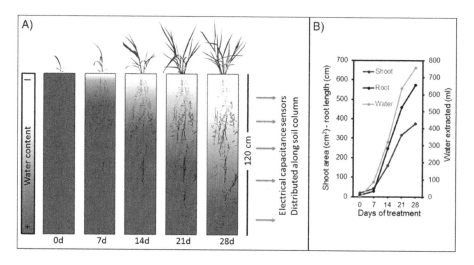

FIGURE 42.1 (See color insert.) Representation of the method to record responses of tropical forages grasses under drought conditions. (A) The method allows simultaneous recording of shoot and root growth and water uptake over a period of time. The electrical capacitance sensors are distributed down the soil profile to estimate water content at different soil profiles, while a weighing scale records the whole weight of the soil column to calculate gravimetric soil water content. (B) Actual data from 1A.

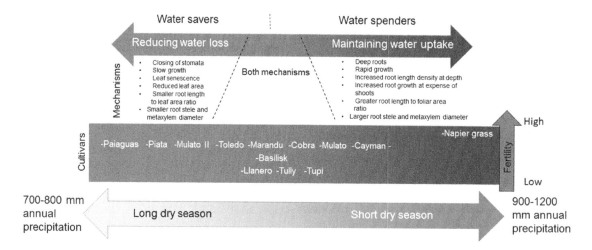

FIGURE 42.2 (See color insert.) Different strategies of water use and associated mechanisms among a number of tropical forage grass cultivars that are being used by farmers. A fine interplay exists between the acquisition of water by roots in drying soil and water loss through transpiration. These two components tend to act simultaneously. Nonetheless, a certain strategy of water use tends to supersede over the other one. The mode of water use was evaluated in conjunction with contrasting soil fertility levels for targeting of forage grasses to different patterns of precipitation and soil fertility levels.

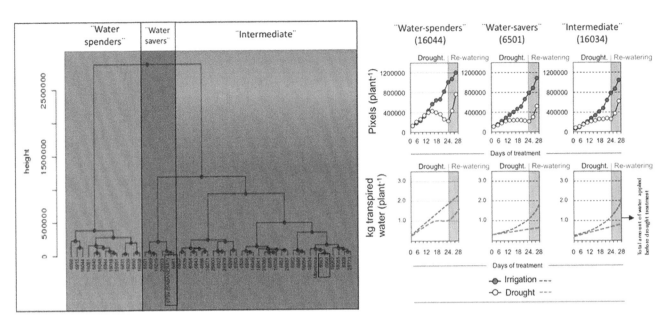

FIGURE 42.3 (See color insert.) An example of hierarchical clustering of models of water use in 50 genotypes of *Panicum maximum*. Examples of three different genotypes—with their accession numbers from CIAT Genetic Resources Program—showing differences in their water use strategies are shown on the right-hand side of the image. The responses of these genotypes based on accumulated growth (pixel-based) and transpired water over a drought period and recovery period are also shown.

FIGURE 42.4 (See color insert.) A hand-propelled vehicle where a number of sensors can be mounted (e.g., cameras). A) an image recorded with an off-the-shelf camera; B) a processed image showing NDVI values; C) a thermal image; D) the propelled vehicle showing just the frame and gathering spectroradiometric data from different genotypes of *Brachiaria*; E) the propelled vehicle covered with a fabric to diffuse light for even illumination and no shadows when capturing images; a laptop computer is connected to the sensors for viewing, triggering and capturing data; F) a top view of the covered vehicle, with an open space to mount an off-the-shelf camera. NDVI, Normalized Difference Vegetation Index; Green healthy plants show values of NDVI close to 1, whereas chlorotic ones and bare soil, show NDVI values close to 0.

42.4.2.2 Field Conditions

Two phenotyping methods are currently in use (Figures 42.4 and 42.5). The simplest of them is a wheeled buggy—or phenocart—which is a hand-propelled vehicle where a number of sensors can be mounted (e.g., cameras, GPSs, spectroradiometers) (Figure 42.4). This with the aim to perform comparative studies and to record responses from a time series. The second method is through the use of unmanned aerial vehicles

(UAVs) (Figure 42.5). UAV technology with its high flexibility and low costs provide an appropriate sampling method for monitoring tropical forage grasses at a high spatiotemporal scale. Consumer UAVs (e.g., DJI Phantom) are highly portable with lightweight which makes them ideal for travel to remote areas and also for hilly/mountainous environments. However, the use of consumer UAVs are mostly restricted for good weather conditions (no rain or heavy winds) and for targeted

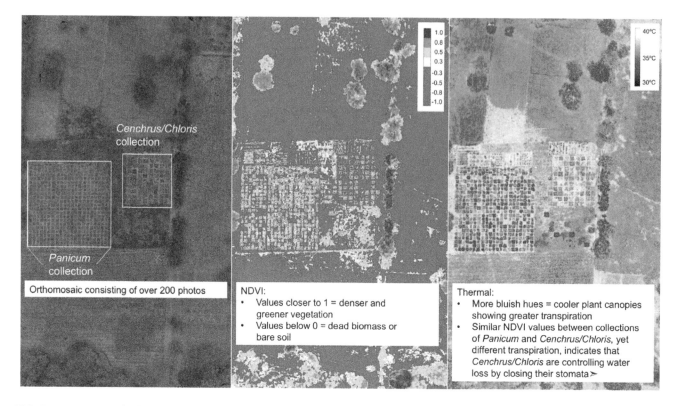

FIGURE 42.5 (See color insert.)   An example of images collected from an UAV showing from left to right: RGB (normal), NDVI values and canopy temperatures of a collection of *Panicum maximum*, *Cenchrus ciliaris*, and *Chloris gayana* germplasm accessions (a total of 153 germplasm accessions from CIAT's Genetic Resources Program and the International Livestock Research Institute, ILRI). The use of UAVs allows simultaneous assessment of a large number of genotypes.

monitoring on relatively small areas (~20 ha). A number of sensors can be mounted to the UAVs (e.g., multispectral and thermal cameras) that can be used for phenotyping purposes.

42.4.2.3   Data Handling

One of the concerns regarding studies of responses to drought in tropical forage grasses has been: 1) repeated failure to describe the experiments, 2) the inconsistency in the terminology used among studies, and 3) difficulty to check results. In response to that, CIAT currently: 1) uses the recommended minimum and extended description of experiments conducted under greenhouse or field conditions (Poorter et al., 2012), 2) makes use of ontologies that provide standard concepts (http://www.cropontology.org/ontology/CO_345/Brachiaria), and 3) saves information in a centralized server (http://octopus.ciat.cgiar.org/) for later dissemination through an open web application (https://dataverse.org/about). The latter with the aim to make data available to other researchers.

42.5   FUTURE OUTLOOK

There are several issues that ought to be sorted out in the short-term to further increase the understanding of responses and coping mechanisms in tropical forage grasses under drought stress. This, in turn, will contribute to their better utilization for improved livestock production in different drought-prone environments.

42.5.1   Increasing Phenotyping Efforts of the Root System

Tropical forage grasses are mostly sown on marginal lands (Rao, 2014). As such, identifying tropical forage grasses with enhanced soil exploration and resource capture abilities are likely to improve yields when these are limited by drought but also with low soil fertility (Rao et al., 2016). A bigger effort to phenotype the root system holds promise for a second green revolution (Lynch, 2007). A set of tools are nowadays available for the non-destructive 2D and 3D analysis of the root system. These tools include 3D reconstructions using: electrical resistance tomography, electromagnetic inductance, ground penetrating radar, magnetic resonance imaging, positron emission tomography, X-Ray computed tomography (Atkinson et al., 2019) or 2D viewing using the more "humble" rhizotron, shovelomics, and the soil core break methods. The latter and more humble methods can be easily established and practiced, as digital imaging using different sensors are no longer cost prohibitive. Furthermore, image analysis packages are getting more accessible and powerful. The current challenge is to develop systems for non-destructive root phenotyping to accurately reflect and capture the root system architecture. Some of these methods have been deployed at CIAT (Figure 42.6). Genetic gains in improving forage grasses are important targets for plant breeding programs, and these gains could be enhanced by identifying root traits that contribute to improved plant performance.

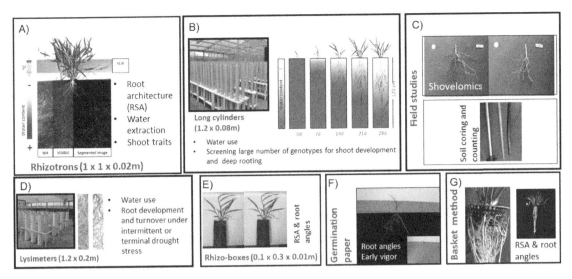

FIGURE 42.6 (See color insert.) Methods available at CIAT to study roots. A) Large rhizotrons that are used to estimate soil water content and root distribution and root angles of roots. This is also used to estimate canopy temperature (using a FLIR system). B) system as the one described in Figure 42.1; C) shovelomics and soil coring; D) large lysimeters to perform long term studies; E, F, and G) methods using rhizoboxes, germination paper, and basket method to study root angles.

42.5.2 RECONCILING PHENOTYPIC INFORMATION FROM GREENHOUSE AND FIELD CONDITIONS

Information gathered from studies conducted under greenhouse and field conditions must be integrated. Albeit phenotypic information under greenhouse conditions might seem irrelevant to some breeders and agronomists, many traits of interest are not easy to record with precision under field conditions. Combining data from both greenhouse and field conditions should provide relevant information to quantify the genotype x environment interactions to set priorities for genetic enhancement of key plant traits in stressful environments.

ACKNOWLEDGMENTS

This work was done as part of the CGIAR Research Program on Livestock. We thank all donors that globally support our work through their contributions to the CGIAR system.

BIBLIOGRAPHY

Araus JL and Kefauver SC. 2018. Breeding to adapt agriculture to climate change: affordable phenotyping solutions. *Current Opinion in Plant Biology* 45:237–247.

Araus JL, Slafer GA, Royo C, and Serret MD. 2008. Breeding yield potential and stress adaptation in cereals. *Critical Reviews in Plant Science* 27:377–412.

Atkinson JA, Pound MP, Bennet MJ, and Wells DM. 2019. Uncovering the hidden plants using new advances in root phenotyping. *Current Opinion in Biotechnology* 55:1–8.

Attere AF. 1992. Conservation and utilization of germplasm in East and Southern Africa: an overview. *Dinteria* 23:19–30.

Baruch Z. 1994a. Responses to drought and flooding in tropical forage grasses. I. Biomass allocation, leaf growth and mineral nutrients. *Plant Soil* 164:87–96.

Baruch Z. 1994b. Responses to drought and flooding in tropical forage grasses. II. Leaf water potential, photosynthesis rate and alcohol dehydrogenase activity. *Plant Soil* 164:97–105.

Baruch Z and Fernández DS. 1993. Water relations of native and introduced C4 grasses in a Neotropical Savanna. *Oecologia* 96:179–185.

Baruch Z, Ludlow MM, and Davies R. 1985. Photosynthesis responses of native and introduced grasses from Venezuelan savannas. *Oecologia* 67:288–293.

Baruch Z and Jackson RB. 2005. Responses of tropical native and invader C4 grasses to water stress, clipping and increased atmospheric CO_2 concentration. *Oecologia* 145:522–532.

Blum A. 2010. *Plant Breeding for Water-Limited Environments.* Springer Publishing, New York, NY. p. 272.

Boyer JS. 1982. Plant productivity and environment. *Science* 218:443–448.

Cardoso JA, Pineda M, Jimenez JC, Vergara M, and Rao IM. 2015. Contrasting strategies to cope with drought conditions by two tropical forage grasses. *AoB Plants* plv107.

Cai S, Chen G, Wang Y, Marchant DB, Wang Y, Tang Q, Dai F, Hills A, Franks PJ, Nevo E, Soltis DE, Soltis PS, Sessa E, Wolf PG, Xue D, Zhang G, Pogson BJ, Blatt MR, and Chen Z. 2017. Evolutionary conservation of ABA signaling for stomata closure. *Plant Physiology* 174:732–747.

Carney JA. 2010. Landscapes and places of memory: African diaspora research and geography. In: *The African Diaspora and the Disciplines*, T. Olaniyan and Sweet JH. (eds.), Indiana University Press. Bloomington, IN and Indianapolis, IN, pp. 101–118.

Carrow RN. 1996. Drought resistance aspects of turfgrasses in southeast: root-shoot responses. *Crop Science* 36:687–694.

Chaves MM, Maroco JP, and Pereira JS. 2003. Understanding plant responses to drought—from genes to whole plant. *Functional Plant Biology* 30:239–264.

Crausbay SD, Ramirez AR, Carter S, Cross MS, Hall K, Bathke DJ, Betancourt JL, Colt S, Cravens AE, Dalton MS, Dunham JB, Hay LE, Hayes MJ, McEvoy J, McNutt CA, Moritz MA, Nislow KH, Raheem N, and Sanford T. 2017. Defining ecological drought for the twenty-first century. *Bulletin of the American Meteorological Society* 98:2543–2550.

Edwards EJ, Osborne CP, Strömberg CAE, and Smith SA. 2010. C4 grass consortium. The origins of grasslands: integrating evolutionary and ecosystem science. *Science* 328:587–591.

Guenni O, Marín D, and Baruch Z. 2002. Responses to drought of five *Brachiaria* species. I. Biomass production, leaf growth, root distribution, water use and forage quality. *Plant Soil* 243:229–241.

Guenni O, Baruch Z, and Marin D. 2004. Responses to drought of five *Brachiaria* species. II. Water relations and leaf gas exchange. *Plant Soil* 258:249–260.

Guenni O, Gil JL, Baruch Z, Márquez L, and Núñez C. 2006. Respuestas al déficit hídrico en especies forrajeras de *Brachiaria* (Trin.) Griseb. (Poaceae). *Interciencia* 31:505–511.

Herrero M, Thornton PK, Gerber P, and Reid RS. 2009. Livestock, livelihoods and the environment: understanding the trade-offs. *Current Opinion in Environmental Sustainability* 1:111–120.

Herrero M, Thornton PK, Notenbaert AM, Wood S, Msangi S, Freeman HA, and Rosengrant M. 2010. Smart investments in sustainable food production: revisiting mixed-crop livestock systems. *Science* 327:822–825.

Hofer RM. 1991. Root hairs. In: *Roots: The Hidden Half*, Waidel Y, Eshel A, and Kafkaki U, (eds.). Marcel Dekker, New York, NY, pp. 129–148.

Houle D, Govindajaru DR, and Omholt S. 2010. Phenomics: the next challenge. *Nature Reviews Genetics* 11:855–866.

Hsiao TC. 1973. Plant responses to water stress. *Annual Review of Plant Physiology* 24:519–570.

Huang H, Duncan RR, and Carrow RN. 1997. Drought-resistance mechanisms of seven warm-season turfgrasses under surface soil drying: I. Shoot response. *Crop Science* 37:1858–1863.

Humphreys LR. 1981. *Environmental Adaptation of Tropical Pasture Plants*. MacMillan Publishing, London.

Humphreys LR. 1994. *Tropical Forages: Their Role in Sustainable Agriculture*. Wiley, Harlow, UK. p. 430.

IPCC. 2014. Climate change 2014: Impacts, adaptation, and vulnerability. Part A: Global and sectoral aspects. Contribution of Working Group II to the fifth assessment report of the intergovernmental panel on climate change. Field CB, Barros VR, Dokken DJ, Mach KJ, Mastrandrea MD, Bilir TE, Chatterjee M, Ebi KL, Estrada YO, Genova RC, Girma B, Kissel ES, Levy AN, MacCracken S, Mastrandrea PR, and White LL, (eds.), Cambridge University Press, Cambridge, UK, and New York, NY, p. 1132.

Kirkegaard JA, Lilley JM, Howe GN, and Graham JM. 2007. Impact of subsoil water use on wheat yield. *Australian Journal of Agricultural Research* 58:303–315.

Kretschmer AE Jr and Pitman WD. 2001. Germplasm resources of tropical forage grasses. In: *Tropical Forage Plants: Development and Use*. Sotomayor-Ríos A and Pitman WD, (ed.). CRC Press, Boca Ratón, FL.

Lawson T and Blatt MR. 2014. Stomatal size, speed and responsiveness impact on photosynthesis and water use efficiency. *Plant Physiology* 164:1556–1570.

Levitt, J. 1980. Chapter 4. Drought avoidance. In: *Responses of Plants to Environmental Stresses*. Levitt J, (ed.). Volume. 2, Academic Press, New York, NY, pp. 93–103.

Loka D, Harper J, Humphreys M, Gasior D, Gwynn-Jones D, Scullion J, Doonan J, Kingston-Smith A, Dodd R, Wang J, Chadwick D, Hill P, Jones D, Mills G, Hayes F, and Robinson D. 2018. Impacts of abiotic stresses on the physiology and metabolism of cool-season grasses: A review. *Food and Energy Security* 8(1): DOI:10.1002/fes3.152.

Ludlow MM and Ng TT. 1976. Effect of water deficits on carbon dioxide exchange and leaf elongation rate of *Panicum maximum* var. trichoglume. *Australian Journal of Plant Physiology* 3:401–413.

Ludlow MM, Ng TT, and Ford CW. 1980. Recovery after water stress of leaf gas exchange in *Panicum maximum* var. trichoglume. *Australian Journal of Plant Physiology* 7:299–313.

Ludlow MM, Fisher MJ, and Wilson JR. 1985. Stomatal adjustment to water deficits in three tropical grasses and a tropical legume grown in controlled conditions and in the field. *Australian Journal of Plant Physiology* 12:131–150.

Lynch JP. 2007. Roots of the second green revolution. *Australian Journal of Botany* 55:493–512.

Miles JW, do Valle CB, Rao IM, and Euclides VPB. 2004. Brachiariagrasses. In: Warm-*Season (C4)* Grasses, Moser L, Burson B, and Sollenberger LE (eds.). ASA-CSSA-SSSA, Madison, WI, pp. 745–783.

Ng TT, Wilson JR, and Ludlow MM. 1975. Influence of water stress on water relations and growth of a tropical (C4) grass, *Panicum maximum* var. trichoglume. *Australian Journal of Plant Physiology* 2:581–595.

Overholt WA and Franck AR. 2017. The invasive legacy of forage introductions into Florida. *Natural Areas Journal* 37:254–264.

Parsons JJ. 1972. Spread of African pasture grasses to the American tropics. *Journal of Range Management* 25:12–17.

Perez-Sanz F, Navarro PJ, and Egea-Cortines M. 2017. Plant phenomics: an overview of image acquisition technologies and image data analysis algorithms. *GigaScience* 6: DOI:10.1093/gigascience/gix092.

Perry B and Sones K. 2007. Poverty reduction through animal health. *Science* 315:333–334.

Passioura JB.1983. Roots and drought resistance. *Agricultural Water Management* 7:265–280.

Passioura JB. 1991. Soil structure and plant growth *Soil Research* 29:717–728.

Passioura JB. 2007. The drought environment: physical, biological and agricultural perspectives. *Journal of Experimental Botany* 58:113–117.

Poorter H, Fiorani F, Stitt M, Schurr U, Finck A, Gibon Y, Usadel B, Munns R, Atkin A, Tardieu F, and Pons TL. 2012. The art of growing plants for experimental purposes: a practical guide for the plant biologist. *Functional Plant Biology* 39:821–838.

Rao IM. 2014. Advances in improving adaptation of common bean and *Brachiaria* forage grasses to abiotic stresses in the tropics. In: *Handbook of Plant and Crop Physiology*, Pessarakli M (ed). Third Edition. CRC Press, Taylor and Francis Group, New York, NY, pp. 847–889.

Rao IM, Miles JW, Beebe SE, and Horst WJ. 2016. Root adaptations to soils with low fertility and toxicities. *Annals of Botany* 118:593–605.

Richards R and Pasioura JB. 1989. A breeding program to reduce diameter of the major xylem vessels in the seminal roots of wheat and its effect on grain yield in rain-fed environments. *Australian Journal of Agricultural Research* 40:943–950.

Sanders GJ and Arndt SK. 2012. Osmotic adjustment under drought conditions. In: *Plant Responses to Drought Stress: From Morphological to Molecular Features*. Aroca R. (ed.). Springer-Verlag, Berlin, pp. 199–229.

Schultze-Kraft R, Rao IM, Peters M, Clements RJ, Bai C, and Guodao L. 2018. Tropical forage legumes for environmental benefits: an overview. *Tropical Grasslands—Forrajes Tropicales* 61–14.

Subrahmanyam VP. 1967. Incidence of continental drought. WMO/IHD Report No. 2. Geneva.

Thornton PK, Kruska RL, Henniger N, Kristjanson PM, Reid RS, Atieno F, Odero A, and Ndegwa T. 2002. *Mapping Poverty and Livestock in the Developing World*. International Livestock Research Institute, Nairobi,. p. 124.

Wasson AP, Richards RA, Chatrath R, Misra SC, Sai Prasard SV, Rebetzke GJ, Kirkegaard JA, Cristopher J, and Watt M. 2012. Traits and selection strategies to improve root systems and water uptake in water-limited wheat crops. *Journal of Experimental Botany* 63:3485–3498.

Wilhite DA and Glantz MH. 1985. *Understanding the Drought Phenomenon: The Tole of Definitions*. Drought Mitigation Center Faculty Publications, National Drought Mitigation Center, Lincoln, NE, p. 20.

Wilson JR and Ludlow MM. 1983a. Time trends for change in osmotic adjustment and water relations of leaves of *Cenchrus ciliarus* during and after water stress. *Australian Journal of Plant Physiology* 10:15–24.

Wilson JR and Ludlow MM.1983b. Time trends of solute accumulation and the influence of potassium fertilizer on osmotic adjustment of water-stressed leaves of three tropical grasses. *Australian Journal of Plant Physiology* 10:523–537.

Wilson JR and Ng TT. 1975. Influence of water stress on parameters associated with herbage quality of *Panicum maximum* var. trichoglume. *Australian Journal of Agricultural Research* 26:127–136.

Wilson JR, Ludlow MM, Fisher MJ, and Schulze ED. 1980. Adaptation to water stress of leaf water relations of four tropical forage species. *Australian Journal of Plant Physiology* 7:207–220.

43 Drought Resistance of Common Bean
Water Spending and Water Saving Plant Ideotypes

Jose A. Polania and Idupulapati M. Rao

CONTENTS

43.1 INTRODUCTION

Common bean (*Phaseolus vulgaris* L.) is the most important food legume in the tropics of Latin America and East, Central, and Southern Africa. It belongs to the family Fabaceae, and it has two gene pools (Mesoamerican and Andean) based on their centers of origin from Central and South America, respectively (Gepts and Debouck, 1991). These gene pools differ in seed size and color, in the protein phaseolin, and in morphological and molecular characteristics (Beebe, 2012). The crop is grown by smallholder farmers in Latin America and East Africa, where it is often exposed to unfavorable conditions such as drought, low soil fertility, and heat (Beebe et al., 2008). It is an inexpensive source of protein and calories for small farmers in countries with endemic poverty (Rao, 2014).

Bean yields are affected by various biotic and abiotic stress factors, but disease is the main constraint on bean production. Among abiotic stress limitations, drought can reduce yields by between 10% and 100%. About 60% of the bean production regions are affected by drought, the second most important factor in yield reduction after diseases (Rao, 2014; Thung and Rao, 1999). Climate change can be a significant threat to bean production by smallholders due to the individual or combined stress factors of drought and heat, which can reduce yield and quality and lead to restricted geographic adaptation (Beebe et al., 2011). Much of the production of beans by small farmers is done under rainfed conditions. In rainfed agriculture, the variation in available moisture for crop growth is a major determinant of the adaptation or the good performance of the bean varieties. This variation in soil moisture has an unpredictable component depending on rainfall patterns, which determine the availability of water at different crop growth stages, and on the farming system, which determines the planting time and therefore, the development of the crop (Fischer et al., 2003). According to rainfall patterns and the bean crop cycle, three types of drought stress can be characterized: an early drought that occurs during vegetative growth; an intermittent drought, which implies different periods of drought during vegetative and reproductive growth; and terminal drought, which occurs from flowering to harvest.

The development of bean varieties resistant to drought and heat stress conditions through breeding is a useful strategy to ensure food security in marginal areas to confront climate change. Understanding the physiological basis of yield limitations will contribute to the development of phenotyping tools in support of plant breeding (Araus et al., 2002; Girdthai et al., 2009; Mir et al., 2012). A useful plant trait or mechanism as a phenotyping tool must exhibit enough genetic variability, correlation with yield, and higher heritability, and its evaluation must be fast, easy, and cheap (Araus et al., 2002; Jackson et al., 1996).

In the case of common bean, physiological traits or mechanisms related to drought adaptation are diverse, and these traits depend on the environments where the crop is grown and the type of drought stress that prevails in the target environment (Polania et al., 2016a). Physiological studies for improving drought resistance generally focus on three key processes: (i) acquiring a greater amount of water by the root system from the soil profile to facilitate transpiration; (ii) acquiring more carbon (biomass) in exchange for the water transpired by the crop; and (iii) increased mobilization of accumulated carbon to the harvestable economic product (Condon et al., 2004), resulting in the identification of key traits that are associated with improved drought resistance. The purpose of this chapter is to report on recent advances in characterizing phenotypic differences in drought resistance in common bean genotypes and in classifying the genotypes as water spending or water saving plant ideotypes.

43.2 SHOOT TRAITS RELATED TO IMPROVED DROUGHT RESISTANCE

Different studies have been carried out with the objective of identifying shoot traits related to drought adaptation in common beans; which range from non-destructive determinations such as stomatal conductance, SPAD chlorophyll meter readings (SCMR), photosynthetic efficiency, canopy temperature, and phenological traits to destructive determinations such as leaf area index, canopy biomass, dry matter partitioning indices, natural abundance of stable carbon and nitrogen isotopes, and nutrient and carbohydrate/sugar concentrations.

43.2.1 CHLOROPHYLL CONTENT AND PHOTOSYNTHETIC EFFICIENCY

The results of the different traits evaluated under field and greenhouse conditions have highlighted some of these as promising and others as less so (Polania et al., 2016a, 2017a). A widely used trait in plant physiology is SCMR, but this trait did not show a good relation to grain yield under drought stress in different field studies. However, drought stress increased SPAD chlorophyll content in the leaves. This increase is because under drought stress conditions, the leaf expansion is reduced, which results in a higher chlorophyll concentration per unit leaf area. In the case of common bean, SCMR allows discrimination between non-stress and stress treatments, but its use as a selection tool for drought resistance is uncertain. A similar story has been reported with photosynthetic efficiency, for which experimental results under field conditions have shown phenotypic differences between water stress treatments, for example, irrigated vs. drought treatment, but no clear differences have been found between genotypes given the same treatment, even under drought conditions, in relationship with grain yield. This is also considered to be an uncertain trait for the selection of genotypes adapted to drought in breeding programs.

43.2.2 EFFECTIVE OR EFFICIENT WATER USE

Traits related to water status and transpiration, such as stomatal conductance, canopy temperature, and carbon isotope discrimination (CID), have been evaluated under field and greenhouse conditions (2016a, Polania et al., 2017a) for their usefulness as selection criteria. The results have indicated that these could serve as promising traits in the identification of genotypes and mechanisms related to drought resistance in beans; but each has advantages and disadvantages. Stomatal conductance provides an instantaneous reading of the extent of opening of stomata, but its use in the evaluation of a large number of genotypes is limited. Canopy temperature depression measured with infrared devices provides an indirect measure of transpiration, and its use in the evaluation of various genotypes is easy. One of the advantages of measuring leaf CID rather than stomatal conductance is that the measurement of CID integrates the process over long periods of gas exchange and its interaction during crop development (Easlon et al., 2014). The use of a grain sample for CID determination makes more sense in the case of common bean under terminal drought stress, because it is taken at the time of maturity and would have more integrated effect of gas exchange from a critical and important crop growth stage which is grain filling. A relationship between stomatal conductance and grain CID under drought conditions has been observed, with differences in performance among some genotypes (Polania et al., 2016a). Stomatal conductance and leaf or grain CID are useful measurements to estimate differences in water use efficiency (WUE), but they depend on different processes. CID depends on variation in photosynthetic biochemistry, conductance of CO_2 to the leaf interior and the chloroplasts, or a combination of these (Easlon et al., 2014; Seibt et al., 2008).

Regarding water use, common beans exhibited phenotypic differences in grain CID, leaf stomatal conductance, canopy biomass, and grain yield under drought stress (Polania et al., 2016a, 2017a). Based on these phenotypic differences, the genotypes are classified into two groups: water savers and water spenders. This grouping facilitates targeting genotypes to specific agro-ecological niches. In bush bean under non-severe droughts or in non-arid environments, it has been observed that there is a positive relationship among grain CID, root length density, and grain yield. This indicates that plants under drought stress generate deeper roots and therefore, access more water, resulting in increased stomatal conductance (Hall, 2004; Polania et al., 2017a, 2012; Sponchiado et al., 1989; White et al., 1990; White, 1993). This increased water use is associated with increased accumulation of carbon and plant growth. The genotypes with these strategies are classified as the water spenders ideotype, and their drought resistance is associated with superior effective use of water (EUW), which was attributed to a deeper and vigorous root system, higher canopy biomass, and better photosynthate remobilization to pod and grain production. These water spending genotypes should be useful for cultivation in areas exposed to intermittent drought stress in Central America, South America, and Africa, particularly in agro-ecological regions where rainfall is intermittent during the season and in soils that can store a greater amount of available water deep in the soil profile.

The drought-resistant genotypes that are classified as the water savers ideotype presented lower values of grain CID and a negative relationship with both root vigor and grain yield. Thus, the water saver genotypes under drought stress generate fewer roots and therefore, use less water, resulting in reduced stomatal conductance and moderate gas exchange and plant growth (Polania et al., 2016a). The water saver genotypes presented higher values of WUE combined with a shallow root system, reduced water loss and plant growth, and better photosynthate partitioning, resulting in a moderate to lower level of grain yield under severe drought stress. These genotypes may be more suitable for semiarid to dry environments, dominated by the terminal type of drought stress in Central America, Africa, northern Mexico, and northeast Brazil.

43.2.3 Remobilization of Photosynthates

However, independently of the water use strategy (spender or saver), the improved harvest index (HI) or enhanced mobilization of photosynthates to grain production plays an essential role in the success of superior genotypes under drought stress. Several studies in common bean have shown that increased photosynthate mobilization to pod and seed formation contributes to a better grain yield under drought as well as low soil fertility stress conditions (Beebe et al., 2013, 2008; Chaves et al., 2018; Devi et al., 2013; Mukankusi et al., 2018; Polania et al., 2017b, 2016a; Rao et al., 2017, 2013; Rao, 2014). Two dry matter partitioning indices that have been useful in characterizing phenotypic differences in photosynthate mobilization under drought stress are the pod partitioning index (PPI), which indicates the extent of mobilization of assimilates from the vegetative structures to pod formation, and the pod harvest index (PHI), which indicates the extent of mobilization of assimilates from the podwall to grain formation and grain filling (Assefa et al., 2013; Beebe et al., 2013; Chaves et al., 2018; Polania et al., 2017b, 2016a, 2016c; Rao et al., 2017, 2013). The relevance of photosynthate remobilization for improved drought resistance in common bean was also demonstrated by biochemical analysis. Research using drought-tolerant and susceptible bean genotypes and techniques such as ^{14}C-labeling to quantify sugar accumulation and partitioning showed that high grain yield under terminal drought stress is associated with an efficient carbon mobilization from leaves to pods and to seeds (Cuellar-Ortiz et al., 2008; Rosales et al., 2012). Different evaluations under field and greenhouse conditions indicated that bean genotypes resistant to terminal drought showed superior mobilization of total non-structural carbohydrates from stems to pods and grain filling (Polania et al., unpublished data). These results showed a closer relationship that may exist between the mobilization of carbon and mobilization of sugar; especially, the hypothesis that a better mobilization of photosynthates is controlled by a better mobilization of sucrose to developing grains. Sucrose can regulate the metabolism of carbon (C) in the source organs and the expansion and cell division in sink organs. Sucrose is the main carbohydrate transported over long distances through the phloem from the source to the sink in most plants (Li et al., 2017). The production of sucrose and its mobilization are affected under drought stress conditions, and this stress also causes significant changes in the expression of sucrose transporter genes (SUT), indicating the importance of the key role of sucrose and its transporters in improving drought resistance in common bean.

43.2.4 Sink Strength: Pod and Seed Formation

A major effect of drought stress on grain legumes is to reduce yield components such as pod number per area (PNA) and seed number per area (SNA) (Assefa et al., 2015; Polania et al., 2017b; Rao et al., 2017). Seed number per pod has been identified as a useful criterion for selection for improving drought resistance because of its higher heritability and contribution to genetic gain (Ramirez-Vallejo and Kelly, 1998). Seed number per pod is determined by the potential of ovules formed by the plant, pollination processes, and the capacity of the plant to transport carbohydrates for the development of embryos (Farooq et al., 2016; Polania et al., 2017b). The results shown in Table 43.1 illustrate that even under conditions of optimal supply of water, the bean plant sacrifices one or two ovules for the grain formation within a pod; and under drought conditions, the plant can sacrifice the formation of half of its ovules in seeds. However, drought-resistant genotypes maintain the formation of an adequate number of seeds per pod, and this ability, combined with a good ability to mobilize photoassimilates from vegetative structures, can result in sustaining superior pod formation per area compared with drought-sensitive genotypes (Table 43.1). This greater mobilization of sugars to pod formation is also evident in the rapid pod growth after the pollination of genotypes that are resistant to drought, such as in INB 841 and SEN 56 (Table 43.1).

The selection of genotypes that have greater sink strength, as reflected in greater values of PNA and SNA, is required to increase grain yield under drought conditions. Pod and seed formation is dependent on the plant's water status and the supply of photoassimilates imported from source leaves through the phloem and also from the remobilization of stored assimilates in the stem tissues (Blum, 2009; Rao et al., 2017). A key challenge to improve bean yield under drought stress is to identify the physiological processes and the genes responsible for imposing limitations on carbohydrate supply to pod and seed development.

43.2.5 Differences between Andean and Mesoamerican Gene Pools

A marked difference in drought resistance has been reported between the Andean and Mesoamerican gene pools of common bean (Polania et al., 2016c). The Mesoamerican gene pool presented a higher level of drought resistance than the Andean gene pool. A significant part of this difference may

TABLE 43.1

Pod Elongation Rate, Number of Ovules per Pod, Seed Number per Pod, and Pod Number per Area (m²) of Four Drought-resistant Bean Genotypes Grown under Well-watered and Terminal Drought Stress Conditions in 2017 at CIAT, Palmira, Colombia

Genotype	Initial Pod Elongation Rate (mm day⁻¹)	Number of Ovules per Pod	Seed Number per Pod under Well-Watered Conditions	Seed Number per Pod Under Drought Conditions	Pod Number per m² under Well-Watered Conditions	Pod Number per m² under Drought Conditions
ALB 91	10.0	6	5	3	324	83
INB 841	12.3	7	5	5	290	160
SEN 56	13.3	7	5	4	300	200
SER 16	10.1	6	5	4	320	190

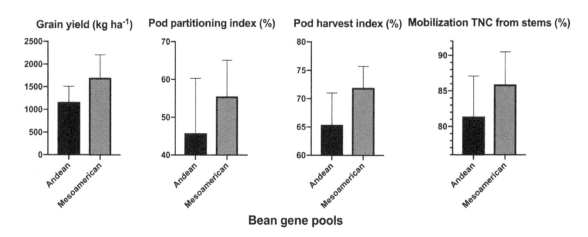

FIGURE 43.1 Differences in grain yield, pod partitioning index (PPI), pod harvest index (PHI) and mobilization of total non-structural carbohydrates (TNC) from stems in the two gene pools of common beans under terminal drought stress conditions. Results from 18 Andean and 18 Mesoamerican genotypes under drought conditions in 2017 at CIAT, Palmira, Colombia.

be due to evolutionary events; historically, the Mesoamerican gene pool could have been more exposed to stress by drought than the Andean gene pool. Research also indicated that the Andean gene pool has a yield potential similar to that of the Mesoamerican gene pool to improve resistance to drought. In general, the ability to mobilize photosynthates was greater with Mesoamerican genotypes than with Andean genotypes, as evidenced by the greater values of PPI, PHI, and total non-structural carbohydrate (TNC) remobilization (Figure 43.1). The sensitivity to drought stress in Andean compared with Mesoamerican genotypes is due to poor performance in the mobilization of these reserves to pod and grain production (Polania et al., 2016c). Breeding programs for Andean beans could focus on selecting the best-performing materials that combine greater values of shoot biomass with greater mobilization of these reserves to pod development and grain filling.

43.2.6 CONTRIBUTION FROM SYMBIOTIC NITROGEN FIXATION

Common bean can supply at least part of its nitrogen (N) requirement through symbiotic nitrogen fixation (SNF). However, compared with other legumes, beans have lower

SNF capacity (Hardarson et al., 1993; Hardarson, 2004; Peoples et al., 2009). Although both abiotic and biotic stress factors affect SNF capacity, SNF is highly sensitive to drought (Barbosa et al., 2018; Devi et al., 2013; Polania et al., 2016b; Ramaekers et al., 2013). The symbiosis is based on the carbon supply from the plant to the rhizobium, which provides fixed N to the plant. Under drought stress, the reduced net photosynthesis decreases the supply of photosynthates and sucrose to the nodules, resulting in lower values of nitrogen fixed (Gonzalez et al., 1995; Sassi et al., 2008). Additionally, due to the evolution process, some genotypes of common bean with adaptation to drought and some genotypes of drought-resistant tepary bean (*Phaseolus acutifolius*) drastically reduce their SNF ability under drought, suggesting an internal control mechanism of SNF ability. This internal SNF control serves for an efficient use of carbon and sugars, thereby decreasing carbon supply to the nodules, leading to superior mobilization of this carbon to pod and grain formation and an increased N use efficiency of acquired N from soil (Polania et al., 2016b). Despite the fact that SNF ability can be drastically affected by drought stress, results have shown that some drought-resistant Mesoamerican lines, classified as water spenders, combined superior grain yield and better SNF ability under drought stress. This response is due to a better source/sink relationship, using the acquired N more efficiently to produce grain, and greater remobilization of both C and N to grain.

43.3 ROOT TRAITS RELATED TO IMPROVED DROUGHT RESISTANCE

43.3.1 ROOT–SHOOT RELATIONSHIPS

Root evaluations are usually complex, expensive, or very labor-intensive. However, several methodologies, both *in situ* and under controlled conditions, have been developed. Evaluations of root traits have shown the contribution of deep rooting to improved drought resistance through the greater acquisition of water (Polania et al., 2017a, 2017b, 2012, 2009; Rao, 2014; Sponchiado et al., 1989; White and Castillo, 1992). For improved performance under drought stress, an increase in water extraction capacity and crop growth must be accompanied by an improved HI. Superior remobilization of photosynthates to pod and grain formation is essential for the success of genotypes under drought stress. The strategic combination of different shoot and root traits seems to be the key to improving the resistance of common bean to drought. The roots and shoots have a complex relationship; the shoot provides the root with carbon and certain hormones, and the root provides the shoot with water, nutrients, and also hormones. To increase grain yield through better plant growth under both optimal and drought stress conditions, the root system must be able to supply water and nutrients to the new plant growth without sequestering too many photoassimilates from the shoot (Bingham, 2001). Genotypes with differential root system architecture should be used for the development of drought-resistant lines with improved root architecture for each specific agro-ecological condition (Heng et al., 2018). The identification of the root traits that are more suitable to different agro-ecological conditions of the crop will play an important role in the development of new varieties adapted to different types of drought stress.

43.3.2 DEEP ROOTING GENOTYPES

The results from different trials using a simple methodology with small plastic soil cylinders as lysimeters to evaluate phenotypic differences in root vigor (Butare et al., 2012, 2011; Polania et al., 2017b, 2017a, 2009) showed the genotypic diversity in root system characteristics that are present in common bean and their relationship with resistance to drought stress. This root phenotyping strategy also contributed to identifying root characteristics that are related to two major plant ideotypes, water savers and water spenders. The genotypes classified as water spenders combined superior grain production under drought stress with a vigorous and deeper root system. The deep rooting genotypes combine different architectural and anatomical phenotypes, such as reduced density of lateral roots and greater loss of roots that do not contribute to water capture (e.g., roots in dry topsoil), that permit the plant to allocate a greater amount of resources to deeper roots, resulting

in greater capture of water from deeper soil layers (Lynch, 2018). The vigorous and deeper root system of these genotypes allows the plant to access available water from deeper soil layers and to continue the processes of gas exchange and carbon accumulation, combined with a better photosynthate remobilization, resulting in a better grain yield under drought stress. The strategy of these genotypes, with deeper roots and better water extraction capacity, is to support the rate of photosynthesis and the accumulation of water-soluble carbohydrates in the stem and their subsequent remobilization to grain filling (Lopes and Reynolds, 2010).

43.3.3 SHALLOW ROOTING GENOTYPES

By exploring the diversity in root vigor in common beans (Polania et al., 2017a, 2017b), it was possible to identify some characteristics that are related to water saving ability. These water saving genotypes combine higher grain yield under drought stress with a slow growing and shallower root system. These water saving genotypes, with a strategy of water conservation and higher WUE, combine this ability with a better remobilization of photosynthates to pod and grain formation. The strategy of these water saver genotypes can be complemented with shoot traits related to conserving water at the vegetative stage, such as lower leaf conductance and a smaller leaf canopy, that would make more water available for reproductive growth and grain filling, resulting in a better grain yield under terminal drought stress conditions (Araújo et al., 2015; Zaman-Allah et al., 2011).

A poor root system can limit the optimal plant development and grain production under drought stress. The drought-susceptible bean genotypes showed a lower level of root production, with a lower rate of root growth at a shallower soil depth under drought conditions (Polania et al., 2017a, 2017b). However, the relationship between root system characteristics and shoot growth can be complex; a genotype with a vigorous and deeper root system could fail in its performance under drought stress due to excessive vegetative growth and poor grain production. A very vigorous root system in a plant that is inefficient at assimilating CO_2 would make the roots another sink competing for photoassimilates and could increase the sensitivity to drought stress. A vigorous and deeper root system with rapid growth is useful but not enough to confer resistance to drought in common bean. What is needed is the strategic combination of traits, such as a better root system combined with the ability to remobilize photosynthates from vegetative structures to the pods and subsequently, to grain production (Beebe et al., 2014; Polania et al., 2016a, 2017a; Rao, 2014).

43.3.4 ROOT MORPHOLOGY AND SYMBIOTIC NITROGEN FIXATION

The response of SNF is related to the root morphology. Different research results have shown that better SNF ability under drought stress is correlated with an increased presence of a thicker root system (Polania et al., 2017a). This relationship is perhaps due to an increased carbon supply to nodules

TABLE 43.2

Root and Shoot Traits Related to the Water Saving Ideotype and the Water Spending Ideotype Proposed for Targeting Improved Common Bean Genotypes to Drought-prone Agro-ecological Zones

Ideotypes	Water Saver Ideotype	Water Spender Ideotype
Root and shoot traits	• Intermediate to shallow rooting system • Intermediate root growth rate and penetration ability • Fine root system • Lower SNF ability • Earliness • High water use efficiency • Reduced transpiration rate • Less carbon isotope discrimination • Limited leaf area and canopy biomass development • Reduced sink strength • Superior photosynthate remobilization to pod and grain formation	• Vigorous and deep rooting system • Rapid root growth rate and penetration ability • Thicker root system • Moderate SNF ability • Earliness • Effective use of water • Moderate transpiration rate • More carbon isotope discrimination • Moderate canopy biomass accumulation • Moderate sink strength • Superior photosynthate remobilization to pod and grain formation
Targeting to specific agro-ecological niches	Zones with terminal drought stress and soils with lower capacity to store available water deep in the soil profile	Zones with intermittent drought stress and soils that can store a greater amount of available water deep in the soil profile

Source: Adapted from Polania, J., et al., *Theor. Exp. Plant Physiol.*, 29, 2017.

under drought stress from the stored carbohydrates in thicker roots, and a large root diameter is correlated with greater sink strength. Also, Polania et al. (2017a) found a positive relationship between the proportion of fine roots and mineral N uptake from the soil, highlighting the importance of a fine root system to acquiring mineral N from the soil. In general, the genotypes classified as water spenders have a more vigorous root system, which additionally allows them to maintain higher levels of SNF, while the genotypes classified as water savers, with a finer root system, have lower levels of SNF, combined with an increased nitrogen use efficiency to produce grain.

43.4 SHOOT AND ROOT TRAITS CONTRIBUTING TO WATER SPENDING OR WATER SAVING PLANT IDEOTYPES COPING WITH DROUGHT STRESS

We propose two ideotypes of bean plants, water spenders and water savers, to cope with drought stress, and our proposal is based on the results of phenotypic evaluation of different shoot and root traits for the past few years (Polania et al., 2017a, 2017b, 2016a, 2016b). Classification of these two ideotypes will help to target bean genotypes to different agro-ecological niches with specific types of soils and rainfall patterns. The water spender genotypes should be useful for cultivation in areas exposed to intermittent drought stress with soils that can store a greater amount of available water deep in the soil profile. The water saver genotypes may be more suitable for farmers in semiarid to dry environments dominated by the terminal type of drought stress. The main morpho-physiological characteristics of the water spender and water saver genotypes are summarized in Table 43.2.

43.5 CONCLUSIONS AND FUTURE PERSPECTIVES

Different plant traits contribute to the characterization of the genotypes as either water savers or water spenders. Traits related to root vigor include visual rooting depth, total root biomass and root length, root volume, and root hair development. Traits related to water use and plant development include carbon isotope discrimination, stomatal conductance, canopy temperature depression, leaf area index, and canopy biomass. Traits related to sink strength include PNA, SNA, 100 seed weight, and seed number per pod. Traits related to photosynthate remobilization include pod partitioning index, pod harvest index, HI, and stem biomass reduction from mid-pod fill growth stage to harvest time. Pod harvest index and carbon isotope discrimination are useful selection criteria for improving drought resistance because of their simplicity in measurement and their correlation with grain yield under both irrigated and drought stress conditions. Further research work is needed to characterize the physiological and molecular mechanisms contributing to phenotypic variation in photosynthate remobilization to grain under different types and intensities of drought stress, both individually and together with other abiotic stress factors, particularly under heat and low soil fertility stress conditions.

ACKNOWLEDGMENTS

The authors acknowledge the support of the Bill and Melinda Gates Foundation (BMGF), the United States Agency for International Development (USAID), and the CGIAR research program on Grain Legumes for financial support. We thank the bean breeding and physiology teams at the International

Center for Tropical Agriculture (CIAT), Colombia for their contribution to this work.

REFERENCES

Araújo, S.S., Beebe, S., Crespi, M., Delbreil, B., Gonzalez, E.M., Gruber, V., Lejeune-henaut, I., Link, W., Monteros, M.J., Rao, I.M., Vadez, V., and Patto, M.C., 2015. Abiotic stress responses in legumes: strategies used to cope with environmental challenges. CRC. *Crit. Rev. Plant Sci.* 34,237–280.

Araus, J.L., Slafer, G.A., Reynolds, M.P., and Royo, C., 2002. Plant breeding and drought in C3 cereals: What should we breed for? *Ann. Bot.* 89,925–940.

Assefa, T., Beebe, S., Rao, I.M., Cuasquer, J., Duque, M.C., Rivera, M., Battisti, A., and Lucchin, M., 2013. Pod harvest index as a selection criterion to improve drought resistance in white pea bean. *F. Crop. Res.* 148,24–33.

Assefa, T., Wu, J., Beebe, S., Rao, I.M., Marcomin, D., and Claude, R.J., 2015. Improving adaptation to drought stress in small red common bean: phenotypic differences and predicted genotypic effects on grain yield, yield components and harvest index. *Euphytica* 303, 477–489.

Barbosa, N., Portilla, E., Buendia, H.F., Raatz, B., Beebe, S., and Rao, I., 2018. Genotypic differences in symbiotic nitrogen fixation ability and seed yield of climbing bean. *Plant Soil* 1–17.

Beebe, S., 2012. Common bean breeding in the tropics. *Plant Breed. Rev.* 36,357–426.

Beebe, S., Ramirez, J., Jarvis, A., Rao, I.M., Mosquera, G., Bueno, J.M., and Blair, M.W., 2011. Genetic improvement of common beans and the challenges of climate change, In: *Crop Adaptation to Climate Change.* pp. 356–369.

Beebe, S., Rao, I.M., Blair, M.W., and Acosta-Gallegos, J.A., 2013. Phenotyping common beans for adaptation to drought. *Front. Physiol.* 4, 1–20.

Beebe, S., Rao, I.M., Cajiao, C., and Grajales, M., 2008. Selection for drought resistance in common bean also improves yield in phosphorus limited and favorable environments. *Crop Sci.* 48, 582–592.

Beebe, S., Rao, I.M., Devi, M., and Polania, J., 2014. Common beans, biodiversity, and multiple stress: challenges of drought resistance in tropical soils. *Crop Pasture Sci.* 65, 667–675.

Bingham, I.J., 2001. Soil-root-canopy interactions. *Ann. Appl. Biol.* 138, 243–251.

Blum, A., 2009. Effective use of water (EUW) and not water-use efficiency (WUE) is the target of crop yield improvement under drought stress. *F. Crop. Res.* 112, 119–123.

Butare, L., Rao, I.M., Lepoivre, P., Cajiao, C., Polania, J., Cuasquer, J., and Beebe, S., 2012. Phenotypic evaluation of interspecific recombinant inbred lines (RILs) of *Phaseolus* species for aluminium resistance and shoot and root growth response to aluminium-toxic acid soil. *Euphytica* 186, 715–730.

Butare, L., Rao, I.M., Lepoivre, P., Polania, J., Cajiao, C., Cuasquer, J., and Beebe, S., 2011. New genetic sources of resistance in the genus Phaseolus to individual and combined aluminium toxicity and progressive soil drying stresses. *Euphytica* 181, 385–404.

Chaves, N., Polania, J., Muñoz, C., Rao, I.M., and Beebe, S., 2018. Caracterización fenotípica por resistencia a sequía terminal de germplasma de frijol común. *Agron. Mesoam.* 29, 1–17.

Condon, A.G., Richards, R.A., Rebetzke, G.J., and Farquhar, G.D., 2004. Breeding for high water-use efficiency. *J. Exp. Bot.* 55, 2447–2460.

Cuellar-Ortiz, S.M., De La Paz Arrieta-Montiel, M., Acosta-Gallegos, J.A., and Covarrubias, A.A., 2008. Relationship between carbohydrate partitioning and drought resistance in common bean. *Plant Cell Environ.* 31, 1399–1409.

Devi, M., Sinclair, T.R., Beebe, S., and Rao, I.M., 2013. Comparison of common bean (Phaseolus vulgaris L.) genotypes for nitrogen fixation tolerance to soil drying. *Plant Soil* 364, 29–37.

Easlon, H.M., Nemali, K.S., Richards, J.H., Hanson, D.T., Juenger, T.E., and McKay, J.K., 2014. The physiological basis for genetic variation in water use efficiency and carbon isotope composition in Arabidopsis thaliana. *Photosynth. Res.* 119, 119–129.

Farooq, M., Gogoi, N., Barthakur, S., Baroowa, B., Bharadwaj, N., Alghamdi, S.S., and Siddique, K.H.M., 2016. Drought stress in grain legumes during reproduction and grain filling. *J. Agron. Crop Sci.* doi.

Fischer, K.S., Lafitte, R., Fukai, S., Atlin, G., and Hardy, B., 2003. *Breeding Rice for Drought-prone Environments.* International Rice Research Institute, Los Baños (Philippines).

Gepts, P. and Debouck, D., 1991. Origin, domestication, and evolution of the common bean (Phaseolus vulgaris L.), In: *Common Beans: Research for Crop Improvement*, pp. 7–53.

Girdthai, T., Jogloy, S., Kesmala, T., Vorasoot, N., Akkasaeng, C., Wongkaew, S., Holbrook, C.C., and Patanothai, A., 2009. Relationship between root characteristics of peanut in hydroponics and pot studies. *Crop Sci.* 50, 159–167.

Gonzalez, E.M., Gordon, A.J., James, C.L., and Arrese-Igor, C., 1995. The role of sucrose synthase in the response of soybean nodules to drought. *J. Exp. Bot.* 46, 1515–1523.

Hall, A.E., 2004. Comparative ecophysiology of cowpea, common bean and peanut, In: H.T, N., A, B. (Eds.), *Physiology and Biotechnology Integration for Plant Breeding.* Marcel Dekker Inc, New York, pp. 271–325.

Hardarson, G., 2004. Enhancement of symbiotic nitrogen fixation in grain legumes: selected results from the FAO/IAEA program, In: Serraj, R. (ed.), *Symbiotic Nitrogen Fixation: Prospects for Enhanced Application in Tropical Agriculture.* Oxford & IBH Publishing, New Delhi, pp. 163–171.

Hardarson, G., Bliss, F.A., Cigales-Rivero, M.R., Henson, R.A., Kipe-Nolt, J.A., Longeri, L., Manrique, A., Peña-Cabriales, J.J., Pereira, P.A., Sanabria, C.A., and Tsai, S.M., 1993. Genotypic variation in biological nitrogen fixation by common bean. *Plant Soil* 152, 59–70.

Heng, Y., Manish, R., Babu, V., Lijuan, Z., Pengyin, C., Pajeev K.V., and Henry T.N., 2018. Genetic diversity of root system architecture in response to drought stress in grain legumes. *J. Exp. Bot.* 69, 3267–3277.

Jackson, P., Robertson, M., Cooper, M., and Hammer, G., 1996. The role of physiological understanding in plant breeding; From a breeding perspective. *F. Crop. Res.* 49, 11–37.

Li, G., Pan, J., Cui, K., Yuan, M., Hu, Q., Wang, W., Mohapatra, P.K., Nie, L., Huang, J., and Peng, S., 2017. Limitation of unloading in the developing grains is a possible cause responsible for low stem non-structural carbohydrate translocation and poor grain yield formation in rice through verification of recombinant inbred lines. *Front. Plant Sci.* 8, 1–16.

Lopes, M.S. and Reynolds, M.P., 2010. Partitioning of assimilates to deeper roots is associated with cooler canopies and increased yield under drought in wheat. *Funct. Plant Biol.* 37, 147–156.

Lynch, J., 2018. Rightsizing root phenotypes for drought resistance. *J. Exp. Bot.* 69, 3279–3292.

Mir, R.R., Zaman-Allah, M., Sreenivasulu, N., Trethowan, R.M., and Varshney, R.K., 2012. Integrated genomics, physiology and breeding approaches for improving drought tolerance in crops. *Theor. Appl. Genet.* 125, 625–645.

Mukankusi, C., Raatz, B., Stanley Nkalubo, Berhanu, F., Binagwa, P., Kilango, M., Williams, M., Enid, K., Chirwa, R., and Beebe, S., 2018. Genomics, genetics and breeding of common bean in Africa: A review of tropical legume project. *Plant Breed.* 1–14.

Peoples, M.B., Brockwell, J., Herridge, D.F., Rochester, I.J., Alves, B.J.R., Urquiaga, S., Boddey, R.M., Dakora, F.D., Bhattarai, S., Maskey, S.L., Sampet, C., Rerkasem, B., Khan, D.F., Hauggaard-Nielsen, H., and Jensen, E.S., 2009. The contributions of nitrogen-fixing crop legumes to the productivity of agricultural systems. *Symbiosis* 48, 1–17.

Polania, J., Poschenrieder, C., Beebe, S., and Rao, I.M., 2016a. Effective use of water and increased dry matter partitioned to grain contribute to yield of common bean improved for drought resistance. *Front. Plant Sci.* 7, 1–10.

Polania, J., Poschenrieder, C., Beebe, S., and Rao, I.M., 2016b. Estimation of phenotypic variability in symbiotic nitrogen fixation ability of common bean under drought stress using 15N natural abundance in grain. *Eur. J. Agron.* 79, 66–73.

Polania, J., Poschenrieder, C., Rao, I., and Beebe, S., 2017a. Root traits and their potential links to plant ideotypes to improve drought resistance in common bean. *Theor. Exp. Plant Physiol.* 29, 143–154.

Polania, J., Rao, I.M., Beebe, S., and Garcia, R., 2009. Root development and distribution under drought stress in common bean (Phaseolus vulgaris L.) in a soil tube system. *Agron. Columbiana* 27, 25–32.

Polania, J., Rao, I.M., Cajiao, C., Grajales, M., Rivera, M., Velasquez, F., Raatz, B., and Beebe, S.E., 2017b. Shoot and root traits contribute to drought resistance in recombinant inbred lines of MD 23–24 × SEA 5 of common bean. *Front. Plant Sci.* 8, 1–18.

Polania, J., Rao, I.M., Cajiao, C., Rivera, M., Raatz, B., and Beebe, S., 2016c. Physiological traits associated with drought resistance in Andean and Mesoamerican genotypes of common bean (Phaseolus vulgaris L.). *Euphytica* 210, 17–29.

Polania, J., Rao, I.M., Mejía, S., Beebe, S., and Cajiao, C., 2012a. Características morfo-fisiológicas de frijol común (Phaseolus vulgaris L.) relacionadas con la adaptación a sequía. *Acta Agron.* 61, 197–206.

Polania, J., Rao, I.M., Mejía, S., Beebe, S.E., and Cajiao, C., 2012b. Morpho-physiological characteristics of common bean (Phaseolus vulgaris L.) related to drought adaptation. *Acta Agron.* 61.

Ramaekers, L., Galeano, C.H., Garzón, N., Vanderleyden, J., and Blair, M.W., 2013. Identifying quantitative trait loci for symbiotic nitrogen fixation capacity and related traits in common bean. *Mol. Breed.* 31, 163–180.

Ramirez-Vallejo, P. and Kelly, J.D., 1998. Traits related to drought resistance in common bean. *Euphytica* 99, 127–136.

Rao, I.M., 2014. Advances in improving adaptation of common bean and Brachiaria forage grasses to abiotic stresses in the tropics, In: M. Pessarakli (Ed.), *Handbook of Plant and Crop Physiology.* CRC Press, Taylor and Francis Group, Boca Raton, FL, pp. 847–889.

Rao, I.M., Beebe, S., Polania, J., Ricaurte, J., Cajiao, C., Garcia, R., and Rivera, M., 2013. Can tepary bean be a model for improvement of drought resistance in common bean? *Afr. Crop Sci. J.* 21, 265–281.

Rao, I.M., Beebe, S.E., Polania, J., Grajales, M., Cajiao, C., Ricaurte, J., Garcia, R., and Rivera, M., 2017. Evidence for genotypic differences among elite lines of common bean in the ability to remobilize photosynthate to increase yield under drought. *J. Agric. Sci.* 155, 857–875.

Rosales, M.A., Ocampo, E., Rodríguez-Valentín, R., Olvera-Carrillo, Y., Acosta-Gallegos, J.A., and Covarrubias, A.A., 2012. Physiological analysis of common bean (Phaseolus vulgaris L.) cultivars uncovers characteristics related to terminal drought resistance. *Plant Physiol. Biochem.* 56, 24–34.

Sassi, S., Gonzalez, E.M., Aydi, S., Arrese-Igor, C., and Abdelly, C., 2008. Tolerance of common bean to long-term osmotic stress is related to nodule carbon flux and antioxidant defenses: Evidence from two cultivars with contrasting tolerance. *Plant Soil* 312, 39–48.

Seibt, U., Rajabi, A., Griffiths, H., and Berry, J.A., 2008. Carbon isotopes and water use efficiency: Sense and sensitivity. *Oecologia* 155, 441–454.

Sponchiado, B.N., White, J.W., Castillo, J.A., and Jones, P.G., 1989. Root growth of four common bean cultivars in relation to drought tolerance in environments with contrasting soil types. *Exp. Agric.* 25, 249–257.

Thung, M. and Rao, I.M., 1999. Integrated management of abiotic stresses, In: Singh, S.P. (ed.), *Common Bean Improvement in the Twenty-First Century.* Springer Netherlands, Kimberly, USA, pp. 331–370.

White, J.W., 1993. Implications of carbon isotope discrimination studies for breeding common bean under water deficits., In: *Stable Isotopes and Plant Carbon-Water Relations.* Academic Press, San Diego, CA, pp. 387–398.

White, J.W. and Castillo, J.A., 1992. Evaluation of diverse shoot genotypes on selected root genotypes of common bean under soil water deficits. *Crop Sci.* 32, 762–765.

White, J.W., Castillo, J.A., and Ehleringer, J.R., 1990. Associations between productivity, root growth and carbon isotope discrimination in Phaseolus vulgaris under water deficit. *Aust. J. Plant Physiol.* 17, 189–198.

Zaman-Allah, M., Jenkinson, D.M., and Vadez, V., 2011. Chickpea genotypes contrasting for seed yield under terminal drought stress in the field differ for traits related to the control of water use. *Funct. Plant Biol.* 38, 270–281.

44 Moringa and Tamarind
Potential Drought-Tolerant Perennial Crops

Satya S.S. Narina, Christopher Catanzaro, and Anwar H. Gilani

CONTENTS

44.1 INTRODUCTION

The family *Fabaceae* is a large family of dicotyledonous plants well known for the fixation of atmospheric nitrogen in the soil by forming root nodules in association with rhizobial bacteria. Some plants in this family, such as tamarind, produce commercially valuable edible fruits (indehiscent legumes or pods) and seeds rich in protein and nutrients beneficial to human health. Fenugreek, also in the family *Fabaceae*, is grown for its leaves, seeds, and sprouts. The leaves are used as a fresh or dried herb or as a fresh vegetable, the seeds as a spice, and the sprouts as microgreens. Leaves and fruits of species in the family *Moringaceae* are used to contribute seed protein and essential nutrients for human and livestock, and to enrich soil nutrients, purify water and provide antipathogen activity (Saini et al., 2016; Tian et al., 2015; Yisehak et al., 2011). The crops included herein, from the plant families *Fabaceae* and *Moringaceae*, have proven sustained economic impacts as

healthy and nutritious foods for human and livestock, as drugs for the pharmaceutical industry, and as various other agricultural products. The nutritional contributions of these crops and various information on taxonomic, morphological, and biochemical aspects of the crops and compounds produced by them are provided herein Tables 44.1–8 and in Figures 44.1–5, as necessary for some products and compounds with brief descriptions of their importance in this chapter.

Although tamarind (*Tamarindus indica* L.) and Fenugreek (*Trigonella foenum-graecum* L.) are both members of the family *Fabaceae*, tamarind is a tree while fenugreek is an annual plant. The *Moringa oleifera* Lam., known as moringa or drumstick tree or horseradish tree and *Moringa stenopetala* (Baker f.) Cufod., known as cabbage tree, are members of the family *Moringaceae*. Tamarind and moringa are grown for their fruits and also for their leaves which are used as a vegetable. Both moringa and tamarind are of tropical origin, with deep taproot systems impacting drought tolerance and thus

TABLE 44.1

A Brief Overview of the Edible Vegetable Resources of Fenugreek, Tamarind, and Moringa

Common Name / Scientific Name	Number of Other Species and Genera that Belong to the Same Family	Center of Origin and Major Center's of Diversity	Plant Parts and Their Uses
Fenugreek / *Trigonella foenum-graecum* L.	260 species and 18 genera	**Center of Origin**: Asia, India, the Mediterranean region **Major Centers of Diversity**: Argentina, West and South Asia, Turkey, Yemen, Abyssinia, Moschino, Sicily, Tuscany, Morocco, Transcaucasia, Afghanistan, China-Iran region, Australia, North America, North Africa, North, and Central Europe and India	The leaf is used as salad, steam cooked with dahls, as a condiment for non-vegetarian dishes and biriyani. The seed is used as aromatic spice and condiment. The seed extract is used as hair softener, body cleanser, and to stop diarrhea.
Moringa/*Moringa oleifera* L.	Single genus with 13 species	**Center of Origin**: Africa, sub-Himalayan tracts of India, Pakistan, Bangladesh, and Afghanistan **Major Centers of Diversity**: Africa, Australia, America, Europe, Indonesia, Pakistan, Sudan, South Asia	The leaf is used as salad, cooked with dahls, and to make cereal bars. Spicy moringa leaf powder is mixed with cereals, leaf flour for snacks in Africa. Leaves, flowers, and fruits are used as a vegetable in Pakistan. The fruit is used in curries, sambar, and soups in India. Seed and bark are water purifiers in Asia and Africa.
Tamarind/*Tamarindus indica* L.	A monotypic genus with single species	**Center of Origin:** India, Africa, and Madagascar. **Major Centers of Diversity**: Indian subcontinent, Central and South America, Africa, Nigeria, China, Europe, Turkey, Bangladesh.	Leaf and mature, sweet, unripe fruit used as salad. Both leaf and fruit used for preparing dahls, soups, and curries. Seed, fruit and seed flour for snacks. In some areas, flowers along with young leaves are used as a vegetable.

Sources: Acharya, (2008); Diallo et al., (2007); Faizi et al., (1994a,1994b,1995 and 1998); Kumar et al., (2017); Verma et al., (2013).

enabling cultivation in arid, semi-arid tropical, and subtropical climates. Within the United States of America (USA), it grows in the southern states, and especially in south Florida.

In limited water situations of crop cultivation, it is a necessity to conserve water resources. One objective of the research review is to draw the public attention to the potential benefits of genetic improvement, cultivation, and adaptation of perennial tree sources for leafy vegetables (tamarind and moringa) for drought-prone regions to replace intensively irrigated annual leafy vegetables like fenugreek.

44.1.1 History, Nomenclature, and Taxonomy of Moringa

The family *Moringaceae* has a single genus, *Moringa*, containing 13 species. Olson (2002) reported species diversity and distribution of the genus *Moringa* along with their countries of origin, *M. arborea* Verdc. (Kenya), *M. rivae* Chiov., and *M. stenopetala* Cufod. (Kenya and Ethiopia), *M. borziana* Mattei and *M. longituba* Engl. (Kenya and Somalia), *M. pygmaea* Verdc. (Somalia), *M. ruspoliana* Engl. (Kenya, Ethiopia, and Somalia), *M. ovalifolia* Dinter and A. Berger (Namibia and Angola), *M. drouhardii* Jum., and *M. hildebrandtii* Engl. (Madagascar), *M. peregrina* (Forssk.) Fiori (Arabia, Red Sea,

and Dead Sea), *M. concanensis* (Dalzell and A. Gibson), and *M. oleifera* Lam. (Sub-Himalayan tracts of South Asia).

Moringa oleifera is native to South Asia, the most commonly cultivated species of moringa and highly cross-pollinated. It is the most studied species within the genus, with anti-tumor, antipyretic, anti-pileptic, anti-inflammatory, anti-ulcer, anti-plasmodic, diuretic, antifungal, and antibacterial, antioxidant, antidiabetic, cholesterol lowering, and hepatoprotective properties, and is valued for its nutritional, medicinal, and pharmacological properties in South Asia (Anwar et al., 2007). It is referred to as mother's best friend for the rural poor of the tropics, subtropics, South Asia and southern Ethiopia due to its ability to increase milk production in nursing mothers. *Moringa stenopetala*, identified with many of the same properties as *M. oleifera*, is native to northeast tropical Africa and India. *M. stenopetala* is widely cultivated in southern parts of Ethiopia. It has vernacular names such as aleko, aluko, halako, haleko, kallanki, shechada, shiferaw, and telahu.

M. oleifera is a tropical, deciduous tree with significant ornamental, medicinal, industrial, food, fodder, and fuel values for both human and livestock use. India is the largest producer with 1.1–1.3 mt of pods from 38,000 ha. The state of Andhra Pradesh (AP) is the major contributor in production and area followed by the states of Karnataka and Tamil Nadu in India. The tree can tolerate drought and heat and slightly tolerates frost. It grows best in a temperature range of 25–35°C in slightly acidic to alkaline soils of pH 5–9. Major regions of its cultivation are South Asia, Northeast and Southwest Africa, Central Europe, Australia, and North and South America (Saini et al., 2016).

M. oleifera is a tall (4–10 m) tree with a wide crown, highly branched with hollow, soft stems, and compound leaves oppositely arranged with 2–6 pinnae bearing obovate light to dark green leaflets, producing bisexual flowers in panicles which are yellow, white, or pink (Gandji et al., 2018; Saini et al., 2016). The fruit is a green pod with groves holding 6 to 12 winged (for wind dispersal) seeds that are brownish-black (Figures 44.1 and 44.2). It is propagated by seed in Sudan and vegetatively in South Asia, Indonesia, and West Africa (Yisehak et al., 2011).

The morphological features of *M. stenopetala* are to those of *M. oleifera*. It is slightly taller (6–12 m) with white to pale gray or silvery smooth bark, softwood, strongly branched crown, sometimes with several trunks. The leaves grow up to 55 cm long, bi- or tri-pinnate, with about five pairs of pinnae and 3–9 elliptic to ovate leaflets per pinna. The inflorescence is pubescent and densely flowered with long (50 cm) panicles. The flowers are very fragrant with cream flushed pink sepals, white, pale yellow or yellow-green petals, white filaments, and yellow anthers. The ovary is ovoid and densely hairy. Pods are elongate, reddish with grayish bloom having grooved valves (Mekonnen, 2002; Nirina et al., 2014).

Both species grow well in sandy, well-drained soils, in both wet and dry regions with low to medium rainfall (500–1,400 mm) at 300–2,500 m above mean sea level (msl). *M. stenopetala* can grow at a temperature range of 24–30°C. Morphological observations revealed similarities to the families *Brassicaceae* and *Capparidaceae* (Yisehak et al., 2011) while observations from genome sequencing revealed similarities to the family *Caricaceae* (Tian et al., 2015), so its taxonomic position remains somewhat unclear.

Yisehak et al. (2011) reported that in southern Ethiopia, having a haleko (*M. stenopetala*) in the backyard is an indication of economic status and ability to feed one's family during the times of famine or drought. This serves as evidence that culture plays a role in the conservation and utilization of locally important tree species.

44.1.2 History, Nomenclature, and Taxonomy of Tamarind

Tamarind is a tree species cultivated mainly for its fruit, leaf, and seed having food, nutritional, medicinal (ayurvedic), pharmaceutical, and industrial values (Table 44.2). India is the world's largest producer with a production of 30,000 t/ha and the largest exporter of tamarind pulp followed by Thailand (SBI, 2018). John (1990) reported its common occurrence in frost-free zones of tropical and subtropical countries and tropical parts of temperate countries at near sea level to 1,500 m above msl (Table 44.1). Tamarind has been in cultivation since 400 BC in Egypt, between 370–287 BC in east Africa

Moringa nursery and tree farm   Moringa flower and fruit   Fruiting Moringa tree

Tamarind seedling   Tamarind flower   Fruits on Tamarind tree

Fenugreek farm   flowering fenugreek   Fruiting fenugreek   Fenugreek seeds

FIGURE 44.1 (See color insert.) Images of morphological and taxonomic features of nutritionally useful parts of moringa, tamarind, and fenugreek.

¹Spicy leaf Powder   ¹Dahl   Moringa Sambar   ²Tamarind block   Tamarind rice

Tamarind Chutney   Tamarind Pickle   Moringa pickle

Tamarind pulp   Tamarind seed   Tamarind fruit   Moringa seed kernel

Sweets and snacks

FIGURE 44.2 (See color insert.)   Images of food or food products made from moringa and tamarind. (¹Dahls and spicy powders of leaves from all crops look alike; ²tamarind block used for soups and rice.)

as per Theophrastus' descriptions of tamarind, and between 1200 and 200 BC as per Indian Brahmasamhita scriptures. Tamarind is also known as "date of India," "assam jawa," and "amlika," indicating its presence in ancient times in India (NAS, 1979; Santosh et al., 2011).

Orwa et al. (2009) and Santosh et al. (2011) reported that globally accepted trade name was "tamarind," although a number of vernacular names are used in different parts of South Asia. It is called imli, or chinta pandu in the state of AP. Tamarind (*Tamarindus indica* L.) is a member of the family *Fabaceae*, the third largest family of flowering plants and belongs to the subfamily *Caesalpinioideae* which is divided into five to nine tribes or groups of genera based on morphological characters. In Bentham's classification (Pettigrew and Watson, 1977), tamarind belongs to the tribe of the *Amherstieae* Benth. which again comprises of 25 genera, 21 from tropical Africa, 2 from tropical America, and 2 from Asia (Léonard, 1957). There are three relic Afro-Asiatic genera within the group namely *Humboldtia*, *Tamarindus*, and *Amherstia*. Both *Tamarindus* and *Amherstia* are more derived, with zygomorphic and showy flowers and stamen filaments connate in a sheath, but have many differences in the floral structure, leaves, fruits, and seeds. *Tamarindus* is said to have some resemblance to *Heterostemon* Desf. from the upper Amazon region of South America (Diallo et al., 2007). The genus *Tamarindus* is monotypic taxon, having a single species. The species has two botanical synonyms, *T. occidentalis* Gaertn. and *T. officinalis* Hook. (Santosh et al., 2011). The most commonly cultivated red fruited variety in India is the intra-specific taxon *T. indica var. rhodocarpa* (John, 1990). The introductions and selections like "Makham Waan," "Manila Sweet," "PKM-1," "Prathisthan," "Urigam," "Tumkur," and "Yogeshwari," besides cultivars with red and

brown pulp types are the most commonly cultivated tamarind varieties in India (Geetha, 1995; John, 1990).

Tamarind is an evergreen, tropical tree, slow growing but tall (15–30 m) at maturity under favorable conditions, with a deep taproot system (Usha, 1990). The main trunk is solid covered with dark grayish brown bark, sometimes hairy, profusely branched, forming a huge central crown. The leaves are glabrous, paripinnate, alternately arranged with a rounded apex and asymmetrical base with an entire margin. Each pinna has 10 to 18 pairs of oppositely arranged leaves that fold during night time. Flowers are 2.5 cm in diameter, hermaphroditic (bisexual) with three stamens and five petals, arranged in yellowish clusters with pink or orange venation (Figure 44.1). The fruit is a berry covered with a greenish-brown shell enclosing a hard, stony, dicotyledon seed with a brown seed coat when matured. Unripe and immature fruits and seeds are green in color (Mowobi et al., 2016).

Only 45% of tamarinds are planted for fruit pulp in Africa and South Asia, with 53% planted as shade trees or on farm boundaries, and use for the remaining 3% unknown. The majority are intercropped with other crops or trees to restore agroclimatic conditions (Edifa-Othieno et al., 2017).

44.1.3 History, Nomenclature, and Taxonomy of Fenugreek

Fenugreek is native to South Asia and Mediterranean regions with 260 known species (Verma et al., 2013; Zandi et al., 2015). It is the oldest known leafy vegetable and seed spice (Basu, 2006). Among which *T. foenum-graecum* is a widely cultivated fenugreek species throughout the world (Zandi et al., 2017), Table 44.1. It is a self-pollinating dicotyledonous plant with branched stems, trifoliate leaves, and yellowish

TABLE 44.2
Major Biochemicals Identified in Various Plant Parts of Fenugreek, Tamarind and Moringa

Name of the Vegetable Crop/Edible Plant Part	Biochemical Compounds/Phytochemical Compounds
Fenugreek **Edible plant parts:** Leaves, seeds **Sources:** Verma et al., (2013); Basu, (2006); Zandi et al., (2015) and Zandi et al., (2017).	**The seed has the following compounds:** **Carbohydrates** (45–60%) as mucilaginous fiber (galactomannans); glycosides such as cholesterol, n-alkanes, nicotinic acid, sitosterol, in minute quantities; sesquiterpenes (0.015%) known as volatile oils. **Proteins** (20–30%) enriched in tryptophan and lysine. **Lipids** (5–10%) or fixed oil. **Minerals**: Calcium and iron. **Vitamins**: Vitamin A, B1, and C in minute quantities from leaf and seed sprouts. **Secondary metabolites:** **Amino acids**: arginine, histidine, 4-hydroxy isoleucine (0.09%), and lysine. **Steroid saponins** (0.6–1.7%): Diosgenin, graecunins, neotigogenin, protogracillin, protodioscin, sarsapogenin, smilagenin, tigogenin, trigofoenosides, trineoside, and yamogenin are major steroidal sapogenins. Gitogenin, neotogenin, and yuccagenin are minor sapogenins or dihydroxy steroidal sapogenins. **Alkaloids of Pyridine type** (0.2–0.38%): Trigonelline. **Choline** (0.5%) and other phytochemicals like carpaine, fenugreekine, and gentianine; **Flavonoids**: Afroside, apigenin, kaempferol, iso-vitexin, iso-quercetin, luteolin, orientin, quercetin, and vitexin. **Isoflavonoid phytoalexins** are maackiain, medicarpin, sativan, and vesitol. **Seed, seed sprouts, and leaves**: Minerals, mostly iron and calcium, had antioxidant, anti-cancerous secondary metabolites (stated above) and antimicrobial compounds.
Tamarind **Edible plant parts**: Leaves, seeds, fruits, fruit pulp, root, flowers, tree bark **Sources:** Santosh et al., (2011); Kuru, (2014). **Reference**: Figure 44.5	**Fruit pulp**: Most of the essential **amino acids** (β-alanine, proline, phenylalanine, leucine, serine), except tryptophan; **carbohydrates** – arabinose, galactose, glucose, xylose, cinnamates, citric and uronic acid, dietary fiber, invert sugar, nicotinic acid,1-malic acid, pectin, pipecolic acid, **volatile oils** (geraniol, limonene), potassium, lipids, proteins, **polyphenols** and **flavonoids**, **vitamins** B3, C and E (α, β, γ –tocopherols), folates; **Secondary metabolites (from mucilage)**: Polyphenols: Flavan-3-ols (catechin, epicatechin), flavonoids (vitexin, iso-vitexin), procyanidin and **triterpenes** [orientin (8-c-β-D-glucopyranosyl-3',4',5,7-tetrahydroxyflavone), iso-orientin (6-substituted luteolin analog)]. **Aqueous pulp extract**: Saponins (2.2%), alkaloids (4.32%) and glycosides (1.59%). **Bark**: Glycosides, lipids, peroxidases, saponins, and tannins. **Seeds**: Amino acids, carbohydrates, cardenolide, protein in mature seeds and antioxidants in green seeds. **Seed extract**: Polyphenol compounds like anthocyanidin, epicatechin, 2-hydroxy- dihydroxy acetophenone, methyl 3,4-dihydroxybenzoate, 3,4-dihydroxy phenylacetate, tannin, and oligomeric pro-anthocyanidins **Leaves**: Antioxidants, calcium, glycosides, iron, orientin, iso-orientin, 1-malic acid, peroxidase. tannins, vitamin C, vitexin, and iso-vitexin.
Moringa **Edible plant parts:** Leaves, seeds, fruits, root, flowers, tree bark, immature pods **Sources:** Anwar et al., (2007); Faizi et al., (1994a); Faizi et al., (1994b) and Faizi et al., (1998). **Reference**: Figure 44.4	**Moringa plant**: β-sitosterol, caffeoylquinic acid, glycosides, kaempferol, quercetin, and zeatin. **Stem**: Alkaloids such as moringine and moringinine **Purified whole gum exudate from _M. oleifera_**: L-arabinose, -galactose, -glucuronic acid, -rhamnose, -mannose, -xylose and degraded gum polysaccharide containing galactose, glucuronic acid, and mannose. **Leaf**: Glycosides (thiocarbamate, carbamate, nitrile groups), antioxidants (ascorbic acid, flavonoids, phenolics, and carotenoids), protein, vitamin C, calcium, and potassium. **Flowers**: Amino acids, D-glucose, kaempferol, kaempferitrin, rhamnetin, sucrose, traces of alkaloids, quercetin, iso-quercitrin, and wax. **Fruit (pod)**: amino acids, β-carotene, minerals, protein, various phenolics and vitamins. **Pods (seed, coat, and pulp)**: O-[2'-hydroxy-3'-(2"-heptenyloxy)] propyl undecanoate, O-ethyl-4-[(α-L-rhmanosyloxy)-benzyl] carbamate, methyl p-hydroxybenzoate, β-sitosterol, and p-hydroxybenzaldehyde **Seeds**: Benzyl isothiocyanate

white flowers producing golden yellow seeds (Acharya et al., 2010, Figure 44.1; Petropoulos, 2002). Fenugreek is a natural nitrogen fixer (gram-negative *Rhizobium* spp.), so it can grow easily on well-drained loamy, or sandy soils in tropical, and subtropical climates as an intercrop, forage crop, or rotation crop to replenish soil nitrogen levels (Acharya et al., 2008). Numerous species of *Trigonella* (*T. foenum-graecum*, *T. balansae*, *T. corniculata*, *T. maritima*, *T. spicata*, *T. occulta*, *T. polycerata*, *T. calliceras*, *T. cretica*, *T. caerulea*, *T. lilacina*, *T. radiata*, *T. spinosa*) were observed to have medicinal and pharmaceutical properties. Because fenugreek is a well-studied and established crop for nutritional uses, not much is discussed about its history, importance, and uses or improvement in this chapter.

44.2　NUTRITIONAL QUALITY OF EDIBLE PORTIONS OF THREE CROPS WITH HEALTH VALUES

44.2.1　Moringa

The leaves and fruit of moringa are rich source of many phytochemicals (Figure 44.4), antioxidants, phenolics, several nutrients, and minerals (calcium (Ca) and iron (Fe)), fiber, vitamins (A, B, and C), amino acids (methionine- and cysteine-rich protein in seed), and polyunsaturated fatty acids (Table 44.2, Anwar et al., 2007). Antioxidants present in the leaves of moringa extend the shelf life of the processed ready-to-eat food products to which the leaves are added (Anwar et al., 2007). The leaves of *M. stenopetala* add a mustardy taste when used as a supplement to major staple foods, and they have been screened for potential compounds and for

pharmaceutical product development (Mekonnen, 2002; Olson, 2002). The green immature and unripe mature pods are eaten pickled and in several vegetarian items after cooking (Figure 44.2, Kumar et al., 2017; Yisehak et al., 2011). The nutritional and medicinal uses of moringa plant parts or products are described in Table 44.3.

The bioproducts (leaf extract, seed cakes) and residual wastes from plant parts are used to replenish soil fertility, as antimicrobial agents against harmful soil pathogens and as a naturally available potential plant growth regulator to protect crop plants from drought stress (Figure 44.3, Yasmeen et al., 2013). Moringa was recognized as having novel health constituents (Figure 44.4, Anwar et al., 2007; Faizi et al., 1994a). Nouman et al. (2018) reported that moringa leaf extract (MLE) was more economical and environmentally friendly for farmers to use than synthetic growth regulator, benzyl amino purine (BAP) (Figure 44.3).

TABLE 44.3
Nutritional and Medicinal Uses of *Moringa*[a] Plant Parts

Name of the Plant Parts/Its Products	Nutritional and Medicinal Use (Food/Food Products)
	Nutritional Uses:
Leaves (Source of minerals; food – Dahl, Pittu, Kukufa)	Calcium (Ca), Iron (Fe) and other trace elements-supplement to major staple food; Leaves + pigeonpea (Dahl); Leaves + split rice (Pittu); leaves + kukufa (cereal dish made with maize and sorghum).
Flowers (Source of Nectar)	Source of nectar for honey bees, source of honey production.
	Leaves and flowers were used in Africa to make tea.
Fruits-Pod (food)	Pickled, Curries (pod + onion + spices) mixed with major cereal rice grain; Eaten in Kenya (Africa), south Asia-India, Europe, the Americas, other growing regions by making curries, soups, and sambar.
Seeds (food)	Eaten as a snack; Rich in edible oil used locally for cooking and lubricant.
Seed, root concoction	Water purifier-cleans mud water; water pollution control by flocculating turbid, clay flood waters.
Powdered seed	Used to purify honey without boiling, to purify sugarcane juice.
Seed Cake	Nitrogen-rich, used as fertilizer.
Gum produced from Cut tree	Used in calico printing, condiment, and medicinal use.
Pulp	Used for making newsprint and writing paper.
Wood	Used in making blue dye in Jamaica and Senegal; Soft spongy wood used for poor firewood.
Bark	Used for rope making.
Bark extracts	Flavoring soups
Bark and gum	Used for extraction of tannins.
	Medicinal Uses:
Leaves (anti-infection and pain reliever)	Leaves boiled in water cure malaria, hypertension, and stomach pain; to expel the retained placenta in women.
Leaf extract (Antifungal, antibacterial, and antibiotic activity)	Remedy for leprosy; used against *Staphylococcus aureus*, *Salmonella typhi* (the causal organism of typhoid), Shigella (the causal organism of dysentery) and *Candida albicans* (a fungus that causes candidiasis or thrush).
Seeds (anti-cancerous)	Glucoconringiin (3%) and O- (rhamnopyranosyloxy) benzyl glucosinolate (19%) of the dry mass of seeds are anti-cancerous.
Seeds	Active substances in cotyledons of the seed have antimicrobial properties that help in separation of gram-positive and gram-negative bacteria separately along with the flocculants in contaminated waters.
Bark	Chewing bark is a treatment for coughs.
Roots (cure malaria)	Roots chopped and mixed with water are used to treat severe cases of malaria.
Root extracts	Cured infective stages of *Trypanosoma brucei* (a parasite that causes sleeping sickness in horse, cattle, and sheep in tropical Africa), *T. cruzi* (causal agent of Chagas disease in South America) and amastigotes of *L. donovani*.
Smoke liberated from burning moringa	Used to treat epilepsy and malaria.
Seeds	Less expensive bio-sorbent for removal of cadmium from aqueous media.

Sources: Gandji et al., (2018); Nirinia et al., (2014); Yisehak et al., (2011).

[a] *moringa*: *M. oleifera* and *M. stenopetala* have similar uses, and thus uses are not differentiated by species.

Moringa leaf extract (a) and Powder (b) Moringa animal feed   Tamarind husk

FIGURE 44.3 (See color insert.)   Images of products used in livestock and other sectors of agriculture.

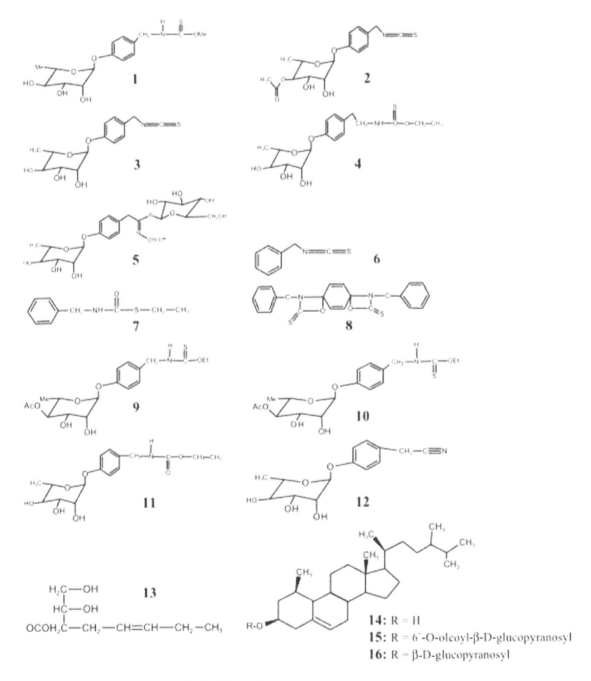

FIGURE 44.4   Structure of selected phytochemicals in moringa.

(Source: Adapted from Anwar et al., 2007).

44.2.2 TAMARIND

Tamarind leaf, seed, and fruit are an easily accessible, inexpensive or free food resource for people in developing countries where tamarind is found. The edible plant portions consumed provides proteins, many amino acids, minerals, vitamins, and phytochemicals to the diets of under-nourished humans (Table 44.2 and 44.4). Tamarind was referred to as a natural blueprint for the development of drugs and phytomedicines for the treatment of diseases (Edifa-Othieno et al., 2017). Various parts of tamarind were identified as containing unique compounds with the potential to provide human health

TABLE 44.4

Nutritional and Medicinal Uses of Tamarind Plant Parts

Name of the Plant Part/Its Product	Nutritional and Medicinal Use (Food/Food Products)
	Nutritional Uses:
Tender leaves	Cooked with dahl or shrimp or onion or any vegetable to eat with rice.
Fresh shoots and tender leaves	Excellent source of vitamin B.
Fresh leaves	An animal feed for goats.
Immature green pods	Eaten fresh or boiled with dahl to eat with cereals or millets.
Immature fruit pulp concentrate	Eaten as sauce.
Mature ripe fruit from a sweet variety	Eaten as a snack.
Ripe fruit pulp extract	Source of potassium, a cold beverage, or can be mixed with sugar or honey for taste.
Ripe fruit pulp	Boiled with potato chips (Ateso) adds flavor and preserves consistency, pulp + sauce added to meat to enhance the taste.
Pulp	Added to sauces, nutritive drink, to flavor local drinks and preservation of foods.
Fruit	Iron bioavailability.
	Medicinal Uses:
Fruit and pulp	Muscle relaxation via calcium channel blockage; fruit extract decreases plasma fluoride concentration and reduces fluoride induced liver and kidney damages. Tamarind fruit extract is an effective drinking water cleaning agent for fluorine, nickel, and lead toxicities. Protects skin from the sun's ultraviolet damage.
Fruit and seed	Regulatory effect on neutrophils due to polyphenols; seed polysaccharide used in eye drops due to its mucoadhesive properties.
Seed extract	Antidiabetic-Diabetes mellitus type 1 and 2 caused by damage due to chronic inflammation of pancreatic β-cell island, blood glucose regulation, reversal damage to pancreatic tissue. Increased white blood cells and thrombocytes.
Decoction from matured leaves, bark, roots or fruits	To treat livestock diseases such as smallpox (Kawali / Dhopadhola), treat abdominal pain (soft parts of bark, roots), constipation (fruit), diarrhea (leaves).
Aqueous pulp and bark extract	Antibacterial activity against *Staphylococcus aureus*, *E coli*, *Pseudomonas aeruginosa*, *Salmonella paratyphi*, *Salmonella typhi*, *Bacillus subtilis*, *B. pseudomallei*, *K. pneumoniae* due to presence of lupeol content.
Fruit	Potential antifungal agent against *Aspergillus niger*, *Candida albicans*.
Plant extract	Antiviral against watermelon mosaic virus, cowpea mosaic virus, tobacco mosaic virus and anti-nematocidal properties against *Bursaphelenchus xylphilus* and molluscicidal properties due to saponin content against *Bulinus trancatus*.
Fruit pulp extract	Used as a cooling agent during fevers in human, relieve pains (pregnant women), treat constipation (easy bowl movement), reduce secondary bacterial infections.
Fruit extract + lime juice or honey or milk or date spice or camphor	Laxative, carminative due to high malic and tartaric acid and potassium contents.
Pulp (digestive role)	As a remedy for biliousness or bile disorder.
Leaves, seeds and raw fruit or partially ripe fruit pulp	Used as an anti-scorbutic, heal inflammations, to treat asthma, cough, sore throat by hydrolysis of phospholipids, due to the presence of polyphenols and flavonoids.
Fruit pulp + salt	Used to treat rheumatism, to alleviate sunstroke, alcoholic intoxication.
Ends of small branches	Chewed to make durable toothbrushes.
Decoction from bark	Treatment of asthma and abdominal upsets.
Tamarind kernel powder	Textile, confectionary, cosmetics, and pharmacy.
Testa	Dyeing and tanning industry.
Seed extract	Nerve repair (xyloglucan), treats peptic ulcers due to the presence of polyphenolic compounds, antioxidants, and tannins; anti-cancerous due to antioxidant enzyme induction by availability of polyphenol compounds from seed extracts.
Leaves	Antimalarial, antiparasitic, anti-apoptotic and liver protective effects, ameliorates fluoride toxicity obtained by toxic drinking water.
Seed and leaves	Antioxidative (dried seeds), hypolipidemic, antiatherosclerotic, immunomodulatory effect.

Sources: Edifa-Othieno et al., (2017); Santosh et al., (2011); Kuru, (2014); SBI, (2018).

benefits, as presented in Tables 44.2 and 44.4 (Kuru, 2014; Santosh et al., 2011). Tamarind fruit juice or concentrate is used as a supplement to tomato or lime juice in south Asian cuisine due to its high levels of acidity (12.2%) and used as a sweet snack due to the presence of 30.2% total soluble solids (Figure 44.2; Usha, 1990).

Tamarind pulp extract has a market demand globally as a food additive (natural preservative, coolant, and flavoring agent), pharmaceutical (antifungal, antiseptic, antimicrobial, antioxidant) and cosmetic (bleaching, skin hydrating agent), and fragrance oil with sweet to sour essence (used in soaps, candles, air fresheners, bath oils, laundry products, etc.). The tamarind pulp extract markets for the industry are divided into five regions, namely North America, Latin America, Europe, Asia Pacific, Middle East, and Africa (FMI, 2018).

44.2.3 FENUGREEK

Fenugreek has been an annual forage legume, traditional spice, and ayurvedic medicinal crop on the Indian subcontinent for centuries, yet it has little share in global markets. It is recommended for cultivation in dry arid and semi-arid regions of Asia, Africa, and Latin America. It is the costliest vegetable to reach the poor in developing countries (Basu, 2006). The nutritional and medicinal values of fenugreek are briefly outlined in Table 44.5.

Consumption of nutritionally rich herbs, spices, vegetables, fruits, and use of herbal medicines were observed in the ancient cultures of Asia, Africa, Europe, and the Americas for nutrition and healthy living. Plant-based ayurvedic medicines are gaining popularity globally as dietary supplements due to their anti-cancerous and cardioprotective properties (Peter, 2006). Fenugreek seed and leaves are a rich source of a wide diversity of medicinally rich phytochemicals, which include steroidal saponins (diosgenin), fenugreekine (alkaloid), galactomannan (carbohydrate), and 4-hydroxy isoleucine (amino acid), as reported by Zandi et al. (2017). Galactomannan is a major polysaccharide consisting of galactose and mannose in 1:1 ratio. Diosgenin [(25R)-spirost-5-en-3β-ol] is a major steroidal sapogenin. Trigonelline is a methyl betaine derivative of nicotinic acid and a major alkaloid component of fenugreek. Trigonelline has hypoglycemic, hypolipidemic, neuroprotective, antimigraine, sedative, memory-improving, antibacterial, antiviral, and anti-tumor activities, and it has been shown to reduce diabetic auditory neuropathy and platelet aggregation. It acts by affecting β cell regeneration, insulin secretion, activities of enzymes related to glucose metabolism, reactive oxygen species, axonal extension, and neuron excitability (Zhou et al., 2012). Biochemical compounds with potential health and nutritive values, identical to those in fenugreek leaves, were identified in plant parts of tamarind and moringa as well. Some of the same compounds have been isolated from all three crops which were presented in Table 44.2 along with other major biochemical compounds that are crop specific. Therefore, the cultivation of tamarind and moringa is more beneficial in drought-prone areas and soils with limited underground water reserves compared to highly irrigated fenugreek.

TABLE 44.5
Nutritional and Medicinal Uses of Fenugreek Plant Parts

Name of Plant Part/Its Product	Nutritional and Medicinal use (Food/Food Products)
	Nutritional Uses:
Leaf	Fresh vegetable, aromatic spice. Phytochemicals in leaf and seed reduce blood sugar and cholesterol in both animals and humans.
Seed	Spice, condiment, for favoring due to its aroma.
Leaf and seed powder	Feed for livestock and poultry. Food for human consumption as vegetable dishes (dhal, soups, and vada by mixing with legumes) to eat with major cereals.
Seed oil	Flavoring for canned foods and syrups.
Entire plant	Livestock feed to goats and ruminants.
	Medicinal Uses:
Leaf and seed	Control or reduce blood sugar and blood cholesterol levels in both human and livestock.
Leaf and seed (seed has steroid – diosgenin)	Fight against a number of diseases (diabetes, cancer, microbial infections).
Seed sprouts and leaves	Anti-cancerous, cardioprotective.
Seed soaked in water or seed powder mixed with buttermilk	A refreshing drink in summer with antidiuretic properties.
Hydrated fenugreek seed paste	Used to remove pain from injuries and heals skin diseases, muscle, bone, and neural pains, as an expectorant, a cooling agent and hair softener, to expel spines removed from the feet.

Sources: Zandi et al., (2015); Verma et al., (2013); Basu., (2006).

44.3 INFLUENCE OF DROUGHT ON WATER USE EFFICIENCY, HEALTH, AND NUTRITIONAL CONSTITUENTS OF TAMARIND AND MORINGA

The plant's growth, in semi-arid and arid soils, is naturally affected by extreme drought resulting in altered photosynthesis, respiration, translocation, ion uptake, carbohydrates, nutrient metabolism, and hormones. Water stress in plants reduces plant-cell water potential and turgor leading to vegetative growth inhibition and reproductive failure followed by accumulation of abscisic acid (ABA), and osmolytes like proline causing wilting. Drought also results in stomatal closure, limits gaseous exchange, reduces transpiration, arrests carbon assimilation rates, and produces negative effects on mineral nutrition (uptake and transport of nutrients) and metabolism causing a reduced leaf area and altered photosynthetic partitioning and synthesis of new proteins and mRNAs associated with drought response (Lisar et al., 2012).

Crops with the ability to thrive and complete their vegetative and reproductive growth during the periods of low moisture availability are said to be drought tolerant. The crops or cultivars with deep root systems, highly bound water in their tissues, reduced leaf area, leaf surfaces with sunken stomata, thick, waxy cuticles, and pubescence, and the ability to adapt to shallow soils, rocky gravelly and undulated wastelands provide evidence of hidden drought tolerance capabilities (Pareek and Sharma, 1991; Sharma et. al., 2013). Moringa and tamarind are two tropical perennial crops which possess all these superior drought-tolerant traits.

Pareek and Sharma, (1991) also reported that foliar feeding of arid plants with nutrients nitrogen (0.5–2.0% urea), zinc (0.05–1.0% zinc sulfate), and boron (0.05–1.0% borax) was observed to be beneficial. Tamarind and moringa do not require foliar feeding in cases of severe drought. An evaluation of underutilized vegetable and field crop species as nutritious foods in resource-poor areas was initiated as early as the 1990s by analyzing the nutritional composition of their edible plant parts (Salvi and Katewa, 2016). Some of these findings were also presented in this chapter for moringa and tamarind.

44.3.1 Moringa

Moringa oleifera is one among the six non-conventional leafy vegetables evaluated for mineral composition (Barminas et al., 1998). Moringa was recommended for cultivation in sandy soils, barren lands, and dry climatic regions with no irrigation source, and for green fodder in areas with large livestock population. It is intercropped with legume vegetables in hortipastoral systems. This crop combination was reported to be advantageous in arid zones (300–500 mm annual rainfall) to avoid the risks of monoculture, to generate net income, to promote efficient use of natural resources, and to improve agroclimatic conditions (Sharma et al., 2013).

Young plants of moringa maintained high water use efficiency (WUE) and high relative water content (RWC) with increased photosynthetic activity and increased activity of antioxidant enzymes when subjected to recurring water stress at 50 days after germination under osmotic potential of −0.3 and −0.4 MPa and observed high rates of photosynthetic and recovery rates during rehydration (Rivas et al., 2012). Young moringa seedlings with a water supply reduced (50% drought) for one month had reduced plant height (−34%), leaf number (−16%), total plant biomass (−60%), shoot dry mass (−37%) and root dry mass (−21%) (Al-Zahrani and Ibrahim, 2018).

Moringa leaves at 100 and 40% soil field capacity had a decrease in growth, chlorophyll a and b, total phenolic contents, antioxidant activities, crude protein, and mineral contents in comparison with those at 75% field capacity. Foliar application of growth regulators on moringa improved the contents of gallic acid, p-coumaric acid, and sinapic acid in leaves. Increased root growth was observed when the plants were sprayed with ascorbic acid (50 mg/l) at 40% field capacity. Increased contents of p-hydroxybenzoic acid and caffeic acid, crude protein, calcium, magnesium, and phosphorous was observed when sprayed on salt-affected wheat cv. Sehar-2006 with MLE (3.3%) and BAP (50 mg/l) at 40% and 75% field capacity (Nouman et al., 2018; Yasmeen et al., 2013).

Improved phenolic acid contents, in general, protects crop plants from drought stress besides other phytochemicals, flavonoids, and isothiocyanates. At a 40% field capacity level, MLE increased total phenolic content to 621 µg g⁻¹, a 57% increase in comparison to untreated plants (Nouman et al., 2018).

The positive correlation of growth parameters like shoot length, number of leaves, root length, and number of roots was observed with an increased level of antioxidants (superoxide dismutase, peroxidase, and catalase) in moringa leaves under drought stress (Nouman et al., 2016; Nouman et al., 2018). The phenolic acid contents improved the nutritional quality of moringa leaves under abiotic stress by protecting from oxidative damage (Nouman et al., 2018). Drought reduced 2S albumin gene expression in leaves and roots of 20-day old seedlings (Al-Zahrani and Ibrahim, 2018).

In another study, thermally unstable kaempferol, glucosides, glucoside malonates, and quercetin were also observed as major constituents of leaves out of 12 flavonoids identified in moringa cultivars (Coppin et al., 2013). Results suggest that these cultivars of moringa might be useful for cultivation in subtropical and temperate zones, to study anti-cancerous properties of flavonoids and to develop moringa cultivars with improved contents of health protective flavonoids in leaves.

44.3.2 Tamarind

In tamarind, low soil moisture availability results in premature leaf fall, shoot tip burning, and abortion, fruit and flower drop to maintain the internal plant water balance and thus avoid drought. Tamarind also exhibits sympodial branching to produce flowers for a long period of time (March to August) in an attempt to avoid unfavorable climatic conditions (Usha, 1990). Tamarind seedlings, meanwhile, have xeromorphic characteristics to minimize water loss. These include small leaf size, small stomata, and trichomes to tolerate drought. Seedlings also tolerate drought through leaf reduction, twig dieback, nonfoliar photosynthesis, and fine root formation. The short- and long-term drought stress experiments revealed a high rate of seedling mortality and slow recovery. Intensive field research on tamarind seedlings is necessary to reduce dry season mortality, increase productivity, and promote efficient use of limited water resources (Van den Bilcke et al., 2013).

Monoculture plantations rely on the identification of superior and use of elite trees without genetic erosion. Genotypes of tamarind which are morphologically similar were determined to be unrelated at the molecular level (Kumar et al., 2015). Therefore, more research is needed to identify tamarind germplasm for useful traits beneficial for both the producers and consumers of tamarind.

Cultivation of leafy vegetables in drought-prone areas is a challenging task whether the crop is an annual (fenugreek) or perennial (moringa and tamarind). Previously, efforts were focused on increasing the yield of annual leafy vegetables through intensive irrigation or fertigation. However, based on the work done over the past three decades, it is evident that further work is needed in germplasm improvement of leafy greens and fruit products, particularly in resource-limited agro-climatic zones.

44.4 ACHIEVEMENTS IN UTILIZING NUTRITIONAL AND HEALTH CONSTITUENTS

Development of molecular and biochemical markers for the selection and introduction of new cultivars of moringa and tamarind with desirable traits in leaves, seed, and fruit is in progress. Excellent progress was made in tissue culture and grafting to produce high-quality planting material. However, some propagation is not easily applicable to growers. A review of the literature yielded no studies on screening of these crops for variability (pod, seed, or leaf) for tolerance to abiotic stress. However, foods (vegetables) and food products (nutrition bars and snacks) with improved nutritional values are available to consumers from these two crops (Figure 44.2).

44.4.1 MORINGA

Moringa is a true diploid (2n=28) and has two chromosomes with a genome size of 1.2 pg with a genetic variation for quantitatively inherited traits. Though the genome is small compared to that of rice (*Oryza sativa* L.) it is compact, and able to control its comparatively fast growth, rapid cell proliferation, high seed production and adaptation to arid and semi-arid environments (Ramachandran et al., 1980). The small size of the genome offers in-depth analysis and genetic improvements of traits through transgenics.

Out of 300 *M. oleifera* genotypes analyzed, Ganesan et al. (2014) identified three main flowering colors of yellow, white, or pink, 19 reproducible SSR markers, and had no observed geographical isolation between genotypes collected from northern and southern parts of India. Gandji et al., (2018) generated DNA-based polymorphic markers using random amplified polymorphic DNA (RAPD), amplified fragment length polymorphisms (AFLPs), and expressed sequence tags-simple sequence repeat markers (EST-SSRs) for two varieties developed for pod production, PKM 1 and PKM 2.

High-quality draft genome reported by Tian et al. (2015) revealed 91.78% of the estimated genome size containing 19,465 protein-coding genes. By comparative analysis of the gene sequence data sets with those of grape (*Vitis vinifera* L.), pigeonpea (*Cajanus cajan* (L.) Millsp.), papaya (*Carica papaya* L.), and apple (*Malus* x *domestica* Borkh.), the authors identified 10,215 shared genes with functional similarities. Of them, 1,777 tRNA genes were coded with a markedly high protein synthesis ability compared to those identified in papaya (388) and grape (600), and this observation led to the consideration of placing *M. oleifera* in the order *Brassicales*.

In the moringa genome, out of 43 transcription factors identified, five copies were positively selected for the WRKY transcription factor, having roles against abiotic stress, including cold, heat, water deficiency, excessive salt, nutrient starvation, and a variable light condition; two copies of the AP2-EREBP transcription factor having impact on hormone, sugar, and redox signaling in response to cold and drought stresses and four copies under positive selection for the C2H2 transcription factor with roles in defense responses and various other physiological processes (Tian et al., 2015).

A total of 133 stress-related heat shock proteins (HSPs) were characterized as HSP 70, HSP 40 (j-proteins), HSP 60 (chaperons), HSP 90, or small HSPs. Twenty-nine copies of BAK 1 (BRI 1 associated receptor kinase 1) gene, which helps to combat cold, heat, and drought stress, were identified in the brassinosteroid pathway. Brassinosteroids are compounds that regulate cell elongation and cell division. Two copies of the STM 2 gene were involved in growth and membrane integrity by balancing the ratio of campesterol to sitosterol (Tian et al., 2015; Table 44.6).

Moringa leaf consumption contributes to human health and nutrition because it has a rich source of secondary metabolites such as flavonoids, glucosinolates, and phenolics. Förster et al. (2015) reported that ecotype has a strong influence on yield, drought tolerance, and secondary metabolite content. They identified two ecotypes, TOT4880 (origin USA) and TOT7267 (origin India), with both superior growth performance and high secondary metabolite production, making them ideal food crop cultivars.

TABLE 44.6
Cultivars of Moringa and Their Biochemical and Genomic Markers Related to Abiotic Stress, and Cultivars of Tamarind

Vegetable Crop (Ploidy)	Potential Cultivars	Biochemical Markers Identified	Genomic Markers Identified with a Role in Abiotic Stress Tolerance
Moringa (2n=28) Source: Tian et al., (2015)	PKM 1, PKM 2	Heat Shock Proteins – HSP70, 40, 60, and 90;	WRKY transcription factor for abiotic stress tolerance
		BAK 1 for cold and drought stress tolerance	AP2-EREBP transcription factor for cold and drought stress tolerance
		STM 2 for membrane integrity	C2 H2 transcription factor for defense and other physiological processes
Tamarind (2n=24) Source: Geetha,(1995) and John, (1990)	Cultivars with red and brownish pulp types, Introductions and Selections	–	–

Minor variations in polyphenolics composition, antioxidant activity, and content of selected nutrients in the leaves of seven cultivars of *M. oleifera* ("Tumu," "Sunyaw," "Kumasi," "Techiman," "China," "Pakistan Black," and "Pakistan White") were found in Pakistan, and were recommended for cultivation in southern Asia (Nouman et. al., 2016) where the climate is purely tropical with high average mean temperatures and humidity. Nouman et al. (2016) identified "Pakistan Black" and "Techiman" as more nutritious cultivars from Pakistan.

44.4.1.1 Utilizing Nutritional Constituents of Moringa

M. oleifera leaves are rich in essential amino acids, proteins, minerals, and vitamins. Moringa seed powder is used as a fish food, leaves are used as animal feed (Figure 44.3), and seeds (kernels) are fried for human consumption to get a sweet to bitter peanut-like taste rich in protein (Figure 44.2). *M. peregrina* is a desert species, and its seeds are a source of edible oil rich in vitamin A, used medicinally as a diuretic, rubefacient, and astringent. Leaves are rich in biologically active carotenoids and vitamin C, which are antioxidants that prevent damage from free radicals. Dry leaves of *M. peregrina* have much less (83 mg) vitamin C than fresh leaves (5,500 mg). The amount of vitamin C in fresh leaves of *M. peregrina* is much more than that in fresh leaves of *M. oleifera* (220 mg) (Asghari et al., 2015).

Extracted *M. oleifera* seed oil has a good oxidative stability and revealed an iodine value of 68.63; refractive index (40°C), 1.4571; density (24°C), 0.903 2g cm⁻³; saponification value, 181.4; unsaponifiable matter, 0.74%; acidity (as oleic acid) 0.81%; and color (1-in. cell) 1.28 R + 31.00 Y. Tocopherol (α, γ, and δ) contents of the oil amounted to 140.5, 63.18 and 61.70 mg kg⁻¹, respectively, and were reduced considerably after degumming. The major sterol components of the oil were β-sitosterol (46.16%), campesterol (17.59%), stigmasterol (18.80%), and Δ-5 avenasterol (9.26%). The wild *M. oleifera* seed oil was found to contain oleic acid up to 73.22%, followed by palmitic, stearic, behenic, and arachidic acids at 6.45, 5.50, 6.16, and 4.08%, respectively (Anwar and Rashid, 2007; Table 44.2).

Rajesh et al. (2016) observed an improved nutritional profile at a 20% level of substitution through pretreatment of moringa leaf flour (MLF). His team developed a ready-to-eat convenience food product through a 20% level of substitution with MLF for good functional and nutritional properties. The improved MLF-based product had high levels of protein (21.6 g/100 g) and dietary fiber (14.8 g/100 g) and low-fat content (3.7 g/100 g). The high-performance liquid chromatography (HPLC) analysis of phenolics revealed MLF containing chlorogenic and gallic acids, pretreated MLF having p-coumaric, caffeic, and gallic acids, and both MLF and pretreated MLF having flavonoids (catechin, kaempferol, rutin, and luteolin).

Leaves were mixed with fruit pastes of high energy foods like tamarind and banana in fixed proportions. One hundred grams of tamarind and banana will provide 252.99 and 205.51 cal, respectively, while 100 g of dry matter from moringa leaves provide 94.37% dry matter, 25 g protein, 55.43%

carbohydrates, and 366.61 cal. The enriched fruit pastries of tamarind with moringa leaf (354.34 cal) and banana with moringa leaf (371.80 cal) provided the same amount of energy as moringa leaf (366.61 cal) and a higher energy compared to pure tamarind and banana fruit pastries, but with an extended shelf life of six months without any bacterial contamination and with a higher nutritional value contributed by combination of fruit and leaf components (Nirina et al., 2014).

The digestibility of the proteins, fats, and the existence of metal chelators (e.g., phytates, oxalates), and protease inhibitors warrants further research with regard to the nutrient bioavailability of moringa plant parts (Asghari et al., 2015). Cultivars of *M. oleifera* or *M. stenopetala* may be improved by increasing the amounts of bioavailable and biologically active nutrients. For example, the amounts of vitamin A in *M. oleifera* and *M. stenopetala* may be genetically altered to be comparable to those in *M. peregrina*. Information currently available on seed nutrient analysis, nutrient availability, and physicochemical properties of leaves are presented in Table 44.7). No data on moringa fruit pulp was found. Nutritional data varied due to the range of species used for biochemical analysis at different locations. The variability observed among species could be used to formulate plans for improvement of cultivars for specific soil types and various forms of abiotic stress.

44.4.1.2 Utilizing Health Constituents of Moringa

Biologically active chemical compounds with pharmaceutical value in South Asia have been identified in a number of species of moringa (Tables 44.1 and 44.5). A number of studies conducted in Pakistan on different parts of *M. oleifera* showed the presence of a wide variety of active constituents with blood pressure lowering and antispasmodic activities mediated possibly through Ca⁺⁺ antagonistic properties (Faizi et al., 1994b; Faizi et al., 1995; Gilani et al., 2004). The hypotensive properties are attributed to the bioactive thiocarbamate glycosides niazimin and niazicin in moringa leaves (Figure 44.4; Faizi et al., 1994a). The presence of amide and thio-amide groups in these bioactive thiocarbamates are responsible for reduced blood pressure. The thiocarbamate glycosides are synthesized by the addition of methanol or ethanol or isothiocyanates present within the moringa plant. The hydrolysis of thiocarbamates results in the biogenesis of carbamates. Similarly, the origin of nitrile glycosides was due to the degradation of glucosinolates, which are the progenitors of organic cyanides and isothiocyanates. All these glycosides contain rhamnose as the sugar residue and are naturally acetylated (Faizi et al., 1995; Figure 44.4).

Thus, leaves of *M. oleifera* were observed with hypotensive, diuretic, anti-fertility, anti-inflammatory, and antispasmodic properties due to the presence of novel glycosides (Faizi et al., 1994b; Faizi et al., 1995). These authors also reported that the carbonyl group of carbamoyl moiety is essential for the platelet inhibition activity in rats; O-carbamoyl taxol analogs were patented in 1993 as neoplasm inhibitors. Faizi et al. (1994a) reported that moringa was used as a cardiac and circulatory tonic with antiseptic properties, and the hypotensive

TABLE 44.7

Nutritional Composition of Edible Parts of Moringa per 100 Grams Fresh Weight

Component	Leaf	Seed
Energy (Cal)	216.1	–
Carbohydrates (g)	9.1–25.27	–
Dry matter (g)	57.19	–
Total protein (g) – raw material	15.83	*23
Total protein (g) – dry material	28.69	31.65
Lipid (g)	5.75	–
Fat (g)	1.7	–
Fiber (g)	2.1	7.54
Fixed seed oil (%)		34.80–*42–54)
Ash (g)	10	6.53
Electrolytes/Minerals		
Sodium (mg)	7.5	–
Potassium (mg)	6–1324 (*900.2)	*572
Magnesium (mg)	35.1–42	
Calcium (mg)	99.1 (*764.8)	1164.8
Iron (mg)	1.3	
Manganese (mg)	0.119	
Phosphorous (mg)	70.8	
Sodium (mg)	70	
Zink (mg)	0.85	
Vitamins		
Vitamin C (mg)	8.6 (*83.00mg)	14.00
Vitamin A (µg)	80 (*6.80mg)	24.80
Thiamine (B1) mg	0.103	–
Ascorbic acid (mg)	220	–
Riboflavin B2 (mg)	0.112	–
Niacin B3 (mg)	1.5	–
Pantothenic acid B5	0.48 (mg)	–
Vitamin B6 (mg)	0.129	–
Folic acid B9 (µg)	41	–
Essential amino acids (µg/ml)		
Threonine	36.77	–
Valine	22.1	–
Methionine	2.13	–
Leucine	20.50	–
Isoleucine	31.8	–
Phenyl alanine	36.8	–
Histidine	30.88	–
Lycine	27.67	–
Arginine	21.45	–

Source: Anwar and Rashid (2007), *Asghari et al., (2015) (*M. peregrina*), Abbas et al., (2018).

properties of isothiocyanate (structures 2 and 3 in Figure 44.4) and the thiocarbamate glycosides niaziminin A and niaziminin B (structures 9 and 10 in Figure 44.4).

Since 1961, a number of medicinal uses of moringa have been identified including cardiac and circulatory stimulants found in the leaves, roots, seed, stem bark, gum, fruit, flowers, and immature pods (Anwar et al., 2007). It was also reported that moringa fruit reduced the serum cholesterol, phospholipids, triglycerides, low-density lipoproteins, very low-density lipoprotein cholesterol to phospholipid ratio, atherogenic index lipids, and the reduced lipid profile in hypercholesteremic rabbits.

The cytotoxic and anti-hypertensive effects of aqueous and ethanolic extracts of whole pods and its parts like seeds, coat, and pulp in moringa were analyzed and identified as displaying similar activity to pods and seeds at a dose of 30 mg/kg. The cytotoxic effects are due to the presence of 4-α-L-rhamnosyloxybenzyl isothiocyanate in seeds, supporting the anti-tumor and antibacterial activities (Faizi et al., 1998).

An exploratory study was conducted recently by researchers for pharmaceutical product development from the most widely grown moringa species in southern Ethiopia, *M. stenopetala*. The leaves and seeds of *M. stenopetala* were used as herbal medicine in areas where the human disease visceral leishmaniasis or kala-azar prevails, which is caused by *Leishmania* parasites. It was reported that anti-cancerous compounds that are identical to glucosinolate compounds in the mustard family were identified in *M. stenopetala* (Nirina et al., 2014). Degradation products of glucosinolates have also been shown to interfere with some metabolic pathways, including the synthesis of thyroid hormones, which warrants further research. Information on the fresh plant parts consumed and products made from moringa are listed in Tables 44.2 and 44.3 for future research investigations.

Genome sequence information generated in the future will be highly helpful in moringa crop and cultivar improvement. This will make moringa a viable perennial crop and lower its cultivation costs in comparison with traditional annual crops (Tian et al., 2015). The United States Department of Agriculture (USDA) is one of the resources which provides two distinct lines of moringa, one that forms underground tuberous roots and the other with the typical taproot. Though there were reports on the existence of variability and germplasm resources, information on how to obtain available moringa germplasm is not readily available online. Genetic research and its potential for the development of improved cultivars, particularly the development of stress tolerant cultivars, is an immediate research need (Leon et al., 2015).

44.4.2 Tamarind

With regard to vegetative propagation methods, softwood grafting, semi-hardwood cuttings, hardwood cuttings, bud grafting, cleft grafting, approach grafting, and air layering were successful. More than 80% graft take (success) was reported with tamarind (Pathak et al., 1991). In tamarind, patch budding was highly successful (82.8%) in August with a maximum number of sprouted buds (21 buds) and under low polyethylene tunnel conditions (15 buds) compared to 19.9% success when budded in October (5 buds) and open conditions (11 buds) at 120 days after budding (Patel, 2016).

Shoot tips, roots, stem, and auxiliary buds from seedlings grown in-vitro were used for micropropagation with a Murashige and Skoog (MS) basal medium containing growth hormones indole-3-acetic acid (IAA) and other growth media to generate new plantlets (Rema et al., 1997; SBI, 2018). Mowobi et al., (2016) also reported using MS basal medium containing growth hormones for callus induction in cotyledon, leaf, and root explants of tamarind and observed success in shoot production with 0.2- 0.05 mg/l of kinetin (6-furfuryl amino purine), 0.5 mg/l of BAP and 0.1 mg/l of naphthalene acetic acid (NAA), indicating positive response with another method of propagation.

Unlike fenugreek and moringa, tamarind can grow readily in drought-prone areas but is very sensitive to frost. The classification of *T. indica* germplasm by early, mid, and late flowering types would help in breeding tamarind cultivars to introduce in various climatic zones under different stresses (Usha, 1990). Genetic diversity analysis using RAPD markers among 36 genotypes of *T. indica* revealed three main clusters with a maximum genetic distance of 9.6% between genotypes. These clusters are 1) genotypes predominantly characterized by brown to dark brown colored fruit pulp, 2) genotypes with straight fruits, and 3) genotypes with semi-curved to curved fruits (Algabal et al., 2011). High levels of intrapopulation genetic diversity was observed in ten populations of tamarind collected from Asia (India and Thailand), Africa (Burkina Faso, Senegal, Kenya, and Tanzania), and from three islands (Madagascar, Réunion, and Guadeloupe) when analyzed using RAPD markers by Diallelo et al. (2007).

The late flowering types of tamarind are best suited for cultivation in USDA Hardiness Zones 9–11, which include the warmer portions of California, Arizona, Alabama, Mississippi, New Mexico, Louisiana, Texas, and Florida (https://planthardiness.ars.usda.gov/PHZMWeb/). Germplasm introduction and evaluation trials will help to enhance cold hardiness, create variability in available genetic resources, and enable increased production of tamarind for various purposes. The germplasm of tamarind identified with drought adaptability and self-incompatibility traits would be useful to fit underutilized croplands for intercrop or multilayered cropping and for breeders to develop breeding lines with frost resistance, ornamental and nutritional values (Narina and Catanzaro, 2018). Afforestation by planting tamarind would be helpful to reduce biodiversity loss due to desertification, deforestation, and erosion (Bello and Gada, 2015). Creation of variability by induced mutation at various growth stages might be beneficial in tamarind to select for improved traits. Screening of vegetatively propagated trees in commercial plantations for visible variability might help in the selection of individuals for pure lines or ecotypes with traits of proven quality and adaptability to adverse environmental conditions.

44.4.2.1 Utilizing Nutritional Constituents of Tamarind

Tamarind is a perennial fruit crop which also has potential as a leafy green vegetable of excellent nutrition (Narina and Catanzaro, 2018). Tamarind fruit is a sweetish sour snack contributing titratable acidity (18.52 g tartaric acid/100 g of pulp) and total soluble solids (44°Brix). Phytochemicals identified in

pulp were alkaloids (4.32%), saponins (2.2%), and glycosides (1.59%), which might be used to reduce losses in processed foods, caused in part by food-borne pathogens. Alkaloids contain complex heterocyclic ring structures and inhibited growth of microbial organisms. Similarly, glycosides are nonvolatile and lack fragrance, but cleavage of glycosidic bonds yields the organic compound aglycone in the extracted pulp, which acts against many microbial organisms (Abukakar et al., 2008).

The aqueous pulp extract of tamarind showed antibacterial activity in a decreasing order of sensitivity as follows: *Staphylococcus aureus* > *Escherichia coli* > *Pseudomonas aeruginosa*. A linear increase occured in the inhibition zone (mm) with a proportionate increase in the concentration of pulp extract (up to 180 mg/ml) using the disk diffusion method (Abukakar et al., 2008).

The effects of various processing methods on the nutritional value, antinutritional compounds, biological value, and protein quality of dehulled seeds of tamarind were evaluated (Vadivel and Pugalenthi, 2010). Data on the mineral and nutritional composition of the various edible parts of tamarind plant parts is presented in Table 44.8.

44.4.2.2 Utilizing Health Constituents of Tamarind

The edible plant parts of tamarind have been evaluated for their antifungal, antiviral, and antibacterial properties. The leaves, seeds, and fruit pulp and their aqueous extracts are rich in flavonoids, polyphenolic compounds, and other phytochemicals which help in spasmolytic, anticancer, antimicrobial, antiparasitic, antifungal, antiviral, and anti-nematocidal effects, wound healing, as well as reduction in dysentery, diarrhea, fever, malaria, and respiratory problems in humans (Tables 44.3 and 44.5). Biochemicals such as triterpenes (lupanone and lupeol), flavones (orientin), and luteolin analogs extracted from fruit pulp and leaves, have a crucial role in health as detailed by Soemardji (2007). Tamarind fruit pulp has vitamin B3, C, and E (α-tocopherol, 93.16 µg, β-tocopherol, 10.89 µg, and γ-tocopherol, 4.73 µg per 100 mg of fruit pulp), folates (5-methyltetrahydrofolate, 41.87 µg, tetrahydrofolate, 12.21 µg, and 5-formyltetrahydrofolate, 5.27 µg per 100 g of pulp), and many useful carbohydrates, secondary metabolites, flavonoids, and polyphenolic compounds (Table 44.2; Figure 44.5).

Seed extracts have immunomodulatory effects (improved immune system) by increasing phagocytosis, inhibiting leukocyte migration and decreasing cell proliferation. The presence of epicatechin in the seed extract of tamarind increases the fecal excretion of total fatty acids, neutral and acidic sterols and has a hypolipidemic effect. Tamarind seed and fruit are suggested as nutritional support for patients with high blood cholesterol levels. Tamarind leaves are associated with liver protective effects by causing membrane stabilization and decreasing glutathione consumption. Tamarind fruit extract regulated lipid metabolism decreased plasma leptin levels and showed weight reduction, hypolipidemic properties, and reversal of hepatosteatosis (Kuru, 2014).

TABLE 44.8
Nutritional Composition of Edible Parts of Tamarind per 100 Grams Fresh Weight

Principle Constituent (Units)	Raw Fruit or Pulp	Percent of RDA from Raw Fruit	Leaf	Flowers
Energy (Kcal)	239.00	12	–	–
Carbohydrates (g)	62.50	40	18.2	–
Invert sugars (g)	30–41	–	–	–
Protein (g)	2.80	5	5.8	0.45
Total fat (g)	0.60	3	2.1	1.54
Cholesterol	0	0	–	–
Dietary fiber (g)	5.10	13	1.9	1.5
Vitamins				
Folates (µg)	14	3.5	–	–
Niacin (mg)	1.938	12	4.1	1.14
Pantothenic acid (mg)	0.143	3	–	–
Ascorbic acid (mg)	0.7–3.0	–	3.0	13.8
Tartaric acid (mg)	8–23.8	–	–	–
Pyridoxine (mg)	0.066	5		
Thiamin (mg)	0.428	36	0.24	0.072
Vitamin A (IU)	30,000	1	–	–
Vitamin C (mg)	3.5	6	–	–
Vitamin E (mg)	0.100	<1	–	–
Vitamin K (µg)	2.800	2	–	–
Electrolytes				
Sodium (mg)	28	2	–	–
Potassium (mg)	628	13	–	–
Minerals				
Calcium (mg)	74	7	101	35.5
Copper (mg)	0.86	9.5	–	–
Iron (mg)	2.80	35	5.2	1.5
Magnesium (mg)	92	23	–	–
Phosphorous (mg)	113	16	140	45.6
Selenium (µg)	1.30	2		
Zinc (mg)	0.10	1	–	–
Phytonutrients				
β Carotene (µg)	18	–	–	–
β Cryptoxanthin (µg)	0	–	–	–
Lutein-Zeaxanthin (µg)	0	–	–	–

Sources: Kuru, 2014; Singh et al., 2007, United State Department of Agriculture – nutrient database; RDA-Recommended Dietary Allowance.

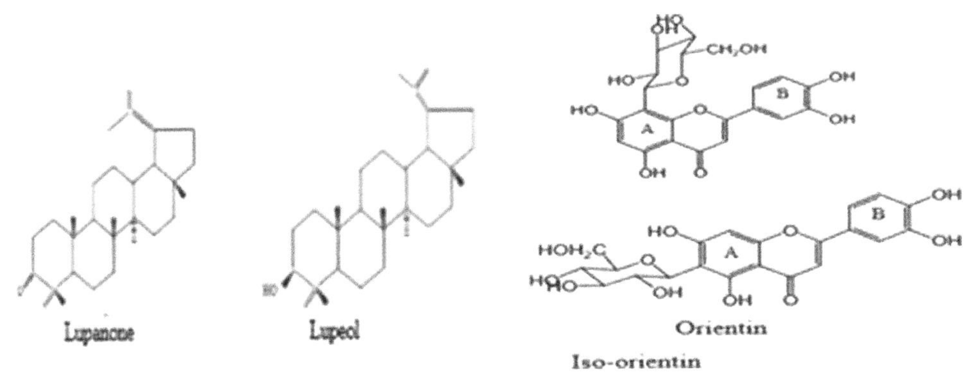

FIGURE 44.5 Chemical structure of the triterpenes, lupanone, and lupeol and the flavones, orientin and iso-orientin in tamarind. (Source: Adapted from Soemardji, 2007).

44.5 PRODUCTION CONSTRAINTS FOR MORINGA, TAMARIND, AND FENUGREEK

44.5.1 MORINGA

Growers listed the following production problems with moringa: livestock damage, theft, limited seed supply, inefficient cultivation practices resulting in inherent low productivity and poor marketing. In some tropical areas in Africa: lack of knowledge of production practices and marketing, competition for land with other food crops, lack of quality seed, pests and disease incidence limited the cultivation (Mudyiwa et al., 2013). Information on how to obtain supplemental income using different production methods to obtain the various saleable products is lacking, as production was confined to traditional areas. Anwar et al. (2007) reported a need to continue research in the anti-cancerous properties of moringa in humans and to identify antipathogenic compounds for humans, livestock, and crops. Moringa flowers are very attractive, so forms with profuse flowering have potential to be sold as ornamental trees. In addition to the above constraints, environmental stresses such as drought, salinity, and poor soil nutrient conditions limit production.

44.5.2 TAMARIND

Tamarind trees are not commercially grown outside of orchards and plantations because of limited land area, long maturation duration, and low monetary values (Edifa-Othiano et al., 2017). Tamarind trees are planted outside on either side of the main roads in South Asia, besides commercial plantations (Narina and Catanzaro, 2018). The foliage of tamarind is very bright, with a potential ornamental use trait for private gardens and parks. Lack of early maturing varieties with frost tolerance, biotic resistance, good fruit quality suitable to machine harvesting, and shortage of quality propagation materials, might be a cause for its limited cultivation in certain agroecological conditions of South Asia and other tropical regions. Therefore, improvement of cultivars with agronomic, ornamental, nutritional and health values and with tolerance to biotic and abiotic stresses will be helpful for future tamarind cultivation.

44.5.3 FENUGREEK

There are no constraints reported for cultivation, production, or marketing of fenugreek. Intensive use of irrigation, fertilizers, and pesticides was reported with the cultivation of fenugreek in South Asia. Typically, low risks to human health due to anticipated pesticide residues and can be eliminated through organic production. Information on fenugreek is limited herein due to the focus on moringa and tamarind as perennial crops suitable for cultivation in tropical regions with limited water supplies.

44.6 RELATIVE COMPARISON AND ECONOMIC ADVANTAGES

Commercial cultivation of annual vegetables, especially leaf and fruit vegetables in commercial farms of the USA, India,

and the rest of the world was involved with intensive use of fertilizers, growth regulators, pesticides, and fungicides starting from sowing until harvest (Ozores-Hampton et al., 2017). The intensive use of various forms of chemicals during production might have a hazardous residual impact on produce that needs costly postharvest processing. These intensive inputs are intended to increase yields, but may also increase production costs, which are passed on to the consumer.

The perennial crops, tamarind and moringa, have relative advantages in terms of pest and disease resistance, providing quality vegetable and fruit products with fewer expenses after orchard establishment, which may yield organic quality fruit and vegetable product at a lower cost.

No state in the USA produces perennial tropical leafy vegetables, though the land area is huge compared to that in developing countries. The possible reasons could be sufficiency of annual leafy vegetable production for domestic consumption or low preference by many for vegetables in comparison with meat or processed foods.

Out of the few vegetables grown in the Midwest and southeastern states of the USA, relatively few leafy vegetables (cabbage, lettuce, spinach, collards, kale, turnip greens) were cultivated with production of 150–200 acres on average in each state (http://hort.purdue.edu/newcrops), with a domestic sale price ranging from $1.75 to $2.50 per pound or per unit (Annual Report, 2015).

The net return per year from a single tamarind tree in South Asia was $220 from fruit, and more than $1,000 from leaf, bark, root, fiber, and from secondary use of byproducts for other medicinal purposes (Usha, 1990). Leaf yield of moringa is 10,000–50,000 kg/ha per year for trees at 1 m × 1 m spacing. 1–5 kg leaf and 230 pods per tree/year. Pod yield of 19 kg/tree/year is equivalent to 31,000 kg/ha/year at 2.5 m × 2.5 m tree spacing. In India, the net return from a kilogram (2.2 lb) moringa products of the leaf, tender pods, flower, seed, and fuelwood are $0.15, $0.40, $0.22, $3.70, and $0.40, respectively (Kumar et al., 2017). Therefore, the net annual return from a single moringa or tamarind tree is $1,000 or more.

Therefore, underutilized plants with drought tolerance and nutritional value should be explored further to develop crop cultivars for overcoming protein-energy malnutrition, for balanced nutrition, biofortification, and to develop transgenic plants. Recently reviews of more than 350 wild and underutilized edible species were reported with their potential use in nutrition and health in India (Gandji et al., 2018; Salvi and Katewa, 2016). Efforts to evaluate beneficial secondary metabolites, phytochemicals, and other nutritional factors in these underutilized plant species, elimination of any undesirable toxic antinutritional compounds and enhancing the ornamental use traits through breeding efforts may be resourceful.

Little genomic research has been conducted on *M. oleifera* (one report by Tian et al., 2015) on the influence of temperature stress on the chemical composition of leaves. *M. stenopetala* was reported to have resistance to pests and disease (Orwa et al., 2009) while *M. peregrina* was reported to have high vitamin A and C content (Asghari et al., 2015). Therefore, the *M. stenopetala* can be used as a rootstock for commercially high yielding cultivars of *M. oleifera* and *M. peregrina* with improved nutritional quality and to adapt to areas with disease and pest incidence and other environmentally unfavorable conditions. This scientific research investment is useful to reduce grower's investment in controlling biotic and abiotic stresses.

The tamarind and moringa crops have potential to be used more extensively in ornamental horticulture due to their aesthetic, bright colored, beautiful flower panicles, and dense foliage, and the ecological values provided by these trees in urban and community forests, an added economic advantage to any country who adapts these crops.

Households in South Asia, India or Africa with average family incomes typically cannot afford fresh vegetables from farmers markets or supermarkets. Typically, calorific requirements are met with inexpensive, less healthy alternatives.

Therefore, based on the information presented, greater utilization of moringa and tamarind as crops with nutritive, pharmaceutical, horticultural, and economic values may help to meet the demands of growing health-conscious populations, provide feed for livestock and poultry, increase efficiency of agricultural production in tropical and subtropical regions, feed human populations with these fresh foods at affordable prices results in increased profits for producers.

44.7 CONCLUSIONS AND FUTURE RESEARCH THRUST

Due to the rapidly increasing population globally, and human and animal dependence on fresh and edible plant resources for health and nutritional values, shifting to more perennial fruit and leafy vegetable crops with drought tolerance and nutritive values will improve the sustainability of food production. Collection, phenotypic, and genotypic characterization of world accessions for useful plant traits by adopting advanced genomic technology is at its initial stage of investigation in both crops. Therefore, it is a necessity to identify the genotypes with economically important traits, screening for variability and understanding how to manipulate the genetic diversity of tamarind and moringa for abiotic stress tolerance and for the cultivation of forms with improved traits. As reported by Narina and Catanzaro (2018), wider cultivation of these two perennial crops in USDA Plant Hardiness Zones 9–11 in the USA would be advantageous to conduct research on crop improvement for potential consumer and industrial markets for apiculture, food, fuel, fodder, horticulture, lipids, medicine, pectins, timber, and tannins.

Future research in moringa is necessary for vegetative propagation techniques like grafting, tissue culture, genome sequencing, breeding, and genetic improvement of the available species on identification of ecotypes, development of drought stress tolerant cultivars or hybrids for industrial value with higher percentage of seed proteins, oil, bioactive compounds with medicinal values and with wide adaptability to diverse agro-climates and marginal lands.

Further research is necessary into genome sequencing to identify the potential drought-tolerant traits in cultivars with high nutritive values and breeding for the development of stress tolerant cultivars to extend cultivation into marginal lands, to identify self-incompatible and cytoplasmic male sterile lines, to develop potential inbred or pure lines with homogeneity with biotic stress tolerance and intensive field research on seedling growth, and random screening of a tamarind stand (as an annual or ratoon or perennial) in commercially feasible growing locations to study tamarind's adaptability to limited water and poor soil conditions.

Both tamarind and moringa were explored and identified with potential health benefits by several scientists. It is the job of agricultural scientists to provide nutrients in a naturally fresh and bioavailable form with health values. There is a true need to increase the use of these crops due to their proven health, ornamental, and nutritional values by introducing their cultivation in non-traditional areas.

ACKNOWLEDGMENT

The authors would like to thank Dr. Toktam Taghvai, Asst. Research Prof. (Plant and Soil Science), Agricultural Research Station, Virginia State University, Petersburg, VA for her valuable contribution in peer reviewing this book chapter.

REFERENCES

Abbas, R.K., Elsharbasy, F.S., and A.A. Fadlelmula. 2018. "Nutritional values of *Moringa oleifera*, total protein, amino acids, vitamins, minerals, carbohydrates, total fat, crude fiber under semi-arid conditions of Sudan." *J. Microb. Biochem. Technol.*, 10(2):56–58.

Abukakar, M.G., Ukwuani, A.N., and R.A. Shehu. 2008. "Phytochemical screening and antibacterial activity of *Tamarindus indica* pulp extract." *Asian. J. Biochem.*, 3(2):134–138.

Acharya, S.N., Thomas, J.E., Prasad, R., and S.K. Basu. 2010. Chapter 19: "Diseases of fenugreek (*Trigonella foenumgraecum* L.) and control measures with special emphasis on fungal diseases." In: *Management of Fungal Pathogens: Current Trends and Progress*, Arya A.P. and Perelló A.E. (eds.). CABI, Nosworthy Way, Wallingford, Oxon, UK. Pages: 245–262.

Acharya, S.N., Thomas, J.E., and S.K. Basu. 2008. "Fenugreek (*Trigonella foenum-graecum* L.) an alternative crop for semi-arid regions of North America." *Crop Sci.*, 48: 841–853

Algabal, A.Q.A.Y., Papanna, N., and L. Simon. 2011. "Estimation of genetic variability in tamarind (*Tamarindus indica* L.) using RAPD markers." *Int. J. Plant Breed.*, 5(1):10–16.

Allen, O.N., and E.K. Allen. 1981. *The Leguminosae: A Source Book of Characteristics, Uses, and Nodulation*. Madison, WS, University of Wisconsin Press. Page 812.

Al-Zahrani, A.A., and A.H. Ibrahim. 2018. "Changes in 2S albumin gene expression in *Moringa oleifera* under drought stress and expected allergenic reactivity in silico analysis." *Theor. Exp. Plant Physiol.*, 30(1):19–27.

Annual Report. 2015. Department of Horticulture, Virginia Tech – Hessler DSF Farm. [Accessed 2018 June 23]. https://www.hort.vt.edu/People/HesslerDSFAnnualReport2015.pdf.

Anwar, F., and U. Rashid. 2007. "Physico-chemical characteristics of *Moringa oleifera* seeds and seed oil from a wild provenance of Pakistan." *Pak. J. Bot.*, 39(5):1443–1453.

Anwar, F., Latif, S., Ashraf, M., and A.H. Gilani. 2007. "*Moringa oleifera*: A food plant with multiple medicinal uses." *Phytoth. Res.*, 21:17–25.

Asghari, G., Palizban, A., and B. Bakhshaei. 2015. "Quantitative analysis of nutritional components in leaves and seeds of the Persian *Moringa peregrina* (Forssk.) Fiori." *Pharmacognosy Res.*, 7(3):242–248.

Barminas, J.T., Charles, M., and D. Emmanuel. 1998. "Mineral composition of non-conventional leafy vegetables." *Plant Foods Human Nutr.*, 53:29–36.

Basu, S.K. 2006. "Seed production technology for fenugreek (*Trigonella foenum-graecum* L.) in the Canadian Priaries." MSc Thesis submitted to the School of Graduate Studies at the University of Lethbridge in partial fulfillment of the requirements for the degree of Master of Science in the Department of Biological Sciences, University of Lethbridge, Lethbridge, Alberta, Canada in 2006 March.

Baye-Niwah, C., and P.M. Mapongmetsem. 2014. "Seed germination and initial growth in *Moringa oleifera* Lam. 1785 (Moringaceae) in Sudano-sahelian zone." *Int. Res. J. Plant Sci.*, 5(2):23–29.

Bello, A.G., and Z.Y. Gada. 2015. "Germination and early growth assessment of *Tamarindus indica* L. in Sokoto State, Nigeria." Hindawi publishing corporation, *Int. J. Forestry Res.*, 2015:5 Article ID 634108.

Coppin, J.P., Yanping, X., Hong, C., Min-Hsiung, P., Chi-Tang, H., Rodolfo, J., James, E.S., and W. Qingli. 2013. "Determination of flavonoids by LC/MS and anti-inflammatory activity in *Moringa oleifera*." *J. Functional Foods*, 5(4):1892–1899.

Diallo, B.O., Joly, H.I., McKey, D., Hossaert-Mckey, M., and M.H. Chevallier. 2007. "Genetic diversity of *Tamarindus indica* populations: Any clues on the origin from its current distribution?" *Afr. J. Biotechnol.*, 6(7):853–860. ISSN 1684-5315. Available online at https://www.academicjournals.org/AJB.

Edifa-Othieno, E., Mughisha, A., Nyeko, P., and J.D. Kabasa. 2017. "Knowledge, attitudes and practices in tamarind (*Tamarindus indica* L.) use and conservation in Eastern Uganda." *J. Ethnobiol. Ethnomed.*, 13:5.

Faizi, S., Siddiqui, B.S., Saleem, R., Siddiqui, S., Aftab, K., Shaheen, F., and A.U.H. Gilani. 1998. "Hypotensive constituents from the pods of *Moringa oleifera*." *Planta Medica*, 64:225–228.

Faizi, S., Siddiqui, B.S., Saleem, R., Siddiqui, S., Aftab, K. and A.U.H. Gilani. 1995. "Fully acetylated carbamate and hypotensive thiocarbamate glycosides from *Moringa oleifera*." *Phytochemistry*, 38(4):957–963. Article ID 0031-9422(94)00729-2.

Faizi, S., Siddiqui, B.S., Saleem, R., Siddiqui, S., Aftab, K. and A.U.H. Gilani. 1994a. "Novel hypotensive agents, Niazimin A, Niazimin B, Niazicin A and Niazicin B from *Moringa oleifera*: Isolation of naturally occurring carbamates." *J. Chem. Soc.*, 3035–3040.

Faizi, S., Siddiqui, B.S., Saleem, R., Siddiqui, S., Aftab, K., and A.U.H. Gilani.1994b. "Isolation and structure elucidation of new nitrile and mustard oil glycosides from Moringa oleifera and their effect on blood pressure." *J. Natu. Prod.*, 57(9):1256–1261.

FMI. 2018. Future market insights: Tamarind – extracts. REP-GB-2127. [Accessed 2018 June 23] www.futuremarketinsights.com/reports/tamarind-extract-market.

Förster, N., Ulrichs, C., Schreiner, M., Arndt, N., Schmidt, R., and I. Mewis. 2015. "Ecotype variability in growth and secondary metabolite profile in Moringa oleifera: impact of sulfur and water availability." *J. Agric. Food Chem.*, 63(11):2852–2861.

Gandji, K., Chadare, F.J., Idohou, R., Salako, V.K., Assogbadjo, A.E., and R.L. Glee Kakai. 2018. "Status and utilization of *Moringa oleifera* Lam: a review." *Afr. Crop Sci. J.*, 26(1):137–156.

Ganesan, S.K., Singh, R., Choudhury, D.R., Bharadwaj, J., Gupta, V., and A. Singode. 2014. "Genetic diversity and population structure study of drumstick (*Moringa oleifera* Lam.) using morphological and SSR markers." *J. Ind. Crops Prod.*, 60:316–325.

Geetha, V.1995. "Urigam tamarind – unique variety." *Spice India*, 8:8.

Gilani, A.H., Aftab, K., Suria, A., Faizi, S., Siddiqui, S., Saleem, R., and B.S. Siddiqui. 1994. "Pharmacological studies on hypotensive and spasmolytic activities of pure compounds from *Moringa oleifera*." *Phytother. Res.*, 8:87–91.

John, A.P. 1990. Tamarind. A review report from Centre for Energy and Environmental Research, University of Puerto Rico, Rio Piedras, Puerto Rico, in collaboration with USDA-forest service. SO-ITF-SM-30 June. www.ieer.org. [Accessed 2018 June 23].

Kumar, Y., Thakur, T.K., Sahu, M.L., and A. Thakur. 2017. "A multifunctional wonder tree: *Moringa oleifera* Lam open new dimensions in field of agroforestry in India." *Int. J. Cur. Microbio. Appl. Sci.*, 6(8):229–235. ISSN: 2319-7706.

Kumar, M., Ponnuswami, V., Rajamanickam, C., and T.L. Preethi. 2015. "Assessment of genetic diversity in tamarind (*Tamarindus indica* L) using random amplified polymorphic markers." *SAARC J. Agri.*, 13(1):27–36.

Kuru, P. 2014. "*Tamarindus indica* and its health-related effects." *Asian Pac. J. Trop. Biomed.*, 4(9):676–681.

Léonard, J. (1957). "Genera des Cynometreae et des *Amherstieae africaines* (Leguminosae-Caesalpinioideae)." *Mémoire Académie Royale Belgique*, 30:1–314.

Leone, A., Spada, A., Battezzati, A., Schiraldi, A., Aristil, J., and S. Bertoli. 2015. "Cultivation, genetic, ethnopharmacology, phytochemistry and pharmacology of *Moringa oleifera* leaves: an Overview." *Int. J. Mol. Sci.*, 16(6):12791–12835.

Lisar, Y.S.S., Motafakkerazad, R., Hossain, M.M., and M.M.I. Rahman. 2012. Introductory Chapter: "Water Stress in Plants: Causes, effects and responses." *In: Water Stress*, M.M.I Rahman and H. Hasegawa (eds.). InTech, Janeza Trdine 9, 51000 Rijeka, Croatia. Pages 1–14978-953-307-963-9, Digitally available online at https://www.researchgate.net/publication/236160968.

Mekonnen, Y. 2002. Chapter 10: "The multipurpose moringa tree: Ethiopia." In: Hand Book on Examples of the development of pharmaceutical products from medicinal plants. Case study prepared for the Institute of Pathobiology, Addis Ababa University, P.O. Box number 1176, Addis Ababa, Ethiopia. Pages: 111–118.

Mowobi, G.G., Osuji, Ch., Salisu, A., and F.M. Yahaya. 2016. "In vitro regeneration of tamarind (*Tamarindus indica* L.) explants." *J. Environ. Life Sci.*, 1(1):26–31. ISSN 2456-6179.

Mudyiwa, S.M., Gadzirayi, C.T., Mupangwa, J.F., Gotosa, J., and T. Nyamugure. 2013. "Constraints and opportunities for cultivation of Moringa oleifera in the Zimbabwean small holder growers." *Int. J. Agr. Res. Innov. Tech.*, 3(1):12–19. ISSN 2224-0616. [Accessed 2018 June 22] http://www.ijarit.webs.com.

Narina, S.S.S. and C.J. Catanzaro. 2018. "Tamarind (*Tamarindus indica* L.), an underutilized fruit crop with potential nutritional value for cultivation in the United States of America: a review." *Asian Food Sci. J.*, 5(1):1–15. Article number AFSJ.43611.

NAS. 1979. "Tropical legumes: A resource for the future." National Academy of Sciences, Washington, DC. 332p.

Nirina, H.A., Louisette, R., and R. Felamboahangy. 2014. "Nutritional quality of fruit pastes enriched with *Moringa oleifera* leaves." *Int. J. Appl. Sci. Technol.*, 4(5):163–175.

Nouman, W., Olson, M.E., Gull, T., Zubair, M., Basra, S.M.A., Qureshi, M.K., Sultan, M.T., and M. Shaheen. 2018. "Drought affects size, nutritional quality, antioxidant activities and phenolic acids pattern of *Moringa oleifera* Lam." *J. Appl. Bot. Food Qual.*, 91:79–87.

Nouman, W., Anwar, F., Gull, T., Newton, A., Rosa, E., and R. Dominguez-Perles. 2016. "Profiling of polyphenolics, nutrients and antioxidant potential of germplasm's leaves from seven cultivars of *Moringa oleifera* Lam." *Ind. Crops Prod.*, 83:166–176. [Accessed 2018 April 20] http://dx.doi.org/10.1016/j.indcrop.2015.12.032.

Ozores-Hampton, M., Kinessary, R., Raid, R.N., Joseph, W.N., Beuzelin, J., and C.F. Miller. 2017. Chapter 9: "Leafy Vegetable Production." In: *Vegetable Production Handbook for Florida*. IFAS Extension, University of Florida. Pages: 107–134.

Olson, M.E. 2002. "Combining data from DNA sequences and morphology for a phylogeny of Moringaceae (Brassicales)." *Syst. Bot.*, 27:55–73.

Orwa, C., Mutua, A., Kindt, R., Jamnadass, R., and S. Anthony. 2009. "Agroforestry Database: A tree reference and selection guide." version 4.0 [Accessed 2018 May 20] (http://www.worldagroforestry.org/sites/treedbs/treedatabases.asp)

Pareek, O.P., and S. Sharma. 1991. "Fruit trees for arid and semi-arid lands." *Indian Farming*, 41:25–30.

Patel, D.K. (2016). "Effect of season and growing environment on success of patch budding in tamarind (*Tamarindus indica* L.)." MSc Thesis submitted to Department of Fruit Science, ASPEE College of Horticulture and Forestry, Navasari Agricultural University, Navasari, Gujarat, 2016 July. Registration No.2020214028.

Pathak, R.K., Ojha, C.M., and R. Dwivedi. 1991. "Adopt patch budding for quicker multiplication in tamarind." *Indian Hortic.*, 36(2):17.

Peter, K.V. 2006. "Handbook of herbs and spices." In: Food Science, Technology and Nutrition. Volume 3. Woodhead Publishing Limited, Abington Hall, Abington, Cambridge CB1 6AH, England. Pages 1000. [Accessed 2018 March 20] www.woodheadpublishing.com.

Petropoulos, G.A. 2002. *Fenugreek: The Genus Trigonella*. Taylor and Francis, London and New York, NY. 1127p.

Pettigrew, C.J., and L. Watson. 1977. "On the classification of *Caesalpinioideae*." *Taxon*, 26:57–64.

Rajesh, D., Yadahally, N.S., and B. Sila. 2015. "Processing effects on bioactive components and functional properties of *Moringa* leaves: Development of a snack and quality evaluation." *J. Food Sci. Technol.*, 53(1): 649–657.

Ramachandran, C., Peter, K.V. and P.K. Gopalakrishnan. 1980. "Drumstick (*Moringa oleifera*): a multipurpose Indian vegetable." *Econ. Bot.*, 34:276–283.

Rema, J., Krishnamoorthy, B., and P.A. Mathew. 1997. "Vegetative propagation of major tree species: a review." *J. Spices Aromatic Crops*, 6(2):87–105.

Rivas, R., Oliveira, M.T., and M.G. Santos. 2012. "Three cycles of water deficit from seed to young plants of *Moringa oleifera* woody species improves stress tolerance." *Plant Physiol. Biochem.*, 63:200–208.

Salvi, J., and S.S. Katewa. 2016. "A review: underutilized wild edible plants as a potential source of alternative nutrition." *Int. J. Bot. Studies*, 1(4):32–36. ISSN 2455-541X.

Saini, R.K., Sivanesan, I., and Y-S Keum. 2016. "Phytochemicals of Moringa oleifera: a review of their nutritional, therapeutic and industrial significance." *Biotech.*, 6(2):203.

Santosh, S.B., Aditya, G., Jitendra, N., Gopal, R., and P.J. Alok. 2011. "*Tamarindus indica*: extent of explored potential." *Pharmacogn. Rev.*, 5(9):73–81.

Sharma, S.K., Singh, R.S., and R. Raghava. 2013. "Arid horticulture: an overview." *Ann. Arid Zone*, 52(3–4):251–264.

Singh, J.S., Batra, V.K., Singh, S.K., and T.J. Singh. 2012. "Diversity of underutilized vegetable crop species in north-east India with special reference to Manipur: a review." *NeBIO*, 3(2):87–95.

SBI. 2018. Spices Board of India, Ministry of Commerce and Industry, Govt. of India. [Accessed 2018 June 23] http://www.indianspices.com/spice-catalog/tamarind.

Soemardji, A.A. 2007. "*Tamarindus indica* L. or Asam jawa: The sour and sweet and useful." [Accessed 2016 May 16] https://www.researchgate.net/publication/242285214.

Suresh, V., and P. Jansirani. 2017. "Cutting Management and Application of Bio-Stimulants on Leaf Yield and Quality of Curry Leaf (*Murraya koenigii* Spreng.)." *Int. J. Curr. Microbiol. App. Sci.*, 6(7):2181–2187.

Tian, Y., Zeng, Y., Zhang, J., Yang, C.G., Yan, L., Wang, X.J., Shi, C.Y., Xie, J., Dai, T.Y., Peng, L., Zeng, H.Y., Xu, A.N., Huang, Y.W., Zhang, J.J., Ma, X., Dong, Y., Hao, S.M., and J. Sheng. 2015. "High quality reference genome of drumstick tree (*Moringa oleifera* Lam.), a potential perennial crop." *Sci. China Life Sci.*, 58:627–638.

Usha, K. 1990. "Studies on dynamics of vegetative and reproductive growth in tamarind (*Tamarindus indica* L.)." PhD Thesis submitted for the award of degree PhD in Horticulture to Univ. Agric. Sci., Bangalore. [Accessed 2018 May 20] www.google.com.

Verma, M.K., Prasad, M., and R.K. Arya. 2013. "Grain yield and quality improvement in fenugreek: a Review." *Forage Res.*, 39(1):1–9.

Vadivel, V., and M. Pugalenthi. 2010. "Evaluation of nutritional value and protein quality of an under-utilized tribal food legume." *Indian J. Trad. Knowl.*, 9(4):791–797.

Van den Bilcke, N., Simbo, D.J. and R. Samson. 2013. "Water relations and drought tolerance of young African tamarind (*Tamarindus indica* L.) trees." *S. Afr. J. Bot.*, 88:352–360.

Yasmeen, A., Basra, S.M.A., Farooq, M., Rehman, H., Hussain, N., and H.R. Athar. 2013. "Exogenous application of moringa leaf extract modulates the antioxidant enzyme system to improve wheat performance under saline conditions." *Plant Growth Regulation*, 69:225–233.

Yisehak, K., Solomon, M., and M. Tadelle. 2011. "Contribution of Moringa (*Moringa stenopetala*, Bac.), a highly nutritious vegetable tree for food security in south Ethiopia: a Review." *Asian J. Appl. Sci.*, 4:477–488.

Zandi, P., Basu, S.K., Bazrkar Khatibani, L., Balogun, M., Aremu, M.O., Sharma, M., Kumar, A., Sengupta, R., Li, X., Li, Y., Tashi, S., Hedi, A., and CetzalIxW.. 2015. "Fenugreek (*Trigonella foenum-graecum* L.) seed: A review of physiological and biochemical properties and their genetic improvement." *Acta Physiologia. Plantarum.*, 37:1714.

Zandi, P., Basu, S.K., CetzalIx, W., Kordrostami, M., Chalaras, S.K., and L.B. Khatibai. 2017. Chapter 12: "Fenugreek (*Trigonella foenum-graecum L.*): an important medicinal and aromatic crop." *INTECH Open Science*, Pages: 207–225. [Accessed 2018 July 1] http://dx.doi.org/10.5772/66506.

Zhou, J., Chan, L., and S. Zhou. 2012. "Trigonelline: A plant alkaloid with therapeutic potential for diabetes and central nervous system disease." *Curr. Med. Chem.*, 19(21):3523–3531.

45 Relationship of Medicinal Plants and Environmental Stresses
Advantages and Disadvantages

Amir Hossein Saeidnejad

CONTENTS

45.1 INTRODUCTION

The earliest records regarding the use of medicinal plants are obtained from Mesopotamian civilizations and are as old as 2600 BC (Gurib-Fakim, 2006). History shows that plants with medicinal properties were in use in the Assyrian, Babylonian, Chinese, Greek, and Hebrew civilizations (Humayun, 2007). People in rural areas have always used native plants and herbs as medicines, partly due to the wide gap of educational, research, and health facilities between rural and urban areas and partly because of socioeconomic issues (Shinwari and Khan, 2000). Today, some medicinal plants need to be cultivated at commercial scales to meet the ever-increasing demands of the pharmaceutical industry. In the mid-twentieth century, the use of synthetic chemical compounds for therapeutic purposes became widespread. However, the worth of medicinal plants came under the spotlight again when researchers failed to produce comparatively affordable and safe medicines, and thus, herbal medicine experienced a revival, especially in Western society.

As sessile organisms, plants live in constantly changing environments that are often unfavorable or stressful for growth and development. These adverse environmental conditions include biotic stress, such as pathogen infection and herbivore attack, and abiotic stress, such as drought, heat, cold, nutrient deficiency, and an excess of salt or toxic metals such as aluminum, arsenate, and cadmium in the soil. Drought, salt, and temperature stresses are major environmental factors that affect the geographical distribution of plants in nature, limit plant productivity in agriculture, and threaten food security (Zhu, 2016).

It is well established that the accumulation of natural products strongly depends on the growing conditions (Ballhorn et al., 2011). It is also well known that various environmental conditions, such as the temperature, the light regime, the nutrient supply, etc., strongly influence the synthesis and accumulation of secondary plant products. Consequently, much more severe environmental influences, such as typical stress situations, which are known to strongly impact on the entire general metabolism, also must alter the secondary metabolism. By now, a tremendous amount of information on the impact of various biological stresses, e.g., pathogen or herbivore attack, on elicitation of the synthesis of natural products is available (Hartmann, 2007; Namdeo, 2007; Wink, 2010). Further on, many papers deal with the impact of abiotic stresses on the secondary metabolism; nonetheless, knowledge about the related biological background is still limited (Selmar et al., 2017).

In most of the few existing studies dealing with the influence of stress on the accumulation of secondary plant products, only one stress factor has been comprehensively investigated, i.e., by comparatively quantifying the content of equivalent natural products in stressed and unstressed plants without being aware that various interactions between numerous factors may have occurred. In this manner, any increase in light intensity may be correlated with elevated leaf temperatures, and a decreased water potential adjusted to induce drought stress frequently also leads to higher salt concentrations in the soil. Moreover, elevated temperatures also affect transpiration and therefore, plant water use. Because of these considerations, many studies are not fully conclusive. Nonetheless, after a thorough review of the literature, certain deductions about the effects of single factors on the accumulation of secondary plant products can be made (Selmar and Kleinwachter, 2013).

45.2 MEDICINAL PLANT FUNCTION UNDER DROUGHT STRESS

A wide range of experiments has shown that plants exposed to drought stress accumulate higher concentrations of secondary metabolites. Such enhancement is mostly seen in all kinds of secondary metabolites, including phenols and numerous terpenes and also nitrogen-containing substances, such as alkaloids. It is therefore obvious that drought stress frequently enhances the concentration of secondary plant products. However, this stress-related increase in natural product concentrations does not mean that the rate of biosynthesis of natural products in the plants has increased, since drought stress also reduces growth and biomass production in most plants. Accordingly, a simple explanation for this effect is that "Plants suffering drought stress, in principle, synthesize and accumulate the same amounts of natural products as under well-watered conditions, but due to the reduction in overall plant biomass, product concentration on a fresh or dry weight basis is enhanced!" To obtain a clear picture of the impact of drought stress on natural product biosynthesis, the total amount of secondary plant products has to be calculated. For such an assessment, however, in addition to the determination of the secondary plant product concentration in stressed and well-watered plants, there also needs to be an estimation of the entire biomass of the plants under both stressed and

unstressed conditions. Unfortunately, in most of the studies published so far, few or no data on the total biomass of the plants analyzed have been published. This lack of information is generally due to the fact that in the related studies, only certain plant parts (e.g. roots, leaves, or seeds) had been analyzed, whereas the total content of natural products on a whole plant basis was not the central focus of the studies. However, in a few studies, the total content of secondary plant products has been given or could be calculated from the data presented (Selmar and Kleinwachter, 2013).

45.2.1 VEGETATIVE GROWTH

It is totally acknowledged that under drought stress, the growth and biomass production of plants are reduced. There are plenty of studies with a focus on this issue. For instance, when different drought stress levels were applied to *Cassia obtusifolia* L., there were significant reductions in plant growth and seed yield attributes of *C. obtusifolia*, which were more pronounced with greater severity of drought (Xue et al., 2018).

The influence of soil water stress on plant height and fresh and dry weight of Iranian *Satureja hortensis* L. was also investigated, and the results showed that greater soil water stress decreased plant height and total fresh and dry weight (Baher et al., 2002). Another experiment was conducted on two varieties, rosea and alba, of *Catharanthus roseus* plants with two watering treatments, 100% and 60% of field capacity (FC), to understand the effects of water deficit on early growth, biomass allocation, and photosynthetic pigment responses. It was found that there were significant differences in early growth and dry matter accumulation between the two varieties. The root length, shoot length, total leaf area, and fresh and dry weights were significantly reduced under water stress treatments (Abdul Jaleel et al., 2008). When the effect of water availability on plant growth and seed yield in *C. obtusifolia* was investigated, the results showed that different levels of drought stress resulted in significantly different reductions in plant growth and seed yield attributes, and these were more pronounced with greater severity of drought. Notably, weak drought stress (70% FC in pot and in field) did not significantly decrease the seed yield, which resulted in the biggest harvest index among other treatments (Xue et al., 2018).

45.2.2 ESSENTIAL OIL QUANTITY AND QUALITY

The effect of drought stress on essential oil compounds and their relationship with the expression levels of the studied genes were investigated in basil (*Ocimum basilicum* L.). The results showed that drought stress increased the amounts of methylchavicol, methyleugenol, β-myrcene, and α-bergamotene. The maximum amounts of these compounds were observed at 50% FC. Real-time polymerase chain reaction (PCR) analysis revealed that severe drought stress (50% FC) increased the expression level of chavicol O-methyltransferase (CVOMT) and eugenol O-methyltransferase (EOMT) by about 6.46 and 46.33 times, respectively, whereas that of CAD relatively remained unchanged. The expression level of 4CL and C4H was reduced under drought stress conditions. These findings also demonstrated that changes in the expression levels of CVOMT and EOMT are significantly correlated with methylchavicol and methyleugenol content (Mandoulakani et al., 2017). It was also reported that chitosan (Naderi et al., 2014) and salicylic acid (Zarei et al., 2015) increase the expression level of CVOMT.

45.2.3 PHENOLIC COMPOUNDS

In *Hypericum brasiliense* plants grown under drought stress, both the concentration and the total amount of the phenolic compounds are drastically enhanced in comparison with the control plants (de Abreu and Mazzafera, 2005). Despite the fact that the stressed *H. brasiliense* plants were considerably smaller, the product of biomass and concentration of the related phenolics yield 10% increase of the total amount of these natural products. In the same manner, in stressed peas (*Pisum sativum*), the overall amount of anthocyanins (product of biomass and anthocyanin concentration) is about 25% higher than in plants cultivated under standard conditions (Nogués et al., 1998): although the biomass of the stressed pea plants is only about one-third that of the control plants, the massive increase in the concentration of phenolic compounds still resulted in a real increase of anthocyanins in the stressed plants. Also, Jaafar et al. (2012) reported that not only the concentration but also the overall production of total phenolics and flavonoids per plant is enhanced in plants suffering drought stress, although the explicit data on biomass per plant are not displayed by the authors. In contrast, the overall yield of flavonoids was nearly the same when the plants were grown either under drought stress or under well-watered, non-stress conditions. In stressed red sage plants (*Salvia miltiorrhiza*), the overall content of furoquinones is slightly lower in plants grown under water deficiency than that of the well-watered controls, although drought stress caused a significant increase of their concentration (Liu et al., 2011).

45.2.4 TERPENOIDS

There are only a few reports available that soundly document drought stress–related increase in the total amount of terpenoids per plant. In sage (*Salvia officinalis*), drought stress results in a massive increase in the concentration of monoterpenes, which easily overcompensates for the reduction in biomass (Nowak et al., 2010). As a result, the entire amount of monoterpenes synthesized in sage plants suffering moderate drought stress is significantly higher than that of the well-watered controls. Yet, in parsley (*Petroselinum crispum*), the drought stress–related concentration enhancement of essential oils in the leaves is more or less completely counterbalanced by the related loss in biomass, resulting in nearly the same overall contents of essential oils in drought-stressed and well-watered plants (Petropoulos et al., 2008). In contrast, in drought-stressed catmint and lemon balm plants, the slight increase in the concentrations of monoterpenes could

not compensate for the stress-related detriment of growth and biomass. Accordingly, the overall content of terpenoids in drought-stressed plants of *Melissa officinalis* and *Nepeta cataria* is lower than that in the corresponding well-watered controls (Manukyan, 2011).

45.2.5 PROLINE CONTENT

In response to different stresses, plants accumulate large quantities of different types of compatible solutes (Serraj and Sinclair, 2002). Compatible solutes are low–molecular weight, highly soluble organic compounds that are usually non-toxic at high cellular concentrations. These solutes provide protection to plants from stress by contributing to cellular osmotic adjustment, reactive oxygen species (ROS) detoxification, protection of membrane integrity, and enzyme/protein stabilization (Ashraf and Foolad, 2007). These include proline, sucrose, polyols, trehalose, and quaternary ammonium compounds (QACs) such as glycine betaine, alanine betaine, proline betaine, and pipecolate betaine. In many plant species, proline accumulation under stress conditions has been correlated with stress tolerance, and its concentration has been shown to be generally higher in stress-tolerant than in stress-sensitive plants (Fougère et al., 1991). Its accumulation normally occurs in the cytoplasm, where it functions as a molecular chaperone stabilizing the structure of proteins, and its accumulation buffers cytosolic pH and maintains cell redox status. It has also been proposed that its accumulation may be part of a stress signal influencing adaptive responses (Petrusa and Winicov, 1997).

During the investigation of drought stress tolerance in *Withania somnifera* (a valuable, endangered medicinal plant), it was found that proline content increased with increasing level of drought stress. The results also showed that the cells were not much affected by drought conditions, which might have been due to the fact that proline helps the cells to cope with the water stress, as it balances the osmotic potential of cells (Sharma et al., 2018). This study is also supported by the study of Sabir et al. (2012), who reported a sudden enhancement in the proline content when calluses were treated with a higher concentration of $CaCl_2$ (200 mM). The response of German chamomile (*Matricaria chamomilla*) to drought stress along with organic fertilizer was also evaluated, and a decrease in water availability under a low-irrigation regime enhanced the leaf proline content, although the leaf proline content increase due to vermicompost was less affected by water deficiency and low irrigation, which increased the content of leaf proline. In all irrigation treatments, the leaf proline content was significantly higher (Salehi et al., 2016). In *Thymus daenensis*, the accumulation of proline under drought stress conditions was also considerable, and plants that experienced severe stress accumulated 53.2% and 27.0% more proline than those under non-stress conditions (Bistgani et al., 2017).

Although proline is mostly known as a regulatory or signaling molecule to activate multiple responses that are part of the adaptation process (Claussen, 2005), there is also some evidence that proline accumulation may not change during drought stress conditions. For instance, leaves of *matricaria chamomilla* grown under different levels of watering had a similar proline content at almost all levels. After rewatering, the synthesis of proline in leaves from all treatments restored the initial content (Pirzad et al., 2011). In fact, a linear relationship could be observed between the proline content and the hydric deficit produced on the plants at different times during this process. This result is easy to explain, considering that proline accumulation may result from both induction of proline biosynthesis and/or inhibition of its oxidation (Hong et al., 2000). Moreover, there are previous reports that the activation of pyrroline-5-carboxylate synthetase (an enzyme from proline biosynthesis) as well as the inhibition of proline dehydrogenase (an enzyme from proline degradation) is more rapid in a fast drought than in a slow drought (Sanada et al., 1995).

It is also necessary to consider that apart from all the well-established ecological functions, the drought stress–related increase in the biosynthesis of highly reduced natural products might also have relevance as an additional mechanism for energy dissipation (Wilhelm and Selmar, 2011).

45.3 SALINITY STRESS AND MEDICINAL PLANT INTERACTIONS

Environmental stress is a major area of scientific concern, because it constrains plants as well as productivity. This situation has been further worsened by anthropogenic activities. Therefore, there is much scientific pressure on researchers to enhance plant productivity under environmental stress. Salinity stress appears to be a major constraint to plant and crop productivity.

Although the effects of salt stress on crops have been investigated widely and well studied, in the case of medicinal plants, there is a lack of information. The molecular mechanisms of salt tolerance and secondary metabolism in these commercially important crops have not been investigated as they have in other crops. Thus, comparing the responses of medicinal plants to salt stress with some other important crops has great value. Salt stress significantly affects the production of essential oils and the constituents of medicinal plants; thus, the investigation of the mechanisms of salt tolerance in medicinal plants has great importance. The creation of salt-tolerant medicinal plants will lead to increased production of raw materials for drugs, flavors, fragrances, and spices all over the world.

Tolerance against salt stress needs profound changes in gene expression, which are accompanied by changes in composition of the plant transcriptome, metabolome, and proteome. Changes in gene expression at the transcript level cannot exactly show the changes at the protein level. This reflects the high importance of the plant proteome, since proteins are directly involved in the plant stress response (Kosova et al., 2011).

45.3.1 GERMINATION AND PRIMARY GROWTH AFFECTED BY SALINITY

Salt stress affects medicinal plants at different physiological stages. One of the most salt-sensitive growth stages, which is severely inhibited by increasing salinity, is the seed germination stage (Sosa et al., 2005). Seed germination of *Ocimum basilicum* (Miceli et al., 2003), *Petroselinum hortense* (Ramin, 2005), sweet marjoram (Ali et al., 2007), and *Thymus maroccanus* (Belaqziz et al., 2009) showed a significant decrease under salt stress.

Water uptake is crucial for the seed germination process; however, the presence of NaCl in the surrounding medium creates a negative water potential, resulting in retardation of water imbibition by seeds. As a consequence, various metabolic perturbations are induced by salinity stress, comprising slow or lower mobilization of reserve nutrients, the arrest of cell division, and decreased the synthesis of nucleic acid, thus causing injury to germinating embryos. This delays the time of radicle emergence and hence, increases the time needed for germination, thus affecting the seed vigor. The reason for the increased lag period could be attributed to the osmotic effect or accumulation of specific ions up to a toxic level (Katembe et al., 1998). This observation was based on a report that deteriorated seeds require more time for metabolic repair before the process of seed germination can begin (Matthews et al., 2011).

Another stage that is negatively influenced by salinity is seedling growth. It has been previously reported that seedling growth of *Thymus maroccanus* (Belaqziz et al., 2009), basil (Ramin, 2005), chamomile, and marjoram (Ali et al., 2007) was severely decreased under salt stress. Slow or lower mobilization of reserve foods, suspending cell division, and enlarging and injuring hypocotyls, which are induced by salt stress, have been proposed as the main reasons for these effects (Said-Al Ahl and Omer, 2011).

45.3.2 EFFECTS ON PRODUCTION

Growth and biomass production are also usually affected by saline conditions. Higher salinity has been reported to induce changes in protein structure and increase cytoplasmic RNAase activity, leading to a decrease in DNA synthesis and creating many cellular menaces to the activity required for development processes in plants (Niu et al., 1995). In an experiment, seeds of *Lepidium sativum* L., *Linum usitatissimum* L., *Nigella sativa* L., *Plantago ovata* Forssk, and *Trigonella foenum-graecum* L. were grown under different NaCl concentrations. Various concentrations of salt had a highly significant effect on the survival age, plant height, the number of branches, shoot fresh and dry weight, root fresh and dry weight, and root moisture content. The number of leaves also varied significantly. However, leaf length and shoot moisture content exhibited non-significant differences. Differences among the test species for all the parameters under consideration were also highly significant (Muhammad and Hussain, 2010). When the effects of saline water on the quality and quantity of fennel was investigated, it was observed that plant height, leaf, and bulb weight dropped by 33%, 49%, and 71%, respectively, with a change from 0.7 to 12 dS m^{-1} salinity treatments. It was also interesting that dry matter content and total soluble solids increased with rising salinity. In fact, with the shift from the control to the more saline treatment, the dry matter in leaves and bulbs increased by 15 and 13%, respectively, and total soluble solids increased by 17%. Moreover, the increase in salinity favored the production of flattened bulbs (Gucci et al., 2014). *Salvia miltiorrhiza*, an important traditional Chinese herbal medicine, was also exposed to salt stress, and the results indicated that salinity treatment significantly inhibited the accumulation of dry matter, although it did not affect the growth of *Salvia miltiorrhiza* in a morphological sense. In fact, the plant height and root length remained virtually unchanged with all the salt treatments (Gengmao et al., 2014). In contrast, the seedlings of cumin under saline conditions showed different responses, and salinity influenced the length of roots and shoots of seedlings. Root length decreased significantly with salinity stress treatments. A significant reduction of 74% and 90% was observed at 80 and 100 mM NaCl, respectively, compared with seedlings grown in control conditions. Apart from growth, salinity also influenced the root morphology of seedlings grown at different NaCl concentrations. Root hair decreased gradually with increasing NaCl concentration, becoming completely absent at 100 mM, and short radicles emerged (Pandey et al., 2015).

45.3.3 SALT STRESS AND PRIMARY PRODUCTS

Oil content and fatty acid synthesis are considered to be important for the salt stress tolerance of plants. It is generally known that they are influenced by salt stress, which modifies fatty acid composition. In *Ricinus communis*, salt stress significantly reduced oil yield in roots, but it was enhanced in shoot tissues (Elfeky and Abbas, 2008). It was also reported that the total fatty acid content of *Coriandrum sativum* leaves decreased significantly due to salinity, and the content of α-linolenic and linoleic acids was reduced (Neffati and Marzouk, 2008).

Carbohydrates are also affected by saline conditions. A significant reduction was observed in fennel plants under salinity stress (Abd El-Wahab, 2006). Several explanations could be presented, including nutritional imbalance, hyperosmotic stress, and reduced photosynthesis (Nou et al., 1995). There are also reports demonstrating enhancement in carbohydrate content under salinity stress in *Salvia officinalis* (Hendawy and Khalid, 2005) and *Satureja hortensis* (Najafi et al., 2010). This may be related to the important role of carbohydrates in turgor maintenance in plants, especially under saline conditions.

Protein content in *Catharanthus roseus* was reduced as a result of salt stress (Osman et al., 2007). An increase in free amino acid content in plants under saline conditions was also reported, which is due to the degradation of intact proteins (Roychoudhury et al., 2011). This implies that salt stress induces protein aggregation within cells and reduces the soluble protein fraction (Ali et al., 2007). Depressed synthesis of crude proteins under salinity stress was also observed in *Achillea fragrantissima* (Abd EL-Azim and Ahmed, 2009). Stimulation of protein synthesis correlated with the degree of salinization has also been observed. Such a spike in protein accumulation might be to provide nitrogen in a stored form that can be reused during the post-stress recovery phase in plants (Roychoudhury and Chakraborty, 2013).

45.4 HEAVY METAL STRESS, A TWO-SIDED ISSUE IN MEDICINAL PLANTS

Unprecedented bioaccumulation and biomagnification of heavy metals in the environment have become a dilemma for all living organisms, including plants. There are two kinds of metals found in soils, which are referred to as *essential micronutrients* for normal plant growth (such as Fe, Mn, and Zn) and *nonessential elements* with unknown biological and physiological functions (such as Cd, Sb, and Pb) (Rascio and Navari-Izzo, 2011). The essential elements play a pivotal role in the structure of enzymes and proteins. Plants require them in tiny quantities for their growth, metabolism, and development; however, the concentration of both essential and nonessential metals is the single most important factor in the growing process of plants, so that their presence in excess can lead to the reduction and inhibition of growth in plants (Zengin and Munzuroglu, 2005). At the toxic level, heavy metals could harm plant functional processes, hindering the functional groups of important cellular molecules, superseding or disrupting the functionality of essential metals in biomolecules, such as pigments or enzymes, and adversely affecting the integrity of the cytoplasmic membrane, resulting in the repression of vital events in plants, such as photosynthesis, respiration, and enzymatic activities (Farid et al., 2013).

On the other hand, considering medicinal and aromatic plants, the environmental impact of toxic metal pollution and the accompanying health effects remain an area of great concern. Although trace amounts of some heavy metals are beneficial to human health, their presence beyond certain thresholds tends to be injurious, causing acute or chronic poisoning (Qelik and Oehlenschlager, 2007). Plants have the capacity to accumulate heavy metals in concentrations much higher than their concentrations in the environment. This ability to bioaccumulate and hence, bioconcentrate metals in plants increases their potential harmful effects (Rao and KumarMeena, 2011). When consumed beyond a certain threshold, lead (Pb) can increase blood pressure, accompanied by debilitating effects on key organs such as the kidney and the brain. Cadmium (Cd) poisoning is associated with a number of respiratory disorders, renal failure, and cardiovascular problems. Even though it is an essential mineral, an overdose of zinc (Zn) can cause fever, nausea, and general weakness. Although iron deficiency causes anemia, too much iron is particularly dangerous in children and could cause gastrointestinal and skin problems (Baby et al., 2010).

One continuing problem in protecting consumers of plant-based medicines is that permissible levels of all heavy metals in herbal medicine have not yet been standardized by governmental regulatory entities. Moreover, there are few existing limit tests for heavy metal contents of medicinal plants or permissible limits for essential dietary minerals in most medicinal plants. The dearth of such limits hamstrings the development of medicinal plant research and delays the release of either new or improved versions of medicinal plants or their components. There are immense discrepancies between countries regarding regulatory requirements to ensure the safety and quality of plant-based products. Several regulations have already been established worldwide for medicinal plants, such as the US Pharmacopoeia (USP), Italian Pharmacopoeia (FUI), and European Pharmacopoeia (Ph. Eur.). Moreover, there are legal frameworks at national and regional levels that are designed to regulate the quality of plant-based products (Sarma et al., 2012).

Although trace amounts of some heavy metals are beneficial to human health, their presence beyond certain thresholds tends to be injurious, causing acute or chronic poisoning. Plants have the capacity to accumulate heavy metals in concentrations much higher than their concentrations in the environment. This ability to bioaccumulate and hence, bioconcentrate metals in plants increases their potential harmful effects (Rao and KumarMeena, 2011). When consumed beyond a certain threshold, lead (Pb) can increase blood pressure, accompanied by debilitating effects to key organs such as the kidney and the brain. Cadmium (Cd) poisoning is associated with a number of respiratory disorders, renal failure, and cardiovascular problems. Even though it is an essential mineral, an overdose of zinc (Zn) can cause fever, nausea, and general weakness. Although iron deficiency causes anemia, too much iron is particularly dangerous in children and could cause gastrointestinal and skin problems (Baby et al., 2010).

45.4.1 Effects on Growth and Productivity

The growth and development of medicinal plants can be adversely affected by high concentrations of heavy metals. A significant reduction in plant biomass, root and shoot length, and the content of proteins, sugars, chlorophyll, and carotenoids in *Phyllanthus amarus* was observed due to increased levels of chromium (Cr) (Rai and Mehrotra, 2008). A significant decrease in plant growth and activity of photosystem II associated with an enhancement in proline content was observed in Indian mustard (*Brassica juncea*) exposed to Cd^{2+} (Sharmila et al., 2016). In another study, *Monochoria hastata* plants exposed to Cd showed several visible toxicity symptoms, such as withering, chlorosis, and falling of leaves (Baruah et al., 2017). When *Allium sativum* L. was exposed to different levels of Cd, a substantial reduction in growth was observed (Jiang et al., 2001). The relative toxicity of heavy metals may depend on the soil properties, the plant species, the age of the plant, and the heavy metal species (Shaw and Rout, 2002).

Depending on the heavy metal concentration and also the medicinal plant species, some plants may become acclimatized to the heavy metal stress and gradually resume normal growth. For instance, in *Catharanthus roseus* seedlings exposed to high levels of $CdCl_2$ and $PbCl_2$, senescence of the lower leaves and extensive chlorosis were observed. However, the affected plants gradually became acclimatized, and the chlorophyll content was almost comparable with that in plants grown under normal conditions (Pandey et al., 2007).

Heavy metal uptake by medicinal plants may interfere with the uptake of macronutrients such as nitrogen, phosphorus, and potassium and so cause deficiency of these main nutrient elements. During a study of the effect of Cd on *Withania somnifera*, NPK deficiency was observed at higher doses of Cd along with stunting of growth, chlorosis, and necrosis (Mishra et al, 2014).

It is also interesting that some heavy metals can moderate the toxic effects of another heavy metal. For example, mustard (*Brassica juncea*) plants exposed to elevated levels of Cd exhibited reduced biomass, pigment content, and relative water content (RWC). However, supplementation of selenium (Se) neutralized the negative effects of Cd and increased biomass, pigment content, and RWC (Ahmad et al., 2016).

45.4.2 Effects on Secondary Metabolite Production

Heavy metal contamination may affect the essential oil chemical composition of medicinal plants and change the quality of the natural plant products. Plants exposed to heavy metal stress show differential responses in the synthesis and accumulation of secondary metabolites. These changes vary from negative effects in a few species (Pandey et al., 2007) to stimulatory effects that result in enhanced metabolite production in other species (Eman et al., 2007). Heavy metals are mobile within plants, and because of this mobility, they may reduce the biosynthesis of active constituents in different plant components. Such effects may result from loss or inactivation of specific essential enzymes or from damage to nonessential biosynthetic processes, such as those involved in production of secondary metabolites. Ultimately, heavy metals may reduce the synthesis and accumulation of key bioactive plant molecules (Pandey et al., 2007). For example, *Hypericum perforatum* seedlings grown in a medium supplemented with 25 or 50 mM Ni lost the capacity to produce or accumulate hyperforin and demonstrated a 15/20-fold decrease in the concentration of pseudohypericin and hypericin (Murch et al., 2003). Increases in heavy metal–induced secondary metabolite biosynthesis have been reported to occur in some medicinal plant species. Induction of phenolic compound biosynthesis, in response to Ni, Al, and Cu toxicity, has been noted in *Phyllanthus tenellus* (Michalak, 2006). An increase in phenolic levels correlated with increased enzyme activity was associated with phenolic compound metabolism, suggesting supplementary synthesis of phenolics under heavy metal stress.

Medicinal plants may lose their capability to synthesize active components under high concentrations of heavy metals. Active components of certain medicinal plants that are tolerant to the heavy metal stress could remain unchanged. After 10 days' exposure of 4-week-old chamomile (*Matricaria chamomilla* L.) plants to low (3 μM) and high (60 and 120 μM) Cd concentrations, it was observed that the concentration of herniarin was not affected by Cd, while its precursors (Z)- and (E)-2-β-d-glucopyranosyloxy-4-methoxycinnamic acids (GMCAs) increased significantly at all the Cd levels tested. Cadmium had no significant effect on umbelliferone, a stress-related metabolite of chamomile (Kovacik et al., 2006). The chemical composition of essential oil of *Mentha crispa* was also affected by high concentrations of Pb, and the concentration of its major component (carvone) varied from 39.3% for plants grown in non-contaminated soil (control) to 90% for all cultivations in Pb-contaminated soils (Sa et al., 2015).

In a greenhouse study, *O. basilicum* L. and *M. spicata* L. were planted under different concentrations of Pb, Cu, and Cd The essential oil composition of aerial parts was analyzed by gas chromatography (GC) and gas chromatography-mass spectrometry (GC-MS). There were significant quantitative variations in the essential oil of *O. basilicum* grown in all heavy metal–amended soils. In the case of *M. spicata*, however, no significant changes were observed in essential oil composition for plants treated with heavy metals (Kunwar et al., 2015). This is a typical example where various plant species have differential responses to metals, highlighting the necessity to choose the right medicinal plant when dealing with a specific metal contamination in soil.

45.4.3 INTERACTION OF HEAVY METALS AND PATHOGENS

It has recently been suggested that plants absorb high concentrations of metals from the substrate as a self-defense mechanism against pathogens and herbivores. This metal defense hypothesis is among the most attractive proposals for the "reason to be" of metal hyperaccumulator species. On a molecular basis, metal defense against biotic stress seems to imply common and/or complementary pathways of signal perception, signal transduction, and metabolism. There is much less information on the influence of excess heavy metals on plant–pathogen relationships. The availability of high levels of potentially toxic metal ions can have a positive or negative impact, or no impact, on biotic stress development. Plant protection by metals requires one or several of the following conditions: the metal is more toxic to the pathogen or herbivore than to the plant; the metal hampers the virulence of the pathogen or herbivore; and/or the metal increases the resistance of the plant to the biotic stress factor (Poschenrieder et al., 2006).

According to the elemental defense hypothesis, metal accumulation within plant tissues can contribute to self-defense against biotic stress. The hypothesis was initially formulated based on an observation that fewer insects feed on Ni hyperaccumulators (Martens and Boyd, 1994). Further investigations have also reported some cases where high levels of Ni, Zn, Cd, or Se have provided effective protection against fungi or even snails and viruses. A recent determination of toxicity thresholds of different metals in the diamondback moth fed with artificial diets containing metal salts has revealed that Cd, Mn, Ni, Pb, and Zn can be toxic to this folivore even at concentrations below accumulator levels (Coleman et al., 2005). However, extending the elemental defense hypothesis to non-hyperaccumulators should be considered with care, because metals given as inorganic salts in an artificial diet can be much more toxic than metals bound by organic ligands inside plant tissues.

A metal-based defense might avoid the need for energy-demanding organic defenses. Relatively few studies on the influence of metal hyperaccumulation and organic plant defenses against pathogens or herbivores are available. One could assume that because of the presence of an elemental defense mechanism, organic defenses could be lowered in leaves but not in roots exposed to pathogens living in metal-rich soil, because they would have evolved high levels of metal tolerance (Ghaderian, et al., 2000). Lower levels of glucosinolate have been found in the leaves of *Thlaspi caerulescens* hyperaccumulating Zn, whereas increasing the supply of Zn enhanced glucosinolate levels (Tolra et al., 2001). High concentrations of phenolic compounds, histidine, or organic acids seem to be constitutive traits of metal hyperaccumulators (Davis et al., 2001; Kramer, 2005). Among these strong metal chelators, only phenolics have an established role in plant defense against biotic stress. The costs derived from metal chelation and compartmentation, and the role of secondary metabolites in plant development, make it difficult to determine quantitative relationships between growth and defense benefits derived from metal hyperaccumulation (Matyssek et al., 2005).

45.5 TEMPERATURE STRESS

Temperature stress may lead to a number of physiological, biochemical, and molecular changes in plant metabolism, such as protein denaturation or perturbation of membrane integrity. Many of these changes can alter the secondary metabolite concentrations in the plant tissues, which are often used as an indicator of stress injury in the plant (Zobayed et al., 2005). High-temperature (35 °C) treatment increased the leaf total peroxidase activity together with an increase in hypericin, pseudohypericin, and hyperforin concentrations in the shoot tissues of St. John's Wort (Zobayed et al., 2005). Also, an exponential increase in a variety of volatile organic compounds, with a linear increase in temperature, has been described in a range of plant species (Sharkey and Loreto, 1993; Sharkey and Yeh, 2001). Cold stress has been shown to stimulate an increase in phenolic production and subsequent incorporation into the cell wall. In particular, levels of anthocyanins increase following cold stress and are thought to protect plants against this effect (Pennycooke et al., 2005). Ncube et al. (2011) regarded the high levels of total phenolic compounds obtained during the winter season in their study as being consistent with this fact, which supports similar findings from previous studies (Pennycooke et al., 2005).

45.6 BIOTIC STRESSES

Plants are the targets of continuous attempted attacks by biotic stress factors such as fungi, viruses, and herbivores. Nonetheless, disease is the exception rather than the rule because of non-host and host resistance. To prosper, the attack must overcome the diverse defense strategies that plants deploy against the invader (Abdul Jaleel et al., 2008). This requires the interaction of a susceptible host and a virulent pathogen.

A third indispensable factor to complete the so-called *disease triangle* is a conducive environment. Climate, soil properties, competition, and human activity are among the most relevant environmental factors determining disease intensity.

On average, each kind of crop plant can be affected by a hundred or more plant diseases. Some pathogens affect only one variety of a plant. Other pathogens affect several dozen or even hundreds of species of plants. Plant diseases are grouped according to the cause of the disease; on this basis, plant diseases are classified as infectious plant diseases (caused by fungi, prokaryotes, parasitic higher plants and green algae, viruses and viroids, nematodes, and protozoa) and noninfectious plant diseases caused by too low or too high a temperature, lack or excess of soil moisture, lack or excess of light, lack of oxygen, air pollution, nutrient deficiencies, soil acidity or alkalinity (pH), etc.

The defense mechanisms of plants include many interactions between the pathogen and the plant, which leads to the production of secondary metabolites as a response. These produced compounds are usually classified into three main groups: phytoalexins, phytoanticipins, and signaling molecules. The phytoanticipins are produced before the attack of pathogens against the host plant. The phytoalexins are low–molecular weight molecules, which accumulate after the exposure of the plant to the pathogen (Cheynier et al., 2013).

Among these secondary metabolites, phenolic compounds play a major role in the chemical defense system of plants under biotic stresses (Kim et al., 2008). They are also known for their participation in transduction pathways and signal transport.

In fact, when the plant is attacked by a pathogen, a complicated immune system recognizes signals from injured cells and responds by activating effective immune responses through a signaling cascade and interactions with phytohormones such as salicylic acid. Nevertheless, the complexity of the plant–microbe interaction can represent different adaptive mechanisms in plants (Zhi-lin et al., 2007).

During the fungal attack, the plant epidermal cells tend to resist penetration by forming papillae that strengthen the cell wall, attempting to stop penetration by the microorganism and also accumulating fungitoxic substances such as the phenolic compounds and ROS (Zeyen et al., 2002).

Another way of avoiding attack by organisms is by the increase of specific enzymes, such as polyphenol oxidase (PPO). PPO activity increases in plants that are wounded, infected by pathogens, or attacked by other types of pests (Vanitha et al., 2009).

A wide range of substances, such as alkaloids, are synthesized in response to the plant defense against microorganism attacks. The alkaloids are important secondary compounds involved in the chemical defense in many plants. For example, compounds such as trigonelline, castanospermine, and camptothecin are increased due to reactions with arbuscular mycorrhizal fungus inoculum (Jia et al., 2016).

Modification in the content and composition of the essential oil is one of the consequences observed in plants attacked by herbivores. Studies show that terpenes, as an important secondary metabolite, are related to the plant defense system against biotic stress. In fact, they show antifungal, antimicrobial, and antiherbivore properties (Bekele and Hassanali, 2001). The conjugation of compounds to minimize the stressor effect of the environment caused by microorganism attack is also described in many research works. For example, we can cite compounds such as phenolamides (also named hydroxycinnamic acid amides), which are combinations of phenolic portions, such as coumaric and ferulic acid, with polyamines or aromatic deaminated amino acids. These compounds have specific functions in the plant's development and defense responses, either as an intermediate product or as the final product (Bassard et al., 2010). When there is a biotic stress, e.g., caused by a fungus, these substances are synthesized in the plant cell wall, increasing the plant's resistance to degradation by the hydrolytic enzymes produced by the pathogen. Additionally, these compounds show antimicrobial activity, acting directly on the pathogen or disease (Bassard et al.,2010; Mikkelsen et al., 2015).

45.7 WEEDS AND MEDICINAL PLANTS

Weeds are generally recognized to be responsible for a reduction in harvested plant quality, but apart from this very general statement, limited attention has been devoted to evaluate in detail the effects of competing weeds on the various plant traits that combine to form plant quality. The requirement for removing weeds in the cultivation of medicinal plants is an almost established principle. Moreover, the absence of weeds from medicinal plant products is obligatory for their high quality to be declared, irrespective of whether they come from cultivation or wild collecting.

In such conditions, weed management is a major constraint. As with all commonly grown crops, also in medicinal and aromatic plants, weeds function as crop competitors, create problems for mechanized harvest, and may alter the end quality when mixed with the harvested product. The well-known interference of weeds has additional relevance for medicinal and aromatic plants for several reasons. First, the synthesis of secondary metabolites in plants is linked to many genetic and environmental factors (Sangwan et al., 2001). Buyers often grade such plants according to their specific quality features, which are primarily determined by their content of essential oils or other secondary metabolites, which in their turn, can be reduced in the presence of weeds (Carrubba and Catalano, 2009). Although few specific experiments have been conducted, the presence of weeds is therefore believed to exert a significant effect on plant metabolic pathways, in this way acting negatively on the market end value of such crops.

The following are the various types of losses caused by weeds to medicinal plants:

1. Weeds cause nutrient losses by competition for essential nutrients.
2. Weeds reduce the yield-contributing character by creating space competition above and below the ground surface.

3. Weeds cause loss of soil moisture.
4. Weeds reduce the quality of medicinal plants due to adulteration/mixing of weeds during post-harvest processing. The root of kans grass (*Cyperus rotundus*) also reduces the quality of root crops such as *Asparagus*, *Chlorophytum*, etc.
5. Weeds drastically reduce the yield of medicinal plants as well as farm income.
6. A high intensity of weeds leads to slow germination and initial growth, wider row spacing, and slow lateral spread, which causes tremendous loss in productivity as well as quality (Upadhyay et al., 2011).

Under the different growth conditions of crops, seed yield decreases were found to range from 34% in garden cress to more than 90% in coriander, fennel, and psyllium (Shehzad et al., 2011). Also, yield losses in biomass and herbage yields were high: they ranged from about 30% in cornmint to 80–90% and more in coriander, fennel, and psyllium (Singh and Saini, 2008). These figures do not vary when the harvestable part is a fraction of total plant biomass: yield losses have been calculated from 51% to 64% in flowers of saffron (Norouzzadeh et al., 2007), more than 20% in bulbs of tassel hyacinth (Bonasia et al., 2012), 49–55% in leaves of rose-scented geranium, and 75% in shoots of sage (Kothari et al., 2002).

In some aspects, the outcome of crop–weed interaction is similar to that assessed for inter-species competition, and the effect of weed removal is similar to that obtained by reducing the plant population: weeding allows bigger plants to be obtained, with a higher number of ramifications and consequently, more flowers and seeds. In *Nigella sativa*, weed control increased yield and yield components such as number of plants, number of branches, number of capsules, number of seeds per capsule, seed weight, and total biomass (Nadeem et al., 2013; Seyyedi et al., 2016); in *Lepidium sativum*, the removal of weeds resulted in higher plants, number of branches, and number of seeds per plant (Shehzad et al., 2011).

Under severe competition, plants are said to grow taller and thinner (Ballaré and Casal, 2000). An increase in height when grown in the presence of weeds is typically observed in medicinal plants. This increase in height, however, not being associated with an increase in size, in most cases does not determine any increase in biomass yield.

Weeds affect the quality of medicinal plants through different aspects. First and foremost, the botanical identity of plants always matters to the consumers, since there is a correlation between the chemical characteristics and the botanical uniformity of medicinal plants (Zhang et al., 2010). Hence, this negative effect needs to be overcome by minimizing the amount of weeds in the whole biomass of the collected plant (Upadhyay et al., 2011). Another issue is the negative effects of the presence of weeds during the growth and production of medicinal plants.

It has been shown that weeds negatively change the amount and chemical composition of the essential oil of medicinal plants. Available reports related to this issue need to be carefully interpreted. The essential oil content was reported to decrease by 20–28.6% in leaves of unweeded rose-scented geranium (Kothari et al., 2002) and coriander (Pouryousef et al., 2015), but conversely, an increase in essential oil was found in basil (Sarrou et al., 2016) and in fatty oil from milk thistle (Zheljazkov et al., 2006). When the oil yield per unit area is considered, a decrease is generally found, although probably, this is due to the strong influence of the decrease of biomass yield. The quantitative variations of essential oil yields, however, are not the only issue, and apart from them, the occurrence of weeds may also alter the chemical essential oil profile of herbs (Rajeswara Rao et al., 2007). The simplest case is when weeds are mixed with the plant material to be distilled: although weeds do not yield any essential oil, their presence may alter the quality features of crop essential oil, conferring on it an unwanted off-flavor (Rajeswara Rao et al., 2007). Considerable change in chemical composition of essential oils as a consequence of weeds has been well documented. In basil, the highest yield of linalool (63.88%) and consequently, of the oxygenated monoterpenes (75.05%) was achieved by weeding treatment (Sarrou et al., 2016). In milk thistle, a decrease in silymarin content was reported as a consequence of weed presence (Zheljazkov et al., 2006). It is also necessary to note that there are some other reports claiming that no considerable difference was observed when comparing the cultivation of medicinal plants with or without control of the weeds.

A wide range of methods, including chemical, mechanical, and agronomical techniques, are applied to obtain a favorable result in weed management in medicinal plants. Considering that the details of these methods are not in the spotlight of this chapter, the author is not focusing on them. But in summary, if the phytotoxic effects of chemical compounds can be avoided, the use of some of them is shown to have no significant effect on seed or essential oil yields or essential oil quality. However, interest in chemical weed control in medicinal and aromatic plants is low because, increasingly, their production involves cultivation using organic methods, which, according to international regulations, prohibit the use of chemicals. For example, when different nonchemical methods for weed management were applied in coriander, fennel, and psyllium cultivation, the data obtained showed high sensitivity of these plants to the presence of weeds. As a consequence, seed yield was 40–90% lower than in the untreated plots (Carrubba and Militello, 2013).

Integrated weed management is a weed population management system that uses all the suitable techniques in a compatible manner to reduce the weed population and maintain weeds at levels below those causing economic injury. No single method is able to control weeds in a satisfactory manner, which creates the problem of resistance in weeds. Use of the same control methods leads to a buildup of tolerant weeds. It is uneconomical to eradicate all weeds, because some may provide food and shelter to insect predators and predators that reduce other pests. Therefore, if we apply all the control measures in integrated form, then weeds can be effectively minimized and controlled.

45.8 GENE EXPRESSION PATTERNS

The quality and quantity of secondary metabolites of medicinal plants strongly depend on environmental conditions. Studies on the effects of abiotic stress on medicinal plants through gene expression analyses are gaining importance, since this knowledge will allow the production of secondary metabolites to be enhanced.

A broad range of variations is effective during the process of production of secondary metabolites in medicinal plants. Gene expression analysis could shed some light on the mechanism of interaction between secondary metabolite production and environmental stresses.

The differential expression, i.e., up- and downregulation, of the genes responsible for increasing the metabolites, has also been analyzed, and the influence of abiotic stresses as signals of secondary metabolism has been reviewed (Ramakrishna and Ravishankar, 2011). Several reports are available on the expression analysis of genes responsible for the development of abiotic stress tolerance in medicinal plants. The biosynthesis of terpenoids in *Cistus creticus* has been reported (Pateraki and Kanellis, 2010). The trichomes of leaves of *C. creticus* excrete a resin rich in several labdane-type diterpenes with verified *in vitro* and *in vivo* cytotoxic and cytostatic activities against human cancer cell lines. It has been demonstrated that the expression pattern of genes related to terpenoid biosynthesis depends on tissues as well as developmental stages. One of the important medicinal plants, *Withania somnifera*, has much therapeutic potential, including sedative, narcotic, thyroid stimulation, anti-inflammatory, hypnotic, anti-stress, general tonic, diuretic, antimicrobial, and anti-tumor activities (Aslam et al., 2017). In *W. somnifera*, the expression analysis of different categories of genes, such as osmoregulation (Δ1-pyrroline-5-carboxylase synthase [Δ1P5CS]), detoxification (glutathione S-transferase [GST] and superoxide dismutase [SOD]), signal transduction (serine/threonine kinase [STK] and phosphoserine phosphatase [PSP]), metabolism (alcohol dehydrogenase [AD] and lactate dehydrogenase [LD]), and transcription regulation (heat shock protein [HSP], MYB, and WRKY) was considered to show their enhanced responses under drought stress condition (Sanchita et al., 2015). In *Rehmannia glutinosa*, the expression response of RghBNG gene has been studied in the presence of different abiotic stresses. The results suggest its role in growth, development, and response to plant growth regulators and abiotic stresses (Zhou et al., 2015). In *Artemisia annua*, on exposure to chilling, heat shock, or ultraviolet (UV) light, the transcription levels of amorpha-4,11-diene synthase (ADS) and cytochrome P450 monooxygenase (CYP71AV1) genes were upregulated when compared with control plants (Yin et al., 2008). The genes related to chilling and freezing tolerance in *Hippophae rhamnoides* L., a Himalayan medicinal plant, have been analyzed by the deep serial analysis of gene expression (SAGE) technique (Chaudhary and Sharma, 2015). It is reported that this study may provide important gene resources to be exploited for the development of stress-tolerant crop plants in the future. The AP2/ERF transcription factor (TF), isolated from *Papaver*

somniferum producing benzylisoquinoline alkaloids, showed overexpression in response to wounding, ethylene, methyl jasmonate, and ABA treatments (Mishra et al., 2015). Another study on *P. somniferum* was carried out to analyze the WRKY family of TFs, which is responsible for the increased accumulation of narcotine and papavarine in response to wounding in the walls of green pods prior to the exudation of latex (Mishra et al., 2013). The study of expression profiling of the same gene against multiple abiotic stresses is important. These multiple stresses affect the gene simultaneously for different time periods of stress. Studies on the effects of multiple stresses are generally limited to crop plants only because of the interest in the development of stress-tolerant varieties.

45.9 PROTEOMIC ANALYSIS AND PRODUCTION OF SECONDARY METABOLITES

Proteomics provides a promising approach for studying secondary metabolism in plants and plant cells. Unfortunately, the natural yield of secondary metabolites in medicinal plants is generally low, and the biochemistry of the biosynthesis of these compounds is complicated and poorly understood. Cell suspension cultures and metabolic engineering are among strategies to increase the yield of these commercially important chemicals. Overexpression of rate-limiting enzymes that are involved in the biosynthesis of these compounds has been used for this purpose (Verpoorte et al.,1999). Therefore, it is necessary to identify proteins involved in the biosynthesis of secondary metabolites. Enzyme isolation and characterization by current approaches is time consuming and troublesome; thus, the proteomic approach is faster and more complete, and by using this technology, we are able to identify regulatory and transport proteins as well as enzymes.

The medicinal plant *Catharanthus roseus* has been used as the best-studied model system for secondary metabolite production (Verpoorte et al., 1997). This plant produces some effective anti-tumor drugs, vinblastine and vincristine, which are alkaloid compounds. However, alkaloid yields in suspension cell cultures are generally too low to allow commercialization. Jacobs et al. (2000) have used two-dimensional gel electrophoresis for proteomic investigations of alkaloid production in *C. roseus*. The influence of zeatin and 2,4-dichlorophenoxyaceticacid (2,4-D) on protein patterns and alkaloid accumulation of *C. roseus* in a proteomic approach showed that proteins that were decreased by 2,4-D and increased by zeatin may have a direct function as an enzyme or an indirect role as a regulatory or transport protein in alkaloid biosynthesis. A 28 kDa polypeptide that was increased by zeatin showed a close correlation with alkaloid production (Jacobs et al., 2000).

45.10 REACTIVE OXYGEN SPECIES AND SECONDARY METABOLITES

Depending on their duration and severity, environmental stresses will inevitably result in oxidative damage due to the overproduction of ROS, also called active oxygen species (AOS) or reactive oxygen intermediates (ROI), which are the result of the partial reduction of atmospheric O_2. There are basically four forms of cellular ROS, singlet oxygen, superoxide radical, hydrogen peroxide, and the hydroxyl radical, each with an oxidizing potential. ROS can be extremely reactive, especially singlet oxygen and the hydroxyl radical, and unlike atmospheric oxygen, they can oxidize multiple cellular components such as proteins, lipids, DNA, and RNA. Unrestricted oxidation of the cellular components will ultimately cause cell death (Mittler, 2002).

ROS seem to have a dual effect under stress conditions depending on their overall cellular amount. If kept at relatively low levels, they are likely to function as components of a stress signaling pathway, triggering stress defense/acclimation responses (Vranova et al, 2002). However, when they reach a certain level of phytotoxicity, ROS become extremely deleterious, initiating uncontrolled oxidative cascades that damage cell membranes and other cellular components, resulting in oxidative stress and eventually, cell death (Dat et al, 2000).

Antioxidants provide protection for living organisms from damage caused by uncontrolled production of ROS. To protect cells against oxidative damage by free radicals, an antioxidant system, including SOD, catalase, glutathione peroxidase, and glutathione reductase enzymes, has evolved in aerobic organisms (Majumdar et al., 2010). Under conditions of elevated ROS production, or when the antioxidant system is compromised, cells are unable to scavenge the free radicals efficiently, leading to ROS accumulation.

It has been demonstrated that many naturally occurring compounds possess notable activity as radical scavengers and lipid peroxidation inhibitors. In addition to plant extracts, numerous naturally occurring compounds are useful as antioxidants, ranging from α-tocopherol and β-carotene to plant antioxidants such as phenolic compounds, alkaloids, and organic sulfur compounds (Mohamed et al., 2006). A large number of experiments have been carried out concerning the antioxidant activity of several plant extracts. The results of these experiments reveal that the activity is due to several secondary metabolites, especially phenolic compounds.

Therefore, the production and accumulation of secondary metabolites and antioxidants have a dual role in medicinal plants. They are able to protect cellular structures from damage by oxidative stress. On the other hand, the accumulation of these compounds is more highlighted in medicinal plants, since they are mostly used for their medical and therapeutic effects. This becomes more prominent when it is noted that antioxidant supplements such as butylated hydroxyanisole (BHA) and butylated hydroxytoluene (BHT) have been suspected of being responsible for liver damage and carcinogenesis (Ashokkumar et al., 2008).

45.11 STRESS COMBINATIONS

It has been known for a long time that often it is the simultaneous occurrence of several abiotic stresses, rather than a particular stress condition, that is most lethal to plants. Surprisingly, the co-occurrence of different stresses is rarely addressed by molecular biologists who study plant acclimation. Recent studies have revealed that the response of plants to a combination of two different abiotic stresses is unique and cannot be directly extrapolated from the response of plants to each of the different stresses applied individually (Mittler, 2006).

It was highlighted that when two or more factors co-occur, their effects are sometimes additive, while in other cases, the influence of one factor has priority (Gouinguene and Turlings, 2002). For instance, higher alkaloid levels were recorded for *Achnatherum inebrians* plants cultivated under salt and drought stress, with levels of ergonovine being higher than those of ergine (Zhang et al., 2011). The concentrations of both alkaloids increased over the life-span of the plant growth period. The influence of multiple stress factors commonly experienced by plants in field environments is often interactive, implying that the combined effect of various stresses is more diverse. Knowledge of how multiple stresses affect secondary metabolite accumulation in plants will provide more information to evaluate the biological roles of these metabolites in mitigating stress, provide criteria for describing their optimum yields and quality, and hence, provide a means of ensuring quality in phytomedicine.

45.12 CONCLUSIONS

Generally, stress causes a reduction in the quality and quantity of yield in agricultural crops. But in the case of medicinal and aromatic plants, the issue is somewhat controversial, and stress has been found to enhance both qualitative and quantitative yield. In this chapter, different aspects of the effects of environmental stresses on medicinal plants were analyzed. Quality, which is a very substantial part of medicinal plants' properties, is, indeed, grown out in the field. It must, however, be pointed out that plant responses to environmental cues are species specific. A multidisciplinary approach to this theme, combining ecology, biochemistry, and molecular physiology, would have great potential to advance this field and unravel the extent to which plant–environment interactions contribute to phytomedicine. The timing of harvesting and/or postharvest handling of plant material also has an influence on the quality of secondary metabolites in medicinal plant extracts. To sum up, the interactions of environmental stresses and medicinal plants need to be interpreted in a careful way. The advantages and positive effects of stress conditions under precise management could be applied in the cultivation and production of medicinal plants.

BIBLIOGRAPHY

Abd EL-Azim, W.M., and S.T. Ahmed. 2009. Effect of salinity and cutting date on growth and chemical constituents of *Achillea fragratissima* Forssk, under Ras Sudr conditions. *Res J Agr Biol Sci* 5:1121–1129.

Abd El-Wahab, M.A. 2006. The efficiency of using saline and freshwater irrigation as alternating methods of irrigation on the productivity of *Foeniculum vulgare* Mill subsp. vulgare var. vulgare under North Sinai conditions. *Res J Agr Biol Sci* 2(6):571–577.

Abdul Jaleel, C., P. Manivannan, G.M.A. Lakshmanan, M. Gomathinayagam, and M. Panneerselvam. 2008. Alterations in morphological parameters and photosynthetic pigment responses of *Catharanthus roseus* under soil water deficits. *Colloids Surf B Biointerf* 61:298–303.

Ahmad, P., E.A. Allah, A. Hashem, M. Sarwat, and S. Gucel. 2016. Exogenous application of selenium mitigates cadmium toxicity in *Brassica juncea* L. (Czern & Cross) by up-regulating antioxidative system and secondary metabolites. *J Plant Growth Regul* 35(4):936–950.

Ali, R.M., H.M. Abbas, and R.K. Kamal. 2007. The effects of treatment with polyamines on dry matter, oil and flavonoid contents in salinity stressed chamomile and sweet marjoram. *Plant Soil Environ* 53:529–543.

Ali, R.M., S.S. Elfeky, and H. Abbas. 2008. Response of salt stressed *Ricinus communis* L. to exogenous application of glycerol and/or aspartic acid. *J Biol Sci* 8(1):171–175.

Ashokkumar, D., V. Thamilselvan, G.P. Senthilkumar, U.K. Mazumder, and M. Gupta. 2008. Antioxidant and free radical scavenging effects of *Lippia nodiflora*. *Pharm Biol* 46:762–771.

Ashraf, M., and M.R. Foolad. 2007. Roles of glycine betaine and proline in improving plant abiotic stress resistance. *Environ Exp Bot* 59:206–216.

Aslam, S., N.I. Raja, M. Hussain, et al. 2017. Current status of *Withania somnifera* (L.) Dunal: an endangered medicinal plant from Himalaya. *Am J Plant Sci* 8:1159–1169.

Baby, J., J.S. Raj, and E.T. Biby. 2010. Toxic effect of heavy metals on aquatic environment. *Int J Biol Chem Sci* 4:939–952.

Baher, Z., M. Mirza, M. Ghorbanli, and M.B. Rezaii. 2002. The influence of water stress on plant height, herbal and essential oil yield and composition in *Satureja hortensis* L. *Flavour Fragr J* 17(4):275–277.

Ballaré, C.L., and J.J. Casal. 2000. Light signals perceived by crop and weed plants. *Field Crops Res* 67:149–160.

Ballhorn, D.J., S. Kautz, M. Jensen, S. Schmitt, M. Heil, and A.D. Hegeman. 2011. Genetic and environmental interactions determine plant defenses against herbivores. *J Ecol* 99:313–326.

Baruah, S., M.S. Bora, P. Sharma, P. Deb, and K.P. Sarma. 2017. Understanding of the distribution, translocation, bioaccumulation and ultrastructural changes of *Monochoria hastate* plant exposed to cadmium. *Water Air Soil Pollut* 228(1):17.

Bassard, J.E., P. Ullmann, F. Bernier, and D. Werck-Reichhart. 2010. Phenolamides: bridging polyamines to the phenolic metabolism. *Phytochemistry* 71:1808–1824.

Bekele, J., and A. Hassanali. 2001. Blend effects in the toxicity of the essential oil constituents of *Ocimum kilimandscharicum* and *Ocimum kenyense* (*Labiateae*) on two post-harvest insect pests. *Phytochemistry* 57:385–391.

Belaqziz, R., A. Romane, and A. Abbad. 2009. Salt stress effects on germination, growth and essential oil content of an endemic thyme species in Morocco (*Thymus maroccanus* Ball.). *J Appl Sci Res* 5:858–863.

Bistgani, Z., A. Siadat, A. Bakhshandeh, A. Ghasemi Pirbalouti, and M. Hashemi. 2017. Interactive effects of drought stress and chitosan application on physiological characteristics and essential oil yield of *Thymus daenensis* Celak. *Crop J* 5:407–415.

Bonasia, A., G. Conversa, C. Lazzizera, P. La Rotonda, and A. Elia. 2012. Weed control in lampascione- *Muscari comosum* (L.) Mill. *Crop Prot* 36:65–72.

Carrubba, A., and C. Catalano. 2009. Essential oil crops for sustainable agriculture—a review. In *Sustainable Agriculture Reviews: Climate Change, Intercropping, Pest Control and Beneficial Microorganisms, Sustainable Agriculture Reviews 2*. ed. Lichtfouse, E. Springer, Berlin, 137–188.

Chaudhary, S., and P.C Sharma. 2015. DeepSAGE based differential gene expression analysis under cold and freeze stress in sea-buckthorn (*Hippophae rhamnoides* L.). *PLoS One* 10:e0121982.

Cheynier, V., G. Comte, K.M. Davies, et al. 2013. Plant phenolics: recent advances on their biosynthesis, genetics, and ecophysiology. *Plant Physiol Biochem* 72:1–20.

Claussen, W. 2005. Proline as a measure of stress in tomato plants. *Plant Sci* 168:241–248.

Coleman, C.M., R.S. Boyd, and M.D. Eubanks 2005. Extending the metal defense hypothesis: dietary metal concentrations below hyperaccumulator levels could harm herbivores. *J Chem Ecol* 31: 1669–1681.

Cucci, G., G. Lacolla, F. Boari, and V. Cantore. 2014. Yield response of fennel (*Foeniculum vulgare* Mill.) to irrigation with saline water. *Acta Agr Scand B Soil Plant Sci* 64(2):129–134.

Dat, J., S. Vandenabeele, E. Vranová, M. Van Montagu, D. Inzé, and F. Van Breusegem. 2000. Dual action of the active oxygen species during plant stress responses. *Cel Molec Life Sci* 57:779–795.

Davis, M.A., S.G. Pritchard, R.S. Boyd, and S.A. Prior. 2001. Developmental and induced responses of nickel-based and organic defences of the nickel-hyperaccumulating shrub, *Psychotria douarrei*. *New Phytol* 150:49–58.

De Abreu, I.N., and P. Mazzafera. 2005. Effect of water and temperature stress on the content of active constituents of *Hypericum brasiliense* Choisy. *Plant Physiol Biochem* 43:241–248.

Eman, A., N. Gad, and N.M. Badran. 2007. Effect of cobalt and nickel on plant growth, yield and flavonoids content of *Hibiscus sabdariffa* L. *Aust J Basic Appl Sci* 1:73–78.

Farid, M., M.B. Shakoor, A. Ehsan, S. Ali, M. Zubair, and M.S. Hanif. 2013. Morphological, physiological and biochemical responses of different plant species to Cd stress. *Int J Chem Biochem Sci* 3:53–60.

Fougère, F., D. Le Rudulier, and J.G. Streeter. 1991. Effects of salt stress on amino acid, organic acid, and carbohydrate composition of roots, bacteroids, and cytosol of alfalfa (*Medicago sativa* L.). *Plant Physiol* 96:1228–1236.

Gengmao, Z., S. Quanmei, H. Yu, L. Shihui, and W. Changhai. 2014. The physiological and biochemical responses of a medicinal plant (*Salvia miltiorrhiza* L.) to stress caused by various concentrations of NaCl. *PLoS One* 9(2):e89624.

Ghaderian, Y.S.M., A.J.E. Lyon, and A.J.M. Baker. 2000. Seedling mortality of metal hyperaccumulator plants resulting from damping off by *Pythium* spp. *New Phytol* 146:219–224.

Gouinguene, S.P., and T.C.J. Turlings. 2002. The effects of abiotic factors on induced volatile emissions in corn plants. *Plant Physiol* 129:1296–1307.

Gurib-Fakim, A. 2006. Medicinal plants: traditions of yesterday and drugs of tomorrow. *Mol Aspects Med* 27:1–93.

Hamayun, M. 2007. Traditional uses of some medicinal plants of Swat Valley, Pakistan. *Ind J Trad Know* 6(4):636–641.

Hartmann, T. 2007. From waste products to ecochemicals: fifty years research of plant secondary metabolism. *Phytochemistry* 68:2831–2846.

Heidari, M., and S. Sarani. 2012. Growth, biochemical components and ion content of Chamomile (*Matricaria chamomilla* L.) under salinity stress and iron deficiency. *J Saudi Soc Agric Sci* 11(1):37–42.

Hendawy, S.F., and K.A. Khalid. 2005. Response of sage (*Salvia officinalis* L.) plants to zinc application under different salinity levels. *J Appl Sci Res* 1:147–155.

Hong, Z., K. Lakkineni, Z. Zhang, and D. Verma. 2000. Removal of feedback inhibition of $\Delta 1$-pyrroline-5-carboxylate synthetase results in increased proline accumulation and protection of plants from osmotic stress. *Plant Physiol* 122:1129–1136.

Jaafar, H.Z., M.H. Ibrahim, and N.F. Mohamad Fakri. 2012. Impact of soil field water capacity on secondary metabolites, phenylalanine ammonialyase (PAL), maliondialdehyde (MDA) and photosynthetic responses of Malaysian kacip fatimah (*Labisia pumila* Benth). *Molecules* 17(6):7305–7322.

Jacobs, D.I., R. VanderHeijden, and R. Verpoorte. 2000. Proteomics in plant biotechnology and secondary metabolism research. *Phytochem Anal* 11:277–287.

Jia, M., L. Chen, H.L. Xin, et al. 2016. A friendly relationship between endophytic fungi and medicinal plants: a systematic review. *Front Microbiol* 7:1–14.

Jiang, W., D. Liu, and W. Hou. 2001. Hyperaccumulation of cadmium by roots, bulbs and shoots of garlic (*Allium sativum* L.). *Bioresour Technol* 76(1):9–13.

Katembe, W.J., I.A. Ungar, and J.P. Mitchell. 1998. Effect of salinity on germination and seedling growth of two Atriplex species (Chenopodiaceae). *Ann Bot* 82:167–175.

Kim, J.H., B.W. Lee, F.C. Schroeder, and G. Jander. 2008. Identification of indole glucosinolate breakdown products with antifeedant effects on *Myzus persicae* (green peach aphid). *Plant J* 54:1015–1026.

Kosova, K., P. Vitamva, I.T. Prasil, and J. Renaut. 2011. Plant proteome changes under abiotic stress contribution of proteomics studies to understanding plant stress response. *J Proteom* 74:1301–1322.

Kothari, S.K., C.P. Singh, and K. Singh. 2002. Weed control in rose-scented geranium (*Pelargonium* spp). *Pest Manage Sci* 58:1254–1258.

Kováčik, J., J. Tomko, M. Bačkor, and M. Repčák. 2006. *Matricaria chamomilla* is not a hyperaccumulator, but tolerant to cadmium stress. *Plant Growth Regul* 50(2–3):239–247.

Kramer, U. 2005. Phytoremediation: novel approaches to cleaning uppolluted soils. *Curr Opin Biotechnol* 16:133–141.

Kunwar, G., C. Pande, G. Tewari, C. Singh, and G.C. Kharkwal. 2015. Effect of heavy metals on terpenoid composition of *Ocimum basilicum* L. and *Mentha spicata* L. *J Essent Oil Bear Pl* 18(4):818–825.

Liu, H., X. Wang, D. Wang, Z. Zou, and Z. Lianga. 2011. Effect of drought stress on growth and accumulation of active constituents in *Salvia miltiorrhiza* Bunge. *Ind Crop Prod* 33:84–88.

Majumdar, S., S. Bhattacharya, and P.K. Haldar. 2010. Comparative in vitro free radical scavenging activity of some indigenous plants. *Int J Pharm Tech Res* 2:1046–1049.

Mandoulakani, B.A., E. Eyvazpour, and M. Ghadimzadeh. 2017. The effect of drought stress on the expression of key genes involved in the biosynthesis of phenylpropanoids and essential oil components in basil (*Ocimum basilicum* L.). *Phytochemistry* 139:1–7.

Manukyan, A. 2011. Effect of growing factors on productivity and quality of lemon catmint, lemon balm and sage under soilless greenhouse production: I. Drought stress. *Med Aromat Plant Sci Biotechnol* 5(2):119–125.

Martens, S.N., and R.S. Boyd. 1994. The ecological significance of nickel hyperaccumulation – a plant-chemical defense. *Oecologia* 98:379–384.

Matthews, S., E. Beltrami, R. El-Khadem, et al. 2011. Evidence that time for repair during early germination leads to vigour differences in maize. *Seed Sci Technol* 39:501–509.

Matyssek, R., R.Agerer, and D. Ernst. 2005. The plant's capacity in regulating resource demand. *Plant Biol* 7:560–580.

Miceli, A., A. Moncada, and F. D'Anna. 2003. Effect of water salinity on seeds-germination of *Ocimum basilicum* L., *Eruca sativa* L. and *Petroselinum hortense* Hoffm. *Acta Hortic* 609:365–370.

Michalak, A. 2006. Phenolic compounds and their antioxidant activity in plants growing under heavy metal stress. *Pol J Environ Stud* 15:523–530.

Mishra, S., V. Triptahi, and S. Singh. 2013. Wound induced transcriptional regulation of benzylisoquinoline pathway and characterization of wound inducible PsWRKY transcription factor from *Papaver somniferum*. *PLoS One* 8, e52784.

Mishra, B., R.S. Sangwan, S. Mishra, J.S. Jadaun, F. Sabir, and N.S. Sangwan. 2014. Effect of cadmium stress on inductive enzymatic and nonenzymatic responses of ROS and sugar metabolism in multiple shoot cultures of Ashwagandha (*Withania somnifera* Dunal). *Protoplasma* 251(5):1031–1045.

Mishra, S., U.J. Phukan, V. Tripathi, D.K. Singh, S. Luqman, and R.K. Shukla. 2015. PsAP2 an AP2/ERF family transcription factor from *Papaver somniferum* enhances abiotic and biotic stress tolerance in transgenic tobacco. *Plant Mol Biol* 89:173–186.

Mittler, R. 2002. Oxidative stress, antioxidants and stress tolerance. *Trends Plant Sci* 7:405–410.

Mittler, R. 2006. Abiotic stress, the field environment and stress combination. *Trends Plant Sci* 11(1):9–15.

Mohamed, S.A., F.A. Marzouk, M.A. Moharram, A.M. Mohamed, and E.A. Gamal Eldeen. 2006. Anticancer and antioxidant tannins from *Pimenta dioica* leaves. *Z Natur Forsch* 62:526–536.

Muhammad, Z., and F. Hussain. 2010. Vegetative growth performance of five medicinal plants under NaCl salt stress. *Pak J Bot* 42(1):303–316.

Murch, S.J., K. Haq, H.P.V. Rupasinghe, and P.K. Saxena. 2003. Nickel contamination affects growth and secondary metabolite composition of St. John's wort (*Hypericum perforatum* L.). *Environ Exp Bot* 49:251–257.

Nadeem, M.A., A. Tanveer, T. Naqqash, A.J. Jhala, and K. Mubeen. 2013. Determining critical weed competition periods for black seed. *J Anim Plant Sci* 23(1):216–221.

Niu, X., R.A. Bressan, P.M. Hasegawa, and J.M. Pardo. 1995. Ion homeostasis in NaCl stress environments. *Plant Physiol* 109:735–742.

Norouzzadeh, S., M. Abbaspoor, and M. Delghandi. 2007. Chemical weed control in saffron fields of Iran. *Acta Hort* 739:119–122.

Naderi, S., B. Fakheri, and S. Esmailzadeh Bahabadi. 2014. Increasing of chavicol Omethyl transfrase gene expression and catalase and ascorbate peroxidase enzymes activity of *Ocimum basilicum* by chitosan (in Persian). Crop Biotech 6:1–9.

Najafi, F., R.A. Khavari-Nejad, and M.S. Ali. 2010. The effects of salt stress on certain physiological parameters in summer savory (*Satureja hortensis* L.) plants. *J Stress Physiol Biochem* 6(1):13–21.

Namdeo, A.G. 2007. Plant cell elicitation for production of secondary metabolites: a review. *Pharmacogn Rev* 1:69–79.

Ncube, B., J.F. Finnie, and J. Van Staden. 2011. Seasonal variation in antimicrobial and phytochemical properties of frequently used medicinal bulbous plants from South Africa. *S Afr J Bot* 77:387–396.

Neffati, M., and B. Marzouk. 2008. Changes in essential oil and fatty acid composition in coriander (*Coriandrum sativum* L.) leaves under saline conditions. *Ind Crops Prod* 28:137–142.

Nogués, S, D.J. Allen, J.I.L Morison, and N.R. Baker. 1998. Ultraviolet-B radiation effects on water relations, leaf development, and photosynthesis in droughted pea plants. *Plant Physiol* 117:173–181.

Nou, X., R.A. Bressan, P.M. Hasegawa, and J.P. Pardo. 1995. Iron homeostasis in NaCl stress environments. *Plant Physiol* 109:735–742.

Nowak, M., R. Manderscheid, H.J. Weigel, M. Kleinwächter, and D. Selmar. 2010. Drought stress increases the accumulation of monoterpenes in sage (*Salvia officinalis*), an effect that is compensated by elevated carbon dioxide concentration. *J Appl Bot Food Qual* 83:133–136.

Osman, M.E.H., S.S. Elfeky, K. Abo El-Soud, and A.M. Hasan. 2007. Response of *Catharanthus roseus* shoots to salinity and drought in relation to vincristine alkaloid content. *Asian J Plant Sci* 6:1223–1228.

Pandey, S., K. Gupta, and A.K. Mukherjee. 2007. Impact of cadmium and lead on *Catharanthus roseus* – a phytoremediation study. *J Environ Biol* 28:655–662.

Pandey, S., M.K. Patel, A. Mishra, and B. Jha. 2015. Physio-biochemical composition and untargeted metabolomics of cumin (*Cuminum cyminum* L.) make it promising functional food and help in mitigating salinity stress. *PLoS One* 10(12):e0144469.

Pateraki, I., and A.K. Kanellis. 2010. Stress and developmental responses of terpenoid biosynthetic genes in *Cistus creticus* subsp creticus. *Plant Cell Rep* 29:629–641.

Pennycooke, J.C., S. Cox, and C. Stushnoff. 2005. Relationship of cold acclimation, total phenolic content and antioxidant capacity with chilling tolerance in petunia (Petunia×hybrida). *Environ Exp Bot* 53:225–232.

Petropoulos, S.A., D. Daferera, M.G. Polissiou, and H.C. Passam. 2008. The effect of water deficit stress on the growth, yield and composition of essential oils of parsley. *Sci Hortic Amsterdam* 115:393–397.

Petrusa, L.M., and I. Winicov. 1997. Proline status in salt tolerant and salt sensitive alfalfa cell lines and plants in response to NaCl. *Plant Physiol Biochem* 35:303–310.

Pirzad, A., M.R. Shakiba, S. Zehtab-Salmasi, A. Mohammadi, R. Darvishzadeh, and A. Samadi. 2011. Effect of water stress on leaf relative water content, chlorophyll, proline and soluble carbohydrates in *Matricaria chamomilla*. *J Med Plant Res* 5(12):2483–2488.

Pouryousef, M., A.R. Yousefi, M. Oveisi, and F. Asadi. 2015. Intercropping of fenugreek as living mulch at different densities for weed suppression in coriander. *Crop Prot* 69:60–64.

Qelik, U., and J. Oehlenschlager. 2007. High contents of cadmium, lead, zinc and copper in popular fishery products sold in Turkish supermarkets. *Food Control* 18:258–261.

Rajeswara Rao, B.R., D.K. Rajput, K.P. Sastry, A.K. Bhattacharya, R.P. Patel, and S. Ramesh. 2007. Effect of crop-weed mixed distillation on essential oil yield and composition of five aromatic crops. *J Essent Oil Bearing Plants* 10(2):127–132.

Ramakrishna, A., and G.A. Ravishankar. 2011. Influence of abiotic stress signals on secondary metabolites in plants. *Plant Signal Behav* 6:1720–1731.

Ramin, A.A. 2005. Effects of salinity and temperature on germination and seedling establishment of sweet basil (*Ocimum basilicum* L.). *J Herbs Spices Med Plants* 11:81–90.

Rao, M.M., and A.G. Kumar-Meena. 2011. Detection of toxic heavy metals and pesticide residue in herbal plants which are commonly used in the herbal formulations. *Environ Monit Assess* 181:267–271.

Rascio, N., and F. Navari-Izzo. 2011. Heavy metal hyperaccumulating plants: how and why do they do it? And what makes them so interesting? *Plant Sci* 180(2):169–181.

Roychoudhury, A., S. Basu, and D.N. Sengupta. 2011. Amelioration of salinity stress by exogenously applied spermidine or spermine in three varieties of indica rice differing in their level of salt tolerance. *J Plant Physiol* 168:317–328.

Roychoudhury, A., and M. Chakraborty. 2013. Biochemical and molecular basis of varietal difference in plant salt tolerance. *Annu Rev Res Biol* 3:422–454.

Sá, R.A., R.A. Sá, O. Alberto, et al. 2015. Phytoaccumulation and effect of lead on yield and chemical composition of *Mentha crispa* essential oil. *Desalin. Water Treat.* 53(11):3007–3017.

Sabir, F., RS. Sangwan, R. Kumar, and N.S. Sangwan. 2012. Salt stress-induced responses in growth and metabolism in callus cultures and differentiating in vitro shoots of Indian ginseng (*W. somnifera* Dunal). *J Plant Growth Regul* 31:537–548.

Salehi, A., H. Tasdighi, and M. Gholamhoseini. 2016. Evaluation of proline, chlorophyll, soluble sugar content and uptake of nutrients in the German chamomile (*Matricaria chamomilla* L.) under drought stress and organic fertilizer treatments. *Asian Pac J Trop Biomed* 6(10):886–891.

Sanada, Y., H. Ueda, K. Kuribayashi, et al. 1995. Novel light-dark change of proline levels in halophyte (*Mesembryanthemum crystallinum* L.) and glycophytes (*Hordeum vulgare* L. and *Triticum aestivum* L.): Leaves and roots under salt stress. *Plant Cell Physiol* 36:965–970.

Said-Al Ahl, H.A.H., and E.A. Omer. 2011. Medicinal and aromatic plants production under salt stress. *Herbapol* 57:72–87.

Sangwan, N.S., A.H.A. Farooqi, F. Shabih, and R.S. Sangwan. 2001. Regulation of essential oil production in plants. *Plant Growth Regul* 34:3–21.

Sarma, H., S. Deka, H. Deka, and R.R. Saikia. 2012. Accumulation of heavy metals in selected medicinal plants. *Rev Environ Contam Toxicol* 214:63–86.

Sarrou, E., P. Chatzopoulou, T.V. Koutsos, and S. Katsiotis. 2016. Herbage yield and essential oil composition of sweet basil (*Ocimum basilicum* L.) under the influence of different mulching materials and fertilizers. *J Med Plants Stud* 4(1):111–117.

Selmar, D. 2008. Potential of salt and drought stress to increase pharmaceutical significant secondary compounds in plants. *Landbauforsch Volk* 58:139–144.

Selmar, D., and M. Kleinwachter. 2013. Stress enhances the synthesis of secondary plant products: the impact of stress-related over-reduction on the accumulation of natural products. *Plant Cell Physiol* 54(6):817–826.

Selmar, D., M. Kleinwächter, S. Abouzeid, M. Yahyazadeh, and M. Nowak. 2017. The impact of drought stress on the quality of spice and medicinal plants. In *Medicinal Plants and Environmental Challenges*, ed. Ghorbanpour, M. and A. Varma. Springer, Cham.

Serraj, R., and T.R. Sinclair. 2002. Osmolyte accumulation: can it really help increase crop yield under drought conditions? *Plant Cell Environ* 25:333–241.

Seyyedi, S.M., P. Rezvani Moghaddam, and M.N. Mahallati. 2016. Weed competition periods affect grain yield and nutrient uptake of Black Seed (*Nigella sativa* L.). *Hortic Plant J* 2(3):172–180.

Sharkey, T.D., and F. Loreto. 1993. Water-stress, temperature, and light effects on the capacity for isoprene emission and photosynthesis of Kudzu leaves. *Oecologia* 95:328–333.

Sharkey, T.D., and S.S. Yeh. 2001. Isoprene emission from plants. Annual review in plant physiology. *Plant Mol Biol* 52:407–436.

Sharma, D., C. Kumar Singh, D. Shankhdhar, and S.C. Shankhdhar. 2018. Evaluation of biochemical markers for managing drought stress tolerance in a valuable endangered medicinal plant *Withania somnifera*. *J Pharmacogn Phytochem* 7(1):203–206.

Sharmila, P., P.K. Kumari, K. Singh, N.V. Prasad, and P. Pardha-Saradhi. 2016. Cadmium toxicity-induced proline accumulation is coupled to iron depletion. *Protoplasma* 254(2):763–770.

Shehzad, M., A. Tanveer, M. Ayub, K. Mubeen, N. Sarwar, M. Ibrahim, and I. Qadir. 2011. Effect of weed-crop competition on growth and yield of garden cress (*Lepidium sativum* L.). *J Med Plants Res* 5(26):6169–6172.

Shinwari, M.I., and M.A. Khan. 2000. Folk use of medicinal herbs of Margalla Hills National Park, Islamabad. *J Ethnopharm* 69:45–56.

Singh, M.K., and S.S. Saini. 2008. Planting date, mulch, and herbicide rate effects on the growth, yield, and physicochemical properties of menthol mint (*Mentha arvensis*). *Weed Technol* 22(4):691–698.

Singh, R., A. Mishra, S.S. Dhawan, P.A. Shirke, M.M. Gupta, and A. Sharma. 2015. Physiological performance, secondary metabolite and expression profiling of genes associated with drought tolerance in *Withania somnifera*. *Protoplasma* 252:1439–1450.

Sosa, L., A. Lianes, H. Reinoso, M. Reginato, and V. Luna. 2005. Osmotic and specific ion effect on the germination of *Prosopis strombulifera*. *Ann Bot* 96:261–267.

Tolra, R., C. Poschenrieder, R. Alonso, D. Barceló, and J. Barceló. 2001. Influence of zinc hyperaccumulation onglucosinolates in *Thlaspi caerulescens*. *New Phytol* 151:621–626.

Upadhyay, R.K., H. Baksh, D.D. Patra, S.K. Tewari, S.K. Sharma, and R.S. Katiyar. 2011. Integrated weed management of medicinal plants in India. *Int J Med Aromat Plants* 1(2):51–56.

Vanitha, S.C., S.R. Niranjana, and S. Umesha. 2009. Role of phenylalanine ammonia lyase and polyphenol oxidase in host resistance to bacterial wilt of tomato. *J Phytopathol* 157:552–557.

Verpoorte, R., R. van der Heijden, and P.R.H. Moreno. 1997. Biosynthesis of terpenoid indole alkaloids in *Catharanthus roseus* cells. In *The Alkaloids*, ed. Cordell, G.A. Academic Press, San Diego, CA, 221–299.

Verpoorte, R., R. van der Heijden, H.J.G. tenHoopen, and J. Memelink. 1999. Metabolic engineering of plant secondary metabolite pathways for the production of chemicals. *Biotechnol Lett* 21:467–479.

Vranova, E., D. Inzé, and F. Van Breusegem. 2002. Signal transduction during oxidative stress. *J Exp Bot* 53:1227–1236.

Wink, M. 2010. Introduction: biochemistry, physiology and ecological functions of secondary metabolites. In *Biochemistry of Plant Secondary Metabolism*. ed. Wink, M. Wiley-Blackwell, 1–19.

Xue, J., S. Zhou, W. Wang, et al. 2018. Water availability effects on plant growth, seed yield, seed quality in *Cassia obtusifolia* L., a medicinal plant. *Agr Water Manage* 195:104–113.

Yin, L., C. Zhao, Y. Huang, R.Y. Yang, and Q.P. Zeng. 2008. Abiotic stress-induced expression of artemisinin biosynthesis genes in *Artemisia annua* L. *Chin J Appl Environ Biol* 14:1–5.

Zarei, H., B.A. Fakheri, S. Esmaeilzadeh, and M. Solouki. 2015. Increasing of chavicol Omethyl transferase gene expression (CVOMT) and methylchavicol value of basil (*Ocimum basilicum*) by salicylic acid. *J Bio Environ Sci* 6:46–53.

Zengin, F.K., and O. Munzuroglu. 2005. Effects of some heavy metals on content of chlorophyll, proline and some antioxidant chemicals in bean (*Phaseolus vulgaris* L.) seedlings. Acta Biol Crac Ser *Bot* 47(2):157–164.

Zeyen, R.J., W.M. Kruger, M.F. Lyngkjaer, and T.L.W. Carver. 2002. Differential effects of D-mannose and 2-deoxy-D-glucose on attempted powdery mildew fungal infection of inappropriate and appropriate Gramineae. *Physiol Mol Plant Pathol* 61:315–323.

Zhang, B., Y. Peng, Z. Zhang, et al. 2010. GAP Production of TCM Herbs in China. *Planta Med* 76:1948–1955.

Zhang, X.X., C.J. Li, and Z.B. Nan. 2011. Effects of salt and drought stress on alkaloid production in endophyte-infected drunken horse grass (*Achnatherum inebrians*). *Biochem Syst Ecol* 39: 471–476.

Zheljazkov, V.D. and N.K. Nedkov. 2006. Herbicides for weed control in blessed thistle (*Silybum marianum*). *Weed Technol* 20:1030–1034.

Zhi-lin, Y., D. Chuan-chao, and C. Lian-qing. 2007. Regulation and accumulation of secondary metabolites in plant-fungus symbiotic system. *Afr J Biotechnol* 6:1266–1271.

Zhou, Y., Y. Zhang, J. Wei, et al. 2015. Cloning and analysis of expression patterns and transcriptional regulation of RghBNG in response to plant growth regulators and abiotic stresses in *Rehmannia glutinosa*. *Springerplus* 4:60.

Zhu, J.K. 2016. Abiotic stress signaling and responses in plants. *Cell* 167(2):313–324.

Zobayed, S.M.A., F. Afreen, and T. Kozai. 2005. Temperature stress can alter the photosynthetic efficiency and secondary metabolite concentrations in St. John's Wort. *Plant Physiol Biochem* 43:977–984.

46 The Role of Beneficial Elements in Mitigation of Plant Osmotic Stress

Maryam Mozafariyan Meimandi, Noémi Kappel, and Mohammad Pessarakli

CONTENTS

46.1 INTRODUCTION

The growth and yield of plants are affected by multiple factors, such as biotic and abiotic stresses, i.e., salinity, water stress, high and low extreme temperatures, heavy metals, and nutrient stress, which may cause a reduction in crop productivity.

Salinity has affected about 33% of the irrigated agricultural lands in the world. Salinity stress influences seed germination and vegetative and reproductive growth stages; causes ion toxicity, osmotic stress, and nutrient imbalance in crops (Noreen et al., 2018); changes various physiological and metabolic processes, depending on the severity and duration of the stress; and ultimately, inhibits crop production (Gupta and Huang, 2014) due to the high osmotic potential of the soil solution, which results in a water deficit. The high concentration of Na^+ and Cl^- causes ion toxicity with the consequence of secondary stresses such as nutritional deficiency and oxidative stress (Yue et al., 2012).

Drought is another major stress that affects plant growth and production in arid and semi-arid regions worldwide. Drought stress may result from a real water deficit in the soil or from excessive salinity of the root zone. A variety of physiological, biochemical, and morphological characteristics are changed under drought stress. Water stress causes stomatal closure and increased respiration, resulting in reduced nutrient uptake.

The harmful effect of stress in plants could be ameliorated by beneficial elements. Crop fertilization with trace elements along with other nutrients can reduce stress in plants. This chapter provides a review of the role of the trace elements in mitigating osmotic stresses.

Seventeen elements are considered as plant essential nutrients. Carbon (C), hydrogen (H), and oxygen (O) are derived from air or water, and others are obtained from soil or nutrient solutions. Nutrient elements that are required in considerable quantities, known as *macronutrients*, include nitrogen (N), phosphorus (P), potassium (K), calcium (Ca), magnesium (Mg), and sulfur (S). Most of the nutrient elements are required in small quantities (*microelements*), including iron (Fe), zinc (Zn), copper (Cu), manganese (Mn), molybdenum (Mo), chlorine (Cl), boron (B), and nickel (Ni). Some elements are not recognized as essential nutrients for plants but have beneficial effects on plant growth and yield. Silicon (Si), cobalt (Co), selenium (Se), and sodium (Na) (Na only for halophytic plants) are notable beneficial elements.

46.2 PHYSIOLOGICAL FUNCTIONS OF BENEFICIAL ELEMENTS IN PLANTS

46.2.1 COBALT (CO)

Cobalt (Co), a transition element, is an essential micronutrient for animals as a component of several enzymes, co-enzymes, and vitamin B_{12} (Palit et al., 1994). It has similar chemical properties to nickel (Ni) and enters the plasma membrane through the same carriers (Kaur et al., 2016). The range of world mean concentrations of Co in surface soils is between 4.5 and 12 mg kg^{-1}, being highest for heavy loamy soils and lowest for light sandy and organic soils (Kabata-Pendias and Mukherjee, 2007), and it is present in plants at 0.1–10 ppm on a dry weight basis (Palit et al., 1994).

Cobalt in the nutrient or soil solution is very easily available to plants and is absorbed from the soil by passive transport. Its uptake is dependent on soil pH, and it is more available to plants as pH decreases. Cobalt, as a beneficial element, can improve plant growth and metabolism, depending on its concentration (Palit et al., 1994). The only physiological role of Co in plants is fixation of molecular nitrogen in the root nodules of leguminous plants (Bond and Hewitt, 1962).

46.2.2 SELENIUM (SE)

Selenium (Se) has properties similar to those of sulfur (Freeman et al., 2006) and is an essential element for humans (Hawrylak-Nowak, 2008; Lin et al., 2012), but it is not yet considered to be an essential element for higher plants (Terry et al., 2000). Typical Se contents in soil range between 0.2 and 5 mg kg^{-1}, with an average of around 0.4 mg kg^{-1} (Kopsell and Kopsell, 2006). Plants absorb Se as selenite (SeO_3^{2-}) and selenate (SeO_4^{2-}) at different rates (Christine et al., 2011). Selenium improves plant growth through the accumulation of starch in the chloroplast (Pennanen et al., 2002). Improved plant growth due to Se was reported in tomato (Mozafariyan et al., 2015) and cucumber (Haghighi et al., 2016). Se increases plant stress tolerance through enzymatic and non-enzymatic antioxidant defense mechanisms.

46.2.3 SILICON (SI)

Silicon (Si) is the second most abundant element on the surface of the earth and as a beneficial element, positively influences plant growth and yield. Si is accumulated in many plants at up to 10% on a dry weight basis (Epstein, 1994). Although Si is abundant in soil, most of it cannot be absorbed directly by plants. This element is available as silicic acid (H_4SiO_4), which moves through transpiration in the xylem and is slightly soluble. Temperature, soil pH, the presence of cations and organic compounds in solution, and water conditions affect silicon accumulation in plants through the formation of soluble silicic acid (Liu et al., 2003). In the early 1950s in Japan, silicate materials were commonly used as fertilizers. Ma and Takahashi (2002) fertilized rice with Si and found a slight increase in the panicle number and a 17% increase in yield.

Silicon can improve plant growth, root development, and hull formation in rice; speed up fruit maturation of citrus, crop quality, and yield; and also induce defense reactions to plant diseases and alleviate stress in plants (Snyder et al., 2007). Increased plant growth in strawberry due to enhanced tissue elasticity and symplastic water volume was observed by Emadian and Newton (1989). Miyake and Takahashi (1983) reported that Si increased the growth and yield of cucumber and decreased wilt disease damage.

46.3 BENEFICIAL ELEMENTS IN MITIGATION OF PLANT OSMOTIC STRESS

46.3.1 COBALT (CO)

Addition of cobalt (Co) to the growth media increased plant growth, nodulation, and leghemoglobin concentration as well as elevating catalase (CAT) activity in legume plants, causing an increase in stress tolerance (Marschner, 1995). In pea, the application of Co (8 mg) in soil increased growth, plant nutrient level, nodule number and weight, and seed pod yield and quality (Gad, 2006). Similarly, Co supplementation had beneficial effects on root growth, yield quality as starch, sugars, L-ascorbic acid, and contents of N, P, K, Mn, Zn, and Cu in sweet potato (Gad and Kandil, 2008).

Tosh et al. (1979) showed that Co supplementation postponed the senescence of lettuce plants by arresting the decline of chlorophyll, protein, RNA, and to a smaller extent, DNA. Beside et al. (1979) reported that Co application increased plant tolerance to drought stress. Moreover, Palit et al. (1994) reported that Co improved biotic stress resistance in medicinal plants by increased accumulation of alkaloids. Also, through inhibition of ethylene production, Co delayed leaf senescence in plants (Palit et al., 1994).

Gad (2005) showed that Co supplementation at a concentration of 7.5 mg L^{-1} under salt stress conditions increased tomato growth and plant nutrient levels, and had beneficial effects on some other physiological parameters. Cobalt application to the growth medium caused an increase in abscisic acid (ABA) level in shoots and roots of tomato (Gad, 2005), and ABA prevented stomata from reopening and induced proline accumulation, which resulted in increased plant resistance to salt stress (Gad, 2005). Li et al. (2005) showed that Co application inhibited ethylene production, alleviated antioxidant enzyme activities, and decreased polyamine content during osmotic stress in potato, protecting the plant under stress conditions.

46.3.2 SELENIUM (SE)

Some studies have confirmed that selenium (Se) at low concentrations can exert beneficial effects on plants, especially under stress conditions (Mozafariyan et al., 2016, 2017). Se contents in soil range between 0.2 and 5 mg kg^{-1}, with an average of around 0.4 mg kg^{-1} (Kopsell and Kopsell, 2006). Recently, several researchers have reported beneficial effects of Se on plants under abiotic stress; for instance, salinity (Mozafariyan et al., 2016), drought (Hajiboland et al., 2014), temperature stress (Hawrylak-Nowak et al., 2010), and heavy metal stress (Mozafariyan et al., 2014). Selenium can directly and indirectly influence the generation of reactive oxygen species (ROS) under stress conditions.

The major effects of Se toward abiotic stress tolerance are the improvement of photosynthesis, antioxidant defense, and higher uptake of water and nutrients (Hawrylak-Nowak et al., 2018). Beneficial effects of Se in different plants under salt stress have been reported by several researchers, i.e., sorrel (Kong et al., 2005), cucumber (Hawrylak-Nowak, 2009), rapeseed (Hasanuzzaman and Fujita, 2011), melon (KeLing et al., 2013), canola (Hashem et al., 2013), lettuce (Hawrylak-Nowak, 2015), tomato (Diao et al., 2014; Mozafariyan et al., 2016), and maize (Jiang et al., 2017).

The results of Hawrylak-Nowak (2009) in cucumber showed that Se improved the growth rate, photosynthetic pigments, and proline accumulation under NaCl stress conditions. In a recent study, Mozafariyan et al. (2016) performed a hydroponic experiment on tomato plants under salt stress;

they demonstrated that exogenous application of Se alleviated the adverse effects of salinity stress. They found that Se application at a concentration of 10 µM was more effective than at a concentration of 5 µM, especially under severe salinity stress, by improving plant water balance and the integrity of the cell membranes.

Selenium protected the cell membranes against lipid peroxidation, enhanced the stability of cucumber plants grown under salt stress conditions, and resulted in plant tolerance to stress conditions (Walaa et al., 2010). In rapeseed plants, exogenous Se applications (25 µM) enhanced antioxidant defense and the methylglyoxal detoxification system and caused tolerance to salt stress (Hasanuzzaman and Fujita, 2011). Exogenous selenite application at low concentration increased the activities of superoxide dismutase (SOD) and peroxidase (POX), improved the accumulation of water-soluble sugars, and stimulated the growth of sorrel under salt stress conditions (Kong et al., 2005).

Improving drought stress damage by Se is related to adjustment of the water status of plants (Hawrylak-Nowak et al., 2018) and increasing root length and diameter (Hajiboland et al., 2014). Kuznetsov et al. (2003) indicated that Se through root system, more than economical use of water in the transpiration, caused an increased water uptake capacity under water deficit conditions in spring wheat. Moreover, the antioxidant role of Se in plants caused significant tolerance in plants under stress conditions. Increasing the activity of some antioxidant enzymes (ascorbate peroxidase [APX], CAT, and glutathione peroxidase [GPX]) and inhibition of lipid peroxidation in olive trees by foliar application of Se improved plant drought tolerance (Proietti et al., 2013). In wheat, Nawaz et al. (2014) and Hajiboland et al. (2014) reported that Se application improved plant growth and increased drought resistance.

Hawrylak-Nowak et al. (2010) applied Se (2.5–20 µM) to cucumber under low-temperature stress and observed increased proline content in leaves and decreased lipid peroxidation in roots. In wheat, Chu et al. (2010) reported that Se supplementation caused a reduction in both ROS production and lipid peroxidation, while anthocyanins, flavonoids, phenolic compounds, and the activity of POX and CAT increased.

Foliar application of Se decreased membrane damage in sorghum by improving antioxidant defense under heat stress conditions (Djanaguiraman et al., 2010). Similarly, the activity of monodehydroascorbate reductase (MDHAR), dehydroascorbate reductase (DHAR), glutathione reductase (GR), GPX, CAT, Gly I, and Gly II was increased in rapeseed by Se applications under high-temperature stress conditions (Hasanuzzaman et al., 2014). A similar trend was reported by Iqbal et al. (2015), who found that CAT and APX activities and anthocyanin, carotenoid, and L-ascorbic acid contents increased in wheat under high-temperature stress and thereby, inhibited grain yield reduction.

Hawrylak Nowak et al. (2018) concluded that Se enhanced tolerance against abiotic stress through improving several physiological processes, including water uptake, nutrient uptake and balance, antioxidant defense, prevention of photosynthetic pigments, maintenance of stomatal conductance,

regulation of transpiration, regulation of osmolytes, upregulation of the glyoxalase system, reduction of methylglyoxal toxicity, and maintenance of cellular water content.

46.3.3   SILICON (SI)

Silicon (Si) can alleviate salt, metal toxicity, nutrient imbalance, drought, radiation, high-temperature, freezing, and UV stress during different growth stages. Most of these beneficial effects are attributed to Si deposition in the cell walls of roots, leaves, stems, and hulls (reviewed by Ma and Yamaji, 2006). The mechanisms of Si-mediated alleviation of drought and salinity stress are explained by Rizwan et al. (2015) as decreasing Na uptake and increasing K, modification of gas exchange attributes (increase in photosynthesis), increasing antioxidant defense (decrease in oxidative damage), and modification of osmolytes and phytohormones.

During seed germination, Si application improved seed germination in tomato under drought stress (Haghighi et al., 2013; Shi et al., 2014).

Si treatment positively affected photosynthetic and transpiration rate, dry weight, root properties, and water potential under drought stress conditions in rice and wheat (Chen et al., 2011; Gong et al., 2006, 2005; Nolla et al., 2012). Drought stress reduced total dry matter, relative water content, and chlorophyll content while increasing proline and electrolyte leakage in maize, and Si supplementation improved drought stress damage (Kaya et al., 2006). The authors reported that water stress decreased K and Ca concentrations in maize, and Si decreased these losses. In potato, Si treatment stimulated proline content and increased Si concentration in leaves under water deficit conditions (Crusciol et al., 2009). Also, decreasing total sugar, protein, and increasing tuber weight and yield were observed by Crusciol et al. (2009) under water stress.

Farooq et al. (2009) explained that transpiration rate and stomatal conductance are important characteristics that influence plant water relations. Mikiciuk and Mikiciuk (2009) reported that foliar application of Si improved photosynthesis and reduced water losses from transpiration in strawberry under deficit irrigation. Transpiration from the leaves occurs mainly through the stomata and partly through the cuticle. As Si is deposited beneath the cuticle of the leaves, forming a Si–cuticle double layer, the transpiration through the cuticle may decrease due to Si deposition (reviewed by Ma, 2004).

Si application increased salt tolerance in many crops, i.e., tomato (Haghighi and Pessarakli, 2013), zucchini (Savvas et al., 2009), and rice (Gong et al., 2006). During the seed germination stage, Si increased the germination percentage of tomato (Haghighi et al., 2012) and rice (Bybordi, 2015) under salt stress conditions.

Plants have complex antioxidant defense systems (enzymatic and non-enzymatic) to scavenge ROS. Antioxidant enzymes, including SOD, CAT, guaiacol peroxidase, APX, DHAR, and GR, among others, are influenced by Si applications under stress conditions. Al-Aghabary et al. (2004) reported that salinity caused an increase in H_2O_2 in cucumber leaves, and Si decreased the salt-induced production of H_2O_2 and improved photosynthesis rates. Si treatments influence SOD, POD, CAT, and GR activities under stress conditions. Increasing SOD activity and decreasing malondialdehyde (MDA) concentration in barley leaves under salt stress was reported by Liang (1999). Zhu et al. (2004) found that Si significantly decreased ELP, H_2O_2, and thiobarbituric acid reactive substances (TBARS) content and significantly enhanced the activities of SOD, GPX, APX, DHAR, and GR in salt-stressed cucumber leaves.

Salinity results in an increase in Na^+ and Cl^- accumulation and a decrease in certain other cations, such as K^+ and Ca^{2+}. Foliar application of Si decreased Na and Cl in okra (Abbas et al., 2015) and grapevine (Abdalla, 2011) under salt stress conditions. Improvement in the K^+/Na^+ ratio under salt stress conditions was reported by Li et al. (2015) in tomato, Khoshgoftarmanesh et al. (2014) in cucumber, and Xu et al. (2015) in aloe. Many researchers have shown that Si treatment increased the uptake and translocation of mineral elements in Egyptian clover (Abdalla, 2011), faba bean (Hellal et al., 2012), canola (Farshidi et al., 2012), and tomato (Li et al., 2015) and caused plant tolerance to stress conditions.

The beneficial effect of Si on photosynthetic pigment was observed in tomato (Al-Aghabary et al., 2004). In a study conducted by Romero-Aranda et al. (2006) in rice plants treated with Si and grown under salt stress conditions, Si increased the stomatal conductance. In tomato, Si improved the fresh and dry weights of plants, root volume, chlorophyll concentration, photosynthetic rate, mesophyll conductance, and plant water use efficiency under salt stress conditions (Haghighi and Pessarakli, 2013). Si application alleviated salt damage by protection of the photosynthetic apparatus through decreasing Na^+ uptake and increasing K^+ uptake in salt-stressed plants (reviewed by Zhu and Gong, 2014).

In addition, Si supplementation can enhance plant resistance to stress conditions by adjusting the levels of solutes and phytohormones, including proline, glycine betaine, and carbohydrates. Application of Si under salt stress conditions decreased proline content in soybean (Lee et al., 2010) and sorghum (Yin et al., , 2016).

46.4 CONCLUSIONS

Crop fertilization with trace elements (Si, Co, and Se) along with other nutrients can reduce stress in plants. Se and Si enhanced tolerance against abiotic stress through improving several physiological processes, including water uptake, nutrient uptake and balance, antioxidant defense, prevention of photosynthetic pigments, maintenance of stomatal conductance, and regulation of transpiration.

ACKNOWLEDGMENTS

This research was supported by the Higher Education Institutional Excellence Program (1783-3/2018/FEKUTSRAT) awarded by the Ministry of Human Capacities within the framework of water-related researches of Szent István University.

"This publication is prepared in number EFOP-3.6.1-16-2016-00016. The specialized of the SZIE Campus of Szarvas Research and Training Profile with intelligent specialization in the themes of water management, hydroculture, pecision mechanical engineering, alternative crop production."

BIBLIOGRAPHY

Abbas T., Balal R.M., Shahid M.A., Pervez M.A., Ayyub C.M., Aqueel M.A., Javaid M.M. (2015). Silicon-induced alleviation of NaCl toxicity in okra (*Abelmoschus esculentus*) is associated with enhanced photosynthesis, osmoprotectants and antioxidant metabolism. *Acta Physiol. Plant.* 37:1–15.

Abdalla M.M. (2011). Beneficial effects of diatomite on growth, the biochemical contents and polymorphic DNA in *Lupinus albus* plants grown under water stress. *Agric. Biol. J. North Am.* 2:207–220.

Al-Aghabary K., Zhu Z., Shi Q. (2004). Influence of silicon supply on chlorophyll content, chlorophyll fluorescence, and antioxidative enzyme activities in tomato plants under salt stress. *J. Plant Nutr.* 27(12):2101–2115.

Bond G., Hewitt E.J. (1962). Cobalt and the fixation of nitrogen by root nodules of Alnus and Casuarina. Nat. Lond. 195:94–95.

Bybordi A. (2015). Influence of exogenous application of silicon and potassium on physiological responses, yield, and yield components of salt-stressed wheat. *Commun. Soil Sci. Plant Anal.* 46(1):109–122.

Chen W., Yao X., Cai K., Chen J. (2011). Silicon alleviates drought stress of rice plants by improving plant water status, photosynthesis and mineral nutrient absorption. *Biol. Trace Elem. Res.* 142:67–76.

Christine A.Z., Messikommer R.E., Caspar W. (2002). Choice feeding of selenium-deficient laying hens affects diet selection, selenium intake and body weight. *J. Plant Nutr.* 132:3411–3417.

Chu J., Yao X., Zhang Z. (2010). Responses of wheat seedlings to exogenous selenium supply under cold stress. *Biol. Trace Elem. Res.* 136:355–363.

Crusciol, C.A.C., Pulz L.A., Lemos L.B., Soratto R.P., Lima G.P.P. (2009). Eff ects of silicon and drought stress on tuber yield and leaf biochemical characteristics in potato. *Crop Sci.* 49:949–954.

Diao M., Ma L., Wang J., Cui J., Fu A., Liu H. (2014) Selenium promotes the growth and photosynthesis of tomato seedlings under salt stress by enhancing chloroplast antioxidant defense system. J. Plant Growth Reg. 33:671–682.

Djanaguiraman M., Prasad P.V.V., Sëppanen M. (2010). Selenium protects sorghum leaves from oxidative damage under high temperature stress by enhancing antioxidant defense system. *Plant Physiol. Biochem.* 48:999–1007.

Emadian S.F., Newton R.J. (1989). Growth enhancement of loblolly pine (*Pinus taeda* L.) seedlings by silicon. *J. Plant Physiol.* 134(1):98–103.

Epstein E. (2001). Silicon in plants: Facts vs. concepts. *Stud. Plant Sci.* 8:1–15.

Farooq M., Wahid A., Kobayashi N., Fujita D., Basra S. M. A. (2009). Plant drought stress: Effects, mechanisms and management. *Agron. Sustain. Dev.* 29:185–212.

Farshidi M., Abdolzadeh A., Sadeghipour H.R. (2012). Silicon nutrition alleviates physiological disorders imposed by salinity in hydroponically grown canola (*Brassica napus* L.) plants. *Acta Physiol. Plant.* 34:1779–1788.

Freeman J.L., Quinn C.F., Marcus M.A., Fakra S., Pilon-Smits E.A.H. (2006). Selenium tolerant diamondback moth disarms hyperaccumulator plant defense. *Curr. Biol.* 16:2181–2192.

Gad N. (2005). Interactive effect of cobalt and salinity on tomato plants I- Growth and mineral composition as affected by cobalt and salinity. Res. J. Agric. Biol. Sci. Pak. 1(3):261–269.

Gad N. (2006). Increasing the efficiency of nitrogen fertilization through cobalt application to pea plant. *Res. J. Agric. Biol. Sci.* 2(6):433–442.

Gad N., Kandil H. (2008). Response of sweet potato (*Ipomoea batatas* L.) plants to different levels of cobalt. *Aust. J. Basic Appl. Sci.* 2(4):949–955.

Gong H.J., Randall D.P., Flowers T.J. (2006). Silicon deposition in the root reduces sodium uptake in rice (*Oryza sativa* L.) seedlings by reducing bypass flow. *Plant Cell Environ.* 29:1970–1979.

Gong H.J., Zhu X.Y., Chen K.M., Wang S.M., Zhang C.L. (2005). Silicon alleviates oxidative damage of wheat plants in pots under drought. *Plant Sci.* 169:313–321.

Gupta B., Huang B. 2014. Mechanism of salinity tolerance in plants: Physiological, biochemical, and molecular characterization. *Int. J. Genom.* 130:2129–2141.

Haghighi M., Afifipour Z., Mozafarian M. (2012). The effect of N–Si on tomato seed germination under salinity levels. *J. Biol. Environ. Sci.* 6:87–90.

Haghighi M., Pessarakli M. (2013). Influence of silicon and nano-silicon on salinity tolerance of cherry tomatoes (*Solanum lycopersicum* L.) at early growth stage. *Sci. Hort.* 161(24):111–117.

Haghighi M., Sheibanirad A., Pessarakli M. 2016. Effects of selenium as a beneficial element on growth and photosynthetic attributes of greenhouse cucumber. *J. Plant Nutr.* 39(10):1493–1498.

Haghighi M., Teixeira da Silva J.A., Mozafariyan M., Afifipour Z. (2013). Can Si and nano-Si alleviate the effect of drought stress induced by PEG in seed germination and seedling growth of tomato? *J. Minerva Biotechnol.* 25(1):17–22.

Hajiboland R., Sadeghzadeh N., Sadeghzadeh B. (2014). Effect of Se application on photosynthesis, osmolytes and water relations in two durum wheat (*Triticum durum* L.) genotypes under drought stress. *Acta Agric. Slov.* 103:167–179.

Hasanuzzaman M., Fujita M. (2011). Selenium pretreatment upregulates the antioxidant defense and methylglyoxal detoxification system and confers enhanced tolerance to drought stress in rapeseed seedlings. *Biol. Trace Elem. Res.* 143:1758–1776.

Hasanuzzaman M., Nahar K., Alam M.M., Fujita M. (2014). Modulation of antioxidant machinery and the methylglyoxal detoxification system in selenium-supplemented *Brassica napus* seedlings confers tolerance to high temperature stress. *Biol. Trace Elem. Res.* 161:297–307.

Hashem H.A., Hassanein R.A., Bekheta M.A., El-Kady F.A. (2013). Protective role of selenium in canola (*Brassica napus* L.) plant subjected to salt stress. *Egypt J. Exp. Biol. (Bot.)* 9:199–211.

Hawrylak-Nowak B. (2008). Effect of selenium on selected macronutrients in maize plants. *J. Biol. Trace Elem. Res.* 13:513–519.

Hawrylak-Nowak B. (2009). Beneficial effects of exogenous selenium in cucumber seedlings subjected to salt stress. *Biol. Trace Elem. Res.* 132:259–269.

Hawrylak-Nowak B., Hasanuzzaman M., Matraszek-Gawron R. (2018). Mechanisms of selenium-induced enhancement of abiotic stress tolerance in plants. In: *Plant Nutrients and Abiotic Stress Tolerance*, pp. 269–295.

Hawrylak-Nowak B., Matraszek R., Pogorzelec M. (2015). The dual effects of two inorganic selenium forms on the growth, selected physiological parameters and macronutrients accumulation in cucumber plants. *Acta Physiol. Plant* 37:41.

Hawrylak-Nowak B., Matraszek R., Szymańska M. (2010). Selenium modifies the effect of short-term chilling stress on cucumber plants. *Biol. Trace Elem. Res.* 138:307–315.

Hellal F.A., Abdelhameid M., Abo-Basha D.M., Zewainy R.M. (2012). Alleviation of the adverse effects of soil salinity stress by foliar application of silicon on faba bean (*Vicia faba* L.). *J. Appl. Sci. Res.* 8:4428–4433.

Jiang C., Zu C., Lu D., Zheng Q., Shen J., Wang H., Li D. (2017). Effect of exogenous selenium supply on photosynthesis, Na+ accumulation and antioxidative capacity of maize (*Zea mays* L.) under salinity stress. *Sci. Rep.* 7:42039.

Kabata-Pendias A., Mukherjee A.B. (2007). *Trace Elements in Soils and Plants.* Springer, Berlin.

Kaur S., Kaur N., Siddique K.H.M., Nayyar H. (2016). Beneficial elements for agricultural crops and their functional relevance in defence against stresses. *Arch. Agron. Soil Sci.* 62(7):905–920.

Kaya C., Tuna L., Higgs D. (2006). Effect of silicon on plant growth and mineral nutrition of maize grown under water-stress conditions. *J. Plant Nutr.* 29:1469–1480.

KeLing H., Ling Z., JiTao W., Yang Y. (2013). Influence of selenium on growth, lipid peroxidation and antioxidative enzyme activity in melon (*Cucumis melo* L.) seedlings under salt stress. *Acta Soc. Bot. Pol.* 82:193–197.

Khoshgoftarmanesh A.H., Khodarahmi S., Haghighi, M. (2014). Effect of silicon nutrition on lipid peroxidation and antioxidant response of cucumber plants exposed to salinity stress. *Arch. Agron. Soil Sci.* 60:639–653.

Kong L., Wang M., Bi D. (2005). Selenium modulates the activities of antioxidant enzymes, osmotic homeostasis and promotes the growth of sorrel seedlings under salt stress. Plant Growth Reg. 45:155–163.

Kopsell D.A., Kopsell D.E. (2007). Selenium. In: Barker AV, Pilbeam D (eds) *Handbook of Plant Nutrition.* CRC Press, Taylor & Francis Group, Boca Raton, pp 515–550.

Kuznetsov V.V., Kholodova V.P., Kuznetsov V.V., Yagodin B.A. (2003). Selenium regulates the water status of plants exposed to drought. *Dokl. Biol. Sci.* 390:266–268.

Lee S.K., Sohn E.Y., Hamayun M., Yoon J.Y., Lee I.J. (2010). Effect of silicon on growth and salinity stress of soybean plant grown under hydroponic system. *Agroforest. Syst.* 80:333–340.

Li H., Zhu Y., Hu Y., Han W., Gong H. (2015). Beneficial effects of silicon in alleviating salinity stress of tomato seedlings grown under sand culture. *Acta Physiol. Plant.* 37:1–9.

Liang, Y. (1999). Effects of silicon on enzyme activity and sodium, potassium and calcium concentration in barley under salt stress. *Plant Soil* 209:217.

Lin L., Zhou W.H., Dai H.X., Cao F.B., Zhang G.P., Wu F.B. (2012). Selenium reduces cadmium uptake and mitigates cadmium toxicity in rice. *J. Hazard. Mater.* 235:343–351.

Liu J.G., Li K.Q., Xu J.K., Liang J.S., Lu X.L., Yang J.C., Zhu Q.S. (2003). Interaction of Cd and five mineral nutrients for uptake and accumulation in different rice cultivars and genotypes. Field Crop Res. 83:271–281.

Ma J.F. (2004). Role of silicon in enhancing the resistance of plants to biotic and abiotic stresses. *Soil Sci. Plant Nutr.* 50(1):11–18.

Ma J.F., Takahashi E. (2002). *Soil, Fertilizer, and Plant Silicon Research in Japan.* Elsevier Science, Amsterdam.

Ma J.F., Yamaji N. (2006). Silicon uptake and accumulation in higher plants. *Trends Plant Sci.* 11(8):392–397. Epub 2006 Jul 12.

Marschner H. (1995). *Mineral Nutrition of Higher Plants*, 2nd ed. Academic Press, London, 889 p.

Mikiciuk G., Mikiciuk M. (2009). The influence of foliar application of potassium and silicon fertilizer on some physiological features of strawberry (*Fragaria ananassa* Duch.), variety Elvira. *Ann. UMCS Sect. E* 64(4):19–27.

Miyake Y., Takahashi E. (1983). Effect of silicon on the growth of solution-cultured cucumber plant. *Soil Sci. Plant Nutr.* 29(1):71–83.

Mozafariyan M., Hawrylak-Nowak B., Kamelmanesh M.M. (2016). Ameliorative effect of selenium on tomato plants grown under salinity stress. *Arch. Agron. Soil Sci.* 62:1368–1380.

Mozafariyan M., Pessarakli M., Saghafi K. (2017). Beneficial effect of selenium on chlorophyll content, relative water content and some morphological characters of tomato (*Lycopersicum esculentum*) under hydroponic conditions. *J. Plant Nutr.* 40:139–144.

Nawaz F., Ashraf M.Y., Ahmad R., Waraich E.A. (2013). Selenium (Se) seed priming induced growth and biochemical changes in wheat under water deficit conditions. *Biol. Trace Elem. Res.* 151:284–293.

Nolla A., de Faria R.J., Korndoerfer G.H., Benetoli da Silva T.R. (2012). Effect of silicon on drought tolerance of upland rice. *J. Food Agric. Environ.* 10:269–272.

Palit S., Sharma A., Talukder G. (1994). Effects of cobalt on plants. *Bot. Rev.* 60:149.

Pennanen A., Xue T., Hartikainen H. (2002). Protective role of selenium in plant subjected to severe UV irradiation stress. *J. Appl. Bot.* 76:66–76.

Proietti P., Nasini L., Del Buono D., D'Amato R., Tedeschini E., Businelli D. (2013). Selenium protects olive (*Olea europaea* L.) from drought stress. *Sci. Hort.* 164:165–171.

Rizwan M., Ali S., Ibrahim M., Farid M., Adrees M., Bharwana S.A., Zia-ur-Rehman M., Qayyum M.F., Abbas F. (2015). Mechanisms of silicon-mediated alleviation of drought and salt stress in plants: A review. *Environ. Sci. Pollut. Res. Int.* 22:15416–15431.

Romero-Aranda M.R., Jurado O., Cuartero J. (2006). Silicon alleviates the deleterious salt effect on tomato plant growth by improving plant water status. *J. Plant Physiol.* 163:847–855.

Savvas D., Karapanos I., Tagaris A., Passam H.-C. (2009). Effects of NaCl and silicon on the quality and storage ability of zucchini squash fruit. *J. Hort. Sci. Biotechnol.* 84(4):381–386.

Shi Y., Zhang Y., Yao H., Wu J., Sun H., Gong H. (2014). Silicon improves seed germination and alleviates oxidative stress of bud seedlings in tomato under water deficit stress. *Plant Physiol. Biochem.* 78:27–36.

Snyder G.H., Martichenkov V.V., Datnoff L.E. (2007). Silicone. In: Barker AV, Pilbean DJ (eds) *Handbook of Plant Nutrition.* CRC Press, Taylor and Francis, New York, pp. 551–568.

Tarabrin V.P., Teteneva T.R. 1979. Presowing treatment of seeds and its effect on the resistance of wood plant seedlings against drought. *Sov. J. Ecol.* 10:204–211.

Terry N., Zayed A.M., De Souza M.P., Tarun A.S. (2000). Selenium in higher plants. *Annu. Rev. Plant Physiol. Plant Mol. Biol.* 51:401–432.

Tosh S., Choudhuri M.A., Chatterjee S.K. (1979). Retardation of lettuce leaf senescence by cobalt ions. *Indian J. Exp. Biol.* 17:1134–1136.

Walaa A.E., Shatlah M.A., Atteia M.H., Sror H.A.M. 2010. Selenium induces antioxidant defensive enzymes and promotes tolerance against salinity stress in cucumber seedlings (*Cucumis sativus*). *Arab. Univ. J. Agric. Sci.* 18:65–76.

Xu C.X., Ma Y.P., Liu Y.L. (2015). Effects of silicon (Si) on growth, quality and ionic homeostasis of aloe under salt stress. *S. Afr. J. Bot.* 98:26–36.

Yin L., Wang S., Tanaka K., Fujihara S., Itai A., Den X., Zhang S. (2016). Silicon-mediated changes in polyamines participate in silicon-induced salt tolerance in *Sorghum bicolor* L. *Plant Cell Environ.* 39:245–258.

Yue Y.,. Zhang M., Zhang J., Duan L., Li Z. (2012). *SOS1* gene overexpression increased salt tolerance in transgenic tobacco by maintaining a higher K+/Na+ ratio. *J. Plant Physiol.* 169(3):255–261.

Zahra N., Mahmood S., Raza Z.A. (2018). Salinity stress on various physiological and biochemical attributes of two distinct maize (*Zea mays* L.) genotypes. *J. Plant Nutr.* 41(11):1368–1380.

Zhu Y., Gong H. (2014). Beneficial effects of silicon on salt and drought tolerance in plants. *Agron. Sustain. Dev.* 34:455.

Zhu Zh., Wei G., Li J., Qian Q., Yuab, J. (2004). Silicon alleviates salt stress and increases antioxidant enzymes activity in leaves of salt-stressed cucumber (*Cucumis sativus* L.). *Plant Sci.* 167(3):527–533.

47 The Role of Grafting Vegetable Crops for Reducing Biotic and Abiotic Stresses

Maryam Mozafarian Meimandi and Noémi Kappel

CONTENTS

47.1 THE PRINCIPLE OF PLANT GRAFTING

Grafting is the connection of two or more pieces of plant tissue which are forced to develop a vascular connection and grow as a single plant (Savvas et al. 2010). Grafting methods have been carried out especially with fruit trees for thousands of years and recently in vegetables. Nowadays, in modern horticulture, grafting is used for many purposes: to have dwarf trees and shrubs, to strengthen plants' resistance to diseases, to retain varietal characteristics, to adapt varieties to unfavorable soil or climatic conditions, to ensure pollination, to propagate certain species that can be propagated in no other way, etc.

47.1.1 IMPORTANCE AND USE OF GRAFTED VEGETABLES

Grafted vegetables have been used commercially in both the *Solanaceae* family (tomato, eggplant, and pepper) and *Cucurbitaceae* family (watermelon, melon, cucumber, pumpkin, and bitter gourd) and rarely in other vegetables such as artichoke and common bean onto vigorous and disease resistance rootstocks (Bie et al. 2017).

The first self-grafting was used to produce a large gourd fruit size as reported in the 1920s in a Chinese and Korean book, but nowadays the cultivated area of a grafted vegetable seedling has increased to reduce biotic and abiotic stress damages for both open fields and protected cultivation. This method is used to improve production and resource efficiency of vegetables (Gaion, Braz, and Carvalho 2018).

Vegetable grafting is a useful technique for several purposes including:

- Improving fruit quality
- Increasing yield (the most important index for the producer)
- Increasing tolerance of abiotic stress in the plant (salt, drought, temperature)
- Reducing disease damage

The efficacy of grafting to control such stress factors can be summarized as follows:

- The grafted plants have a stronger root system.
- Water and nutrient uptake of the grafted plant is more intense.
- Enhanced photosynthetic activity and improved water absorption characterize the grafted plants.
- A stronger antioxidative defense system is formed due to the grafting.
- Increased hormonal signal transmission in the grafted plants.

Therefore, it is important to consider these physiological changes in the cultivation of a grafted plant since this is due to the fact that a different technology has to be applied in relation to self-rooted plant cultivation.

Despite all the advantages, some problems of grafting still exist; for instance, labor and technique essential for grafting process (A. Kumar and Sanket 2017), the price of grafted seedlings and automated grafting machines (Tsaballa et al. 2013).

Grafting involves four phases: (1) selection of rootstock and scion species, (2) creation of a graft union by physical manipulation, (3) healing of the union, and (4) acclimation of the grafted plant.

47.2 ABIOTIC STRESS REDUCTION DUE TO GRAFTING

There is a gap between actual to potential yield due to biotic and abiotic stress conditions in farming and even in modern greenhouses (Peleg et al. 2011). One-third of the yield gap is related to biotic stress including nematodes and soil-borne pathogens, and other gaps are for abiotic stress including salinity, drought, flooding, heavy metal, temperature, and nutrient deficiencies or toxicities (Rouphael, Kyriacou, and Colla 2018).

One environment-friendly technique for avoiding or reducing losses in production caused by stress would be to graft the produced variety onto tolerant rootstock. Vegetable grafting is considered as a rapid alternative to slow breeding method for increasing plant resistance against abiotic stress (Flores et al. 2010; Penella et al. 2017).

According to the review by Rouphael, Kyriacou, and Colla (2018), the reason for resistance in grafted plants against biotic and abiotic stress was related to improving photosynthesis activity and water relation, increasing nutrient and water uptake, stronger antioxidative defense, long-distance movement of mRNAs, small RNAs and proteins, and also a more vigorous root system.

47.2.1 TEMPERATURE STRESS

One of the most important climatic factors that affects plant growth and yield is temperature. The range above or below the optimum temperature (depending on the duration and severity) cause stress and lead to physiological, morphological, and biochemical changes in shoot and root (Colla, Pérez-Alfocea, and Schwarz 2017). The temperature threshold for the growth of pepper, eggplant, cucumber, tomato, and melon is about 8–12°C (Schwarz, Rouphael, and Venema 2010). Also, root growth optimum temperature is slightly lower than for shoot growth and it is for tomato, 22–26°C; eggplant, 22–30°C; pepper, 25–30°C; cucumber, 19–23°C; melon, 20–32°C; and watermelon, 33–37°C (Rouphael, Kyriacou, and Colla 2017).

Using the heat or cold tolerance rootstocks with the more extensive root system and wide optimum temperature can protect plants, which grow under low or high-temperature stress. Several grafting in Cucurbitaceae and Solanaceae families did not have an influence on plant growth and fruit yield under low-temperature stress. Also, grafting doesn't improve plant tolerance against harmful stress automatically (Schwarz, Rouphael, and Venema 2010).

Fig leaf gourd (Cucurbita ficifolia Bouché) with an optimal root temperature of 15°C (6°C lower compared to cucumber) and bur cucumber (Sicos angulatus L.) used as rootstock for cucumber. Recently, Shibuya et al. (2007) grafted cucumber onto squash (Cucurbita moschata Duch) and observed that it could tolerate low temperatures in comparison with self-grafted plants.

According to previous studies by Venema et al. (2008), Keatinge et al. (2014), and Okimura et al. (1986), cold tolerance rootstocks LA 1777 (Solanum habrochaites), 'KNVF' (the interspecific hybrid of Solanum lycopersicum × Solanum habrochaites) and chill-tolerant lines from backcrossed progeny of Solanum habrochaites LA 1778 × Solanum lycopersicum cv. T5 alleviate suboptimal temperatures in tomato for different scions.

In watermelon, Davis et al. (2008) used a Shin-tosa-type, an interspecific squash hybrid (Cucurbita maxima Duch × Cucurbita moschata Duch), and in tomato a Torvum vigor (Okimura et al. 1986) as a rootstock for preventing the negative effects of suboptimal temperatures.

In comparison with tomato, eggplant is tolerant to high soil temperatures, so it can be used as rootstock for tomato (Abdelmageed and Gruda 2009). According to Wang et al. (2007), grafting onto heat tolerant eggplant increased growth period and fruit yield of tomato at 28°C.

47.2.2 SALINITY STRESS

Salinity is one of the most important abiotic stresses which reduces both plant growth and yield especially in arid and semi-arid regions, and most vegetables are sensitive to salinity even at low electrical conductivity. The low water potential of the root medium, the toxic effect of ion Na^+, Cl^- and SO_4^{2-}, and the imbalance of nutrients caused harmful damages to the plants (Colla et al. 2010). Grafting is a beneficial way of reducing losses in the production of Solanaceae and Cucurbitaceae families under NaCl stress conditions. Rootstock, scion and their interactions influence a wide range of physiological and morphological characters (Colla et al. 2010).

Root systems (root density, length, number of root hair, and surface area) play an important role in the tolerance of grafted vegetables against salinity stress (Colla et al. 2010). Zhu et al. (2008), Huang et al. (2009), and Yetisir and Uygur (2010) reported that the decreasing of root properties were smaller in a grafted plant under saline stress conditions. So better root growth of grafted vegetables in comparison with non-grafted may result in better plant growth under salt stress.

Several researchers explained that higher accumulation of proline, sugar, and antioxidant, and lower accumulation of Na^+ and Cl^- in the leaves of grafted plants cause an increased tolerance under stress conditions (Fernández-García, Martínez, and Carvajal 2004; Estan et al. 2005; JuanM. Ruiz et al. 2005; López-Gómez et al. 2007; GORETA et al. 2008). A higher water use efficiency (WUE) indicated higher salt tolerance which may reduce the absorption of ions and also alleviate drought stress induced by salt stress (Colla et al. 2010).

Salinity has impacts on both the vegetative and reproductive stages of plants. During the reproductive stage, water and nutrient uptake will improve due to the rootstock's vigorous root system; production enhancement of endogenous-hormones and improvement of scion vigor can increase grafted fruit yield (Zijlstra, Groot, and Jansen 1994; J.M. Ruiz et al. 1997). For instance, transferring cytokinins from root to shoot improved tomato productivity by reducing flower abortion and increasing fruit size under salt stress conditions (Albacete, Martínez-Andújar, and Pérez-Alfocea 2014).

Several salt tolerance rootstocks were suggested for tomato: wild tomato species including *Solanum pimpinellifolium*, *Solanumperuvianum*, *Solanum cheesmaniae*, *Solanum habrochaites*, *Solanum chmielewskii*, and *Solanumpennellii* (Rao et al. 2013); Pera (Santa-Cruz et al. 2002) and Radja, Volgogradskij, Pera, and Volgogradskij × Pera (Santa-Cruz et al. 2002). He et al. (2009) grafted Chinese commercial tomato hybrids (Hezuo 903) onto Chinese commercial tomato rootstock (Zhezhen No. 1) under 0 to 150 mM NaCl; they observed that photosynthesis characteristics improved in grafted plants compared to the control and self-graft.

Liu et al. (2007) and Wei et al. (2007) found that grafting eggplant cv. Suqiqie onto Torvum vigor (*Solanum torvum* Swartz) resists when grown under saline stress conditions. Grafting common eggplant onto wild eggplant rootstock Tuolubamu (*Solanum torvum*) alleviates salt injury stress through plant growth, proline content, enzyme activity, enhancing photosynthesis, increasing N and P, reduction of f Na^+/K^+ value, and a balanced absorption of Ca, Mg, Cu, Fe, Zn and Mn (Qian, Chen, and Cui 2013).

Penella et al. (2017) grafted pepper onto different rootstocks from the COMAV gene bank at the UPV university (Valencia, east Spain): one from *Capsicum chinense* Jacq., one from *Capsicum baccatum* L. var. pendulum, and one from *Capsicum annuum* L. and showed a higher marketable yield in grafted plants than non-grafted and grafted onto commercial rootstock (Antinema) due to a high photosynthesis level under stress conditions.

As reported by Santa-Cruz et al. (2002) grafting watermelon cv. Crimson Tide onto *Cucurbita maxima* Duchand and *Lagenaria siceraria* (Molina) Standl. rootstocks caused higher plant growth in comparison with non-grafted plants under saline conditions. Another experiment by Goreta et al. (2008) demonstrated that shoot weight and leaf area reduction were lower in watermelon when grafted onto Strongtosa rootstock (*Cucurbita maxima* Duch × *Cucurbita moschata* Duch) than non-grafted plants.

In cucumbers, plant growth and yield (cv. Jinchun No. 2) alleviated in grafted plants onto bottle gourd rootstock Chaofeng 8848 (*Laganaria siceraria* (Molina) Standl.) compared to ungrafted plants under salinity stress (Huang et al. 2005).

47.2.3 DROUGHT STRESS

In arid and semi-arid regions, water sources are limited, and water resource security is one of the most important issues. Drought resistance is a complex quantitative trait controlled by many genes, and it is a challenge for breeders (Kumar et al. 2017). One rapid way of reducing water loss damages and increasing water use efficiency is grafting (García-Sánchez et al. 2007).

Sanchez-Rodriguez et al. (2012) grafted tomato cv. Zarina (drought tolerant) and Josefina (drought sensitive) onto themselves and results showed that drought tolerant rootstock increased fruit number per plant as well as sugar and mineral nutrient. As eggplant root systems uptake more water than tomato, it would be a good rootstock under water deficit (Schwarz, Rouphael, and Venema 2010). Several researchers reported that grafting tomato to tolerant rootstocks increased root growth, nutrient uptake and hydraulic conductivity (Cantero-Navarro et al. 2016) and resulted in more efficient stomatal control, enhancing polyamine content and antioxidant enzyme activity, increasing proline and K^+ with better osmotic adjustment and nitrate assimilation (Altunlu and Gul 2012; Nilsen et al. 2014; Poudyala, Khatria and Uptmoora 2015).

Rouphael et al. (2008) used commercial hybrid rootstock 'PS 1313' for mini watermelons and demonstrated that fruit yield in grafted plants was bigger than in non-grafted ones under water stress due to higher N, K, and Mg increased water absorption, and higher CO_2 assimilation. Liu et al. (2016) reported that grafting cucumber onto luffa caused decreasing stomatal density at the stage of reduced transpiration, increasing ABA accumulation and sensitivity of stomatal movement to ABA and also improvement of antioxidant enzyme activities under drought stress. Strong and higher root growth in pumpkin as rootstock for 'Charleston Gray' watermelon (Poor 2015); tomato cv. Zarina (Sánchez-Rodríguez et al. 2014) and an introgression line IL 'LA3957' (Poudyala, Khatria, and Uptmoora 2015) as rootstock for tomato showed better resistance against drought stress. Increasing uptake and translocation of mineral nutrients through deep and vigorous root systems in watermelon (Rouphael et al. 2008) tomato (Sánchez-Rodríguez et al. 2014) and pepper (Penella et al. 2014) caused an improving drought tolerance in grafted vegetables. Kumar et al., (2017) concluded that faster induction of hormone accumulation especially ABA, higher root hydraulic conductance, reduction in stomata density, better osmoregulation, higher antioxidant activities could increase the resistance of grafted plants.

47.2.4 NUTRIENT STRESS (DEFICIENCY OR TOXICITY)

Many studies revealed that grafting increased absorption and transportation of nutrients to shoot relatives with non-grafted plants (Rouphael et al. 2008; Savvas, Rouphael, and Schwarz 2010). A recent study (Martínez-Andújar et al. 2017) showed that tomato rootstocks improve plant growth and phosphor use efficiency under low P nutrients through hormonal and nutritional mechanisms.

Potassium (K) deficiency in watermelon has a negative effect and grafting onto rootstock with high K^+ uptake can improve both plant growth and yield (Savvas et al. 2010; Huang et al. 2013; Zhong et al. 2018). Zhong et al. (2018) performed an experiment on watermelon cv. Zaojia 8424 grafted onto the rootstock Yongshi (*Citrullus lanatus* (Thumb.) Matsum & Nakai.), 'Jingxinzhen No.1' (*Lagenaria siceraria* (Molina) Standl.), and 'Qingyanzhen No.1' (*Cucurbita maxima* Duch × *Cucurbita moschata* Duch) subjected to low K⁺ concentrations. According to the result of this experiment, grafting increased the quality (total soluble solids, sucrose, vitamin C, lycopene, β-carotene) and total yield of watermelon fruit under deficit conditions. Also, the result of fruit transcriptome analysis showed that the rootstock decreased the sensitivity of watermelon at the transcriptome level under K⁺ stress. Improved K⁺ absorption was observed by Qi et al. (2006) who grafted melon onto pumpkin 'Shengzhen'; Rouphael et al. (2008) grafted mini watermelon onto pumpkin, and Zhu et al. (2008) grafted cucumber onto *Cucurbita moschata* Duch.

47.2.5 HEAVY METAL STRESS

Heavy metals such as Cd, Cu, Zn, Ni, Co, Cr, Pb, V, and As are important environmental pollutants and their toxicity is a problem for an ecological, nutritional, and environmental reason. Besides several other management practices, grafting of scion cultivar onto an appropriate rootstock provides a suitable solution. The use of bottle gourd and pumpkin rootstock enhanced the plant growth and V stress tolerance of watermelon (Azher Nawaz et al. 2018).

Excessive levels of Cu cause morphological and physiological disorders in plants, and cucumber plants, considered as a Cu-sensitive crop, grafted onto the commercial rootstock 'Shintoza' (*Cucurbita maxima* Duch × *Cucurbita moschata* Duch), improved crop performance due to inhibition of Cu accumulation in shoot and resulted in better nutritional status (Rouphael et al. 2008).

Cadmium (Cd) is toxic heavy metal for all organisms. Grafting eggplant onto *Solanum torvum* reduced Cd concentration in the fruits (Savvas et al. 2010). Mori, Murata, and Uraji (2009) indicated that it reduces Cd translocation in fruit related to the process of xylem loading in a grafted plant onto *Solanum torvum* in comparison with a self-grafted plant. Gene expression in the scion influences the root and affects Cd uptake and translocation (Si et al. 2010). Similarly, Edelstein and Ben-Hur (2007) explained that there is lower heavy metal content in the fruit of a grafted plant related to the root system.

Savvas, et al., (2010) examined the effect of grafting cucumber onto different rootstocks under Cd and Ni stress. They grafted cucumber cv. 'Creta' onto four commercial *C. maxima* Duch × *C. moschata* Duch rootstocks ('Power,' 'TZ-148,' 'Ferro,' and 'Strong Tosa'). Grafting onto rootstock 'Power' restricted Cd concentration in the fruit due to the genotype and improved K, Zn, and Mn content in the fruit under non-stress conditions. Also, Ni uptake was restricted by self-grafting in the fruit and old leaves.

47.3 DISEASE RESISTANCE DUE TO GRAFTING

Pests and diseases have a significant effect on vegetable production. Due to the special cultivation conditions, the primary reason for grafting vegetables has been to provide resistance to soil-borne diseases. Grafting can increase the resistance to more than 10 fungal, bacterial and soil-borne pests.

47.3.1 NEMATODES

Several resistant cultivars to root-knot nematodes (*Meloidogyne incognita*, *Meloidogyne javanica*, *Meloidogyne aernaria*) are available. Hence, sensitive, commercial varieties can be grafted onto nematode resistance rootstocks (Keatinge et al. 2014).

Burelle and Rosskopf (2011) reported that grafting tomato 'Florida 47' onto three tomato rootstocks (TX301, Multifort, and Aloha) controlled nematode galling and reduced the numbers of infective *Meloidogyne incognita* juveniles in the soil and roots compared to non-grafted plants. Similarly, commercial tomato rootstock (Multifort) reduced root galling inoculation with *Meloidogyne javanica* (Rumbos, Khah, and Sabi 2011). Dong et al. (2007) used Beaufort, Energy, and He-man as root-knot nematode-resistant rootstocks for tomato scion cv. FA189. A similar result was observed by Wang, Kong, and Yang (2008) who grafted tomato variety 'Rainbow 101' onto eggplant rootstocks (*Solanum integrifolium* and *Solanum sisymbriifolium*).

Grafting onto *Solanum* spp. wild relatives *Solanum torvum* and *Solanum sisymbriifolium* controlled nematodes in eggplant (Rahman et al. 2002). Grafting eggplant scion (cv. Bonica) onto tomato rootstock (Brigeor) protected the plant completely against root-knot nematodes, corky root rot, and some protection were observed for *Verticillium* wilt in summer and winter (Ioannou 2001). Other studies by Gisbert, Prohens, and Nuez (2011) evaluated the yield quality of eggplant fruits under nematodes infected soil. The grafting eggplant cv. 'Black Beauty' onto *Solanum torvum*, *Solanum macrocarpon* and crosses between *Solanum incanum* and *Solanum aethiopicum* resulted in minimal galling, good compatibility, good yield, and quality, especially by *Solanum incanum*.

47.3.2 DISEASES

Bacterial wilt (*Ralstonia solanacearum*) causes the death of all plants, especially in the soil culture and during continuous cultivation. Controlling bacterial wilt through grafting methods was established by Peregrine and Ahmad (1982).

Using tomato rootstock cv. Hawaii 7996 (VI043614) showed effective resistance in bacterial wilt-infected fields by tomato (Cardoso et al. 2012). The results of Zhang et al. (2010) showed that grafting tomato cv. 'Zaoguan 30' onto wilt-resistant tomato rootstocks (LA2701, LA3202, and LA3526) inhibited bacterial wilt pathogens damage. Eggplant rootstocks 'EG 203' (Palada and Wu 2007); 'Cheong Gang,' 'BHN 1054,' and 'BHN 998' (McAvoy et al. 2012); interspecific hybrid tomato rootstocks (*Solanum lycopersicum* × *Solanum habrochaites*) (Rivard et al. 2010); tomato rootstocks 'Dai Honmei' and 'RST-04-105-T' (Rivard et al. 2012) controlled bacterial wilt in tomato grafted plants. McAvoy et al. (2012) grafted tomato cv. 'BHN 602' onto 'Cheong Gang,' 'BHN 1054,' 'BHN 998' rootstocks and itself in open field and in greenhouse conditions and indicated less bacterial wilt than control and self-grafted plants.

The Asian Vegetable Research and Development Center (AVRDC) classified a wide range of eggplant rootstock resistant to bacterial wilt (i.e., TS3, TS60A, TS89, TS36, Terong Lalap, TS75, Hijau Besar, Terung Hijau Kecil, and Terong Hijau). Dabirian, Inglis, and Miles (2017) demonstrated that the Super Shintoza rootstock reduced the microsclerotia incidence, and increased tolerance to *Verticillium* wilt in watermelon.

47.4 PLANT MATERIAL AND ROOTSTOCK SELECTION

Various rootstocks have been examined from the existing cultivars in each crop, and growers try to select a suitable rootstock according to their specific requirement (Lee et al. 2010).

47.4.1 CUCURBITACEAE ROOTSTOCKS

Due to a high demand in the off-season, Cucurbit plants cultivate under inadequate conditions (cold, wet, dry, etc.) in some countries and also continued cropping cause salinity, pests, and soil-borne diseases (Davis et al. 2008). Grafting on *Cucurbitaceae* rootstocks is one of the best methods for avoiding these problems. Rootstocks have a severe impact on the scion through improving plant growth and yield, and tolerance to biotic and abiotic stresses. In this part, some tolerant rootstock for the *Cucurbitaceae* family with their features is reviewed.

Increasing tolerance to *Fusarium* resistance, *Phytophthora* blight, *Verticillium* wilt, *Phomopsis* rot, root-knot nematodes, viruses, and cold tolerance are the most important aims of watermelon grafting (Davis et al. 2008). For instance, bottle gourd (*Lagenaria siceraria* L), squash (*Cucurbita moschata* Duch), and pumpkin (*Cucurbita pepo* L.) induced resistance to *Fusarium* and low-temperature tolerance with vigorous root systems. Bottle gourd, *Cucurbita* spp., *Citrullus* spp., interspecific hybrids between *Cucurbita maxima* Duch and *Cucurbita moschata* Duch, and wild watermelon (*Citrullus lanatus* var. *citroides*) have high compatibility with watermelon.

Grafting watermelon onto interspecific hybrid rootstocks decreased flesh quality, increased firmness (depending on the scion, increasing firmness can be an advantage or a disadvantage) and better shelf life and field holding abilities (King et al. 2010).

Increasing cold tolerance and resistance to *Fusarium* wilt is the primary objective of cucumber grafting. Cucumber rootstocks included: *Cucurbita ficifolia* cv. Heukjong (it has low temperature and disease resistance), *Cucurbita moschata* cv. Butternut, Unyoung #1, Super Unyoung (improving fruit quality and *Fusarium* tolerance), *Cucurbita maxima* × *C. moschata* cv. Shintos wa, Keumtozwa, Fero RZ, 64-05 RZ, Gangryuk Shinwha (*Fusarium* and low-temperature tolerance), *Sicyos angulatus* cv. Andong (*Fusarium*, high soil moisture, nematode, and low-temperature tolerance), *Cucumis matuliferus* cv. NHRI-1 (*Fusarium* and nematode tolerance, weak temperature resistance) and luffa rootstock for heat and drought tolerance (Liu et al. 2016).

The primary reason for melon grafting is resistance to soil-borne pathogens: *Monosporascus cannonballus*, *Fusarium* *oxysporum* f. sp. *melonis* (Fom), and *Stagonosporopsis* spp. and root-knot nematodes (Gaion, Braz, and Carvalho 2018). *Cucurbita moschata*, *Cucurbita maxima* × *Cucurbita moschata*, *Cucurbita pepo*, *Cucumis melo* cv. Kangyoung, Keonkak, and Keumgang (*Fusarium* tolerance and fruit quality improvement) and African cucumber (*Fusarium* and nematode tolerance, low temperature and high soil moisture resistance) are used as melon rootstocks.

47.4.2 SOLANACEAE ROOTSTOCKS

Solanaceae family is one of the most important sources of food for human health. However, biotic and abiotic stress reduced their growth and productivity. In this section, some resistance rootstocks for tomato, eggplant, and pepper are reviewed.

The reasons for tomato grafting and looking for different rootstocks are extending the harvest season in the greenhouse, disease resistance in non-heated soil under a plastic tunnel, control of bacterial wilt and brown root rot in open field production, and also abiotic stress tolerance. Respectively, bacterial wilt and brown root rot occur in warm and cold temperature, so it is necessary to select specific rootstock for each environment (Lee et al. 2010).

Some tomato rootstocks are included: tomato genotypes, interspecific hybrids, *Solanum lycopersicon*, *Solanum habrochaites* cv. 'Maxifort' and 'Beaufort' (resistant to *Tomato Mosaic Virus*, *Fusarium* root rot and *Fusarium* crown rot, corky root, *Verticillium*, and nematodes). Recently, some tomato rootstock 'AR-9704' (Fernández-García, Martínez, and Carvajal 2004) and 'Maxifort' (Semiz and Suarez 2015) were introduced as salt tolerance rootstocks.

Control of soil-borne diseases, *Verticillium* wilt, bacterial wilt, *Fusarium* wilt, and root-knot nematodes is the purpose of eggplant grafting (King et al. 2010). Wild species (*Solanum integrifolium*) was the first rootstock for eggplant with highly resistant *Fusarium* wilt and bacterial wilt (Tachibana 1994). Other eggplant rootstocks are tomato hybrid rootstocks, *Solanum torvum* and *Solanum sisymbriifolium* (Bletsos, Thanassoulopoulos, and Roupakias 2003), interspecific hybrid rootstocks *Solanum melongena* × *Solanum incanum* and *Solanum melongena* × *Solanum aethiopicum* (Gisbert, Prohens, and Nuez 2011), tomato interspecific hybrids *Solanum lycopersicon* × *Solanum habrochaites* (Johnson, Inglis, and Miles 2014; Giuffrida et al. 2015; Miles, Wimer, and Inglis 2015). Eggplant grafted onto *Solanum torvum* and *Solanum sisymbriifolium* provide a high level of resistance to bacterial wilt, *Fusarium* wilt, *Verticillium* wilt, and root-knot nematode (Bletsos, Thanassoulopoulos, and Roupakias 2003; Gaion, Braz, and Carvalho 2018). Moreover, grafting eggplant onto tomato interspecific hybrids 'Beaufort' F1 and 'He-Man' improved plant growth under non-saline conditions but was not affected under salt stress (Giuffrida et al. 2015). In other experiments by Johnson, Inglis, and Miles (2014) and Miles, Wimer, and Inglis (2015) reported that grafting eggplant cv. 'Epic' onto 'Beaufort' rootstock increased resistance to *Verticillium* wilt.

For controlling *Phytophthora capsica*, nematodes and abiotic stress resistance, pepper are usually grafted onto

Capsicum annuum and other genera of the *Solanaceae* family (King et al. 2010). Pepper has been grafted less than other *Solanaceae* plants (tomato and eggplant) due to limited compatibility with other species (Gaion, Braz, and Carvalho 2018).

Totally, as biotic and abiotic stress change rapidly, plant breeders face the challenge to explore new rootstocks germplasms (King et al. 2010). There is a wide range of rootstocks, but they need research in select match rootstocks for special scion and environment (Behera and Singh 2002).

47.5 CONCLUSIONS

The ancient Greeks used grafting procedures, and Chinese sources from the same period refer to grafting as a propagation method used in the Far East. Grafting was probably taught to gardeners by nature. In contact with the cambiums of close-up branches, natural extinction was created. By observing this phenomenon, gardeners in ancient times were able to experiment with grafting. In ancient Rome, the grafting of fruit trees was an ordinary method.

Nowadays, the use of grafted vegetables is increasing worldwide, led by the needs of new emerging growers: such as leaving crop rotation requirements, the benefits of methyl bromide; demand for more stress-resistant crops; more developed roots, better conditions, increased plant growth vigor; higher yields and better yield quality; better demand for water and nutrient utilization.

The primary aim of the grafting of vegetable species is to protect the soil against pests and pathogens in cases where genetic and chemical solutions are not available to overcome these problems. By placing a susceptible scion on a resistant rootstock, some cultivated varieties can be resistant, saving the long and expensive breeding work. In addition, grafting can increase resistance to abiotic stress (temperature tolerance, extreme moisture conditions, salt stress), improve yields, produce more efficient water and nutrient use, extend the harvest period, and influence crop quality. The choice of species used as rootstock is made according to these goals.

ACKNOWLEDGMENTS

This research was supported by the Higher Education Institutional Excellence Program (1783-3/2018/FEKUTSRAT). The financial support was awarded by the Ministry of Human Capacities within the framework of water-related researches of Szent István University. This publication is prepared in number EFOP-3.6.1-16-2016-00016. The specialized of the SZIE Campus of Szarvas Research and Training Profile with intelligent specialization in the themes of water management, hydroculture, precision mechanical engineering, alternative crop production.

REFERENCES

Abdelmageed, A.H.A. and N. Gruda. 2009. Influence of Grafting on Growth, Development and Some Physiological Parameters of Tomatoes under Controlled Heat Stress Conditions. *European Journal of Horticultural Science* 47: 16–20.

Albacete, A.A., C. Martínez-Andújar, and F. Pérez-Alfocea. 2014. Hormonal and Metabolic Regulation of Source–Sink Relations under Salinity and Drought: From Plant Survival to Crop Yield Stability. *Biotechnology Advances* 32 (1): 12–30.

Altunlu, H. and A. Gul. 2012. Increasing Drought Tolerance of Tomato Plants by Grafting. *Acta Horticulture* 960: 183–190.

Behera, T.K. and Narendra Singh. 2002. Inter-Specific Crosses between Eggplant (*Solanum Melongena* L.) with Related Solanum Species. *Scientia Horticulturae* 95 (1–2): 165–172.

Bie, Z., M.A. Nawaz, Y. Huang, J-M. Lee, and G. Colla. 2017. Importance and Use of Vegetable Grafting. In: Colla, G., F.P. Alfocea, and D. Schwarz, eds. *Vegetable Grafting, Principles and Practices*. CABI Publishing, UK, https://cab.presswarehouse.com/sites/stylus/resrcs/chapters/1780648979_1stChap.pdf.

Bletsos, F., C. Thanassoulopoulos, and D. Roupakias. 2003 Effect of Grafting on Growth, Yield, and Verticillium Wilt of Eggplant. *HortScience* 38: 183–186.

Burelle N.K. and E.N. Rosskopf. 2011. Microplot evaluation of rootstocks for control of Meloidogyne incognita on grafted tomato, muskmelon, and watermelon. *Journal of Nematology* 43: 166–171.

Cantero-Navarro, E., R. Romero-Aranda, R. Fernández-Munoz, C. Martínez-Andújara, F. Pérez-Alfoceaa, and A. Albacete. 2016. Improving agronomic water use efficiency in tomato by rootstock-mediated hormonal regulation of leaf biomass. *Plant Science* 251: 90–100.

Cardoso, S.C., A.C.F. Soares, S. Brito Ade, A.P. dos Santos, F.F. Laranjeira, L.A. de Carvalho. 2012. Evaluation of Tomato Rootstocks and its Use to Control Bacterial Wilt Disease. *Ciências Agrarias, Londrina* 33: 595–604.

Colla, G., F. Pérez-Alfocea, and D. Schwarz, eds. 2017. *Vegetable Grafting: Principles and Practices*. Wallingford: CABI.

Colla, G., Y. Rouphael, C. Leonardi, and Z. Bie. 2010. Role of Grafting in Vegetable Crops Grown under Saline Conditions. *Scientia Horticulturae*. 127 (2): 147–155. Elsevier. Accessed September 28, 2018. https://www.sciencedirect.com/science/article/pii/S0304423810003705.

Colla, G., Y. Rouphael, M. Cardarelli, A. Salerno, and E. Rea. 2010. The Effectiveness of Grafting to Improve Alkalinity Tolerance in Watermelon. *Environmental and Experimental Botany* 68 (3): 283–291.

Dabirian, S., D. Inglis, and C.A. Miles. 2017. Grafting Watermelon and Using Plastic Mulch to Control Verticillium Wilt Caused by *Verticillium Dahliae* in Washington. *HortScience* 52 (3): 349–356.

Davis, A.R., P. Perkins-Veazie, Y. Sakata, S. López-Galarza, J.V. Maroto, S.-G. Lee, Y.-C. Huh, Z. Sun, A. Miguel, S.R. King, R. Cohen, and J.-M. Lee. 2008. Cucurbit Grafting. *Critical Reviews in Plant Sciences* 27 (1): 50–74.

Dong, D., Z. Cao, X. Wang, J. Hu, and M.L. Gullino. 2007. Effect of Nematode Resistant Rootstocks on Growth Characteristics and Yields of Tomato. *Acta Horticulturae Sinica* 34: 1305–1308.

Estan, M.T., M.M. M.-R., F. Perez-Alfocea, T.J. Flowers, and M.C. Bolarin. 2005. Grafting Raises the Salt Tolerance of Tomato through Limiting the Transport of Sodium and Chloride to the Shoot. *Journal of Experimental Botany* 56 (412): 703–712.

Fernández-García, N., V. Martínez, and M. Carvajal. 2004. Effect of Salinity on Growth, Mineral Composition, and Water Relations of Grafted Tomato Plants. *Journal of Plant Nutrition and Soil Science* 167 (5): 616–622.

Flores, F.B., P. Sanchez-Bel, M.T. Estañ, M.M. Martinez-Rodriguez, E. Moyano, B. Morales, J.F. Campos, J.O. Garcia-Abellán, M.I. Egea, N. Fernández-Garcia, F. Romojaro, and M.C. Bolarín. 2010. The Effectiveness of Grafting to Improve Tomato Fruit Quality. *Scientia Horticulturae* 125 (3): 211–217.

Gaion, L.A., L.T. Braz, and R.F. Carvalho. 2018. Grafting in Vegetable Crops: A Great Technique for Agriculture. *International Journal of Vegetable Science* 24 (1): 85–102.

García-Sánchez, F., J.L. Jifon, M. Carvajal, and J.P. Syvertsen. 2002. Gas Exchange, Chlorophyll and Nutrient Contents in Relation to Na and Cl Accumulation in 'Sunburst' Mandarin Grafted on Different Rootstocks. *Plant Science* 162: 705–712.

García-Sánchez, F., J.P. Syvertsen, V.G., P. Botía, and J.G. Perez-Perez. 2007. Responses to Flooding and Drought Stress by Two Citrus Rootstock Seedlings with Different Water-Use Efficiency. *Physiologia Plantarum* 130 (4): 532–542.

Gisbert, C., J. Prohens, and F. Nuez. 2011. Performance of Eggplant Grafted onto Cultivated, Wild, and Hybrid Materials of Eggplant and Tomato. *International Journal of Plant Production* 5(4). www.ijpp.info.

Giuffrida, F., C. Cassaniti, M. Agnello, and C. Leonardi. 2015. Growth and Ionic Concentration of Eggplant as Influenced by Rootstocksunder Saline Conditions. *Acta Horticulturae* 1086 (June): 161–166.

Goreta, S., V. Bucevic-Popovic, G. Vuletin Selak, M. Pavela-Vrancic, and S. Perica. 2008. Vegetative Growth, Superoxide Dismutase Activity and Ion Concentration of Salt-Stressed Watermelon as Influenced by Rootstock. *The Journal of Agricultural Science* 146 (06): 695.

He, Y., Z. Zhu, J. Yang, X. Ni, and B. Zhu. 2009. Grafting Increases the Salt Tolerance of Tomato by Improvement of Photosynthesis and Enhancement of Antioxidant Enzymes Activity. *Environmental and Experimental Botany* 66 (2): 270–278.

Huang, Y., R. Tang, Q. Cao, and Z. Bie. 2009. Improving the Fruit Yield and Quality of Cucumber by Grafting onto the Salt Tolerant Rootstock under NaCl Stress. *Scientia Horticulturae* 122 (1): 26–31.

Huang, Y., J. Zhu, A. Zhen, L. Chen, and Z. Bie. 2005. Organic and Inorganic Solutes Accumulation in the Leaves and Roots of Grafted and Ungrafted Cucumber Plants in Response to NaCl Stress. *Journal of Food, Agriculture & Environment* 7 (2): 703–708. www.world-food.net.

Ioannou, N. 2001. Integrating Soil Solarization with Grafting on Resistant Rootstocks for Management of Soil-borne Pathogens of Eggplant. *Journal of Horticultural Science and Biotechnology* 76: 396–401.

Johnson, S., D. Inglis, and C. Miles. 2014. Grafting Effects on Eggplant Growth, Yield, and Verticillium Wilt Incidence. *International Journal of Vegetable Science* 20 (1): 3–20.

Keatinge, J.D.H., L.-J. Lin, A.W. Ebert, W.Y. Chen, J.d'A. Hughes, G.C. Luther, J.-F. Wang, and M. Ravishankar. 2014. Overcoming Biotic and Abiotic Stresses in the Solanaceae through Grafting: Current Status and Future Perspectives. *Biological Agriculture & Horticulture* 30 (4): 272–287.

King, S.R., A.R. Davis, X. Zhang, and K. Crosby. 2010. Genetics, Breeding and Selection of Rootstocks for Solanaceae and Cucurbitaceae. *Scientia Horticulturae* 127 (2): 106–111.

Kumar, A. and K. Sanket. 2017. Grafting of Vegetable Crops as a Tool to Improve Yield and Tolerance against Diseases: A Review. *International Journal of Agricultural Science* 9 (13): 4050–4056.

Kumar, P., Y. Rouphael, M. Cardarelli, and G. Colla. 2017. Vegetable Grafting as a Tool to Improve Drought Resistance and Water Use Efficiency. *Frontiers in Plant Science* 8 (June): 1130.

Lee, J.-M., C. Kubota, S.J. Tsao, Z. Bie, P. Hoyos Echevarria, L. Morra, and M. Oda. 2010. Current Status of Vegetable Grafting: Diffusion, Grafting Techniques, Automation. *Scientia Horticulturae* 127 (2): 93–105.

Liu, Z.-L., Y.-L. Zhu, C.-M. Hu, G.-P. Wei, L.-F. Yang, and G.-W. Zhang. 2007. Effects of NaCl Stress on the Growth, Antioxidant Enzyme Activities and Reactive Oxygen Metabolism of Grafted Eggplant. *The Journal of Applied Ecology* 18 (3): 537–541.

Liu, S., H. Li, X. Lv, G.J. Ahammed, X. Xia, J. Zhou, K. Shi, T. Asami, J. Yu, and Y. Zhou. 2016. Grafting Cucumber onto Luffa Improves Drought Tolerance by Increasing ABA Biosynthesis and Sensitivity. *Nature* 6 (20212): 1–14.

López-Gómez, E., M.A.San Juan, P. Diaz-Vivancos, J. Mataix Beneyto, M.F. García-Legaz, and J.A. Hernández. 2007. Effect of Rootstocks Grafting and Boron on the Antioxidant Systems and Salinity Tolerance of Loquat Plants (*Eriobotrya Japonica* Lindl.). *Environmental and Experimental Botany* 60 (2): 151–158.

Martínez-Andújar, C., J.M. Ruiz-Lozano, I.C. Dodd, A. Albacete, and F. Pérez-Alfocea. 2017. Hormonal and Nutritional Features in Contrasting Rootstock-Mediated Tomato Growth under Low-Phosphorus Nutrition. *Frontiers in Plant Science* 08 (April): 533.

McAvoy, T., J.H. Freeman, S.L. Rideout, S.M. Olson, and M.L. Paret. 2012. Evaluation of Grafting Using Hybrid Rootstocks for Management of Bacterial Wilt in Field Tomato Production. *HortScience* 47: 621–625.

Miles, C., J. Wimer, and D. Inglis. 2015. Grafting Eggplant and Tomato for Verticillium Wilt Resistance. *Acta Horticulturae* 1086 (June): 113–118.

Mori, I.C., Y. Murata, and M. Uraji. 2009. Integration of ROS and Hormone Signalling. In: del Rio L.A., A. Puppo, eds. *Reactive Oxygen Species in Plant Signalling*. Berlin: Springer-Verlag, pp. 25–42.

Nawaz, M.A., C. Chen, F. Shireen, Z. Zheng, Y. Jiao, H. Sohail, M. Afzal, M. Imtiaz, M.A. Ali, Y. Huang, and Z. Bie. 2018. Improving Vanadium Stress Tolerance of Watermelon by Grafting onto Bottle Gourd and Pumpkin Rootstock. *Plant Growth Regulation* 85: 41–56.

Nilsen, E.T., J. Freeman, R. Grene, and J. Tokuhisa. 2014. A Rootstock Provides Water Conservation for a Grafted Commercial Tomato (*Solanum lycopersicum* L.) Line in Response to Mild-drought Conditions: A Focus on Vegetative Growth Andphotosynthetic Parameters. *PLoS ONE* 9: e115380.

Okimura, M., Matsuo, S., Arai, K., and Okitsu, S. 1986. Influence of soil temperature on the growth of fruit vegetable grafted on different stocks. *Bull. Veg. & Ornam. Crops Res. Stn. Japan*, Series C 9: 43–58 (in Japanese with English summary).

Palada, M.C. and D.L. Wu. 2009. Grafting Sweet Peppers for Production in the Hot-Wet Season. International Cooperators Guide. Pub. No.09-722-e. AVRDC (Asian Vegetable Research and Development Center), Taiwan.

Peleg, Z., M. Reguera, E. Tumimbang, H. Walia, and E. Blumwald. 2011. Cytokinin-Mediated Source/Sink Modifications Improve Drought Tolerance and Increase Grain Yield in Rice under Water-Stress. *Plant Biotechnology Journal* 9 (7): 747–758.

Penella, C., S.G. Nebauer, A. San Bautista, S. López-Galarza, and Á. Calatayud. 2014. Rootstock Alleviates PEG-Induced Water Stress in Grafted Pepper Seedlings: Physiological Responses. *Journal of Plant Physiology* 171 (10): 842–851.

Penella, C., S.G. Nebauer, S. López-Galarza, A. Quiñones, A. San Bautista, and Á. Calatayud. 2017. Grafting Pepper onto Tolerant Rootstocks: An Environmental-Friendly Technique Overcome Water and Salt Stress. *Scientia Horticulturae* 226 (December): 33–41.

Peregrine, W.T.H. and K.B. Ahmad. 1982. Grafting – A Simple Technique for Overcoming Bacterial Wilt in Tomato. *Tropical Pest Management* 28: 71–76.

Poudyala, D., Laxman K., and R. Uptmoora. 2015. An Introgression of *Solanum habrochaites* in the Rootstock Improves Stomatal Regulation and Leaf Area Development of Grafted Tomatoes under Drought and Low Root-Zone-Temperatures. *Advances in Crop Science and Technology* 03 (03): 175.

Qian, Z.W., H.L. Chen, and Y.L. Cui. 2013. Effects of Grafting on Yield and Mineral Elements of Eggplants with Seawater Cultivation. *Chin Veg.* 2: 58–65.

Qi, H.Y., T.L. Li, Y.F. Liu, and D. Li. 2006. Effects of Grafting on Photosynthesis Characteristics, Yield, and Sugar Content in Melon. *Journal of Shenyang Agricultural University* 37: 155–158.

Rahman, M.A., M.A. Rashid, M.M. Hossain, M.A. Salam, and A.S.M.H. Masum. 2002. Grafting Compatibility of Cultivated Eggplant Varieties with Wild Solanum Species. *Pak. J. Biol. Sci.* 5: 755–757.

Rao, E.S., P. Kadirvel, R.C. Symonds, and A.W. Ebert. 2013. Relationship between Survival and Yield Related Traits in *Solanum pimpinellifolium* under Salt Stress. *Euphytica* 190 (2): 215–228.

Rivard, C.L., S. O'Connell, M.M. Peet, R.M. Welker, and F.J. Louws. 2012. Grafting tomato to manage bacterial wilt caused by Ralstonia solanacearum in the southeastern United States. *Plant Disease* 96: 973–978.

Rivard, C.L., O. Sydorovych, S. O'Connell, M.M. Peet, and F.J. Louws. 2010. An economic analysis of two grafted tomato transplant production systems in the United States. *HortTechnology* 20: 794–803.

Rouphael, Y., M. Cardarelli, E. Rea, and G. Colla. 2008. Grafting of Cucumber as a Means to Minimize Copper Toxicity. *Environmental and Experimental Botany* 63 (1–3): 49–58.

Rouphael, Y., M. Cardarelli, G. Colla, and E. Rea. 2008. Yield, Mineral Composition, Water Relations, and Water Use Efficiency of Grafted Mini-watermelon Plants under Deficit Irrigation. *Hortscience* 43: 730–736.

Rouphael, Y., M.C. Kyriacou, and G. Colla. 2018. Vegetable Grafting: A Toolbox for Securing Yield Stability under Multiple Stress Conditions. *Frontiers in Plant Science* 8 (January): 2255.

Rouphael, Y., M.C. Kyriacou, and G. Colla. 2017. Vegetable Grafting: A Toolbox for Securing Yield Stability under Multiple Stress Conditions. *Frontiers in Plant Science* 8: 2255.

Ruiz, J.M., A. Belakbir, I. López-Cantarero, and L. Romero. 1997. Leaf-Macronutrient Content and Yield in Grafted Melon Plants: A Model to Evaluate the Influence of Rootstock Genotype. *Scientia Horticulturae* 71 (3–4): 227–234.

Ruiz, J.M., B. Blasco, R.M. Rivero, and L. Romero. 2005. Nicotine-Free and Salt-Tolerant Tobacco Plants Obtained by Grafting to Salinity-Resistant Rootstocks of Tomato. *Physiologia Plantarum* 124 (4): 465–475.

Rumbos, C.I., E.M. Khah, N. Sabi. 2011. Response of Local and Commercial Tomato Cultivars and Rootstocks to Meloidogyne Javanica Infestation. *Australian Journal of Crop Science* 5: 1388–1395.

Sánchez-Rodríguez, E., R. Leyva, C. Constán-Aguilar, L. Romero, and J.M. Ruiz. 2012. Grafting under Water Stress in Tomato Cherry: Improving the Fruit Yield and Quality. *Ann. Appl. Biol.* 161: 302–312.

Sánchez-Rodríguez, E., R. Leyva, C. Constán-Aguilar, L. Romero, and J.M. Ruiz. 2014. How Does Grafting Affect the Ionome of Cherry Tomato Plants under Water Stress? *Soil Science and Plant Nutrition* 60 (2): 145–155.

Santa-Cruz, A., M.M. Martinez-Rodriguez, F. Perez-Alfocea, R. Romero-Aranda, and M.C. Bolarin. 2002. The Rootstock Effect on the Tomato Salinity Response Depends on the Shoot Genotype. *Plant Science* 162 (5): 825–831.

Savvas, D., Y. Rouphael, D. Schwarz, G. Colla, Y. Rouphael, and D. Schwarz. 2010. Amelioration of Heavy Metal and Nutrient Stress in Fruit Vegetables by Grafting. *Scientia Horticulturae* 127 (2): 156–161.

Schwarz, D., Y. Rouphael, and J.H. Venema. 2010. Grafting as a Tool to Improve Tolerance of Vegetables to Abiotic Stresses: Thermal Stress, Water Stress and Organic Pollutants. *Scientia Horticulturae* 127 (2): 162–171.

Semiz, G.D. and D.L. Suarez. 2015. Tomato Salt Tolerance: Impact of Grafting and Water Composition on Yield and Ion Relations. *Turkish Journal of Agriculture and Forestry* 39: 876–886.

Shibuya, T., A. Tokuda, R. Terakura, K. Shimizu-Maruo, H. Sugiwaki, Y. Kitaya, and M. Kiyota. 2007. Short-Term Bottom-Heat Treatment during Low-Air-Temperature Storage Improves Rooting in Squash (*Cucurbita moschata* Duch.) Cuttings Used for Rootstock of Cucumber (*Cucumis sativus* L.). *Journal of the Japanese Society for Horticultural Science* 76 (2): 139–143.

Si, Y., F. Dane, A. Rashotte, K. Kang, and N. K. Singh. 2010. Cloning and Expression Analysis of the Ccrboh Gene Encoding Respiratory Burst Oxidase in Citrullus Colocynthis and Grafting onto *Citrullus Lanatus* (Watermelon). *Journal of Experimental Botany* 61 (6): 1635–1642.

Tachibana, S. 1982. Comparison of Root Temperature on the Growth and Mineral Nutrition of Cucumber Cultivars and Figleaf Gourd. *Journal of the Japanese Society for Horticultural Science* 51: 299–308.

Tsaballa, A., C. Athanasiadis, K. Pasentsis, I. Ganopoulos, I. Nianiou-Obeidat, and A. Tsaftaris. 2013. Molecular Studies of Inheritable Grafting Induced Changes in Pepper (Capsicum Annuum) Fruit Shape. *Scientia Horticulturae* 149 (January): 2–8.

Venema, J.H., B.E. Dijk, J.M. Bax, P.R. van Hasselt, and J.T.M. Elzenga. 2008. Grafting Tomato (*Solanum lycopersicum*) onto the Rootstock of a High-Altitude Accession of *Solanum habrochaites* Improves Suboptimal-Temperature Tolerance. *Environmental and Experimental Botany* 63 (1–3): 359–367.

Wang, S., R. Yang, J. Cheng, J. Zhao. 2007. Effect of Rootstocks on the Tolerance to High Temperature of Eggplants under Solar Greenhouse during Summer Season. *Acta Horticulturae* 761: 357–360.

Wang, S.H., Y. Kong, and R. Yang. 2008. Studies on Rootstock Screening and Resistance to Root-knot Nematodes for Grafted Tomato. *Chin Veg.* 12: 24–27.

Wei, G.P., Y.L. Zhu, Z.L. Liu, L.F. Yang, and G.W. Zhang. 2007. Growth and Ionic Distribution of Grafted Eggplant Seedlings with NaCl Stress. *Acta Botanica Boreali-Occidentalia Sinica* 27: 1172–1178 (in Chinese with English summary).

Wei, G., Y. Zhu, Z. Liu, L. Yang, and G. Zhang. 2007. Growth and Ionic Distribution of Grafted Eggplant Seedlings with NaCl Stress. https://europepmc.org/abstract/cba/639334.

Yetisir, H. and V. Uygur. 2010. Responses of Grafted Watermelon Onto Different Gourd Speciesto Salinity Stress. *Journal of Plant Nutrition* 33 (3): 315–327.

Zhang, Z.-K., S.-Q. Liu, A.-Q. Hao, and S.-H. Liu. 2010. Grafting Increases the Copper Tolerance of Cucumber Seedlings by Improvement of Polyamine Contents and Enhancement of Antioxidant Enzymes Activity. *Agricultural Sciences in China* 9: 985–994.

Zhong, Y., C. Chen, M.A. Nawaz, Y. Jiao, Z. Zheng, X.g Shi, W. Xie, Y. Yu, J. Guo, S. Zhu, M. Xie, Q. Kong, F. Cheng, Z. Bie, and Y. Huang. 2018. Using Rootstock to Increase Watermelon Fruit Yield and Quality at Low Potassium Supply: A

Comprehensive Analysis from Agronomic, Physiological and Transcriptional Perspective. *Scientia Horticulturae* 241 (November): 144–151.

Zhu, J., Z. Bie, Y. Huang, and X. Han. 2008. Effect of Grafting on the Growth and Ion Concentrations of Cucumber Seedlings under NaCl Stress. *Soil Science and Plant Nutrition* 54 (6): 895–902.

Zijlstra, S., S.P.C. Groot, and J. Jansen. 1994. Genotypic Variation of Rootstocks for Growth and Production in Cucumber; Possibilities for Improving the Root System by Plant Breeding. *Scientia Horticulturae* 56 (3): 185–196.

48 Why Root Morphology is Expected to Be a Key Factor for Crop Salt Tolerance

Uwe Schleiff

CONTENTS

48.1 INTRODUCTION

48.1.1 BACKGROUND

Around the world, there is increasing pressure on agriculture, horticulture, and landscape greening to replace good quality water with 'unconventional water' of lower qualities such as brackish, drainage, and/or treated wastewaters, especially in (semi)arid areas. The area of irrigated land using more or less treated sewage waters is estimated to be at least 3.5 million ha, China not included (Hamdy 1995; Qadir et al. 2010). Annual economic loss caused by soil salinity is estimated to be some 12 billion US$ globally (FAO 2018). It is estimated that about 10 million ha of irrigated land is abandoned yearly, due to adverse effects of irrigation management, mainly from secondary salinization (Pessarakli and Szabolcs 2011).

There are many components involved that limit plant growth under saline conditions. These include depletion of nutrients (e.g., Potassium, Calcium, Nitrate) and specific ion toxicities (e.g., Sodium, Magnesium, Chloride, Sulfate), which can affect growth at the cell level, specific organs, and entire plants (Oster 1999; Munns and Gilliham 2015). But with respect to irrigated field grown crops, yields are negatively affected by salt-induced falls in the water supply to plants, which is the most critical growth factor. Water supply to shoots becomes critical when the availability of soil water to roots is limited by the decrease of the total water potential (Ψt) of the soil. In saline soils, root water uptake does not only decrease because of dropping soil matric water potential (Ψm) as happens in non-saline soils but also salts dissolved in soil water decrease soil osmotic water potential (Ψo), which additionally stunts plant growth (Wadleigh and Ayers 1946; Kirkham 2005).

Different plans vary significantly in their tolerance of soil salinity. In irrigated agriculture, for practical applications, a rating of soil salinity on crop growth utilizes ECe measurements (electrical conductivity EC in dS/m) of soil saturation extracts (ECe). The advantage of the ECe-based rating is the simplicity of measurement and its close relationship to the osmotic water potential of the soil solution, which is a genuine growth restricting factor, but more difficult to measure (Richards 1954; Rhoades et al. 1992). For a wide range of relevant soil conditions, the well-proven rule is that an EC of the soil water of 3.3 dS/m corresponds to a Ψo of 1.0 bar (= −0.1 MPa resp. −100 kPa).

As shown in Table 48.1, crops that include fodder halophytes can grow in a soil salinity that exceeds an ECe of 50 dS/m. Yields of the most sensitive crops (e.g., *Phaseolus vulgaris*, *Pisum sativum*, etc.) begin to decline at ECe-values >1.5 dS/m and can start failing at ECe-values ~8 dS/m. Salt tolerant crops (e.g., *Hordeum vulgare*, *Beta vulgaris*, etc.) can tolerate ECe-values up to 10 dS/m, without yield loss and will collapse at an ECe of around 32 dS/m (Maas and Hoffmann 1977; Ayers and Westcot 1985; Rhoades et al.

1992). Halophytes can even survive ECe-values exceeding 50 dS/m, at least in moist soils.

48.1.2 Outlining of the Proposed Concept on Crop Salt Tolerance Research

In saline soils, the availability of soil water to roots is the most limiting growth factor to plants. Thus the basis of salt tolerance rating in irrigated agriculture focuses on soil water salinity in the root zone, with respect to plant growth. The sketch presented in Figure 48.1 is our attempt to combine principles of the current concept on salt tolerance rating, as developed by the USDA (Maas and Hoffmann 1977), with our recently proposed concept that includes impacts of transpiration driven processes occurring between two water applications, caused from the exclusion of salts by roots.

As sketched on the left-hand side of Figure 48.1, USDA-rating focuses on the assumption that under controlled irrigation there is the development of a vertical increase of salts in the rooted soil profile, which follows continuous decreases in rooting densities and root water uptake. This increase in soil water salinity downwards into the soil profile arises from the continuous drop of soil water percolating downwards and

TABLE 48.1
'Threshold' and 'Zero Yield' Soil Salinity of Average Root Zone (ECe) for Field Grown Crops of Different Salt Tolerance (Supplemented)

Crop Sensitivity	Threshold[a] (ECe in dS/m)	Zero Yield (ECe in dS/m)	Crop Examples
Sensitive	0–1.5	<8	Bean, pea, onion, carrot, leek
Moderately sensitive	1.5–3	8–16	Maize, broad bean, alfalfa, tomato
Moderately tolerant	3–6	16–24	Sorghum, soybean, wheat, rape, beet
Tolerant	6–10	24–32	Barley, cotton, sugar beet, date palm
Very tolerant	>10	50–80	Spec. of Atriplex, Agropyron, Kochia,
No plant growth		>50–80	(depending on moisture conditions)

[a] Soil salinity at which yield loss commences.

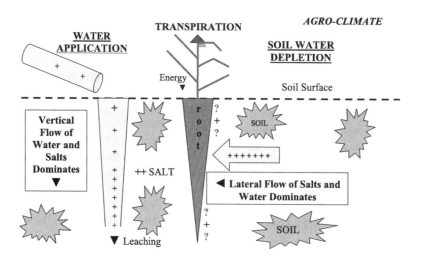

FIGURE 48.1　Sketch to describe fundamental flows of water and salts in the rooted soil layer under controlled brackish water irrigation.

affects leaching of dissolved salts. Soil water salinity is lowest at the soil surface, immediately following watering when the soil water content is highest, and the salinity of the soil water is most diluted, down to the salinity level of the applied water. Downwards into the soil profile, the salt concentration of the soil water increases steadily, as water extraction by roots decreases the amount of water flowing downwards thereby leaving less soil water to leach salts. Soil water salinity is greatest at the bottom of the root zone, where root water uptake ceases. The objective of controlled irrigation management using saline waters is to achieve crop specific root zone salinity levels through optimal watering. Salt leaching is the key activity for controlling soil water salinity. After watering, conditions for root water uptake and water supply to shoots improve and enable high growth rates. In general, when between 10 and 30% of produced assimilates are directed to supply root systems, plant water supply is also affected (York et al. 2016).

The right-hand side of Figure 48.1 indicates impacts between waterings when plant roots deplete soil water to cover the transpiration demand of the shoot. A severe shortcoming in the USDA's salt tolerance ratings is that they do not account for lateral water and salt movement (Schleiff 2011, 2013a, 2013b). When soil water content drops below field capacity, vertical water flows cease and lateral flows of soil water and salts directed at roots dominate. Firstly, when water is taken up from the soil fraction nearest to roots, the rhizospheric soil (= part of rooted soil volume directly altered by roots, as compared to the bulk soil, where the root meets the soil), a gradient of soil matric water potential (Ψm) between rhizospheric and bulk soil develops. This gradient is expected to generate a flow of soil solution from the bulk soil into the rhizospheric soil to cover the demand for plant transpiration. In saline soils, simultaneous increases of rhizospheric soil water salinity occur, as most salts located near the roots are excluded from root uptake (Barber 1979; Schubert and Laeuchli 1990; Hamza and Aylmore 1992). The increase of salinity in the rhizospheric soil solution arises from two causes: first, decreasing soil water contents causes an inverse increase of salt concentration, and second salts imported from the bulk soil accumulate in the soil fraction near roots. Little is known about this process and how it relates to root water uptake and crop salt tolerance, especially in the case of the combined impact of decreasing osmotic and matric soil water potentials. In any case, the combined effect of simultaneously decreasing soil water potentials decreases the water supply to shoots stimulating shoot wilting, reducing assimilation, and simultaneously degrading energy supply of roots (Schleiff and Schaffer 1984). To the right side of the shoot, shown in Figure 48.1, younger leaves often suffer more from insufficient water uptake than older leaves, as they are less adapted to saline growth conditions. The agro-climatic conditions that surround the shoot (temperature, wind, light, air humidity, etc.) are mentioned, as they affect transpiration and wilting significantly.

Introduction of the USDA concept was, doubtless, a great step for improving controlled irrigation of crops of different salt tolerance, but this concept has a long list of limitations.

Maas and Hoffman (1977), for example, identified a whole set of factors, e.g., soil fertility, climatic conditions, differences in varieties, growth stages, irrigation techniques, and root characteristics that produce significant deviations in collected critical values, due to mostly unknown and/or neglected impacts. We will focus on the relation between root traits and root surrounding soil fractions affecting the water supply of crops and thus salt tolerance under saline soils conditions. Model calculations presented in the following section will show the need to improve our understanding of dynamic processes happening at the soil/root interface, between roots and soil, in particular, when roots differ in their morphology.

48.2 SIMPLIFIED MODEL CALCULATIONS

48.2.1 PRINCIPALS OF LATERAL WATER AND SALT DISTRIBUTION AROUND ROOTS

The principal concept for expanding our understanding of the processes that occur under brackish water irrigation, in the rooted soil layer, are presented in Figures 48.2 and 48.3. In both sketches root surrounding soil is divided into three lateral fractions of increasing distance from the soil surface:

- The soil fraction inside the rhizocylinder, which is in immediate contact with the root surface and penetrated with root hairs, when such exist (rhizospheric soil volume)
- The soil fraction of the transition zone, which surrounds the rhizocylinder and where a sharp gradient of water and salinity, between bulk and rhizospheric soil water can occur within hours or less (ion diffusion)
- The soil fraction remotest from the root surface, and not directly affected by roots, the bulk soil

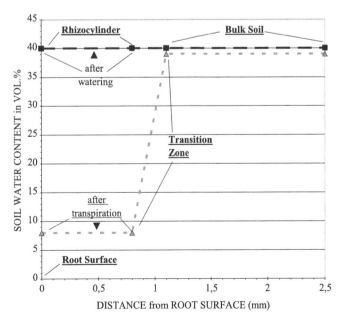

FIGURE 48.2 Lateral soil water gradients around irrigated roots as affected by watering and transpiration (Schleiff 2013b).

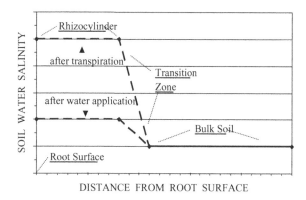

FIGURE 48.3 Lateral gradients of easily soluble salts around irrigated roots, as affected by watering and transpiration (Schleiff 2013b).

Figure 48.2 shows the impact of a watering upon the lateral water distribution surrounding roots and results of plant transpiration upon soil water contents. Immediately after a watering, soil water contents are highest at all three soil fractions (around 'field capacity': the broken line). We assume that in this phase, when root water uptake occurs directly from the rhizocylinder, shoot water supply is at its greatest and can afford high transpiration and growth rates for hours. Consequently, a significant drop of soil water content (soil water matric potential Ψm) within the rhizocylinder volume, causes a significant water potential gradient between rhizocylinder and bulk soil. Between both fractions, there develops a transition zone, where the gradients balance. Plant growth decreases as rhizospheric soil water depletes and root water uptake predominantly depends upon the flow of soil water from the bulk soil outside the rhizocylinder.

The sketch presented in Figure 48.3 outlines this concept of the lateral distribution of salts around roots of transpiring shoots, as plant growth occurs under controlled water management in saline soils. The salinity of the soil water (ECsw on Y-axis) relates to the water of the soil fractions that surround the roots at increasing distances (X-axis; mm): (a) the soil water of the rhizocylinder, in direct contact with the root surface and intensively penetrated by root hairs (broken line); (b) the bulk soil water, which is the remote soil fraction and not directly affected by roots (the continuous line); (c) the water of the transition zone that surrounds the rhizocylinder and where a sharp gradient in salinity between bulk and rhizospheric soil water fraction occurs (dotted line: ion diffusion).

The lower lines show the lateral salt distribution immediately after watering, as the downward flow of soil water ceases. Rhizospheric soil water salinity is at its lowest, but under controlled water management, is still higher than in bulk soil. This is due to the fact that prior to watering, rhizocylinder soil water salinity is expected to be significantly higher when compared to bulk soil salinity.

The impact of the succeeding, intensive plant transpiration and the exclusion of salts from root uptake are so significant that within hours a relatively rapid increase of rhizospheric soil water salinity develops that exceeds the bulk soil water salinity by multiple factors.

We expect the build-up of such gradients, during transpiration, will differ significantly among crops and depend upon plant growth rate, soil and root characteristics, but this build-up usually develops in a few hours. Soil water salinity in the rhizocylinder will be lowest just after watering, due to high water contents and some leaching effects. However, under well-controlled irrigation management, we expect the salinity outside the rhizocylinder to be lower than the salinity inside the rhizocylinder because salt that leaches inside the rhizocylinder starts from a much higher level of soil salinity (Schleiff 1981). The transpiration demand of growing plants is primarily met by root water uptake from the soil fraction nearest to the roots, the rhizocylinder. Thus, plant transpiration and salt exclusion by the roots produce increasing soil water salinity at simultaneously decreasing soil water content. In addition, saline soil water from the bulk soil travels into the rhizocylinder and replenishes its water loss, thereby raising soil water salinity even further. Salinity gradients between rhizocylinder and bulk soil differ greatly and can amount to several hundred percent. Hence, there develops a transition zone between both soil fractions, where the balance between both concentrations occurs. The next section will apply this concept to conditions of root water uptake when plant roots differ in their morphological traits growing in the soil of various textures.

48.2.2 Calculated Effect of Root Morphology on Rhizospheric Soil Water

It is our goal to calculate the quantitative amounts of soil water (total soil water contents) contained in the rhizospheric soil volume, by roots of different morphology and that grow in soils of different textures and soil water contents at 'field capacity' (SWC in Vol.%). It is assumed that the availability of this water fraction to plant roots is in principle much higher ('readily plant available' soil water) than soil water from outside the rhizocylinder and therefore of special importance. In Figure 48.4, we distinguish between roots of varying radii of rhizospheric soil volume. The length of root hairs that form rhizospheric soil volumes can range from 0.25 and 1.5 mm (Jungk 2002). Additionally, we consider soils of various total SWCs in the range of 20 to 40 Vol.% (e.g., sandy to loamy/silty soils). The rhizospheric soil volumes formed by a given root system can be calculated using the following formula:

$$V_{soil} = \pi * R^2 * RL \qquad (48.1)$$

where V_{soil} is the volume (cm³) of the rhizocylinder; π is the circle constant ~3,14; R is the average radius of the rhizocylinder formed by root hairs (in cm); and root length (RL) is the length of the active roots (in cm). The volume of the rhizocylindric soil water (V_{water} in cm³) is calculated from V soil multiplied by the soil water content, as given in the formula (48.2):

$$V_{water} = V_{soil} * (Vol.\%/100) \qquad (48.2)$$

As shown on the Y-axis in Figure 48.4, the total amount of soil water in rhizospheric soil volumes can vary in an extremely large range, running from 0.39 to 28 ml/10 m root length (e.g., a

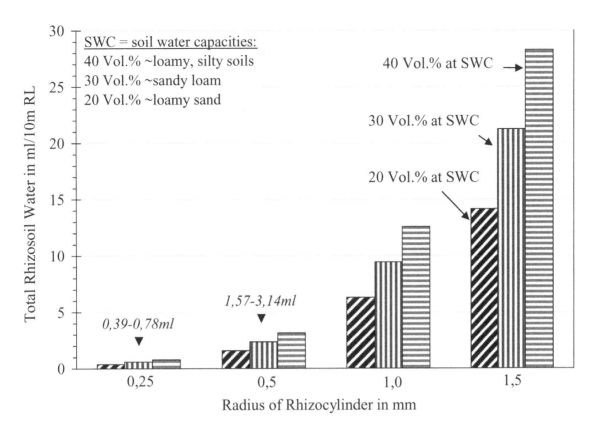

FIGURE 48.4 (See color insert.) Total amount of rhizosoil water as related to soil texture and root morphology (SWC = soil water capacity, RL = root length).

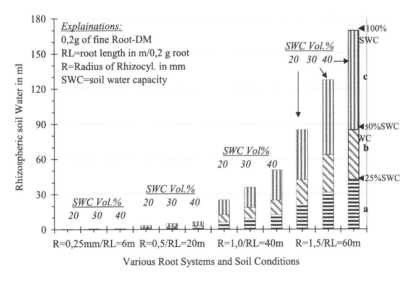

FIGURE 48.5 (See color insert.) Rhizospheric soil water fractions (ml) in various soils around roots of different morphology.

70-fold range). The amount of water is lowest (0.39 ml/10 m RL) when plants grow in light textured soils (20 Vol.% SWC) and are equipped with root hairs that form a small rhizocylinder volume (radius of 0.25 mm). On the other hand, the amount of rhizospheric soil water is highest (28 ml/10 m RL) when plants grow in loamy/silty soils (40 Vol.% SWC) and root hairs form a large rhizocylinder volume (radius of 1.5 mm).

However, it is not usually the total amount of the rhizospheric soil water (SWC) that is relevant for plant water supply, but fractions that differ in their contribution to plant growth and will be discussed later (Figure 48.5). The volume differences of the plant available rhizospheric soil water can be calculated, using the following equation (48.3), an application of the formulas (48.1) and (48.2):

$$V_{\text{rhizowater}} = \pi * R^2 * RL * (\text{Vol.}\%/100) \qquad (48.3)$$

where V is the water volume of the rhizospheric soil fraction in ml; л is the circle constant ~3,14; R is the radius of the rhizocylinder with values from 0,025 and 0.15 cm; RL in cm between 6 and 60 m/0.2 g root dry matter (DM); and the soil water content SWC (Vol.%/100). We anticipate that the assumed conditions for roots and soils cover the range of most field conditions.

The examples shown in Figure 48.5 clearly show that there are potentially large differences in the amounts of total rhizospheric soil water (full length of bars). Rhizospheric soil water is nearly negligible (0.24 to 0.47 ml/0.2 g root-DM) when the rhizospheric soil volume is too small due to short root hairs (0.25 mm) and shortness of roots (RL = 6 m length, e.g., onions). We assume it is a lateral flow of soil water from outside the small rhizocylinder that dominates the water supply of plants under such conditions.

The volume of the rhizospheric soil water increases tremendously when roots are longer and equipped with root hairs. Roots of many *Gramineae* are significantly longer (e.g., 40 m/0.2 g root-DM) and at least partly equipped with root hairs of about 1.0 mm length (Figure 48.5). This results in a rhizospheric soil volume of 126 cm³, which may contain between 25 ml (20 Vol.%SWC) and 50 ml total soil water (40 Vol.%SWC) depending on the soil texture (SWC).

However, it is not the total soil water that is relevant for crop growth, but only a fraction of it. Per the scale on the right-hand side of Figure 48.5, we can divide the SWC of the various soils into three fractions of different plant availabilities: (a) the fraction 0–25% (horizontal lined), which contributes very little to nothing to the plant water supply of; (b) the fraction 25 to 50% of SWC (diagonal lined), which enables most field crops to survive periods of water shortage between waterings; and (c) the fraction 50 to 100% (~'field capacity') of SWC (vertical lined), the fraction most relevant for achieving high growth rates of crops ('readily plant available'). The 'non-available' water fraction, (a), may vary from 0.06 ml/0.2 g root-DM, when R of the rhizocylinder is 0.25 mm, RL is only 6 m, and SWC is low at 20 Vol.%, to 42.4 ml/0.2 g root-DM, when R is 1.5 mm, RL reaches 60 m/0.2 g root-DM, at a SWC of 40 Vol.%. The (b)-fraction of lower plant availability (25 to 50% SWC) varies in the same ranges, as those calculated for fraction (a). This fraction is assumed to supply plants in periods of water stress, and can secure low growth rates and survival of plants. Fraction (c) is that part of SWC, most important to achieve high growth rates of plants, as required for an efficient irrigated crop production. Under the conditions assumed in Figure 48.5, the amounts of readily root available soil water may vary from 0.12 ml (R = 0.25; RL = 6 m; SWC = 20 Vol.%) to 84.8 ml (R = 1.5; RL = 60 m; SWC = 40 Vol.%).

These model calculations offer a convincing thesis concerning the importance of interactions between root and soil, and improves our understanding of quantitative plant watering. These simplified calculations do not depict the complexity of all the processes involved in root water uptake, but they foreshadow big potential and the need to develop adequately needed experimental designs (Carminati et al. 2017). Also, the volume of rhizopheric soil water is of high relevance for

the development of roots surrounding salinity, produced by plant transpiration, which will be analyzed in Section 48.2.4.

48.2.3 SUPPLY OF TRANSPIRING SHOOTS FROM RHIZOSPHERIC SOIL WATER

The characteristic trait of rhizospheric soil fractions is that they are densely penetrated by the finest root hairs, which enable roots to absorb water and other growth factors (e.g., nutrients) directly and effectively from the rhizocylinder. It is the special value of this soil fraction that the supply of plants with growth factors is only slightly dependent upon time-consuming transport processes inside the soil. Consequently, we understand the rhizospheric soil water to be that part of the total soil water that is taken up most easily and provides the shoot with sufficient water to achieve the highest growth rates.

The following calculations attempt to quantify the potentially large differences that exist between roots of different morphology, and that provide transpiring shoots with readily available soil water for DM production. These calculations follow equation (48.4), where the shoot-DM (Y) produced from the rhizospheric soil water (W) is calculated from W (in ml/root), and the plant-specific water productivity or transpiration coefficient (TC in ml/g DM):

$$Y[\text{mg shoot-DM}] = W[\text{ml rhizowater}] / TC[300\ \text{ml}] \quad (48.4)$$

The calculation considers 50% of the rhizospheric SWC of a medium textured soil (SWC of 30 Vol.%), i.e., 15 Vol.% rhizosoil water to gain high shoot-DM growth rates (see fraction 'c' in Figure 48.5). The TC of plants differs significantly among species and their growth stages, ranging for example from 200 to 600 ml/g DM shoots and even higher, depending on various growth conditions (e.g., climate, soil fertility, soil water, salinity). For our model calculation, we chose a relatively favorable TC of 300 ml/g shoot-DM (= 0.3ml/mg shoot), a typical value for efficient growth conditions of many soil-grown crops.

As shown in the examples of Figure 48.6, the potential shoot-DM production from the rhizospheric soil water increases as root length per unit shoot-DM (Z-axis) and the radius of the rhizocylinder (X-axis) increases. It is lowest (~1.0 mg shoot-DM) when the active RL is only a 10 m/g shoot at a rhizocylinder radius of only 0.25 mm. It is highest when RL reaches a 30 m/g shoot at a rhizocylinder radius of 1.5 mm (~100 mg shoot-DM). Thus the model calculations confirm our expectation that there must be a significant impact of root morphology on root water uptake, plant water supply, and growth.

48.2.4 TIME SPAN REQUIRED TO DEPLETE RHIZOSPHERIC SOIL WATER

The rhizospheric soil water fraction is the most important fraction for achieving the highest growth rates of shoots. Thus the rhizospheric soil water deserves special attention. The

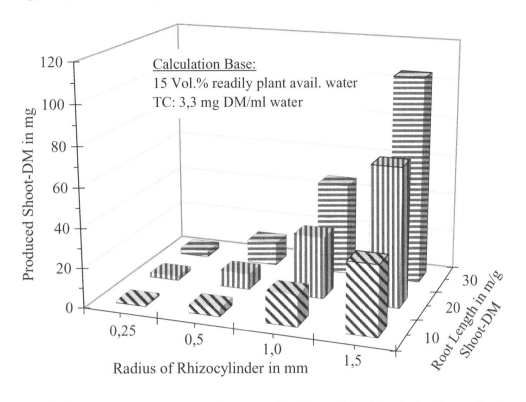

FIGURE 48.6 (See color insert.) Shoot-DM produced from the readily plant available rhizospheric soil water fraction of a medium textured soil (15% Vol at 50% SWC) by roots of different morphology and length (transpiration coeff. TC 300ml/g DM shoot).

objective of the following calculations is to obtain a quantitative idea on the time this soil water fraction provides water for the high water uptake rates of roots, when root systems that differ in their morphological traits are exposed to a medium textured soil containing 15 Vol.% of readily plant available soil water (see 100 to 50% of SWC, 'c' in Figure 48.5).

The time, T in hours, that plants can meet their water requirement for transpiration is calculated from the rhizospheric soil water (SWC/2 in ml) and its productivity per unit of time (20 mg shoot-DM/hour). To simplify the calculation, we assume a constant growth rate (20 mg/hour referred to an initial shoot-DM of 1g/plant) for this short growth period of a few hours:

$$T \ (\text{in hours}) = Y\left[\text{mg DM prod.}\right] / \text{GR}\left[20 \text{ mg DM/h}\right]$$

Example taken from

Figure .48.6 to 48.7:5.25 h = 105 mg/20 mg/h

(48.5)

In Figure 48.7, the calculated time span of covering the transpiration water demand from the readily available rhizospheric soil water depends heavily on root characteristics. Time is extremely low, covering only 3 minutes of transpiration demand, when the rhizocylinder R is small (0.25 mm) and only 10 m RL/g shoot-DM are involved. The time span reaches 5 hours when R is 1.5 mm, and active RL is 30 m, about 100-fold higher than at R of 0.25 mm. We conclude that differences in root morphology may play an important role in water supply for shoots. We take the results of our model

calculations as a clear indication that differences in root traits play a significant role in the understanding of plant water supply from soils. However, we also expect significant differences between model calculations and experimental measurements, as the nature of water uptake will likely prove to be more complex, e.g., interactions occurring between roots, etc.

48.2.5 Increase of Rhizospheric Soil Water Salinity by Root Water Depletion

Calculations presented in the previous section clearly show the importance of root morphological traits for the availability of soil water to roots and water supply of shoots growing under non-saline soil conditions. The objective of the following model calculations is to evaluate the impact of soil water depletion on rhizospheric soil water salinity when roots are exposed to saline soils and additionally differ in their morphology. For decades, it has been well-known that under saline growth conditions it is the salinity of the soil water (soil osmotic water potential Ψo) that reduces the availability of soil water to plants, due to restricted water uptake rates by roots (USDA 1954). But the course of soil water salinities in the root surrounding soil fraction during periods of water depletion is still a black box being extremely difficult to measure directly. We offer a calculation procedure that can provide relevant salinity ranges.

The curves presented in Figure 48.8 are calculated from formula (48.6), where the impact of water uptake, from a medium textured soil (30 Vol.% SWC), by roots (in ml/10 m root length) of various morphologies (R from 0.25 to 1.5 mm),

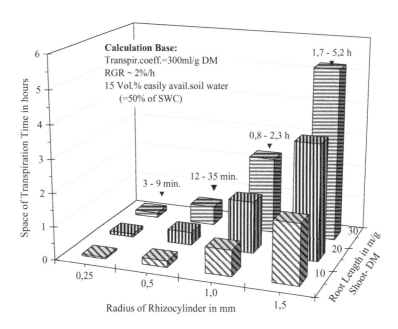

FIGURE 48.7 Time span (h)to meet transpiration demand of shoots is met from the rhizospheric soil water (50% of SWC) of different roots and soils and (RGR = Relative Growth Rate in 20 mg/(g DM*h)).

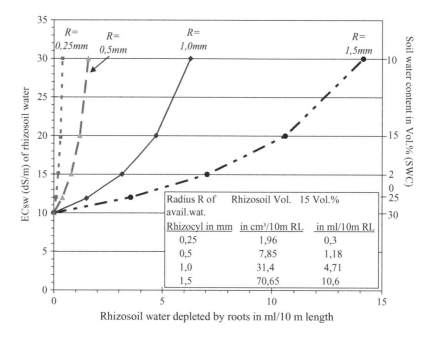

FIGURE 48.8 Impact of water depletion by roots of different morphology on rhizospheric soil water salinity.

on the course of the rhizospheric soil water salinity ECrh (Y-axis), is demonstrated:

$$ECrh = \rightarrow ECini * \left(Wini/Wrh \right)$$

(48.6)

15 and 10 Vol.%

ECrh (salinity of rhizospheric soil water) is calculated from ECini (initial ECrh at SWC after watering; e.g., 10 dS/m) multiplied by the quotient Wini/Wrh, where Wini is the initial soil water content (e.g., 30 Vol.%) and Wrh the rhizospheric soil water

content, after water depletion (e.g., 10 Vol.%). Ergo the radius R, formed by roots of various morphologies and the rhizospheric soil volume are factors of great importance for the course of root surrounding soil water salinity the roots are exposed to. For example, water uptake of only 1.18 ml/10 m RL causes an ECrh of 20 dS/m at an SWC of 15 Vol.%, when the root forms a small rhizocylinder R of only 0.5 mm. But for a large R of 1.5 mm, the identical ECrh of 20 dS/m is achieved after only 10.6 ml/10m RLdepletion (in comparison to an R of 0.5 mm about 10-fold higher). We conclude, conditions for root water uptake significantly improve, when roots form a larger rhizocylinder volume.

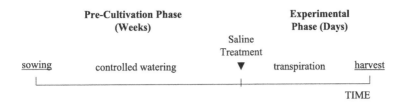

FIGURE 48.9 Vegetation technique procedure (VeTe and Schleiff 1987a, 2013a).

The presented model calculations clearly show an urgent need for a deepened understanding of comparative, quantitative processes occurring between roots of different morphology and the rhizospheric soil fraction to improve the growth of field grown crops under saline soil conditions. In the following section, we offer an experimental design with the potential to verify, quantitatively, the principal considerations of the presented model calculations.

48.3 BASICS OF EXPERIMENTAL DESIGN TO DETERMINE ROOT WATER UPTAKE FROM THE RHIZOCYLINDER AND OUTSIDE OF IT

48.3.1 CONCEPTUAL APPROACH

Under controlled irrigation, crops growing in saline soils, a simultaneous decrease soil water contents, and an increase of soil water salinity of the rooted soil reduce plant transpiration and plant growth (Schleiff and Schaffer 1984; Oster 1999). Earlier we discussed the evaluation of salt tolerance under the USDA concept, which is based on the impact of vertical salt distribution in the rooted soil layer. But besides a build-up of a vertical salinity gradient during periods of soil water depletion, simultaneously there is also a build-up of lateral soil salinity gradients around roots. Thus, water uptake by roots occurs during rapid changes in soil water salinity (rhizospheric soil water) that significantly differs from the salinity determined for the average soil of the rooted layer.

The results of the model calculation presented in Figure 48.8 clearly show a significant impact of root morphology for the process of rhizospheric salt accumulation that affects water uptake by roots, shoot water supply, and crop salt tolerance. Thus, to understand the quantitative impact of processes happening around roots, on their water uptake, we need to develop a methodology for measuring these factors. We propose an experimental set-up with the potential to simultaneously measure the following parameters and provide these data sets:

- ECsw (dS/m) of soil water and soil osmotic water potential (Ψo)
- Total soil water potential ($\Psi t = \Psi m + \Psi o$)
- Shoot and root dry matter weight of plants with roots of different morphology
- Transpiration rate and relative growth rate (RGR) of shoot-DM

Experiments using technical tools (e.g., micro-tensiometers, micro-suction-cups, CTR) for direct and simultaneous measurements of the required parameters proved unpromising, in particular when it came to quantifying impacts of lower soil water contents on plants (Vetterlein et al. 2004; Roose et al. 2016). We anticipate our vegetation technique (VeTe) has the potential to fill this gap and provides missing data via the continuous measuring of pot water losses, which serves as a basis for several calculations. In Figure 48.9, we show our VeTe, which basically consists of two growth periods, a pre-cultivation and an experimental phase. Each of which is described, in detail, in the following sections.

48.3.2 PRE-CULTIVATION PHASE FOR ACHIEVING HOMOGENEOUS PLANTS

The objective of the pre-cultivation phase is to raise plants with homogeneous masses of the shoot- and root-DM by the end of this phase. Depending upon the research objective, and/or in order to achieve different rooting densities (very high, high, or moderate) of the soil, plants are cultivated in pots of different volumes (e.g., 250, 500, and 1000 ml). The impact of different soil textures (e.g., light, moderate, or heavy textured) can be studied, when pots are filled with the different soils for simultaneous study.

As already shown in previous papers (Schleiff 1987a, 1987b), the big advantage of our pre-cultivation phase is that we can start the experimental phase with uniform plants (e.g., in a growth chamber). In particular, we control watering during the pre-cultivation phase. For rape plants, the standard deviation of shoot-DM of five replicates was 2% (707 mg DM/shoot +−15 mg) and for the root-DM 11% (183 mg DM/root +−21 mg), whereas with less controlled watering the deviation for the shoot-DM was 16% (Schleiff 1987a). Similar results were obtained for wheat plants, under controlled watering.

The homogeneity of our plants at the commencement of the experimental phase offers the chance to reduce the number of replicates and instead to increase the number of treatments, which helps us to research more complex questions and reduces our workload. In our approach, we obtain data on various impacts of factors, such as soil water contents and soil water salinity, soil textures, and rooting densities on water supply, and growth of various crops/varieties, which we require to quantitatively improve our understanding of crop salt tolerance. Our VeTe can provide the required data by continual monitoring of pot water losses after saline waterings until we observe severe wilting. The specific objectives of an experiment determine the length of the experimental phase that follows the pre-cultivation phase, which may last from a few hours to a couple of days.

48.3.3 Examples for Planning Saline Treatments and Experimental Set-Ups

Detailed planning for the experimental phase is an important pre-assumption to match the complexity of our approach and the attainment of a reliable set of data. The experimental set-up follows two basic objectives: first, the quantitative study of the impact of increasing soil salinity in rhizospheric soil fractions on the water uptake by roots of different morphology, directly from the 'readily' plant available fraction. The second objective focuses on the impact of the soil fraction that at the commencement of the experimental phase is located outside the rhizocylinder. During periods of soil water depletion, non-rhizospheric soil fraction can contribute to an improved root water uptake by two mechanisms: (1) soil water explored by the growth of roots into the non-rhizospheric soil becomes rhizospheric soil water during a depletion phase. This contribution may be significant in cases when the energy supply of the root system is high, and roots are able to form a large rhizocylinder volume per unit energy supply from the shoot. Carminati et al. (2017) have highlighted the importance of root hairs for the water supply of crops. (2) The non-rhizospheric soil contribution to the water supply is assumed to emerge from a gradient of soil matric water potential between rhizospheric and non-rhizospheric soil fractions. Basically, we assume water flow from non-rhizospheric soil to rhizospheric soil, which follows the lower water content (lower soil matric potential Ψm) of soil near to the roots. But the water flow along the decreasing Ψm is most probably also affected by the osmotic gradient between the non-rhizospheric and rhizospheric fractions, which we expect to decrease the matric flow. We cannot fully exclude ion diffusion as a contributing factor in the dilution of root surrounding soil water, although its impact on facilitating root water uptake will be small. So, we do not confront this low-grade facilitation here.

48.3.3.1 Experimental Set-Up to Study the Impact of Rhizospheric Soil Volumes

The objective of this experimental set-up is to quantify the variation of water uptake patterns by roots from rhizospheric soil fractions at different levels of salinity, which continuously increase during periods of soil water depletion. For demonstration and analysis of the impact of root morphology, we selected rape (large rhizocylinder volume per unit root-DM) and leek (small rhizocylinder volume) as model plants. So rape was pre-cultivated for 5 weeks and leek for 10 weeks in soil volumes of 300 ml/plant. Rape achieved a rooting density of 0.5 g/300 ml soil volume (1.6 mg/ml soil) and leek 0.7 g root-DM (2.3 mg/ml soil). A typical shoot-DM for rape was 0.83 g/plant and for leek 1.3 g/plant by the end of pre-cultivation, prior to our treatment of these plants with saline waters. In rape, we estimate half of the total root-DM forms fine roots that includes hairs; the other half forms coarse roots (taproot

and main branches), which we assume to be much less efficient at the immediate depletion of soil resources.

The objective of saline water applications at the outset of our experiment is to raise the SWC to 100% pot water capacity (PWC), which, for different soils, can range from less than 20 Vol.% (sandy soil) to more than 40 Vol.% (loamy, silty soil, clayey soils). In the silt soil, used in most of our experiments, water depletion was down to 25% PWC (= 30 ml/300 ml soil volumes (SV) at 10 Vol.%) before watering at the outset the experimental phase. Then, the watering required to achieve a PWC of 100% was 90 ml (= 120 ml/300 ml SV at 40 Vol.%).

The experimental phase began with the application of saline waters that achieved water salinities of the rhizospheric soil to cover wide ranges of salt concentrations. An example recommended for rape and leek is shown in Table 48.2. But the salt concentrations selected for specific crops can vary significantly from our estimated salt tolerance for each crop. Experiments with crops of known higher salt tolerance may require higher salinity levels than crops known to be less tolerant to salinity.

Calculation of ECsw at the outset from the salinity of the applied water (ECw), diluted by the low salinity (ECre neglected) of the soil water before watering (e.g., 25% PWCre), is shown in equation (48.7). The osmotic water potential Ψo of the soil water is obtained with equation (48.8) using this constant 0.36 (Richards 1954).

$$ECsw = ECw * \left[\left(100\% \; PWC - 25\% \; PWCre^{1)} \right) / 100 \right]$$

Example: 8.25 dS/m = 11 dS/m * 0.75   (48.7)

[1)] PWCre: remaining soil water before saline watering

$$\Psi o \; [bar] = ECsw [dS/m] * 0.36$$

Example: 3.0 bar = 8.3 dS/m * 0.36 = −0.3 MPa = −300 kPa

(48.8)

TABLE 48.2

Example of Planning an Experimental Set-up to Compare Water Uptake from Rhizospheric Soils (e.g., 300 ml Densely Rooted Soil Volume) by Roots of a Small (e.g., Leek) and a Large (e.g., Rape) Rhizocylinder Volume Using Leek and Rape as Model Crops

Salinity Treatment ECw dS/m	Calculated Initial Salinity ECsw of Rhizosoil	
	Small (e.g., leek)	Large (e.g., rape)
(Non-Saline)	–	–
Low: 5,5	4,2/(1,5)[a]	4.2/(1,5)
Medium: 11	8.3/(3,0)	8.3/(3,0)
High: 21	15.8/(5,7)	15.8/(5,7)
Very High: 30	22.5/(8,1)	22.5/(8,1)

[a] in brackets: corresponding osmotic soil water potentials OP in bar.

To simplify the calculation, we assume a low salinity (< 2 dS/m) for the remaining soil water fraction upon the termination of the pre-cultivation phase that can be neglected for further calculations of the soil water salinity.

48.3.3.2 Experimental Set-Up to Study Impacts of Non-Rhizospheric Soil Fractions

Model calculations presented in Figures 48.5 through 48.8 indicate significant potentials of root morphologic traits, soil textures, and soil salinity, related to the availability of rhizospheric soil water for roots, growth, and salt tolerance of plants. However, the root water uptake is not limited to the rhizospheric soil fraction only, but contributions from outside the rhizospheric soil fraction, as already sketched in Figures 48.1 and 48.2, can be most relevant. Impacts from interactions between rhizospheric and non-rhizospheric soil water fractions and the continuous expansion of rhizospheric soil volume are anticipated as affecting root water uptake. We expect quantitatively relevant impacts, related to differences in root growth and root morphology for the various plants. Suitable technical equipment (i.e., microtensiometers, micro cups, TDR) for reliable direct and quantitative measurement of water uptake by roots that expand into the non-rhizospheric soil fraction of different salinity levels are currently not very promising. Hence we propose an experimental set-up that is a modification of our VeTe and provides this missing data (Table 48.3).

The experimental set-up proposed in Table 48.3 advances the VeTe we developed earlier, as we studied the impacts of rhizospheric soil salinity on shoot water supply (Schleiff 1987a, 2013a). Pre-cultivated potted plants having various SV that contain 250, 500, and 1000 ml/pot are employed. Pre-cultivating plants in soils of varying volumes produces rooted

soils that differ in rooting densities and also produces plants with homogeneous shoot and root mass (DM base). Further details on the pre-cultivation phase were described earlier (Schleiff 2013a). Upon termination of the pre-cultivation phase, all plants should have consumed similar amounts of water to achieve a very similar root- and shoot-DM. A typical example for the outcome of the pre-cultivation phase, before starting the experimental phase, is given in Figure 48.10 for rape, where an amount of 280 ml/pot transpired water produced 0.96 g/plant shoot-DM and 0.45 g/plant root-DM. Leaf surface of an example plant was ~142 cm² at the outset of the experimental phase but increased until harvesting in the non-saline treatment (SV–1.0) to 226 cm².

The experimental phase commences with the salinization of the rooted soils by application of 50 ml/pot for the SV-250, 100 ml/pot for the SV-500, and 200 ml/pot for the 1000 ml/pot treatment to achieve SWCs of 30 Vol.%. The applied waters were salinized with 75 mM/l and 150 mMol/l to obtain ECw's of about 8 dS/m resp. 16 dS/m, and initial ECsw's of ~6 dS/m resp. ~12 dS/m at the experimental start. Nutritional status of all plants was optimal. In order to prevent any Calcium problems, some gypsum was added to the water prophylactic.

The course of soil water salinity ECsw during the experimental phase was determined from the soil water contents, which were measured bi-hourly from the pot water losses according to the formula in equation (48.9):

$$Ya = Yi * Xi/Xa \qquad (48.9)$$

where Ya is the actual ECsw (dS/m) at the time of weighing the pot; Yi is the initial ECsw after watering at the start of the experiment (hour 0) when plants started to transpire; Xi (Vol.%) is the soil water content SWC at the start; and Xa is the SWC at the actual moments when the pot water losses were measured.

TABLE 48.3

Planning an Experimental Set-up for Quantitative Comparison of Water Uptake from the Rhizospheric and Non-Rhizospheric Soil Fractions (SV = Soil Volume in ml; RD = Rooting Density in Length cm/ml Soil; Ecw of Water Applied at the Outset of Experimental Phase)

| | Crop/Variety A | | |
Salinity Treatment	SV-250 ml RD[a]	SV-500 ml RD[b]	SV-1000 ml RD[c]
(Non-saline) (e.g., ECw 0.3 dS/m)	very high	high	moderate
Non–Low (e.g., ECw 3 dS/m)	very high	high	moderate
Low–Mod (e.g., ECw 8 dS/m)	very high	high	moderate
Mod–High (e.g., ECw 15 dS/m)	very high	high	moderate

[a] rhizospheric soil vol. only.
[b] rhizosoil plus transition zone.
[c] rhizo-soil plus bulk soil vol.

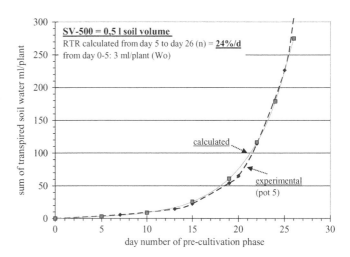

FIGURE 48.10 Relation between calculated and experimentally determined soil water losses (RTR = ml/(g DM*d)) during typical pre-cultivation period of young rape plants, here grown in a soil volume of 500 ml (Schleiff 2013a).

The value for Yi, the initial ECsw at the experimental start, was calculated accordingly:

$$Yi = ECw * \left[(SWCa/SWCt) \right] \rightarrow SWCa/SWCt = 0.67 \quad (48.10)$$

where ECw is the EC (dS/m) of the water applied at the experimental start; SWCt (30 Vol.%) is the soil water content achieved after watering; and SWCa ((SWCt) 30 Vol.% minus(SWCre) 10 Vol.% = (SWCa) 20 Vol.%) is the amount of water required to raise the SWC from an initial value of 10 Vol.% (SWCre before watering) to an SWCt of 30 Vol.%. The salinity of the soil water before watering was considered as negligible.

Example for watering a pot with ECw of 11.0 dS/m from 10 to 30 Vol.%:

$$Yi = 11 \text{ dS/m} * (20 \text{ Vol.}\% / 30 \text{ Vol.}\%) = 7.3 \text{ dS/m} \quad (48.11)$$

Following earlier model calculations (Figure 48.5), we expected that for rape plants there would be a significant contribution coming from the non-rhizospheric soil fraction to cover the transpiration water demand during a water depletion period. For a direct comparison of any two crops or varieties, simultaneous pre-cultivation and saline treatment are recommended.

48.4 PRESENTATION AND DISCUSSION OF SOME OUTSTANDING RESULTS

Critical evaluation of full data sets from experiments of this complexity overburdens the scope of this paper. Instead, the next sections focus on selected results that appear most relevant to improving understanding of the dynamics between root and soil under saline conditions.

Section 48.4.1 focuses on the uptake of rhizospheric soil water under saline conditions, when roots significantly differ in their morphologies. Young leek and rape plants were taken as our model plants because their roots are known to form different rhizospheric soil volumes due to significant morphological differences (Jungk et al. 1982).

Section 48.4.2 focuses on the potential contribution of non-rhizospheric soil fraction that can improve plant water supply and in consequence crop salt tolerance. Growing roots can expand into the soil fraction outside the rhizocylinder and, thereby increase rhizospheric soil volume. It is obvious that roots differ greatly in their ability to explore water the non-rhizospheric soil fraction, even under saline conditions. To test the ability of crops to profit from the water outside the rhizocylinder, we took rape as a model plant. From research on nutrient uptake by roots, we know that their roots may expand relatively rapidly and explore non-rhizospheric soil nutrients and soil water (Jungk 2002). Consequently, we expect that rape plants can provide us with significant data about the non-rhizospheric soil water as a contributing factor to improve plant water supply and crop salt tolerance.

48.4.1 IMPACT OF RHIZOSPHERIC SOIL VOLUMES

48.4.1.1 Patterns of Soil Water Extraction and Transpiration

In Figure 48.11, the experimental phase for 'medium salinity treatment' (see Table 48.2) commenced with watering the well-rooted soils of pre-cultivated plants (leak and rape) with water of an ECw 11.0 dS/m (= 100 mmol/l NaCl). At the experimental commencement root-DM was 390 mg/plant for rape and 520 mg/plant for leek; shoot-DM for rape 750 mg and for leek 1150 mg. Shoot/root ratios (DM based) for both plants were similar, 1.9 for rape and 2.2 for leek.

In this experiment, all soil volume was considered as rhizospheric (250 cm³/pot). Plants were then exposed to constant lighting in a growth chamber for up to 42 hours. Plant transpiration was measured bi-hourly by weighing pot water losses. The initial soil water content after watering was about 40 Vol.% (= 100% PWC) for both plants, the initial soil water salinity ECsw 8.2 dS/m (Ψo = ~3.0 bar). Temporal development of pot water losses (Vol.%) and soil water salinity (ECsw) varied greatly between plants, up to the termination, when both plant species strongly wilted. The total water uptake of rape was 95 ml/pot and terminated at 7 Vol.% soil water, which was near to the PWP (permanent wilting point), and soil water salinity ECsw approached 40 dS/m (Figure 48.11). Simultaneously total water uptake by leek was only 42 ml/pot and was already negligibly low at 25 Vol.% soil water (~75% of PWC) when ECsw was only about 14 dS/m. It is clearly shown that pot water loss and plant transpiration is much higher by rape as compared to leek plants (Figure 48.11).

We conclude that the experimental set-up can quantitatively reflect the anticipated differences in water depletion from saline, rhizospheric soil fractions of roots that carry variant morphological traits.

48.4.1.2 Root Water Uptake as Related to Soil Water Content and Salinity

The combined impact of decreasing soil water content (soil matric water potential Ψm) and increasing soil water salinity (soil osmotic water potential Ψo) around roots, respectively, in the rhizosphere can limit root water uptake during periods of soil water depletion from saline soils (Schleiff and Schaffer 1984; Homaee and Schmidhalter 2008). However, only little is known about root water uptake from a saline rhizosphere, when roots of different morphology are employed (Schleiff 2008, 2013). Detailed knowledge of this aspect is required for the development of sound views on options to improve the salt tolerance of crops (see Figure 48.4).

In Figure 48.12, we plot pot water losses during the experiment (= water extraction by roots) at decreasing soil water contents and increasing the salinity of soil water. For both plant species, the initial soil water content was 100% PWC (SWC 40 Vol.%) with a soil water salinity of ~8 dS/m (potential osmotic Ψo of 2.9 bar resp. –0.29 MPa) at the outset of the experiment. Over a period of 38 hours, terminating with strong wilting of both plants, the rape extracted nearly all soil water down to about 10 Vol.%, while soil water salinity

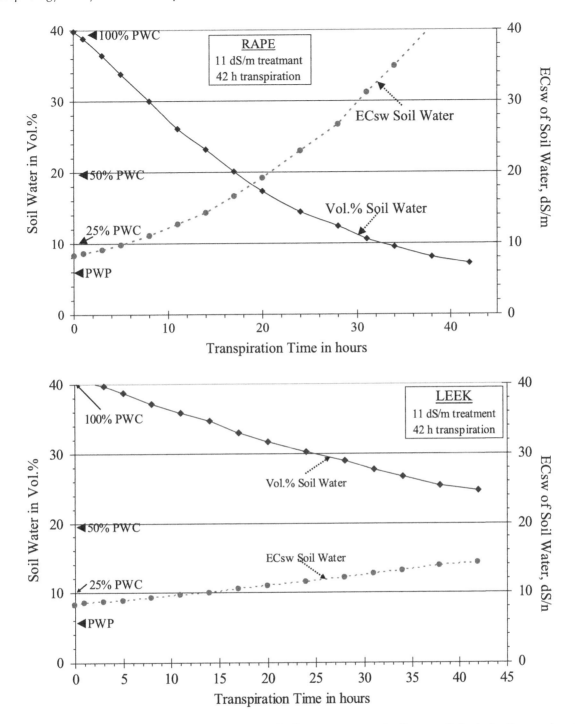

FIGURE 48.11 Effect of transpiration time on the development of soil water contents and soil water salinity during the experimental phase of 42 hours for rape and leek plants.

(ECsw) increased to 35 dS/m (Ψo of 12.5 bar). Water uptake rates were highest with 9.3 ml h^{-1} g^{-1} root-DM at the experiment commencement and had decreased to 2.3 ml h^{-1} g^{-1} root-DM at the experiment's termination.

The pattern of soil water extraction by leek roots differed greatly from that of rape plants. At the outset, when the soil water content was highest, and soil water salinity lowest, soil water extraction was 2.6 ml h^{-1} g^{-1} root-DM or only 28% of the water uptake rate of rape roots. Depletion of soil water due to plant wilting already ceased when soil water contents were

yet relatively high at about 25 Vol.%, while soil water salinity was relatively low at 14 dS/m (Ψo of 5.0 bar). By the termination of the experiment, the water uptake rate of leek had dropped to 1.4 ml h^{-1} g^{-1} root-DM or 54% of the initial uptake. It is obvious that there are significant differences in soil water depletion from saline soils when plants have roots that differ in morphology. Despite higher root and shoot masses for leek, water depletion was significant in comparison to rape.

We outlined earlier that in saline soils it is the osmotic soil water potential of the rooted layer that is most crucial for root

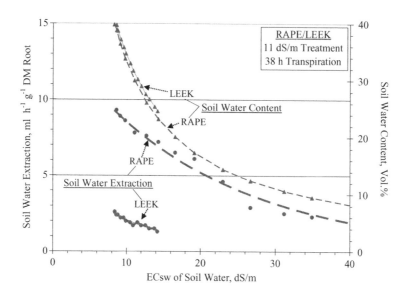

FIGURE 48.12 Water uptake by roots of young leek (salt sensitive) and rape plants (moderately tolerant) from simultaneously decreasing soil water contents and increasing soil water salinity.

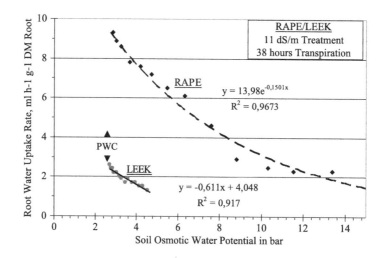

FIGURE 48.13 Water uptake by roots of young leek and rape plants when related to the development of the soil osmotic water potential.

water uptake. But crops differ greatly in their ability to 'live' with saline soil water. Our VeTe offers the chance to compare root water uptake rates of both species when roots are exposed to increasing Ψo. In Figure 48.13, we related their uptake rates to the soil osmotic water potentials after 11 dS/m watering. The highest RWU for rape roots was 9.3 ml h^{-1} g^{-1} root-DM when they were exposed to Ψo of –2.9 MPa, but leek roots achieved an RWU of only 2.6 ml h^{-1} g^{-1} root-DM, which is ~70% less than for rape. When we consider uptake rates of around 2 ml h^{-1} g^{-1} root-DM as inadequate for any relevant growth, in the case of leek Ψo-values exceeding 3 bar (–0.3 MPa) were already most critical, but for rape only around 10 bar (–1.0 MPa).

In conclusion, these results provide a significant hint that water uptake from saline soils closely relates to the morphological traits of roots and therefore also plays an important role in crop salt tolerance.

48.4.1.3 Root Water Uptake from the Rhizospheric Soil Fraction as Related to the Total Soil Water Potential

The total water potential Ψt of soils consisting of Ψm and Ψois is often a useful criterion for evaluating the water stress to crops. We apply this criterion to our experiment, where roots of different traits were exposed to rhizospheric soils of a low and a moderate level of soil salinity. In the low saline treatment (Figure 48.14), plants were watered with water of 3 dS/m. After watering, roots of both plants were exposed to a PWC of around 100% at a Ψt of –0.18 MPa (for leek) resp. –0.2 MPa (for rape). Water uptake rates were highest at the outset of the experimental phase but differed greatly among species. Initial water uptake by rape roots (18.2 ml h^{-1} g^{-1} root-DM) was three times that of leek roots (6.1 ml h^{-1} g^{-1} root-DM). Water uptake by rape ceased, when Ψt had dropped

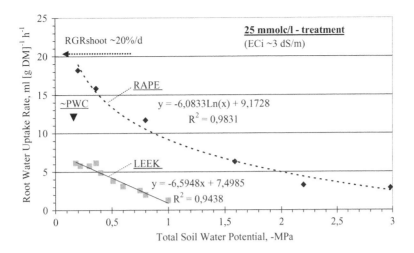

FIGURE 48.14   Water uptake by roots, related to total soil water potential (–MPa) from a low saline soil

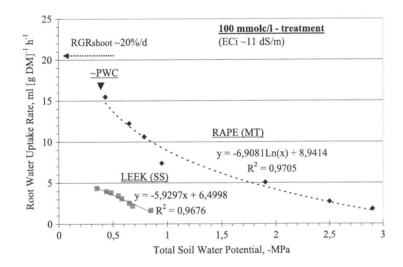

FIGURE 48.15   Water uptake by roots, related to total soil water potential (–MPa) of moderate soil salinity.

to a value of –2.0 MPa, while in case of leek a Ψt near to –1.0 MPa already ceased water uptake.

In the moderate salinity treatment (Figure 48.15), plants were irrigated with water of 11 dS/m salinity. At the experimental outset, when PWC was around 100%, plant roots were exposed to significantly lower Ψt-values of –0.43 MPa for rape and –0.35 MPa for leek. In comparison with the low saline treatment, the highest water uptake rate was 15% lower for rape (15.5 ml h^{-1} g^{-1} root-DM) and 30% even lower for leek (4.3 ml h^{-1} g^{-1} root-DM). Water uptake rates were insignificant (<2 ml h^{-1} g^{-1} root-DM), when Ψt had dropped to about –0.8 MPa for leek and –2.5 MPa for rape, when strongly wilting was reached.

48.4.1.4   Resistances to Plant Water Supply at the Soil/Root Interface

Movement of water through the soil–plant–atmosphere continuum follows the principals of Ohm's law (R=U/I), where resistance (R) is related to the soil water potentials (U) divided through the water uptake rates (I) of roots (Feddes

1981; Kirkham 2005). This relationship is an accepted tool for quantifying evaluation water uptake from soils. Low R-values at the soil/root interface correspond to a good water supply for plants, higher R-values to a critical water supply.

We apply this correspondence to our evaluation of crop water supply under saline soil conditions, when the total soil water potential (Ψt) is formed by integrating the matric (Ψm) with the osmotic (Ψo) soil water potentials.

It is obvious, there is a clear impact of Rt (=total soil/root resistance) on water uptake rates by roots, but there are also significant differences among species, at least if grown under saline soils conditions (Figure 48.16). At the outset, when PWC was ~100% and transpiration began, root water uptake for rape was about 350% higher (9.3 ml h^{-1} g^{-1} root-DM) in comparison to leek roots (2.6 ml h^{-1} g^{-1} root-DM), at low Rt-values of <1 for both plants. Subsequent root water uptake yielded decreasing soil water potentials (Ψm and Ψo) for both crops. In the case of rape roots, the higher water uptake rates yielded significantly increasing Rt-values of 5–10, at which plants wilted strongly. At all Rt-values, water uptake by rape

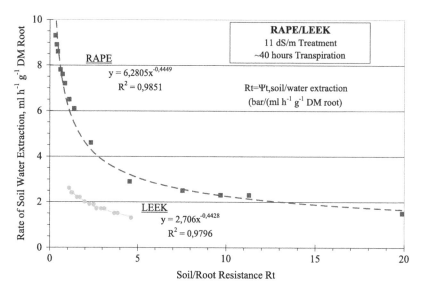

FIGURE 48.16   Water extraction by roots of rape (MT) and leek (SS) related to resistances at the soil/root interface in a saline soil.

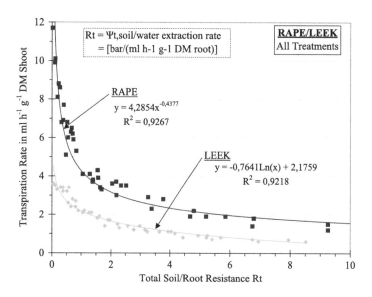

FIGURE 48.17   Transpiration of rape (MT) and leek (SS) shoots in relation to the soil resistance at soil/root interface (Rt), over a wide range of salinity treatments (0, 25, 50, 100, 150, 200, 250 meq/l NaCl-treatments; unpublished data).

roots was significantly higher by comparing with leek roots. In our concept, it is the difference in morphological traits of plant roots that contributes essentially to this result.

We also tested this concept by exposing rape and leek plants to a wide range of salinity treatments to compare non-saline treatment with heavy saline treatments (applications from 0 to 250 meq/l NaCl-water; Figure 48.17). The lowest Rt-values of 0.02 (rape) and 0.05 (leek) corresponded with the highest transpiration rates of 11.7 ml h^{-1} g^{-1} shoot-DM (rape) resp. 3.6 ml h^{-1} g^{-1} shoot-DM (leek), in the non-saline treatments. Ultimately, plant wilting was associated with Rt-values >2. At low soil salinity (application of 25 meq/l NaCl-water), Rt-values rose considerably. For rape/leek the Rt-values increased to 0.11/0.29, with slightly reduced transpiration rates of 10.1 ml h^{-1} g^{-1} shoot-DM (rape) resp. 3.4 ml h^{-1} g^{-1} DM (leek), as PWC were close to 100%. Plant wilting occurred at Rt-values of >6 for rape resp.

>3 for leek. Leek transpiration rates around 3.0 ml h^{-1} g^{-1} shoot-DM corresponded to Rt of ~0.3, but for rape a much higher Rt-value of 2.2 occurred. In conclusion calculation of soil/root-resistances can serve as a useful tool for analyzing quantitative processes at soil/root interfaces for roots having variant morphological traits most relevant to our understanding growth of crops under brackish water irrigation.

48.4.2   IMPACTS OF THE NON-RHIZOSPHERIC SOIL FRACTIONS

The amount of soil water directly and readily available to plants from the rhizospheric soil fraction is estimated to meet the transpiration demand of well-growing crops for a few hours at most (see Figure 48.7). This hints at the value of another source of water supply to plants, contributed by the non-rhizospheric soil fraction. This supplemental water source can be rendered plant available by two means, (1) from roots that grow into the bulk soil and (2) from a flow of water from outside the rhizospheric soil, reliant on a gradient of the soil water potential to roots. Model calculations presented earlier (Figure 48.4 and 48.5) estimate the quantities of soil water gained from additional root penetration (growth) into non-rhizospheric soil in the ranges from less than 1 ml/10 m RL to nearly 30 ml, depending upon soil characteristics and root traits. Thus, soil volumes explored by growing roots are expected to contribute significantly to plant water supply, both under non-saline and saline growth conditions. At present, the measurement of root water uptake from the different root fractions by instruments (e.g., electrodes, tensiometers) is to be extremely difficult, perhaps even impossible. But instead of an instrumental solution, we propose the modification of our VeTe to achieve data on the impact of non-rhizospheric soil fractions upon plant water supply. In the following sections, we present initial results obtained from our VeTe, when contributions of the non-rhizospheric soil fractions are included. We selected young rape as our test plant, as its root system is well-known for its ability to explore growth factors (i.e., nutrients) from the non-rhizospheric soil fraction more efficiently than many other crops (Jungk 2002). But we shall not present and analyze all experimental data collected with our VeTe, as this will overwhelm the scope of this section.

A better understanding of interactions between roots and their surrounding soil fractions is important for discovering ways to improve salt tolerance of soil-grown crops, especially with regard to improve brackish water irrigation techniques and breeding of salt tolerant crops.

48.4.2.1 Patterns of Water Extraction from Soils at Various Rooting Densities

In the experiments presented in Section 48.4.1, we focused on the analysis of root water uptake from the rhizospheric soil fraction of roots that differed in their morphological traits.

Plant roots were exposed to varying salinity levels, but all to a very small soil volume of 0.25 l/pot only, which assured soils of a very high rooting density. It is the objective of the next section to present results from a modified VeTe, which offers the chance to obtain quantitative data on the contribution of the non-rhizospheric soil fractions to supply plants with soil water, at varying soil salinity. As already described in Section 48.3.3.2, the modification concerns the pre-cultivation of plants in different SV (0.25 l, 0.5 l, 1.0 l) to achieve rooted soils of different rooting densities for saline treatments. In the treatment SV-0.25 l roots are exposed to the rhizospheric soil fraction only, but in the treatments SV-0.5 l and SV-1.0 l they are additionally exposed to non-rhizospheric soil volumes.

48.4.2.1.1 Patterns of Water Extraction from Non-Saline Soil

Roots of rape seedlings establish a root system, whose morphological traits favor the effective uptake of nutrients from the rhizospheric soil fraction and also from the non-rhizospheric soil (Jungk 2002). Therefore the objective of our next experiment using rape plants growing in pots of increasing soil volumes (decreasing rooting densities) is to tabulate the course of root water uptake under non-saline conditions. Principally our model calculations (Figures 48.4 and 48.5) show that rape roots are able to explore water from the non-rhizospheric soil fraction such that plant water supply will not significantly suffer, once the 'readily available' water from the rhizospheric soil fraction is depleted.

In Figure 48.18, over the first 5 to 6 hours after watering, the amount of soil water extracted (triangles) reaches about 40 ml in all treatments (SV of 0.25 l, 0.5 l, 1.0 l). But SWC (dots) did not yet drop below 15 Vol.% (50% PWC). Therefore, the first phase of 5–6 hours, the 'readily plant available water' of the rhizospheric soil exclusively met the water requirements of plants in all treatments. It seems that the non-rhizospheric soil water was hardly involved in the water supply of the plants at all.

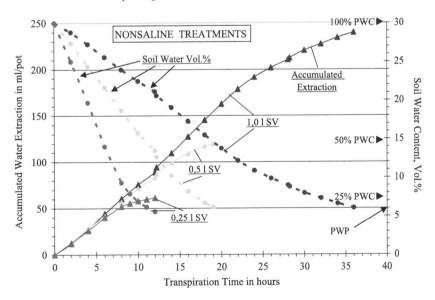

FIGURE 48.18 Time required for plant transpiration to affect the course of soil water contents and accumulated water uptake from non-saline soil, with different rooting densities (from SV of 0.25 l, 0.5 l, 1.0 l).

In the next phase, from 6 to 12 hours after watering, significant differences in the course of soil water extraction and SWCs among treatments appear. In this 6 hour-phase, the volume of extracted water increased relatively little, to a mere 60 ml, in the highest rooting density (SV-0.25 l) that involved the rhizospheric soil fraction only. The drop of SWC to about 6 Vol.% corresponds to the PWP that terminated water extraction. The lower rooting densities, when non-rhizospheric soil water was involved, produced a significantly higher accumulated water uptake. Concurrently, water extraction in the treatment 0.5 l-SV was 80 ml and even 85 ml at 1.0 l-SV, associated with SWCs of 13 Vol.% (SV-0.5 l) resp. 21 Vol.% (SV-1.0 l) in the soil average.

At 12–19 hours after watering, water extraction increased, at SV-0.5 l treatment from 80 to 120 ml, equivalent to an SWC of 6 Vol.% (PWP). Thus, at the SV-0.5 l level (high rooting density) the root system succeeded in exploring the total rhizospheric and non-rhizospheric soil water. Simultaneously, water extraction in treatment SV-1.0 l accumulated up to 160 ml that yielded a SWC of 16 Vol.%.

The transpiration period ran from 19 to 36 hours of transpiration, as plants strongly wilted and the experiment ended. The total water uptake approached 240 ml at an SWC of 6 Vol.%, corresponding to the PWP. We conclude that rape roots prove very efficient in their ability to explore water from the non-rhizospheric soil to meet transpiration demands of the shoot, at least at a lower level. This experimental set-up can be applied with other plants to test the efficiency of their roots to explore non-rhizospheric soil water, which is important to maintain growth rates under decreasing soil water conditions.

48.4.2.1.2   Comparison of Water Extraction from Soils Having Various Salinity Levels and Decreasing Rooting Densities

Reduced water uptake by roots is the most critical factor for plant growth in saline soils resp. crop salt tolerance. But, in

Table 48.1, plants differ greatly in their responses to increased soil salinity. To understand, quantitatively, we initially will analyze experiments on rape plants, exposed to increasing soil salinity and the resultant water uptake occurring in soils of divergent rooting densities.

Treatment at 'very high rooting density': 0.25 l soil volume

Figure 48.19 focuses on the impact of transpiration time on soil water content and water uptake by a root system at a very high rooting density and all soil is considered rhizospheric (SV-0.25 l). During the first 8 hours of transpiration, soil water content drops nearly linear but reaches different endpoints. In non-saline soil, SWC dropped to 9 Vol.%, but it fell much less in treatments of 8 dS/m (to 14 Vol.%) resp. 16 dS/m (to 18 Vol.%). Correspondently, the first 8 hours accumulated water extraction proved highest, under non-saline conditions (52 ml = 100%), but fell to 77% (40 ml) resp. 58% (30 ml) in the 8 dS/m resp. 16 dS/m treatments.

Water uptake in the non-saline treatment terminated after 12 hours transpiration when soil water content reached the PWP (6 Vol.%), and the total amount of depleted soil water was at 61 ml. In the 8 dS/m treatment, the total amount of extracted soil water also achieved 61 ml, but plant transpiration time nearly doubled, reaching 22 hours. Under the 16 dS/m treatment, plant transpiration lasted 24 hours before strong wilting set in, and water extraction terminated at an SWC of 22 Vol.%. In this treatment, not only did the SWC stop the water uptake, but in addition rhizospheric soil water salinity rose to 20 dS/m (~–0.7 MPa). Small jumps in the progression curve indicate the interruption of lighting at nighttime, e.g., after 12 hours transpiration.

Treatment at 'high rooting density': 0.5 l soil volume

In Figure 48.20, we plot the course of SWCs and water extraction at various salinity levels, when plant roots were exposed to an increased soil volume (0.5 l-SV) that exceeded the volume of the rhizospheric soil fraction. We expected

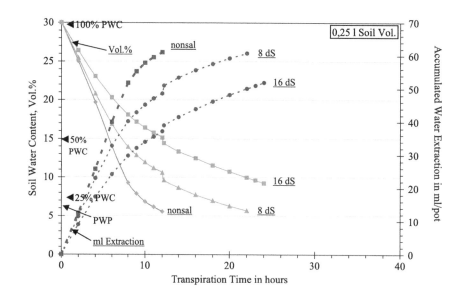

FIGURE 48.19   Transpiration time as affecting the course of soil water extraction and SW from the rhizospheric soil (SV-0.25 l: very high rooting density) after applying waters of different salinities (0.3, 8.0 and 16.0 dS/m).

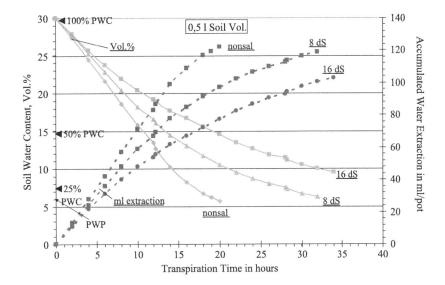

FIGURE 48.20 Transciption time as affecting the course of soil water extraction and SWC from the fraction 'rhizospheric soil inclusive transition zone' (SV-0.5 l: high rooting density; see Figure 48.3) after applying waters of different salinities (0.3, 8.0, and 16.0 dS/m).

this treatment to yield water uptake gains from the soil water fraction outside the rhizospheric soil. And, for rape roots, the gains should prove relatively important because of their morphological traits (see Figure 48.5).

In the non-saline treatment of 0.5 l-SV, plants extracted 72 ml of soil water during the first 10 hours of transpiration, 14 ml (~20%) more than in the 0.25 l-SV treatment (58 ml). The simultaneous decrease in the SWCs differed greatly too. For the treatment of 0.5 l-SV, the SWC dropped from 30 to 15.5 Vol.%, but at 0.25 l-SV treatment from 30 to ~6.5 Vol.% resulted, which is a value very close to the PWP. Soil water extraction terminated after 20 hours after ~125 ml of the soil water was taken up at 6 Vol.% SWC. Thus, for both rooting densities, the total amount of plant available soil water was taken up, which shows a significant contribution of the non-rhizospheric soil fraction to plant water supply.

At 8 dS/m, total water extraction was slightly affected by salinity, achieving 120 ml at an SWC of 6.5 Vol.% and took the plant 32 hours instead of 20 hours in the non-saline treatment. Salinity hardly affected the total amount of plant available soil water, but lower rates of water uptake by roots increased the time needed to extract all the soil water, which was significantly prolonged by 60%. This also shows that all the soil volume (rhizospheric fraction and transition zone) contributed to the water supply of rape plants, but uptake rates decreased with time due to falling PWCs and simultaneously increasing soil water salinity.

After watering with 16 dS/m, the total amount of extracted soil water dropped to 100 ml, and a transpiration time of 34 hours produced a final SWC of 9 Vol.%. By comparison with the SV-0.25 l treatment (when roots were exposed to the rhizospheric soil fraction only), access to water of the non-rhizospheric transition zone (in the 0.5 l-SV treatment) contributed to a prolonged water depletion of plants for some 10 hours. We conclude, for this treatment, it was not only the matric soil water potential Ψm that finally stopped root water uptake,

but the osmotic soil water potential Ψo also played a significant role in uptake termination.

Treatment at 'moderate rooting density': 1.0 l soil volume

In the SV-1.0 l treatment, plant roots were exposed to the largest soil volume of this experiment, including the rhizospheric and non-rhizospheric soil fractions. In the non-saline treatment, the PWP of the soil stopped root water uptake, once a total of 240 ml was extracted after 36 hours of transpiration (Figure 48.21).

Under 8 dS/m, simultaneous water extractions achieved 185 ml/pot (23% less than non-saline) and under 16 dS/m 130 ml only (46% less), which corresponded SWCs of 11.5 Vol.% in the 8 dS/m resp. 17 Vol.% in the 16 dS/m treatment at the experimental end. So, for both saline treatments, it was not just the Ψm of the soil that limited soil water extraction, but soil water salinity (Ψo) was also strongly involved. The non-saline treatment plants wilted strongly, after 36 hours of transpiration, but under saline conditions plant transpiration continued at a low level, without stopping, even after 40 hours.

48.4.2.2 Impact of Plant Transpiration on Development on Soil Water Contents and Soil Water Salinity

Plant transpiration and the exclusion of ions from root uptake are the driving forces that cause the build-up of salinity gradients between the rhizospheric and non-rhizospheric soil fractions. Consequently, salt concentrations near the roots of transpiring plants are expected to be significantly higher than in bulk soil. In this section, we analyze the course of soil water salinities, when roots are exposed to decreasing rooting densities of soils. Application of equation (48.9) on the various treatments of the VeTe offers a chance to calculate the impact of pot water losses upon the soil water losses (PWC) and soil water salinities (ECsw) that roots are exposed to during water depletion (Figure 48.22 through 48.24).

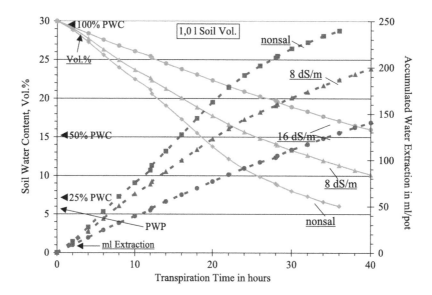

FIGURE 48.21 Effects of transpiration time on soil water extraction and SWC (Vol.%) from soils at moderate rooting density (SV 1.0 l: rhizospheric plus non-rhizospheric bulk soil) after watering by different salinities (0.3, 8.0, and 16.0 dS/m).

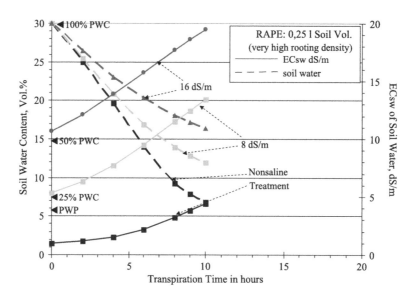

FIGURE 48.22 Impact of transpiration time on the course of soil water contents and salinity of the rhizospheric soil water (SV-0.25 l) after applying the water of different salinities.

Treatment at 'very high rooting density' (SV 0.25 l)

In this treatment, where total soil volume was rhizospheric soil, it took the plant, in the non-saline treatment, 10 hours to deplete a total amount of plant available soil water (58 ml depletion) from a water content of 30 Vol.% down to the PWP of 6 Vol.% (Figure 48.22). Concurrently, the SWC in the 8 dS/m treatment fell to 12 Vol.% (45 ml depletion), and in the 16 dS/m to 17 Vol.% (34 ml depletion) only. Increases of soil water salinity in the non-saline treatment are minimal (< 5 dS/m at PWP). But in the 8 dS/m treatment, soil water salinity increased from 5 to 18 dS/m (~350%) and in the 16 dS/m treatment from 11 dS/m to 19.5 dS/m (~80%). These results indicate a strong impact from soil water salinity upon the availability of the rhizospheric soil water.

Treatment at 'high rooting density' (SV-0.5 l)

In the SV-0.5 l treatment, it took the non-salinated plant 18 hours to deplete the total amount of plant available soil water (116 ml depletion; SWC of 6 Vol.%) down to the PWP, at which plants wilted strongly (Figure 48.23). It shows that roots of rape could continuously explore all non-saline soil water, without restrictions. At that point of time, when the non-saline treatment was depleted, the SWCs of the saline treatments had dropped to 11.5 Vol.% (8 dS/m treatment) resp. 14 Vol.% (16 dS/m treatment) only. In both saline treatments, salinity prolonged the period of water depletion considerably, and was negligibly low, as SWCs dropped to 7.5 Vol.% (8 dS/m treatment: after 112 ml depletion) resp. 12 Vol.% (16 dS/m treatment: after 92 ml depletion) and ended in both treatments

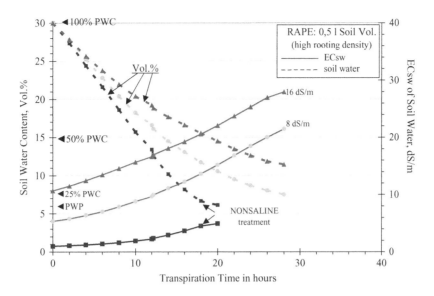

FIGURE 48.23 Impact of transpiration time on the course of soil water contents and soil water salinity at high rooting density (0.5 l-SV) after applying the water of different salinities.

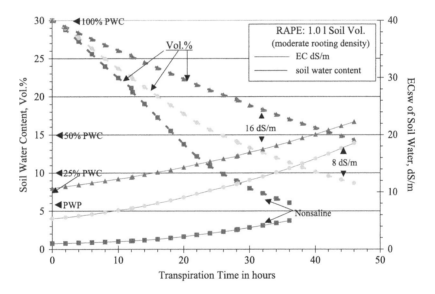

FIGURE 48.24 Impact of transpiration time on the course of soil water contents and soil water salinity at moderate rooting density (1.0 l-SV) after applying water of different salinities.

after 28 hours only. Comparisons of soil water losses between non-saline and saline treatments show a salt-induced decline of water uptake by plants, which can contribute to a prolongation of plant transpiration and thus help the plant to survive periods of critical availability of soil water. In this experimental phase, soil water salinity increased from an initial 5.5 dS/m to 22 dS/m (for 400%; 8 dS/m treatment) resp. from 11 dS/m to 28 dS/m (increase of 255%; 16 dS/m treatment) by the end of experiment, when plants wilted.

Treatment at 'moderate rooting density' (SV-1.0 l)

The course of SWCs and ECsw under lower rooting densities was measured at SV-1.0 l (Figure 48.24). Under non-saline conditions, it took the plant 36 hours to deplete all plant available soil water (PWP at 6 Vol.%; 239 ml depletion).

Simultaneous (after 36 hours) the SWCs of the saline treatments dropped to about 11 Vol.% (8 dS/m treatment) resp. 17 Vol.% (16 dS/m treatment) only. As predicted from earlier model calculations (Figures 48.4 and 48.5) these results confirm that roots of rape are most effective to explore soil resources such as soil water from the non-rhizospheric soil fractions, too. In both salinity treatments, root water uptake became negligible after some 45 hours of transpiration, when PWCs had dropped to 8.5 Vol.% (8 dS/m treatment: 210 ml depletion) resp. 14 Vol.% (16 dS/m treatment: 154 ml depletion). Parallel soil water salinities (ECsw) increased with time from 4 to 18 dS/m (8 dS/m treatment) resp. from 8 dS/m to 22 dS/m (16 dS/m treatment) for 450 and 270% respectively. We conclude that, in case of rape, the water of a relatively large soil volume is available to roots, because of its effective

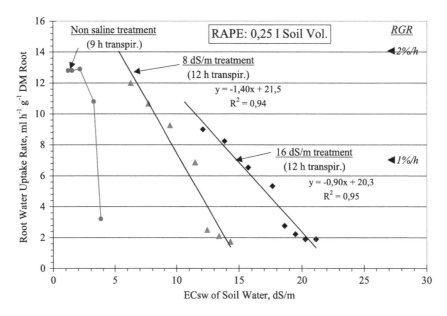

FIGURE 48.25 Water uptake rates of roots related to soil water salinity (ECsw) at very high rooting density (all soil volume 'rhizospheric').

root system. For comparison, further experiments with plant roots of other morphological traits are recommended.

48.4.2.3 Root Water Uptake as Related to Soil Water Salinity at Varied Rooting Densities

Soil water salinity in the rooted soil layer limits root water uptake and consequently plant growth. Plant roots evidence big differences in their uptake of water from soils of increasing soil water salinities. Earlier we saw that very little is known about the quantitative effects of increasing soil water salinity when plant roots differed in their morphological traits. As presented in Figure 48.8, root morphology is expected to be most relevant for the ability of roots to explore water from both soil fractions, the rhizospheric and the non-rhizospheric. This section presented some significant results on water uptake of rape roots when roots were exposed to various levels of soil water salinity at different rooting densities.

Water uptake at 'very high rooting density' (SV 0.25 l)

When plants were exposed to the non-saline rhizospheric soil fraction only, with no external contribution of non-rhizospheric soil (Figure 48.25), water uptake achieved rates between 11 to 13 ml h^{-1} g^{-1} root-DM, during the first 8 hours of transpiration. Water uptake fell sharply after 8 hours, to less than 3 ml h^{-1} g^{-1} root-DM, at an SWC below 8 Vol.%.

In the 8 dS/m treatment, soil water salinity increased from ~5 to ~14 dS/m, after SWC had dropped from 30 Vol.% (100% PWC) to ~11 Vol.% during 12 hours of plant transpiration. By comparing to the non-saline treatment, initial water uptake was slightly reduced to 12 ml h^{-1} g^{-1} root-DM. The total transpired water dropped from 55 ml during 9 hours (non-saline) to 48 ml during 12 hours (8 dS/m), which was an average of 4 ml h^{-1} g^{-1} root-DM over all the transpiration time (8 dS/m treatment).

After watering with 16 dS/m, rhizospheric soil water salinity was ~11 dS/m (at 30 Vol.% SWC) at the outset and increased

after 12 hours of transpiration to 21 dS/m at 15 Vol.% SWC (Figure 48.25). The total amount of transpired water fell to 37 ml only, some 33% lower than in the non-saline treatment and 23% less than in the 8 dS/m treatment. The initial water uptake rate of 9 ml h^{-1} g^{-1} root-DM at ECsw 11 dS/m was 31% less by comparing to the non-saline treatment (13 ml). Water uptake was already negligible (<2 ml h^{-1} g^{-1} root-DM), after 12 hours transpiration, when ECsw had dropped to ~20 dS/m at SWC of 15 Vol.%. The average RWU was 3.1 ml h^{-1} g^{-1} root-DM, which was the lowest value of all treatments.

Treatment at 'high rooting density' (SV-0.5 l)

In the 'high rooting density' treatment, we assume a full contribution by rhizospheric soil water and, in addition, an impact from the soil fraction in direct contact with the rhizospheric soil fraction, which surrounds the rhizocylinder volume (see Figure 48.2: 'transition zone').

After a non-saline treatment, a plant took 18 hours to deplete 116 ml soil water and to reduce SWC from 30 Vol.% at the outset to 6.6 Vol.% (near to PWP), after water uptake terminated. At the outset, water uptake rates for 'very high' and 'high rooting density' reached almost identical values of 13 ml h^{-1} g^{-1} root-DM (Figures 48.25 and 48.26). The average water uptake rates at both levels of rooting density occurring before plant wilting did not significantly differ with values of 6.1 (SV-0.25 l) and 6.4 ml h^{-1} g^{-1} root-DM (SV-0.5 l). A sudden drop of RWU to below 2.0 ml h^{-1} g^{-1} root-DM occurred when SWC fell under 6.5 Vol.%.

Under 'irrigation,' with water of 8 dS/m, the initial soil water salinity reached an ECsw of 5.3 dS/m (SWC of 30 Vol.%). After depleting 106 ml soil water during 24 hours of plant growth, the ECsw rose up to 18.5 dS/m at 8.5 Vol.% SWC (Figure 48.26). The initial RWU near 100% PWC achieved 11 ml h^{-1} g^{-1} root-DM, which was about 15% less than in the non-saline treatment. When comparing the RWUs of the 8 dS/m treatments during the transpiration phase, the average RWU was about

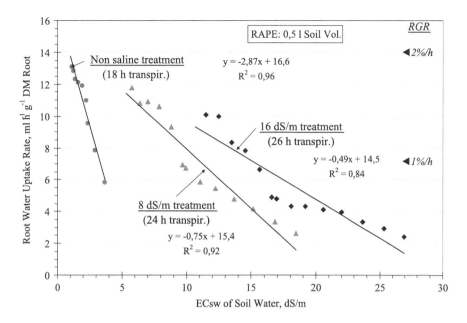

FIGURE 48.26 Water uptake rates for roots, related to soil water salinity (ECsw) at high rooting density, where small part of soil volume is non-rhizospheric (transition zone).

10% higher in SV-500 ml (106 ml/24 hours = 4.4 ml/hours) than in the SV-250 ml (48 ml/12 hours = 4.0 ml/hours). RWU nearly stopped at ECsw of 13 to 14 dS/m in the SV-250 ml treatment (exclusive rhizospheric soil), but only near 19 dS/m in the SV-500 ml treatment, when some non-rhizospheric soil of the 'transition zone' was involved.

When the plant was watered with 16 dS/m, initial ECsw climbed to 10.6 dS/m after watering (Figure 48.26). Plant transpiration during the following 26 hours caused an increase of ECsw to 27 dS/m (Ψo 9.7 bar) on average and a fall of SWC to 11.9 Vol.% when RWU dropped to critical values of less than 2.5 ml h^{-1} g^{-1} root-DM. Total water uptake reached 90 ml/26 hours transpiration, which was 3.5 ml/hours on average and about 20% lower than in the non-saline treatment. The initial RWU at 100% SWC was 10ml h^{-1} g^{-1} root-DM, which was 23% less than under non-saline conditions, but only 5% less than under 8 dS/m watering. Compared with SV-0.25 l treatment ECsw values exceeding 20 dS/m appeared less critical for RWU, which is due to the larger root surrounding soil volume.

Water uptake at 'moderate rooting density' (SV-1.0 l)

It is the objective of the SV-1.0 l treatment to quantitatively analyze the importance of the non-rhizospheric soil fractions for plant water supply, when roots have depleted the 'readily plant available' rhizospheric soil water, and plant water supply depends on availability of soil water outside the rhizocylinder volume (see Figure 48.2: transition zone plus bulk soil). As rape roots are known to form comparatively large soil volume per unit root dry matter, we expect a significant contribution of the bulk soil to ensure an adequate water supply of rape under non-saline and saline soils conditions.

In the non-saline treatment, it took the plant 30 hours of transpiration to deplete a total amount of 220 ml soil water

until the plant wilted, which was on average 7.3 ml h^{-1} g^{-1} root-DM (Figure 48.27). The SWC fell from 30 to 8 Vol.%. At the experimental outset, RWU reached 15 ml h^{-1} g^{-1} root-DM, which was the highest value of all treatments, 2 ml higher (15%) than in both non-saline treatments of lower root densities (SV-0.25 l and SV-0.5 l). RWU fell below 3 ml h^{-1} g^{-1} root-DM when the SWC approached 6 Vol.%. The final leaf area of the plant reached 236 cm².

After watering with 8 dS/m, roots were exposed to an initial ECsw of 5.3 dS/m at a SWC of 30 Vol.% (Figure 48.27). After plant transpiration of 213 ml, the SWC decreased to 8.6 Vol.%, and the ECsw increased to 18.5 dS/m. The initial RWU reached 12 ml h^{-1} g^{-1} root-DM, which was the same value than in both smaller soil volumes, but about 25% lower (3 ml) than in the SV-1.0 l non-saline treatment. On average RWU reached 4.6 ml h^{-1} g^{-1} root-DM, which was somewhat higher than at higher root densities (SV-0.25 l and SV-0.5 l). The lowest RWU of <3 ml h^{-1} g^{-1} root-DM was related to SWC of 8.6 Vol.% and an ECsw of 18.5 dS/m at experimental end.

When the water of 16 dS/m was used, the initial ECsw reached 10.6 dS/m at 30 Vol.% SWC. After 42 hours of plant transpiration (146 ml), the SWC fell to 15.4 Vol.%, and ECsw rose to 20.8 dS/m (~7.5 bar). The initial RWU was 9.5 ml h^{-1} g^{-1} root-DM, which was about 35% lower than in the non-saline treatment and 20% lower than in the 8 dS/m treatment. In average of the 42 hours of transpiration, RWU was nearly 3.5 ml h^{-1} g^{-1} root-DM, which was comparable to the value of the less saline SV-05 l treatment. In the case of this test plant, we observed that plants had not yet wilted when harvested, as there was still a low water uptake of about 4 ml h^{-1} g^{-1} root-DM. We think that this uptake rate was just enough to protect the shoot from wilting. We conclude that, in this treatment, it was not the soil matric water potential (SWC of 15.4

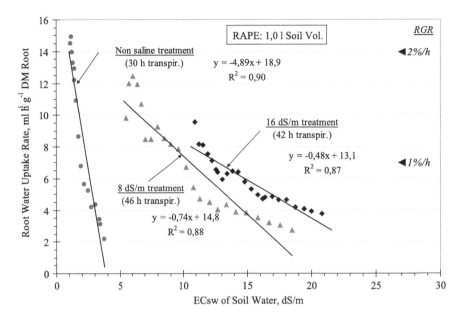

FIGURE 48.27 Water uptake rates of roots as related to the soil water salinity (ECsw) at moderate rooting density (higher part of soil volume is non-rhizospheric).

Vol.%) that was critical for root water uptake, but it was the combined effect with the soil osmotic water potential (~7.5 bar). Water uptake fell to a value that was probably too low for additional shoot growth, but still high enough to maintain the turgor pressure of leaves. We assume that it is an advantage for rape roots to form a large rhizocylinder volume when they explore soil water from the less saline soil outside the rhizospheric soil fraction. This ability improves salt tolerance of irrigated crops.

48.5 SUMMARY AND CONCLUSIONS

When crops grow under brackish water irrigation or in saline soils, their roots are exposed to soil water containing easily soluble salts (Riley and Barber 1970; Schleiff 1981; Homae and Schmidhalter 2008). Soil water salinity harms plant growth due to several impacts, but for most crops water supply is by far the most critical growth factor. Based on this finding, the Salinity Laboratory/USDA in Riverside was the leading institution when the internationally applied concept to control soil salinity in irrigated soils by adequate leaching of salts from the root zone of specific crops was developed (Ayers and Westcot 1985). This concept for rating the salt tolerance of crops considers the impact of vertical salt distribution in the rooted soil layer, but it does not yet consider the impact of the lateral distribution of salts in the root surrounding soil that develops through transpiring plants during periods of soil water depletion after crop watering. Uptake of soil water by roots usually by far exceeds the uptake of salts dissolved in the soil water. Therefore, salts excluded from root uptake accumulate in the soil volume directly contacting roots, the rhizospheric soil fraction, where roots meet soil. Consequently, shoot transpiration and root water uptake can be considered as driving forces to develop lateral salinity

gradients between the highly saline fraction near to roots (the volume of rhizocylinder penetrated by root hairs) and the less saline bulk soil, which is remote from the root surface and not directly root affected. Between both fractions exists a transition zone, which contributes to balance the gradients.

Before presenting our vegetation technique to experimentally determine the effect of salts accumulated in the rhizospheric soil water, we perform some model calculations to achieve a quantitative idea on impacts of root morphologic traits on rhizospheric soil water conditions. In a first step, we will calculate the amounts of water enclosed in the rhizospheric soil volumes when roots differ in their morphology and soils in their texture. In soils around roots forming a small rhizocylinder (radius R of 0.25 mm), the total water amount is <1.0 ml/10 m RL only. On the other hand, when roots form a large rhizocylinder (R of 1.5 mm), they are directly surrounded by 14 ml (light textured soil) to 28 ml/10 m RL (loamy/silty soil). About half of this water can be taken as easily plant available. These significant differences of the rhizospheric soil water (factor up to 70) can significantly affect the salt accumulation process when it is related to the amount of soil water taken up by roots. It is estimated that the plant available water of the non-saline rhizospheric soil fraction can cover the transpiration demand of a well transpiring plant for a period of few minutes only (R = 0.25 mm) to about 5 hours (R = 1.5 mm) depending from soil type and root characteristics.

In saline soils, water depletion from the rhizospheric soil fraction and simultaneous salt exclusion by the roots causes an increase of rhizospheric soil water salinity. During a period of root water uptake, we expect increases that differ significantly among roots, especially when related to the amount of depleted soil water. Model calculations based on an initial soil water salinity of 10 dS/m at 'field capacity' soon after watering resulted in a rhizospheric soil water salinity of 30 dS/m (ECsw), after depleting only <0.5 ml/10 m root length, when roots formed a very small rhizocylinder volume of about 2 cm³/10 m root length (R = 0.25 mm). But when roots formed a large rhizocylinder volume of about 70 cm³/10 m root length (R = 1.5 mm), the critical ECsw of 30 dS/m (Ψo ~10 bar) was achieved after depletion of 14 ml/10 m root length, which is about 30-fold the value for the small volume.

After plants have taken up the readily plant available water from rhizospheric soil fraction, further root water uptake predominantly occurs from the exploration of bulk soil water, from outside the volume of the rhizocylinder. Principally two components contribute to profit roots from the bulk soil water during soil water depletion, that is the flow of soil water to the root surface following a gradient of the soil matric water potential, and the exploration of bulk soil water by root growth, a continuous increase of rhizospheric soil volume. We expect that plant roots differ significantly in their efficiency to explore bulk soil water. To our knowledge, it is not yet clear how soil salinity affects the flow of soil water to roots quantitatively.

Against this background, we offer a new vegetation technique (VeTe) that has the potential to improve our understanding on the complex and dynamic processes occurring between roots and soils, and their quantitative impact on root water uptake and crop salt tolerance. It is the basic strategy of our research concept to divide the rooted soil layer into three lateral, root surrounding soil fractions: (a) the rhizospheric soil, which is in immediate contact with the root (rhizocylinder); (b) the bulk soil, which is remote from the root surface and nearly unaffected from roots; (c) and the transition zone between, which balances existing gradients.

Basically our VeTe consist of two phases, a pre-cultivation phase of few weeks to produce soils of varying rooting densities, very high, high, and moderately, and an experimental phase when the pre-cultivated plants are salinized. In the pre-cultivated treatment 'very high' rooting density all soil volume (i.e., 250 ml SV) is considered as rhizospheric. In the treatment 'high' rooting density (i.e., 500 ml SV) roots are exposed to the rhizospheric soil and additional soil of the transition zone, which surrounds the rhizospheric soil. At 'moderate' rooting density (i.e., 1.0 l SV) roots grow under conditions where they can grow into the bulk soil to cover their demands on growth factors if they have the capacity.

Our experimental design requires plants of the homogeneous shoot and root masses at the outset of the experimental phase when pots of all treatments are supplied with saline waters to achieve different salinities of the soil water and roots are exposed to 100% PWC. Depending on the salt tolerance of the specific crop and soil conditions (i.e., texture), salt concentrations of the applied water in the range from 25 to 250 meq/l can be recommended to study most non-halophytic crops. The watered pots are exposed to well-controlled optimal growth conditions in a climate chamber. During the following hours and days, pot water losses of all treatments are determined bi-hourly by pot weighing until plant wilting. The measured water losses (ml/pot) serve to determine the courses

of soil water contents and soil matric water potentials, soil water salinity, soil osmotic and total soil water potential, relative shoot transpiration, and root water uptake rates. Earlier experiments using maize, sugar beets, wheat, and barley have already indicated that an endless accumulation of salts in the rhizospheric soil water could not occur, but there is a critical value (1981; 1982b, 1983). This value is determined by the water potential of leaves, which cannot be surpassed. For example, for young maize plants, the rhizospheric salinity cannot exceed an osmotic potential in the range of –0.8 to –0.9 MPa (Schleiff 1982a). Rhizospheric soil matric and osmotic water potentials of the same value did not affect root water uptake to the same degree, but decreasing osmotic water potentials depressed water supply of many crops much less than decreasing soil matric water potentials, at least in light textured soils (Schleiff and Schaffer 1984; Schleiff 1986).

The impact of the morphological traits of roots for water uptake from saline soils was investigated using rape and leek as test plants, both grown in 250 ml soil volumes to achieve very high rooting densities (rhizospheric soil only). Rape is known to form a large volume of the rhizocylinder, while leek forms a small volume per unit root-DM. Shoots of both plants are similar in their response to saline growth conditions by osmotic adoption of leaves to a certain degree. When roots of both plants were exposed to a silty soil, holding at 100% PWC saline soil water of ~9 dS/m (Ψo ~2.9 bar), water uptake rates differed significantly. The uptake rate of rape roots (large rhizocylinder volume; 9.3 ml h^{-1} g^{-1} root-DM) was ~350% higher compared to roots of leek (small rhizocylinder; 2.6 ml h^{-1} g^{-1} root-DM). We conclude that the morphology of plant roots is the most promising item for extending our understanding of the salt tolerance of soil-grown crops.

The amounts of soil water of the rhizospheric soil fractions differ significantly. Depending on soil type and root traits, this fraction can meet transpiration demand of well-growing crops from a few minutes to a few hours (i.e., 5 hours) at most. This hints at the prominent value of non-rhizospheric soil water for an adequate plant water supply. Our VeTe was developed to quantitatively determine the significance of rhizospheric soil water salinity for water supply, and salt tolerance can be modified to include the impact of non-rhizospheric soil water by growing young rape plants in soils of increasing soil volumes to achieve soils of decreasing rooting densities (rhizospheric soil (SV 0.25 l); rhizospheric plus non-rhizospheric transition zone (SV 0.5 l); rhizospheric plus bulk soil (SV 1.0 l)). Results showed that the roots of young rape plants are most efficient in the exploration of saline and non-saline soil water from outside of the rhizospheric soil fraction. Water uptake from the SV-1.0 l (moderate rooting density) of non-saline and moderately saline (8 dS/m) treatments ended at equal SWC around the PWP of 6 Vol.%, but in the saline treatment it took the plant 32 hours of transpiration instead of 20 hours (non-saline) to deplete soil water. Thus soil salinity can stretch the time of plant transpiration, which can help plants to survive periods of critical water supply, but RGR is lower. On the other hand, soil salinity can significantly reduce the amount of plant available soil water. Under the 16 dS/m treatment root water uptake nearly ceased already at SWCs of 12 Vol.% (SV-0.5 l) resp. 14 Vol.% (SV-1.0 l), when ECsw values exceeded 22 dS/m.

For future research, we propose the execution of experiments using plants of different root traits and salinized soils of different textures and rooting densities to learn more about dynamic processes happening at the soil/root interface and their relevance for salt tolerance of irrigated crops. We expect valuable suggestions for breeding crops of improved salt tolerance, especially when related to brackish water irrigation.

GLOSSARY

DM: the dry matter of shoot or root mass in g or mg
ECw, ECsw: electrical conductivity of water resp. soil water in dS/m
MT: moderately salt tolerant crop
Ψm, Ψo, Ψt: matric, osmotic, total soil water potential
PWC: pot water capacity in Vol.%
PWP: permanent wilting point
resp.: respectively
R: radius of rhizocylinder in mm or cm
RL: root length in m or cm
Rt, Rm, Ro: total, matric or osmotic soil resistance
RTR: relative transpiration rate of a shoot in ml/(g DM*hour)
RGR: the relative growth rate of the shoot in mg/(g DM/hour) or %/hour or %/d
RWU: root water uptake rate in ml h^{-1} g^{-1} root-DM
SS: salt-sensitive crop
SWC: soil water content in Vol.%
TC: transpiration coefficient in ml/g shoot-DM
VeTe: vegetation technique

ACKNOWLEDGMENT

I want to thank my friend Jim Hattab of Houston, Texas, for critically reviewing my 'German' English

REFERENCES

Ayers R.S. 1977. Quality of water for irrigation. *J. Irrig. Drain. Div. ASCE.* 103(IR2), 135–154.

Ayers R.S., Westcot D.W. 1985. Water quality for agriculture. FAO Rome, Irrigation and Drainage *Paper 29 Rev.* 1, 13–58.

Barber S.A. 1979. Growth requirements for nutrients in relation to demand at the root surface. In: *The Soil–Root Interface.* Harley J.L. and Russell R.S. (eds). Academic Press, London, pp. 5–20.

Carminati A., Passioura J.B., Zarebanadkouki M., Delhaize E. 2017. Root hairs enable high transpiration rates in drying soils. *New Phytol.* doi.

FAO. 2018. Salt-affected soils. Available at: http://www.fao.org/s oils-portal/soil-management/management-of-some-problem-soils/salt-affected-soils/more-information-on-salt-affected-soils/en/

Feddes R.A. 1981. Water use models for assessing root zone modification. In: *Modifying the Root Environment to Reduce Crop Stress.* American Society of Agriculture and Engineering, MI, pp. 347–390.

Hamdy A. 1995. Saline water use and management for sustainable agriculture in the Mediterranean Region. In: *Workshop On Farm Sustainable Use of Saline Water in Irrigation: Mediterranean Experience*, INAT, CIHEAM/IAM-B, CRGR, Hammamet, Tunisia.

Hamza M.A., Aylmore L.A.G. 1992. Soil solute concentration and water uptake by single lupin and radish plant roots. II. Driving forces and resistances. *Plant Soil.* 145, 197–205.

Homaee M., Schmidhalter U. 2008. Water integration by plants root under non-uniform soil salinity. *Irrig. Sci.* 27, 83–95.

Jungk A., Claassen N., Kuchenbuch R. 1982. Potassium depletion of the soil-root interface in relation to soil parameters and root properties. Proceedings of the 9th International Plant Nutrition Colloquium, Coventry, pp. 250–255.

Jungk A.O. 2002. Dynamics of nutrient movement at the soil–root interface. In: *Plant Roots – The Hidden Half.* 3rd edition. Waisel Y., Eshel A., and U. Kafkafi (eds). Marcel Dekker, Inc., New York, NY, Basel, Hongkong, pp. 587–616.

Kirkham M.B. 2005. *Principles of Soil and Plant Water Relations.* Elsevier Academic Press, Amsterdam, Boston, Heidelberg, pp. 207–239; 341–356.

Maas E.V., Hoffmann G.J. 1977. Crop salt tolerance – Current assessment. *J. Irrig. Drain. Div.* 103(IR 2), 115–134.

Munns R., Gilliham M. 2015. Salinity tolerance of crops – What is the cost? *New Phytol.* 208(3), 668–673.

Oster J.D. 1999. Use of marginal quality waters for irrigation. Proceedings of the Irrigation Management and Saline Conditions; Regional Symposium, JUST, Irbid, Jordan, pp. 1–16.

Pessarakli M., Szabolcs I. 2011. Soil salinity and sodicity as particular plant/crop stress factors. In: *Handbook of Plant and Crop Stress.* 3rd edition. Pessarakli M. (ed). Taylor & Francis Group, Chapter 1, pp. 3–21.

Qadir M., Quillerou E., Nangia V., Murtaza G., Singh M., Tomas R.J., Drechsel P., Noble A.D.. 2014. Economics of salt-induced land degradation and restoration. *Nat. Resour. Forum.* 38, 282–295.

Rhoades J.D., Kandiah A., Mashali A.M. 1992. The use of saline waters for crop production. *FAO Irrig. Drain.* 48, Rome, IV, 23–69.

Richards L.A. 1954. Diagnosis and improvement of saline and alkali soils. *USDA Handb.* 60, 7–18.

Riley D., Barber S.A. 1970. Salt accumulation at the soybean (*Glycin max.* L. Merr.) root-soil interface. *Soil Sci. Soc. Amer. Proc.* 34, 154–155.

Roose T., Keyes S.D., Daly K.R, Carminati A., Otten W., Vetterlein D., Peth S. 2016. Challenges in imaging and predictive modeling of rhizosphere processes. *Plant Soil*; Springerlink.com.

Schleiff U. 1981. Osmotic potentials of roots of onions and their rhizospheric soil solutions when irrigated with saline drainage waters. *Agric. Water Manage.* 3, 317–323.

Schleiff U. 1982a. An experimental approach to determine water uptake of roots from rhizospheric soil solutions of different salt concentrations. Proceedings of the 9th International Plant Nutrition Colloquium, Coventry, England, Commonwealth Agricultural Bureaux, pp. 567–581.

Schleiff, U. 1982b. Maximum salt concentration in the rhizospheric soil solution of young sugar beets independence from the salt adaptation of their shoots. *Plant Soil.* 66, 397–404.

Schleiff, U. 1983. Water uptake of barley roots from rhizospheric soil solution of different salt concentrations. *Irrig. Sci.* 4, 177–189.

Schleiff, U. 1986. Water uptake by barley roots as affected by the osmotic and matric potential in the rhizosphere. *Plant Soil.* 94, 143–147.

Schleiff U. 1987a. A vegetation technique to study the water uptake by roots from salinized rhizospheric soils. *Z. Pflanzenernähr. Bodenk.* 150, 139–146.

Schleiff U. 1987b. A comparison of transpiration rates of young rape plants from salinized soils of different texture. *Z. Pflanzenernähr. Bodenk*. 150, 403–440.

Schleiff U. 2011. The forgotten link in improving crop salt tolerance research under brackish irrigation – Lateral soil salinity gradients around roots. In: *Handbook of Plant and Crop Stress*. 3rd edition. Pessarakli M. (ed). Taylor & Francis Group, Chapter 45, pp. 1145–1152.

Schleiff U. 2013a. Soil-based vegetation technique to quantify effects of rhizospheric soil osmotic and matric water potentials on crop salt tolerance. *J. Agron. Crop Sci*. 198(6), 94–105.

Schleiff U. 2013b. Mechanistic approach to understand crop salt tolerance under brackish irrigation. In: *Messung, Monitoring und Modellierung von Prozessen im System Boden-Pflanze-Atmosphaere*, 16–17 November 2012, Leipzig, Helmholtzzentrum; DBG Print-Archiv 2013.

Schleiff U., Schaffer G. 1984. The effect of decreasing soil osmotic and soil matric water potential in the rhizosphere of a loamy and a sandy soil on the water uptake rate of wheat roots. *Z. Acker. Pflanzenbau*. 153, 373–384.

Schubert S., Laeuchli, A. 1990. Sodium exclusion mechanisms at the root surface of two maize cultivars. *Plant Soil*. 123, 205–209.

Vetterlein D., Kuhn K., Schubert S., Jahn R. 2004. Consequences of sodium exclusion for the osmotic potential in the rhizosphere – Comparison of two maize cultivars differing in Na+ uptake. *J. Plant Nutr. Soil Sci*. 167, 337–344.

Wadleigh C.H., Ayers R.S. 1946. Growth and biochemical composition of bean plants as conditioned by soil moisture tension and salt concentration. *Plant Physiol*. 20, 106–132.

York L.M., Carminati A., Mooney S.J., Ritz K., Bennett M.J. 2016. The holistic rhizosphere: Integrating zones, processes, and semantics in the soil influenced by roots. *J. Exp. Bot*.

49 Improving Plant Yield and Quality under Normal and Stressful Conditions by Modifying the Interactive Signaling and Metabolic Pathways and Metabolic Interaction Networks

Omid Askari-Khorasgani and Mohammad Pessarakli

CONTENTS

ABBREVIATIONS

$O_2{}^{\bullet-}$	hydroxyl radical
$^1O_2$	singlet oxygen
5-ALA	5-aminolevulinic acid
ABA	abscisic acid
ALMT	aluminum-activated malate transporter
AMDH	aminoaldehyde dehydrogenase
AOA	amino-oxyacetate
AsA	ascorbic acid
AUX	auxin
BEE1	BRASSINOSTEROID ENHANCED EXPRESSION1
BR	brassinosteroid
BTH	benzothiadiazole
bZIP	basic leucine zipper
CAB	chlorophyll a/b-binding protein
CaM	calmodulin
cAMP	cyclic adenosine monophosphate
CK	cytokinin
COMT	*caffeic acid 3-O-methyltransferase*
DHA	dehydroascorbate
ETH	ethylene

FaSAMDC	*Fragaria ananassa S-adenosyl methionine decarboxylase*
GABA	γ-aminobutyric acid
GABA-T	GABA transaminase
GAD	glutamate decarboxylase
GAs	gibberellins
GB	glycine betaine
GFR	G2-LIKE FLAVONOID REGULATOR
GRAS	generally recognized as safe
GSH	glutathione
GSSG	glutathione disulfide
H_2O_2	hydrogen peroxide
$HO_2{}^{\bullet}$	hydroperoxyl
IAA	indole-3-acetic acid
JA	jasmonic acid
MAPKs	mitogen-activated protein kinases
MeJA	methyl jasmonate
MG	methylglyoxal
mODC	mouse *ornithine decarboxylase*
$O_2{}^{\bullet-}$	superoxide anion
O_3	ozone
$ONOO^-$	peroxynitrite
PAs	polyamines

PIF	PHYTOCHROME INTERACTING FACTOR
PRRs	pattern-recognition receptors
PS	photosystem
PTMs	post-translational modifications
RbcS	ribulose 1,5-bisphosphate carboxylase/oxygenase small subunit
RES	reactive electrophilic species
RLKs	receptor-like kinases
RLPs	receptor-like proteins
RNAi	RNA interference
RNS	reactive nitrogen species
RO•	alkoxyl
ROO•	peroxyl
ROS	reactive oxygen species
SAMDC	*S-adenosyl methionine decarboxylase*
SlCAT9	*Solanum lycopersicum* cationic amino acid transporter 9
SlNAGS1	*Solanum lycopersicum N-acetyl-l-glutamate synthase1*
SNAT	serotonin *N*-acetyltransferase
SNOs	*S*-nitrosothiols
SPDS	*spermidine synthase*
SSA	succinic semialdehyde
TCA	tricarboxylic acid

49.1 INTRODUCTION

The escalating demographic trend in population, which is expected to increase from 7.7 billion in 2019 to approximately 9.7 billion in 2050 (Prospects World Population Projections, 2019) and the negative effects of climate change (Clapp et al., 2018) are the two major concerns that put food security at risk. The integrated effects of environmental stresses, which can be exacerbated by climate change, disturb not only plant yield but also quality attributes (Clapp et al., 2018; Kizildeniz et al., 2018). Therefore, agricultural management and breeding strategies that maintain plant yield and quality under constantly changing environmental conditions are of critical importance to meet the global demand and food quality standards. Governed by plant genetic and environmental cues, the interplay reactions between redox, hormone, and chemical signals regulate all aspects of plant growth and development cycles by regulating cell signaling, gene expression, metabolism (such as enzyme activity and protein–protein interactions), homeostasis, and post-translational modifications (PTMs) under normal and stressful conditions (Couée et al., 2006; Xia et al., 2015; Askari-Khorasgani, 2018; Askari-Khorsgani et al., 2018; Brunetti et al., 2018; Charpentier, 2018; Demidchik and Shabala, 2018; Foyer et al., 2018; Manishankar et al., 2018; Mhamdi and Van Breusegem, 2018). Redox (reduction-oxidation) status and its effects on plant cell cycles and stress responses are dependent on the concentrations, combinations, and interplay reactions of the reactive electrophilic species (RES) involved in reduction–oxidation reactions, including reactive nitrogen species (RNS), reactive oxygen species (ROS), and methylglyoxal (MG) molecules (del Río et al., 2006; Hoque et al., 2016;

Askari-Khorasgani and Pessarakli, 2019a). ROS include free radicals such as superoxide anion ($O_2^{•-}$), hydroxyl (•OH), hydroperoxyl ($HO_2^•$), peroxyl ($ROO^•$), and alkoxyl ($RO^•$) radicals, and the air pollutant ozone (O_3) as well as non-radical molecules, such as hydrogen peroxide (H_2O_2), singlet oxygen ($^1O_2$), etc. (Kangasjärvi and Kangasjärvi, 2014; Mhamdi and Van Breusegem, 2018; Ortega-Villasante et al., 2018). The main RNS include peroxynitrite ($ONOO^-$), produced via the interaction between $O_2^{•-}$ and nitric oxide ($NO^•$), and *S*-nitrosothiols (SNOs), produced by the reaction of the free radical $NO^•$ with thiol groups (Corpas and Palma, 2018). On stress perception, redox, hormone, and chemical signaling cascades activate plant defense mechanisms by stimulating phytometabolites to maintain cell homeostasis. Phytometabolites such as flavonoids, γ-aminobutyric acid (GABA), polyamines (PAs), proline, glutamate, and other regulatory metabolites and metabolically active nutrient compounds, particularly Ca^{2+}, C, and N (Majumdar et al., 2016; Brunetti et al., 2018; Zhang et al., 2018a), interactively react with each other and redox/hormone/chemical signals and thereby, regulate all aspects of plant cell physiological processes, including growth, development, fruit quality, and stress responses (Demidchik et al., 2018; Manishankar et al., 2018). Depending on plant genotype, mild stresses such as light intensity and mild drought stress can activate plant cell signaling pathways and thereby, stimulate the synthesis of plant bioactive compounds with health-promoting effects not only in plants but also in herbivores such as humans and animals (Brunetti et al., 2018; Fraga et al., 2018; Askari-Khorasgani and Pessarakli, 2019b). The knowledge of the interacting metabolically active molecules, cell signaling pathways, and homeostasis and the associated gene regulation and stress-induced metabolic reprogramming can be used in breeding programs to optimize plant quality, such as marketability and nutritional values and yield under normal and unfavorable environmental conditions. In this regard, this chapter aims to discuss the critical roles of regulating metabolic interaction networks in improving plant quality, and yield under normal and stressful conditions.

49.2 REGULATORY ROLES OF SIGNALING MOLECULES, THEIR INTERACTIVE PATHWAYS, AND INTERPLAY REACTIONS ON PLANT METABOLITES AND, CONSEQUENTLY, PHYSIOLOGICAL RESPONSES UNDER NORMAL AND STRESSFUL CONDITIONS

Environmental cues are primarily sensed by a diverse family of plant cell surface-localized pattern-recognition receptors (PRRs), including receptor-like kinases (RLKs) and receptor-like proteins (RLPs) containing extracellular ligand-binding domains. PRRs are synthesized in the endoplasmic reticulum and transported to the plasma membrane via the secretory pathway (Nejat and Mantri, 2017) and/or the cytoplasm (Chen et al., 2013). Accordingly, the analysis of chemical/hormone/

redox signals, PRR-mediated signal perception, transduction, and function provides insight into understanding the functional characterization of plant signals and their integrative effects on gene expression, protein function, and physiological responses in association with internal and external stimuli. Direct mass spectrometric detection of cysteine redox modifications is not usually possible due to their low abundance and frequent instability. Most redox proteomics strategies exploit the intrinsic chemical reactivity of thiols or cysteine PTMs to introduce labels enabling enrichment (via biotinylation or thiol-reactive resin). Quantification, including of PTM site occupancy, is becoming increasingly important, and most redox proteomic approaches lend themselves well to isotope-based comparative quantification; for example, exploiting labeling steps to introduce isotopic moieties. Quantification methods can be broadly divided into two categories: those that exploit the intrinsic chemistry of redox PTMs to introduce a label directly at the modification site and indirect methods that label free thiols and, then, apply selective reduction to release specific cysteine PTM sites for differential labeling (Foyer et al., 2018). Colorimetric analysis and fluorescence probes have been used for the detection of ROS (Ortega-Villasante et al., 2018). Depending on plant genotype, growth, and phenological stage, the interplay of different stressors may suppress or activate plant signaling pathways and thereby, control metabolic reprogramming to, in turn, activate plant defense responses by, for example, the synthesis of plant bioactive compounds, osmoprotectants, and/or antioxidants involved in the regulation of cell water, redox, hormone, and chemical homeostasis, and thereby, physiological processes, affecting stress tolerance, yield, and quality attributes (Brunetti et al., 2018; Fraga et al., 2018; Kizildeniz et al., 2018). In this context, this chapter provides information regarding the regulatory effects of phytometabolites and their interactions on yield and fruit quality under normal and stressful conditions.

49.3 REGULATORY ROLES OF FLAVONOID CONTENT AND COMPOSITION ON PLANT GROWTH, FRUIT QUALITY, AND YIELD UNDER DIFFERENT ENVIRONMENTAL CONDITIONS

Among the bioactive compounds, flavonoids act as antioxidants and play essential roles in regulating cell homeostasis, particularly redox and hormone homeostasis, and thereby, enhancing tolerance to both biotic and abiotic stresses (Nakabayashi et al., 2014; Brunetti et al., 2018). Besides promoting fruit quality and stress responses in plants, flavonoids possess strong nutraceutical and pharmaceutical properties and, hence, health-promoting effects in humans. Flavonoids can indirectly improve the yield and quality of plants by promoting symbiosis. There is a consensus that flavonols, particularly derivatives of quercetin, can improve water and nutrient uptake by improving the relationships between plant and symbionts, such as nitrogen-fixing bacteria and mycorrhiza. Flavonoid-mediated hormone and redox homeostasis promote

nodule organogenesis by inhibiting auxin (AUX) transport to the shoot during nodule development (Brunetti et al., 2018). Depending on plant species and genotype, growth and phenological stages, environment and stress conditions, and plant–environment interactions, agrochemicals (such as sucrose, chitosan, and selenium), (bio)fertilizers (such as humic acid and calcium nitrate), biostimulants (such as seaweed extracts [such as alginate] and pomegranate peel extract), symbiotic relations, hormones (such as benzothiadiazole [BTH], ABA, ethylene [ETH], AUX, and methyl jasmonate [MeJA]), agronomical and cultural practices, and their crosstalk effects can promote the synthesis of plant bioactive compounds, such as flavonoids, and their composition and content (Böttcher et al., 2010, 2013; Prakongkha et al., 2013; Liu et al., 2014; Champa et al., 2015; Neto et al., 2017; Olivares et al., 2017; Nair et al., 2018; Torres et al., 2018; Zhu et al., 2018). Accordingly, during breeding programs and agronomical management, the effects of all these contributing factors and their relationships should be taken into account to obtain high-quality plants with optimal performance.

The crosstalk signals induced by the interplay between environmental cues, including ambient temperature, high solar irradiation (either in the presence or in the absence of UV radiation), and internal cues such as redox, hormone, and chemical signals (for example, NaCl and Cu^{2+}) regulate the synthesis of enzymatic and non-enzymatic antioxidants (e.g., flavonoids, GABA, glutathione [GSH], alkaloids, and hydroxycinnamic acid) to differentially protect plants against stress conditions. High temperature, above 35 °C, restricts enzyme activity and thus, the synthesis of plant bioactive compounds such as flavonoids (Brunetti et al., 2018; Zhang et al., 2018a; Askari-Khorasgani and Pessarakli, 2019c). An experiment with pea plants (*Pisum sativum* "Meteor") showed that solar blue light is more effective than UV-A and UV-B radiation in inducing the synthesis of flavonoids and other phenolic compounds and increasing the ratio of quercetin to kaemferol derivatives (Siipola et al., 2014). Excluding heat stress, at a specific solar irradiation and more effectively with blue light, flavonoid synthesis may occur at the expense of hydroxycinnamate synthesis. While flavonoids primarily act as antioxidants involved in free radical scavenging, the protective roles of hydroxycinnamic acids against UV irradiation are more prominent during photoprotection (Siipola et al., 2014; Brunetti et al., 2018; Zhang et al., 2018a; Askari-Khorasgani and Pessarakli, 2019c). Hence, a balance between hydroxycinnamic acids and flavonoids would be beneficial when light stress is combined with other stresses. The effect of flavonoid composition and content on fruit quality and stress responses varies depending on plant species, genotype, and use purposes (for example, between red and white wine and table grapes). Affected by plant–environment interactions, stress conditions (such as stress severity, repetition, duration, combination, time, light wavelength, etc., and their interactions with plant characteristics such as genotype, age, phenological stage, rootstock–scion combination, etc.) play determinant roles in regulating the composition and contents of phytometabolites, such as flavonoids and amino acids, which may possess health-promoting effects in plants and humans (Zhang et al., 2018a; Askari-Khorasgani and Pessarakli, 2019d). The manipulation of signaling pathways, genotype selection, rootstock–scion combination, and regulation of transcription factors and/or genes can be used as effective strategies to regulate plant metabolites and thereby, yield, quality, and resilience (Askari-Khorasgani, 2018; Prinsi et al., 2018; Askari-Khorasgani and Pessarakli, 2019d). Hence, in addition to OMICS analysis, understanding the relationships between internal and external cues, such as the relationships between gene transcription, metabolite synthesis, and redox/hormone/chemical homeostasis under growing conditions at different phenological stages and day/night cycles, is critical to correctly interpret the data. This would demonstrate which metabolite composition is effective under specific environmental conditions to enhance plant adaptation and tolerance and to optimize plant quality and yield, facilitating improvements in breeding programs.

49.4 INTERACTIONS BETWEEN POLYAMINES, ETHYLENE, AMINO ACIDS (PARTICULARLY, γ-AMINOBUTYRIC ACID) REGULATE CELL HOMEOSTASIS AND, THEREBY, PLANT YIELD, FRUIT QUALITY, AND STRESS RESPONSES

Polyamines (PAs), which are ubiquitous aliphatic amines and biogenic regulators, are differentially involved in many physiological processes, including growth, development, senescence, and stress responses (Guo et al., 2018). Accordingly, depending on plant genotype and environmental conditions, a specific composition of PAs can induce positive regulatory effects on plant growth, yield, and quality parameters, such as fruit maturation, ripening, softening, senescence, longevity, nutrient composition, titratable acidity, and total soluble solids (Majumdar et al., 2016; Guo et al., 2018; Wannabussapawich and Seraypheap, 2018; Zarei et al., 2018). As illustrated in Figure 49.1, ornithine is a key metabolite sitting at a crossroads of several interactive pathways, including PAs, GABA, proline, and glutamate synthesis, in plants. Their interacting networks regulate plant growth, development, and tolerance against multiple biotic and abiotic stresses (Majumdar et al., 2016; Zarei et al., 2018). Proline and arginine are synthesized independently of GABA and PAs (Majumdar et al., 2016). GABA is recognized as a GRAS (generally recognized as safe) compound and exhibits high potential to improve preharvest and postharvest plant physiological responses under normal and stressful conditions, modulating cell homeostasis and phytometabolites and, thereby, plant growth, quality, and yield. GABA can be elicited in different ways: for example, by applying exogenous GABA treatment, manipulating its biosynthetic pathways, and/or regulating its precursors (Sheng et al., 2017). Proline can be used as a source of GABA synthesis, while glutamate, ornithine, and arginine can contribute to both PA and GABA synthesis (Majumdar et al., 2016; Zarei et al., 2018). Although knowledge of their molecular interactions

(Figure 49.1) has increased in recent years, their interactive activities with other metabolites, which might be under the influence of a wide range of internal and external factors and their ratios, are not yet well understood. Constitutive overexpression of a *Solanum lycopersicum N-acetyl-l-glutamate synthase1* (*SlNAGS1*) gene in *Arabidopsis* resulted in overproduction of ornithine and citrulline without significant influence on arginine accumulation, which resulted in improving seed germination and tolerance to drought and salt stresses. With similar phenotypic traits compared with the wild type, overexpressing mouse *ornithine decarboxylase* (*mODC*) plants were more effective in biomass production and quality traits, which could be attributed to the increased total N, C, chlorophyll, PA, and GABA contents, hormone homeostasis (particularly, indole-3-acetic acid [IAA], ABA, and ETH), reduced respiration and water loss, and, consequently, improved photosynthetic capacity (Majumdar et al., 2016; Guo et al., 2018). In strawberry, comparing PA compositions, the exogenous application of putrescine resulted in delayed fruit ripening, while spermidine and, more notably, spermine promoted fruit ripening (Guo et al., 2018). Thus, putrescine would be more effective than spermidine and spermine to promote postharvest quality characteristics of fruits by delaying senescence-related physiological processes and prolonging their life span. For example, putrescine promoted the postharvest quality characteristics of green pepper (*Capsicum annum* L.) by extending its shelf life, preventing vitamin C, chlorophyll, and color degradation, promoting antioxidant activities, and reducing the adverse physiological changes caused by mechanical injury (Wang et al., 2018). *Fragaria ananassa S-adenosyl methionine decarboxylase* (*FaSAMDC*)-overexpressing fruits affected the genes involved in PA biosynthetic pathways and, thereby, promoted fruit ripening by increasing the conversion of putrescine to spermidine and particularly, spermine, whereas fruit ripening was delayed by *FaSAMDC* silencing in RNA interference (RNAi) strawberries, most likely as a result of reduced conversion of putrescine to spermine (Guo et al., 2018). As with phenolic compounds and flavonoid composition (Brunetti et al., 2018; Zhang et al., 2018a), the integrative effects of PAs with other phytometabolites, such as hormones (particularly, ABA, IAA, and ETH) and amino acids, are determinant of plant physiological responses, yield, and quality attributes, depending on plant species and genotype. For example, in strawberry, spermine synthesis promoted fruit quality by promoting softening, as well as anthocyanin and sugar accumulation by increasing the ABA to IAA and ETH ratio, while promoting IAA and inhibiting ETH signaling (Guo et al., 2018), which could possibly be mediated by the influence of spermine on redox signaling and homeostasis in addition to gene and molecular interaction networks. Unlike strawberry, in tomato, spermidine was more effective in improving fruit ripening than spermine (Guo et al., 2018), although the interactive effects of plant genotype with other internal factors, such as other hormone and nutrient compositions, and external cues, such as agronomical practices, harvesting time, and stress conditions, should not be excluded. While at early fruit

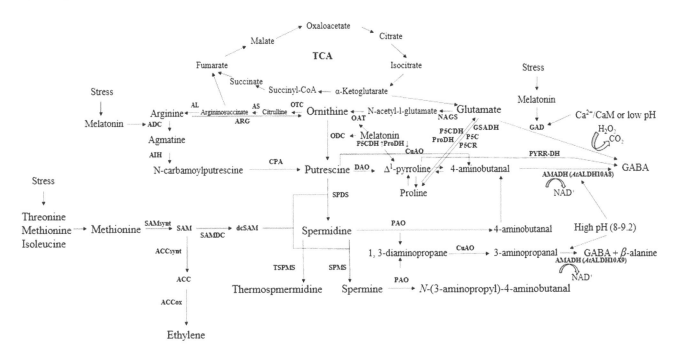

FIGURE 49.1 Interactive synthetic pathways and metabolic interaction networks of polyamines, ethylene, melatonin, and γ-aminobutyric acid. TCA, tricarboxylic acid; GAD, glutamate decarboxylase (EC 4.1.1.15); NAGS, *N*-acetyl-l-glutamate synthase (EC 2.3.1.1); OAT, ornithine-d-aminotransferase (EC 2.6.1.13); ODC, ornithine decarboxylase (EC 4.1.1.17); SPDS, spermidine synthase (EC 2.5.1.16); TSPMS, thermospermine synthase (EC 2.5.1.79); SPMS, spermine synthase (EC 2.5.1.22); CuAO, copper-containing amine oxidase (EC 1.4.3.22); OTC, ornithine transcarbamylase (EC 2.1.3.3); AS, argininosuccinate synthase (EC 6.3.4.5); AL, argininosuccinate lyase (EC 4.3.2.1); ADC, arginine decarboxylase (EC 4.1.1.19); AIH, agmatine iminohydrolase (EC 3.5.3.12); CPA, *N*-carbamoyl putrescine amidohydrolase (EC 3.5.1.53); DAO, EC 1.4.3.6; SAM, *S*-adenosyl-L-methionine; dcSAM, decarboxylated *S*-adenosyl-L-methionine; SAMS, *S*-adenosyl-L-methionine synthetase (EC 2.5.1.6); SAMDC, *S*-adenosylmethionine decarboxylase (EC 4.1.1.50); ACC, 1-aminocyclopropane-1-carboxylic acid; ACCsynt, ACC synthase (EC 4.4.1.14); ACCox, ACC oxidase (EC 1.14.17.4); P5C, Δ¹-pyrroline-5-carboxylate; P5CDH, P5C dehydrogenase (EC 1.2.1.88, formerly EC 1.5.1.12); P5CR, P5C reductase (EC 1.5.1.2); PYRR-DH, pyrroline dehydrogenase (EC 1.2.1.88); AMDH, aminoaldehyde dehydrogenase (EC 1.2.1.19); ProDH, proline dehydrogenase (EC 1.5.5.2, formerly 1.5.99.8); GSADH, glutamate-5-semialdehyde dehydrogenase (EC 1.2.1.41); GABA, γ-aminobutyric acid; PAO, polyamine oxidase (EC 1.5.3.11).

development stages, a higher GA₃ + IAA to ABA ratio stimulates cell expansion and thereby, fruit enlargement, the metabolic interaction network of ABA (higher ABA to IAA ratio), ETH, sugars, organic acids, amino acids, phenolic compounds (particularly, flavonoids), and PAs play essential roles in regulating fruit ripening and quality characteristics at later stages, the metabolic pathways of which can be promoted under specific mild stress conditions either naturally or by human interference (Bünger-Kibler and Bangerth, 1982; Guo et al., 2018; Askari-Khorasgani and Pessarakli, 2019b,d). Preharvest mild drought stress conditions, which can promote MeJA, ABA, and sugar accumulation and flavonoid biosynthesis, and postharvest controlled conditions, particularly temperature, which can prevent flavonoid degradation and excessive enzyme activities and ETH accumulation, can promote fruit quality by regulating metabolic interaction networks. In addition to preharvest quality attributes, metabolic and hormone homeostasis, principally the composition of PAs, sugars, MeJA, amino acids, enzymatic and non-enzymatic antioxidant activities, and specifically GABA are determinants of postharvest quality attributes and longevity (Palma et al., 2014; Guo et al., 2018; Askari-Khorasgani and Pessarakli, 2019b,d). During plant growth and fruit development, glutamate homeostasis is essential for N metabolism, and thus, the overconversion of

glutamate/ornithine to putrescine may disturb amino acid ratios and, consequently, cause N and protein deficiency as well as C:N, NO, and NO₃⁻ imbalance, which are crucially involved in biomass production and stress responses (Majumdar et al., 2016). The fact that high C:N ratio is a limiting factor for PA (particularly, putrescine) synthesis, but not GABA, suggests that there might be other sources for GABA synthesis, such as glutamate, proline, melatonin, ornithine, and/or arginine (Majumdar et al., 2016; Aghdam and Fard, 2017). As illustrated in Figure 49.1, Zarei et al.'s (2016) study demonstrated that overexpression of NAD⁺-dependent aminoaldehyde dehydrogenase (AMDH) genes encoding 4-aminobutanal can be used to stimulate GABA production and, consequently, enhance tolerance to multiple stresses, such as drought, salinity, and chilling (Zarei et al., 2016). It can be hypothesized that when N is limited, N-rich compounds, such as pre-accumulated proline (Figure 49.1), redirection of ornithine to putrescine, and/or GABA-mediated N uptake and assimilation, by improving cell homeostasis and indirectly replenishing C reservoirs, can restore GABA synthesis. In return, GABA contributes to redox homeostasis, its catabolism promotes the synthesis of amino acids such as glutamate, aspartate, alanine, serine, and proline and organic acids, and thereby, it reloads C and N reservoirs by, for example, regulating 14-3-3 genes and aluminum-activated malate transporter (ALMT) and, consequently, nutrient uptake to provide energy for maintaining the activity of the tricarboxylic acid (TCA) cycle and photosynthetic performance by enhancing ATP level and electron transport flux in the photosynthetic apparatus, optimizing phytonutrient composition (e.g., by modifying the composition of PAs, amino acids, and organic acids) and performance under normal and particularly stress conditions (Snowden et al., 2015; Majumdar et al., 2016; Sheng et al., 2017; Carillo, 2018; Kalhor et al., 2018; Podlesakova et al., 2018). ALMT proteins are activated by specific anions (such as malate, sulfate, CO_2/HCO^-, NO_3^-, and Cl^-), Al^{3+}, amino-oxyacetate (AOA) (1 mM) (an inhibitor of glutamate decarboxylase [GAD] and GABA transaminase [GABA-T]), and vigabatrin (100 µM) (a GABA-T inhibitor) to transport both ions (such as malate) and GABA efflux from roots, modulating physiological responses such as C metabolism and GABA-mediated cell homeostasis, particularly under stress conditions. Phenolic compounds such as flavonoids and PAs (particularly, putrescine) can promote plant physiological responses directly, by regulating cell homeostasis, or indirectly, by improving symbiotic relationships (Salloum et al., 2018). Depending on stress conditions and plant genotype, different types of metabolites such as PAs accumulate in different cell, tissue and organ types (Lopez-Gomez et al., 2014), so the regulation of phytometabolites according to plant genotype and environmental conditions would be an effective strategy to boost plant performance, particularly under stressful conditions. Thus, the tune balance between metabolite compositions, such as phenols–GABA–PA synthesis in root, has the potential to promote root morphology and plant–microbiome symbiotic relations, which will require more attention in the future. The elicitation of GABA can promote plant growth and yield as well as the pre- and postharvest quality (e.g., by enhancing nutrient composition as well as inhibiting browning and decay incidence) of fruits by regulating cell homeostasis, such as signaling molecules; ion homeostasis; nutrient composition, such as C, organic acids, and amino acids; and antioxidant activities (Sheng et al., 2017; Gao et al., 2018). However, at a specific C deficiency level (which can disrupt proline, glutamate, putrescine, arginine, and GABA synthesis, but may have insignificant effects on spermidine and its derivatives), the extra N might not be sufficient to mitigate the negative impact of C deficiency on metabolic pathways (Majumdar et al., 2016). Since their relations can be strongly affected by different internal and external interacting factors, more studies are still required to develop strategies for optimizing their metabolic pathways under stressful conditions. Under a climate change scenario, the elevated CO_2 may inhibit C depletion, but the disruptive effects of heat and light stress on enzyme activities and, thereby, cell homeostasis and nutrient composition may affect their relationships and physiological responses, which will require further studies. When the knowledge of their interactive activities and pleiotropic functions is complete, manipulation of their biosynthetic pathways and their homeostasis, for example, by converting ornithine, glutamate, and

ETH to PAs and GABA, can be used to promote plant cell physiological processes and, thereby, yield, fruit quality, and stress responses (Majumdar et al., 2016). For example, overexpression of *spermidine synthase* (*SPDS*) and *S-adenosyl methionine decarboxylase* (*SAMDC*) genes was effective in improving the quality of tomato fruits by positively regulating a high ratio of spermidine and spermine to putrescine, sugars and ETH production, and carotenoid and lycopene accumulation, which consequently promoted ripening processes. The pre- and postharvest exogenous application of PAs can also promote fruit quality characteristics by, for example, promoting cell expansion and fruit development (enlargement) and aroma in grapevine and oil-palm, reducing ETH synthesis and, thereby, fruit softening in peach, and extending shelf life in mango (*Mangifera indica*), peach (*Prunus persica*), plum (*Prunus domestica*), and apple (*Malus domestica*) fruits. However, controversial results have been reported (Guo et al., 2018), which can be ascribed to a large number of internal and external interacting factors, such as plant genotype and environmental (normal and stressful) conditions, harvesting time, agronomical management, the interactions between nutrient compositions, etc. Affected by plant–environment interactions, disparate and variable stress conditions differentially affect cell homeostasis, and, hence, taking into consideration their relationships, ratios, cellular transport, compartmentalization and availability, and circadian rhythm as well as cell, tissue, and organ type and age, future experiments should be conducted under different environmental conditions to provide sufficient information concerning their relationships and effects on plant cell physiological responses.

49.5 INTERACTIVE PHYTOMETABOLITES, INCLUDING MELATONIN, γ-AMINOBUTYRIC ACID, POLYAMINES, PROLINE, AMINO ACIDS, FLAVONOIDS, GLYCINE BETAINE, AND PHYTOHORMONES, REGULATE PLANT PHYSIOLOGICAL RESPONSES AND THEREBY, PLANT YIELD AND QUALITY UNDER NORMAL AND STRESSFUL CONDITIONS

After perception, stress-induced interactive (redox/hormones/chemicals) signaling pathways generate a Ca^{2+} signature, which can be sensed and transduced by Ca^{2+}-binding proteins, such as calmodulin (CaM) (Scholz et al., 2015). At the early fruit development stage, the activated Ca^{2+}/CaM signals activate some of the genes encoding cytosolic glutamate decarboxylase (GAD) (EC 4.1.1.15) enzymes at neutral (around 7.3) pH; their optimum activity is at pH 5.8, and they can also be induced in a Ca^{2+}-independent manner (Snedden et al., 1995). Then, the activated GAD enzymes induce GABA synthesis, mainly in the cytosol, during fruit formation. The cytosolic GABA is then transported into the vacuole via specific tonoplast transporters, such as Uga4, and mainly into the mitochondria via GABA permease (GABP), peaking when the fruit has reached its maximum volume (Uemura et al., 2004; Carillo, 2018; Podlesakova et al., 2018).

During the initial phase of fruit development, as observed in tomato, tonoplast GABA transporters, such as cationic amino acid transporter proteins, transfer vacuole-localized GABA into the cytosol. For example, in tomato, *Solanum lycopersicum* cationic amino acid transporter 9 (SlCAT9) transports GABA from the vacuole to the cytosol. Then, during fruit ripening, the cytosolic GABA stoichiometrically declines by transfer into the mitochondria for metabolism into glutamate and aspartate, followed by their export into the cytosol and uptake into the vacuole to participate in osmotic adjustment and turgor pressure regulation, particularly under hyperosmotic stress. Studies indicate that that the conversion of GABA into its cytosolic glutamate and aspartate by-products provides N content and regulates vacuole and cytosolic pH to modulate physiological responses, particularly under stress conditions, which can promote the nutrient balance of fruits from the beginning of fruit development until the end of postharvest storage life (Snowden et al., 2015; Podlesakova et al., 2018). Since cold stress causes cytosolic acidification, GABA

synthesis can alleviate chilling injury by consuming H⁺ and preventing acidification (Aghdam and Fard, 2017), as GABA carries a net positive charge (7.3) at the acidic pH of the vacuole, while glutamate (3.2) and aspartate (2.8) carry a net negative charge at the near-neutral pH of the cytosol (Snowden et al., 2015). The mitochondrial GABA is also catabolized to succinic semialdehyde (SSA) by GABA-T to provide a substrate for both the TCA cycle and the electron transport chain required for photosynthesis (Figure 49.2). Additionally, the GABA-mediated conversion of glyoxylate to glycine can regulate photorespiration (Figure 49.2). Stress-induced signals can downregulate the expression of genes involved in GABA metabolism, such as GABA-T, and, consequently, disturb cell homeostasis and nutrient balance, disrupting physiological and, thus, morphological processes such as root and/or fruit development (Snowden et al., 2015; Podlesakova et al., 2018). Overall, it can be hypothesized that improvements in vacuole-localized GABA content and its conversion to cytosolic amino acids would have a prominent role in promoting both pre- and postharvest quality of fruits, while the cytoplasmic GABA is mainly required to optimize the preharvest quality of fruits,

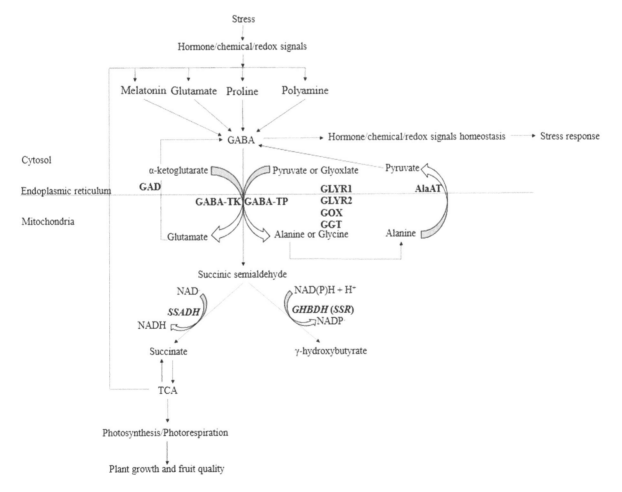

FIGURE 49.2 Stress-induced metabolic (melatonin, glutamate, proline, polyamine) pathways involved in γ-aminobutyric acid synthesis and the consequent metabolic reprogramming to optimize plant growth, quality, and stress responses.

GABA, γ-aminobutyric acid; GAD, glutamate decarboxylase (EC 4.1.1.15); GABA-TK, GABAtransaminase α-ketoglutarate; GABA-TP, GABAtransaminase pyruvate; SSADH, succinic semialdehyde dehydrogenase (EC 1.2.1.16); GHBDH, γ-hydroxybutyrate dehydrogenase (EC 1.1.1.61); SSR, succinic semialdehyde reductase (EC 1.1.1.61); TCA, tricarboxylic acid.

particularly during fruit development and specifically under stress conditions. Yet, the influence of GABA compartmentalization, for example, by manipulation of GABA transporters, on physiological responses, particularly under different stress conditions, remains to be investigated. Considering the influence of melatonin on the stimulation of GABA synthesis, it can be hypothesized that stress-induced melatonin (Aghdam and Fard, 2017), activation of its biosynthetic pathways and/ or exogenous applications of melatonin have the potential to induce *GAD* genes independently of Ca^{2+} signals, or there might be other regulatory pathways, the mechanism(s) of which remain to be investigated. Analysis of the relationships between plant small RNAs and GAD activities under different environmental conditions can also provide more insights into their regulatory pathways.

The metabolite interaction networks of GABA and N sources are illustrated in Figures 49.1 and 49.2. So far, the relationships between the activity of these regulatory metabolites and plant performance, quality, and tolerance have been extensively studied, but with many contradictory results (Podlesakova et al., 2018). By looking at their metabolic pathways, it can be hypothesized that the differential responses could be attributed to a large number of interacting internal factors, such as plant species, genotype, growth stage, age, and cell, tissue, and organ type, as well as metabolite composition and content and external factors such as stress conditions and agricultural practices (Askari-Khorasgani and Pessarakli, 2019b,c,d). A fine-tuned balance between the composition of organic acids and sugars and in some plants, such as berries, flavonoids and amino acids is of critical importance in determining the pre- and postharvest quality characteristics, particularly flavor, taste, and nutritional quality, of fruits (Sheng et al., 2017; Askari-Khorasgani and Pessarakli, 2019d).

Melatonin (N-acetyl-5-methoxy-tryptamine) and serotonin (5-hydroxytryptamine) are small indoleamine derivatives of the amino acid L-tryptophan that are produced in plants, humans, and some animals (Figure 49.3). Phytomelatonin is involved in rhythmic and cyclic processes, such as chronoregulation, photomorphogenesis, skotomorphogenesis (growth in darkness), and seasonal and senescence processes, as well as modulation of reproductive development, control of root and shoot organogenesis, maintenance of plant tissues, and responses to biotic and abiotic stresses (Erland and Saxena, 2018; Hwang and Back, 2018). Melatonin is involved in innate plant immunity and DNA protection and repair. Studies indicate that melatonin promotes the plant immune system by regulating hormone (particularly, salicylic acid [SA]), redox (particularly, NO, but also H_2O_2 and O_2), and sugar homeostasis, signaling, and thereby, their integrative mode of action (Shi et al., 2015; Hardeland, 2016; Lee et al., 2017; Majidinia et al., 2017; Hernández-Ruiz and Arnao, 2018; Lee and Back, 2018; Liang et al., 2018; Zhang et al., 2018b). Coordinately, the pleiotropic indoleamines melatonin and serotonin act as signaling molecules with hormone-like activities [respectively, similar to auxin–cytokinin (AUX–CK) activities] to modulate cell homeostasis and signaling and, consequently, interact with phytohormones such as SA, CK, ABA, jasmonic

acid (JA), gibberellins (GAs), ETH, and brassinosteroid (BR) to regulate a wide range of physiological processes, spanning reproduction, germination, vegetative growth, morphogenesis, and responses to both biotic and abiotic stresses (Erland et al., 2018; Erland and Saxena, 2018; Hernández-Ruiz and Arnao, 2018; Hwang and Back, 2018). As illustrated in Figure 49.3, studies indicate that upregulation of *caffeic acid 3-O-methyltransferase* (*COMT*) during the day and downregulation at night can induce the synthesis of melatonin, ferulic acid, and scopolin as well as methylated anthocyanins (the last even under low light conditions). Scopolin enhances tolerance to cold stress, while methylated anthocyanins promote antioxidant capacity to enhance tolerance to multiple stresses (Erland and Saxena, 2018; Hwang and Back, 2018; Askari-Khorasgani and Pessarakli, 2019b,c). Another important gene involved in melatonin synthesis, *serotonin N-acetyltransferase2* (*SNAT2*), (Figure 49.3) is upregulated at night, while *SNAT1* is irresponsive to light changes and, consequently, upregulates the *DWARF4*, *D11*, and *RAVL1* genes responsible for BR synthesis to maintain skotomorphogenesis (Hwang and Back, 2018). In addition to enhancing tolerance to low light, BRs can activate the signaling and biosynthetic pathways (briefly, by optimizing GA-AUX signaling to maintain PHYTOCHROME INTERACTING FACTOR [PIF] activity) to enhance tolerance to heat and UV-B, which may negatively or positively affect flavonoid synthesis (Hwang and Back, 2018; Askari-Khorasgani and Pessarakli, 2019c). In tea (*Camellia sinensis* L.), exogenous BR induced production of endogenous NO and, consequently, flavonoids (Li et al., 2017a), the mechanism of which can vary depending on plant genotype and stress conditions. In *Arabidopsis*, both *Arabidopsis g2-like flavonoid regulator* (*gfr*) and *brassinosteroid enhanced expression1* (bee1) mutants accumulated less quercetin and scopolin and more anthocyanin at low temperature compared with the control plants, the mechanism of which could be regulated by SA/JA antagonistic effects on the phenylpropanoid pathway (Petridis et al., 2016; Askari-Khorasgani and Pessarakli, 2019c). Since quercetins and scopolin can enhance tolerance to cold stress (Askari-Khorasgani and Pessarakli, 2019c), studies on the influence of melatonin and GABA on *GFR/BEE1* activities and metabolite composition can provide more insights into their regulatory roles in inducing cold acclimation. Therefore, the metabolic interaction networks between BR, SA, and JA, which can be modulated by phytometabolites such as GABA and melatonin, can regulate flavonoid composition and content, depending on environmental conditions and plant genotype. Since flavonoids can differentially suppress melatonin biosynthetic pathways (Lee et al., 2018), BRs may also indirectly affect melatonin synthesis by regulating flavonoid composition and content (Li et al., 2017a), which remains to be studied in different plants under different environmental conditions. Overall, it can be hypothesized that redox/chemical/phytohormone signals regulate melatonin synthesis and, in turn, may negatively or positively modulate metabolite interaction networks such as BRs–flavonoids and redox status to enhance tolerance to different stresses such as low and high light conditions,

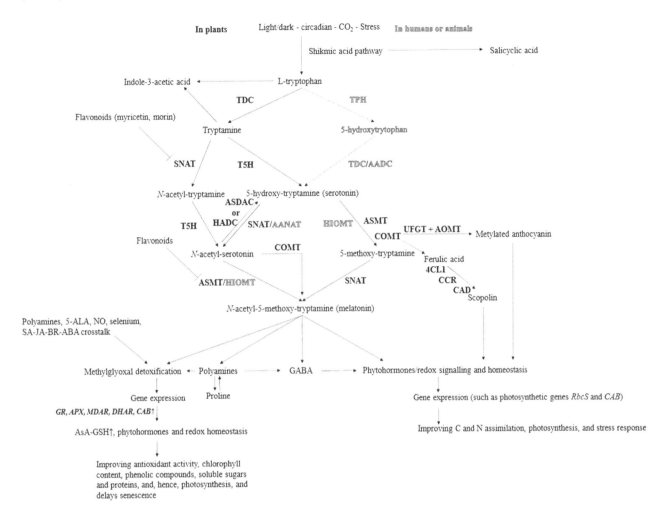

FIGURE 49.3 Metabolite interaction networks involved in serotonin and melatonin synthesis and the consequent activation of signaling molecules, γ-aminobutyric acid, methylglyoxal, phytohormones, and redox to control plant yield, quality, and stress responses. TDC, tryptophan decarboxylase; T5H, tryptamine-5-hydroxylase; TPH, tryptophan hydroxylase; AADC, aromatic-L-amino-acid decarboxylase; SNAT, serotonin-N-acetyltransferase; ASDAC, N-acetylserotonin deacetylase; HADC, histone deacetylase; HIOMT, hydroxyindole-O-methyltransferase; COMT, caffeic acid O-methyltransferase; ASMT, N-acetyl-serotonin methyltransferase; UFGT, UDP glucose:flavonoid 3-O-glucosyltransferase; AOMT, anthocyanin O-methyltransferase; 4CL1, 4-coumarate:CoA ligase 1; CCR, cinnamoyl-CoA reductase; CAD, cinnamyl alcohol dehydrogenase; GABA, γ-aminobutyric acid; 5-ALA, 5-aminolevulinic acid; NO, nitric oxide; SA, salicylic acid; JA, jasmonic acid; BR, brassinosteroid; ABA, abscisic acid; GR, glutathione reductase; APX, ascorbate peroxidase; MDAR, monodehydroascorbate reductase; DHAR, dehydroascorbate reductase; CAB, chlorophyll a/b-binding protein; AsA, ascorbate; GSH, glutathione; RbcS, ribulose 1,5-bisphosphate carboxylase/oxygenase small subunit.

heat, and low temperatures. Yet, more studies are required to demonstrate the influence of melatonin on different phytometabolites in different plants under different environmental conditions. In plants such as red wine grapes, where flavonoids play essential roles in determining fruit quality, the negative or positive effects of melatonin and hormones such as BRs on flavonoids under normal and stressful conditions should be taken into account to choose the best strategies for obtaining high-quality fruits.

In plants, manipulation of enzymatic activities linked with both humans or animals and plants can be used to regulate melatonin synthesis and, hence, stress responses (Wang et al., 2014). Melatonin enhances plant productivity, fruit nutritional quality, and stress tolerance by promoting GABA-mediated signaling and, thereby, redox and hormone homeostasis and

signaling, gene expression such as the *RbcS* (*ribulose-1,5 bisphosphate carboxylase/oxygenase small subunit* or Rubisco) and *CAB* (*chlorophyll a/b binding protein*) photosynthetic genes, photosynthetic efficiency, and carbohydrate metabolism, as well as plant pre- and postharvest quality attributes and yield (Hasan et al., 2015; Erland et al., 2017; Li et al., 2017b; Fan et al., 2018; Zhang et al., 2018b). Melatonin can be used as an effective preharvest and postharvest elicitor to promote plant quality, yield, and tolerance to multiple stresses by promoting cell homeostasis (particularly, redox homeostasis) and inducing the synthesis of bioactive phytochemicals such as PAs (putrescine, spermine, and spermidine), phenolic compounds such as flavonoids, enzymatic antioxidants, GABA, lycopene, and ascorbic acid (AsA) (Aghdam and Fard, 2017). Melatonin enhanced C assimilation in control wheat and the chlorophyll b–deficient wheat mutant *ANK32B* in response to elevated CO_2 via boosting the activities of ATPase and sucrose synthesis and maintaining a relatively high level of total chlorophyll content (Li et al., 2017b). The fact that melatonin regulates redox, hormone, and GABA homeostasis and signaling, and root morphogenesis as well as sugar, C, and N metabolism, particularly under stress conditions (Sulieman and Schulze, 2010), could be a sign of its potential for improving symbiotic relationships and performance (Askari-Khorasgani, 2018), which remains to be investigated in the future. Exogenous application of melatonin delayed the senescence of apple and kiwifruit leaves by increasing AsA and GSH content and soluble sugar and protein content, maintaining high glutathione:glutathione disulfide (GSH:GSSG) and ascorbic acid:dehydroascorbate (AsA:DHA) ratios (better to say GSH/(GSH + GSSG) and AsA/(AsA + DHA)), and promoting chlorophyll content, nutrient homeostasis, photosynthesis and, thereby, growth and productivity under stressful conditions (Wang et al., 2012; Liang et al., 2018). Thus, it can be hypothesized that in addition to modulating GABA signaling, melatonin can enhance tolerance to multiple stresses by methylglyoxal (MG) detoxification and signaling as the result of regulating AsA-GSH and phytohormones (particularly, JA). Melatonin-induced MG homeostasis promotes plant adaptation and stress responses by regulating cell (particularly, redox) homeostasis and stress signaling. It can be hypothesized that the regulatory roles of melatonin on gene expression could be mediated by the influence of MG on activating mitogen-activated protein kinases (MAPKs) and ABA-inducible basic leucine zipper (bZIP) transcription factors as well as transmitting stress signaling (for example, by inducing second messengers such as Ca^{2+}, ROS, cyclic adenosine monophosphate [cAMP], and, consequently, ABA and K^+) (Tan et al., 2012; Wang et al., 2012; Fan et al., 2018; Askari-Khorasgani and Pessarakli, 2019a). Melatonin is implicated in skotomorphogenesis by modulating BR biosynthesis in rice plants (Hwang and Back, 2018). Whether the melatonin-induced BR synthesis promotes skotomorphogenesis downstream of GABA signaling and/or MG signaling remains to be studied. Since BR can maintain physiological responses under combined heat and light stresses (Askari-Khorasgani and Pessarakli, 2019c), melatonin-mediated BR biosynthesis (Hwang and Back, 2018) may induce such responses, the mechanism of which remains to be investigated.

Studies indicate that the metabolized CO_2 might produce melatonin at the expense of reducing substrate for SA, flavonoids, and/or IAA synthesis. Melatonin may induce flavonoid synthesis by upregulating genes involved in flavonoid synthesis (Liang et al., 2018), and in turn, flavonoids can downregulate the genes involved in melatonin synthesis. The evaluation of metabolic interaction networks involved in the synthesis of phytometabolites such as melatonin is necessary for the correct interpretation of their influence on plant cell physiological responses. The exogenous application of melatonin on heat- and salinity-stressed tomato (*Solanum lycopersicon* cv. Boludo, Monsanto, Torre Pacheco, Murcia, Spain) demonstrated the efficacy of melatonin in improving photosynthesis

(Martinez et al., 2018), which could be ascribed to the influence of melatonin on modulating MG (Askari-Khorasgani and Pessarakli, 2019a) or GABA signaling (Carillo, 2018). Melatonin improved transpiration rate and, consequently, alleviated heat stress damage to tomato by lowering leaf temperatures, improving ROS detoxification, and thereby, preventing irreversible damage to the protein core of photosystem I (PSI) and PSII and maintaining photosynthetic pigments (Martinez et al., 2018). The regulatory roles of melatonin (particularly, phytomelatonin) in regulating plant signaling molecules, particularly MG and GABA, remain to be fully explored. Future studies are still required to elucidate the influence of hormone/chemical/redox signaling on the activation of melatonin synthetic pathways and the associated phenotypic traits under different stress conditions.

Glycine betaine (GB) is a quaternary organic compound that regulates plant growth, yield, nutrient composition, and stress responses (particularly, osmotic stress) by modulating diverse plant metabolic and physiological responses under normal and multiple environmental stress conditions (Rasheed et al., 2018). GB mainly accumulates in young leaves independently of N content under prolonged stress conditions, but high light suppresses GB and glutamate synthesis. Proline is preferentially accumulated at the onset of stress in older leaves when N is sufficient, while at low N concentration, glutamate can be used as a proline substrate. Proline together with GB induces a strong protective role against drought and salinity stresses. As illustrated in Figures 49.1 and 49.2, glutamate can be used as a precursor of proline under stress conditions such as low nitrate and high light; however, with sufficient N content, depending on the stress conditions, PAs, glutamate, and proline can mitigate the adverse effects of stress by inducing GABA synthesis in a calcium- and pH-dependent manner (Carillo, 2018). The manipulation of genes encoding PA (Nahar et al., 2016), GB, proline, (Hossain and Fujita, 2010), JA (Shan et al., 2018), SA (Alam et al., 2013), and trehalose (Aldesuquy and Ghanem, 2015; Ibrahim and Abdellatif, 2016) can be used to promote stress responses by, for example, detoxifying MG. Similar to the effects of melatonin, 5-aminolevulinic acid (5-ALA), selenium (sodium selenate), PAs, NO, PA–NO crosstalk, and a tune balance between phytohormones, particularly, SA–JA–BR–ABA, can promote MG detoxification and, thereby, plant cell antioxidant capacity by maintaining high GSH:GSSG and AsA:DHA ratios, chlorophyll content, nutrient homeostasis, and photosynthesis, leading to higher productivity under normal and stressful conditions (Askari-Khorasgani and Pessarakli, 2019a). GB stabilizes the structure and efficiency of PSII by regulating the activity of RuBisCo activase, promoting ion homeostasis (particularly, by promoting root morphology, reducing Na and Cl accumulation, and promoting K and Ca^{2+} uptake and transport), water status, chlorophyll concentration, efficient energy trapping, electron transport chain and antioxidant activities, and optimizing plant yield, quality, and nutritional value (for example, by increasing phenols, flavonoids, and anthocyanins in hawthorn fruit during storage at low temperature) under various stress conditions (Khalid et al., 2015; Rasheed et al.,

2018; Razavi et al., 2018). Similar observations have also been reported by applying 5-ALA (Naeem et al., 2010). A joint application of GB and 5-ALA may increase their efficiency compared with single applications, which remains to be investigated. Both 5-ALA (Li et al., 2016) and GB (, Razavi et al. 2018) have strong potential to promote physiological characteristics and nutrient composition of fruits during postharvest storage. Overall, GABA, melatonin, 5-ALA, and GB have the potential to promote both pre- and postharvest quality attributes of produce.

49.6 CONCLUDING REMARKS

Because of the synergistic and antagonistic effects between metabolic compounds and their differential influences on physiological responses under the interacting internal and external cues, the influence of manipulating not only a specific metabolic compound, but also the whole metabolic network should be considered during breeding programs to obtain stress-tolerant plants with high-quality indices. The manipulation of signaling and metabolic pathways (for example, GABA and melatonin) may positively or negatively affect the metabolic relationships (for example, flavonoid composition and content), which thereby affect plant qualities such as marketability and nutraceutical values, yield, and stress tolerance. Hence, the analysis of plant metabolic interaction networks in different plants under specific environmental conditions is of critical importance during breeding programs so as to obtain plants with optimal performance and quality attributes.

REFERENCES

Aghdam, M. S., Fard, J. R. 2017. Melatonin treatment attenuates postharvest decay and maintains nutritional quality of strawberry fruits (Fragaria×anannasa cv. Selva) by enhancing GABA shunt activity. *Food Chemistry*, 221, 1650–1657.

Alam, M. M., Hasanuzzaman, M., Nahar, K., Fujita, M. 2013. Exogenous salicylic acid ameliorates short-term drought stress in mustard (Brassica juncea L.) seedlings by up-regulating the antioxidant defense and glyoxalase system. *Australian Journal of Crop Science*, 7 (7), 1053–1063.

Aldesuquy, H., Ghanem, H. 2015. Exogenous salicylic acid and trehalose ameliorate short term drought stress in wheat cultivars by up-regulating membrane characteristics and antioxidant defense system. *Journal of Horticulture*, 2 (2), 139.

Askari-Khorasgani, O. 2018. Regulation of trehalose/sucrose-nonfermenting-related protein kinase1 and its signaling pathways involved in drought stress responses. *MedCrave Online Journal of Cell Science & Report*, 5 (3), 39–41.

Askari-Khorsgani, O., Flores, F., Pessarakli, M. 2018. Plant signaling pathways involved in stomatal movement under drought stress conditions. *Advances in Plants and Agriculture Research*, 8(3), 290–297.

Askari-Khorasgani, O., Pessarakli, M. 2019a. Manipulation of plant methylglyoxal metabolic and signaling pathways for improving tolerance to drought stress. *Journal of Plant Nutrition*, 42 (10), 1268–1275.

Askari-Khorasgani, O., Pessarakli, M. 2019b. Fruit quality and nutrient composition of grapevines - a review. *Journal of Plant Nutrition*, In press.

Askari-Khorasgani, O., Pessarakli, M. 2019c. Relationship between signalling and metabolic pathways of grapevines under temperature and light stresses – a review. *Journal of Plant Nutrition*, In press.

Askari-Khorasgani, O., Pessarakli, M. 2019d. Grapevine selection for improving nutrient content and composition and the associated quality indices – a review *journal of Plant Nutrition*, In press.

Böttcher, C., Burbidge, C. A., Boss, P. K., Davies, C. 2013. Interactions between ethylene and auxin are crucial to the control of grape (Vitis vinifera L.) berry ripening. *BMC Plant Biology*, 13 (1), 222.

Böttcher, C., Keyzers, R. A., Boss, P. K., Davies, C. 2010. Sequestration of auxin by the indole-3-acetic acid-amido synthetase GH3-1 in grape berry (Vitis vinifera L.) and the proposed role of auxin conjugation during ripening. *Journal of Experimental Botany*, 61 (13), 3615–3625.

Brunetti, C., Fini, A., Sebastiani, F., Gori, A., Tattini, M. 2018. Modulation of phytohormone signaling: a primary function of flavonoids in plant–environment interactions. *Frontiers in Plant Science*, 9 (1042).

Bünger-Kibler, S., Bangerth, F. 1982. Relationship between cell number, cell size and fruit size of seeded fruits of tomato (Lycopersicon esculentum Mill.), and those induced parthenocarpically by the application of plant growth regulators. *Plant Growth Regulation*, 1 (3), 143–154.

Carillo, P. 2018. GABA Shunt in Durum Wheat. *Frontiers in Plant Science*, 9 (100).

Champa, W., Harindra, A., Gill, M. I. S., Mahajan, B. V. C., Arora, N. K. 2015. Preharvest salicylic acid treatments to improve quality and postharvest life of table grapes (Vitis vinifera L.) cv. flame seedless. *Journal of Food Science and Technology*, 52 (6), 3607–3616.

Charpentier, M. 2018. Calcium signals in the plant nucleus: origin and function. *Journal of Experimental Botany*, 69 (17), 4165–4173.

Chen, L.-J., Wuriyanghan, H., Zhang, Y.-Q., Duan, K.-X., Chen, H.-W., Li, Q.-T., Lu, X., He, S.-J., Ma, B., Zhang, W.-K., Lin, Q., Chen, S.-Y., Zhang, J.-S. 2013. An S-domain receptor-like kinase, OsSIK2, confers abiotic stress tolerance and delays dark-induced leaf senescence in rice. *Plant Physiology*, 163 (4), 1752–1765.

Clapp, J., Newell, P., Brent, Z. W. 2018. The global political economy of climate change, agriculture and food systems. *The Journal of Peasant Studies*, 45 (1), 80–88.

Corpas, J. F., Palma, M. J. 2018. Assessing nitric oxide (NO) in higher plants: an outline. *Nitrogen*, 1 (1), 12–20.

Couée, I., Sulmon, C., Gouesbet, G., El Amrani, A. 2006. Involvement of soluble sugars in reactive oxygen species balance and responses to oxidative stress in plants. *Journal of Experimental Botany*, 57 (3), 449–459.

del Río, L. A., Sandalio, L. M., Corpas, F. J., Palma, J. M., Barroso, J. B. 2006. Reactive oxygen species and reactive nitrogen species in peroxisomes. production, scavenging, and role in cell signaling. *Plant Physiology*, 141 (2), 330–335.

Demidchik, V., Maathuis, F., Voitsekhovskaja, O. 2018. Unravelling the plant signalling machinery: an update on the cellular and genetic basis of plant signal transduction. *Functional Plant Biology*, 45 (2), 1–8.

Demidchik, V., Shabala, S. 2018. Mechanisms of cytosolic calcium elevation in plants: the role of ion channels, calcium extrusion systems and NADPH oxidase-mediated 'ROS-Ca²⁺ Hub'. *Functional Plant Biology*, 45 (2), 9–27.

Erland, L. A. E., Saxena, P. K. 2018. Melatonin in plant morphogenesis. *In Vitro Cellular & Developmental Biology - Plant*, 54 (1), 3–24.

Erland, L. A. E., Saxena, P. K., Murch, S. J. 2018. Melatonin in plant signalling and behaviour. *Functional Plant Biology*, 45 (2), 58–69.

Erland, L. A. E., Shukla, M. R., Singh, A. S., Murch, S. J., Saxena, P. K. 2017. Melatonin and serotonin: Mediators in the symphony of plant morphogenesis. *Journal of Pineal Research*, 64 (2), e12452.

Fan, J., Xie, Y., Zhang, Z., Chen, L. 2018. Melatonin: A multifunctional factor in plants. *International Journal of Molecular Sciences*, 19 (5), 1528.

Foyer, C. H., Wilson, M. H., Wright, M. H. 2018. Redox regulation of cell proliferation: Bioinformatics and redox proteomics approaches to identify redox-sensitive cell cycle regulators. *Free Radical Biology and Medicine*, 122, 137–149.

Fraga, C. G., Oteiza, P. I., Galleano, M. 2018. Plant bioactives and redox signaling: (–)-Epicatechin as a paradigm. *Molecular Aspects of Medicine*, 61, 31–40.

Gao, H., Wu, S., Zeng, Q., Li, P., Guan, W. 2018. Effects of exogenous γ-aminobutyric acid treatment on browning and foodborne pathogens in fresh-cut apples. *Postharvest Biology and Technology*, 146, 1–8.

Guo, J., Wang, S., Yu, X., Dong, R., Li, Y., Mei, X., Shen, Y. 2018. Polyamines regulate strawberry fruit ripening by abscisic acid, auxin, and ethylene. *Plant Physiology*, 177 (1), 339–351.

Hardeland, R. 2016. Melatonin in plants – Diversity of levels and multiplicity of functions. *Frontiers in Plant Science*, 7 (198).

Hasan, M. K., Ahammed, G. J., Yin, L., Shi, K., Xia, X., Zhou, Y., Yu, J., Zhou, J. 2015. Melatonin mitigates cadmium phytotoxicity through modulation of phytochelatins biosynthesis, vacuolar sequestration, and antioxidant potential in *Solanum lycopersicum* L. *Frontiers in Plant Science*, 6 (601).

Hernández-Ruiz, J., Arnao, B. M. 2018. Relationship of melatonin and salicylic acid in biotic/abiotic plant stress responses. *Agronomy*, 8(4), 33.

Hoque, T. S., Hossain, M. A., Mostofa, M. G., Burritt, D. J., Fujita, M., Tran, L.-S. P. 2016. Methylglyoxal: An emerging signaling molecule in plant abiotic stress responses and tolerance. *Frontiers in Plant Science*, 7 (1341).

Hossain, M. A., Fujita, M. 2010. Evidence for a role of exogenous glycinebetaine and proline in antioxidant defense and methylglyoxal detoxification systems in mung bean seedlings under salt stress. *Physiology and Molecular Biology of Plants: An International Journal of Functional Plant Biology*, 16 (1), 19–29.

Hwang, O. J., Back, K. 2018. Melatonin is involved in skotomorphogenesis by regulating brassinosteroid biosynthesis in rice plants. *Journal of Pineal Research*, 65 (2), e12495.

Ibrahim, H. A., Abdellatif, Y. M. R. 2016. Effect of maltose and trehalose on growth, yield and some biochemical components of wheat plant under water stress. *Annals of Agricultural Sciences* 61 (2), 267–274.

Kalhor, M. S., Aliniaeifard, S., Seif, M., Asayesh, E. J., Bernard, F., Hassani, B., Li, T. 2018. Enhanced salt tolerance and photosynthetic performance: Implication of ɤ-amino butyric acid application in salt-exposed lettuce (*Lactuca sativa* L.) plants. *Plant Physiology and Biochemistry*, 130, 157–172.

Kangasjärvi, S., Kangasjärvi, J. 2014. Towards understanding extracellular ROS sensory and signaling systems in plants. *Advances in Botany*, 2014, 10.

Khalid, A., Zafar, Z. U., Akram, A., Hussain, K., Manzoor, H., Al-Qurainy, F., Ashraf, M. A. 2015. Photosynthetic capacity of canola (*Brassica napus* L.) plants as affected by glycinebetaine under the salt stress. *Journal of Applied Botany and Food Quality*, 88 (1), 77–86.

Kizildeniz, T., Pascual, I., Irigoyen, J. J., Morales, F. 2018. Using fruit-bearing cuttings of grapevine and temperature gradient greenhouses to evaluate effects of climate change (elevated CO_2 and temperature, and water deficit) on the cv. red and white Tempranillo. Yield and must quality in three consecutive growing seasons (2013–2015). *Agricultural Water Management*, 202, 299–310.

Lee, H. Y., Back, K. 2018. Melatonin induction and its role in high light stress tolerance in *Arabidopsis thaliana*. *Journal of Pineal Research*, 65 (3), e12504.

Lee, K., Hwang, O. J., Reiter, R. J., Back, K. 2018. Flavonoids inhibit both rice and sheep serotonin *N*-acetyltransferases and reduce melatonin levels in plants. *Journal of Pineal Research*, 65 (3), e12512.

Lee, K., Lee, H. Y., Back, K. 2017. Rice histone deacetylase 10 and Arabidopsis histone deacetylase 14 genes encode *N*-acetylserotonin deacetylase, which catalyzes conversion of *N*-acetylserotonin into serotonin, a reverse reaction for melatonin biosynthesis in plants. *Journal of Pineal Research*, 64 (2), e12460.

Li, X., Zhang, L., Ahammed, G. J., Li, Z. X., Wei, J. P., Shen, C., Yan, P., Zhang, L. P., Han, W. Y. 2017a. Nitric oxide mediates brassinosteroid-induced flavonoid biosynthesis in *Camellia sinensis* L. *Journal of Plant Physiology*, 214, 145–151.

Li, X., Brestic, M., Tan, D.-X., Zivcak, M., Zhu, X., Liu, S., Song, F., Reiter, R. J., Liu, F. 2017b. Melatonin alleviates low PS I-limited carbon assimilation under elevated CO_2 and enhances the cold tolerance of offspring in chlorophyll b-deficient mutant wheat. *Journal of Pineal Research*, 64 (1), e12453.

Li, Y., Li, Z., Wang, L. 2016. Applications of 5-aminolevulinic acid on the physiological and biochemical characteristics of strawberry fruit during postharvest cold storage. *Ciência Rural*, 46 (12), 2103–2109.

Liang, D., Shen, Y., Ni, Z., Wang, Q., Lei, Z., Xu, N., Deng, Q., Lin, L., Wang, J., Lv, X., Xia, H. 2018. Exogenous melatonin application delays senescence of kiwifruit leaves by regulating the antioxidant capacity and biosynthesis of flavonoids. *Frontiers in Plant Science*, 9 (426).

Liu, Z., Shi, M.-Z., Xie, D.-Y. 2014. Regulation of anthocyanin biosynthesis in *Arabidopsis thaliana* red *pap1-D* cells metabolically programmed by auxins. *Planta*, 239 (4), 765–781.

Lopez-Gomez, M., Cobos-Porras, L., Hidalgo-Castellanos, J., Lluch, C. 2014. Occurrence of polyamines in root nodules of *Phaseolus vulgaris* in symbiosis with *Rhizobium tropici* in response to salt stress. *Phytochemistry*, 107, 32–41.

Majidinia, M., Sadeghpour, A., Mehrzadi, S., Reiter, R. J., Khatami, N., Yousefi, B. 2017. Melatonin: A pleiotropic molecule that modulates DNA damage response and repair pathways. *Journal of Pineal Research*, 63 (1), e12416.

Majumdar, R., Barchi, B., Turlapati, S. A., Gagne, M., Minocha, R., Long, S., Minocha, S. C. 2016. Glutamate, ornithine, arginine, proline, and polyamine metabolic interactions: the pathway is regulated at the post-transcriptional level. *Frontiers in Plant Science*, 7 (78).

Manishankar, P., Wang, N., Köster, P., Alatar, A. A., Kudla, J. 2018. Calcium signaling during salt stress and in the regulation of ion homeostasis. *Journal of Experimental Botany*, 69 (17), 4215–4226.

Martinez, V., Nieves-Cordones, M., Lopez-Delacalle, M., Rodenas, R., Mestre, T. C., Garcia-Sanchez, F., Rubio, F., Nortes, P. A., Mittler, R., Rivero, R. M. 2018. Tolerance to stress combination in tomato plants: new insights in the protective role of melatonin. *Molecules*, 23 (3), 535.

Mhamdi, A., Van Breusegem, F. 2018. Reactive oxygen species in plant development. *Development*, 145 (15), dev164376.

Naeem, M. S., Jin, Z. L., Wan, G. L., Liu, D., Liu, H. B., Yoneyama, K., Zhou, W. J. 2010. 5-Aminolevulinic acid improves photosynthetic gas exchange capacity and ion uptake under salinity stress in oilseed rape (*Brassica napus* L.). *Plant and Soil*, 332 (1), 405–415.

Nahar, K., Hasanuzzaman, M., Alam, M. M., Rahman, A., Suzuki, T., Fujita, M. 2016. Polyamine and nitric oxide crosstalk: antagonistic effects on cadmium toxicity in mung bean plants through upregulating the metal detoxification, antioxidant defense and methylglyoxal detoxification systems. *Ecotoxicology and Environmental Safety*, 126, 245–255.

Nair, M. S., Saxena, A., Kaur, C. 2018. Effect of chitosan and alginate based coatings enriched with pomegranate peel extract to extend the postharvest quality of guava (*Psidium guajava* L.). *Food Chemistry*, 240, 245–252.

Nakabayashi, R., Yonekura-Sakakibara, K., Urano, K., Suzuki, M., Yamada, Y., Nishizawa, T., Matsuda, F., Kojima, M., Sakakibara, H., Shinozaki, K., Michael, A. J., Tohge, T., Yamazaki, M., Saito, K. 2014. Enhancement of oxidative and drought tolerance in Arabidopsis by overaccumulation of antioxidant flavonoids. *The Plant Journal*, 77 (3), 367–379.

Nejat, N., Mantri, N. 2017. Plant immune system: crosstalk between responses to biotic and abiotic stresses the missing link in understanding plant defence. *Current Issues in Molecular Biology*, 23, 1–16.

Neto, F. J. D., Junior, A. P., Borges, C. V., Cunha, S. R., Callili, D., Lima, G. P. P., Roberto, S. R., Leonel, S., Tecchio, M. A. 2017. The exogenous application of abscisic acid induce accumulation of anthocyanins and phenolic compounds of the 'Rubi' grape. *American Journal of Plant Sciences*, 8 (10), 2422–2432.

Olivares, D., Contreras, C., Muñoz, V., Rivera, S., González-Agüero, M., Retamales, J., Defilippi. B. G. 2017. Relationship among color development, anthocyanin and pigment-related gene expression in 'Crimson Seedless' grapes treated with abscisic acid and sucrose. *Plant Physiology and Biochemistry*, 115, 286–297.

Ortega-Villasante, C., Burén, S., Blázquez-Castro, A., Barón-Sola, Á., Hernández, L. E. 2018. Fluorescent in vivo imaging of reactive oxygen species and redox potential in plants. *Free Radical Biology and Medicine*, 122, 202–220.

Palma, F., Carvajal, F., Jamilena, M., Garrido, D. 2014. Contribution of polyamines and other related metabolites to the maintenance of zucchini fruit quality during cold storage. *Plant Physiology and Biochemistry*, 82, 161–171.

Petridis, A., Döll, S., Nichelmann, L., Bilger, W., Mock, H.-P. 2016. Arabidopsis thaliana G2-LIKE FLAVONOID REGULATOR and BRASSINOSTEROID ENHANCED EXPRESSION1 are low-temperature regulators of flavonoid accumulation. *New Phytologist*, 211(3), 912–925.

Podlesakova, K., Ugena, L., Spichal, L., Dolezal, K., De Diego, N. 2018. Phytohormones and polyamines regulate plant stress responses by altering GABA pathway. *New Biotechnology*.

Prakongkha, I., Sompong, M., Wongkaew, S., Athinuwat, D., Buensanteai, N. 2013. Changes in salicylic acid in grapevine treated with chitosan and BTH against *Sphaceloma ampelinum*, the causal agent of grapevine anthracnose. *African Journal of Microbiology Research*, 7 (7), 557–563.

Prinsi, B., Negri, A. S., Failla, O., Scienza, A., Espen, L. 2018. Root proteomic and metabolic analyses reveal specific responses to drought stress in differently tolerant grapevine rootstocks. *BMC Plant Biology*, 18 (1), 126.

Prospects World Population Projections. 2019. Prospects World Population Projections. Available: https://www.worldometers.info/world-population/

Rasheed, R., Iqbal, M., Ashraf, M. A., Hussain, I., Shafiq, F., Yousaf, A., Zaheer, A. 2018. Glycine betaine counteracts the inhibitory effects of waterlogging on growth, photosynthetic pigments, oxidative defence system, nutrient composition, and fruit quality in tomato. *The Journal of Horticultural Science and Biotechnology*, 93 (4), 385–391.

Razavi, F., Mahmoudi, R., Rabiei, V., Aghdam, M. S., Soleimani, A. 2018. Glycine betaine treatment attenuates chilling injury and maintains nutritional quality of hawthorn fruit during storage at low temperature. *Scientia Horticulturae*, 233, 188–194.

Salloum, M. S., Menduni, M. F., Benavides, M. P., Larrauri, M., Luna, C. M., Silvente, S. 2018. Polyamines and flavonoids: key compounds in mycorrhizal colonization of improved and unimproved soybean genotypes. *Symbiosis*, 76 (3), 265–275.

Scholz, S. S., Reichelt, M., Vadassery, J., Mithöfer, A. 2015. Calmodulin-like protein CML37 is a positive regulator of ABA during drought stress in *Arabidopsis*. *Plant Signaling & Behavior*, 10 (6), e1011951.

Shan, C., Wang, T., Zhou, Y., Wang, W. 2018. Hydrogen sulfide is involved in the regulation of ascorbate and glutathione metabolism by jasmonic acid in *Arabidopsis thaliana*. *Biologia Plantarum*, 62 (1), 188–193.

Sheng, L., Shen, D., Luo, Y., Sun, X., Wang, J., Luo, T., Zeng, Y., Xu, J., Deng, X., Cheng, Y. 2017. Exogenous γ-aminobutyric acid treatment affects citrate and amino acid accumulation to improve fruit quality and storage performance of postharvest citrus fruit. *Food Chemistry*, 216, 138–145.

Shi, H., Chen, Y., Tan, D. X., Reiter, R. J., Chan, Z., He, C. 2015. Melatonin induces nitric oxide and the potential mechanisms relate to innate immunity against bacterial pathogen infection in *Arabidopsis*. *Journal of Pineal Research*, 59 (1), 102–108.

Siipola, S. M., Kotilainen, T., Sipari, N., Morales, L. O., Lindfors, A. V., Robson, T. M., Aphalo, P. J. 2014. Epidermal UV-A absorbance and whole-leaf flavonoid composition in pea respond more to solar blue light than to solar UV radiation. *Plant, Cell & Environment*, 38 (5), 941–952.

Snedden, W. A., Arazi, T., Fromm, H., Shelp, B. J. 1995. Calcium/calmodulin activation of soybean glutamate decarboxylase. *Plant Physiology*, 108 (2), 543–549.

Snowden, C. J., Thomas, B., Baxter, C. J., Smith, J. A. C., Sweetlove, L. J. 2015. A tonoplast Glu/Asp/GABA exchanger that affects tomato fruit amino acid composition. *The Plant Journal*, 81 (5), 651–660.

Sulieman, S., Schulze, J. 2010. Phloem-derived γ-aminobutyric acid (GABA) is involved in upregulating nodule N2 fixation efficiency in the model legume Medicago truncatula. *Plant, Cell & Environment*, 33 (12), 2162–2172.

Tan, D.-X., Hardeland, R., Manchester, L. C., Korkmaz, A., Ma, S., Rosales-Corral, S., Reiter, R. J. 2012. Functional roles of melatonin in plants, and perspectives in nutritional and agricultural science. *Journal of Experimental Botany*, 63 (2), 577–597.

Torres, N., Goicoechea, N., Carmen Antolín, M. 2018. Influence of irrigation strategy and mycorrhizal inoculation on fruit quality in different clones of Tempranillo grown under elevated temperatures. *Agricultural Water Management*, 202, 285–298.

Uemura, T., Tomonari, Y., Kashiwagi, K., Igarashi, K. 2004. Uptake of GABA and putrescine by UGA4 on the vacuolar membrane in *Saccharomyces cerevisiae*. *Biochemical and Biophysical Research Communications*, 315 (4), 1082–1087.

Wang, L., Zhao, Y., Reiter, R. J., He, C., Liu, G., Lei, Q., Zuo, B., Zheng, X. D., Li, Q., Kong, J. 2014. Changes in melatonin levels in transgenic 'Micro-Tom' tomato overexpressing ovine *AANAT* and ovine *HIOMT* genes. *Journal of Pineal Research*, 56 (2), 134–142.

Wang, P., Yin, L., Liang, D., Li, C., Ma, F., Yue, Z. 2012. Delayed senescence of apple leaves by exogenous melatonin treatment: toward regulating the ascorbate-glutathione cycle. *Journal of Pineal Research*, 53 (1), 11–20.

Wang, Y., Zhou, F., Zuo, J., Zheng, Q., Gao, L., Wang, Q., Jiang, A. 2018. Pre-storage treatment of mechanically-injured green pepper (*Capsicum annuum* L.) fruit with putrescine reduces adverse physiological responses. *Postharvest Biology and Technology*, 145, 239–246.

Wannabussapawich, B., Seraypheap, K. 2018. Effects of putrescine treatment on the quality attributes and antioxidant activities of 'Nam Dok Mai No.4' mango fruit during storage. *Scientia Horticulturae*, 233, 22–28.

Xia, X.-J., Zhou, Y.-H., Shi, K., Zhou, J., Foyer, C. H., Yu, J.-Q. 2015. Interplay between reactive oxygen species and hormones in the control of plant development and stress tolerance. *Journal of Experimental Botany*, 66 (10), 2839–2856.

Zarei, A., Trobacher, C. P., Shelp, B. J. 2016. Arabidopsis aldehyde dehydrogenase 10 family members confer salt tolerance through putrescine-derived 4-aminobutyrate (GABA) production. *Scientific Reports*, 6 (1), 35115.

Zarei, A., Trobacher, C. P., Shelp, B. J. 2018. Corrigendum: *Arabidopsis* aldehyde dehydrogenase 10 family members confer salt tolerance through putrescine-derived 4-aminobutyrate (GABA) production. *Scientific Reports*, 8, 46967.

Zhang, X., Ding, X., Ji, Y., Wang, S., Chen, Y., Luo, J., Shen, Y., Peng, L. 2018a. Measurement of metabolite variations and analysis of related gene expression in Chinese liquorice (*Glycyrrhiza uralensis*) plants under UV-B irradiation. *Scientific Reports*, 8 (1), 6144.

Zhang, J., Li, D., Wei, J., Ma, W., Kong, X., Rengel, Z., Chen, Q. 2018b. Melatonin alleviates aluminum-induced root growth inhibition by interfering with nitric oxide production in *Arabidopsis*. *Environmental and Experimental Botany*, 161, 157–165.

Zhu, Z., Zhang, Y., Liu, J., Chen, Y., Zhang, X. 2018. Exploring the effects of selenium treatment on the nutritional quality of tomato fruit. *Food Chemistry*, 252, 9–15.

Section X

Beneficial Aspects of Stress on Plants/Crops

50 Beneficial Effects of Various Environmental Stresses on Vegetables and Medicinal Plants for the Production of High Value-Added Plants

Satoru Tsukagoshi and Wataru Yamori

CONTENTS

50.1 INTRODUCTION

Environmental factors such as light, temperature, humidity, CO_2 concentration, and the composition of nutrient solutions influence plant growth and have a direct impact on biochemical pathways (Yamori et al., 2009, 2010, 2011, 2014, 2015, 2016; Yamori and Shikanai, 2016), and these factors affect the metabolism of both primary and secondary products (e.g., Joshi et al., 2017; Lu et al., 2015, 2017; Maneejantra et al., 2016; Tewolde et al., 2016). Plant factories are plant production facilities that provide growers with the advantage of controlling the various environmental factors, mentioned above, thus enabling the control of plant growth and development (Kozai et al., 2015; Yamori et al., 2014). The ability to control environmental factors using technology in plant factories has allowed producers to maintain close control of their production systems.

Plant factory technology is currently being used for the commercial production of leafy vegetables, herbs, and transplants in Asian regions, including Japan, Korea, China, Taiwan, and Thailand (Hu et al., 2014; Kozai, 2013). This chapter describes the production of functional vegetables and medicinal plants that is achieved by (1) controlling the composition of the hydroponic nutrient solution and (2) optimizing light quality, based on the science behind the response of plants to various environmental stresses. Understanding plant responses to various environmental stress would be necessary to optimize the environmental control to improve the plant biomass production and desired functional ingredients in plant factories.

50.2 CHARACTERISTICS OF HYDROPONICS SUITABLE FOR PRODUCING HIGH SUGAR CONTENT FRUIT AND HIGH FUNCTIONAL VEGETABLES

In plant factories, hydroponic systems have been commonly used to control the nutrient and water supply for optimizing plant growth. In hydroponic agriculture, water is enriched with nutrients, creating a perfectly balanced, pH adjusted nutrient solution. The hydroponic system can increase or decrease the electrical conductivity (EC) of a nutrient solution or even the supply of a specific nutrient, as desired. For example, the sugar content of tomato fruit can be increased by increasing EC, since a nutrient solution with a high EC suppresses the water uptake of plants. In addition, the functional improvement of hydroponic crops, such as the production of low nitrate (NO_3^-) spinach, or low potassium (K^+) lettuce, is now achieved by reducing the macro-nutrient supply at a specific growth stage. To produce these functional vegetables, it is necessary to alter the nutrient condition immediately after the plants have reached a certain size. Changing the growth conditions for a short period of time is only possible in hydroponic cultivation systems. Thus, hydroponics is generally considered to be suitable for the production of high sugar content fruit and high functional vegetables.

50.2.1 Production of High Sugar Content Tomatoes Using a Solution with High EC

The fruit sugar content of tomatoes (*Solanum lycopersicum* L.) is increased by water stress (e.g., Nahar and Gretzmacher, 2002; Sato et al., 2006; Zushi et al., 2005) In soil culture, plants are usually subjected to water stress because of limited irrigation. However, it is difficult to maintain a uniform water condition over the entire cultivation area. In such cases, the fruit sugar content becomes unstable, and the roots are easily damaged by dry soil. In addition, when the water supply is insufficient, the nutrient supply is also insufficient, because plants absorb nutrients when they are dissolved in water. So, in soil culture, it is generally difficult to produce tomatoes with high sugar content, while avoiding the exhaustion of the plant body and physiological disorders.

On the other hand, a solution with a high EC is generally used in hydroponics to induce water stress in plants. The high EC is provided by increasing the nutrient concentrations in the solution or by adding NaCl in the solution. To increase the sugar content of tomato fruit, previous studies indicate that an EC value of 5 ~ 8 dS m⁻¹ around the root zone appears to be effective (Sakamoto et al., 1999; Wu and Kubota, 2008). With hydroponics, the grower can maintain this condition uniformly for all the plants in the production site, and the extent of the stress can be easily regulated depending on the plant status. However, even with hydroponics, it is difficult to achieve long-term or year-round tomato production while imposing stress, since stress inevitably influences plant growth.

From this viewpoint, "a low truss with a high-density tomato production system" could be an effective way to produce high sugar content tomato fruit (Figure 50.1). With this cultivation method, tomato plants are placed 10 ~ 20 cm apart on a cultivation bed, which is basically a nutrient film

technique (NFT) system, and only 1 ~ 3 fruit trusses are set. The grower can provide strong stress from a relatively early growth stage because only a few trusses are the target as regards to improving the quality, and it is not necessary to consider later growth. The general method for producing high sugar content tomatoes using this system is 1) a nutrient solution at a standard concentration (EC 1.2 ~ 2.4 dS m⁻¹) is supplied when seedlings are placed on the bed, 2) the EC of the nutrient solution is gradually increased after the flowers emerge, 3) a nutrient solution EC of 6 ~ 8 dS m⁻¹ is achieved at the first fruit set, and 4) a high EC solution is maintained until the end of the harvest (Johkan et al., 2014).

A more effective way to increase sugar content involves the combination of the cultivation method and pot culture (Figure 50.2). The specially made pot has a water-permeable root-barrier sheet at the bottom, as shown in Figure 50.2. The seedling is transplanted into a pot filled with a substrate such as granulated rockwool, or coconut fiber. The EC of the nutrient solution is gradually increased from the time the flowers emerge to the first fruit set, as described previously. The solution basically moves upwards in the pot from the bottom to the surface, and only the water evaporates from the surface. As a result, the solution is often concentrated in the substrate, and the EC around the root exceeds the EC of the supplied solution. In addition to this high EC, it is hard for plants to absorb water since the root mass is restricted by the pot. Thus, the plants are more effectively subjected to water stress.

However, in any case, it is necessary to be aware of the possibility of the frequent occurrence of blossom-end rot, extreme reduction in fruit weight, and root damage caused by highly-concentrated salts around the roots.

50.2.2 Production of Functional Vegetables by Controlling the Composition of the Hydroponic Nutrient Solution

One of the methods described below is chosen to reduce the specific ion content in the edible part of hydroponically grown

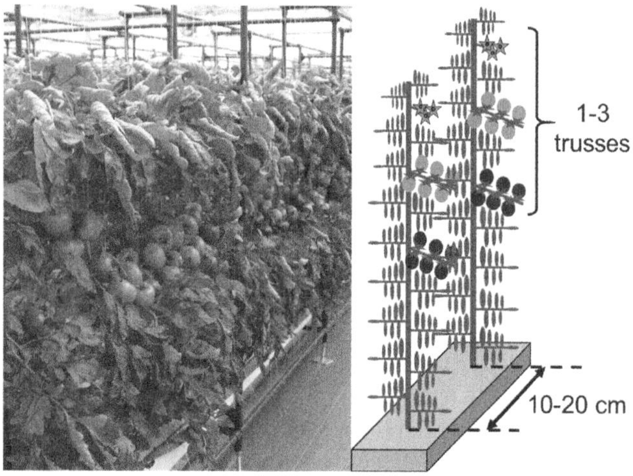

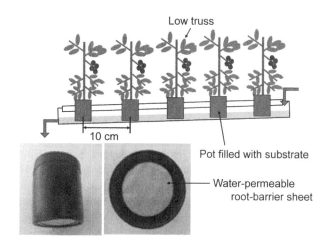

FIGURE 50.1 (See color insert.) Low truss with a high-density tomato production system.

FIGURE 50.2 Combination of low truss with the high-density tomato production system and pots filled with substrate.

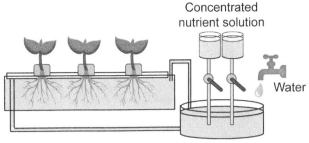

Set the targeted nutrient concentration lower
than the concentration taken up by plants

FIGURE 50.3 Basic theory behind reducing the inorganic ingredient (I). Set the targeted nutrient concentration lower than the concentration taken up by plants.

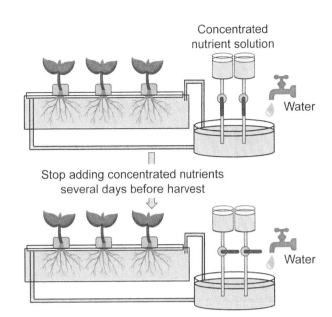

Stop adding concentrated nutrients
several days before harvest

FIGURE 50.4 Basic theory behind reducing the inorganic ingredient (II). Stop adding concentrated nutrients several days before harvest.

These crops are harvested in a couple of days

Solution tank & supply pump for 2 or 3 planting beds

FIGURE 50.5 (See color insert.) Commercial production system for spinach with low NO_3^- concentration.

vegetables, particularly leafy vegetables, but a combination of the methods can also be adopted.

The first method is to cultivate the vegetables from the start in a solution that its target ion concentration is low (Figure 50.3). It is possible to estimate the amounts of nutrient uptake (L) and water uptake (mmol) by plants. The ratio of the amounts of nutrient and water uptake is usually called "n/w." When the target ion concentration is lower than n/w, the ion concentration gradually decreases as cultivation proceeds, and the ion content in the plant tissue also decreases. However, it is possible to employ this method when using a re-circulating hydroponic system. In addition, when there is a relatively large quantity of nutrient solution per plant in the cultivation system, the effect of a low nutrient concentration hardly appears because the total amount of the target ion that is supplied is too much even if the concentration is low (total supplied amount = solution volume x concentration). The effect also hardly appears when water and concentrated nutrients are frequently supplied.

The second method is to terminate the supply of concentrated nutrient and to continue to supply only water to the solution tank from several days to 2 weeks before harvest (Figure 50.4). Although the supply of concentrated nutrient is halted while only water is supplied, the plants continue to absorb the nutrient solution, leading to the dilution of the nutrient solution concentration. As the nutrient solution concentration decreases, the ion content in the plant tissue also decreases (Tsukagoshi et al., 1999). With this method, the entire ion concentration in the solution tank decreases. However, the degree of the decrease depends on the type of the ion, since the metabolism and translocation differ depending on the ion. In addition, the cultivation system must be subdivided into plural units (containing plural solution tanks) for year-round production (Figure 50.5).

The third method is to move the plants from cultivation beds with a standard solution to cultivation beds with a specific solution that contains no target ions from several days to 2 weeks before harvest (Figure 50.6). Although moving planting panels and plants takes time and requires physical labor, it is possible to terminate the supply of target ions to the plants just after the moving work is complete (Ogawa et al., 2007). In

addition, the treatment is much easier than the second method since only two cultivation systems are needed, namely one for cultivating the plants until they reach a specific size and the other for treatment.

50.2.2.1 Practical Method for Low NO_3^- Vegetable Production

Plants accumulate excess nitrogen that is not used for growth as nitrate in vacuoles (Granstedt and Huffaker, 1982). This is considered to be a reserve pool that is utilized during nitrogen starvation. Although nitrate is not very toxic, it has been considered harmful to human health and hazardous when ingested, because it is converted to nitrite in the gastrointestinal tract and then the nitrite leads to the formation

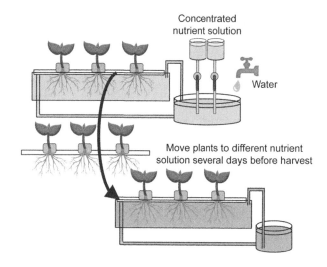

FIGURE 50.6 Basic theory behind reducing the inorganic ingredient (III). Move plants to a different nutrient solution containing no target ions several days before harvest.

TABLE 50.1

Formula of the Nutrient Solution Used for Low NO$_3^-$ Spinach Production

NH$_4$-N	NO$_3$-N	PO$_4$-P	K	Ca	Mg
		(mmol L^{-1})			
1.3	16	4	12	4	4

TABLE 50.2

Effect of Restriction of the Nutrient Supply on the Growth and Quality in Spinach

Days after Nutrient Supply Was Stopped (Days)	Shoot Fresh Weight (g plant^{-1})	NO$_3^-$ conc. in Leaves (ppm)	Ascorbic Acid Conc. (mg per 100 g Fresh Weight)
0	12.8	3,232	21.4
2	12.6	2,870	20.3
4	12.3	2,342	35.3
6	10.8	1,781	31.7

Cultivar: Joker. Experimental period: September 29 ~ November 29, 1997.

of nitrosamines, which are potent carcinogens (Sohar and Domoki, 1980). However, recently, the general view has been changing, and it is now considered that NO$_3^-$ intake from vegetables is of low risk to human health (European Food Safety Authority, Ministry of Agriculture, Forestry and Fisheries). In this section, we disregard the effect of NO$_3^-$ on human health and summarize a cultivation strategy for reducing NO$_3^-$ in plants.

More than 90% of the nitrogen-containing nutrients in most nutrient solutions, including Hoagland's solution, are in the form of NO$_3$-N. To prevent the accumulation of an excess of nitrate in vegetables, they should be fed a suitable amount of nitrogen nutrient that is exhausted during their growth. The nitrate concentration in leafy vegetables can be reduced by 1) eliminating NO$_3^-$ in the nutrient solution or 2) replacing it with NH$_4^+$ a few weeks before harvest.

50.2.2.2 Restriction When Feeding NO$_3^-$ in Nutrient Solution

The combined method is shown in Figure 50.1 and Figure 50.2 is adopted to produce vegetables with a low NO$_3^-$ concentration. The concentration (n/w) of the NO$_3^-$ uptake by spinach obtained from a previous experiment is approximately 17 mmol L^{-1} and so the concentration of NO$_3^-$ in the nutrient solution, namely the "supplied concentration," is presumed to be set at 16 mmol L^{-1} (Table 50.1). The nutrient solution with this composition is supplied to the cultivation system, and supply of just the nutrient is halted several days before harvest. In addition to the gradual reduction in the NO$_3^-$ concentration in the solution caused by the gap between the uptake and the supply concentration, the rapid reduction in the ion concentration that results when the nutrient supply is stopped before the harvest can effectively produce low NO$_3^-$ spinach (Tsukagoshi et al., 1999).

Although the degree of the NO$_3^-$ reduction depends on the cultivation season and for how many days the nutrient supply is stopped, the leaf NO$_3^-$ concentration of the spinach

fell from 3,200 ppm to around 2,500 ppm by stopping the nutrient supply for 2 ~ 4 days (Table 50.2, Tsukagoshi et al., 1999). The NO$_3^-$ concentration decreases more if the nutrient is withdrawn for a larger number of days. However, it is necessary to examine the most suitable period during which to stop fertilization under each cultivation condition in order to avoid leaf etiolation and a reduction in fresh weight.

The complete elimination of NO$_3^-$ from the nutrient solution 1 week before harvest resulted in a 56% decrease in the NO$_3^-$ concentration in celery leaves compared with those grown in the regular nutrient solution (Martignon et al., 1994). When plants were deprived of nitrogen nutrient, they were able to reuse the NO$_3^-$ accumulated in vacuoles for growth, and as a result, the accumulated NO$_3^-$ decreased (Santamaria et al., 1998). Eliminating nitrogen nutrient completely 7–10 days before harvest appears to reduce the fresh mass of lettuce leaves at harvest (Santamaria et al., 1998).

50.2.2.3 Replacing NO$_3^-$ with NH$_4^+$ in the Nutrient Solution

Replacing some of the NO$_3^-$ in the nutrient solution with NH$_4^+$ can reduce the NO$_3^-$ concentration of vegetables without reducing the yield (Saigusa and Kumazaki, 2014). In general, vegetables cultivated in fields, such as lettuce, spinach, rocket, chicory, and endive, prefer NO$_3^-$ to NH$_4^+$, and feeding them with NH$_4^+$ is considered to cause stunting (Ohashi-Kaneko, 2015). However, some of these plant species that absorb NO$_3^-$ preferentially, including lettuce, can grow vigorously in a nutrient solution in which the main source of nitrogen is NH$_4^+$, provided the pH is kept strictly at 5.5. The biomass in lettuce plants is almost the same as that of lettuce plants cultured in the conventional nutrient solution (Morituguet al., 1980; Moritugu and Kawasaki, 1982, 1983). When plants take up NH_4^+, they excrete protons into the nutrient solution to maintain their internal electrical conductivity balance. This reduces the pH of the nutrient solution (Tsukagoshi, 2002), and plants sensitive to acidic conditions are damaged (Troelstra et al., 1990). However, spinach and Chinese cabbage cultured in a solution in which the main source of nitrogen is NH_4^+ are stunted even when the pH is held constant, and they seem to be sensitive to both NH_4^+ and pH (Moritugu and Kawasaki, 1983).

In leaf lettuce, the NO_3^- content was very low without a reduction in yield when the available nitrogen status in the nutrient solution was cut by 1/5 (1,375 ppm vs. 6,450 ppm for plants fed with the conventional nutrient solution), with the NO_3^-/NH_4^+ molar ratio kept at 1 from transplanting to harvest (Saigusa and Kumazaki, 2014). When chicory plants were transferred from 4 to 1 mM NO_3^- plus 3 mMNH_4^+ 6 days before harvesting, the NO_3^- content in leaves decreased by 58% without a decrease in fresh mass compared with growth with no change in the nutrient solution (Santamaria et al., 1998).

50.2.2.4 Practical Method for Low K⁺ Vegetable Production

Patients suffering from kidney disease or those undergoing dialysis cannot excrete excess K^+ in their body fluids. So, their intake of K^+ is often restricted to 1,500 mg a day. Therefore, there has been a strong need for vegetables with low K^+ concentrations for the patients' meals. A method for cultivating vegetables with low K^+ was established for spinach by Ogawa et al. (2007), developed by Ogawa et al. (2012) and this spinach is now commercially produced using hydroponics. Generally, a plant grown under low K^+ conditions has a lower K^+ content in the leaves (Asao et al., 2013; Terry and Ulrich, 1973). Thus, the production technique is very similar to that for low NO_3^- spinach, namely, the limitation of K^+ supply during the cultivation and/or withdrawal of K^+ from the nutrient solution 1 ~ 2 weeks before harvest.

There is a great difference between low NO_3^- and low K^+ vegetable production. The NO_3^- concentration decreases if the supply from the root stops because it is consumed by amino acid synthesis in leaves. On the other hand, K^+ exists in an ion form inside the plant tissue and is not consumed by the synthetic reaction of the plant. To produce low K^+ lettuce, the plant growth must continue after stopping the supply of K^+ to dilute the K^+ concentration in the plant tissue in addition to restricting the absorption (Figure 50.7). Thus, the solution, which aims to decrease the K^+ content in the plant, should contain some nitrogen to help the plant grow after the withdrawal of K^+.

The K^+ concentration of spinach with a low K^+ decreased to about 24% (165 mg K^+ per 100 g fresh weight; Ogawa et al., 2007) of that in spinach supplied with the conventional nutrient solution (MEXT, 2015) without any significant reduction in fresh weight. Lettuce (*Lactuca sativa* L.) with a low K^+ is

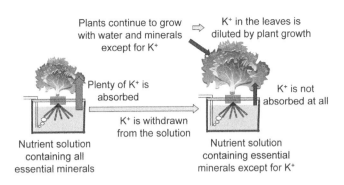

FIGURE 50.7 Strategy for producing leafy vegetables with a low K^+ concentration.

already on the market produced in Japan by plant factories with artificial lighting (e.g., Fujitsu Home & Office Services Limited, Dr. Vegetable, Aizufujikako Co., Ltd.). In the commercial production of lettuce with a low K^+, the plants are raised in a complete nutrient solution or a solution with a relatively low K^+ concentration until the plants reach a certain size. They are then moved to the solution containing no K^+ and left for 10 ~ 14 days. In this case, the K^+ content in the leaves fell below 100 mg K^+ per 100 g fresh weight.

50.2.2.5 Trials to Produce Low K⁺ Fruit Vegetables

The production of low K^+ fruit vegetables is considered to be very difficult, compared with that of low K^+ leaf vegetables (Besford and Maw, 1975). The restriction of K^+ supply and the reduction in fruit K^+ content is very difficult for various reasons; 1) a sufficient supply of K^+ is needed to ensure the vegetative growth required for normal flower and fruit development, 2) K^+ influences fruit quality, particularly the acid content, 3) K^+ easily moves among plant tissue and a large amount of K^+ is translocated into fruit from other tissue very easily during the fruit development stage (Figure 50.8). In other words, the extreme restriction of K^+ supply suppress the plant growth, but the halfway restriction of the K^+ supply does not affect the fruit K^+ content.

In melon fruit (*Cucumis melo* L.), the fruit K^+ content was successfully reduced to around 60% compared with the standard value of K^+ content of melon (340 mg K^+ per 100 g fresh weight) (MEXT, 2015) without affecting the sugar and acid content of the fruit (Asao et al., 2013). Successful production was realized by transferring plants from a half-strength Enshi formula nutrient solution (KNO_3 404 mg L^{-1}, $Ca(NO_3)_2 \bullet 4H_2O$ 472 mg L^{-1}, $NH_4H_2PO_4$ 76 mg L^{-1}, $MgSO_4 \bullet 7H_2O$ 246 mg L^{-1}) to a nutrient solution containing 1/16 KNO_3 or without K^+ during the fruit enlargement period.

Low K^+ tomato fruit has also been successfully produced (Tsukagoshi et al., 2016). It was considered more difficult to produce low K^+ tomato fruit than low K^+ melon, since the tomato has plural fruit trusses on a shoot, and a truss includes plural fruits, whereas a melon has only one fruit per shoot. With medium-sized tomatoes in a low truss with a high-density tomato production system, the seedlings were first grown in a half-strength Enshi formula nutrient solution, and then KNO_3 was withdrawn when the third truss flowered. Although

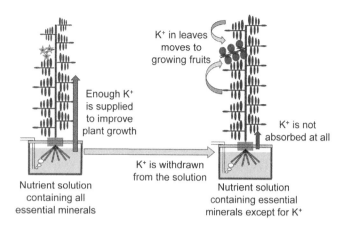

FIGURE 50.8 Strategy for producing fruit and vegetables with a low K⁺ concentration.

TABLE 50.3

Effect of Different Amount of K⁺ Supply on the Yield and Fruit Quality in Cherry-Type Tomato

Amount of K⁺ Supply Before Anthesis – After Anthesis (g week⁻¹ plant⁻¹)	Yield (g plant⁻¹)	Brix (%)	Acid Content (%)	K⁺ Content in Fruit (mg per 100 g Fresh Weight)
0.3–0.6	685	7.5	0.47	152
0.15 – 0.3	505	8.1	0.40	143
0.08–0.15	527	8.2	0.31	108
0.04 – 0.08	470	6.7	0.33	76
0.03–0.05	394	6.0	0.27	61

Cultivar: Carol 10. Experimental period: October 9, 2014 ~ February 6, 2015.

the extent of the reduction in K⁺ depended on the cultivars, the fruit K⁺ content decreased to 100 ~ 140 mg per 100 g fresh weight (Tsukagoshi et al., 2016). Those values corresponded to 34 ~ 50% compared with the standard value for K⁺ content (290 mg per 100 g fresh weight) (MEXT, 2015). However, the acidity of the fruit also decreased with low K⁺ treatment. The problem remained of reducing the fruit K⁺ content while as far as possible maintaining the taste of the fruit.

Furthermore, research was also carried out using a cherry-type tomato (Tsukagoshi and Aoki, 2018). In this experiment, quantitative K⁺ management was adopted to regulate the K⁺ supply more precisely. The solution is usually managed using the EC value, and this method is called "concentration management." Under this condition, some ions such as NO_3^-, $H_2PO_4^-$ and K⁺ tend to be absorbed rapidly by plants, resulting in a large quantity of such ions being absorbed. In addition, the total quantity of the nutrient supply depends on the actual solution volume in the cultivation system. In a contrasting approach called "quantitative management," nutrient is added to the nutrient solution tank several times a day, once a day, once a week, or at other intervals. For example, when the plant absorbs 1 g of K⁺ (potassium) per day, nutrient containing 7 g of K⁺ is added to the solution tank once a week. The cherry-type tomato plants were supplied with 0.04 or 0.08 g of K⁺ for a week per plant before the first flower opened, and then with 0.08 or 0.15 g of K⁺ after flowering, and topped (i.e., the shoot tip was removed) above the second truss leaving just two leaves (Table 50.3, Tsukagoshi and Aoki, 2018). The fruit K⁺ content decreased to 26 ~ 33 % compared with the standard value (290 mg per 100 g fresh weight). In addition, the fruit K⁺ content decreased to 150 mg per 100 g fresh weight when 0.3–0.6 g of K was supplied. This result suggests that quantitative management is a very effective way to produce low K⁺ vegetables because K is a nutrient with which "luxurious absorption" can happen easily by employing concentration management. However, because extreme K⁺ limitation has a large influence on yield and quality, particularly on the acid content, it is important to set the target value of the K⁺ content, and then plan the amount of K⁺ to be supplied.

Both low K⁺ melon and low K⁺ tomato have already been commercialized, but production is extremely limited. In addition, the production of low K⁺ strawberries has been examined, but it has not yet been commercialized, since the flowering period is longer than tomato, and long-term K⁺ control to balance the vegetative and reproductive growth is very difficult to achieve. Further research and the development of new approaches are needed to commercialize low K⁺ strawberries.

50.3 IMPROVING THE QUALITY OF VEGETABLES AND MEDICINAL PLANTS BY CONTROLLING NUTRIENT SOLUTION AND LIGHT QUALITY

The functional ingredients in plants are mainly products of secondary metabolism and include essential oils, polyphenols, carotenoids, anthocyanins, and alkaloids. The secondary metabolite ingredients in plants function as defensive chemical substances. For example, when plants are exposed to biotic stress, such as being eaten by insects or animals, the synthesis of essential oils is promoted (Koseki, 2004). The production of secondary metabolites is also promoted by abiotic stress, such as high temperature, low temperature, dryness, and UV light irradiation. Recently, to develop a novel cultivation system, ways to improve the functional ingredients in vegetables and medicinal plants have been explored by controlling the nutrient solution and light quality.

Secondary metabolites in perilla plants, such as perillaldehyde and rosmarinic acid, reportedly have the potential to prevent disease, particularly due to their anti-allergic and anti-inflammatory properties (Ito et al., 2011; Makino et al., 2003; Sanbongi et al., 2004; Zhang et al., 2008). Recently, functional ingredients in both red and green perilla (*Perilla frutescens* (L.)) have been obtained by optimizing the light intensity and nutrient solution (Lu et al., 2017). The plants were cultivated for 3 weeks with an EC of 1.2 d Sm⁻¹ of OAT Agrio A formula (N:P:K:Ca:Mg = 18.2:1.7:8.6:4.1:1.5 mmol L⁻¹ in a standard concentration) and then subjected to different PPFD (100,

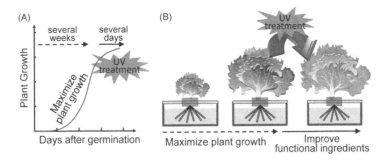

FIGURE 50.9 Improving the quality of vegetables and medicinal plants by controlling light quality.

200, and 300 μmol m^{-2} s^{-1}) with cool white fluorescent lamps combined with different EC levels (1.0, 2.0, and 3.0 d Sm^{-1}) for 5 weeks. The perillaldehyde concentration was largely unaffected by any nutrient or light treatment; however, the rosmarinic acid concentration was significantly increased with a nutrient solution with a lower EC and a higher light intensity.

Cultivation techniques that enhance functional ingredients by controlling the light environment have also been developed for leafy vegetables, herbs, and medicinal plants. For example, when red leaf lettuce plants are cultivated under irradiation with mixtures of red and blue light, there is an increase in the leaf anthocyanin content and the expression of genes encoding several enzymes related to the anthocyanin synthesis pathway as the blue light ratio increases (Ooshima et al., 2013; Shoji et al., 2010). Moreover, it has been reported that anthocyanin synthesis was induced in leaves of red leaf lettuce by UV-B (Park et al., 2007) and UV-A (Voipio and Autio, 1995). Also, Ebisawa et al. (2008) showed that supplementary irradiation with UV-B or UV-A during a dark period increased the anthocyanin content in the leaves of red leaf lettuce compared with those receiving no supplementary irradiation. In addition, supplementation with UV-B increased the limonene and l-menthol concentrations in the leaves of Japanese mint plants (Hikosaka et al., 2010), levels of several essential oils including eugenol and linalool in leaves of sweet basil (Johnson et al., 1999) and rosmarinic acid in leaves of red perilla (Iwai et al., 2010). It has also been reported that supplementary irradiation with UV-B could affect the functional ingredients in shoots as well as in roots. Glycyrrhizin, the major bioactive component of *Glycyrrhiza uralensis*, is widely used as a natural sweetener. UV-B radiation improved the production of glycyrrhizin in the root tissues of *G. uralensis* (Afreen et al., 2006). Most of these researches indicate that high concentrations of functional ingredients can be achieved within a short production period. Since the growth of medicinal plants is generally slow, it is essential to grow them at a fast growth rate until near the end of the cultivation cycle, and then provide environmental stresses several days before harvest to sharply enhance the production of the medicinal components (Figure 50.9).

50.4 CONCLUSIONS

Progress is underway on cultivation methods for functional vegetables and medicinal plants by controlling the composition of the hydroponic nutrient solution and the quality of the lighting. The production of functional ingredients in plant factories has several advantages: 1) plants can be grown under sterile, standardized conditions and are free from biotic and abiotic contamination, 2) uniform plant growth with consistent plant material can be achieved, and 3) consistent biochemical profiles can be achieved with the uniform production of metabolites. Closed systems can ensure that optimum environmental conditions are achieved with which to maximize biomass production and facilitate the induction of stresses. The manipulation of environmental factors such as light and the composition of nutrient solutions are the key techniques with which to promote the accumulation of desired functional ingredients. Plant production using this technology is not only a part of agriculture but also of a new plant-based industry that combines molecular biology and environmental engineering.

REFERENCES

Afreen, F., Zobayed, S., and Kozai, T. 2006. Spectral quality and UV-B stress stimulate glycyrrhizin concentration of Glycyrrhiza uralensis. *Plant Physiol. Biochem.* 43, 1074–1081.

Asao, T., Asaduzzaman, M., Mondal, M., Tokura, M., Adachi, F., Ueno, M., Kawaguchi, M., Yano, S., and Ban, T. 2013. Impact of reduced potassium nitrate concentrations in nutrient solution on the growth, yield and fruit quality of melon in hydroponics. *Sci. Hort.* 164, 221–231.

Besford, R.T. and Maw, G.A. 1975. Effect of potassium nutrition on tomato plant growth and fruit development. *Plant Soil.* 42, 395–412.

Bornman, J.F., Reuber, S., Cen, Y.P., and Weissenbock, G. 1997. Ultraviolet radiation as a stress factor and the role of protective pigments. In: Lusden, P. (Ed.), *Plants and UV-B. Responses to Environmental Change.* Society for Experimental Biology Seminar Series 64. Cambridge University Press, Cambridge, pp. 157–168.

Ebisawa, M., Shoji, K., Kato, M., Shimomura, K., Goto, F., and Yoshihara, T. 2008. Effect of supplementary lighting of UV-B, UV-A, and blue light during night on growth and coloring in red-leaf lettuce (in Japanese). *J. Sci. High Technol. Agric.* 20, 158–164.

Granstedt, R.C. and Huffaker, R.C. 1982. Identification of the leaf vacuole as a major nitrate storage pool. *Plant Physiol.* 70, 410–413.

Hikosaka, S., Ito, K., and Goto, E., 2010. Effects of ultraviolet light on growth, essential oil concentration, and total antioxidant capacity of Japanese mint. *Environ. Control Biol.* 48, 185–190.

Hu, M. C., Chen, Y.H., and Huang, L.C. 2014. A sustainable vegeta- ble supply chain using plant factories in Taiwanese markets: A Nash-Cournot model. *Int. J. Prod. Econ.* 152, 49–56.

Ito, N., Nagai, T., Oikawa, T., Yamada, H., and Hanawa, T. 2011. Antidepressant like effect of l-perillaldehyde in stress- induced depression-like model mice through regulation of the olfactory nervous system. *Evid. Based Complement. Alternat. Med.* 2011, 512697.

Iwai, M., Ohta, M., Tsuchiya, H., and Suzuki, T. 2010. Enhanced accumulation of caffeic acid, rosmarinic acid and luteolin- glucoside in red perilla cultivated under red diode laser and blue LED illumination followed by UV-A irradiation. *J. Funct. Foods.* 2, 66–70.

Johkan, M., Nagatsuka, A., Yoshitomi, A., Nakagawa, T., Maruo, T., Tsukagoshi, S., et al. 2014. Effect of moderate salinity stress on the sugar concentration and fruit yield in single-truss, high- density tomato production system. *J. Japan. Soc. Hort. Sci.* 83, 229–234.

Johnson, C.B., Kirby, J., Naxakis, G., and Pearson, S., 1999. Substantial UV-B mediated induction of essential oils in sweet basil (*Ocimum basilicum* L.). *Phytochemistry.* 51, 507–510.

Joshi, J., Zhang, G., Shen, S., Supaibulwatana, K., Watanabe, C.K.A., and Yamori, W. 2017. A combination of downward lighting and supplemental upward lighting improves plant growth in a closed plant factory with artificial lighting. *Hortscience.* 52, 831–835.

Koseki, Y., 2004. Secondary metabolites and plant defense (in Japanese). In: Nishitani, K., Shimazaki, K. (Eds.), *Plant Physiology* (Written in Japanese: Shokubutsu seirigaku). Baifukan, Tokyo, pp. 282–312.

Kozai, T. 2013. Resource use efficiency of closed plant production system with artificial light: concept, estimation and applica- tion to plant factory. *Proc. Jpn. Acad. Ser. B Phys. Biol. Sci.* 89: 447–461.

Kozai, T., Niu, G., and Takagaki, M. (Eds.). 2015. *Plant Factory: An Indoor Vertical Farming System for Efficient Quality Food Production*. Academic Press, Amsterdam.

Lu, N., Bernardo, E.L., Tippayadarapanich, C., Takagaki, M., Kagawa, N., and Yamori, W. 2017. Growth and accumulation of secondary metabolites in Perilla as affected by photosyn- thetic photon flux density and electrical conductivity of the nutrient solution. *Front. Plant Sci.* 8, 708.

Lu, N., Nukaya, T., Kamimura, T., Zhang, D., Kurimoto, I., Takagaki, M., Maruo, T., Kozai, T., and Yamori, W. 2015. Control of vapor pressure deficit (VPD) in greenhouse enhanced tomato growth and productivity during the winter season. *Sci. Hort.* 197, 17–23.

Makino, T., Ono, T., Matsuyama, K., Nogaki, F., Miyawaki, S., Honda, G., et al. 2003. Suppressive effects of Perilla frute- scens on IgA nephropathy in HIGA mice. *Nephrol. Dial. Transpl.* 18, 484–490.

Maneejantra, N., Tsukagoshi, S., Lu, N., Supaibulwatana, K., Takagaki, M., and Yamori, W. 2016. A quantitative analysis of nutrient requirements for hydroponic spinach (*Spinacia olera- cea* L.) production under artificial light in a plant factory. *J. Fertil. Pestic.* 7, 170.

Martignon, G., Casarotti, D., Venezia, A., Sciavi, M., and Malorgio, F. 1994. Nitrate accumulation in celery as affected by grow- ing system and N content in the nutrient solution. *Acta Hortic.* 361, 583–589.

Ministry of Education, Culture, Sports, Science and Technology (MEXT). 2015. *Standard Tables of Food Composition in Japan – 2015* (7th revised edition). http://www.mext.go.jp/en/policy/sci ence_technology/policy/title01/detail01/1374030.htm.

Moritugu, M. and Kawasaki, T. 1982. Effect of solution pH on growth and mineral uptake in plants under constant pH condi- tion. Ber. Ohara Inst. Landwirtsch. *Forsch. Okayama Univ.* 18, 77–92.

Moritugu, M. and Kawasaki, T. 1983. Effect of nitrogen source on growth and mineral uptake in plants under nitrogen restricted culture condition. Ber. Ohara Inst. Landwirtsch. *Forsch. Okayama Univ.* 18, 145–158.

Moritugu, M., Suzuki, T., and Kawasaki, T. 1980. Effect of nitrogen sources upon plant growth and mineral uptake. 1. Comparison between constant pH and conventional culture method. *JSSSPN.* 51, 447–456.

Nahar, K. and Gretzmacher, R. 2002. Effect of water stress on nutrient uptake, yield and quality of tomato (*Lycopersicon esculentum* Mill.) under subtropical conditions. *Die Bodenkultur.* 53, 45–51.

Ogawa, A., Taguchi, S., and Kawashima, C. 2007. A cultivation method of spinach with a low potassium content for patients on dialysis. *Jpn. J. Crop Sci.* 76, 232–237 (in Japanese with English abstract).

Ogawa, A., Udzuka, K., Toyofuku, K., Ikeda, T. 2012. Japan patent: P2012-183062A.

Ohashi-Kaneko, K. 2015. Functional components in leafy vegetables. In: Kozai, T., Niu, G., and Takagaki, M. (Eds.), *Plant Factory: An Indoor Vertical Farming System for Efficient Quality Food Production*. Academic Press, Amsterdam, pp. 177–185.

Ooshima, T., Hagiya, K., Yamaguchi, T., and Endo, T. 2013. Commercialization of LED used plant factory (written in Japanese). Nishimatsu Construction's Technical Research Institute Report 36, 1–6.

Park, J.S., Choung, M.G., Kim, J.B., Hahn, B.S., Kim, J.B., Bae, S.C., Roh, K.H., Kim, Y.H., Cheon, C.I., Sung, M.K., and Cho, K.J. 2007. Genes up-regulated during red coloration in UV-B irradiated lettuce leaves. *Plant Cell Rep.* 26, 507–516.

Saigusa, M. and Kumazaki, T. 2014. Nitrate concentration and trial of decreasing nitrate concentration in factory vegetables (in Japanese). In: Takatsuji, M. and Kozai, T. (Eds.), *Important Problem and Measure Against Plant Factory Management* (written in Japanese: Shokubutsu kojo keiei no juuyo kadai to taisaku). Johokiko Co., Ltd., Tokyo, pp. 208–214.

Sakamoto, Y., Watanabe, S., Nakashima, T., and Okano, K. 1999. Effects of salinity at two ripening stages on the fruit quality of single-truss tomato grown in hydroponics. *J. Hortic. Sci. Biotechnol.* 74, 690–693.

Sanbongi, C., Takano, H., Osakabe, N., Sasa, N., Natsume, M., Yanagisawa, R., et al. 2004. Rosmarinic acid in Perilla extract inhibits allergic inflammation induced by mite allergen in a mouse model. *Clin. Exp. Allergy.* 34, 971–977.

Santamaria, P., Elia, A., Parente, A., and Serio, F. 1998. Fertilization strategies for lowering nitrate content in leafy vegetables: chicory and rocket salad cases. *J. Plant Nutr.* 21, 1791–1803.

Sato, S., Sakaguchi, S., Furukawa, H., and Ikeda, H. 2006. Effects of NaCl application to hydroponics nutrient solution on fruit characteristics of tomato (*Lycopersicon esculentum* Mill.). *Sci. Hortic.* 109, 248–253.

Shoji, K., Goto, E., Hashida, S., Goto, F., and Yoshihara, T. 2010. Effect of red light and blue light on the anthocyanin accu- mulation and expression of anthocyanin biosynthesis gene in red-leaf lettuce (in Japanese). *J. Sci. High Technol. Agric.* 22, 107–113.

Sohar, J. and Domoki, J. 1980. Nitrite and nitrate in human nutrition. *Bibl. Nutr. Dieta.* 29, 65–74.

Terry, N. and Ulrich, A. 1973. Effects of potassium deficiency on the photosynthesis and respiration of leaves of sugar beet. *Plant Physiol.* 51, 783–786.

Tewolde, F.T., Lu, N., Shiina, K., Maruo, T., Takagaki, M., Kozai, T., and Yamori, W. 2016. Nighttime supplemental LED interlighting improves growth and yield of single-truss tomatoes by enhancing photosynthesis in both winter and summer. *Front. Plant Sci.* 7, 448.

Troelstra, S.R., Wagenaar, R., Smant, W. 1990. Growth responses of plantago to ammonium nutrition with and without pH control: comparison of plants precultivated on nitrate or ammonium. In: van Beusichem, M.L. (Ed.), *Plant Nutrition – Physiology and Applications. Developments in Plant and Soil Sciences 41.* Springer, Netherlands, pp. 39–43.

Tsukagoshi, S. 2002. pH (in Japanese). In: Japan Greenhouse Horticulture Association (Ed.), *New Manual of Hydroponics* (Written in Japanese: Yoeki Saibai no shin-manuaru). Seibundo-Shinkosha, Tokyo, pp. 180–183.

Tsukagoshi, S. and Aoki, M. 2018. New method of nutrient management in hydroponics and the application for production of low potassium tomato fruit. *J. Food Process Technol.* doi.

Tsukagoshi, S., Hamano, E., Hohjo, M., and Ikegami, F. 2016. Hydroponic production of low-potassium tomato fruit for dialysis patients. *Int. J. Veg. Sci.* 22, 451–460.

Tsukagoshi, S., Maruo, T., Ito, T., Fuso, H., and Okabe, K. 1999. Effect of withdrawal of NO3−N or all nutrient from the NFT System prior to harvest on the growth, NUMBER3 content in the spinach plant and the final mineral concentration in the nutrient solution. *J. Japan Soc. Hort. Sci.* 68, 1022–1026 (in Japanese with English abstract).

Voipio, I. and Autio, J. 1995. Responses of red-leaved lettuce to light intensity, UV-A radiation and root zone temperature. *Acta Hortic.* 399, 183–187.

Wu, M. and Kubota, C. 2008. Effects of high electrical conductivity of nutrient solution and its application timing on lycopene, chlorophyll and sugar concentrations of hydroponic tomatoes during ripening. *Sci. Hortic.* 116, 122–129.

Yamori, W. 2016. Photosynthetic response to fluctuating environments and photoprotective strategies under abiotic stress. *J. Plant Res.* 129, 379–395.

Yamori, W., Hikosaka, K, and Way, D.A. 2014. Temperature response of photosynthesis in C3, C4 and CAM plants: temperature acclimation and temperature adaptation. *Photosynth. Res.* 119, 101–117.

Yamori, W., Makino, A., and Shikanai, T. 2016. A physiological role of cyclic electron transport around photosystem I in sustaining photosynthesis under fluctuating light in rice. *Sci. Rep.* 6, 20147.

Yamori, W., Nagai, T., and Makino, A. 2011. The rate-limiting step for CO_2 assimilation at different temperatures is influenced by the leaf nitrogen content in several C3 crop species. *Plant Cell Environ.* 34, 764–777.

Yamori, W., Noguchi, K., Hikosaka, K., and Terashima, I. 2009. Cold-tolerant crop species have greater temperature homeostasis of leaf respiration and photosynthesis than cold-sensitive species. *Plant Cell Physiol.* 50, 203–215.

Yamori, W., Noguchi, K., Hikosaka, K., and Terashima, I. 2010. Phenotypic plasticity in photosynthetic temperature acclimation among crop species with different cold tolerances. *Plant Physiol.* 152, 388–399.

Yamori, W. and Shikanai, T. 2016. Physiological functions of cyclic electron transport around photosystem I in sustaining photosynthesis and plant growth. *Annu. Rev. Plant Biol.* 67, 81–106.

Yamori, W., Shikanai, T., and Makino, A. 2015. Photosystem I cyclic electron flow via chloroplast NADH dehydrogenase-like complex performs a physiological role for photosynthesis at low light. *Sci. Rep.* 5, 13908.

Yamori, W., Zhang, G., Takagaki, M., and Maruo, T. 2014. Feasibility study of rice growth in plant factories. *J. Rice Res.* 2, 119.

Zhang, Y., Seeram, N. P., Lee, R., Feng, L., and Heber, D. 2008. Isolation and identification of strawberry phenolics with antioxidant and human cancer cell antiproliferative properties. *J. Agric. Food Chem.* 56, 670–675.

Zushi, K., Matsuzoe, N., Yoshida, S., and Chikushi, J. 2005. Comparison of chemical composition contents of tomato fruit grown under water and salinity stresses. *J. SHITA.* 17, 128–136 (in Japanese with English abstract).

Index

Printed and bound by CPI Group (UK) Ltd, Croydon, CR0 4YY

17/10/2024

01775702-0001